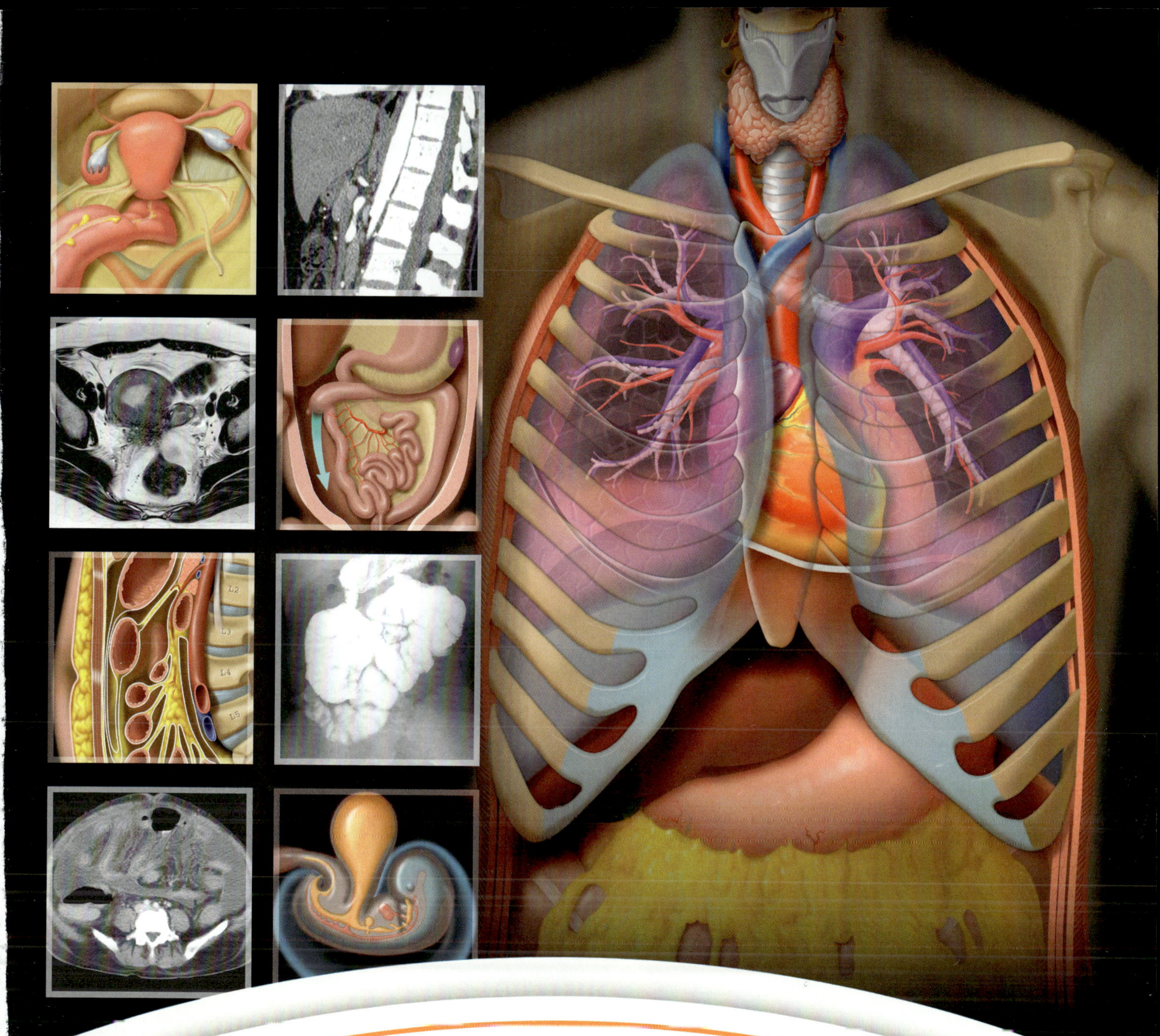

THIRD EDITION

IMAGING ANATOMY

Chest, Abdomen, Pelvis

Raman | Rosado-de-Christenson | Zaheer
Martínez-Jiménez | Fananapazir | Garrana
Rogers | Foster

ELSEVIER

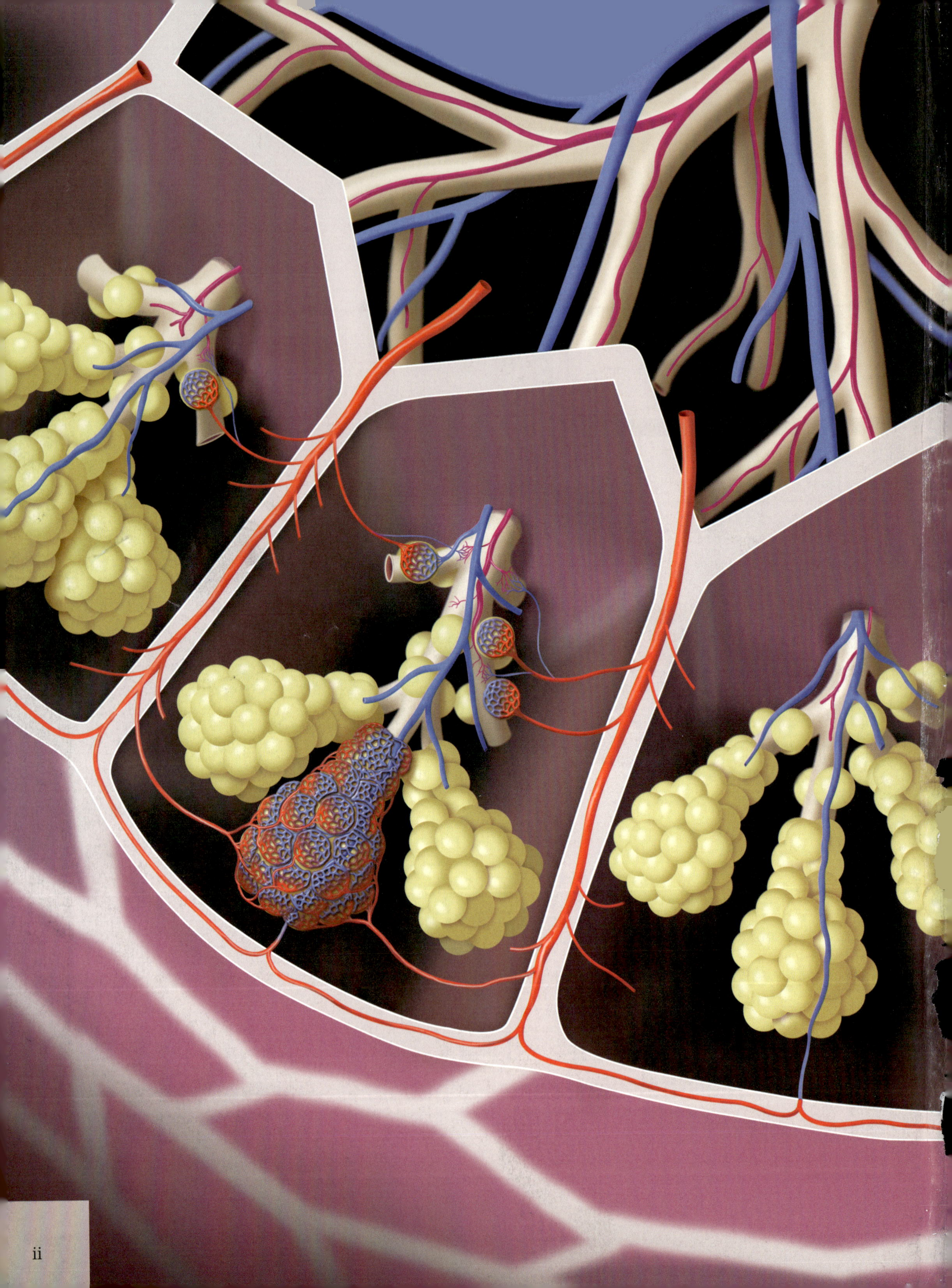

IMAGING ANATOMY

Chest, Abdomen, Pelvis

THIRD EDITION

Siva P. Raman, MD
Bay Imaging Consultants
Walnut Creek, California

Melissa L. Rosado-de-Christenson, MD, FACR, FAAWR
Attending Radiologist
Division of Cardiothoracic Imaging
Department of Medical Imaging
Banner – University Medical Group Tucson
Professor of Medical Imaging
The University of Arizona College of Medicine – Tucson
Tucson, Arizona

Atif Zaheer, MD, FSAR
Professor of Radiology, Oncology and Medicine
The Johns Hopkins University School of Medicine
Baltimore, Maryland

Santiago Martínez-Jiménez, MD
Department of Radiology
Saint Luke's Hospital of Kansas City
Professor of Radiology
University of Missouri-Kansas City School of Medicine
Kansas City, Missouri

Ghaneh Fananapazir, MD, FSAR, FSRU, FSABI
Professor of Radiology
University of California, Davis
Sacramento, California

Sherief H. Garrana, MD
Assistant Professor of Radiology
Department of Radiology
Saint Luke's Hospital of Kansas City
University of Missouri-Kansas City School of Medicine
Kansas City, Missouri

Douglas Rogers, MD
Associate Professor
Department of Radiology and Imaging Sciences
University of Utah School of Medicine
Salt Lake City, Utah

Bryan R. Foster, MD
Professor
Department of Radiology
Oregon Health & Science University
Portland, Oregon

Additional Contributors

Rania Farouk El Sayed, MD, PhD
Michael P. Federle, MD, FACR
Akram M. Shaaban, MBBCh
Paula J. Woodward, MD

Elsevier
1600 John F. Kennedy Blvd.
Ste 1800
Philadelphia, PA 19103-2899

IMAGING ANATOMY: CHEST, ABDOMEN, PELVIS, THIRD EDITION

ISBN: 978-0-443-11800-5

Previous edition copyrighted 2017.

Library of Congress Control Number: 2023943574

Printed in Canada by Friesens, Altona, Manitoba, Canada

Last digit is the print number: 9 8 7 6 5 4 3 2 1

Dedications

To my loving wife, Janani, my two children, Jay and Mina, and my parents, Raghu and Visali, who have supported me in all my endeavors and who have provided me with invaluable guidance in every facet of my life.

SPR

To my family, but especially my husband, Paul, whose love, encouragement, and support have seen me through this and so many other projects. To my chest imaging coauthors, Santiago Martínez-Jiménez and Sherief Garrana, from whom I continue to learn every day. To my division chief, colleague, and friend, Dr. Kavitha Yaddanapudi, for warmly welcoming me to the University of Arizona family and providing the academic environment that allowed me to complete this work.

MLR

For the one whose love and inspiration have been my compass, this book is lovingly dedicated to my mother, and also to the brave souls who navigated the intricate corridors of anatomy, defying doubts and discovering the marvels that make us human.

AZ

While working on this book, one of the most important people in my life passed away. Fanny, you were truly the epitome of revolution! I am so thankful that you were around for me. Thank you for staying. Buen viento y buena mar!

SM

To my son, Sameh (the listener): Thank you for your patience and willingness to listen to me; you have lived up to your name. To my daughter, Nura (light): Thank you for your happiness and spirit that brighten my life; you too have lived up to your name. To my wife, Melody: Thank you for taking me as your most challenging coaching client; you are the best at it!

GF

To my coauthors, Melissa Rosado-de-Christenson and Santiago Martínez-Jiménez, who continue to be at the very core of my support system, personally and professionally. I am truly privileged and forever grateful for your mentorship and friendship.

SHG

I dedicate this to individuals committed to safeguarding universities as havens of open inquiry and constructive dialogue, particularly with regard to the anatomical and physiological aspects in my area of subspecialty interest of gynecology.

DR

To Nolan and Bennett, who are endlessly curious, and to Sarah, who makes it all possible.

BRF

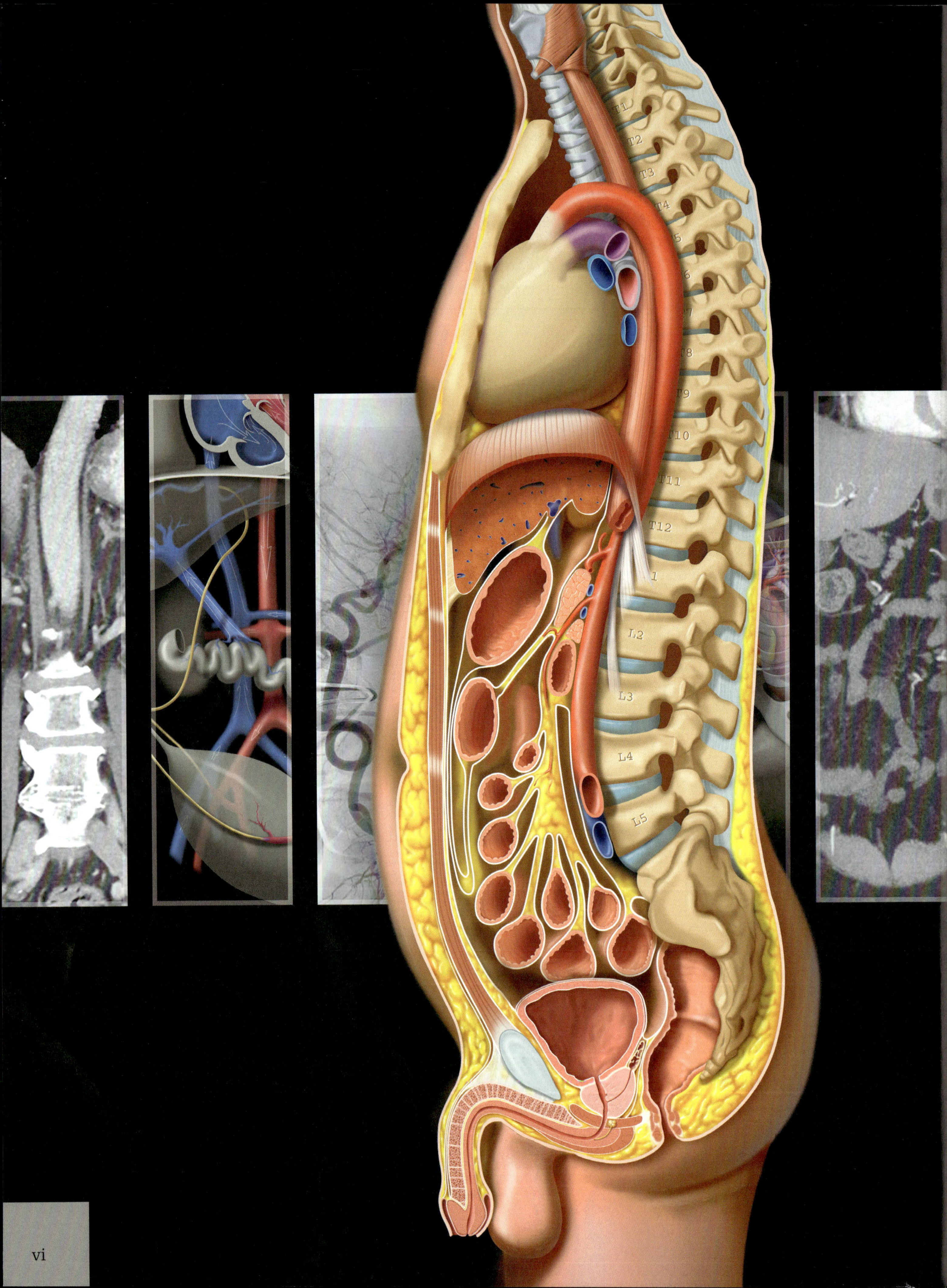
T1
T2
T3
T4
T8
T9
T10
T11
T12
L2
L3
L4
L5

Preface

The evolution of the study of anatomy has been a cumulative effort spanning centuries and involving contributions from various cultures and civilizations. From the early observations of ancient Egyptians and Greeks to the remarkable achievements of the Muslim anatomists during the Islamic Golden Age to the Renaissance physicians, the quest to understand the complex organization of our bodies has captivated scholars throughout the ages. Pioneers like Galen, Ibn Sina, Ibn al-Nafis, Leonardo da Vinci, and Andreas Vesalius laid the groundwork for anatomic knowledge, revealing the intricacies of the human form through meticulous dissections and detailed illustrations.

In the modern era, the remarkable development of advanced machines using computed tomography, magnetic resonance imaging, and ultrasound has had a profound impact on the study of anatomic complexities. These technologies visualize organs, vessels, and tissues with unprecedented clarity, revolutionizing medical diagnostics, research, and education. Their noninvasive and detailed visualization of internal structures has nearly supplanted the traditional reliance on dissectional anatomy.

The third edition of this book features a well-organized layout and rich illustrations, facilitating easy access to vital information. The templated format, concise text, and cutting-edge images assist clinicians in swiftly identifying clinical entities within each anatomic region. This approach deepens understanding of normal anatomy while enabling accurate recognition of pathologic conditions.

We express heartfelt gratitude to the authors for their invaluable contributions to the field of anatomy. Their commitment to disseminating knowledge through captivating visuals will inspire future generations of students, researchers, and health care professionals. They continue the legacy of the "House of Wisdom" by furthering the quest for knowledge and understanding of the human body.

Atif Zaheer, MD, FSAR
Professor of Radiology, Oncology and Medicine
The Johns Hopkins University School of Medicine
Baltimore, Maryland

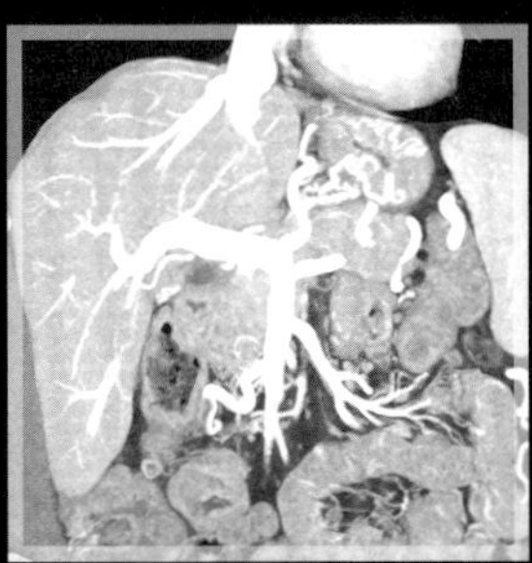
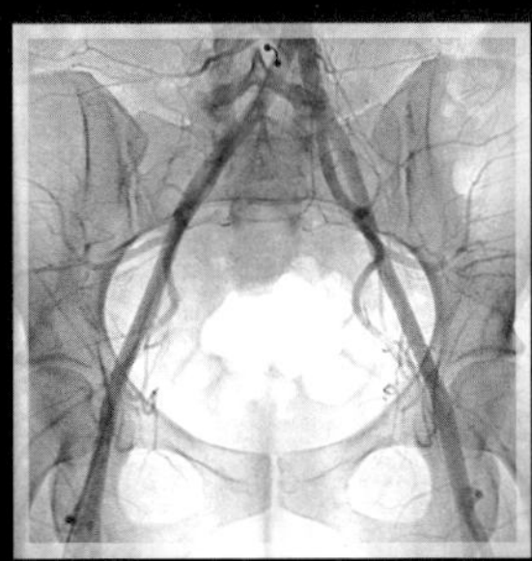
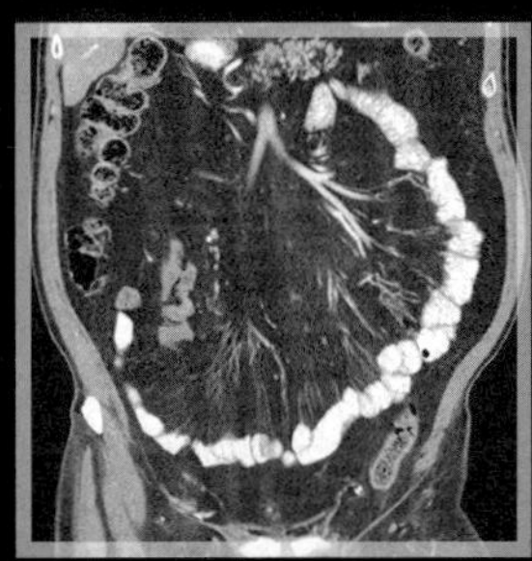
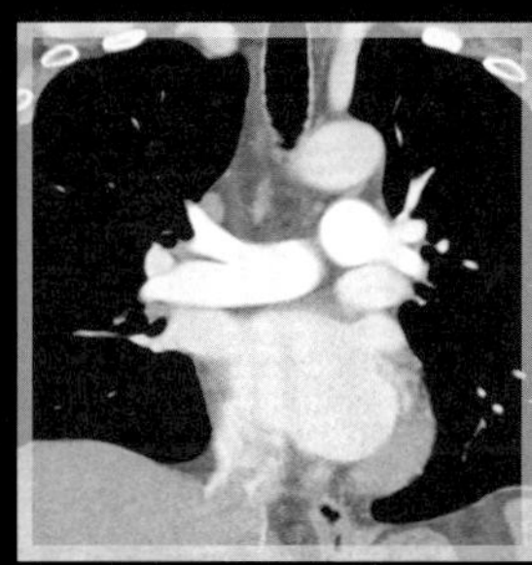

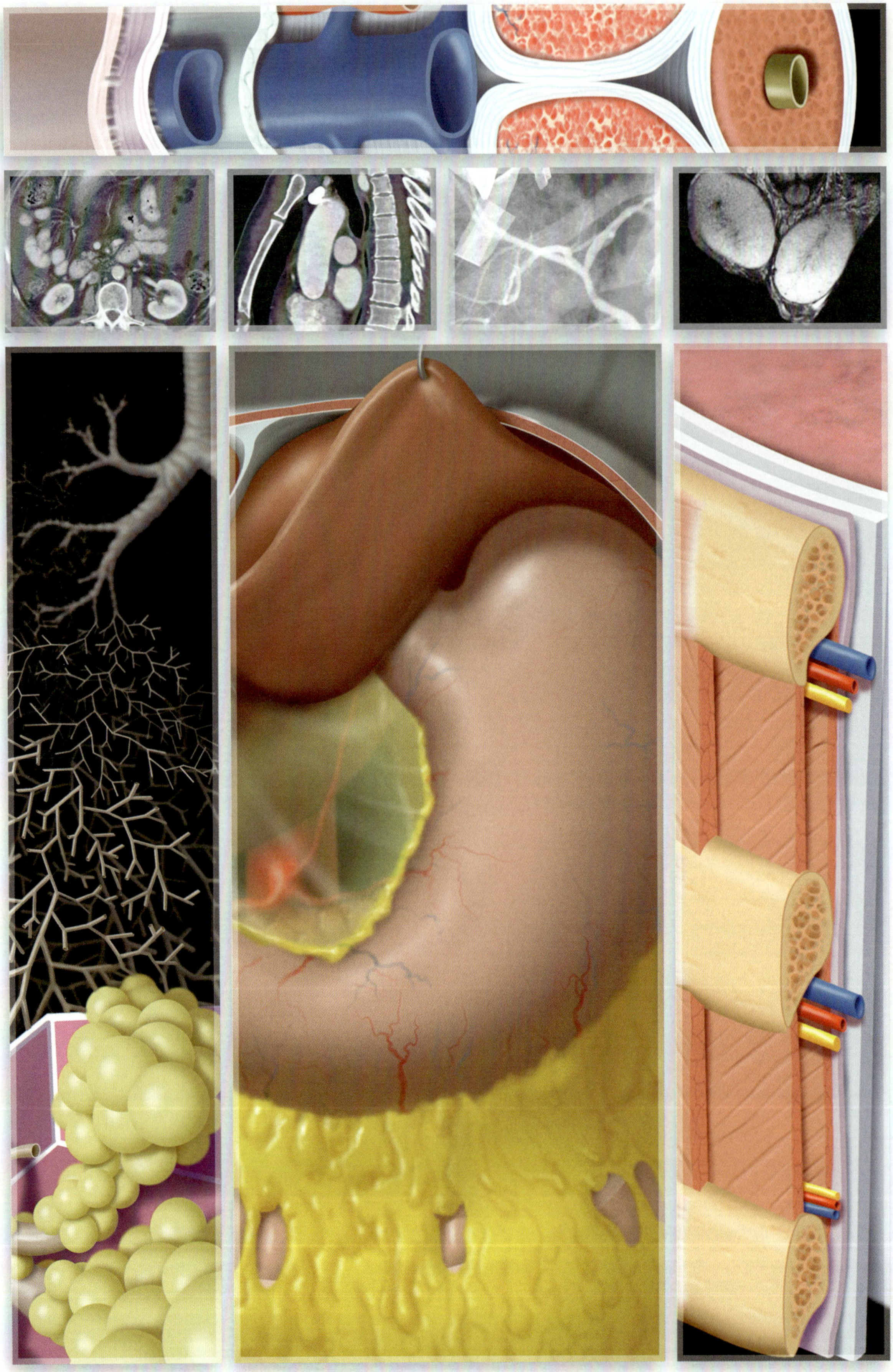

Acknowledgments

LEAD EDITOR
Terry W. Ferrell, MS

LEAD ILLUSTRATOR
Laura C. Wissler, MA

TEXT EDITORS
Arthur G. Gelsinger, MA
Rebecca L. Bluth, BA
Nina Themann, BA
Megg Morin, BA
Kathryn Watkins, BA
Shannon Kelly, MA

ILLUSTRATIONS
Lane R. Bennion, MS
Richard Coombs, MS

IMAGE EDITORS
Jeffrey J. Marmorstone, BS
Lisa A. M. Steadman, BS

ART DIRECTION AND DESIGN
Cindy Lin, BFA

PRODUCTION EDITORS
Emily C. Fassett, BA
John Pecorelli, BS

ELSEVIER

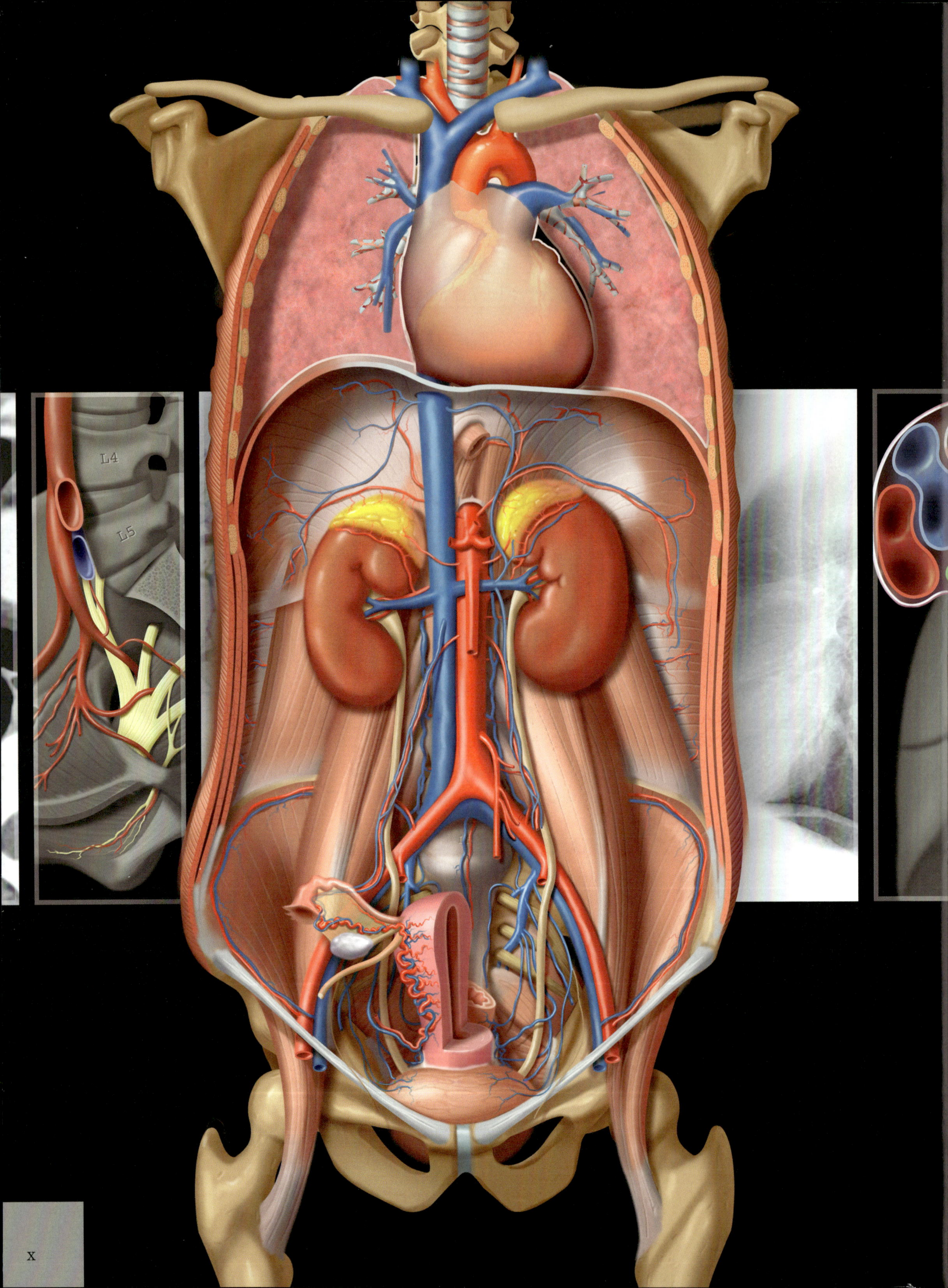
L4
L5

Sections

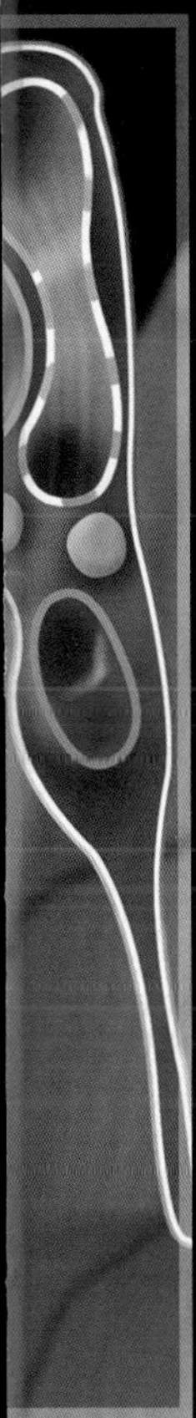
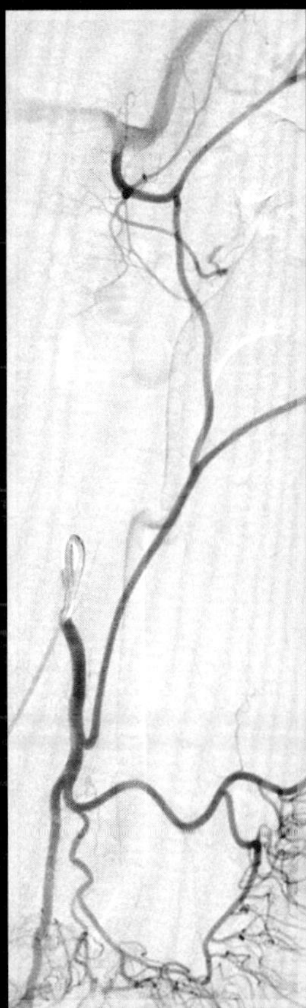

SECTION 1:
Chest

SECTION 2:
Abdomen

SECTION 3:
Pelvis

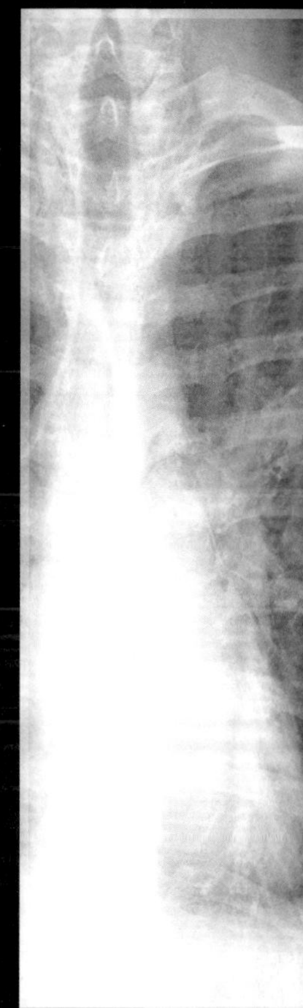
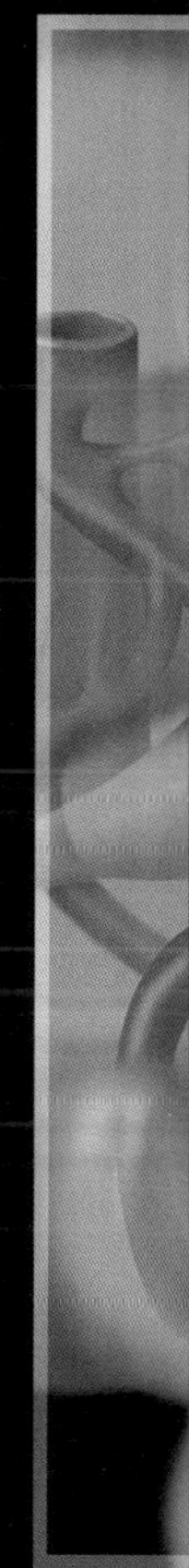

TABLE OF CONTENTS

SECTION 1: CHEST

SECTION 2: ABDOMEN

SECTION 3: PELVIS

MALE

TABLE OF CONTENTS

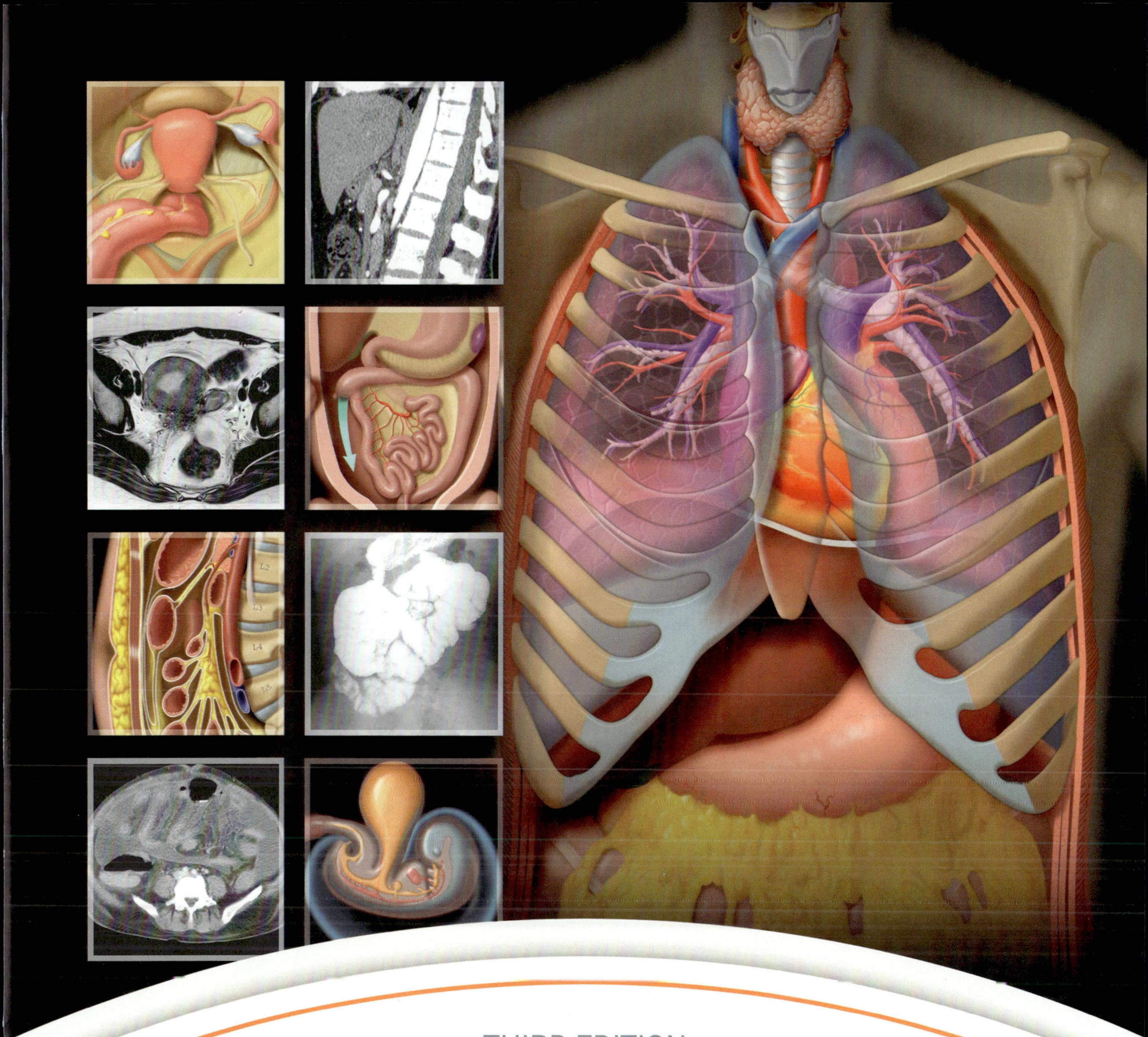

THIRD EDITION

IMAGING ANATOMY
Chest, Abdomen, Pelvis

Raman | Rosado-de-Christenson | Zaheer
Martínez-Jiménez | Fananapazir | Garrana
Rogers | Foster

ELSEVIER

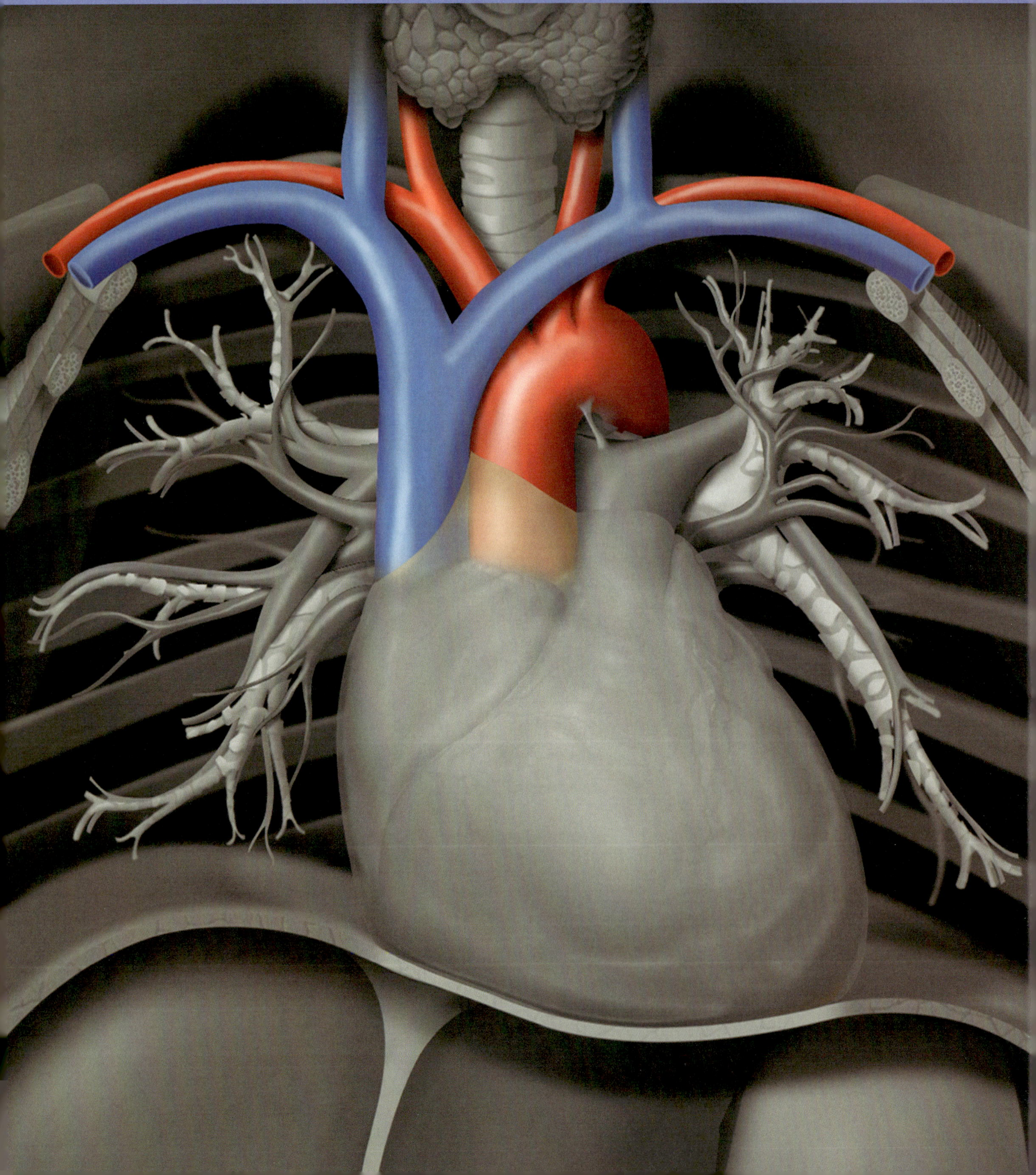

SECTION 1
Chest

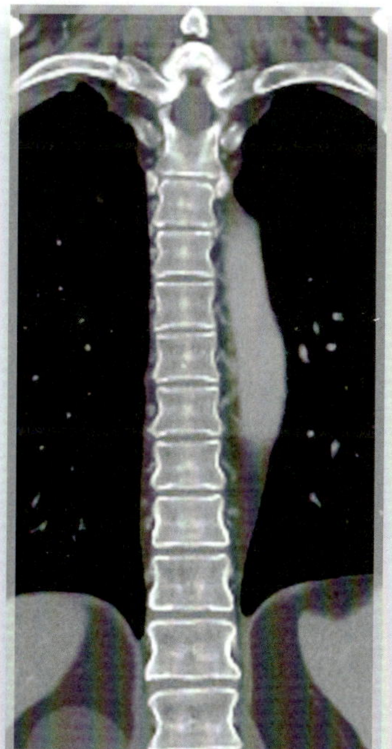

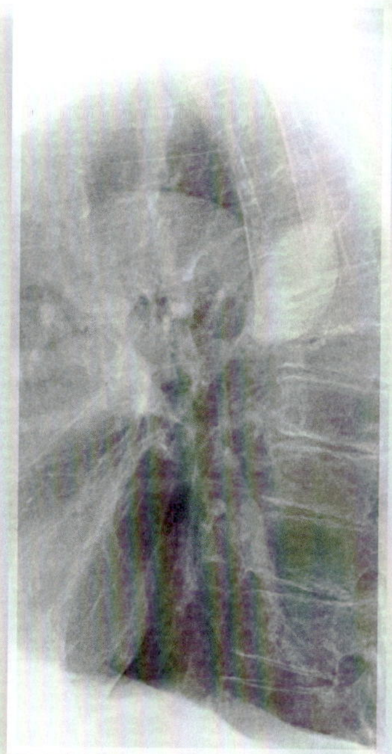

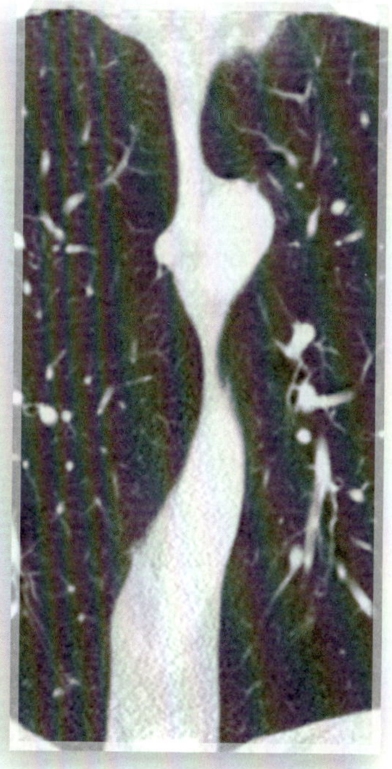

GENERAL ANATOMY AND FUNCTION

Chest Wall

- Anatomy
  - Spine
  - Sternum
  - Ribs
  - Clavicles
  - Scapulae
  - Muscles
  - Nerves and vessels
  - Skin and subcutaneous fat
- Function
  - Protects lungs, cardiovascular structures, and intrathoracic organs
  - Participates in bellows-like process of respiration

Pleura

- Anatomy
  - Thin, continuous membrane
  - Parietal pleura: Lines nonpulmonary surfaces
  - Visceral pleura: Lines pulmonary surfaces and fissures
  - Pleural space: Potential space between pleural surfaces
- Function
  - Production and absorption of normal pleural fluid
    - Lubrication of pleural surfaces
    - Facilitation of lung motion during respiration
  - Clearance of abnormal pleural fluid

Airways

- Anatomy
  - Trachea
  - Mainstem bronchi
  - Lobar bronchi
  - Segmental/subsegmental bronchi
  - Bronchioles
  - Small airways and alveoli
- Function
  - Gas transport and exchange during respiration
  - Protective mechanisms against foreign particles
    - Mucociliary escalator
    - Cough reflex
  - Gas exchange at alveolar-capillary interface

Heart and Great Vessels

- Anatomy
  - Venae cavae and great veins
  - Right atrium
  - Right ventricle
  - Pulmonary arteries
  - Capillary network
  - Pulmonary veins
  - Left atrium
  - Left ventricle
  - Thoracic aorta and branch vessels
- Function
  - Pump action for systemic and pulmonary circulations
  - Transport of deoxygenated blood to capillary-alveolar interface
  - Transport of oxygenated blood to tissues

CHEST RADIOGRAPHY

Standard Chest Radiography

- Imaging study of choice for initial assessment of cardiopulmonary disease
- PA and lateral chest radiography
  - Orthogonal views (i.e., at right angles to each other)
  - Analysis of orthogonal views for precise anatomic localization of imaging abnormalities

Standard Radiographic Positioning

- Upright patient
- Full inspiration and breath hold near total lung capacity
- No rotation or motion
- Positioning to minimize overlying osseous structures
- Area of interest closest to image receptor (IR)
- Radiographic technique
  - High kVp to increase latitude (shades of gray)
  - Source-to-image receptor distance (SID): 72 inches to minimize magnification
  - Central x-ray beam centered on thorax
  - Beam collimation to include outer chest wall

Radiographic Projections

- **Posteroanterior (PA) chest radiography**
  - Term PA: Describes PA direction of x-ray beam traversing chest toward IR
  - Anterior chest against IR
  - Head vertically positioned and chin on top of grid device
  - Dorsal wrists on hips and elbows rotated anteriorly to move scapulae laterally
  - Shoulders moved caudally and squarely against IR to bring clavicles below lung apices
- **Left lateral chest radiography**
  - Term **left lateral**: Denotes that left lateral chest wall is against IR
  - x-ray beam traverses chest from right to left toward IR
  - Arms above head to move upper extremities away from anterior/apical lungs and mediastinum
- **Anteroposterior (AP) chest radiography**
  - Term AP: Describes AP direction of x-ray beam traversing chest toward IR
  - Supine and bedside (portable) radiography and imaging of sitting and semiupright patients
    - Neonates, infants, and young children
    - Traumatized, critically ill, and bed-ridden patients
    - Debilitated and unstable patients
  - Distinctive features
    - Magnification of anterior structures (heart and mediastinum) farthest from IR; shorter SID
    - Clavicles course horizontally and partially obscure lung apices
    - Ribs assume horizontal course
- **Lateral decubitus chest radiography**
  - Recumbent position with right or left side down
  - Elevation of chest on radiolucent support
  - Frontal radiograph (AP or PA) with horizontal x-ray beam
  - Indications
    - Evaluation of pleural fluid in dependent pleural space (x-ray beam tangential to fluid-lung interface)

- Evaluation of air in nondependent pleural space (x-ray beam tangential to visceral pleura-air interface)
- **Apical lordotic** (AP or PA) **chest radiography**
  - Superior angulation of x-ray beam from horizontal plane of 15-20°
  - Distinctive features
    - Clavicles and 1st anterior ribs project above apices
    - Ribs course horizontally
    - Magnification (foreshortening) of mediastinum
  - Indications
    - Radiographic visualization of lung apices, upper mediastinum, and thoracic inlet
    - Enhanced visualization of minor fissure on frontal radiographs in suspected middle lobe atelectasis
- Expiratory radiography
  - Evaluation of air-trapping (e.g., aspirated foreign body)
  - Evaluation of pneumothorax (limited value)
    - No clear difference in sensitivity or specificity for diagnosis

Radiographic Interpretation

- Assessment of patient's identity and location of right/left markers
- Assessment of entire thorax
  - **Frontal radiography**
    - Inclusion of all thoracic structures from larynx to lung bases
    - Full inspiration
      - Hemidiaphragm intersected by 5th-7th anterior ribs at midclavicular line
  - **Lateral radiography**
    - Inclusion of AP extent of chest wall
    - Inclusion of apical lung and posterior costodiaphragmatic sulci
- Assessment of **appropriate radiographic positioning**
  - No rotation
    - Spinous process of T3 (posterior structure) centered between medial clavicles (anterior structures) on frontal chest radiography
    - Superimposition of right and left ribs posterior to thoracic vertebrae on lateral chest radiography
  - Medial aspects of scapulae lateral to lungs on frontal chest radiography
  - Arms above thorax without superimposition on lung and mediastinum on lateral chest radiography
- **Appropriate exposure**
  - Assessment of peripheral pulmonary vasculature
  - Visualization of pulmonary vessels through heart and diaphragm on frontal radiography
  - Visualization of vertebrae through heart on frontal radiography
- Systematic evaluation
  - Assessment of multiple superimposed structures and tissues
  - Assessment of all visible structures including portions of neck, shoulders, and upper abdomen
  - Anatomic localization of abnormalities and associated findings
  - Comparison to prior studies
- Challenges
  - Evaluation of retrocardiac lung
  - Evaluation of retrodiaphragmatic lung
  - Evaluation of apical lung

Radiographic Densities

- 5 radiographic densities
  - **Air**
  - **Water** (fluid, blood, soft tissue)
  - **Fat**
  - **Mineral or calcium** (bone)
  - **Metal** (contrast, metallic portions of medical devices, metallic foreign bodies)
- **Silhouette sign**
  - Intrathoracic process (mass, consolidation, pleural fluid) that abuts mediastinum or diaphragm and obscures visualization of their borders on radiography
  - Critical for radiographic diagnosis of
    - **Atelectasis**
    - **Consolidation**
    - **Pleural effusion**

COMPUTED TOMOGRAPHY

General Concepts

- Imaging based on x-ray absorption by tissues with differing atomic numbers
- Display of differences in x-ray absorption in cross-sectional format
- Good spatial and contrast resolution
- Enhanced visualization of structures of different tissue attenuation based on display of wide range of Hounsfield unit (HU) measurements
  - **Window width**: Number of HU displayed
  - **Window level**: Median (center) HU
  - Window width and level for assessment of thoracic CT
    - **Lung window** (width of 1,500 HU; level of -650 HU)
      - Evaluation of lungs, airways, and air-containing portions of gastrointestinal tract
      - Identification of pneumomediastinum, pneumoperitoneum, and soft tissue gas
    - **Soft tissue (mediastinal) window** (width of 400 HU; level of 40 HU)
      - Evaluation of vascular structures and soft tissues of mediastinum and chest wall
    - **Bone window** (width of 2,000 HU; level of 400 HU)
      - Evaluation of skeletal and calcified structures
      - Evaluation of metallic objects and medical devices

Conventional CT

- Identification, localization, and characterization of abnormalities detected on radiography
- Anatomic localization/assessment of lesions in preparation for CT-guided biopsy/drainage

Contrast-Enhanced CT

- Administration of intravenous contrast
  - Evaluation of vascular structures/abnormalities
  - Distinction of vascular structures from adjacent soft tissues (e.g., hilar lymph nodes, masses)
  - Determination of lesion/tissue enhancement
- Administration of enteric contrast
  - Evaluation of gastrointestinal perforations/leaks

CT Angiography

- Vascular imaging
  - Timing of contrast bolus
  - Imaging of specific vascular structures
    - CT pulmonary angiography for evaluation of pulmonary thromboembolic disease
    - CT aortography for evaluation of traumatic aortic injury or aneurysm
    - CT aortography for evaluation of acute aortic syndrome: Baseline unenhanced images required prior to contrast administration to document intramural hematoma

High-Resolution CT

- Technique
  - Thin sections to minimize partial volume effects
  - High-resolution reconstruction algorithm
- Indications
  - Evaluation of diffuse lung disease, including idiopathic interstitial pneumonias and fibrosing interstitial lung disease
  - Evaluation of patients with unexplained dyspnea and normal chest radiographs
- Special techniques
  - Prone imaging for evaluation of peripheral basilar lung disease
  - Expiratory imaging for identification/assessment of small airways disease/constrictive bronchiolitis

Postprocessing Techniques

- Multiplanar reformatted images
  - Multiplanar evaluation of axially oriented structures and abnormalities
  - Assessment of anatomic location of lung lesions in relation to interlobar fissures
  - Evaluation of chest wall and mediastinal involvement by adjacent pulmonary lesions
  - Coronal/sagittal imaging: Lesion measurement/assessment of apicobasilar distribution of diffuse lung disease
- Maximum-intensity projection (MIP)
  - Evaluation of vascular structures
  - Identification of pulmonary nodules
- Minimum-intensity projection (minIP)
  - Evaluation of central airways, cystic/lucent lung lesions, air-trapping
- Surface-rendered displays
- Volume-rendered techniques for problem solving and education: Virtual bronchoscopy

MAGNETIC RESONANCE IMAGING

General Concepts

- Application of radiofrequency to excite protons within magnetic field
- Detection of signal emitted by nuclei as they relax to their original alignment with generation of images depicting their spatial distribution
- Advantages of MR
  - Excellent contrast resolution/increased soft tissue contrast
  - Multiplanar imaging
  - Intrinsic vascular "contrast"

Techniques/Sequences

- Traditional/anatomic: Evaluation of anatomy and pathology (e.g., signal intensity characterization, contrast enhancement)
- Functional (e.g., diffusion weighted)
- Cine based

Indications

- Imaging of heart and great vessels
- Assessment of vascular integrity without contrast
- Diagnostic imaging of mediastinum, hila, thymus
- Assessment of chest wall and diaphragm
- Pulmonary imaging: Emerging role of anatomic/quantitative/functional imaging
  - Children, pregnant women, young patients (exclusion of pulmonary thromboembolism, serial imaging)

ANGIOGRAPHY

Pulmonary Angiography

- Venous catheterization of pulmonary arteries
  - Assessment/management of selected cases of thromboembolic disease
  - Evaluation of congenital and acquired pulmonary vascular abnormalities

Aortography

- Arterial catheterization of thoracic aorta
  - Evaluation/management of traumatic aortic and great vessel injury
  - Assessment of congenital arterial vascular anomalies

Bronchial Artery and Intercostal Arteriography

- Bronchial arteries
  - Diagnosis and treatment of hemoptysis
- Intercostal arteries

OTHER CHEST IMAGING MODALITIES

Nuclear Medicine

- **Ventilation-perfusion scintigraphy**
  - Diagnosis of pulmonary thromboembolic disease
    - Study of choice in pregnant patients with normal chest radiographs
  - Assessment of pre- and posttreatment lung perfusion
- **Positron-emission tomography (PET)**
  - Determination of metabolic activity of lesions
  - Staging of malignant neoplasms
  - Use of integrated PET/CT and PET/MR

Ultrasound

- Evaluation of **pleural effusion**
  - Free vs. loculated
  - Thoracentesis/biopsy planning and thoracostomy tube placement
- Assessment of diaphragmatic motion in suspected paralysis

CHEST OVERVIEW

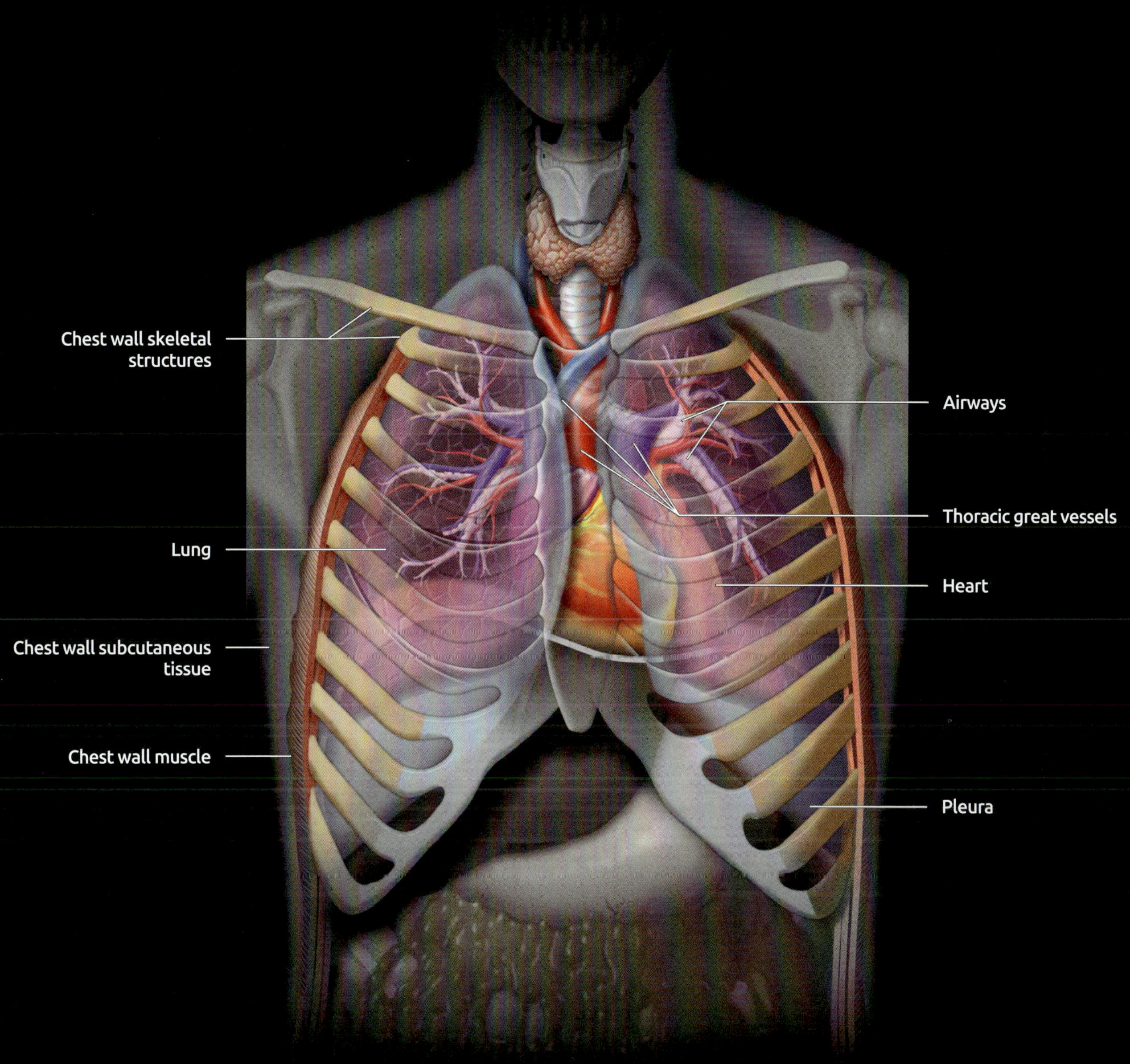

Graphic shows the complex and diverse structures and organs of the thorax. The chest wall skeletal and soft tissue structures surround and protect the primary organs of respiration, the thoracic cardiovascular system, and the proximal/upper gastrointestinal tract. The apposed visceral and parietal pleurae create a potential space that normally contains a small amount of fluid, which lubricates the pleural surfaces and reduces friction during respiratory motion. The airways deliver oxygen to the alveolar-capillary interface and carry carbon dioxide out to the environment. The heart, pulmonary vessels, and systemic vessels deliver deoxygenated blood to the capillary-alveolar interface and oxygenated blood to the body's organs and tissues.

PA CHEST RADIOGRAPHY

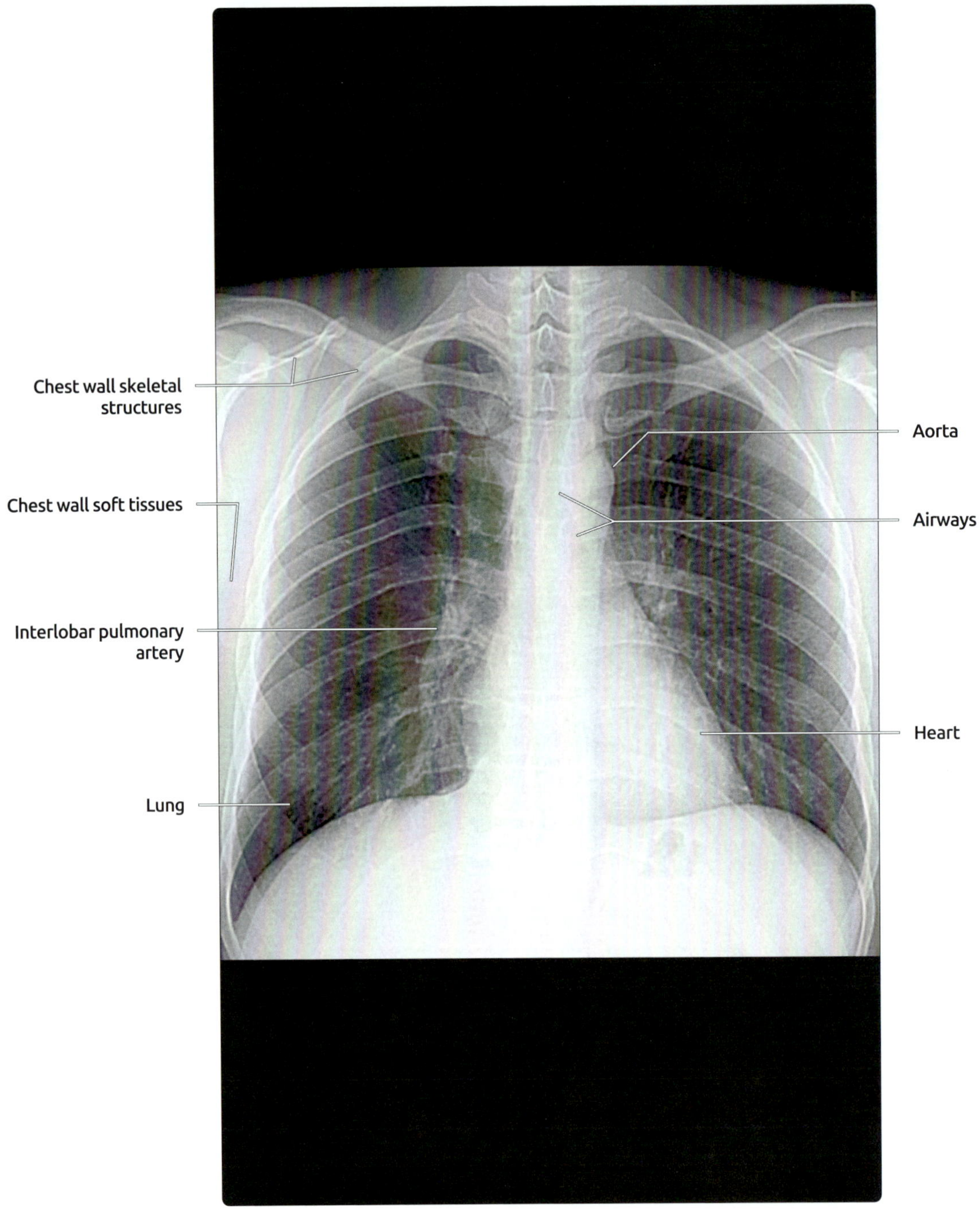

Normal posteroanterior (PA) chest radiograph helps illustrate inherent challenges regarding radiographic interpretation. Chest radiography displays a wide range of structures and tissue types with significant superimposition of structures of different radiographic densities. Portions of the lung may be obscured by overlying mediastinal soft tissues and skeletal structures. Attention to radiographic image quality is of paramount importance for identification and accurate diagnosis of subtle abnormalities.

LATERAL CHEST RADIOGRAPHY

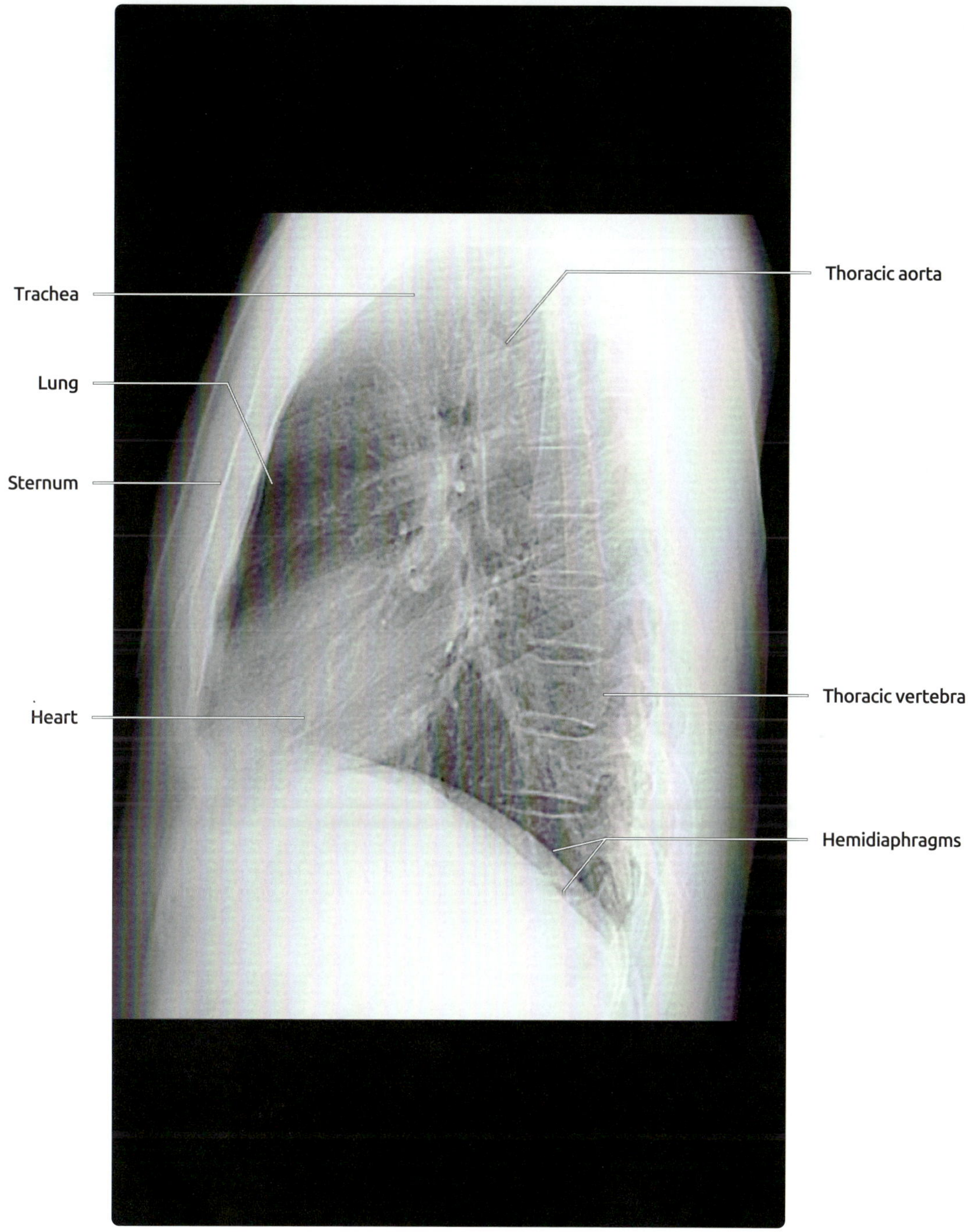

Lateral chest radiography is orthogonal (at 90°) to PA chest radiography. The lateral chest radiograph is a complementary radiographic projection that allows visualization of the retrocardiac left lower lobe and the retrodiaphragmatic lung bases and allows evaluation of the thoracic vertebrae. As on the PA chest radiograph, multiple structures of various radiographic densities are superimposed and must be evaluated in a systematic manner. Evaluation of both PA and lateral chest radiographs allows identification, anatomic localization, and characterization of thoracic abnormalities and the formulation of an appropriate differential diagnosis.

PA CHEST RADIOGRAPHY, POSITIONING AND COLLIMATION

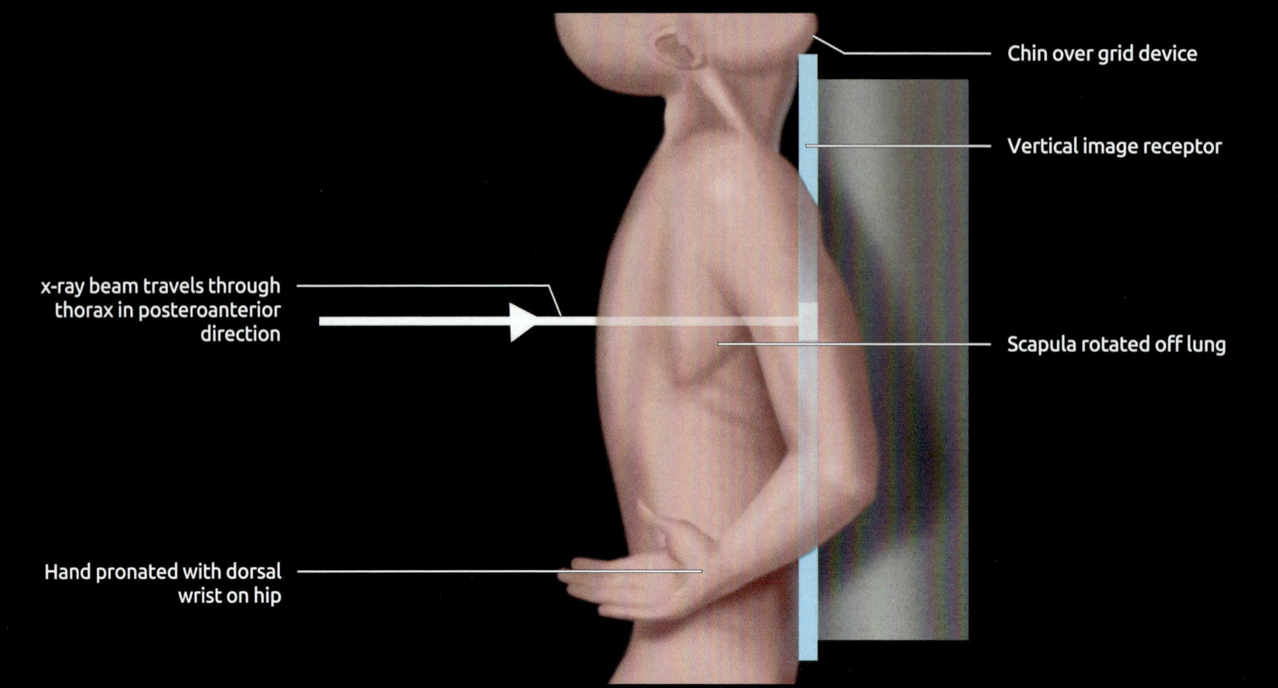

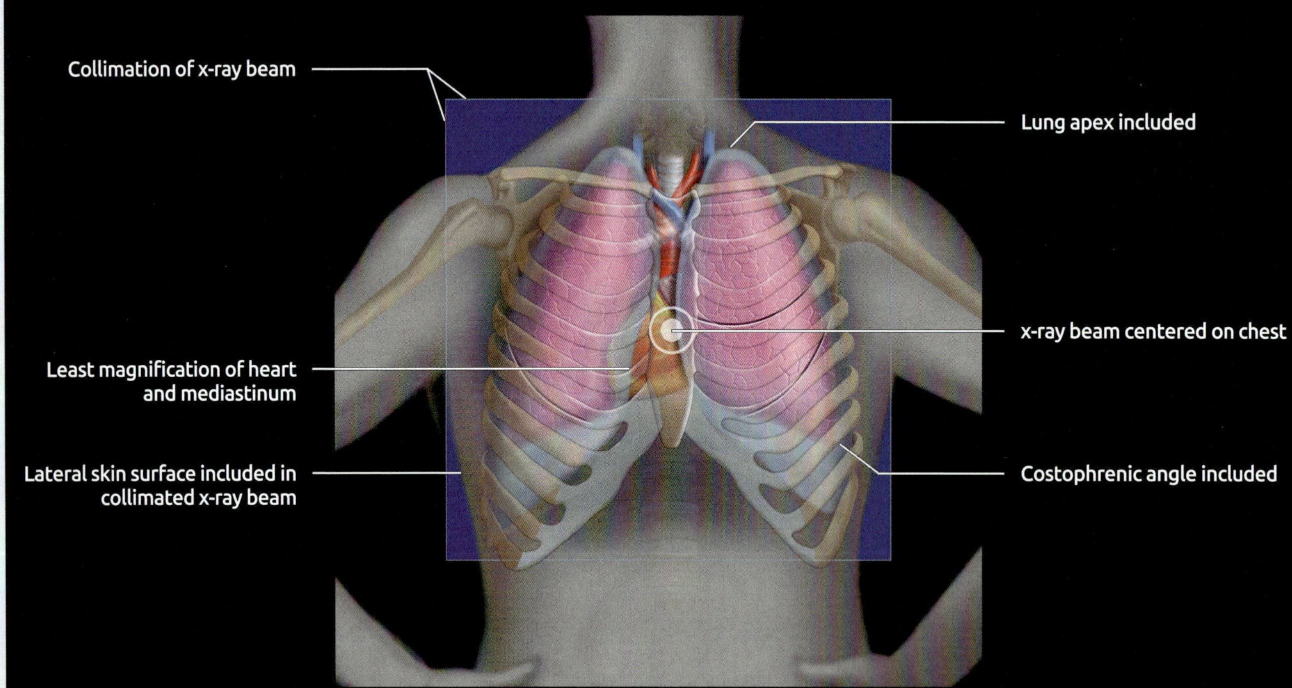

(Top) *Graphic shows proper patient positioning for PA chest radiography. The patient is upright with the anterior chest placed against the vertically oriented image receptor, the chin rests over the top of the device, the arms are flexed anteriorly with the dorsal aspects of the hands on the hips, and the shoulders are internally rotated to move the scapulae off the lungs. The x-ray beam travels through the patient in a PA direction.* **(Bottom)** *Graphic shows proper PA chest radiographic collimation. The white target sign shows the centering of the x-ray beam in the midportion of the thorax. The blue overlay represents the collimated x-ray beam that extends from the cervical airway superiorly to below the costophrenic angles inferiorly and includes the left and right skin surfaces. The anterior structures of the thorax (shown in color) are closest to the image receptor and experience the least degree of magnification.*

PA CHEST RADIOGRAPHY

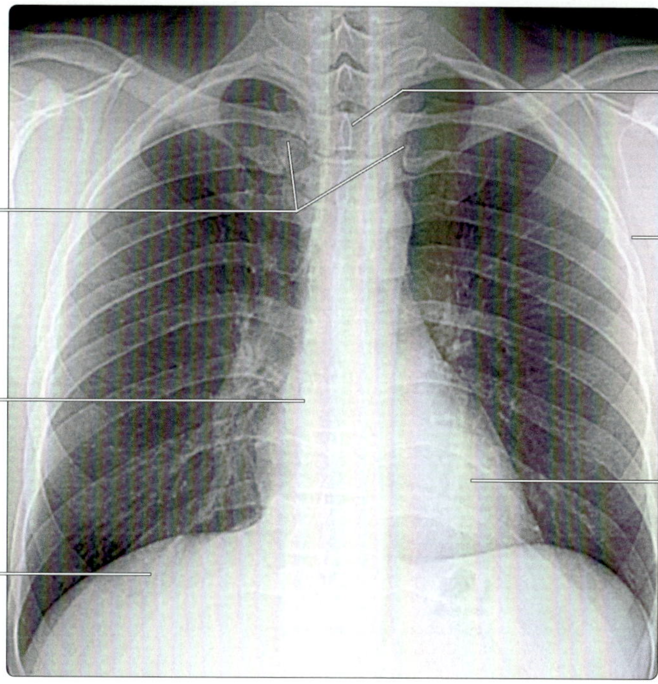

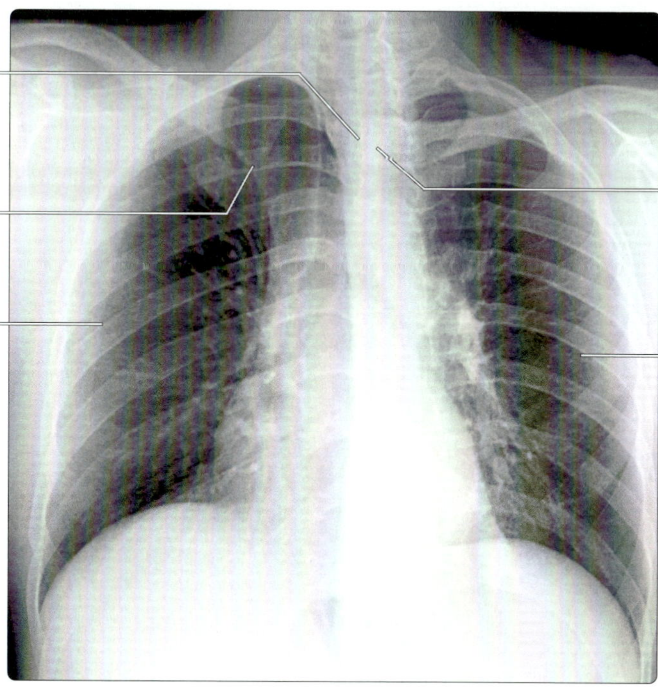

(Top) *Well-positioned, well-exposed normal PA chest radiograph shows that the scapulae are rotated off the lungs and the spinous process of T3 is equidistant from the bilateral medial clavicles. Proper collimation spans from the cervical trachea superiorly to below the bilateral costophrenic angles inferiorly and includes the lateral aspects of the chest wall. Optimal exposure allows visualization of the vertebral bodies (visible through the mediastinum) and peripheral pulmonary vessels, including those in retrocardiac and retrodiaphragmatic locations.* **(Bottom)** *Poorly positioned PA chest radiograph due to marked rotation to the right shows that the left medial clavicle overlies the spinous process of T3, and the right medial clavicle is displaced to the right of midline and projects over the right lung apex. A portion of the right scapula projects over the right lateral lung. Increased opacity of the left hemithorax results from x-ray penetration of a greater thickness of left-sided chest wall soft tissues, when compared to the right, due to rotation.*

LEFT LATERAL CHEST RADIOGRAPHY, POSITIONING AND COLLIMATION

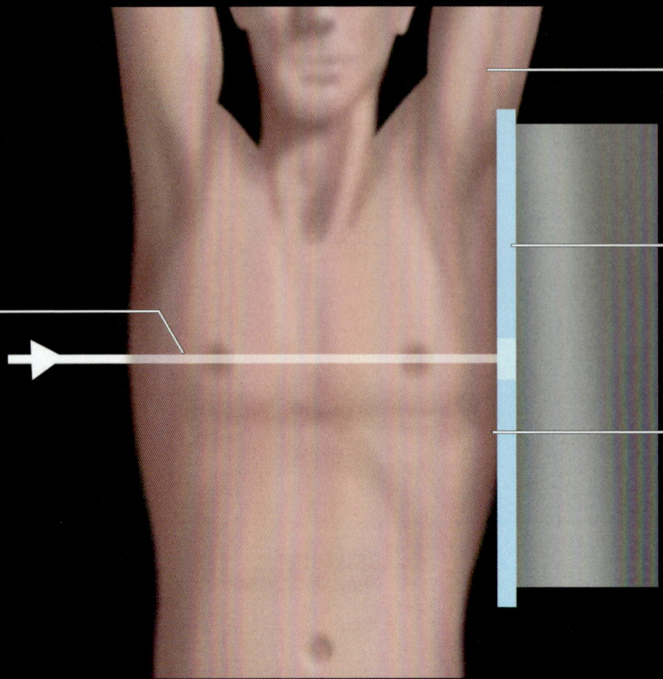

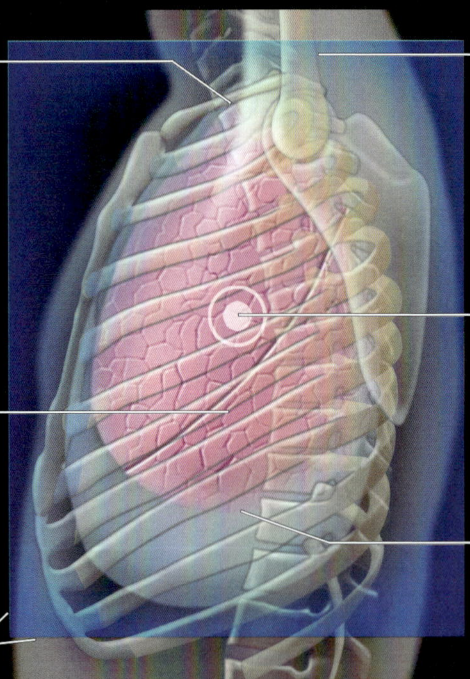

(Top) *Graphic shows proper positioning for left lateral chest radiography. The patient is upright with the left lateral chest placed against the vertically oriented image receptor. The patient's arms are extended upward to provide unobstructed visualization of the upper lung zones. The x-ray beam travels through the patient from right to left.* **(Bottom)** *Graphic shows proper left lateral chest radiographic collimation. The white target sign shows centering of the x-ray beam in the midportion of the thorax. The blue overlay represents the collimated x-ray beam that extends from the cervical airway superiorly to below the costophrenic angles inferiorly and includes the anterior and posterior skin surfaces. The structures of the left thorax (shown in color) are closest to the image receptor and experience the least degree of magnification.*

LATERAL CHEST RADIOGRAPHY

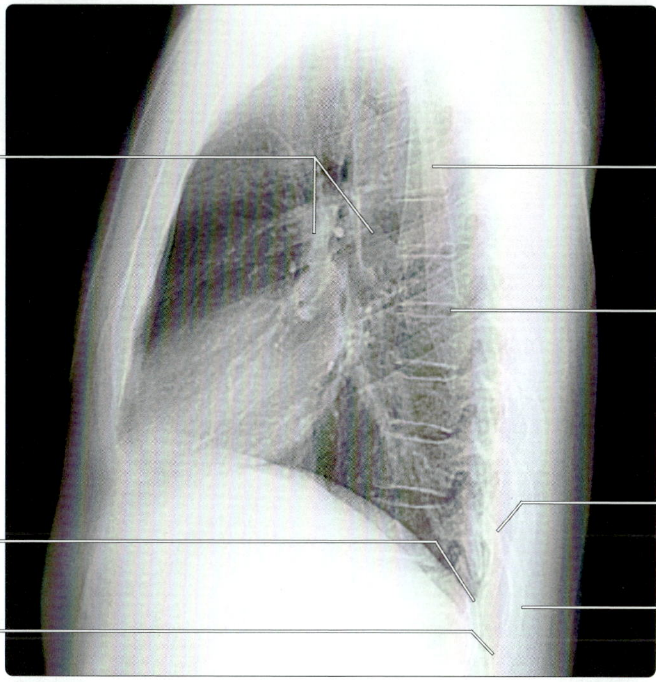

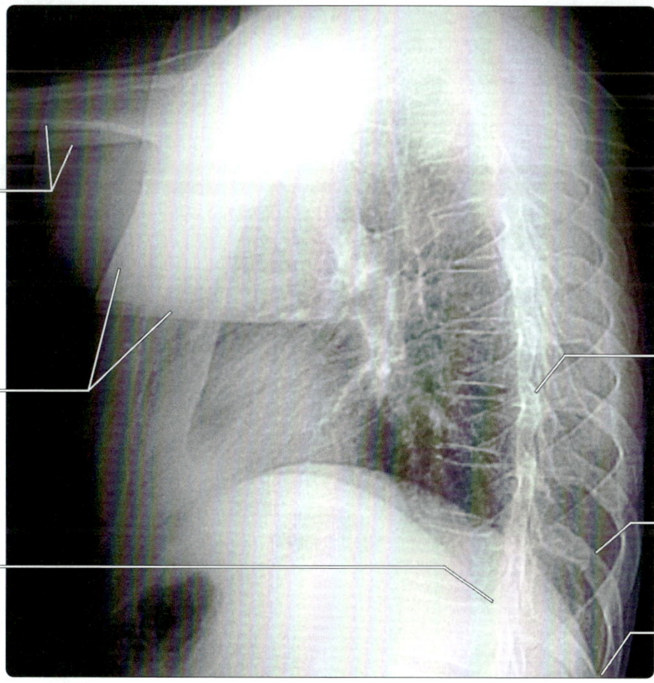

(Top) *Well-positioned, well-exposed normal left lateral chest radiograph shows that the upper extremities are not visible, the hila are centrally located in the thorax, the thoracic intervertebral disks are visible, and the posterior aspects of the ribs are nearly superimposed and project behind the thoracic vertebrae. There is minimal magnification of the left posterior ribs, which appear sharper and smaller compared to the right posterior ribs. Proper collimation allows inclusion of the lung apices, posterior costophrenic angles, and anterior and posterior skin surfaces.* **(Bottom)** *Poorly positioned left lateral chest radiograph shows that the skeletal and soft tissue structures of the bilateral upper extremities obscure the anterior aspects of the lungs and mediastinum. Rotation precludes superimposition of the posterior ribs. The right posterior ribs appear larger and project behind the left posterior ribs. The right costophrenic angle projects posterior to the left costophrenic angle.*

AP CHEST RADIOGRAPHY, POSITIONING AND COLLIMATION

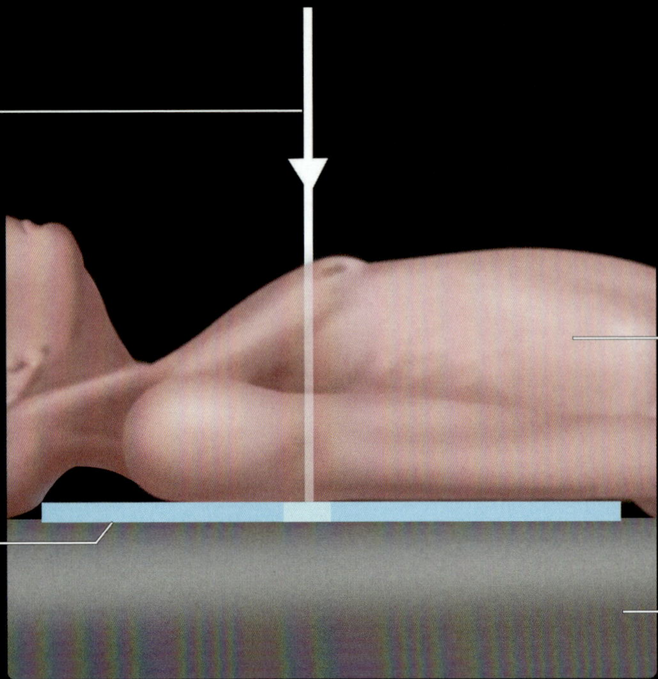

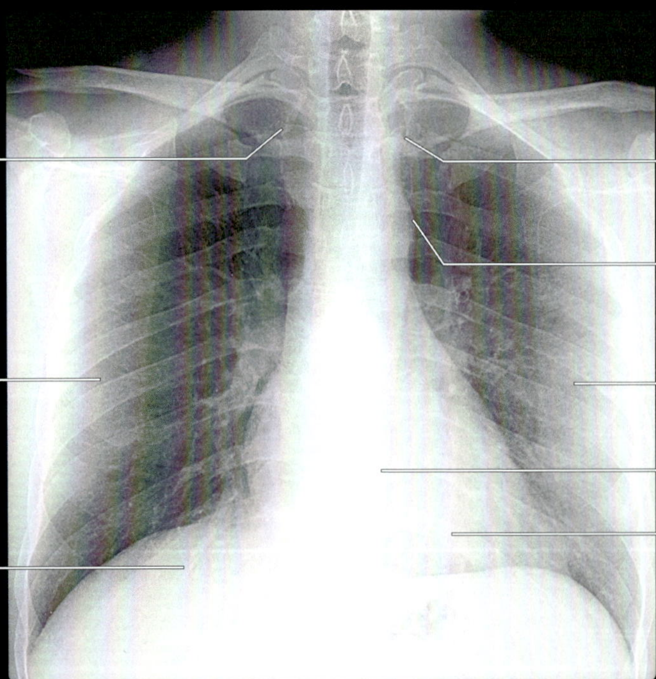

(Top) *Graphic shows proper positioning for supine anteroposterior (AP) chest radiography. The patient's back is against the horizontally oriented radiographic cassette, and the upper extremities are by the patient's sides. Internal rotation of the shoulders will minimize the degree of superimposition of the scapulae on the lateral aspects of the lungs. The x-ray beam travels through the patient in an AP direction. The heart and anterior chest structures are farthest from the cassette and experience some degree of magnification.* **(Bottom)** *Normal AP chest radiograph shows that the heart and great vessels appear mildly magnified. The clavicles show a horizontal course, and their medial portions overlie the lung apices. The medial scapulae project over the lateral aspects of the lungs. Note that exposure factors and collimation are optimal with visualization of retrocardiac and retrodiaphragmatic skeletal and vascular structures.*

PORTABLE AP CHEST RADIOGRAPHY, TRAUMA AND INTENSIVE CARE

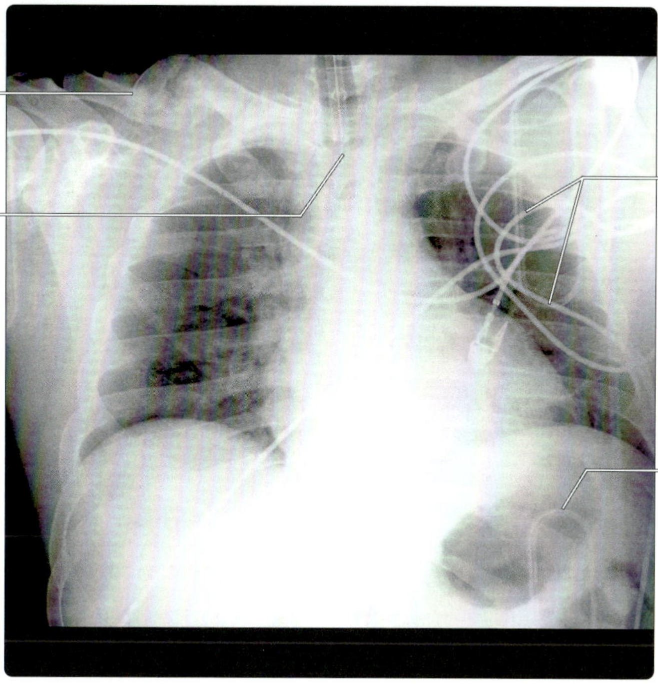

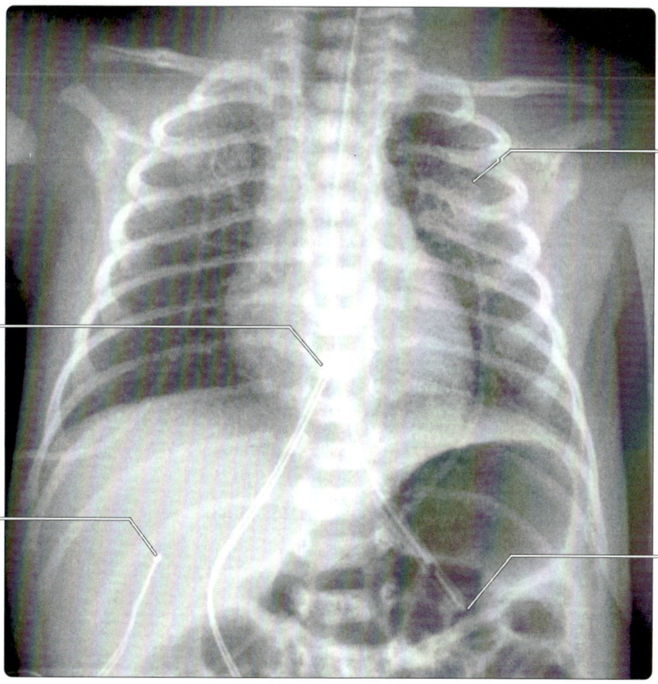

(Top) *Supine bedside (portable) AP chest radiograph of a patient involved in a motor vehicle collision shows features of AP chest radiography in trauma victims. Portable radiographs are used for imaging debilitated, seriously ill, and traumatized patients. AP chest radiographs in the setting of trauma are often compromised by technical factors related to overlying radiopaque monitoring and stabilizing devices. However, they allow a rapid assessment of the integrity of the thoracic structures and the position of life support devices.* **(Bottom)** *Supine AP chest radiograph shows a 1-day-old infant born at 31-weeks gestation with mild respiratory distress syndrome. Portable radiography is optimal for imaging neonates and infants, particularly those who are seriously ill due to congenital abnormalities &/or prematurity. Portable chest radiography allows rapid assessment of life support devices for proper positioning as well as evaluation of the lungs, cardiothymic silhouette, pleural spaces, thoracic skeleton, and upper abdomen.*

PA AND AP CHEST RADIOGRAPHY

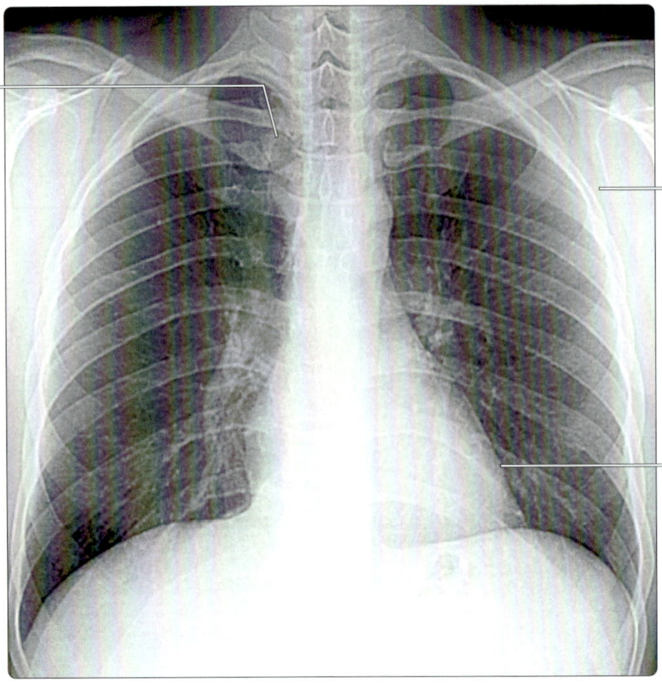

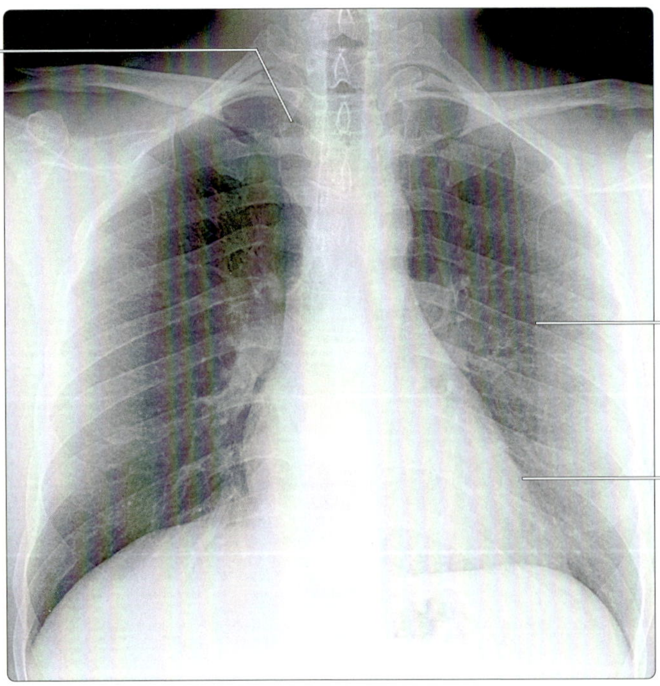

(Top) *First of 2 normal chest radiographs of the same patient is a PA chest radiograph showing that the heart and mediastinum, which are closest to the image receptor, undergo the least degree of magnification. The medial clavicles curve inferiorly and do not overlie the upper lung apices. The scapulae are rotated laterally and do not obscure the lateral aspects of the lungs.* **(Bottom)** *AP chest radiograph of the same patient shows that the heart and mediastinum appear slightly larger, as they are farthest from the image receptor and undergo some degree of magnification. The clavicles exhibit a horizontal course, and their medial aspects overlie the upper lung apices. The medial portions of the scapulae overlie the lateral aspects of the lungs.*

INSPIRATORY AND EXPIRATORY CHEST RADIOGRAPHY

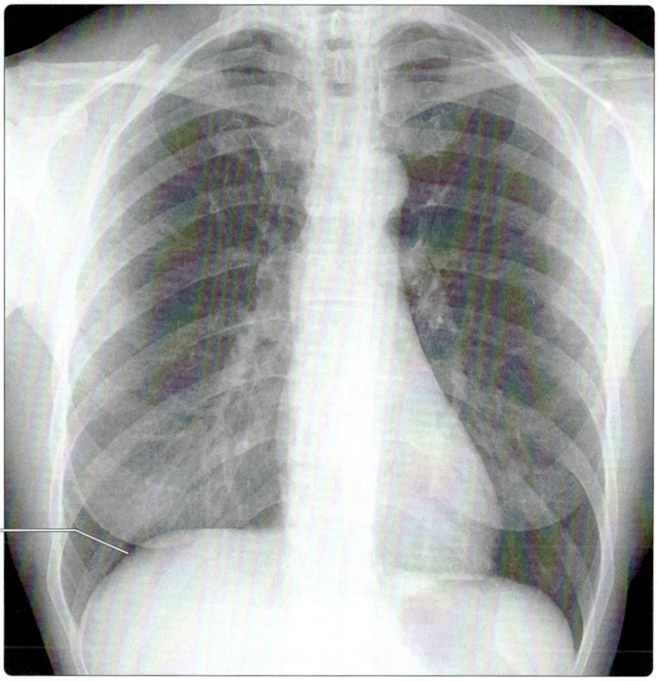

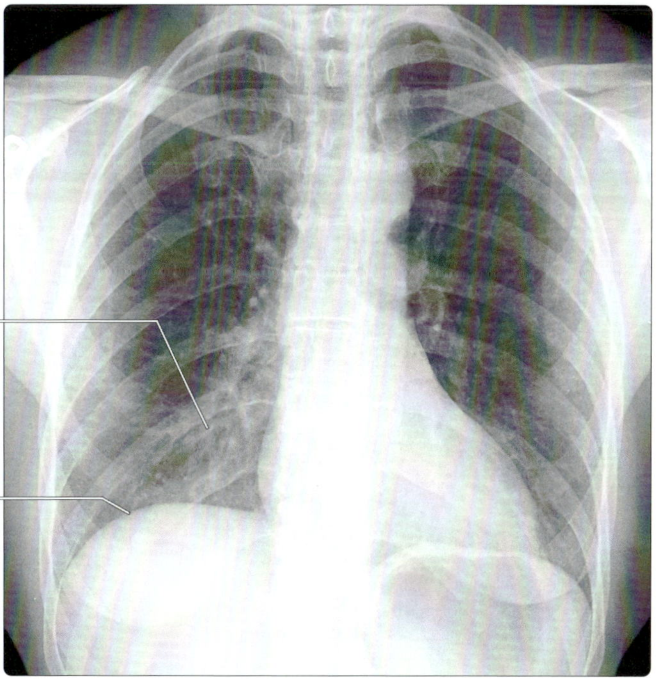

(Top) *First of 2 normal chest radiographs of the same patient is a PA chest radiograph obtained at full inspiration showing optimal visualization of the lung bases and the retrocardiac and retrodiaphragmatic lung. Note that the right 7th anterior rib intersects the right hemidiaphragm at the midaxillary line, consistent with normal lung volume at full inspiration.* **(Bottom)** *PA chest radiograph of the same patient obtained at end expiration shows decreased lung volume and mild diaphragmatic elevation. Note that the right 5th anterior rib now intersects the right hemidiaphragm at the midaxillary line. Note mildly increased pulmonary opacity in the right lung base due to expiratory imaging, which may be misinterpreted as early airspace disease.*

LATERAL DECUBITUS CHEST RADIOGRAPHY, POSITIONING AND COLLIMATION

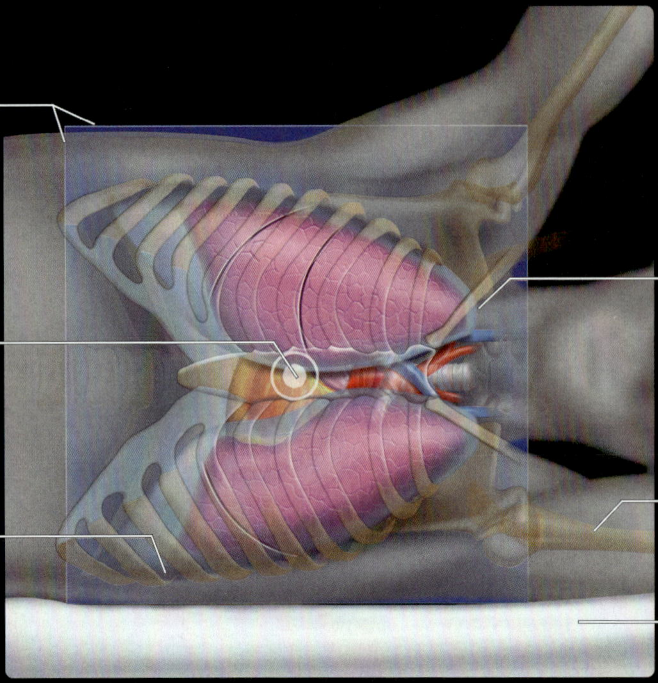

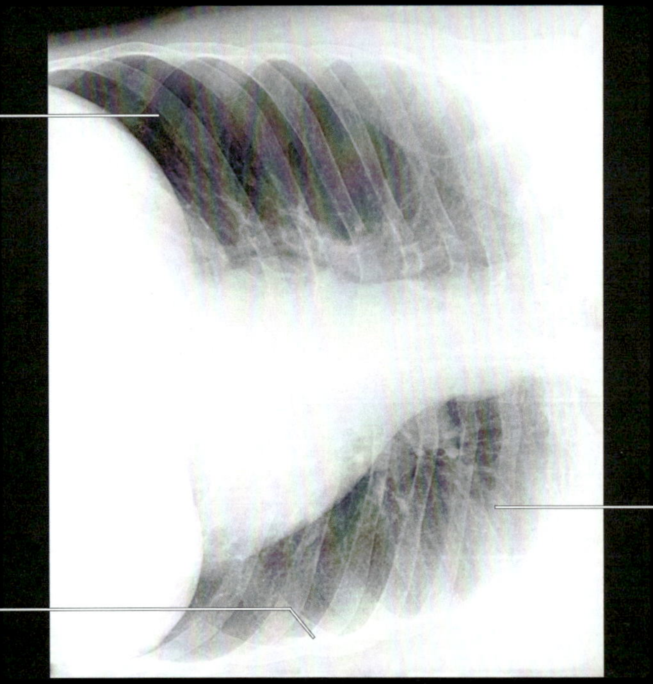

(Top) *Graphic shows proper lateral decubitus PA chest radiographic collimation. The white target sign shows the centering of the x-ray beam in the midthorax. The blue overlay represents the collimated x-ray beam that extends from the cervical airway superiorly to below the costophrenic angles inferiorly and includes the left and right skin surfaces. The thorax is elevated on a radiolucent pad to ensure inclusion of the dependent pleura and chest wall. The anterior structures of the chest (shown in color) are closest to the image receptor and experience the least magnification.* **(Bottom)** *Normal left lateral decubitus radiograph shows a larger lung volume in the nondependent right lung and volume loss manifesting as increased density in the dependent left lung. There is good visualization of the dependent left lung and pleura and the nondependent right lung and pleura. Decubitus radiography is helpful in the assessment of pleural disease and lung volume abnormalities.*

APICAL LORDOTIC CHEST RADIOGRAPHY, POSITIONING

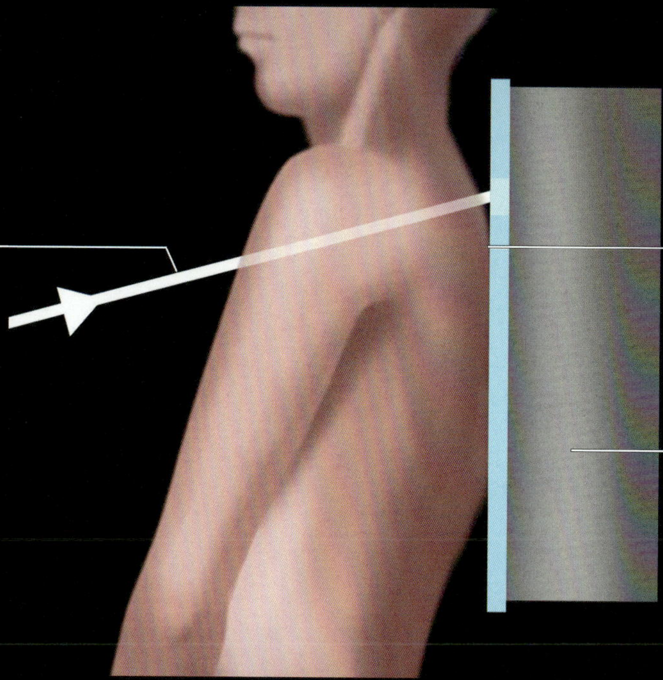

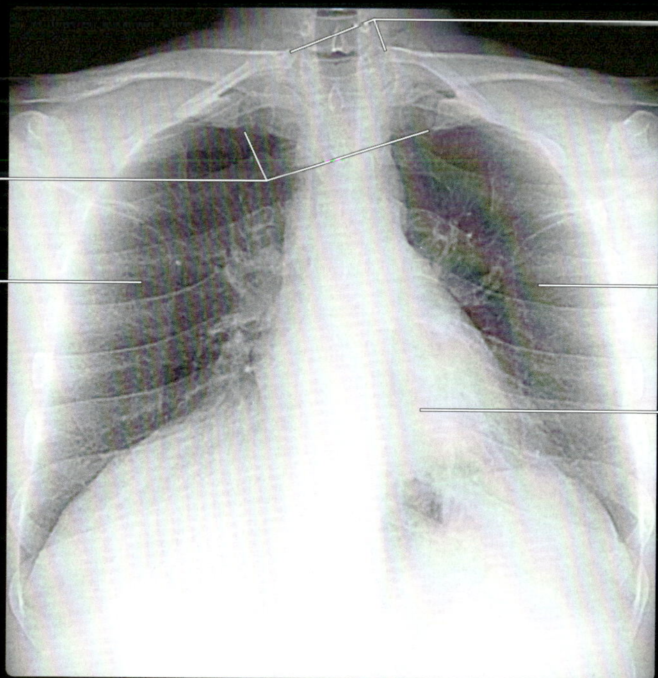

(Top) *Graphic shows proper positioning for AP apical lordotic chest radiography. The patient is upright with the posterior shoulders placed against the vertically oriented image receptor, and the arms are internally rotated to move the scapulae away from the lungs. The x-ray beam travels through the patient from anterior to posterior, is centered at the sternal manubrium, and is oriented superiorly at a 20° angle from the horizontal plane.* **(Bottom)** *Normal apical lordotic chest radiograph projects the medial aspects of the clavicles above the lung apices. Note that the apices are partly obscured by the anterior aspects of the 1st ribs and their costochondral articulations in this case. The mediastinum is foreshortened and mildly magnified. The scapulae overlie a significant portion of the bilateral lateral lungs.*

RADIOGRAPHIC DENSITIES

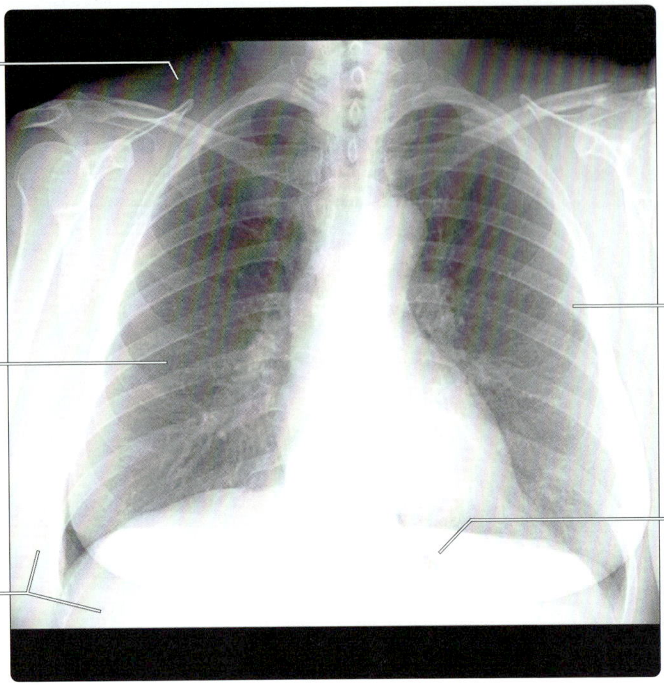

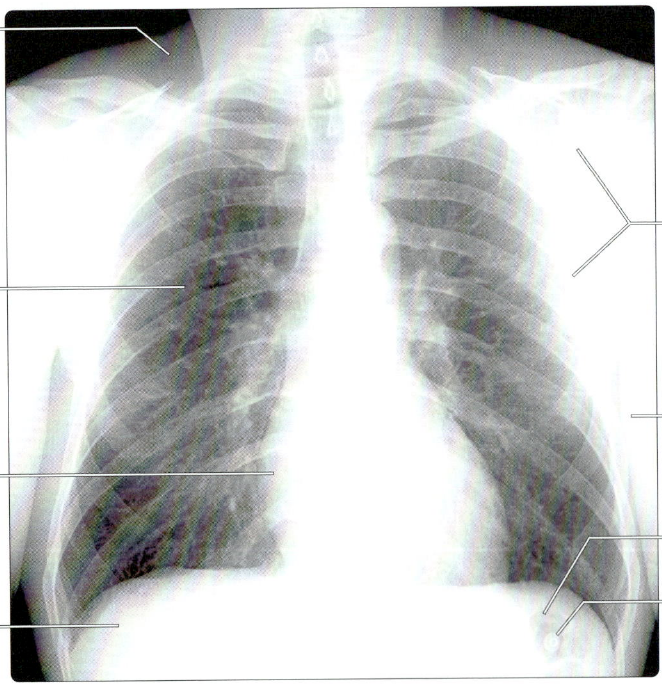

(Top) *PA chest radiograph shows the 4 radiographic densities. Air is present in the lungs bilaterally and within the gastric air bubble. Water (or soft tissue) density is seen in the mediastinum, abdomen, and subcutaneous tissues. Fat density is visible in the subcutaneous fat of the supraclavicular regions. Calcium or mineral density is noted in the skeletal structures.* **(Bottom)** *PA chest radiograph shows the 4 radiographic densities. Air density is present in the lungs and in bowel gas. Water density is seen in the mediastinum, abdominal viscera, and subcutaneous soft tissues. Fat is more difficult to demonstrate in this thin patient but is present in the normal subcutaneous supraclavicular regions. Calcium or mineral density is represented by the skeletal structures. Metal, a 5th radiographic density, is represented by an overlying metallic snap on the patient's gown.*

RADIOGRAPHIC DENSITIES, MEDICAL DEVICES

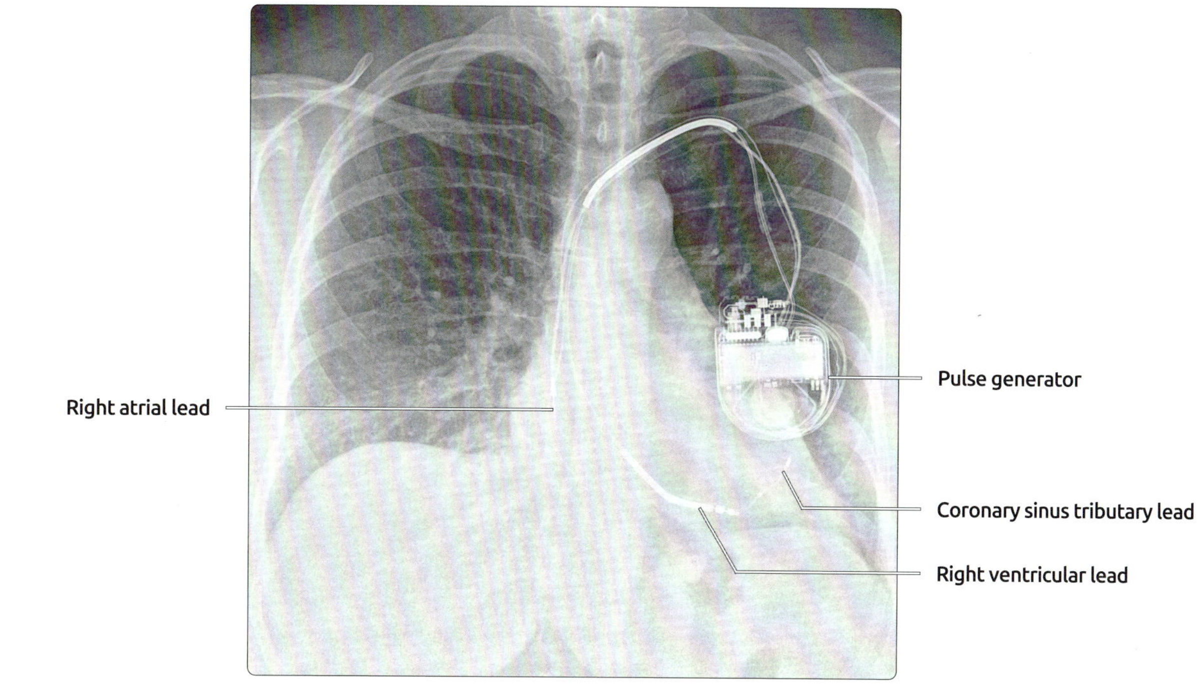

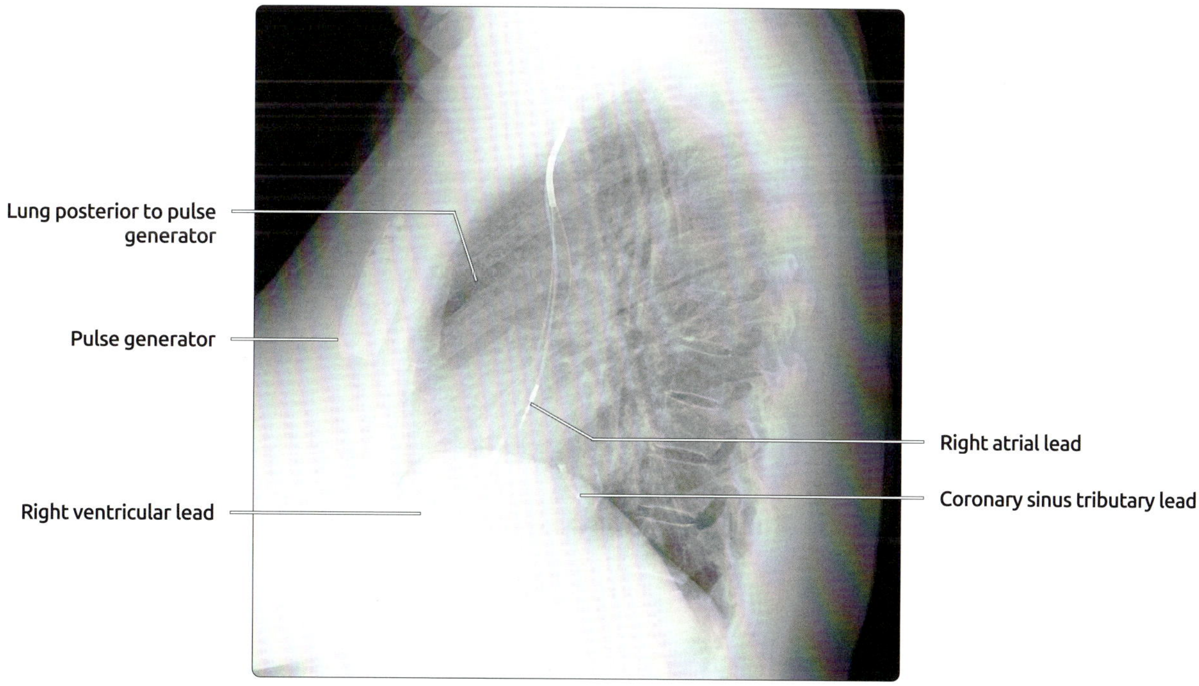

(Top) *PA chest radiograph of an asymptomatic patient shows a left pectoral dual-chamber biventricular transvenous pacer/implantable cardioverter defibrillator (ICD) with lead tips in the right atrium, right ventricle, and a coronary sinus tributary. The metallic portions of the pulse generator obscure visualization of the left midlung zone.* **(Bottom)** *Lateral chest radiograph of the same patient shows atrial and ventricular leads as well as a 3rd lead in a tributary of the coronary sinus. The left lung posterior to the pulse generator is now visible, although superimposed on the right lung. Metallic portions of medical devices enable radiographic assessment for proper placement and positioning.*

ANATOMIC LOCALIZATION WITH ORTHOGONAL RADIOGRAPHY

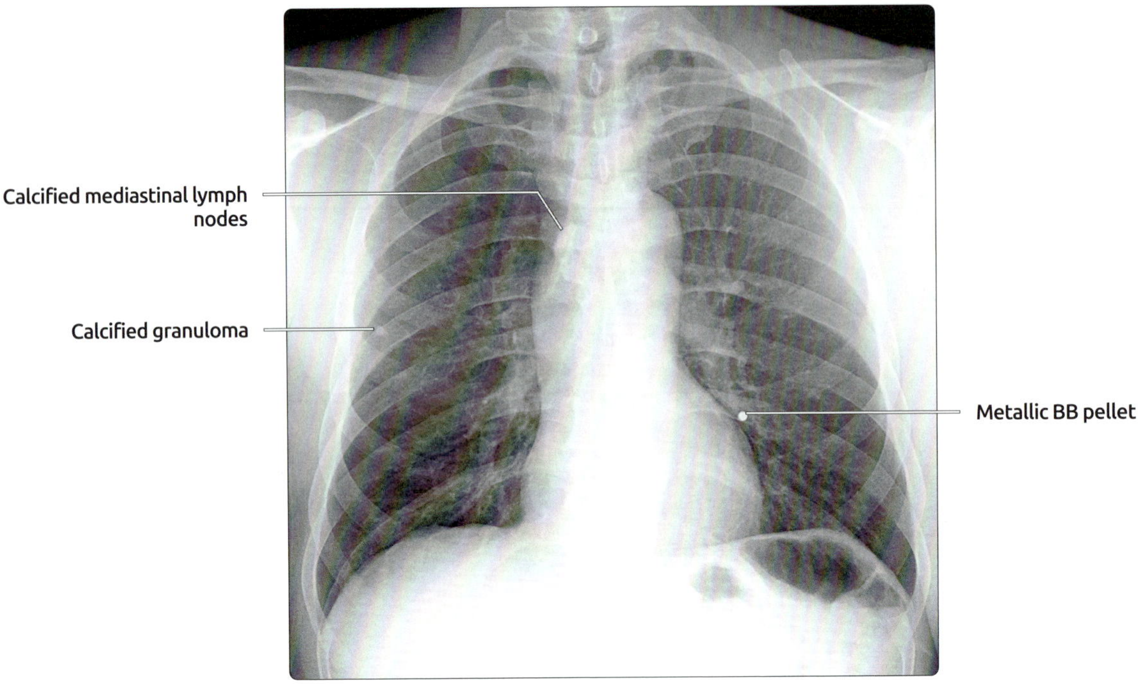

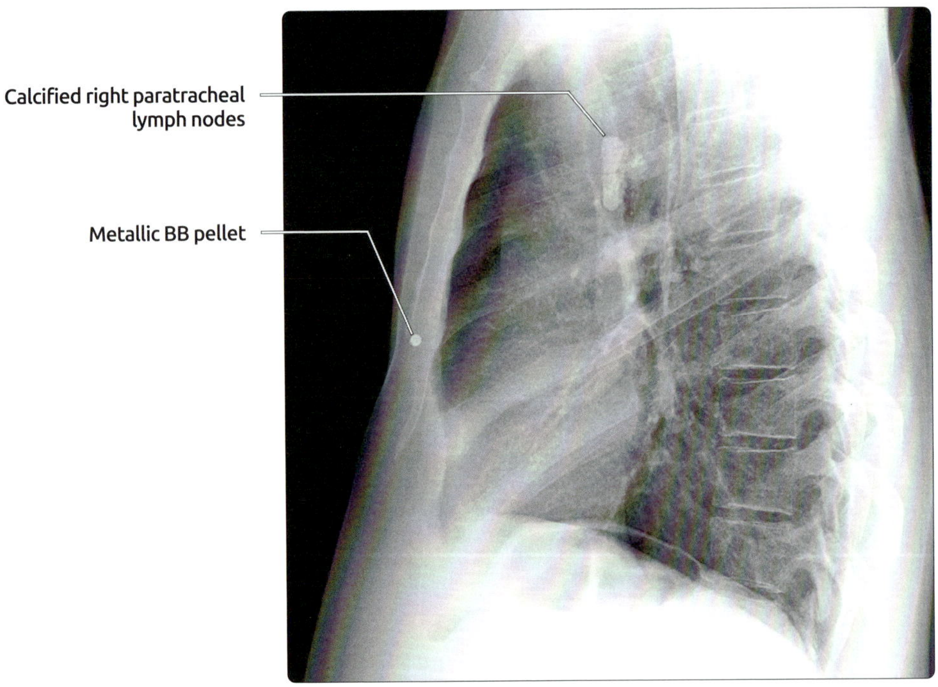

(Top) *First of 2 chest radiographs of an asymptomatic 57-year-old man is shown. PA chest radiograph shows a BB pellet projecting over the left lower lung zone adjacent to the left cardiac border. Note the calcified right midlung zone granuloma and the calcified right upper mediastinal lymph nodes. Note the difference in radiographic density between the metallic BB pellet and the calcified right lung granuloma.* **(Bottom)** *Lateral chest radiograph (orthogonal to the PA chest radiograph) shows that the BB pellet projects over the lower sternal body and anterior to the lung, thus allowing accurate anatomic localization of the BB pellet to the left anterior chest wall. The small granuloma is poorly visualized, but the calcified lymph nodes can be localized to the paratracheal region. PA and lateral chest radiography allows anatomic localization of imaging abnormalities.*

SILHOUETTE SIGN, LEFT LOWER LOBE AIRSPACE DISEASE

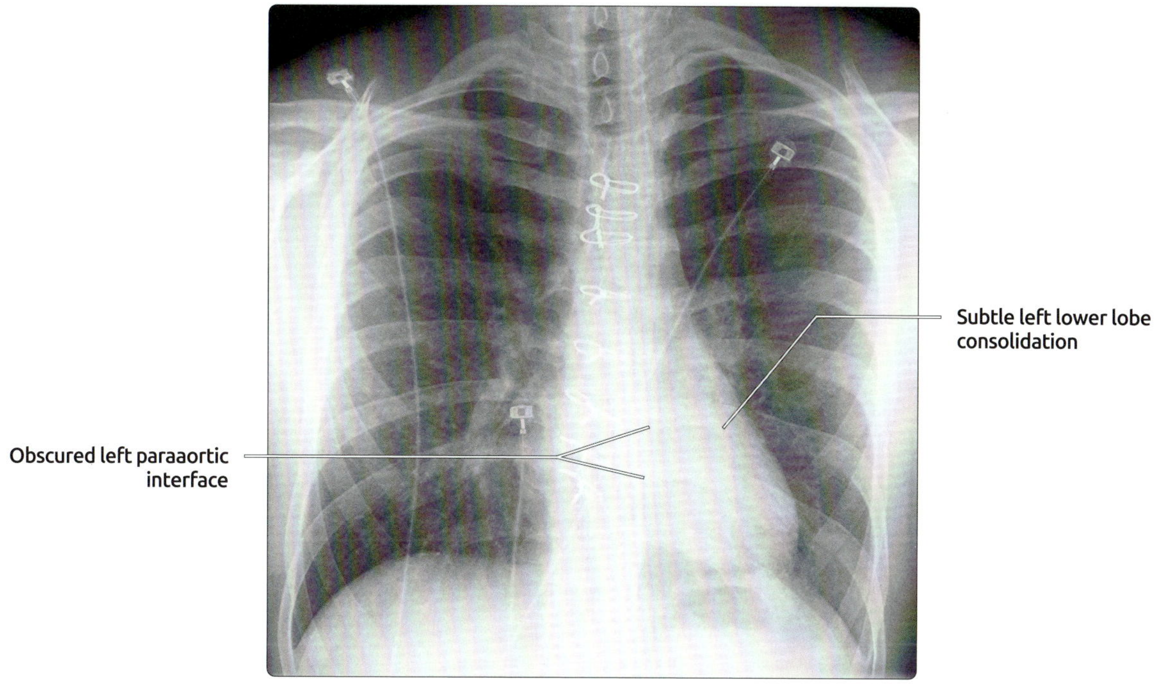

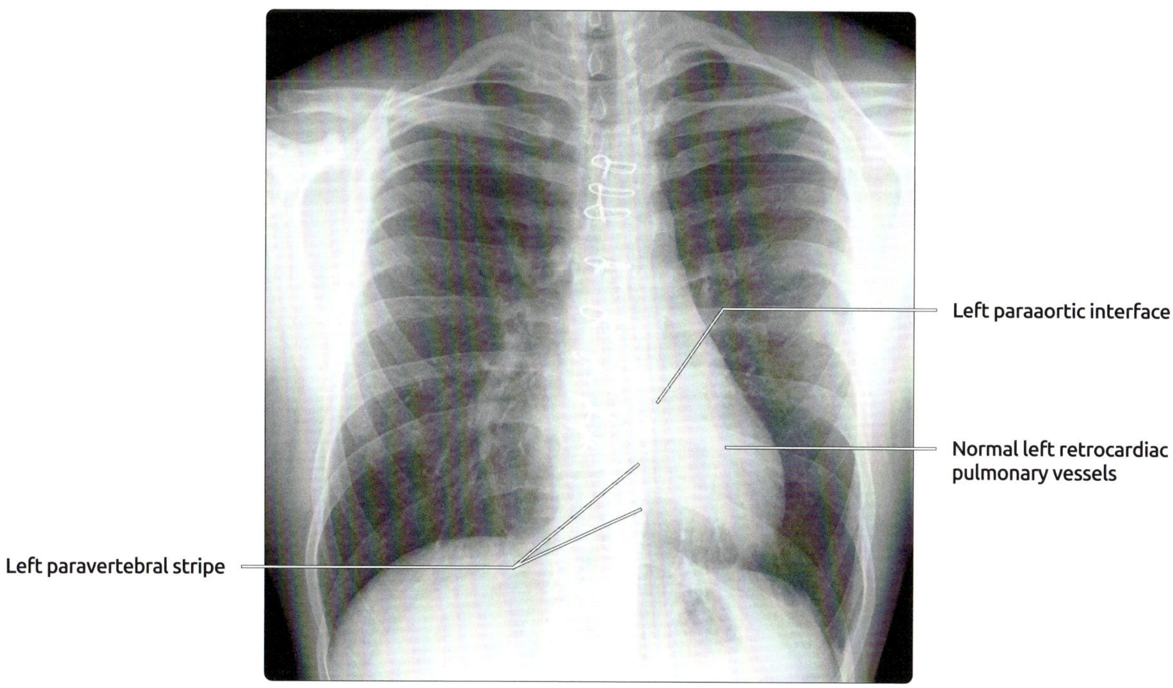

(Top) *First of 2 PA chest radiographs of a 20-year-old man who presented with cough and fever shows a subtle left retrocardiac opacity that obscures the left retrocardiac pulmonary vasculature and portions of the paraaortic interface and the left paravertebral stripe, consistent with consolidation secondary to left lower lobe pneumonia.* **(Bottom)** *Comparison PA chest radiograph of the same patient obtained 2 months earlier shows a normal appearance of the left lower lobe with visualization of the left retrocardiac pulmonary vasculature, the left paraaortic interface, and the left paravertebral stripe. This case illustrates the value of understanding the normal radiographic densities and the normal mediastinal stripes and interfaces for recognition of the silhouette sign as an indicator of pulmonary pathology and the importance of comparison with prior studies for the diagnosis of sometimes subtle radiographic abnormalities.*

SILHOUETTE SIGN, LEFT LOWER LOBE AIRSPACE DISEASE

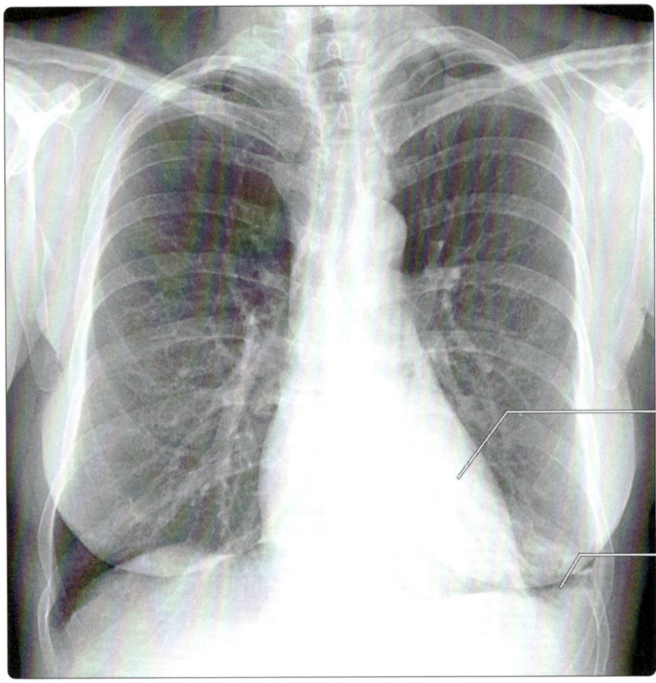

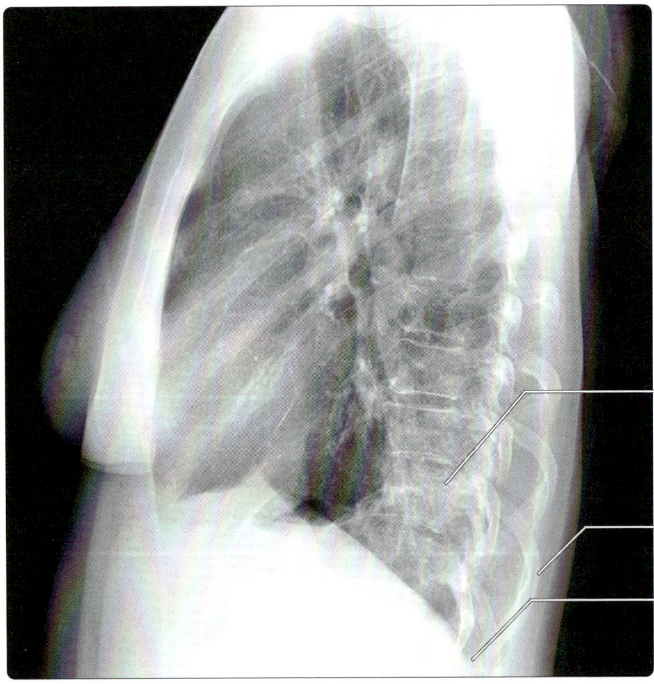

(Top) *First of 2 chest radiographs of a 45-year-old woman with left lower lobe pneumonia is shown. PA chest radiograph shows a subtle retrocardiac left basilar airspace opacity that obscures the lateral aspect of the left hemidiaphragm. The alveolar air in the left lower lobe has been replaced by an inflammatory process of water density that produces the silhouette sign.* **(Bottom)** *Lateral chest radiograph of the same patient shows that the consolidation is located in the posterior basilar segment of the left lower lobe, produces the radiographic spine sign (increasing opacity over the lower thoracic spine), and obscures the posterior left hemidiaphragm. The right hemidiaphragm is visualized in its entirety, as there is no right lower lobe airspace disease. The left lateral chest radiograph facilitates confident identification of the right ribs (and the normal ipsilateral right hemidiaphragm) by virtue of their magnification, as they are farthest from the image receptor.*

SILHOUETTE SIGN, MIDDLE LOBE AIRSPACE DISEASE

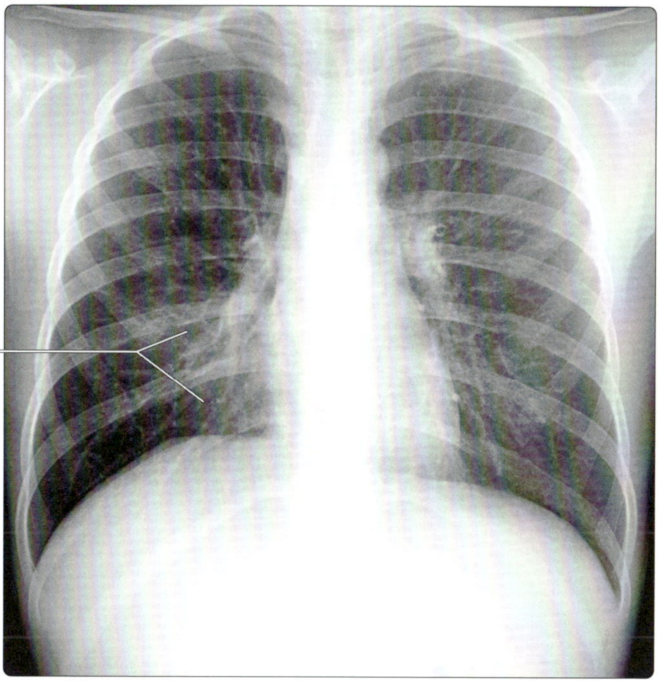

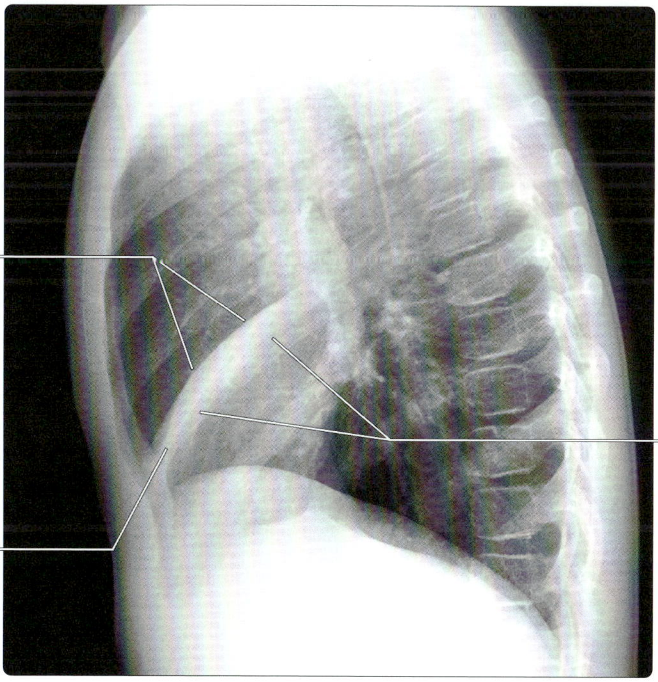

(Top) *First of 2 chest radiographs of a 10-year-old asthmatic boy with middle lobe atelectasis is shown. PA chest radiograph shows airspace opacity in the medial aspect of the right lower lung zone, which obscures the right cardiac border. The location of the process can be inferred by noting obscuration of the right cardiac border while the adjacent medial right hemidiaphragm is well visualized. Atelectasis has resulted in evacuation of alveolar air from the middle lobe producing the silhouette sign.* **(Bottom)** *Lateral chest radiograph shows a band-like opacity that projects over the heart and represents the atelectatic middle lobe. Posteroinferior displacement of the minor fissure and anterosuperior displacement of the inferior aspect of the right major fissure are typical of middle lobe volume loss and help distinguish atelectasis from consolidation.*

DECUBITUS CHEST RADIOGRAPHY, EVALUATION OF PLEURAL DISEASE

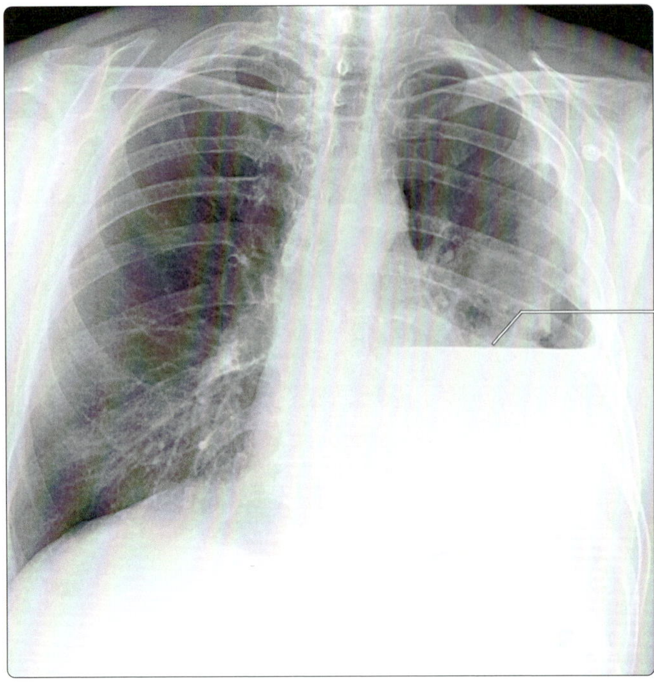

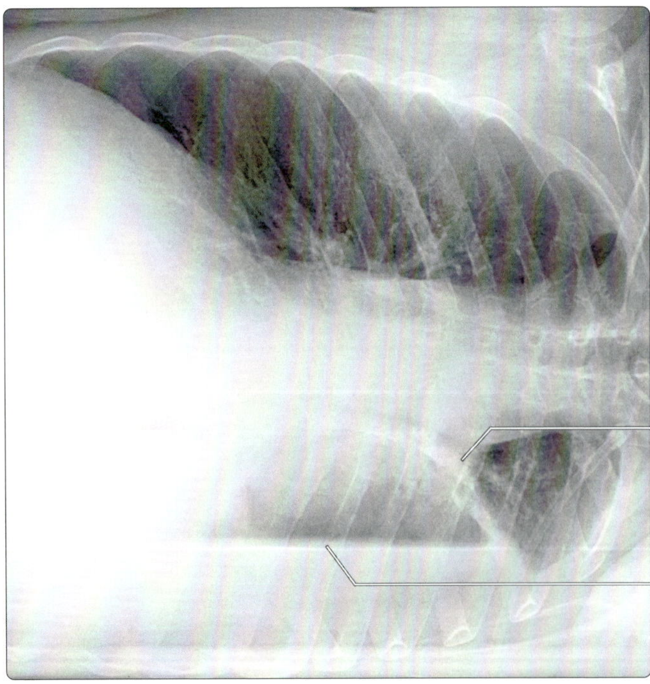

(Top) *First of 2 chest radiographs of a patient with a left empyema and a bronchopleural fistula is shown. The PA chest radiograph shows a large air-fluid level that spans the width of the left hemithorax.* **(Bottom)** *Left lateral decubitus chest radiograph of the same patient shows a discrepant length of the air-fluid level, which appears longer than on the PA chest radiograph, correlating with the elongate shape of the air and fluid collection and allowing its anatomic localization to the left pleural space. Note the thick medial wall of the air and fluid collection. The findings are characteristic of a loculated pleural effusion. The presence of air indicates a communication with the tracheobronchial tree (bronchopleural fistula), and the findings are diagnostic of a complicated empyema. In this case, the findings on the PA and lateral decubitus radiographs allow localization of the abnormality to the pleural space and differentiation from lung parenchymal disease.*

LORDOTIC CHEST RADIOGRAPHY, EVALUATION OF APICAL LESION

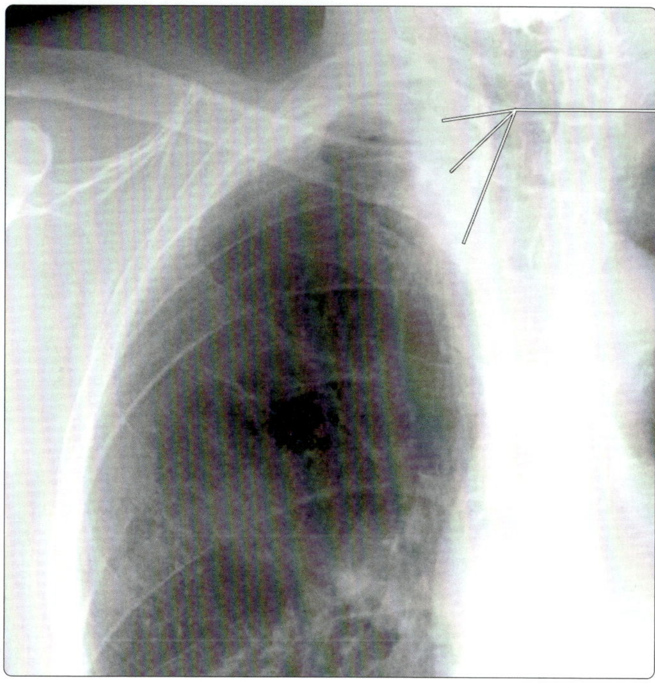

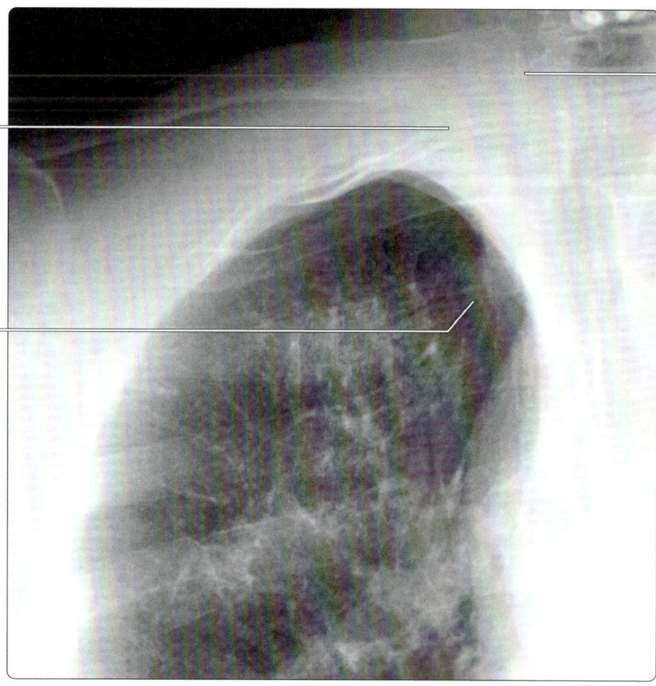

(Top) *First of 2 chest radiographs of a patient with a right apical mass is shown. PA chest radiograph coned down to the right apex demonstrates an abnormal irregular apical soft tissue mass and thickening of the medial aspect of the right apical pleura.* **(Bottom)** *AP apical lordotic chest radiograph coned down to the right upper lobe allows visualization of the medial aspect of the right apical lung by projecting the right medial clavicle and the right 1st anterior rib above the lung apex. The spiculated lateral border of this right apical lung cancer is now visible.*

CROSS-SECTIONAL ANATOMY

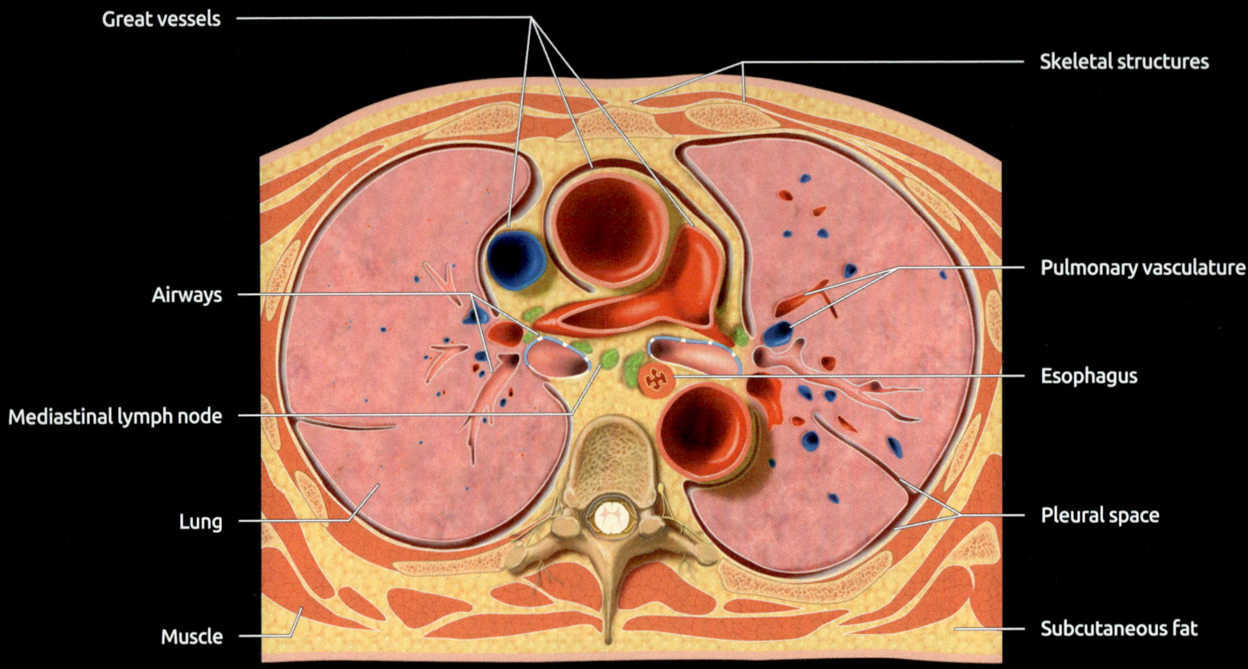

Graphic shows the cross-sectional appearance of the midthorax and illustrates the manner of visualization and assessment of various organs and tissues in cross section. Cross-sectional imaging allows assessment of the organs, structures, and tissues of the thorax. The soft tissues of the chest wall consist of skin, subcutaneous fat, and chest wall muscles. Together with the skeletal structures, the soft tissues of the chest wall surround and protect the thoracic cavity and its internal organs and tissues. The apposed pleural surfaces form the potential pleural space. The pulmonary arteries and veins course through the lungs. The mediastinal fat, mediastinal vascular structures, esophagus, central tracheobronchial tree, and lymph nodes are also depicted.

AXIAL CHEST CT, WINDOW AND LEVEL SETTINGS

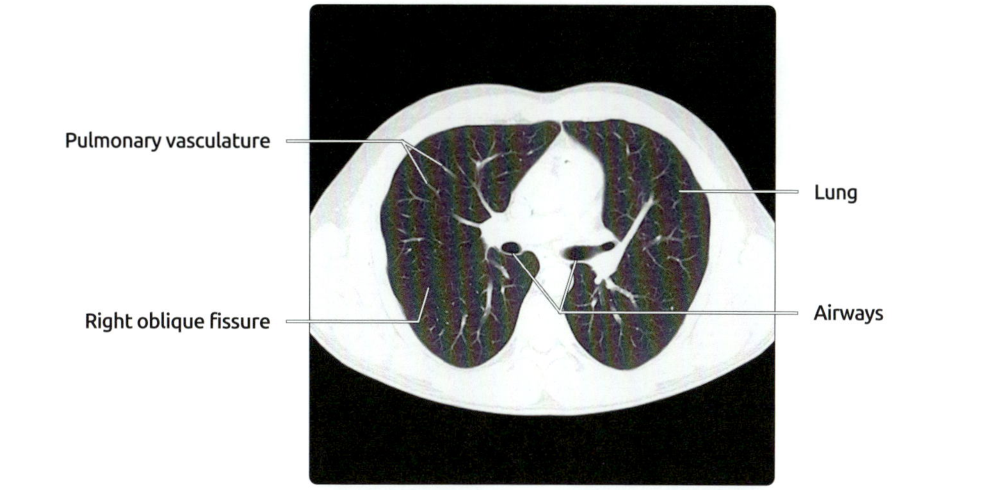

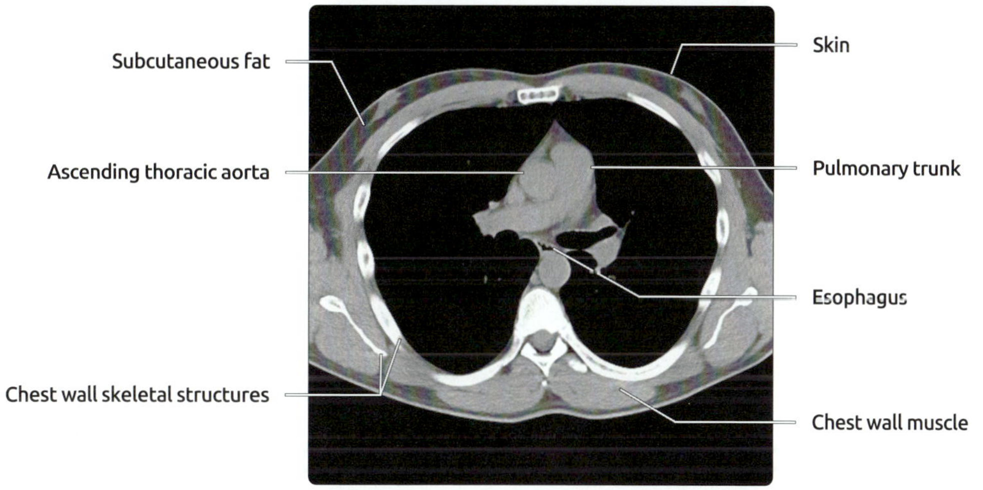

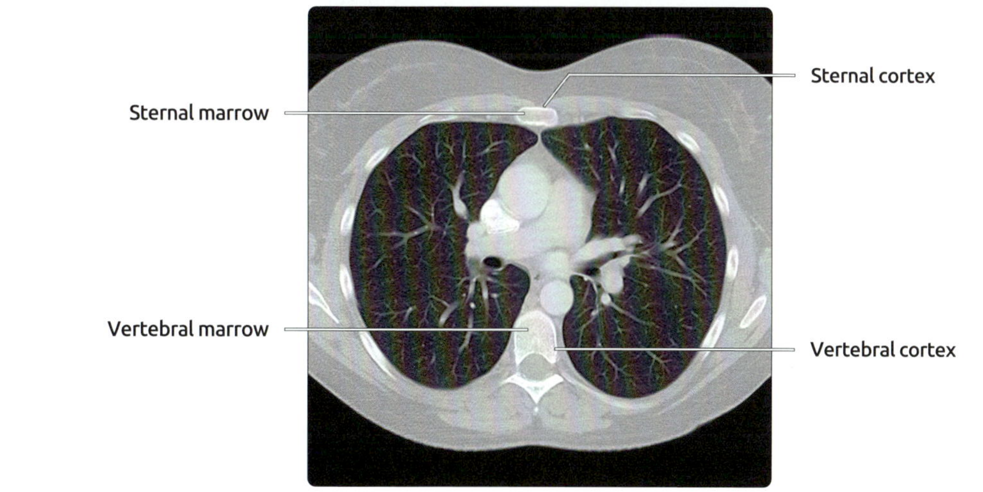

(Top) *Axial CECT (lung window) allows evaluation of the lungs, pulmonary vasculature, central tracheobronchial tree, and pleural surfaces. The mediastinum and chest wall are poorly evaluated in this window setting.* **(Middle)** *Axial NECT (mediastinal window) allows evaluation of the soft tissue structures of the mediastinum and the soft tissues and skeletal structures of the chest wall. The pulmonary parenchyma, pleura, and central tracheobronchial tree are not well evaluated.* **(Bottom)** *Axial CECT (bone window) allows assessment of the skeletal structures with visualization of their cortices and marrow spaces. Note improved skeletal visualization when compared to the soft tissue window settings used on the prior image. This window setting is also useful for evaluation of calcifications and metallic medical devices.*

CHEST NECT AND CECT

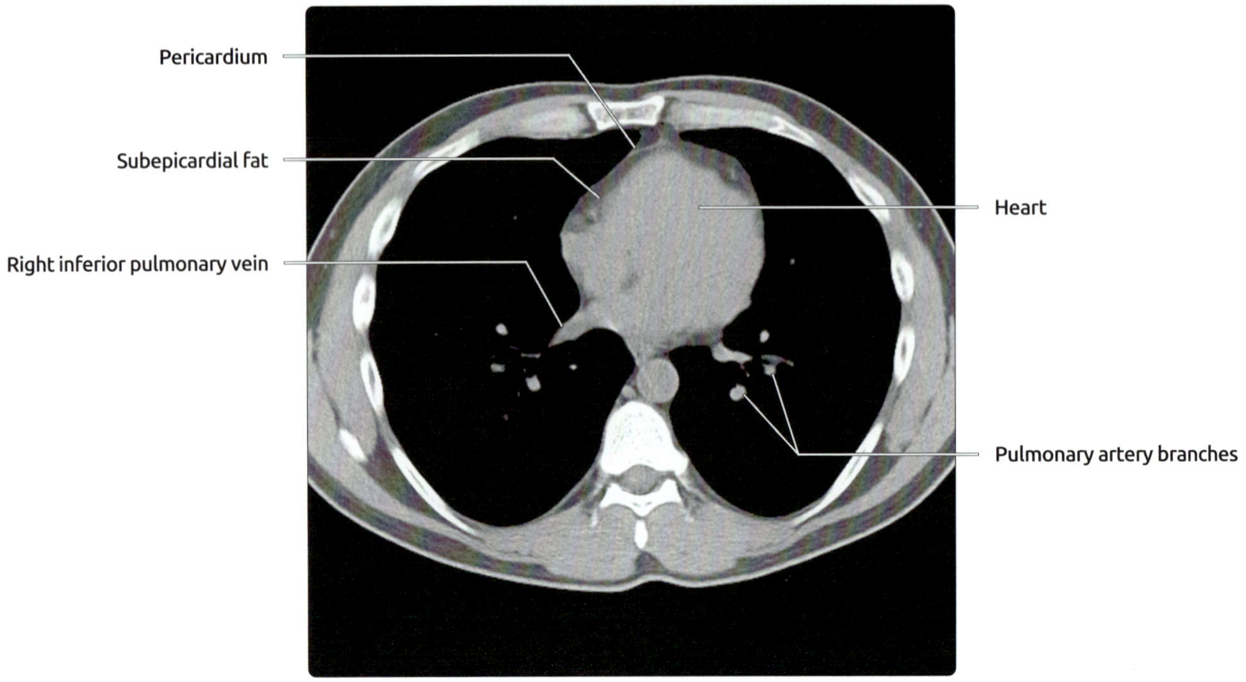

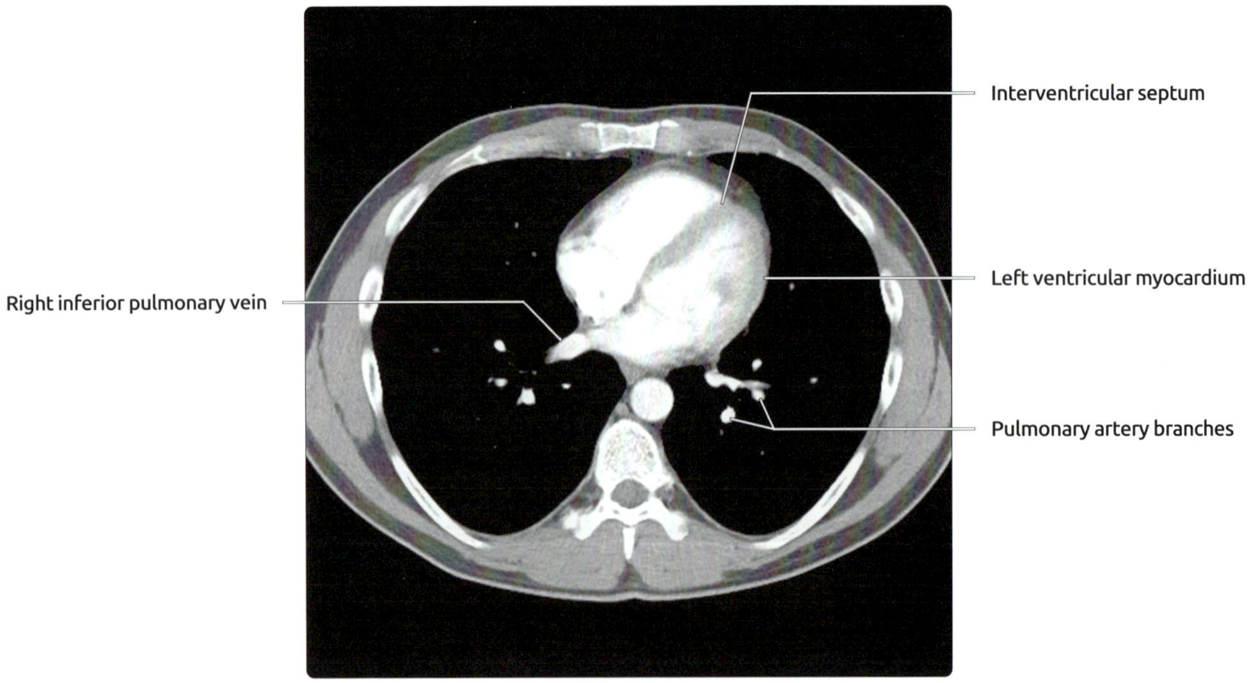

(Top) *First of 2 normal chest CT images through the heart of the same patient is shown. Axial NECT (mediastinal window) shows the heart, pericardium, and intervening subepicardial fat. The inferior pulmonary veins are visible bilaterally. Note that individual cardiac chambers cannot be completely resolved without the use of intravenous contrast.* **(Bottom)** *Axial CECT (mediastinal window) shows the excellent visualization of the inferior pulmonary veins, pulmonary arteries, and cardiac chambers afforded by the use of intravenous contrast. The interventricular septum and the left ventricular myocardium are well demonstrated against the contrast-filled cardiac chambers.*

CONVENTIONAL CT AND HRCT

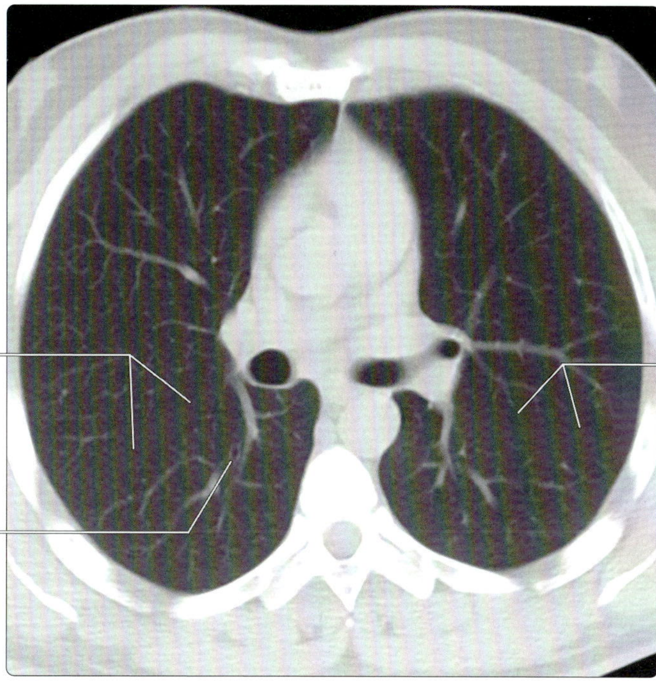

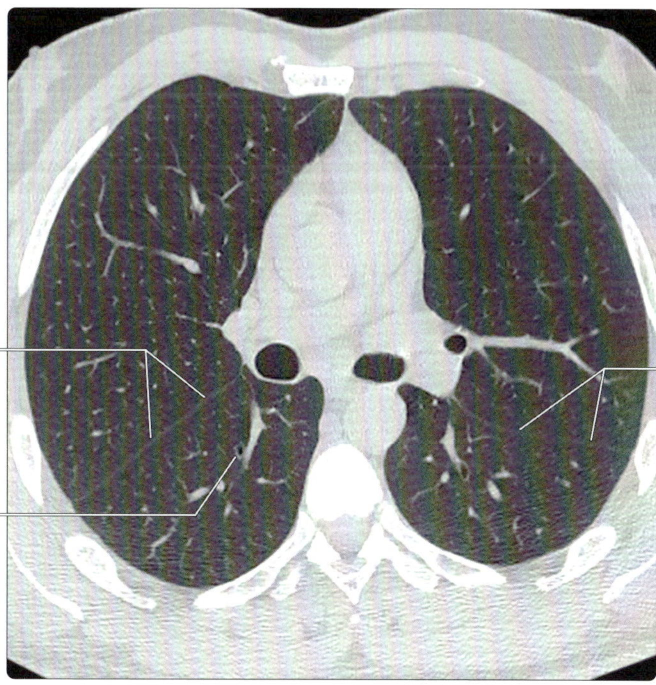

(Top) *First of 2 images of a normal CECT is shown. Conventional axial CECT (lung window) with 5-mm slice thickness shows adequate visualization of the lung parenchyma, pulmonary vasculature, and tracheobronchial tree. The major fissures are visible as avascular bands coursing obliquely through the lungs.* **(Bottom)** *Axial HRCT (lung window) with 1.25-mm slice thickness at the same level as the previous image shows markedly improved visualization of pulmonary detail. The major fissures now manifest as distinct thin soft tissue attenuation lines. There is improved visualization of the airway walls and sharper outlines of the pulmonary vessels.*

SUPINE AND PRONE HRCT

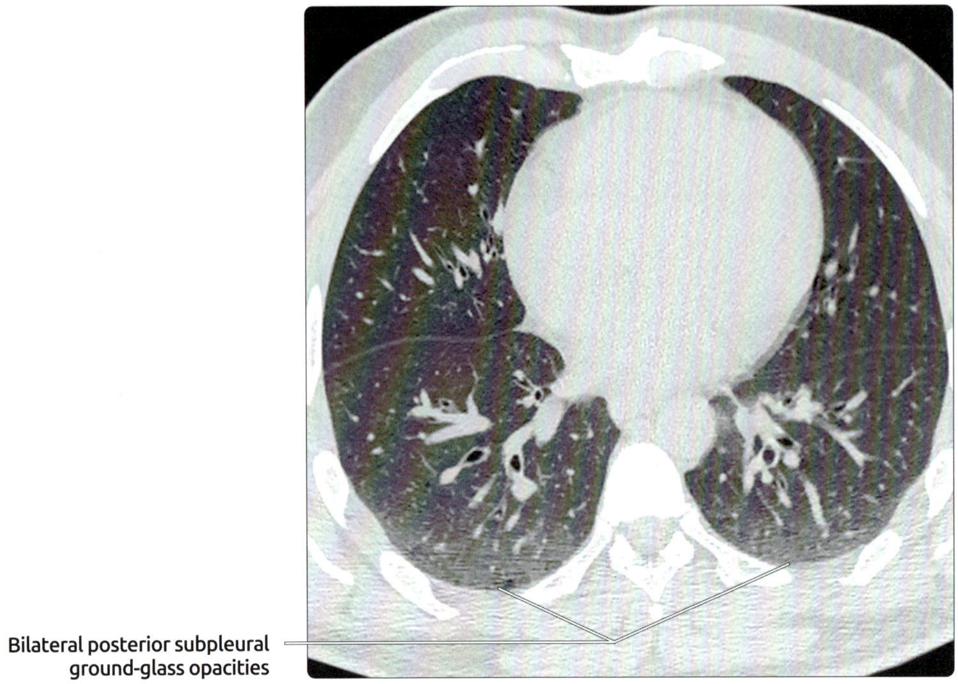

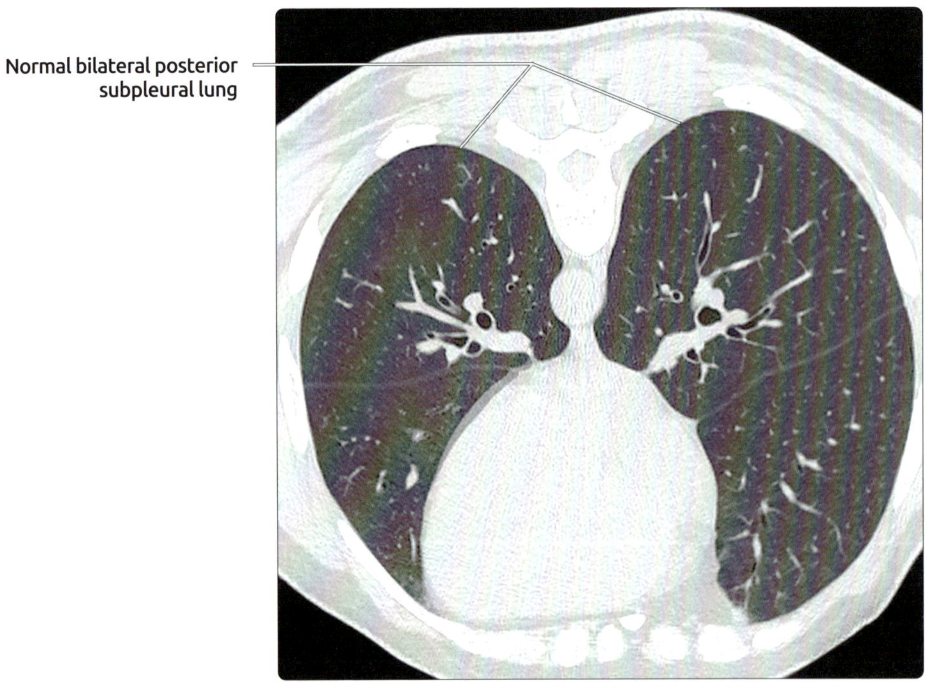

(Top) *First of 2 normal high-resolution CT images of a patient with mild dyspnea is shown. Supine axial HRCT (lung window) shows bilateral posterior subpleural ground-glass opacity. There is no reticulation, architectural distortion, or other abnormality.* **(Bottom)** *Prone axial HRCT (lung window) of the same patient shows clearance of previously demonstrated basilar subpleural ground-glass opacities and confirms that they are related to dependent or supine atelectasis.*

INSPIRATORY AND EXPIRATORY HRCT

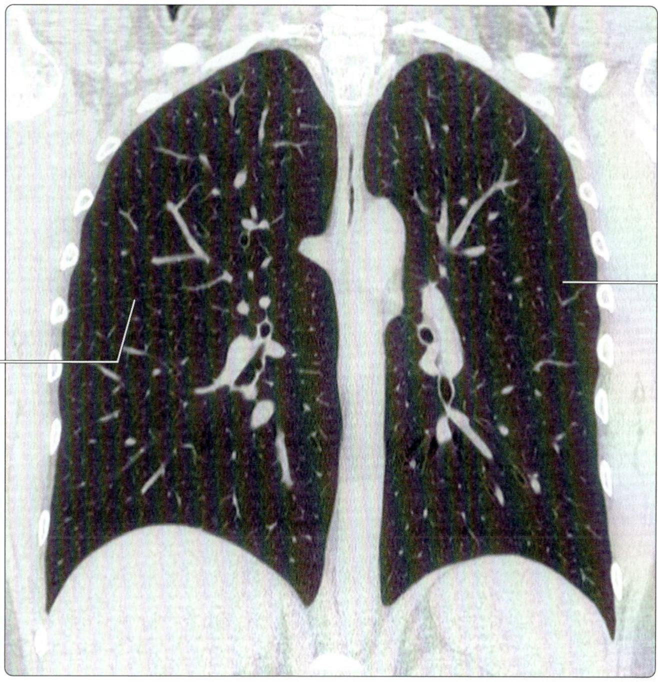

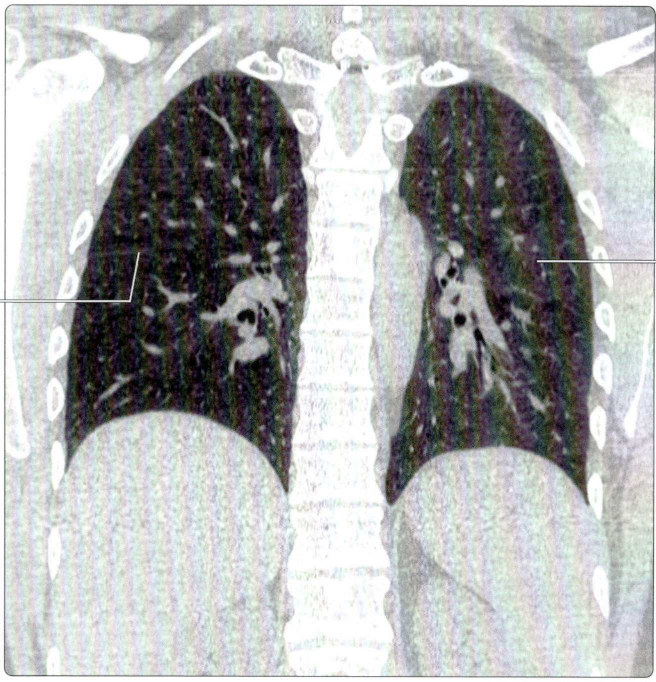

(Top) *First of 2 images from a normal high-resolution chest CT is shown. Inspiratory coronal HRCT (lung window) shows normal homogeneous pulmonary attenuation. Note excellent visualization of vascular structures, central bronchi, and interlobar fissures.* **(Bottom)** *Coronal expiratory HRCT (lung window) at the same level as the previous image shows elevation of the hemidiaphragms, decreased lung volume, and increased attenuation of the lung parenchyma, more pronounced on the left in this case. Although lung attenuation is heterogeneous, there is no evidence of expiratory air-trapping.*

CHEST CT, MAXIMUM- AND MINIMUM-INTENSITY PROJECTIONS

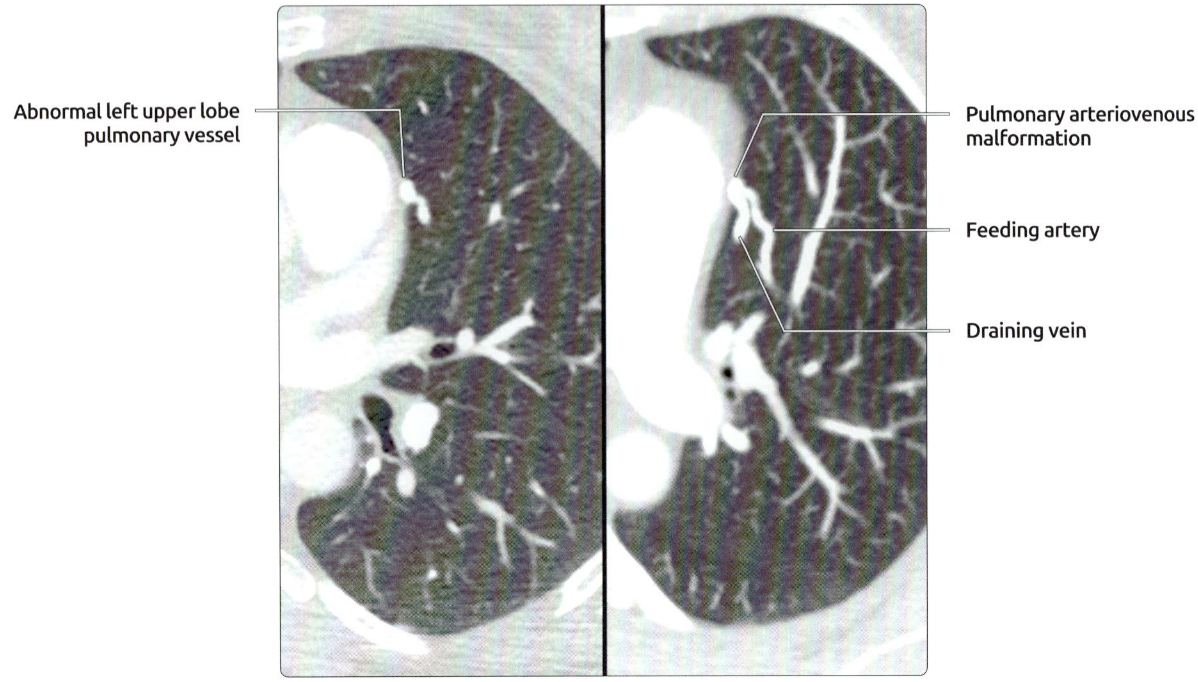

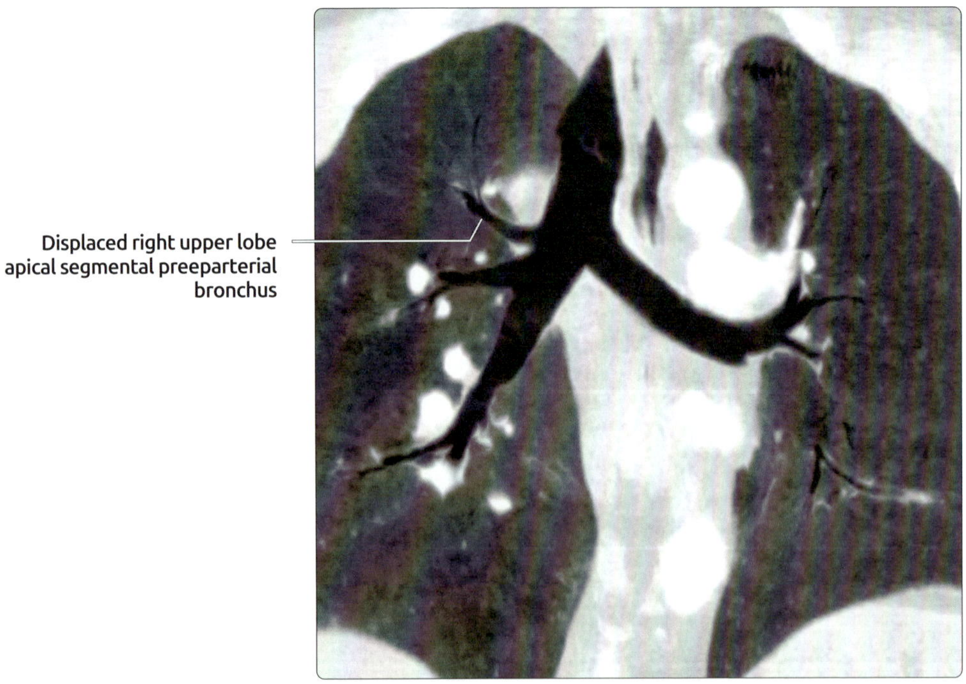

(Top) *Composite image with axial CECT (left) and oblique axial CECT maximum-intensity projection (MIP) image (right) (lung window) shows a left upper lobe abnormal pulmonary vessel. MIP image allows a confident diagnosis of pulmonary arteriovenous malformation and visualization of the feeding artery and draining vein. MIP images are useful for depicting vascular lesions and help increase the conspicuity of small pulmonary nodules.* **(Bottom)** *Coronal oblique CECT (lung window) minimum-intensity projection (MinIP) image demonstrates the branching pattern of the central tracheobronchial tree. Note a displaced apical segmental preeparterial bronchus arising from the upper aspect of the right mainstem bronchus and coursing cephalad to supply the apical segment of the right upper lobe. MinIP images are useful for demonstrating the central airways, cystic/lucent abnormalities, and pulmonary air-trapping.*

CORONAL AND SAGITTAL CHEST CT ANGIOGRAPHY

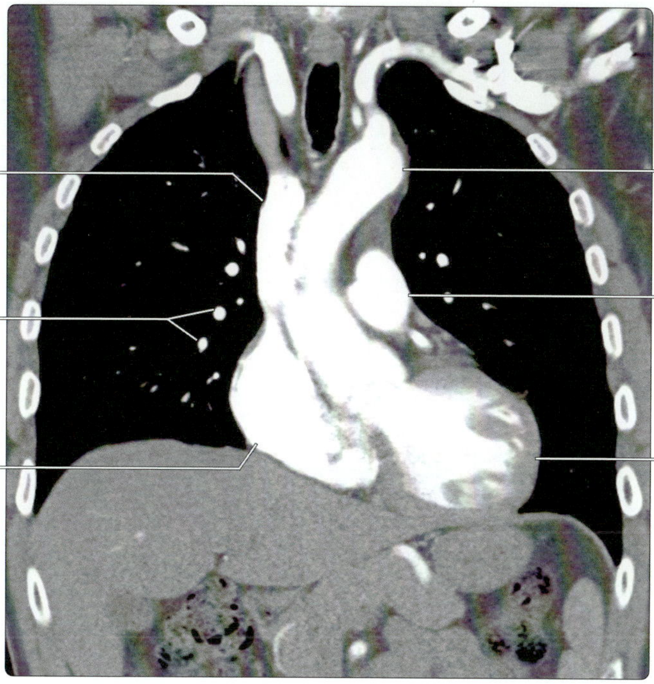

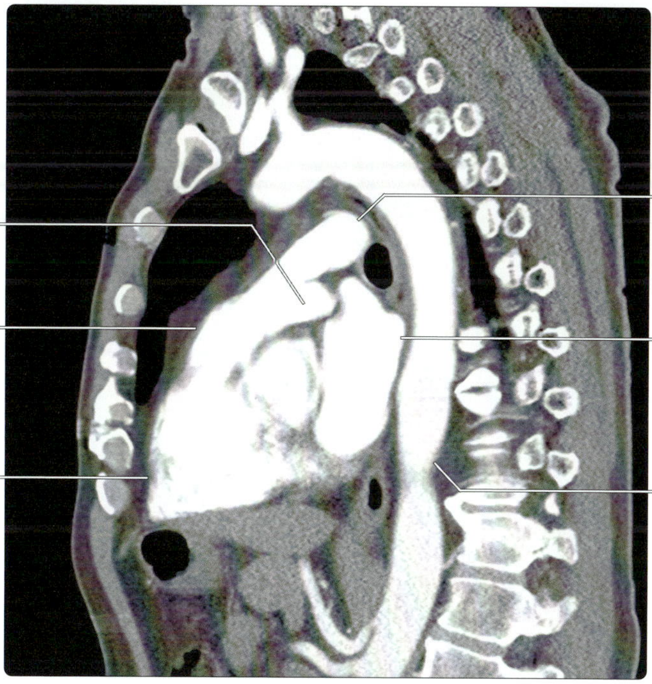

(Top) *First of 2 images from a normal CT angiogram of the chest is shown. Coronal CECT (mediastinal window) allows visualization and evaluation of the cardiac chambers and the lumina of the thoracic great vessels. CT angiography also allows assessment of the peripheral pulmonary vasculature for exclusion of pulmonary thromboembolic disease.* **(Bottom)** *Sagittal CECT (mediastinal window) allows visualization of the descending thoracic aorta and its branches. Note visualization of the right ventricle, right ventricular outflow tract, and right and left pulmonary arteries. Multiplanar reformatted images of CT angiograms allow exquisite visualization of the vascular lumina and cardiac chambers for evaluation of luminal enlargement, endoluminal thrombus or tumor, vascular disruption, and congenital anomalies.*

CHEST CT VOLUME RENDERING

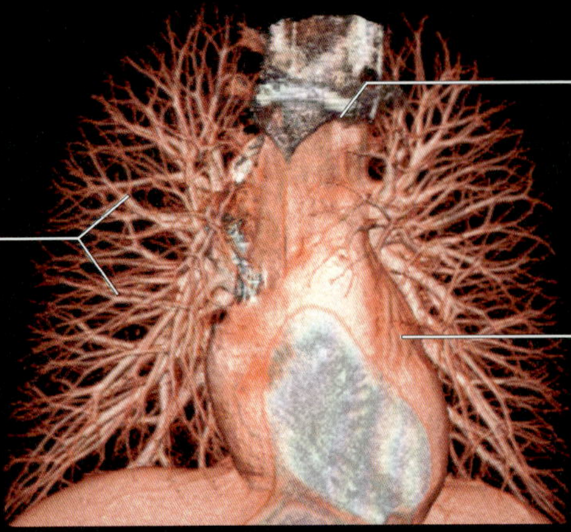

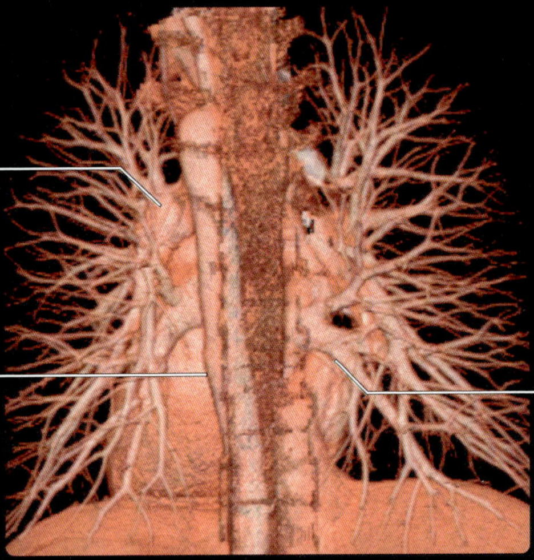

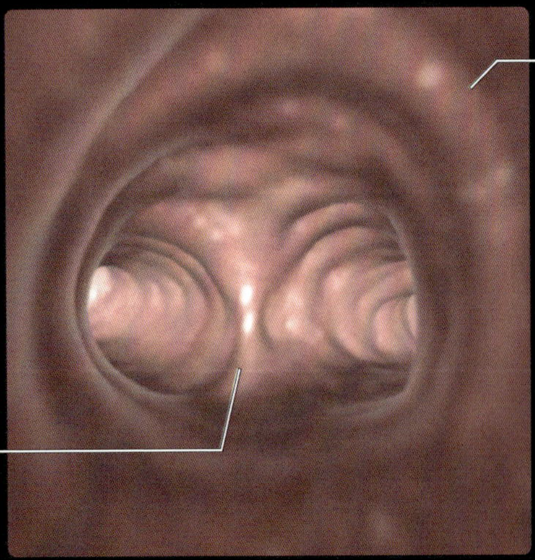

(Top) *First of 2 volume-rendered CT images of the chest is shown. Coronal volume-rendered CT of the anterior chest shows the pulmonary vasculature and the superior extent of the pericardial reflections.* **(Middle)** *Coronal volume-rendered CT of the posterior chest allows distinction of pulmonary veins from pulmonary arteries and shows portions of the descending thoracic aorta.* **(Bottom)** *Virtual bronchoscopy of the normal central airways shows the tracheal carina and its bifurcation into bilateral mainstem bronchi. Note the depiction of the tracheal lumen and the visualization of individual tracheal cartilages.*

AXIAL, CORONAL, AND SAGITTAL CHEST MR

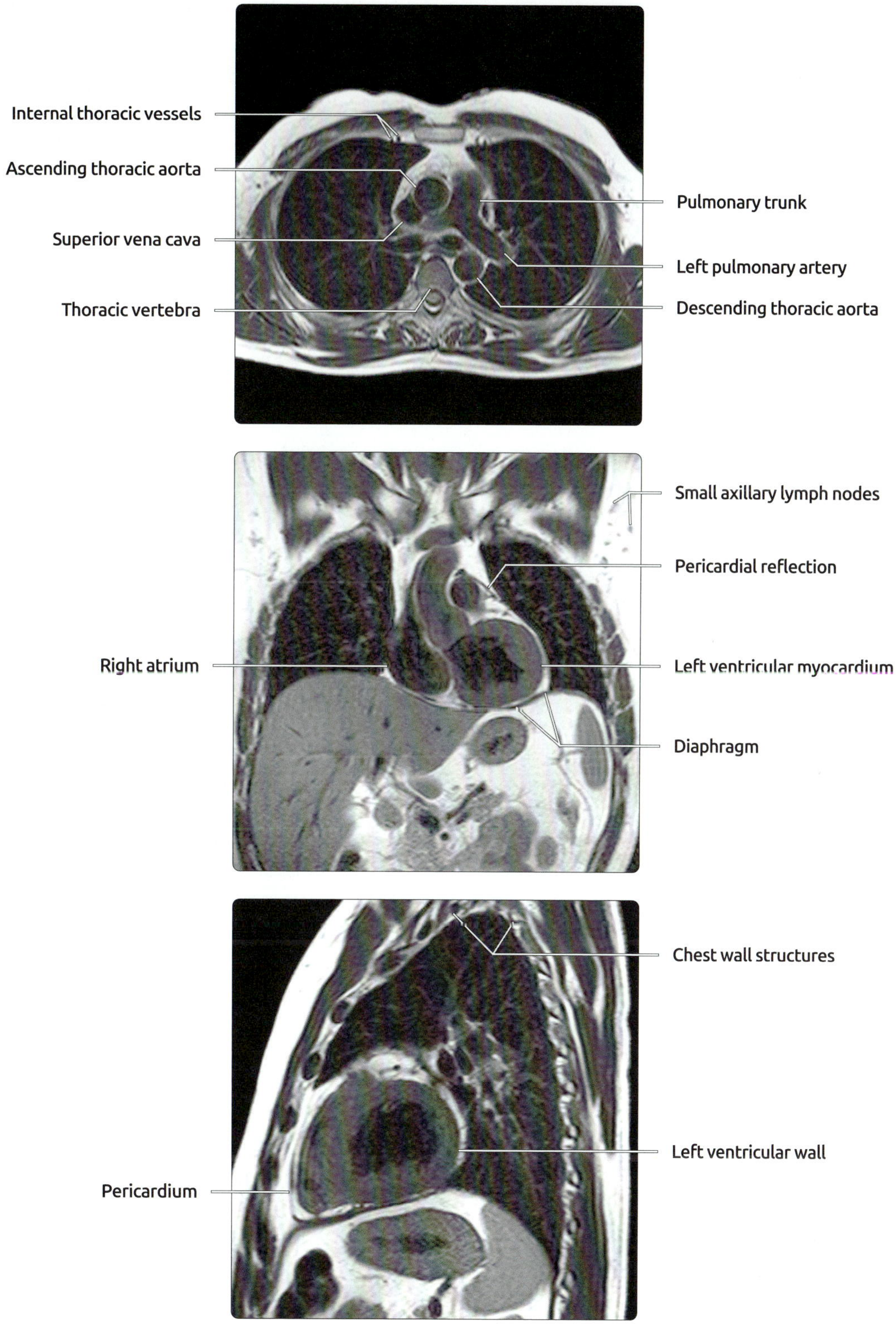

(Top) *Axial cardiac-gated T1 MR shows the value of MR in the evaluation of thoracic vascular structures without iodinated contrast. White blood sequences are also used for this purpose.* **(Middle)** *Coronal cardiac-gated T1 MR shows outstanding visualization of the mediastinum and chest wall, particularly the heart, great vessels, diaphragm, and chest wall structures, including axillary lymph nodes. Note the excellent visualization of the pericardium, cardiac chambers, and left ventricular myocardium.* **(Bottom)** *Sagittal cardiac-gated T1 MR shows exquisite visualization of the various chest wall tissues. The heart and pericardium are nicely visualized. Thoracic MR is often used as a problem-solving modality for assessment of mediastinal and chest wall tumor involvement and is particularly useful for staging apical tumors. MR is also useful in the diagnosis of mediastinal lesions, including congenital cysts and thymic lesions. MR is the imaging modality of choice for evaluating thoracic neurogenic neoplasms.*

CT PULMONARY ANGIOGRAPHY, ACUTE PULMONARY EMBOLI

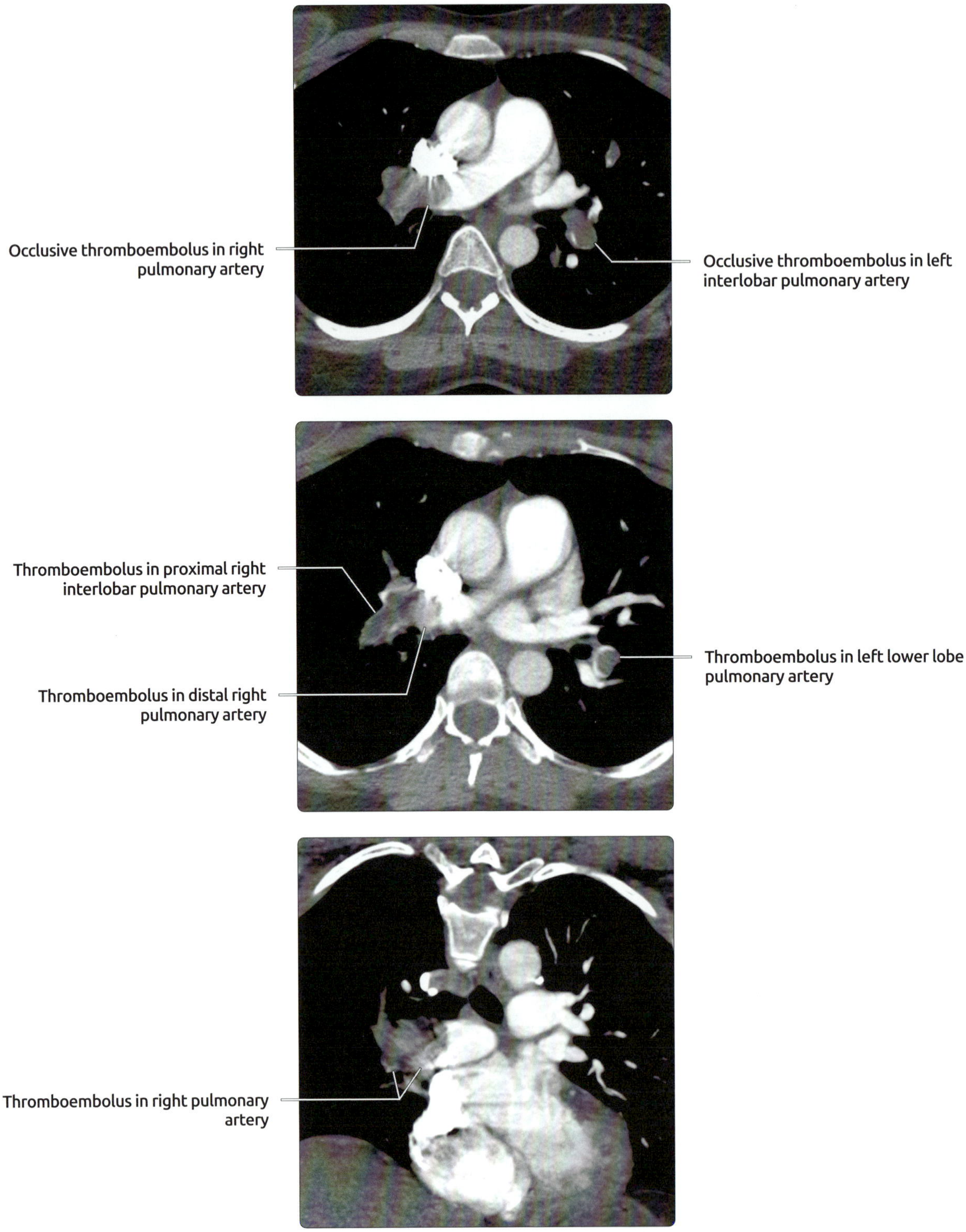

(Top) *Axial CECT (mediastinal window) of a 30-year-old woman who developed severe dyspnea 2 weeks after elective abdominoplasty and breast implant surgery shows large burden bilateral pulmonary thromboembolic disease with occlusive thrombus in the right pulmonary artery and left interlobar pulmonary artery.* **(Middle)** *Axial CECT (mediastinal window) of the same patient shows a large pulmonary thromboembolus in the right pulmonary artery with propagation into the right interlobar pulmonary artery as well as thrombus in the left lower lobe pulmonary artery.* **(Bottom)** *Coronal CECT (mediastinal window) shows the large occlusive thromboembolus in the right pulmonary artery. The patient was evaluated for pharmacomechanical thrombolysis.*

PULMONARY ANGIOGRAPHY AND THROMBOLYSIS, ACUTE PULMONARY EMBOLI

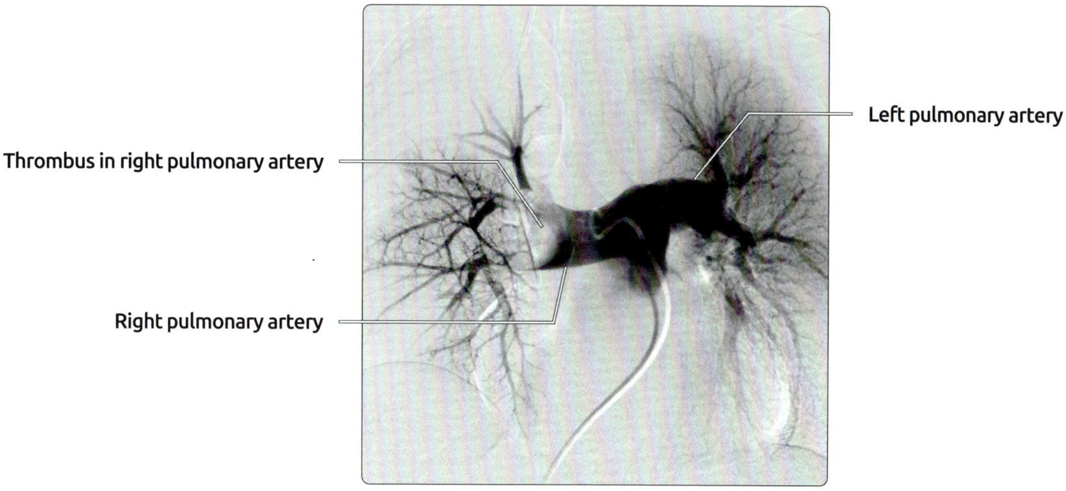

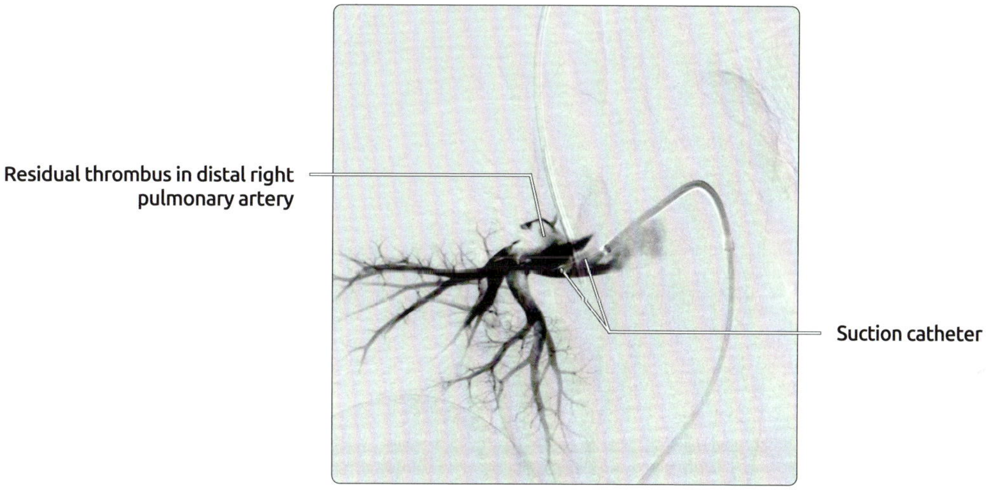

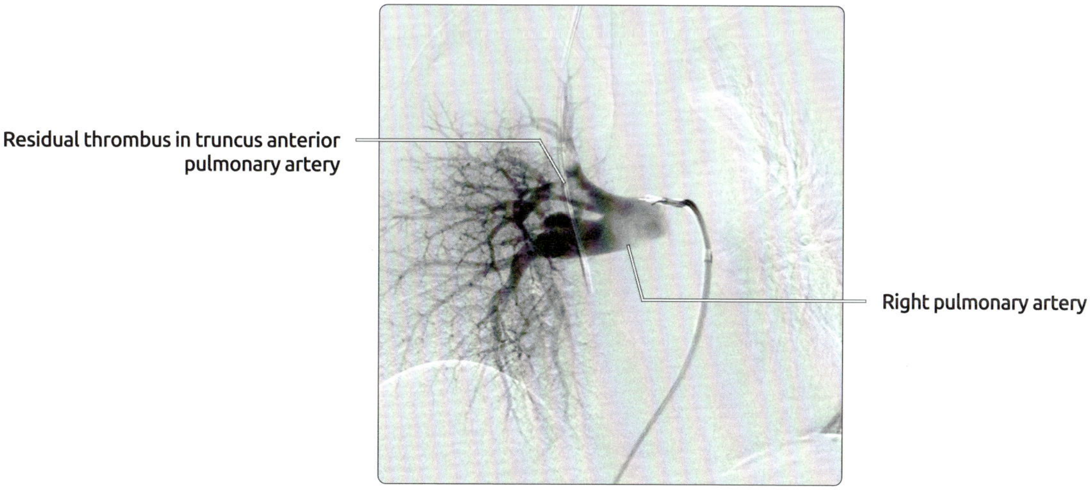

(Top) *AP digital subtraction angiography (DSA) of the pulmonary arteries of the same patient confirms a large, nearly occlusive thrombus in the right pulmonary artery.* **(Middle)** *AP DSA of the same patient shows that a suction catheter has been introduced into the distal aspect of the right pulmonary artery for mechanical suction thrombectomy and intermittent tissue plasminogen activator administration. Note the decreased thrombus burden.* **(Bottom)** *AP DSA of the same patient at the end of the procedure shows significant decrease in thrombus burden with residual thrombus in the truncus anterior pulmonary artery. The pulmonary artery pressure at the beginning of the procedure was 32 mm Hg and decreased to 18 mm Hg by the end of the procedure.*

CHEST RADIOGRAPHY AND CT, LUNG CANCER

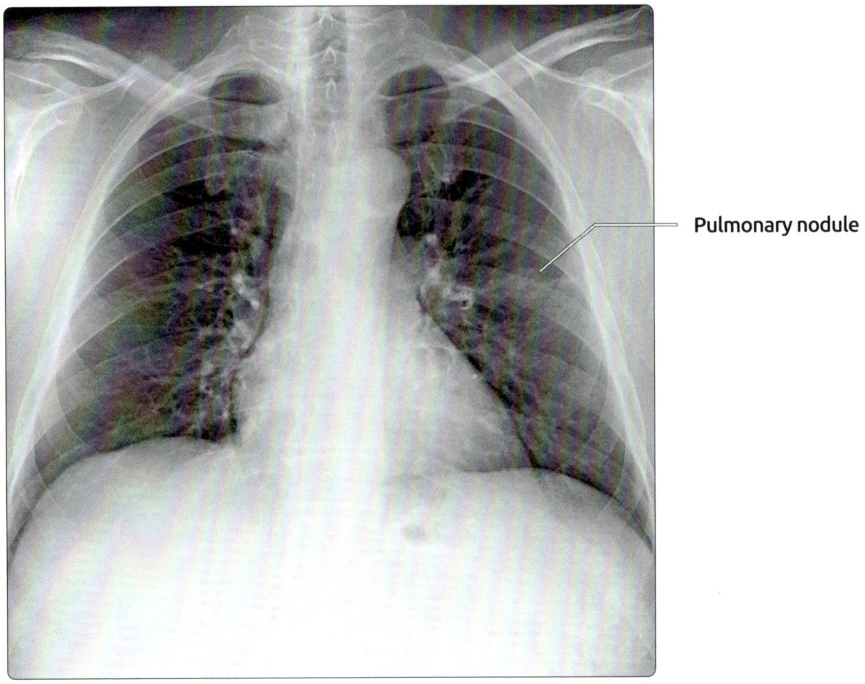

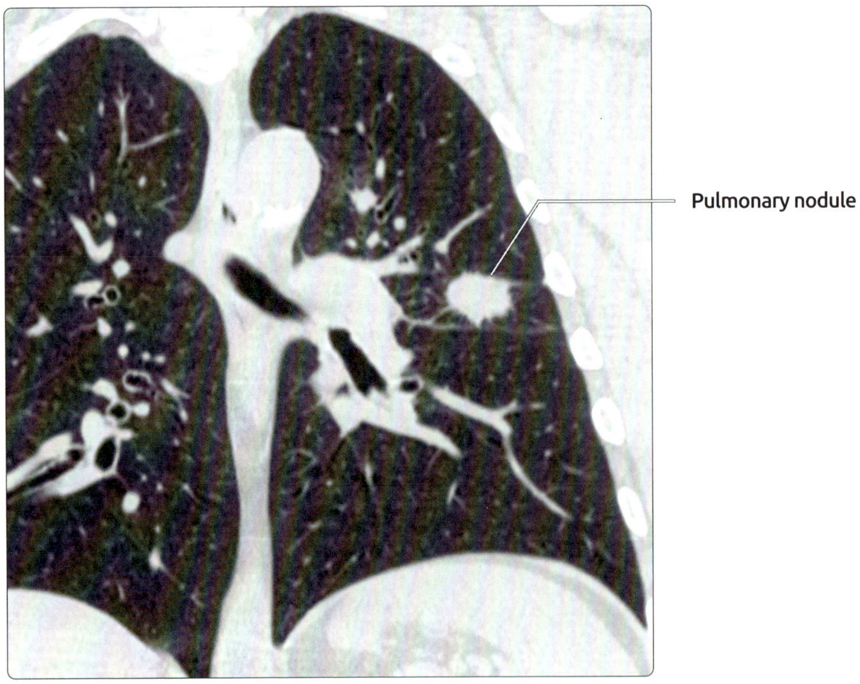

(Top) *PA chest radiograph of an asymptomatic 64-year-old man shows an incidentally discovered pulmonary nodule projecting over the left midlung zone.* **(Bottom)** *Coronal NECT (lung window) of the same patient shows that the nodule is located in the left upper lobe and exhibits spiculated borders and pleural tags. These findings are highly suspicious for primary lung cancer.*

CHEST CT AND FDG PET/CT, LUNG CANCER

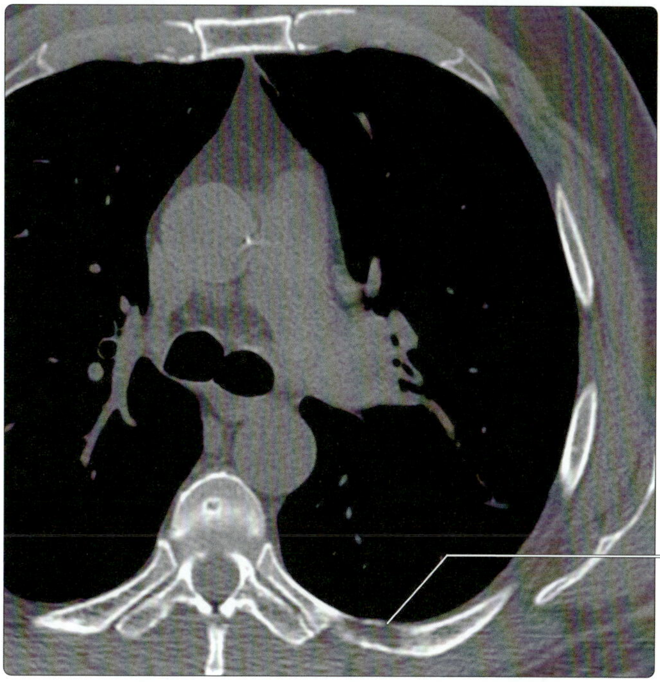

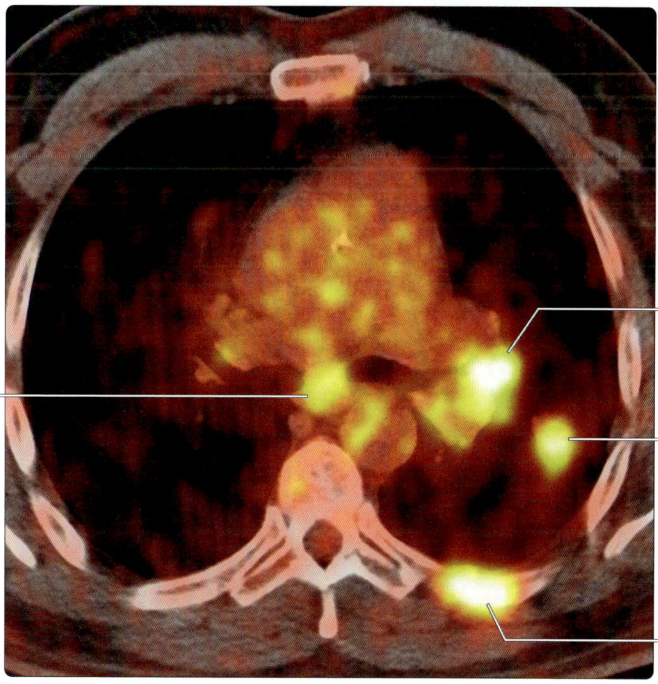

(Top) *Axial NECT (bone window) of the same patient shows a lytic lesion in a left posterior rib, which is highly concerning for neoplastic involvement by hematogenous metastatic disease.* **(Bottom)** *Axial fused FDG PET/CT of the same patient demonstrates that the left upper lobe nodule is intensely FDG avid. In addition, left hilar and subcarinal lymph nodes also exhibit intense FDG avidity, consistent with lymph node metastases. Finally, the lytic left posterior rib lesion seen on CT also exhibits FDG avidity, consistent with hematogenous metastatic disease. FDG PET/CT is useful in the clinical staging of patients with primary lung cancer. In this case, biopsy of the bone lesion allowed diagnosis and surgical-pathologic staging of this advanced stage IV primary lung cancer.*

CHEST RADIOGRAPHY AND VENTILATION-PERFUSION SCINTIGRAPHY

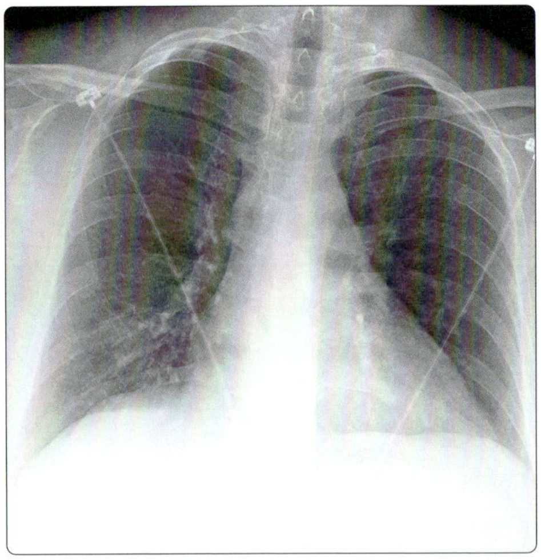

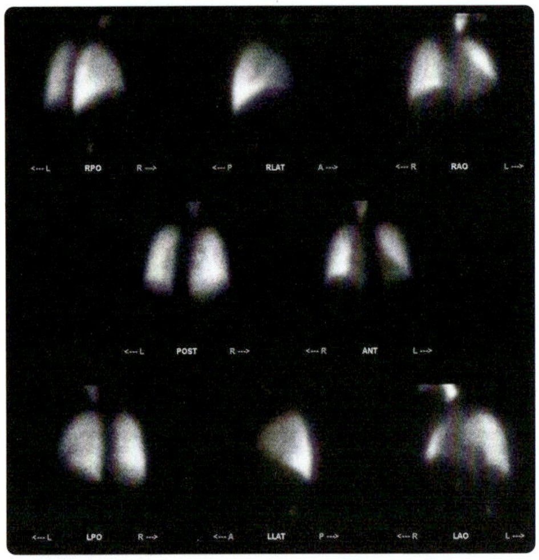

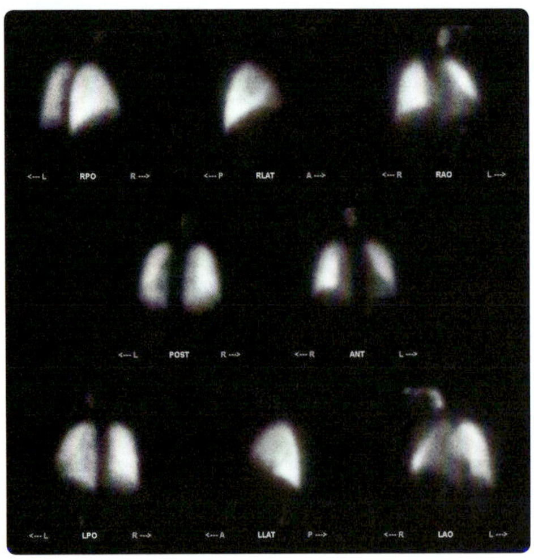

(Top) *AP chest radiograph of a 35-year-old pregnant woman with suspected pulmonary thromboembolism is normal. Chest radiography should be the 1st radiation-associated imaging study in pregnant patients with suspected pulmonary thromboembolic disease. Because the chest radiograph was normal, a ventilation-perfusion nuclear medicine scan was ordered for further assessment.* **(Middle)** *Ventilation scan performed after administration of Tc-99m DTPA aerosol shows normal distribution of activity throughout the lungs.* **(Bottom)** *Perfusion scan performed after intravenous administration of Tc-99m MAA shows no perfusion defects and normal homogeneous distribution of activity throughout the lungs, excluding pulmonary thromboembolism. In a pregnant patient with suspected pulmonary thromboembolism and a normal chest radiograph, ventilation-perfusion scintigraphy or only perfusion scintigraphy should be the next diagnostic imaging study, rather than CT pulmonary angiography.*

CHEST CT AND ULTRASONOGRAPHY, EMPYEMA

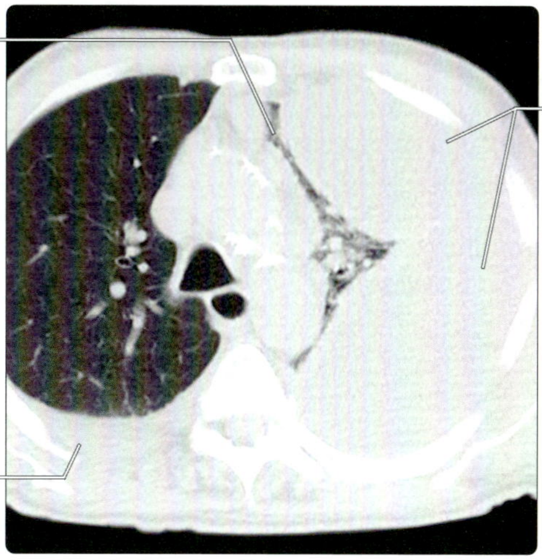

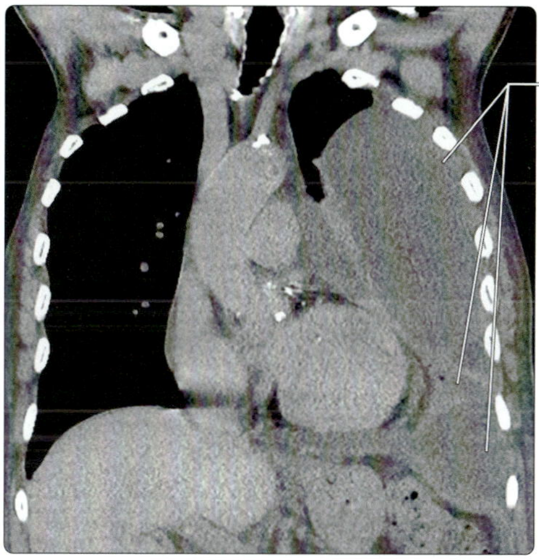

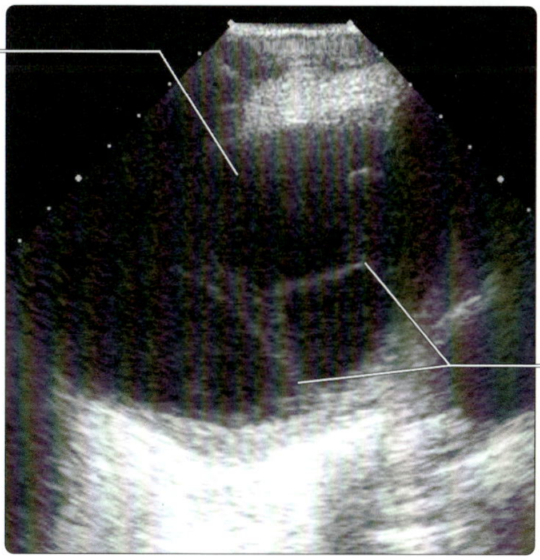

(Top) *Axial NECT (lung window) of an 86-year-old critically ill man who presented with dyspnea, fever, and leukocytosis shows a large, multiloculated left pleural effusion that produces marked relaxation atelectasis of the left lung. Note a small free right pleural effusion.* **(Middle)** *Coronal NECT (mediastinal window) of the same patient shows a large, multiloculated left pleural effusion concerning for empyema, given the imaging findings and the clinical presentation.* **(Bottom)** *Transverse thoracic ultrasound of the same patient shows a multiloculated left pleural effusion with multifocal echogenic septations. Ultrasound was performed for image guidance during chest tube placement. Two chest tubes were placed in the largest residual fluid collection, and the patient improved significantly. However, because of persistent leukocytosis, the patient eventually required thoracotomy and decortication.*

Lung Development

GENERAL CONCEPTS

Overview of Lung Development

- Larynx and trachea
  - Origin of primitive larynx from laryngotracheal groove
  - Formation of **tracheal bud** and primitive trachea
  - Separation of primitive trachea from foregut (developing esophagus)
- Bronchi
  - Tracheal bud branches into 2 primitive bronchial buds
  - **Bronchial buds** are precursors of bilateral **mainstem bronchi**
- Lungs
  - Sequential branching of primitive bronchi
  - Formation of distinct pulmonary lobes
  - Formation of distinct pulmonary segments
- Distal airways and lung parenchyma
  - Interaction of endodermal and mesodermal elements allows normal lung development
  - Continued branching of primitive airways
  - Progressive vascularization of surrounding mesenchyme
  - Development of **alveolar-capillary interfaces**
  - Postnatal airway development and maturation

Lung Developmental Stages

- Embryonic stage
- Pseudoglandular stage
- Canalicular stage
- Saccular stage
- Alveolar stage

Concurrent Developmental Processes

- Pulmonary vessels
  - Pulmonary arteries form along developing airways
  - Vasculogenesis within primitive mesenchyme to form capillary network
  - Pulmonary vein and lymphatic development along segmental boundaries
- Pleura
  - Developing lungs protrude into coelomic cavity
  - Separation of pericardial cavity, pleural cavities, and peritoneal cavity
  - Pleural investment of lungs within bilateral hemithoraces
- Diaphragm: Fusion of 4 embryonic components
  - Septum transversum: Primordium of central tendon; semicircular, separates heart from liver; persistent open pericardioperitoneal canal on either side of esophagus
  - Pleuroperitoneal membranes: Fuse with dorsal mesentery of esophagus, complete partition between chest and abdominal cavities; primordial diaphragm
  - Dorsal mesentery of esophagus: Forms median portion of diaphragm, courses on either side of aorta
  - Muscular ingrowth from lateral body walls

LUNG DEVELOPMENTAL STAGES

Embryonic Stage (26 Days to 6 Weeks)

- **Laryngotracheal groove**
  - Develops 26-28 days after fertilization
  - Arises caudal to primitive pharynx and paired 4th pharyngeal pouches
  - Longitudinal growth
  - Endoderm-derived structures
    - Tracheobronchial and pulmonary epithelium
    - Tracheobronchial glands
  - Mesoderm-derived structures
    - Tracheobronchial cartilage
    - Tracheobronchial connective tissue
    - Tracheobronchial smooth muscle
- **Respiratory diverticulum or lung bud**
  - Develops 4 weeks after fertilization
  - Pouch-like outgrowth from caudal aspect of **laryngotracheal groove**
  - Caudal growth
  - Invested in mesoderm-derived splanchnic mesenchyme
- **Tracheal bud**
  - Globular enlargement of distal lung bud
  - Caudal growth from primordial pharynx
  - Proximal communication with foregut through primordial laryngeal inlet
- **Tracheoesophageal septum**
  - Longitudinal **tracheoesophageal folds** form on either side of developing tracheal bud
  - Medial growth of bilateral tracheoesophageal folds
  - Formation of **tracheoesophageal septum** from midline fusion of **tracheoesophageal folds**
  - Separation of primitive trachea from developing esophagus
- **Primary bronchial buds and branches**
  - Branching of primitive tracheal bud into right and left branches (5th week after fertilization)
    - **Right bronchial bud**: Longer and vertically oriented
    - **Left bronchial bud**: Shorter and horizontally oriented
  - Branching of primary bronchial buds into 2 primitive lobar bronchi
    - **Right superior lobar bronchus**: Right upper lobe bronchus
    - **Right inferior lobar bronchus**: Primitive bronchus intermedius
      - Right middle lobe bronchus
      - Right lower lobe bronchus
    - **Left superior lobar bronchus**: Left upper lobe bronchus
    - **Left inferior lobar bronchus**: Left lower lobe bronchus
  - Branching of primitive lobar bronchi into primitive segmental bronchi

Pseudoglandular Stage (6-16 Weeks)

- **Microscopic morphology**
  - Gland-like appearance of lung
  - Formation of tracheobronchial cartilages, mucus glands, and cilia by 13 weeks after fertilization
  - Primitive airways lined by endoderm-derived columnar epithelium
  - Primitive airways surrounded by mesoderm-derived mesenchymal tissue
  - Absent alveolar-capillary interface
- **Important events**
  - Formation of all major airway elements
    - Bronchial development complete to level of terminal bronchioles

- All bronchopulmonary segments formed by 7 weeks after fertilization
- **Physiologic implications**
  - Respiration not possible
  - No possibility of extrauterine survival

Canalicular Stage (16-28 Weeks)

- **Microscopic morphology**
  - Continued enlargement of primitive airway lumina
  - Continued thinning of airway epithelium
  - Epithelial differentiation into **type 1 and type 2 cells**
  - Airways separated by reduced but significant mesenchymal tissue
- **Important events**
  - Continued vascularization of lung
  - Continued development of primitive airways
    - **Terminal bronchioles** give rise to 2 or more **respiratory bronchioles**
    - **Respiratory bronchioles** give rise to 3-6 **alveolar ducts**
    - Development of small number of **terminal saccules**
  - Lamellar inclusions within type 2 pneumocytes in terminal saccules with potential for surfactant production
- **Physiologic implications**
  - Limited surfactant production
  - Respiration is possible in late canalicular stage
  - Possibility of neonatal survival with intensive care and appropriate life support

Saccular Stage (28-36 Weeks)

- **Microscopic morphology**
  - Continued airway differentiation
  - Continued thinning of airway epithelium
  - Some capillaries abut and bulge into developing alveoli
  - Terminal sacs begin to approach morphology of adult alveoli
- **Important events**
  - Development of increasing numbers of **terminal saccules**
  - Establishment of primitive **alveolar-capillary interface**
  - Increased potential for **surfactant production**
- **Physiologic implications**
  - Respiration with adequate gas exchange is possible
  - Survival of premature neonates with appropriate life support

Alveolar Stage (36 Weeks to 8 Years)

- **Microscopic morphology**
  - Continued thinning of epithelial lining of terminal sacs
  - Formation of primordial **alveoli**
  - Adjacent capillaries bulge into terminal saccules
- **Important events**
  - Continued development of distal airways
  - Primordial alveoli form along respiratory bronchioles and terminal saccules
  - Development of thin **alveolar-capillary membrane**
- **Physiologic implications**
  - Presence of nearly mature alveolar-capillary interface
  - Adequate surfactant production
  - Respiration possible without external support

IMPORTANT FACTORS FOR NORMAL PRENATAL LUNG DEVELOPMENT

Volume Requirements

- Adequate **intrathoracic volume** required for normal pulmonary development
- Factors predisposing to **pulmonary hypoplasia**
  - **Intrathoracic** masses/lesions that produce mass effect on developing lung
    - Pleural effusion
    - Neonatal chylothorax
    - Neoplasms: Teratoma, neuroblastoma
    - **Space-occupying thoracic congenital anomalies**
      - Pulmonary sequestration
      - Congenital pulmonary airway malformation
    - **Congenital diaphragmatic hernia**
      - Complex diaphragm development may result in congenital defects and diaphragmatic hernia
      - Herniated abdominal content into thorax acts as space-occupying thoracic mass
      - Inhibition of lung development → pulmonary hypoplasia
  - **Thoracic wall** abnormalities
    - Skeletal dysplasias
  - **Extrathoracic** masses/lesions producing mass effect
    - Renal dysplasia, urinary outflow obstruction
    - Ascites
  - **Intrauterine volume reduction**
    - Oligohydramnios

Circulatory Requirements

- Circulation affects pulmonary development
- Abnormal circulation may result in pulmonary hypoplasia

NEONATAL LUNG

Clearance of Retained Fluid

- External clearance: Mouth, nasal cavity
- Internal clearance: Lymphatics, capillaries

1st Breath

- Diaphragmatic contraction
- Role of **surfactant**
  - Alveolar expansion results in surfactant discharge by **type 2 pneumocytes**
  - Decreased surface tension of remaining intraalveolar fluid
  - Increased surfactant activity with decreased surface area
  - **Prevention of alveolar collapse** during expiration
- Pulmonary vascular changes
  - Fluid-filled lungs result in high resistance of pulmonary circulation
    - Small portion of cardiac output goes to lungs prior to birth
  - Pulmonary expansion with 1st breath
    - Vascular vasodilatation
    - Increased pulmonary blood flow

Postnatal Lung Development

- 24 million terminal sacs and alveoli at birth
  - 300 million terminal sacs and alveoli in adult life

- 5-fold increase of alveolar numbers within 1st year of life
- Formation of 95% of adult alveoli by 8 years of age

DISORDERS OF LUNG IMMATURITY

Retained Fetal Lung Fluid Syndrome

- Synonyms: Transient tachypnea of newborn, wet lung disease
- Neonatal respiratory distress in term or preterm infants
  - Duration: 24-72 hours
- Imaging
  - Normal lung volume
  - Small pleural effusions
  - Increased interstitial opacities
  - Resolution within 1 or 2 days
- Treatment
  - Supportive management
  - Supplemental oxygen
  - Continuous positive airway pressure (CPAP)

Respiratory Distress Syndrome

- Impaired capacity of **type 2 pneumocytes** to replenish **surfactant** prior to 36 weeks of gestation
- Symptoms occur shortly after birth in premature infants
- Imaging
  - Low lung volume
  - Opaque lung parenchyma with air bronchograms
  - Volume loss related to surfactant insufficiency
- Treatment
  - Supplemental oxygen; CPAP
  - Endotracheal intubation
  - Administration of exogenous surfactant
  - Typical improvement by age 3-4 days

DISORDERS RELATED TO TRACHEOBRONCHIAL MORPHOLOGY

Central Tracheobronchial Tree

- Horizontal course of left mainstem bronchus
- Vertical course of right mainstem bronchus
- Propensity for aspiration into right-sided airways
  - Oral secretions
  - Ingested material
  - Foreign bodies

DEVELOPMENTAL DISORDERS

Abnormal Tracheoesophageal Development

- **Tracheal aplasia/agenesis**
  - Abnormal course and anomalous ventral location of tracheoesophageal folds in early embryonic development
    - Interruption of proximal tracheal lumen
  - Foregut derived bronchi may communicate with each other
  - Foregut derived bronchi may arise directly from esophagus
- **Esophageal atresia with tracheoesophageal fistula**
  - Abnormal course and anomalous dorsal location of tracheoesophageal folds in early embryonic development
    - Interruption of primitive proximal esophagus
  - Tracheoesophageal fistula with communication between distal trachea and distal esophageal segment inferred by presence of gastric &/or bowel gas on radiography
- **Esophageal atresia without tracheoesophageal fistula**
  - Abnormal course and anomalous dorsal location of tracheoesophageal folds in early embryonic development
    - Interruption of primitive esophagus
  - Absence of tracheoesophageal fistula inferred by gasless abdomen on radiography
- **Tracheoesophageal fistula**
  - Failure of complete midline fusion of tracheoesophageal folds
  - Abnormal communication between proximal trachea and esophagus

Abnormalities of Foregut Budding

- **Extralobar sequestration**
  - Anomalous supernumerary foregut bud induces primitive mesenchyme to form lung parenchyma
    - Supernumerary lung tissue (sequestration) develops in thorax, diaphragm, or abdomen
    - No communication with normal tracheobronchial tree
    - Systemic blood supply
  - Neonatal respiratory distress from mass effect &/or pulmonary hypoplasia
  - Imaging
    - Homogeneous soft tissue mass typically located in left hemithorax
    - Systemic arterial supply
- **Bronchogenic cyst**
  - Anomalous supernumerary foregut bud that does not induce development of lung parenchyma
    - Mediastinal mass in subcarinal location
    - Unilocular, thin-walled cyst
  - Symptomatic or asymptomatic older children and young adults
  - Imaging
    - Subcarinal middle/visceral mediastinal spherical mass
    - Thin-walled cystic lesion with fluid attenuation content
    - May exhibit soft tissue attenuation secondary to high-attenuation fluid content

Abnormalities of Distal Airway Development

- **Bronchial atresia**
  - Postulated focal in utero vascular compromise of short segment of distal bronchial wall
    - Focal bronchial atresia
    - Near normal distal airway development
    - Formation of mucocele distal to atretic segment
  - Asymptomatic adults
  - Imaging
    - Round, tubular or branching opacity representing mucocele distal to atretic airway
    - Surrounding hyperlucent lung parenchyma

EARLY LUNG DEVELOPMENT

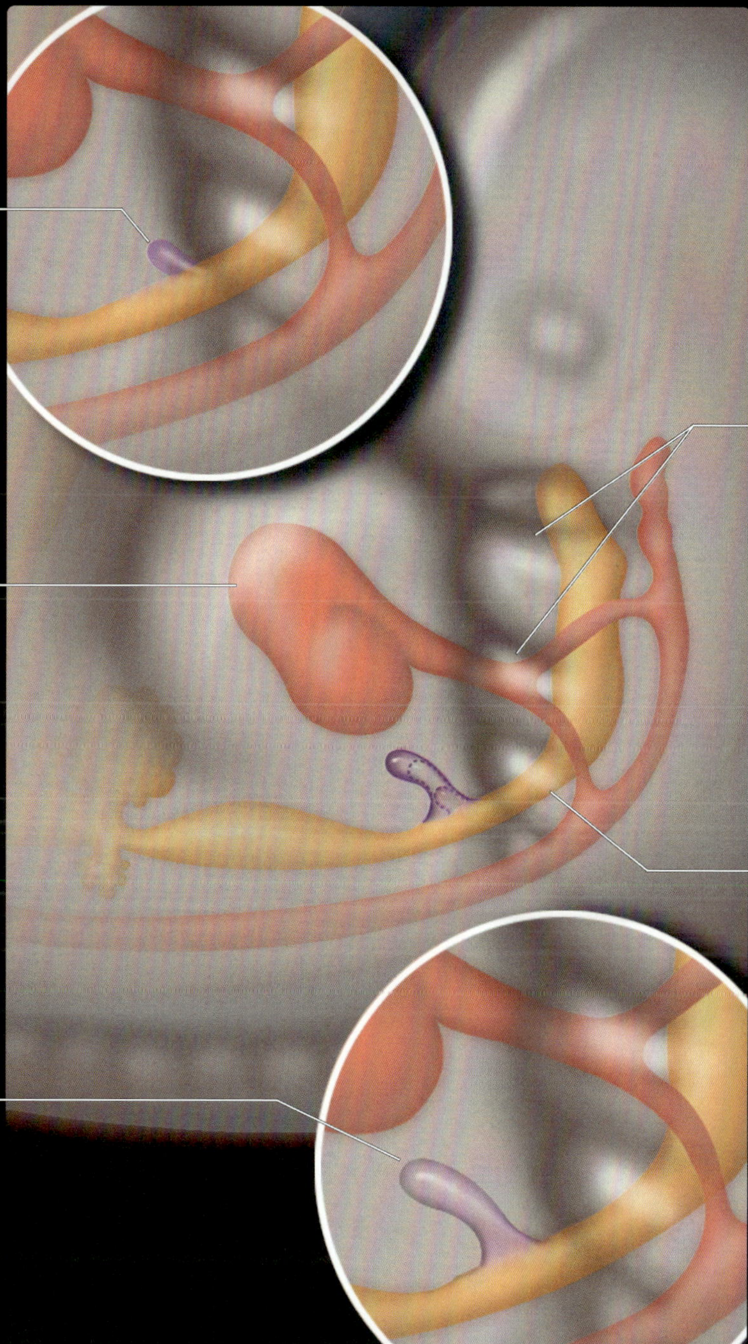

Graphic shows the development of the primitive lung. The respiratory diverticulum (top left inset) arises from the laryngotracheal groove near the primordial esophagus caudal to the 4th pharyngeal pouches. The sequential evolution of the respiratory diverticulum into the tracheal bud (bottom right inset) is shown. The tracheal bud will then divide into 2 outpouchings (primary bronchial buds), which will progressively differentiate into the right and left bronchial tree. Note the close relationship of the developing tracheobronchial tree with the primitive foregut or primordial esophagus.

EMBRYONIC STAGE

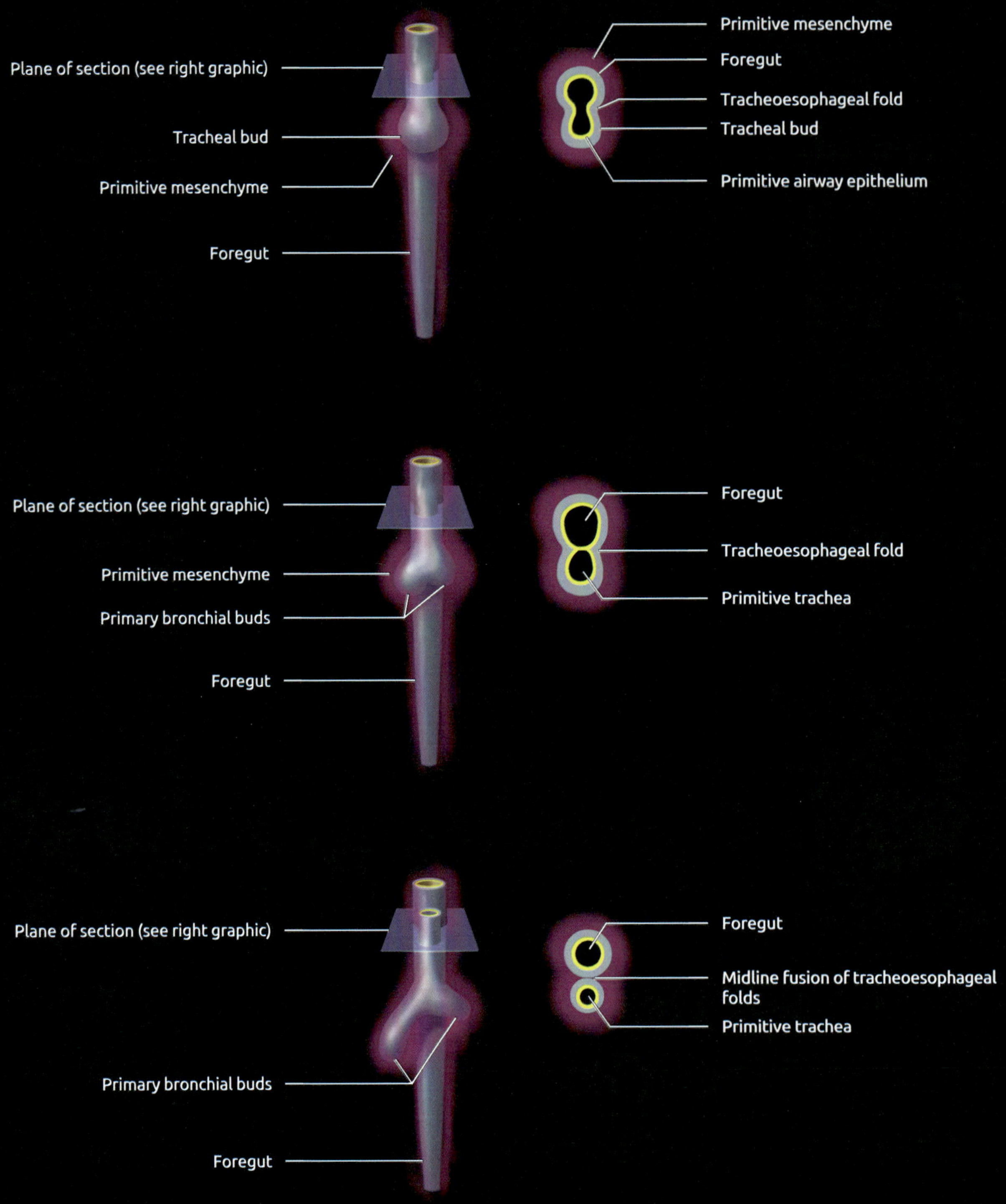

(Top) *First of 3 graphics shows the development of the tracheobronchial tree. The tracheal bud is a ventral outpouching of the primitive foregut lined by endoderm-derived epithelium and surrounded by mesoderm-derived mesenchyme. The axial plane of section (right graphic) shows communication between the tracheal bud and the primitive foregut as well as formation of bilateral longitudinal tracheoesophageal folds.* **(Middle)** *Graphic shows early branching of the tracheal bud into primary bronchial buds, which will eventually form the right and left mainstem bronchi. The longitudinal tracheoesophageal folds (right graphic) continue to migrate medially to fuse in the midline, thus, separating the trachea from the esophagus.* **(Bottom)** *Graphic shows further development of the primary bronchial buds with elongation and vertical growth. Note that the right bronchial bud is vertically oriented, and the left follows a more horizontal course. The tracheoesophageal folds (right graphic) have fused in the midline to completely separate the trachea from the esophagus.*

EMBRYONIC STAGE

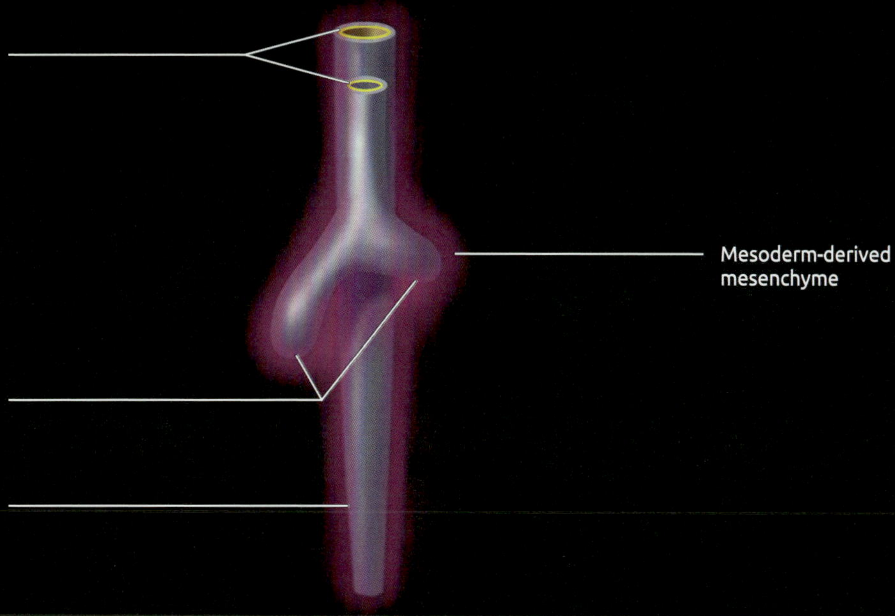

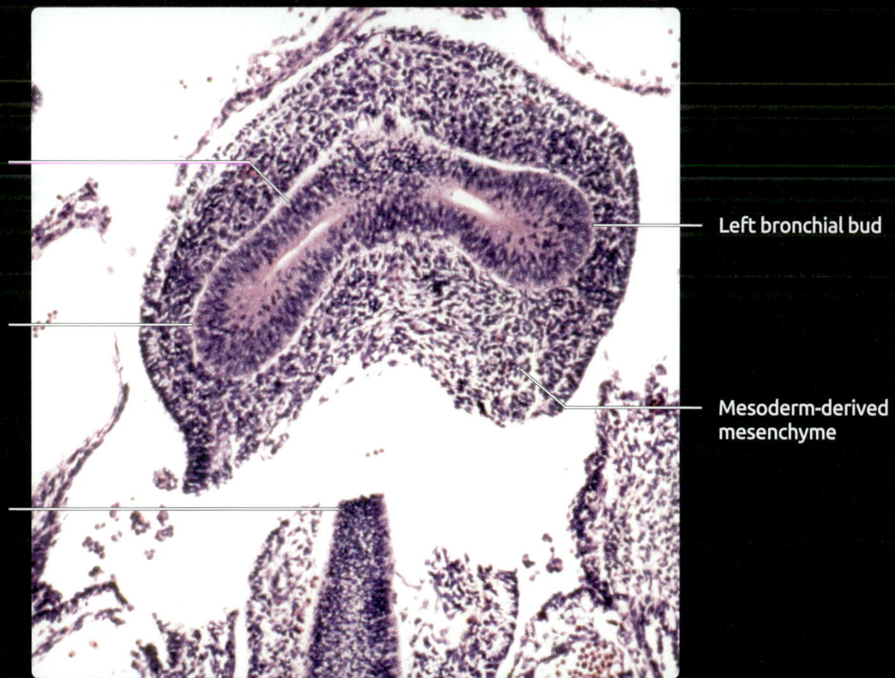

(Top) *Graphic shows the development and morphology of the bronchial buds as they invaginate into the primitive mesenchyme. The bronchial buds are the precursors of the bilateral mainstem bronchi and determine their anatomic orientation in adulthood.* **(Bottom)** *Photomicrograph (hematoxylin and eosin stain) of a coronal section through the primitive thorax of a 4-mm embryo shows the bilateral bronchial buds invaginating into the surrounding primitive mesenchyme. Contact between the developing tracheobronchial tree and the primitive mesenchyme induces development of the surrounding lung parenchyma. The primitive foregut is located dorsal to the plane of section. Note that the right bronchial bud is larger and courses in a more vertical direction than the left bronchial bud. These structures are the precursors of the central tracheobronchial tree. (Courtesy J. Thomas Stocker, MD.)*

EMBRYONIC AND PSEUDOGLANDULAR STAGES

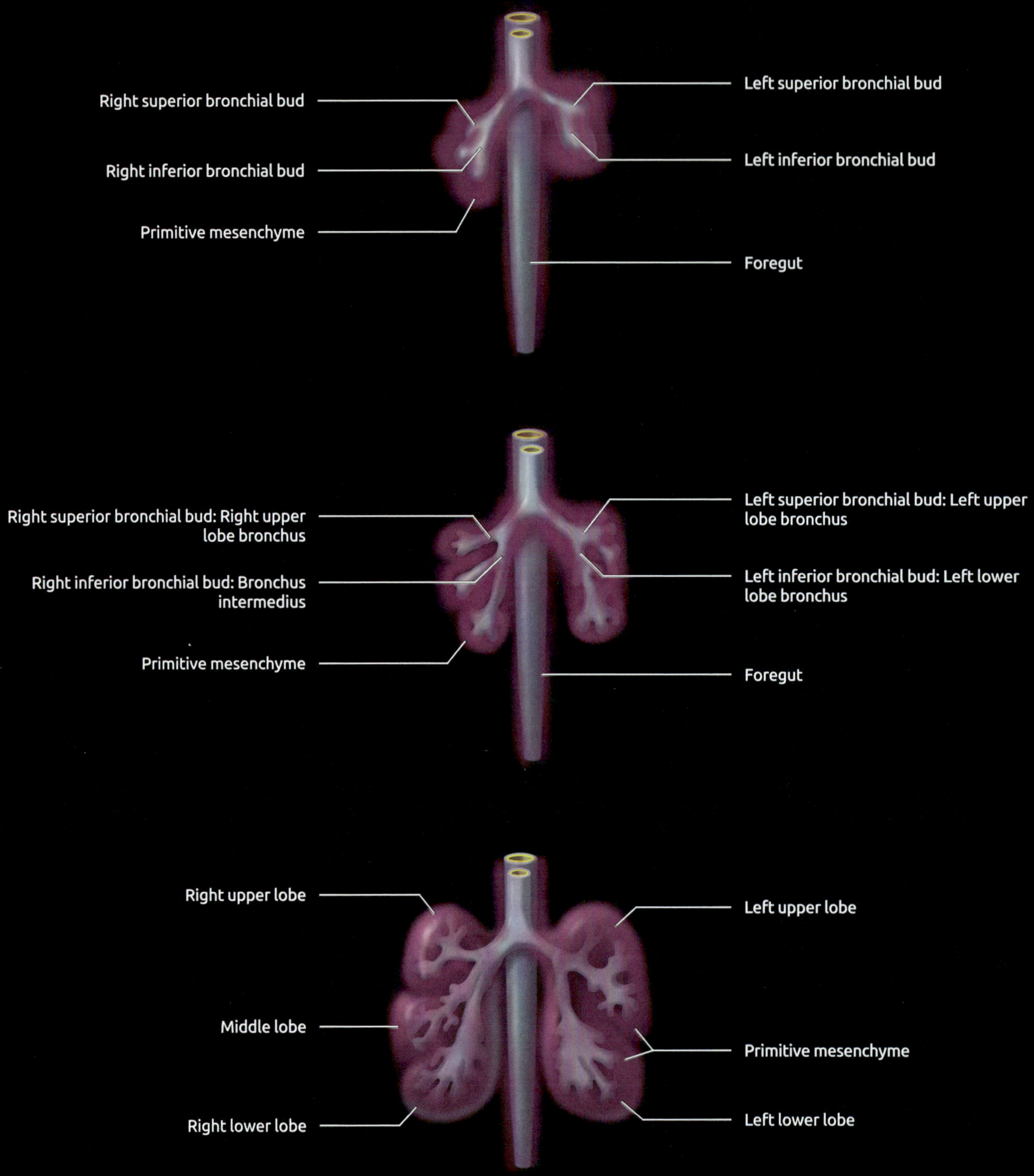

(Top) *Graphic shows the developing tracheobronchial tree at 28 days of gestation. The right and left bronchial buds begin to divide into right and left superior and inferior bronchial buds. The developing tracheobronchial tree is surrounded by primitive mesenchyme.* **(Middle)** *Graphic shows the developing tracheobronchial tree at 42 days of gestation with continued elongation and branching of the bronchial buds to form rudimentary lobar bronchi. Further growth and branching of the distal primitive airways form rudimentary segmental bronchi. The rudimentary bronchus intermedius gives rise to primitive middle and right lower lobe bronchi.* **(Bottom)** *Graphic shows the tracheobronchial development at 56 days of gestation with continued branching of the primitive airways. The primitive mesenchyme surrounds the developing airways forming rudimentary lung lobes.*

PSEUDOGLANDULAR STAGE

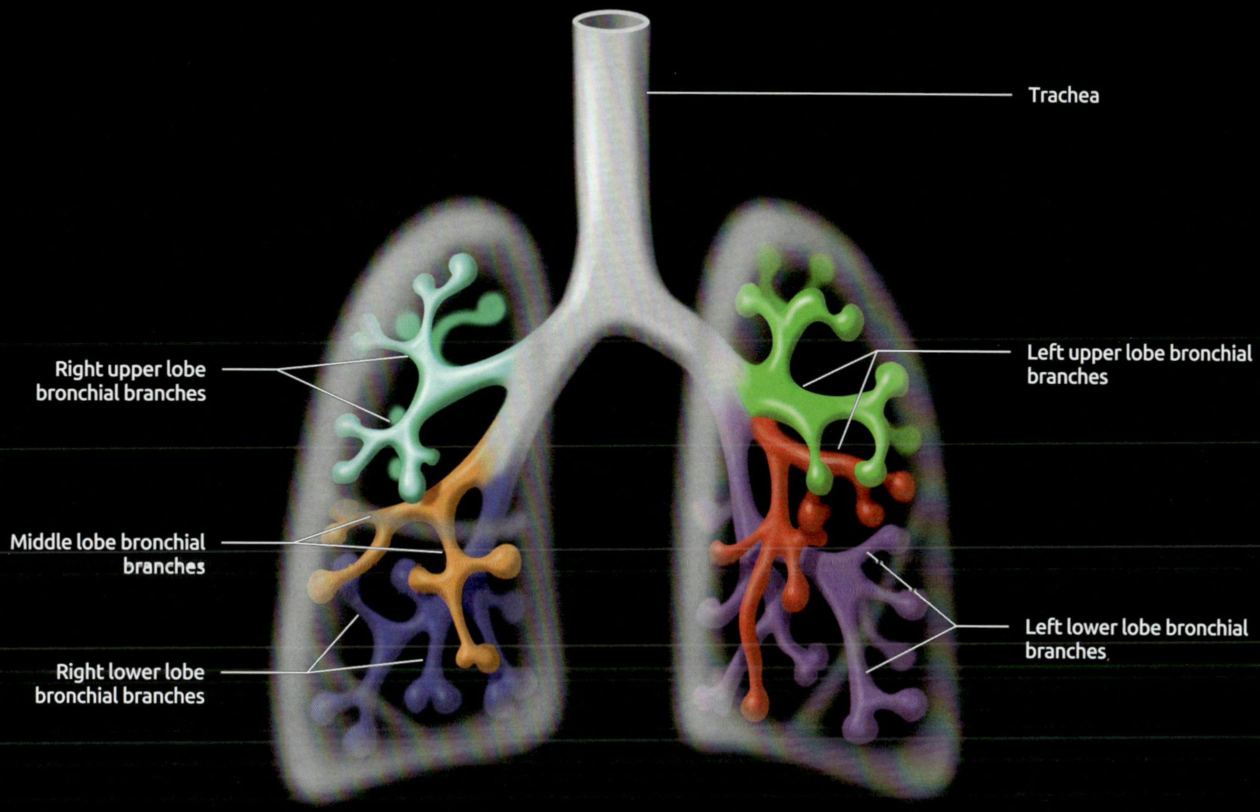

Graphic shows tracheobronchial development at ~ 10 weeks of gestation. Note airway differentiation into rudimentary lobar bronchial branches (shown in different colors) and segmental bronchial branches. Note that the green and red bronchial branches represent different portions of the primitive left upper lobe. Recognizable lung lobes are present. By the end of the pseudoglandular stage of lung development, all major elements of the airways are formed. The interaction between the developing tracheobronchial tree and the surrounding primitive mesenchyme induces the development of lung parenchyma.

PSEUDOGLANDULAR AND CANALICULAR STAGES

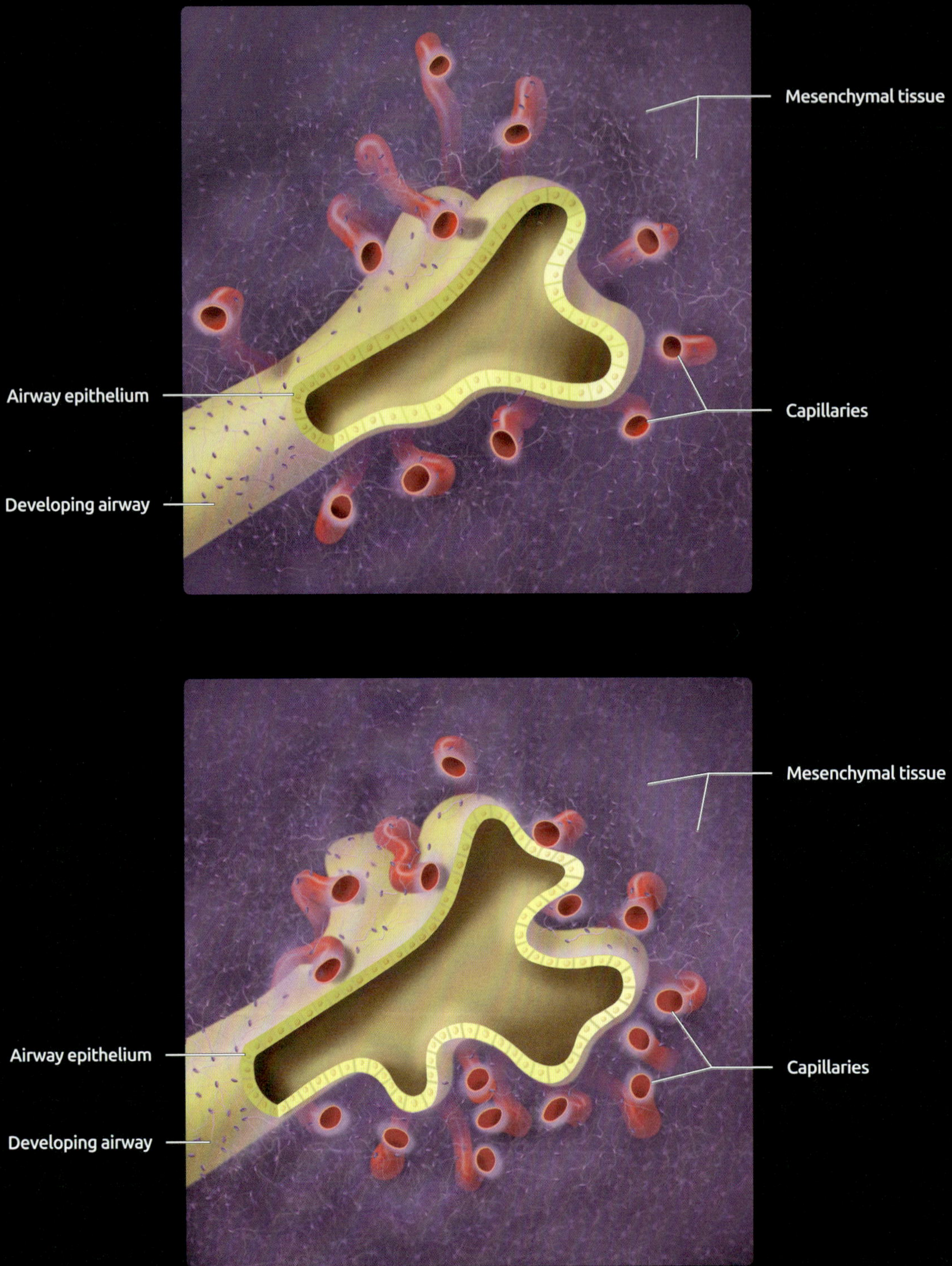

(Top) *Graphic shows a schematic of the primitive airway in the pseudoglandular stage of development (6-16 weeks). The airways are blind-ending tubules. There is no alveolar-capillary interface, as mesenchymal tissue separates the thick-walled primitive airway from the developing pulmonary capillaries. Respiration is not possible.* **(Bottom)** *Graphic shows a schematic of the primitive airway in the canalicular stage of development (16-28 weeks). The airway lumen has enlarged, and the airway epithelium has thinned. There is an increased number of vessels within the primitive mesenchyme, and some of the vessels abut the airway wall. Respiration is possible at the end of this stage of lung development, but these infants require intensive care and support for survival.*

SACCULAR AND ALVEOLAR STAGES

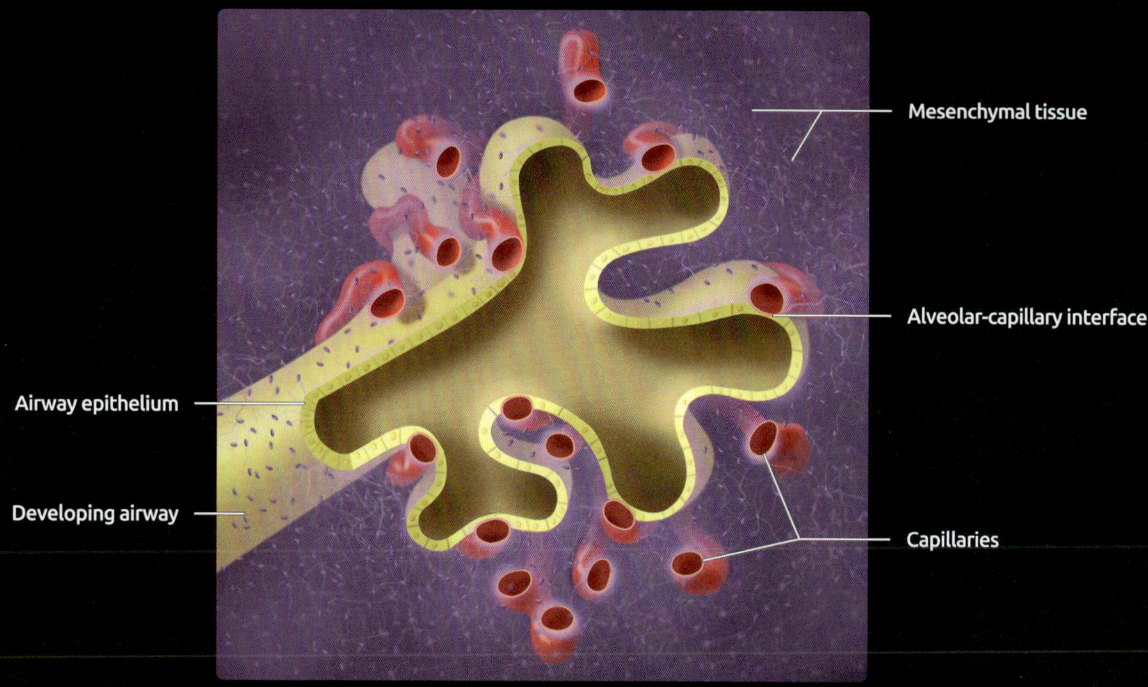

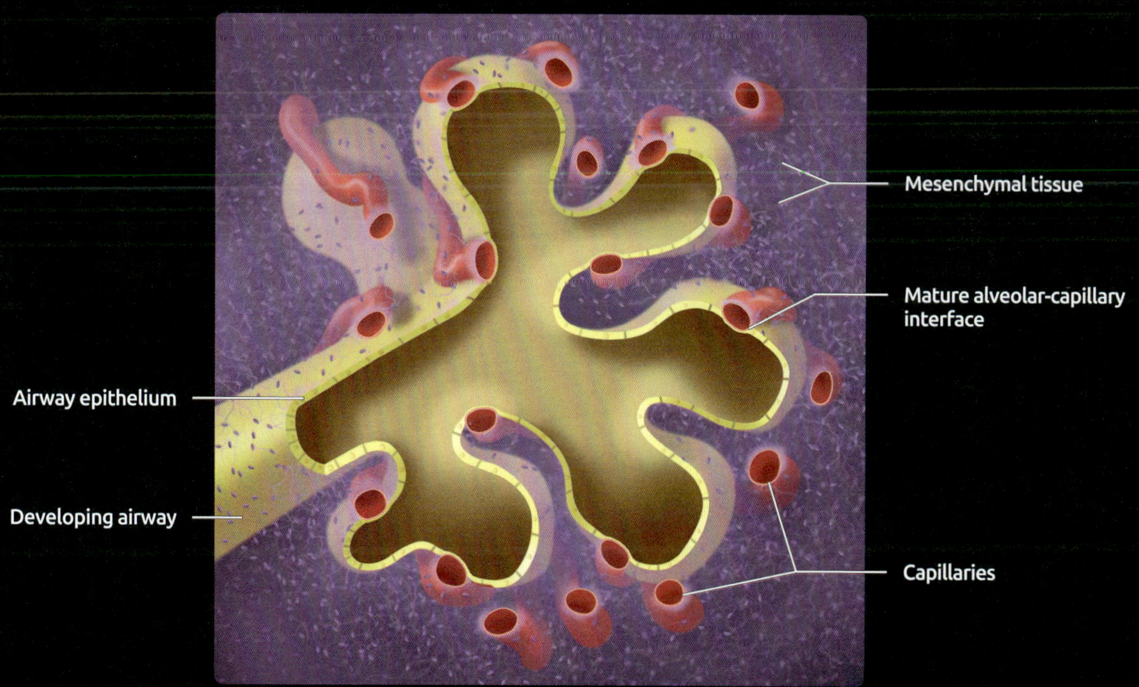

(Top) *Graphic shows a schematic of the developing airway in the saccular stage of lung development (28-36 weeks). The airway lumen continues to enlarge. The lining epithelium continues to thin. More numerous capillaries abut the walls of the primitive airways, and some bulge into the airway lumina. The alveolar-capillary interface continues to mature. Respiration is possible, and many infants born at this stage of pulmonary development survive with proper medical management and support.* **(Bottom)** *Graphic shows a schematic of the developing airway in the alveolar stage of pulmonary development (36 weeks to 8 years). The airway lumen continues to enlarge, and there is less surrounding mesenchymal tissue. The airway epithelium is thin, and many capillaries bulge into the airway lumina, establishing mature alveolar-capillary interfaces. Airway development continues after birth and into childhood.*

PSEUDOGLANDULAR, CANALICULAR, AND ALVEOLAR STAGES

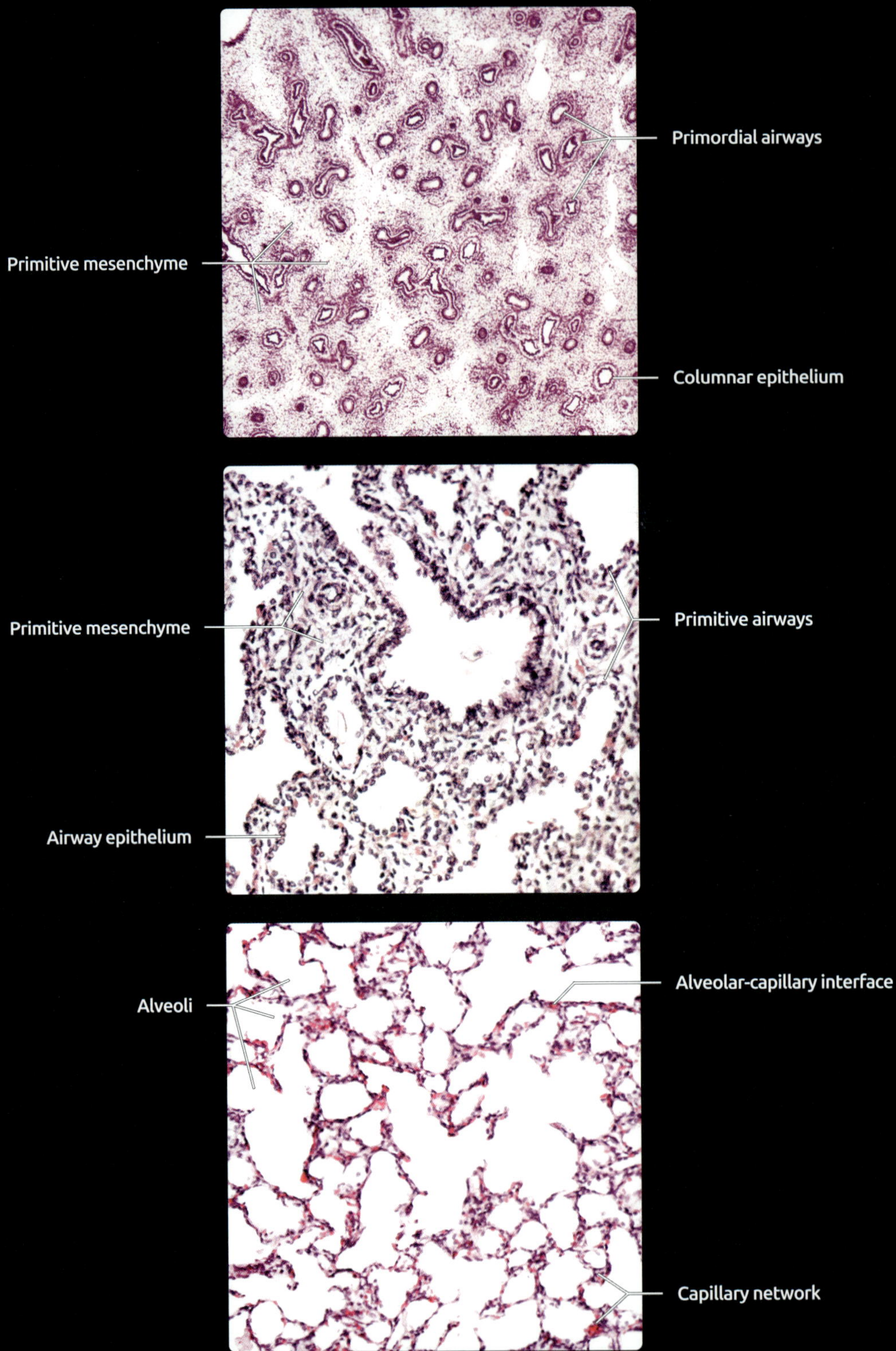

(Top) *Photomicrograph of fetal lung (hematoxylin and eosin stain) at 10 weeks of gestation in the pseudoglandular stage of lung development shows the gland-like microscopic morphology of the lung. The primordial airways are lined by columnar epithelium and separated by a significant quantity of primitive mesenchyme. (Courtesy J. Thomas Stocker, MD.)* **(Middle)** *Photomicrograph of fetal lung (hematoxylin and eosin stain) at 24 weeks of gestation in the canalicular stage of lung development shows the enlarging airway lumina. Portions of the airway epithelium have thinned, and there is less intervening primitive mesenchyme. (Courtesy J. Thomas Stocker, MD.)* **(Bottom)** *Photomicrograph (hematoxylin and eosin stain) of a normal lung at 38 weeks of gestation in the alveolar stage of lung development demonstrates the formation of adult-like alveoli and alveolar-capillary interfaces that allow the process of gas exchange enabling respiration. Lung development continues into infancy and early childhood. (Courtesy J. Thomas Stocker, MD.)*

NEONATAL LUNG

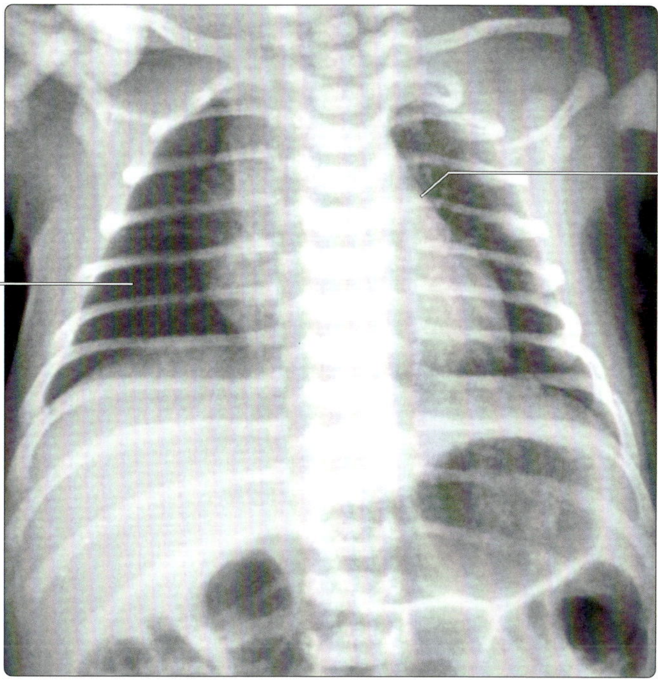

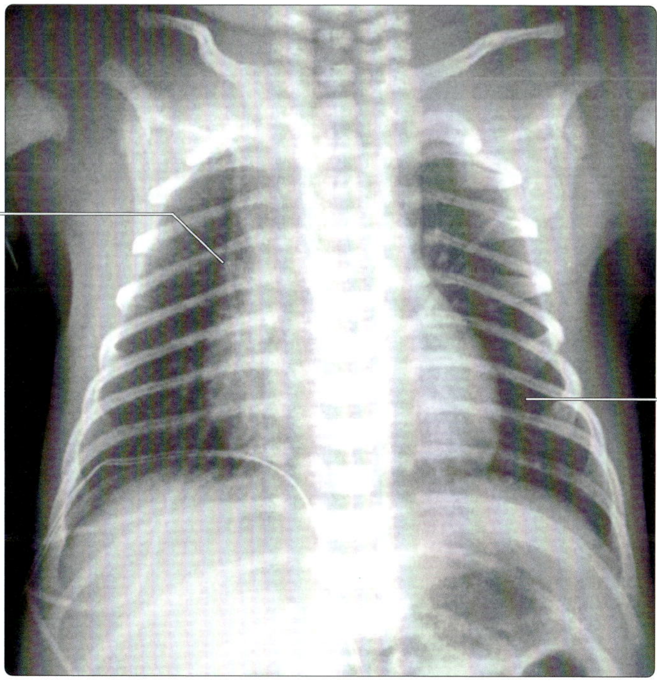

(Top) *AP chest radiograph of a symptomatic term infant imaged shortly after birth for evaluation of a holosystolic murmur shows normal lung volume and a normal cardiothymic silhouette. The patient was diagnosed with mild aortic stenosis.* **(Bottom)** *AP chest radiograph of a mildly symptomatic premature infant born at 33 weeks of gestation shows normal lung volume and a normal cardiothymic silhouette. The patient responded to conservative management and required no further imaging.*

RETAINED FETAL LUNG FLUID SYNDROME

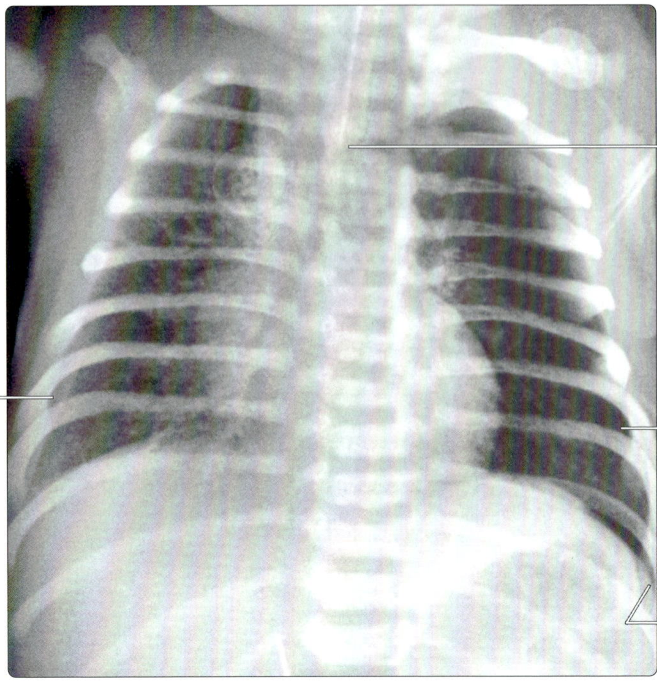

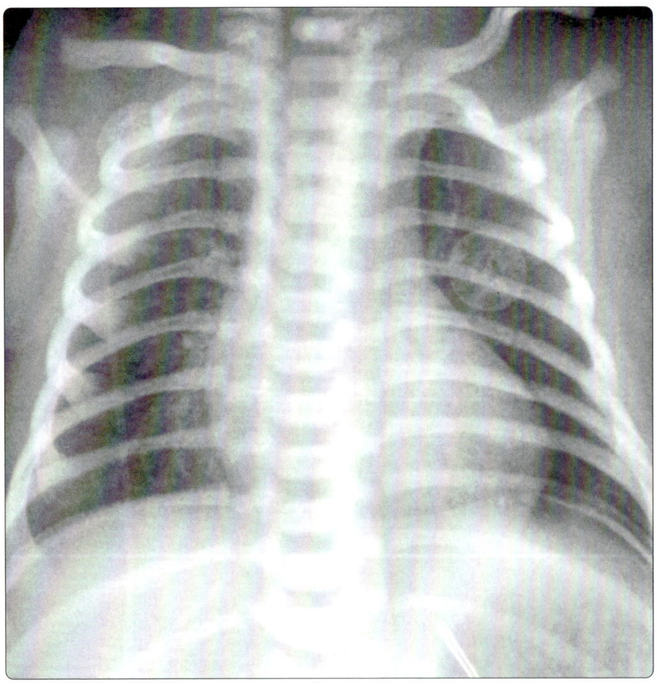

(Top) *First of 2 AP chest radiographs of a term infant with retained fetal lung fluid syndrome who presented with tachypnea and respiratory distress at birth shows coarse linear opacities in the perihilar regions and a small right pleural effusion. A small left pneumothorax manifests with a subtle visible pleural line and the deep sulcus sign. The endotracheal tube is appropriately positioned.* **(Bottom)** *AP chest radiograph of the same patient obtained 48 hours later shows that the patient has been extubated. The lungs are clear and exhibit normal volume. The pleural effusion and pneumothorax have resolved. The rapid resolution of abnormalities is consistent with suspected retained fetal lung fluid syndrome or transient tachypnea of the newborn. While the etiology of this condition is not completely understood, lung immaturity is thought to play a role in its pathogenesis.*

RESPIRATORY DISTRESS SYNDROME

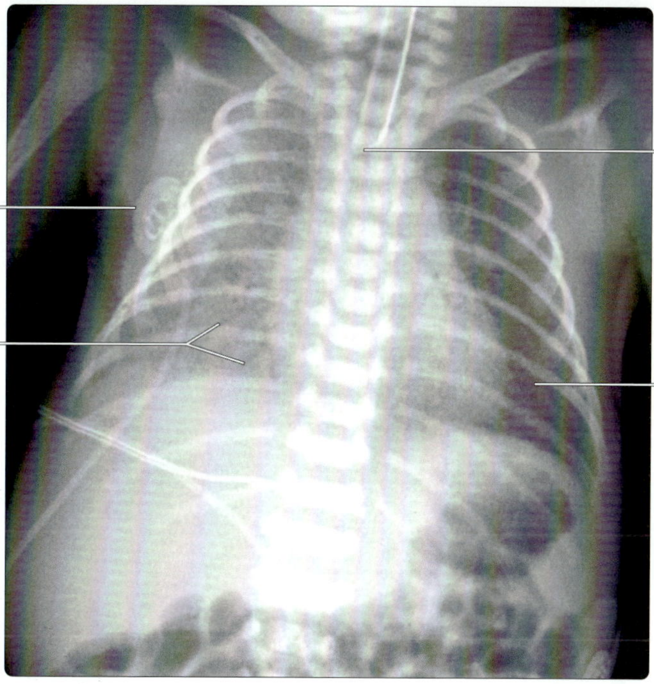

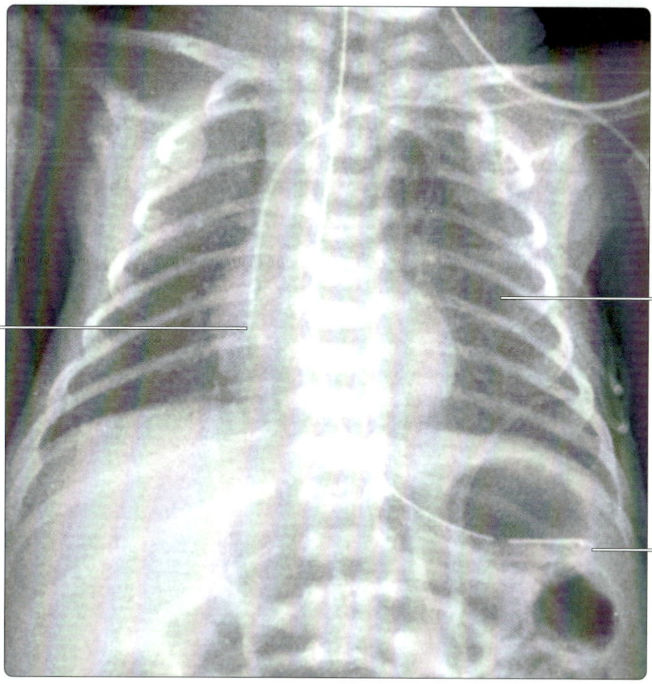

(Top) *First of 2 AP chest radiographs of a preterm infant born at 26 weeks of gestation who developed respiratory distress syndrome shows asymmetric granular pulmonary opacities and basilar air bronchograms. The pulmonary opacities result from a combination of adhesive atelectasis and retained fluid. There is no pleural effusion. The endotracheal tube is appropriately positioned.* **(Bottom)** *AP chest radiograph of the same patient obtained at 10 days of age shows that the patient has been extubated. The lung volume has improved with minimal residual fine granular opacities. An orogastric tube and a peripherally inserted central catheter (PICC) are noted. Approximately 60-80% of preterm infants born before 28 weeks of gestation develop respiratory distress syndrome.*

MORPHOLOGY AND ORIENTATION OF MAINSTEM BRONCHI

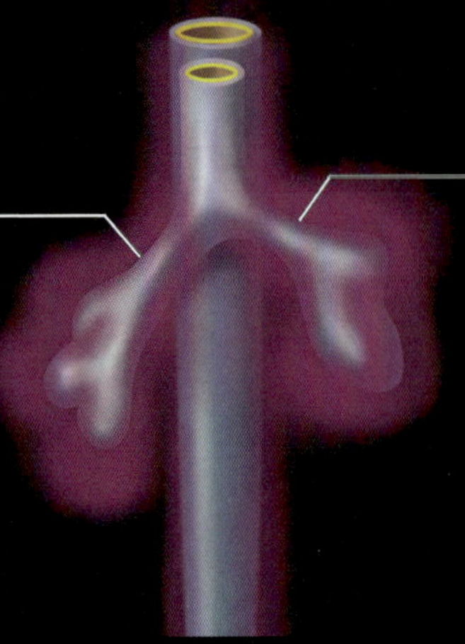

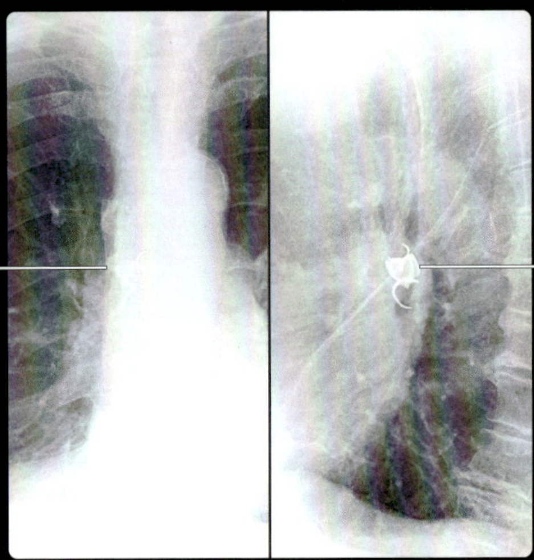

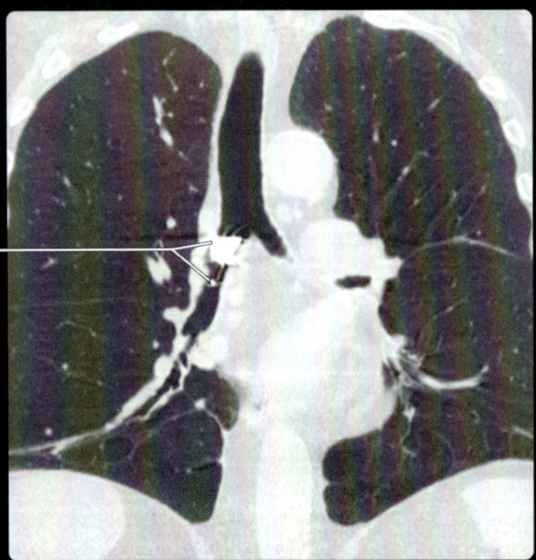

(Top) *Graphic shows the morphology and orientation of the central airways at 28 days of gestation. Note that the primordial right mainstem bronchus is more vertically oriented and larger than the left. As a result, the right tracheobronchial tree is more susceptible to aspiration of foreign bodies and secretions.* **(Middle)** *Composite image with PA (left) and lateral (right) chest radiographs of a 92-year-old man who aspirated a dental crown shows a metallic foreign body within the right mainstem bronchus.* **(Bottom)** *Coronal NECT (lung window) of the same patient shows the aspirated dental crown in the right mainstem bronchus. Given the more vertical orientation of the right tracheobronchial tree, aspirated foreign bodies and secretions exhibit a predilection for lodging in the right-sided airways.*

EMBRYONIC STAGE, TRACHEAL ATRESIA

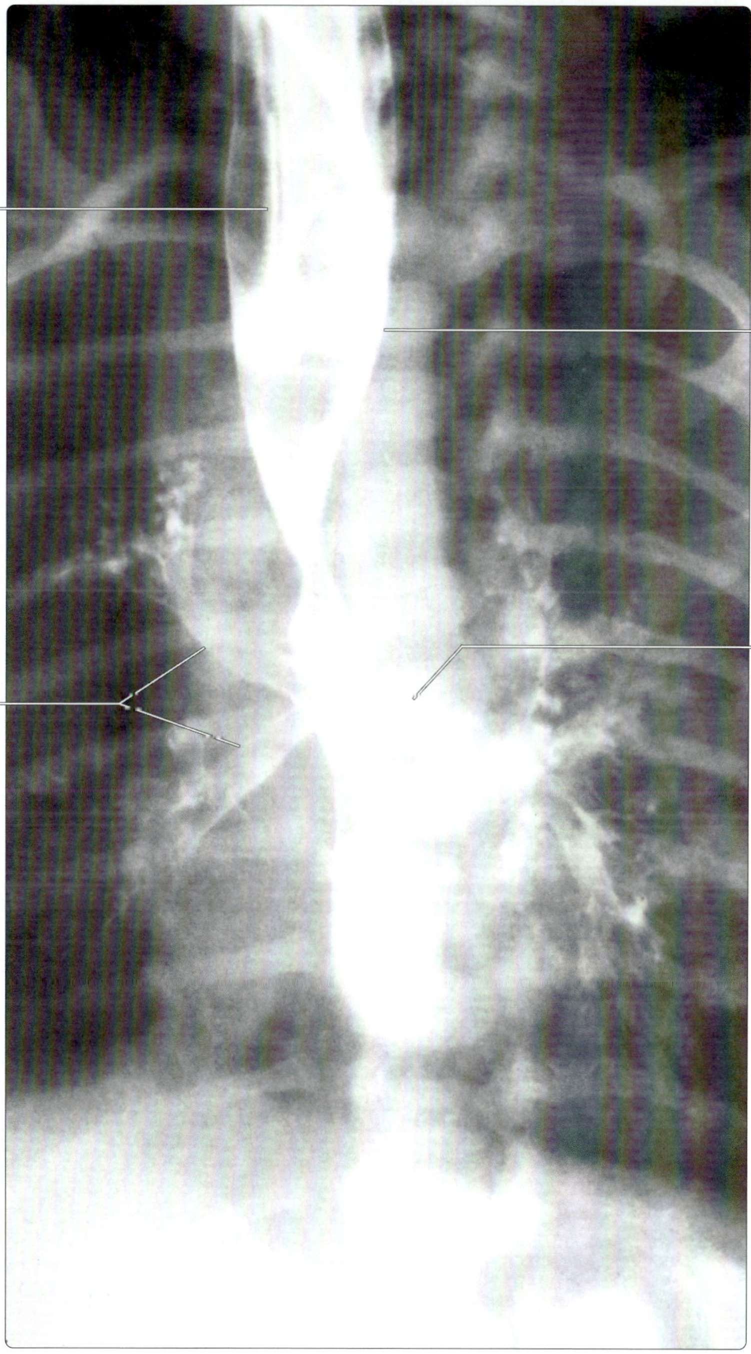

AP postmortem radiograph of an infant with tracheal atresia shows contrast administered via an endoluminal catheter, which opacifies the esophagus and the anomalous right and left bronchi that arise from the lateral esophageal walls. Tracheal atresia is thought to result from abnormal ventral location of the primitive tracheoesophageal folds, which interrupt the lumen of the developing trachea. Because the bronchi are foregut-derived structures, they maintain their primitive communication with the esophagus.

EMBRYONIC STAGE, TRACHEOESOPHAGEAL FISTULA AND ESOPHAGEAL ATRESIA

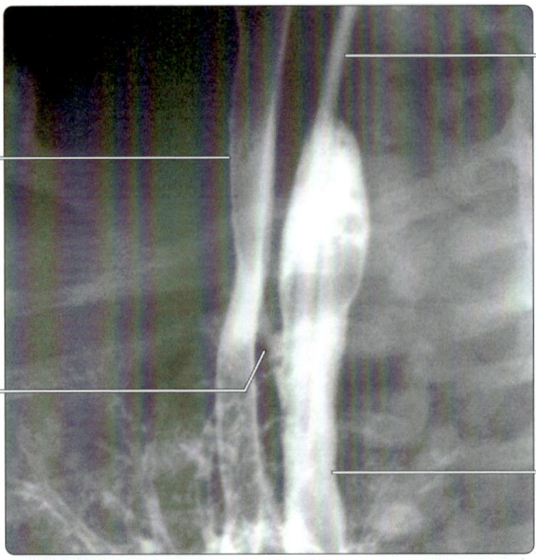

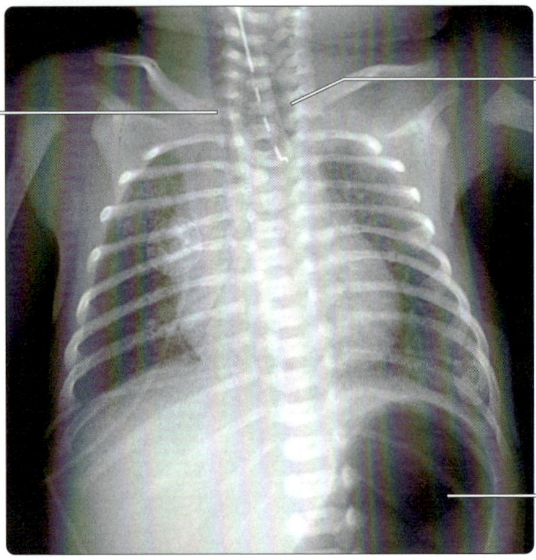

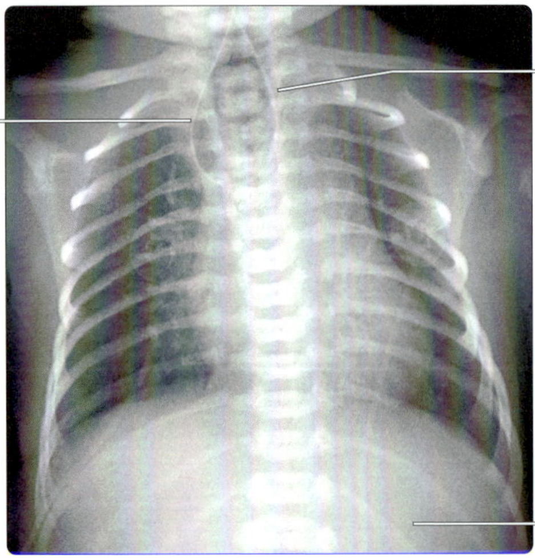

(Top) *Esophagram of an infant with recurrent pulmonary infections and congenital tracheoesophageal fistula shows an esophageal catheter used for contrast administration and an abnormal communication between the anterior esophagus and the posterior trachea, diagnostic of tracheoesophageal fistula, thought to result from failure of midline fusion of the primitive tracheoesophageal folds.* **(Middle)** *AP chest radiograph of a newborn with esophageal atresia shows a catheter in the atretic air-filled proximal esophagus. Gastric gas indicates a coexistent tracheoesophageal fistula. An upper thoracic hemivertebra and duodenal atresia (not shown) were consistent with VATER association. Esophageal atresia is thought to result from abnormal dorsal location of the tracheoesophageal folds, which interrupt the normal development of the esophageal lumen.* **(Bottom)** *AP chest radiograph of a neonate with esophageal atresia shows a catheter coiled in the air-distended atretic proximal esophagus. Absence of gastric or bowel gas indicates absence of a coexistent tracheoesophageal fistula.*

EMBRYONIC STAGE, EXTRALOBAR SEQUESTRATION

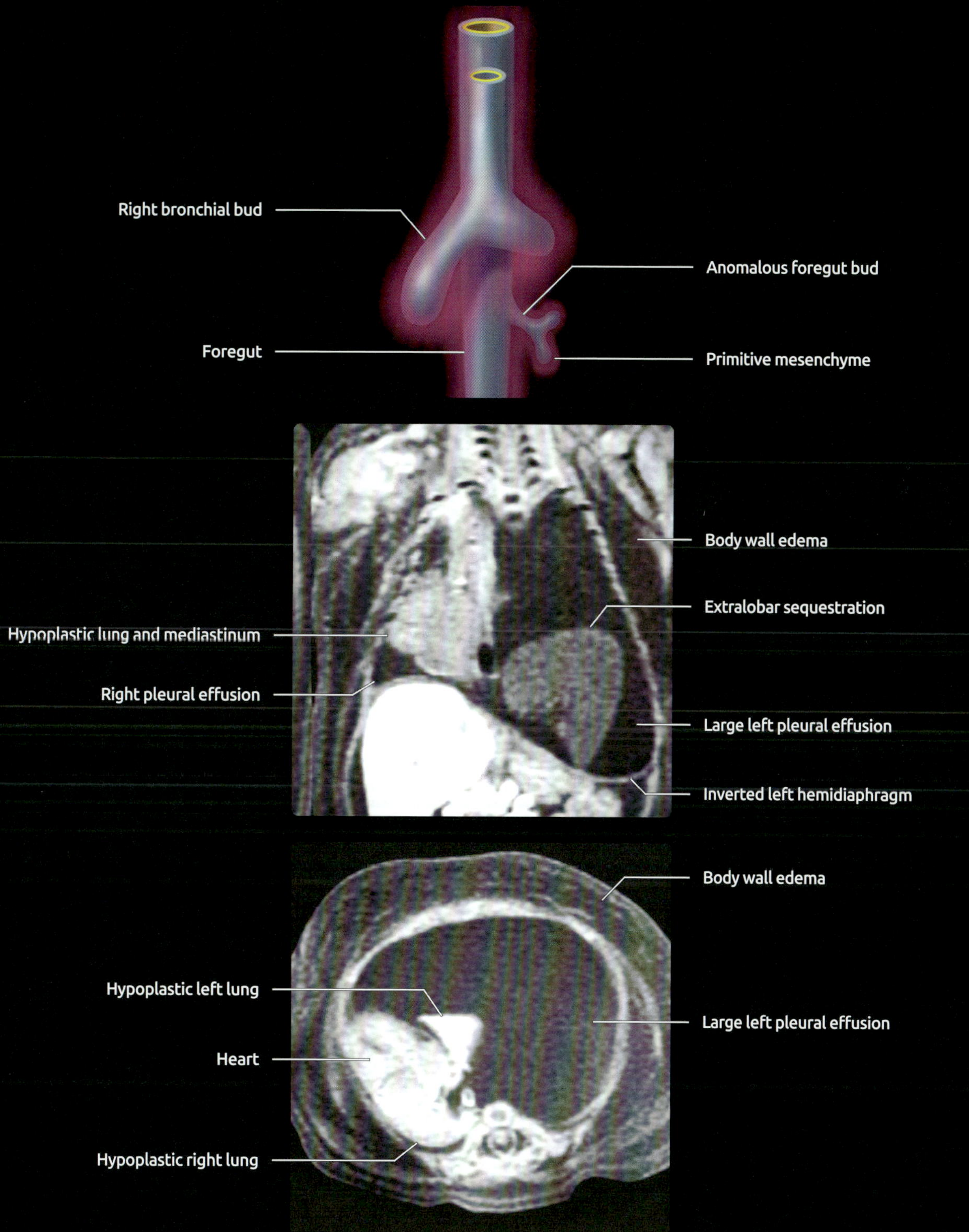

(Top) *Graphic shows the postulated pathogenesis of extralobar sequestration. A supernumerary foregut bud comes in contact with adjacent primitive mesenchyme and induces the development of lung parenchyma. The foregut communication typically involutes with resultant sequestered lung with a systemic blood supply.* **(Middle)** *Postmortem coronal T1 MR shows a large left extralobar sequestration, a large left pleural effusion that produces inversion of the left hemidiaphragm, and a small right pleural effusion. There is resultant severe pulmonary hypoplasia with most of the lung and the mediastinum located in the right hemithorax. Associated abdominal ascites and body wall edema are consistent with fetal hydrops.* **(Bottom)** *Axial T1 MR shows a large left pleural effusion, severe pulmonary hypoplasia, and marked body wall edema. Normal lung development requires adequate intrathoracic volume. In this case, mass effect from the extralobar sequestration and the large left hydrothorax precluded normal lung development and resulted in pulmonary hypoplasia.*

EMBRYONIC STAGE, BRONCHOGENIC CYST

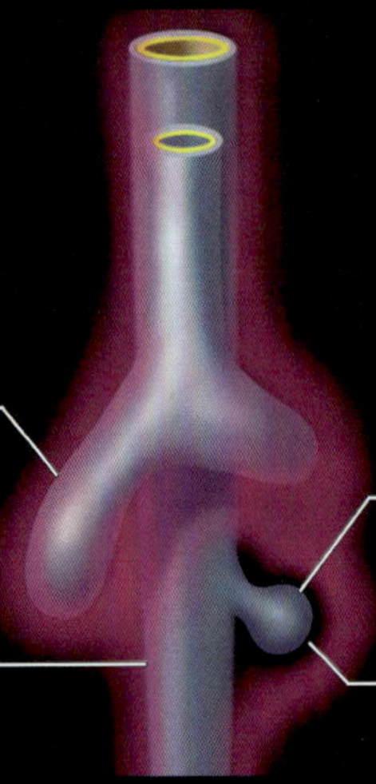

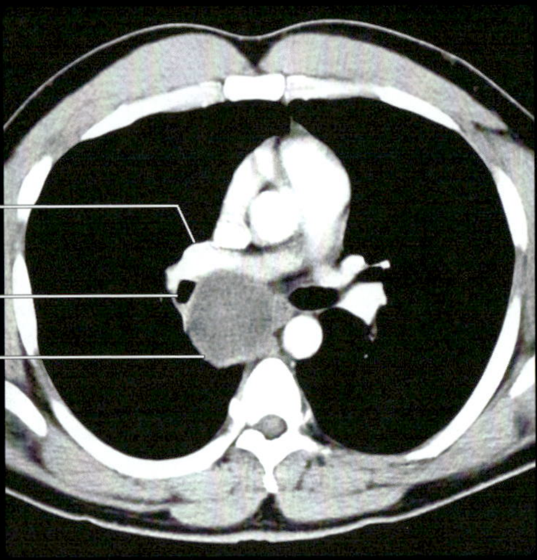

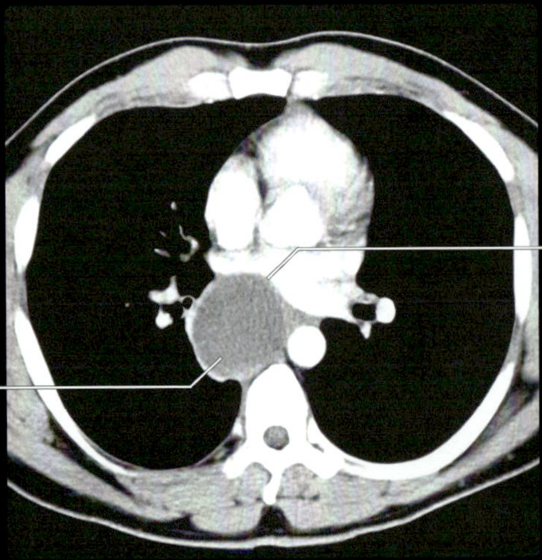

(Top) *Graphic shows the proposed pathogenesis of bronchogenic cyst. An anomalous supernumerary bud from the primitive foregut fails to induce the formation of surrounding lung parenchyma. The original communication with the foregut typically involutes resulting in a blind-ending pouch or cyst. Because the anomalous bud develops in the same way as the primitive central airways, the cyst wall contains bronchial wall components, hence, the term bronchogenic.* **(Middle)** *Axial CECT (mediastinal window) of a young man with a bronchogenic cyst shows a subcarinal thin-walled spherical cyst with water attenuation contents that produces mass effect on the right pulmonary artery and the bronchus intermedius.* **(Bottom)** *Axial CECT (mediastinal window) of the same patient shows the spherical subcarinal cystic lesion with nonenhancing water attenuation contents and a thin enhancing cyst wall. The lesion produces mass effect on the left atrium. The morphology and location of the lesion are characteristic of bronchogenic cyst.*

PSEUDOGLANDULAR STAGE, BRONCHIAL ATRESIA

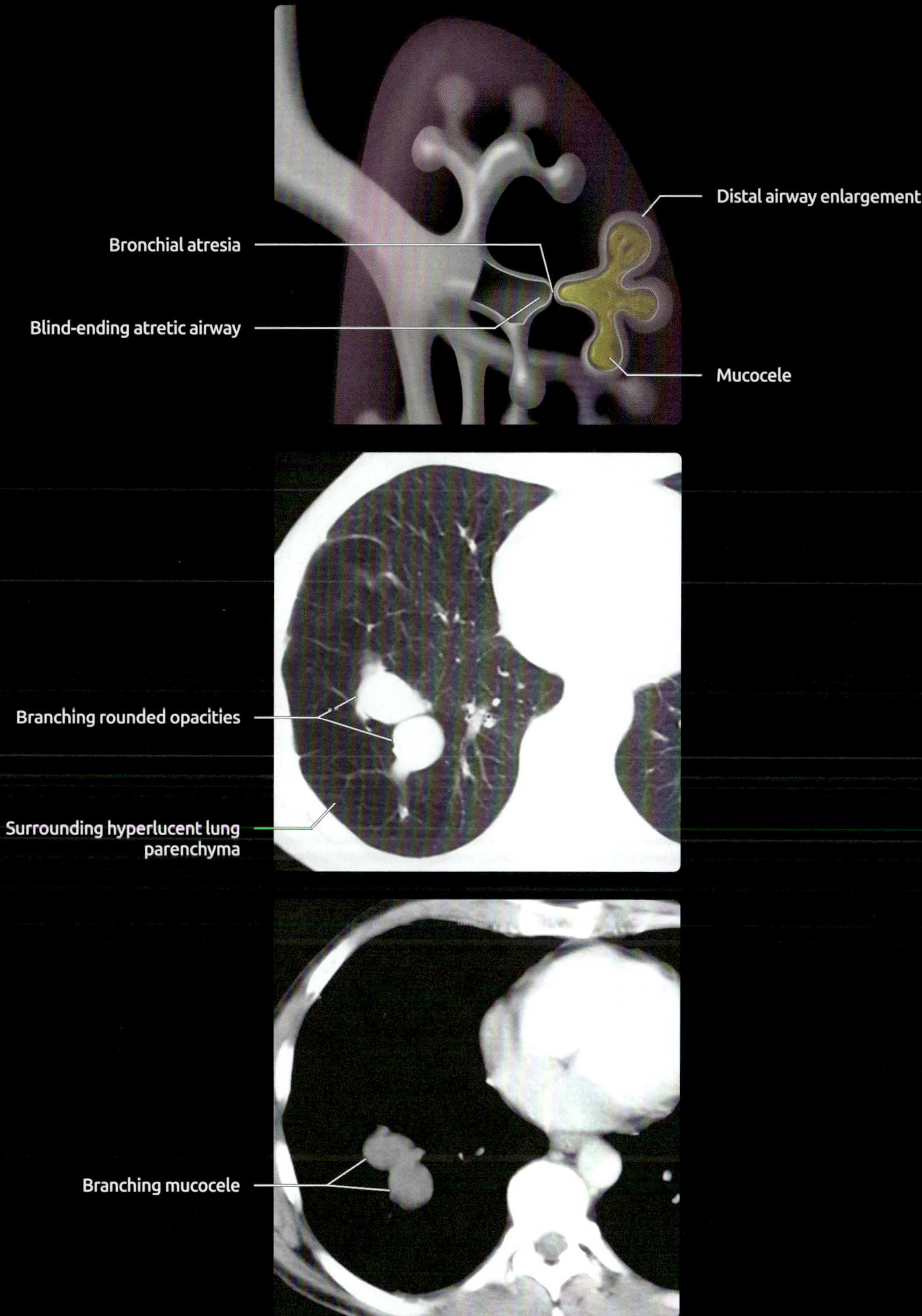

(Top) *Graphic shows the proposed pathogenesis of bronchial atresia, which is postulated to occur during the pseudoglandular stage of lung development when all the elements of the tracheobronchial tree are present. Bronchial atresia may result from in utero vascular compromise of a portion of the airway wall. The atretic bronchus ends blindly at the point of atresia. Distal airway secretions result in the formation of a bronchocele or mucocele, which enlarges the distal airway lumen. After birth, surrounding air-trapping produces hyperlucent lung parenchyma.* **(Middle)** *Axial CECT (lung window) of a young man with bronchial atresia shows right lower lobe central branching rounded opacities surrounded by hyperlucent lung. Air distal to the atretic bronchus is thought to reach the affected lung via collateral air drift.* **(Bottom)** *Axial CECT (mediastinal window) of the same patient shows nonenhancing right lower lobe branching opacities that represent the mucocele that forms distal to the point of atresia. The findings are diagnostic of bronchial atresia.*

NORMAL DIAPHRAGMATIC DEVELOPMENT

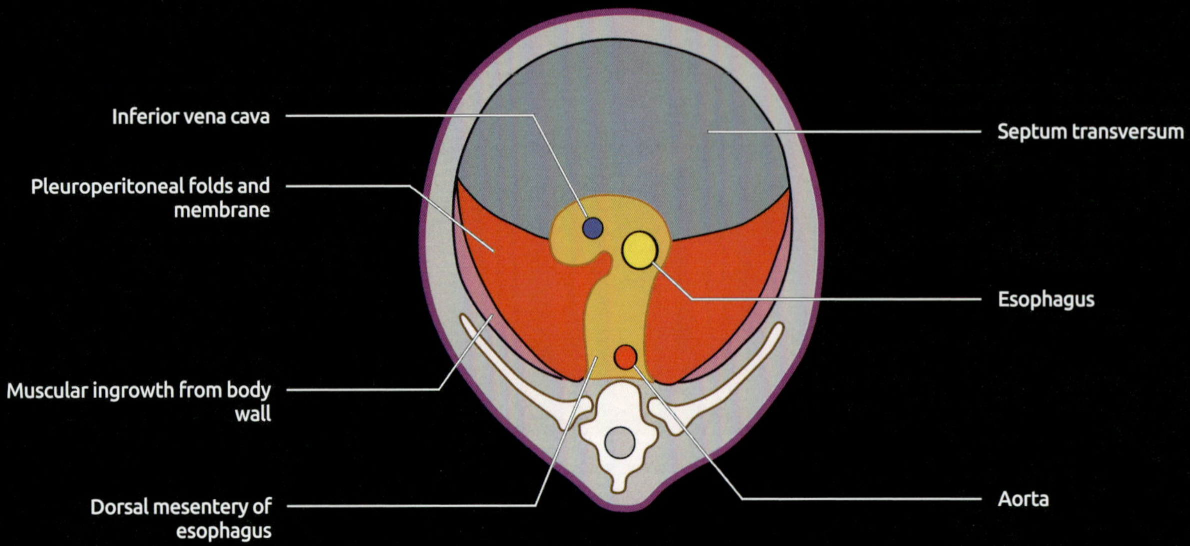

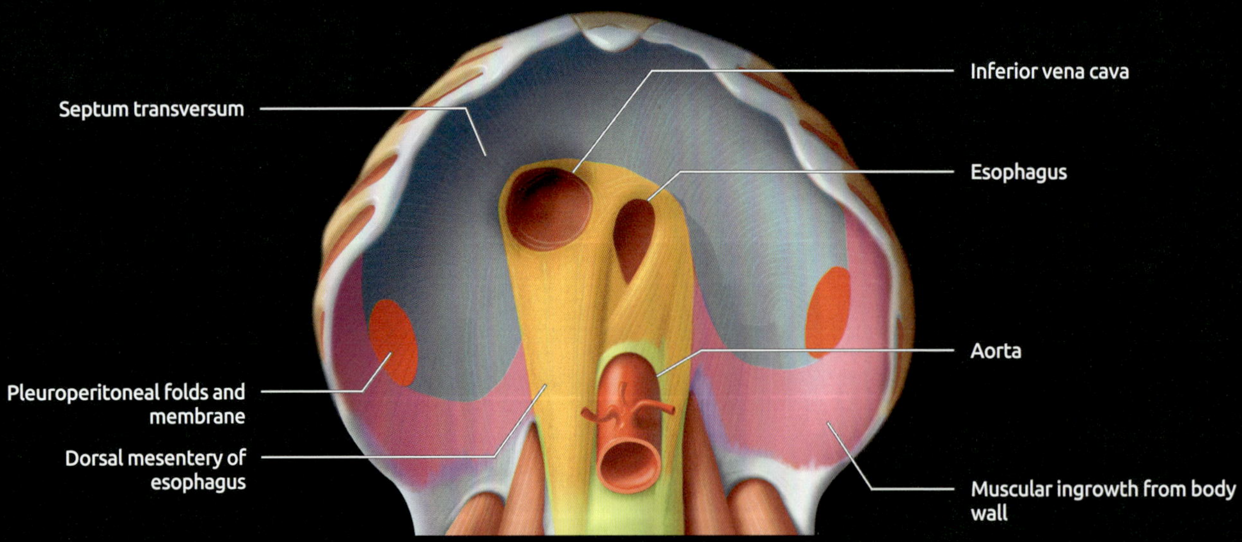

(Top) *Graphic depicts a transverse section of a fetus at 12 weeks of gestation and illustrates the complexity of diaphragmatic development from 4 separate embryonic components, the septum transversum, the dorsal mesentery of the esophagus, the pleuroperitoneal folds and membranes, and the muscular ingrowths from the body wall. These 4 components fuse together to form the primordial diaphragm and achieve closure of the pleuroperitoneal canal.* **(Bottom)** *Graphic shows an inferior view of the diaphragm at birth and the relative contributions from the 4 embryonic diaphragmatic components (shaded in color to correspond to the previous graphic). Abnormalities of fusion &/or agenesis of portions of the embryonic diaphragmatic components may result in congenital diaphragmatic hernia in which abdominal contents herniate into the thorax through diaphragmatic defects of varying size.*

CONGENITAL DIAPHRAGMATIC HERNIA

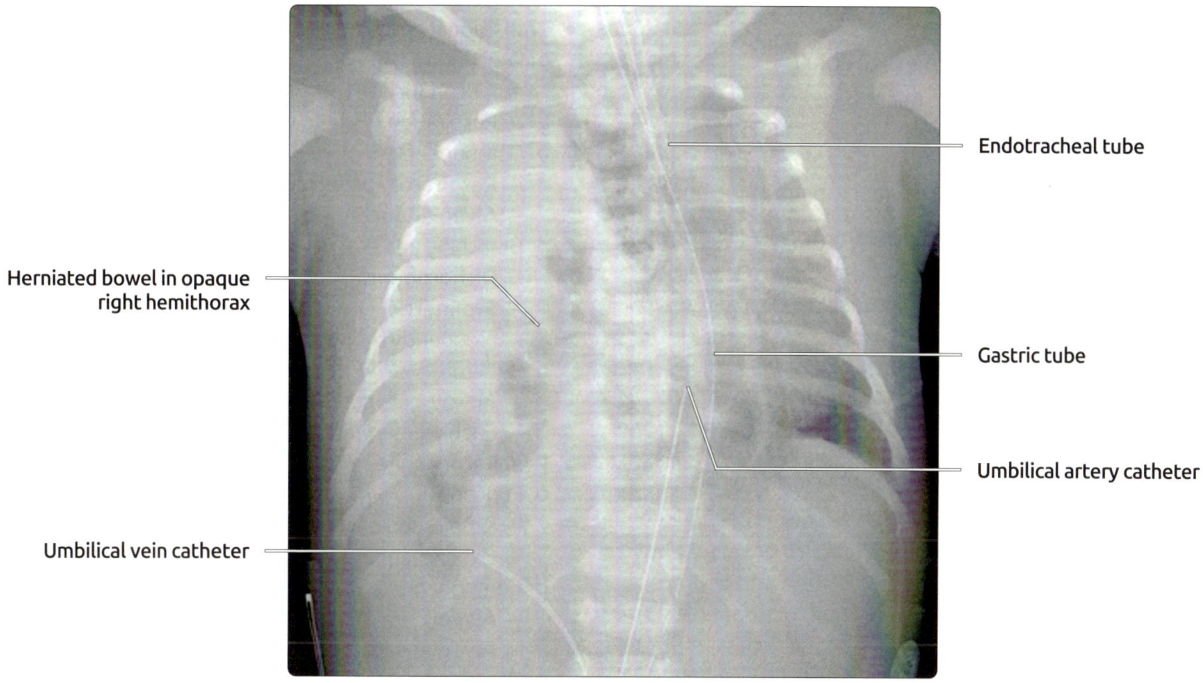

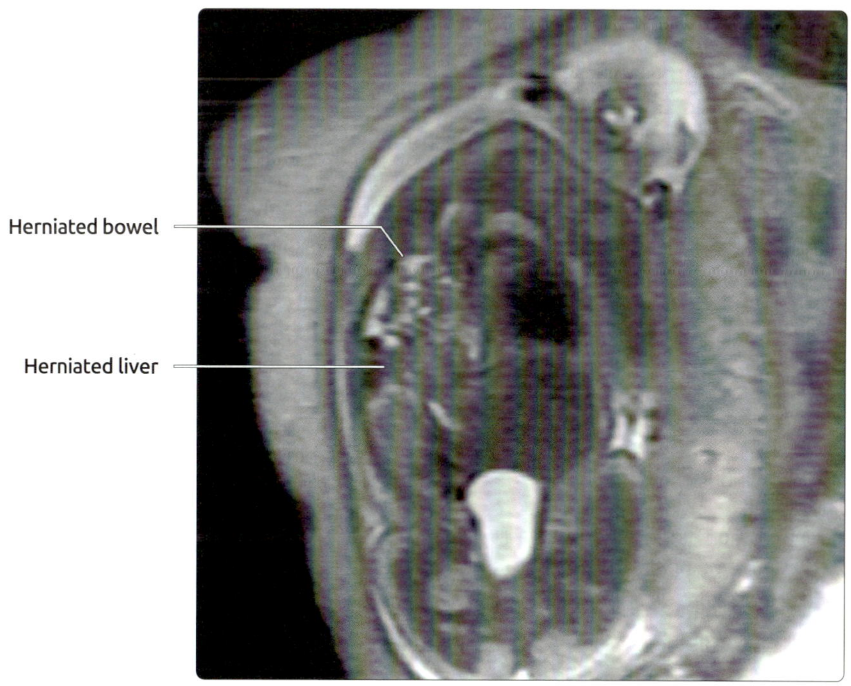

(Top) *AP chest radiograph of a newborn obtained shortly after birth shows a nearly opaque right hemithorax that contains air-filled bowel, consistent with congenital diaphragmatic hernia. Note mediastinal shift to the left and displacement of the gastric tube and umbilical artery catheter to the left. The umbilical vein catheter is displaced to the right within herniated liver.* **(Bottom)** *Coronal fetal MR of the same patient shows herniation of both bowel and liver into the right hemithorax through a right-sided congenital diaphragmatic hernia. Herniated bowel produces mass effect on the developing lung. Normal lung development requires normal intrathoracic volume, and abnormalities that reduce or compromise intrathoracic volume and exert mass effect on the fetal lung are associated with abnormal lung development and may result in pulmonary hypoplasia.*

- Most often seen in apices, anteriorly, and near mediastinal pleura
- Location may be inferred by identification of septal veins
- Lobular bronchioles at lower limits of thin-section CT resolution
  - Not normally visible
  - Lobular bronchiole measures ~ 1 mm in diameter
  - Visibility correlates with wall thickness
  - Location may be inferred by identification of centrilobular artery

Acinus

- Normal acini are not visible on imaging
- Experimental filling of single acinus results in rosette appearance and progresses to spherical opacity
  - Acinar/airspace nodule

ANATOMY-RELATED IMAGING ABNORMALITIES

Centrilobular and Tree-in-Bud Micronodules

- Often related to cellular bronchiolitis
  - Rarely due to vascular causes (e.g. excipient lung disease)
- Etiologies
  - Infectious
    - Acute: Bacteria, viruses
    - Chronic: Mycobacteria (tuberculosis and nontuberculous), fungi
  - Lymphoproliferative: Follicular bronchiolitis
  - Inflammatory: Respiratory bronchiolitis
  - Allergic: Hypersensitivity pneumonitis
  - Diffuse aspiration bronchiolitis
  - Vascular tree-in-bud
    - Excipient lung disease
      - □ Rare but more common in IV drug users or patients with long-term indwelling catheters
      - □ Illicit intravenous injection of drugs intended for oral use which contain talc, cellulose, or crospovidone
    - Tumor embolism
- CT features
  - Small micronodules
    - Variable attenuation
    - Size range from few mm to 1 cm
  - Centrilobular location
    - Situated 5-10 mm from pleural surface, typically spare interlobar fissures
  - Tree-in-bud opacities
    - Clustered centrilobular micronodules and branching opacities
    - Resemble tree in bud
- Pathologic correlation
  - **Inflammatory** (cellular) **bronchiolitis**
    - Inflammation/infiltration of centrilobular bronchioles
    - Involvement of surrounding interstitium and alveoli

Centrilobular Low Attenuation: Centrilobular Emphysema

- CT features
  - Foci (3-10 mm) of centrilobular low attenuation
  - Low attenuation surrounding centrilobular artery
  - Imperceptible walls
- Pathologic correlation
  - **Centrilobular** (**proximal acinar**, **centriacinar**) **emphysema**
    - Involves proximal acinus
    - Distention and destruction of respiratory bronchioles
    - Enlarged airspace in central acinus with relatively normal distal acinus
  - Most severe involvement of upper lobes and superior segments of lower lobes

Lobular Low Attenuation: Panlobular Emphysema

- CT features
  - Diffuse extensive areas of low attenuation
  - Reduced size of pulmonary vessels
- Pathologic correlation
  - **Panlobular** (**panacinar**) **emphysema**
    - Affects entire acinus and all acini in secondary pulmonary lobule
    - Diffuse or lower lobe predominance
    - Association with α-1-antitrypsin deficiency

Lobular Low Attenuation: Paraseptal Emphysema

- CT features
  - Juxtapleural cystic areas near interlobular septa, large vessels, and bronchi
  - Frequent association with centrilobular emphysema
- Pathologic correlation
  - Affects periphery of pulmonary acinus and subpleural secondary pulmonary lobules
  - Enlarged alveolar ducts
  - Predominant upper lobe involvement
  - Association with **bullous disease**

Lobular Low Attenuation: Constrictive Bronchiolitis

- CT features
  - Mosaic attenuation
  - Patchy distribution
  - **Air-trapping** on expiratory CT
- Pathologic correlation
  - Concentric narrowing of membranous bronchioles by fibrosis with resultant air flow obstruction
    - **Mosaic attenuation**
    - **Air-trapping** on expiratory imaging

Acinar Nodules

- CT features
  - 6- to 10-mm fluffy nodular opacities
- Pathologic correlation
  - Infection
    - Inflammation of terminal and respiratory bronchioles
    - Spared distal air spaces
    - Airway dissemination of infection
      - □ **Tuberculosis**
      - □ Early **varicella pneumonia**
  - Edema
    - Alveolar extravascular water
  - Hemorrhage
    - Alveolar blood products
  - Malignancy
    - Invasive mucinous adenocarcinoma

OVERVIEW OF AIRWAY STRUCTURE

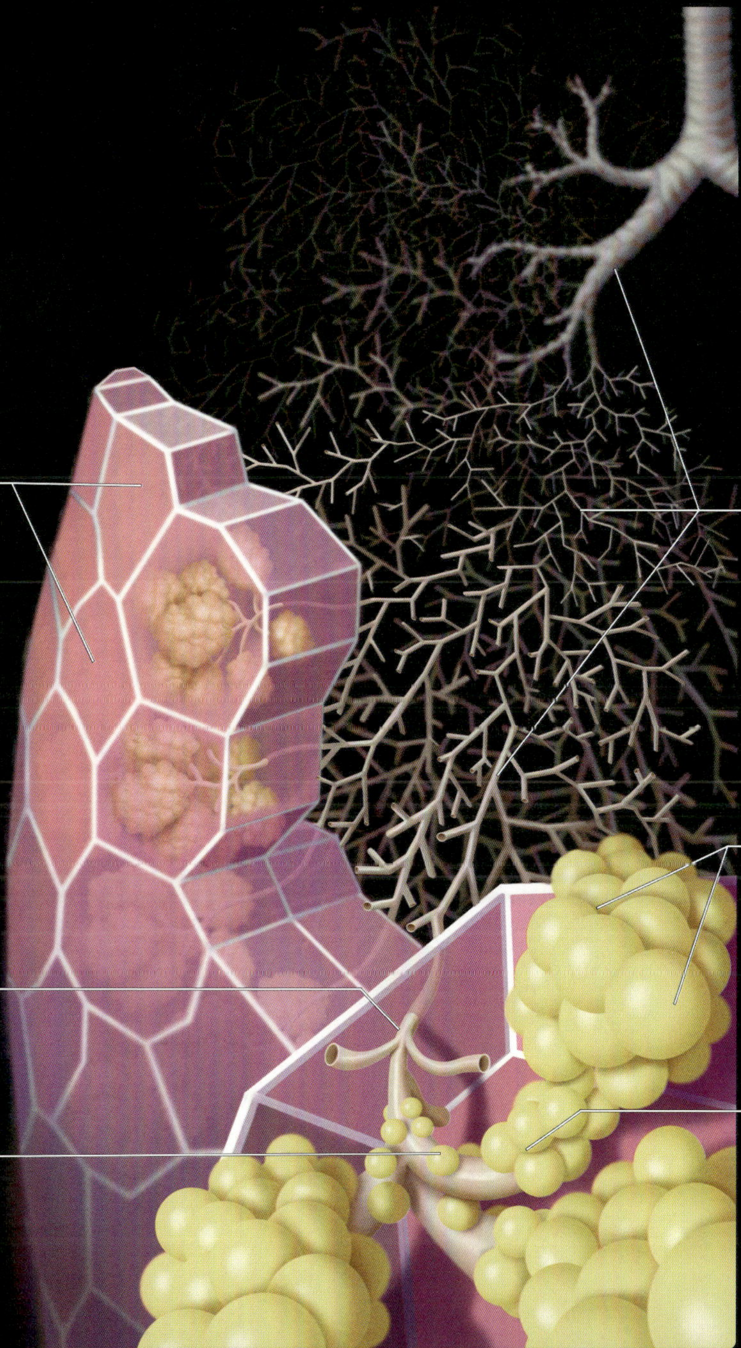

Graphic representation shows 24 airway generations branching dichotomously from the trachea to the most distal airways that compose the secondary pulmonary lobule. The secondary pulmonary lobule is the smallest unit of lung surrounded by connective tissue septa and has a polyhedral shape. Each secondary pulmonary lobule contains the distal branches of a lobular bronchiole and its accompanying pulmonary (lobular) artery. The acinus is composed of the airways distal to a terminal bronchiole with each secondary pulmonary lobule containing 12 or fewer acini. The terminal bronchiole gives rise to 2 or 3 respiratory bronchioles, which in turn give rise to 3 alveolar ducts, each terminating in alveolar sacs and alveoli. The respiratory bronchiole is characterized by its alveolated wall. The alveolar duct wall is covered with alveoli. The alveolar sacs terminate in clustered alveoli.

ANATOMY OF BRONCHI, BRONCHIOLES, AND ACINI

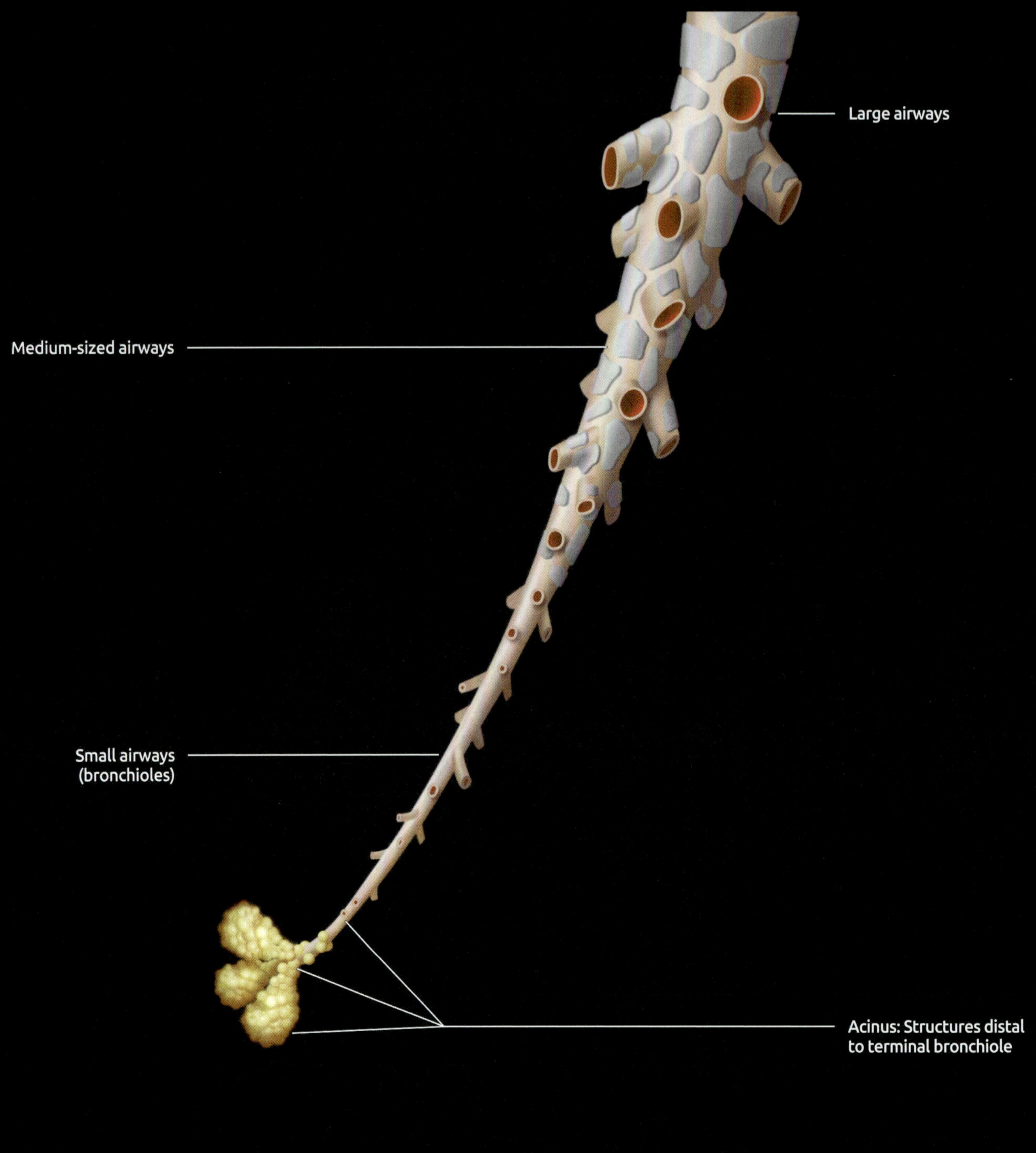

Graphic telescoping depiction of an airway illustrates the decreasing size of the different airway types and the structural changes of the airway wall with a decreasing number and size of cartilage plates. Cartilage plates are present in large and medium-sized airways (trachea and bronchi) but gradually become smaller and less numerous in the medium-sized bronchi. The walls of small airways (bronchioles) do not contain cartilage. Distal clusters of alveoli and alveolar sacs form acini, the pulmonary functional unit of gas exchange. The acinus is defined as the airways, vessels, and supporting structures distal to a terminal bronchiole.

MICROSCOPIC STRUCTURE OF TRACHEA, BRONCHI, AND BRONCHIOLES

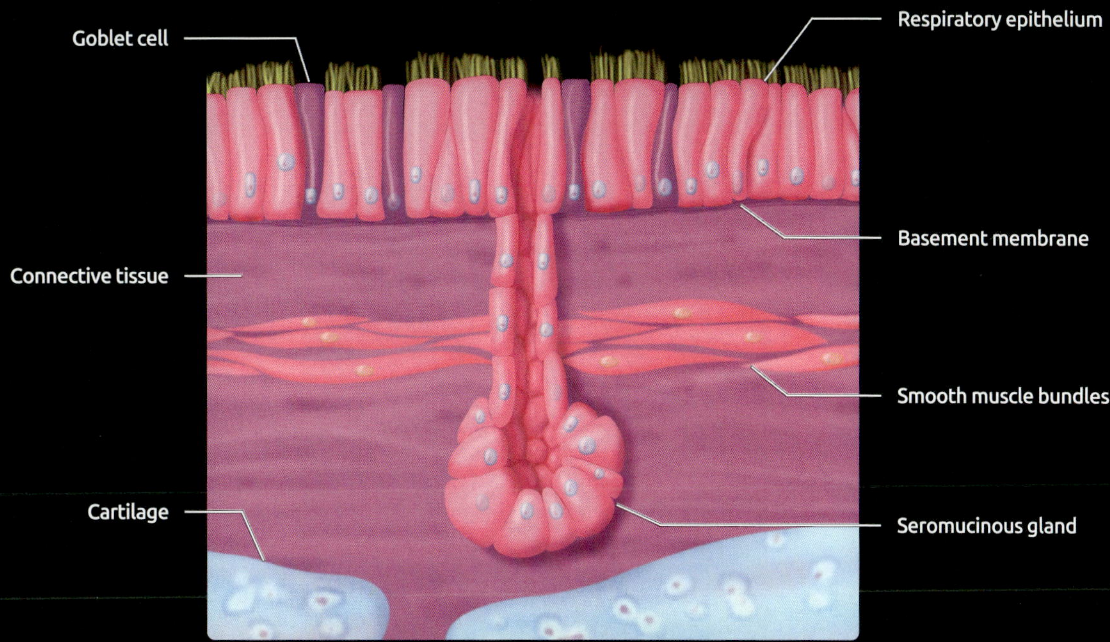

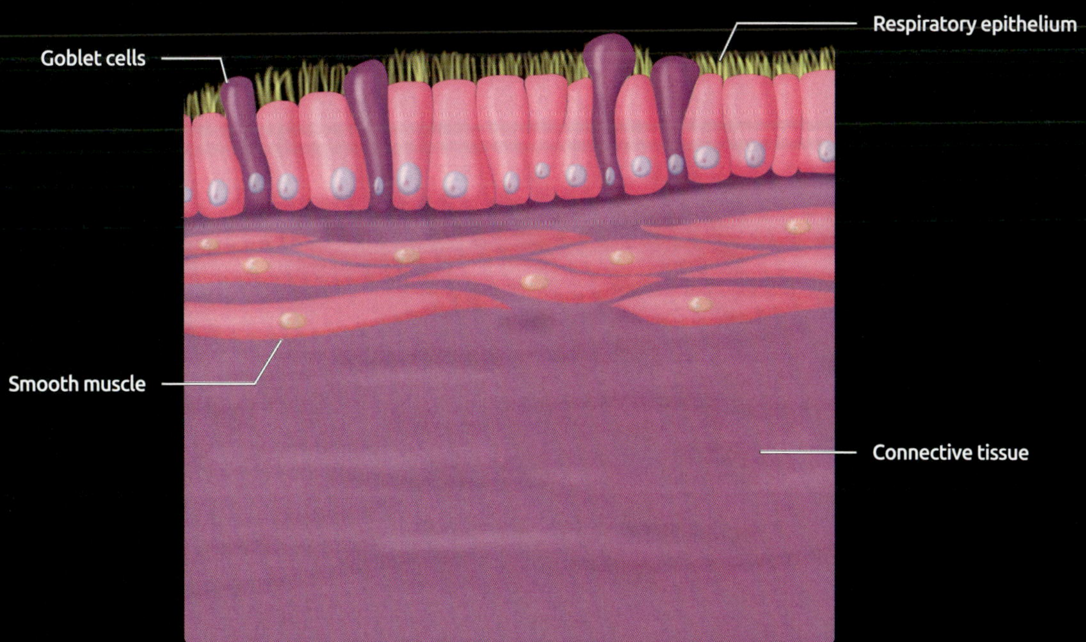

(Top) *Graphic shows the microscopic structure of the large cartilaginous airways. These airways are lined by pseudostratified ciliated columnar (respiratory) epithelium that rests on a basement membrane. The cilia participate in the mucociliary escalator that pushes overlying mucin in a cephalad direction and provides clearance of secretions and particulate matter. Submucosal loose connective tissue beneath the basement membrane contains smooth muscle bundles and seromucinous glands. Cartilage plates are found beneath the submucosal layer.* **(Bottom)** *Graphic shows the microscopic structure of the bronchioles that are lined by respiratory epithelium. Goblet cells contribute to airway mucus production and are interspersed between columnar ciliated cells. Smooth muscle bundles are spirally arranged in the submucosa. There are no cartilage plates or bronchial glands.*

LARGE AIRWAYS, RADIOGRAPHY AND CT

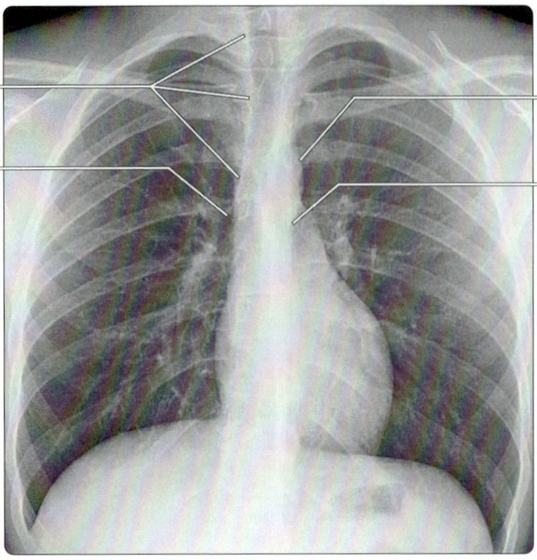

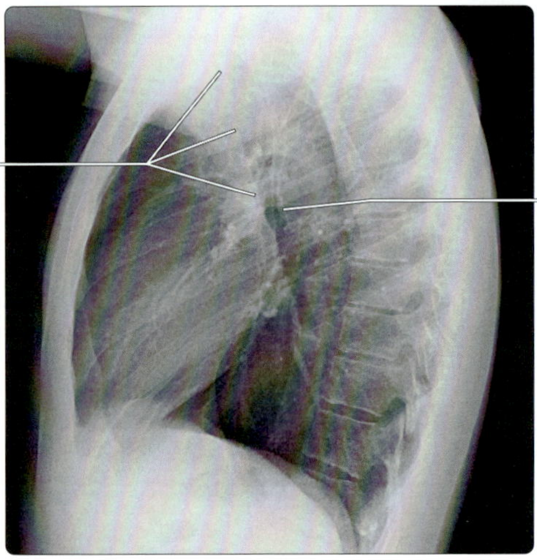

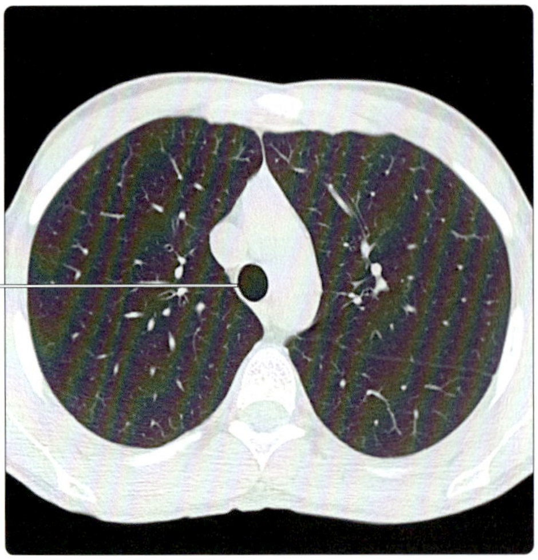

(Top) *PA chest radiograph of a normal individual shows a normal configuration of the central airways. The aortic arch causes a mild indentation on the distal left tracheal wall and mildly displaces the trachea to the right. Despite the fact that the central airways are well visualized on radiography, focal central stenosis and central endoluminal lesions are often missed. The peripheral airways are not well visualized on radiography, although indirect and nonspecific findings, such as micronodules, bronchial wall thickening, and bronchiectasis, may be visible.* **(Middle)** *Lateral chest radiograph of the same individual shows the radiographic appearance of the central airways. In spite of the size of the central airways and their visibility, chest radiography has a low sensitivity for the detection of central airway pathology. Familiarity with the normal radiographic anatomy, morphology, and location of the central airways is critical for radiographic identification of subtle airway abnormalities, such as stenosis, endoluminal lesions, and abnormal displacement.* **(Bottom)** *Axial NECT (lung window) of the same individual obtained during inspiration shows a normal rounded tracheal configuration.*

LARGE AIRWAYS, CT

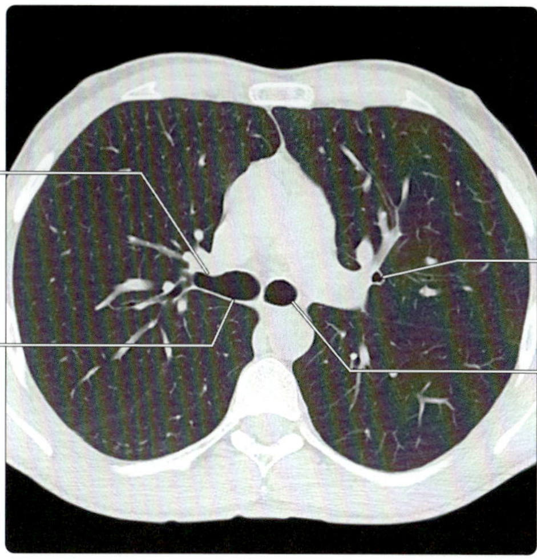

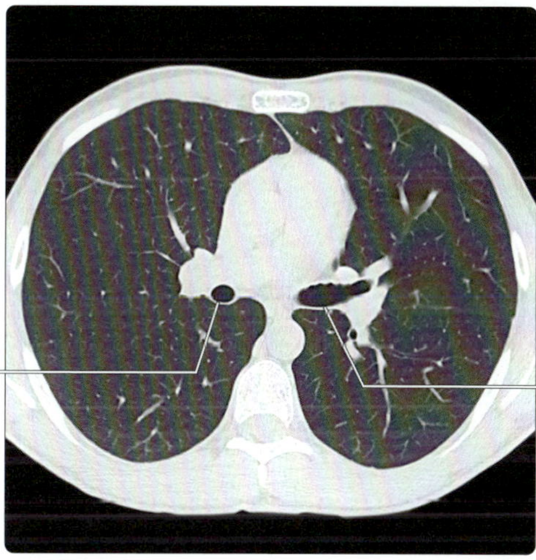

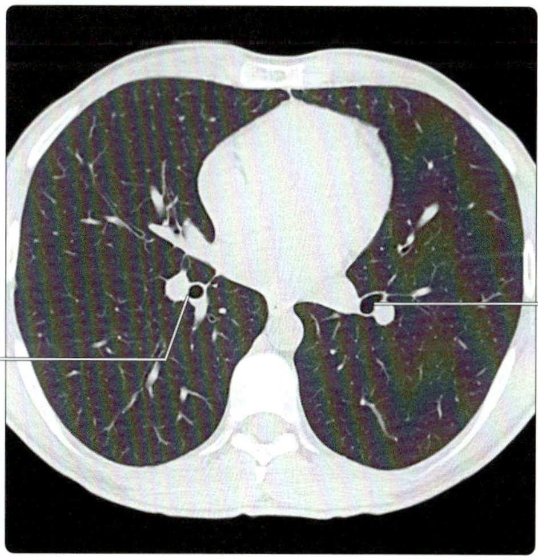

(Top) *Axial NECT (lung window) of the same individual obtained slightly below the carina shows the normal right and left mainstem bronchi. Note the right upper lobe bronchus, which is the 1st branch of the right mainstem bronchus.* **(Middle)** *Axial NECT (lung window) of the same individual obtained at a more caudal level shows the bronchus intermedius, which will give rise to the middle and right lower bronchi. Note the normal distal left mainstem bronchus.* **(Bottom)** *Axial NECT (lung window) of the same individual obtained at a more caudal level shows the truncus basalis portions of the right lower and left lower bronchi, which will give rise to the bilateral basilar segmental bronchi. The small airways distal to the muscular bronchioles are not visible with current imaging modalities.*

LARGE AIRWAYS, CT

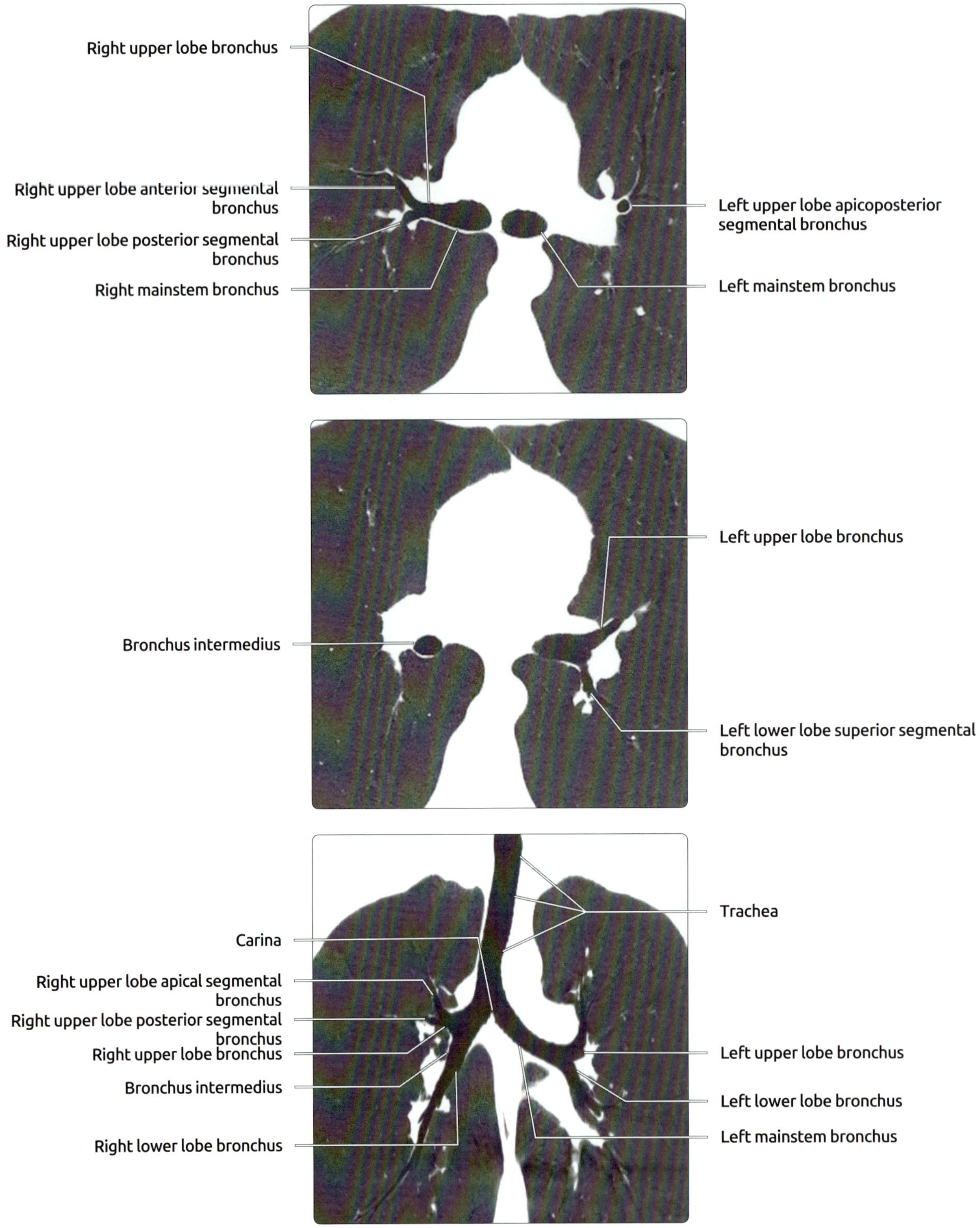

(Top) *Axial NECT (lung window) minIP reformatted image of the same individual is shown. minIP reformatted images highlight low-attenuation structures and lesions in the lung, particularly the airways, and may be helpful in understanding normal airway anatomy. In this case, 2 segmental branches of the right upper lobe bronchus (anterior and posterior segmental bronchi) and a segmental branch of the left upper lobe bronchus (apicoposterior segmental bronchus) are visible on a single image.* **(Middle)** *Axial NECT (lung window) minIP reformatted image of the same individual at a more caudal level shows the left upper lobe bronchus, the bronchus intermedius, and the left lower lobe superior segmental bronchus on the same image.* **(Bottom)** *Coronal NECT (lung window) minIP reformatted image of the same individual shows the normal coronal anatomy of the trachea and the central bronchi, which correlates with the course and morphology of the central airways on radiography.*

LARGE AIRWAYS, CT

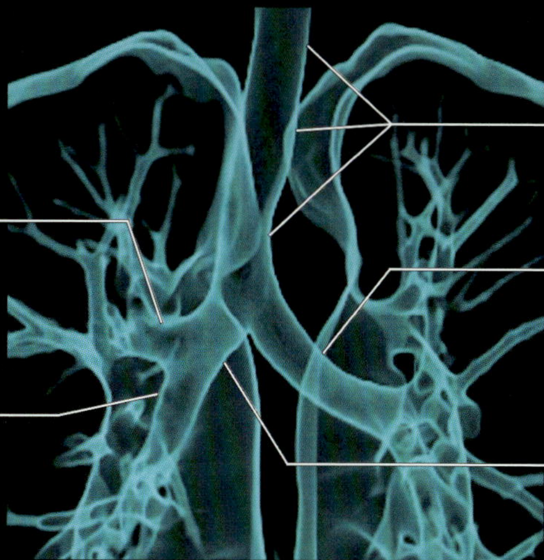

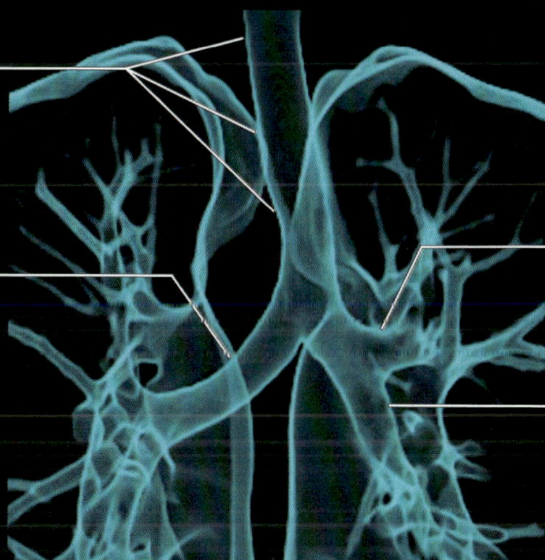

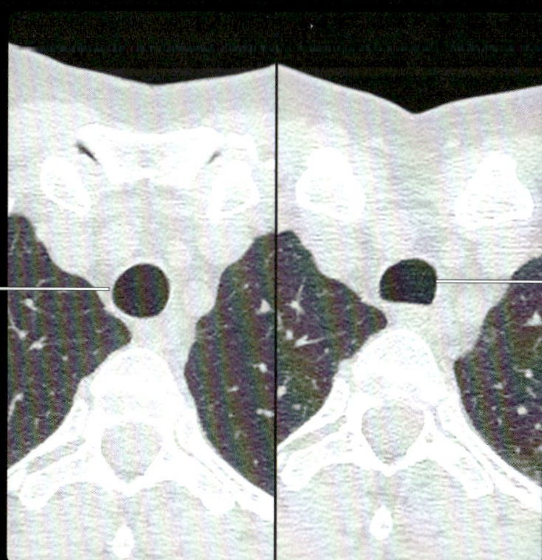

(Top) *Coronal NECT anterior view 3D surface-rendered reformatted image of the same individual shows the anatomy of the normal central airways. While 3D surface-rendered reformatted images are not usually helpful for diagnostic purposes, they may be useful to clinicians in planning endoscopic and surgical interventions.* **(Middle)** *Coronal NECT posterior view 3D surface-rendered reformatted image of the same individual shows a posterior view of the normal central airways. Evaluation of anterior and posterior views of the airways may be useful in planning endoscopic and surgical interventions.* **(Bottom)** *Composite image with axial NECT (lung window) obtained at inspiration (left) and at expiration (right) shows normal tracheal changes related to the dynamics of respiration. The normal trachea exhibits a round morphology on inspiration. During expiration, the posterior membranous trachea flattens due to changes in intrathoracic pressure and the absence of tracheal cartilages in the posterior tracheal wall. Excessive expiratory tracheal collapse is often abnormal and likely related to tracheomalacia.*

LARGE AIRWAY CARTILAGE CALCIFICATION, RADIOGRAPHY AND CT

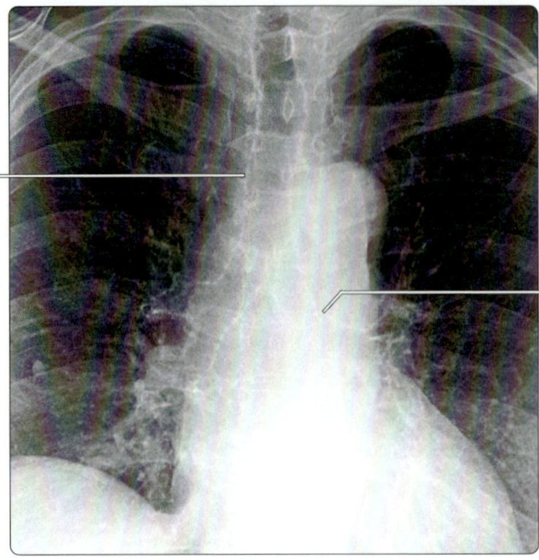

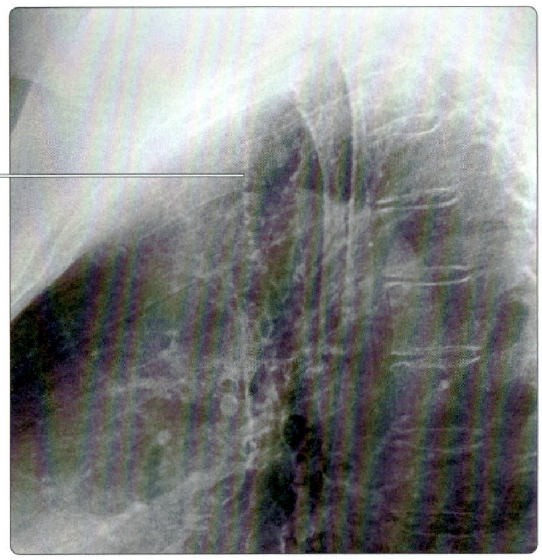

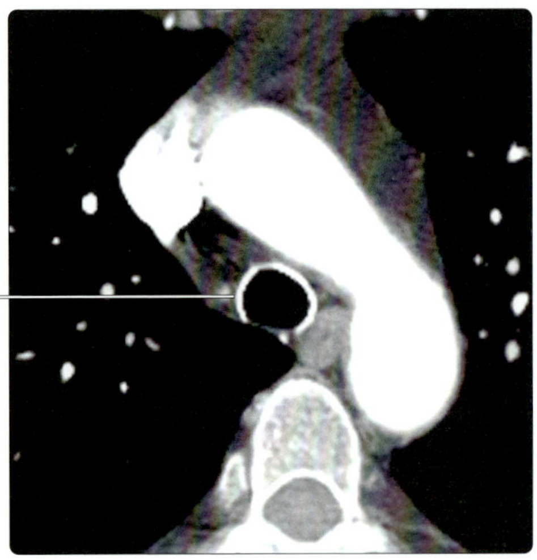

(Top) *PA chest radiograph of an older adult patient with normal tracheal calcification shows calcified tracheal and bronchial cartilages manifesting as thin white lines along the airway walls.* **(Middle)** *Lateral chest radiograph of the same patient shows discontinuous and corrugated calcification of the anterior tracheal wall, which correlates with calcification of individual C-shaped tracheal cartilages.* **(Bottom)** *Axial CECT (mediastinal window) of the same patient shows continuous calcification of an individual C-shaped tracheal cartilage and absence of calcification along the posterior membranous tracheal wall, which is devoid of cartilage.*

CARTILAGE CALCIFICATION AND LARGE AIRWAY DILATATION, CT

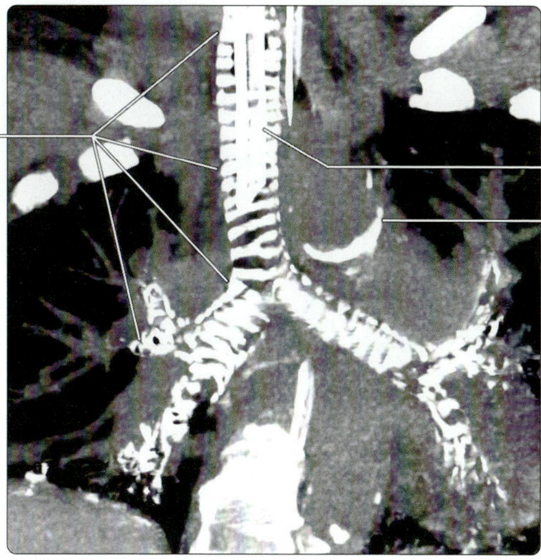

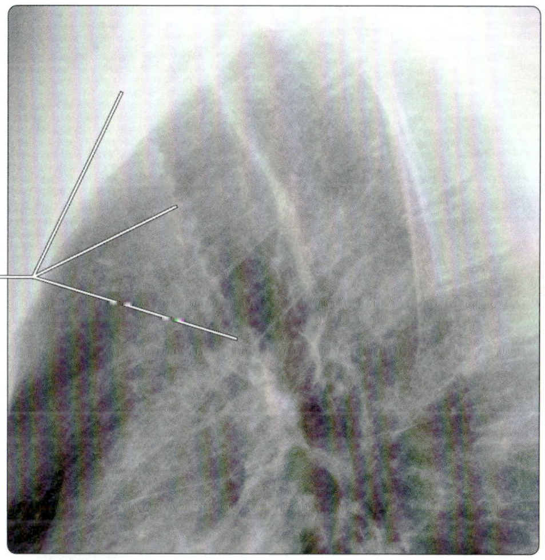

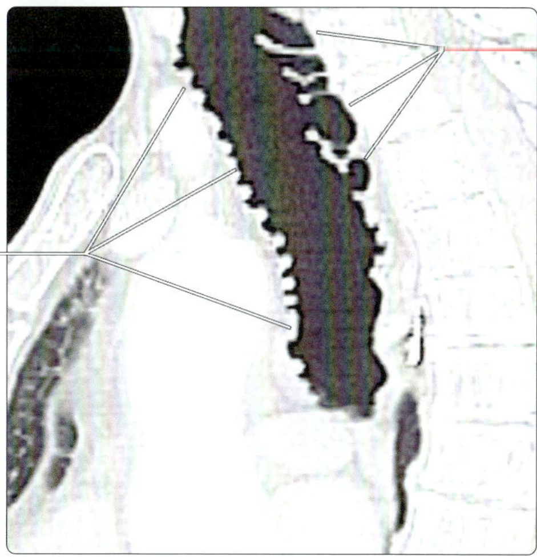

(Top) *Coronal NECT (mediastinal window) MIP reformatted image of a patient with a history of long-term use of warfarin shows extensive and exuberant calcifications of the tracheobronchial cartilaginous rings. Although tracheobronchial cartilage calcification may occur as a consequence of aging, it is also described in the context of long-term warfarin use.* **(Middle)** *Lateral chest radiograph of a patient with tracheobronchomegaly due to Mounier-Kuhn syndrome shows marked tracheal dilatation and a corrugated contour of the anterior tracheal wall.* **(Bottom)** *Sagittal CECT (lung window) of the same patient shows marked tracheal dilatation with associated anterior and posterior mural corrugation and numerous posterior tracheal diverticula, which represent diverticular outpouchings of the airway mucosa.*

SMALL AIRWAY STRUCTURE AND SECONDARY PULMONARY LOBULE

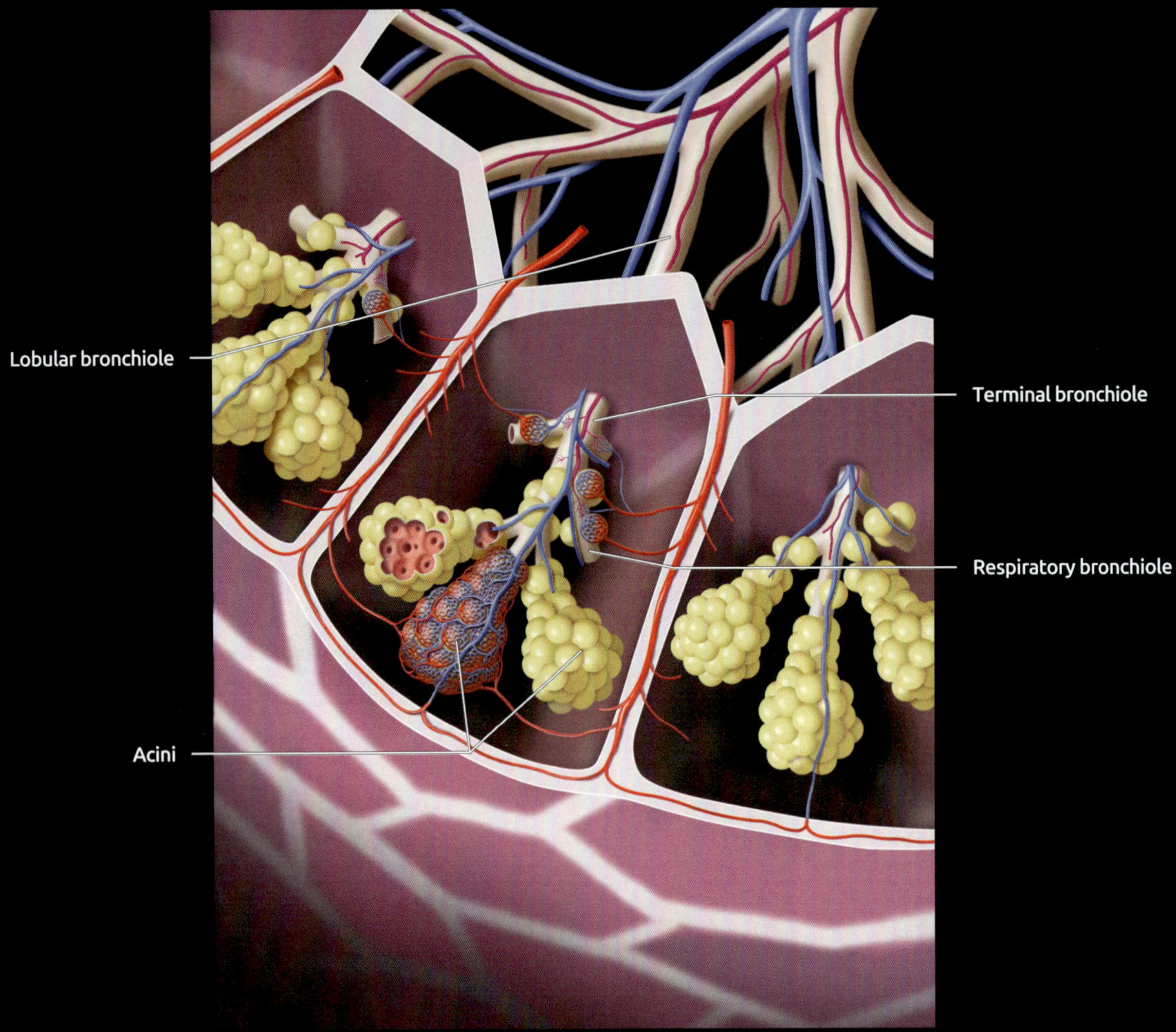

Graphic shows the structure of a portion of a pulmonary acinus within the secondary pulmonary lobule. The acinus is composed of the structures distal to a terminal bronchiole. The terminal bronchiole is the last conducting airway and gives rise to 2 or more respiratory bronchioles, which are transitional airways characterized by their alveolated walls. Each respiratory bronchiole gives rise to 3 alveolar ducts, which are airways lined by alveoli. Alveolar ducts terminate in alveolar sacs and alveoli. Pores of Kohn provide communication between adjacent alveoli. Canals of Lambert provide communication between various airways within an acinus as well as between adjacent acini and facilitate collateral ventilation.

SMALL AIRWAY STRUCTURE, MICROSCOPY

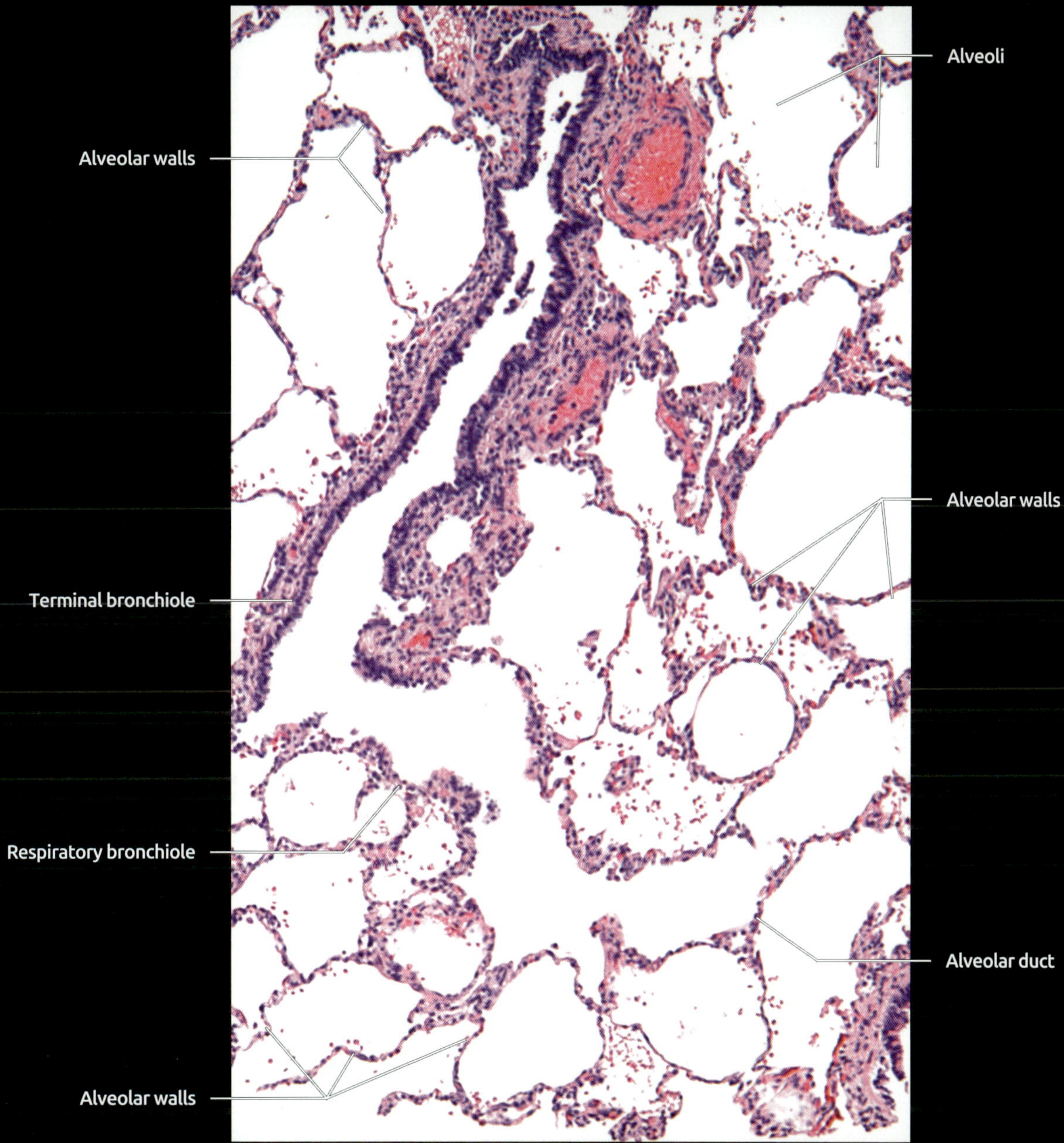

High-power photomicrograph (hematoxylin and eosin stain) shows the microscopic structure of the small airways. The terminal bronchiole is the last purely conducting airway and gives rise to respiratory bronchioles, which are characterized by their partially alveolated walls. The alveolar duct is a tubular space lined by alveoli.

ALVEOLAR STRUCTURE

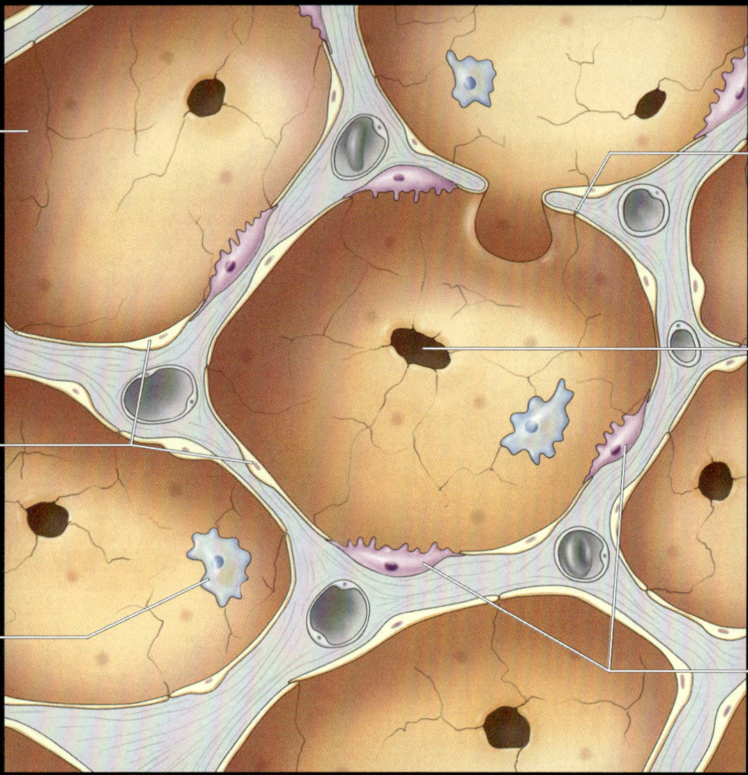

Graphic shows the microscopic structure of the alveoli. The alveolar-capillary interface is the principal site of respiration where the alveoli come in contact with the rich capillary network of the lung. Inspired oxygen is delivered to the capillaries and carbon dioxide is delivered to the airways. The type 1 pneumocyte is a flat cell with tight cell junctions that lines the alveolar surface. The tight cell junctions prevent permeability of fluid into the alveolar space. The type 2 pneumocyte is a larger polygonal cell that produces surfactant and processes circulating vasoactive substances. The alveolar macrophage is a migratory cell that forms part of the defense mechanisms of the lung.

ALVEOLAR STRUCTURE, MICROSCOPY

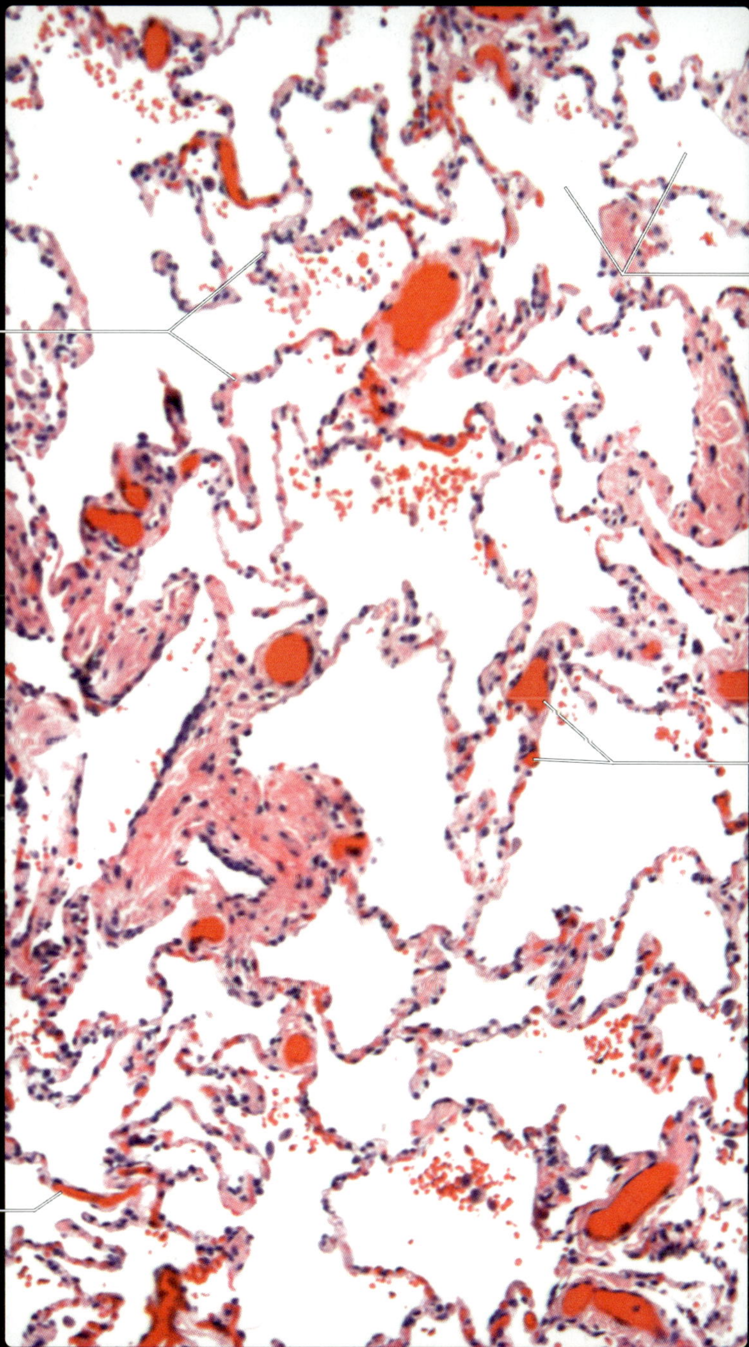

High-power photomicrograph (hematoxylin and eosin stain) shows the microscopic structure of the alveoli. The flat, thin nature of the alveolar pneumocytes facilitates gas exchange with adjacent capillaries at the alveolar capillary interface. The 300 million alveoli in the human lung provide an enormous surface area (70 m^2) for gas exchange.

TREE-IN-BUD OPACITIES, TERMINOLOGY

(Top) *Photograph of a leafless tree in bud shows the origin of the term tree-in-bud opacities used to describe the CT findings of cellular bronchiolitis. The small tree branches correspond to dilated centrilobular bronchioles filled with inflammatory cells &/or fluid on HRCT; the round buds correspond to peribronchiolar inflammatory tissue. (Courtesy Gerald F. Abbott, MD.)* **(Bottom)** *Photograph of a tree during early spring at the Hospital de Sant Pau in Barcelona shows leafless buds and early leaves. The term tree-in-bud refers to the morphologic resemblance of cellular bronchiolitis with leafless buds. The early leaves are equivalent to the formation of acinar opacities, as centrilobular micronodules grow and become less defined.*

CENTRILOBULAR AND TREE-IN-BUD OPACITIES, CT

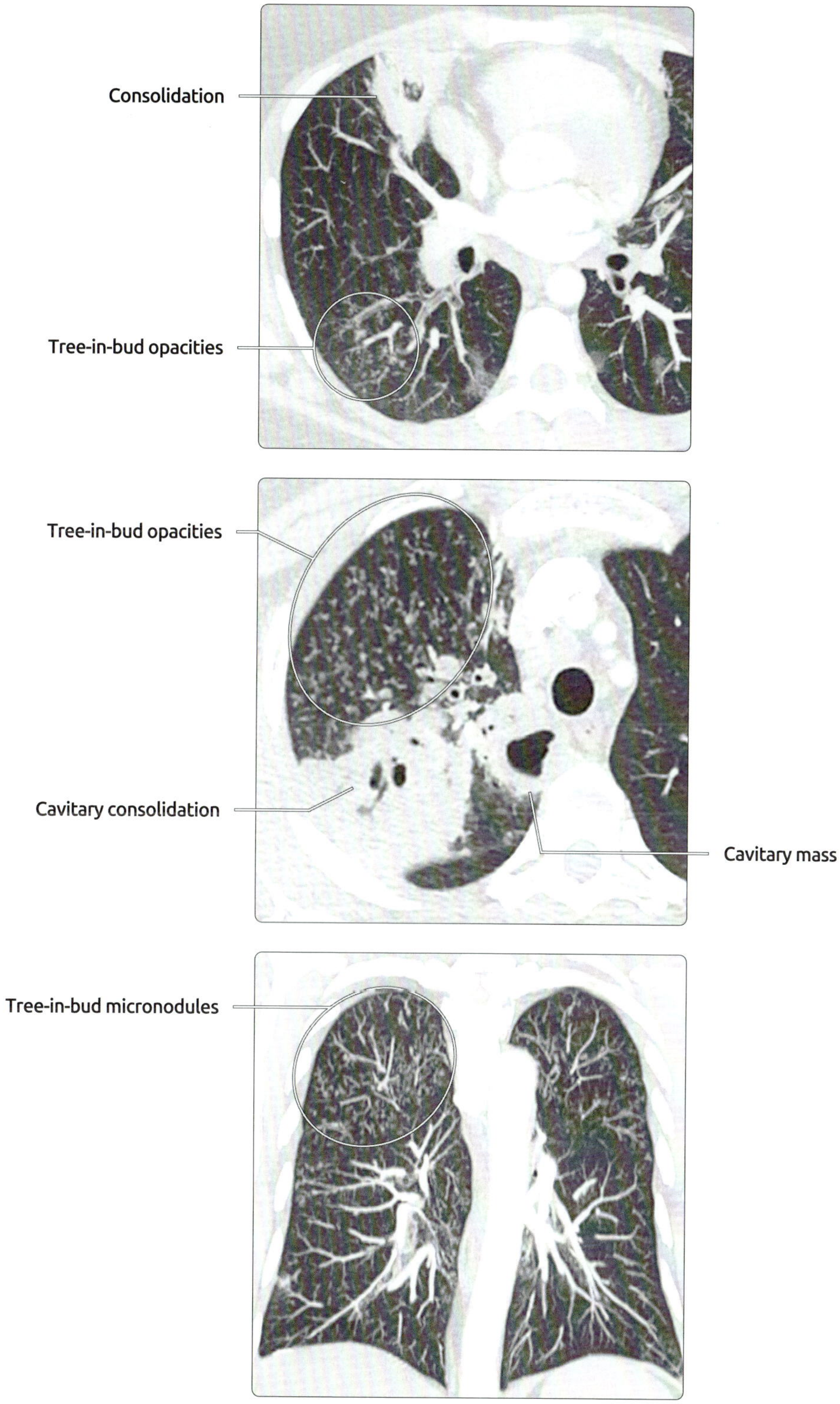

(Top) *Axial CECT (lung window) of a patient with pneumonia due to Mycoplasma pneumoniae shows coexisting areas of consolidation and tree-in-bud opacities. Mycoplasma pneumonia often manifests with cellular bronchiolitis in the early stages of infection.* **(Middle)** *Axial CECT (lung window) of a patient with the postprimary pattern of tuberculosis shows a cavitary consolidation and a cavitary mass in the fright upper lobe and extensive adjacent centrilobular micronodules and tree-in-bud opacities. The pulmonary micronodules spare the subpleural lung, consistent with their centrilobular location within the secondary pulmonary lobule.* **(Bottom)** *Coronal CECT (lung window) MIP reformatted image of a patient with rheumatoid arthritis and follicular bronchiolitis shows extensive bilateral upper lobe predominant centrilobular micronodules and tree-in-bud opacities.*

CENTRILOBULAR LOW ATTENUATION, RADIOGRAPHY AND CT

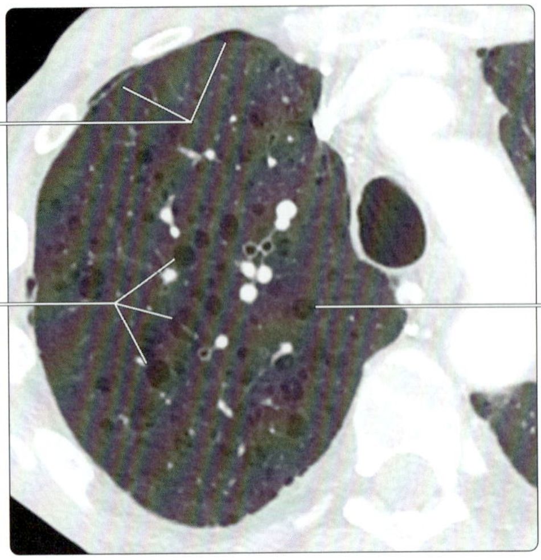

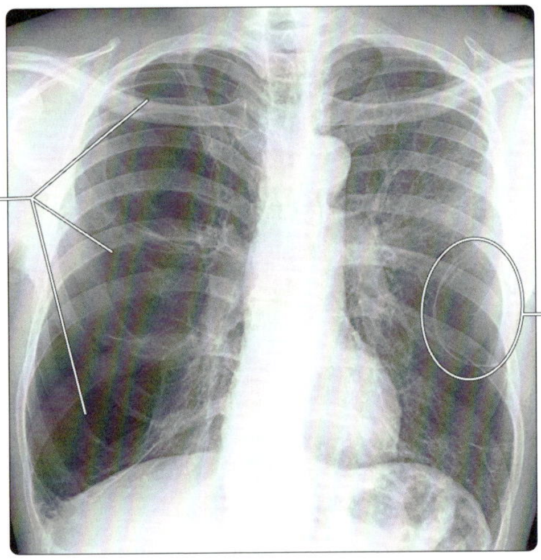

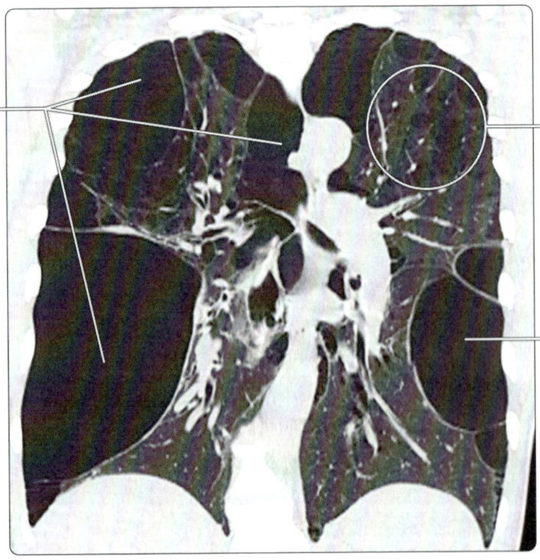

(Top) *Axial CECT (lung window) of a patient with centrilobular emphysema shows rounded centrilobular foci of low attenuation without perceptible walls, characteristic of centrilobular emphysema. Centrilobular emphysema is typically differentiated from cystic lung disease by the absence of a perceptible peripheral wall and the presence of the so-called central-dot sign, which represents the central lobular artery of the secondary pulmonary lobule.* **(Middle)** *PA chest radiograph of a patient with severe right > left bilateral bullous disease shows characteristic large cystic areas devoid of intrinsic vascular structures that produce mass effect on the adjacent lung.* **(Bottom)** *Coronal NECT (lung window) of the same patient shows that the cystic changes noted on radiography correspond to numerous large bilateral bullae, which coexist with upper lobe predominant centrilobular emphysema.*

LOBULAR LOW ATTENUATION, RADIOGRAPHY AND CT

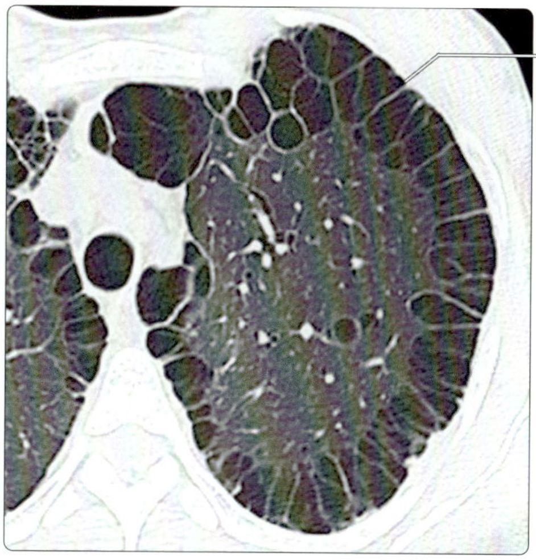

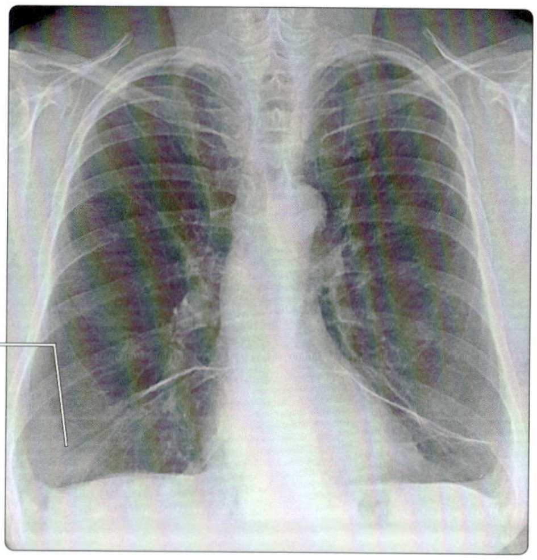

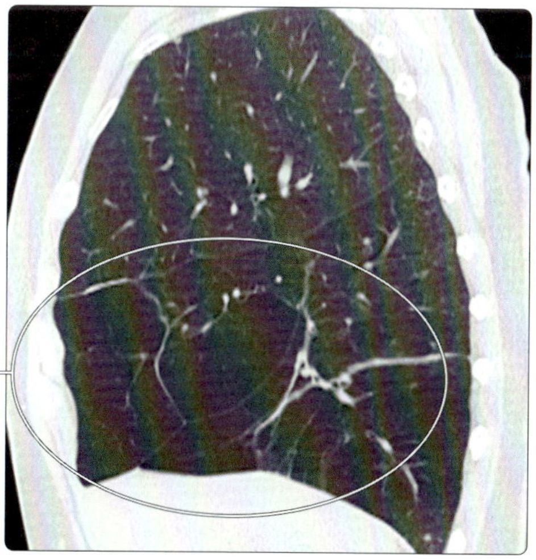

(Top) *Axial NECT (lung window) of a patient with extensive paraseptal emphysema is shown. Paraseptal emphysema manifests as a single subpleural layer of ovoid lucencies with intervening smooth walls (interlobular septa) and no intrinsic spared lung. Paraseptal emphysema characteristically affects the periphery of the pulmonary acinus and subpleural secondary pulmonary nodules.* **(Middle)** *PA chest radiograph of a patient with alpha-1 antitrypsin deficiency and panlobular emphysema shows hyperlucency and vascular attenuation predominantly involving the lower lobes.* **(Bottom)** *Sagittal NECT (lung window) of the same patient shows extensive areas of basilar predominant panlobular emphysema involving the middle and right lower lobes with relative sparing of the upper lobes. Panlobular emphysema affects the entirety of the secondary pulmonary lobule and typically manifests with hyperlucent lung and paucity of intrinsic vascular markings.*

LOBULAR LOW ATTENUATION, CT

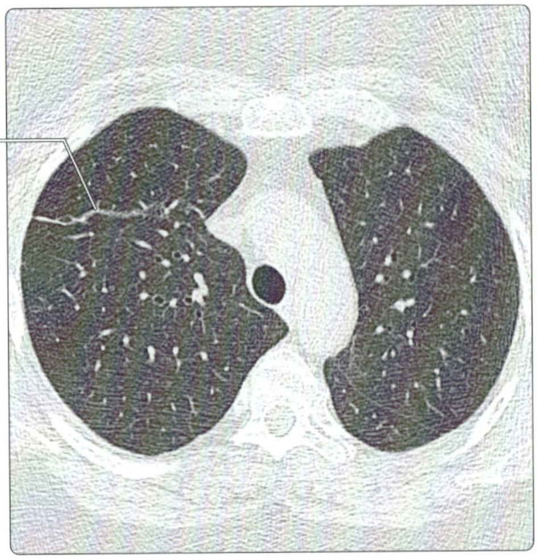

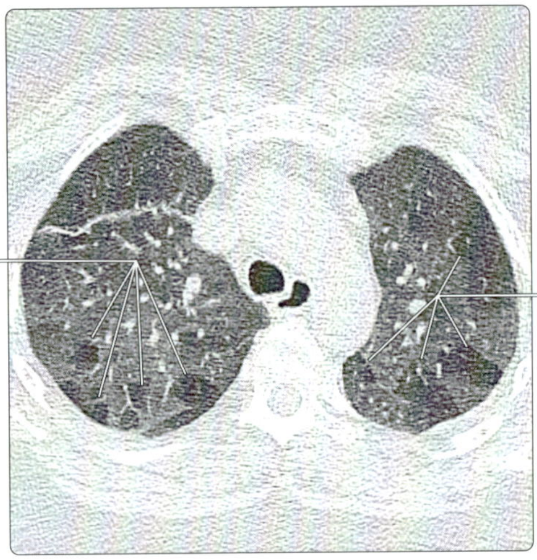

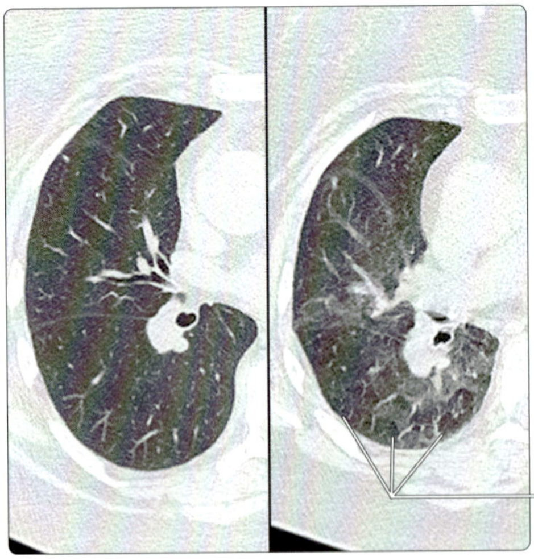

(Top) *Axial HRCT (obtained at inspiration) of a patient with rheumatoid arthritis and constrictive bronchiolitis shows relatively normal lung attenuation bilaterally with an area of subsegmental atelectasis in the right upper lobe.* **(Middle)** *Axial HRCT (obtained at expiration) of the same patient shows scattered areas of lobular hypoattenuation, consistent with expiratory air-trapping. The normal lung attenuation should increase homogeneously during expiratory imaging.* **(Bottom)** *Composite image with axial NECT (lung window) obtained during inspiration (left) and during expiration (right) of a patient with Sjögren syndrome and constrictive bronchiolitis shows normal relatively homogeneous lung attenuation during inspiration and patches of air-trapping on expiration.*

ACINAR NODULES, CT

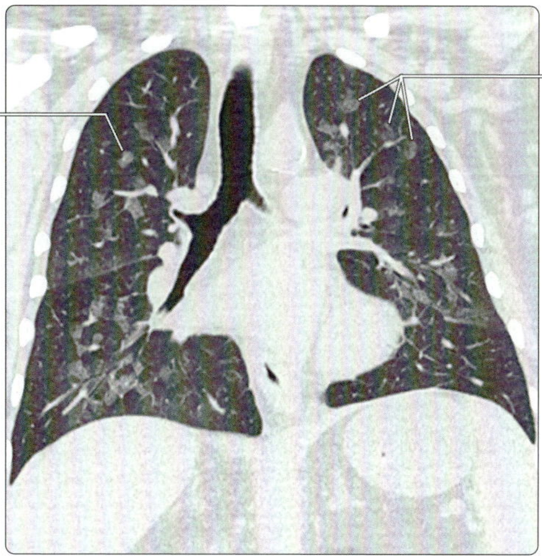

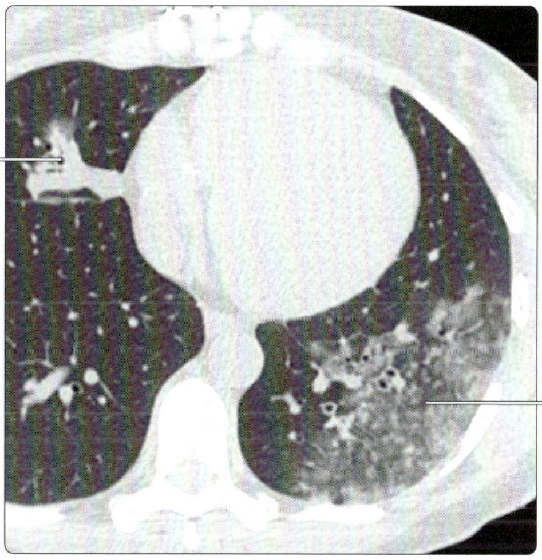

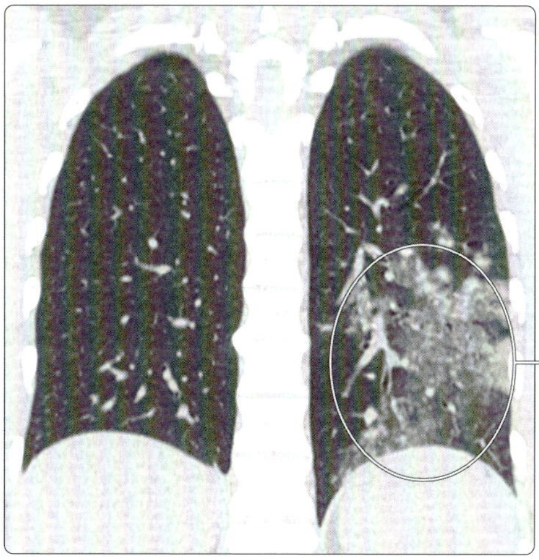

(Top) *Coronal NECT (lung window) of a patient with cardiogenic pulmonary edema shows profuse acinar nodules and ground-glass acinar opacities bilaterally that represent extravascular alveolar water. Pulmonary alveolar hemorrhage may appear similar.* **(Middle)** *Axial NECT (lung window) of a patient with a history of lung transplantation and pneumonia shows extensive left lower lobe acinar opacities and a middle lobe peribronchovascular consolidation.* **(Bottom)** *Coronal NECT (lung window) of the same patient shows profuse left lower lobe acinar opacities with intervening patchy ground-glass opacities.*

OVERVIEW OF PULMONARY VASCULAR STRUCTURE

Arteries

- Dual arterial supply
  - Pulmonary arteries
  - Systemic arteries (bronchial arteries)
- **Pulmonary arterial system**
  - Pulmonary circulation
  - Conducts blood from right ventricle to capillary-alveolar interface
- **Bronchial arterial system**
  - Systemic circulation
  - Conducts blood from aorta to airway walls, vessel walls, and visceral pleura

Veins

- Dual venous drainage with pulmonary and systemic components
- **Pulmonary venous system**
  - Pulmonary circulation
  - Conducts blood from alveolar-capillary interface to left atrium
  - Venous drainage of lung
- **True bronchial venous system**
  - Conducts blood from perihilar bronchi and vessels to azygos venous system
  - Venous drainage of walls of central vessels and tracheobronchial tree

Other Systemic Vessels

- Sources of collateral blood supply to lung from systemic circulation
- Pulmonary ligament arteries
  - Aortic branches located in pulmonary ligament
  - Supply esophageal plexus
  - Supply medial visceral pleura
  - May supply adjacent lung parenchyma
    - Potential source of systemic blood supply to intralobar sequestrations
- Subclavian and axillary arteries
- Intercostal arteries
- Inferior phrenic arteries

Lymphatics

- Vascular channels
- Collection and transport of lymphatic fluid
- Conduit of lymphatic fluid toward hila

Relationship of Vascular Structures

- Anatomy
  - **Bronchoarterial bundle**: Single connective tissue sheath that contains
    - Pulmonary arteries
    - Bronchi
    - Bronchial arteries
  - **Lymphatics** located along
    - Bronchi to level of respiratory bronchioles
    - Pulmonary arteries
    - Pulmonary veins
    - Interlobular and connective tissue septa
    - Visceral pleural connective tissue
- **Vascular anastomoses**
  - **Pulmonary to bronchial arteries**
    - Communication between pulmonary and systemic circulations
    - Located in walls of larger airways
    - Functionally closed in normal individuals
    - May become patent in disease states that produce pulmonary artery compromise
  - **Bronchial arteries to pulmonary veins**
    - Communication between systemic and pulmonary circulations
    - Communication between capillary bed of bronchial wall and pulmonary veins
  - **Pulmonary veins to lymphatics**
    - Communication with arterial channels
    - Communication with venous channels

GENERAL ANATOMY AND FUNCTION

Anatomy

- **Pulmonary circulation**
  - Pulmonary arteries
  - Capillary network
  - Pulmonary veins
- **Systemic circulation**
  - Bronchial arteries
  - Nonbronchial systemic arteries
  - True bronchial veins
- **Pulmonary lymphatics**
  - Complex network of vascular channels
    - Reservoir lymphatics: Broad, ribbon-like
    - Conduit lymphatics: Tubular
    - Saculotubular lymphatics: Plexiform complex around vessels and bronchi
  - Bronchus-associated lymphoid tissue (BALT)
  - Intrapulmonary lymph nodes
  - Intrapulmonary peribronchial lymph nodes

Function

- **Pulmonary circulation**
  - **Pulmonary arteries**
    - Collection of deoxygenated blood from right cardiac chambers
    - Conduit of deoxygenated blood to pulmonary capillary-alveolar interface
  - **Pulmonary alveolar-capillary network**: Gas exchange
  - **Pulmonary veins**
    - Collection of oxygenated blood from alveolar-capillary network and conduit of oxygenated blood to left atrium and systemic circulation
    - Collection of blood from capillary beds of bronchial/bronchiolar walls, vessel walls, visceral pleura, and conduit to left atrium
- **Systemic circulation**
  - **Bronchial arteries**
    - Conduit of oxygenated blood from aorta to airway walls, vessel walls, and visceral pleura
  - **True bronchial veins**
    - Conduit of deoxygenated blood from central vessel/bronchial walls to azygos system

- **Lymphatics**
  - Lymphatic fluid drainage toward hila
  - Lymphatic drainage to abdominal lymph nodes

VASCULAR STRUCTURE

Pulmonary Arteries

- General concepts
  - **Pulmonary trunk** arises from right ventricle
  - Dichotomous branching into right and left pulmonary arteries
  - Pulmonary artery branches
    - **Lobar**
    - **Segmental**
    - **Subsegmental**
- Physiology
  - **Low-pressure pulmonary circulation**
  - Minimal flow through lung apices
    - Rapid increase in cardiac output accommodated through recruitment of closed capillaries in underperfused upper lung zones
  - Muscular arteries constrict in response to hypoxia
    - **Ventilation-perfusion matching** to prevent hypoxemia
- Microscopic anatomy
  - Proximal large arteries: Elastic arteries
  - Distal small arteries
    - Transition to muscular arteries at level of bronchioles
    - External elastic lamina
    - Muscular media
    - Internal elastic lamina
  - Smallest arteries: Loss of smooth muscle in vessel wall
    - Muscle in media thins as artery size decreases
    - Single elastic lamina distally
    - Difficult distinction between arteries and arterioles due to varying amounts of muscle in vessel walls
      - Term "arteriole" not considered applicable to pulmonary circulation
- Imaging anatomy
  - Pulmonary arteries course alongside bronchi
    - Medial to bronchi in upper lobes
    - Lateral to bronchi in middle lobe, lingula, and lower lobes
  - Accessory branches directly penetrate lung
  - **Lobular arteries** of secondary pulmonary lobule
    - Typically dominant lobular branch in center of lobule (centrilobular)
    - Manifest as central dot-like opacities within 1 cm of pleural surfaces

Capillary Network

- Anatomy
  - Origin from distal most pulmonary arteries
  - Surround alveoli of respiratory bronchioles, alveolar ducts, alveolar sacs, and alveoli
  - Respiratory surface with enormous surface area
    - ~ 300 million alveoli
    - ~ 1,000 capillaries per alveolus
    - Surface area of 70 m^2
- Microscopic anatomy
  - Lined by flat endothelial cells on basement membrane
  - Loose endothelial cell junctions
    - Permit passage of low-molecular-weight proteins
- Imaging anatomy
  - Below resolution of CT

Pulmonary Veins

- General concepts
  - Right and left pulmonary veins
  - Sources of venous drainage
    - Efferent alveolar capillaries
    - Bronchial/vascular wall capillary networks
    - Visceral pleura
- Microscopic anatomy
  - Small pulmonary veins indistinguishable from small arteries
  - Mural smooth muscle cells and elastic lamina
- Imaging anatomy
  - Pulmonary veins travel in periphery of lung units
    - Acinus
    - Lobule
    - Segment
  - Pulmonary vein branches
    - Septal veins in periphery of secondary pulmonary lobule
      - Visible along interlobular septa; help define boundaries of secondary pulmonary lobule
      - 0.5 mm in diameter
      - Arcuate or branching structures identified 1.0-1.5 cm from pleural surface
    - Veins in periphery of subsegments and segments
    - Veins in periphery of lobes
    - Superior pulmonary veins
      - Drainage of upper lobes and middle lobe
    - Inferior pulmonary veins
      - Drainage of lower lobes
    - Variable anatomy of central pulmonary veins at anastomoses with left atrium

Bronchial Arteries

- General concepts
  - Variable origin from systemic circulation
    - Left bronchial arteries originate directly from aorta
    - Right bronchial arteries may originate directly from aorta but typically have shared origin with another artery (usually intercostal artery)
    - Orthotopic: Origin from proximal descending aorta
    - Ectopic: Origin from inferior aortic arch, distal descending aorta, subclavian artery, brachiocephalic trunk, internal thoracic artery
  - Vascular supply to
    - Subepithelial capillary plexus along airway length from trachea to terminal bronchioles
    - Peribronchial, perivascular connective tissue
    - Vessel walls: Great vessels, pulmonary arteries, and veins
    - Paratracheal, subcarinal, hilar, intrapulmonary lymph nodes
    - Visceral pleura
    - Esophagus
  - Numerous anastomoses with other systemic mediastinal vessels

- Travel in close relationship to (within) walls of bronchi and bronchioles
- Microscopic anatomy
  - Muscular media
    - High-pressure systemic circulation
  - Prominent internal elastic lamina
  - No external elastic lamina
- Imaging anatomy
  - Typically inapparent
  - Visualization of central bronchial arteries in several disease states with collateral circulation to lung

Bronchial Veins

- General concepts
  - True bronchial veins only in perihilar regions
  - Drainage to azygos and hemiazygos systems
- Imaging anatomy
  - Generally not visible on imaging studies

Pulmonary Lymphatics

- General concepts
  - Diffuse and complex pulmonary lymphatic network
  - Lymphatic channels along
    - Bronchi
      - To level of respiratory bronchioles
    - Vessels
      - Pulmonary arteries
      - Pulmonary veins
    - Connective tissue septa
      - Interlobular septa
    - Intralobular interstitium
    - Visceral pleura
- Microscopic anatomy
  - Thin endothelium
  - Endoluminal valves direct flow toward hila
  - Not normally visible on histologic specimens
    - May become conspicuous in certain disease states
- Imaging anatomy
  - Generally not visible on imaging studies

VASCULAR STRUCTURE OF SECONDARY PULMONARY LOBULE

Lobular Pulmonary Artery

- Each lobular pulmonary artery supplies single lobule
- Courses alongside distal airway
- Ramifies as alveolar capillary network
- Imaging
  - Lobular artery: Central dot-like structure measuring ~ 1 mm in diameter
  - Intralobular acinar arteries: Vessels as small as 0.2 mm in diameter resolved on thin-section CT

Pulmonary Vein

- Intralobular pulmonary veins drain into septal pulmonary veins
  - Septal veins are visible on CT and measure ~ 0.5 mm in diameter
  - Located ~ 5-10 mm from arteries
- Each lobular (septal) pulmonary vein drains 2 or more adjacent lobules
- Tributaries from
  - Alveolar capillary network
  - Bronchial artery-derived airway wall (subendothelial) capillary network
  - Visceral pleura capillaries

Bronchial Artery

- Courses within airway wall
- Ramifies as airway wall (subepithelial) capillary network

Lymphatics

- Peribronchovascular lymphatics
  - Alongside lobular pulmonary arteries and airways
- Perilobular lymphatics
  - Along interlobular septa and pulmonary veins
- Visceral pleural lymphatics
- Intercommunication via anastomotic channels
  - Peribronchovascular to perilobular
  - Peribronchovascular to pleural

ANATOMY-RELATED IMAGING ABNORMALITIES

Random Pulmonary Micronodules

- Common etiologies: **Hematogenous metastases**, **miliary infection**
- CT features
  - Nodules randomly and evenly distributed in relation to structures of secondary pulmonary lobule
  - Some nodules may exhibit specific relationship to small pulmonary arteries
    - Helpful finding to suggest hematogenous mechanism of dissemination

Perilymphatic Pulmonary Micronodules

- Common etiologies
  - **Sarcoidosis**
  - **Lymphangitic carcinomatosis**
- CT features
  - Nodules located along lymphatic channels of secondary pulmonary lobule
    - **Peribronchovascular**
    - **Interlobular septal**
    - **Subpleural**
    - **Centrilobular**

Centrilobular Pulmonary Micronodules

- Common etiologies
  - **Cellular bronchiolitis** (infection, aspiration, follicular, etc.)
  - Excipient lung disease
- CT features
  - Nodules limited to center of secondary lobule (i.e., centrilobular)
  - Characteristically fissures and subpleural lung

ANATOMY OF SECONDARY PULMONARY LOBULE ARTERIES

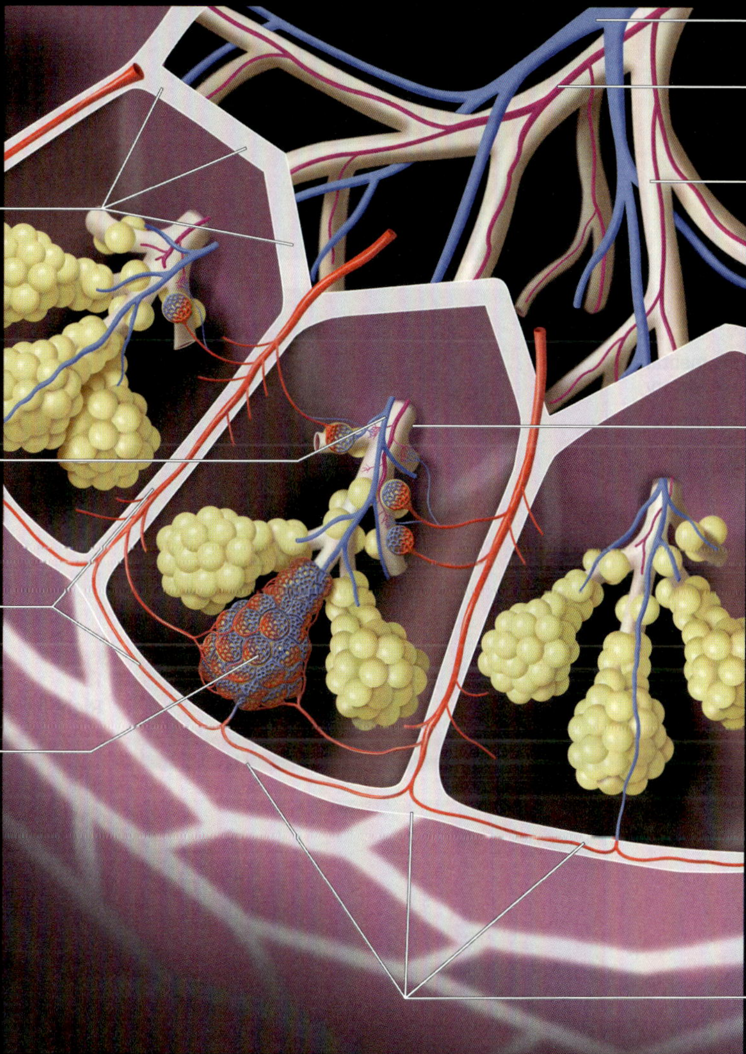

Graphic shows the complexity of the vascular components of the secondary pulmonary lobule, which reproduces (in miniature) the structural morphology of the lung with respect to the organization of vascular structures, airways, and lymphatics. The pulmonary arteries travel alongside the airways, supply the capillary network that surrounds the alveoli, and bring deoxygenated blood to the capillary-alveolar interface. The bronchial arteries travel within the airway walls and supply the capillary networks within the walls of the airways.

PULMONARY AND BRONCHIAL ARTERIES, MICROSCOPY

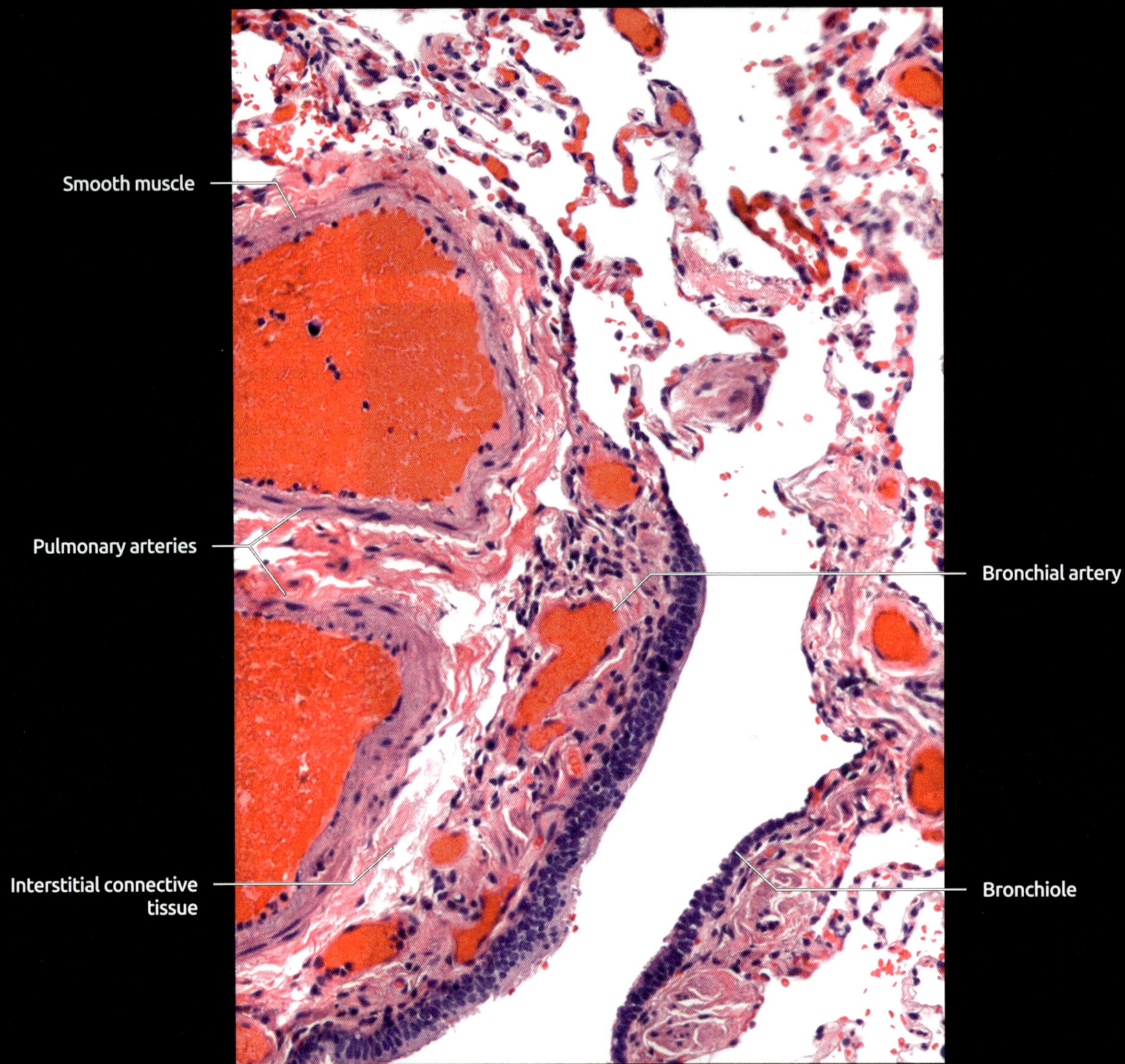

High-power photomicrograph (H&E stain) of normal lung demonstrates a bronchiole lined by pseudostratified cuboidal epithelium. The adjacent pulmonary arteries that travel alongside the bronchus within the bronchovascular sheath are surrounded by intervening loose interstitial connective tissue. The muscular pulmonary arteries exhibit circularly oriented smooth muscle located between their internal and external elastic laminae. Bronchial arteries are intimately related to their corresponding airway walls and are often described as traveling within them.

PULMONARY ARTERIES, CT

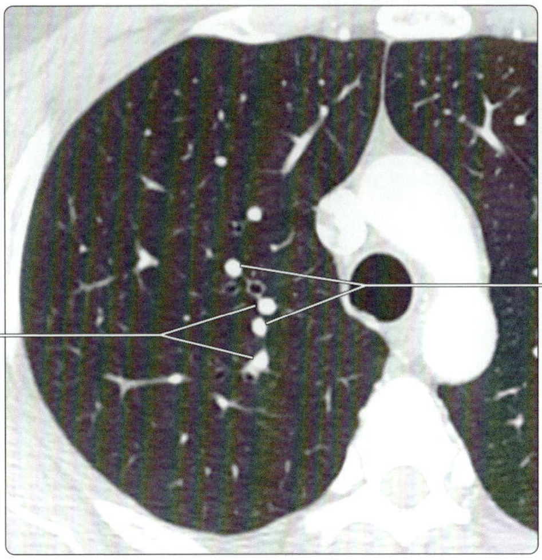

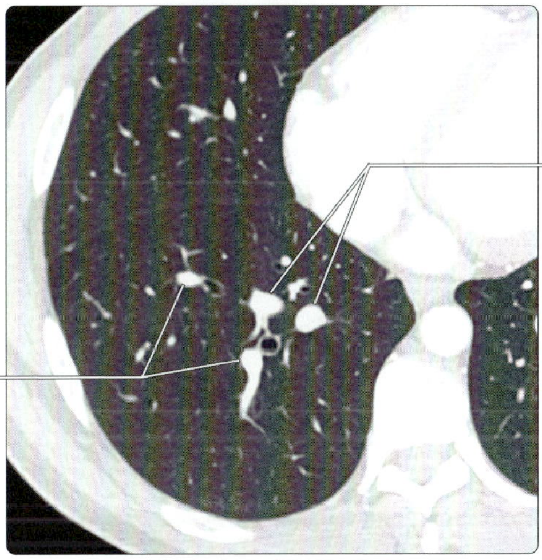

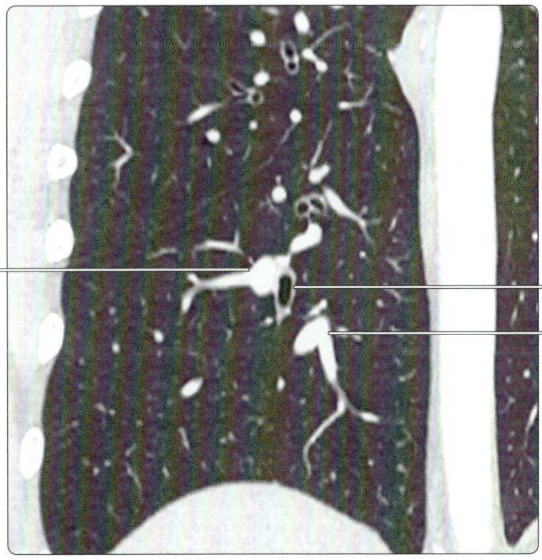

(Top) *Axial CECT (lung window) of a normal subject obtained through the level of the aortic arch shows the relationship of the pulmonary arteries and veins to the adjacent airways. The right upper lobe pulmonary arteries course alongside bronchi in the bronchovascular sheath and are located medial to them. The pulmonary veins are located lateral to adjacent bronchi.* **(Middle)** *Axial CECT (lung window) of the same individual shows the right lower lobe pulmonary arteries located lateral to their adjacent bronchi. The pulmonary veins exhibit a more horizontal course along connective tissue septa and are located medial to the adjacent bronchi. In clinical practice, scrolling through sequential stacks of images permits tracing the peripheral vessels to their central portions to definitively distinguish pulmonary arteries from pulmonary veins.* **(Bottom)** *Coronal CECT (lung window) of the same individual shows the relationship of right lower lobe pulmonary arteries and veins to adjacent bronchi. The right lower lobe pulmonary artery branch is located lateral to the corresponding bronchus, while the pulmonary vein is located medially.*

ANATOMY OF CAPILLARY-ALVEOLAR INTERFACE

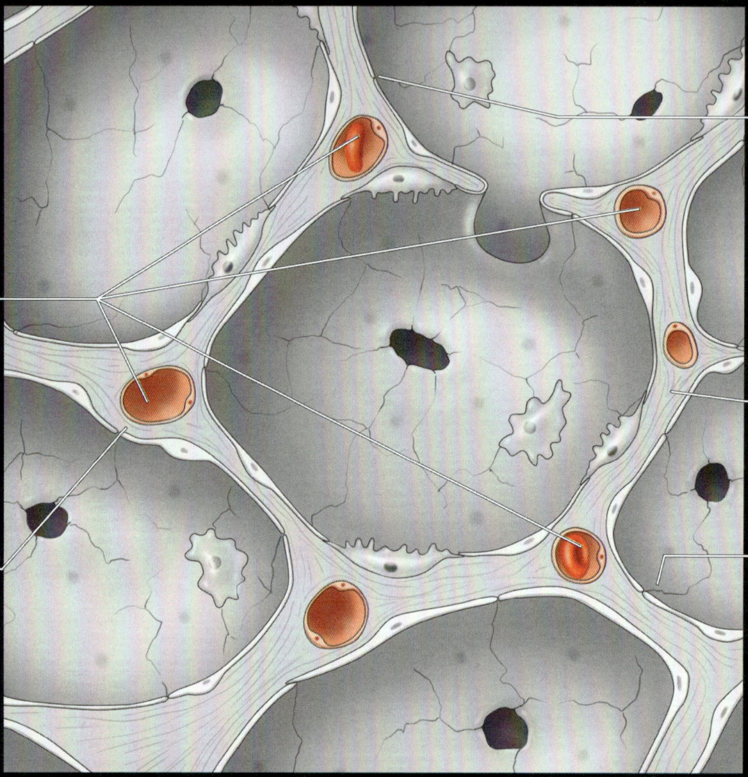

Graphic shows the relationship of the capillary network to the alveoli. The capillary-alveolar interface is the site of gas exchange. The large number of capillaries surrounding each alveolus provides an enormous surface area for respiration. The thin vascular endothelium and the thin alveolar wall cells facilitate the process of gas exchange. Loose endothelial cell junctions permit passage of intravascular low-molecular-weight substances.

CAPILLARY-ALVEOLAR INTERFACE, MICROSCOPY

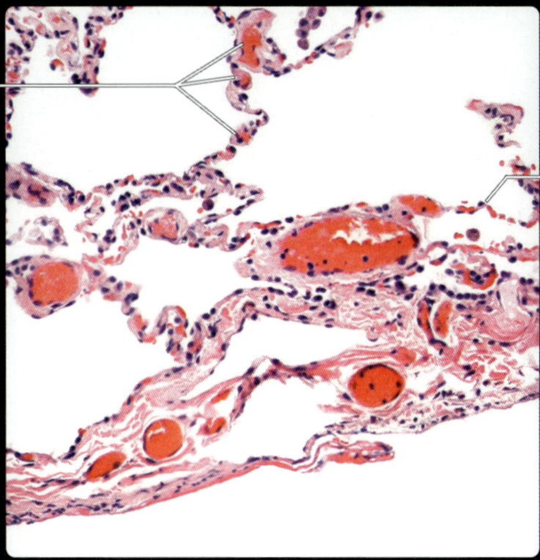

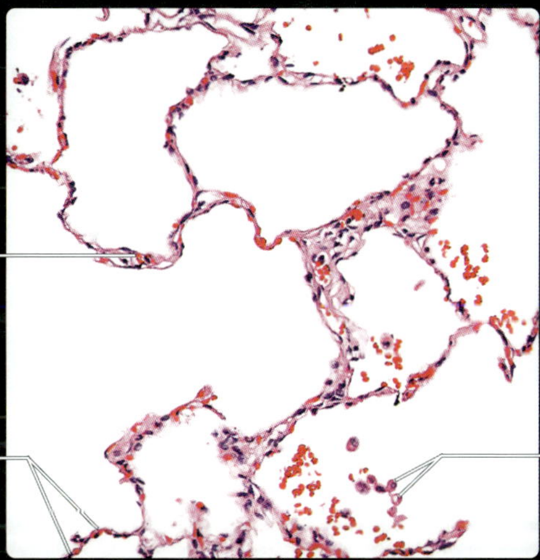

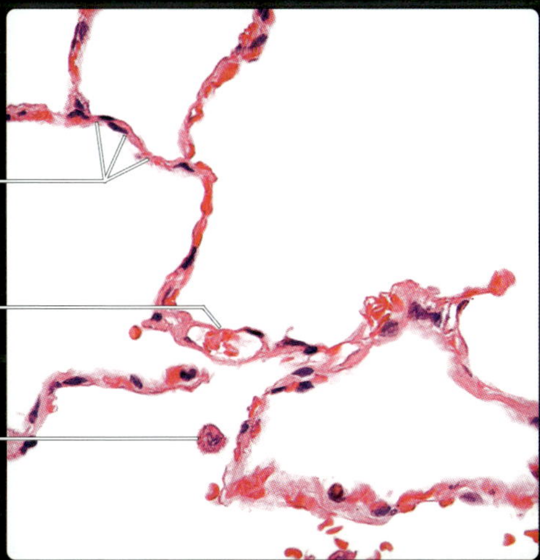

(Top) *Intermediate-power photomicrograph (H&E stain) of normal lung shows the thin alveolar walls and the numerous capillaries associated with them. The capillaries and other vascular structures are easily identified, as they contain red blood cells in their lumina. There are ~ 300 million alveoli and 1,000 capillaries per alveolus. The enormous number of capillaries associated with the numerous alveoli provides an immense surface area for respiration.* **(Middle)** *High-power photomicrograph (H&E stain) shows the alveolar walls and numerous capillaries identified by red blood cells in their lumina. Small intraalveolar macrophages are not uncommon normal findings. (From DP: Normal Histology.)* **(Bottom)** *High-power photomicrograph (H&E stain) at very high magnification demonstrates thin alveolar walls and alveolar wall capillaries, identified by their open lumina containing red blood cells. These alveolar-capillary interfaces make respiration possible. Note the single alveolar macrophage. (From DP: Normal Histology.)*

PULMONARY VEINS, CT

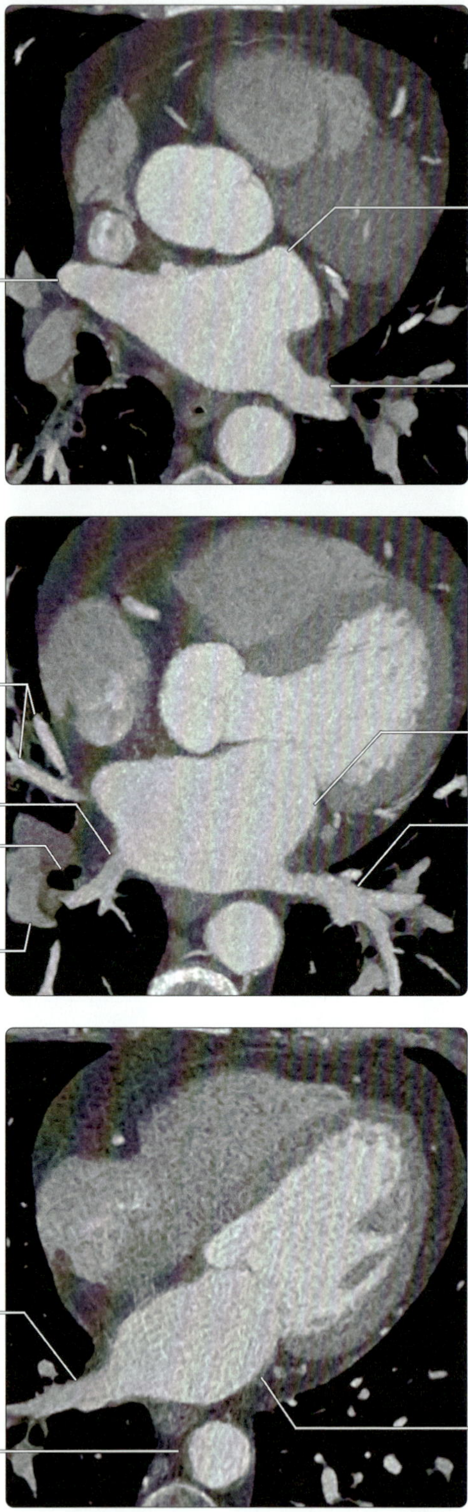

(Top) *Axial CECT (mediastinal window) of a normal subject shows the right and left superior pulmonary veins. There are characteristically 2 left and 2 right pulmonary veins, although the number and morphology of the central pulmonary veins may be variable.* **(Middle)** *Axial CECT (mediastinal window) of the same subject shows the bilateral inferior pulmonary veins coursing horizontally to drain into the left atrium and located medial to lower lobe bronchi. Note right lower lobe pulmonary artery branches located lateral to the right lower lobe bronchi.* **(Bottom)** *Axial CECT (mediastinal window) of the same subject shows the right inferior pulmonary vein, which follows a horizontal orientation as it courses into the left atrium. Detailed description of the anatomy of the pulmonary veins in relationship to the left atrium is critical for left atrial ablation and surgical planning.*

PULMONARY VEINS, CT

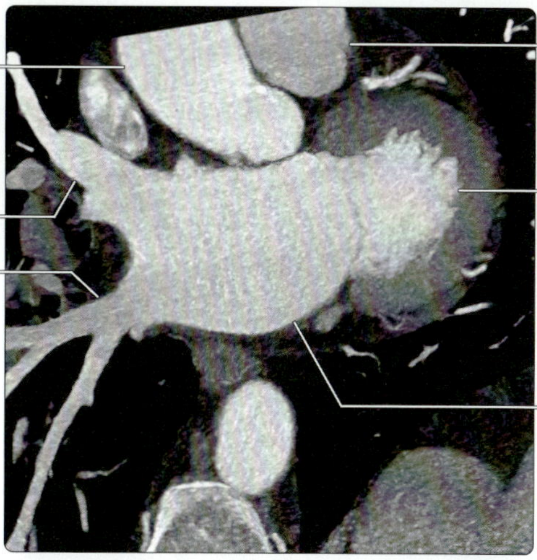

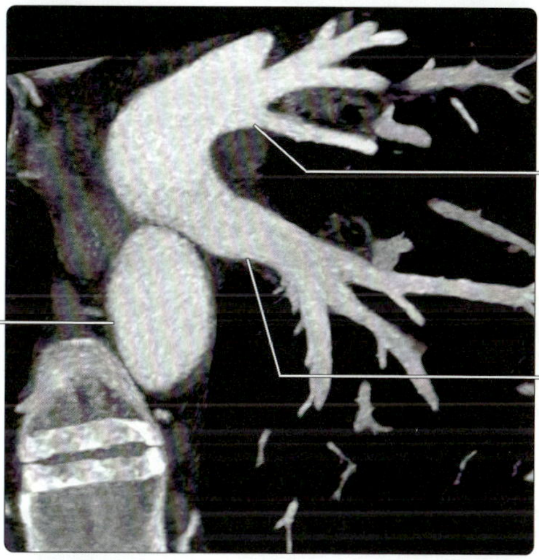

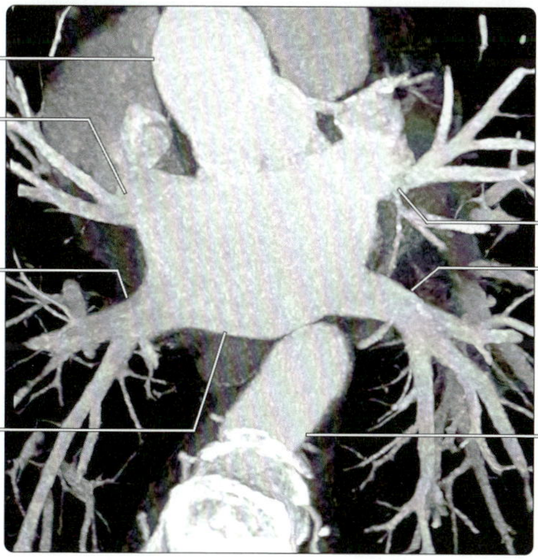

(Top) *Coronal oblique CECT (mediastinal window) MIP reformatted image of the same individual shows the right superior and right inferior pulmonary veins as they course medially to drain into the left atrium.* **(Middle)** *Coronal oblique CECT (mediastinal window) MIP reformatted image of the same individual shows the left superior and left inferior pulmonary veins as they course medially to drain into the left atrium.* **(Bottom)** *Coronal oblique CECT (mediastinal window) MIP reformatted image of the same individual shows all 4 pulmonary veins. While these types of reformations may have little value for the radiologist in understanding the anatomy of the pulmonary veins, they may be of value to clinicians and surgeons for procedural planning, especially in cases that exhibit anatomic variants.*

ANATOMY OF SECONDARY PULMONARY LOBULE VEINS

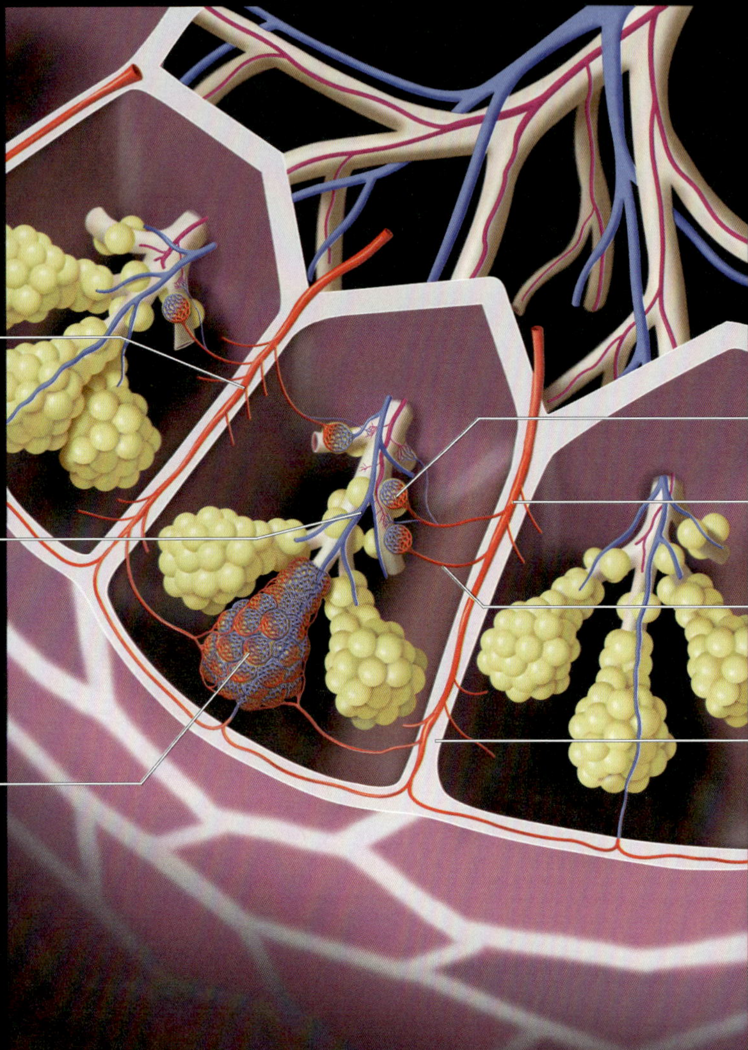

Graphic shows the vascular anatomy of the secondary pulmonary lobule. The veins are found in the interlobular septa that define the boundaries of the secondary pulmonary lobule. Intralobular pulmonary veins arise from the capillary network that surrounds the alveoli and drains into the septal veins. The septal pulmonary veins drain > 1 secondary pulmonary lobule. The pulmonary veins also drain the subendothelial capillary network of the airway walls and the visceral pleura.

SECONDARY PULMONARY LOBULE, CT

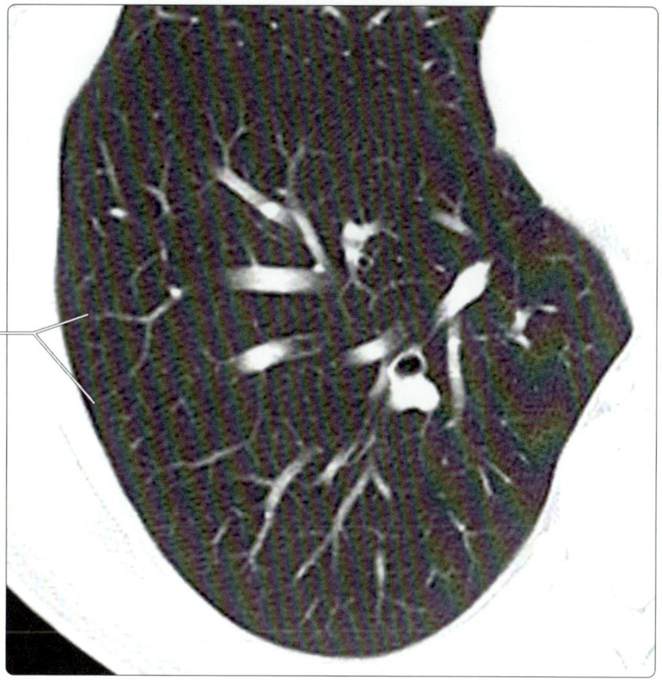

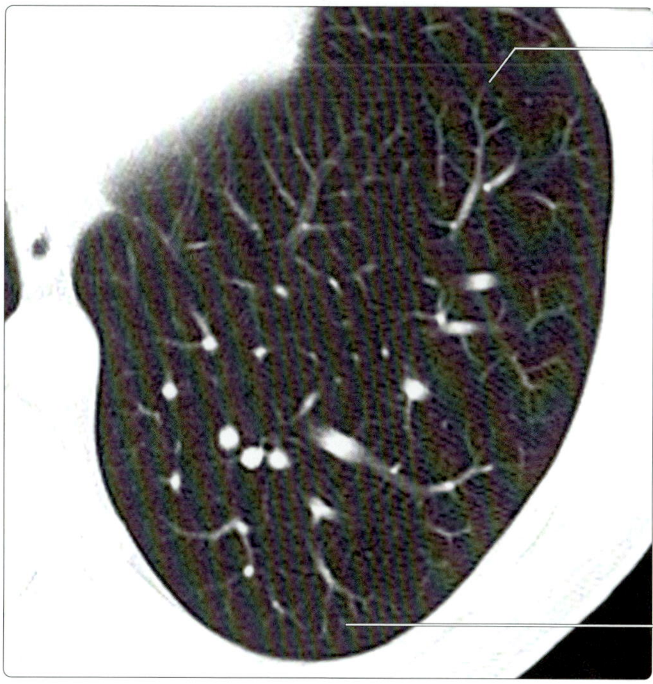

(Top) *First of 2 axial images of a normal contrast-enhanced chest CT shows the vessels of the secondary pulmonary lobule. Axial CECT (lung window) coned down to the right lower lobe shows the peripheral pulmonary vein branches that demarcate the boundaries of the secondary pulmonary lobule. Note that the interlobular septa are not visible in normal subjects, but their location can be inferred by identification of the normal septal veins.* **(Bottom)** *Axial CECT (lung window) coned down to the left lower lobe shows septal pulmonary veins outlining the boundaries of the secondary pulmonary lobules. The vessels in the center of the space bound by the septal veins represent the central lobular arteries of the secondary pulmonary lobules. The normal secondary pulmonary lobule is not readily visible on CT, but knowledge of the normal pulmonary vascular anatomy allows identification of its vascular components.*

SECONDARY PULMONARY LOBULE, CT

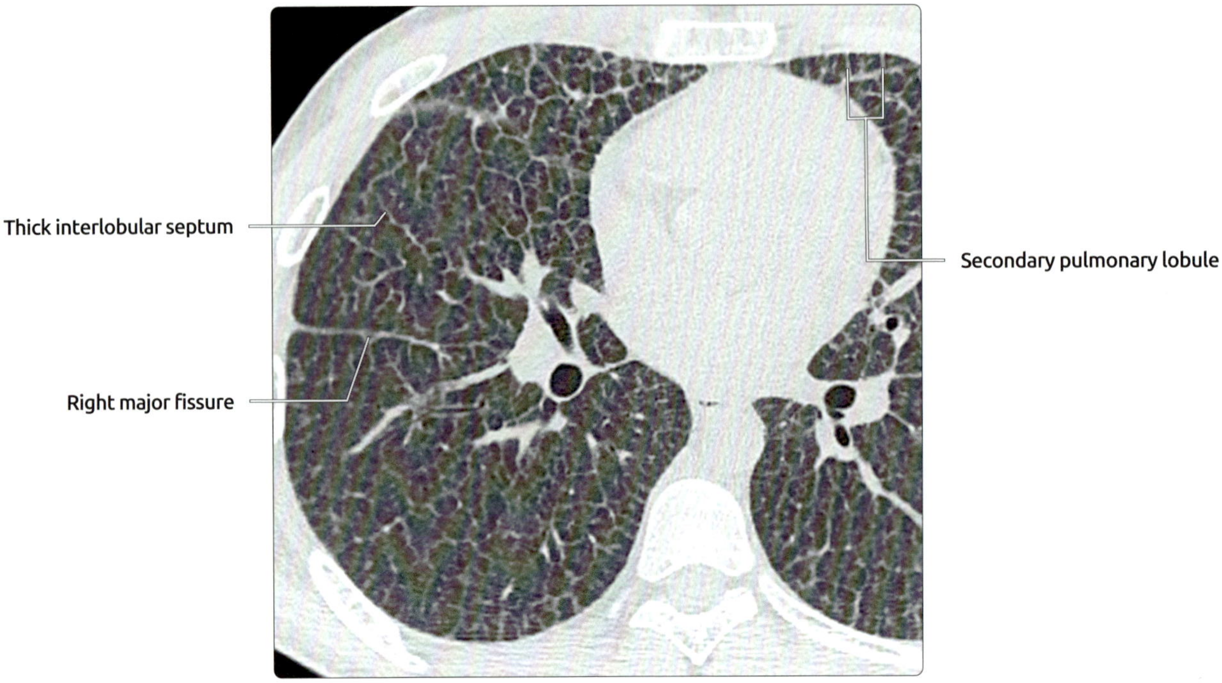

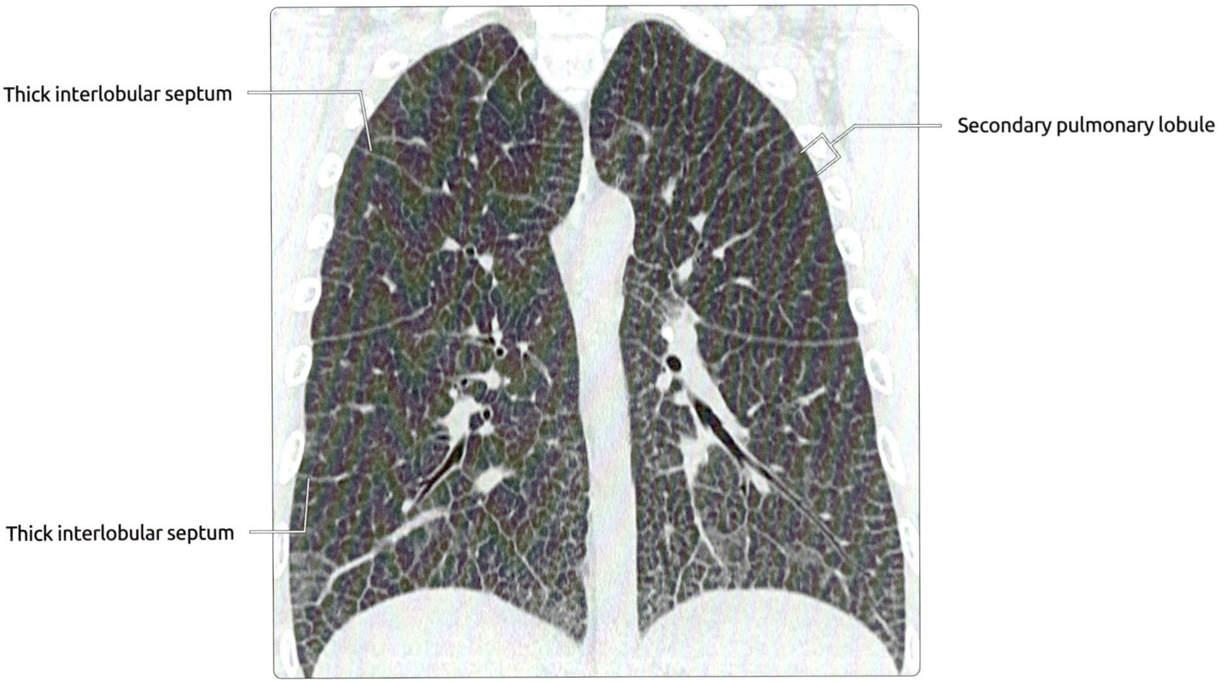

(Top) *Axial NECT (lung window) of a patient with Niemann-Pick disease shows diffuse delineation of virtually each and every secondary pulmonary lobule, which manifest as back-to-back polygonal structures. The normal secondary pulmonary lobule is not visible, as its boundaries are below the resolution of CT and only inferred through identification of normal pulmonary veins that course along its periphery and pulmonary arteries that course along its center.* **(Bottom)** *Coronal NECT (lung window) of the same patient shows diffuse delineation of nearly all secondary pulmonary lobules. In Niemann-Pick disease, there is abnormal deposition of lipids in the brain, spleen, liver, lungs, and bone marrow. Pulmonary involvement is characterized by accumulation of Niemann-Pick cells in the alveolar septa, pleura, and bronchial walls, which highlight the peripheral boundaries of the secondary pulmonary lobule as well as the centrilobular areas.*

ANATOMY OF PULMONARY LYMPHATICS

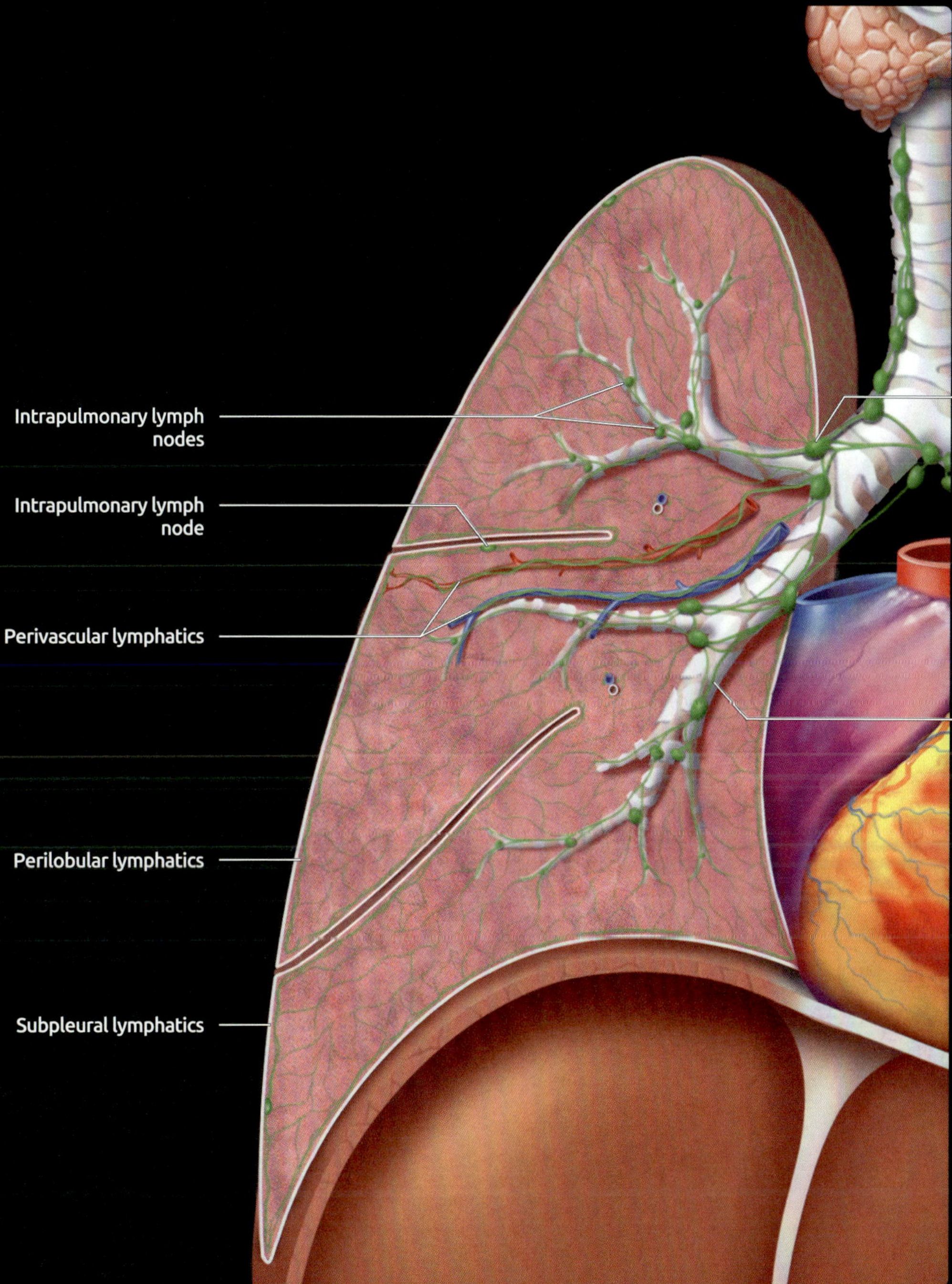

Graphic shows the complex lymphatic network of the lungs. The pulmonary lymphatics are found along bronchi, vessels, and in the subpleural connective tissue. Collecting lymphatics also course within the pulmonary connective tissue septa. The lymphatic network becomes organized as small intrapulmonary lymph nodes that typically occur at the bifurcations of large airways.

ANATOMY OF SECONDARY PULMONARY LOBULE LYMPHATICS

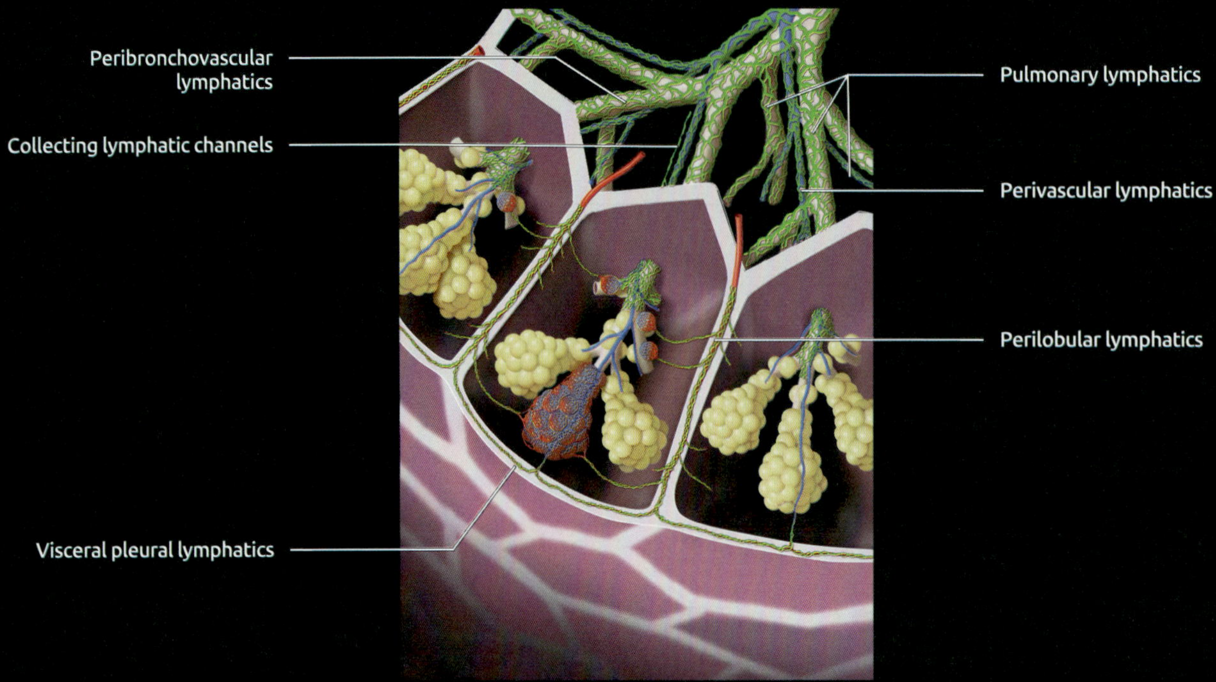

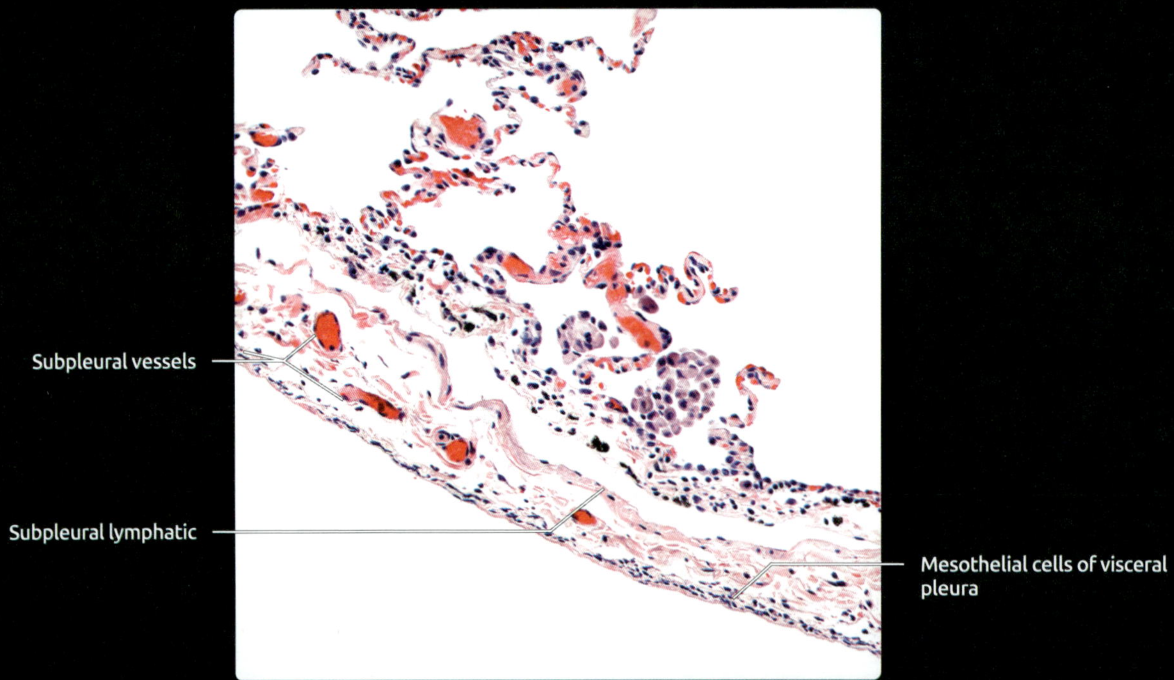

(Top) *Graphic shows the complex network of lymphatic channels within the secondary pulmonary lobule. Peribronchovascular lymphatics course along the airway walls to the level of the respiratory bronchioles. Perilobular lymphatics course in the interlobular septa and surround the pulmonary veins. Subpleural lymphatics are also illustrated. Pulmonary lymphatics are numerous and complex interconnecting vascular structures. Understanding the structure and anatomic location of the lymphatics of the secondary pulmonary lobule is important in identifying a perilymphatic distribution of disease on thin-section CT.* **(Bottom)** *High-power photomicrograph (H&E stain) shows a visceral pleural (subpleural) lymphatic channel surrounded by loose connective tissue and pulmonary veins coursing in the visceral pleura.*

RANDOM NODULES, CT

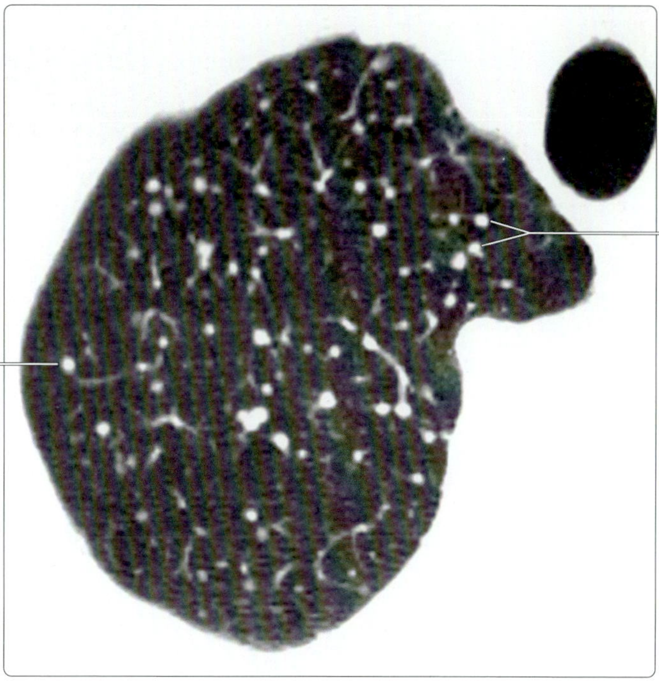

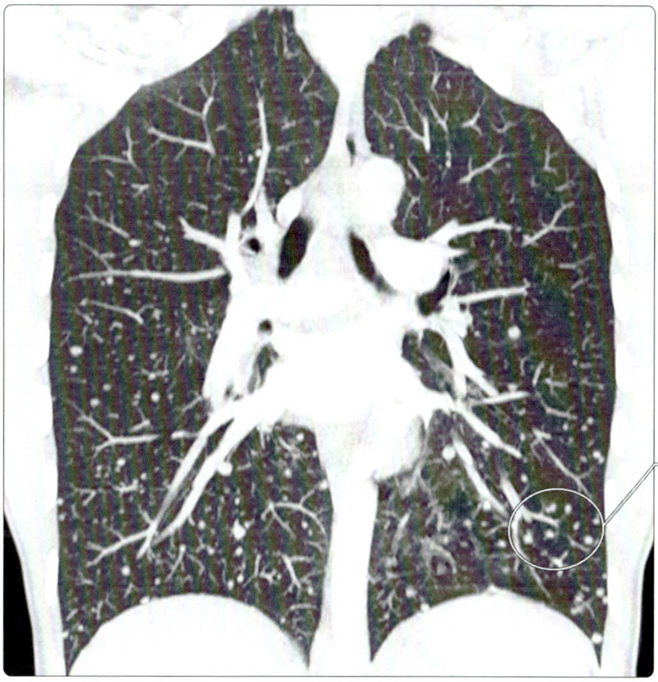

(Top) *Axial NECT (lung window) coned down to the right upper lobe of a patient with hematogenous pulmonary metastases from colon cancer demonstrates tiny, spherical, well-defined metastatic pulmonary nodules that exhibit a random distribution with respect to the structures of the secondary pulmonary lobule. Some nodules occur at the end of small pulmonary arteries indicating the hematogenous route of dissemination of the metastases.* **(Bottom)** *Coronal CECT (lung window) of a patient with metastatic papillary thyroid carcinoma shows numerous randomly distributed small well-defined nodules bilaterally. Note that the nodules are more numerous in the lower lobes in keeping with the dominant pulmonary blood flow to the lower lobes.*

RANDOM NODULES, CT

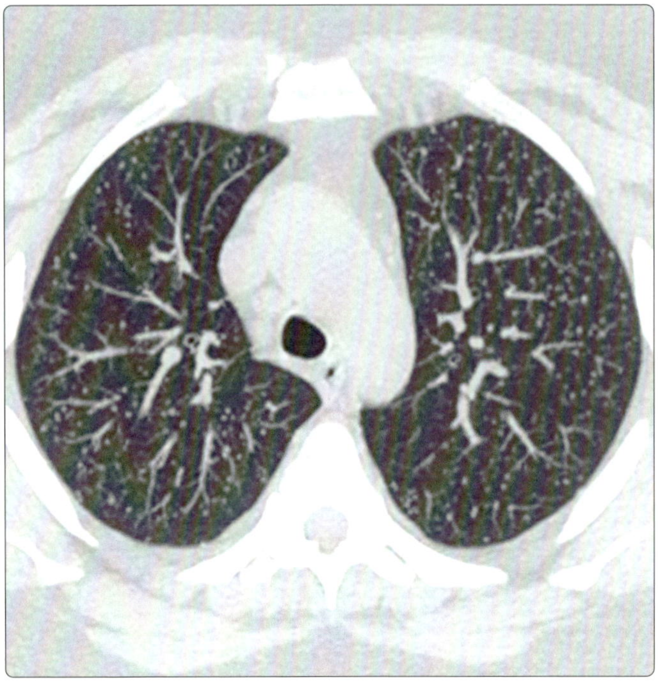

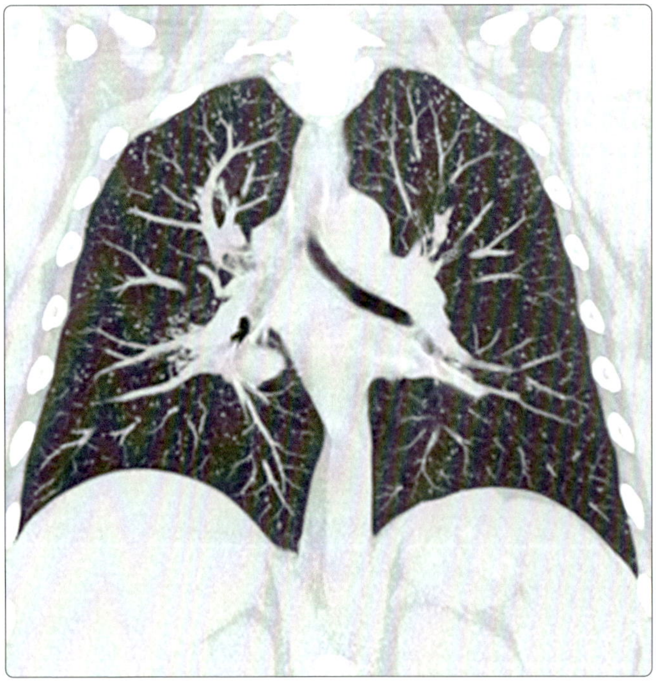

(Top) *Axial NECT (lung window) MIP reformatted image of a patient with hematogenous dissemination of histoplasmosis shows small, multifocal, well-defined pulmonary micronodules that exhibit a random distribution with respect to the structures of the secondary pulmonary lobule. Note that there is not a predilection for involvement of the pleura, although several subpleural nodules are present.* **(Bottom)** *Coronal NECT (lung window) MIP reformatted image of the same patient shows the random distribution of multiple miliary pulmonary nodules characteristic of hematogenous dissemination of infection.*

RANDOM NODULES, CT

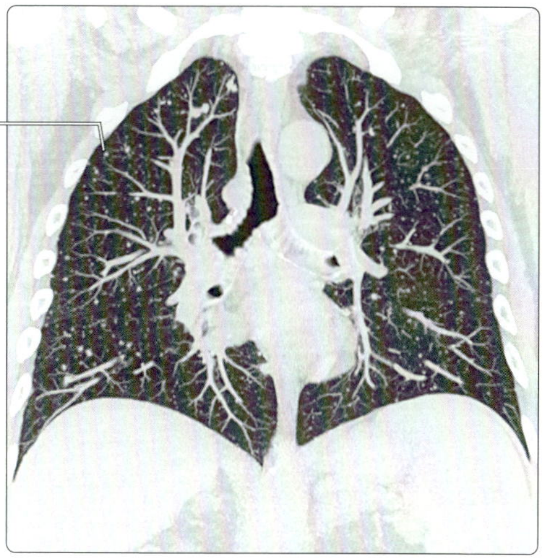

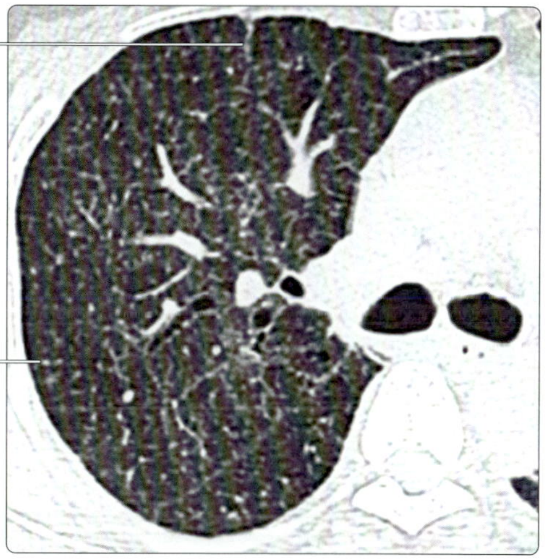

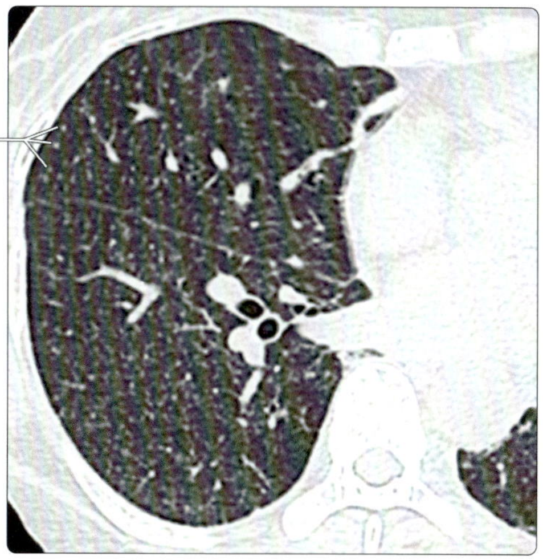

(Top) *Coronal CECT (lung window) MIP reformatted image of a patient with prostate cancer treated with intravesical Bacillus Calmette-Guérin (BCG) who developed miliary pulmonary BCG shows multifocal bilateral pulmonary micronodules (some of which exhibit an angiocentric distribution) randomly distributed throughout the lungs, consistent with hematogenous dissemination of BCG.* **(Middle)** *Axial HRCT of a patient with miliary tuberculosis shows multifocal pulmonary miliary micronodules that exhibit a random distribution throughout the lung. In this case, some of the nodules are located in the subpleural lung parenchyma and along interlobular septa.* **(Bottom)** *Axial HRCT of the same patient obtained through the right lower and middle lobes shows profuse micronodules that exhibit a random distribution. Some nodules are located in the subpleural aspect of the right major fissure. Although the mode of disease dissemination is hematogenous, it is difficult to associate the nodules to small pulmonary vessels in this case.*

PERILYMPHATIC NODULES, CT

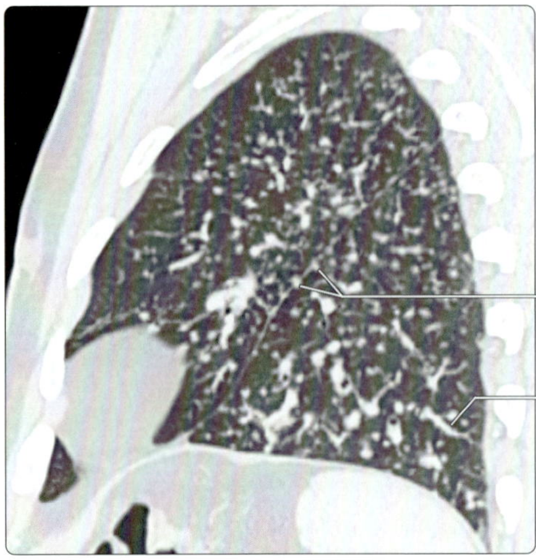

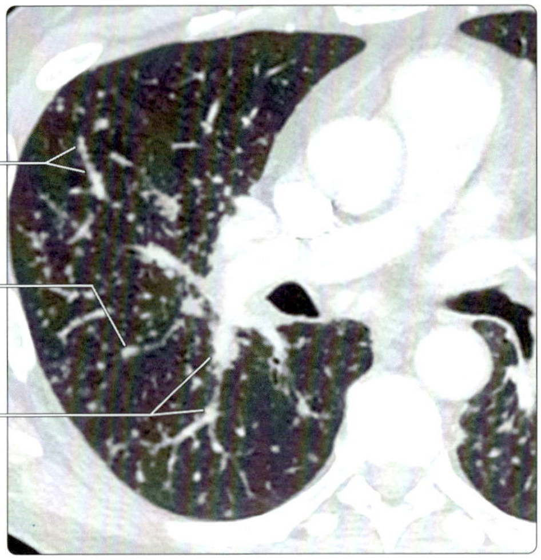

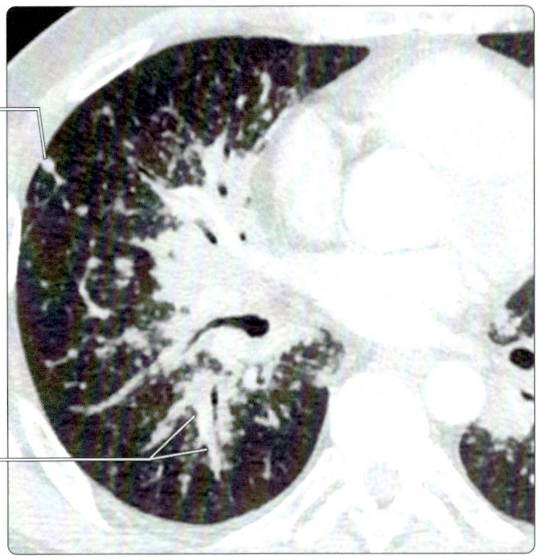

(Top) *Sagittal CECT (lung window) MIP reformatted image of a patient with sarcoidosis shows profuse perilymphatic pulmonary micronodules representing sarcoid granulomas. The nodules follow the anatomic distribution of the pulmonary lymphatics and are located along vascular structures, interlobular septa, and pleura. Note profuse nodularity along the major fissure, consistent with a perilymphatic distribution.* **(Middle)** *Axial CECT (lung window) of a patient with sarcoidosis shows multifocal pulmonary micronodules that follow a perilymphatic distribution. Note the beaded appearance of pulmonary vessels and the right major fissure resulting from micronodules located along peribronchovascular and subpleural lymphatics.* **(Bottom)** *Axial CECT (lung window) of a patient with sarcoidosis demonstrates coalescent mass-like opacities predominantly affecting the central aspects of the lung, surrounding and encasing central bronchi and vessels. Note irregular nodular thickening of bronchial walls characteristic of the perilymphatic distribution of sarcoidosis. Silicosis and coal worker's pneumoconiosis may produce similar findings.*

PERILYMPHATIC NODULES, CT

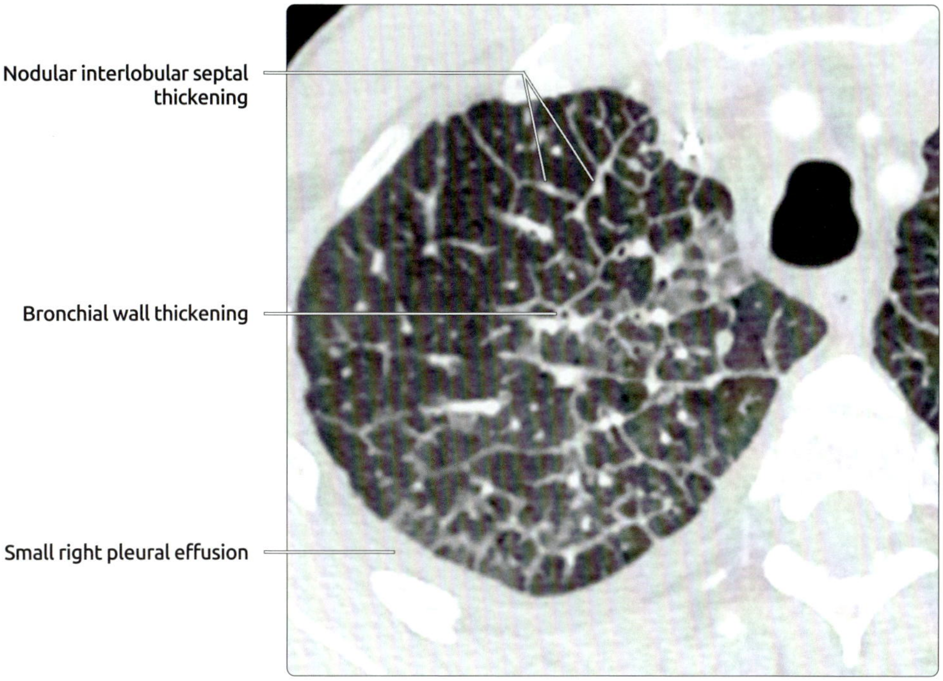

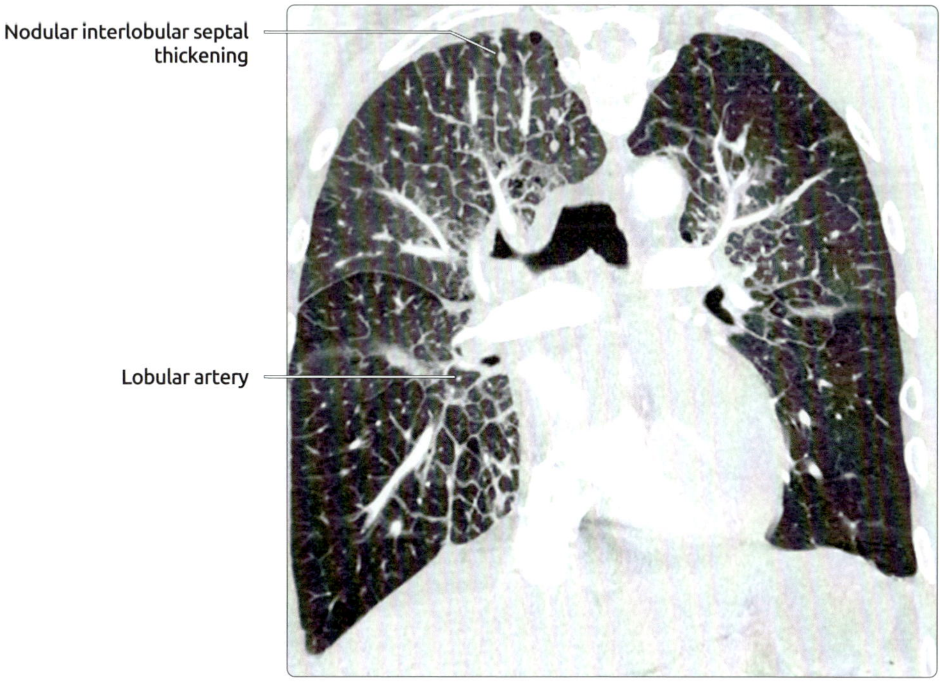

(Top) *Axial CECT (lung window) of a patient with lymphangitic carcinomatosis in the context of advanced prostate cancer shows profuse, smooth, and nodular thickening of the interlobular septa and bronchial wall thickening in a characteristic perilymphatic distribution. Lymphangitic carcinomatosis is characterized by tumor in pulmonary lymphatics and desmoplasia of the surrounding interstitium.* **(Bottom)** *Coronal CECT (lung window) of the same patient shows asymmetric distribution of smooth and nodular thickening of the interlobular septa, which demarcate the boundaries of the secondary pulmonary lobules. The central dot-like structures within several secondary pulmonary lobules represent the lobular arteries.*

OVERVIEW OF PULMONARY INTERSTITIUM

Embryology

- Embryologic remnant of splanchnopleuric mesenchymal bed
  - Site of airway and vessel ingrowth during lung morphogenesis

Anatomy

- Continuum of loose connective tissue
  - Extends from pulmonary hila to visceral pleura
    - Anchored at hila
    - Under tension by negative visceral pleural (intrapleural) pressure

Microscopic Features

- Fine reticulin fibers
- Fine elastin fibers
- Coarser collagen fibers
- Most prominent around large bronchovascular structures

Components

- Matrix components
  - Fibrous network of collagen and elastin fibers
    - Collagen fibers: Cannot be stretched or extended
    - Elastin fibers: Can be extended
- Cellular components: Fibroblasts, mast cells, tissue macrophages, lymphocytes
- Continuous epithelial and endothelial basement membranes
  - Form barriers defining outer boundaries of interstitium

Function

- Provides lung with structural integrity
- Permits lung deformation during respiration

INTERSTITIAL FIBER NETWORK

3 Subdivisions Forming Continuum

- **Axial**
  - Peribronchovascular
  - Centrilobular
- **Parenchymal** (**intralobular**)
- **Peripheral**
  - Subpleural
  - Interlobular (septal)

Axial

- **Peribronchovascular**: Surrounds bronchi and arteries at hila and central lungs
- **Centrilobular**: Peripheral continuation of interstitium; serves as support for bronchioles and arterioles at center of secondary pulmonary lobule (i.e., centrilobular)

Parenchymal (Intralobular)

- Fine network of very thin connective tissue fibers within alveolar walls
- Between alveolar and capillary basement membranes
- Supports structures of secondary pulmonary lobule
- Connects axial and peripheral interstitia

Peripheral

- **Subpleural**: Interstitium immediately beneath pleural mesothelium
- **Interlobular**: Interstitium distributed along periphery of secondary pulmonary lobule

IMAGING OF NORMAL INTERSTITIUM

Chest Radiography

- Interstitium is not normally visible
- Peribronchovascular interstitium may be visible as thin bronchial walls
- Centrilobular and intralobular interstitia not normally visible
- Interlobular interstitium not normally visible
- Subpleural interstitium visible along interlobar fissures (normal thin appearance)

Thin-Section CT

- Peribronchovascular interstitium may be visible as thin bronchial walls
- Centrilobular and intralobular interstitia not normally visible
- Subpleural interstitium visible along interlobar fissures (normal thin appearance)
- Occasional thickened interlobular septa (septal lines) may be normal
  - Typically located along peripheral pulmonary veins

IMAGING FINDINGS OF ABNORMAL INTERSTITIUM

Chest Radiography

- **Peribronchovascular**
  - Peribronchial cuffing
    - Bronchial wall thickening when visualized end on
    - Most apparent in perihilar regions
  - Peribronchovascular (perihilar) micronodules
- **Centrilobular**
  - Nonspecific opacities; difficult determination of whether they represent airspace disease, airway nodules, or interstitial involvement
- **Interlobular**: Septal lines
  - Kerley B lines
- **Subpleural**: Thickening of interlobar fissures

Thin-Section CT

- **Peribronchovascular**
  - Peribronchial cuffing
  - Peribronchovascular micronodularity
  - Perivascular cuffing or micronodules (on contrast-enhanced studies)
- **Centrilobular**
  - Centrilobular micronodules; airway micronodules (i.e., cellular bronchiolitis) may be indistinguishable from vascular micronodules
- **Intralobular**
  - Nonspecific ground-glass opacities due to volume averaging as structures are beyond limits of resolution of thin-section CT
  - Differentiation of ground-glass opacities arising in abnormal interstitium from alveolar filling may not be possible
  - Fine reticular opacities

- **Interlobular**
  - Septal lines (i.e., Kerley B lines)
    - Perpendicular to and in contact with pleura
    - 1.5-2.0 cm in length, 1-2 mm in width
    - Smooth: Pulmonary edema, lymphangitic carcinomatosis (often nodular)
    - Nodular: Lymphangitic carcinomatosis, sarcoidosis
    - Irregular: Fibrosis
  - Kerley A lines
    - Straight linear opacities in upper lung zones
      - 2-6 cm in length, 1-3 mm in width
    - Extend radially from hila toward upper lungs
    - Extend toward pleura but do not reach pleural surface
  - Kerley C lines
    - Branching linear opacities at lung bases
    - Fine and net-like
    - Represent Kerley B lines seen en face
- **Subpleural**
  - Fissural thickening
  - Fissural micronodularity

ADDITIONAL IMAGING FINDINGS OF ABNORMAL INTERSTITIUM

Reticular and Nodular Opacities

- **Reticular opacities**
  - Multiple intersecting irregular lines with net-like appearance
    - Fine: < 3 mm thick
    - Medium: 3-10 mm thick
    - Coarse: > 10 mm thick
- **Reticulonodular opacities**
  - Perceived combination of lines and dots; often artifactual
  - Superimposition of lines/reticulation may mimic nodules
  - Superimposition of nodules may mimic reticulation
- **Honeycombing** (end-stage lung disease)
  - Indicates extensive lung fibrosis
  - Thick-walled, air-filled cysts: 3 mm to 3 cm
  - Cysts share walls, occur in several layers in subpleural lung
  - Associated with traction bronchiectasis/bronchiolectasis resulting from lung fibrosis
    - Fibrous tissue produces outward traction on bronchial walls; resultant irregular bronchial dilatation
    - Traction bronchiolectasis involves small airways in peripheral lung
    - Associated with reticular opacities, parenchymal distortion, and honeycombing

Perihilar Haze

- Interstitial edema surrounding bronchoarterial bundles; indistinguishable from alveolar filling
- Blurred vascular borders
- Best appreciated after comparison to prior normal radiographs

ANATOMY-RELATED IMAGING ABNORMALITIES

Interstitial Pulmonary Edema

- **Radiography**
  - Axial interstitium: Conspicuous bronchovascular bundles
    - Blurred bronchial and vascular margins
    - Peribronchial cuffing when visualized end on
  - Peripheral interstitium: Conspicuity and thickening of interlobar fissures
    - Conspicuous minor fissure on frontal chest radiography
    - All fissures conspicuous on lateral chest radiography
    - Kerley lines (A &/or B)
- **CT**
  - Axial interstitium: Thickening of bronchovascular bundles
  - Peripheral interstitium: Smooth thickening of interlobular septa

Lymphangitic Carcinomatosis

- **Radiography**
  - Axial interstitium: Smooth and nodular thickening of bronchovascular bundles; Kerley A lines
  - Peripheral interstitium: Smooth and nodular thickening of interlobar fissures; Kerley B lines
- **CT**
  - Axial interstitium: Smooth and nodular thickening of bronchovascular bundles
    - Centrilobular nodules, peribronchovascular thickening
  - Peripheral interstitium: Smooth &/or nodular thickening of interlobular septa and interlobar fissures

Sarcoidosis

- **Radiography**
  - Axial interstitium: Thickening of bronchovascular bundles
    - Reticular and nodular opacities may emanate from hila
    - Predominant involvement of mid- and upper lung zones
  - Peripheral interstitium: Thickening and nodularity of interlobular septa and interlobar fissures
- **CT**
  - Axial interstitium: Small nodules along bronchovascular bundles
    - Thickening of bronchovascular bundles
  - Peripheral interstitium: Small nodules along interlobular septa and interlobar fissures
    - Small subpleural nodules

Interstitial Fibrosis

- **Radiography**
  - Peripheral interstitium: Reticular opacities, honeycombing
    - Subpleural and basilar predominant
  - Parenchymal interstitium: Fine basilar reticular opacities
- **CT**
  - Peripheral interstitium: Irregular thickening of interlobular septa and interlobar fissures
  - Parenchymal interstitium: Intralobular reticular opacities, honeycombing
  - Associated traction bronchiectasis

ANATOMY OF AXIAL, PARENCHYMAL, AND PERIPHERAL INTERSTITIUM

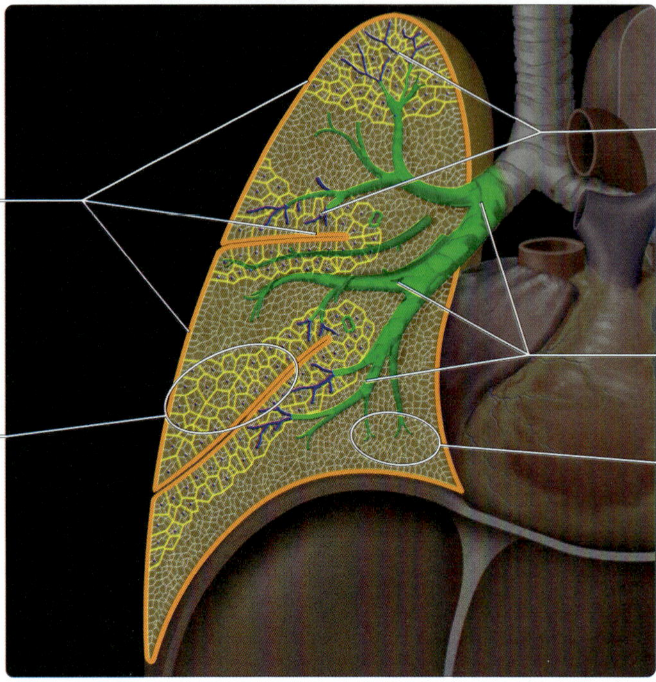

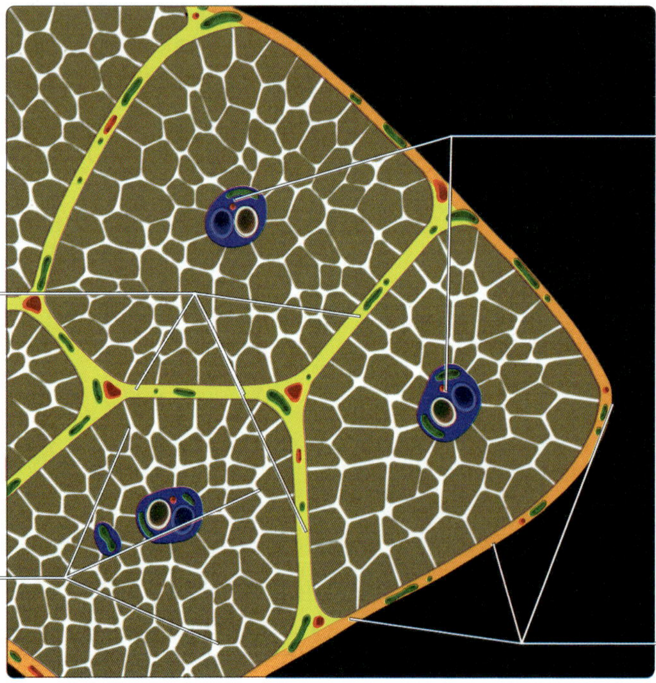

(Top) *Graphic shows all the components of the interstitial network. The axial (peribronchovascular) interstitium extends along the bronchovascular structures from the hilum to the lung periphery. Once it reaches the secondary pulmonary lobule this axial interstitium is referred as the centrilobular interstitium. The peripheral (subpleural/interlobular septal) interstitium extends along the subpleural regions, including the interlobar fissures, and contiguously along the interlobular septa with extension to the pulmonary hila along the pulmonary veins and lymphatics. The fine interstitial network of the parenchymal (intralobular) interstitium is distributed throughout the lung.* **(Bottom)** *Coned down graphic shows the different interstitial compartments, including the axial (centrilobular) interstitium extending along the center of the secondary pulmonary lobule, the peripheral (subpleural/interlobular septal) along the periphery of the secondary lobule, and the parenchymal (intralobular) interstitium between central and peripheral interstitia.*

ANATOMY OF AXIAL, PARENCHYMAL, AND PERIPHERAL INTERSTITIUM

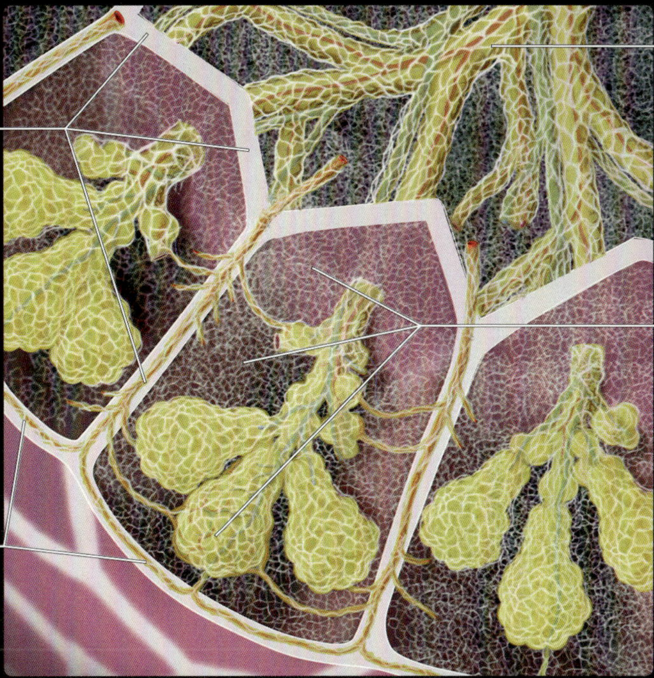

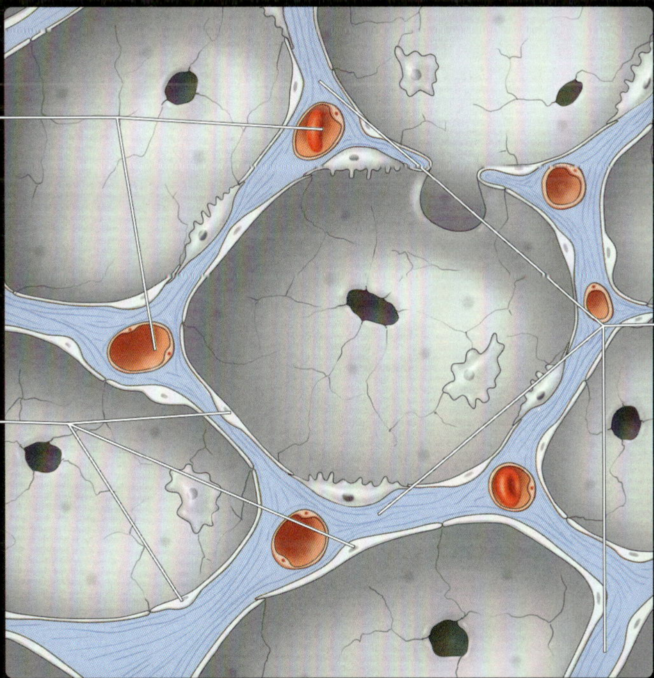

(Top) *Graphic shows the axial, parenchymal and peripheral interstitium of the secondary pulmonary lobule. The peripheral interstitium extends along the subpleural regions and along the interlobular septa and extends along the pulmonary veins and lymphatics toward the pulmonary hilum. The most distal portions of the axial interstitial sheath are shown along the bronchovascular structures as they enter secondary pulmonary lobules. The parenchymal interstitium forms a meshwork around clusters of alveoli and alveolar sacs. A continuous interstitial fiber network within the secondary pulmonary lobule extends to the interlobular septa.* **(Bottom)** *The parenchymal (intralobular) interstitium is interposed between the capillary endothelium and the alveolar epithelium and is located between the peripheral and the axial interstitium and provides structural support to the alveolar wall.*

MICROSCOPY OF NORMAL PULMONARY INTERSTITIUM

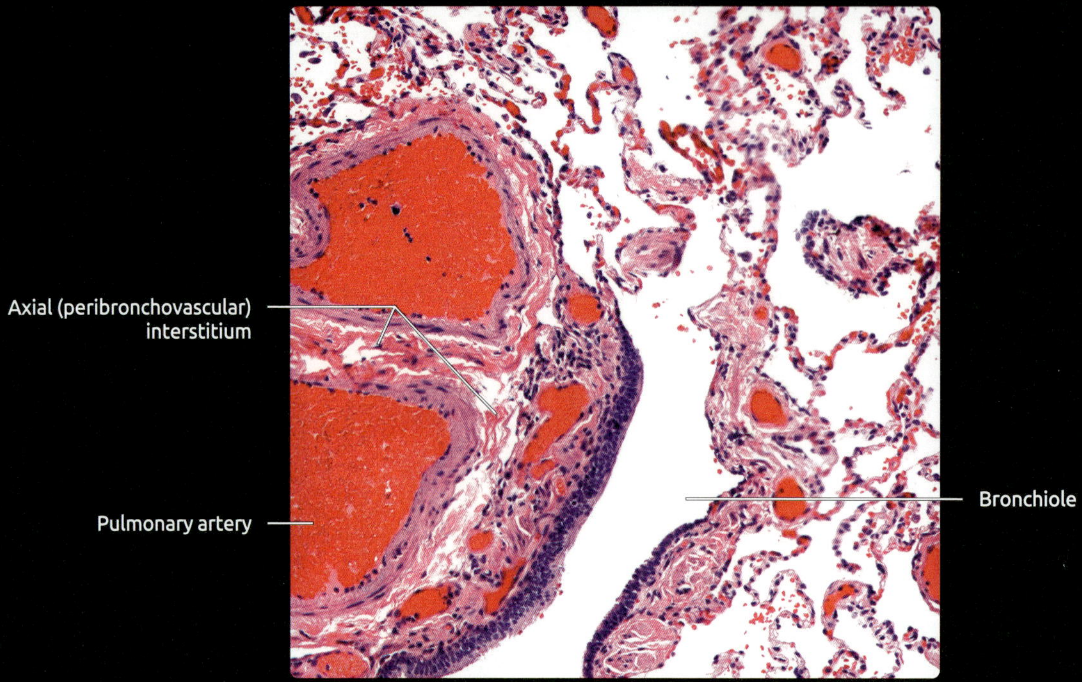

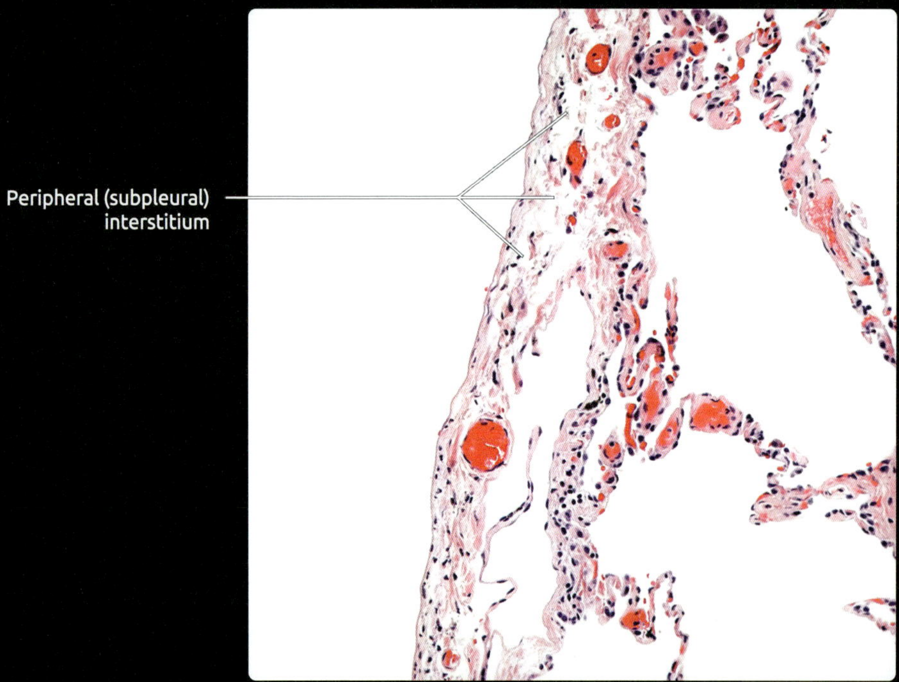

(Top) *High-power photomicrograph (H&E stain) shows the fine connective tissue elements that compose the axial (peribronchovascular) interstitium that surrounds the pulmonary arteries and the airways.* **(Bottom)** *High-power photomicrograph (H&E stain) shows the peripheral (subpleural) interstitium. Fine collagen and elastin fibers occupy the subpleural region and surround subpleural vessels and lymphatics.*

PULMONARY INTERSTITIUM, RADIOGRAPHY AND CT

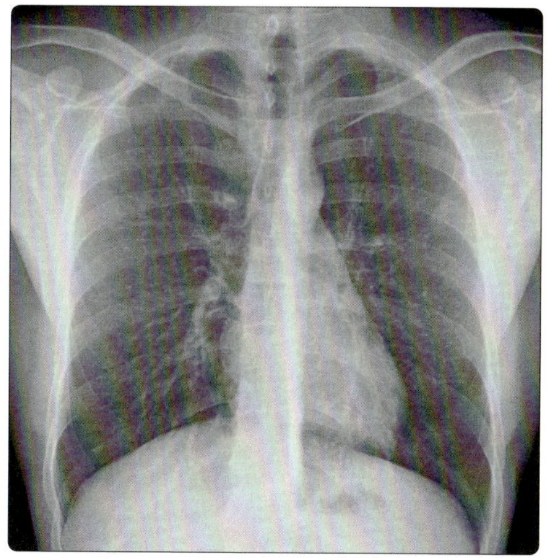

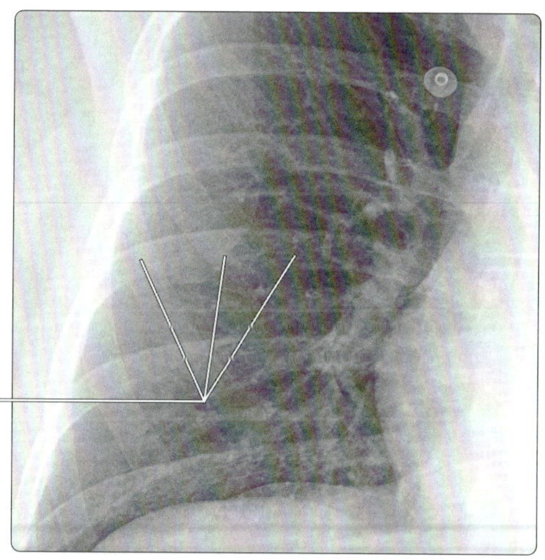

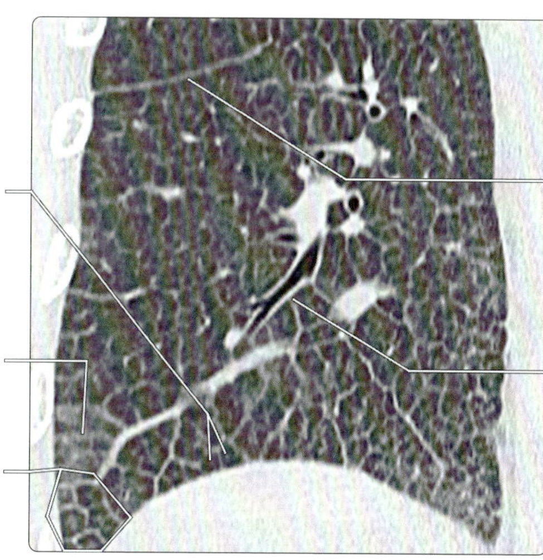

(Top) *PA chest radiograph of a normal subject shows a normal appearance of the lungs. The pulmonary interstitium is not visible on normal chest radiographs.* **(Middle)** *PA chest radiograph of a normal subject coned down to the right lung shows the normal appearance of the minor (horizontal) fissure, which manifests as an inconspicuous thin line. The normal minor fissure and its adjacent subpleural interstitium are visualized due to the adjacent normal aerated lung.* **(Bottom)** *Coronal HRCT of a patient with Niemann-Pick disease allows visualization of all the interstitial compartments (not normally visible) due to diffuse abnormal interstitial accumulation of lipid-laden macrophages. In this case, there is thickening of the subpleural interstitium along the interlobar fissures, the peripheral interlobular interstitium along the interlobular septa, and the peribronchovascular interstitium with resultant thickening of the bronchial walls. Ill-defined nonspecific ground-glass opacities likely correlate with involvement of the intralobular interstitium.*

OVERVIEW OF LUNGS

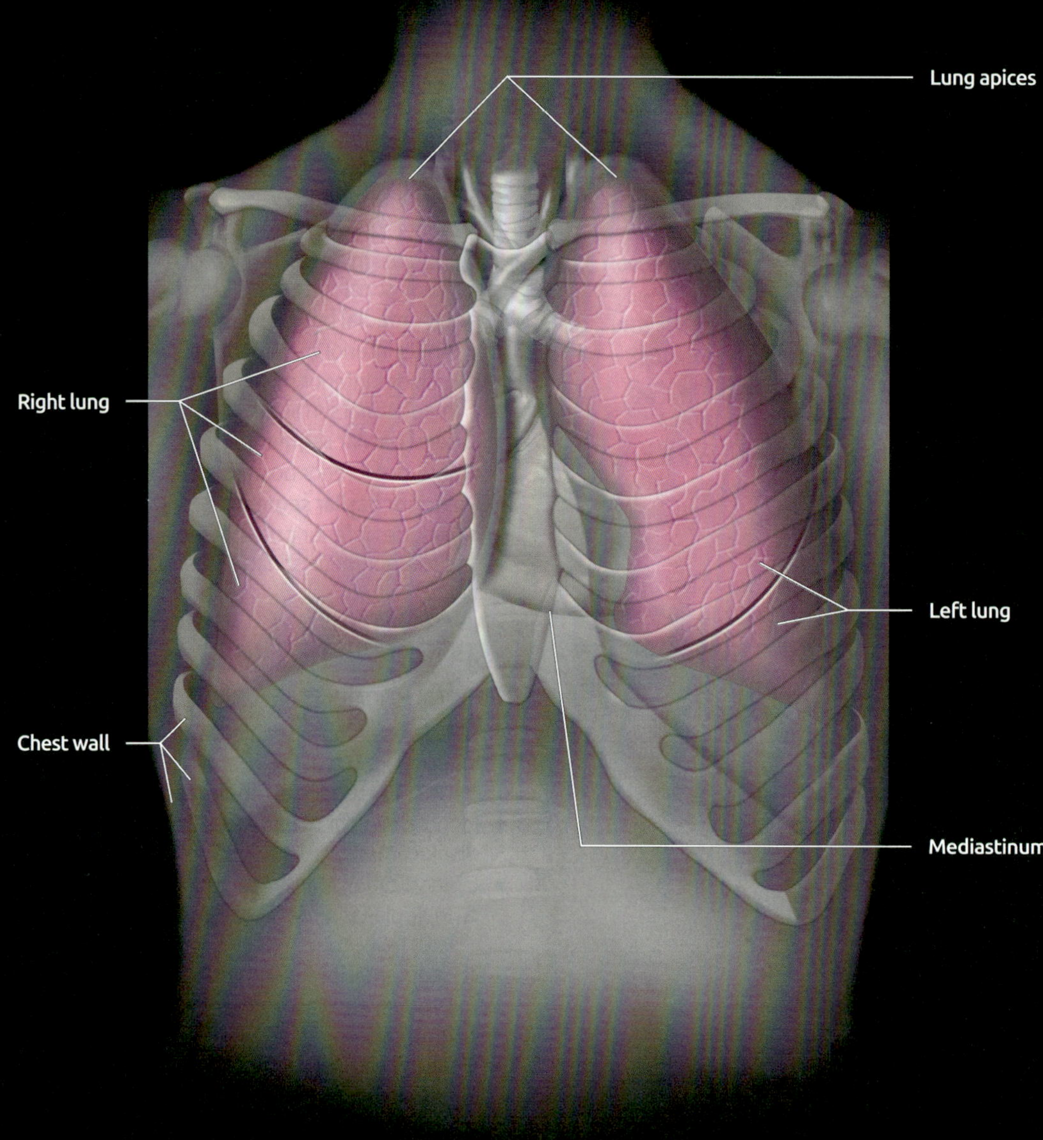

Graphic shows an overview of the lung anatomy. The lungs are surrounded by the pleura and the osseous and soft tissue structures of the chest wall. The 2 lungs are located on either side of the mediastinum. Each lung is freely mobile within its surrounding pleural space and is attached to the mediastinum at the hilum and the pulmonary ligament. The lung apices project above the medial clavicles and anterior 1st ribs and extend toward the root of the neck. The right lung has 3 lobes and is larger than the left lung. The left lung is smaller than the right lung and has 2 lobes.

SURFACE ANATOMY OF ANTERIOR AND POSTERIOR LUNGS

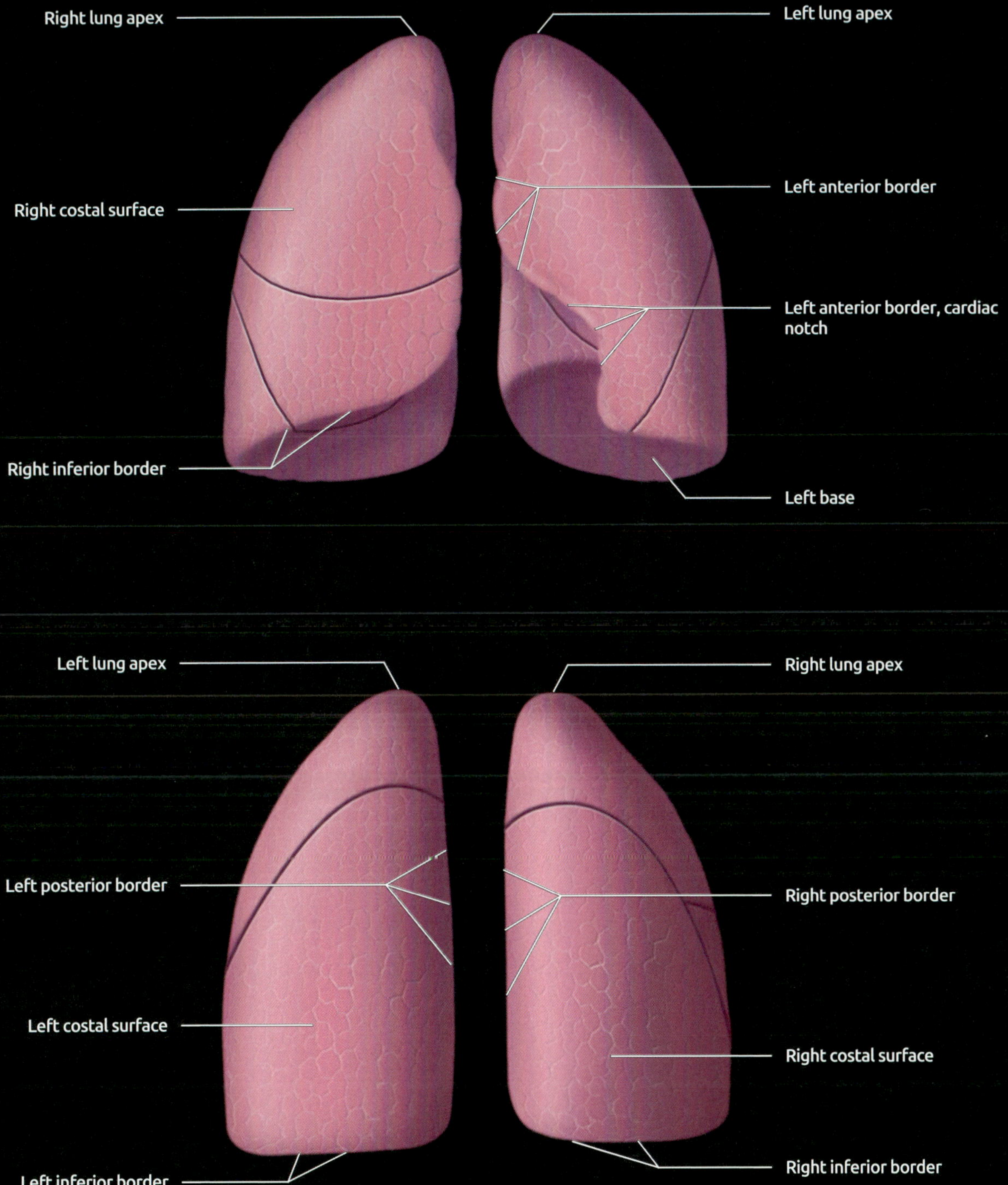

(Top) *Graphic depicts the surface anatomy of the anterior lungs. The shape of each lung resembles that of a half cone. The surface anatomy of each lung is characterized by an apex, a base, 2 surfaces, and 3 borders. The apices represent the uppermost extent of the lungs. The anterior borders separate the anterior costal surfaces from the mediastinal (medial) surfaces. Note the arcuate morphology of the inferior aspect of the left anterior border, also called the cardiac notch. The costal surfaces are adjacent to the chest wall, and the mediastinal surfaces are adjacent to the mediastinum. The inferior borders separate the costal surfaces from the lung bases.* **(Bottom)** *Graphic depicts the surface anatomy of the posterior lungs. The inferior lung borders separate the costal surfaces from the corresponding lung bases. The posterior borders separate the costal surfaces of the lungs from the corresponding mediastinal (medial) surfaces.*

SURFACE ANATOMY OF LATERAL LUNGS

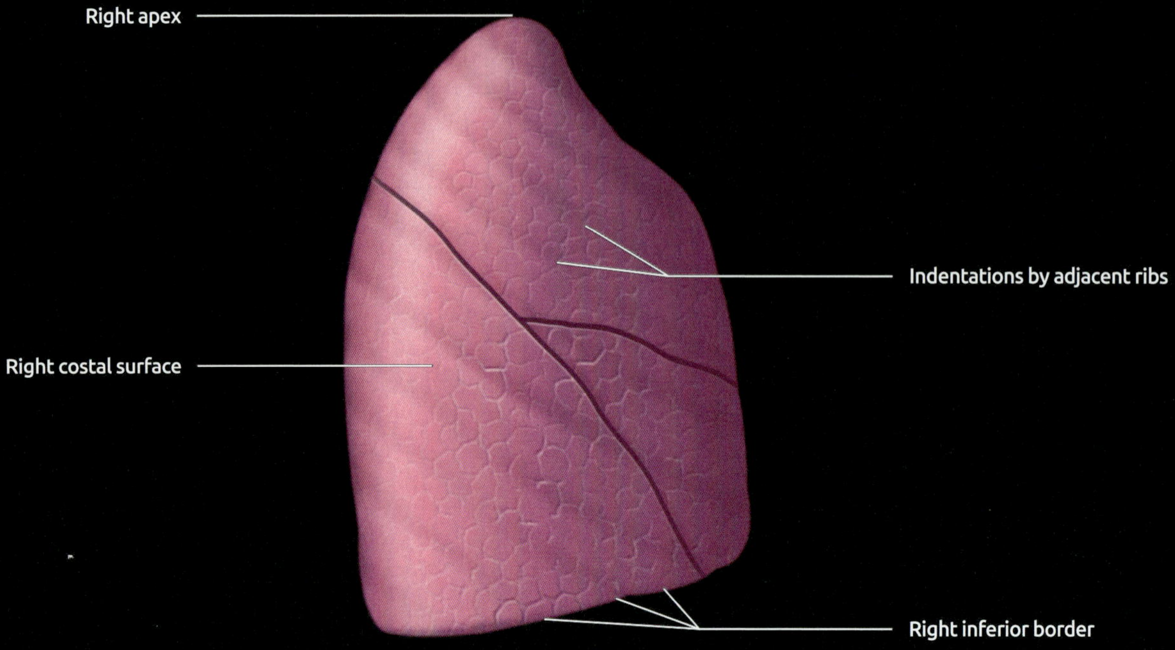

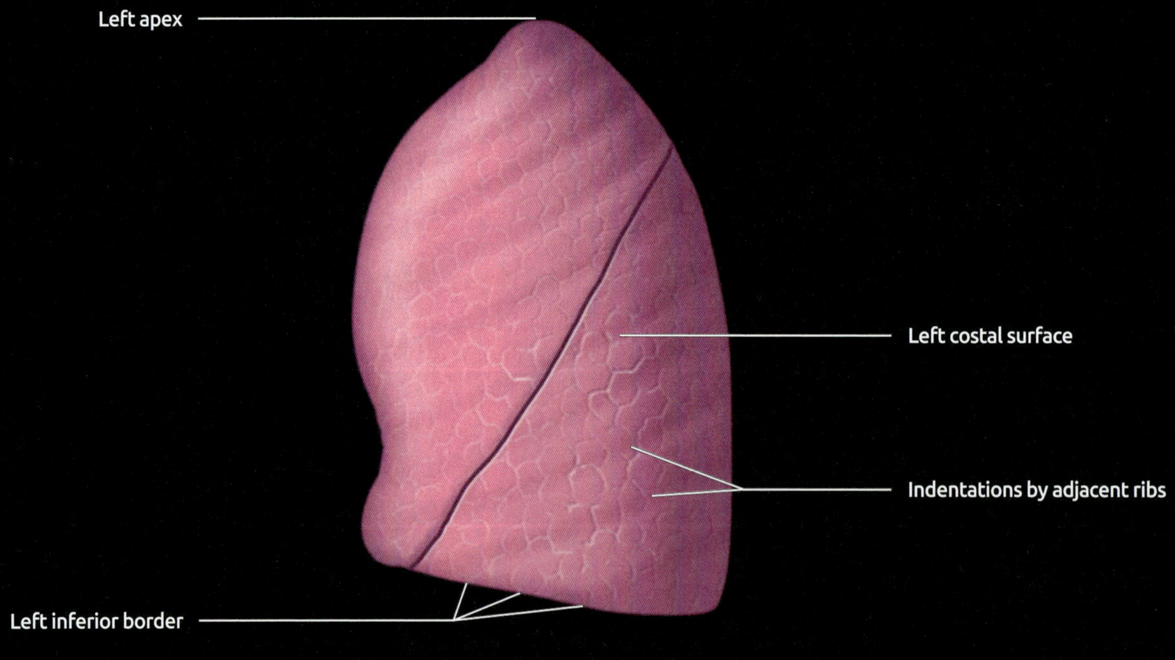

(Top) *Graphic depicts the lateral surface anatomy of the right lung. The lateral lung surface forms part of the costal surface named for the adjacent ribs (and intercostal spaces), which produce obliquely oriented indentations on the lung parenchyma. The inferior border separates the lateral costal surface from the lung base.* **(Bottom)** *Graphic depicts the lateral surface anatomy of the left lung. The lateral surface forms part of the costal surface named for the adjacent ribs (and intercostal spaces), which produce obliquely oriented indentations on the lung. The inferior border separates the costal surface from the lung base.*

SURFACE ANATOMY OF MEDIAL LUNGS

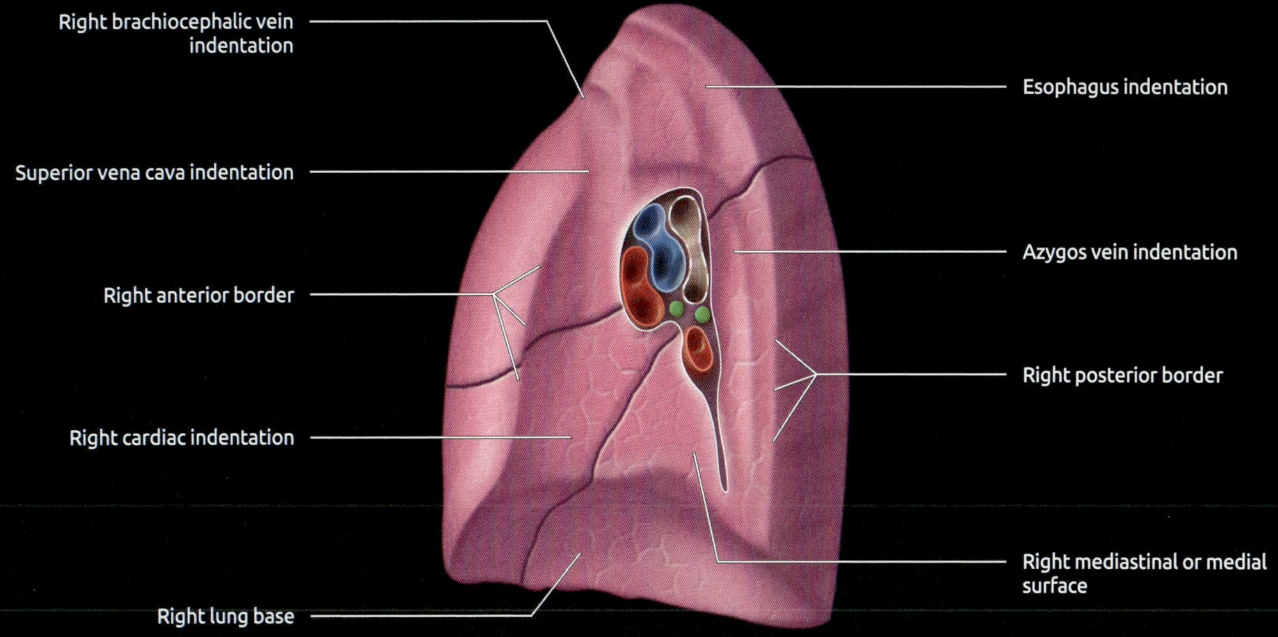

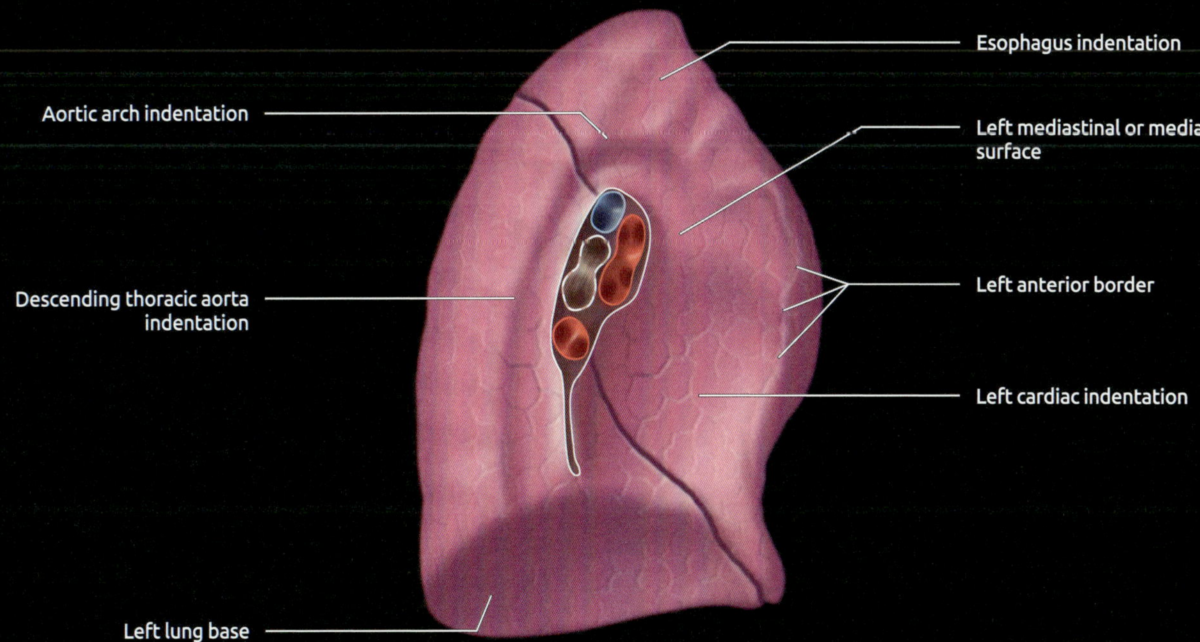

(Top) *Graphic depicts the surface anatomy of the medial right lung. The mediastinal or medial surface is concave and exhibits indentations produced by adjacent mediastinal structures, including vessels and organs. The right cardiac indentation is produced predominantly by the right atrium. The anterior border separates the mediastinal surface from the costal surface. The right hilum is located centrally on the mediastinal surface.* **(Bottom)** *Graphic depicts the surface anatomy of the medial left lung. The mediastinal or medial surface exhibits indentations produced by adjacent mediastinal structures. The left cardiac indentation is predominantly produced by the left ventricle. The descending thoracic aorta indents the left lower lobe and correlates with visualization of the retrocardiac descending aorta on frontal chest radiography. The hilum is located centrally in the mediastinal or medial surface.*

LUNG LOBES

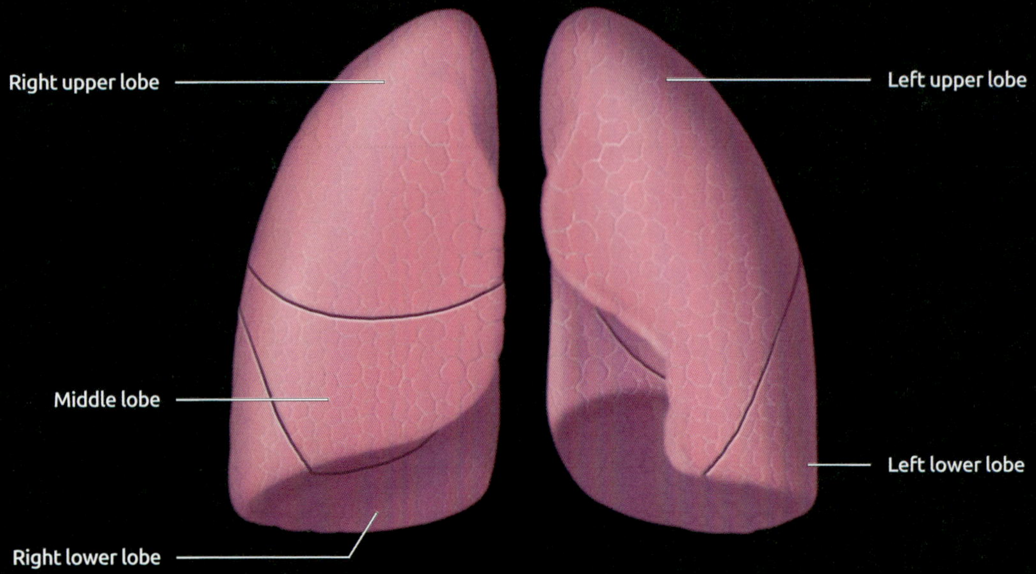

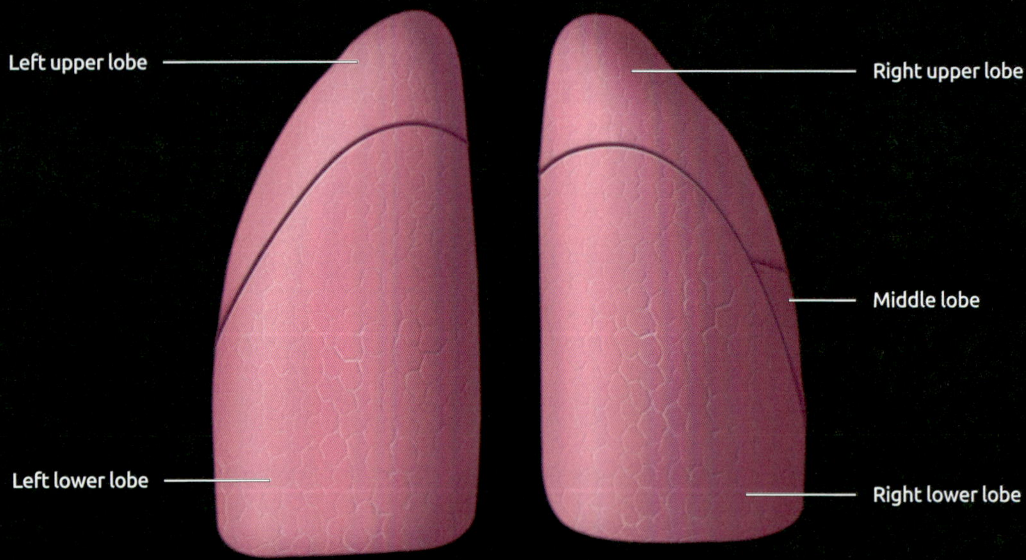

(Top) *Graphic depicts the anatomy of the anterior lungs and their lobes. The right lung has 3 lobes and is larger than the left. The left lung has 2 lobes. Visceral pleura lines each of the lobes, which are demarcated by the interlobar fissures.* **(Bottom)** *Graphic depicts the anatomy of the posterior lungs and the lung lobes. The lobar boundaries are demarcated by the interlobar fissures.*

LUNG LOBES

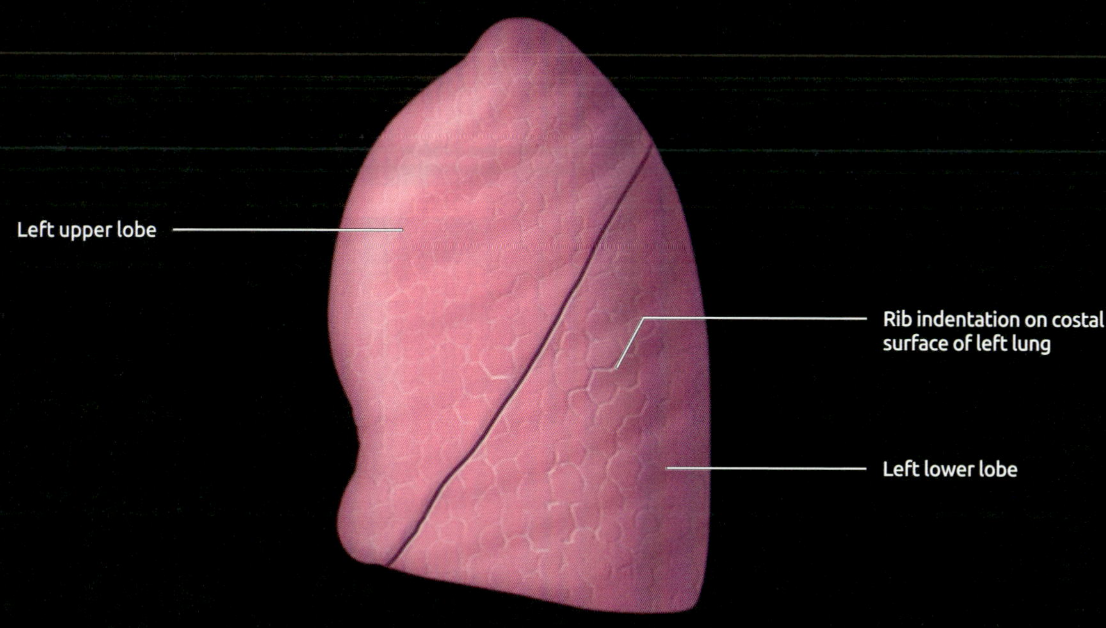

(Top) *Graphic shows the lateral aspect of the costal surface of the right lung, which exhibits linear rib indentations, and the location of the 3 right lung lobes demarcated by the interlobar (major and minor) fissures. The right upper lobe and middle lobe occupy the anterior right lung. The right lower lobe occupies the mid- and inferior aspects of the posterior right lung.* **(Bottom)** *Graphic shows the lateral aspect of the costal surface of the left lung, which exhibits rib indentations, and the location of the 2 left lung lobes, which are separated by the left major fissure. The left upper lobe occupies the anterior left lung. The left lower lobe occupies the mid- and inferior aspects of the posterior left lung.*

LOBAR AND SEGMENTAL LUNG ANATOMY, AXIAL CT

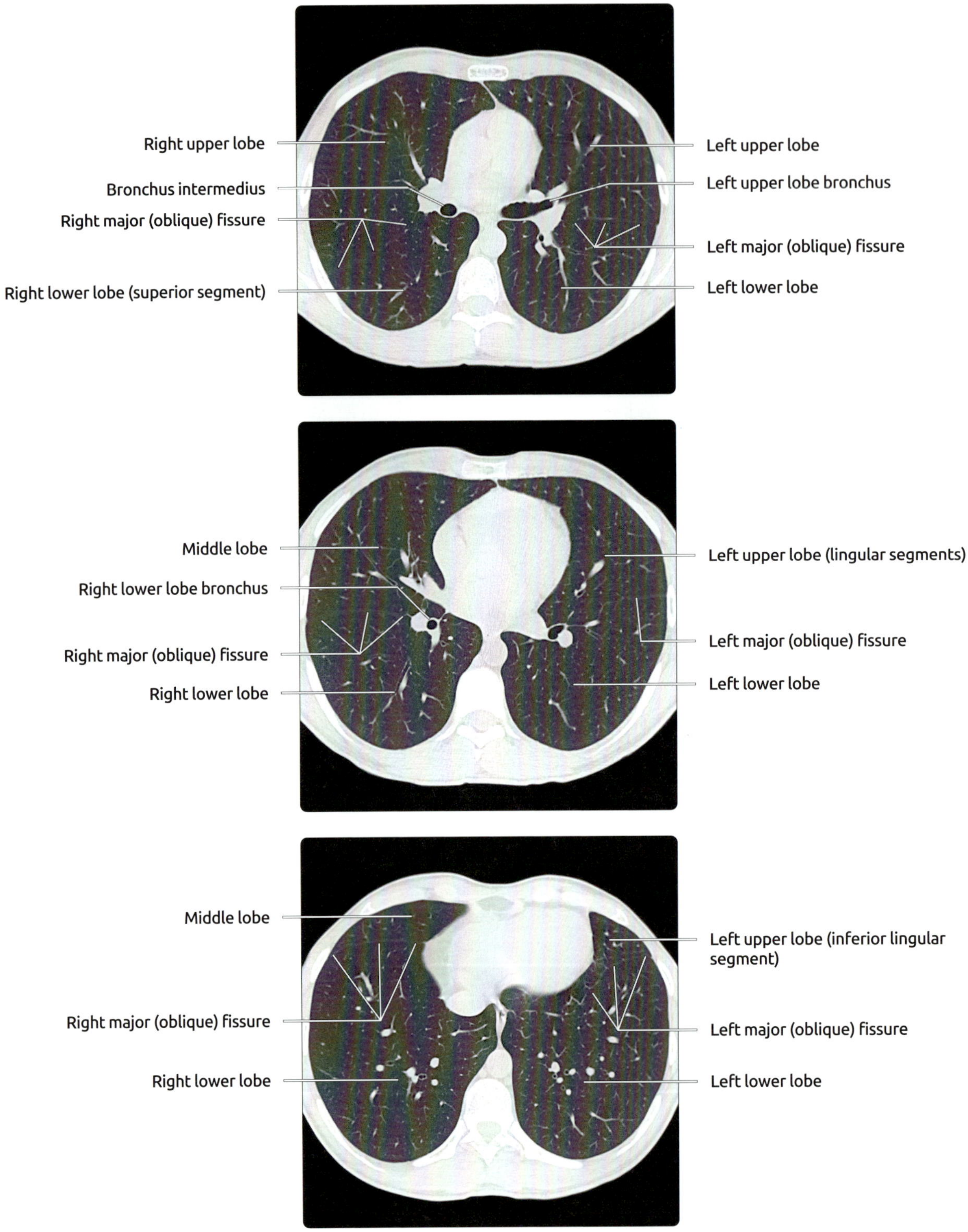

(Top) *Axial NECT (lung window) of the same patient shows the lobar and segmental anatomy of the lungs based on the corresponding bronchial and fissural anatomy.* **(Middle)** *Axial NECT (lung window) of the same patient shows the normal lobar, segmental, bronchial, and fissural anatomy of the lungs. Identification of such landmarks allows the radiologist to provide specific information regarding the anatomic location of imaging abnormalities.* **(Bottom)** *Axial NECT (lung window) of the same patient shows the normal lobar and segmental anatomy of the lungs. At the inferiormost portions of the hemithoraces, most of the visualized lung corresponds to the lower lobes and their basilar segments.*

LOBAR LUNG ANATOMY, CORONAL CT

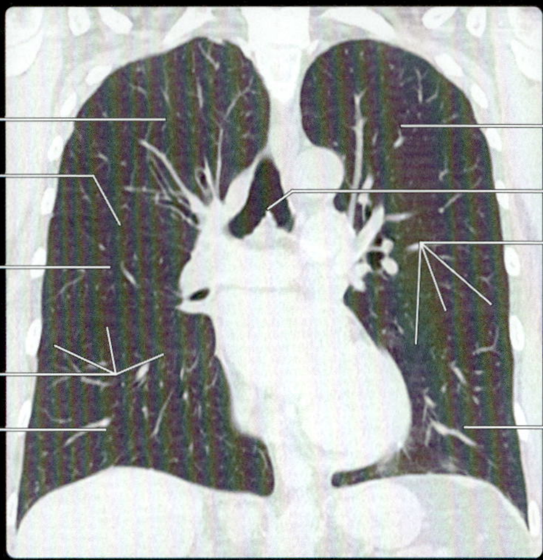

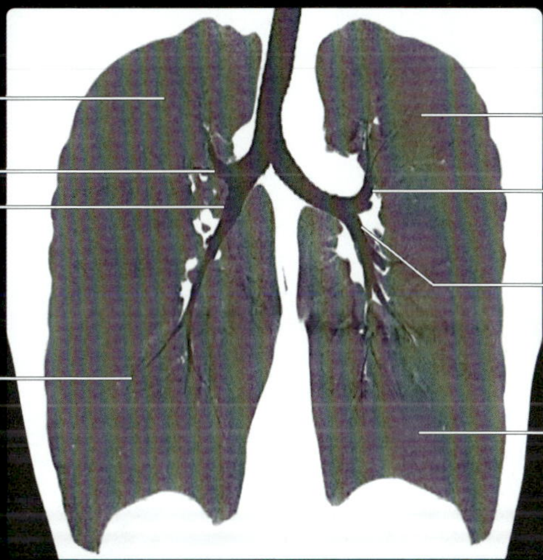

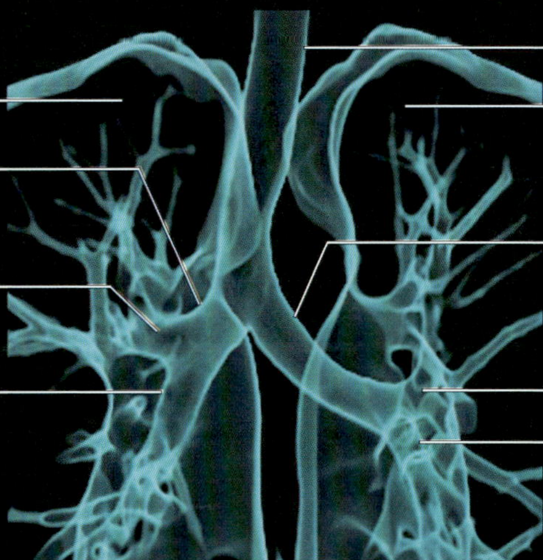

(Top) *Coronal reformatted NECT (lung window) of a normal individual shows the normal coronal lung anatomy at the level of the tracheal carina. Familiarity with tracheobronchial anatomy and the normal distribution of the interlobar fissures in several imaging planes is critical for correct anatomic localization of imaging abnormalities, especially those that may require intervention. Reformatted images may be helpful for accurate lesion localization.* **(Middle)** *Coronal reformatted NECT (minIP image) of the same patient shows the normal anatomy of the central airways at the level of the tracheal carina. Note that the middle lobe bronchus is not visualized as it courses anteriorly from the bronchus intermedius and is not depicted at this level.* **(Bottom)** *Coronal 3D surface-rendered image of an NECT of the chest of a normal individual shows the normal morphology of the central airways. Knowledge of the normal tracheobronchial anatomy of the lung is critical for accurate lesion localization to a specific lung lobe &/or segment.*

LOBAR AND SEGMENTAL LUNG ANATOMY, AXIAL AND SAGITTAL CT

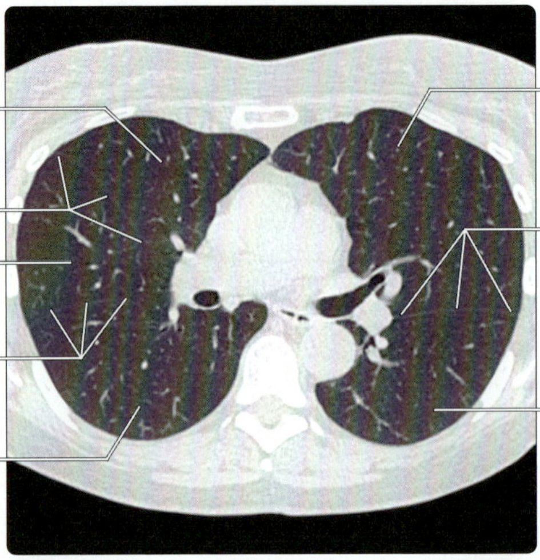

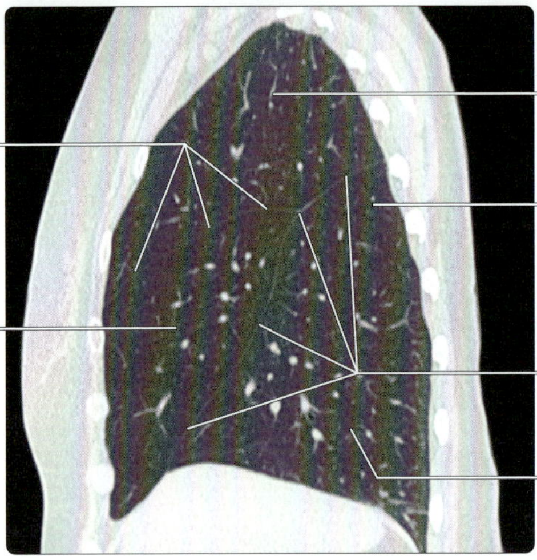

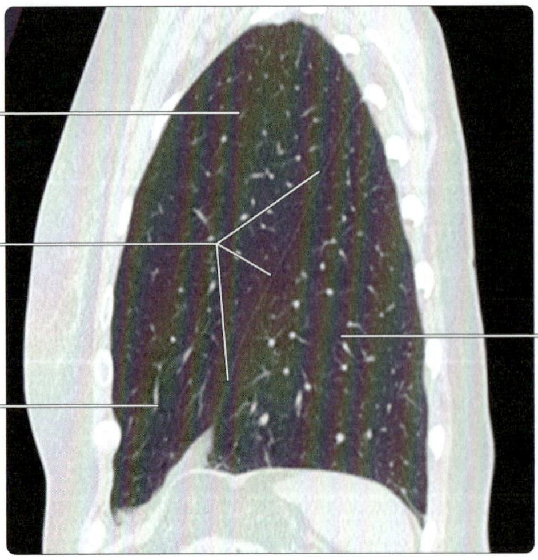

(Top) *Axial NECT (lung window) of a normal individual shows the normal morphology and orientation of the minor and the bilateral major fissures. Recognition of the interlobar fissures allows identification of lung lobes on CT and accurate lesion localization.* **(Middle)** *Sagittal NECT (lung window) of the same patient shows the normal morphology and orientation of the minor and right major fissures in the sagittal plane. Recognition of normal fissural anatomy on multiplanar CT allows accurate identification of the lung lobes. Interlobar fissures separate the 3 right pulmonary lobes (i.e., upper, middle, and lower).* **(Bottom)** *Sagittal NECT (lung window) of the same patient shows the normal morphology and orientation of the left major fissure, which separates the left upper lobe from the left lower lobe. The lingular segments are within the left upper lobe. Typically, there is not a minor fissure on the left, but an accessory left minor fissure may occur as an anatomic variant, producing separation between the lingular segments and the rest of the left upper lobe. Identification of a left minor fissure may be relevant for surgical planning in selected cases.*

LOBAR AND SEGMENTAL LUNG ANATOMY, AXIAL AND SAGITTAL CT

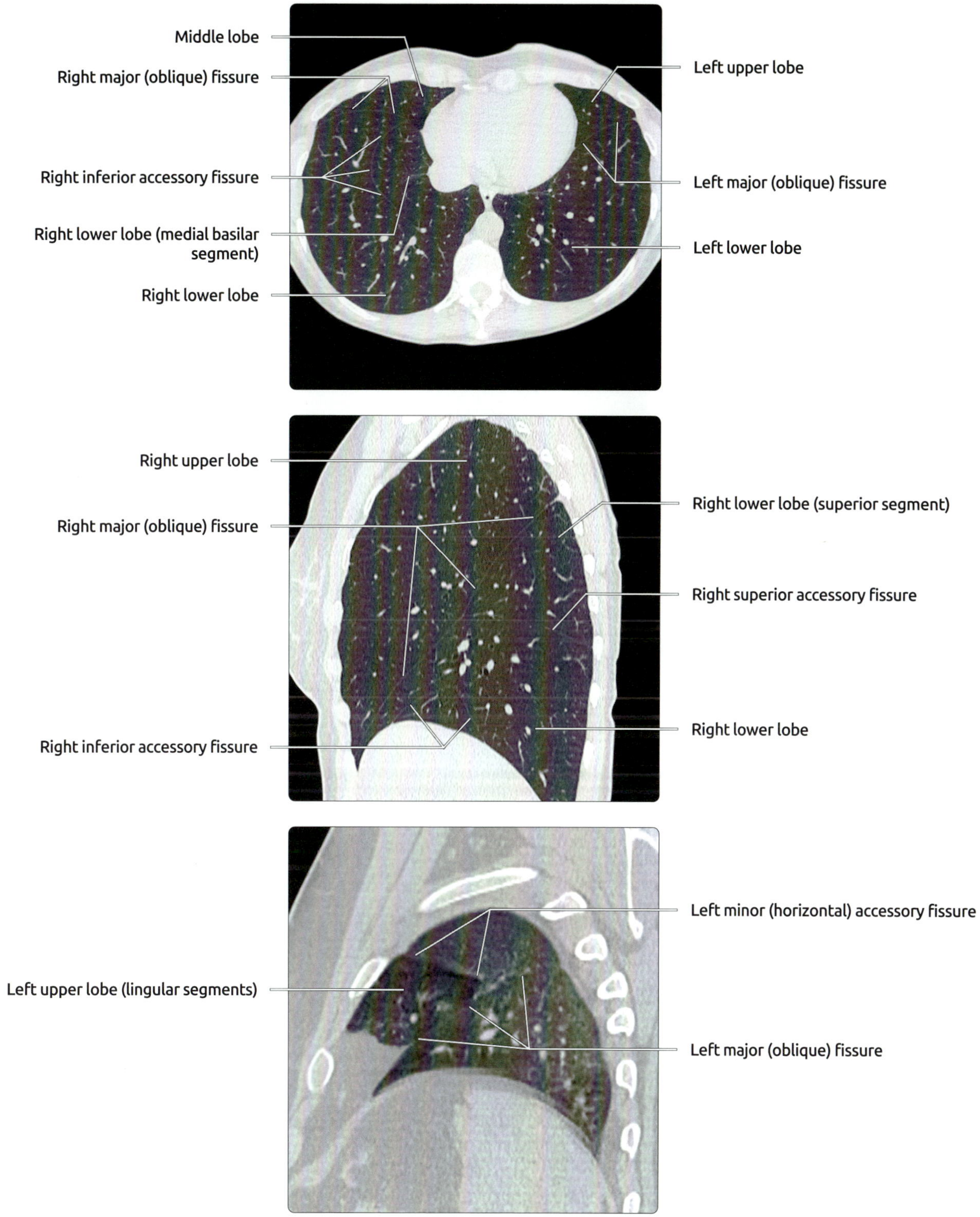

(Top) *Axial NECT (lung window) of a normal individual shows the right inferior accessory fissure, an anatomic variant of no clinical significance. The right inferior accessory fissure separates the right lower lobe medial basilar segment from the remaining right lower lobe basilar segments.* **(Middle)** *Sagittal NECT (lung window) of the same patient shows the right inferior accessory fissure and an additional right superior accessory fissure. The right superior accessory fissure separates the right lower lobe superior segment from the right lower lobe basilar segments.* **(Bottom)** *Sagittal expiratory NECT (lung window) of a normal subject shows an accessory left minor fissure, an anatomic variant that separates the left upper lobe lingular segments from the remaining left upper lobe segments.*

RIGHT UPPER LOBE ANATOMY, RADIOGRAPHY AND CT

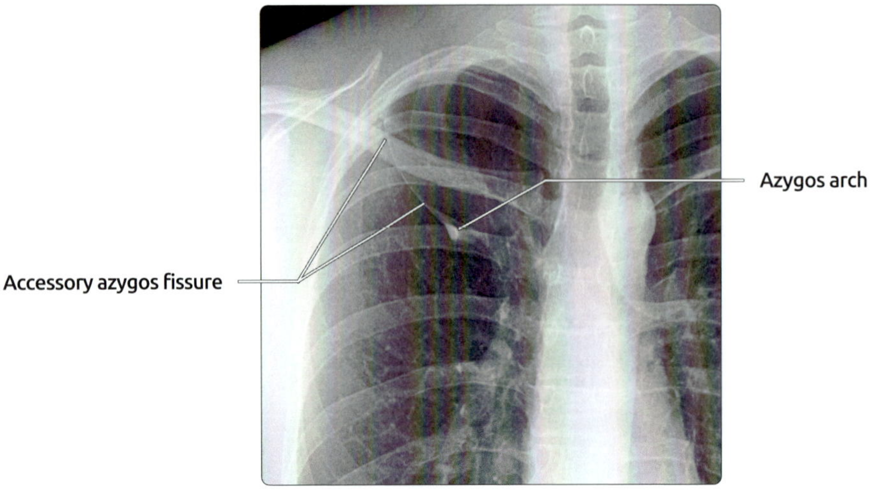

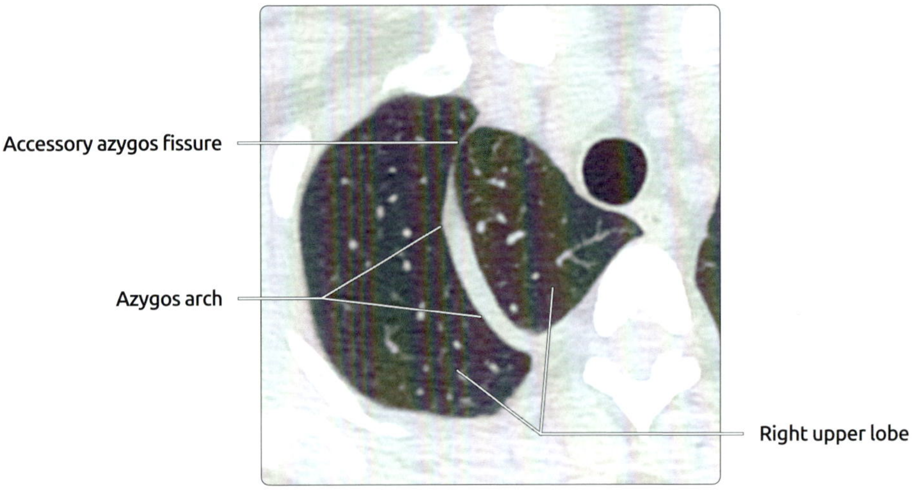

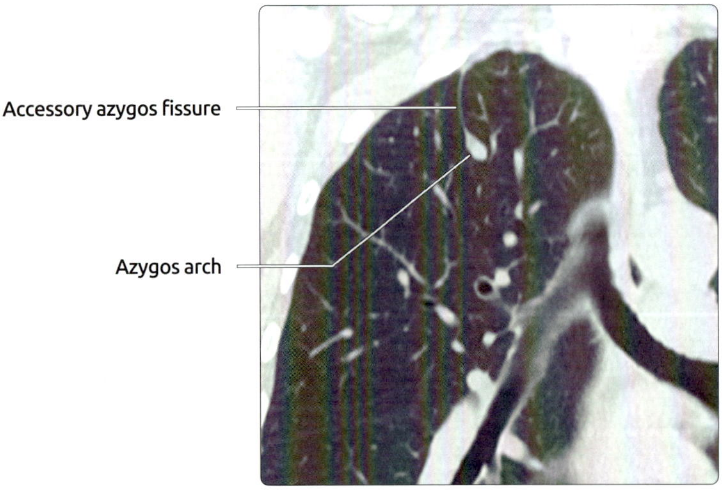

(Top) *PA chest radiograph of a normal subject shows an accessory azygos fissure, an anatomic variant that contains the azygos arch with its classic teardrop configuration. Note absence of the azygos arch from its normal location at the right tracheobronchial angle. The accessory azygos fissure subdivides the right upper lobe. The term "azygos lobe" is a misnomer, as by definition, a lung lobe is defined by its supplying bronchus, rather than by the fissural anatomy.* **(Middle)** *Axial NECT (lung window) of a normal individual with an accessory azygos fissure shows the arcuate morphology of the azygos arch contained within the fissure. Rarely, subjects with an accessory azygos fissure who sustain a right pneumothorax may exhibit a so-called "empty" azygos fissure due to migration of the azygos arch to its normal location secondary to lung collapse.* **(Bottom)** *Coronal NECT (lung window) of the same patient shows the morphology of the accessory azygos fissure and the azygos azygos arch within the right upper lobe. Note the classic teardrop configuration of the azygos arch, corresponding to its appearance on frontal chest radiography.*

RIGHT LUNG SEGMENTAL ANATOMY

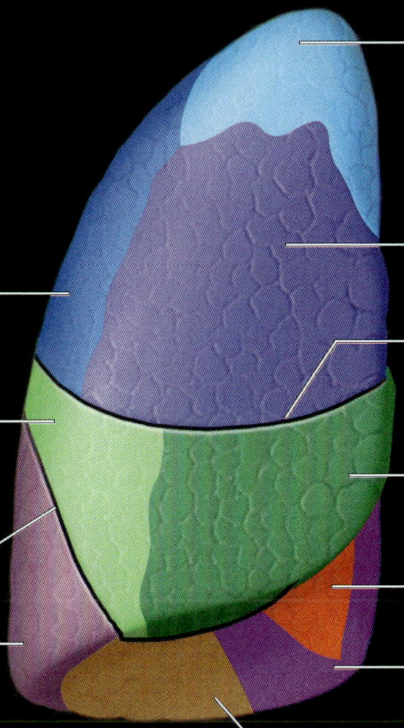

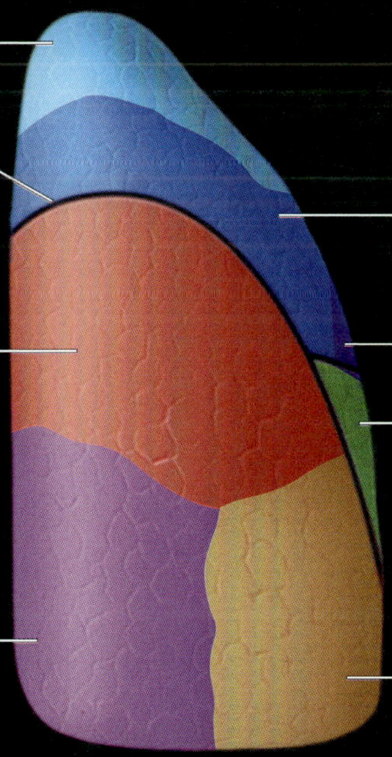

(Top) *First of 4 graphics depicts the segmental anatomy of the right lung and shows the anterior costal surface and right lung base. The 3 right upper lobe segments are named after their respective segmental bronchi (apical, anterior, and posterior). The right upper lobe anterior segment abuts the minor fissure. The middle lobe has a medial segment that abuts the right atrium and a lateral segment that abuts the lower aspect of the right major fissure.* **(Bottom)** *Graphic depicts the posterior costal surface of the right lung. The right upper lobe posterior segment abuts the posterosuperior aspect of the right major fissure. The right lower lobe superior segment occupies the apical aspect of the right lower lobe. The right lower lobe basilar segments are located inferior to the superior segment. The right lower lobe lateral basilar segment lobe is located lateral to the right lower lobe posterior basilar segment.*

RIGHT LUNG SEGMENTAL ANATOMY

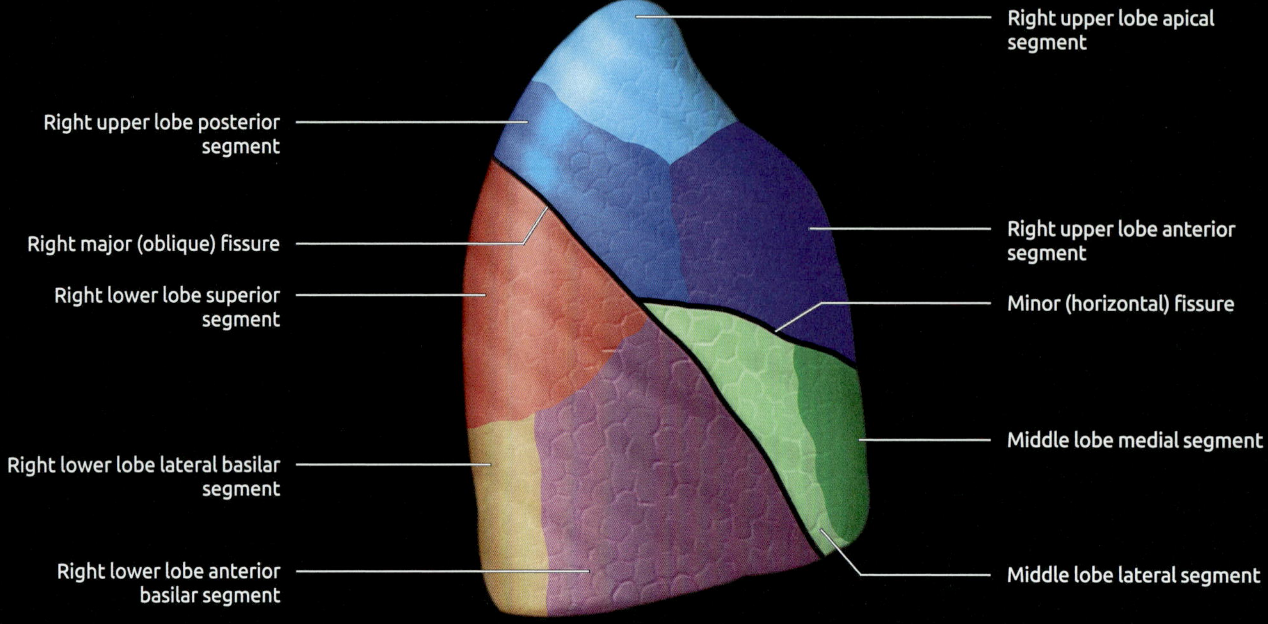

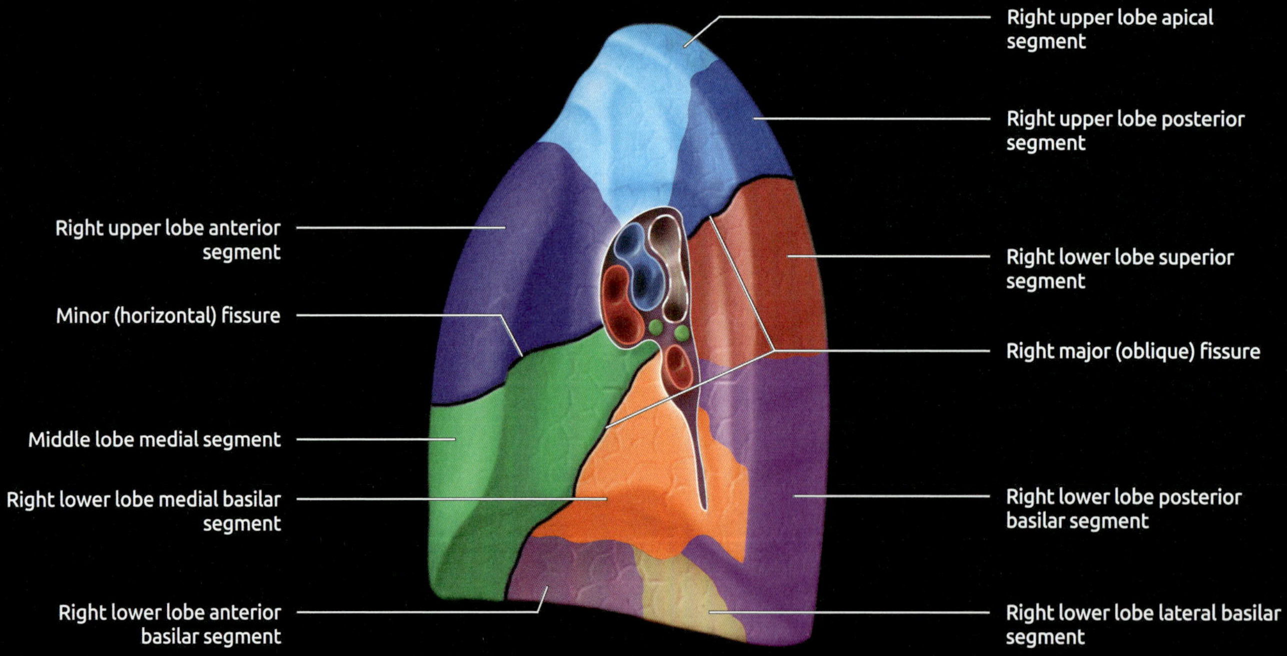

(Top) *Graphic depicts the lateral costal surface of the right lung and shows the location of the right upper lobe segments (posterior, apical, and anterior) corresponding to their anatomic positions within the lobe. The anterior segment of the right upper lobe abuts the minor fissure. The medial segment of the middle lobe is located anterior to the lateral segment. Note the relationship between the anterior basilar and lateral basilar right lower lobe segments located below the superior segment of the right lower lobe.* **(Bottom)** *Graphic depicts the mediastinal or medial surface of the right lung and shows the anatomic locations of the right upper lobe segments. The anterior segment abuts the minor fissure. The posterior segment abuts the superior aspect of the major fissure. The medial segment of the middle lobe abuts the right atrium. Note the relationship between the medial and posterior basilar segments of the right lower lobe situated below the superior segment and their relationship to the major fissure.*

LEFT LUNG SEGMENTAL ANATOMY

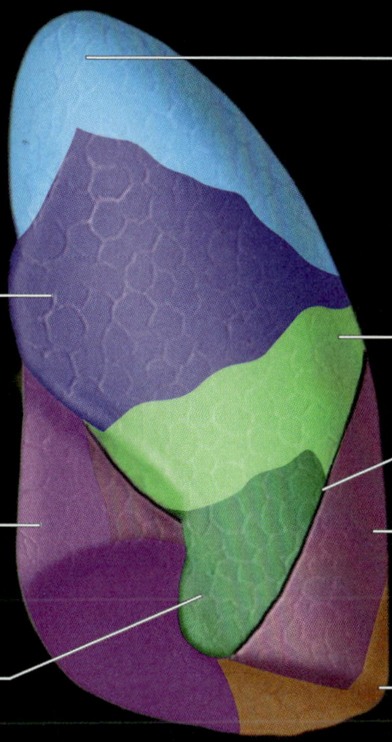

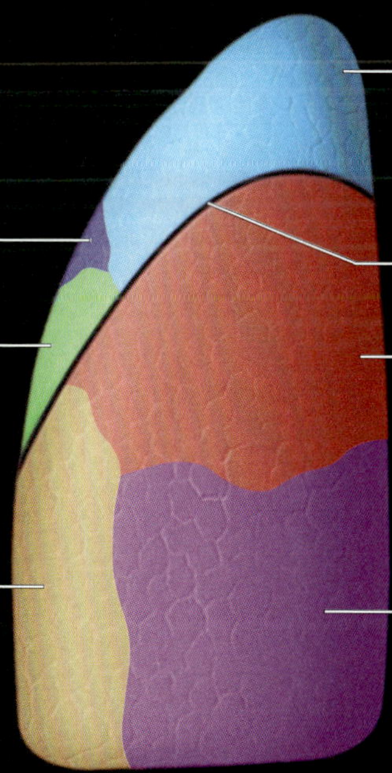

(Top) *First of 4 graphics depicts the segmental anatomy of the left lung and shows the anterior costal surface of the left lung and the 4 segments of the left upper lobe. The apicoposterior segment forms the left lung apex and is located above the anterior segment, which in turn is located above the lingular segments. The superior and inferior lingular segments abut the superior and inferior aspects of the left cardiac border, respectively. Note the anterior and medial location of the anteromedial basilar segment of the left lower lobe and the posteromedial location of the posterior basilar segment.* **(Bottom)** *Graphic depicts the posterior costal surface of the left lung and shows the apicoposterior segment of the left upper lobe above the posterior superior aspect of the left major fissure. The superior segment of the left lower lobe also abuts the posterior superior left major fissure. The posterior basilar segment of the left lower lobe is located medial to the left lower lobe lateral basilar segment.*

LEFT LUNG SEGMENTAL ANATOMY

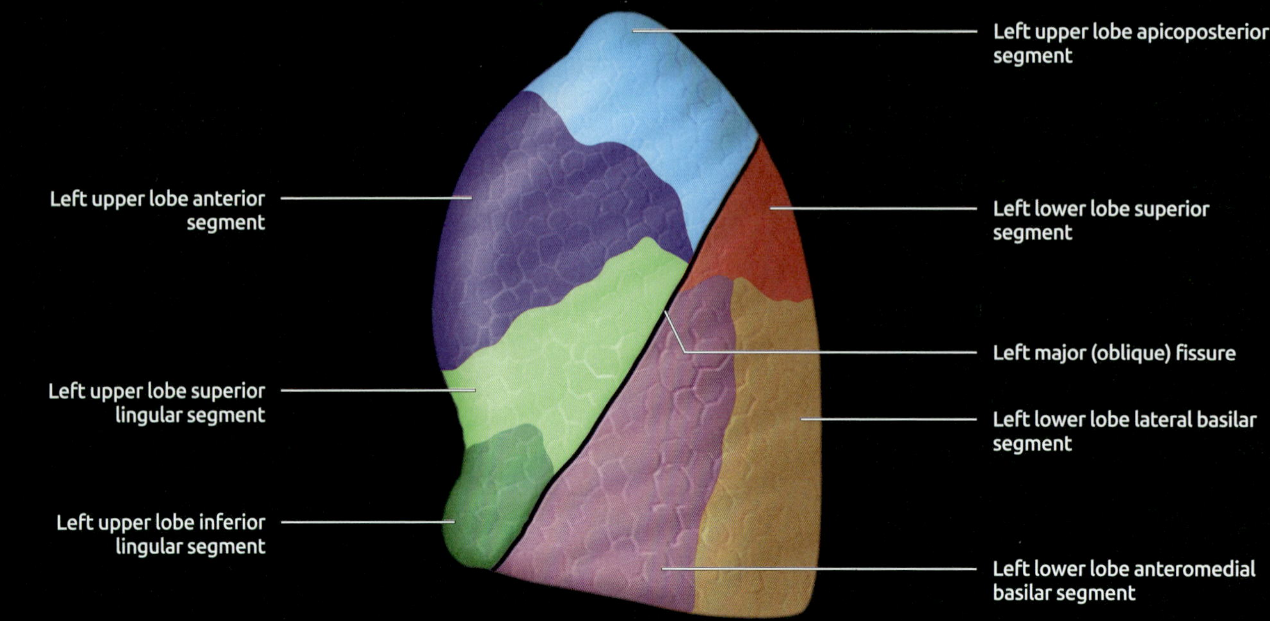

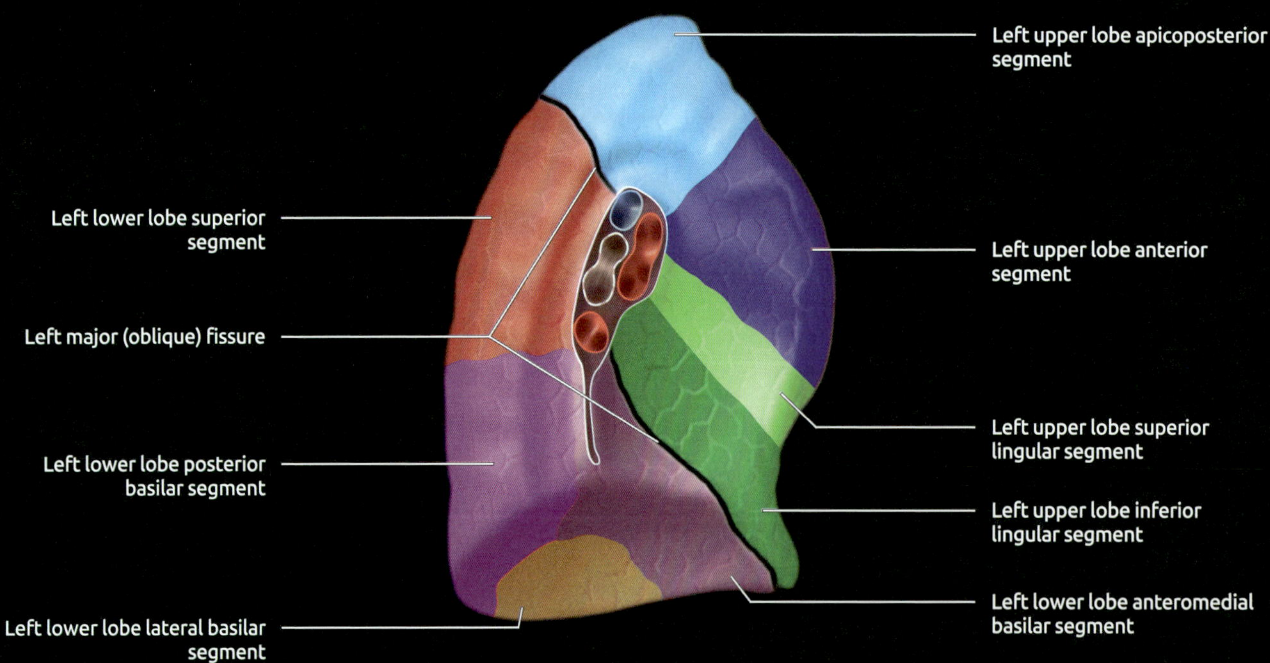

(Top) *Graphic depicts the lateral costal surface of the left lung and shows that the apicoposterior and anterior segments occupy the corresponding regions of the left upper lobe. The superior lingular segment abuts the midportion of the left major fissure, and a small portion of the inferior lingular segment abuts the inferior aspect of the left major fissure. Note the anterior location of the anteromedial basilar segment of the left lower lobe with respect to the lateral basilar segment. The apex of the left lower lobe is formed by the superior segment.* **(Bottom)** *Graphic depicts the mediastinal or medial surface of the left lung and shows the relative locations of the left upper lobe segments (apicoposterior, anterior, superior lingular, and inferior lingular). The superior segment of the left lower lobe abuts the superior aspect of the left major fissure and is situated above the basilar segments. Note the relationship between the posterior and anteromedial basilar segments of the left lower lobe.*

LUNG PERFUSION SCINTIGRAPHY

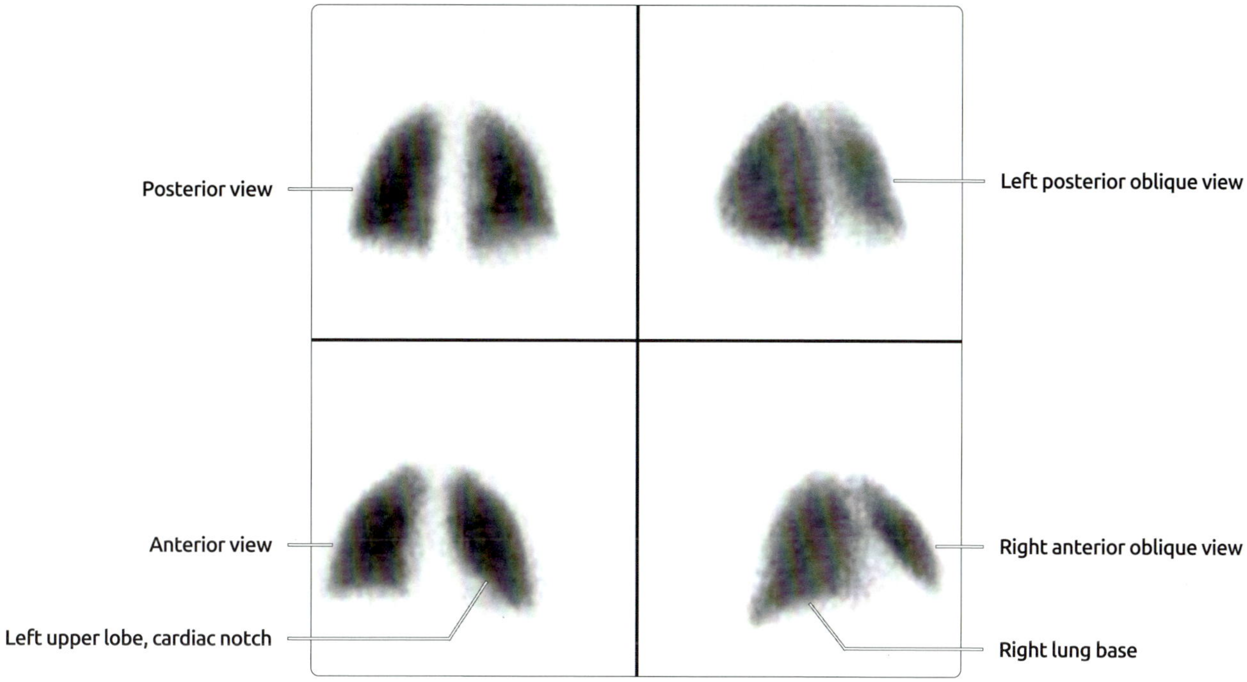

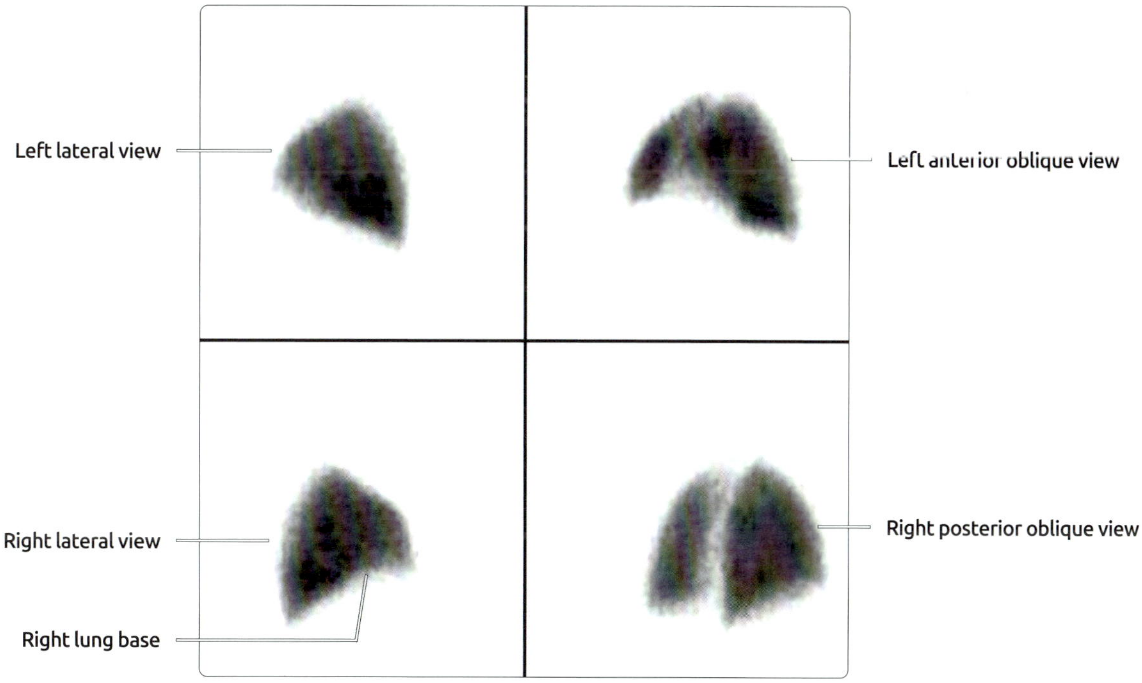

(Top) *Images from a normal perfusion lung scan show homogeneous distribution of activity throughout the pulmonary lobes. Understanding the segmental anatomy of the lungs in various projections is crucial for accurate interpretation of ventilation-perfusion scintigraphy. Note the inferior extent of the lungs on the posterior view. The cardiac notch of the left upper lobe is nicely depicted on the anterior view. The right anterior oblique view shows the concave morphology of the right lung base.* **(Bottom)** *Images from a normal perfusion lung scan of the same subject demonstrate homogeneous distribution of activity throughout the lungs, indicating normal pulmonary perfusion and no evidence of thromboembolic disease. The left lateral view shows relative absence of activity in the anterior inferior left lung corresponding to the cardiac notch of the left anterior lung border. The right lateral view shows the concave morphology of the right lung base. The right posterior oblique view shows the inferior extent of the posterior right lower lung.*

LOBAR AND SEGMENTAL LUNG ANATOMY, AXIAL CT

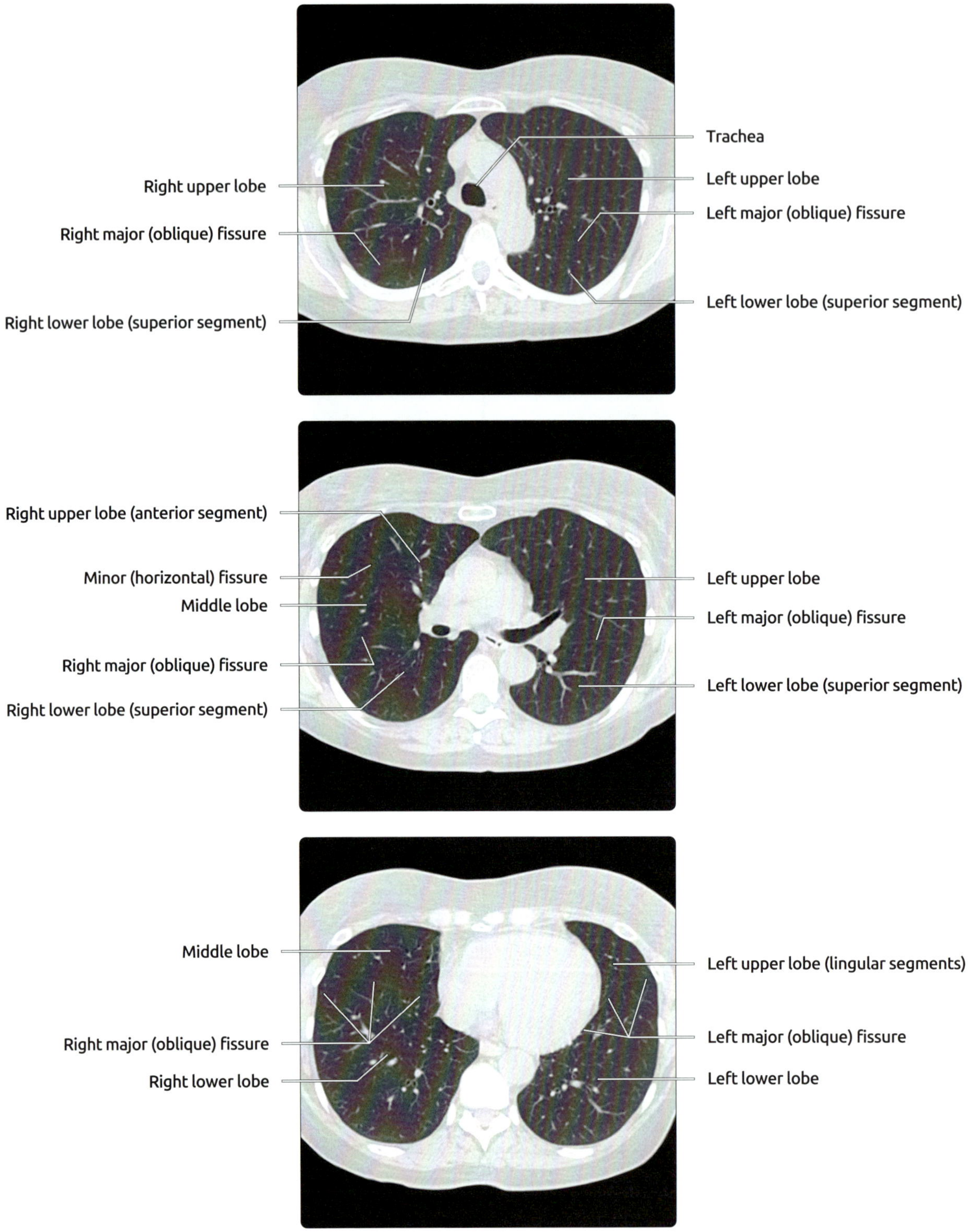

(Top) *Axial NECT (lung window) of a normal individual shows the lobar and segmental anatomy of the lungs. Identification of the interlobar fissures on CT allows confident identification of the lung lobes and segments.* **(Middle)** *Axial NECT (lung window) of a normal individual shows the lobar and segmental anatomy of the lungs. Identification of the interlobar fissures allows accurate localization of the lung lobes on CT.* **(Bottom)** *Axial NECT (lung window) of a normal individual shows the lobar and segmental anatomy of the lungs.*

LOBAR CONSOLIDATION, RADIOGRAPHY

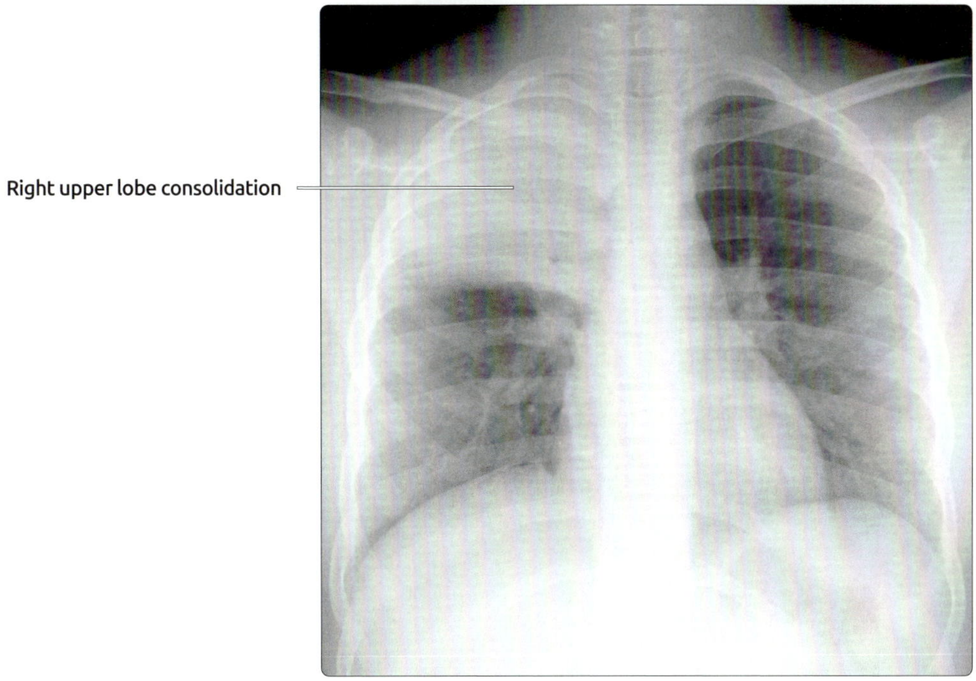

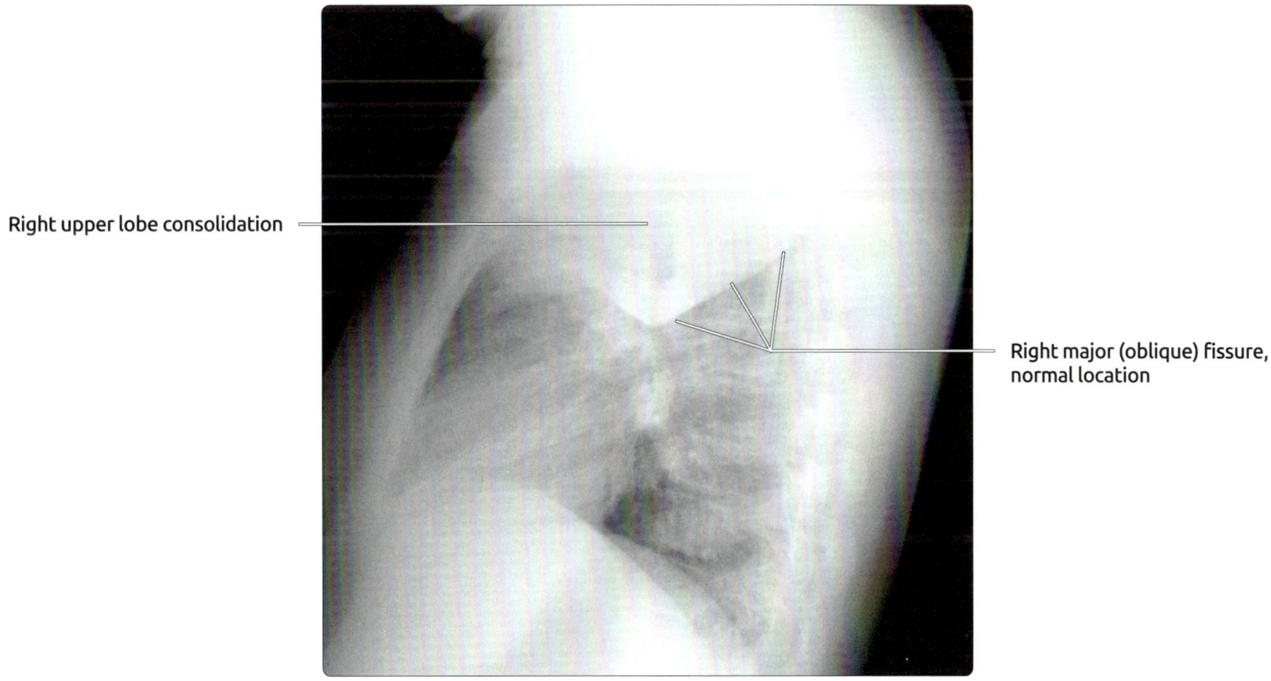

(Top) *PA chest radiograph of a patient with right upper lobe pneumonia shows complete opacification of the right upper lobe with intrinsic air bronchograms due to pulmonary consolidation.* **(Bottom)** *Lateral chest radiograph of the same patient shows complete right upper lobe opacification due to consolidation. Note that the posterior aspect of the right upper lobe is sharply demarcated by the superior aspect of the right major (oblique) fissure.*

LOBAR ATELECTASIS, RADIOGRAPHY AND CT

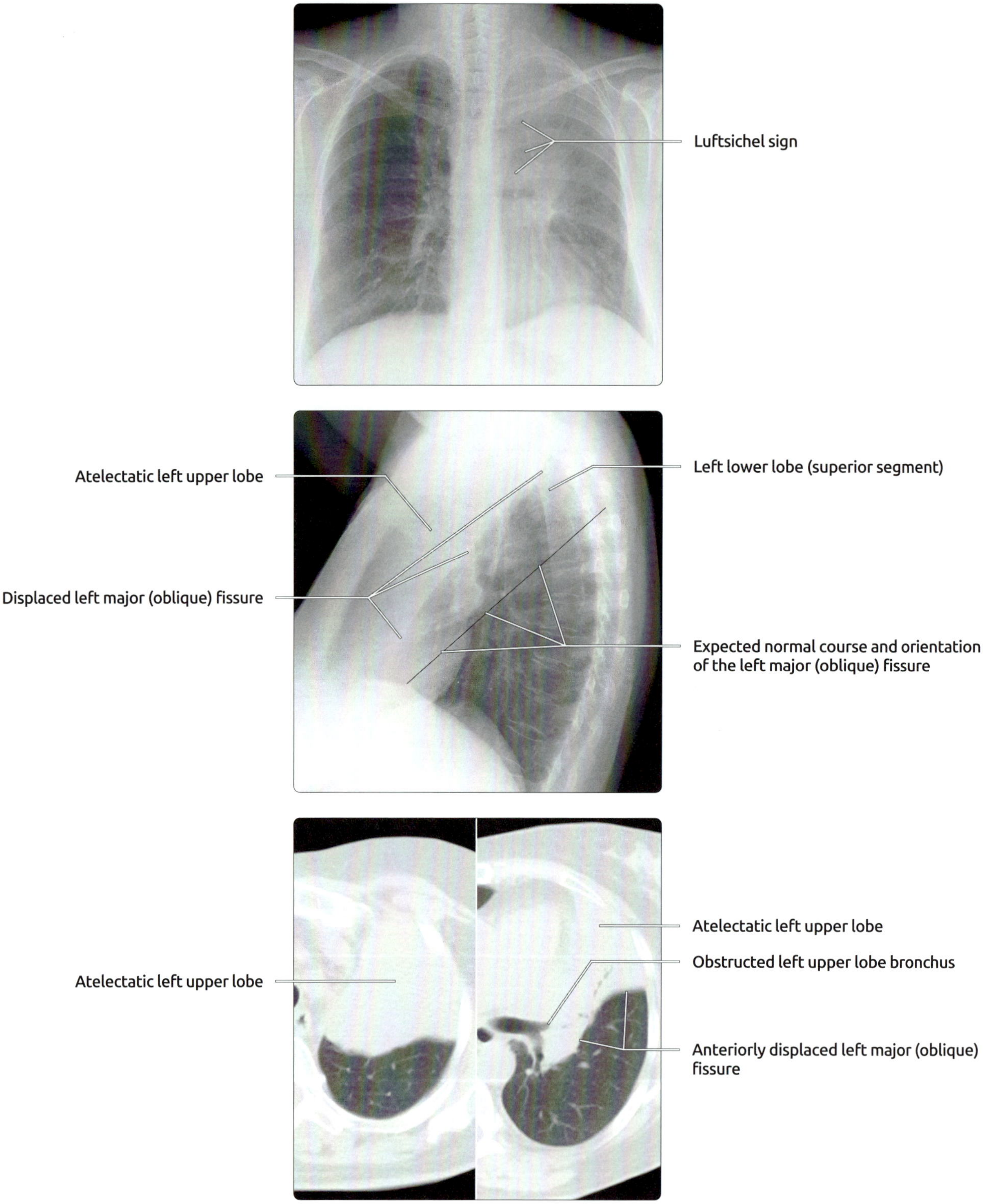

(Top) *PA chest radiograph of a patient with a central obstructing primary lung cancer shows complete left upper lobe atelectasis manifesting with a veil-like opacity in the left upper lung zone with obscuration of the adjacent left mediastinal contour. Note the sickle-shaped lucency adjacent to the aortic arch, the so-called Luftsichel sign. In spite of the degree of atelectasis, displacement of the left major fissure is not appreciated on frontal radiography.* **(Middle)** *Lateral chest radiograph of the same patient shows marked anterior displacement of the left major fissure adjacent to a completely atelectatic left upper lobe. An oblique black line shows the expected normal course and location of the left major fissure. The Luftsichel sign is produced by superior migration of the overexpanded left lower lobe superior segment, which abuts the aortic arch.* **(Bottom)** *Composite image with axial NECT (lung window) of the same patient shows the atelectatic left upper lobe, abnormal anterior displacement of the left major fissure, and obstruction of the left upper lobe bronchus by the central lung cancer.*

LOBAR CONSOLIDATION, RADIOGRAPHY AND CT

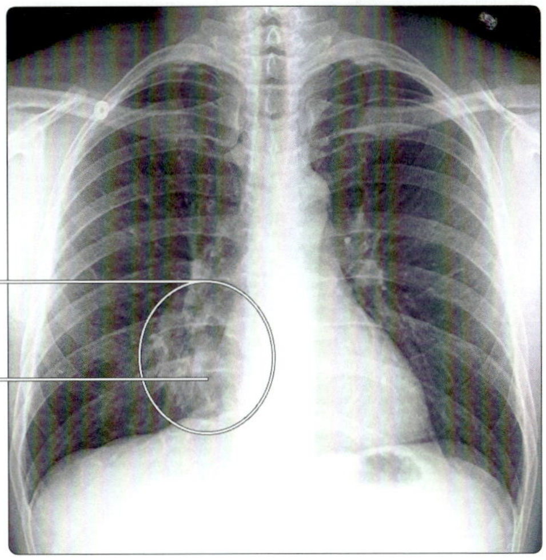

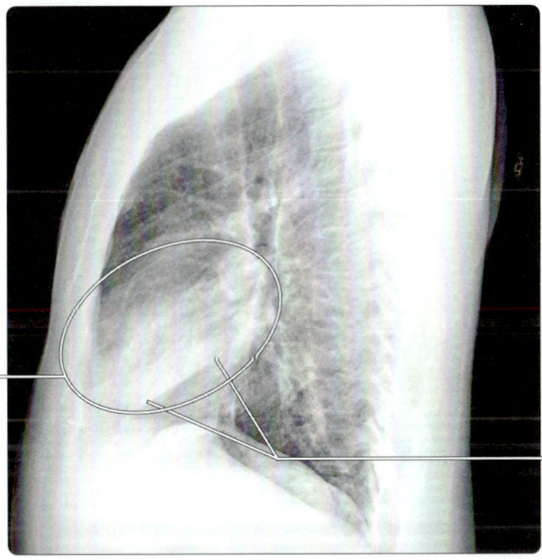

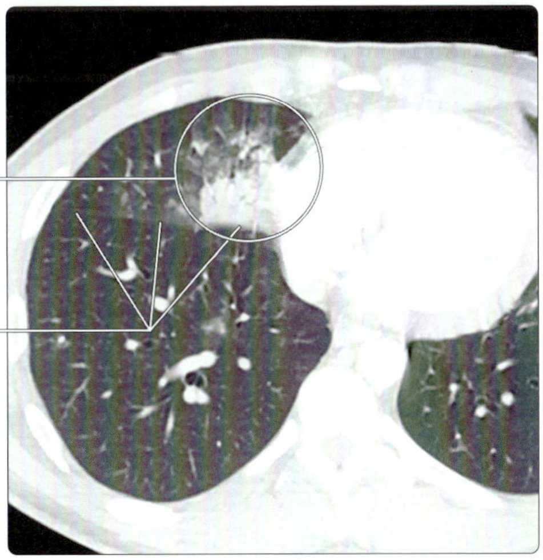

(Top) *PA chest radiograph of a patient with middle lobe pneumonia shows an ill-defined opacity of the medial right lower lung zone with obscuration of the adjacent right cardiac border produced by opacification of the medial segment of the middle lobe. Similarly, opacification of the lingular segments of the left upper lobe would result in obscuration of the left cardiac border.* **(Middle)** *Lateral chest radiograph of the same patient shows pulmonary consolidation involving the expected anatomic location of the middle lobe that outlines the normal course of the inferior aspect of the right major fissure. Regardless of the side of involvement, a lower lobe consolidation is always located posterior to the ipsilateral major or oblique fissure.* **(Bottom)** *Axial CECT (lung window) of the same patient shows heterogeneous consolidations that predominantly affect the medial segment of the middle lobe. Identification of the adjacent right major or oblique fissure allows localization of the abnormality to the middle lobe.*

LESION LOCALIZATION, RADIOGRAPHY

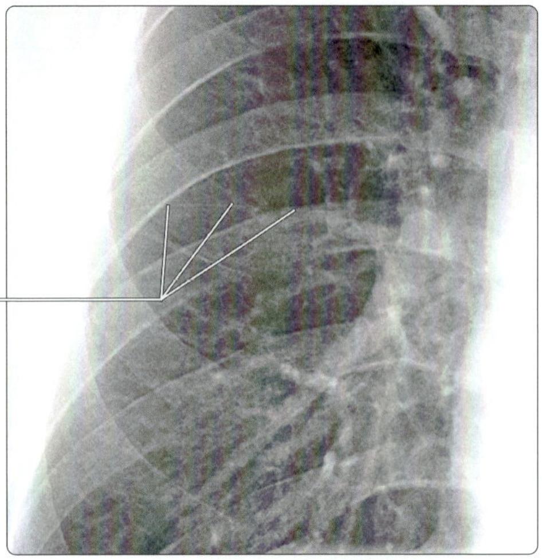

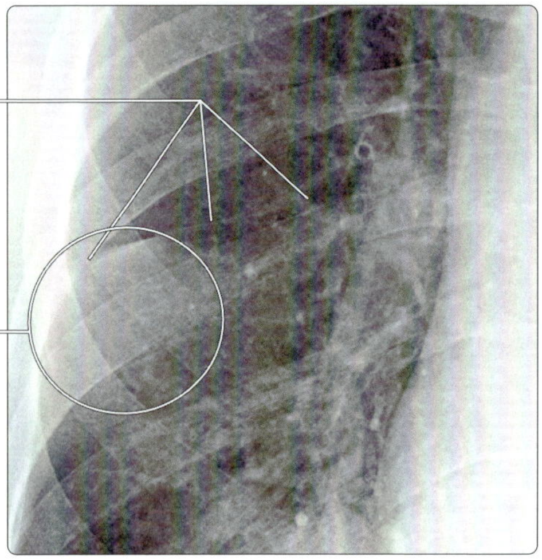

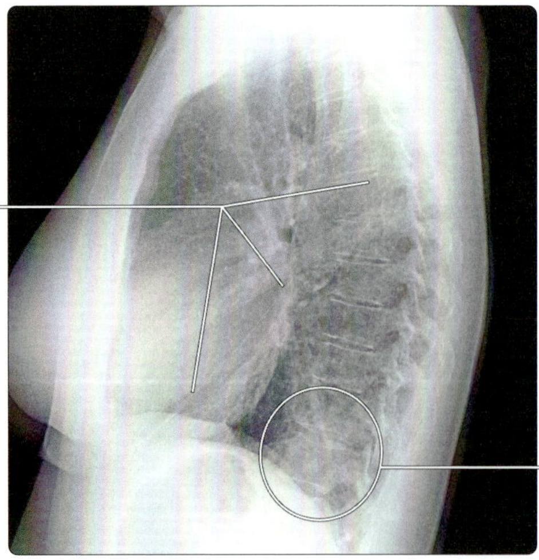

(Top) *PA chest radiograph coned down to the right midlung shows the normal morphology and orientation of the minor or horizontal fissure, which is frequently identified on frontal chest radiographs. The minor fissure anatomically separates the right upper lobe anterior segment from the adjacent middle lobe. Displacement of the minor fissure from its normal position should suggest adjacent volume loss.* **(Middle)** *PA chest radiograph coned down to the right lung shows a right midlung consolidation located inferior and lateral to the minor fissure, consistent with pneumonia involving the middle lobe lateral segment. Familiarity with normal radiographic anatomy allows accurate anatomic localization of radiographic abnormalities.* **(Bottom)** *Lateral chest radiograph of a patient with pneumonia shows the expected orientation of the major fissure, which anatomically separates the upper lobe from the lower lobe. The major fissure is not typically visible on frontal radiography unless it is abnormally displaced. Note the right lower lobe consolidation that obscures the posterior right hemidiaphragm and produces the so-called radiographic spine sign.*

SEGMENTAL CONSOLIDATION, RADIOGRAPHY

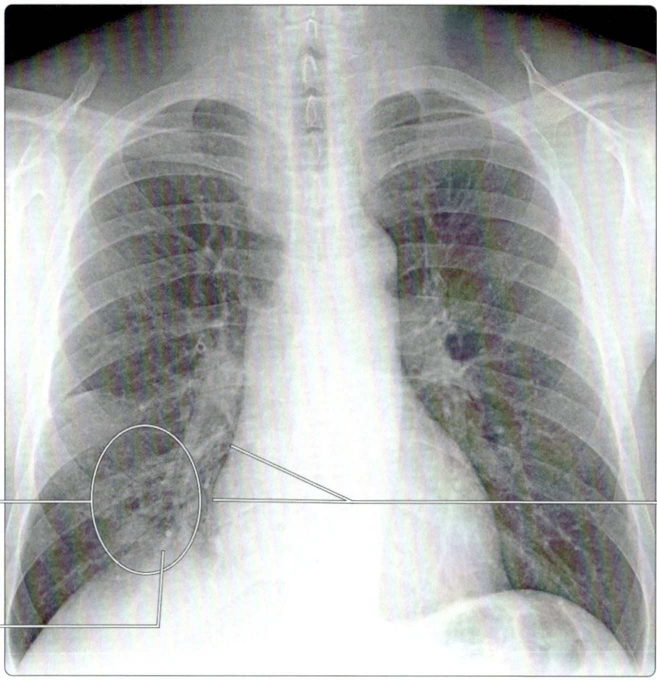

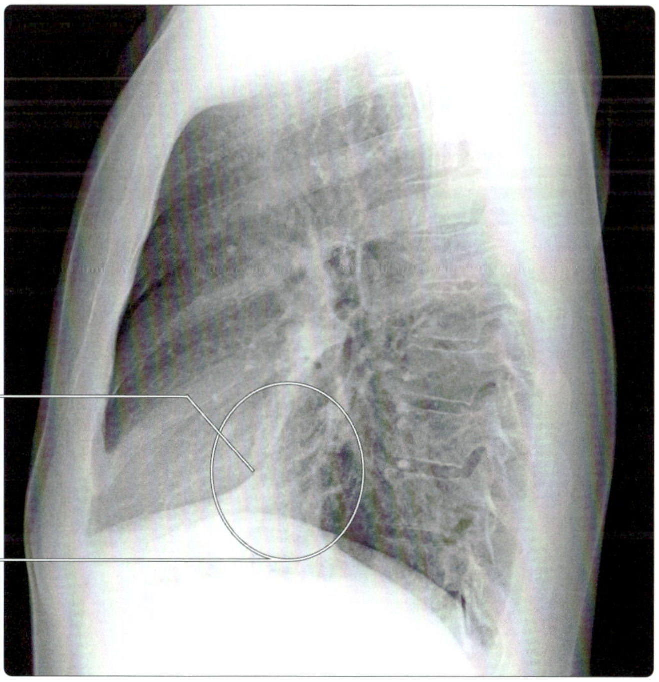

(Top) *PA chest radiograph of a patient with pneumonia involving the right lower lobe anterior basilar segment shows pulmonary opacity that involves the medial inferior right lung, which obscures the medial aspect of the right hemidiaphragm and allows localization of the abnormality to the right lower lobe. Note that the consolidation does not obscure the right cardiac border (as would be seen in middle lobe pneumonia, as the middle lobe is not in anatomic contact with the diaphragm).* **(Bottom)** *Lateral chest radiograph of the same patient shows consolidation of the anterior basilar segment of the right lower lobe anatomically located posterior to the right major fissure. Note partial obscuration of the adjacent right hemidiaphragm.*

SEGMENTAL CONSOLIDATION, RADIOGRAPHY

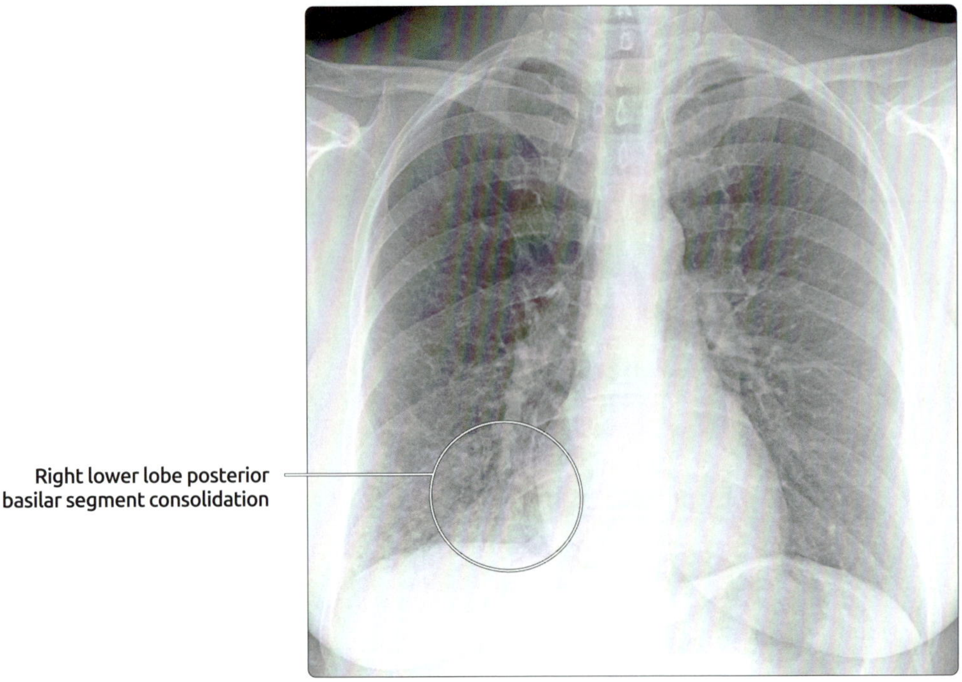

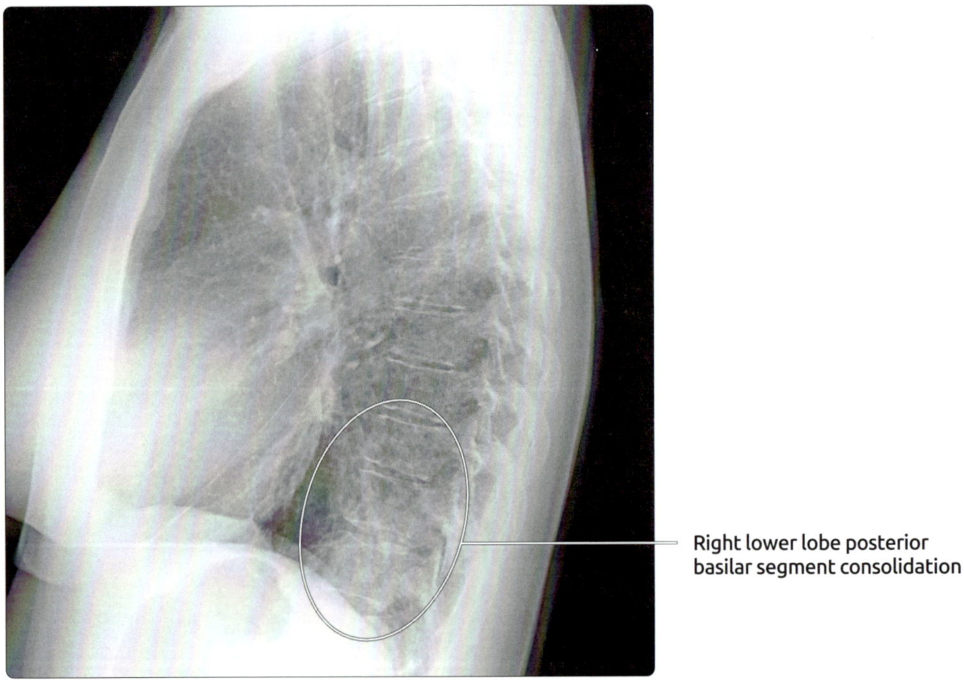

(Top) *PA chest radiograph of a patient with pneumonia involving the posterior basilar segment of the right lower lobe shows pulmonary opacity involving the inferior medial aspect of the right lung without obscuration of the right cardiac border or the adjacent right hemidiaphragm.* **(Bottom)** *Lateral chest radiograph of the same patient shows the right lower lobe consolidation manifesting with the so-called radiographic spine sign. Under normal circumstances, the lower thoracic spine vertebral bodies should be more lucent than the upper thoracic vertebral bodies. Consolidation of a lower lobe posterior basilar segment results in a denser appearance of the lower thoracic vertebral bodies (i.e., spine sign).*

SEGMENTAL CONSOLIDATION, RADIOGRAPHY

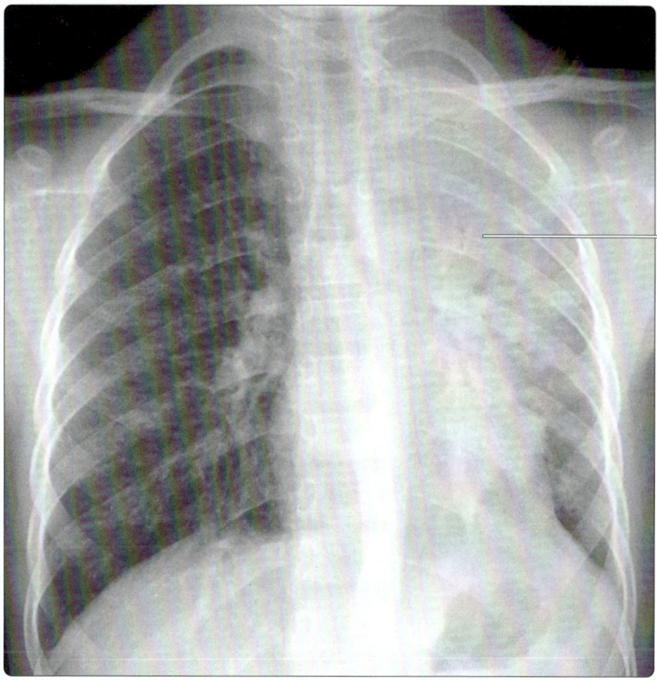

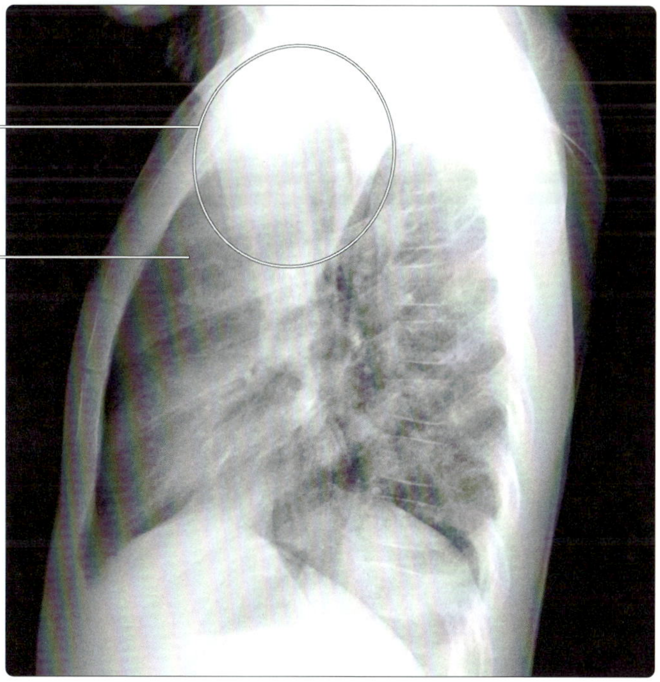

(Top) *PA chest radiograph of a patient with left upper lobe pneumonia involving the apicoposterior segment shows dense consolidation of the left upper lung with intrinsic air bronchograms. Note that a portion of the left cardiac border is still visible.* **(Bottom)** *Lateral chest radiograph of the same patient shows consolidation of the apicoposterior segment of the left upper lobe and sparing of the left upper lobe anterior segment. Unlike the right upper lobe, which has 3 segments (i.e., anterior, posterior, and apical), the left upper lobe has 4 segments (i.e., anterior, apicoposterior, superior lingular, and inferior lingular).*

SEGMENTAL CONSOLIDATION, RADIOGRAPHY AND CT

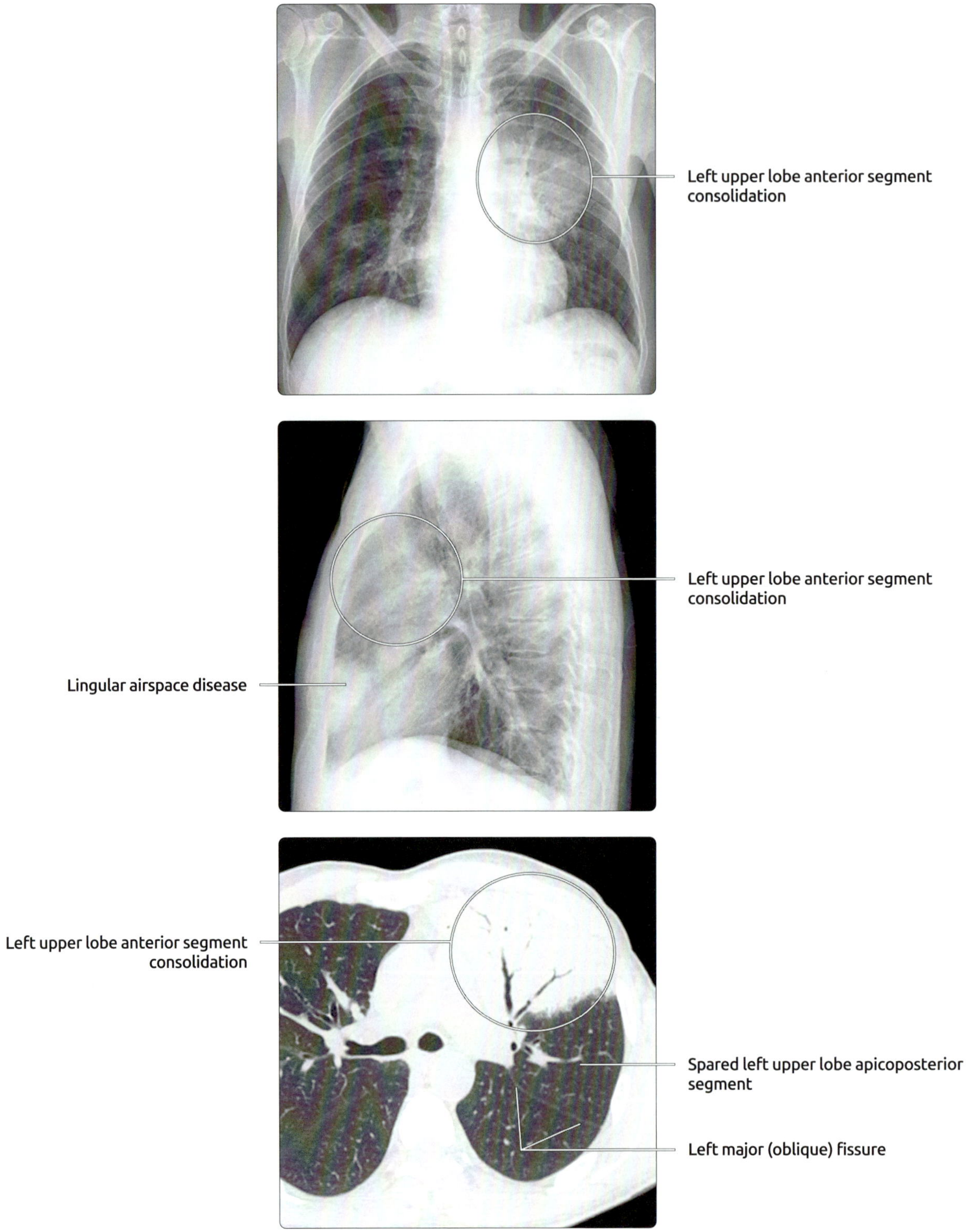

(Top) *PA chest radiograph of a patient with left upper lobe pneumonia shows consolidation that predominantly involves the anterior segment of the left upper lobe.* **(Middle)** *Lateral chest radiograph of the same patient shows extensive consolidation that involves of the anterior segment of the left upper lobe. Note involvement of portions of the left upper lobe lingular segments.* **(Bottom)** *Axial CECT (lung window) of the same patient shows extensive consolidation with intrinsic air bronchograms involving the anterior segment of the left upper lobe.*

SEGMENTAL CONSOLIDATION, RADIOGRAPHY

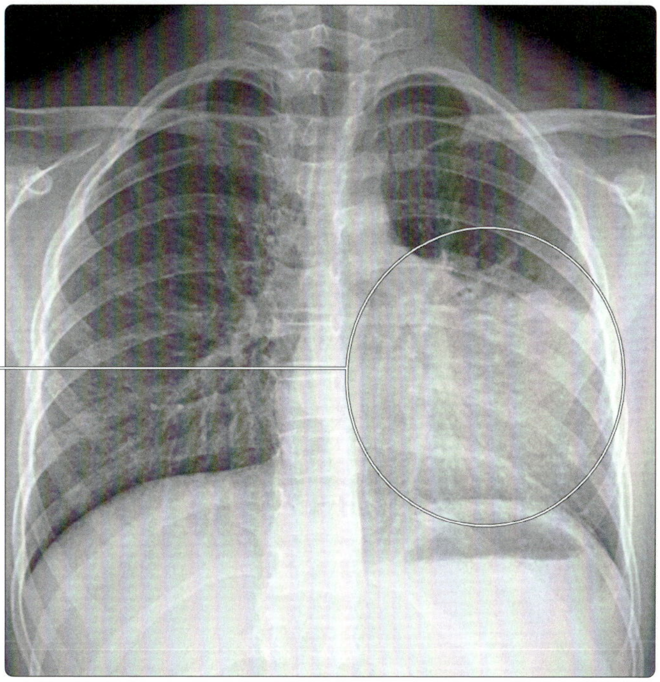

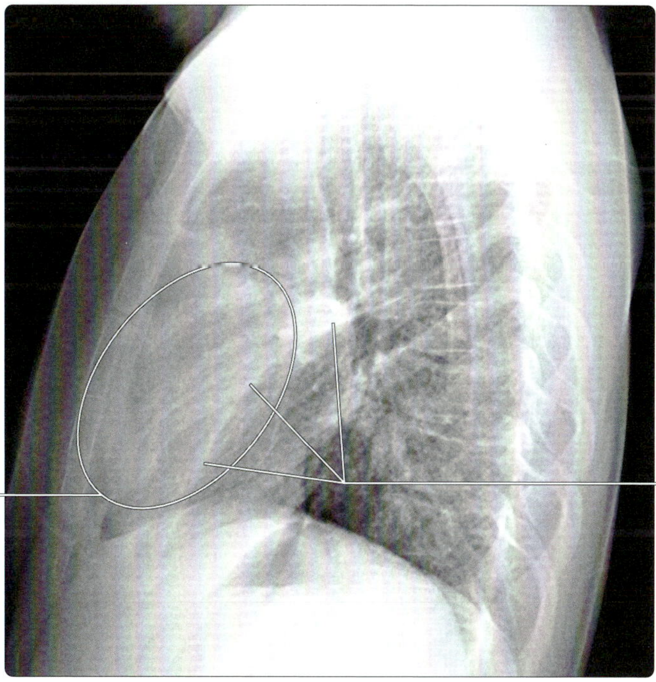

(Top) *PA chest radiograph of a patient with left upper lobe pneumonia involving predominantly the lingular segments shows pulmonary opacity that completely obscures the left cardiac border.* **(Bottom)** *Lateral chest radiograph of the same patient shows consolidation involving the lingular segments of the left upper lobe. The left upper lobe lingular segments are always located anterior to the inferior aspect of the left major or oblique fissure.*

SEGMENTAL CONSOLIDATION, RADIOGRAPHY

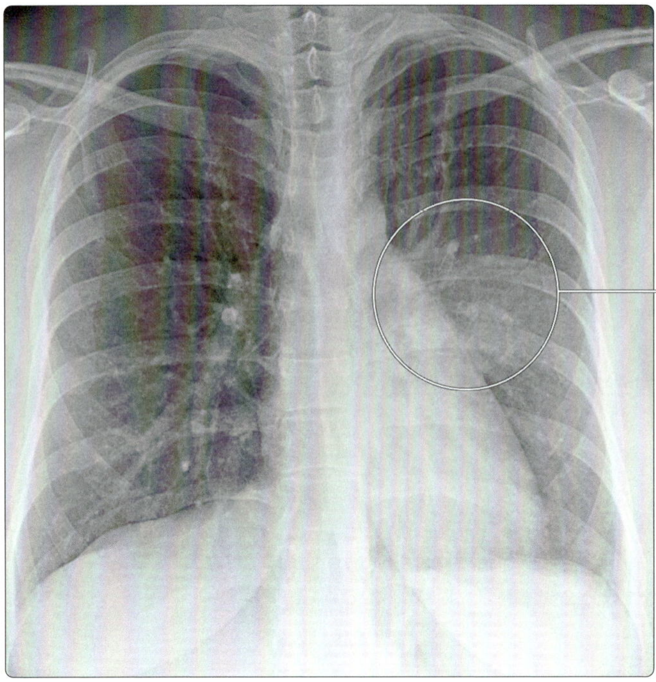

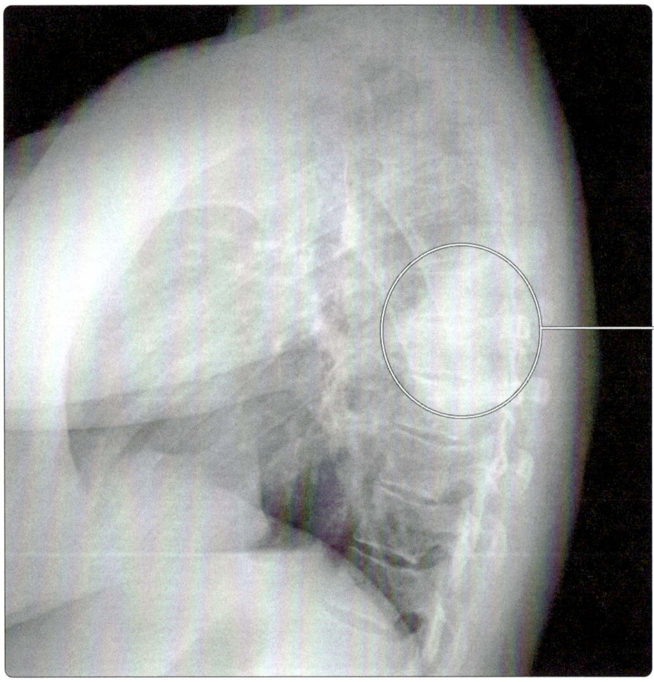

(Top) *PA chest radiograph of a patient with left lower lobe pneumonia shows an ill-defined left midlung zone opacity that does not obscure the left cardiac border. However, it is difficult to determine whether the airspace disease involves the anterior segment of the left upper lobe or the superior segment of the left lower lobe on frontal chest radiography alone.* **(Bottom)** *Lateral chest radiograph of the same patient shows consolidation of the superior segment of the left lower lobe. The superior segment of the lower lobe is located posterior and inferior to the ipsilateral major or oblique fissure.*

SEGMENTAL CONSOLIDATION, RADIOGRAPHY

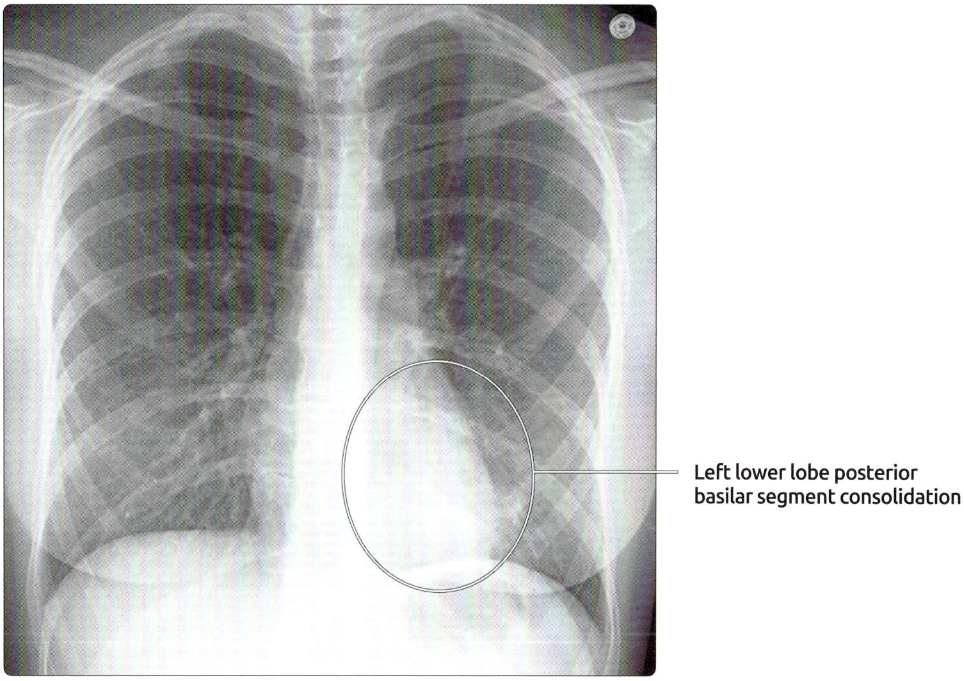

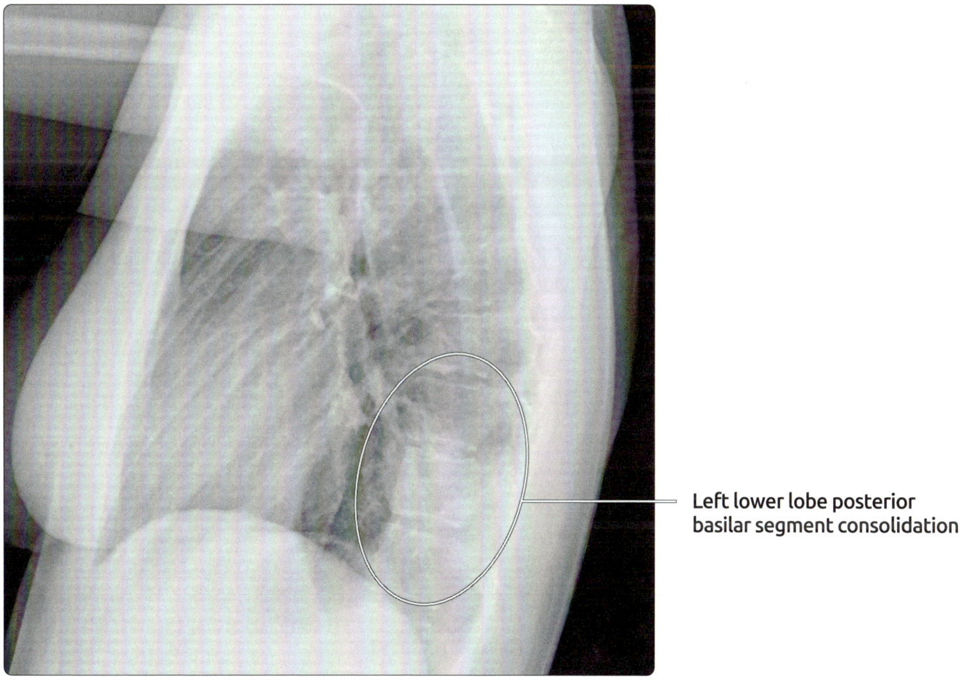

(Top) *PA chest radiograph of a patient with left lower lobe pneumonia involving the posterior basilar segment shows ill-defined retrocardiac hazy opacities.* **(Bottom)** *Lateral chest radiograph of the same patient shows extensive consolidation of the left lower lobe posterior basilar segment that obscures the posterior left hemidiaphragm and exhibits the so-called radiographic spine sign. Normally, the inferior thoracic vertebral bodies appear more lucent than the superior thoracic vertebral bodies on lateral chest radiography. In this case, the inferior thoracic vertebral bodies appear denser than the superior thoracic vertebral bodies due to the overlying consolidation.*

TERMINOLOGY

Definitions

- **Hilum**: Small point of attachment of a seed to its base or support (plural: **Hila**)
- **Lung hilum**
  - Point of connection between lung and mediastinum
  - Central area between mediastinum medially and lung laterally through which bronchi, vessels, and other structures course into lung

GENERAL ANATOMY AND FUNCTION

Anatomy

- **Hilum** located on central mediastinal lung surface
- Hilar structures
  - **Pulmonary arteries**
  - **Pulmonary veins**
  - **Mainstem** and **lobar bronchi**
  - **Bronchial arteries/veins**
  - **Nerves**
  - **Lymph nodes**
  - **Lymphatics**
- Hilar boundaries
  - Mediastinal pleural reflections
  - Inferior extension as **pulmonary ligament**

Function

- May play role in stabilizing lung
- **Pulmonary ligament**
  - May play role in stabilizing lower lobe
  - May accommodate cephalocaudad motion of hilar structures during respiration

HILAR ANATOMY

Right Hilum

- Structures
  - Right mainstem bronchus and branches
  - Right pulmonary artery and branches
  - Right pulmonary veins
  - Bronchopulmonary lymph nodes: Hilar and interlobar
- Anatomic location
  - Right mainstem bronchus located posteriorly and superiorly
    - Right upper lobe bronchus origin above right pulmonary artery: **Eparterial bronchus**
  - Right pulmonary artery, anterior to right mainstem bronchus
  - Right superior pulmonary vein, anterior to right pulmonary artery
  - Right inferior pulmonary vein, inferior to right hilum

Left Hilum

- Structures
  - Left mainstem bronchus and branches
  - Left pulmonary artery and branches
  - Left pulmonary veins
  - Bronchopulmonary lymph nodes: Hilar and interlobar
- Relationships
  - Left mainstem bronchus, located posteriorly in midhilum
  - Left pulmonary artery above left mainstem bronchus
    - Left upper lobe bronchus origin below left pulmonary artery: **Hyparterial bronchus**
  - Left superior pulmonary vein
    - Anterior to left mainstem bronchus and left pulmonary artery
    - Posterior to left atrial appendage
  - Left inferior pulmonary vein
    - Inferior to left hilum

RADIOGRAPHY OF HILA

General Concepts

- Hilar components: Pulmonary arteries, pulmonary veins, central bronchi, lymph nodes
- Pulmonary arteries provide greatest contribution to hilar opacity seen on radiography
- Lateral radiography
  - Right hilar vasculature located anteriorly
  - Left hilar vasculature located posteriorly

Frontal (PA/AP) Radiography

- **Superior right hilum**
  - Right ascending (i.e., truncus anterior) pulmonary artery and branches located medially
  - Right superior pulmonary vein located laterally
    - Forms lateral margin of superior portion of right hilum
  - Apical segmental and anterior segmental right upper lobe bronchi located centrally
- **Hilar angle**
  - Angle formed by right superior pulmonary vein as it obliquely crosses right descending (i.e., interlobar) pulmonary artery
  - Shallow concavity in midright hilum
- **Inferior right hilum**
  - Right interlobar pulmonary artery
    - Lateral aspect of inferior right hilum
    - Measures 15-16 mm in thickness
  - Bronchus intermedius
    - Located medial to right interlobar pulmonary artery
  - Right inferior pulmonary vein located below right hilum
- **Superior left hilum**
  - Left pulmonary artery
    - Courses above left mainstem bronchus
    - Upper aspect of superior left hilum
  - Left superior pulmonary vein located laterally
  - Anterior and apicoposterior segmental left upper lobe bronchi located centrally
- **Inferior left hilum**
  - Left interlobar pulmonary artery located lateral to left lower lobe bronchus
  - Left inferior pulmonary vein located below left hilum
- **Hilar opacity**
  - Greatest contribution from **pulmonary arteries** and **superior pulmonary veins**
  - Minimal contribution from bronchial walls, lymph nodes, and surrounding tissues
- **Hilar height**
  - Left hilum higher than right in **97%** of cases
  - Equal hilar height in **3%** of cases
  - Right hilum never higher than left in normal subjects

Lateral Radiography

- Airways
  - **Right upper lobe bronchus**
    - Superior round or ovoid lucency
    - Visible in **50%** of cases
    - Circumferential visualization of bronchial walls suggests abnormal adjacent soft tissue
  - **Left mainstem/left upper lobe bronchus**
    - Horizontal portions manifest as inferior round or ovoid lucency
    - Visible in **75%** of cases; visualization enhanced by surrounding water density vessels
  - **Intermediate stem line**
    - Vertical thin linear opacity formed by posterior walls of right **mainstem** bronchus and bronchus **intermedius**
    - Normal thickness ≤ **3 mm**
    - Visible in **95%** of cases
    - Outlined by endoluminal air and retrobronchial aerated lung abutting azygoesophageal recess
    - Courses through posterior/middle 1/3 of left upper lobe bronchus
  - **Anterior wall of bronchus intermedius**
    - Edge posterior to right hilar vascular opacity
  - **Anterior wall of left lower lobe bronchus**
    - Merges superiorly with left upper lobe bronchus
    - Visible in **45%** of cases
  - **Middle lobe bronchus**
    - May be visible below right hilar vascular opacity
  - **Anterior wall of right lower lobe bronchus**
    - Visible in **15%** of cases
  - **Left retrobronchial line**
    - Formed by posterior walls of left mainstem bronchus and left lower lobe bronchus
    - Normal thickness ≤ **3 mm**
    - Terminates at origin of left lower lobe superior segmental bronchus
    - Shorter and posterior to intermediate stem line
- Vessels
  - **Right hilar vascular opacity**
    - Anterior to hilar airways
    - Posteriorly located right truncus anterior and interlobar arteries
    - Anteriorly located right superior pulmonary vein
    - Surrounding areolar tissue and lymph nodes
    - Right pulmonary artery is intramediastinal and does not contribute to right hilar vascular opacity
  - **Left hilar vascular opacity**
    - **Left pulmonary artery** forms arcuate opacity superior and posterior to **left upper lobe bronchus**
    - Superior aspect of left pulmonary artery visible in **95%** of cases
  - **Inferior hilar window**
    - Avascular zone in anterior inferior hilum
    - Inferior hilar window soft tissue should suggest lymphadenopathy or mass

CT OF HILA

6 Characteristic Axial Levels

- **Supracarinal region/trachea**
  - Right upper lobe apical segmental artery medial to apical segmental bronchus
  - Right apical vein lateral to apical segmental bronchus
  - Left upper lobe apicoposterior segmental bronchus and artery
  - Left apical and anterior veins anteromedial to bronchus and artery
- **Carina/right upper lobe bronchus**
  - Right upper lobe anterior and posterior segmental bronchi
  - Right upper lobe ascending (truncus anterior) pulmonary artery anterior to right mainstem bronchus
  - Right upper lobe anterior segmental artery medial to anterior segmental bronchus
  - Lateral location of right superior pulmonary vein
  - Left upper lobe apicoposterior segmental bronchus and artery, lateral to left pulmonary artery
  - Left superior pulmonary vein located anteromedial to bronchus and artery
- **Proximal bronchus intermedius/left upper lobe bronchus**
  - Bronchus intermedius covered anterolaterally by horizontal and vertical portions of interlobar artery
  - Right superior pulmonary vein abuts junction of horizontal and descending interlobar artery
    - Characteristic **elephant head** and **trunk** morphology
  - Distal left mainstem and left upper lobe bronchi
  - Left interlobar artery posterolateral to bronchi
- **Bronchus intermedius/lingular bronchi**
  - Bronchus intermedius
  - Right interlobar artery lateral to bronchus intermedius
  - Proximal lingular and left lower lobe superior segmental bronchi
  - Left interlobar artery lateral to bronchial bifurcation
- **Middle lobe bronchus**
  - Origins of middle lobe and right lower lobe superior segmental bronchi (latter may arise slightly above this level)
  - Middle lobe artery lateral to middle lobe bronchus
  - Right interlobar artery lateral to bifurcation of bronchus intermedius into middle and lower lobe bronchi
  - Anteromedial right superior pulmonary vein
  - Left interlobar artery lateral to left lower lobe truncus basalis bronchus
- **Basilar lower lobe bronchi/inferior pulmonary veins**
  - Right lower lobe medial basilar segmental bronchus anterior to right inferior pulmonary vein
  - Right lower lobe anterior, lateral, and posterior basilar segmental bronchi lateral and posterior to right inferior pulmonary vein
  - Left lower lobe anteromedial basilar segmental bronchus anterior to left inferior pulmonary vein
  - Left lower lobe lateral and posterior basilar segmental bronchi posterior to left inferior pulmonary vein

HILAR LYMPH NODES

Normal Lymph Nodes

- Radiography: Not normally visible
- CT: Small ovoid/rounded soft tissue nodules ± fatty hila
  - Conspicuous at right pulmonary artery bifurcations and at middle lobe, left upper lobe, and lingular bronchial bifurcations
  - Optimal visualization on CECT
  - Identification of subtle calcification
  - Short-axis measurement ≤ **1 cm**
- MR: Distinction of lymph nodes from adjacent vessels
  - Not sensitive to calcification

International Association for the Study of Lung Cancer (IASLC) Lymph Node Map

- Specific designations for assessment of lung cancer lymph node metastases
  - 14 stations; 7 lymph node zones; right (R) or left (L)
- **Hilar lymph nodes (station 10)**
  - Adjacent to mainstem bronchi and hilar vessels
  - Station 10R: Bound superiorly by lower rim of azygos arch
  - Station 10L: Bound superiorly by upper rim of pulmonary artery
  - 10R and 10L: Bound inferiorly by interlobar regions
- **Interlobar lymph nodes (station 11)**
  - Between origins of lobar bronchi
  - Station 11R
    - Station 11Rs: Between right upper lobe bronchus and bronchus intermedius
    - Station 11Ri: Between middle and right lower lobe bronchi
  - Station 11L: Between left upper and lower lobe bronchi

HILAR SIGNS

Hilum Overlay Sign

- Visualization of pulmonary artery through abnormal hilar convexity on frontal radiography
- Abnormality is not in hilum
  - Abnormality located anterior or posterior to hilum
  - Classically described for anterior mediastinal masses

Hilum Convergence Sign

- Pulmonary artery branches arise from lateral margin of abnormal hilar convexity on frontal radiography
  - Radiographic evidence of pulmonary artery enlargement

ANATOMY-RELATED IMAGING ABNORMALITIES

Alteration of Hilar Height

- Typically seen with loss of lung volume
  - Inferior hilar displacement with lower lobe volume loss
  - Superior hilar displacement with upper lobe volume loss

Increased Hilar Density

- **Calcification**
  - **Granulomatous disease**
    - Punctate, irregular, or diffuse
    - Association with calcified pulmonary granulomas
  - **Pneumoconiosis** (**silicosis**, **coal worker's pneumoconiosis**)
    - ± peripheral (**eggshell**) calcification
    - Association with upper/posterior lung zone silicotic nodules ± calcification
  - **Metastases**
    - Bone-forming primary malignant neoplasms ± calcified hilar lymph node/pulmonary metastases
- **Hilar Mass**: **Lung Cancer**
  - Asymmetric hilar density on radiography
    - CT for exclusion of central/hilar mass
  - Central/hilar mass ± bronchial obstruction
    - Typical of **squamous cell carcinoma**
  - Hilar/mediastinal lymphadenopathy
    - Typical of **small cell carcinoma**
  - CT and PET/CT for staging
    - Assessment of primary neoplasm (T)
    - Assessment of lymph node involvement (N)
    - Assessment of metastatic disease (M)
- **Hilar Mass**: **Metastatic Disease**
  - Primary lung cancer
  - Extrathoracic malignancy
  - Lymphoma

Bilateral Hilar Enlargement: Pulmonary Arteries

- **Pulmonary hypertension**
  - Pulmonary trunk width > 3 cm on CT
  - Pulmonary trunk diameter > diameter of adjacent ascending aorta

Bilateral Hilar Enlargement: Lymph Nodes

- **Nonneoplastic lymphadenopathy**
  - **Sarcoidosis**
    - Bilateral symmetric hilar lymphadenopathy
    - Associated mediastinal lymphadenopathy (paratracheal, aortopulmonary window)
    - Associated perilymphatic lung nodules/micronodules
- **Neoplastic lymphadenopathy**
  - **Metastases**
  - **Lymphoma**

Unilateral Hilar Enlargement: Pulmonary Artery

- **Pulmonic stenosis**
  - Asymptomatic adults
  - Poststenotic dilatation of left pulmonary artery related to direction of high-velocity jet of blood through stenotic valve

Unilateral Hilar Enlargement: Lymph Nodes

- Neoplastic lymphadenopathy
  - Lung cancer lymph node metastases
  - Metastases from other primary malignancies
  - Lymphoma

Thick Intermediate Stem Line

- Thickness > 3 mm on lateral radiography
- Etiologies
  - Interstitial edema
  - Hilar/perihilar mass/lymphadenopathy

OVERVIEW OF HILA

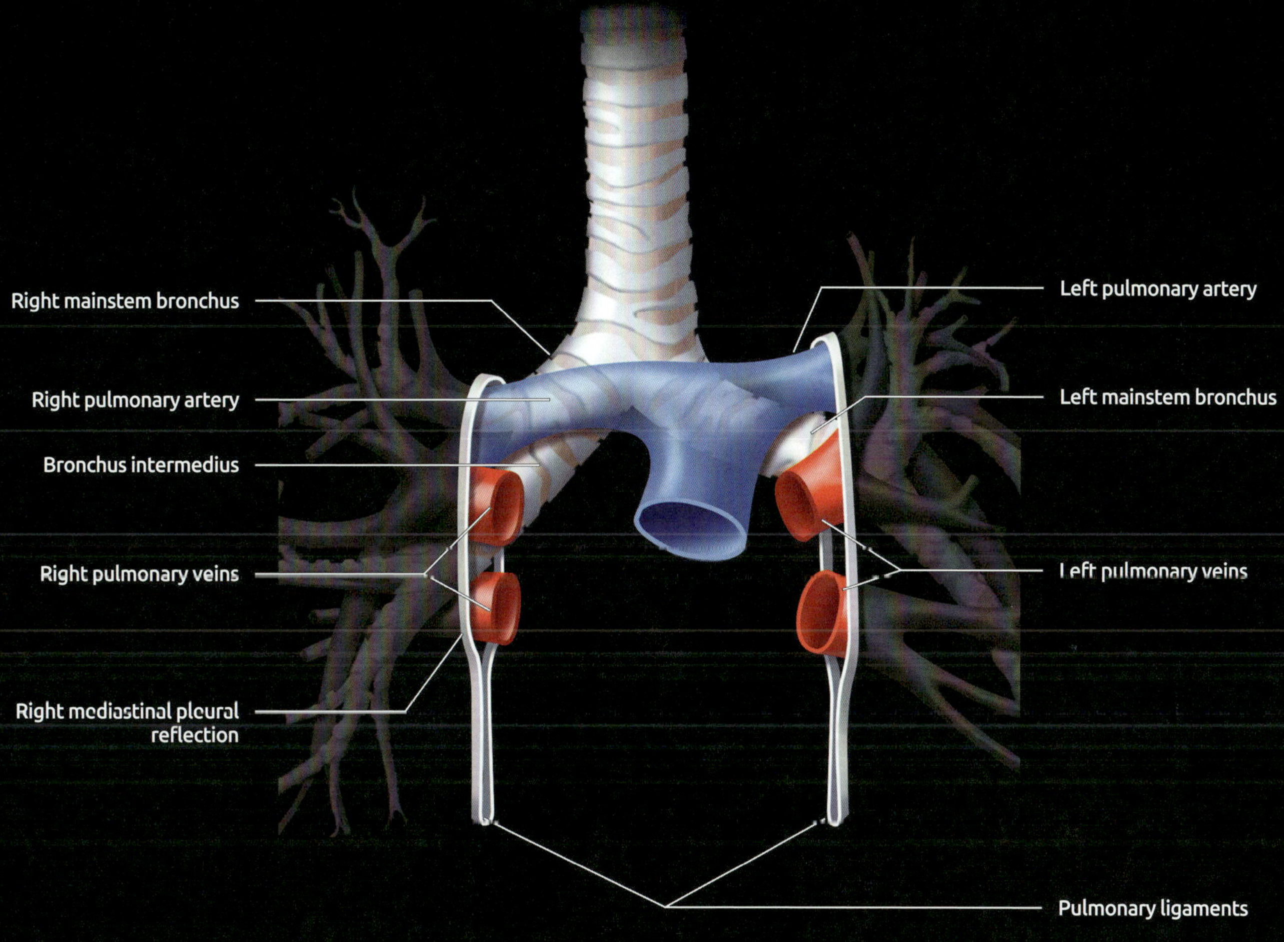

Graphic shows the normal anatomy of the pulmonary hila. The hila constitute the "roots" of the lungs and are surrounded by the mediastinal pleural reflections, which extend inferiorly as the pulmonary ligaments. The hila are the sites through which airways, vessels, and other structures travel between the mediastinum and the bilateral adjacent lungs. This graphic shows the central bronchi, pulmonary arteries, and pulmonary veins, which are the principal components of the hila.

RIGHT HILUM

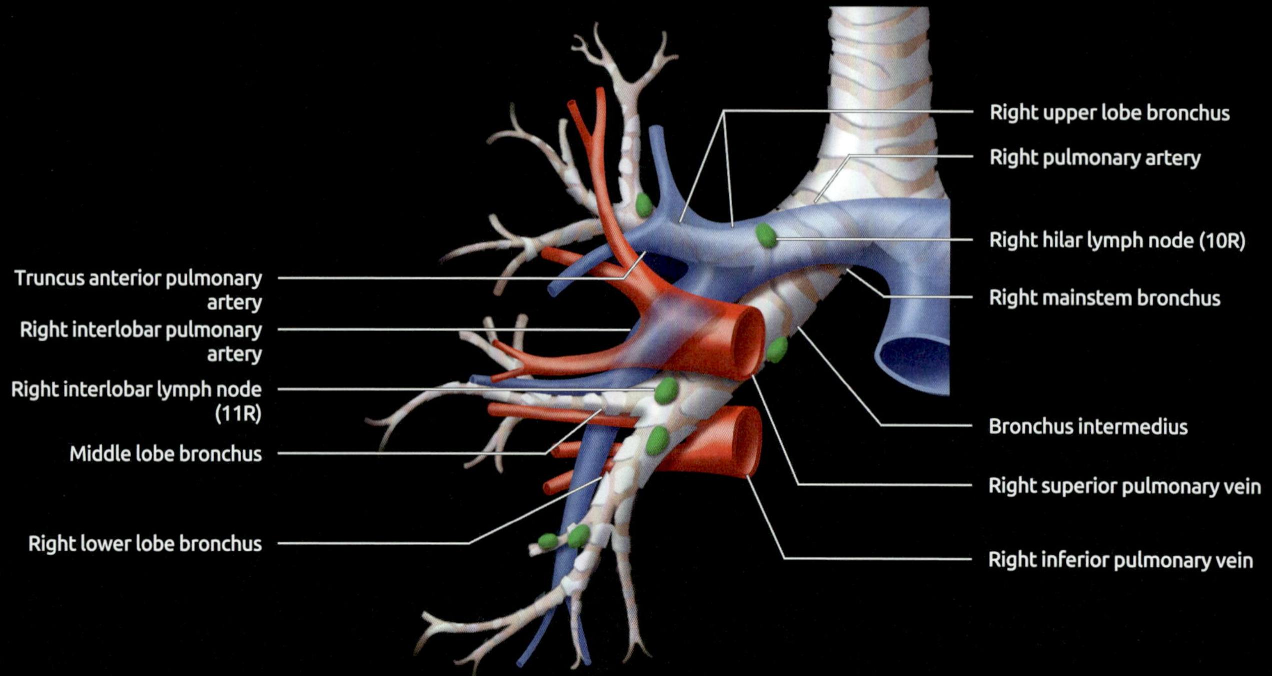

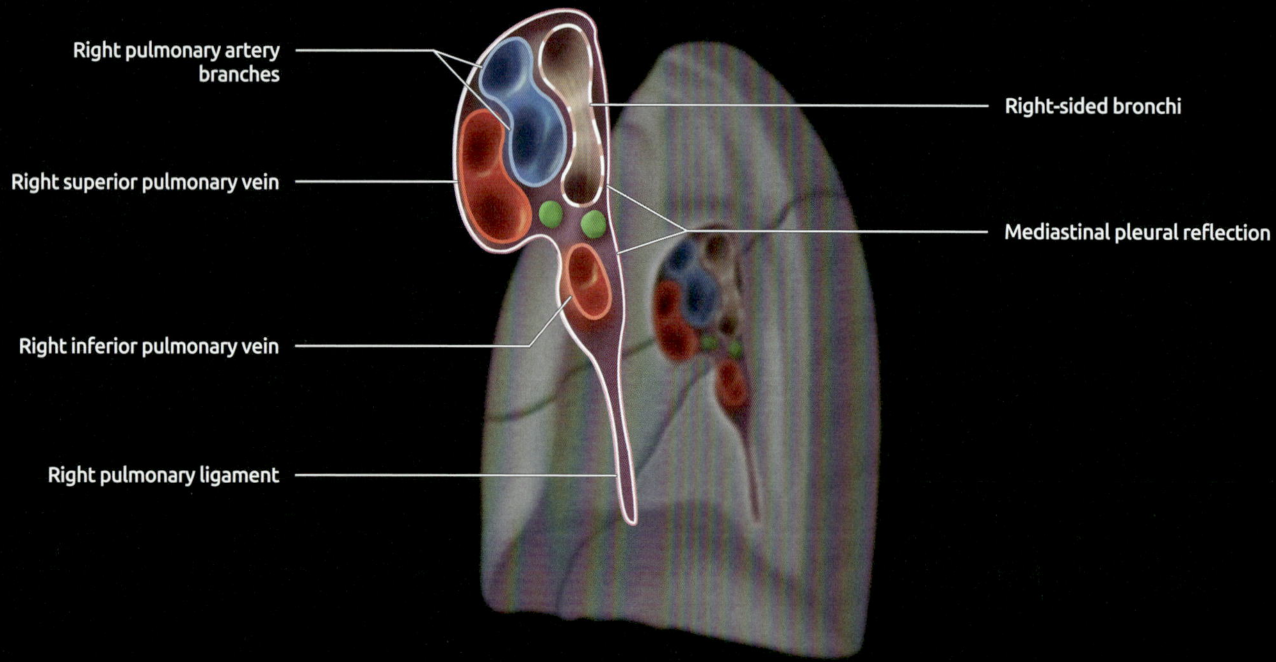

(Top) *First of 2 graphics illustrates the normal right hilar anatomy. Anterior view shows the pulmonary arteries, which course along their respective homologous bronchi. The right upper lobe bronchus is eparterial because it originates above the right pulmonary artery. The right pulmonary artery has ascending (truncus anterior) and descending (interlobar) branches. The pulmonary veins follow a horizontal course relative to that of the pulmonary arteries. Normal lymph nodes in the hilar and interlobar (10R and 11R, respectively) International Association for the Study of Lung Cancer (IASLC) lymph node stations are shown in green.* **(Bottom)** *Graphic shows a sagittal or medial view of the right hilum and highlights its central location on the mediastinal lung surface and the relative locations of its various components. The right hilum contains the right superior pulmonary vein, right pulmonary artery, and right-sided bronchi (from anterior to posterior). The right inferior pulmonary vein is inferior to the hilum proper. Normal hilar lymph nodes are depicted in green.*

LEFT HILUM

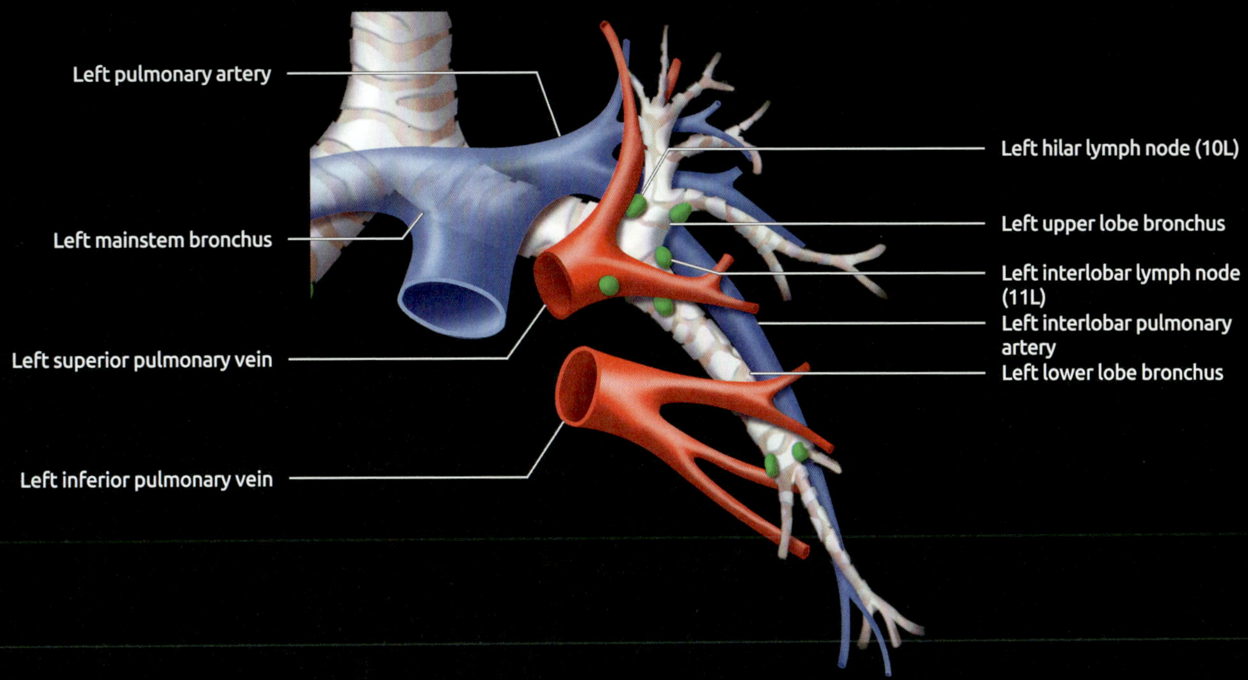

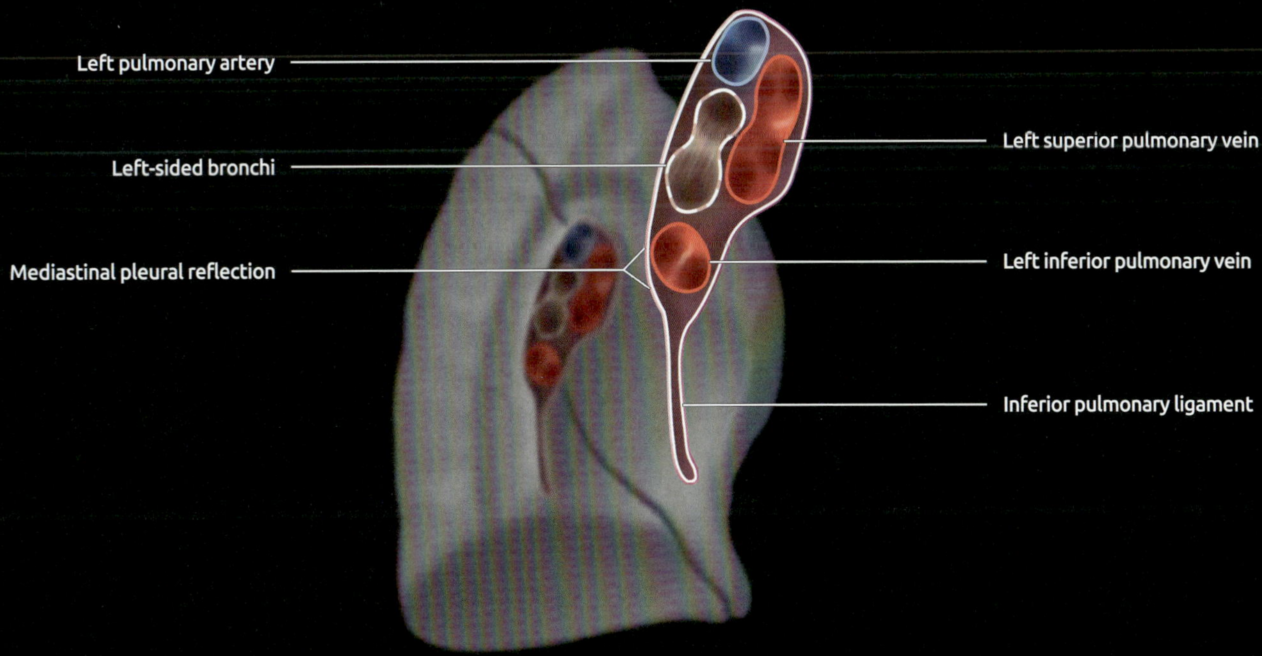

(Top) *First of 2 graphics illustrates the normal left hilar anatomy. Anterior view shows that the left pulmonary artery courses over the left mainstem bronchus. The left upper lobe bronchus is hyparterial because it originates below the left pulmonary artery. The left interlobar pulmonary artery courses along the posterolateral aspect of the left lower lobe bronchus. Note the anterior location of the pulmonary veins. Normal lymph nodes in the hilar and interlobar (10L and 11L, respectively) IASLC lymph node stations are shown in green.* **(Bottom)** *Graphic shows a sagittal or medial view of the left hilum and highlights its central location on the mediastinal lung surface and the relative locations of its various components. The left-sided bronchi are situated posteriorly within the midleft hilum. The pulmonary artery is located above the left mainstem bronchus. The left superior pulmonary vein is located anterior to the left bronchi. The left inferior pulmonary vein is located inferior to the hilum proper.*

RADIOGRAPHY OF NORMAL HILA, RADIOGRAPHY

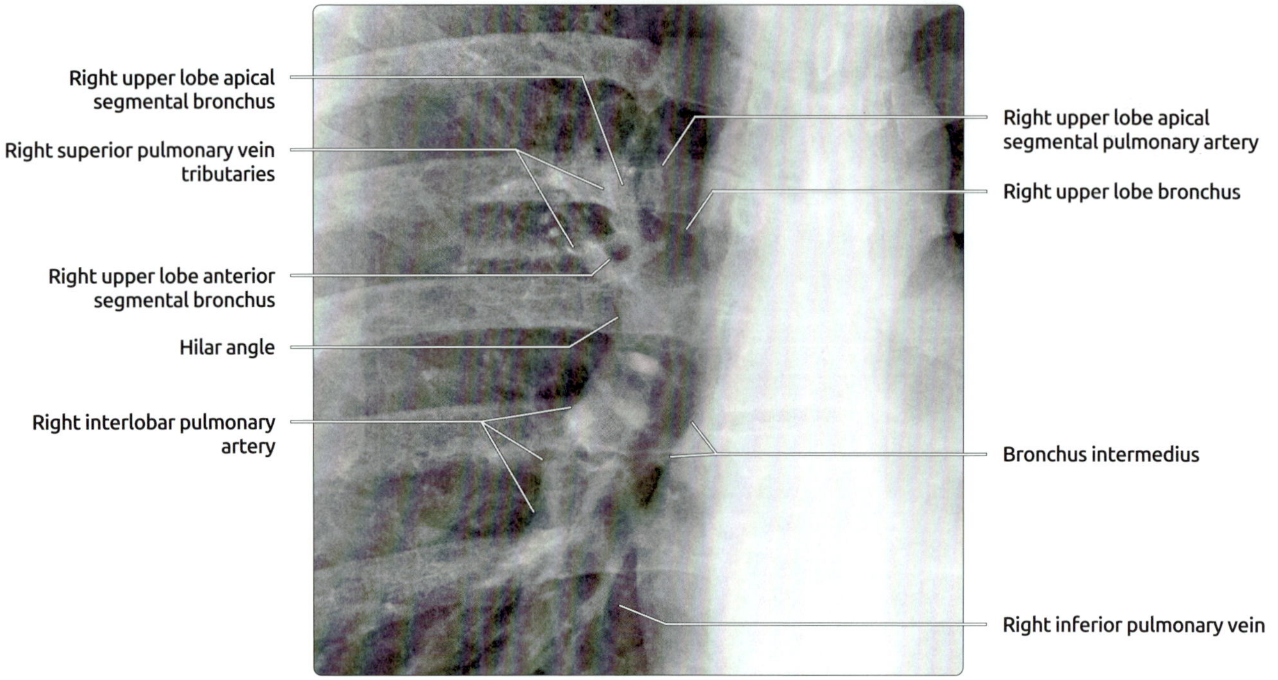

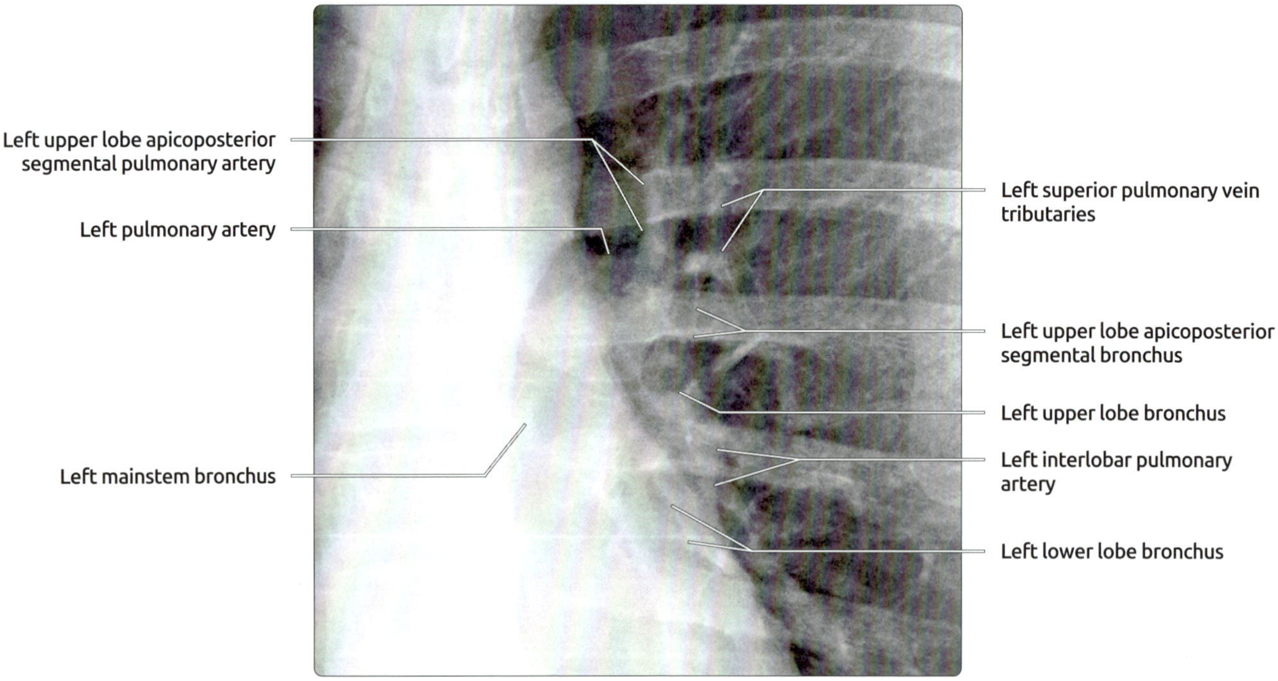

(Top) *PA chest radiograph coned down to the right hilum shows that the lateral border of the upper right hilum is formed by tributaries of the right superior pulmonary vein, while the medial border is formed by the apical segmental branch of the truncus anterior pulmonary artery. Segmental branches of the right upper lobe bronchus may be visible, as in this case. The hilar angle is a shallow concavity located where the right superior pulmonary vein crosses the right interlobar pulmonary artery. The inferior aspect of the right hilum is formed by the right interlobar pulmonary artery, which is located lateral to the bronchus intermedius.* **(Bottom)** *PA chest radiograph coned down to the left hilum shows that the left pulmonary artery courses above the left mainstem bronchus. The upper aspect of the left hilum is formed by the left superior pulmonary vein and its tributaries laterally and the left upper lobe apicoposterior segmental pulmonary artery medially. The left upper lobe bronchus and its branches may be visible in the upper left hilum. The left interlobar pulmonary artery is located lateral to the left lower lobe bronchus.*

NORMAL HILA, CT

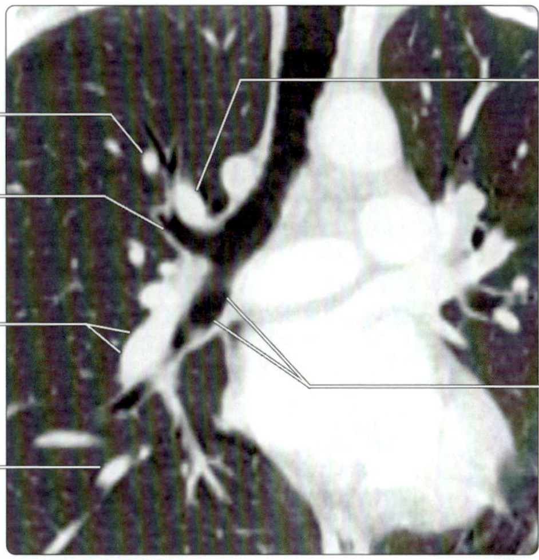

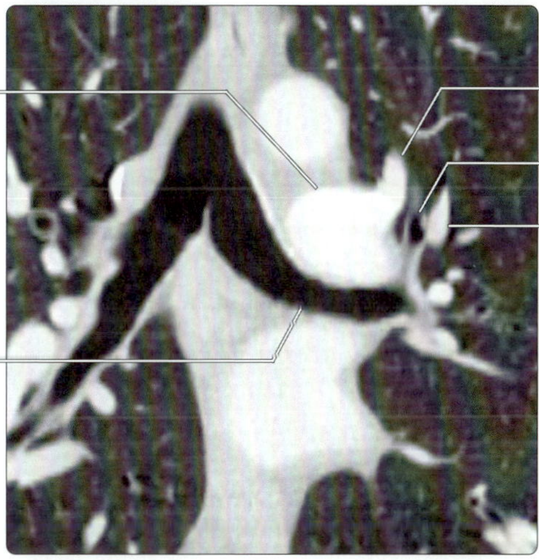

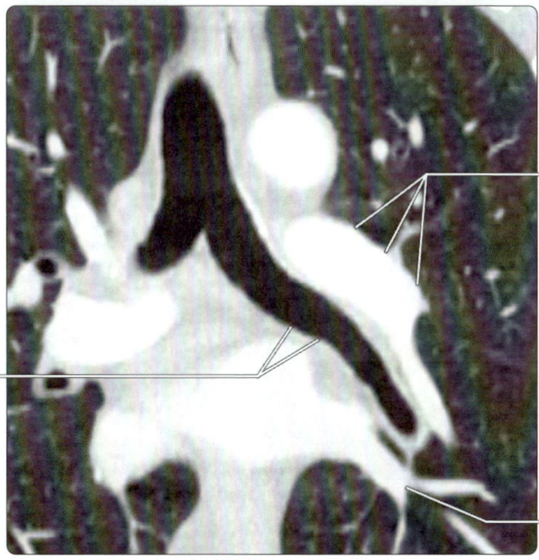

(Top) *Coronal oblique CECT (lung window) shows the upper aspect of the right hilum formed by tributaries of the right superior pulmonary vein laterally and branches of the truncus anterior pulmonary artery medially on either side of the right upper lobe bronchus and its branches. The right interlobar pulmonary artery and bronchus intermedius form the inferior aspect of the right hilum.* **(Middle)** *Coronal oblique CECT (lung window) shows that the upper aspect of the left hilum is formed by the left superior pulmonary vein laterally and the apicoposterior segmental pulmonary artery medially, located on either side of the left upper lobe bronchus and its branches. The left pulmonary artery courses above the left mainstem bronchus.* **(Bottom)** *Oblique coronal CECT (lung window) shows the inferior left hilum formed by the left interlobar pulmonary artery laterally and the left lower lobe bronchus medially. The left inferior pulmonary vein is located below the left hilum.*

NORMAL HILA, RADIOGRAPHY

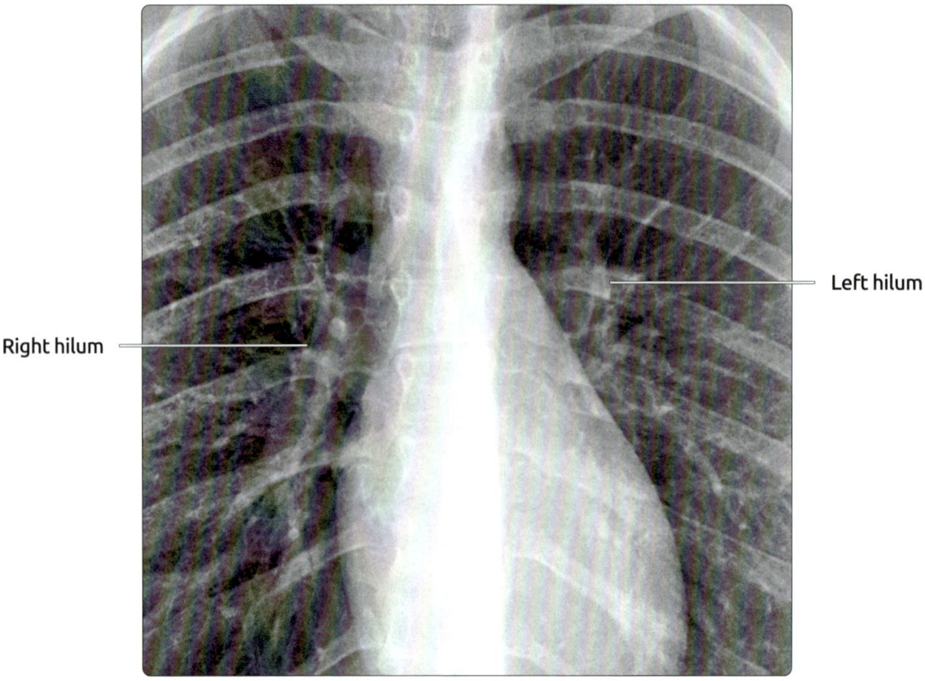

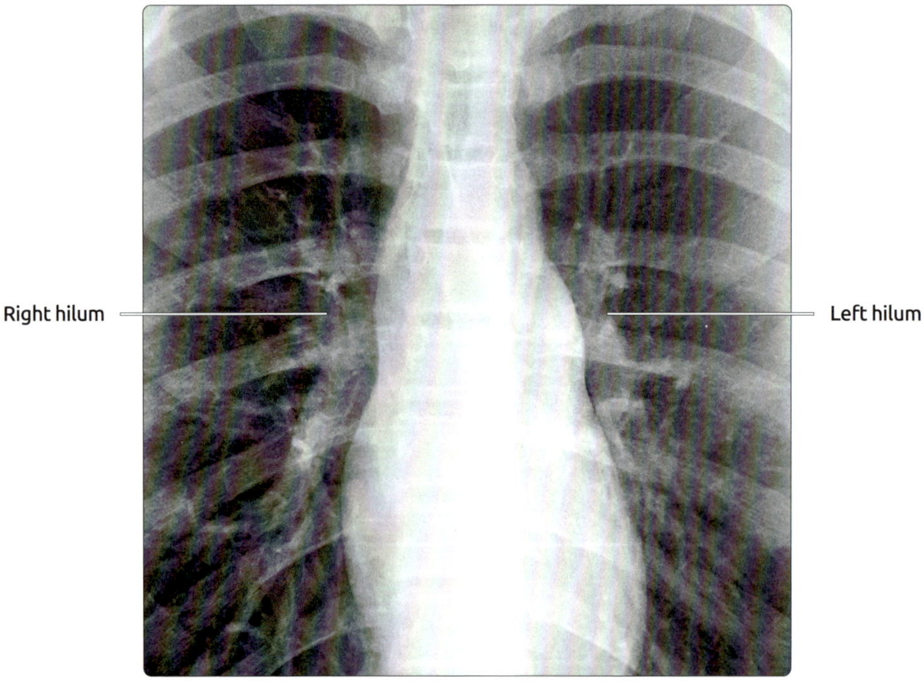

(Top) *PA chest radiograph coned down to the hila shows the normal and typical relationship of the right and left hila. Most of the hilar density visible on radiography is attributable to the pulmonary arteries and superior pulmonary veins. There is normally a mild asymmetry in hilar height with the right hilum located slightly lower than the left, as in this case. This hilar configuration is seen in ~ 97% of normal subjects.* **(Bottom)** *PA chest radiograph coned down to the hila shows normal bilateral hilar structures. Note that in this case the bilateral hila are located at roughly the same height. This hilar configuration is seen in ~ 3% of normal subjects. Abnormal hilar relationships are helpful clues to the presence of volume loss.*

NORMAL HILA, GRAPHIC AND LATERAL RADIOGRAPHY

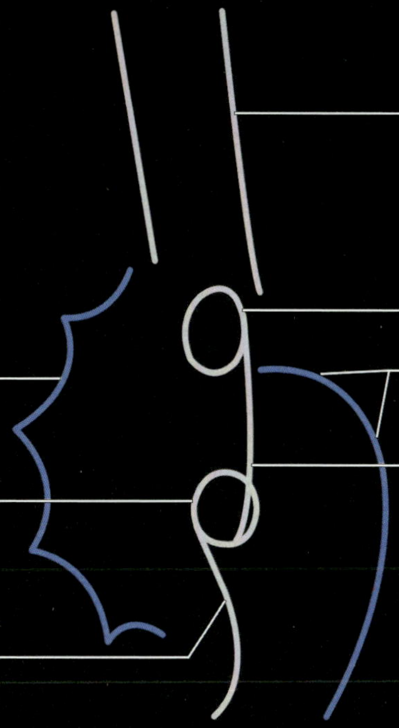

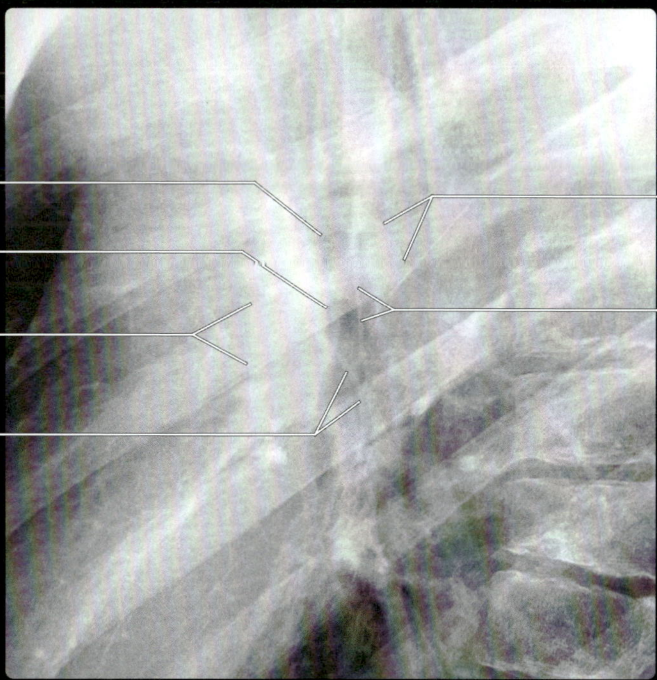

(Top) *Graphic provides a schematic of the normal hilar anatomy, as seen on lateral chest radiography. Two roughly vertical parallel gray lines represent the trachea. Two gray circles (one superior, one inferior) represent the right upper lobe bronchus and left mainstem/left upper lobe bronchi, respectively. A gray line drawn from the posterior wall of the superior circle intersects the posterior aspect of the inferior circle and represents the intermediate stem line formed by the posterior walls of the right mainstem bronchus superiorly and the bronchus intermedius inferiorly. The arcuate gray line below the inferior gray circle represents the anterior wall of the left lower lobe bronchus. The anterior and posterior blue lines depict the right hilar vascular opacity and the left pulmonary artery, respectively.* **(Bottom)** *Lateral chest radiograph coned down to the hila shows the right hilar vascular opacity anteriorly and the left pulmonary artery posteriorly. The right upper lobe bronchus and left mainstem/upper lobe bronchi are visible in 45% and 75% of normal subjects, respectively. Note the normal intermediate stem line, which measures up to 3 mm in thickness.*

NORMAL HILA, LATERAL RADIOGRAPHY

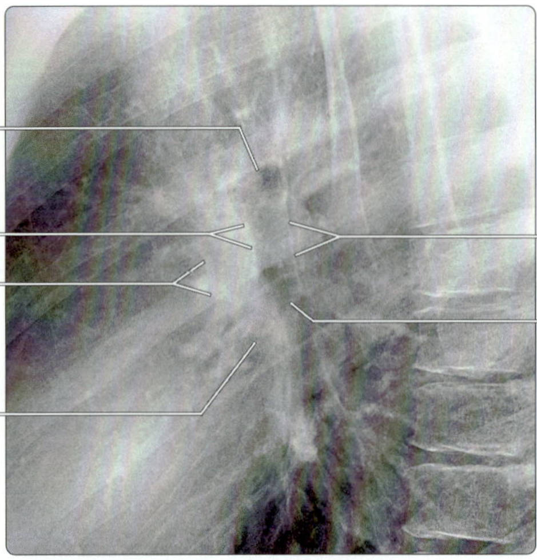

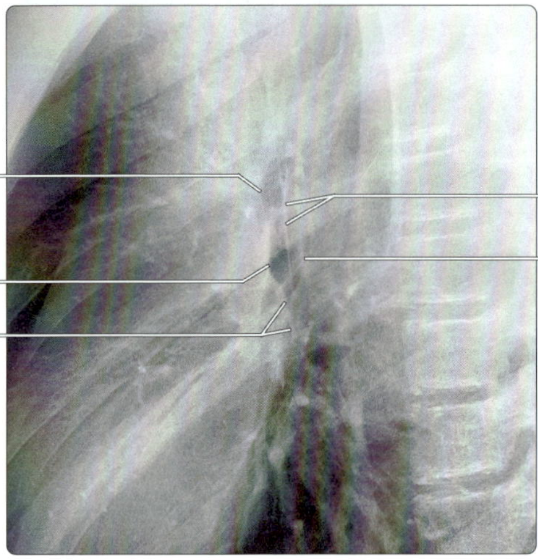

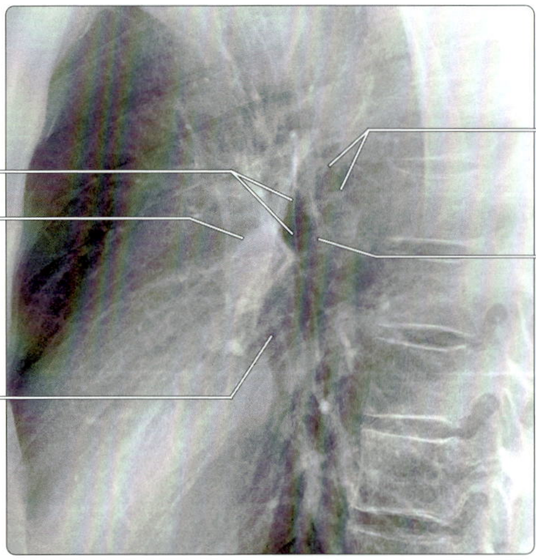

(Top) *First of 3 lateral chest radiographs coned down to the hila shows various normal anatomic landmarks. The right hilar vascular opacity outlines the anterior wall of the bronchus intermedius. The intermediate stem line corresponds to the posterior walls of the right mainstem bronchus and the bronchus intermedius. The left mainstem/left upper lobe bronchi manifest as a round or oval lucency in the inferior hilum. The inferior hilar window is a relatively avascular zone in the anterior inferior hilum.* **(Middle)** *Lateral chest radiograph shows the right upper lobe bronchus superiorly and the left mainstem/left upper lobe bronchi inferiorly. An arcuate interface inferior to the left mainstem/left upper lobe bronchi represents the anterior wall of the left lower lobe bronchus. Note the normal thin intermediate stem line.* **(Bottom)** *Lateral chest radiograph coned down to the hila shows the right hilar vascular opacity anteriorly and the left pulmonary artery posteriorly. Note the normal inferior hilar window and the normal thin intermediate stem line.*

NORMAL HILA, SAGITTAL CT

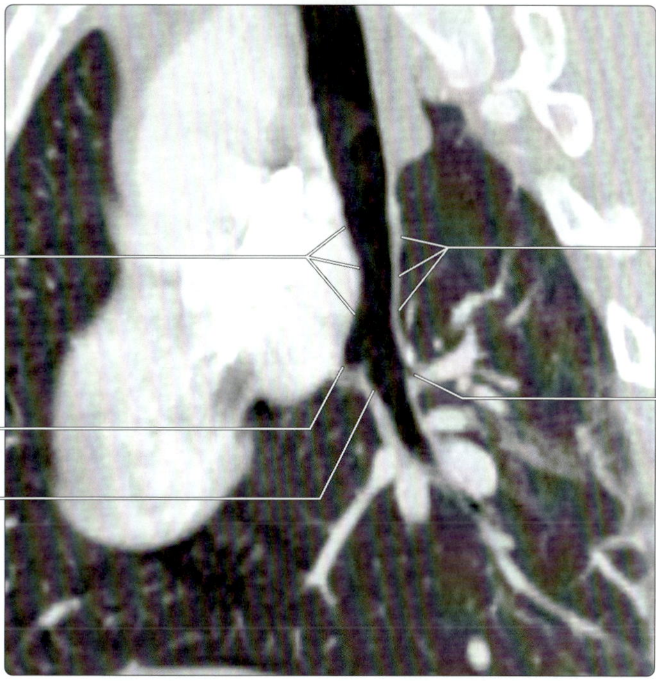

(Top) *Sagittal oblique CECT (lung window) of the right hilum correlates with the anatomic landmarks seen on lateral chest radiography. The right hilar vascular opacity outlines the anterior wall of the bronchus intermedius. The intermediate stem line is formed by the posterior walls of the right mainstem bronchus and bronchus intermedius and normally measures up to 3 mm in thickness. The bronchus intermedius branches into the middle and right lower lobe bronchi.* **(Bottom)** *Sagittal CECT (lung window) of the left hilum correlates with the anatomic landmarks seen on lateral chest radiography. The left superior pulmonary vein is located anterior to the central airways. The left pulmonary artery courses above the left mainstem bronchus. The left inferior pulmonary vein is located inferior to the hilum.*

NORMAL HILA, AXIAL CT

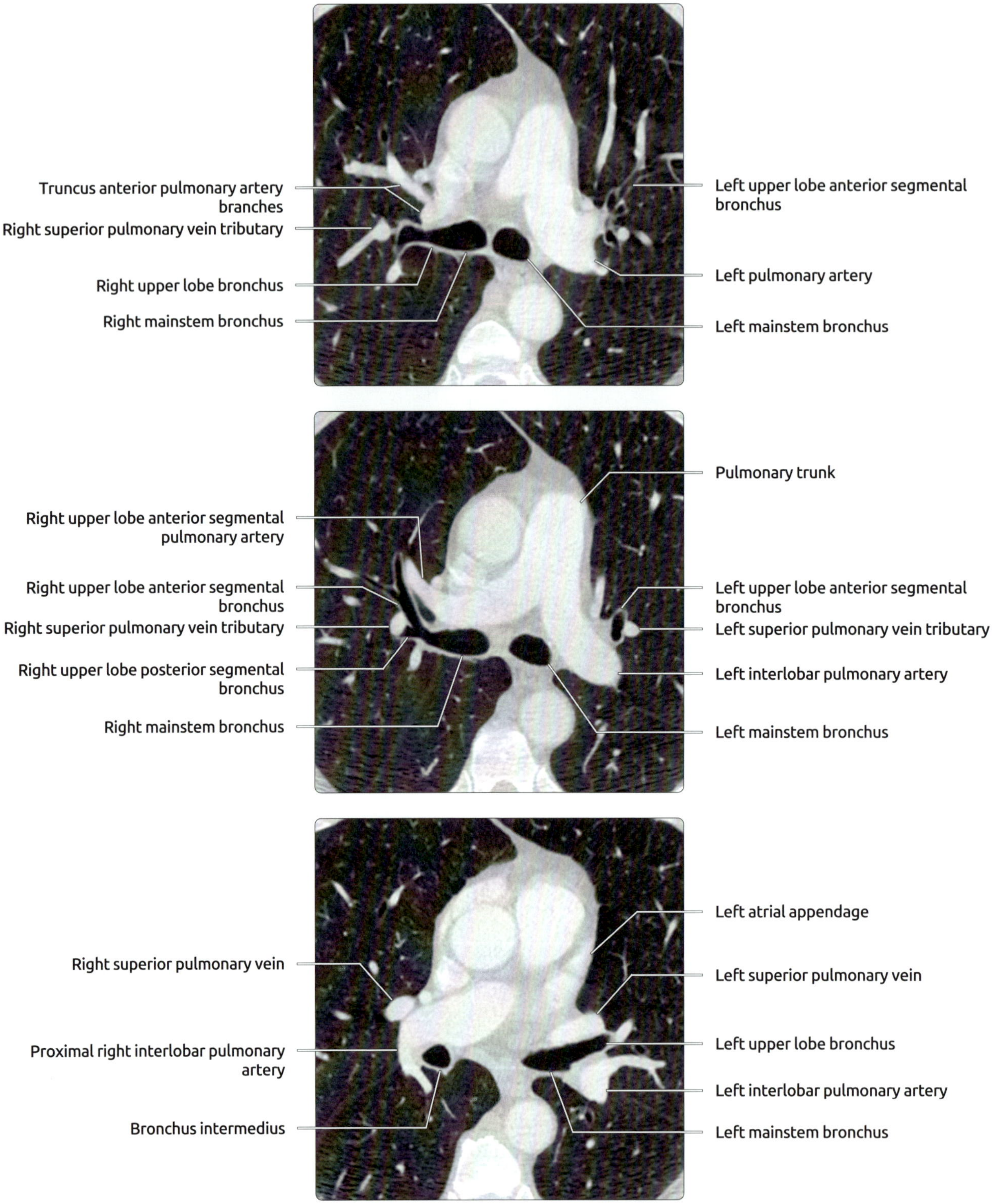

(Top) *First of 6 axial CECT images (lung window) shows the normal hilar anatomy. The truncus anterior pulmonary artery is located anteromedial to the right upper lobe bronchus, and the right superior pulmonary vein is located laterally. The left pulmonary artery courses over the left mainstem bronchus, and its branches are medial to the left upper lobe bronchi.* **(Middle)** *Axial CECT (lung window) shows branches of the truncus anterior pulmonary artery located anterior and medial to the right upper lobe bronchi. The right superior pulmonary vein is located lateral to the upper lobe bronchi. The same relationship exists on the left where branches of the left upper lobe pulmonary artery are medial to the airways, and tributaries of the left superior pulmonary vein are lateral to the airways.* **(Bottom)** *Axial CECT (lung window) shows the right superior pulmonary vein located anterolateral to the right pulmonary artery at the level of the bronchus intermedius, where the proximal right interlobar pulmonary artery exhibits a characteristic elephant head and trunk configuration. The left superior pulmonary vein lies anterior to the left upper lobe bronchus and posterior to the left atrial appendage.*

NORMAL HILA, AXIAL CT

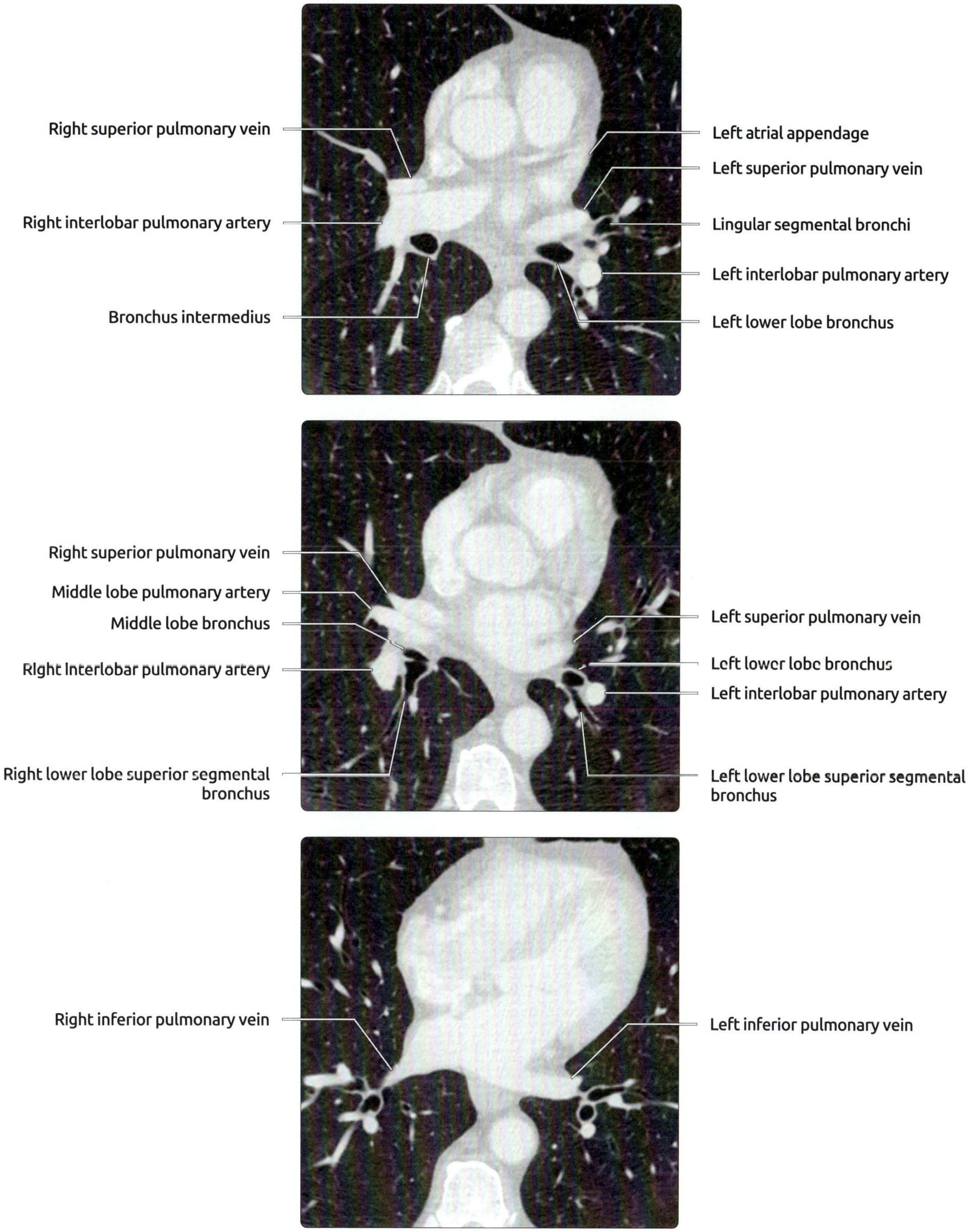

(Top) *Axial CECT (lung window) shows the right superior pulmonary vein coursing anterior to the right interlobar pulmonary artery. Note the thin posterior wall of the bronchus intermedius outlined by the adjacent right lower lobe. The left interlobar pulmonary artery lies posterolateral to the left lower lobe bronchus. The left superior pulmonary vein is located anterior to the left-sided bronchi and immediately posterior to the left atrial appendage.* **(Middle)** *Axial CECT (lung window) shows the origins of the bilateral lower lobe superior segmental bronchi, which arise roughly at the same axial level as the middle lobe bronchus.* **(Bottom)** *Axial CECT (lung window) shows the bilateral inferior pulmonary veins, which are located below the respective right and left hila.*

NORMAL HILA, CORONAL CT

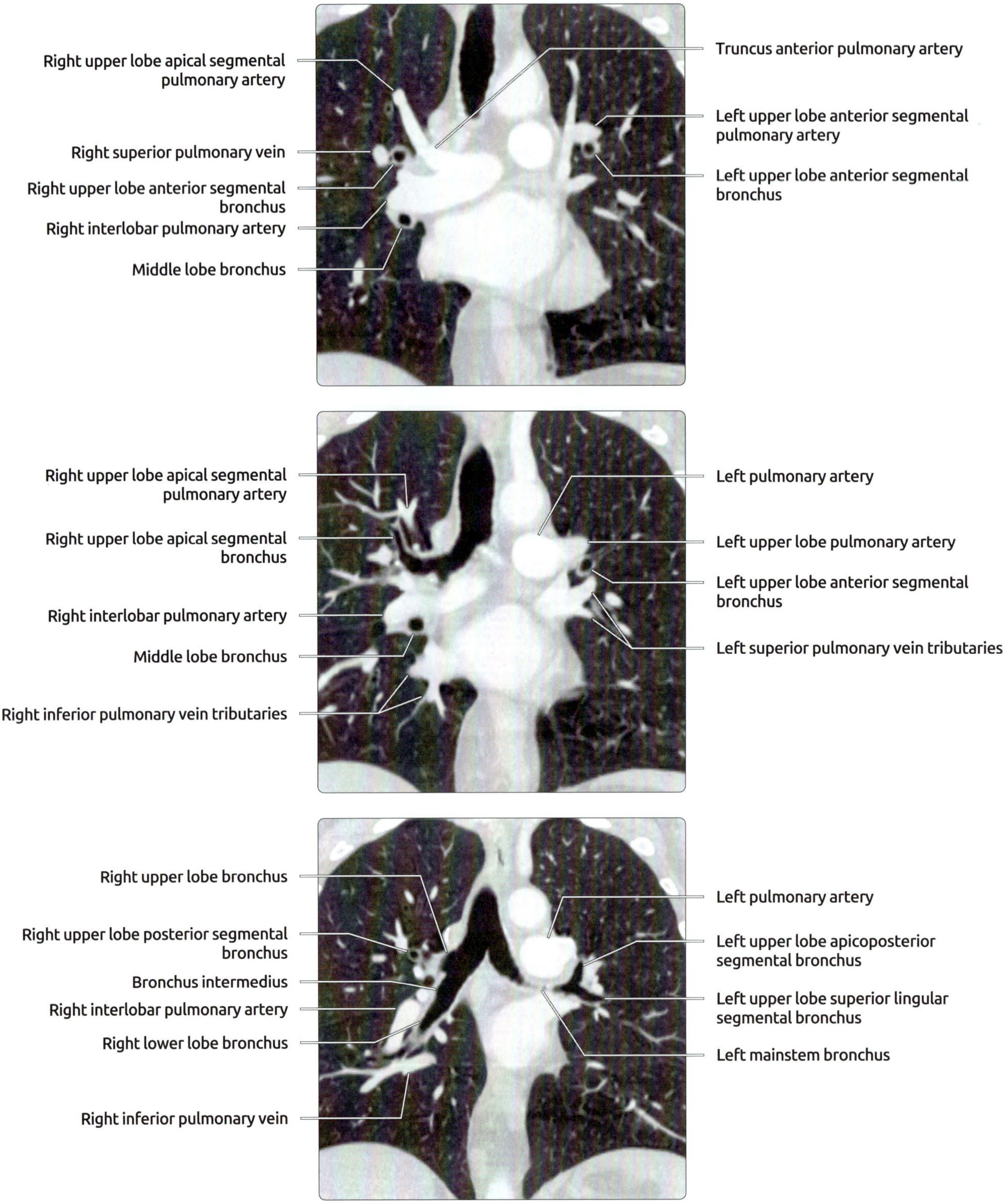

(Top) *First of 6 coronal CECT images (lung window) obtained anterior to the carina shows the normal anatomy of the upper hila. The upper right hilum is formed by branches of the truncus anterior pulmonary artery medially and the right superior pulmonary vein and its tributaries laterally. These vessels are located on either side of the right upper lobe bronchus and its branches. Likewise, branches of the left upper lobe pulmonary artery lie medial to branches of the left upper lobe bronchus in the upper aspect of the left hilum.* **(Middle)** *Coronal CECT (lung window) shows the anatomy of the bilateral upper hila. The pulmonary artery branches form the medial margins of the upper hila and the superior pulmonary veins form the lateral margins. The central airways are located between the adjacent pulmonary veins and pulmonary arteries.* **(Bottom)** *Coronal CECT (lung window) obtained at the carina shows the left pulmonary artery coursing posteriorly above the left mainstem bronchus. Note the right interlobar pulmonary artery coursing alongside the bronchus intermedius and the right lower lobe bronchus and forming the inferior aspect of the right hilum.*

NORMAL HILA, CORONAL CT

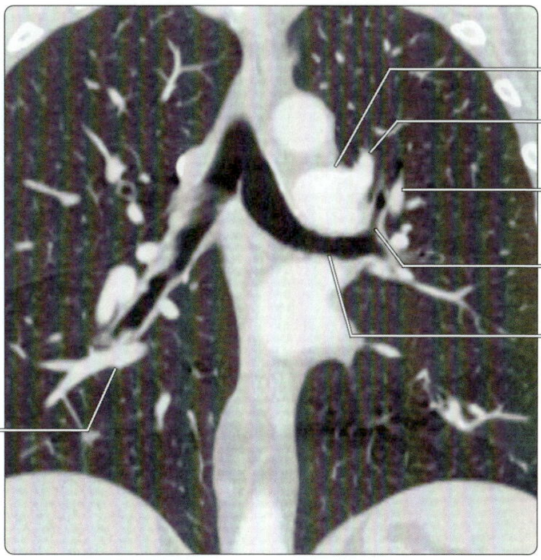

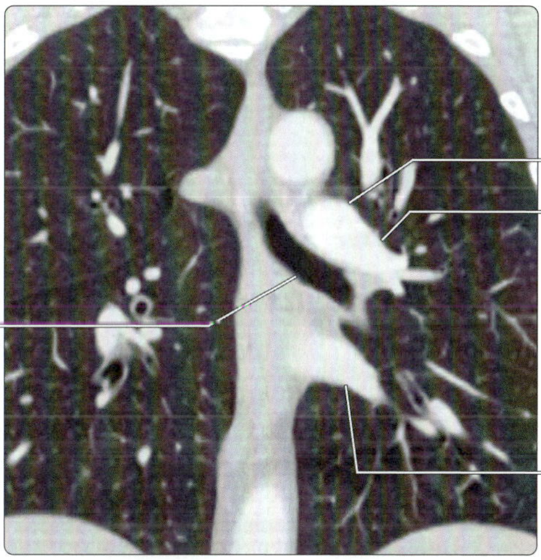

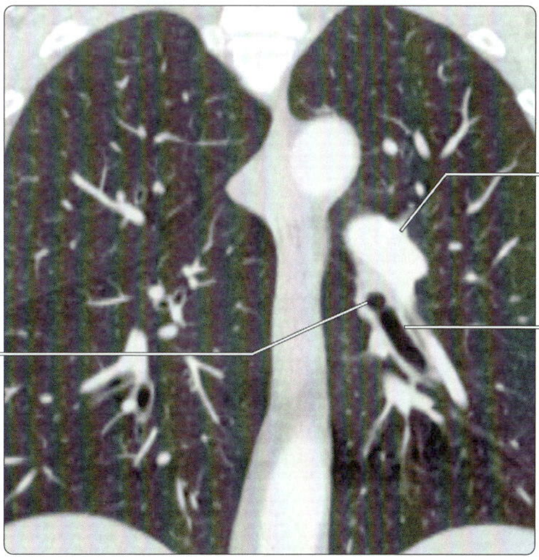

(Top) *Coronal CECT (lung window) shows the anatomy of the upper left hilum. The left pulmonary artery courses posteriorly immediately above the left mainstem bronchus. The left upper lobe apicoposterior segmental pulmonary artery, apicoposterior segmental bronchus, and a tributary of the left superior pulmonary vein (from medial to lateral) contribute to the left hilar opacity. Note that the right inferior pulmonary vein is located below the right hilum.* **(Middle)** *Coronal CECT (lung window) shows the upper aspect of the left interlobar pulmonary artery coursing posterolateral to the left lower lobe bronchus. The left inferior pulmonary vein is located inferior to the left hilum.* **(Bottom)** *Coronal CECT (lung window) shows the left interlobar pulmonary artery coursing posterolateral to the left lower lobe bronchus. The 1st branch of the left lower lobe bronchus is the superior segmental bronchus (seen end on). The left lower lobe truncus basalis bronchus continues inferiorly to give rise to the left lower lobe basilar segmental bronchi.*

NORMAL RIGHT HILUM, GRAPHIC, SAGITTAL CT, AND SAGITTAL MR

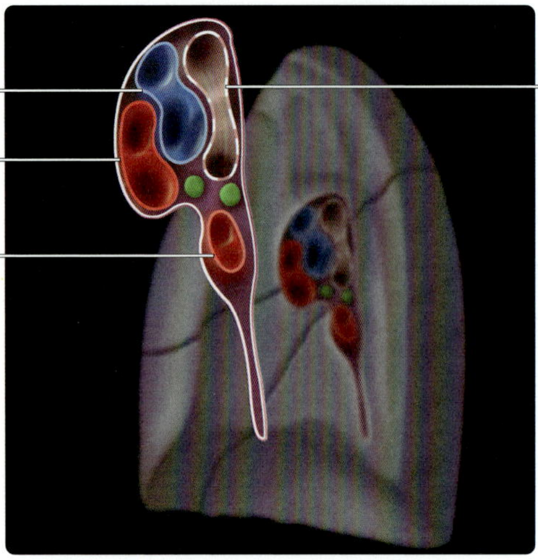

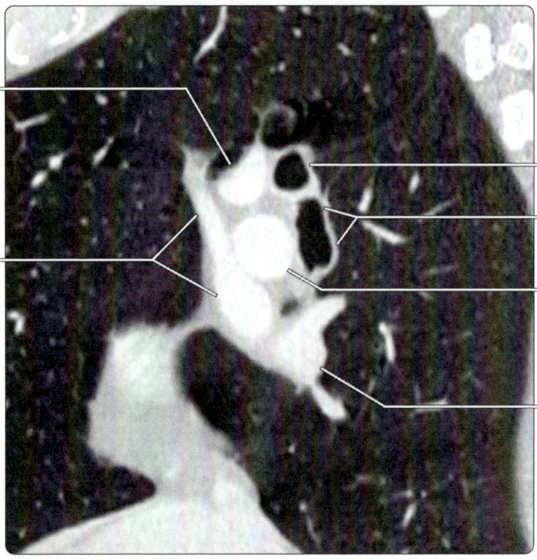

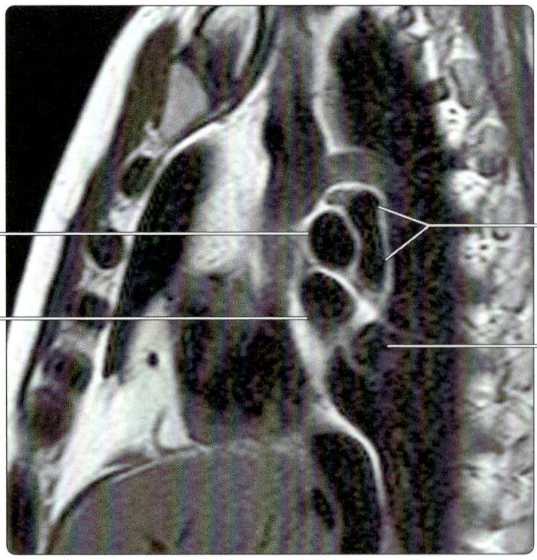

(Top) *Graphic correlates with normal sagittal CT and MR of the right hilum and shows the normal relationships of the right hilar vessels to the right-sided bronchi. The right-sided bronchi are the most posterior structures in the upper right hilum, and the superior pulmonary vein is the most anterior. The pulmonary artery is located centrally between the superior pulmonary vein and the bronchi, and the inferior pulmonary vein is located inferior to the hilum.* **(Middle)** *Sagittal CECT (lung window) of the right hilum shows the relationship of the airways and vascular structures depicted in the graphic. The right superior pulmonary veins are located anterior to the pulmonary arteries, which in turn are located anterior to the airways. The right inferior pulmonary vein is located inferior to the hilum.* **(Bottom)** *Sagittal T1 MR through the right hilum shows the anterior and inferior locations of the pulmonary veins and the posterior location of the airways. The right pulmonary artery is located in the central hilum. The right inferior pulmonary vein is located inferior to the hilum.*

NORMAL LEFT HILUM, GRAPHIC, SAGITTAL CT, AND SAGITTAL MR

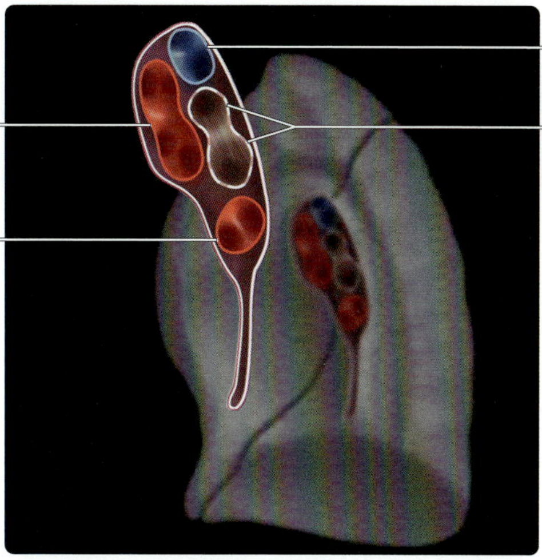

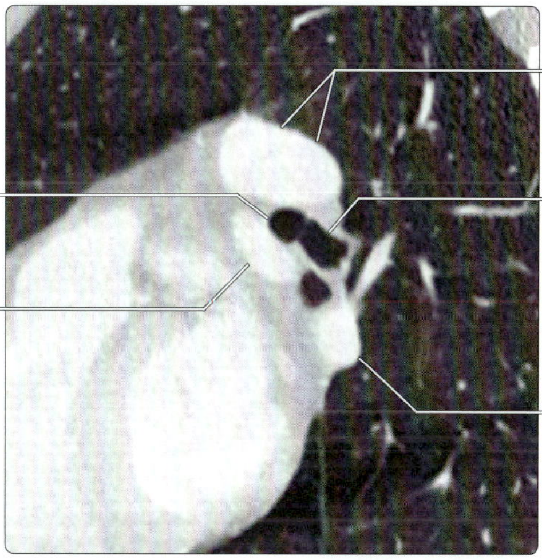

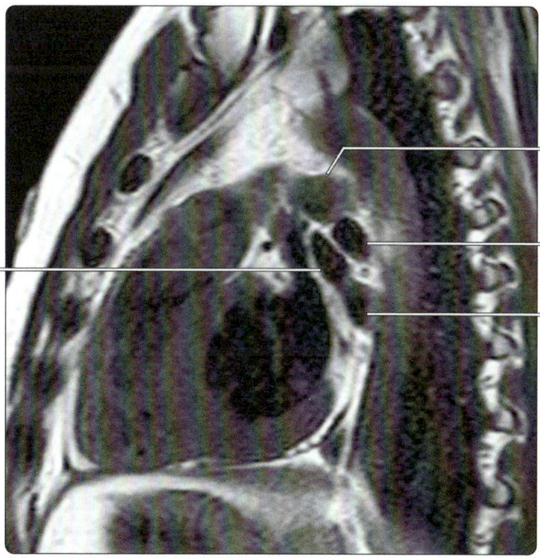

(Top) *Graphic correlates with sagittal CT and MR of the normal left hilum and shows the normal relationships of the left hilar vessels to the left-sided bronchi. The pulmonary veins are located anteriorly and inferiorly, and the left-sided bronchi are located posteriorly. The left pulmonary artery is located superiorly and courses above the left mainstem bronchus.* **(Middle)** *Sagittal CECT (lung window) shows the anatomy of the normal left hilum. The left pulmonary artery courses above the left mainstem bronchus and will continue caudally as the left interlobar pulmonary artery. The left superior pulmonary vein is located anterior to the airways. The left inferior pulmonary vein is located inferior to the hilum.* **(Bottom)** *Sagittal T1 MR through the left hilum shows the anatomic relationships shown in the graphic and the CT. The left pulmonary vein courses above the left mainstem bronchus. The left superior pulmonary vein is located anterior to the airways, and the left inferior pulmonary vein is located below the left hilum.*

NORMAL HILA, AXIAL MR

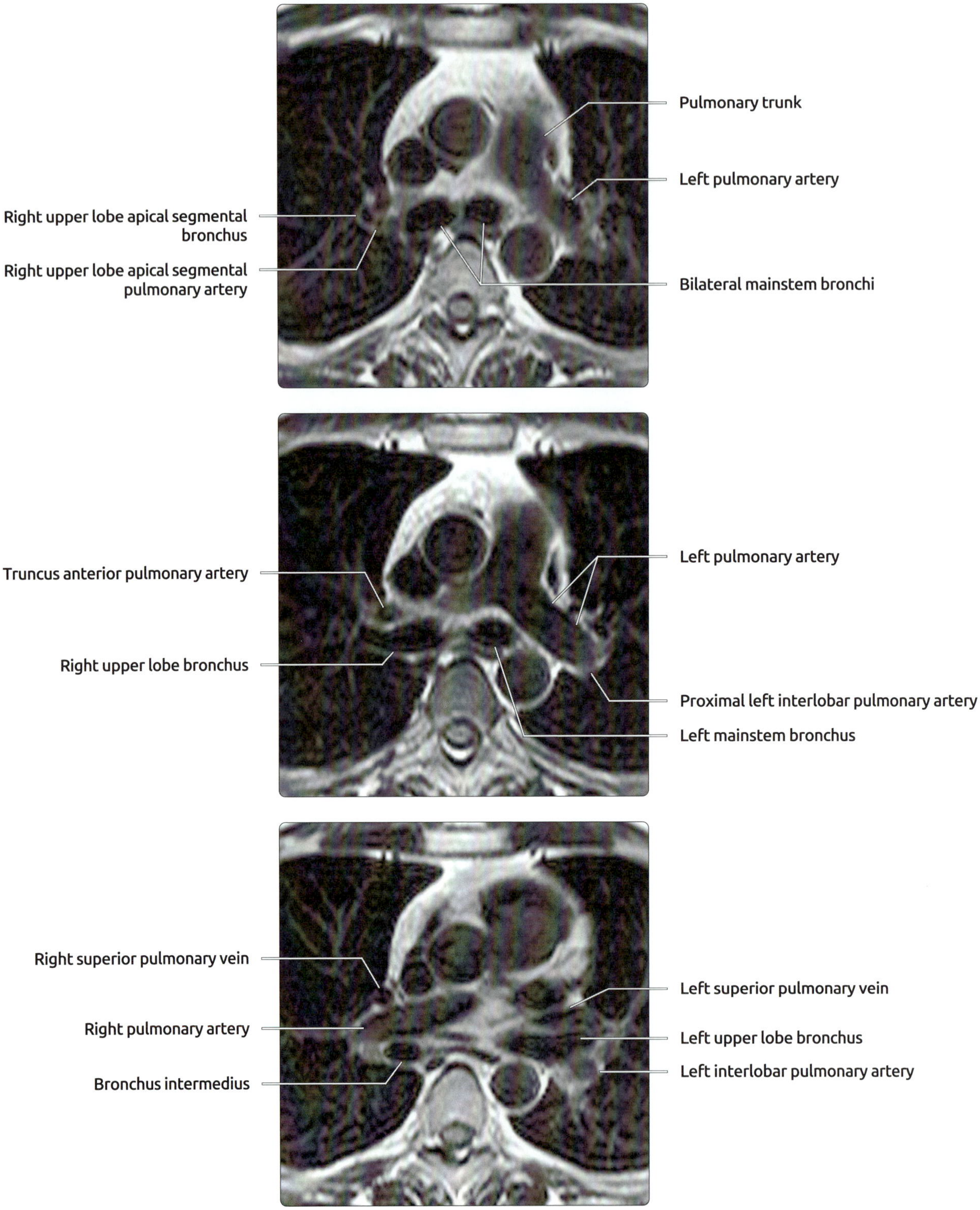

(Top) *First of 3 normal axial T1 MR images shows the normal hilar anatomy. The left pulmonary artery courses over the left mainstem bronchus (not visible in this plane of section). The right upper lobe apical segmental bronchus is identified with its corresponding pulmonary artery.* **(Middle)** *Axial T1 MR just below the carina demonstrates the horizontal portion of the left pulmonary artery, which courses above the left mainstem bronchus (not visible in this plane of section). The left pulmonary artery gives rise to the left interlobar pulmonary artery. The right upper lobe bronchus and the anteriorly located truncus anterior pulmonary artery are seen in the right hilum.* **(Bottom)** *Axial T1 MR at the level of the bronchus intermedius shows the horizontal portion of the right pulmonary artery coursing anterior to the bronchus intermedius, and the left interlobar pulmonary artery is located posterolateral to the left upper lobe bronchus. Note the anterior location of the bilateral superior pulmonary veins.*

NORMAL HILA, CORONAL MR

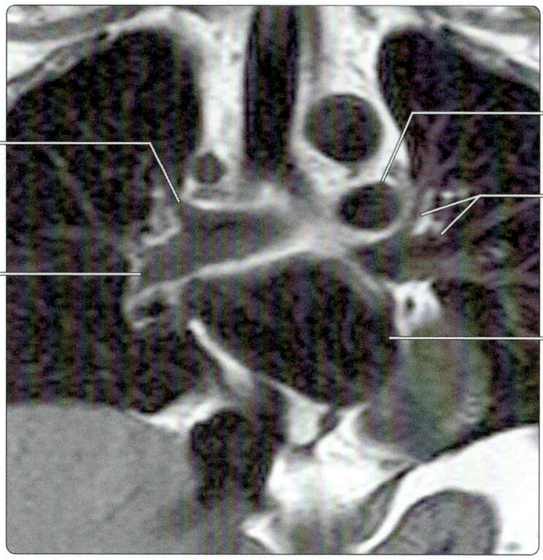

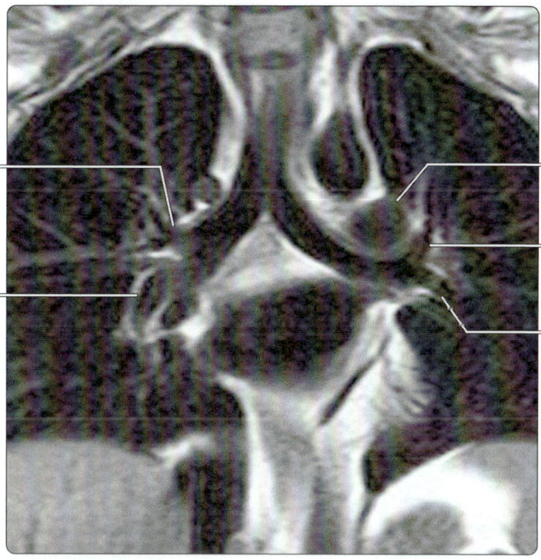

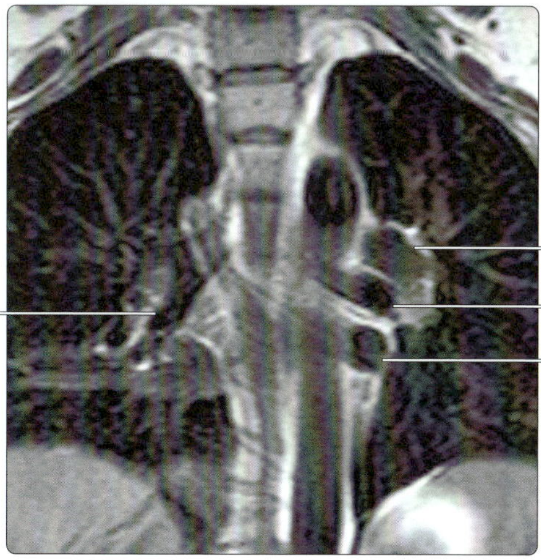

(Top) *First of 3 coronal T1 MR images shows the normal anatomy of the hila. Note the horizontal portion of the right pulmonary artery and its ascending and descending branches. The left superior pulmonary veins course into the left atrium below the left pulmonary artery.* **(Middle)** *Coronal T1 MR through the carina demonstrates the relationship of the bilateral pulmonary arteries to their ipsilateral bronchi. The left pulmonary artery courses above the left mainstem bronchus. The right upper lobe bronchus is eparterial, as it originates above the ipsilateral right pulmonary artery.* **(Bottom)** *Coronal T1 MR through the bronchus intermedius and the posterior left mainstem bronchus demonstrates the posterior aspect of the left pulmonary artery as it begins to course inferiorly to become the left interlobar pulmonary artery. Note the horizontal orientation of the left inferior pulmonary vein.*

ABNORMAL HILAR HEIGHT, RADIOGRAPHY

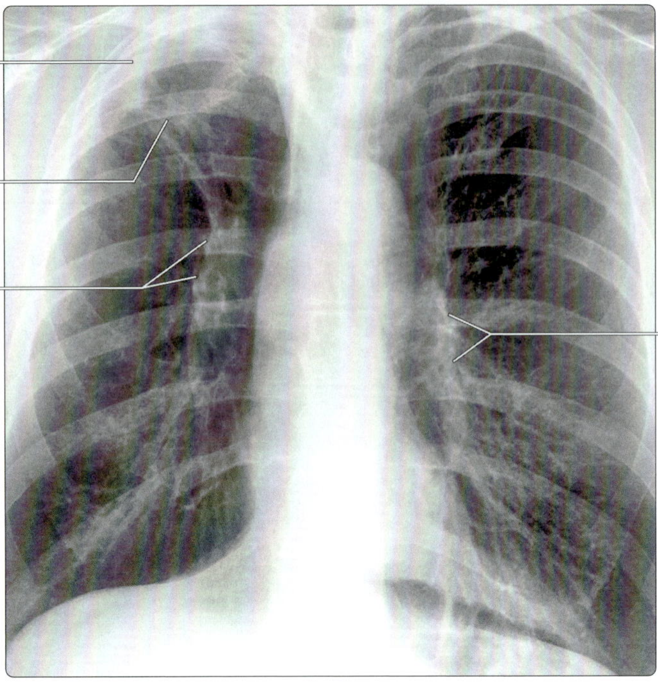

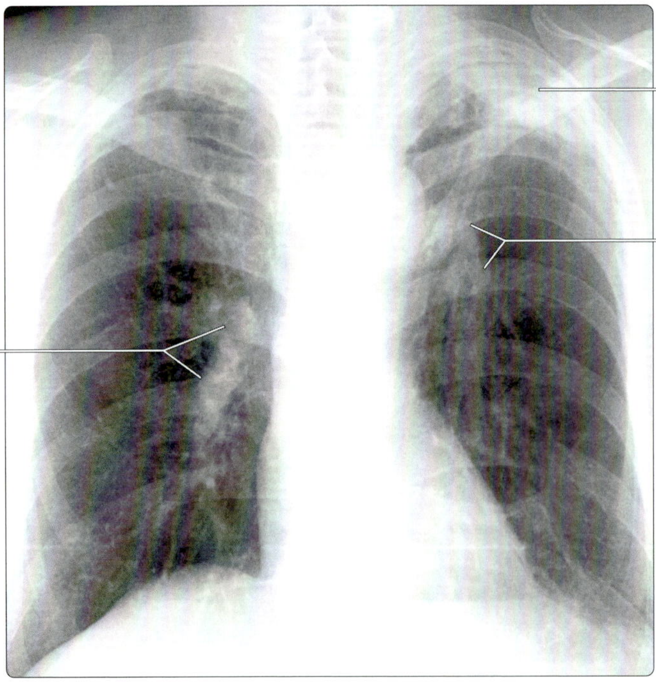

(Top) *PA chest radiograph of an asymptomatic 60-year-old man with a remote history of tuberculosis demonstrates abnormal elevation of the right hilum secondary to right upper lobe volume loss from right apical fibrosis manifesting with right apical opacities and adjacent right apical pleural thickening. The right hilum is normally located slightly lower than the left.* **(Bottom)** *PA chest radiograph of a 57-year-old man with asymmetric left apical pleural thickening and left upper lobe fibrosis secondary to prior tuberculosis demonstrates elevation and distortion of the left hilum. Although the left hilum is typically higher than the right, the asymmetry of hilar height is not normally this pronounced. The right hilum is of normal size and is in normal position.*

HILAR AND INTERLOBAR LYMPH NODES, CT

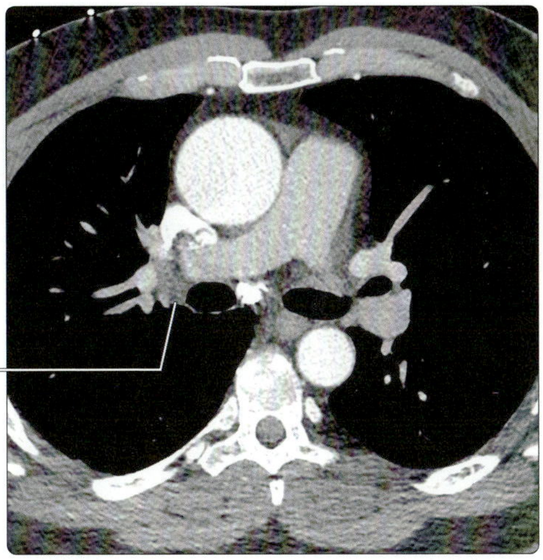

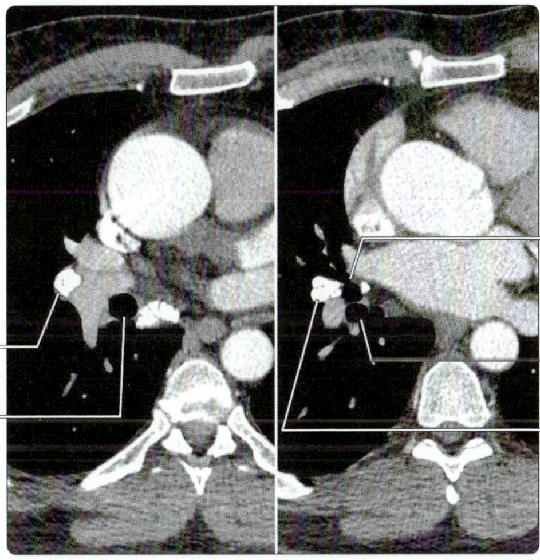

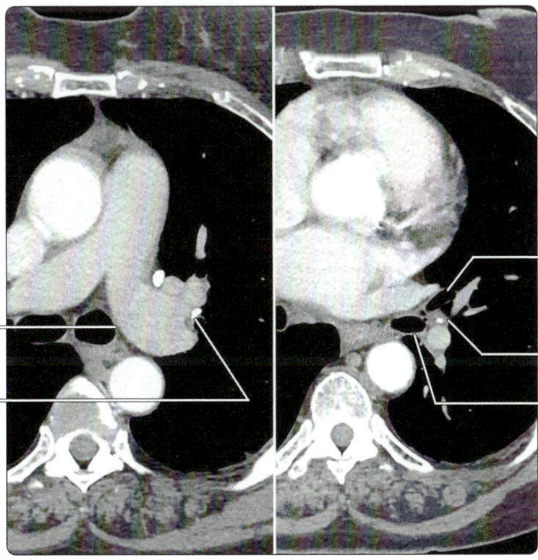

(Top) *First of 2 axial CECT images of the same patient shows the anatomic locations of right hilar and interlobar lymph nodes according to the IASLC lymph node map. Axial CECT (mediastinal window) shows a normal-sized right hilar lymph node designated as 10R based on the IASLC lymph node map.* **(Middle)** *Composite image with axial CECT (mediastinal window) shows the anatomic locations of normal-sized calcified right interlobar lymph nodes designated as 11Rs (left) and 11Ri (right), respectively. Station 11Rs is located between the right upper lobe bronchus and the bronchus intermedius. Station 11Ri is located between the middle and right lower lobe bronchi.* **(Bottom)** *Composite image with axial CECT (mediastinal window) shows the anatomic locations of left hilar (left) and interlobar (right) lymph nodes, designated as 10L and 11L, respectively, according to the IASLC lymph node map.*

HILUM OVERLAY SIGN, RADIOGRAPHY

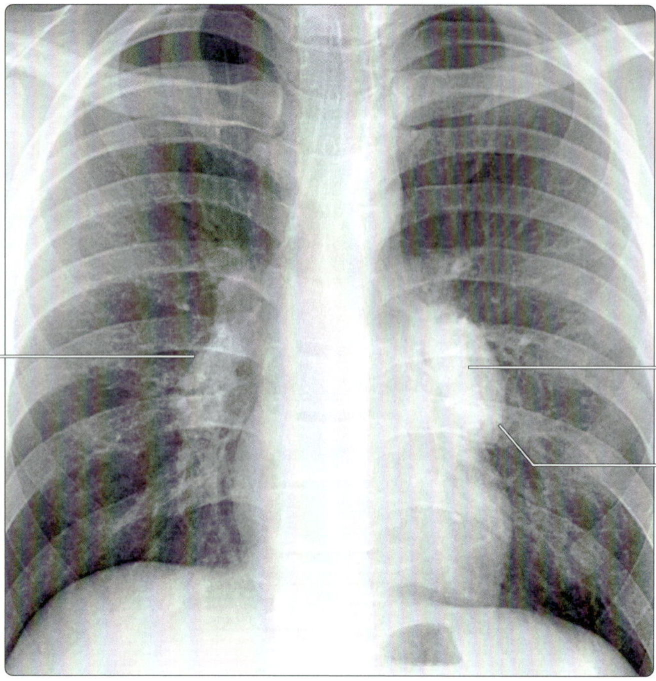

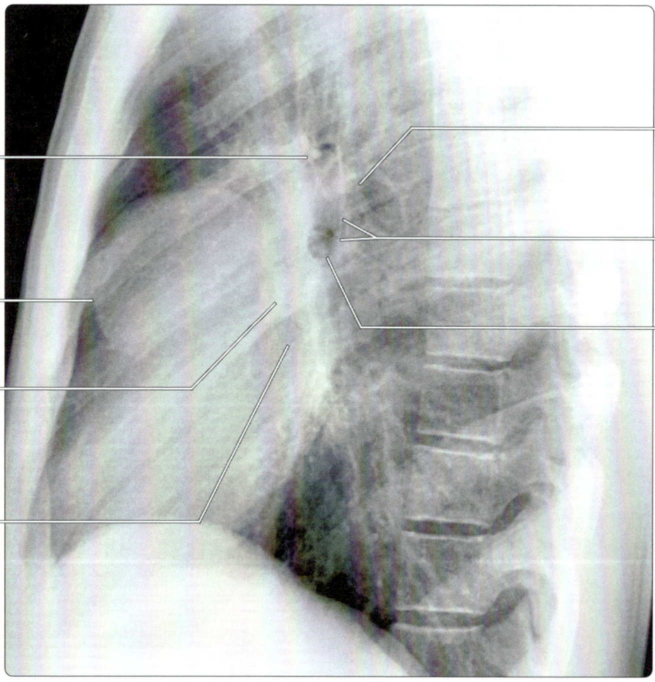

(Top) *First of 2 chest radiographs of a 23-year-old man with mediastinal T-cell lymphoma demonstrates the hilum overlay sign. PA chest radiograph shows an ovoid left mediastinal mass that projects over the left hilum. The left hilum and the left interlobar pulmonary artery are visible through the soft tissue mass, indicating that the mass is separate from the hilum. This finding illustrates the hilum overlay sign, which is classically associated with anterior mediastinal masses.* **(Bottom)** *Lateral chest radiograph of the same patient confirms the anterior mediastinal location of the mass, completely separate from the left hilum. The left pulmonary artery and the bilateral hila are normal. The right hilar vascular opacity is noted anterior to the hilar airways. Other normal hilar landmarks, such as the normal bilateral upper lobe bronchi, the intermediate stem line, and a normal inferior hilar window, are also visible.*

HILUM CONVERGENCE SIGN, RADIOGRAPHY

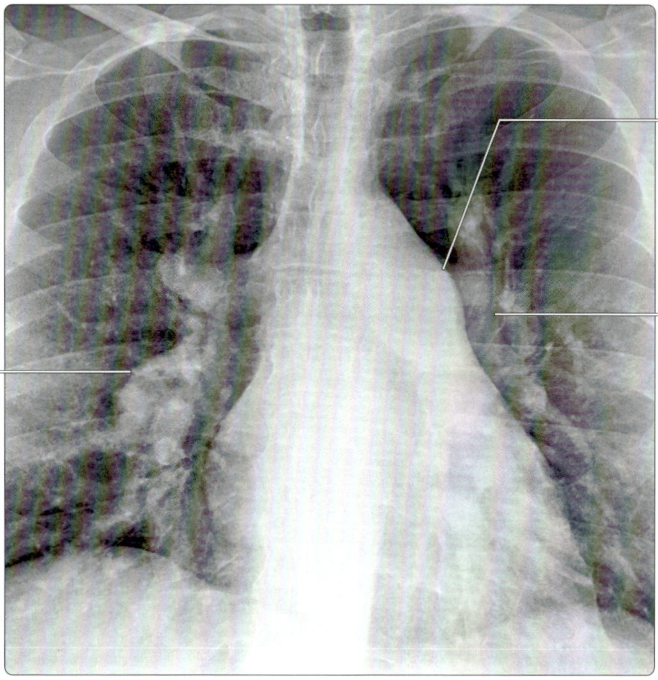

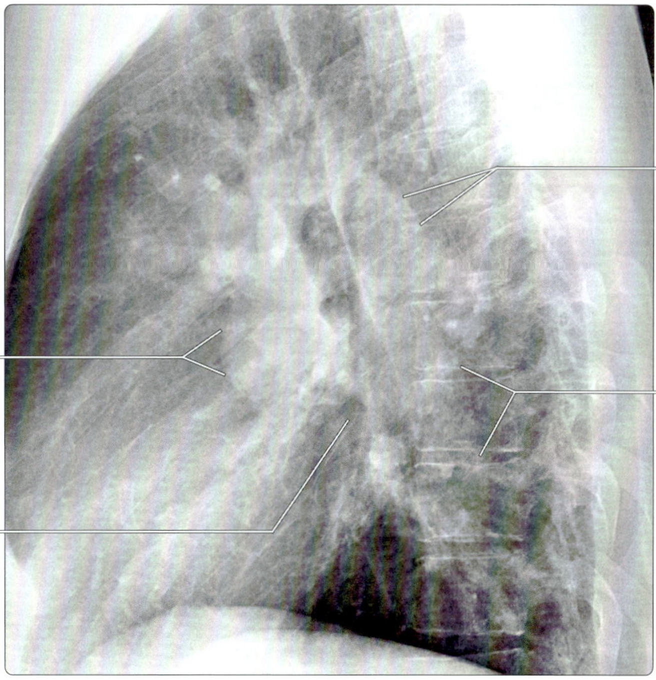

(Top) *First of 2 chest radiographs of a 42-year-old man shows findings consistent with pulmonary hypertension. PA chest radiograph shows bilateral hilar enlargement and enlarged pulmonary vessels coursing directly to the bilateral hila demonstrating the hilum convergence sign. The hilar enlargement is secondary to enlarged central pulmonary arteries. The markedly enlarged pulmonary trunk supports the diagnosis of pulmonary hypertension.* **(Bottom)** *Lateral chest radiograph of the same patient confirms marked bilateral hilar enlargement with a large right hilar vascular opacity and a large left pulmonary artery. Enlarged pulmonary artery branches course directly into the left hilum. Note the normal inferior hilar window.*

INCREASED HILAR DENSITY: CALCIFICATION, RADIOGRAPHY AND CT

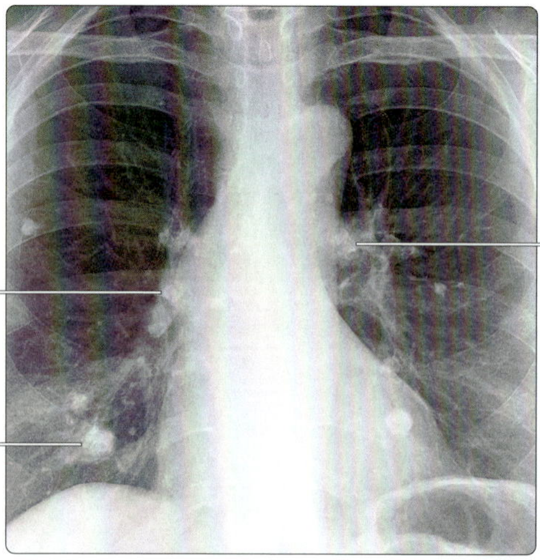

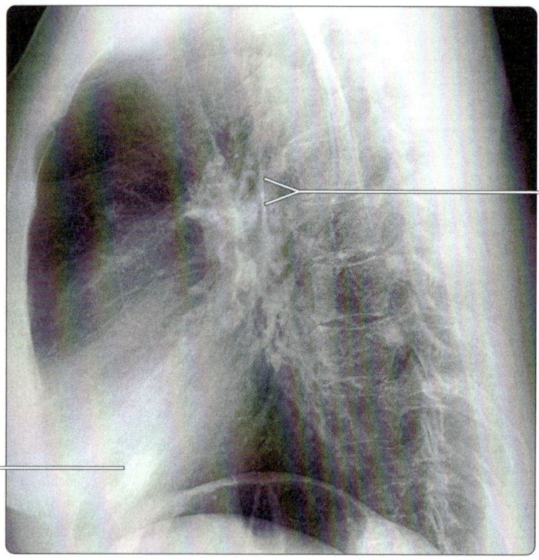

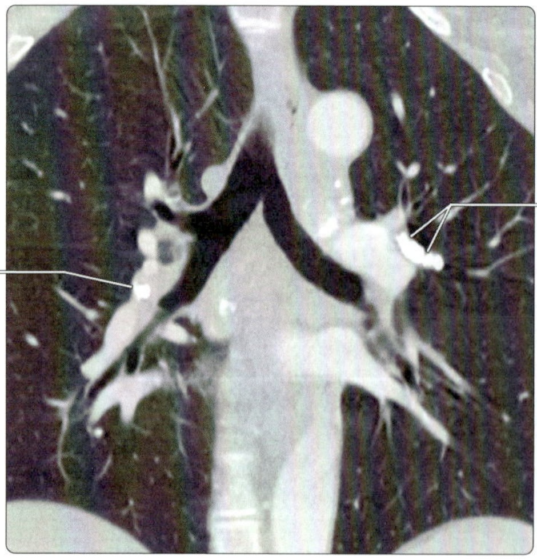

(Top) *PA chest radiograph of an asymptomatic 60-year-old woman with known prior histoplasmosis shows multifocal calcified pulmonary granulomas. The hila are of normal size and are in normal position. Note the bilateral small nodular heterogeneous hilar calcifications, consistent with calcified hilar lymph nodes.* **(Middle)** *Lateral chest radiograph of the same patient shows bilateral calcified nodules superimposed on the hila. However, the hilar anatomy and relationships remain normal, and the intermediate stem line exhibits normal thickness.* **(Bottom)** *Coronal CECT (lung window) of the same patient demonstrates calcifications in hilar and interlobar lymph nodes, consistent with known remote histoplasmosis affecting station 11R and 10L lymph nodes.*

INCREASED HILAR DENSITY: HILAR MASS, RADIOGRAPHY AND CT

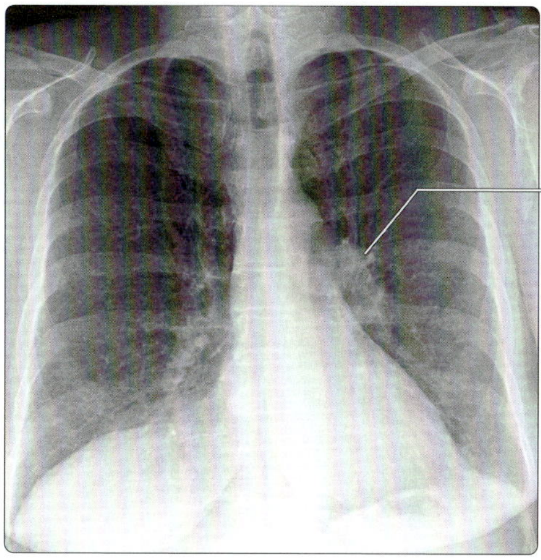

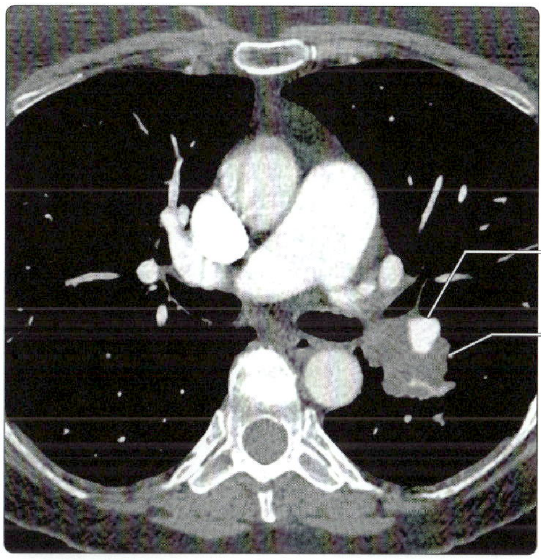

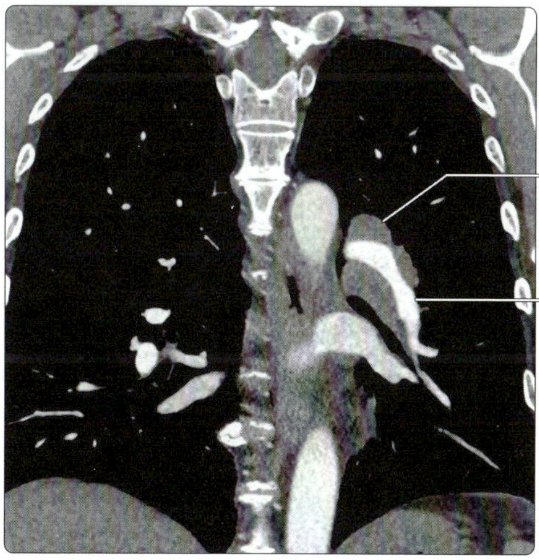

(Top) *PA chest radiograph of a 59-year-old woman smoker who presented with weight loss shows unilateral left hilar enlargement secondary to a large left hilar mass.* **(Middle)** *Axial CECT (mediastinal window) of the same patient shows left hilar lymphadenopathy that partially encases the left interlobar pulmonary artery and corresponds to the radiographic abnormality.* **(Bottom)** *Coronal CECT (mediastinal window) of the same patient shows left hilar lymphadenopathy encasing the distal left pulmonary artery and the left interlobar pulmonary artery. Biopsy confirmed the diagnosis of small cell lung carcinoma. Liver metastases (not shown) were also confirmed at presentation. Primary lung cancer, lymphoma, and metastatic disease may all produce unilateral hilar enlargement due to lymphadenopathy.*

BILATERAL HILAR ENLARGEMENT: PULMONARY ARTERIES, RADIOGRAPHY

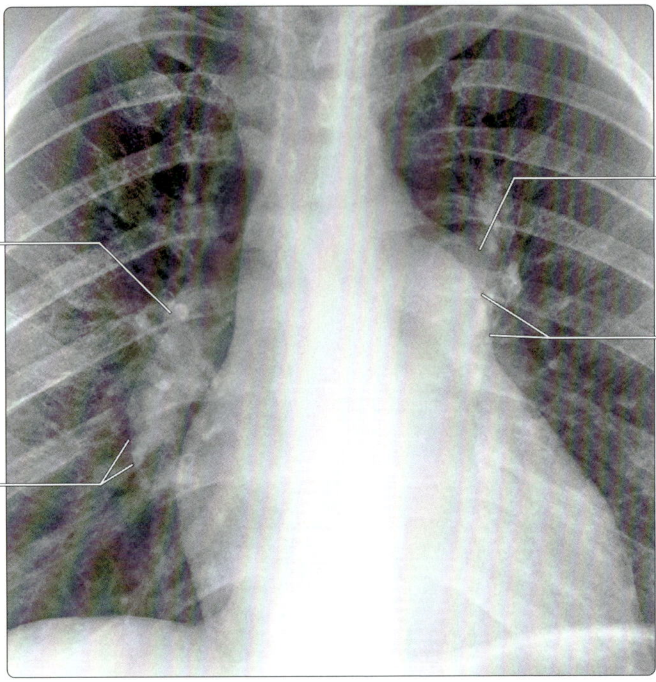

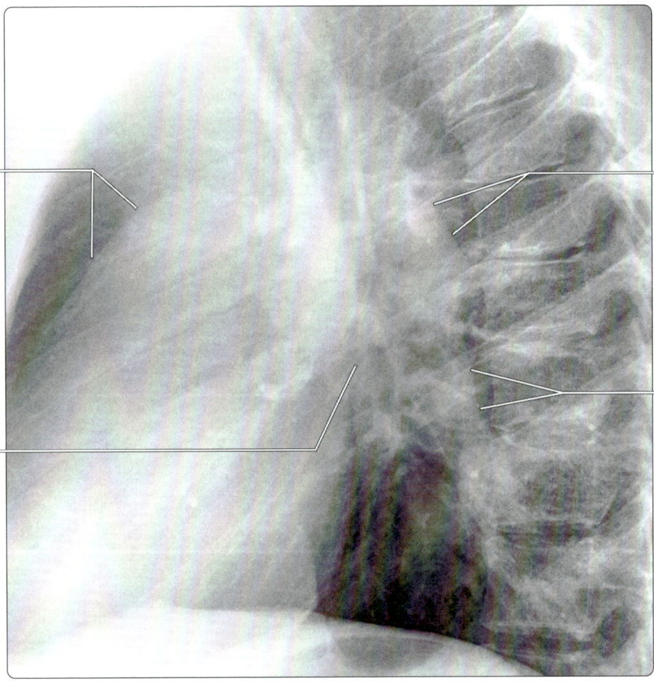

(Top) *PA chest radiograph of a 31-year-old man with known pulmonary hypertension shows bilateral hilar enlargement associated with enlargement of the pulmonary trunk. Large vessels course to the bilateral enlarged hila, consistent with the hilum convergence sign. There are no other abnormalities to suggest lymphadenopathy or mass.* **(Bottom)** *Lateral chest radiograph of the same patient shows marked enlargement of the left pulmonary artery and the pulmonary trunk and a conspicuous right hilar vascular opacity. Note the absence of abnormal soft tissue in the inferior hilar window. All these observations help differentiate lymphadenopathy from vascular disease as the etiology of bilateral hilar enlargement.*

BILATERAL HILAR ENLARGEMENT: LYMPHADENOPATHY, RADIOGRAPHY

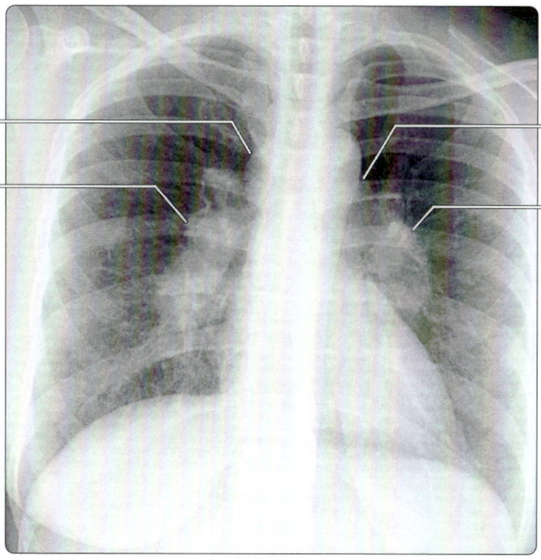

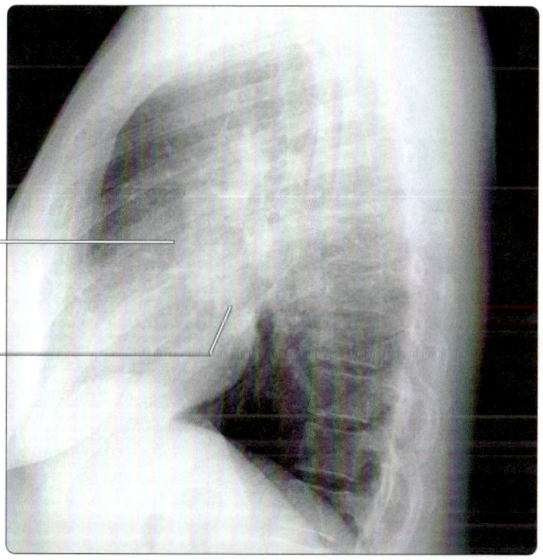

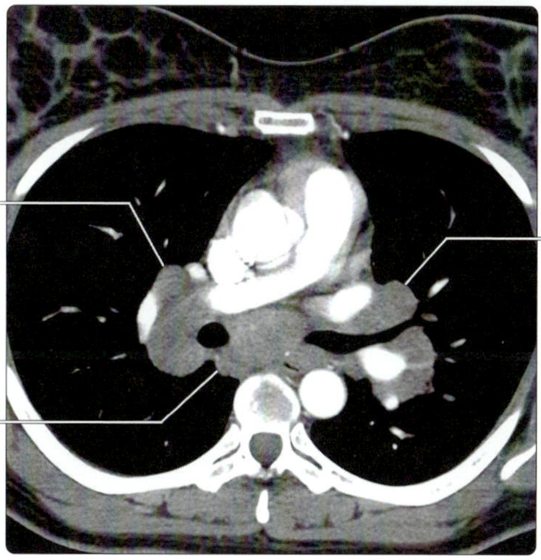

(Top) *PA chest radiograph of a young woman with sarcoidosis shows marked bilateral relatively symmetrical hilar lymphadenopathy. Note associated right paratracheal and aortopulmonary window lymphadenopathy. Although the radiographic appearance is characteristic, sarcoidosis is a diagnosis of exclusion, and other infectious and neoplastic etiologies should always be considered.* **(Middle)** *Lateral chest radiograph of the same patient demonstrates bilateral hilar enlargement due to lymphadenopathy. Note abnormal density filling the inferior hilar window, consistent with subcarinal lymphadenopathy producing the radiographic doughnut sign.* **(Bottom)** *Axial CECT (mediastinal window) of the same patient confirms bilateral hilar lymphadenopathy as well as subcarinal lymphadenopathy, which manifests as abnormal subcarinal soft tissue that produces an abnormal convexity at the azygoesophageal recess. The diagnosis of sarcoidosis was confirmed based on correlation of clinical, imaging, and histologic findings.*

UNILATERAL HILAR ENLARGEMENT: ENLARGED LEFT PULMONARY ARTERY, RADIOGRAPHY

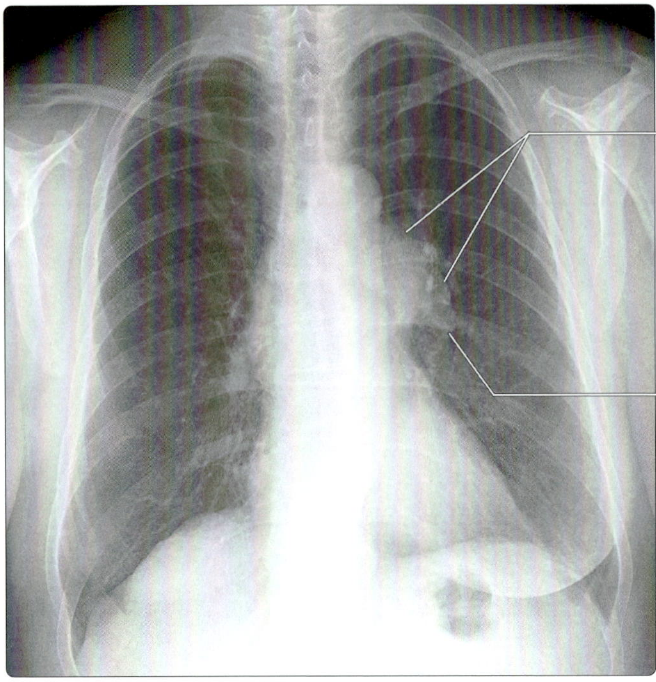

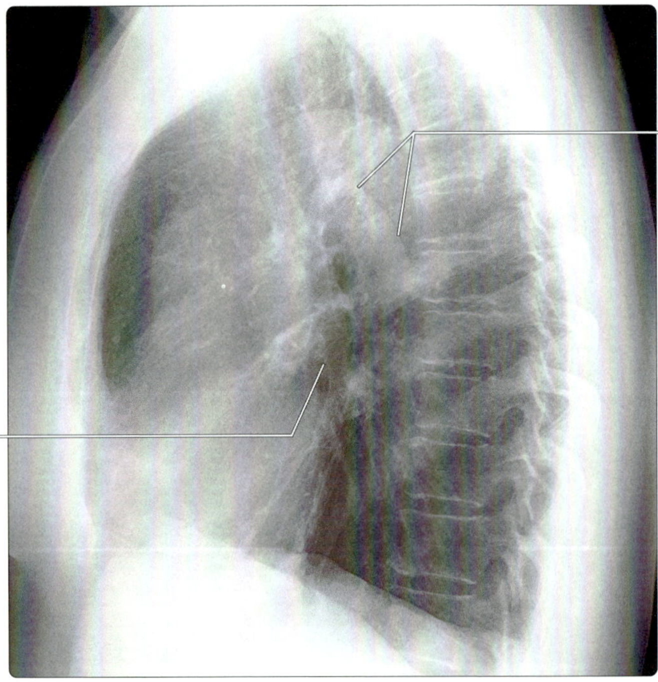

(Top) *PA chest radiograph of a 69-year-old woman with pulmonic stenosis shows unilateral enlargement of the left hilum. Note that enlarged pulmonary artery branches course to the left hilum producing in the so-called hilum convergence sign and suggest a vascular etiology for the radiographic abnormality.* **(Bottom)** *Lateral chest radiograph of the same patient shows an enlarged left pulmonary artery, a normal inferior hilar window, and no other hilar or mediastinal abnormality. The finding of unilateral left pulmonary artery enlargement should suggest the diagnosis of pulmonic stenosis.*

UNILATERAL HILAR ENLARGEMENT: ENLARGED LEFT PULMONARY ARTERY, CT

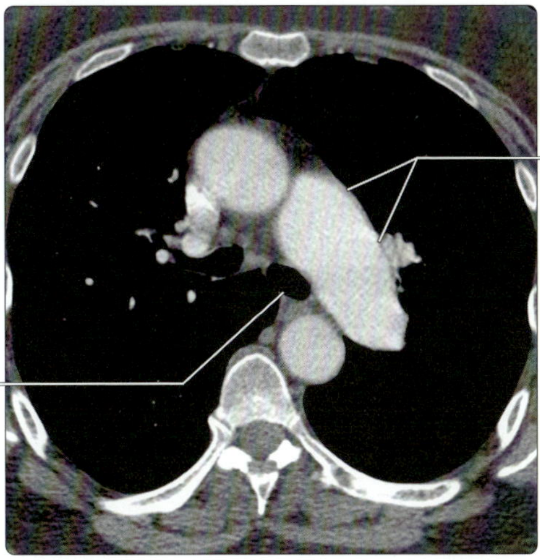

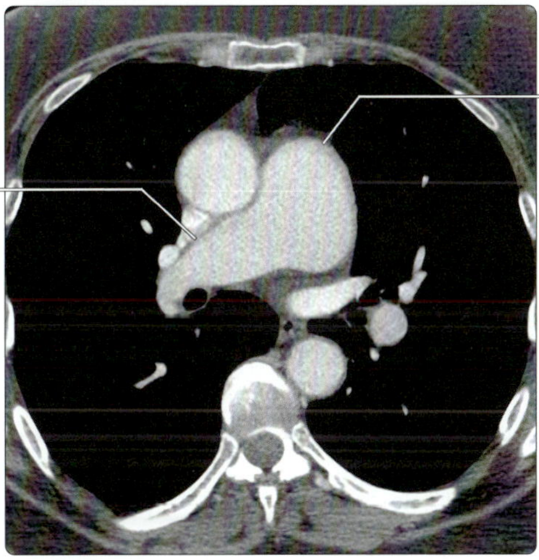

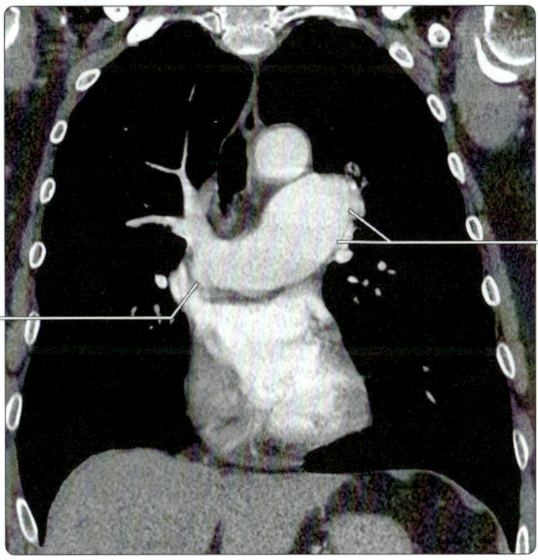

(Top) *Axial CECT (mediastinal window) of the same patient shows marked enlargement of the left pulmonary artery as it crosses over the left mainstem bronchus.* **(Middle)** *Axial CECT (mediastinal window) of the same patient shows marked enlargement of the pulmonary trunk and a normal right pulmonary artery. Unilateral enlargement of the left pulmonary artery associated with enlargement of the pulmonary trunk should suggest the diagnosis of pulmonic stenosis.* **(Bottom)** *Coronal CECT (mediastinal window) of the same patient shows marked enlargement of the left pulmonary artery and normal size of the right pulmonary artery. The diagnosis of pulmonic stenosis was confirmed on echocardiography.*

UNILATERAL HILAR ENLARGEMENT AND THICK INTERMEDIATE STEM LINE, LUNG CANCER, RADIOGRAPHY AND CT

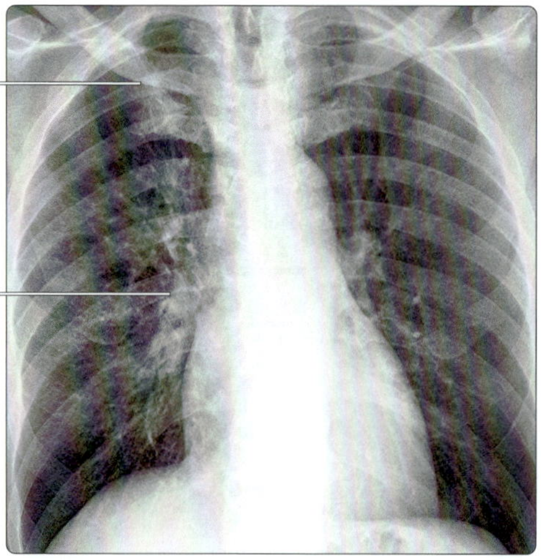

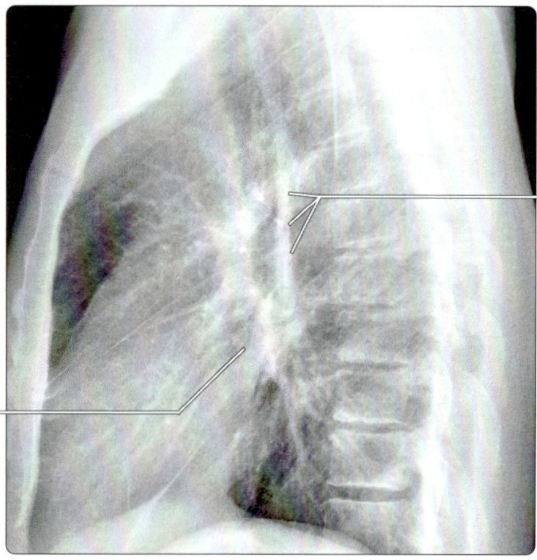

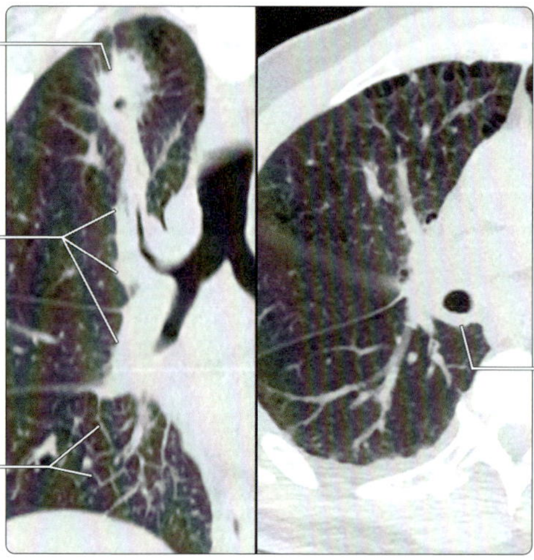

(Top) *PA chest radiograph of a 68-year-old man who presented with cough, dyspnea, and weight loss shows mild volume loss in the right lung, an asymmetric right apical nodular opacity, asymmetric increased density in the right hilum, and increased right lung interstitial markings.* **(Middle)** *Lateral chest radiograph of the same patient demonstrates abnormal thickening of the intermediate stem line and abnormal density in the inferior hilar window. The constellation of findings is highly concerning for advanced lung cancer.* **(Bottom)** *Composite image with coronal (left) and axial (right) NECT (lung window) of the same patient demonstrates a cavitary right upper lobe spiculated nodule and right hilar enlargement secondary to infiltrating soft tissue that encases the central tracheobronchial tree and thickens the posterior wall of the bronchus intermedius. Thick interlobular septa in the right lower lobe are consistent with lymphangitic carcinomatosis in this case of advanced pulmonary adenocarcinoma.*

THICK INTERMEDIATE STEM LINE, PULMONARY EDEMA, RADIOGRAPHY

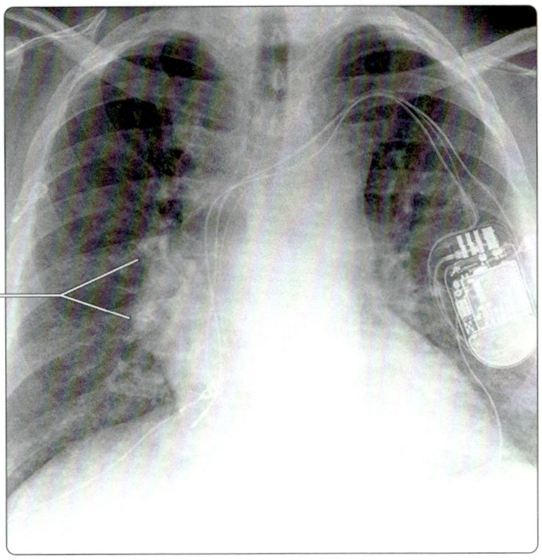

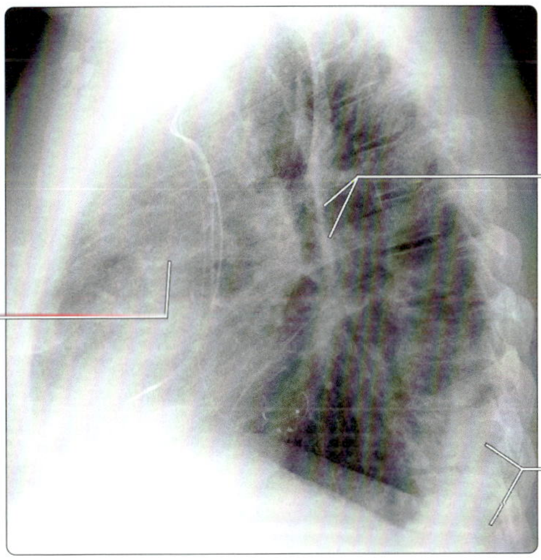

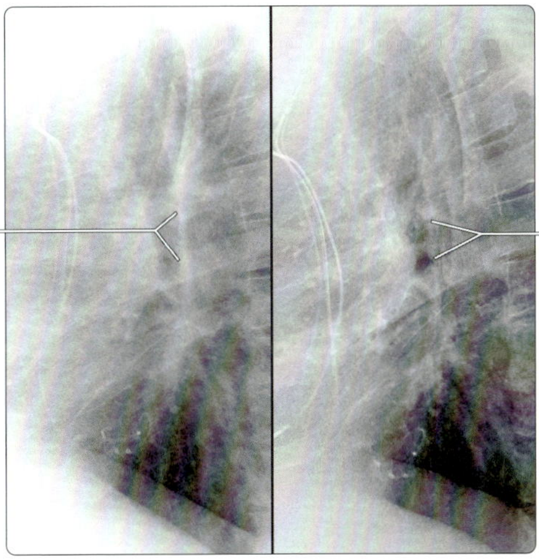

(Top) *PA chest radiograph of a 71-year-old man who presented with dyspnea secondary to interstitial pulmonary edema shows a left pectoral dual-chamber biventricular pacer, cardiomegaly, and bilateral perihilar haze, consistent with the suspected clinical diagnosis of pulmonary edema.* **(Middle)** *Lateral chest radiograph of the same patient shows findings of pulmonary edema with indistinct pulmonary artery margins, thick interlobar fissures, and small bilateral pleural effusions. Note the abnormally thick intermediate stem line.* **(Bottom)** *Composite image with lateral chest radiographs of the same patient with (left) and without (right) pulmonary edema shows that the intermediate stem line is virtually imperceptible in the absence of pulmonary edema. Interstitial edema and resultant extravascular lung water in the axial interstitium are common nonneoplastic causes of thickening of the intermediate stem line.*

NOMENCLATURE

Jackson-Huber Nomenclature (1943)

- Bronchial tree anatomy nomenclature based on lobar and segmental bronchial branches
- Spurs (septa along bronchial orifice edges at bifurcation points): Landmarks of airway origins

System of Boyden (1961)

- Right or left (**R**/**L**) segmental bronchus designation (**B**); followed by sequential number (e.g., B1, B2, etc.)

OVERVIEW OF AIRWAY ANATOMY

Endoscopy Nomenclature

- Differs from Boyden nomenclature
  - Anterior segmental upper lobe bronchi: B2 according to Boyden; B3 on endoscopy
  - Posterior segmental upper lobe bronchi: B3 according to Boyden; B2 on endoscopy

Trachea

- Divides into right and left mainstem bronchi

Mainstem Bronchi

- Right mainstem bronchus
- Left mainstem bronchus

Lobar and Segmental Bronchi

- Right upper lobe bronchus
  - Apical segmental bronchus (B1)
  - Posterior segmental bronchus (B2)
  - Anterior segmental bronchus (B3)
- Middle lobe bronchus
  - Lateral segmental bronchus (B4)
  - Medial segmental bronchus (B5)
- Right lower lobe bronchus
  - Superior segmental bronchus (B6)
  - Truncus basalis gives origin to basilar segmental bronchi
    - Medial basilar segmental bronchus (B7)
    - Anterior basilar segmental bronchus (B8)
    - Lateral basilar segmental bronchus (B9)
    - Posterior basilar segmental bronchus (B10)
- Left upper lobe bronchus
  - Apical-posterior (apicoposterior) segmental bronchus (B1+2)
  - Anterior segmental bronchus (B3)
  - Superior lingular segmental bronchus (B4)
  - Inferior lingular segmental bronchus (B5)
- Left lower lobe bronchus
  - Superior segmental bronchus (B6)
  - Truncus basalis gives origin to basilar segmental bronchi
    - Anteromedial basilar segmental bronchus (B7+8)
    - Lateral basilar segmental bronchus (B9)
    - Posterior basilar segmental bronchus (B10)

TRACHEAL ANATOMY

General Anatomy

- Midline structure, 10-12 cm long; intrathoracic portion 6-9 cm long
- Smooth left mural indentation by aortic arch and mild rightward deviation
- Extends caudally from inferior aspect of cricoid cartilage (level of C6 vertebral body)
- Bifurcates at carina (level of T5 vertebral body) into right and left mainstem bronchi
- Normal tracheal diameters
  - Coronal: 13-25 mm (men); 10-21 mm (women)
  - Sagittal: 13-27 mm (men); 10-23 mm (women)

Relationships

- Thyroid gland abuts anterolateral upper trachea above thoracic inlet
- Esophagus lies posterior to trachea, between trachea and thoracic vertebrae
- Aorta lies along anterior and left lateral trachea
- Brachiocephalic artery lies along anterior and right lateral trachea
- Right upper lobe adjacent to right lateral tracheal wall

RIGHT BRONCHIAL ANATOMY

Right Mainstem Bronchus

- Origin anterior to esophagus; inferolateral course; posterior to right pulmonary artery
- Relatively short; more vertical than left mainstem bronchus; more prone to foreign body aspiration
- Divides into right upper lobe bronchus and bronchus intermedius

Right Upper Lobe Bronchus

- Origin from lateral right mainstem bronchus at or just below carina; cephalad to left upper lobe bronchus
- **Eparterial** (i.e., "situated above artery"); refers to relationship with adjacent right pulmonary artery
- Courses horizontally and laterally (1-2 cm) before branching
- Posterior wall is important anatomic landmark; in direct contact with aerated lung (< 3 mm thick)
  - Thickened by edema, neoplasm, lymphadenopathy
  - Posteromedial azygos vein may mimic thickening
- Right upper lobe bronchial segments
  - Apical segmental bronchus
    - 1st branch with cephalad (vertical) course
    - Circular lucency on axial cross section, arises from distal upper aspect of right upper lobe bronchus
  - Posterior and anterior segmental bronchi
    - Typically horizontal course, parallel to axial plane
    - Anterior segmental bronchus courses anteriorly
    - Posterior segmental bronchus courses posteriorly and cephalad

Bronchus Intermedius

- Origin immediately below right upper lobe bronchus
- Oblique course (3-4 cm); directly posterior to right pulmonary artery
- Posterior wall abuts aerated lung; normally thin, uniform thickness (< 3 mm)
  - Thickening: Edema, lymphadenopathy, tumor infiltration
- Branches into middle lobe and right lower lobe bronchi
- Obstruction: Combined middle and right lower lobe volume loss/pneumonia

Middle Lobe Bronchus

- Origin from anterolateral bronchus intermedius; at or near right lower lobe bronchus origin; origins separated by spur
- Courses anterolaterally, caudally, and obliquely
- Middle lobe segmental bronchi
  - Lateral segmental bronchus
    - More horizontal course than medial segmental bronchus; visible over greater length
  - Medial segmental bronchus
    - Equal size of medial and lateral segmental bronchi in 50% of cases; medial segmental bronchus larger than lateral segmental bronchus in most others

Right Lower Lobe Bronchus

- Superior segmental bronchus: Posterior origin from short proximal portion of right lower lobe bronchus
- Right lower lobe bronchus continues for 5-10 mm as truncus basalis and branches into 4 basilar segmental bronchi
  - Medial, anterior, lateral, and posterior basilar segmental bronchi arise (on axial CT) in counterclockwise order
  - Identified by relative position to each other and course toward respective lobar segments

LEFT BRONCHIAL ANATOMY

Left Mainstem Bronchus

- Origin anterior to esophagus; inferolateral course
- Longer, more horizontal than right mainstem bronchus; less prone to aspiration
- Divides into left upper lobe and left lower lobe bronchi

Left Upper Lobe Bronchus

- Origin from left mainstem bronchus; bifurcates or trifurcates
- **Hyparterial** (i.e., "situated below artery"); refers to relationship with adjacent left pulmonary artery
- Typically branches into superior and lingular divisions
- Left upper lobe bronchial segments
  - Superior portion divides into apicoposterior and anterior segmental bronchi
  - Inferior (lingular) portion courses obliquely, inferiorly and anterolaterally; analogous to middle lobe bronchus
    - Divides into superior and inferior lingular segmental bronchi

Left Lower Lobe Bronchus

- Superior segmental bronchus: Posterior origin from short proximal portion of left lower lobe bronchus; at or near level of lingular bronchus
- Left lower lobe bronchus continues as truncus basalis and branches into 3 basilar segmental bronchi
  - Anteromedial, lateral, and posterior basilar segmental bronchi arise (on axial CT) in clockwise order
  - Identified by relative position to each other and course toward respective lobar segments

VARIANTS OF BRONCHIAL ANATOMY

General Concepts

- Numerous branching anomalies described; may involve entire lung or specific lobes
- Typically asymptomatic and diagnosed incidentally
- Symptoms usually related to associated congenital anomalies, bronchial obstruction/stenosis, recurrent atelectasis/infection
- Preoperative recognition may be important for endotracheal intubation/surgical planning
- Displaced vs. supernumerary
  - Displaced: Anomalous bronchus + absence of normal corresponding bronchus that supplies affected lung
  - Supernumerary: Anomalous bronchus + coexisting normal bronchus

Tracheal Bronchus

- Anomalous bronchus origin directly from lateral tracheal wall
  - Term formerly included upper lobe (lobar or segmental) bronchi arising from trachea or mainstem bronchi
- Most frequently right sided (0.1-1.3% of adults); typically displaced
  - May be left sided or bilateral
- "Pig" bronchus: Upper lobe bronchus displaced to trachea (term "pig" due to corresponding normal anatomy in swine)

Right Preeparterial Bronchus

- Anomalous bronchus directed to right upper lobe with origin from right mainstem bronchus above level of right upper lobe bronchus
- Nearly twice as frequent as right tracheal bronchus
- Most reported cases are displaced apical segmental bronchi

Right Posteparterial Bronchus

- Anomalous bronchus directed to right upper lobe with origin below level of right upper lobe bronchus

Left Eparterial Bronchus

- Anomalous bronchus directed to left upper lobe with origin from left mainstem bronchus above level where left pulmonary artery crosses over left mainstem bronchus
- Most common left-sided bronchial anomaly
- Most reported cases are displaced bronchi

Left Prehyparterial Bronchus

- Anomalous bronchus directed to left upper lobe with origin from left mainstem bronchus between level where left pulmonary artery crosses over left mainstem bronchus and origin of left upper lobe bronchus

Left Posthyparterial Bronchus

- Anomalous bronchus directed toward left upper lobe with origin below level of left upper lobe bronchus

Accessory Cardiac Bronchus

- Supernumerary bronchus; up to 0.5% prevalence in adults
- Origin from medial right mainstem bronchus or bronchus intermedius
- Caudal course toward mediastinum and heart (hence "cardiac" designation); typically blind ending but may supply rudimentary lung tissue/lobule

Bronchial Situs

- Predictor of visceroatrial situs (position/morphology of atria, lungs, liver, stomach, spleen relative to midline)
- Situs solitus: Usual arrangement of bronchi, atria, viscera
  - Right upper lobe eparterial bronchus
  - Left upper lobe hyparterial bronchus

- Situs inversus: Mirror image of situs solitus
  - Right upper lobe hyparterial bronchus
  - Left upper lobe eparterial bronchus
  - Association with Kartagener syndrome: Situs inversus, chronic sinusitis, bronchiectasis
- Right isomerism: Bilateral right sidedness
  - Bilateral eparterial bronchi, bilateral trilobed lungs
  - Bilateral morphologic right atria, asplenia, midline liver, variable gastric location
  - High association with congenital heart disease
- Left isomerism: Bilateral left sidedness
  - Bilateral hyparterial bronchi, bilateral bilobed lungs
  - Bilateral morphologic left atria, polysplenia, midline liver, variable gastric location
  - Less frequent association with congenital heart disease

IMAGING OF AIRWAYS

Radiography

- Tracheal air column visible on PA and lateral chest radiography; smooth, parallel luminal borders
- Right paratracheal stripe
  - Thin (1- to 4-mm) stripe corresponding to right lateral tracheal wall on PA chest radiography
  - Extends from clavicle to azygos arch at right tracheobronchial angle
  - Visible in 2/3 of normal subjects
  - Thickening suggests lymphadenopathy or mass
- Bronchoarterial (BA) pairs may be visible in perihilar region
- Normal bronchi not visible in mid to peripheral lung

CT

- Optimal visualization/assessment of airways
- BA pairs: Bronchi paired with respective pulmonary arteries
  - Bronchi and accompanying pulmonary arteries ~ equal in diameter
  - BA ratio: Internal luminal bronchial diameter divided by outer diameter of adjacent pulmonary artery
    - Normal BA ratio: 0.65-0.70
- Expiratory air-trapping in small airways disease
- Virtual bronchoscopy simulates visual airway assessment through bronchoscope
  - 3D internal surface rendering of spiral CT data

ANATOMY-RELATED IMAGING ABNORMALITIES

Foreign Body Aspiration

- Foreign bodies typically lodge in right or left mainstem bronchi; less commonly in trachea or lobar bronchi
- Predilection for aspiration into right mainstem bronchus due to its more vertical course

Saber Sheath Trachea

- Deformity limited to intrathoracic trachea
- Coronal diameter ≤ 2/3 sagittal diameter at same level
- Typically men > 50 years of age; association with chronic obstructive pulmonary disease
  - Usually of no clinical significance

Tracheal Narrowing/Stenosis

- Reduction of normal diameter by > 10%
- Etiologies
  - Trauma, intubation, tracheostomy
  - Extrinsic compression: Thyromegaly, mass, vascular abnormalities
  - Infection
  - Neoplasia
  - Systemic disorders: Granulomatosis with polyangiitis, relapsing polychondritis, amyloidosis

Tracheomalacia

- Abnormal compliance of tracheal wall and supporting cartilage; usually apparent during forced expiratory imaging

Endoluminal Tumor

- Tracheal neoplasms: Squamous cell carcinoma, adenoid cystic carcinoma, mucoepidermoid carcinoma, papillomatosis
- Bronchial neoplasms: Lung cancer, carcinoid (frequent bronchial relationship), metastasis, hamartoma, papillomatosis
- Associated abnormalities
  - Airway stenosis/obstruction leading to atelectasis, pneumonia, mucoid impaction, abscess formation
    - Persistent atelectasis/pneumonia: Assess feeding bronchi with CT for exclusion of airway neoplasm
    - Combined (atelectasis/pneumonia) of middle and right lower lobes suggests involvement of bronchus intermedius

Mounier-Kuhn (Tracheobronchomegaly)

- Rare; unknown etiology
- Dilated trachea and central bronchi; corrugated contours from mucosal prolapse between cartilaginous rings

Bronchiectasis

- Chronic, irreversible bronchial dilatation, usually due to inflammation (transient airway dilatation described in pneumonia and atelectasis)
- BA ratio > 1
- Reid classification based on gross, bronchographic, and CT/HRCT appearances
  - Cylindrical (mild)
    - Parallel walls; radiographic tram-track sign
    - Lack of normal tapering; signet-ring sign on CT
  - Varicose (moderate)
    - Irregular luminal diameter; alternating luminal constriction and dilatation
    - Beaded bronchial morphology
  - Cystic/saccular (severe)
    - Balloon-like; > 1 cm in diameter; reduced number of bronchial divisions
- Traction bronchiectasis from retractile interstitial fibrosis; corkscrew morphology
- Imaging
  - Optimal assessment with CT/HRCT
  - Demonstration of bronchial dilatation ± bronchial wall thickening
  - Identification of mucoid impaction of bronchiectatic airways: Nodular/tubular &/or branching endoluminal opacities

OVERVIEW OF AIRWAY ANATOMY

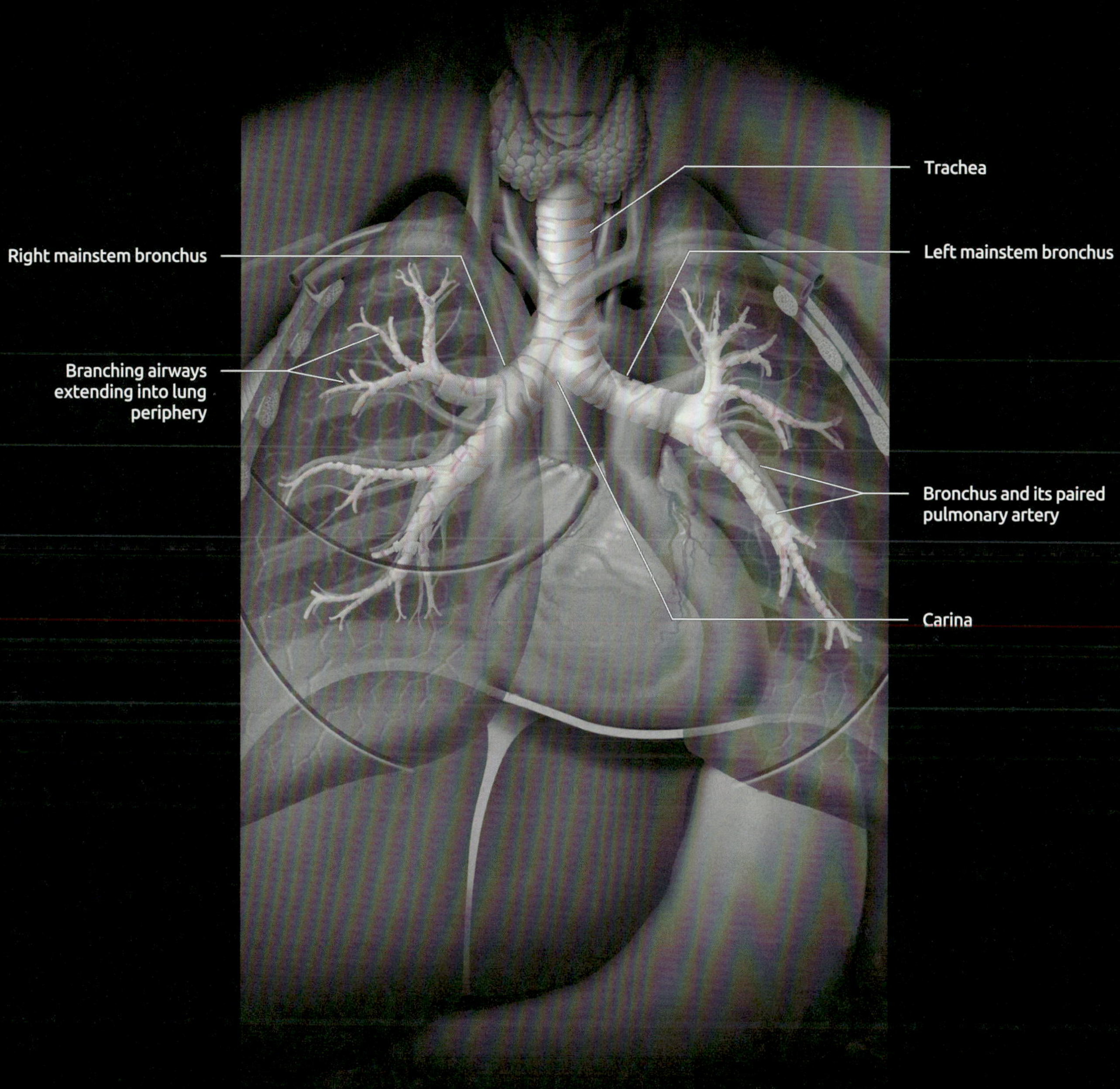

Overview of the airways shows the tracheobronchial tree extending bilaterally from the tracheal bifurcation at the carina. Lobar and segmental bronchi emanate from the right and left mainstem bronchi as they branch, taper, and extend into the lung periphery. Each airway beyond the pulmonary hila is accompanied by its paired, homologous pulmonary artery.

ANTERIOR VIEW, SEGMENTAL BRONCHI

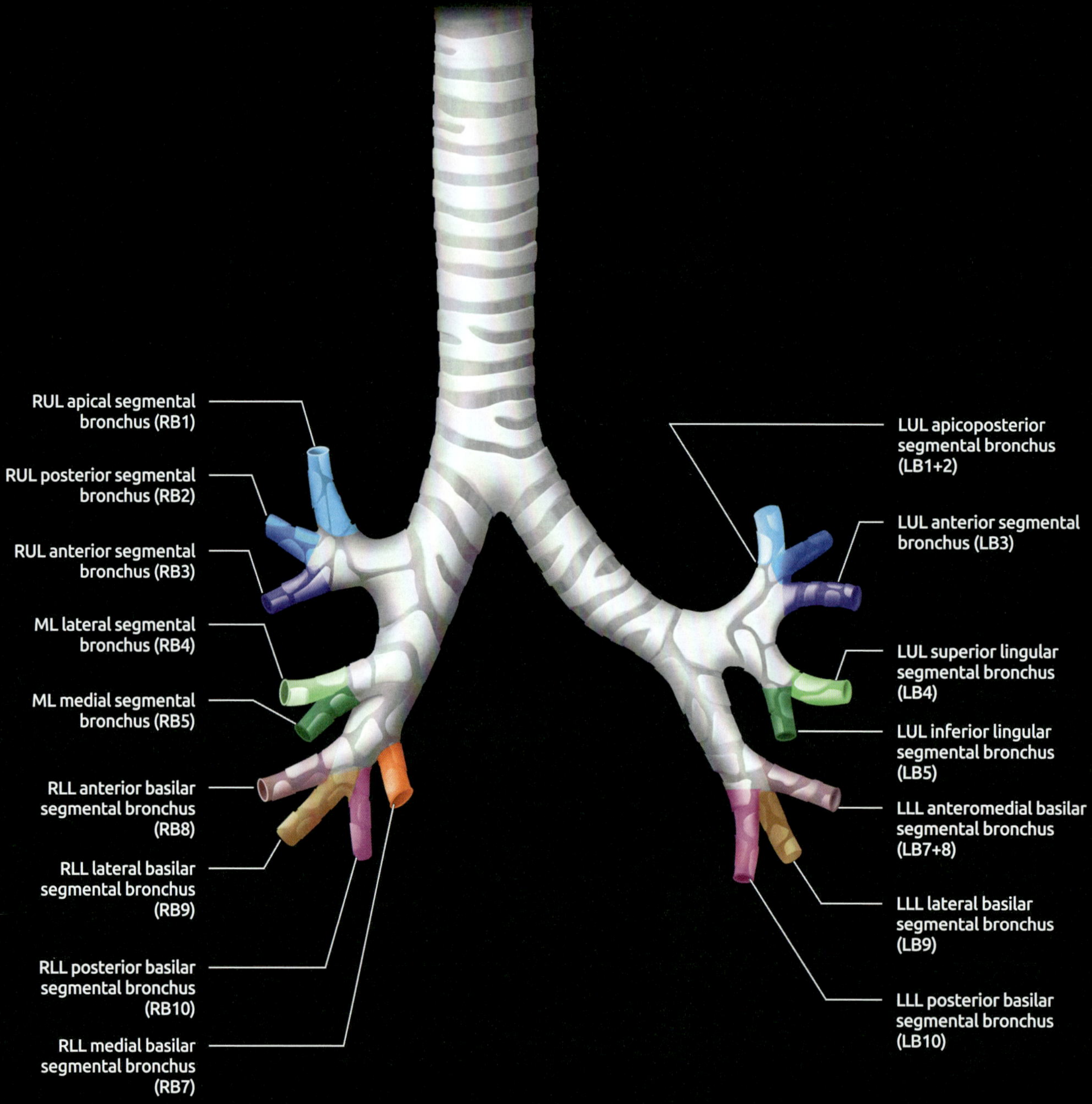

Graphic shows an anterior view of the tracheobronchial tree and depicts the color-coded segmental bronchi that correspond to color-coded lung segments. The segmental bronchi are labeled using the endoscopic adaptation of the Boyden nomenclature. Right upper lobe (RUL), middle lobe (ML), right lower lobe (RLL), left upper lobe (LUL), and left lower lobe (LLL) abbreviations are used to designate the right and left lobar bronchi.

POSTERIOR VIEW, SEGMENTAL BRONCHI

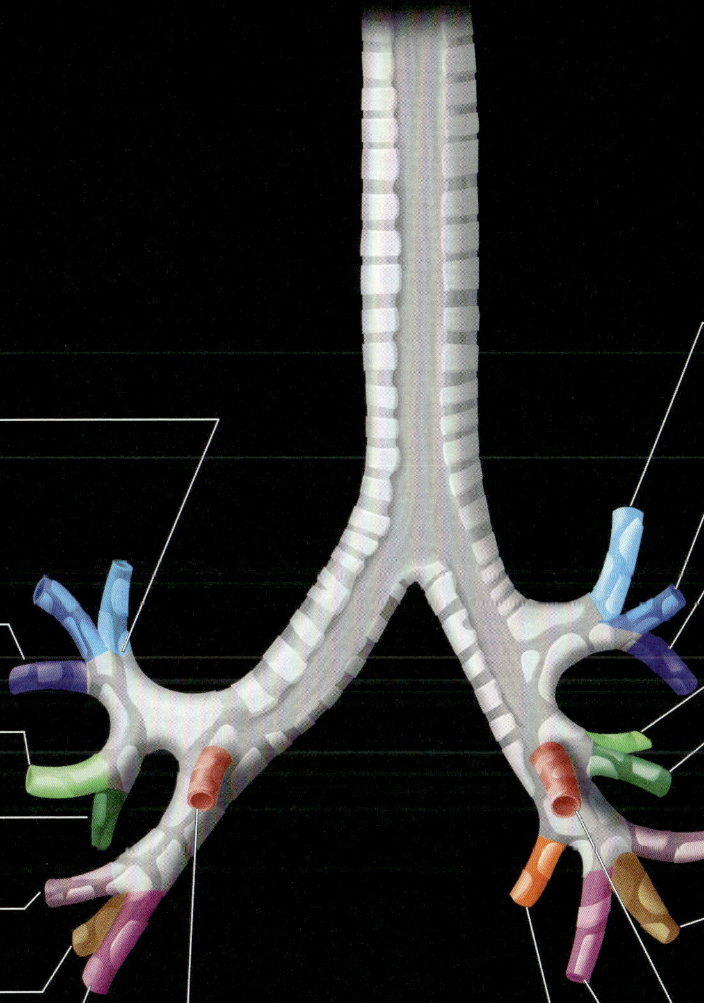

Graphic shows a posterior view of the tracheobronchial tree and depicts the color-coded segmental bronchi that correspond to color-coded lung segments. The segmental bronchi are labeled using the endoscopic adaptation of the Boyden nomenclature. Right upper lobe (RUL), middle lobe (ML), right lower lobe (RLL), left upper lobe (LUL), and left lower lobe (LLL) abbreviations are used to designate the right and left lobar bronchi.

CENTRAL AIRWAYS, RADIOGRAPHY

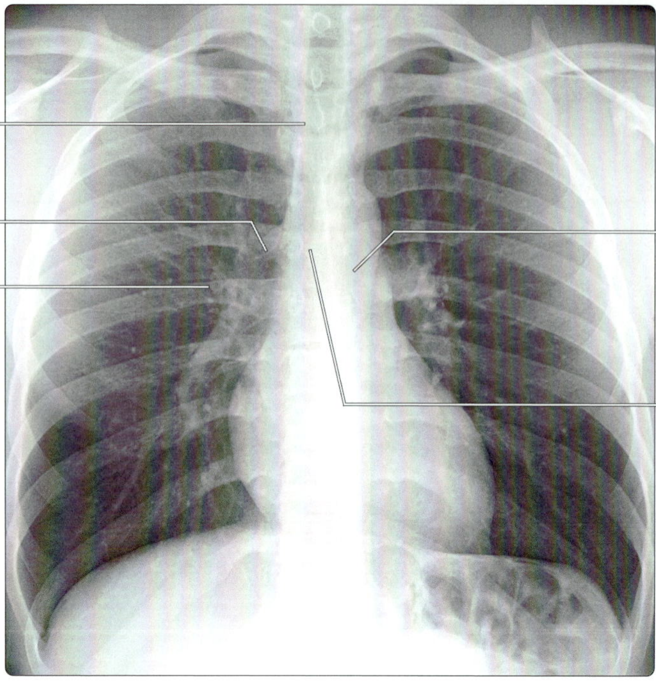

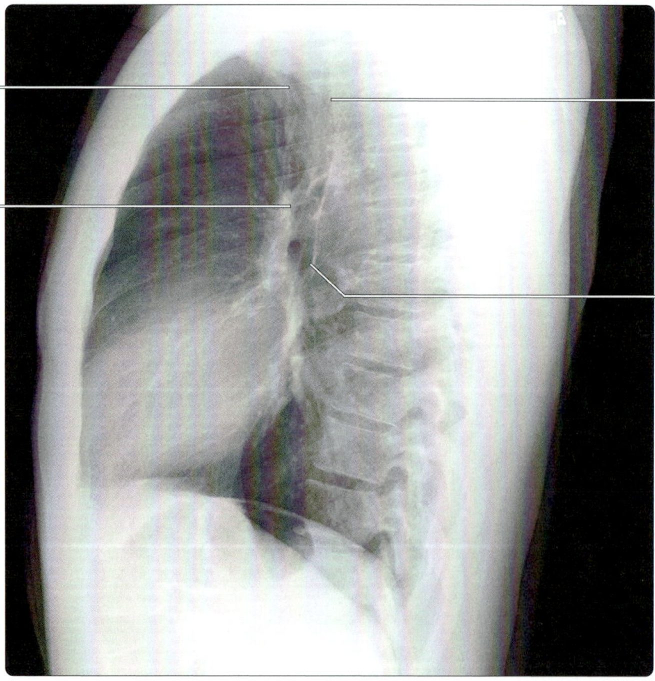

(Top) *Normal PA chest radiograph demonstrates the midline tracheal air column. The trachea bifurcates at the carina into the right and left mainstem bronchi. The anterior segmental bronchus of the right upper lobe is frequently visible end on on PA chest radiographs and manifests as a thin-walled lucency located near the right hilum. Small peripheral airways are not normally visible on radiography.* **(Bottom)** *Normal lateral chest radiograph of the same patient demonstrates the tracheal air column and the characteristic ovoid lucencies formed by the right upper lobe bronchus and the left upper lobe bronchus seen end on. The right upper lobe bronchus is located cephalad to the left upper lobe bronchus.*

CENTRAL AIRWAYS, AXIAL AND CORONAL CT

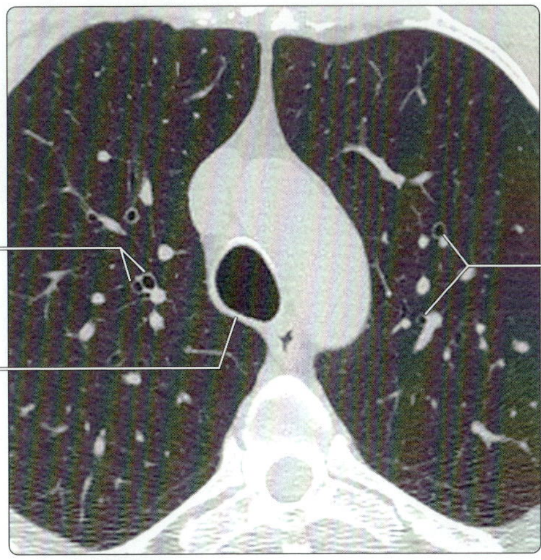

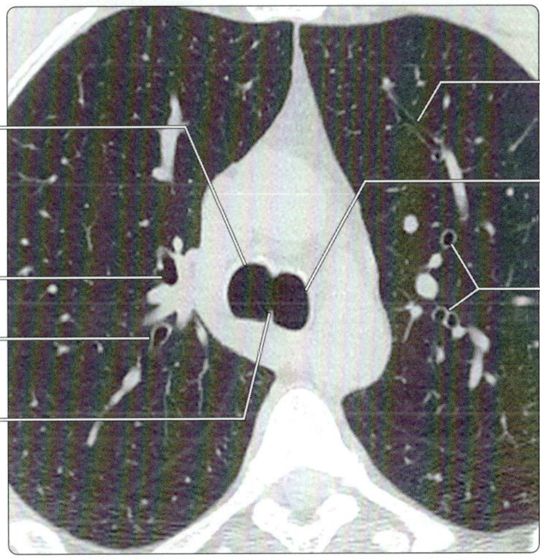

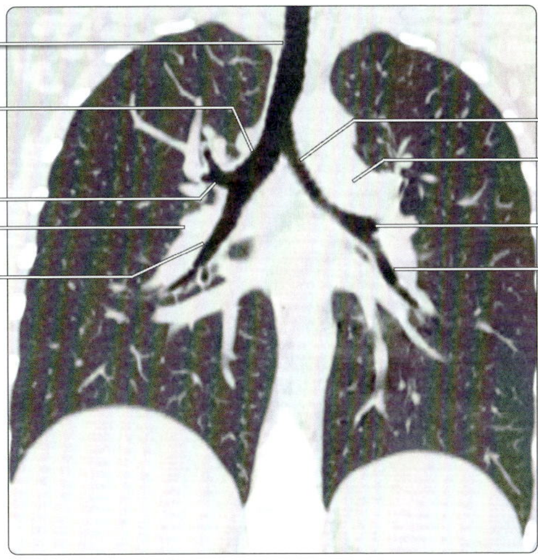

(Top) *Coned-down axial NECT (lung window) shows the trachea, a midline mediastinal air-filled structure with well-defined luminal margins. The posterior membranous tracheal wall bows posteriorly on scans obtained during full-suspended inspiration resulting in a nearly circular tracheal morphology. Subsegmental branches of the right upper lobe apical segmental bronchus and the left upper lobe apicoposterior segmental bronchus are also visible in cross section.* **(Middle)** *Coned-down axial NECT (lung window) of the same patient demonstrates the midline carina where the trachea bifurcates into right and left mainstem bronchi. Segmental branches of the right upper lobe bronchus and subsegmental branches of the left upper lobe bronchus are visible at this level.* **(Bottom)** *Coronal NECT (lung window) shows the relatively vertical course of the right mainstem bronchus compared to the more oblique course of the left mainstem bronchus. The right upper lobe bronchus is eparterial (above ipsilateral pulmonary artery). The left upper lobe bronchus is hyparterial (below ipsilateral pulmonary artery). The truncus basalis bronchi give rise to the lower lobe basilar segmental bronchi.*

BRONCHOARTERIAL RATIO

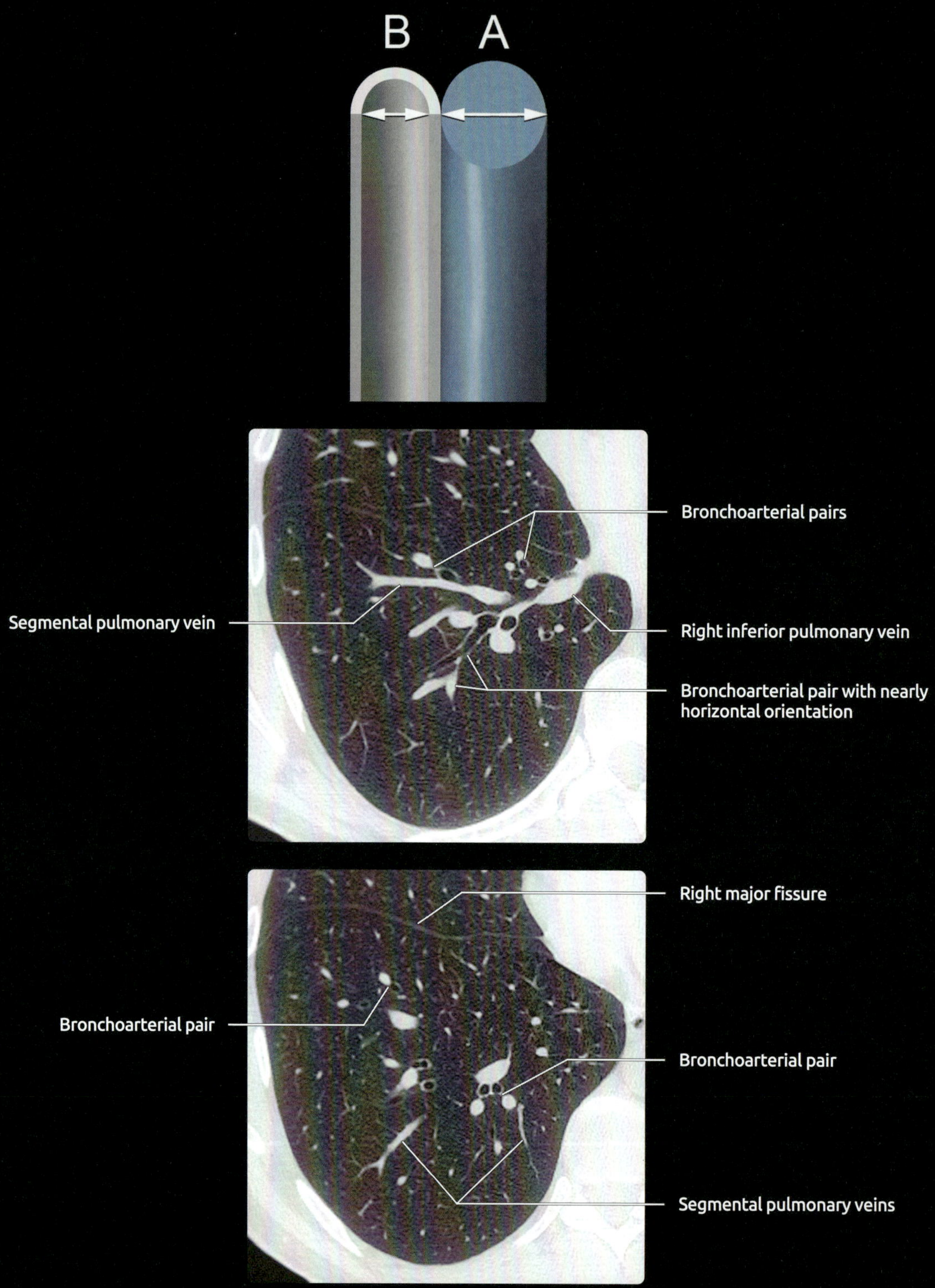

(Top) *Graphic demonstrates the normal bronchoarterial ratio (BA), determined by dividing the inner luminal diameter of an airway by the outer diameter of its accompanying (homologous) pulmonary artery. The BA ratio in normal individuals ranges from 0.65-0.70.* **(Middle)** *Axial HRCT (lung window) coned down to the right lower lobe demonstrates several normal bronchoarterial pairs, at least 1 of which exhibits a horizontal orientation. Note the relatively horizontal course of a right lower lobe segmental pulmonary vein as it courses toward the right inferior pulmonary vein.* **(Bottom)** *Axial HRCT (lung window) coned down to the right lower lobe demonstrates the normal oblique vertical course of lower lobe bronchoarterial pairs seen in cross section, each composed of a bronchus and its adjacent (homologous) pulmonary artery. Note the relatively horizontal orientation of pulmonary vein tributaries as they course toward the left atrium.*

SEGMENTAL ANATOMY, RIGHT LUNG

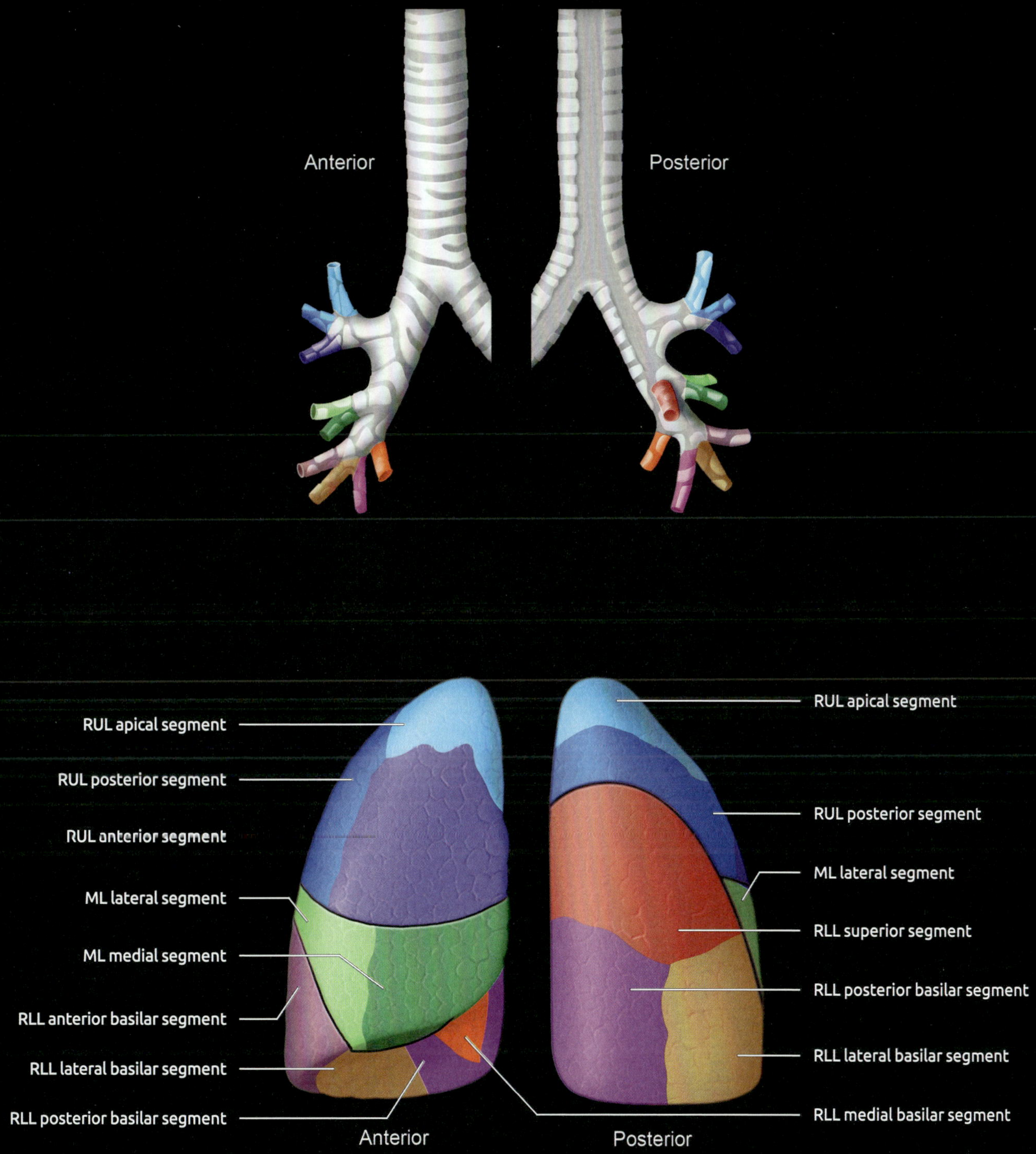

(Top) *Graphic shows anterior (left) and posterior (right) views of the trachea and the right tracheobronchial tree and depicts color-coded segmental bronchial origins that correspond to color-coded lung segments.* **(Bottom)** *Graphic shows anterior (left) and posterior (right) views of the right lung and shows the lung segments of each lobe color coded to correspond to the segmental bronchial origins depicted in the previous graphic.*

SEGMENTAL ANATOMY, LEFT LUNG

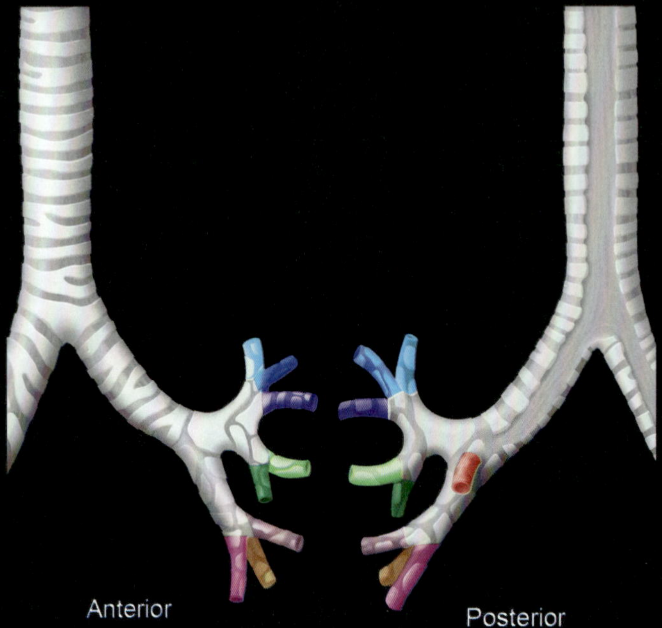

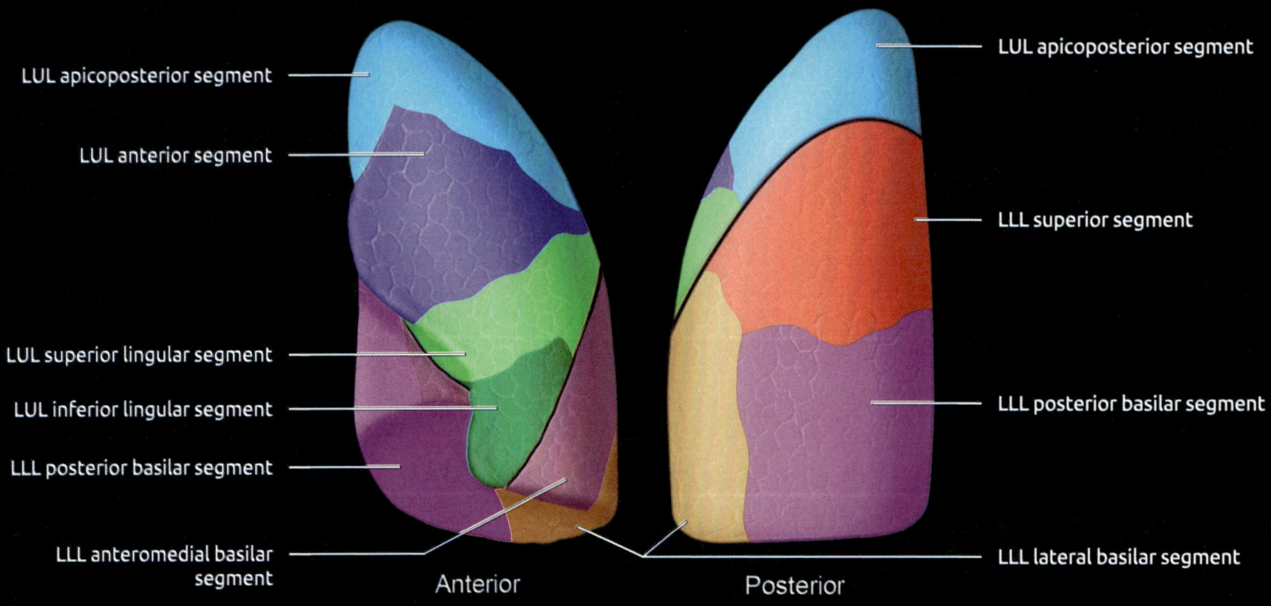

(Top) *Graphic depicts anterior (left) and posterior (right) views of the trachea and the left tracheobronchial tree. Segmental airway origins are color coded to correspond to lung segments.* **(Bottom)** *Graphic shows anterior (left) and posterior (right) views of the left lung with the lung segments of each lobe color coded to correspond to the segmental bronchial origins depicted in the previous graphic.*

LOBAR AND SEGMENTAL BRONCHI, AXIAL CT

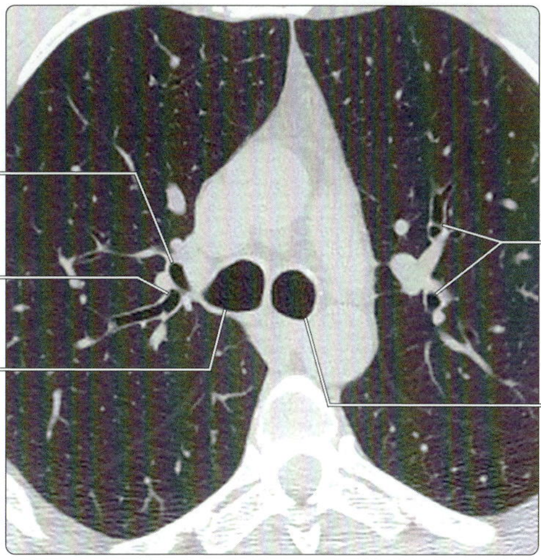

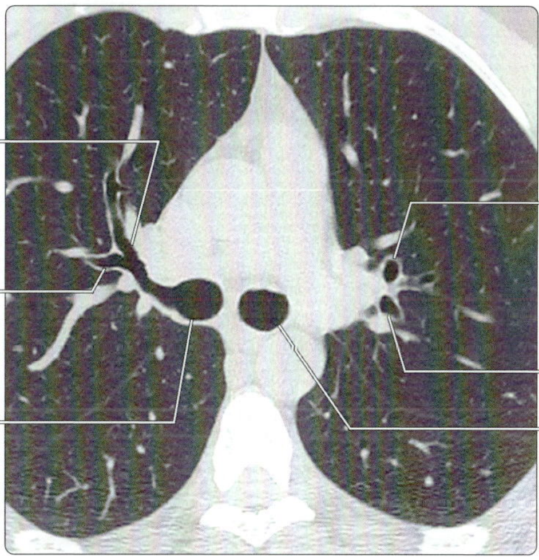

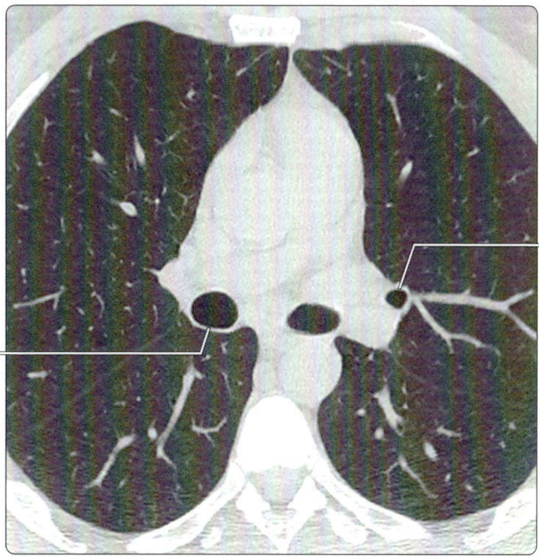

(Top) *First of 9 axial NECT (lung window) images shows the bilateral mainstem bronchi, the right upper lobe apical segmental and posterior segmental bronchi, and subsegmental branches of the left upper lobe apicoposterior segmental bronchus. The right upper lobe apical segmental bronchus typically manifests as a nearly circular lucency located just cephalad to the superior wall of the right upper lobe bronchus. The right upper lobe posterior segmental bronchus follows a nearly horizontal orientation.* **(Middle)** *Axial NECT (lung window) just below the carina shows the characteristic horizontal orientation of the right upper lobe anterior segmental and posterior segmental bronchi. Note the apical and posterior subsegmental branches of the left upper lobe apicoposterior segmental bronchus.* **(Bottom)** *Axial NECT (lung window) below the right upper lobe bronchus shows the bronchus intermedius and its thin uniform posterior wall that normally measures < 3 mm in thickness. Note the vertical orientation of the left upper lobe apicoposterior segmental bronchus, which manifests as a circular lucency in axial cross section.*

LOBAR AND SEGMENTAL BRONCHI, AXIAL CT

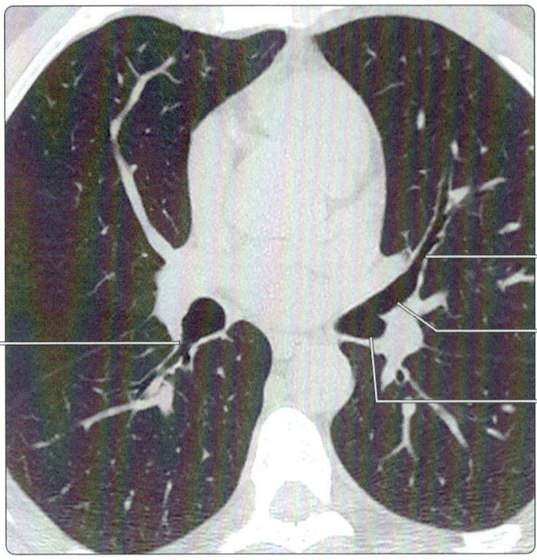

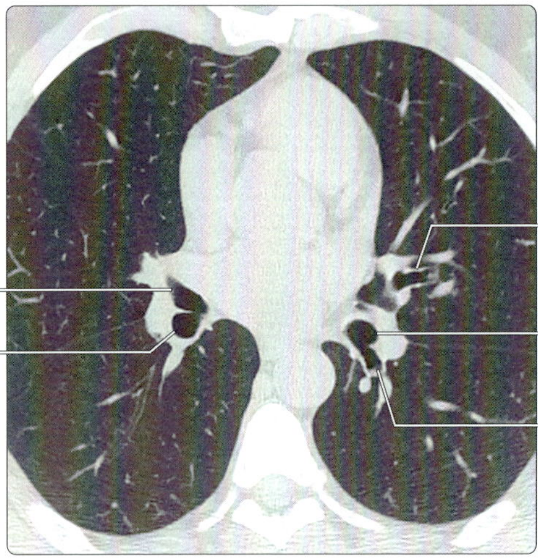

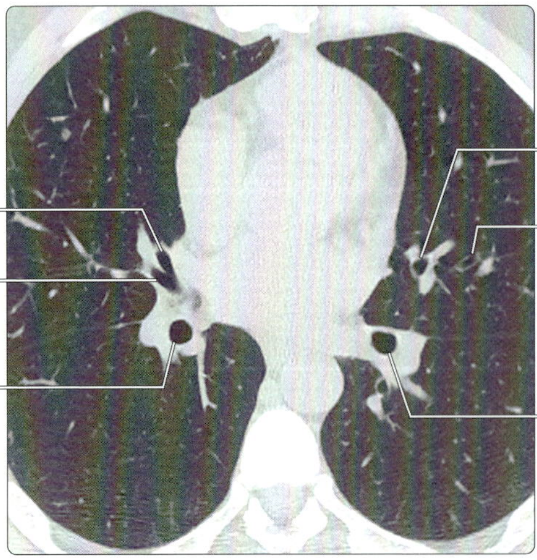

(Top) *Axial NECT (lung window) shows the right lower lobe superior segmental bronchus, which arises from the posterior wall of the right lower lobe bronchus and courses horizontally. Note the bifurcation of the left mainstem bronchus into left upper lobe and left lower lobe bronchi and the nearly horizontal course of the left upper lobe anterior segmental bronchus.* **(Middle)** *Axial NECT (lung window) shows the anterior course of the proximal middle lobe bronchus and the adjacent right lower lobe truncus basalis bronchus. Note the left upper lobe inferior lingular segmental bronchus and the left lower lobe superior segmental bronchus. The latter originates from the posterior wall of the left lower lobe bronchus and courses posteriorly and horizontally.* **(Bottom)** *Axial NECT (lung window) shows the bifurcation of the middle lobe bronchus into medial and lateral segmental bronchi. Note the left upper lobe superior and inferior lingular segmental bronchi and the bilateral lower lobe truncus basalis bronchi, which will give rise to the lower lobe basilar segmental bronchi.*

LOBAR AND SEGMENTAL BRONCHI, AXIAL CT

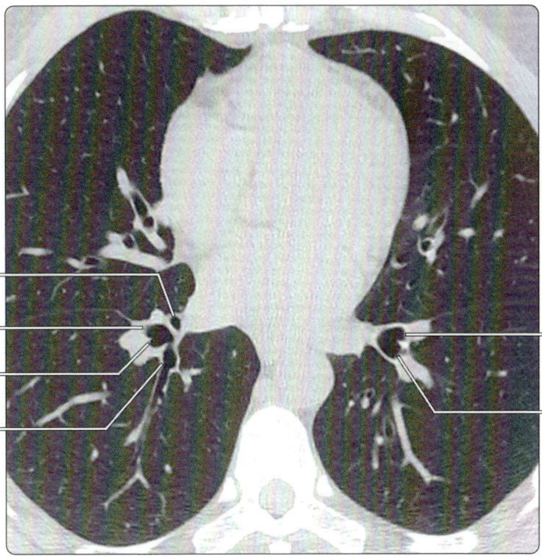

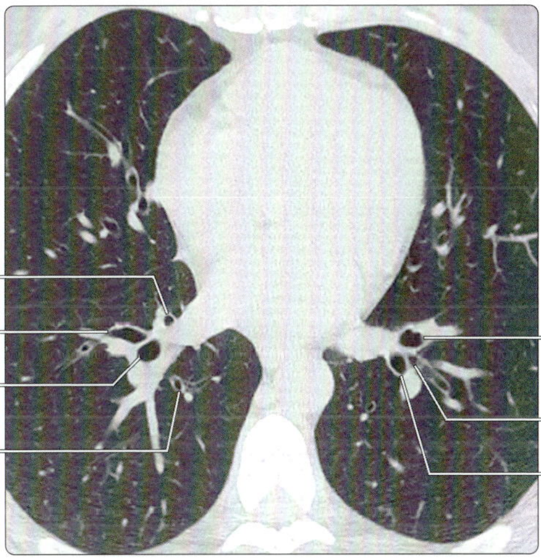

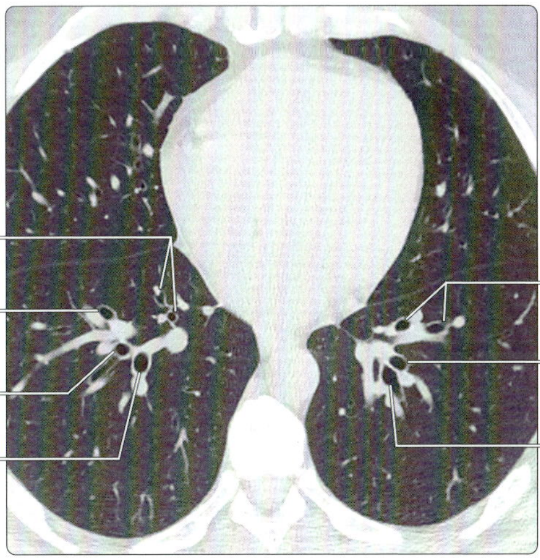

(Top) *Axial NECT (lung window) shows the basilar segmental bronchi, which originate from the distal lower lobe truncus basalis bronchi. The right lower lobe medial, anterior, lateral, and posterior basilar segmental bronchi arise in counterclockwise order and course toward their respective lung segments. The left lower lobe anteromedial, lateral, and posterior basilar segmental bronchi arise in clockwise order from the distal left lower lobe truncus basalis bronchus. Each bronchus is accompanied by its paired, homologous, segmental pulmonary artery.* **(Middle)** *Axial NECT (lung window) demonstrates the bilateral lower lobe basilar segmental bronchi as they begin to diverge and course peripherally toward their respective basilar lung segments. Note the left lower lobe anteromedial, lateral, and posterior basilar segmental bronchi.* **(Bottom)** *Axial NECT (lung window) shows bilateral basilar segmental and subsegmental bronchi. The right lower lobe medial, anterior, lateral, and posterior basilar segmental bronchi arise in counterclockwise order on axial CT, while the left lower lobe anteromedial, lateral, and posterior basilar segmental bronchi arise in clockwise order.*

LOBAR AND SEGMENTAL BRONCHI, CORONAL CT

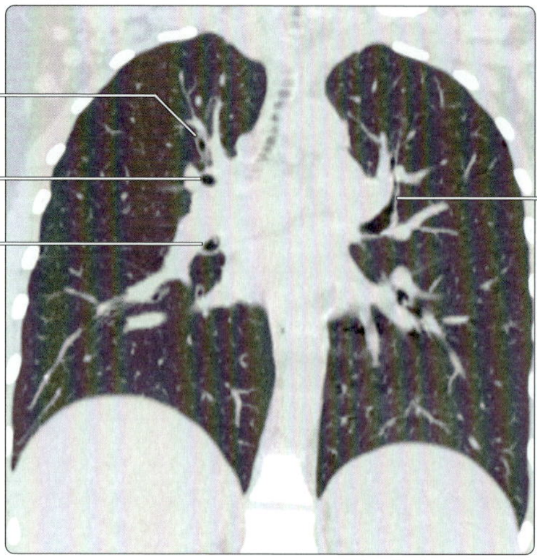

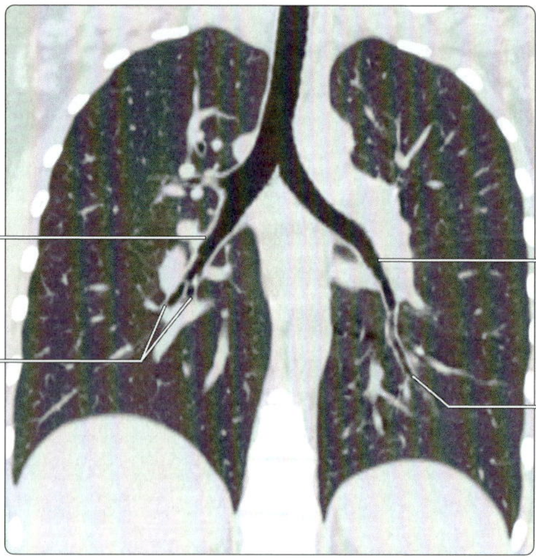

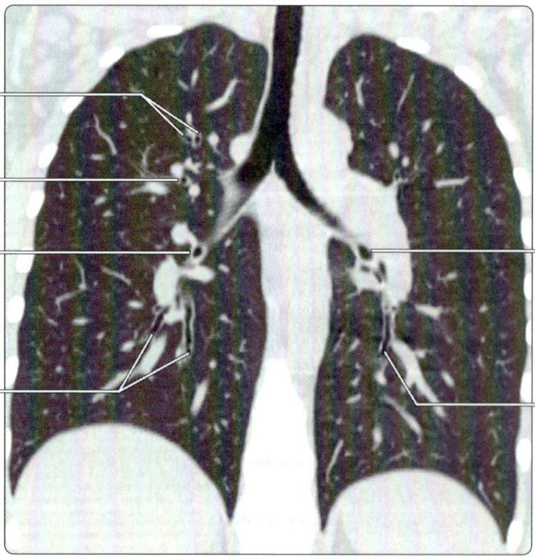

(Top) *First of 3 oblique coronal NECT (lung window) images of the same patient obtained just anterior to the trachea and carina shows the right upper lobe anterior segmental bronchus and the middle lobe bronchus end on due to their anterior origins and horizontal course. Note the cephalad course of the right upper lobe apical segmental bronchus and the left upper lobe apicoposterior segmental bronchus.* **(Middle)** *Oblique coronal NECT (lung window) shows the bilateral lower lobe truncus basalis bronchi that give rise to the bilateral basilar segmental bronchi.* **(Bottom)** *Oblique coronal NECT (lung window) obtained along the posterior aspect of the carina demonstrates the bilateral lower lobe superior segmental bronchi seen end on due to their posterior origins and horizontal course as they arise from the bilateral lower lobe bronchi. Note the bilateral basilar segmental bronchi as they course inferiorly after originating from the bilateral lower lobe truncus basalis bronchi.*

VIRTUAL BRONCHOSCOPY

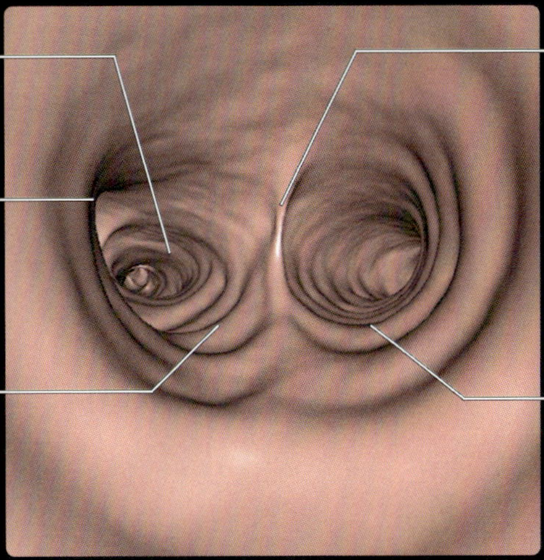

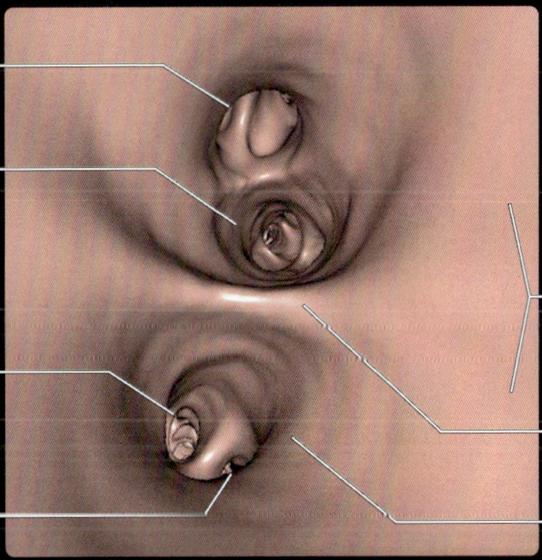

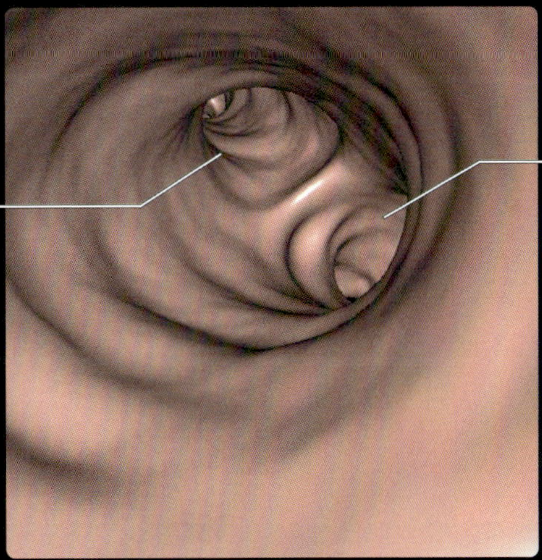

(Top) *First of 3 normal virtual bronchoscopy images is shown. Each image is aligned with the anterior aspect of the patient at the bottom of the image. The midline carina marks the bifurcation of the distal trachea to form the right and left mainstem bronchi.* **(Middle)** *Virtual bronchoscopy image obtained from the perspective of the distal aspect of the bronchus intermedius shows the soft tissue ridge that demarcates the origins of the lower lobe bronchus (above) and the middle lobe bronchus (below) and manifests on CT as a thin septum or spur.* **(Bottom)** *Virtual bronchoscopy image obtained from the perspective of the left mainstem bronchus shows its bifurcation into the left upper lobe bronchus and the left lower lobe bronchus. Virtual bronchoscopy is a noninvasive imaging technique that may provide useful information for thoracic surgeons and pulmonary medicine physicians prior to direct airway assessment with bronchoscopy.*

VARIANTS OF BRONCHIAL ANATOMY, GRAPHIC AND CT

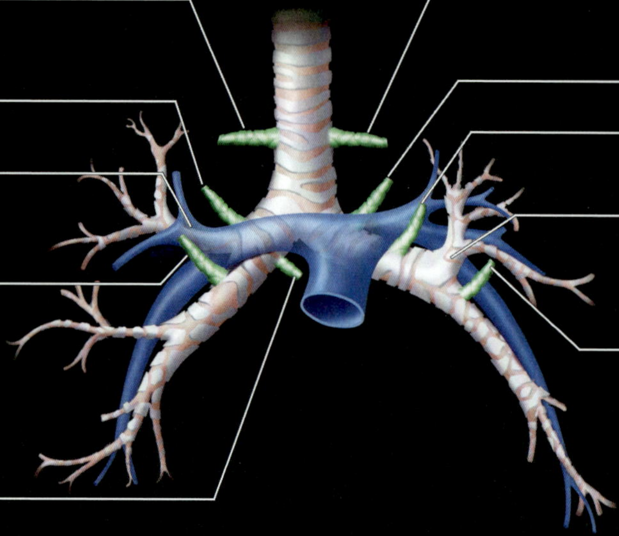

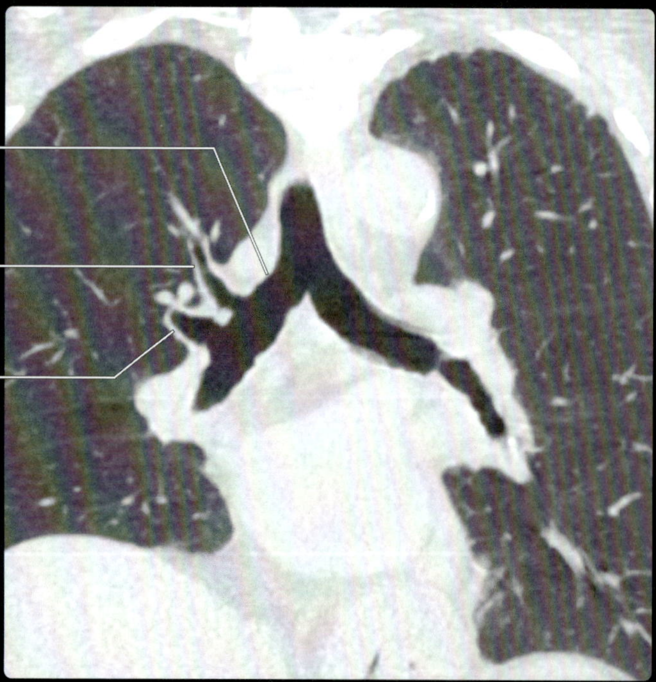

(Top) *Graphic demonstrates variants of bronchial anatomy (shown in green) in relationship to the right upper lobe eparterial bronchus and the left upper lobe hyparterial bronchus. Anomalous bronchi may be either displaced or supernumerary. The most commonly encountered variant of bronchial anatomy is the displaced right upper lobe preeparterial apical segmental bronchus, an anomalous bronchus that arises from the lateral aspect of the right mainstem bronchus above the right upper lobe eparterial bronchus. The later typically only exhibits anterior and posterior segmental bronchial branches.* **(Bottom)** *Oblique coronal NECT (lung window) shows a displaced right upper lobe preeparterial bronchus that arises from the lateral wall of the right mainstem bronchus and supplies the apical segment of the right upper lobe. The right upper lobe bronchus only gives rise to anterior and posterior segmental bronchi (not shown on this plane of imaging).*

TRACHEAL BRONCHUS, CT

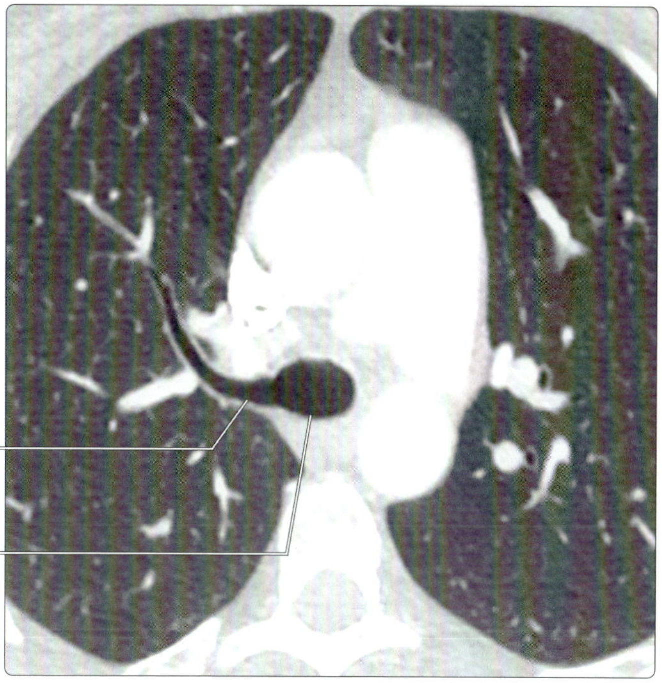

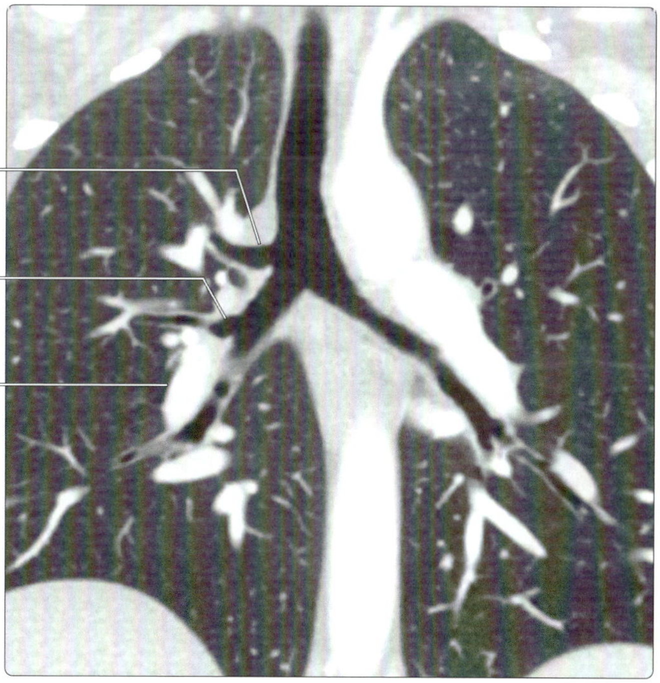

(Top) *Axial CECT (lung window) of a patient with a displaced right upper lobe apical segmental tracheal bronchus is shown. The anomalous bronchus arises from the lateral wall of the trachea. Right tracheal bronchus is much more frequent than the much rarer left tracheal bronchus or bilateral tracheal bronchi.* **(Bottom)** *Oblique coronal CECT (lung window) of the same patient shows that the displaced right upper lobe apical segmental right tracheal bronchus arises from the distal right lateral tracheal wall and courses cephalad to supply the right upper lobe apical segment. The right upper lobe bronchus only gave origin to anterior and posterior segmental bronchi, as the apical segmental bronchus was "displaced" to the right lateral tracheal wall.*

ACCESSORY CARDIAC BRONCHUS, CT

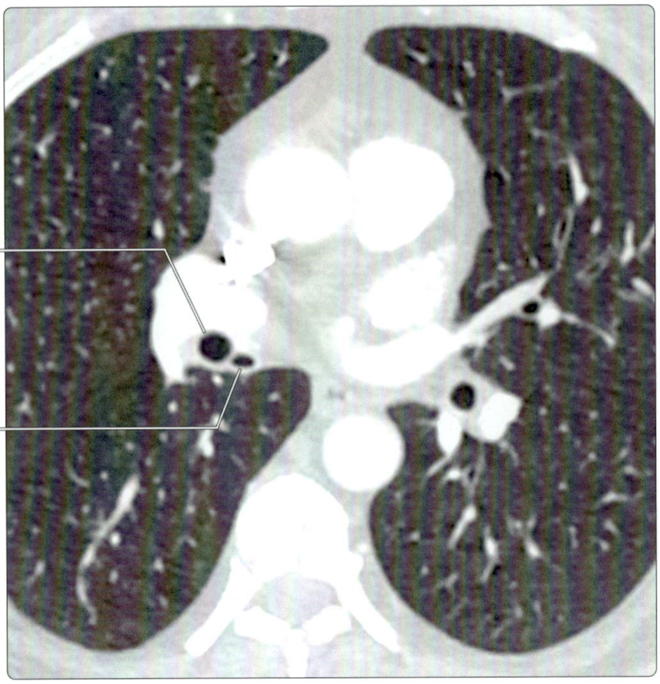

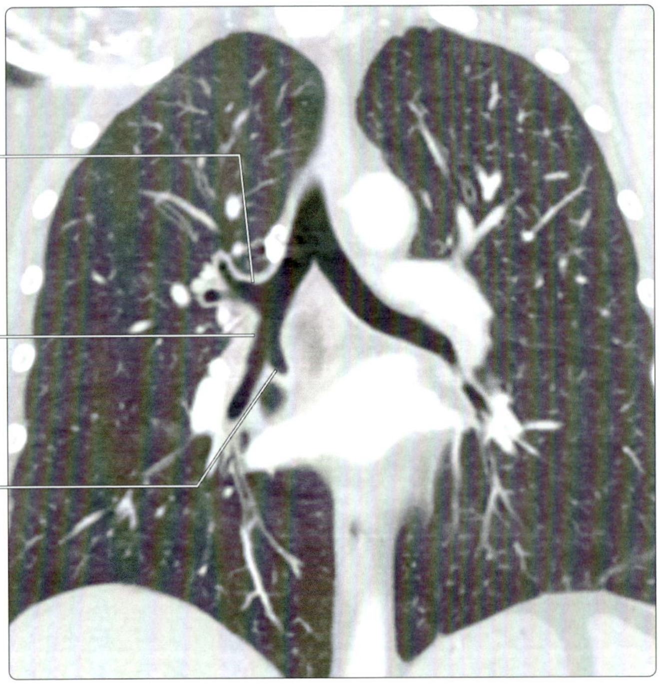

(Top) *Axial CECT (lung window) of a patient with an incidentally diagnosed accessory cardiac bronchus shows that the anomalous bronchus arises from the medial wall of the bronchus intermedius and is blind ending in the mediastinum.* **(Bottom)** *Oblique coronal CECT of the same patient shows the accessory cardiac bronchus originating from the medial aspect of the bronchus intermedius and ending blindly in the mediastinum. While the accessory cardiac bronchus is typically blind ending, it may sometimes supply rudimentary lung tissue in the right lower lobe.*

SITUS SOLITUS, RADIOGRAPHY

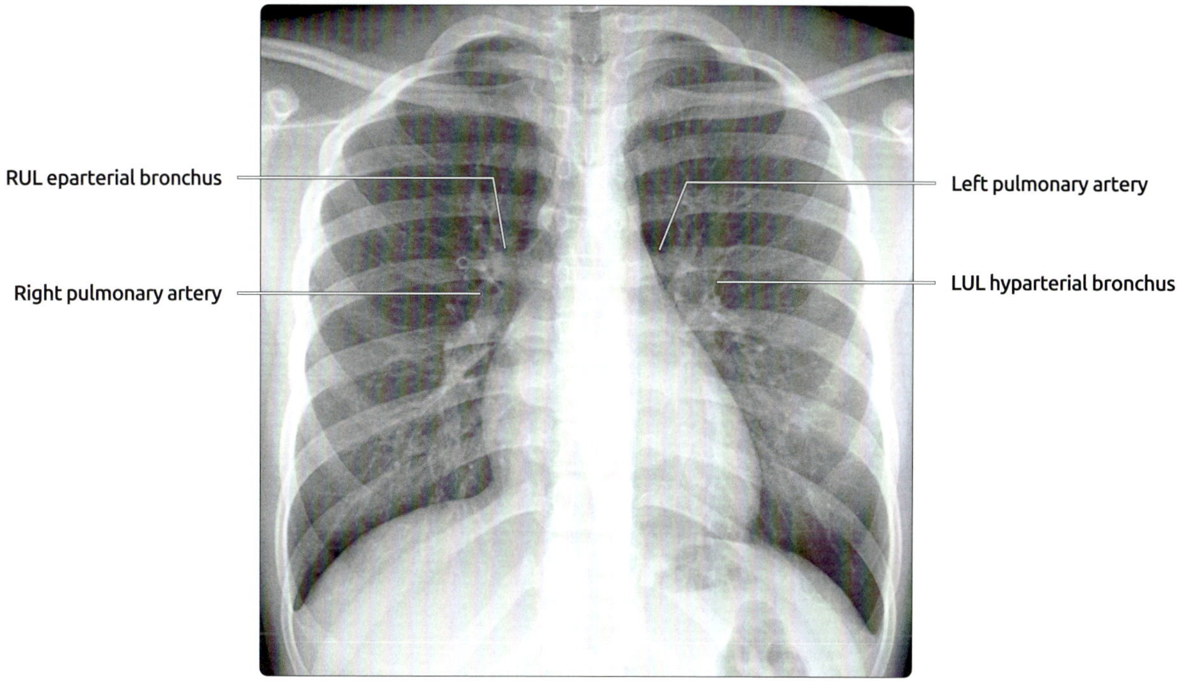

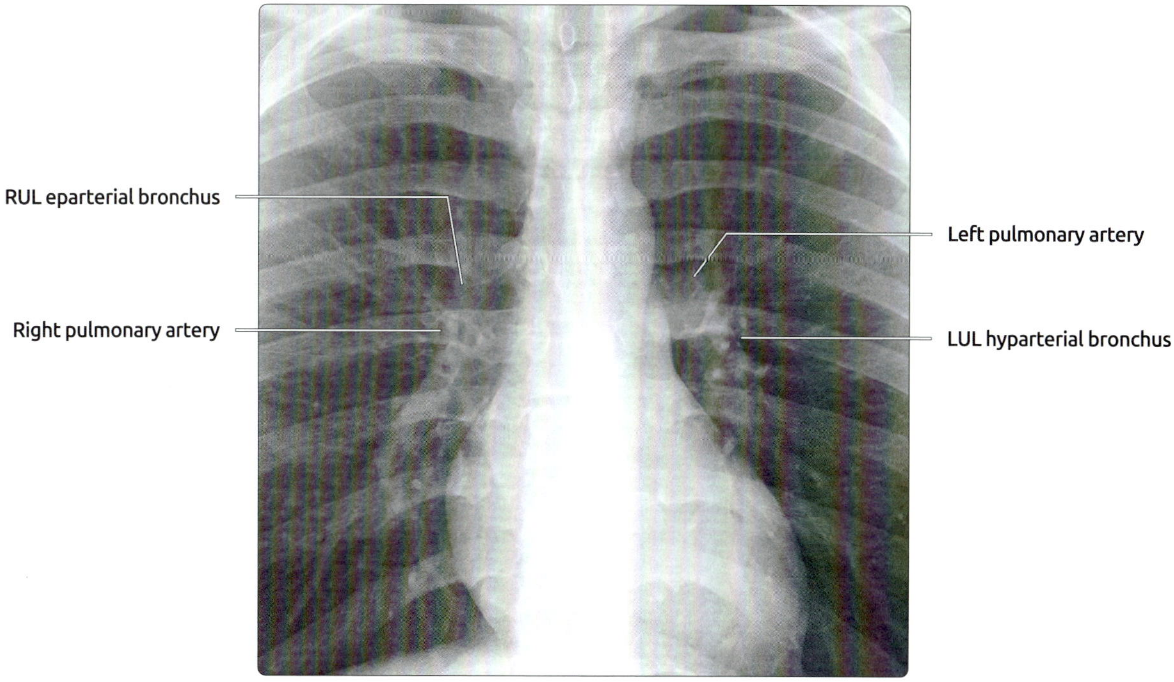

(Top) *Normal PA chest radiograph shows the normal situs solitus tracheobronchial morphology seen in most patients. The right upper lobe bronchus is eparterial and arises above the level of the ipsilateral right pulmonary artery. The left upper lobe bronchus is hyparterial and arises below the level of the ipsilateral left pulmonary artery. Note presence of levocardia, a left aortic arch, a left gastric bubble, and a right major hepatic lobe.* **(Bottom)** *Normal coned-down PA chest radiograph demonstrates the normal situs solitus bronchial morphology. The right upper lobe bronchus is eparterial, and the right lung is trilobed. The left upper lobe bronchus is hyparterial, and the left lung is bilobed.*

SITUS INVERSUS, RADIOGRAPHY AND CT

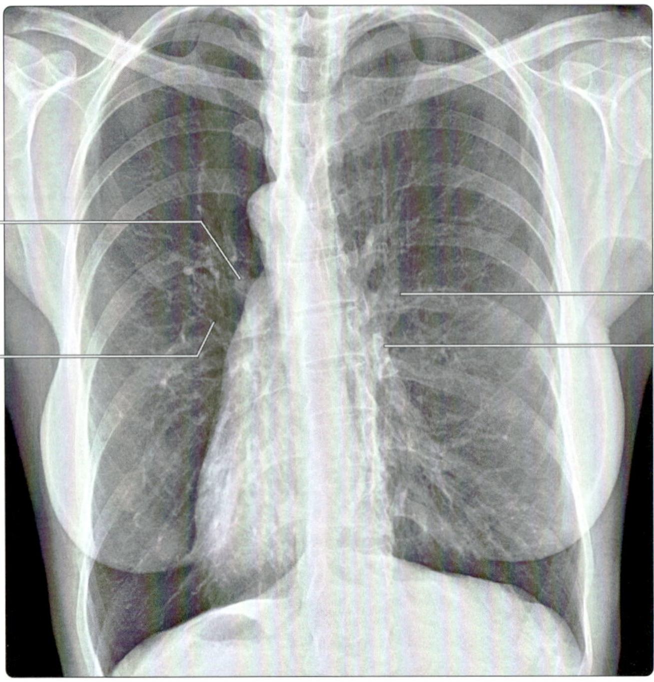

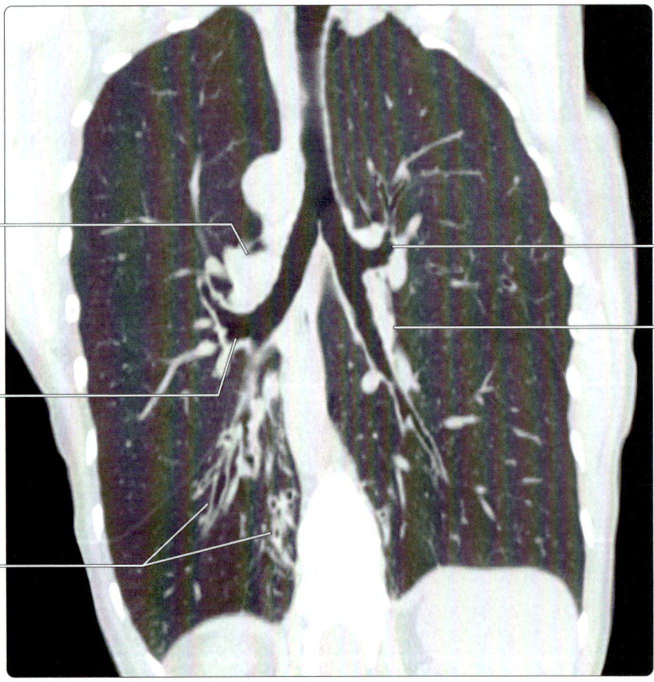

(Top) *PA chest radiograph of a patient with situs inversus shows that the right upper lobe bronchus is hyparterial arising below the ipsilateral right pulmonary artery, and the left upper lobe bronchus is eparterial arising above the ipsilateral left pulmonary artery. Note the right aortic arch, dextrocardia, a right-sided gastric bubble, and a left-sided major hepatic lobe.* **(Bottom)** *Oblique coronal NECT (lung window) of the same patient better demonstrates the right upper lobe hyparterial (arising below the ipsilateral pulmonary artery) bronchus and the left upper lobe eparterial (arising above the ipsilateral pulmonary artery) bronchus characteristic of situs inversus. The right lung is bilobed, and the left lung is trilobed. Note bilateral right greater than left lower lobe bronchiectasis and bronchial wall thickening in this patient with Kartagener syndrome.*

LEFT ISOMERISM, RADIOGRAPHY AND CT

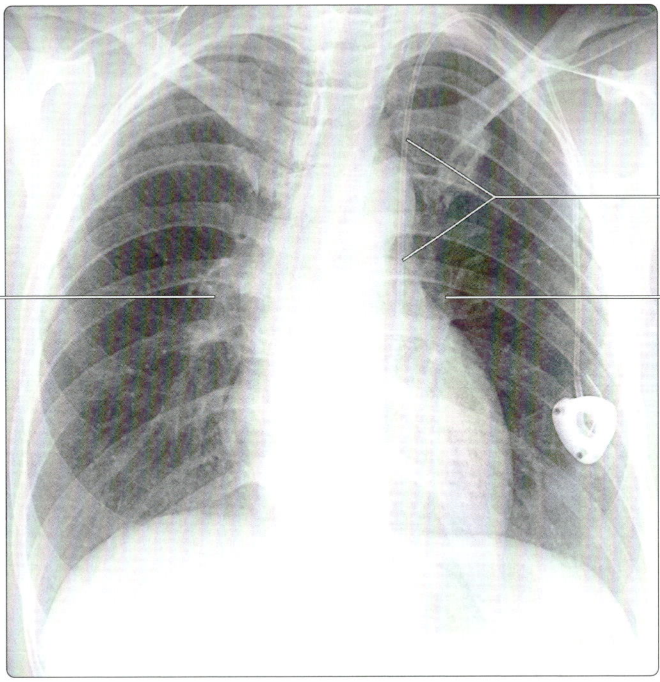

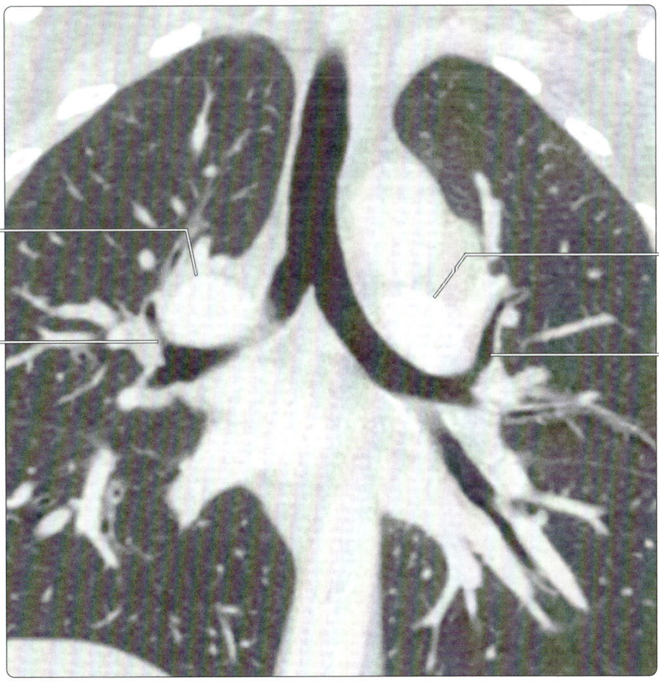

(Top) *PA chest radiograph of a patient with left isomerism (bilateral left sidedness) shows bilateral hyparterial bronchi and a persistent left superior vena cava diagnosed based on the course of the left pectoral port catheter. In this case, there are levocardia, a left aortic arch, a left gastric bubble, and a right major hepatic lobe. Persistent left superior vena cava and azygos continuation of the inferior vena cava are frequently associated with left isomerism.* **(Bottom)** *Coronal CECT (lung window) of the same patient optimally demonstrates the symmetric tracheobronchial morphology of the bilateral hyparterial bronchi (arising distal to the corresponding ipsilateral pulmonary arteries as the latter cross over the mainstem bronchi) in this case of left isomerism.*

ASPIRATED FOREIGN BODY AND SABER SHEATH TRACHEA, RADIOGRAPHY AND CT

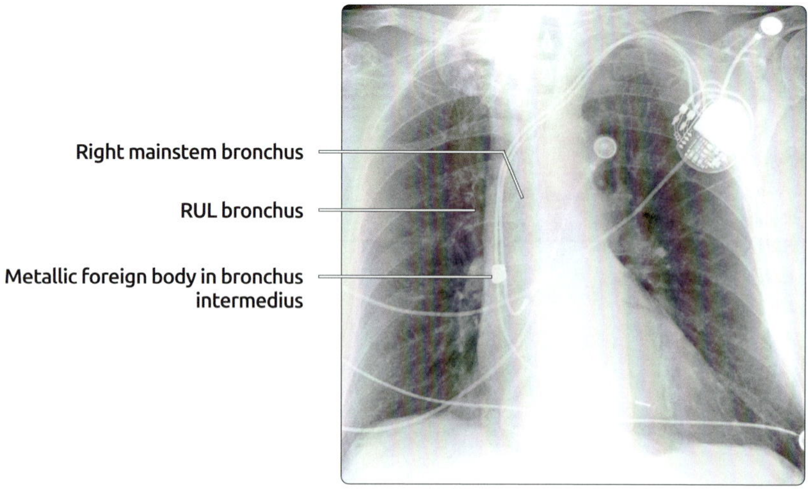

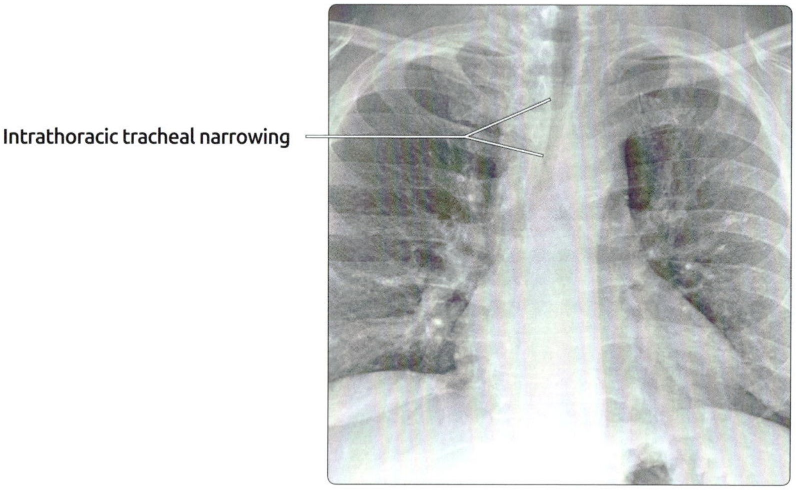

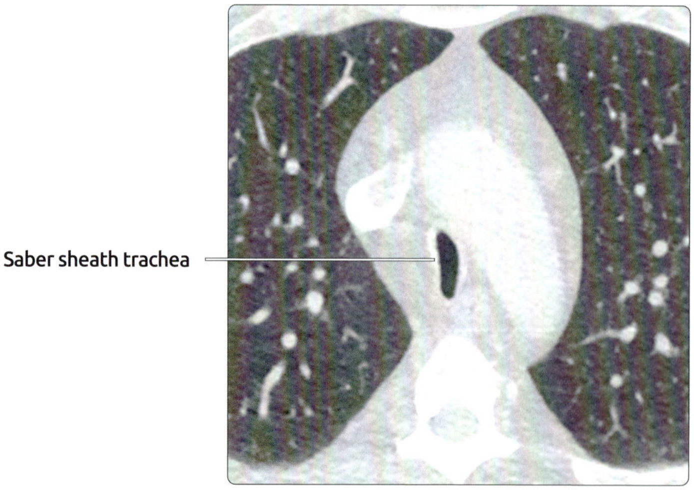

(Top) *AP chest radiograph of a patient who aspirated a tooth crown shows the metallic foreign body lodged in the bronchus intermedius. Foreign body aspiration more frequently affects the right tracheobronchial tree compared to the left due to the more vertical course of the right mainstem bronchus.* **(Middle)** *PA chest radiograph of an asymptomatic man with chronic obstructive pulmonary disease shows diffuse narrowing of the intrathoracic trachea.* **(Bottom)** *Axial CECT (lung window) of the same patient demonstrates an abnormal morphology of the intrathoracic trachea characterized by marked coronal narrowing and a normal sagittal caliber, confirming the diagnosis of saber sheath trachea. Saber sheath trachea typically affects men over the age of 50 years, is associated with chronic obstructive pulmonary disease, and usually has no clinical significance.*

TRACHEAL STENOSIS, RADIOGRAPHY AND CT

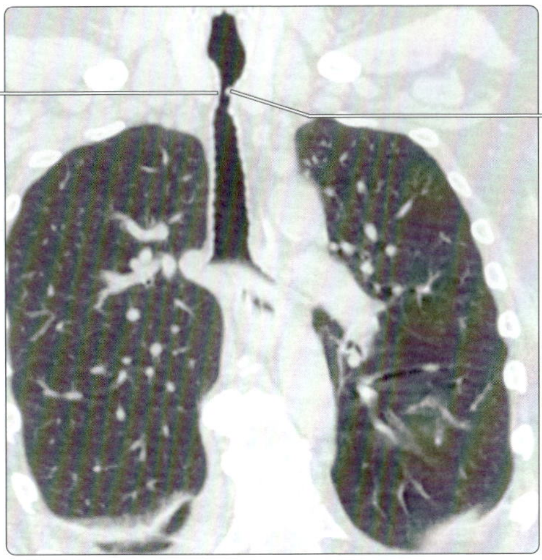

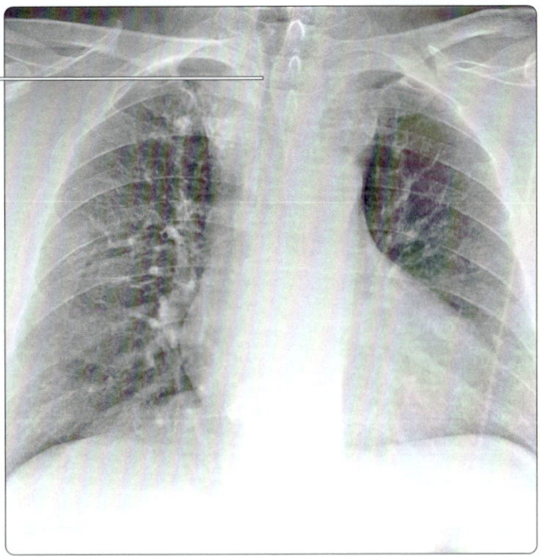

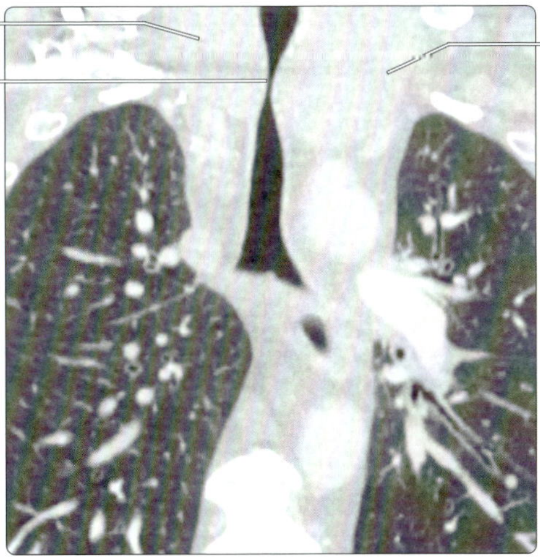

(Top) *Oblique coronal NECT (lung window) of a patient who developed stridor following a prolonged hospital stay shows marked upper tracheal stenosis and nodular left tracheal mural thickening secondary to granulation tissue. Postintubation tracheal stenosis is the most common type of acquired tracheal stenosis.* **(Middle)** *AP chest radiograph of a patient with a known large thyroid goiter with intramediastinal extension who presented with stridor shows marked smooth stenosis of the upper intrathoracic trachea associated with mass effect on the left cervical trachea, widening of the upper mediastinum, and the radiographic cervicothoracic sign.* **(Bottom)** *Oblique coronal CECT (lung window) of the same patient shows smooth stenosis of the upper intrathoracic trachea secondary to a large thyroid goiter with intramediastinal extension. Tracheal stenosis may be secondary to extrinsic compression by thyromegaly, mass, or a vascular abnormality.*

TRACHEAL STENOSIS SECONDARY TO MALIGNANCY, CT

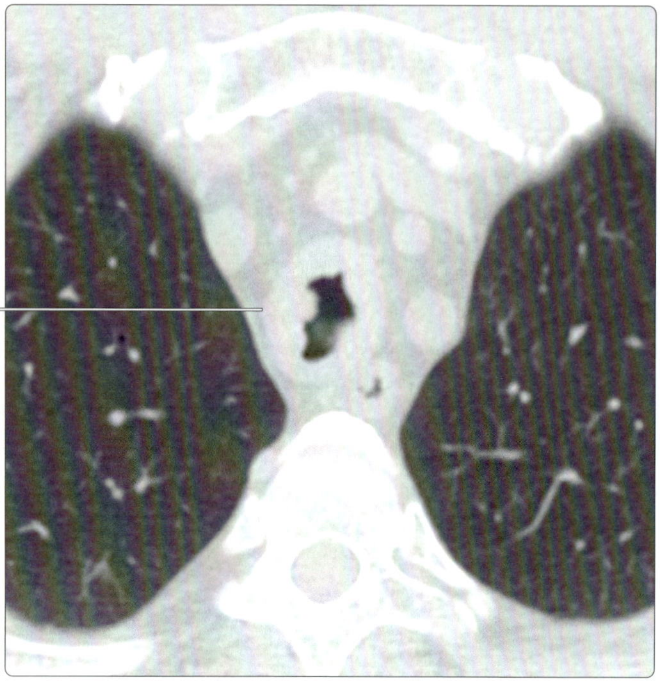

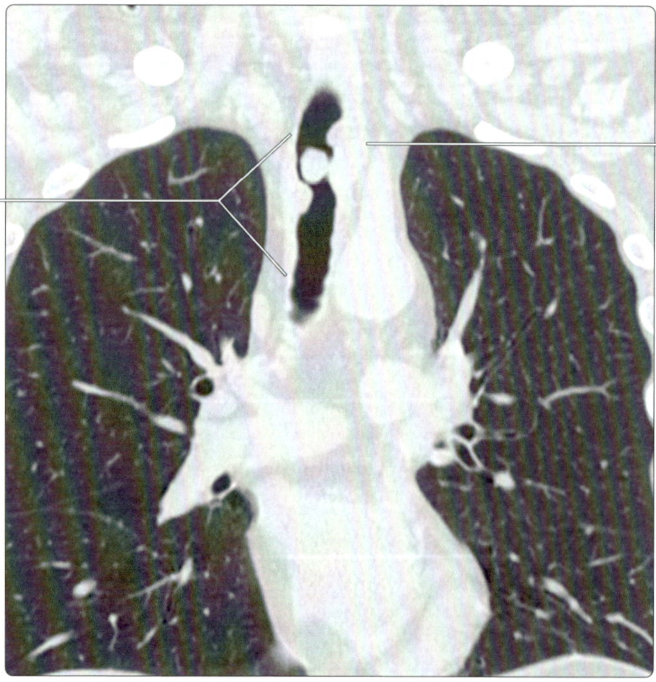

(Top) *Axial NECT (lung window) of a man who presented with stridor shows marked tracheal stenosis with circumferential nodular mural thickening, which is highly concerning for tracheal malignancy.* **(Bottom)** *Oblique coronal NECT (lung window) of the same patient shows marked upper tracheal stenosis and demonstrates the longitudinal extent of the asymmetric nodular tracheal wall thickening, which is valuable information in preoperative planning. Biopsy confirmed adenoid cystic carcinoma, considered the most common primary malignant tracheal neoplasm.*

MOUNIER-KUHN SYNDROME AND BRONCHIECTASIS, CT

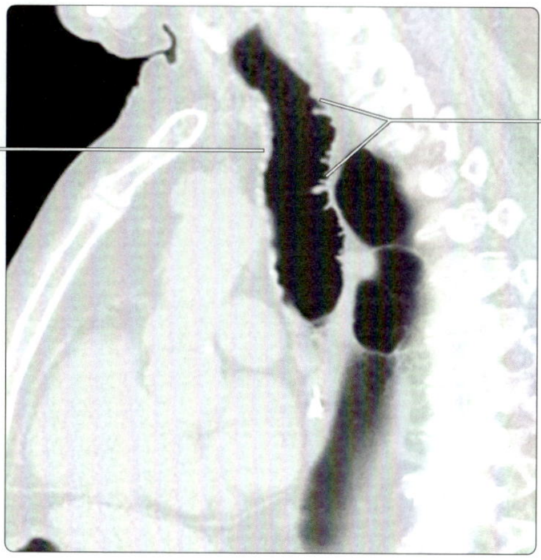

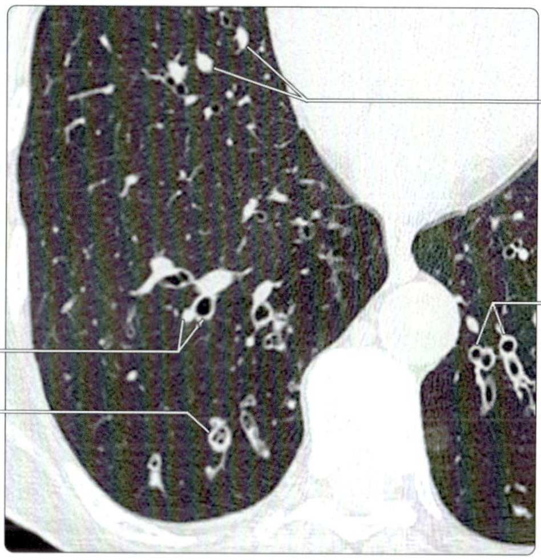

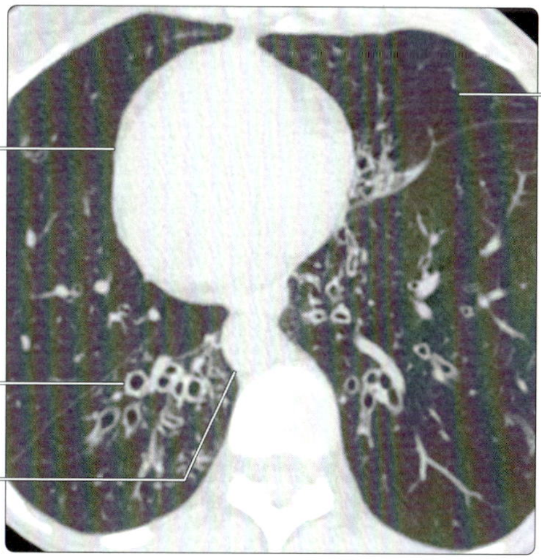

(Top) *Sagittal NECT (lung window) of a patient who presented with cough and dyspnea shows marked tracheal dilatation. Note the corrugated appearance of the tracheal wall, which results from airway mucosal prolapse between adjacent tracheal cartilage rings. The findings are characteristic of Mounier-Kuhn syndrome.* **(Middle)** *Axial HRCT (lung window) of a patient with a history of recurrent pneumonia demonstrates multifocal cylindrical bronchiectasis, bronchial wall thickening, and endoluminal content representing mucoid impaction of bronchiectatic airways. Note mosaic attenuation of the lung parenchyma likely related to superimposed constrictive bronchiolitis.* **(Bottom)** *Axial NECT (lung window) of a patient with situs inversus and Kartagener syndrome shows bronchiectasis and bronchial wall thickening involving predominantly the right lower lobe, but also the left lower lobe and the left middle lobe, due to recurrent pulmonary infection secondary to primary ciliary dyskinesia. Note the presence of dextrocardia, a right-sided descending thoracic aorta, a right bilobed lung, and a left trilobed lung.*

PULMONARY ARTERIES

Function

- Conduit of deoxygenated blood to capillary-alveolar interface

Pulmonary Trunk

- Originates from right ventricular outflow tract
  - Anterior and to left of ascending aorta
- Posterosuperior course toward the left, posterior to ascending aorta
- Proximal aspect contained within pericardium
  - Common serous pericardial sheath contains proximal pulmonary trunk and ascending thoracic aorta

Right Pulmonary Artery and Branches

- **Right pulmonary artery**
  - Longer and larger than left pulmonary artery
  - Horizontal intramediastinal course
  - Relationships
    - Posterior to ascending aorta, superior vena cava, and right superior pulmonary vein
  - Branches
    - Ascending trunk or truncus anterior, supplies right upper lobe
    - Descending or interlobar pulmonary artery, supplies middle and right lower lobes
    - Peripheral pulmonary artery branches
      - Located medial to bronchi in upper lobes
      - Located lateral to bronchi in middle and lower lobes

Left Pulmonary Artery and Branches

- **Left pulmonary artery**
  - Shorter and smaller than right pulmonary artery
  - Slightly posterior to right pulmonary artery
  - Courses above left mainstem bronchus
  - Branches
    - Short ascending branch, supplies left upper lobe apicoposterior and anterior segments
    - Descending or interlobar pulmonary artery, supplies left upper lobe superior and inferior lingular segments and left lower lobe
    - Peripheral pulmonary artery branches
      - Located medial to bronchi in left upper lobe anterior and apicoposterior segments
      - Located lateral to bronchi in lingula and lower lobe

PULMONARY VEINS

Function

- Conduit of oxygenated blood from capillary-alveolar interface to left heart chambers

Embryologic Anatomy

- Left atrium partially derived from primitive common pulmonary vein
- Atrial expansion with incorporation of primitive common pulmonary vein into left atrial chamber
- 2 pairs of superior and inferior pulmonary veins with frequent variations in number
  - 70% of population has 4 pulmonary veins
  - Underincorporation of pulmonary veins into atrium
    - More common on left
    - Common trunk for superior and inferior pulmonary veins in 12-25% of subjects
  - Overincorporation of pulmonary veins into atrium resulting in supernumerary/accessory veins
    - More common on right
  - Frequent variations in central anastomosis of middle lobe pulmonary vein
    - Drains into right superior pulmonary vein in 53-69% of normal subjects
    - Drains directly into left atrium in 17-23% of normal subjects
    - Drains into inferior pulmonary vein in 3-8% of normal subjects
- Clinical significance
  - Identification of anatomic variants
  - Preablation imaging of patients with atrial fibrillation

General Anatomy

- Pulmonary veins course along peripheral aspects of pulmonary subsegments, segments, and lobes
- Superior pulmonary veins course obliquely into left atrium
- Inferior pulmonary veins course obliquely into left atrium
- **Right pulmonary veins**
  - Superior pulmonary vein
    - Drains upper and middle lobes
  - Inferior pulmonary vein
    - Drains lower lobe
- **Left pulmonary veins**
  - Superior pulmonary vein
    - Drains upper and lower divisions of left upper lobe
  - Inferior pulmonary vein
    - Drains lower lobe

PULMONARY ARTERIES AND VEINS

General Concepts

- Nomenclature system generally based on Boyden nomenclature of tracheobronchial tree and Jackson and Huber segmental and subsegmental anatomy
  - Not frequently used in clinical practice
  - Right (R), left (L); arteries (A); veins (V)
  - Numerical descriptors analogous to those of bronchial nomenclature
- Typical branching patterns
  - 10 segmental pulmonary arteries, 2nd-order branches
  - 20 subsegmental pulmonary arteries, 3rd-order branches
  - Subsegmental pulmonary arteries, 4th-order branches; dichotomous division of segmental arteries
  - Subsequent dichotomous divisions into 5th- and 6th-order branches

Right Upper Lobe Pulmonary Arteries

- Ascending trunk (truncus anterior)
  - Apical segmental artery (RA1)
    - Posteriorly located apical and anterior subsegmental branches
  - Anterior segmental artery (RA2)
    - Laterally located posterior and anterior subsegmental branches

- Posterior segmental artery (RA3)
  - Originates from interlobar pulmonary artery
  - Laterally located apical and posterior subsegmental branches

Middle Lobe Pulmonary Arteries

- Origin from right interlobar pulmonary artery
  - May arise as single vessel
  - May arise as 2 distinct vessels
  - Origin slightly superior to origin of right lower lobe superior segmental artery
- Lateral segmental artery (RA4)
  - Posterior and anterior subsegmental branches
- Medial segmental artery (RA5)
  - Superior and inferior subsegmental branches

Right Lower Lobe Pulmonary Arteries

- Right lower lobe pulmonary artery
  - Continuation of interlobar pulmonary artery distal to origin of middle lobe artery(ies)
  - Superior segmental artery (RA6)
    - Posterior origin; combined medial/superior and lateral subsegmental branches
  - Common basilar artery
    - Gives rise to basilar segmental arteries
  - Medial basilar segmental artery (RA7)
    - Anterolateral and anteromedial subsegmental branches
  - Anterior basilar segmental artery (RA8)
    - Lateral and basilar subsegmental branches
  - Lateral basilar segmental artery (RA9)
    - Lateral and basilar subsegmental branches
  - Posterior basilar segmental artery (RA10)
    - Laterobasilar and mediobasilar subsegmental branches

Left Upper Lobe Pulmonary Arteries

- Greater number of separate subsegmental branches
- Upper division
  - Independent origin of apical and posterior segmental arteries, unlike bronchial divisions
  - Apical segmental artery (LA1)
  - Anterior segmental artery (LA2)
    - Lateral and anterior subsegmental arteries
  - Posterior segmental artery (LA3)
- Lower division
  - Single lingular artery gives rise to 2 segmental arteries
  - Superior lingular segmental artery (LA4)
    - Posterior and anterior subsegmental branches
  - Inferior lingular segmental artery (LA5)
    - Superior and inferior subsegmental branches

Left Lower Lobe Pulmonary Arteries

- Superior segmental artery (LA6)
  - Posterior origin, superior to that of lingular artery
  - Superomedial and lateral subsegmental branches
- Common basilar artery
  - Gives rise to basilar segmental arteries
- Anteromedial basilar segmental artery (LA7+8)
  - Anterior, medial, lateral, and basilar subsegmental branches
- Lateral basilar segmental artery (LA9)
  - Lateral and basilar subsegmental branches
- Posterior basilar segmental artery (LA10)
  - Lateral and basilar subsegmental branches

Right Upper Lobe Pulmonary Veins

- Apical segmental vein (RV1)
  - Apical and anterior subsegmental tributaries
- Anterior segmental vein (RV2)
  - Inferior and superior tributaries
- Posterior segmental vein (RV3)
- RV1, RV2, and RV3 join to form large right superior pulmonary vein located anterior to interlobar pulmonary artery

Middle Lobe Pulmonary Veins

- Lateral segmental vein (RV4)
- Medial segmental vein (RV5)
- Form common trunk that typically courses below middle lobe bronchus and joins right superior pulmonary vein

Right Lower Lobe Pulmonary Veins

- Superior segmental vein (RV6)
  - Drains into right inferior pulmonary vein
- Medial basilar segmental vein (RV7)
- Anterior basilar segmental vein (RV8)
- Lateral basilar segmental vein (RV9)
- Posterior basilar segmental vein (RV10)
- Common basilar vein drains right lower lobe basilar segments

Right Top Pulmonary Vein

- Anatomic variant
- Pulmonary vein enters left atrium superomedial to right superior pulmonary vein
- Drains superior segment of right lower lobe &/or posterior segment of right upper lobe

Left Upper Lobe Pulmonary Veins

- Upper division
  - Apicoposterior segmental vein (LV1+3)
  - Anterior segmental vein (LV2)
- Lower division
  - Superior lingular segmental vein (LV4)
  - Inferior lingular segmental vein (LV5)

Left Lower Lobe Pulmonary Veins

- Superior segmental vein (LV6)
- Anteromedial basilar segmental vein (LV7+8)
- Lateral basilar segmental vein (LV9)
- Posterior basilar segmental vein (LV10)

IMAGING OF PULMONARY ARTERIES

Radiography

- **Pulmonary trunk**
  - Prominent in normal children, adolescents, young adults
- **Right interlobar pulmonary artery**
  - Lateral to bronchus intermedius on frontal radiography
  - Normal transverse measurements
    - 15 mm in women;16 mm in men

CT

- **Pulmonary trunk**
  - Normal transverse diameter up to 29 mm; may normally be up to 33-34 mm
  - Measured in axial plane of bifurcation perpendicular to vascular long axis
- **Interlobar pulmonary artery**
  - Normal transverse diameter up to 16.8 mm (13.0 ± 1.9 mm)

Angiography

- Assessment of endoluminal integrity and luminal size
- Visualization of branching patterns and distal arterial tree (including capillary bed)
- Visualization of abnormal vascular connections
- Interventional procedures: Thrombectomy, embolization, foreign body retrieval

ANATOMY-RELATED IMAGING ABNORMALITIES

Pulmonary Venous Hypertension

- **Vascular equalization** (mean pulmonary venous pressure: 13-15 mm Hg)
  - Earliest radiographic manifestation; equal size of upper and lower lobe vessels
- **Vascular redistribution** (mean pulmonary venous pressure: 15-18 mm Hg)
  - Upper lobe vessels more dilated than lower lobe vessels

Pulmonary Hypertension

- Mean pulmonary artery pressure: > 25 mm Hg at rest; > 30 mm Hg during exercise
- Enlarged pulmonary trunk: Diameter > 2.9-3.4 cm
- Pulmonary trunk:ascending aorta diameter ratio > 1
- Exclusion of idiopathic dilatation of pulmonary trunk

Arteriovenous Malformation

- **Congenital pulmonary arteriovenous malformation**
  - Direct communication between pulmonary artery(ies) and vein(s) without intervening capillary bed
  - Hereditary hemorrhagic telangiectasia (Rendu-Osler-Weber syndrome)
  - Imaging
    - Single or multiple nodular opacity(ies) with feeding and draining vessel(s)
    - Arteriography: Exclusion of multifocal disease, embolotherapy
- **Hepatopulmonary syndrome**
  - Acquired pulmonary arteriovenous malformations in patients with end-stage liver disease; hepatic dysfunction, hypoxemia, pulmonary vascular dilatation
  - Imaging
    - Dilated peripheral lower lobe pulmonary vessels extending to pleura
    - Direct visualization of abnormal arteriovenous connections

Meandering Pulmonary Vein

- Anomalous course of pulmonary vein
- Drainage into left atrium
- Typically incidental finding in asymptomatic patient

Anomalous Pulmonary Venous Return

- **Partial anomalous pulmonary venous return**
  - Left upper lobe venous drainage to anomalous vertical vein coursing along left superior mediastinum and draining into left brachiocephalic vein; absent normal left superior pulmonary vein at hilum
  - Right upper lobe venous drainage to superior vena cava; associated with **sinus venosus atrial septal defect**
- **Scimitar syndrome** (**congenital venolobar syndrome**)
  - Partial anomalous right lung pulmonary venous drainage into inferior vena cava, right atrium, or portal veins
  - Associations: Right pulmonary hypoplasia, congenital heart disease, bronchopulmonary malformations, systemic arterial supply to right lung, diaphragmatic abnormalities
  - Imaging
    - Small right lung volume
    - Visualization of right scimitar vein and identification of anomalous venous drainage
    - Assessment of associated anomalies

Pulmonary Vein Varix

- Congenital or acquired enlargement of central pulmonary vein or veins
- Associated with chronic elevation of left atrial pressures
- Imaging: Nodular/mass-like opacity on radiography that corresponds to enlarged pulmonary vein on CT/MR

Pulmonary Artery Aneurysm/Pseudoaneurysm

- Etiologies: Pulmonary hypertension, collagen vascular disease (Marfan, Takayasu, Behçet), infection, iatrogenic
- Aneurysmal dilatation, absence of normal luminal tapering of pulmonary artery

Pulmonary Thromboembolic Disease

- Common disease with significant morbidity/mortality
- Symptomatic patients; postoperative/bed-ridden patients, malignancy
- Acute thromboembolic disease
  - Acute pulmonary trunk dilatation; pulmonary hypertension
  - Visualization of occlusive/nonocclusive filling defects within contrast-opacified pulmonary arteries
  - Assessment of right heart strain as prognostic indicator
- Chronic thromboembolic disease
  - Pulmonary hypertension
  - Partial/complete filling defects in vascular lumina with eccentric locations and irregular borders
  - Recanalized/calcified emboli

Foreign Body Embolism

- Catheter embolism
  - Shearing by introducer needle during catheter manipulation
  - Pulmonary arteries commonly affected
  - Abnormal catheter/catheter fragment location on imaging
- Cement embolism (polymethylmethacrylate)
  - Vertebroplasty and kyphoplasty
    - Tubular metallic densities in vertebral venous plexi
  - Tubular and branching metallic densities in peripheral pulmonary arteries

OVERVIEW OF PULMONARY ARTERIES AND VEINS

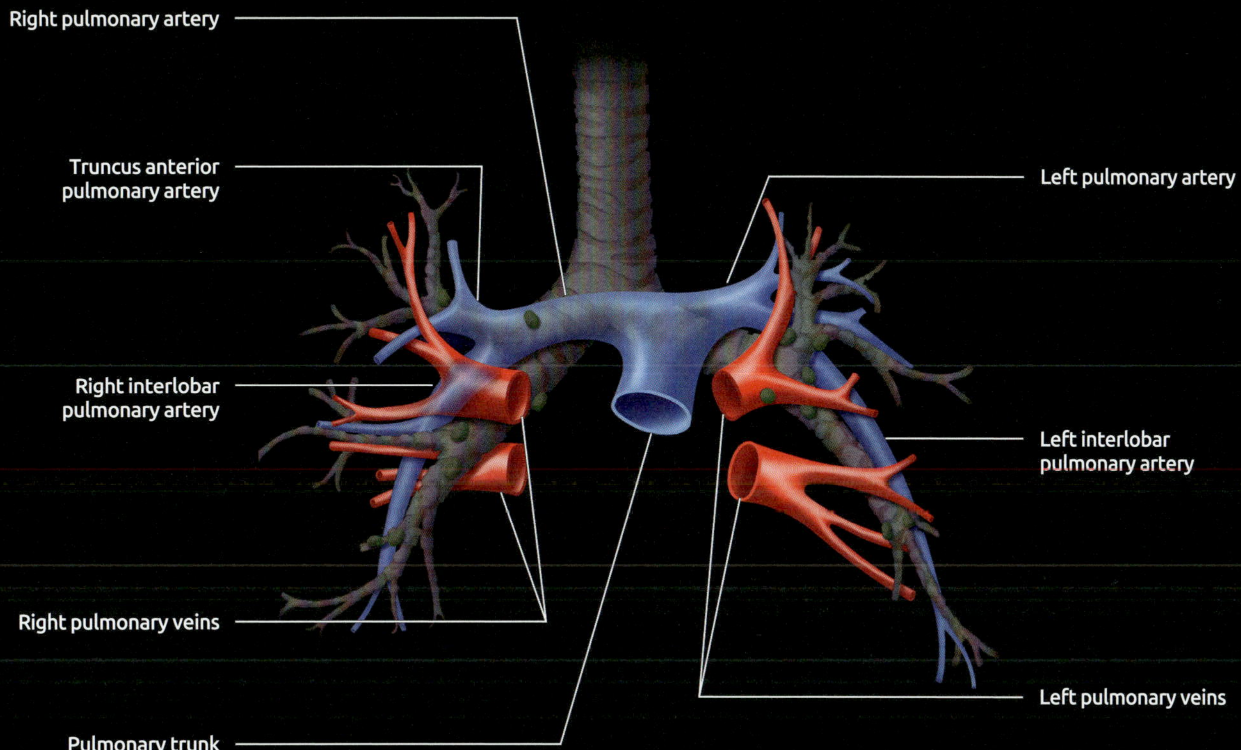

Graphic depicts the anatomy and relationships of the pulmonary vessels. The pulmonary trunk arises from the right ventricular outflow tract and gives rise to right and left pulmonary arteries. The pulmonary arteries (in blue) accompany the bronchi and carry deoxygenated blood to the capillary-alveolar interface. The pulmonary veins (in red) travel along the periphery of the pulmonary units (secondary pulmonary lobules, pulmonary segments, and pulmonary lobes) and carry oxygenated blood from the alveolar-capillary interface to the left atrium and the left-sided circulation. Pulmonary vein tributaries drain into the bilateral paired pulmonary veins that in turn drain into the left atrium.

NORMAL PULMONARY ARTERIES AND VEINS, RADIOGRAPHY

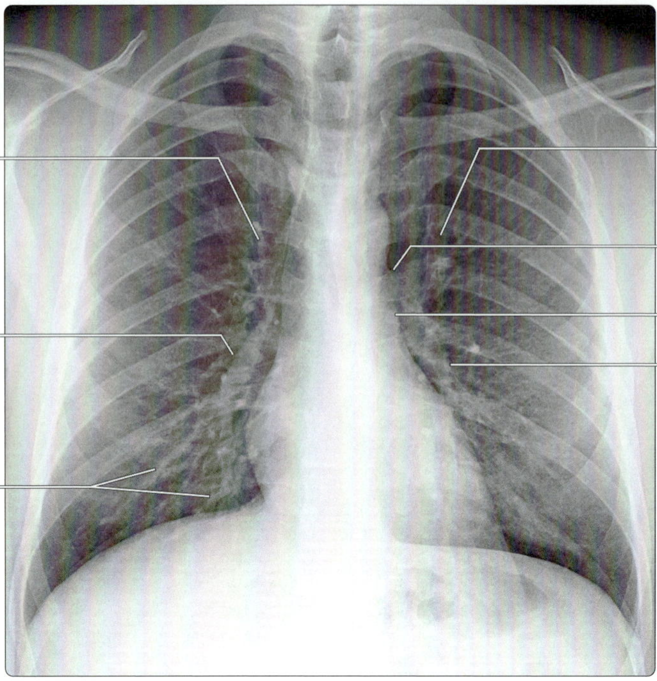

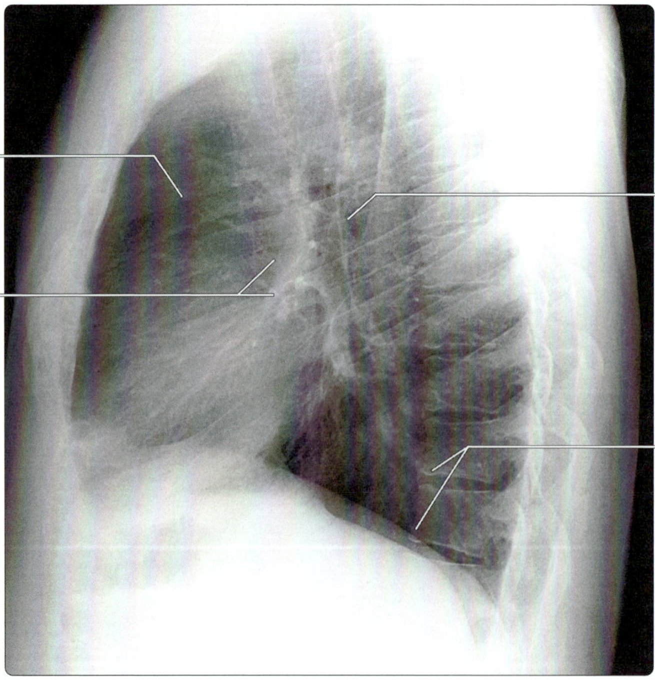

(Top) *First of 2 normal chest radiographs demonstrates the radiographic appearance of the normal pulmonary vessels. PA chest radiograph shows that the pulmonary trunk is partly visible as a midleft mediastinal interface. The left pulmonary artery courses over the left mainstem bronchus. The right pulmonary artery is intramediastinal and is not visible on radiography. The bilateral interlobar (descending) pulmonary arteries are typically evident on normal frontal chest radiographs (measure up to 16.8 mm transversely) as are several normal central pulmonary artery branches and pulmonary vein tributaries. The superior pulmonary veins are typically visible in the bilateral suprahilar central lungs. Pulmonary arteries and veins are least perceptible in the subpleural lung periphery.* **(Bottom)** *Lateral chest radiograph of the same patient shows the normal pulmonary vessels. The pulmonary trunk is barely apparent in this normal patient. The right hilar vascular opacity is produced by summation of hilar pulmonary veins and arteries. The left pulmonary artery is visible in most normal subjects. Pulmonary vessels are least apparent in the lung periphery.*

PULMONARY TRUNK, PA CHEST RADIOGRAPHY

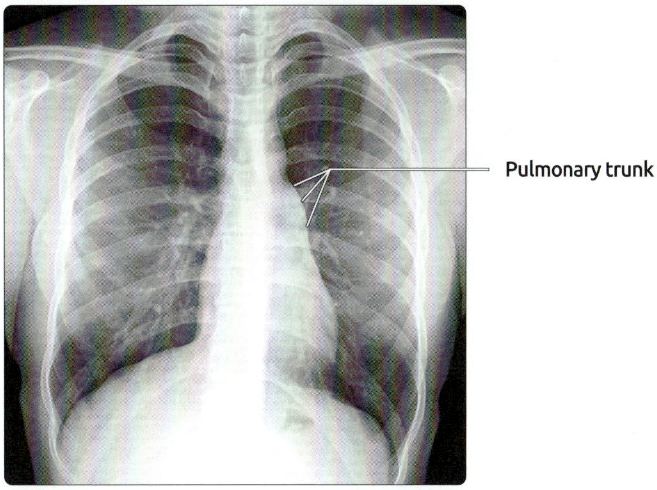

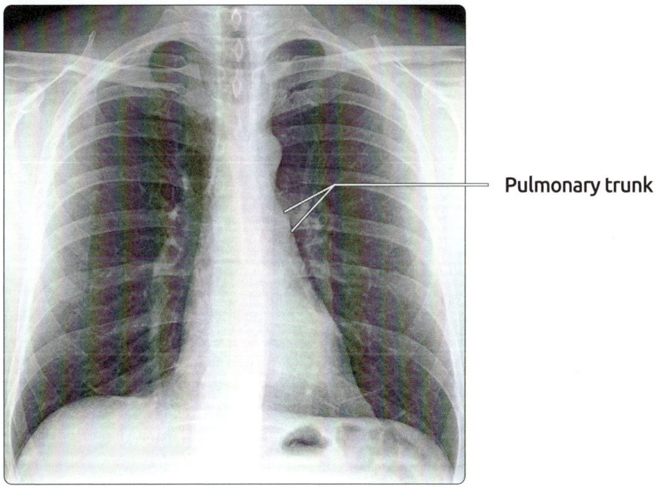

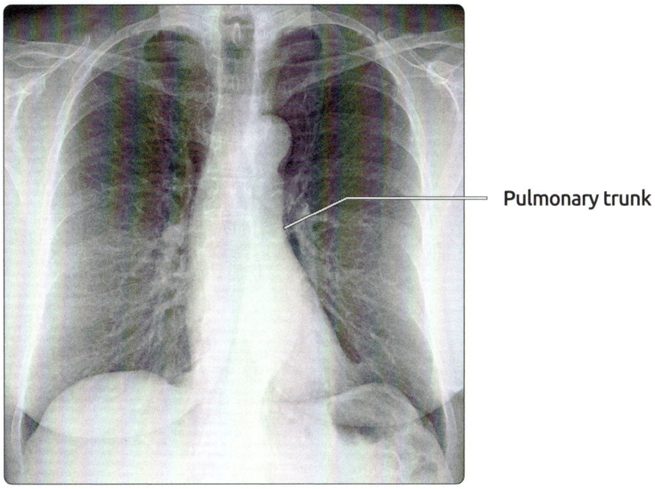

(Top) *PA chest radiograph of a 20-year-old woman shows normal prominence of the pulmonary trunk, which manifests as a midleft mediastinal convexity. The normal pulmonary trunk may be prominent in chest radiographs of children, adolescents, and young adults.* **(Middle)** *PA chest radiograph of a 36-year-old man shows a normal cardiomediastinal silhouette. Note absence of convexity and low conspicuity of the normal pulmonary trunk in this case.* **(Bottom)** *PA chest radiograph of a 60-year-old woman shows a normal heart size and aortic tortuosity. Note the straight to minimally concave contour of the pulmonary trunk mediastinal interface, characteristic of normal adult subjects.*

ANATOMY OF PULMONARY ARTERIES AND VEINS

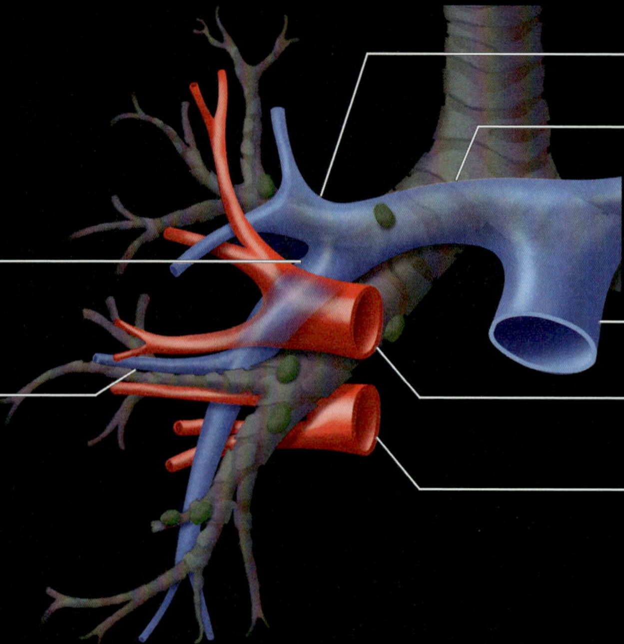

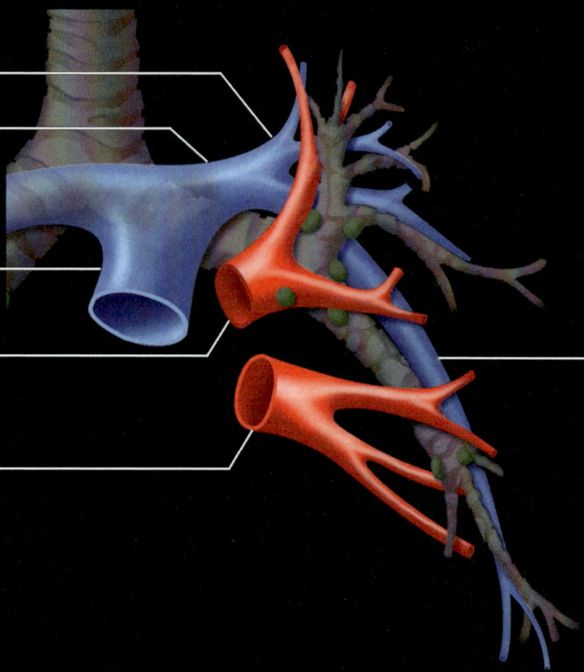

(Top) *First of 2 graphics depicting the anatomy of the pulmonary arteries and veins is shown. Anterior view of the right pulmonary vasculature shows the right pulmonary artery arising from the pulmonary trunk and bifurcating into ascending (truncus anterior) and descending (interlobar) branches. Pulmonary artery branches generally course medial to the bronchi in the right upper lobe and lateral to the bronchi in the middle and lower lobes. The right superior pulmonary vein courses anterior to the right upper lobe bronchi and right pulmonary artery branches. The right inferior pulmonary vein courses posterior to the bronchi inferior to the right hilum.* **(Bottom)** *Anterior view of the left pulmonary vessels shows that the pulmonary trunk gives rise to the left pulmonary artery and its branches, which course along the bronchi. The left pulmonary artery courses posteriorly and over the left mainstem bronchus. Pulmonary artery branches generally course medial to the left upper lobe apicoposterior and anterior segmental bronchi and lateral to the lingular and lower lobe bronchi. The pulmonary veins course anterior to the bronchi to anastomose with the left atrium.*

PULMONARY ARTERIES, ANGIOGRAPHY

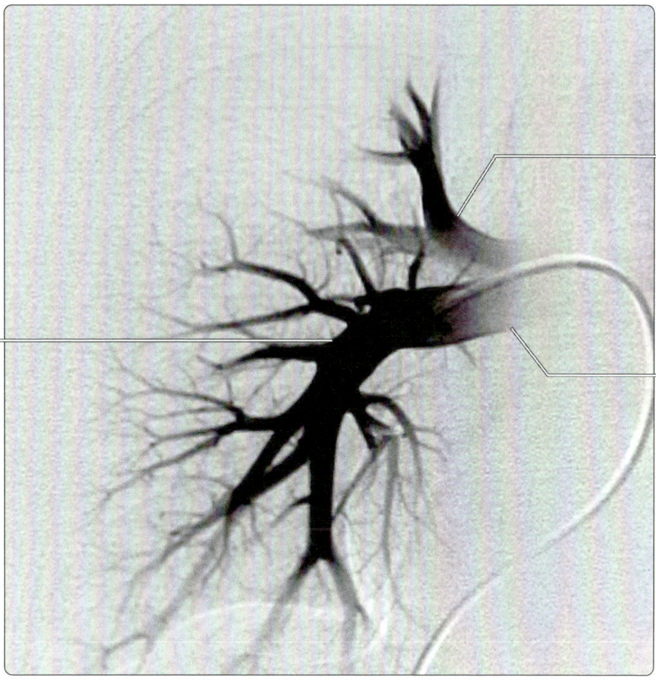

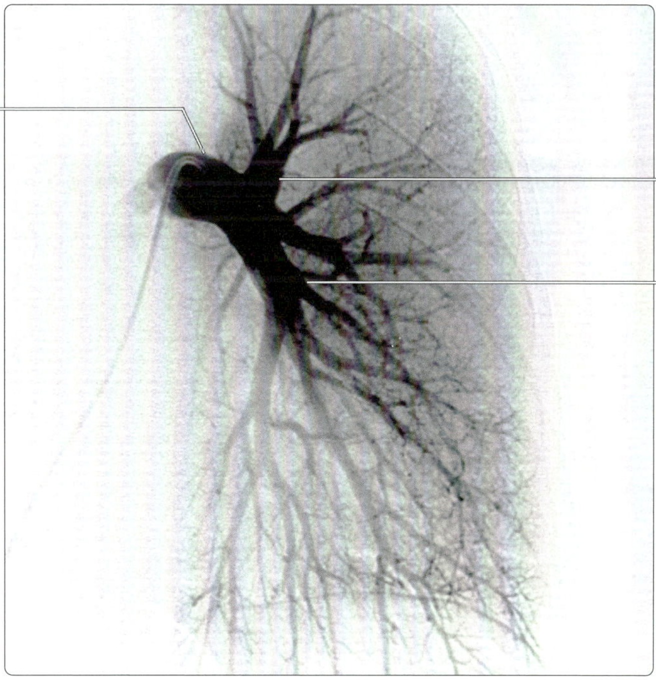

(Top) *First of 2 images from a normal pulmonary arteriogram is shown. Right pulmonary arteriogram shows an endovascular catheter that courses from the inferior vena cava through the right atrium, right ventricle, pulmonary trunk, into the right pulmonary artery and terminates in the proximal right interlobar pulmonary artery. Note the horizontal course of the right pulmonary artery, the superior course of the truncus anterior pulmonary artery, and the arcuate/inferior vertical course of the right interlobar pulmonary artery.* **(Bottom)** *Left pulmonary arteriogram shows an endovascular catheter coursing from the right atrium through the right ventricle and pulmonary trunk to terminate in the distal left pulmonary artery. The left pulmonary artery is shorter and smaller than the right, courses posteriorly over the left mainstem bronchus, and gives rise to the left upper lobe and left interlobar pulmonary arteries.*

PULMONARY ARTERIES AND VEINS, AXIAL CT

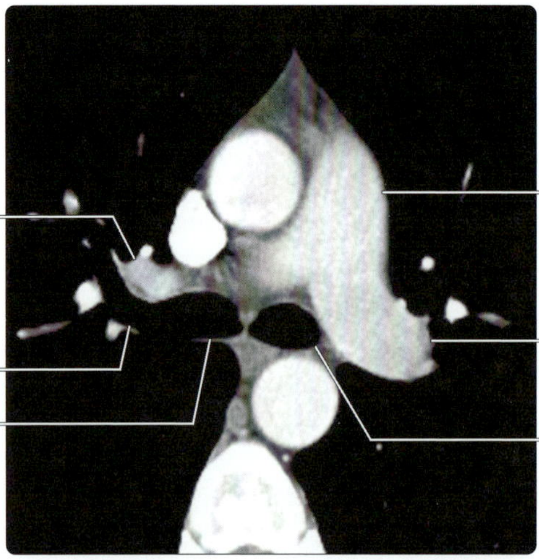

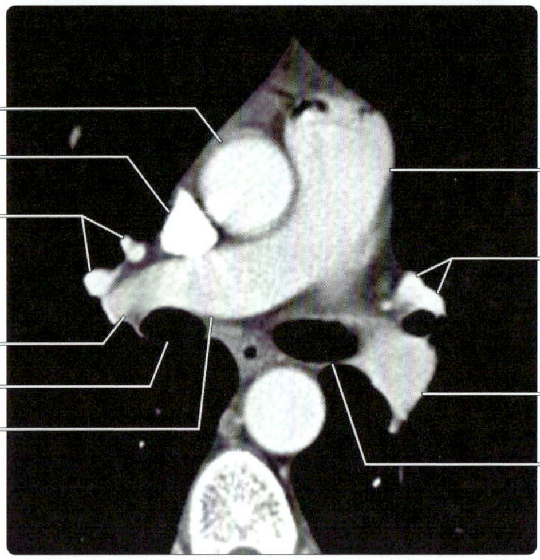

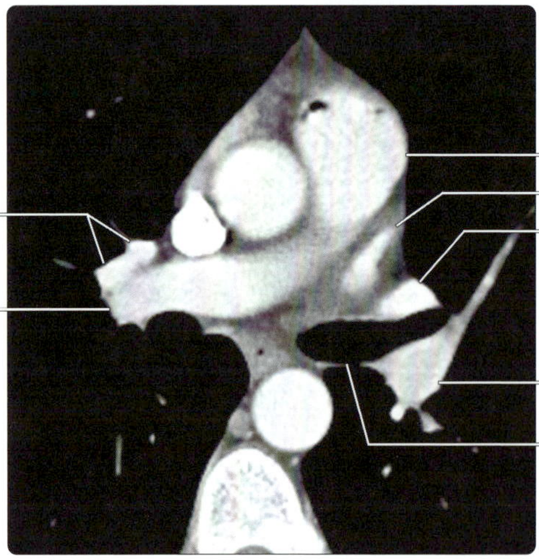

(Top) *First of 6 normal contrast-enhanced axial CT images shows the normal anatomy of the central pulmonary arteries. Axial CECT (mediastinal window) shows that the left pulmonary artery is located cephalad to the right pulmonary artery and courses posteriorly over the left mainstem bronchus. The truncus anterior pulmonary artery is located anterior to the right upper lobe bronchus.* **(Middle)** *Axial CECT (mediastinal window) shows the pulmonary trunk, which courses posteriorly and is located to the left of the ascending thoracic aorta. The left mainstem bronchus is immediately below the left pulmonary artery, which continues vertically as the left interlobar pulmonary artery. The right pulmonary artery follows a horizontal intramediastinal course behind the ascending thoracic aorta and superior vena cava.* **(Bottom)** *Axial CECT (mediastinal window) shows that pulmonary trunk is the most anteriorly located great vessel in the axial plane. The bilateral superior pulmonary veins are located anterior to the bronchi and pulmonary arteries. The normal left superior pulmonary vein is consistently located just behind the left atrial appendage.*

PULMONARY ARTERIES AND VEINS, AXIAL CT

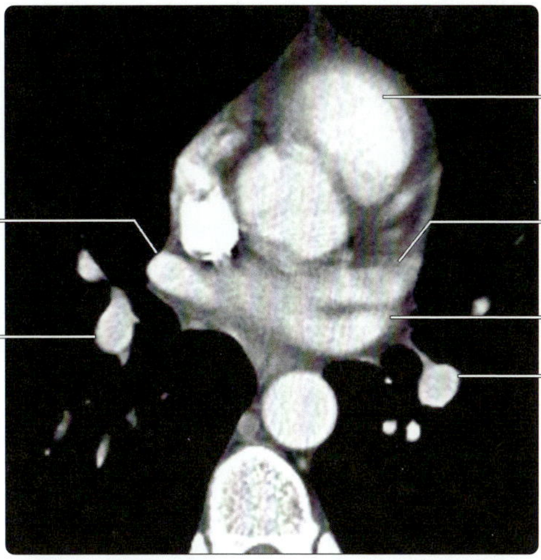

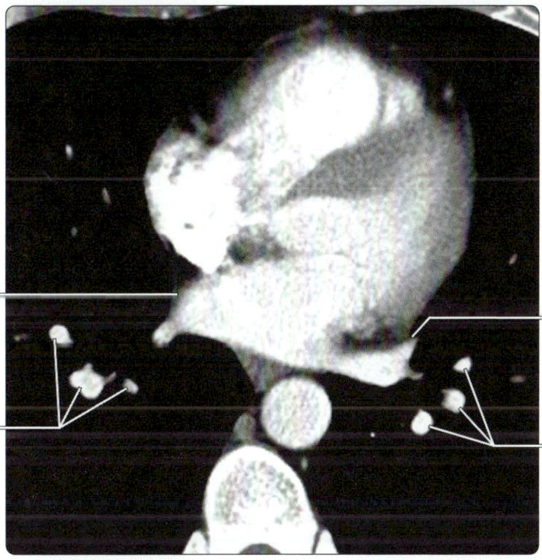

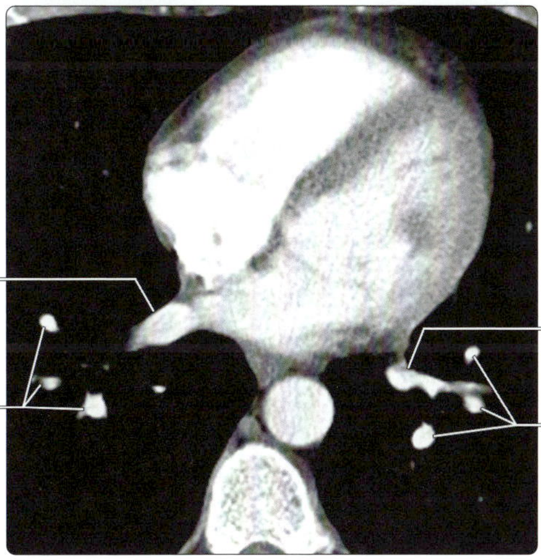

(Top) *Axial CECT (mediastinal window) shows the consistent relationship between the left atrial appendage anteriorly and the left superior pulmonary vein posteriorly and inferiorly. The right superior pulmonary vein is located anterior to the bronchus intermedius and its branches. The bilateral interlobar pulmonary arteries are located posterolateral to their accompanying bronchi.* **(Middle)** *Axial CECT (mediastinal window) shows the left inferior pulmonary vein and the bilateral basilar segmental lower lobe pulmonary arteries. The arteries follow a relatively vertical course with respect to the more horizontal course of the pulmonary veins.* **(Bottom)** *Axial CECT (mediastinal window) through the inferior aspect of the left atrium demonstrates the bilateral inferior pulmonary veins and their tributaries, which follow a nearly horizontal course into the left atrium. The bilateral lower lobe basilar segmental pulmonary arteries follow a vertical course toward the lung bases.*

PULMONARY VEINS, ANGIOGRAPHY

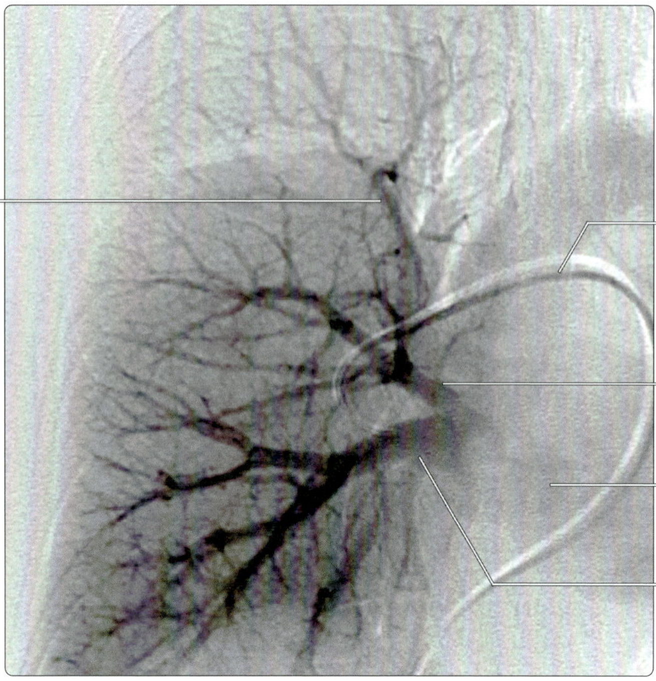

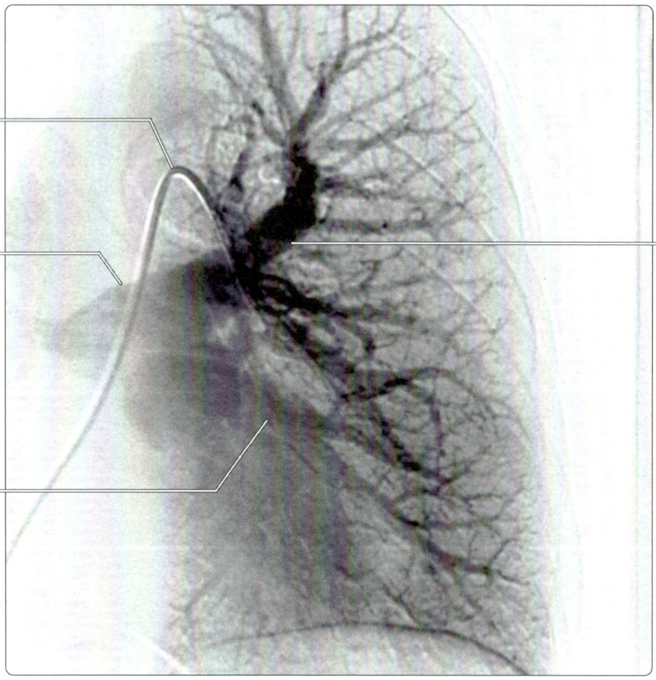

(Top) *First of 2 images from a normal pulmonary arteriogram shows the normal anatomy of the pulmonary veins. AP view of the venous phase of a right pulmonary arteriogram shows an arterial catheter that courses through the right atrium, right ventricle, pulmonary trunk, and right pulmonary artery with its tip in the right interlobar pulmonary artery. This delayed image shows the morphology of the right pulmonary veins as they course medially to anastomose with the left atrium. In this case, the middle lobe pulmonary vein drains into the right inferior pulmonary vein. The right superior pulmonary vein is not well opacified.* **(Bottom)** *AP view of the venous phase of a left pulmonary arteriogram demonstrates the left pulmonary veins draining into the left atrium. The pulmonary artery catheter courses through the right atrium, right ventricle, pulmonary trunk, and left pulmonary artery to terminate in the left interlobar pulmonary artery. The normal left superior pulmonary vein is well opacified. The left inferior pulmonary vein is less clearly visualized.*

PULMONARY ARTERIES AND VEINS, CORONAL CT

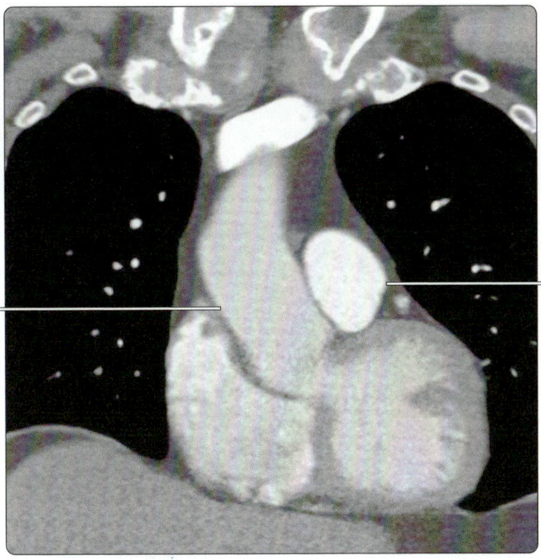

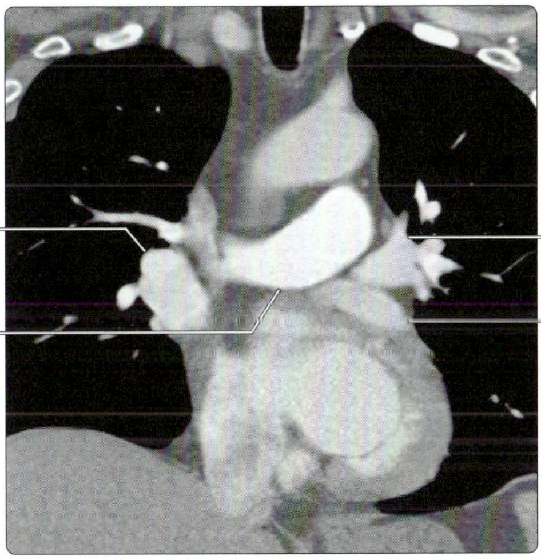

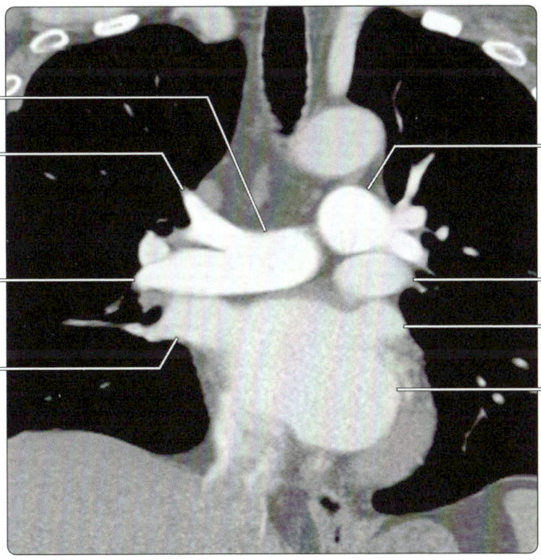

(Top) *First of 6 normal contrast-enhanced chest CT images of the same patient presented from anterior to posterior shows the normal anatomy of the central pulmonary arteries and veins. Coronal CECT (mediastinal window) shows the pulmonary trunk coursing posteriorly and to the left of the ascending thoracic aorta.* **(Middle)** *Coronal CECT (mediastinal window) shows the right pulmonary artery and its horizontal intramediastinal course. The bilateral superior pulmonary veins course obliquely into the left atrium and are anteriorly located structures. The left superior pulmonary vein is located posterior to the left atrial appendage, a characteristic anatomic relationship between these 2 structures.* **(Bottom)** *Coronal CECT (mediastinal window) shows the bifurcation of the right pulmonary artery into truncus anterior and interlobar branches. The left pulmonary artery courses posteriorly and bifurcates into upper lobe and interlobar branches. The right superior pulmonary vein is visualized as it anastomoses with the left atrium.*

PULMONARY ARTERIES AND VEINS, CORONAL CT

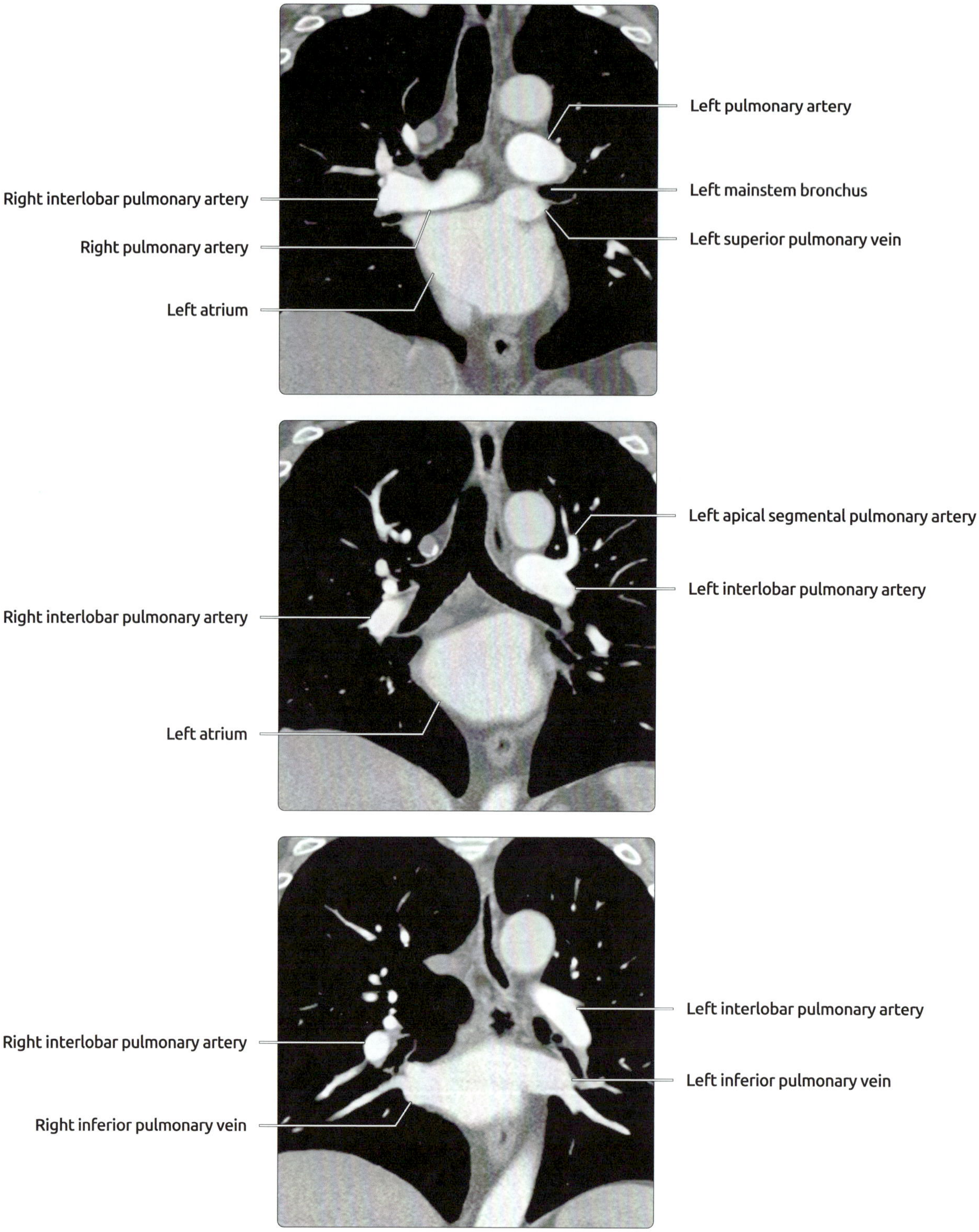

(Top) *Coronal CECT (mediastinal window) shows that the left pulmonary artery is located cephalad to the right pulmonary artery and courses posteriorly over the left mainstem bronchus. The lateral aspect of the horizontal portion of the right pulmonary artery gives rise to the right interlobar pulmonary artery.* **(Middle)** *Coronal CECT (mediastinal window) shows the bilateral interlobar pulmonary arteries and the left apical segmental pulmonary artery arising as a branch of the left upper lobe pulmonary artery.* **(Bottom)** *Coronal CECT (mediastinal window) through the posterior aspect of the left atrium shows the bilateral inferior pulmonary veins anastomosing with the left atrium. The bilateral interlobar pulmonary arteries and their relative vertical course with respect to the relative horizontal course of the pulmonary veins are also shown.*

PULMONARY VEIN VARIANTS, CORONAL CT

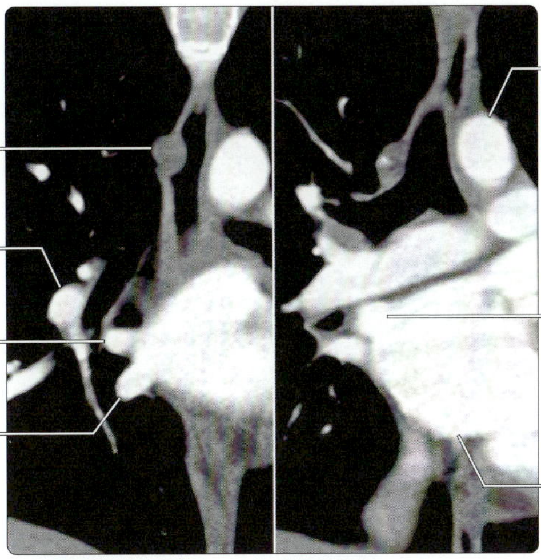

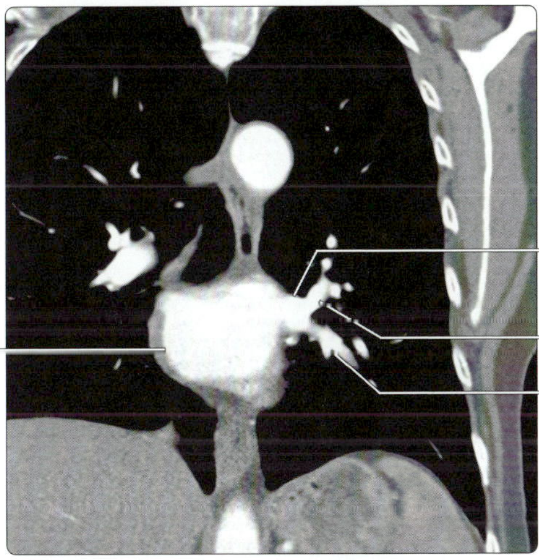

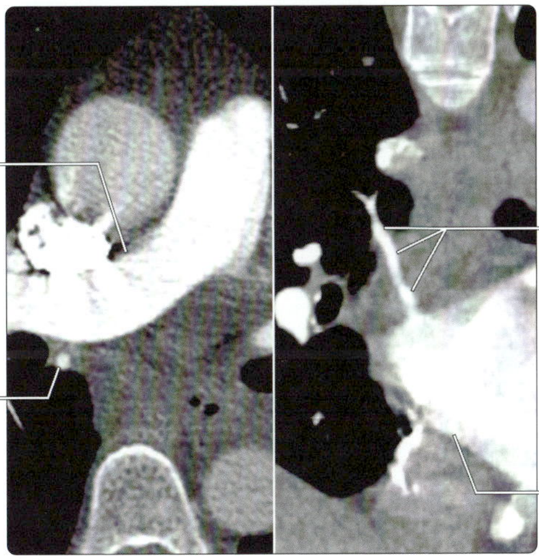

(Top) *Composite image with coronal CECT (mediastinal window) through the right side of the left atrium demonstrates a middle lobe pulmonary vein that drains directly into the left atrium between the ostia of the right superior and right inferior pulmonary veins. Supernumerary pulmonary veins are more common on the right and result from overincorporation into the left atrium during embryogenesis.* **(Middle)** *Coronal CECT (mediastinal window) through the posterior aspect of the left atrium shows that the left superior and inferior pulmonary veins share a common trunk that drains into the left atrium, the result of underincorporation of these vessels into the left atrium during embryogenesis.* **(Bottom)** *Composite image with axial (left) and coronal oblique (right) CECT (mediastinal window) of the same patient shows the anatomic variant known as the right top pulmonary vein. In this case, a small vein drains the posterior segment of the right upper lobe and courses vertically to enter the left atrium superomedially to the right superior pulmonary vein.*

RIGHT PULMONARY ARTERIES AND VEINS, AXIAL CT

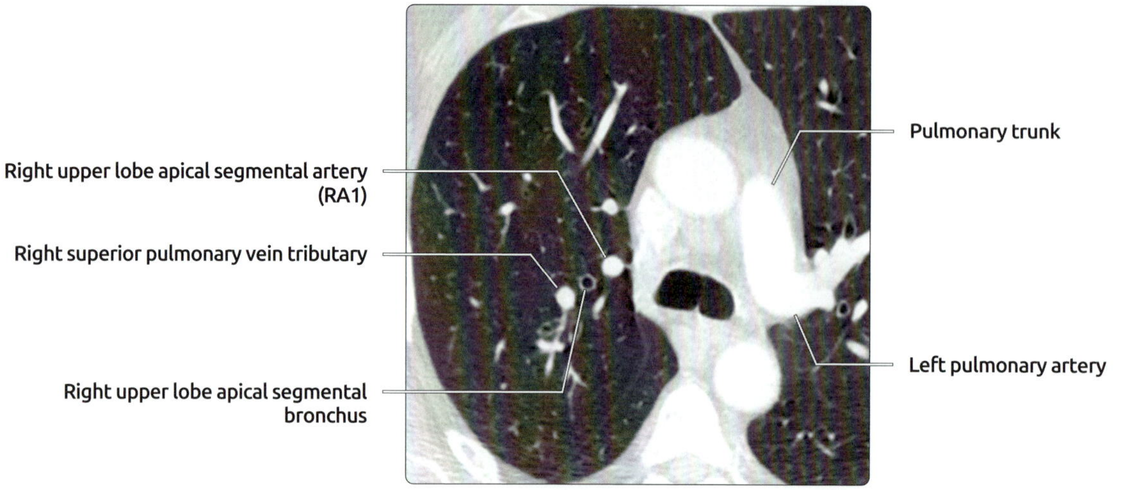

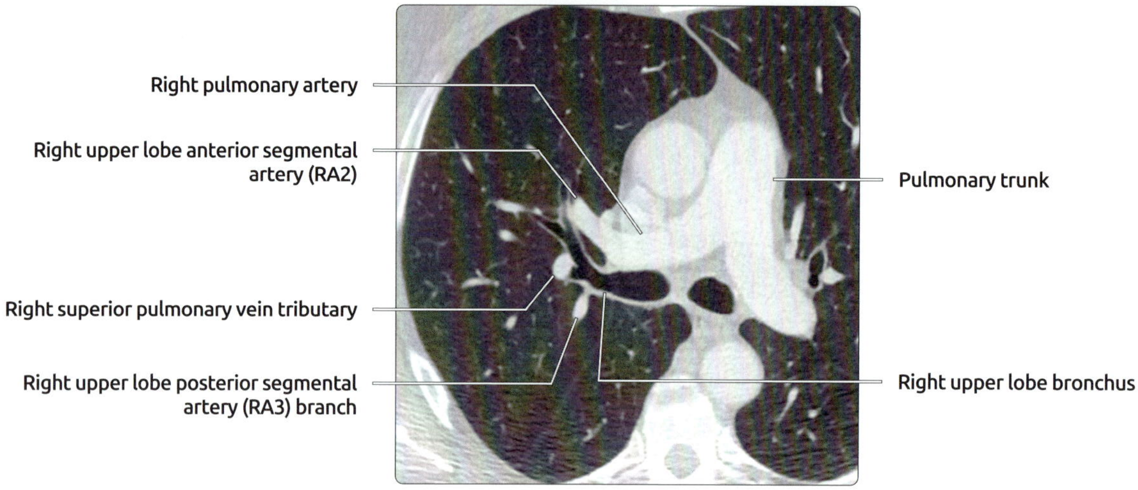

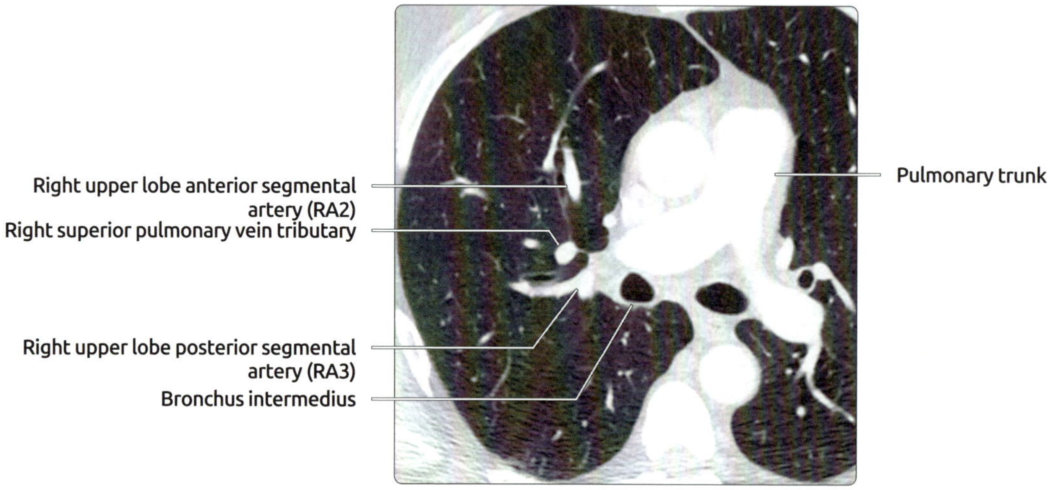

(Top) *First of 9 axial contrast-enhanced chest CT images through the right lung of the same patient shows the normal anatomy of the pulmonary arteries and veins presented from cephalad to caudad. Axial CECT (lung window) shows the right upper lobe apical segmental pulmonary artery (RA1) located medial to the right upper lobe apical segmental bronchus. A tributary of the right superior pulmonary vein is located lateral to the right upper lobe apical segmental bronchus.* **(Middle)** *Axial CECT (lung window) through the right upper lobe bronchus shows the right upper lobe anterior segmental (RA2) pulmonary artery and a subsegmental branch of the right upper lobe posterior segmental (RA3) pulmonary artery. A tributary of the right superior pulmonary vein is located lateral to the bronchi.* **(Bottom)** *Axial CECT (lung window) through the bronchus intermedius shows the proximal right upper lobe posterior segmental pulmonary artery (RA3), which typically arises from the right interlobar pulmonary artery.*

RIGHT PULMONARY ARTERIES AND VEINS, AXIAL CT

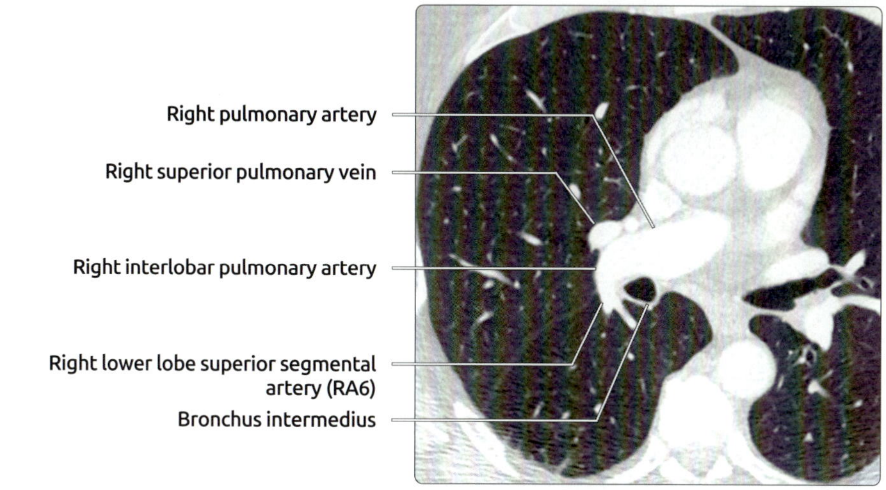

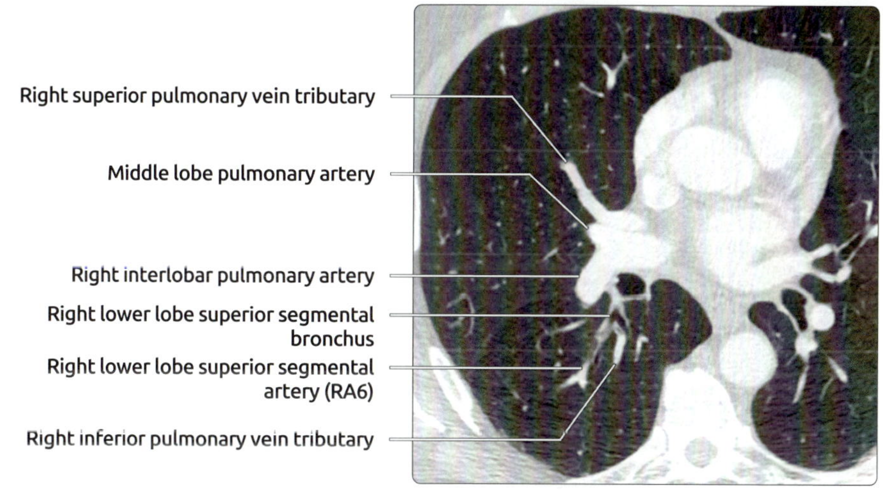

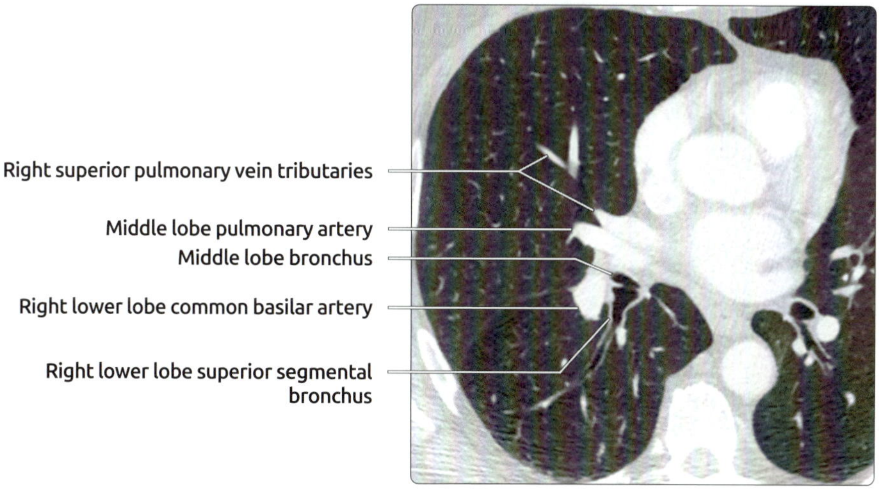

(Top) *Axial CECT (lung window) through the bronchus intermedius shows the origin of the right lower lobe superior segmental pulmonary artery (RA6) arising from the right interlobar pulmonary artery. The right superior pulmonary vein is located immediately anterior to the right pulmonary artery at this level.* **(Middle)** *Axial CECT (lung window) obtained near the origin of the right lower lobe superior segmental bronchus shows the proximal middle lobe pulmonary artery and the distal right lower lobe superior segmental pulmonary artery (RA6), which typically arises just cephalad to its corresponding segmental bronchus. The middle lobe pulmonary artery arises as a single trunk from the right interlobar pulmonary artery.* **(Bottom)** *Axial CECT (lung window) through the right lower lobe superior segmental bronchus shows the middle lobe pulmonary artery prior to its bifurcation as well as the right lower lobe common basilar pulmonary artery, which gives rise to the basilar segmental pulmonary arteries.*

RIGHT PULMONARY ARTERIES AND VEINS, AXIAL CT

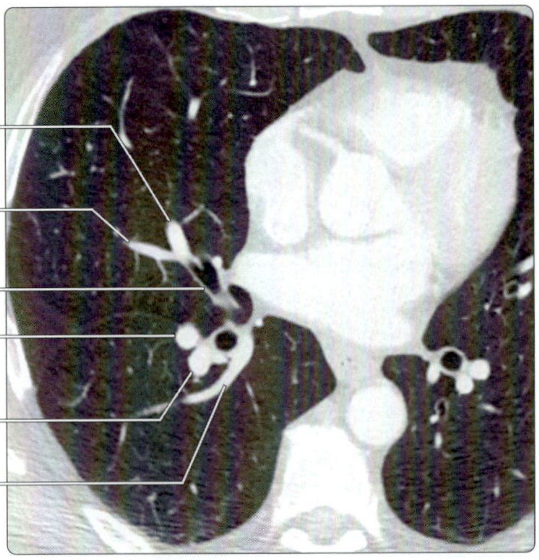

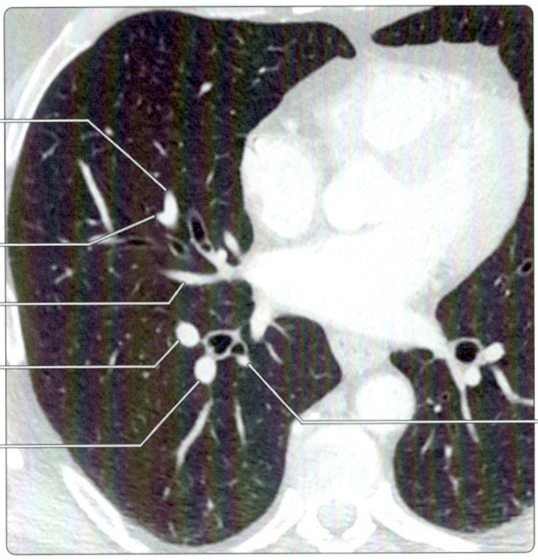

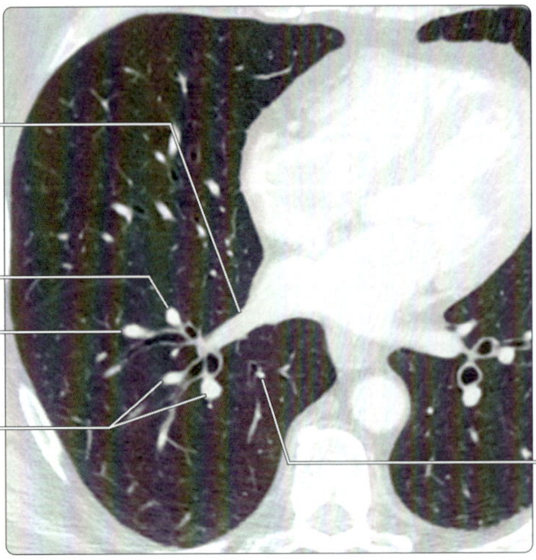

(Top) *Axial CECT (lung window) through the middle lobe bronchus shows the bifurcation of the middle lobe pulmonary artery into lateral (RA4) and medial (RA5) segmental pulmonary arteries. The right lower lobe pulmonary artery is the caudal continuation of the right interlobar pulmonary artery after the takeoff of the middle lobe pulmonary artery. After giving off the right lower lobe superior segmental pulmonary artery (RA6), it continues as the right lower lobe common basilar artery prior to giving rise to the basilar segmental pulmonary arteries.* **(Middle)** *Axial CECT shows the bifurcation of the middle lobe pulmonary artery into medial (RA5) and lateral (RA4) segmental pulmonary arteries. Note the basilar segmental right lower lobe pulmonary arteries. In this case, there is a separate middle lobe pulmonary vein.* **(Bottom)** *Axial CECT shows the right lower lobe basilar segmental pulmonary arteries located lateral to the corresponding bronchi: The medial (RA7), anterior (RA8), lateral (RA9), and posterior (RA10) basilar segmental pulmonary arteries. The left inferior pulmonary vein courses horizontally and is located medial to the bronchi.*

LEFT PULMONARY ARTERIES AND VEINS, AXIAL CT

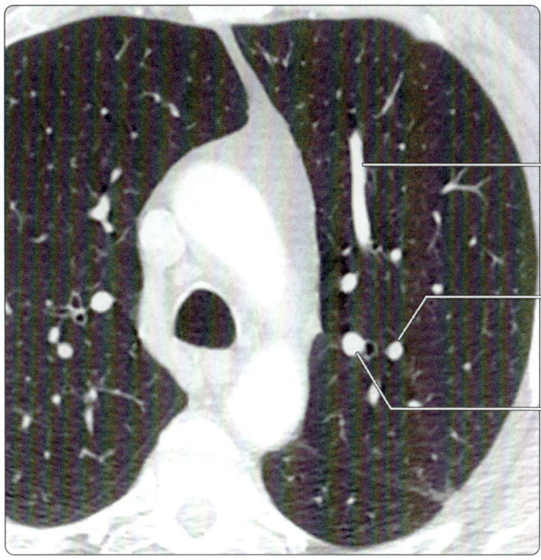

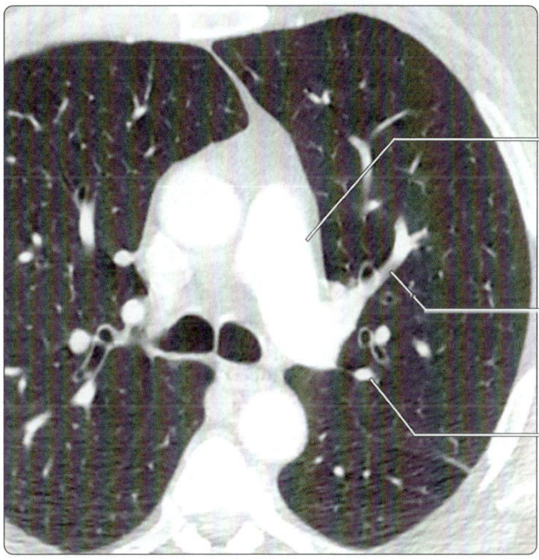

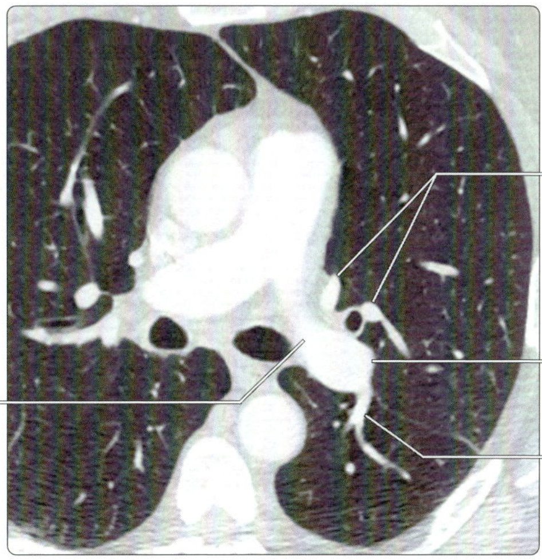

(Top) *First of 9 axial contrast-enhanced chest CT images of the same patient shows the normal anatomy of the left pulmonary arteries and veins presented from cephalad to caudad. Axial CECT (lung window) through the lower trachea shows the medially located left upper lobe apical segmental pulmonary artery (LA1) and the laterally located tributaries of the left superior pulmonary vein. Note the horizontal course of a subsegmental branch of the left upper lobe anterior segmental pulmonary artery (LA2).* **(Middle)** *Axial CECT (lung window) through the carina shows the left upper lobe anterior segmental pulmonary artery (LA2) coursing anteriorly and the left upper lobe posterior segmental pulmonary artery (LA3).* **(Bottom)** *Axial CECT (lung window) shows the left pulmonary artery coursing posteriorly directly above the left mainstem bronchus. Note the origin of the left lower lobe superior segmental pulmonary artery (LA6) from the left interlobar pulmonary artery and the anteriorly located tributaries of the left superior pulmonary vein.*

LEFT PULMONARY ARTERIES AND VEINS, AXIAL CT

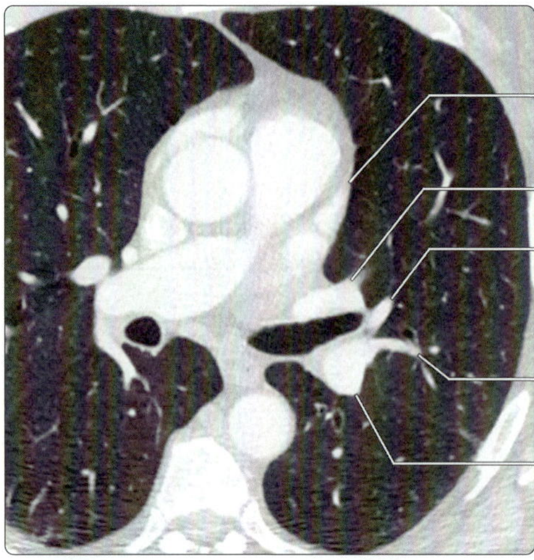

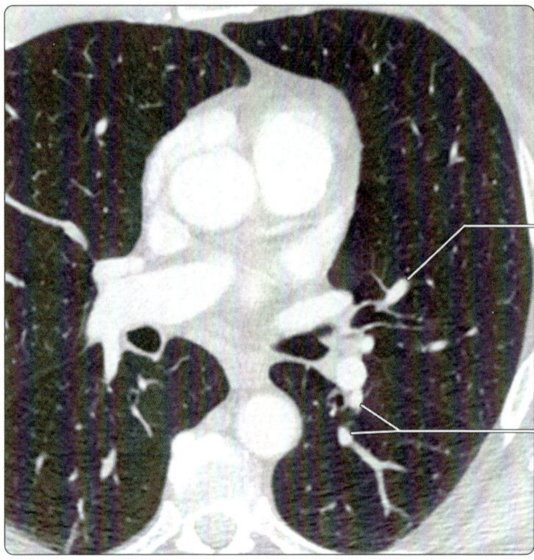

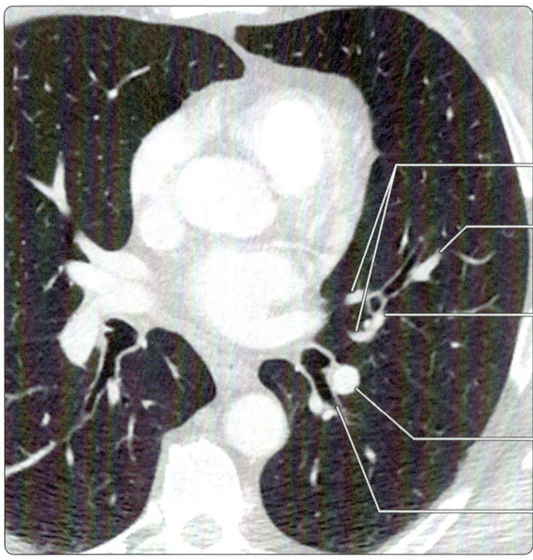

(Top) *Axial CECT (lung window) through the left mainstem and left upper lobe bronchi shows 2 left upper lobe superior lingular segmental pulmonary arteries (LA4). The left superior pulmonary vein is also shown located posterior to the left atrial appendage.* **(Middle)** *Axial CECT (lung window) shows the left upper lobe superior lingular segmental pulmonary artery (LA4) adjacent to the left upper lobe superior lingular segmental bronchus. Note a tortuous left lower lobe superior segmental pulmonary artery (LA6) arising from the left interlobar pulmonary artery.* **(Bottom)** *Axial CECT (lung window) shows tributaries of the left superior pulmonary vein, the distal left upper lobe superior lingular segmental pulmonary artery (LA4), and the proximal left upper lobe inferior lingular segmental pulmonary artery (LA5). Note the left lower lobe common basilar pulmonary artery, which gives rise to the basilar segmental pulmonary arteries.*

LEFT PULMONARY ARTERIES AND VEINS, AXIAL CT

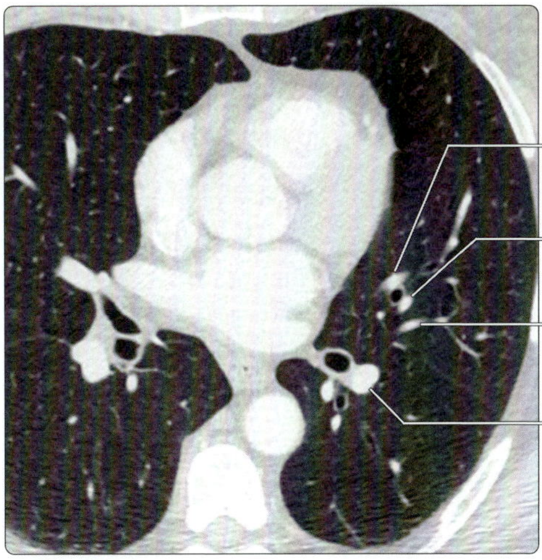

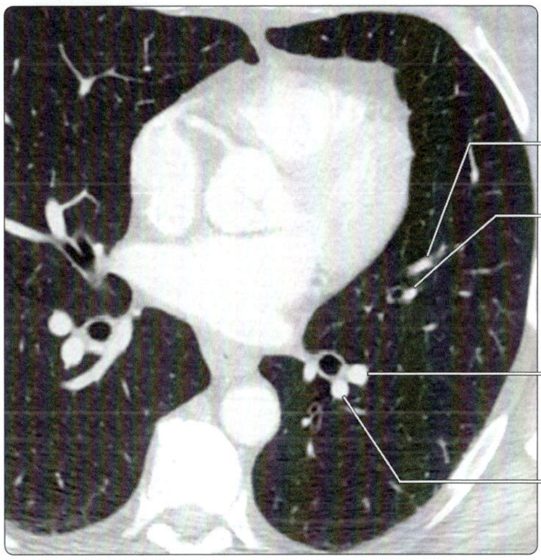

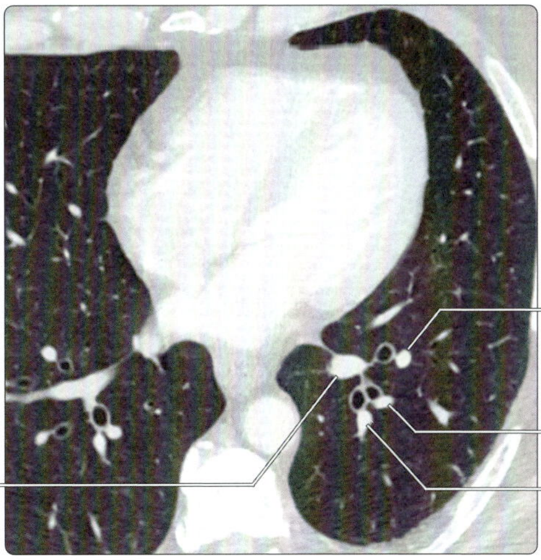

(Top) *Axial CECT (lung window) shows the proximal left upper lobe inferior lingular segmental pulmonary artery (LA5) coursing lateral to its corresponding bronchus. The corresponding pulmonary vein courses medial to the bronchus.* **(Middle)** *Axial CECT (lung window) through the left lower lobe truncus basalis bronchus shows the anteriorly located left lower lobe anteromedial basilar segmental pulmonary artery (LA7+8) and the common trunk for the left lower lobe lateral and posterior basilar segmental pulmonary arteries.* **(Bottom)** *Axial CECT (lung window) shows the left lower lobe basilar segmental bronchi and the basilar segmental pulmonary arteries. These are (clockwise from anterior to posterior) the left lower lobe anteromedial basilar segmental (LA7+8), the lateral basilar segmental (LA9), and the posterior basilar segmental (LA10) pulmonary arteries.*

PULMONARY VENOUS HYPERTENSION AND INTERSTITIAL EDEMA, RADIOGRAPHY

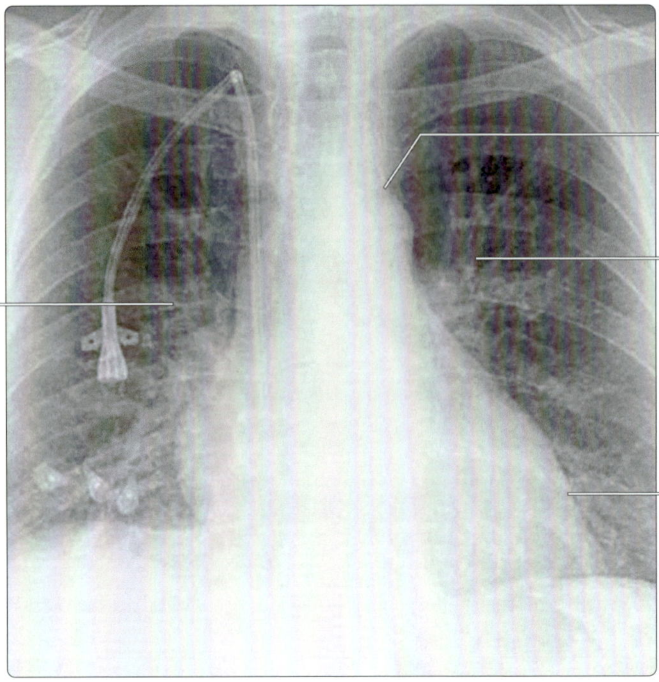

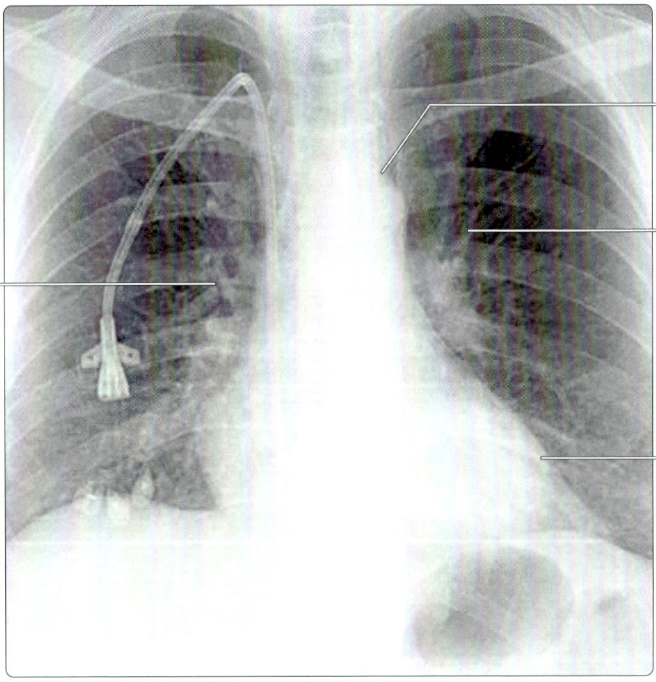

(Top) *First of 2 images of a 52-year-old man who developed dyspnea after an autologous bone marrow transplant is shown. PA chest radiograph demonstrates redistribution of pulmonary vascular flow to the upper lung zones and indistinctness of the pulmonary vessels and the hilar angle. There is cardiomegaly and a wide vascular pedicle. The findings indicate the presence of pulmonary venous hypertension and interstitial edema.* **(Bottom)** *Follow-up PA chest radiograph of the same patient demonstrates resolution of vascular redistribution, visualization of distinct central vascular margins, improved visualization of the hilar angle, a normal vascular pedicle, and resolution of previously demonstrated cardiomegaly and interstitial edema.*

PULMONARY ARTERY MEASUREMENTS AND PULMONARY HYPERTENSION, CT

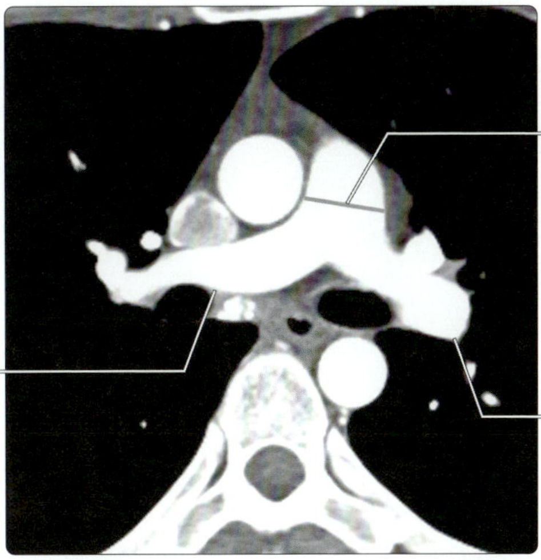

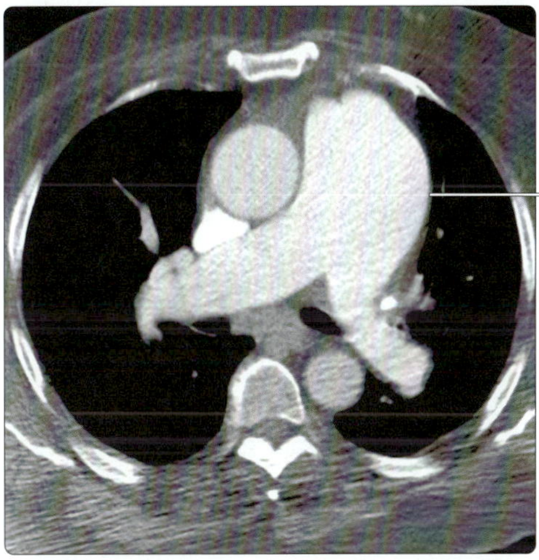

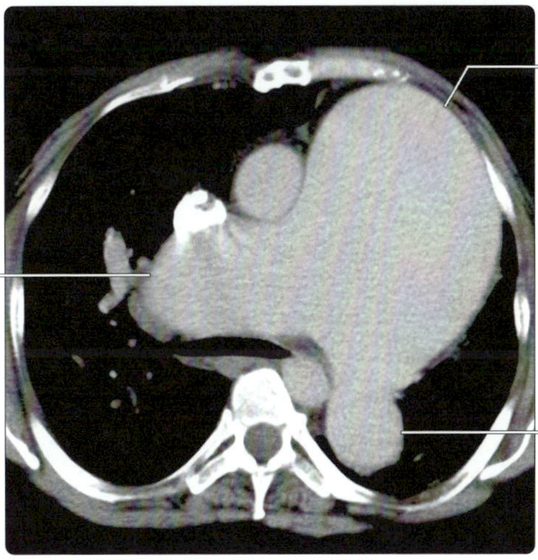

(Top) *Axial gated cardiac CECT (mediastinal window) shows the method for measuring the pulmonary trunk (gray line) on CT. The pulmonary trunk width is measured perpendicular to the vascular long axis at the level of the bifurcation.* **(Middle)** *First of 2 axial contrast-enhanced chest CT images through the pulmonary trunks of 2 different patients is shown. Axial CECT (mediastinal window) of a 74-year-old woman with pulmonary hypertension and multiple prior episodes of pulmonary thromboembolic disease shows marked enlargement of the pulmonary trunk, which measures 5.7 cm.* **(Bottom)** *Axial CECT (mediastinal window) of a 51-year-old woman with severe pulmonary hypertension shows massive enlargement of the pulmonary trunk and bilateral pulmonary arteries. The pulmonary trunk measures 7.4 cm, and the pulmonary artery pressure was 56 mm Hg. Pulmonary hypertension is defined as pulmonary artery pressure > 25.0 mm Hg at rest and > 30 mm Hg during exercise.*

CONGENITAL PULMONARY ARTERIOVENOUS MALFORMATION, RADIOGRAPHY, CT, ANGIOGRAPHY

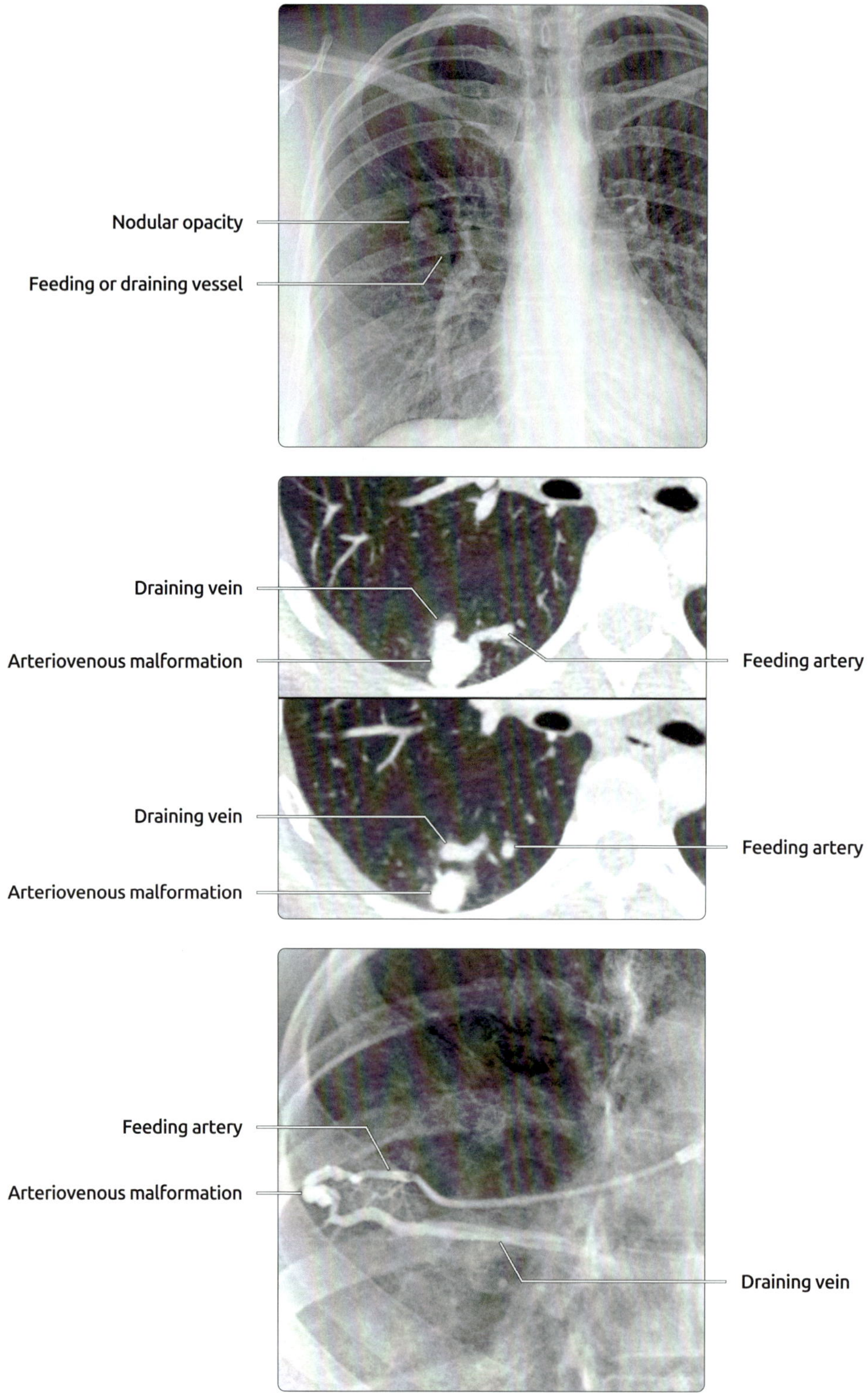

(Top) *PA chest radiograph of an asymptomatic 28-year-old woman shows a right upper lung zone nodule with an associated tubular opacity, which courses obliquely toward the right hilum, highly suggestive of a pulmonary arteriovenous malformation.* **(Middle)** *Composite image with axial CECT (lung window) of the same patient shows that the pulmonary nodule corresponds to a right lower lobe pulmonary arteriovenous malformation. The tubular opacity noted on radiography corresponds to a feeding or draining vessel.* **(Bottom)** *Selective injection from a pulmonary arteriogram into the feeding artery of an arteriovenous malformation shows the smaller caliber feeding artery and the larger caliber draining vein. An arteriovenous malformation provides a direct communication between the pulmonary arterial and venous systems without an intervening capillary bed and produces a right-to-left shunt.*

HEPATOPULMONARY SYNDROME, CT

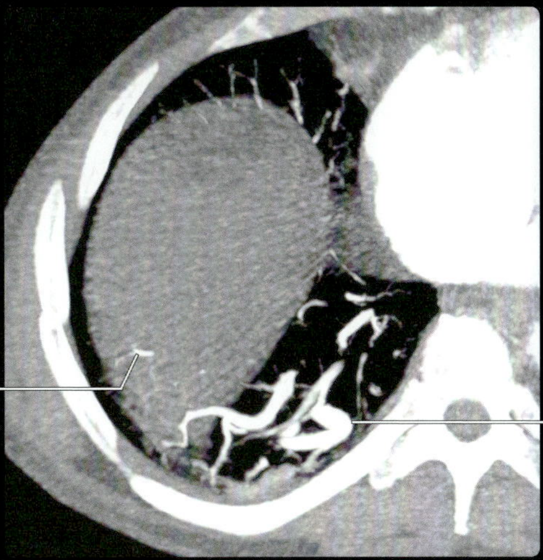

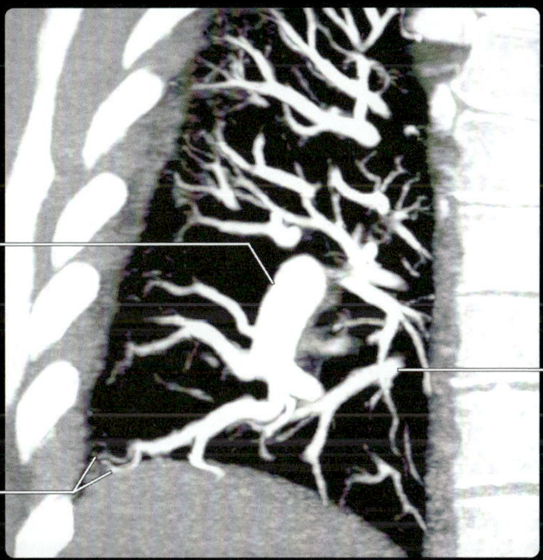

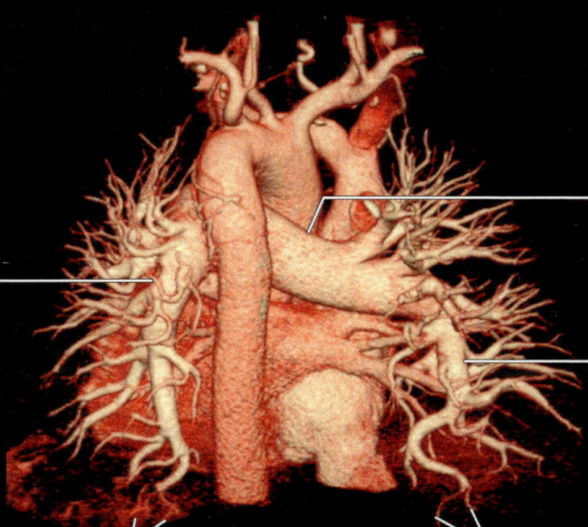

(Top) *First of 3 contrast-enhanced chest CT images of a 48-year-old man with end-stage liver disease and hepatopulmonary syndrome is shown. Axial CECT (mediastinal window) MIP reformatted image through the right lung base shows markedly dilated right lower lobe pulmonary arteries and veins and peripheral acquired arteriovenous communications.* **(Middle)** *Coronal CECT (mediastinal window) MIP reformatted image of the same patient demonstrates marked enlargement of the right pulmonary arteries and veins. Peripheral arteriovenous malformations or communications are visible in the right lung base.* **(Bottom)** *Coronal posterior volume-rendered CT through the thorax of the same patient shows enlarged pulmonary arteries and enlarged bilateral interlobar pulmonary arteries. Subpleural arteriovenous communications and dilated peripheral pulmonary vessels are also visible.*

MEANDERING PULMONARY VEIN, CT

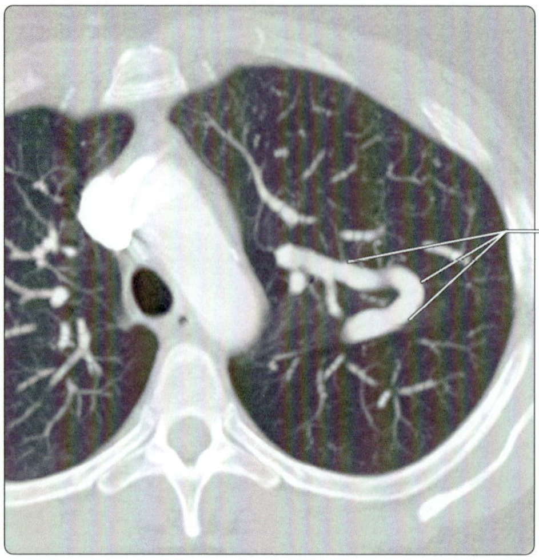

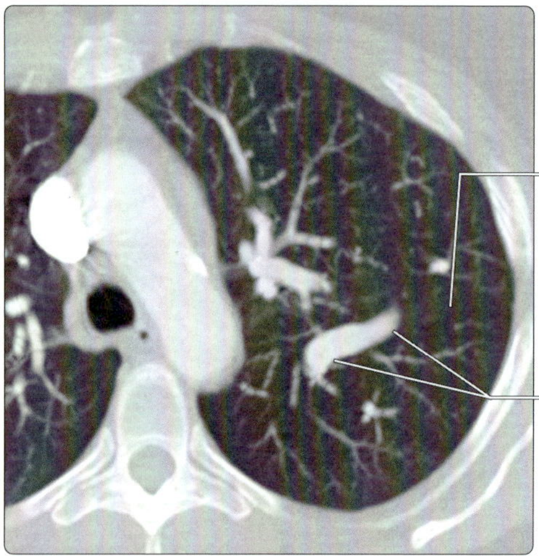

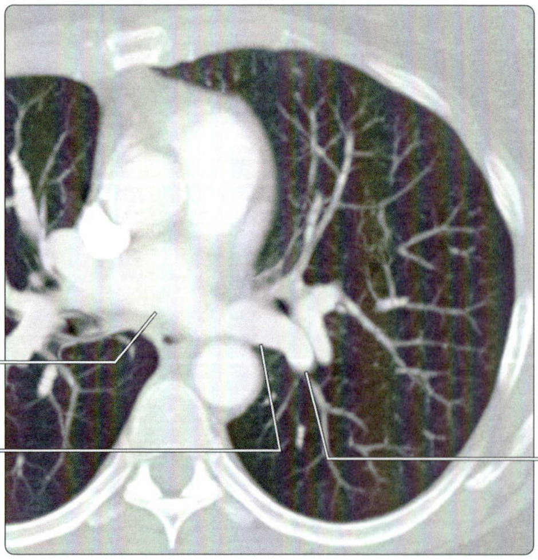

(Top) *Axial CECT (lung window) MIP reformatted image of a patient evaluated because of a left upper lobe indeterminate lung nodule seen on chest radiography (not shown) shows that the radiographic abnormality corresponds to a dilated tortuous pulmonary vein in the apicoposterior segment of the left upper lobe. The anomalous pulmonary vein drained into the left inferior pulmonary vein after following a tortuous and meandering course through the left lung.* **(Middle)** *Axial CECT (lung window) MIP reformatted image of the same patient shows the unusual course of the left upper lobe meandering pulmonary vein as it courses into the left lower lobe superior segment.* **(Bottom)** *Axial CECT (lung window) MIP reformatted image of the same patient shows that the meandering pulmonary vein drains into the left atrium via the left inferior pulmonary vein.*

PARTIAL ANOMALOUS PULMONARY VENOUS RETURN, CT

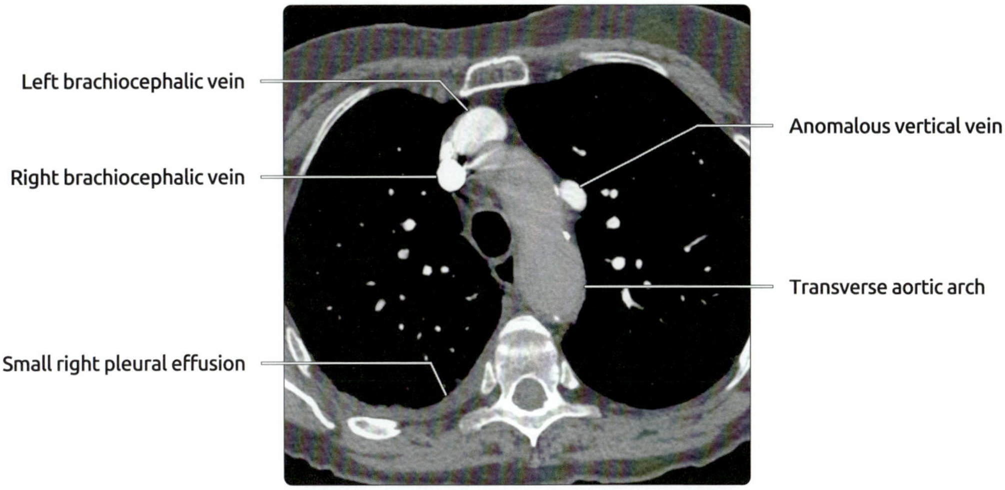

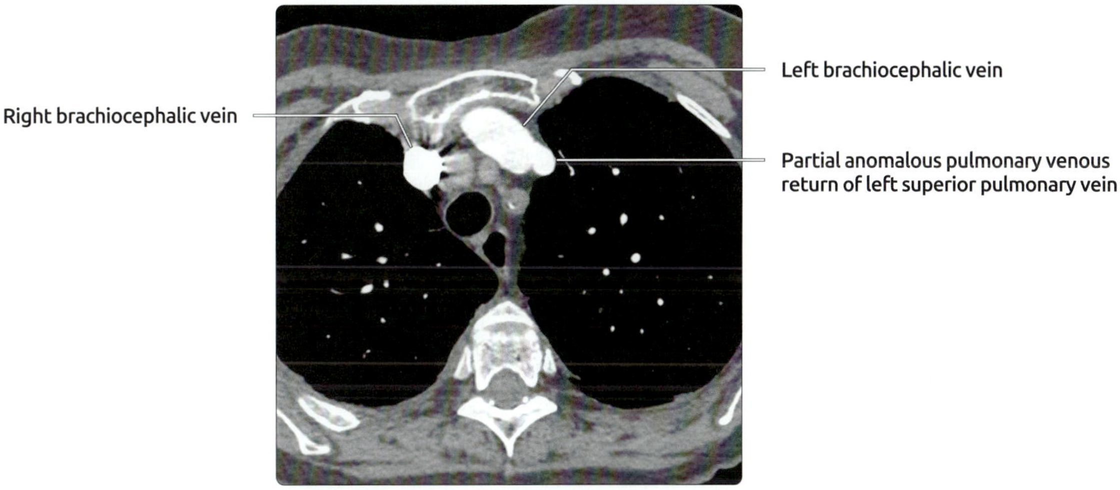

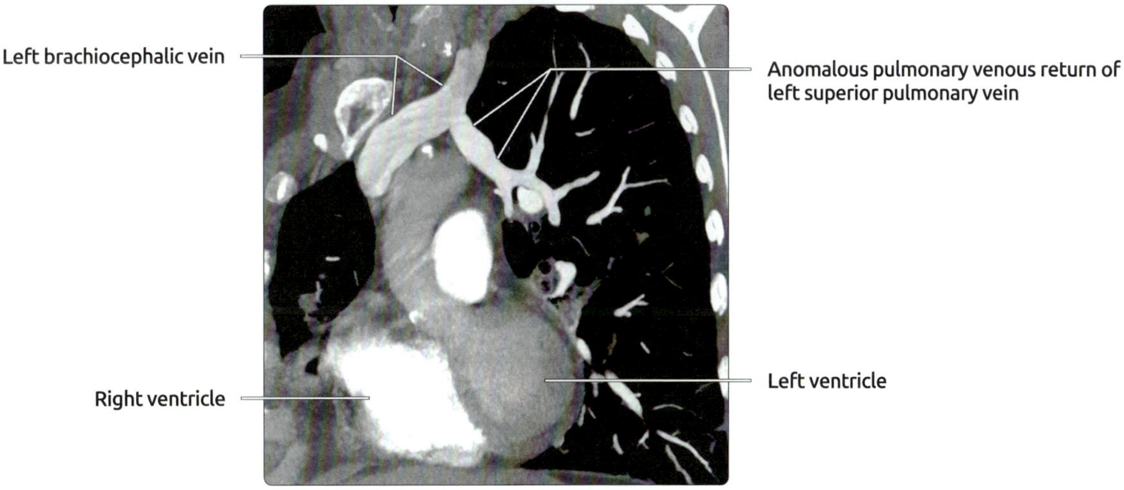

(Top) *Axial CECT (mediastinal window) of a patient with pulmonary hypertension shows an anomalous vessel that courses vertically in the left mediastinum and abuts the adjacent transverse aortic arch. Note the small right pleural effusion.* **(Middle)** *Axial CECT (mediastinal window) of the same patient shows that the anomalous vessel is a vertical vein that courses cephalad to drain into the left brachiocephalic vein. The findings are diagnostic of partial anomalous pulmonary venous return of the left upper lobe venous drainage, which constitutes a left-to-right shunt.* **(Bottom)** *Coronal oblique CECT (mediastinal window) MIP reformatted image of the same patient shows the anomalous course of the left superior pulmonary vein along the left mediastinum and its drainage into the left brachiocephalic vein. Note right ventricular enlargement, consistent with pulmonary hypertension secondary to the left-to-right shunt.*

PARTIAL ANOMALOUS PULMONARY VENOUS RETURN, CT

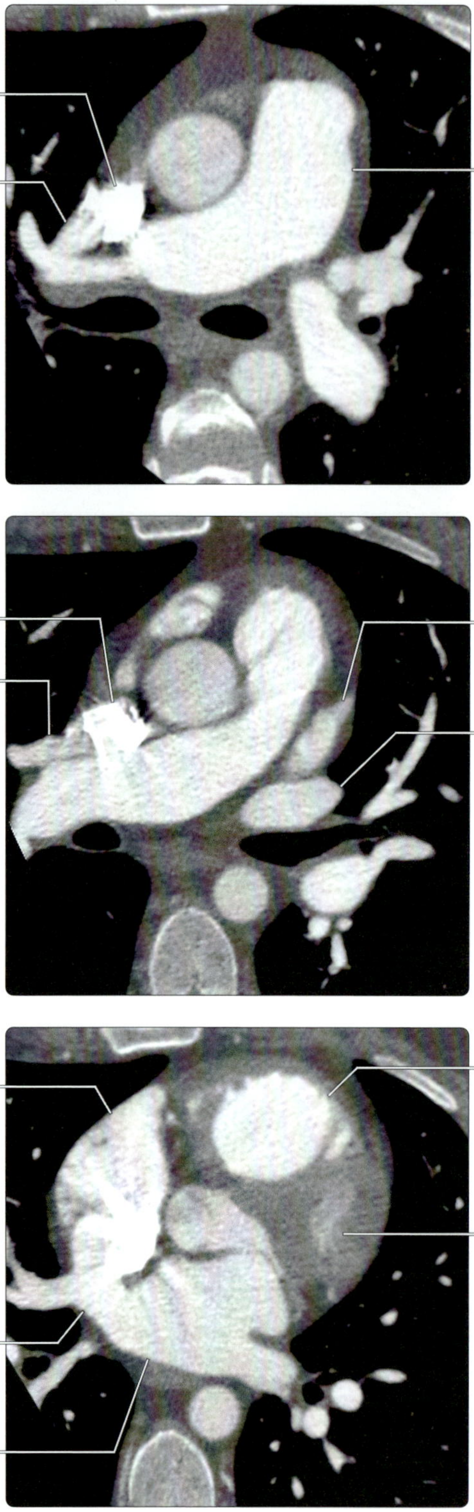

(Top) *Axial CECT (mediastinal window) of a patient with pulmonary hypertension secondary to an intracardiac left-to-right shunt shows enlargement of the pulmonary trunk, consistent with known pulmonary hypertension. Note partial anomalous pulmonary venous return of a right pulmonary vein to the superior vena cava.* **(Middle)** *Axial CECT (mediastinal window) of the same patient shows partial anomalous pulmonary venous return of a right superior pulmonary vein to the superior vena cava.* **(Bottom)** *Axial CECT (mediastinal window) of the same patient shows a sinus venosus atrial septal defect. Partial anomalous pulmonary venous return of the right superior pulmonary vein is strongly associated with atrial septal defect of the sinus venosus type.*

SCIMITAR SYNDROME, RADIOGRAPHY AND CT

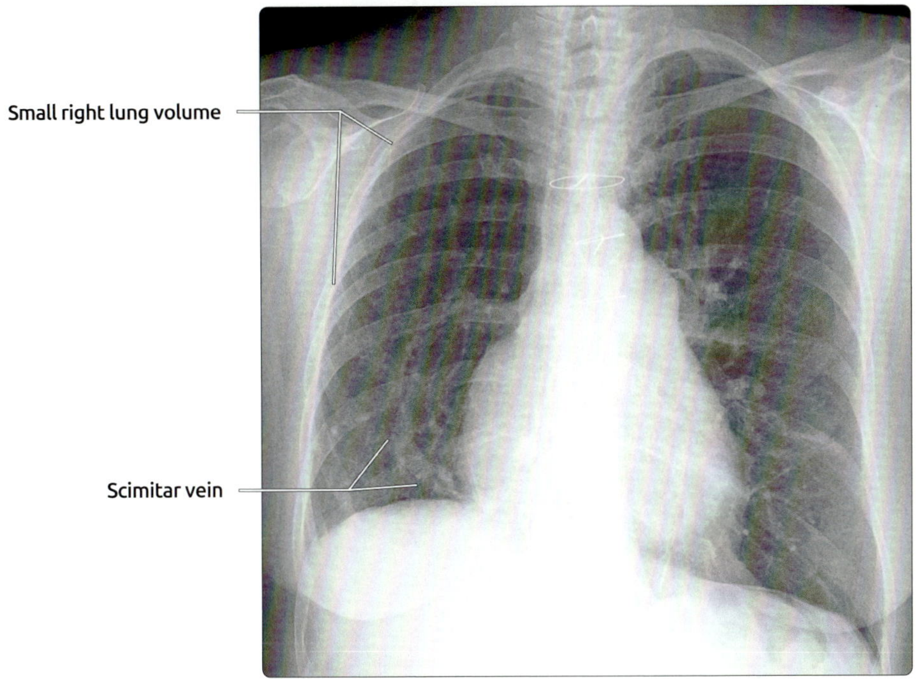

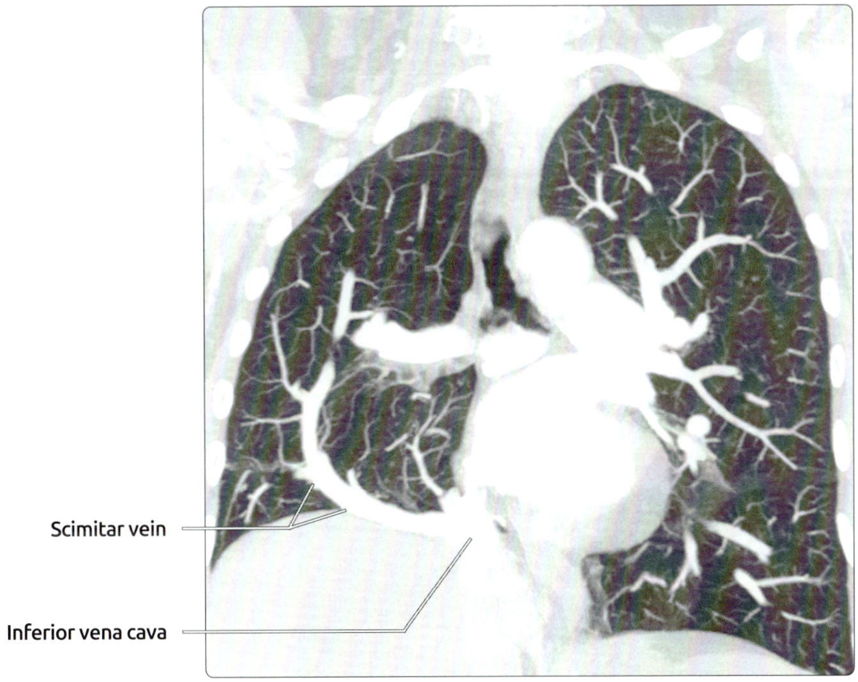

(Top) *PA chest radiograph of an adult patient status post remote surgical correction of congenital heart disease shows low right lung volume and an associated anomalous arcuate vessel in the right mid- and lower lung zones that courses toward the inferior aspect of the right mediastinum. The anomalous vessel resembles a scimitar (Turkish sword). The findings are consistent with congenital venolobar or scimitar syndrome.* **(Bottom)** *Coronal CECT (lung window) MIP reformatted image of the same patient shows a low right lung volume and the anomalous scimitar vein that drains into the inferior vena cava in this patient with scimitar syndrome, a type of partial anomalous pulmonary venous return.*

PULMONARY VEIN VARIX, RADIOGRAPHY AND CT

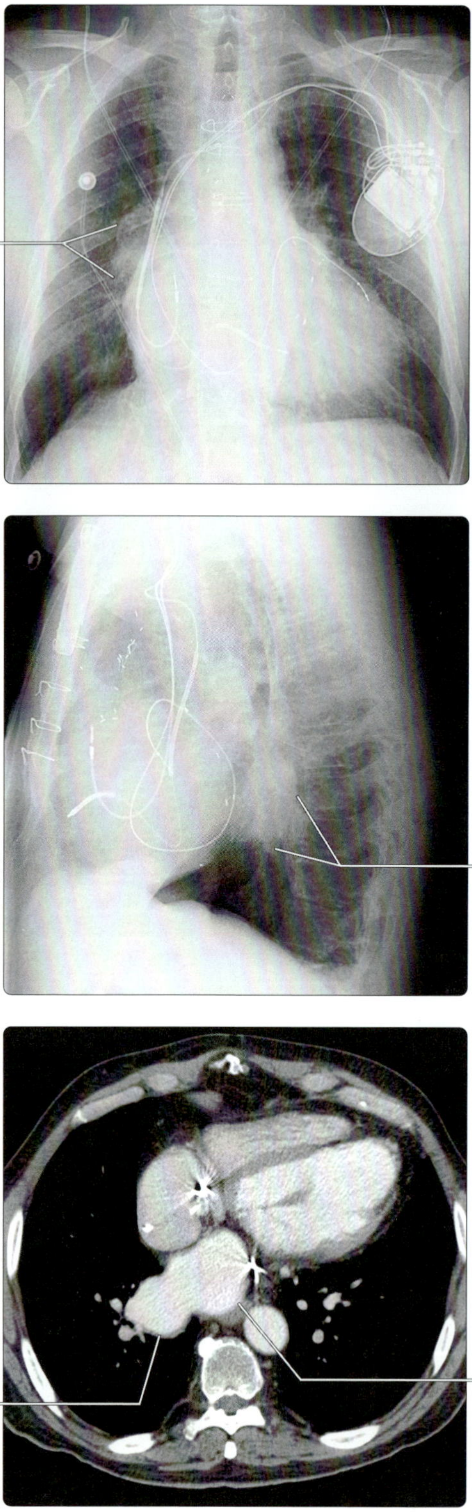

(Top) *PA chest radiograph of a patient with chronic heart failure shows a left pectoral dual-chamber biventricular pacer/implantable cardioverter defibrillator, cardiomegaly, a wide vascular pedicle, and mild interstitial edema. Note apparent enlargement of the right hilum.* **(Middle)** *Lateral chest radiograph of the same patient shows that the abnormality noted on PA chest radiography corresponds to a mass-like opacity that projects posterior to the heart and abuts the region of the left atrium.* **(Bottom)** *Axial CECT (mediastinal window) of the same patient shows that the mass-like radiographic abnormality corresponds to a large right inferior pulmonary vein that drains into an enlarged left atrium. The findings are consistent with pulmonary vein varix.*

PULMONARY ARTERY ANEURYSM OR PSEUDOANEURYSM, CT

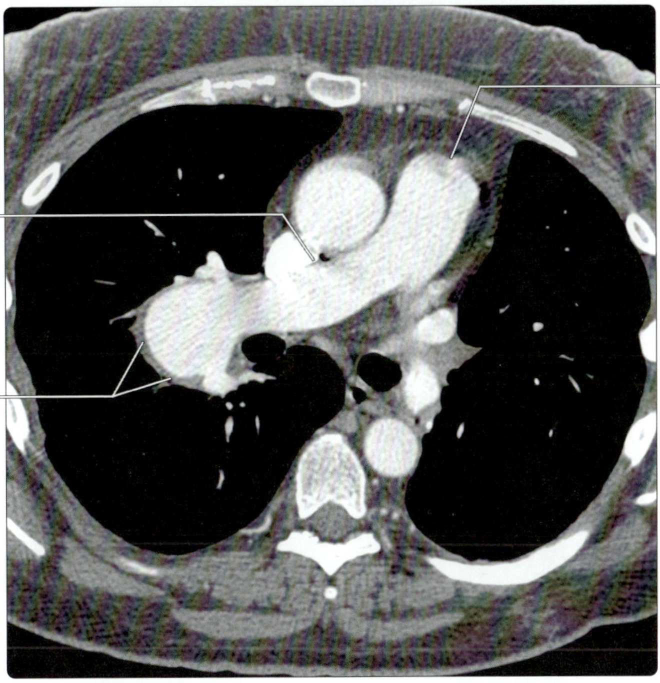

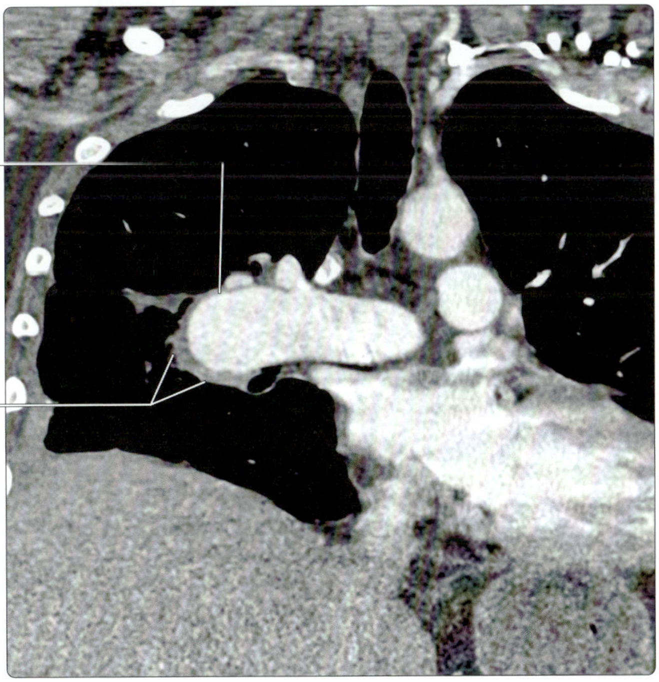

(Top) *Axial CT (mediastinal window) of a patient who presented with signs and symptoms of infection and a right hilar mass on radiography (not shown) shows marked spherical enlargement of the right interlobar pulmonary artery, consistent with a pulmonary artery aneurysm or pseudoaneurysm, clinically presumed mycotic in etiology.* **(Bottom)** *Coronal CECT (mediastinal window) of the same patient shows the large aneurysm or pseudoaneurysm of the right interlobar pulmonary artery and peripheral mural thickening of the aneurysm.*

ACUTE PULMONARY THROMBOEMBOLIC DISEASE, CT

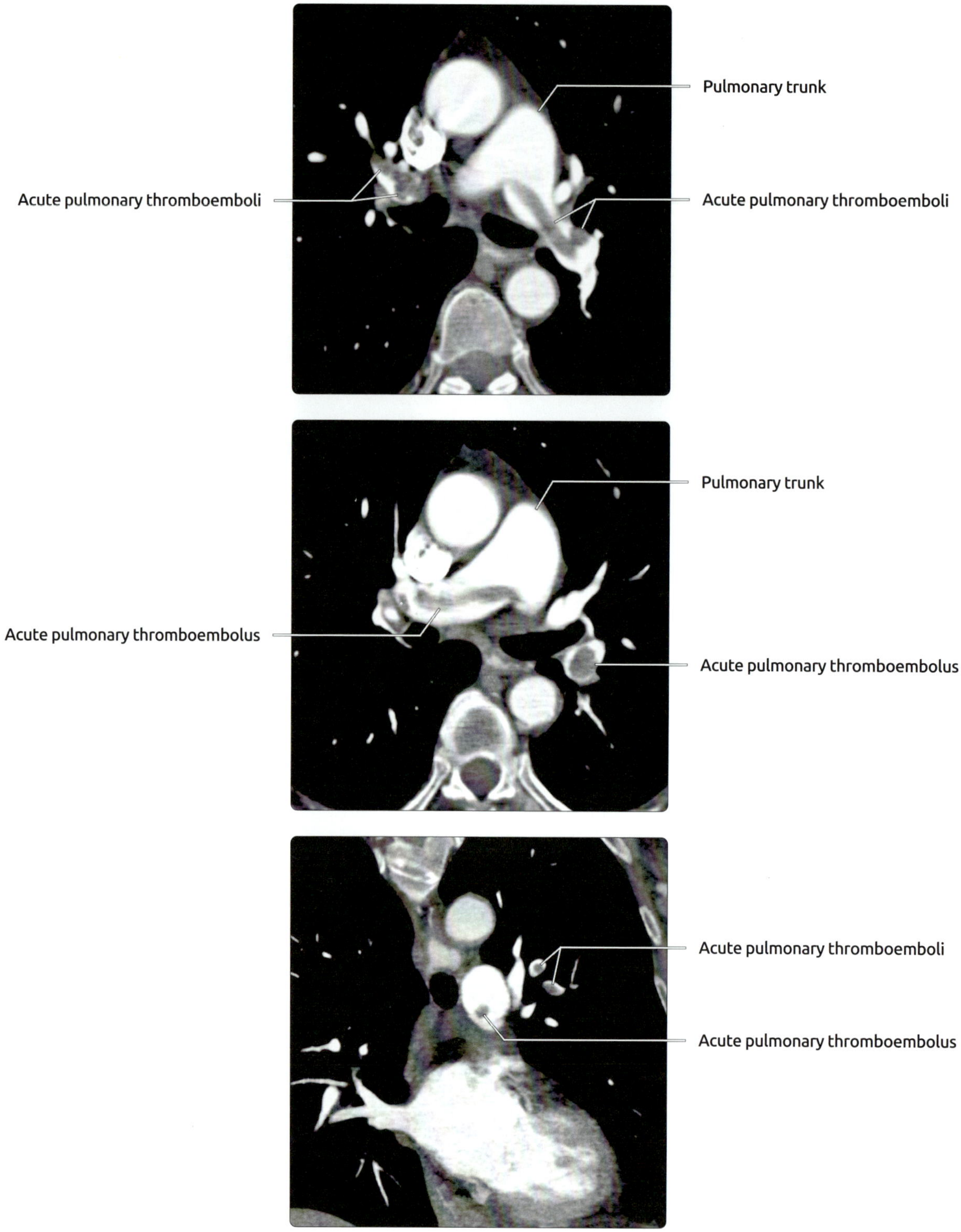

(Top) *First of 3 images from a CT pulmonary angiogram of a 67-year-old woman with acute pulmonary thromboembolic disease is shown. Axial CECT (mediastinal window) shows pulmonary thromboemboli within the left pulmonary artery that extend into the proximal left interlobar pulmonary artery. A large nearly occlusive pulmonary thromboembolus is also noted within an enlarged truncus anterior pulmonary artery.* **(Middle)** *Axial CECT (mediastinal window) through the pulmonary trunk of the same patient shows low-attenuation tubular filling defects with saddle morphology involving the bilateral proximal pulmonary arteries. Note the large left greater than right pulmonary thromboemboli in the interlobar pulmonary arteries. The pulmonary trunk is enlarged, consistent with pulmonary hypertension.* **(Bottom)** *Coronal oblique CECT (mediastinal window) through the left pulmonary artery of the same patient shows endoluminal centrally located acute thromboemboli in the left pulmonary artery and in branches of the left upper lobe anterior segmental pulmonary artery.*

CHRONIC PULMONARY THROMBOEMBOLIC DISEASE, CT

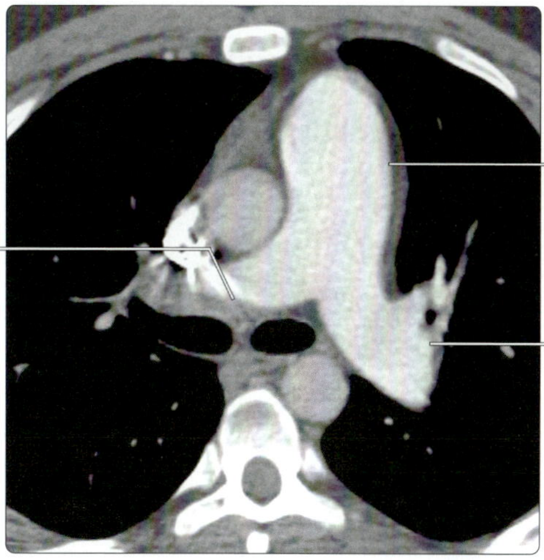

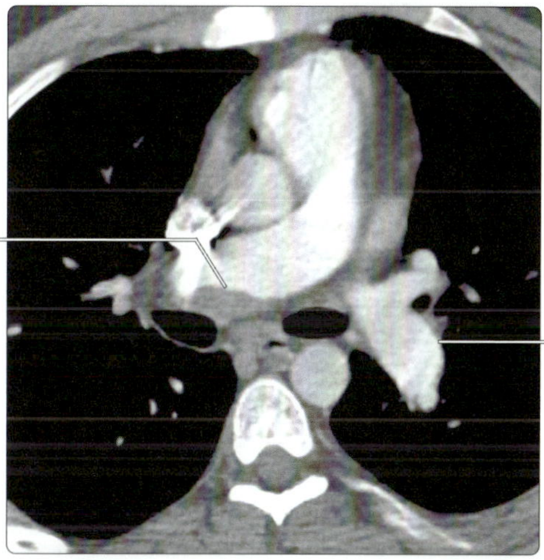

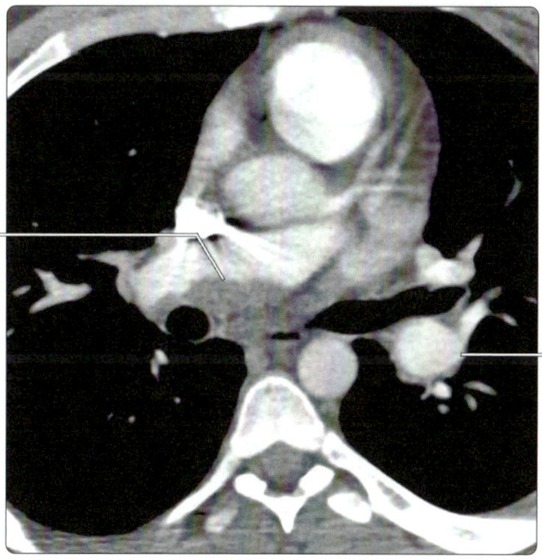

(Top) *First of 3 axial images from a contrast-enhanced chest CT of a 20-year-old man with pulmonary hypertension (pulmonary artery pressure of 51 mm Hg) secondary to chronic pulmonary thromboembolism is shown. Axial CECT (mediastinal window) through the pulmonary trunk shows diffuse pulmonary artery enlargement. The pulmonary trunk measured 3.5 cm. Note the peripheral filling defect in the posterior right pulmonary artery, consistent with chronic pulmonary thromboembolism.* **(Middle)** *Axial CECT (mediastinal window) of the same patient shows eccentric thrombus in the posterior lumen of the right pulmonary artery and enlargement of the left interlobar pulmonary artery.* **(Bottom)** *Axial CECT (mediastinal window) through the root of the aorta shows the posterior eccentric soft tissue filling defect in the right pulmonary artery that extends into the proximal right interlobar pulmonary artery, consistent with chronic pulmonary thromboembolic disease. Note the enlarged left interlobar pulmonary artery, consistent with pulmonary hypertension.*

FOREIGN BODY EMBOLISM, RADIOGRAPHY AND ANGIOGRAPHY

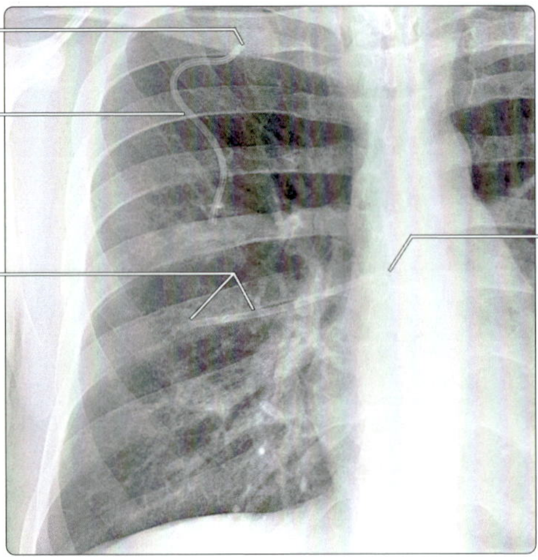

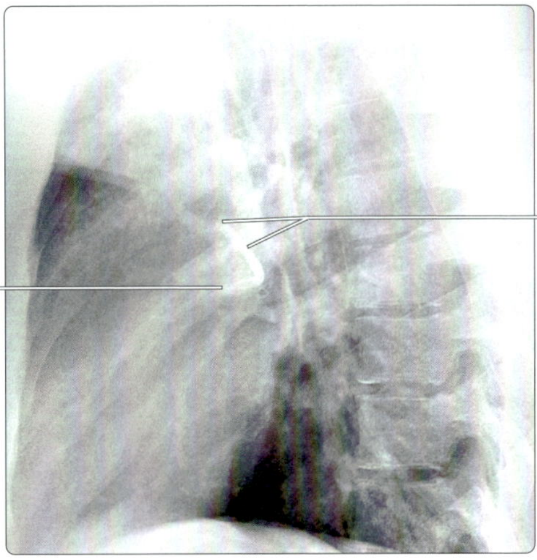

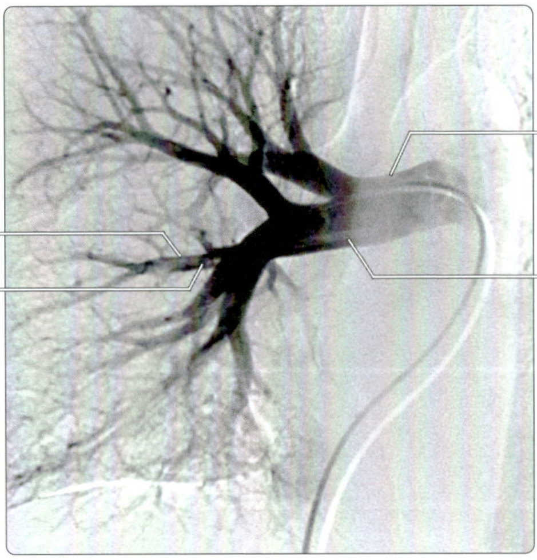

(Top) *First of 3 images of a 44-year-old man with a fractured right subclavian central catheter, which embolized to the middle lobe pulmonary artery, is shown. PA chest radiograph coned down to the right lung shows the site of catheter fracture and the embolized catheter fragment within the right pulmonary artery and branches. The proximal aspect of the fractured catheter fragment is in the intramediastinal (horizontal) right pulmonary artery.* **(Middle)** *Coned-down lateral chest radiograph of the same patient demonstrates the fractured catheter fragment projecting over the right hilar vascular opacity within the proximal right pulmonary artery and extending into the middle lobe pulmonary artery.* **(Bottom)** *Right pulmonary arteriogram of the same patient performed prior to catheter retrieval shows the location of the fractured catheter fragment with its proximal portion in the right pulmonary artery and its distal portion in the middle lobe pulmonary artery.*

FOREIGN BODY EMBOLISM, RADIOGRAPHY AND CT

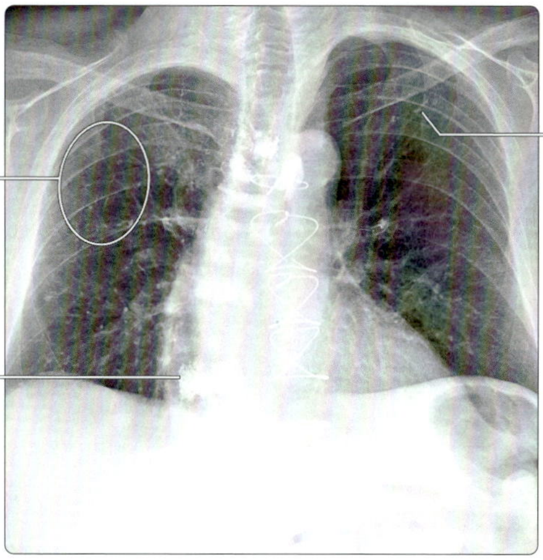

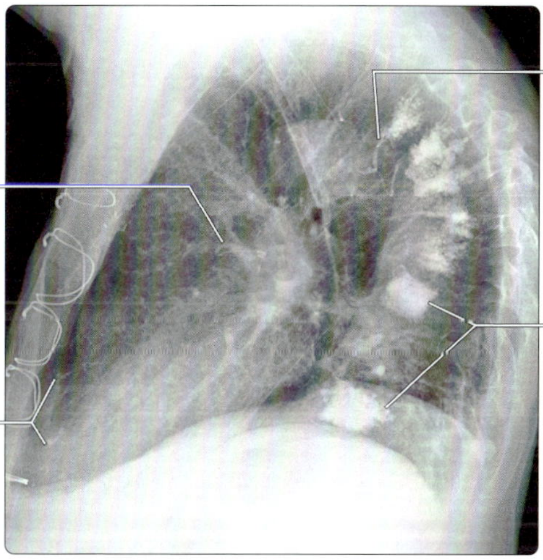

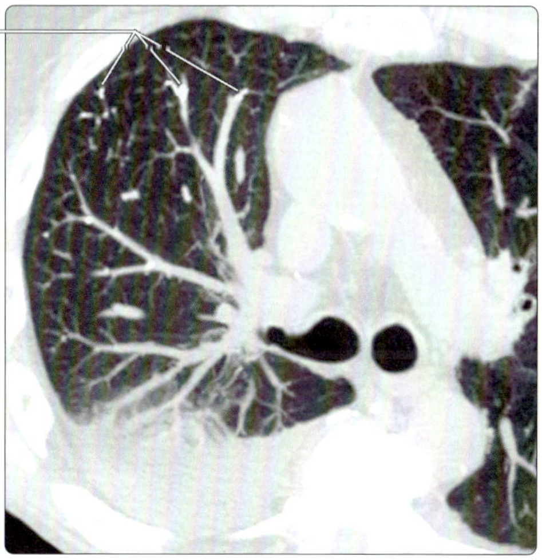

(Top) *PA chest radiograph of an 83-year-old woman status post multiple thoracic spine vertebroplasties shows multifocal bilateral small linear and rounded metallic density foreign bodies in the lungs, many of which exhibit branching configurations.* **(Middle)** *Lateral chest radiograph of the same patient again demonstrates multifocal thoracic spine vertebroplasties and multiple thin tubular and branching metallic densities representing embolized polymethylmethacrylate in the paravertebral venous plexus and in pulmonary artery branches.* **(Bottom)** *Axial NECT (lung window) of the same patient demonstrates peripheral intraarterial metallic densities with branching configurations, consistent with embolized polymethylmethacrylate into the pulmonary circulation as a complication of vertebroplasty.*

GENERAL ANATOMY AND FUNCTION

Anatomy

- Systemic arteries
  - Thoracic aorta
  - Brachiocephalic trunk (artery)
    - Right common carotid artery
    - Right subclavian artery: Gives rise to right vertebral and right internal mammary (thoracic) arteries
  - Left common carotid artery
  - Left subclavian artery: Gives rise to left vertebral and left internal mammary (thoracic) arteries
- Systemic veins
  - Right brachiocephalic vein
  - Left brachiocephalic vein
  - Azygos system
    - Azygos vein
    - Hemiazygos vein
    - Accessory hemiazygos vein
  - Superior vena cava
  - Inferior vena cava

Function

- Systemic arteries
  - Conduit and delivery of oxygenated blood from left heart to tissues
- Systemic veins
  - Conduit and delivery of deoxygenated blood from tissues to right heart

ANATOMY OF SYSTEMIC VESSELS

Systemic Arteries

- **Thoracic aorta**
  - Anatomic divisions
    - Ascending thoracic aorta
    - Aortic arch (transverse aorta)
    - Descending thoracic aorta
  - **Ascending thoracic aorta**
    - Originates distal to left ventricular outflow tract
    - Origin and proximal aspect within pericardium
    - Courses anterosuperiorly and slightly to the right
    - 2 anatomic portions: Aortic root and tubular ascending aorta
    - Aortic root (sinuses of Valsalva and sinotubular junction)
    - Sinotubular junction: Well-defined angle that occurs between aortic root and tubular portion of ascending thoracic aorta
    - Tubular ascending thoracic aorta (from sinotubular junction to aortic arch)
    - Branches: Coronary arteries
    - Relationships
      - Central location with respect to other vascular structures
      - Anterior to left atrium, posterior to right ventricle
      - Medial to right atrium and left ventricle
  - **Aortic arch**
    - Relationships
      - Ascends anterior to right pulmonary artery and tracheal carina
      - Apex on left side of distal trachea
      - Descends posterior to left hilum
    - Ligamentum arteriosum
      - Remnant of ductus arteriosus connecting superior aspect of left pulmonary artery to inferior aspect of aortic arch
      - Connects pulmonary artery to aortic arch during embryonic/fetal circulation, allowing blood to bypass lungs in utero; closes after birth
    - Branches: Brachiocephalic artery, left common carotid artery, left subclavian artery
  - **Descending thoracic aorta**
    - Descends on left of midline
    - Exits thorax through aortic hiatus posterior to diaphragm
    - Branches
      - Bronchial arteries (variable number) supply right and left mainstem bronchi
      - Posterior intercostal arteries
      - Pericardial, esophageal, mediastinal, superior phrenic, and subcostal branches
- **Brachiocephalic artery** (**trunk**)
  - Relationships
    - 1st branch of aortic arch
    - Origin posterior to sternal manubrium and left brachiocephalic vein, anterior to trachea
    - Ascends along right side of trachea toward right sternoclavicular joint, where it bifurcates
  - Branches
    - Right common carotid artery
      - Supplies right head and neck
    - Right subclavian artery
      - Supplies right upper extremity and upper thorax
      - Gives rise to right vertebral and right internal mammary (thoracic) arteries
    - Thyroid ima artery: Supplies thyroid gland
- **Left common carotid artery**
  - Relationships
    - 2nd branch of aortic arch
    - Origin posterior to sternal manubrium, posterior and to the left of brachiocephalic trunk
    - Ascends anterior to left subclavian artery
    - Courses anterior and to the left of trachea
- **Left subclavian artery**
  - Relationships
    - 3rd branch of aortic arch
    - Ascends lateral to left aspect of trachea and left common carotid artery
    - Gives rise to left vertebral and left internal mammary (thoracic) arteries

Systemic Veins

- **Azygos system**
  - Drains posterior chest/abdominal walls and mediastinum
  - May drain lower body in cases of inferior vena cava obstruction
  - **Azygos vein**
    - Potential collateral pathway between superior and inferior venae cavae in cases of venous obstruction
    - Drains posterior chest wall and abdominal wall

  - Ascends on right side of anterior aspects of T12-T5 thoracic vertebral bodies
  - Arches anteriorly at T4 over right mainstem bronchus to drain into posterior aspect of superior vena cava
  - Tributaries
    - Posterior intercostal veins; including right superior intercostal vein formed by 2nd, 3rd, and 4th intercostal veins
    - Hemiazygos and accessory hemiazygos veins
    - Mediastinal, esophageal, pericardial, and right bronchial veins
- **Hemiazygos vein**
  - Ascends on left side of anterior T12-T9 vertebrae
  - Posterior to thoracic aorta up to T9 vertebral body
  - Crosses to right to join azygos vein
  - Tributaries
    - Inferior 4 or 5 left posterior intercostal veins
    - Esophageal and mediastinal veins
- **Accessory hemiazygos vein**
  - Descends on left side of anterior aspects of T4-T8 vertebral bodies
  - Frequently connects to azygos, hemiazygos, and left superior intercostal veins

- **Vena cavae**
  - **Superior vena cava**
    - Formed by anastomosis of right and left brachiocephalic veins
    - Distal portion partially invested by pericardium
    - Drains into right atrium
    - Tributary: Azygos vein
  - **Inferior vena cava**
    - Courses through diaphragm to drain into right atrium
    - Partially invested by pericardium
- **Brachiocephalic veins**
  - **Right brachiocephalic vein**
    - Formed by anastomosis of right subclavian and right internal jugular veins
    - Origin posterior to medial right clavicle
    - Vertical course to form superior vena cava
    - Tributaries
      - Vertebral vein
      - 1st posterior intercostal vein
      - Internal thoracic vein
      - Inferior thyroid and thymic veins
  - **Left brachiocephalic vein**
    - Formed by anastomosis of left subclavian and left internal jugular veins
    - Origin posterior to medial left clavicle
    - Crosses midline, courses to the right to anastomose with the left brachiocephalic vein and form the superior vena cava
    - Tributaries
      - Vertebral vein
      - Left superior intercostal vein
      - Inferior thyroid vein
      - Internal thoracic vein
      - Thymic and pericardial veins
- **Left superior intercostal vein**
  - Tributaries
    - 2nd and 3rd left intercostal veins
    - Left bronchial veins
    - Left pericardiacophrenic vein
  - Courses horizontally along left side of aortic arch
    - Lateral to left vagus nerve
    - Medial to left phrenic nerve
  - Drains into left brachiocephalic vein
  - May connect with accessory hemiazygos vein

IMAGING OF NORMAL GREAT VESSELS

Radiography

- **PA chest radiograph**
  - **Superior vena cava interface**
    - Right upper mediastinum
    - Visualization of lateral border of superior vena cava above azygos arch and below thoracic inlet as it abuts mediastinal surface of right upper lobe
  - **Azygos arch**
    - Right upper mediastinum
    - Ovoid opacity at right tracheobronchial angle
    - Normal width of up to 10 mm
  - **Left subclavian artery interface**
    - Left upper mediastinum
    - Visualization of lateral border of left subclavian artery above aortic arch as it abuts mediastinal surface of left upper lobe
  - **Vascular pedicle**
    - Width of great vessels arising from heart
    - Measured from vertical line drawn from where superior vena cava interface crosses right mainstem bronchus to point where left subclavian artery arises from aortic arch: Normal width of up to **58 mm**
  - **Aortic arch interface**
    - Left upper mediastinum
    - Focal convexity at left tracheobronchial angle with mass effect on distal left tracheal wall
  - **Left superior intercostal vein** (**aortic nipple**)
    - Left upper mediastinum
    - Rounded or triangular bulge of aortic arch contour; seen in < 5% of normal subjects
  - **Left paraaortic interface**
    - Left mid- and lower mediastinum
    - Lateral wall of descending aorta visible from aortic arch to diaphragm, abuts left mediastinal lung surface
- **Lateral chest radiograph**
  - **Ascending thoracic aorta**
    - May be visible in superior aspect of mediastinum, particularly in patients with aortic ectasia
  - **Proximal descending thoracic aorta**
    - Visible in most individuals posterior to distal trachea
  - **Inferior vena cava**
    - Convex, straight, or concave interface posteroinferior to left ventricle

CT/MR

- **Supraaortic mediastinum**
  - Great veins anterior and lateral to great arteries
  - Aortic branches surround anterolateral trachea
  - **Brachiocephalic veins**
    - Right brachiocephalic vein: Shorter than left, vertical course

- Left brachiocephalic vein: Longer than right, horizontal or oblique course
  - **Brachiocephalic artery**
    - Anterior to trachea, right of midline
  - **Left common carotid artery**
    - Posterior and to the left of brachiocephalic artery
    - Smallest great artery at this level
  - **Left subclavian artery**
    - Posterolateral or lateral to trachea
- **Aortic arch level**
  - **Aortic arch**
    - Proximal arch is anterior and to the right of trachea
    - Courses posteriorly and to the left
    - Posterior arch is anterolateral to vertebral bodies
    - Proximal ascending thoracic aorta: Average diameter: 3.6 cm (range: 2.4-4.7 cm)
    - Ascending thoracic aorta proximal to arch: Average diameter: 3.5 cm (range: 1.6-3.7 cm)
  - **Superior vena cava**
    - Ovoid or round morphology
    - Courses vertically anterior and to right of trachea
- **Pericarinal and subcarinal regions**
  - **Azygos arch**
    - Courses anteriorly at right tracheobronchial angle
    - Drains into posterior aspect of superior vena cava
  - **Azygos vein**
    - Vertical course along right anterolateral vertebrae
  - **Hemiazygos vein**
    - Courses vertically along left anterolateral vertebral bodies posterior to descending thoracic aorta
  - **Descending thoracic aorta**
    - Courses along posterolateral aspect of spine
    - Mid descending thoracic aorta: 2.5-cm diameter (range: 1.6-3.7 cm)
    - Distal descending thoracic aorta: 2.4-cm diameter (range: 1.4-3.3 cm)

ANATOMY-RELATED IMAGING VARIANTS AND ABNORMALITIES

Aortic Arch Anomalies

- **Bovine arch**
  - Left common carotid artery origin from brachiocephalic artery
  - Most common great vessel anomaly, seen in up to 20% of population
- **Aberrant right subclavian artery**
  - Seen in 0.4-2.3% of population, typically asymptomatic but may produce dyspnea/dysphagia
  - Last aortic branch arising distal to left subclavian artery, courses cephalad posteriorly and to the right, usually behind trachea and esophagus
  - Diverticulum of Kommerell, focal dilatation of proximal aberrant subclavian artery, seen in 60% of cases
- **Anomalous left vertebral artery**
  - Direct origin from aortic arch, seen in 10% of population
- **Right aortic arch**
  - Seen in 0.1-0.2% of population
  - Mirror image great vessel branching associated with congenital heart disease
  - Non-mirror image great vessel branching, associated with aberrant left subclavian artery
- **Double aortic arch**
  - Seen in 0.05-0.30% of population; may produce stridor/dysphagia or may be asymptomatic
  - Most common complete vascular ring
  - Right arch gives off right subclavian and right common carotid arteries
    - Larger, superior, and posterior compared to left arch
  - Left arch gives off left common carotid and left subclavian arteries

Thoracic Aortic Aneurysm

- Aortic dilatation > 50% or > 1.5x normal diameter
- Imaging
  - Radiography
    - Ascending aortic/aortic arch aneurysm: Often not visible
    - Descending aorta aneurysm: Focal or diffuse contour abnormality of left paraaortic interface
    - ± peripheral/curvilinear calcification
    - Rupture: Wide mediastinum, pleural effusion
  - CT/MR
    - Aortic dilatation ± mural calcification
      - Saccular or fusiform morphology
    - Crescentic/mural high attenuation; consider contained/impending rupture
- Etiologies
  - Atherosclerosis
  - Infectious (mycotic)
  - Cystic medial necrosis
    - Effacement of sinotubular angle may be early sign of aortic dilatation (common in Marfan syndrome)

Bronchial Artery Hypertrophy

- Normal diameter is 1.5 mm; hypertrophy: > 2-mm diameter
- Visualization should suggest pulmonary artery compromise
- Source of bleeding in 90% of patients with hemoptysis

Venous Anomalies

- **Persistent left superior vena cava**
  - Seen in 1-3% of population, usually asymptomatic
  - Arises from anastomosis of left subclavian and left jugular veins
    - Typically drains into coronary sinus
    - May coexist with right superior vena cava, occasional venous connection (bridging vein) between 2 superior venae cavae
    - Vertical course, anomalous vessel anterior to left superior pulmonary vein
  - Association with left isomerism
- **Azygos continuation of inferior vena cava**
  - Infrahepatic interruption of inferior vena cava with azygos continuation
  - Enlarged azygos vein, may mimic lymphadenopathy
  - Association with left isomerism

OVERVIEW OF THORACIC SYSTEMIC VESSELS

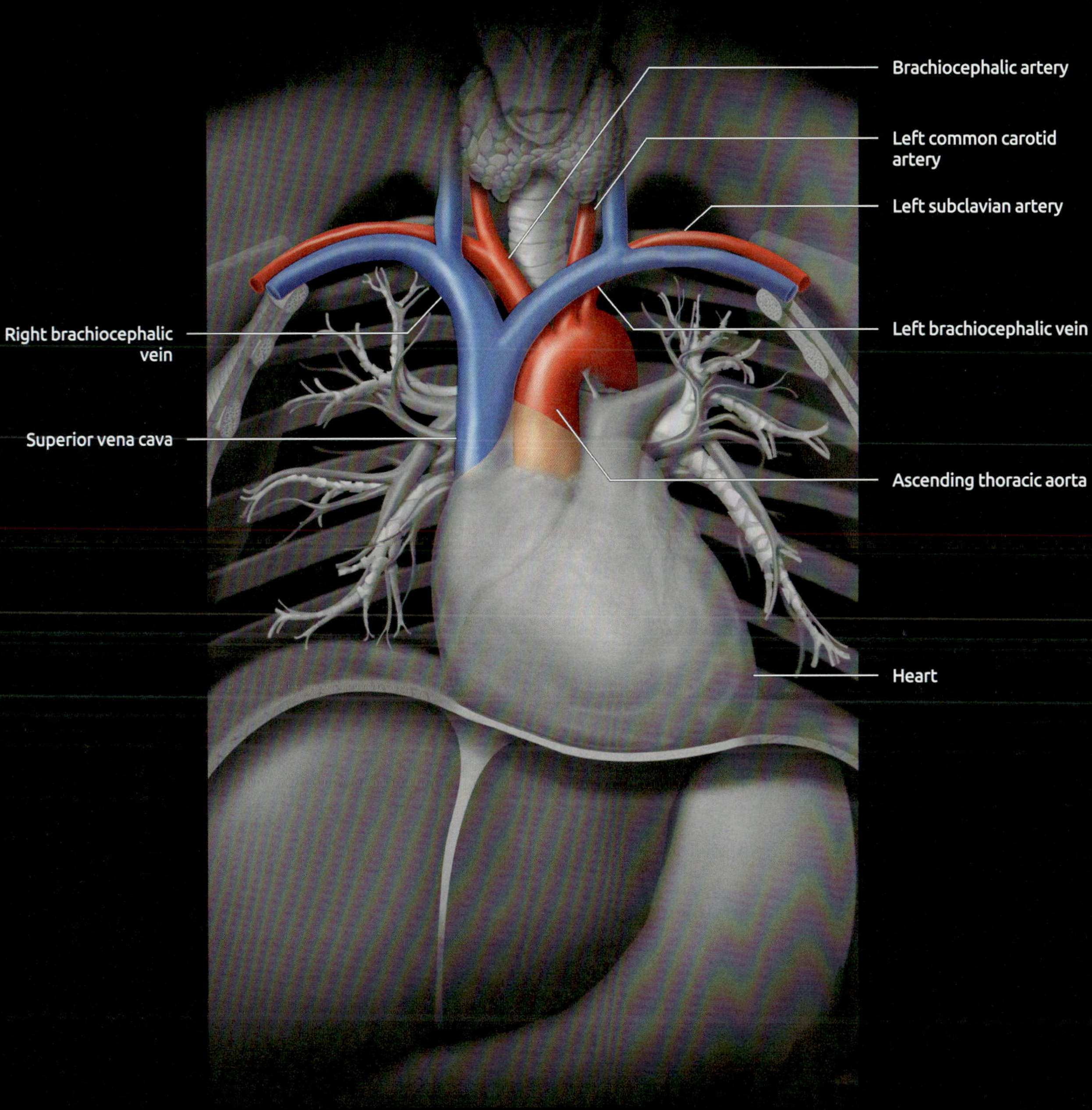

Graphic depicts the anatomy of the principal systemic thoracic great vessels and their relationship to the adjacent thoracic organs and structures. The venae cavae transport deoxygenated blood to the right heart chambers, which in turn deliver it to the alveolar-capillary interface via the pulmonary arteries. The superior vena cava is formed by the anastomosis of the bilateral brachiocephalic veins. The thoracic aorta transports oxygenated blood from the left heart chambers to the body. The 3 main branches of the thoracic aorta, the brachiocephalic, left common carotid, and left subclavian arteries, supply the upper extremities, the thorax, and the head and neck.

ANATOMY OF SYSTEMIC VESSELS

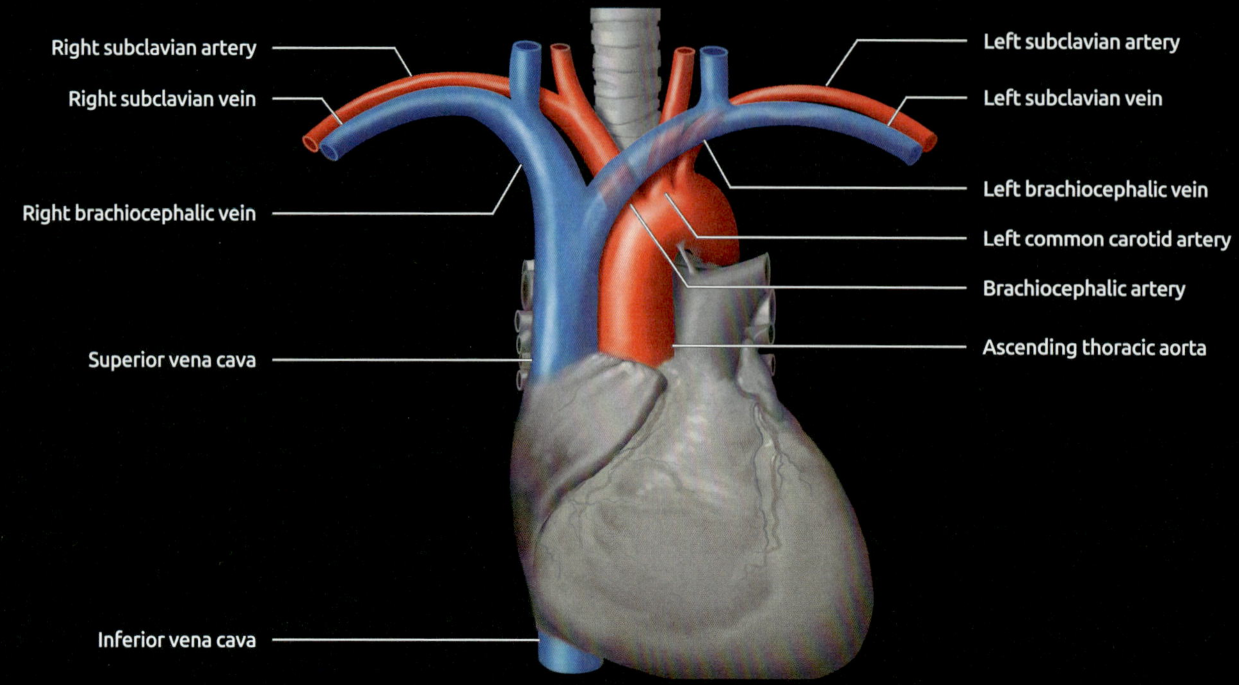

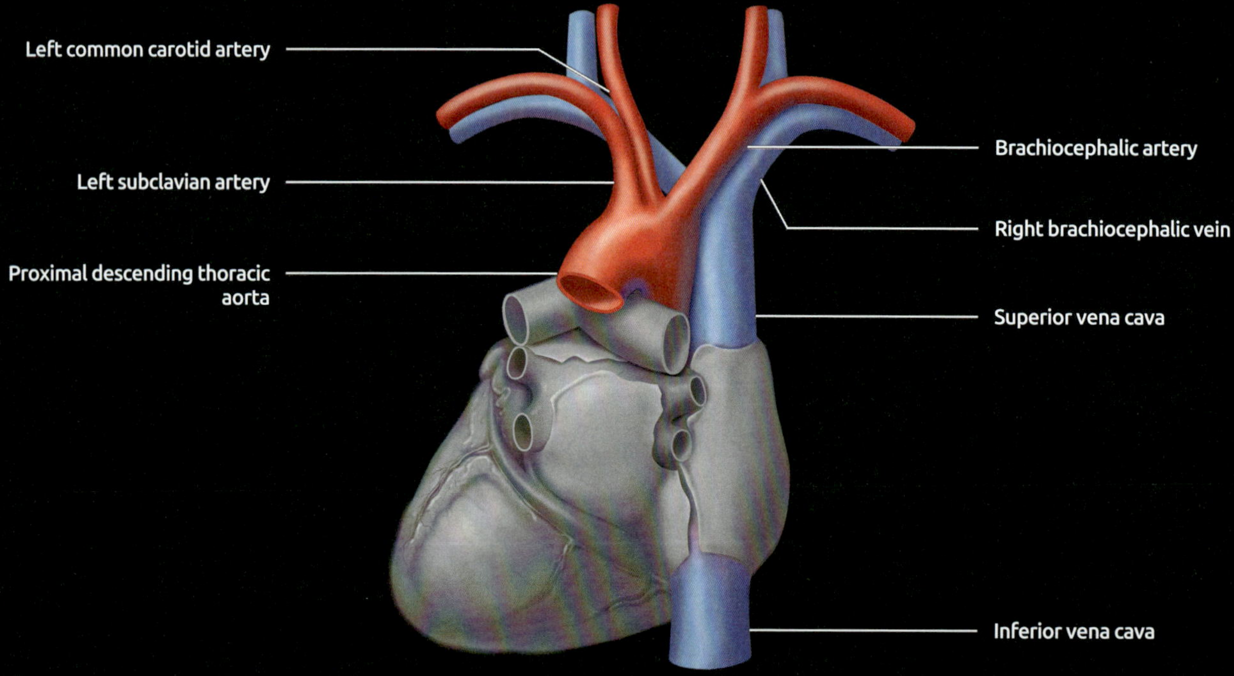

(Top) *First of 2 graphics illustrates the anatomy of the thoracic systemic great vessels and their relationship to the heart and the pulmonary vessels. Anterior view shows the superior vena cava, which is formed by the anastomosis of the right and left brachiocephalic veins, located anterior to the branches of the aortic arch. The venae cavae drain into the right atrium. The proximal thoracic aorta is located centrally with respect to the surrounding vascular structures. Its ascending portion courses superiorly. The aortic arch courses posteriorly and to the left and gives off the brachiocephalic, left common carotid, and left subclavian arteries in sequence.* **(Bottom)** *Posterior view shows the anatomy of the thoracic great vessels. The superior and inferior venae cavae course into the right atrium. The branches of the aortic arch are located behind the anteriorly located thoracic great veins.*

ANATOMY OF SYSTEMIC ARTERIES

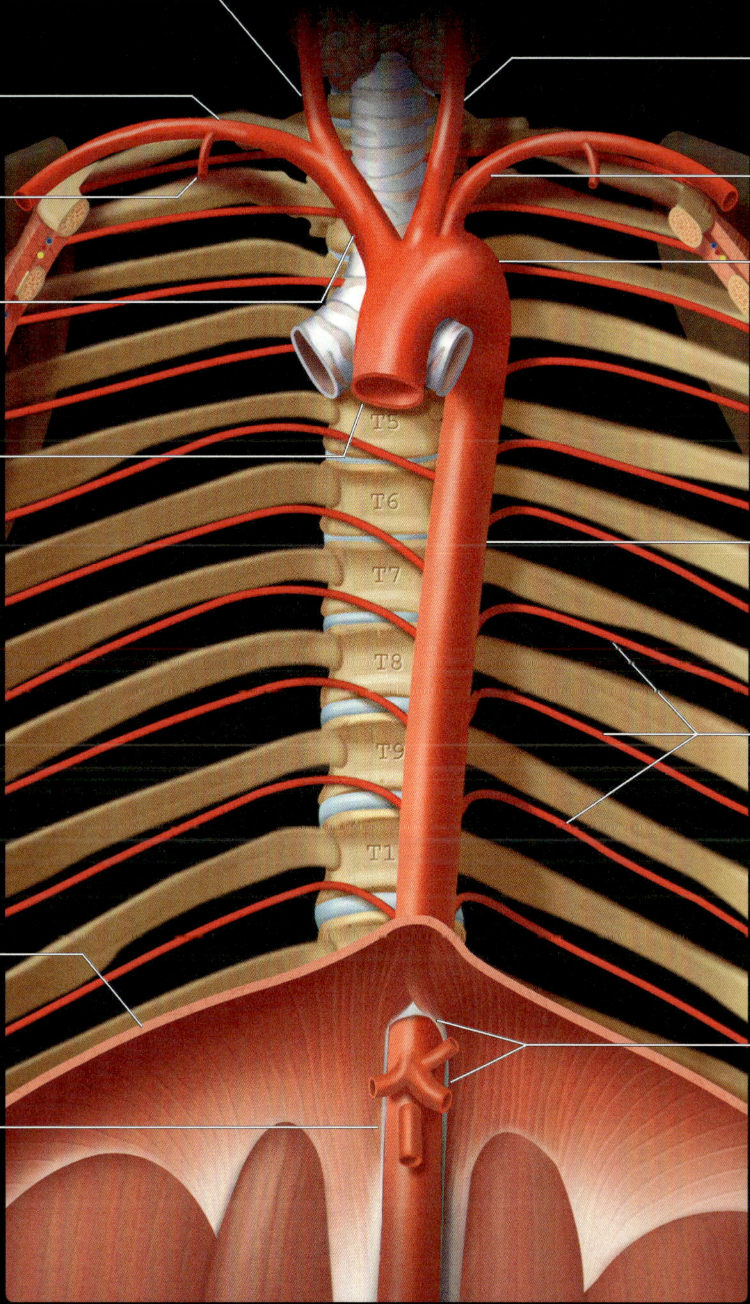

Graphic shows the anatomy of the thoracic aorta and its branches. The thoracic aorta has 3 major portions: Ascending thoracic aorta, aortic arch, and descending thoracic aorta. The ascending thoracic aorta courses superiorly and slightly toward the right. The aortic arch courses posteriorly and toward the left over the left tracheobronchial angle. The 1st branch of the aortic arch is the brachiocephalic artery followed by the left common carotid and left subclavian arteries. These vessels supply the upper thorax, upper extremities, and the head and neck. The brachiocephalic artery gives rise to the right common carotid and right subclavian arteries. The subclavian arteries give rise to the bilateral vertebral arteries and to the bilateral internal thoracic (or internal mammary) arteries that supply the anterior chest wall. The descending thoracic aorta gives rise to paired intercostal arteries.

ANATOMY OF SYSTEMIC VEINS

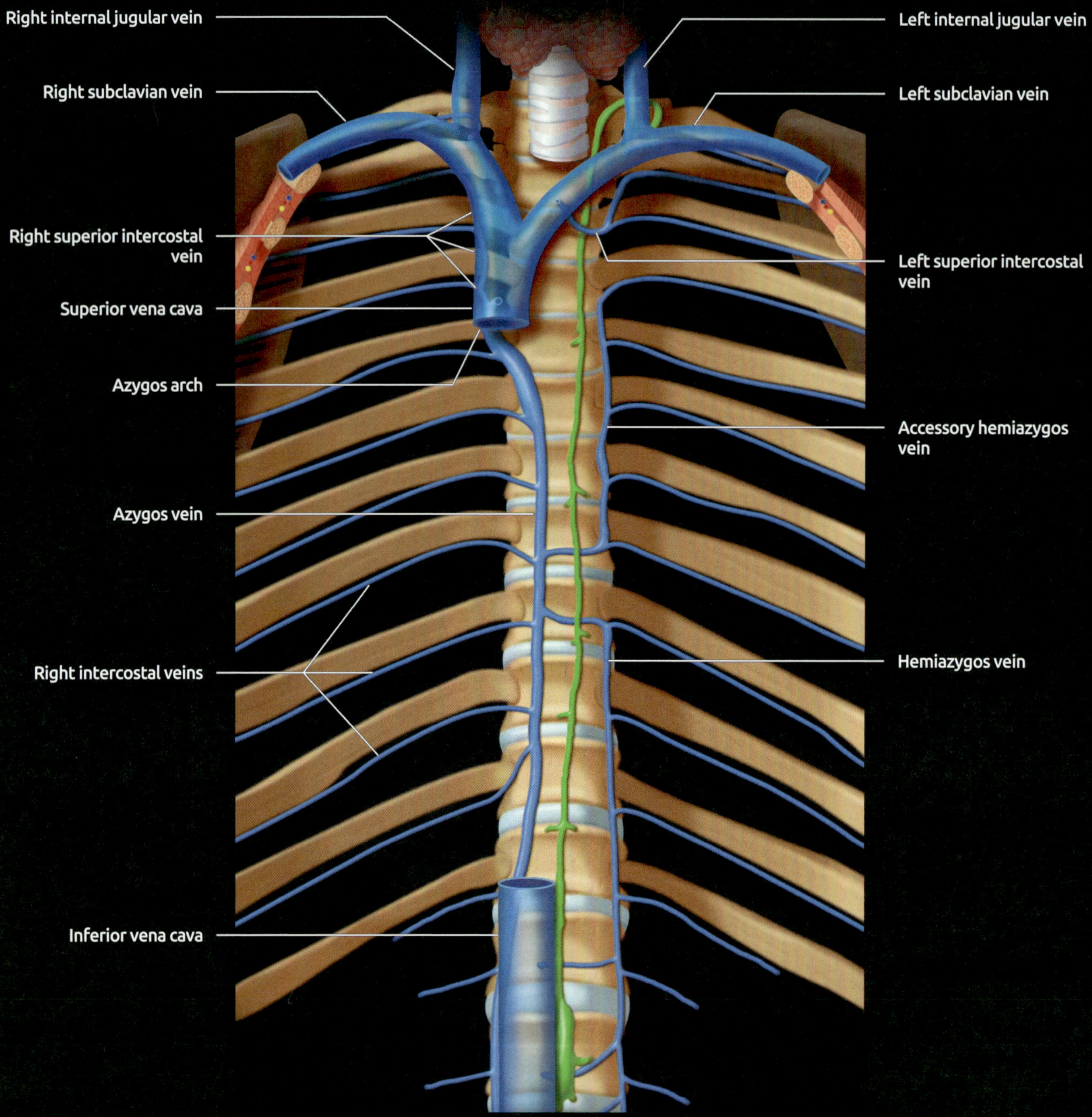

Graphic shows the anatomy of the thoracic systemic veins. The superior vena cava is formed by the anastomosis of the right and left brachiocephalic veins, which in turn are formed by the anastomosis of the internal jugular and subclavian veins. The chest wall is drained by intercostal veins that course along the undersurfaces of the ribs and drain into the azygos vein on the right and the hemiazygos and accessory hemiazygos veins on the left. Venous anastomoses connect the azygos and hemiazygos veins. The 2nd and 3rd left intercostal veins drain into the left superior intercostal vein that courses along the lateral aspect of the aortic arch to drain into the posterior aspect of the left brachiocephalic vein. The right superior intercostal vein, shown behind the superior vena cava and right brachiocephalic vein, receives the 2nd, 3rd, and 4th right intercostal veins.

ANATOMY OF SYSTEMIC VEINS

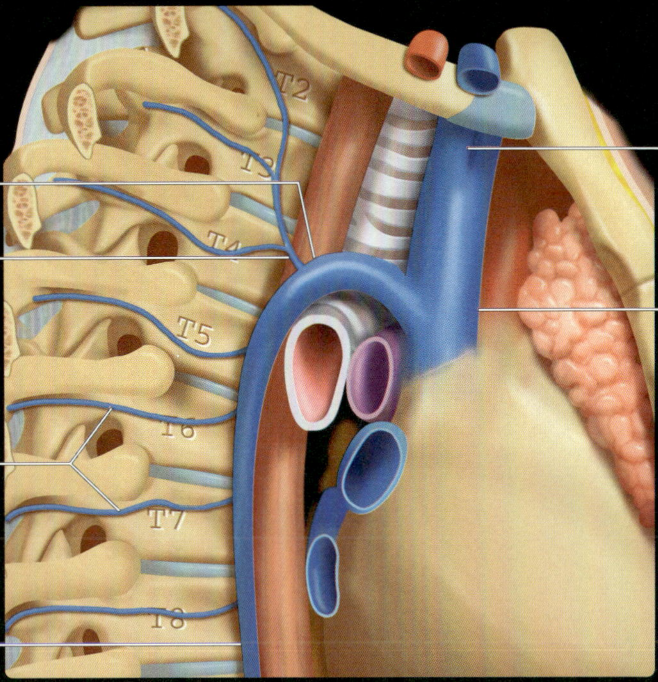

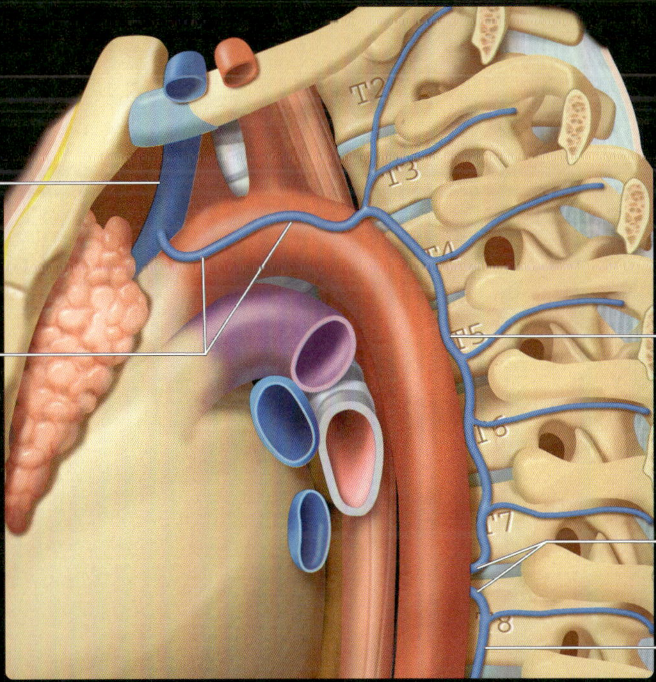

(Top) *First of 2 graphics shows the anatomy of the azygos venous system. Graphic of the right mediastinal surface shows the intercostal veins draining into the azygos vein. The azygos arch courses over the right mainstem bronchus at the right tracheobronchial angle to drain into the posterior aspect of the superior vena cava. The right superior intercostal vein drains into the azygos vein.* **(Bottom)** *Graphic of the left mediastinal surface shows the intercostal veins draining into accessory hemiazygos and hemiazygos veins, which communicate with the azygos vein through venous anastomoses. The left superior intercostal vein receives tributaries from the 2nd and 3rd left intercostal veins and courses along the lateral aspect of the aortic arch to drain into the left brachiocephalic vein. The left superior intercostal vein may be conspicuous in a small percentage of normal subjects and in patients with venous obstruction and collateral vessels. In such cases, it may be visible on frontal chest radiography as the aortic nipple.*

SYSTEMIC VESSELS, RADIOGRAPHY

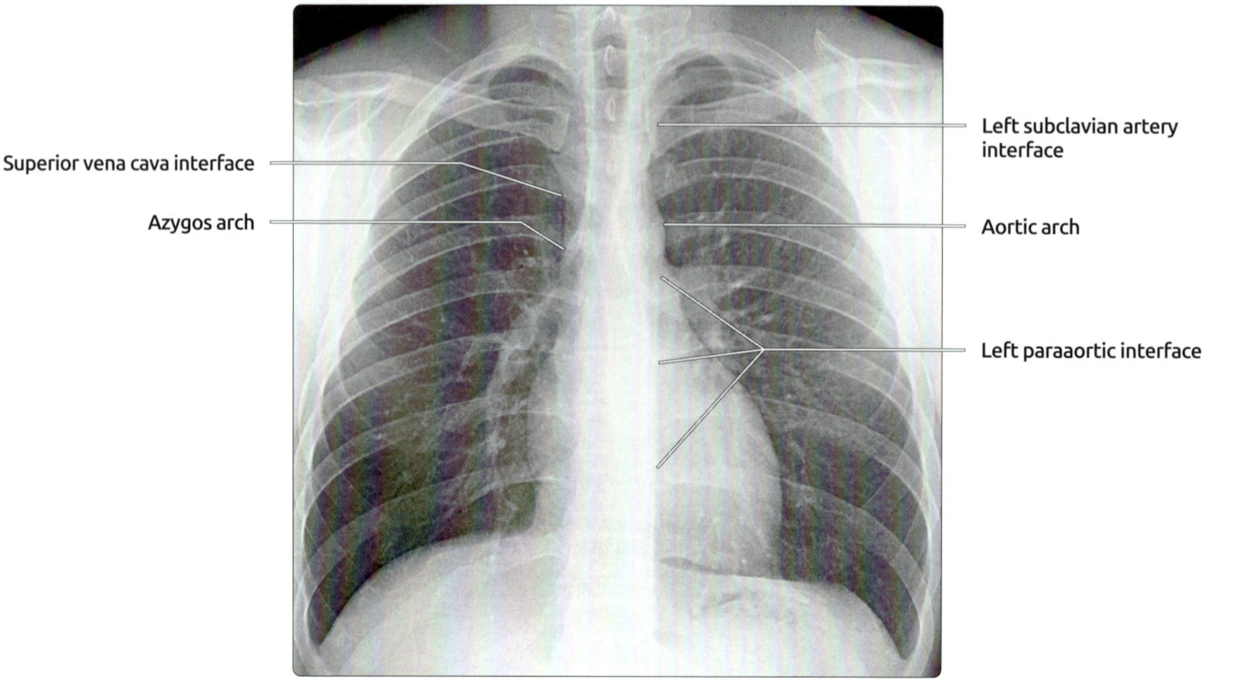

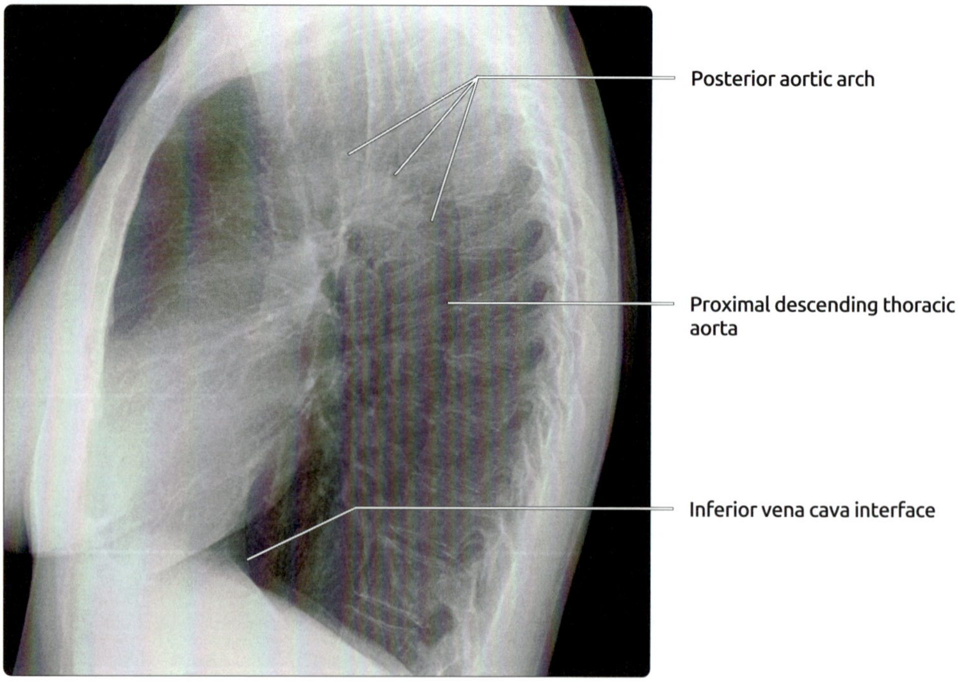

(Top) *PA chest radiograph of a normal individual shows the radiographic anatomy of the systemic vessels. The superior vena cava interface results from contact between the superior vena cava and the mediastinal surface of the right upper lobe. The azygos arch is visible at the right tracheobronchial angle. The aortic arch is located at the left tracheobronchial angle and produces mild mass effect on the distal left tracheal wall. The left subclavian artery interface is visible cephalad to the aortic arch. The left paraaortic interface is visible through the mediastinum and is produced by contact between the lateral aspect of the descending thoracic aorta and the mediastinal surface of the left lung.* **(Bottom)** *Lateral chest radiograph of a normal individual shows the posterior aspect of the aortic arch, the proximal aspect of the descending thoracic aorta, and the inferior vena cava interface.*

SYSTEMIC VESSEL INTERFACES, RADIOGRAPHY

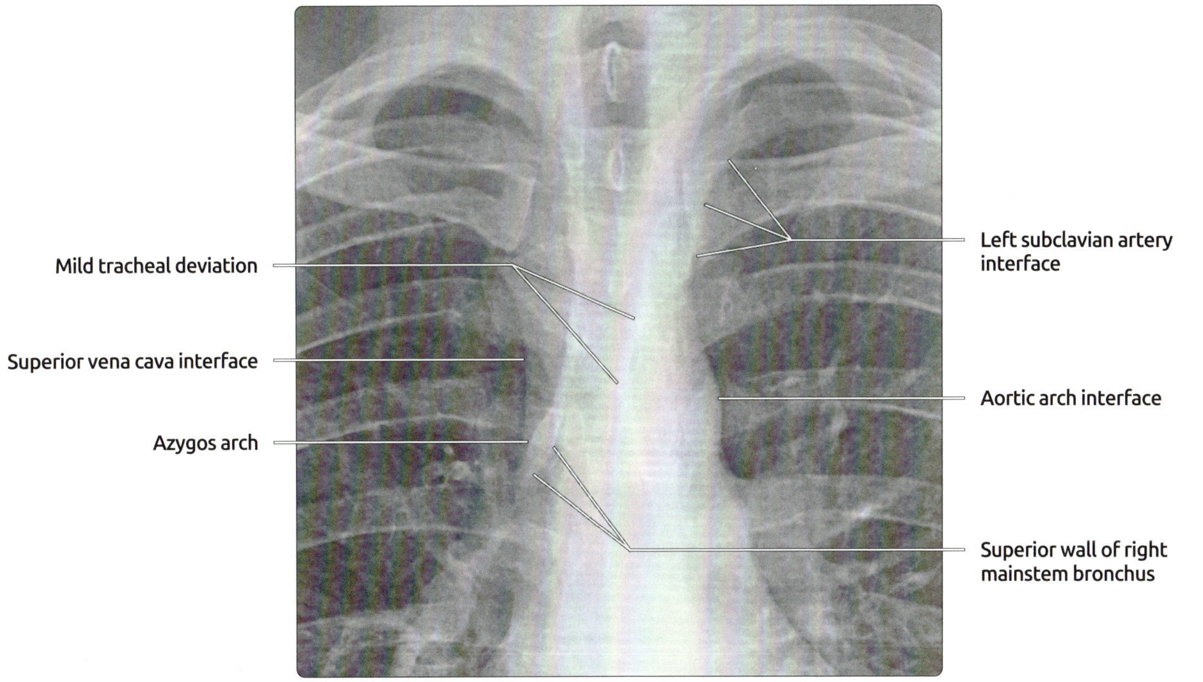

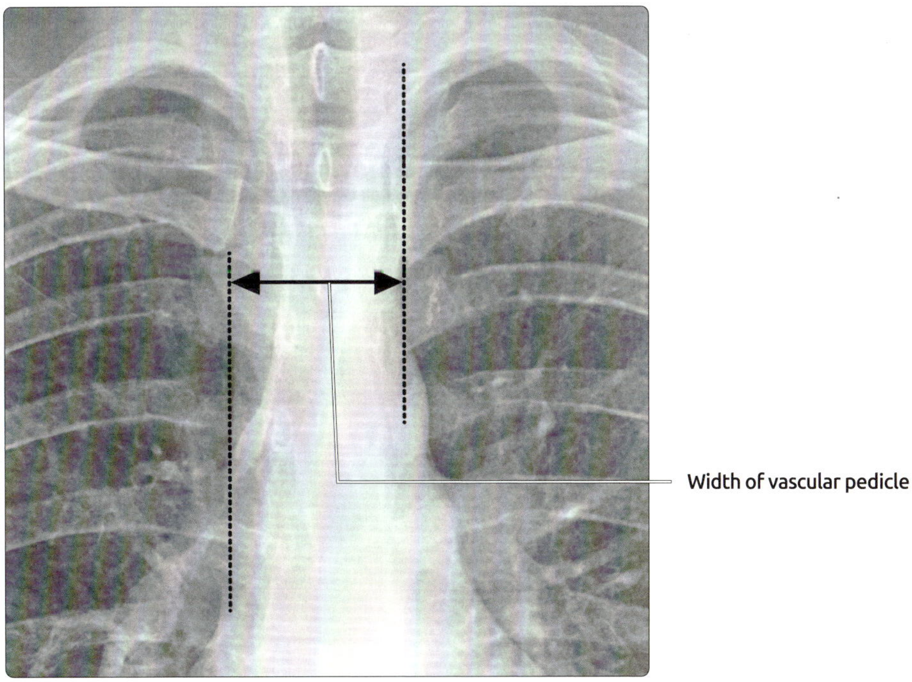

(Top) *PA chest radiograph of a normal individual coned down to the upper thorax shows the superior vena cava interface and the aortic arch interface in the right and left upper mediastinum, respectively. The left subclavian artery interface is visible in the left upper mediastinum, where the vessel abuts the left mediastinal lung surface. Note the mild rightward tracheal deviation produced by the aortic arch.* **(Bottom)** *PA chest radiograph of the same subject shows the method for measuring the vascular pedicle. The vascular pedicle is measured from a vertical line drawn from where the superior vena cava interface crosses the right mainstem bronchus to the origin of the left subclavian artery from the aortic arch. The normal width of the vascular pedicle is up to 58 mm.*

SYSTEMIC VESSELS, CORONAL CT

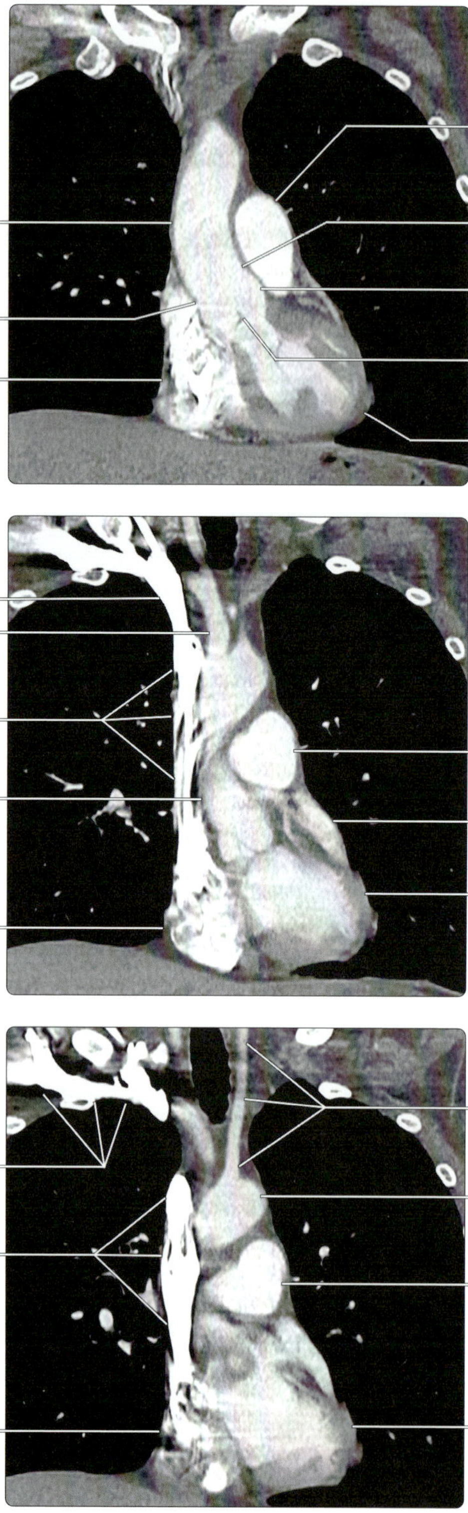

(Top) *Coronal CECT (mediastinal window) of a normal individual shows the ascending thoracic aorta composed of the aortic root, which is bulbous in morphology, and the tubular portion. The sinotubular junction is located between the aortic root and the tubular ascending thoracic aorta.* **(Middle)** *Coronal CECT (mediastinal window) of the same individual shows the right brachiocephalic vein as it drains into the superior vena cava. The ascending thoracic aorta is partially visualized. The brachiocephalic artery or trunk arises as the 1st branch of the aortic arch.* **(Bottom)** *Coronal CECT (mediastinal window) of the same individual shows the left common carotid artery arising from the aortic arch as its 2nd branch.*

SYSTEMIC VESSELS, CORONAL CT

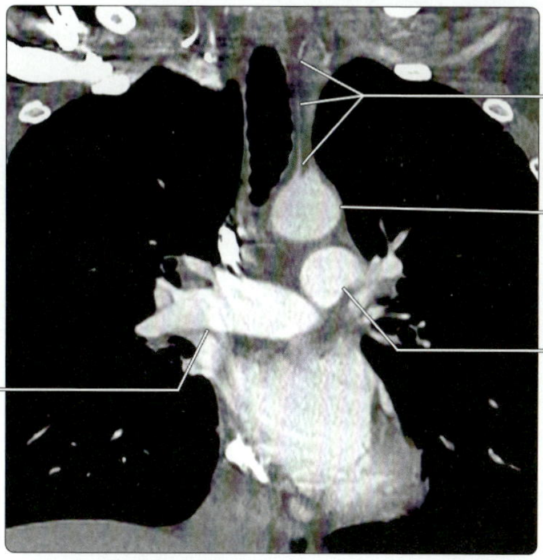

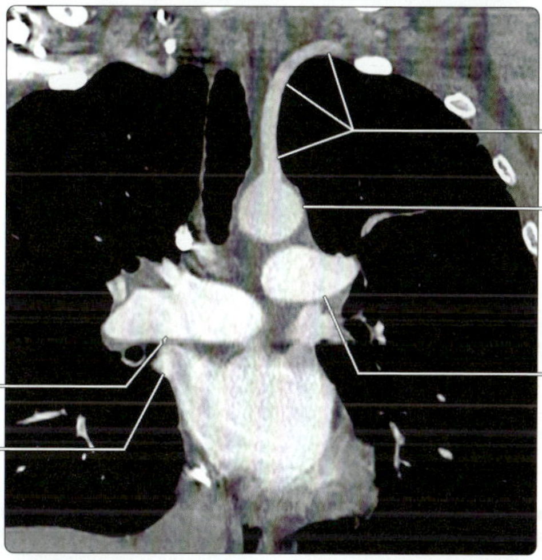

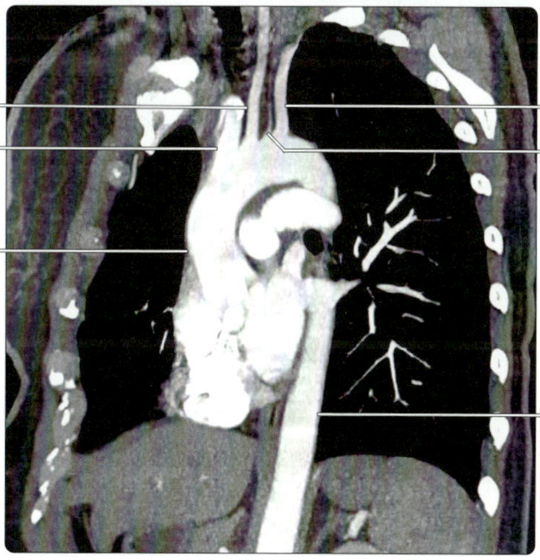

(Top) *Coronal CECT (mediastinal window) of the same individual shows the left vertebral artery arising directly from the aortic arch, a common anatomic variant. More frequently, the left vertebral artery arises from the ipsilateral left subclavian artery.* **(Middle)** *Coronal CECT (mediastinal window) of a normal individual shows the left subclavian artery, the last arterial branch of the aortic arch.* **(Bottom)** *Coronal oblique CECT (mediastinal window) (candy cane reformation) of the same individual shows the ascending thoracic aorta and its branch vessels: The brachiocephalic artery or trunk, the left common carotid artery, the anomalous left vertebral artery, and the left subclavian artery.*

SYSTEMIC VEINS, CT AND 3D SURFACE-RENDERED IMAGES

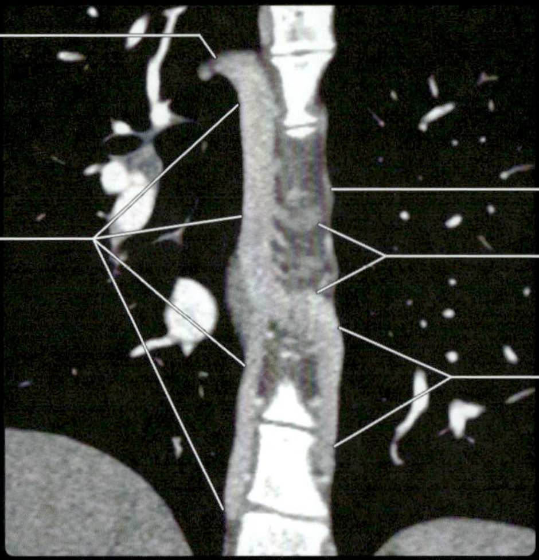

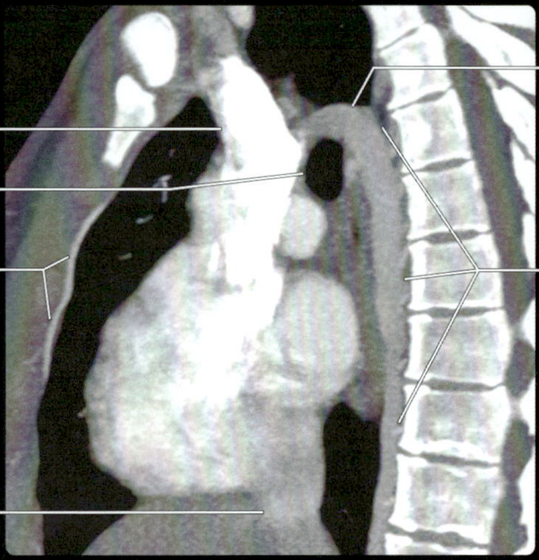

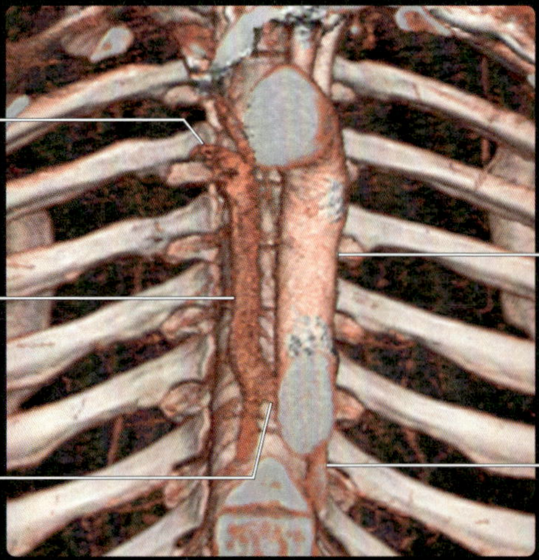

(Top) *Coronal CECT (mediastinal window) of a normal individual shows portions of the azygos vein and azygos arch. The azygos vein courses vertically along the right mediastinum and contributes to the azygoesophageal recess interface. The accessory hemiazygos and hemiazygos veins are also visible. Note anastomoses of the azygos vein with the hemiazygos and accessory hemiazygos veins.* **(Middle)** *Sagittal oblique CECT (mediastinal window) MIP reformatted image shows the normal course and morphology of the azygos vein. Note the azygos arch, which courses over the right mainstem bronchus to drain into the dorsal aspect of the superior vena cava. The superior and inferior vena cavae drain into the right atrium. Note the right internal thoracic vein coursing along the anterior chest wall.* **(Bottom)** *Coronal 3D surface-rendered CECT through the azygos vein shows its vertical course in the right mediastinum and a portion of the azygos arch. The hemiazygos vein is partially obscured by the descending thoracic aorta. Note the anastomosis between the azygos and hemiazygos veins.*

THORACIC AORTA AND BRANCH VESSELS, MR

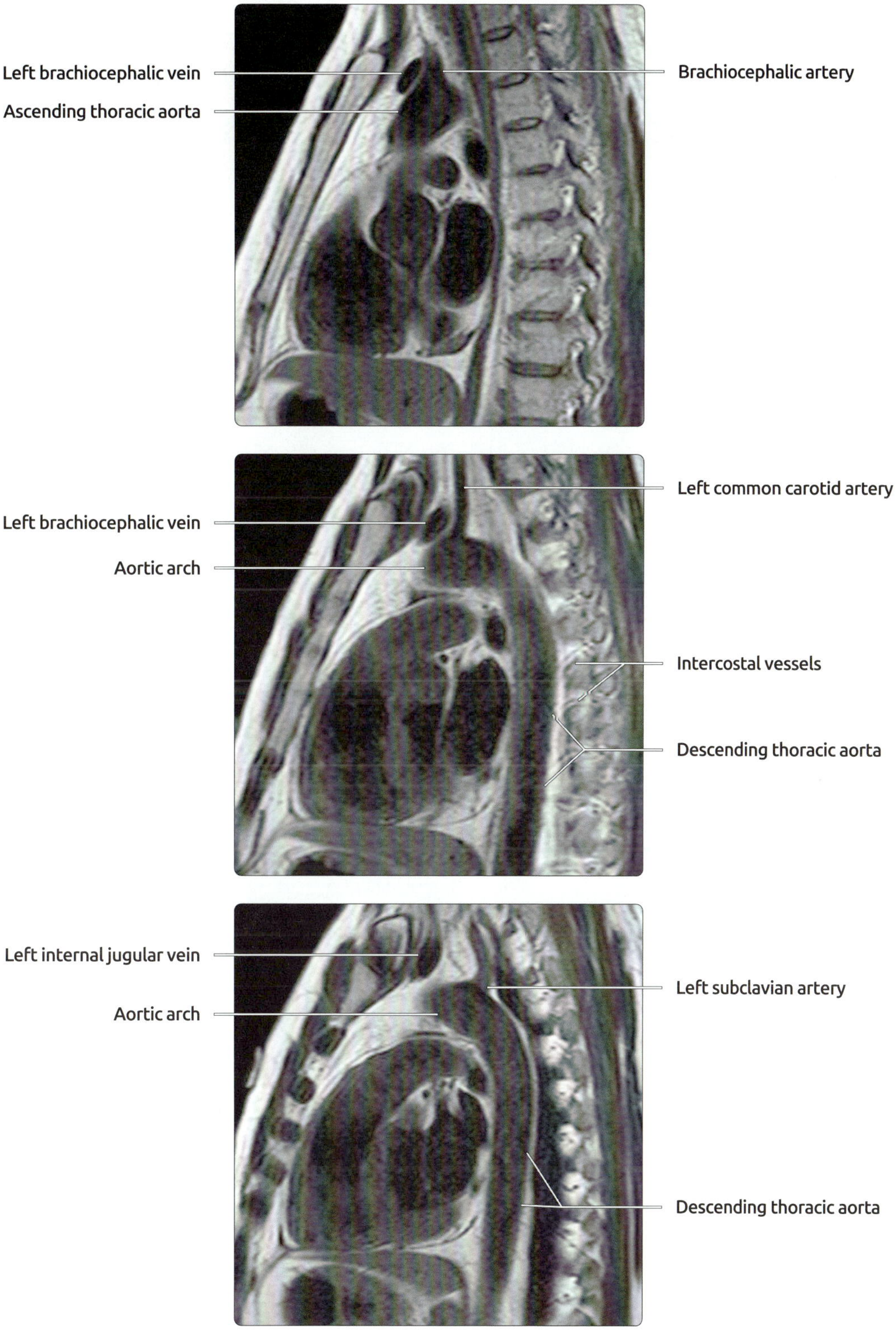

(Top) *Sagittal T1 MR of a normal subject shows the distal ascending thoracic aorta, the proximal aortic arch, and the origin of the brachiocephalic artery. Note the anterior location and horizontal course of the left brachiocephalic vein imaged in cross section.* **(Middle)** *Sagittal T1 MR of the same subject shows the midportion of the aortic arch and demonstrates the origin and course of the left common carotid artery, the 2nd branch of the aortic arch. The descending thoracic aorta and some of its intercostal branches are also visible.* **(Bottom)** *Sagittal T1 MR of the same normal subject shows the distal aspect of the aortic arch and demonstrates the origin of the left subclavian artery, the last branch of the aortic arch. The descending thoracic aorta is also visualized.*

AZYGOS VEIN, MR

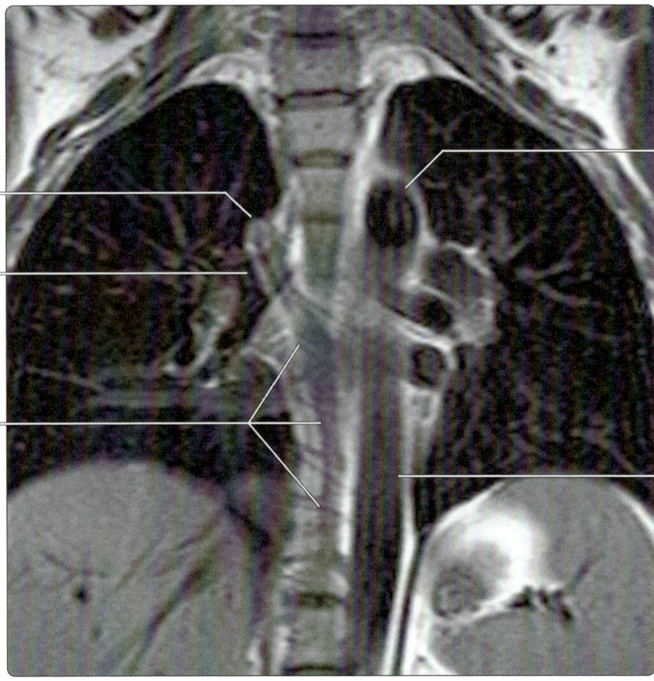

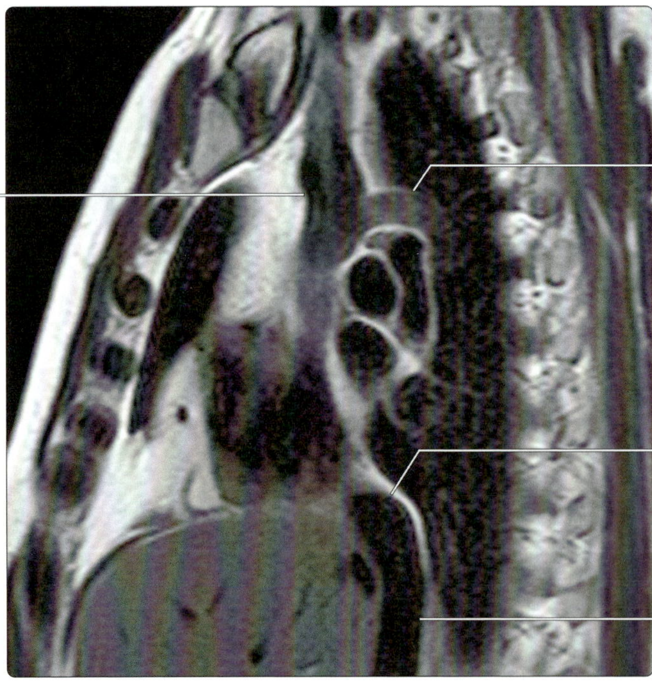

(Top) *Coronal T1 MR of the same normal subject shows the anatomy of the azygos vein. Note the oblique vertical course of the azygos vein in the right retrocardiac mediastinum and the anatomic basis for the morphology of the azygoesophageal recess seen on radiography. The azygos arch courses over the right mainstem bronchus. The descending thoracic aorta is also visualized coursing into the abdomen at the aortic hiatus.* **(Bottom)** *Sagittal T1 MR of the same normal subject shows the normal morphology of the azygos arch as it courses over the right mainstem bronchus to drain into the posterior aspect of the superior vena cava. The intrahepatic and intrathoracic portions of the inferior vena cava are also seen.*

LEFT SUPERIOR INTERCOSTAL VEIN, RADIOGRAPHY AND CT

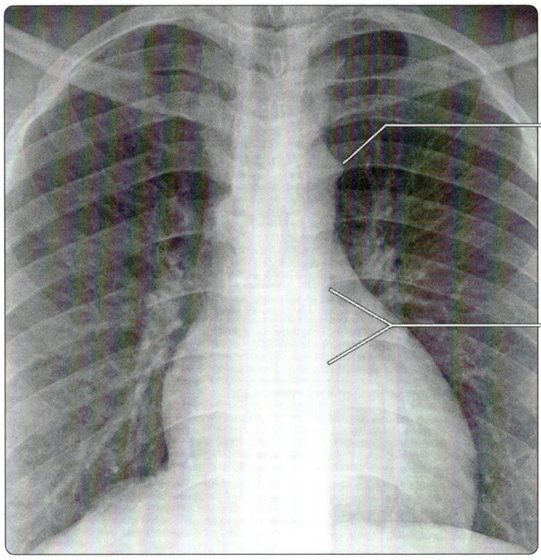

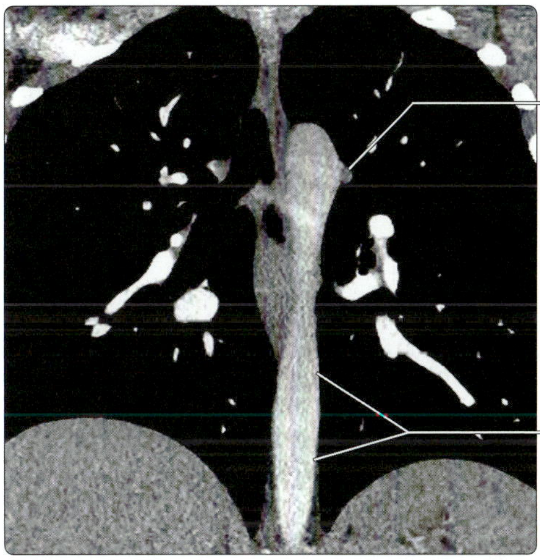

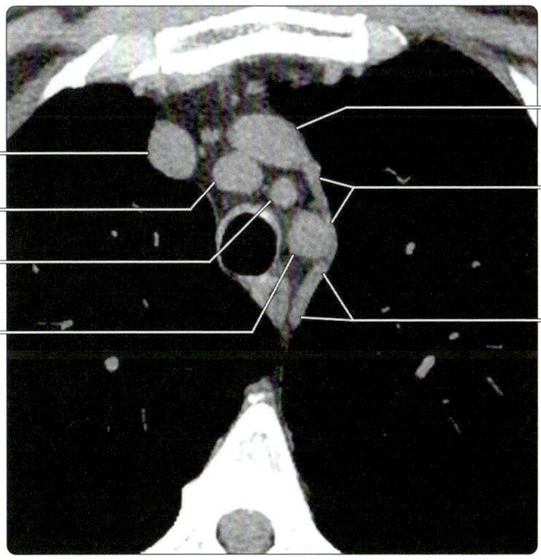

(Top) *PA chest radiograph of a normal individual shows a small nodular bulge at the superolateral aspect of the aortic arch interface that corresponds to the left superior intercostal vein as it courses horizontally and lateral to the aortic arch. This structure, also known as the aortic nipple based on its morphology, is seen in 5% of normal subjects.* **(Middle)** *Coronal CECT (mediastinal window) of the same individual shows the left superior intercostal vein seen in cross section coursing along the lateral aspect of the aortic arch and producing the anatomic correlate for the radiographic aortic nipple.* **(Bottom)** *Axial NECT (mediastinal window) of the same individual shows the anatomic basis for the aortic nipple seen on radiography. The left superior intercostal vein courses along or slightly cephalad (as in this case) to the lateral aspect of the aortic arch and may manifest as a small nodular bulge of the aortic arch on frontal radiography. The left superior intercostal vein receives the 2nd and 3rd left intercostal veins, usually anastomoses with the accessory hemiazygos vein, and drains into the left brachiocephalic vein.*

AORTOGRAPHY

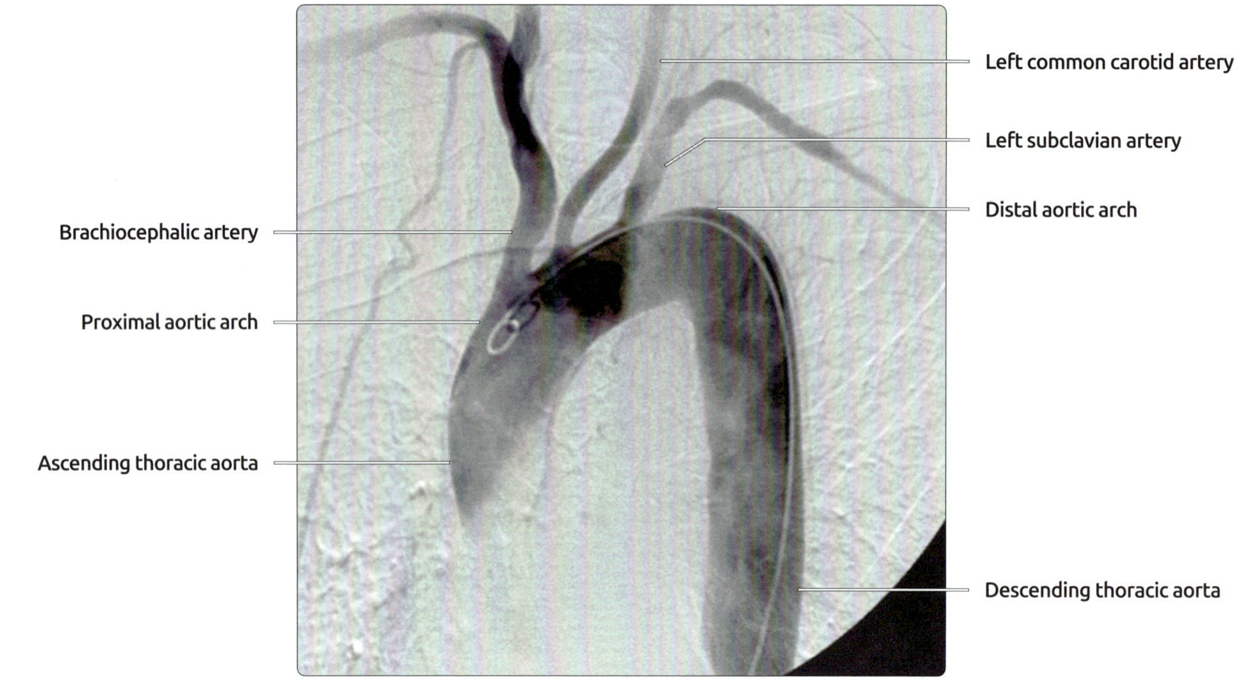

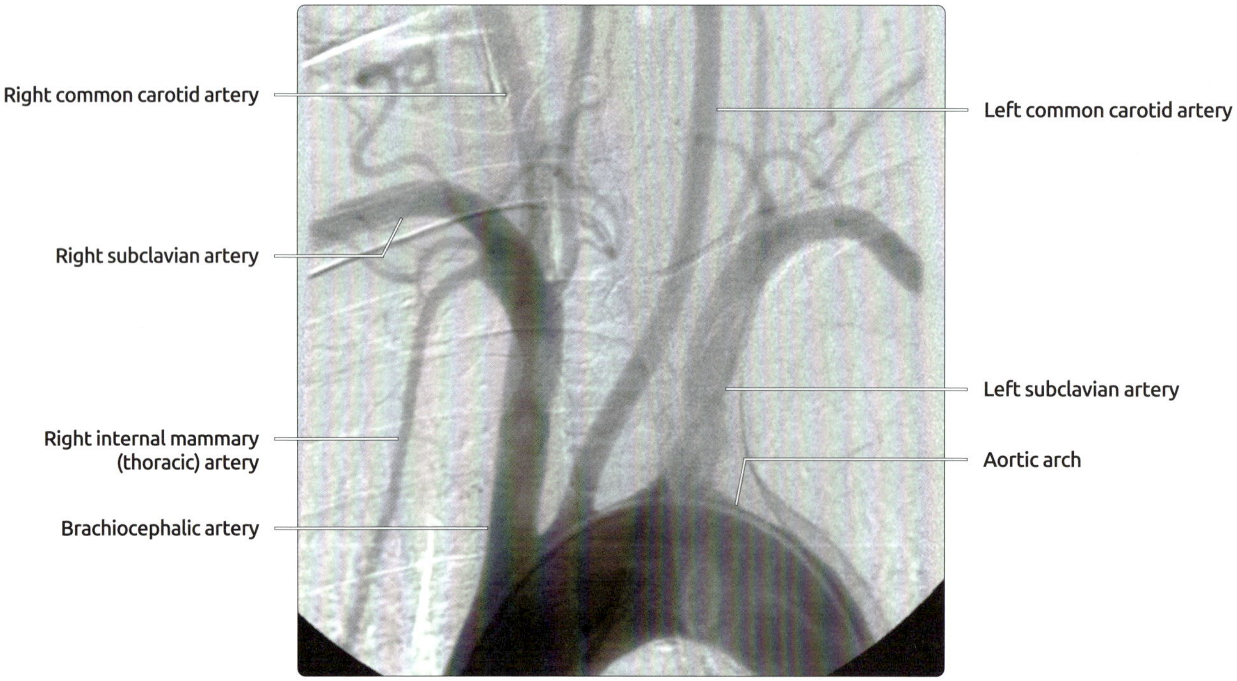

(Top) *Left anterior oblique thoracic aortogram shows the anatomy of the aortic arch and its branches. Contrast injected into the distal ascending thoracic aorta opacifies the aortic arch and its branches: The brachiocephalic artery, followed by the left common carotid artery, and the left subclavian artery. The irregular contour of the descending thoracic aorta and the luminal narrowing of the left subclavian artery are secondary to atherosclerosis.* **(Bottom)** *Left anterior oblique thoracic aortogram shows the normal branching pattern of the aortic arch. The 1st branch is the brachiocephalic artery, which gives rise to the right subclavian and right common carotid arteries. The left common carotid and left subclavian arteries are the 2nd and 3rd aortic arch branches, respectively. The right internal mammary (or internal thoracic) artery, a branch of the right subclavian artery, is also visualized.*

BRONCHIAL ARTERIES, GRAPHIC AND CT

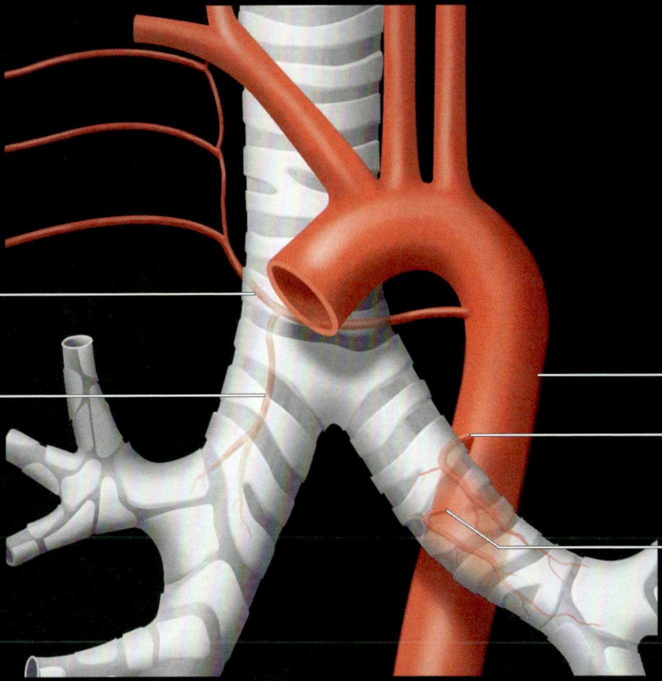

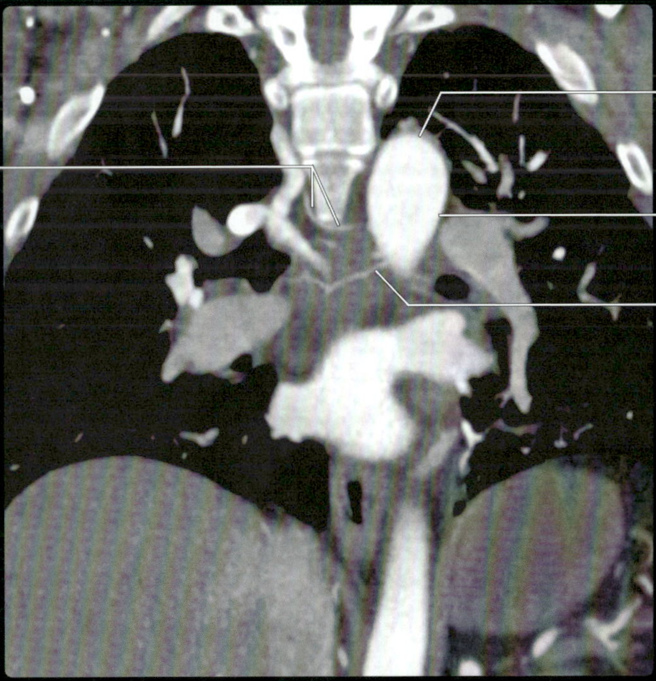

(Top) *Graphic depicts a common variation of the anatomy of the bronchial arteries. In this graphic, the right bronchial artery arises from the right intercostal artery and supplies the wall of the right mainstem bronchus. Superior and inferior left bronchial arteries arise from the descending thoracic aorta and supply the wall of the left mainstem bronchus.* **(Bottom)** *Coronal CECT (mediastinal window) of a patient with abdominal arterial thrombosis (not shown) shows partial visualization of the bronchial arteries. Several small bronchial artery branches arise from the descending thoracic aorta. One of the bronchial artery branches courses anteriorly to supply the right mainstem bronchus. A right intercostal artery is also shown.*

ANATOMIC VARIANTS OF BRONCHIAL ARTERIES

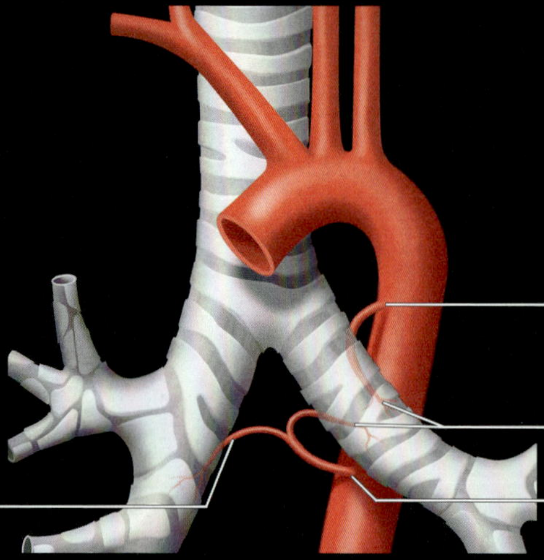

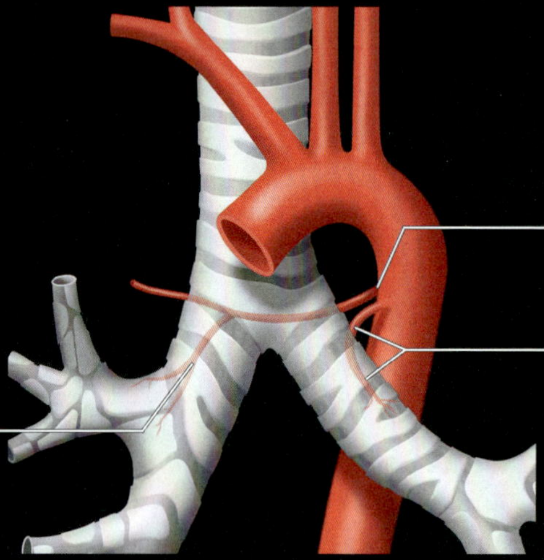

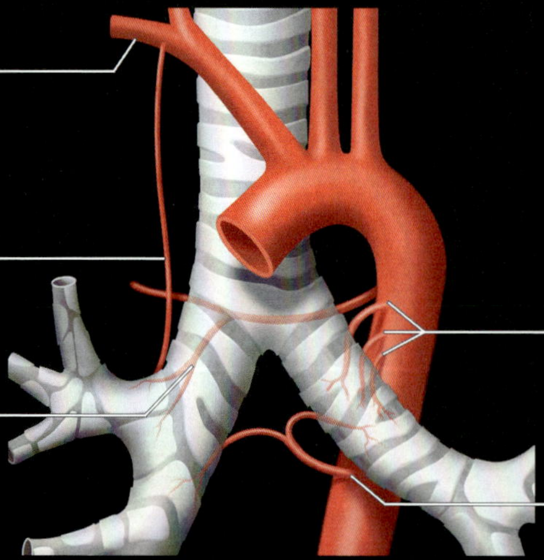

(Top) *First of 3 graphics illustrates anatomic variations of the bronchial arteries. In this graphic, the left superior bronchial artery supplies the left mainstem bronchus. The left inferior bronchial artery also supplies the left mainstem bronchus but gives off a right bronchial artery that supplies the right mainstem bronchus.* **(Middle)** *Graphic shows single bronchial arteries arising from the descending thoracic aorta to supply the right and left mainstem bronchi.* **(Bottom)** *Graphic shows several supernumerary bronchial arteries supplying the right and left mainstem bronchi. There is also an inferior bronchial artery with branches to the right and left mainstem bronchi and a right bronchial artery originating from the right subclavian artery.*

HYPERTROPHIED BRONCHIAL ARTERIES, CT

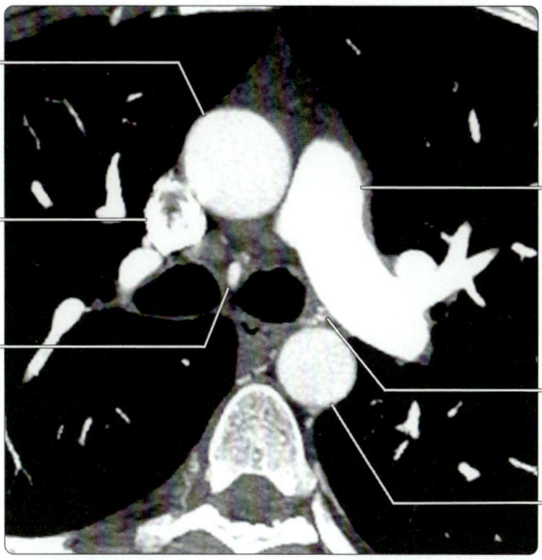

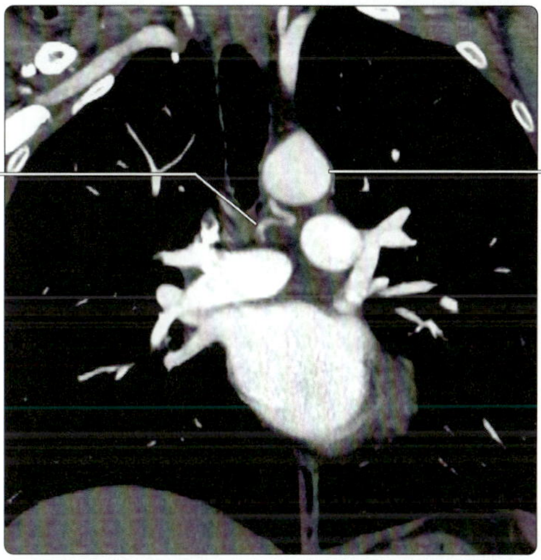

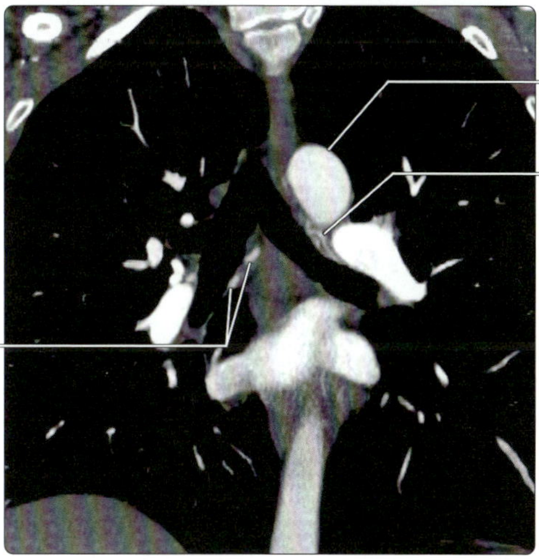

(Top) *Axial CECT (mediastinal window) MIP reformatted image of a 65-year-old man who presented with hemoptysis shows hypertrophied bronchial arteries visualized in the mediastinum and located adjacent to the bilateral mainstem bronchi.* **(Middle)** *Coronal CECT (mediastinal window) of the same patient shows a hypertrophied bronchial artery coursing toward the right mainstem bronchus. Note that the right bronchial artery originates from the inferior aspect of the posterior aortic arch.* **(Bottom)** *Coronal CECT (mediastinal window) of the same patient shows hypertrophied right bronchial artery branches and smaller left bronchial artery branches supplying the walls of the right and left central tracheobronchial tree, respectively.*

HYPERTROPHIED BRONCHIAL ARTERIES, CT

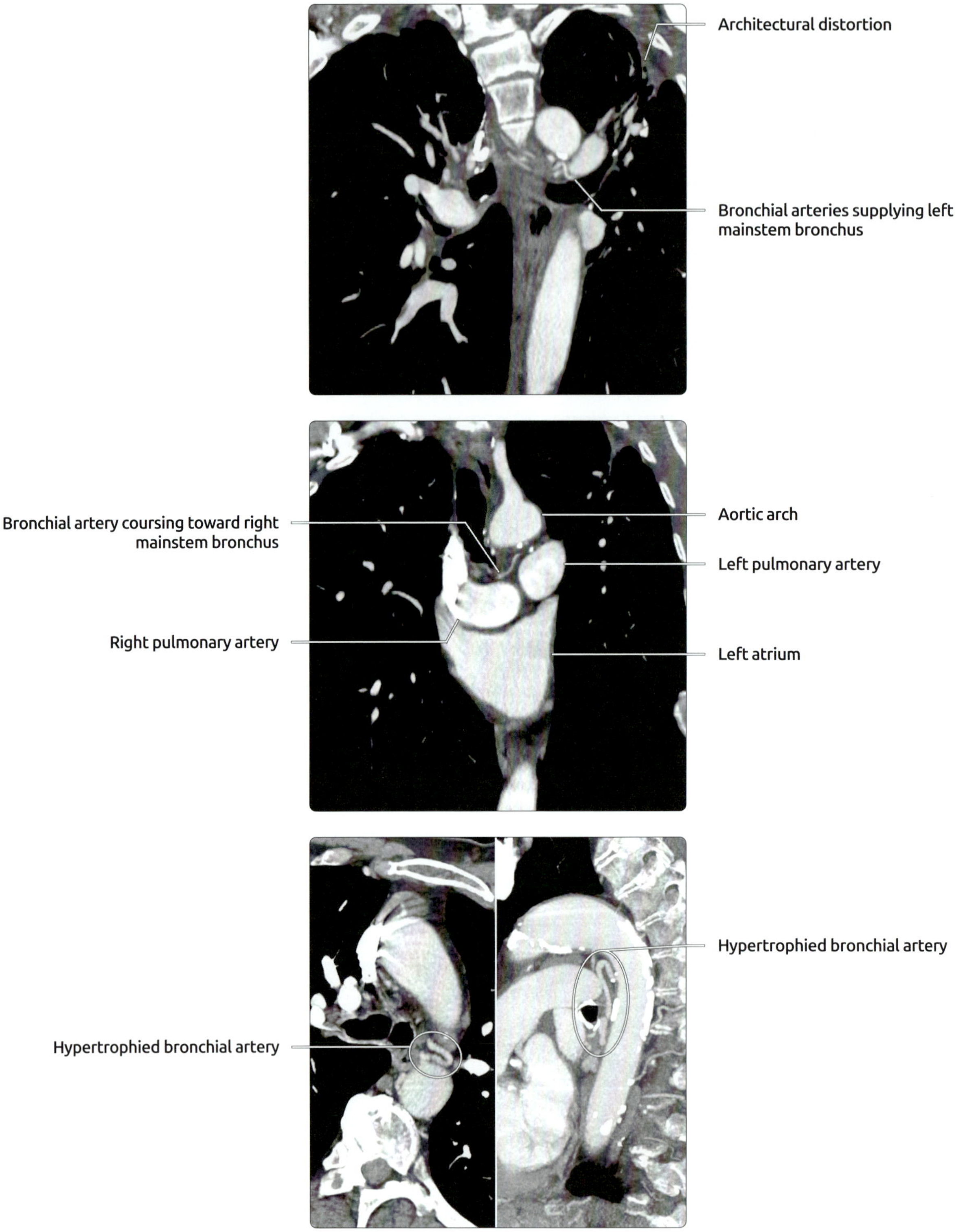

(Top) *Coronal NECT (mediastinal window) of the same patient shows hypertrophied bronchial arteries coursing toward the left mainstem and left upper lobe bronchi. Note bilateral upper lobe volume loss and architectural distortion secondary to chronic pulmonary infection.* **(Middle)** *Coronal oblique CECT (mediastinal window) of the same patient shows a bronchial artery coursing toward the right mainstem bronchus and its branches. Visualization of hypertrophied bronchial arteries should suggest pulmonary artery compromise. Bronchial artery hypertrophy may be identified in patients presenting with hemoptysis.* **(Bottom)** *Composite image with axial (left) and sagittal (right) CECT (mediastinal window) shows hypertrophied bronchial arteries in patients with chronic pulmonary thromboembolic disease (left) and bronchiectasis (right), respectively. Hypertrophied bronchial arteries occur in chronic processes that impair the pulmonary circulation, including pulmonary hypertension, Takayasu arteritis, certain chronic infections (e.g., tuberculosis and aspergilloma), and a variety of congenital conditions.*

ANOMALOUS LEFT VERTEBRAL ARTERY, CT

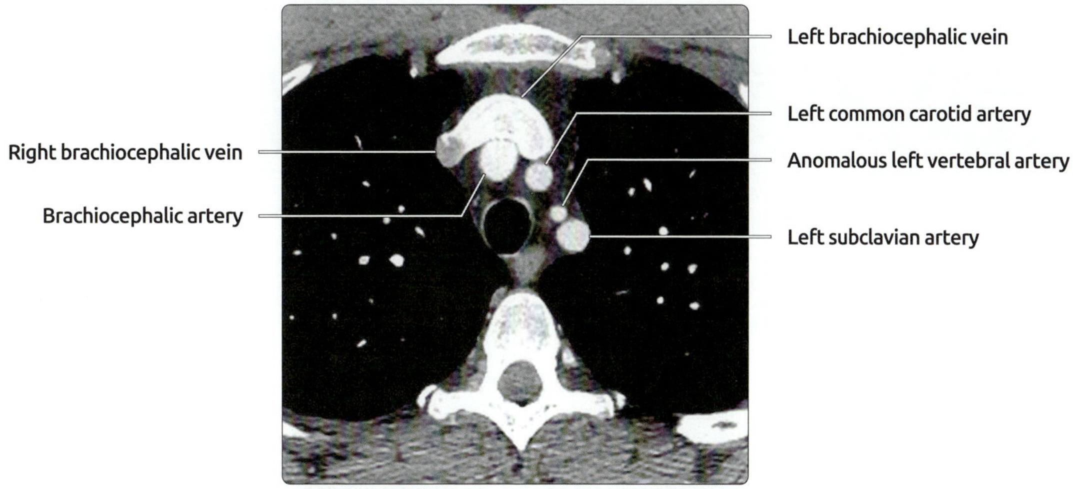

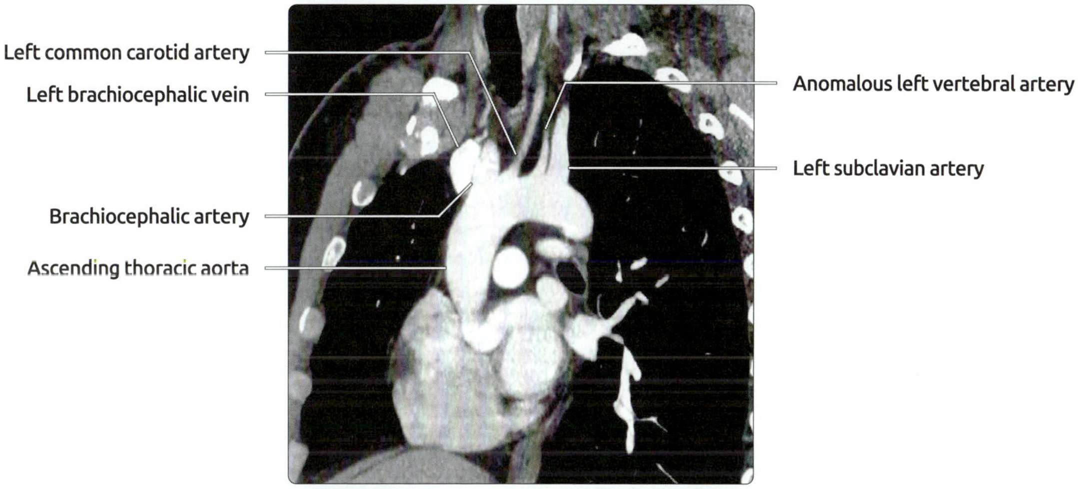

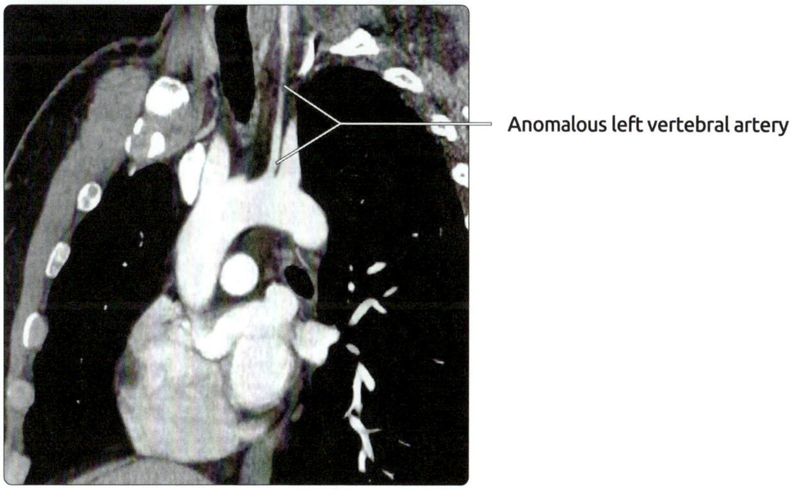

(Top) *Axial CECT (mediastinal window) of a normal individual shows an incidental anomalous left vertebral artery, which arises directly from the aortic arch between the left common carotid and left subclavian arteries.* **(Middle)** *Coronal oblique CECT (mediastinal window) of the same patient confirms that the aortic arch gives rise to 4 distinct branches: The brachiocephalic artery, the left common carotid artery, the anomalous left vertebral artery, and the left subclavian artery.* **(Bottom)** *Coronal oblique CECT (mediastinal window) of the same patient shows the course of the anomalous left vertebral artery from its origin in the aortic arch to its distal portion in the soft tissues of the left neck. Anomalous left vertebral artery is found in up to 2.3% of the population.*

ABERRANT RIGHT SUBCLAVIAN ARTERY, GRAPHIC AND RADIOGRAPHY

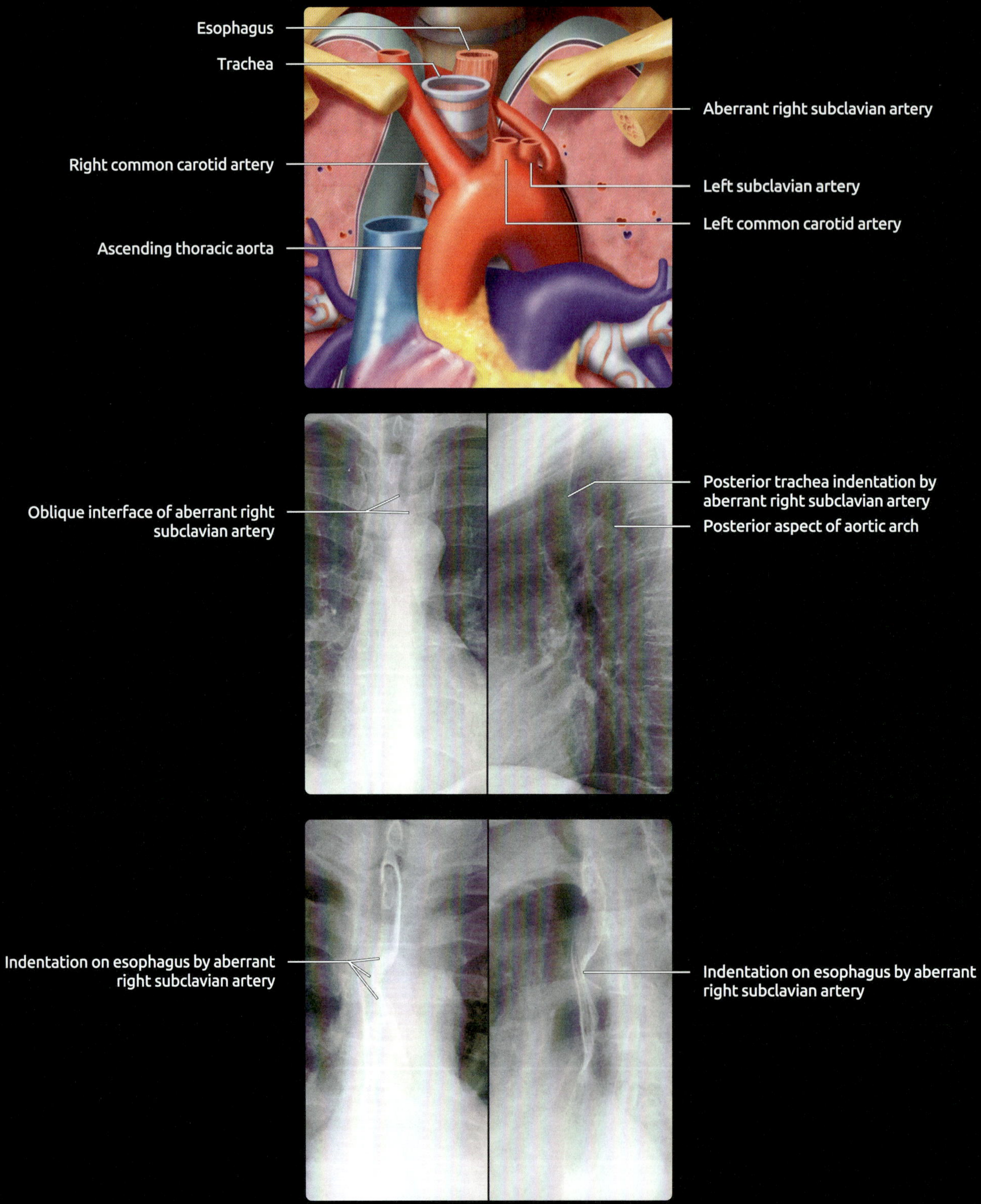

(Top) *Graphic depicts the anatomy of the aberrant right subclavian artery, which arises as the last branch of the left aortic arch. The anomalous vessel courses obliquely, typically behind the esophagus and trachea, to supply the right upper extremity. Mass effect on the esophagus and trachea may produce symptoms.* **(Middle)** *Composite image with PA (left) and lateral (right) chest radiographs of a patient who presented with mild dysphagia shows an oblique interface near the trachea on the PA chest radiograph and mass effect on the posterior tracheal wall near the aortic arch on the lateral chest radiograph. These findings are consistent with an aberrant right subclavian artery in this patient with a left aortic arch.* **(Bottom)** *Composite image with AP (left) and right posterior oblique (right) esophagram of the same patient shows an indentation on the esophagus at the level of the aortic arch caused by an aberrant right subclavian artery. The aberrant vessel courses behind the trachea and esophagus and may result in symptoms, as in this case.*

BOVINE AORTIC ARCH, AORTOGRAPHY AND CT

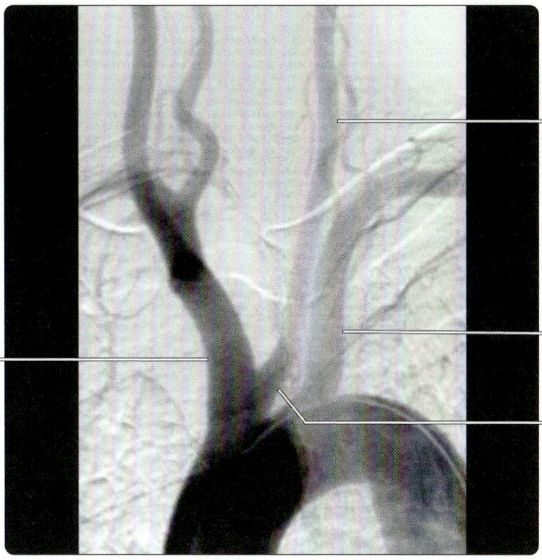

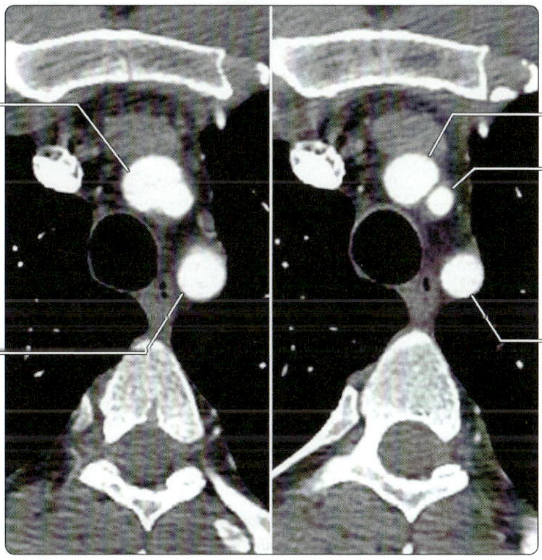

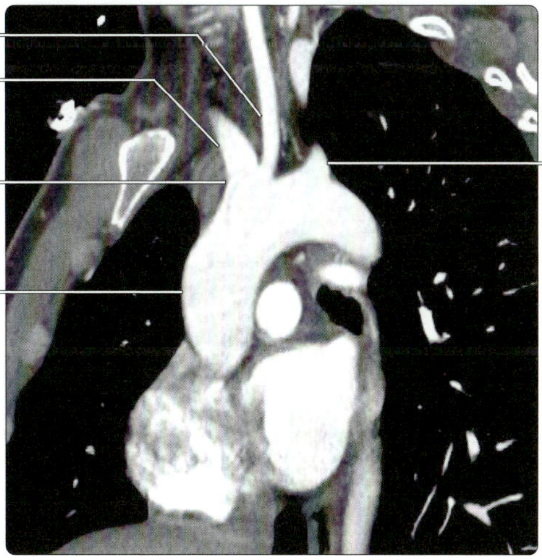

(Top) *Left anterior oblique aortogram of a 59-year-old man demonstrates the so-called bovine aortic arch, a common anatomic variant. The left common carotid artery arises from the proximal aspect of the brachiocephalic artery. The left subclavian artery arises independently from the aortic arch.* **(Middle)** *Composite image with axial CECT (mediastinal window) through the origin of the aortic arch branches shows that the brachiocephalic and left common carotid arteries arise from a common trunk. The left subclavian artery arises independently from the aortic arch.* **(Bottom)** *Coronal oblique CECT (mediastinal window) MIP reformatted image of the same patient shows the proximal aspects of the aortic arch branches. The brachiocephalic and left common carotid arteries share a common trunk. The left subclavian artery arises independently from the aortic arch. The bovine aortic arch is a common anatomic variant seen in up to 20% of the population.*

ABERRANT RIGHT SUBCLAVIAN ARTERY, CT

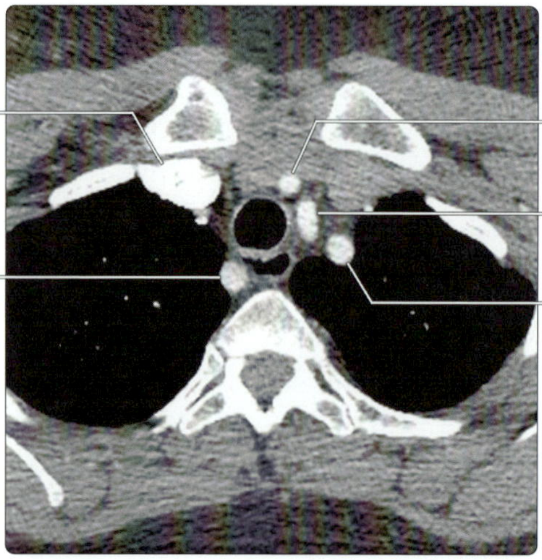

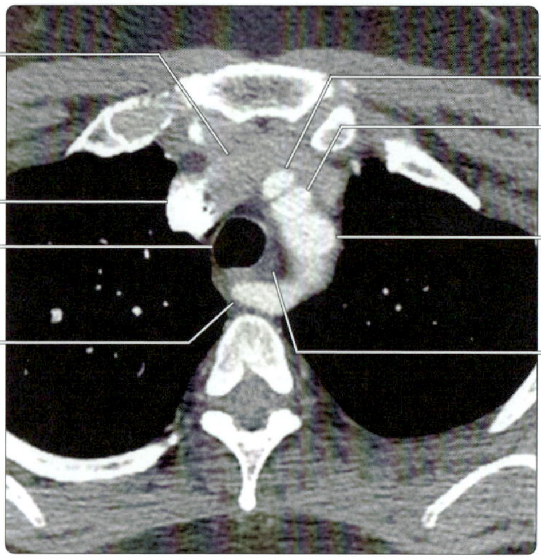

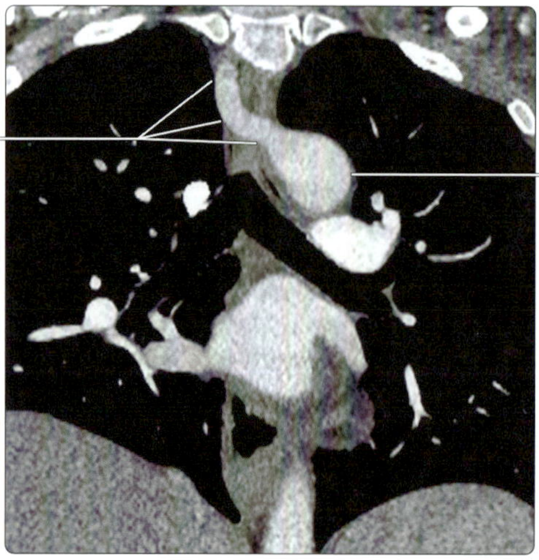

(Top) *Axial CECT (mediastinal window) of the same patient shows the aberrant right subclavian artery after having coursed behind the esophagus and trachea. The right common carotid artery arises independently from the aortic arch. The aortic arch branches arise in the following order: Right common carotid, left common carotid, left subclavian, and aberrant right subclavian arteries.* **(Middle)** *Axial CECT (mediastinal window) of the same patient obtained through the roots of the aortic arch branches shows the anomalous course of the aberrant right subclavian artery behind the trachea and esophagus. Note the potential for esophageal compression, which may result in symptoms.* **(Bottom)** *Coronal CECT (mediastinal window) of the same patient shows the anomalous course of the aberrant right subclavian artery, which correlates with the abnormalities noted on radiography and esophagram.*

RIGHT AORTIC ARCH WITH NON-MIRROR IMAGE BRANCHING, CT

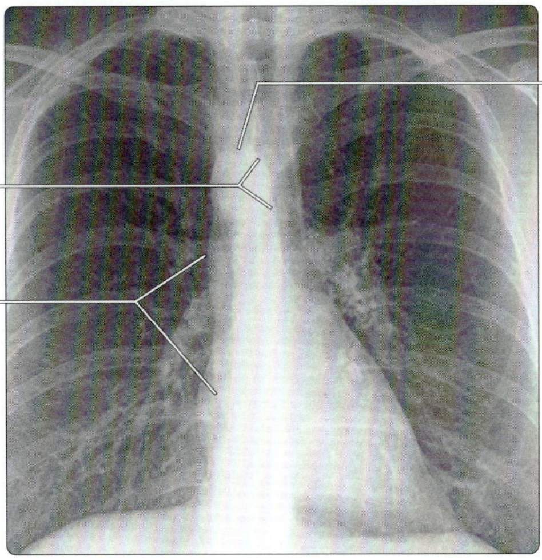

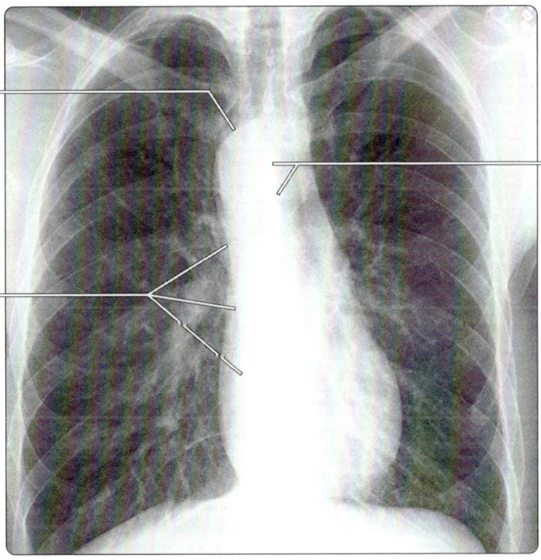

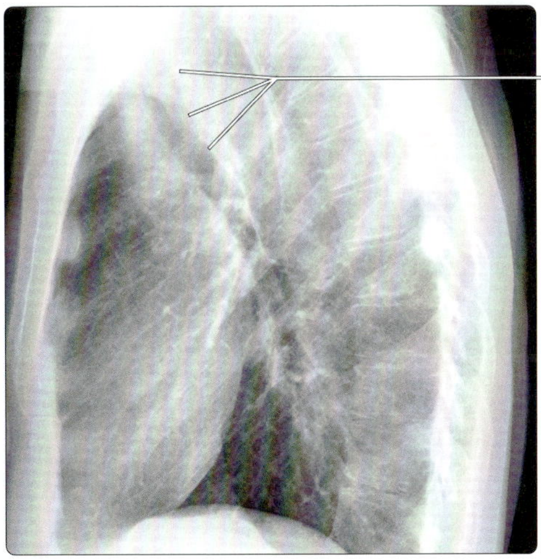

(Top) *PA chest radiograph of an asymptomatic adult shows an incidental right aortic arch. The aortic arch produces mild mass effect on the right lower lateral tracheal wall. Note the right paraaortic interface and no left paraaortic interface, consistent with a right-sided descending thoracic aorta.* **(Middle)** *PA chest radiograph of an asymptomatic 47-year-old man shows a right aortic arch. The aortic arch produces mild mass effect on the right lower lateral tracheal wall. There is a right paraaortic interface and no left paraaortic interface, consistent with a right-sided descending thoracic aorta.* **(Bottom)** *Lateral chest radiograph of a patient with a right aortic arch shows bowing of the posterior wall of the upper trachea, consistent with an aberrant left subclavian artery and indicating that the right aortic arch exhibits non-mirror image branching. Therefore, there is a low likelihood of associated congenital heart disease in this patient.*

RIGHT AORTIC ARCH WITH NONMIRROR IMAGE BRANCHING, CT

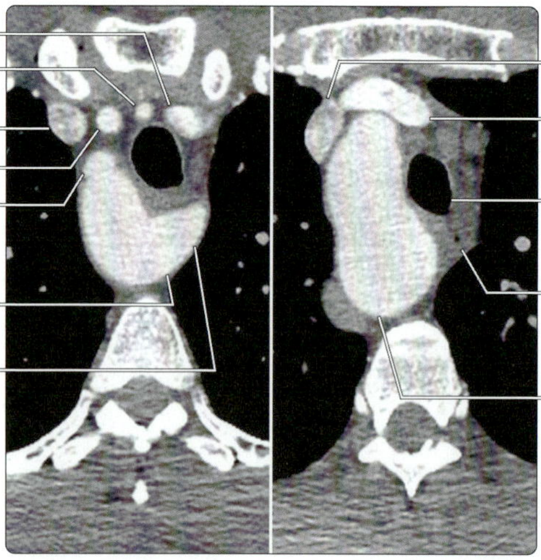

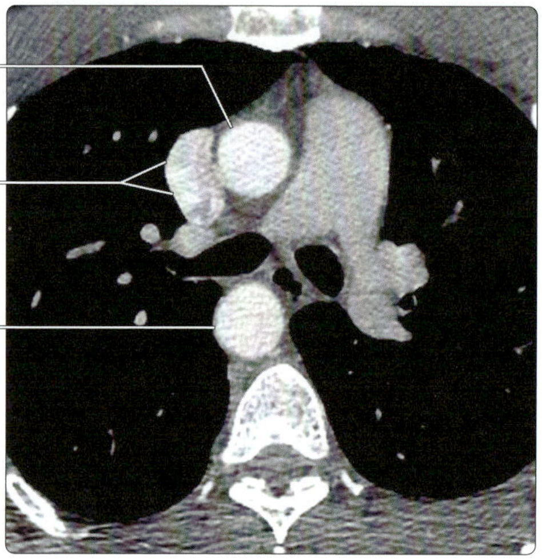

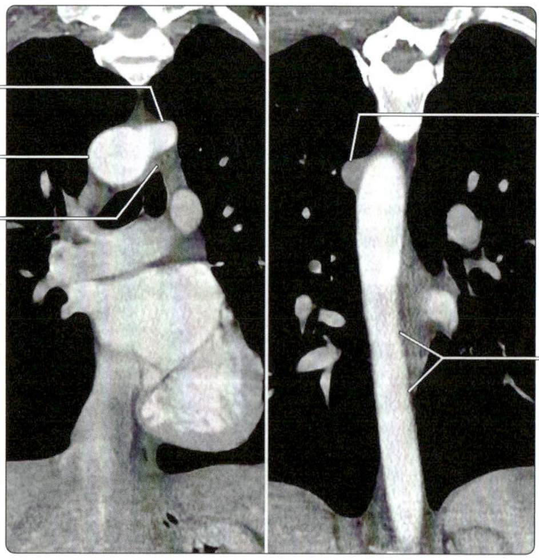

(Top) *Composite image with axial CECT (mediastinal window) of the same patient shows the non-mirror image branching pattern of the right aortic arch with an aberrant left subclavian artery arising (as the last aortic arch branch) from a diverticulum of Kommerell, which produces mild mass effect on the posterior tracheal wall (left). The right aortic arch produces mild mass effect on the right lateral tracheal wall (right).* **(Middle)** *Axial CECT (mediastinal window) of the same patient shows the descending thoracic aorta located to the right of midline, which correlates with the presence of a right paraaortic interface on radiography.* **(Bottom)** *Composite image with coronal CECT (mediastinal window) of the same patient shows the retrotracheal, retroesophageal course of the aberrant left subclavian artery, which arises as the last aortic branch from the posterior aspect of the aortic arch (left). Note the right-sided descending thoracic aorta and the azygos arch coursing lateral to the upper descending thoracic aorta (right).*

DOUBLE AORTIC ARCH, CT

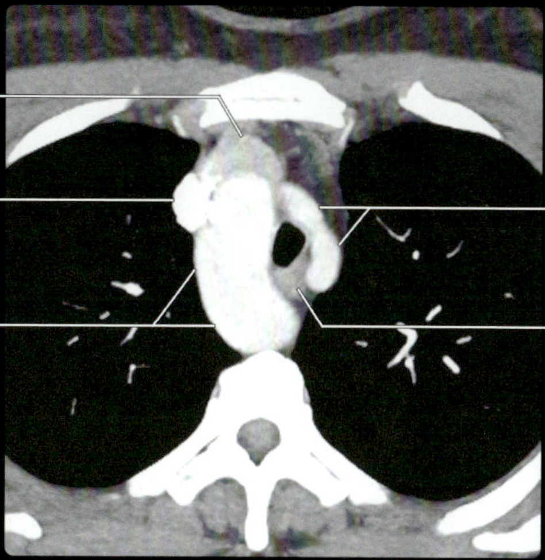

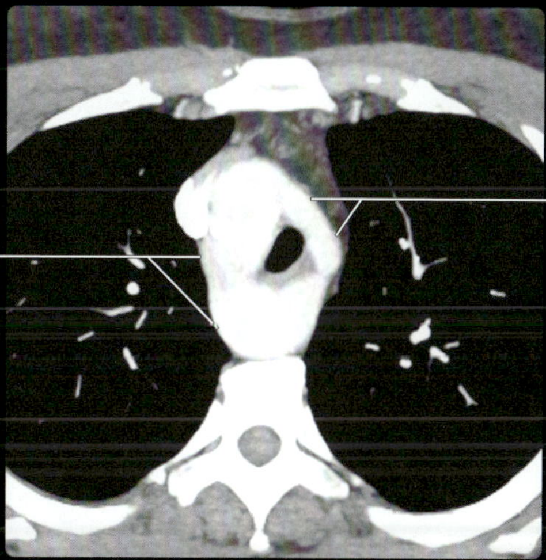

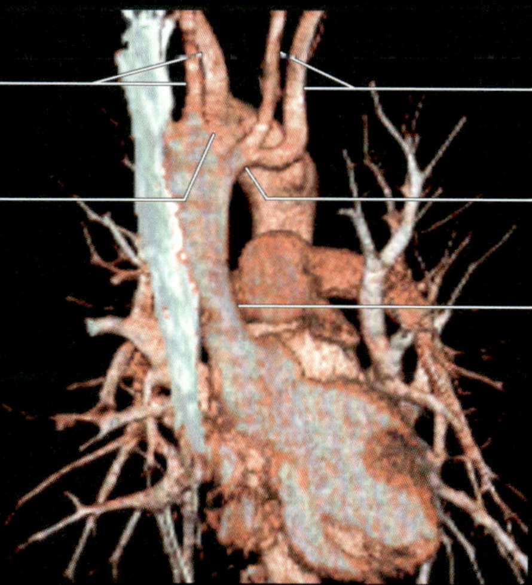

(Top) *Axial CECT (mediastinal window) of a 50-year-old woman with a double aortic arch who presented with chest pain shows that the right aortic arch is higher and larger than the left aortic arch. This anomaly produces a complete vascular ring, and affected patients may present with stridor &/or dysphagia.* **(Middle)** *Axial CECT (mediastinal window) of the same patient shows the midportion of the left aortic arch and demonstrates the vascular ring that surrounds the trachea and esophagus and produces mild tracheal narrowing.* **(Bottom)** *Coronal 3D surface-rendered CECT demonstrates the larger right aortic arch and the smaller left aortic arch. Note that each arch gives off separate common carotid and subclavian arteries.*

AORTIC ANEURYSM, RADIOGRAPHY AND CT

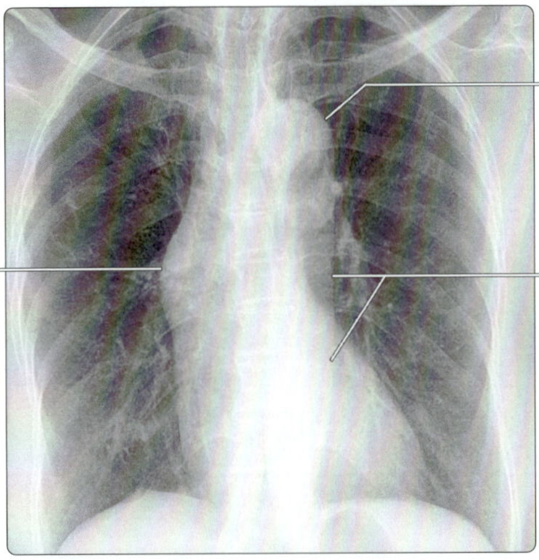

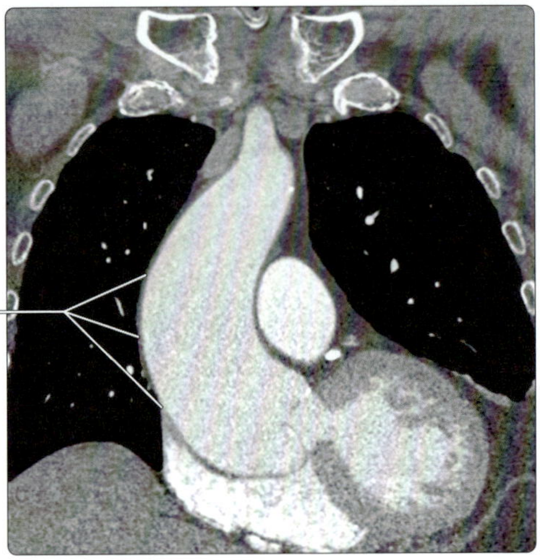

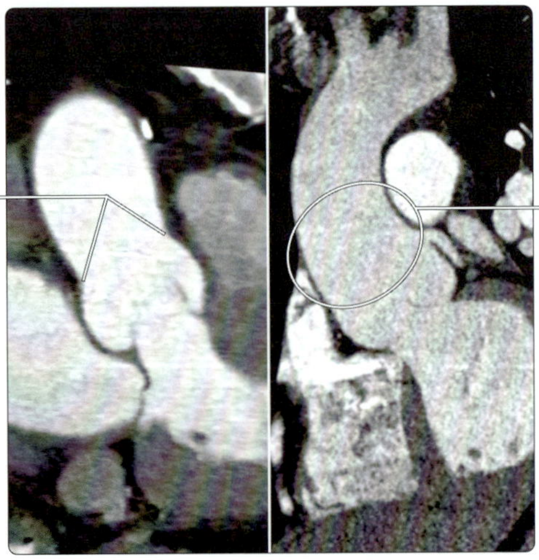

(Top) *PA chest radiograph of a 28-year-old woman with Marfan syndrome shows an enlarged ascending thoracic aorta secondary to medial degeneration, which produces a laterally convex interface in the midportion of the right mediastinum. Note conspicuity and ectasia of the aortic arch and the descending thoracic aorta.* **(Middle)** *Axial CECT (mediastinal window) of a 68-year-old man with an ascending thoracic aorta aneurysm shows aneurysmal dilatation of the ascending thoracic aorta, which produces convexity of the right midmediastinal interface and correlates with the radiographic abnormalities seen in patients with enlargement of the ascending thoracic aorta.* **(Bottom)** *Composite image with coronal CECT (mediastinal window) through the aortic root of a normal individual (left) and of a patient with Marfan syndrome and annuloaortic ectasia (right) shows a normal well-defined sinotubular junction that separates the aortic root from the tubular ascending thoracic aorta (left) and effacement of the sinotubular junction (right), which is described as an early sign of ascending thoracic aorta dilatation in patients with Marfan syndrome.*

AORTIC ANEURYSM, RADIOGRAPHY AND CT

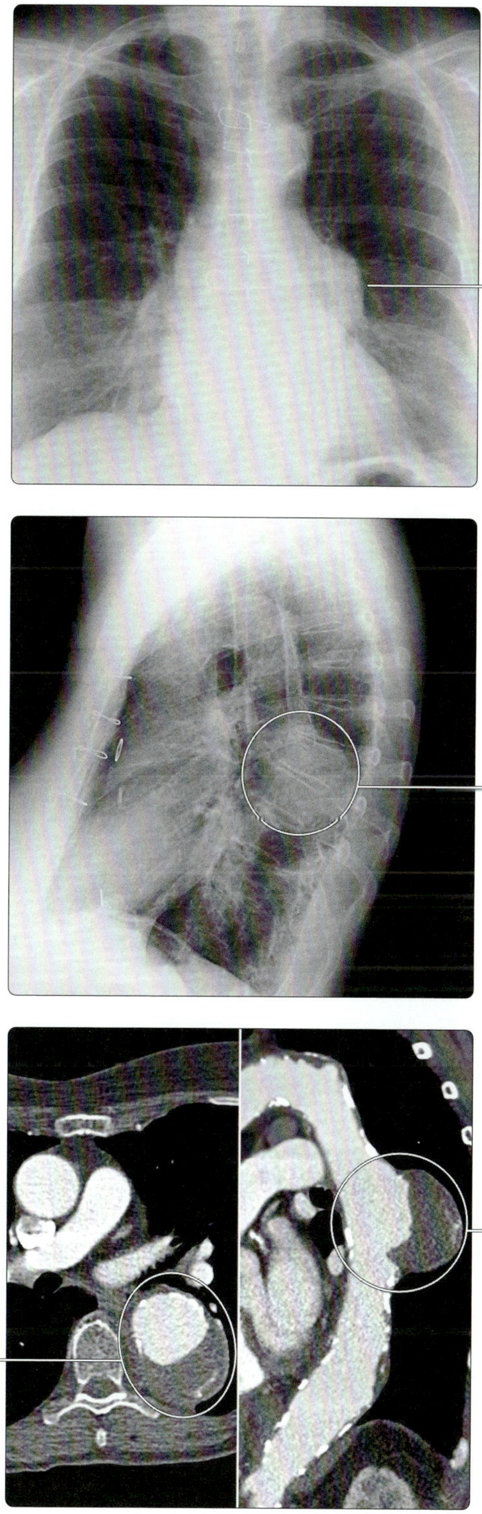

(Top) *PA chest radiograph of a patient with a saccular aneurysm involving the descending thoracic aorta shows abnormal convexity overlying the left hilum manifesting as a focal mediastinal mass.* **(Middle)** *Lateral chest radiograph of the same patient shows a mass-like opacity overlying the midthoracic spine, which corresponds to the saccular aneurysm of the descending thoracic aorta.* **(Bottom)** *Composite image with axial (left) and oblique sagittal (right) CECT (mediastinal window) of the same patient shows focal dilatation of the descending thoracic aorta (left), which appears more saccular in nature on multiplanar imaging (right). Note extensive atherosclerotic changes of the aorta. Atherosclerosis is a common etiology of thoracic aorta aneurysms.*

PERSISTENT LEFT SUPERIOR VENA CAVA, RADIOGRAPHY

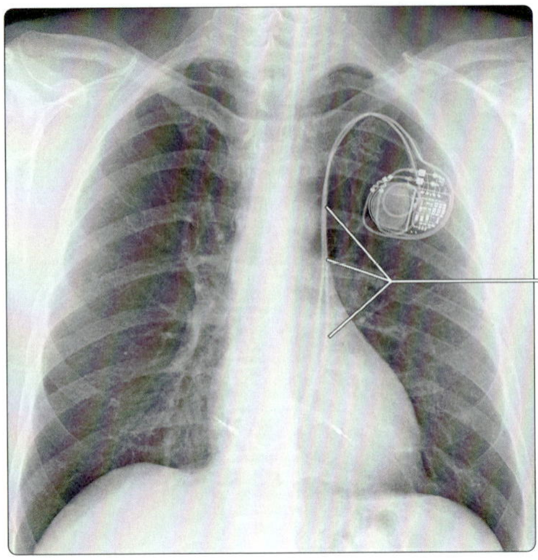

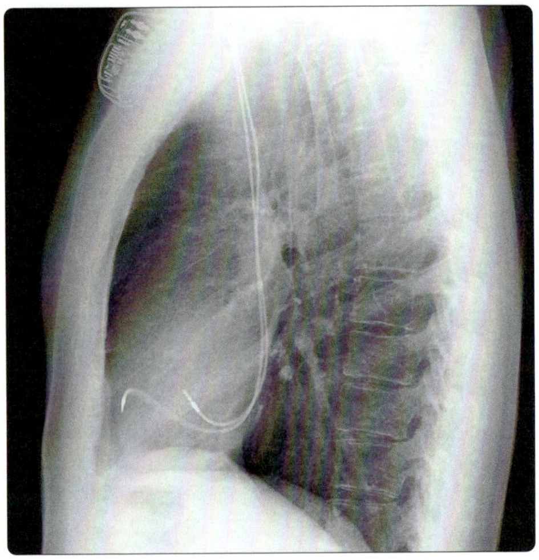

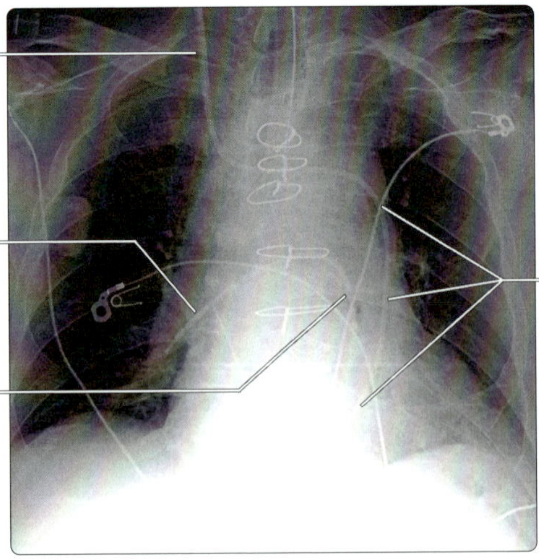

(Top) *PA chest radiograph of an asymptomatic 50-year-old man with a left pectoral dual-chamber pacemaker with leads in the right atrium and right ventricle is shown. The pacer leads course vertically along the left mediastinum within a persistent left superior vena cava to reach the right atrium via the coronary sinus. The right ventricular lead then courses through the tricuspid valve to reach the right ventricle.* **(Middle)** *Lateral chest radiograph of the same patient shows appropriately positioned atrial and ventricular pacer leads. Persistent left superior vena cava is seen in 1-3% of the population, often in completely asymptomatic patients. In this case, the diagnosis was made based on the course of the cardiac pacemaker leads along the left mediastinum.* **(Bottom)** *AP chest radiograph of a patient with persistent left superior vena cava who underwent placement of a right internal jugular pulmonary artery catheter shows that the catheter exhibits an unexpected course that corresponds to the anatomy of a persistent left superior vena cava. The catheter courses into the right atrium (via the coronary sinus), the right ventricle, and terminates in the distal right interlobar pulmonary artery.*

PERSISTENT LEFT SUPERIOR VENA CAVA, RADIOGRAPHY AND CT

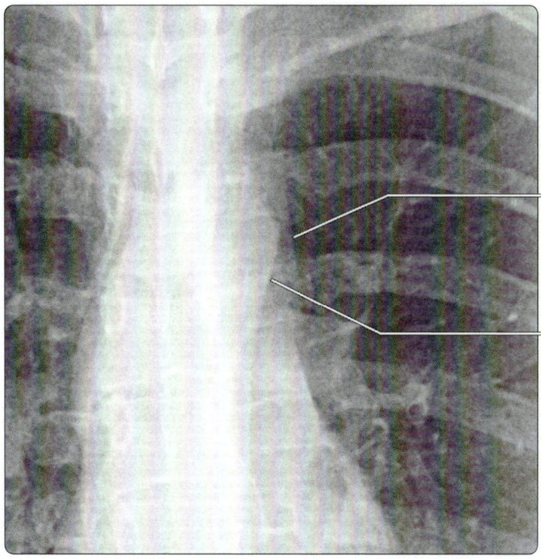

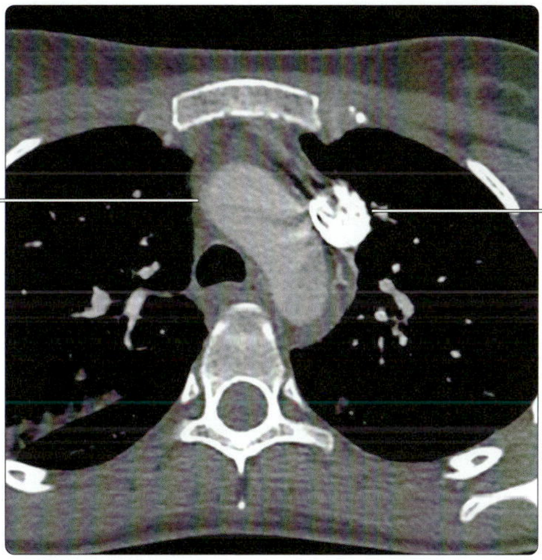

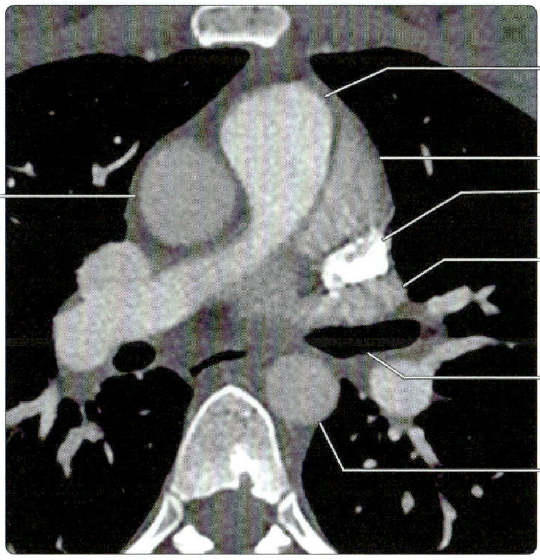

(Top) *PA chest radiograph of a patient who sustained trauma shows an abnormal contour of the left superior mediastinum characterized by lateralization of the aortic-pulmonary reflection, which manifests as a vertical interface located lateral to the aortic arch. This finding was secondary to a persistent left superior vena cava but is also described with mediastinal fat deposition, lymphadenopathy, and vascular abnormalities, such as partial anomalous pulmonary venous return (i.e., vertical vein).* **(Middle)** *Axial CECT (mediastinal window) of the same patient shows dense opacification of the persistent left superior vena cava, which accounted for the radiographic abnormality.* **(Bottom)** *Axial CECT (mediastinal window) of the same patient shows the persistent left superior vena cava coursing vertically along the left mediastinum directly anterior to the left superior pulmonary vein. In patients with left-sided partial anomalous pulmonary venous return, there is absence of the left superior pulmonary vein in its expected location posterior to the left atrial appendage.*

PERSISTENT LEFT SUPERIOR VENA CAVA, CT

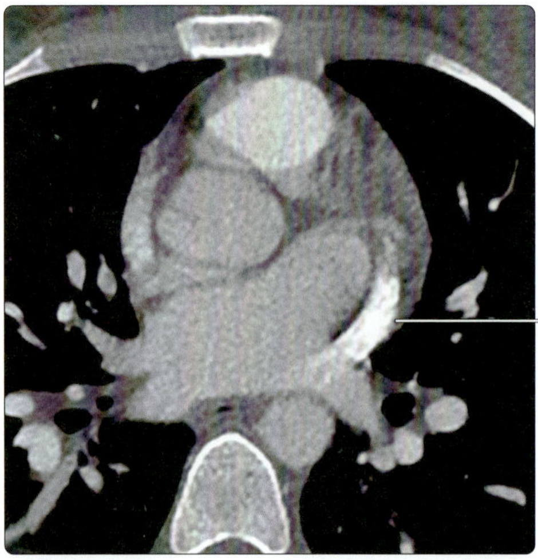

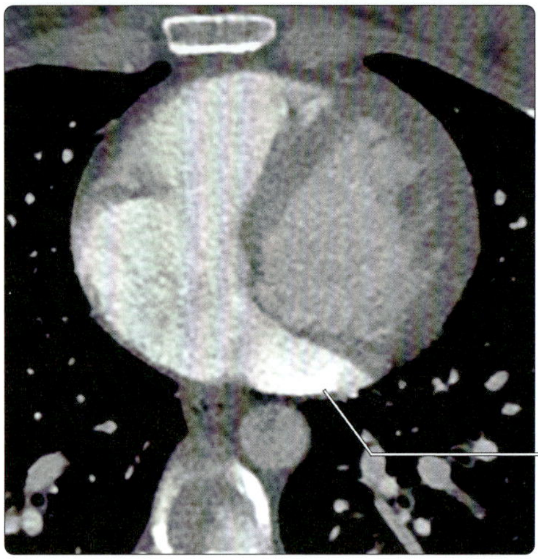

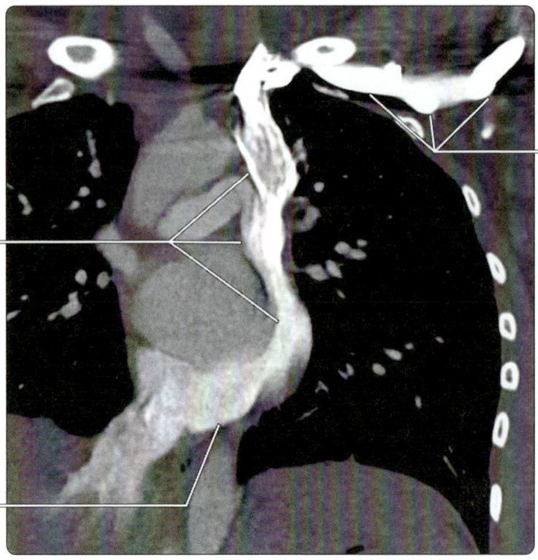

(Top) *Axial CECT (mediastinal window) of the same patient shows opacification of the distal aspect of the persistent left superior vena cava.* **(Middle)** *Axial CECT (mediastinal window) of the same patient shows a dilated coronary sinus due to increased flow from the persistent left superior vena cava. The coronary sinus may not be dilated if there is a concomitant right superior vena cava.* **(Bottom)** *Coronal CECT (mediastinal window) of the same patient shows the entire course of the persistent left superior vena cava and the dilated coronary sinus. In this case, there was no coexistent right superior vena cava.*

AZYGOS CONTINUATION OF INFERIOR VENA CAVA, RADIOGRAPHY AND CT

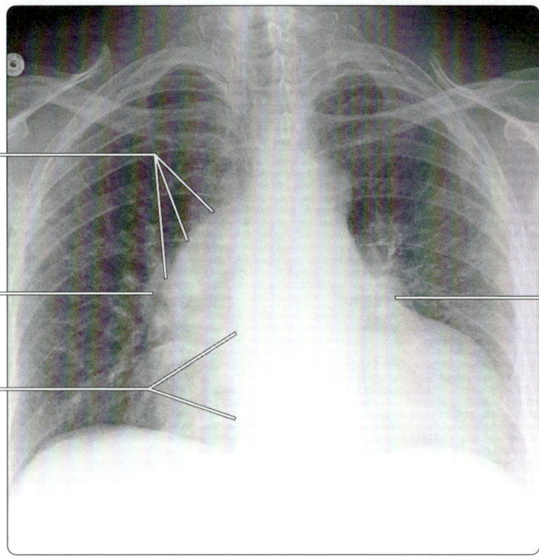

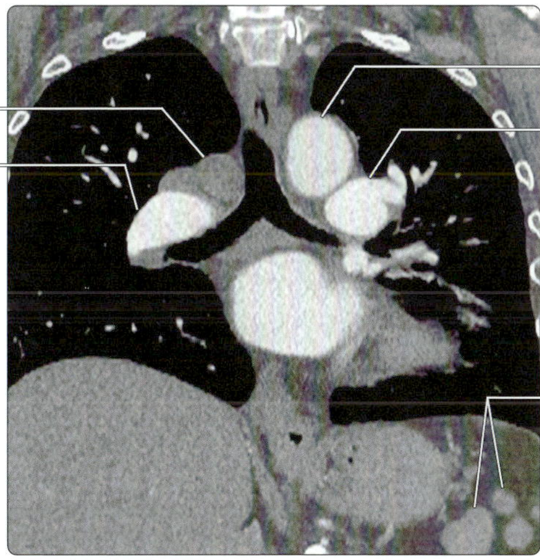

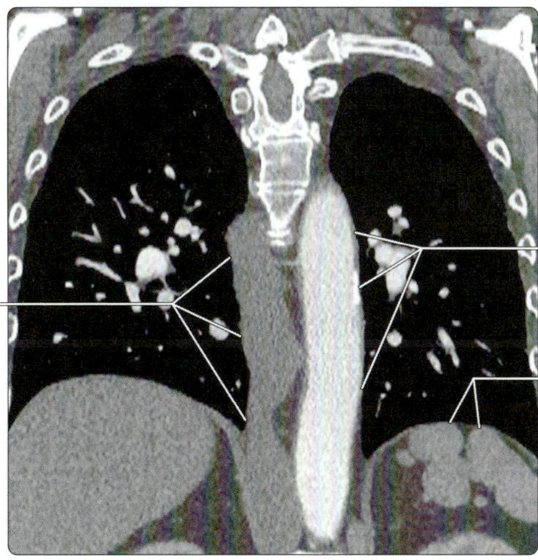

(Top) *PA chest radiograph of a 70-year-old woman with left isomerism shows a markedly dilated azygos arch and a vertical retrocardiac interface that corresponds to a dilated azygos vein. Note bilateral hyparterial bronchi, cardiomegaly, and vascular redistribution due to pulmonary venous hypertension.* **(Middle)** *Coronal CECT (mediastinal window) of the same patient shows bilateral hyparterial bronchi with the bilateral upper lobe bronchi arising distal to or below the ipsilateral pulmonary arteries. Note the enlarged unenhanced azygos arch and polysplenia in the left upper quadrant of the abdomen.* **(Bottom)** *Coronal CECT (mediastinal window) of the same patient shows the vertical course of the unenhanced dilated azygos vein that carries deoxygenated blood from the abdomen into the superior vena cava through a dilated azygos arch. Azygos continuation of the inferior vena cava may occur as an isolated condition or in association with left isomerism.*

PERSISTENT LEFT SUPERIOR VENA CAVA AND HEMIAZYGOS CONTINUATION OF INFERIOR VENA CAVA, CT

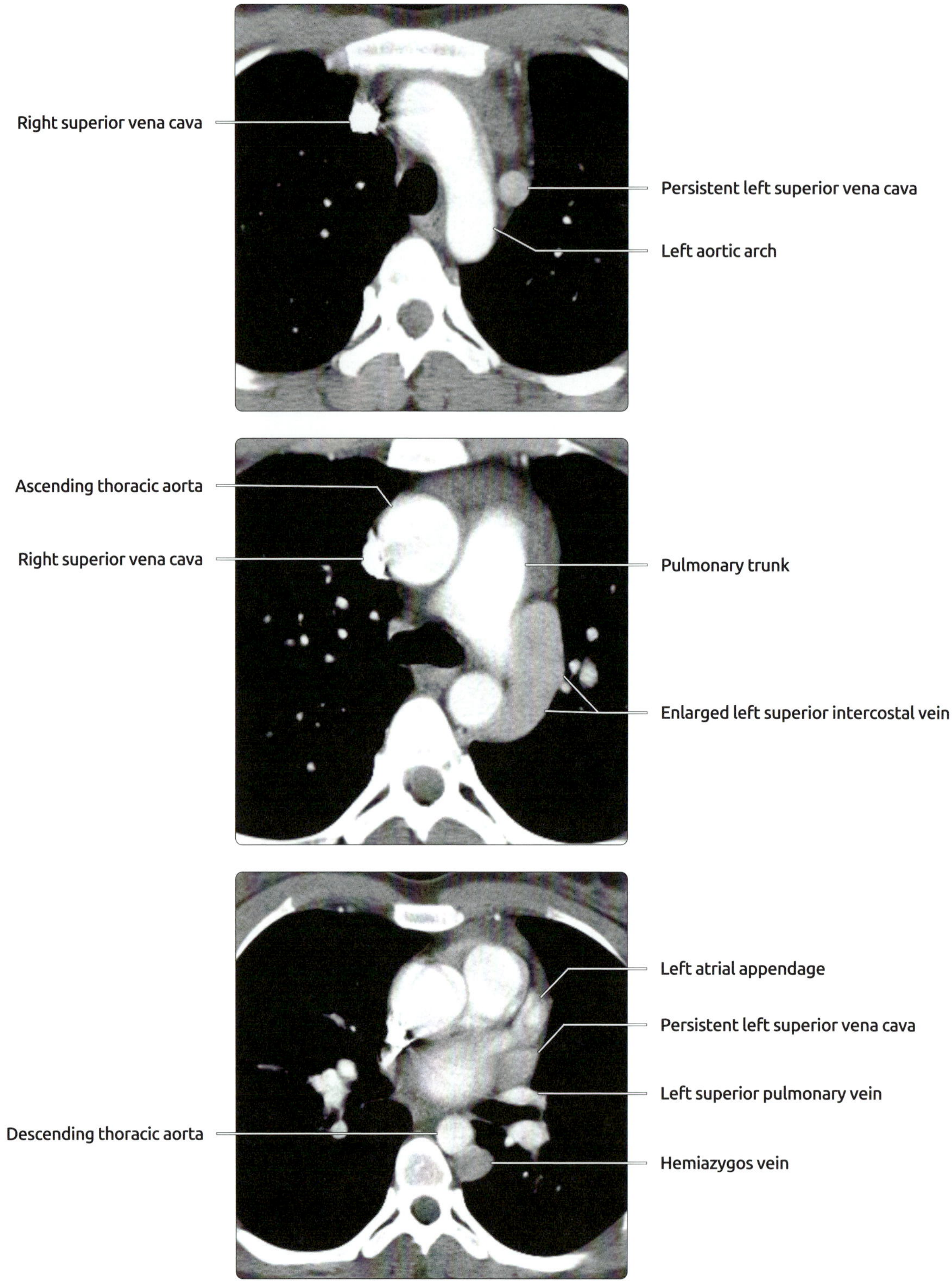

(Top) *Axial CECT (mediastinal window) of a 32-year-old man with left isomerism shows a persistent left superior vena cava manifesting as a vertically oriented vessel lateral to the aortic arch. A right superior vena cava is also present without an intervening bridging vein between the venae cavae.* **(Middle)** *Axial CECT (mediastinal window) of the same patient shows an enlarged left-sided venous structure coursing horizontally along the left superior mediastinum, consistent with an enlarged left superior intercostal vein draining into the posterior aspect of the left superior vena cava.* **(Bottom)** *Axial CECT (mediastinal window) of the same patient shows the persistent left superior vena cava coursing vertically between the left atrial appendage anteriorly and the left superior pulmonary vein posteriorly to drain into the coronary sinus. The enlarged hemiazygos vein courses alongside the descending thoracic aorta.*

PERSISTENT LEFT SUPERIOR VENA CAVA AND HEMIAZYGOS CONTINUATION OF INFERIOR VENA CAVA, CT

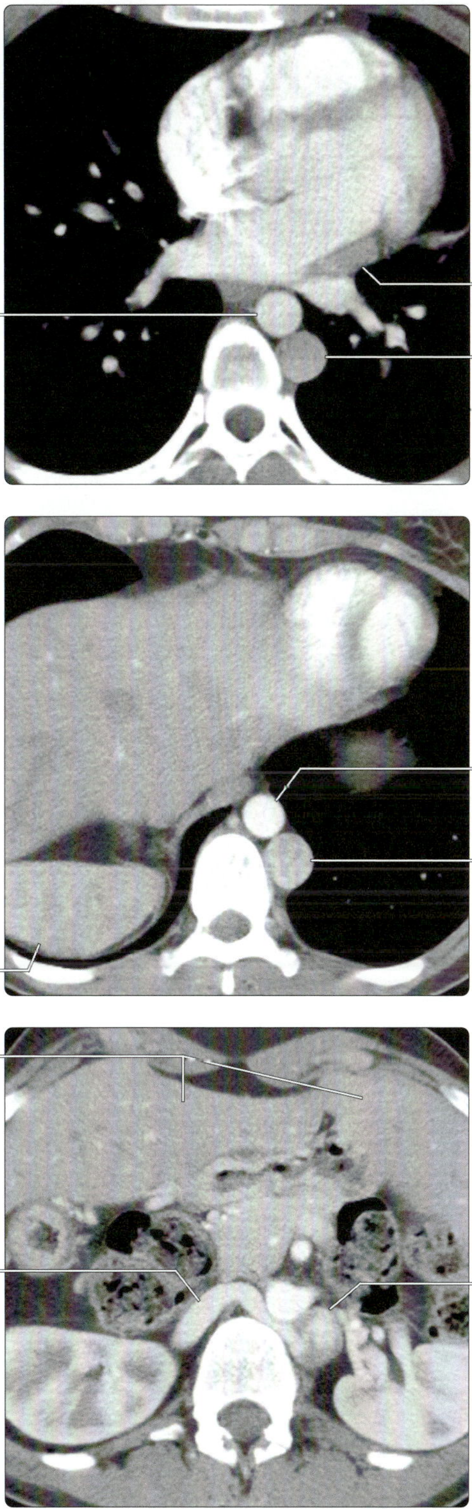

(Top) *Axial CECT (mediastinal window) of the same patient shows the distal aspect of the persistent left superior vena cava as it courses medially to drain into the coronary sinus. A dilated hemiazygos vein courses vertically posterior to the descending thoracic aorta along the left paravertebral space.* **(Middle)** *Axial CECT (mediastinal window) of the same patient obtained through the inferior thorax/upper abdomen shows imaging findings of isomerism. The spleen is on the right side. The dilated hemiazygos vein courses posterior to the descending thoracic aorta and along the left paravertebral space.* **(Bottom)** *Axial CECT (mediastinal window) of the same patient obtained through the upper abdomen shows a horizontal liver and the right renal vein draining into a left-sided inferior vena cava. In patients with this anomaly, abdominal deoxygenated blood reaches the right atrium by sequentially draining into the inferior vena cava, the hemiazygos vein, the persistent left superior vena cava, and finally into the coronary sinus to reach the right atrium.*

GENERAL ANATOMY AND FUNCTION

Anatomy

- Lymphatic vessels
  - Lymphatic capillaries: Blind ending; lined by single layer of overlapping endothelial cells
    - More permeable than vascular capillaries
  - Lymphatic collecting vessels: Thin 3-layer wall, thinner compared to veins but more numerous valves
    - Interposed by lymph nodes at intervals
    - Superficial and deep sets
  - Bilateral thoracic and lower neck lymphatic trunks
    - Jugular, subclavian, intercostal, bronchomediastinal
- Lymphatic ducts
  - Thoracic duct
  - Right lymphatic duct
- Lymph nodes: Small, round or ovoid; along lymphatic vessel pathways
  - Superficial and deep groups
  - Afferent vessels enter lymph node periphery; efferent vessels emerge from lymph node hilum

Function

- Maintenance of blood fluid volume: Reabsorption of excess interstitial fluid (returned to systemic venous circulation)
- Transport of dietary lipids: Lacteals drain to larger lymphatic vessels and eventually to systemic veins
- Lymphocyte development; immune response
- Lymphatic fluid transport: Lymphatic capillaries join to form lymphatic collecting vessels → filtering lymph nodes → lymphatic trunks → lymphatic ducts → systemic venous system

CENTRAL LYMPHATIC SYSTEM ANATOMY

Lymphatic Ducts

- **Thoracic duct**
  - Largest lymphatic vessel, drains 75% of total lymphatic fluid volume; does not drain upper right hemithorax, right upper extremity, or right neck
  - Origin at aortic hiatus between T12-L2 as cisterna chyli (sac-like tubular structure formed by confluence of subdiaphragmatic lymphatic trunks)
  - Ascends in visceral mediastinum to the right of descending thoracic aorta, close to esophagus, typically crosses midline near T4-T6
  - Ascends superiorly, arches above left clavicle, drains into left subclavian vein near confluence with left internal jugular vein; referred to as **left venous angle**
  - Receives multiple tributaries from lungs, heart, esophagus, intercostal vessels, and diaphragm
  - Highly variable anatomy; classic pattern described in 40-60% of cases
- **Right lymphatic duct**: Highly variable anatomy
  - < 2 cm long; drains remaining 25% of lymphatic fluid volume; drains right hemithorax, right upper extremity, and right head/neck
  - Formed by union of right jugular, right subclavian, and right bronchomediastinal trunks; mirrors thoracic duct as it joins venous system at confluence of right subclavian and right internal jugular veins; referred to as **right venous angle**

DRAINAGE PATHWAYS AND LYMPH NODES

Anterior Chest Wall

- Skin lymphatic drainage
  - Superficial lymphatic vessels converge in subcutaneous space, take lymph back to axillary lymph nodes
  - Deep lymphatic system: Drains to intercostal, parasternal/internal mammary, and diaphragmatic lymph nodes
- Parasternal/internal mammary lymph nodes: Anterior intercostal spaces, adjacent to internal mammary vessels
  - Drain breasts, deep epigastrium, deep anterior chest wall, and cranial liver surface
  - Lymph node efferent lymphatic vessels anastomose with tracheobronchial and brachiocephalic lymphatic vessels to form bilateral bronchomediastinal trunks
    - Frequent drainage to junction of ipsilateral internal jugular and subclavian veins
    - Right bronchomediastinal trunk may join right lymphatic duct and left thoracic duct
- Diaphragmatic lymph nodes: Thoracic (superior) diaphragm surface
  - Anterior, middle, and posterior groups
    - Anterior diaphragmatic: Anterior phrenic, paracardiac, cardiophrenic lymph nodes
      - Lower prevascular mediastinal fat, anterolateral to heart, above diaphragm
      - Drain convex liver surface and anterior diaphragm
      - Efferent lymphatic vessels drain to internal mammary lymph nodes
    - Middle diaphragmatic (juxtaphrenic) lymph nodes
      - Adjacent to inferior vena cava (IVC) near where phrenic nerves enter diaphragm
      - May occur within fibrous pericardium anterior to distal IVC
      - Drain middiaphragm; those on the right also drain convex liver surface
      - Efferents drain to posterior mediastinal lymph nodes
    - Posterior diaphragmatic (retrocrural) lymph nodes
      - Posterior to diaphragmatic crura
      - Drain posterior diaphragm, posterior (visceral and paravertebral) mediastinum, lumbar spine
      - Communication between posterior mediastinal and lateral aortic lymph nodes

Posterolateral Chest Wall

- Intercostal lymph nodes: At costovertebral junctions along thoracic vertebrae
  - Drain posterolateral chest wall
  - Efferents from caudal 4-7 intercostal spaces anastomose as single trunk that courses inferiorly into proximal thoracic duct
- Posterior diaphragmatic (retrocrural) lymph nodes

Intrathoracic and Supraclavicular Lymph Nodes (Lower Respiratory Tract)

- Intrathoracic respiratory system: Lower 1/2 of trachea (from inferior T1 border) and bronchial branches, lungs, and pleurae
- Supraclavicular, tracheobronchial (mediastinum and lungs), and hilar lymph nodes

- Lymphatic drainage to pulmonary, bronchopulmonary/hilar, tracheobronchial, and paratracheal lymph nodes
- Drainage of lungs and pleurae: Superficial to deep
  - Parietal pleura drains through stomata to lymphatics in intercostal spaces, then anteriorly to internal mammary lymph nodes and posteriorly to intercostal lymph nodes
  - Visceral pleura drains via submesothelial lymphatic network that spans lung periphery and extends into lung parenchyma, via deep pulmonary plexus to hilar lymph nodes
  - Superficial and deep subpleural plexuses have few communications (one being at hilar region)
  - Efferent vessels of superficial plexus follow lung borders and fissure margins to terminate in bronchopulmonary lymph nodes
  - Intrapulmonary lymph nodes: Along fissures, subpleural lung, bronchovascular bundles; small (average diameter < 10 mm), smooth margins; oval, lentiform, triangular
- **International Association for the Study of Lung Cancer (IASLC) Lymph Node Map**
  - Facilitates reliable assessment of lymph node metastases for staging lung cancer: 14 lymph node stations, 7 lymph node zones
- **IASLC lymph node zones** and **stations**
  - **Supraclavicular zone**: **Station 1**
    - **Station 1R**: Right low cervical, supraclavicular, sternal notch lymph nodes
      - Upper border: Lower margin of cricoid cartilage
      - Lower border: Bilateral clavicles and upper manubrium border at midline
      - Boundary between 1R and 1L: Midline of trachea
    - **Station 1L**: Left low cervical, supraclavicular, sternal notch lymph nodes
      - Upper border: Lower margin of cricoid cartilage
      - Lower border: Bilateral clavicles and upper manubrium border at midline
  - **Upper zone** (**superior mediastinal**): **Stations 2**, **3**, and **4**
    - **Station 2R**: Right upper paratracheal lymph nodes
      - Upper border: Right lung apex and pleural space, upper manubrium border at midline
      - Lower border: Intersection of caudal margin of brachiocephalic vein with trachea
      - Boundary between 2R and 2L: Left lateral wall of trachea
    - **Station 2L**: Left upper paratracheal lymph nodes
      - Upper border: Left lung apex and pleural space, upper manubrium border at midline
      - Lower border: Superior border of aortic arch
    - **Station 3A**: Prevascular lymph nodes
      - Upper border: Apex of chest
      - Lower border: Carina
      - Anterior border: Posterior aspect of sternum
      - Posterior border: Anterior superior vena cava border on right, left carotid artery on left
    - **Station 3P**: Retrotracheal lymph nodes (posterior to trachea)
      - Upper border: Apex of chest
      - Lower border: Carina
    - **Station 4R**: Right lower paratracheal lymph nodes
      - Upper border: Intersection of caudal margin of brachiocephalic vein with trachea
      - Lower border: Lower border of azygos arch
      - Boundary between 4R and 4L: Left lateral tracheal border
    - **Station 4L**: Left lower paratracheal lymph nodes
      - Upper border: Upper margin of aortic arch
      - Lower border: Upper rim of left pulmonary artery
  - **Aortopulmonary zone**: **Stations 5** and **6**
    - **Station 5**: Subaortic lymph nodes
      - Lateral to ligamentum arteriosum
      - Upper border: Lower border of aortic arch
      - Lower border: Upper rim of left pulmonary artery
    - **Station 6**: Paraaortic lymph nodes
      - Lymph nodes anterior and lateral to ascending aorta and aortic arch
      - Upper border: Upper border of aortic arch
      - Lower border: Lower border of aortic arch
  - **Subcarinal zone**: **Station 7**
    - **Station 7**: Subcarinal lymph nodes
      - Upper border: Carina
      - Lower border: Lower border of bronchus intermedius on right, upper border of lower lobe bronchus on left
  - **Lower zone** (**inferior mediastinal**): **Stations 8** and **9**
    - **Station 8**: Paraesophageal lymph nodes
      - Adjacent to esophagus (right or left of midline)
      - Upper border: Lower border of bronchus intermedius on right, upper border of lower lobe bronchus on left
      - Lower border: Diaphragm
    - **Station 9**: Pulmonary ligament lymph nodes
      - Upper border: Inferior pulmonary vein
      - Lower border: Diaphragm
  - **Hilar/interlobar zone** (**pulmonary**): **Stations 10** and **11**
    - **Station 10**: Hilar lymph nodes
      - Immediately adjacent to mainstem bronchi and hilar vessels, including proximal pulmonary veins and pulmonary trunk; right (10R) and left (10L) divisions
      - Upper border: 10R, lower rim of azygos arch; 10L, upper rim of pulmonary artery
      - Lower border: Bilateral interlobar regions
    - **Station 11**: Interlobar lymph nodes
      - Between origins of lobar bronchi; right (11R) and left (11L) divisions
      - 11R: **11Rs** (between right upper lobe bronchus and bronchus intermedius), **11Ri** (between middle lobe and right lower lobe bronchi)
  - **Peripheral zone** (**pulmonary**): **Stations 12**, **13**, and **14**
    - **Station 12**: **Lobar lymph nodes**
      - Adjacent to lobar bronchi; right (12R) and left (12L) divisions
    - **Station 13**: **Segmental lymph nodes**
      - Adjacent to segmental bronchi; right (13R) and left (13L) divisions
    - **Station 14**: **Subsegmental lymph nodes**
      - Adjacent to subsegmental bronchi; right (14R) and left (14L) divisions
- Miscellaneous thoracic lymph nodes (not in IASLC map)
  - Internal mammary, axillary, subpectoral, intercostal, diaphragmatic

- Nonregional for lung cancer staging, regional for staging other primary thoracic malignancies (e.g., pleural mesothelioma, esophageal cancer)

Heart

- 3 plexuses: Subepicardial, myocardial, and subendocardial spaces
  - Subepicardial plexus receives efferents from subendocardial and myocardial plexuses, then gives rise to left and right cardiac collecting trunks
  - Left trunks travel superiorly in anterior interventricular groove receiving afferents from ventricles → tracheobronchial lymph nodes
  - Right trunks receive tributaries from right heart border, diaphragmatic surface of right ventricle, right atrium → left brachiocephalic lymph node

Esophagus

- Esophageal lymph nodes
  - Cervical esophagus: Drains to paratracheal and deep cervical lymph nodes
  - Thoracic esophagus: Drains to paraesophageal and subcarinal mediastinal lymph nodes
  - Abdominal esophagus: Drains to left gastric lymph nodes

Thymus

- No afferent lymphatic vessels leading to thymus
- Efferent lymphatics terminate in internal mammary, brachiocephalic, and tracheobronchial lymph nodes

IMAGING

Radiography

- Limited sensitivity and specificity
- Abnormal mediastinal lines, stripes, and interfaces &/or hilar enlargement may indicate lymphadenopathy
- Kerley B lines (interlobular septal thickening)
- Pleural effusions

Conventional CT/MR

- Identification of lymphadenopathy, septal thickening
- Pleural/pericardial effusions/thickening

Conventional Lymphangiography (Oil Based)

- Preferred technique for imaging lymphatics
- Excellent visualization of lymphatics and localization of foci of obstruction or disruption/leak
- Simultaneous interventional treatment

CT and MR Lymphangiography

- CT: Injection of standard iodinated contrast into inguinal lymph nodes under ultrasound or fluoroscopy guidance
  - CT fluoroscopy: Monitor contrast flow to cisterna chyli
  - Chest CT to assess for lymphatic leak
- MR lymphangiography with or without contrast
  - Noncontrast MR; heavily T2-weighted sequences
  - Contrast MR; gadolinium injection of inguinal lymph nodes

Lymphoscintigraphy With SPECT/CT

- Evaluation of known or suspected lymphatic disorders; localization of sentinel lymph nodes in cancer staging
- Intradermal injection of technetium 99m sulfur colloid or Tc-tilmanocept in extremities
- Demonstration of thoracic duct and main lymph chains
  - Better identification and localization of lymphatic leaks achieved with SPECT/CT

ANATOMY-RELATED IMAGING ABNORMALITIES

Common Causes of Thoracic Lymphatic Disruption

- Thoracic surgery: Esophagectomy, congenital heart defect repair, coronary artery bypass grafting, lung transplant
- Malignancy: Metastatic disease, lymphoma
- Heart failure, renal failure, cirrhosis

Uncommon Causes of Thoracic Lymphatic Disruption

- Vascular/lymphatic malformations
- Genetic syndromes: Congenital (pulmonary) lymphangiectasis, lymphangiomatosis, lymphangioleiomyomatosis, Waldmann disease, *EPHB4*-associated central conducting lymphatic anomaly (CCLA)
- Autoimmunity: Systemic lupus erythematosus, rheumatoid arthritis, scleroderma, dermatomyositis
- Secondary lymphatic disorders: Pulmonary lymphatic perfusion syndrome, secondary lymphangiectasis

Imaging Manifestations of Lymphatic Disruption

- **Lung**
  - Smooth interlobular septal thickening: Interstitial edema &/or dilated lymphatics
    - Volume overload, heart failure, renal failure
    - Septal nodularity may represent congested lymphatics and increased vascular permeability
  - Nodular interlobular septal thickening
    - Lymphangitic carcinomatosis, sarcoidosis
    - Rapid onset suggests interstitial edema rather than malignancy
  - Septal thickening from other lymphatic abnormalities
  - Lymphatic flow reversal and contrast intravasation into pulmonary lymphatics
- **Pleura** and **pericardium**
  - Chylous effusion (pleural or pericardial)
    - Traumatic (including postsurgical) vs. nontraumatic
    - Large, free, or loculated
    - Recurrent chylous effusion common in malignancy
  - Nonchylous effusion (transudative)
  - Pleural/pericardial thickening
    - Chronic or recurrent effusions
    - Proliferative lymphatic disorders (lymphangiomatosis)
  - Increased extrapleural fat: Chronic pleural effusion
- **Mediastinum**
  - Thoracic duct abnormalities
    - Focal disruption or leak, duct dilatation, missing segment, plexiform thoracic duct
  - Lymphatic malformations: Focal or diffuse
    - 10% in mediastinum, usually superior and prevascular
    - Often insinuate between mediastinal compartments without invasion or mass effect
  - Lymphadenopathy
    - Etiologies: Infections, inflammatory conditions, malignancy, autoimmunity, pulmonary edema
    - Pattern of metastatic lymphadenopathy correlates with anatomic lymphatic drainage pathways

OVERVIEW OF LYMPHATICS

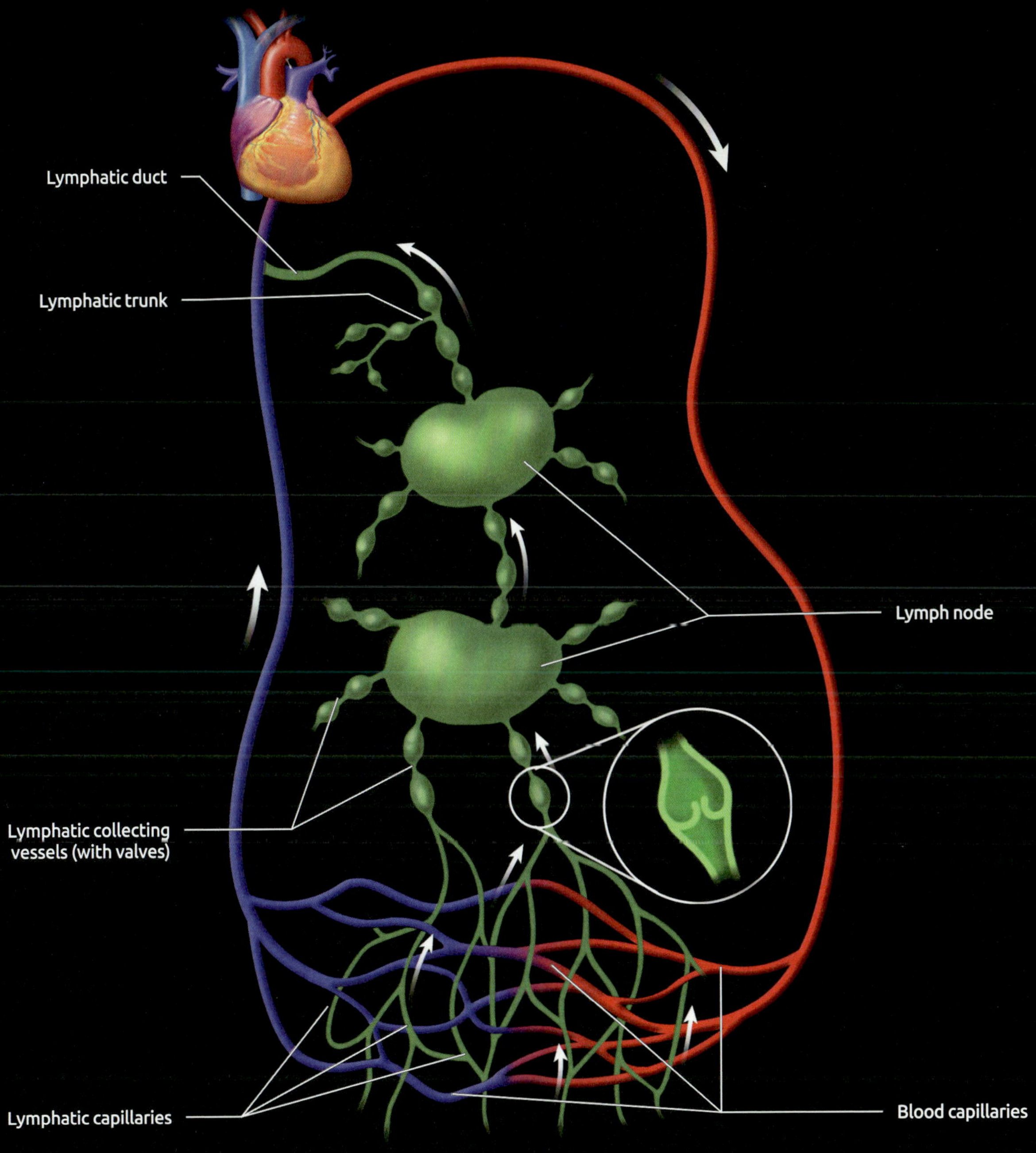

Graphic presents a broad overview of the lymphatic system. Lymphatic vessels include lymphatic capillaries and lymphatic collecting vessels that drain into superficial and deep interposed lymph nodes. Lymph nodes drain into lymphatic collecting vessels, which in turn drain into lymphatic trunks, and then ducts, including the thoracic duct (that drains 75% of the total lymphatic fluid volume) and the right lymphatic duct (that drains the remaining 25% of the lymphatic fluid volume). The thoracic duct and the right lymphatic duct ultimately drain into the systemic venous system. The lymphatic system is responsible for maintaining the blood fluid volume, the transport of dietary lipids, and the development of lymphocytes that play an important role in the immune response.

OVERVIEW OF LYMPHATIC DRAINAGE

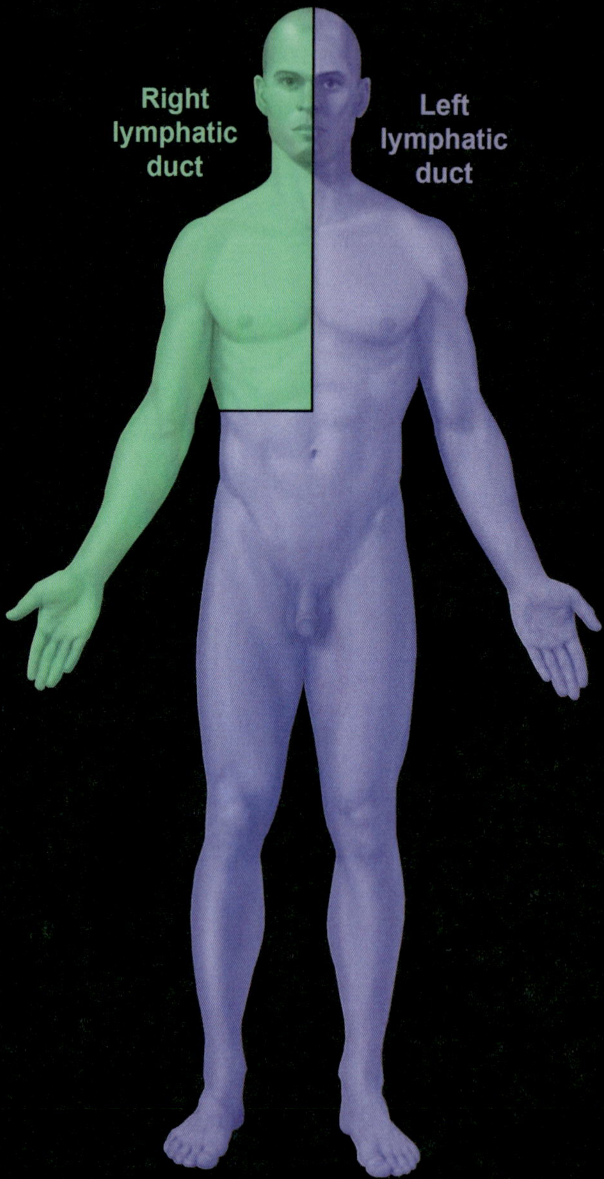

Graphic illustrates the relative contributions of the thoracic duct and the right lymphatic duct to the body's lymphatic drainage. The thoracic duct is the largest lymphatic vessel and drains ~ 75% of the total lymphatic fluid volume (outlined in purple). The right lymphatic duct is responsible for the rest of the lymphatic drainage (~ 25%) and drains the right hemithorax, the right upper extremity, and the right head and neck (outlined in green).

ANATOMY OF THORACIC DUCT

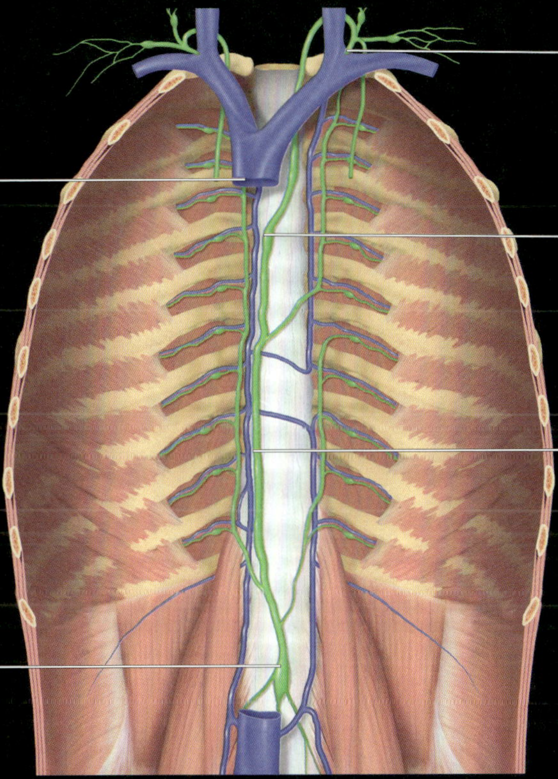

Graphic shows the thoracic duct, which arises from the cisterna chyli near the L2 vertebra. The thoracic duct courses cephalad to the right of midline along the anterior spine, crosses to the left at the upper thoracic spine (near the T4-T6 thoracic vertebral bodies), and continues cephalad to the left of midline to drain near the left internal jugular vein/left subclavian vein anastomosis, referred to as the left venous angle. The thoracic duct receives tributaries from the lungs, the heart, the esophagus, the intercostal vessels, and the diaphragm.

ANATOMY OF THORACIC DUCT VARIANTS

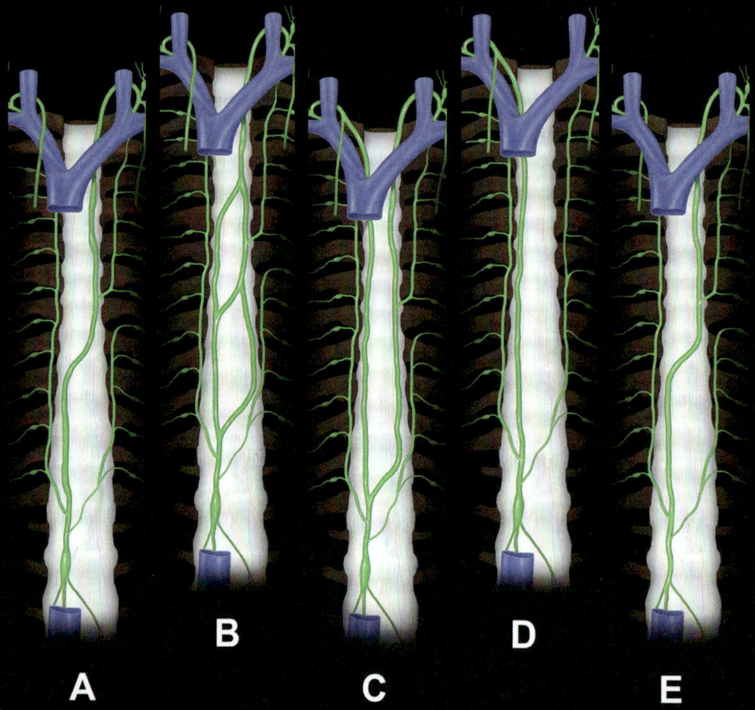

Graphic illustrates variations in thoracic duct anatomy. In the classically described pattern, the thoracic duct courses cephalad along the anterior thoracic spine to the right of midline and crosses over to the left near the T4-T6 thoracic vertebrae to drain into the left venous angle, a pattern described in 40-60% of cases (A). The thoracic duct (10-40% of cases) has multiple channels &/or partial duplication (B). A rare variant (1-1.5% of cases) is a duplicated thoracic duct, terminating bilaterally in the neck (C). A right-sided thoracic duct that does not cross midline and terminates in the right neck occurs in 2-6% of cases (D). The cisterna chyli may be absent or have a plexiform configuration (or may be associated with other configurations) in 50-85% of cases (E).

INTRATHORACIC LYMPHATIC DRAINAGE

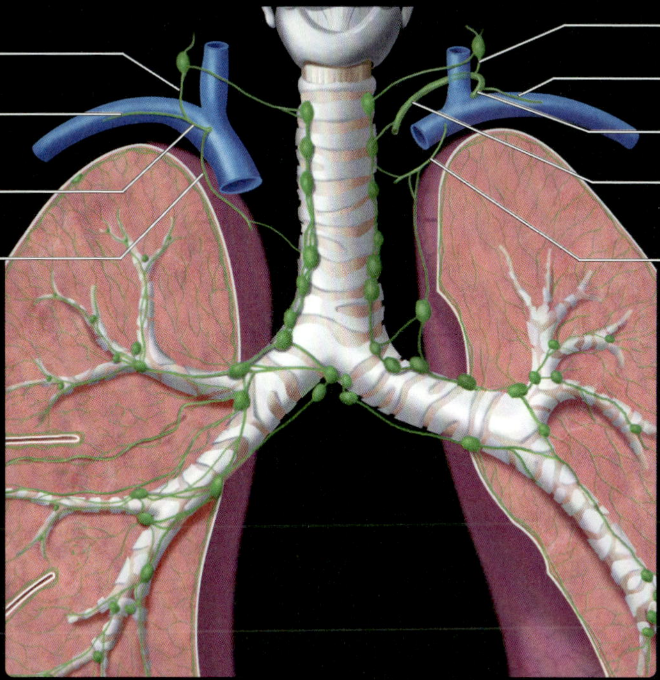

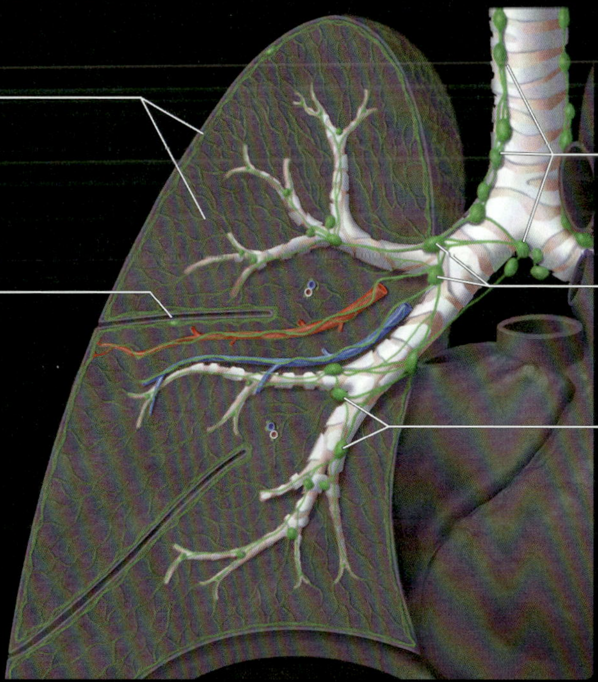

(Top) *Graphic shows the lymphatic drainage of the thorax. The right bronchomediastinal lymphatic trunk joins with the right subclavian and jugular trunks as the right lymphatic duct, which drains into the proximal right brachiocephalic vein. The thoracic duct typically drains into the posterior aspect of the junction of the left internal jugular and subclavian veins, known as the "left venous angle". The left bronchomediastinal lymphatic trunk usually drains into the left brachiocephalic vein but sometimes may drain into the thoracic duct.* **(Bottom)** *Graphic depicts the normal anatomy of the pulmonary lymphatic vessels. These small structures are not visible on normal imaging studies. They course centripetally toward the hilum and form aggregates of lymphoid tissue or lymph nodes. Although lymph nodes typically occur at bronchial bifurcations, they may also be located elsewhere throughout the lung. Bronchopulmonary lymph nodes and hilar lymph nodes are considered intrapulmonary in the International Association for the Study of Lung Cancer (IASLC) lymph node map.*

PULMONARY LYMPHATIC DRAINAGE

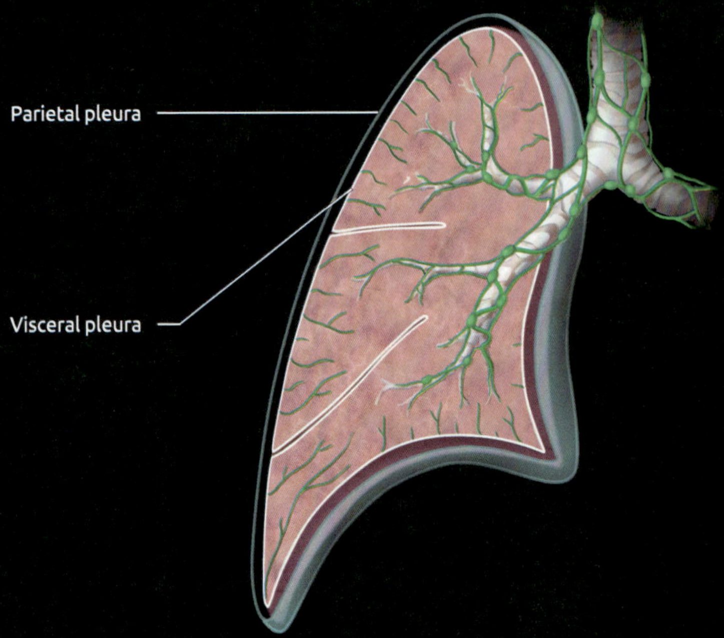

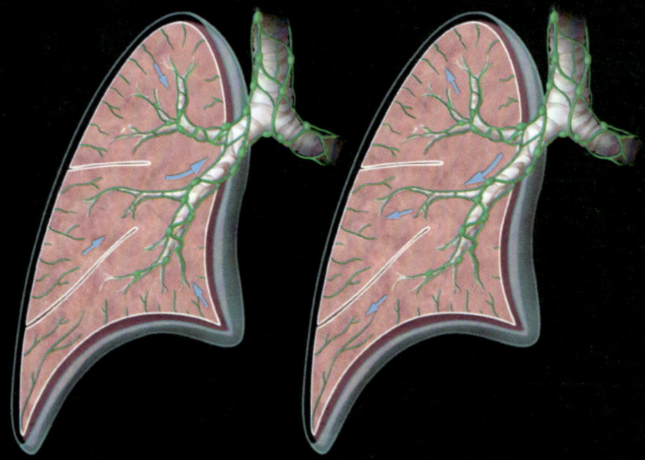

(Top) *Graphic shows the anatomy of the pulmonary lymphatics (green lines), which course along the subpleural regions, the interlobular septa, the central aspect of the secondary pulmonary lobule, and along the bronchovascular structures to drain to the ipsilateral hilum and subsequently to the mediastinum. Parietal pleura drains through stomata to lymphatics in intercostal spaces, then anteriorly to internal mammary lymph nodes and posteriorly to intercostal lymph nodes. Visceral pleura drains via submesothelial lymphatic network that spans lung periphery and extends into lung parenchyma, via deep pulmonary plexus to hilar lymph nodes.* **(Bottom)** *Graphic shows normal (left) and abnormal (right) lymphatic flow patterns. Normal flow is typically centripetal from the lung periphery toward the hilum and mediastinum. Abnormal lymphatic flow may exhibit reversal from the normal pattern with lymph flowing centrifugally and away from the hilum toward the lungs.*

DIAPHRAGM DRAINAGE

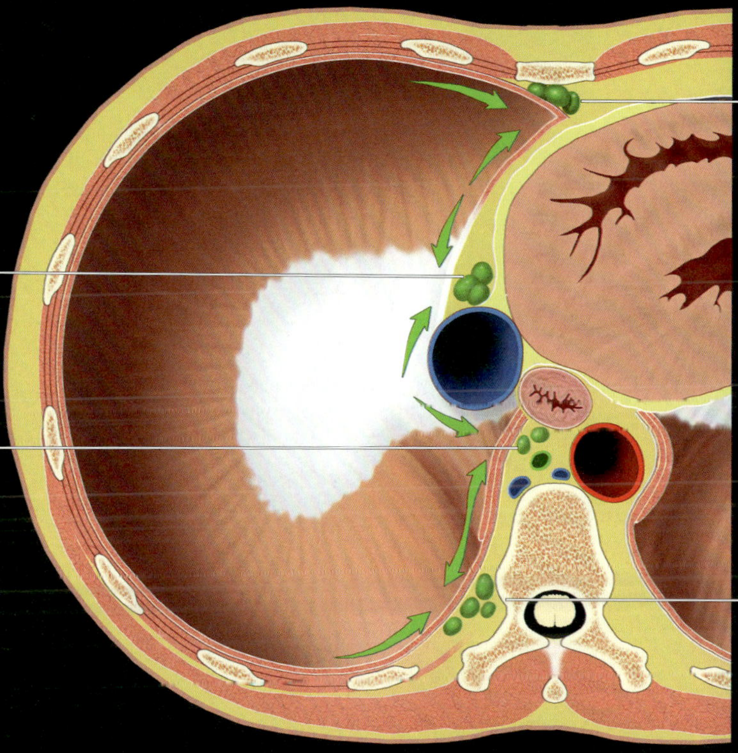

Graphic shows normal lymphatic drainage pathways at the level of the diaphragm. The anterior lower pleura, hemidiaphragm, and convex liver surface drain to anterior and middle diaphragmatic lymph nodes. The posterior lower pleura and hemidiaphragm drain into posterior intercostal, paravertebral, and retrocrural lymph nodes. As none of these lymph nodes drain the lung, they are not included in the IASLC lymph node map.

UPPER THORACIC DUCT, LYMPHANGIOGRAPHY

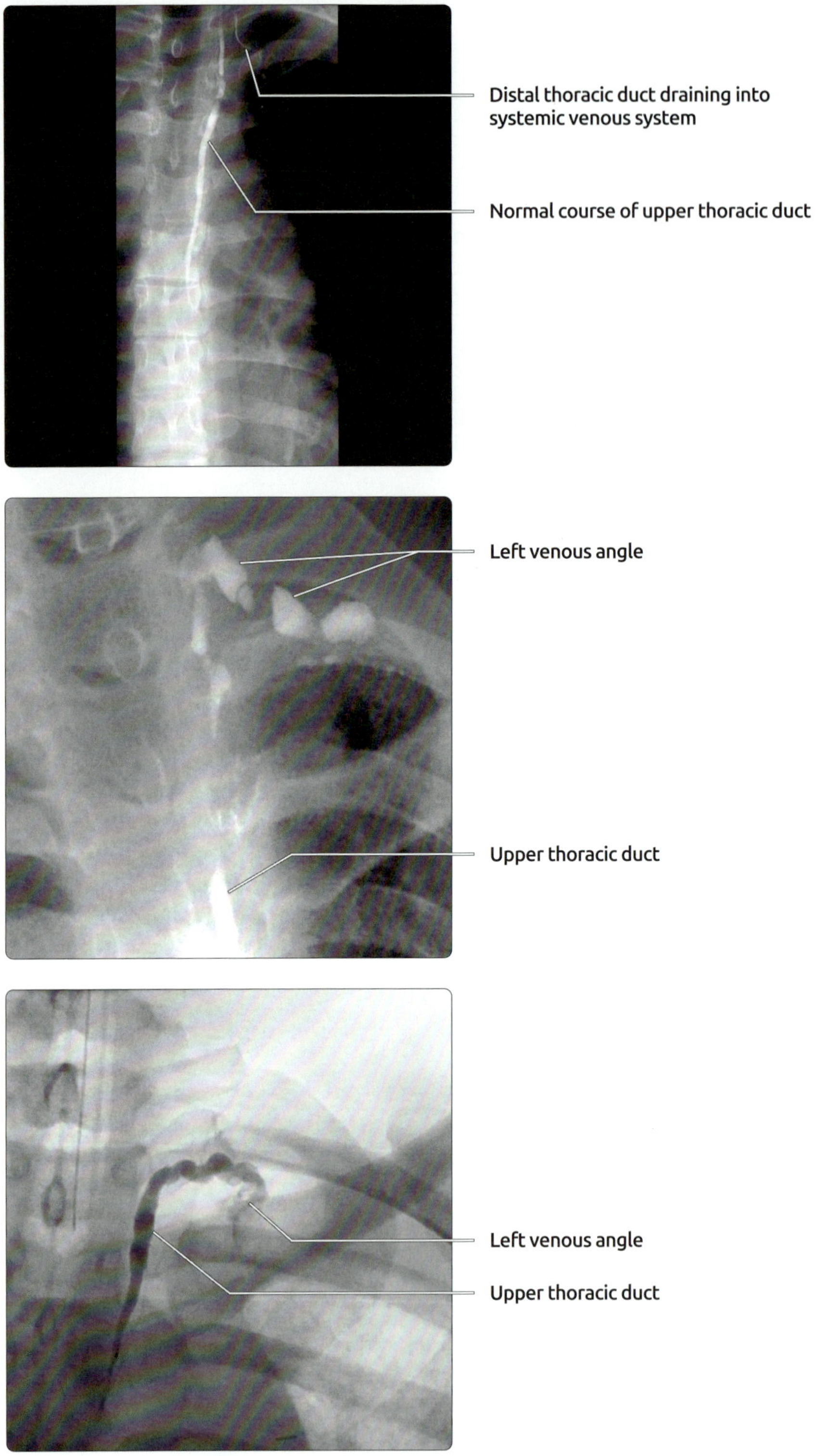

(Top) *AP chest radiograph from a lymphangiogram (coned down to the left upper hemithorax) shows the normal course of the distal thoracic duct. The upper aspect of the thoracic duct courses within the left mediastinum and drains into the left subclavian vein near its confluence with the left internal jugular vein. (From DI: Chest.)* **(Middle)** *AP chest radiograph from a lymphangiogram (coned down to the left lung apex) shows the termination of the thoracic duct at the left venous angle, in the lower left neck. (From DI: Chest.)* **(Bottom)** *AP fluoroscopic spot image from a lymphangiogram shows the normal anatomic course of the upper thoracic duct in the upper left mediastinum and its termination at the left venous angle in the left lower neck.*

THORACIC DUCT, CT

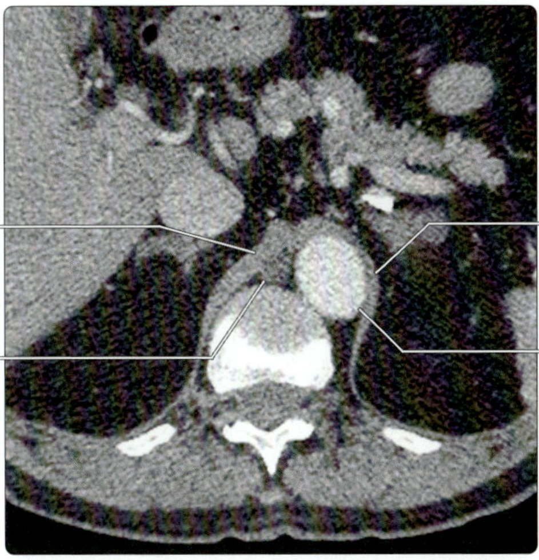

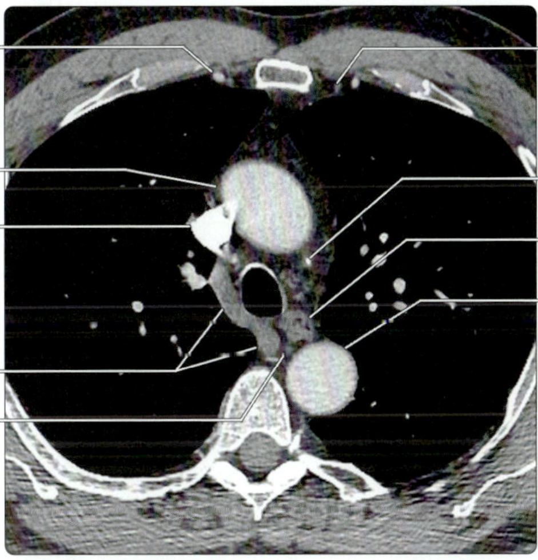

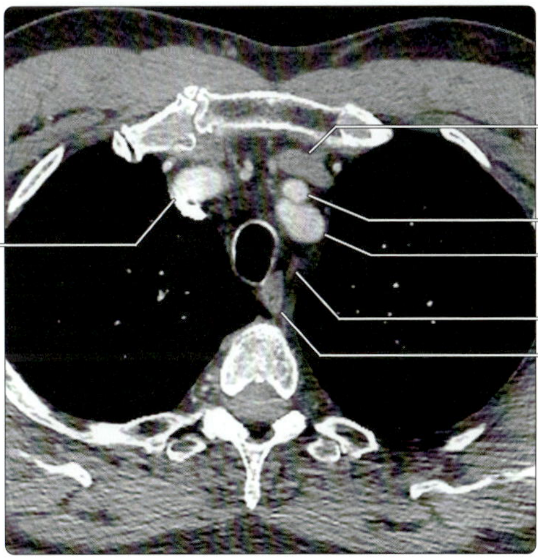

(Top) *First of 3 axial CECT (mediastinal window) images shows the normal appearance and anatomic course of the thoracic duct, beginning as the cisterna chyli at the aortic hiatus and coursing cephalad in the mediastinum. The normal cisterna chyli manifests as a sac-like tubular structure with simple fluid attenuation.* **(Middle)** *Axial CECT (mediastinal window) of the same patient obtained near the level of the T5 vertebral body demonstrates the thoracic duct coursing in the left side of the visceral mediastinum, just after crossing the midline. Note the close association of the thoracic duct with the adjacent esophagus.* **(Bottom)** *Axial CECT (mediastinal window) of the same patient obtained through the upper mediastinum near the proximal aspects of the aortic arch branch vessels demonstrates the thoracic duct coursing in the upper left mediastinum. The thoracic duct will continue its cephalad course to drain into the venous system at the left venous angle.*

THORACIC DUCT, CT

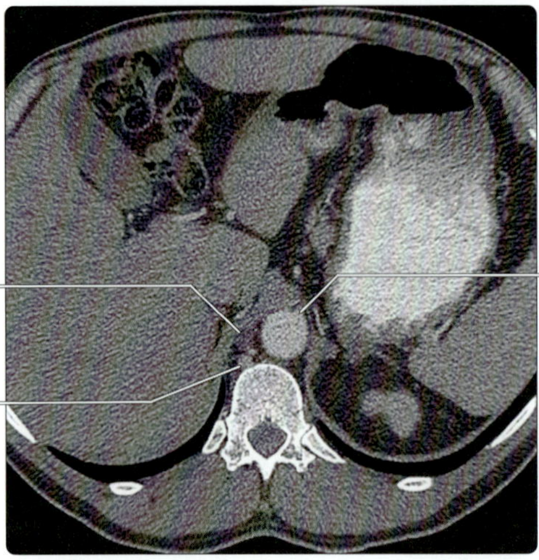

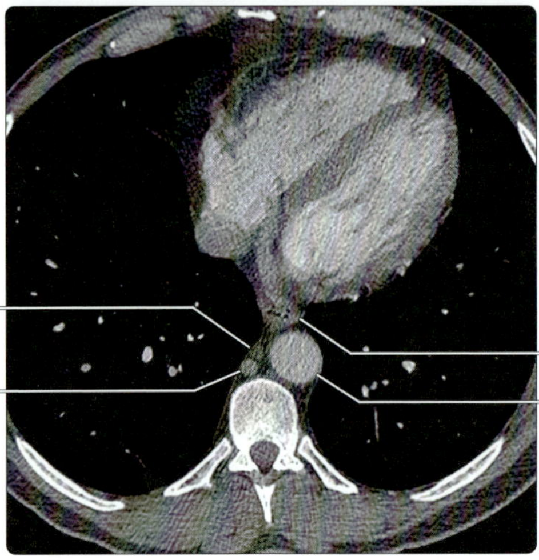

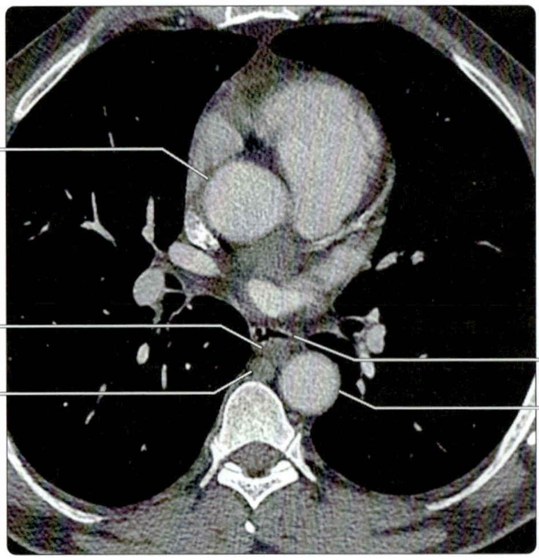

(Top) *First of 8 CECT images of a patient with renal cell carcinoma is shown. Axial CECT (mediastinal window) shows the cisterna chyli visible at its expected location adjacent to the right diaphragmatic crus.* **(Middle)** *Axial CECT (mediastinal window) shows a slightly dilated lower thoracic duct located slightly to the right of midline and coursing along the descending thoracic aorta and anterior to the azygos vein.* **(Bottom)** *Axial CECT (mediastinal window) shows a dilated midthoracic duct as it crosses the midline toward the left side. Note the close association with the adjacent esophagus, which is located just anterior to the thoracic duct.*

THORACIC DUCT, CT

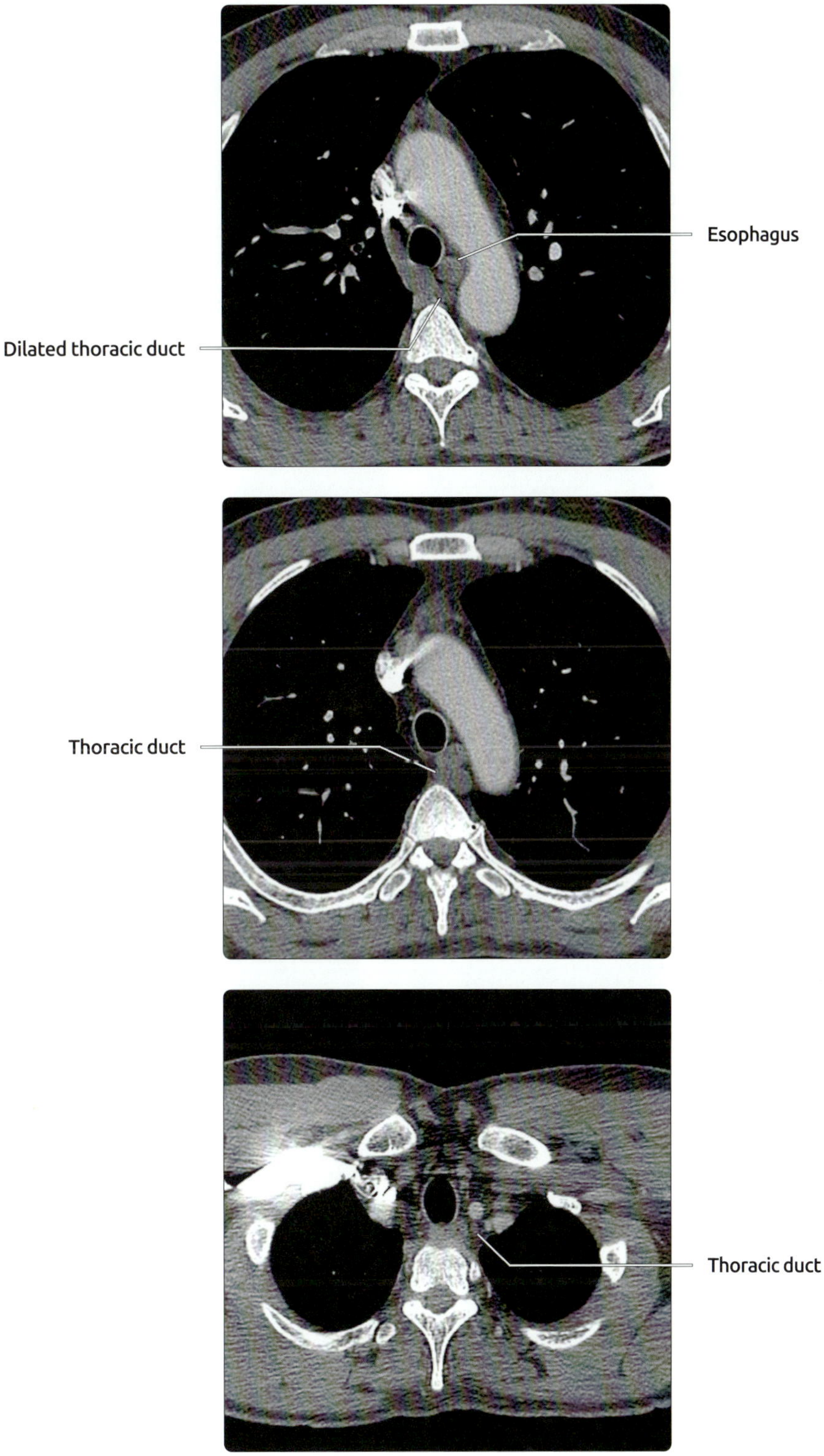

(Top) *Axial CECT (mediastinal window) at the level of the aortic arch shows the dilated thoracic duct coursing to the left of midline, immediately posterior to the esophagus.* **(Middle)** *Axial CECT (mediastinal window) at the upper aspect of the aortic arch demonstrates the markedly dilated thoracic duct coursing just to the left of midline.* **(Bottom)** *Axial CECT (mediastinal window) through the upper mediastinum shows the thoracic duct coursing superiorly and to the left, toward its terminus at the left venous angle in the left lower neck.*

THORACIC DUCT, CT

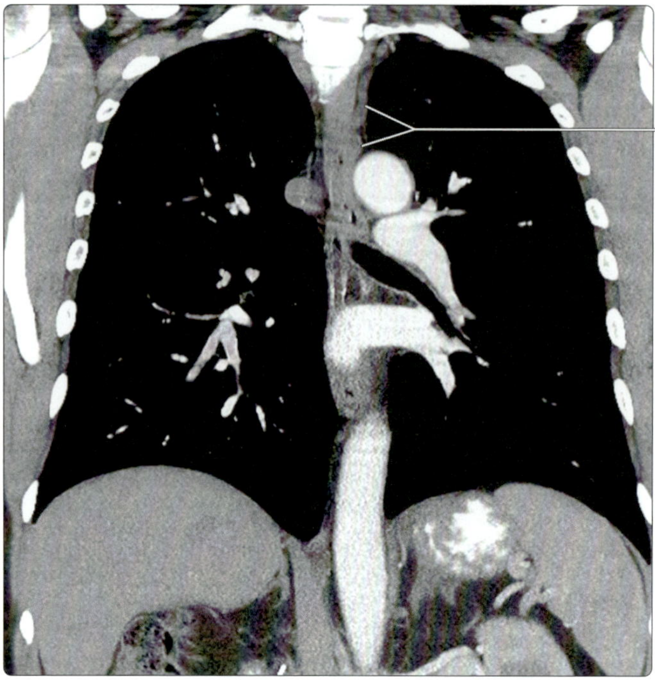

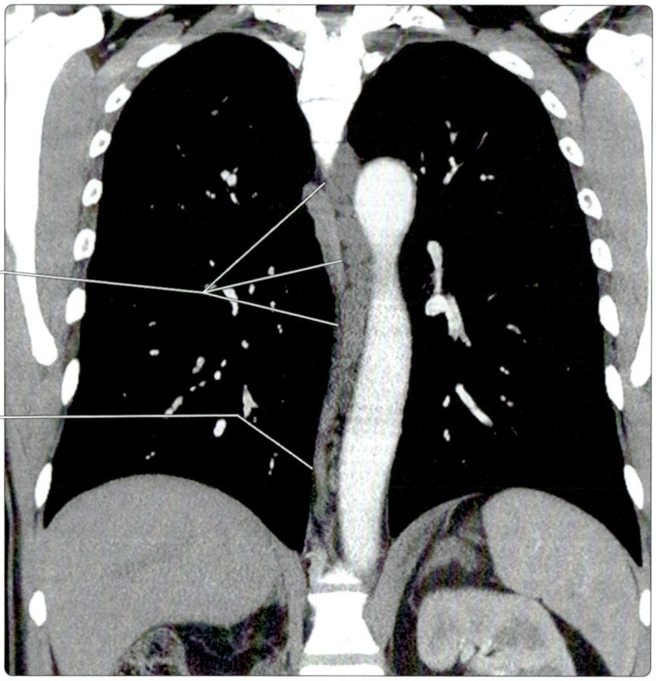

(Top) *Coronal CECT (mediastinal window) shows the dilated upper thoracic duct coursing to the left of midline.* **(Bottom)** *Coronal CECT (mediastinal window) shows the cephalad course of the dilated and varicoid thoracic duct as it crosses midline from right to left.*

DIAPHRAGMATIC LYMPH NODES, CT

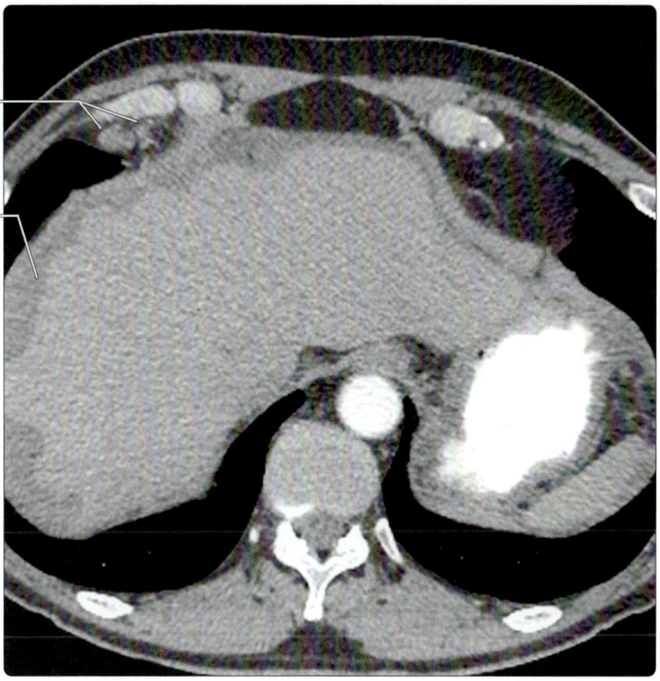

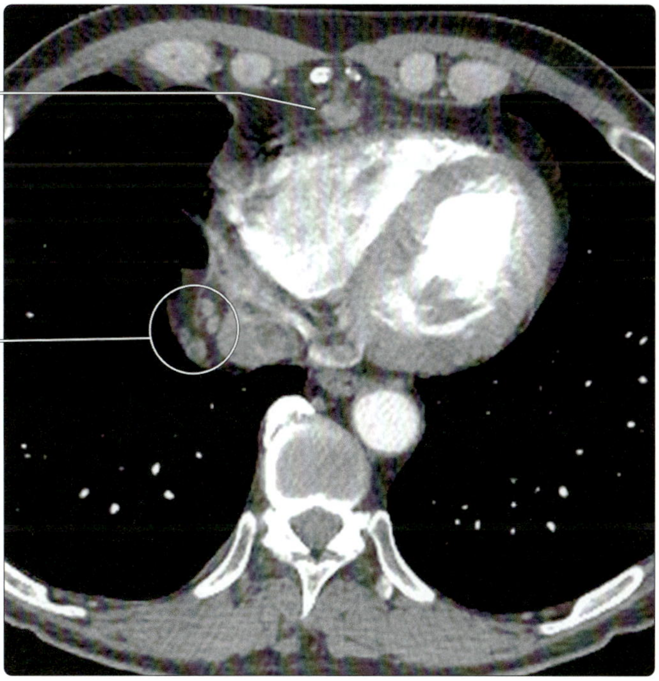

(Top) *Axial CECT (mediastinal window) of a patient with metastatic colon adenocarcinoma demonstrates enlarged right anterior diaphragmatic lymph nodes, compatible with metastases, and a small amount of upper abdominal ascites. The anterior diaphragmatic lymph nodes drain the convex liver surface and the anterior diaphragm.* **(Bottom)** *Axial CECT (mediastinal window) of the same patient shows enlarged metastatic right middle diaphragmatic lymph nodes and right anterior diaphragmatic/cardiophrenic lymph nodes. These lymph nodes drain the convex liver surface and the anterior and middiaphragm.*

RETROCRURAL, INTERNAL MAMMARY, AND INTRAPULMONARY LYMPH NODES, CT

Retrocrural lymphadenopathy

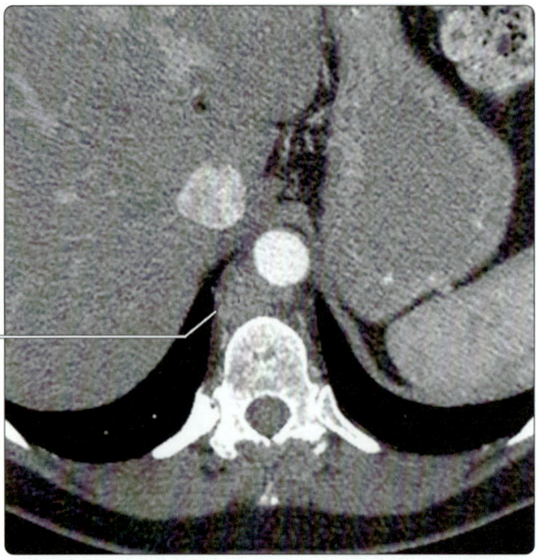

Right internal mammary lymphadenopathy

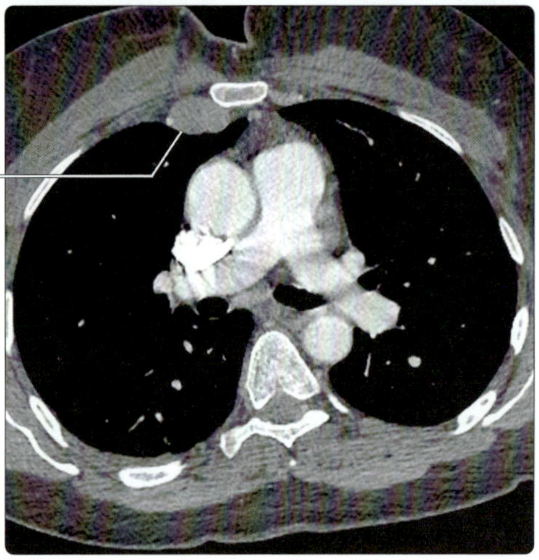

Intrapulmonary lymph node

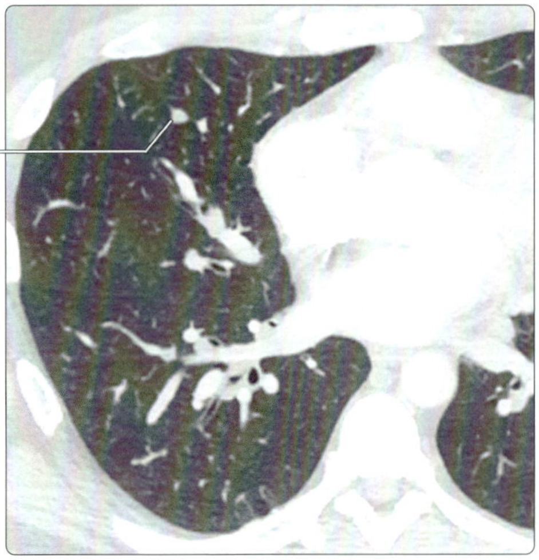

(Top) *Axial CECT (mediastinal window) of a patient with metastatic cervical adenocarcinoma demonstrates enlarged coalescent retrocrural or posterior diaphragmatic lymphadenopathy. These lymph nodes drain the posterior diaphragm, the posterior mediastinum, and the lumbar spine.* **(Middle)** *Axial CECT (mediastinal window) of a patient with advanced breast cancer shows metastatic right internal mammary lymphadenopathy. These lymph nodes drain the breast, the deep epigastrium, the deep anterior chest wall, and the cranial liver surface.* **(Bottom)** *Axial CECT (lung window) shows a small intrapulmonary lymph node located in the middle lobe along the minor fissure. Normal intrapulmonary lymph nodes are small (average diameter < 10 mm) with smooth margins and are oval, lentiform, or triangular-shaped. They are frequently seen along the interlobar fissures and subpleural lung.*

INTERNATIONAL ASSOCIATION FOR THE STUDY OF LUNG CANCER LYMPH NODE MAP

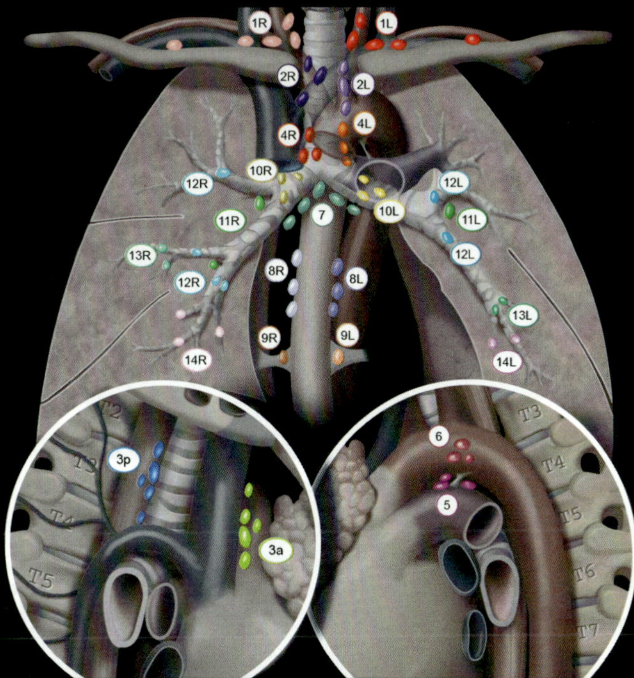

Right

Low Cervical Nodes
- 1R. R. Low cervical
  Supraclavicular
  Sternal notch nodes

Superior Mediastinal Nodes
- 2R. R. Upper paratracheal
- 4R. R. Lower paratracheal

Inferior Mediastinal Nodes
- 8R. R. Paraesophageal
- 9R. R. Pulmonary ligament

Pulmonary Nodes
- 10R. R. Hilar
- 11R. R. Interlobar
- 12R. R. Lobar
- 13R. R. Segmental
- 14R. R. Subsegmental

Left

Low Cervical Nodes
- 1R. R. Low cervical
  Supraclavicular
  Sternal notch nodes

Superior Mediastinal Nodes
- 2L. L. Upper paratracheal
- 4L. L. Lower paratracheal

Inferior Mediastinal Nodes
- 8L. L. Paraesophageal
- 9L. L. Pulmonary ligament

Pulmonary Nodes
- 10L. L. Hilar
- 11L. L. Interlobar
- 12L. L. Lobar
- 13L. L. Segmental
- 14L. L. Subsegmental

Aortic Nodes
- 5. Subaortic
- 6. Paraaortic
- 7. Subcarinal

Superior Mediastinal Nodes
- 3a. Prevascular
- 3p. Retrotracheal

(Top) *Graphic demonstrates the lymph node stations typically affected by metastatic lung cancer as defined by the IASLC. The IASLC lymph node map defines 7 lymph node zones and 14 different lymph node stations, which may be further divided into right and left lymph node stations, as appropriate.* **(Bottom)** *Graphic demonstrates the specific lymph node stations depicted in the graphic of the IASLC lymph node map. The 14 lymph node stations are divided into 7 zones, including low cervical/supraclavicular, superior mediastinal, aortopulmonary, subcarinal, inferior mediastinal, hilar/interlobar, and peripheral or pulmonary zones.*

INTERLOBULAR SEPTAL THICKENING, CT

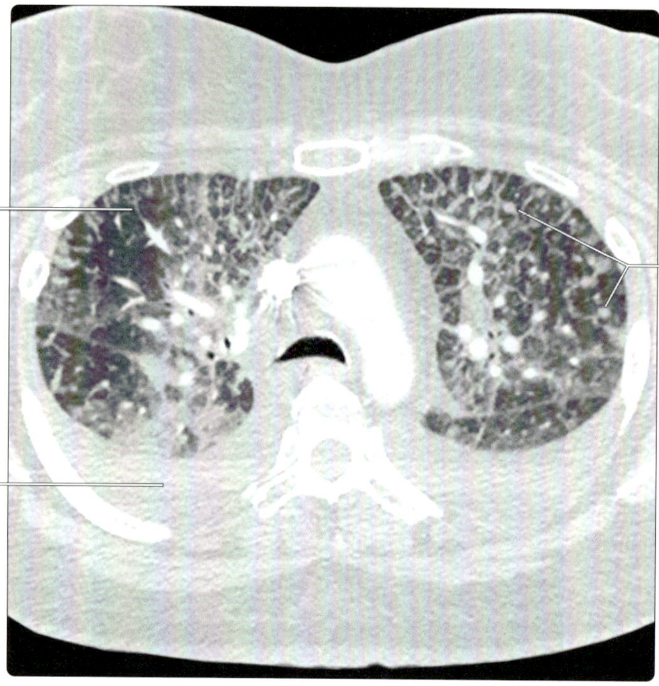

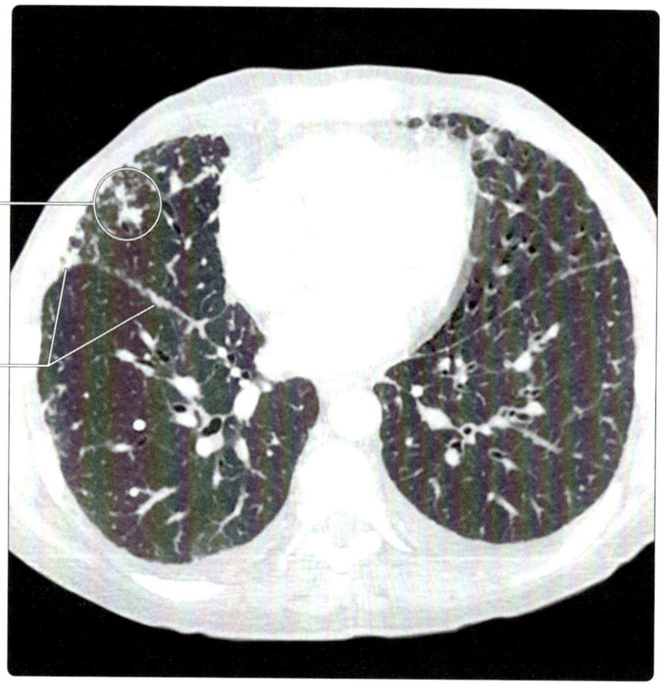

(Top) *Axial CECT (lung window) of a patient with heart failure and diffuse interstitial and alveolar pulmonary edema demonstrates ill-defined bilateral ground-glass opacities and diffuse interlobular septal thickening. While smooth interlobular septal thickening predominates, there are areas of nodular septal thickening likely representing congested pulmonary lymphatics.* **(Bottom)** *Axial CECT (lung window) of a patient with metastatic breast cancer demonstrates areas of nodular interlobular septal thickening and nodular thickening of the interlobar fissures, consistent with lymphangitic carcinomatosis. Note small bilateral pleural effusions.*

LYMPHATIC MALFORMATION, CT AND MR

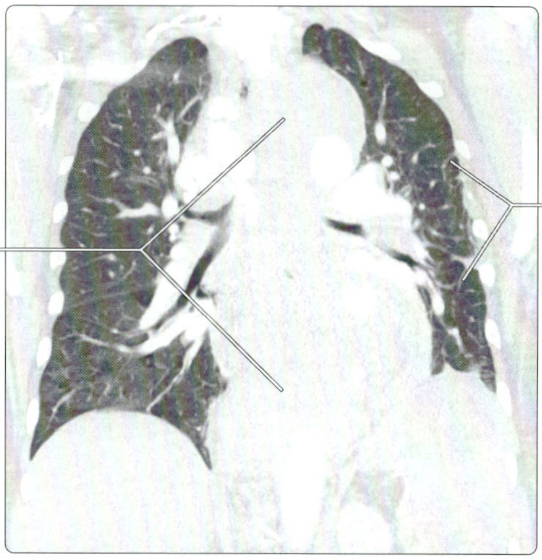

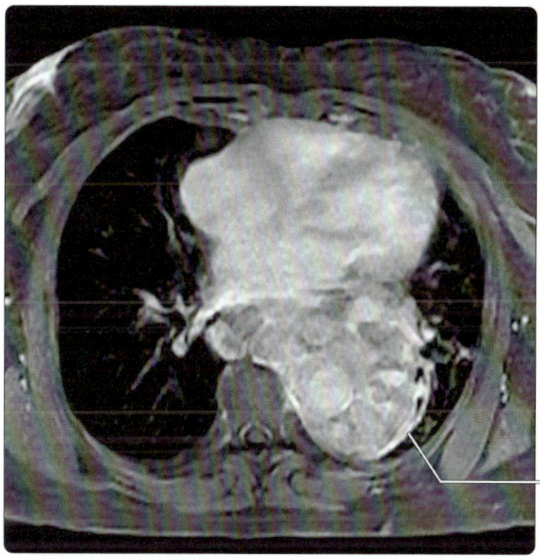

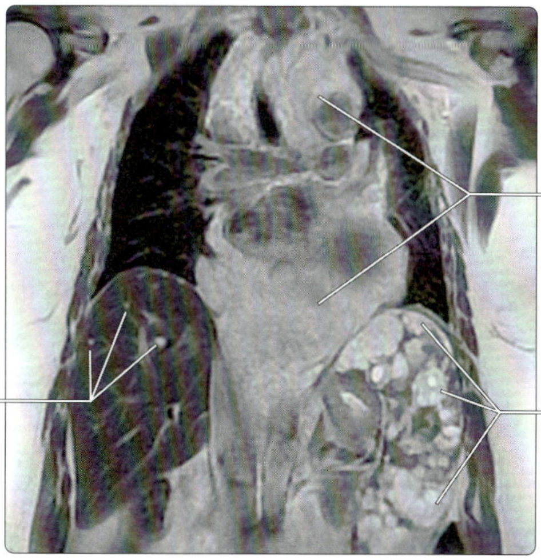

(Top) *Coronal CECT (lung window) of a patient with lymphangiomatosis demonstrates bilateral interlobular septal thickening secondary to lymphangiectasis. Note the large infiltrative mediastinal mass that encases central vascular structures and the central tracheobronchial tree.* **(Middle)** *Axial T1 fat-suppressed contrast-enhanced MR of the same patient demonstrates progressive tubular/serpiginous enhancement within a large venolymphatic malformation, which involves the visceral and paravertebral mediastinal compartments, and communicates with the azygos and hemiazygos venous systems as well as with left-sided pulmonary veins.* **(Bottom)** *Coronal T2 MR of the same patient demonstrates the diffusely infiltrative lobulated mediastinal venolymphatic malformation with T2-hyperintense signal. Multiple T2-hyperintense cystic lesions are present within the enlarged spleen, compatible with lymphatic malformations. Small T2-hyperintense foci in the liver could represent additional lymphatic malformations or cysts.*

LYMPHATIC DISRUPTION, CT AND RADIOGRAPHY

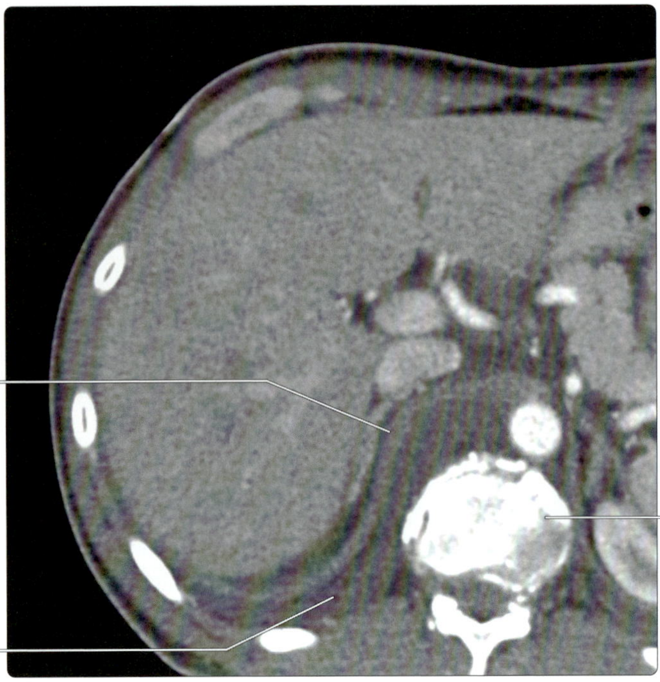

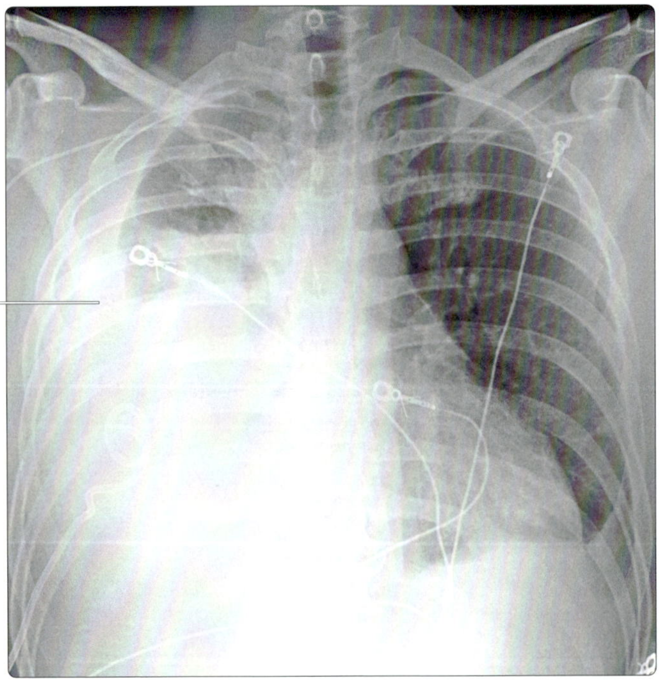

(Top) *Axial CECT (mediastinal window) of a man who sustained a traumatic fracture of the T10 thoracic vertebra, with associated injury to the adjacent cisterna chyli and thoracic duct, shows lateral displacement of the right diaphragmatic crus and a simple fluid attenuation right chylous pleural effusion. (From DI: Chest.)* **(Bottom)** *AP chest radiograph of the same patient obtained 4 days later shows a large right chylous pleural effusion despite catheter drainage and low-fat diet. Chylous pleural effusions are typically indistinguishable from pleural effusions of other etiologies. (From DI: Chest.)*

LYMPHATIC DISRUPTION, LYMPHANGIOGRAPHY

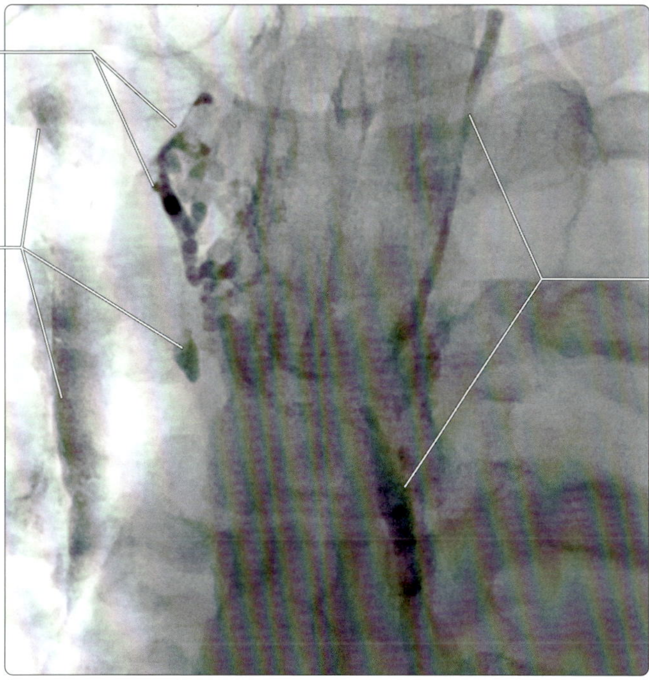

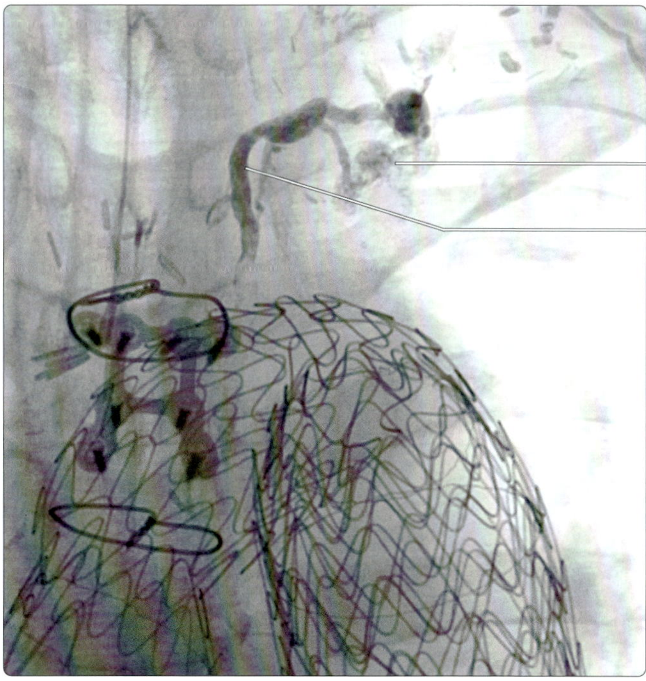

(Top) *AP fluoroscopic spot image acquired during lymphangiography of a patient who recently underwent thymectomy demonstrates right-sided chyle leak, secondary to iatrogenic injury of right-sided mediastinal lymphatics.* **(Bottom)** *AP fluoroscopic spot image acquired during lymphangiography of a patient who recently underwent surgical and endovascular thoracic aortic aneurysm repair demonstrates active extravasation of contrast from the thoracic duct terminus at the left venous angle.*

LYMPHATIC DISRUPTION, RADIOGRAPHY

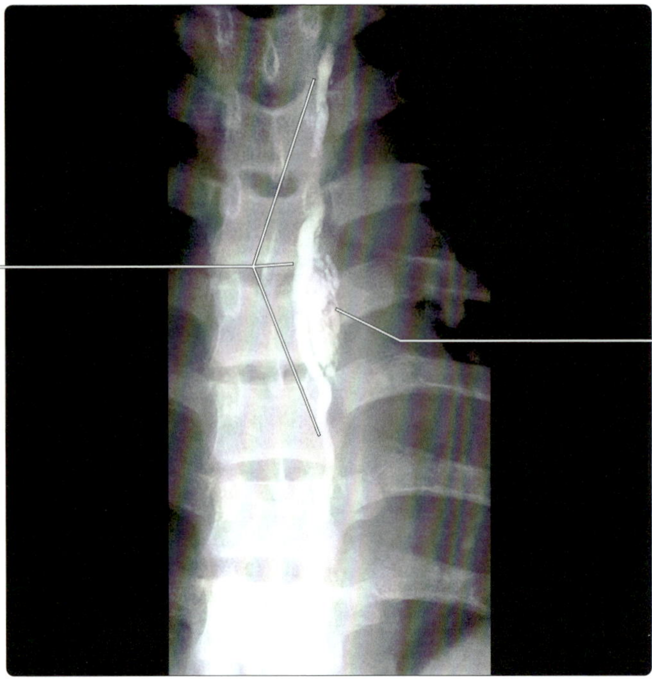

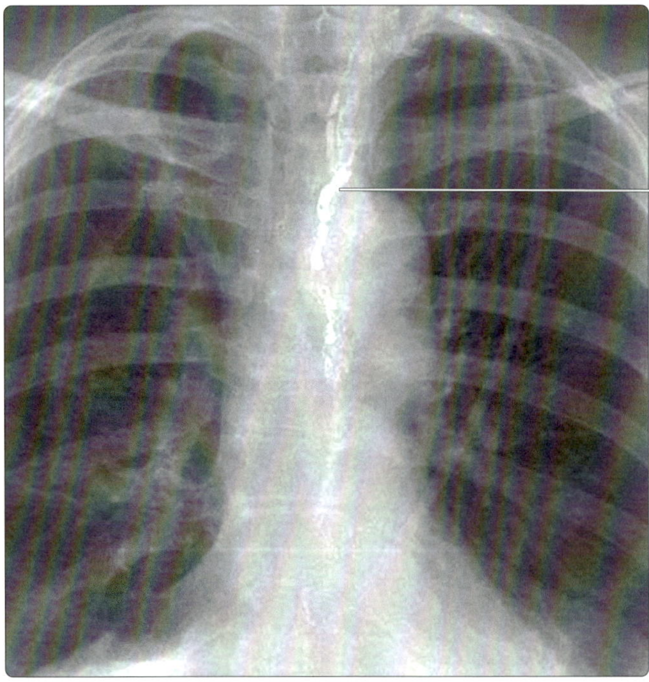

(Top) *AP chest radiograph from a lymphangiogram coned down to the left mediastinum shows a thoracic duct tear manifesting with extravasation of contrast. (From DI: Chest.)* **(Bottom)** *PA chest radiograph shows embolization coils along the left mediastinum following repair of a thoracic duct tear. (From DI: Chest.)*

LYMPHATIC DISRUPTION, CT

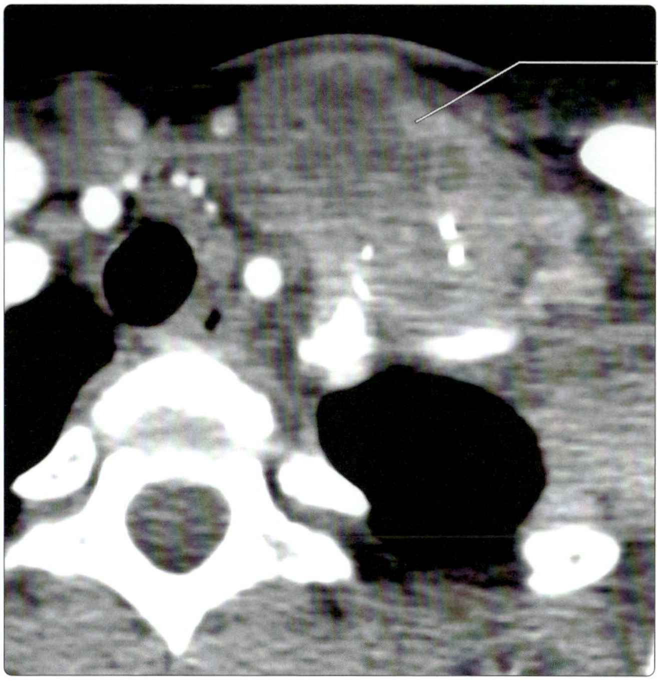

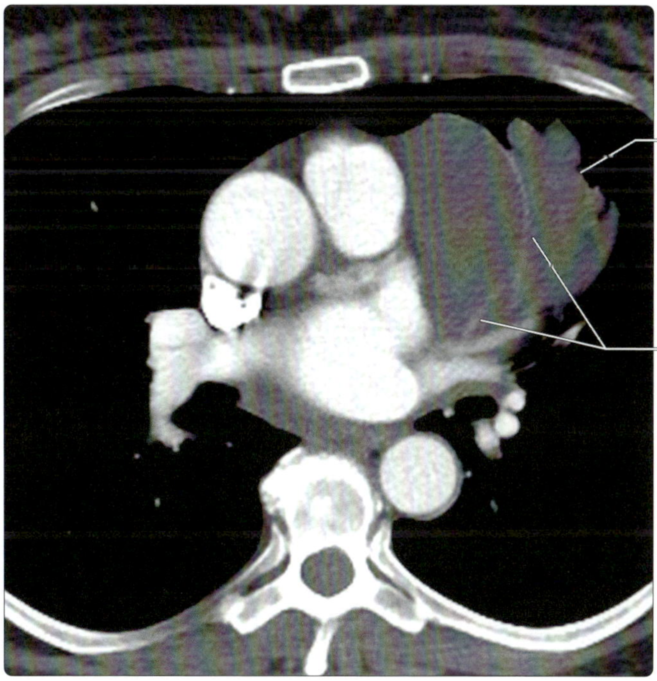

(Top) *Axial CECT (mediastinal window) of a patient with a right aortic arch (not shown) status post surgical ligation of a diverticulum of Kommerell complicated by thoracic duct injury shows a resultant left anterior chest wall and left lower neck chyloma confirmed after drainage. Thoracic duct ligation was performed for treatment.* **(Bottom)** *Axial CECT (mediastinal window) of a patient with a mediastinal lymphatic malformation shows a left prevascular mediastinal multilocular fluid attenuation cystic lesion with intrinsic septations, which encases the left tracheobronchial tree and extends into the adjacent left upper lobe. The lesion produces mild mass effect on the mediastinal structures.*

GENERAL ANATOMY AND FUNCTION

General Anatomy

- Parietal pleura
- Visceral pleura
- Fissures
  - Standard interlobar fissures
    - Right major (oblique) fissure
    - Minor (horizontal) fissure
    - Left major (oblique) fissure
  - Accessory fissures
    - Accessory azygos fissure
    - Accessory left minor fissure
    - Superior accessory fissure
    - Inferior accessory fissure

Pleural Structure

- Continuous (visceral and parietal) surface epithelium and underlying connective tissue
- Visceral pleura adheres to pulmonary surfaces
- Parietal pleura covers chest wall, diaphragm, and mediastinal surfaces
- Visceral and parietal pleurae form pleural spaces
  - Potential spaces; contain small amount of serous fluid
- Combined thickness of visceral and parietal pleurae and fluid-containing pleural space is < 0.5 mm

Function

- Visceral pleura directly apposes and slides freely over parietal pleura during respiration
- Inspiration: Muscles of respiration and diaphragm increase intrathoracic volume
  - Create negative pressure within pleural space and lung
  - Resultant lung expansion causes reduction in intraalveolar pressure
    - Prompts conduction of air through upper respiratory tract and airways into alveoli

Pleural Space

- Potential space; normally contains 2-10 mL of fluid
- Fluid production capacity of 100 mL/hr; fluid absorption capacity of 300 mL/hr
- Fluid flux: Parietal pleura capillaries to pleural space, absorption by microscopic parietal pleural stomata

PLEURAL ANATOMY

Parietal Pleura

- General concepts
  - Covers nonparenchymal surfaces; lines thoracic cavities
  - Costal portion extends along ribs and intercostal spaces
    - Diaphragmatic portions cover hemidiaphragms
    - Mediastinal portions cover mediastinum
- Blood supply and drainage
  - Supply from adjacent chest wall (intercostal, internal mammary, diaphragmatic arteries)
  - Drainage to bronchial veins (diaphragmatic pleural drainage to inferior vena cava and brachiocephalic trunk)
- Innervation
  - Intercostal nerves (costal and peripheral diaphragmatic pleura) and phrenic nerves (mediastinal and central diaphragmatic pleura)
  - Irritation of costal or peripheral diaphragmatic pleura refers pain along intercostal nerves to thoracic or abdominal wall
  - Irritation of mediastinal or central diaphragmatic pleura refers pain to lower neck and shoulder
- Histology
  - Single layer of parietal mesothelial cells over loose, fat-containing areolar connective tissue
  - Bound externally by endothoracic fascia

Visceral Pleura

- General concepts
  - Covers pulmonary surfaces and fissures
- Blood supply and drainage
  - Supply by systemic bronchial arteries; drainage by pulmonary and bronchial veins
  - Lymphatic drainage to deep pulmonary plexus within interlobar and peribronchial spaces toward hilum
- Innervation
  - Visceral afferent nerves that course along bronchial vessels; lacks pain fibers
- Histology
  - Mesothelial layer, thin connective tissue layer, chief layer of connective tissue, vascular layer, limiting lung membrane (connected to chief layer by collagen and elastic fibers)
    - Single layer of flat mesothelial cells separated by basal lamina from underlying lamina propria composed of loose connective tissue

Pleural Reflections

- Pulmonary ligament
  - Inferior extension of double layer of mediastinal pleura
- Costodiaphragmatic recesses
  - Caudal pleural extension beyond inferior lung border
  - Costal and diaphragmatic pleurae separated by narrow slit, costodiaphragmatic recess
  - ~ 5 cm below inferior lung border during quiet inspiration; caudal extent at ~ 12th rib posteromedially

Interlobar Fissures

- General concepts
  - Complete fissures extend from lung surface to hilum
  - Incomplete fissures fail to extend to hilum
    - Allow parenchymal communication and collateral air drift between adjacent lobes
    - Frequency: 12.5-73% (major fissures); 60-90% (minor fissure); more frequent on right than on left
- Major (oblique) fissures
  - Originate posteriorly, near T5 vertebral body; left major fissure originates near T4 in 75% of individuals
  - Terminate along anterior diaphragmatic pleural surface, 3-4 cm posterior to anterior chest wall
  - Right major fissure separates right upper and middle lobes from right lower lobe
  - Left major fissure separates left upper lobe from left lower lobe
  - Change in contour from upper portion (concave anterior aspect) to lower portion (convex anterior aspect); termed propeller-like morphology
- Minor (horizontal) fissure
  - Separates upper middle lobe from right upper lobe

- Incomplete in > 80% of individuals

Accessory Fissures

- General concepts
  - Clefts of varying depth in outer surface of lung; occur in 22-32% of individuals
- Accessory azygos fissure
  - Right sided
  - Failure of normal migration of azygos vein to tracheobronchial angle
  - Invaginated visceral and parietal pleurae (4 pleural layers) in medial aspect of right lung apex
- Accessory left minor fissure
  - Separates lingula from remainder of left upper lobe
  - Frequency of 8-18%; rarely detected on frontal chest radiography (frequency of 1.6%)
- Superior accessory fissure
  - Separates superior segment of lower lobe from basilar segments
  - Horizontal or oblique orientation
- Inferior accessory fissure
  - Incompletely separates lower lobe medial basilar segment from rest of basilar segments
    - Basis for juxtaphrenic peak sign seen in some cases of upper lobe volume loss
  - Right more common than left
  - Frequency: 5-10% on chest radiography, 16-21% on chest CT, 30-50% of anatomic specimens

IMAGING

Radiography

- Major (oblique) fissures
  - Not normally seen on frontal chest radiography
  - Portions of major fissures typically visible on lateral chest radiography
  - Identified by continuity with respective hemidiaphragm
- Minor (horizontal) fissure
  - Visible in 50-80% of PA chest radiographs as horizontal linear opacity at or near anterior 4th rib
  - Variable contour
  - Contacts lateral chest wall near axillary portion of right 6th rib
  - Anterior and lateral portions curve gently downward
  - Ends medially at interlobar pulmonary artery; never crosses right hilar opacity
- Accessory azygos fissure
  - Thin, curvilinear opacity; inferomedial course from right lung apex toward hilum
  - Characteristic teardrop configuration of inferior end (azygos arch)
- Accessory left minor fissure
  - Thin linear opacity
  - More cephalad and oblique than right minor fissure
- Superior accessory fissure
  - Projects below and medial to minor fissure on frontal chest radiography
- Inferior accessory fissure
  - Thin linear opacity; extends obliquely from medial hemidiaphragm toward hilum on frontal chest radiography
  - Occasionally visible on lateral chest radiography

CT

- Nonfissural normal pleura not visible by CT/HRCT
- Intercostal stripe
  - Normal thin linear opacity (1-2 mm thick)
  - Overlies intercostal spaces; connects inner aspects of ribs
  - 2 layers of pleura, extrapleural fat, endothoracic fascia, and innermost intercostal muscle
  - Disappears on inner aspects of ribs (innermost intercostal muscle absent); may be mimicked by intercostal veins in paravertebral region
- Major (oblique) fissures
  - Thin linear opacities
  - Curvilinear in upper and lower thorax (concave and convex anterior aspects, respectively)
  - May appear as thin ground-glass opacity bands when oblique to axial plane
- Minor (horizontal) fissure
  - Variable appearance
    - Thin, curvilinear opacity
    - Ground-glass opacity
    - Area devoid of vessels (roughly triangular)

MR

- Normal pleura not visible

Ultrasound

- Normal pleura not visible
- Role in imaging of abnormal pleura

ANATOMY-RELATED IMAGING ABNORMALITIES

Pleural Effusion

- Categorized as transudate or exudate
  - Based on pleural fluid analysis
  - Differentiation by Light criteria
    - Transudates not associated with pleural disease
      - Systemic disease: Heart failure, pericardial disease, cirrhosis, pregnancy, hypoalbuminemia, overhydration, renal failure
    - Exudates indicate pleural disease
      - Pneumonia, empyema, tuberculosis, neoplasm, pulmonary embolism, collagen vascular disease
- Radiography
  - Blunt costophrenic angle (focal opacification with meniscus-shaped upper border)
    - Upright frontal chest radiography
      - Blunt lateral costophrenic angle correlates with at least 200 mL of pleural fluid
      - Obscured hemidiaphragm if > 500 mL of pleural fluid
    - Upright lateral chest radiography: Blunt posterior costophrenic angle correlates with at least 75 mL of pleural fluid
    - Lateral decubitus chest radiography: Detection of > 10 mL of pleural fluid
    - Supine frontal chest radiography: Increased density of affected hemithorax, veil-like opacity, apical cap, thick paravertebral stripe

- Small to moderate free pleural effusion: Basilar pleural fluid with meniscus-shaped upper border
  - Lateral aspect more cephalad than medial portion
- Large pleural effusions may opacify hemithorax, displace mediastinum, invert hemidiaphragm
- Subpulmonic pleural effusion
  - Accumulation of fluid in subpulmonic pleural space
  - Lateral displacement of dome or peak of "pseudohemidiaphragm" on frontal chest radiography
  - Flattening, elevation of undersurface of posterior lung on lateral radiography; flattening of inferior lung contour anterior to major fissure; sharp fissural downward angulation (resembles profile of Rock of Gibraltar)
  - Left subpulmonic pleural effusion > 2 cm distance between aerated lung base and superior border of gastric air bubble
- Pseudotumor
  - Fissural or interlobar pleural fluid
  - Most associated with cardiac decompensation
  - Associated with heart failure; spontaneous resolution after treatment
  - Fissural elliptical/lenticular opacity; minor fissure most commonly affected
  - Tapered peripheral margins as they merge with affected fissure
- Loculation associated with exudative pleural effusion; limited by pleural adhesions
  - Fixed, nonmobile, mass-like opacity; typically elliptical or lenticular
  - May exhibit incomplete border sign: Sharply defined on radiography when surface is tangential to x-ray beam; ill-defined when imaged en face
- Empyema
  - Infected pleural effusion
  - Association with pneumonia
  - Loculated pleural effusion; intrinsic gas suggests bronchopleural fistula

- CT
  - May not distinguish transudate from exudate
  - Most effusions near water attenuation (20-30 HU) regardless of cause; range: 0-100 HU
  - Small effusions in posterior costodiaphragmatic recess; concave/meniscoid anterior margin
  - Large effusions track anteriorly and cephalad; may extend into fissures
  - Empyema
    - Elliptical or lenticular loculated fluid collection
      - Intrinsic air suggests bronchopleural fistula in absence of prior intervention
      - Involvement of adjacent chest wall suggests empyema necessitatis
    - Nondependent location, sharp demarcation from adjacent lung
    - Compression/displacement of adjacent lung
    - **Split pleura** sign
      - Fluid bound by smooth, uniform thick visceral and parietal pleurae
      - Thick pleura may enhance with contrast
    - **Pomegranate sign**: Formation of air bubbles along hydropneumothorax air-fluid level; resemble pomegranate cut surface; indicative of exudate
- MR
  - Transudate: T2 hyperintense, no enhancement, absence of septa
  - Exudate: Lower T2 signal intensity, heterogeneity, enhancement, septations
  - Identification of associated pleural nodules/masses indicative of malignancy
- Ultrasound
  - Guide for pleural fluid analysis/pleural biopsy
  - Assessment of echogenicity and morphology; visualization of echogenic visceral pleura line
  - Transudate: Anechoic
  - Exudate: Anechoic or echogenic ± septations
  - Identification of associated pleural nodules/masses indicative of malignancy

Pneumothorax

- Air or gas within pleural space
  - Spontaneous primary: Association with blebs and bullae, no underlying diffuse lung disease
  - Spontaneous secondary: Association with diffuse/cystic/cavitary lung disease
  - Traumatic
  - Iatrogenic
- Imaging
  - Radiography
    - Thin visceral pleural line adjacent to air in pleural space; apical pleural line in upright position
    - Variable appearance based on degree of volume loss, presence of tension, patient position
      - Deep sulcus sign: Lucency extending to costodiaphragmatic sulcus in supine position
  - CT: High sensitivity for small pneumothoraces

Pleural Thickening

- Nonneoplastic
  - Pleural plaques: Bilateral, discontinuous, predominantly along 6th-8th ribs ± Ca++, diaphragmatic pleura, spare apices and costophrenic sulci
  - Diffuse pleural fibrosis: Continuous; differentiation from malignant pleural disease
  - Fibrothorax: Localized or diffuse; frequent calcification
  - Healed/organized hemothorax: Frequent calcification
  - Thoracic splenosis: Multifocal pleural nodules/masses; left pleura; splenectomy
- Neoplastic
  - Localized fibrous tumor: Pleural nodule or mass of variable size
  - Pleural metastases, lung cancer with direct pleural involvement, malignant pleural mesothelioma: Malignant pleural effusion ± solid pleural nodules/masses; circumferential nodular pleural thickening, involvement of mediastinal pleura

OVERVIEW OF PLEURAL ANATOMY

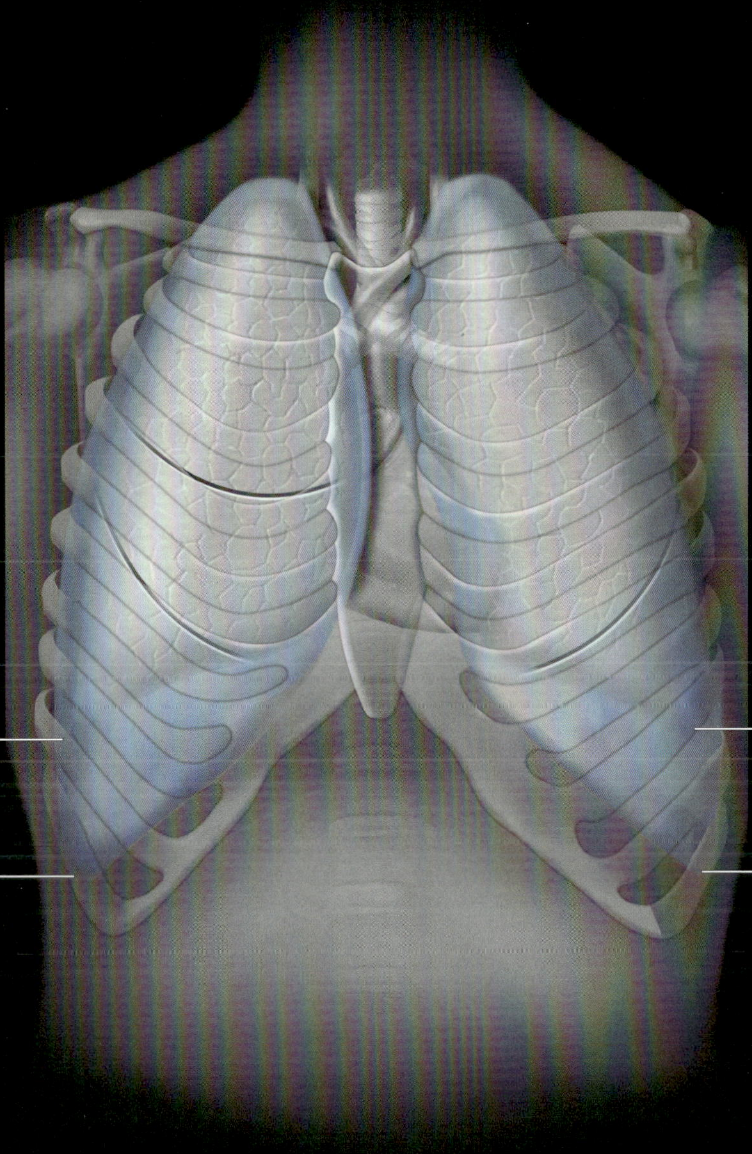

Graphic shows the anatomy of the pleura, a continuous thin layer divided into visceral and parietal pleurae. The visceral pleura covers the pulmonary surfaces and extends into the interlobar fissures. The parietal pleura covers the nonpulmonary surfaces of the thoracic cavity and extends more caudally than the inferior aspects of the lungs. The inferior reflection of the parietal pleura extends within the costophrenic sulci to the level of the upper kidneys. The apposition of the visceral and parietal pleurae results in a potential space that normally contains a small amount of serous fluid.

CORONAL ANATOMY OF PLEURAL REFLECTIONS

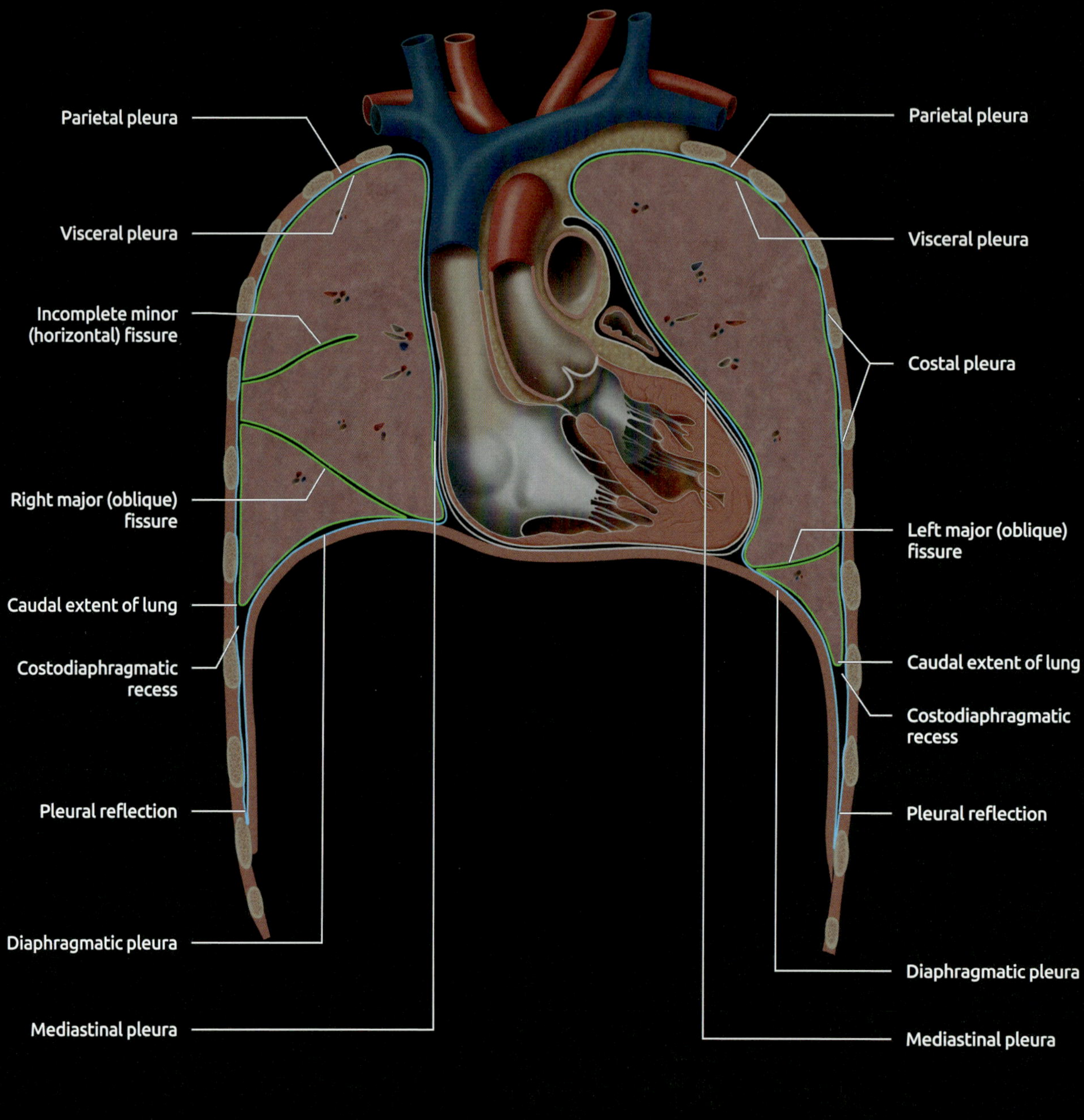

Graphic shows the extent and distribution of the pleura as visualized in the coronal plane. The visceral pleura covers the surfaces of both lungs and forms interlobar fissures that may be complete or incomplete in their extension to the hila. The parietal pleura lines the nonpulmonary surfaces of the bilateral thoracic cavities and may be designated by its location as costal, diaphragmatic, or mediastinal pleura. Inferiorly, the parietal pleura extends deeply into the costodiaphragmatic recesses where the costal and diaphragmatic pleurae are in apposition. The apposition of the visceral and parietal pleurae results in the potential bilateral pleural spaces, which contain a small amount of pleural fluid.

SAGITTAL ANATOMY OF PLEURAL REFLECTIONS

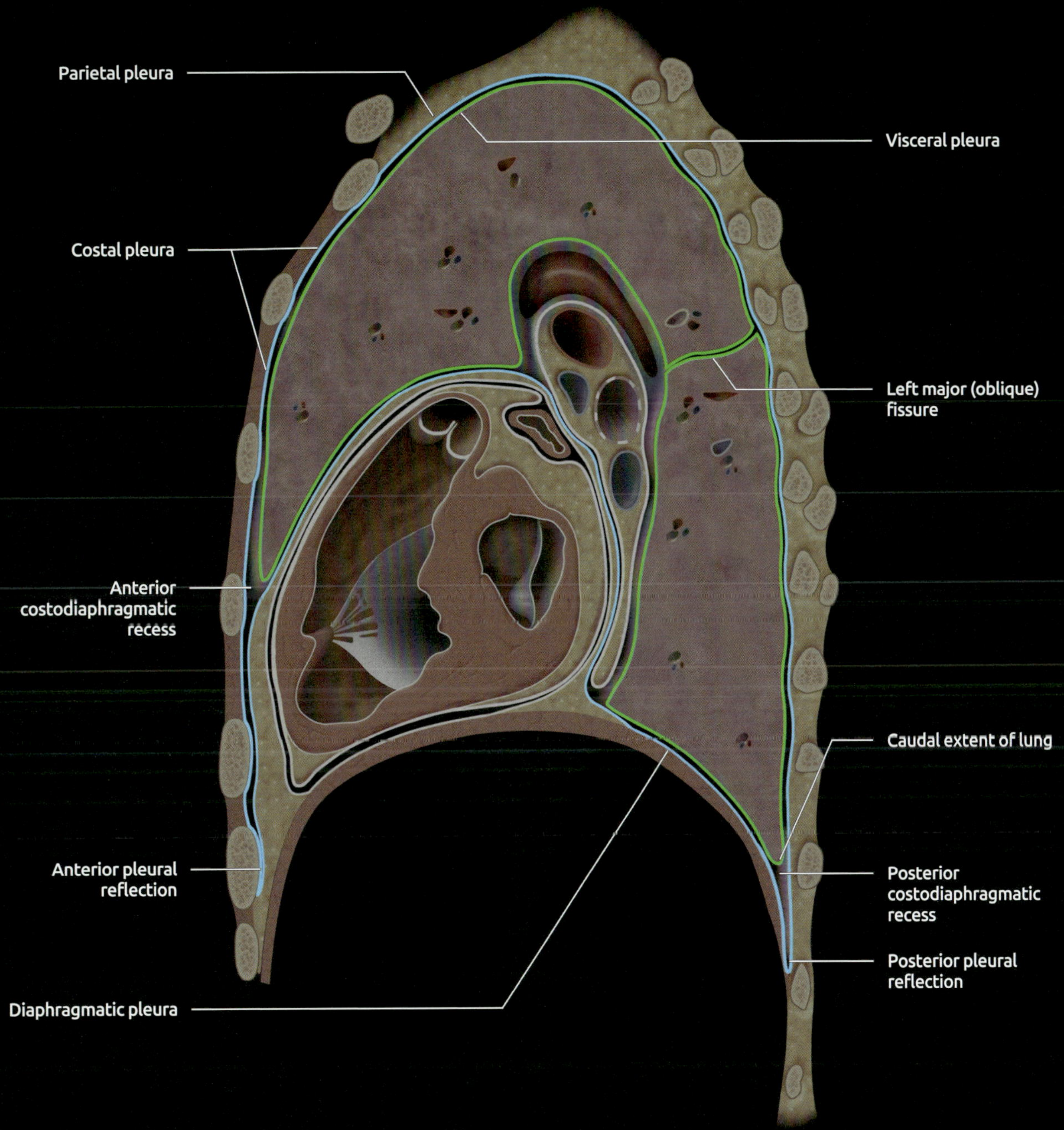

Graphic shows the extent of the parietal and visceral pleurae as visualized in the sagittal plane at the left midclavicular zone. Notice the depth of the pleural reflections and their capacity to "hide" small amounts of pleural fluid.

ANATOMY OF STANDARD AND ACCESSORY FISSURES

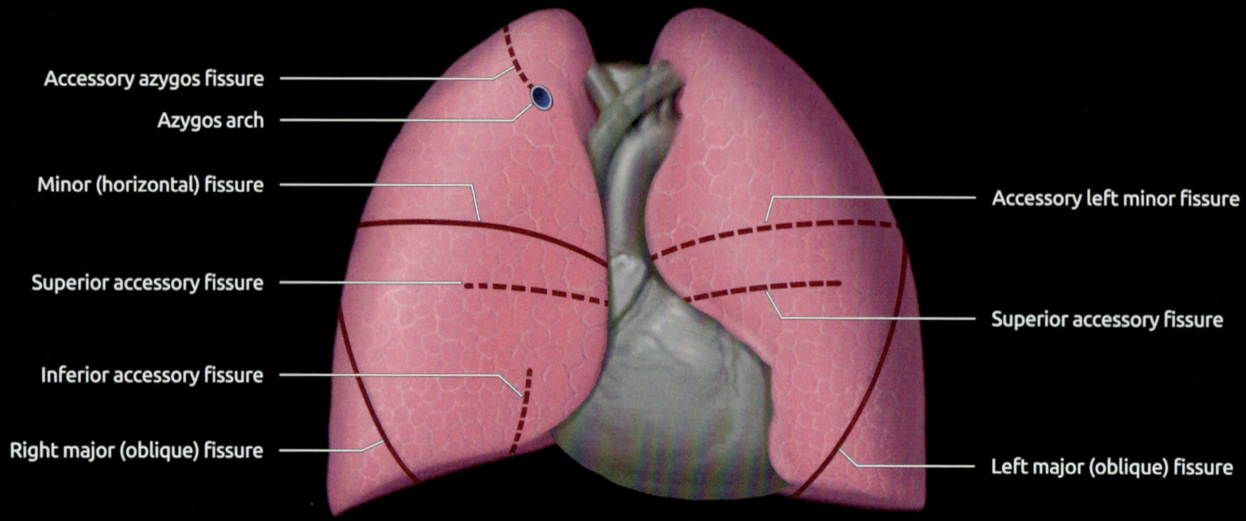

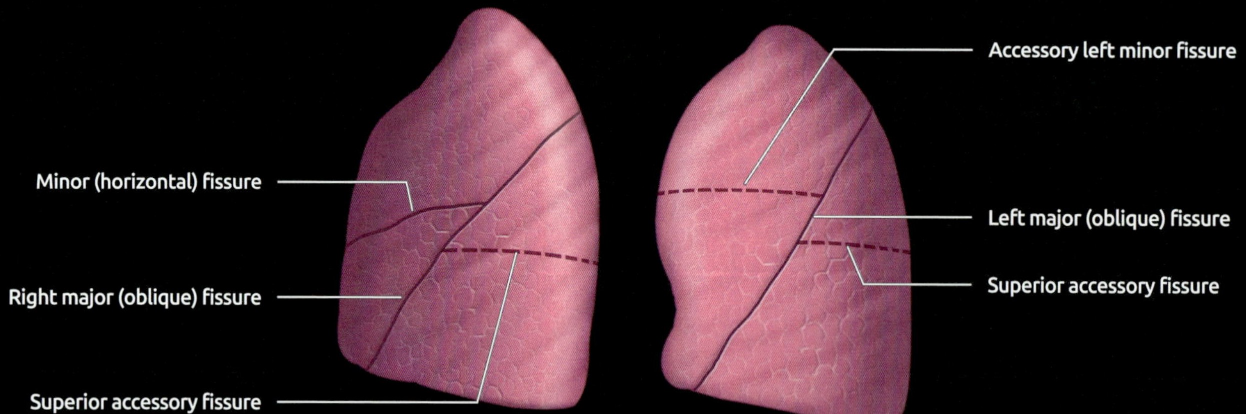

(Top) *Graphic shows a frontal view of the standard interlobar fissures depicted as solid lines and the most common accessory fissures depicted as dashed lines. The standard interlobar fissures include the bilateral major (oblique) fissures and the minor (horizontal) fissure. The accessory fissures include the accessory azygos fissure, accessory minor fissure, superior accessory fissure, and the inferior accessory fissure.* **(Bottom)** *Graphic shows the costal surfaces of the bilateral lungs and depicts both standard (as solid lines) and accessory (as dashed lines) fissures. The minor (horizontal) fissure is typically slightly convex superiorly, and its anterior aspect courses inferiorly. The bilateral major (oblique) fissures course obliquely from posterior to anterior toward the basilar diaphragmatic pleurae.*

STANDARD INTERLOBAR FISSURES, RADIOGRAPHY

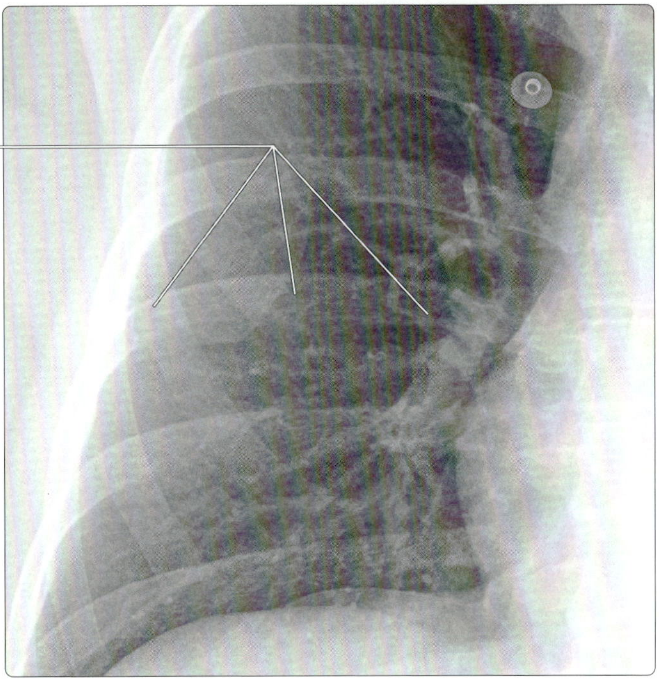

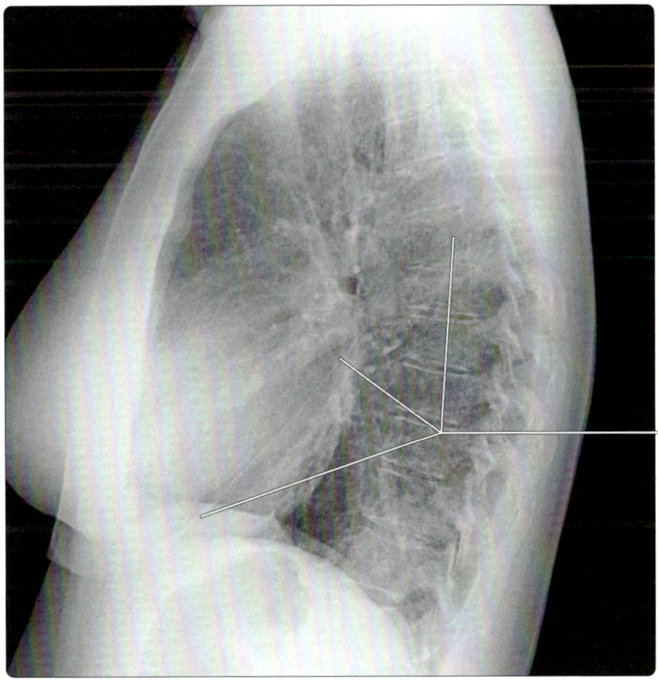

(Top) *PA chest radiograph, coned down to the right lung, shows a subtle thin, curvilinear opacity overlying the posterolateral aspect of the right 7th rib that represents the normal minor or horizontal fissure. The major or oblique fissures are not usually visible on frontal chest radiography.* **(Bottom)** *Lateral chest radiograph shows the characteristic trajectory and orientation of the major (oblique) fissure, which follows an oblique course from posterior to anteroinferior and divides the left upper lobe from the left lower lobe and the right upper and middle lobes from the right lower lobe. In this case, only the left major or oblique fissure is visible, as the right major fissure is obscured by airspace disease. A good strategy to differentiate the right major or oblique fissure from the left is to follow the course of each fissure to the corresponding hemidiaphragm.*

STANDARD INTERLOBAR FISSURES, CT

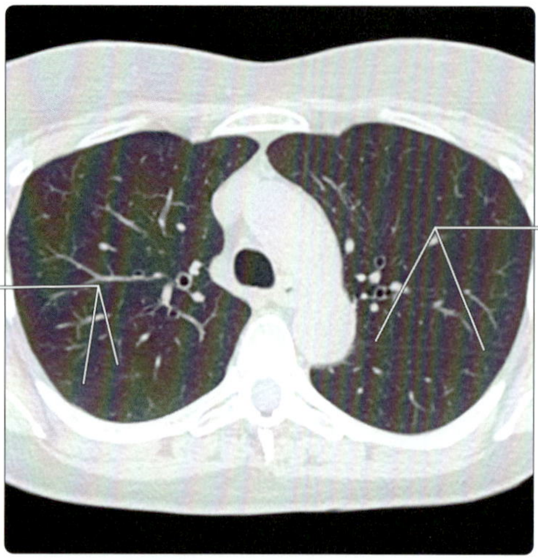

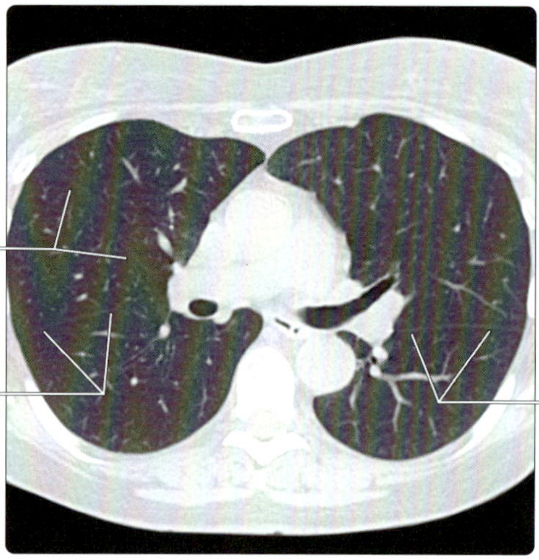

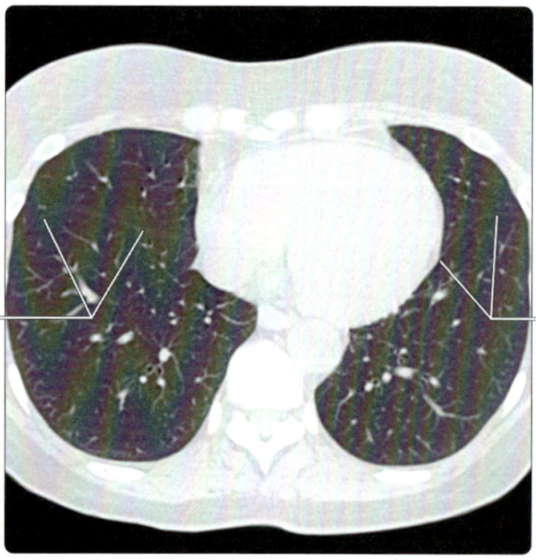

(Top) *Axial NECT (lung window) of a normal individual shows the right and left major or oblique fissures in the upper thorax. The normal upper major fissures manifest as thin curvilinear opacities with straight to slightly convex anterior aspects. Note that while the left major fissure manifests as a thin line, the right major fissure manifests as an ill-defined band.* **(Middle)** *Axial NECT (lung window) of the same individual shows the midportions of the major or oblique fissures near the level of the hila. The major fissures manifest as thin linear opacities that are nearly straight in configuration with a mild anterior convexity. The minor or horizontal fissure manifests as a thin linear band-like opacity.* **(Bottom)** *Axial NECT (lung window) of the same individual shows the curvilinear contours of the major or oblique fissures, which are now anteriorly convex. This shift in configuration along the vertical axis of the major or oblique fissures has been described as a propeller-like morphology. Alterations of this normal shift in configuration may indicate volume loss.*

MINOR FISSURE, RADIOGRAPHY

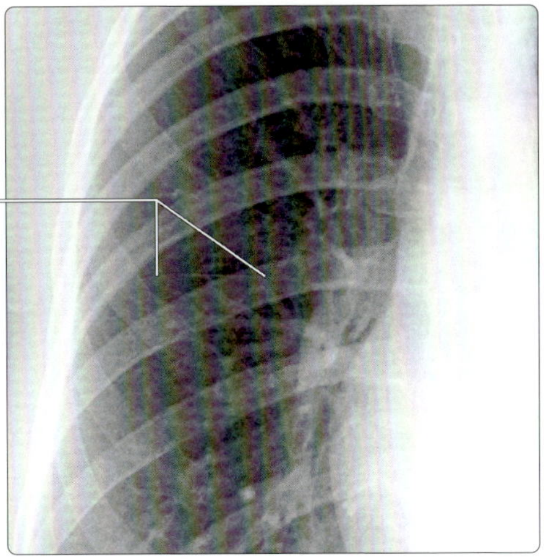

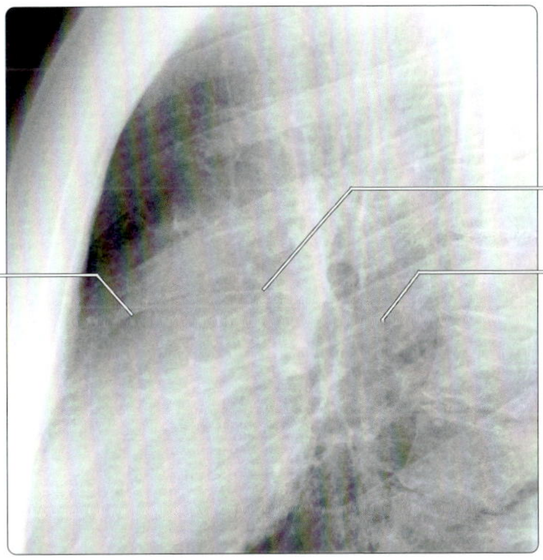

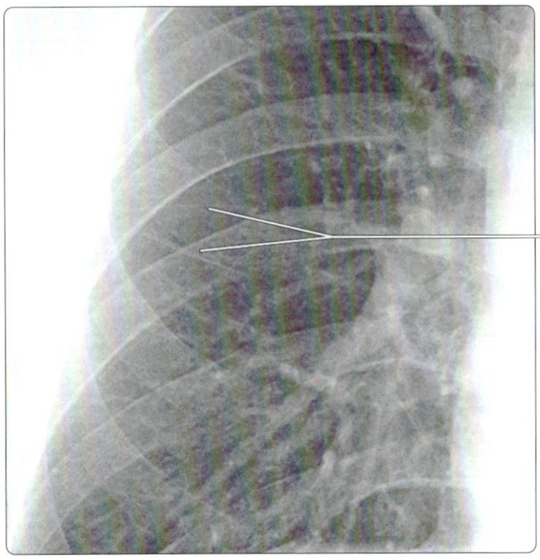

(Top) *PA chest radiograph coned down to the right lung shows the anatomy of the minor fissure, which manifests as a thin, linear opacity that extends across the midright hemithorax, originating near the axillary portion of the right 6th rib and ending medially at the right interlobar artery. Note that the linear opacity never crosses the hilar vessels.* **(Middle)** *Lateral chest radiograph shows that the anterior and posterior portions of the minor fissure exhibit a characteristic gentle downward curve.* **(Bottom)** *PA chest radiograph (coned down to the right lung) of a normal subject shows the minor or horizontal fissure manifesting as 2 adjacent and roughly parallel thin linear opacities. Because of its curving configuration, the minor fissure often manifests as a double line on frontal radiography, produced by the x-ray beam encountering tangential components of the fissure at 2 locations along its course.*

MAJOR AND MINOR FISSURES, CT

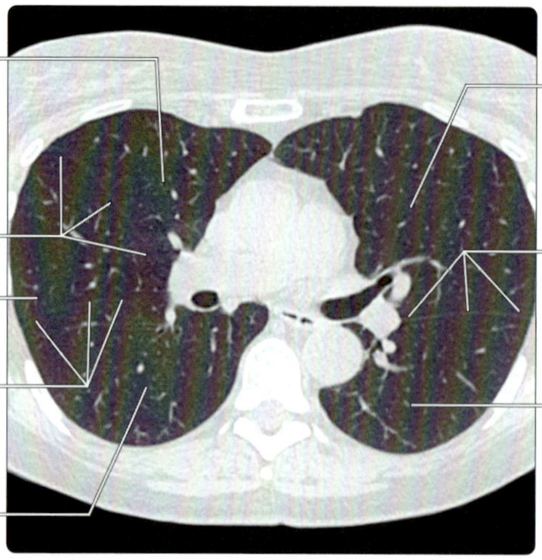

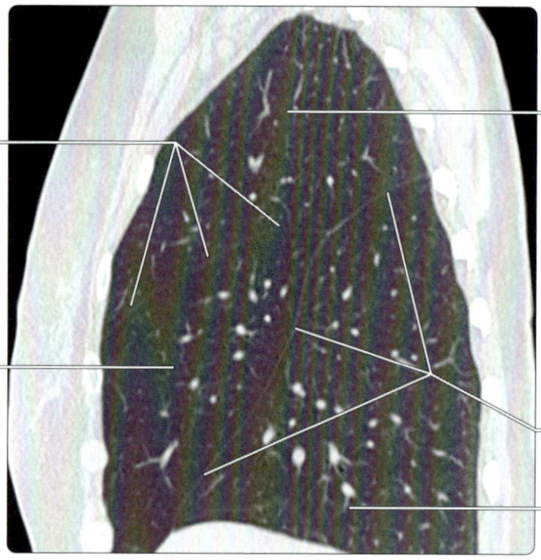

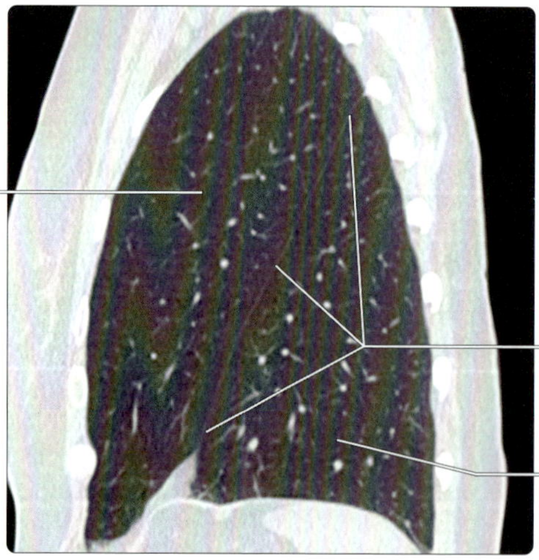

(Top) *Axial NECT (lung window) of a normal individual shows the horizontal or minor fissure as a thin linear opacity that separates the the right upper lobe from the middle lobe. Likewise, the major or oblique fissures separate the upper lobes (and middle lobe on the right) from the lower lobes.* **(Middle)** *Sagittal NECT (lung window) through the right lung of the same individual shows the normal course and orientation of the right major or oblique and minor or horizontal fissures. The right major or oblique fissure separates the right upper and middle lobes from the right lower lobe.* **(Bottom)** *Sagittal NECT (lung window) through the left lung of the same individual shows the normal course and orientation of the left major or oblique fissure. The left major fissure separates the left upper lobe from the left lower lobe.*

INCOMPLETE FISSURES, GRAPHIC AND CT

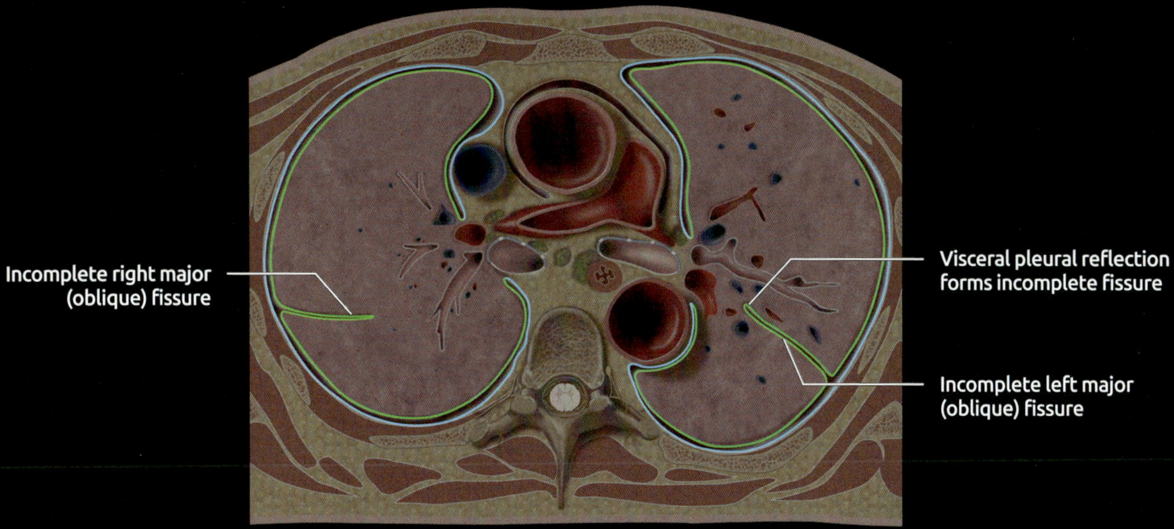

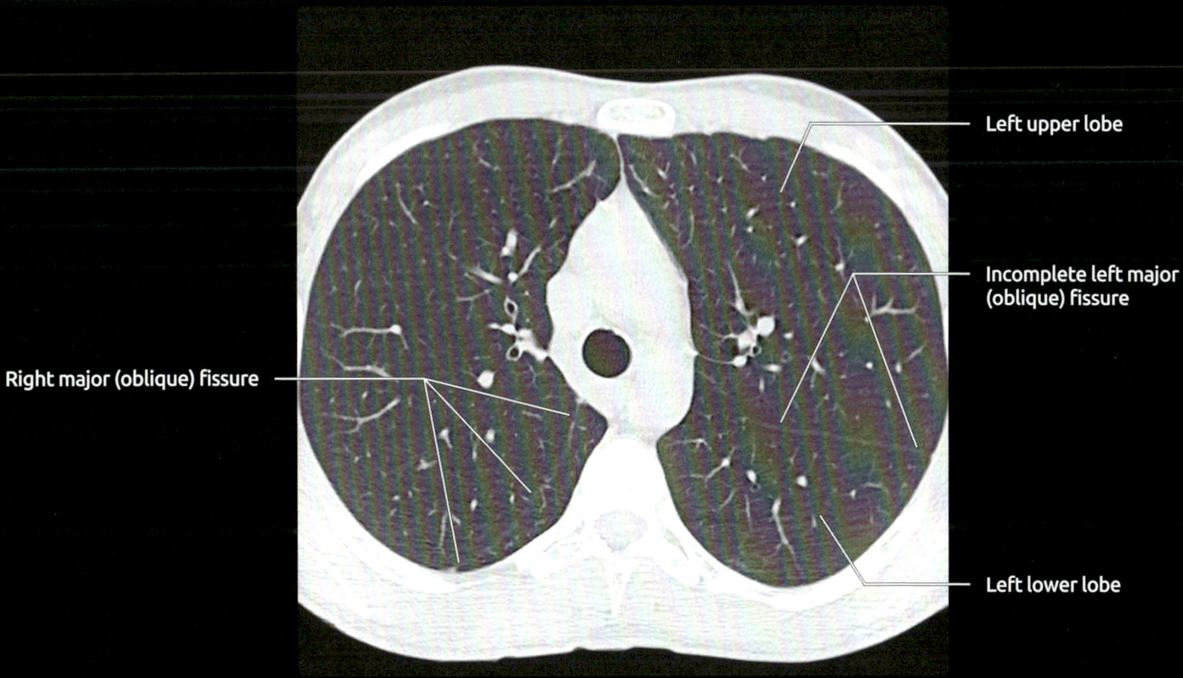

(Top) *Graphic shows an axial view of the lung and demonstrates bilateral incomplete major or oblique fissures that extend inward from the costal lung surface but fail to reach the hila. The visceral pleura reflects upon itself at the medial termination of each incomplete fissure.* **(Bottom)** *Axial NECT (lung window) of a normal individual shows an incomplete left major or oblique fissure and a complete right major or oblique fissure. The incomplete left major or oblique fissure allows parenchymal communication between the left upper lobe and the left lower lobe, in this case.*

ACCESSORY AZYGOS FISSURE, GRAPHICS AND RADIOGRAPHY

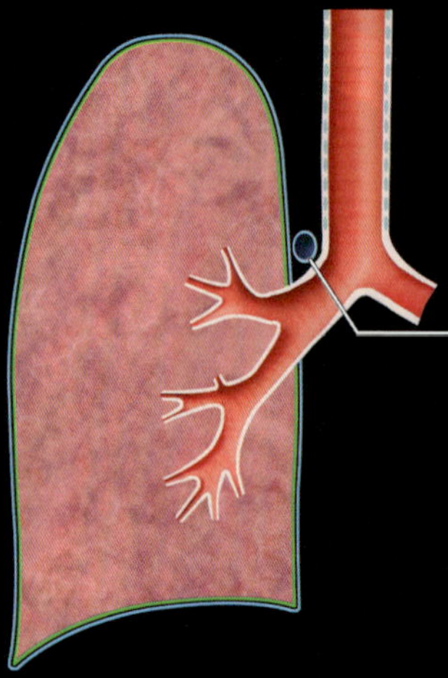

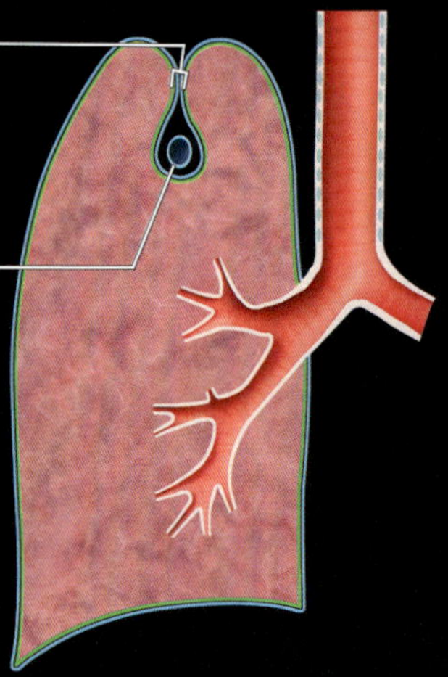

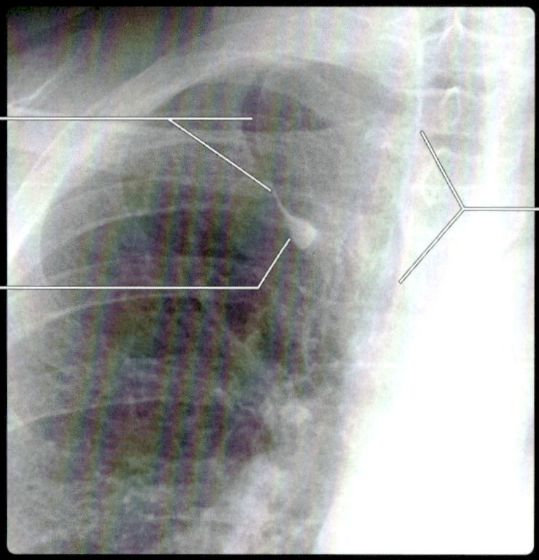

(Top) *Graphic shows a coronal view of the normal position of the azygos arch in the right tracheobronchial angle that results from normal migration of the primitive posterior cardinal vein to form the azygos vein.* **(Middle)** *Graphic shows the embryologic basis of the accessory azygos fissure. Early migration of the primitive posterior cardinal vein results in an abnormal course of the azygos arch, which invaginates inferiorly into the lung apex forming an accessory azygos fissure, composed of 4 layers of pleura, which encase the azygos arch.* **(Bottom)** *PA chest radiograph (coned down to the right lung apex) of a normal individual shows an accessory azygos fissure manifesting as a thin curvilinear opacity extending inferomedially from the right lung apex. The teardrop-shaped opacity at the inferior aspect of the accessory fissure represents the azygos arch. Note the absence of the azygos arch in the right tracheobronchial angle.*

ACCESSORY AZYGOS FISSURE, RADIOGRAPHY AND CT

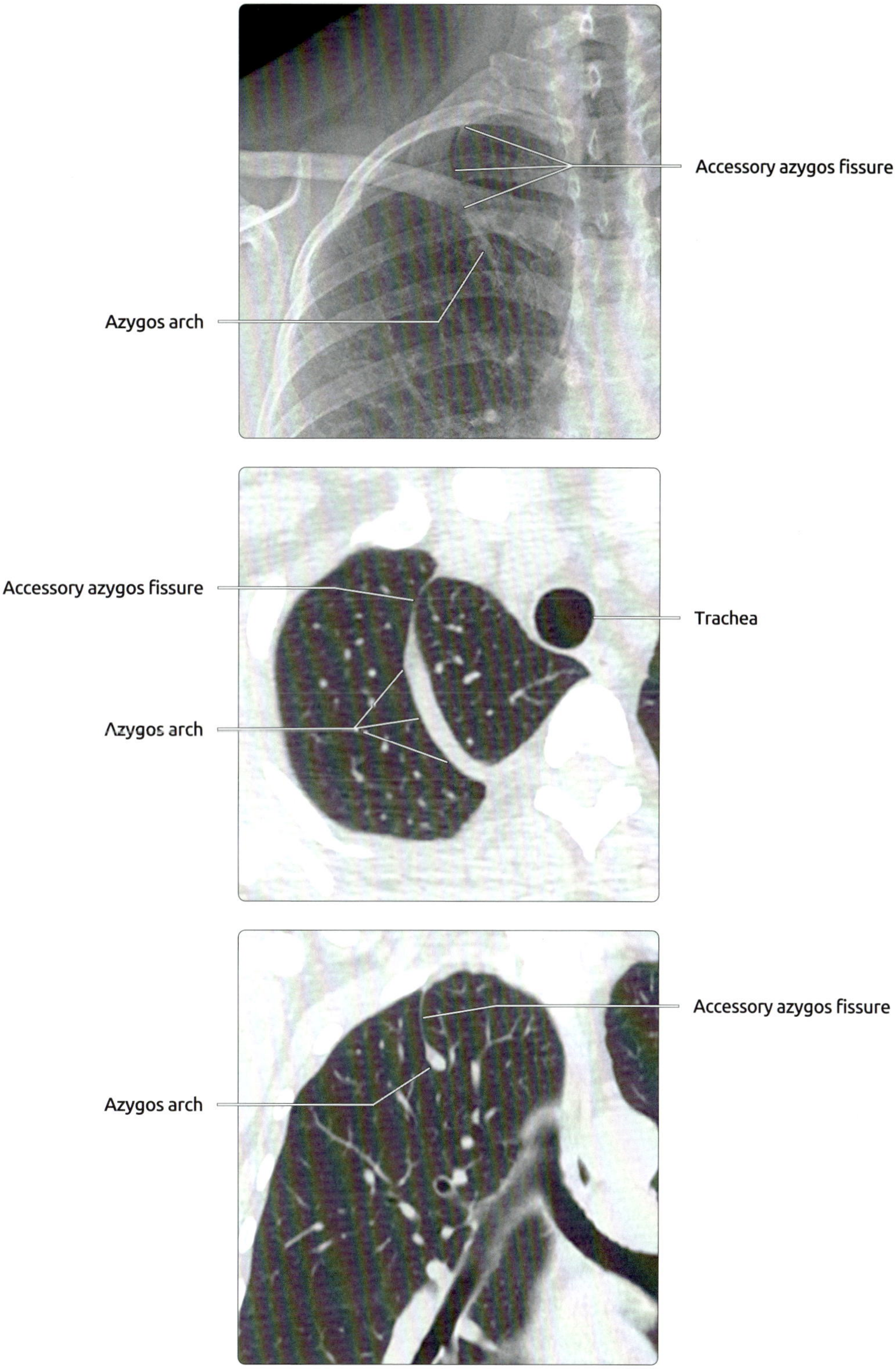

(Top) *PA chest radiograph (coned down to the right lung apex) of a normal subject with an accessory azygos fissure shows the characteristic thin curvilinear opacity, which courses inferomedially. The linear thickening of the inferior aspect of the accessory azygos fissure corresponds to the azygos arch, which typically exhibits a teardrop shape.* **(Middle)** *Axial NECT (lung window) of the same patient shows the course of the azygos arch along the most inferior aspect of the accessory azygos fissure.* **(Bottom)** *Coronal NECT (lung window) of the same patient shows the characteristic teardrop configuration of the azygos arch, which correlates with its typical appearance on frontal chest radiography.*

PLEURAL EFFUSION, RADIOGRAPHY

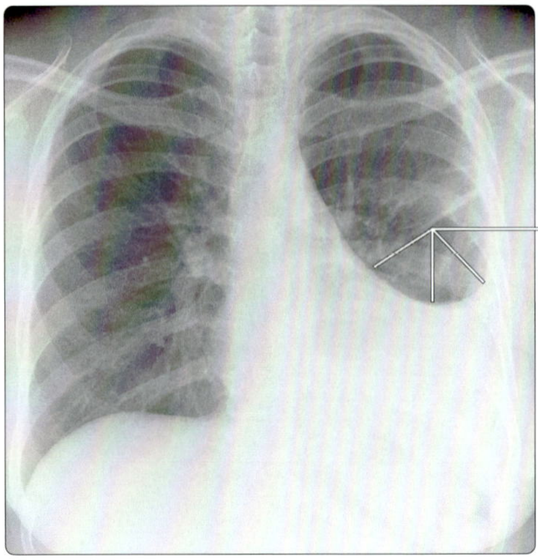

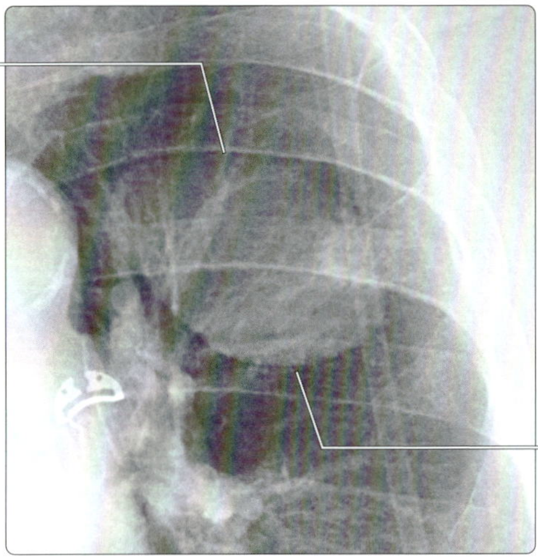

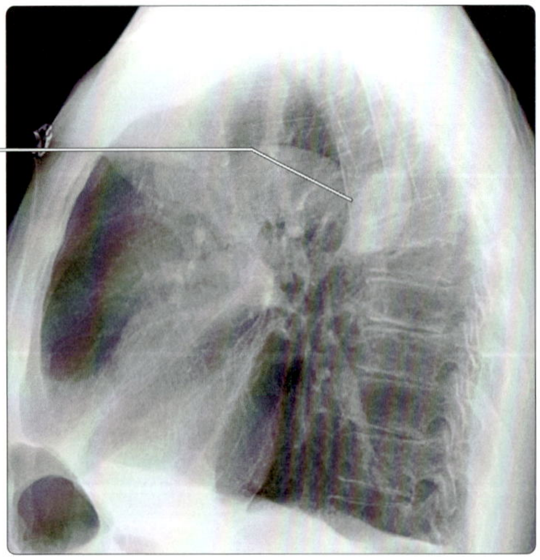

(Top) *PA chest radiograph of a patient with autoimmunity who presented with dyspnea shows a moderate left pleural effusion that exhibits a characteristic meniscus morphology or meniscus sign, present in most free pleural effusions on upright chest radiography.* **(Middle)** *PA chest radiograph (coned down to the left apex) of a patient with chronic heart failure shows a small loculated pleural effusion located in the upper aspect of the left major fissure. The loculated pleural effusion exhibits the incomplete border sign characterized as the presence of both sharply defined and ill-defined margins associated with the same lesion and allows placement of the lesion in an extrapulmonary location.* **(Bottom)** *Lateral chest radiograph of the same patient shows a fusiform opacity within the upper aspect of the left major fissure, consistent with intrafissural loculated pleural fluid. This is frequently referred to as a "phantom" or "vanishing" tumor or pseudotumor. This abnormality is typically seen in the setting of chronic heart failure and often disappears after appropriate treatment, thus accounting for the "phantom" descriptor.*

PLEURAL EFFUSION, CT

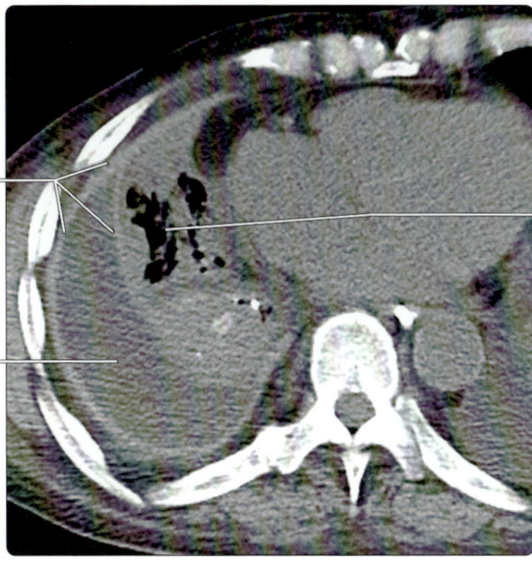

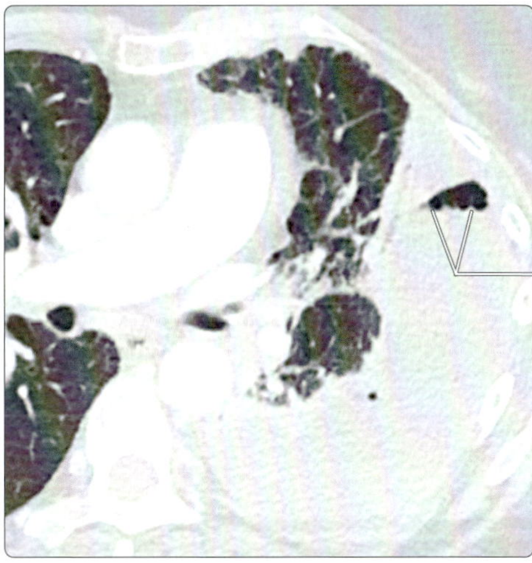

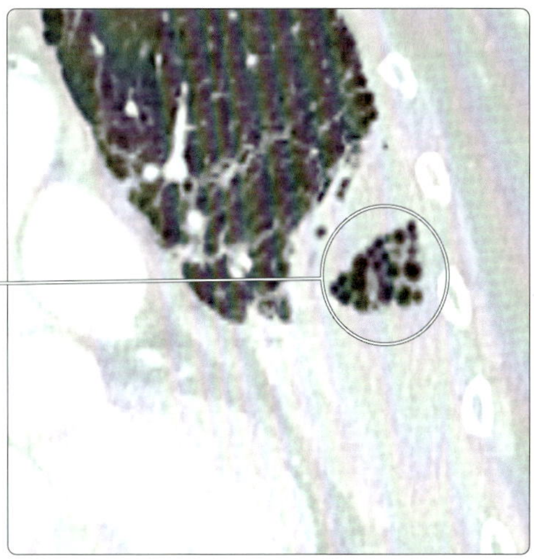

(Top) *Axial NECT (mediastinal window) of a patient with empyema secondary to necrotizing pneumonia shows a loculated right pleural effusion with marked thickening of the parietal and visceral pleurae, the so-called split pleura sign, which is common in empyema and chronic pleural disease. The split pleura sign was initially described on contrast-enhanced CT but can be seen in unenhanced CT, as in this case.* **(Middle)** *Axial CECT (lung window) of a patient with a left empyema complicated by bronchopleural fistula shows gas in the left pleural space forming an air-fluid level. Note the presence of air bubbles along the length of the air-fluid level.* **(Bottom)** *Coronal NECT (lung window) of the same patient shows the grouped air bubbles along the air-fluid level that result in the so-called pomegranate sign. This sign is often associated with exudative pleural effusion.*

PNEUMOTHORAX, RADIOGRAPHY AND CT

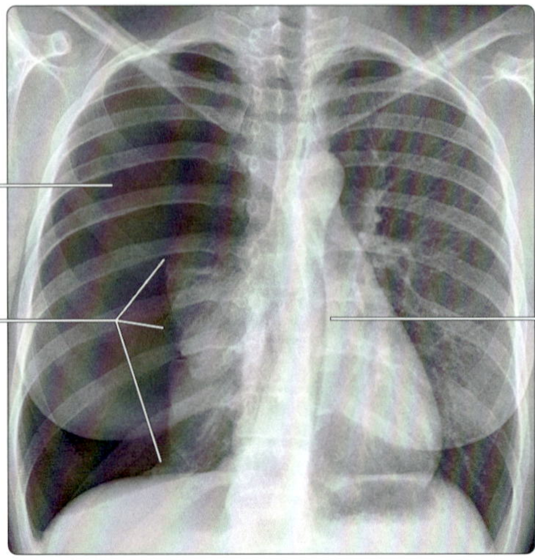

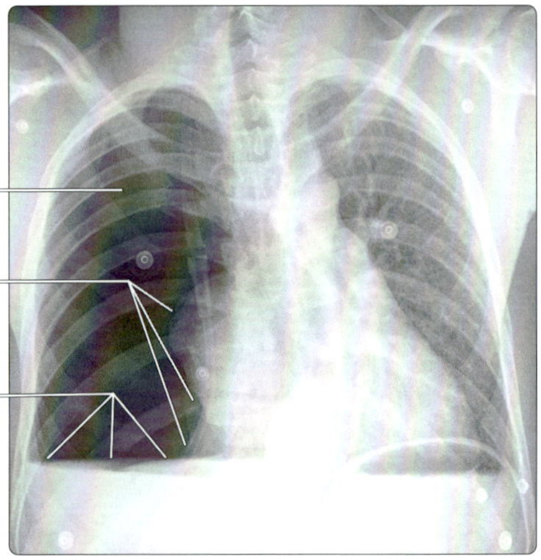

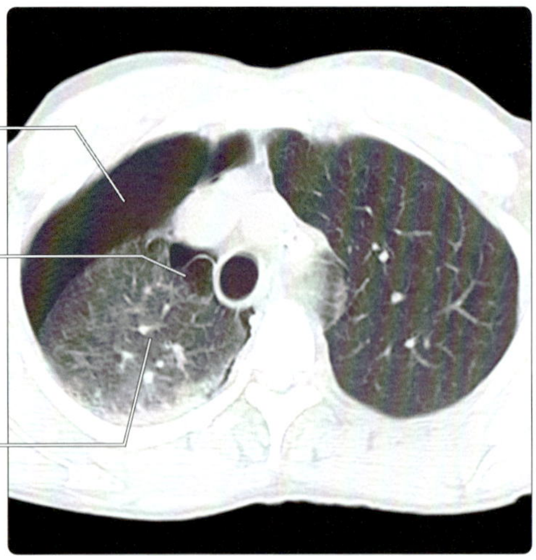

(Top) *PA chest radiograph of a patient with a right-sided primary spontaneous pneumothorax shows marked atelectasis of the right lung and mass effect on and contralateral shift of the cardiomediastinal silhouette.* **(Middle)** *PA chest radiograph of a patient with a primary spontaneous pneumothorax shows a large right hydropneumothorax with near-complete atelectasis of the right lung, a large right basilar pleural air-fluid level, and contralateral shift of the cardiomediastinal silhouette. The findings are concerning for tension physiology.* **(Bottom)** *Axial NECT (lung window) of a patient with a primary spontaneous pneumothorax shows a moderate right anterior pneumothorax with partial atelectasis of the right lung. Note right apical bullae, likely responsible for the pneumothorax.*

NONNEOPLASTIC PLEURAL THICKENING, RADIOGRAPHY AND CT

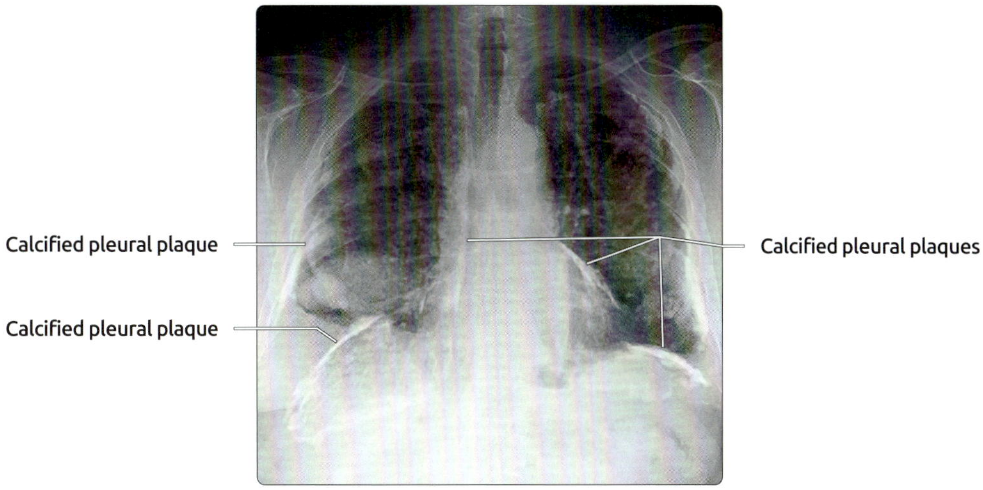

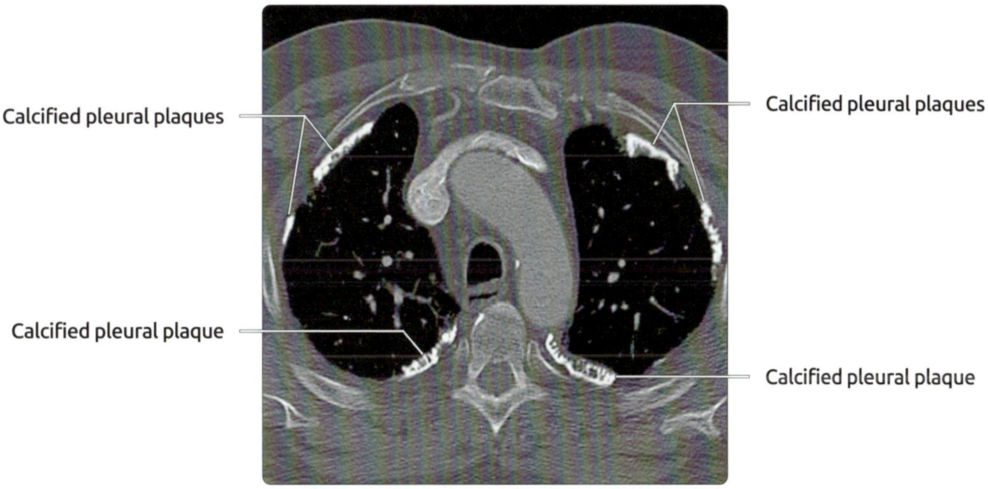

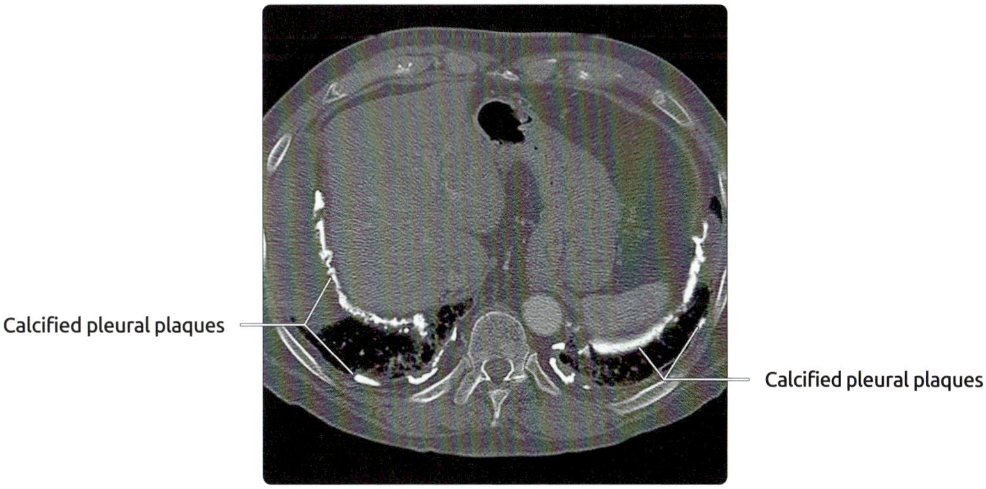

(Top) *PA chest radiograph of a patient with a history of occupational asbestos exposure shows multiple ill-defined calcified irregular lesions overlying the bilateral hemithoraces. Note involvement of the bilateral diaphragmatic pleural surfaces. Note dense vertical band of pleural calcification along the right paravertebral region.* **(Middle)** *Axial CECT (bone window) of the same patient at the level of the aortic arch shows multifocal, discontinuous areas of pleural thickening with calcification. Several plaques extend along the inner aspects of the ribs manifesting as calcified thickening of the costal parietal pleura.* **(Bottom)** *Axial CECT (bone window) of the same patient shows calcified pleural plaques along the diaphragmatic pleurae bilaterally, a common finding in asbestos-related pleural plaques.*

NONNEOPLASTIC PLEURAL THICKENING, RADIOGRAPHY AND CT

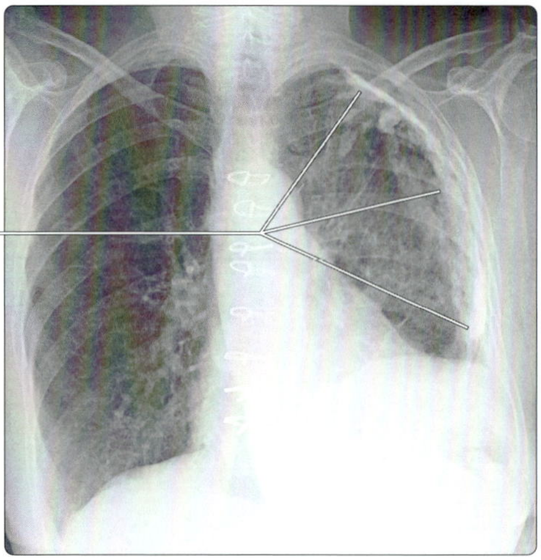

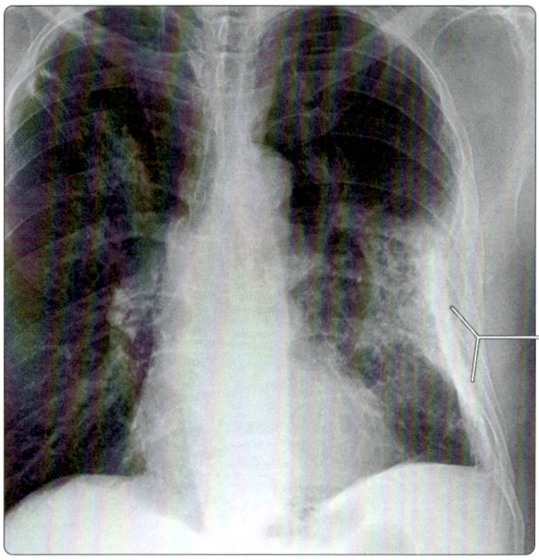

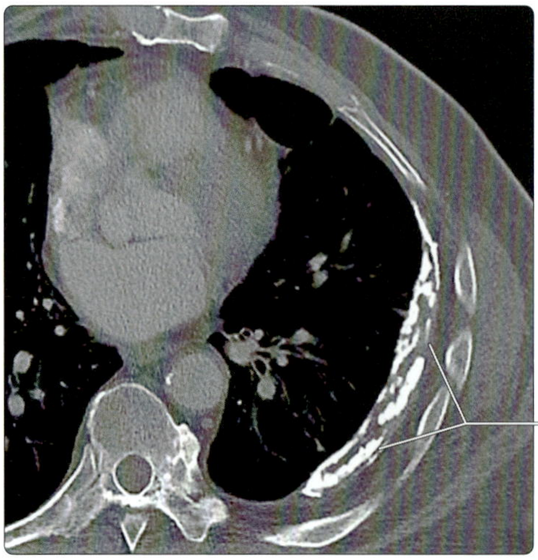

(Top) *PA chest radiograph of patient with a left fibrothorax due to a remote left postsurgical hemothorax shows extensive calcified left pleural thickening with associated low left lung volume. Untreated hemothorax is known to cause pleural organization that often leads to extensive calcification. Fibrothorax may be associated with impaired ventilation of the affected lung.* **(Middle)** *PA chest radiograph of a patient with a previous history of left tuberculous empyema shows marked left pleural thickening and calcification with resultant left fibrothorax.* **(Bottom)** *Axial CECT (bone window) of the same patient shows the thickened and calcified left pleura surrounding tiny residual pleural fluid &/or noncalcified thickening consistent with fibrothorax.*

NEOPLASTIC PLEURAL THICKENING, GRAPHIC, RADIOGRAPHY, AND CT

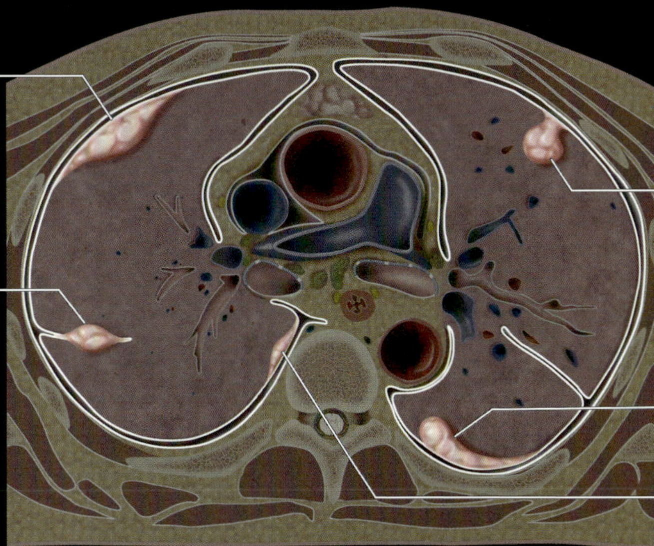

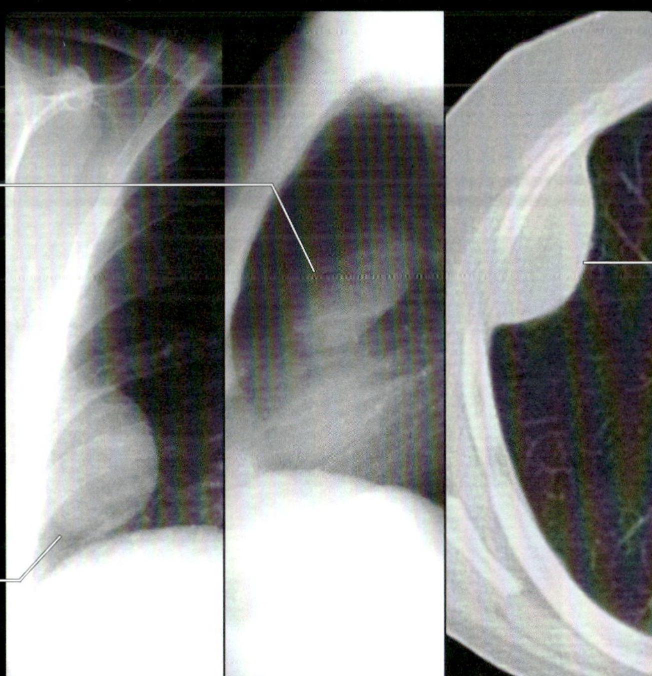

(Top) *Graphic shows the variable shapes and configurations of pleural masses. Pleural masses along fissures may be fusiform, while those occurring along peripheral pleural surfaces may be symmetrically or asymmetrically lenticular with obtuse or acute angles at their interface with the adjacent pleura. Lung parenchymal lesions may invade the adjacent pleura.* **(Bottom)** *Composite image with PA (left) and lateral (middle) chest radiographs and axial NECT (lung window) (right) through the same region shows the characteristic lentiform shape of a pleural mass with obtuse margins at the interface of the mass with the adjacent pleura. When the x-ray beam hits the lesion tangentially, a sharp interface with the adjacent lung results. When the x-ray beam hits the lesion en face, an ill-defined border results. This is the radiographic basis for the so-called incomplete border sign.*

- Posterior mediastinum
  - Structures posterior to middle mediastinum
- **Mediastinal compartments according to Fraser, Müller, Colman, and Paré**
  - Based on lateral chest radiograph
  - Localization of lesions as **predominantly** within a compartment
    - Addresses multicompartment abnormalities
  - **Anterior mediastinum**
    - Identical to anterior mediastinum of **Felson**
  - **Middle-posterior mediastinum**
    - Posterior to anterior mediastinum, anterior to thoracic vertebrae
  - **Paravertebral region**

CT Mediastinal Compartments: International Thymic Malignancy Interest Group (ITMIG)

- General concepts
  - CT-based mediastinal divisions; also applicable to MR imaging
  - International multidisciplinary consensus
  - 3 mediastinal compartments extending from thoracic inlet superiorly to diaphragm inferiorly
    - Prevascular, visceral, paravertebral
- **Prevascular compartment**
  - Bound by sternum anteriorly and anterior pericardial margin posteriorly
  - Contents: Thymus, lymph nodes, left brachiocephalic vein, mediastinal fat
- **Visceral compartment**
  - Bound by anterior pericardial margin anteriorly and by visceral-paravertebral boundary line posteriorly
  - **Visceral-paravertebral compartment boundary line**: Imaginary vertical line "drawn" along anterior thoracic vertebrae, 1 cm posterior to their anterior margins
  - Contents: Pericardium, heart, aorta, vena cavae, pulmonary trunk, intrapericardial pulmonary arteries and veins, thoracic duct, trachea, carina, esophagus, lymph nodes, mediastinal fat
- **Paravertebral compartment**
  - Bound by visceral-paravertebral compartment boundary line anteriorly and imaginary vertical lines "drawn" along posterior chest wall at lateral margins of thoracic vertebral transverse processes

THYMUS

Anatomy

- Bilobed encapsulated organ
- Closely related to central great vessels and pericardium
- Age-related changes
  - Involution beginning after puberty
  - Fatty infiltration after 40 years of age

Imaging

- Not visible radiographically in normal adults
- Conspicuous in normal infants and children
  - **Thymic sail sign**, morphologic appearance of nautical sail (5% of infants)
  - **Thymic wave sign**, thymic indentations by anterior ribs
- CT
  - Quadrilateral morphology in infancy and childhood
  - Triangular morphology in late childhood and adulthood
  - Fat replacement beginning at 25 years, usually complete by 40 years of age
  - Size (thymic lobe thickness)
    - Children ≤ 5 years; average thickness: **1.4 cm**
    - Adults ≤ 20 years; maximum thickness: **1.8 cm**
    - Adults > 20 years; maximum thickness: **1.3 cm**

ANATOMY-RELATED IMAGING ABNORMALITIES

General Concepts

- Mediastinal masses may be subtle on radiography
- Identification relies on knowledge of radiographic anatomy
  - Alterations of normal mediastinal contours, lines, stripes, and interfaces should suggest mediastinal pathology
- ~ 10% of mediastinal abnormalities are of vascular etiology

Systematic Approach to Imaging Diagnosis of Mediastinal Abnormalities

- Frontal radiography: Determination of presence of mediastinal mass
- Characterization of abnormality as focal or diffuse
  - Focal abnormalities should suggest primary lesions
  - Diffuse abnormalities should suggest lymphadenopathy
- Lateral radiography: Localization to mediastinal compartment
- Identification of associated abnormalities: Calcification, pleural effusion, osseous abnormalities
- Formulation of focused differential diagnosis after correlation of imaging findings with demographic and clinical information
- Further assessment with contrast-enhanced chest CT (MR if posterior mediastinal location)
  - Localization to CT mediastinal compartment and space
  - Associated imaging features: Attenuation, lymphadenopathy, mass effect, local invasion
- Formulation of final differential diagnosis &/or diagnosis
- Management recommendation: Additional diagnostic imaging; transthoracic/surgical biopsy; excision

Mediastinal Abnormalities

- **Pneumomediastinum**: Air in mediastinum
  - Spontaneous due to elevated lung pressure with airway rupture; esophageal perforation; iatrogenic due to traumatic instrumentation
- **Mediastinal enlargement**
  - Focal mediastinal mass
    - Primary neoplasm: Benign or malignant
    - Congenital cyst: Typically unilocular and subcarinal
    - Vascular lesion: Vascular enhancement, continuity with adjacent vascular lumen
    - Glandular enlargement: Thymus or thyroid
    - Herniations: Mediastinal abdominal contents
  - Diffuse mediastinal enlargement
    - Lymphadenopathy involving multiple lymph node stations: Granulomatous disease, malignant neoplasia
    - Lymphadenopathy with nodal coalescence: Malignant neoplasia
    - Lipomatosis

OVERVIEW OF MEDIASTINUM

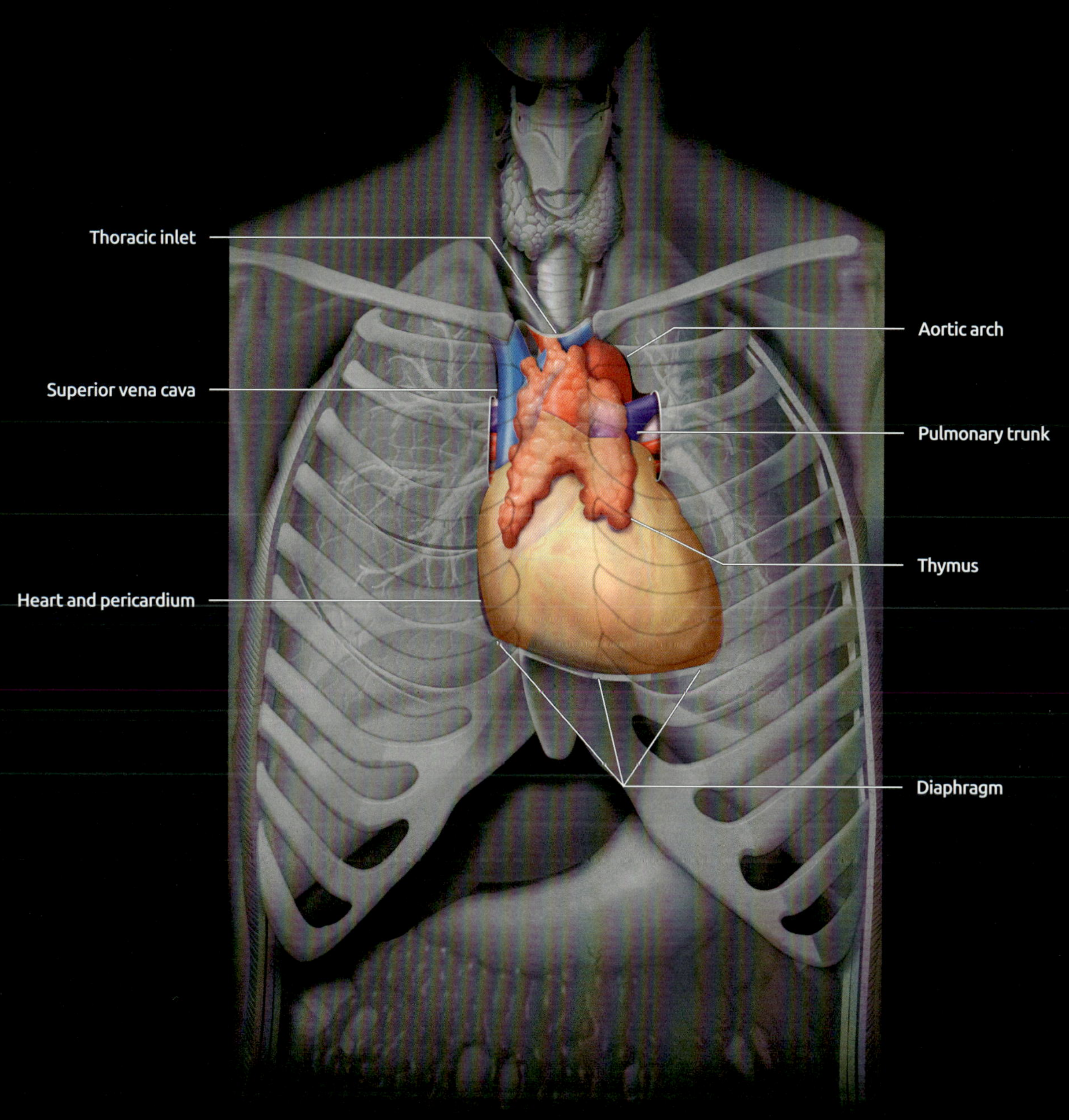

Graphic demonstrates the location of the mediastinum with respect to the adjacent thoracic organs and structures. The mediastinum is the space between the bilateral lungs and pleurae. It contains the heart, the pericardium, the thymus, the aerodigestive tract, the central portions of the thoracic great arteries and veins, nerves and ganglia, lymph nodes and lymphatics, and mediastinal fat. The mediastinum extends from the thoracic inlet superiorly to the diaphragm inferiorly and from the sternum anteriorly to the thoracic vertebrae posteriorly.

ANATOMY OF MEDIASTINUM AND PARAVERTEBRAL REGION

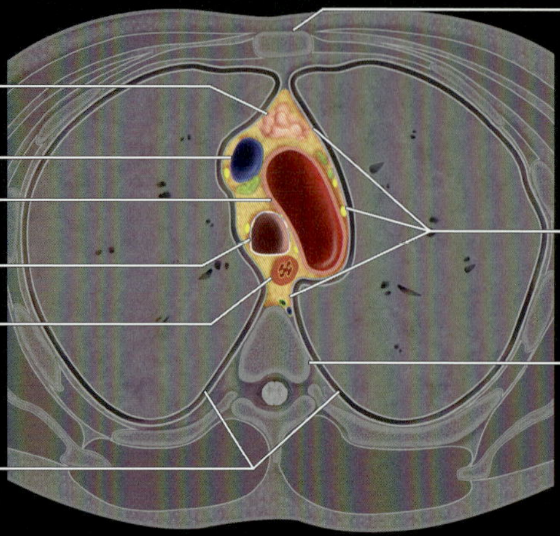

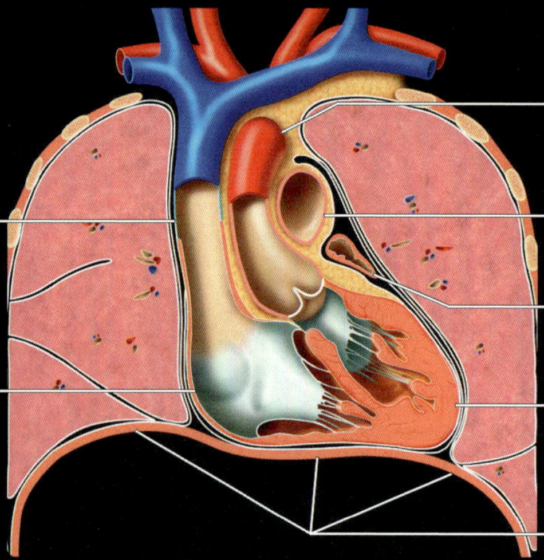

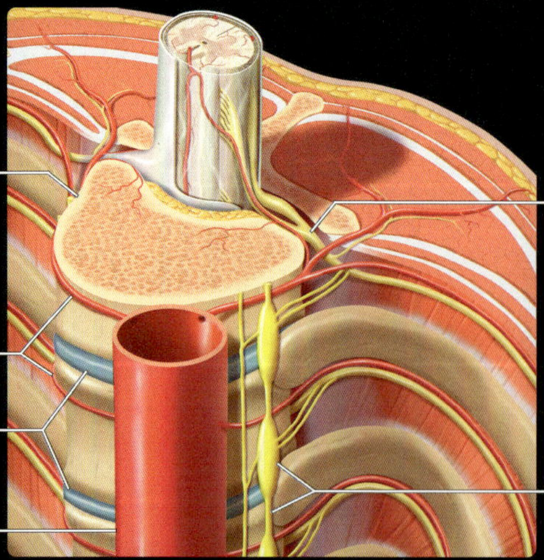

(Top) *Graphic shows the axial cross-sectional anatomy of the mediastinum (shown in color) at the level of the aortic arch. By definition, the mediastinum is the space between the bilateral lungs and pleurae. It is bound anteriorly by the sternum and posteriorly by the thoracic vertebrae.* **(Middle)** *Graphic shows the coronal cross-sectional anatomy of the mediastinum located between the bilateral lungs and pleurae and extending from the thoracic inlet superiorly to the diaphragm inferiorly. Note the anatomic locations of the superior vena cava, right atrium, aortic arch, pulmonary trunk, left atrial appendage, and left ventricle, which form the mediastinal contours seen on frontal chest radiography.* **(Bottom)** *Graphic shows the left paravertebral region, which contains vessels, lymph nodes and lymphatics, nerves, and fat. Peripheral nerves course through the paravertebral region as they exit the neuroforamina. The sympathetic trunk and ganglia are oriented vertically along the bilateral paravertebral regions. Although the paravertebral region is not part of the mediastinum proper, it is often included in imaging classifications of the mediastinal compartments.*

ANATOMY OF MEDIASTINAL NERVES

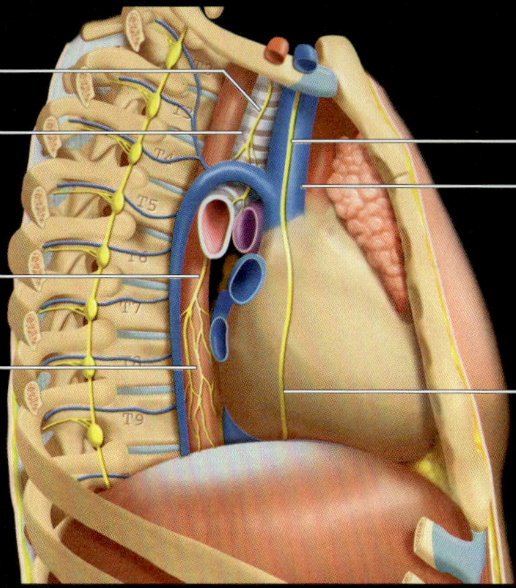

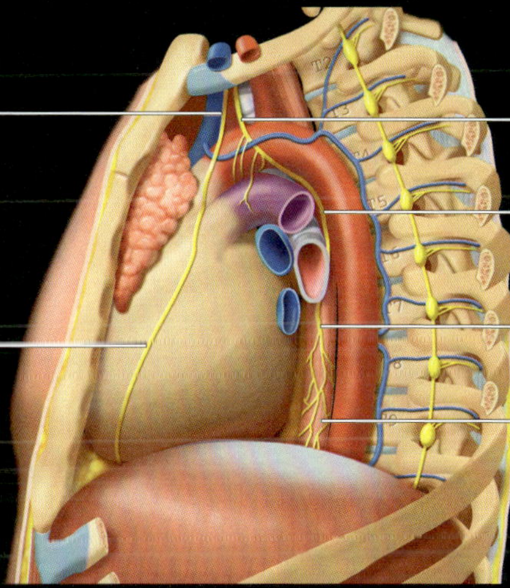

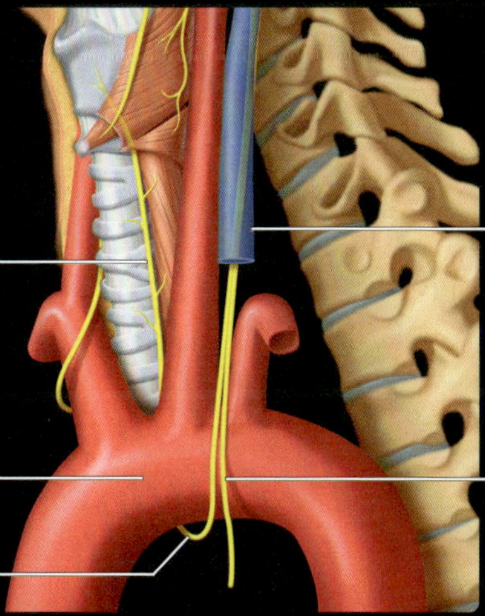

(Top) *First of 3 graphics depicting the anatomy of the mediastinal nerves is shown. The right vagus nerve courses inferiorly along the right lateral trachea posterior to the right hilum and along the right lateral esophageal wall. The right phrenic nerve courses inferiorly lateral to the right brachiocephalic vein and superior vena cava, anterior to the right hilum, and along the right pericardium to enter the abdomen along the inferior vena cava.* **(Middle)** *The left vagus nerve courses inferiorly along the lateral aspect of the aortic arch and continues caudally coursing posterior to the left hilum along the left lateral esophageal wall. The left phrenic nerve courses inferiorly, anterior to the left hilum along the left pericardium and pierces the diaphragm to enter the abdomen.* **(Bottom)** *The left recurrent laryngeal nerve is a branch of the left vagus nerve that follows an arcuate course under the aortic arch and continues superiorly along the left lateral tracheal wall to supply the larynx.*

MEDIASTINAL CONTOURS, RADIOGRAPHY AND CT

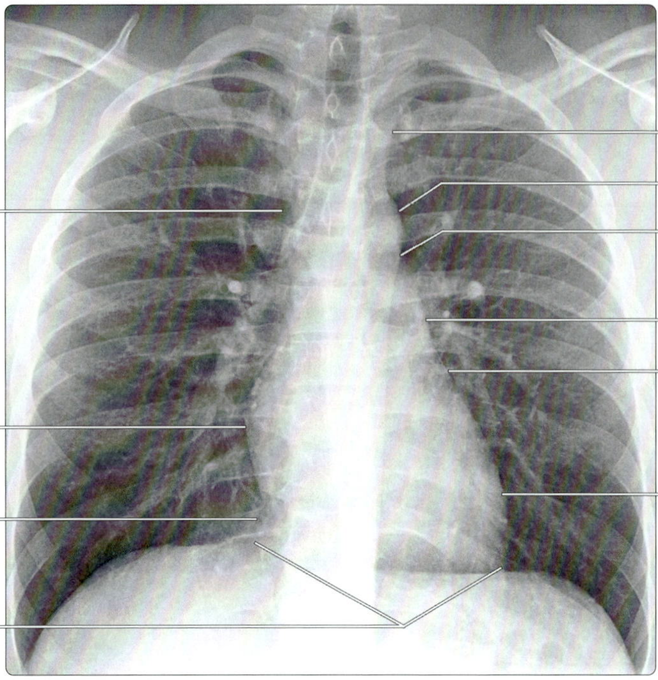

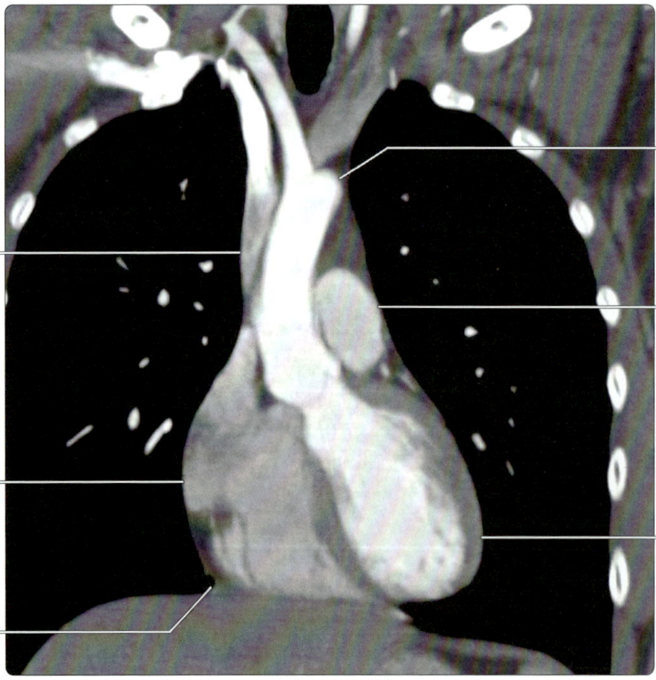

(Top) *Posteroanterior (PA) chest radiograph shows the normal mediastinal contours. The right mediastinal contour (from superior to inferior) is formed by the superior vena cava interface and the right atrium (which forms the right cardiac border). The inferior vena cava may be visible below the right atrium. The left mediastinal contour (from superior to inferior) is formed by the left subclavian artery interface, the aortic arch, the aortopulmonary window, the pulmonary trunk, the left atrial appendage, and the left ventricle. The left ventricle accounts for most of the left cardiac border. The right ventricle is not border forming on frontal chest radiography. Note the bilateral cardiophrenic angles.* **(Bottom)** *Coronal CECT (mediastinal window) shows some of the anatomic structures that form the radiographic mediastinal contours. The superior vena cava and right atrium form most of the right mediastinal contour. The aortic arch, the pulmonary trunk, the left atrial appendage (not shown on this plane of section), and the left ventricle form most of the left mediastinal contour.*

MEDIASTINAL CONTOURS, RADIOGRAPHY AND CT

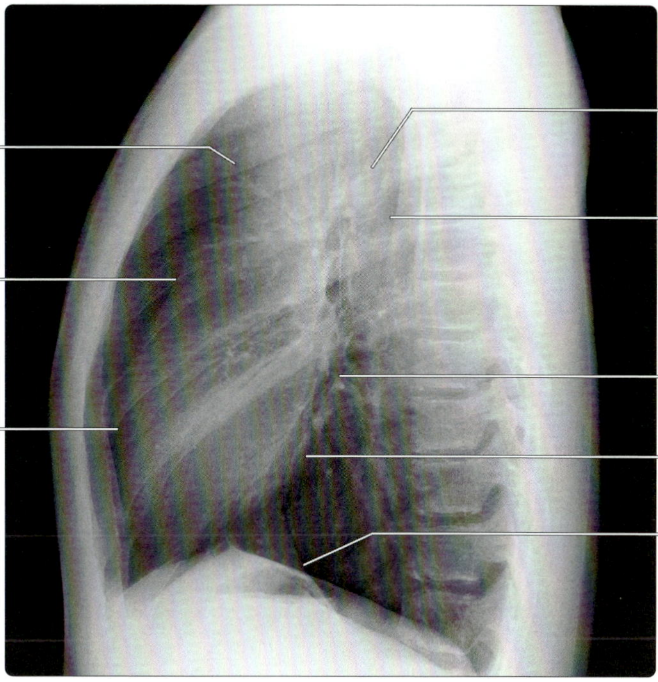

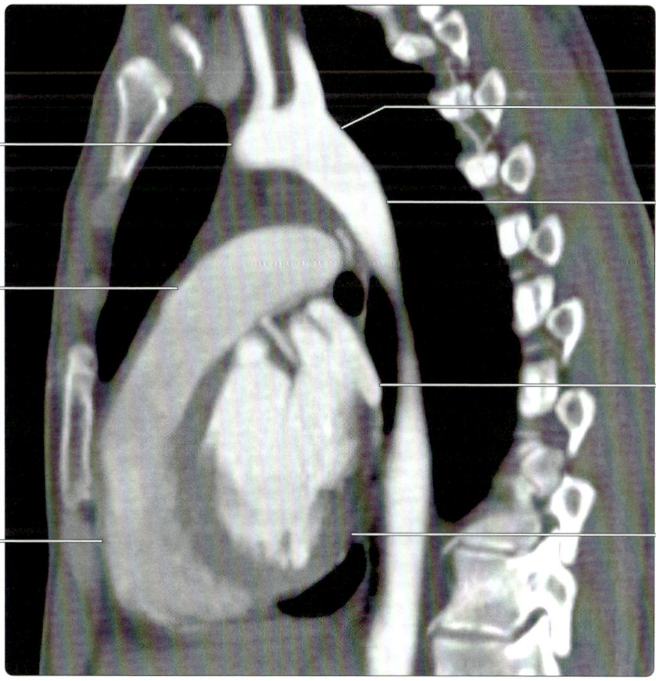

(Top) *Lateral chest radiograph shows the normal mediastinal contours. The anterior mediastinal contour (from superior to inferior) is formed by the pulmonary trunk and the right ventricle. A portion of the distal ascending thoracic aorta may be visible anteriorly and superiorly. The posterior mediastinal contour (from superior to inferior) is formed by the aortic arch, the proximal descending thoracic aorta, the left atrium, the left ventricle, and the inferior vena cava. The right atrium is not border forming on lateral chest radiography.* **(Bottom)** *Sagittal CECT (mediastinal window) shows the anatomic structures that form the mediastinal contours seen on lateral chest radiography. The anterior mediastinal contour (from superior to inferior) is formed by the distal ascending thoracic aorta, the pulmonary trunk, and the right ventricle. The posterior mediastinal contour (from superior to inferior) is formed by the distal aspect of the aortic arch/proximal descending thoracic aorta, the left atrium, and the left ventricle. The inferior vena cava is not visible on this plane of section.*

AORTOPULMONARY WINDOW AND AORTIC-PULMONARY STRIPE, RADIOGRAPHY AND CT

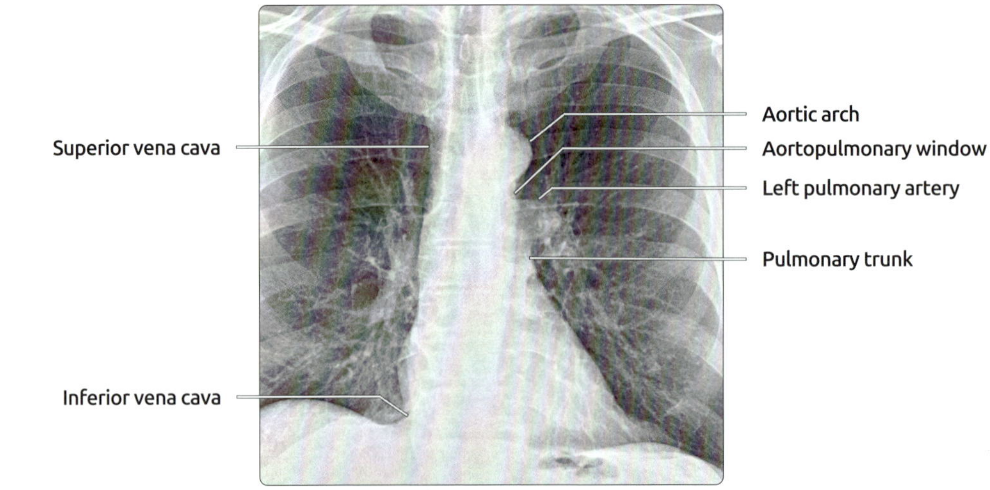

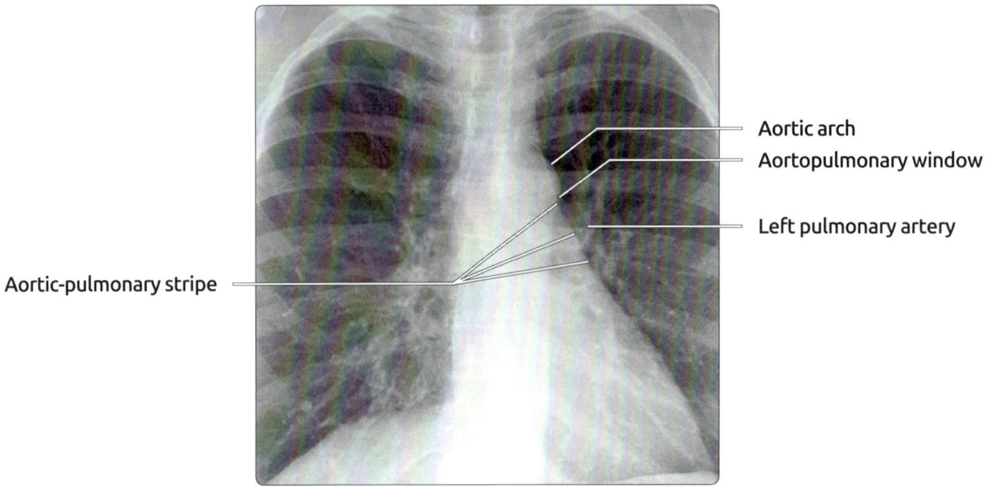

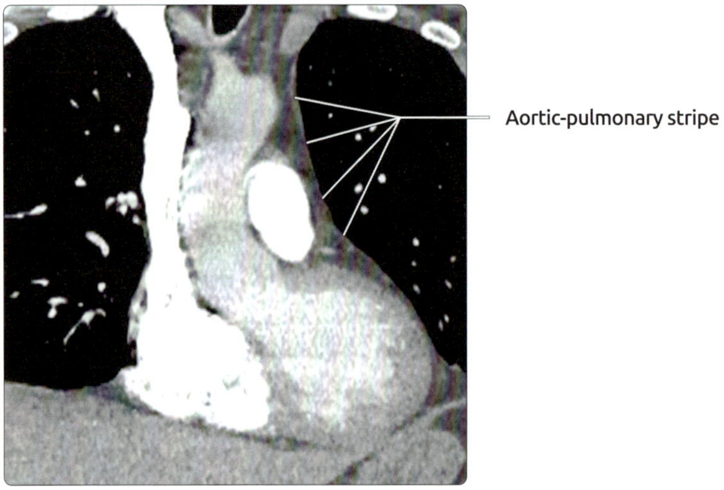

(Top) *PA chest radiograph demonstrates the laterally concave aortopulmonary window interface formed by contact of the left lung with the mediastinal interface located just below the aortic arch and just above the left pulmonary artery. Lateral convexity of the aortopulmonary window may be secondary to mediastinal lymphadenopathy or mass.* **(Middle)** *PA chest radiograph demonstrates the aortopulmonary window and the aortic-pulmonary stripe. The aortic-pulmonary stripe or reflection is located anterior to the aortopulmonary window and extends from the aortic arch to the left mainstem bronchus where it continues inferiorly to merge with the left cardiac border.* **(Bottom)** *Coronal CECT (mediastinal window) shows the anatomic basis of the aortic-pulmonary stripe or reflection located anterior to the aortic arch and the left pulmonary artery. This mediastinal interface is normally straight or laterally convex. Mediastinal fat and anomalous vasculature may produce lateralization of the aortic-pulmonary stripe and must be distinguished from lymphadenopathy or mass.*

ANTERIOR JUNCTION LINE, RADIOGRAPHY AND CT

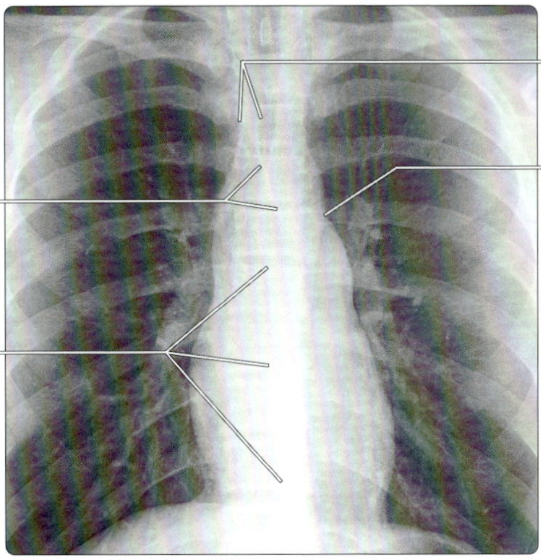

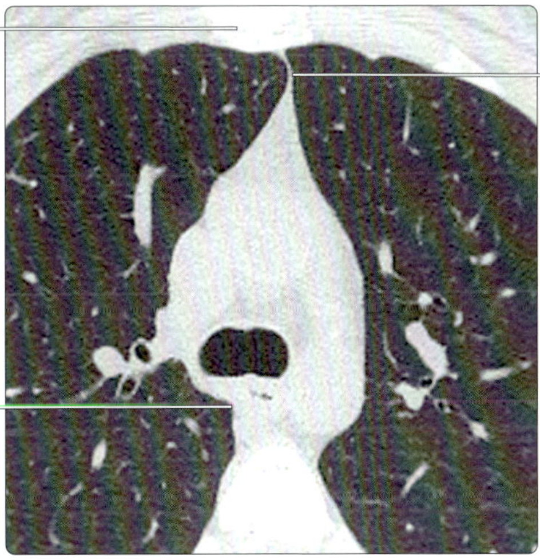

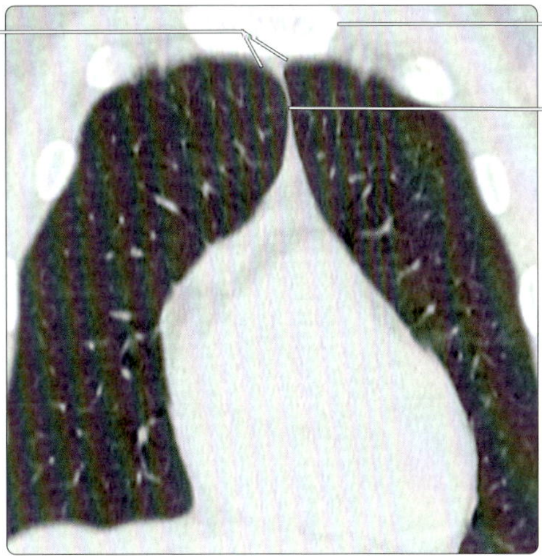

(Top) *PA chest radiograph shows the anterior junction line formed by anterior contact between the bilateral upper lobes and their 4 pleural layers (visceral and parietal pleurae) posterior to the sternum. The anterior junction line courses obliquely from cephalad to caudad from the manubrium sternum superiorly and follows a leftward downward course. The upper aspect of the anterior junction line may exhibit a triangular morphology known as the anterior mediastinal triangle.* **(Middle)** *Axial NECT (lung window) shows the cross-sectional anatomy of the anterior junction line, formed by anterior contact of the bilateral upper lobes and their 4 pleural layers (visceral and parietal pleurae) behind the sternum. The anterior junction line may manifest as an anterior junction stripe due to intervening mediastinal fat.* **(Bottom)** *Coronal NECT (lung window) of the same patient shows the anterior junction line. Divergence of the upper lung surfaces at the upper anterior junction line produces a triangular morphology known as the anterior mediastinal triangle.*

POSTERIOR JUNCTION LINE, RADIOGRAPHY AND CT

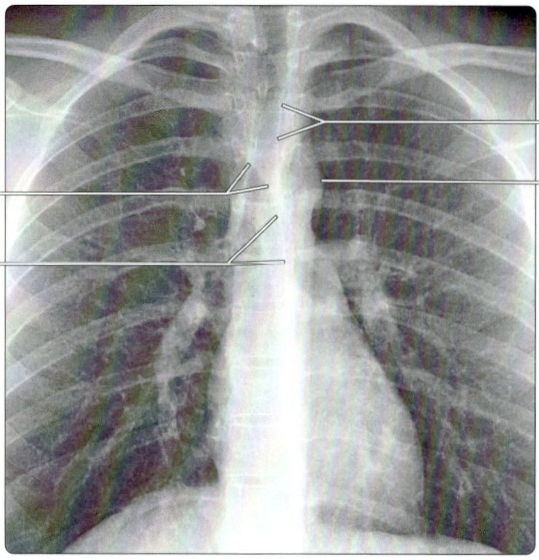

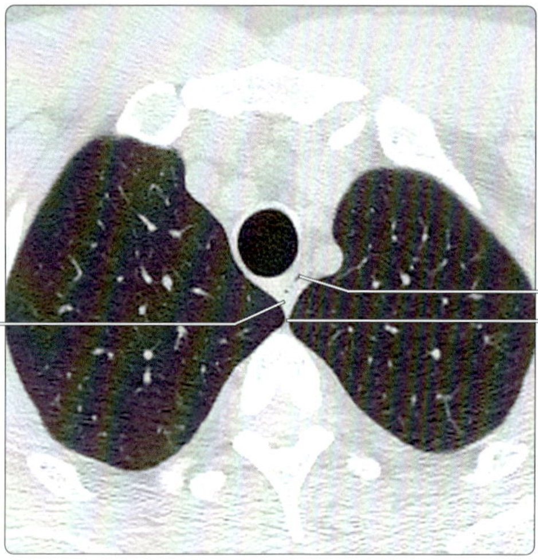

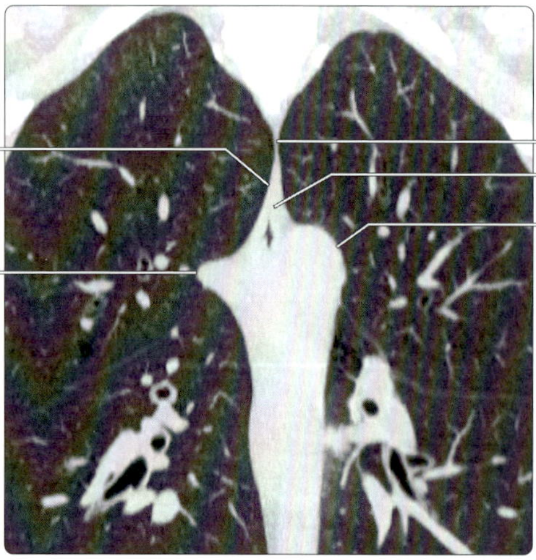

(Top) *PA chest radiograph shows the posterior and anterior junction lines and the anterior mediastinal triangle, located cephalad to the upper anterior junction line. The posterior junction line is formed by contact of the bilateral upper lobes and their 4 pleural layers behind the upper mediastinum and anterior to the thoracic vertebrae. The posterior junction line terminates inferiorly at the superior aspect of the aortic arch or the midportion of the azygos arch.* **(Middle)** *Axial NECT (lung window) shows the anatomic basis of the posterior junction line formed by contact of the bilateral upper lobes and their corresponding pleural layers posterior to the upper mediastinum and anterior to the thoracic vertebrae. Intervening mediastinal fat may result in a posterior junction stripe.* **(Bottom)** *Coronal NECT (lung window) shows contact of the posterior right and left upper lobes behind the upper mediastinum to form the posterior junction line. The posterior junction line usually terminates at the level of the aortic arch or at the level of the azygos arch. Intervening mediastinal fat may produce a posterior junction stripe.*

PARATRACHEAL STRIPES, RADIOGRAPHY AND CT

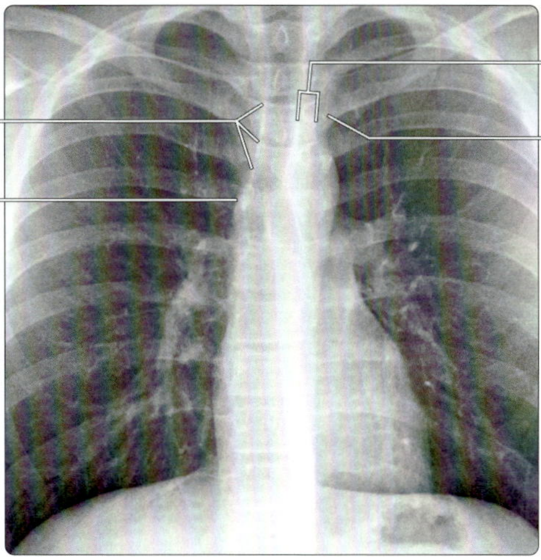

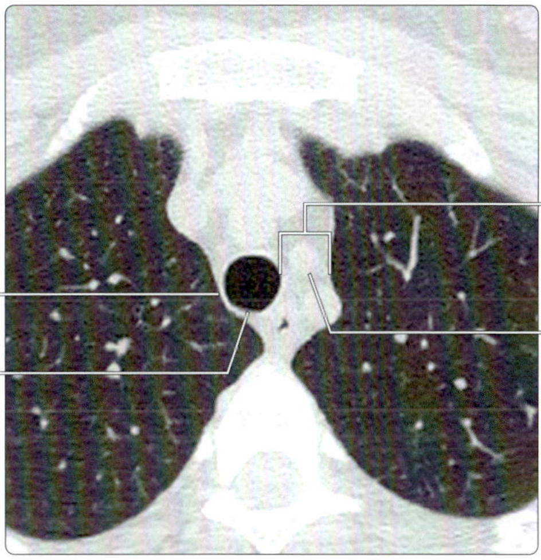

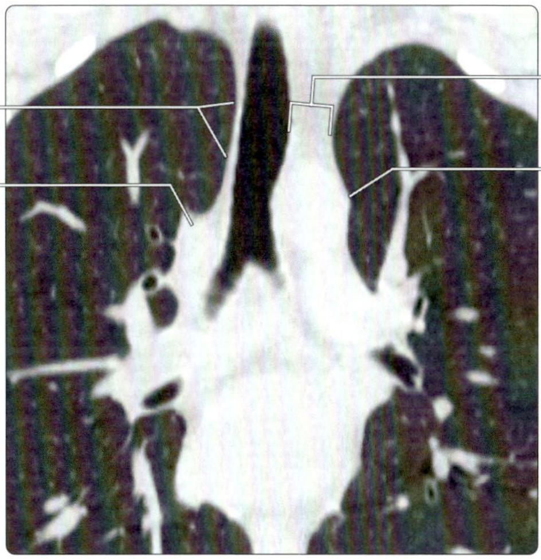

(Top) *PA chest radiograph shows the right and left paratracheal stripes. The right paratracheal stripe is formed by contact between the right upper lobe (and the adjacent 2 pleural layers) and the right lateral tracheal wall. It extends from the thoracic inlet superiorly to the azygos arch inferiorly and normally measures up to 4 mm in thickness. The left paratracheal stripe is formed by contact between the left upper lobe and the left lateral tracheal wall and its adjacent tissues. It extends from the thoracic inlet superiorly to the aortic arch inferiorly and normally exhibits variable thickness.* **(Middle)** *Axial NECT (lung window) shows the thin right paratracheal stripe formed by contact between the right lung and the right lateral tracheal wall. The left paratracheal stripe exhibits variable thickness and includes mediastinal tissues and structures (in this case, the left common carotid artery).* **(Bottom)** *Coronal NECT (lung window) shows the thin right paratracheal stripe and the thicker left paratracheal stripe. The right and left paratracheal stripes extend from the thoracic inlet superiorly to the azygos arch and the aortic arch inferiorly, respectively.*

AZYGOESOPHAGEAL RECESS, RADIOGRAPHY AND CT

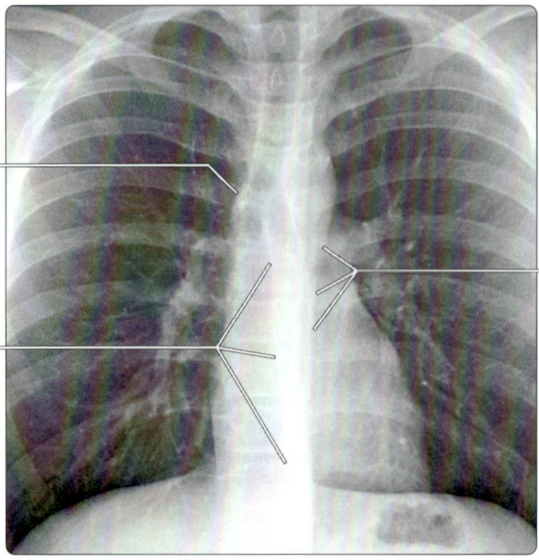

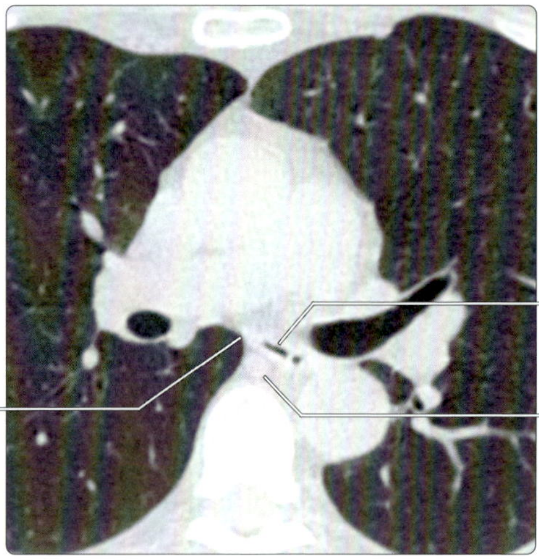

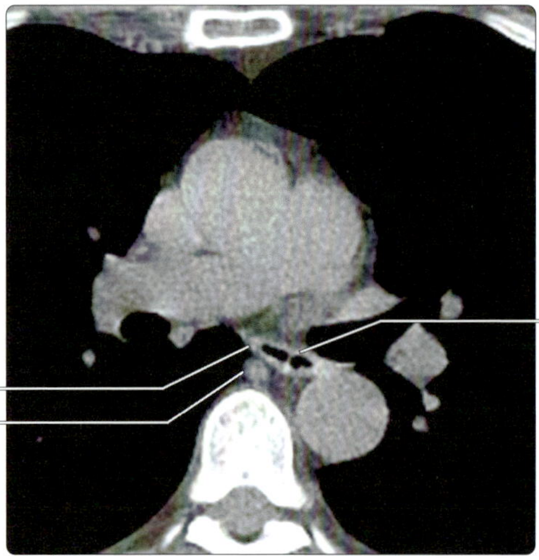

(Top) *PA chest radiograph shows the azygoesophageal recess, a vertically oriented interface formed by contact of the right lung with the adjacent retrocardiac mediastinum along the vertically oriented esophagus and azygos vein. It extends from the azygos arch superiorly to the diaphragm inferiorly and often exhibits a gentle reverse C or reverse S configuration.* **(Middle)** *Axial NECT (lung window) illustrates the normal laterally concave azygoesophageal recess interface formed by contact between the right lower lobe and the retrocardiac mediastinum adjacent to the vertically oriented esophagus and azygos vein.* **(Bottom)** *Axial NECT (mediastinal window) illustrates the normal concave appearance of the azygoesophageal recess, which follows the vertical course of the azygos vein and the esophagus.*

PLEUROESOPHAGEAL STRIPES, RADIOGRAPHY AND CT

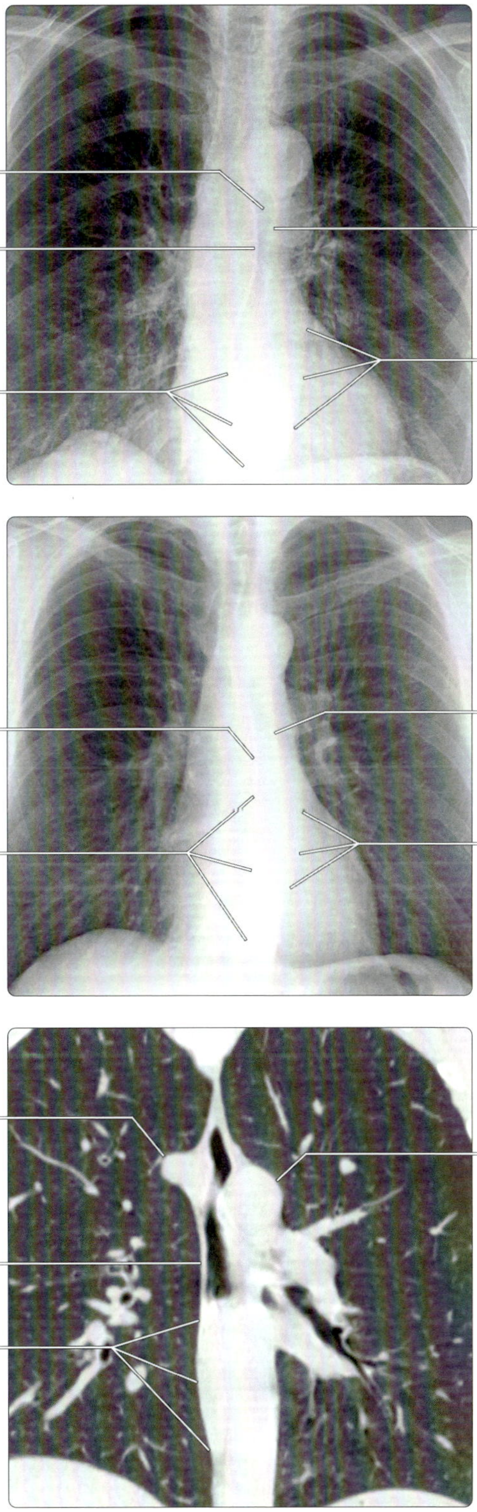

(Top) *PA chest radiograph shows the inferior aspect of the azygoesophageal recess. The upper azygoesophageal recess in this case manifests as a right pleuroesophageal stripe formed by the presence of air within the esophagus. Note the contralateral left pleuroesophageal stripe.* **(Middle)** *PA chest radiograph shows the normal course and configuration of the azygoesophageal recess. Air in the esophagus produces a right pleuroesophageal stripe along the course of the azygoesophageal recess. A left pleuroesophageal stripe is also visible secondary to contact between the left lower lobe and the left esophageal wall. The pleuroesophageal stripes are characteristically visualized immediately inferior to the aortic arch.* **(Bottom)** *Coronal NECT (lung window) of the same patient shows the normal azygoesophageal recess, which extends from the azygos arch superiorly to the diaphragm inferiorly. Air within the esophageal lumen results in visualization of a right pleuroesophageal stripe, which is typically visualized just below the aortic and azygos arches.*

RETROSTERNAL STRIPE, RADIOGRAPHY AND CT

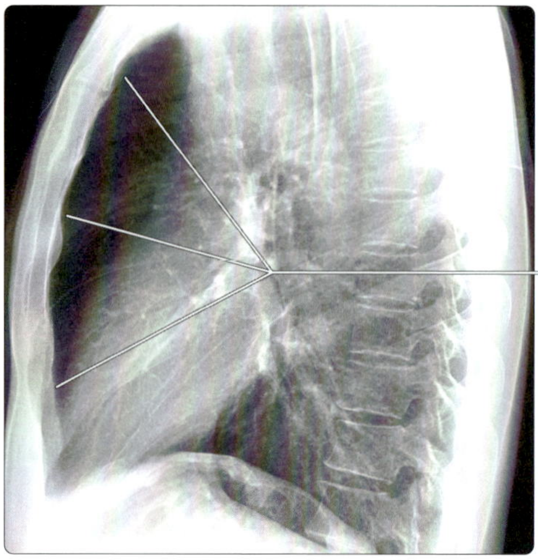

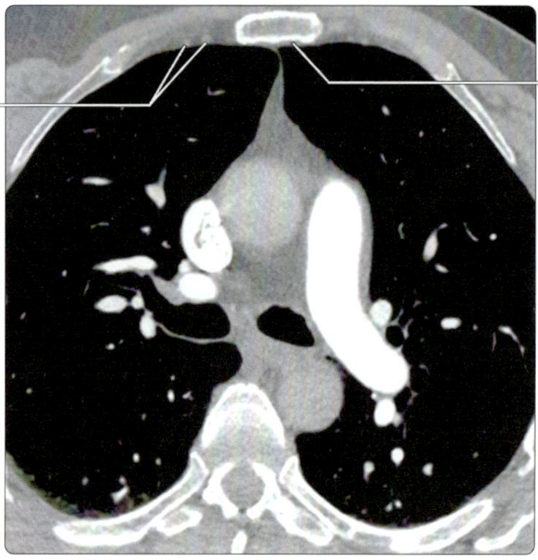

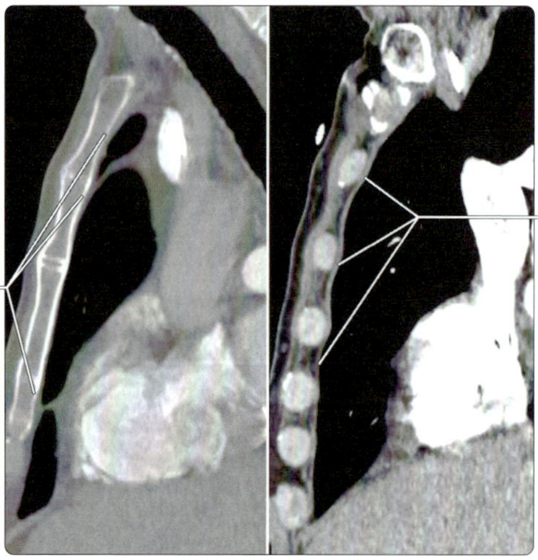

(Top) *Lateral chest radiograph shows the retrosternal stripe formed by contact between the bilateral anterior lungs and the retrosternal soft tissues. The retrosternal stripe includes fat, lymph nodes, and internal thoracic vessels, exhibits a variable thickness, and normally measures up to 7 mm.* **(Middle)** *Axial CECT (bone window) demonstrates the anatomic basis for the radiographic retrosternal stripe. The retrosternal stripe is formed by the normal internal thoracic arteries and veins and surrounding retrosternal fat.* **(Bottom)** *Composite sagittal CECT in bone window (left) and mediastinal window (right) shows the anatomic basis of the retrosternal stripe with contributions from retrosternal fat deposition and the internal thoracic arteries and veins, respectively.*

POSTERIOR TRACHEAL STRIPE, RADIOGRAPHY AND CT

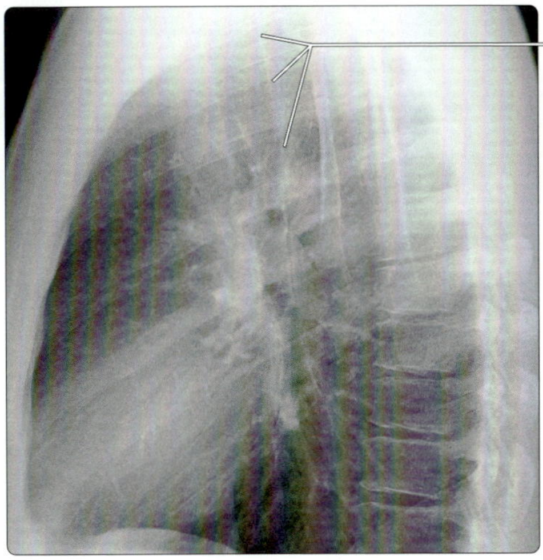

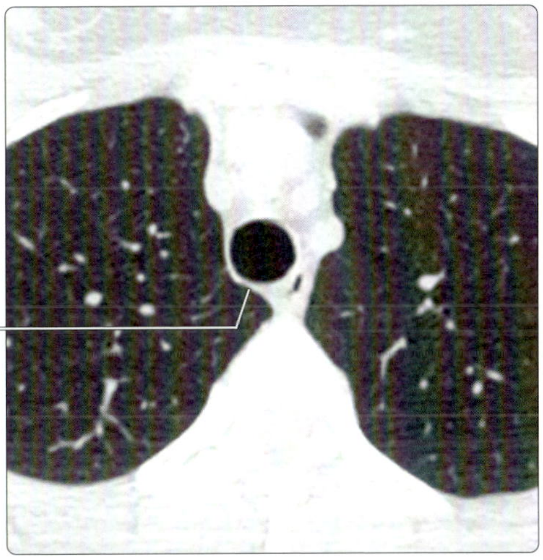

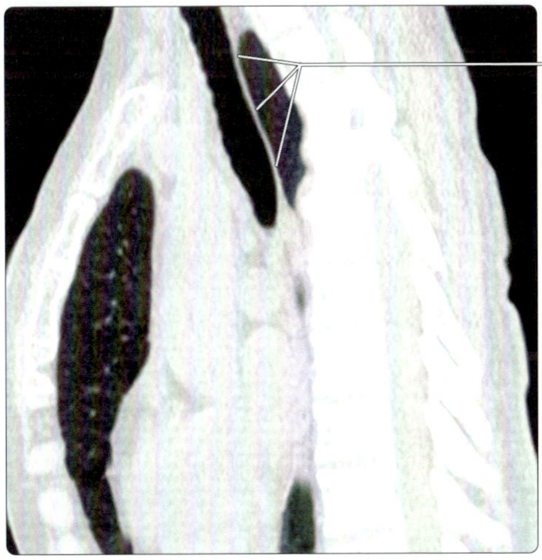

(Top) *Lateral chest radiograph demonstrates the normal posterior tracheal stripe formed by contact between the lung and the posterior tracheal wall. The posterior tracheal stripe extends from below the thoracic inlet superiorly to above the tracheal carina inferiorly.* **(Middle)** *Axial NECT (lung window) demonstrates the anatomic basis for the radiographic posterior tracheal stripe that results from contact between the lung and the posterior tracheal wall. The posterior tracheal stripe may be thickened by portions of the esophagus.* **(Bottom)** *Sagittal NECT (lung window) shows the posterior tracheal stripe, which represents the posterior tracheal wall and adjacent tissues outlined by air in the tracheal lumen anteriorly and aerated lung posteriorly. The posterior tracheal stripe normally includes the posterior tracheal wall, a small amount of mediastinal fat, and occasionally portions of the esophagus.*

ANATOMY OF MEDIASTINAL SPACES

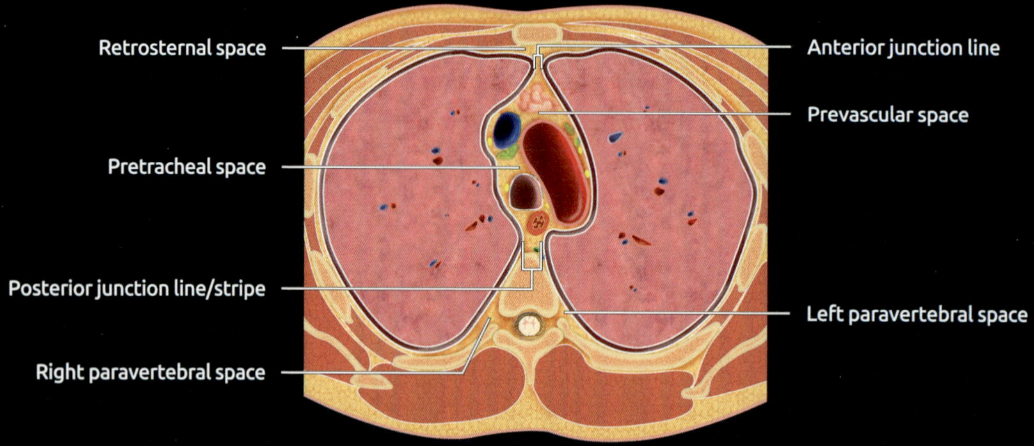

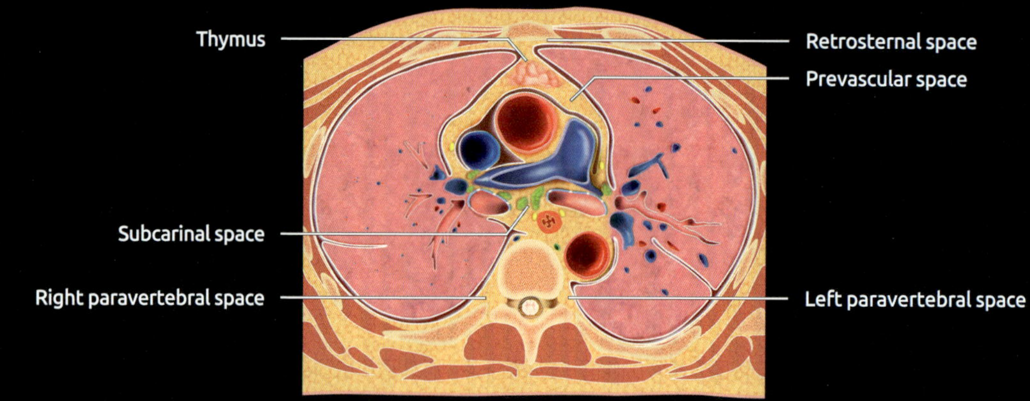

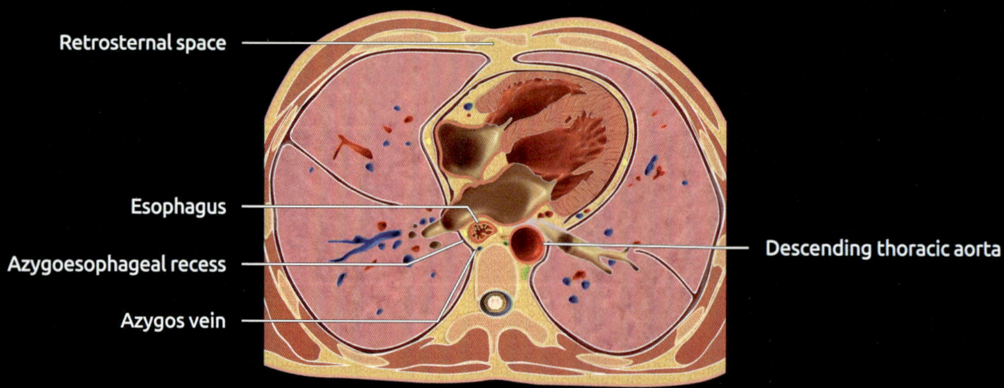

(Top) *Graphic shows the mediastinal spaces at the level of the aortic arch. The retrosternal space is located immediately behind the sternum. The prevascular space is located anterior to the aorta and superior vena cava and contains the thymus and mediastinal fat. The pretracheal space is located anterior to the trachea and posterior to the superior vena cava. The paravertebral spaces as well as the anterior and posterior junction lines/stripes are also shown.* **(Middle)** *Graphic shows the bifurcation of the pulmonary trunk and the prevascular space that contains the thymus and surrounding mediastinal fat. Note the subcarinal space, which contains mediastinal fat and lymph nodes (in green) and the bilateral paravertebral spaces.* **(Bottom)** *Graphic at the level of the 4-chamber heart shows the normally concave azygoesophageal recess, which courses along the length of the esophagus and the azygos vein. Note the normal retrosternal space located immediately posterior to the sternum.*

MEDIASTINAL SPACES, AXIAL CT

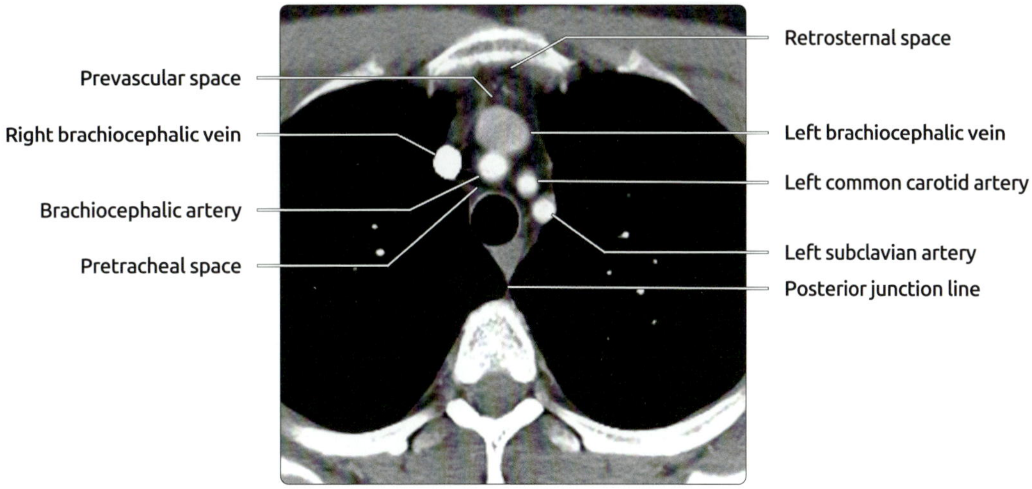

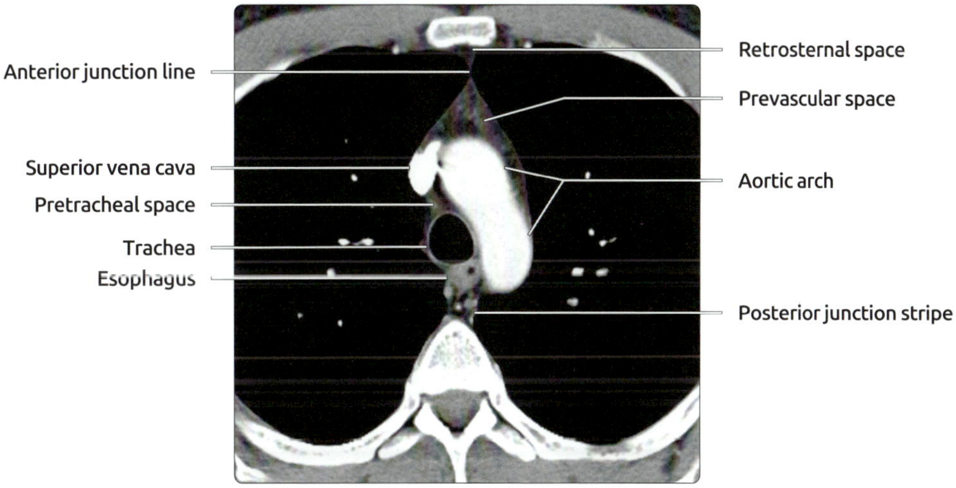

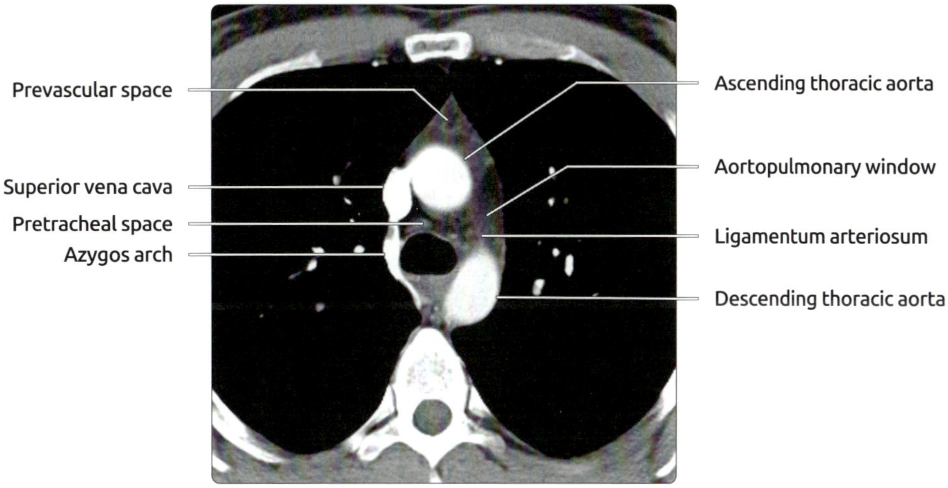

(Top) *First of 6 axial CECT (mediastinal window) shows the upper aspect of the mediastinum. The left brachiocephalic vein courses anterior to the brachiocephalic, left common carotid, and left subclavian arteries. The right brachiocephalic vein is also shown. The prevascular space is located anterior to the thoracic great vessels, and the retrosternal space is located immediately posterior to the sternum.* **(Middle)** *Axial CECT (mediastinal window) at the level of the aortic arch shows the prevascular space located anterior to the aorta and superior vena cava, which contains fat and thymic tissue. The pretracheal space is bound anteriorly by the superior vena cava and posteriorly by the trachea.* **(Bottom)** *Axial CECT (mediastinal window) at the level of the azygos arch shows the prevascular space located anterior to the ascending thoracic aorta and superior vena cava. The pretracheal space is located anterior to the trachea. The aortopulmonary window is located between the aortic arch superiorly and the left pulmonary artery (not shown) inferiorly and contains the ligamentum arteriosum and mediastinal fat.*

MEDIASTINAL SPACES, AXIAL CT

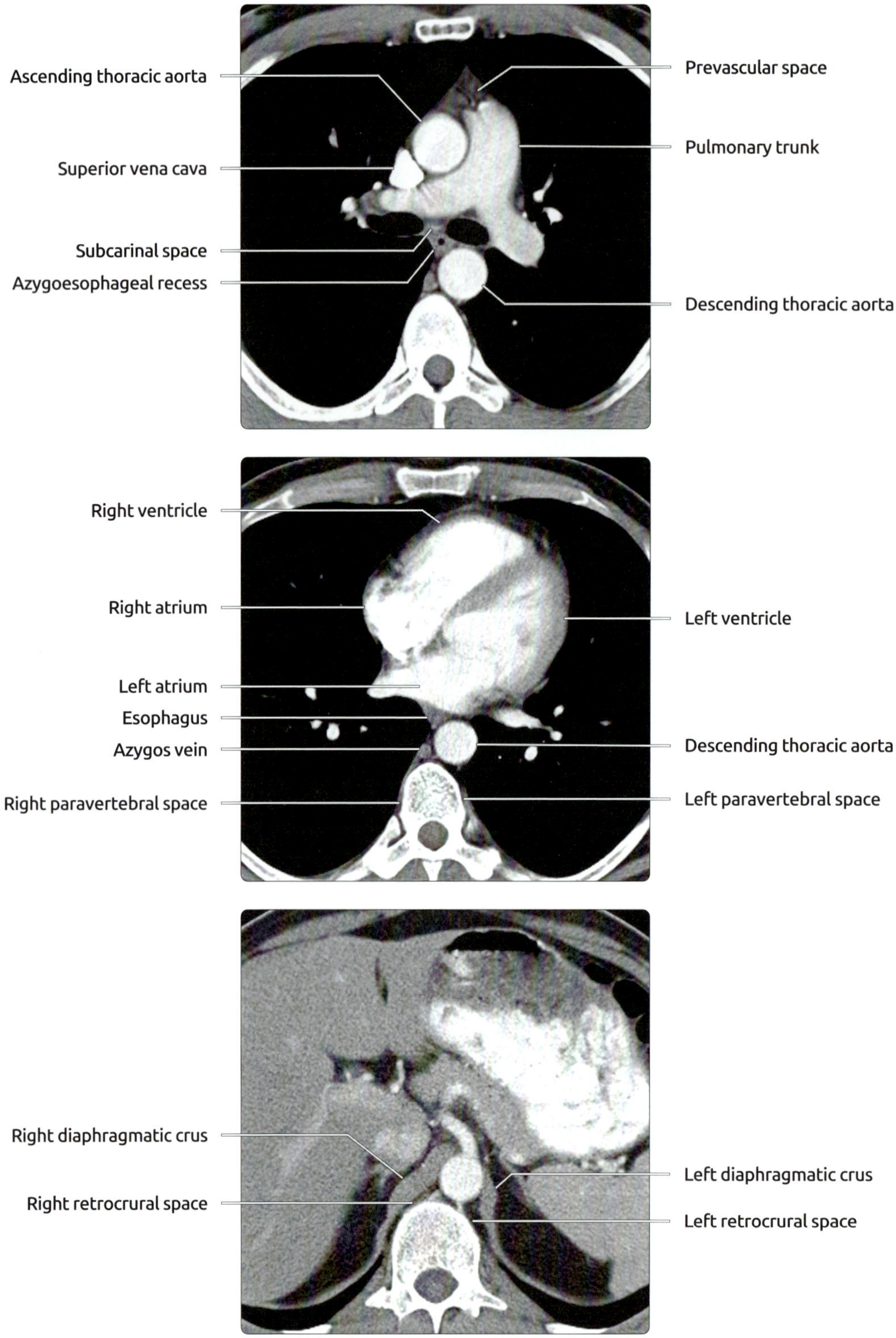

(Top) *Axial CECT (mediastinal window) at the level of the bifurcation of the pulmonary trunk shows the prevascular space located anterior to the great vessels and the subcarinal space located just below the tracheal carina and medial to the mainstem bronchi.* **(Middle)** *Axial CECT (mediastinal window) at the level of the heart shows the bilateral paravertebral spaces, which typically contain paravertebral fat, blood vessels, and nerves.* **(Bottom)** *Axial CECT (mediastinal window) shows the bilateral retrocrural spaces located medial to the bilateral diaphragmatic crura. The retrocrural spaces are generally considered intrathoracic in location.*

MEDIASTINAL SPACES, CORONAL MR

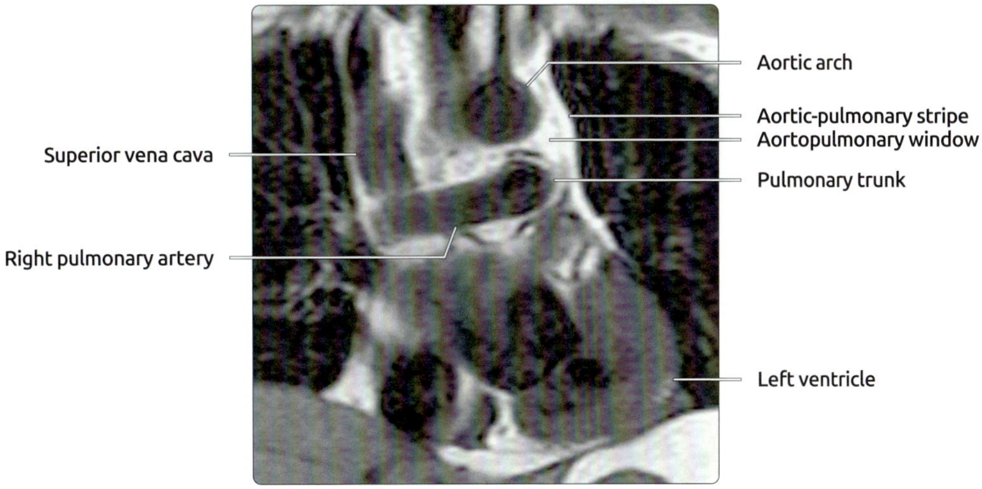

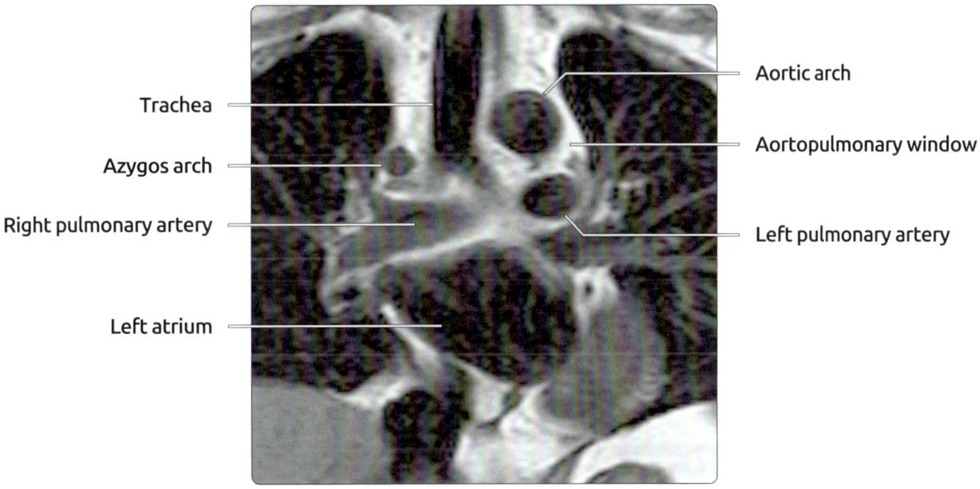

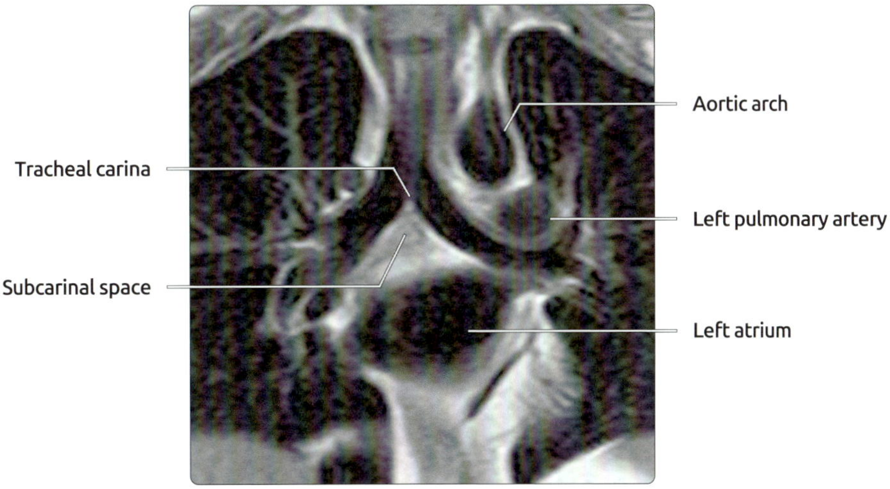

(Top) *First of 3 coronal T1 MR images through the mediastinum demonstrates the normal mediastinal spaces. Coronal image through the right pulmonary artery demonstrates the aortopulmonary window located between the aortic arch superiorly and the pulmonary trunk and left pulmonary artery inferiorly. The aortopulmonary window is filled with high signal intensity mediastinal fat.* **(Middle)** *Coronal T1 MR through the distal trachea demonstrates the posterior aspect of the aortopulmonary window. Small foci of intermediate signal intensity within the high-signal mediastinal fat correspond to normal mediastinal lymph nodes.* **(Bottom)** *Coronal T1 MR through the carina demonstrates the normal subcarinal space located between the tracheal bifurcation superiorly and the left atrium inferiorly.*

MEDIASTINAL COMPARTMENTS, ANATOMIC AND RADIOGRAPHIC

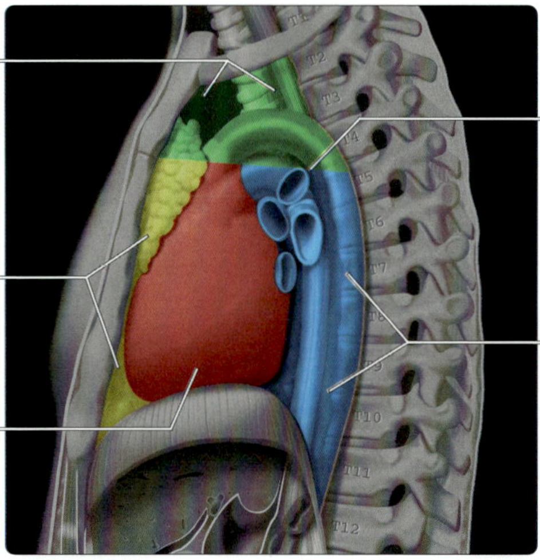

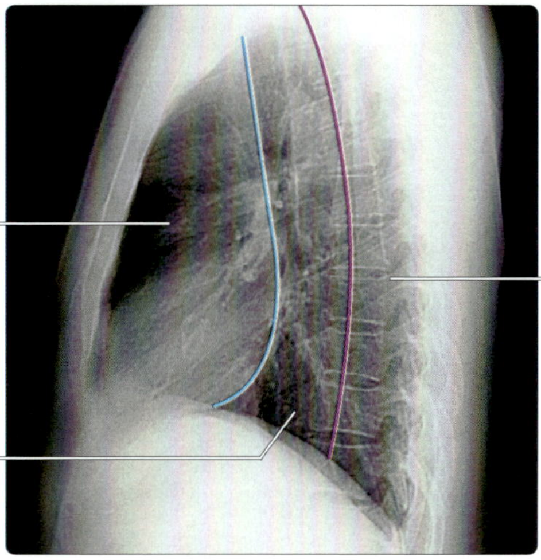

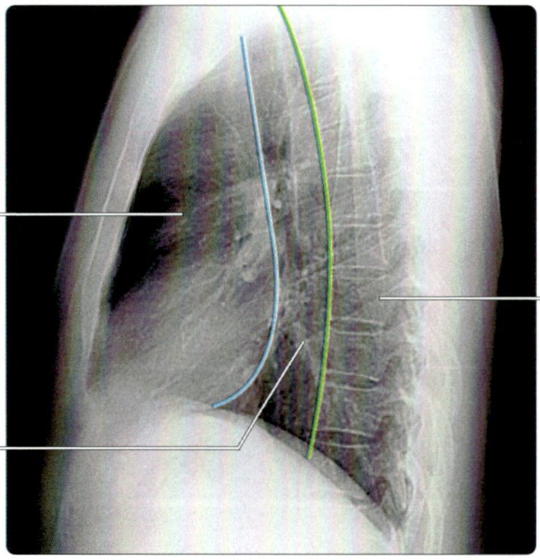

(Top) *Graphic shows the anatomic mediastinal compartments. The superior mediastinum is located above an imaginary line between the sternomanubrial junction and the T4-T5 disk. Anterior, middle (defined by the fibrous pericardium), and posterior compartments are located below the superior mediastinum. The paravertebral region is not in the mediastinum and therefore not included as an anatomic mediastinal compartment.* **(Middle)** *Lateral chest radiograph shows the mediastinal compartments according to Felson. The anterior mediastinum is located anterior to a line drawn along the anterior trachea and posterior heart. The posterior mediastinum is located behind a line that connects points located 1 cm behind the anterior vertebral bodies. The middle mediastinum is located between the anterior and posterior compartments.* **(Bottom)** *Lateral chest radiograph shows the mediastinal compartments according to Fraser et al. The anterior mediastinum is identical to that described by Felson. The paravertebral region is located behind a line drawn along the anterior vertebrae. The middle-posterior mediastinum is located between the anterior mediastinum and the paravertebral region.*

CT MEDIASTINAL COMPARTMENTS, ITMIG

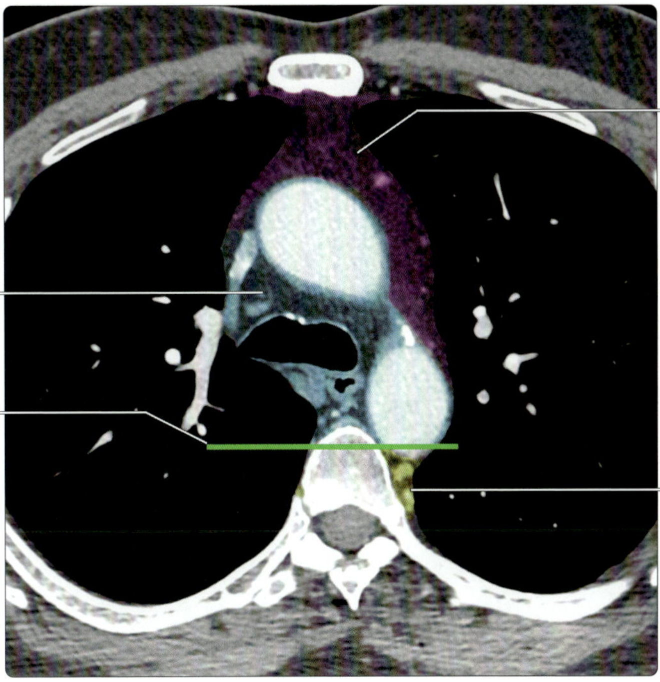

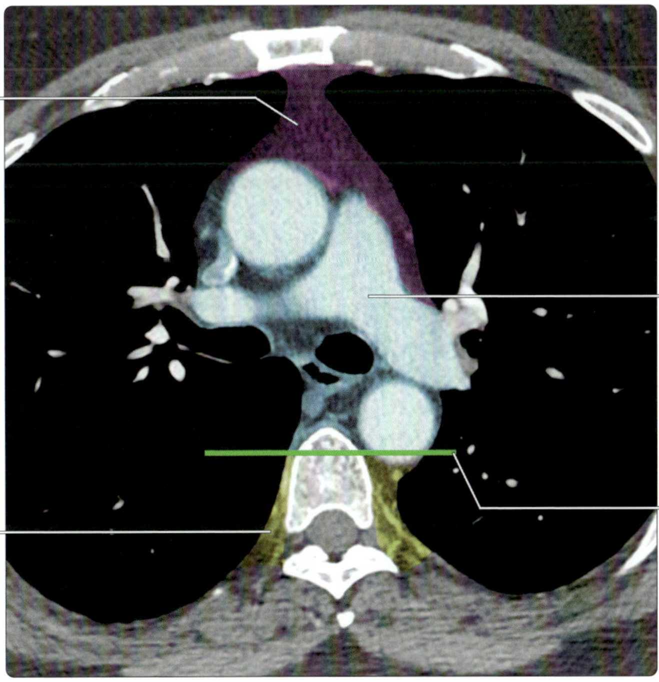

(Top) *The International Thymic Malignancy Interest Group (ITMIG) mediastinal compartments are based on CT imaging. Axial CECT (mediastinal window) shows the prevascular compartment in magenta, the visceral compartment in blue, and the paravertebral compartment in yellow. The horizontal green line drawn 1 cm posterior to the anterior vertebral margin represents the visceral-paravertebral compartment boundary line that separates the visceral and paravertebral compartments. The prevascular compartment wraps around the anterolateral aspects of the thoracic great vessels.* **(Bottom)** *Axial CECT (mediastinal window) at the level of the bifurcation of the pulmonary trunk shows the prevascular (magenta), visceral (blue), and paravertebral (yellow) ITMIG mediastinal compartments. The visceral and paravertebral compartments are separated by the horizontal green line that represents the visceral-paravertebral compartment boundary line. Note that the prevascular compartment wraps around the anterolateral aspects of the thoracic great vessels.*

CT MEDIASTINAL COMPARTMENTS, ITMIG

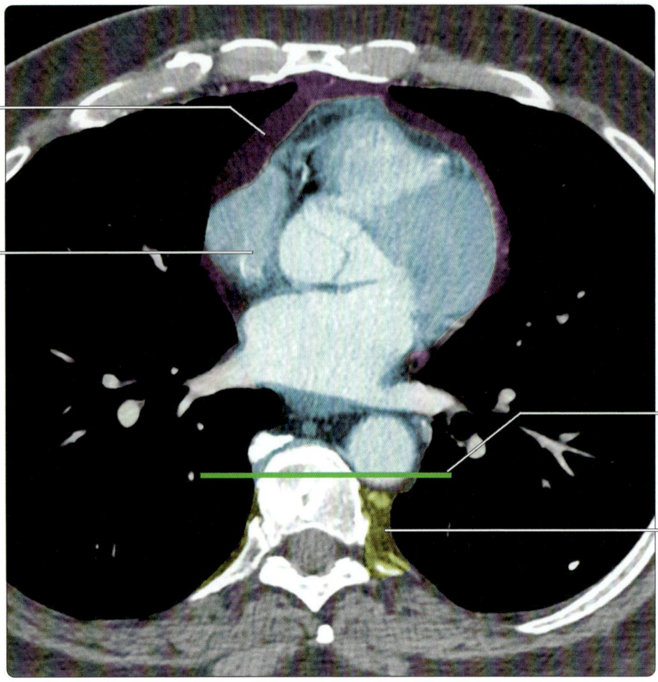

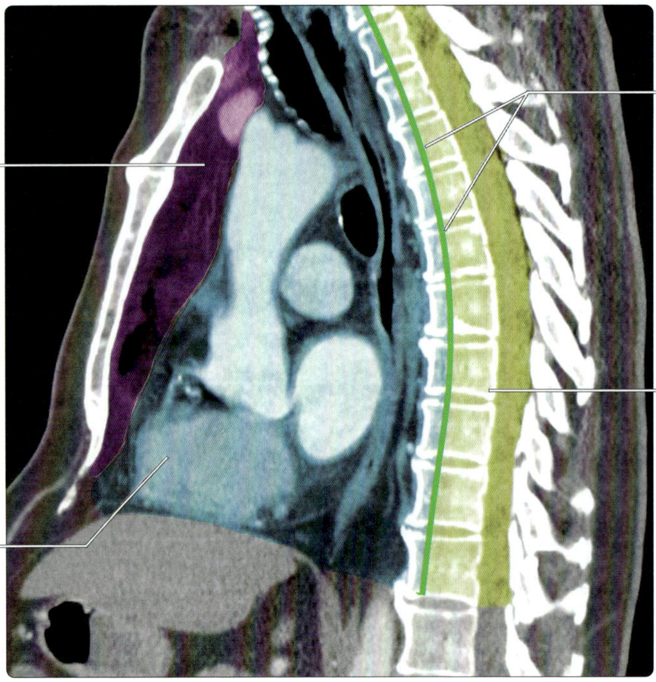

(Top) *Axial CECT (mediastinal window) illustrates the ITMIG mediastinal compartments. The prevascular compartment (magenta) wraps around the anterolateral aspects of the heart and pericardium. The visceral compartment is shown in blue. The paravertebral compartment is shown in yellow. The horizontal green line represents the visceral-paravertebral compartment boundary line that separates the visceral and paravertebral compartments.* **(Bottom)** *Sagittal CECT (mediastinal window) illustrates the ITMIG mediastinal compartments. The prevascular compartment is shown in magenta, the visceral compartment is shown in blue, and the paravertebral compartment is shown in yellow. Note the vertically oriented green visceral-paravertebral compartment boundary line. The ITMIG mediastinal compartments are CT-based mediastinal divisions established based on an expert international multidisciplinary consensus.*

APPROACH TO IMAGING DIAGNOSIS OF MEDIASTINAL ABNORMALITIES

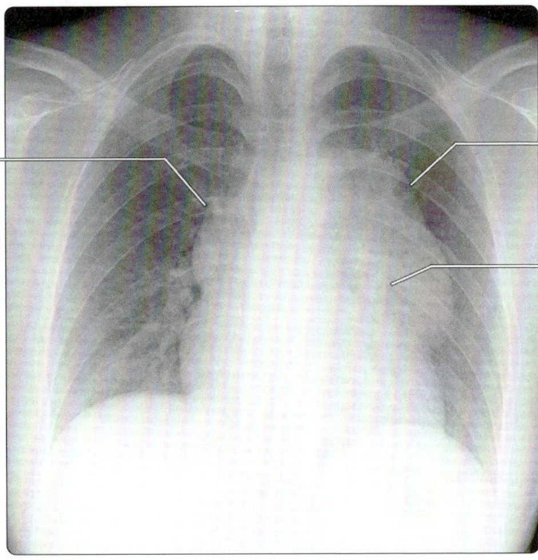

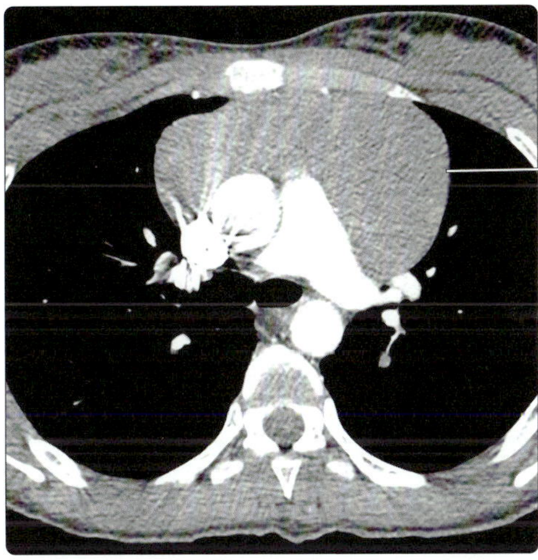

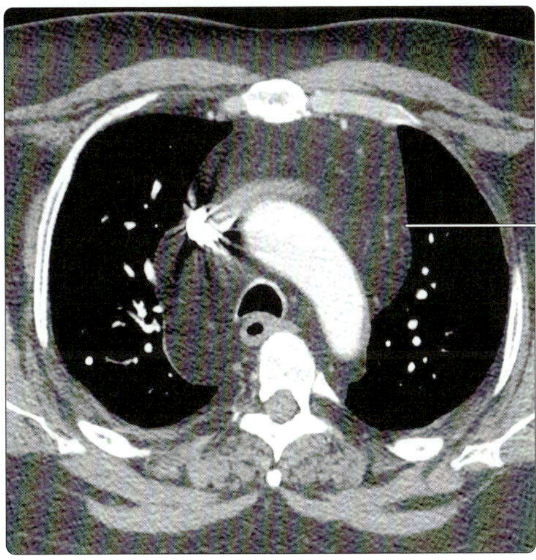

(Top) *AP chest radiograph of a 27-year-old woman with constitutional symptoms shows diffuse left greater than right mediastinal enlargement with lobulated contours. Although a lateral chest radiograph was not obtained, the presence of the hilum overlay sign is consistent with an anterior mediastinal location of the large mass suggestive of mediastinal lymphadenopathy. Based on morphology and location of the mass and the patient's demographic and clinical presentation, the diagnosis of mediastinal lymphoma is favored.* **(Middle)** *Axial CECT (mediastinal window) of the same patient shows the large homogeneous prevascular mediastinal mass that represented nodal coalescence. Mediastinal biopsy confirmed the diagnosis of Hodgkin lymphoma.* **(Bottom)** *Axial CECT (mediastinal window) of an asymptomatic 71-year-old man with diffuse mediastinal enlargement on chest radiography (not shown) shows that the diffuse mediastinal abnormality is secondary to diffuse unencapsulated mediastinal fat, consistent with and diagnostic of mediastinal lipomatosis.*

GENERAL ANATOMY AND FUNCTION

Anatomy

- Cardiac chambers
  - Right atrium, right ventricle, left atrium, left ventricle
- Cardiac valves
  - Atrioventricular (AV) valves: Tricuspid, mitral
  - Semilunar valves: Pulmonic (or pulmonary), aortic
- Cardiac structure
  - **Epicardium**: Serous visceral pericardium
  - **Myocardium**: Specialized cardiac muscle that forms atrial and ventricular walls
  - **Endocardium**: Thin layer of cells, lines inner cardiac chambers, participates in cardiac contraction

Function

- Pump action
  - Delivery of deoxygenated blood to alveolar-capillary interface
  - Delivery of oxygenated blood to tissues
- Cardiac conduction system
  - **Sinoatrial (SA) node**
    - Cardiac pacemaker; atrial contraction
    - Right atrium: Superior end of crista terminalis at superior vena cava orifice
  - **AV node**
    - Right atrium: Interatrial septum, near coronary sinus orifice, near tricuspid valve septal cusp attachment
    - Receives impulse generated at SA node, propagates it to ventricles, produces ventricular contraction
  - **AV bundle**
    - Continuation of AV node along interventricular septum; divides into right and left bundle branches
    - **Right bundle branch**
      - Continues along right ventricular side of interventricular septum to reach subendocardial Purkinje fibers
    - **Left bundle branch**
      - Continues along left ventricular side of interventricular septum to apex to reach subendocardial Purkinje fibers

CARDIAC SHAPE AND ORIENTATION

Shape

- Pyramidal

Orientation

- Position analogous to pyramid on its side
- Apex oriented anteriorly, inferiorly, and to the left

CARDIAC SURFACES

Posterior Surface (Base)

- Quadrilateral shape
- Faces posteriorly
- Structures: Left atrium, small portion of right atrium, central great veins

Apex

- Faces anteriorly, inferiorly, and to the left
- Structures: Inferolateral left ventricle

Anterior Surface

- Faces anteriorly
- Structures: Right ventricle (anterior free wall), portions of right atrium and left ventricle

Diaphragmatic Surface

- Faces inferiorly
- Rests on diaphragm
- Structures: Left ventricle, portion of right ventricle

Right Pulmonary Surface

- Faces right lung
- Structures: Right atrium

Left Pulmonary Surface

- Faces left lung
- Structures: Left ventricle, portion of left atrium

CARDIAC BORDERS (MARGINS)

Upper Border

- Not well appreciated on imaging

Inferior (Acute) Border

- Edge between anterior and diaphragmatic surfaces
- Structures: Right ventricle, portion of left ventricle

Obtuse (Left) Border

- Curved morphology
- Between anterior and left pulmonary surfaces
- From left atrial appendage to apex
- Structures: Left ventricle, portion of left atrial appendage

Right Border

- Right atrium; between anterior and right pulmonary surfaces

CARDIAC SULCI OR GROOVES

General Features

- Heart divided into chambers
- Internal cardiac partitions demarcate chamber boundaries
- Sulci: External grooves related to internal partitions

Atrioventricular (Coronary) Sulcus or Groove

- Surrounds heart, separates atria from ventricles
- Structures: Right coronary/circumflex branch of left coronary artery, small cardiac vein, great cardiac vein, coronary sinus

Anterior and Posterior Interventricular Sulci or Grooves

- Separate the ventricles
- **Anterior interventricular sulcus or groove**
  - Anterior heart surface
  - Structures: Left anterior descending (LAD) coronary artery, anterior interventricular vein
- **Posterior interventricular sulcus or groove**
  - Diaphragmatic heart surface
  - Structures: Posterior descending (interventricular) coronary artery, middle cardiac vein (posterior interventricular vein), ± distal wraparound LAD
- Anterior and posterior interventricular sulci continue inferiorly to right of apex

CARDIAC CHAMBERS

Right Atrium

- Forms right cardiac surface and portion of anterior surface
- Receives deoxygenated blood from
  - Superior vena cava: Superior posterior right atrium
  - Inferior vena cava: Inferior posterior right atrium
  - Coronary sinus: Inferior posterior right atrium
    - Thebesian valve: Prevents reflux of blood back into coronary sinus from right atrium
  - Thebesian veins: Small veins that drain subendocardium
  - Anterior cardiac vein: Venous return from anterior right ventricle
- Blood exits through AV tricuspid valve
- Structures
  - Compartmentalized by external sulcus terminalis cordis
    - From right of superior vena cava to right of inferior vena cava
  - Compartmentalized by internal crista terminalis
    - Smooth ridge that begins at atrial roof anterior to superior vena cava orifice and extends to anterior lip of inferior vena cava orifice
  - Sinus of venae cavae, posterior to **crista terminalis**
  - Right atrium proper
    - Anterior to crista terminalis; wall covered by ridge-like pectinate muscles
    - Right atrial appendage (auricle): Broad opening, triangular shape
  - Vascular orifices
    - Orifice of superior vena cava
    - Orifice and valve (eustachian valve or ridge) of inferior vena cava
    - Coronary sinus
  - Interatrial septum
    - Fossa ovalis: Depression in septum above orifice of inferior vena cava
    - Limbus of fossa ovalis: Margin of fossa ovalis
    - Fossa ovalis marks location of primitive foramen ovale, which allows oxygenated blood to enter left atrium and bypass lungs in utero

Right Ventricle

- Forms anterior cardiac surface and small portion of diaphragmatic surface
- Structures
  - Conus arteriosus: Smooth-walled right ventricular infundibulum or outflow tract
  - Right ventricular inflow tract lined by trabeculae carneae forming ridges and bridges
  - Papillary muscles: Trabeculae carneae attached to ventricular surface and chordae tendineae; connect chordae tendineae to free edges of tricuspid valve
    - Anterior papillary muscle: Largest, arises from anterior ventricular wall
    - Posterior papillary muscle: Some chordae tendineae arise directly from ventricular wall
    - Septal papillary muscle: Most inconsistent
  - Septomarginal trabecula or **moderator band**
    - Forms bridge between lower interventricular septum and base of anterior papillary muscle

Left Atrium

- Forms base or posterior cardiac surface
- Structures
  - Posterior or inflow portion: Smooth walled, receives pulmonary veins
  - Anterior or outflow portion: Continuous with left atrial appendage, lined by pectinate muscles
    - Left atrial appendage: Narrow opening; finger-like morphology
  - Interatrial septum
    - Contains valve of foramen ovale, which prevents blood from passing from left atrium to right atrium
    - Valve of foramen ovale may provide passage between atria during cardiac instrumentation

Left Ventricle

- Contributions to anterior, diaphragmatic, and left pulmonary cardiac surfaces; forms cardiac apex
- Thickest myocardium
- Structures
  - Fine, delicate trabeculae carneae
  - Papillary muscles larger than those in right ventricle
    - Anterior papillary muscle
    - Posterior papillary muscle
  - Interventricular septum
    - Thick, muscular portion forms major part of septum
    - Membranous portion

CARDIAC SKELETON

Anatomy

- Annulus fibrosus: 4 fibrous rings between atria and ventricles
  - Points of origin of atrial (superior) and ventricular (inferior) myocardium
  - Maintains integrity of orifices
  - Traversed by AV bundle
- Surrounds AV valves and aortic/pulmonary trunk orifices
- Interconnecting fibrous tissue
  - Right fibrous trigone: Between aortic and right AV rings
  - Left fibrous trigone: Between aortic and left AV rings

CARDIAC VALVES

Tricuspid Valve

- 3 leaflets (or cusps) attached to fibrous ring
  - Anterior, septal, posterior
- Anatomy of right AV valve
  - Free margins attached to chordae tendineae
    - Leaflets continuous with each other at their bases forming commissures
  - Chordae tendineae arise from 2 papillary muscles and attach to each leaflet
- Separated from pulmonic valve by muscular ridge (crista supraventricularis)

Pulmonic (Pulmonary) Valve

- 3 semilunar leaflets (or cusps)
  - Left, right, anterior
- Anatomy of semilunar leaflets (or cusps)
  - Free edges project into lumen of pulmonary trunk, forming sinuses

- Each cusp
  - Thick, central focus: Nodule of semilunar cusp
  - Thin, lateral portion: Lunule of semilunar cusp

Mitral Valve

- 2 leaflets attached to fibrous ring
  - Anterior and posterior
    - 3 subdivisions each (A1, A2, A3 and P1, P2, P3; 1 is superior)
- Anatomy of left AV valve
  - Cusps continuous with each other at commissures
  - Chordae tendineae attach papillary muscles to free borders of cusps
- In fibrous continuity with aortic valve
- Mitral annulus
  - Tissue juncture of left atrium, left ventricle, and mitral leaflets
  - D-shaped structure

Aortic Valve

- 3 semilunar cusps
  - Right, left, and noncoronary (or posterior)
- 3 sinuses of Valsalva
  - Right coronary artery originates from right sinus
  - Left coronary artery originates from left sinus
  - Noncoronary sinus always faces interatrial septum

IMAGING

Radiography

- Analysis of cardiac borders and surfaces
  - Right cardiac border: Right atrium
  - Left cardiac border: Left atrial appendage and left ventricle
  - Anterior cardiac surface: Right ventricle
  - Posterior cardiac surface: Left atrium and left ventricle
- Variations in cardiac morphology
  - Infancy: Prominent cardiothymic silhouette
  - Childhood, adolescence, and young adulthood: Prominent pulmonary trunk
  - Adulthood: Progressive left ventricular configuration with dominant left-sided structures and concavity of upper left cardiac border
- Analysis of cardiac size
  - **Cardiothoracic ratio**
    - Maximum transverse cardiac diameter:transverse thoracic diameter ratio
      - □ Normal cardiothoracic ratio ≤ **0.55**
    - Influenced by rotation, lung volume, projection
  - Analysis of individual chamber enlargement
- Analysis of abnormal cardiac density/calcification

CT/MR Anatomy

- **Right atrium**
  - Right atrial appendage (trabeculated) anterior and superior to right ventricle
  - Crista terminalis: Vertical ridge in right atrium extending from superior to inferior venae cavae
- **Left atrium**
  - Most superior and posterior cardiac chamber
  - Left atrial appendage (trabeculated) anterior and superior to left ventricle
  - Smooth muscle ridge at junction of left atrial appendage and central left superior pulmonary vein: Coumadin ridge
- **Interatrial septum**
  - Thin structure, difficult identification on CT
  - Increased visibility with fat deposition
    - Fat spares fossa ovalis and may allow its identification
- **Right ventricle**
  - Most anterior cardiac chamber, anterior heart surface
  - Heavy trabeculations, thin wall
    - **Moderator band**: Connects anterior papillary muscle to interventricular septum near right ventricular apex; contains right bundle branch
  - Anterior, posterior, and septal papillary muscles
- **Left ventricle**
  - Posterior and diaphragmatic cardiac surfaces
  - Thicker than right ventricle, less trabeculated
  - Anterior and posterior papillary muscles
- **Interventricular septum**
  - Thicker than interatrial septum
- **Valves**
  - Imaging in longitudinal and perpendicular planes
  - Thin, low-attenuation/-signal structures on CECT and bright-blood cardiac MR, respectively
  - Assessment of function, morphology, calcification

CT/MR

- Assessment of cardiac chambers, valves, myocardium
  - Size, morphology, wall thickness, calcification, function
- **Short-axis view**
  - Cross section through short axis of left ventricular cavity and bodies of papillary muscles
- **Paraseptal long-axis (2-chamber) view**
  - Displays left atrial and ventricular chambers; mitral valve evaluation
- **4-chamber view**
  - Displays 4 cardiac chambers and AV valves
- **Left ventricular outflow tract view**
  - Displays left atrium, left ventricle, left ventricular outflow tract, and aorta
- **Right ventricular outflow tract view**
  - Displays right atrium, right ventricle, right ventricular outflow tract, and pulmonary trunk

ANATOMY-BASED CARDIAC AND VALVULAR ABNORMALITIES

Cardiomegaly and Chamber Enlargement

- Increased cardiothoracic ratio
- Analysis of specific chamber enlargement

Myocardial Calcification

- Sequelae of myocardial infarction
- May be associated with ventricular aneurysm

Annular and Valvular Calcification

- Mitral annulus calcification: Involvement of mitral valve ring, C- or J-shaped morphology on radiography
- Valvular calcification
  - Aortic stenosis: Congenital (bicuspid or bicommissural valve), degenerative (fibrocalcific), rheumatic heart disease
  - Mitral stenosis: Rheumatic heart disease

OVERVIEW OF HEART

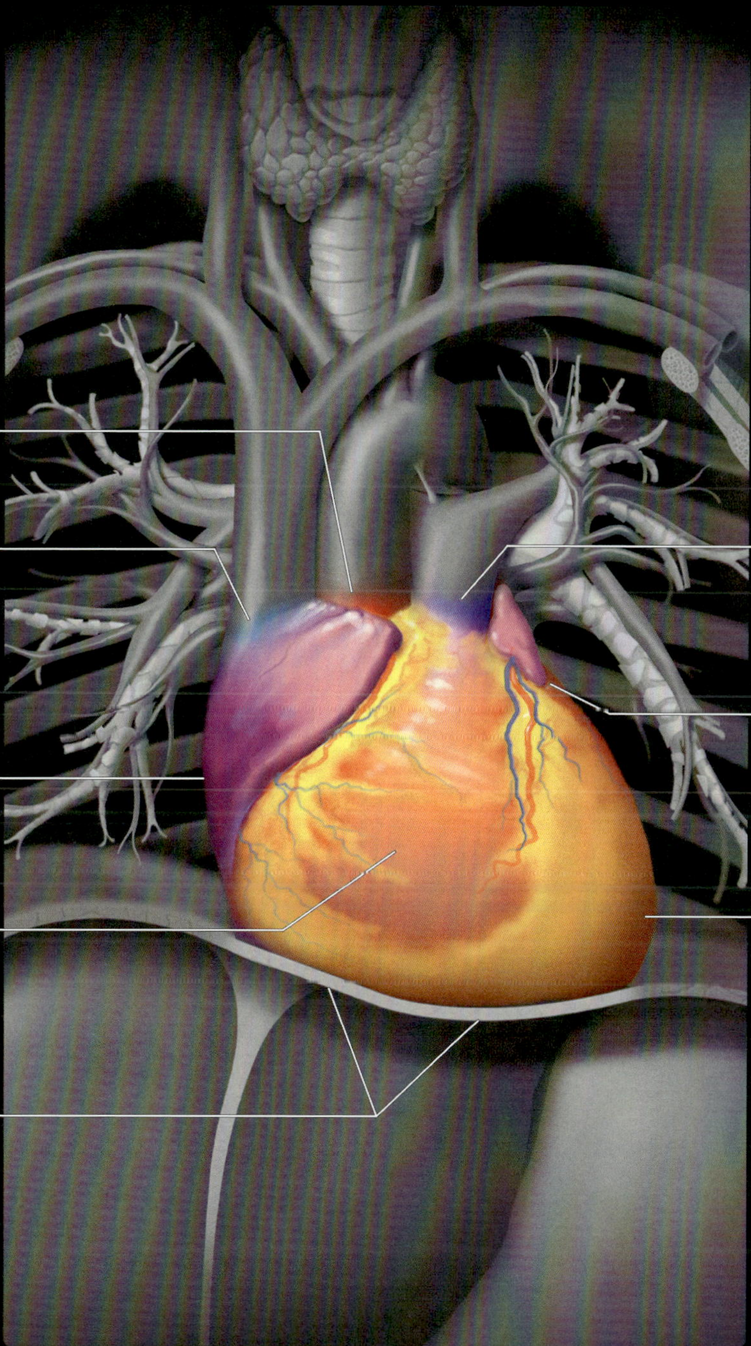

Graphic shows the location of the heart with respect to the rest of the thoracic structures and organs. The heart is surrounded by the pericardium (not shown) and has a pyramidal shape with its apex oriented inferiorly and toward the left side. The cardiac base faces posteriorly. The 4 heart chambers connect with the great pulmonary and systemic vessels. The heart receives deoxygenated blood from the systemic circulation via the vena cavae. The deoxygenated blood is then pumped into the pulmonary arteries for delivery to the capillary-alveolar interfaces. Oxygenated blood then returns to the heart via the pulmonary veins and is pumped into the systemic circulation via the aorta for delivery to the tissues and organs of the body.

ANATOMY OF HEART SURFACES, MARGINS, AND SULCI

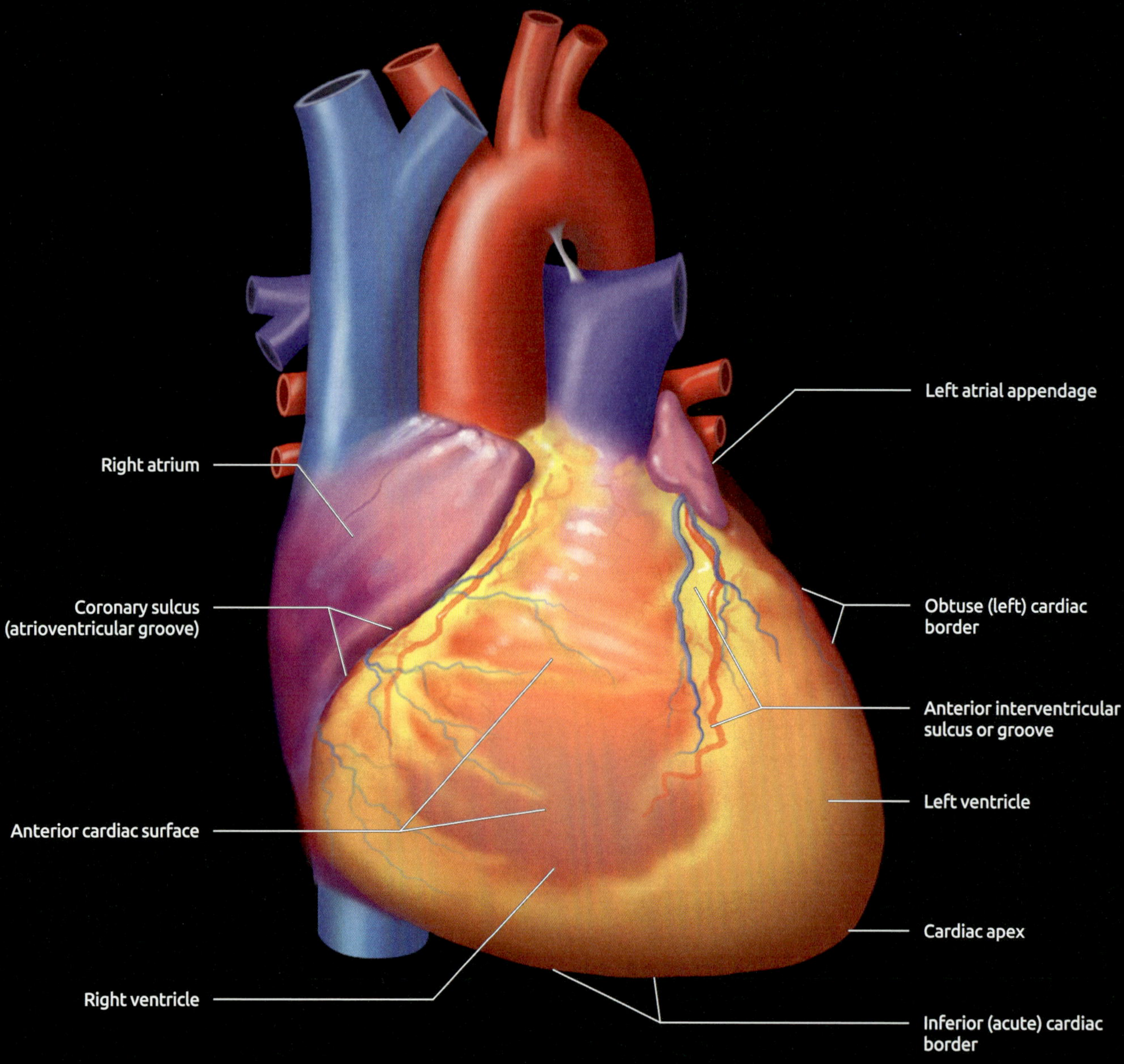

First of 3 graphics illustrating the surface anatomy of the heart is shown. The heart has a pyramidal shape with a base and an apex and has anterior, diaphragmatic, posterior, and right and left pulmonary surfaces. The anterior surface (shown here) is formed by the right ventricle with contributions from the right atrium and the left ventricle. The obtuse (left) cardiac border separates the anterior and left pulmonary cardiac surfaces. The inferior (acute) cardiac border separates the anterior and diaphragmatic cardiac surfaces. External sulci or grooves correspond to the internal partitions that divide the heart into chambers. There are anterior and posterior interventricular sulci or grooves and a coronary sulcus (or atrioventricular groove). The coronary sulcus is circumferential and separates the atria from the ventricles. The anterior and posterior interventricular sulci or grooves separate the ventricles.

ANATOMY OF HEART SURFACES, MARGINS, AND SULCI

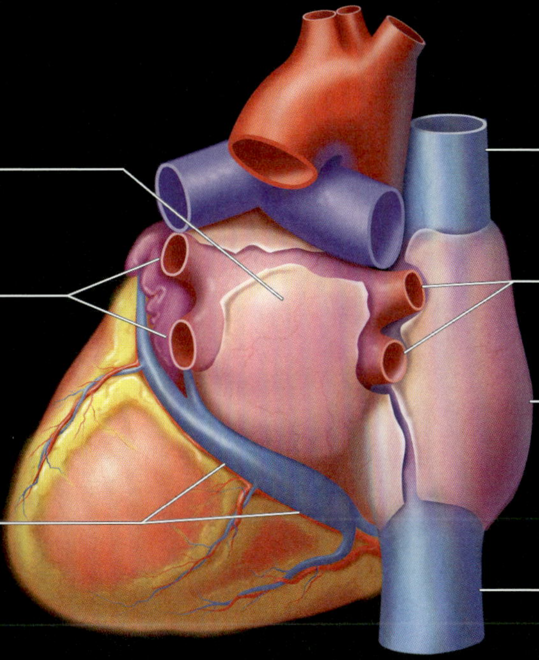

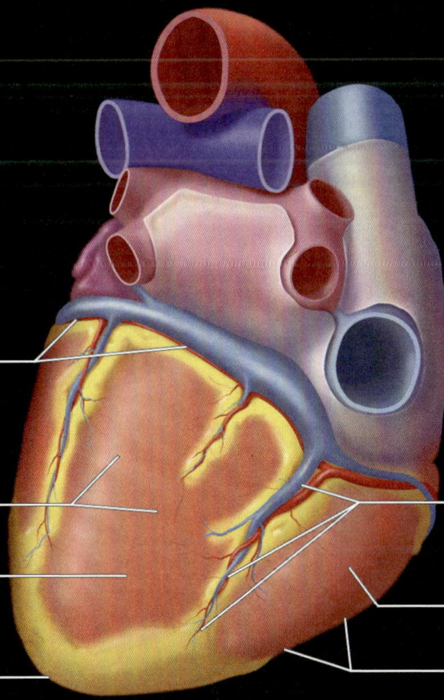

(Top) *Graphic shows the posterior heart surface, or base of the heart, which is formed by the left atrium, a small portion of the right atrium, the paired superior and inferior pulmonary veins, and the superior and inferior venae cavae, which fix the heart base to the pericardium. The atrioventricular groove (or coronary sulcus) is located at the junction of the left atrium and left ventricle and contains the coronary sinus.* **(Bottom)** *Graphic shows the diaphragmatic cardiac surface, which is formed by the right and left ventricles and is separated from the heart base by the coronary sulcus (or atrioventricular groove). The inferior (acute) cardiac border separates the diaphragmatic cardiac surface from the anterior cardiac surface. The posterior interventricular sulcus (or groove) corresponds to the location of the interventricular septum.*

ANATOMY OF ANTERIOR CARDIAC SURFACE AND RIGHT ATRIUM

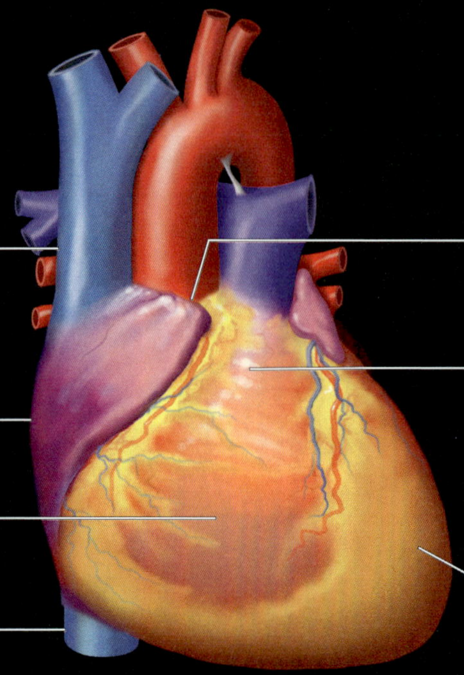

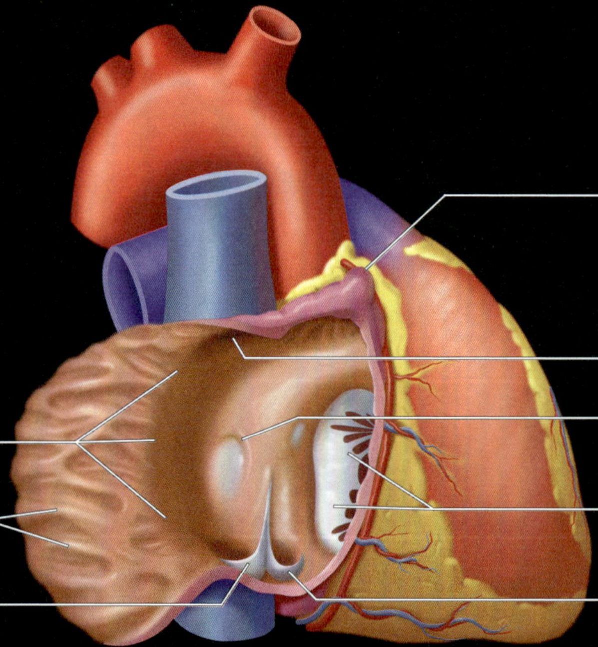

(Top) *First of 3 graphics demonstrating the anatomy of the anterior cardiac surface and the right cardiac chambers is shown. Graphic shows the anterior cardiac surface predominantly formed by the right ventricle and portions of the right atrium and left ventricle. The right atrial appendage also forms part of the anterior cardiac surface.* **(Bottom)** *Graphic shows the internal anatomy of the right atrium. Note the trabeculated portion of the right atrium formed by the pectinate muscles and separated from the smooth portion of the right atrial chamber by the crista terminalis. The posterior smooth portion of the right atrium receives the superior and inferior vena cavae (sinus of the venae cavae) as well as the coronary sinus. The anterior trabeculated portion is known as the right atrium proper. The interatrial septum (located in the smooth portion) contains the fossa ovalis, which marks the location of the embryonic foramen ovale. Note the so-called thebesian valve of the coronary sinus.*

ANATOMY OF RIGHT VENTRICLE

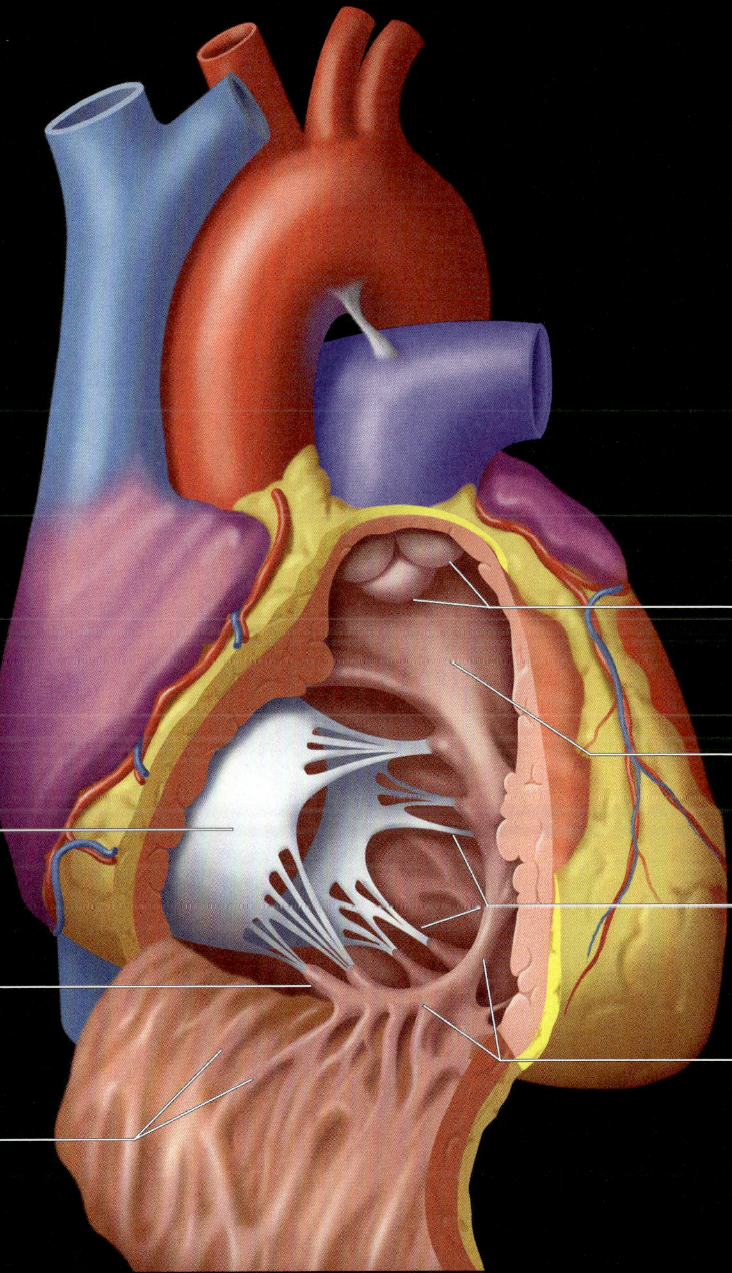

Graphic shows that the internal anatomy of the right ventricle is characterized by trabeculated and smooth areas. The proximal (inflow portion) right ventricle is characterized by the trabeculae carneae, some of which form papillary muscles. Papillary muscles attach to the free edges of the tricuspid (right atrioventricular) valve leaflets via chordae tendineae. There are anterior, septal, and posterior papillary muscles. The septal papillary muscle is not visualized in all subjects. A thick trabecula carnea connects the interventricular septum to the anterior papillary muscle and is known as the moderator band or septomarginal trabecula. Visualization of this structure allows identification of the morphologic right ventricle on cross-sectional imaging studies. The right ventricular outflow tract leads to the pulmonic (or pulmonary) valve and is characterized by its smooth walls.

RIGHT CARDIAC CHAMBERS, AXIAL CT

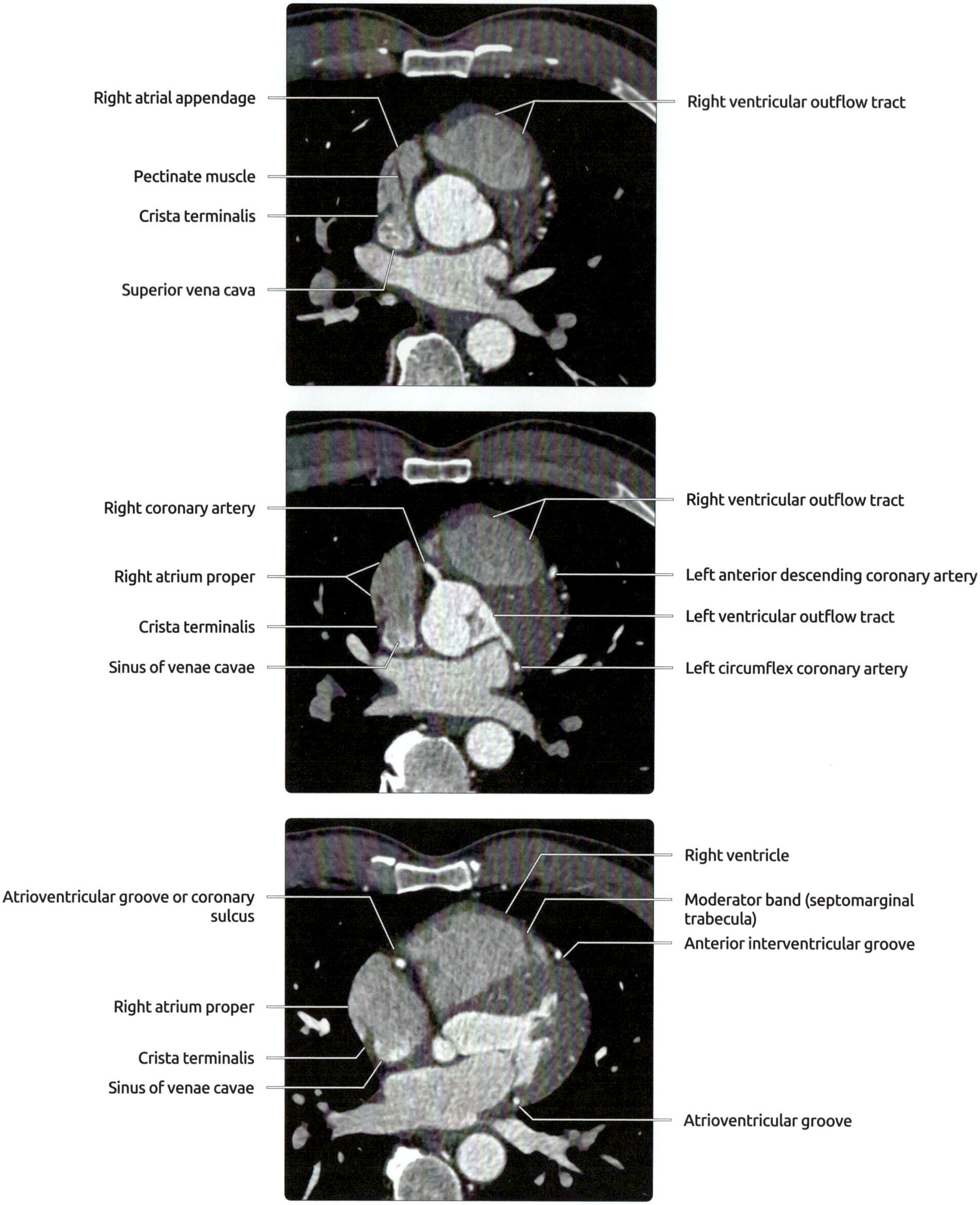

(Top) *First of 6 axial images from a normal gated cardiac CECT of the same patient shows the anatomy of the right cardiac chambers. Axial CECT (mediastinal window) shows the superior aspect of the right atrium and the anteriorly oriented right atrial appendage. Note its trabeculated wall and its junction with the superior vena cava.* **(Middle)** *Axial CECT (mediastinal window) shows the superior vena cava as it anastomoses with the right atrium. The right atrium is located anterior and to the right of the left atrium and the ascending thoracic aorta. The right ventricle is located anterior and to the left of the ascending thoracic aorta. Note coronary arteries coursing within the coronary or atrioventricular grooves, which separate the atria from the ventricles.* **(Bottom)** *Axial CECT (mediastinal window) demonstrates the superior aspects of the right atrium and right ventricle, which are separated by the atrioventricular groove. The crista terminalis courses between the orifices of the venae cavae and separates the right atrium proper from the sinus of the venae cavae.*

RIGHT CARDIAC CHAMBERS, AXIAL CT

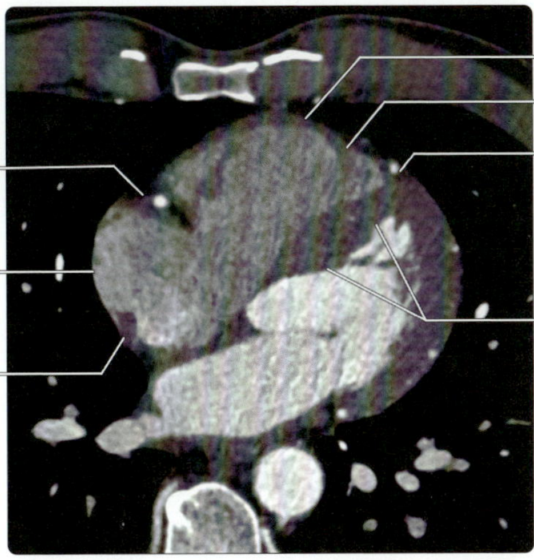

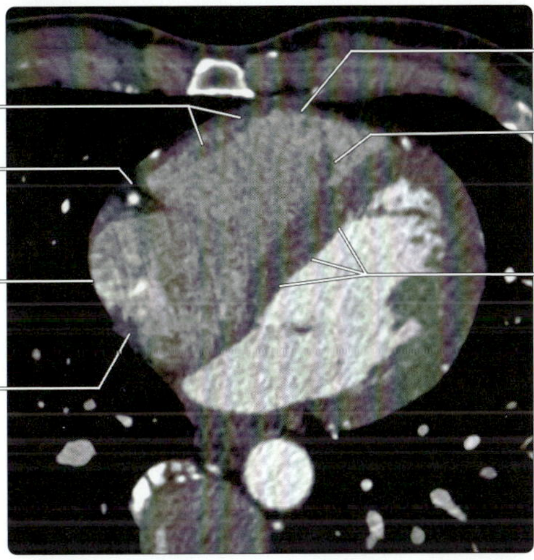

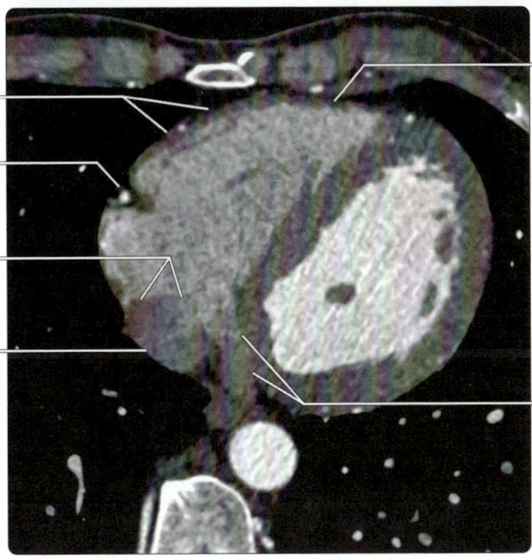

(Top) *Axial CECT (mediastinal window) shows the midportions of the right atrium and right ventricle separated by the atrioventricular or tricuspid valve (poorly visualized on this image) and the atrioventricular groove or coronary sulcus. The crista terminalis is located on the posterolateral wall of the right atrium.* **(Middle)** *Axial CECT (mediastinal window) shows the trabeculae carneae that characterize the internal surface of the free wall of the right ventricle. The septomarginal trabecula (or moderator band) courses from the inferior aspect of the interventricular septum to the base of the anterior papillary muscle. The atrioventricular groove (or coronary sulcus) contains the right coronary artery.* **(Bottom)** *Axial CECT (mediastinal window) shows the junctions of the inferior vena cava and coronary sinus with the right atrium. The atrioventricular groove contains the right coronary artery. Note the fine trabeculations of the anterior right ventricular myocardium produced by the trabeculae carneae. The right ventricular myocardium is very thin when compared with that of the left ventricle.*

ANATOMY OF POSTERIOR CARDIAC SURFACE, CARDIAC BASE, AND LEFT ATRIUM

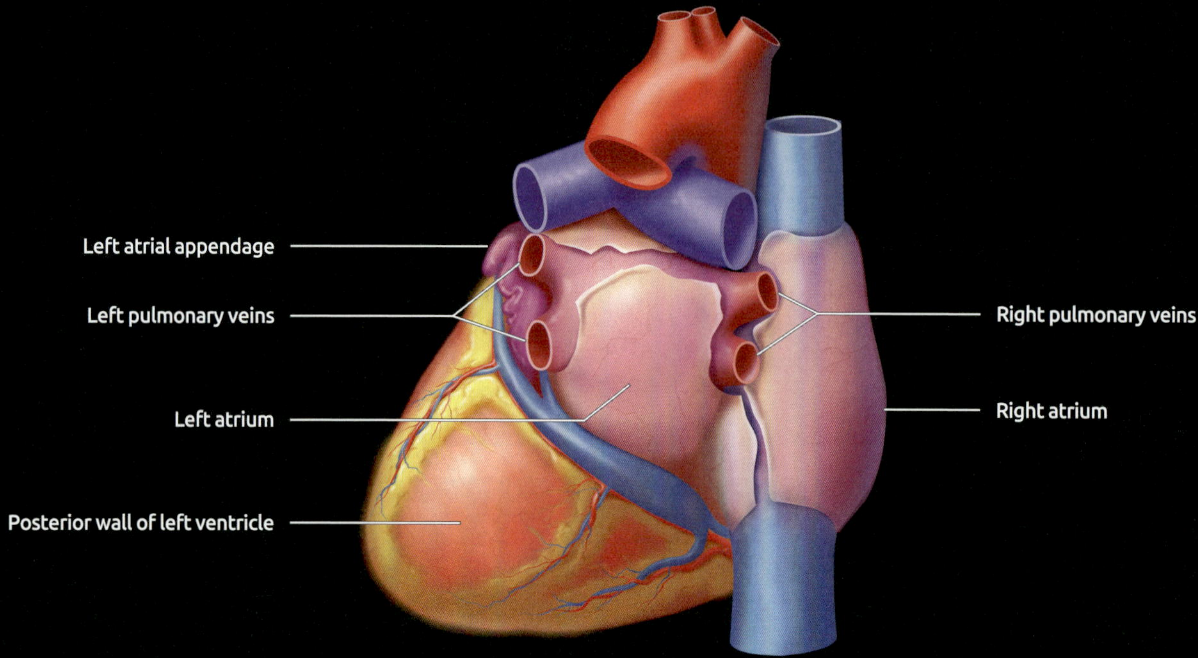

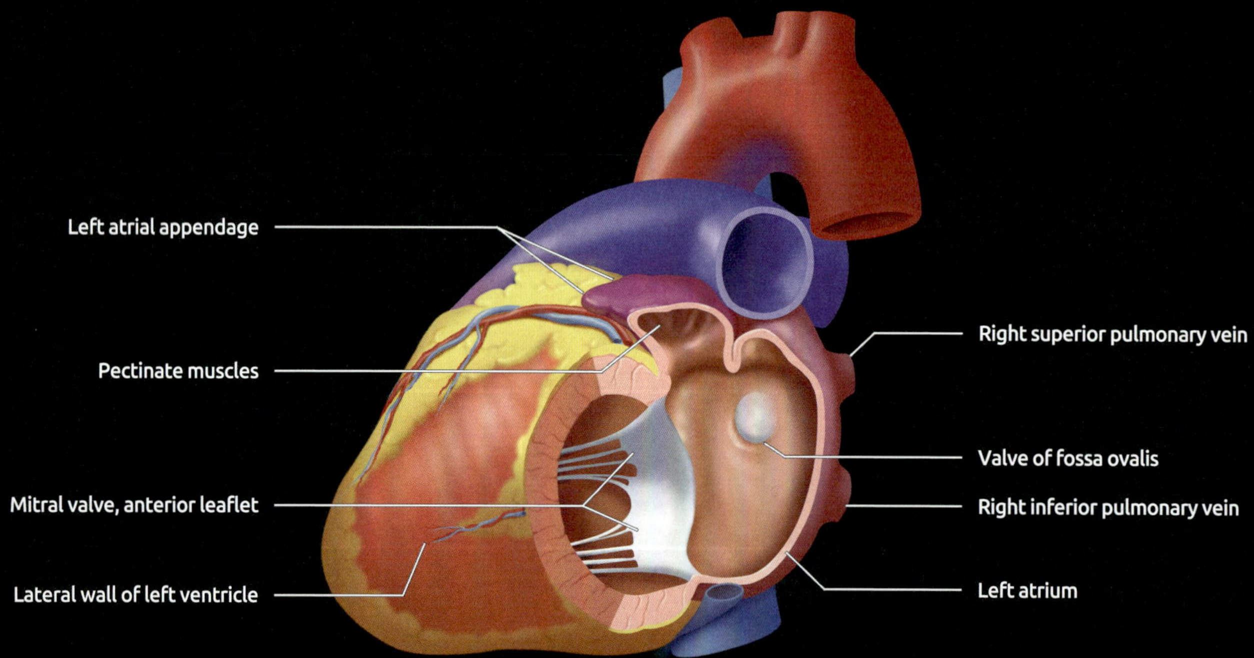

(Top) *First of 3 graphics illustrating the anatomy of the posterior cardiac surface and the left cardiac chambers is shown. Posterior view of the heart shows that the cardiac base is formed primarily by the left atrium and a small portion of the right atrium. The left atrium receives the paired superior and inferior pulmonary veins. The posterior wall of the left ventricle forms most of the diaphragmatic cardiac surface.* **(Bottom)** *Graphic shows the internal anatomy of the left atrium. The left atrium has smooth and trabeculated inner surfaces. The left atrial appendage is characterized by its trabeculated inner surface produced by the pectinate muscles. The interatrial septum is located in the smooth portion of the left atrium. The valve of the fossa ovalis is located on the interatrial septum and prevents passage of blood between the atria.*

ANATOMY OF LEFT VENTRICLE

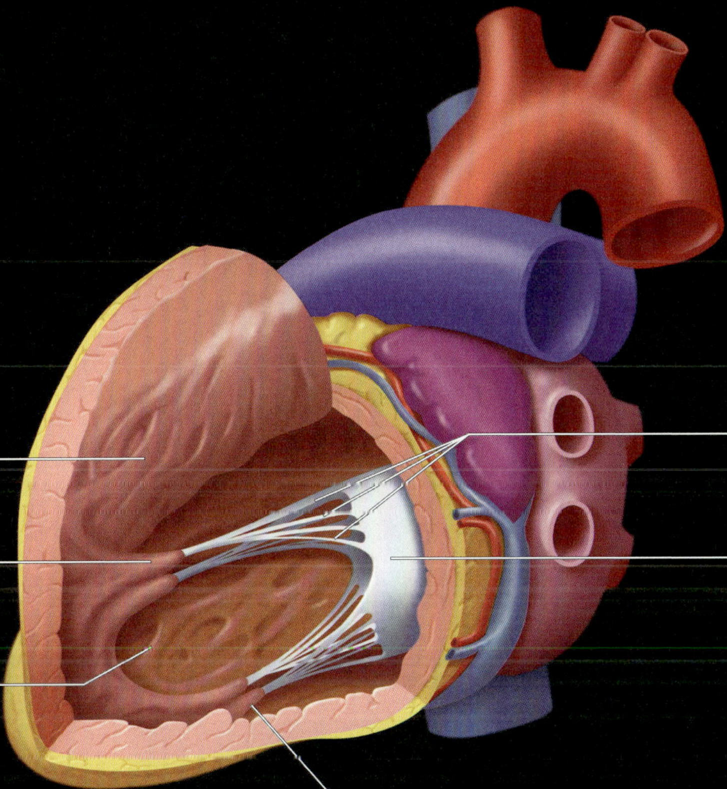

Graphic shows the internal anatomy of the left ventricle. The left ventricle has a thicker myocardium compared to that of the right ventricle but thinner and more delicate trabeculae carneae when compared to those of the right ventricle. The anterior and posterior papillary muscles attach to the anterior and posterior leaflets of the mitral valve, respectively.

LEFT CARDIAC CHAMBERS, AXIAL CT

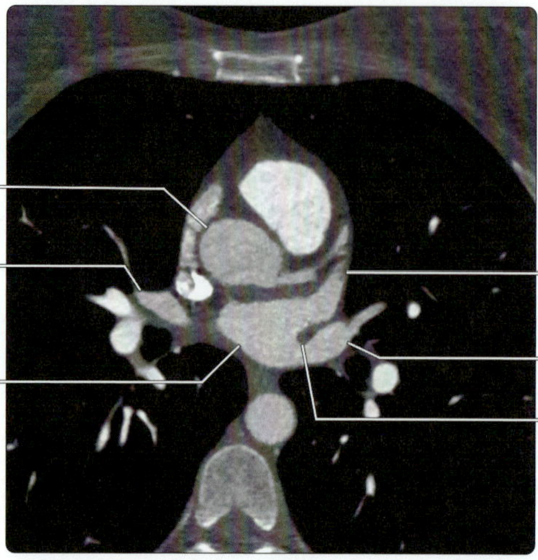

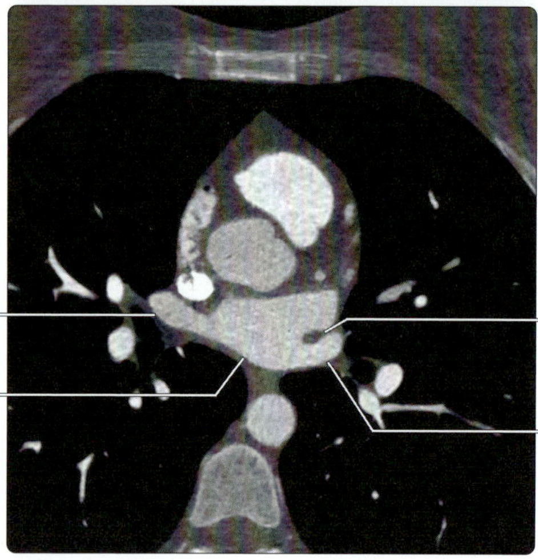

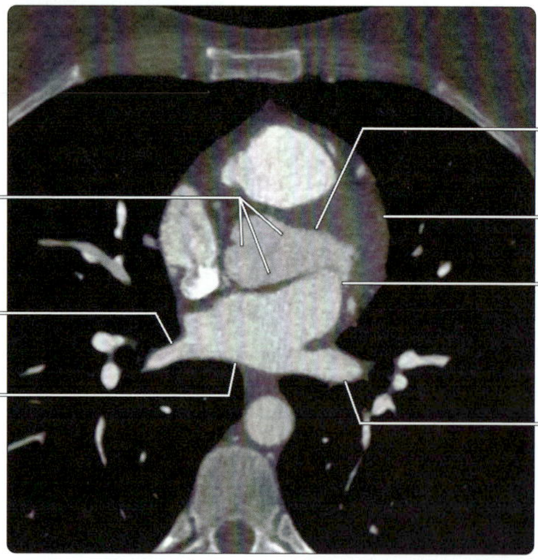

(Top) *First of 6 gated cardiac CECT images of the same patient illustrating the anatomy of the left cardiac chambers is shown. Axial CECT (mediastinal window) demonstrates the relationship between the left atrial appendage and left superior pulmonary vein (always located posterior to the former). A smooth or nodular (Coumadin) ridge is located at the junction of these 2 structures. Note the trabeculated inner wall of the left atrial appendage formed by the pectinate muscles.* **(Middle)** *Axial CECT (mediastinal window) shows the smooth internal wall of the left atrium proper in contrast to the trabeculated wall of the left atrial appendage. Note the Coumadin ridge located between the left superior pulmonary vein and the left atrial appendage. This structure may mimic thrombus on transesophageal echocardiography, hence its name.* **(Bottom)** *Axial CECT (mediastinal window) shows the normal thick left ventricular wall and the smooth wall of the left ventricular outflow tract. The inferior pulmonary veins are seen draining into the left atrium.*

LEFT CARDIAC CHAMBERS, AXIAL CT

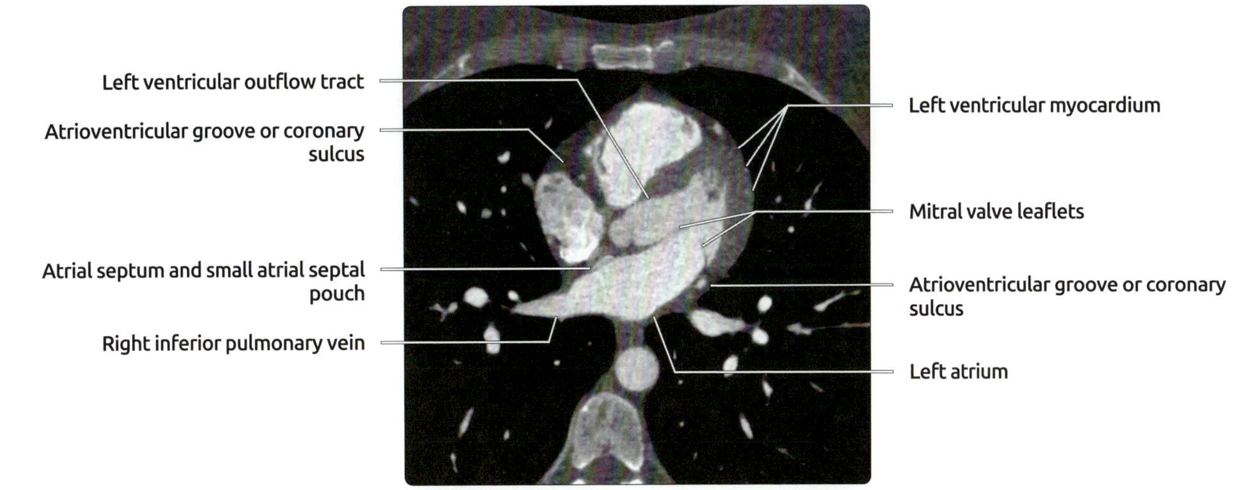

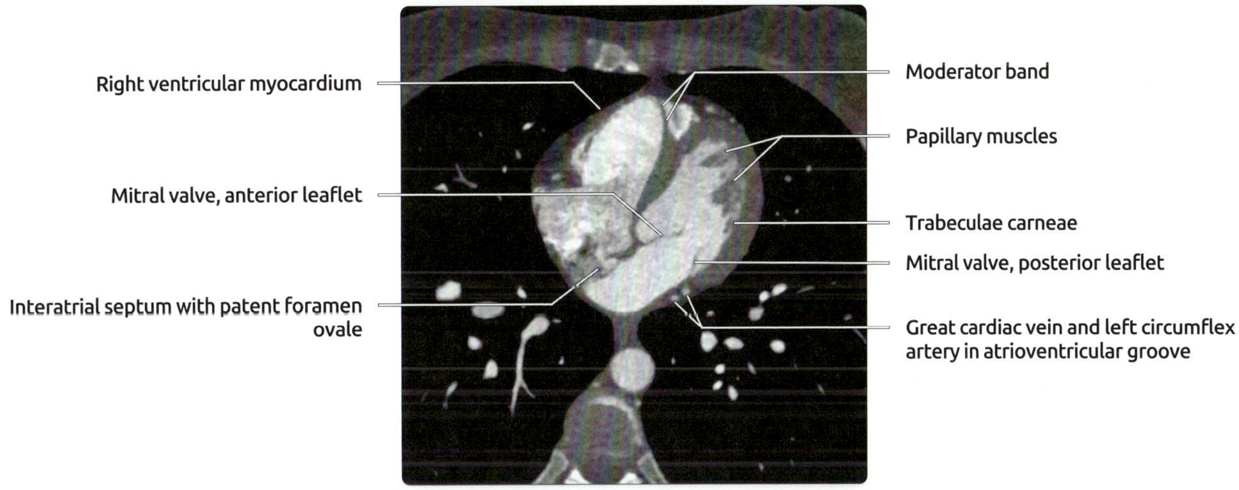

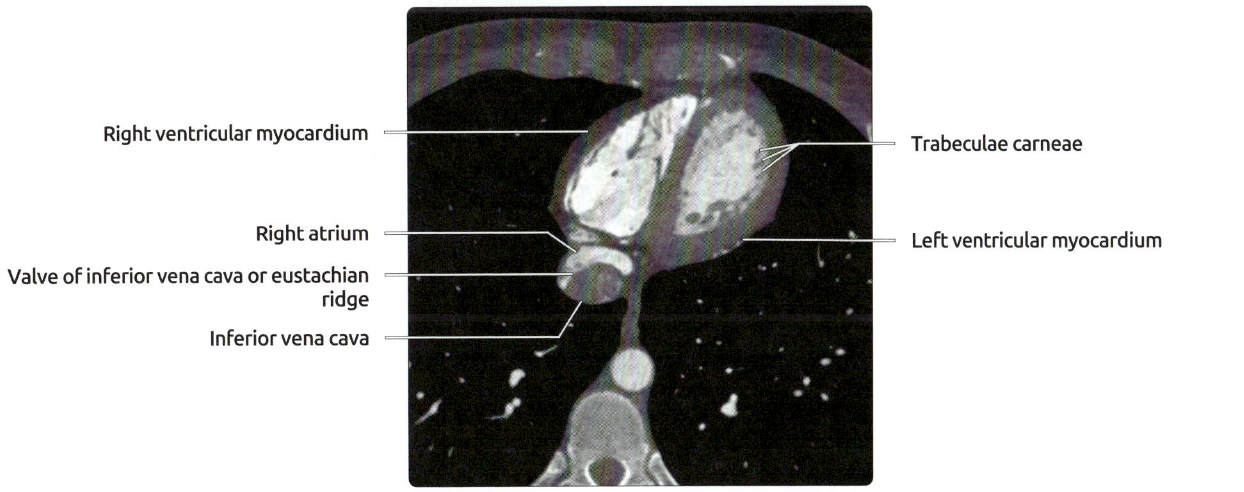

(Top) *Axial CECT (mediastinal window) demonstrates the normal thick left ventricular wall and its irregular inner wall caused by the trabeculae carneae in contrast to the smooth wall of the left ventricular outflow tract. The mitral valve leaflets are thin and open during ventricular diastole. The atrioventricular groove or coronary sulcus separates the atria from the ventricles.* **(Middle)** *Axial CECT (mediastinal window) shows the trabeculated inner wall of the left ventricle secondary to the trabeculae carneae and the papillary muscles. Note the thick left ventricular myocardium in contrast to the relatively thin right ventricular myocardium. The thin anterior and posterior mitral valve leaflets are open during ventricular diastole.* **(Bottom)** *Axial CECT (mediastinal window) shows the inferior left ventricle and its trabeculated inner wall formed by the trabeculae carneae. The inner wall of the right ventricular myocardium is also trabeculated. Note the valve of the inferior vena cava, also called the eustachian ridge or valve.*

HEART, CORONAL CT

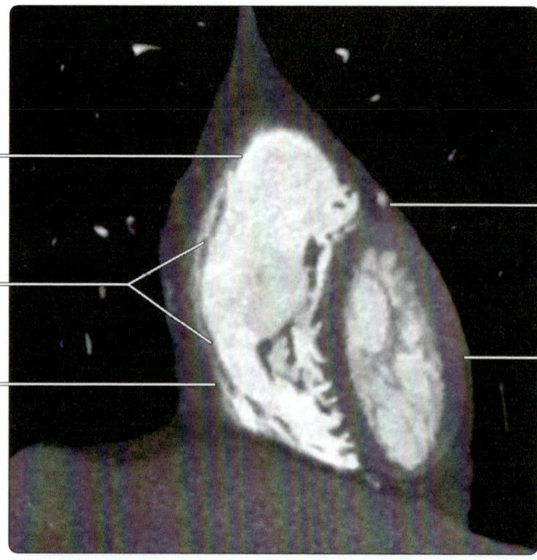

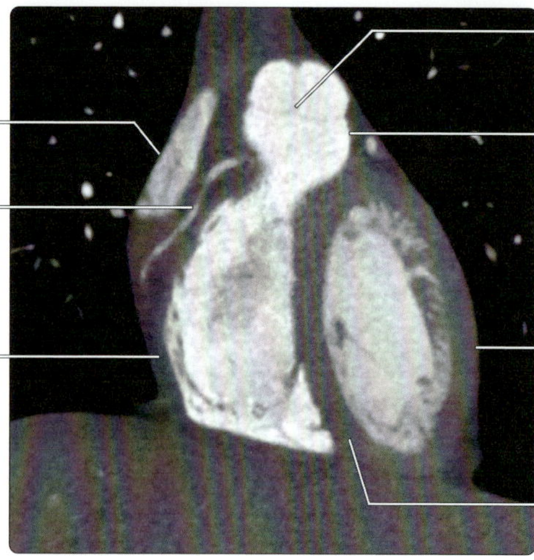

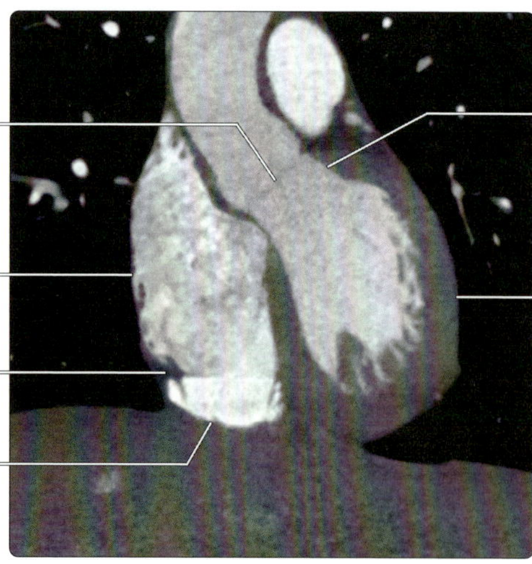

(Top) *First of 6 coronal CECT images of the same patient demonstrating the coronal anatomy of the heart, presented from anterior to posterior, is shown. Coronal CECT (mediastinal window) shows the anteriorly located trabeculated right ventricle leading to the superiorly located and relatively smooth right ventricular outflow tract. The left ventricle forms the left cardiac border.* **(Middle)** *Coronal CECT (mediastinal window) shows the right and left ventricles separated by the interventricular septum. The left ventricle forms the left heart border and has a thicker wall than that of the right ventricle. Note the location of the right atrial appendage. The right ventricle is anterior and to the left of the right atrium.* **(Bottom)** *Coronal CECT (mediastinal window) shows the smooth internal wall of the left ventricular outflow tract compared with the trabeculated inner surface of left ventricle proper. The atrioventricular groove or coronary sulcus separates the right atrium from the right ventricle. The right atrium forms the right heart border on radiography.*

HEART, CORONAL CT

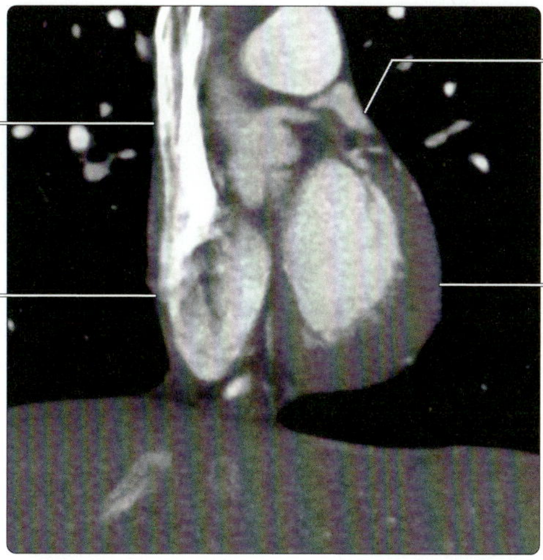

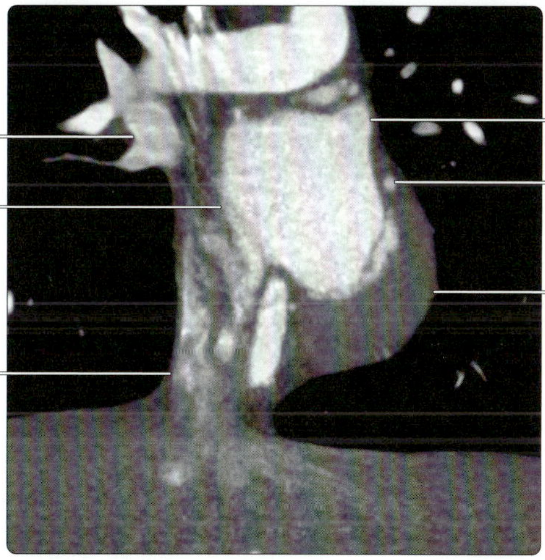

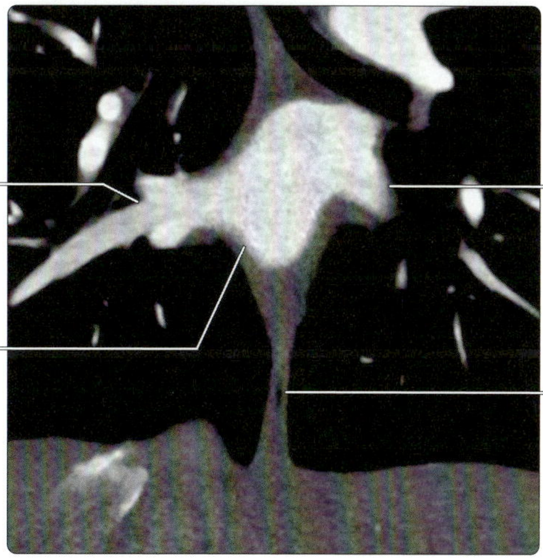

(Top) *Coronal CECT (mediastinal window) through the posterior aspect of the right atrium shows its junction with the superior vena cava. The left atrial appendage is superior to the left ventricle and forms a portion of the obtuse or left cardiac border.* **(Middle)** *Coronal CECT (mediastinal window) through the midportion of the left atrium shows its junction with the left superior pulmonary vein. The junction of the left atrium with the right superior pulmonary vein is not visible in this plane of section.* **(Bottom)** *Coronal CECT (mediastinal window) through the posterior aspect of the left atrium shows the bilateral inferior pulmonary veins coursing obliquely to drain into the left atrium.*

HEART, SAGITTAL CT

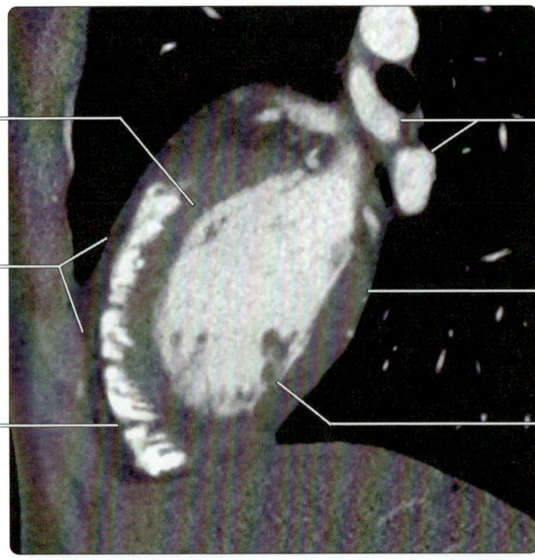

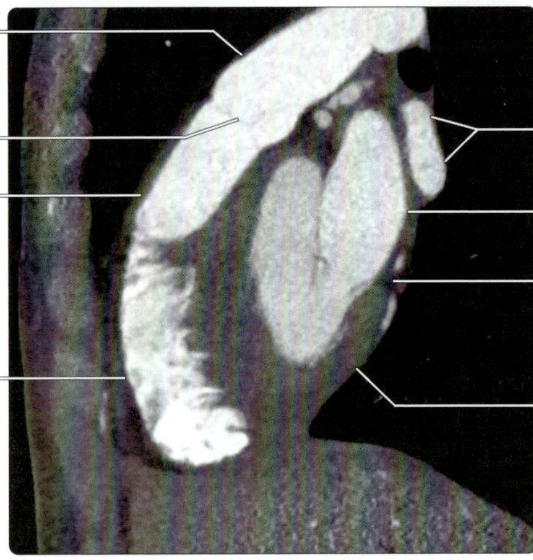

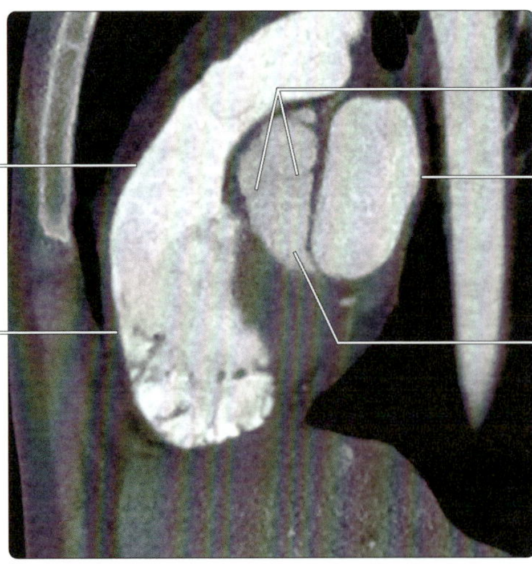

(Top) *First of 6 CECT images of the same patient demonstrating the sagittal anatomy of the heart, presented from left to right, is shown. Sagittal CECT (mediastinal window) shows that the left ventricular myocardium is thicker than the right ventricular myocardium, and the left ventricle exhibits well-defined papillary muscles. The right ventricle forms the anterior cardiac surface.* **(Middle)** *Sagittal CECT (mediastinal window) through the right ventricular outflow tract shows the anterior location of the right ventricle and the posterosuperior course of the right ventricular outflow tract and pulmonary trunk. The left ventricle is located posterior to the right ventricle. The left atrium is located superior to the left ventricle and is separated from it by the atrioventricular groove or coronary sulcus.* **(Bottom)** *Sagittal CECT (mediastinal window) through the midheart shows the central location of the left ventricular outflow tract and the aortic root.*

HEART, SAGITTAL CT

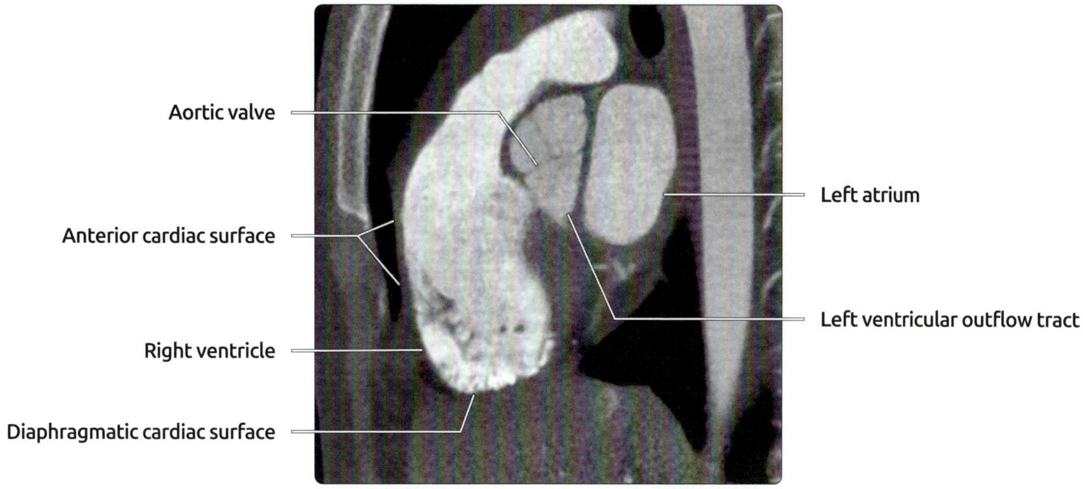

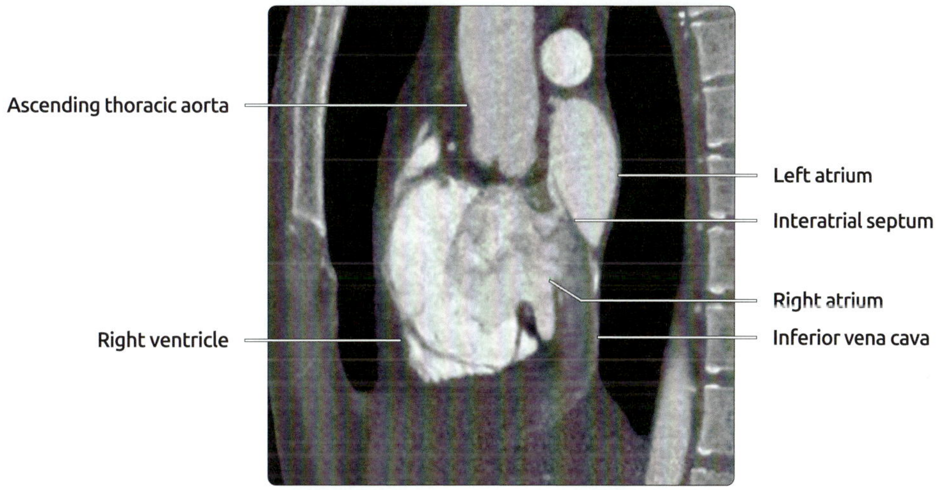

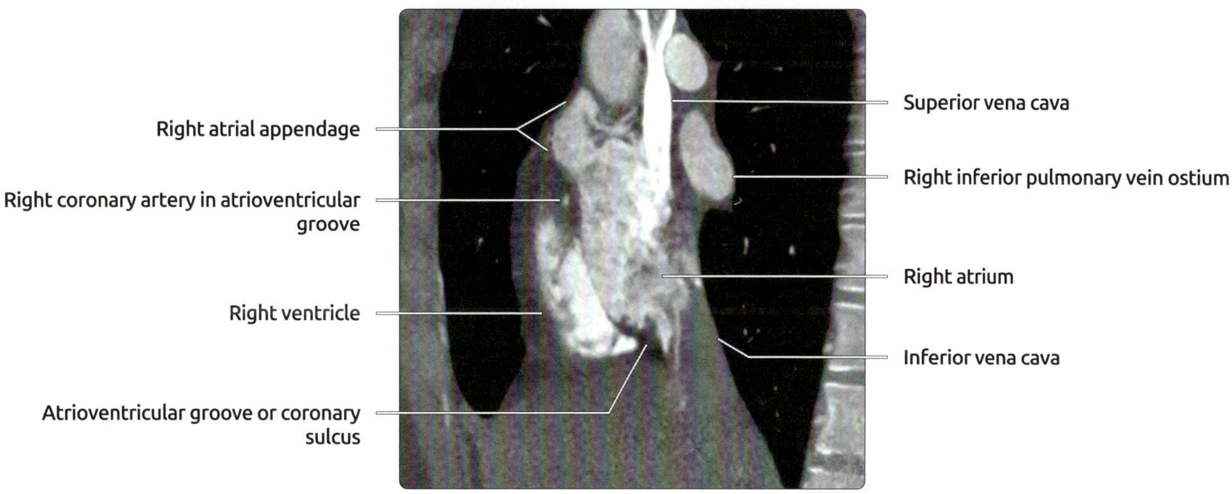

(Top) *Sagittal CECT (mediastinal window) shows the anatomic location of the right ventricle, which forms the anterior cardiac surface and also contributes to the diaphragmatic cardiac surface. The left atrium forms the posterior cardiac surface or base of the heart.* **(Middle)** *Sagittal CECT (mediastinal window) through the right heart chambers shows the posterior location of the right atrium with respect to the right ventricle and its connection with the inferior vena cava.* **(Bottom)** *Sagittal CECT (mediastinal window) through the right side of the heart shows the connection of the venae cavae with the right atrium. The right ventricle is located anteriorly and separated from the right atrium by the atrioventricular groove or coronary sulcus.*

RIGHT ATRIUM, CT

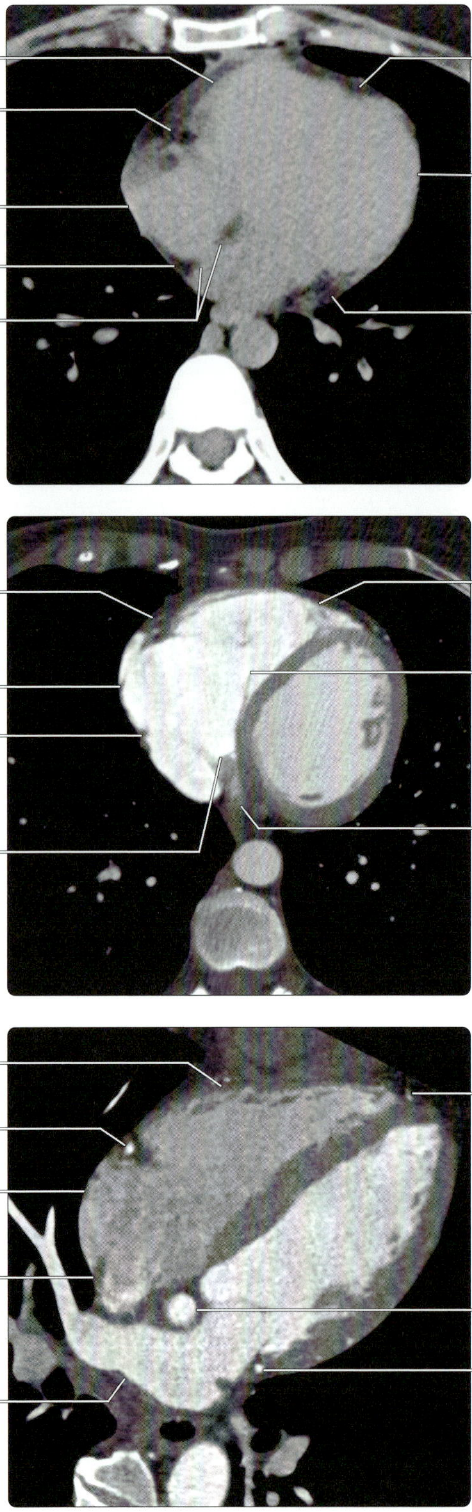

(Top) *Axial NECT (mediastinal window) demonstrates fat in the interatrial septum that spares the fossa ovalis and allows its identification and localization in the axial plane. This appearance may mimic an atrial septal defect.* **(Middle)** *Axial CECT (mediastinal window) shows a dilated right atrium. Note unopacified blood in the coronary sinus. The thebesian valve prevents reflux of blood from the right atrium back into the coronary sinus.* **(Bottom)** *Four-chamber view from a cardiac CECT (mediastinal window) shows the location of the crista terminalis, which must be differentiated from intraatrial thrombus. The crista terminalis is a smooth ridge of tissue that begins at the roof of the right atrium anterior to the superior vena cava orifice and extends inferiorly to the anterior lip of the inferior vena cava orifice.*

LEFT ATRIUM, CT

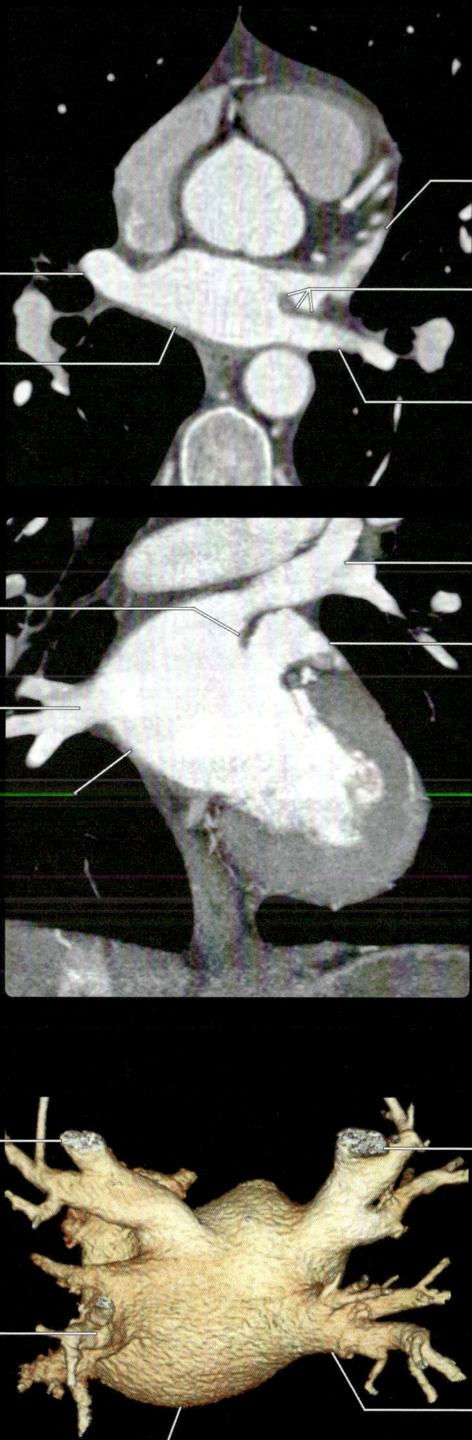

(Top) *Axial CECT (mediastinal window) shows the nodular appearance of the ridge at the junction of the left atrium and the left superior pulmonary vein adjacent to the left atrial appendage (also known as the Coumadin ridge, as it may be mistaken for thrombus at transesophageal echocardiography). Note the constant relationship between the left superior pulmonary vein and the adjacent left atrial appendage. The left atrial appendage is always anterior and inferior to the left superior pulmonary vein.* **(Middle)** *Coronal CECT (mediastinal window) through the left atrial appendage demonstrates the normal soft tissue (Coumadin) ridge that occurs at the junction of the left atrium with the left superior pulmonary vein adjacent to the left atrial appendage. Note that the left atrial appendage is always inferior to the left superior pulmonary vein.* **(Bottom)** *Posterior 3D volume-rendered CT angiogram shows the normal appearance of the pulmonary veins and their relationship to the left atrium.*

ANATOMY OF CARDIAC SKELETON AND HEART VALVES

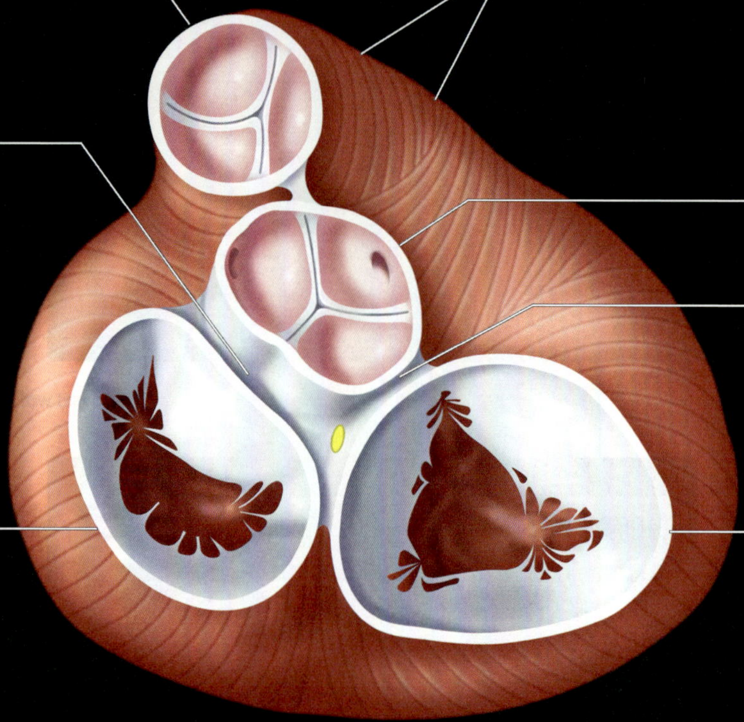

Graphic demonstrates the anatomy of the cardiac skeleton located between the atria and ventricles. The cardiac skeleton consists of thick, fibrous connective tissue and provides support for the valve orifices and the areas of attachment for the valve leaflets. In this graphic, the atria have been "removed" to expose the cardiac skeleton, and the heart valves are seen from above. The 4 fibrous rings that surround the valves are known as the annulus fibrosus. The right fibrous trigone is the connective tissue bridge between the aortic valve and the right atrioventricular (tricuspid) valve rings. The left fibrous trigone is the connective tissue bridge between the aortic valve and the left atrioventricular (mitral) valve rings. The yellow dot represents the atrioventricular bundle seen in cross section as it courses caudally from the atria to the ventricles.

PROSTHETIC AORTIC AND MITRAL VALVES, RADIOGRAPHY

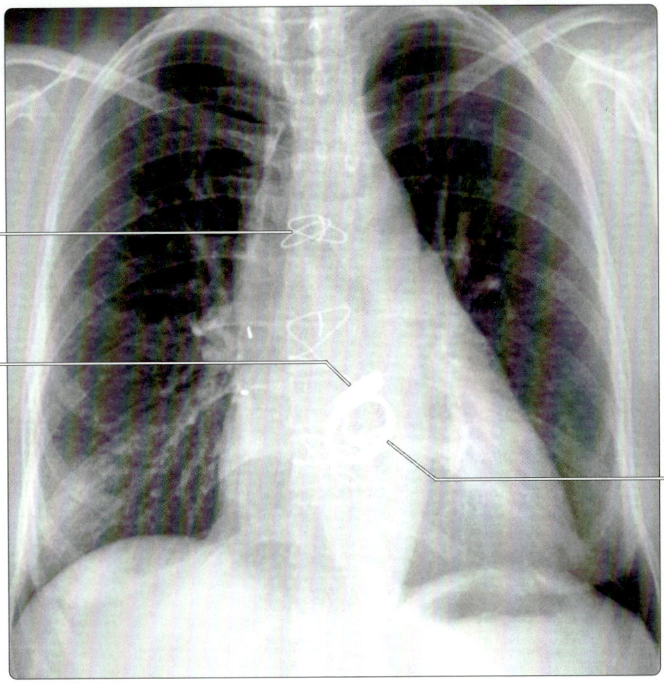

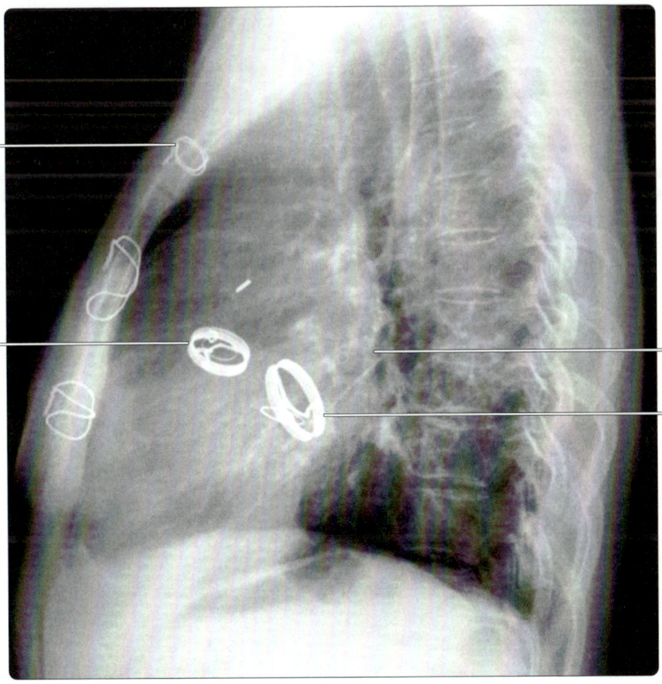

(Top) *PA chest radiograph of a 56-year-old woman status post aortic and mitral valve replacements (Bjork-Shiley mechanical heart valves) for rheumatic heart disease shows the close relationship of the aortic and mitral valves and expected postsurgical changes. The aortic valve prosthesis is located in the center of the heart and is oriented across the long axis of the ascending aorta. The mitral valve prosthesis is located more inferiorly, and its orifice exhibits a more horizontal orientation.* **(Bottom)** *Lateral chest radiograph of the same patient shows the aortic valve prosthesis located in the center of the heart and the more posteriorly located mitral valve prosthesis oriented across the long axis of the left atrioventricular orifice. The close relationship of these 2 prosthetic valves is consistent with the fact that they share a common fibrous annulus at the left fibrous trigone.*

ANATOMY OF ATRIOVENTRICULAR VALVES, GRAPHICS AND CT

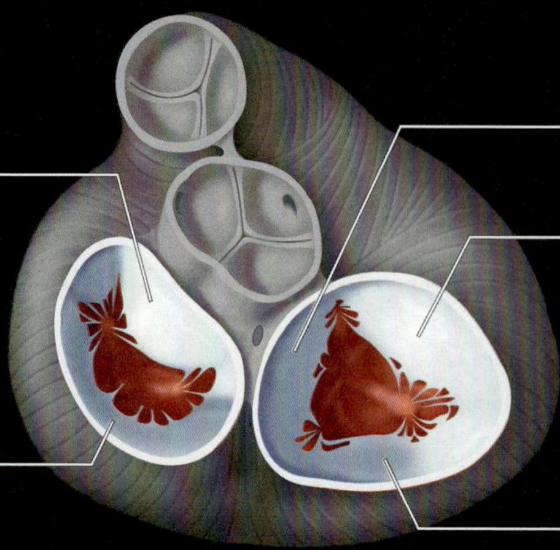

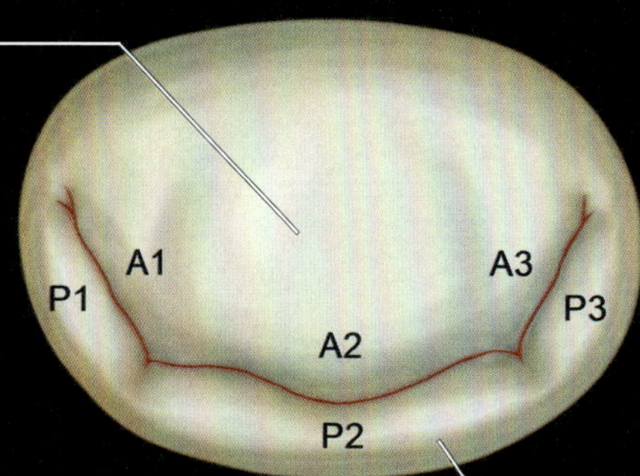

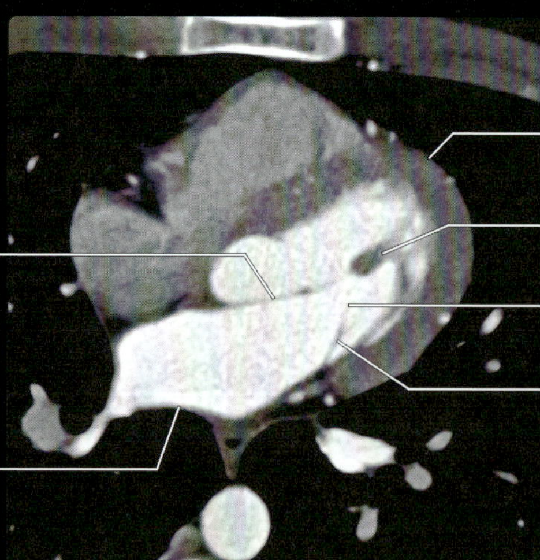

(Top) *Graphic shows the anatomy of the atrioventricular valves. The mitral (or left atrioventricular) valve has 2 leaflets, anterior and posterior, attached by chordae tendineae to the anterior and posterior papillary muscles, respectively. The tricuspid (or right atrioventricular) valve has 3 leaflets: Anterior, posterior, and septal. Chordae tendineae attach the tricuspid valve leaflets to papillary muscles in the right ventricle.* **(Middle)** *Graphic shows the Carpentier alphanumeric nomenclature that divides the mitral valve leaflets into separate anatomic regions. Each leaflet is subdivided into scallops or zones of apposition with highly variable anatomic configurations. The A and P designations apply to the anterior and posterior valve leaflets, respectively. There are 6 different regions: A1, A2, A3, P1, P2, and P3.* **(Bottom)** *Axial gated cardiac CECT (mediastinal window) shows the mitral valve during ventricular filling (diastole). The valve leaflets protrude into the left ventricular lumen and are attached to the papillary muscles by thin chordae tendineae.*

ANATOMY OF SEMILUNAR VALVES, GRAPHIC, CT, AND MR

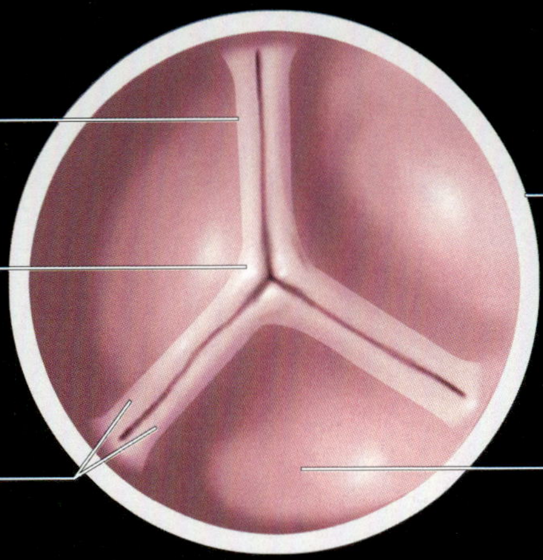

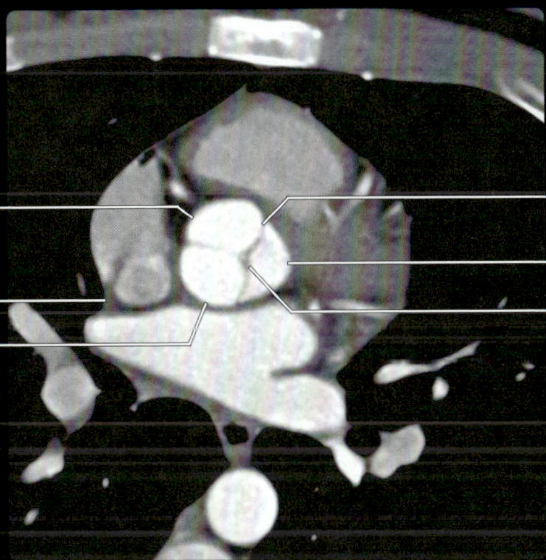

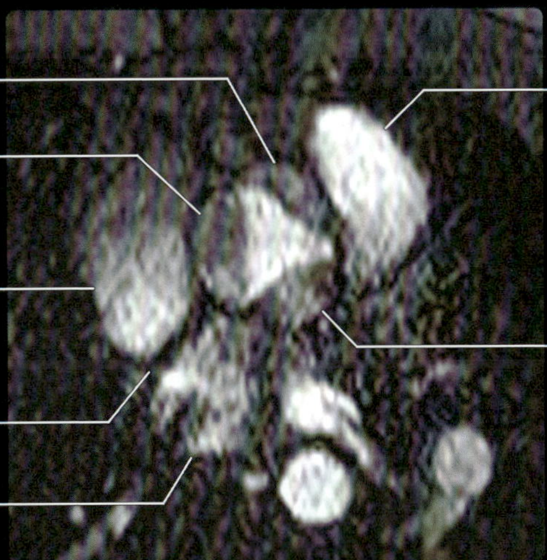

(Top) *Graphic shows the anatomy of the semilunar (pulmonic and aortic) valves, which are tricuspid (with 3 cusps or leaflets) and have free edges without tendinous attachments to the ventricles. The free edges of the cusps project into the vessel lumen during valve closure, forming sinuses. Retrograde blood flow after ventricular contraction forces the valve shut. Antegrade blood flow during ventricular systole forces the valve open. The superior free edge of the valve cusp has a central nodule of the semilunar cusp and a thin lateral free edge, or lunule, of the semilunar cusp.* **(Middle)** *Axial gated cardiac CECT (mediastinal window) shows the morphology of the aortic valve during diastole. Thick nodular foci in the central aspects of the free edges of the cusps represent the nodules of the semilunar cusps.* **(Bottom)** *Axial SSFP MR shows the normal aortic valve during systole. The aortic valve is composed of 3 cusps or leaflets and corresponding sinuses of Valsalva. The noncoronary cusp can always be identified as it faces the interatrial septum.*

ANATOMY OF TRICUSPID AND PULMONIC VALVES

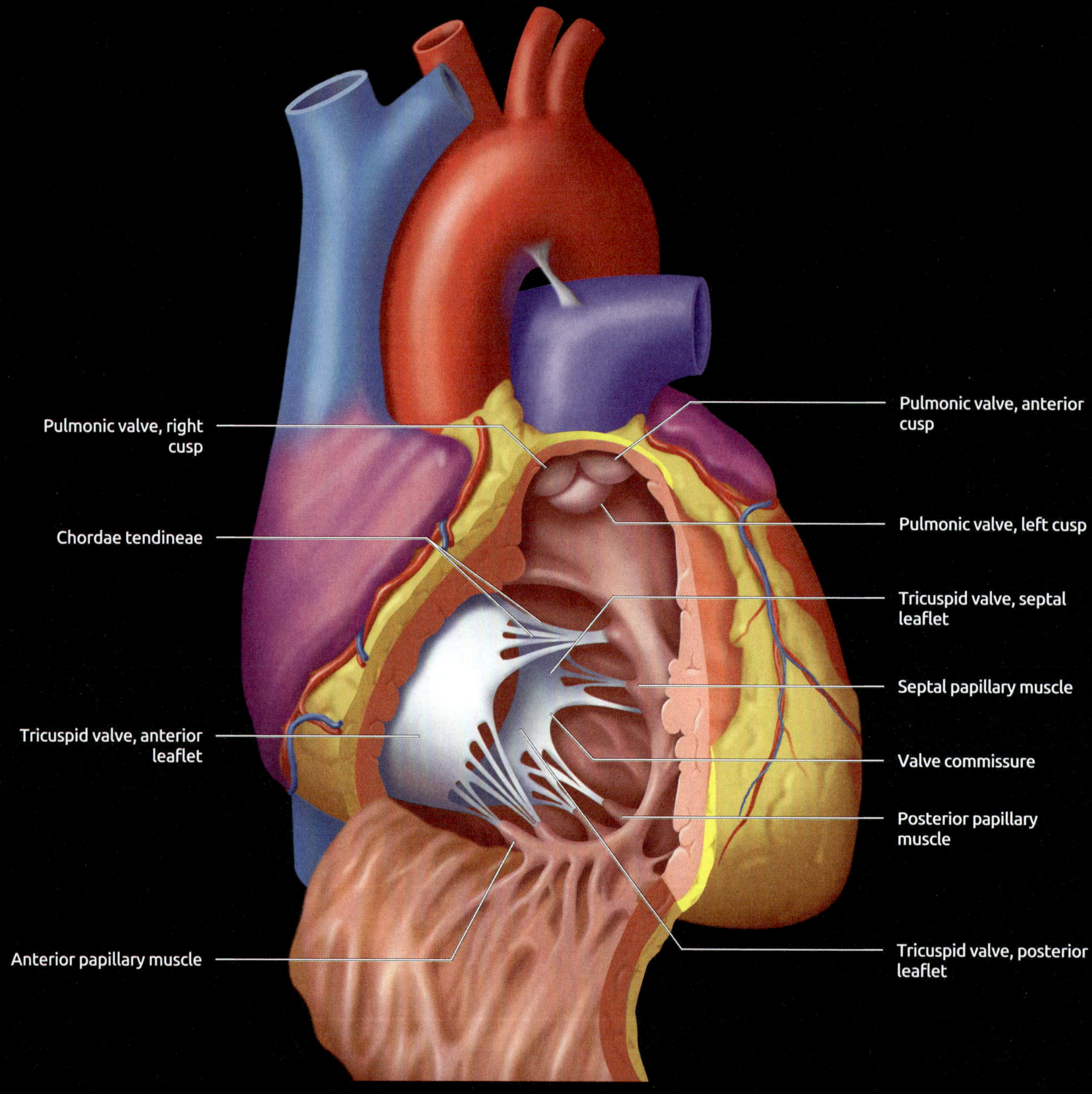

Graphic illustrates the internal anatomy of the right ventricle and the anatomy of the right heart valves. The right atrioventricular valve is the tricuspid valve. It has 3 cusps or leaflets that are attached to the fibrous valve ring and are continuous with each other at the valve commissures. The cusps are named according to their positions: Anterior, septal, and posterior. The free edges of the valves attach to chordae tendineae that, in turn, attach to the tips of papillary muscles that are also named according to their location. The pulmonic valve is a tricuspid semilunar valve located just distal to the right ventricular outflow tract. The free edges of the valve project into the pulmonary trunk, forming sinuses, and coapt at the nodules of the semilunar cusps. There are left, right, and anterior pulmonic valve cusps.

RIGHT HEART VALVES, CT

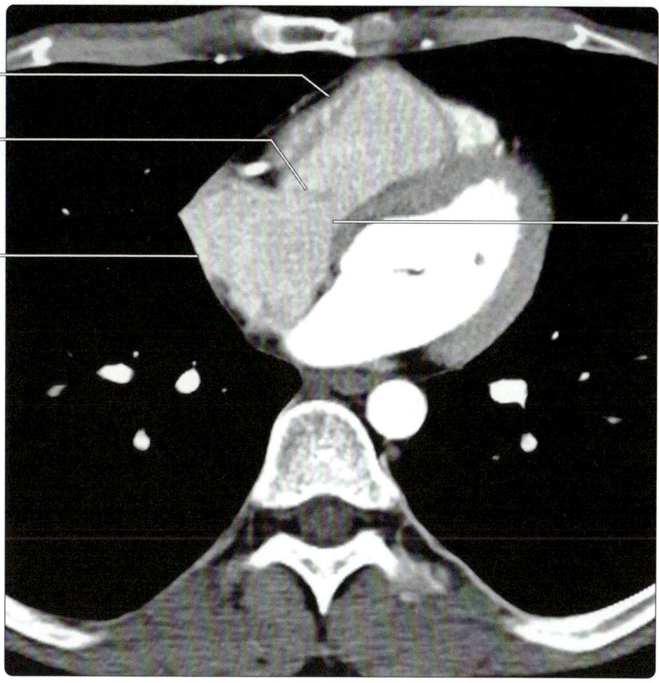

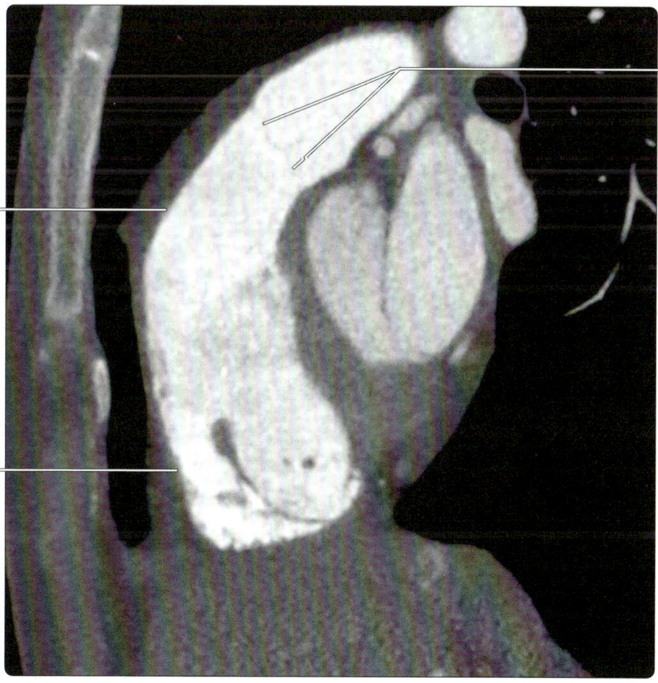

(Top) *First of 2 CECT images (mediastinal window) of the same patient demonstrating the anatomic location of the right heart valves is shown. The normal valves are thin and difficult to visualize on conventional CT. Axial gated cardiac CECT (mediastinal window) shows the tricuspid valve imaged during ventricular systole. The coapted septal and anterior valve cusps manifest as thin, linear soft tissue structures within the contrast-filled right heart chambers.* **(Bottom)** *Oblique sagittal CECT (mediastinal window) through the pulmonary trunk shows the anatomy and location of the pulmonic valve. The pulmonic valve is tricuspid and is located at the apex of the right ventricular outflow tract. This image is obtained during ventricular filling (diastole) and shows 2 of the 3 valve cusps coapted within the vessel lumen. The free valve edges protrude into the vascular lumen and form sinuses.*

ANATOMY OF LEFT HEART VALVES

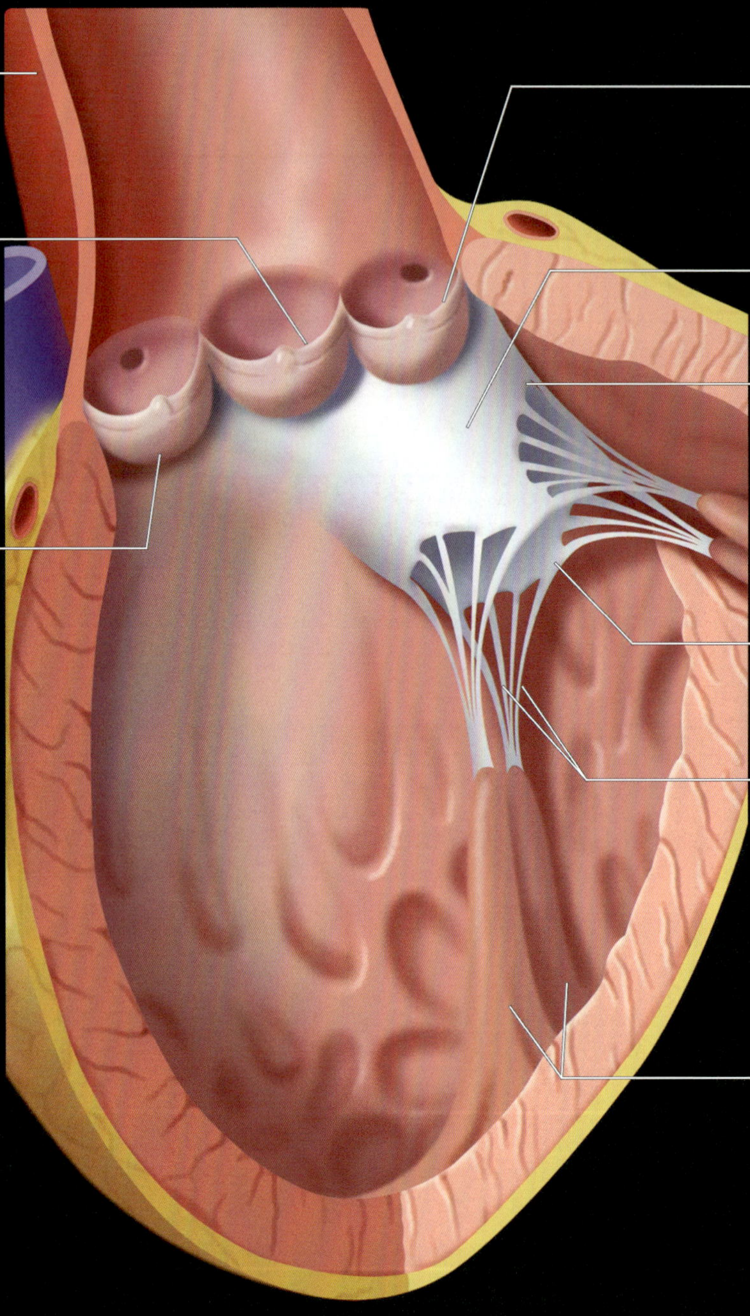

Graphic depicts the close relationship between the aortic and mitral valves. These valves are supported by fibrous valve rings connected by the left fibrous trigone. Thus, the anterior leaflet of the mitral valve is closely related to the left coronary cusp of the aortic valve. The aortic valve right coronary, left coronary, and noncoronary cusps are shown. The semilunar cusps form sinuses during valve closure, which occurs by coaptation of the free cusp edges. Each valve has a fibrous nodule on the central portion of its free edge that is called the nodule of the semilunar cusp. The mitral valve leaflets are continuous along the left atrioventricular fibrous valve ring and are connected at the valve leaflet commissures. The free edges of the mitral valve leaflets attach to the anterior and posterior papillary muscles via chordae tendineae.

ANATOMY OF LEFT HEART VALVES, GRAPHIC AND CT

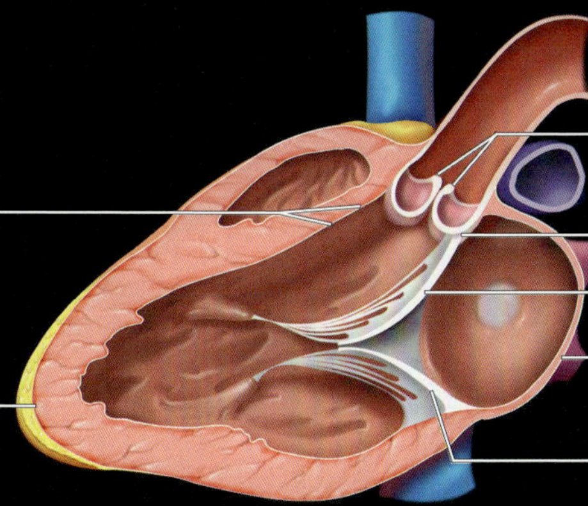

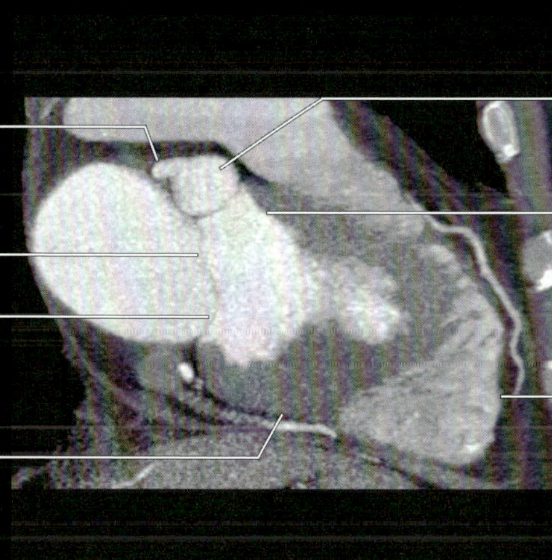

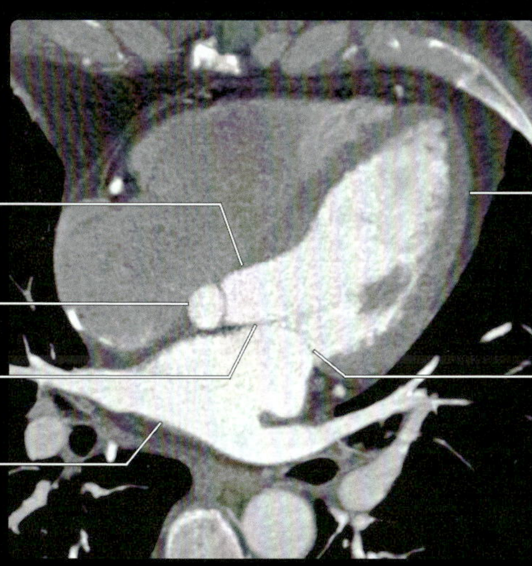

(Top) *Graphic illustrates a 3-chamber (left ventricular outflow tract) view of the heart and the relationship between the mitral and aortic valves. The fibrous rings that support these valves share a common fibrous bridge called the left fibrous trigone. The fibrous bridge forms a connection between the anterior leaflet of the mitral valve and the left coronary cusp of the aortic valve.* **(Middle)** *First of 2 gated cardiac CECT images (mediastinal window) of the same patient demonstrating the anatomic relationship between the aortic and mitral valves is shown. The left coronary cusp of the aortic valve shares a common fibrous attachment with the anterior leaflet of the mitral valve, as shown here.* **(Bottom)** *Four-chamber gated cardiac CECT (mediastinal window) demonstrates the close relationship between the mitral valve and the aortic valve.*

LEFT VENTRICULAR SEGMENTATION

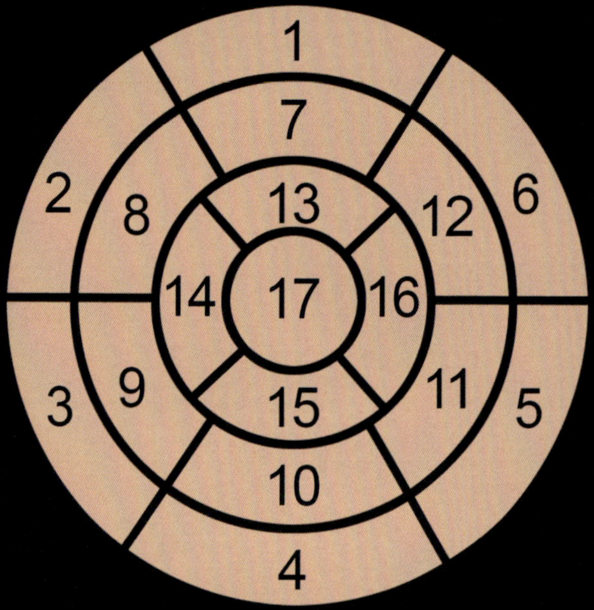

1. Basal anterior
2. Basal anteroseptal
3. Basal inferoseptal
4. Basal inferior
5. Basal inferolateral
6. Basal anterolateral
7. Midanterior
8. Midanteroseptal
9. Midinferoseptal
10. Midinferior
11. Midinferolateral
12. Midanterolateral
13. Apical anterior
14. Apical septal
15. Apical inferior
16. Apical lateral
17. Apex

Graphic shows the 12 standardized myocardial segments that were adopted by the Cardiac Imaging Committee of the Council on Clinical Cardiology of the American Heart Association for tomographic imaging of the left ventricle. The names for the myocardial segments define the location relative to both the long and short axis of the left ventricle. Basal, midcavity, and apical designations localize the segments along the long axis of the left ventricle. With regard to the circumferential location on the bull's-eye view, the basal (segments 1-6) and midcavity (segments 7-12) sections are divided into 6 segments of 60° each. The apical section (segments 13-16) is divided into 4 segments of 90° each. The cardiac apex is segment 17. The attachment of the right ventricular wall to the left ventricle is used to separate the septum from the left ventricular anterior and inferior free walls.

CORONARY ARTERY TERRITORIES

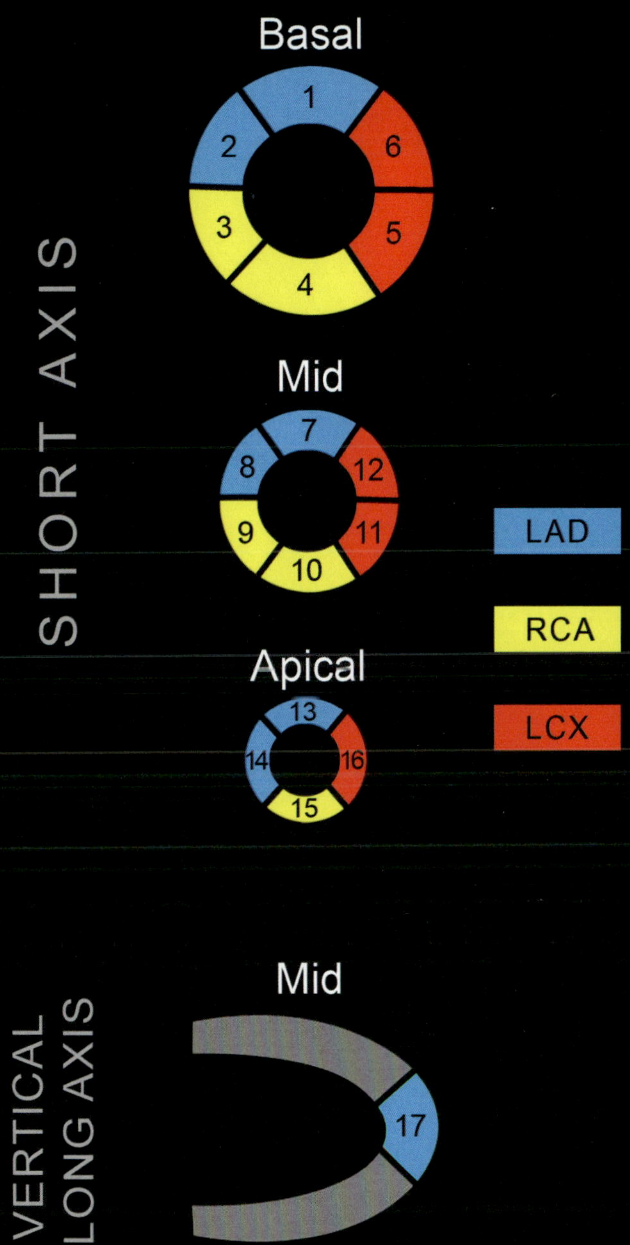

Graphic shows the left ventricular myocardial segments and the distribution of coronary blood flow to these segments. Segmentation of the left ventricular myocardium is used to assess myocardial perfusion, left ventricular function, and coronary anatomy. Myocardial perfusion and function can be assessed using nuclear medicine cardiac SPECT, coronary CTA with fractional flow reserve (FFR), and cardiac MR.

CORONARY ARTERY TERRITORIES

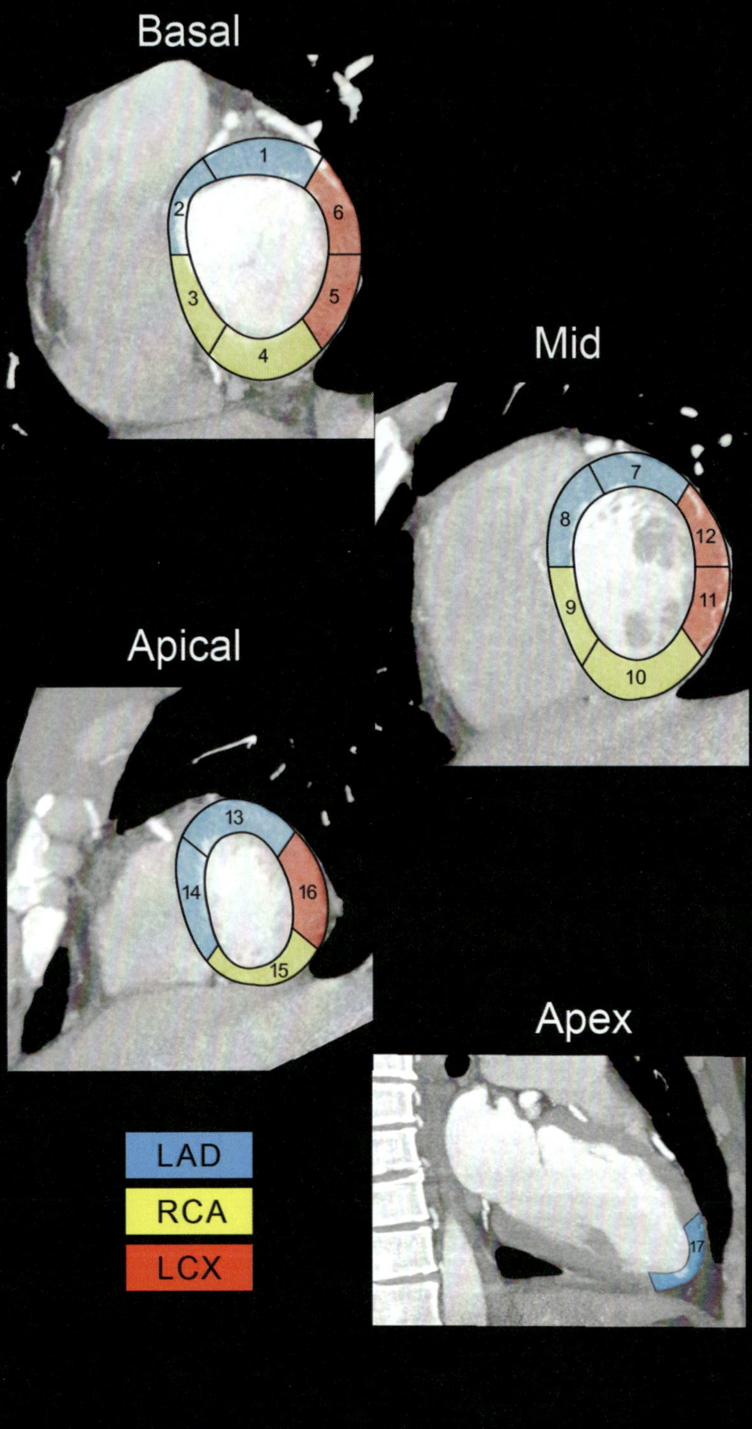

The first 3 images are short-axis CECT views of the left ventricular wall at the basal, mid-, and apical levels. The 4th image is a vertical long-axis image, obtained from the same data set, reformatted to show the apex. Different colors are assigned to show the vascular territories of the coronary arteries. The LAD supplies the anterior interventricular septum, the anterior wall, and in most cases, the apex. On a short-axis image, it usually supplies from 9 to 1 o'clock at the basal and midventricular levels. The LCX supplies the lateral wall, usually from 2 to 4 o'clock. The RCA supplies the inferior wall segments and the posterior interventricular septum, usually from 5 to 8 o'clock. It should be emphasized, however, that great variation in the distribution of coronary blood flow is observed in clinical practice.

CATHETER CORONARY ANGIOGRAPHY

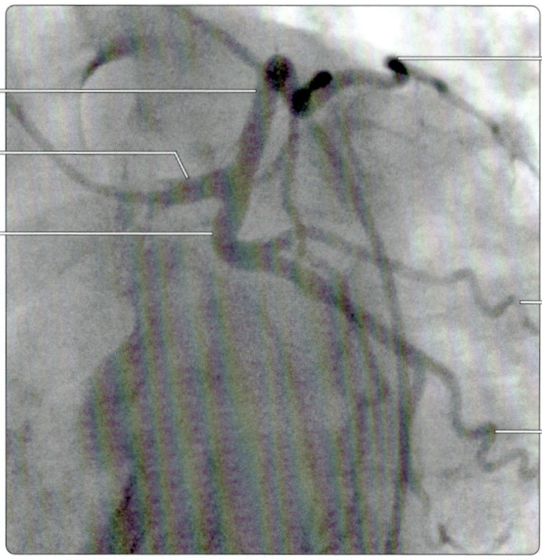

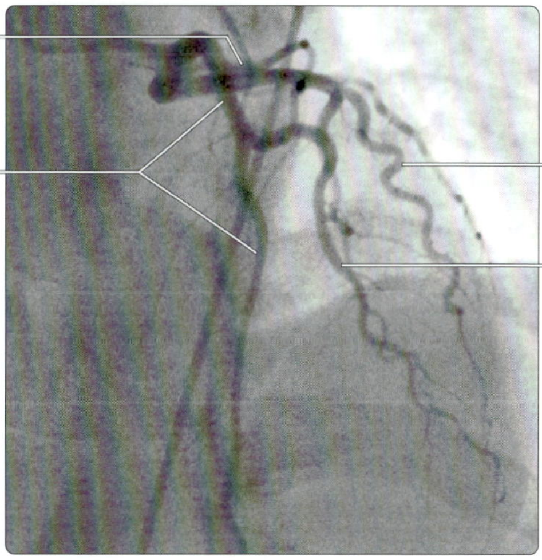

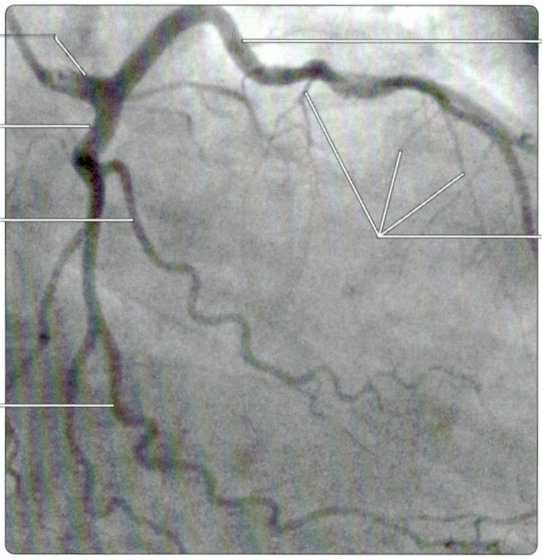

(Top) *First of 4 images shows the left coronary circulation obtained during catheter coronary angiography of a 48-year-old patient who presented with chest pain. Because of the overlap between the branches of the coronary arteries, multiple radiographic projections are obtained to show different segments in greater detail. Coronary artery angiogram, left anterior oblique (LAO) caudal view shows the bifurcation of the left main coronary artery into the LAD and LCX. The LAD is foreshortened on this projection, while the LCX with its obtuse marginal branches is well seen.* **(Middle)** *Coronary artery angiogram LAO cranial projection shows the LAD and its diagonal branch. It is difficult to separate the LCX and its obtuse marginal branches on this projection.* **(Bottom)** *Coronary artery angiogram right anterior oblique (RAO) caudal view shows the origin of the LAD and LCX from the left main coronary artery. The diagonal branch of the LAD projects over the LAD. Note the septal perforator branches of the LAD. Note the obtuse marginal branches of the LCX.*

CATHETER CORONARY ANGIOGRAPHY

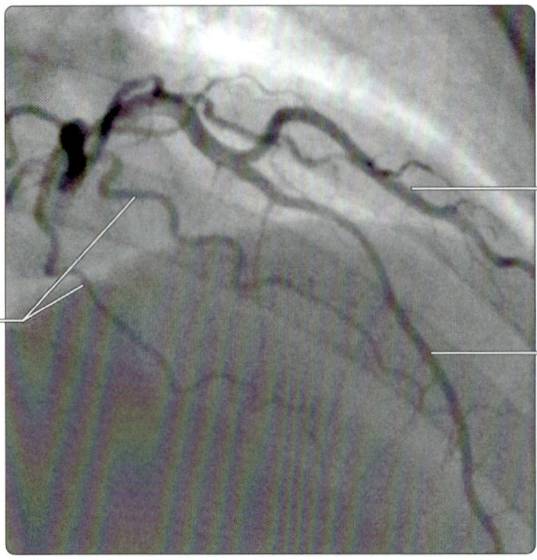

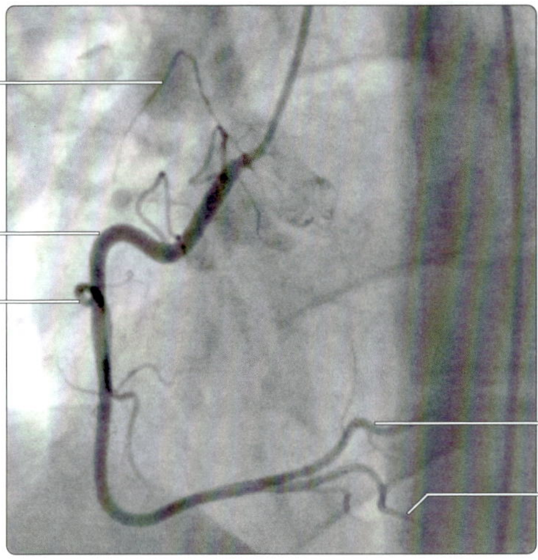

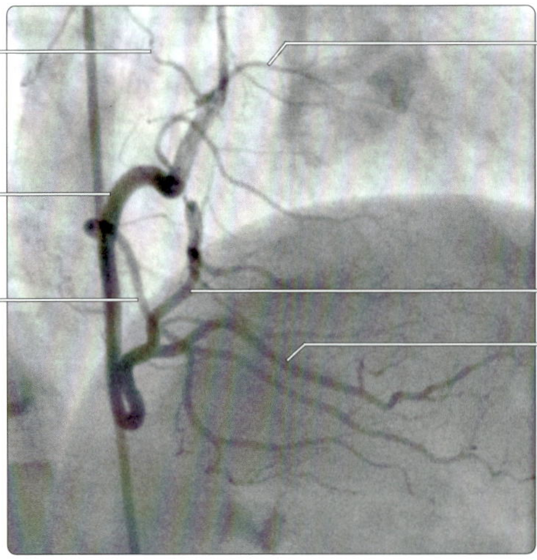

(Top) *Coronary artery angiogram RAO cranial projection separates the LAD from its diagonal branch. The LCX is not well seen, but its obtuse marginal branches are well delineated.* **(Middle)** *First of 2 images obtained during a coronary artery angiogram after injection of the RCA is shown. Coronary artery angiogram LAO cranial view shows the branches of the RCA and its distal bifurcation into the PDA and the PLB. The acute marginal artery is superimposed on the RCA on this projection.* **(Bottom)** *Coronary artery angiogram RAO cranial projection shows the branches of the RCA. The conus artery is usually the 1st branch and supplies the pulmonary outflow tract. The acute marginal artery is separated on this projection from the RCA.*

CORONARY ARTERIES, CT ANGIOGRAPHY

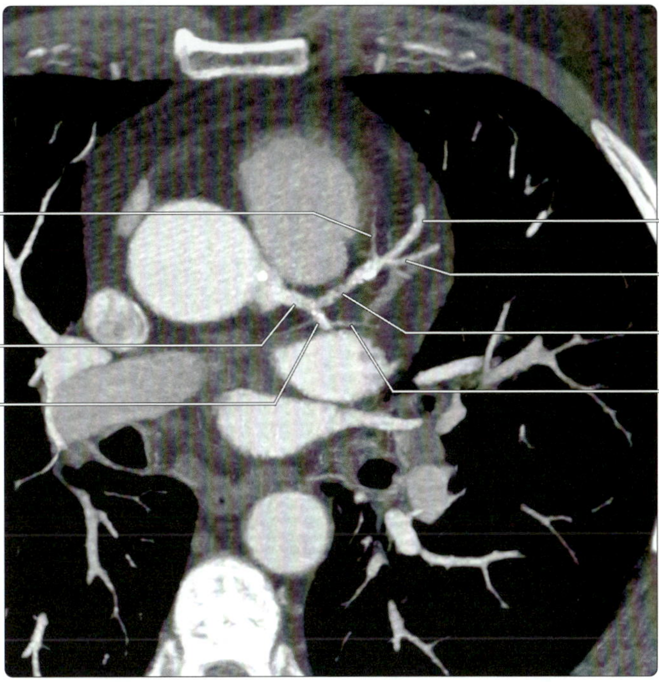

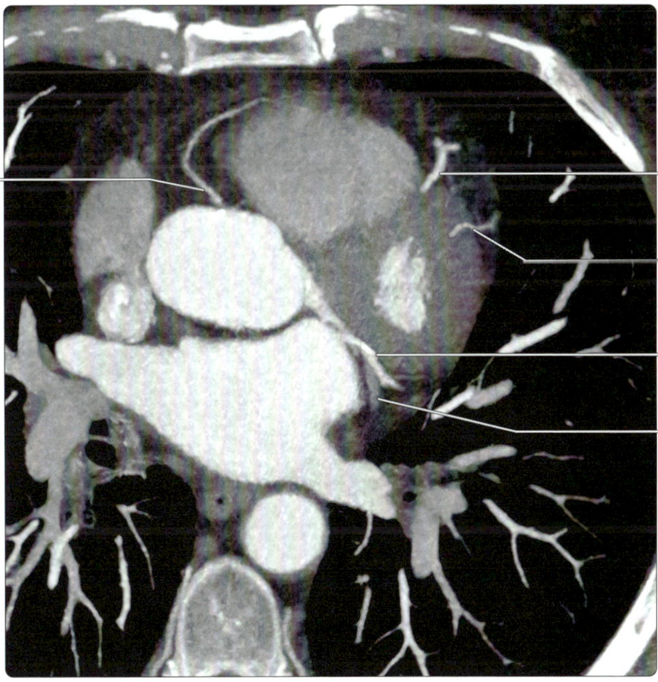

(Top) *Axial CT angiogram MIP reformatted image demonstrates the left main coronary artery arising from the left coronary sinus. The left main coronary artery bifurcates into the LAD and the LCX, which then give rise to their respective branches.* **(Bottom)** *Axial CT angiogram MIP reformatted image demonstrates the conus artery arising directly from the right coronary sinus.*

RIGHT-DOMINANT CORONARY CIRCULATION, CT ANGIOGRAPHY

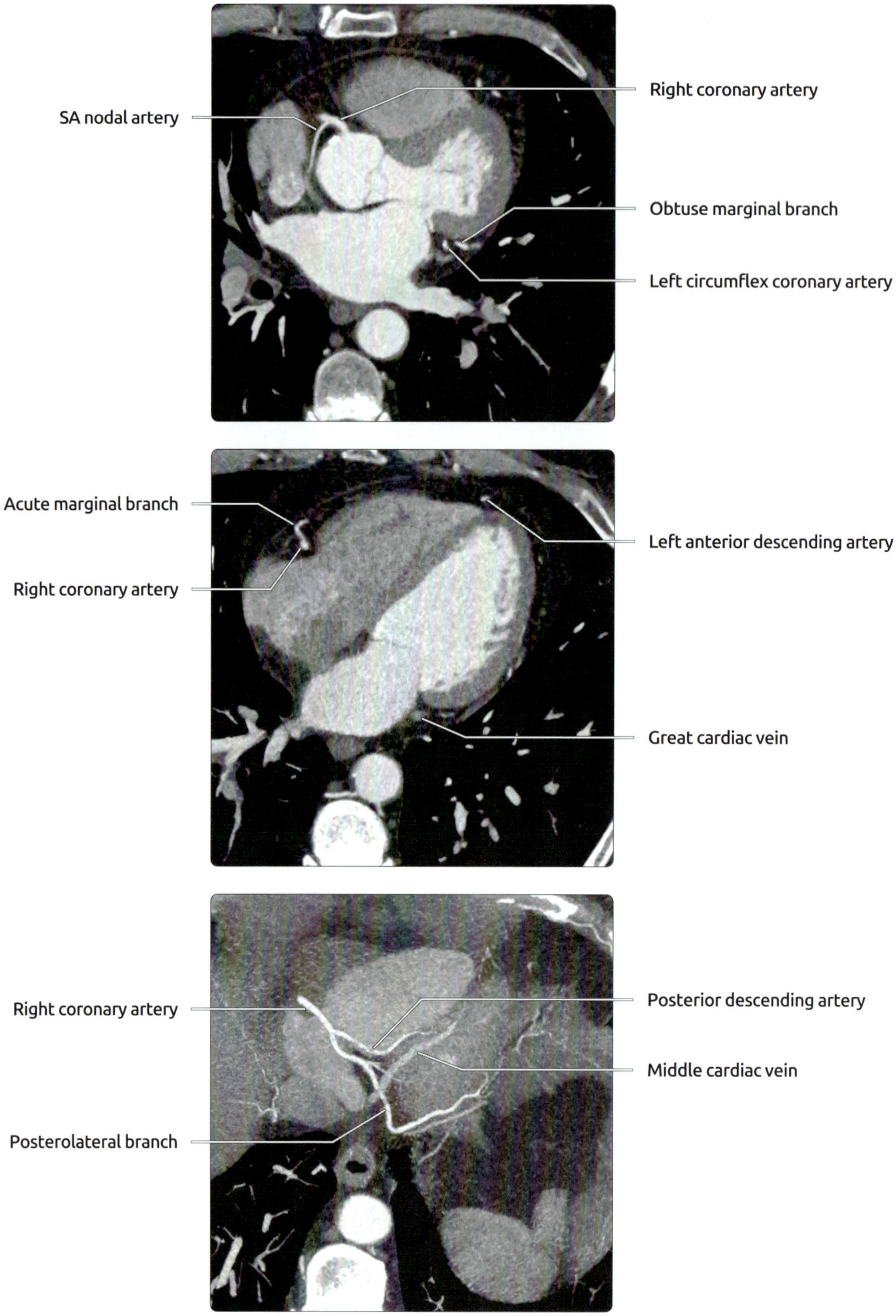

(Top) *Axial CT angiogram MIP reformatted image of the same patient demonstrates the RCA arising from the right coronary sinus. The sinoatrial (SA) nodal artery is the 1st branch off the RCA in this case, as the conus branch had a separate origin from the right coronary sinus.* **(Middle)** *Axial CT angiogram MIP reformatted image demonstrates the RCA coursing in the atrioventricular groove and giving rise to the acute marginal branch, which courses along the lateral wall of the right ventricle.* **(Bottom)** *Axial CT angiogram MIP reformatted image demonstrates the typical terminal branching of the RCA in a right-dominant system, giving rise to the PDA and the PLB.*

CORONARY ARTERIES, CT ANGIOGRAPHY

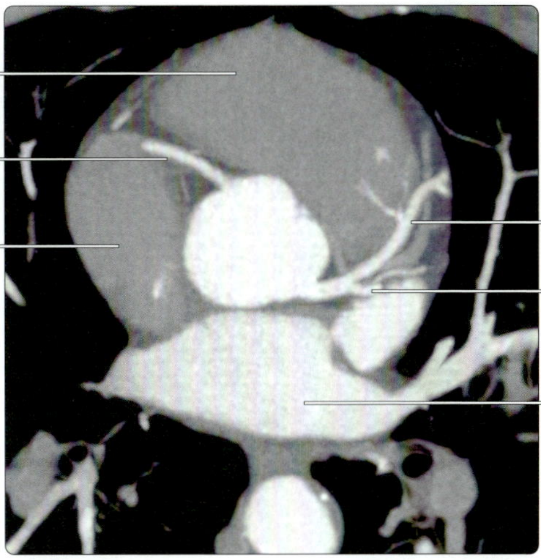

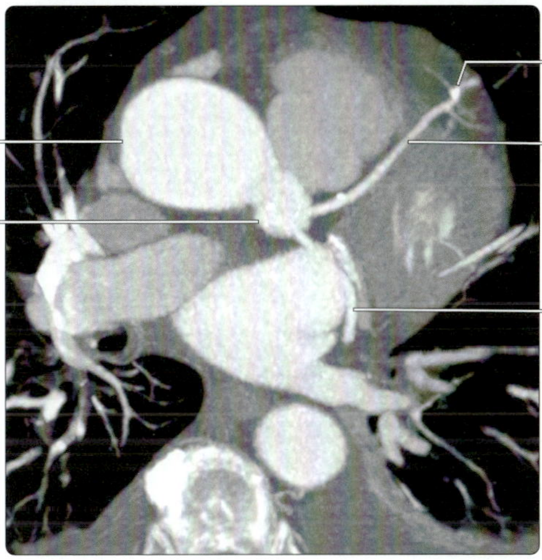

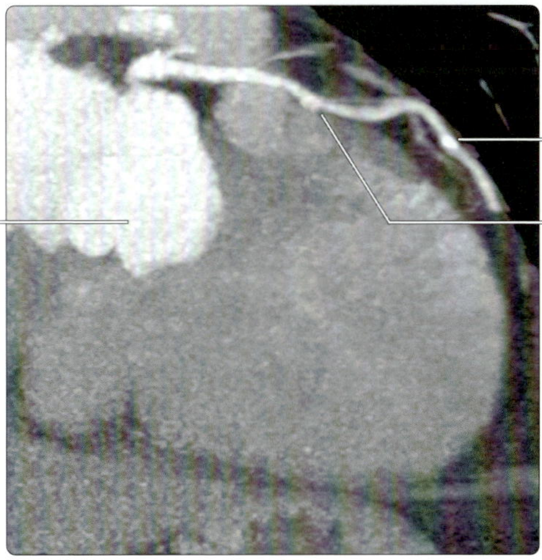

(Top) *First of 3 CT angiogram maximum intensity projection (MIP) reformatted images depicts the origins of both coronary arteries. The RCA arises from the right coronary sinus and courses to the right behind the pulmonary outflow tract to reach the coronary sulcus. The short left main coronary artery divides into the LAD and LCX.* **(Middle)** *Axial oblique CT angiogram MIP reformatted image of the left coronary system of another patient with mild coronary artery disease shows the LAD in the anterior interventricular groove and the LCX in the coronary sulcus. Note the mild calcification of the LAD. The obliquity of the plane of reconstruction makes the left coronary sinus appear aneurysmal and masks the left main coronary artery origin.* **(Bottom)** *CT angiogram coronal oblique thin MIP reformatted image of the same patient delineates the course of the LAD in the anterior interventricular groove and again shows calcifications of the vessel wall.*

CORONARY ARTERIES, CT ANGIOGRAPHY

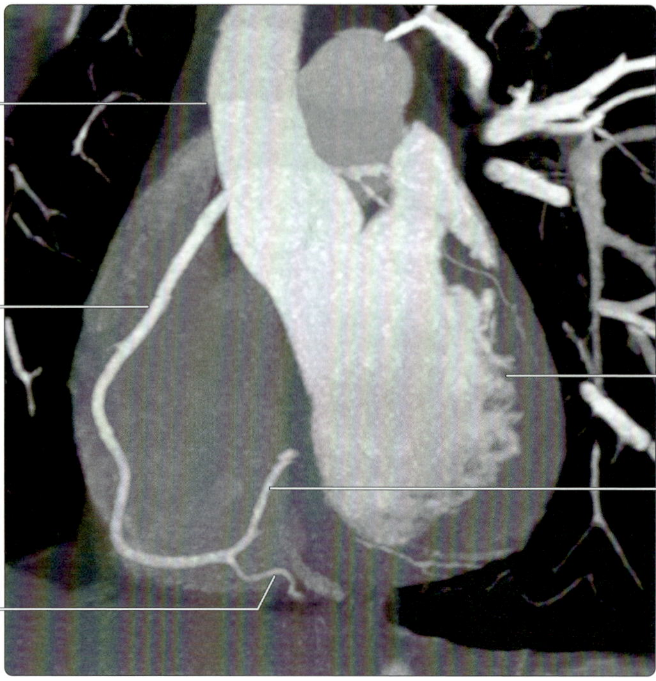

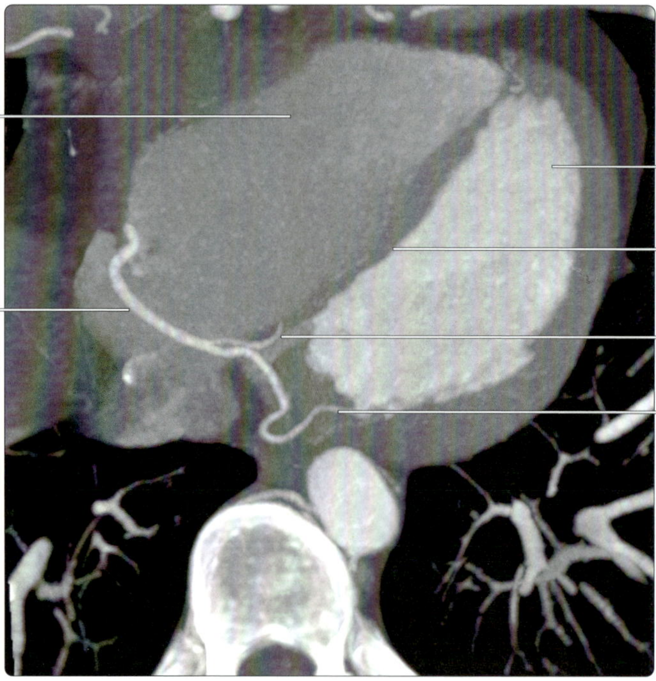

(Top) *CT angiogram coronal oblique MIP reformatted image shows the vertical course of the RCA within the coronary sulcus and its 2 terminal branches: The PDA and the PLB. This is the usual branching pattern in a right-dominant coronary artery circulation. The coronary circulation is right dominant if the RCA supplies the PDA and at least 1 PLB.* **(Bottom)** *CT angiogram axial MIP reformatted image through the base of the heart shows the horizontal course of the RCA as it courses to the posterior interventricular septum before branching into the PDA and PLB. The PDA runs in the posterior interventricular groove. MIP images are useful to delineate the course of an artery but should not be used alone to diagnose arterial stenosis because areas of narrowing can be masked by contrast in the plane of reconstruction.*

CORONARY ARTERY BRANCHING VARIANTS, CT ANGIOGRAPHY

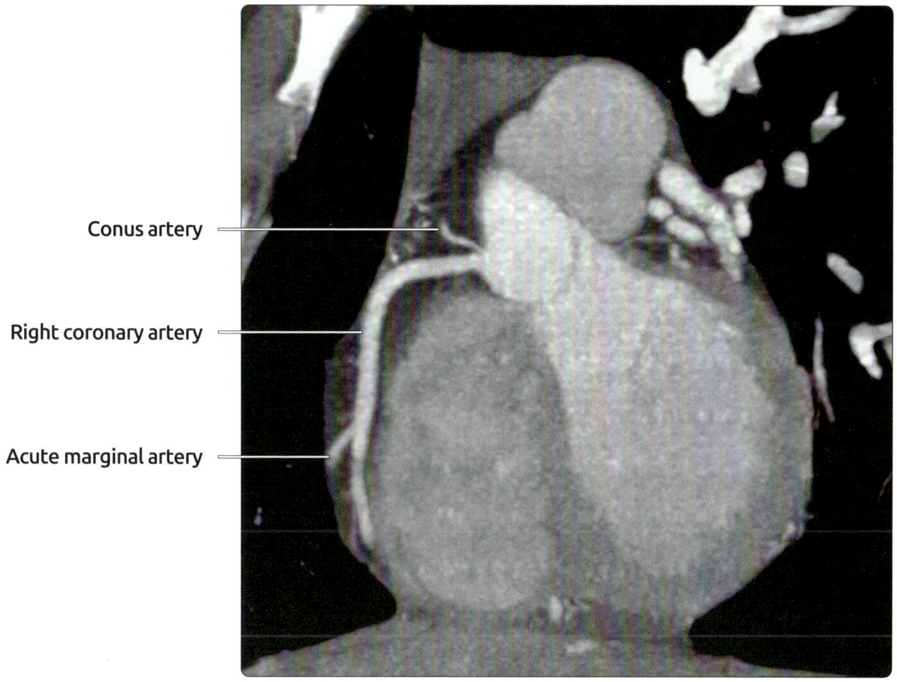

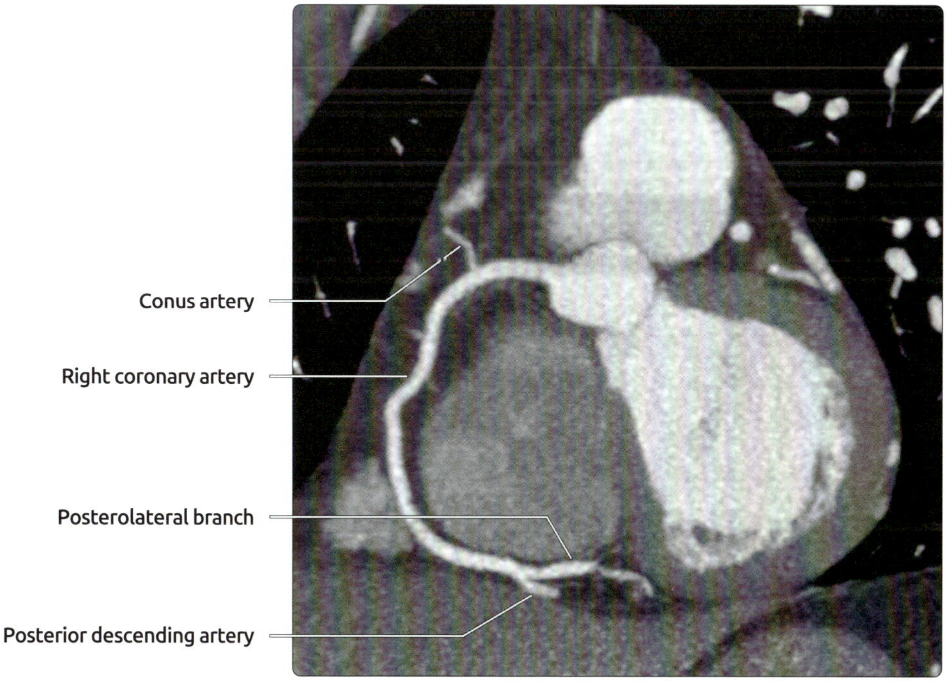

(Top) *CT angiogram coronal oblique MIP reformatted image shows a separate ostium of the conus artery, which arises directly from the right coronary sinus rather than from the RCA.* **(Bottom)** *CT angiogram coronal oblique MIP reformatted image demonstrates the normal course and branching of the RCA. The conus artery is the 1st branch arising from the RCA.*

CORONARY ARTERY BRANCHING VARIANTS, CT ANGIOGRAPHY

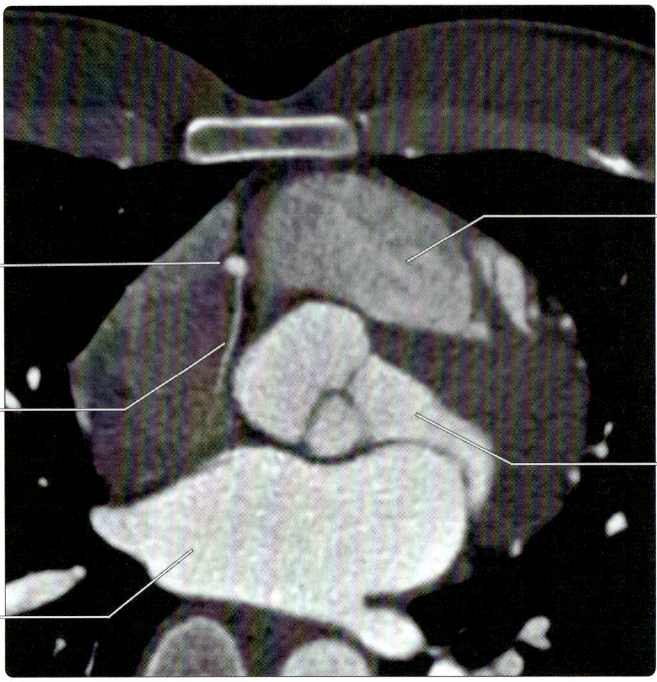

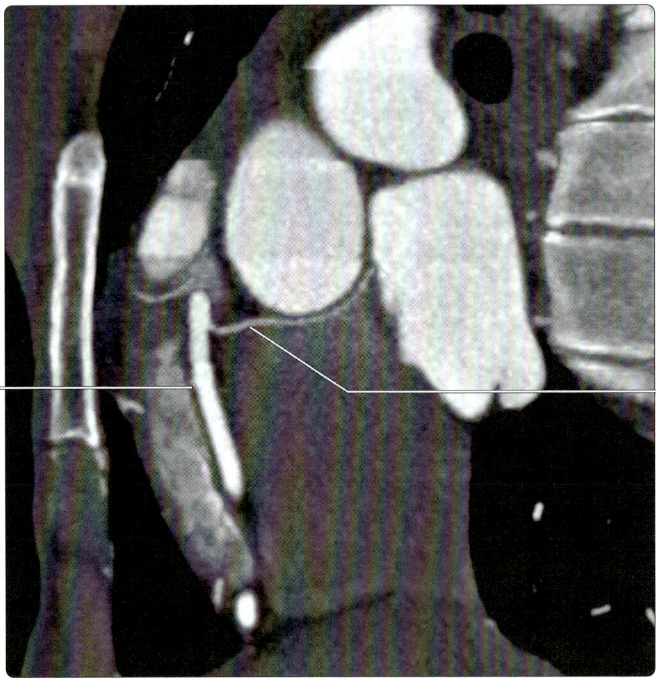

(Top) *CT angiogram axial MIP reformatted image demonstrates the SA nodal artery arising as the 2nd branch off the RCA and coursing posteriorly.* **(Bottom)** *CT angiogram curved MPR image of the same patient demonstrates the SA nodal artery arising from the RCA and coursing posteriorly and superiorly.*

CORONARY ARTERIES, SURFACE-RENDERED CT ANGIOGRAPHY

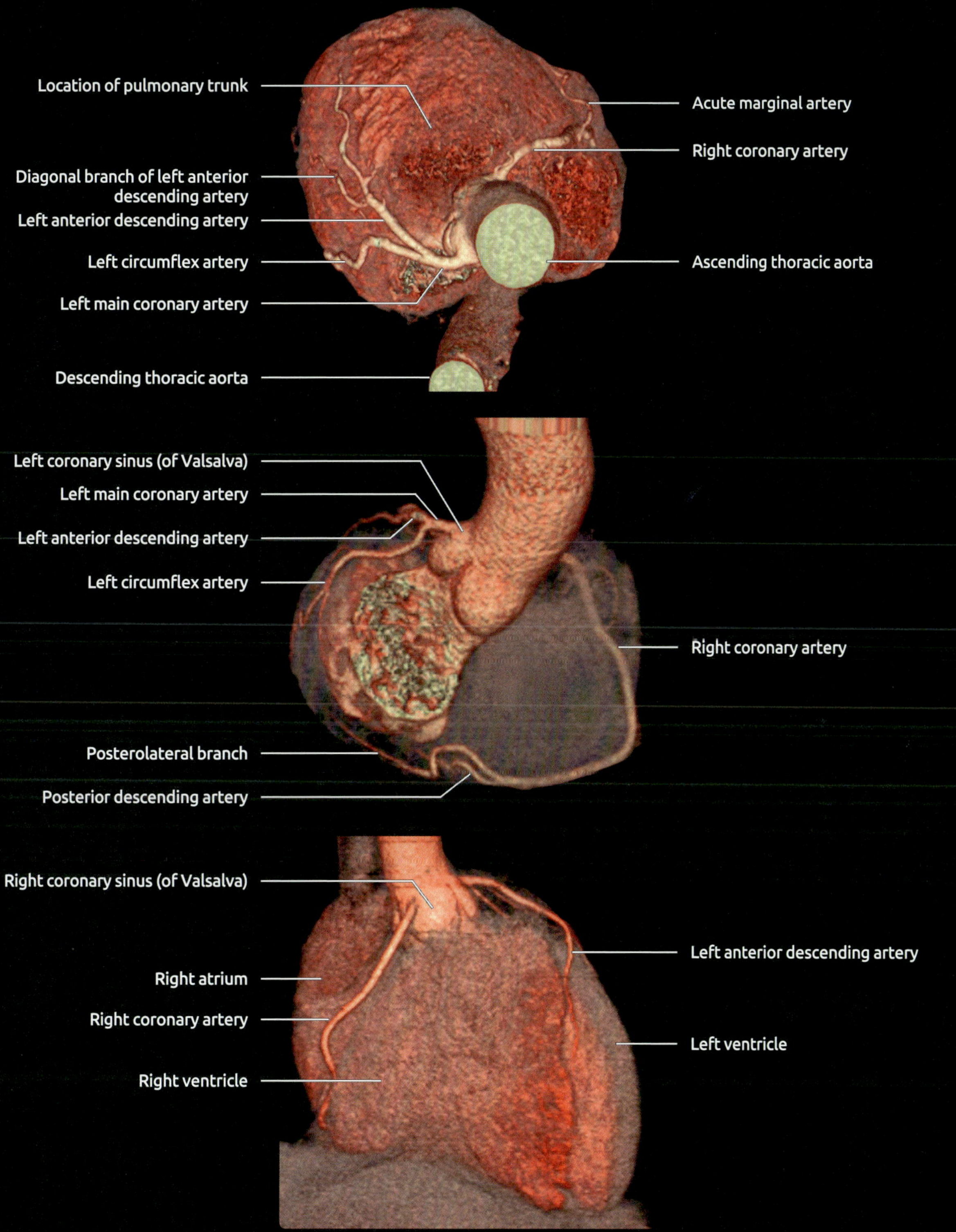

(Top) *First of 3 CT angiogram 3D surface-rendered images of the coronary arteries is shown. This image shows the origins of the coronary arteries viewed from above with the pulmonary trunk digitally removed. The RCA arises from the right coronary sinus. The short left main coronary artery arises from the left coronary sinus and divides into the LAD, which runs in the anterior interventricular groove, and the LCX which runs in the coronary sulcus.* **(Middle)** *CT angiogram posterior oblique 3D surface-rendered image shows the RCA and LCX forming a circle as each runs in its respective atrioventricular groove.* **(Bottom)** *CT angiogram frontal oblique 3D surface-rendered image shows the origin of the RCA from the right coronary sinus. The RCA courses to the right of the pulmonary artery (digitally removed during image post processing) to reach the atrioventricular groove.*

RIGHT-DOMINANT CIRCULATION, SURFACE-RENDERED CT ANGIOGRAPHY

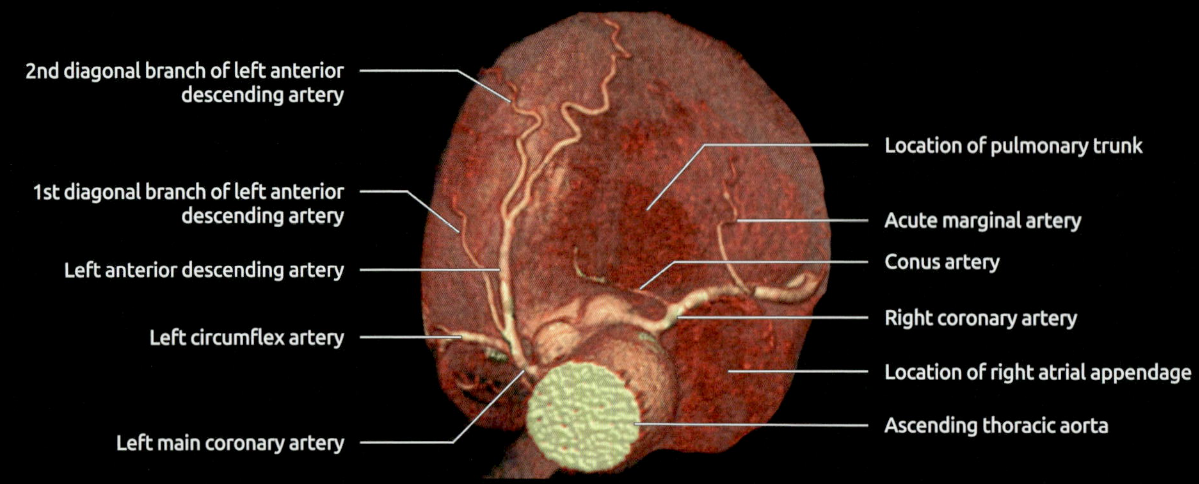

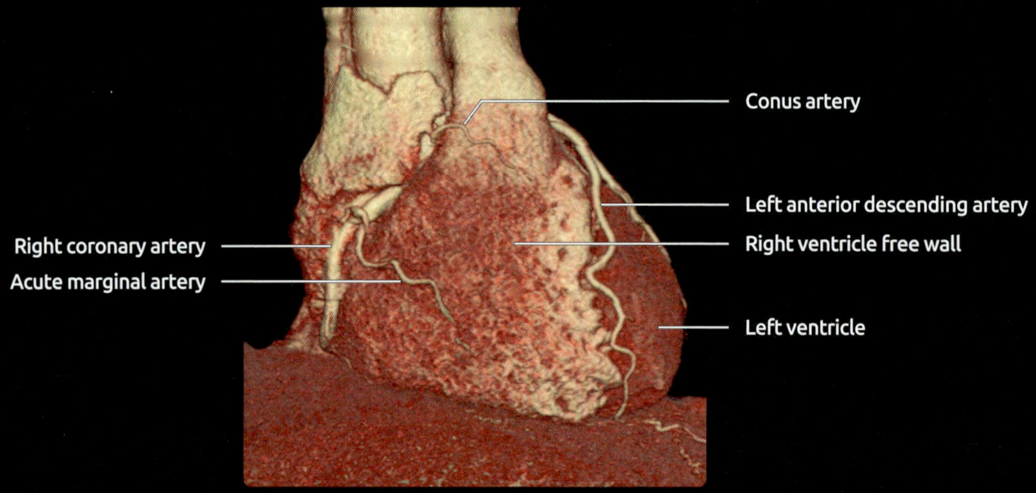

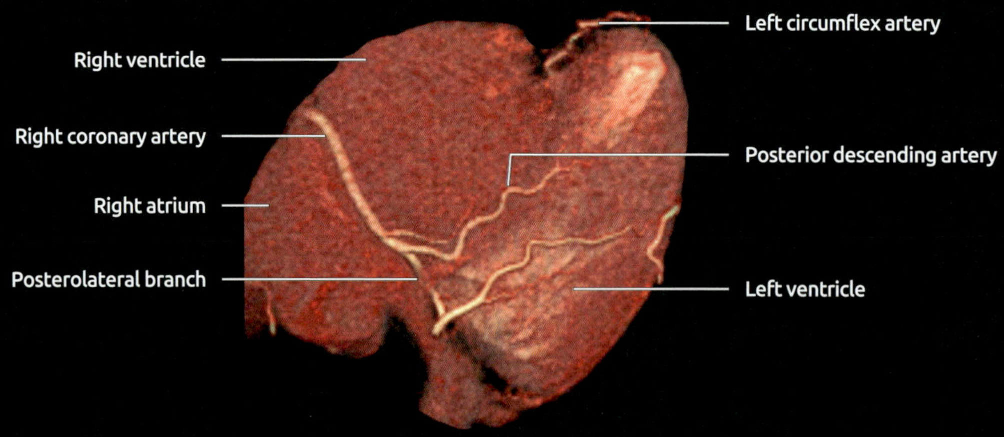

(Top) *First of 3 CT angiogram 3D surface-rendered images of a patient with right-dominant coronary circulation shows the conus artery, usually the 1st branch off the RCA, which supplies the pulmonary outflow tract. The acute marginal artery arises at the junction between the mid- and distal parts of the RCA and supplies the anterior wall of the right ventricle.* **(Middle)** *CT angiogram frontal 3D surface-rendered image shows the conus and acute marginal arteries. During its course in the coronary sulcus, the RCA also gives off multiple branches to the right ventricular free wall.* **(Bottom)** *CT angiogram axial 3D surface-rendered image through the base of the heart shows the 2 terminal branches of the RCA in a right-dominant coronary circulation. The PLB supplies the posterolateral wall of the left ventricle, while the PDA diverges to run in the posterior interventricular groove.*

LEFT-DOMINANT CIRCULATION, SURFACE-RENDERED CT ANGIOGRAPHY

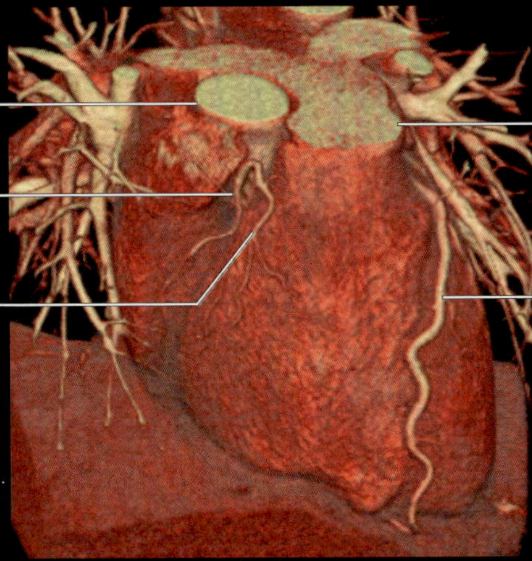

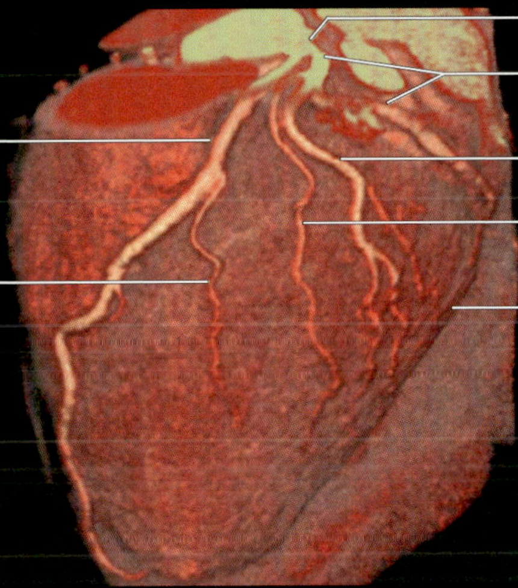

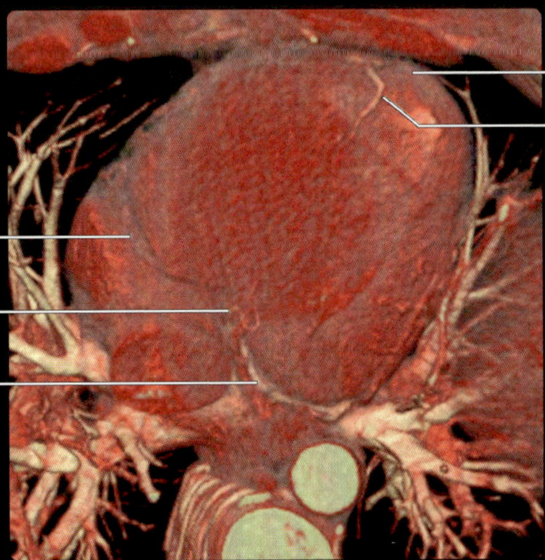

(Top) *First of 3 CT angiogram 3D surface-rendered images of a patient with left-dominant coronary circulation is shown. CT angiogram frontal oblique 3D surface-rendered image shows a common origin of the RCA and the conus artery from the right coronary sinus. The RCA is small and ends before reaching the posterior interventricular septum.* **(Middle)** *CT angiogram frontal oblique 3D surface-rendered image shows the branches of the left coronary artery. The LAD is large and wraps around the cardiac apex. The LCX is also large and runs in the coronary sulcus to reach the posterior interventricular groove.* **(Bottom)** *CT angiogram diaphragmatic view 3D surface-rendered image of the heart shows the LAD wrapping around to supply the cardiac apex and the anterior portion of the interventricular septum. The LCX supplies a small PDA. Note the absence of the RCA in the right coronary sulcus.*

ANOMALOUS CORONARY ARTERIES, CT ANGIOGRAPHY

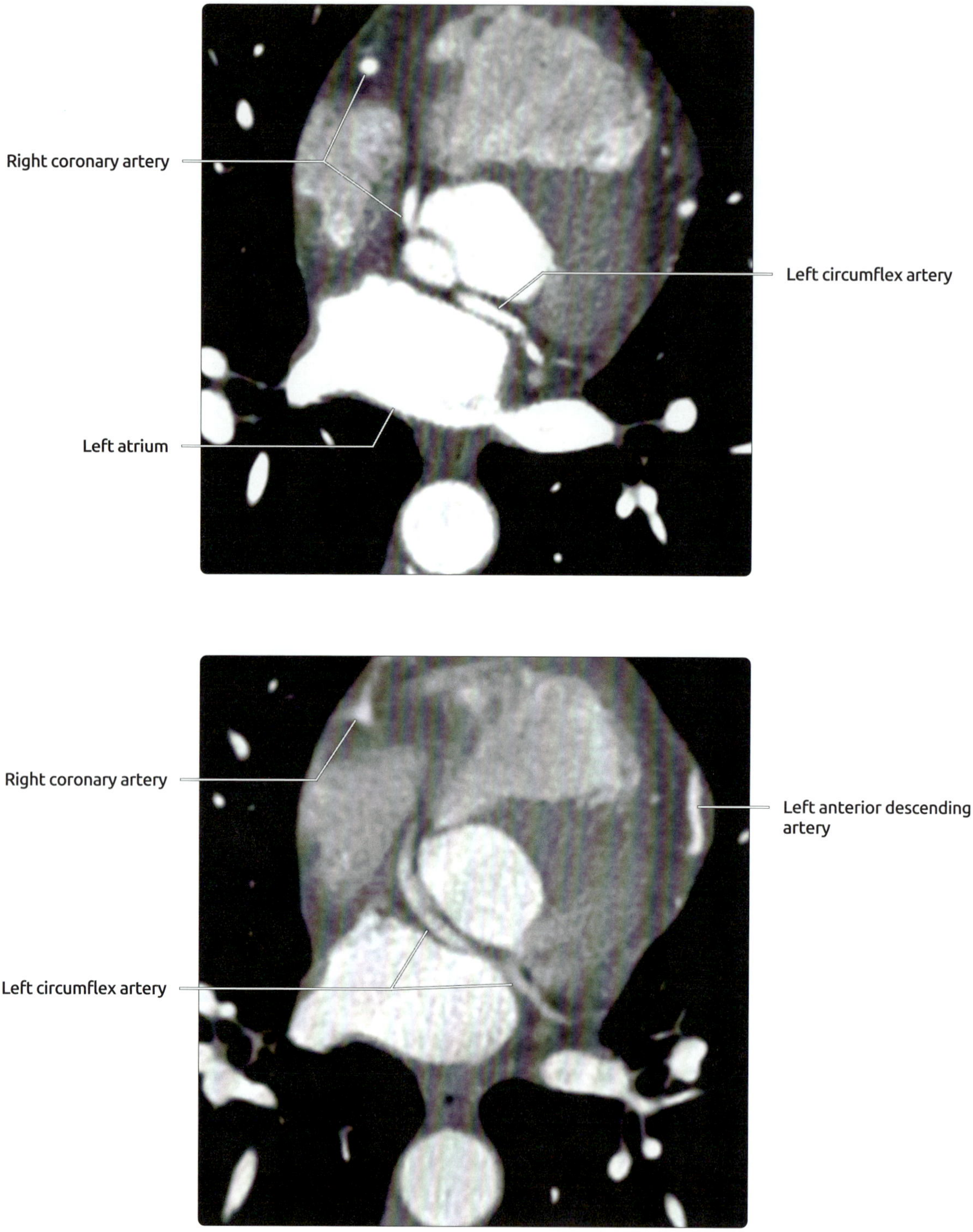

(Top) *Axial CT angiogram MIP reformatted image demonstrates an anomalous left circumflex coronary artery arising from the right coronary sinus with a retroaortic course.* **(Bottom)** *CT angiogram axial MIP reformatted image of the same patient shows the anomalous retroaortic course of the LCX as it extends into the left atrioventricular groove.*

ANOMALOUS CORONARY ARTERIES, CT ANGIOGRAPHY

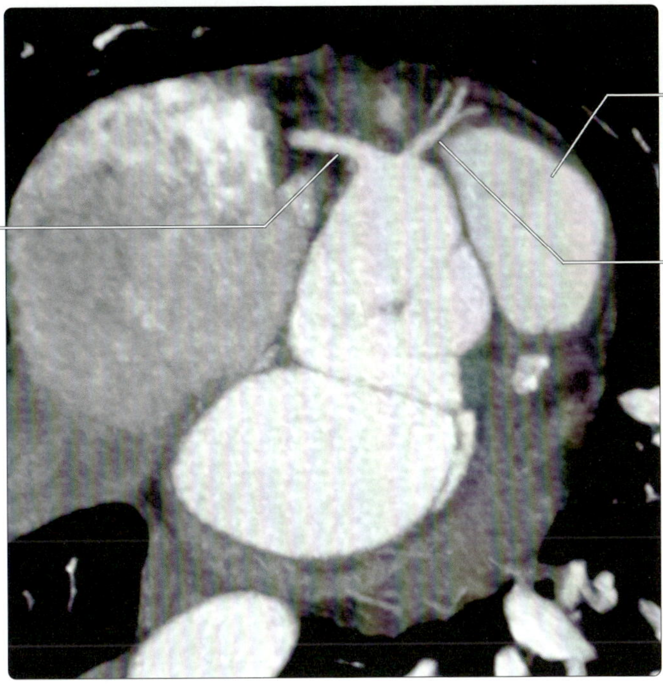

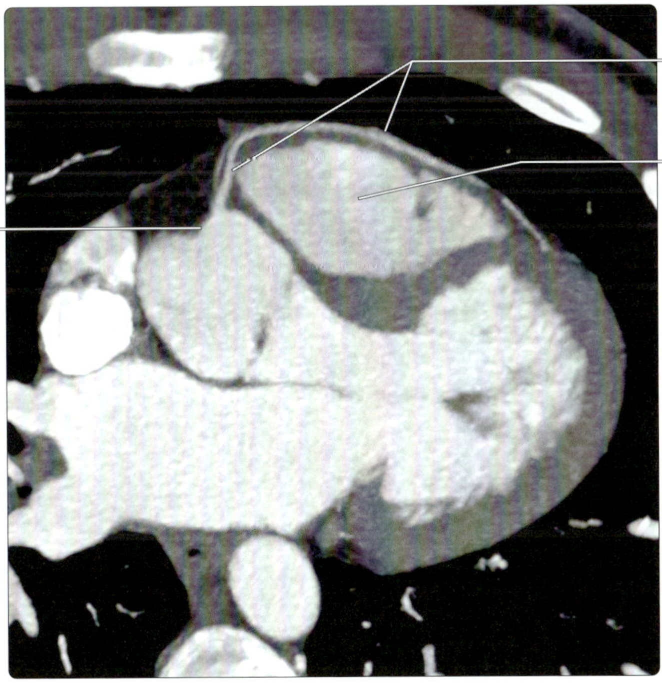

(Top) *CT angiogram axial oblique MIP reformatted image shows an anomalous LAD arising from the right sinus of Valsalva with a prepulmonic course.* **(Bottom)** *CT angiogram curved MIP reformatted image of the same patient demonstrates the prepulmonic course of the anomalous LAD, arising from the right sinus of Valsalva, and coursing anterior to the right ventricular outflow tract prior to descending within the anterior interventricular sulcus.*

ANOMALOUS CORONARY ARTERIES, CT ANGIOGRAPHY

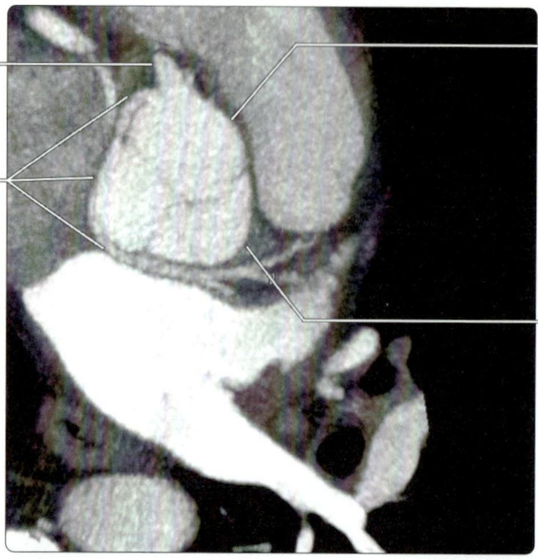

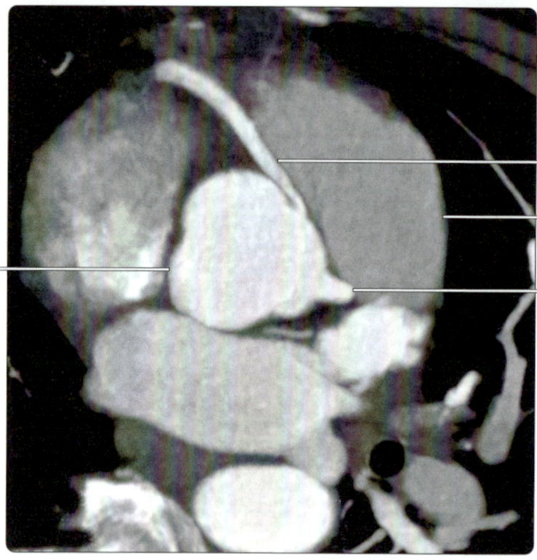

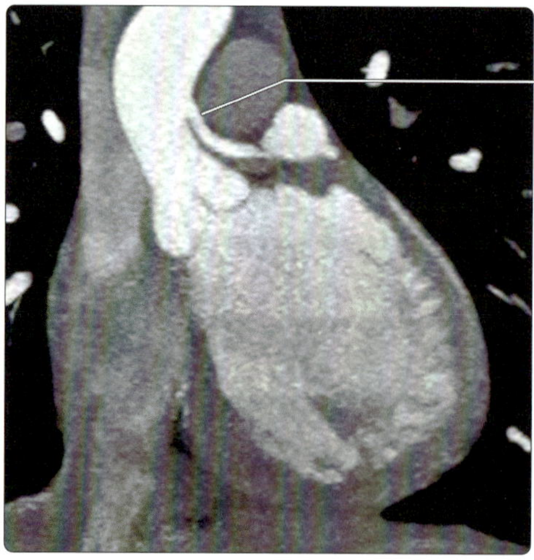

(Top) *CT angiogram MIP reformatted image shows an anomalous origin of the left coronary artery arising from the right coronary cusp. The anomalous artery courses behind the root of the aorta (retroaortic).* **(Middle)** *CT angiogram MIP reformatted image shows an anomalous RCA arising from the left coronary cusp. The anomalous artery has a slit-like origin and courses between the aorta and right ventricular outflow tract (interarterial), 2 reasons implicated in the occurrence of sudden death with this coronary artery anomaly.* **(Bottom)** *CT angiogram coronal oblique MIP reformatted image demonstrates an anomalous high origin of the left main coronary artery, arising above the level of the sinotubular junction of the ascending thoracic aorta.*

CORONARY ARTERY TO SINUS FISTULA, CT ANGIOGRAPHY

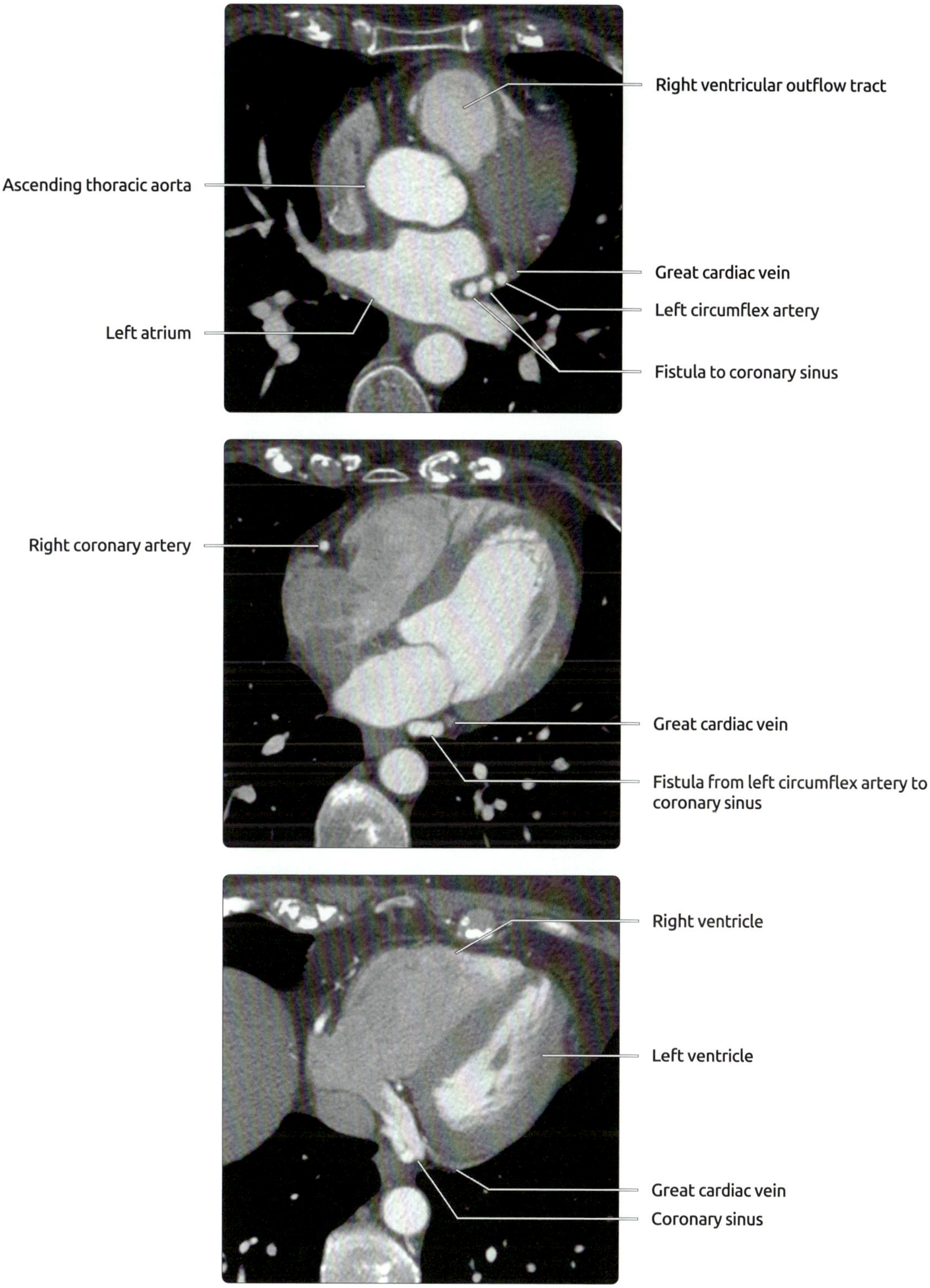

(Top) *First of 3 CT angiogram axial MIP reformatted images of a patient with a left circumflex coronary artery to coronary sinus fistula shows serpiginous vessels that arise from the left circumflex coronary artery.* **(Middle)** *CT angiogram axial MIP reformatted image of the same patient demonstrates a left circumflex coronary artery to coronary sinus fistula.* **(Bottom)** *CT angiogram axial MIP reformatted image of the same patient shows increased enhancement within the coronary sinus in comparison to the poorly opacified adjacent great cardiac vein, secondary to fistulous communication with the LCX.*

MYOCARDIAL BRIDGING, CT ANGIOGRAPHY

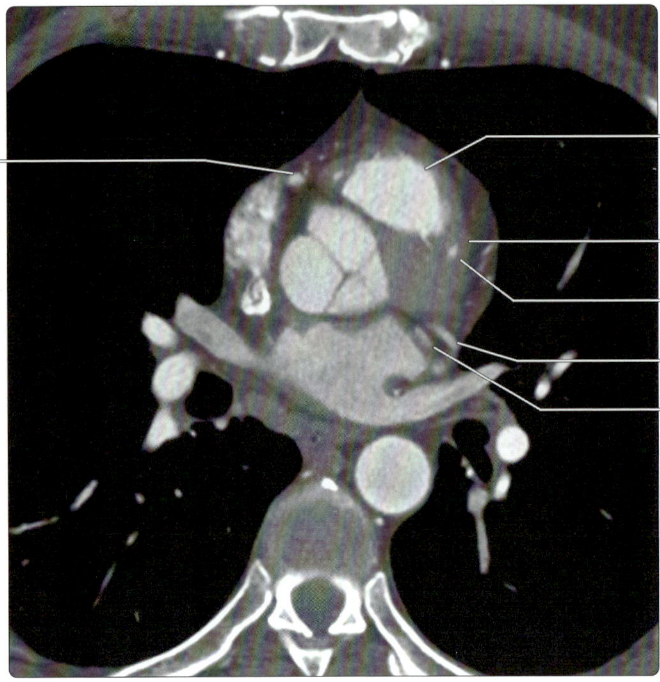

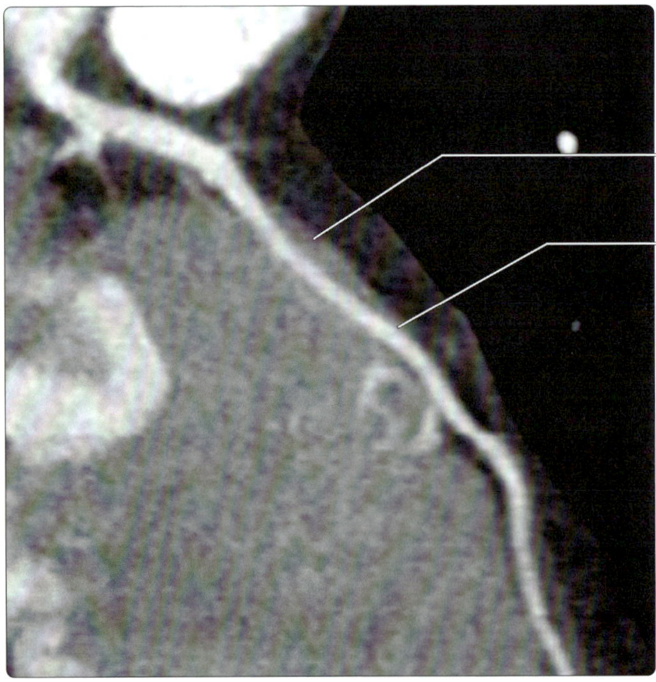

(Top) *Axial CT angiogram demonstrates a segment of myocardial bridging involving the LAD, where it is seen coursing through and surrounded by the left ventricular myocardium.* **(Bottom)** *CT angiogram curved MPR of the LAD demonstrates a short segment of myocardial bridging surrounding the mid-LAD.*

SINGLE RIGHT CORONARY ARTERY, CT ANGIOGRAPHY AND SURFACE-RENDERED CT

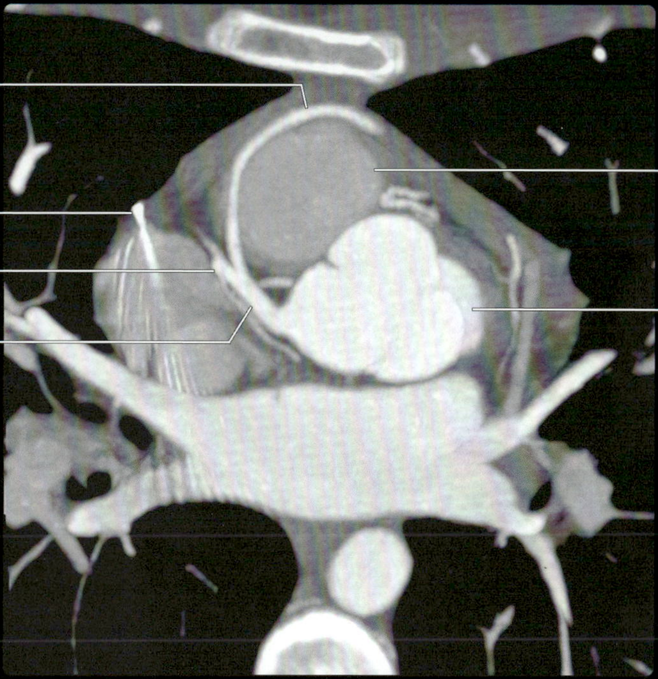

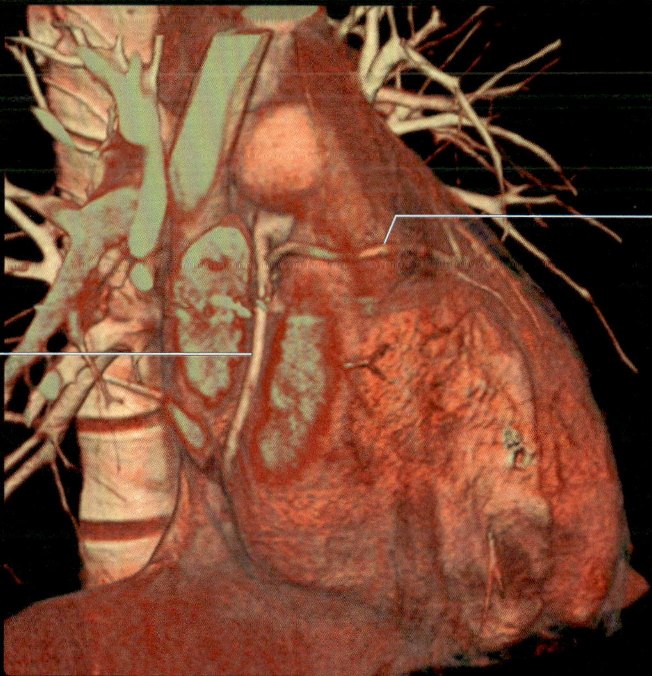

(Top) *CT angiogram axial MIP reformatted image of a patient with D-transposition of the great vessels who underwent arterial switch procedure as an infant shows a single coronary artery arising from the right coronary sinus. This single artery divides into a left coronary artery and an RCA proper, which continues in the expected course of the RCA. The anomalous left coronary artery loops in front of the pulmonary trunk in this patient. A single RCA, with anomalous left coronary artery coursing anterior to the pulmonary trunk, occurs in ~ 10% of patients with D-transposition of great vessels. Note the location of the pulmonary trunk to the right and anterior to the aorta.* **(Bottom)** *CT angiogram 3D surface-rendered image of the same patient shows the course of the anomalous left coronary artery, anterior to the pulmonary trunk.*

ANTERIOR VIEW OF CORONARY VEINS

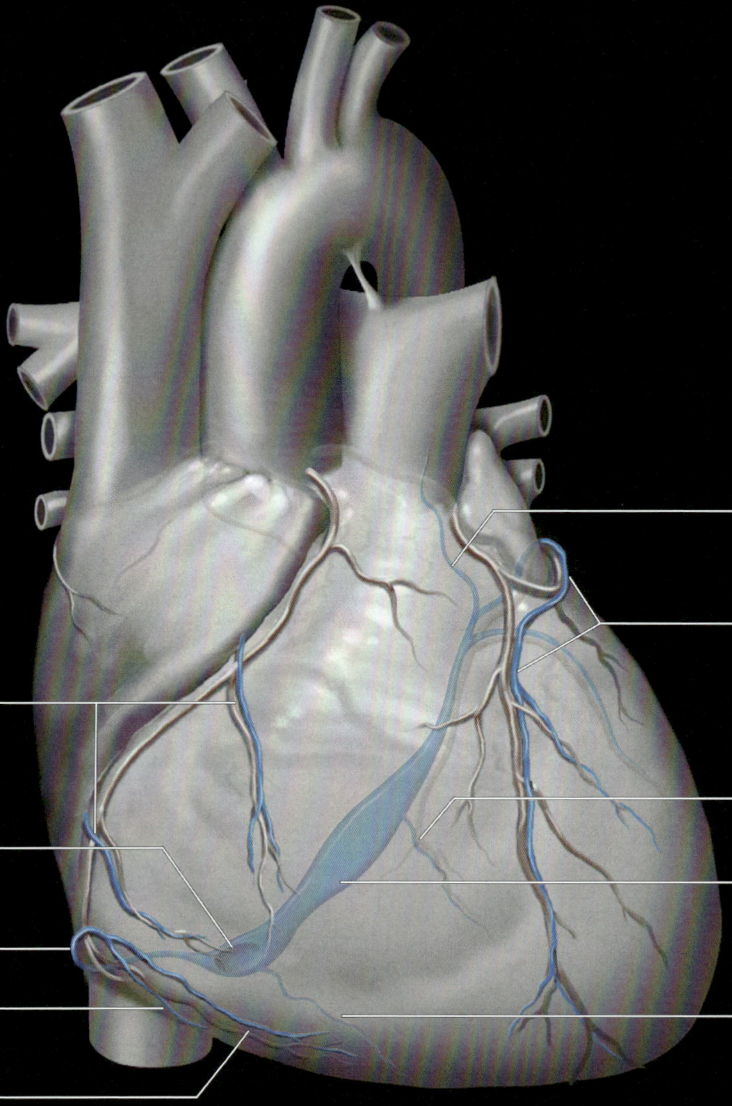

Graphic illustrates the venous drainage of the heart. The great cardiac vein originates at the cardiac apex, runs in the anterior interventricular groove, and curves to the left in the coronary sulcus to drain into the left end of the coronary sinus. The oblique vein of the left atrium runs obliquely on the posterior aspect of the left atrium to merge with the great cardiac vein. The small cardiac vein runs in the coronary sulcus to drain into the right end of the coronary sinus. The middle cardiac vein originates at the cardiac apex, runs in the posterior interventricular groove, and drains into the right end of the coronary sinus. The posterior vein of the left ventricle runs on the diaphragmatic surface of the left ventricle to reach the coronary sinus. The anterior cardiac veins (3 or 4) drain directly into the right atrium. The right marginal vein may join the small cardiac vein or drain directly into the right atrium.

CORONARY SINUS, SURFACE-RENDERED CT ANGIOGRAPHY

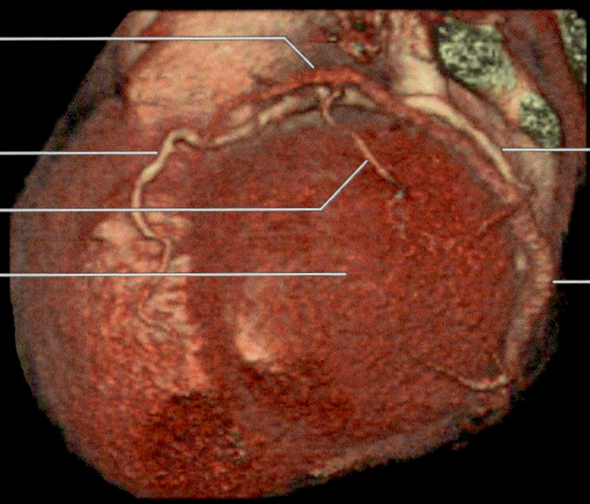

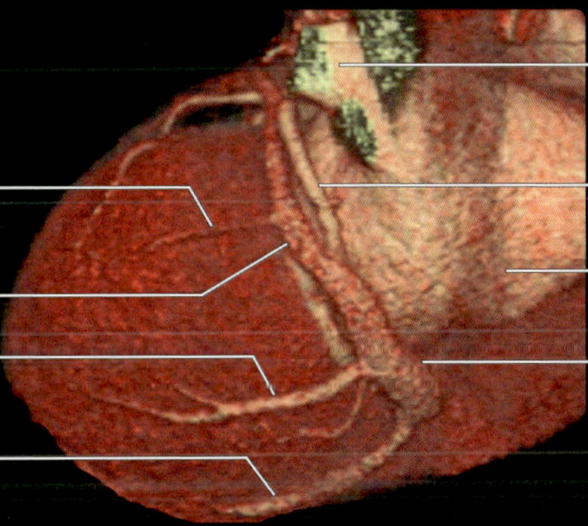

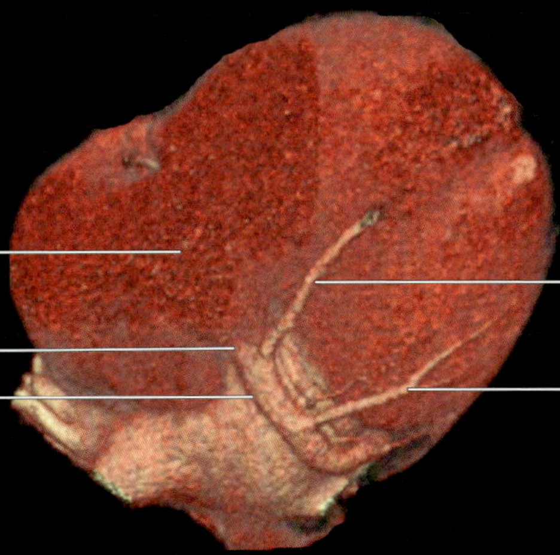

(Top) *First of 3 CT angiogram 3D surface-rendered images of the same patient shows the coronary sinus venous system. CT angiogram 3D surface-rendered left lateral view shows the great cardiac vein as it courses along the LAD in the anterior interventricular groove and curves in the coronary sinus to course along the LCX. The confluence of the great cardiac vein with the oblique vein of the left atrium (not shown) marks the origin of the coronary sinus. Multiple left ventricular veins drain into the great cardiac vein.* **(Middle)** *CT angiogram 3D surface-rendered posterior view of the coronary sinus shows its major tributaries: The great cardiac vein, the middle cardiac vein, and the posterior vein of the left ventricle.* **(Bottom)** *CT angiogram 3D surface-rendered diaphragmatic view shows the coronary sinus as it drains into the right atrium. The middle cardiac vein courses along the PDA in the posterior interventricular groove.*

SINGLE CORONARY ARTERY FROM RIGHT CORONARY SINUS, MR

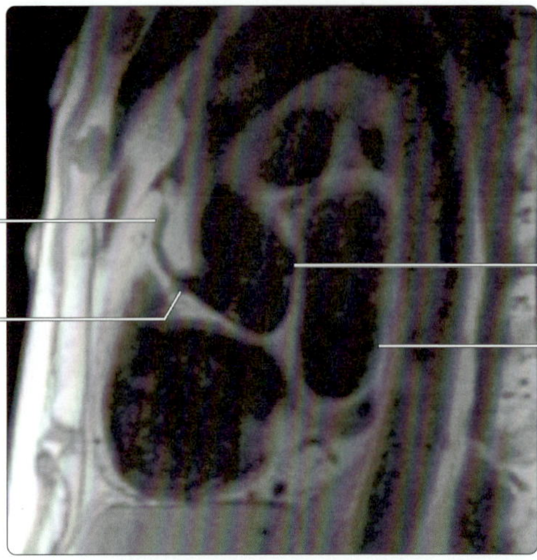

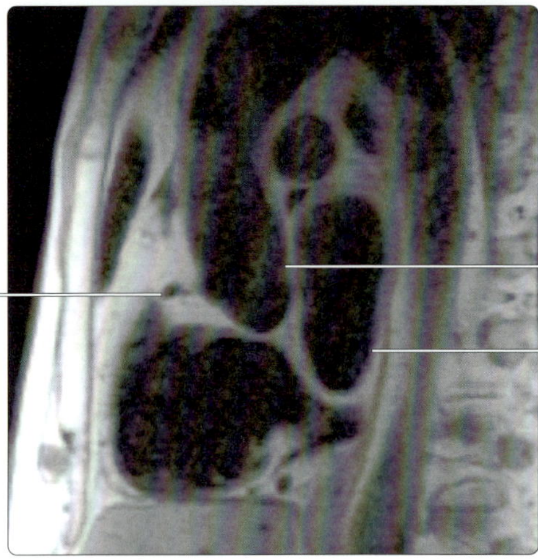

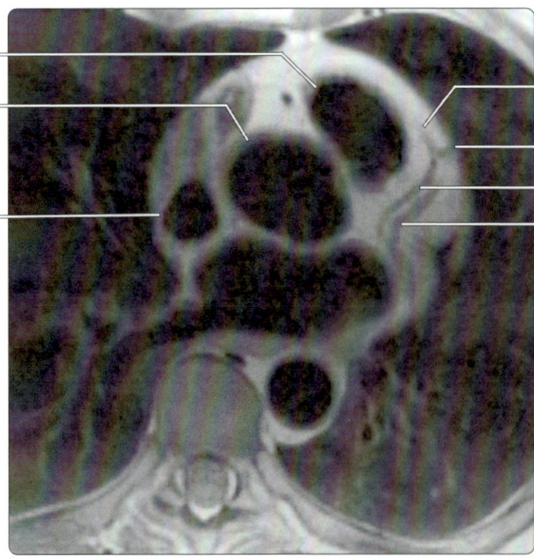

(Top) *First of 3 double IR MR images of a 70-year-old patient with chest pain is shown. Sagittal double IR MR through the root of the aorta shows a single coronary artery segment arising from the right coronary sinus. The left main coronary artery arises from the common trunk and is directed upward.* **(Middle)** *Sagittal double IR MR shows the RCA, which has a normal course.* **(Bottom)** *Axial double IR MR shows the left main coronary artery (after looping anterior to the pulmonary trunk) dividing into the LAD and the LCX. The LAD descends in the anterior interventricular groove, while the LCX, which has a longer course, runs in the upper part of the anterior interventricular groove (where the LAD normally runs) and then curves to assume its normal course in the coronary sulcus.*

BIVENTRICULAR PACING, RADIOGRAPHY

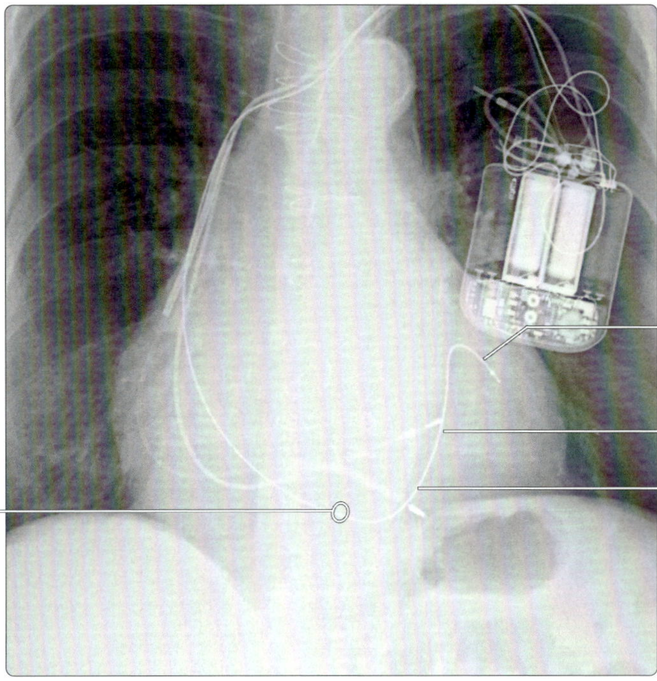

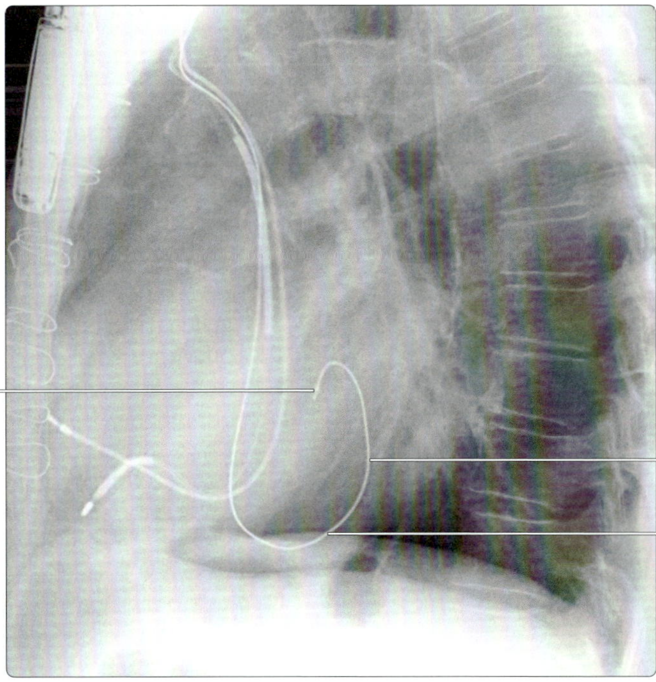

(Top) *First of 2 chest radiographs of a patient with dilated cardiomyopathy and biventricular pacing for cardiac resynchronization therapy is shown. PA chest radiograph shows a left pectoral transvenous left ventricular pacing lead extending posteriorly from the right atrium into the coronary sinus, along the course of the great cardiac vein, and into an anterior ventricular vein. The coronary sinus lead shows the oblique orientation of the coronary sinus in the coronary sulcus.* **(Bottom)** *Lateral chest radiograph shows the coronary sinus lead, which outlines the location and orientation of the coronary sinus. Prolonged conduction in patients with heart failure leads to asynchronous contraction of the ventricles and decreased cardiac output. Biventricular pacing results in synchronization of ventricular contraction and improved cardiac output. Stimulation of the left ventricle is achieved by the coronary sinus lead.*

ANATOMY OF PERICARDIUM

General Concepts

- Definition: Sac-like membrane; surrounds heart and proximal great vessels; boundary between International Thymic Malignancy Interest Group (ITMIG) prevascular and visceral mediastinal compartments
- Function
  - Heart protection; maintenance of normal heart position
  - Limits heart distention and pericardial fluid volume
  - Pericardial space
    - Permits cardiac motion and changes in morphology
  - Pericardial fluid
    - Provides lubrication during cardiac motion
    - Up to 15-50 mL of serous fluid in normal subjects

Fibrous Pericardium

- Cone-shaped, sac-like, tough connective tissue layer
  - Surrounds heart but is not attached to it
  - Lined internally by serous parietal pericardium
  - Continuous with great vessel adventitia
  - Continuous with pretracheal fascia
- Defines **anatomic middle mediastinum**
- Attachments
  - Base attached to central tendinous portion of diaphragm
    - Pericardiophrenic ligament
  - Anterior surface attached to anterior chest wall (posterior aspect of sternum)
    - Superior sternopericardial ligaments
    - Inferior sternopericardial ligaments
- Conduit for neurovascular structures
  - Phrenic nerves
  - Pericardiacophrenic vessels

Serous Pericardium

- Closed sac-like membrane that surrounds heart **within fibrous pericardium**
- Continuous serosal layer with parietal and visceral components
  - Lined by mesothelial cells
- **Serous parietal pericardium**
  - Lines internal surface of fibrous pericardium
- **Serous visceral pericardium**
  - Adherent to heart
    - a.k.a. **epicardium**
    - Covers **subepicardial fat**

Pericardial Reflections

- Superior reflection surrounds
  - Ascending aorta
  - Pulmonary trunk
- Posterior reflection surrounds
  - Superior vena cava
  - Inferior vena cava (IVC)
  - Pulmonary veins

Pericardial Innervation

- Vagus nerves
- Sympathetic trunks
- Phrenic nerves (C3, C4, C5)
  - Somatic innervation of parietal pericardium
  - Supraclavicular region dermatomes; referred shoulder pain in cases of pericardial inflammation

Pericardial Vessels

- Arterial supply
  - Internal thoracic arteries
    - Pericardiacophrenic arteries
    - Musculophrenic arteries
  - Thoracic aorta
    - Superior phrenic arteries
    - Coronary arteries
- Venous drainage
  - Azygos system
  - Internal thoracic veins
    - Pericardiacophrenic veins

PERICARDIAL SPACE AND PERICARDIAL RECESSES

General Concepts

- Serosal pericardial reflections arranged around 2 tube-like structures
  - One "tube" encloses aorta and pulmonary trunk
  - One "tube" encloses superior and inferior venae cavae and pulmonary veins
- **Transverse sinus**
  - Passage located between 2 tube-like structures
- **Oblique sinus**
  - Posterior pericardial extension behind left atrium
- No communication between transverse and oblique sinuses
  - Separated by 2 pericardial reflections

Transverse Sinus

- General
  - Horizontal orientation
  - Posterior to ascending aorta and pulmonary trunk
  - Superior to left atrium
  - Superior and anterior to oblique sinus
  - May mimic aortic dissection or lymphadenopathy
- **Superior aortic recess**
  - Superior extent of transverse sinus
  - Partially surrounds ascending aorta
  - Abuts aorta without intervening tissue plane
  - **Posterior portion**
    - a.k.a. superior pericardial recess/sinus
    - Crescent-shaped fluid collection abutting posterior wall of ascending aorta
    - May extend cephalad into right paratracheal region
    - May mimic lymphadenopathy or foregut cyst
  - **Anterior portion**
    - Triangular-shaped recess with beak-like extension between ascending aorta and pulmonary trunk
    - May mimic aortic dissection
  - **Right lateral portion**
    - Between ascending aorta and superior vena cava
- **Inferior aortic recess**
  - Inferior extension of transverse sinus
  - Crescent-shaped recess between right lateral ascending aorta and right atrium
    - Posterior to aorta
    - Anterior to left atrium

- **Right and left pulmonic recesses**
  - Lateral extensions of transverse sinus
  - **Right pulmonic recess**
    - Inferior to proximal right pulmonary artery
  - **Left pulmonic recess**
    - Inferior to left pulmonary artery, superior to left superior pulmonary vein

Pericardial Cavity

- Gives rise to postcaval and pulmonary venous recesses
- **Postcaval recess**
  - Small fluid-filled recess along posterolateral aspect of superior vena cava
- **Right and left pulmonary venous recesses**
  - Small fluid-filled recesses along lateral heart borders
  - Around superior and inferior pulmonary veins
  - May mimic lymphadenopathy

Oblique Sinus

- **Inferior to transverse sinus**
  - Separated from transverse sinus by double pericardial reflection
  - **No communication between transverse** and **oblique sinuses**
- Superior and posterior to left atrium
- Posterior to right pulmonary artery, medial to bronchus intermedius
  - Sometimes called posterior pericardial recess
- May mimic lymphadenopathy, esophageal pathology, or foregut cyst

IMAGING OF NORMAL PERICARDIUM

General Concepts

- Typically imperceptible on radiography
- **Echocardiography**: Modality of choice for initial evaluation
  - Assessment of pericardial effusion and tamponade
  - Inability to assess entire pericardium or detect pericardial thickening

Cross-Sectional Imaging (CT/MR)

- Normal pericardium thickness of **1-2 mm**
  - Measurement reflects combined thicknesses of fibrous pericardium, serous parietal pericardium, and serous visceral pericardium
  - Serous visceral pericardium (**epicardium**) closely adherent to heart surface
- Optimal pericardial visualization anterior to right ventricle
  - Outlined by subepicardial and mediastinal fat
- Visualization/identification of pericardial sinuses and recesses
  - May mimic lymphadenopathy or aortic dissection
- MR
  - Pericardial visualization in 80% of normal subjects
  - Normal pericardium exhibits low signal on T1- and T2-weighted SE images
  - Outlined by bright signal mediastinal and subepicardial fat
  - Increased thickness near diaphragm; ligamentous insertions

ANATOMY-RELATED IMAGING ABNORMALITIES

Congenital Absence of Pericardium

- **Complete absence**: Left cardiac displacement and posterior cardiac rotation
- **Partial absence**
  - Lung interposed between ascending aorta and pulmonary trunk on CT/MR
  - More common than complete absence, usually affects left pericardium
  - Association with potentially life-threatening left atrial herniation

Pericardial Cyst

- Developmental anomaly; noncommunicating fluid-filled outpouching of **parietal pericardium**
- Typically in **anterior cardiophrenic angle** (70% on right)
- Imaging
  - Smooth, round, ovoid, or teardrop-shaped mass abutting heart; homogeneous fluid attenuation/signal
  - Thin, imperceptible wall; no contrast enhancement

Pericardial Effusion

- Abnormal amount of pericardial fluid; up to 300 mL
- Imaging
  - **Water bottle sign**: Morphology of cardiopericardial silhouette secondary to large pericardial effusion, resembles old-fashioned water bottle
  - **Fat pad sign**: Water density band (> 4 mm) between subepicardial and mediastinal fat stripes on lateral radiography; a.k.a. **Oreo cookie sign**
  - Uncomplicated effusion: Homogeneous water attenuation/signal on CT/MR
  - Hemopericardium: High attenuation; high signal on T1 MR; low signal on gradient-echo cine images
- **Cardiac tamponade**: Pericardial fluid accumulation → significant heart compression → ↓ cardiac output → shock; **clinical diagnosis**

Pericardial Thickening

- Thickness > 4 mm
- Difficult distinction from small pericardial effusion on CT
- **Constrictive pericarditis**
  - Symptomatic patients with ventricular dysfunction
  - Thick, fibrotic pericardium limits ventricular expansion
  - Etiologies: Radiation, heart surgery, infection (tuberculosis), neoplasia, connective tissue disorders
  - CT and MR findings that suggest diagnosis
    - Pericardial thickening and calcification; diffuse, focal, or multifocal
    - Tubular ventricular deformity (particularly right ventricle) and decreased volume
    - Flat or sigmoid (inverted) interventricular septum
    - Dilated atria, coronary sinus, systemic veins

Pneumopericardium

- Air in pericardial space typically from barotrauma, less frequently due to blunt/penetrating trauma
- May coexist with pneumomediastinum &/or pneumothorax
- May produce cardiac tamponade

OVERVIEW OF PERICARDIUM

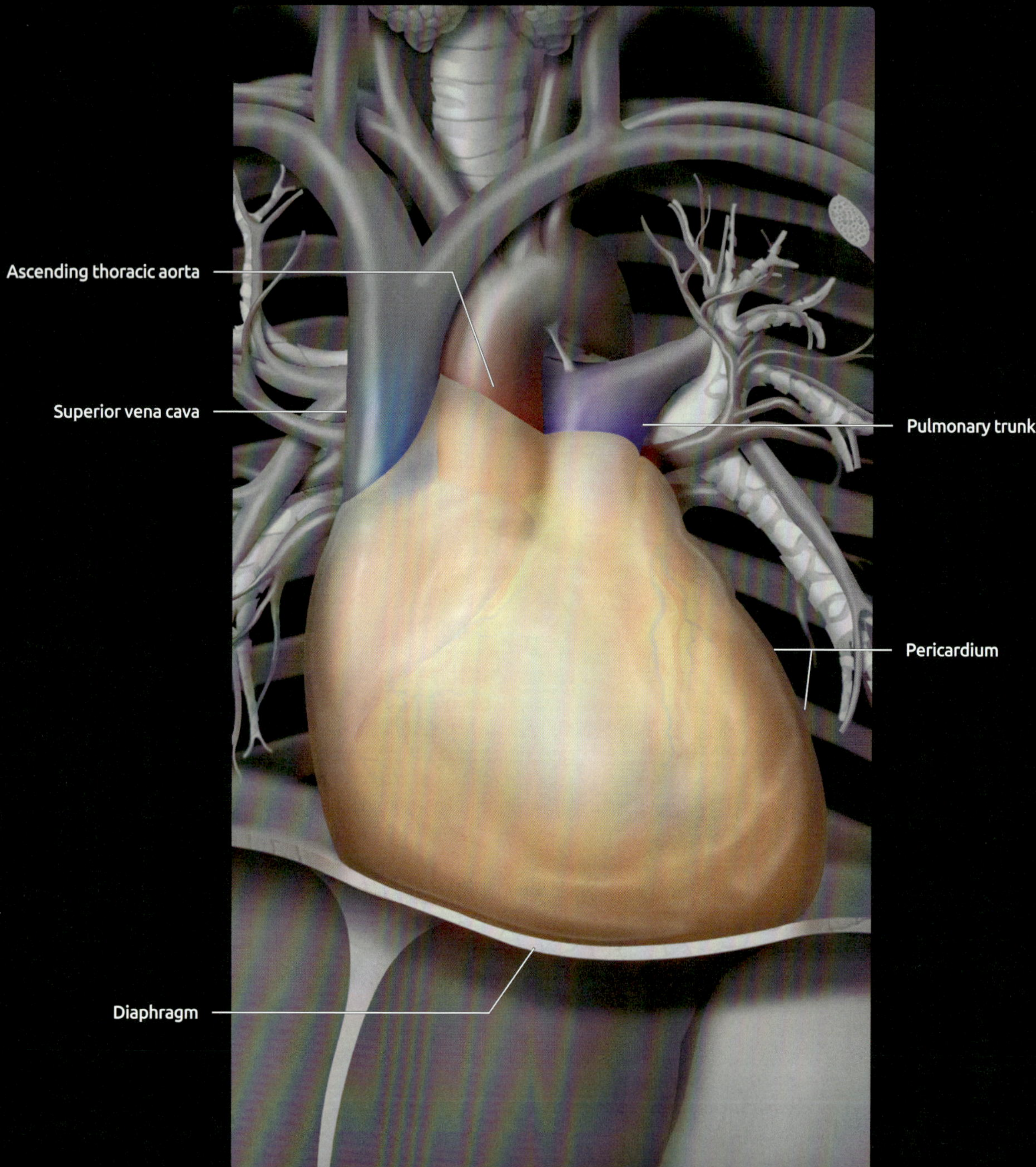

The pericardium is a sac-like membrane that surrounds the heart and the origins of the thoracic great vessels. It has a conical shape with the apex of the cone oriented inferiorly and to the left. The pericardium normally contains up to 15-50 mL of serous fluid that lubricates its apposing surfaces during cardiac motion. The pericardium helps maintain the normal position of the heart in the thorax. Within the pericardial cavity, the heart may move relatively freely and change its morphology, but the pericardium poses some restrictions on cardiac motion and pericardial fluid volume.

ANATOMY OF PERICARDIUM

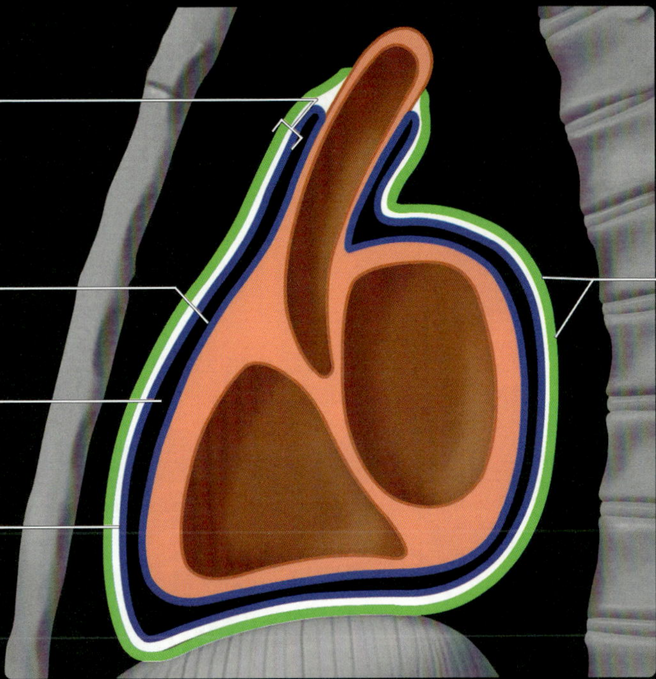

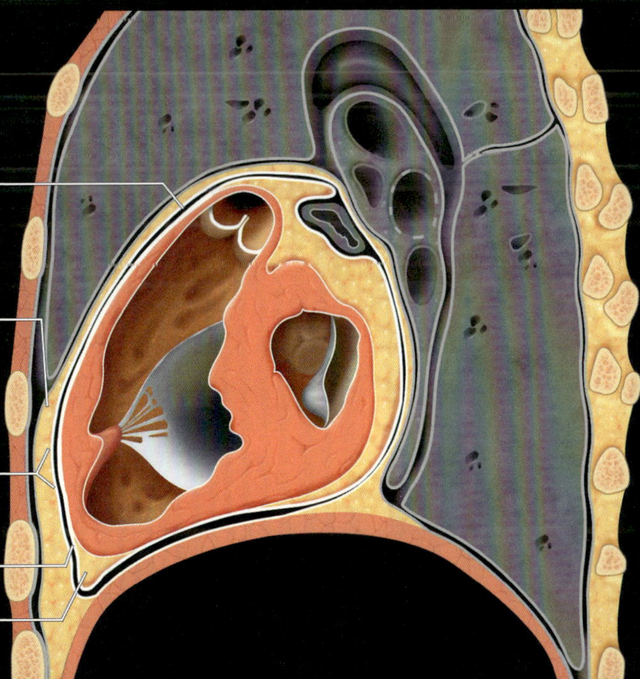

(Top) *Graphic shows the cross-sectional anatomy of the 3 pericardial layers. The fibrous pericardium (outlined in green) surrounds the heart, is composed of tough connective tissue, is continuous with the proximal great vessel adventitia, and defines the anatomic middle mediastinum. The serous pericardium (outlined in blue) surrounds the heart (within the confines of the fibrous pericardium) and is composed of 2 continuous thin serous pericardial layers lined by mesothelial cells. The serous parietal pericardium lines the internal surface of the fibrous pericardium. The serous visceral pericardium (or epicardium) lines the heart. Between the 2 serous pericardial layers is the potential pericardial space that contains a small amount of pericardial fluid (up to 15-50 mL).* **(Bottom)** *Graphic shows the pericardial cavity in sagittal cross section and demonstrates the anatomic location of the pericardial layers with respect to surrounding fat. Visualization of the pericardium on cross-sectional imaging is enhanced by subepicardial fat deep to the serous visceral pericardium (or epicardium) and mediastinal fat surrounding the combined thicknesses of the serous parietal pericardium and fibrous pericardium.*

ANATOMY OF PERICARDIUM

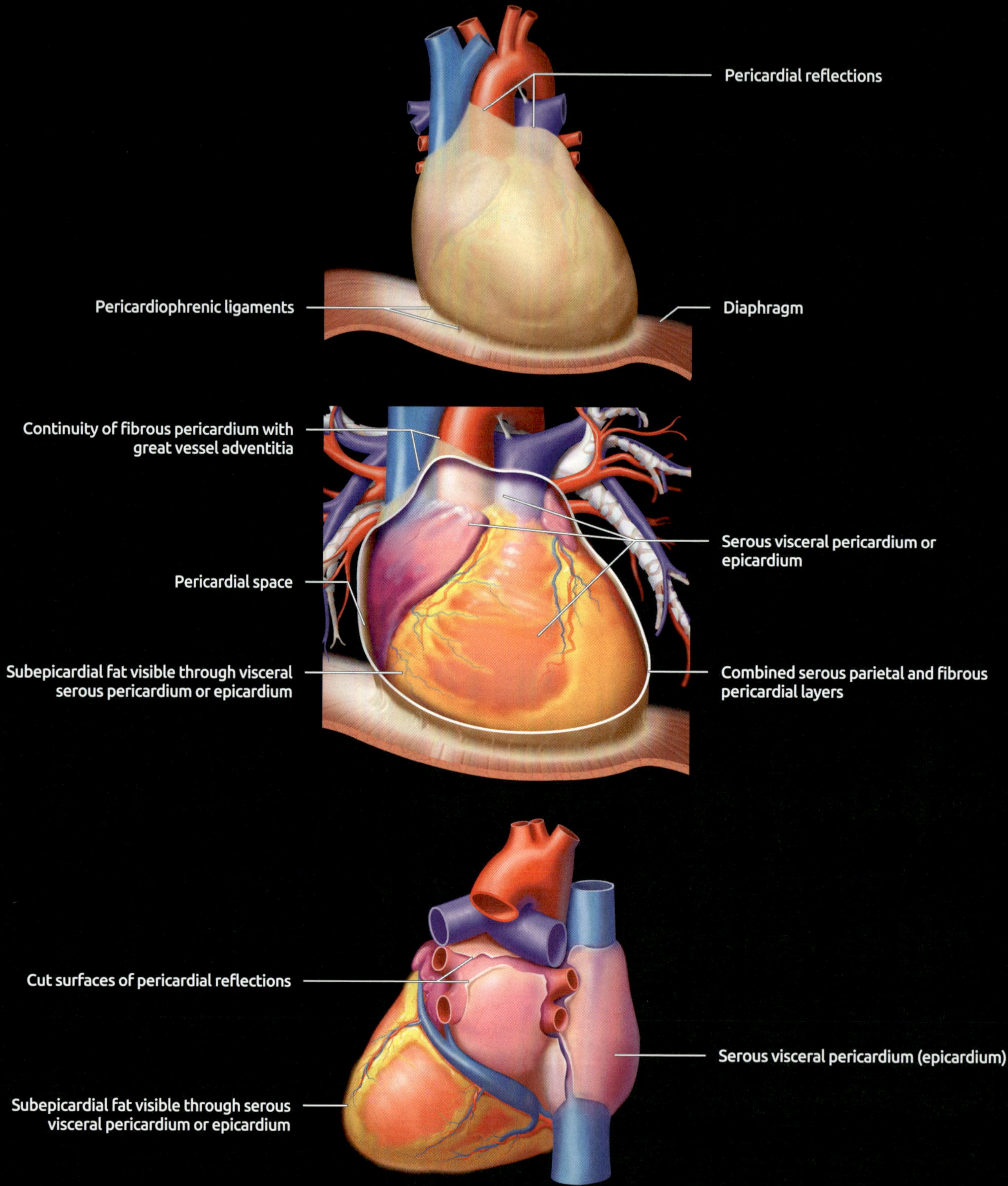

(Top) *Graphic shows the anterior surface anatomy of the fibrous pericardium, which forms the boundaries of the anatomic middle mediastinum. The base of the pericardium is attached to the diaphragm by the pericardiophrenic ligaments and to the chest wall by the sternopericardial ligaments (not shown). These attachments anchor the pericardium, hold the heart and pericardium in place, and help prevent excessive cardiac motion.* **(Middle)** *Graphic shows the pericardial space or cavity after "cutting away" the anteriorly located combined fibrous and serous parietal pericardial layers. The fibrous pericardium is continuous with the thoracic great vessel adventitia and is lined by the serous parietal pericardium, which is continuous with the serous visceral pericardium (epicardium) that covers the heart and its subepicardial fat.* **(Bottom)** *Graphic shows the posterior heart surface after "cutting away" the posterior aspect of the combined fibrous and serous parietal pericardial layers to reveal the posterior visceral serous pericardium and its reflections.*

ANATOMY OF PERICARDIAL RECESSES

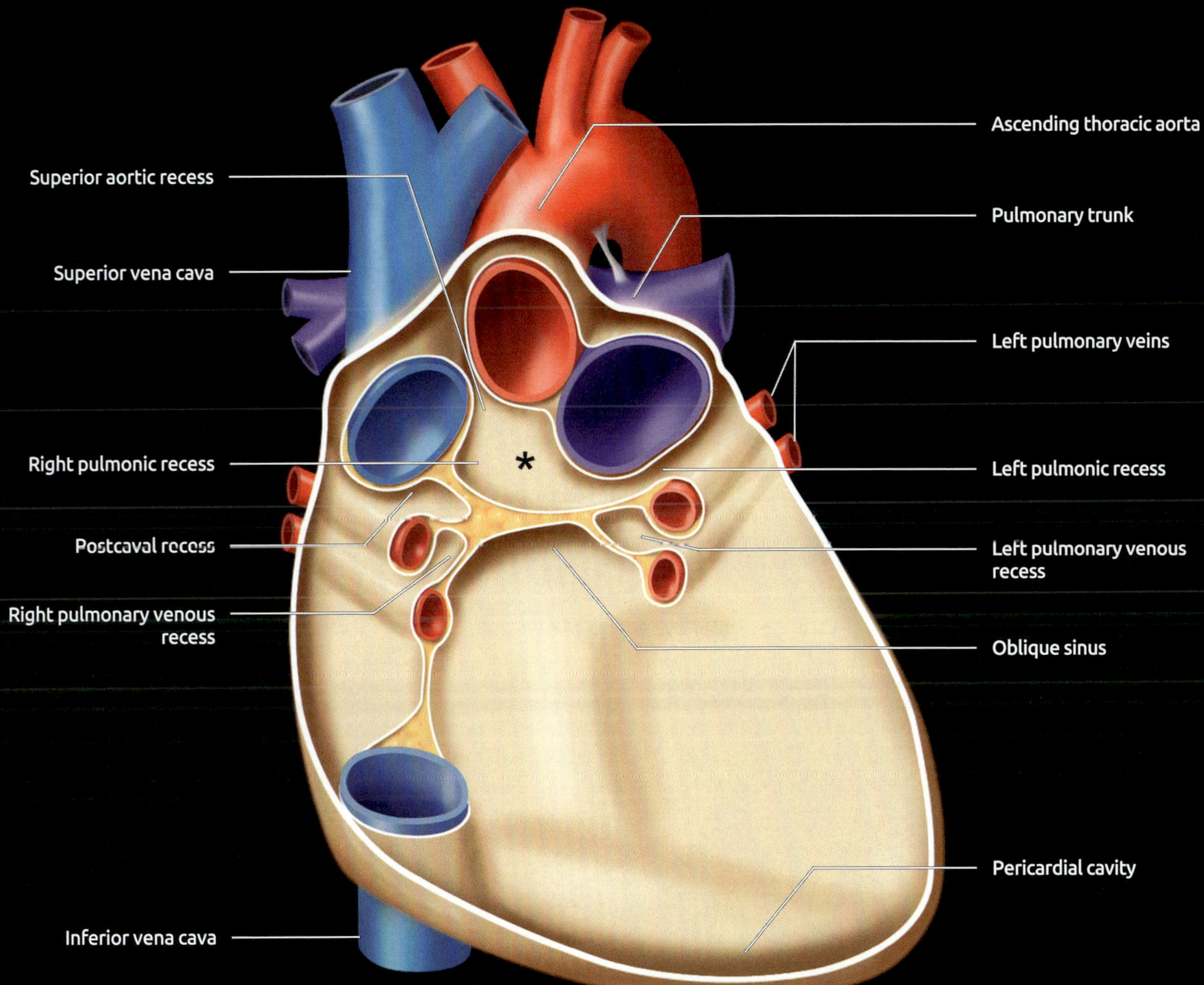

Graphic depicts the pericardial sinuses and recesses after removing the anterior pericardium and the heart. Two tube-like structures, one surrounding the ascending thoracic aorta and pulmonary trunk and the other surrounding the superior and inferior venae cavae and pulmonary veins, are separated by the transverse sinus (*). The transverse and oblique sinuses are separated by 2 pericardial reflections. The superior aortic, inferior aortic, right pulmonic, and left pulmonic recesses arise from the transverse sinus. The oblique sinus is located above and behind the left atrium, does not give rise to recesses, and does not communicate with the transverse sinus. The postcaval, left pulmonary venous, and right pulmonary venous recesses arise from the pericardial cavity proper.

PERICARDIUM, AXIAL CT

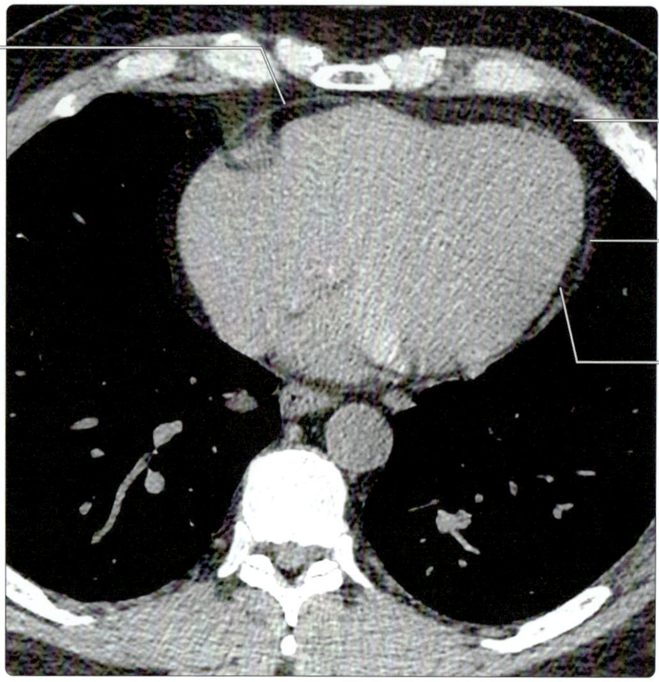

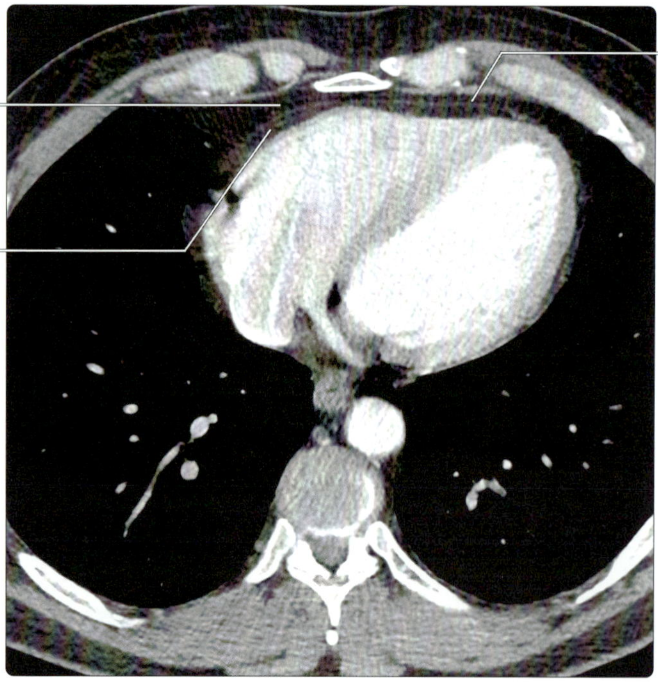

(Top) *Axial NECT (mediastinal window) shows the cross-sectional imaging appearance of the normal pericardium, which manifests as a soft tissue attenuation thin linear structure outlined by subepicardial fat immediately adjacent to the heart and mediastinal fat along its periphery. The thin pericardial line identified on CT represents the combined thicknesses of the fibrous pericardium and the apposing serous parietal and serous visceral pericardial layers. Note that there is optimal visualization of the normal pericardium anteriorly, typically anterior to the right ventricle. The posterior aspect of the normal pericardium is not always visible on cross-sectional imaging.* **(Bottom)** *Axial CECT (mediastinal window) of the same patient shows the normal pericardium manifesting as a thin soft tissue attenuation linear structure outlined by subepicardial and mediastinal fat. Note poor visualization of the posterior aspect of the pericardium. The normal thickness of the pericardium is < 2 mm in most cases but may normally reach up to 4 mm due to variations in normal pericardial fluid volume.*

PERICARDIUM, CORONAL AND SAGITTAL CT

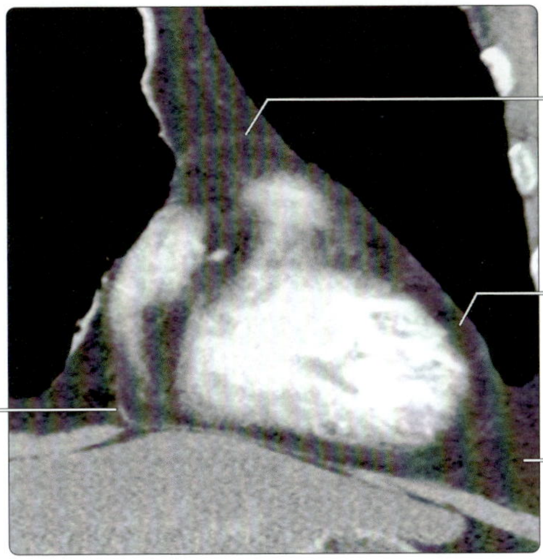

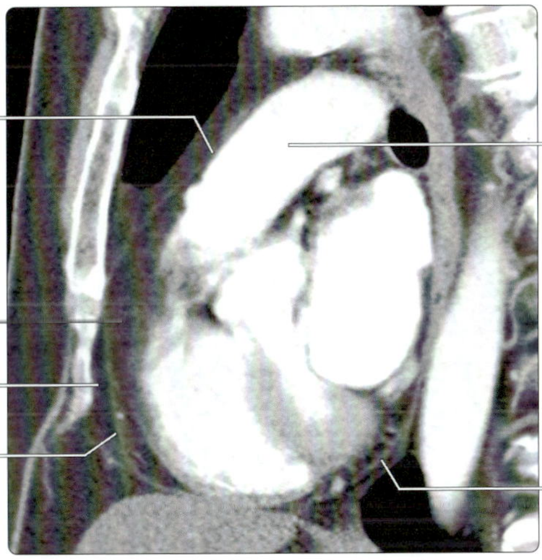

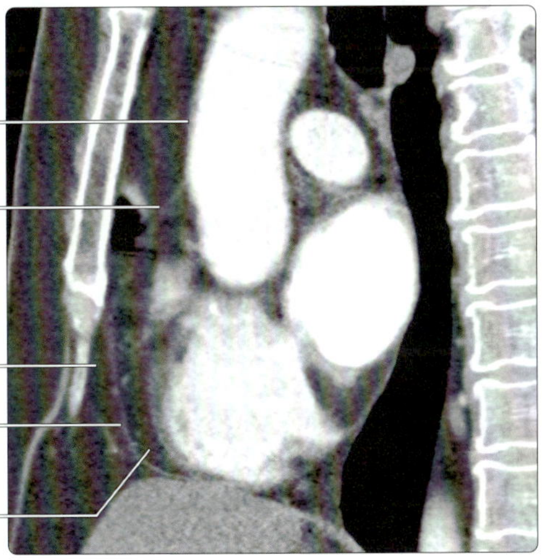

(Top) *First of 3 gated cardiac CECT images of the same patient shows the coronal appearance of the normal pericardium. Coronal gated cardiac CECT (mediastinal window) through the anterior heart shows the normal pericardium manifesting as a soft tissue attenuation thin linear structure. The superior aspect of the pericardium manifests as an ill-defined soft tissue band.* **(Middle)** *Sagittal gated cardiac CECT (mediastinal window) through the pulmonary trunk shows the anterior and posterior aspects of the normal pericardium. The normal pericardium is usually not visible over the atria or left ventricle. Note the superior pericardial reflection at the pulmonary trunk where the fibrous pericardium is continuous with the vascular adventitia.* **(Bottom)** *Sagittal gated cardiac CECT (mediastinal window) through the ascending thoracic aorta demonstrates the normal pericardium and its superior pericardial reflection. The superior pericardium surrounds the proximal thoracic great vessels and is continuous with the vascular adventitia of the ascending thoracic aorta.*

PERICARDIUM, AXIAL AND CORONAL MR

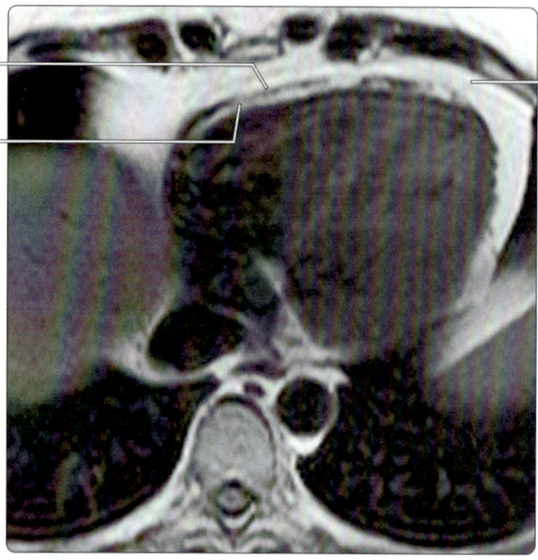

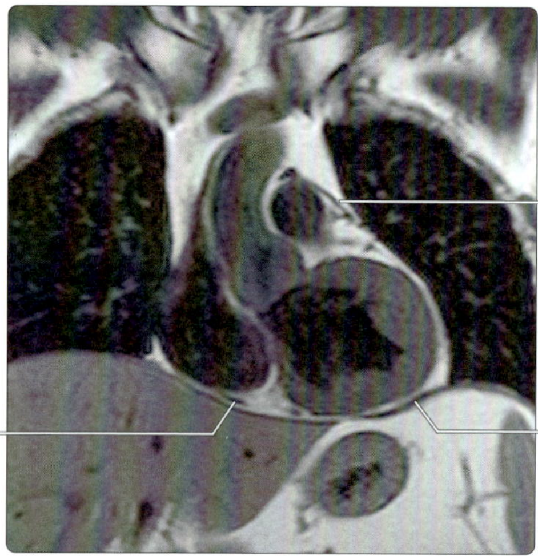

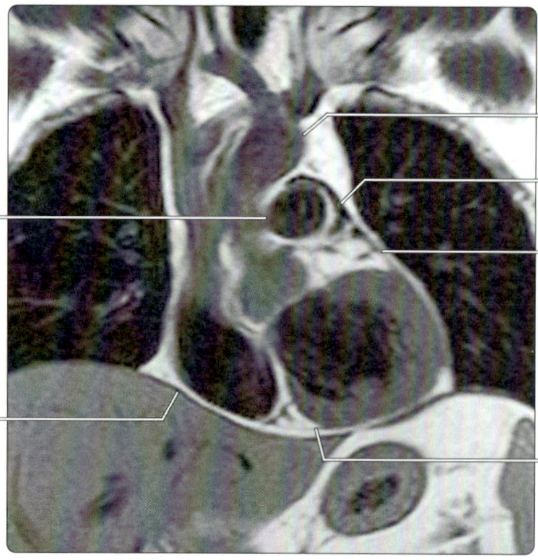

(Top) *First of 6 normal double inversion recovery (IR) MR images through the mediastinum of the same patient shows the normal pericardium. Axial double IR MR through the inferior aspect of the heart shows the normal pericardium manifesting as a thin, low-signal intensity linear structure outlined by high-signal subepicardial fat and mediastinal fat. The pericardium is typically optimally visualized along its anterior surface on both CT and MR imaging.* **(Middle)** *Coronal double IR MR through the aortic root shows the normal pericardium manifesting as a low-signal intensity thin linear structure. Note the superior pericardial reflection that surrounds the proximal pulmonary trunk. The superior aspect of the fibrous pericardium is continuous with the great vessel adventitia.* **(Bottom)** *Coronal double IR MR through the proximal aortic arch shows the thin, low-signal pericardium surrounded by high-signal subepicardial and mediastinal fat. The superior pericardial reflection is visible adjacent to the pulmonary trunk and contains a small amount of fluid located within the left pulmonic and superior aortic pericardial recesses of the transverse sinus.*

PERICARDIUM, SAGITTAL MR

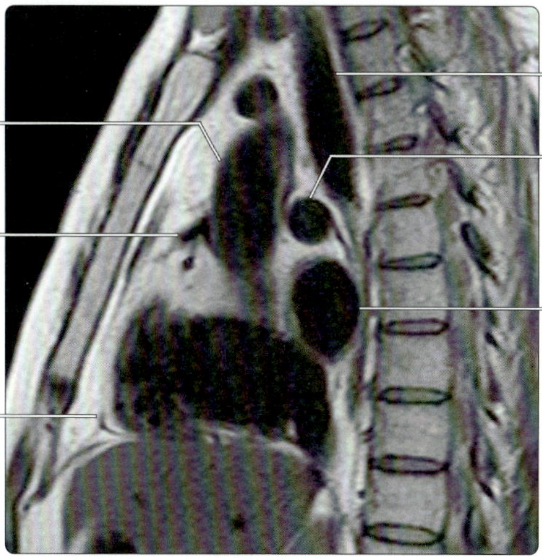

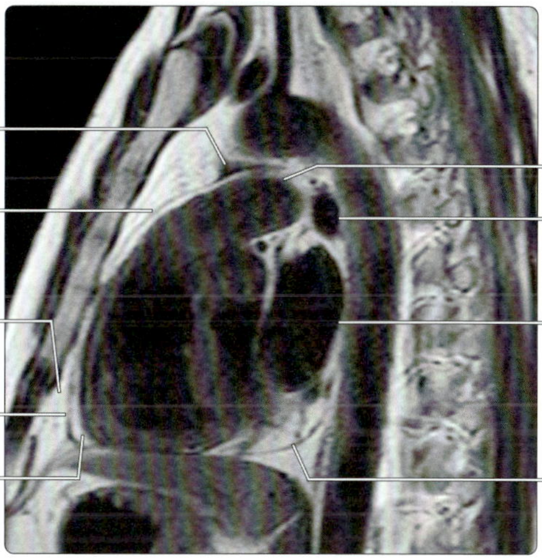

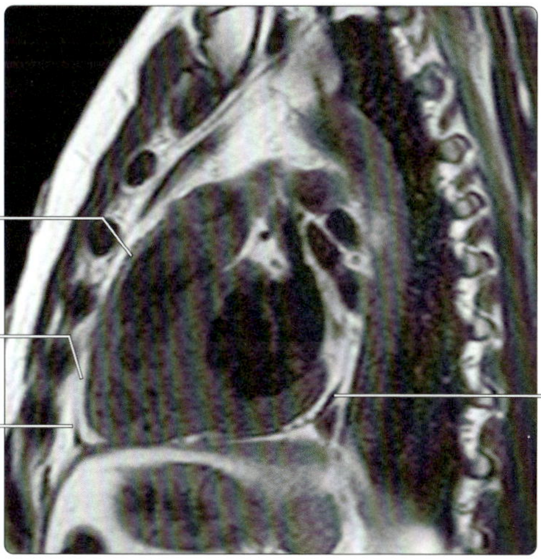

(Top) *Sagittal double IR MR through the ascending thoracic aorta shows the normal pericardium visible anteriorly and inferiorly. The superior pericardial reflection is also visible, as is fluid in the superior aortic recess of the transverse sinus.* **(Middle)** *Sagittal double IR MR through the pulmonary trunk shows the anterior and posterior aspects of the normal pericardium, which manifests as a low-signal intensity thin linear structure surrounded by high-signal fat (subepicardial and mediastinal). The pericardium is not normally visible adjacent to the left atrium. Note the small amount of normal pericardial fluid in the anterior portion of the superior aortic recess of the transverse sinus.* **(Bottom)** *Sagittal double IR MR through the right ventricular outflow tract demonstrates visualization of the normal pericardium. In this case, the posterior pericardium over the left ventricle is well visualized due to a small amount of normal pericardial fluid.*

PERICARDIAL SINUSES, AXIAL CT

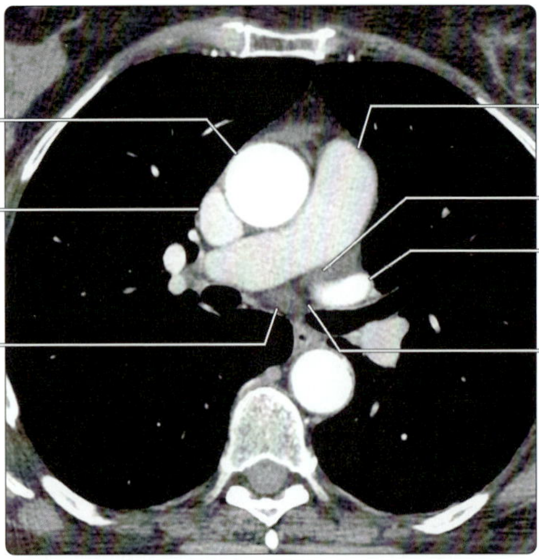

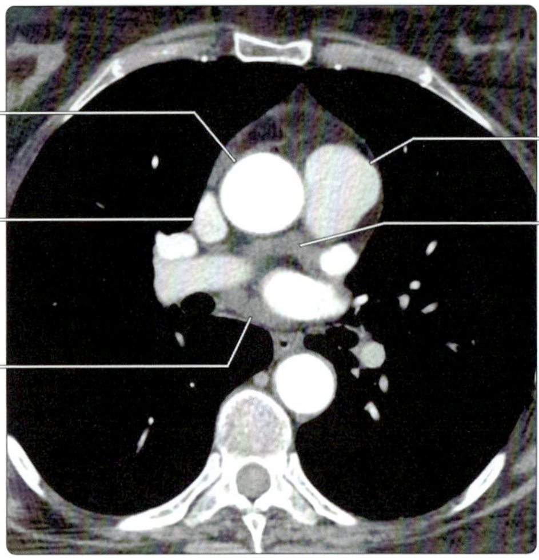

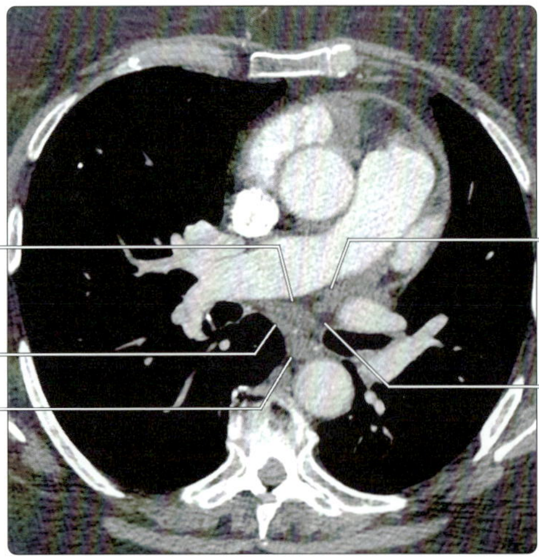

(Top) *Axial CECT (mediastinal window) shows the transverse sinus and the oblique sinus of the pericardium. These sinuses do not communicate with each other and are separated by 2 pericardial reflections that may manifest as a linear fat attenuation septum-like structure, as in this case.* **(Middle)** *Axial CECT (mediastinal window) of the same patient shows the transverse and oblique pericardial sinuses, which are separated by 2 pericardial reflections that manifest as intervening fat attenuation. There is no direct communication between the pericardial sinuses. The transverse sinus gives rise to superior aortic, inferior aortic, and bilateral pulmonic recesses. The oblique sinus does not give rise to any pericardial recess.* **(Bottom)** *Axial CECT (mediastinal window) of a patient with a small-volume high-attenuation pericardial effusion shows the transverse and oblique pericardial sinuses separated by 2 pericardial reflections that manifest as linear fat attenuation. Note the normal mediastinal lymph node located posterior to the oblique sinus. In cases such as this, the combined oblique sinus and adjacent normal lymph node could potentially be misinterpreted as lymphadenopathy.*

PERICARDIAL RECESSES, AXIAL CT

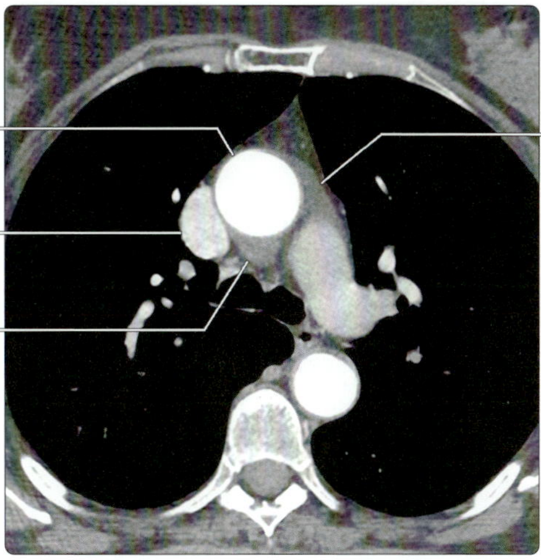

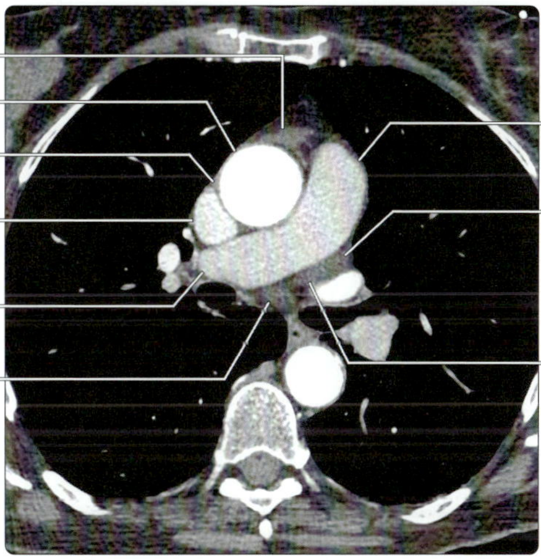

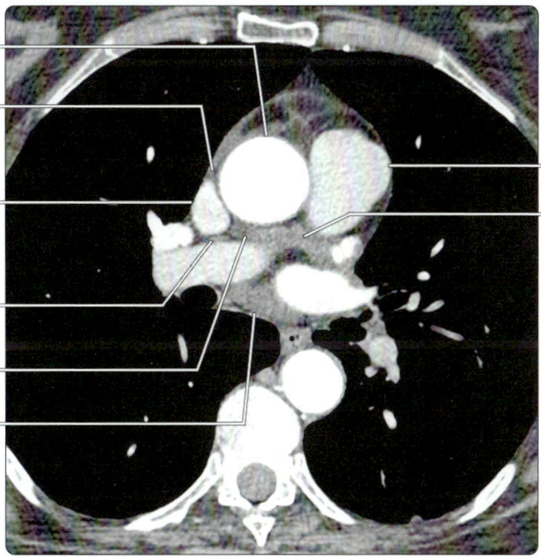

(Top) *Axial CECT (mediastinal window) shows the superior aortic recess, which arises from the transverse sinus (not shown). The superior aortic recess has anterior, posterior, and right lateral portions. The anterior portion is a triangular fluid collection with a beak-like extension between the ascending thoracic aorta and the pulmonary trunk. The posterior portion is crescent-shaped and abuts the posterior wall of the ascending thoracic aorta.* **(Middle)** *Axial CECT (mediastinal window) shows the left pulmonic recess, which is located immediately below the left pulmonary artery and arises from the adjacent transverse sinus. Note the anterior and lateral portions of the superior aortic recess, which also arises from the transverse sinus. The oblique sinus does not give rise to recesses and is separated from the transverse sinus by 2 pericardial reflections.* **(Bottom)** *Axial CECT (mediastinal window) shows the transverse sinus and its right pulmonic recess, which is located below the proximal right pulmonary artery. A small amount of fluid in the postcaval recess, which arises from the pericardial cavity proper, is also visible.*

PERICARDIAL RECESSES; AXIAL AND CORONAL CT

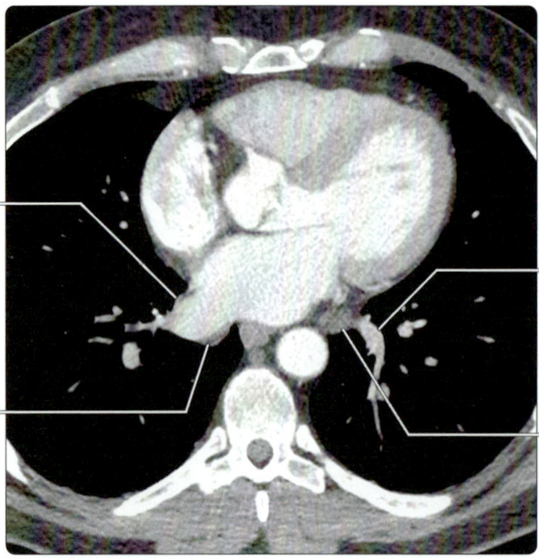

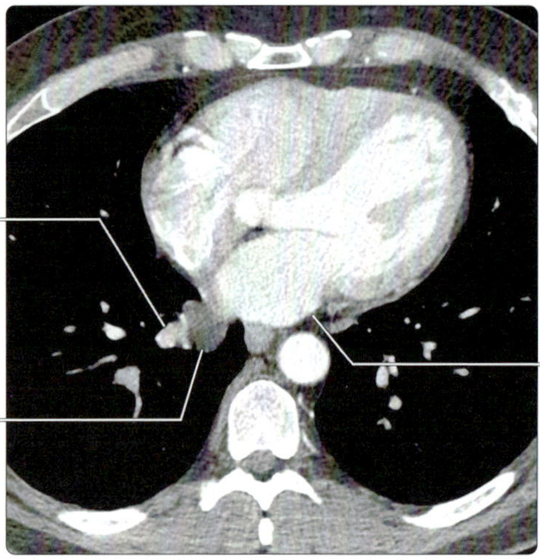

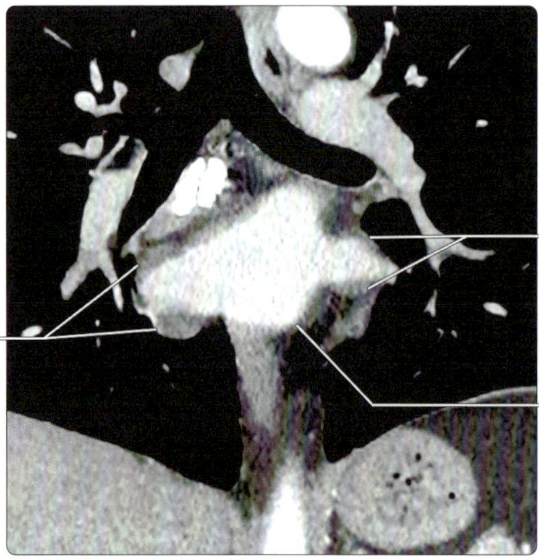

(Top) *First of 3 CECT images of the same patient shows the bilateral pulmonary venous recesses. Axial CECT (mediastinal window) shows small amounts of fluid in the right and left pulmonary venous recesses, which arise from the pericardial cavity proper. These recesses surround the pulmonary veins as they enter the left atrium.* **(Middle)** *Axial CECT (mediastinal window) shows the right pulmonary venous recess partially surrounding the inferior aspect of the right inferior pulmonary vein adjacent to its anastomosis with the left atrium. The right pulmonary venous recess may exhibit a nearly spherical morphology, as in this case, and should not be mistaken for lymphadenopathy.* **(Bottom)** *Coronal CECT (mediastinal window) shows the bilateral pulmonary venous recesses surrounding the bilateral pulmonary veins as they anastomose with the left atrium. Note the spherical morphology of the inferior aspect of the right pulmonary venous recess, which may mimic mediastinal lymphadenopathy.*

PERICARDIAL RECESSES, AXIAL CT

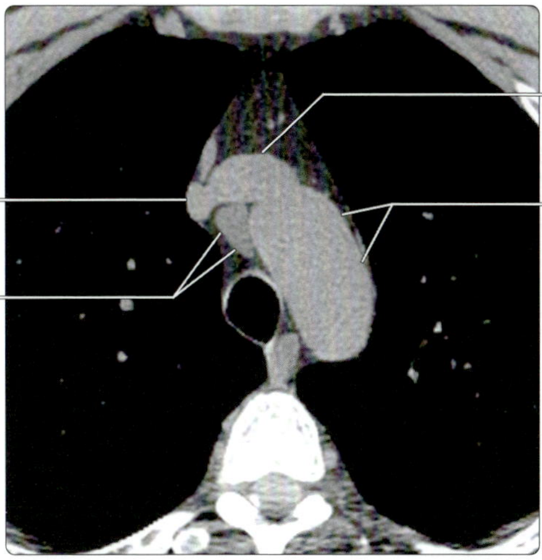

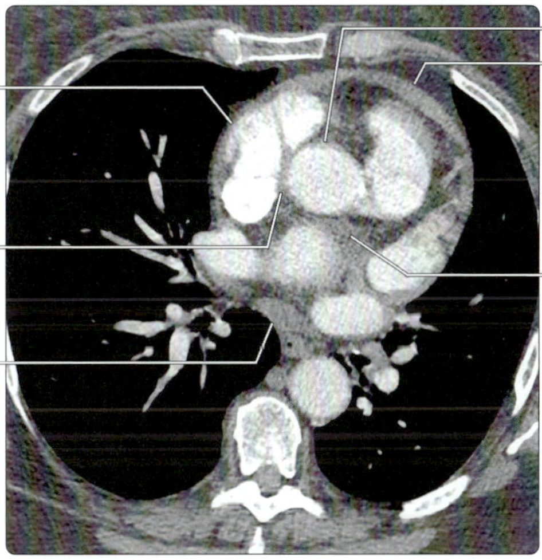

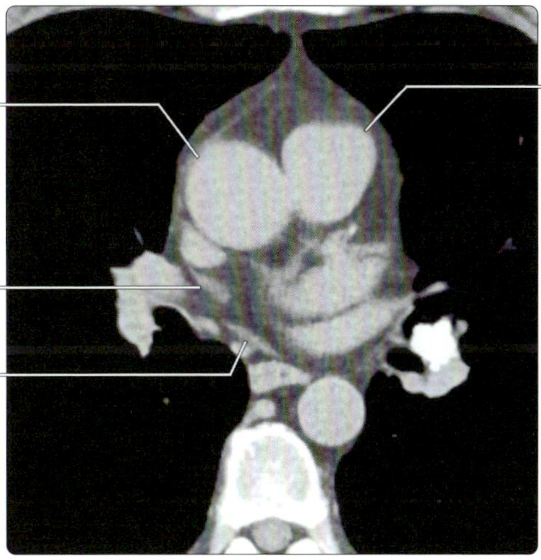

(Top) *Axial NECT (mediastinal window) shows a high-riding posterior portion of the superior aortic recess. Such high-riding recesses may mimic lymphadenopathy and congenital lesions, such as foregut or bronchogenic cysts.* **(Middle)** *Axial CECT (mediastinal window) of a patient with a small pericardial effusion shows the right inferior aortic recess located between the ascending thoracic aorta and the adjacent right atrium. The right inferior aortic recess is an extension of the transverse sinus.* **(Bottom)** *Axial NECT (mediastinal window) shows a small amount of fluid in the postcaval pericardial recess and in the oblique sinus. The postcaval recess arises from the pericardial cavity proper. The oblique sinus does not give rise to recesses.*

CONGENITAL ABSENCE OF PERICARDIUM, RADIOGRAPHY, MR, AND CT

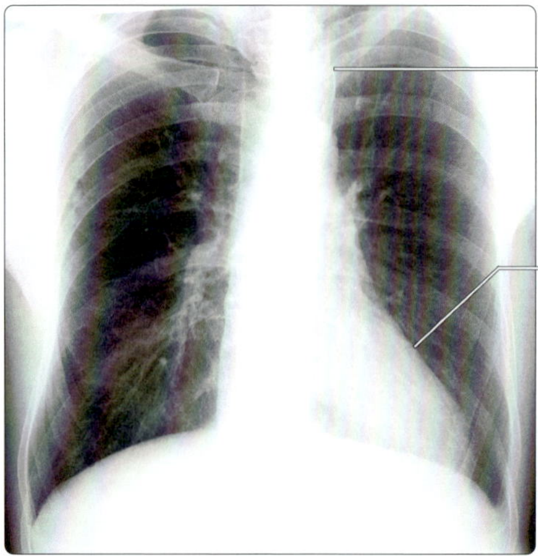

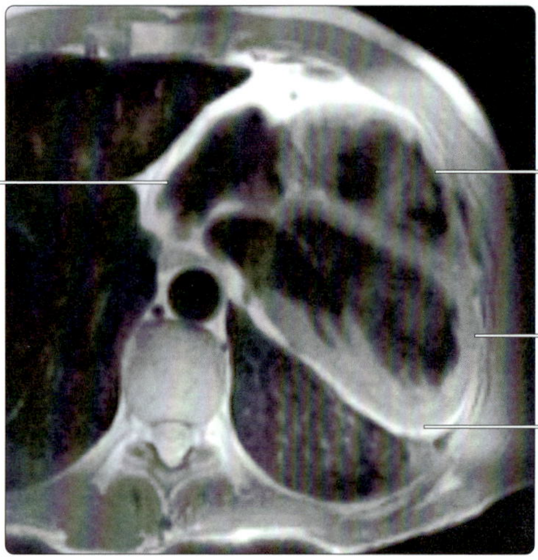

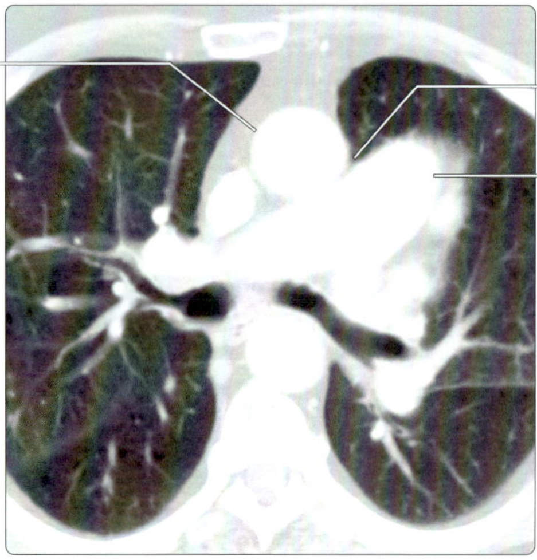

(Top) *First of 2 images of an asymptomatic 22-year-old man with congenital absence of the pericardium is shown. PA chest radiograph shows upper thoracic scoliosis and an unusual configuration of the heart characterized by an elongate morphology and an inferior orientation of its long axis.* **(Middle)** *Axial double IR MR of the same patient shows displacement of the heart to the left and a posterior orientation of the cardiac apex, consistent with the diagnosis of complete congenital absence of the pericardium.* **(Bottom)** *Axial CECT (lung window) of an asymptomatic 84-year-old man evaluated because of an abnormal cardiac configuration seen on chest radiography (not shown) shows interposition of lung between the ascending thoracic aorta and the pulmonary trunk in the expected location of the anterior portion of the superior aortic pericardial recess, indicating partial absence of the pericardium in this area.*

PERICARDIAL CYST, RADIOGRAPHY AND CT

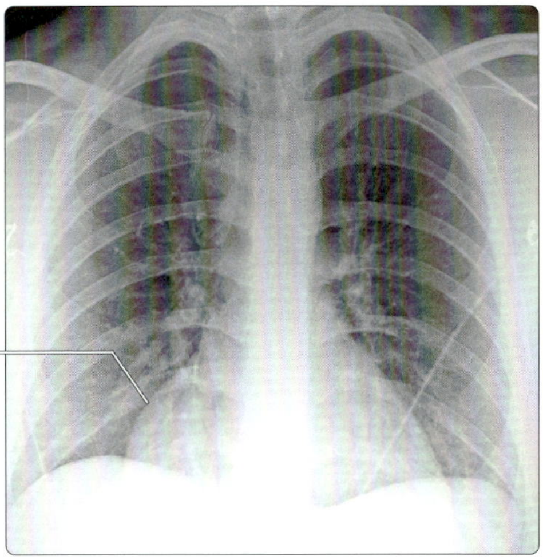

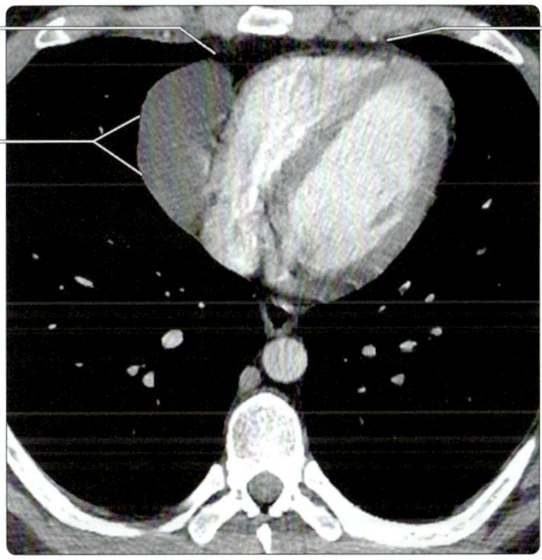

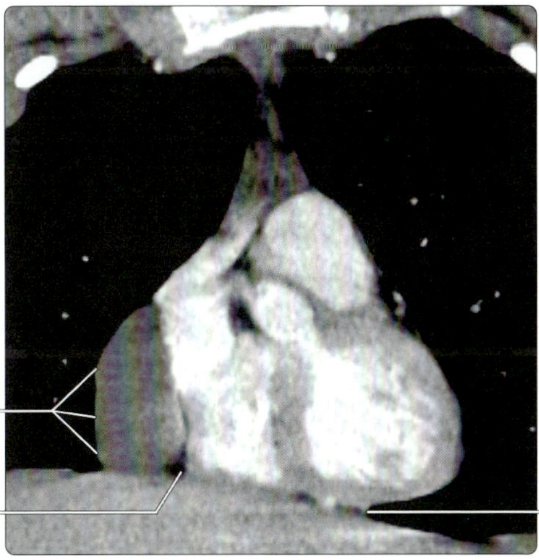

(Top) *PA chest radiograph of a 19-year-old man who presented with chest pain and dyspnea shows an abnormal cardiac contour that suggests cardiomegaly and is accentuated by low lung volume. The right heart border appears enlarged, suggesting the presence of a right cardiophrenic angle mass.* **(Middle)** *Axial CECT (mediastinal window) of the same patient shows that the mass corresponds to an ovoid fluid attenuation cystic lesion with an imperceptible wall located at the right cardiophrenic angle.* **(Bottom)** *Coronal CECT of the same patient shows the ovoid homogeneous water attenuation cystic lesion at the right cardiophrenic angle. The findings are consistent with a pericardial cyst.*

PERICARDIAL EFFUSION, RADIOGRAPHY

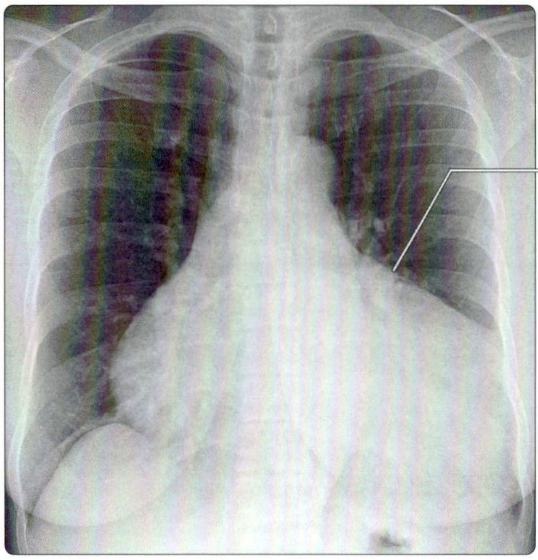

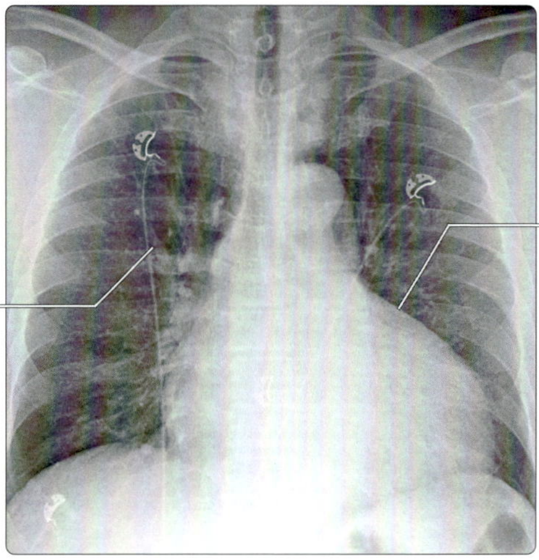

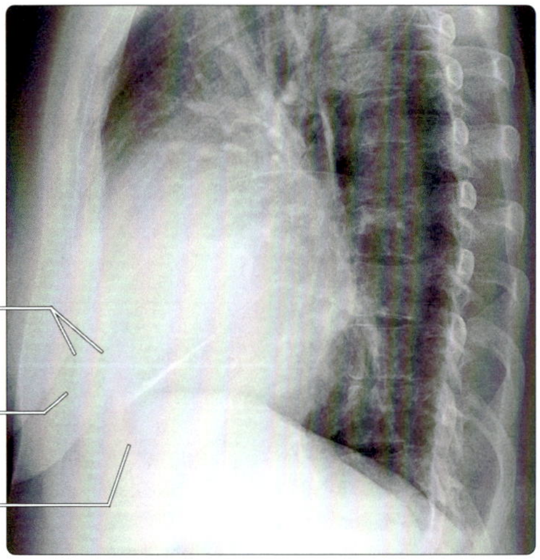

(Top) *PA chest radiography shows a large pericardial effusion manifesting with massive enlargement of the cardiopericardial silhouette and the so-called water bottle sign in which the cardiac silhouette morphologically resembles an old-fashioned water bottle.* **(Middle)** *PA chest radiograph of a patient with a moderate pericardial effusion shows enlargement of the cardiopericardial silhouette and vascular redistribution.* **(Bottom)** *Lateral chest radiograph of the same patient confirms the presence of a moderate pericardial effusion that exhibits the so-called fat pad or Oreo cookie signs. These signs are based on visualization of an arcuate band of water density pericardial fluid of variable thickness (> 4 mm) outlined by surrounding lower density mediastinal fat anteriorly and subepicardial fat posteriorly and located over the anterior inferior heart border on lateral chest radiography.*

PERICARDIAL EFFUSION, CT

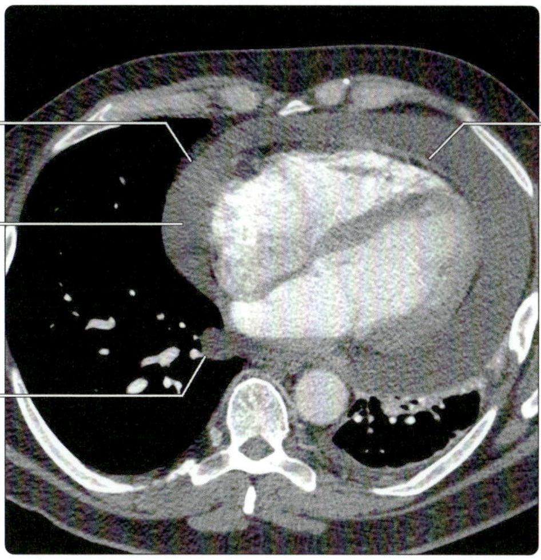

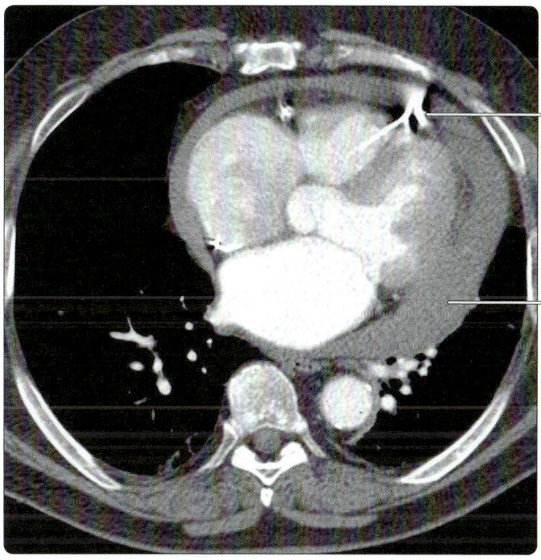

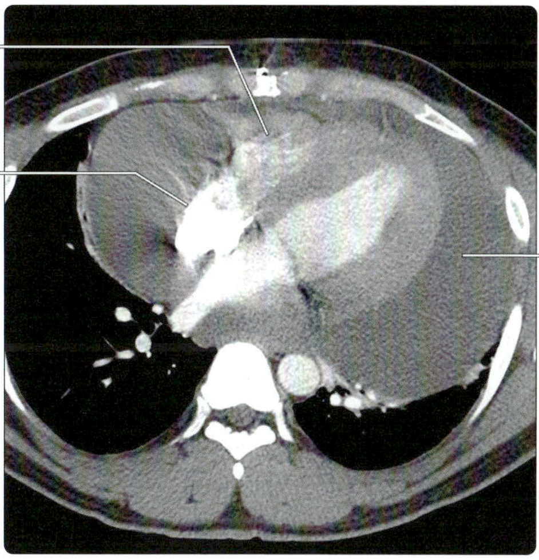

(Top) *Axial CECT (mediastinal window) of a patient with a moderate pericardial effusion shows fluid attenuation material in the pericardial space that separates the serous visceral pericardium (epicardium) from the combined thicknesses of the serous parietal and fibrous pericardial layers. Note fluid in the right pulmonary venous recess and trace bilateral pleural fluid.* **(Middle)** *Axial CECT (mediastinal window) of a patient who developed a symptomatic moderate hemopericardium following placement of a cardiac pacer due to perforation of the right ventricle by the right ventricular lead is shown.* **(Bottom)** *Axial CECT (mediastinal window) of a patient who developed a large circumferential pericardial effusion that produced mass effect on the right ventricle and right atrium is shown. Cardiac tamponade was suspected based on the imaging findings, but tamponade physiology is a clinical diagnosis.*

PERICARDIAL EFFUSION AND THICKENING, CT

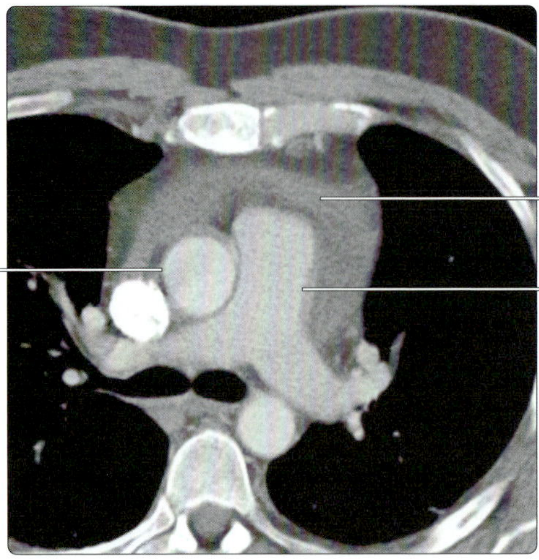

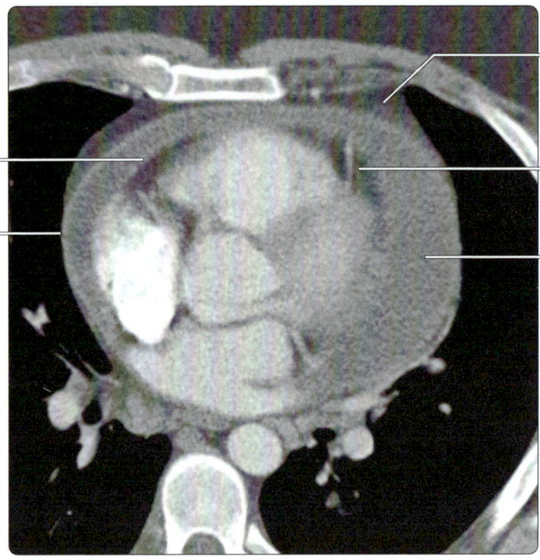

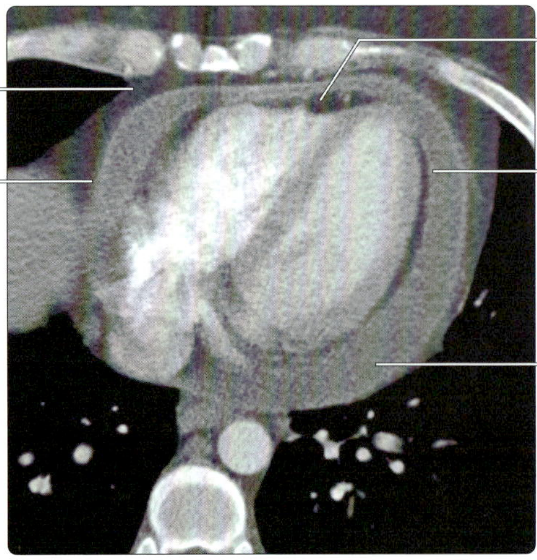

(Top) *First of 3 axial contrast-enhanced chest CT images of a patient with viral pericarditis is shown. Axial CECT (mediastinal window) through the pulmonary trunk shows fluid surrounding the ascending thoracic aorta and the pulmonary trunk in the expected location of the upper anterior pericardial space.* **(Middle)** *Axial CECT (mediastinal window) through the aortic root shows a large pericardial effusion and mild thickening and enhancement of the serous visceral and serous parietal pericardial layers, consistent with inflammation. The presence of the pericardial effusion allows distinction of the serous visceral pericardium or epicardium from the combined thicknesses of the fibrous and serous parietal pericardial layers.* **(Bottom)** *Axial CECT (mediastinal window) through the inferior aspect of the heart demonstrates low-attenuation pericardial fluid between the mildly thickened and enhancing serous visceral pericardium (epicardium) that covers the heart and the subepicardial fat and the combined thicknesses of the serous parietal and fibrous pericardial layers. Enhancement of the pericardial layers is consistent with pericardial inflammation.*

PERICARDIAL THICKENING AND CALCIFICATION, RADIOGRAPHY AND CT

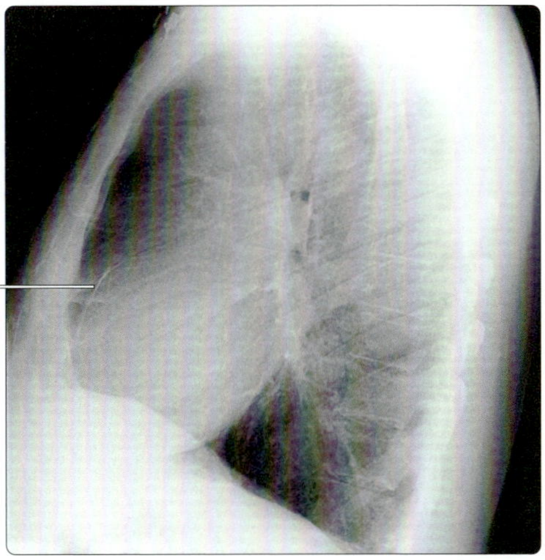

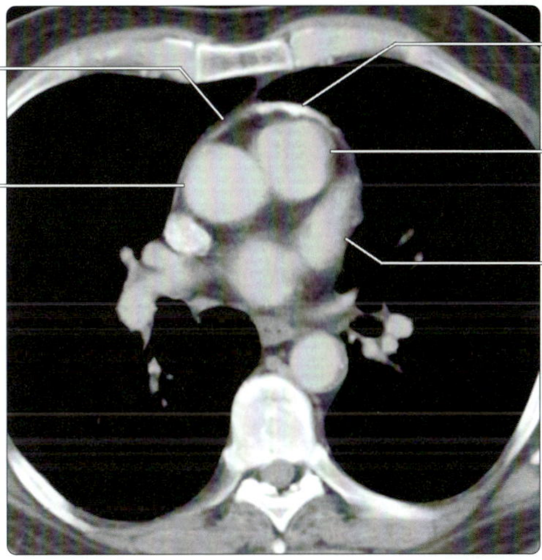

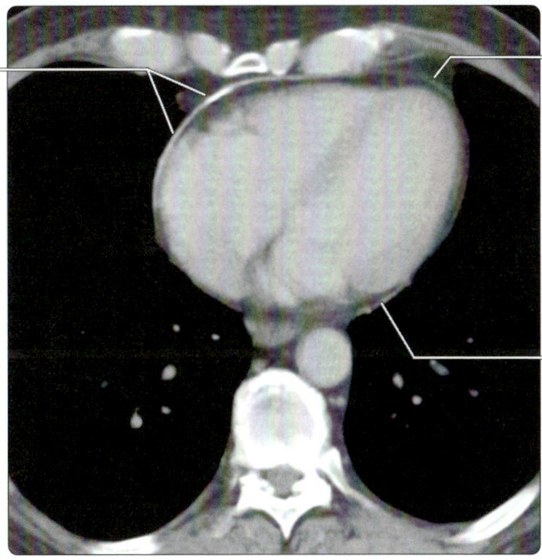

(Top) *Lateral chest radiograph of an asymptomatic 49-year-old man shows anterior thin curvilinear pericardial calcification. The heart size, pulmonary vascularity, and pleural surfaces are normal. Pericardial calcification may result from a variety of conditions, including infection, pneumoconiosis, cardiac surgery, and radiation.* **(Middle)** *First of 2 contrast-enhanced chest CT images of a 60-year-old man with incidentally discovered pericardial calcification is shown. Axial CECT (mediastinal window) demonstrates mild thickening and partial calcification of the anterosuperior aspect of the pericardium.* **(Bottom)** *Axial CECT (mediastinal window) of the same patient shows thickening of a portion of the anterior pericardium with focal calcification. There is no associated pericardial effusion or morphologic abnormality of the underlying cardiac chambers. Portions of the anterior and posterior pericardium demonstrate normal thickness. Pericardial calcification is typically secondary to prior injury, infection, or connective tissue disease but may also be idiopathic.*

CONSTRICTIVE PERICARDITIS, CT

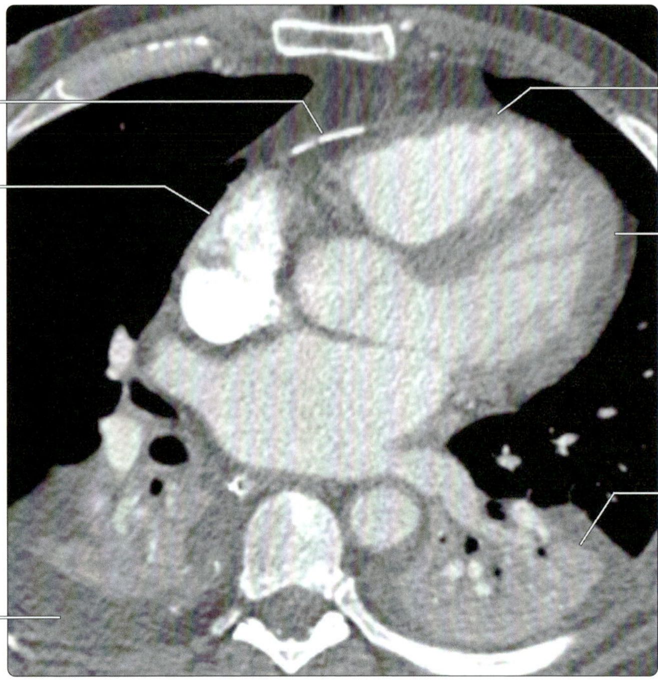

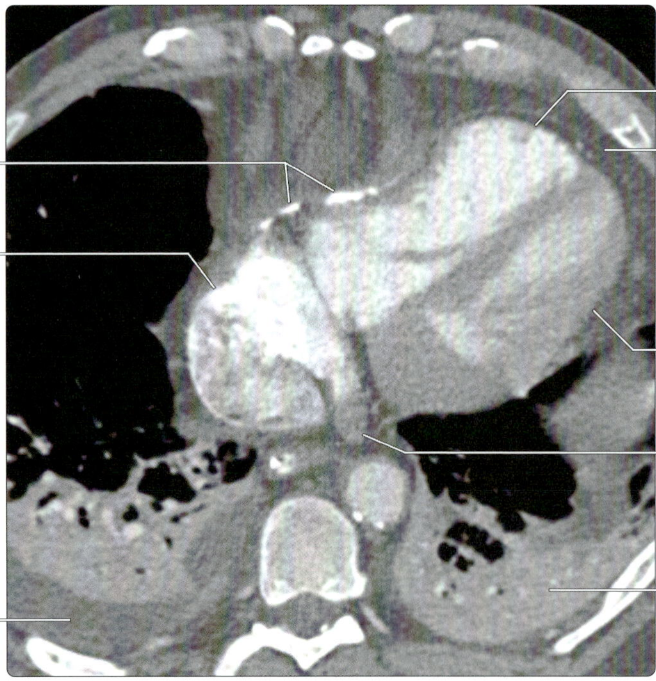

(Top) *Axial CECT (mediastinal window) of a 67-year-old man with diastolic ventricular dysfunction secondary to constrictive pericarditis shows focal linear calcification of the anterior aspect of the pericardium, small bilateral pleural effusions, and bilateral lower lobe relaxation atelectasis.* **(Bottom)** *Axial CECT (mediastinal window) through the inferior aspect of the heart of the same patient shows discontinuous linear calcifications of the anterior pericardium. Note the associated mass effect on the right ventricle, enlargement of the right atrium and the coronary sinus, bilateral lower lobe relaxation atelectasis, and small bilateral pleural effusions. The tubular morphology of the right ventricle is consistent with the diagnosis of constrictive pericarditis, and the presence of pericardial calcification allows differentiation of this condition from restrictive cardiomyopathy.*

PNEUMOPERICARDIUM, RADIOGRAPHY AND CT

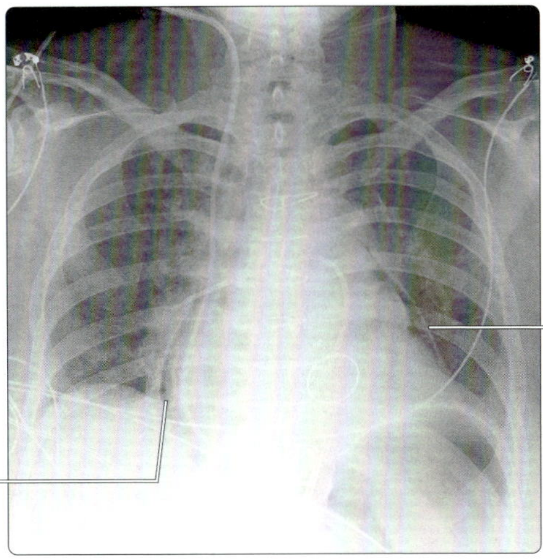

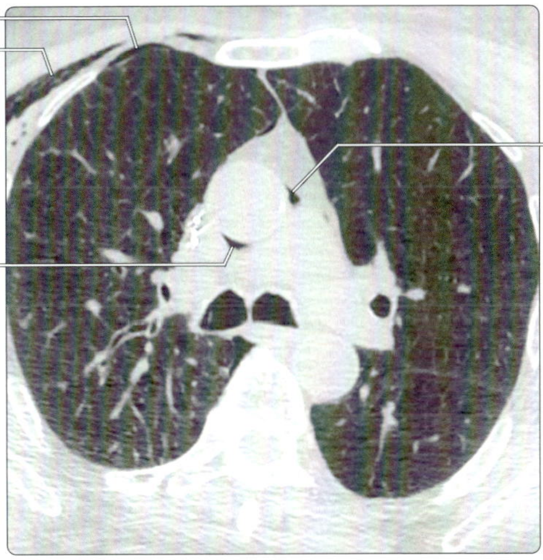

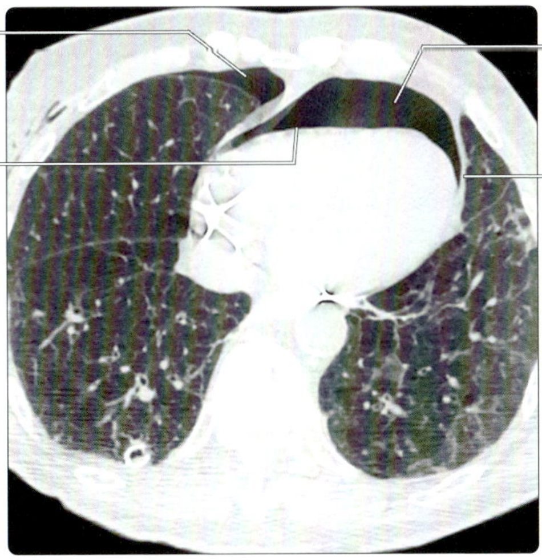

(Top) *AP portable chest radiograph of a patient who is status post cardiac surgery shows posttreatment changes and a small to moderate postsurgical pneumopericardium. Note that the abnormal mediastinal gas is contained by the anatomic boundaries of the pericardium and its superior reflections about the origins of the thoracic great vessels. The combined thicknesses of the fibrous pericardium and the serous parietal pericardium are outlined by air in the pericardial space and alveolar air in the adjacent lung.* **(Middle)** *Axial NECT (lung window) of a patient who sustained thoracic trauma shows a pneumopericardium with air within the anterior and posterior portions of the superior aortic recess. Note the small right pneumothorax and the right chest wall subcutaneous air.* **(Bottom)** *Axial NECT (lung window) of the same patient shows a moderate to large pneumopericardium that produces mass effect on the heart and a small right pneumothorax. A moderate to large pneumopericardium may produce tamponade physiology.*

GENERAL ANATOMY AND FUNCTION

Chest Wall Anatomy

- Skin, subcutaneous fat
- Blood vessels, lymphatics, nerves
- Bone, cartilage
- Musculature
- Endothoracic fascia: Loose fibroelastic connective tissue layer deep to intercostal spaces/ribs; separates them from parietal pleura

Function

- Musculoskeletal cage: Surrounds cardiorespiratory system; affects respiration by expansion and contraction during ventilation

Surface Landmarks

- Suprasternal (jugular) notch: Superior aspect of sternal manubrium, between clavicular heads
- Sternal angle: Landmark for internal thoracic anatomy; anterior projection at costal cartilage of 2nd rib
- Costal margins: Inferior margins of lowest ribs and costal cartilages

CHEST WALL ANATOMY

Skeletal Structures

- **Thoracic vertebrae**
  - 12 vertebrae (T1-T12); normal kyphotic curvature
  - Articular facets for ribs on vertebrae and transverse processes (except T11-T12)
  - Broad laminae and spinous processes (projecting inferiorly) overlap with those of vertebra beneath
- **Shoulder girdle**
  - 3 synovial articulations between clavicle, scapula, and proximal humerus
    - Acromioclavicular, sternoclavicular, glenohumeral joints
  - 1 functional articulation (scapulothoracic "joint")
    - Mobile scapula attached to ribs by musculature
- **Sternum**
  - Flat, broad bone in anterior thoracic wall; 3 components
    - Manubrium: Superior part of sternum
    - Body: Articulates with manubrium superiorly, xiphoid process inferiorly, and bilateral costal cartilages of 2nd-7th ribs
    - Xiphoid process: Variable size, shape, and ossification; articulates with sternal body superiorly
- **Clavicle**
  - Slender, S-shaped; connects sternum to scapula bilaterally
  - Conoid tubercle (coracoid tuberosity): Anatomic variant; bony prominence on inferior surface of lateral 1/3 of clavicle and its articulation with coracoid process; insertion point of conoid ligament
- **Scapula**
  - Large, triangular flat bone; parallel to upper posterior thorax; extends from 2nd-7th ribs
  - Glenoid fossa at glenohumeral joint
- **Ribs**
  - 12 pairs, symmetrically arrayed; numbered in accordance with attached vertebral body
  - True ribs (1-7) attach to sternum by costal cartilages (synovial joints)
  - False ribs (8-10) articulate by costal cartilages with costal cartilage of 7th rib
  - Floating ribs (11-12) do not articulate with sternum or rib costal cartilages; short costal cartilages terminate in abdominal wall musculature
  - Head articulates with demifacets of 2 adjacent vertebral bodies; neck located between head and tubercle of each rib; tubercle articulates with vertebral transverse process
  - Body: Longest part of each rib
  - Angle: Most posterior part
  - Costal groove on inner surface of inferior border; accommodates intercostal neurovascular bundle

Muscles

- **Pectoral**
  - Pectoralis major: Largest muscle in breast and pectoral region; originates from anterior chest wall, sternum, and clavicle; adducts, flexes, and medially rotates arm
  - Pectoralis minor: Deep to pectoralis major; originates from chest wall, inserts onto coracoid process of scapula; stabilizes scapula
- **Intercostal**
  - External: Contained within 11 intercostal spaces; from tubercle of rib to costochondral junction
  - Internal: Middle layer; occupy 11 intercostal spaces; extend from border of sternum to angle of rib
  - Innermost: Forms inner layer of chest wall muscles with subcostalis and transversus thoracis muscles
- **Serratus anterior**
  - Thin muscular sheet; overlies lateral thoracic cage and intercostal muscles; arises from upper 8 ribs; wraps around rib cage; inserts along medial border of anterior scapular surface
- **Back muscles**
  - Superficial extrinsic muscles (connect upper limbs to trunk; limb movement); trapezius, latissimus dorsi, levator scapulae, rhomboids
  - Intermediate extrinsic muscles (superficial respiratory muscles); serratus posterior
  - Deep intrinsic muscles (postvertebral muscles; control posture, vertebral and head movement); splenius, erector spinae, deep transversospinales muscles
- **Rotator cuff muscles**
  - Attachments to scapula; stabilize shoulder joint, performance of arm movements
  - Supraspinatus, infraspinatus, teres minor, subscapularis

Vessels

- **Arteries**
  - Internal thoracic (internal mammary): Branch of subclavian artery; descends posterior to first 6 costal cartilages; supplies upper anterior chest wall (anterior intercostal arteries to 1st-6th intercostal spaces)
- **Veins**
  - Azygos vein receives drainage from posterior intercostal, hemiazygos, and accessory hemiazygos veins
- **Lymphatics**: Chest wall drainage through **thoracic duct**; right upper limb, right face, and right neck drained by **right lymphatic duct**

Soft Tissues

- **Skin** and **subcutaneous tissues**
  - Nipple: Superficial to 4th intercostal space (males and prepubertal females)
- **Nerves**
  - Anterior rami of thoracic spinal nerves (T1-T11) supply skin, tissues of chest wall; form intercostal nerves
  - Intercostal nerves run in costal grooves between internal and innermost intercostal muscles
  - Brachial plexus: Branching network of nerve roots, trunks, divisions, cords, and branches
    - Spinal roots form 3 trunks; behind clavicle, each dividing into anterior and posterior divisions

ANATOMIC REGIONS

Thoracic Inlet

- Opening at superior end of thoracic rib cage; conduit for cervical structures to enter thorax
- Bound by T1 vertebral body, right and left 1st ribs and their costal cartilages, and manubrium of sternum

Thoracic Outlet

- Opening at inferior end of thoracic rib cage; conduit for thoracic structures to exit thorax into abdomen
- Bound by T12 vertebral body, right and left 12th ribs, costal cartilages of 7th-12th ribs, xiphisternal joint

Supraclavicular Region

- Supraclavicular lymph nodes in and around carotid sheath
- Lymph drainage of breast superiorly to supraclavicular and inferior deep lymph nodes, laterally to axillary lymph nodes, medially to parasternal (internal mammary) and mediastinal lymph nodes, inferiorly to diaphragmatic lymph nodes

Axilla

- Pyramidal-shaped space between lateral chest and upper arm; bound by pectoralis muscles (anteriorly); subscapularis, latissimus dorsi, and teres major muscles (posteriorly); convergence of axillary fold muscles (laterally); and clavicle, scapula, and outer border of 1st rib (apex)
- Axillary lymph nodes drain breast, thoracoabdominal wall above umbilicus, and upper arm

Breast and Pectoral Region

- Anterior and superior part of chest; muscles and fascia assist movement of upper limb; mammary glands

IMAGING

Radiography

- Limited capabilities; may detect congenital deformities, soft tissue masses, bone destruction

CT

- CT with multiplanar reformats allows excellent visualization of osseous and soft tissue lesions

MR

- Multiplanar capabilities and advanced pulse sequences optimal for assessment of chest wall soft tissues and bones, particularly for tumor evaluation

ANATOMY-RELATED IMAGING ABNORMALITIES

Congenital and Developmental Abnormalities

- Pectus excavatum (syn. funnel chest): Abnormal growth of costal cartilages; sternal depression/rotation
- Pectus carinatum (syn. pigeon breast): Abnormal growth of costal cartilage; sternal protrusion
- Cervical rib: Supernumerary rib, usually arising from 7th cervical vertebra
- Poland syndrome
  - Uncommon; partial or total absence of pectoralis major and minor muscles; associated malformations of ipsilateral ribs (2-5) and clavicle; congenital absence of ipsilateral breast
- Sternalis muscle: Uncommon accessory chest wall muscle; should not be mistaken for pathologic lesion

Inflammatory and Infectious Disease

- Etiologies
  - Primary chest wall infection is rare
    - Immunosuppression, diabetes mellitus
    - Complication of trauma, surgery
  - Secondary chest wall involvement more common
    - Pulmonary infection (tuberculous, fungal)
    - Pleural empyema (empyema necessitatis)
- Intravenous drug users prone to septic arthritis of sternoclavicular and sternochondral joints
- Other causative organisms
  - *Actinomyces* (dental origin), *Nocardia*, *Streptococcus pneumoniae*
  - Post sternotomy: *Staphylococcus aureus*, *Streptococcus epidermidis*, *Pseudomonas aeruginosa*
- Key imaging features
  - Pulmonary consolidation with involvement of chest wall
  - Chest wall soft tissue mass
  - Osseous destruction, periostitis, osteomyelitis
  - Associated findings: Pleural effusion, bronchopleural fistula, subcutaneous gas

Neoplasia

- Benign
  - Lipoma: Predominant adipose tissue; minimal soft tissue components
  - Smooth pressure erosion of bone implies slow growth (e.g., neurofibroma, schwannoma)
  - Fibrous dysplasia: Monostotic and polyostotic forms
- Malignant
  - Metastatic disease: Lytic or blastic osseous metastases, soft tissue metastases to chest wall, direct invasion of chest wall by primary intrathoracic malignancy
  - Chondrosarcoma: Lytic, expansile, often thick sclerotic border; chondroid calcification (60-75%)
    - Chest wall: Most commonly affects ribs (anteriorly at costochondral junctions) and vertebrae
  - Osteosarcoma: Soft tissue mass with osseous destruction and internal foci of ossification/calcification
  - Multiple myeloma: Commonly manifests as osseous lytic lesions, sometimes with associated soft tissue mass
  - Liposarcoma: Contains fat and prominent soft tissue components; may exhibit aggressive behavior

OVERVIEW OF CHEST WALL

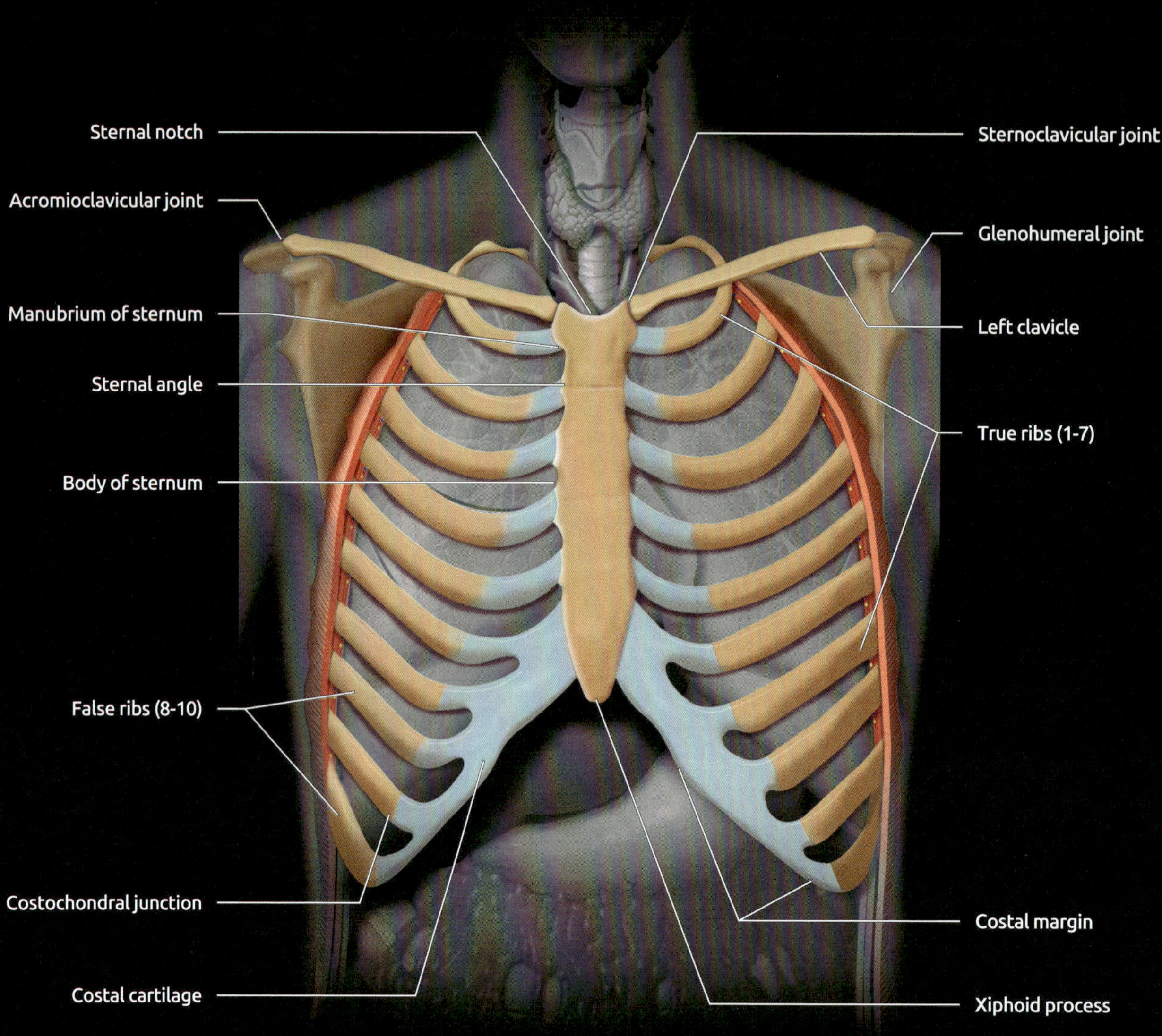

Graphic depicts the chest wall structures, which form a musculoskeletal thoracic cage that surrounds the cardiorespiratory organs. The chest wall structures affect respiration by expanding and contracting during ventilation.

ANATOMY OF CHEST WALL

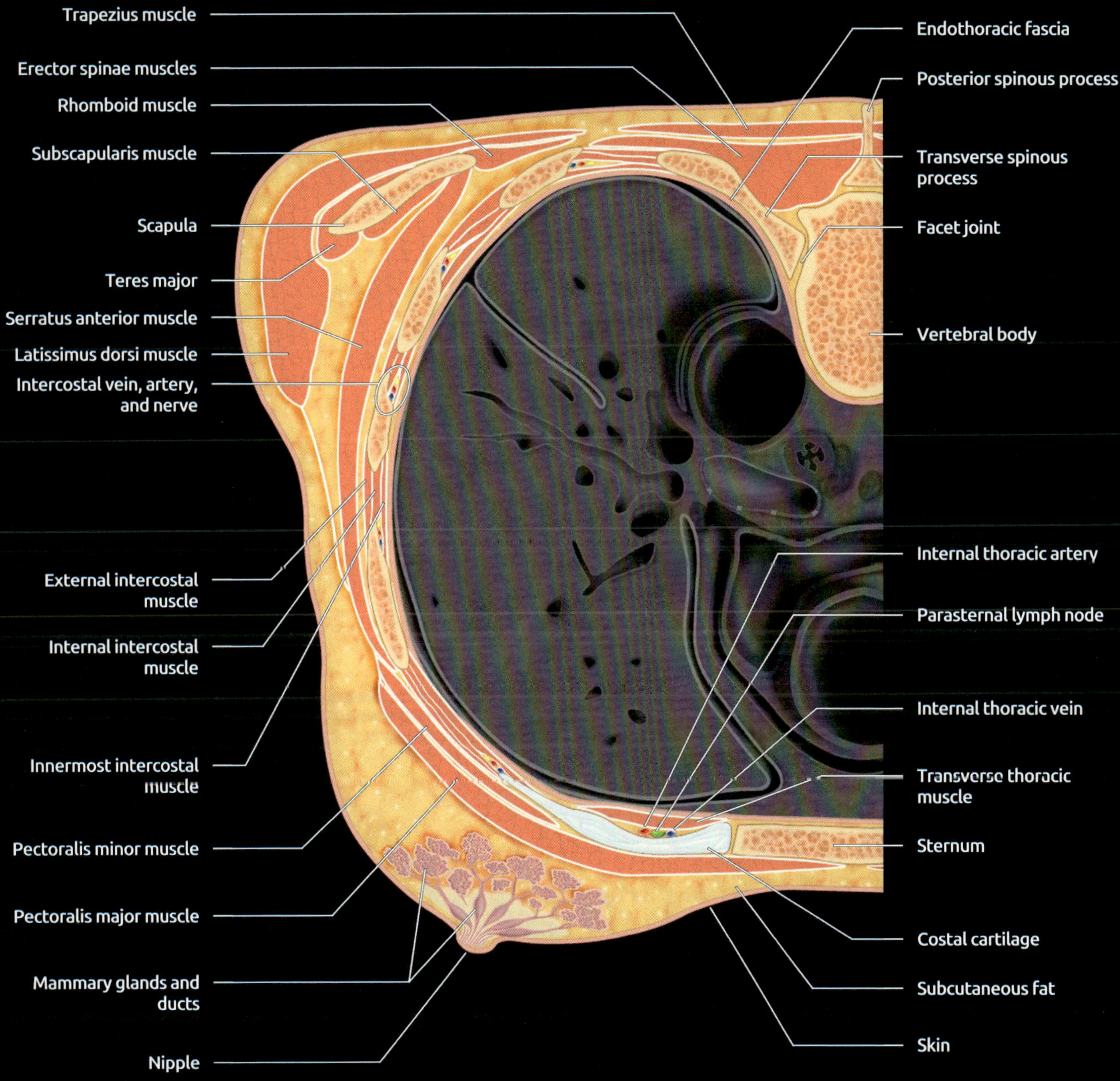

Graphic depicts the chest wall layers as visualized in the axial plane, including skin, subcutaneous fat, blood vessels, lymphatics, muscles, and osseous structures. The innermost layer, the endothoracic fascia, is a fibroelastic connective tissue layer located between the inner aspect of the chest wall and the pleura.

ANATOMY OF THORACIC INLET, SUPRACLAVICULAR, AND AXILLARY REGIONS

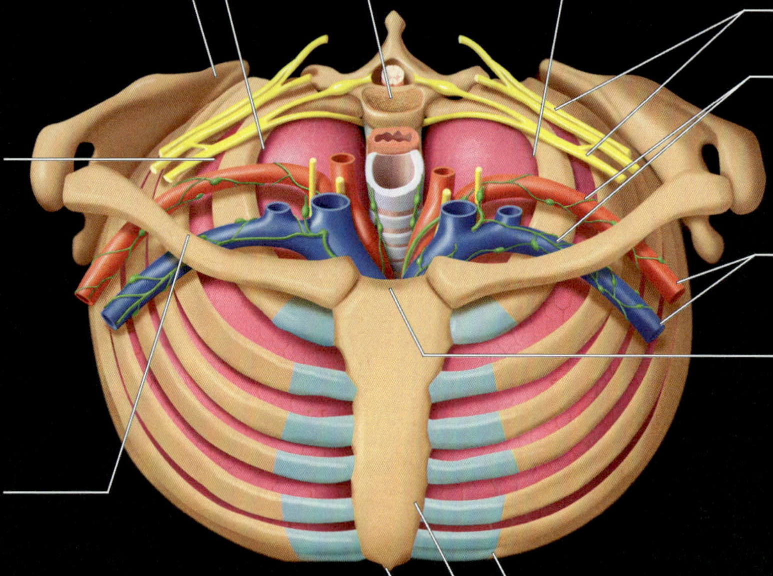

Graphic depicts the thoracic inlet, supraclavicular structures, and the axillary regions. The thoracic inlet is bound by the T1 vertebral body, the right and left 1st ribs and their costal cartilages, and the manubrium of the sternum. The apex of the axillary region is bound by the clavicle, the scapula, and the outer border of the 1st rib. Vascular structures allow blood flow to enter and exit the thorax through the thoracic inlet and join with brachial plexus components to supply the thorax and upper limbs.

INTERCOSTAL REGION, GRAPHIC AND MR

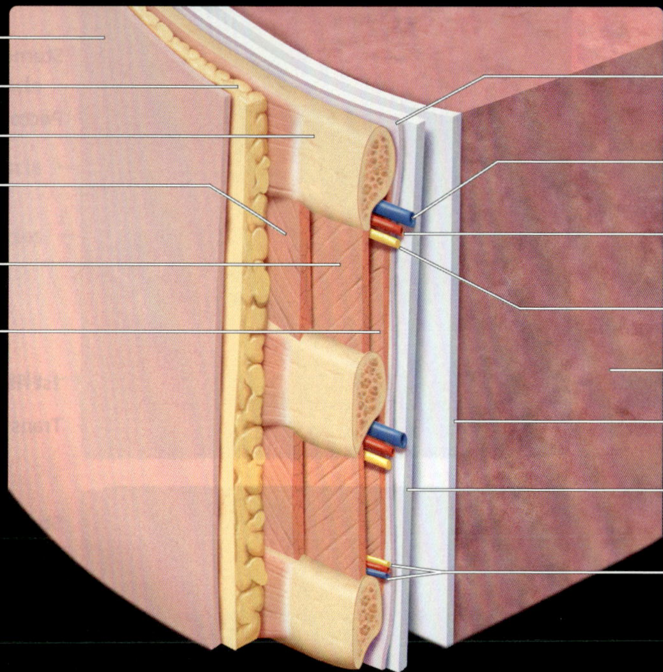

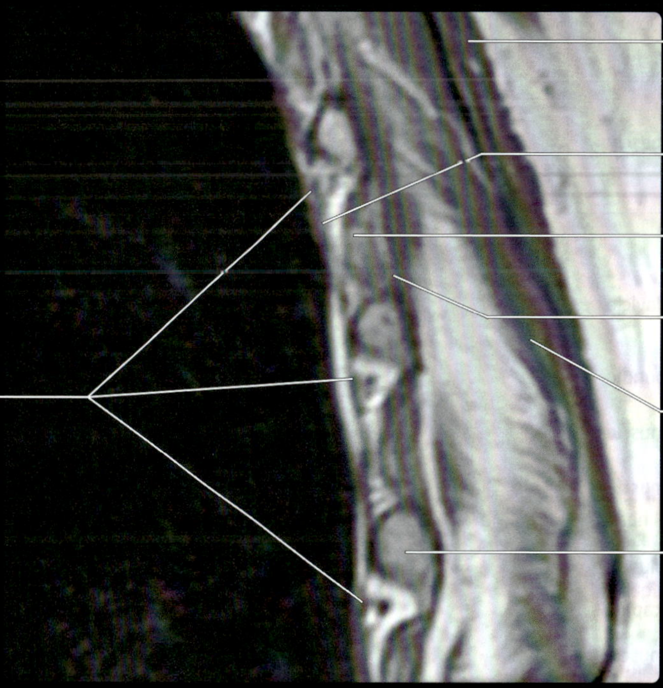

(Top) *Graphic demonstrates details of the intercostal region, showing 3 layers of intercostal muscles (external, internal, and innermost) between the ribs. The costal groove along the inferomedial aspect of each rib accommodates the intercostal neurovascular bundle (vein, artery, and nerve). Small collateral branches of the major intercostal vessels and nerves may be present above the body of the subjacent rib. The endothoracic fascia forms a connective tissue layer between the inner aspect of the chest wall and the costal parietal pleura.* **(Bottom)** *Detail of a sagittal oblique T1 MR through the midchest wall demonstrates the intercostal neurovascular bundles coursing within the costal grooves, along the inferior and inner aspect of the ribs. The neurovascular bundles are located deep to the internal intercostal muscles and superficial to the innermost intercostal muscles.*

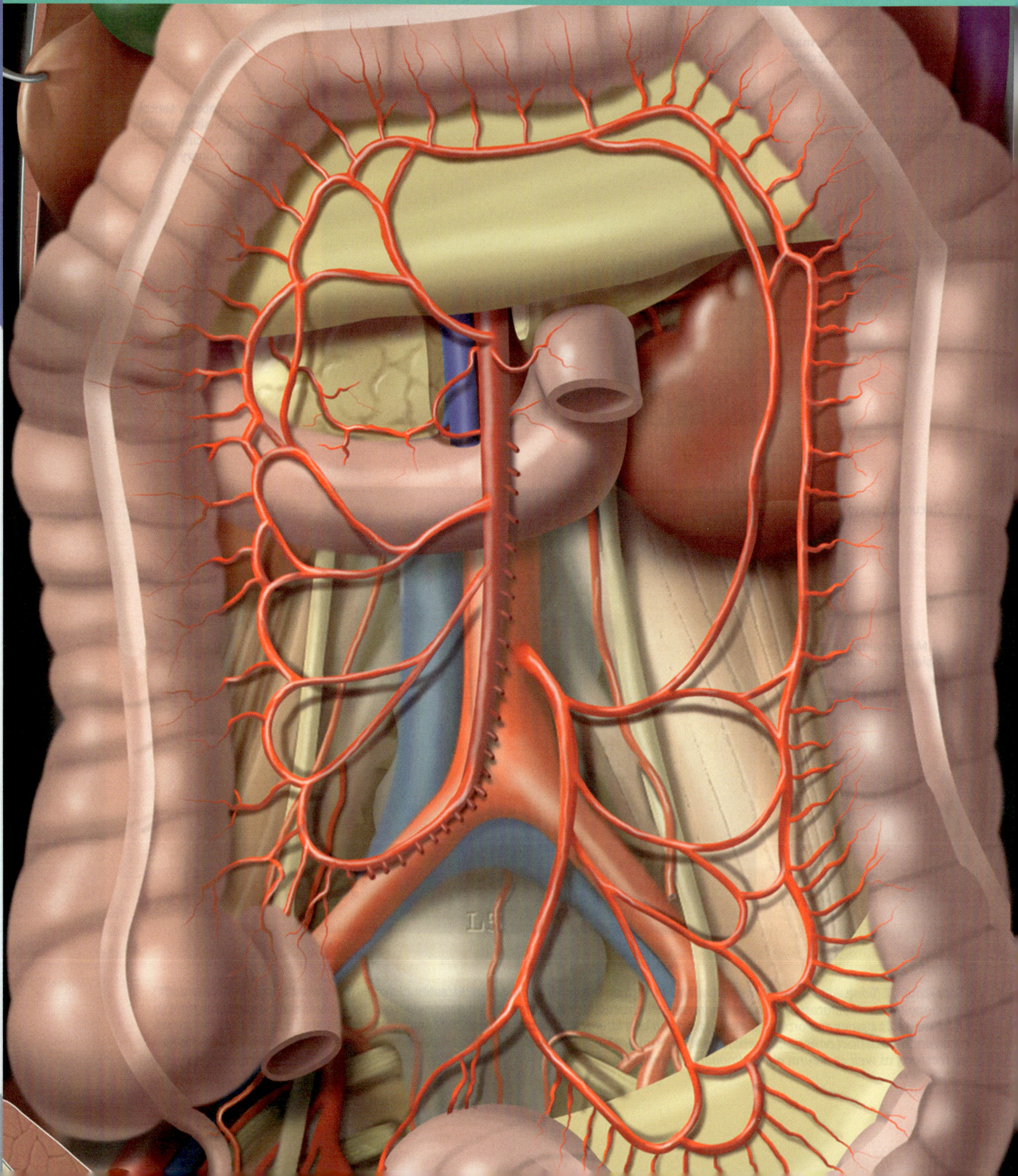

SECTION 2

Abdomen

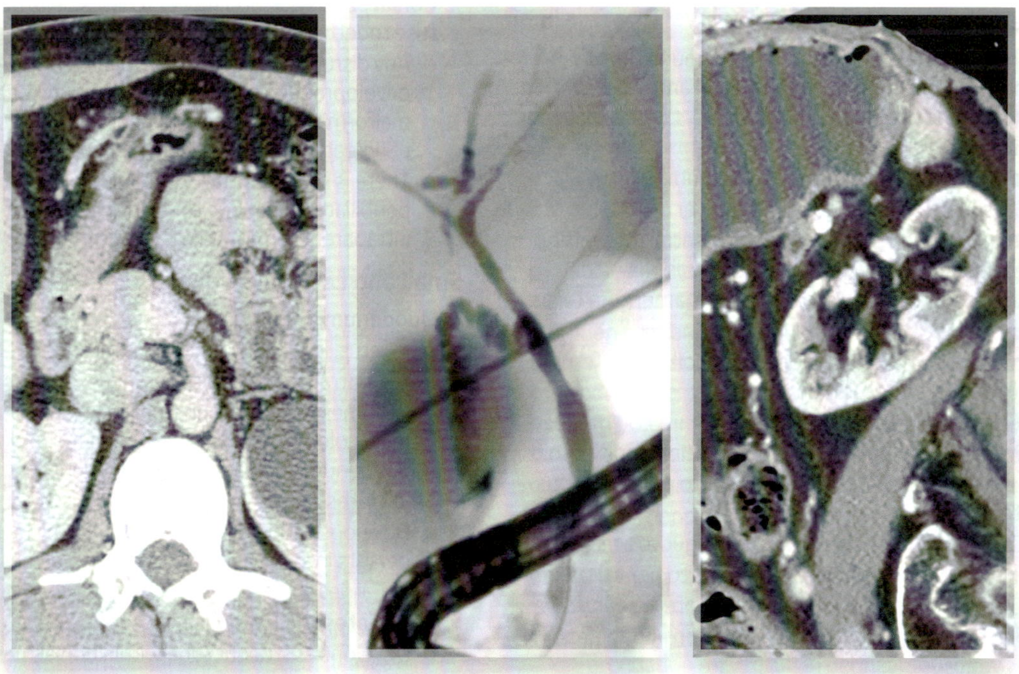

18-DAY EMBRYO (LATERAL)

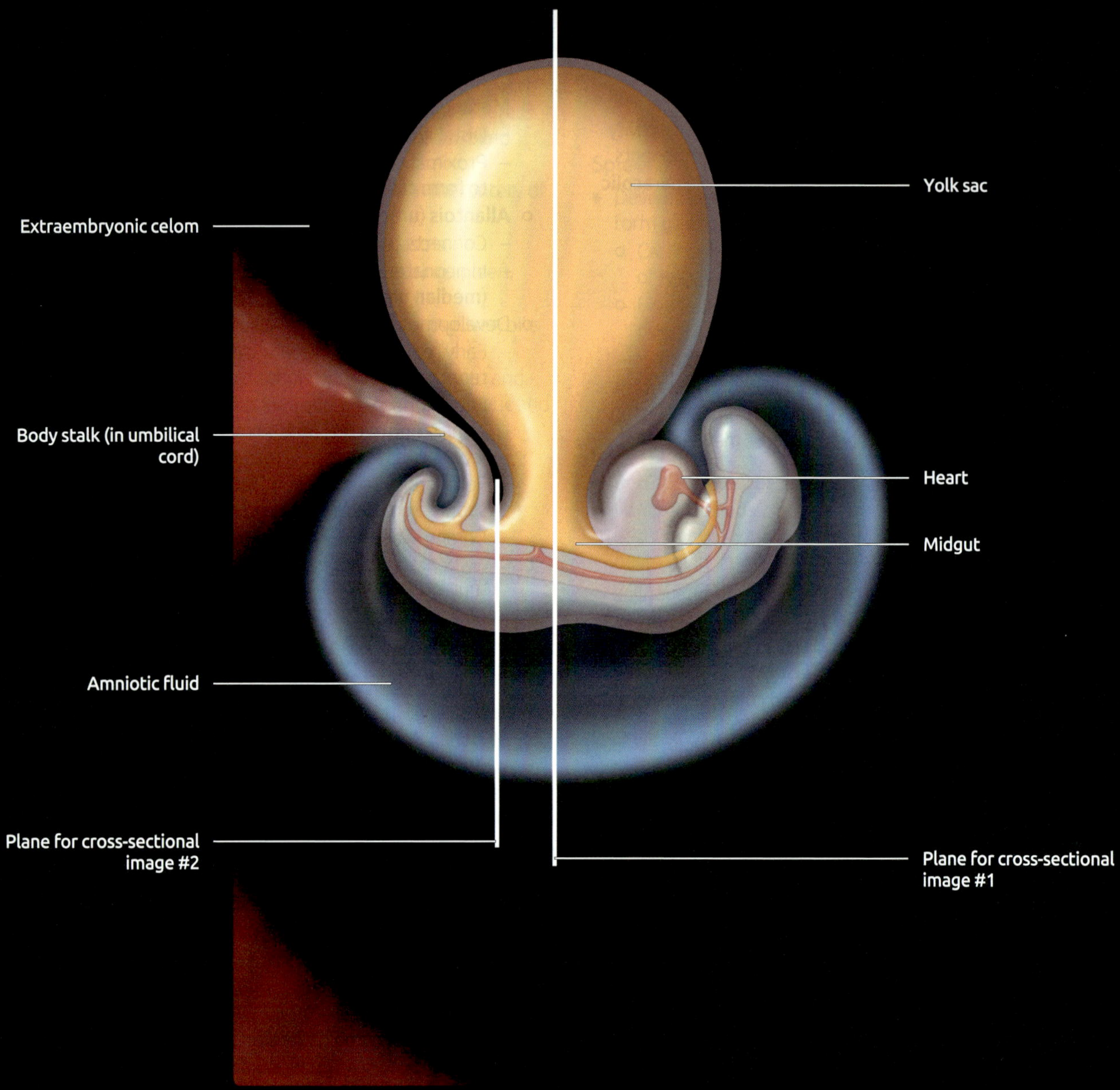

Lateral graphic of an 18-day embryo is shown. The roof of the yolk sac becomes incorporated in the form of a tube, the primitive gut. The cranial end of the tube becomes the foregut, and the caudal end becomes the hindgut.

18-DAY EMBRYO (CROSS SECTION)

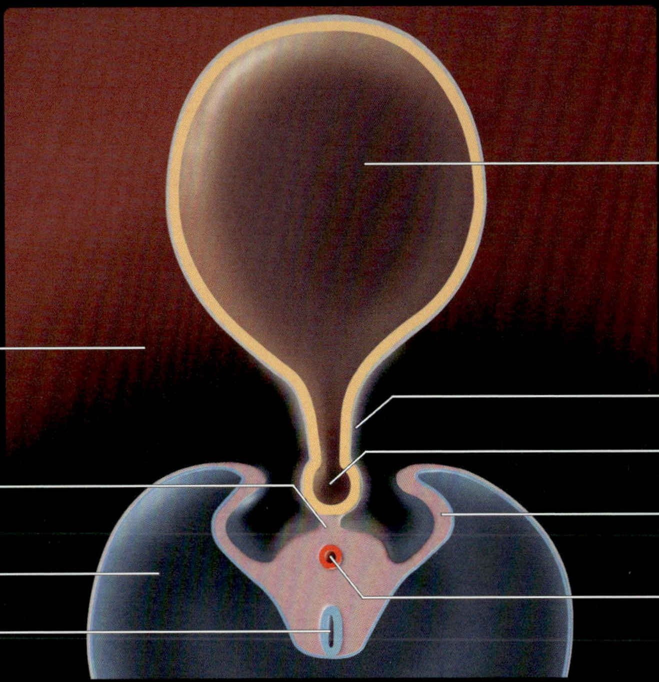

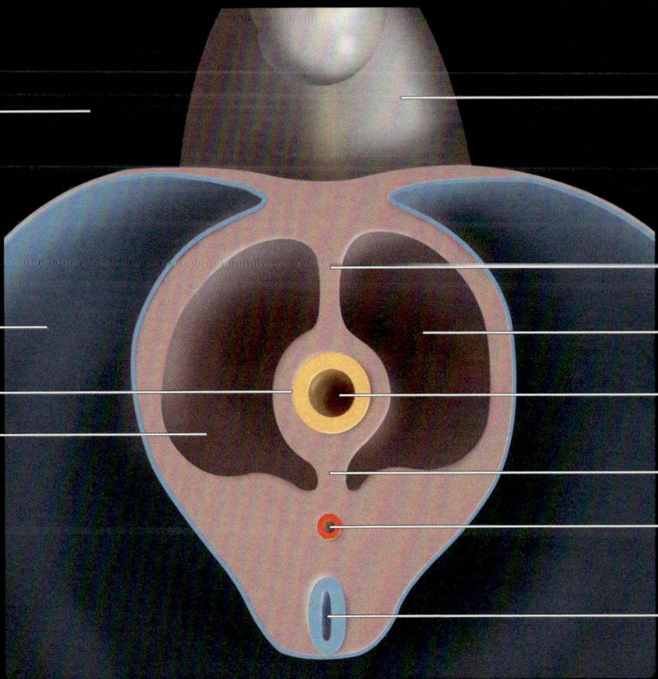

(Top) *Cross-sectional graphic along plane #1, indicated on the lateral 18-day embryo, is shown. The midgut has a wide communication with the yolk sac at this phase.* **(Bottom)** *Cross-sectional graphic along plane #2, indicated on the lateral 18-day embryo, is shown. The gut is suspended by the ventral and dorsal mesenteries.*

4-WEEK EMBRYO (LATERAL)

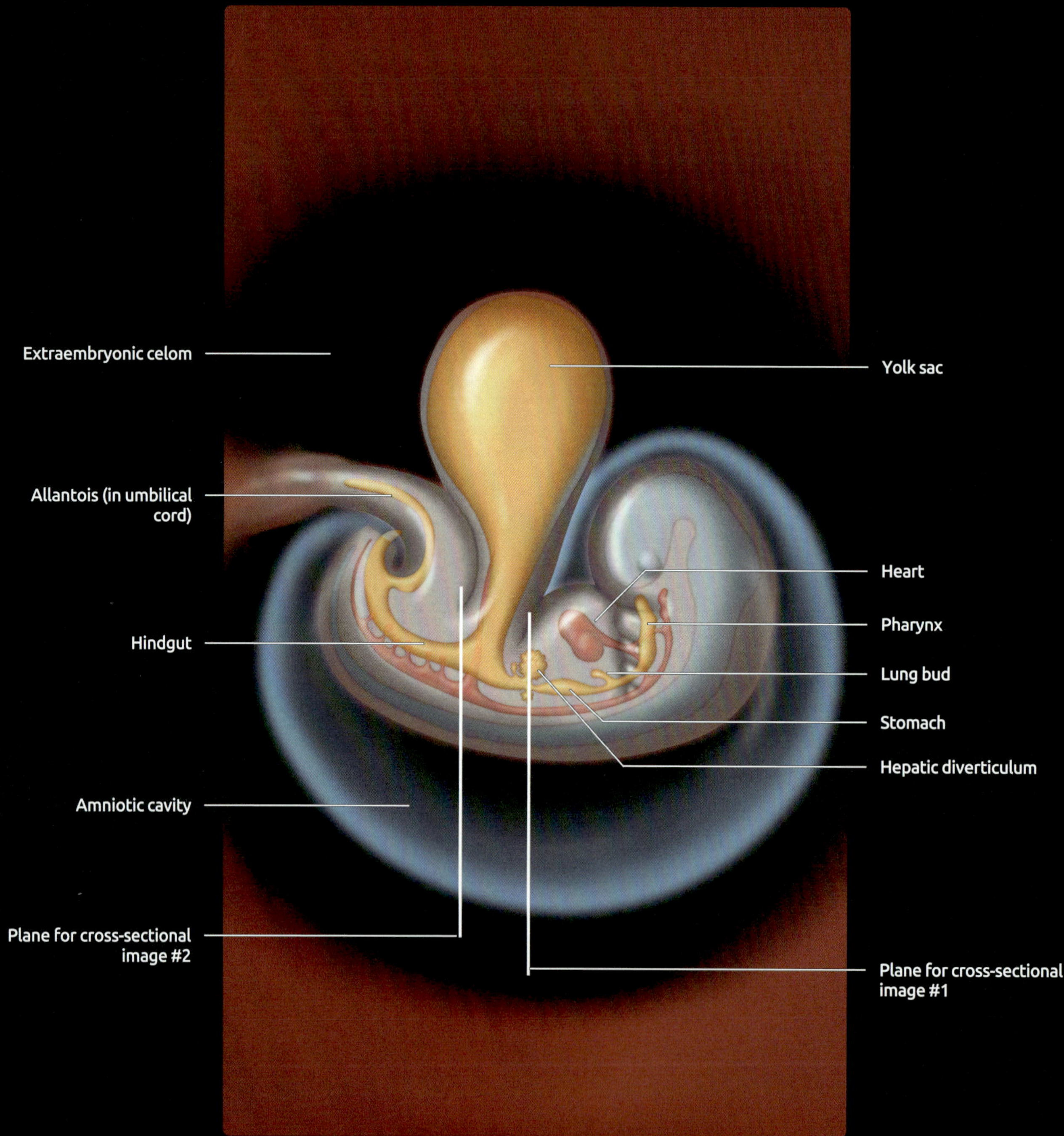

Lateral graphic of a 4-week embryo is shown. The pharynx and lung bud arise from the foregut, along with the stomach. The allantois connects the body stalk to the hindgut. The yolk sac communicates broadly with the primitive gut. Errors in development include communication between the foregut branches, such as a tracheoesophageal fistula.

4-WEEK EMBRYO (CROSS SECTION)

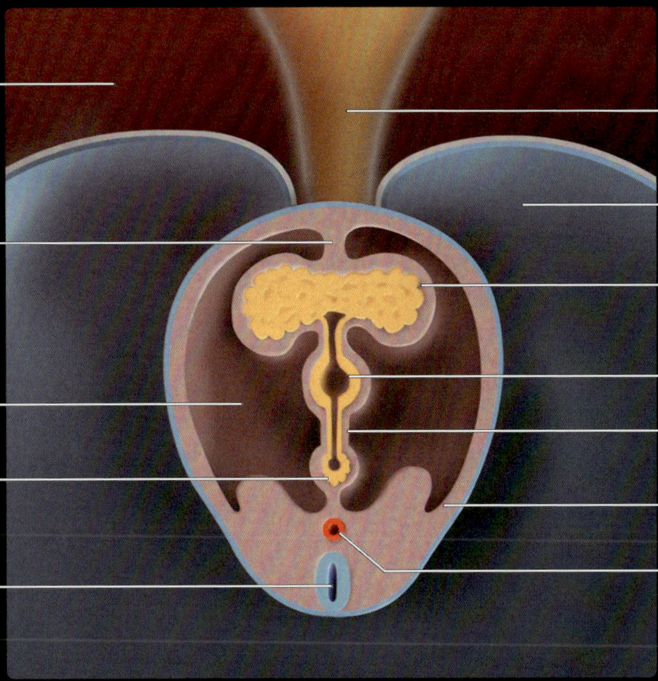

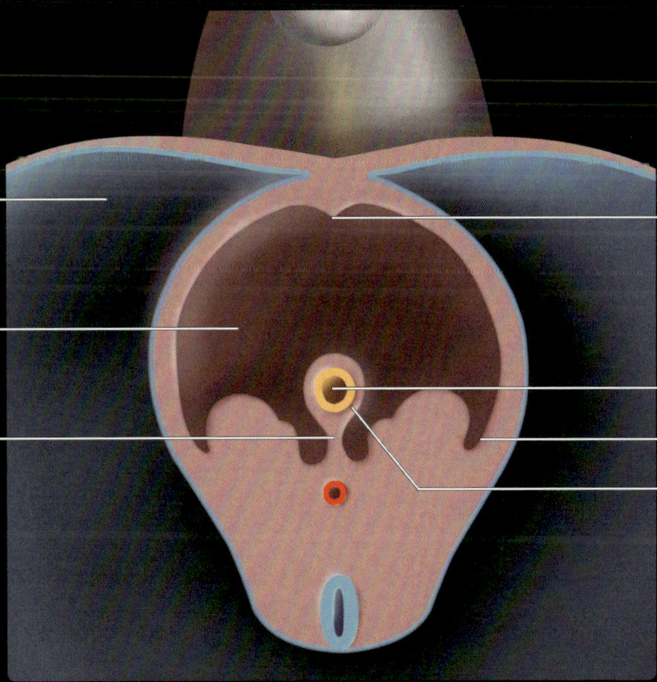

(Top) *Cross-sectional graphic along plane #1, indicated on this lateral 4-week embryo, is shown. The liver arises from a ventral bud of the foregut, whereas the pancreas arises from the dorsal mesentery.* **(Bottom)** *Cross-sectional graphic along plane #2, indicated on this lateral 4-week embryo, is shown. The ventral mesentery begins to disintegrate to allow communication between the right and left sides of the abdominal cavity.*

DEVELOPMENT OF GI TRACT (5 WEEKS)

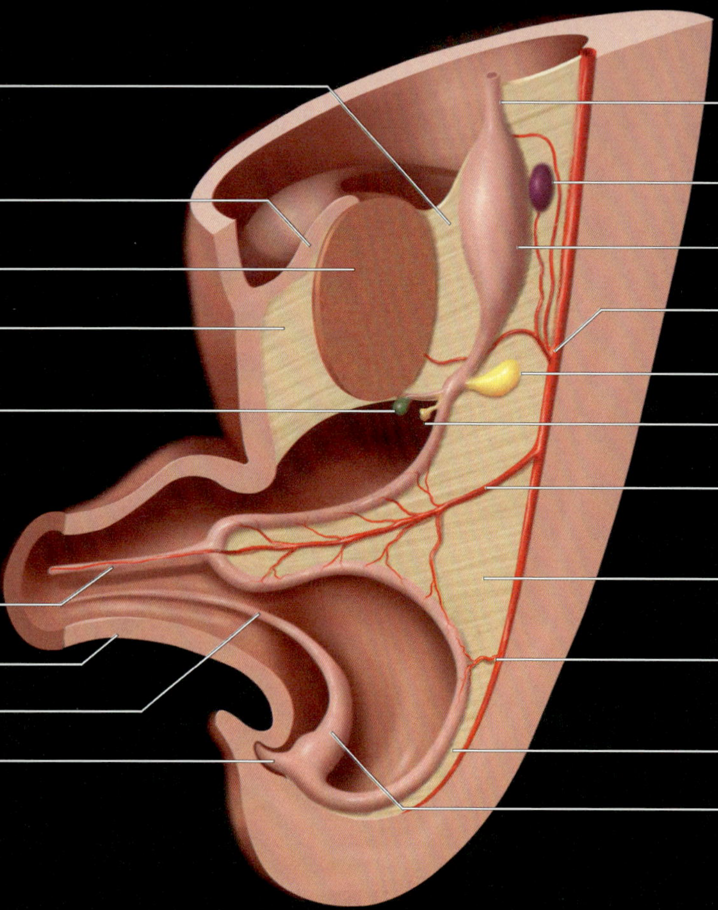

The transverse septum grows as a shelf from the anterior body wall, becoming the ventral part of the diaphragm. The esophagus and stomach begin to develop as distinct structures. During a period of its development, the esophageal lumen is occluded, and it shares a foregut origin with the trachea. Errors in development may result in esophageal atresia or tracheoesophageal fistula. The primary gut begins to elongate along with its dorsal mesentery. The hepatic diverticulum gives rise to the biliary tree and ventral pancreas. The arterial supply to the gut is already defined: Celiac artery (foregut), superior mesenteric artery (midgut), and inferior mesenteric artery (hindgut).

DEVELOPMENT OF GI TRACT (6 WEEKS)

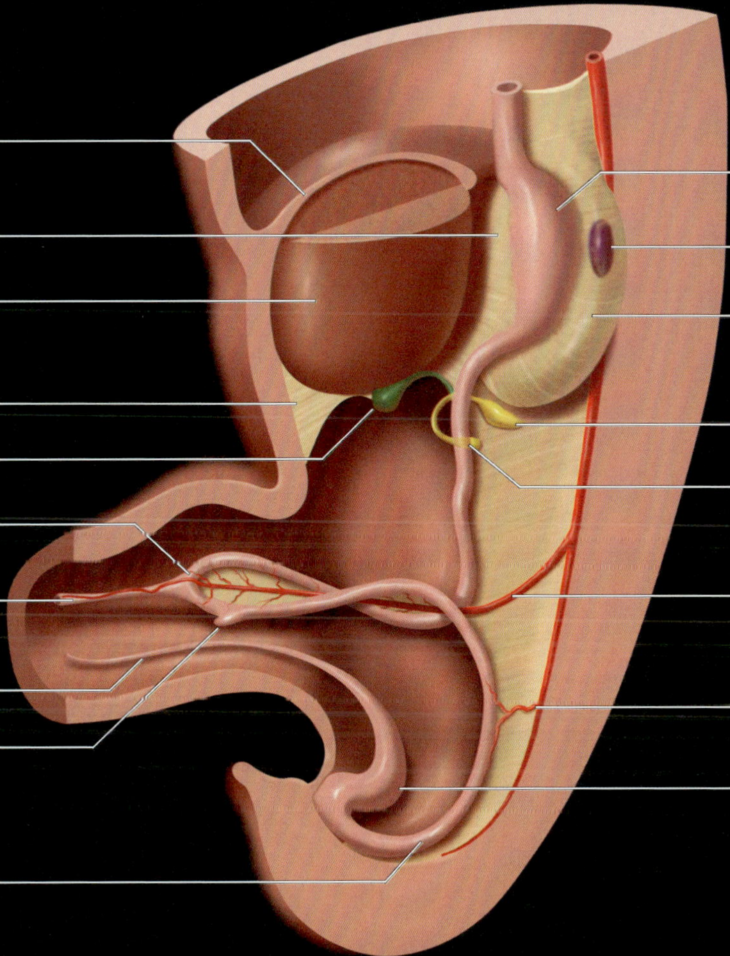

The liver expands within the ventral mesentery, which later disintegrates, leaving only the falciform ligament and the lesser omentum. The common bile duct, portal vein, and hepatic artery traverse the caudal part of the lesser omentum. The primary gut elongates and herniates out into the umbilical cord. The primitive gut and urinary system terminate in the cloaca and connect to the umbilical cord via the yolk stalk and allantois, respectively.

DEVELOPMENT OF GI TRACT (8 WEEKS)

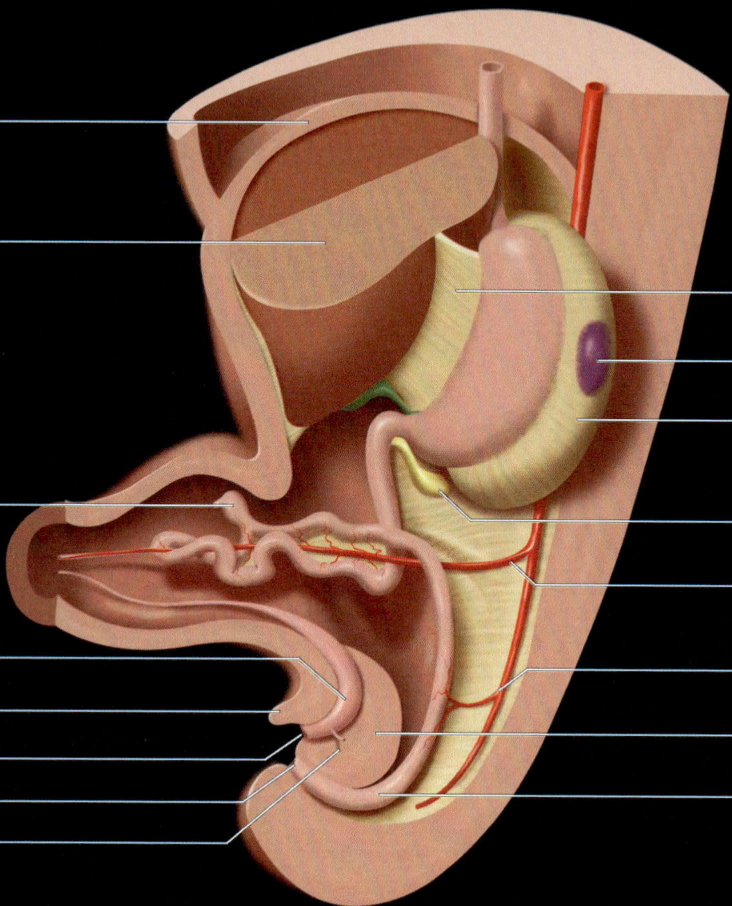

The diaphragm is nearly complete. The liver continues its rapid enlargement. Only the caudal part of the ventral mesentery remains (falciform ligament), allowing the more cephalad portions of the peritoneal cavity to communicate. The dorsal mesogastrium elongates considerably, forming the left and caudal portions of the lesser sac. The gut continues to elongate and rotates counterclockwise around the superior mesenteric artery within the dorsal mesentery. The urogenital sinus has separated from the rectum and anus. Common developmental errors include midgut malrotation, persistent omphalocele, and imperforate anus.

DEVELOPMENT OF LESSER SAC

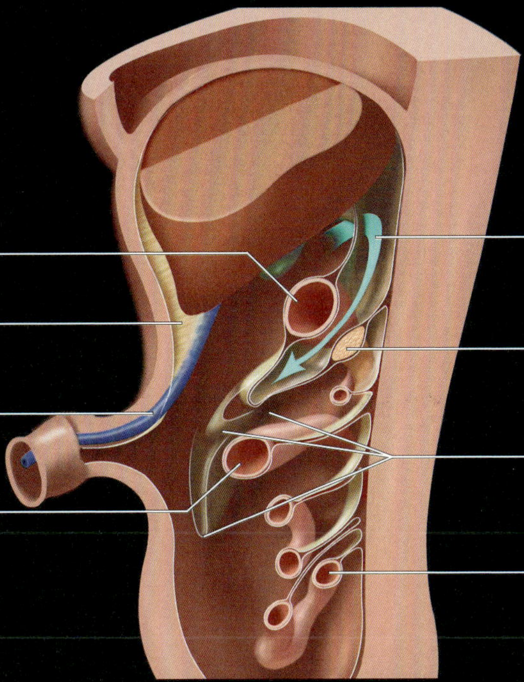

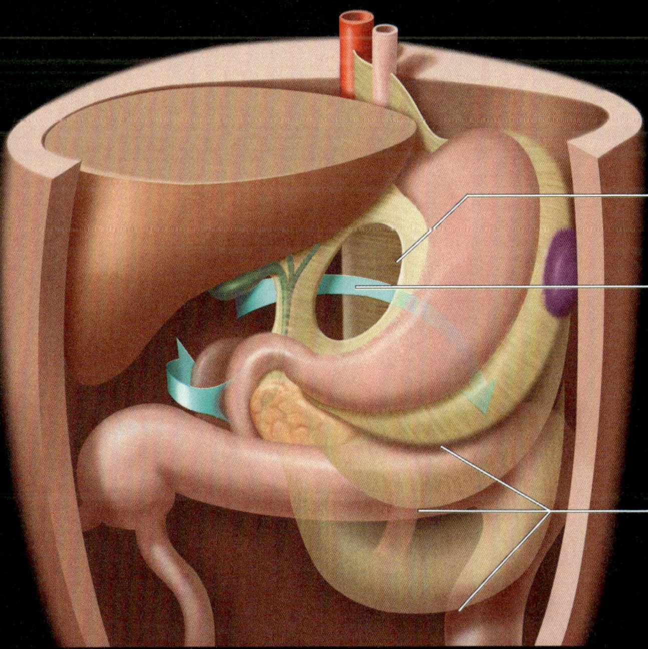

(Top) *The umbilical vein enters the liver along the caudal (free) edge of the falciform ligament. The duodenum and pancreas are intraperitoneal at this point. The leaves of the greater omentum elongate to the left and caudally, expanding the volume of the lesser sac and beginning to cover the transverse colon and small intestine.* **(Bottom)** *The greater peritoneal cavity and lesser sac communicate through the epiploic foramen.*

DEVELOPMENT OF GI TRACT (10-WEEK FETUS)

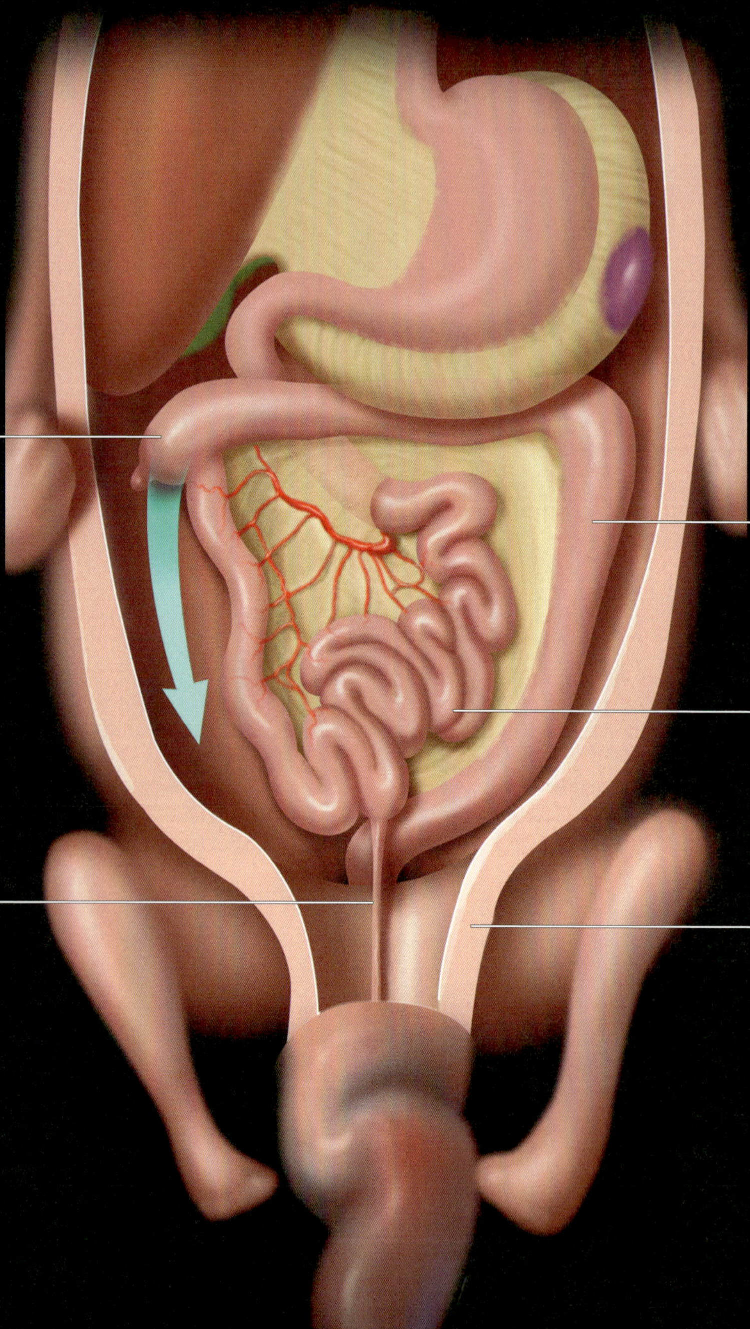

The small intestine has returned to the abdomen, having previously elongated and herniated out into the umbilical cord. The yolk stalk (vitelline duct) is disintegrating, having connected the yolk sac to the primitive gut at the level of the distal small intestine. The cecum is the last part of the gut to return and continues to rotate in a counterclockwise direction until reaching its adult position in the right lower quadrant. The transverse colon lies superficial to the intestine. Errors in development include persistence of a part of the yolk stalk (Meckel diverticulum) and errors of bowel rotation and mesenteric fusion.

DEVELOPMENT OF GI TRACT (16-WEEK FETUS)

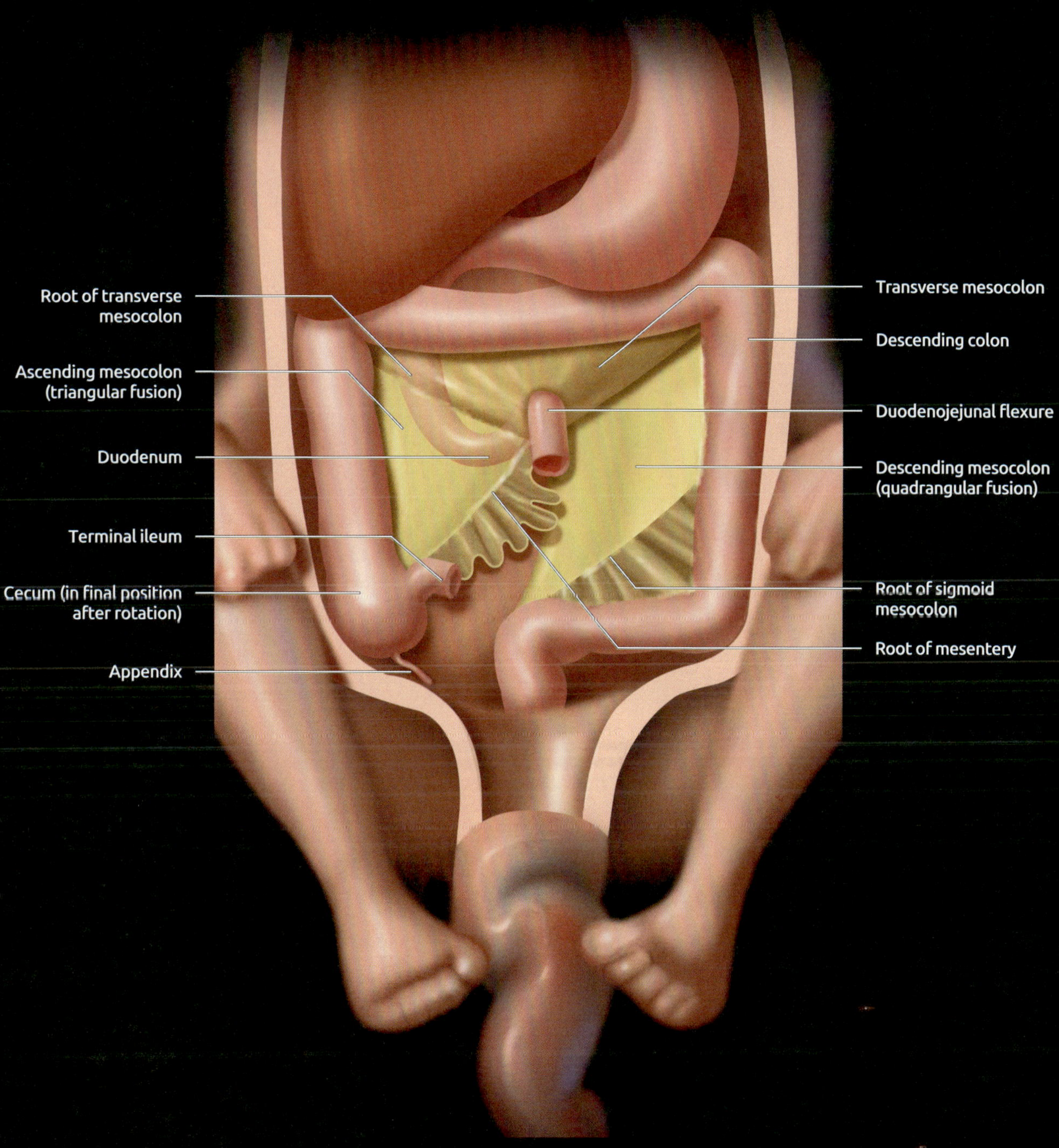

By 4-5 months of gestation, the ascending and descending colon are fixed in a retroperitoneal location by fusion of their mesocolons to the posterior abdominal wall. The fusion of the ascending mesocolon covers the duodenum and pancreas, resulting in their retroperitoneal location. The transverse and sigmoid colon remain intraperitoneal, suspended on a mesentery. The root of the small bowel mesentery is fused to the posterior abdominal wall, but the intestine is suspended on a long, fan-shaped mesentery.

DEVELOPMENT OF HEPATIC VEINS (4- TO 6-WEEK FETUS)

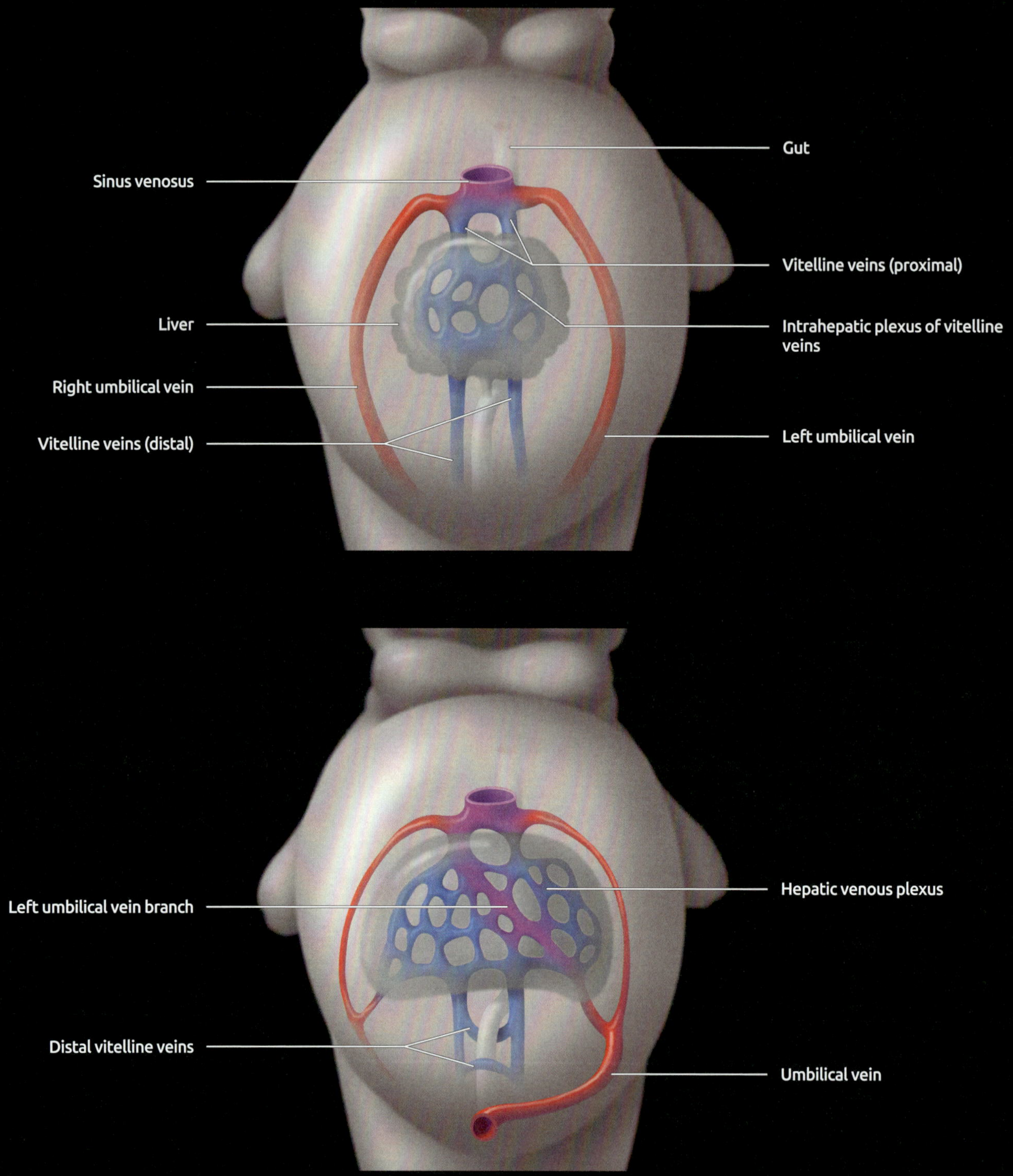

(Top) *The vitelline veins return blood from the yolk sac and branch out within the liver to form the hepatic sinusoids and venous system. They unite again to form the proximal vitelline veins, which join with the (initially) paired umbilical veins to enter the sinus venosus of the heart.* **(Bottom)** *Graphic shows the entire right umbilical vein and much of the left atrophy. The left umbilical vein sends a large branch to the liver, which anastomoses with the plexus derived from the vitelline veins. The extrahepatic (distal) vitelline veins form around the gut, as precursors to the portal venous system.*

DEVELOPMENT OF HEPATIC VEINS (8- TO 12-WEEK FETUS)

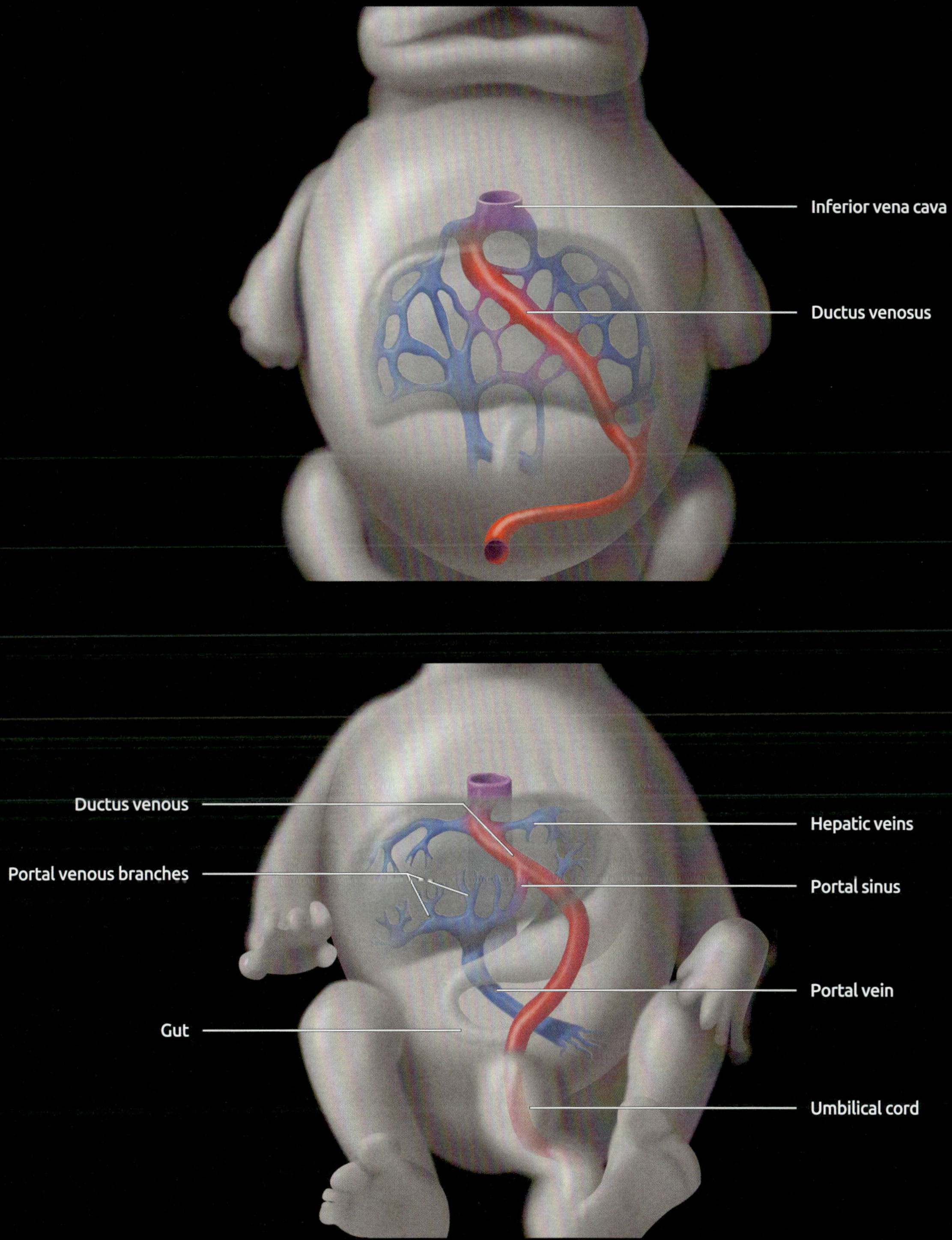

(Top) *The entire right umbilical vein and much of the left have disappeared. The left hepatic branch of the umbilical vein has become the ductus venosus, which bypasses the liver to deliver oxygenated blood from the placenta directly to the heart.* **(Bottom)** *The portal sinus diverts some oxygenated blood to the liver. The proximal parts of the vitelline veins have become the hepatic veins, returning blood from the liver to the heart. The distal parts have developed into the portal venous system, returning blood from the gut to the liver sinusoids.*

PRENATAL ABDOMINAL CIRCULATION

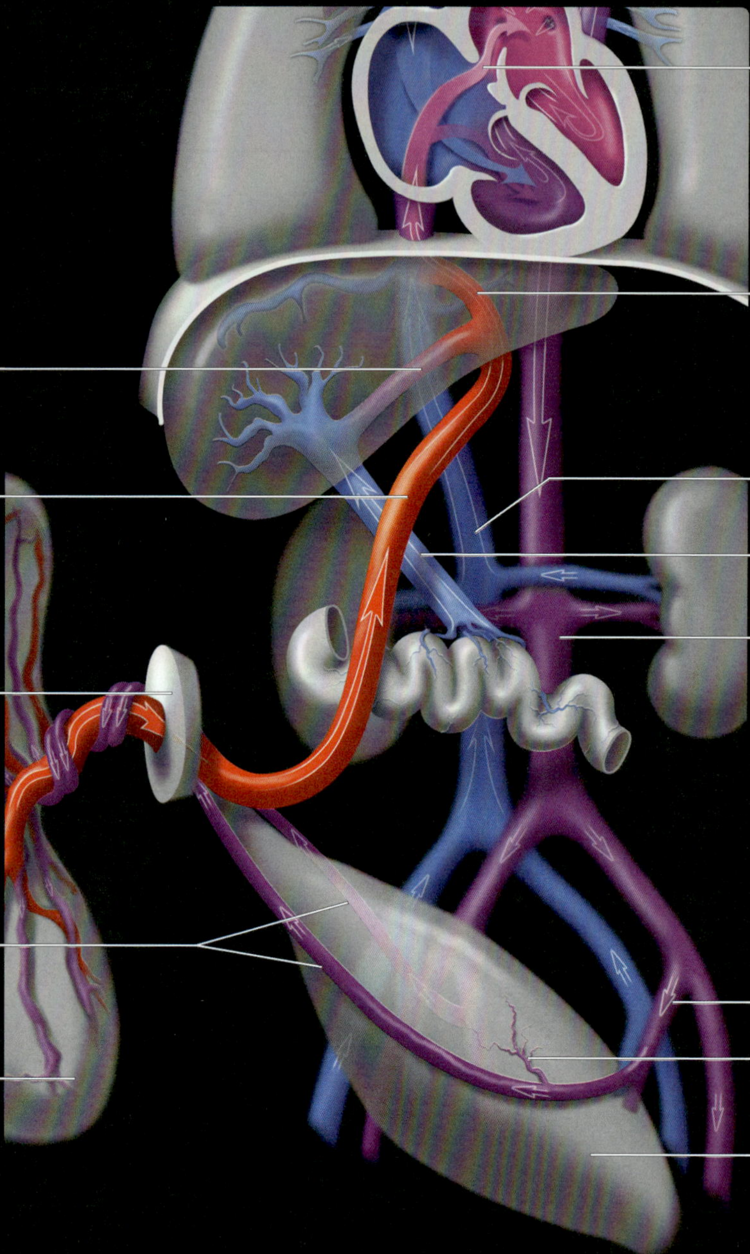

A simplified view of fetal circulation, not drawn to scale, is shown. The placenta sends oxygenated blood and nutrients to the fetus via the umbilical vein. Some oxygenated blood perfuses the liver via reflux through the portal sinus, while most blood bypasses the liver via the ductus venosus, which empties into the inferior vena cava. The patent foramen ovale allows oxygenated blood to pass from the right atrium into the left atrium. The portal vein returns blood from the fetal gut to the liver, and the aorta carries moderately oxygenated blood to the entire body and returns blood to the placenta via the umbilical arteries and branches of the internal iliac arteries. The fetal bladder is large and is perfused by branches of the internal iliac artery.

POSTNATAL CIRCULATION

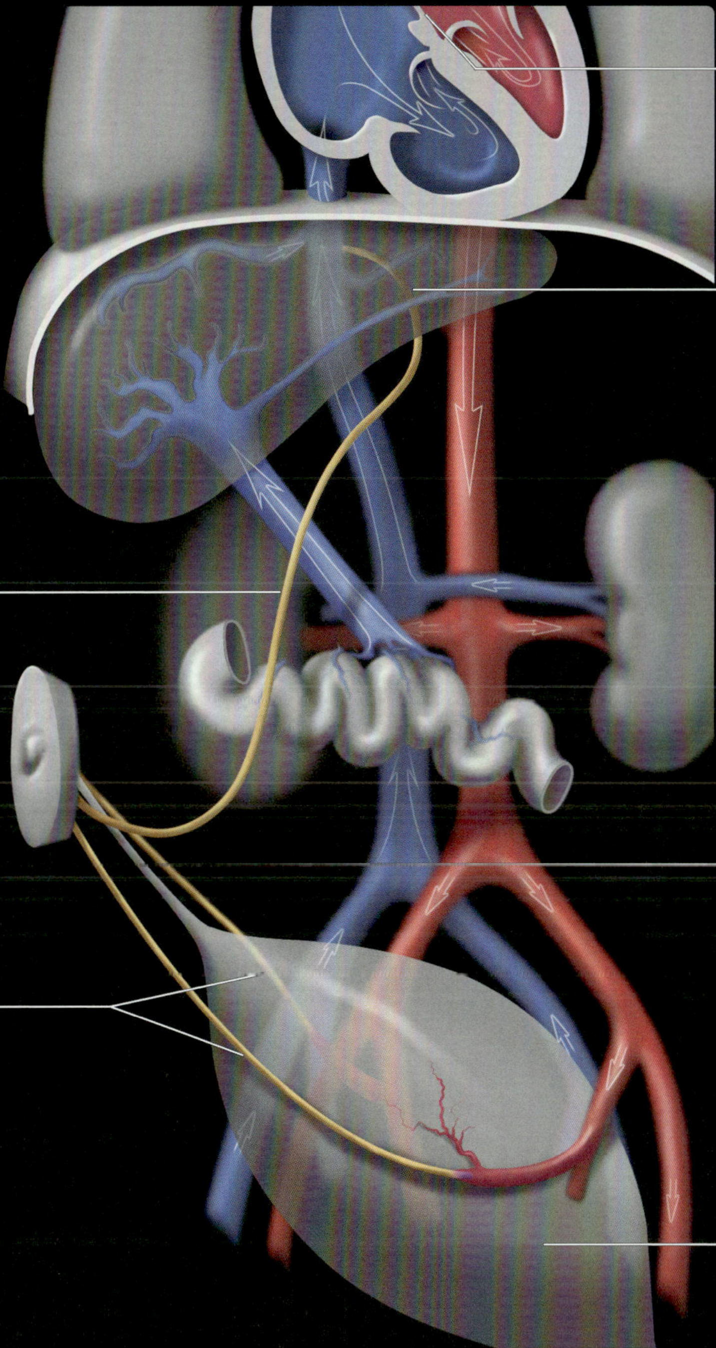

After birth, the umbilical vein is obliterated, becoming the ligamentum teres in the free edge of the falciform ligament. The ductus venosus also closes to become the ligamentum venosum. The foramen ovale closes, as the lungs now return oxygen-rich blood to the left atrium and aorta. The umbilical arteries atrophy to become the lateral umbilical ligaments. The allantois that connected the urinary bladder to the umbilicus closes to become the median umbilical ligament. Failure of this tract to close results in a urachal cyst or diverticulum.

PERITONEAL REFLECTIONS IN PARACOLIC GUTTERS

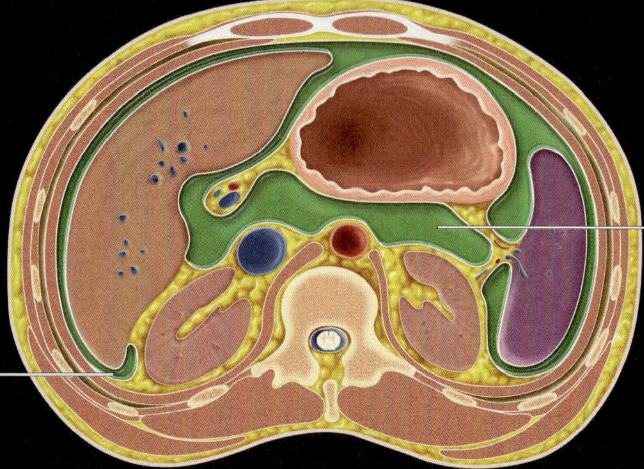

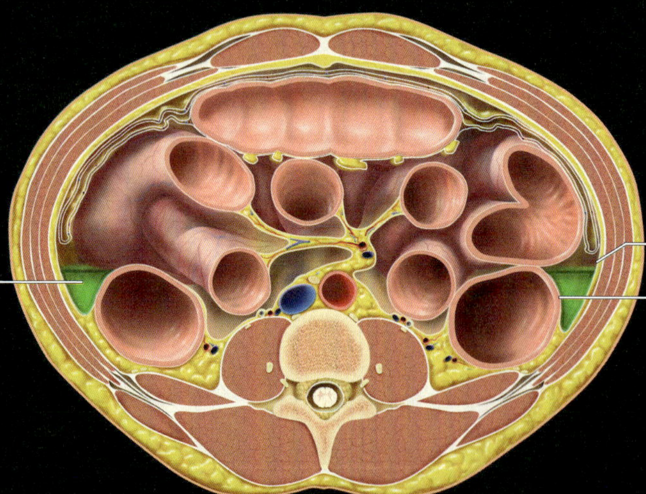

(Top) *Graphic of axial section through the upper abdomen shows ascites (fluid, green) within the subphrenic spaces bilaterally and lesser sac. The hepatorenal fossa (Morison pouch) is the most dependent peritoneal recess in the upper abdomen.* **(Bottom)** *The paracolic gutters are the main conduit for fluid between the upper (supramesocolic) and lower (inframesocolic) parts of the peritoneal cavity.*

SAGITTAL SECTION OF MESENTERIES AND PERITONEAL CAVITY

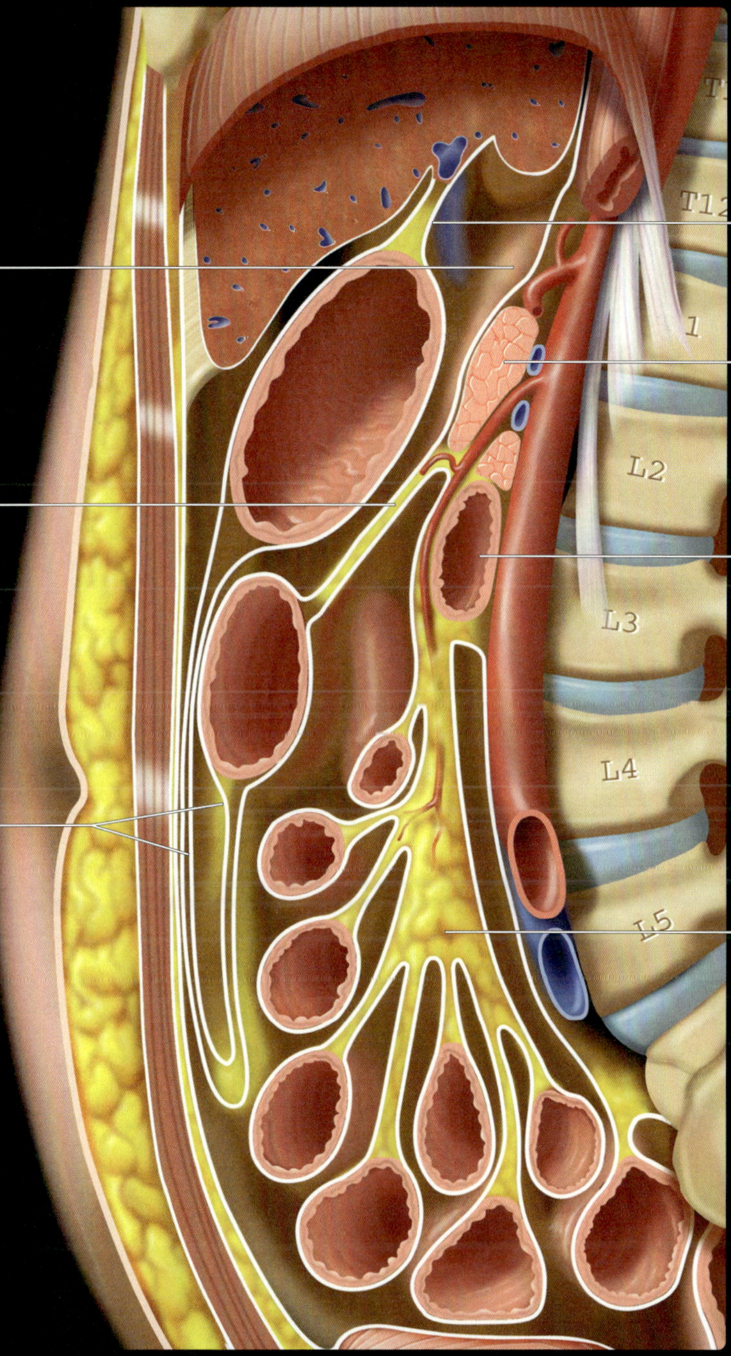

The peritoneal cavity is artificially distended, as if insufflated with air. The peritoneum is represented by a white line covering the entire cavity, lining the inner abdominal wall (parietal peritoneum) and the surface of abdominal organs (visceral peritoneum or serosa). Organs are considered intraperitoneal if they are mostly covered with peritoneum (e.g., intestine) or retroperitoneal if they have no or only partial peritoneal covering. Intraperitoneal and retroperitoneal compartments communicate via the subperitoneal space, between the leaves of the mesenteries. Note how bleeding or inflammation originating in the (retroperitoneal) pancreas can easily spread through the mesocolon and mesentery to affect the bowel. The greater omentum is depicted as a 4-layered fold of peritoneum, although these layers usually fuse together below the transverse colon.

PERITONEAL REFLECTIONS AND SPACES

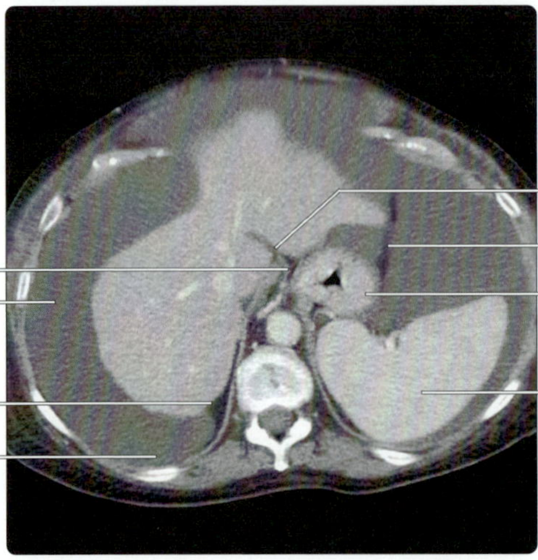

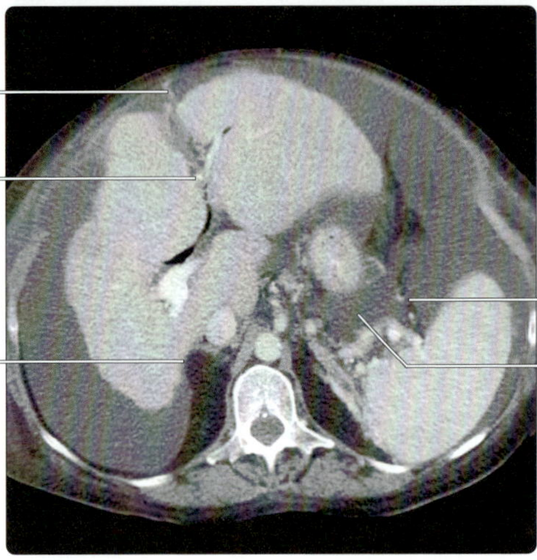

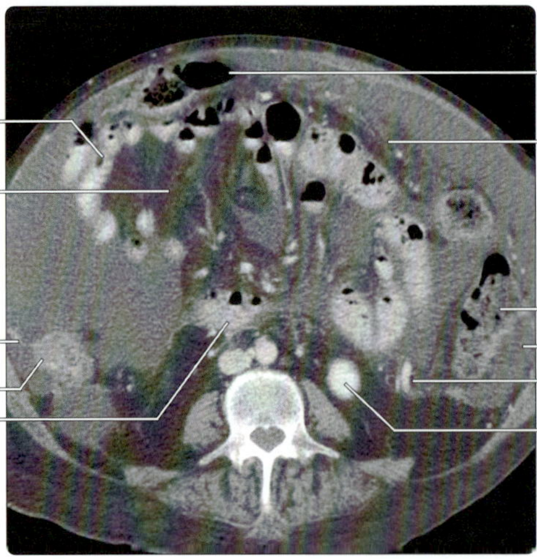

(Top) *First of 6 axial CECT sections of a 60-year-old man with cirrhosis and ascites is shown. The ascites distends the peritoneal cavity and outlines the mesenteries and ligaments. Note the bare area, the nonperitonealized attachment of the liver to the diaphragm, where ascites is excluded. Most mesenteries, omenta, and ligaments contain enough fat to allow them to be identified.* **(Middle)** *Paraumbilical venous collaterals have formed in the falciform ligament due to portal hypertension, reestablishing a fetal site of collateral circulation.* **(Bottom)** *The vessels to the intraperitoneal organs are carried through the various peritoneal reflections (mesenteries, omenta, and ligaments). Vessels to the retroperitoneal organs, including the ascending and descending colon, do not traverse mesenteries.*

PERITONEAL REFLECTIONS AND SPACES

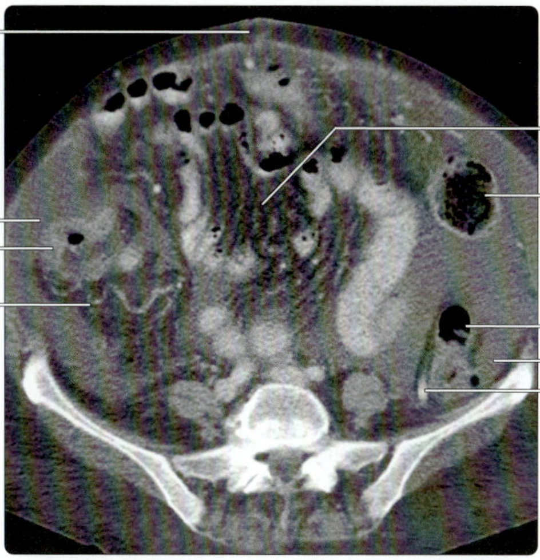

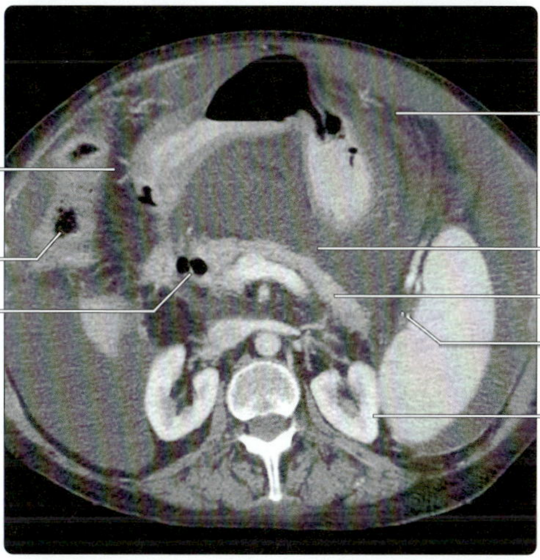

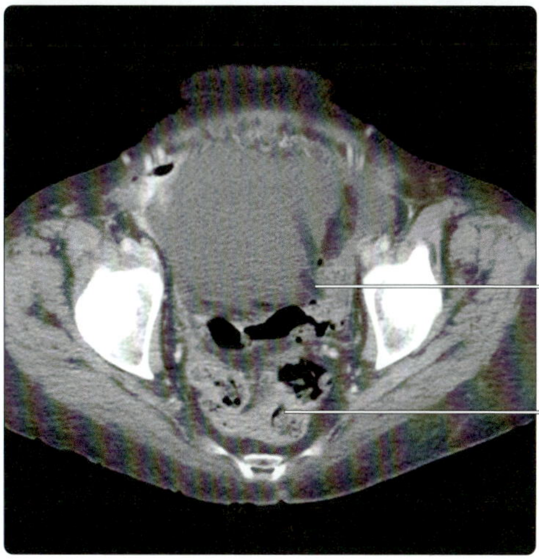

(Top) *The upper and lower portions of the peritoneal cavity "communicate" via the left and right paracolic recesses (gutters), which are formed by reflections of the peritoneum covering the colon and the inside of the anterolateral abdominal wall.* **(Middle)** *The margins of the lesser sac include the stomach, caudate lobe of the liver, gastrosplenic and splenorenal ligaments, pancreas, and the lesser omentum (gastrohepatic and hepatoduodenal ligaments). The intraperitoneal organs (liver, spleen, stomach, intestine, and colon) seem to be suspended and surrounded by ascites (intraperitoneal fluid), while the retroperitoneal pancreas, duodenum, and kidneys are not.* **(Bottom)** *The sigmoid colon is intraperitoneal, suspended on the sigmoid mesocolon through which the inferior mesenteric vessels run. Ascites outlines fatty epiploic appendages, which are prominent in the sigmoid colon. The rectum lies in the pelvic extraperitoneal space and is not surrounded by ascites.*

MESENTERIES AND PERITONEAL SPACES

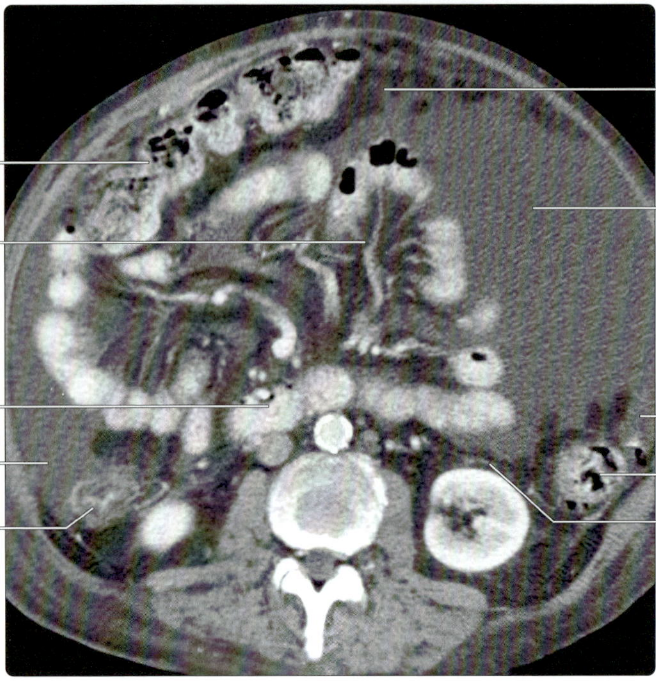

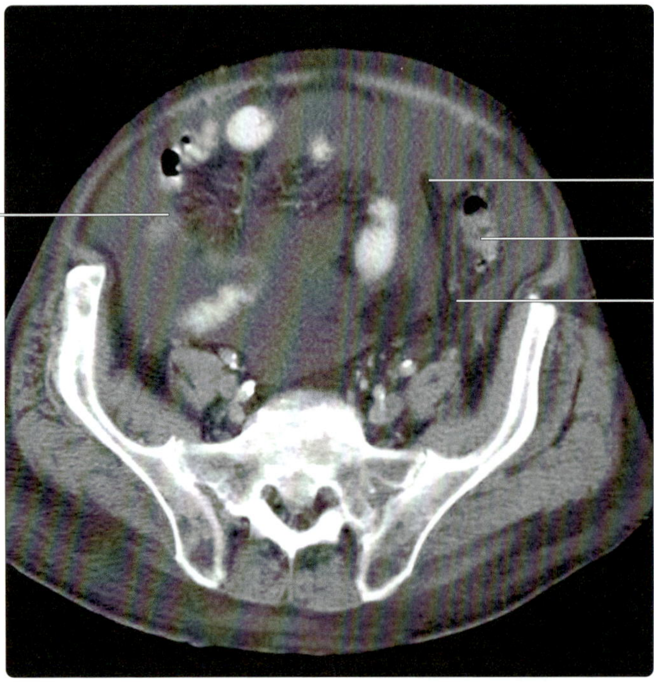

(Top) *First of 5 axial CECT images of a 50-year-old man with cirrhosis and ascites is shown. The relationship between the intraperitoneal and retroperitoneal organs is evident in this patient with ascites. The vessels, nerves, and lymphatics travel through the retroperitoneum and, for the intraperitoneal organs, extend out through the fibrofatty tissue in the mesenteries, between layers of visceral peritoneum.* **(Bottom)** *Note the paracolic gutters, formed by reflections of peritoneum over the ascending and descending colon.*

MESENTERIES AND PERITONEAL SPACES

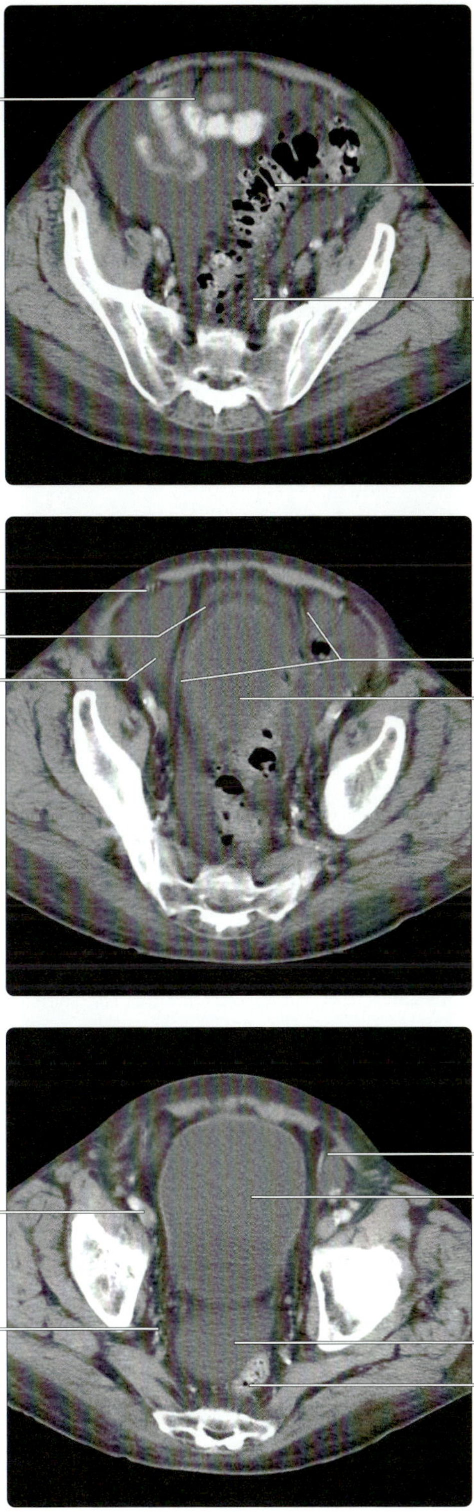

(Top) *Ascites surrounds the intraperitoneal small bowel and sigmoid colon.* **(Middle)** *Ascites outlines the lateral umbilical ligaments (remnants of the umbilical arteries). The main vessels and nerves supplying the anterolateral abdominal wall travel between muscle layers or between the peritoneum and the transversalis fascia, such as the inferior epigastric vessels, which arise from the external iliac vessels in the pelvis.* **(Bottom)** *The rectovesical space is the most dependent recess of the peritoneal cavity.*

RETROPERITONEAL SPACES AND PLANES

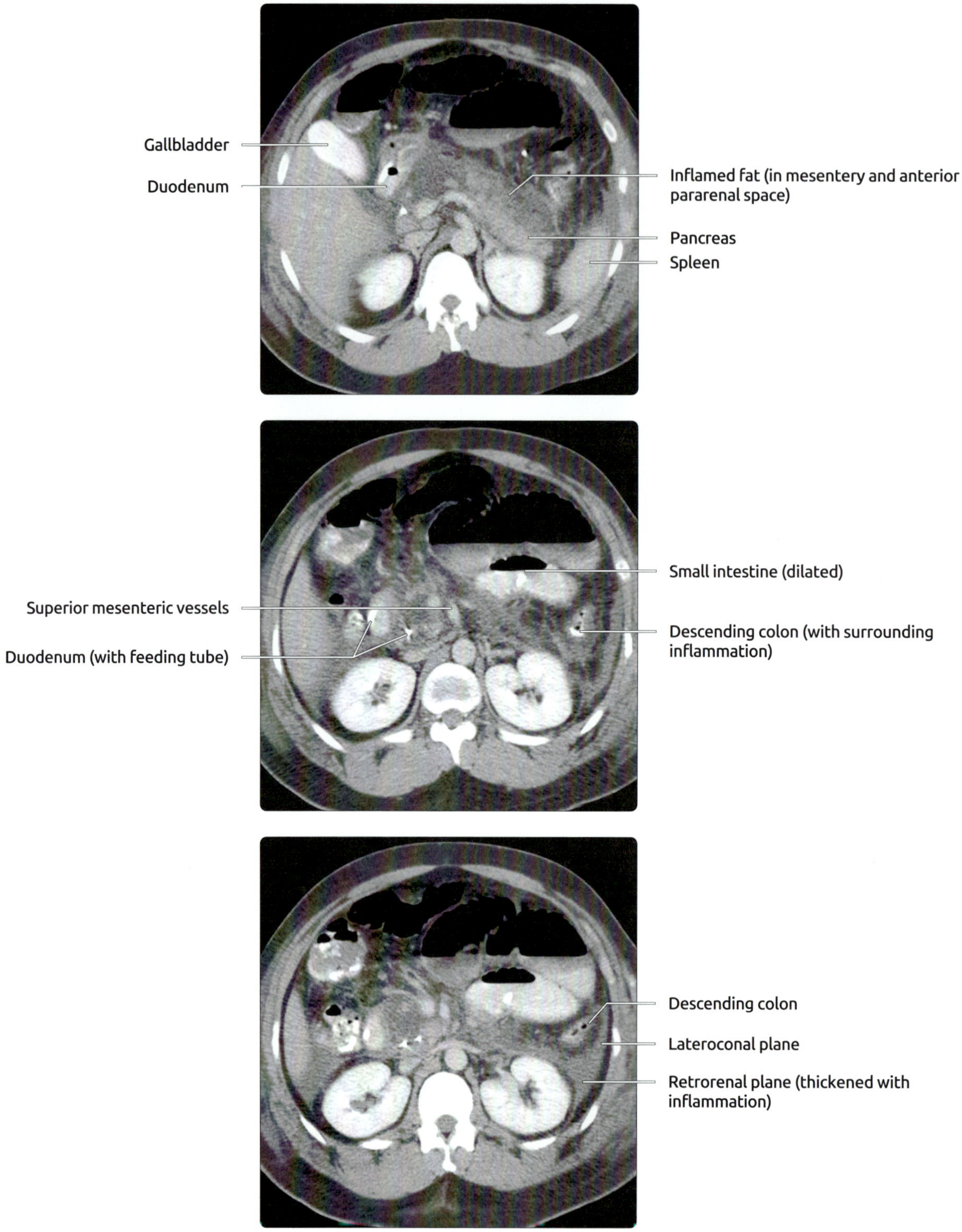

(Top) *First of 6 axial CECT images of a young man with acute pancreatitis is shown. The gallbladder is opacified due to prior administration of contrast material with vicarious excretion. The pancreas is enlarged and enhances heterogeneously due to acute pancreatitis. Inflammation spreads ventrally into the mesentery and laterally throughout the anterior pararenal space, which is bordered by the peritoneum in front and the perirenal fascia behind.* **(Middle)** *Mesenteric involvement is evident by infiltration of the fat surrounding the superior mesenteric vessels. Organs sharing the anterior pararenal space with the pancreas, namely the duodenum, ascending colon, and descending colon, are similarly involved.* **(Bottom)** *The inflammatory process splits apart and thickens the double-layered renal and lateroconal fascia to reveal retrorenal and lateroconal planes.*

RETROPERITONEAL SPACES AND PLANES

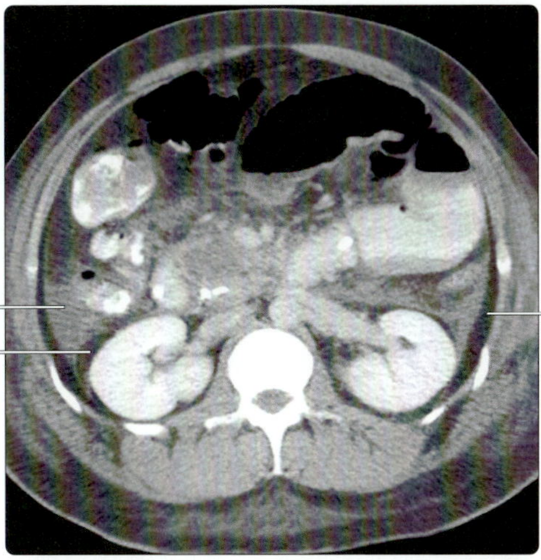

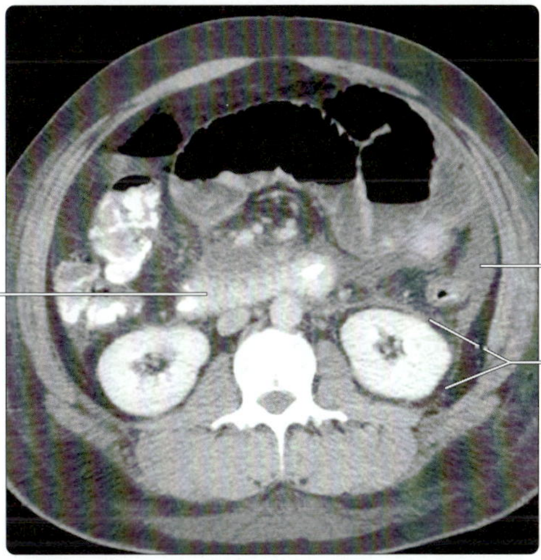

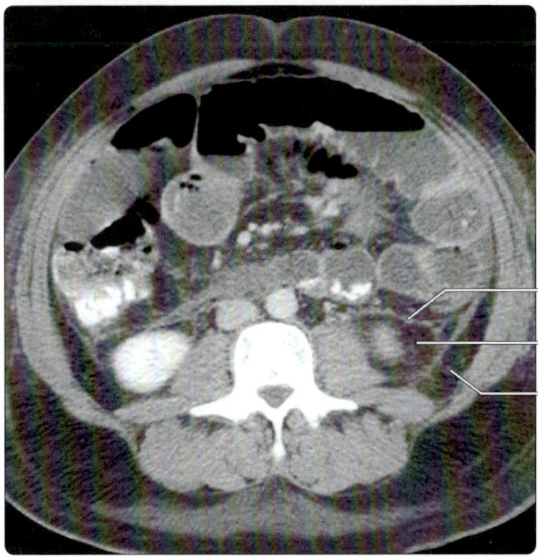

(Top) *The perirenal and posterior pararenal spaces remain uninvolved, as the renal and lateroconal fascia have blocked further spread of inflammation. Intraperitoneal fluid (ascites) in the right paracolic gutter is adjacent to, and difficult to distinguish from, inflammation in the right anterior pararenal space, which surrounds the ascending colon. Inflammation has breached the posterior parietal peritoneum to cause the ascites.* **(Middle)** *Ascites in the left paracolic gutter lies in the recess lateral to the colon. The space behind the colon and the thickening of the perirenal fascia are representative of the retroperitoneal (anterior pararenal space) inflammation.* **(Bottom)** *On this more caudal section, just above the iliac wing, the leaves of the renal fascia approach each other. Caudal to this point, there is only a single abdominal retroperitoneal space, which communicates with the pelvic extraperitoneal spaces (pre- and perivesical).*

COMMUNICATION AMONG EXTRAPERITONEAL SPACES

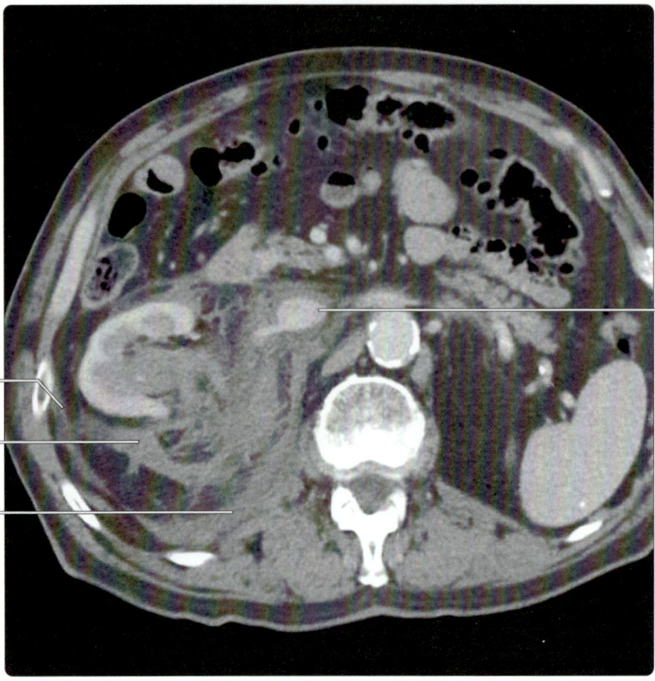

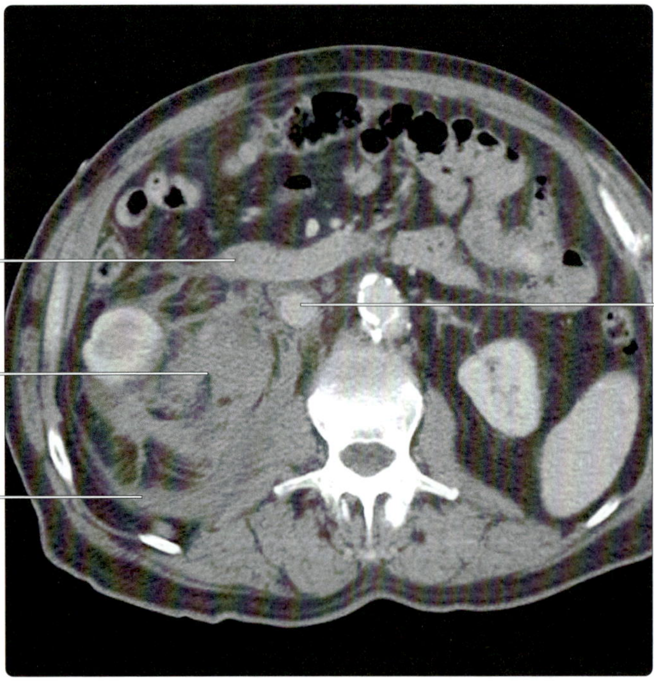

(Top) *First of 6 axial CECT images of an older man (on Coumadin therapy) with a spontaneous perirenal bleed is shown. Blood spreads through the perirenal space along the perirenal septa to "decompress" into the retrorenal interfascial plane.* **(Bottom)** *The perirenal space does not extend to the midline, but the interfascial planes do, allowing hemorrhage to surround the inferior vena cava.*

COMMUNICATION AMONG EXTRAPERITONEAL SPACES

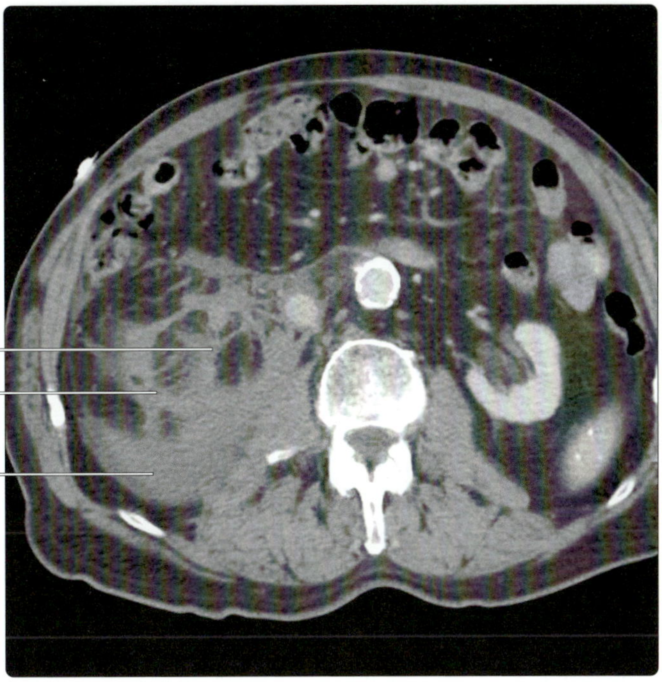

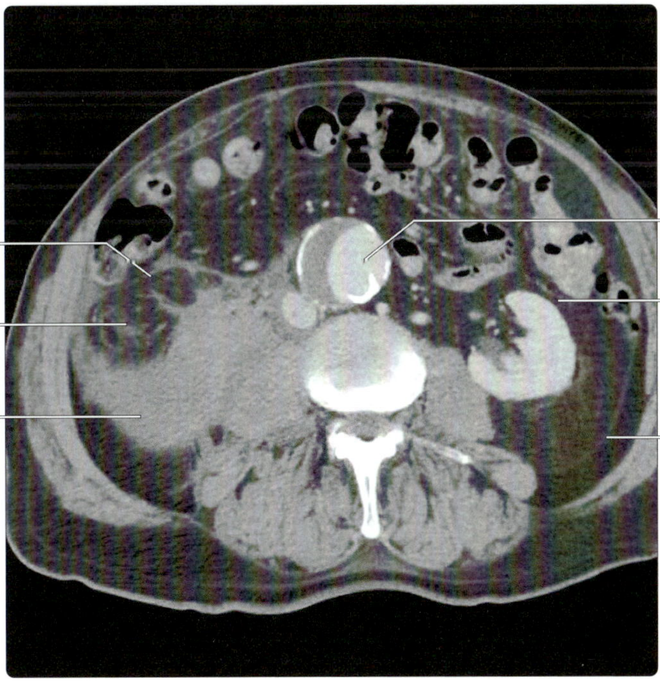

(Top) *The thickened perirenal septa and retrorenal interfascial planes are well defined. The perirenal space decreases in diameter caudally as the leaves of the renal (Gerota) fascia approach each other.* **(Bottom)** *An aortic aneurysm is noted on this image but is not the source of the hemorrhage. Caudal to the perirenal space, there is only a single infrarenal retroperitoneal space.*

COMMUNICATION AMONG EXTRAPERITONEAL SPACES

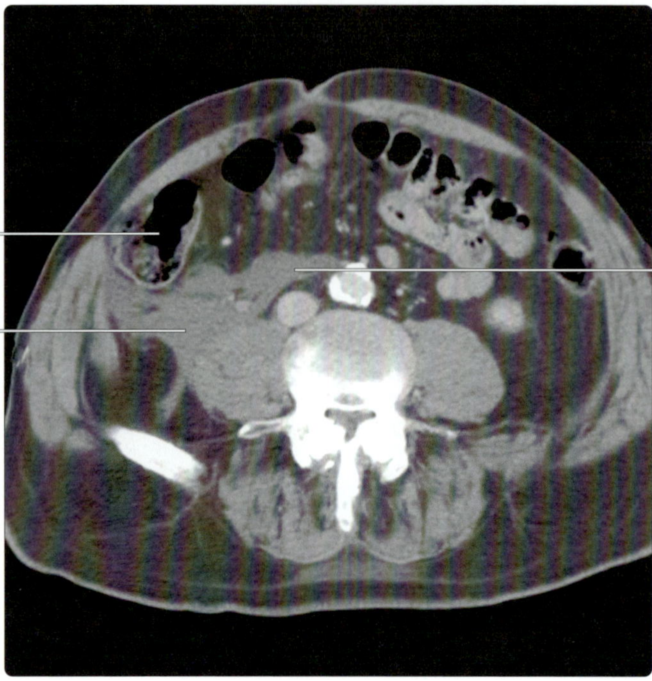

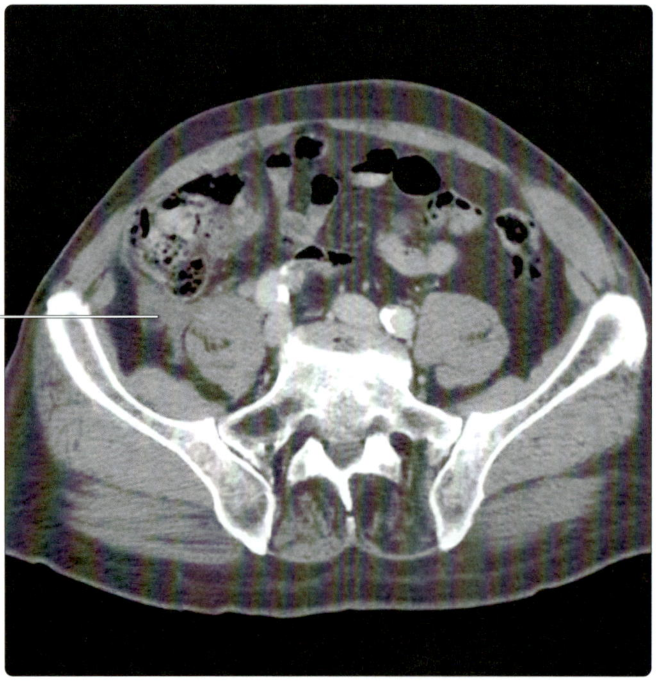

(Top) *The infrarenal retroperitoneal space is contiguous with the psoas muscle and communicates across the midline.* **(Bottom)** *The infrarenal retroperitoneal space communicates caudally with the extraperitoneal pelvic spaces.*

CONGENITAL ANOMALIES

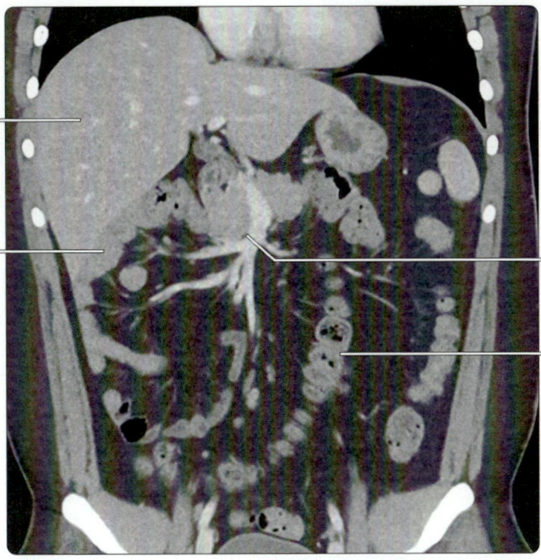

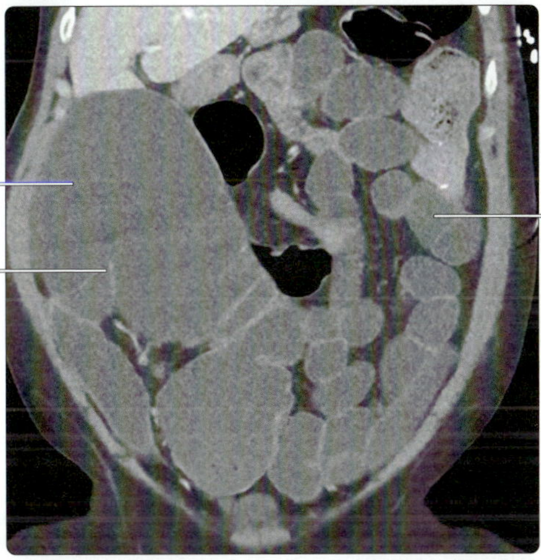

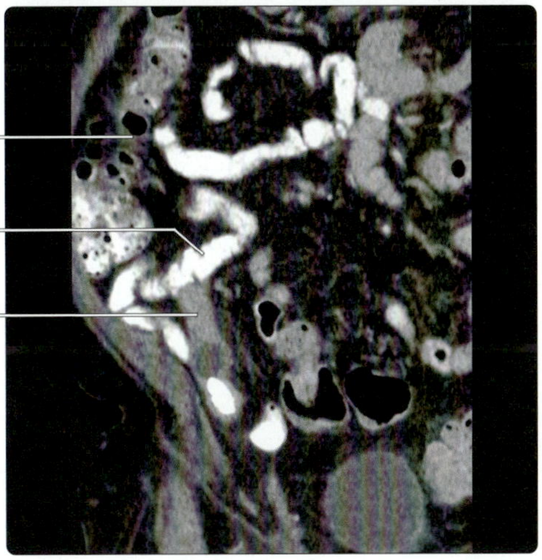

(Top) *Bowl malrotation is also associated with pancreatic contour abnormalities, such as aplasia/hypoplasia of the uncinate process.* **(Middle)** *Coronal reformatted CT in a patient with markedly distended cecum is shown. The cecum is upside down and backward with the ileocecal valve pointed laterally. Coronal-reformatted CT is usually the most definitive study to show the characteristic features of cecal volvulus. Cecal volvulus is a condition where the cecum twists around its own supporting tissue, resulting in bowel obstruction. Abnormal motility of the right colon can occur due to a long congenital cecal mesentery resulting from inadequate fusion between the mesentery and retroperitoneal structures.* **(Bottom)** *Coronal CT shows an incidentally found blind-ending sac arising from the distal ileum, consistent with a Meckel diverticulum, which are a remnant of the omphalomesenteric duct, a structure that connects the developing intestine to the umbilical cord during fetal development. These are seen in ~ 2% of population, located within 2 feet of ileocecal valve, have a length of ~ 2 inches, and are symptomatic usually before 22 years of age.*

DEVELOPMENT OF GU TRACT

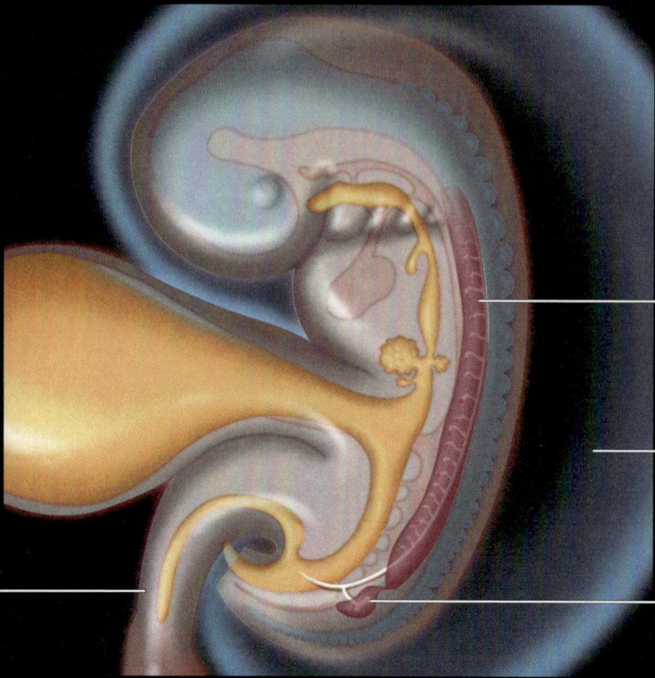

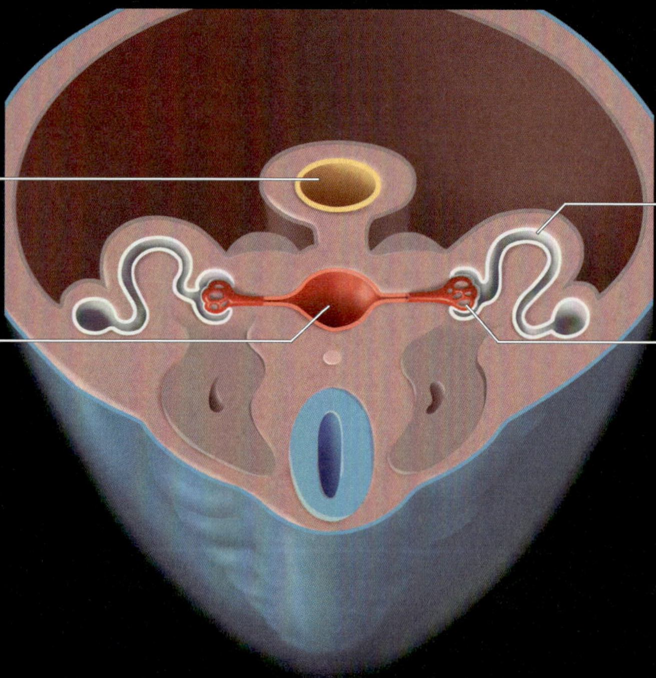

(Top) *The primitive GU system develops in parallel with the GI tract. The pronephron is a transient development in vertebrate animals and degenerates to be replaced by the mesonephron. The mesonephron also degenerates (in mammals) and is replaced by the metanephron, which will be the permanent kidney.* **(Bottom)** *By the 4th fetal week, the mesonephric tubules and duct are formed. Branched vessels from the aorta reach the blind ends of the tubules to form glomeruli. Although these achieve excretory function in the human embryo, they degenerate from the cranial toward the caudal end. The caudal end of the mesonephric duct persists and differentiates into the müllerian and wolffian ducts, which later give rise to the internal genitalia.*

DEVELOPMENT OF GU TRACT (5-WEEK FETUS)

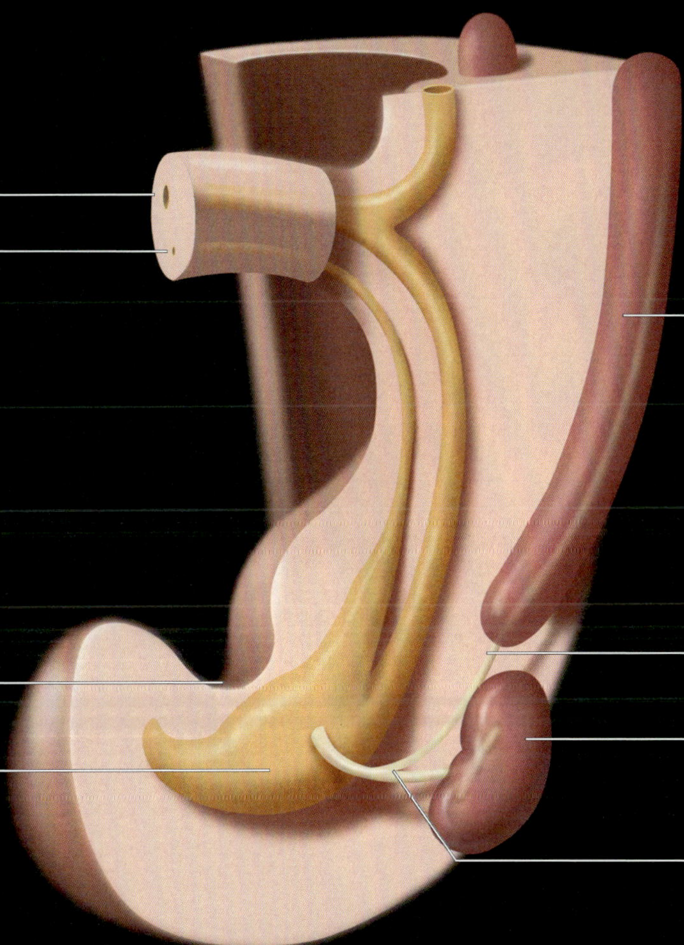

The ureteric bud (metanephric duct) has grown out of the mesonephric duct near its termination in the cloaca and has extended into the metanephrogenic tissue (kidney). Within the kidney, the ureteric bud branches and expands to form calces, and these arborize to form successive generations of collecting ducts. At the 5th fetal week, the hindgut and developing urinary tract both terminate in the blind-ending cloaca.

DEVELOPMENT OF GU TRACT (20-WEEK FEMALE FETUS)

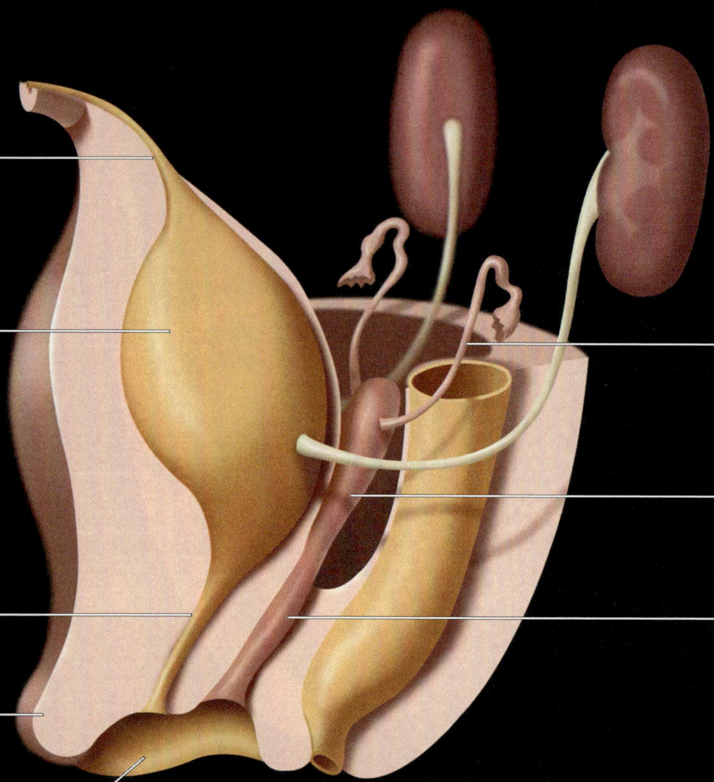

In the female fetus, the lower müllerian ducts unite to form a uterovaginal canal. The unfused portions of the ducts give rise to the fallopian tubes. The distal part of the vagina forms from invagination of a solid mass of cells between the uterovaginal canal and the urogenital sinus. Other parts of the urogenital sinus form the bladder, urethra, and vestibule. Accessory glands derived from the urethra are inconspicuous in most female fetuses as urethral and periurethral glands, whereas these develop into the prostate gland in male fetuses.

KIDNEY ASCENT AND ROTATION (6 WEEKS)

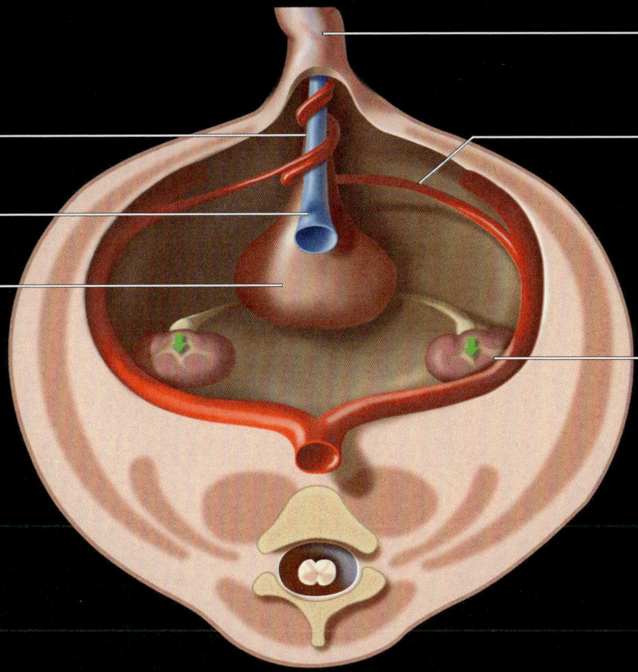

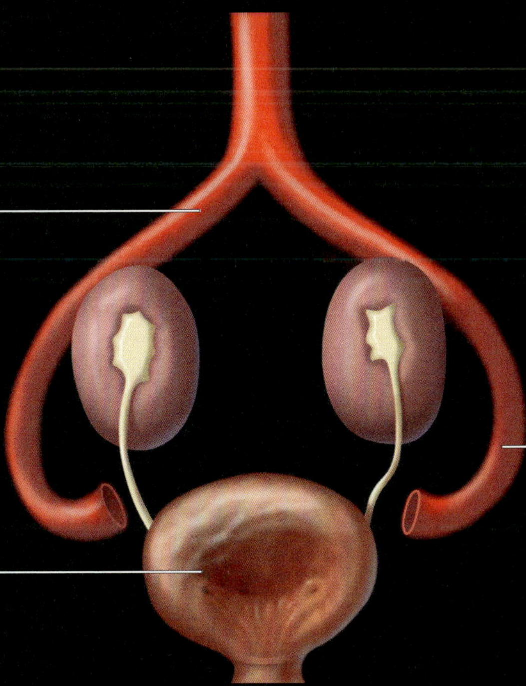

(Top) *The early metanephros (kidneys) lie close together in the pelvis and may even touch each other. This may lead to various anomalies of fusion and ectopic location.* **(Bottom)** *The fetal kidneys face forward.*

KIDNEY ASCENT AND ROTATION (7 WEEKS)

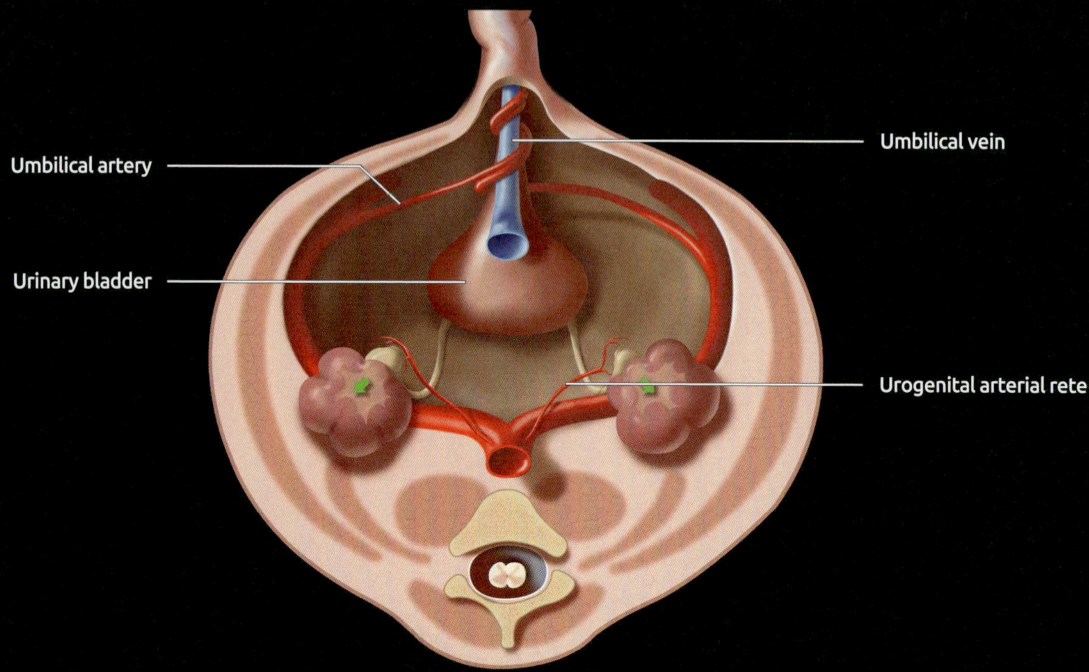

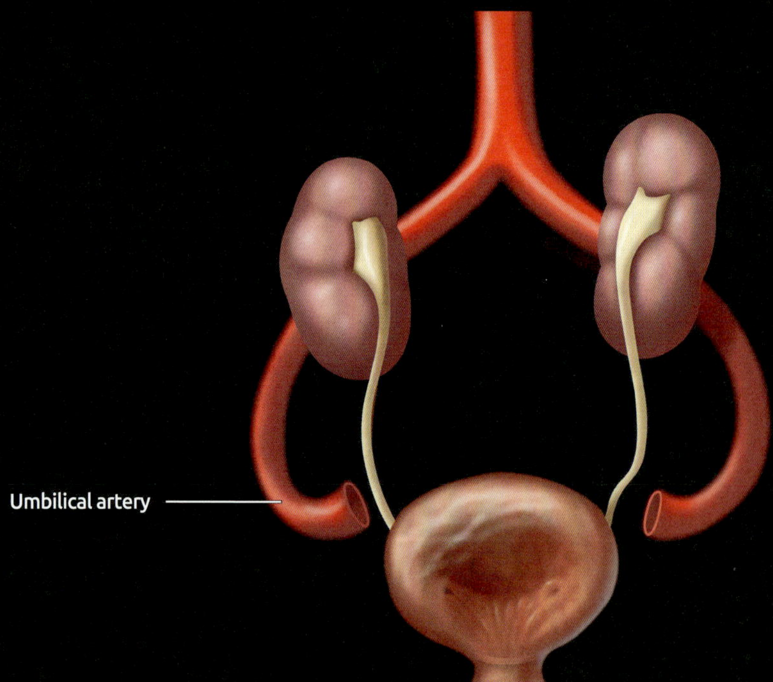

(Top) *The early metanephros (kidney) lies below the aortic bifurcation. Arterial supply is from multiple branches of the aorta, each part of a plexus called the urogenital arterial rete. As the kidneys "ascend" with fetal growth, usually only 1 artery remains to supply each kidney.* **(Bottom)** *At this stage of development, the fetal kidneys face forward. If the fetal kidney fails to ascend normally (remains ectopic), it is likely to retain a multiple arterial supply and fail to rotate medially.*

KIDNEY ASCENT AND ROTATION (9 WEEKS)

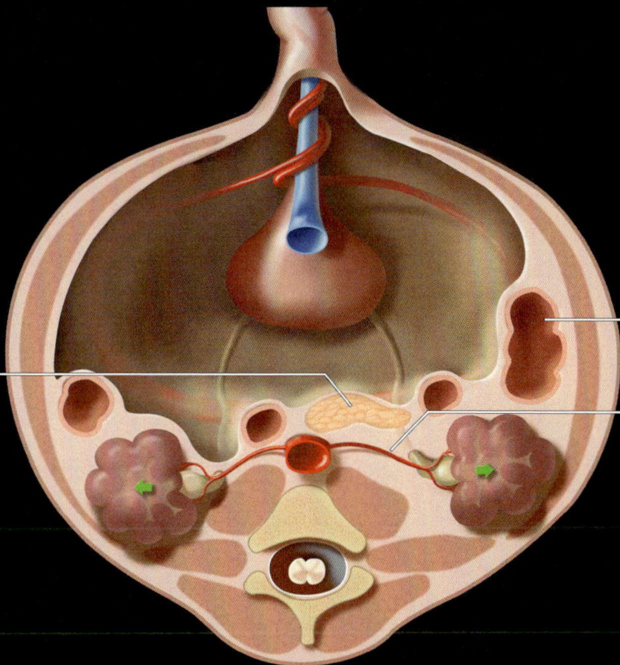

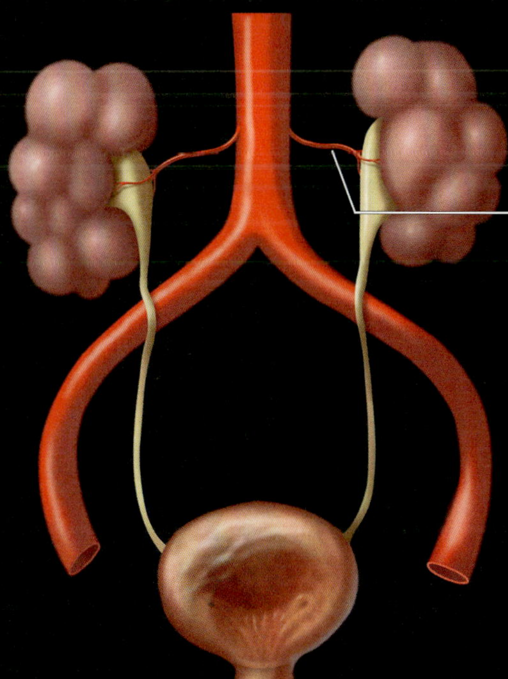

(Top) *By the 9th fetal week, the kidneys have occupied their retroperitoneal location (along with the pancreas and duodenum). Each kidney is usually supplied by a single arterial branch of the aorta.* **(Bottom)** *The kidneys now lie at about the level of the 3rd lumbar vertebra and have rotated along their longitudinal axis so that the renal hila face medially. Fetal lobation remains a distinctive feature, reflecting the lobular origin of the primitive kidney.*

GROSS ANATOMY

Muscles of Anterior Abdominal Wall

- Consists of 3 **flat muscles** (external and internal obliques and transverse abdominal), which are located laterally in abdominal wall, and 1 important **strap-like muscle** (rectus), which has vertical orientation near midline
  - Combination of muscles and **aponeuroses** (sheet-like tendons) act as corset to confine and protect abdominal viscera
  - Actions of anterior abdominal wall muscles
    - Support and protect abdominal viscera
    - Help flex/twist trunk and maintain posture
    - Increase intraabdominal pressure for defecation, micturition, and childbirth
    - Stabilize pelvis during walking and sitting up
- **External oblique muscle**
  - Largest and most superficial of 3 flat muscles, which courses inferomedially in lateral abdominal wall
  - Origin: External surfaces of ribs 5-12
  - Insertion: Linea alba, iliac crest, and pubis via broad aponeurosis, which also forms external inguinal ring
- **Internal oblique muscle**
  - Middle of 3 flat abdominal muscles (deep to external oblique), which runs superomedially
  - Origin: Posterior layer of thoracolumbar fascia, iliac crest, and inguinal ligament
  - Insertion: Ribs 10-12 posteriorly, linea alba via broad aponeurosis, and pubis
- **Transversus abdominis (transverse abdominal) muscle**
  - Innermost of 3 flat muscles (deep to external/internal oblique muscles), which runs in transverse plane
  - Origin: 6 most inferior costal cartilages, thoracolumbar fascia, iliac crest, and inguinal ligament
  - Insertion: Linea alba via broad aponeurosis, pubis
  - **Transversalis fascia** located deep to this muscle
- **Rectus abdominal muscle**
  - Paired muscles on either side of midline (divided by linea alba), which run in vertical plane
    - Lateral border of muscles defines **linea semilunaris**
    - 3-4 fibrous bands extend perpendicularly into rectus (creates visual appearance of 6 pack)
  - Origin: Pubic symphysis and pubic crest
  - Insertion: Xiphoid process and costal cartilages 5-7
- **Pyramidalis** is 2nd small strap muscle anterior to inferior rectus abdominis but not of major clinical importance
- **Rectus sheath**
  - Rectus sheath: Strong fibrous compartment that envelops rectus muscles, formed by aponeuroses of 3 flat muscles
    - Rectus sheath contains **superior** and **inferior epigastric vessels**
  - Most of rectus sheath has anterior (aponeuroses of external/internal oblique) and posterior wall (aponeuroses of internal oblique/transversus abdominis)
    - No posterior wall starting halfway between umbilicus and pubic symphysis (defined as **arcuate line**)
- **Linea alba** is fibrous raphe stretching from xiphoid to pubis, formed by interlacing fibers of flat muscle aponeuroses
  - Forms central anterior attachment for abdominal wall muscles
- **Transversalis fascia**
  - Lies deep to abdominal wall muscles and lines entire abdominal wall, separated from **parietal peritoneum** by layer of **extraperitoneal fat**

Muscles of Posterior Abdominal Wall

- Composed of psoas (major and minor), iliacus, and quadratus lumborum
- **Psoas major**: Long, thick, fusiform muscle lying immediately lateral to vertebral column
  - Origin: Transverse processes and T12-L5 vertebral bodies
  - Insertion: Lesser trochanter of femur (passing behind inguinal ligament)
  - Action: Flexes thigh at hip joint and bends vertebral column laterally
  - 60% have **psoas minor** muscle, which runs anterior to psoas major and originates from T12/L1 vertebrae
- **Iliacus**: Large triangular sheet of muscle lying along lateral aspect of psoas
  - Origin: Superior part of iliac fossa
  - Insertion: Lesser trochanter of femur (after joining with psoas tendon)
  - Action: Iliopsoas muscle flexes thigh
- **Quadratus lumborum**: Thick muscle adjacent to lumbar transverse processes and superficial to psoas muscles
  - Origin: Iliac crest and lumbar transverse processes
  - Insertion: 12th rib and L1-L4 transverse processes
  - Actions: Stabilizes position of thorax and pelvis during respiration and walking and bends trunk to side
- Fascial layer covers posterior abdominal wall muscles and divides them from parietal peritoneum
  - Includes psoas fascia, which covers psoas major, and thoracolumbar fascia, which covers quadratus lumborum and deep back muscles

CLINICAL IMPLICATIONS

Clinical Importance

- **Rectus** and iliopsoas muscle compartments are common sites for **spontaneous bleeding** in coagulopathic patients
  - Absence of posterior wall of rectus sheath below arcuate line allows rectus bleeding to extend into pelvic cavity
- Obesity and lack of exercise result in atrophy of abdominal wall muscles
  - **Pannus** (panniculus): Lax abdominal wall and excess subcutaneous fat may simulate ventral hernia both clinically and on imaging
  - Results in common occurrence of abdominal **wall hernias**, especially following abdominal surgery
- Congenital points of weakness in aponeuroses predispose to other common external hernias
  - **Ventral hernia**: Through congenital defect in aponeuroses forming linea alba in midline
  - **Spigelian hernia**: Occurs lateral to rectus muscle below umbilicus through defect in aponeurosis of internal oblique and transverse abdominal muscles at point where inferior epigastric vessels penetrate rectus sheath
    - External oblique aponeurosis is often intact; hernia sac may be intermuscular
  - **Lumbar hernia**: At congenital point of deficiency just above iliac crest = **inferior lumbar triangle** (**of Petit**)

POSTERIOR ABDOMINAL WALL MUSCLES

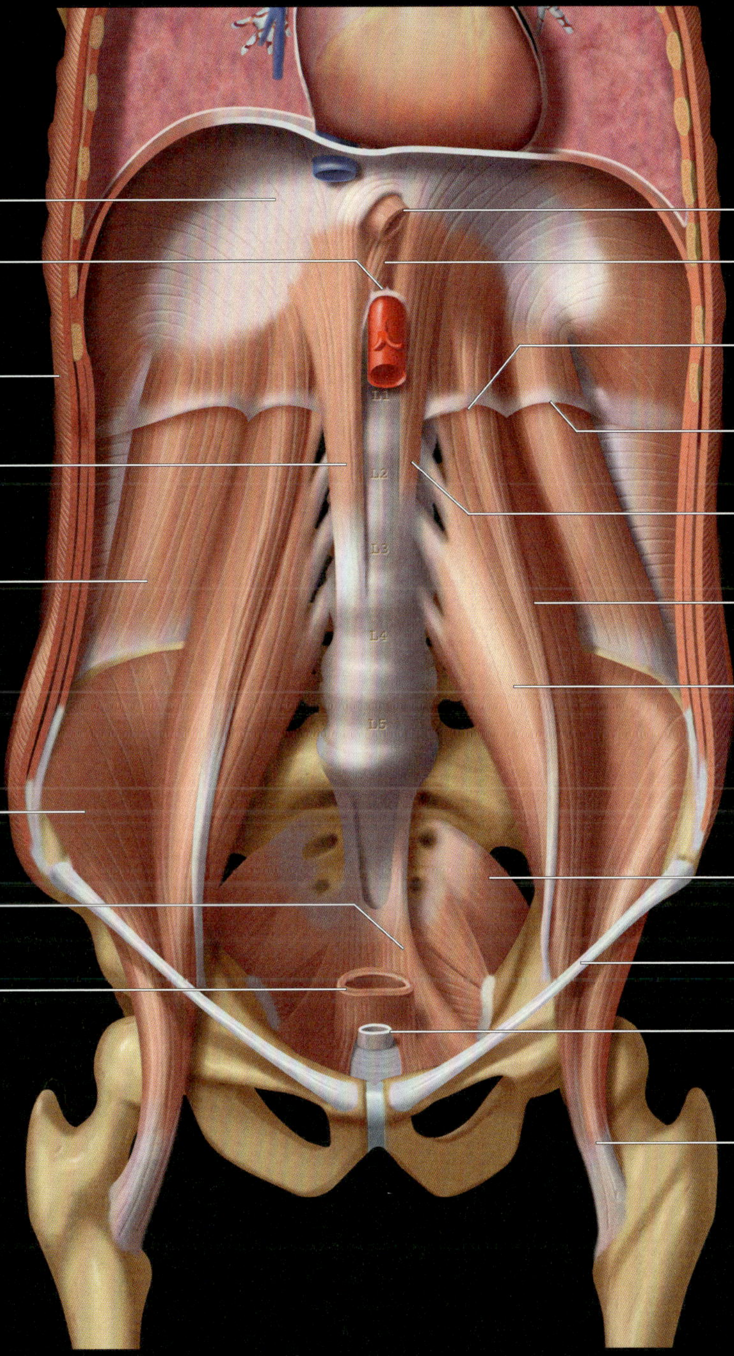

Lumbar vertebrae are covered and attached by the anterior longitudinal ligament. The diaphragmatic crura are closely attached to the lumbar vertebra, as are the origins of the psoas muscles, which also arise from the transverse processes. The iliacus muscle arises from the iliac fossa of pelvis and inserts into the tendon of the psoas major, constituting the iliopsoas muscle, which inserts into the lesser trochanter. The quadratus lumborum arises from the iliac crest and inserts into the 12th rib and transverse processes of the lumbar vertebrae. The diaphragm and transverse abdominal fibers interlace. The psoas and quadratus lumborum pass behind diaphragm under medial and lateral arcuate ligaments.

ANTERIOR WALL MUSCLES

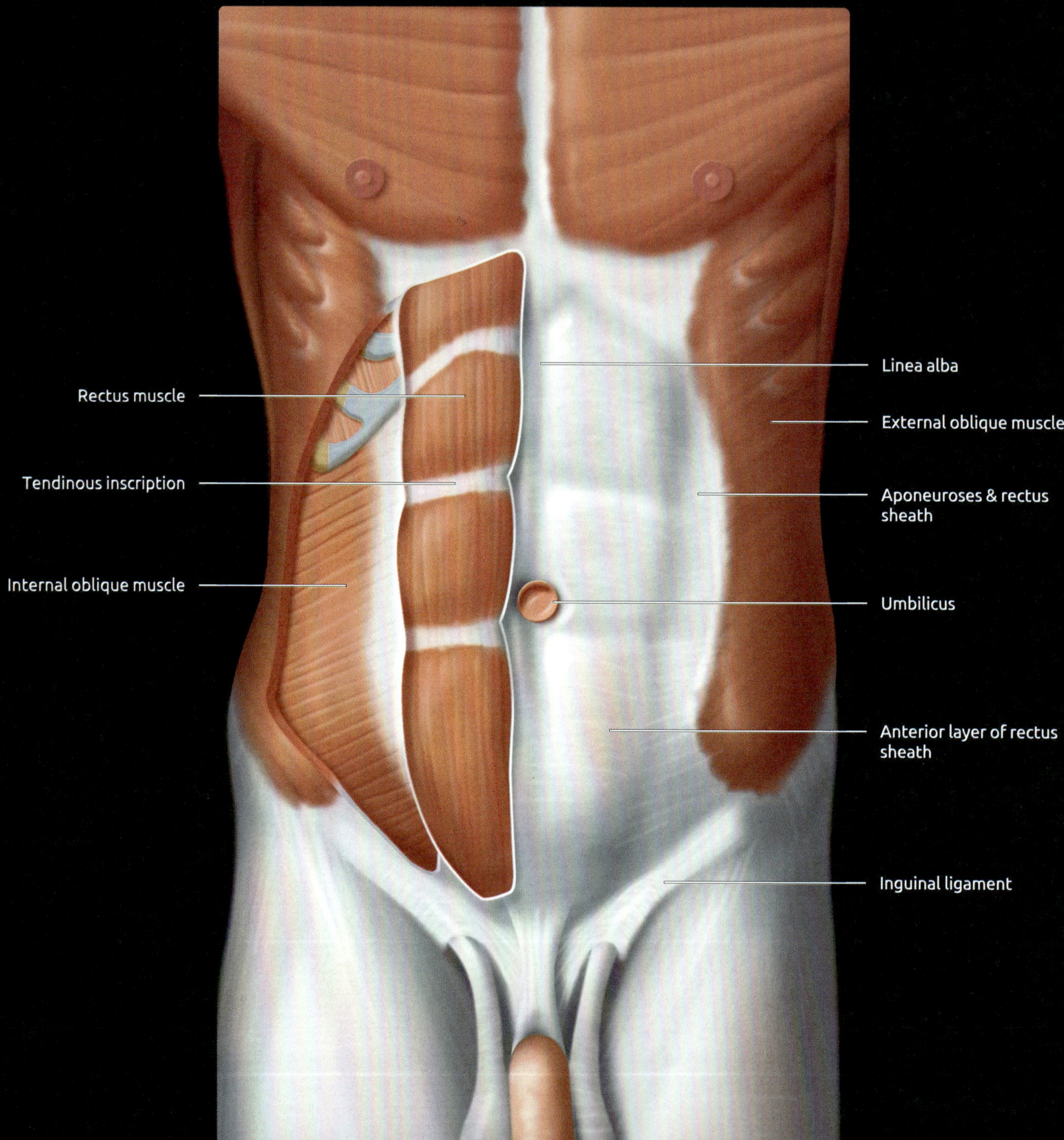

The aponeuroses of the internal and external oblique and transverse abdominal muscles are 2 layered and interweave with each other, covering the rectus muscle and constituting the rectus sheath and linea alba. About midway between the umbilicus and symphysis, the posterior rectus sheath ends, and the transversalis fascia is the only structure between the rectus muscle and parietal peritoneum.

ANTERIOR ABDOMINAL WALL

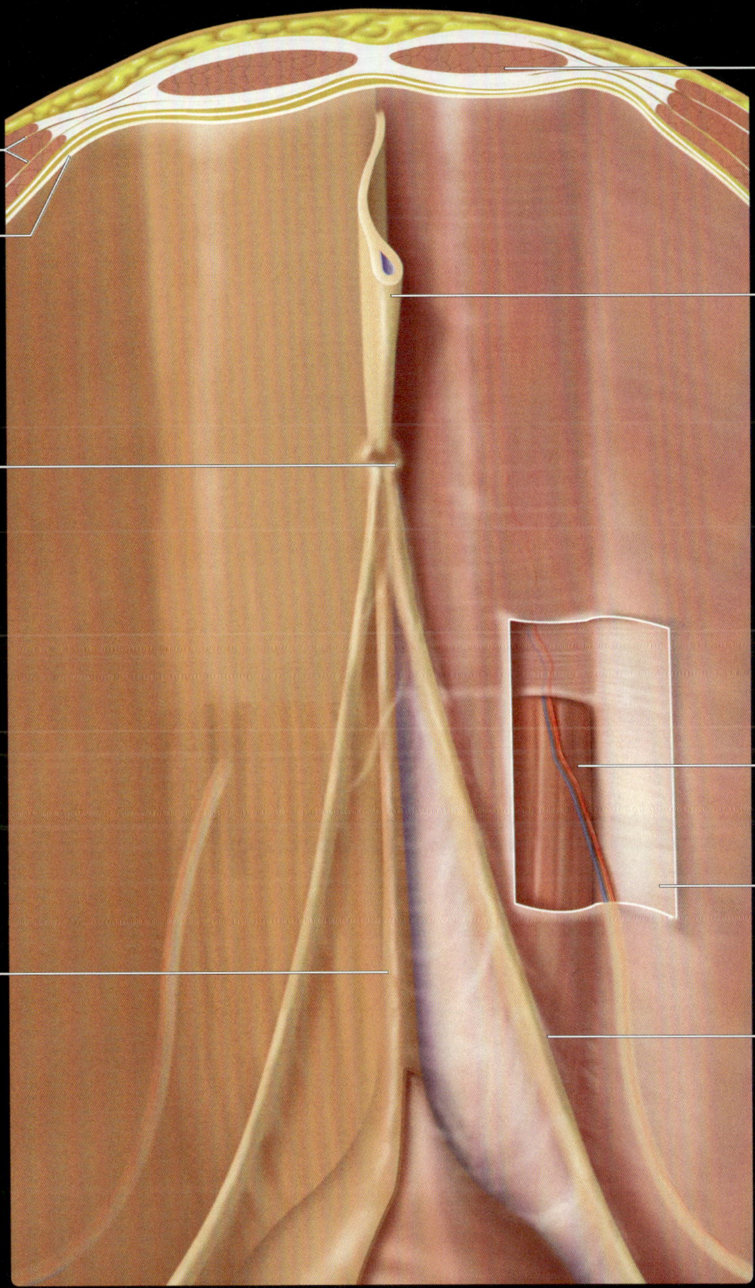

Graphic view of the anterior abdominal wall from the inside is shown. The umbilical ligaments are remnants of the fetal umbilical arteries and allantois (urachus). The falciform ligament is the remnant of the umbilical vein. All of these converge at the umbilicus. Transversalis fascia lines the muscular wall of the abdomen. The epigastric vessels lie between the transversalis fascia and parietal peritoneum.

THORACOLUMBAR FASCIA

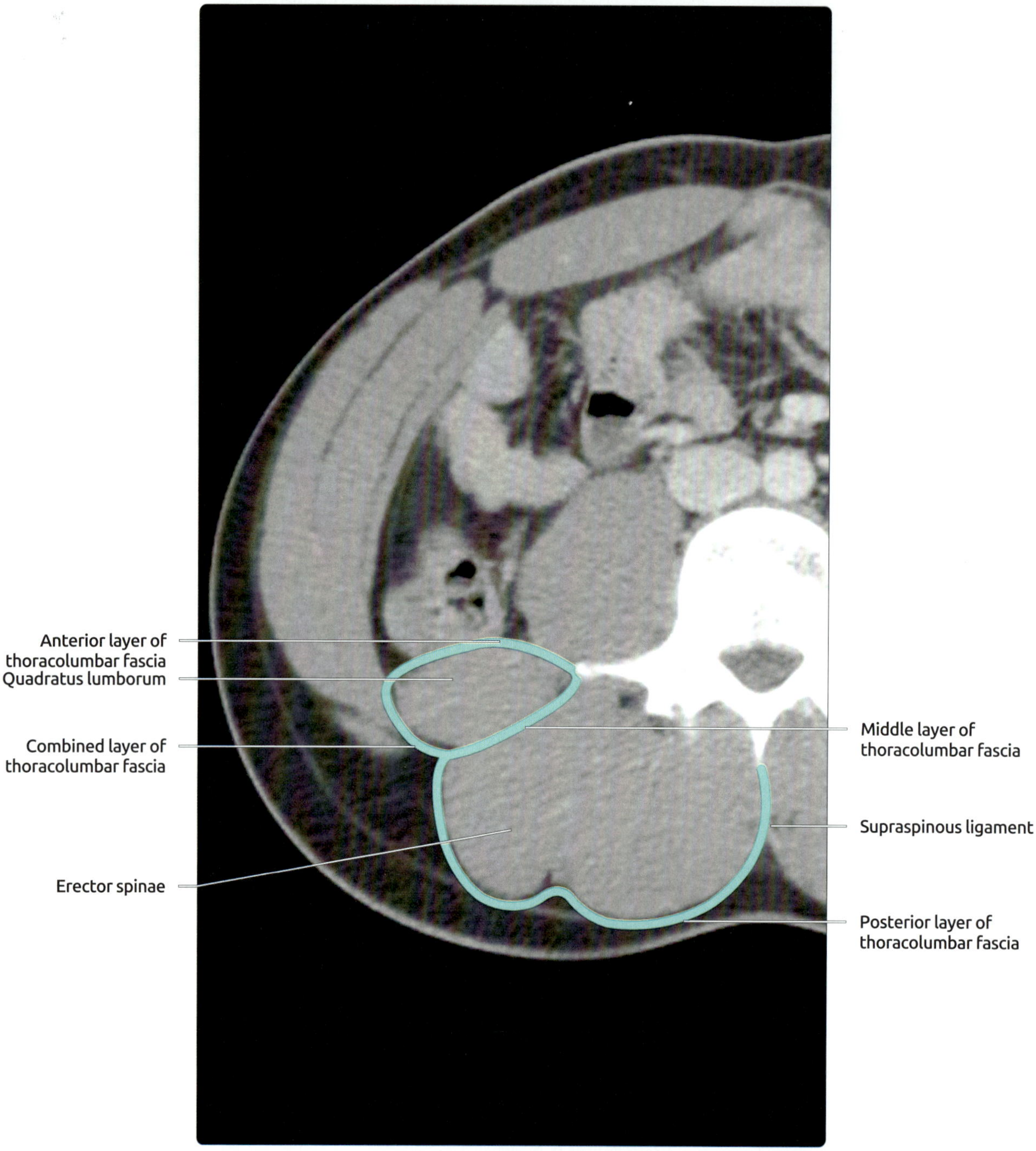

The thoracolumbar (lumbodorsal) fascia is a strong fibrous sheath that encloses the quadratus lumborum and erector spinae muscles. The thoracoabdominal fascia has 3 layers (anterior, middle, and posterior) with the quadratus lumborum lying between the anterior and middle layers, while the erector spinae lies between the middle and posterior layers. The anterior layer is continuous with the transversus abdominis muscle aponeurosis laterally, while the posterior layer is continuous with the aponeuroses of the internal oblique and transversus abdominis muscles.

AXIAL CT, ABDOMINAL WALL MUSCLES

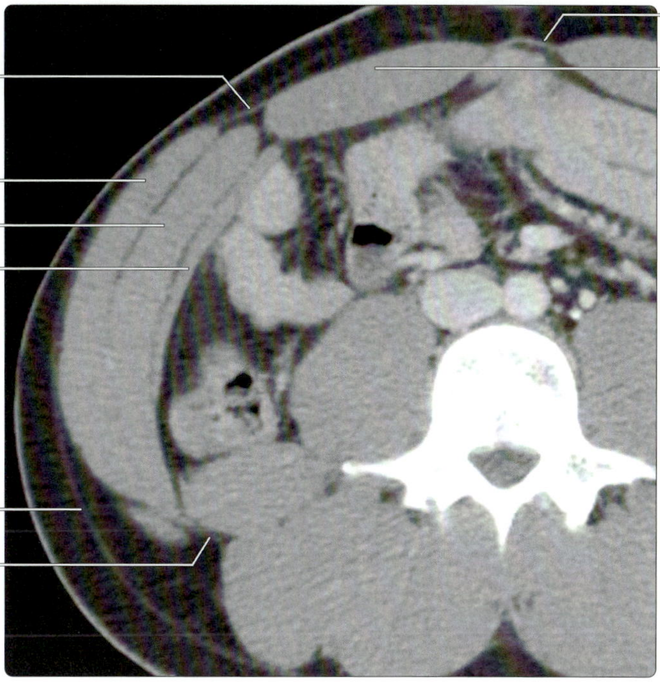

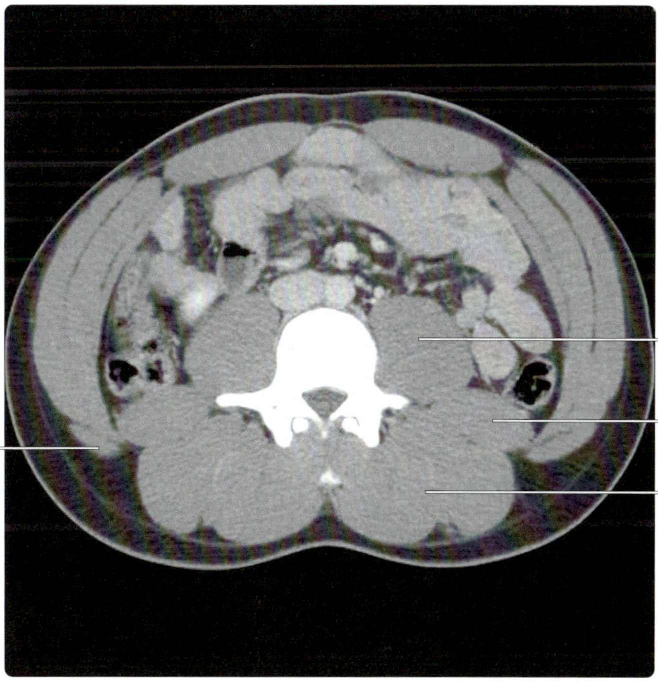

(Top) *First of 2 axial CT sections in a muscular young man is shown. Note that the aponeuroses of the oblique and transverse abdominal muscles surround the rectus muscle as the rectus sheath and then continue to form and insert upon the linea alba.* **(Bottom)** *Aponeuroses of oblique and transverse abdominal muscles blend with and attach to the thoracolumbar fascia, which surrounds the paraspinal and quadratus lumborum muscles.*

AXIAL CT, ABDOMINAL WALL MUSCLES

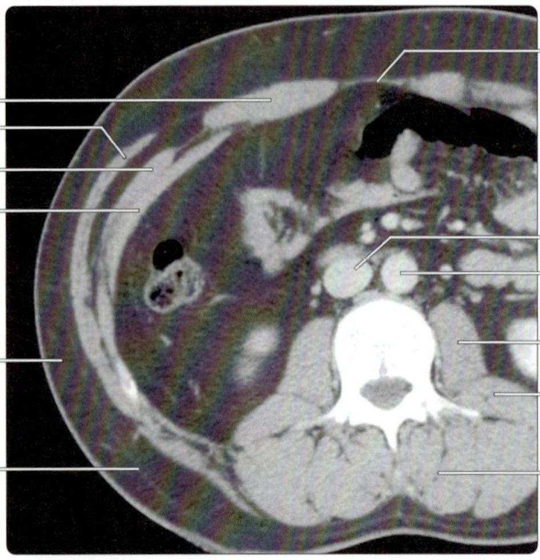

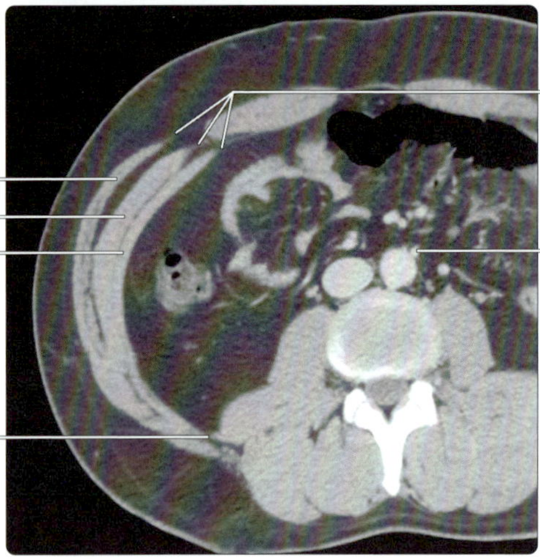

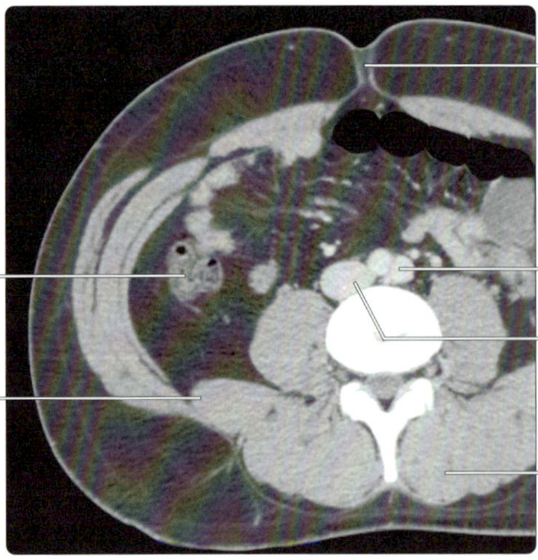

(Top) *First of 8 axial CECT sections in a young man is shown. The rectus sheath is formed by the aponeuroses of the oblique and transverse abdominal muscles, which also form the linea alba, the central anterior attachment for anterior muscles. Subcutaneous tissue has 2 layers, superficial fatty layer (Camper fascia) and deep membranous layer (Scarpa fascia), that continue down to the perineum.* **(Middle)** *Thoracolumbar fascia surrounds the quadratus lumborum and paraspinal muscles, forming an important posterior attachment for the abdominal wall muscles. Aponeuroses are the flat tendons that extend from the muscles.* **(Bottom)** *The origins of the anterior abdominal wall muscles include the iliac crest and thoracolumbar fascia inferiorly. The transversalis fascia lies deep to the abdominal wall muscles and lines the entire wall but is usually not detectable on imaging studies.*

AXIAL CT, ABDOMINAL WALL MUSCLES

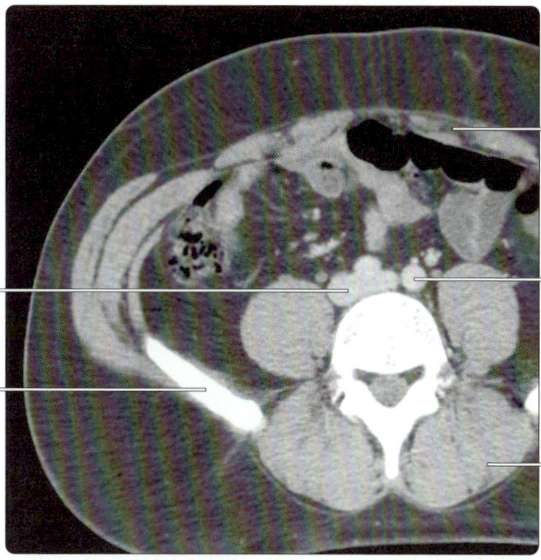

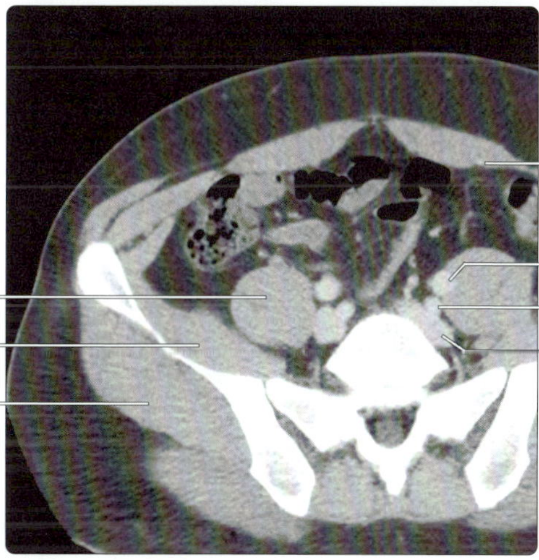

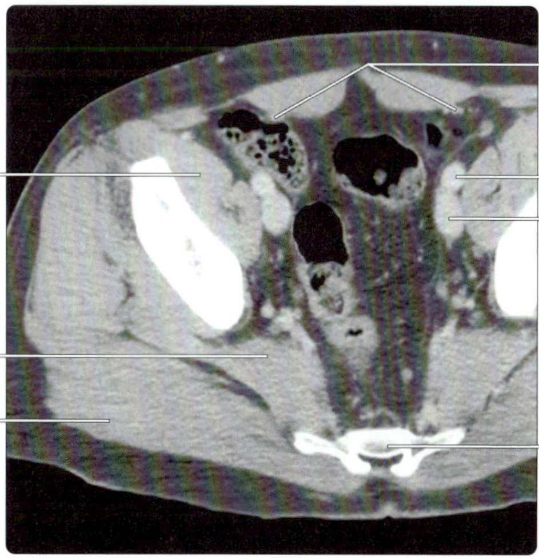

(Top) *Muscle attachments to the iliac crest are evident.* **(Middle)** *The inferior epigastric arteries and veins course along the underside of the rectus muscles and arise from the external iliac vessels low in the pelvis.* **(Bottom)** *The external iliac vessels run along the surface of the iliopsoas muscles.*

AXIAL CT, ABDOMINAL WALL MUSCLES

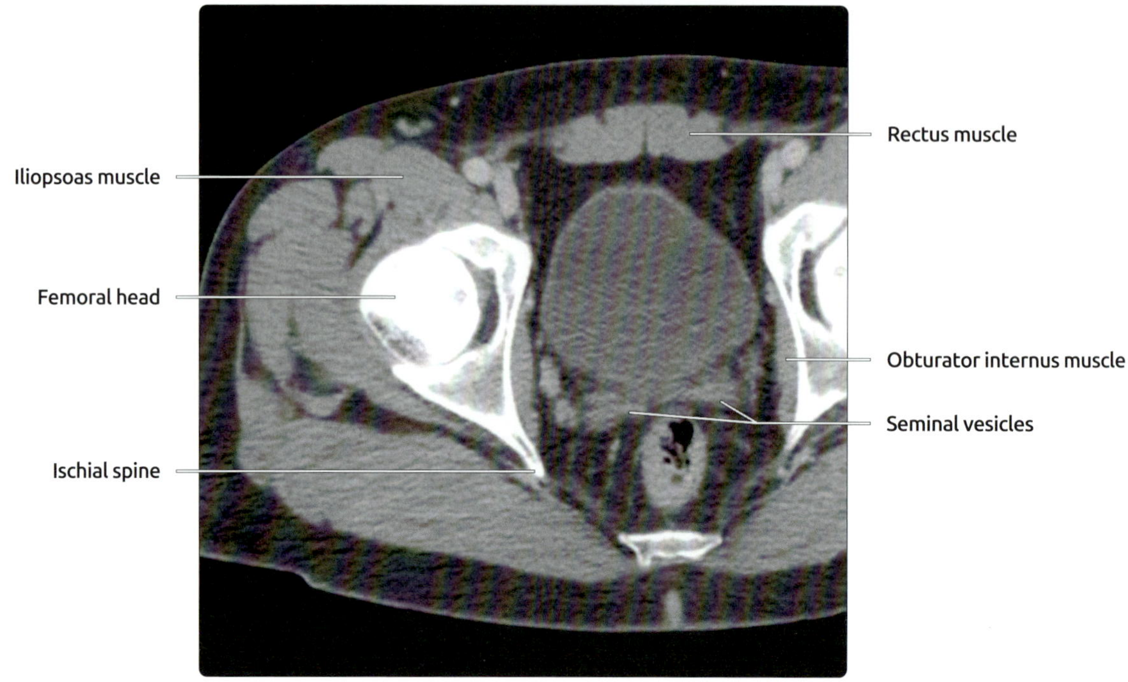

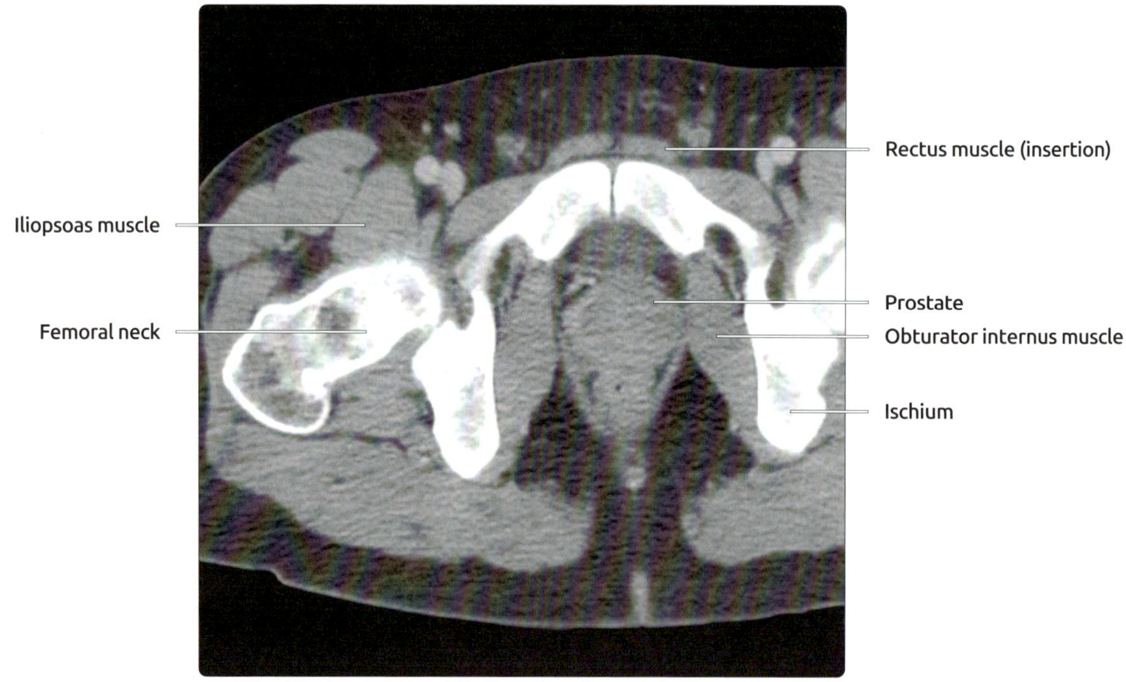

(Top) *The iliopsoas muscle leaves the abdomen, passing under the inguinal ligament.* **(Bottom)** *The iliopsoas muscle passes anterior to the femoral head and neck to insert on the lesser trochanter of the femur. The rectus muscle inserts on the pubis.*

AXIAL MR, ABDOMINAL WALL MUSCLES

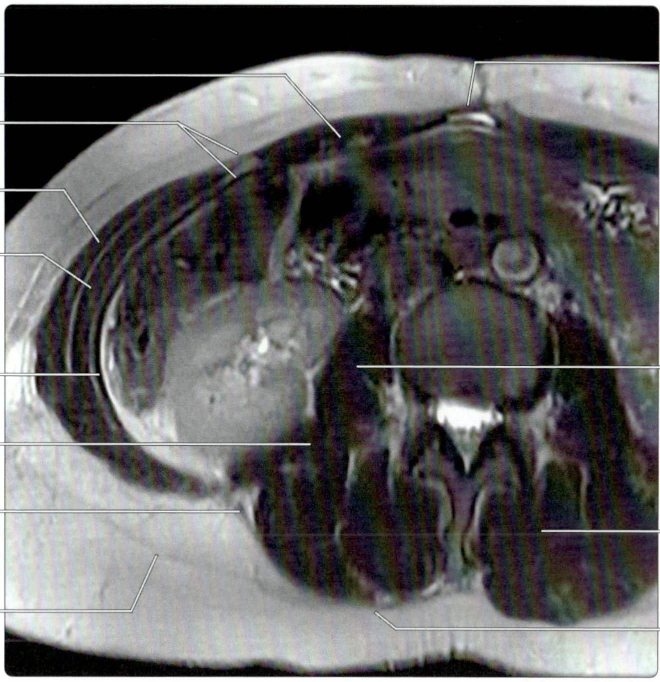

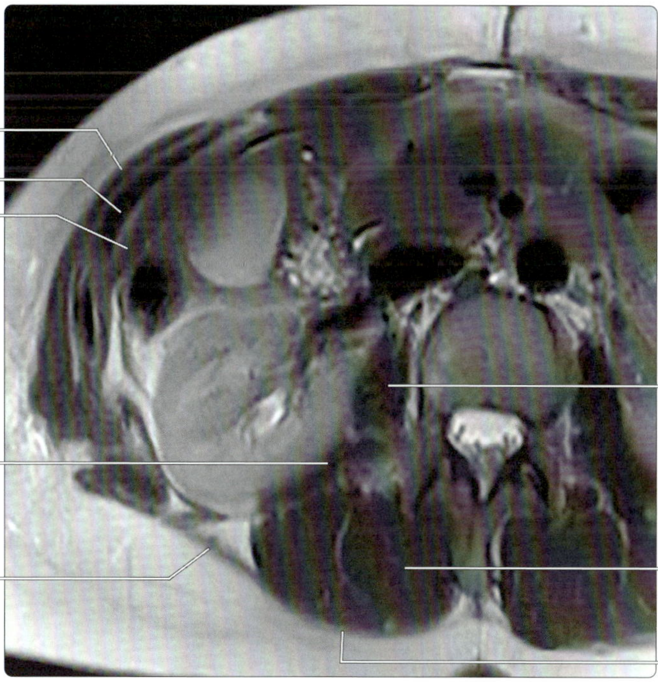

(Top) *Axial T2 HASTE MR (without fat saturation) demonstrates the normal appearance of the internal/external oblique and transverse abdominis muscles. Note the manner in which the aponeuroses of these muscles are nicely visualized on MR and are seen to extend medially to surround the rectus abdominis muscle and subsequently converge at the linea alba.* **(Bottom)** *MR nicely demonstrates the aponeuroses of the oblique and transverse abdominal muscles as they become confluent with the thoracolumbar fascia, which surrounds the paraspinal and quadratus lumborum muscles.*

CORONAL CT, ABDOMINAL WALL, MUSCLES AND VESSELS

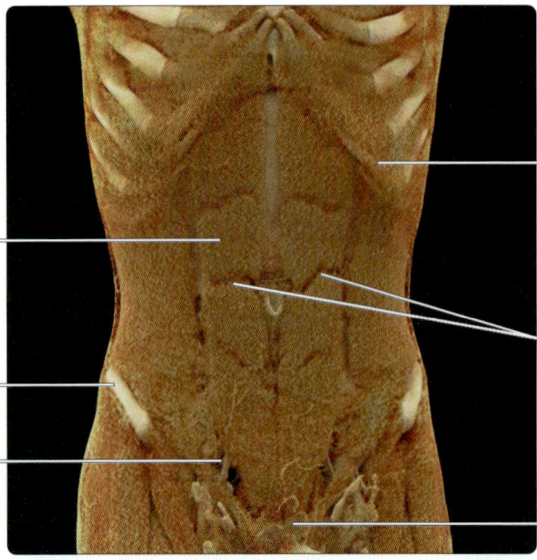

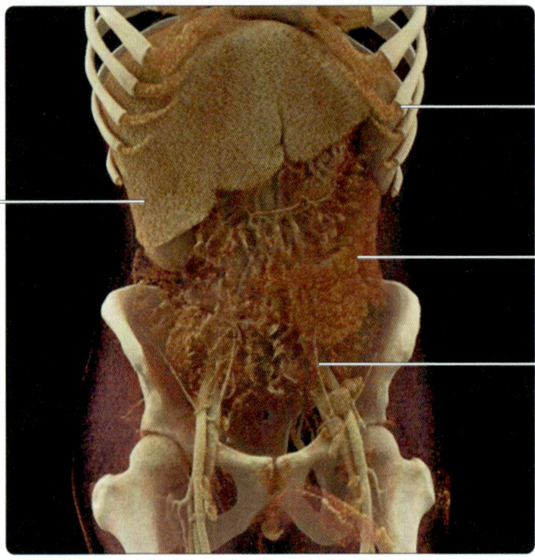

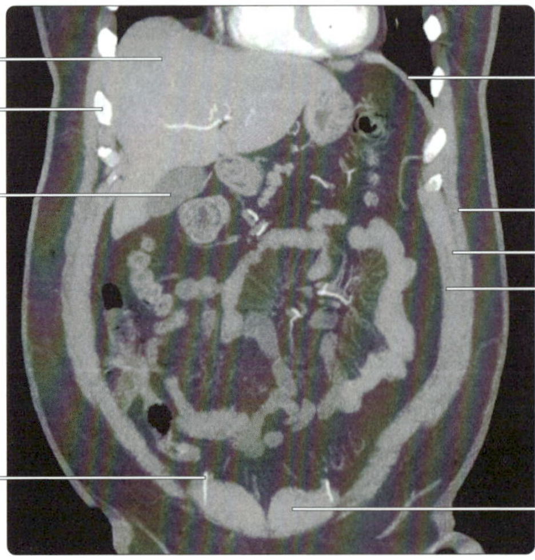

(Top) *Cinematic rendering of CECT images through the abdominal wall is shown. The rectus muscles arise from the pubis and insert into the xiphoid process and costal cartilages 5-7. Cephalad attachments of the oblique and transverse abdominal muscles include the lower ribs and costal cartilages.* **(Middle)** *Cinematic rendering of CECT images after removal of the abdominal wall shows internal structures.* **(Bottom)** *Attachment of the oblique and transverse abdominal muscles to the lower ribs is evident.*

CORONAL CT, ABDOMINAL WALL, MUSCLES AND VESSELS

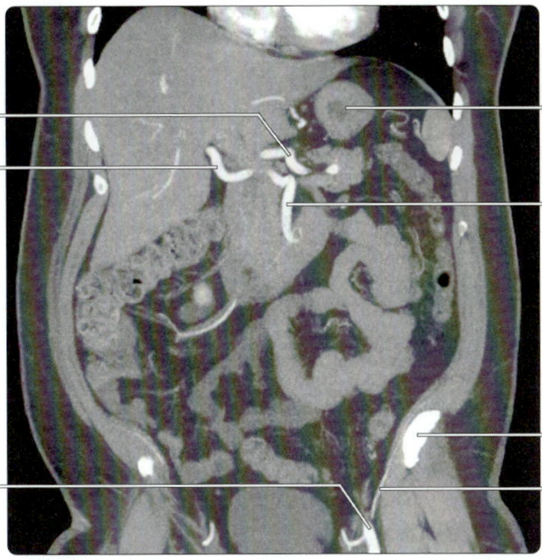

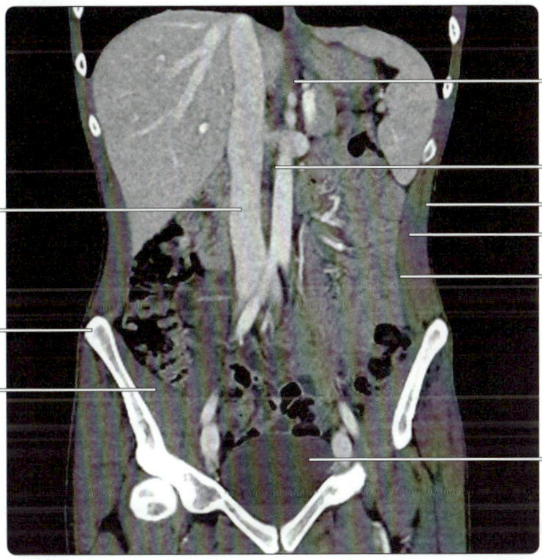

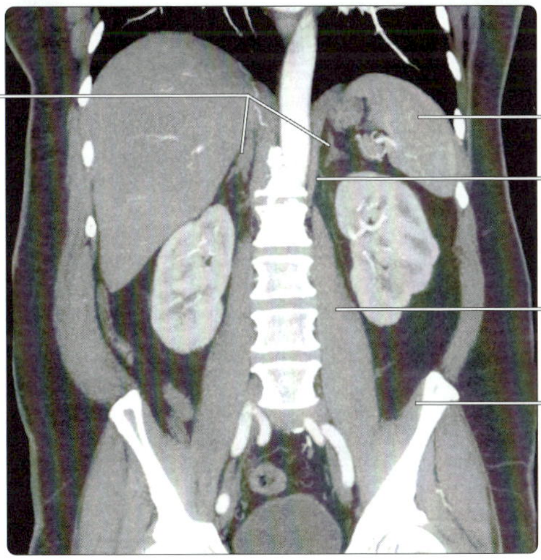

(Top) *Caudal attachment of the oblique and transverse abdominal muscles to the iliac crest is shown.* **(Middle)** *Aorta enters the abdomen behind the median arcuate ligament, the junction of the diaphragmatic crura. Attachment of the oblique and transverse abdominal muscles to the lower ribs is seen.* **(Bottom)** *Psoas muscle arises from the transverse processes and bodies of lumbar vertebrae. After joining the iliacus muscle in the pelvis, it passes beneath the inguinal ligament to enter the thigh and insert on the lesser trochanter of the femur.*

CORONAL CT, ABDOMINAL WALL, MUSCLES AND VESSELS

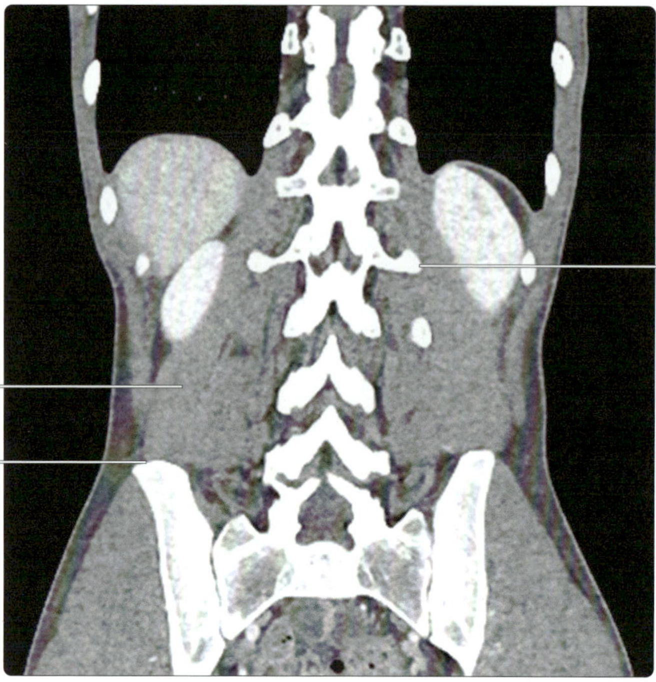

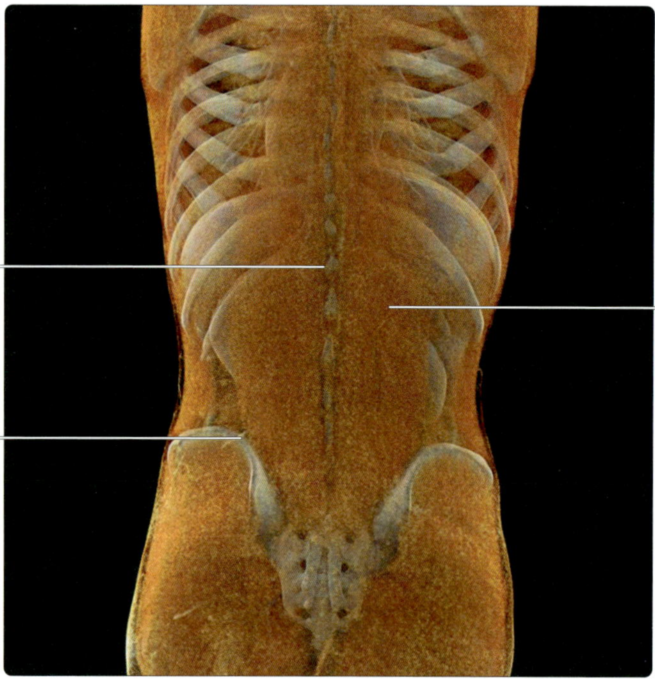

(Top) *The quadratus lumborum muscle arises from the iliac crest, iliolumbar ligament, and transverse processes of lumbar vertebrae. The quadratus lumborum passes behind the diaphragm under the lateral lumbocostal arch (not evident on imaging studies).* **(Bottom)** *Posterior view of the back on CECT cinematic rendered image shows the erector spinae muscles that lie on either side of the spinous processes helping to stabilize the spine.*

CORONAL CT, ABDOMINAL WALL MUSCLES

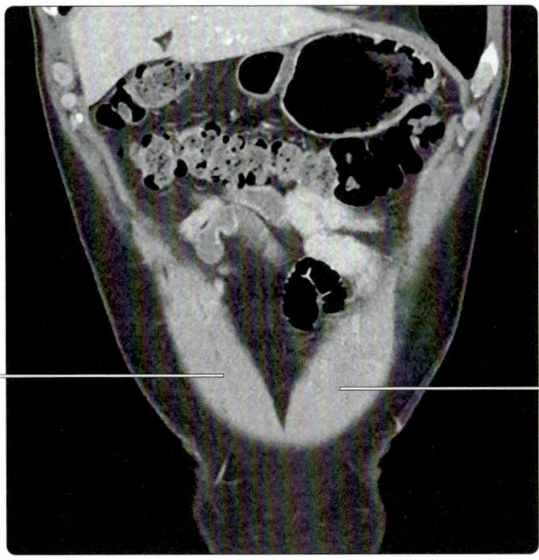

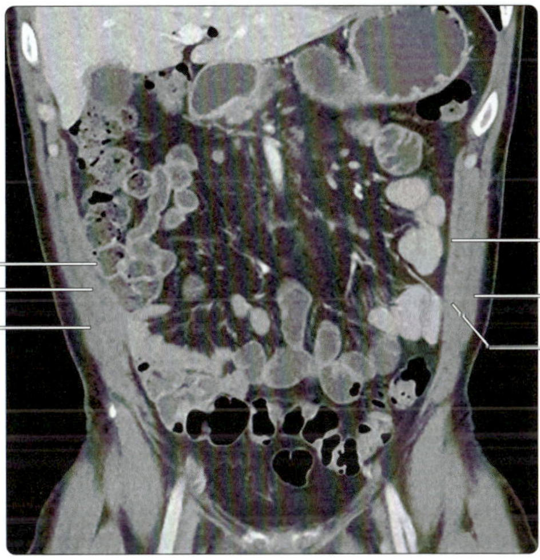

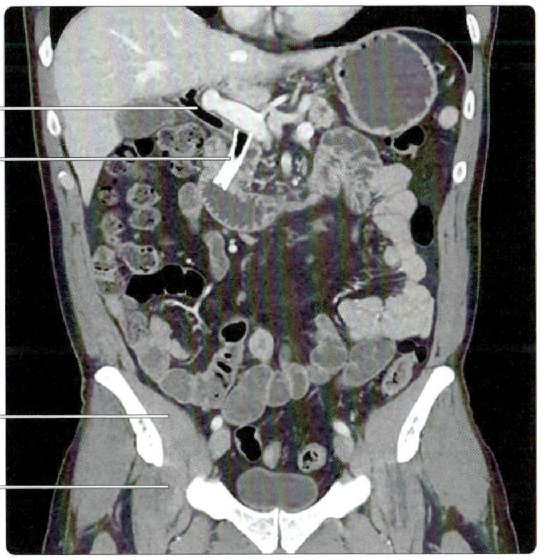

(Top) *First of 8 coronal CECT images demonstrates the normal appearance of the abdominal wall. Note the bilateral rectus muscles converging centrally towards the linea alba. The rectus muscles run in the vertical plane extending more inferiorly to insert on the pubis (not shown).* **(Middle)** *More posterior image demonstrates the 3 "flat" muscles closely layering on top of each other (external oblique, internal oblique, transverse abdominal muscles). Note the attachments of these muscles to the ribs more superiorly.* **(Bottom)** *More posterior image demonstrates the iliacus muscle coming into view. The iliacus muscle fills the iliac fossa and forms the lateral wall of the abdominopelvic cavity. The combined iliacus and psoas (iliopsoas) is seen extending inferiorly into the pelvis to attach to the lesser trochanter of the femur.*

CORONAL CT, ABDOMINAL WALL MUSCLES

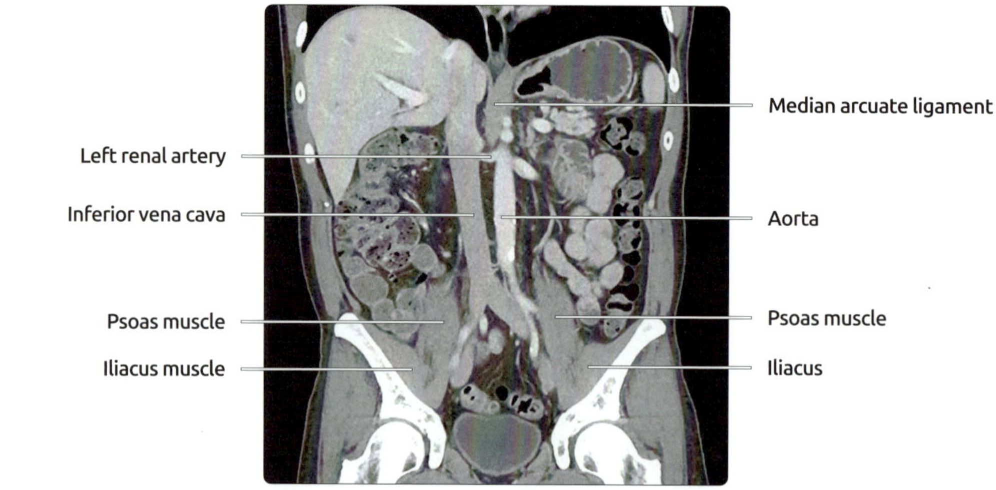

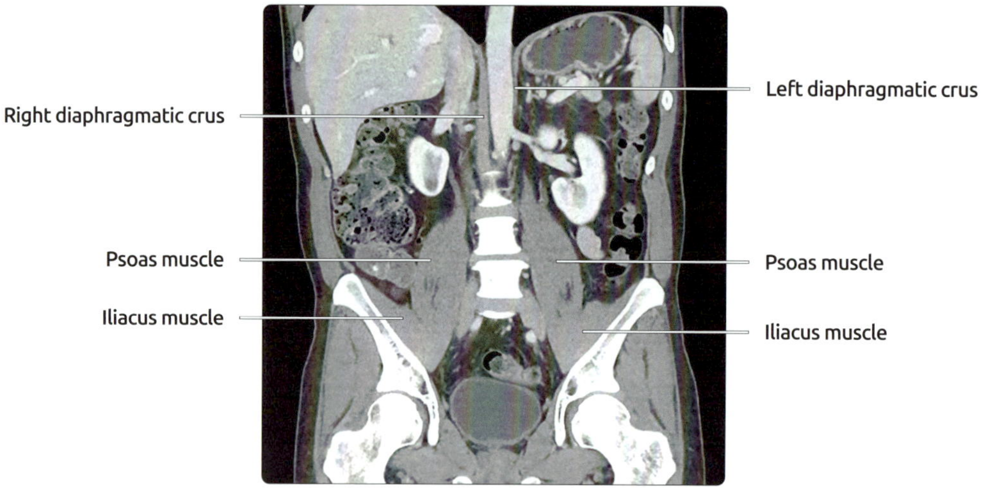

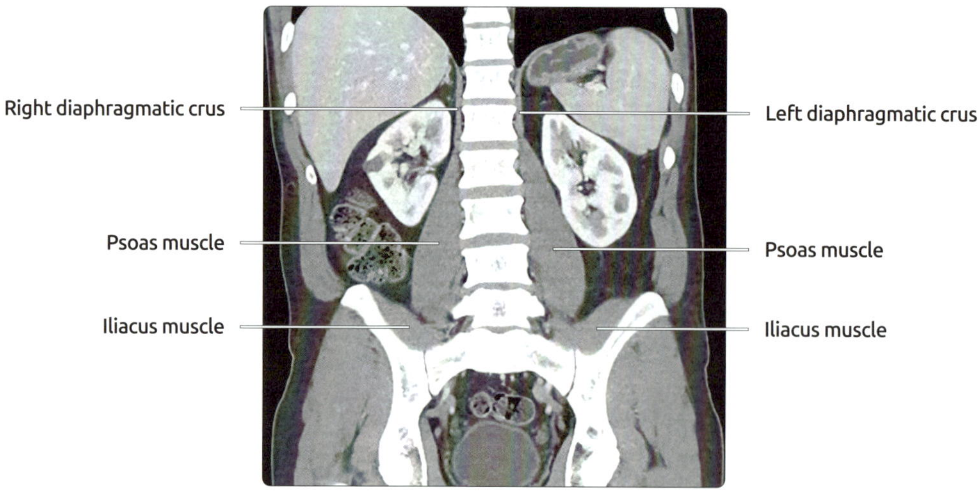

(Top) *Slightly more posterior image demonstrates the psoas muscles coming into view. Note the way in which the psoas and iliacus muscles converge to form the iliopsoas complex, which has a major role in the flexion of the thigh. Note the location of the median arcuate ligament, which represents the convergence of the diaphragmatic crura.* **(Middle)** *Coronal CECT demonstrates a larger portion of the thick psoas muscles bilaterally, which are long, thick muscles lying immediately lateral to the vertebral column and converging distally with the iliacus muscles to form the iliopsoas complex.* **(Bottom)** *The psoas muscles originate from the transverse processes of the T12-L5 vertebral bodies and insert on the lesser trochanter of the femur, playing a major role in the flexion of the thigh and bending the vertebral column laterally.*

CORONAL CT, ABDOMINAL WALL MUSCLES

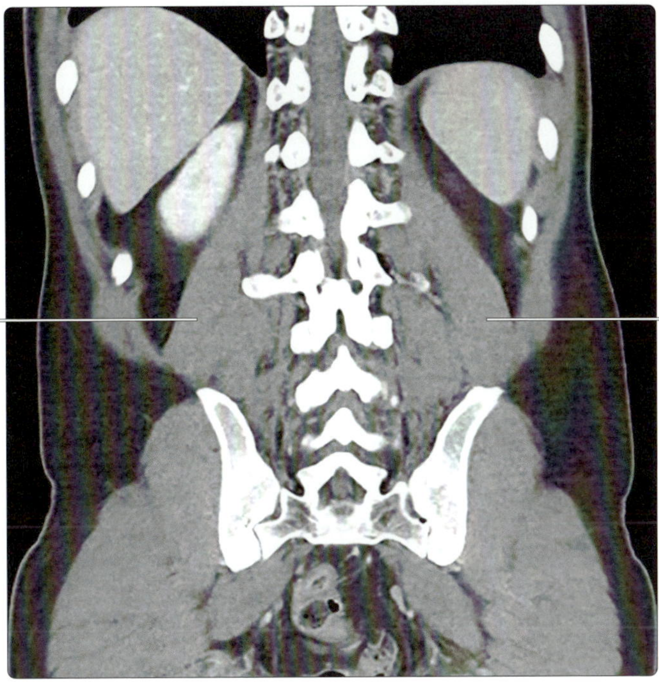

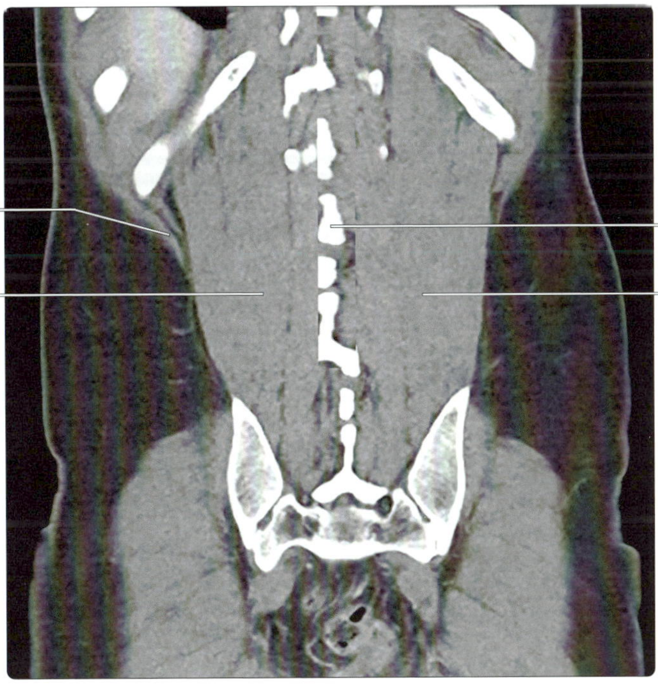

(Top) *Slightly more posterior image brings the quadratus lumborum muscles into view. These muscles are adjacent to the lumbar transverse processes and superficial to the psoas muscles.* **(Bottom)** *Most posterior CECT shows the erector spinae muscles, which play an important role in the stabilization of the spine. This image also shows the flat muscle aponeuroses converging to join the thoracolumbar fascia.*

RECTUS HEMORRHAGE INTO PERIVESICAL SPACE

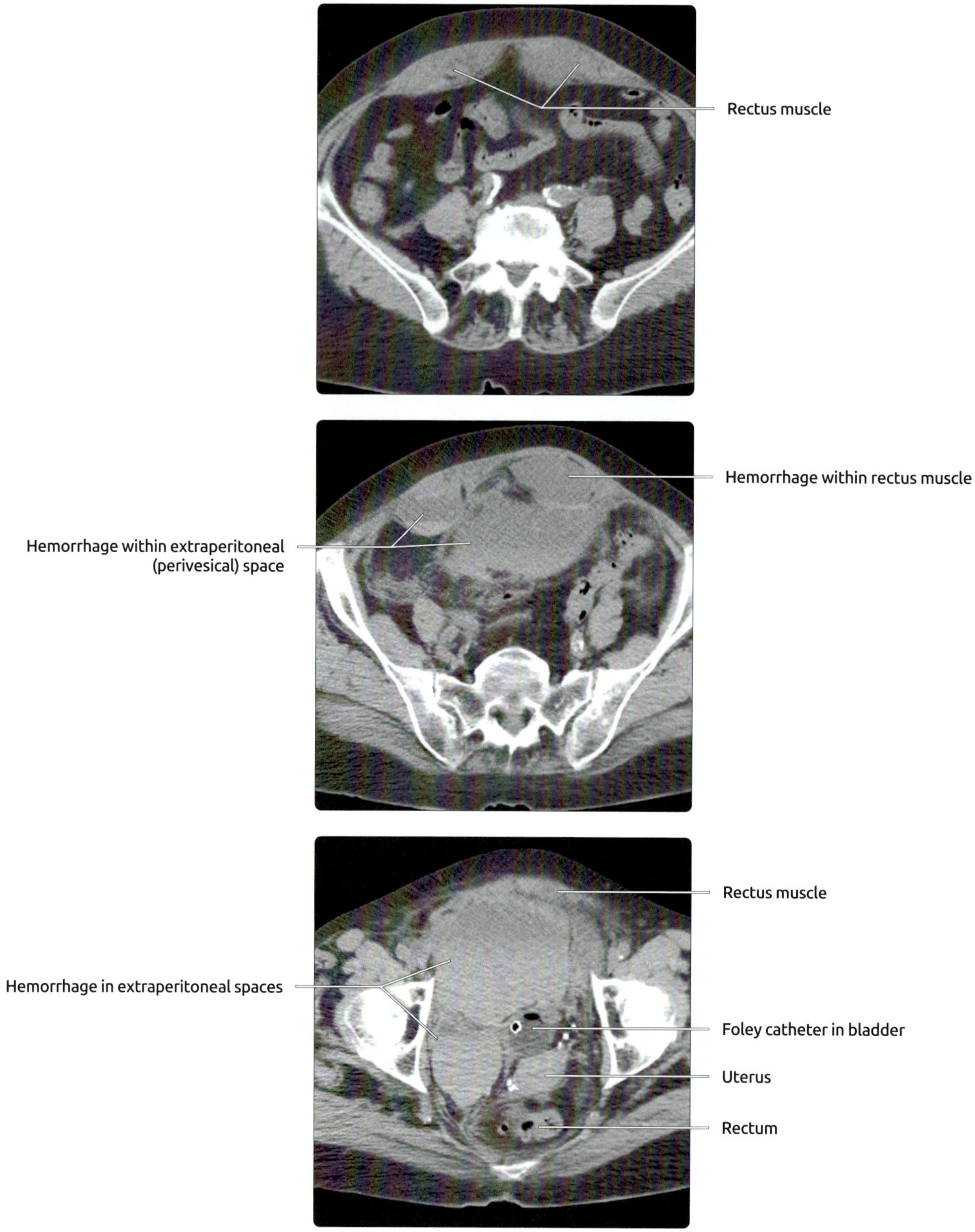

(Top) *First of 3 axial NECT images depicts a spontaneous abdominal wall and extraperitoneal hemorrhage in an older woman receiving anticoagulant medication. The rectus muscles are enlarged due to hemorrhage within the rectus sheath.* **(Middle)** *Hemorrhage has "leaked" through the incomplete posterior layer of the rectus sheath to extend into the extraperitoneal pelvis. Note the cellular-fluid levels within the hemorrhage, called the hematocrit sign, indicative of a coagulopathic hemorrhage.* **(Bottom)** *The abdominal extraperitoneal spaces communicate caudally with the pelvic extraperitoneal spaces, including the perivesical space.*

PANNUS SIMULATING VENTRAL HERNIA

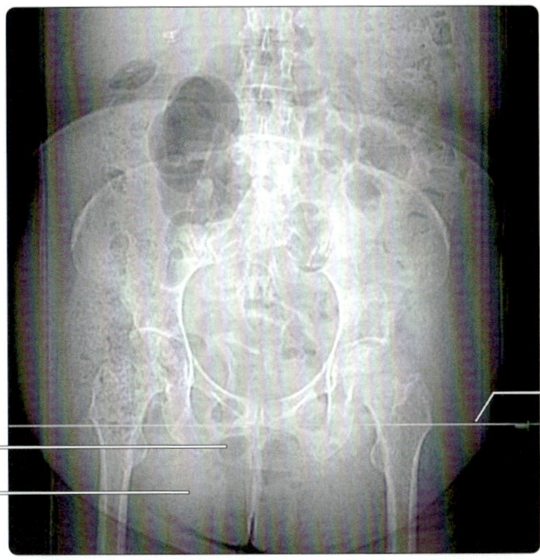

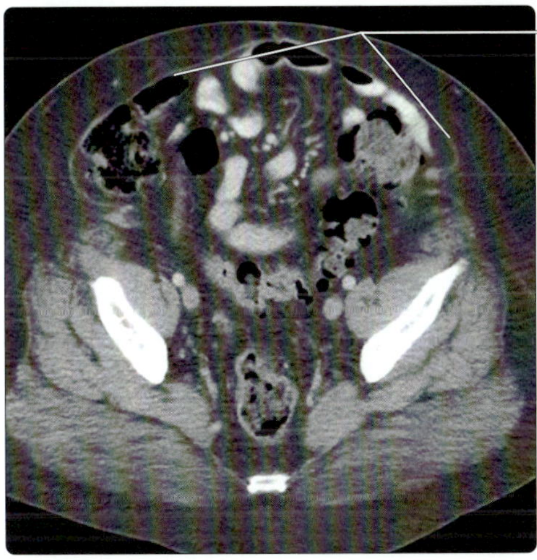

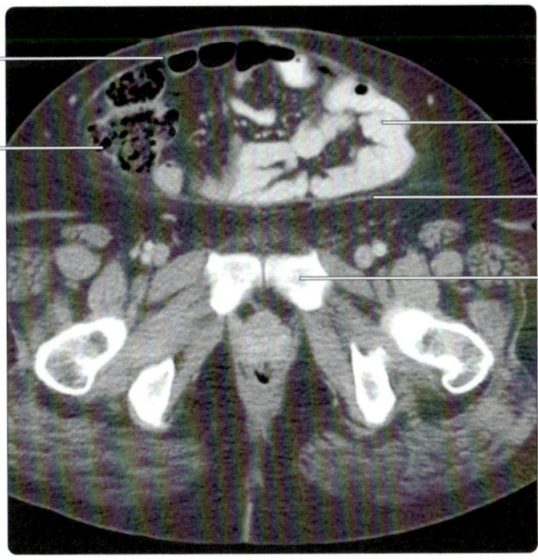

(Top) *An older woman with a protuberant abdomen simulating a ventral hernia is shown. Frontal computed radiograph shows an obese abdomen and bowel gas below the level of the symphysis pubis.* **(Middle)** *First of 2 axial CT sections in the same patient shows that the thin abdominal wall is stretched and elongated, allowing anterior and caudal protrusion of abdominal contents (pannus).* **(Bottom)** *This image shows bowel in the pannus below the symphysis pubis. While these findings could be superficially confused with a hernia, there is no real abdominal wall defect in this case.*

VENTRAL HERNIA, INCARCERATED CAUSING BOWEL OBSTRUCTION

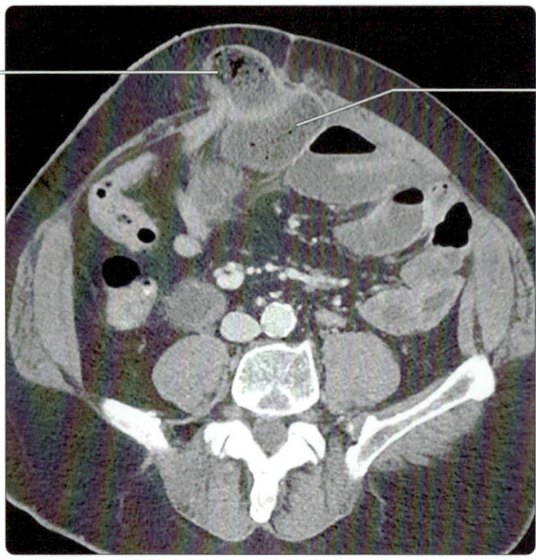

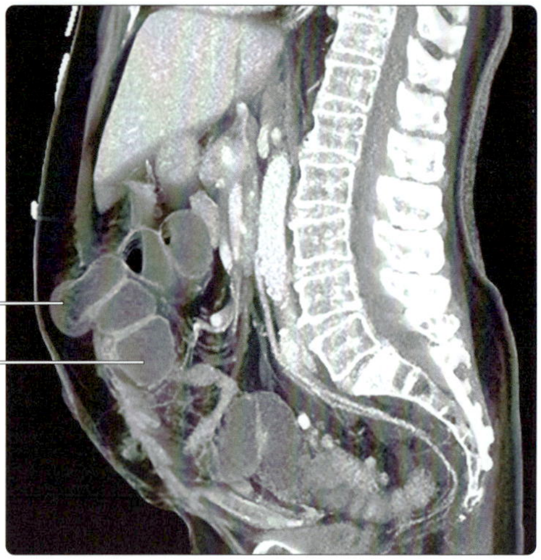

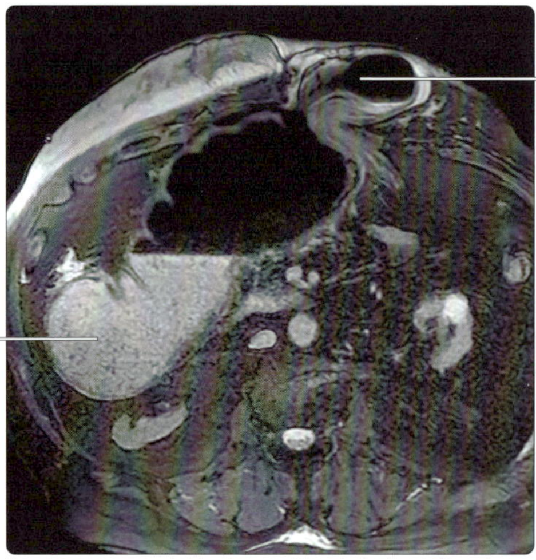

(Top) *Axial CECT demonstrates a small ventral hernia containing a short segment of the small bowel. Note the fecalized appearance of the small bowel within the hernia (containing speckled fecal material) as well as the dilated loops of bowel more proximally within the abdominal cavity, consistent with a small bowel obstruction.* **(Middle)** *Sagittal volume-rendered CECT in the same patient nicely demonstrates the strangulated ventral hernia with the moderately dilated, more proximal small bowel, findings that ultimately required surgical correction.* **(Bottom)** *Axial T2 MR demonstrates a markedly dilated and obstructed tranverse colon with abrupt narrowing as the colon enters a narrow midline ventral hernia.*

INCISIONAL HERNIA CONTAINING COLON

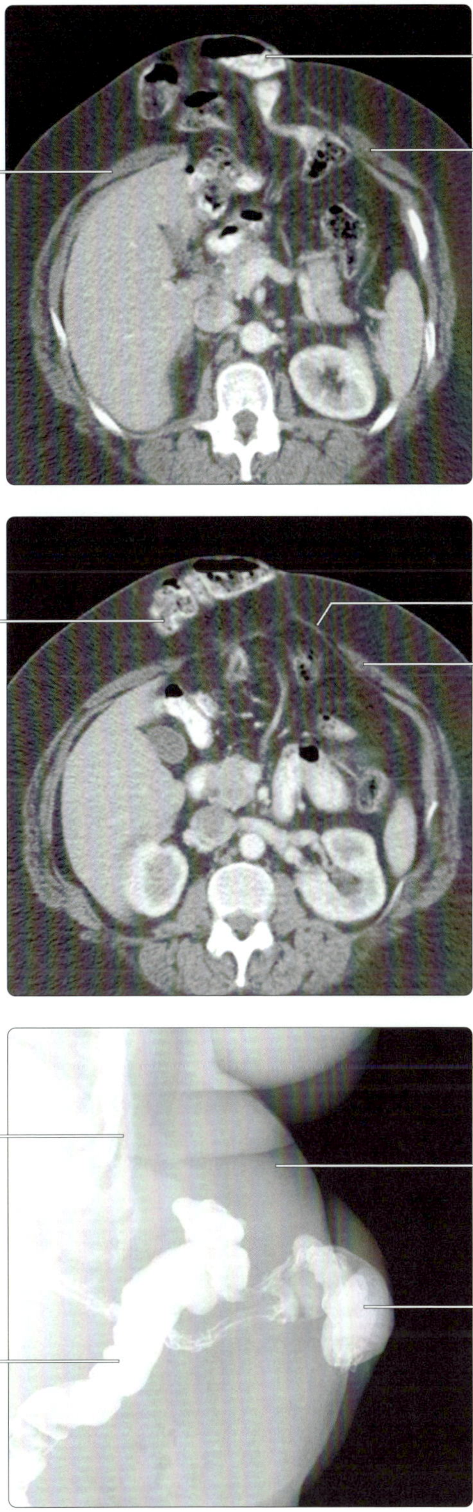

(Top) *First of 2 axial CT images in an older woman shows a ventral incisional hernia following laparotomy. The defect in the anterior abdominal wall (through the linea alba) allows herniation of the colon and omental fat.* **(Middle)** *The rectus muscles, linea alba, and aponeuroses are thin and stretched.* **(Bottom)** *Lateral image from a barium enema in the same patient demonstrates that the colon is herniated into the subcutaneous fat with no overlying muscle.*

INGUINAL HERNIA CAUSING SMALL BOWEL OBSTRUCTION

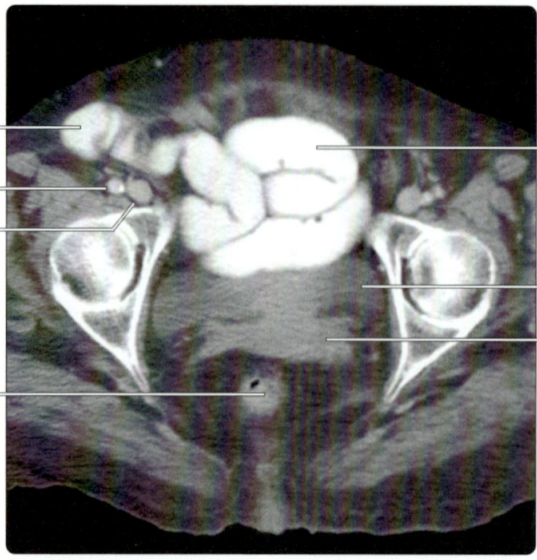

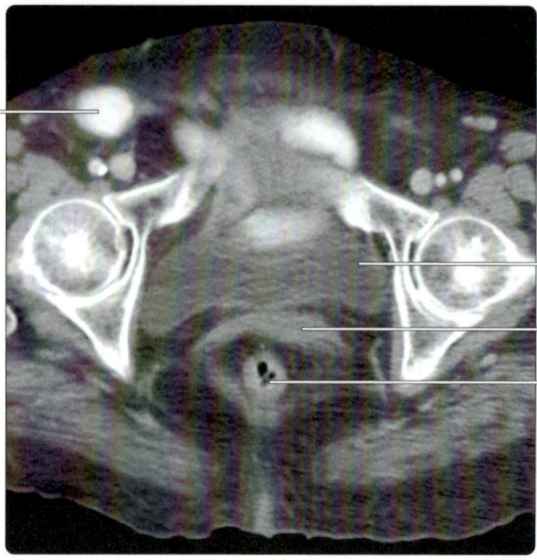

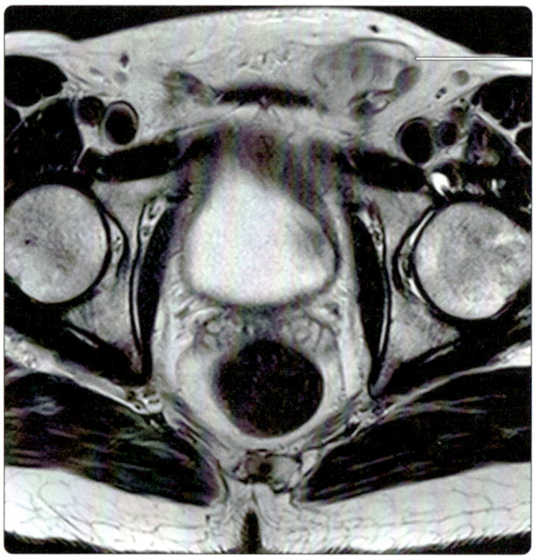

(Top) *First of 2 axial CECT images in an older woman with an inguinal hernia causing small bowel obstruction is shown. Note the dilated small intestine and collapsed rectum.* **(Middle)** *A segment of small intestine has herniated into the inguinal canal and is causing the bowel obstruction.* **(Bottom)** *Inguinal liposarcoma mimicking an inguinal hernia is shown. Note the soft tissue nodularity with the fatty mass. There was no herniation of the pelvic fat into the inguinal canal.*

SPIGELIAN HERNIA CAUSING COLONIC OBSTRUCTION

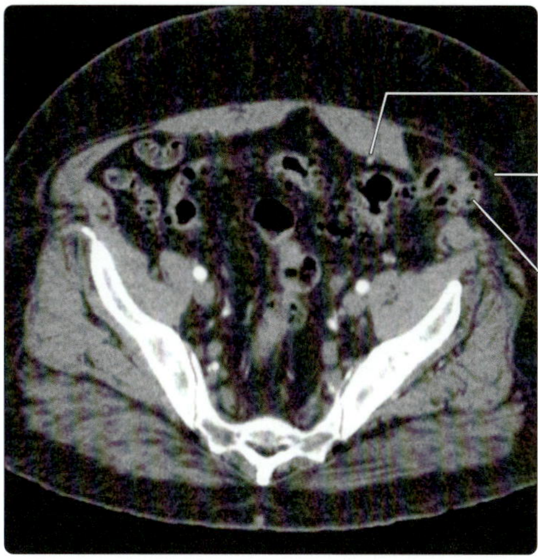

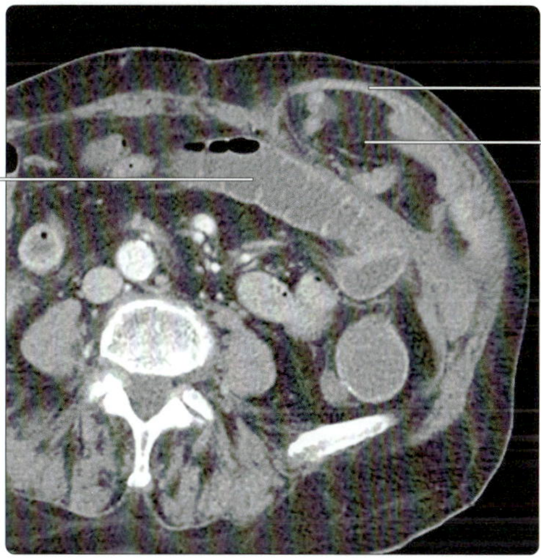

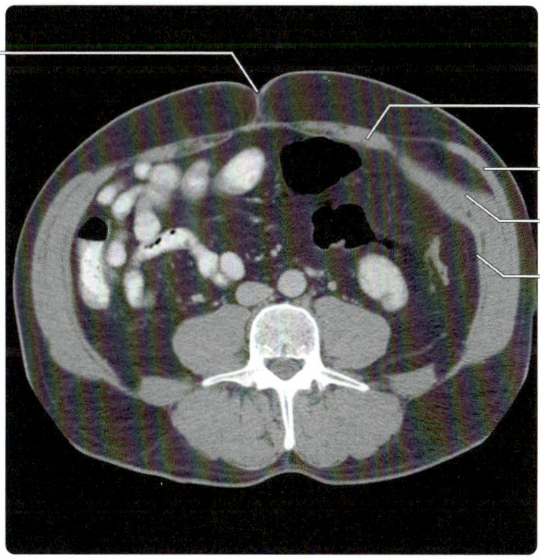

(Top) *Axial CECT demonstrates a left spigelian hernia containing a loop of large bowel covered only by the external oblique muscle aponeurosis. Note the left inferior epigastric artery as it inserts in the left rectus muscle. Almost all spigelian hernias occur at or below the arcuate line. The absence of posterior rectus fascia may contribute to an inherent weakness in this area where the inferior epigastric vessels penetrate the muscle.* **(Middle)** *Axial CECT demonstrates a large spigelian hernia in the left lower quadrant containing a loop of dilated, obstructed small bowel. The hernia is covered only by the external oblique muscle and its aponeurosis.* **(Bottom)** *First of 4 axial CT sections in a patient with a spigelian hernia nicely demonstrates the relationship of the hernia relative to the abdominal wall muscles. Note that the hernia (only partially seen on this image) is arising between the left internal and external oblique muscles.*

SPIGELIAN HERNIA CAUSING COLONIC OBSTRUCTION

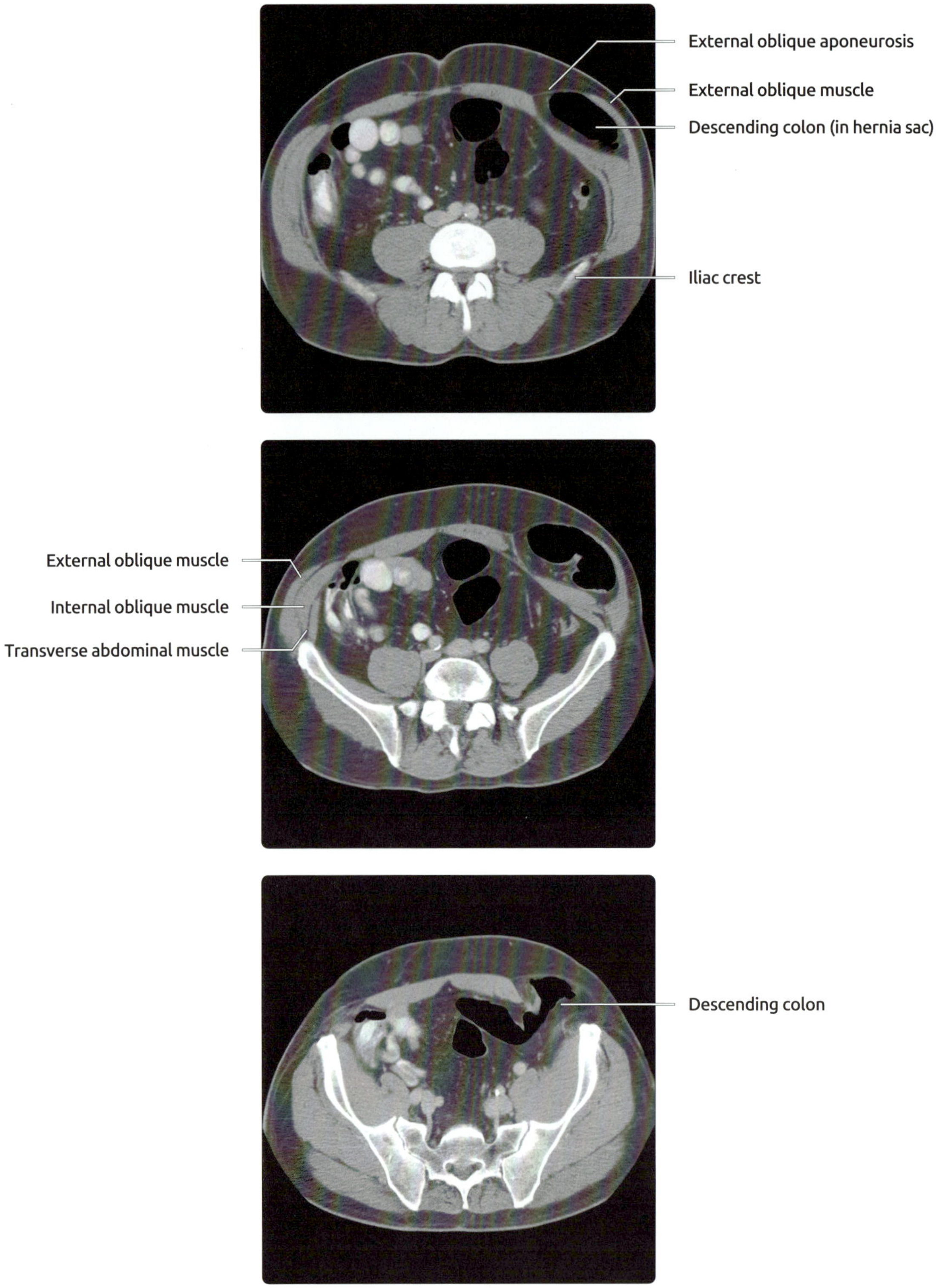

(Top) *More caudal CECT demonstrates that the descending colon has herniated into the subcutaneous fat and is covered only by the external oblique muscle and its aponeurosis.* **(Middle)** *Compare the location of the left-sided spigelian hernia with the intact muscles of the right anterior abdominal wall.* **(Bottom)** *This final, most caudal CECT demonstrates the colon herniating through a defect in the aponeurosis of the internal oblique and transverse abdominal muscles.*

LUMBAR HERNIA

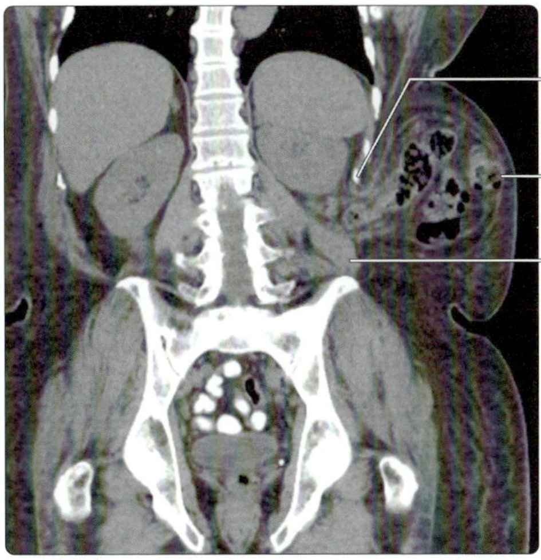

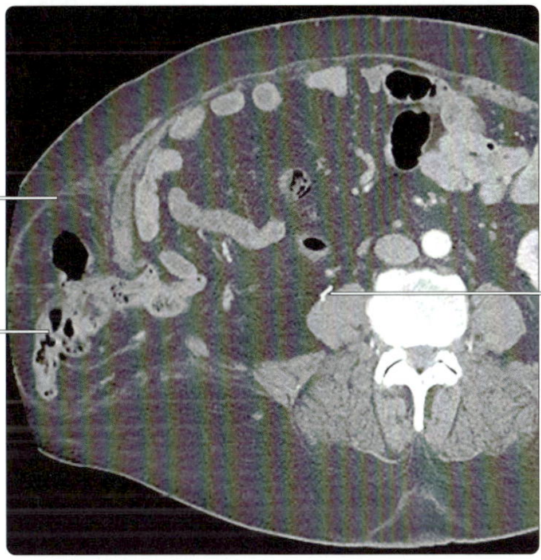

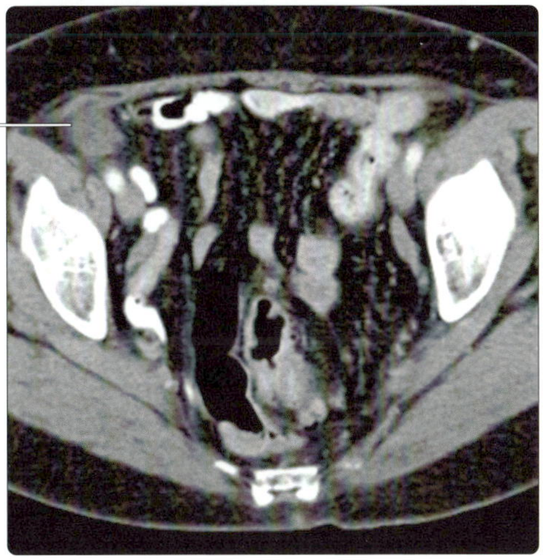

(Top) *Superior lumbar hernia with herniation of bowel loops through the superior lumbar, or Grynfeltt-Lesshaft triangle, is shown. The triangle is formed medially by the quadratus lumborum, laterally by the posterior border of internal abdominal oblique muscle, and superiorly by the 12th rib.* **(Middle)** *Axial CECT demonstrates a large left-sided lumbar hernia containing a portion of the right colon. Note the presence of surgical clips in the right renal fossa. This lumbar hernia was attributable to the patient's prior incision in this location during a nephrectomy.* **(Bottom)** *Patient with a history of right inguinal hernia repair shows a Prolene plug at the site of repair. Hernia plugs can be easily mistaken for lymphadenopathy, and a careful look at patient's history of repair is always helpful.*

GROSS ANATOMY

Overview

- Diaphragm: Sheet of skeletal muscle that forms 2 "domes" along inferior aspect of rib cage
  - Primary functions of diaphragm
    - Diaphragm descends to increase thoracic volume and decrease intrathoracic pressure
    - Chief muscle of respiration
    - Helps circulation of blood by creating reciprocal increases/decreases in thoracic/abdominal pressure
    - Helps raise intraabdominal pressure for defecation, micturition, and childbirth
    - Separates thoracic and abdominal cavities
- Origins/attachments of diaphragm
  - Anterior: **Xiphoid process**
  - Anterolateral: Ribs and costal cartilage margins 7-12
    - Interdigitate with transverse abdominal muscle
  - Posterolateral: Lumbar vertebrae via crura and arcuate ligaments
    - **Diaphragmatic crura** are tendinous structures that blend with anterior longitudinal ligament of vertebral column
    - Right crus arises from L1-L3 vertebral bodies and intervertebral discs and is longer than left
      - Fibers of right crus surround esophageal hiatus and have "sphincter" function, which helps prevent reflux
    - Left crus arises from L1-L2 vertebral bodies and intervertebral discs
  - 3 **arcuate ligaments** give rise to diaphragm posteriorly
    - **Median arcuate ligament** unites both crura
      - Median arcuate ligament passes over anterior surface of aorta (and can compress celiac artery in some individuals)
    - **Medial arcuate ligaments** attach to thoracolumbar fascia over psoas muscles
    - **Lateral arcuate ligaments** attach to thoracolumbar fascia over quadratus lumborum muscles
    - **Vertebrocostal triangle** is thin membrane that may result from failure of arcuate ligament to reach 12th rib
- Muscular portions converge medially as aponeurosis (central tendon)
  - **Central tendon** is fused to inferior aspect of fibrous pericardium but has no osseous attachment
  - 2 "domes" of diaphragm (right and left) are found on either side of fibrous pericardium with right dome slightly higher than left
- Apertures (openings) allow structures to traverse diaphragm
  - **Vena caval foramen** located to right of central tendon
    - Usually located at T8 vertebral level
    - May also transmit right phrenic nerve
  - **Esophageal hiatus** located just left of midline and passes through right crus of diaphragm
    - Usually located at T10 vertebral level
    - Also transmits **vagus nerves** and esophageal branches of **left gastric artery and vein**
  - **Aortic hiatus**
    - Actually passes posterior to diaphragm and median arcuate ligament (**retrocrural space**)
    - Also transmits **thoracic duct**, **azygous/hemiazygos veins**, lymph nodes, and sympathetic trunk
- Arterial supply
  - Primarily via **inferior phrenic arteries**, which usually arise directly from aorta just above celiac artery (or, more rarely, from celiac or renal artery)
  - Smaller arterial branches may also contribute (superior phrenic artery, pericardiophrenic artery, etc.)
- Nervous supply
  - Each 1/2 of diaphragm supplied by phrenic nerve (right or left) formed by cervical plexus (C3-C5 nerve roots)

ANATOMY IMAGING ISSUES

Imaging Recommendations

- Diaphragm is difficult to demonstrate on axial CT or MR and usually better visualized in coronal or sagittal plane
- **Diaphragmatic insertions** on thoracoabdominal wall may take form of thin "slips" of muscle and tendon, rather than smooth sheet of tissue
  - Can be mistaken for nodular peritoneal tumor implants when viewed on axial CT/MR
- Ultrasound shows diaphragm as thin echogenic line

CLINICAL IMPLICATIONS

Clinical Importance

- **Elevation** of **hemidiaphragm** is common
  - May be caused by upward displacement of diaphragm by abdominal mass
  - May be due to diaphragmatic paralysis
    - Phrenic nerve is sole motor supply to diaphragm and can be injured by trauma, thoracic surgery, or thoracic malignancy
    - Paralysis results in permanent elevation and paradoxical movement of hemidiaphragm on forced inspiration
  - May be due to congenital thinning (**eventration**)
- Diaphragm is not always impermeable
  - Ascites, pleural effusion, tumor, infection, or extraluminal air may pass from abdomen into thorax (and vice versa)
  - **Defects in diaphragm** increase with age and emphysema
- Congenital defects in diaphragm are common
  - Posterolateral defect at **vertebrocostal triangle** is known as **Bochdalek hernia** (abdominal fat, bowel, or viscera may herniate into chest)
  - Eventration of diaphragm is thinning or absence of part of diaphragm (left > central tendon > right)
  - Anterior paramedian defect at **sternocostal hiatus** is known as **Morgagni hernia** (right > left)
- Acquired hernias are common
  - **Hiatal hernia**: Widening of esophageal hiatus, allowing part of stomach to herniate into chest
- **Traumatic rupture**: Relatively uncommon, usually resulting from severe blunt or penetrating trauma
  - > 90% are left sided, through apex or posterior portion

OMENTUM AND PERITONEAL REFLECTIONS

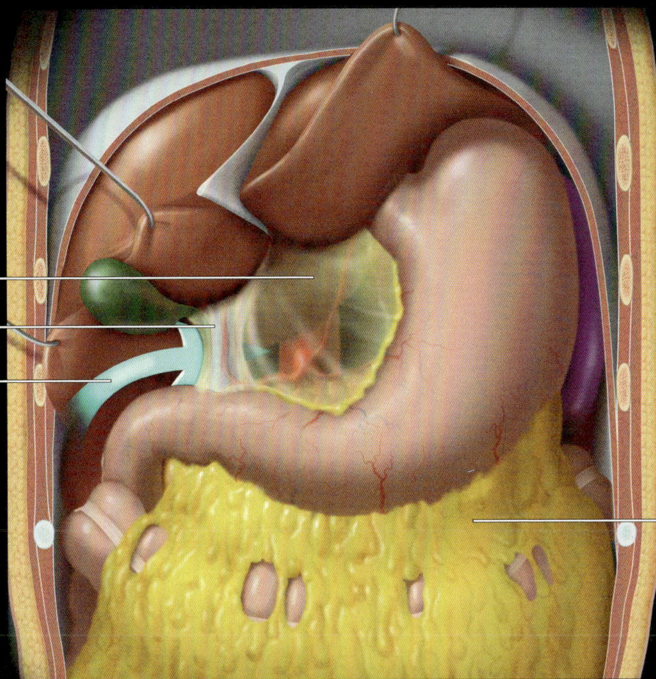

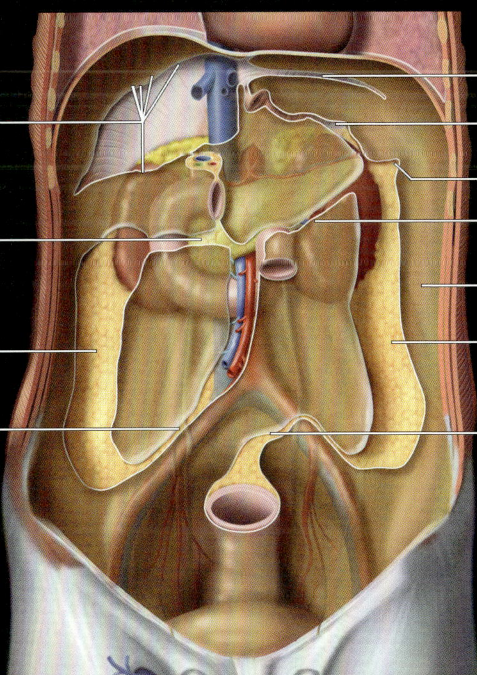

(Top) *The liver has been retracted upward in this graphic, revealing the lesser omentum. The lesser omentum, which is composed of the hepatoduodenal and hepatogastric ligaments, forms part of the anterior wall of the lesser sac and conveys the common bile duct, hepatic artery, portal vein, and gastric arteries. The aorta and celiac artery can be seen through the lesser omentum, as they lie just posterior to the lesser sac. The greater omentum is seen on this graphic as a large fatty apron extending from the stomach downward along the anterior abdomen.* **(Bottom)** *Frontal graphic shows the abdomen with all of the intraperitoneal organs removed. The root of the transverse mesocolon divides the peritoneal cavity into supramesocolic and inframesocolic spaces that communicate only along the paracolic gutters. The coronary and triangular ligaments suspend the liver from the diaphragm. The superior mesenteric vessels traverse the small bowel mesentery whose root crosses obliquely from the upper left to the lower right posterior abdominal wall.*

PERITONEAL SPACES AND REFLECTIONS

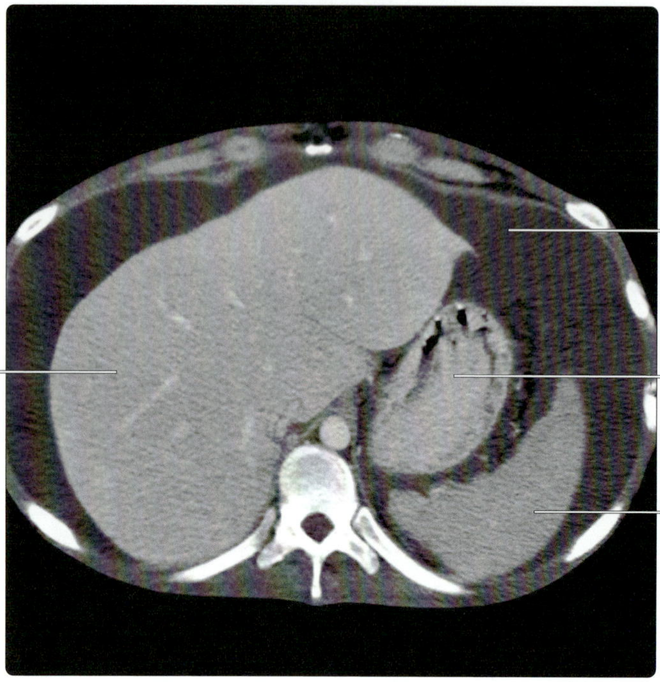

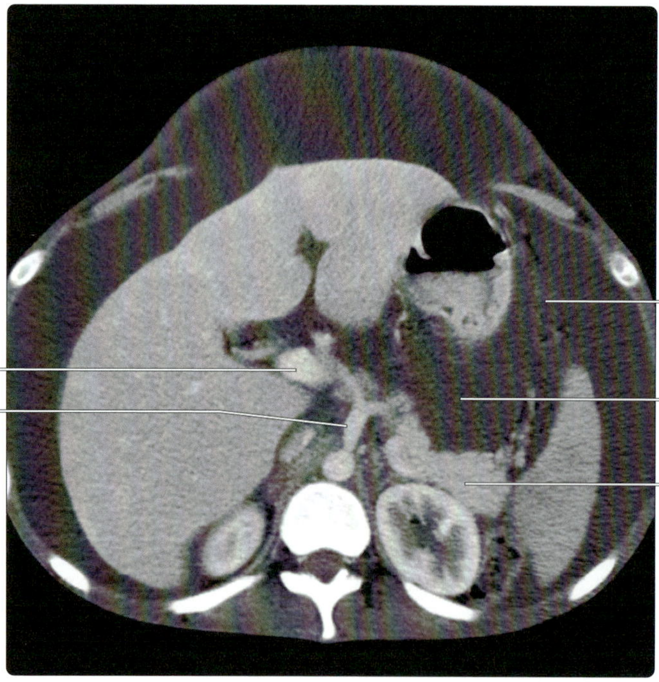

(Top) *First of 4 axial CECT images of a middle-aged man with cirrhosis with delineation of the peritoneal spaces and reflections is shown. Ascites distends the peritoneal cavity in these images, allowing visualization of recesses and peritoneal reflections, which are not normally seen.* **(Bottom)** *The lesser and greater sacs of the peritoneal cavity are distended with ascites. The gastrosplenic ligament and pancreas border the lesser sac, as does the lesser omentum, whose position is marked by the portal vein and celiac trunk.*

PERITONEAL SPACES AND REFLECTIONS

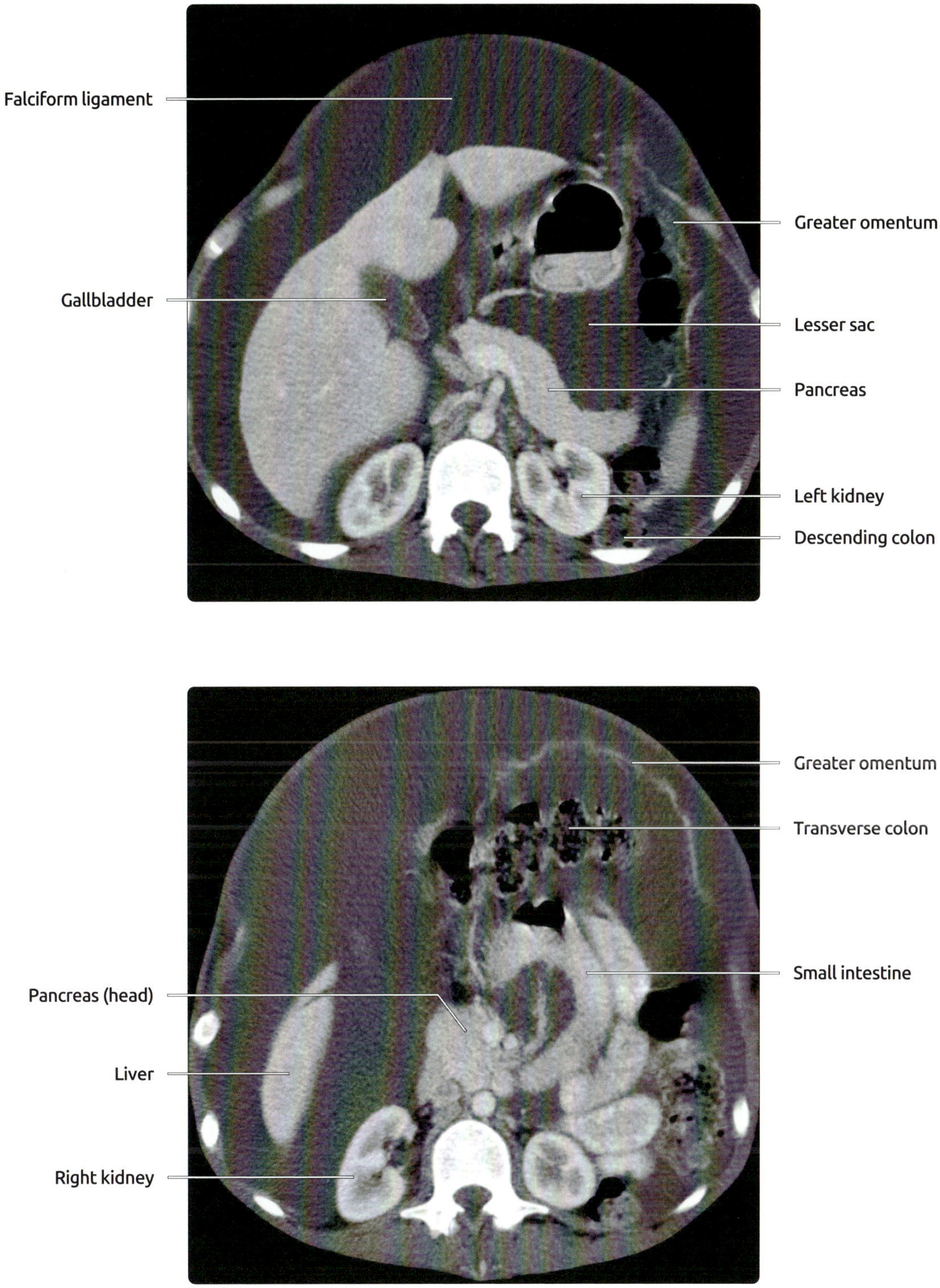

(Top) *The falciform ligament suspends the liver from the anterior abdominal wall. The greater omentum lies between the bowel and the anterior abdominal wall.* **(Bottom)** *The intraperitoneal organs, such as the liver, transverse colon, and small bowel, are suspended within the ascites, while the position of the retroperitoneal organs, such as the kidneys and pancreas, is unaffected by the ascites.*

DISTENDED PERITONEAL CAVITY

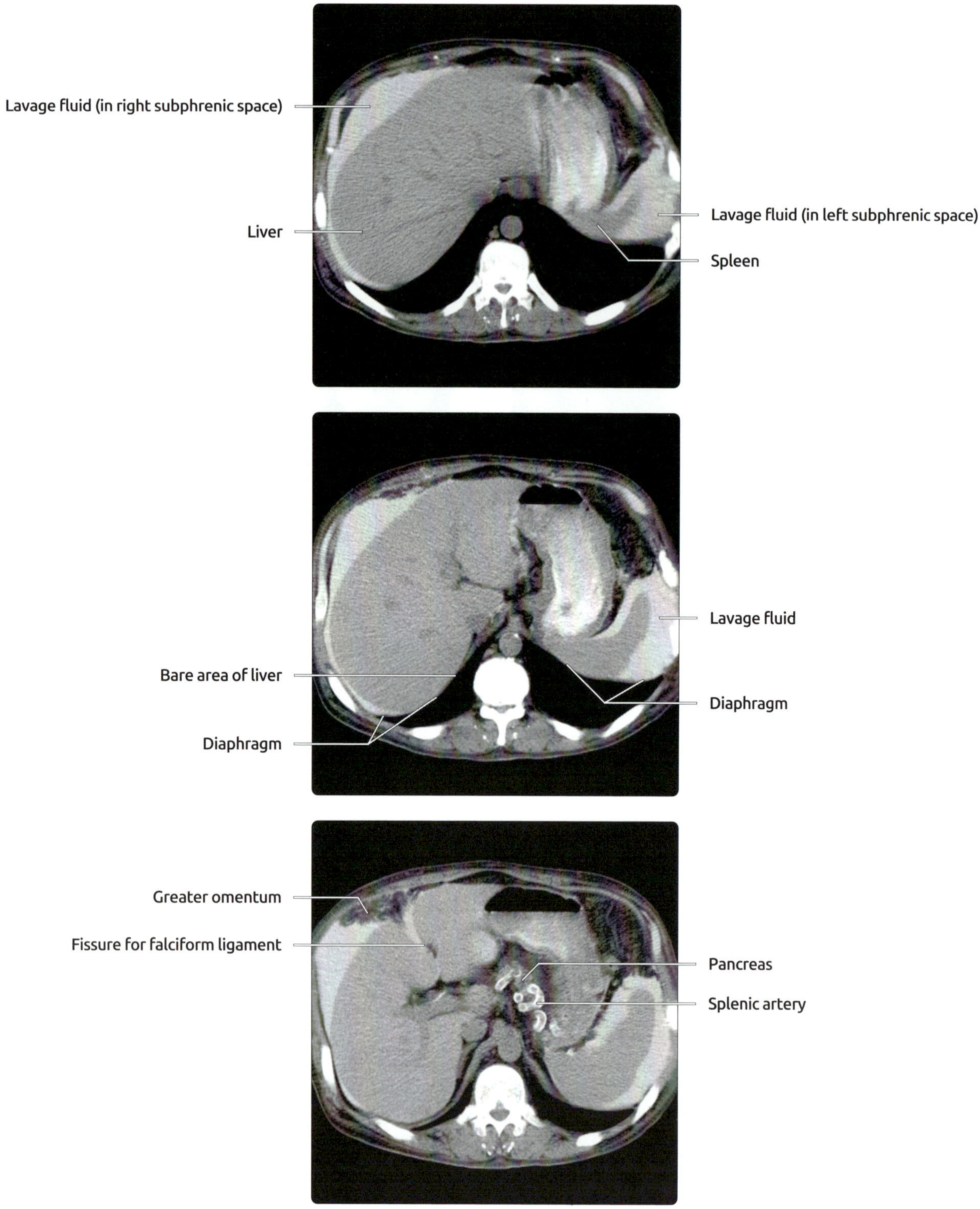

(Top) *First of 6 axial CECT images of an older man with renal failure treated with peritoneal dialysis is shown. Contrast medium was added to the dialysate to identify potential sites of loculated fluid and accounts for the dense appearance of the fluid. Note the intraperitoneal fluid collecting in the subphrenic spaces.* **(Middle)** *Note how the diaphragm suspends the liver and spleen away from the chest wall. The bare area of the liver is in direct contact with the diaphragm but not with the peritoneal cavity, and, accordingly, lavage fluid is not in contact with the bare area.* **(Bottom)** *Fluid invaginates into the fissure for the falciform ligament. Note that there is no fluid within the lesser sac. As a general rule, unless the ascites is tense or of a "local" source (such as a perforated gastric ulcer or pancreatitis), it remains confined to the greater peritoneal cavity and does not pass through the epiploic foramen. The greater omentum "floats" on top of the ascites and is normally of fat density with small internal vessels.*

DISTENDED PERITONEAL CAVITY

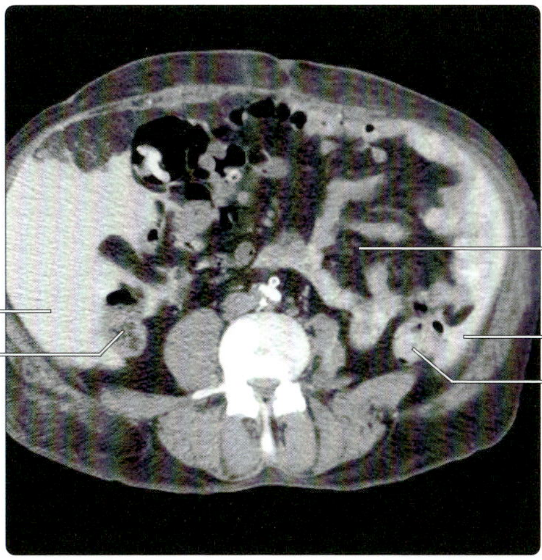

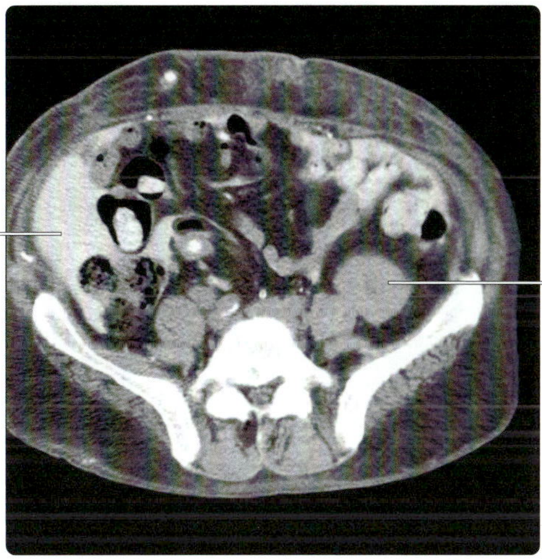

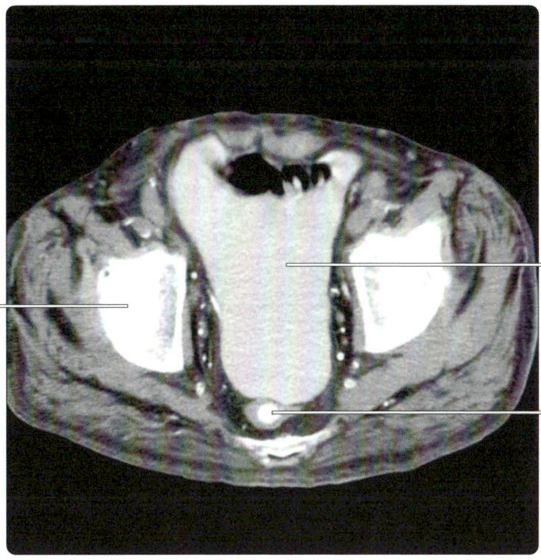

(Top) *The peritoneal fluid is mostly confined to the paracolic gutters at this level.* **(Middle)** *The peritoneal fluid is somewhat loculated in this patient, typical in the setting of chronic peritoneal dialysis, which results in inflammation and scarring of the peritoneum over time. Note the transplanted kidney in the left iliac fossa, which had stopped functioning due to rejection.* **(Bottom)** *The most dependent recess of the peritoneal cavity is in the pelvis, which is distended by dialysis fluid in this patient.*

LESSER SAC (OMENTAL BURSA)

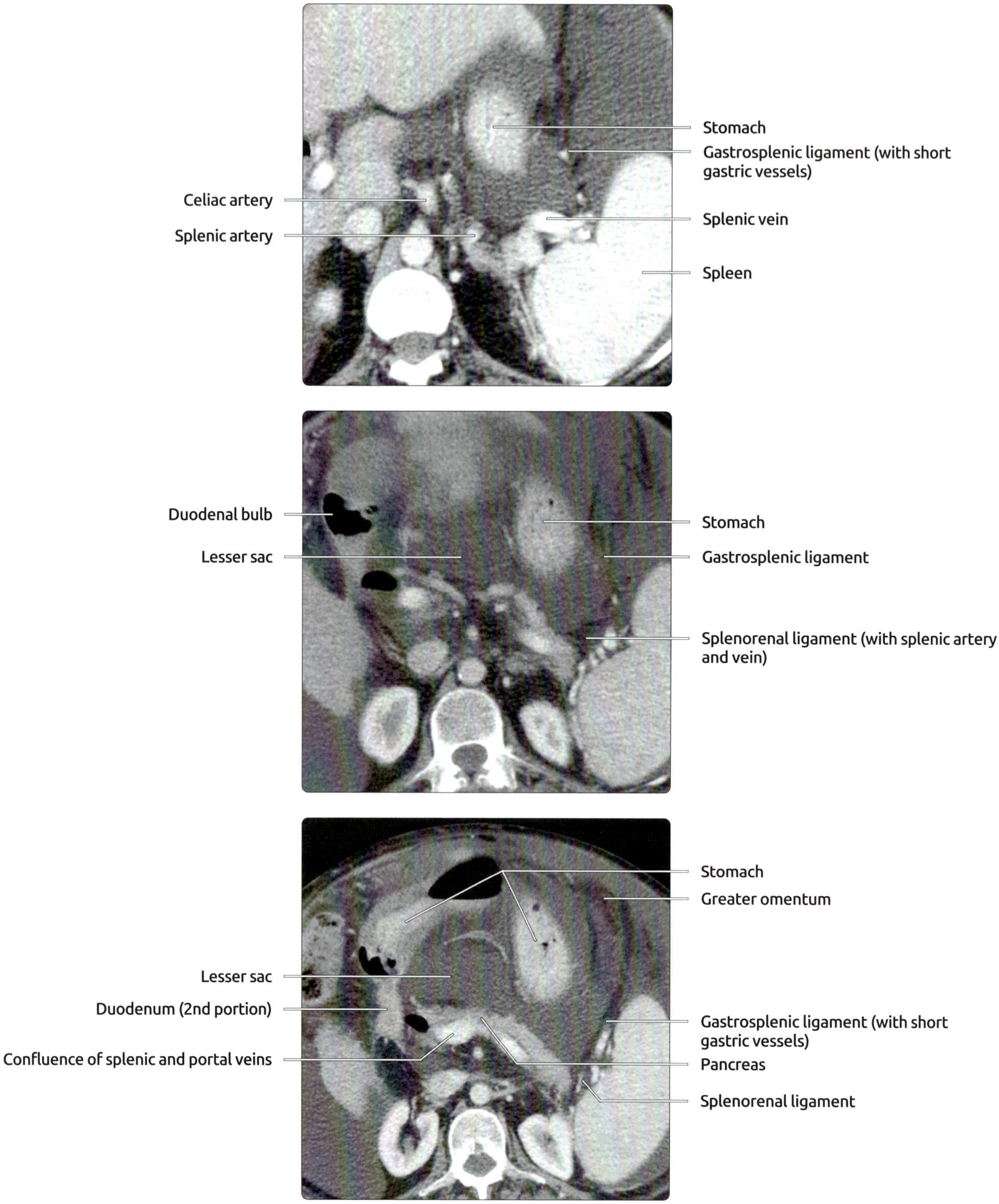

(Top) *First of 3 axial CECT images focusing on the lesser sac is shown. The gastrosplenic ligament connects the stomach to the spleen and carries the short gastric vessels. Abdominal "ligaments" are double-layered folds of peritoneum that connect one viscus to another. They contain fat and transmit vessels, nerves, and lymphatics between the retroperitoneum and the abdominal viscera.* **(Middle)** *The gastrosplenic and splenorenal ligaments form the left anterior and posterior walls of the lesser sac, respectively.* **(Bottom)** *Note the structures abutting the lesser sac, including the stomach anteriorly and the pancreas posteriorly.*

UMBILICAL LIGAMENTS, DELINEATED BY ASCITES

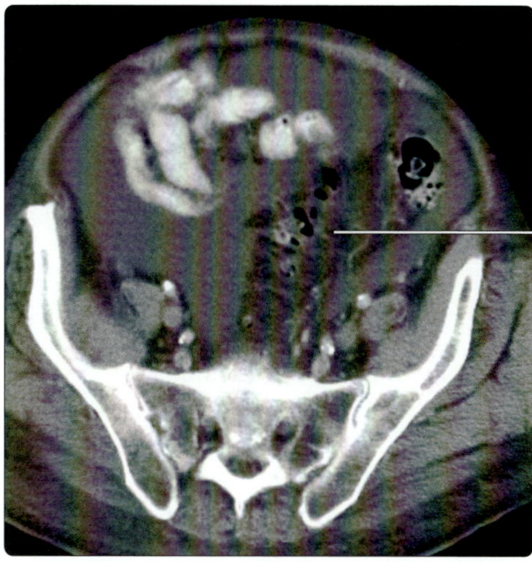

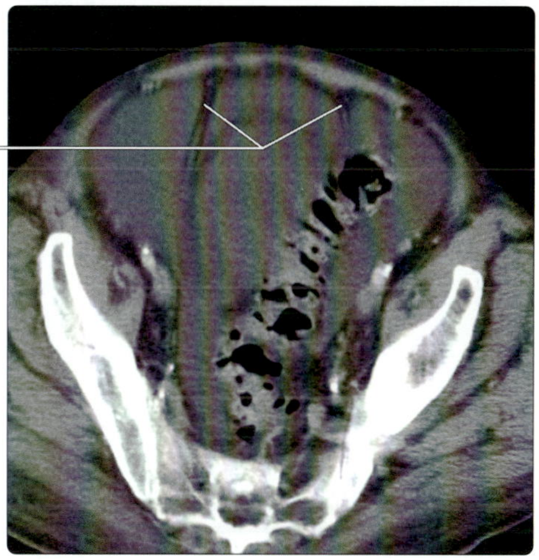

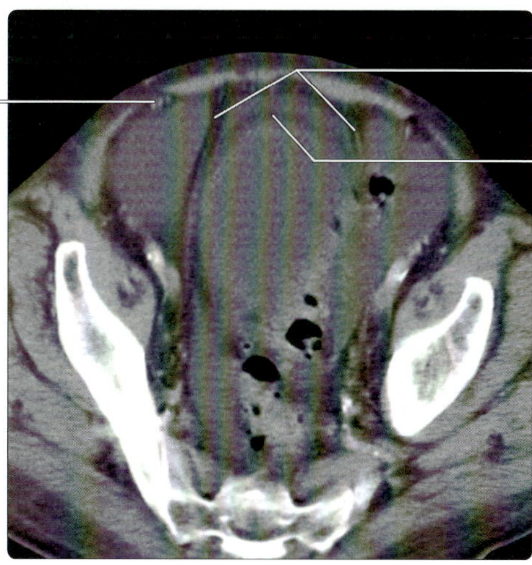

(Top) *First of 3 axial CECT images focuses on the umbilical ligaments. Ascites outlines the sigmoid mesocolon and small bowel loops in this image.* **(Middle)** *The umbilical ligaments are outlined by ascites in this more caudal image. These are the remnants of the fetal umbilical arteries that had connected the internal iliac arteries to the umbilical cord. The peritoneal reflections covering these ligaments are the lateral umbilical folds.* **(Bottom)** *The ascites is not loculated in this patient, but normal structures, such as the umbilical ligaments and the wall of the urinary bladder, may be mistaken for septations within the fluid collection.*

UMBILICAL FOLDS (LIGAMENTS)

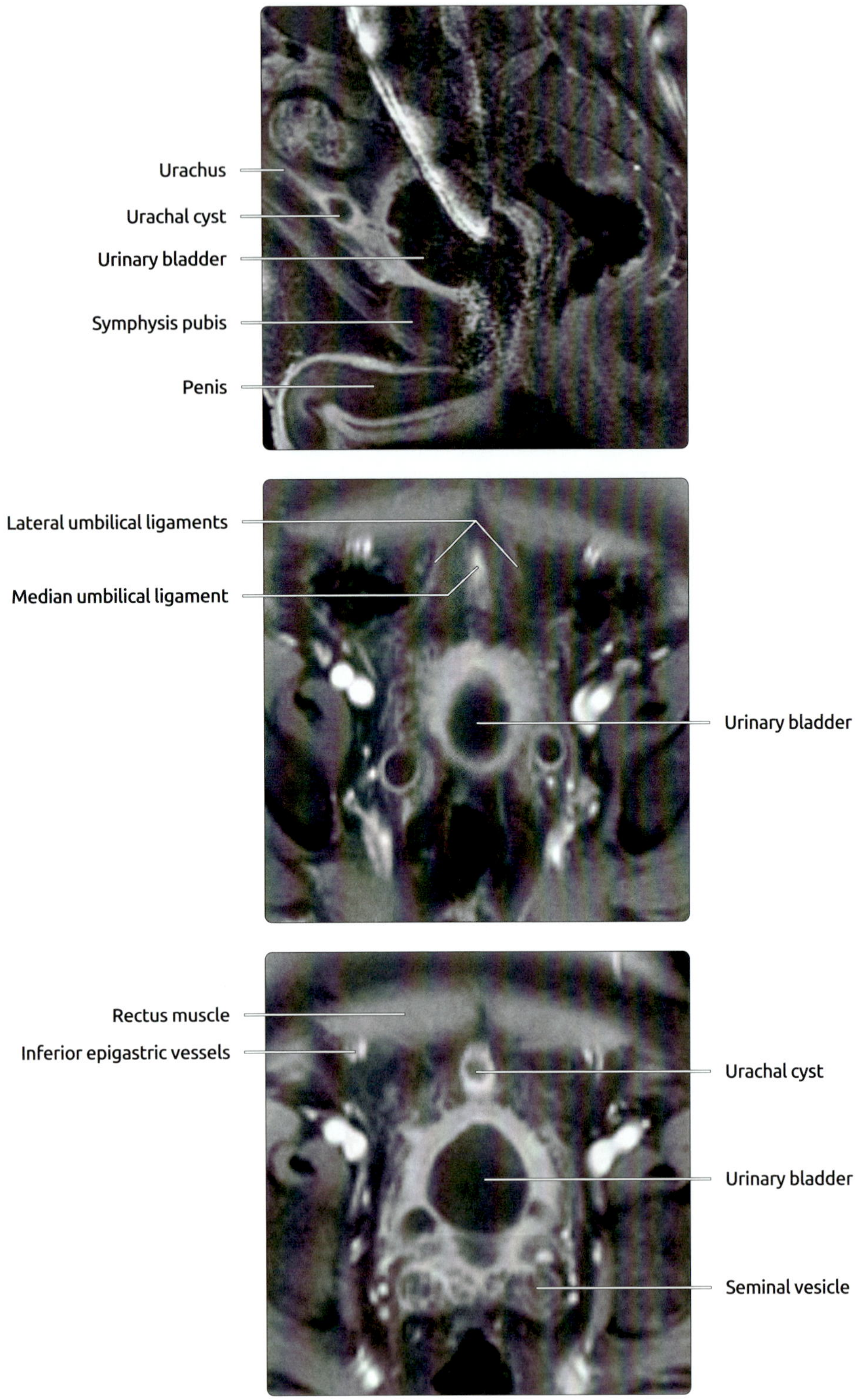

(Top) *Sagittal T1 C+ FS MR of the pelvis shows a linear structure extending from the umbilicus to the dome of the urinary bladder, representing the urachus (median umbilical ligament). This is the fibrous remnant of the allantois and should be completely obliterated after birth. In some individuals, parts of the tract may remain patent leading to a urachal diverticulum or a urachal cyst, as in this patient.* **(Middle)** *The median and lateral umbilical ligaments are evident in this patient on this axial T1 C+ FS MR. Recall that these ligaments are covered with peritoneal reflections called the median and lateral umbilical folds, respectively.* **(Bottom)** *The urachal cyst is evident within the median umbilical fold. The enhancement around the cyst indicates inflammation (infection) of the cyst, which ultimately brought the patient to clinical attention.*

PERITONITIS WITH LOCULATED ASCITES

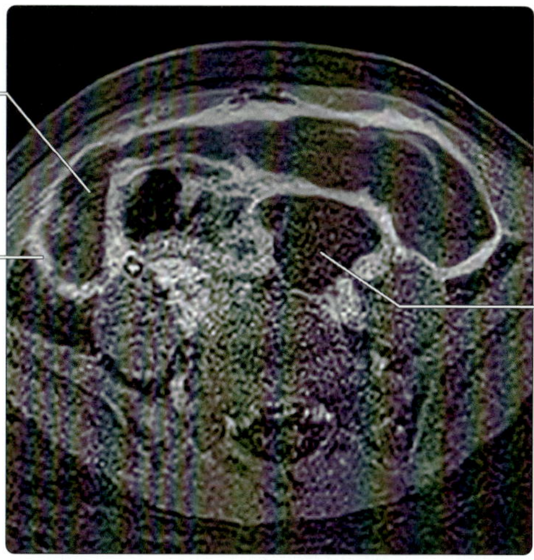

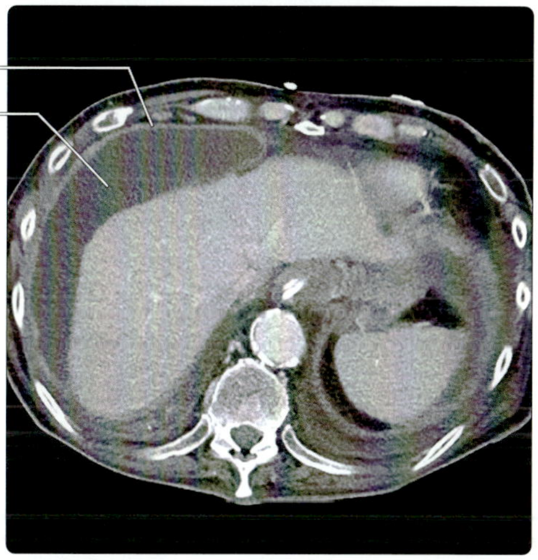

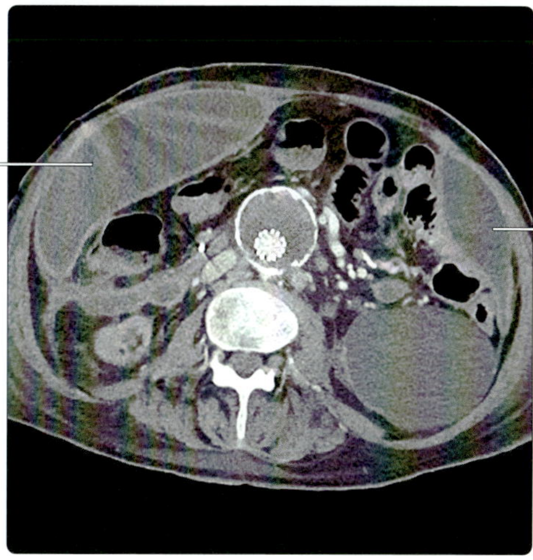

(Top) *Axial T1 C+ FS MR in a patient with perforated appendicitis demonstrates findings of peritonitis with extensive thickening and enhancement of the peritoneum with large loculated ascites and fluid collections. The peritoneal lining should not be visible in normal patients and is typically only visible when thickened by infection, inflammation, or tumor.* **(Middle)** *Axial CECT in a septic patient after surgery demonstrates prominent peritoneal enhancement and thickening with an associated loculated fluid collection along the margin of the liver.* **(Bottom)** *Axial CECT in the same patient demonstrates multiple loculated intraperitoneal fluid collections throughout the abdomen and pelvis, including 2 large collections extending down the paracolic gutters. The peritoneum adjacent to the these collections is thickened and enhancing. This constellation of findings is typical for peritonitis with abscess formation.*

TERMINOLOGY

Abbreviations

- Inferior vena cava (IVC)
- Superior mesenteric artery (SMA)
- Superior vena cava (SVC)
- Inferior mesenteric artery (IMA)

GROSS ANATOMY

Overview

- **Abdominal aorta**
  - Enters abdomen at aortic hiatus in diaphragm (~ T12 level) and bifurcates at ~ L4 level into right and left common iliac arteries
    - Gradually tapers and narrows as its extends inferiorly
    - Lies slightly to left of midline, separated from spine only by anterior longitudinal ligament and lumbar veins
  - Gives rise to arteries in 3 vascular planes
    - **Anterior midline plane**: Unpaired visceral arteries to alimentary tract (**celiac artery**, **SMA**, and **IMA**) and **median sacral artery**
      - **Median** (**middle**) **sacral artery** arises from posterior aspect of distal abdominal aorta and extends to lumbosacral spine and coccyx
      - Terminal branch of IMA is **superior rectal artery**, which extends into deep pelvis toward rectum
    - **Lateral plane**: Paired visceral arteries to urogenital and endocrine organs [**renal**, **adrenal**, and **gonadal arteries** (testicular or ovarian)]
    - **Posterolateral plane**: Paired parietal arteries to diaphragm and body wall (**subcostal**, **inferior phrenic**, and **lumbar arteries**)
      - **Lumbar arteries** are tiny paired vessels arising from posterior aspect of L1-L4, which extend posteriorly and laterally along margin of vertebral body
      - **Inferior phrenic arteries** are 2 small arteries that arise immediately above celiac (less commonly from celiac or renal arteries) and extend upward to supply diaphragm
      - **Subcostal arteries** extend immediately below 12th rib
- **Iliac arteries**
  - **Common iliac arteries**: Extend from aortic bifurcation and bifurcate into internal and external iliac arteries at roughly level of sacroiliac joints
    - Extend inferiorly and laterally along medial border of psoas muscles
    - Ureters course anterior to common iliac arteries and common iliac veins are posterior
    - Give off small unnamed branches to surrounding soft tissues and may give off accessory renal or iliolumbar arteries
  - **External iliac arteries**: Extend from common iliac artery bifurcation into groin (under inguinal ligament) to become **common femoral arteries**
    - Major branches include
      - **Inferior epigastric artery**, which extends medially and upward and anastomoses with superior epigastric artery, providing blood supply to rectus muscles and anterior abdominal wall; linea semilunaris where inferior epigastric vessels penetrate rectus sheath is where **spigelian hernias** occur
      - **Deep circumflex iliac artery**, which extends laterally and upward from external iliac (exactly opposite to inferior epigastric artery origin) toward anterior inferior iliac spine
  - **Internal iliac arteries**: Originate from common iliac artery bifurcation and extend posteromedially to supply blood to pelvic sidewalls, perineum, buttock, medial thigh, and pelvic organs
    - a.k.a. "hypogastric" arteries
    - ~ 4 cm in length and generally larger in diameter compared to external iliac arteries
    - Divide into **anterior** and **posterior divisions** at level of greater sciatic foramen with highly variable branching pattern
    - Anterior division
      - **Umbilical artery** and **superior vesical artery**: Umbilical artery gives rise to superior vesical artery, which travels to cranial aspect of bladder; distal portion of umbilical artery thromboses and regresses after birth and becomes medial umbilical ligament (along internal aspect of anterior abdominal wall)
      - **Inferior vesical arteries**: Supplies prostate, seminal vesicles, and inferior portion of bladder in men
      - **Obturator artery**: Travels in obturator foramen along with obturator nerve and vein
      - **Vaginal artery**: Takes place of inferior vesical artery in women and supplies vagina and inferior bladder
      - **Uterine artery**: Located within broad ligament and anastomoses with ovarian artery
      - **Middle rectal artery**: Supplies distal rectum and frequently anastomoses with superior and inferior rectal arteries
      - **Internal pudendal artery**: Travels with pudendal nerve in greater sciatic foramen toward perineum (providing blood supply to external genitalia) and gives rise to inferior rectal artery
      - **Inferior gluteal artery**: Last branch of anterior division, which is located next to internal pudendal artery in greater sciatic foramen and then extends into gluteal region (below piriformis muscle)
    - Posterior division
      - **Iliolumbar artery**: Travels superiorly and laterally to supply posterior musculature (including psoas muscles)
      - **Lateral sacral artery**: Travels inferiorly and medially to supply branches that enter sacral foramina
      - **Superior gluteal artery**: Largest and terminal branch of posterior division, which exits via greater sciatic foramen and travels laterally into gluteal tissues (superior to piriformis) to supply gluteal musculature via superficial and deep branches
- **IVC**

- Returns poorly oxygenated blood to heart from lower extremities, abdominal wall, and visceral organs (all structures below diaphragm)
  - Blood from gastrointestinal tract passes through portal venous system before entering IVC through **hepatic veins**
- Begins at L5 level with union of **common iliac veins** and exits abdomen via IVC hiatus in diaphragm at ~ T8 level (and subsequently drains into right atrium)
  - Entirely retroperitoneal course with IVC running to right of spine and abdominal aorta
- IVC tributaries correspond to similarly named paired visceral and parietal branches of abdominal aorta
  - **Right adrenal vein**, **left** and **right renal veins**, **hepatic veins**, and **right gonadal vein** drain directly into IVC
  - **Left adrenal and gonadal veins** drain into left renal vein (which then drains into IVC)
  - Paired parietal branches (**inferior phrenic veins**, **lumbar veins**) drain directly into IVC
  - Spleen, pancreas, and small/large bowel do not directly drain into IVC but drain into portal venous system first with venous return ultimately progressing to IVC via hepatic veins
- IVC and SVC are connected through **azygous and ascending lumbar veins**, which are important collateral pathways when SVC or IVC flow is obstructed or slowed
  - Additional collateral pathways include epidural venous plexus and **epigastric veins** (anterior abdominal wall)
  - Azygous continuation of IVC seen in **asplenia and polysplenia syndromes**
  - Duplication of IVC associated with other congenital anomalies, such as congenital heart defects and renal anomalies, such as crossed fused ectopia or circumaortic renal vein
- IVC development has complex embryology
  - Various anomalies are common (~ 10% of population), especially at and below level of renal veins
  - All are variations of persistence/regression of embryologic subcardinal and supracardinal veins

• **Lymphatic system**
- Lymphatic system drains surplus fluid from extracellular spaces and returns it to bloodstream
  - Important function in defense against infection, inflammation, and tumor via lymphoid tissue that is present in lymph nodes, bowel wall, spleen, and thymus
  - Absorbs and transports dietary lipids from bowel to thoracic duct and bloodstream
- Major lymphatic vessels and nodal chains in abdomen/pelvis are closely related to arterial supply, mostly lying along major blood vessels (aorta, IVC, iliac) and having similar names (e.g., common iliac nodes)
- Lymphatic drainage from stomach, liver, spleen, and pancreas drain via **celiac nodes**, while small and large bowel drain via **superior** and **inferior mesenteric nodes**
  - Efferent vessels from these alimentary tract nodes form **intestinal lymphatic trunks**
- Paraaortic nodes drain iliac nodes (which receive lymph from pelvis and lower extremities) and also receive drainage from kidneys, adrenals, and gonads
- **Cisterna chyli** (**chyle cistern**)
  - Formed by confluence of intestinal lymphatic trunks and **right** and **left lumbar lymphatic trunks** (which receive lymph from nonalimentary viscera, abdominal wall, and lower extremities via paraaortic nodes) and may be discrete sac or plexiform convergence
- **Thoracic duct**: Originates from cisterna chyli anterior to L1-L2 level
  - Formed by convergence of main lymphatic ducts of abdomen and ultimately drains lymph from abdomen, lower extremities, left chest, left upper extremity, and left head/neck
  - Ascends through **aortic hiatus** in diaphragm to enter posterior mediastinum
  - Ends by entering junction of **left subclavian** and **internal jugular veins**

• **Major nerves**
- Nerves of anterolateral abdominal wall
  - Arise from anterior rami of spinal nerves T7-L1
  - **Dermatomes** (areas of sensory innervation) resemble oblique stripes around trunk following slope of ribs beginning posteriorly over intervertebral foramen from which spinal nerve exits
  - Nerves run in **neurovascular plane** between transverse abdominal and internal oblique muscles and supply muscles and skin
- Nerves of posterior abdominal wall
  - Sensory and motor fibers come from anterior and posterior rami of spinal nerves T12 and L1-L5 (**lumbar plexus**); lie deep to **psoas muscle**
- Innervation of abdominal viscera
  - **Sympathetic** innervation through **lower thoracic** and **abdominopelvic splanchnic nerves** from spinal cord levels T5-L3 via paraaortic prevertebral ganglia (innervates blood vessels of abdominal viscera; inhibitory to parasympathetic innervation)
  - **Parasympathetic** innervation from **vagus nerve** (esophagus through transverse colon) and **pelvic splanchnic nerve** (descending colon to rectum) **Intrinsic parasympathetic ganglia** within muscular wall of stomach and gut, called **myenteric plexus** (**of Auerbach**) promotes peristalsis and secretion (though secretion is largely controlled hormonally)
  - **Sensory innervation of viscera**: Pain sensation carried by fibers accompanying sympathetic nerves; reflex afferent innervation accompanies **vagus** (parasympathetic) nerves
- Greater sciatic foramen is point of exit for sciatic nerve as it passes through pelvis, positioned beneath piriformis muscle and above superior gemellus muscle, to reach lower limb

CORONAL ABDOMINAL ARTERIES

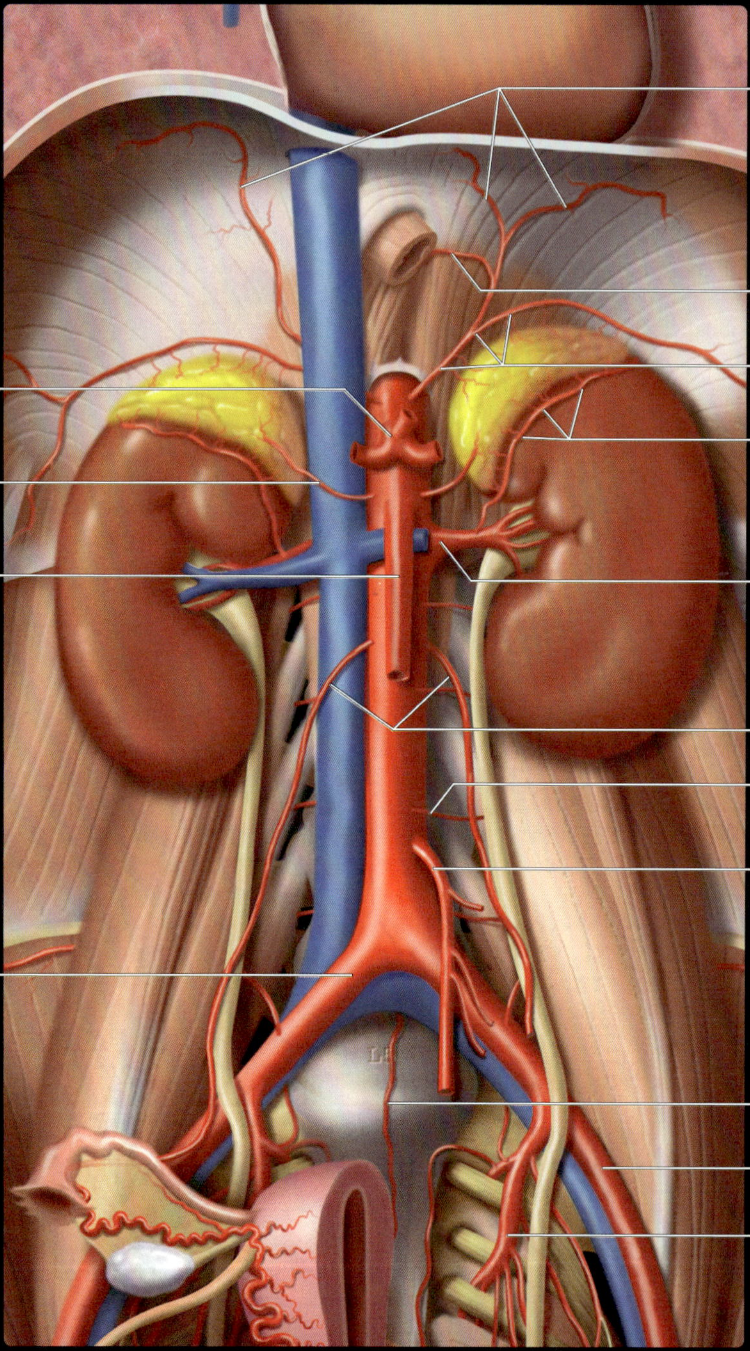

The major arteries to the gastrointestinal tract arise as unpaired vessels from the aortic midline plane and include the celiac, superior mesenteric artery (SMA), and inferior mesenteric artery (IMA). Branches to the urogenital and endocrine organs arise as paired vessels in the lateral plane and include the renal, adrenal, and gonadal (testicular or ovarian) arteries. The diaphragm and posterior abdominal wall are supplied by paired branches in the posterolateral plane, including the inferior phrenic and lumbar arteries (4 pairs, only 1 of which is labeled in this graphic). The anterior abdominal wall is supplied by the inferior epigastric and deep circumflex iliac arteries, both branches of the external iliac artery. The inferior epigastric artery turns superiorly to run in the rectus sheath, where it anastomoses with the superior epigastric artery, a terminal branch of the internal mammary (thoracic) artery.

INTERNAL ILIAC ARTERY

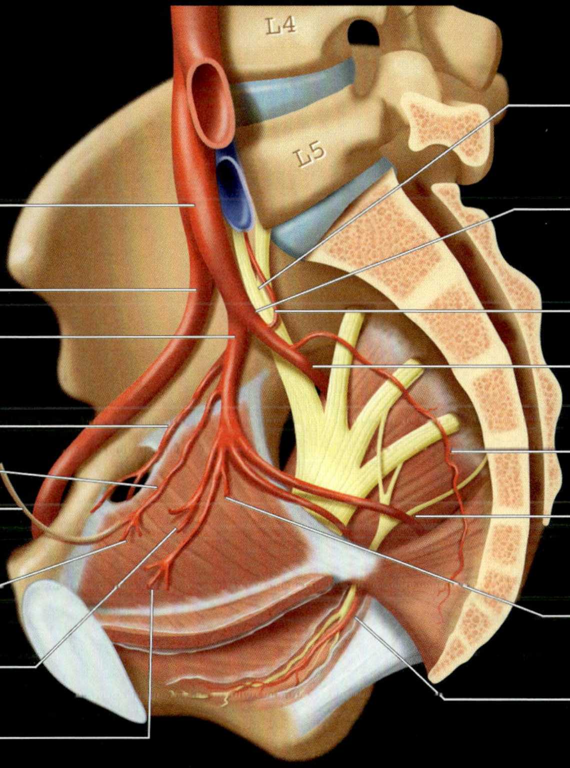

The internal iliac artery divides into anterior and posterior divisions with the branching pattern of individual branches varying considerably. The 3 major branches of the posterior division include iliolumbar artery, lateral sacral artery, and superior gluteal artery, the large terminal branch of the posterior division, which passes posteriorly and runs between the lumbosacral nerve trunk and the anterior ramus of the S1 nerve. There are several branches of the anterior division, which can vary between male and female patients. Major branches of the anterior division include the superior and inferior vesical arteries, uterine artery, middle rectal artery, obturator artery, vaginal artery, and internal pudendal artery. Only the proximal portion of the umbilical arteries remains patent after birth, while the distal portion obliterates, forming the medial umbilical ligaments.

MESENTERIC VASCULAR ANATOMY

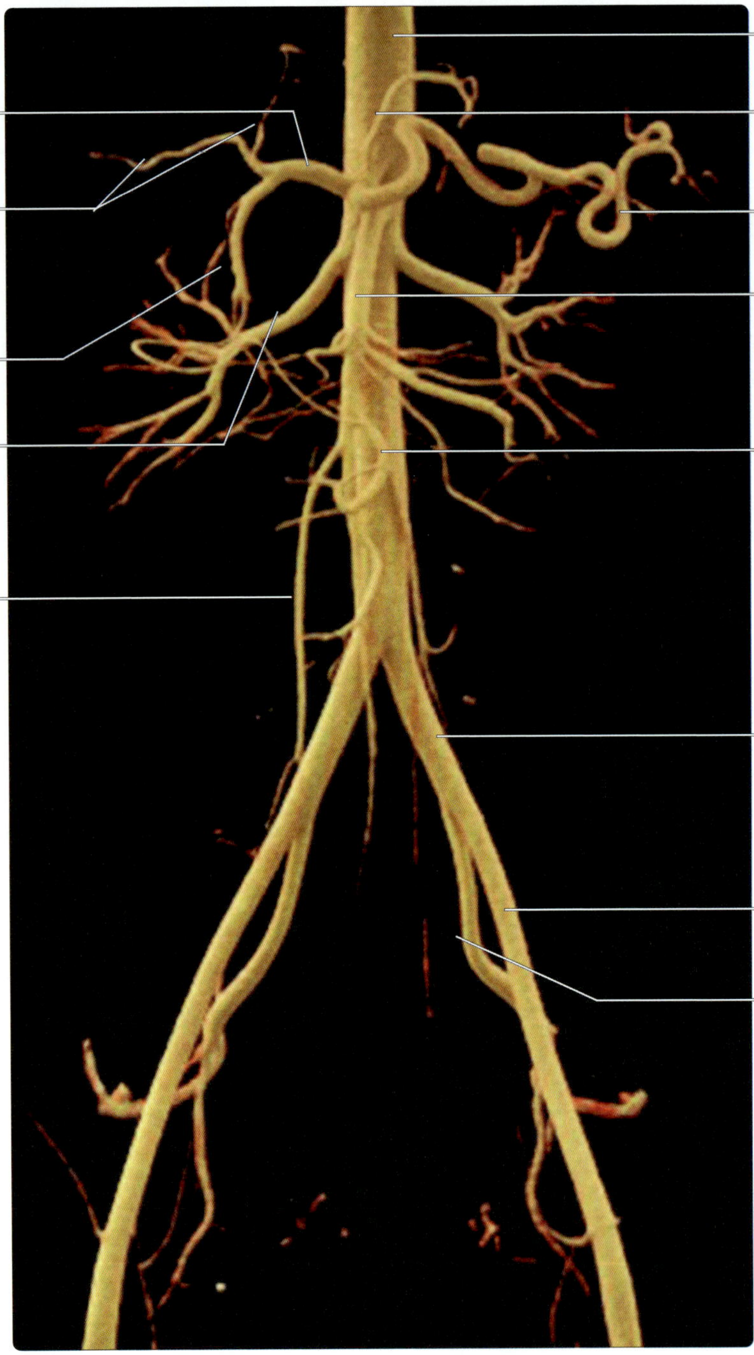

Cinematic-rendered image shows arterial anatomy of the aorta and its branches.

NORMAL CATHETER AORTOGRAM

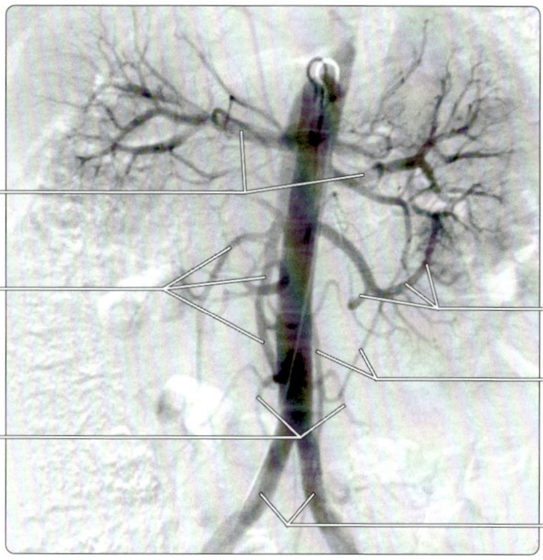

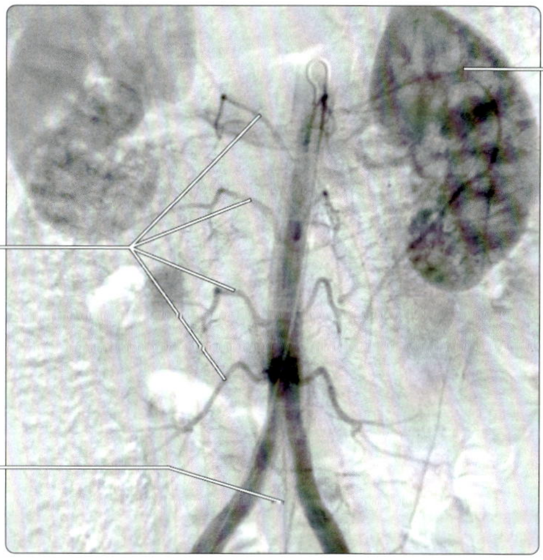

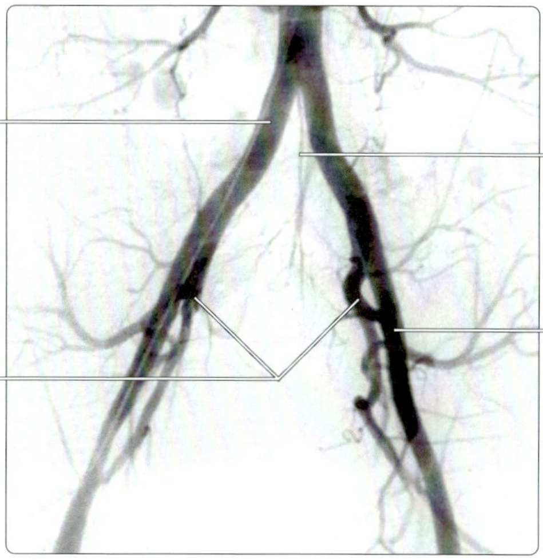

(Top) *First of 3 images from a catheter angiogram of the aorta shows many of the major branches of the aorta overlapped to a degree due to the nonselective (upper aortic) injection and the nontomographic nature of the radiograph.* **(Middle)** *The 4 paired lumbar arteries are well shown in this image arising from the posterolateral abdominal aorta. A 5th lumbar artery may arise from the median sacral or internal iliac artery.* **(Bottom)** *The last branches of the abdominal aorta are the median sacral and common iliac arteries. The latter branch into the external and internal iliac arteries. The internal iliac (hypogastric) artery supplies all of the pelvic viscera and muscles, while the external iliac supplies the anterior abdominal wall (through the deep circumflex iliac and inferior epigastric arteries) before leaving the abdomen (behind the inguinal ligament) to supply the lower extremity.*

ANGIOGRAM, CELIAC ARTERY

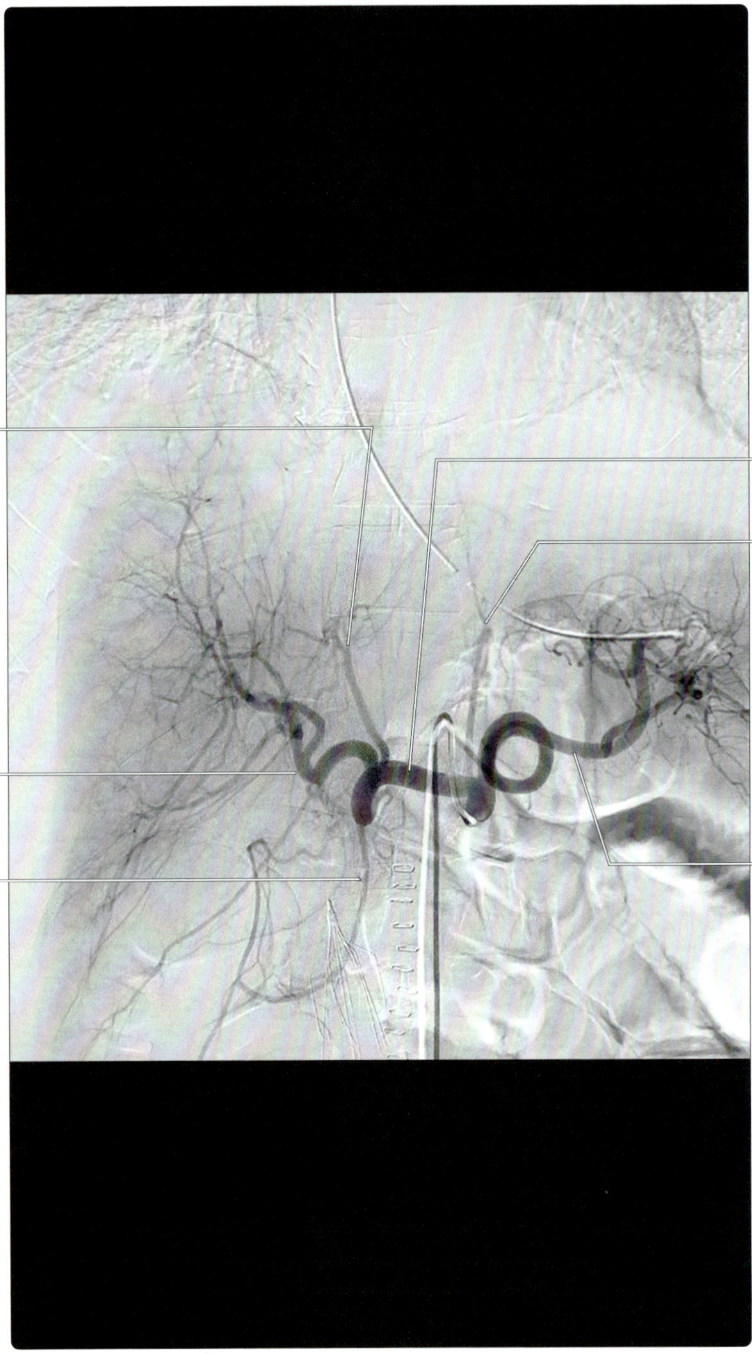

Frontal angiogram after celiac artery injection demonstrates "conventional" anatomy with the celiac artery trifurcating into the splenic artery, common hepatic artery, and left gastric artery. The gastroduodenal artery (GDA) extends downward from the common hepatic artery and divides the common hepatic from the proper hepatic artery.

ANGIOGRAM, SUPERIOR MESENTERIC ARTERY

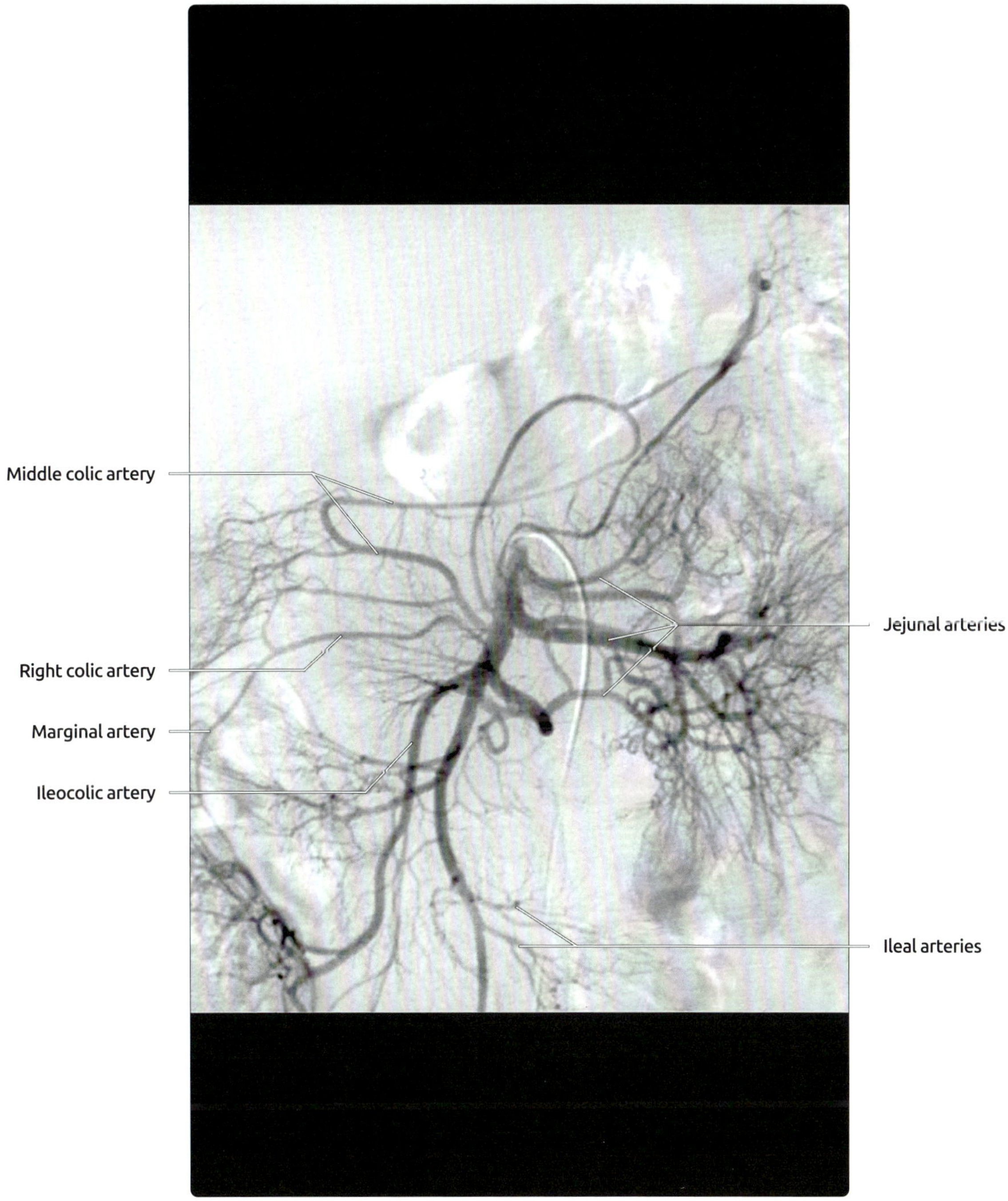

Catheter injection of the SMA shows opacification of all of the arterial branches that supply the entire small bowel, appendix, cecum, ascending colon, and transverse colon. The marginal artery has a course parallel to the entire length of the colon and anastomoses with the marginal artery on the left, which is supplied by branches of the IMA. This is an important collateral pathway that can supply flow to parts of the colon that would otherwise be become ischemic due to occlusion of the main trunk or branches of the SMA or IMA.

ANGIOGRAM, INFERIOR MESENTERIC ARTERY

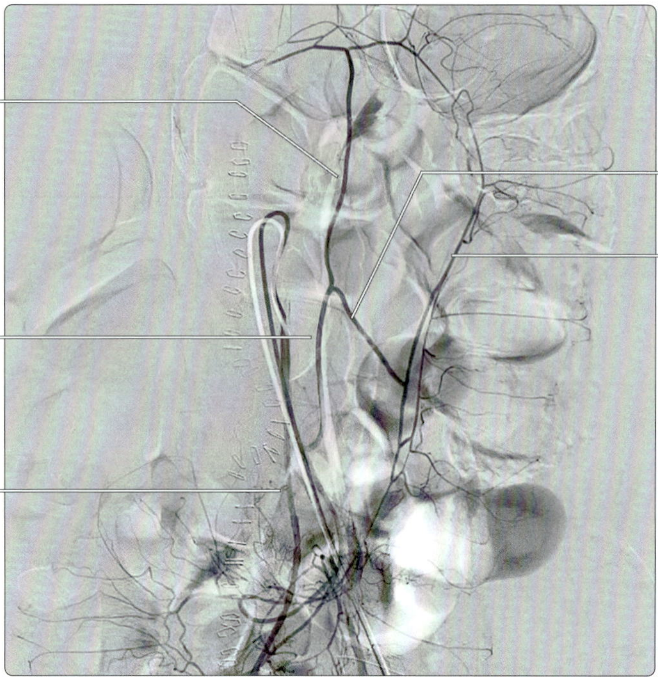

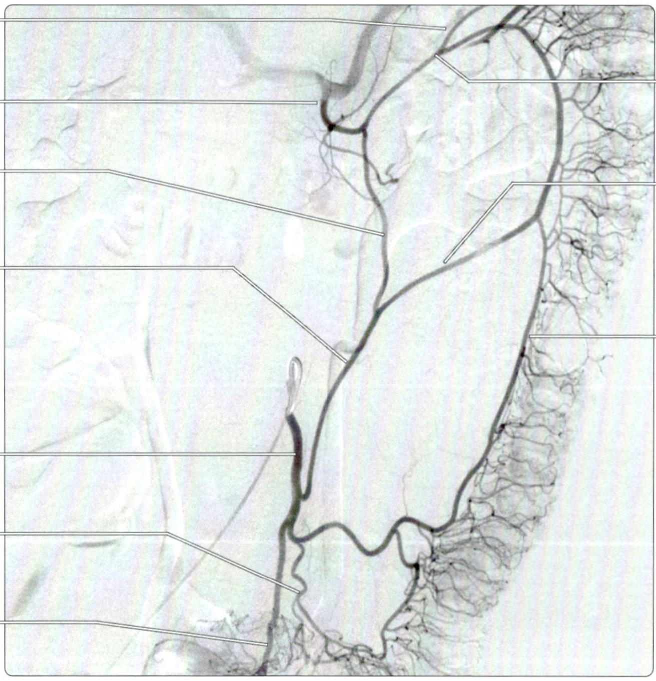

(Top) *Frontal image from an IMA angiogram demonstrates the 1st major branch of the IMA, the left colic artery, which then gives off ascending and descending branches. The IMA is then seen to give off a number of smaller sigmoid branches in the left lower quadrant before continuing into the pelvis as the superior rectal artery.* **(Bottom)** *Frontal image from an IMA angiogram shows left colic artery branches and the marginal artery of Drummond. Contrast reflux shows the dorsal pancreatic artery with the the middle colic artery arising from the dorsal pancreatic artery as a variant. The dorsal pancreatic artery can yield part or all of the middle colic artery in ~ 1% of patients.*

ANGIOGRAM, HEPATIC ARTERY VARIANTS

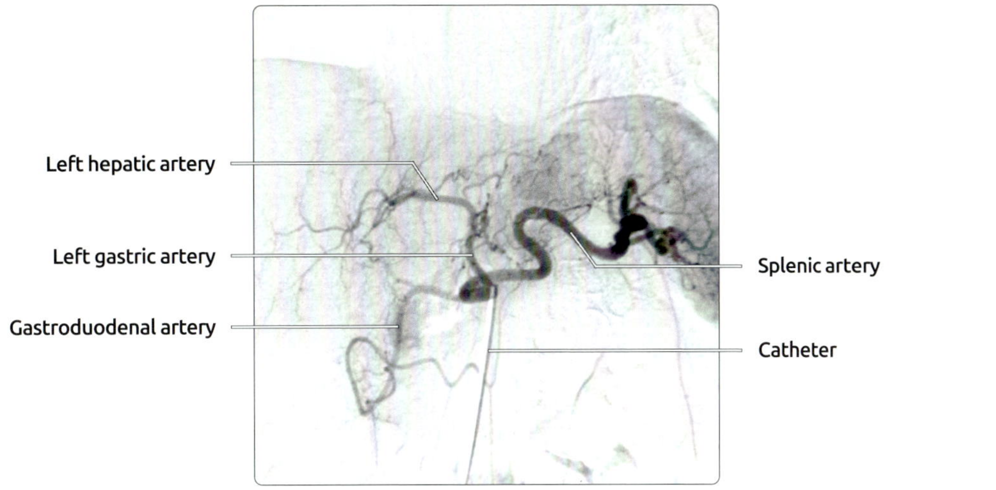

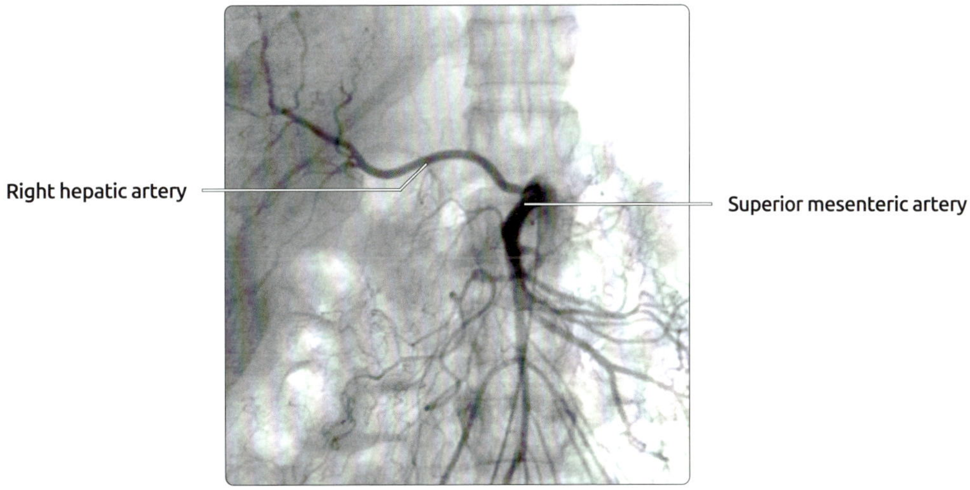

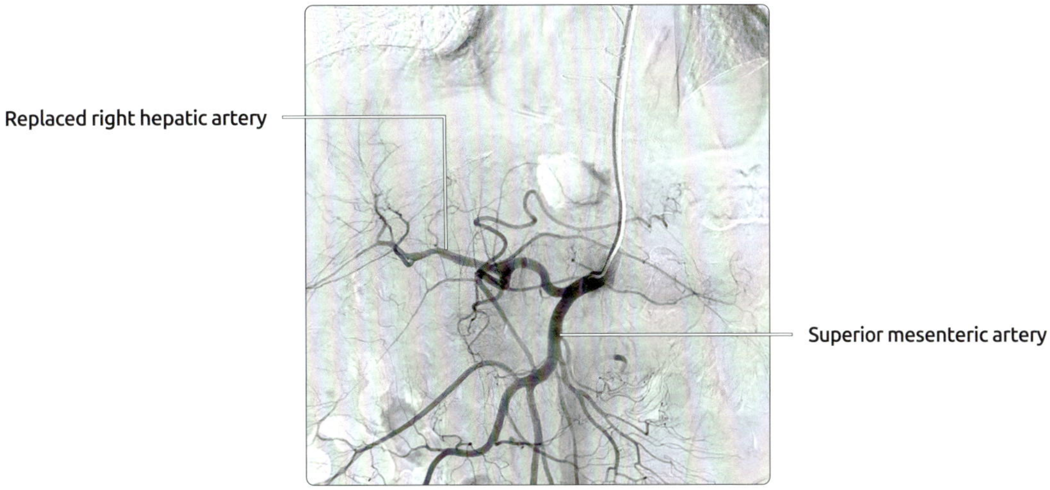

(Top) *Catheter arteriogram with injection of the celiac trunk shows a normal splenic artery, a large left gastric artery that gives off the left hepatic artery, and the GDA arising directly from the celiac trunk. This aberrant origin of the left hepatic artery is referred to as a "replaced" left hepatic artery arising from the left gastric.* **(Middle)** *Catheter injection of the SMA shows the right hepatic artery arising from the SMA.* **(Bottom)** *Frontal SMA angiogram demonstrates the typical orientation of a replaced right hepatic artery arising from the SMA with the artery extending upward and to the right. When viewed on a CT or MR, this replaced artery would be seen extending between the main portal vein and Inferior vena cava (IVC).*

CT, ABDOMINAL AND PELVIC ARTERIAL VASCULATURE

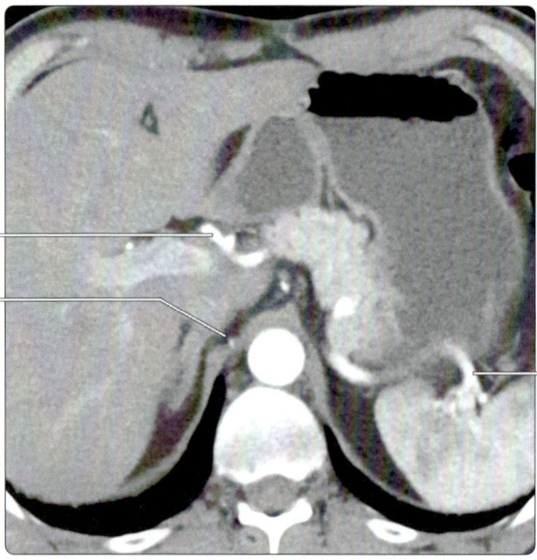

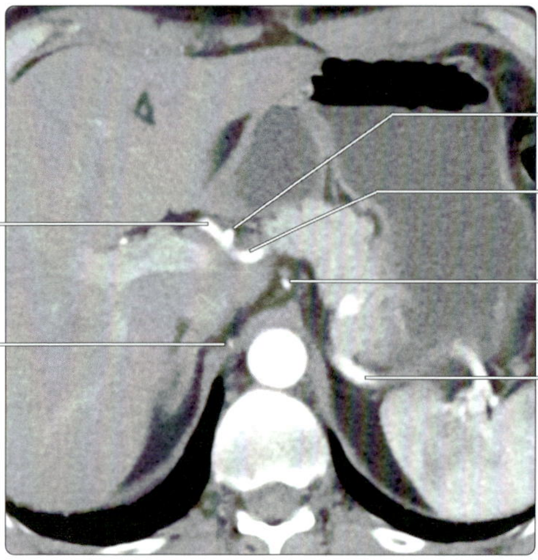

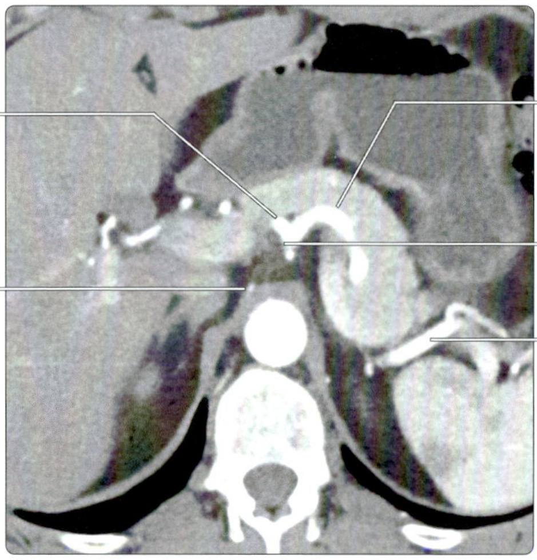

(Top) *First of 15 arterial-phase CT images through the abdomen and pelvis delineating normal arterial anatomy is shown. Axial CECT through the superior abdomen demonstrates a tiny artery running along the right diaphragmatic crus toward the right hemidiaphragm, representing the right inferior phrenic artery. This small artery usually arises from the aorta above the celiac artery but can also arise from the celiac itself or the renal artery.* **(Middle)** *Axial CECT slightly more inferiorly demonstrates the origin of the GDA, which divides the common hepatic artery (a branch of the celiac artery) from the proper hepatic artery. The left gastric artery is seen extending upward in the gastrohepatic ligament.* **(Bottom)** *Axial CECT slightly more inferiorly demonstrates the celiac artery dividing into the common hepatic artery and splenic artery with the left gastric artery (which extends superiorly) being the 3rd branch in patients with "conventional" anatomy.*

CT, ABDOMINAL AND PELVIC ARTERIAL VASCULATURE

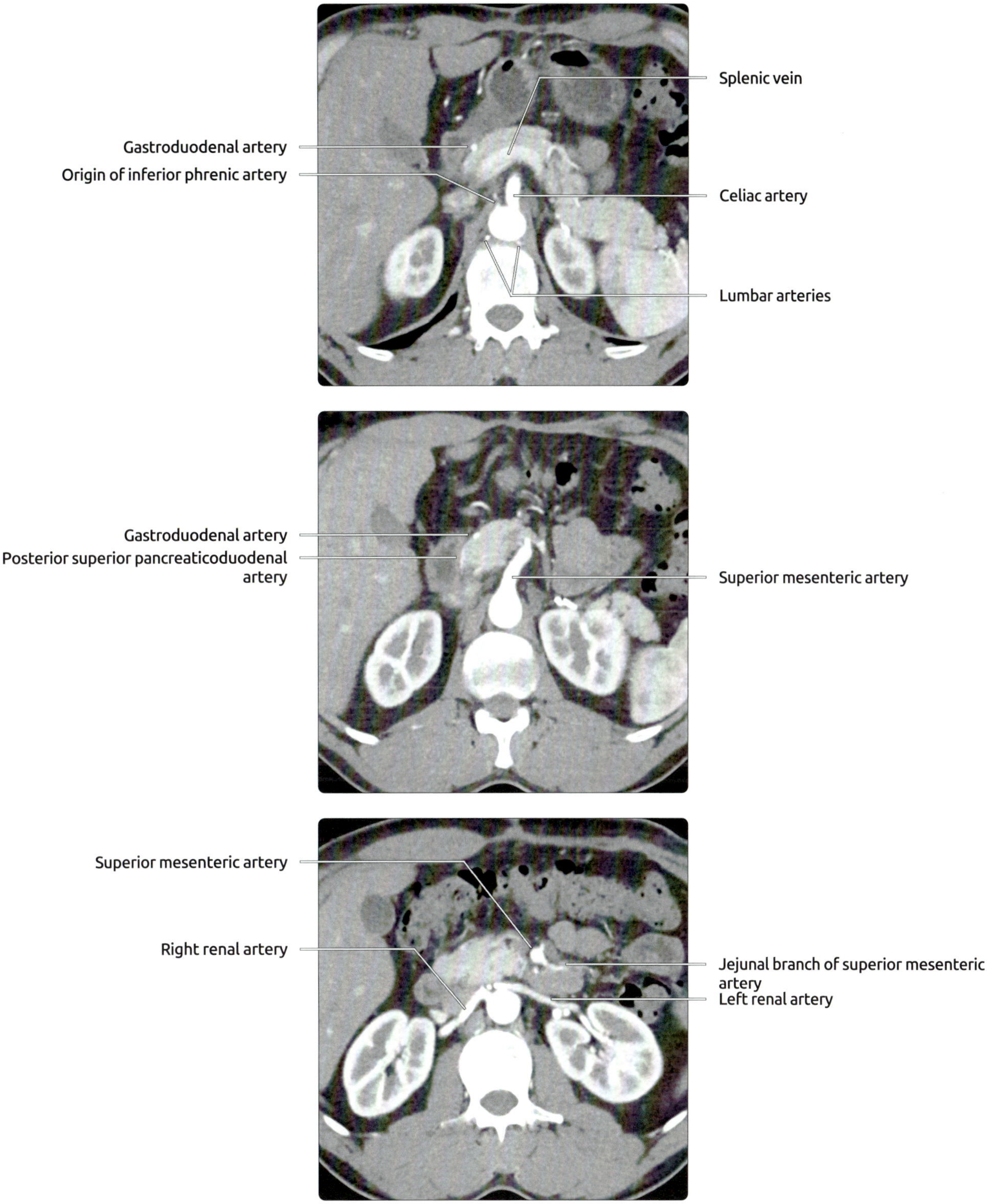

(Top) *Axial CECT demonstrates the GDA extending inferiorly along the anterior aspect of the pancreatic head. Also note the paired lumbar arteries arising along the posterior aspect of the aorta, which are seen from the L1-L4 levels.* **(Middle)** *Axial CECT demonstrates the SMA arising just below the celiac and extending downward into the mesentery. The GDA gives off several branches, including the anterior and posterior superior pancreaticoduodenal arteries and the right gastroepiploic artery more distally.* **(Bottom)** *Axial CECT demonstrates both renal arteries. The SMA is diving downward into the mesentery and gives off jejunal and ileal branches into the left hemiabdomen. The anatomy of the SMA is complex and is often best appreciated using a coronal reformation with 3D reconstruction.*

CT, ABDOMINAL AND PELVIC ARTERIAL VASCULATURE

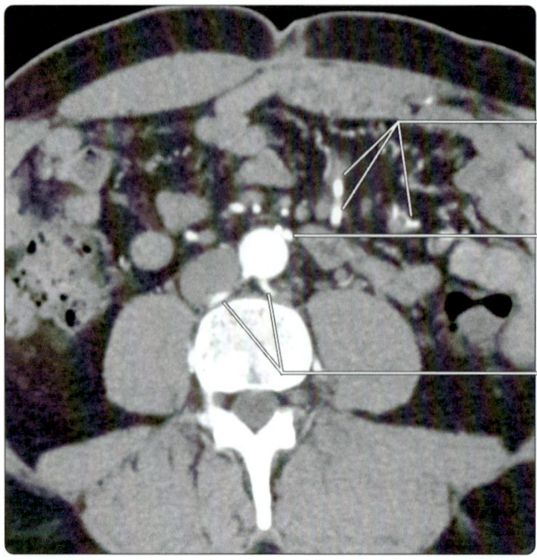

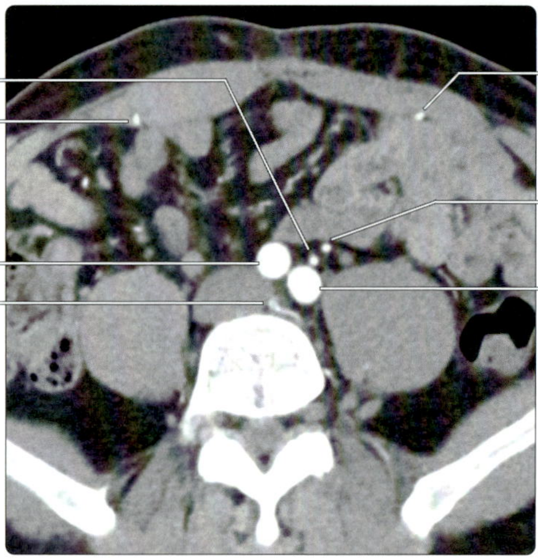

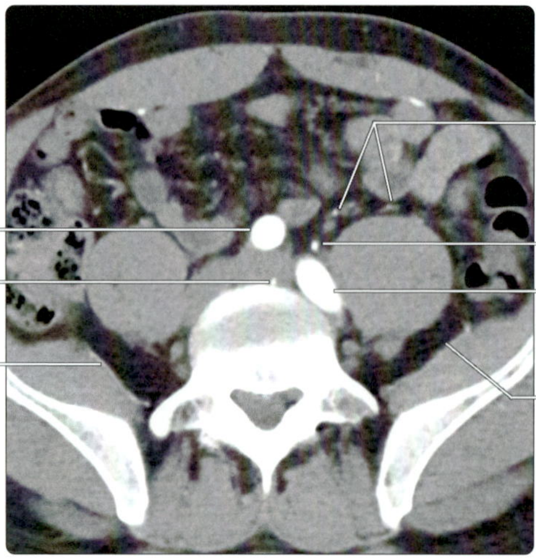

(Top) *Axial CECT demonstrates small jejunal and ileal branches of the SMA extending into the mesentery. The IMA arises from the anterior aspect of the abdominal aorta (usually near the L3 level) and courses downward slightly to the left where it supplies blood to the distal colon (beyond the splenic flexure) and rectum.* **(Middle)** *Axial CECT slightly more inferiorly shows the aortic bifurcation into right and left common iliac arteries. The IMA has begun giving off branches, including the left colic artery and (more inferiorly) sigmoid arteries, which extend into the sigmoid mesocolon. The median sacral artery extends off the posterior aspect of the aortic bifurcation and proceeds inferiorly along the anterior aspect of the spine and scrum.* **(Bottom)** *Axial CECT demonstrates the 1st branch of the posterior division of the internal iliac arteries, the iliolumbar arteries, which extend anteriorly/superiorly toward the anterior/inferior iliac spine. Small sigmoid arteries are visible extending downward toward the sigmoid colon.*

CT, ABDOMINAL AND PELVIC ARTERIAL VASCULATURE

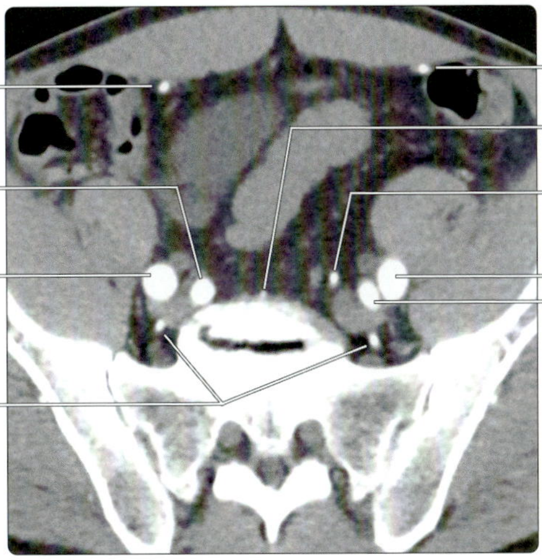

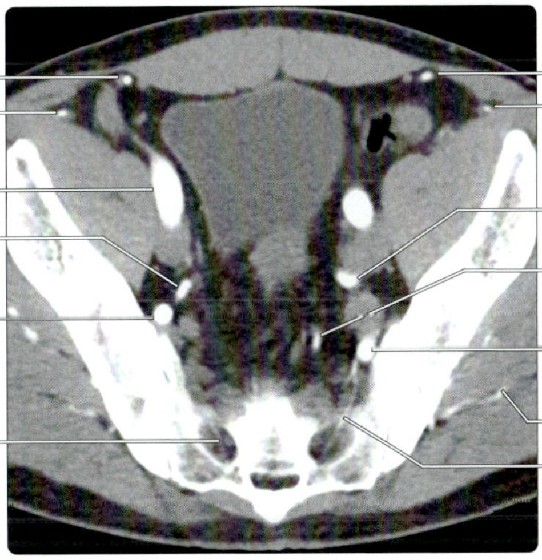

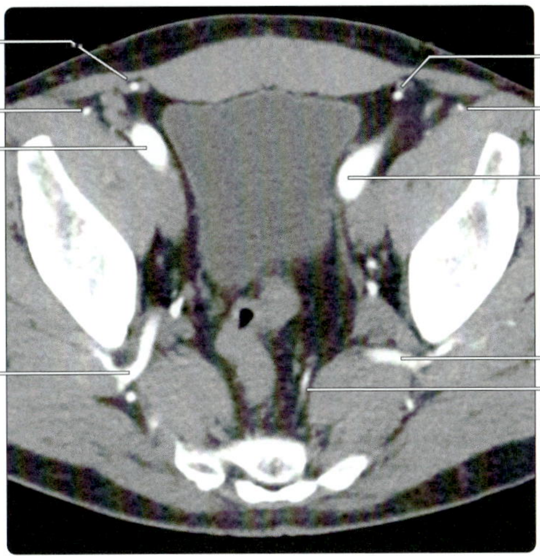

(Top) *The common iliac arteries have bifurcated into the external and internal iliac arteries. The iliolumbar arteries, arising from the posterior division of the internal iliacs, are still visualized. The median sacral artery continues to proceed downward along the anterior aspect of the sacrum. Note the course of the inferior epigastric arteries, branches of the external iliac arteries, along the anterior abdominal wall.* **(Middle)** *Another branch of the posterior division of the internal iliac arteries now comes into view with the lateral sacral arteries extending into the sacral foramina. The 2 branches of the external iliac arteries are visible on this image, the deep circumflex iliac artery (which extends upward and laterally) and the inferior epigastric artery (which extends medially and upward).* **(Bottom)** *The superior gluteal artery is the terminal branch of the posterior division of the internal iliac artery and gives off superficial and deep branches that supply the gluteal muscles. The IMA continues as the superior rectal artery as it proceeds toward the rectum.*

CT, ABDOMINAL AND PELVIC ARTERIAL VASCULATURE

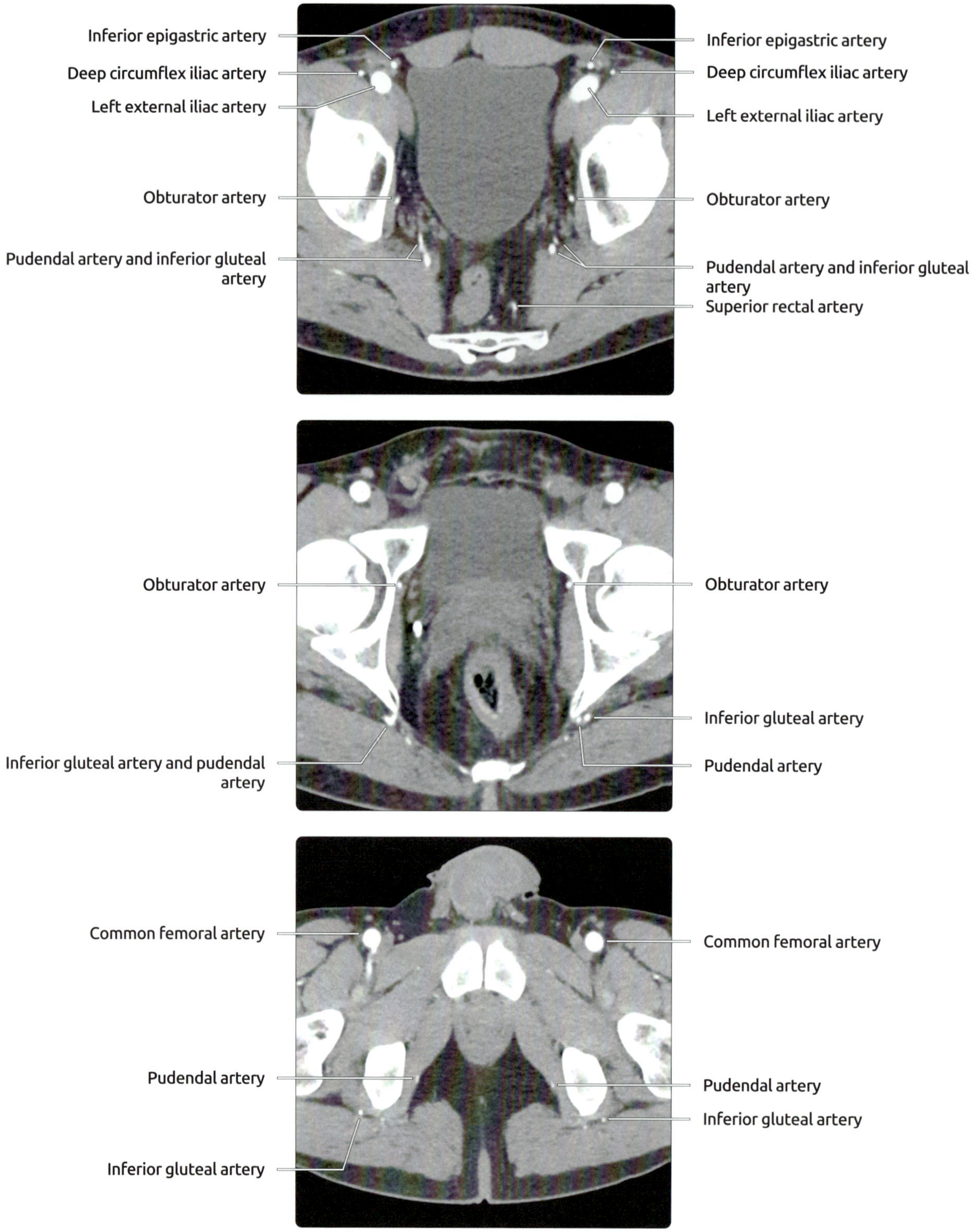

(Top) *A more inferior image demonstrates several branches of the anterior division of the internal iliacs, including the obturator arteries extending along the pelvic sidewalls toward the obturator foramen and the pudendal and inferior gluteal arteries, which proceed together through the greater sciatic foramen below the piriformis muscle.* **(Middle)** *The obturator arteries continue toward the obturator foramen. The inferior gluteal artery begins to extend laterally to supply the gluteal soft tissues and musculature, while the pudendal nerve will progress anteriorly toward the perineum.* **(Bottom)** *Note the anterior course of the pudendal artery and lateral course of the inferior gluteal artery.*

CT, ABDOMINAL AND PELVIC ARTERIAL VASCULATURE

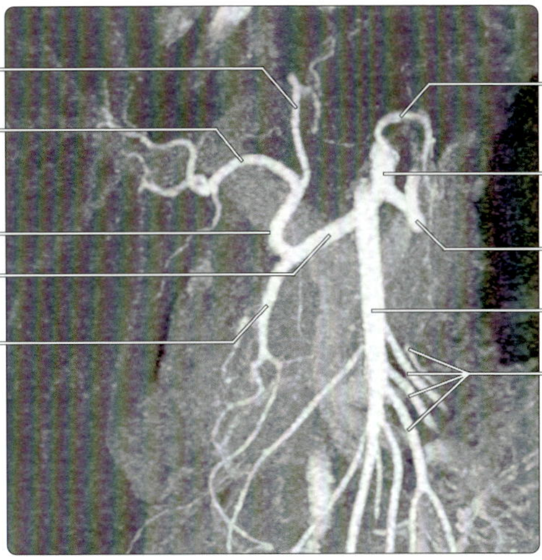

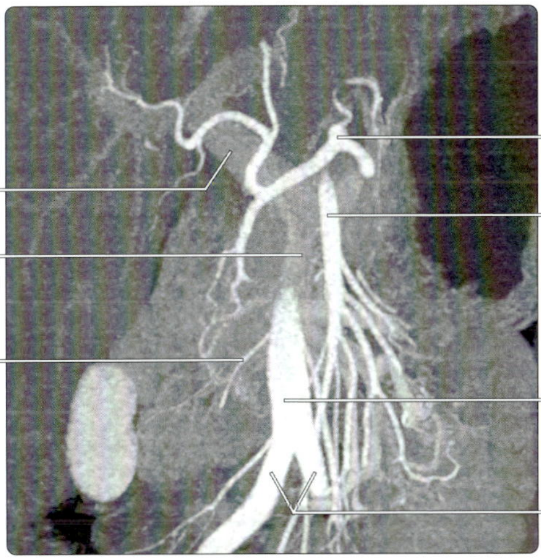

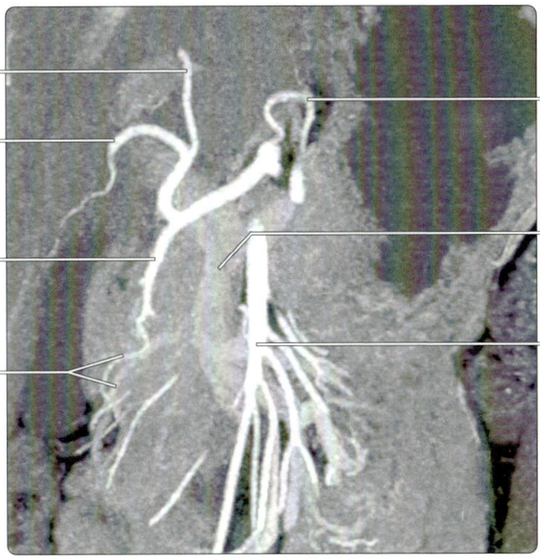

(Top) *First of 3 CECT images with 3D reconstruction is shown. MIP in the coronal plane shows the overlapping branches of the celiac artery and SMA. An arbitrarily set thickness of the plane of reconstruction results in some peripheral portions of the SMA and splenic artery being excluded from the image.* **(Middle)** *A more ventral plane of reconstruction shows peripheral branches of the SMA and also some of the distal aorta and common iliac arteries. The portal vein and its major branches are seen faintly, as these CT images were acquired in the predominantly arterial phase after IV administration of contrast.* **(Bottom)** *A more ventral coronal reconstruction shows additional branches of the SMA and celiac artery. The GDA is seen to give off the superior pancreaticoduodenal arteries, which anastomose with the inferior pancreaticoduodenal arteries that arise from the SMA.*

CT, HEPATIC ARTERY VARIANTS

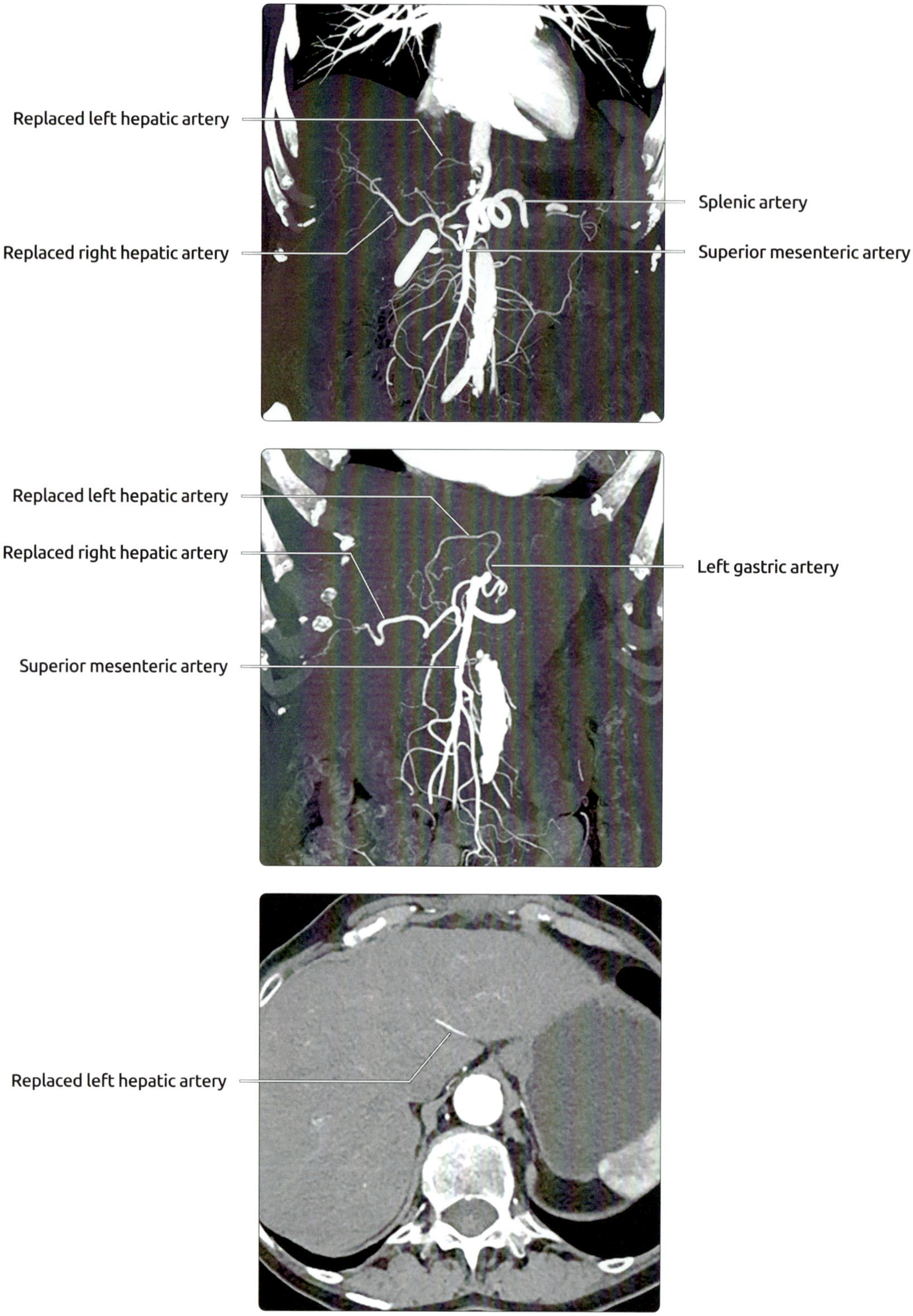

(Top) *Coronal CECT with MIP reconstruction demonstrates a replaced right hepatic artery arising from the SMA as well as a smaller replaced left hepatic artery arising from the left gastric artery, both of which are very common variants, which are quite important to report in patients who will be undergoing hepatobiliary or transplant surgery.* **(Middle)** *Coronal CECT with MIP reconstruction demonstrates the typical courses of replaced right and left hepatic arteries.* **(Bottom)** *Axial CECT demonstrates a replaced left hepatic artery in the fissure of the ligamentum venosum. Any artery present in this fissure must represent an aberrant left hepatic artery (either replaced or accessory), as no artery should normally run in this space.*

CT, HEPATIC ARTERY VARIANTS

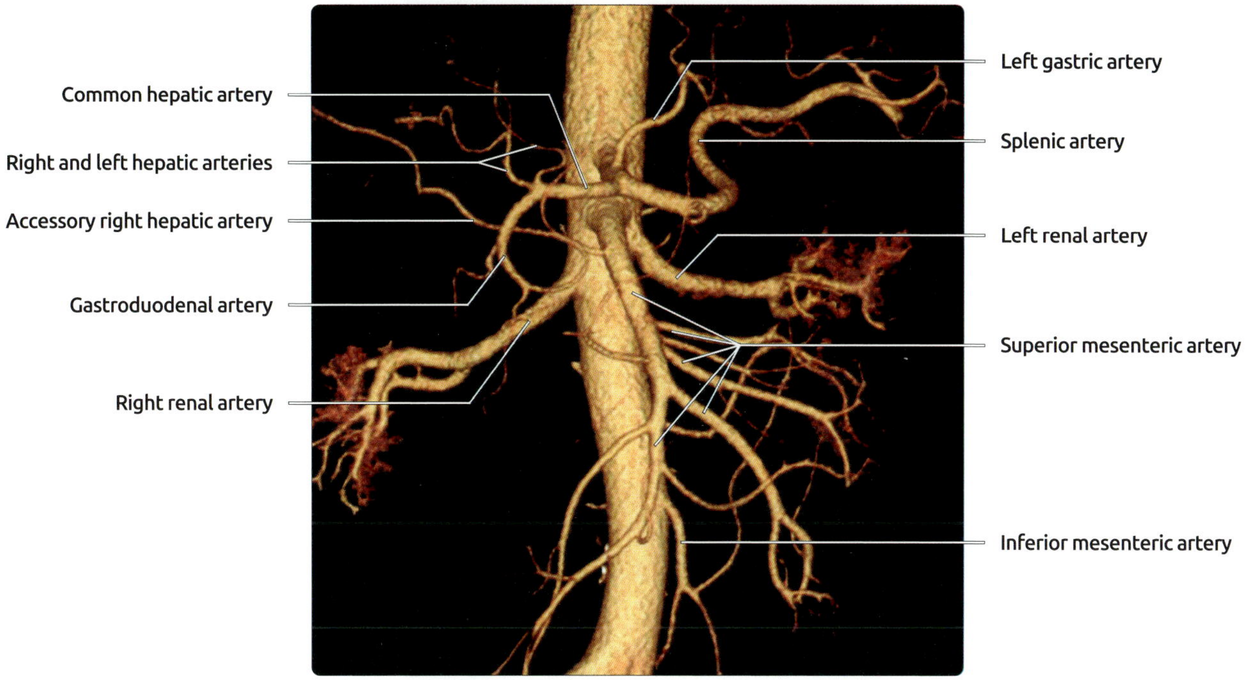

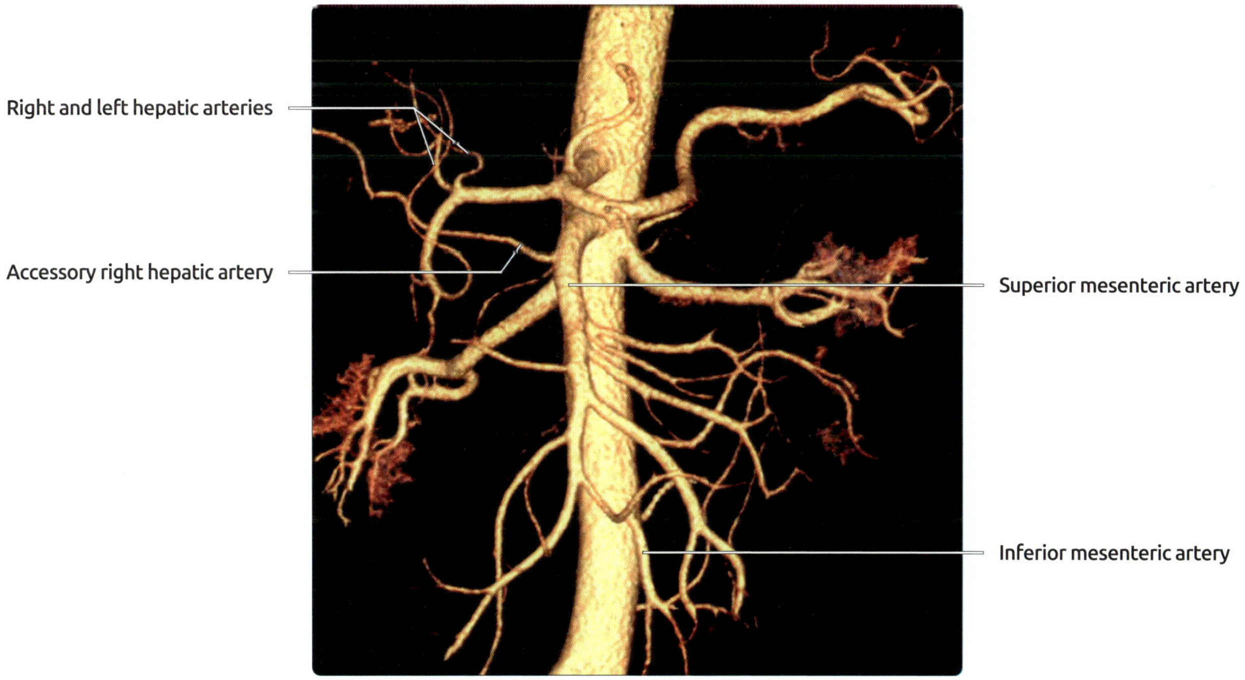

(Top) *First of 2 CECT images with volume-rendered reconstruction demonstrates the conventional origins of all the major abdominal aortic visceral branches. The right hepatic artery, arising from the proper hepatic branch of the celiac artery, is smaller than usual because there is an accessory right hepatic artery arising from the SMA, a common variant.* **(Bottom)** *Oblique volume-rendered reconstruction of the same CTA clearly shows the origin of the accessory right hepatic artery from the SMA.*

CT, RENAL AND HEPATIC ARTERY VARIANTS

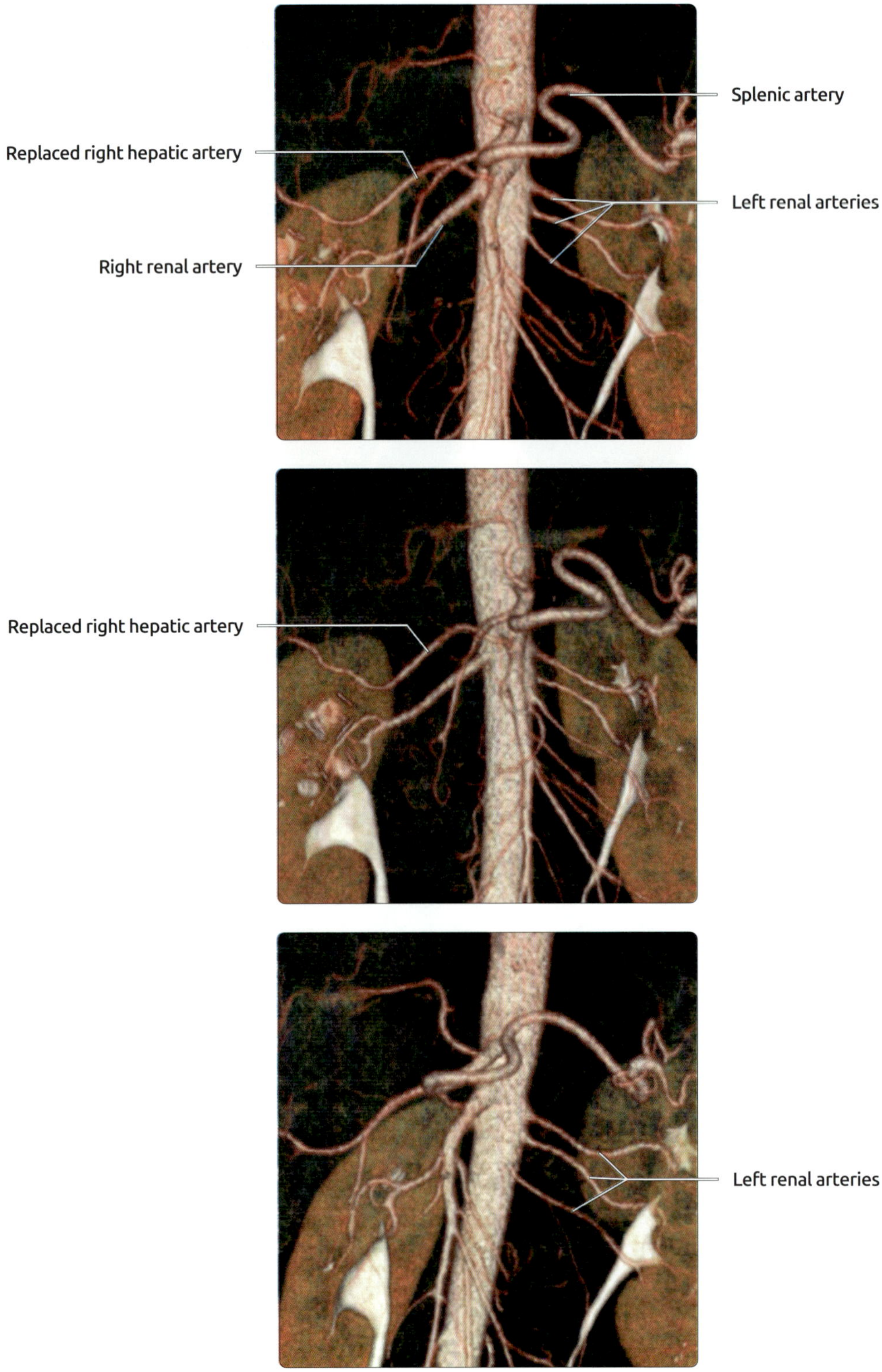

(Top) *First of 3 images of a CT arteriogram with volume-rendered reconstruction is shown. There are 3 left renal arteries, each having a separate origin from the aorta. A single right renal artery is present. The right hepatic artery arises from the SMA, rather than the celiac artery, a common variation called a "replaced" hepatic artery.* **(Middle)** *In this oblique plane, it is easier to recognize the replaced hepatic artery but more difficult to distinguish the multiple left renal arteries from nearby branches of the SMA.* **(Bottom)** *The 3 left renal arteries are best shown in this obliquity. Multiple renal arteries are a quite common anatomic variant, particularly common with accessory arteries supplying the renal poles, and are important to note in patients undergoing renal surgery or transplant donation.*

CT, CELIAC ARTERY VARIANTS

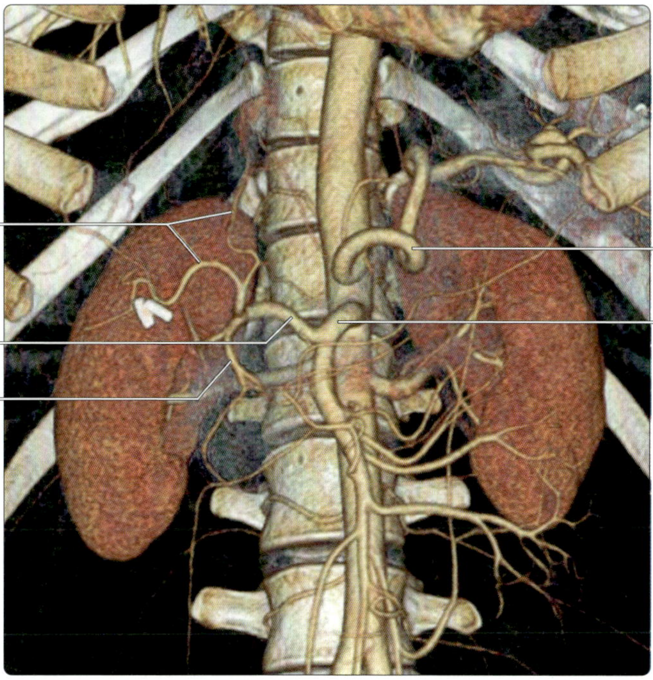

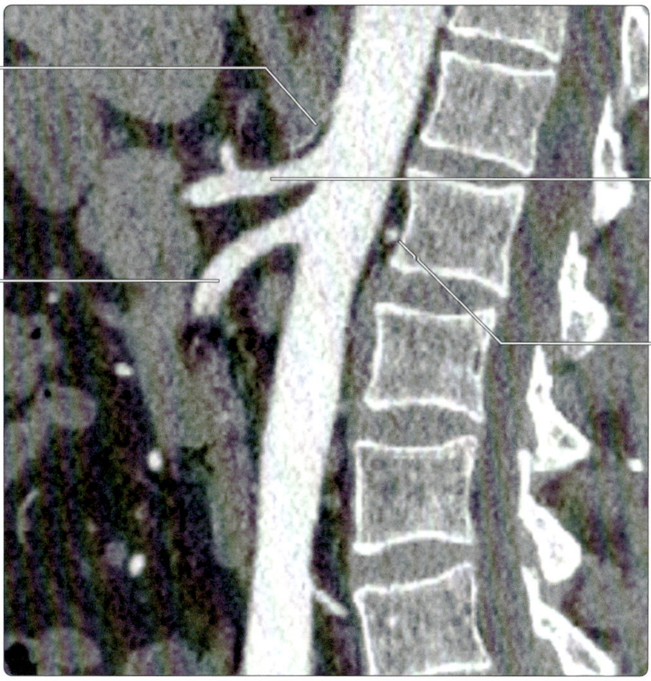

(Top) *Coronal volume-rendered CECT shows the entire common hepatic artery arising from the SMA. The left gastric artery also has a separate origin from the aorta, though difficult to perceive on this image. The "celiac trunk" in this patient consists only of the splenic artery.* **(Bottom)** *Sagittal CECT with MIP reconstruction in the midline shows the small left gastric artery with its separate origin from the aorta. The celiac trunk, essentially the splenic artery in this variation, and the origin of the SMA are shown.*

ABDOMINAL AND PELVIC ARTERIAL AND VENOUS VASCULATURE

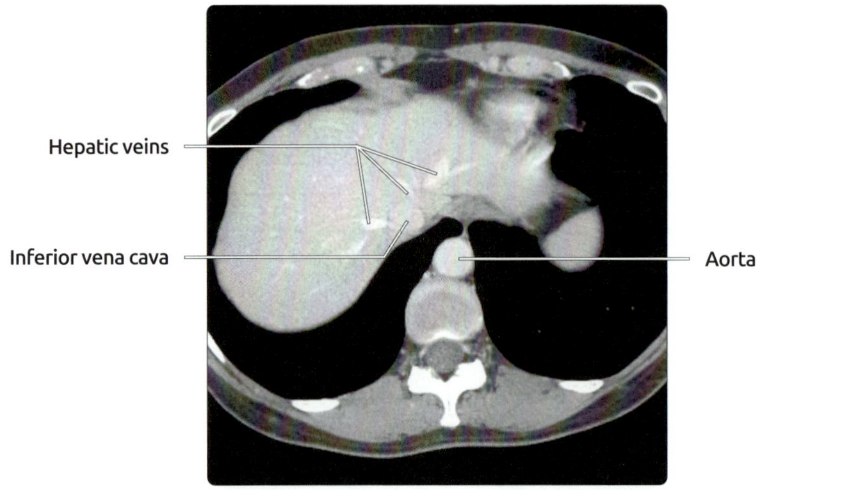

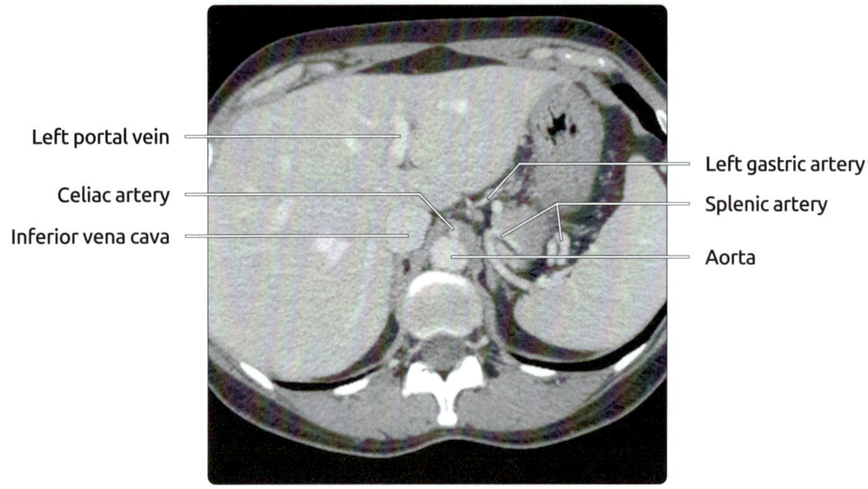

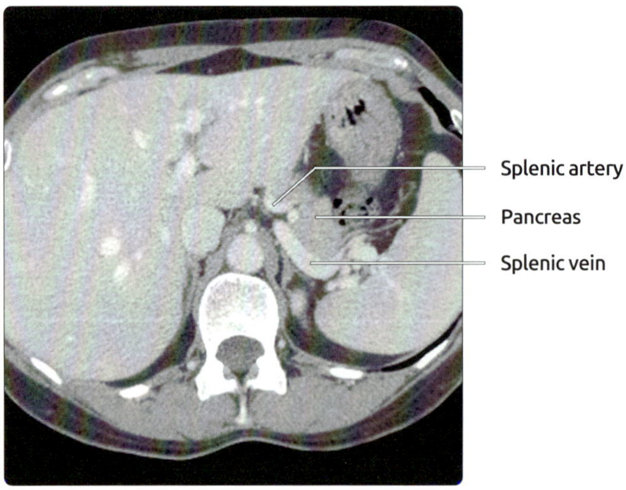

(Top) *First of 12 axial CT sections details the arteries and veins of the abdomen and pelvis. The hepatic veins (usually 3 major branches) join the IVC just below the diaphragm.* **(Middle)** *The celiac artery is the 1st (most cephalic) midline branch of the abdominal aorta. Its 3 major branches are the left gastric, splenic, and common hepatic arteries.* **(Bottom)** *The splenic vein and artery run along the body of the pancreas. The artery is more tortuous (curved) than the vein and curves into and out of the plane of axial sections, while the splenic vein usually lies in a straight horizontal plane running in a groove along the posterior margin of the pancreatic body and tail.*

ABDOMINAL AND PELVIC ARTERIAL AND VENOUS VASCULATURE

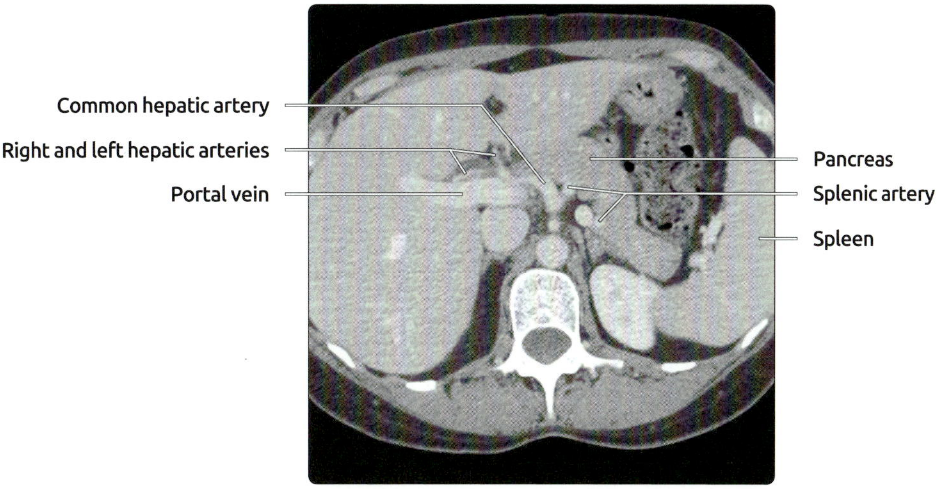

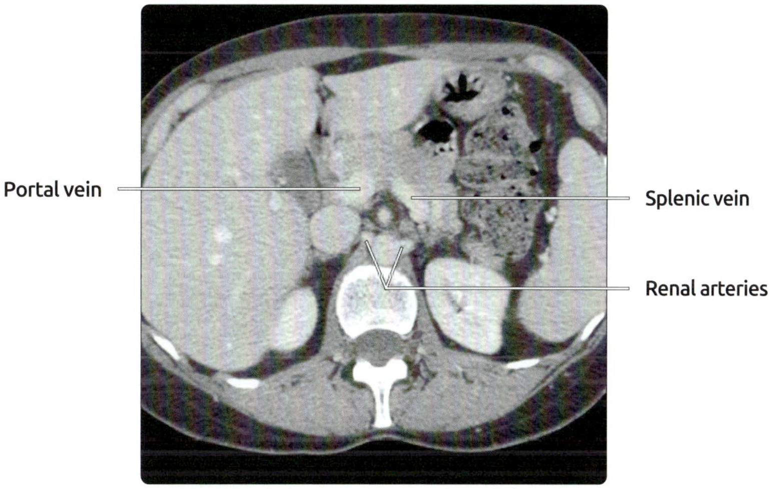

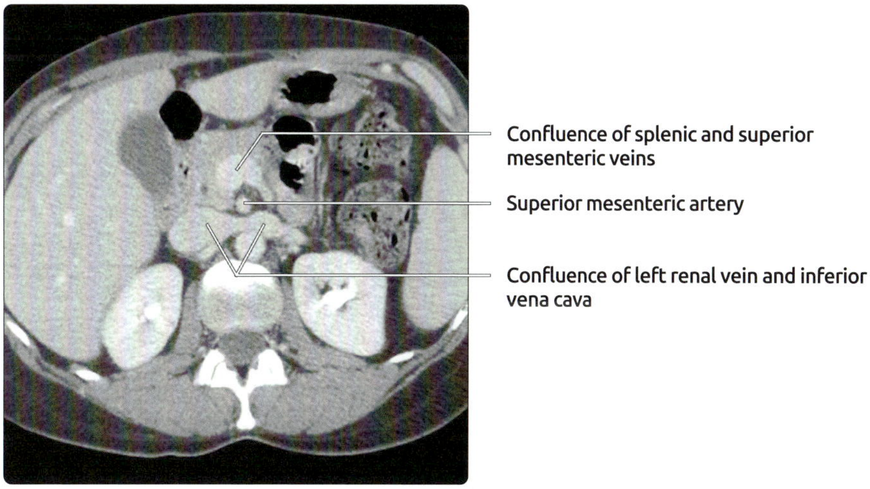

(Top) *The main branches of the celiac axis (with many variations) are the common hepatic, left gastric, and splenic arteries. The hepatic artery is smaller than, and lies ventral to, the portal vein at the porta hepatis.* **(Middle)** *The splenic and superior mesenteric veins join to form the main portal vein. The renal arteries are the 1st major paired lateral branches of the abdominal aorta. The inferior phrenic arteries are the 1st paired lateral branches but are much smaller, usually much less apparent on imaging, and of less clinical importance.* **(Bottom)** *The left renal vein usually crosses in front of the aorta and behind the SMA to join the IVC.*

ABDOMINAL AND PELVIC ARTERIAL AND VENOUS VASCULATURE

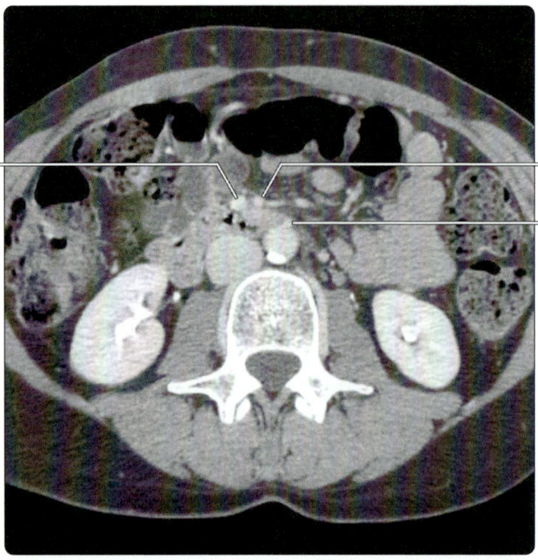

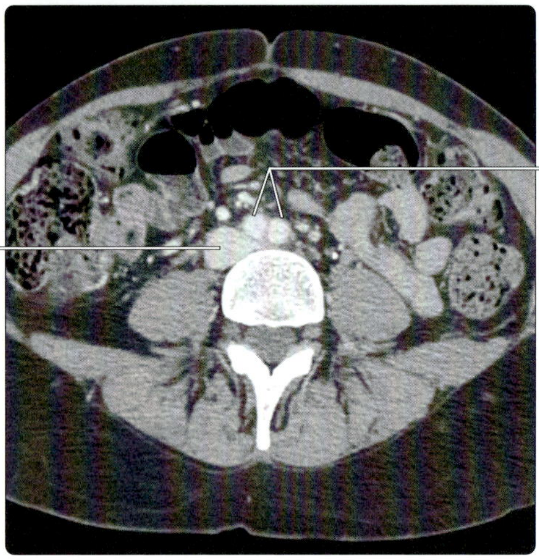

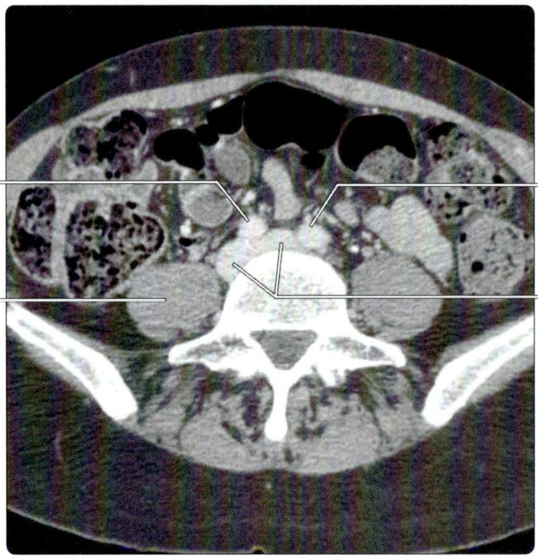

(Top) *The SMA usually lies to the left of and is smaller than the superior mesenteric vein. The IMA arises from the aorta just above the aortic bifurcation and supplies the left side of the colon.* **(Middle)** *The aorta bifurcates into the common iliac arteries at the level of the 4th lumbar vertebra ~ 2 cm higher (more cephalic) than the confluence of the common iliac veins, which join to form the IVC.* **(Bottom)** *The confluence of the common iliac veins is shaped like a peanut and lies behind the right common iliac artery.*

ABDOMINAL AND PELVIC ARTERIAL AND VENOUS VASCULATURE

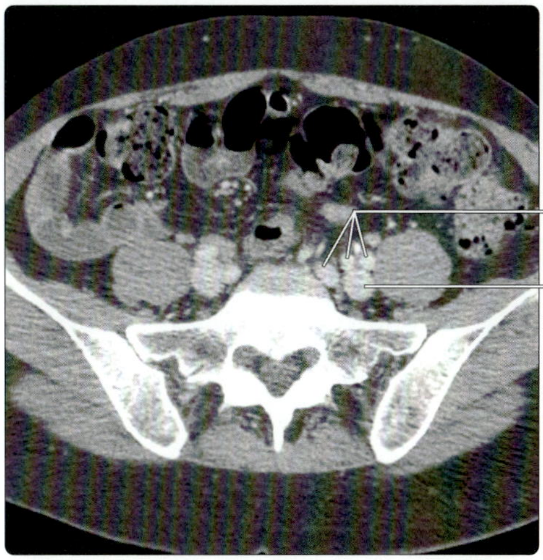

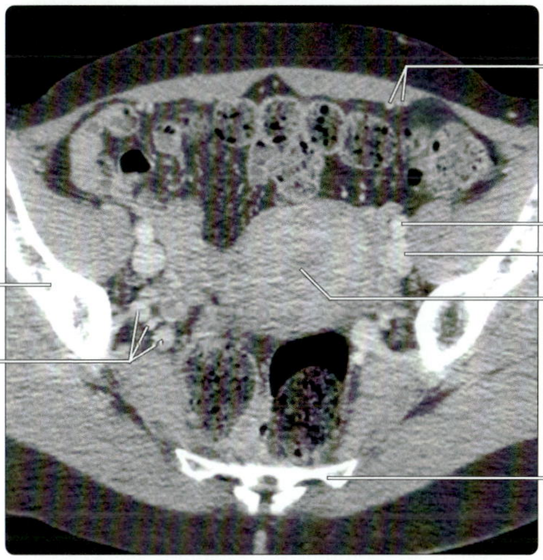

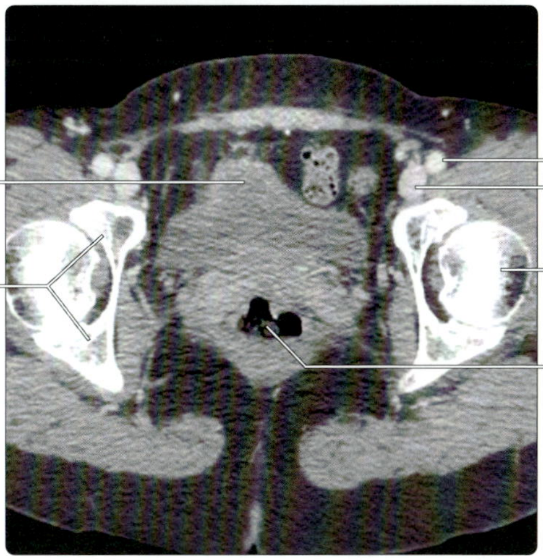

(Top) *The internal iliac artery gives off numerous small branches to pelvic muscles and viscera, only some of which can be consistently identified on CT.* **(Middle)** *The major pelvic branches of the external iliac artery are the inferior epigastric and deep circumflex iliac arteries, which feed the anterior abdominal wall muscles from below and freely anastomose with branches of the superior epigastric and musculophrenic arteries. These are the terminal branches of the internal mammary (thoracic) artery and constitute important collateral pathways in the event of occlusion of the iliac arteries (a relatively common result of atherosclerosis).* **(Bottom)** *As the external iliac artery passes under the inguinal ligament, it becomes the femoral artery, which supplies the lower extremity as well as branches to the anterior abdominal wall and external genitalia. The femoral artery lies lateral to the vein and is smaller in diameter.*

CORONAL ABDOMINAL VEINS

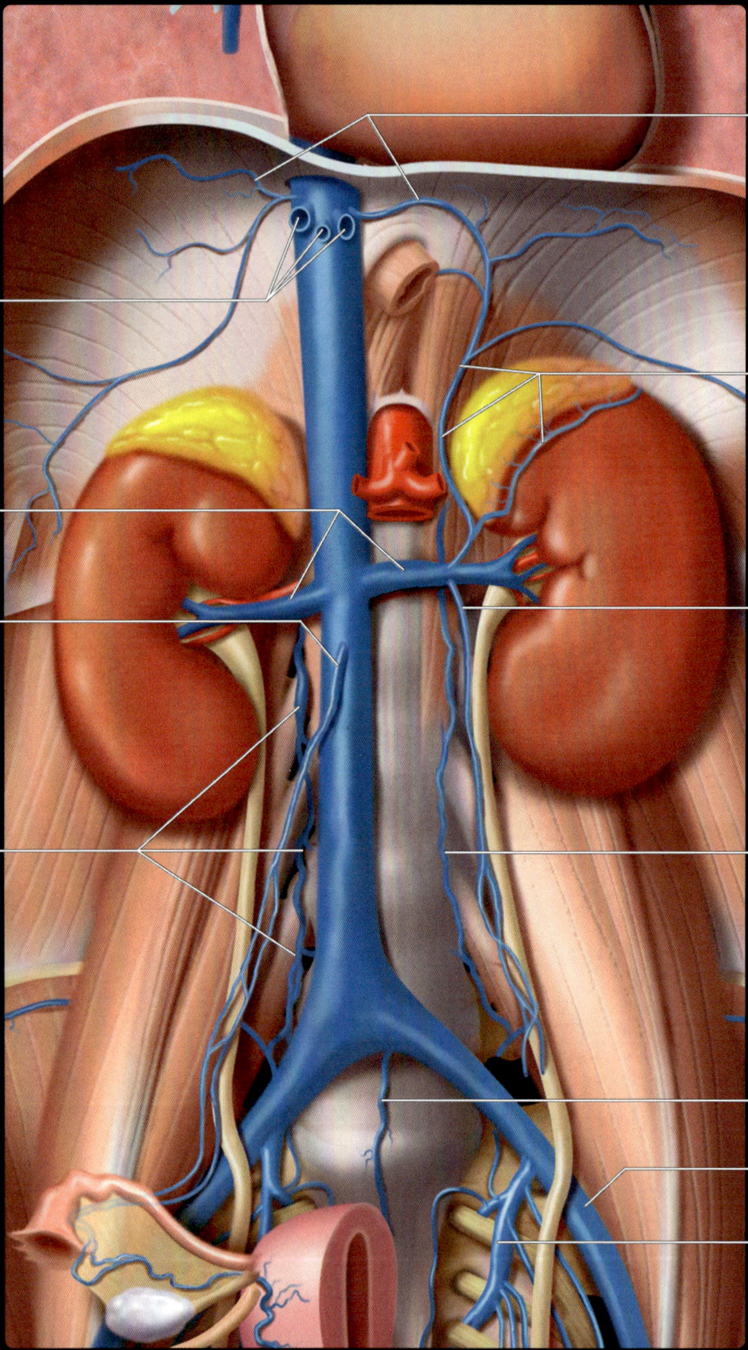

The IVC begins at the L5 level with the confluence of the common iliac veins, which are themselves the result of the confluence of the internal and external iliac veins. Note the ascending lumbar veins, which anastomose freely between the IVC, the azygous and hemiazygos veins, and the renal veins. These are an important pathway for collateral flow in the event of obstruction of the IVC or one of its major tributaries, and these veins play an important role in spread of tumor and infection from the pelvis and spine to the thorax, upper spine, and brain. The right renal vein rarely receives tributaries, while the left receives the gonadal, adrenal, and lumbar veins. The left adrenal vein also anastomoses with the inferior phrenic vein. The hepatic veins return blood from the liver as they join the IVC just below its hiatus in the diaphragm at about the T8 level.

INFERIOR VENA CAVA DUPLICATION AND ANOMALIES

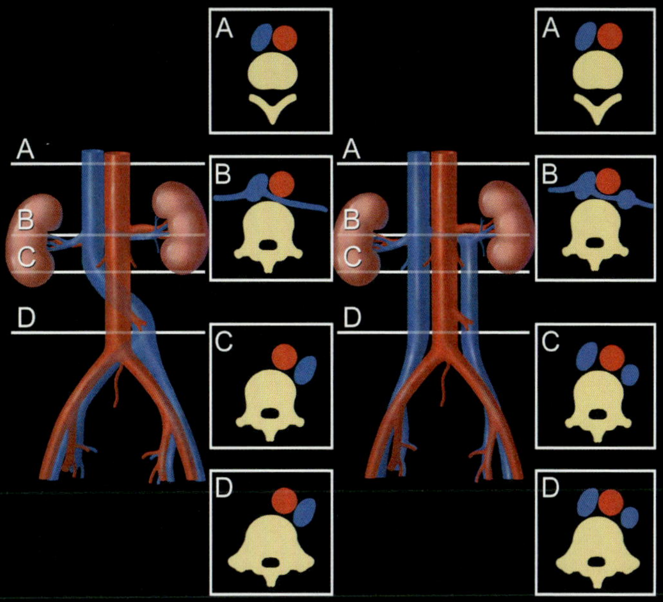

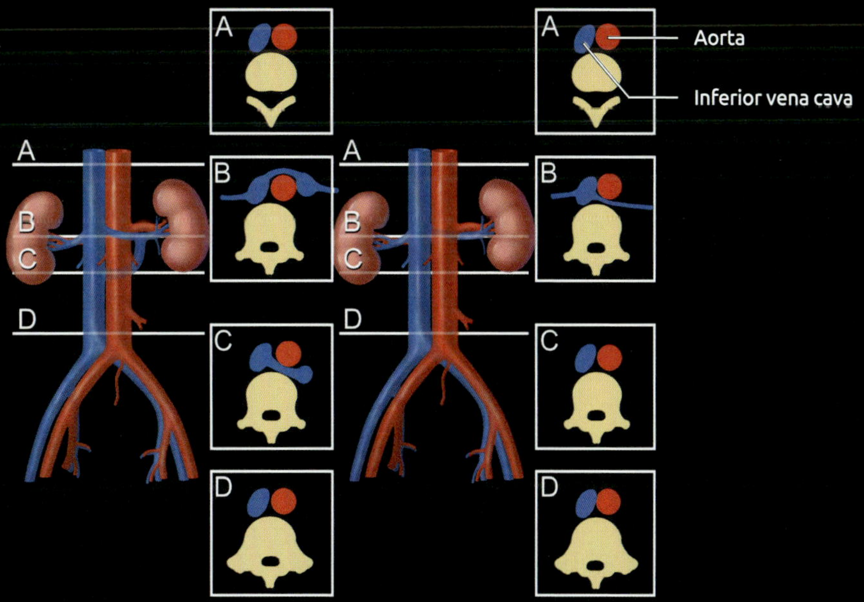

(Top) *Four graphics illustrate common variations of the IVC. The labeled lines on the frontal graphics correspond to the levels of the axial sections. The left graphic shows transposition of the IVC, in which the infrarenal portion of the IVC lies predominantly to the left side of the aorta. A more common anomaly is shown on the right graphic, a "duplication" of the IVC in which the left common iliac vein continues in a cephalad direction without crossing over to join the right iliac vein. Instead, it joins the left renal vein and then crosses over to the right. The suprarenal IVC has a conventional course and appearance.* **(Bottom)** *The left graphic shows a circumaortic left renal vein with the smaller, more cephalic vein passing in front of the aorta and the larger vein passing behind and caudal. The right graphic shows a completely retroaortic renal vein.*

VARIATION, LEFT-SIDED INFERIOR VENA CAVA (TRANSPOSITION)

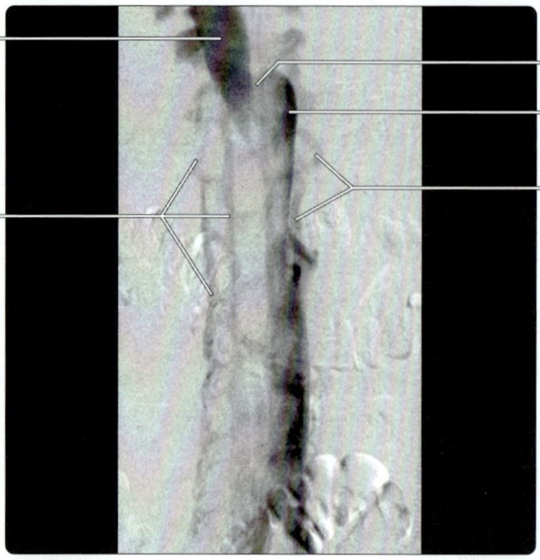

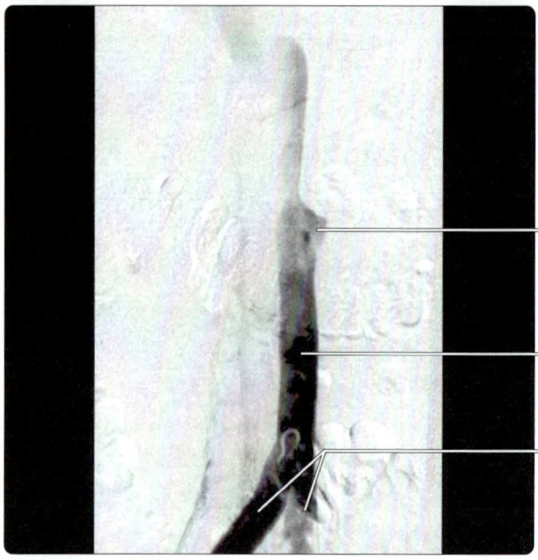

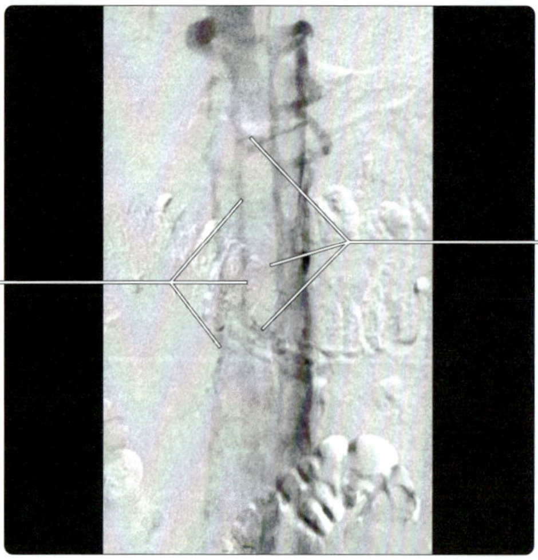

(Top) *First of 3 images from a catheter injection of the IVC shows the IVC lying to the left of midline, an anomalous position. In the midabdomen, it crosses over to the right side behind the aorta, which results in compression of the IVC. The luminal narrowing results in increased flow through collateral venous channels, including the lumbar veins.* **(Middle)** *This frontal image in the same patient shows the common iliac veins and some retrograde opacification of the orifice of the left renal vein.* **(Bottom)** *A later film from the same study shows that contrast material has cleared from most of the IVC, while the collateral veins of the vertebral plexus and ascending lumbar veins remain opacified. Note the numerous connections between the venous plexus, which surrounds the vertebrae and the lumbar veins.*

VARIATION, LEFT-SIDED INFERIOR VENA CAVA (TRANSPOSITION)

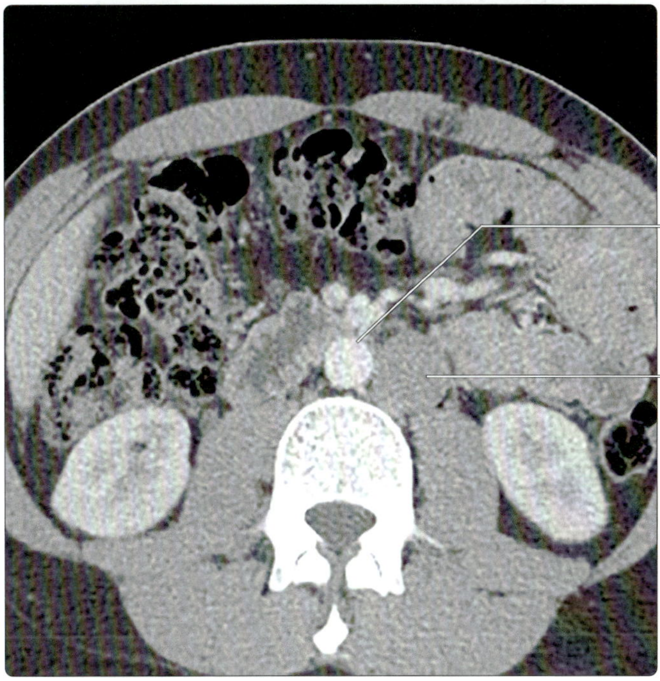

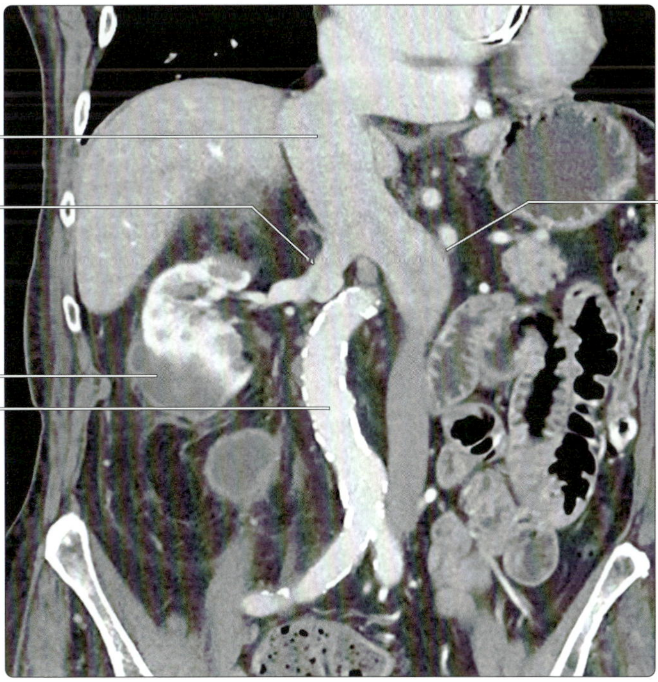

(Top) *Axial CECT demonstrates the IVC lying in an abnormal position immediately to the left (instead of the right) of the aorta.* **(Bottom)** *Coronal CECT in the same patient demonstrates the classic findings of a left-sided IVC with the IVC extending upward in the left abdomen and finally turning into the right abdomen where it joins with the right renal vein. This patient incidentally has a right-sided perinephric abscess.*

VARIATION, INFERIOR VENA CAVA DUPLICATION

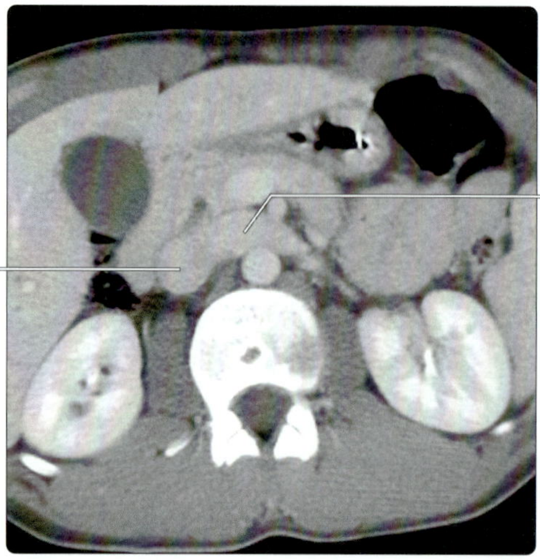

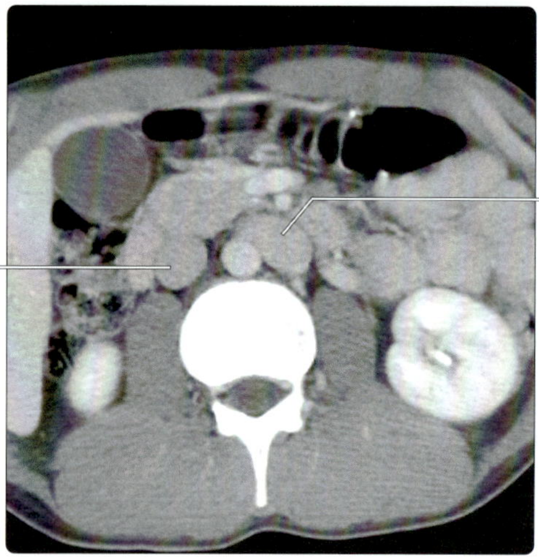

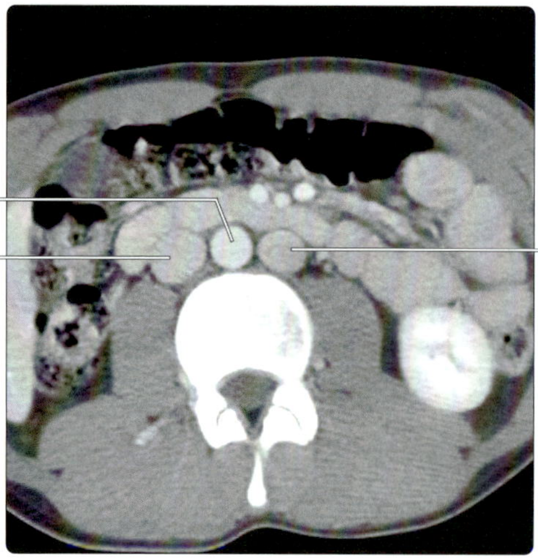

(Top) *First of 5 axial CECT images shows an enlarged left renal vein joining the IVC. From this level cephalad, the IVC had a normal caliber and course lying in its usual site to the right of the aorta.* **(Middle)** *More caudal CECT shows the infrarenal duplicated IVC joining the left renal vein.* **(Bottom)** *Caudal image below the kidneys shows both the left- and right-sided IVC. Failure to recognize this as a cylindrical structure in continuity with the common iliac vein caudally and the left renal vein cephalically might lead to a mistaken diagnosis of a paraaortic mass, such as lymphadenopathy.*

VARIATION, INFERIOR VENA CAVA DUPLICATION

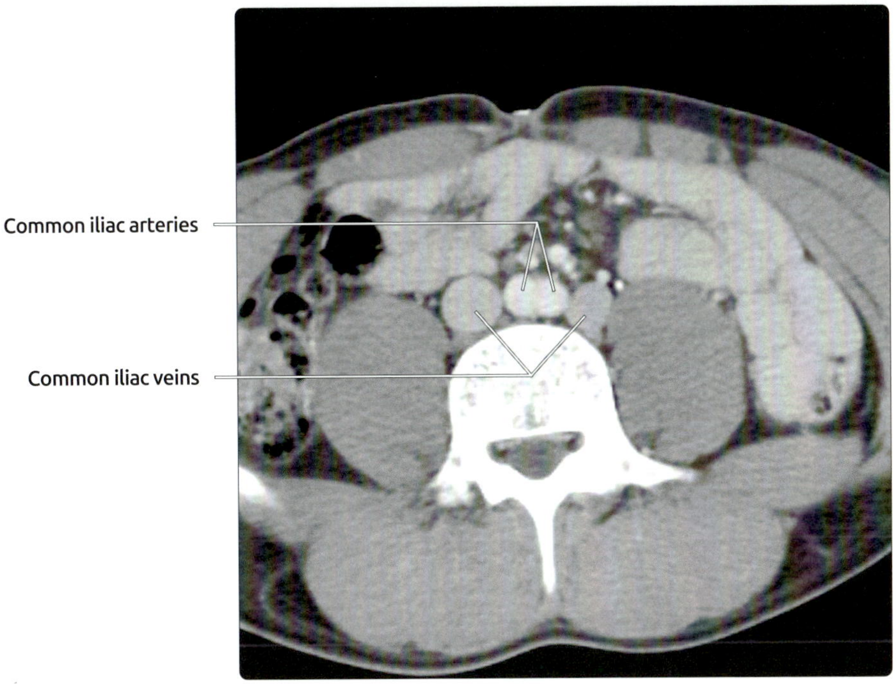

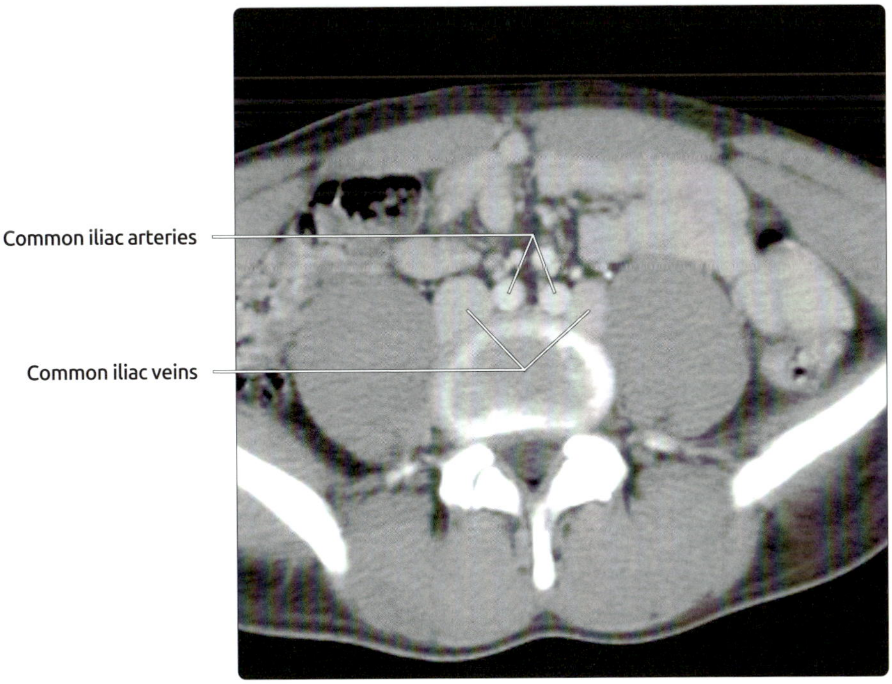

(Top) *More caudal image shows the bifurcation of the aorta into the common iliac arteries. At this level, the left common iliac vein would normally have crossed over to join the right common iliac vein to form the IVC.* **(Bottom)** *A more caudal section shows a normal appearance and location of the common iliac veins.*

VARIATION, CIRCUMAORTIC RENAL VEIN

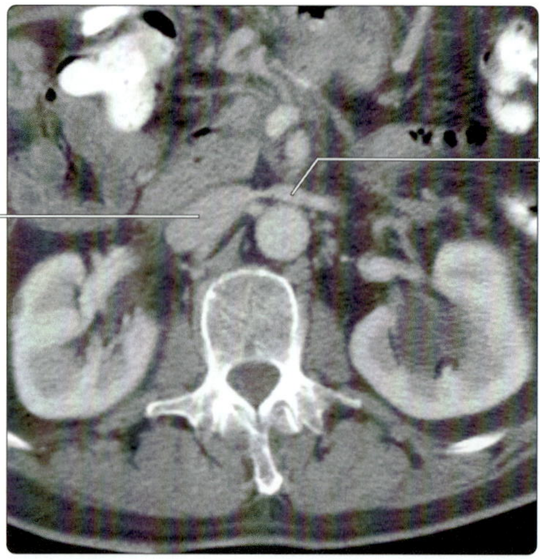

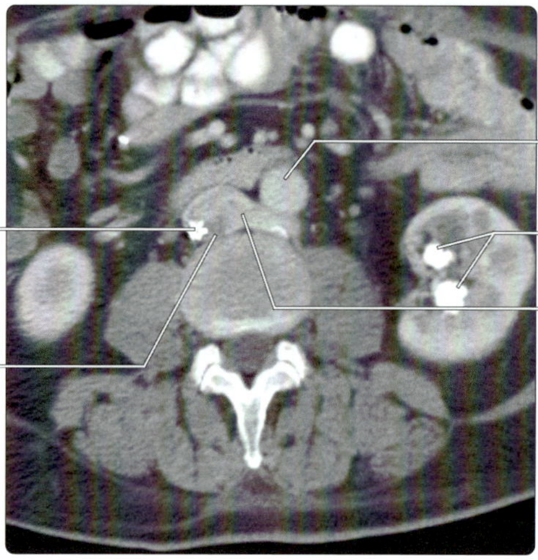

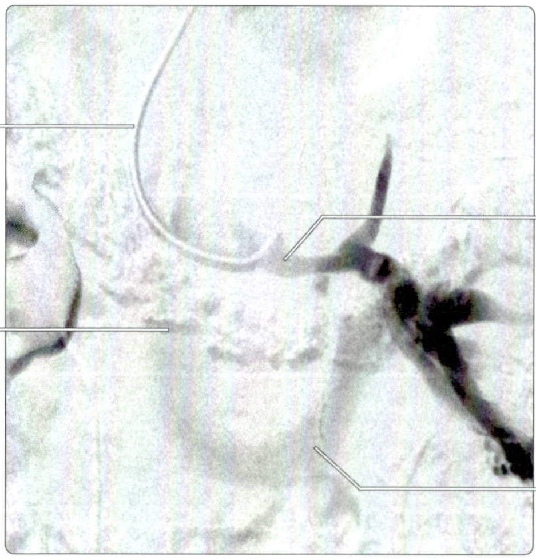

(Top) *Variations of the left renal vein are much more common than those affecting the right vein. One common anomaly is a circumaortic left renal vein in which the preaortic branch passes, as usual, between the aorta and the superior mesenteric vessels. This preaortic branch is smaller and more cephalad than the retroaortic component.* **(Middle)** *The retroaortic branch of the left renal vein is usually larger and lies ~ 1 vertebral body more caudal than the preaortic branch. This anomaly is of importance if surgery of the left kidney is being considered or if interventions on the IVC are planned (such as placement of an IVC filter in this patient with deep vein thrombosis).* **(Bottom)** *In anticipation of placing an IVC filter, a catheter was introduced into the IVC from an arm vein. The tip of the catheter is seen curving to the left and entering a renal vein branch that represents the preaortic branch seen on CT. Note its small size and cephalic position relative to the retroaortic branch, which is also opacified by injection of contrast material.*

VARIATION, AZYGOUS CONTINUATION OF INFERIOR VENA CAVA AND POLYSPLENIA

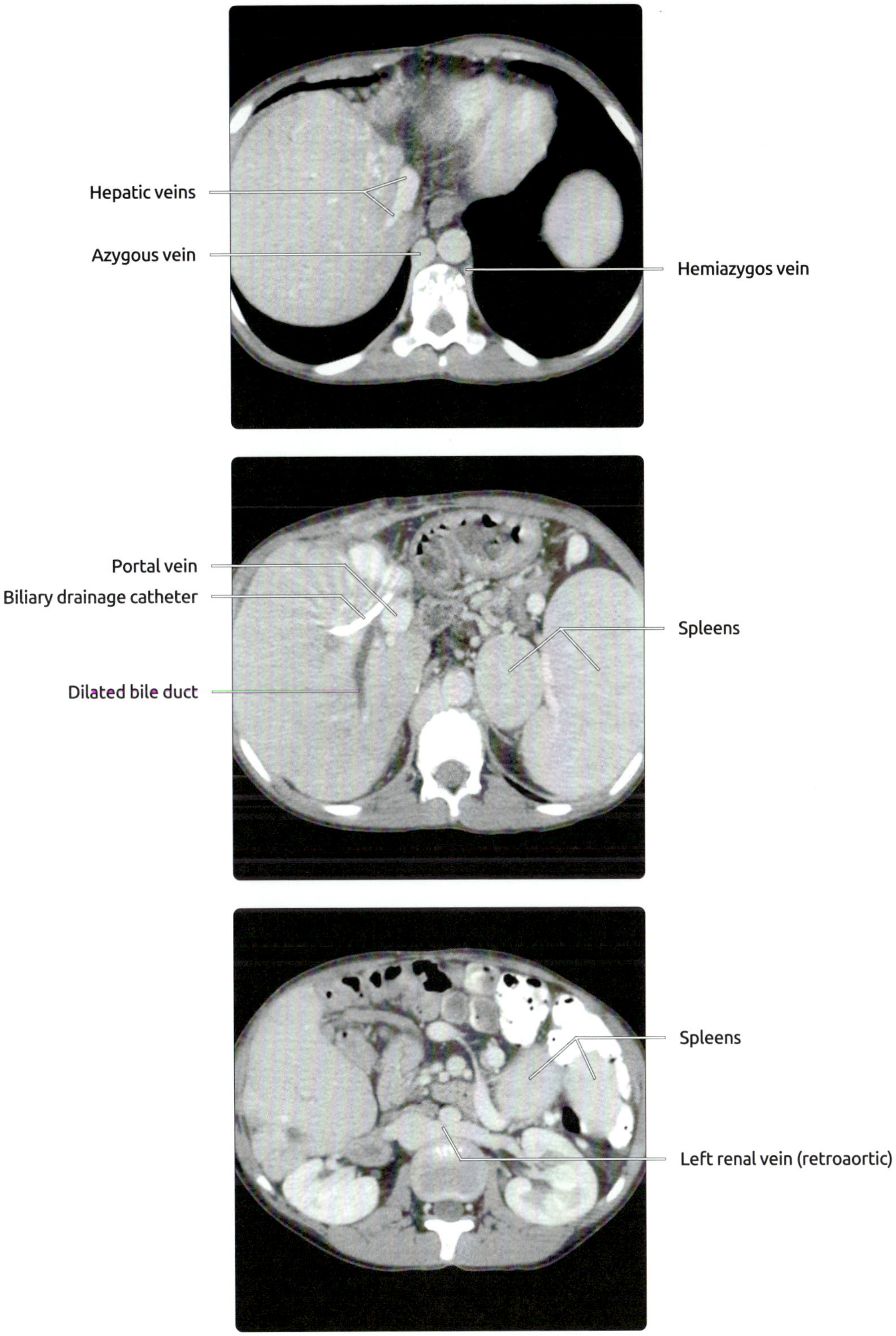

(Top) *First of 3 axial CECT images in a patient with polysplenia syndrome shows an enlarged azygous vein, which serves as the primary route of venous drainage of the abdomen and lower extremities in this condition. The IVC is absent between the level of the renal veins and the hepatic veins, which drain directly into the right atrium.* **(Middle)** *A more caudal CECT shows absence of the retrohepatic IVC. Dilation of the intrahepatic bile ducts is noted along with a biliary drainage catheter. Multiple splenic masses are noted and may number anywhere from 2-16 in patients with this disorder.* **(Bottom)** *An associated vascular anomaly is a retroaortic left renal vein. Above this level, the IVC is absent, and the azygous/hemiazygos and other collateral veins must return blood to the heart.*

VARIATION, ACCESSORY HEPATIC VEIN

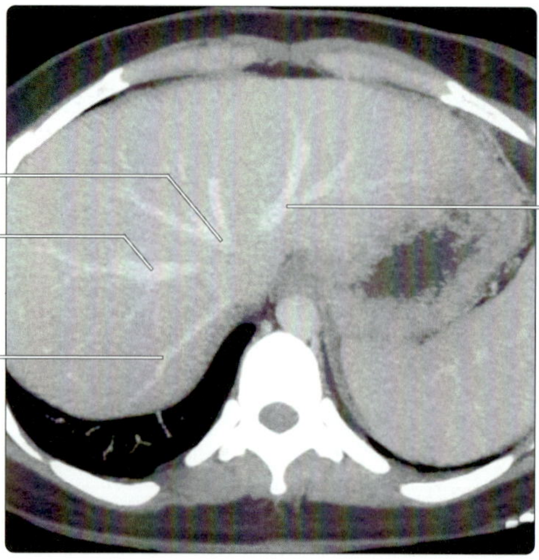

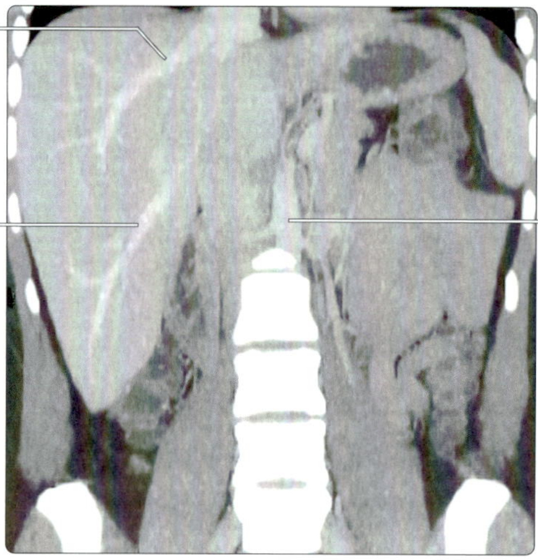

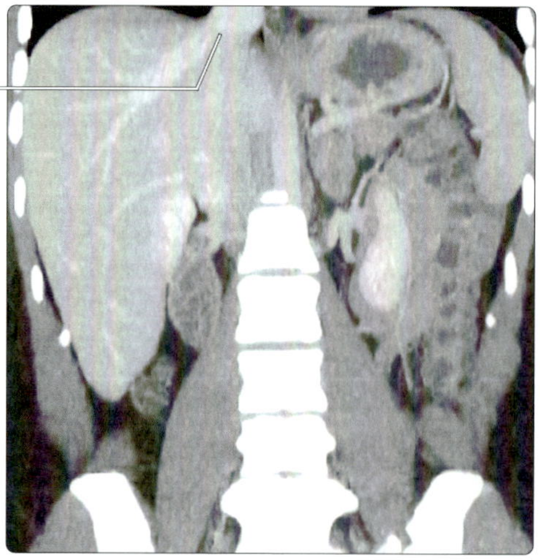

(Top) *First of 3 CECT images shows a variation in hepatic venous drainage into the IVC. Axial CECT with MIP reconstruction shows the 3 major hepatic veins (left, middle, and right), with an additional accessory right hepatic vein, which has a separate entry into the IVC.* **(Middle)** *Coronal CECT shows the long axis of the right and accessory right hepatic veins, which have separate orifices into the IVC. Venous anomalies, such as these, are important to recognize if hepatic surgery (e.g., partial hepatectomy) is being considered.* **(Bottom)** *The 3 major hepatic veins enter the IVC just below the diaphragm. They may have 3 separate orifices or may join together before entering the IVC. Commonly, the left and middle hepatic veins join and empty through a single orifice.*

VARIATION, INTERRUPTED INFERIOR VENA CAVA

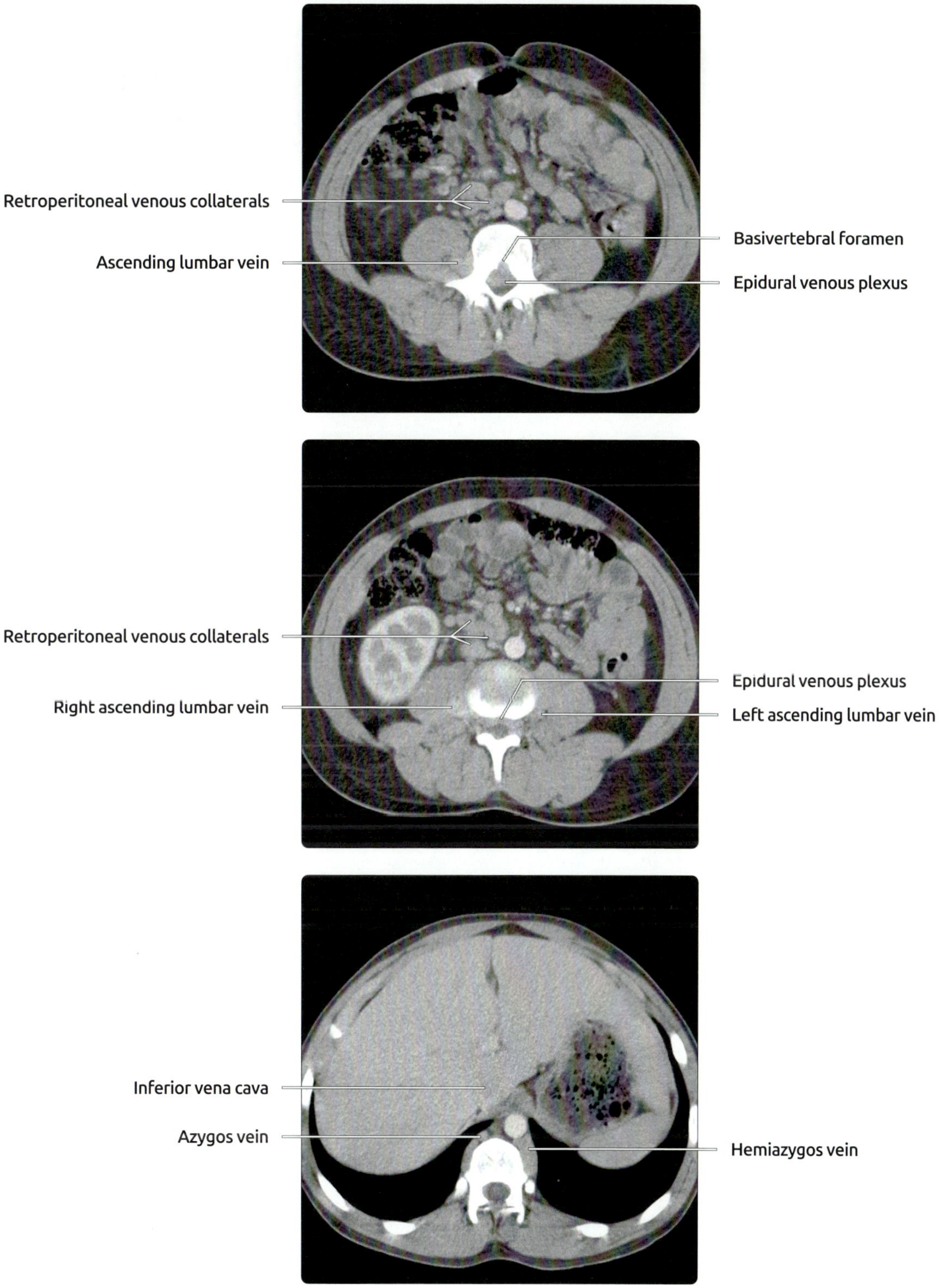

(Top) *Axial CECT in a young patient with interruption of the infrarenal IVC demonstrates that the IVC is absent and is replaced by multiple retroperitoneal venous collaterals. Note the widened basivertebral foramen due to dilatation of the basivertebral vein connecting the dilated epidural plexus to the ascending lumbar veins on both sides of the lumbar vertebral bodies.* **(Middle)** *The left lumbar vein joins a medial root from the left renal vein to form the hemiazygos vein, whereas the right ascending lumbar vein joins a medial root from the IVC to form the azygos vein. Note the dilated ascending lumbar veins.* **(Bottom)** *The hemiazygos vein crosses the midline at T8 to drain into the azygos vein, which drains into the superior vena cava. This is a crucial pathway for drainage of the lower part of the body in patients with IVC obstruction. Note the dilatation of both the azygos and hemiazygos veins carrying blood from the lower part of the body to drain into the superior vena cava.*

CORONAL LYMPH NODES

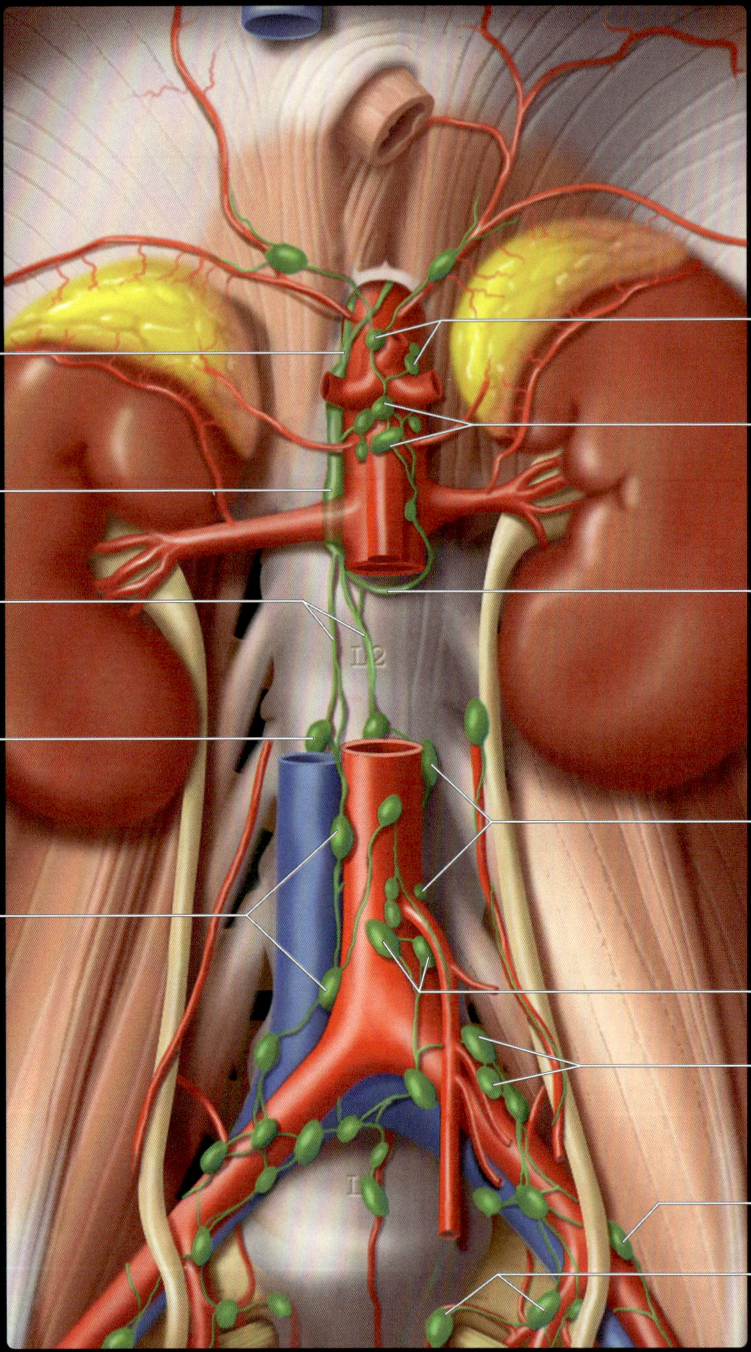

The major lymphatics and lymph nodes of the abdomen are located along, and share the same names as, the major blood vessels, such as the external iliac nodes, celiac nodes, and superior mesenteric nodes. The paraaortic and paracaval nodes are also referred to as the lumbar nodes and receive afferents from the lower abdominal viscera, abdominal wall, and lower extremities. They are frequently involved in inflammatory and neoplastic processes. The lumbar trunks join with an intestinal trunk (at about the L1 level) to form the cisterna chyli, which may be a discrete sac or a plexiform convergence. The cisterna chyli and other major lymphatic trunks join to form the thoracic duct, which passes through the aortic hiatus to enter the mediastinum. After picking up additional lymphatic trunks within the thorax, the thoracic duct empties into the left subclavian or innominate vein.

LYMPHATIC ANATOMY

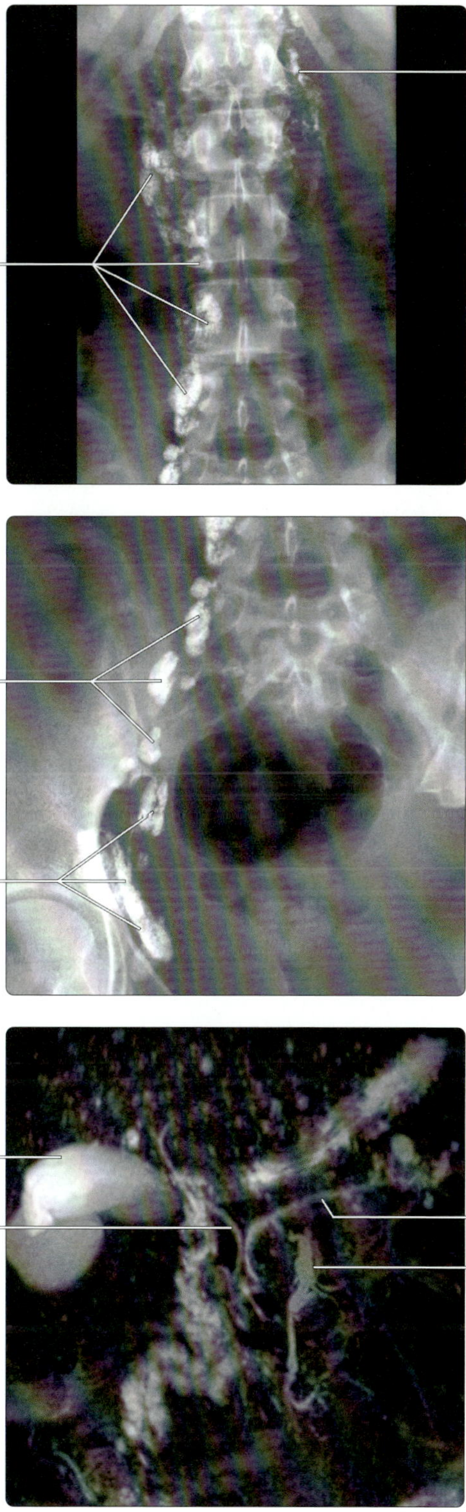

(Top) *First of 3 frontal images from a lymphangiogram in which iodinated oil is slowly infused into the lymphatics of the foot to produce opacification of the lymphatic channels and nodes is shown. Note the subcentimeter (short-axis) diameter of these normal retroperitoneal lymph nodes.* **(Middle)** *Lymphatic channels and lymph nodes parallel the course of major blood vessels and share similar names, such as these common iliac node. With the availability of CT, MR, and PET, lymphangiograms are performed much less frequently than in the past.* **(Bottom)** *The cisterna chyli is a sac-like structure located in the a retroperitoneum anterior to the 2nd lumbar vertebra. It is part of the lymphatic system and acts as a storage area for lymph that receives lymphatic flow from lumbar and intestinal lymphatic trunks and serves as the origin of the thoracic duct. On MR, it is typically seen as a well-defined, fluid-filled structure with a fluid signal intensity on T2 MR images.*

POSTLYMPHANGIOGRAM CT

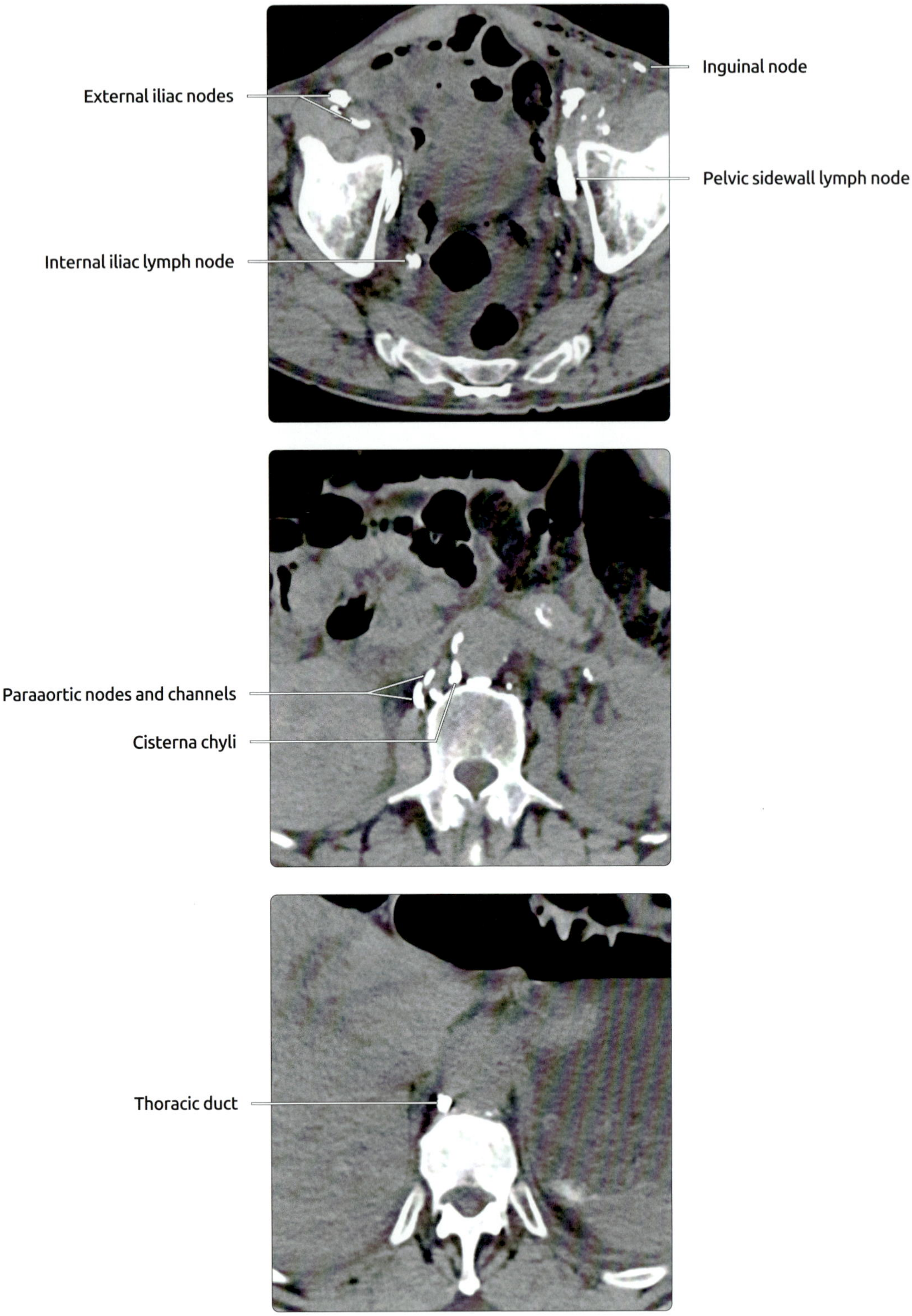

(Top) *Axial NECT acquired shortly after a conventional lymphangiogram was performed shows high-attenuation contrast collecting within inguinal and iliac lymph nodes and lymphatic channels. These nodes and channels run along the course of major vessels and are often similarly named.* **(Middle)** *Axial CT acquired slightly more cephalad now shows multiple paraaortic nodes and lymphatic channels filled with lymphangiogram contrast. The cisterna chyli is seen at this level and is formed by the confluence of intestinal lymphatic trunks and the right and left lumbar lymphatic trunks.* **(Bottom)** *Most caudal image demonstrates high-attenuation contrast material extending upward along the right side of the aorta into the thorax representing the thoracic duct (which is found superior to the L1 level). On routine studies, the cisterna chyli and thoracic duct can sometimes be confused for a node or cyst, particularly when large or distended.*

NODAL ANATOMY OF PELVIS

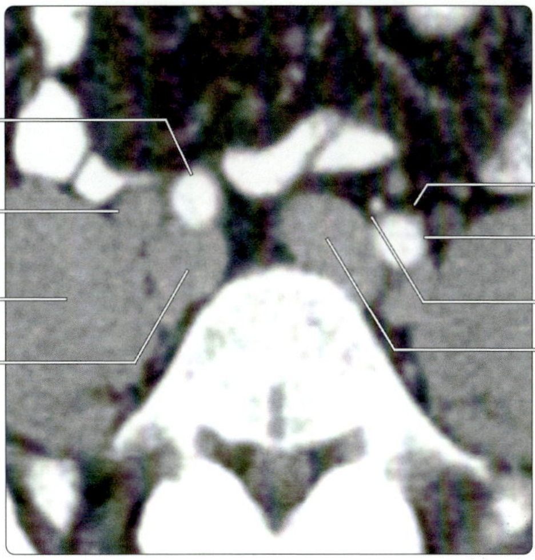

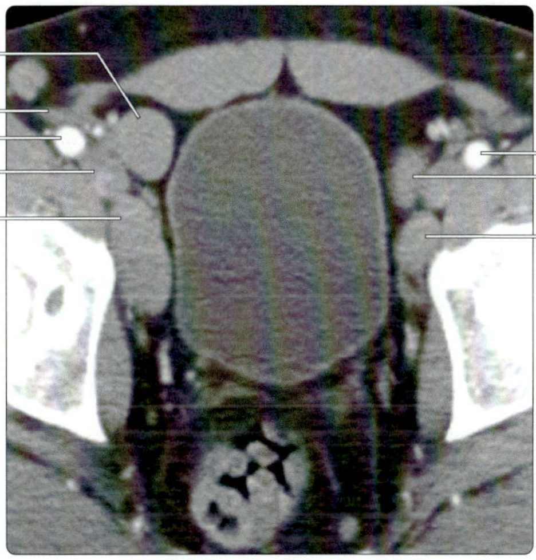

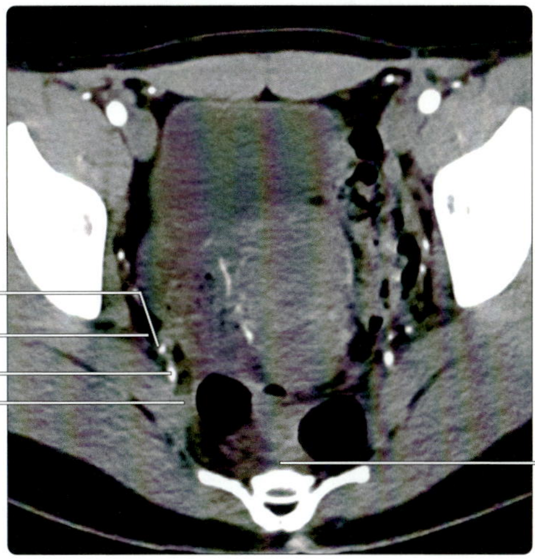

(Top) *Common iliac lymph nodes are located near the common iliac artery and vein and can be divided into 3 subgroups based on their location: Lateral, medial, and middle. The middle subgroup are situated below the aortic bifurcation and above the bifurcation of the common iliac vessels in the lumbosacral fossa and is bordered by the lower lumbar or upper sacral vertebral bodies, the psoas muscle, and the common iliac artery and vein.* **(Middle)** *External iliac lymph nodes are near the external iliac artery and vein and have 3 subgroups: Lateral (lateral to external iliac artery), middle (medial to the artery and lateral to the vein), and medial (medial to the artery, including obturator). The medial subgroup is medial to both vessels and may be posteromedial to the vessels and close to the obturator internus muscle and is sometimes referred to as obturator nodes.* **(Bottom)** *The internal iliac lymph nodes are divided into subdivisions named after adjacent vessels. The subdivisions include lateral sacral nodes, presacral nodes, and anterior internal iliac nodes.*

AUTONOMIC NERVES AND GANGLIA OF ABDOMEN

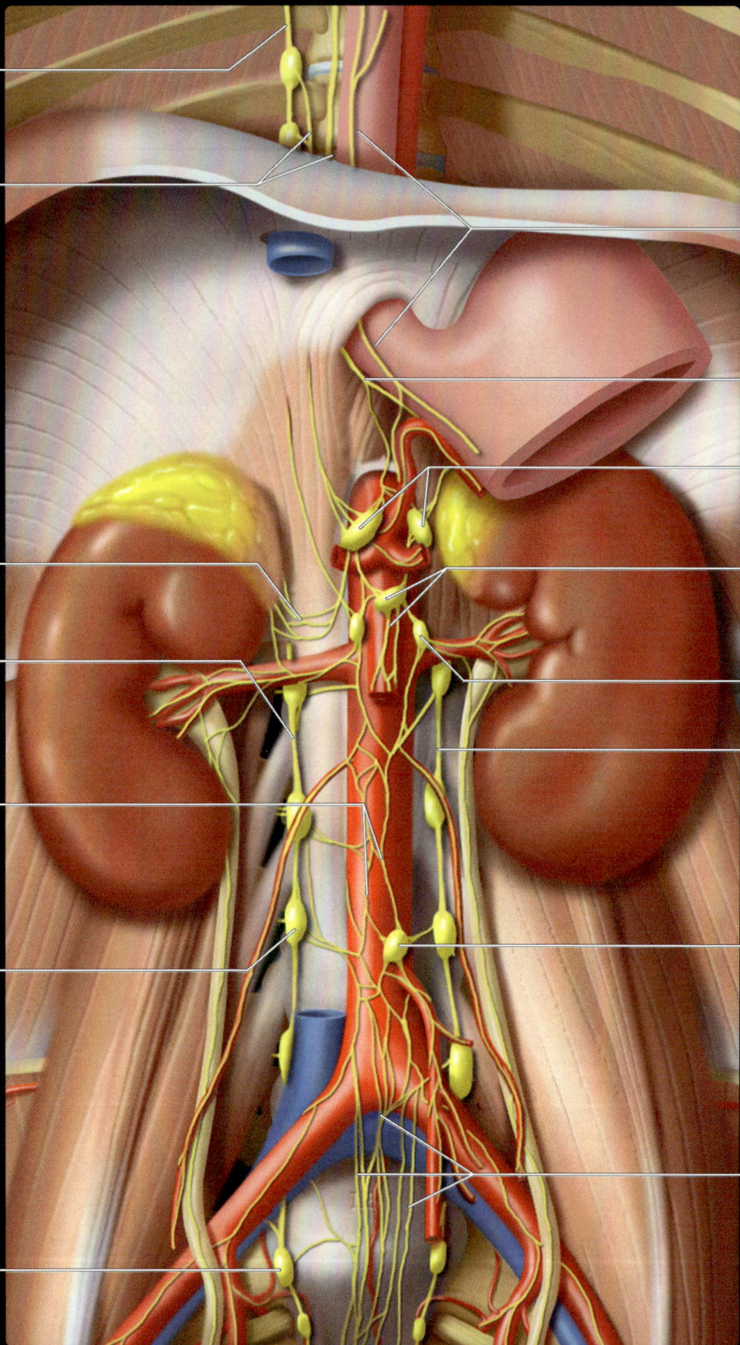

Autonomic innervation of the abdominal viscera comes from several splanchnic nerves and 1 cranial nerve, the vagus (CNX), which deliver presynaptic sympathetic and parasympathetic fibers (respectively) to the aortic plexus and its sympathetic ganglia. Periarterial extensions of these plexuses deliver postsynaptic sympathetic and parasympathetic fibers to the abdominal viscera where intrinsic parasympathetic ganglia occur. The various plexuses are mixed, sharing sympathetic, parasympathetic, and visceral afferent fibers. Thoracic splanchnic nerves are the main source of presynaptic sympathetic fibers to the abdominal viscera. The cell bodies of the postsynaptic sympathetic neurons constitute the prevertebral ganglia that cluster around the roots of the major branches of the abdominal aorta, including the celiac, aortorenal, and superior/inferior mesenteric ganglia.

CELIAC GANGLION AND INVASION BY PANCREATIC CANCER

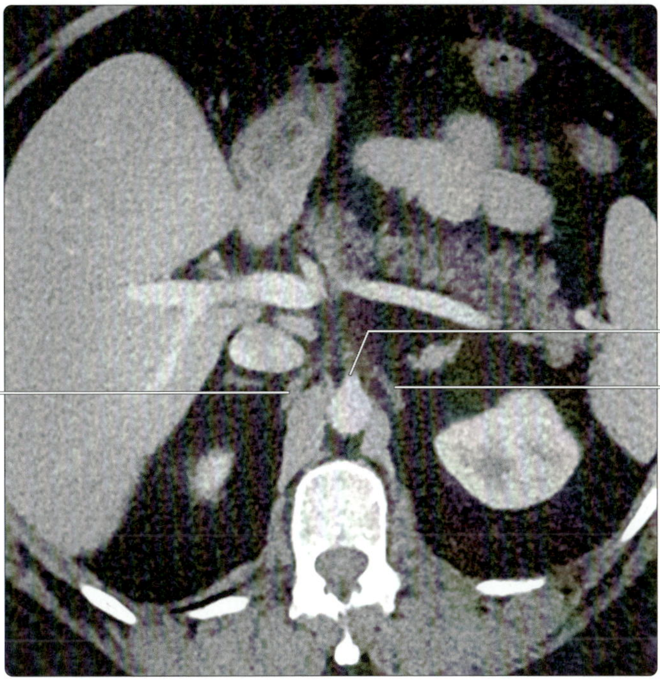

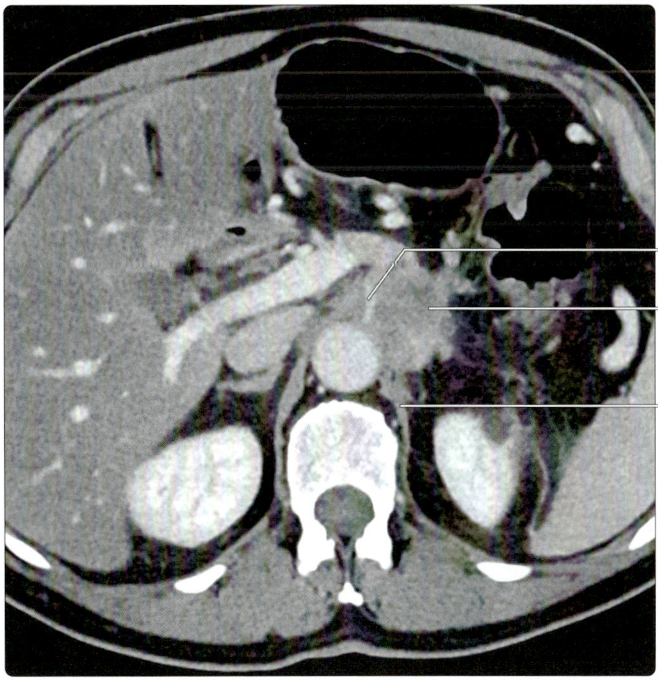

(Top) *Bilateral normal celiac ganglia are observed at the level between the celiac artery and the SMA with a density comparable to that of the liver and spleen or slightly less than the diaphragmatic crura. These may get confused with paraaortic lymph nodes.* **(Bottom)** *Axial CECT in a patient with abdominal pain shows a mass in the pancreatic body representing pancreatic ductal adenocarcinoma. The mass invades the celiac ganglion (not visualized, replaced by tumor).*

TERMINOLOGY

Definitions

- **A ring**: Sporadically imaged indentation of esophageal lumen at cephalic end of lower esophageal sphincter
- **B ring**: Transverse mucosal fold marking esophagogastric junction, often corresponding to mucosal junction between squamous and columnar epithelium

GROSS ANATOMY

Overview

- Esophagus is fibromuscular tube in posterior mediastinum ~ 25 cm long that extends from pharynx to stomach and conveys swallowed material into stomach
  - Enters thorax at ~ T1 level
  - Exits thorax and enters abdomen through **esophageal hiatus** in diaphragm (opening in right crus) at ~ T10 level
  - Physiologic areas of narrowing or constriction
    - At its origin by **cricopharyngeus muscle** (upper esophageal sphincter)
    - By **arch of aorta** (left anterolateral surface of esophagus)
    - By **left main bronchus**
    - By **diaphragm** [and mucosal B ring at gastroesophageal junction (GEJ) ~ 40 cm from incisor teeth]
- Bounded superiorly by upper esophageal sphincter and inferiorly by lower esophageal sphincter
  - **Upper esophageal sphincter** (under conscious control) separates pharynx from esophagus and prevents ingested material from traveling backward into airway
    - Composed of 3 skeletal muscles: Inferior constrictor, cricopharyngeus, and proximal esophagus
  - **Lower esophageal sphincter** (not under conscious control) prevents gastric contents from refluxing into esophagus
    - Composed of skeletal muscle of crural diaphragm and smooth muscle of distal esophagus
    - Essentially synonymous with "esophageal vestibule" or "**phrenic ampulla**"
    - Defined clinically by manometric evidence of high resting tone or pressure
    - Occasionally recognized radiographically as 2- to 4-cm long luminal dilation between esophageal A and B rings
- **Mural anatomy**
  - Internal circular and external longitudinal **muscle layers**
    - Superior 1/3 of esophagus: Voluntary, striated muscle; lower 1/3: Smooth muscle; middle 1/3: Both types
  - Lined by stratified squamous epithelium
  - Lacks serosal lining
- **GEJ**
  - Marked on mucosal surface by **Z line** (1 cm below proximal border of lower esophageal sphincter): Demarcates end of smooth, pearly pink esophageal mucosa and beginning of reddish, textured gastric columnar mucosa
  - GEJ attached to liver at fissure for ligamentum venosum by **gastrohepatic ligament**
  - GEJ often appears thickened on axial imaging (may simulate tumor)
- Attached to diaphragm by **phrenicoesophageal ligament** (collagenous band) tends to weaken and elongate with age, may lead to hiatal hernia
- **Vessels, nerves, lymphatics**
  - Arterial supply
    - Bronchial and esophageal branches of aorta in thorax
    - Left gastric (via celiac) and inferior phrenic arteries in abdomen
    - Cervical esophagus supplied by inferior thyroidal artery
  - Venous drainage through azygous system (systemic) (along with intercostal and bronchial veins) in thorax, left gastric vein (to portal venous system) in abdomen, and inferior thyroidal vein for cervical esophagus
  - Lymphatic drainage
    - Extensive network of submucosal lymphatics along entire course of esophagus
    - Upper 2/3 → posterior mediastinal nodes
    - Lower 1/3 → left gastric and celiac nodes
  - Innervation: Right/left **vagus nerves** and sympathetic trunk

ANATOMY IMAGING ISSUES

Imaging Recommendations

- Best means of evaluating mucosal disease (inflammation, superficial tumor): Double-contrast barium esophagram and endoscopy
- Best test for gastroesophageal reflux disease: Esophagram and pH testing by esophageal probe or capsule
- Best test for stricture: Single-contrast esophagram
- Best test for mass lesion: Esophagram and endoscopy
- Best test for depth of tumor invasion: Endoscopic ultrasound
- Best test for staging esophageal cancer: PET/CT

CLINICAL IMPLICATIONS

Clinical Importance

- **Hiatal hernia** and GERD are extremely common and usually occur together
  - 2 types: Sliding hiatal hernia: GEJ and gastric cardia pass through esophageal hiatus and paraesophageal hernia: Gastric fundus ± other parts of stomach herniate into chest
  - Reflux often causes spasm of esophageal longitudinal muscles, resulting in shortening of esophagus and more herniation
- **Esophageal varices** are common
  - Submucosal veins drain into systemic and portal venous system and constitute potential collateral pathway
  - "**Uphill varices**": More common, due to portal hypertension (cirrhosis), ↑ portal venous pressure → upward venous flow via dilated esophageal collaterals to superior vena cava (SVC)
  - "**Downhill varices**": Obstruction of SVC → downward venous flow via esophageal collaterals to portal vein and inferior vena cava

ESOPHAGUS FROM PHARYNX TO STOMACH

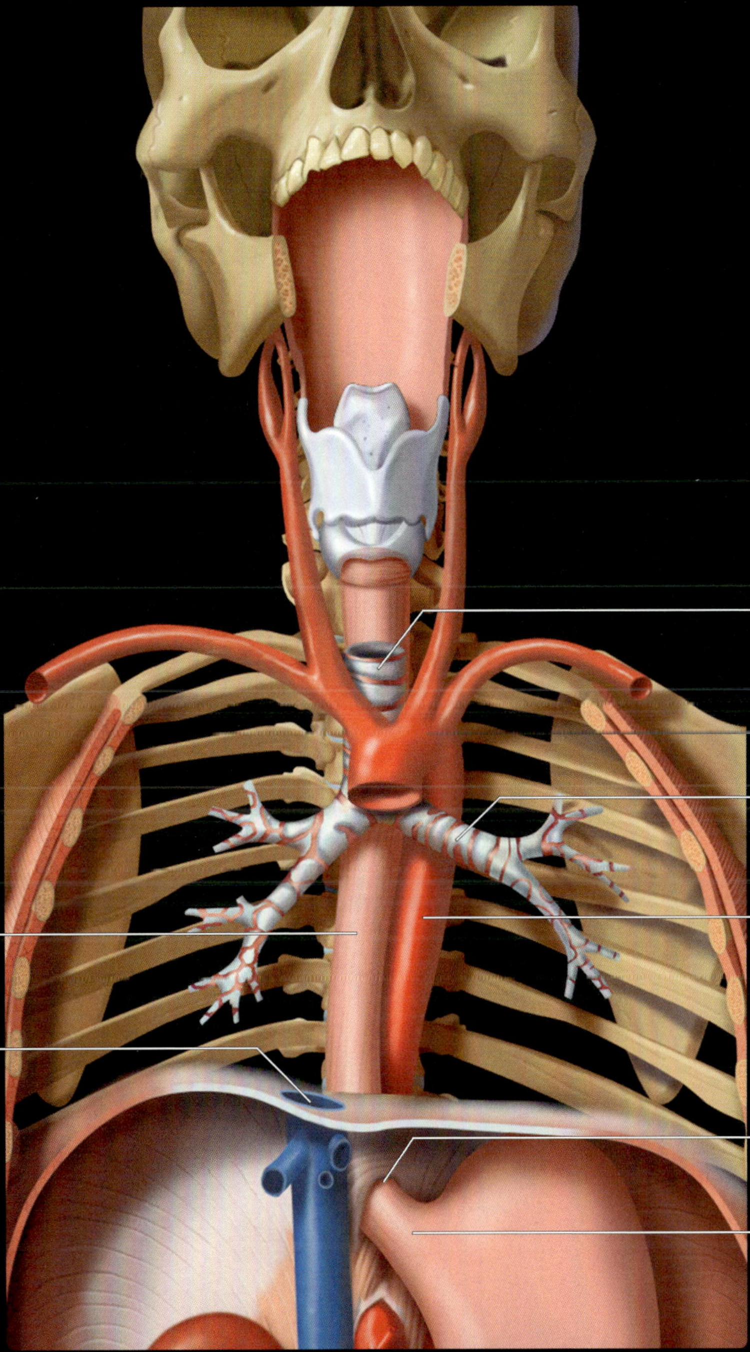

The esophagus enters the thorax at roughly the T1 level posterior and slightly to the left of the trachea. It is usually indented on its left anterolateral surface by the arch of the aorta and the left main bronchus. The esophagus is closely applied to the aorta throughout its course and may be pushed or pulled by aortic abnormalities, such as an aneurysm or ectasia. The esophagus enters the abdomen at approximately the T10 level between fibers of the right crus of the diaphragm. The esophageal hiatus lies more caudal than the hiatus from the inferior vena cava and cephalad to the aortic hiatus.

ESOPHAGUS SAGITTAL RELATIONS

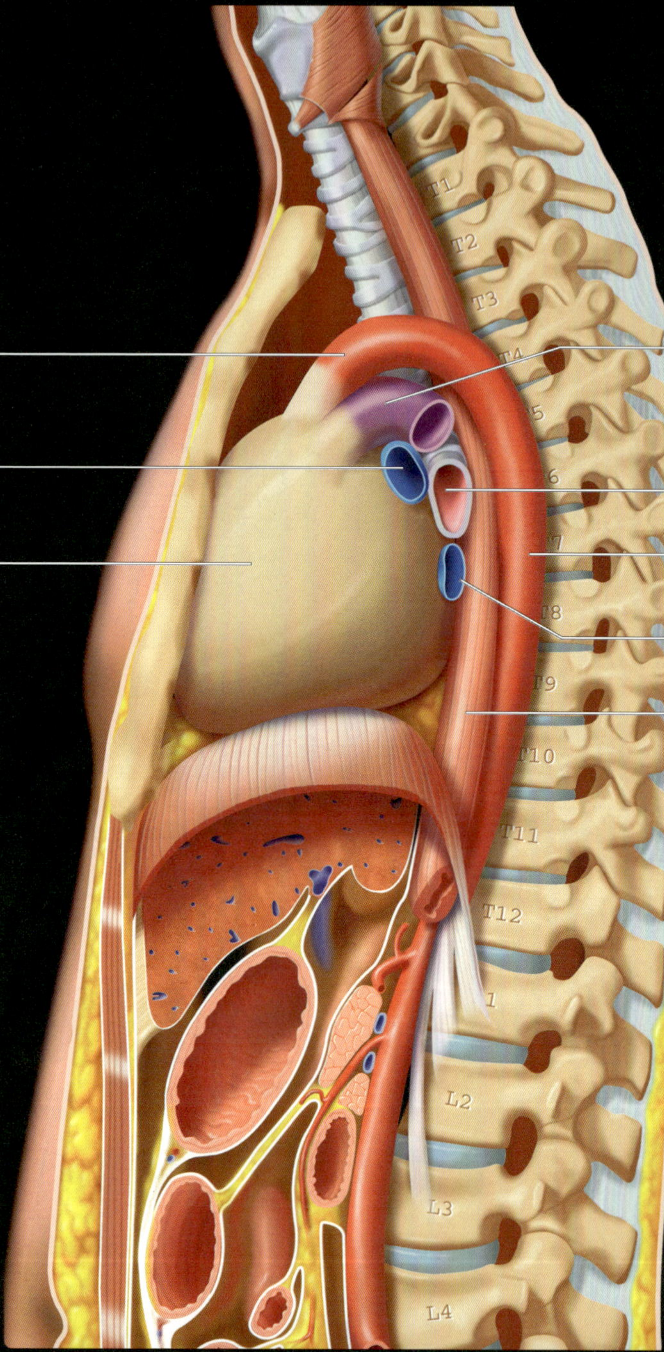

The esophagus is ~ 25 cm long and extends from the level of the cricopharyngeus muscle (at the C5-C6 level) to the gastroesophageal junction (at approximately the T10-T11 level). Note the relationship between the esophagus and adjacent structures, including the heart, which may indent or displace the esophageal lumen.

GASTROESOPHAGEAL JUNCTION AND LYMPHATICS SYSTEM

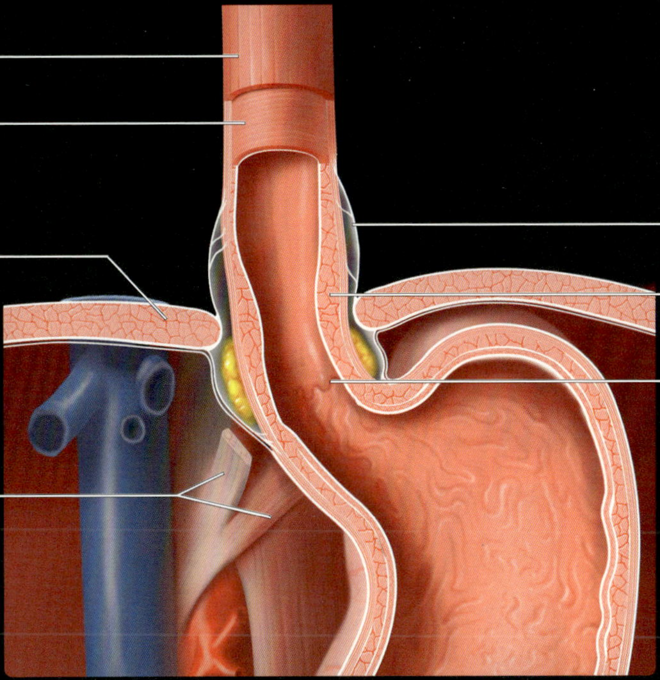

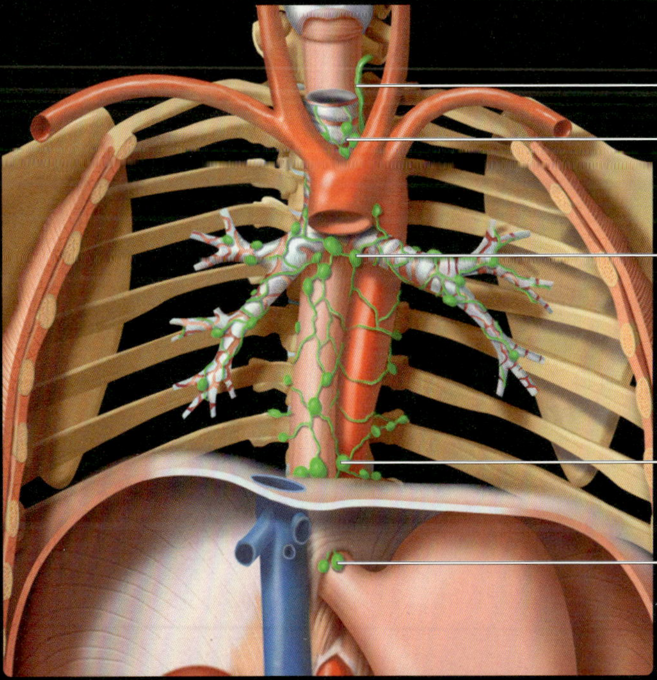

(Top) *The esophageal wall musculature consists of an inner circular layer and an outer longitudinal layer. In the region of the lower esophageal sphincter (LES), the muscle layers are thickened. This, along with diaphragmatic constriction and the angulation of the esophagus as it passes into the abdomen and enters the stomach, helps to prevent reflux of gastric contents. The Z line marks the junction of the esophageal and gastric mucosa.* **(Bottom)** *Lymphatic drainage from the upper part of the esophagus is usually to the paratracheal and posterior mediastinal nodes, while the distal esophagus drains to the diaphragmatic, celiac, and left gastric nodes (although there is considerable overlap). The tracheobronchial nodes lie near the carina and may be responsible for the formation of traction diverticula when they become fibrosed, usually as a result of tuberculosis or histoplasmosis.*

NORMAL IMPRESSIONS ON ESOPHAGUS

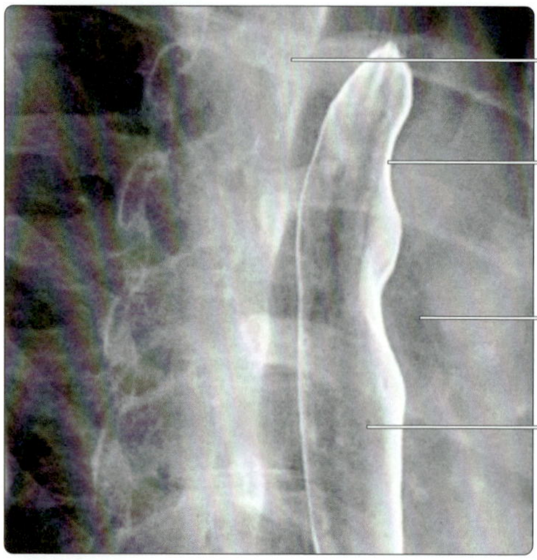

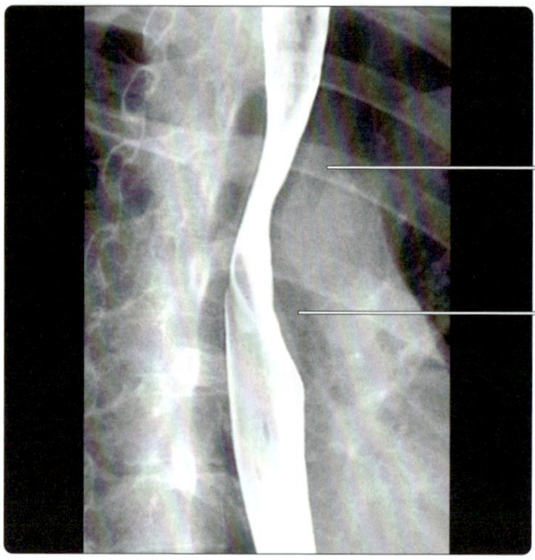

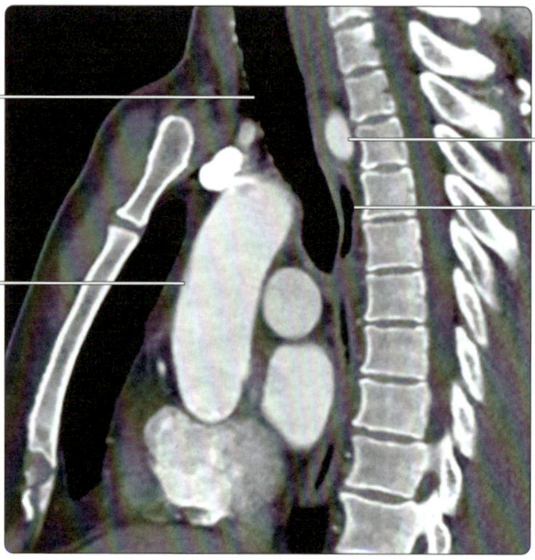

(Top) *First of 2 films from a barium esophagram with air contrast (lumen is distended by gas from ingested gas granules and water) is shown. The esophagus is normal and has a featureless smooth mucosal surface. The left anterior wall of the esophagus is indented by 2 adjacent structures, the arch of the aorta and the left main bronchus.* **(Middle)** *The esophageal lumen appears narrowed between the aortic arch and the left main bronchus, but this is a normal finding. The bronchus is identified as the air-filled tubular structure next to the esophagus.* **(Bottom)** *Aberrant right subclavian artery has a retroesophageal course and causes posterior indentation of the esophagus. This is usually asymptomatic but may sometimes cause dysphagia.*

BARIUM ESOPHAGRAM, ESOPHAGEAL A AND B RINGS

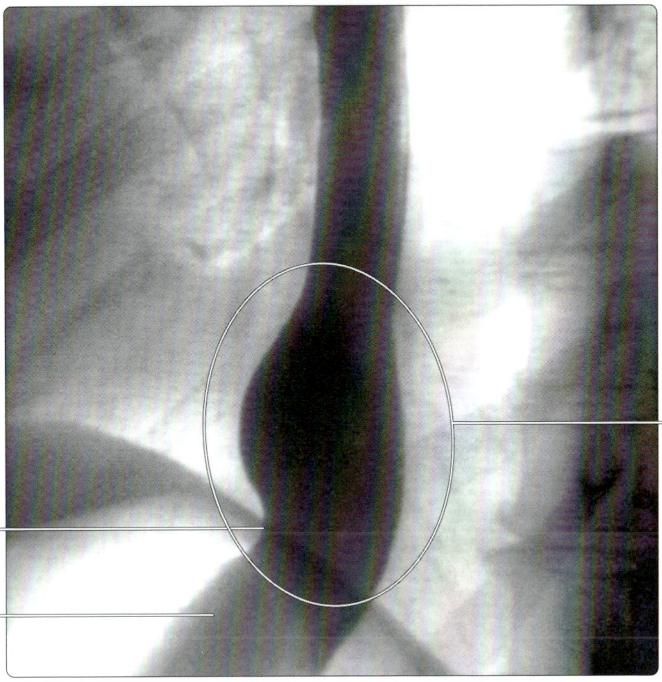

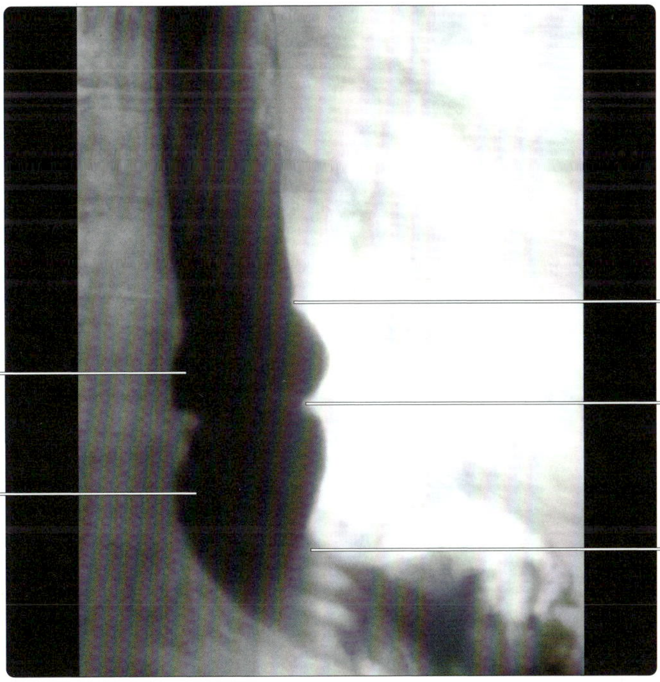

(Top) *First of 2 images from a single-contrast barium esophagram is shown. This image shows smooth dilation of the distal few cm of the esophageal lumen, often referred to as the "phrenic ampulla" or "esophageal vestibule." This is the anatomic correlate to the LES, which is defined manometrically as the zone of increased resting tone or pressure within the esophagus. The LES contributes to the antireflux valve effect at the gastroesophageal junction.* **(Bottom)** *Second esophagram image from the same patient with the film taken during abdominal straining (Valsalva maneuver) is shown. A small hiatal hernia is demonstrated. The A and B rings of the esophagus are well defined and mark the upper and lower limits of the LES, respectively. The B ring marks the gastroesophageal junction, which is above the diaphragm in this patient, indicating a hiatal hernia.*

NORMAL GASTROESOPHAGEAL JUNCTION

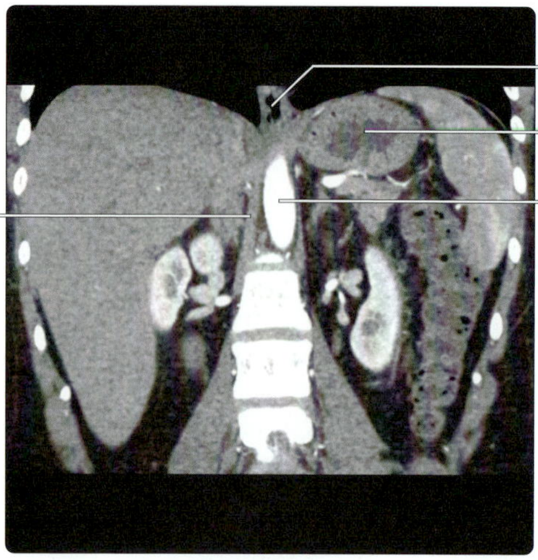

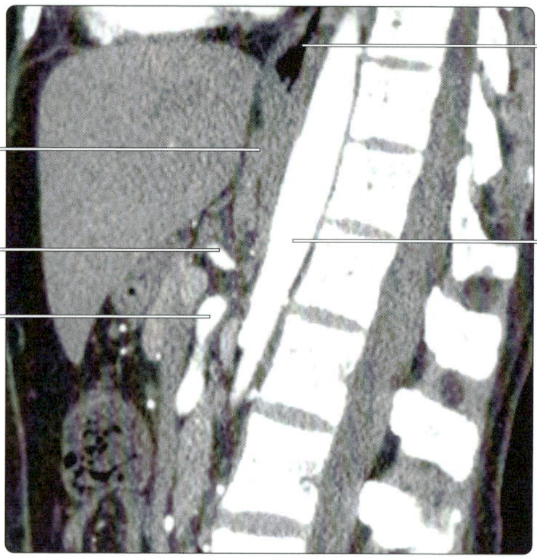

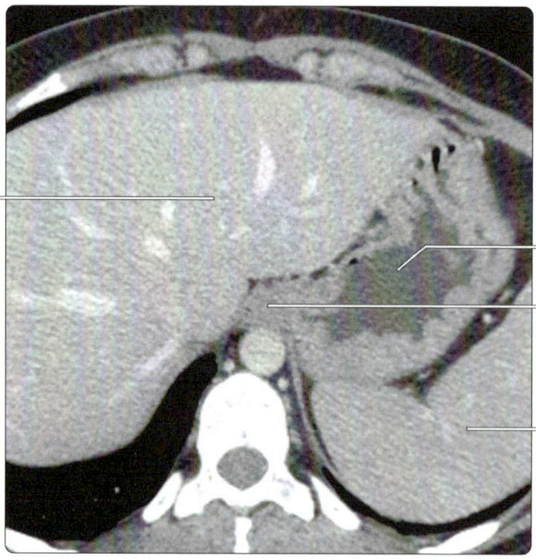

(Top) *First of 3 CT sections through the upper abdomen is shown. Coronal CT shows the esophagus and aorta entering the abdomen. The esophagus enters at approximately the T10 vertebral level between fibers of the right crus of the diaphragm. The aorta enters at approximately the T12 level behind the median arcuate ligament of the diaphragm.* **(Middle)** *Sagittal CECT in the midline shows the esophagus and aorta entering the abdomen. Only portions of the celiac and superior mesenteric arteries are seen in this section, but the origin of the celiac is usually just caudal to the median arcuate ligament, which marks the entry of the aorta into the abdomen.* **(Bottom)** *Axial CECT demonstrates that the gastroesophageal junction usually lies at the same level as the fissure for the ligamentum venosum in the liver and commonly appears thickened relative to the walls of the distal esophagus or proximal stomach.*

GASTROESOPHAGEAL JUNCTION

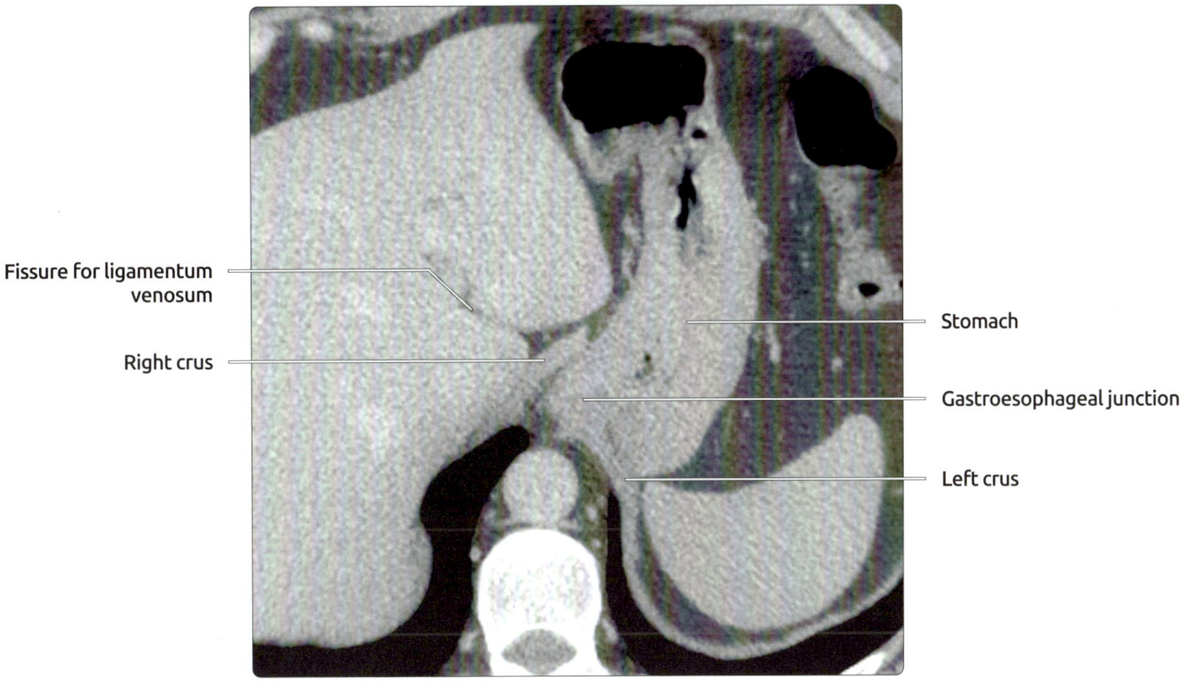

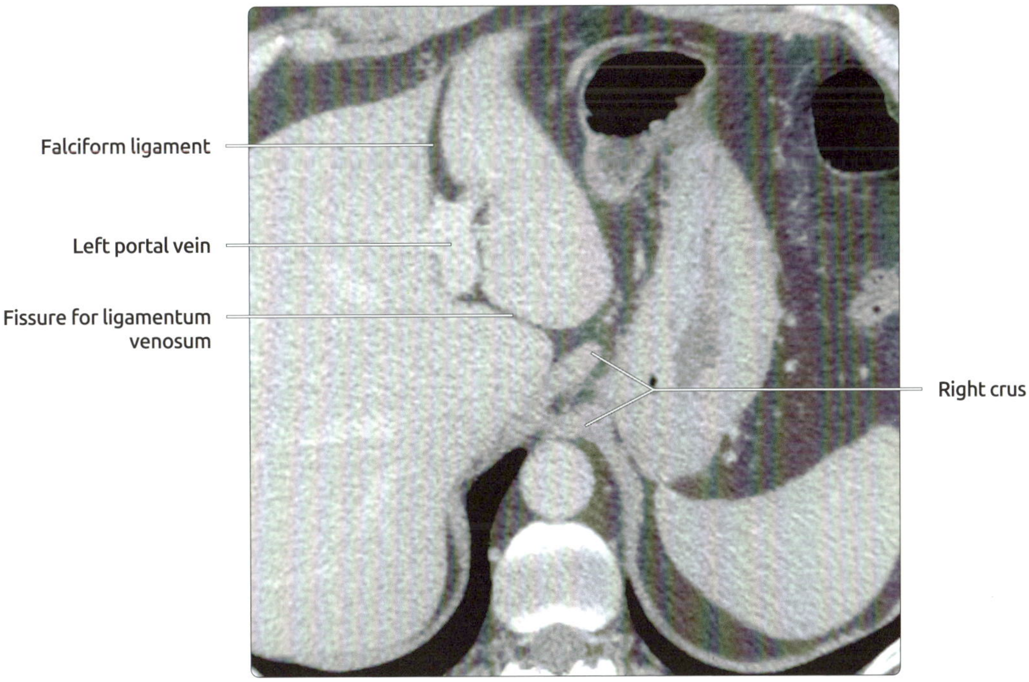

(Top) *First of 2 axial CECT sections through the upper abdomen is shown. The esophagus enters the abdomen through fibers of the right crus, and the gastroesophageal junction is normally at approximately the level of the fissure for the ligamentum venosum.* **(Bottom)** *The decussation (crossing of fibers) of the right crus is well shown on this CT section, allowing the traversal of the esophagus into the abdomen.*

PATULOUS DIAPHRAGMATIC HIATUS WITH HERNIA

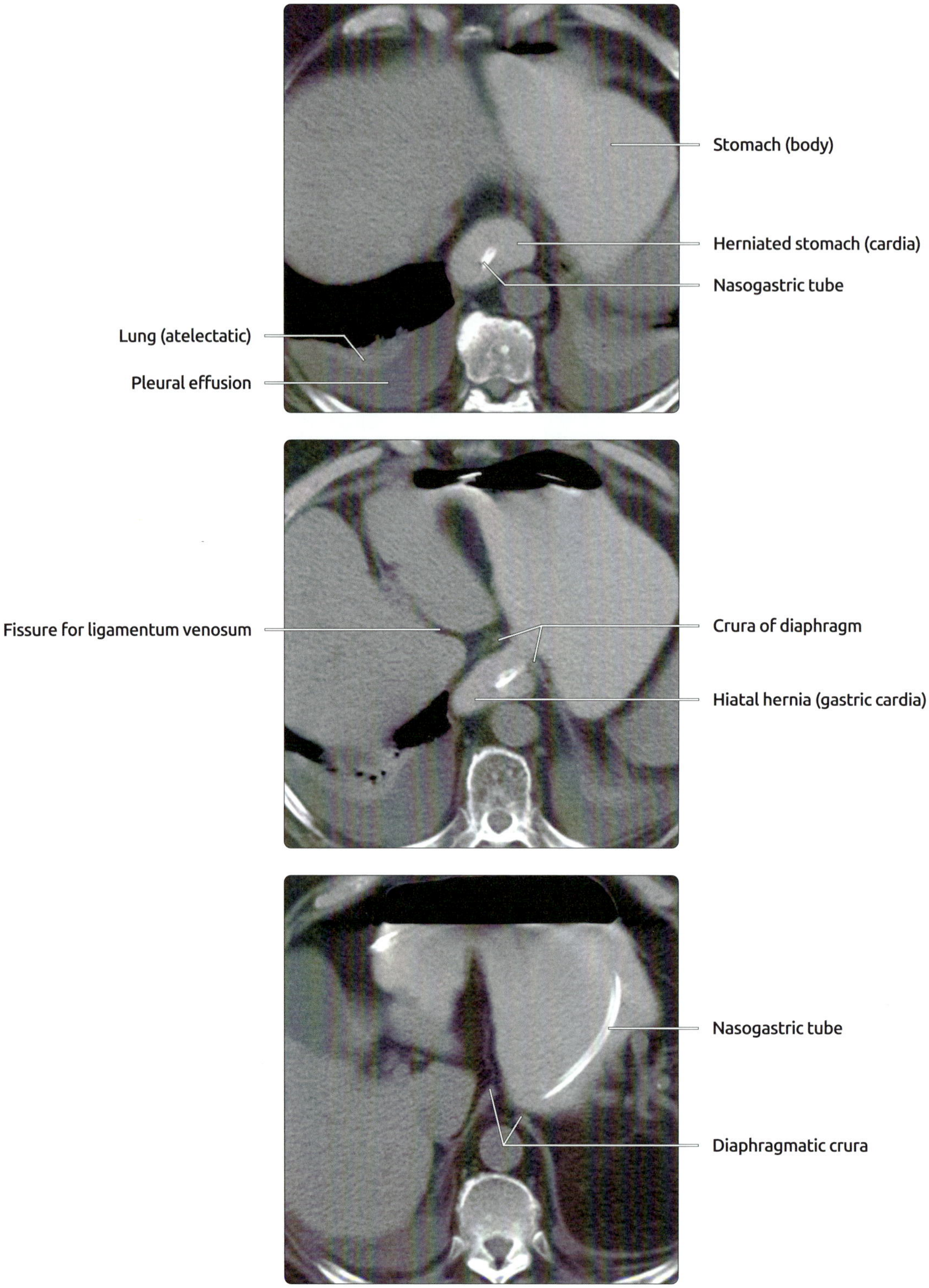

(Top) *First of 3 axial CECT sections in an older man is shown. In this image through the lower chest, a portion of the stomach is visualized, opacified by oral contrast medium.* **(Middle)** *At the level of the fissure for the ligamentum venosum, which should mark the gastroesophageal junction, note the wide opening between the crura of the diaphragm, which has allowed the stomach (gastric cardia) to herniate. This is a sliding hiatal hernia.* **(Bottom)** *Most caudal CT section shows a persistent gap between the crura, which should have joined together in the midline at this level.*

ESOPHAGEAL B RING, HIATAL HERNIA

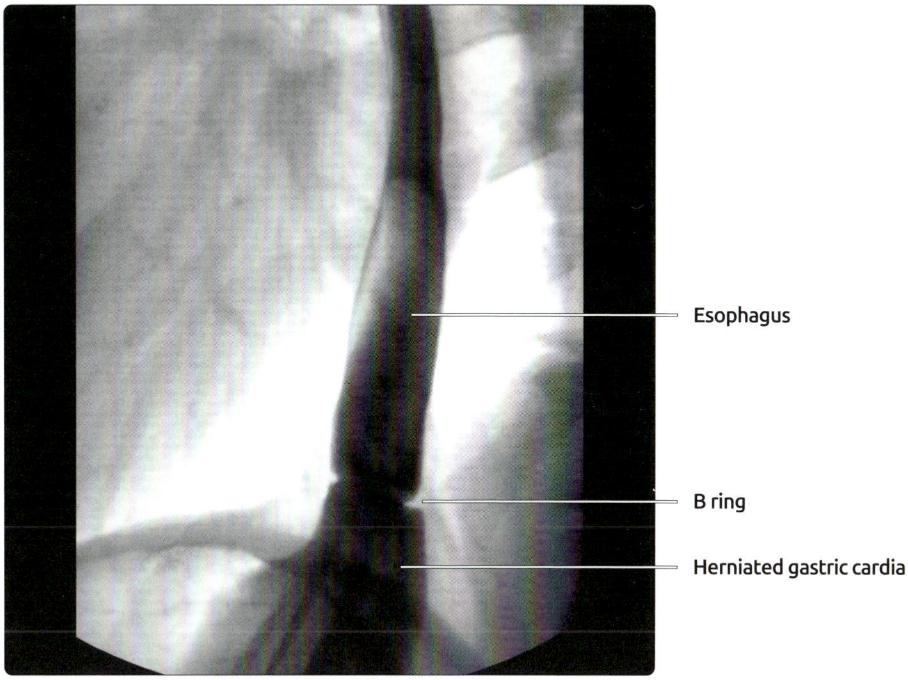

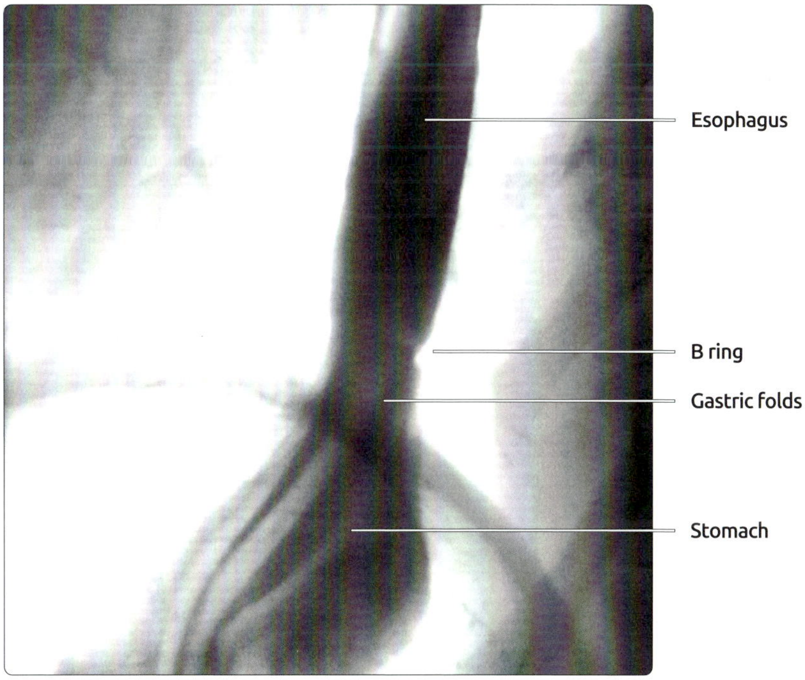

(Top) *First of 2 films from a barium esophagram shows a web-like narrowing at the gastroesophageal junction, a typical feature of the esophageal B ring, which marks the squamocolumnar junction. The B ring is well seen due to the hiatal hernia and good fluoroscopic technique, which has distended the distal esophagus and proximal stomach.* **(Bottom)** *Second esophagram image from the same patient demonstrates that the gastric folds extend above the diaphragm to the B ring, another sign of a hiatal hernia.*

STRICTURE AT B RING (SCHATZKI RING)

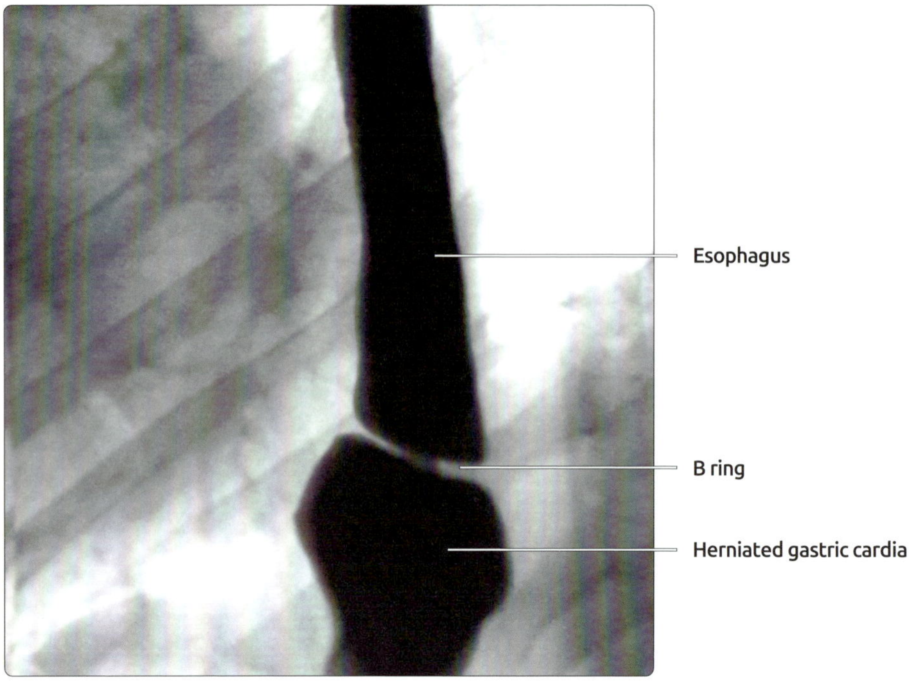

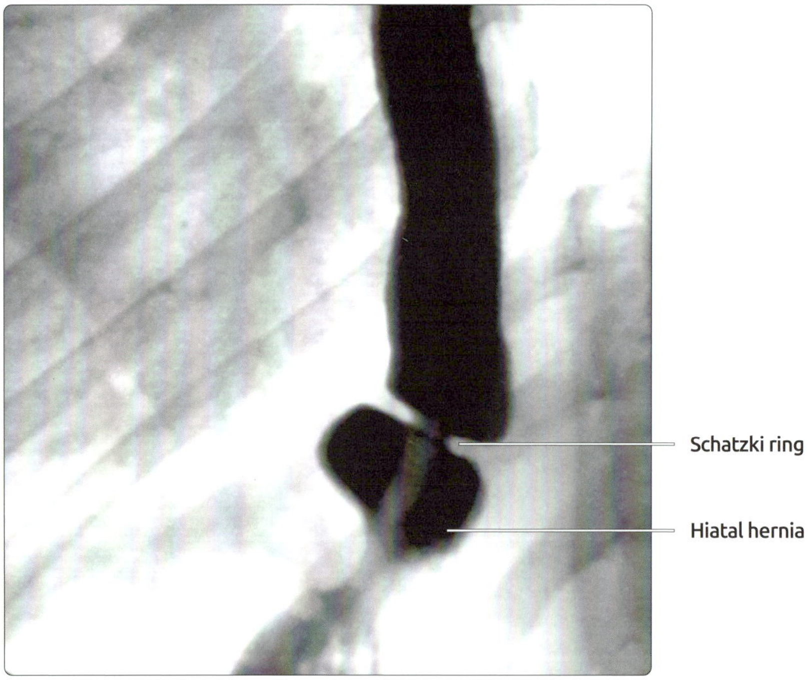

(Top) *First of 2 films from a barium esophagram shows a narrowed lumen through the B ring at the gastroesophageal junction. The lower esophageal ring is thickened, and the lumen is narrowed due to inflammation and scarring, almost always due to gastroesophageal reflux. Note the associated presence of a small hiatal hernia.* **(Bottom)** *When the lumen is narrowed through the B ring, it is sometimes referred to as a Schatzki ring after the radiologist who noted that patients with this finding frequently had complaints of reflux and food "sticking" in their esophagus, sometimes requiring endoscopic treatment.*

CRICOPHARYNGEAL ACHALASIA AND HIATAL HERNIA

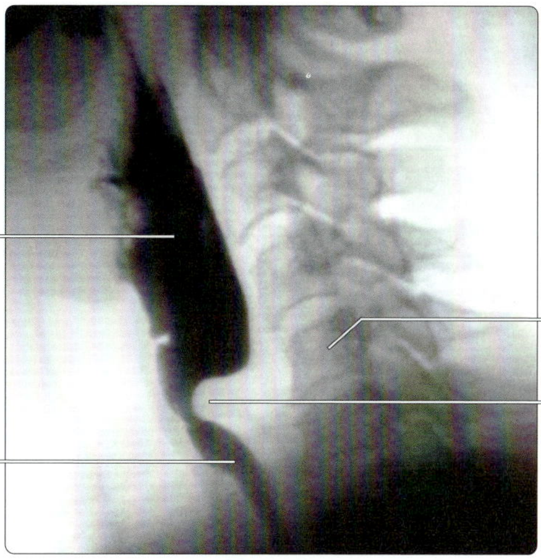

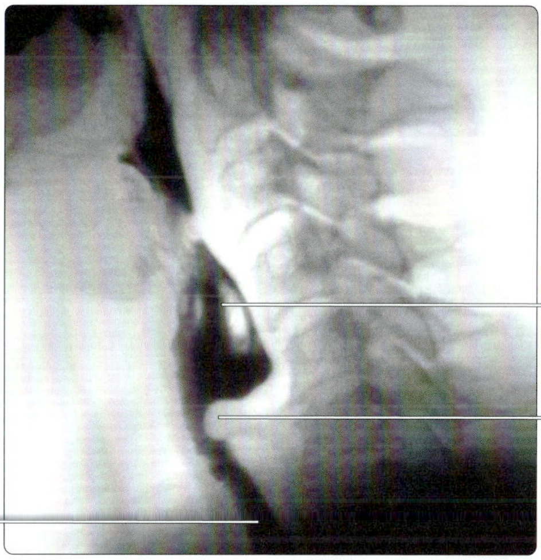

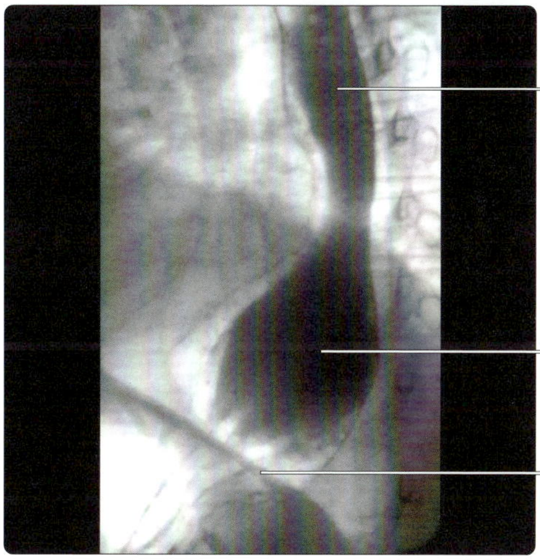

(Top) *First of 3 films from a barium esophagram is shown. This lateral view shows a rounded indentation of the posterior wall of the pharyngoesophageal junction at the level of the disc space between the 5th and 6th cervical vertebrae. The pharynx is distended, indicating impaired passage of the barium. These are typical features of spasm or achalasia (failure to relax) of the cricopharyngeus muscle, which is part of the upper esophageal sphincter.* **(Middle)** *The cricopharyngeus normally relaxes in anticipation of the arrival of a bolus of food. This lateral film shows persistent filling of the pharynx and contraction of the cricopharyngeus muscle after the bolus of barium has passed.* **(Bottom)** *Film of the lower chest in the same patient shows a hiatal hernia. Cricopharyngeal achalasia is often the indirect result of disordered esophageal motility and acid reflux.*

"FELINE" ESOPHAGUS

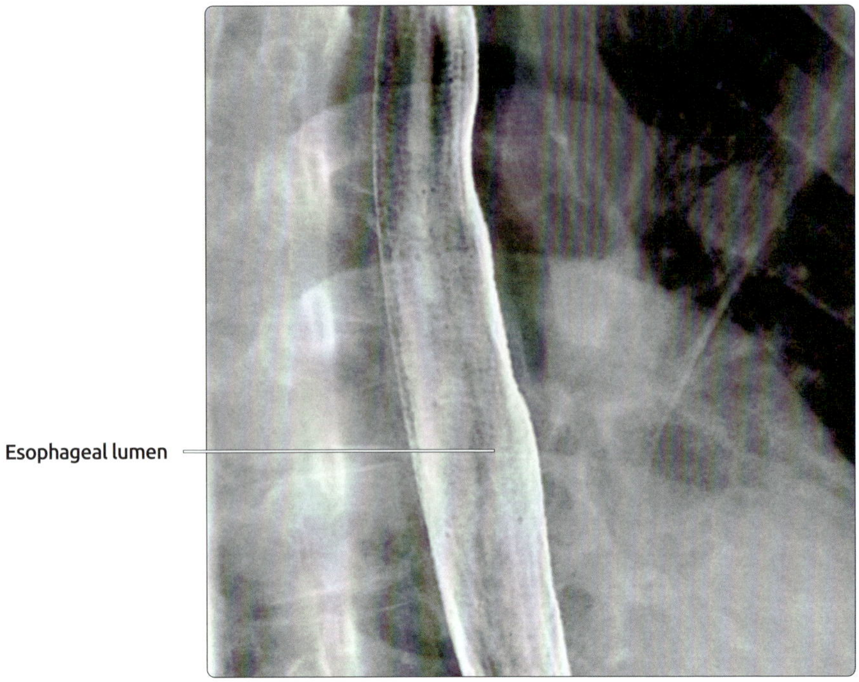

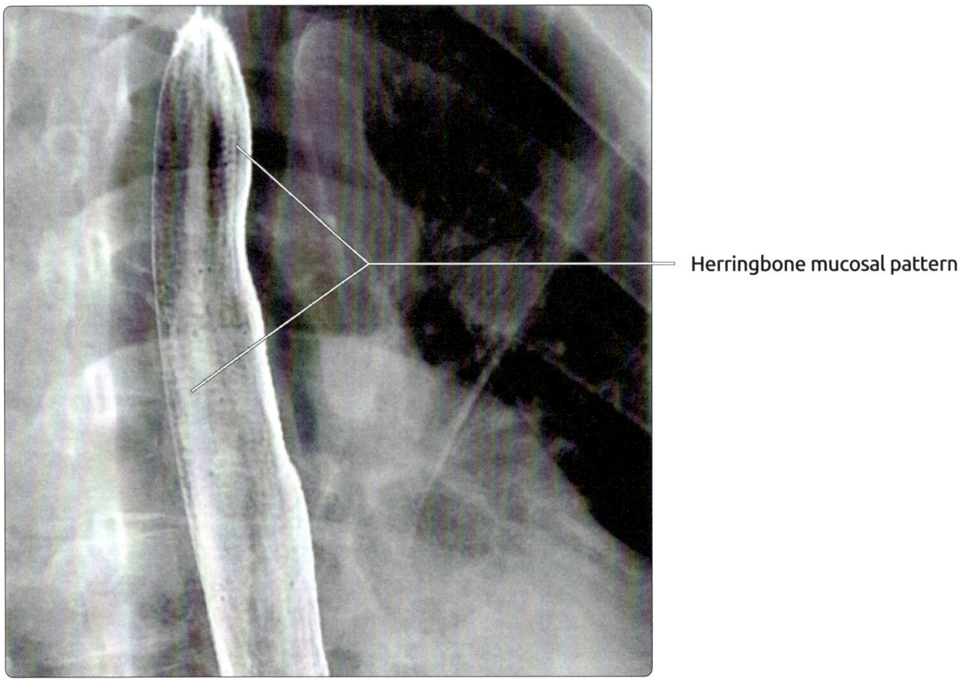

(Top) *Two films from an air-contrast barium esophagram are shown. Both show a peculiar herringbone surface-coating pattern rather than the normal featureless surface pattern of the esophagus. This is a transient finding and represents the result of contraction of the muscularis mucosa and usually results from an episode of reflux, which "irritates" the esophagus.* **(Bottom)** *This herringbone-like pattern of barium coating is an uncommon feature in the human esophagus but is, apparently, a typical feature of the cat esophagus, hence, the descriptive term "feline" esophagus.*

PARAESOPHAGEAL HERNIA

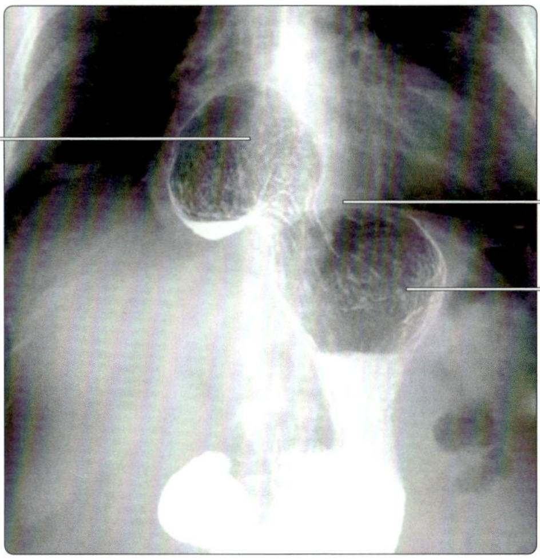

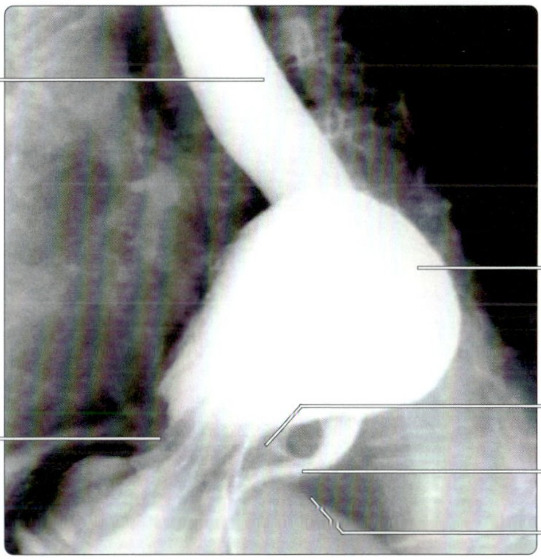

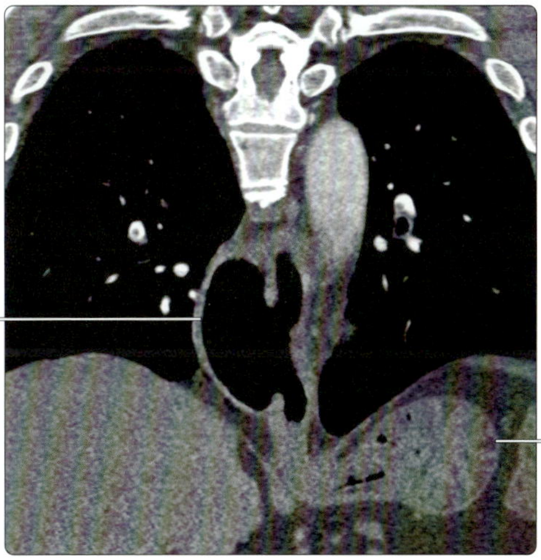

(Top) *First of 2 films from barium upper GI series shows herniation of the gastric fundus as well as the gastric cardia. This constitutes a paraesophageal hernia and is considered a more compelling indication for surgical repair than a simple sliding hiatal hernia, which involves only the cardia.* **(Middle)** *In most paraesophageal hernias, the gastroesophageal junction lies above the diaphragm; these are classified as type III paraesophageal hernia. Those in which the gastroesophageal junction is below the diaphragm are classified as a type II paraesophageal hernia. The term "paraesophageal" refers to the position of the intrathoracic stomach alongside ("para") the esophagus.* **(Bottom)** *Coronal CECT shows a large pulsion diverticulum projecting laterally from the distal esophagus. A pulsion diverticulum is an esophageal saccular protrusion or outpouching that may mimic a hiatal hernia (especially type 3 paraesophageal hernia) on barium study. Absence of gastric folds on barium within the diverticulum and a normal stomach on CT may be helpful clues.*

ESOPHAGEAL CANCER

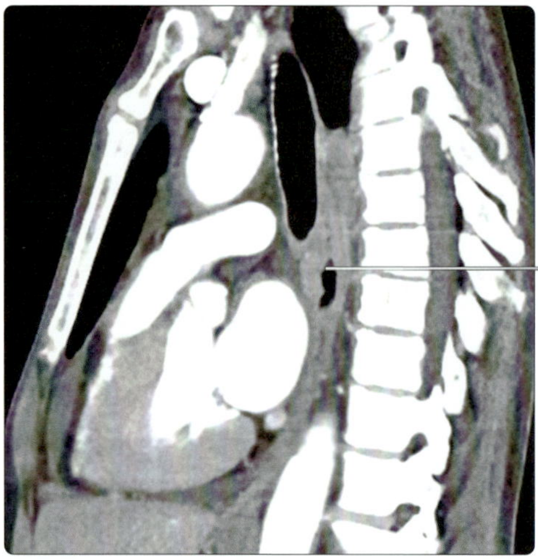

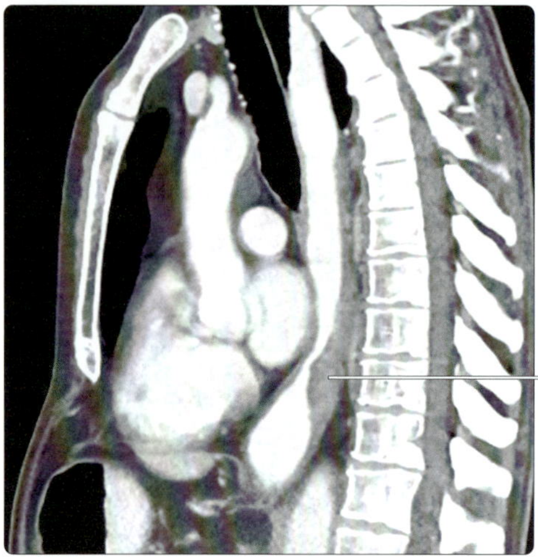

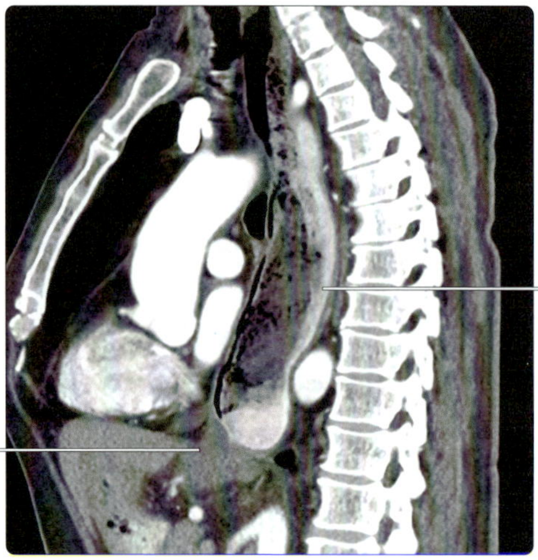

(Top) *A 56-year-old woman presenting with ongoing reflux, dysphagia, emesis, and weight loss is shown. Sagittal contrast-enhanced CT without oral contrast shows circumferential esophageal wall thickening with submucosal nodular enhancement at the level of midthoracic esophagus, concerning for primary esophageal malignancy. The mass is difficult to visualize due to the absence of oral contrast. Pathology showed squamous cell carcinoma.* **(Middle)** *A 62-year-old man with adenocarcinoma of the distal esophagus is shown. Note a broad based mass in the posterior wall that is well visualized due to the presence of oral contrast. Oral contrast administration right before the CT can help better visualization of the esophageal lumen.* **(Bottom)** *A 66-year-old man with progressive dysphagia found to have Barrett esophagus with development of adenocarcinoma is shown. Note the soft tissue mass at the gastroesophageal junction with extensive upstream dilatation of the esophagus containing oral contrast and particulate matter.*

ESOPHAGEAL VARICES

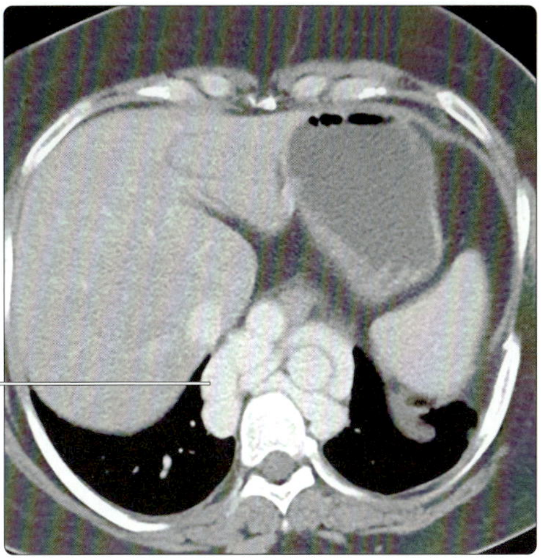

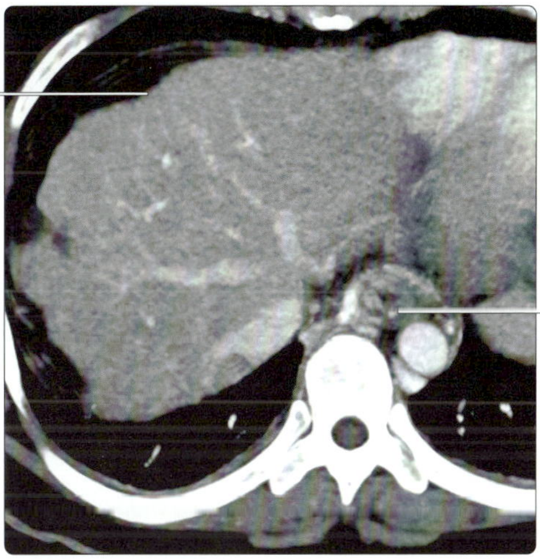

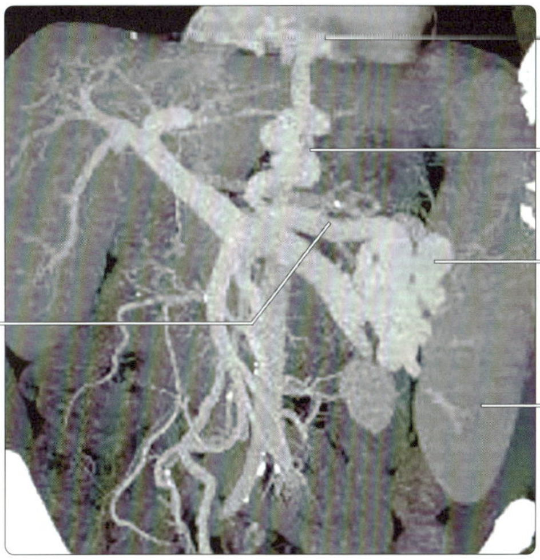

(Top) *Axial CECT in a patient with known chronic liver disease demonstrates extensive varices surrounding the esophagus. These are "uphill" esophageal varices related to portal hypertension and hepatofugal flow of blood seeking to bypass the liver.* **(Middle)** *A patient with hepatic cirrhosis and related "uphill" varices is shown. Note dilated vessels in the paraesophageal location as well as within the wall of the esophagus.* **(Bottom)** *Coronal volume-rendered CECT demonstrates several of the most common sites of collateral formation in a patient with portal hypertension, including esophageal, perigastric, and perisplenic varices. The left renal vein is markedly dilated and communicates with the perisplenic varices, and the splenic vein is compatible with a splenorenal shunt. Note the presence of a markedly enlarged spleen, also on the basis of portal hypertension.*

GROSS ANATOMY

Overview

- **Stomach**: Portion of gastrointestinal tract (between esophagus and duodenum) serving as alimentary reservoir for mixing and enzymatic digestion of food
  - **Cardia**: Surrounds esophageal orifice into stomach (divided from esophagus by lower esophageal sphincter); where lesser and greater curvature meet
  - **Fundus**: Most cephalic part of stomach, located superiorly and to left relative to cardia, touching left hemidiaphragm
  - **Body**: Largest central portion of stomach, serving as principal site of acid production
  - **Antrum**: Prepyloric portion of stomach (vestibule)
  - **Pylorus**: Sphincter opening into duodenum (to right of midline), formed by thickened middle layer of smooth muscle and thin, fibrous septum
    - Pyloric sphincter controls transit of ingested material from stomach into duodenum
  - Superior and inferior borders of stomach known as **lesser curvature** and **greater curvature**, respectively
    - Greater curvature is longer convex margin of stomach (inferior and lateral border) attached to transverse colon by **gastrocolic ligament**
      - **Greater omentum** (double layer of peritoneum) extends downward from greater curvature
    - Lesser curvature is shorter concave margin of stomach (superior) attached to liver via hepatogastric ligament
      - **Lesser omentum** (double layer of peritoneum contiguous with gastric/duodenal peritoneal covering) extends upward from stomach to liver
      - **Incisura angularis** is "notch" in lesser curvature near pylorus used to delineate gastric body from pylorus
    - **Lesser sac** (**omental bursa**) lies posterior to stomach (and anterior to pancreas) and communicates with main peritoneal cavity via **foramen of Winslow**
  - **Mural anatomy**
    - Wall consists of 3 layers of smooth muscle (outermost = longitudinal; middle = circular; inner = oblique) with circular layer thickest
    - **Gastric folds** (**rugae**): Redundant folds of gastric mucosal surface, which are most evident when stomach is empty (especially along greater curvature)
    - Mucosa is columnar epithelium
    - Gastric glands: Produce mucus (which lines and protects gastric surface), pepsinogen (precursor to pepsin), and hydrochloric acid (activates digestive enzymes and assists with breakdown of food)
  - **Vessels**, **nerves**, and **lymphatics**
    - Lesser curvature: Arterial supply via branches of **left and right gastric arteries** (lie in lesser omentum)
    - Greater curvature: Arterial supply via left and right **gastroepiploic** arteries (in greater omentum) and **short gastric arteries**
    - Venous drainage into portal vein through **left and right gastric veins** and **splenic/superior mesenteric veins**
    - Lymphatic drainage along course of arteries (**gastric** and **gastroomental nodes**), then to **celiac nodes** via efferent lymphatic ducts
    - Innervation: Parasympathetic supply via posterior vagal trunks stimulates peristalsis and acid production; sympathetic supply via celiac and splanchnic ganglia and plexus (also carry pain receptor nerves)
- **Duodenum**: Shortest (~ 25 cm), most proximal aspect of small bowel, which extends from gastric pylorus to duodenojejunal junction and forms C loop
  - **Bulb** (1st part): 2 cm; triangular segment suspended by hepatoduodenal ligament; also contains bile duct, portal vein, and hepatic artery
  - **Descending** (2nd part): 8 cm; site of **major papilla of Vater** (entry site of common duct and main pancreatic duct)
  - **Transverse** (3rd part): 8 cm; extends in front of aorta and IVC
  - **Ascending** (4th part): 4 cm; ending at duodenojejunal junction
  - **Mural anatomy**
    - Mucosa, submucosa, circular and longitudinal smooth muscle, and serosa (anterior to duodenum only)
    - Duodenal bulb intraperitoneal; rest is retroperitoneal (with peritoneum only covering anterior surface)
    - **Brunner glands**: Most prominent in proximal duodenum and secrete fluid with mucus and proteolytic enzymes
  - Anatomic relations
    - Duodenojejunal junction usually at ~ L2 level (about same level as pylorus), suspended by **ligament of Treitz** (extension of right crus of diaphragm)
    - 2nd and 3rd portions of duodenum are closely attached to pancreatic head [surgical resection of pancreatic head (**Whipple procedure**) requires resection of duodenum as well]
    - 2nd part of duodenum lies just anterior to hilum of right kidney (inflammation from duodenal perforation may extend into perirenal space)
  - **Vessels** and **nerves**
    - Arterial supply to duodenal bulb via right gastric and gastroepiploic arteries
    - Arterial supply to remainder of duodenum via superior and inferior pancreaticoduodenal arteries
    - Venous drainage of bulb into portal vein (via prepyloric vein), while remainder of duodenum drains into superior and inferior pancreaticoduodenal veins
    - Lymphatics follows arterial supply and drain to celiac/superior mesenteric nodes
    - Innervation: Sympathetic via celiac plexus and parasympathetic via vagal trunks

CLINICAL IMPLICATIONS

Clinical Importance

- Gastric and duodenal ulcers are common
  - Rich vascular supply of stomach and duodenum makes upper GI bleeding common result
- **Gastritis** presents with thickening of gastric folds
- **Gastric malignancies**
  - Rich lymphatic and venous drainage makes nodal and liver metastases common at time of diagnosis
- **Gastric** and **duodenal diverticula** are common and usually asymptomatic but can mimic other pathology

FRONTAL VIEW OF STOMACH AND GASTRIC MUCOSA

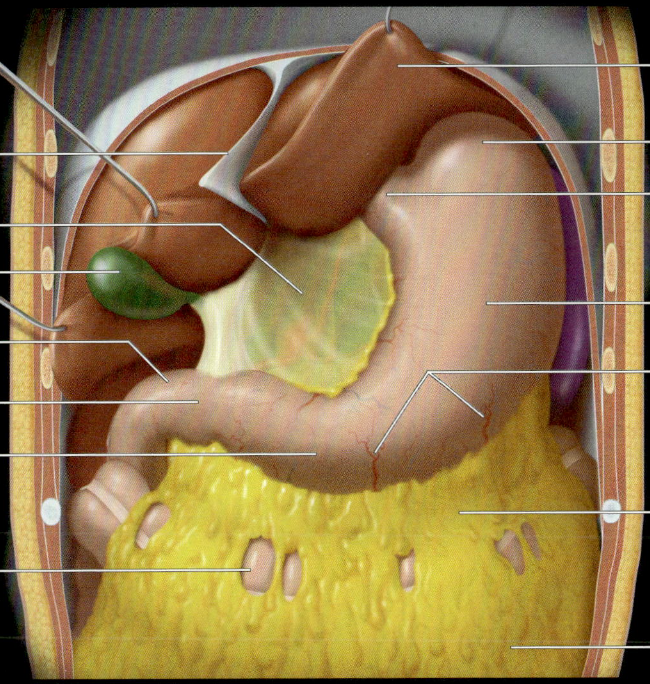

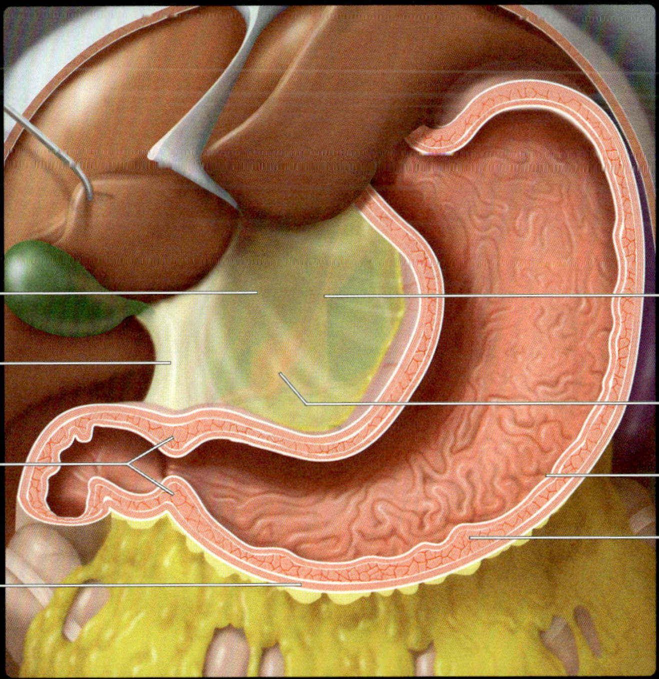

(Top) *The liver and gallbladder have been retracted upward in this graphic. Note that the lesser curvature and anterior wall of the stomach touch the underside of the liver, and the gallbladder abuts the duodenal bulb. The greater curvature is attached to the transverse colon by the gastrocolic ligament, which continues inferiorly as the greater omentum, covering most of the colon and small bowel.* **(Bottom)** *The lesser omentum extends from the stomach to the porta hepatis and can be divided into the broader and thinner hepatogastric ligament and the thicker hepatoduodenal ligament. The lesser omentum carries the portal vein, hepatic artery, common bile duct, and lymph nodes. The free edge of the lesser omentum forms the ventral margin of the epiploic foramen, the passage connecting the lesser sac with the main peritoneal cavity. The celiac artery can be seen through the surface of the lesser sac. Note the layers of gastric muscle with the middle circular layer being the thickest.*

DUODENUM, FRONTAL AND INTERNAL FOLD PATTERN

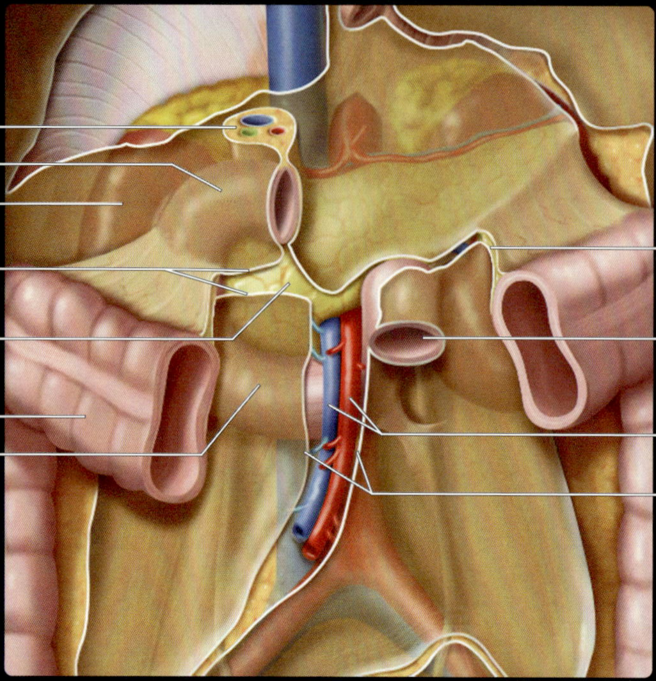

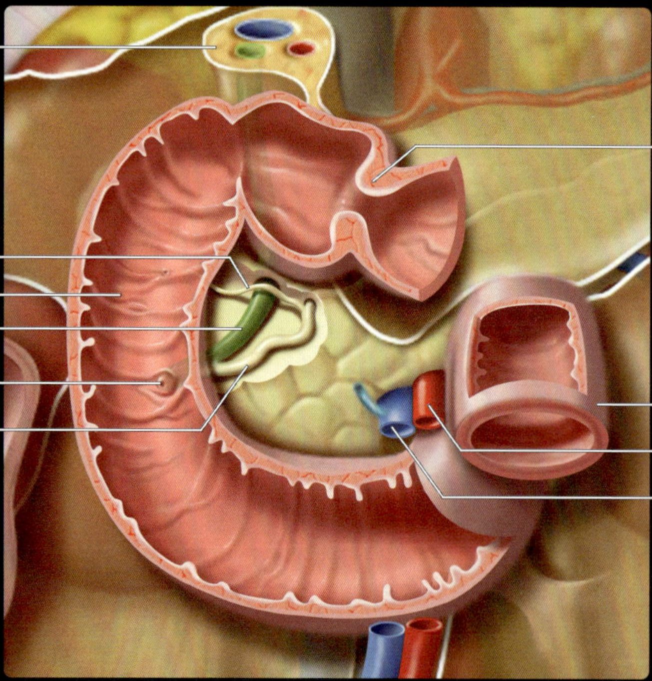

(Top) *The duodenum is retroperitoneal, except for the intraperitoneal duodenal bulb (1st part), while the proximal jejunum is completely intraperitoneal. The hepatoduodenal ligament attaches the duodenum to the porta hepatis and contains the portal triad (bile duct, hepatic artery, portal vein). The root of the transverse mesocolon and mesentery both cross the duodenum. The 3rd portion of the duodenum crosses in front of the aorta and inferior vena cava and behind the superior mesenteric vessels [superior mesenteric artery (SMA) and superior mesenteric vein (SMV)]. The 2nd portion of duodenum is attached to the pancreatic head and lies close to the hilum of the right kidney.* **(Bottom)** *The duodenal bulb is suspended by the hepatoduodenal ligament. The duodenojejunal flexure is suspended by the ligament of Treitz, an extension of the right diaphragmatic crus. The major pancreaticobiliary papilla enters the medial wall of the 2nd portion of the duodenum. The duodenal wall consists of mucosa, submucosa, and 2 muscle layers.*

LYMPHATICS OF STOMACH AND SMALL INTESTINE

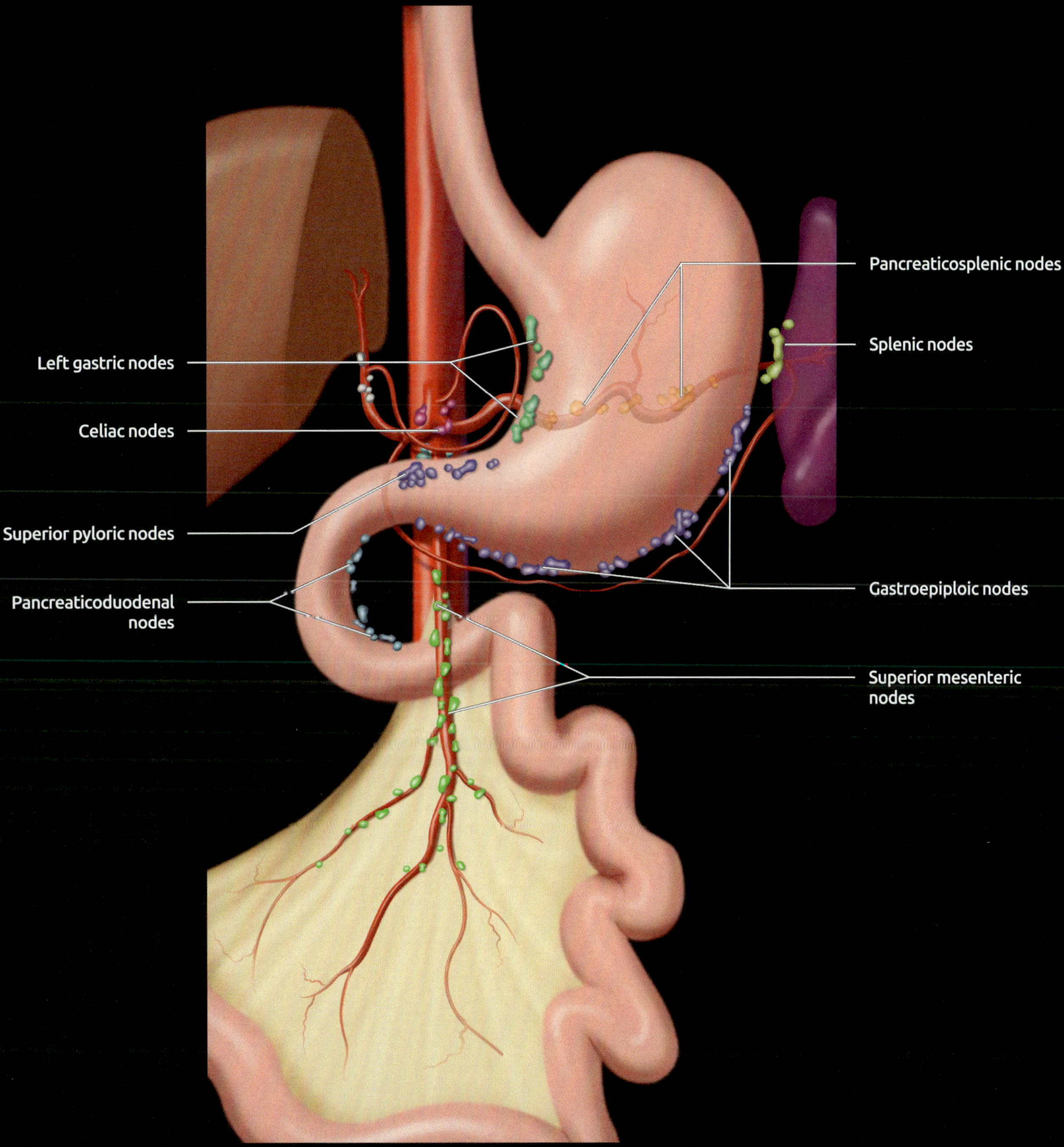

Gastric lymphatic vessels accompany arteries with similar names along the greater and lesser curvatures of the stomach. Lymph from the anterior and posterior surfaces of the stomach drains to the left gastric and gastroepiploic nodes, along with the pancreaticosplenic nodes. The pyloric nodes drain the inferior 1/3 of the stomach, primarily from the lesser curvature, while the distal part of the greater curvature and the duodenum drain to pancreaticoduodenal nodes. All of these nodal groups subsequently drain to the celiac nodes, which are grouped at the base of the celiac artery.

ARTERIES OF STOMACH AND DUODENUM

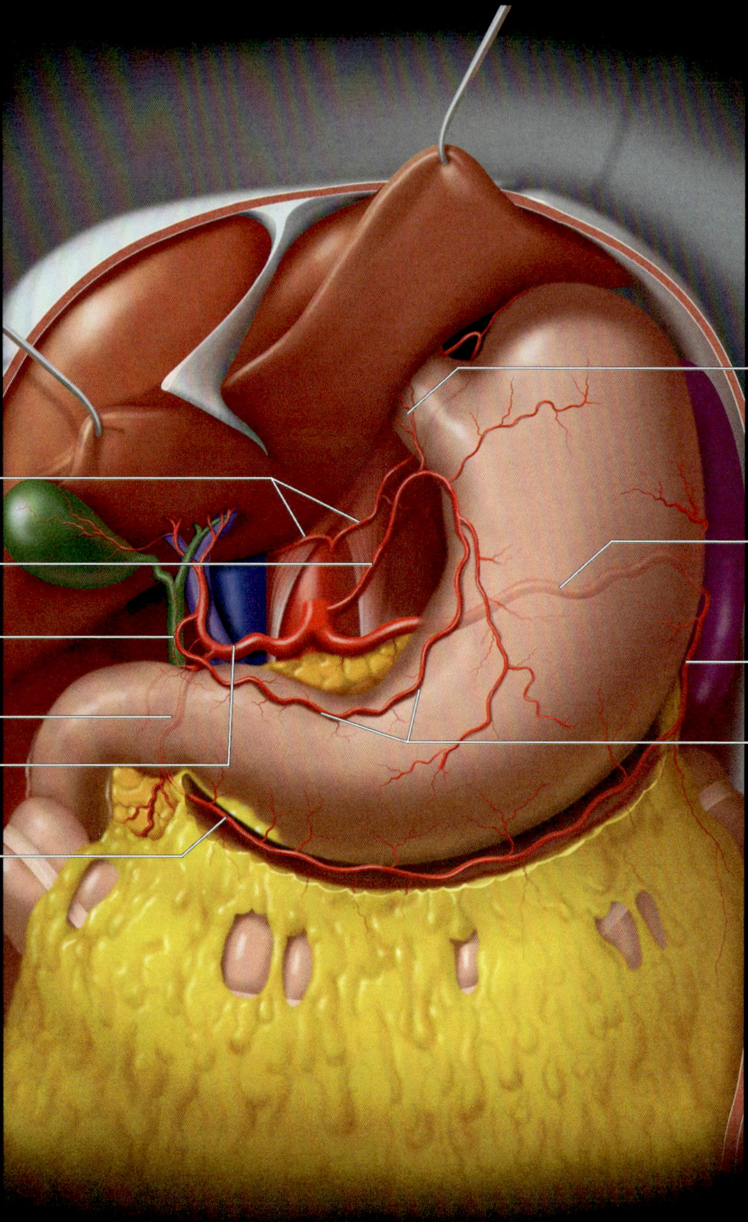

"Conventional" arterial anatomy of the stomach and duodenum (present in only ~ 50% of the population) has the left gastric artery arising from the celiac trunk, supplying the lesser curvature, and anastomosing with the right gastric artery, a branch of the proper hepatic artery. The gastroduodenal artery is the 1st major branch of the common hepatic artery and branches into the superior pancreaticoduodenal and right gastroepiploic arteries. The greater curvature of the stomach is supplied by anastomosing branches of the right and left gastroepiploic arteries with the left gastroepiploic artery arising from the splenic artery. The duodenum and pancreas are supplied by a rich pancreaticoduodenal "arcade" composed of multiple anastomosing branches from the gastroduodenal artery and SMA.

PORTAL VENOUS SYSTEM

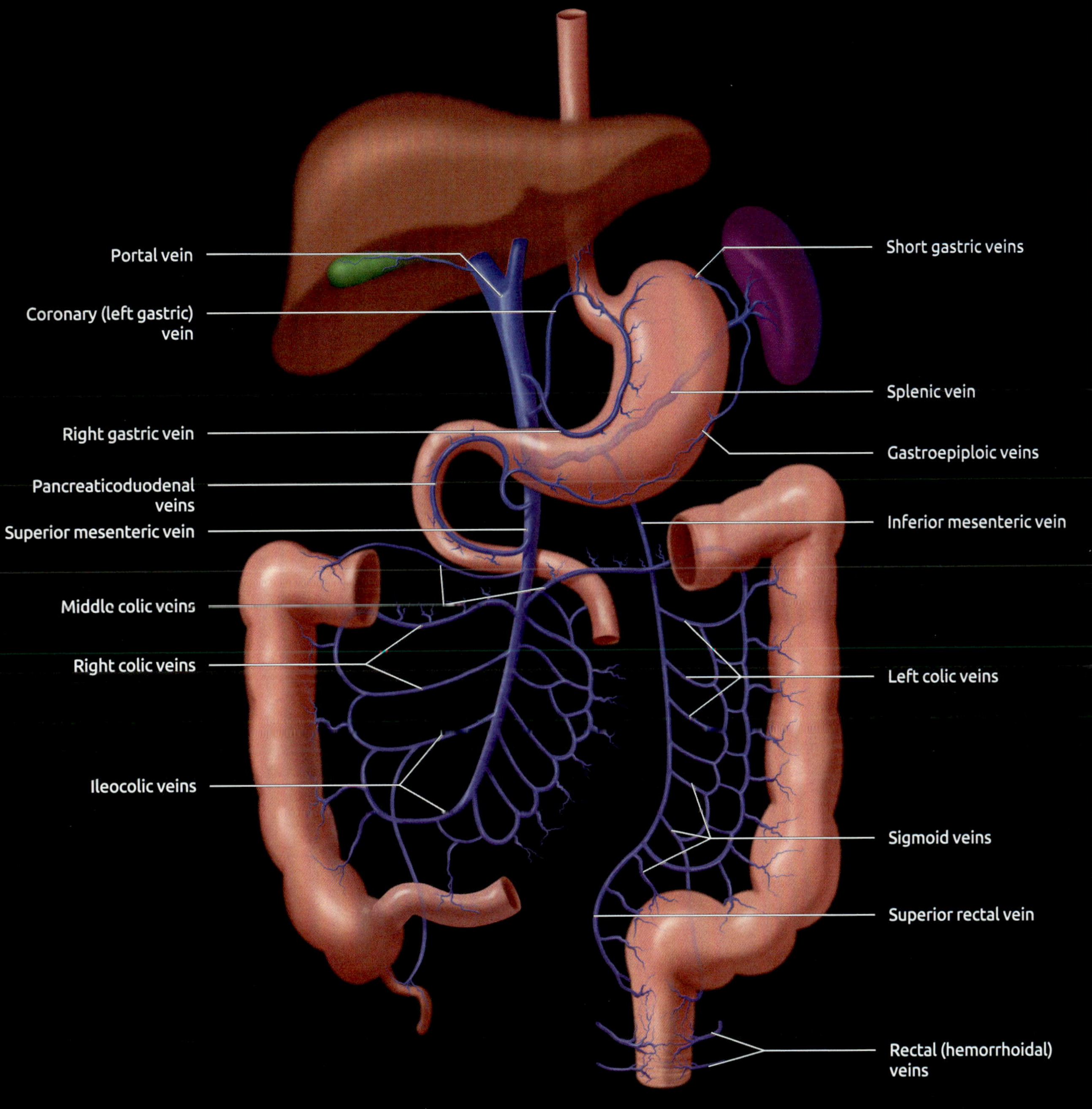

Venous drainage from the stomach, duodenum, small bowel, and colon (up to the splenic flexure) is to the SMV. Pancreatic venous drainage is to the splenic veins and SMV. The descending and sigmoid colon drain through the inferior mesenteric vein (IMV). The splenic vein, SMV, and IMV are the main tributaries of the portal vein. These veins are valveless and freely communicate through numerous collateral pathways, which become important and evident in cases of thrombosis or compression of veins [e.g., splenic vein occlusion by pancreatic cancer, leading to collateral flow through the coronary and short gastric veins (gastric varices)]. There are also extensive potential collaterals between the portal venous and systemic venous circulations that become important in portal hypertension, especially gastroesophageal, parumbilical, and hemorrhoidal varices.

NORMAL STOMACH

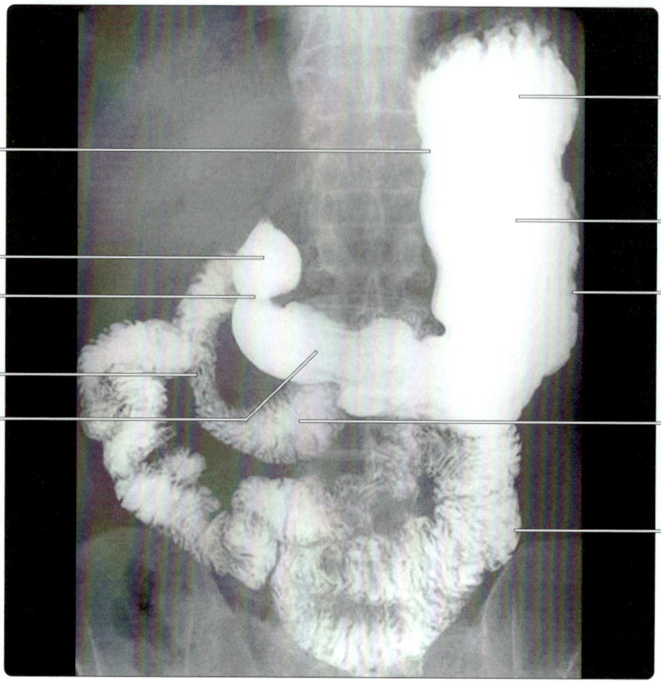

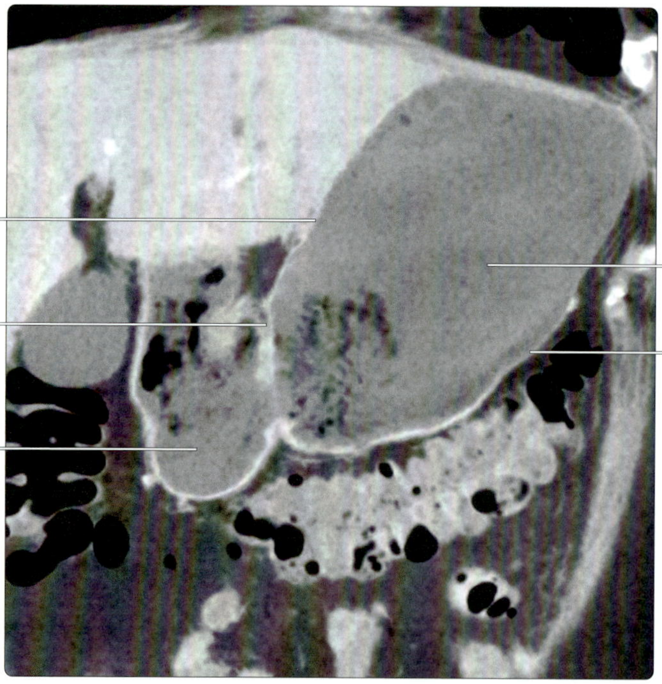

(Top) *Frontal film from a barium upper GI series is shown. The lesser curvature is the concave border, and the greater curvature is the convex border of the stomach. The fundus is the uppermost hood-like portion of the stomach, which intersects with the cardia, where the esophagus enters the stomach, at an acute angle. The body is the main portion, and the antrum is the distal part of the stomach, which empties into the duodenum through the pylorus.* **(Bottom)** *Minimum intensity projection CT of the stomach shows the anatomy.*

UPPER GI, NORMAL STOMACH AND DUODENUM

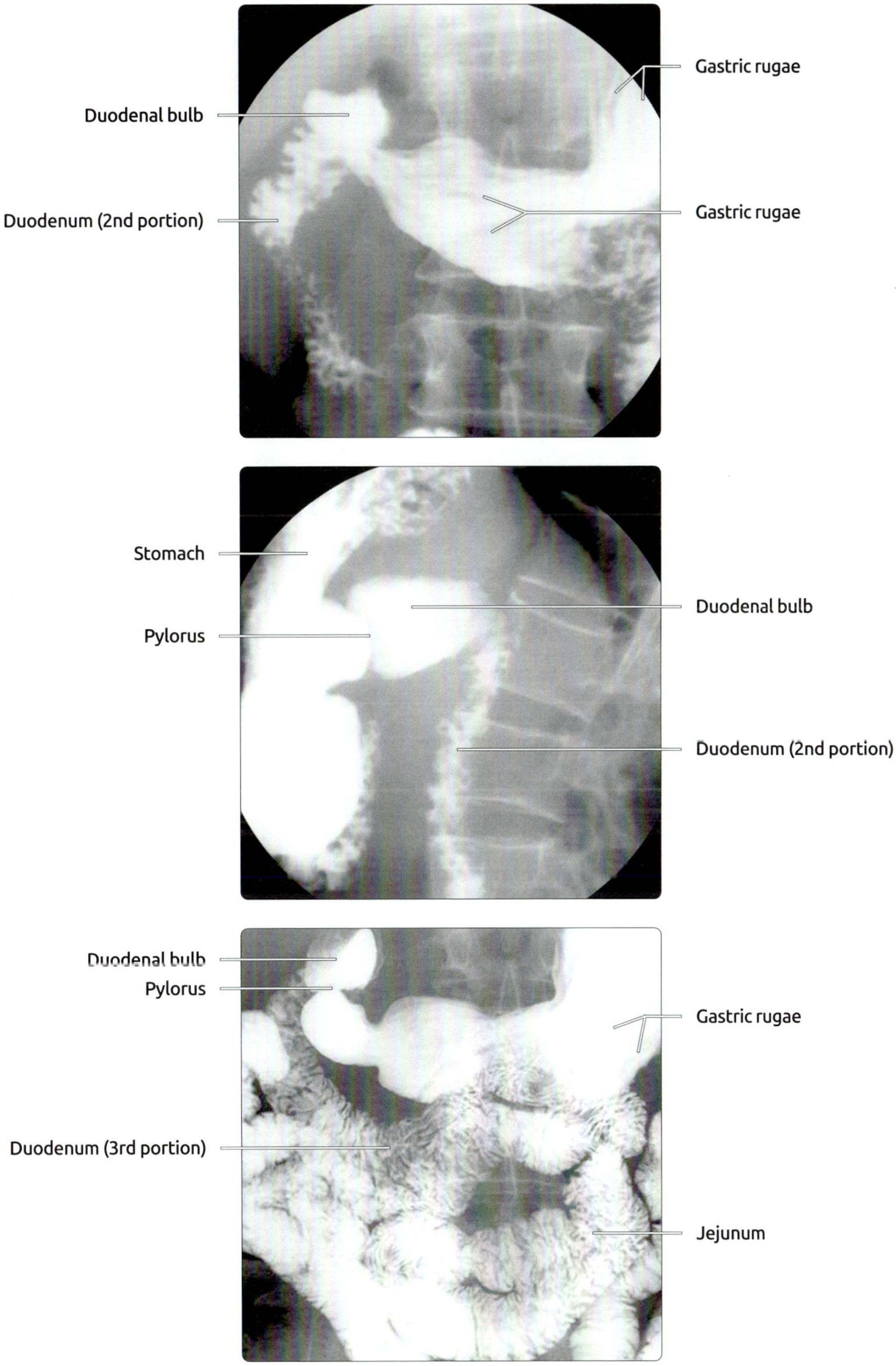

(Top) *First of 3 films from a barium upper GI series shows the distal stomach and duodenum. Note the normal gastric folds, or rugae, as smooth, linear filling defects in the barium pool.* **(Middle)** *Lateral image from an upper GI series shows the anterior and posterior walls of the stomach as well as the pylorus in profile. The duodenal bulb is well distended with its normal triangular shape.* **(Bottom)** *Another image from an upper GI series demonstrates the normal smooth mucosal surface of the duodenal bulb, as opposed to the normal feathery fold pattern of the remaining duodenum and jejunum.*

AXIAL CECT, NORMAL GASTRODUODENAL ANATOMY

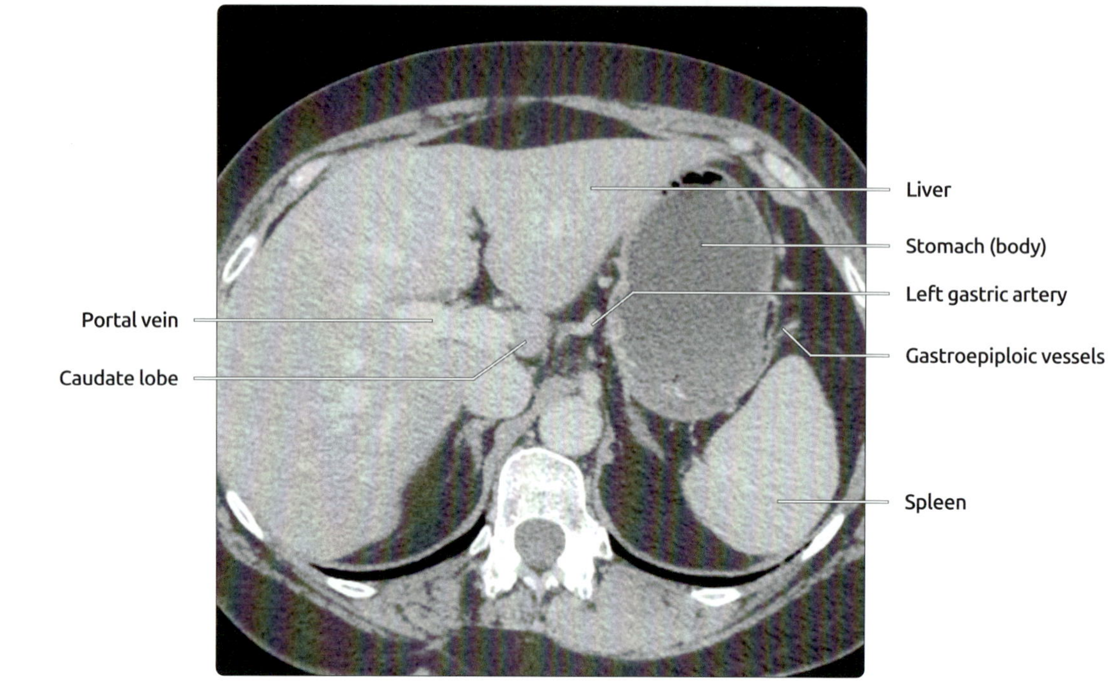

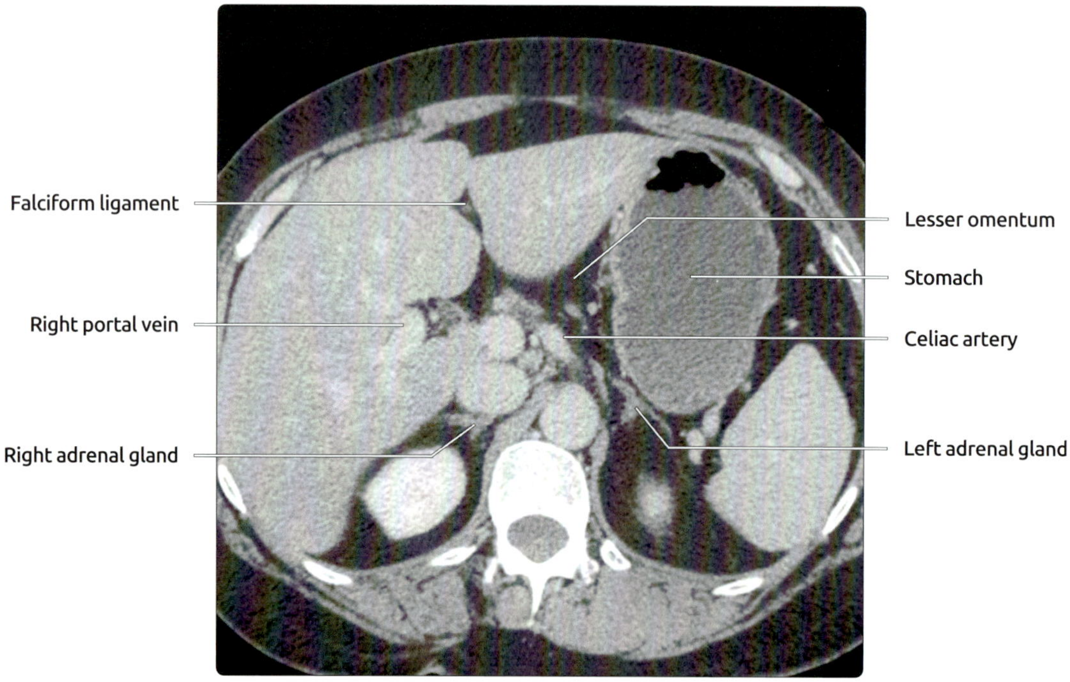

(Top) *First of 4 axial CECT sections shows the normal relationships of the stomach to adjacent organs. Note that the stomach may be compressed by an enlarged liver or spleen.* **(Bottom)** *Note the fat-containing lesser omentum that carries vessels and lymphatics to the stomach and liver.*

AXIAL CECT, NORMAL GASTRODUODENAL ANATOMY

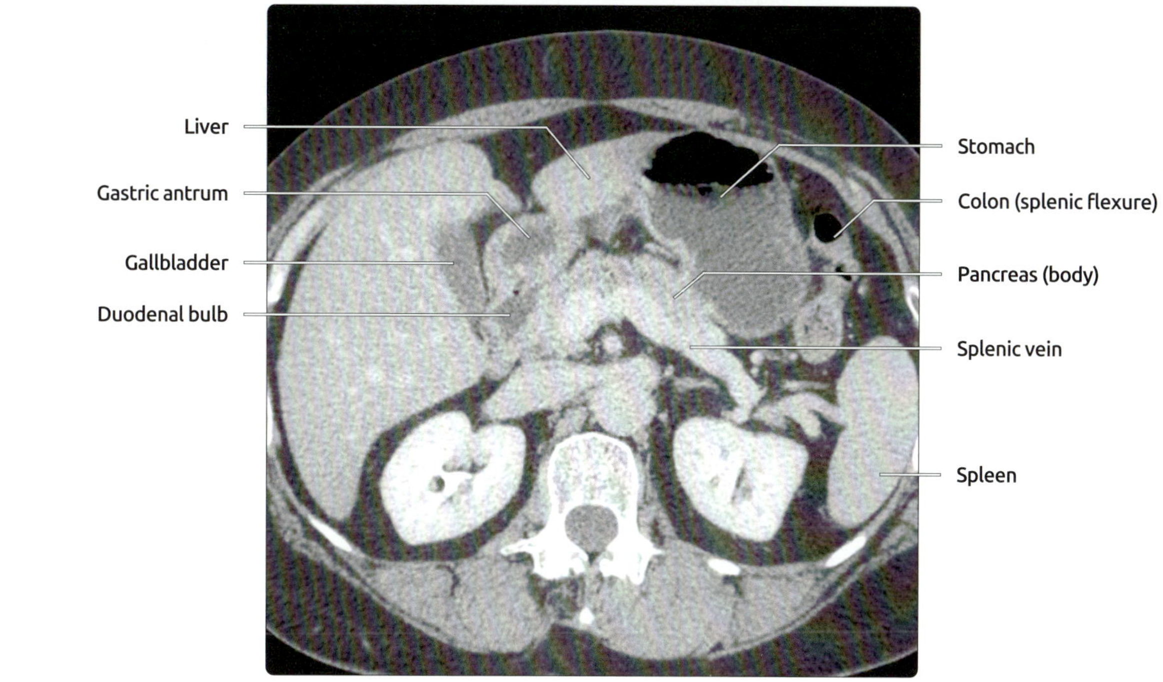

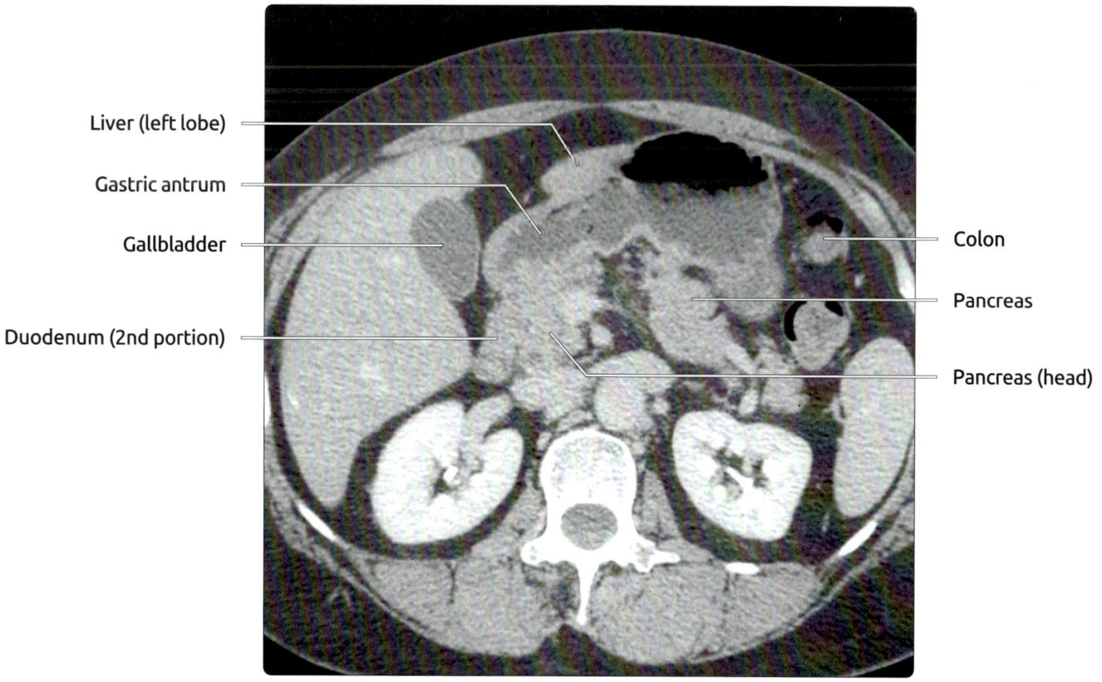

(Top) *The posterior wall of the stomach abuts the pancreas with only the lesser sac lying between them. The gallbladder abuts the gastric antrum and duodenal bulb. The greater curvature touches the splenic flexure of the colon.* **(Bottom)** *The gastric antrum abuts the pancreatic head posteriorly and the gallbladder laterally.*

GASTRIC RUGAE

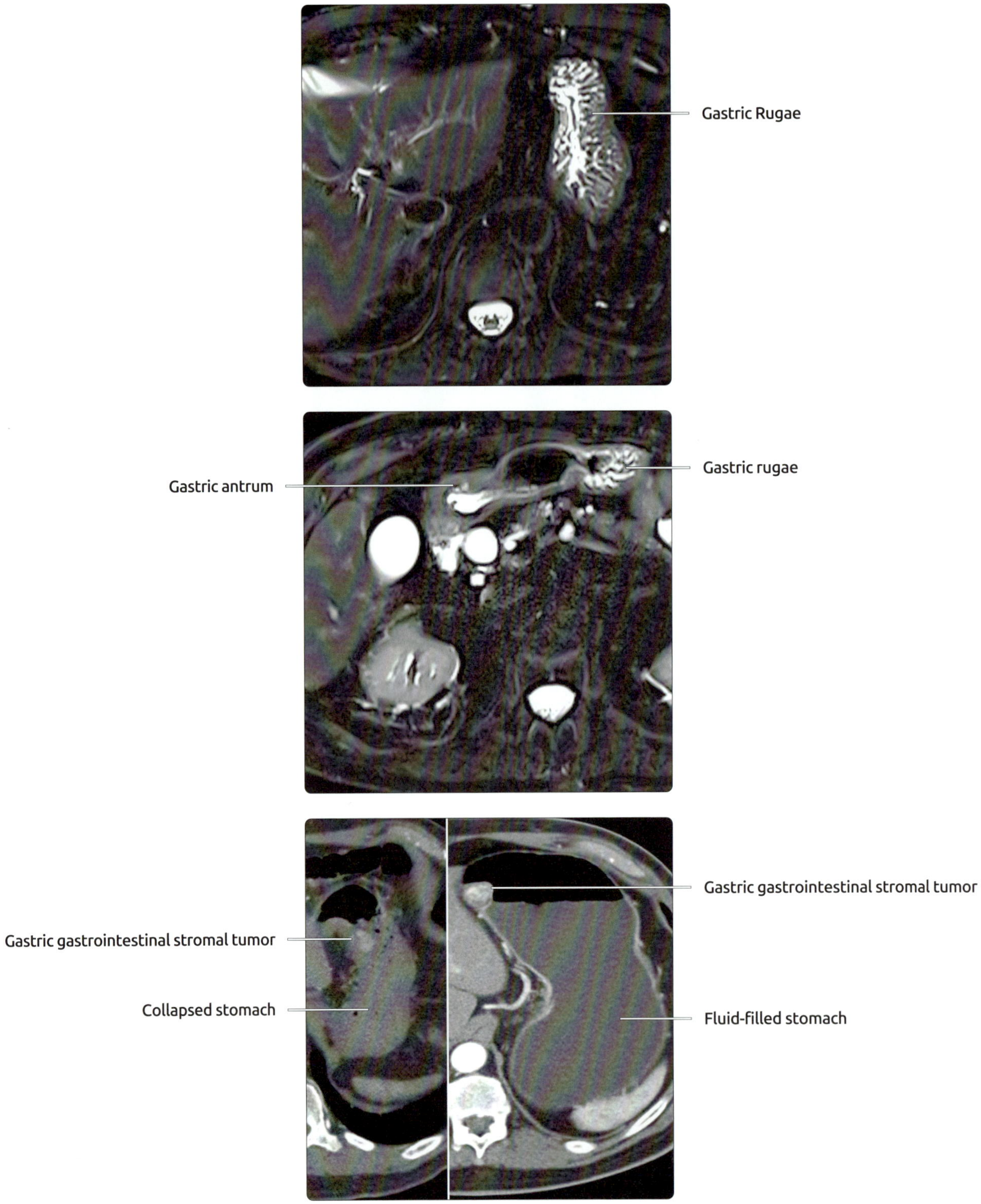

(Top) *Gastric rugae are redundant tubular folds of gastric mucosa are well visualized in parts of the proximal stomach. The thickness of the rugal folds varies with the degree of gastric distention.* **(Middle)** *Gastric folds are well visualized in the body of the stomach. Note the gastric antrum devoid of rugal folds.* **(Bottom)** *Gastric distention is important for the assessment of gastric anatomy. Note that in this patient with gastric gastrointestinal stromal tumor (GIST), the mass may missed easily, as it may be mistaken for food matter in a collapsed stomach (left image). In the same patient, gastric distention was obtained by drinking 2 cups of water right before the CT, and the endophytic mass is well appreciated (right image).*

CATHETER AND CT ANGIOGRAPHY, ARTERIAL VARIATIONS AND VARICES

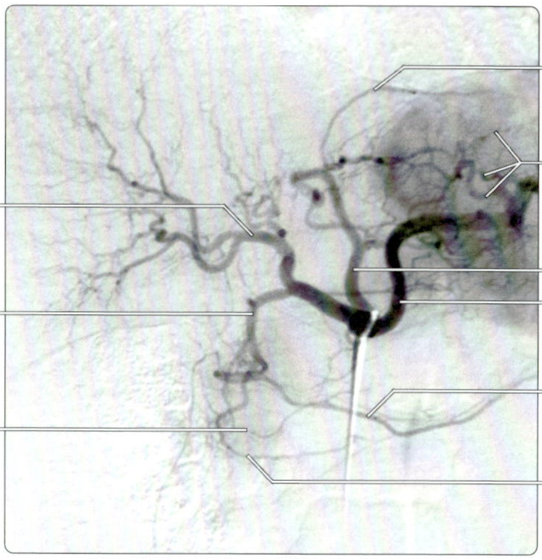

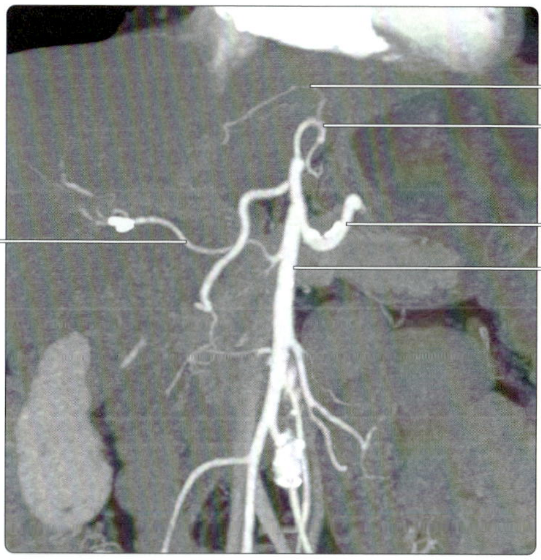

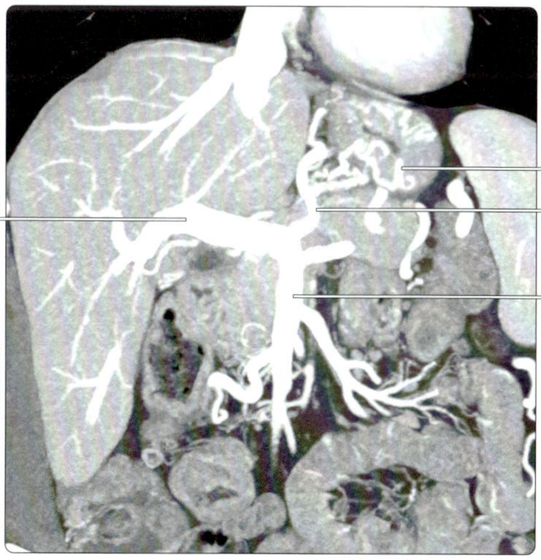

(Top) *Catheter angiogram shows the common hepatic artery arising from the celiac axis, its most common arrangement, and giving rise to the gastroduodenal and right hepatic arteries in this individual.* **(Middle)** *Coronal CECT with MIP reconstruction demonstrates the origin of the left hepatic artery from the left gastric artery, a common variant. Branches of the portal venous system are less well opacified due to the deliberate timing of the CT to show the arteries preferentially. The left hepatic artery is "replaced" to the left gastric artery in ~ 10% of individuals. Replaced right hepatic artery arising from the SMA, the most common hepatic arterial anatomic variant, is also present.* **(Bottom)** *Patient with history of necrotizing pancreatitis complicated by thrombosis of multiple peripancreatic veins (including the splenic vein) presenting with melena is shown. Note multiple gastric varices on the coronal portal venous-phase MIP image.*

ARTERIAL VARIATIONS, SEPARATE ORIGIN OF LEFT GASTRIC ARTERY

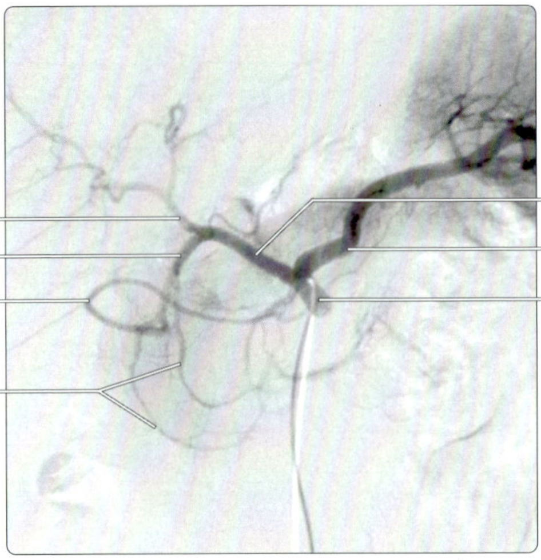

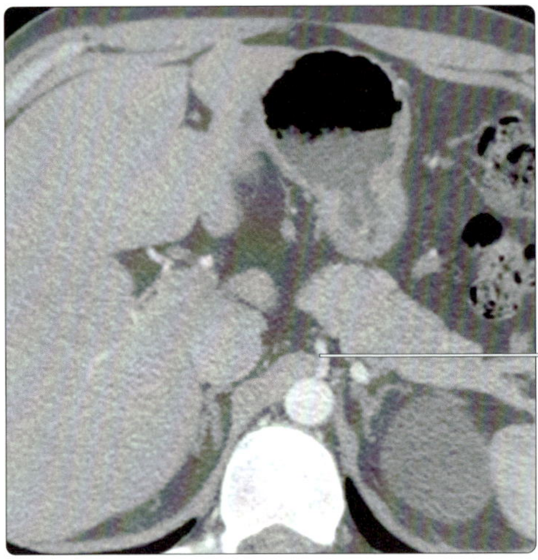

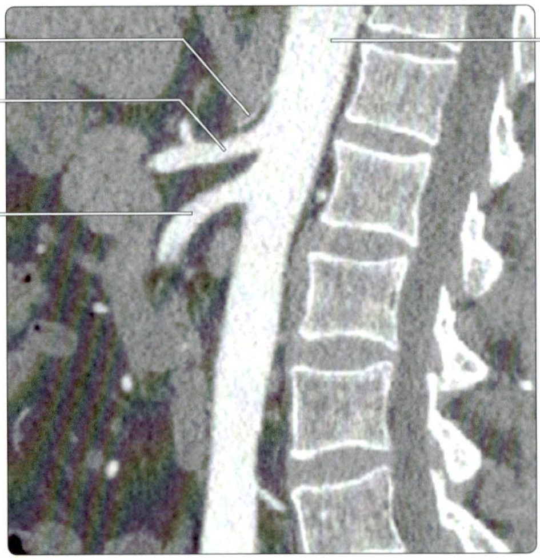

(Top) *Arterial phase of a celiac angiogram shows only the common hepatic and splenic arteries arising from the celiac trunk. The left gastric artery has a separate origin from the aorta in this individual, as shown on the subsequent CT.* **(Middle)** *Axial CECT through the upper abdomen shows the left gastric artery arising separately from the aorta, rather than having its usual origin from the celiac trunk.* **(Bottom)** *Sagittal CECT reconstruction in the same patient shows the origins of the celiac artery and SMA from the proximal abdominal aorta. Barely seen is the small left gastric artery, which has a separate origin from the aorta in this patient.*

BRUNNER GLAND HYPERPLASIA

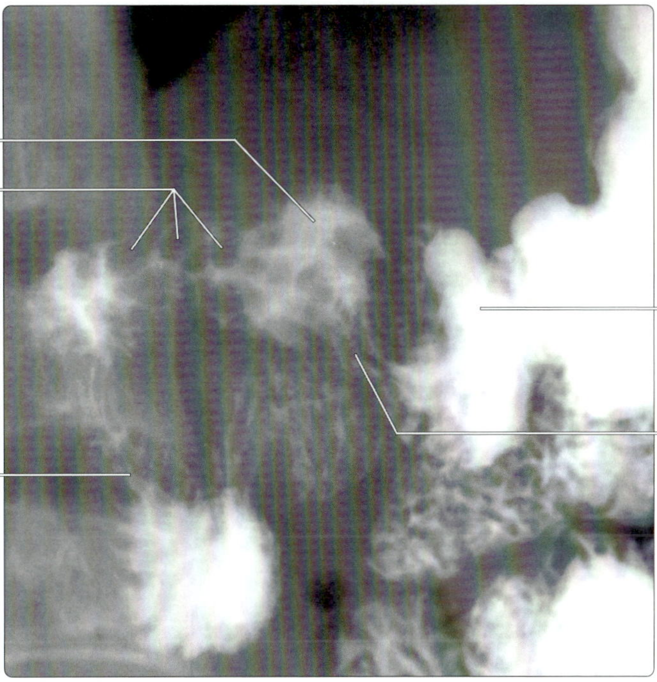

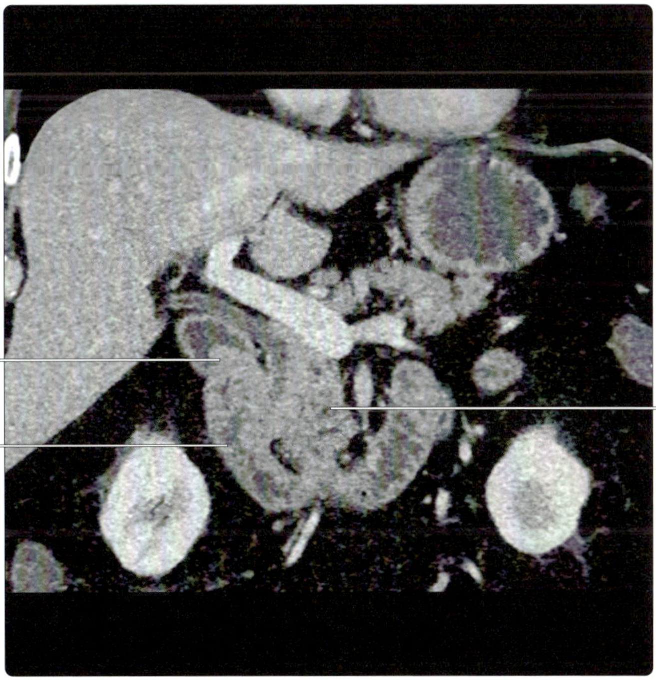

(Top) *Image from an upper GI series shows numerous polypoid filling defects in the duodenal bulb characteristic of hyperplastic Brunner glands. These glands are normal constituents of the duodenal wall, more numerous in the bulb and 2nd portion. They are normally only 1 or 2 mm in diameter and are usually not evident on radiographic studies. The Brunner glands secrete a clear fluid that contains mucus and a weak proteolytic enzyme acting in an acid milieu.* **(Bottom)** *Coronal CT in a patient with abdominal pain shows a duodenal mass. Endoscopy and cytology were indeterminate. Pancreaticoduodenectomy showed chronic inflammation and Brunner gland hyperplasia of the duodenum without cancer.*

PERFORATED DUODENAL ULCER

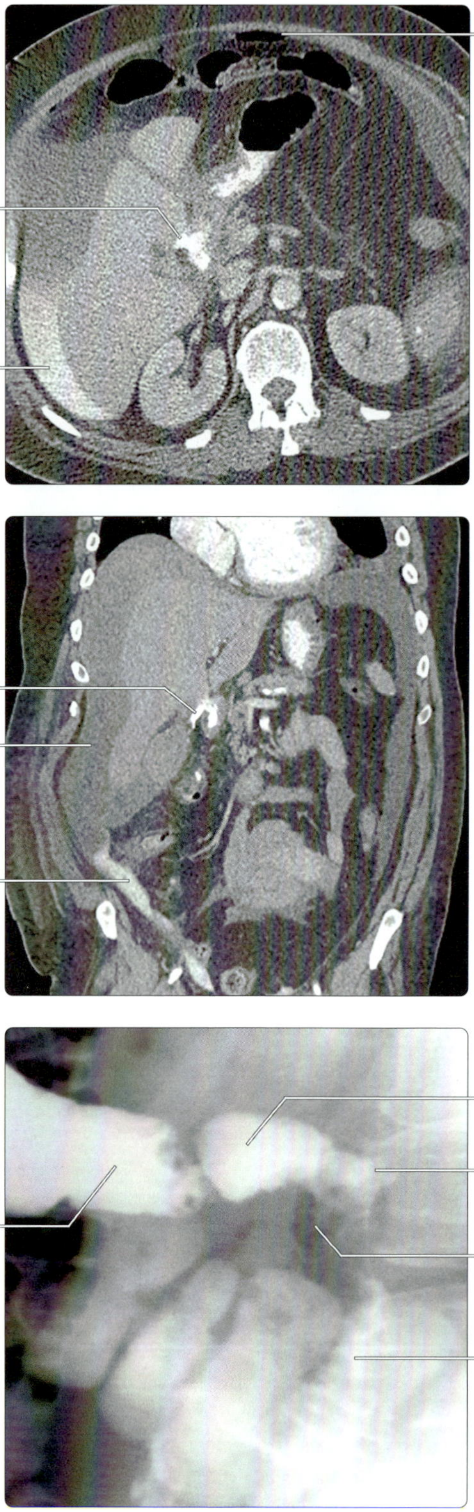

(Top) *Axial CECT in a patient with acute abdominal pain demonstrates large ascites and a small amount of free air in the anterior abdomen. Most importantly, there is extraluminal enteric contrast material from the duodenum as well as layering high-density contrast within the perihepatic ascites.* **(Middle)** *Coronal CECT in the same patient nicely demonstrates extraluminal contrast from the duodenum with associated extensive free ascites (and free contrast material). This was found at surgery to be secondary to a perforated duodenal ulcer.* **(Bottom)** *Oblique film from an upper GI series shows an ulcer at the apex of the duodenum (postbulbar segment that begins the 2nd portion of duodenum). The lumen of the adjacent segment of duodenum is narrowed due to spasm, and extraluminal gas is present. A perforated ulcer was confirmed at surgery. While the duodenal bulb is intraperitoneal, the rest of the duodenum is retroperitoneal, which explains the presence of retroperitoneal inflammation and gas in this case.*

PERFORATED DUODENAL ULCER

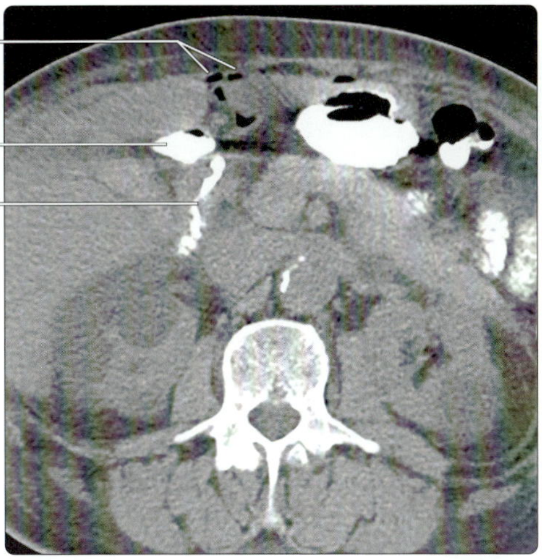

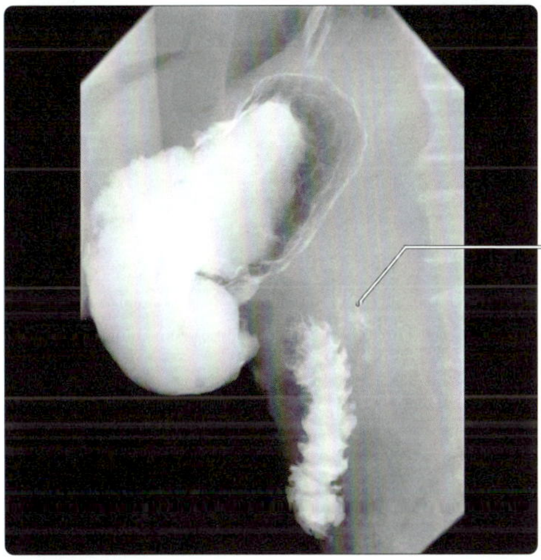

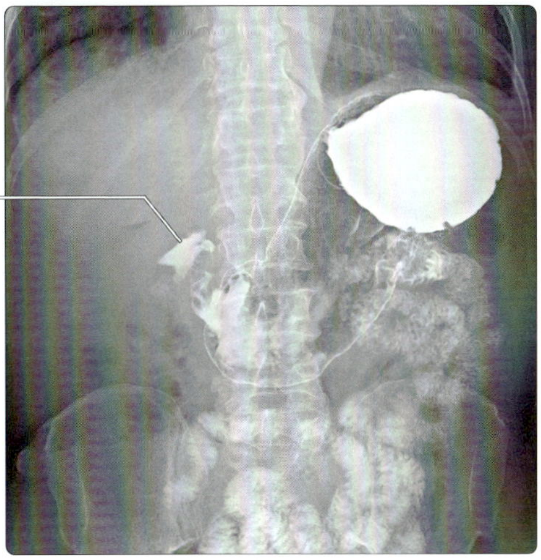

(Top) *Axial NECT in a patient with acute abdominal pain demonstrates extravasation of enteric contrast from the duodenum (with contrast pooling immediately anterior to the duodenal bulb). Note the presence of several foci of free intraperitoneal air.* **(Middle)** *Image from an upper GI series in the same patient demonstrates subtle contrast extravasation from the proximal duodenum, corresponding to the site of extravasation on CT. This was subsequently determined to be secondary to a perforated duodenal ulcer.* **(Bottom)** *Frontal image from an upper GI series in the same patient nicely demonstrates the pooling of ectopic contrast immediately adjacent to the duodenal bulb, corresponding to the collection of contrast also seen on CT.*

PERFORATED DUODENAL ULCER, TISSUE PLANES

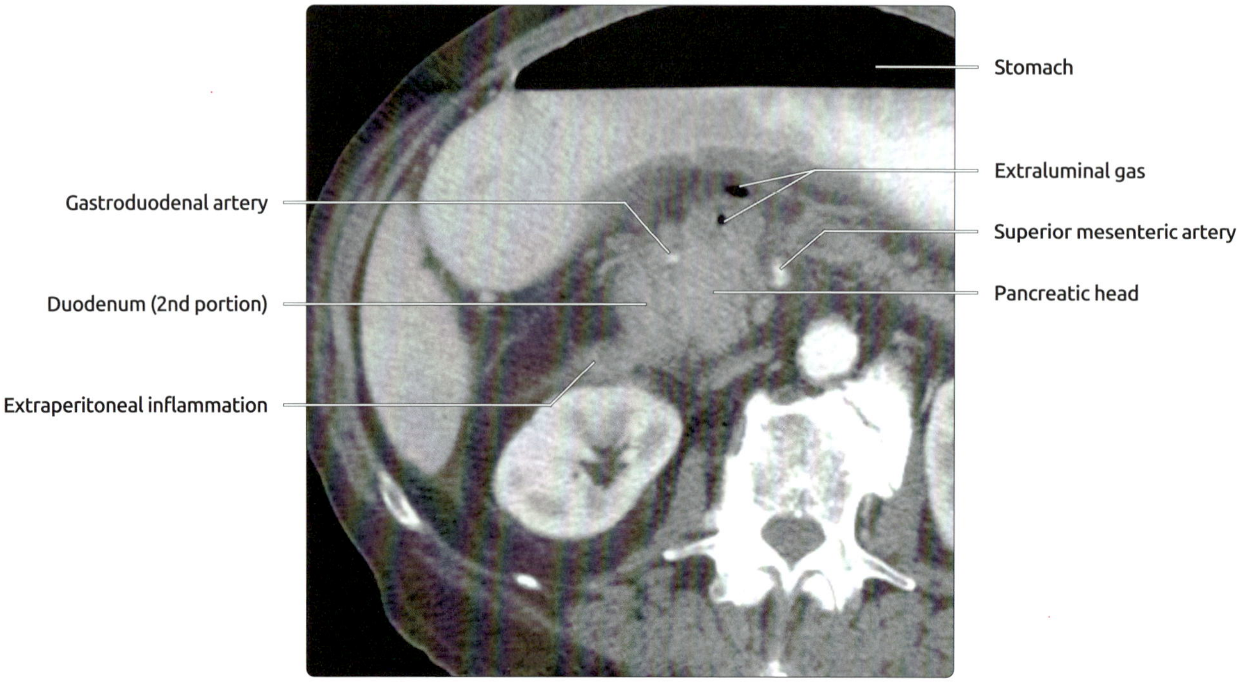

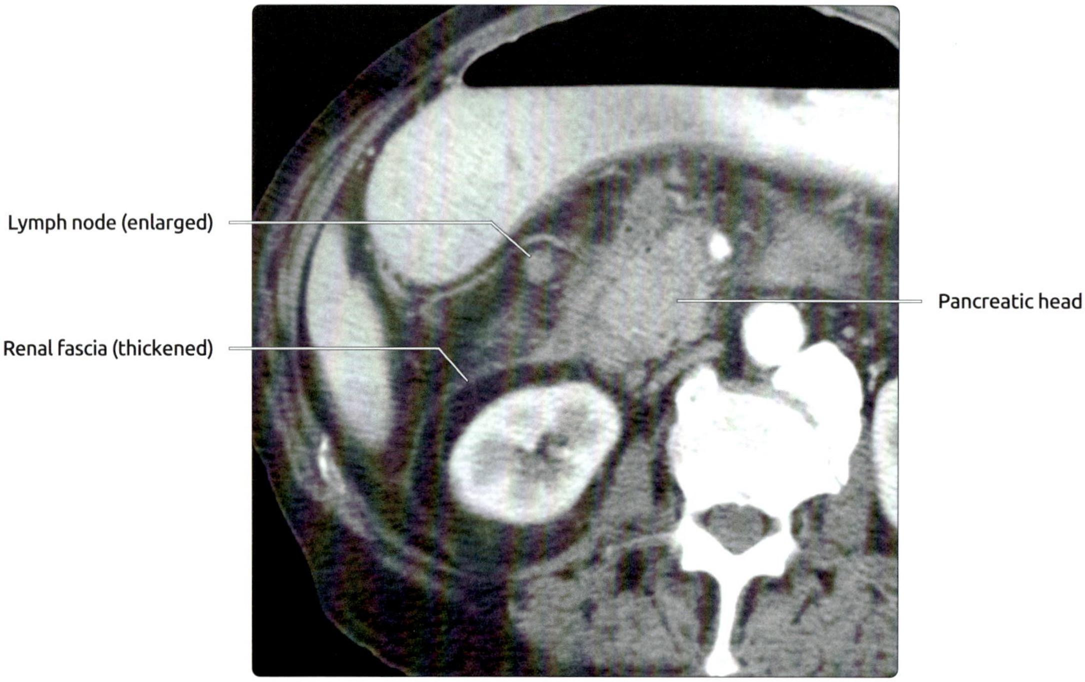

(Top) *Axial CECT in an older man with acute abdominal pain shows extraluminal gas near the 2nd portion of the duodenum and infiltration of the retroperitoneal (anterior pararenal) space on the right. The gastroduodenal artery lies in a plane between the pancreatic head and the 2nd portion of duodenum.* **(Bottom)** *The 2nd portion of the duodenum always lies immediately lateral to the pancreatic head. The lumen of the duodenum is collapsed, and the adjacent retroperitoneal fascial planes and spaces are infiltrated. An adjacent lymph node is enlarged due to inflammation (reactive hyperplasia). A perforated ulcer was subsequently confirmed. Inflammation of the pancreas or duodenum will result in infiltration of the anterior pararenal space and thickening of the renal fascia, which forms the dorsal boundary to this space.*

PERFORATED GASTRIC ULCER

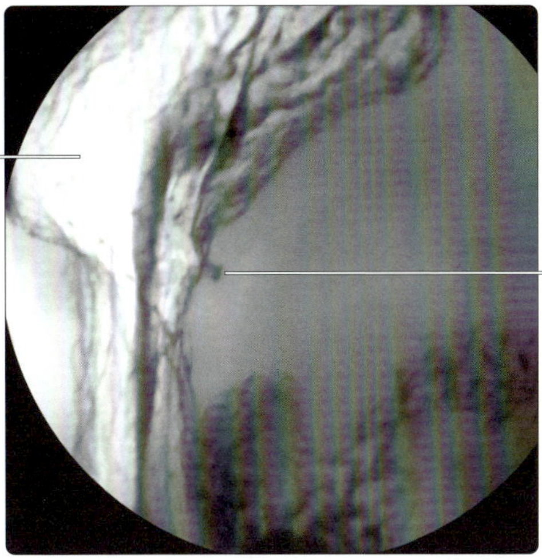

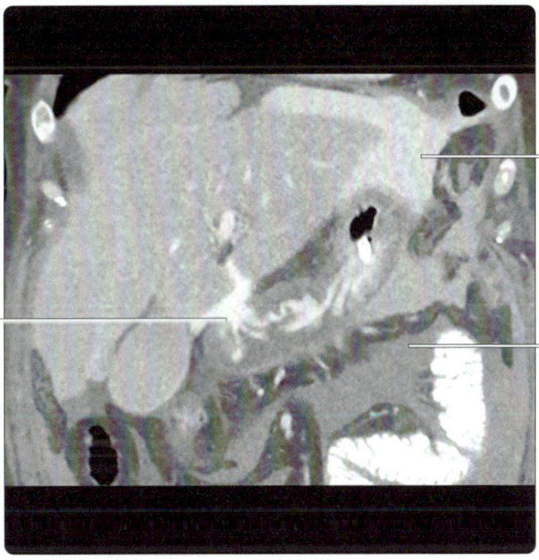

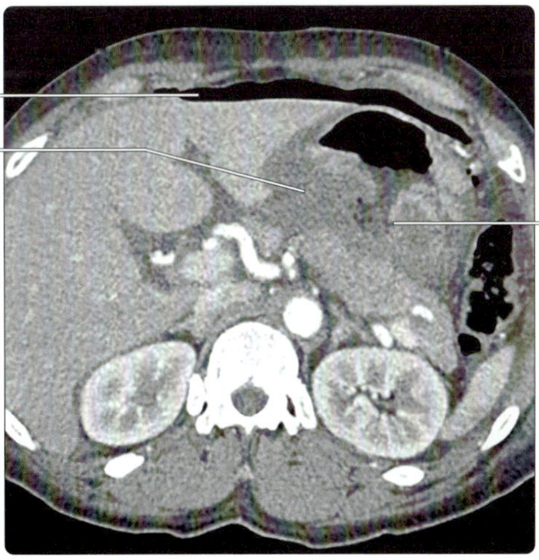

(Top) *Lateral view from an air contrast upper GI series shows a focal outpouching from the posterior wall of the stomach, diagnostic of a benign gastric ulcer. The ulcer was confirmed by endoscopy and resolved with medical therapy. Recall that the lesser sac lies immediately adjacent to the posterior wall of the stomach.* **(Middle)** *Coronal CECT in a patient shows the antral gastric wall defect with extraluminal-positive oral contrast and free fluid. Endoscopy revealed a perforated ulcer.* **(Bottom)** *Axial CECT shows an ulcer in the posterior gastric wall with the surrounding wall demonstrating severe thickening. In addition, there is extensive free air in the upper abdomen. This constellation of findings is virtually diagnostic of a perforated gastric ulcer.*

GASTRIC ADENOCARCINOMA

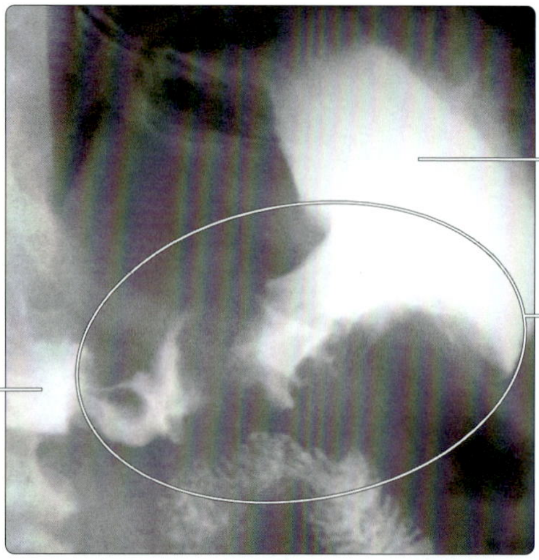

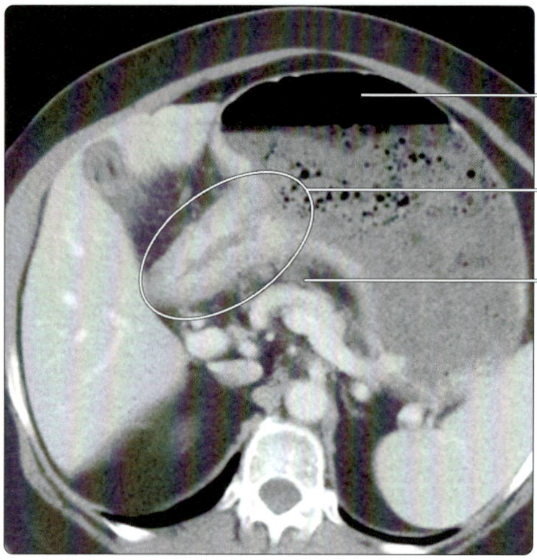

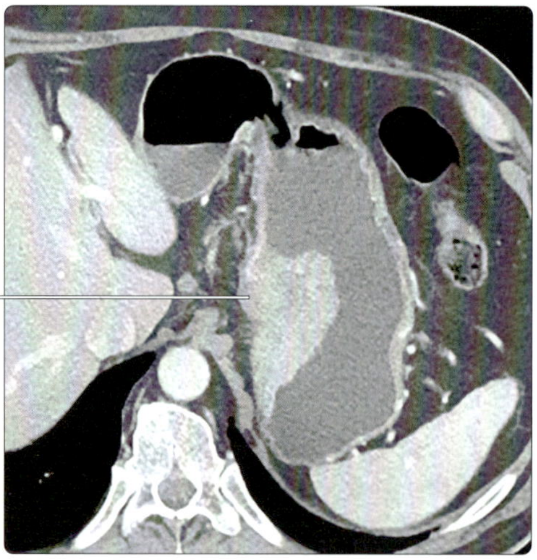

(Top) *Frontal film from an upper GI series in an older man with early satiety and weight loss shows marked irregular thickening of the wall of the stomach throughout the distal body and antrum. There is an abrupt transition between the normally distensible fundus and proximal body and the fixed, nondistensible distal stomach. The duodenal bulb is normal.* **(Middle)** *Axial CECT demonstrates abrupt transition from the dilated proximal stomach with its thin wall to the narrowed lumen and thick wall of the distal stomach. When carcinoma affects the distal stomach, it often results in a rigid, nondistensible condition that obstructs gastric emptying. Spread to upper nodal chains is also typical.* **(Bottom)** *Axial CECT demonstrates a large, polyploid mass arising from the lesser curvature of the stomach, confirmed at endoscopy to represent a gastric adenocarcinoma.*

GASTRIC ADENOCARCINOMA

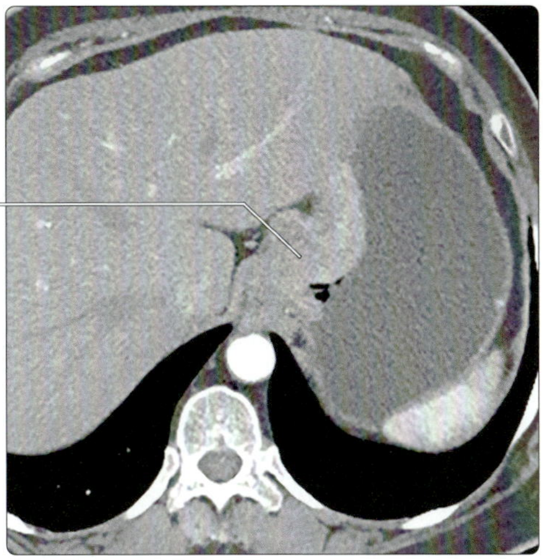

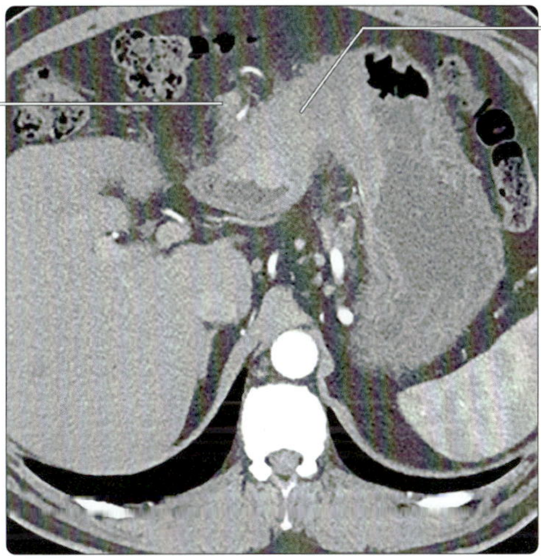

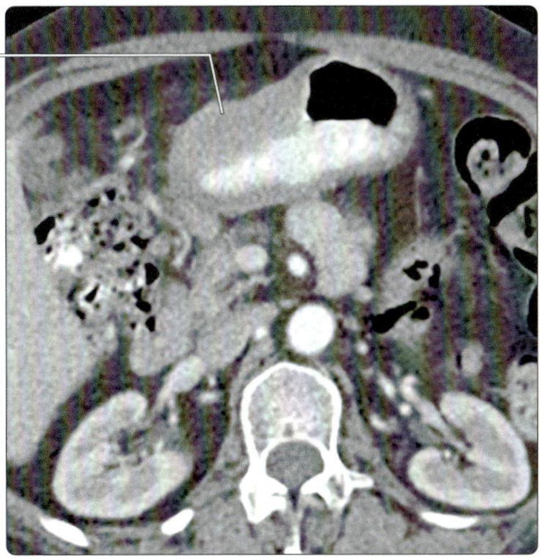

(Top) *Axial CECT demonstrates extensive, mass-like wall thickening along the lesser curvature of the stomach. While gauging wall thickening can sometimes be difficult on CT, particularly when the stomach is not well distended, the presence of focal wall thickening should always raise concern for malignancy.* **(Middle)** *Axial CECT demonstrates severe wall thickening of the distal stomach. Notice the discrete tumor nodule extending into the fat between the stomach and the transverse colon, representing spread of tumor into the gastrocolic ligament. The gastrocolic ligament serves as a potential conduit for the spread of malignancy from the stomach to the adjacent colon.* **(Bottom)** *Axial CECT demonstrates focal wall thickening along the anterior aspect of the gastric antrum/pylorus. While this region can be difficult to evaluate on CT, as a result of pyloric contractions simulating wall thickening, the manner in which the wall thickening in this case is asymmetrically involving the anterior wall should suggest the correct diagnosis of gastric adenocarcinoma.*

GASTRIC CARCINOMA WITH GASTROCOLIC LIGAMENT TUMOR EXTENSION

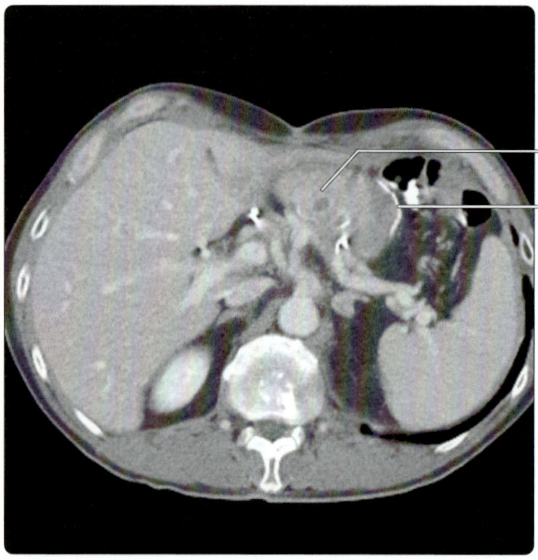

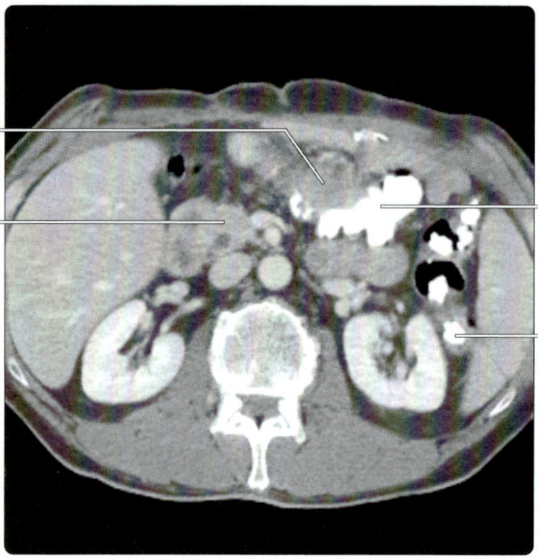

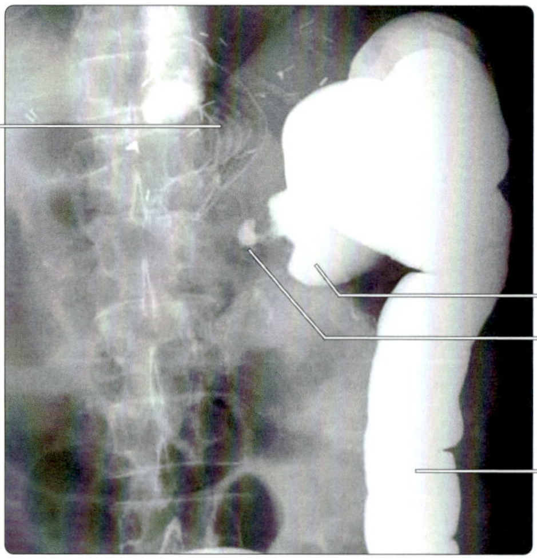

(Top) *First of 3 axial CECT images in a patient with recurrence of gastric carcinoma after prior resection shows a recurrent tumor mass adjacent to a line of sutures that marks the resection margin following partial gastrectomy.* **(Middle)** *More caudal CECT section shows tumor in the gastrocolic ligament, the portion of the greater omentum that connects the greater curvature of the stomach with the transverse colon. The tumor distorts and narrows the lumen of the transverse colon.* **(Bottom)** *Frontal film from a barium enema shows complete obstruction to retrograde flow of barium in the transverse colon. Note the metallic stent that had been inserted into the stomach in an attempt to prevent complete gastric outlet obstruction. This serves as a radiologic marker for the gastric tumor and shows how close the gastric tumor is to the colon, which led to the tumor invasion and obstruction of the colon.*

OTHER GASTRIC MASSES

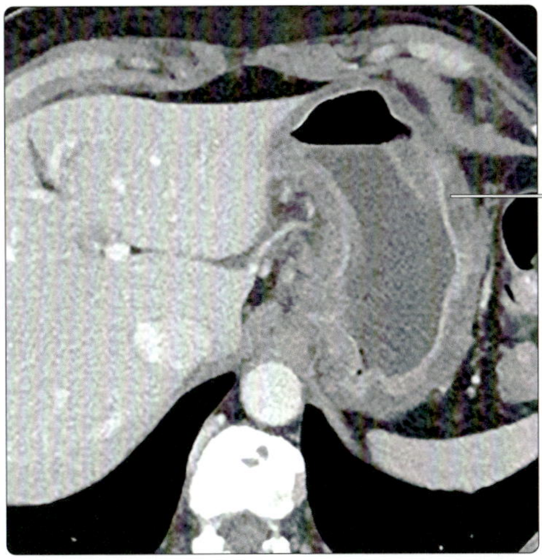

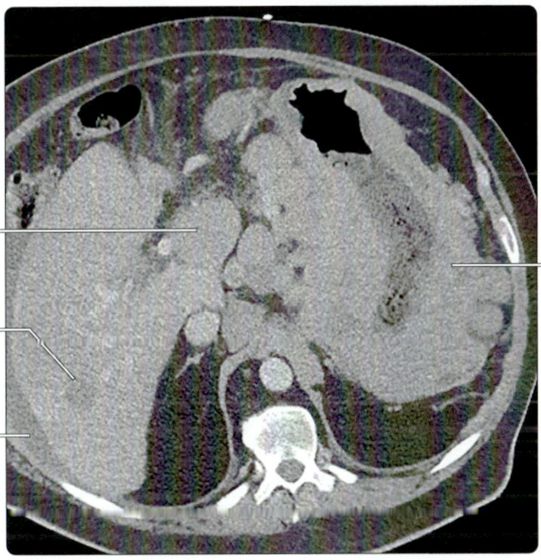

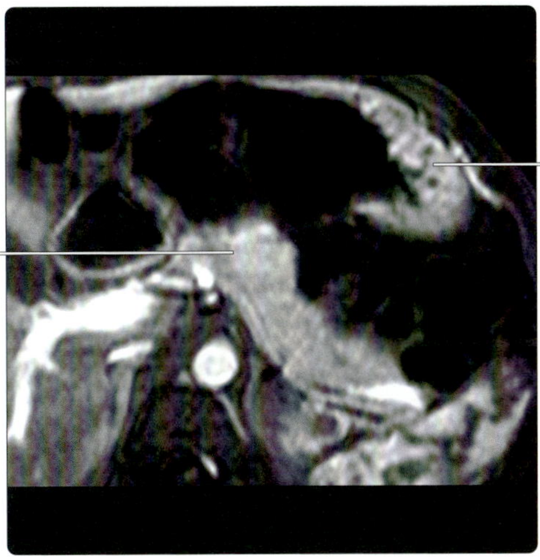

(Top) *Axial CECT in a patient with a history of breast cancer demonstrates diffuse wall thickening of the entire stomach. While diffuse gastric thickening is usually inflammatory or infectious, this turned out to be a case of linitis plastica with diffuse tumor infiltration of the entire gastric wall by metastatic breast cancer.* **(Middle)** *Axial CECT demonstrates extensive circumferential wall thickening throughout the entire stomach along with multiple bulky locoregional lymph nodes (particularly in the gastrohepatic ligament). Also note the subtle hypodense liver lesion in the right lobe. When confronted with this degree of lymphadenopathy, lymphoma must be considered, and this was ultimately proven to represent non-Hodgkin lymphoma.* **(Bottom)** *Arterial-phase postcontrast MR shows thickened gastric folds secondary to gastrin-producing neuroendocrine tumor (gastrinoma) in the pancreatic body.*

DUODENAL ADENOCARCINOMA

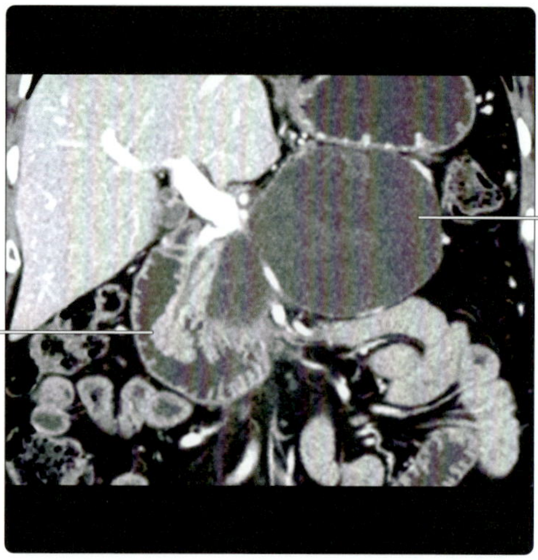

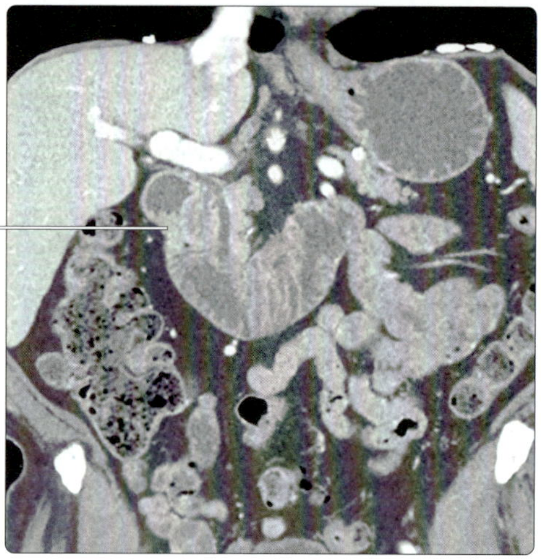

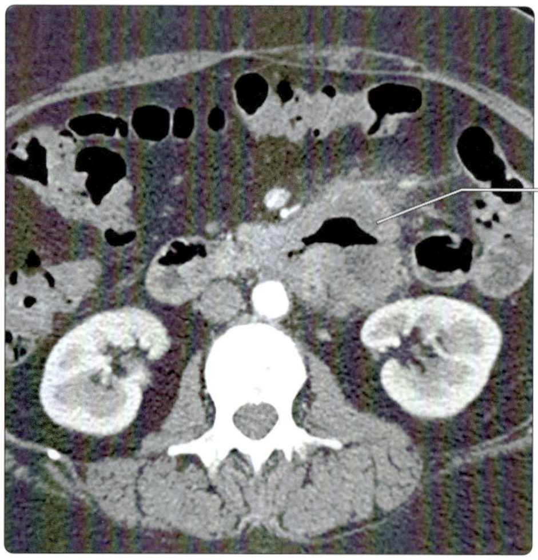

(Top) *Coronal CECT in a middle-aged woman shows acute pancreatitis with pseudocyst formation secondary to an obstructing ampullary mass. The mass was surgically removed, and pathology showed tubulovillous adenoma with high-grade dysplasia.* **(Middle)** *Coronal CECT demonstrates an annular constricting mass in the periampullary duodenum, found to represent a duodenal adenocarcinoma. Compared to other potential malignancies in the region of the ampulla (such as pancreatic adenocarcinoma, distal common bile duct cholangiocarcinoma, or ampullary carcinoma), duodenal adenocarcinoma carries a more favorable prognosis.* **(Bottom)** *Axial CECT demonstrates focal mass-like wall thickening of the transverse duodenum with associated aneurysmal dilatation and ulceration, found to represent a duodenal adenocarcinoma at surgical resection.*

GASTROINTESTINAL STROMAL TUMORS OF STOMACH AND DUODENUM

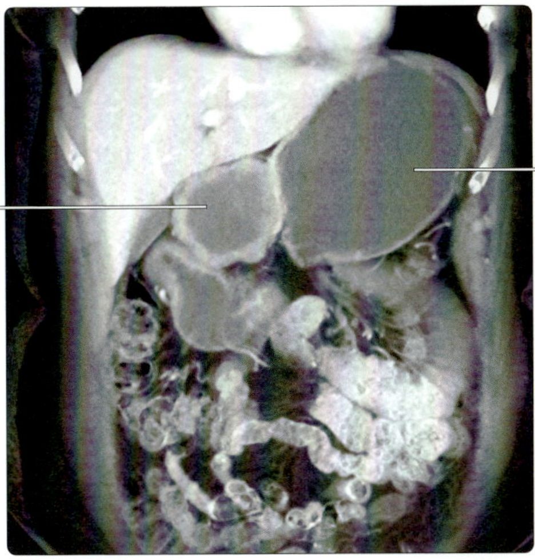

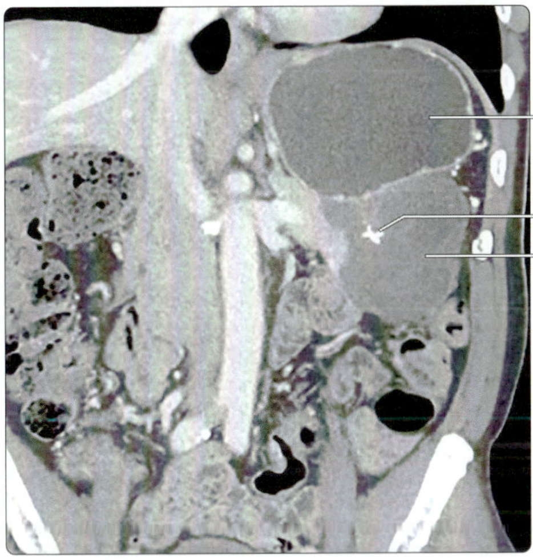

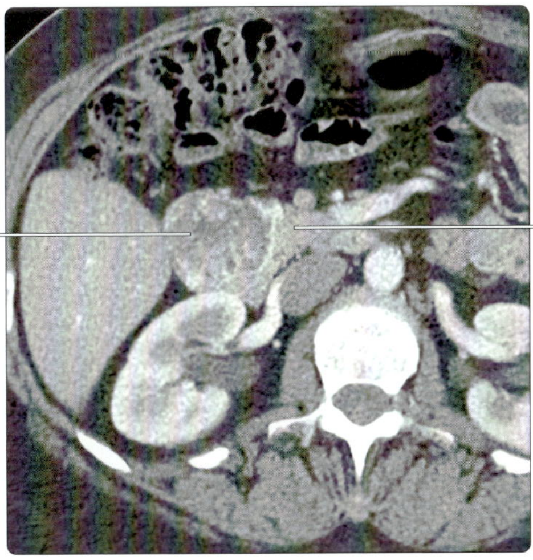

(Top) *Coronal volume-rendered CECT demonstrates a large, centrally necrotic mass exophytically arising from the stomach. GISTs can arise from any part of the gastrointestinal tract and are not uncommonly exophytic.* **(Middle)** *Coronal CECT demonstrates a hypodense mass arising from the inferior aspect of the stomach with a coarse internal calcification. This mass was resected and found to be a GIST at resection.* **(Bottom)** *Axial CECT demonstrates a hypervascular mass arising from the 2nd/3rd portions of the duodenum. While an avidly enhancing mass in this location could be confused with a pancreatic neuroendocrine tumor or duodenal carcinoid tumor, some GISTs can show avid enhancement. This mass was resected, and the diagnosis of GIST was confirmed.*

DUODENAL COMPRESSION, SUPERIOR MESENTERIC ARTERY SYNDROME

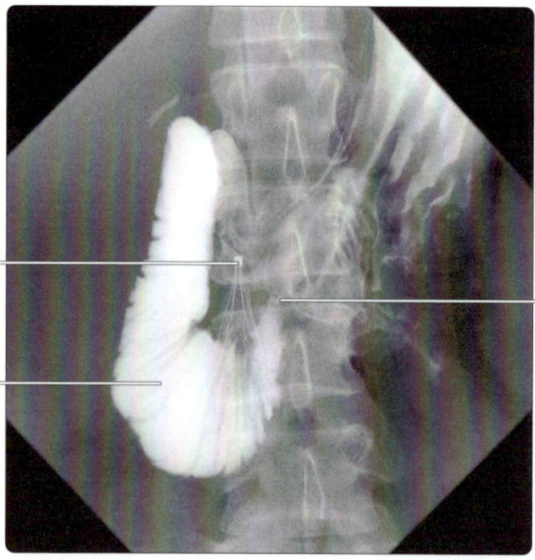

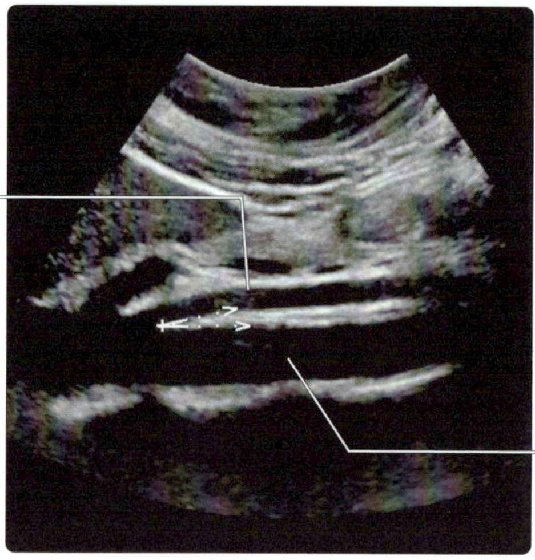

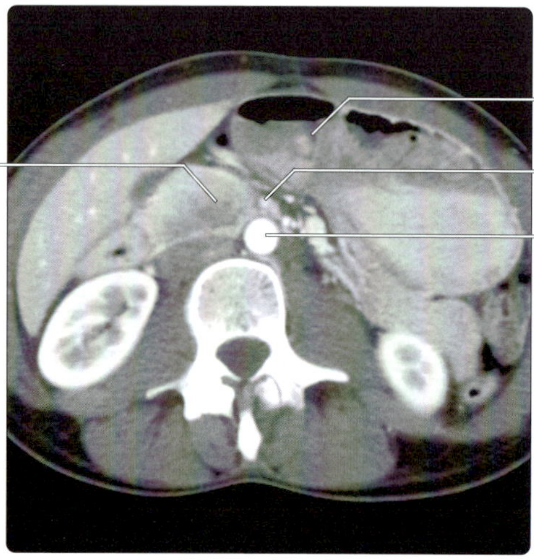

(Top) *Frontal image from an upper GI series in a young female patient with weight loss and chronic nausea demonstrates dilation of the 2nd and 3rd portions of the duodenum with an abrupt vertical, band-like narrowing as the duodenum crosses the midline.* **(Middle)** *Sagittal ultrasound in the same patient demonstrates narrowing of the space between the aorta and SMA. The calipers are measuring the aortomesenteric angle, which is markedly narrowed to 11.9°, a finding strongly suggestive of SMA syndrome, especially given the patient's history.* **(Bottom)** *Axial CECT in a young woman with weight loss demonstrates that as the duodenum crosses between the aorta and the superior mesenteric vessels, it is compressed, and the lumen is markedly narrowed. The duodenum proximal to that point is mildly dilated. This is sometimes referred to as SMA syndrome and is felt to reflect functional partial obstruction of the duodenum between the aorta and SMA.*

AORTOENTERIC FISTULA

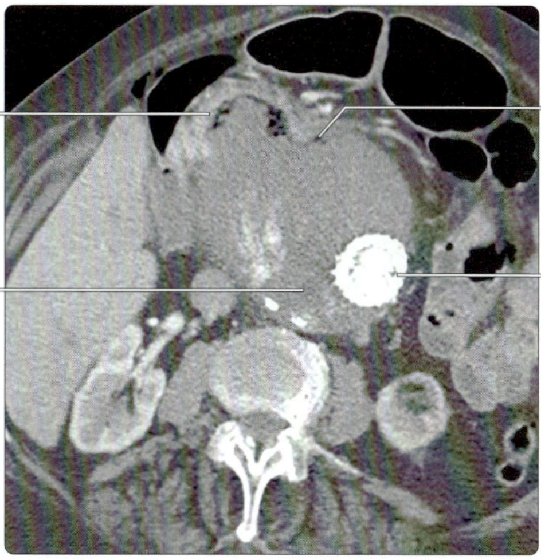

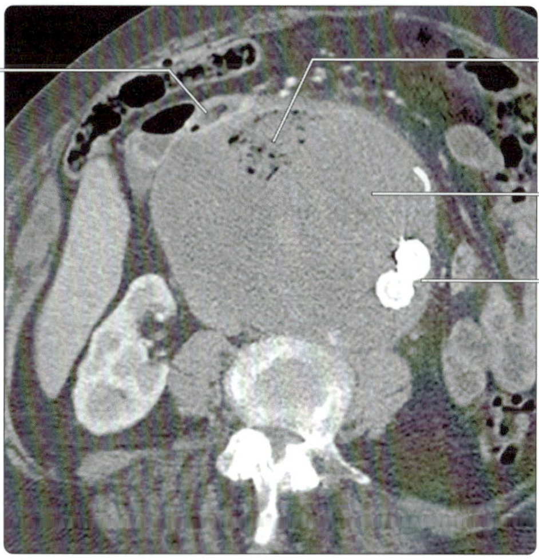

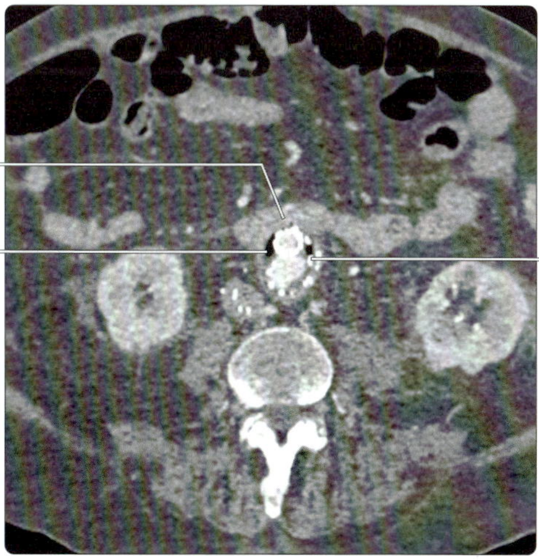

(Top) *Axial CECT demonstrates a large aortic aneurysm treated with stent graft repair. Notice that the aorta and adjacent duodenum are directly adherent to one another without an intervening fat plane. A tiny focus of ectopic gas is identified within the aneurysm sac itself.* **(Middle)** *Axial CECT in the same patient demonstrates a large amount of ectopic gas within the aneurysm sac, a finding virtually diagnostic of an aortoenteric fistula. This patient presented in septic shock and was taken to the operating room for repair. Even with operative treatment, the vast majority of patients with aortoenteric fistula do not survive.* **(Bottom)** *Axial CECT demonstrates ectopic gas within the aorta in a patient who has undergone prior stent graft placement, compatible with an aortoenteric fistula. Primary aortoenteric fistulas are exceedingly uncommon, and this is entity is almost always secondary in patients who have undergone prior aortic intervention.*

GASTRIC DIVERTICULUM

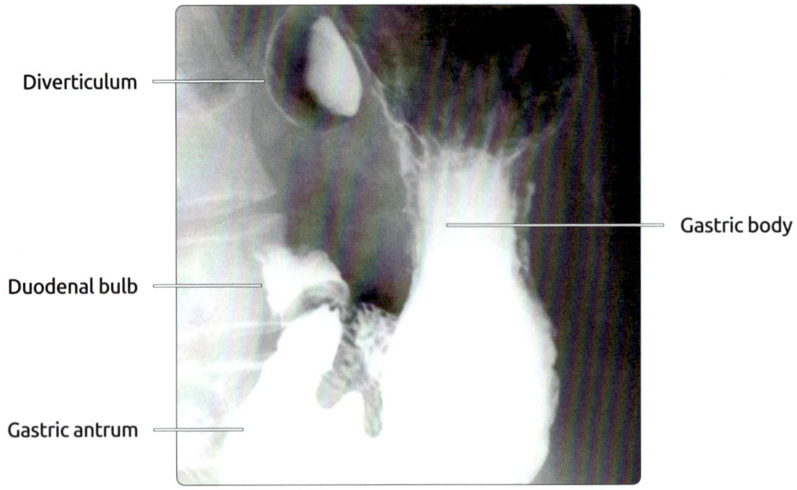

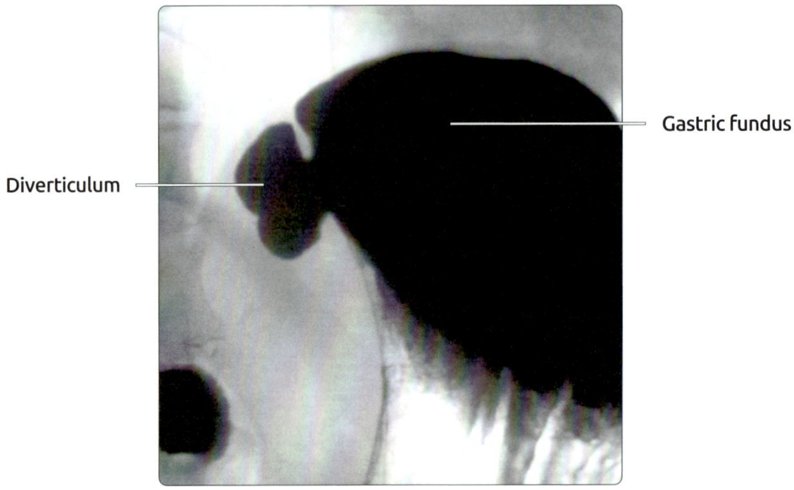

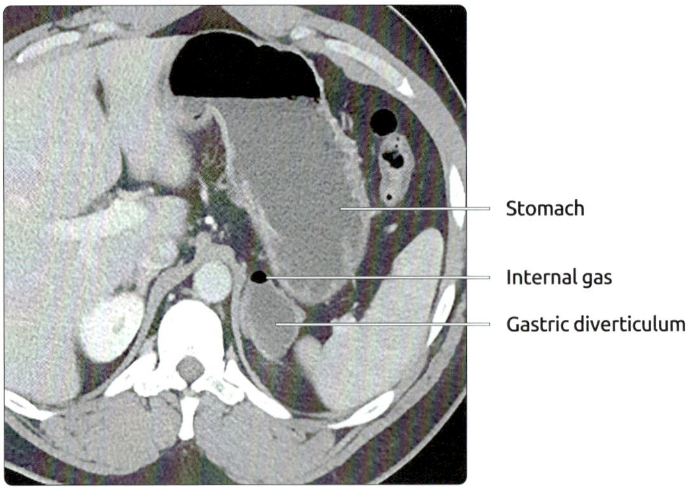

(Top) *Lateral film from an upper GI series demonstrates a gastric diverticulum as an outpouching from the posterior wall of the fundus, just above the gastroesophageal junction.* **(Middle)** *Supine frontal film from an upper GI series demonstrates both the gastric fundus and a gastric diverticulum filling with barium. Note the wide "mouth" of the diverticulum, which usually permits free entry and exit of food and fluid.* **(Bottom)** *Axial CECT demonstrates a cystic "mass" arising in the vicinity of the left adrenal gland. Although this could be superficially mistaken for a low-density adrenal mass or cyst, notice the presence of subtle internal gas layering internally, suggesting communication with the GI tract. This is a classic location and appearance for a gastric diverticulum, and the communication with the stomach can usually be identified with careful examination.*

LARGE DUODENAL DIVERTICULUM AND MIMIC

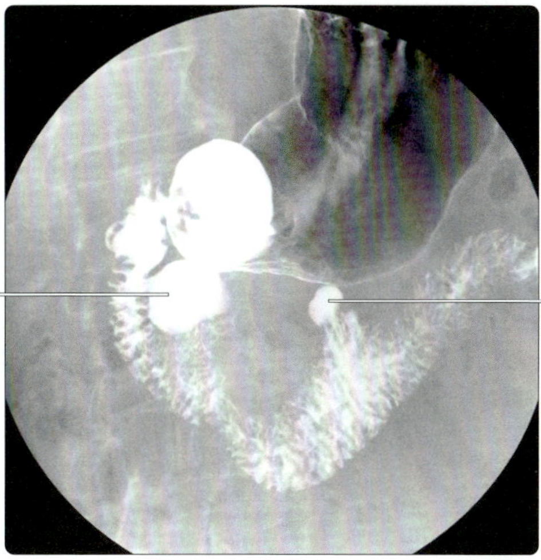

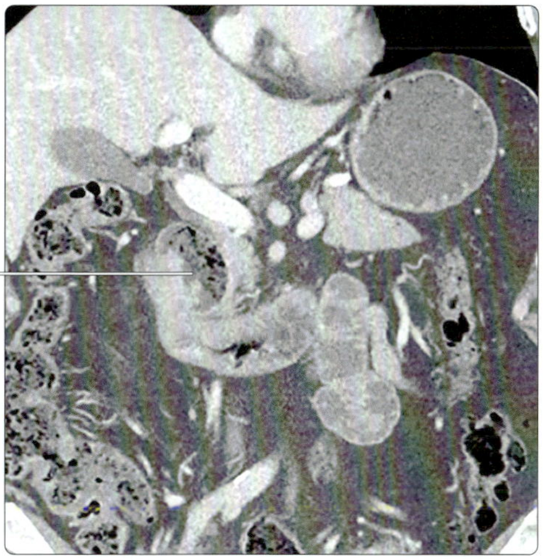

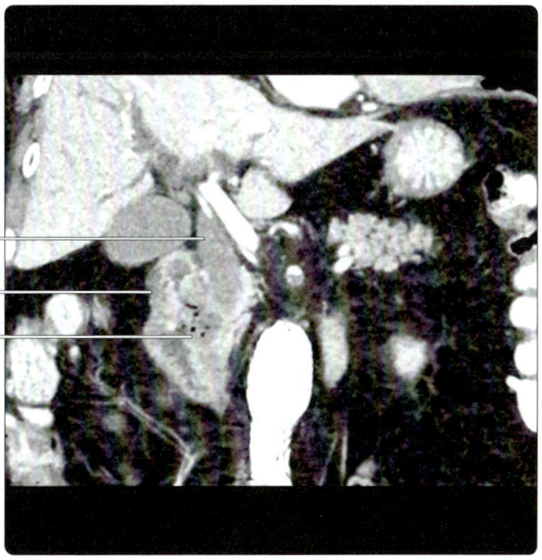

(Top) *Frontal image from an upper GI series demonstrates 2 discrete duodenal diverticula, the 1st arising from the 2nd portion of the duodenum and a 2nd smaller diverticulum arising in the transverse duodenum. These are usually asymptomatic, but may rupture or bleed, and are not uncommonly mistaken for ulcers or masses on various imaging tests.* **(Middle)** *Coronal CECT demonstrates a large duodenal diverticulum arising in the 2nd portion of the duodenum in close proximity to the pancreatic head and ampulla. Although most often asymptomatic, large diverticula can cause mild biliary obstruction due to mass effect and have been reportedly associated with pancreatitis.* **(Bottom)** *A patient with severe abdominal pain is shown. Coronal CT shows a duodenal ulcer with contained perforation confirmed on endoscopy. In some cases, the patient's acute history may the only clue to the correct diagnosis.*

GROSS ANATOMY

Overview

- Mesenteric small bowel is suspended from posterior abdominal wall by fan-shaped **mesentery** and is located within peritoneal cavity

Divisions

- **Jejunum**
  - Begins at **duodenojejunal flexure** (to left of L2 vertebral body)
    - Duodenojejunal flexure is suspended by musculotendinous **ligament of Treitz** (**suspensory ligament of duodenum**), extension of right crus of diaphragm
  - Constitutes ~ 40% of length of small bowel (2-3 m long)
  - Predominantly located in left upper quadrant
  - Thicker, more vascular wall with tall, closely spaced **circular folds** (4-7 folds per in) and fewer **lymphoid follicles** (**Peyer patches**) in submucosa
    - Circular folds also referred to as **valvulae conniventes**, **plicae circulares**, or **folds of Kerckring**
    - Prominence of folds and wall thickness vary by age (more in younger) and degree of bowel distention
  - Wider than ileum (2.5 cm vs. 2 cm)
  - Longer vasa recta and fewer arterial arcades (1-2 cm vs. 3-4 cm for ileum)
- **Ileum**
  - No clear point of transition between jejunum and ileum; gradual transition
  - Constitutes distal 60% of intestine (~ 4 m long)
  - Predominantly located in right lower abdomen and pelvis
  - Has thinner, less vascular wall with shorter, more widely spaced circular folds and more **lymphoid follicles**
  - Vasa recta supplying ileum shorter than in jejunum but with more arterial arcades
  - Ends at **ileocecal valve**
- **Small bowel mesentery**
  - Double fold of peritoneum attached to posterior abdominal wall (root extends from left L2 transverse process to right sacroiliac joint); measures ~ 15 cm
    - Root of small bowel is bare area continuous with anterior pararenal space and contains superior mesenteric artery (SMA) and superior mesenteric vein (SMV)
    - Fans outward to cover entire length of jejunum and ileum
    - All vessels/nerves supplying small bowel lie between layers of **small bowel mesentery**
- **Arterial supply**
  - **SMA** supplies entirety of small bowel distribution
    - Arises from aorta at L1 level, ~ 1 cm caudal to celiac artery, and sends ~ 20 branches to intestine
      - ☐ Ileocolic, right colic, and middle colic arteries extend off right aspect of SMA and jejunal/ileal branches extend off left aspect
    - Arteries unite to form arches/loops (**arterial arcades**) → straight arteries (**vasa recta**)
- **Venous supply**
  - **SMV** drains entirety of small bowel
    - Lies to right of SMA with branches named similarly to SMA branches
    - Unites with **splenic vein** to form **portal vein** posterior to pancreatic neck
- **Lymphatics**
  - Begin within **intestinal villi** (tiny projection of mucous membrane) as **lacteals**
    - Specialized lymphatic vessels that absorb fat from gut
    - Empty milky chyle into **lymphatic plexuses** in intestinal walls → **lymphatic vessels** (in mesentery) → lymph nodes → thoracic duct
  - Draining lymph nodes include **juxtaintestinal nodes** near wall of intestine and **mesenteric lymph nodes** following arterial arcades
- **Nerves** are autonomic with both sympathetic and parasympathetic supply
  - Sympathetic: Via greater and lesser splanchnic nerves of **superior mesenteric plexus**
    - Reduces motility, secretion, and vascularity
  - Parasympathetic: From **posterior vagal trunk** via celiac plexus
    - Increases motility, secretion, vascularity, and digestion
    - Includes sensory branches that can detect stretching and distention but no other pain stimuli; bowel obstruction → distended lumen → spasms of poorly localized crampy abdominal pain (colic)

CLINICAL IMPLICATIONS

Clinical Importance

- **Bowel obstruction** is common clinical problem
  - Usual causes are adhesions and hernias
  - Much more commonly affects small bowel than colon (small bowel is longer, more mobile)
- **Bowel ischemia** is important diagnosis in emergency and critical care settings
  - May result from arterial or venous occlusion (SMA or SMV) or hypoperfusion (e.g., shock, cardiac failure)
  - Imaging pattern can vary depending on underlying cause of ischemia (e.g., arterial vs. venous occlusion)
- **Crohn disease** is one of most common inflammatory abnormalities to affect small bowel with predisposition for involvement of distal ileum
- Tumors of small bowel are very rare with most common malignancies, including **carcinoid**, **adenocarcinoma**, **lymphoma**, and **gastrointestinal stromal tumor**

EMBRYOLOGY

Embryologic Events

- Embryological foregut: Esophagus, stomach, duodenum, liver, biliary system
- Midgut: Small bowel and right side of colon
- Hindgut: Left side of colon and rectum
- Foregut and midgut herniate into early fetal umbilical cord, usually returning to abdominal cavity after 270° counterclockwise rotation and are then fixed into position by modification of mesenteries
- Errors in this process are common (e.g., malrotation and midgut volvulus)

SMALL INTESTINE VESSELS

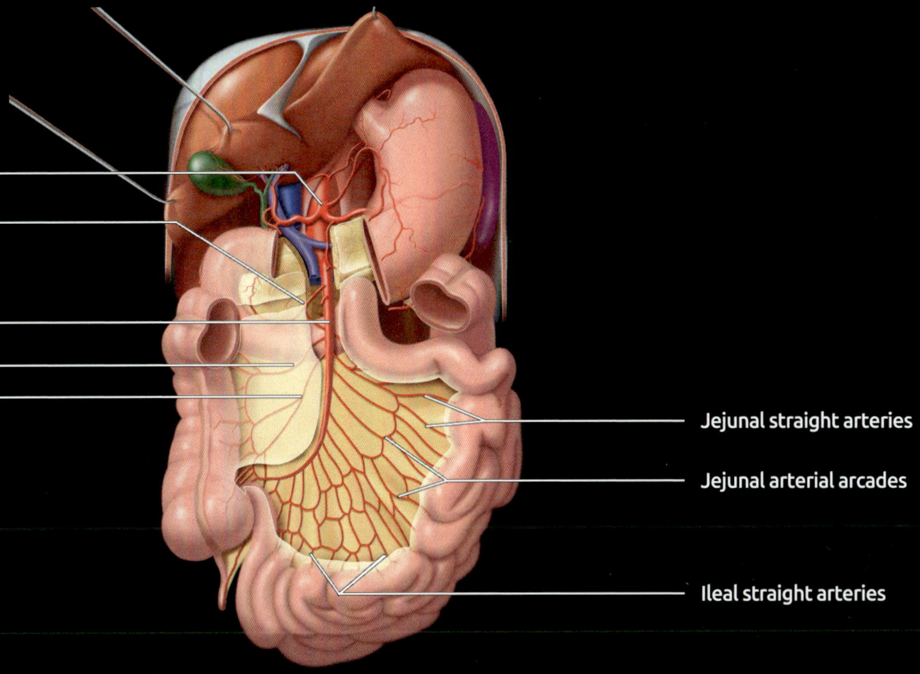

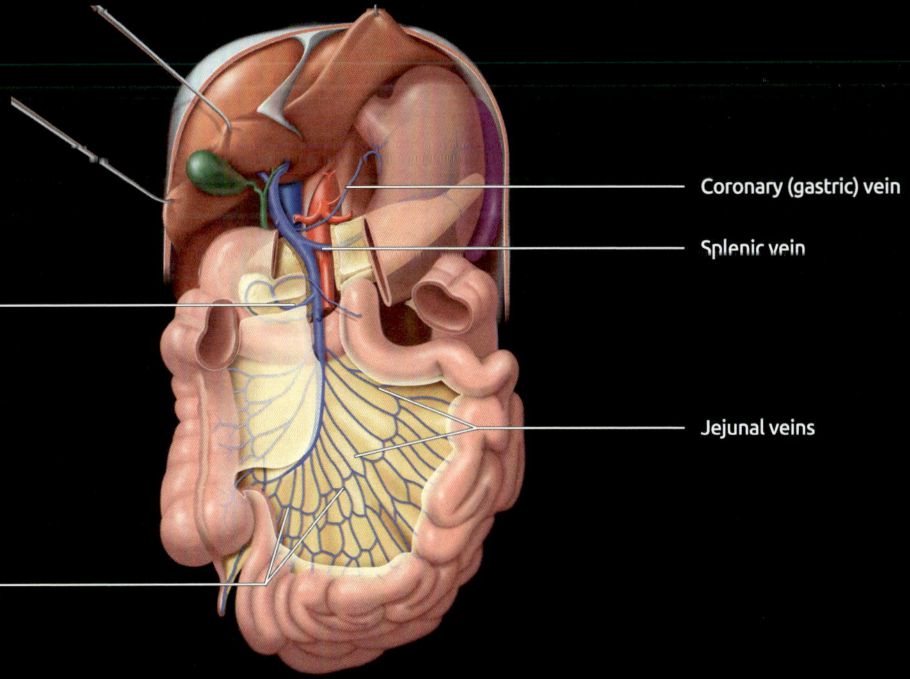

(Top) *The superior mesenteric artery (SMA) supplies the entire small intestine. Arising from the anterior aspect of the aorta at about the L1 vertebral level, its 1st branch is the inferior pancreaticoduodenal artery, which supplies the duodenum and pancreas and anastomoses freely with branches from the celiac trunk. The next branch is the middle colic, supplying the transverse colon, followed by right colic and ileocolic branches. Arising from the convex or left side of the SMA are numerous branches to the jejunum and ileum. Jejunal arteries are generally larger and longer than those of the ileum. After a straight course, the arteries form multiple curvilinear arcades, which form lateral communications between the arteries. Finally, straight arteries (arteriae rectae) extend to and penetrate the wall of the intestine.* **(Bottom)** *In number, point of origin, course, and name, the intestinal veins (tributaries of the superior mesenteric vein) are similar to the arteries.*

JEJUNUM

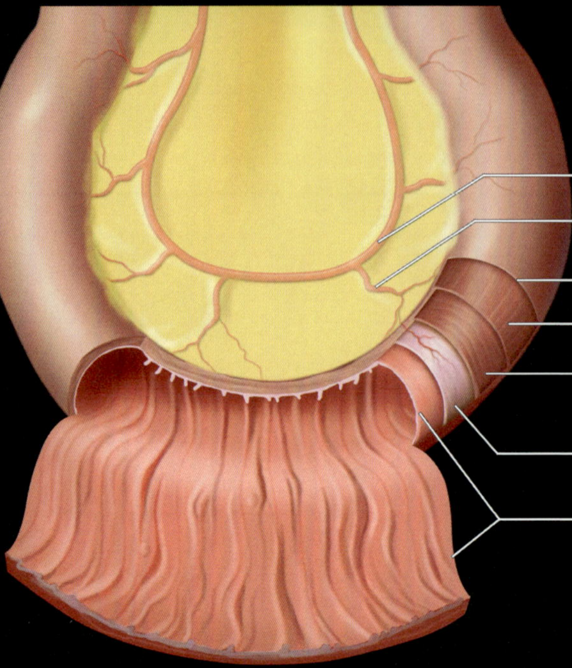

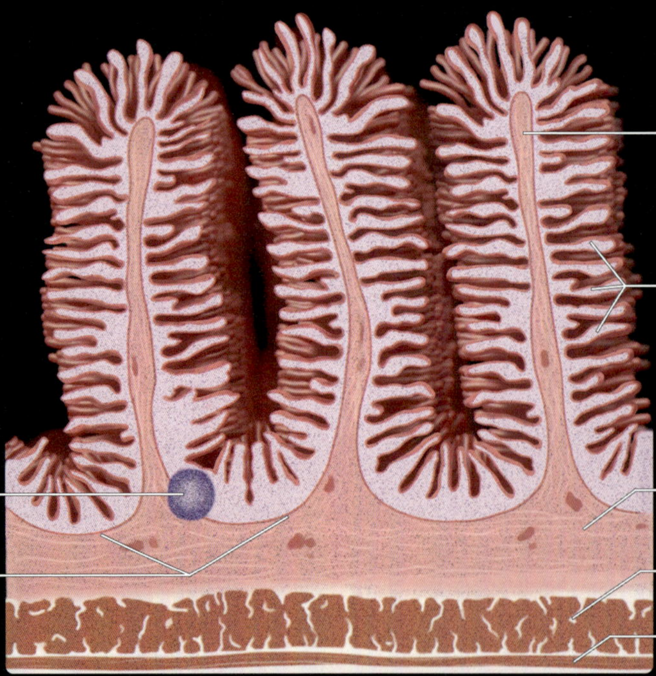

(Top) *There is no sharp point of demarcation between the jejunum and ileum, and both have the same basic structure. The jejunum has a richer vascular supply, thicker wall, and wider lumen. There are 5 layers of the bowel wall. The innermost is the mucosa, the absorptive surface of the gut. The jejunal mucosa is extensively plicated (folded), and these transverse ("circular") folds lie perpendicular to the long axis of the bowel. The other layers are the submucosa, circular muscle, longitudinal muscle, and serosa, the peritoneal lining of the bowel.* **(Bottom)** *The mucosal surface of the jejunum is increased by prominent villi, finger-like projections of mucosa. The muscularis mucosa separates the mucous membrane from the submucosa. The submucosa has a network of capillaries, lymphatics, and a nerve plexus (of Meissner). The jejunum has few and small, discrete lymphoid nodules.*

ILEUM

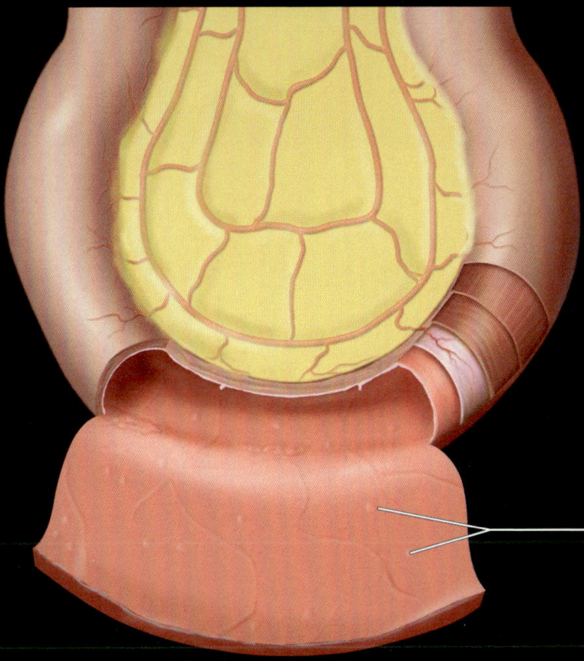

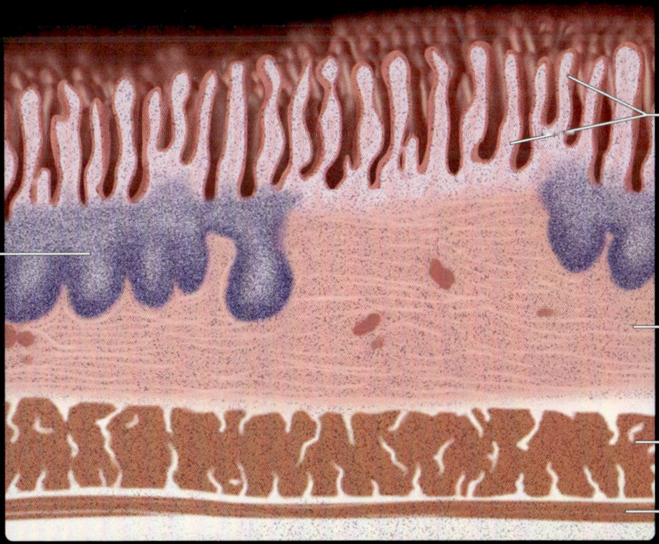

(Top) *The ileum is distinguished by a thinner wall, less vascularity, and less prominent transverse folds and villi than the jejunum. It has the same 5 layers of the bowel wall.* **(Bottom)** *Low-power microscopic view shows a section of ileum. The villi in the ileum are shorter and narrower than in the jejunum, and the transverse folds are much less prominent. They are often not visible on radiographic studies of the intestine. Conversely, submucosal lymphoid follicles become progressively more prominent along the course of the distal small intestine. In the distal ileum, lymphoid follicles may aggregate into macroscopic collections called Peyer patches, which may be evident as submucosal "masses" of a few mm in diameter on barium studies of the small intestine.*

MESENTERY OUTLINED BY ASCITES

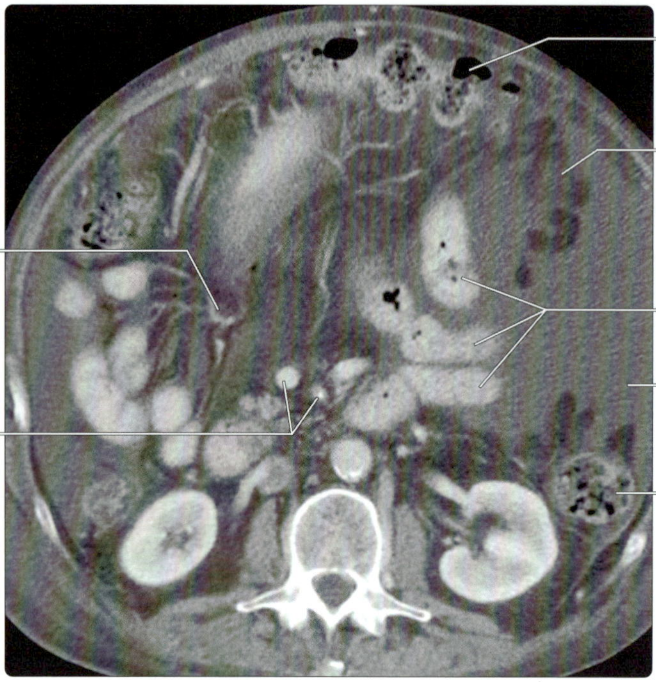

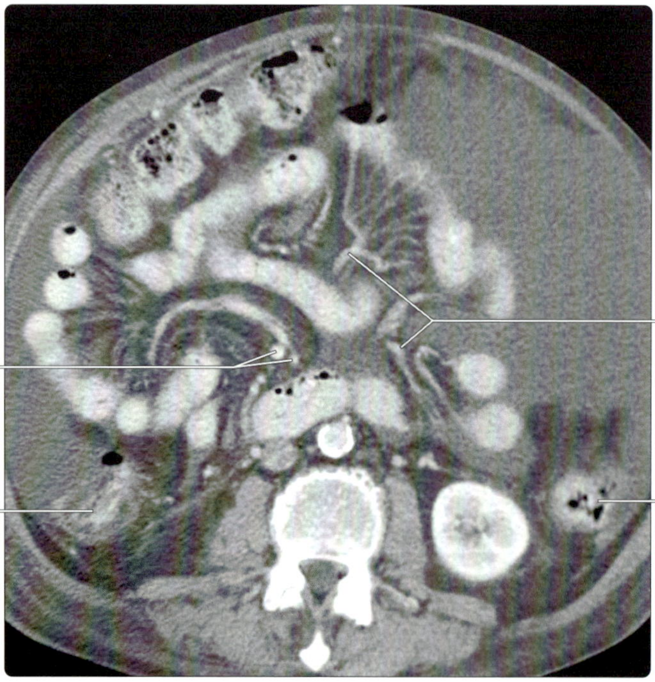

(Top) *First of 2 axial CECT images in a patient with ascites accentuates the small bowel mesentery, as its leaves are separated by the fluid. The mesenteric vessels, nerves, and lymphatics travel to and from the bowel between the layers of the mesentery, surrounded by a layer of fat and loose connective tissue. Note the SMA and vein at the base of the mesentery.* **(Bottom)** *CECT through the right lower quadrant shows the ileocolic vessels supplying the ileum and ascending colon as well as multiple jejunal vessels to the left of the midline. The vessels to the ascending and descending colon travel through the retroperitoneum, rather than through a mesentery.*

MR ENTEROGRAPHY, NORMAL JEJUNUM AND ILEUM

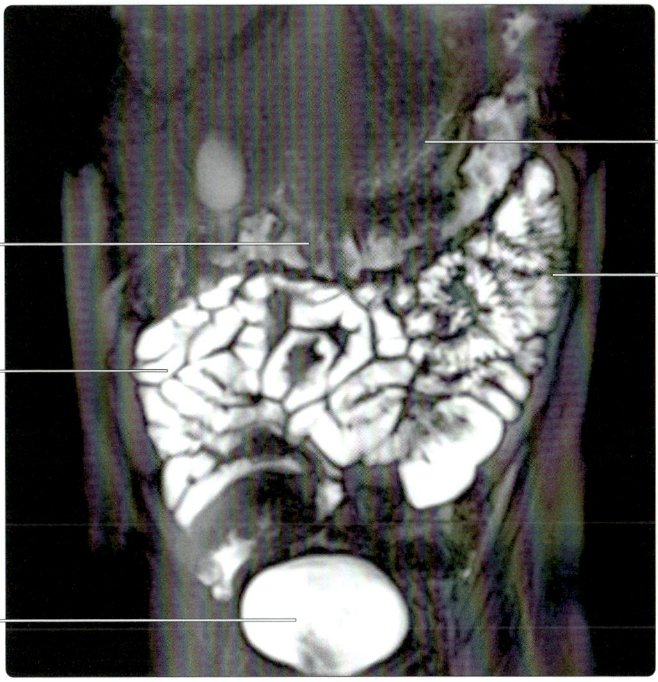

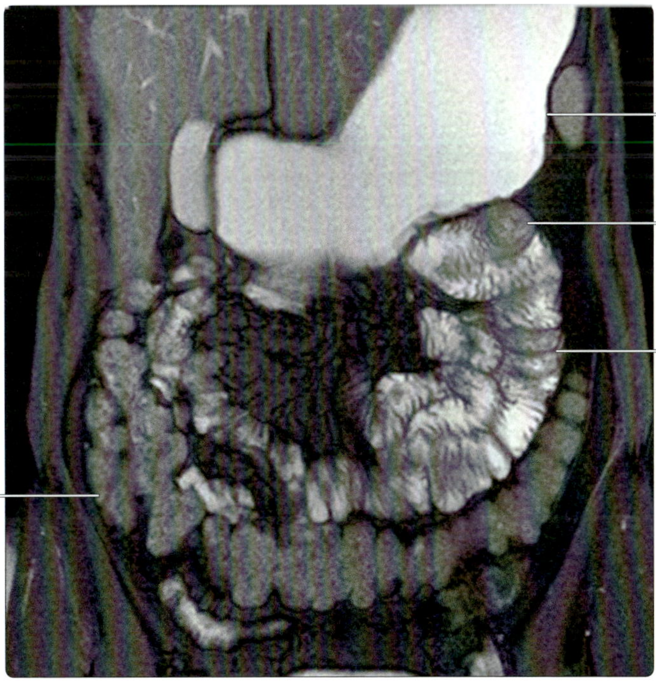

(Top) *Coronal thick-slab HASTE MR from an MR enterography study again demonstrates the normal appearance of the jejunum and ileum. Notice that the folds are far more visible and apparent in the left upper quadrant jejunum compared to the right abdominal ileum. While there is no clear point of transition from the jejunum to the ileum, the jejunum tends to be located in the left upper quadrant, while the ileum is primarily located in the right lower quadrant and pelvis.* **(Bottom)** *Coronal true fast imaging with steady-state free precession (TRUFI) image of the abdomen shows normal bowel loops. Note the normal thickening of the jejunal loops. Incidentally noted transient intussusception can also be appreciated in the proximal jejunum.*

BARIUM STUDY, NORMAL SMALL INTESTINE

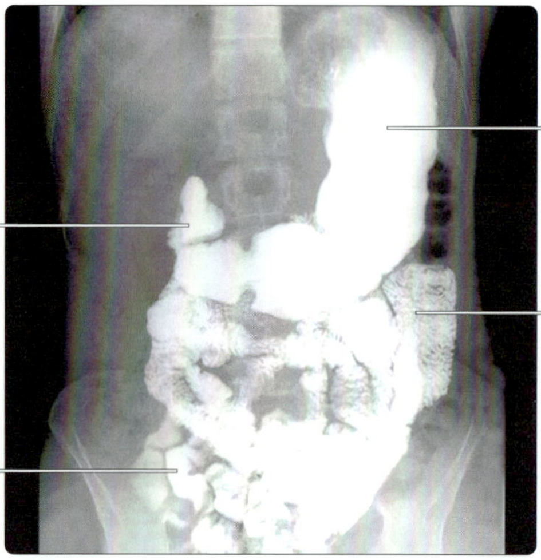

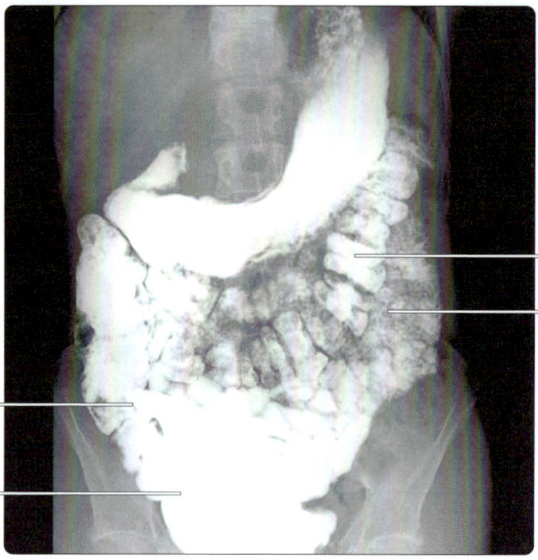

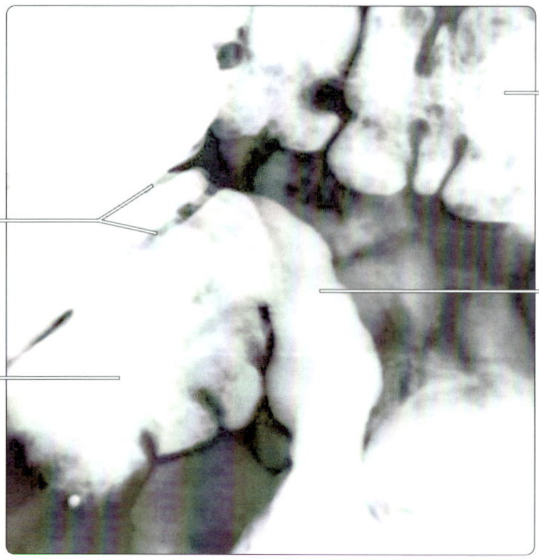

(Top) *First of 3 images from a barium small bowel follow-through shows the normal appearance of the small intestine. Note the position of the jejunum in the left upper quadrant and its prominent, feathery mucosal fold pattern (in the nondistended state). The ileum lies predominantly in the right lower quadrant and has a less prominent fold pattern.* **(Middle)** *Later image from the same series shows barium progressively opacifying the distal small bowel and colon. Note the difference in spacing between the transverse semilunar folds of the colon and the closely spaced, circular folds of the small bowel.* **(Bottom)** *Frontal coned-down "spot" image from the same series shows the terminal ileum, ileocecal valve, and cecum.*

ENTEROCLYSIS, NORMAL SMALL INTESTINE

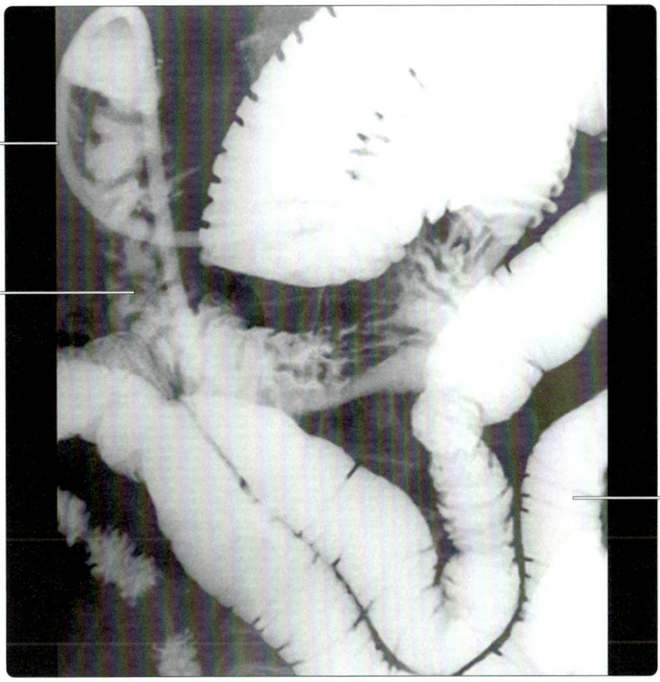

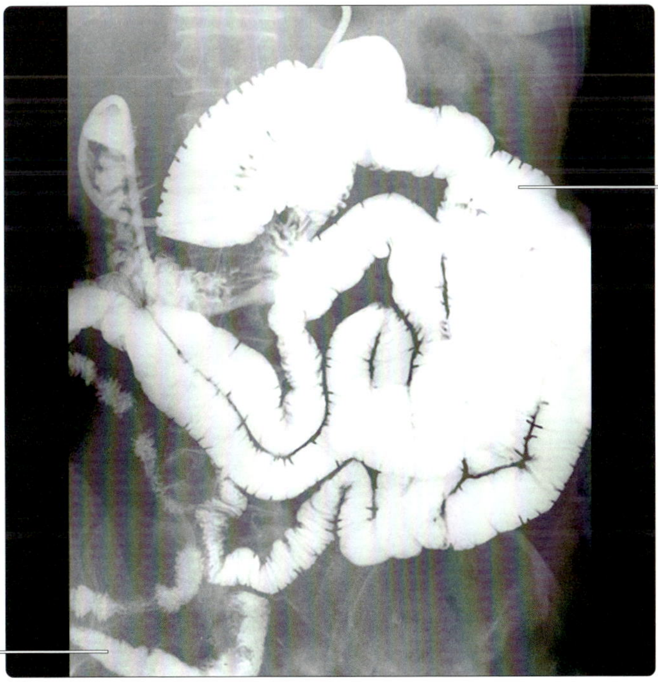

(Top) *First of 2 images from an enteroclysis study in which barium is pumped into the bowel through a nasoenteric tube, the tip of which has been advanced to the duodenojejunal junction under fluoroscopic control, is shown. By bypassing the gastric pylorus and using a pump to infuse barium at ~ 70 mL per minute, the bowel lumen can be distended optimally, allowing much better visualization of the circular folds. Note how thin and evenly spaced the jejunal folds are with ~ 4-7 folds per linear inch in the jejunum and somewhat fewer and less prominent folds in the ileum, as a general rule.* **(Bottom)** *Another image from the same enteroclysis study shows the less prominent fold pattern in the ileum, compared with the jejunum.*

LYMPHOID NODULES

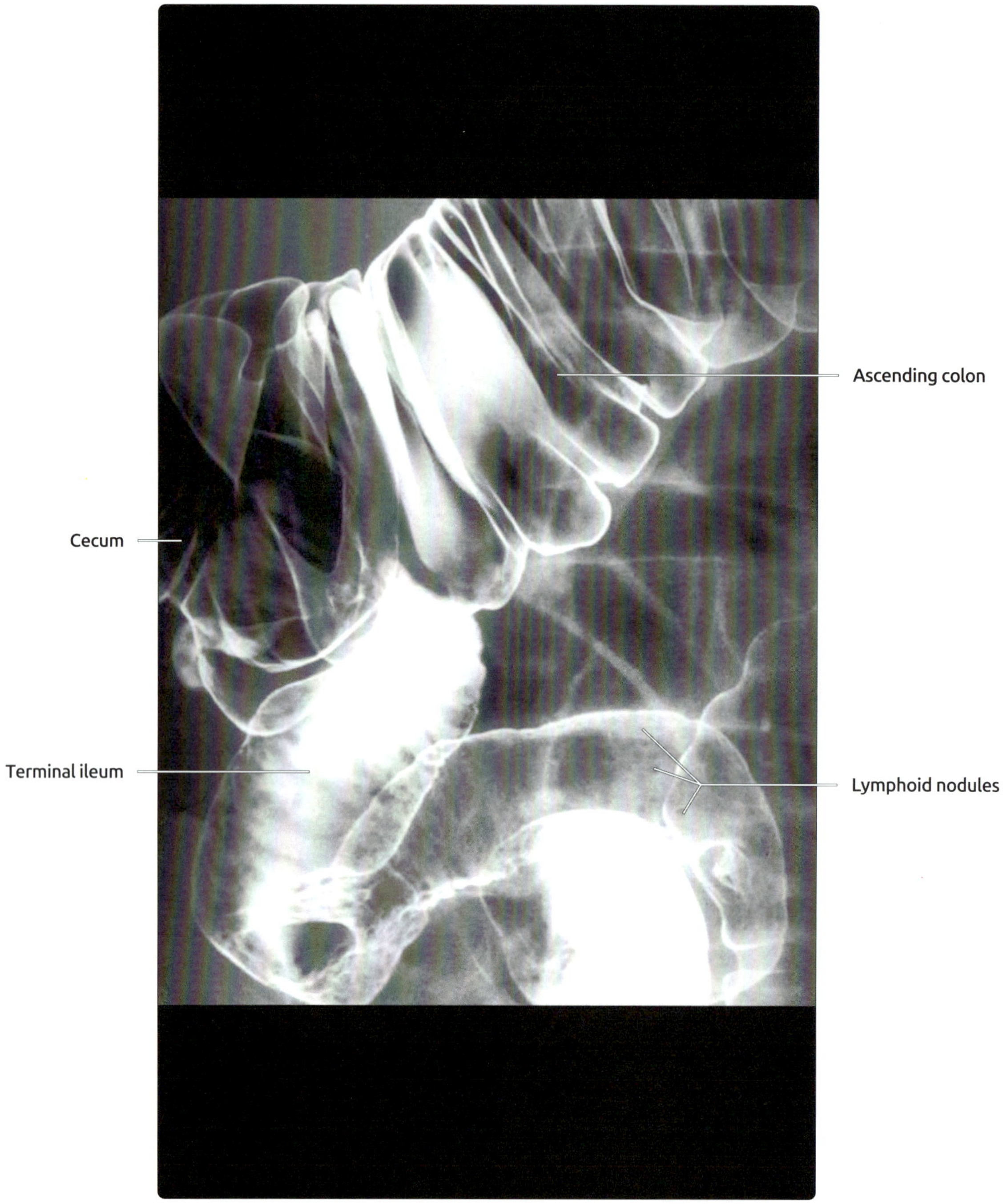

Oblique image from an air-contrast barium enema shows the innumerable small lymphoid nodules (Peyer patches), which are normal aggregates of lymphoid tissue that lie in the submucosa of the bowel wall. In general, these are more prominent in the terminal ileum and in younger patients.

CATHETER ANGIOGRAM, SUPERIOR MESENTERIC ARTERY AND VEIN

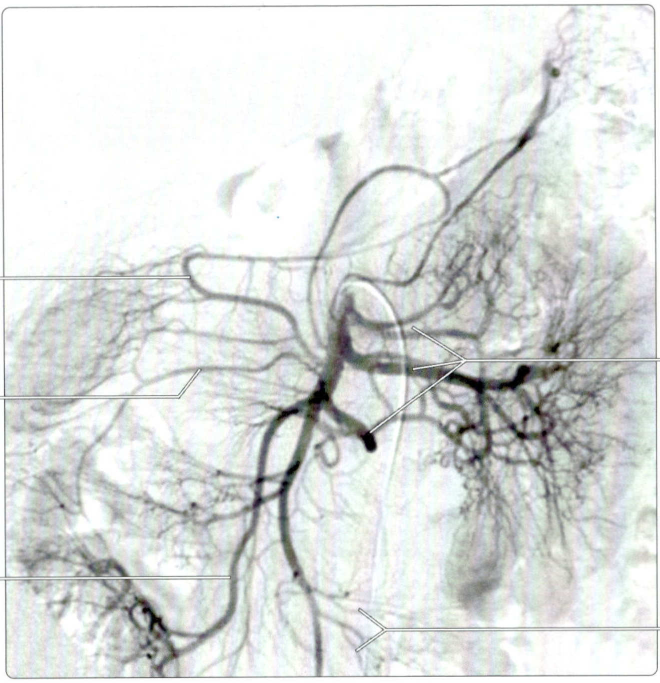

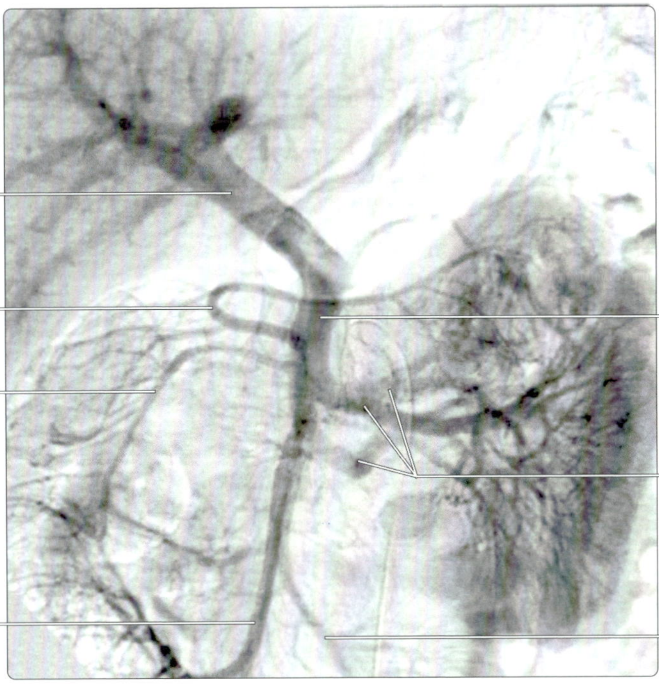

(Top) *First of 2 images from a catheter injection of contrast material into the SMA shows multiple (15-20) arterial branches to the small bowel as well as 3 major branches to the ascending and transverse colon (ileocolic, right colic, and middle colic arteries).* **(Bottom)** *Venous-phase image from the same study shows the superior mesenteric vein and some of its major tributaries as well as the portal vein. The veins parallel the arteries and have similar names.*

CT ANGIOGRAM, SUPERIOR MESENTERIC ARTERY AND VEIN

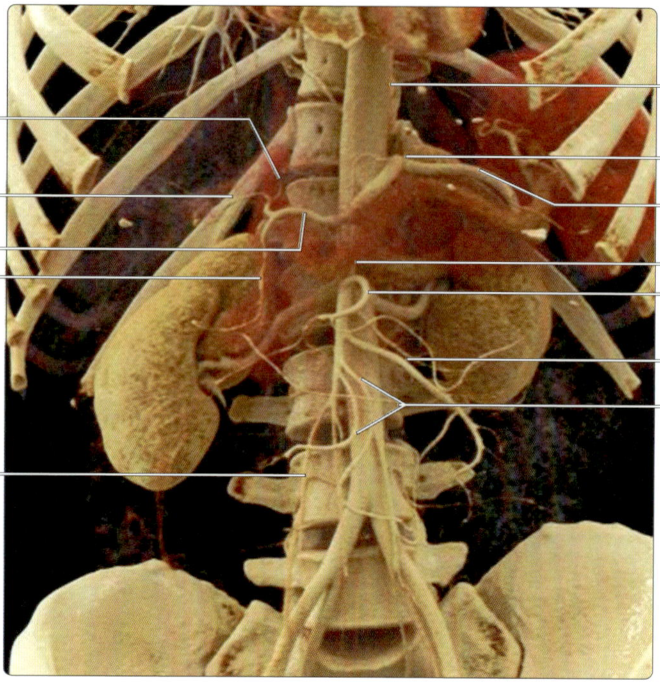

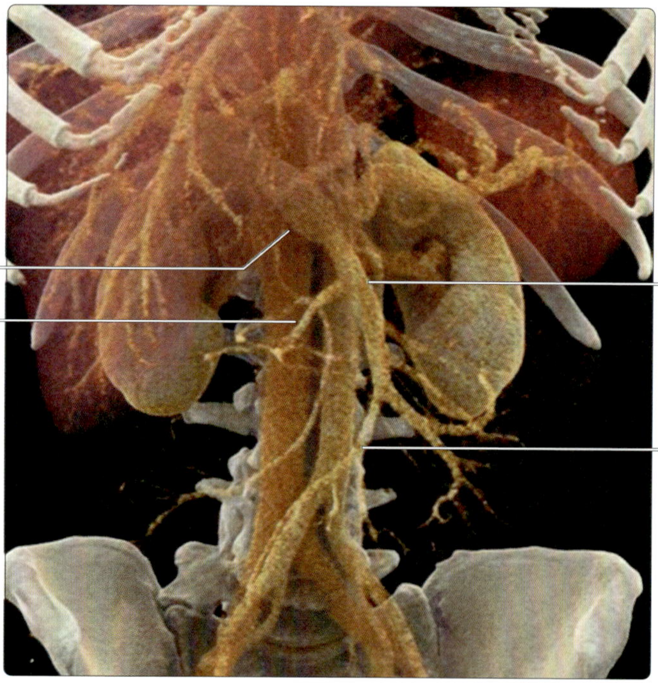

(Top) *Cinematic volume-rendered image of the abdomen shows mesenteric arterial anatomy.* **(Bottom)** *Cinematic volume-rendered image of the abdomen shows mesenteric venous anatomy.*

SMALL BOWEL INFECTIONS

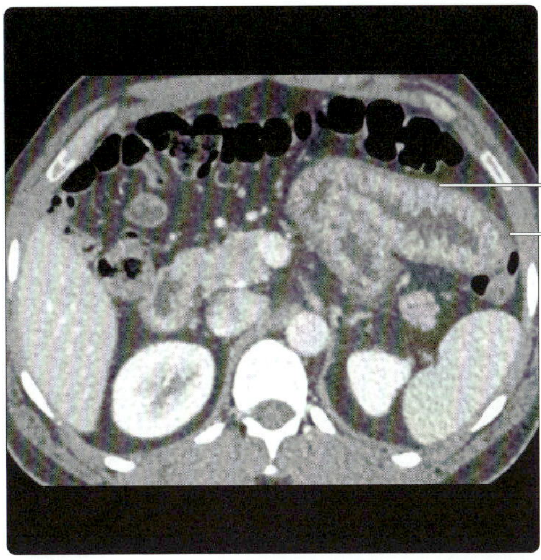

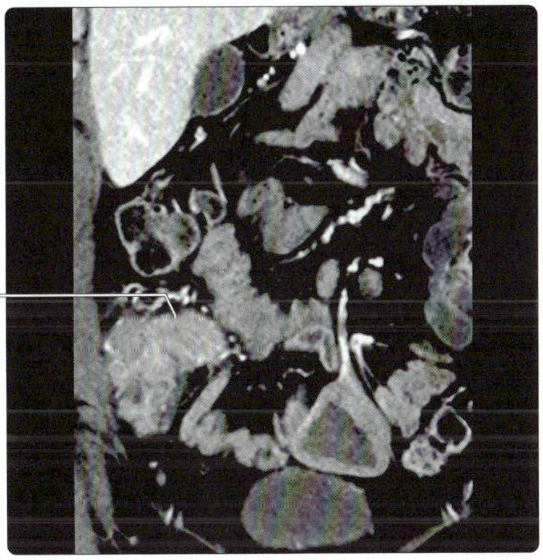

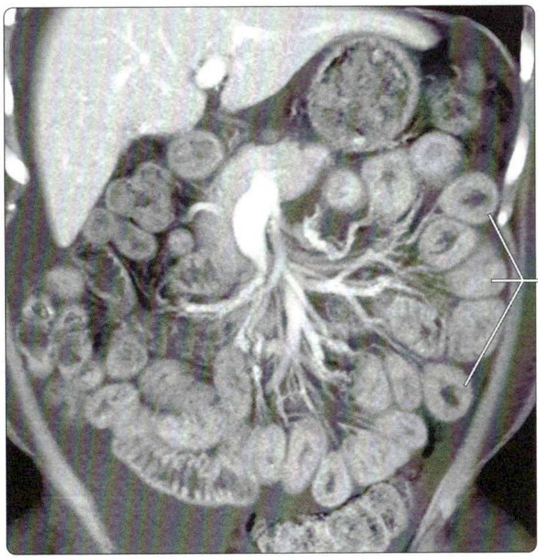

(Top) *CT in an HIV patient with nausea, vomiting, and diarrhea who had a positive stool antigen test for Giardia shows jejunal fold thickening and submucosal edema, consistent with Giardiasis.* **(Middle)** *Coronal CT in a 29-year-old man with severe diarrhea and abdominal pain secondary to salmonella enteritis shows thickening of the terminal ileum for which salmonella has a predilection.* **(Bottom)** *Coronal volume-rendered CECT in an AIDS patient demonstrates extensive wall thickening of the entire small bowel. This appearance persisted over time and was discovered to be secondary to a chronic Mycobacterium avium-intracellulare infection.*

CROHN DISEASE

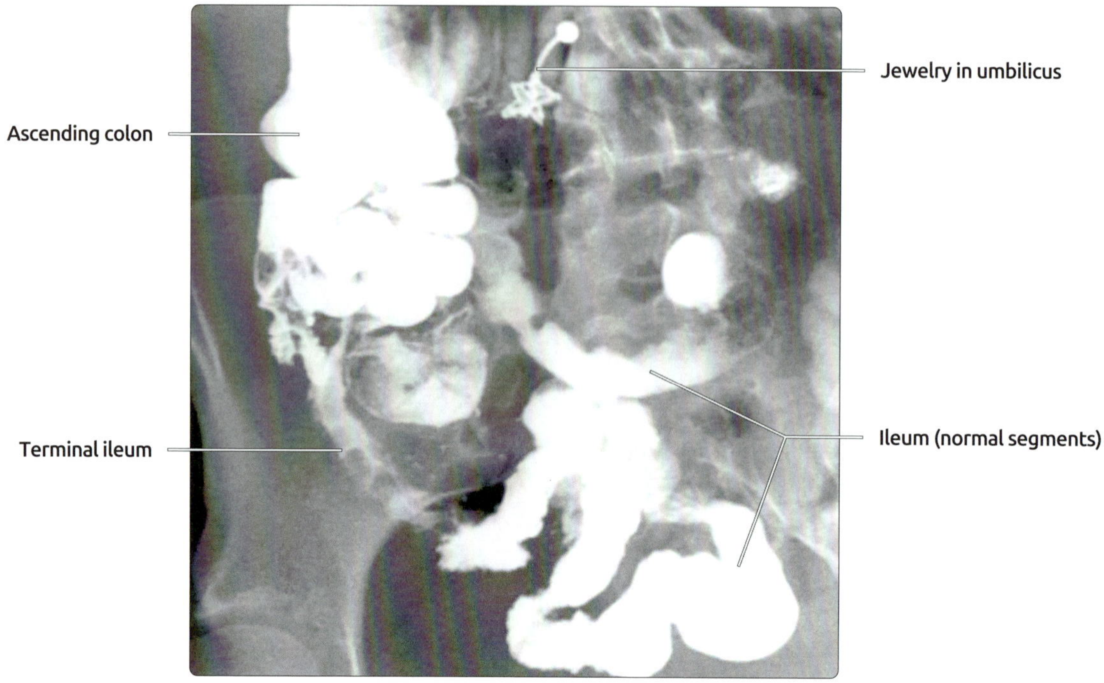

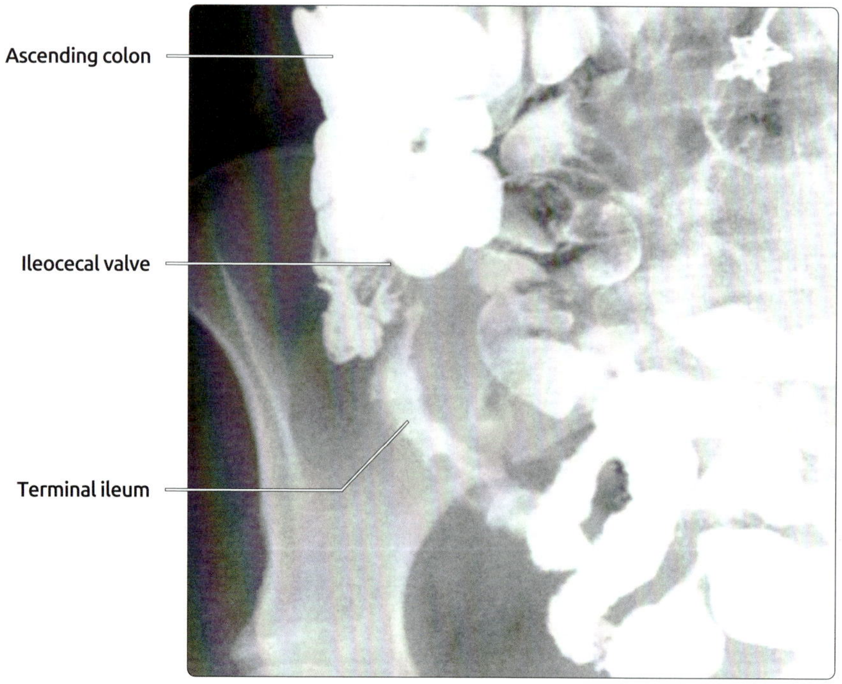

(Top) *First of 2 fluoroscopic images of a young woman with Crohn disease is shown. First image from a barium small bowel follow-through shows irregular nodular thickening of the wall of the terminal ileum.* **(Bottom)** *Another image from the same patient demonstrates that the lumen of the terminal ileum is narrowed, and the wall is irregularly thickened. This abnormal segment of bowel appears straightened and stands away from other segments of bowel, reflecting bowel wall thickening and infiltration of the mesenteric fat near this segment of bowel. The terminal ileum is usually the 1st portion of bowel to be inflamed in patients with Crohn disease, a common inflammatory bowel condition of uncertain etiology. The clinical and radiographic features in this case are typical of this disease.*

CROHN DISEASE

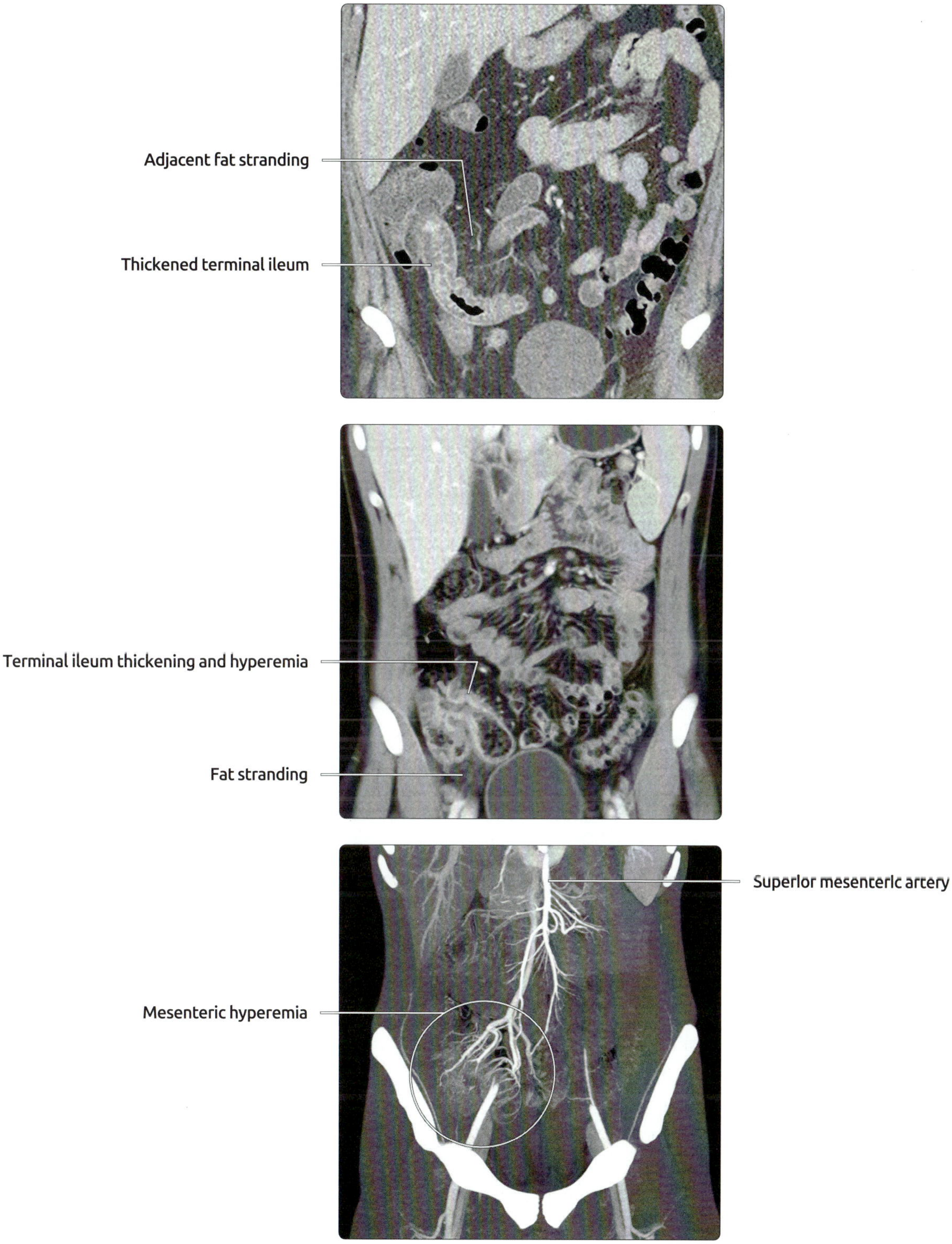

(Top) *Coronal CECT in a young patient demonstrates prominent wall thickening and submucosal edema of the terminal ileum as well as subtle adjacent fat stranding. This is a typical appearance and distribution for Crohn disease with acute terminal ileitis.* **(Middle)** *Coronal CECT with volume-rendered reconstruction demonstrates thickening of the terminal ileum with mucosal hyperemia in a patient with known Crohn disease.* **(Bottom)** *Coronal CECT with MIP reconstruction in the same patient demonstrates the prominent mesenteric vascularity (with engorged vasa recta) supplying this segment of inflamed bowel. Increased mesenteric vascularity can be a useful clue when trying to identify subtle sites of Crohn-related inflammation and may be seen even prior to the development of frank bowel wall thickening.*

CROHN DISEASE

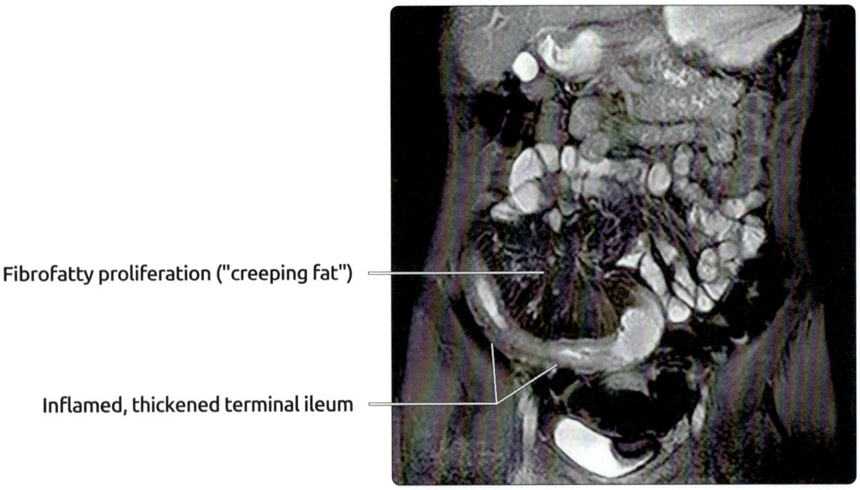

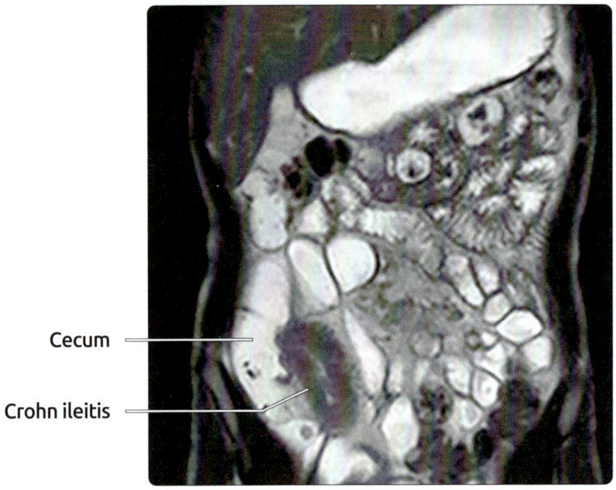

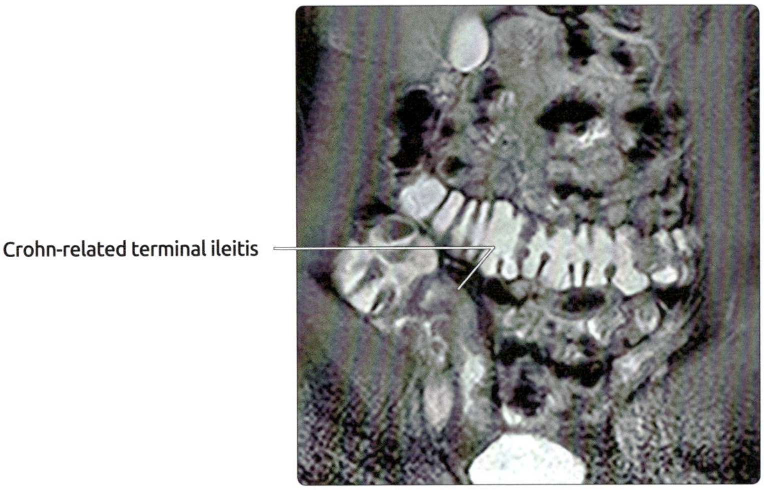

(Top) *Coronal FIESTA from an MR enterography study demonstrates long-segment wall thickening of the terminal ileum with subtle T2 hyperintensity of the ileal wall suggesting mild active Crohn-related inflammation. Notice the extensive fibrofatty proliferation ("creeping fat") in the ileocolic mesentery, likely the sequelae of chronic bouts of inflammation.* **(Middle)** *Coronal FIESTA from an MR enterography study in a patient with Crohn disease demonstrates marked wall thickening of the terminal ileum, compatible with severe Crohn ileitis.* **(Bottom)** *Coronal FIESTA from an MR enterography study demonstrates subtle wall thickening in the terminal ileum compatible with mild Crohn ileitis in this patient with known Crohn disease. MR enterography offers the advantage of lack of ionizing radiation in a patient population that will undoubtedly be imaged repeatedly over the course of their lifetimes. It also offers the advantage of potentially distinguishing chronic changes related to Crohn from true active inflammation.*

CROHN DISEASE WITH COMPLEX FISTULIZING DISEASE

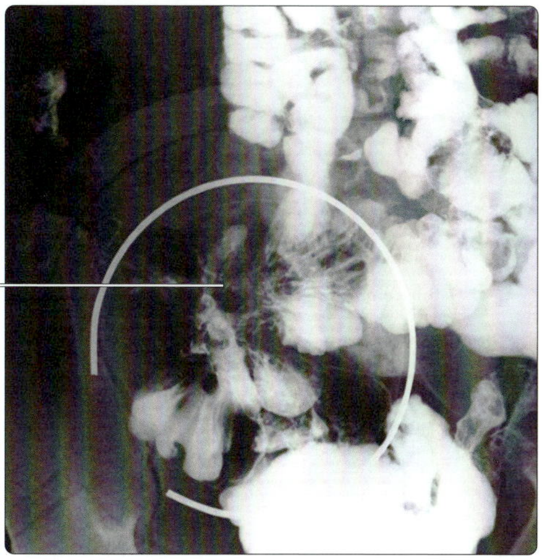

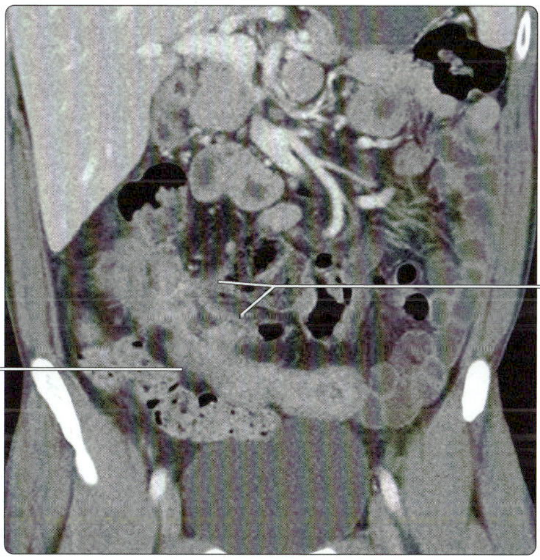

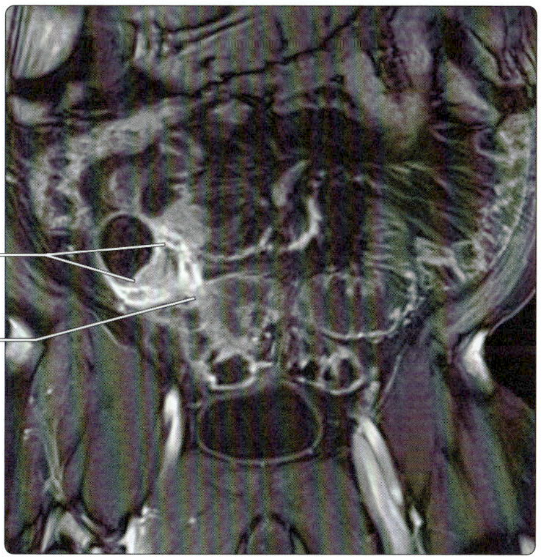

(Top) *Frontal spot compression image from a small bowel follow-through shows a stellate arrangement of the small bowel with acute angulation of loops. The bowel loops converge to a central point and appear to potentially communicate. This appearance is typical of complex fistulizing disease in patients with Crohn disease.* **(Middle)** *Coronal CECT in the same patient demonstrates thickening of the terminal ileum due to Crohn disease. There is communication between the terminal ileum and adjacent bowel loops, compatible with fistulizing disease.* **(Bottom)** *Coronal postcontrast T1 MR in the same patient demonstrates spiculated, tethered bowel loops in the right pelvis with better delineation of the enhancing fistulous tracks between loops of small bowel along with small bowel angulation.*

SMALL BOWEL ISCHEMIA, SUPERIOR MESENTERIC ARTERY OCCLUSION

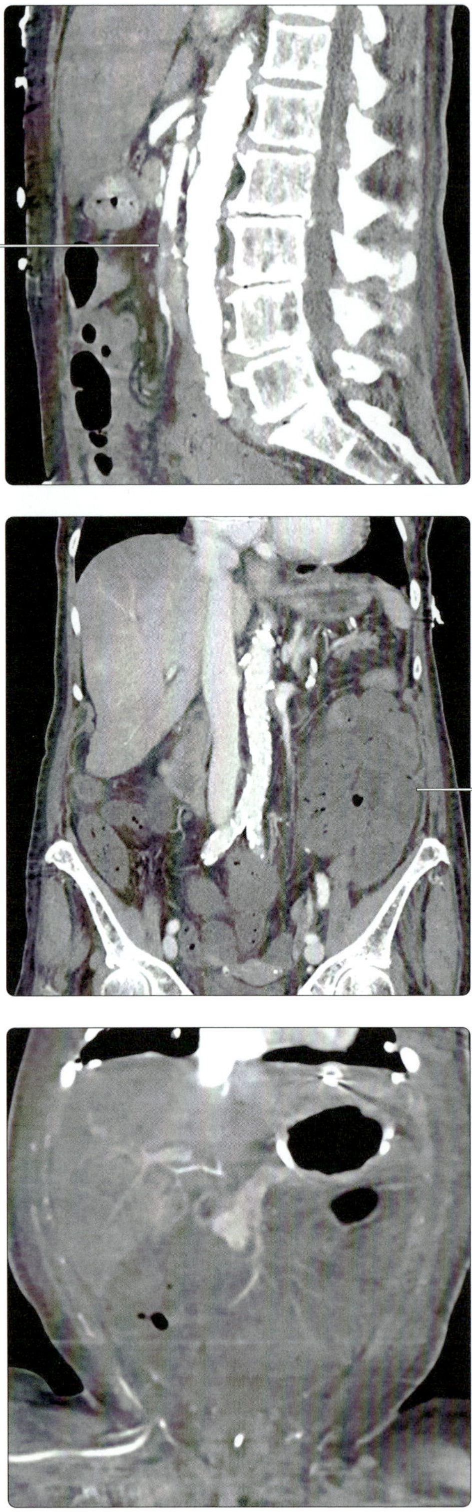

(Top) *Sagittal CECT in the arterial phase in a patient with acute abdominal pain demonstrates essentially occlusive thrombus within the mid- to distal SMA. Notice the presence of quite considerable atherosclerosis elsewhere in the visualized arteries.* **(Middle)** *Coronal CECT in the same patient demonstrates mild dilatation of the small bowel with bowel wall hypoperfusion throughout, particularly evident in left abdominal bowel loops. Bowel ischemia due to arterial occlusion often does not result in significant bowel wall thickening or mesenteric inflammation, and may be apparent only as wall hypoenhancement.* **(Bottom)** *Coronal CECT in a newborn infant (10 days old) with known congenital heart disease and complete thrombosis of the entire aorta and SMA (not shown) demonstrates complete nonenhancement of the entire small bowel. In fact, the degree of bowel nonenhancement is so severe that individual bowel loops are not truly discernible on this image, compatible with extensive small bowel ischemia.*

SMALL BOWEL ISCHEMIA, VENOUS THROMBOSIS

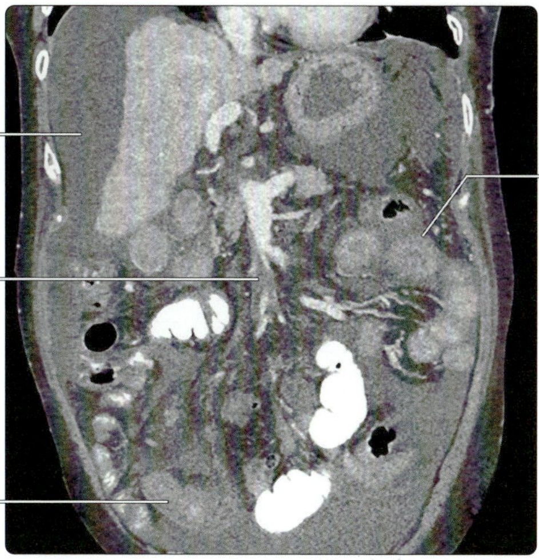

Ascites

Superior mesenteric vein thrombus

Thickened small bowel

Thickened, hyperemic small bowel

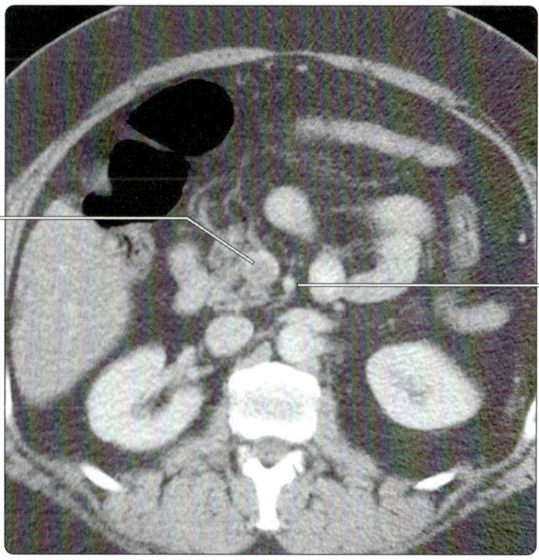

Superior mesenteric vein

Superior mesenteric artery

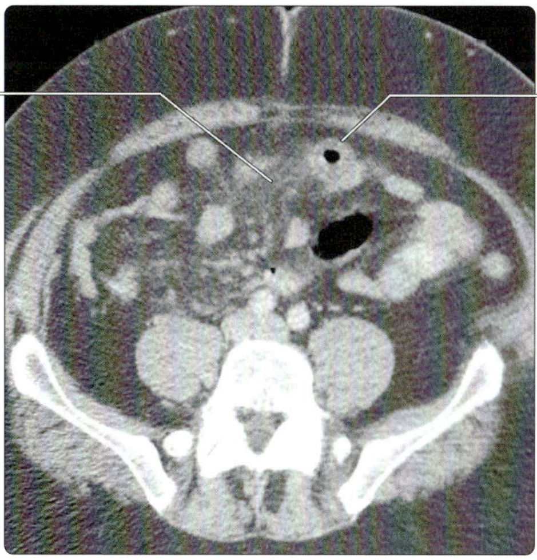

Edematous mesentery

Thick bowel wall

(Top) *Coronal CECT demonstrates nearly occlusive thrombus within the mid- to distal superior mesenteric vein. The presence of multiple thickened, hyperemic loops of small bowel, particularly in the setting of ascites and mesenteric fat stranding/edema, is strongly suggestive of small bowel ischemia due to venous occlusion.* **(Middle)** *Axial CECT in a patient with severe abdominal pain shows a normal-caliber and uniformly enhancing lumen of the SMA, while the lumen of the superior mesenteric vein is distended and partly occluded by a (nonenhancing) clot.* **(Bottom)** *More caudal section shows diffuse edema of the mesentery and a thick-walled segment of small bowel, indicative of some sort of acute injury (in this case, acute bowel ischemia related to the superior mesenteric vein thrombosis). Mesenteric ischemia due to venous occlusion is much more likely to result in bowel wall thickening and significant mesenteric inflammation compared to arterial occlusion.*

SMALL BOWEL ISCHEMIA AND INFARCTION WITH PNEUMATOSIS AND PORTAL VENOUS GAS

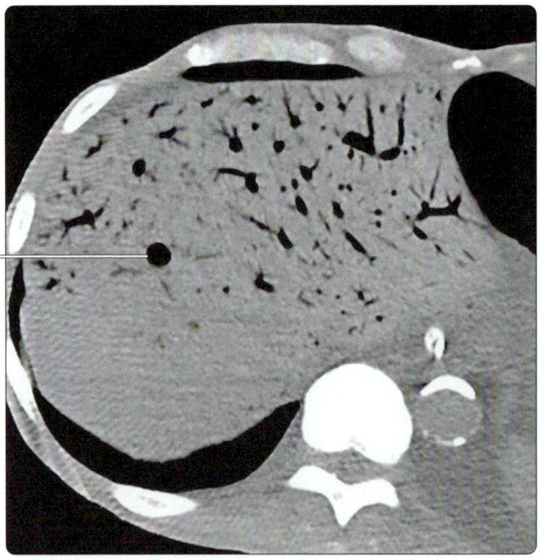

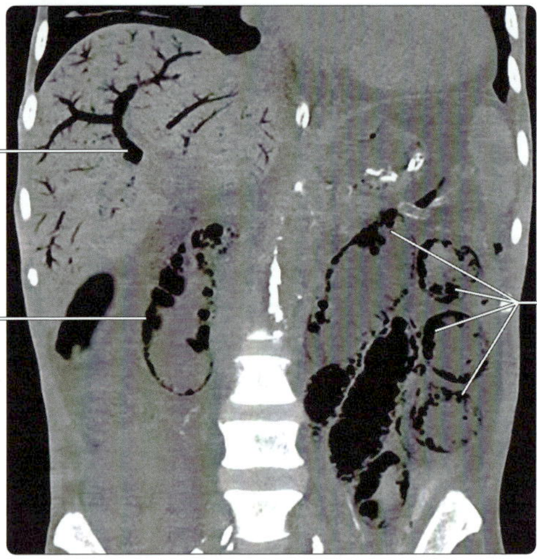

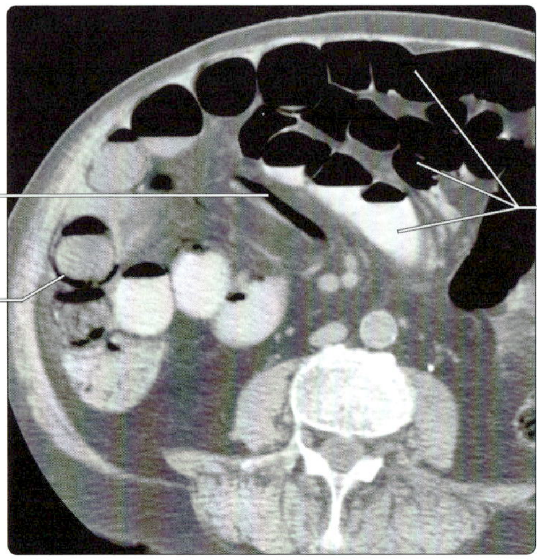

(Top) *Axial NECT in an older adult patient with sepsis and hypotension demonstrates extensive branching gas throughout the liver. Notice that unlike pneumobilia, this gas extends to the periphery of the liver and seems to follow the course of the portal venous system, compatible with portal venous gas.* **(Middle)** *Coronal NECT in the same patient demonstrates extensive portal venous gas as well as multiple dilated small bowel loops with significant intramural gas (i.e., pneumatosis). The patient's lactate level was markedly elevated, and these findings are consistent with mesenteric ischemia and likely small bowel infarction.* **(Bottom)** *Axial NECT shows subtle pneumatosis within the wall of dilated segments of ileum in the right lower quadrant. The jejunum is dilated, but its wall is normal. Gas is present in the mesenteric vein draining the ileal segment with the pneumatosis (intramural gas).*

SMALL INTESTINE OBSTRUCTION

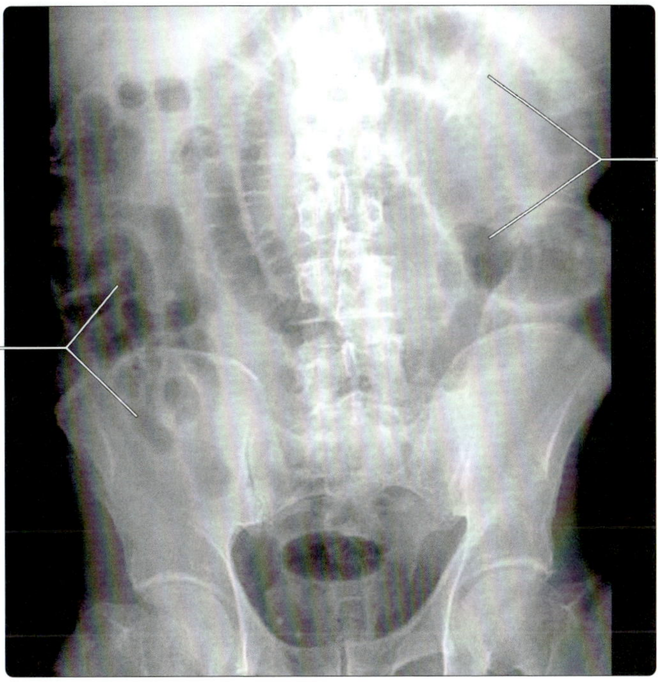

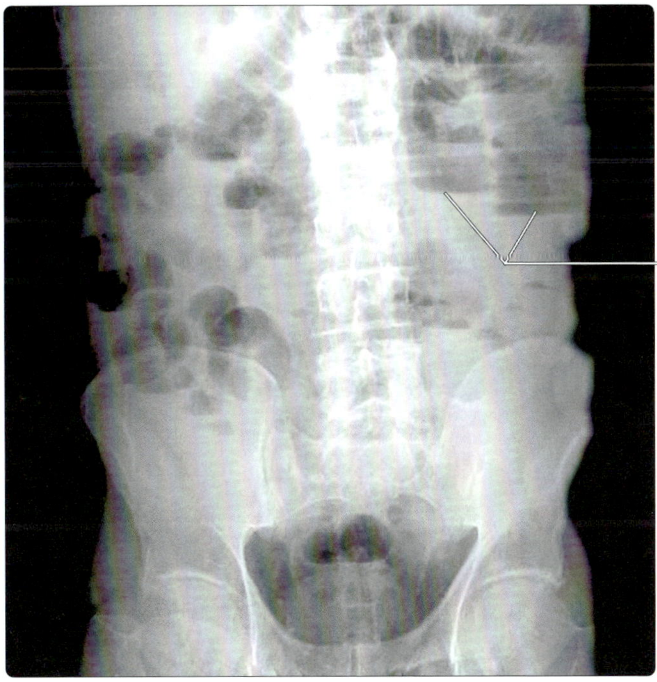

(Top) *First of 4 images of a patient with abdominal distention and crampy pain is shown. A supine image of the abdomen shows dilated segments of proximal small bowel (jejunum) in the left upper quadrant. Distal small bowel and colon are of normal caliber.* **(Bottom)** *An upright image demonstrates multiple air-fluid levels within the dilated segments of bowel, a characteristic feature of small bowel obstruction.*

SMALL INTESTINE OBSTRUCTION

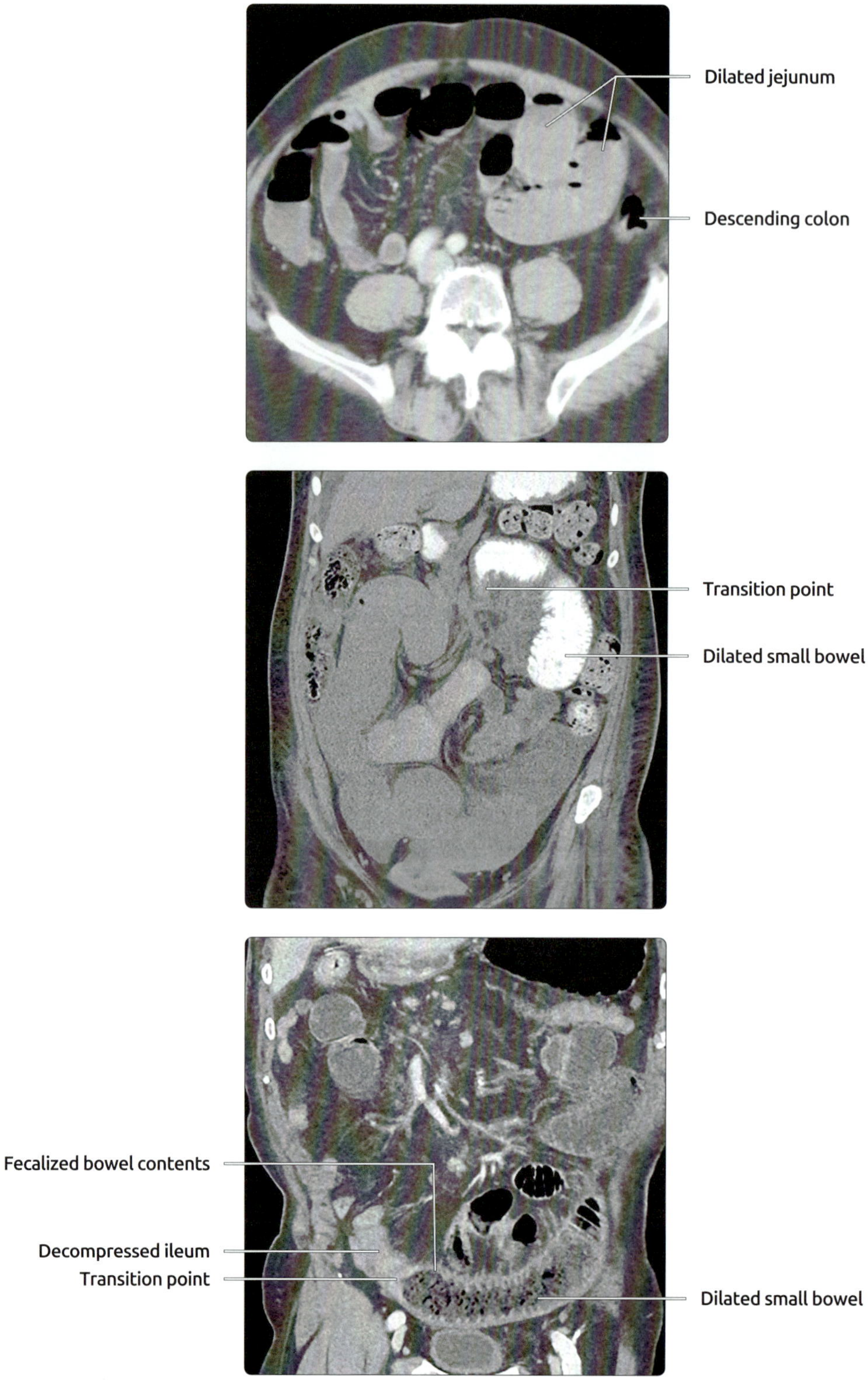

(Top) *Axial CECT section in the same patient shows dilated jejunum with air-fluid levels and "collapsed," normal-caliber ileum and colon.* **(Middle)** *Coronal NECT in a patient with abdominal pain demonstrates dilated small bowel with a discrete transition point to normal-caliber or decompressed bowel, compatible with a small bowel obstruction. Notice that positive oral contrast extends to the level of the transition point without progressing more distally.* **(Bottom)** *Coronal CECT demonstrates dilated small bowel with a discrete transitional point in the right lower quadrant and distally decompressed ileum. The ileum near the transition point appears fecalized, a useful imaging clue when trying to find a transition point in a small bowel obstruction. While no mass is seen on these images, this small bowel obstruction was found at surgery to be secondary to small bowel lymphoma.*

PLAIN RADIOGRAPHS, MALROTATION

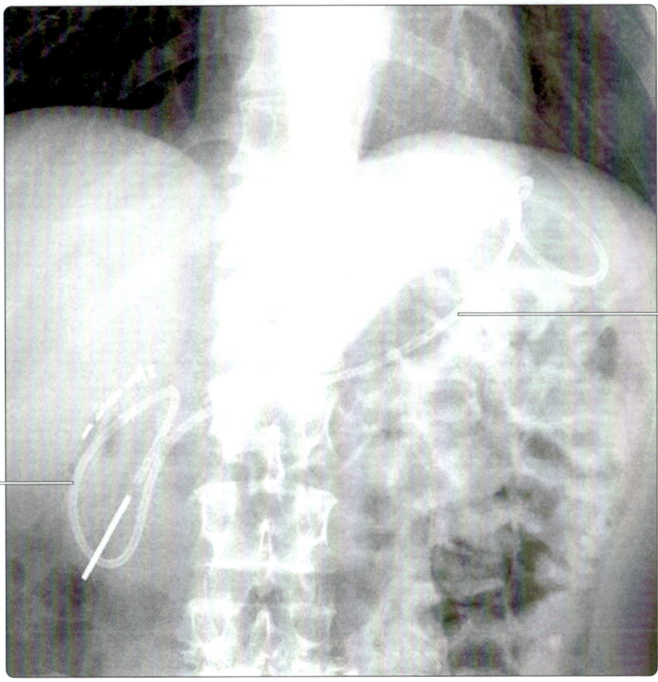

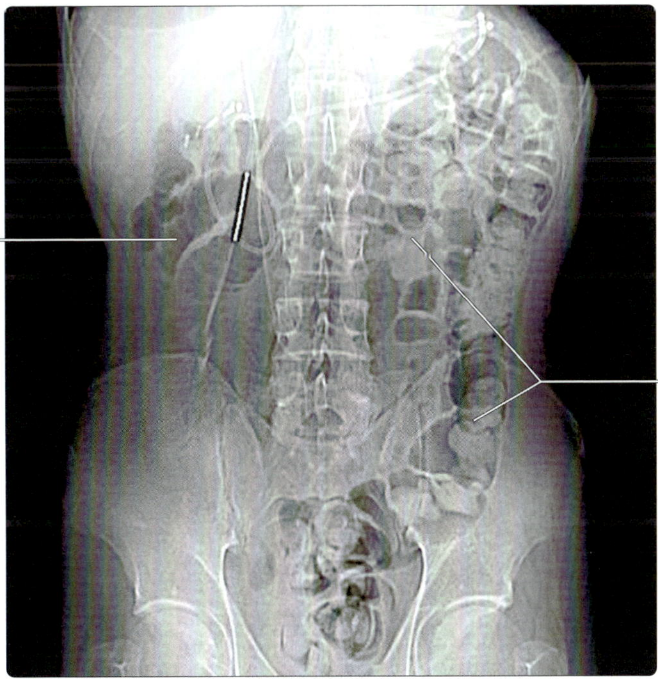

(Top) *First of 5 images shows malrotation. Frontal radiograph shows a feeding tube that has been advanced through the nose and stomach. The course of the tube within the duodenum shows that the duodenum is "looped" upon itself to the right of the midline instead of crossing the midline as the 3rd portion of the duodenum does normally.* **(Bottom)** *Frontal radiograph shows the duodenum and small bowel (marked by the feeding tube and gas) in the right upper quadrant. Gas and stool within the colon suggest that all of the colon lies to the left of midline.*

CT, MALROTATION

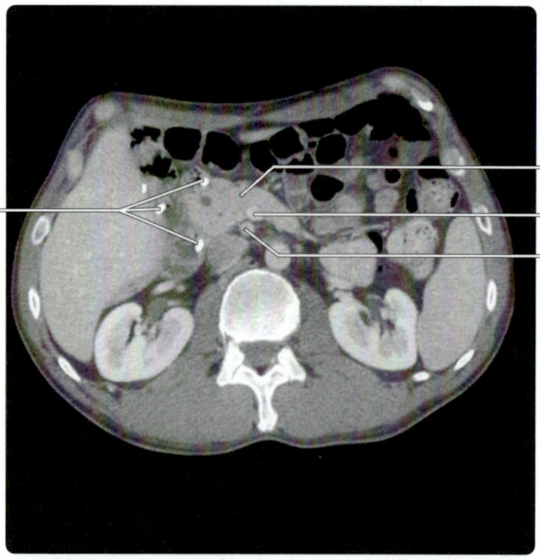

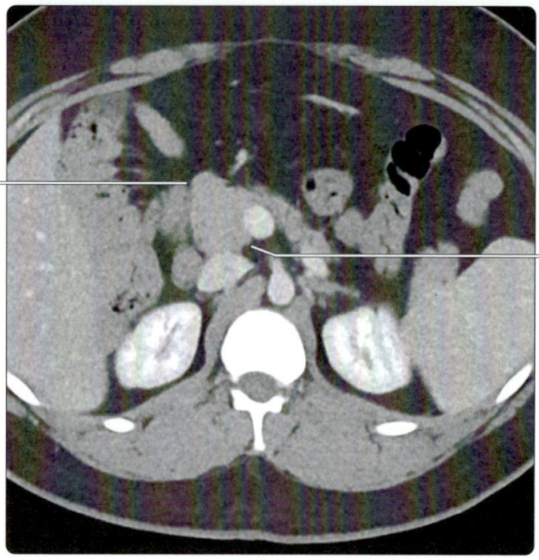

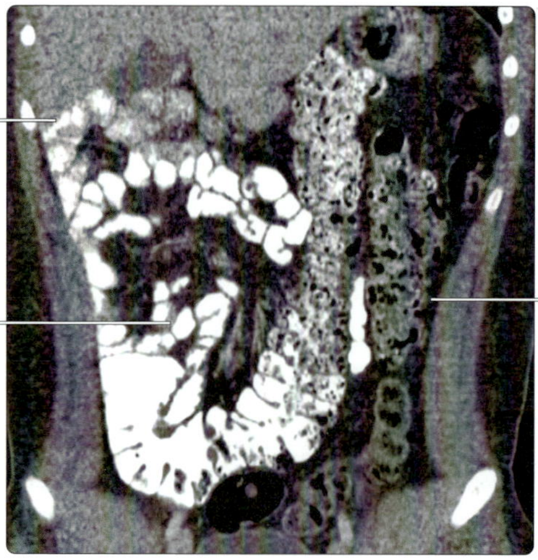

(Top) *Axial CECT in the same patient shows the radiopaque feeding tube coiled within the duodenum, all of which lies to the right of midline. The superior mesenteric vein lies ventral and to the left of the SMA, which is opposite to its usual anatomic relationship.* **(Middle)** *Axial contrast-enhanced image in a 16-year-old boy shows globular enlargement of the pancreatic head and aplasia of the uncinate process. Bowel malrotation is associated with variation in pancreatic anatomy.* **(Bottom)** *Coronal CT in the same patient shows bowel malrotation with large bowel in the left abdomen and jejunal and ileal loops in the right abdomen.*

FLUOROSCOPY, MALROTATION

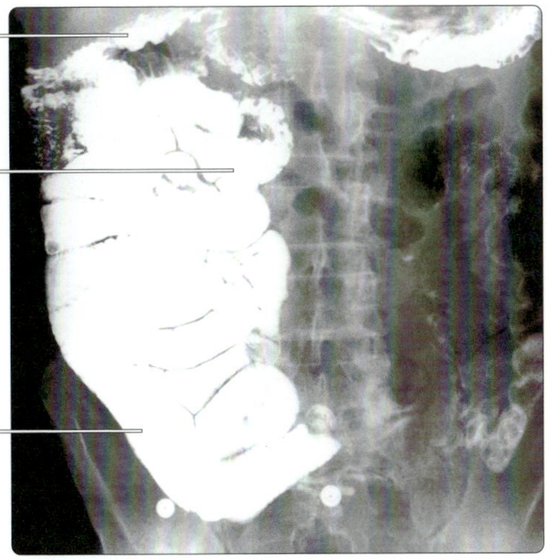

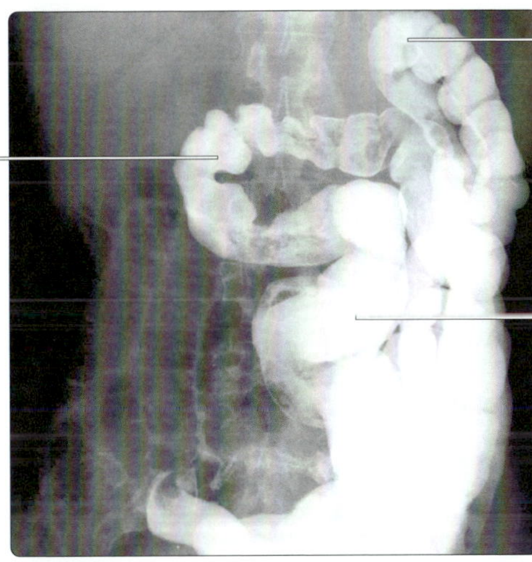

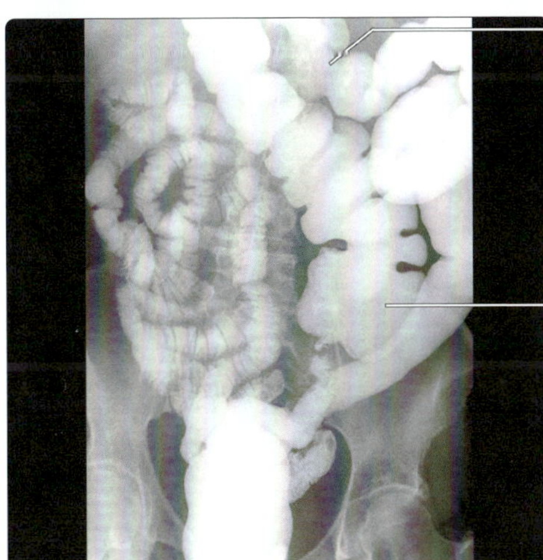

(Top) *First of 2 images from a barium small bowel follow-through study shows essentially all of the small intestine lying to the right of midline. The duodenum never crosses the midline, and the jejunum is not in its usual left upper quadrant location.* **(Middle)** *Delayed image from a small bowel study shows all of the colon lying in the left abdomen. The descending colon, sigmoid colon, and rectum, representing the embryological hindgut, are in normal position. It is the embryologic midgut that has failed to rotate properly on its return from the umbilical cord to the peritoneal cavity, resulting in the abnormal position of the small bowel and right side of the colon. This situation can predispose to volvulus and obstruction of the bowel and its vasculature.* **(Bottom)** *Delayed image from a small bowel study shows all of the small bowel lying to the right of midline, while all of the colon lies to the left. The embryologic midgut has failed to rotate properly, resulting in malrotation of the small intestine and the colon up to the splenic flexure.*

MIDGUT VOLVULUS

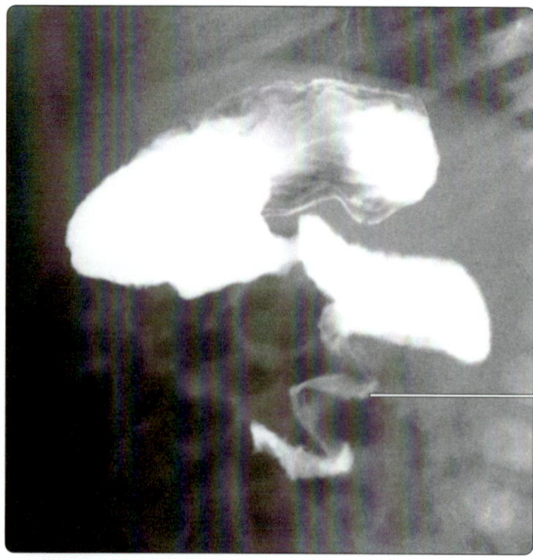

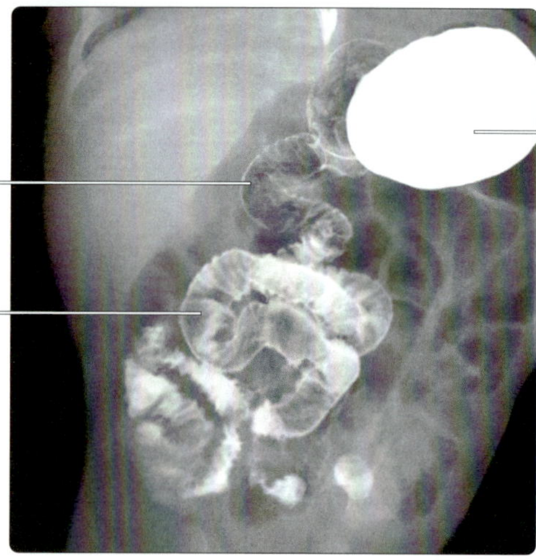

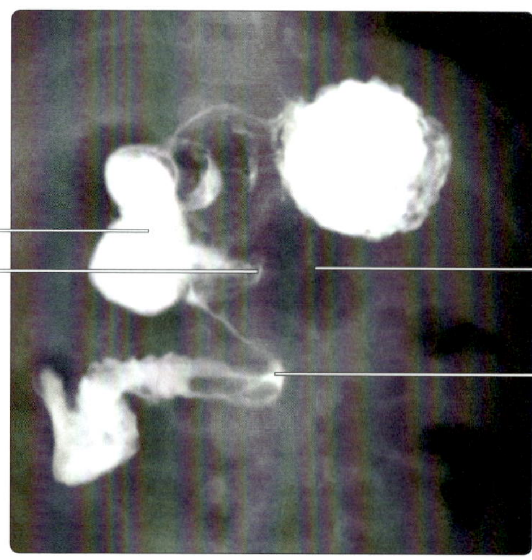

(Top) *Upper GI series image in a young child demonstrates an abnormal appearance of the duodenum and proximal small bowel with the duodenum appearing distended and the jejunum extending downward with an abnormally narrowed, corkscrew appearance, classic for midgut volvulus.* **(Middle)** *Frontal delayed image from the same upper GI series demonstrates that the duodenum does not cross the midline but instead extends down the right abdomen. In addition, the jejunum lies in the right abdomen instead of the left, compatible with malrotation.* **(Bottom)** *Frontal upper GI series image demonstrates a typical example of midgut volvulus with an abnormally positioned duodenaljejunal junction to the right of the pedicles. The typical corkscrew appearance of the proximal small bowel and mildly dilated proximal duodenum in this patient are compatible with the subsequently surgically proven midgut volvulus.*

CONGENITAL DUPLICATION CYST

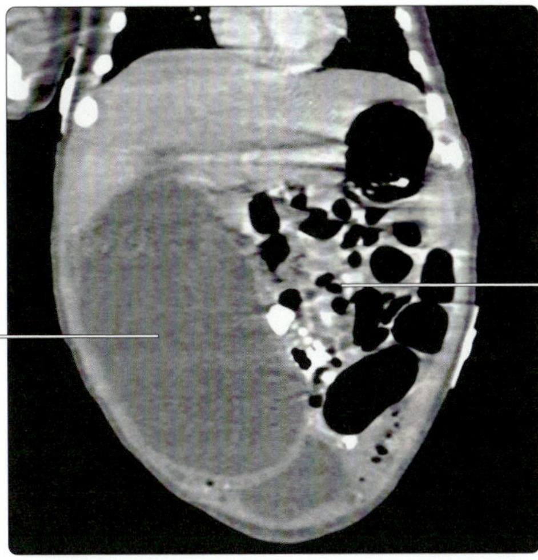

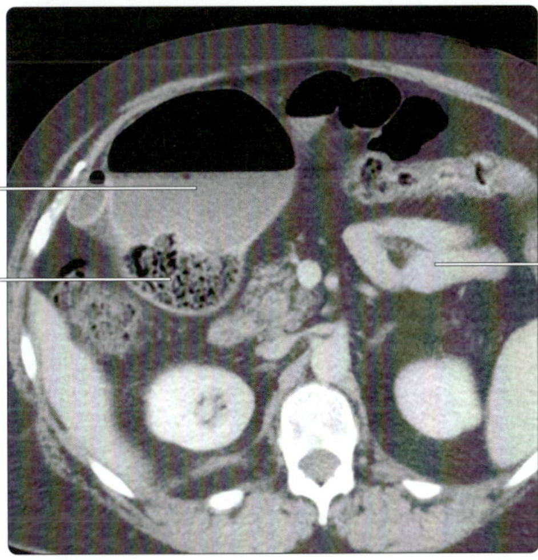

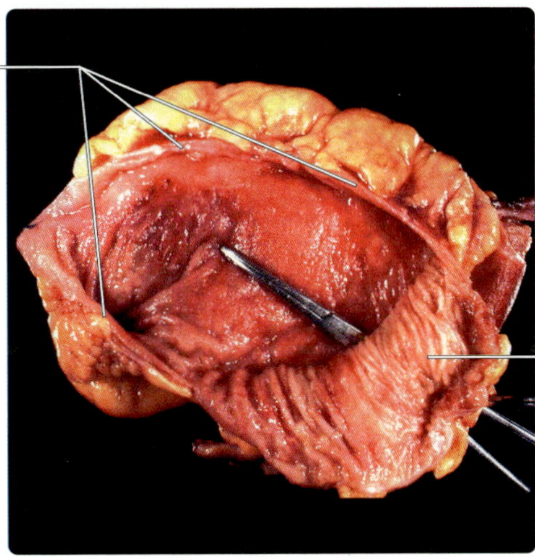

(Top) *Coronal CECT in a newborn baby demonstrates a massive cystic mass in the right abdomen, which was surgically resected and found to be a duplication cyst arising from the small bowel mesentery.* **(Middle)** *Axial CECT of a young man with crampy abdominal pain demonstrates a spherical mass with an air-fluid level adjacent to a dilated segment of small bowel that has particulate matter and gas within it simulating stool, a common feature of small bowel obstruction. The air-fluid level within the mass suggests communication with the adjacent bowel.* **(Bottom)** *Surgical specimen photograph in the same patient shows a probe passing from the small bowel lumen into the cyst, which was lined by intestinal mucosa. Congenital duplications or cysts may occur along the entire length of the alimentary tube. In rare cases, long segments of bowel or colon may be duplicated. These lesions may remain asymptomatic, become infected, or cause bowel obstruction. They may or may not communicate with the lumen of the gut.*

MECKEL DIVERTICULUM

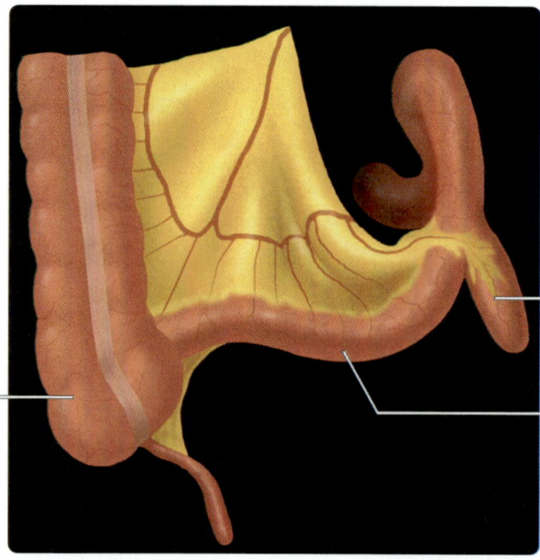

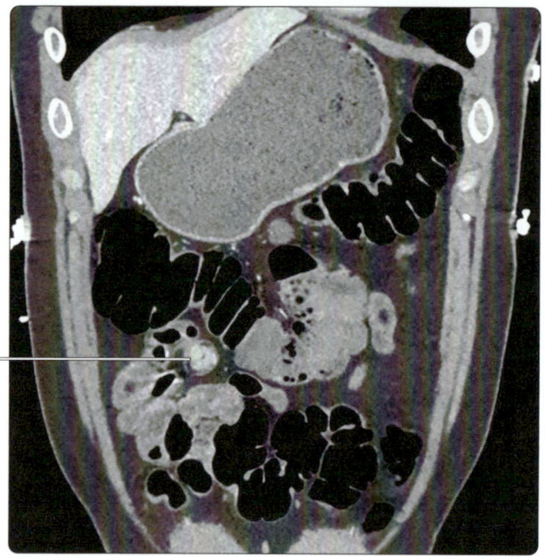

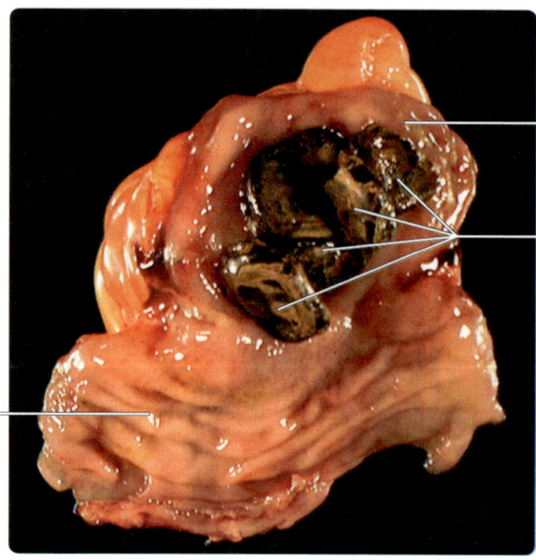

(Top) *First of 3 images demonstrates a Meckel diverticulum. Graphic shows a blind-ending diverticulum arising from the antimesenteric border of the distal ileum, a characteristic appearance of a Meckel diverticulum. This is a remnant of the embryologic omphalomesenteric or vitelline duct, also known as the yolk stalk, which normally connects the fetal gut to the yolk sac. This connection atrophies and disappears soon after birth in most individuals.* **(Middle)** *Coronal CECT demonstrates a bulbous, blind-ending diverticulum arising from the distal small bowel confirmed to represent a Meckel diverticulum at surgery. This young patient was exhibiting signs of lower GI bleeding, and the high-density material within the diverticulum may represent blood products.* **(Bottom)** *Photograph of a resected Meckel diverticulum shows the opened diverticulum and some of the dark brown stones or enteroliths that had formed within it.*

SMALL BOWEL HEMORRHAGE

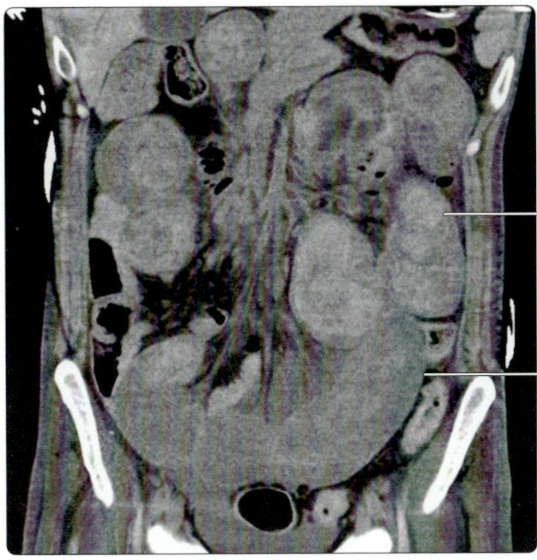

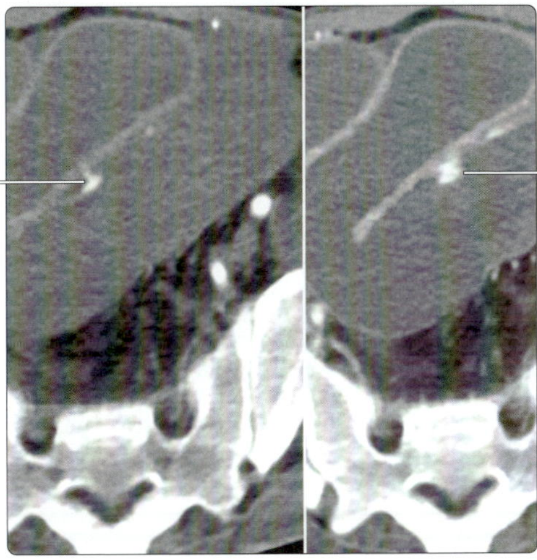

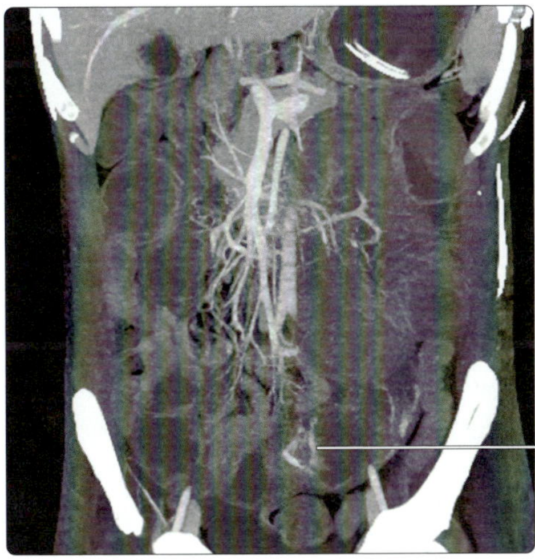

(Top) *Patient with dropping hematocrit is shown. Dilated loops of small bowel are seen containing hyperdense material, most concerning for hemorrhage.* **(Middle)** *Axial arterial-phase image in the same patient shows a focus of contrast blush that increases in size on the venous phase, most consistent with active hemorrhage.* **(Bottom)** *Coronal MIP image in the same patient shows active extravasation of contrast consistent with active bleeding.*

SMALL BOWEL TUMORS

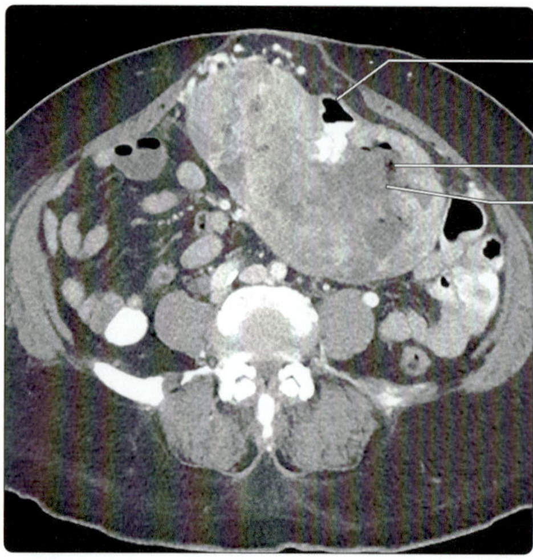

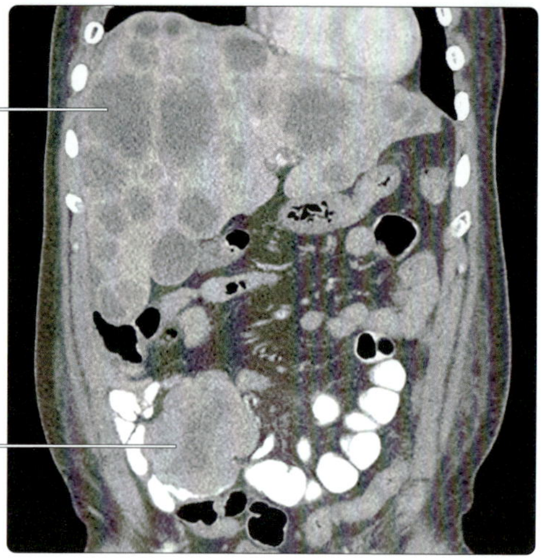

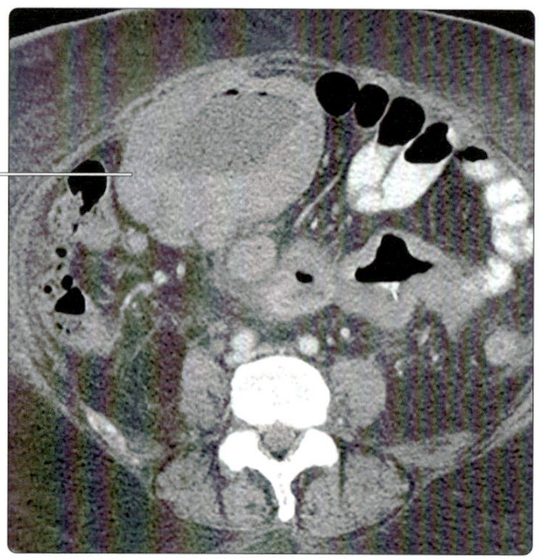

(Top) *Axial CECT demonstrates a large heterogeneous mass in the midabdominal mesentery closely associated with an adjacent small bowel loop. The presence of gas within the mass suggests ulceration and communication with the bowel. Gastrointestinal stromal tumor (GIST) should be a primary consideration when confronted with a large heterogeneous mesenteric mass, as in this case.* **(Middle)** *Coronal CECT demonstrates a large heterogeneous mass abutting small bowel loops in the right lower quadrant as well as multiple liver metastases. The mass was inseparable from the small bowel at resection, and this was found to be a GIST arising from the small intestine. The exophytic nature of the mass can serve as a clue to this diagnosis.* **(Bottom)** *Axial CECT demonstrates an aneurysmally dilated loop of small bowel with thick, irregular peripheral wall thickening. While lymphoma and GIST are both considerations given this appearance, the patient's history of primary melanoma allows the correct diagnosis of a small bowel metastasis in this case.*

SMALL BOWEL TUMORS

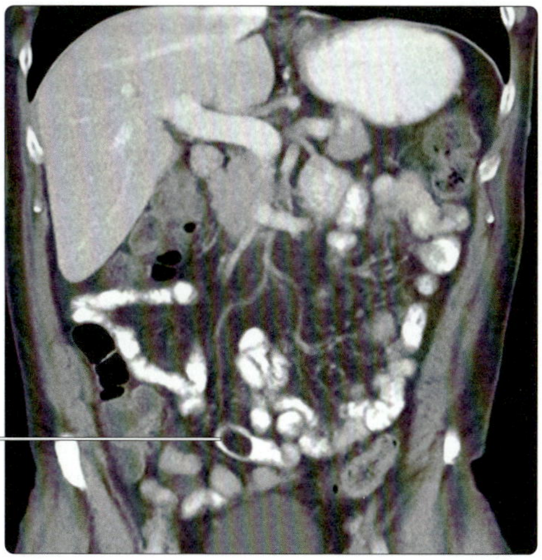

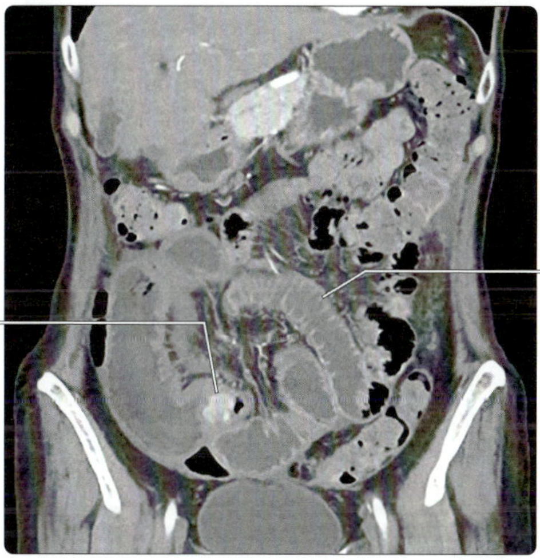

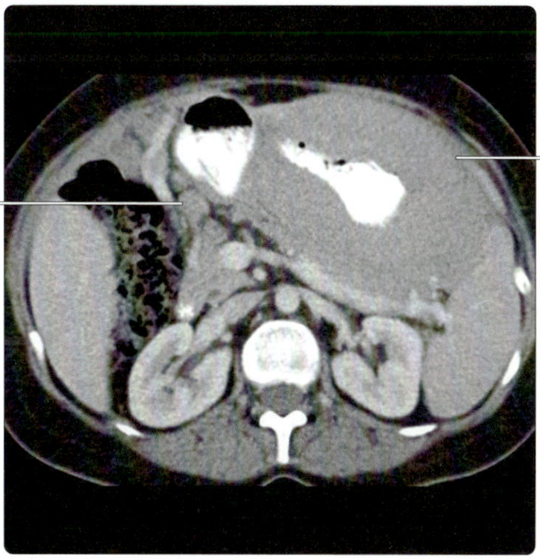

(Top) *A small, well-circumscribed attenuation mass in the small bowel, consistent with a benign lipoma, is shown. Small lipomas (< 1 cm) are usually detected incidentally and may be completely asymptomatic.* **(Middle)** *Coronal CECT in the arterial phase demonstrates a hypervascular mass in the terminal ileum, a classic location for a carcinoid. These lesions tend to be avidly vascular and most apparent in the arterial phase. Notice that the proximal bowel is dilated, compatible with small bowel obstruction secondary to the lesion.* **(Bottom)** *A large, circumferential mass involving the small bowel with aneurysmal dilation and mesenteric lymphadenopathy is shown in a patient with mucosa-associated lymphoid tissue (MALT) lymphoma.*

GROSS ANATOMY

Overview

- **Colon** (**large intestine**): Distalmost portion of gastrointestinal tract responsible for absorption of water and salt from bowel content (chyme) that was not digested and absorbed by small intestine
  - Converts ingested contents into semisolid stool that is stored until defecation occurs

Segments

- **Cecum**: Blind-ending 1st part of colon (~ 7 cm in length), which is contiguous with ascending colon and receives ingested material from distal ileum through ileocecal valve
  - Defined as portion of right colon inferior to ileocecal junction
  - Does not have its own mesentery but shares mesentery with ileum
    - Loosely attached to posterior and lateral abdominal wall via peritoneal (cecal) folds
  - Receives terminal ileum through ileocecal valve
    - Valve lips have submucosal fat content usually evident on CT (which can hypertrophy as normal variant and potentially mimic fat-containing mass)
    - Valve usually prevents reflux of colonic contents into small bowel
  - **Appendix** ("vermiform" appendix)
    - Narrow, blind-ending tubular structure (i.e., diverticulum) measuring 6-15 cm in length and < 7 mm in diameter, which arises from cecum, usually along its posteromedial surface
    - Always arises from tip of cecum but may point and lie in many different locations (2/3 retrocecal)
    - Has its own mesentery (mesoappendix)
  - Both cecum and appendix are intraperitoneal
  - Arterial and venous supply of cecum and appendix: **Ileocolic artery** and **vein** [branch of superior mesenteric artery (SMA) and superior mesenteric vein (SMV)]
- **Ascending colon**
  - Extends from cecum to transverse colon (meets transverse colon at hepatic flexure where it turns 90° to left)
  - Located in retroperitoneum
  - Blood supply: **Right colic artery** (branch of SMA) and **right colic vein** (tributary of SMV)
- **Transverse colon**
  - Extends across abdomen from ascending colon (at hepatic flexure) to descending colon (at splenic flexure), where it makes inferior 90° turn
    - Most variable portion of colon, as it can be quite tortuous and extend into pelvis
    - Transverse colon is intraperitoneal
    - Transverse colon has its own mesentery (**transverse mesocolon**), through which vessels, nerves, and lymphatics course
    - **Greater omentum** drapes downward from transverse colon, and **gastrocolic ligament** connects greater curvature of stomach and transverse colon
  - Arterial supply: **Middle colic artery** (branch of SMA)
  - Venous drainage: SMV
- **Descending colon**
  - Retroperitoneal portion of left colon, which extends from splenic flexure downward and turns medially as it becomes sigmoid colon
  - Arterial supply: **Left colic artery** and **superior sigmoid artery** [branches of inferior mesenteric artery (IMA)]
  - Venous drainage: Inferior mesenteric vein (IMV)
- **Sigmoid colon**
  - Mobile, S-shaped intraperitoneal portion of left colon
    - Quite variable in length, redundancy, and location
    - Has its own long mesentery (**sigmoid mesocolon**)
  - Arterial supply: **Sigmoid arteries** (branches of IMA)
  - Venous drainage: IMV
- **Rectum**
  - Final 15-20 cm of colon that extends from **rectosigmoid junction** (variable position but usually begins at S3 level) down to anal canal
    - Divided into 3 segments: Lower 1/3 (0-5 cm from anal verge), middle 1/3 (5.1-10 cm from anal verge), and upper 1/3 (10.1-15 cm from anal verge)
    - Lies in extraperitoneal pelvis surrounded by perirectal fat (i.e., "mesorectum"), which is bounded by mesorectal fascia
      - Mesorectal fascia defines "circumferential resection margin" utilized surgically when patients undergo total mesorectal excision for rectal cancer (and key landmark when staging rectal cancer using MR)
    - Upper 1/3 covered by peritoneal lining anteriorly and laterally, while middle 1/3 covered by peritoneum anteriorly only; no peritoneal covering for lower 1/3
      - Peritoneal reflections form rectovesical pouch in men and rectouterine pouch (of Douglas) in female patients, both of which are most inferior peritoneal recesses
      - Only upper 1/3 of rectum is intraperitoneal
    - Has several rectal folds (valves) analogous to semilunar folds of colon
    - Has continuous layer of longitudinal muscle, rather than taeniae (separate bands of muscle) seen more proximally in colon
  - Distal rectum expands to store fecal material prior to defecation ("ampulla" of rectum)
  - Arterial supply: Both mesenteric and systemic arterial supply via superior rectal artery (terminal branch of IMA), middle rectal artery (branch of internal iliac artery), and inferior rectal artery (branch of internal pudendal artery via internal iliac artery)
  - Venous drainage: Superior rectal vein (drains into portal venous system), middle rectal vein (systemic drainage), and inferior rectal vein (systemic drainage)
  - Nervous supply
    - Sympathetic supply via lumbar splanchnic nerves and superior/inferior hypogastric plexuses
    - Parasympathetic supply via pelvic splanchnic nerves and inferior hypogastric plexus
  - Lymphatics: Pararectal lymph nodes drain to inferior mesenteric nodes and internal iliac nodes
- **Anus** and **perianal region**
  - Anal canal (~ 4 cm) is extraperitoneal terminal portion of GI tract, which extends from rectum to anal verge
    - Anorectal junction defined by anorectal ring (where puborectalis muscle loops around anus/rectum)

   - Surrounded by internal sphincter (involuntary sphincter) and external sphincter (under voluntary control)
     - Internal sphincter contiguous with circular muscle of rectum
     - External sphincter composed of 3 parts (deep, superficial, and subcutaneous)
   - Primary anorectal musculature includes **levator ani** (primary support for pelvic floor) and **puborectalis muscle** (forms sling around anorectal junction and plays important role in fecal continence)
 - Pectinate line (not visible on imaging) divides superior anal canal (derived from embryonic hindgut) from inferior canal (derived from embryonic proctodeum)
 - Arterial supply: Superior canal supplied by superior rectal artery and inferior canal supplied by inferior rectal artery (both anastomose with middle rectal artery)
 - Venous supply: Internal rectal plexus surrounds anus and drains into superior rectal vein above pectinate line and inferior rectal vein below pectinate line (both of which anastomose with middle rectal vein)
 - Lymphatics: Lymphatics above pectinate line drain into internal iliac nodes, while below pectinate line, they drain into superficial inguinal lymph nodes
 - Nervous supply
   - Internal sphincter supplied by parasympathetic pelvic splanchnic nerves (relaxes sphincter when rectum is distended)
   - External sphincter supplied by inferior rectal nerve (S4 level)
   - Anal canal above pectinate line has visceral supply via inferior hypogastric plexus (only stretch sensitivity), while below pectinate line has somatic supply via inferior rectal nerves (sensitive to pain, temperature, and touch)
- **Paracolic gutters**
  - Right and left paracolic gutters are part of peritoneal cavity and run immediately lateral to right and left colon and extend downward into pelvis
  - Provide pathway for infectious or inflammatory abnormalities (including those originating from right/left colon) in upper abdomen to track into pelvis
  - Right paracolic gutter starts at hepatic flexure and surrounds cecum before extending into pelvis
    - Continuous with right subhepatic and subphrenic spaces
  - Left paracolic gutter starts at splenic flexure and extends into pelvis
    - Separated from left subphrenic space by phrenicocolic ligament
  - Also location for peritoneal spread of metastatic disease
- **Vascular anatomy**
  - Blood supply to colon primarily via SMA (right colic and middle colic arteries) and IMA (left colic and sigmoid arteries)
  - **Marginal artery (of Drummond)** runs along inner margin of colon and provides continuous series of anastomoses between terminal SMA and IMA branches
    - Protects blood supply in cases of stenosis or ↓ flow
    - Least robust at "watershed" zone between SMA and IMA (splenic flexure), making this site most vulnerable to ischemia
  - **Arc of Riolan** is another communication between SMA and IMA (not always present)

Mural (Wall) Anatomy

- Mucosa, submucosa, double-layer muscularis, serosa (for intraperitoneal portions), and submucosa (adventitia for extraperitoneal portions)
- Longitudinal muscle layer not continuous (unlike small bowel) but separated into taeniae (except in rectum)
- Unlike small bowel, mucosa of colon not covered with villous projections
- Submucosa contains numerous discrete **lymphoid follicles** that may be apparent as subtle 3- to 4-mm nodules on double-contrast barium enema exam
- **Taeniae coli**: 3 thickened, flat bands of smooth muscle constituting outer longitudinal layer of smooth muscle
- **Haustra**: Sacculations of colon wall caused by contractions of taeniae, separated by semilunar folds
- **Semilunar folds (plicae semilunares)**
  - Furrows between haustra consisting of mucosa, submucosa, and circular muscle (small bowel folds lack muscle layer)
- **Epiploic (omental) appendages (or appendices)**
  - Subserosal pockets of fat extending from colonic surface

CLINICAL IMPLICATIONS

Clinical Importance

- **Appendicitis**
  - Results from obstruction of appendiceal lumen, leading to appendiceal inflammation
  - Enlarged, inflamed appendix with thick, hyperemic wall and periappendiceal fat stranding on CT
- **Diverticulosis**
  - All parts of colon, except rectum, may develop diverticula (most prevalent in sigmoid colon)
    - Diverticula protrude through weak points in colon wall where nutrient arteries penetrate muscle coat
  - Diverticula may perforate, leading to colonic inflammation (i.e., diverticulitis)
- **Epiploic appendagitis**
  - Colonic appendages may twist and infarct, resulting in clinical symptoms that mimic diverticulitis
- **Colonic volvulus**
  - Sigmoid mesocolon may be long with narrow base of attachment to posterior abdominal wall
    - Predisposes to volvulus (twisting) of colon that often obstructs lumen and compresses vessels, potentially leading to ischemia and perforation
  - Cecum may also be on long mesentery, predisposing to volvulus, obstruction, and ischemia ("cecal volvulus")
- **Rectal carcinoma**
  - Due to systemic and portal venous drainage, may metastasize to systemic sites (lungs, bones, etc.) prior to spreading to liver (whereas carcinoma elsewhere in colon almost always metastasizes to liver first)
  - MR serves as most valuable imaging tool in staging of rectal cancer

COLON AND ITS MESENTERY

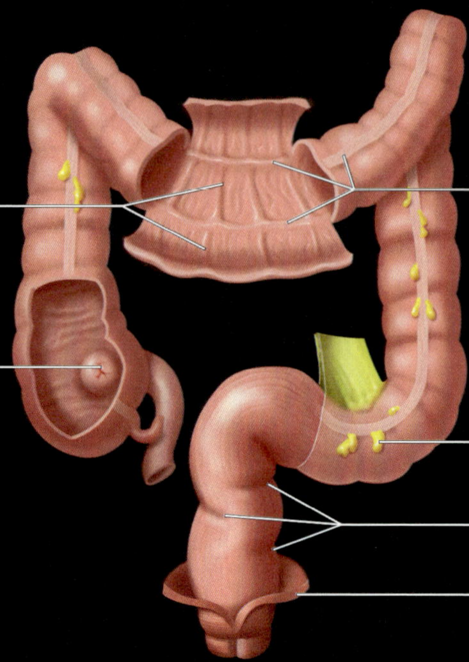

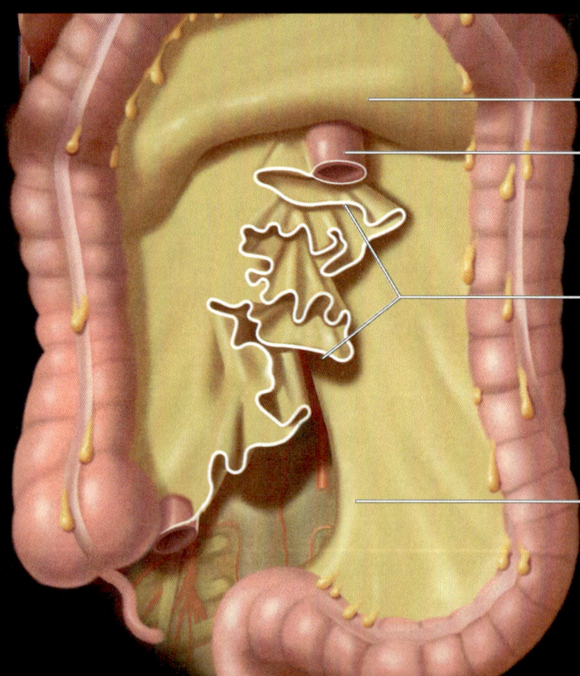

(Top) *Graphic shows a surface and mucosal view of the colon. The appendix and ileocecal valve enter the cecum, the 1st part of the colon, with the ileocecal valve acting as a sphincter to prevent reflux. The semilunar folds lie at right angles to the long axis of the colon and are analogous to the rectal folds (valves). The outpouchings between the folds are the haustra, and the spaces between the transverse folds are wider than in the small intestine. The longitudinal muscle layer of the rectum is continuous, as it is in the small intestine. In the colon, this layer is separated into 3 thickened flat bands of muscle called the taeniae coli.* **(Bottom)** *The ascending and descending colon are largely retroperitoneal, while the transverse and sigmoid colon have a mesentery (mesocolon). This graphic shows the transverse mesocolon and colon reflected upward.*

ARTERIES

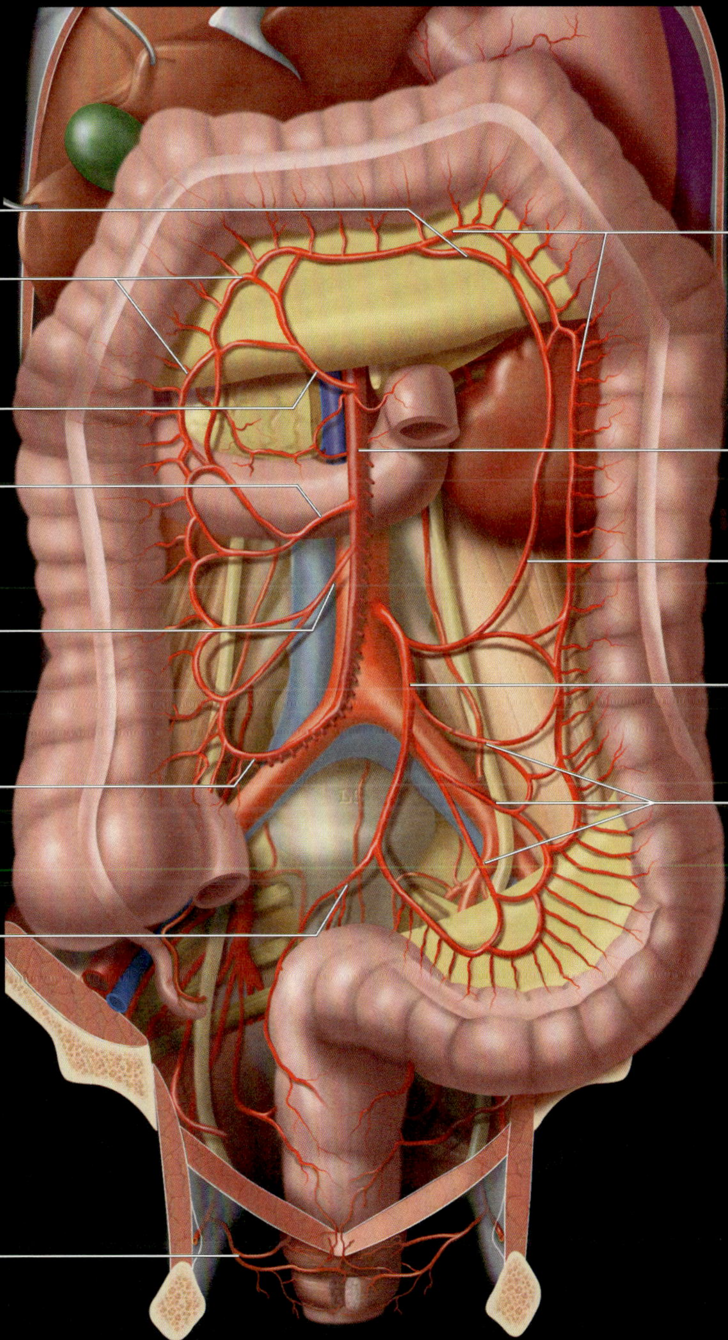

Graphic shows the small intestine has been removed, and the transverse colon has been reflected upward. This shows the conventional depiction of the superior mesenteric artery (SMA) supplying the colon from the appendix through the splenic flexure and the inferior mesenteric artery (IMA) supplying the descending colon through the rectum. The cecum is supplied by the ileocolic branch, the ascending colon by the right colic, and the transverse colon by the middle colic artery. These arterial branches are highly variable, and all are connected by anastomotic arterial arcades and the marginal artery (of Drummond) and arc of Riolan, which also anastomose with branches of the IMA that feed the descending and sigmoid colon. The arcades give off the straight arteries (arteriae rectae) to the colonic wall.

VEINS

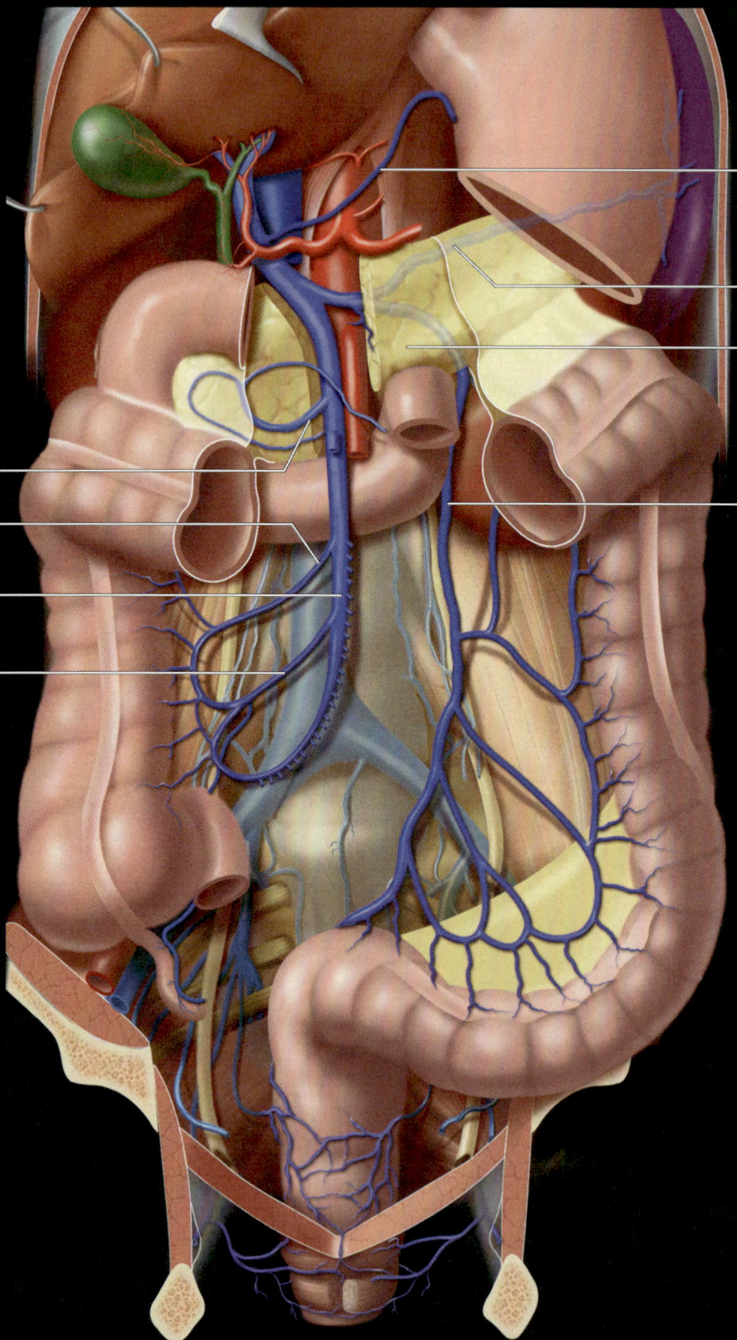

The veins that drain the colon generally follow a similar course and have similar names as the arteries that they accompany. The inferior mesenteric vein (IMV) drains the left colon and empties into the splenic vein or, less commonly, the superior mesenteric vein (SMV) or portal vein. The SMV and IMV lie just deep to the neck of the pancreas.

ILEOCECAL REGION

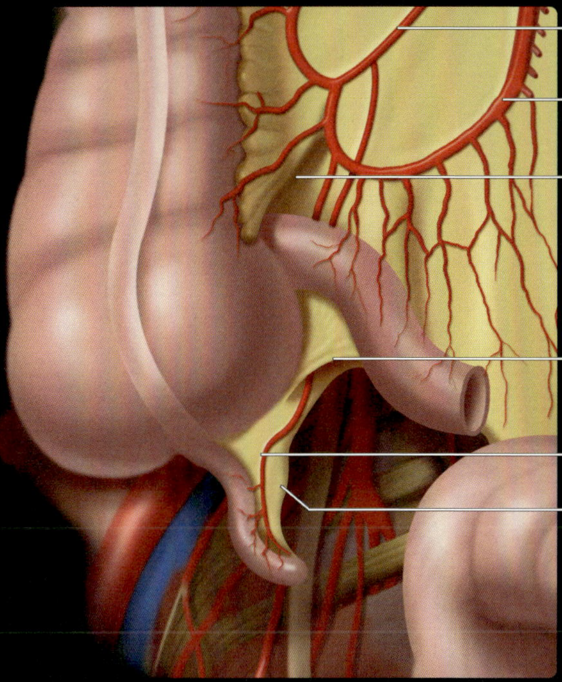

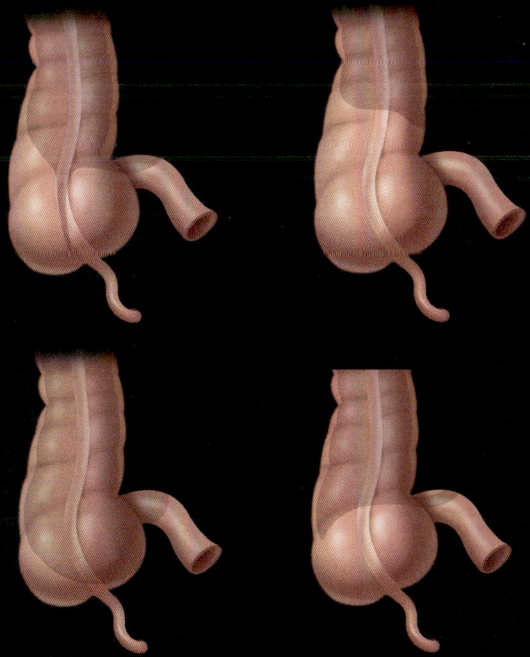

(Top) *The cecum is the blind-ending pouch that lies caudal to the ileocecal valve. It is usually the widest part of the colon and is easily distended with intestinal gas and the fluid contents that empty into it from the small intestine. The cecum usually rests in the right iliac fossa near the external iliac vessels and iliacus muscle but may extend down into the pelvis. Note that the appendix has its own mesentery (mesoappendix) with the appendiceal artery arising from the ileocolic artery.* **(Bottom)** *Graphics show different variations of the posterior peritoneal attachments of the cecum with the dark area representing the retroperitoneal attached segment. Note that the cecum is usually relatively mobile, as is the appendix and terminal ileum, while the ascending colon is usually fixed retroperitoneally. The appendix can, and often does, lie in a retrocecal space. An especially "free" (unattached) cecum and ascending colon predisposes to volvulus and obstruction of this part of the colon.*

LATERAL VIEW, RECTOSIGMOID

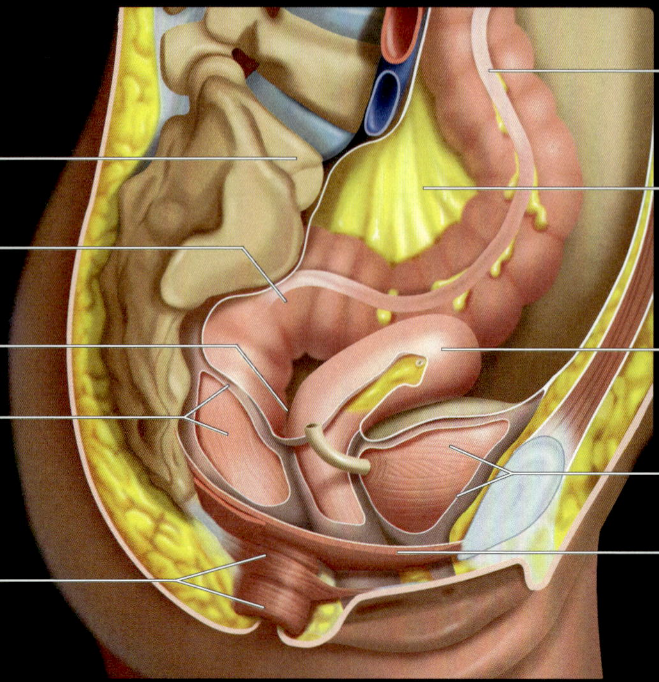

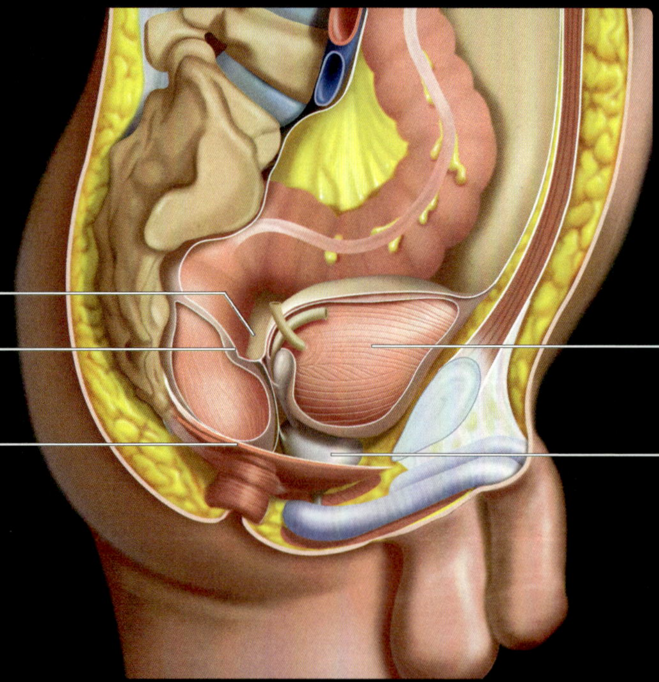

(Top) *The sigmoid colon is on a mesentery, while the rectum is retroperitoneal. The anterior surface of the rectum has a peritoneal covering, which extends deep in the pelvis in women, forming the rectouterine pouch (of Douglas) as it is reflected along the posterior surface of the uterus. The rectum is narrowed as it passes through the pelvic diaphragm and then enters the anal canal with 3 levels of the anal sphincter (deep, superficial, and subcutaneous). The rectum has a continuous external longitudinal coat of muscle, unlike the colon with its discontinuous rows of taeniae.* **(Bottom)** *The rectovesical space (recess) is the most dependent portion of the peritoneal cavity in men. The anterior wall of the rectum has more contiguity with the peritoneal cavity than its lateral or posterior walls.*

RECTAL VEINS AND VALVES

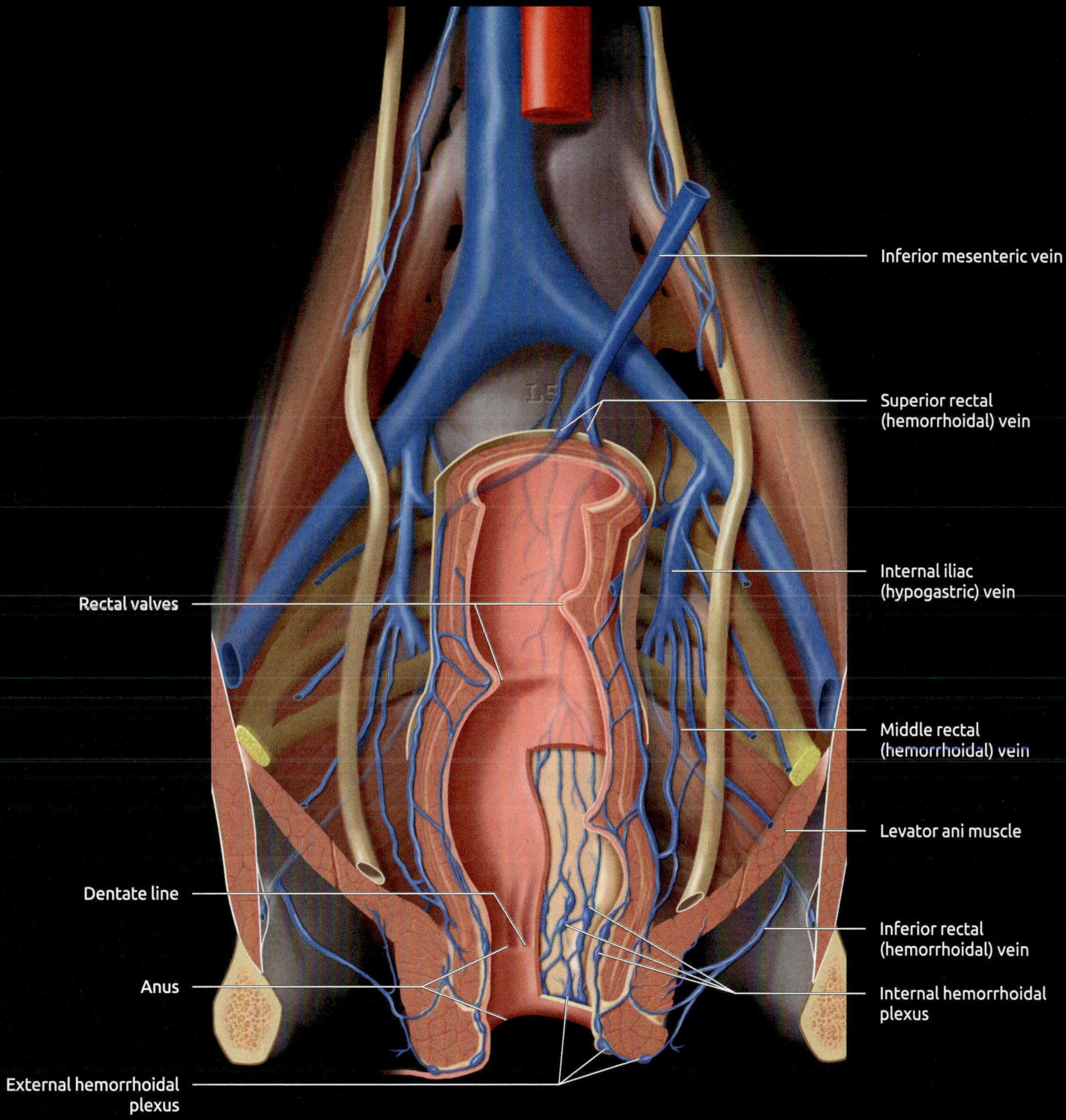

The rectal valves (folds) are crescentic infoldings of the wall that include the mucosa, submucosa, and circular muscle layers. The dentate line marks the squamocolumnar boundary. The veins of the rectum return blood to 2 different systems: The superior rectal vein to the portal system and the middle and inferior rectal veins to the inferior vena cava. These anastomose with each other and have important clinical implications (e.g., patterns of hematologic spread of rectal cancer, perirectal varices in portal hypertension). The external hemorrhoidal plexus lies in the subcutaneous tissue surrounding the anus, while the internal plexus is in the submucosal tissue of the rectum. These veins have no valves and are easily distended due to portal hypertension or chronic straining at stool, resulting in "hemorrhoids," which are common, painful swellings that may bleed &/or thrombose.

BARIUM ENEMA, NORMAL COLON

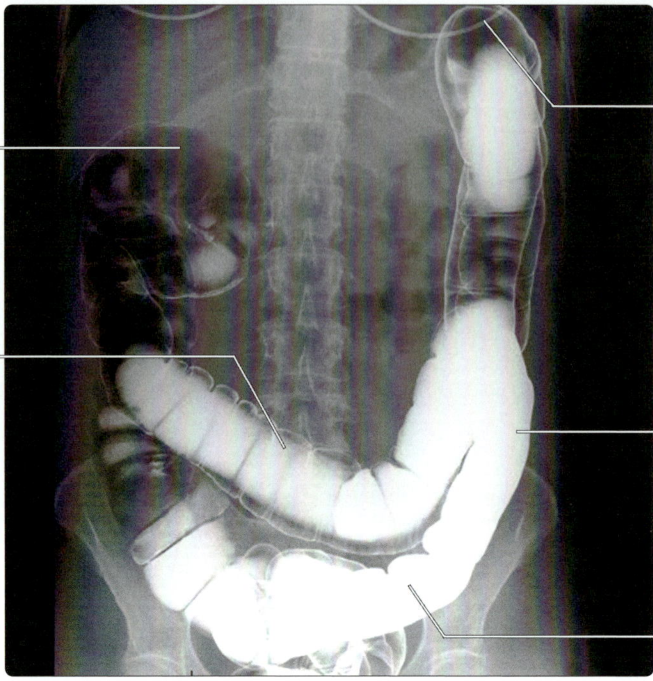

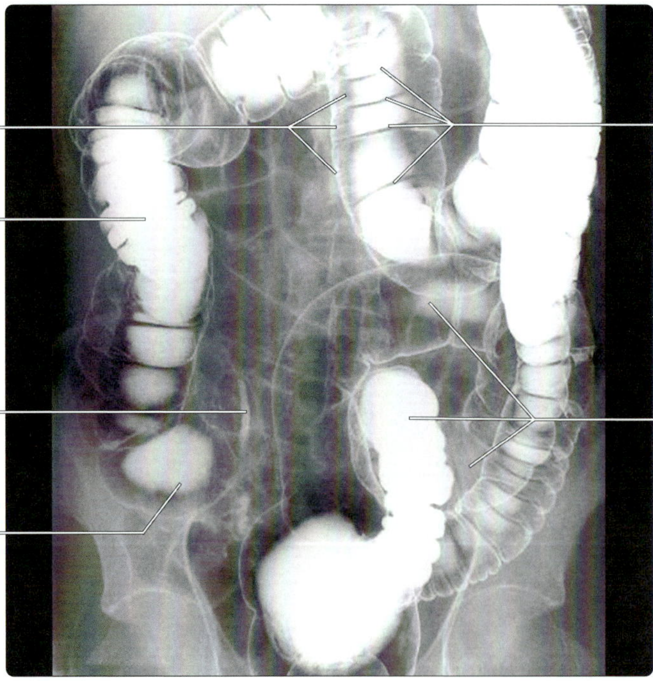

(Top) *First of 5 images from an air-contrast barium enema with the patient in a prone position shows barium pooling in the dependent portions of the transverse and sigmoid colon. Note the "radiologic" hepatic and splenic flexures of the colon, which are the most cephalic portions of the transverse colon. The "anatomic" flexures are the transitions from the retroperitoneal ascending (and descending) colon to the intraperitoneal transverse colon.* **(Bottom)** *In this supine position image, barium pools in the ascending and descending colon as well as the rectum. Note the transverse semilunar colonic folds and the haustra, the saccular outpouchings of the lumen between the folds. Colonic transverse folds are spaced farther apart than those in the small intestine, a useful marker in distinguishing bowel segments on plain radiographs.*

BARIUM ENEMA, NORMAL COLON

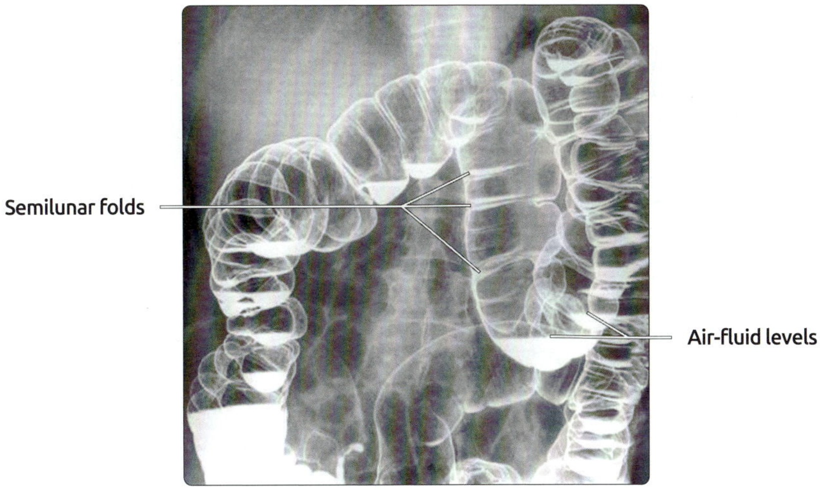

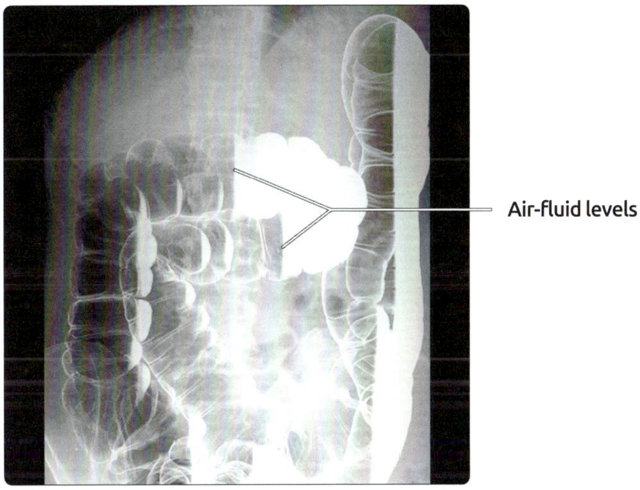

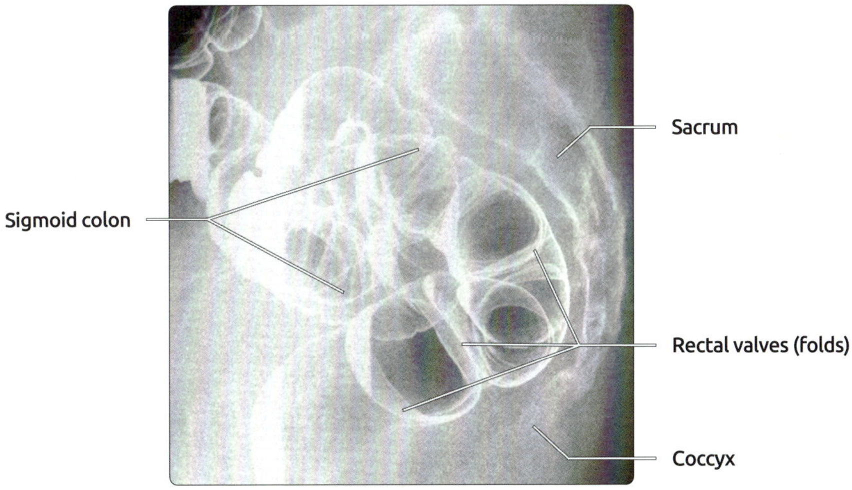

(Top) *Upright image from an air-contrast barium enema shows barium pooling in dependent haustra and curvatures of the colon.* **(Middle)** *A left lateral decubitus image shows barium pooling in the dependent left (descending) colon and within dependent haustra. This patient has a relatively long and curving ("redundant") colon, a common variant of no clinical concern, although this condition makes colonoscopic inspection of the entire colon more difficult.* **(Bottom)** *A cross-table lateral image, with the patient in the prone position, shows barium pooling along the anterior surface of the rectum and colon. Note the rectal valves (folds), which are analogous to the semilunar folds of the colon.*

NORMAL ILEOCECAL VALVE

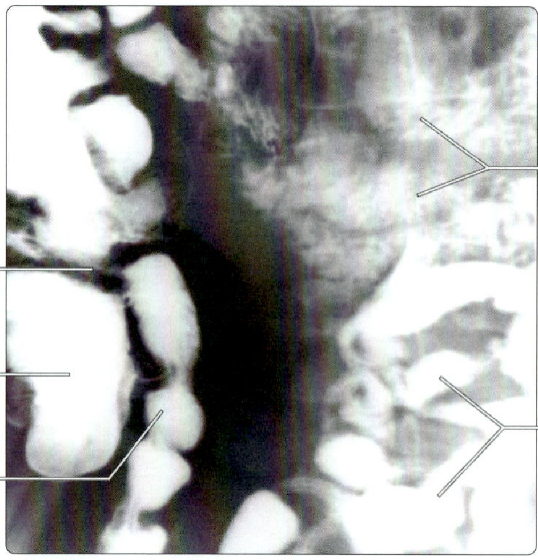

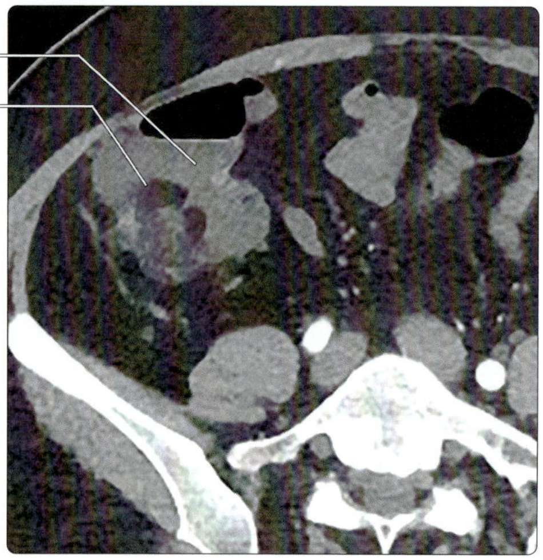

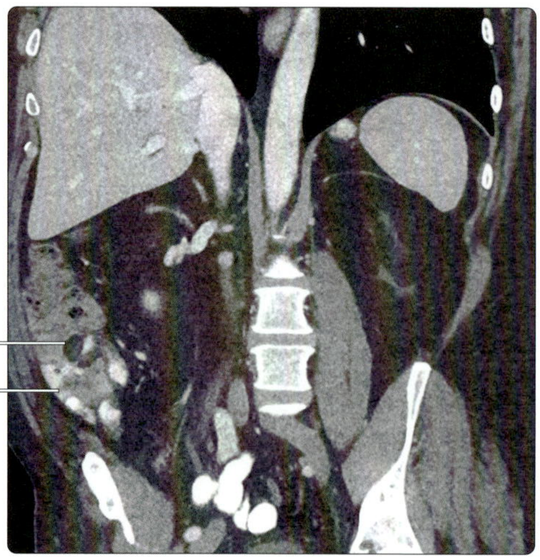

(Top) *Barium small bowel follow-through image shows a normal ileocecal valve. The "lips" of the ileocecal valve compress the terminal ileum as it enters the cecum. In most patients, this prevents reflux of colonic contents into the small intestine.* **(Middle)** *Axial CECT demonstrates normal low attenuation (density) within the "lips" of the ileocecal valve due to fibrofatty tissue. This constitutes a useful anatomic marker for identification of the ileocecal valve on CT sections. In some instances, fatty hypertrophy of the valve might be superficially mistaken for a lipoma in the cecum.* **(Bottom)** *Angled coronal CECT in a patient demonstrates the normal appearance of the ileocecal valve with fat density within the "lips" of the valve.*

NORMAL APPENDIX

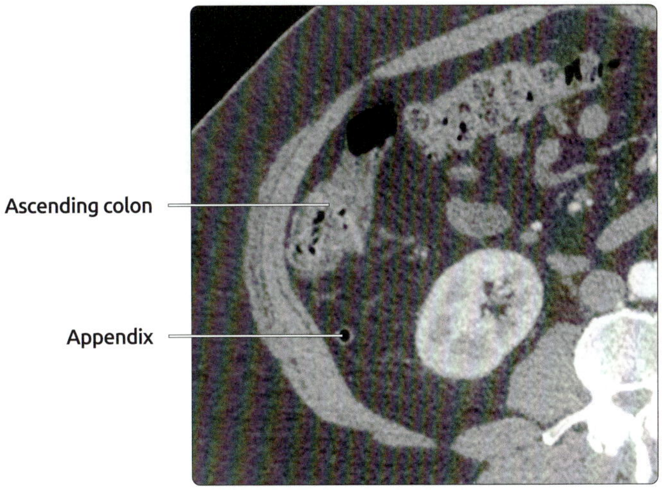

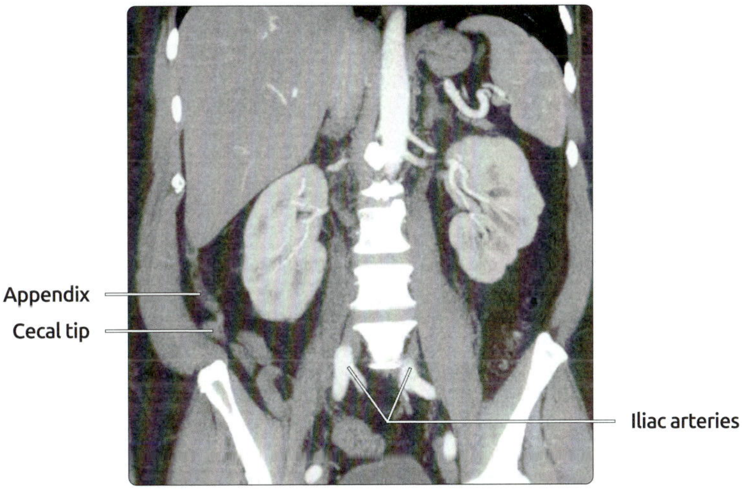

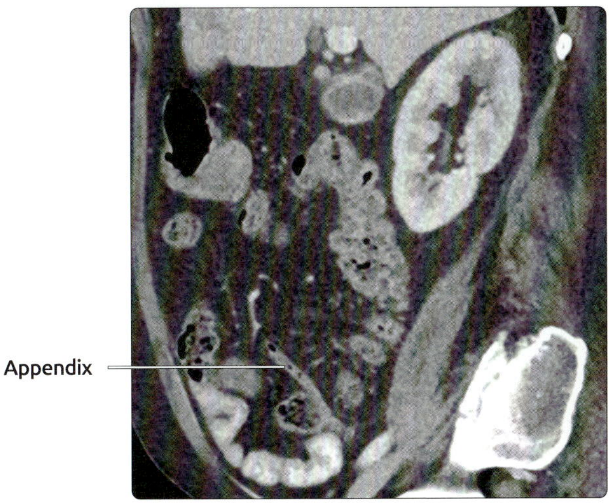

(Top) *The appendix commonly lies behind the cecum and ascending colon, as in this subject. The appendix may be intra- or retroperitoneal. Note the thin wall and gas content of the appendix as well as the homogeneous, noninflamed appearance of the periappendiceal fat planes.* **(Middle)** *Coronal CECT in the same patient demonstrates that the normal appendix has a small amount of fecal content and gas with a luminal diameter < 6 mm and a thin wall. The base of the appendix is seen at the tip of the cecum.* **(Bottom)** *Sagittal image of the abdomen shows the appendix with a diameter of ~ 5 mm and containing a few air bubbles, consistent with a normal appearance. Sagittal view can be helpful in identification of the appendix.*

SUPERIOR MESENTERIC VESSELS

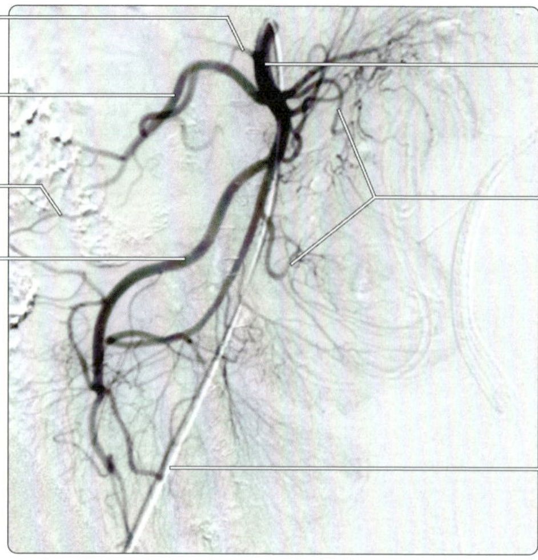

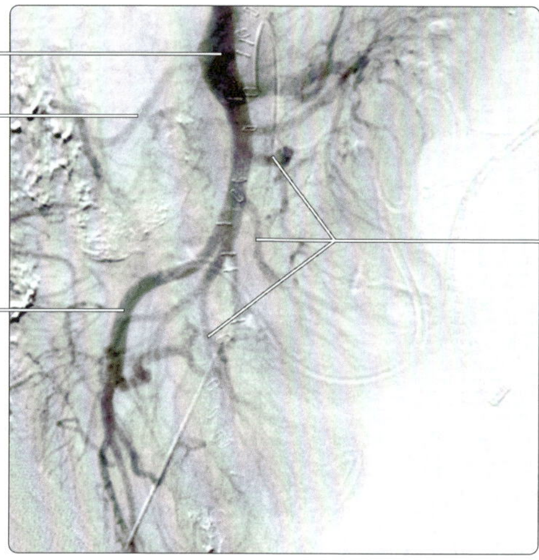

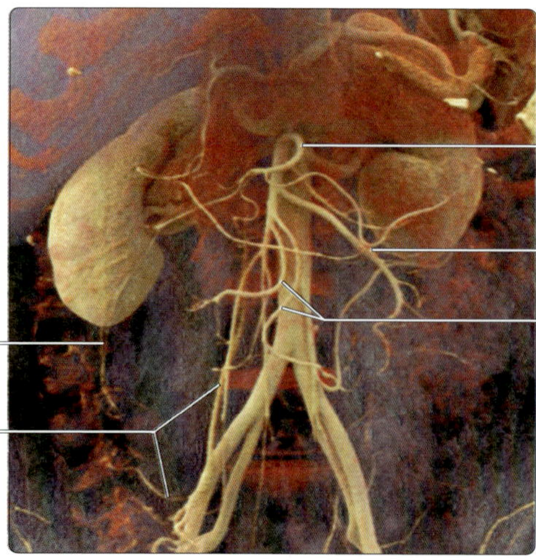

(Top) *First of 2 images from a catheter injection of the SMA shows the jejunal arteries arising from the left or convex side of the SMA and the colic branches from the right side. The middle colic is the 1st colonic branch of the SMA and supplies the transverse colon. The ascending colon is supplied by the right colic, and the cecum, appendix, and ileum are supplied by the ileocolic artery. There are many variations of this pattern and multiple anastomoses between colonic arteries, most notably through the marginal artery, which connects branches of all colonic arteries.* **(Middle)** *Venous-phase angiogram in the same patient shows opacification of the major branches of the SMV. These follow the same course and have similar names as the major arteries. The SMV usually joins with the splenic vein just behind the neck of the pancreas to form the portal vein.* **(Bottom)** *Cinematic volume-rendered image of the abdomen shows mesenteric arterial anatomy.*

INFERIOR MESENTERIC VESSELS

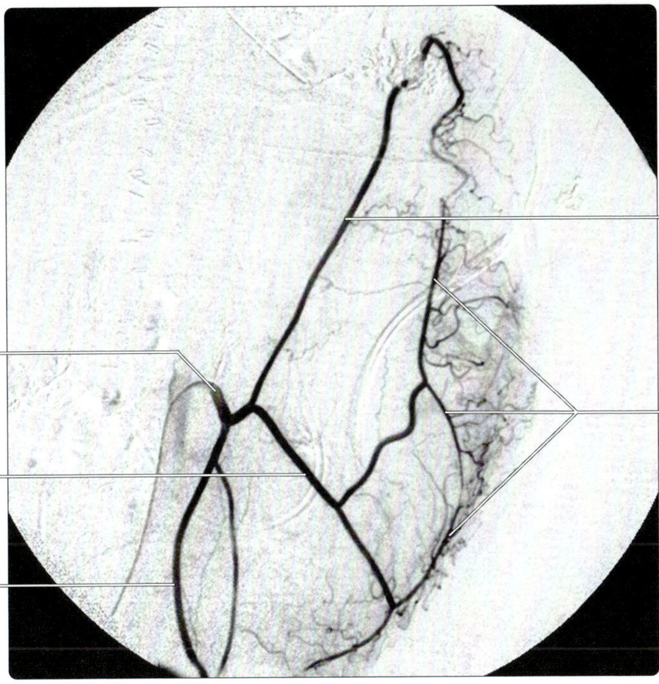

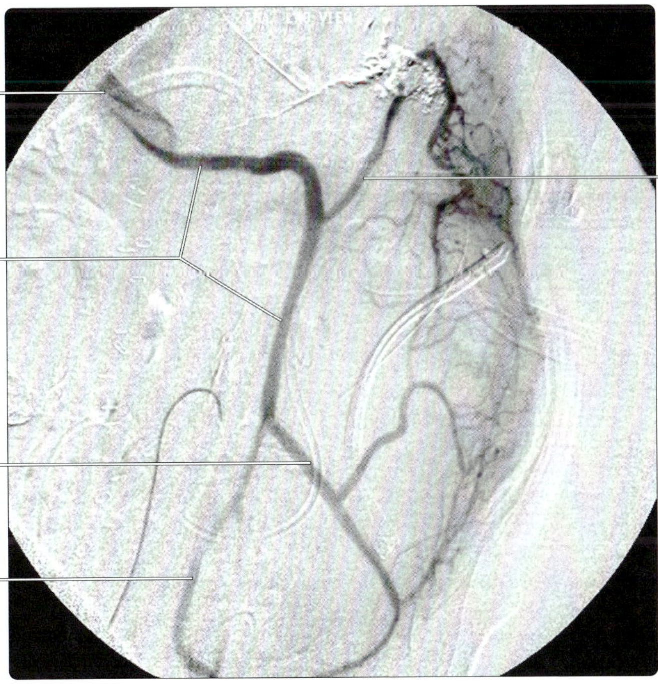

(Top) *First of 2 images from a catheter injection of the IMA shows its major branches, including the left colic (to the descending colon), sigmoid, and rectal arteries. The marginal artery parallels the course of the entire colon, gives rise to the terminal straight branches, and forms an important pathway for collateral flow to segments of the colon. The marginal artery connects branches of the SMA and IMA, and these collateral vessels may be sufficient to maintain colonic viability, even with complete occlusion of the origin of the IMA. The distribution of the SMA and IMA overlap in the "watershed" region of the splenic flexure. In the setting of shock or diminished cardiac output, this region is prone to ischemic injury.* **(Bottom)** *A venous-phase image from the inferior mesenteric angiogram shows the major tributaries of the IMV, which are named similarly to branches of the IMA.*

MESENTERY OUTLINED BY ASCITES

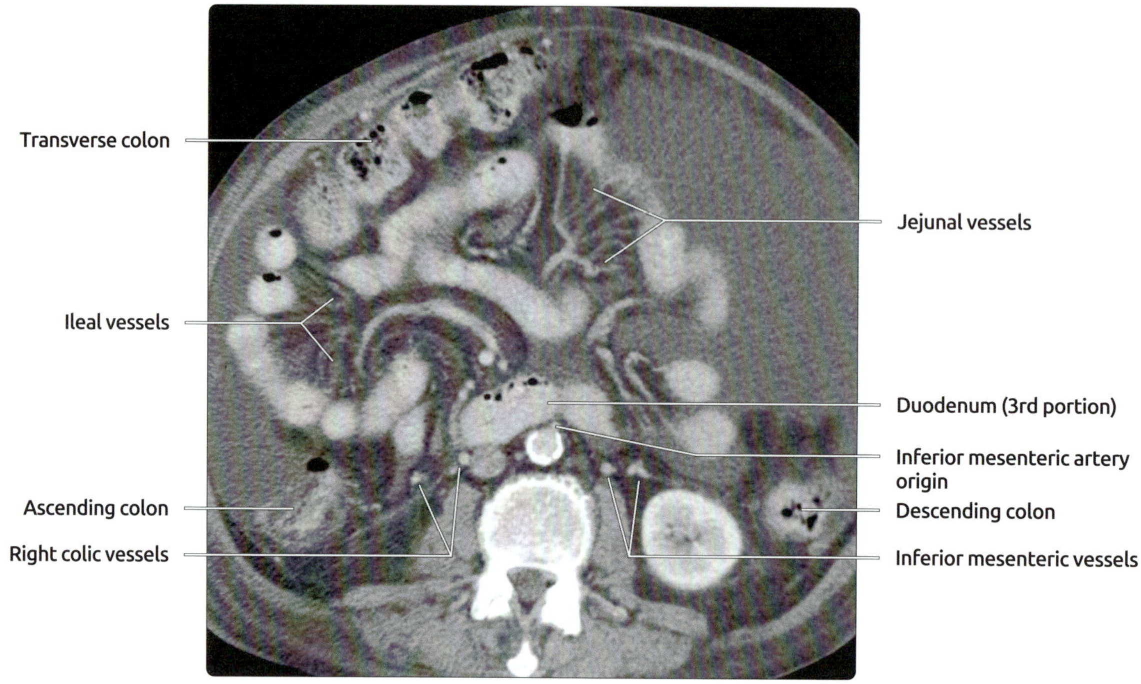

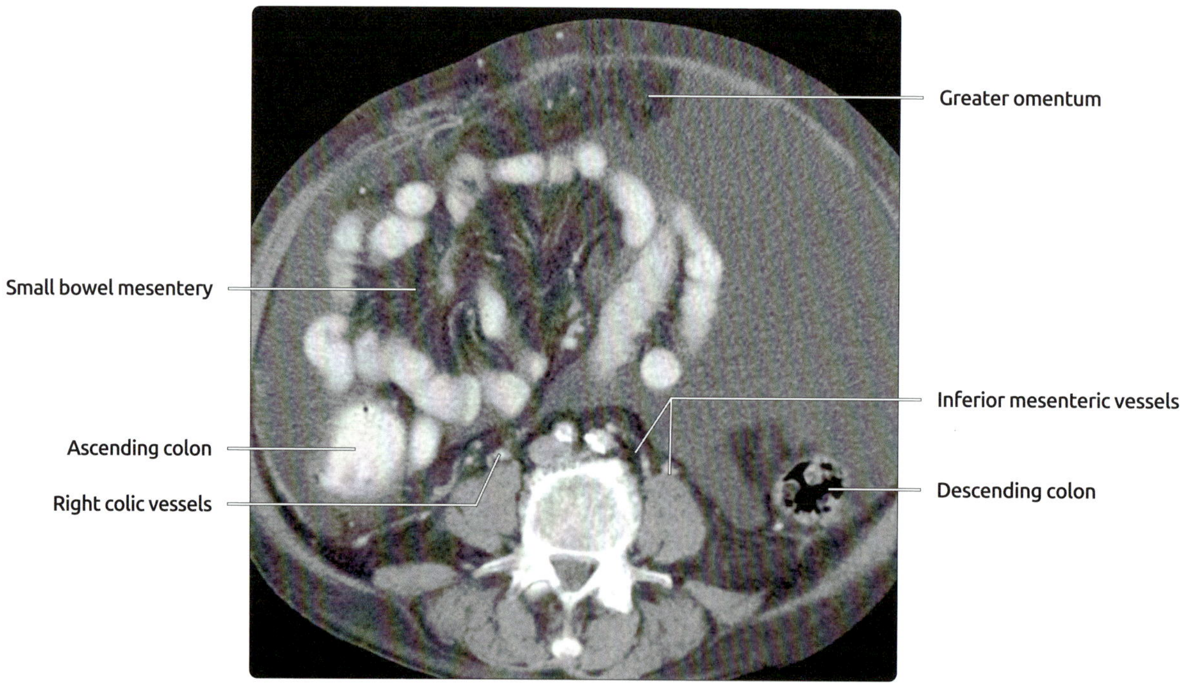

(Top) *First of 5 axial CECT sections in a patient with ascites shows the vascular supply to the bowel. The small bowel and transverse colon receive their supply through the mesentery and transverse mesocolon, respectively, and these are highlighted by mesenteric fat and separated by ascites in this subject. The ascending and descending colon are retroperitoneal, as are their vessels. The origin of the IMA is seen on this section arising from the distal aorta just caudal and dorsal to the 3rd portion of the duodenum.* **(Bottom)** *This image clearly shows the distinction between the intra- and retroperitoneal bowel segments and their blood supplies.*

MESENTERY OUTLINED BY ASCITES

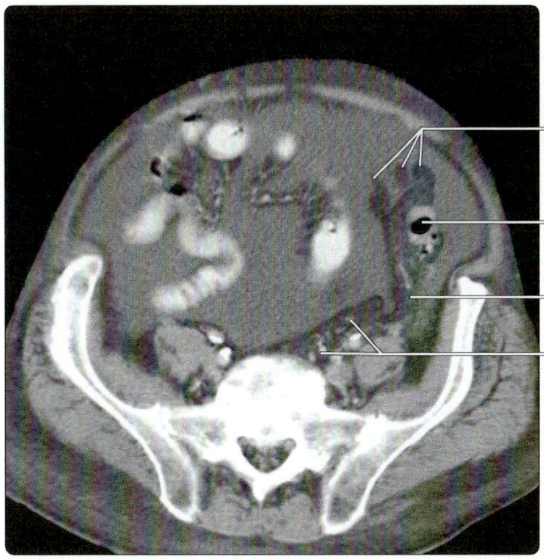

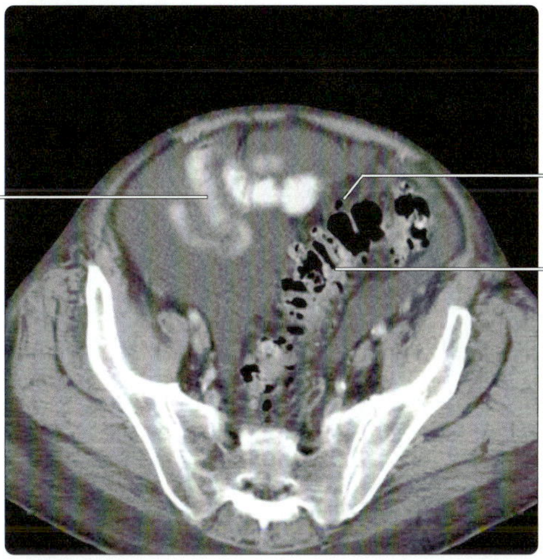

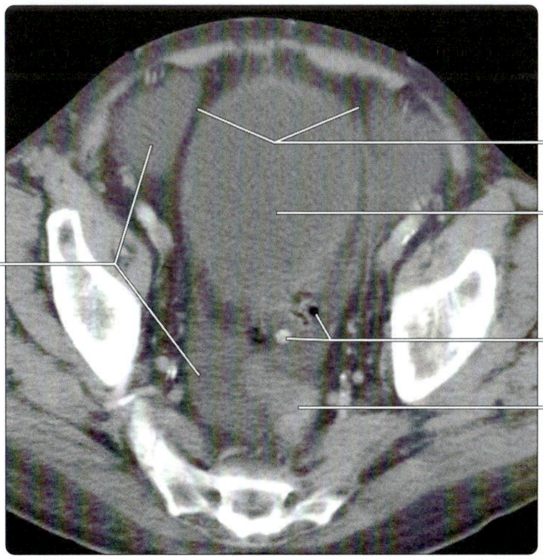

(Top) *More caudal section shows the sigmoid colon on its mesocolon, which carries the sigmoid branches of the IMVs. Note the normal epiploic appendages, vestigial fatty tags on the antimesenteric surface of the colon.* **(Middle)** *The sigmoid colon is mobile on its mesocolon and is surrounded by ascites in this subject. One of many diverticula is identified.* **(Bottom)** *The rectosigmoid junction marks the transition from the intraperitoneal sigmoid colon to the extraperitoneal rectum. Diverticula potentially affect all portions of the colon, especially the sigmoid, but spare the rectum.*

INFERIOR MESENTERIC VESSELS, ASCITES

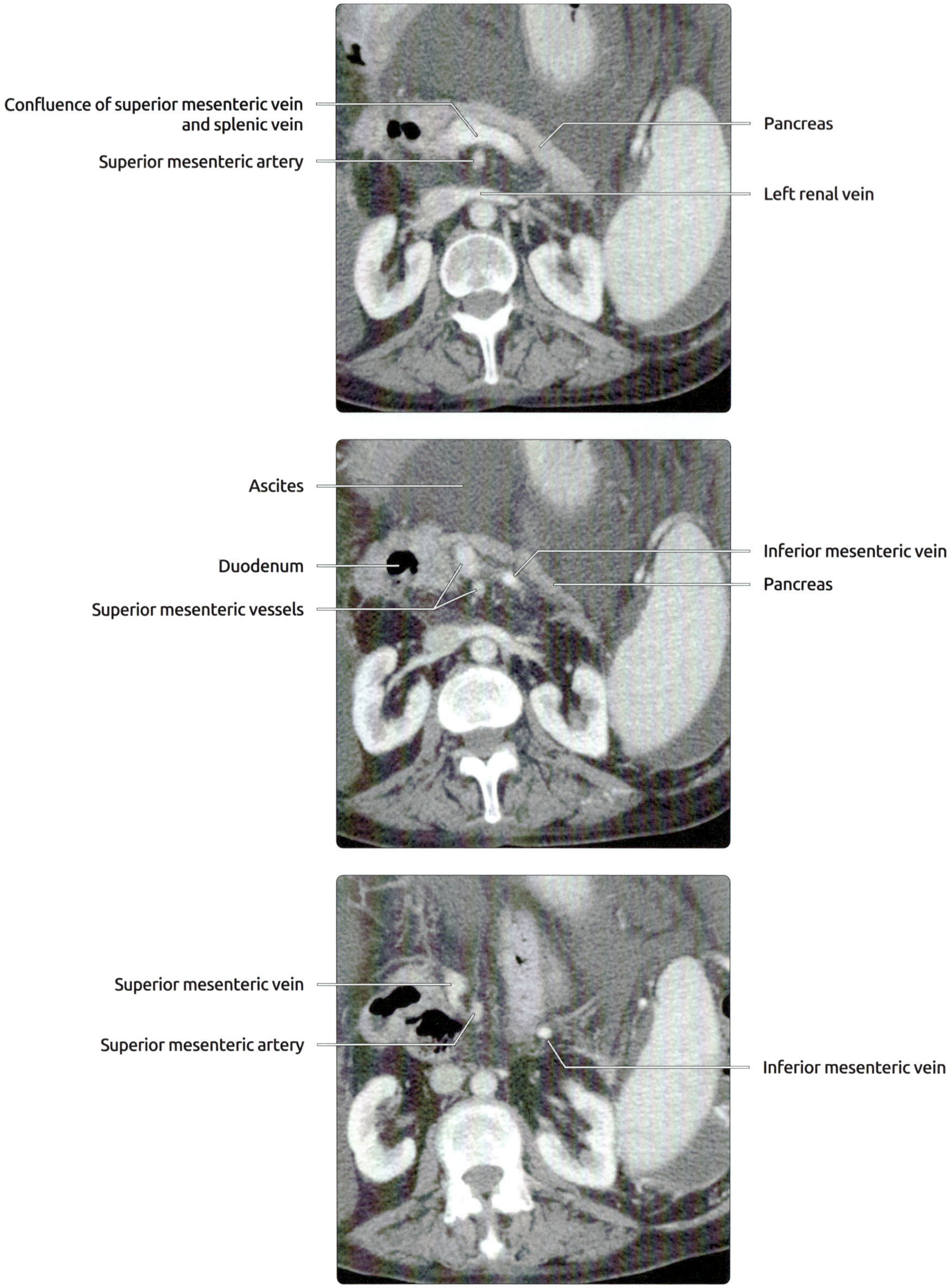

(Top) *First of 6 axial CECT sections in a patient with ascites shows the confluence of the splenic veins and SMV behind the neck of the pancreas, forming the portal vein.* **(Middle)** *A more caudal section shows the confluence of the inferior mesenteric and splenic veins behind the body of the pancreas.* **(Bottom)** *The IMV has a cephalocaudal course, and its small size can make it difficult to identify on many axial CT studies.*

INFERIOR MESENTERIC VESSELS, ASCITES

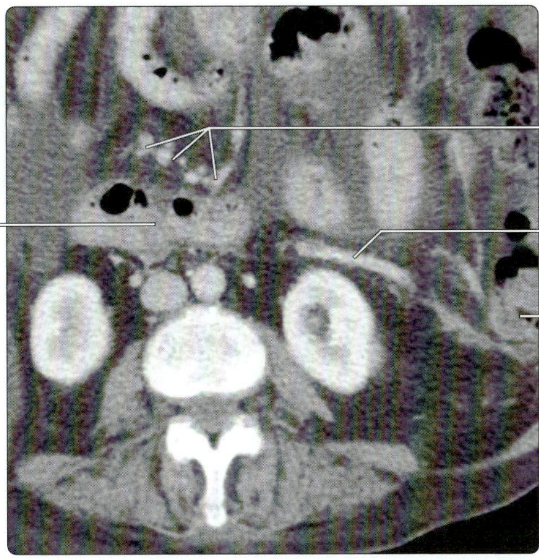

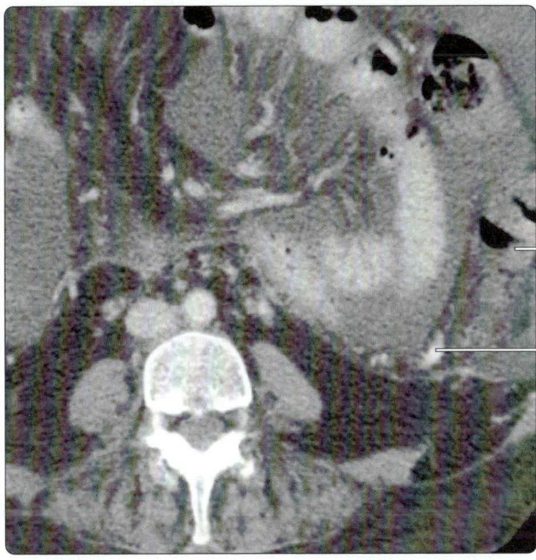

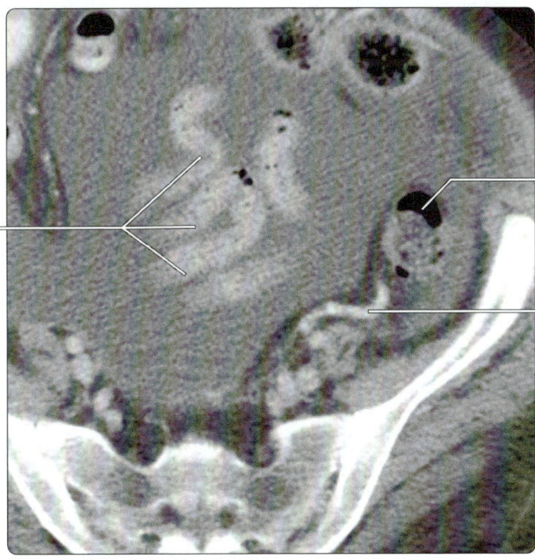

(Top) *A more caudal section shows the left colic vein draining the descending colon.* **(Middle)** *The left colic vein joins with tributaries from the sigmoid colon.* **(Bottom)** *The sigmoid colon and its vessels are seen within the sigmoid mesocolon. Veins are generally larger in caliber and less tortuous than arteries.*

INFERIOR MESENTERIC VESSELS, ASCITES

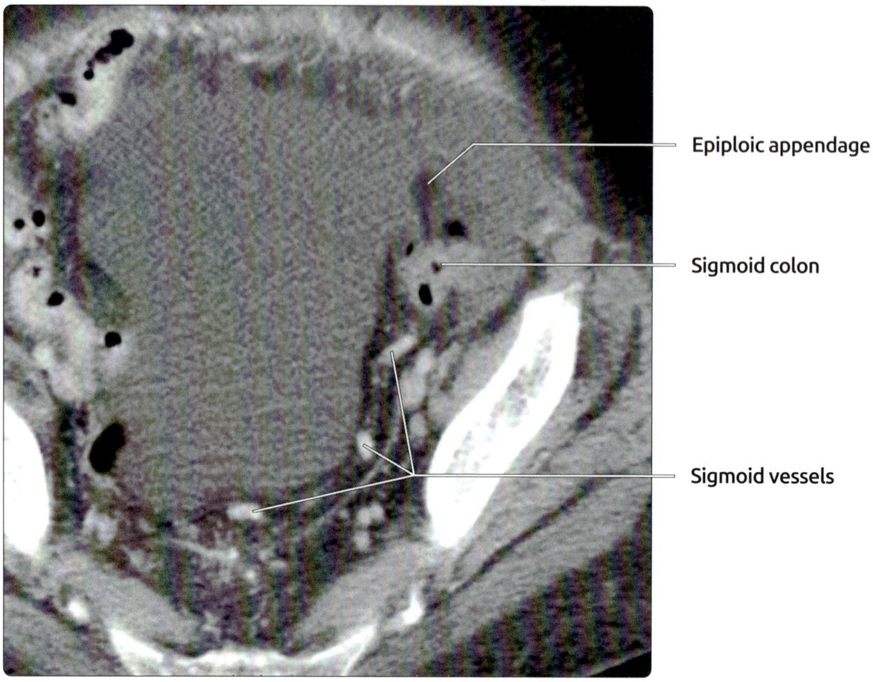

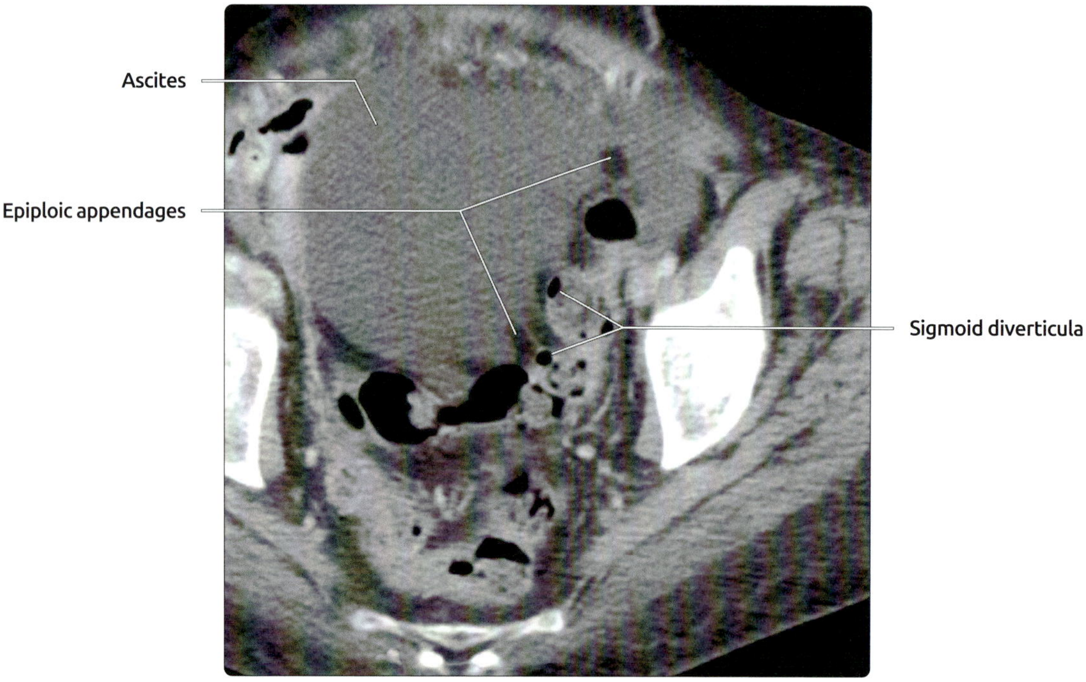

(Top) *First of 2 axial CT sections in a patient with ascites shows the sigmoid colon and its mesocolon, which carries the vessels to and from the colon. Note the epiploic appendages, fatty tags on the antimesenteric border that are especially prominent in the sigmoid colon. These are normal anatomic features that are made more evident by the presence of ascites.* **(Bottom)** *Another common, though abnormal, feature of the sigmoid colon is the presence of diverticula, which are outpouchings of the mucosa and submucosa through the muscle layers of the colonic wall.*

MR, RECTUM

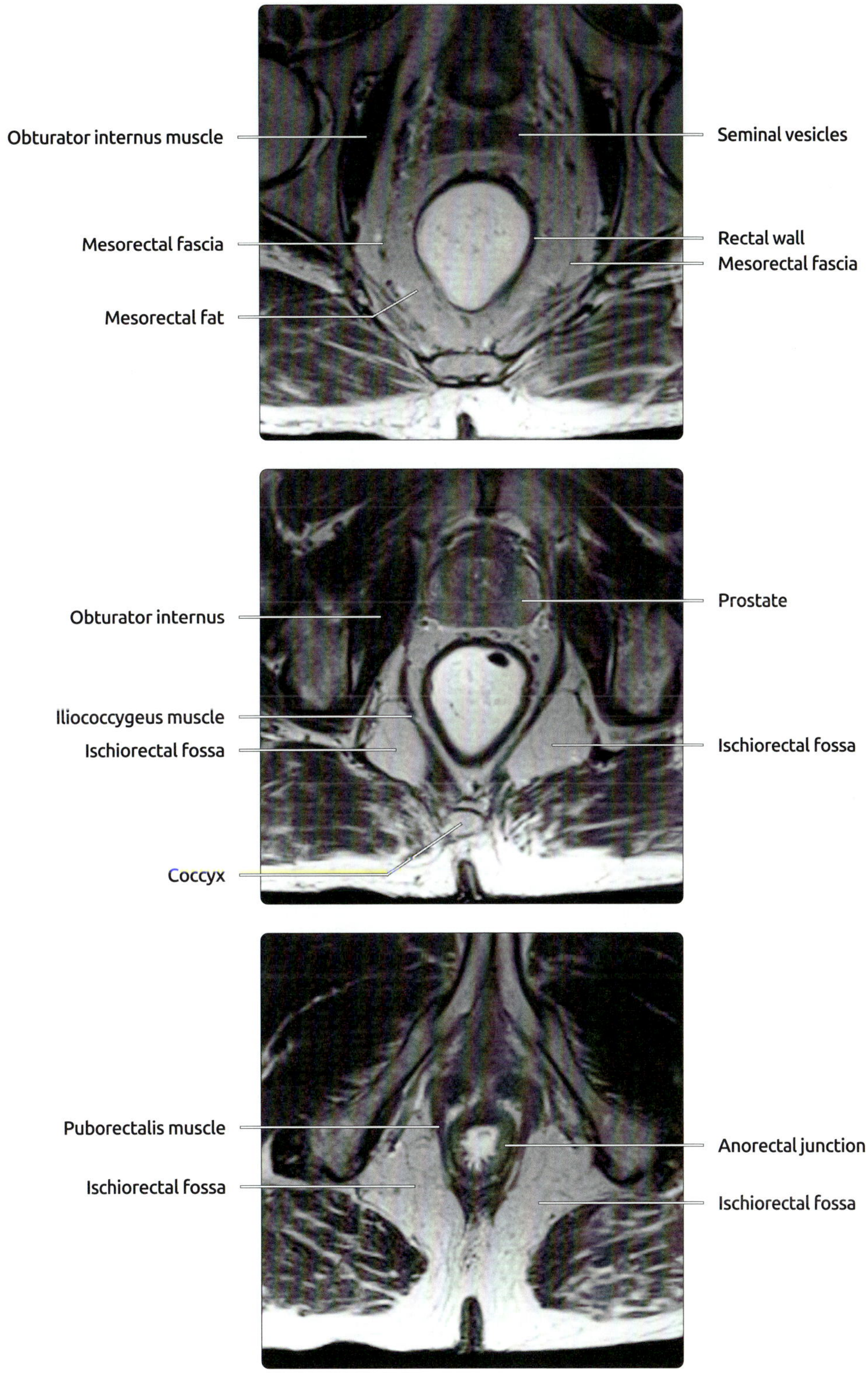

(Top) *Axial T2 MR demonstrates a thin hypointense fascial layer surrounding the rectum containing a layer of perirectal fat. The mesorectal fascia is an important anatomic landmark when performing a total mesorectal excision for rectal cancer, and the distance between a cancer and the mesorectal fascia is an important element of staging rectal cancer using MR.* **(Middle)** *Axial T2 MR at a more caudal level demonstrates the levator ani muscle complex defining the adjacent ischiorectal fossa. The iliococcygeus muscle (1 of the 3 components of the levator ani muscle complex, along with the puborectalis and pubococcygeus muscles) inserts onto the coccyx posteriorly.* **(Bottom)** *Most caudal T2 MR demonstrates the anorectal junction defined by the puborectalis "sling" wrapping around the rectum.*

ANAL SPHINCTERS

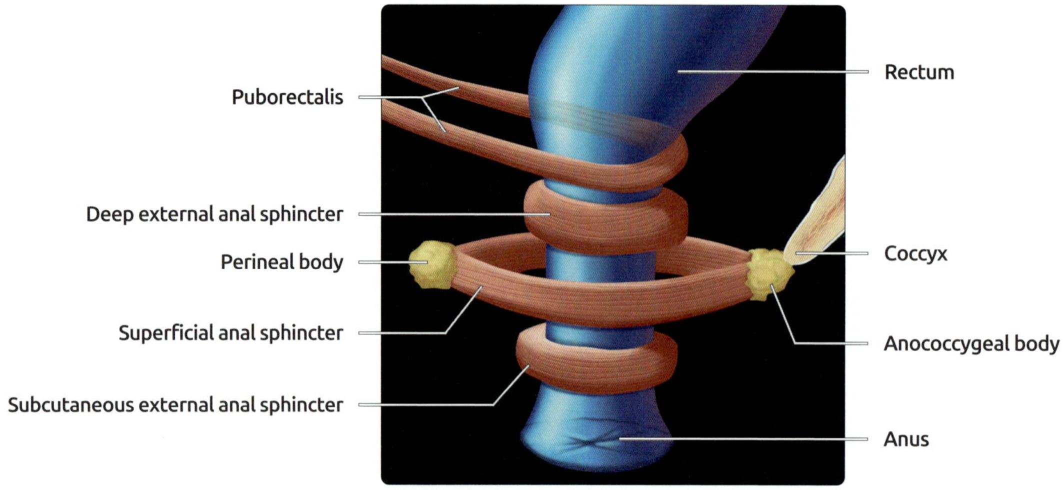

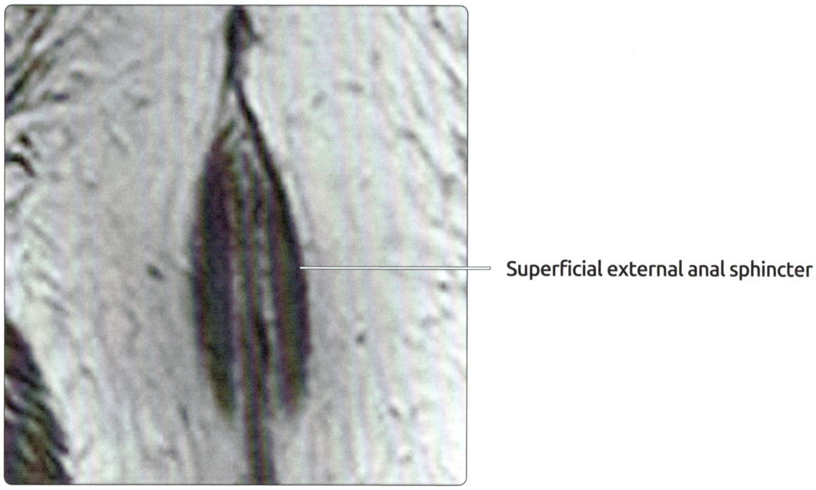

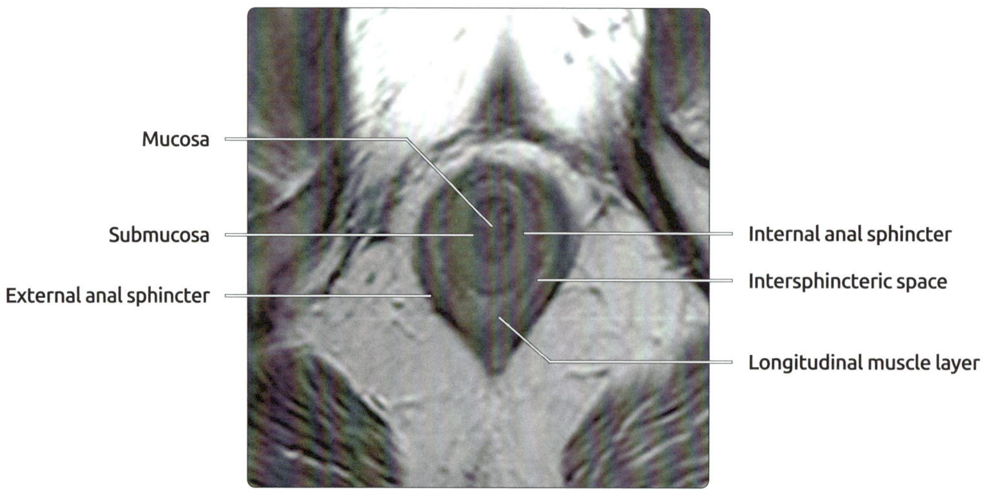

(Top) *The external anal sphincter (EAS) consists of subcutaneous, superficial, and deep parts. For simplicity on MR, the subcutaneous and superficial parts are considered 1 muscle, the superficial EAS (SEAS). The puborectalis muscle is responsible for the creation of the anorectal angulation. (From DI: GYN.)* **(Middle)** *Axial oblique MR shows the SEAS. It is the lowest part of the external sphincter and appears as 2 parallel muscles directed anteriorly to posteriorly. (From DI: GYN.)* **(Bottom)** *The anal sphincter complex shows 4 layers of different signal intensity. The mucosa is the thin folded inner layer of high signal intensity. The submucosa is low signal and has a folded internal contour and smooth outer contour. The internal anal sphincter (IAS) appears as a homogeneous isointense to hyperintense smooth circular band surrounding the anal canal. The longitudinal smooth muscle layer and the fatty intersphincteric space are more prominent at the distal part of the deep EAS.*

ANAL SPHINCTERS

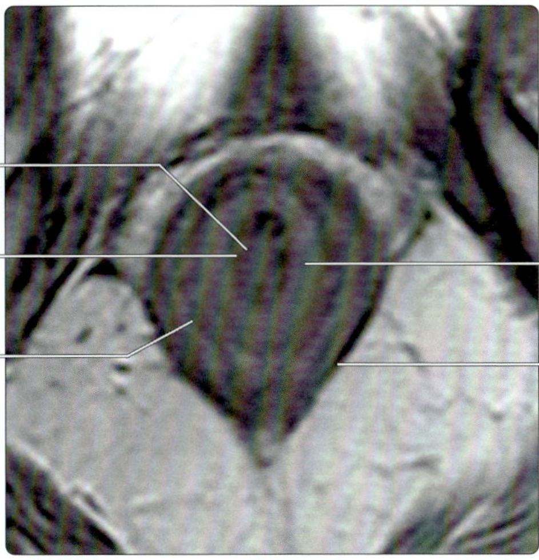

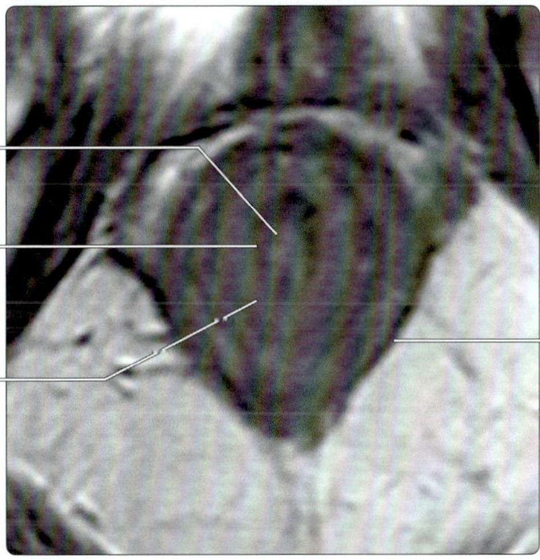

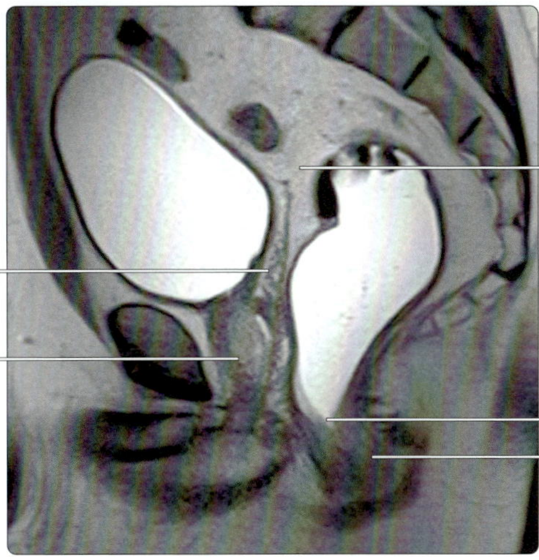

(Top) *Axial oblique BFFE MR of the anal sphincter complex shows the deep EAS (DEAS). The middle 1/3 of the DEAS forms the most prominent part of the sphincter and has a characteristic teardrop shape around the IAS. (From DI: GYN.)* **(Middle)** *Axial oblique BFFE MR shows the anal sphincter complex at the upper 1/3 of the DEAS. The innermost fibers of the EAS are completely circular. The deep part of the external sphincter is thick and encircles the anal canal.* **(Bottom)** *In men, the apex of the seminal vesicle serves as a landmark to identify the peritoneal reflection. In women, the equivalent to the seminal vesicles is the cervix. Rectal cancer is staged as T4a when the tumor involves the peritoneal reflection.*

ANAL SPHINCTERS

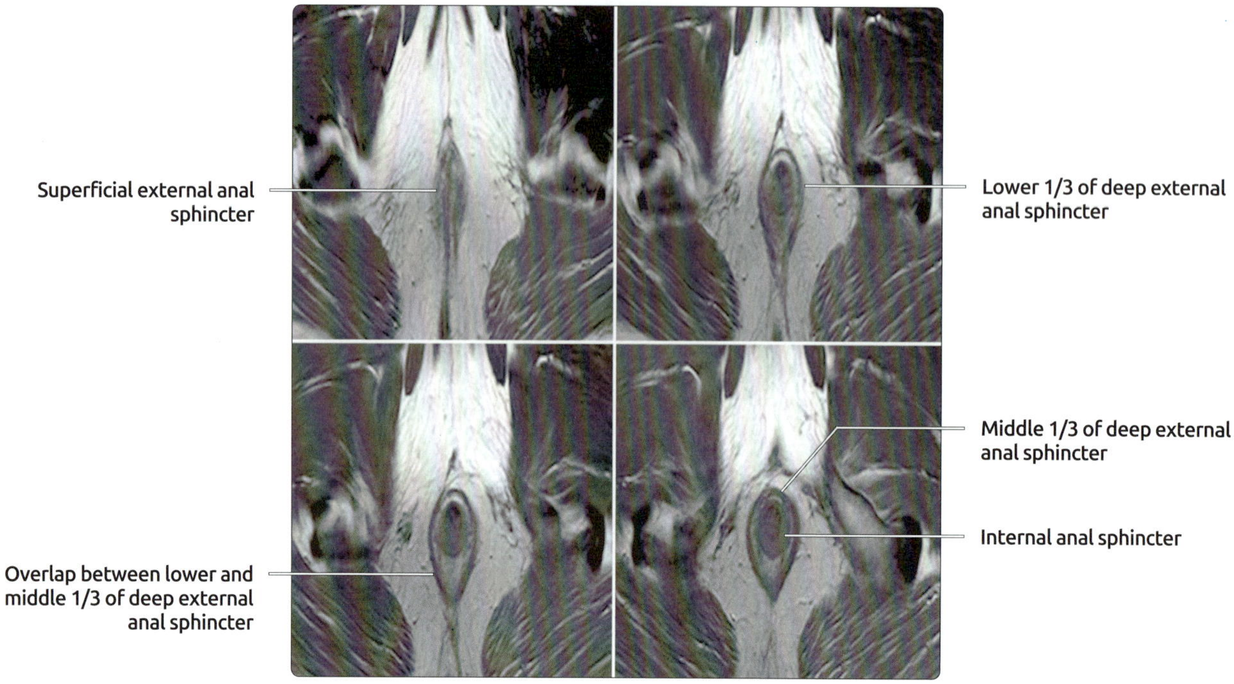

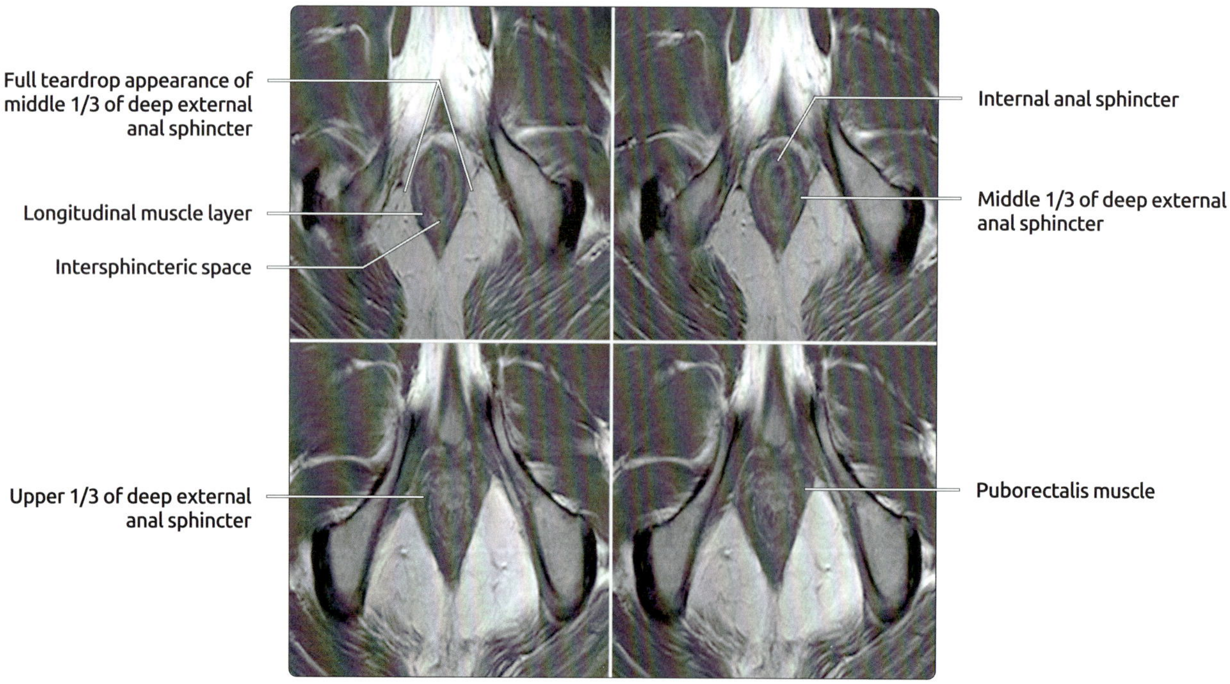

(Top) *Cranial to the level of the SEAS (which can be distinguished by its unique MR configuration) is the DEAS. The DEAS is further subdivided into the lower 1/3, which is the first 3-4 sections below the level of the transversus perinei superficialis muscle. The superficial part of the external sphincter is present, just cranial to the SEAS, middle 1/3, and upper 1/3. (From DI: GYN.)* **(Bottom)** *The middle 1/3 of the DEAS can be identified by the teardrop appearance of the sphincter. The puborectalis muscle is seen just cephalad to the IAS and EAS and does not extend along the ventral portion of the rectal wall. The puborectalis forms a sling along the posterior anorectum. (From DI: GYN.)*

ANAL SPHINCTERS

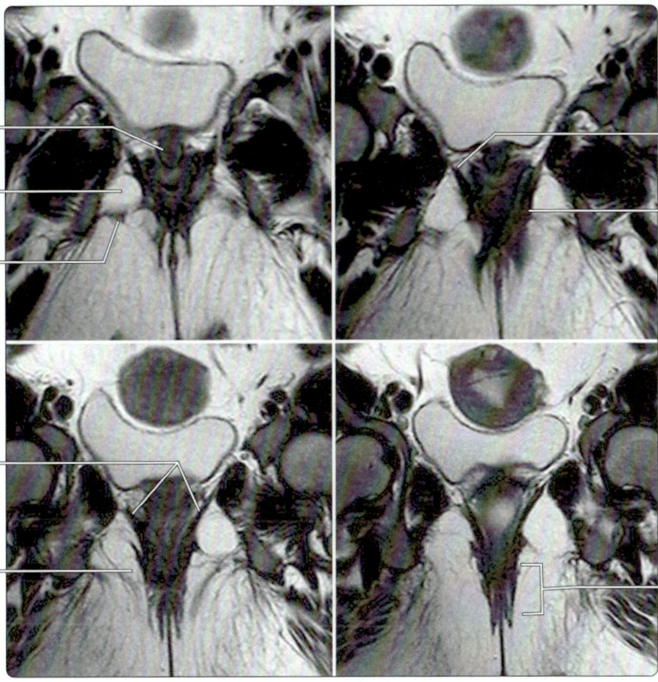

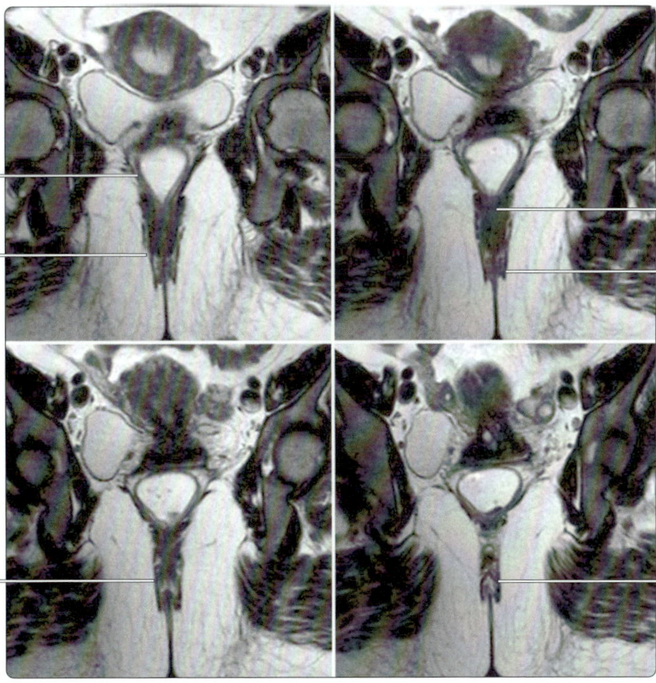

(Top) *Sequential coronal T2-weighted images of a female volunteer show the details of the sphincter complex from anterior to posterior. The upper 2 sections are anterior to the anal canal. The puborectalis muscle is inferolateral to the iliococcygeus muscle. Both muscles run perpendicular to the urogenital diaphragm. Note the anterior extension of the ischioanal space. More posteriorly, the DEAS is visible as a separate muscle. Note the funnel-shaped iliococcygeus muscle and the ischioanal space. (From DI: GYN.)* **(Bottom)** *Sequential coronal T2-weighted images of a female volunteer show the anal sphincter complex. The upper 2 images are at the anal canal proper. The lower part of the anal canal is surrounded by the IAS and the EAS, while its upper part is surrounded by the IAS and puborectalis muscle. The lower 2 images are more posterior and show the cleft between the EAS and the puborectalis muscle. The direction of the fibers differ between the EAS and puborectalis muscle. The iliococcygeus muscle has a more horizontal orientation. (From DI: GYN.)*

RECTUM AND ANAL CANAL

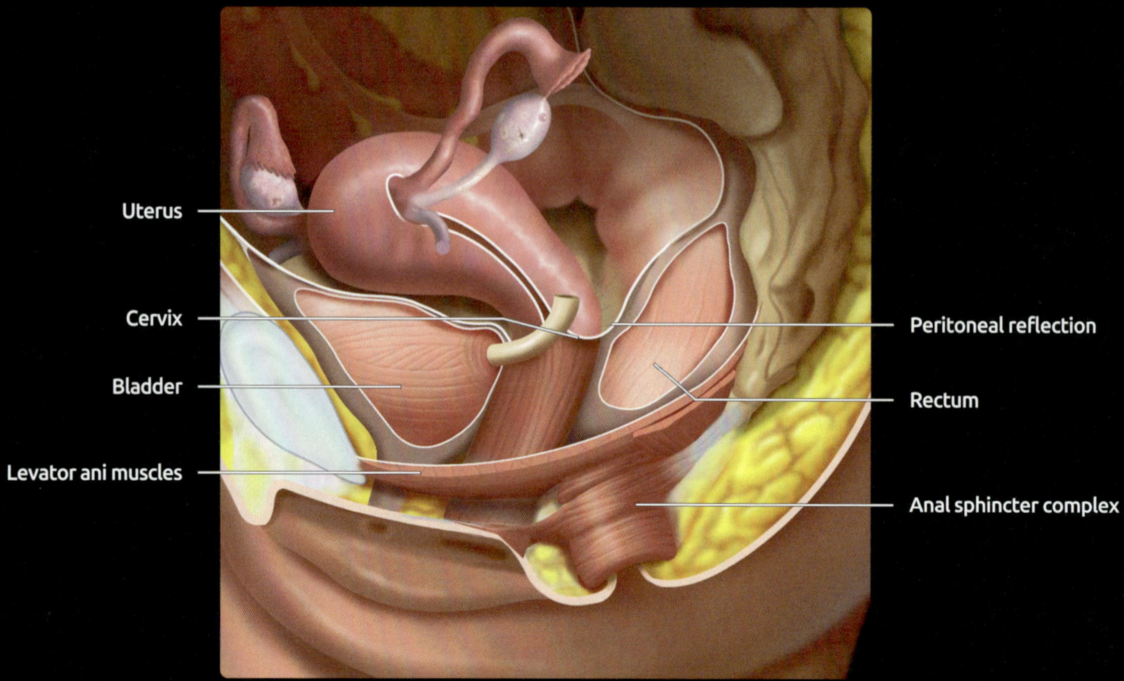

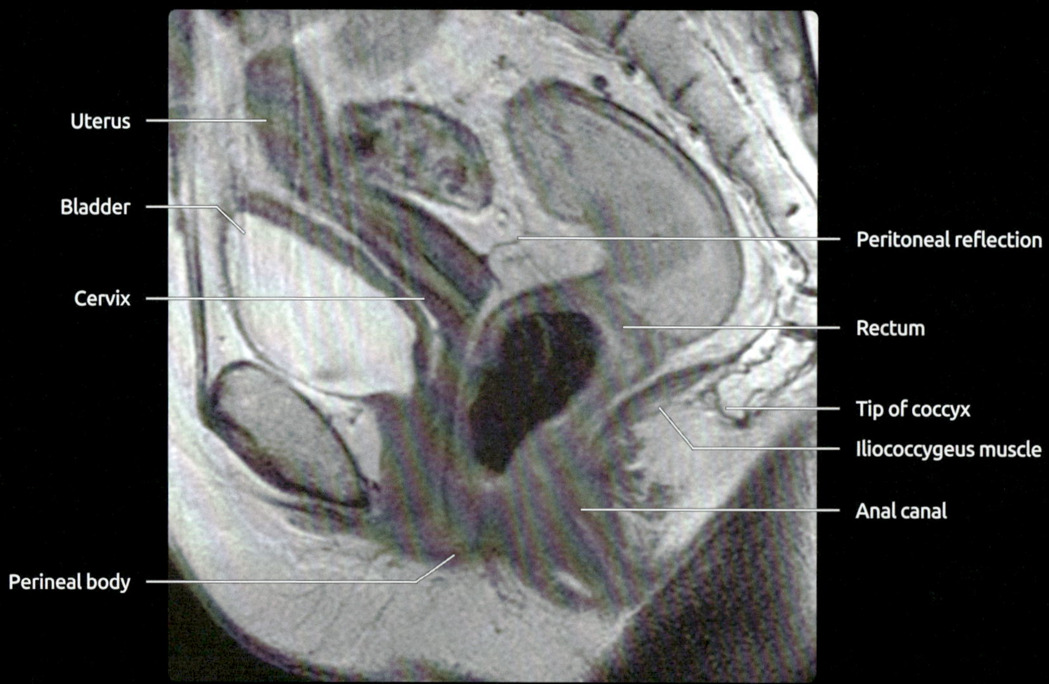

(Top) *Graphic shows the anatomy and subdivisions of the rectum. The intraperitoneal rectum is related anteriorly to the upper vagina and uterus. The extraperitoneal rectum is related anteriorly to the posterior vaginal wall and rectovaginal septum. The inferior rectum has no mesentery but is enveloped in fat and is bordered by the mesorectal fascia (mesorectum). The anal sphincter envelops the anal canal and is composed of several cylindrical layers.* **(Bottom)** *Sagittal T2 MR shows the normal relationships of the anal canal. Anteriorly, the anal canal is related to the perineal body, the lower part of the vagina, and the anovaginal septum. The attachment to the perineal body is important for anal canal support.*

ANAL SPHINCTER COMPLEX

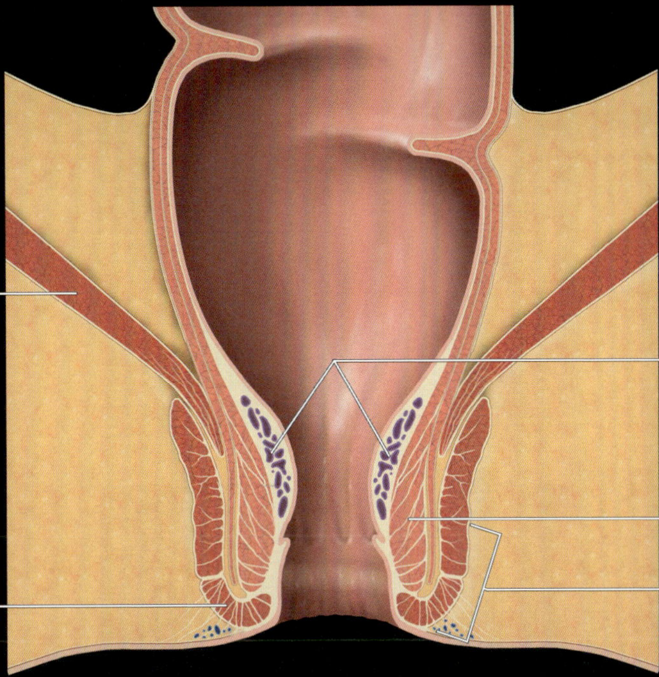

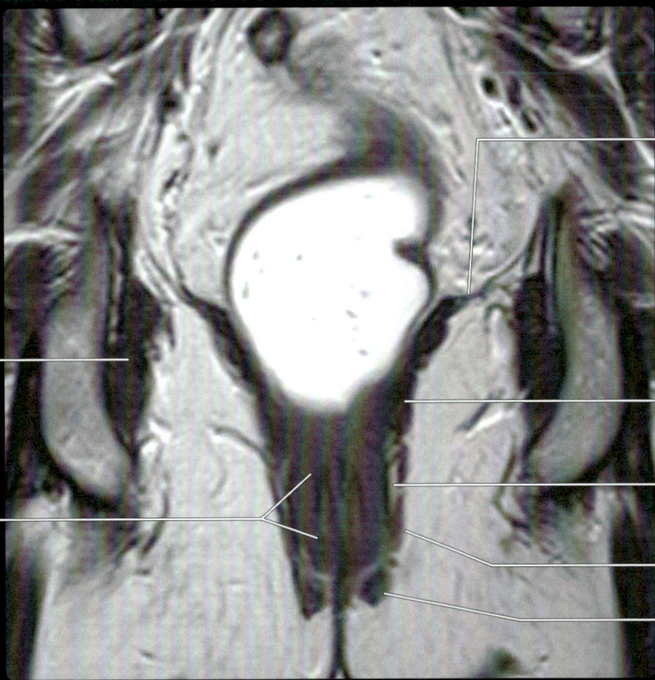

(Top) *The IAS is a continuation of the circular muscle layer of the muscularis propria of the rectum. The longitudinal muscle layer is located within the intersphincteric fat and is a continuation of the longitudinal muscle layer of the muscularis propria of the rectum. The EAS is composed of multiple components and constitutes the outer and inferior part of the anal sphincter complex. The lowermost part of the anal canal is surrounded by the SEAS. At a higher level, the anal canal is surrounded by the IAS, longitudinal muscle layer, and the EAS. The uppermost part is surrounded by the IAS, longitudinal muscle layer, and puborectalis muscle. (From DI: GYN.)* **(Bottom)** *Coronal T2 MR shows the sphincter complex. The EAS envelops the intersphincteric space. The IAS extends from the anorectal junction to ~ 1 cm below the dentate line. The lower muscular part of the anal sphincter is made only of the external sphincter, levator ani, and the transverse perineal muscles. (From DI: GYN.)*

ACUTE APPENDICITIS

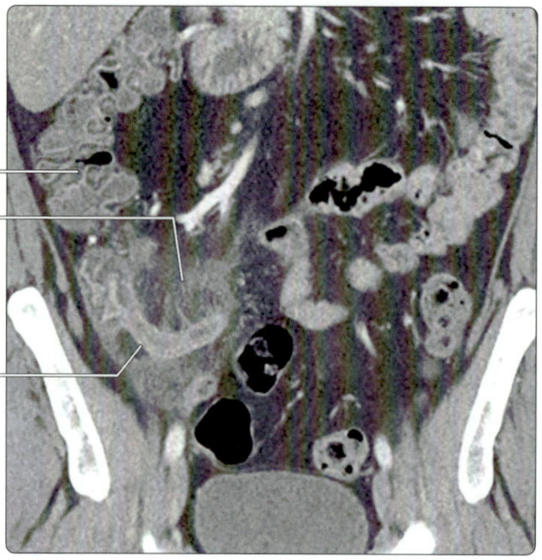

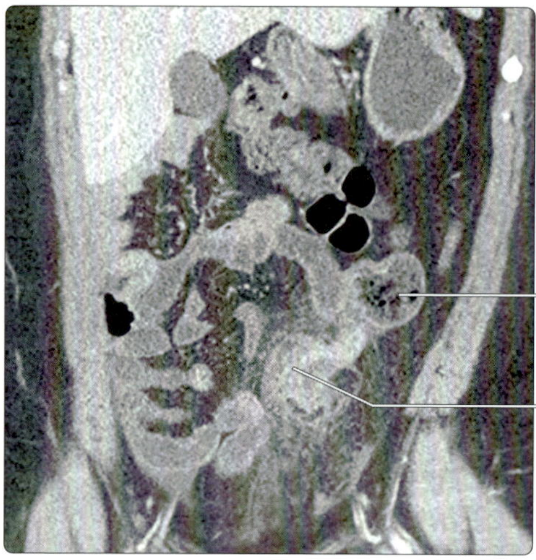

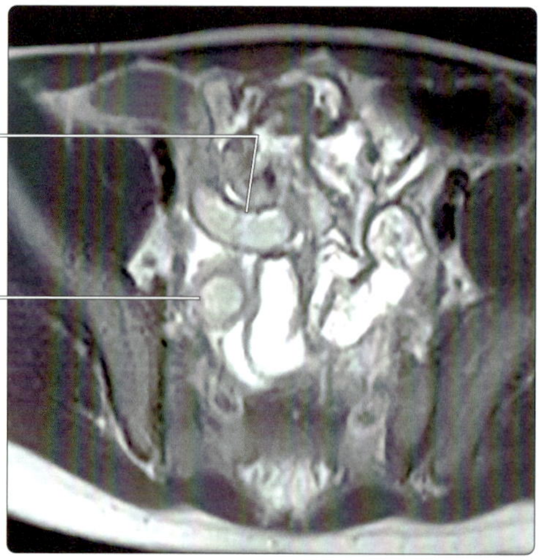

(Top) *Coronal CECT demonstrates a dilated appendix in the right lower quadrant with prominent wall thickening and mucosal hyperemia, as well as extensive periappendiceal fat stranding and inflammation, compatible with acute appendicitis.* **(Middle)** *Coronal CECT demonstrates the appendix extending from the colon in the left lower quadrant, which appears dilated with extensive surrounding fat stranding or inflammation. Notice that the small bowel is entirely located in the right abdomen, while the colon is in the left abdomen. This represents an unusual instance of acute appendicitis in a patient with malrotation.* **(Bottom)** *Axial T2 HASTE MR demonstrates a dilated, fluid-filled appendix in a young child with right lower quadrant pain. Notice that the appendiceal tip is also dilated but demonstrates a thick wall with subtle adjacent fat stranding, findings compatible with acute appendicitis.*

ACUTE APPENDICITIS

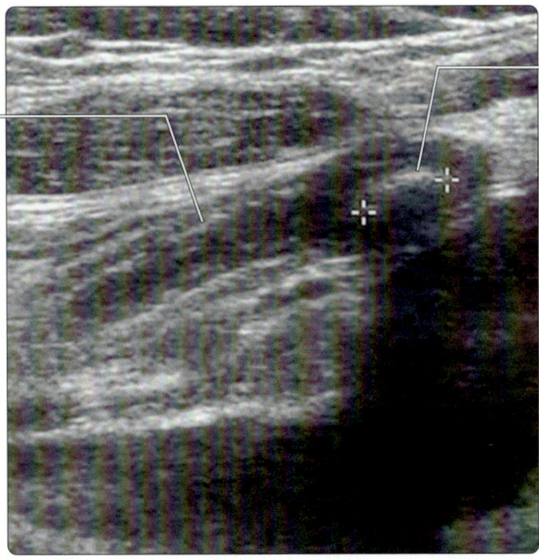

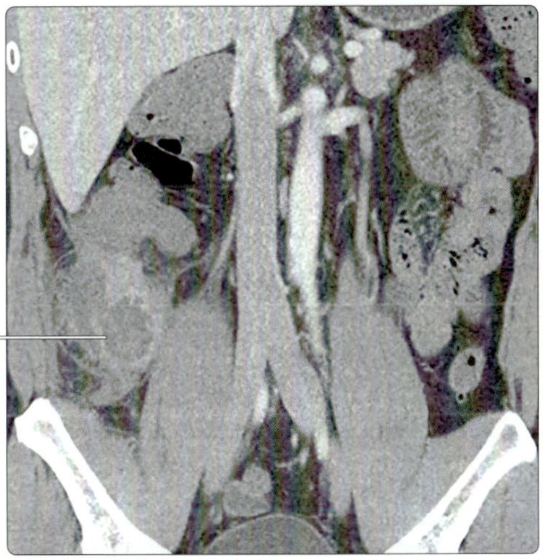

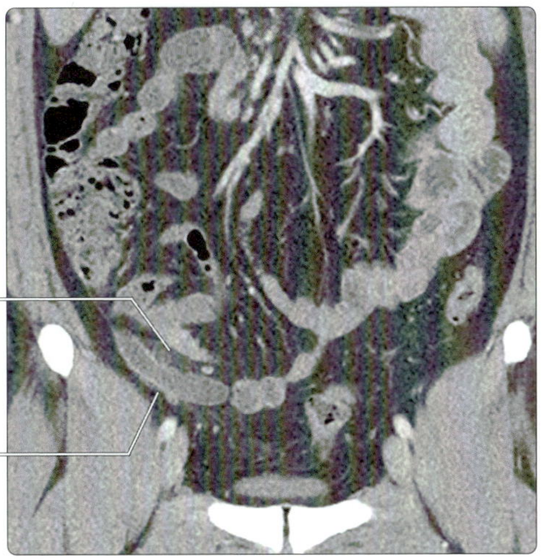

(Top) *Ultrasound demonstrates a dilated appendix with thickening of the submucosal echogenic line as well as a large calcified appendicolith at the level of the appendiceal tip, findings compatible with acute appendicitis in this patient with right lower quadrant pain.* **(Middle)** *Coronal CECT demonstrates a large loculated fluid collection in the right lower quadrant with adjacent fat stranding and edema. The appendix is not visualized in the midst of this extensive inflammation. This constellation of findings is compatible with acute perforated appendicitis. Perforated appendicitis should always be considered when confronted with an abscess in the right lower quadrant.* **(Bottom)** *Coronal CECT demonstrates a dilated appendix with adjacent periappendiceal fat stranding. Notice that the wall of the appendix demonstrates subtle hyperenhancement. This constellation of findings is compatible with acute uncomplicated appendicitis.*

DIVERTICULITIS

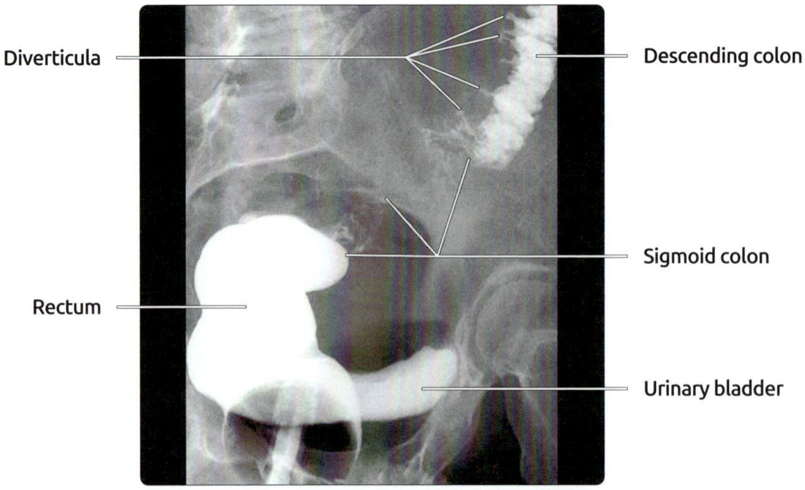

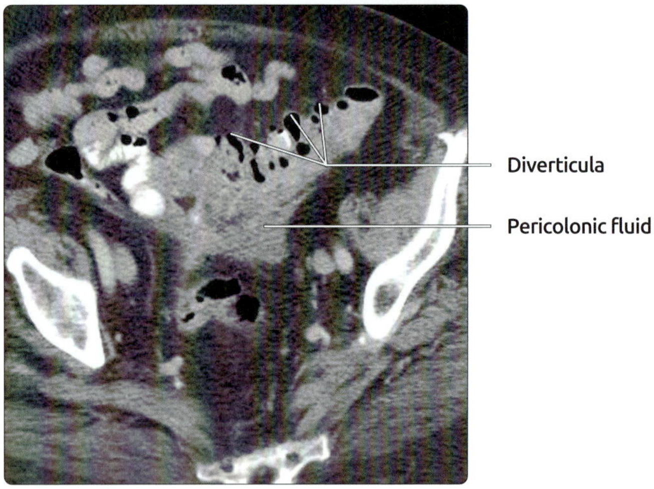

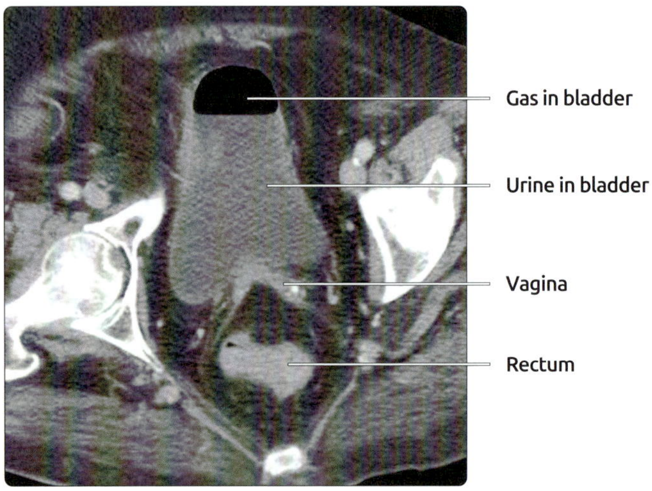

(Top) *Oblique image from a barium enema shows marked spasm and luminal narrowing of the sigmoid colon and opacification of the urinary bladder through a fistula (that is not evident on this image). Numerous diverticula are seen in the descending and sigmoid colon.* **(Middle)** *Axial CECT in the same patient demonstrates extraluminal fluid adjacent to the sigmoid colon due to perforation of the diverticula and inflammation (diverticulitis).* **(Bottom)** *Axial CECT in the same patient demonstrates a gas-fluid level in the urinary bladder indicative of a colovesical fistula.*

DIVERTICULITIS

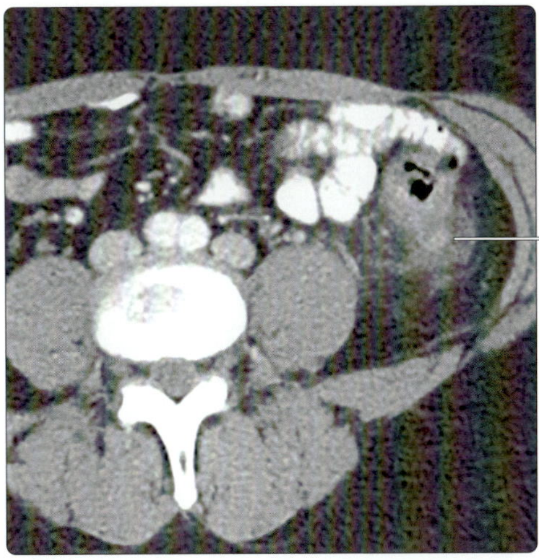

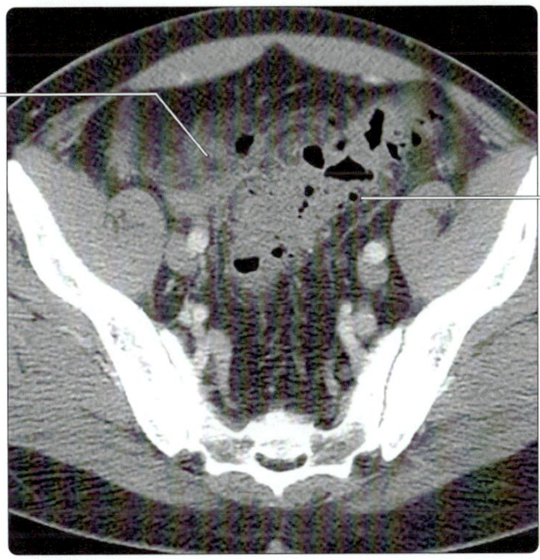

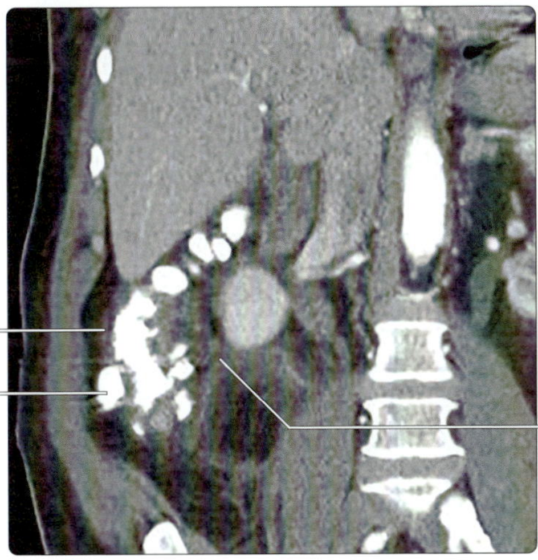

(Top) *Axial CECT demonstrates a focally inflamed diverticulum in the left colon, compatible with acute uncomplicated diverticulitis.* **(Middle)** *Axial CECT demonstrates thickening of the sigmoid colon with extensive pericolonic fat stranding. Several small diverticula are identified in this segment of colon suggesting the correct diagnosis of acute diverticulitis.* **(Bottom)** *Axial CECT demonstrates wall thickening of the cecum with adjacent inflammatory change. Several diverticula are identified in this location, compatible with acute cecal diverticulitis. This location is far less common for both diverticulosis and diverticulitis compared to the left colon.*

EPIPLOIC APPENDAGITIS

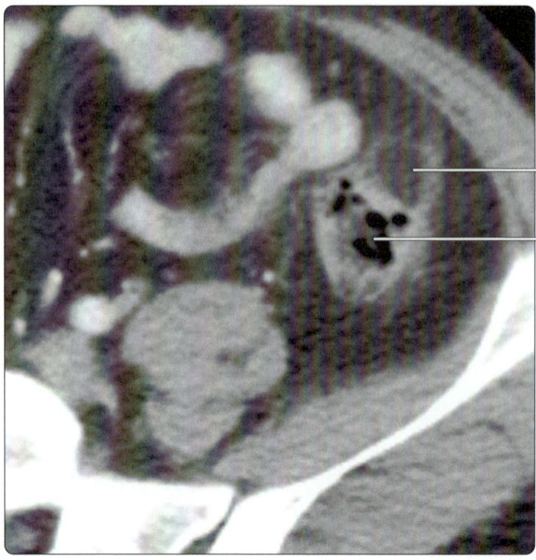

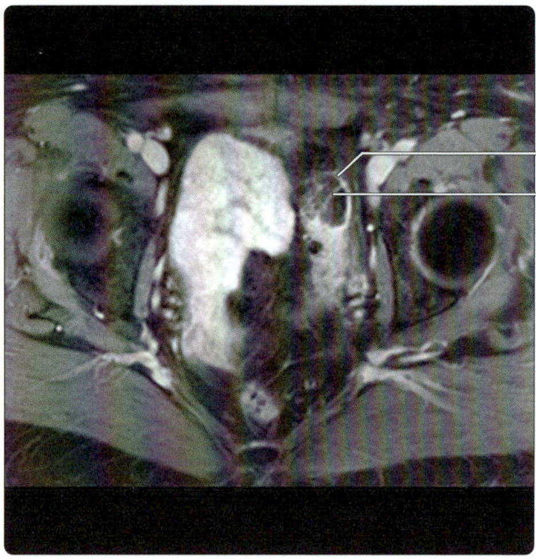

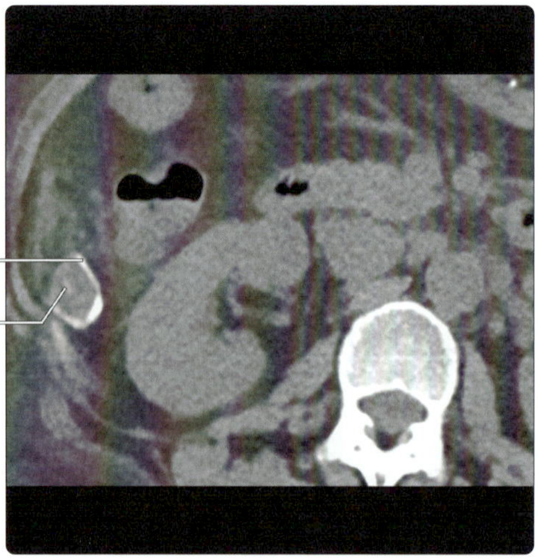

(Top) *Axial CECT in a patient with acute left lower quadrant pain shows inflammation of the fat planes anterior to the descending colon. In the middle of the inflammation is an oval fat density lesion that represents an infarcted epiploic appendage. Recall that the normal epiploic appendage is a tag of fat that extends off the antimesenteric border of the colon.* **(Middle)** *Postcontrast T1 MR in the same patient shows rim enhancement of the lesion, the CT equivalent of the hyperattenuating ring sign. No enhancement is seen in the central thrombosed vein.* **(Bottom)** *A small area of fat attenuation abutting the ascending colon with calcified peripheral rim and calcification of the central dot is consistent with an old infarcted appendage, as seen here. This is a benign entity that does not require any active intervention.*

COLON POLYPS

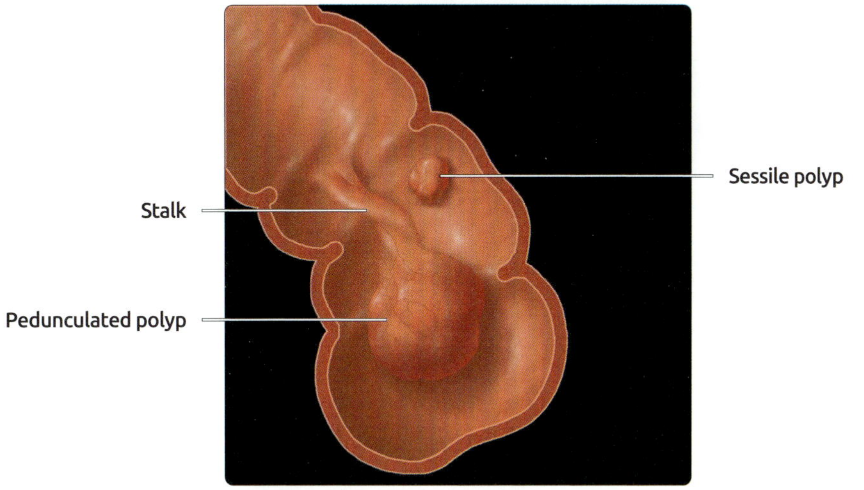

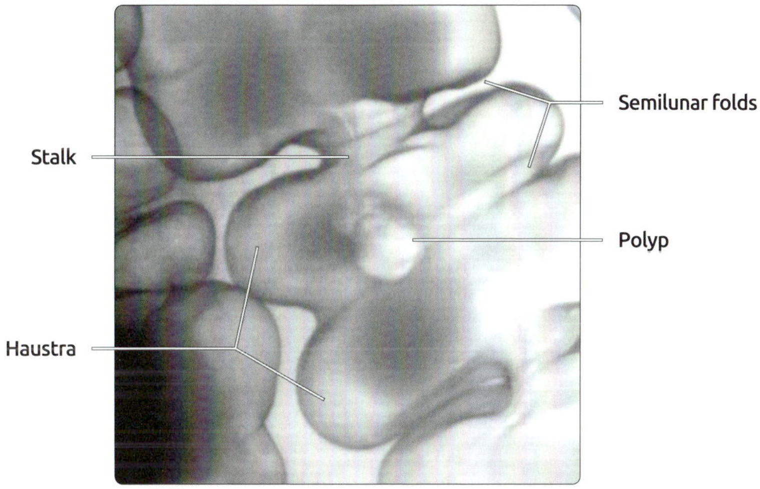

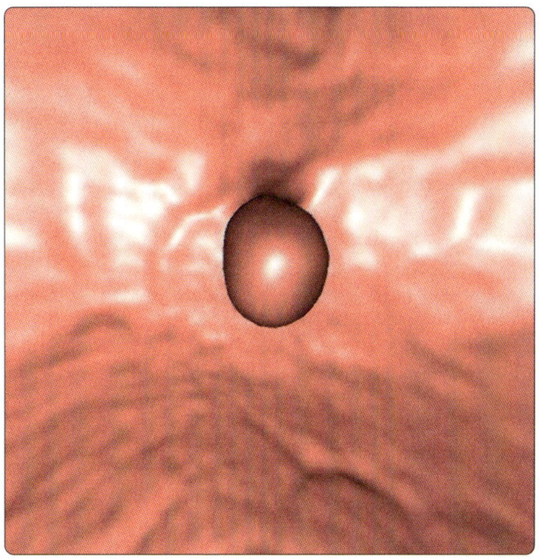

(Top) *Graphic shows 2 common shapes of colonic polyps: A pedunculated polyp on an elongated stalk and a sessile, or relatively flat, polyp. In adults, most polyps are adenomatous (gland forming) and are considered precursors to colon cancer. For this reason, surveillance techniques, such as colonoscopy or CT colonography, are recommended to detect polyps so that they can be removed before undergoing malignant degeneration.* **(Middle)** *"Spot film" from an air-contrast barium enema reveals a spherical pedunculated polyp on a long thin stalk. A polyp of this size (~ 1 cm) is usually benign but should be resected, preferably via colonoscopy.* **(Bottom)** *Volume-rendered CT from a CT colonography study shows a polyp projecting off the rectal wall. CT colonography has largely supplanted barium enema in the screening of patients for colon cancer.*

SIGMOID VOLVULUS

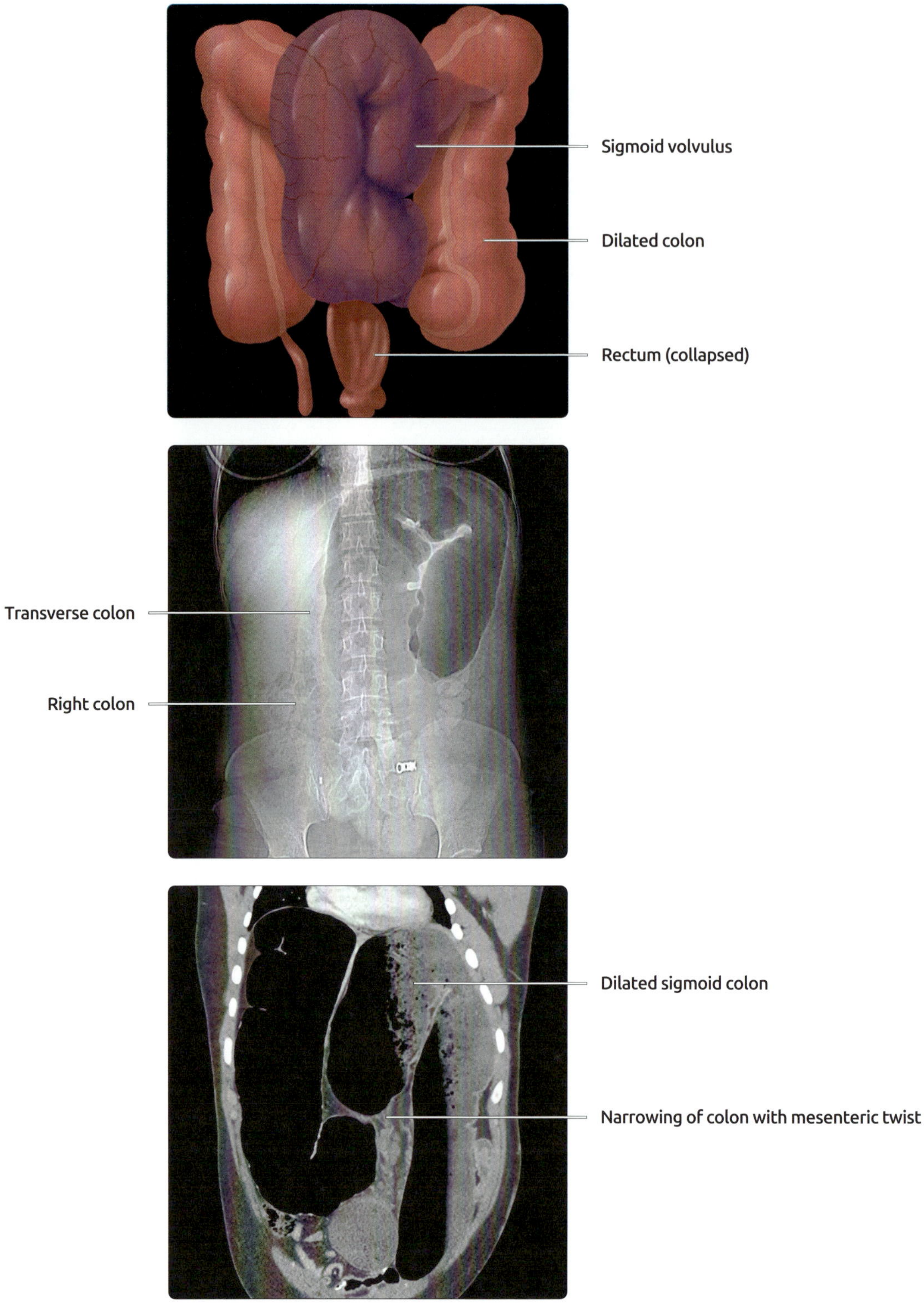

(Top) *Graphic shows a sigmoid volvulus in which the sigmoid colon is twisted around the base of its mesocolon. The sigmoid lumen and blood vessels are occluded, which may lead to ischemia and perforation.* **(Middle)** *Frontal radiograph shows massive distention and elongation of the sigmoid colon, which has a shape likened to a football or coffee bean with the "seam" representing the apposed walls of the sigmoid. The remainder of the colon is collapsed or normal in caliber.* **(Bottom)** *Coronal CECT in the same patient demonstrates the massively dilated sigmoid colon with abrupt narrowing at a site of mesenteric "twisting," compatible with a sigmoid volvulus.*

CECAL VOLVULUS

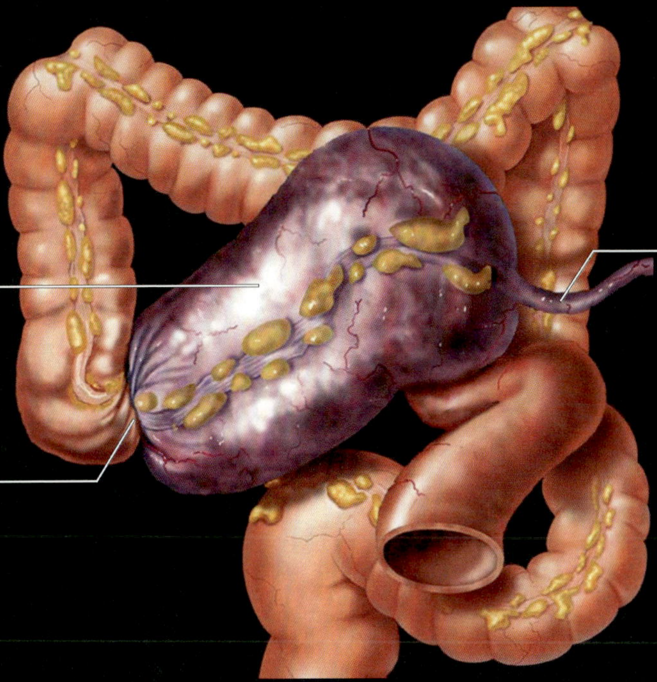

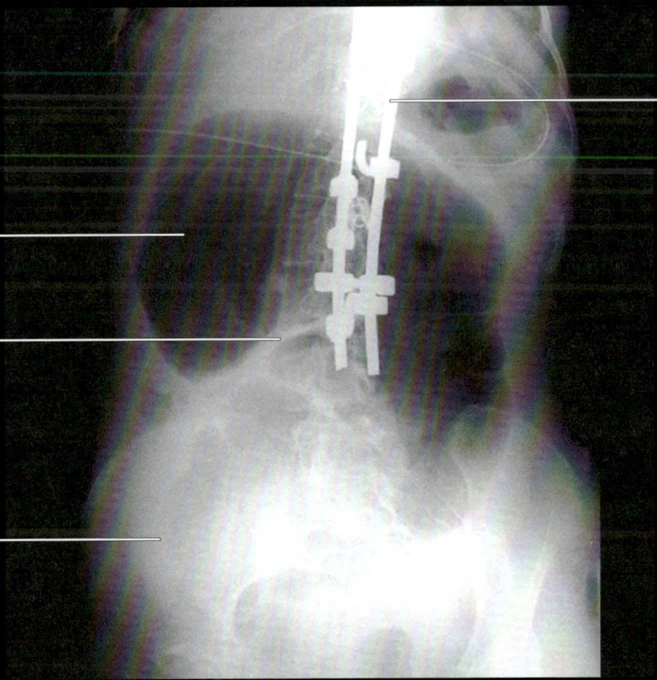

(Top) *Graphic shows a typical cecal volvulus. The ascending colon is usually fixed in a retroperitoneal location. Patients who have an elongated cecum and ascending colon that is on a mesentery are prone to twisting and obstruction of these segments resulting in obstruction of the bowel lumen and blood vessels of the twisted colonic segment. Note that the cecum is dilated and displaced toward the left upper quadrant. The discoloration represents ischemic injury to the bowel.* **(Bottom)** *First of 3 frontal abdominal radiographs of a patient who had a recent spinal fusion shows massive distention of the right side of the colon (technically, the cecum and a portion of the ascending colon, which are inverted and obstructed).*

CECAL VOLVULUS

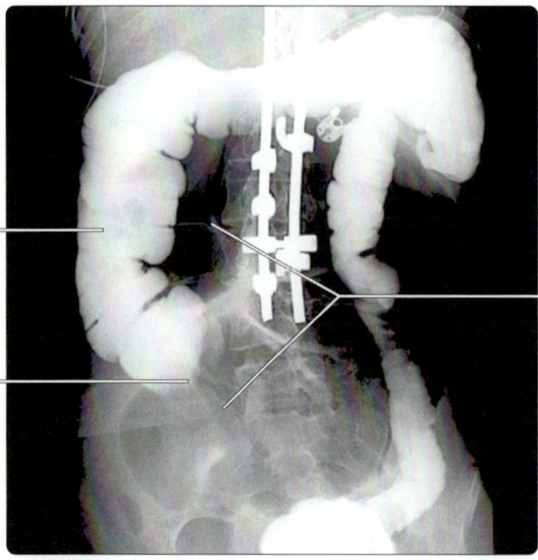

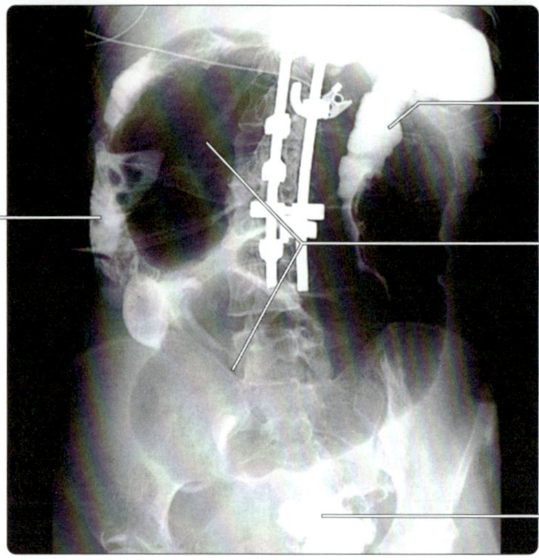

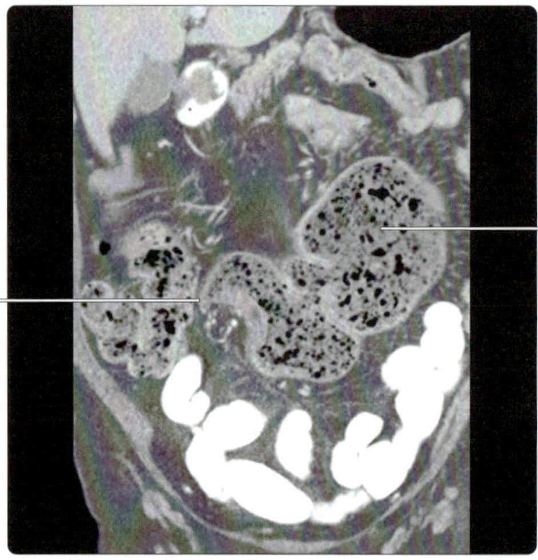

(Top) *Image from a barium enema in the same patient shows retrograde opacification of the normal colon with abrupt obstruction in the ascending colon.* **(Middle)** *"Postevacuation" image in the same patient shows a collapsed normal colon and massive distention of the cecum and the twisted portion of the ascending colon diagnostic of cecal volvulus.* **(Bottom)** *Coronal CT in a patient with abdominal pain due to cecal volvulus shows dilated stool-filled cecum shifted to the midline. There is a beak-like tapering at the level of the volvulus. The distal colon is decompressed.*

CECAL VOLVULUS

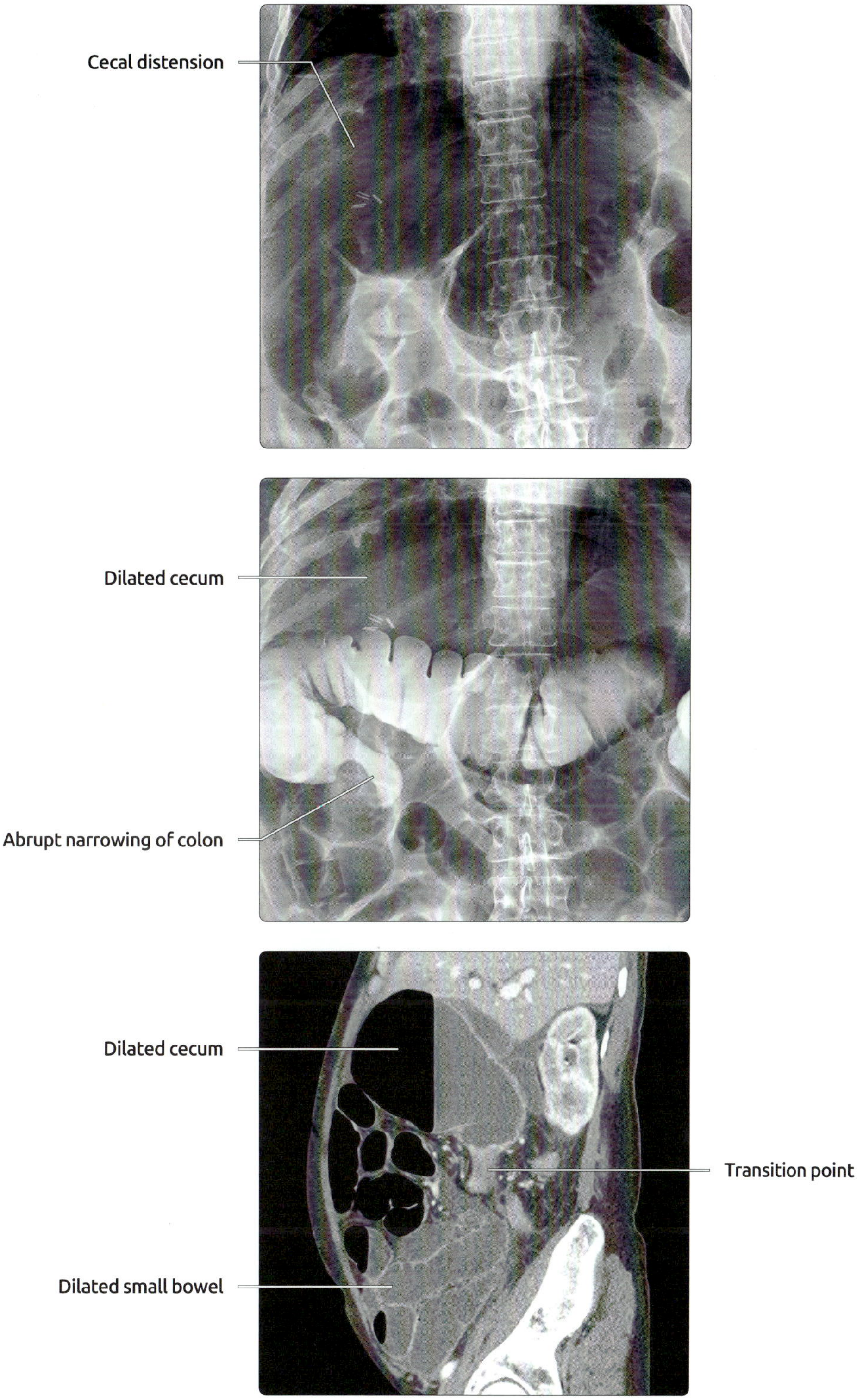

(Top) *Frontal radiograph demonstrates a markedly dilated cecum in a patient with acute abdominal pain.* **(Middle)** *Frontal image from a barium enema in the same patient demonstrates abrupt narrowing of the contrast column at the level of the cecum with no contrast entering the markedly dilated cecum.* **(Bottom)** *Sagittal CECT in the same patient demonstrates twisting of the mesentery with abrupt narrowing of the colon at this level. The findings in this series of images are classic for cecal volvulus. Notice the dilated small bowel due to the obstruction at the level of the cecum.*

COLON CANCER

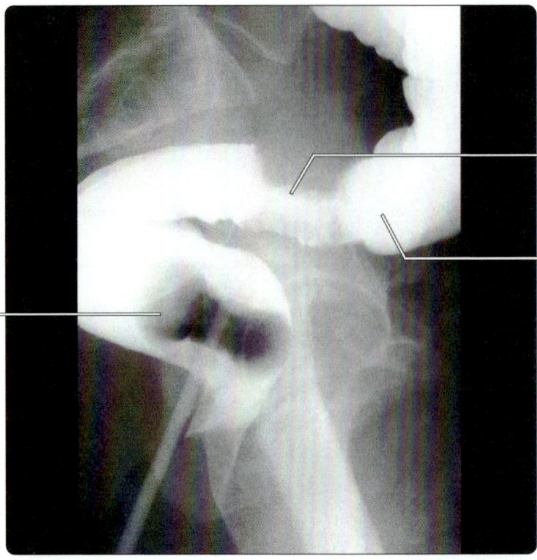

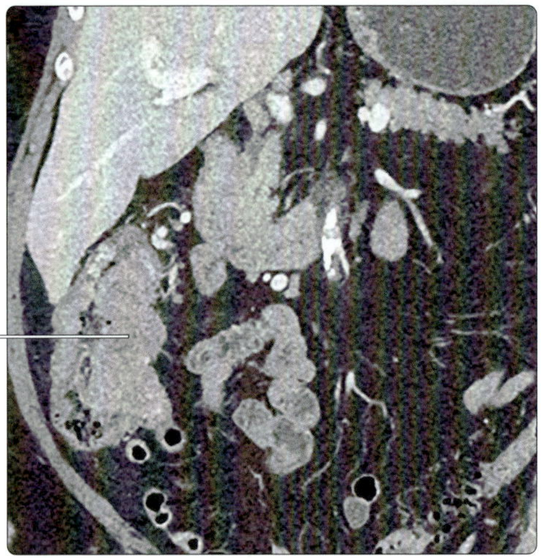

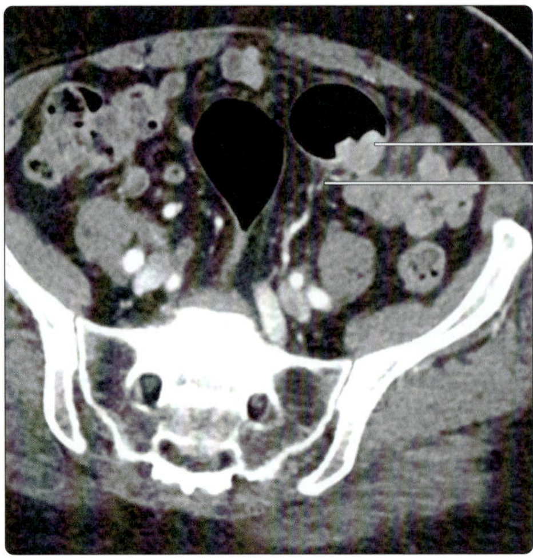

(Top) *A lateral view from a barium enema shows abrupt, irregular narrowing of the sigmoid colon by a carcinoma.* **(Middle)** *Coronal CECT demonstrates irregular nodular wall thickening of the cecum, particularly along its medial margin, confirmed to represent cecal cancer at colonoscopy.* **(Bottom)** *Axial CECT shows a lobulated mass in the sigmoid colon, which may be easily mistaken for stool. Clues to diagnosis are the prominent adjacent vessels, suggestive of neovascularization of the tumor. Biopsies of the sigmoid mass showed poorly differentiated colon cancer.*

PERIANAL FISTULAS

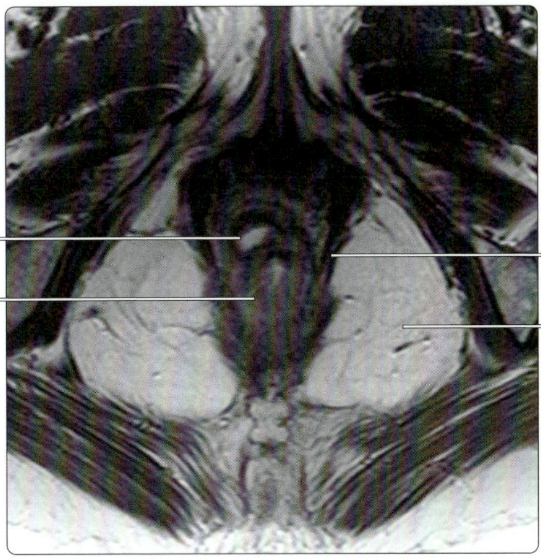

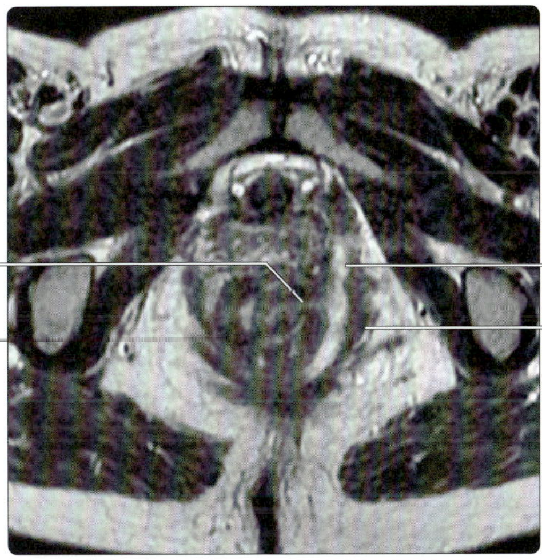

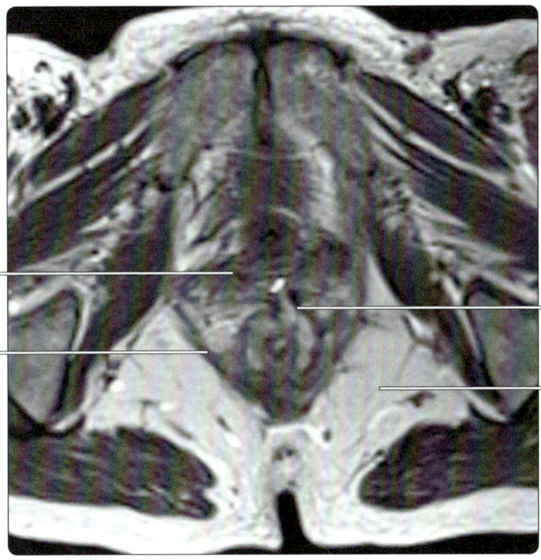

(Top) *MR for the diagnosis and classification of perianal fistulae provides essential information with both preoperative and prognostic value. Intersphincteric fistula are the most common of all perianal fistulas. St James University Hospital grading system is based upon the MR imaging appearance of perianal fistulae and constitutes 6 types: Grade 0: Normal; grade 1: Simple intersphincteric fistula; grade 2: Intersphincteric fistula with a secondary tract or abscess; grade 3: Simple transsphincteric fistulae; grade 4: Complicated transsphincteric fistula with a secondary tract or abscess; grade 5: Complicated abscess with a supra or translevator component.* **(Middle)** *A patient with Crohn disease and a complex transsphincteric perianal fistula traversing the external sphincter is shown.* **(Bottom)** *A patient with history of Crohn disease with a rectovaginal fistula is shown.*

ISCHEMIC COLITIS

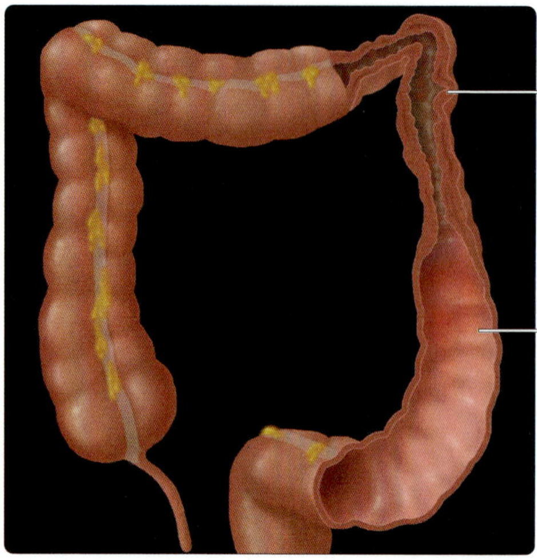

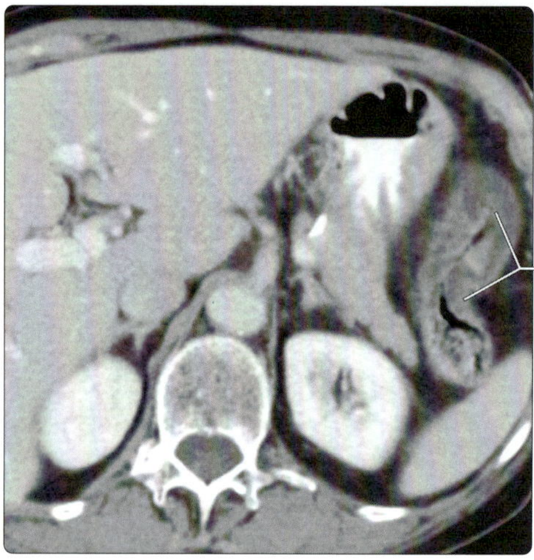

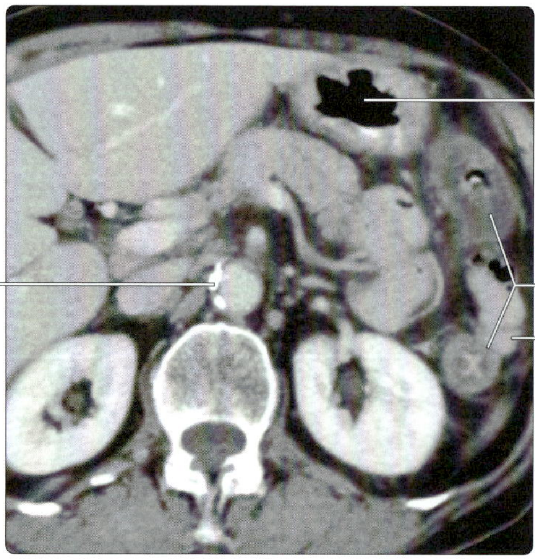

(Top) *Graphic shows the typical effect of hypoperfusion ischemic injury of the colon. The "watershed" region of the splenic flexure, where the distribution of the superior and IMAs intersect, is the most vulnerable segment. Note the luminal narrowing and wall thickening.* **(Middle)** *Axial CECT shows luminal narrowing and submucosal edema within the splenic flexure region of the colon, a classic location for ischemic colitis, as this is a "watershed" area between the SMA and IMA vascular distributions.* **(Bottom)** *The ischemic injury to the colon in this patient extends into the distal transverse and proximal descending colon.*

ISCHEMIC COLITIS

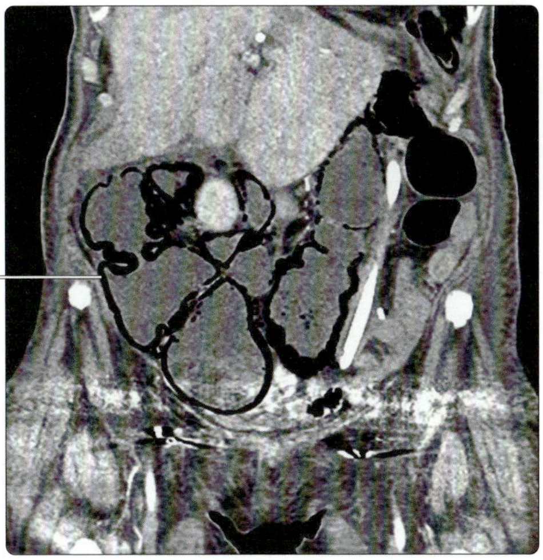

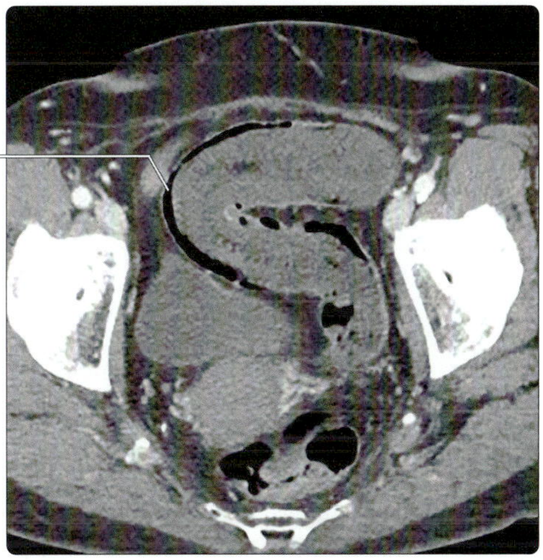

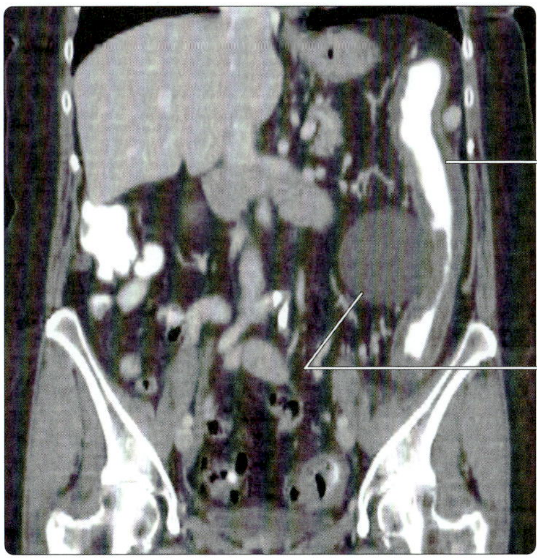

(Top) *Coronal CECT in a hypotensive older adult demonstrates extensive pneumatosis throughout the visualized colon, compatible with ischemic colitis. A pancolonic distribution, as in this case, is unusual for ischemic colitis but reflects the severity of the hypotensive insult to this patient.* **(Middle)** *Axial CECT in an older adult after extensive hepatobiliary surgery demonstrates pneumatosis throughout the sigmoid colon, compatible with ischemic colitis. The sigmoid colon is probably the 2nd most common location for ischemic colitis after the splenic flexure.* **(Bottom)** *Coronal CECT demonstrates the classic distribution of ischemic colitis in a hypotensive patient with isolated wall thickening of the splenic flexure of the colon and the descending colon with sparing of the remainder of the more proximal colon and the rectum.*

RECTAL CANCER

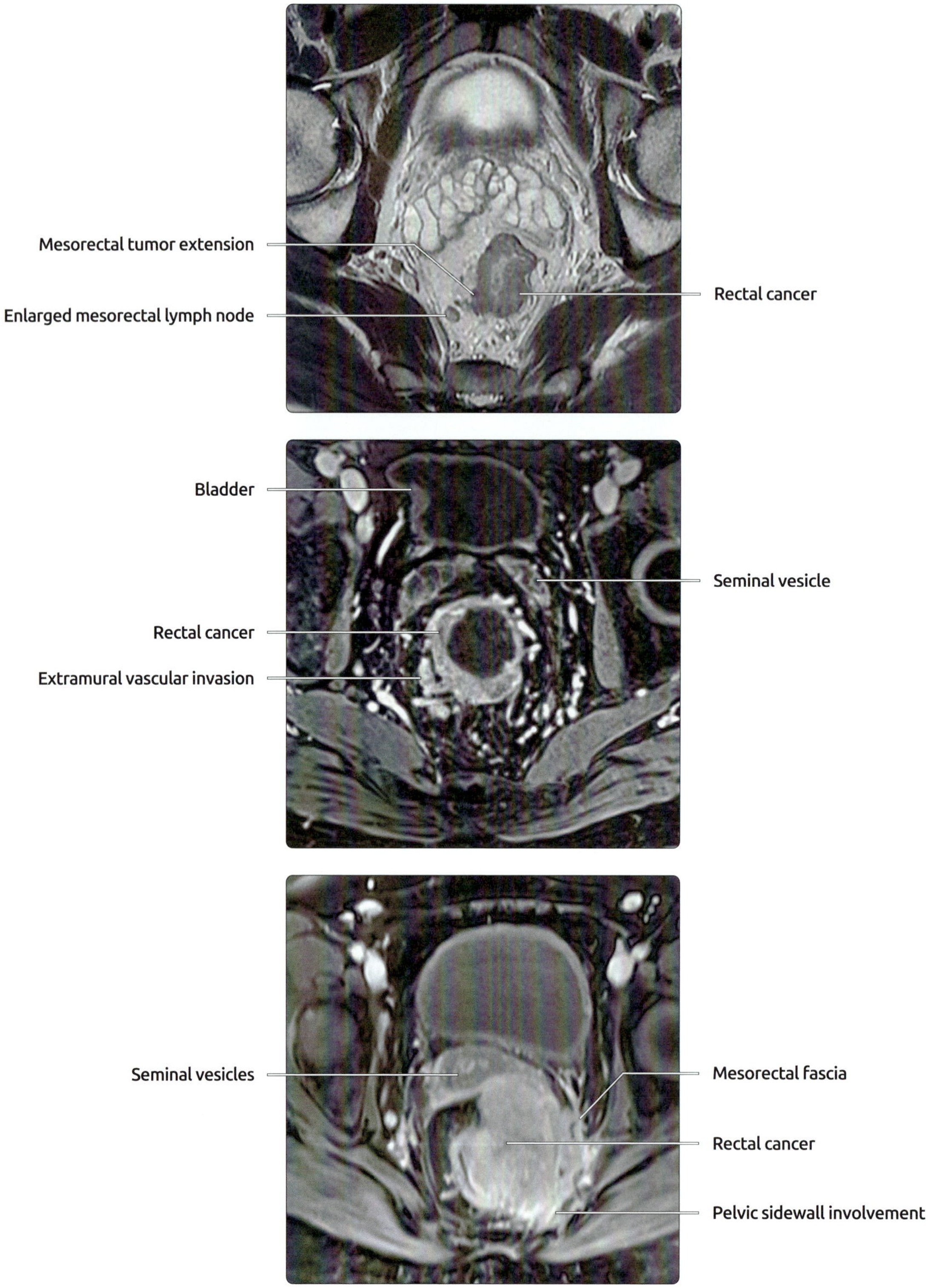

(Top) *Axial T2 MR demonstrates a circumferential mass involving the rectum. Note that there is subtle extension of tumor into the mesorectal fat, particularly evident at the 7 o'clock position. In addition, there is at least 1 enlarged mesorectal lymph node at the 8 o'clock position.* **(Middle)** *A patient with rectal cancer with extramural vascular invasion (EMVI), seen as focal enlargement of the vessel, heterogeneous signal intensity of the tumor, and wall irregularity, is shown. EMVI, along with mucin content, and involvement of the mesorectal fascia are considered poor prognostic factors.* **(Bottom)** *Axial T1 C+ FS MR demonstrates a large enhancing mass arising from the rectum, compatible with a primary rectal cancer. The mass extends anteriorly to involve the seminal vesicles and also extends laterally to the left to involve the left-sided mesorectal fascia and the left pelvic sidewall musculature.*

RECTAL CANCER

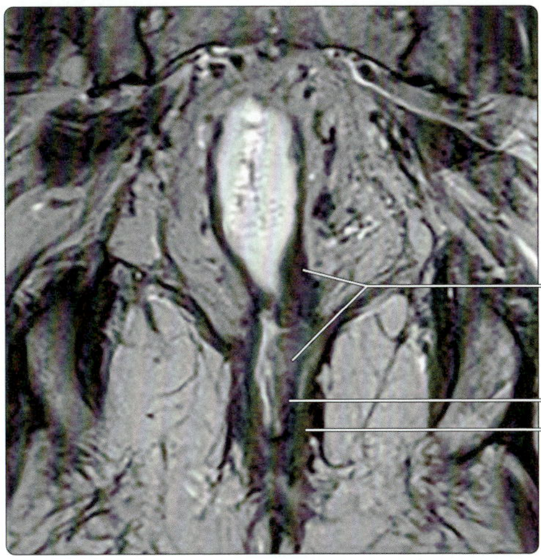

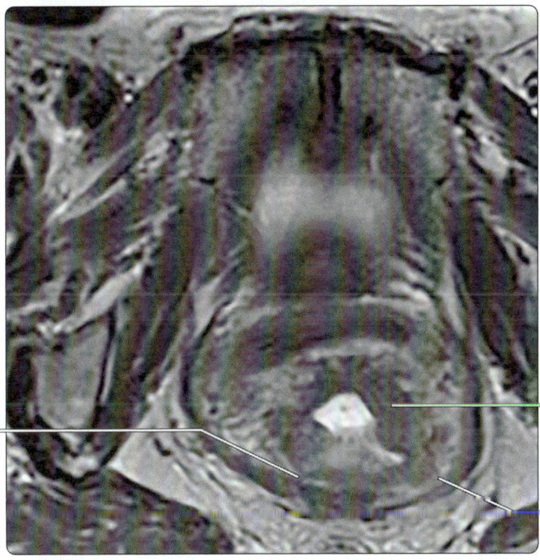

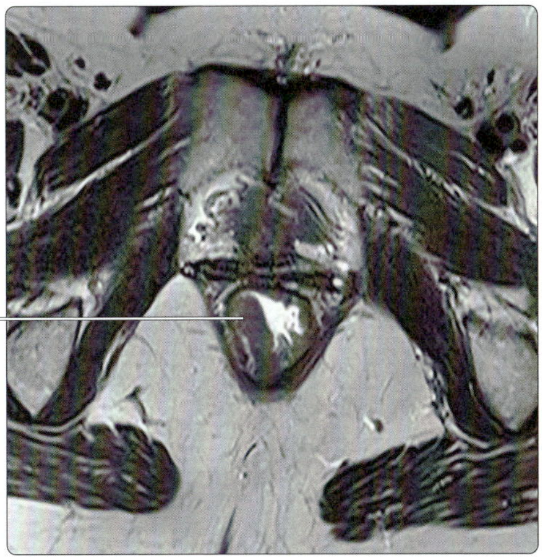

(Top) *Coronal T2 MR demonstrates irregular wall thickening along the left side of the rectum, compatible with the patient's known rectal malignancy. The lesion extends downward to at least involve the internal sphincter on the left with possible involvement of the external sphincter as well.* **(Middle)** *Axial T2 MR demonstrates circumferential, nodular thickening of the rectum, compatible with a primary rectal cancer. The tumor extends into the adjacent mesorectal fat in several locations and also appears to extend outward to involve the mesorectal fascia, a key imaging feature when staging rectal cancer on MR.* **(Bottom)** *Axial T2 MR demonstrates a polyploid mass in the inferior rectum nicely demarcated from the adjacent T2-hypointense wall of the rectum. There is no evidence of tumoral invasion of the mesorectal fat in this case.*

GROSS ANATOMY

Overview

- Spleen is largest lymphatic organ
  - Size is variable with no consensus on absolute size thresholds to determine enlargement/splenomegaly
    - Usually < 12 cm long, 8 cm wide, 5 cm thick
    - Usual volume range: 100-250 cm^3 with mean of 150 cm^3 in adults
    - Volume > 470 cm^3 = splenomegaly
  - Primary functions
    - Manufactures lymphocytes and plasma cells
    - Component of reticuloendothelial system, which filters blood and is involved in phagocytosis (removes damaged red blood cells and platelets)
    - Acts as blood reservoir and can expand or contract in response to changes in blood volume
- **Histology**
  - Soft organ with fibroelastic **capsule** and composed of **pulp** (i.e., substance of spleen)
    - **White pulp**: Lymphoid nodules/tissue primarily located surrounding vasculature
    - **Red pulp**: Sinusoidal spaces containing blood, which are primarily involved in filtration function of spleen and as reservoir for phagocytic white blood cells
    - **Trabeculae**: Extensions of capsule into parenchyma, which carry arterial and venous branches
- **Relations** and **vessels**
  - Surrounded entirely by peritoneum (except at splenic hilum) and suspended by several ligaments
  - Spleen contacts posterior surface of stomach and is connected via **gastrosplenic ligament** (GSL)
    - GSL is left anterior margin of lesser sac
    - GSL carries **short gastric** and **left gastroepiploic arteries** and venous branches to spleen
  - Spleen contacts pancreatic tail and surface of left kidney and is connected to these by **splenorenal ligament** (SRL)
    - SRL carries **splenic arterial** and venous branches to spleen
    - SRL is left posterior margin of **lesser sac**
  - **Splenic artery** usually arises from celiac artery and divides into multiple smaller branches both at splenic hilum and within spleen itself
    - Prior to reaching spleen, splenic artery also provides branches, which extend to pancreas, short gastric branches, and left gastroepiploic artery
  - **Splenic vein** runs in groove along dorsal surface of pancreatic body and tail
    - Receives **inferior mesenteric vein** (IMV) and other smaller veins (left gastroepiploic, short gastric, etc.)
    - Combined splenic vein and IMV join **superior mesenteric vein** to form **main portal vein**

ANATOMY IMAGING ISSUES

Multimodality Imaging Appearance

- NECT: Spleen typically 5-10 HU lower than liver
- Spleen shows **heterogeneous enhancement on** contrast-enhanced CT or MR in **arterial phase**
  - Described as moiré pattern of enhancement, which can be mistaken for pathology (e.g., infarct, laceration)
  - Reflects presence of red and white pulp
  - Spleen should demonstrate homogeneous enhancement on venous- or delayed-phase CT/MR
- MR: Spleen is usually slightly hypointense to liver on T1 MR and hyperintense on T2 MR
- Ultrasound: Spleen is typically isoechoic/hyperechoic relative to liver and hyperechoic to kidney
  - Typically has crescent shape, which can become more rounded as spleen enlarges
  - May normally show notched or indented appearance along its inner border, but outer margin usually smooth
- Spleen is relatively mobile and can change position after resection of adjacent organs (e.g., post nephrectomy)
  - Easily indented/displaced by masses or fluid collections
- Accessory spleen (e.g., splenule) is normal variant found in up to 25% of normal patients (multiple in 10%)
  - Can be confused with pathology (e.g., lymph nodes, carcinomatosis, pancreatic tail mass)
  - Should have identical attenuation, enhancement, MR signal to primary spleen on any imaging modality

CLINICAL IMPLICATIONS

Clinical Importance

- Spleen is **commonly injured** in blunt trauma, especially with fracture of left lower ribs
- Neoplastic involvement of spleen
  - Spleen is commonly involved in Hodgkin/non-Hodgkin **lymphoma** and leukemia
  - Uncommonly involved with other metastatic disease (e.g., metastatic melanoma), especially in absence of widespread metastatic disease elsewhere
  - **Very rarely** site of primary malignancy
- Tail of pancreas contacts splenic hilum and lies in SRL (can be considered intraperitoneal), allowing pancreatic tail pathology to spread into spleen
  - Pancreatitis can extend into splenic hilum and result in **intrasplenic pseudocyst**
- **Splenic infarction** is relatively common cause of acute left upper quadrant pain
  - Appears as sharply marginated, wedge-shaped area of hypoenhancement abutting splenic capsule

EMBRYOLOGY

Embryologic Events

- Originates from dorsal mesogastrium during 5th fetal week
- Embryologic rotation of stomach leads to spleen (usually) moving to its normal location in left upper quadrant
  - Usually fixed into place by peritoneal reflections linking it to diaphragm, abdominal wall, kidney, and stomach

Practical Implications

- Splenic ligaments may be congenitally lax or can weaken over time
  - May allow migration of spleen into other locations in abdomen (i.e., "**wandering spleen**"), resulting in splenic torsion or infarction
- **Asplenia** and **polysplenia**
  - Rare congenital conditions, which can be associated with other cardiovascular anomalies, situs inversus, etc.

SPLEEN AND ITS RELATIONS

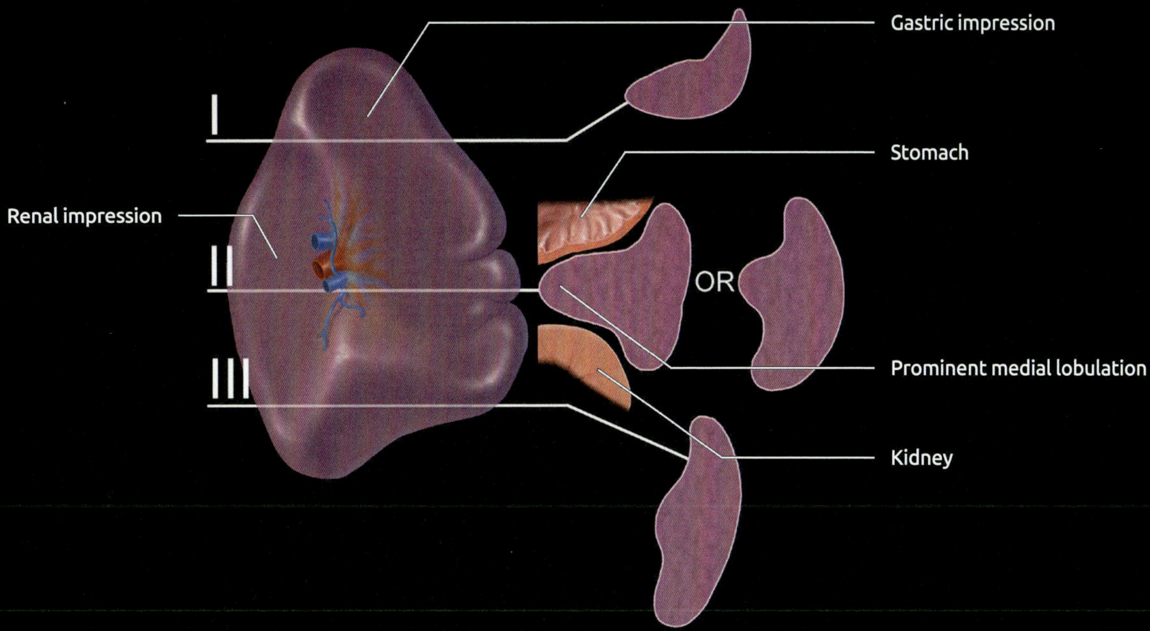

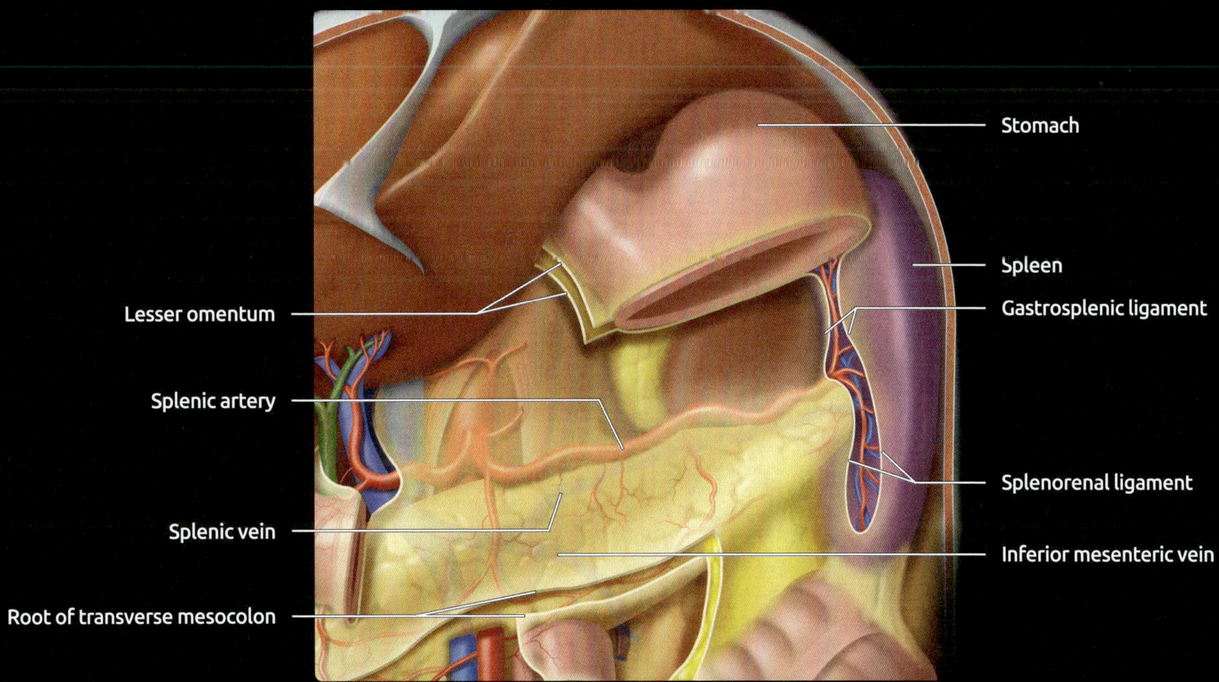

(Top) *Graphic shows the medial surface of the spleen and representative axial sections at 3 levels through the parenchyma. The spleen is of variable shape and size, even within the same individual, varying with states of nutrition and hydration. It is a soft organ that is easily indented by adjacent organs. The medial surface is often quite lobulated, as it is interposed between the stomach and the kidney.* **(Bottom)** *The liver is retracted upward, and the stomach is transected to reveal the pancreas and spleen. The splenic artery and vein course along the body of the pancreas, and the tail of the pancreas lies within the splenorenal ligament. The gastrosplenic ligament carries the short gastric and left gastroepiploic vessels to the stomach and spleen. The splenic vein receives the inferior mesenteric vein and joins the superior mesenteric vein behind the neck of the pancreas to form the portal vein.*

AXIAL CT, NORMAL SPLEEN

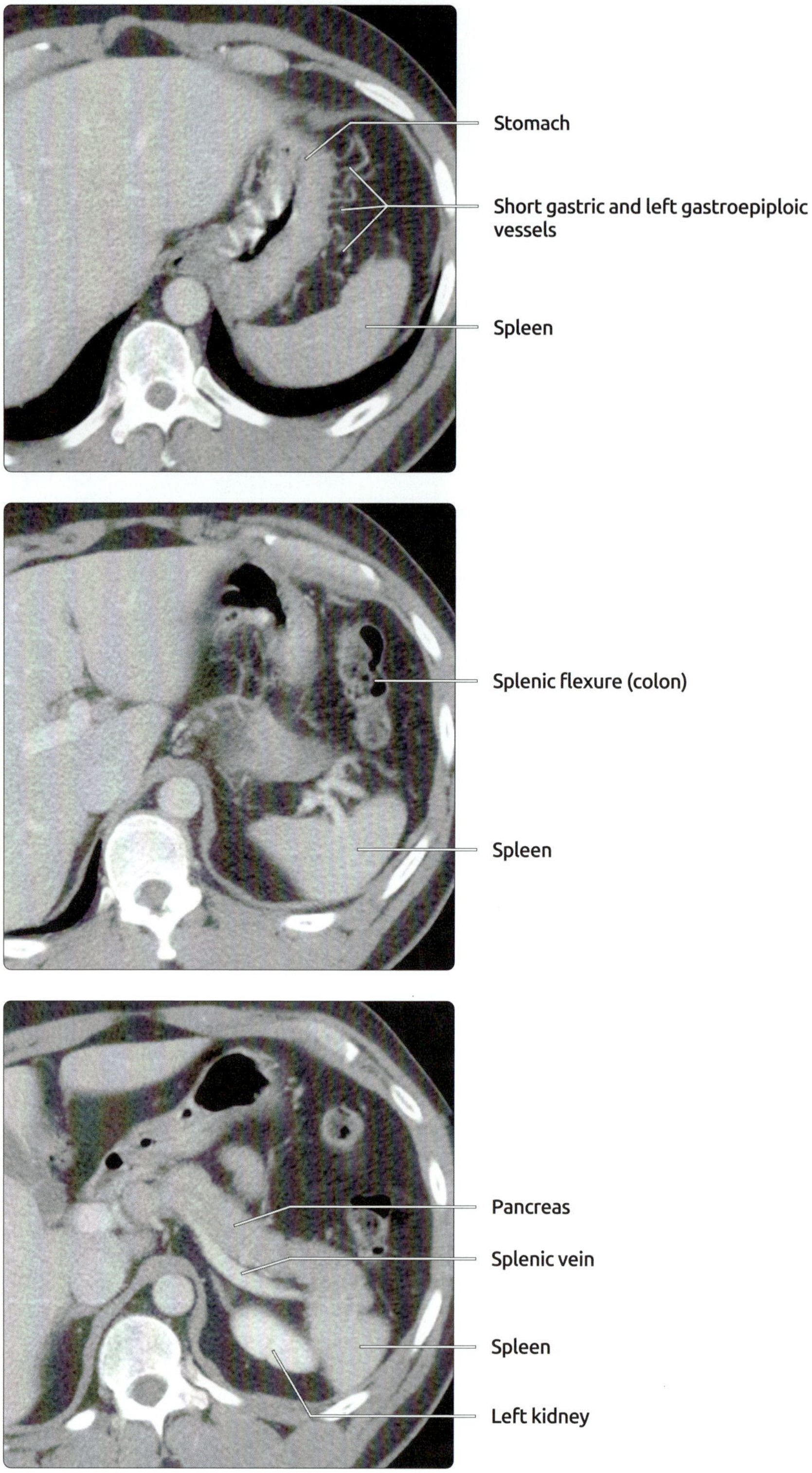

(Top) *First of 3 axial CT sections shows a normal spleen abutting the left hemidiaphragm. The short gastric and left gastroepiploic vessels lie in the gastrosplenic ligament and supply both the stomach and spleen.* **(Middle)** *Note the relation between the spleen and the left ("splenic") flexure of the colon.* **(Bottom)** *The body of the pancreas courses along the splenic vein, and the tail of the pancreas inserts into the splenic hilum. Note the relation of the spleen to the left kidney.*

CORONAL CT, NORMAL SPLEEN

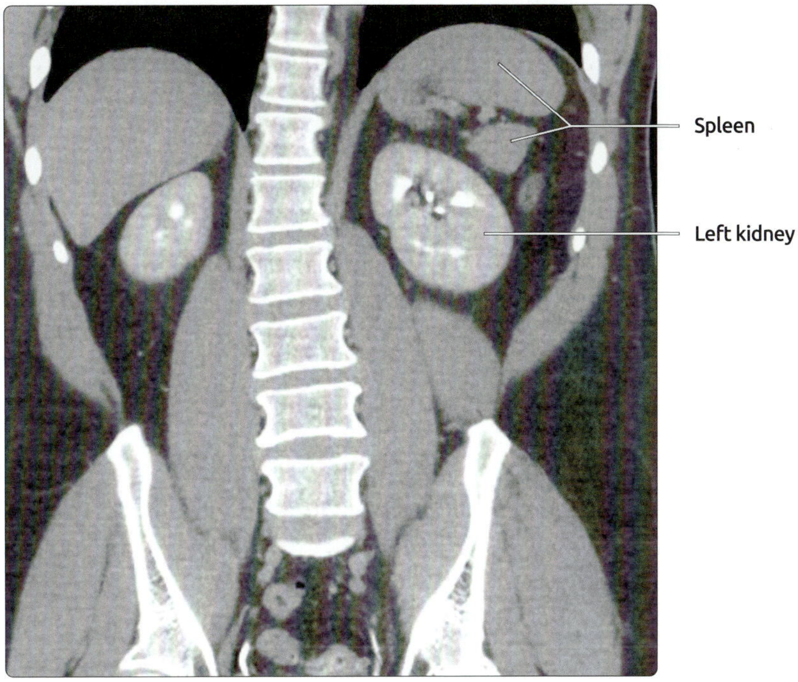

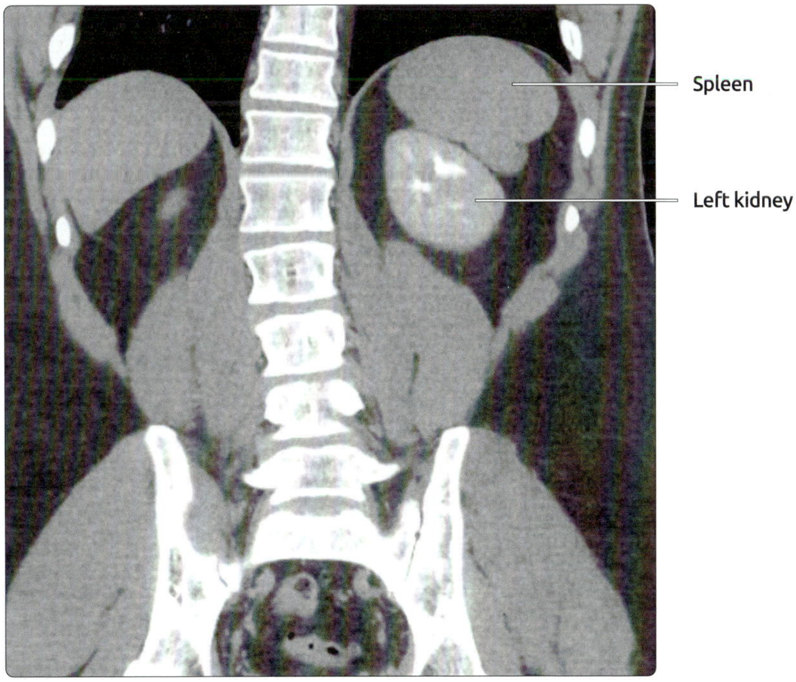

(Top) *First of 2 coronal CT images shows a normal spleen in its subdiaphragmatic location. Note the normal fatty clefts in the splenic parenchyma. Infoldings of the splenic capsule into the parenchyma are called trabeculae, and these carry the branches of the splenic vessels.* **(Bottom)** *Note the close relation between the spleen and the left kidney.*

NORMAL SPLEEN

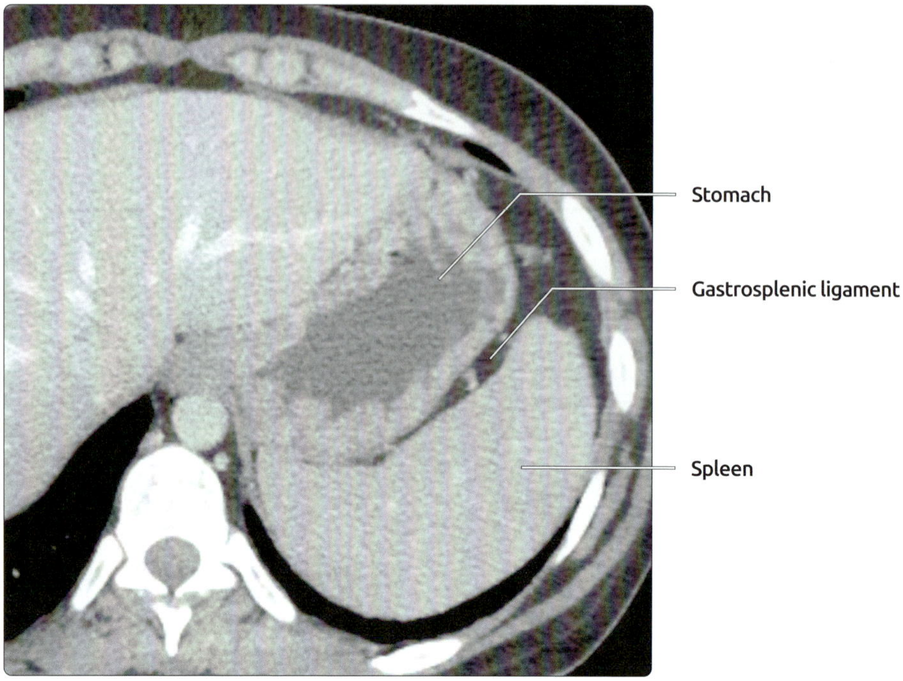

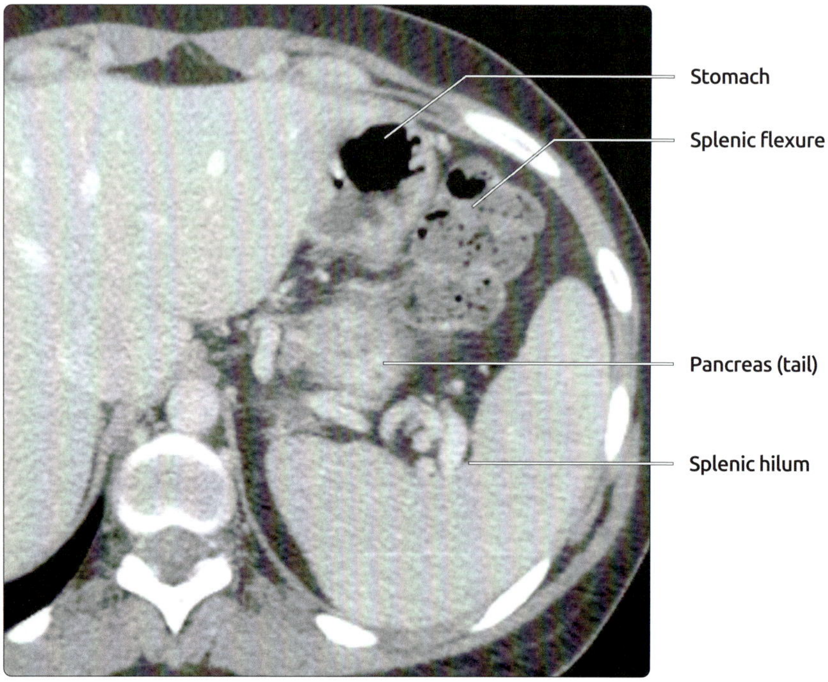

(Top) *First of 5 axial CT sections shows a normal spleen and its relations. Note the fat-containing gastrosplenic ligament, which carries the short gastric and left gastroepiploic vessels to the stomach and spleen.* **(Bottom)** *The tail of the pancreas courses with the splenic vessels through the splenorenal ligament and lies in the splenic hilum. The tail of the pancreas is the only intraperitoneal portion of this organ. Processes that affect the pancreatic tail, including pancreatitis and cancer, can easily extend into the splenic hilum and invade the splenic parenchyma.*

NORMAL SPLEEN

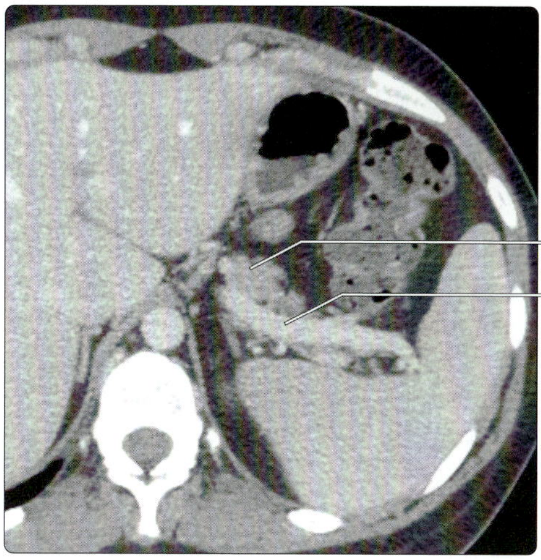

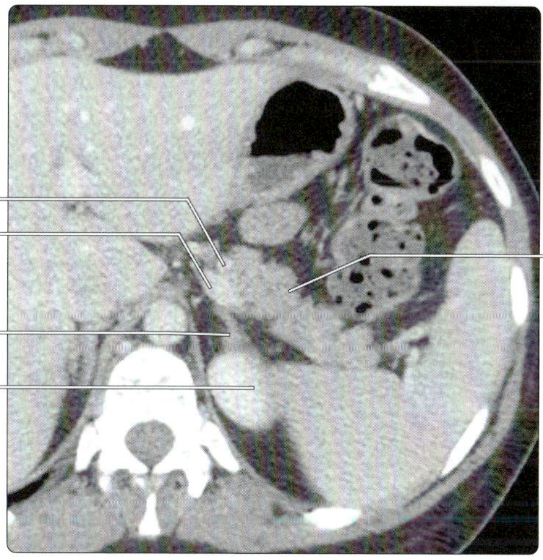

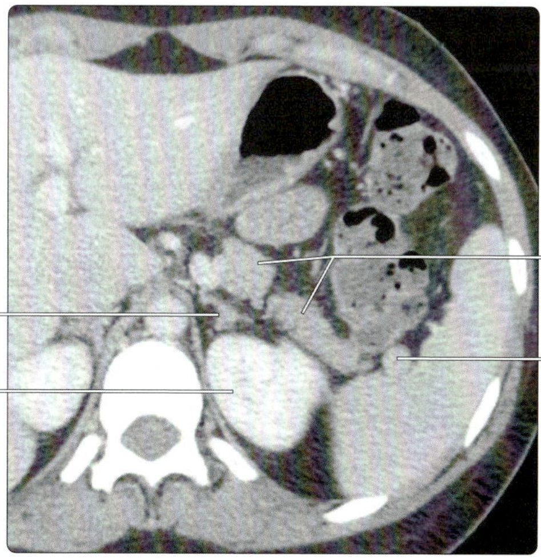

(Top) *The splenic vein runs in a transverse plane in a groove in the dorsal surface of the pancreas. This is a reliable anatomic marker to help identify the pancreatic body.* **(Middle)** *The splenic artery has a much more circuitous route, especially in older individuals, and frequently moves into and out of the axial plane of CT sections.* **(Bottom)** *Accessory splenic tissue is found in up to 30% of the population. As in this example, it is usually seen as a small, spherical "mass" in the splenic hilum.*

HETEROGENEOUS SPLENIC ENHANCEMENT

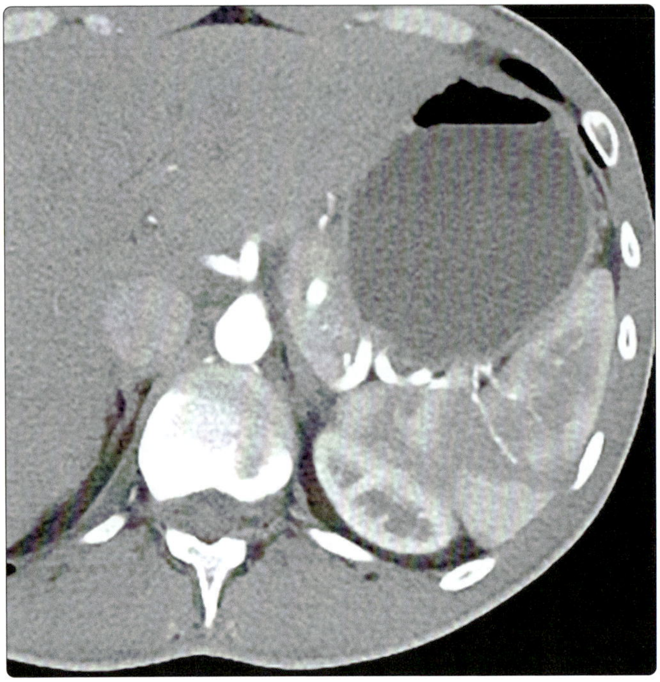

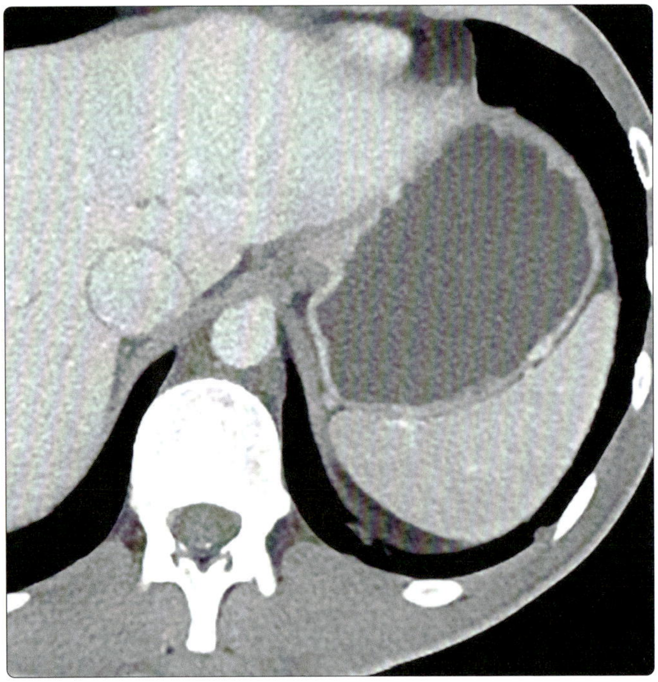

(Top) *First of 2 axial CT sections through a normal spleen is shown. This image was obtained in the arterial phase of enhancement (~ 35 seconds after the intravenous injection of contrast) and shows heterogeneous enhancement (moiré pattern) of the splenic parenchyma. This reflects the unique histology of the spleen with cords of tissue and sinusoidal spaces, rather than a capillary bed, as is found in most organs.* **(Bottom)** *A repeat CT section after a 75-second delay shows homogeneous enhancement of the splenic parenchyma. Splenic enhancement should be homogeneous in the venous or delayed phases, although heterogeneous enhancement may persist slightly longer than usual in patients with portal hypertension.*

SPLENIC LIGAMENTS AND VESSELS

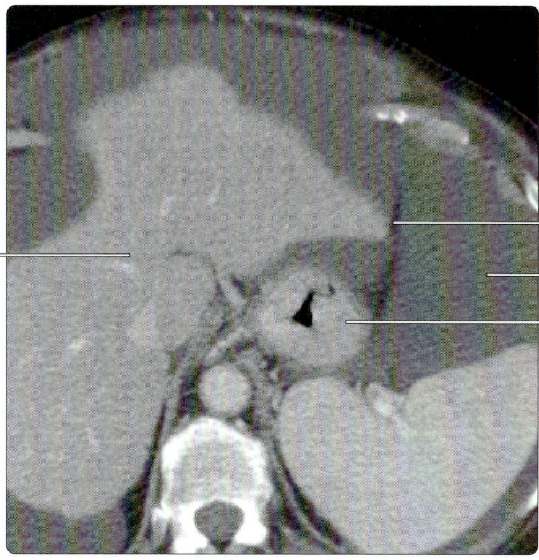

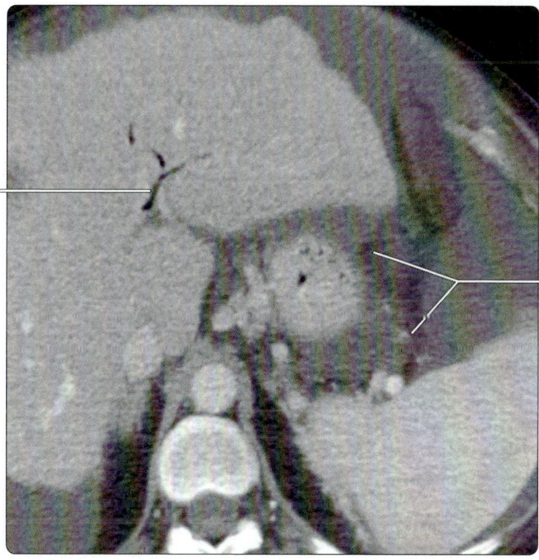

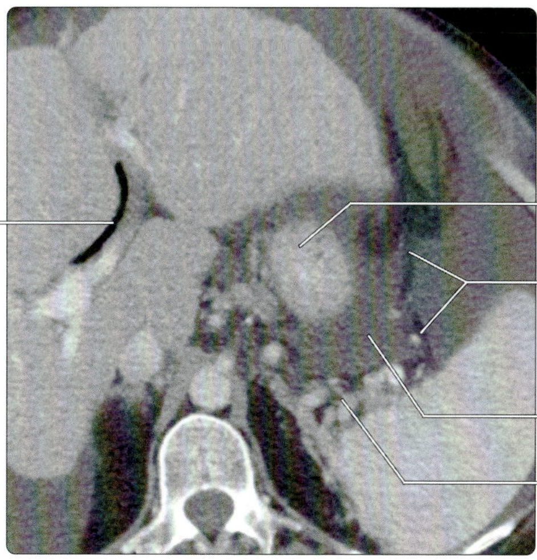

(Top) *First of 5 axial CT sections in a man with ascites that helps to accentuate the peritoneal ligaments that carry vessels to and from the spleen is shown. Axial image shows ascites surrounding the liver and spleen.* **(Middle)** *The gastrosplenic ligament is the peritoneal reflection that connects the stomach and spleen. It carries the short gastric and left gastroepiploic vessels to both organs.* **(Bottom)** *The gastrosplenic and splenorenal ligaments form the left lateral margin of the lesser sac and convey blood vessels, nerves, and lymphatics to the spleen.*

SPLENIC LIGAMENTS AND VESSELS

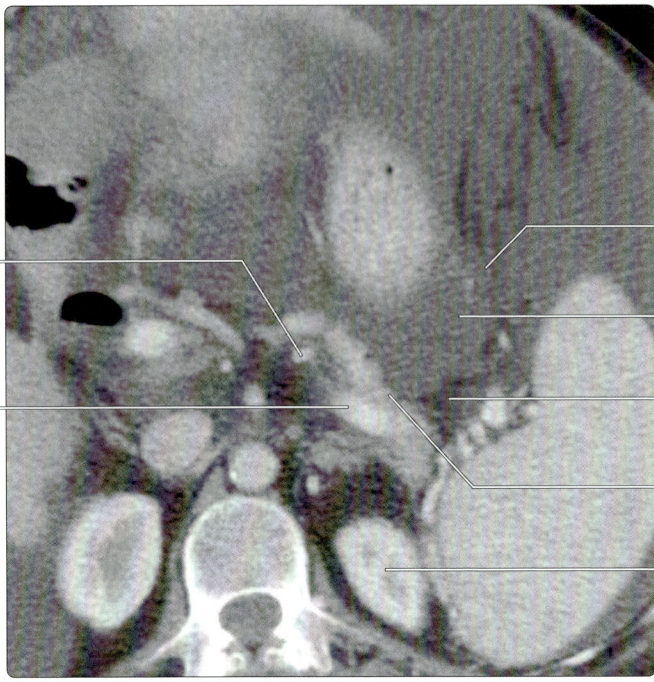

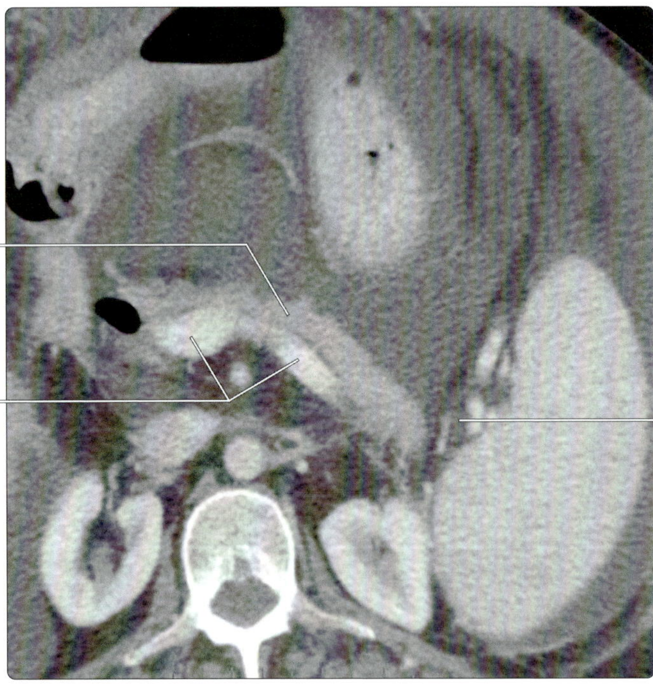

(Top) *The (main) splenic artery and vein course along the pancreas and enter the spleen through the splenorenal ligament.* **(Bottom)** *The splenic vein runs parallel to the body of the pancreas and lies in a groove along its dorsal margin.*

AXIAL T2 MR, NORMAL SPLEEN

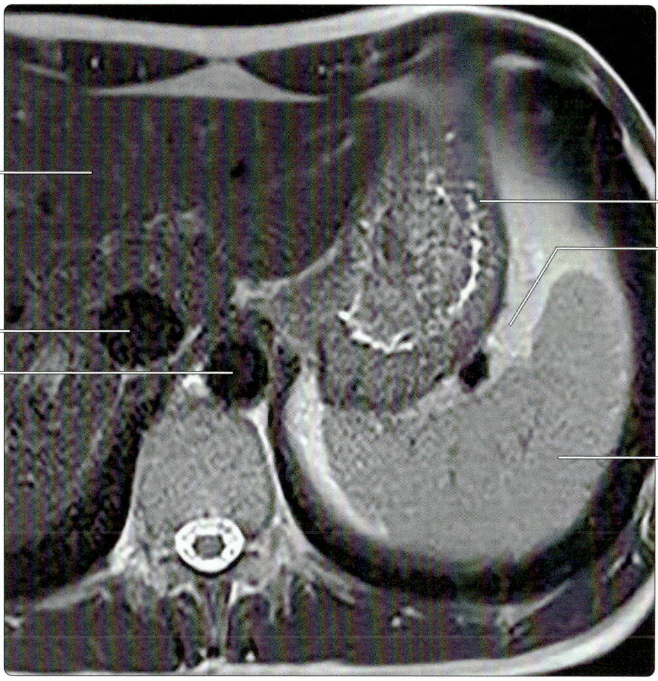

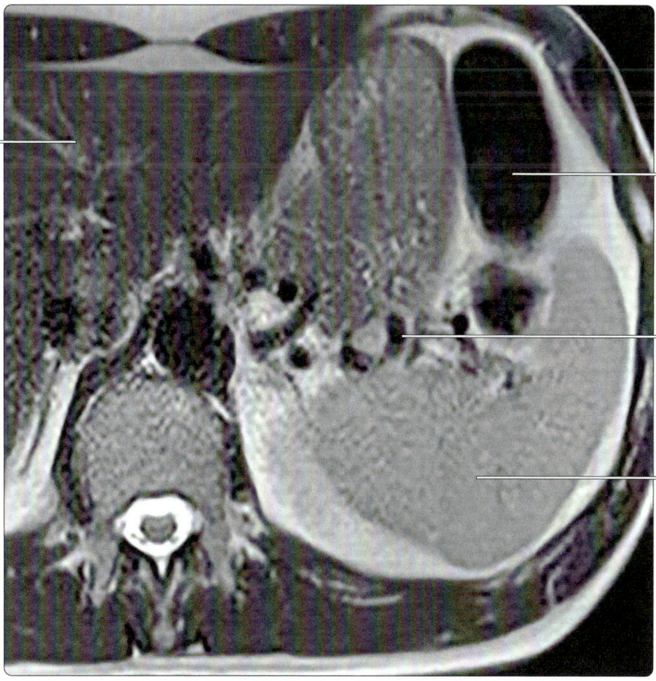

(Top) *First of 5 axial T2 HASTE MR images demonstrates a normal spleen abutting the left hemidiaphragm. The gastrosplenic ligament, the fatty attachment between the stomach and spleen, carries the short gastric and left gastroepiploic vessels to the stomach and spleen. Note that the spleen is higher in signal on T2 MR compared to the liver.* **(Bottom)** *Axial T2 HASTE MR at a slightly more caudal level demonstrates the tortuous flow void of the splenic artery coursing above the pancreas into the splenic hilum. The splenic artery has a very variable course (unlike the more predictable course of the splenic vein). Also note the relationship of the spleen with the adjacent splenic flexure of the colon more anteriorly.*

AXIAL T2 MR, NORMAL SPLEEN

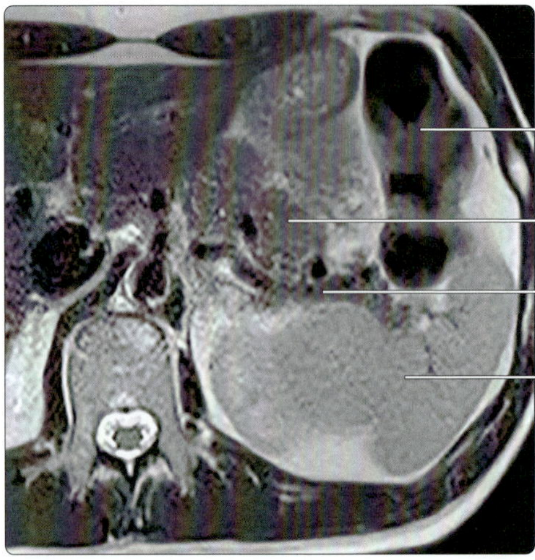

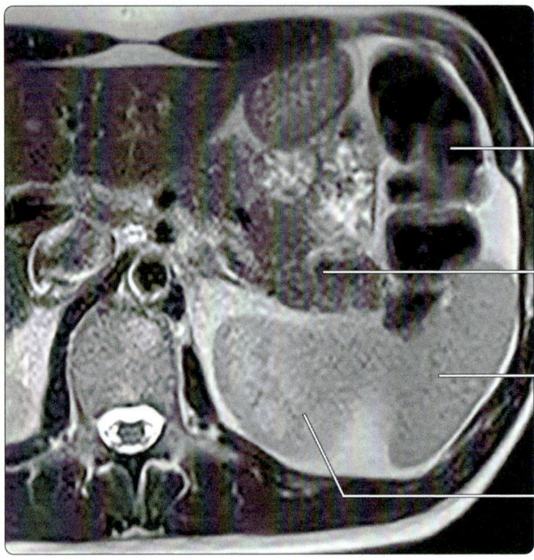

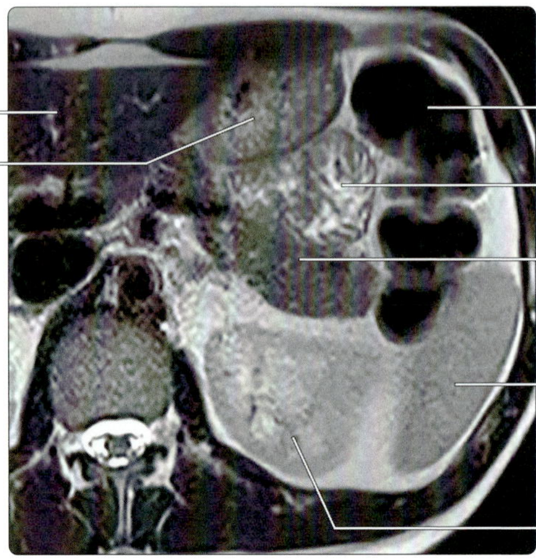

(Top) *More caudal axial T2 HASTE MR demonstrates the pancreatic tail extending into the splenic hilum. The pancreatic tail is intraperitoneal, allowing inflammatory or neoplastic processes to spread into the splenic hilum and spleen. The splenic vein flow void courses in a groove posterior to the pancreatic body and tail.* **(Middle)** *Axial T2 HASTE MR at a more inferior level demonstrates the relationship of the spleen with the immediately adjacent left kidney.* **(Bottom)** *Final axial T2 HASTE MR demonstrates the close anatomic relationship of the inferior aspect of the spleen with the immediately adjacent splenic flexure of the colon, left kidney, and pancreatic tail.*

ANGIOGRAM, SPLENIC VESSELS

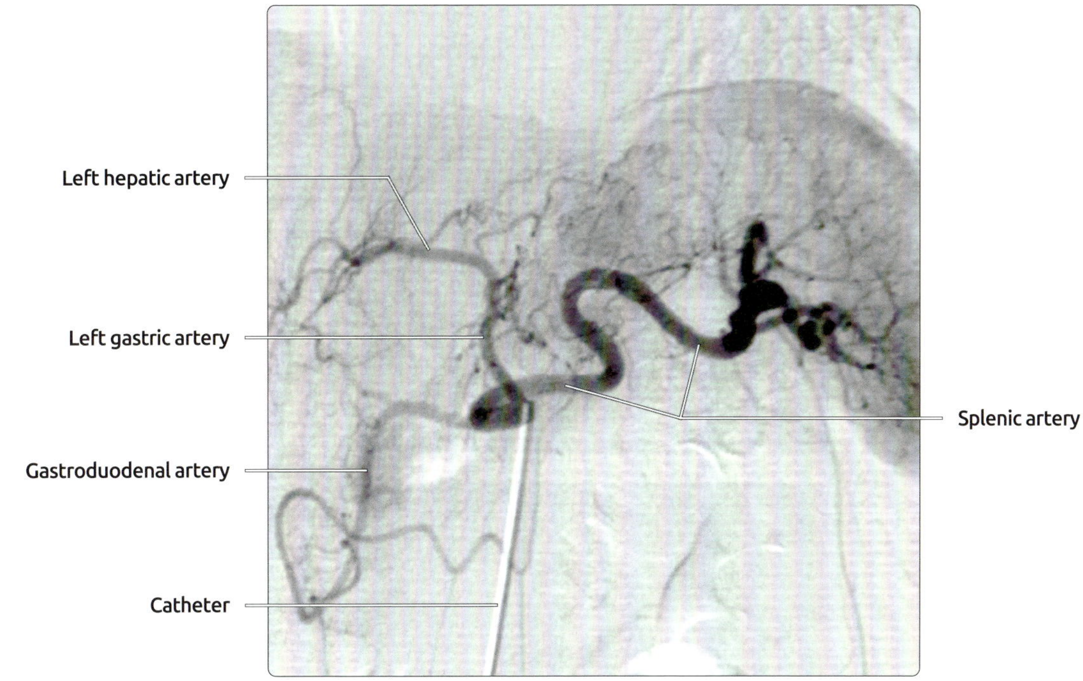

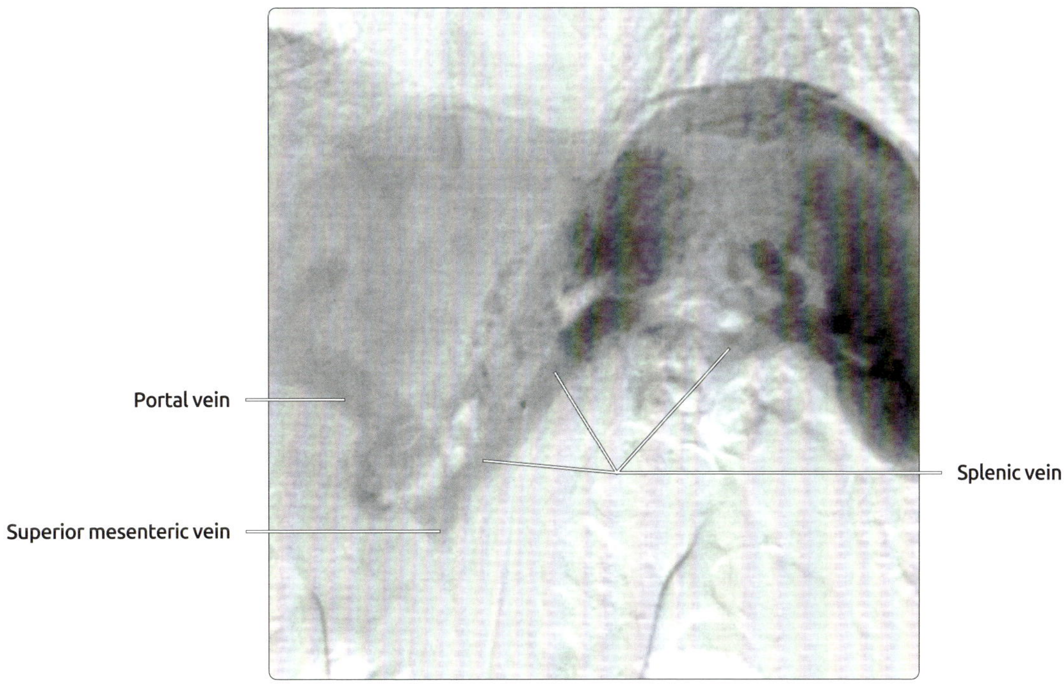

(Top) *A catheter injection of the celiac artery shows the typically tortuous course of the splenic artery. This individual exhibits a common variant in which the right hepatic artery arose from the superior mesenteric artery and is, therefore, not opacified on this injection.* **(Bottom)** *The splenic vein joins the superior mesenteric vein to form the portal vein. The inferior mesenteric vein, draining the left side of the colon, usually joins the splenic vein just prior to its confluence with the superior mesenteric vein.*

LYMPHOMA

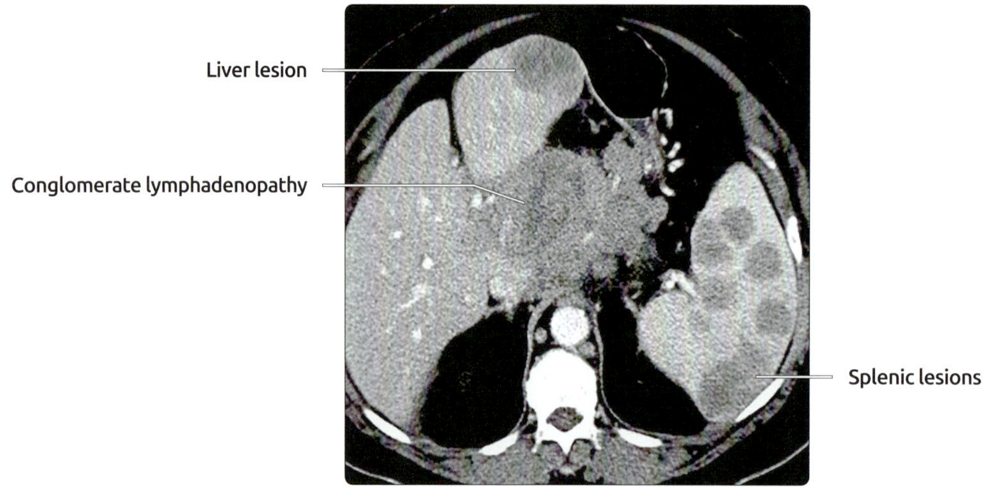

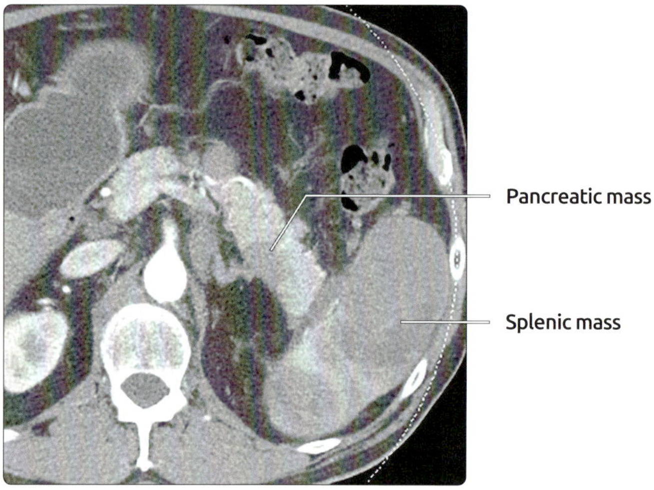

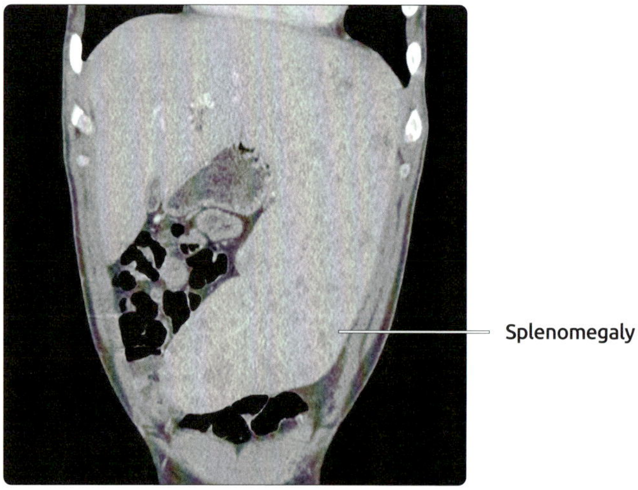

(Top) *Axial CECT in a patient with non-Hodgkin lymphoma demonstrates conglomerate lymphadenopathy in the upper abdomen and numerous solid hypodense masses in the spleen as well as a similar-appearing solid hypodense mass in the liver.* **(Middle)** *Axial CECT in a patient with non-Hodgkin lymphoma demonstrates extensive infiltration of the spleen with multiple sizable hypodense masses as well as a similar-appearing hypodense mass in the pancreas.* **(Bottom)** *Coronal CECT demonstrates a markedly enlarged spleen with innumerable tiny hypodense foci, an appearance strongly suggestive of splenic lymphomatous involvement in this patient with known non-Hodgkin lymphoma.*

SPLENIC VEIN OCCLUSION, CANCER

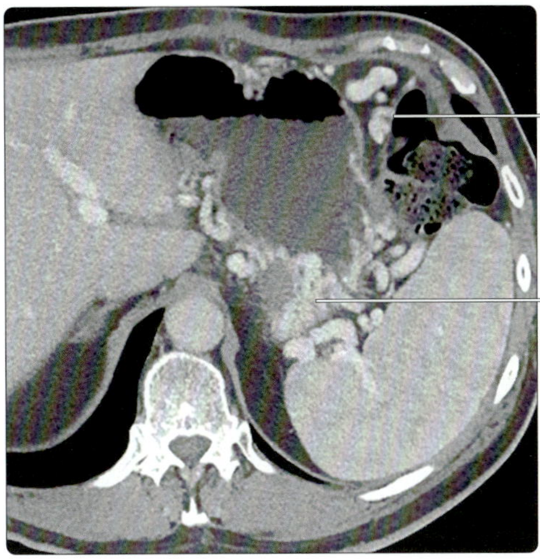

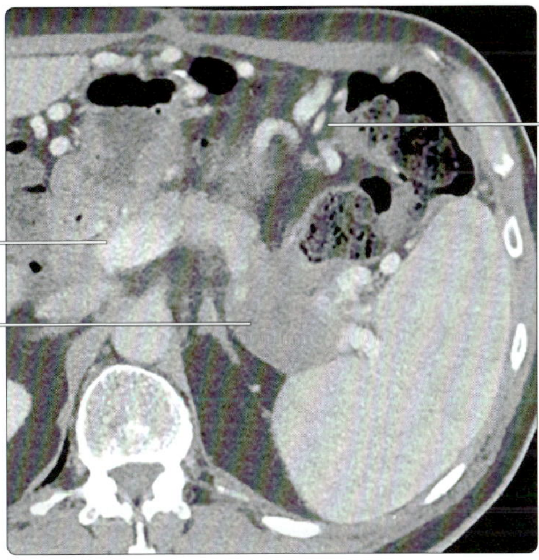

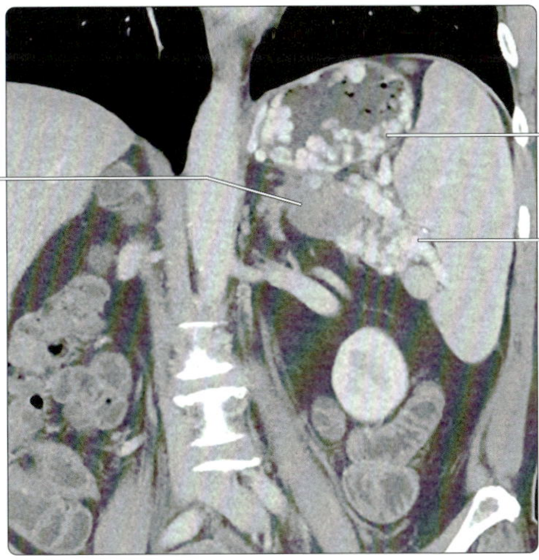

(Top) *Axial CECT from a patient with pancreatic adenocarcinoma of the pancreatic tail shows enlarged perigastric varices, which are collateral veins connecting with the short gastric veins and splenic vein tributaries, as well as gastroepiploic varices coursing around the greater curvature of the stomach, a classic set of findings in splenic vein occlusion.* **(Middle)** *The splenic vein should be opacified and visible on this section but is instead occluded by a pancreatic tail adenocarcinoma. Instead, collateral perigastric and gastroepiploic varices return blood to the portal vein via the left gastric (coronary) vein. Perigastric varices in the absence of cirrhosis and portal hypertension are indicative of splenic vein occlusion, usually due to pancreatic carcinoma or chronic pancreatitis.* **(Bottom)** *Coronal CECT in the same patient nicely demonstrate the characteristic pattern of varices involving the stomach and near the splenic hilum that are seen in the setting of splenic vein occlusion.*

INTRASPLENIC PSEUDOCYST

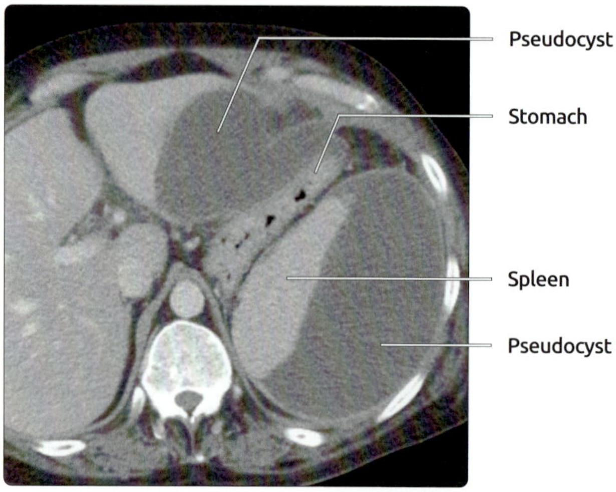

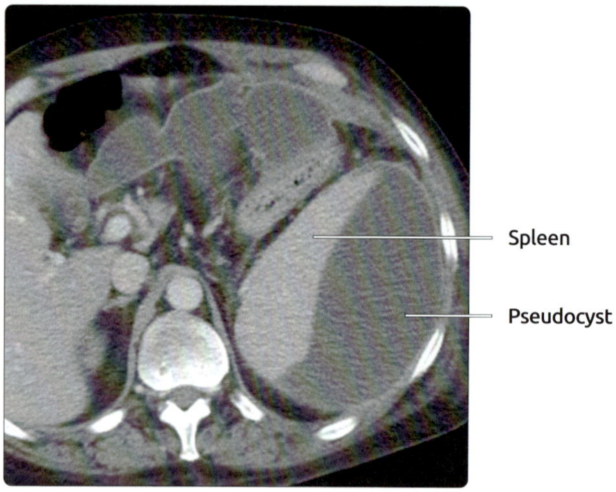

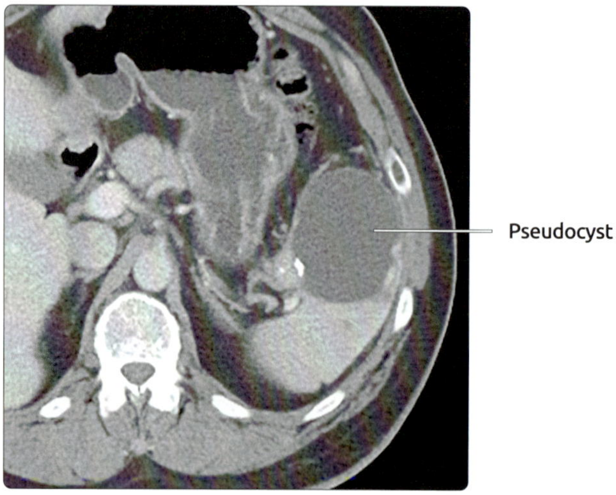

(Top) *Axial CECT in a patient with recurrent pancreatitis shows the stomach compressed between 2 pseudocysts. One of these is within the splenic capsule and causes marked compression of the splenic parenchyma.* **(Middle)** *More caudal section in the same patient again shows the encapsulation of the splenic pseudocyst and its effect on the splenic parenchyma.* **(Bottom)** *Axial CECT in a patient with a history of pancreatitis demonstrates a cystic lesion invaginating into the spleen with some peripheral calcification, representing a pseudocyst. Pseudocysts can occur in a multitude of different locations, but the splenic hilum and surrounding spleen are relatively common locations.*

ACUTE SPLENIC INFARCTION

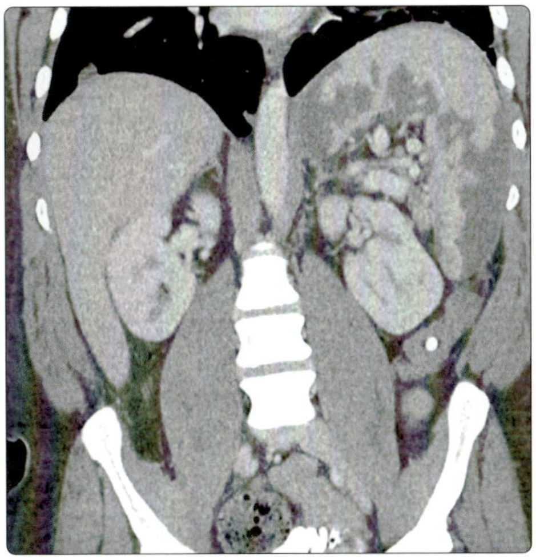

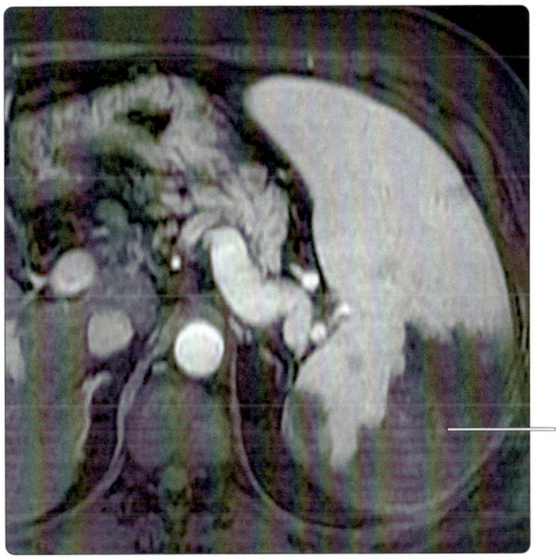

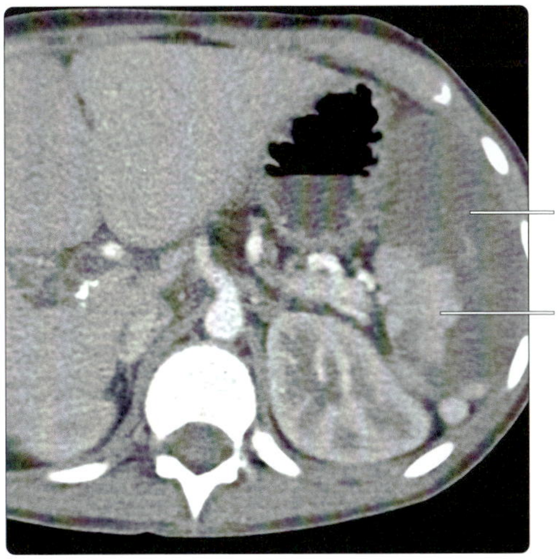

(Top) *Coronal CECT demonstrates multiple wedge-shaped areas of hypoenhancement extending out to the capsular surface of the spleen, compatible with acute splenic infarcts in this patient with acute abdominal pain and underlying sickle cell disease.* **(Middle)** *Axial T1 C+ MR demonstrates a large splenic infarct with hypoenhancing infarcted spleen juxtaposed against enhancing viable splenic parenchyma.* **(Bottom)** *Axial CECT demonstrates extensive infarction of the spleen with the infarcted, nonenhancing portions of the spleen clearly appearing distinct from the enhancing, viable portions.*

SPLENIC INFARCTION AND CYST

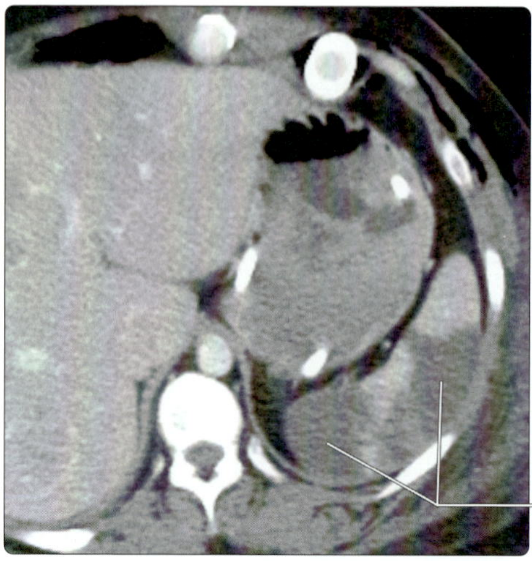

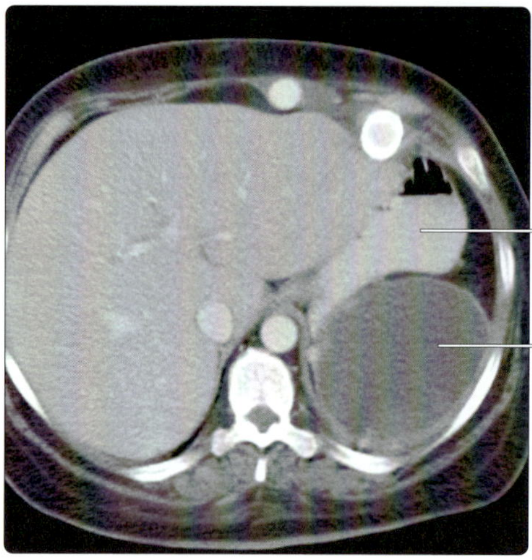

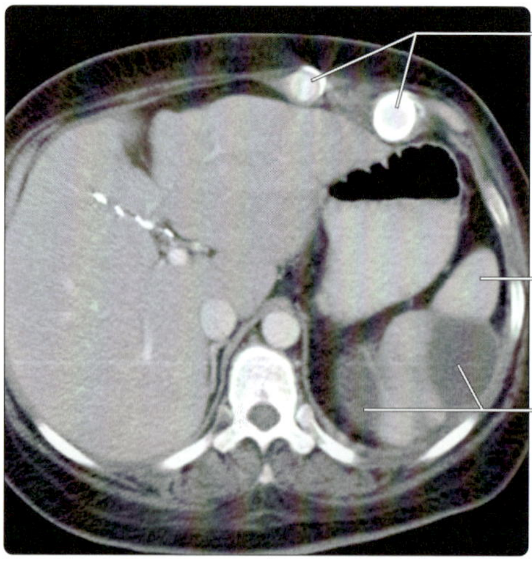

(Top) *First of 3 axial CECT images of a patient with heart failure, a ventricular assist device, and acute abdominal pain shows wedge-shaped regions of nonenhancing splenic parenchyma that extend to the capsular surface, characteristic of acute splenic infarctions.* **(Middle)** *CT obtained 7 days later shows interval development of splenic cysts near water density lesions that may develop after splenic infarction or other forms of injury.* **(Bottom)** *Additional splenic cysts now occupy the regions of the spleen that were infarcted.*

CHRONIC SPLENIC INFARCTION IN SICKLE CELL DISEASE

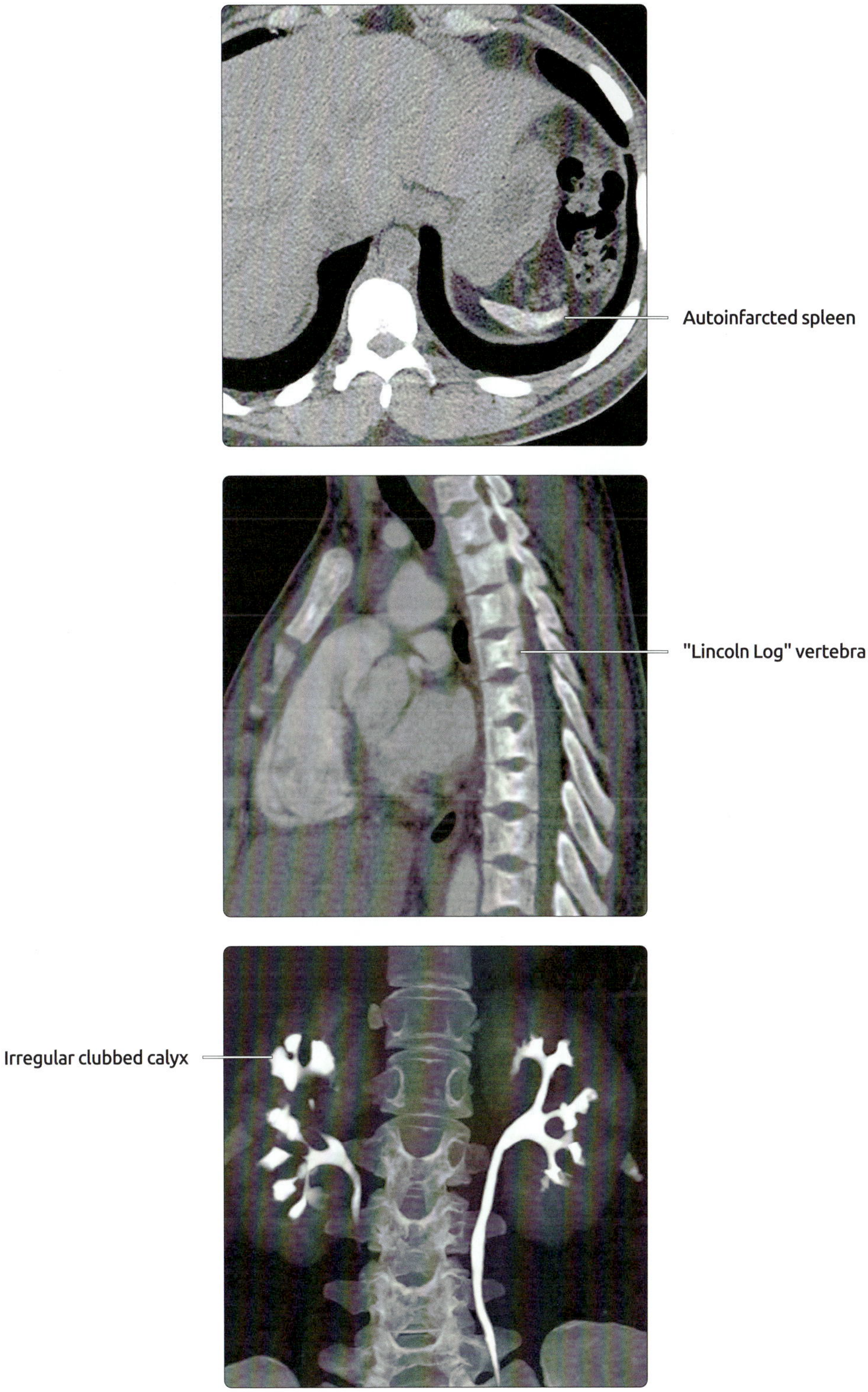

(Top) *Axial NECT demonstrates a markedly atrophic, small, calcified spleen in the left upper quadrant, a classic example of an "autoinfarcted" spleen in sickle cell disease.* **(Middle)** *Sagittal CECT demonstrates diffusely sclerotic, coarsened bones as well as multiple endplate depressions producing classic "Lincoln Log" vertebral bodies, common osseous features of sickle cell disease.* **(Bottom)** *Coronal CT urogram maximum intensity projection image demonstrates bilaterally deformed, irregular-appearing calyces (particularly on the right), representing papillary necrosis. Papillary necrosis is not an uncommon manifestation of sickle cell disease in the kidneys.*

ENLARGED ACCESSORY SPLEENS

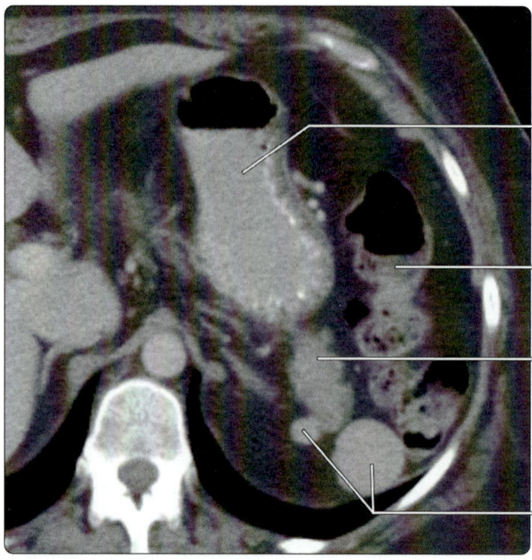

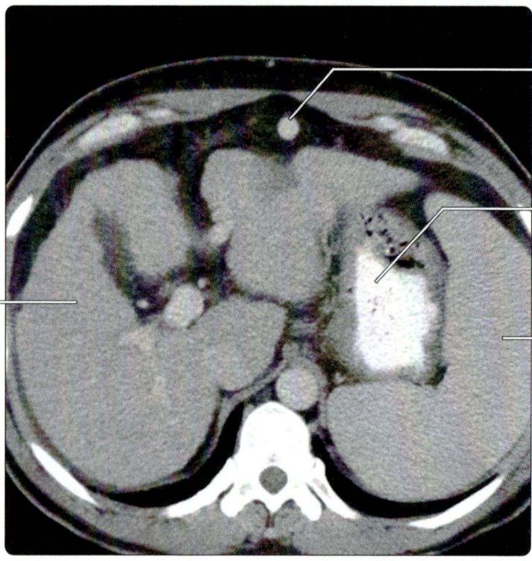

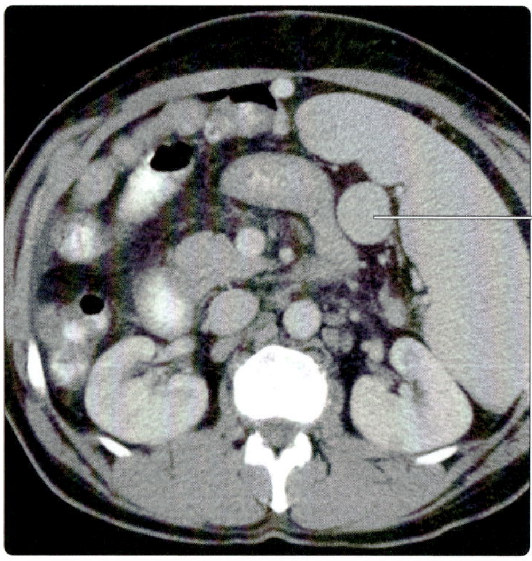

(Top) *Axial CECT demonstrates 2 left upper quadrant masses that represent enlarged accessory spleens that have hypertrophied to compensate for the surgical removal of the "main" spleen. A similar situation may occur after trauma, and many small pieces of splenic tissue may implant throughout the peritoneal cavity and acquire a new blood supply, developing into multiple splenic masses. This latter condition is known as "splenosis."* **(Middle)** *First of 2 axial CECT images shows a cirrhotic, nodular liver with evidence of portal hypertension, including varices (enlarged parumbilical vein) and splenomegaly.* **(Bottom)** *Axial CECT demonstrates a spherical mass near the splenic hilum representing an enlarged accessory spleen. Processes that result in splenomegaly may also cause increased size of accessory spleens that may be present.*

ACCESSORY SPLEEN, INTRAPANCREATIC

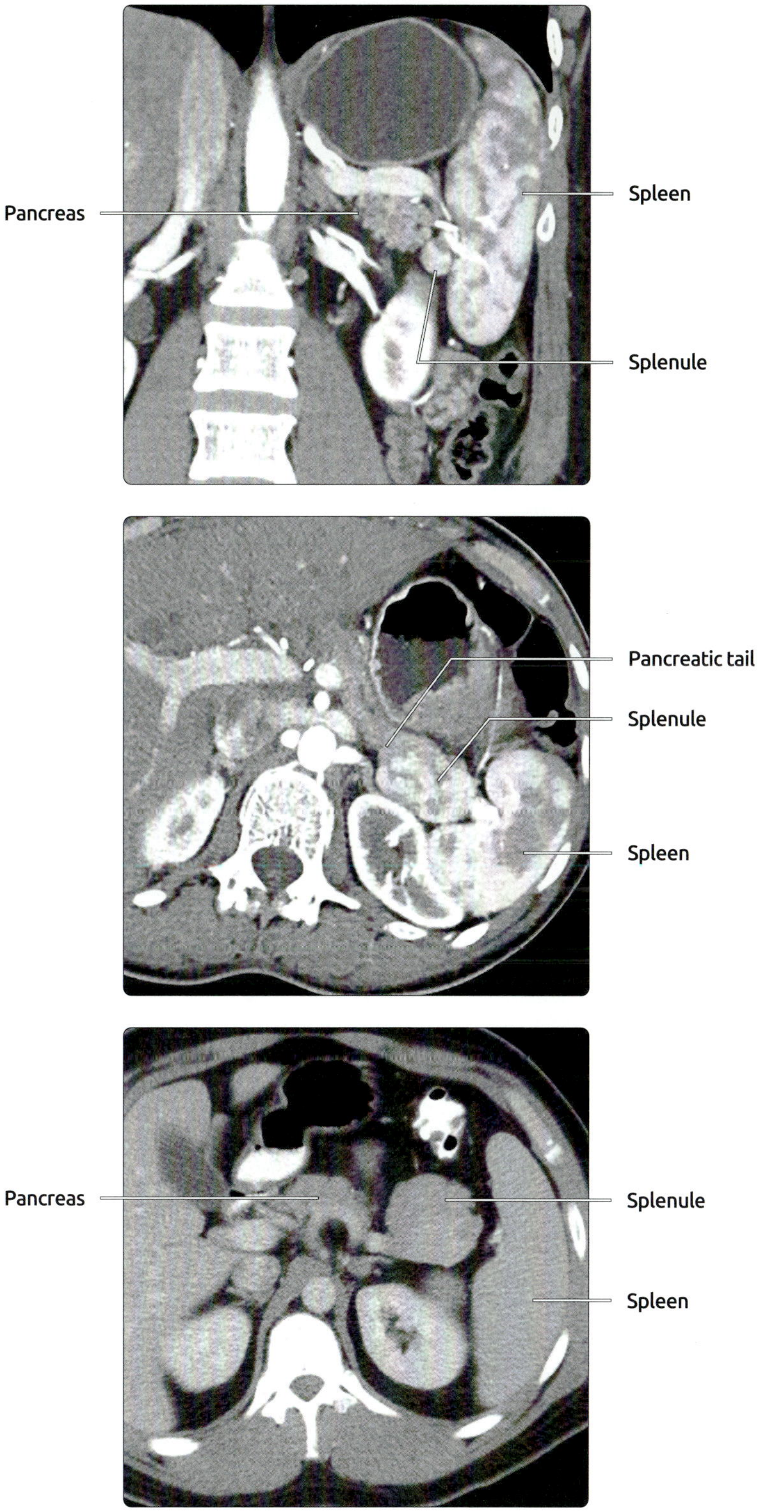

(Top) *Coronal CECT demonstrates a soft tissue "mass" abutting the pancreatic tail. Notice that this "mass" demonstrates identical enhancement to the adjacent main spleen, clearly allowing one to distinguish a benign splenule (as in this case) from a primary pancreatic tail neoplasm (such as a pancreatic neuroendocrine tumor).* **(Middle)** *Axial CECT in the arterial phase demonstrates a soft tissue mass at the level of the splenic hilum intimately associated with the pancreatic tail. Notice that the enhancement of this mass is identical to that of the spleen, both of which demonstrate the moiré pattern of enhancement that is so characteristic of splenic tissue in the arterial phase.* **(Bottom)** *Axial CECT in the venous phase demonstrates a soft tissue "mass" near the pancreatic tail with enhancement similar to the spleen. This was unfortunately thought to represent a pancreatic mass and underwent resection (where the diagnosis of a splenule was confirmed).*

WANDERING SPLEEN

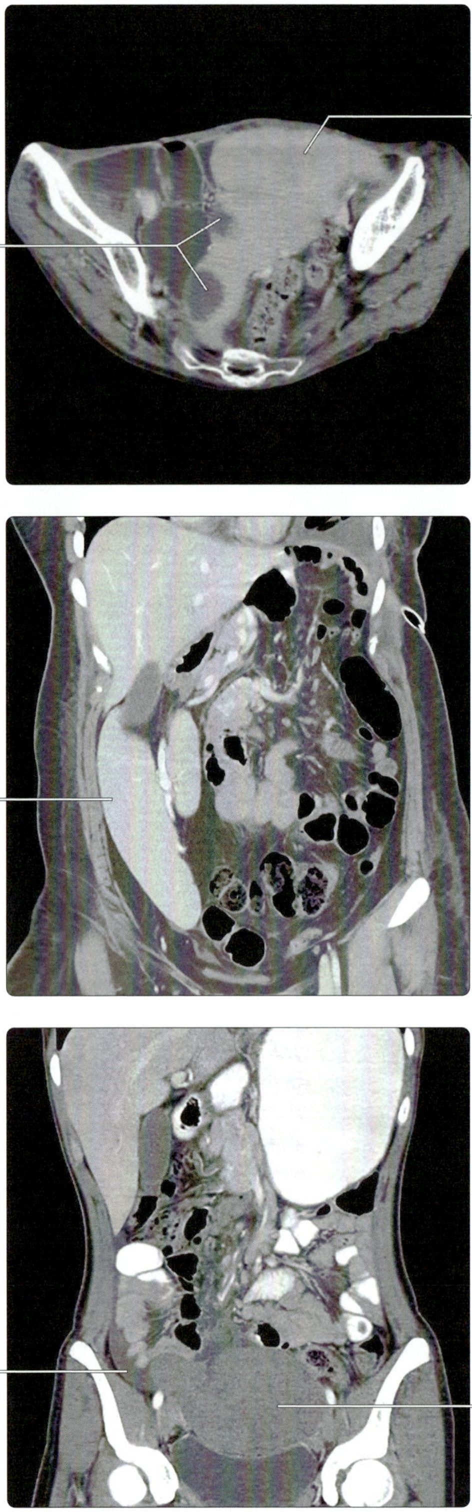

(Top) *Axial CECT demonstrates a large "mass" in the pelvis with areas of cystic change. No splenic tissue was evident in the upper abdomen (not shown). This pelvic mass was proved to be a "wandering spleen," which was on a long mesentery allowing it to descend into the pelvis. Low-density peripheral lesions represent acute splenic infarctions.* **(Middle)** *Coronal CECT demonstrates the markedly abnormal position of a "wandering" spleen located in the right lower quadrant. In this case, despite the abnormal position of the spleen, the patient was asymptomatic, although its location would clearly put it at risk for torsion and subsequent infarction.* **(Bottom)** *Axial CECT demonstrates a large, hypoenhancing mass in the pelvis of a young patient presenting with severe lower abdominal pain. Notice the absence of the spleen in the left upper quadrant. The patient was taken to surgery as a result of her symptomatology, and this large mass was found to be an infarcted wandering spleen.*

POLYSPLENIA SYNDROME

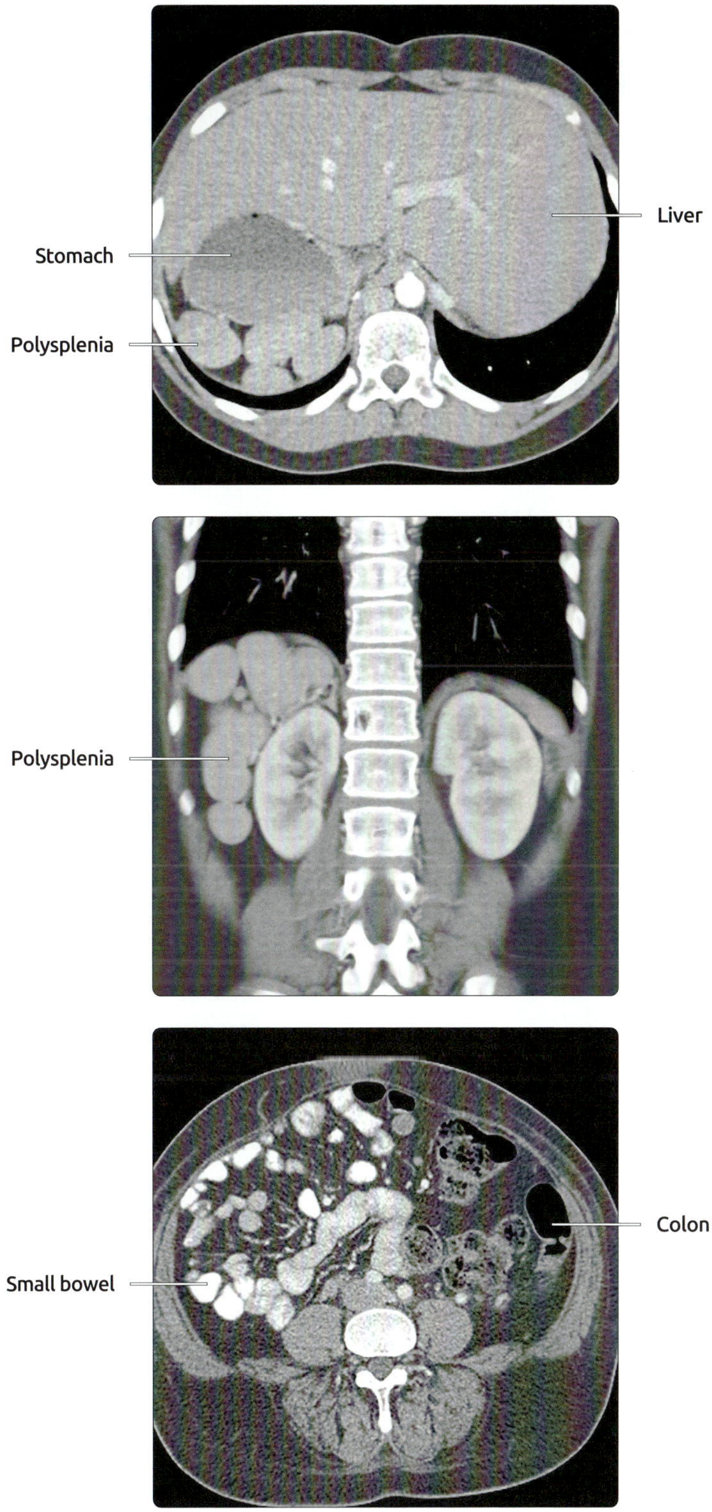

(Top) *Axial CECT shows the stomach and liver located on the "wrong" side of the patient, along with multiple spleens in the right upper quadrant.* **(Middle)** *Coronal CECT in the same patient demonstrates that multiple splenic "masses" are present on the right side of the abdomen, compatible with the patient's diagnosis of polysplenia.* **(Bottom)** *Axial CECT in another patient with polysplenia demonstrates malrotation with the small bowel entirely in the right abdomen and the colon entirely in the left abdomen. Polysplenia can be associated with a number of congenital abnormalities (including malrotation).*

GROSS ANATOMY

Overview

- Liver is largest internal body organ (average weight of 1,500 g) with variety of functions, including processing of nutrients absorbed from GI tract (via portal vein), storage of glycogen, and secretion of bile
  - Gross anatomic relationships
    - Anterior and superior surfaces (i.e., diaphragmatic surface) are smooth and convex and contact diaphragm
    - Posterior and inferior surfaces (i.e., visceral surface) are indented by colon, stomach, right kidney, right adrenal gland, duodenum, inferior vena cava (IVC), and gallbladder
  - Covered by visceral peritoneum except at **gallbladder fossa**, **porta hepatis**, and "**bare area**"
    - "**Bare area**": Posterior and superior surface where liver abuts diaphragm and is not covered by peritoneum
      - Located to right of IVC and contacts IVC and undersurface of diaphragm
    - **Porta hepatis**: Site of entry/exit of portal vein, hepatic artery, and bile duct
      - Located immediately anterior to caudate lobe along undersurface of liver
  - Covered by thick fibrous capsule (**Glisson capsule**)
- Vascular supply: Liver has unique dual blood supply via portal vein and hepatic arteries
  - **Portal vein**
    - Accounts for 75-80% of blood supply to liver
    - Carries nutrients from bowel, hepatotrophic hormones from pancreas, and oxygen (contains 40% more oxygen than systemic venous blood but significantly less oxygen than hepatic artery)
  - **Hepatic artery**
    - Accounts for 20-25% of blood supply to liver
    - Richer in oxygen than portal venous blood but does not contain same nutrients seen in portal venous system
    - Liver is less dependent than biliary tree on hepatic arterial blood supply
    - Usually arises from celiac artery, but variations in origin of hepatic artery are very common, such as replaced arteries arising from superior mesenteric artery or left gastric artery
  - **Hepatic veins**
    - Usually 3 (right, middle, and left), although there are many variations possible, and accessory veins are common
    - Return blood to IVC at **confluence of hepatic veins** just below diaphragm and entrance of IVC into heart
  - **Portal triad**
    - At all levels of size and subdivision, branches of hepatic artery, portal vein, and bile ducts travel together
    - Blood flows into hepatic sinusoids from interlobular branches of hepatic artery and portal vein → hepatocytes (detoxify blood and produce bile) → bile enters bile ducts and blood enters central veins → hepatic veins
- Segmental anatomy of liver
  - Older, more traditional division of liver into 4 segments (right anterior and posterior, left medial and lateral) has been supplanted by **Couinaud classification**, which divides liver into 8 segments
    - More clinically and surgically relevant classification, as each individual segment receives secondary or tertiary branch of hepatic artery and portal vein and is drained by its own bile duct (intrahepatic) and hepatic vein branch
    - Each segment is functionally independent and can be individually resected
    - Hepatic artery, portal vein, and bile duct segmental branches enter center of each segment, while hepatic veins are at periphery of segments (and divide segments)
    - Liver divided in vertical planes by 3 hepatic veins (hepatic vein planes), while portal vein bifurcation divides liver in horizontal plane (portal plane)
      - Middle hepatic vein divides right and left hepatic lobes (**Cantlie line**), while right hepatic vein divides right lobe into anterior and posterior parts and left hepatic vein divides left lobe into medial and lateral parts
    - Segments are numbered in clockwise order, as if looking at anterior surface of liver
    - Because of variations of vascular and biliary branching within liver (common), it is frequently impossible to designate precisely boundaries between hepatic segments on imaging studies
    - Right lobe usually slightly larger than left (~ 60% of total liver volume)
  - Segment I: Caudate lobe
    - Unique segment located around intrahepatic IVC, which has independent portal triads and direct hepatic venous drainage to IVC (via small veins)
  - Left lobe
    - Lateral superior = segment II (located to left of left hepatic vein and above portal plane)
    - Lateral inferior = segment III (located to left of left hepatic vein and below portal plane)
    - Medial superior = segment IVA (located to right of left hepatic vein and above portal plane)
    - Medial inferior = segment IVB (located to right of left hepatic vein and below portal plane)
  - Right lobe
    - Anterior inferior = segment V (located to right of middle hepatic vein and below portal plane)
    - Posterior inferior = segment VI (located to right of right hepatic vein and below portal plane)
    - Posterior superior = segment VII (located to right of right hepatic vein and above portal plane)
    - Anterior superior = segment VIII (located to right of middle hepatic vein and above portal plane)
- Major ligaments (double layers of peritoneum, which fix liver to adjacent structures)
  - **Falciform ligament**: Extends from anterior surface of liver to anterior abdominal wall
    - Separates right and left **subphrenic peritoneal recesses** (between liver and diaphragm)
    - Marks plane separating medial and lateral segments of left hepatic lobe

- Carries **round ligament** (**ligamentum teres**), fibrous remnant of obliterated umbilical vein
  - **Coronary ligaments**: Connect superior surface of liver to diaphragm (formed by anterior and posterior layers of triangular ligaments uniting together)
    - Bare area of liver located between 2 leaves of coronary ligament (to right of IVC)
    - Posterior layer of right coronary ligament a.k.a. hepatorenal ligament
  - **Triangular ligaments**: Connect superior surface of liver to diaphragm
  - **Lesser omentum**: Hepatoduodenal ligament and hepatogastric ligament (2 constituents of lesser omentum) connect liver with duodenum and stomach, respectively
  - Liver also fixed by fibrous tissue to IVC posteriorly
- Perihepatic recesses (spaces surrounding liver where fluid can collect)
  - **Morison pouch (hepatorenal recess)**: Most dependent portion of peritoneal cavity in supine patient between right hepatic lobe and kidney
  - **Subphrenic space**: Space between liver and diaphragm divided into right and left subphrenic spaces by falciform ligament
  - **Subhepatic space**: Space between undersurface of liver and adjacent transverse colon
- Nervous supply
  - Sympathetic and parasympathetic supply via hepatic nerve plexus (which runs in portal triads)
    - Sympathetics via celiac plexus and parasympathetics via anterior and posterior vagal trunks
- Normal multimodality imaging appearance
  - CT
    - Normal liver parenchyma on NECT is ~ 10 HU > spleen (important relationship when trying to diagnose hepatic steatosis)
    - Liver parenchyma on CECT far more variable making determination of "normal" vs. steatosis much more difficult
    - No absolute threshold for hyperdense liver (i.e., hemochromatosis, amiodarone effect, etc.), but liver often > 80 HU on NECT or markedly hyperdense compared to spleen
  - US
    - Normal liver should be equal or slightly echogenic relative to adjacent right kidney
    - Measurement from liver dome to right hepatic lobe tip often used in assessment of hepatomegaly with 15.5 cm used as threshold for hepatomegaly (questionable validity)
    - Liver surface/capsule should be smooth and regular when viewed using high-frequency transducer with surface nodularity suggesting cirrhosis/liver fibrosis
    - Portal veins have thick, echogenic wall, as opposed to thin wall of hepatic veins
  - MR
    - Normal liver slightly hyperintense on T1WI to muscle and kidneys
    - Normal liver hypointense to spleen, kidneys, and pancreas on T2WI and DWI
- Embryology
  - Liver develops as ventral diverticulum from junction of foregut and midgut
    - Biliary tree and gallbladder also share this origin from ventral diverticulum

CLINICAL IMPLICATIONS

Clinical Importance

- Advances in hepatic surgery (tumor resection, transplantation) make it essential to depict lobar and segmental anatomy, volume, blood supply, and biliary drainage as accurately as possible
  - Combination of standard multiplanar imaging and 3D reconstructions (CT, MR, US) may be necessary for optimal preoperative planning
  - "Invasive" imaging studies (catheter angiography and percutaneous transhepatic or endoscopic cholangiography) are now usually not necessary due to image quality of CT and MR/MRCP
  - Segmental Couinaud anatomy is critical for guiding surgical treatment with variety of surgeries (segmentectomy, bisegmentectomy, hepatectomy, extended right/left hepatectomy) based on segmental anatomy
    - Nonanatomic liver resections not adhering to segmental anatomy termed wedge resection
- **Liver metastases are common** and much more common that primary liver malignancies
  - Primary carcinomas of gastrointestinal tract, such as colon, pancreas, and stomach malignancies are most common
  - Portal venous drainage usually results in liver being initial site of metastatic spread from these tumors
  - Arterial-phase imaging critical when evaluating liver for potential metastases from hypervascular primary tumors (e.g., pancreatic neuroendocrine tumor, carcinoid, renal cell carcinoma, etc.), as small metastases may be virtually invisible on portal venous phase
- Most common primary liver malignancy is **hepatocellular carcinoma** (a.k.a. **hepatoma**)
  - Very common malignancy worldwide and most commonly seen in patients with underlying liver cirrhosis as result of chronic viral hepatitis or alcohol abuse
  - Primary liver malignancies (particularly hepatocellular carcinoma) can invade hepatic vasculature (especially portal venous system)
  - Arterial-phase imaging critical for detection of small tumors, and MR generally considered superior to CT for assessment of cirrhotic liver
- Benign liver lesions are incredibly common and, in most instances, can be easily distinguished from malignant lesions based on characteristic enhancement patterns on multiphase imaging
  - Most common benign liver lesions include cysts, hemangioma, focal nodular hyperplasia, and hepatic adenoma
  - Focal fat or focal fatty sparing are common confused for focal liver "masses" and can be easily diagnosed in difficult cases using chemical-shift MR

HEPATIC VISCERAL SURFACE

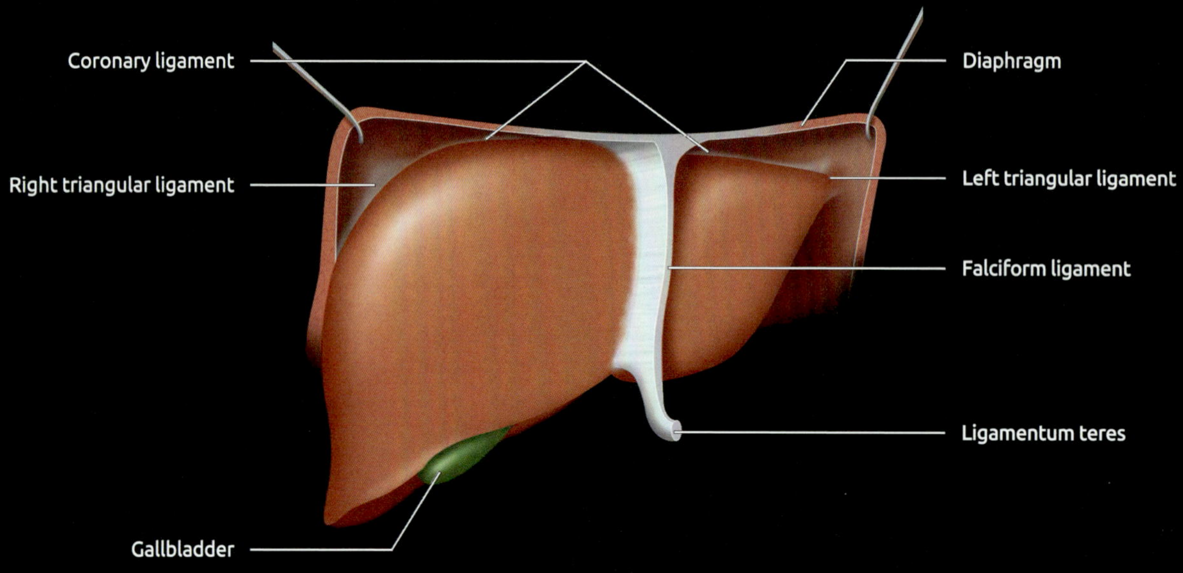

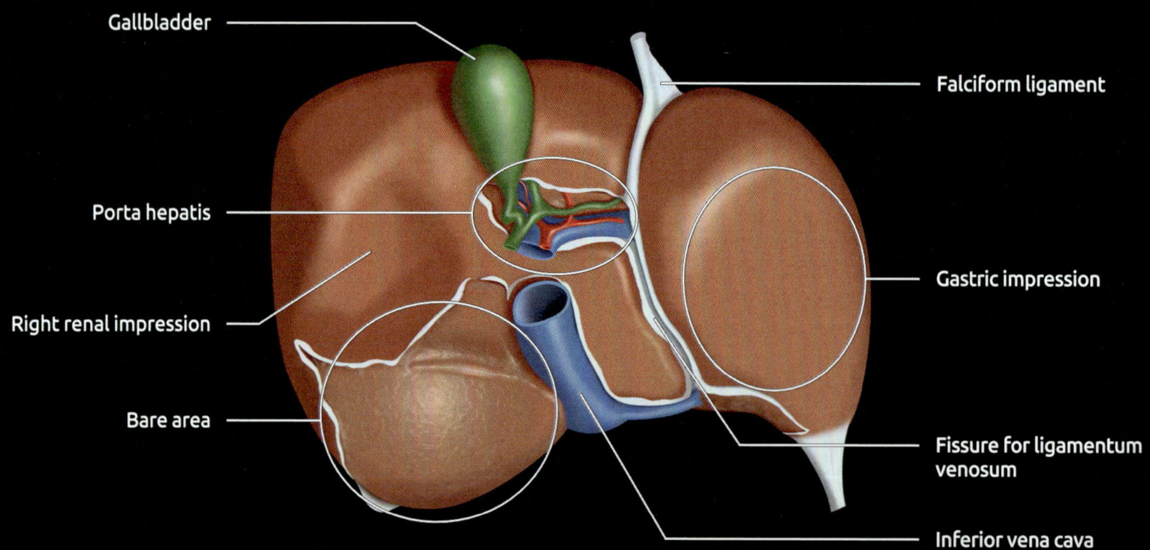

(Top) *The anterior surface of the liver is smooth and molds to the diaphragm and anterior abdominal wall. Generally, only the anterior/inferior edge of the liver is palpable on physical exam. The liver is covered with peritoneum, except for the gallbladder bed, porta hepatis, and the bare area. Peritoneal reflections form various ligaments that connect the liver to the diaphragm and abdominal wall, including the falciform ligament, the inferior edge of which contains the ligamentum teres, and the obliterated remnant of the umbilical vein.* **(Bottom)** *Graphic shows the liver inverted, somewhat similar to the surgeon's view of the upwardly retracted liver. The structures in the porta hepatis include the portal vein (blue), hepatic artery (red), and the bile ducts (green). The visceral surface of the liver is indented by adjacent viscera. The bare area is not easily accessible.*

HEPATIC ATTACHMENTS AND RELATIONS

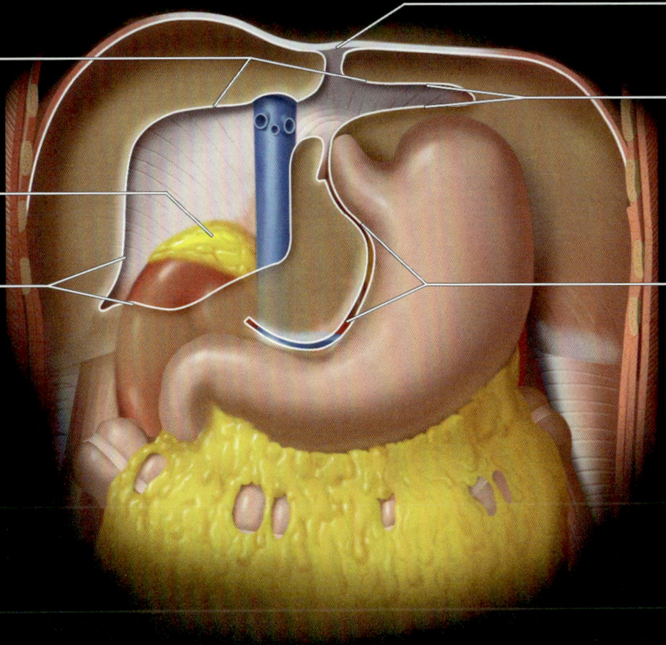

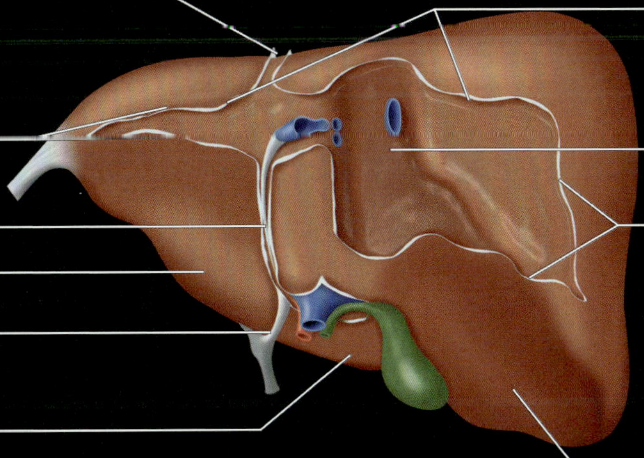

(Top) *The liver is attached to the posterior abdominal wall and diaphragm by the left and right triangular ligaments and the coronary ligaments. The falciform ligament attaches the liver to the anterior abdominal wall. The bare area is in direct contact with the right adrenal gland, right kidney, and inferior vena cava (IVC).* **(Bottom)** *Posterior view of the liver shows the ligamentous attachments. While these may help to fix the liver in position, abdominal pressure alone is sufficient to keep the liver in place, as evidenced by orthotopic liver transplantation, after which the ligamentous attachments are lost without the liver shifting position. The diaphragmatic peritoneal reflection is the coronary ligament whose lateral extensions are the right and left triangular ligaments. The falciform ligament separates the medial and lateral segments of the left lobe.*

HEPATIC VESSELS AND BILE DUCTS

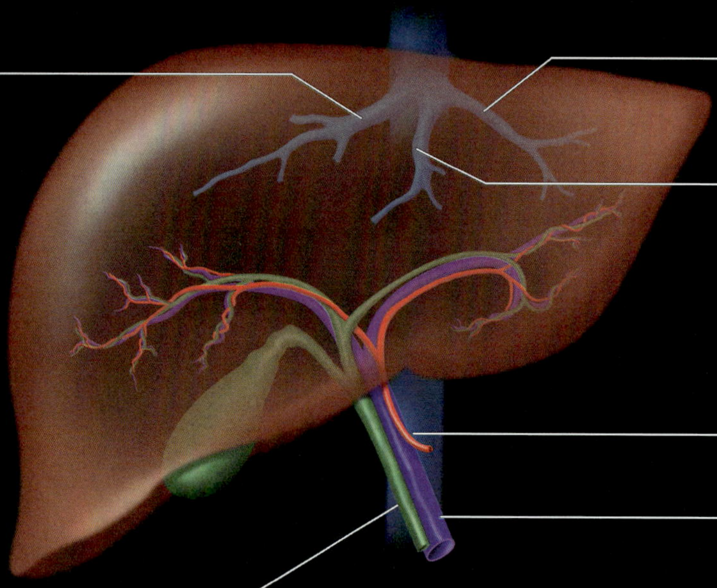

Graphic emphasizes that at every level of branching and subdivision, the portal veins, hepatic arteries, and bile ducts course together, constituting the "portal triad." Each segment of the liver is supplied by branches of these vessels. Conversely, hepatic venous branches lie between hepatic segments and interdigitate with the portal triads but never run parallel to them.

AXIAL CT, NORMAL LIVER

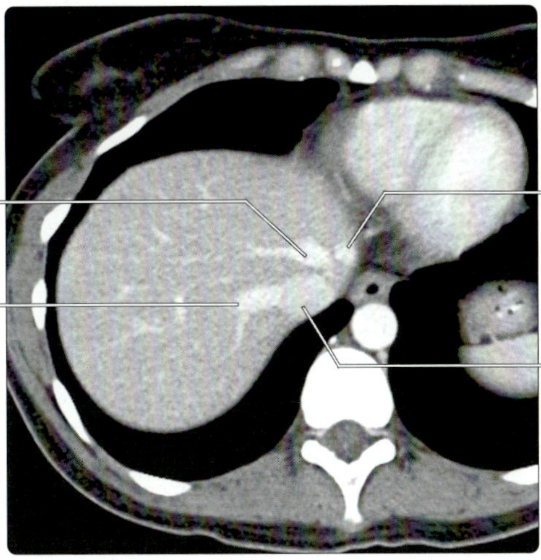

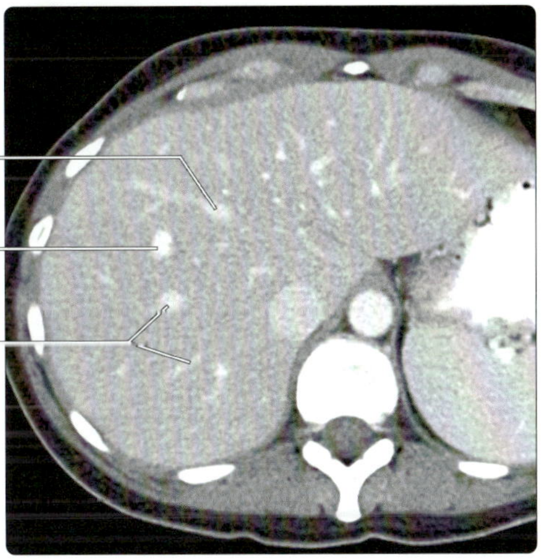

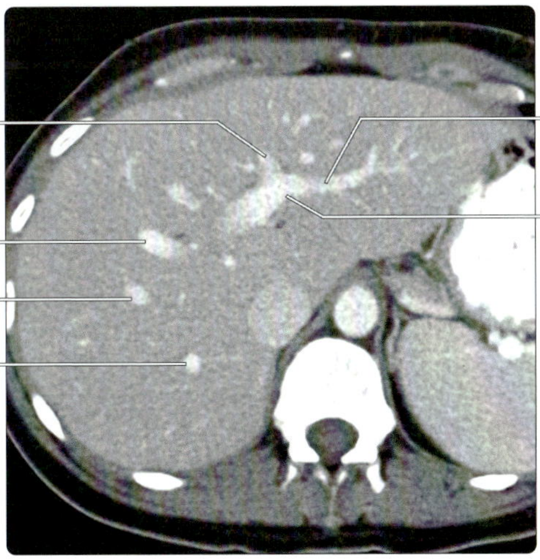

(Top) *First of 9 axial CT images shows the confluence of the hepatic veins with the IVC just below the diaphragm and the entrance of the IVC into the right atrium.* **(Middle)** *At this level, the portal veins run in a predominantly cephalocaudal direction and bisect the angle made by the hepatic veins. Portal veins generally lie within hepatic segments, while hepatic veins lie between segments.* **(Bottom)** *The horizontal plane defined by the left portal vein divides the lateral segment into segment 2 (above the vein) and 3 (below the vein).*

AXIAL CT, NORMAL LIVER

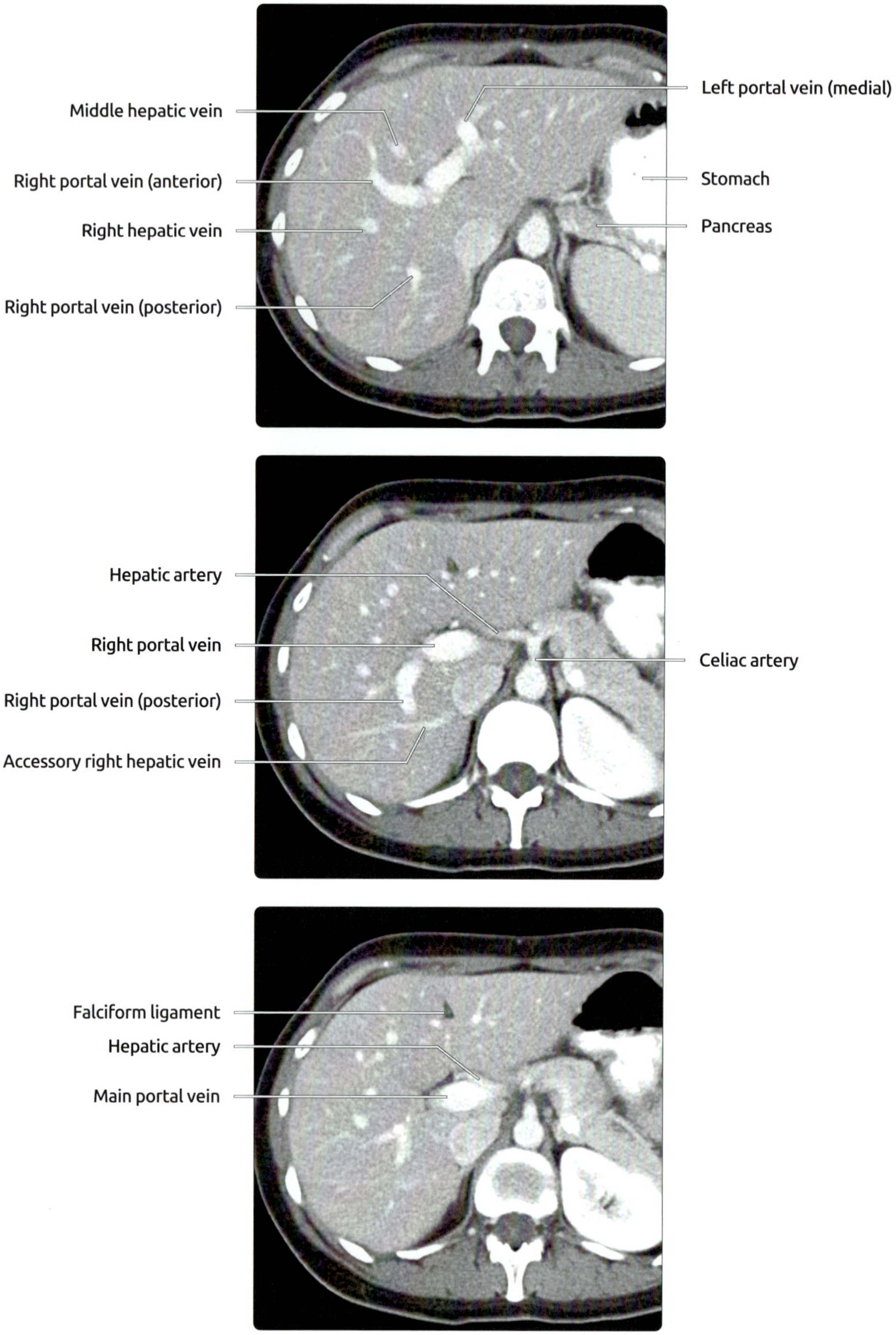

(Top) *The plane of section through the long axis of the right portal vein divides segments 7 and 8 (above) from segments 5 and 6 (below). Note the relationships between the liver, stomach, and pancreas.* **(Middle)** *The hepatic artery arises conventionally in this subject from the celiac artery. There is an accessory right hepatic vein that drains directly into the IVC caudal to the confluence of the other hepatic veins.* **(Bottom)** *Note the fissure for the falciform ligament, which separates the medial and lateral segments of the liver (segment 4 from segments 2 and 3).*

AXIAL CT, NORMAL LIVER

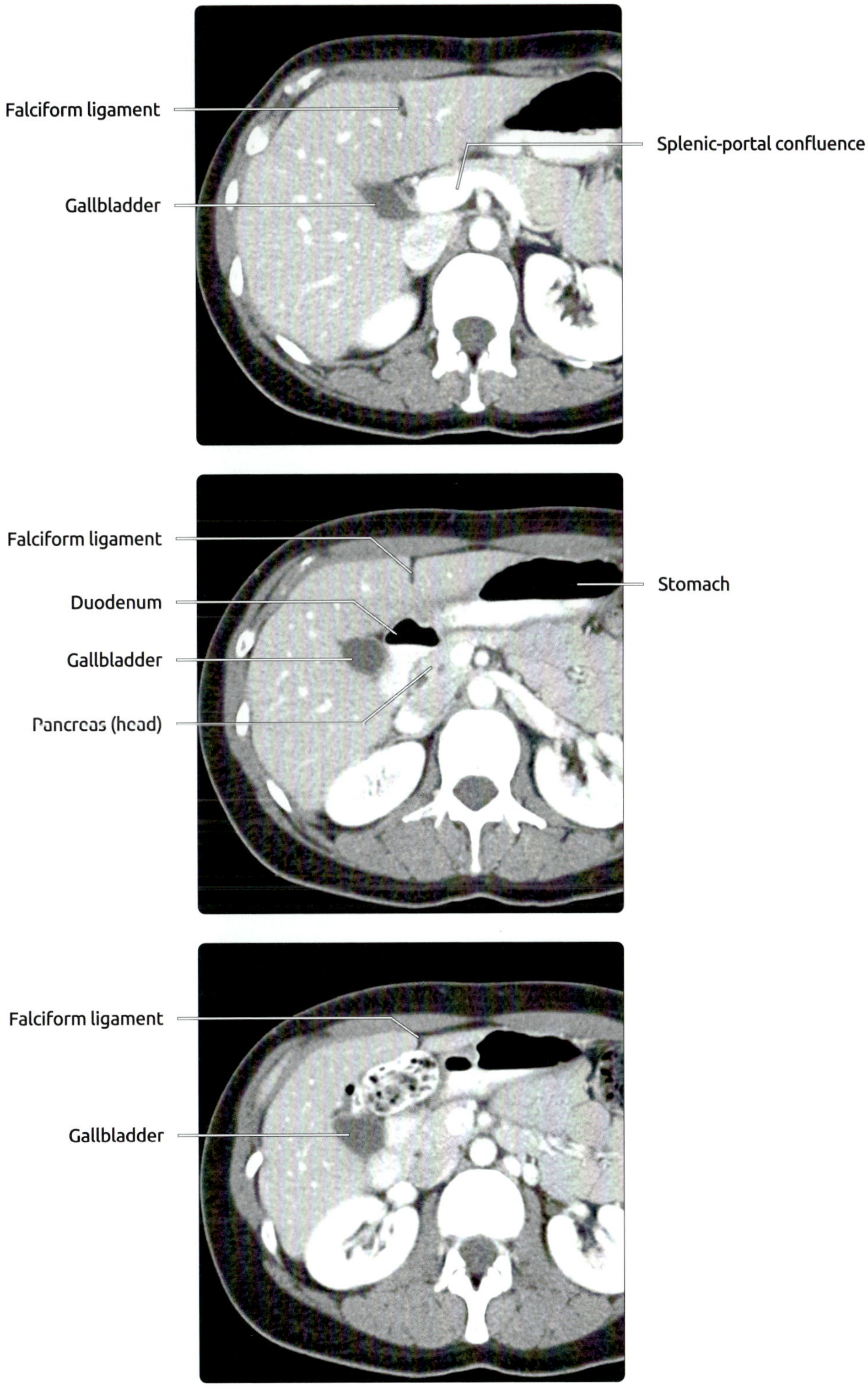

(Top) *A vertical plane through the falciform ligament separates the medial and lateral segments of the left lobe, while a vertical plane through the gallbladder fossa and middle hepatic vein separates the right and left lobes.* **(Middle)** *Note the relationships between the liver and adjacent organs.* **(Bottom)** *The right lobe of the liver extends much more caudally than the left, occasionally, even into the pelvis. The considerable variability of the shape of the liver makes it difficult to identify hepatomegaly by physical examination alone, especially since only the ventral, inferior edge of the liver can be palpated.*

PORTAL VENOUS SYSTEM

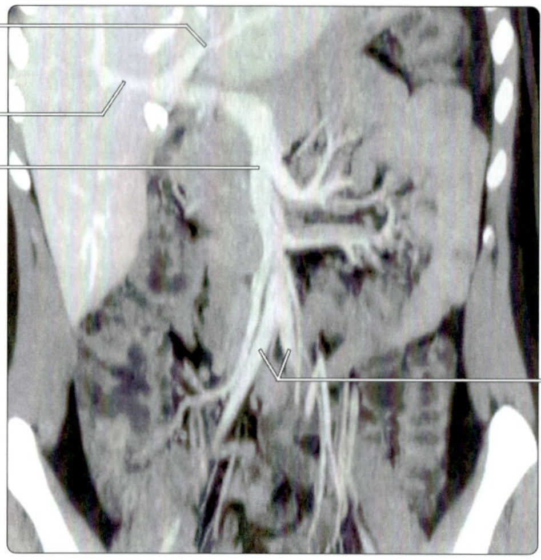

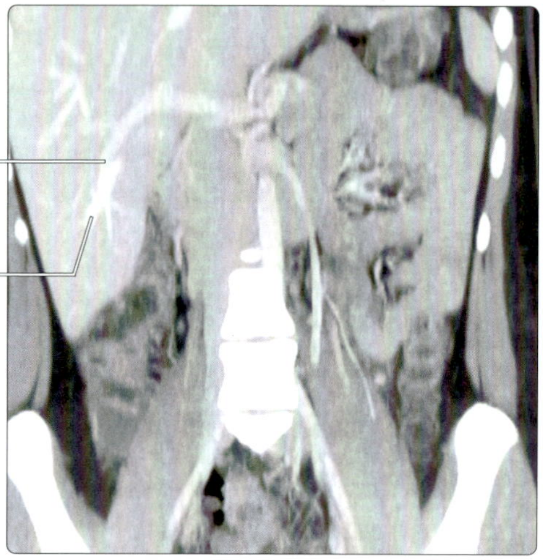

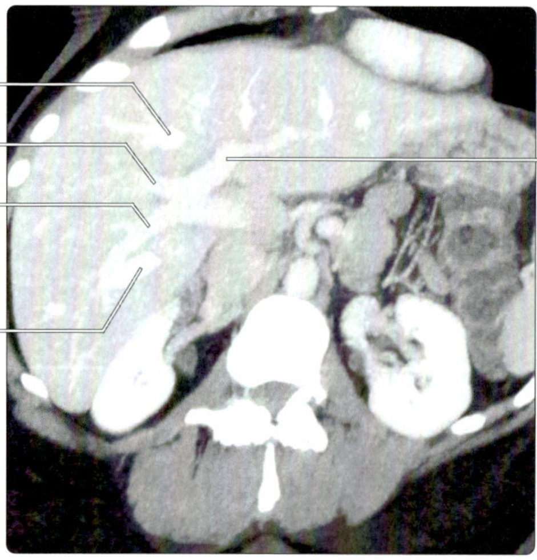

(Top) *First of 3 CT images emphasizes the portal venous system. This coronal reformation of a CT in the venous phase of imaging shows the superior mesenteric vein and its branches as well as the portal vein. The (arbitrary) thickness and plane of the section exclude the splenic vein and include the iliac arteries.* **(Middle)** *Coronal plane through the posterior segments of the liver shows the hepatic artery and portal vein branches to segment 6.* **(Bottom)** *Thick reconstructed axial section shows a trifurcation pattern of the portal vein in which the left portal vein and the anterior and posterior branches of the right portal vein all arise directly from the main portal vein.*

LIVER SEGMENTAL ANATOMY

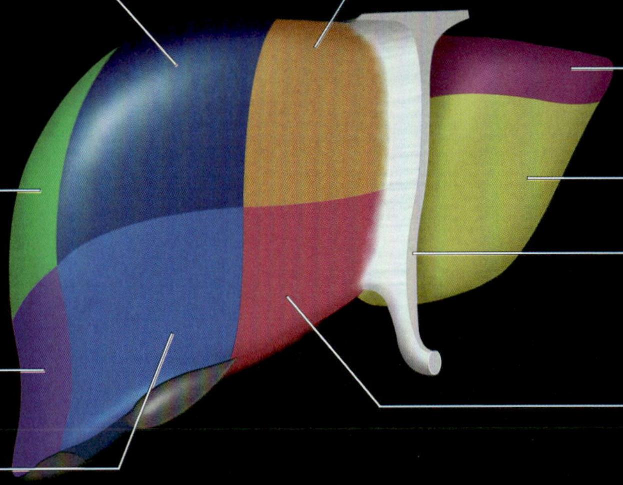

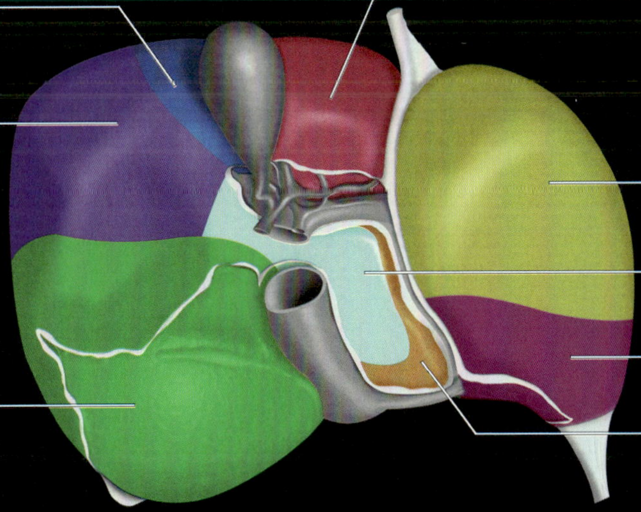

(Top) *First of 2 graphics demonstrates the segmental anatomy of the liver in a somewhat idealized fashion. Segments are numbered in a clockwise direction starting with the caudate lobe (segment I), which cannot be seen on this frontal view. The falciform ligament divides the lateral (segments II and III) from the medial (segments IVa and IVb) left lobe. The horizontal planes separating the superior from the inferior segments follow the course of the right and left portal veins. An oblique vertical plane through the middle hepatic vein, gallbladder fossa, and IVC divides the right and left lobes.* **(Bottom)** *Posterior view of the liver shows that the caudate is entirely posterior, abutting the IVC, ligamentum venosum, and porta hepatis. A plane through the IVC and gallbladder divides the left and right lobes.*

LIVER SEGMENTAL ANATOMY

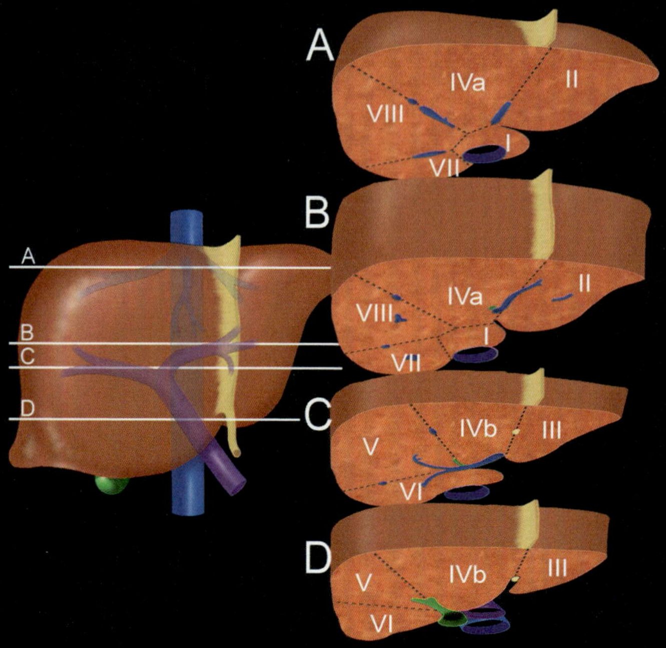

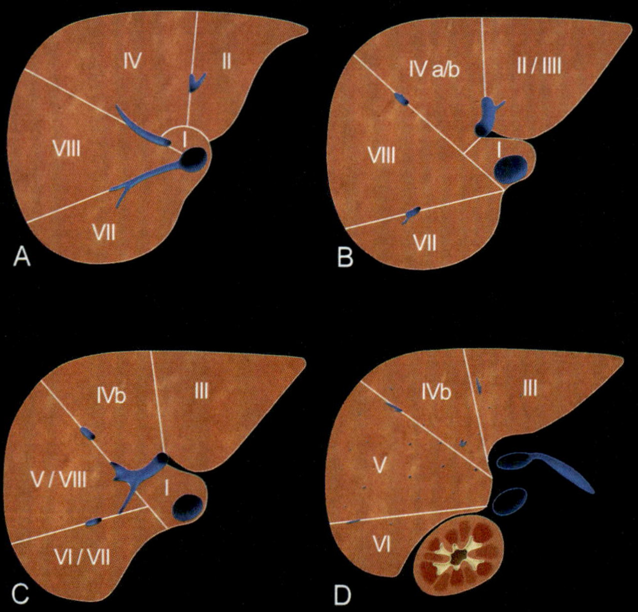

(Top) *Graphic demonstrates the division of the Couinaud segments of the liver at 4 different levels of the liver. The Couinaud segments are defined by the hepatic veins (hepatic vein plane) and portal veins (portal vein plane).* **(Bottom)** *Another graphic focuses on the segments at each of the 4 levels shown in the previous graphic with graphic A being the most inferior (at the level of the portal-splenic confluence) and graphic B being the most superior (at the level of the IVC-hepatic vein confluence). The left hepatic vein divides segments II and III from IVa and IVb, which the plane of the left portal vein divides the superior segments of the left lobe (II and IVa) from the inferior segments (III and IVb). The middle hepatic vein divides segments IVa/b from V and VIII, while the right hepatic vein divides segments V and VIII from VI and VII. The right portal vein divides the superior segments of the right lobe (VII and VIII) from the inferior segments (V and VI).*

LIVER SEGMENTAL ANATOMY

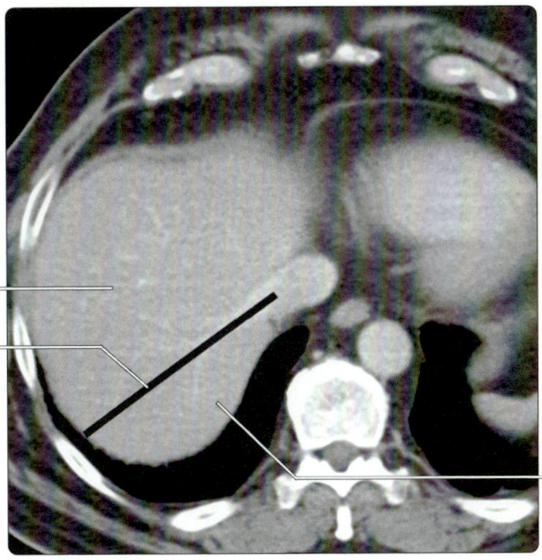

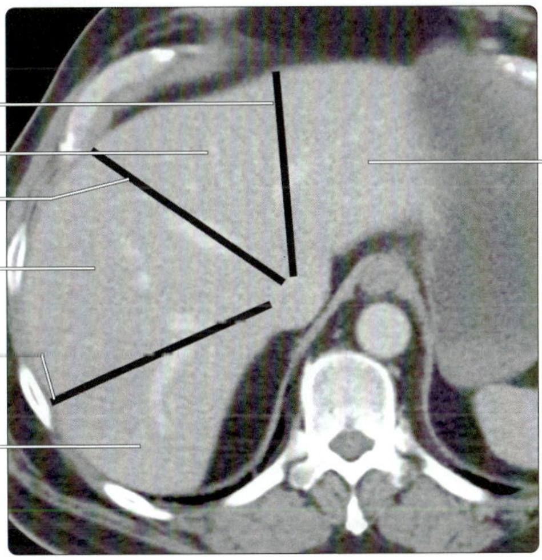

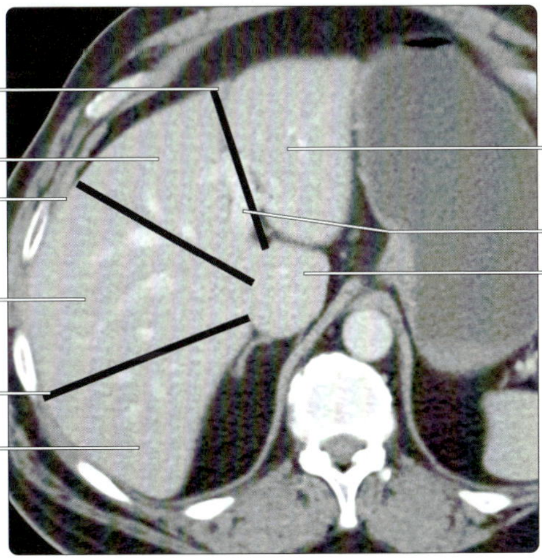

(Top) *Axial CECT through the superior liver demonstrates the right hepatic vein coursing through the right lobe, dividing segments VII and VIII. The right hepatic vein plane also divides segments V and VI below the level of the portal veins.* **(Middle)** *A slightly more caudal image (still above the portal vein plane) demonstrates the left hepatic vein plane (line extrapolated through the plane of the left hepatic vein) dividing segments II and IVa of the liver, the middle hepatic vein plane (line extrapolated through the middle hepatic vein) dividing segments IVa and VIII, and the right hepatic vein plane (line extrapolated through the middle hepatic vein) dividing segments VII and VIII.* **(Bottom)** *More caudal level now demonstrates the portal veins coming into view, which divide the liver into superior and inferior segments. The portal vein plane demarcates the division between segments II and III, IVa and IVb, V and VIII, and VI and VII.*

LIVER SEGMENTAL ANATOMY

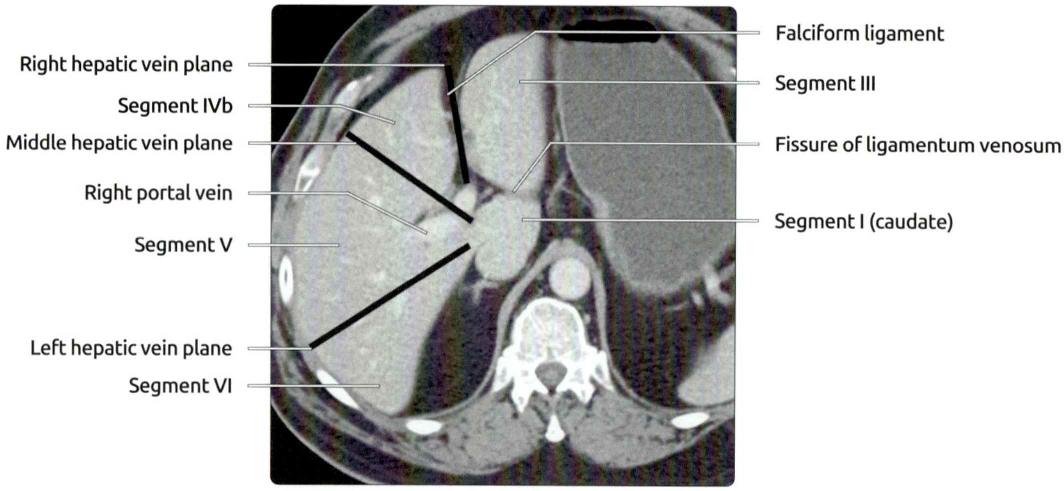

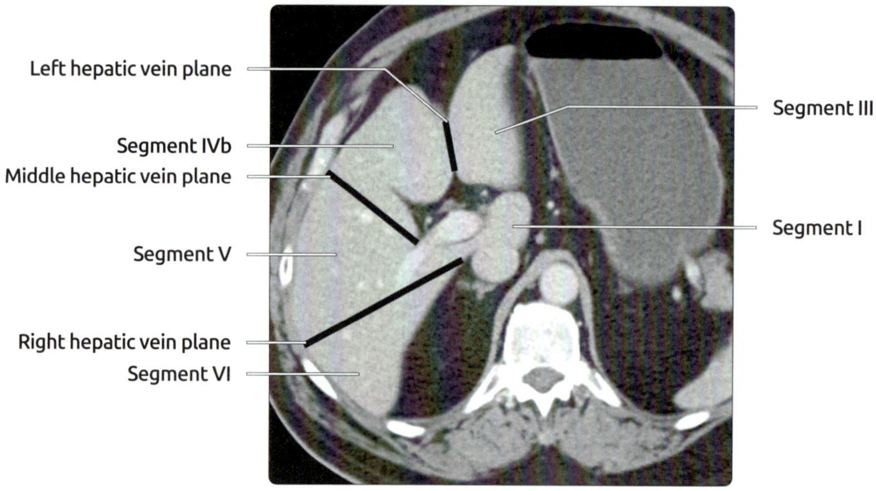

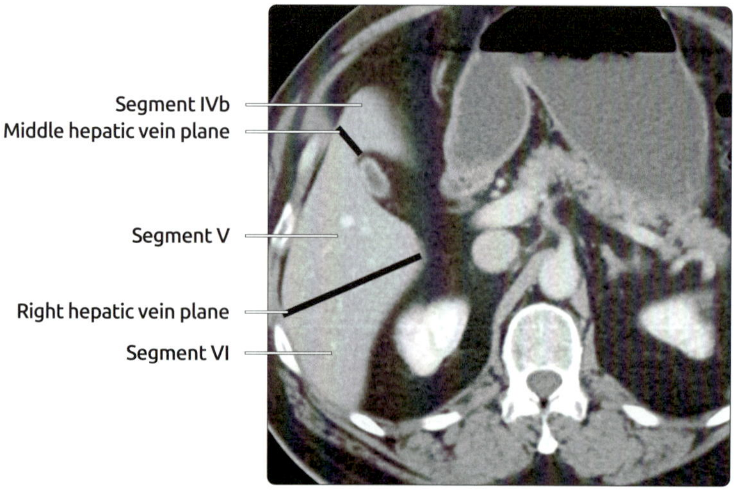

(Top) *Axial CECT below the portal vein plane now demonstrates segments III, IVb, VI, and VII. Note the location of the caudate lobe surrounding the IVC and posterior to the fissure of the ligamentum venosum.* **(Middle)** *Axial CECT more inferiorly continues to demonstrate the inferior segments of the liver, including III, IVb, V, and VI.* **(Bottom)** *Final image through the inferior liver now shows that most of the left hepatic lobe has gone out of view in this patient, leaving only the inferior segments of the right hepatic lobe (V and VI) defined by the middle and right hepatic veins.*

FALCIFORM LIGAMENT

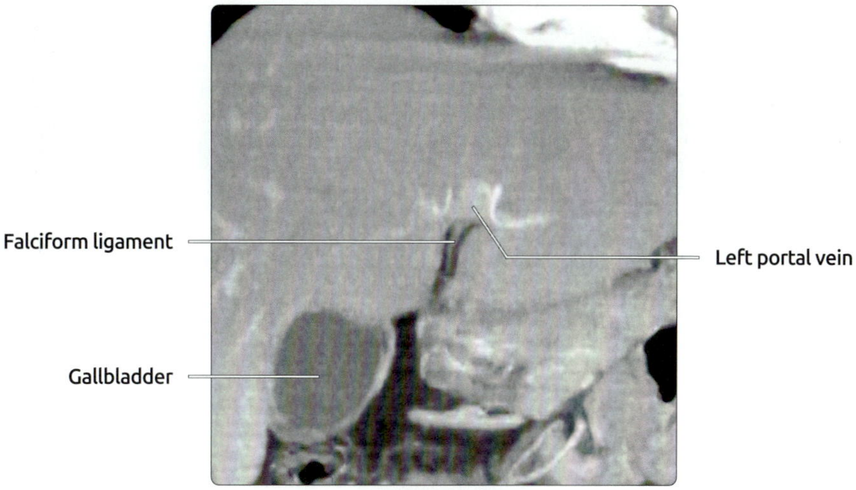

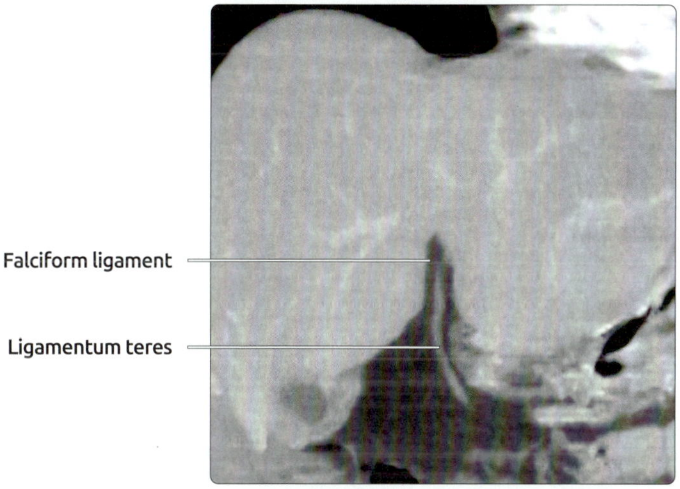

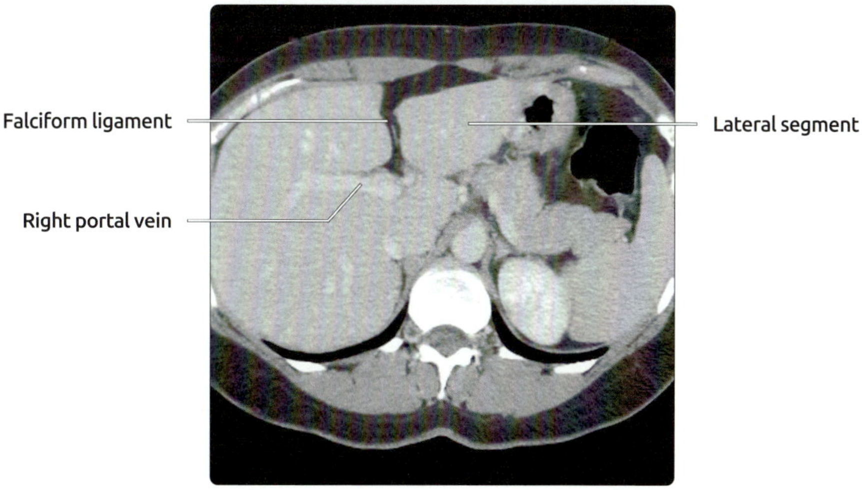

(Top) *First of 3 CECT images shows the falciform ligament in various planes. Coronal reformation through the middle of the liver shows the falciform ligament dividing the medial and lateral segments of the left lobe. The ascending portion of the left portal vein also lies in the falciform ligament.* **(Middle)** *A more anterior coronal section shows the ligamentum teres, the obliterated remnant of the umbilical vein, which lies in the falciform ligament before leaving the anteroinferior surface of the liver to extend to the umbilicus.* **(Bottom)** *This axial CECT section shows the ligamentum teres within the falciform ligament.*

HEPATIC ARTERIAL ANATOMY OVERVIEW

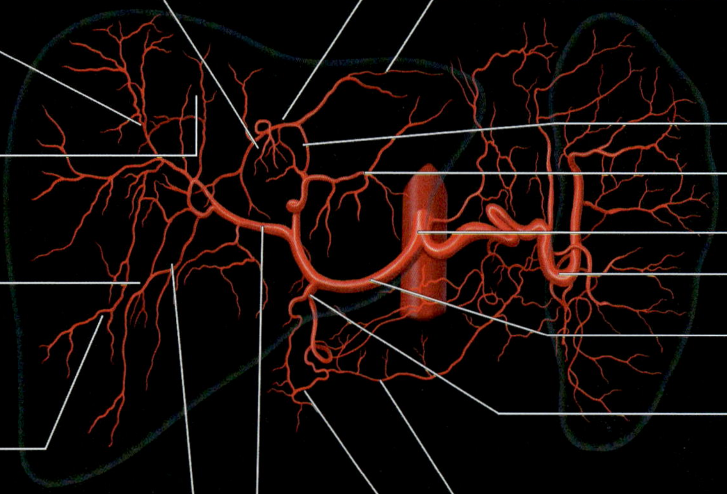

Graphic demonstrates the conventional hepatic arterial supply to the liver. The celiac artery arises at roughly the T12 level before dividing into the common hepatic artery, left gastric artery, and splenic artery. The common hepatic artery gives off the gastroduodenal artery inferiorly and becomes the proper hepatic artery, which then divides into the right and left hepatic arteries at the liver hilum. The left hepatic artery courses superiorly and slightly to the left before giving off branches to segments II-IV. In some instances, the segment IV artery may arise directly from the proper hepatic artery and is then termed the middle hepatic artery. The right hepatic artery divides into anterior and posterior branches, which take a upward vertical course and horizontal course, respectively. The anterior branch gives off arteries supplying segments V and VIII, while the posterior branches supplies segments VI and VII. Segment I (caudate) is typically supplied by small branches of the right or left hepatic arteries (or both).

ARTERIOGRAM, CELIAC AND HEPATIC ARTERIES

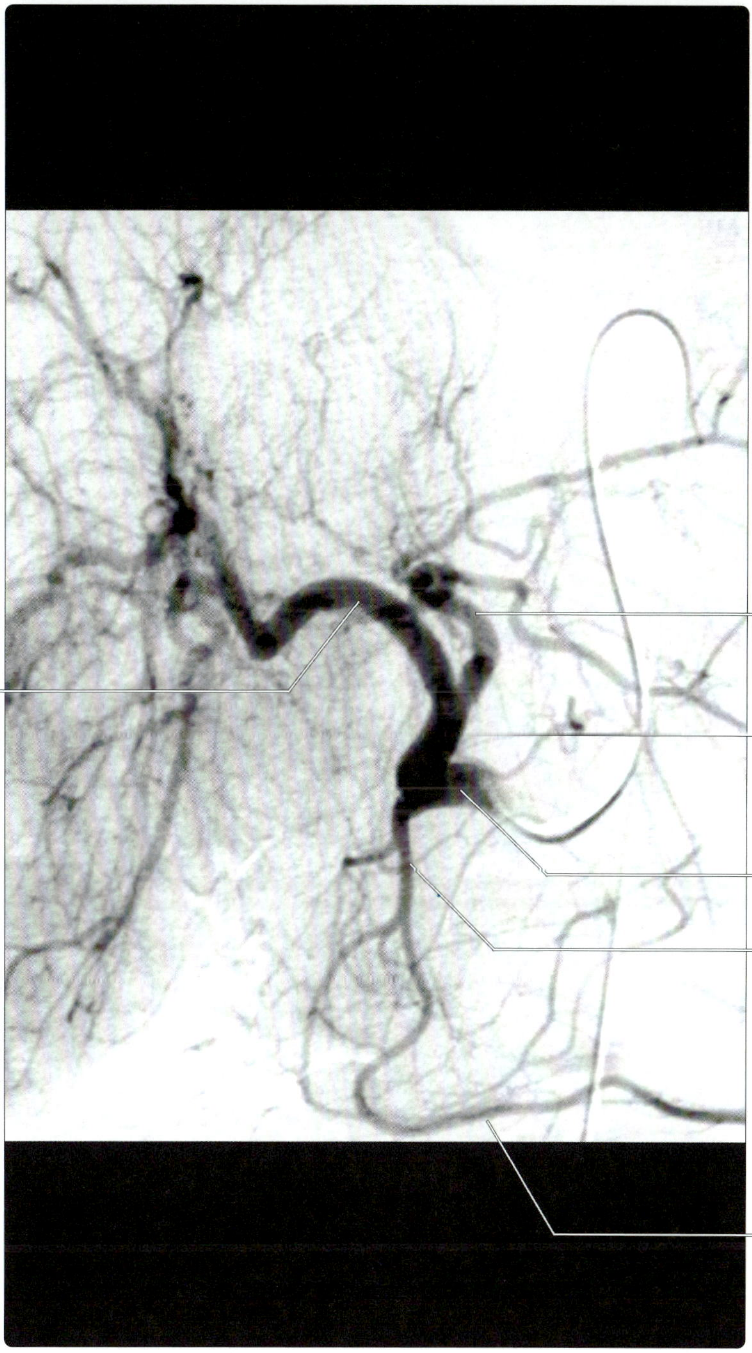

Selective catheterization of the common hepatic artery demonstrates conventional arterial anatomy in which all the hepatic arteries arise from the celiac axis. Variations in hepatic arterial supply are very common and important to recognize, especially if partial hepatic resection is being considered.

HEPATIC ARTERY ANGIOGRAM

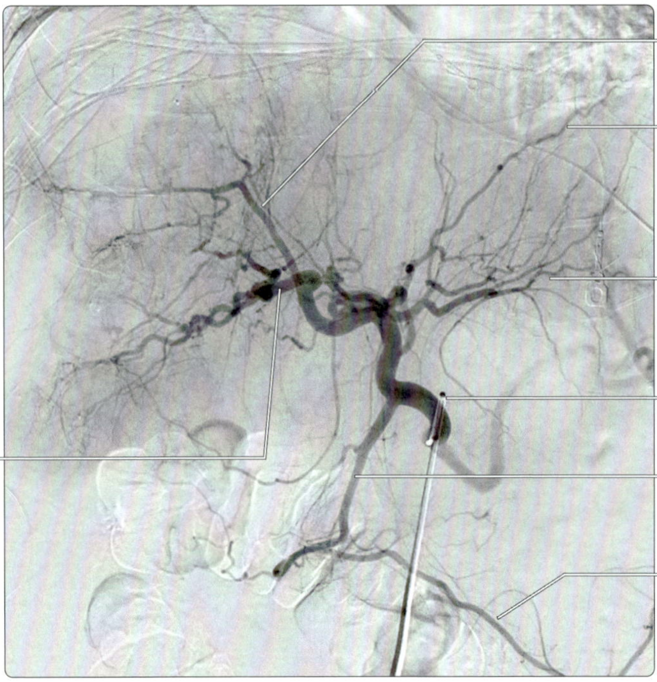

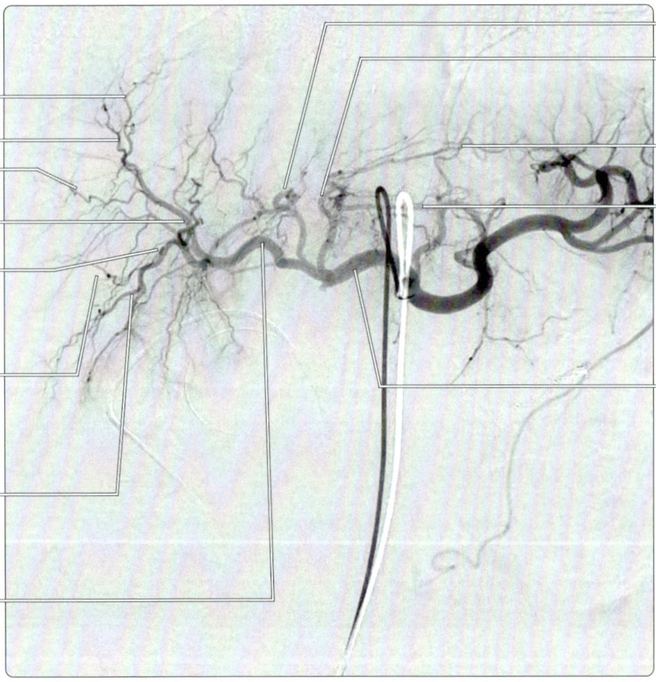

(Top) *Frontal angiogram demonstrates the typical orientations of the anterior and posterior divisions of the right hepatic artery, which parallel those of the bile ducts with the anterior division extending upward in a vertical fashion, while the posterior division extends laterally and horizontally. Also note the course of segmental branches supplying segments II and II arising from the left hepatic artery.* **(Bottom)** *Hepatic artery angiogram demonstrates the presence of a "middle" hepatic artery arising directly from the proper hepatic artery with a trifurcation configuration (right, left, and middle arising from proper hepatic artery). Segment IV can be supplied either from a segmental branch extending rightward from the left hepatic artery or from a middle hepatic artery directly arising from the proper hepatic artery. Also note the "arch" of the left hepatic artery, a characteristic turn of the left hepatic artery as it overrides the umbilical portion of the left portal vein and then gives off the branches to segments II and III to the left.*

CELIAC AND HEPATIC ARTERIES

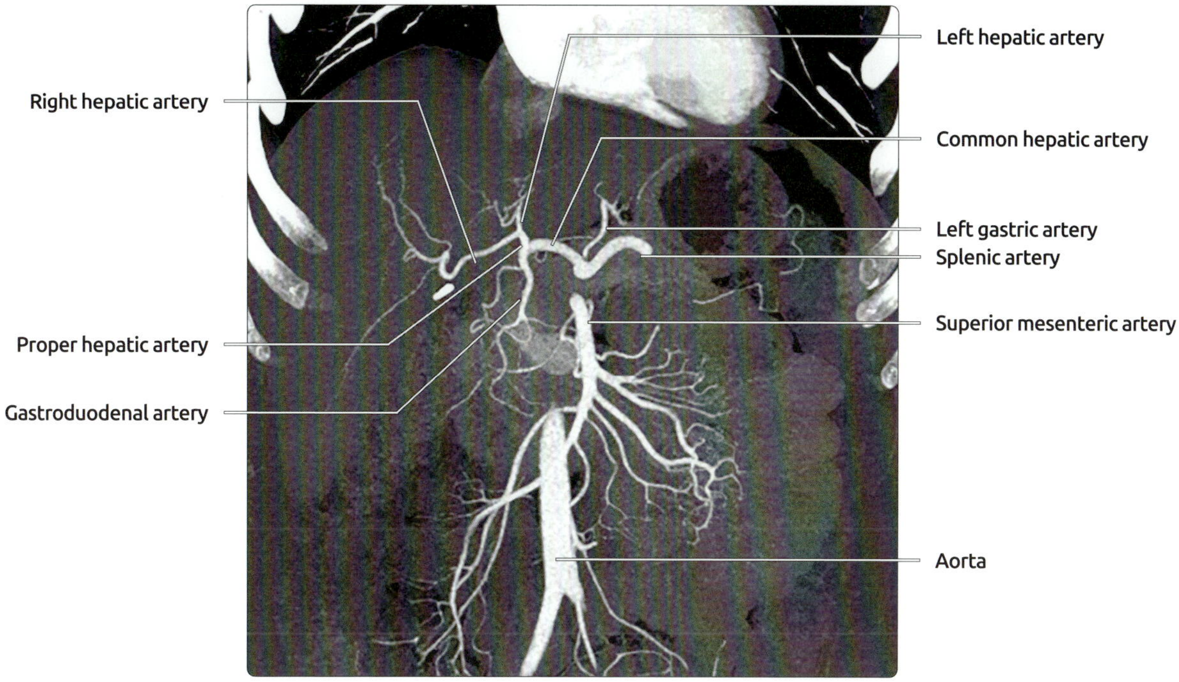

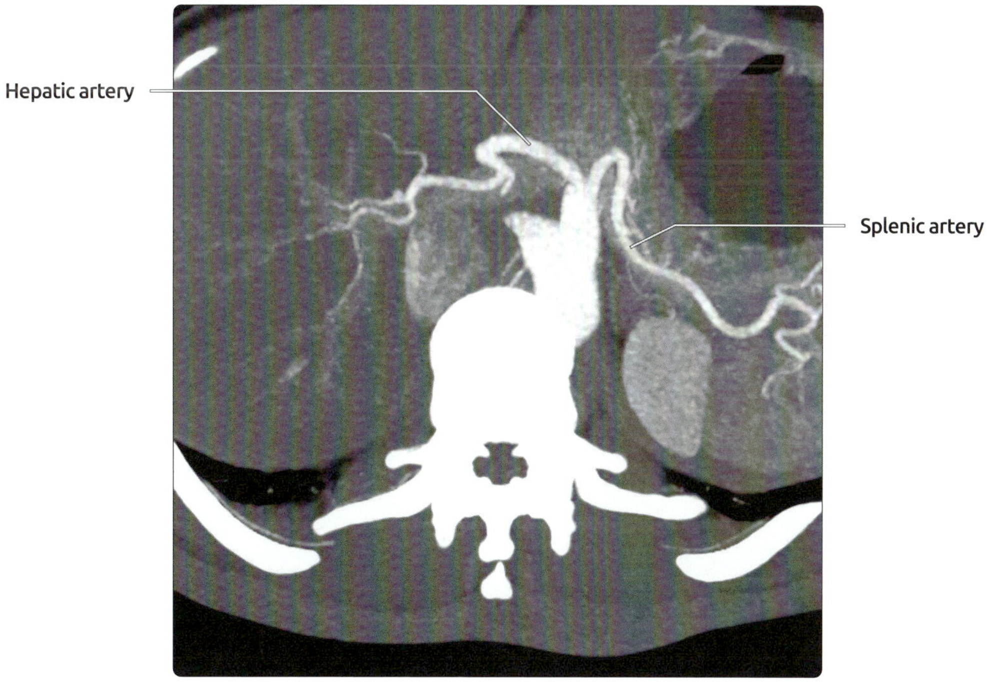

(Top) *Coronal CECT with MIP reconstruction demonstrates conventional hepatic arterial anatomy. This coronal reformation shows both hepatic arteries arising from the proper hepatic artery which, in turn, arises from the common hepatic artery.* **(Bottom)** *Axial CECT with MIP reconstruction shows the hepatic and splenic arteries arising from the celiac artery.*

HEPATIC AND PORTAL VEINS

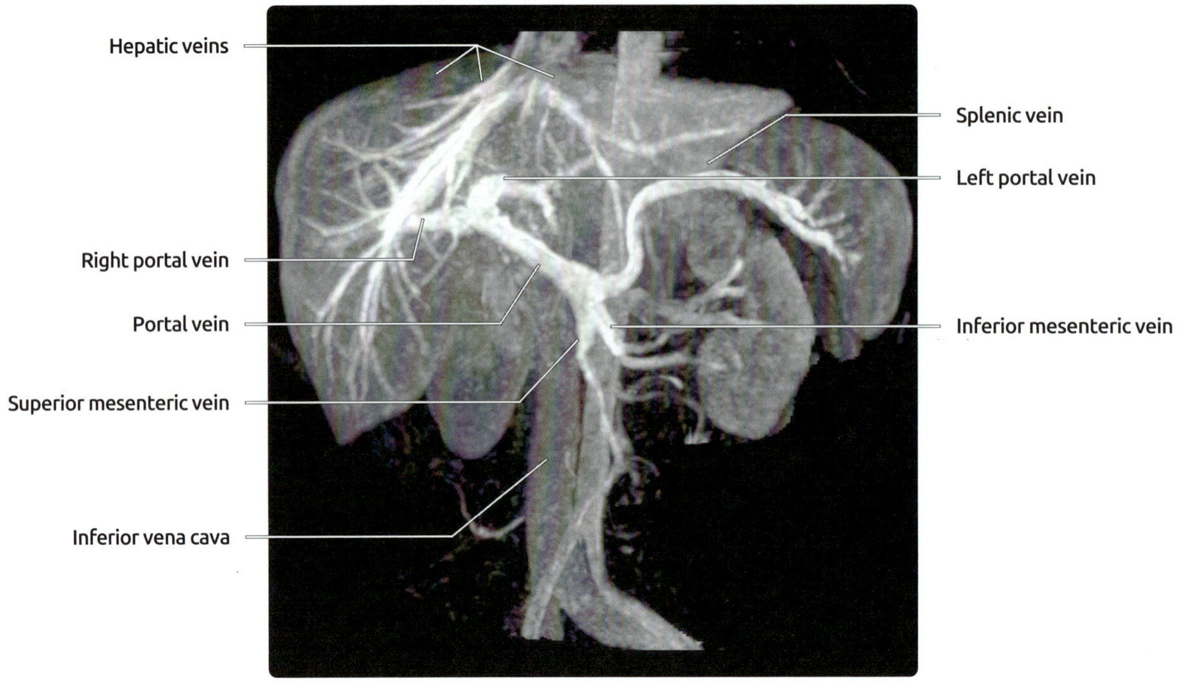

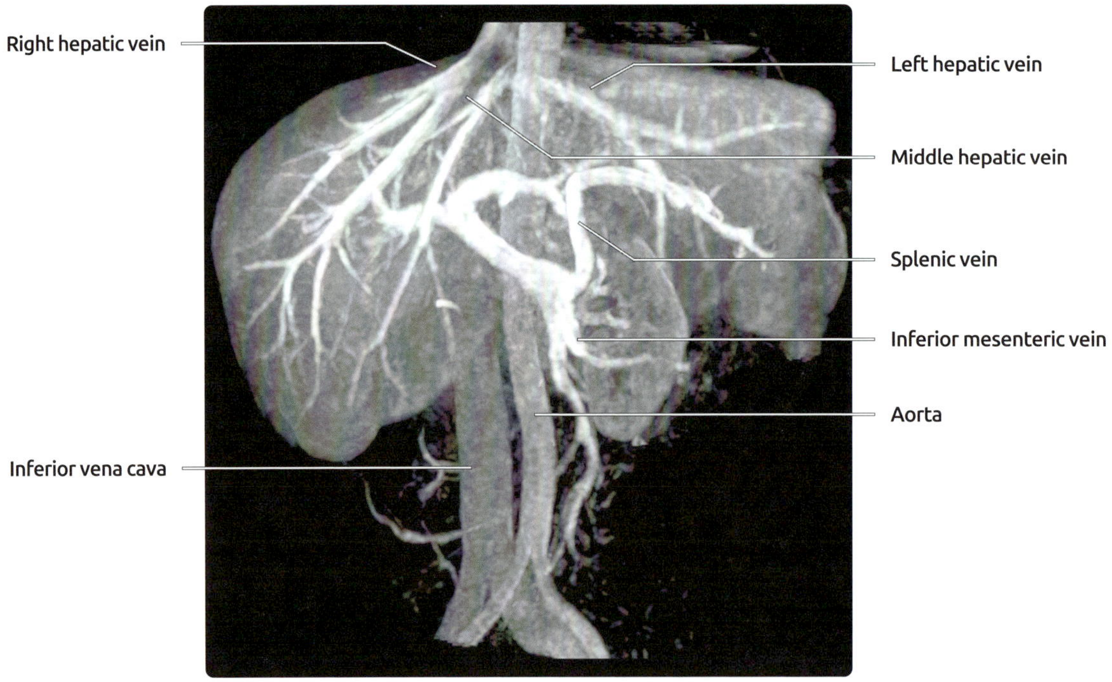

(Top) *First of 2 coronal MIP images from a contrast-enhanced MRA shows the major divisions and tributaries of the hepatic and portal veins. Only a few branches of the superior mesenteric vein are included in this plane of section, making it appear artifactually small.* **(Bottom)** *In this subject, the inferior mesenteric vein joins the confluence of the superior mesenteric and splenic veins. Some of the intravenous contrast medium is still circulating through the arteries, resulting in partial opacification of the aorta.*

MULTIPLANAR CT, NORMAL VESSELS

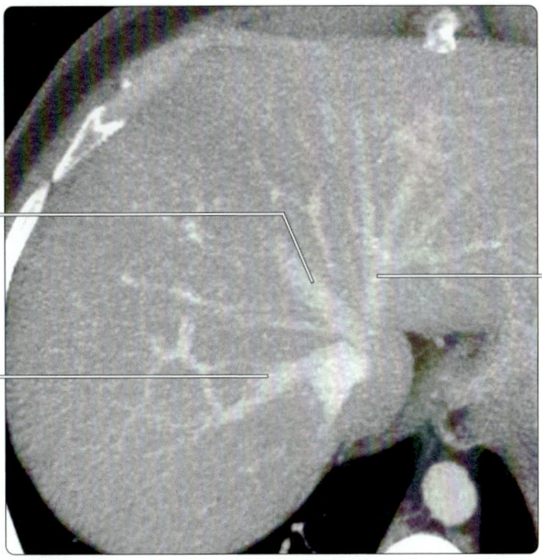

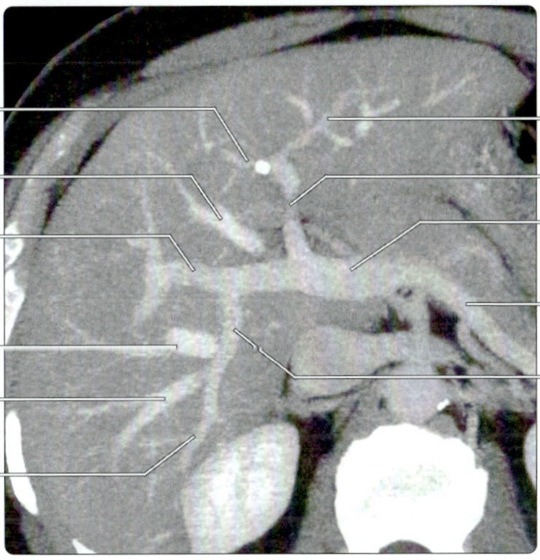

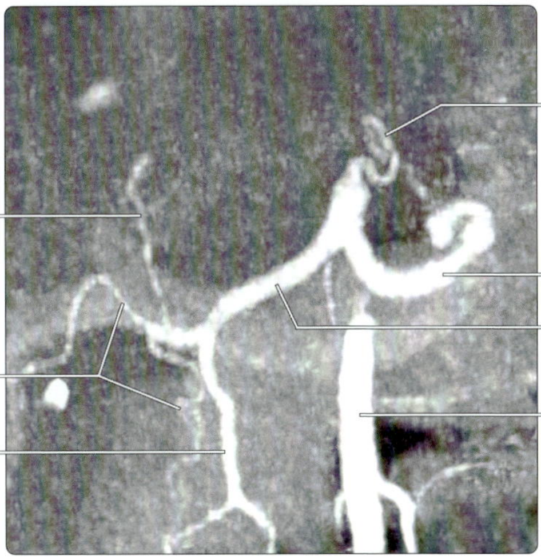

(Top) *First of 6 CECT sections through a normal liver is shown. This thick axial plane reconstruction shows the confluence of the hepatic veins.* **(Middle)** *Another thick axial reconstructed image shows the intrahepatic portal and hepatic veins.* **(Bottom)** *Coronal CECT with MIP reconstruction in the arterial phase shows the origins of the celiac and superior mesenteric arteries. The hepatic arteries arise conventionally from branches of the celiac trunk.*

MULTIPLANAR CT, NORMAL VESSELS

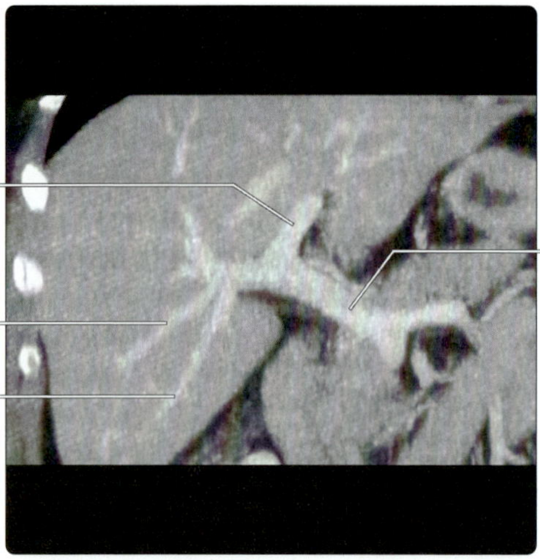

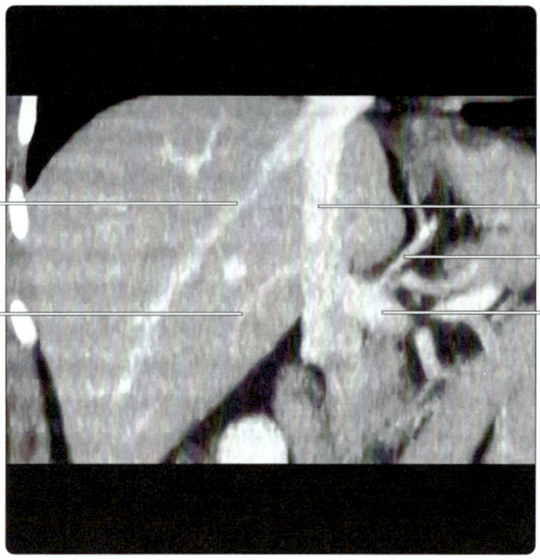

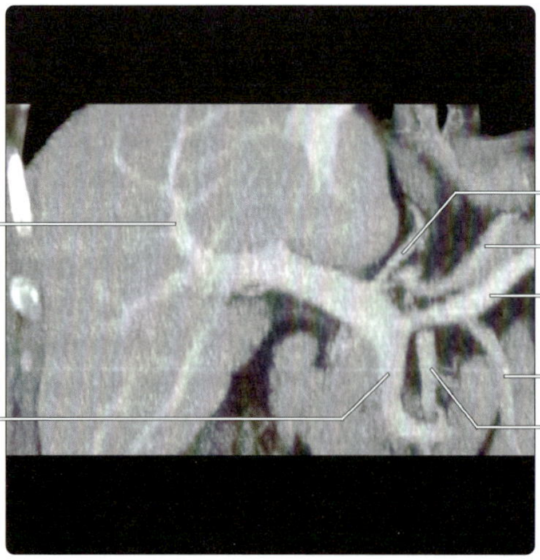

(Top) *A coronal image from the portal venous phase of enhancement shows the major branches of the portal vein.* **(Middle)** *Coronal plane image shows the entry of the right hepatic vein into the IVC. A small accessory right hepatic vein is also present. The thick plane of reconstruction artifactually suggests that the portal vein is entering the IVC.* **(Bottom)** *Coronal image includes the major tributaries of the portal vein, including the superior and inferior mesenteric veins and the splenic vein. Recirculation of contrast medium results in residual enhancement of some arteries that also lie in this plane of section.*

AXIAL T2 FS MR

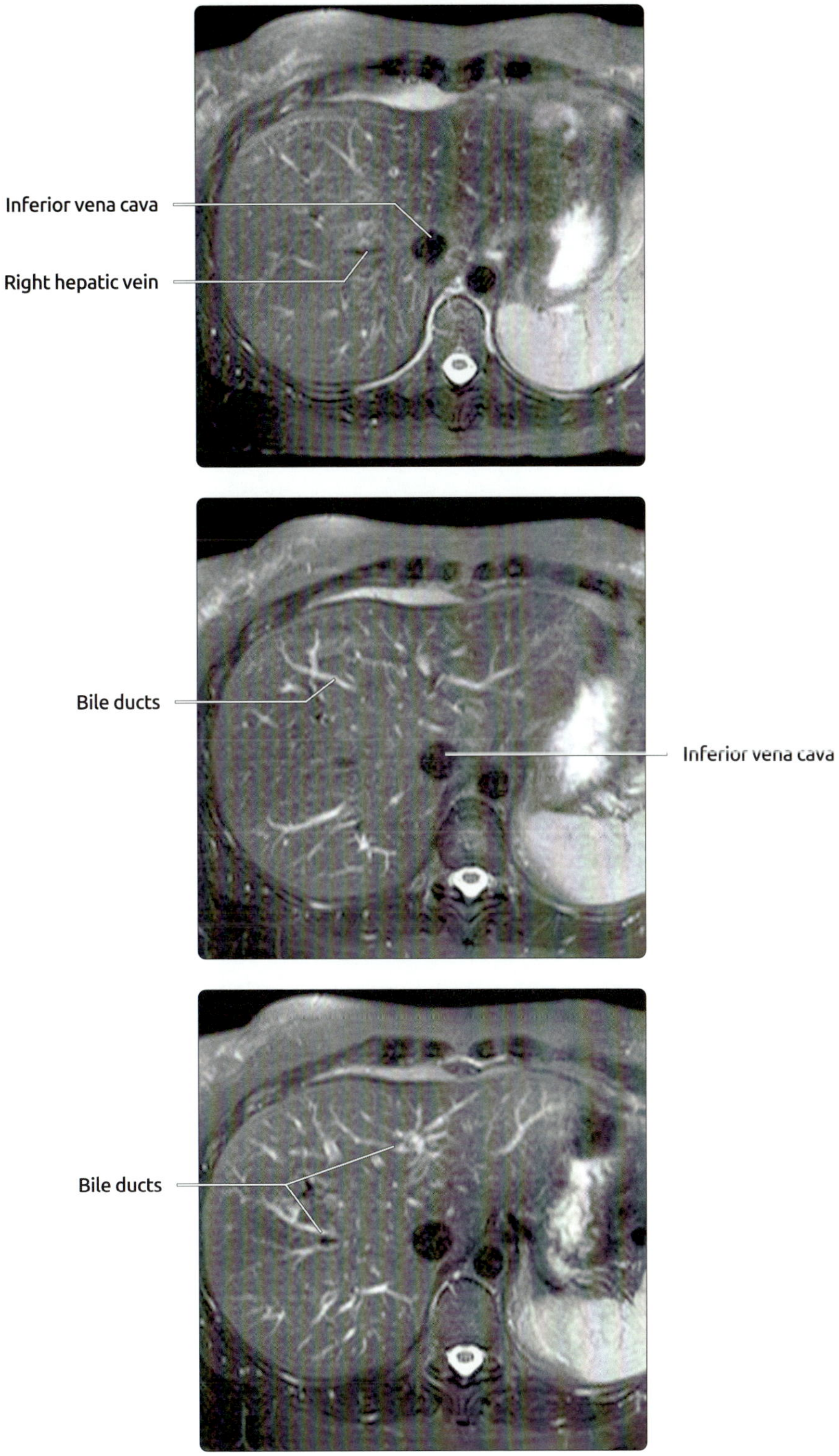

(Top) *First of 9 axial T2 fat-suppressed MR images shows relatively low signal from the liver parenchyma. Flowing blood appears very dark, while static fluid, such as bile and spinal fluid, appears quite bright.* **(Middle)** *The branching pattern of the intrahepatic bile ducts is evident.* **(Bottom)** *A more caudal section shows the bile ducts becoming larger as they approach the porta hepatis.*

AXIAL T2 FS MR

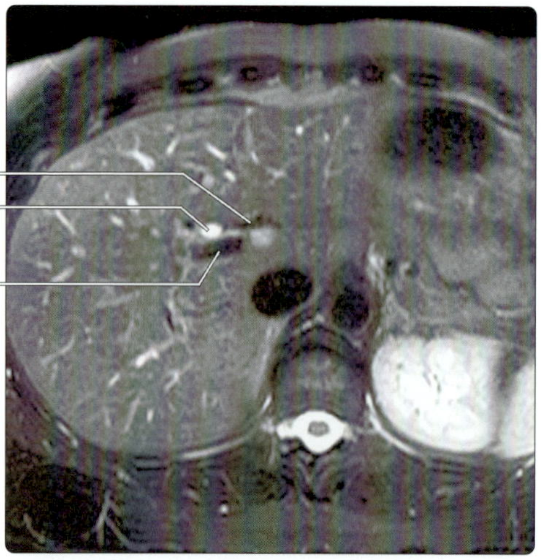

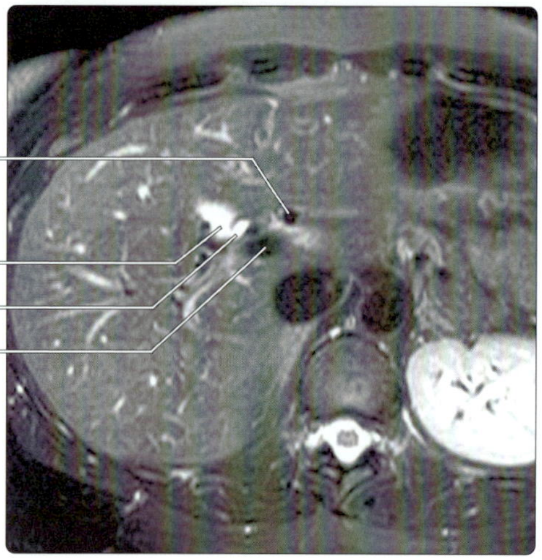

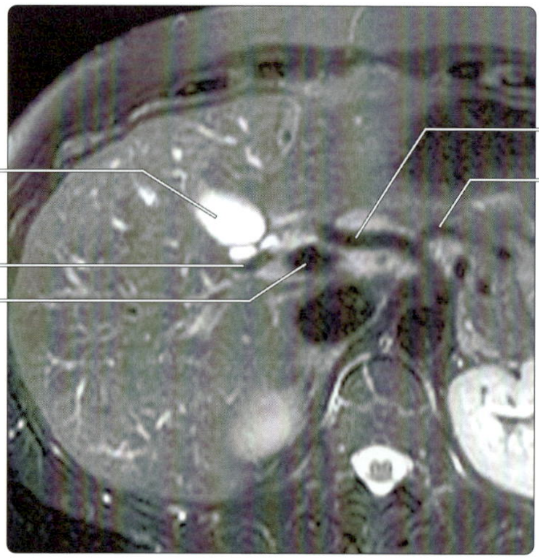

(Top) *More caudal section shows the usual relation between the bile duct, portal vein, and hepatic artery near the porta hepatis.* **(Middle)** *The neck of the gallbladder and cystic duct are seen on this section.* **(Bottom)** *Section through the porta hepatis shows the hepatic artery and portal vein entering the liver.*

AXIAL T2 FS MR

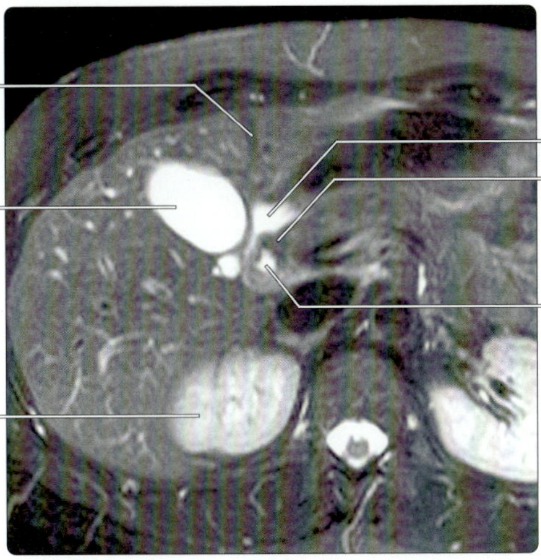

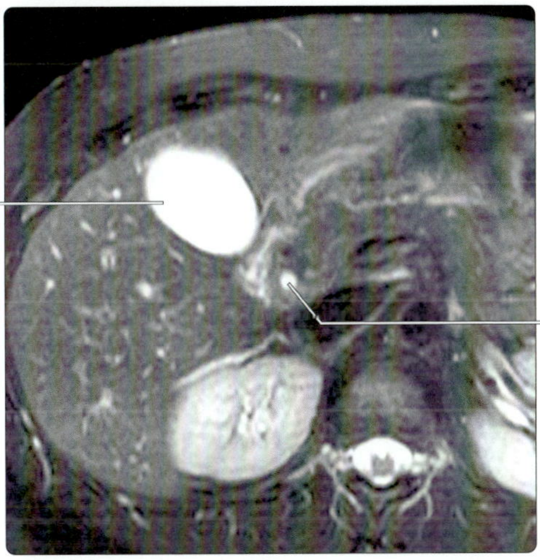

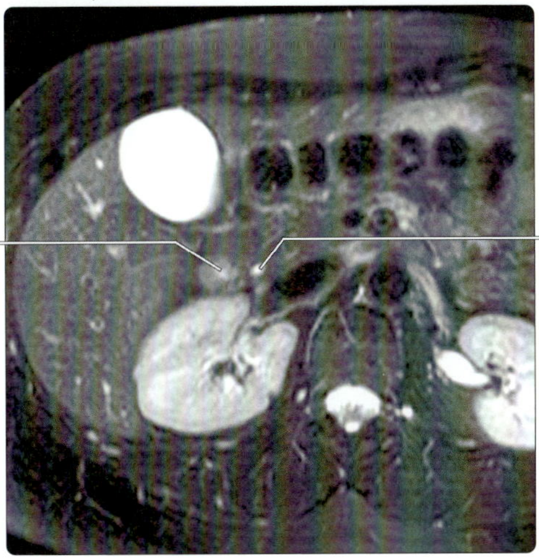

(Top) *The common duct leaves the liver, descends medial to the 2nd portion of the duodenum, and enters the pancreatic head.* **(Middle)** *The vertical course of the bile duct within the pancreatic head is seen well on this image.* **(Bottom)** *The bile duct is seen just proximal to its confluence with the pancreatic duct and their entry into the pancreaticobiliary ampulla in the duodenum.*

AXIAL T1 MR, HEPATIC ARTERIAL PHASE

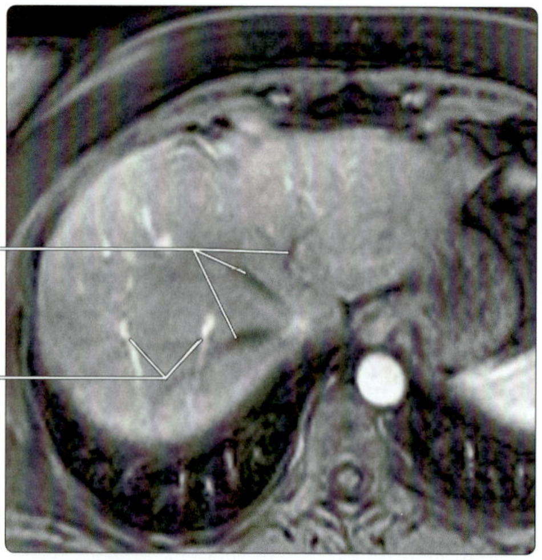

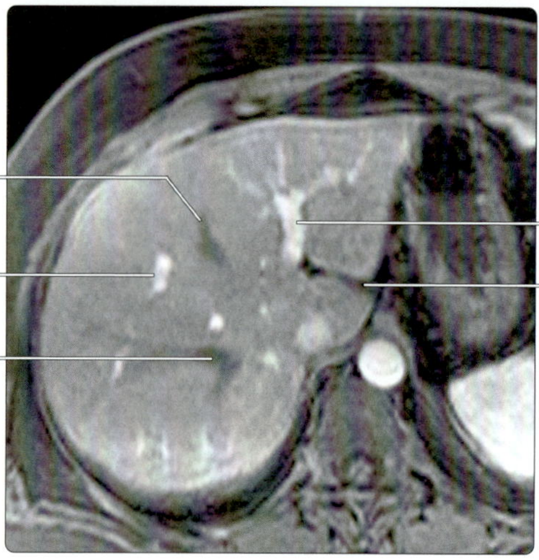

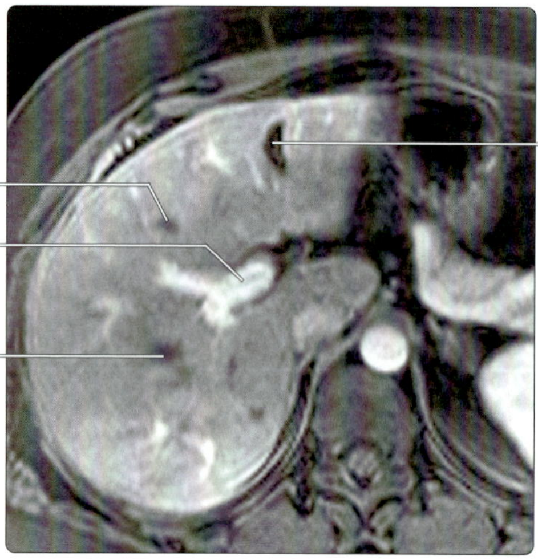

(Top) *First of 6 axial T1 MR images obtained during the arterial (portal venous inflow) phase of contrast enhancement shows the hepatic veins as dark, unenhanced structures, while the aorta, arteries, and portal venous branches are bright due to contrast enhancement.* **(Middle)** *The portal and hepatic venous branches are well seen on this section.* **(Bottom)** *A more caudal section again shows the opacified portal vein branches and the unopacified hepatic veins.*

AXIAL T1 MR, HEPATIC ARTERIAL PHASE

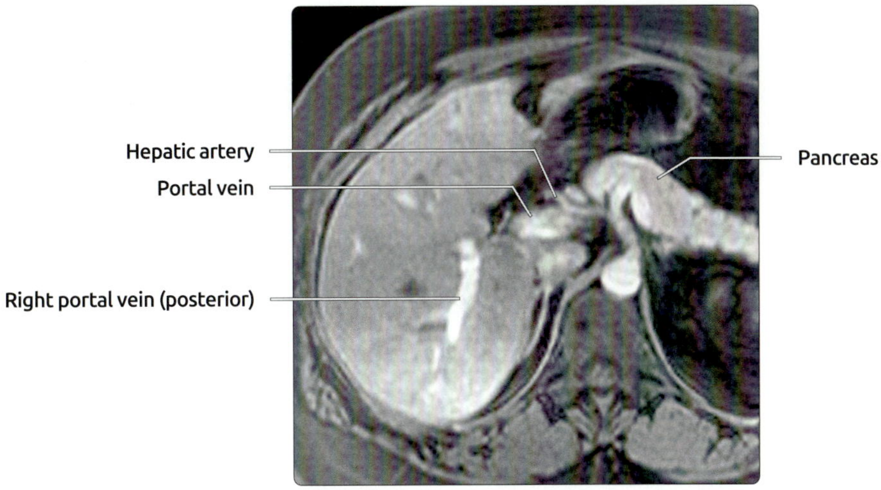

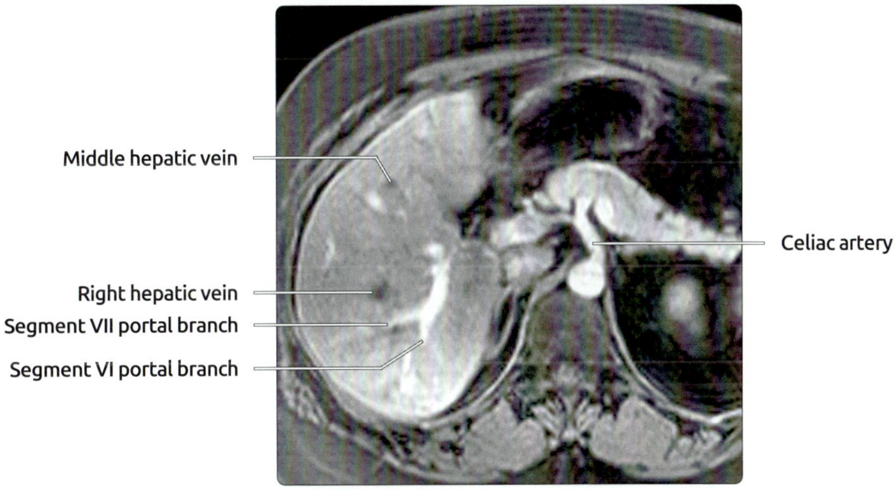

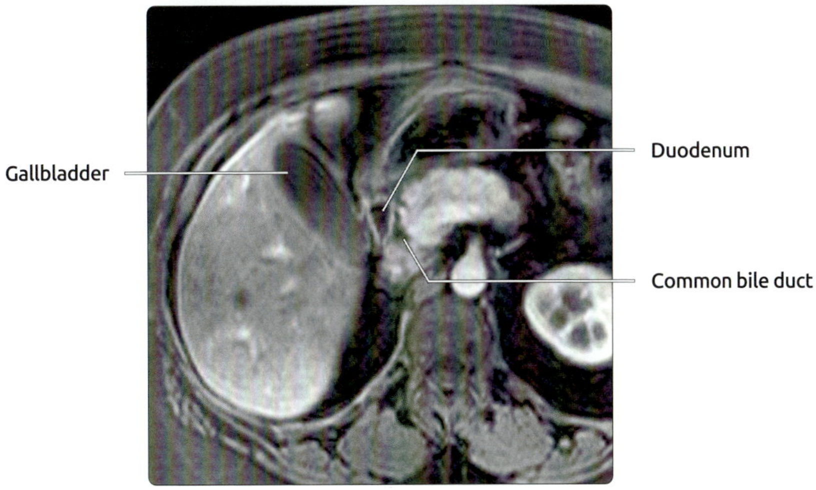

(Top) *The pancreas, arteries, and portal veins are well opacified on this image.* **(Middle)** *Note the intrasegmental position of the portal vein branches, while the hepatic veins lie between segments (in general).* **(Bottom)** *On these T1 images, static fluid-containing structures, such as the gallbladder and duodenum, appear dark (low signal).*

NORMAL AXIAL T1 MR, VENOUS PHASE

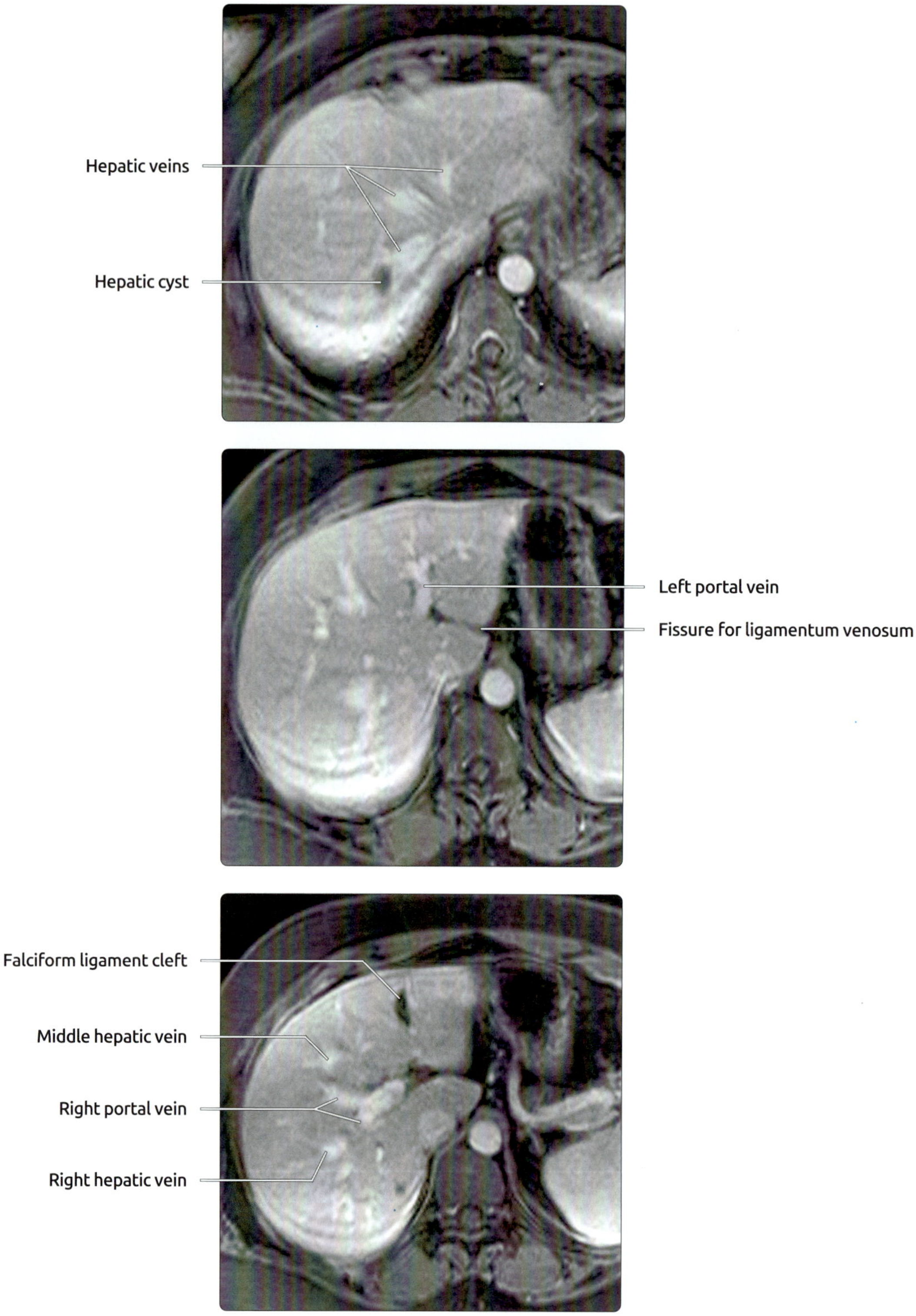

(Top) *First of 3 axial T1 MR images obtained during the hepatic venous phase of enhancement, when all the vascular structures in the liver are of bright signal due to contrast enhancement, is shown. A small cyst is seen adjacent to the right hepatic vein.* **(Middle)** *The portal and hepatic vein branches are all enhanced.* **(Bottom)** *The fissures and paths of the major veins are well identified, aiding in the definition of hepatic segmental anatomy.*

US, NORMAL LIVER

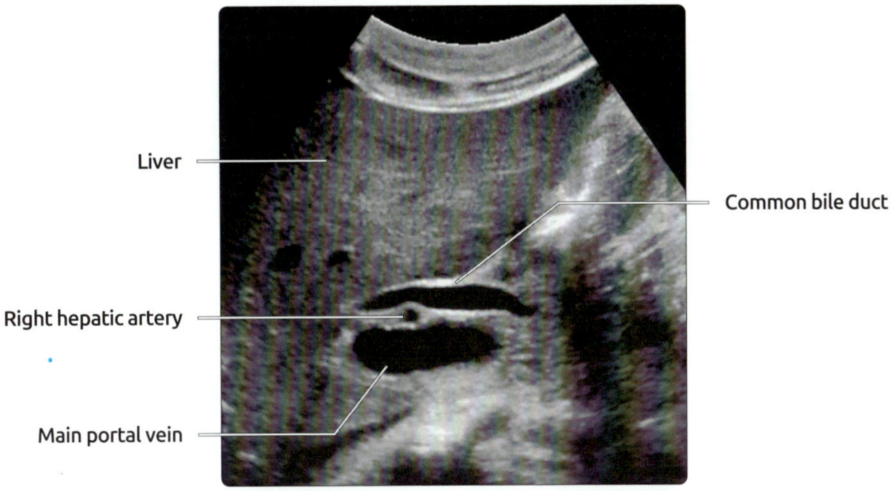

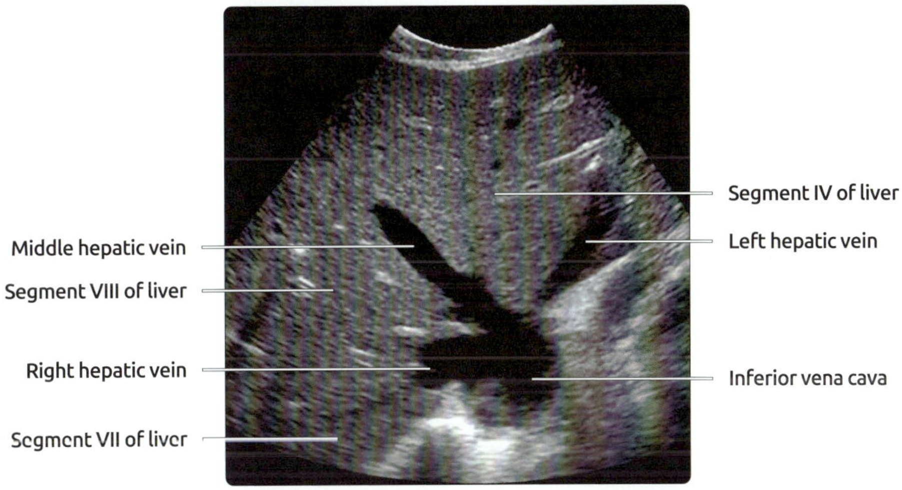

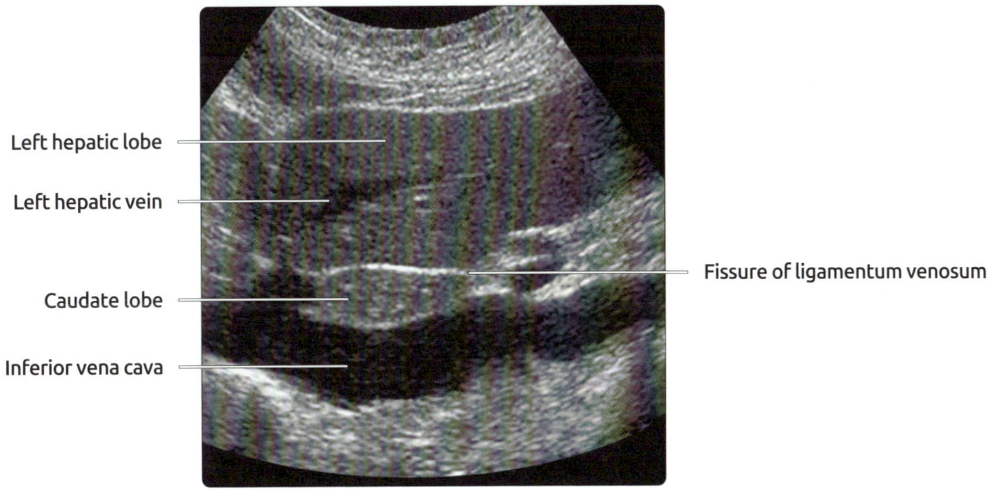

(Top) *Sagittal US demonstrates the characteristic relationship of the bile duct, portal vein, and hepatic artery in the porta hepatis. The common bile duct consistently runs anterior to the main portal vein in the porta hepatis, an important anatomic relationship when trying to identify the common duct. The right hepatic artery is seen in cross section in this plane.* **(Middle)** *Transverse US demonstrates the characteristic appearance of the hepatic veins converging on the IVC in axial cross section. This appearance has been described as the bunny sign, as the hepatic veins resemble the ears of a rabbit.* **(Bottom)** *Sagittal US demonstrates the fissure of the ligamentum venosum, dividing the caudate lobe from the left hepatic lobe. Replaced or accessory left hepatic arteries arising from the left gastric artery would be visualized in this fissure. The left hepatic vein is seen coursing toward the IVC.*

US, NORMAL LIVER

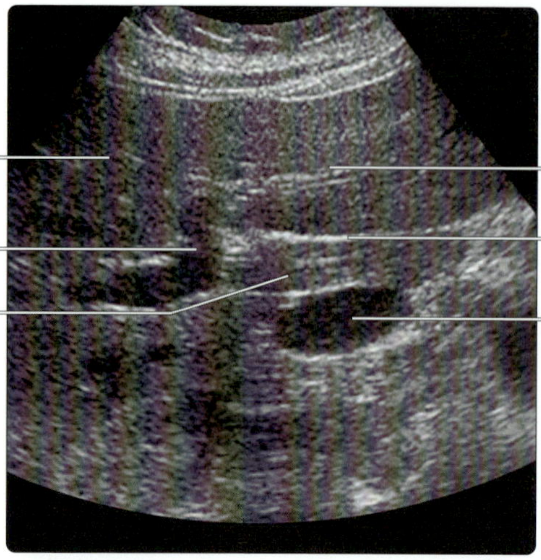

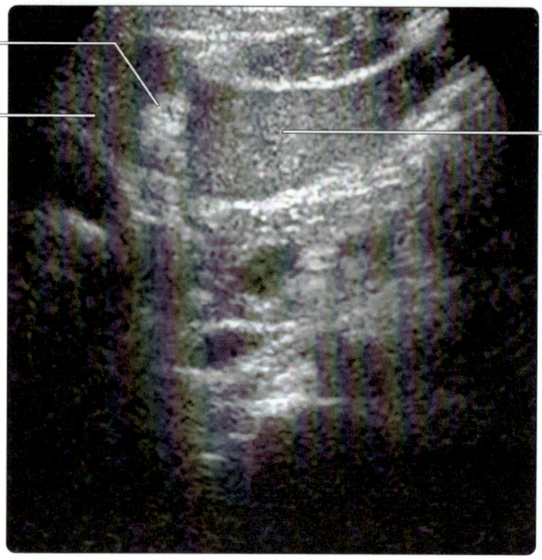

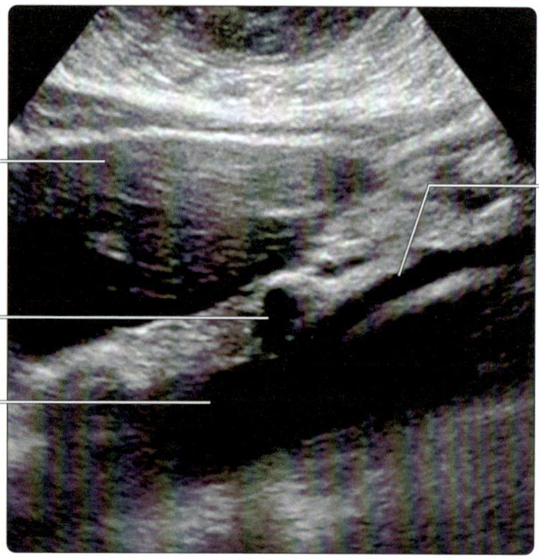

(Top) *The fissure of the ligamentum venosum is again visualized, this time in the transverse plane. Note the relationship of the left portal vein, medial and lateral segments of the left lobe, fissure of the ligamentum venosum, caudate, and IVC.* **(Middle)** *Transverse US demonstrates a normal appearance of the falciform ligament, an echogenic structure that has not uncommonly been mistaken for an echogenic mass. Segments II and III are seen lateral to the falciform, while segment IVa/b is seen medially.* **(Bottom)** *Sagittal US demonstrates the classic relationship of the celiac and superior mesenteric arteries in the sagittal plane akin to the appearance of these vessels on a sagittal CT. The superior mesenteric artery is seen diving inferiorly into the mesentery.*

SONOGRAPHY, NORMAL ANATOMY

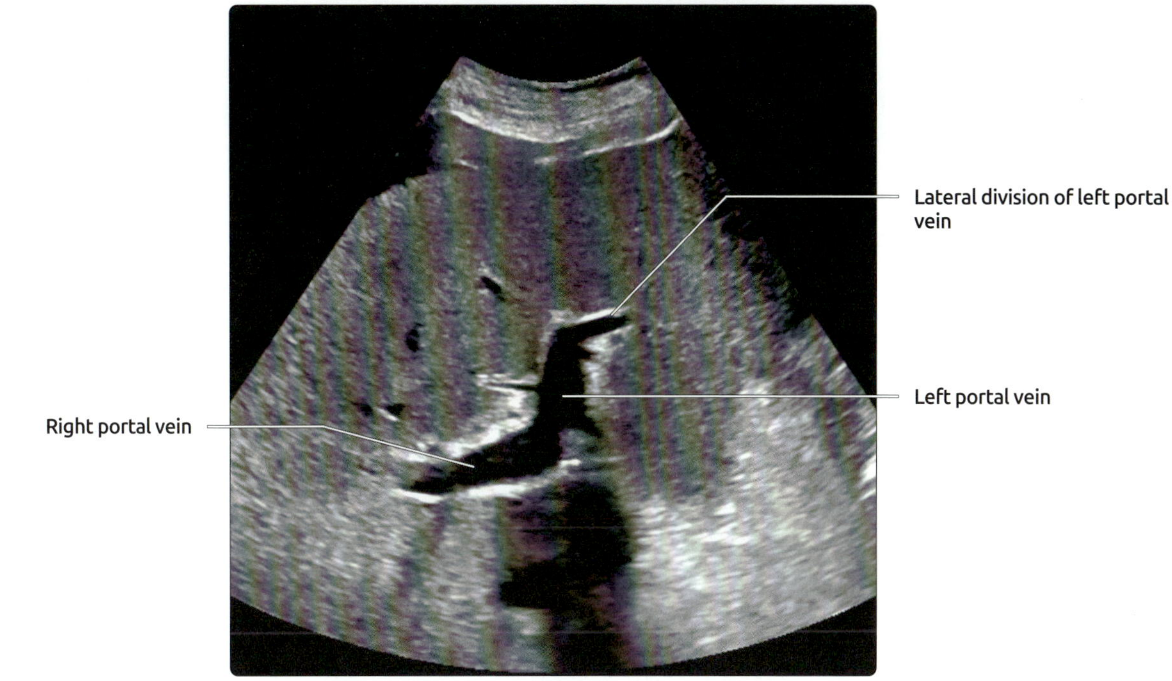

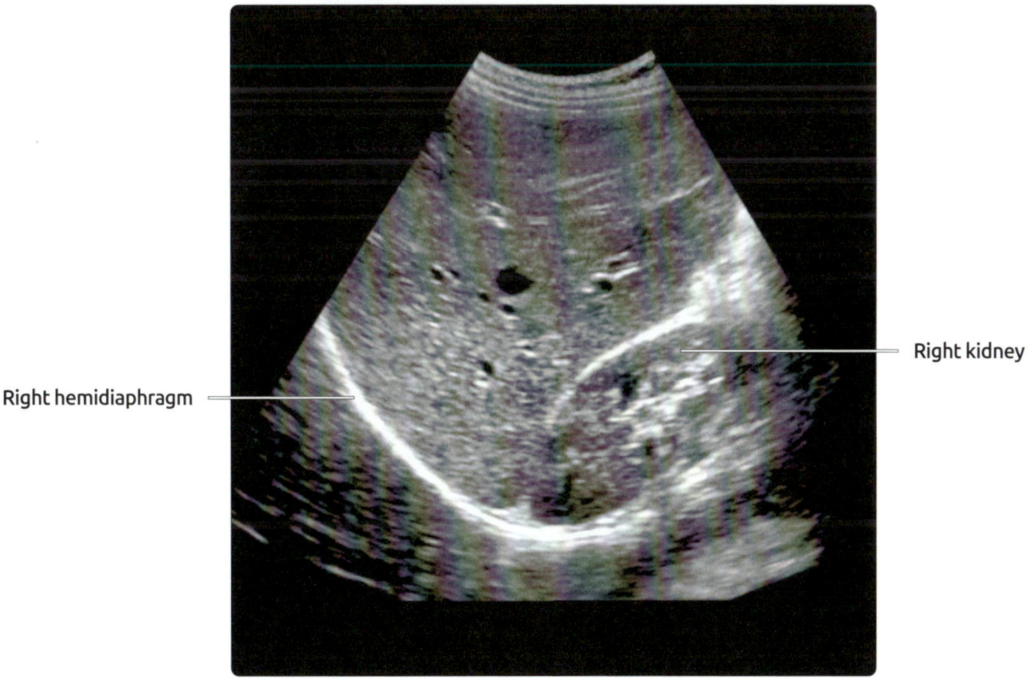

(Top) *Transverse US demonstrates the normal right and left portal veins, both of which have the thick echogenic wall that is characteristic of the portal venous system (and allows portal veins to be easily differentiated from hepatic veins). The left portal vein divides into medial and lateral branches, while the right portal vein divides into anterior and posterior branches.* **(Bottom)** *Sagittal US through the liver shows its smooth and homogeneous echogenicity with interspersed vessels and small intrahepatic bile ducts. Notice in this image that the right kidney and the liver parenchyma have roughly similar echogenicities. The relationship in echogenicity between these 2 structures is critical in determining whether the liver is abnormally echogenic (suggesting steatosis) or whether the kidney is abnormally echogenic (suggesting chronic medical renal disease).*

CONGENITAL HYPOPLASIA OF SEGMENTS

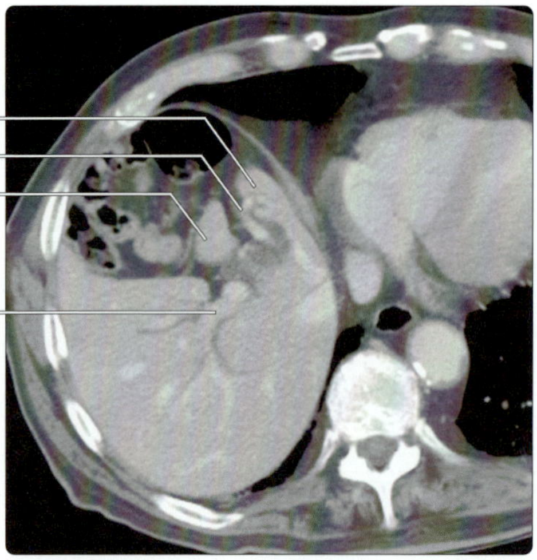

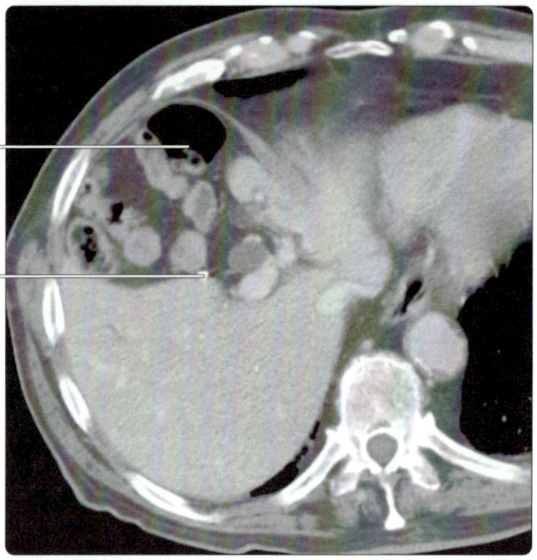

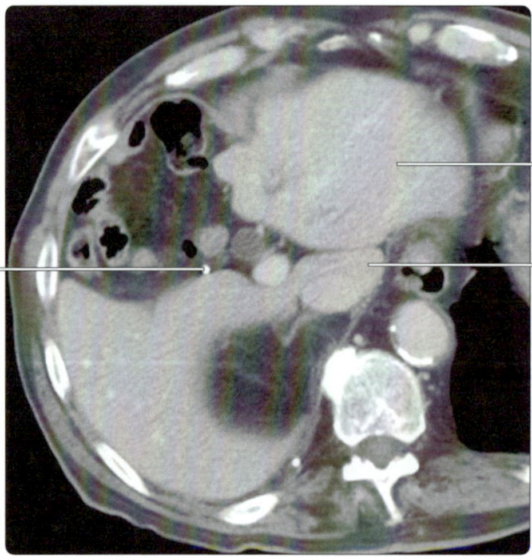

(Top) *First of 3 axial CECT images shows the colon in the right subphrenic region normally occupied by the anterior and medial segments of the liver. This patient is asymptomatic and has no history of any surgical resection or liver disease. Note the relative absence of tissue lateral to the left portal vein, which lies in the falciform ligament (boundary between the medial and lateral segments). The anterior branch of the right portal vein is absent.* **(Middle)** *A surgical clip is present from a prior cholecystectomy during which this congenital hypoplasia of hepatic segments was also noted.* **(Bottom)** *Congenital hypoplasia usually affects the anterior and medial segments, though any portion of the liver may fail to develop normally. This is one of the common causes of so-called interposition of the colon between the liver and the right hemidiaphragm.*

HEPATIC ARTERIAL VARIANTS

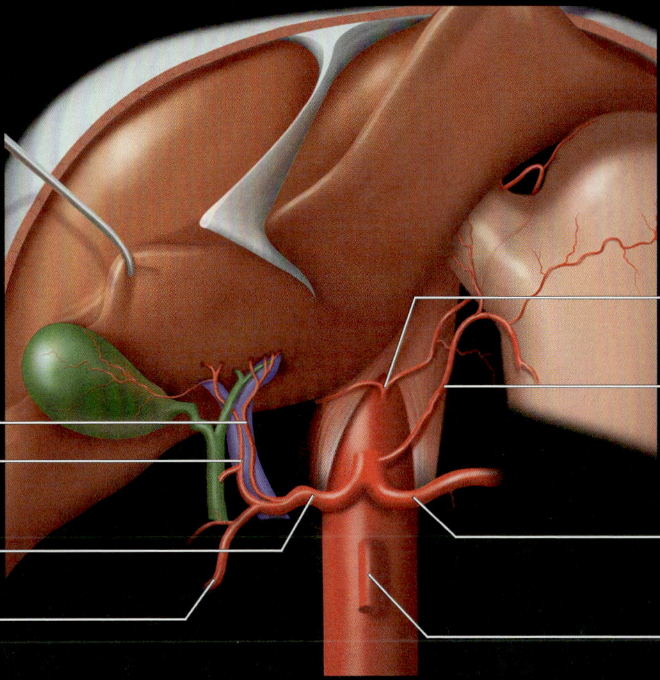

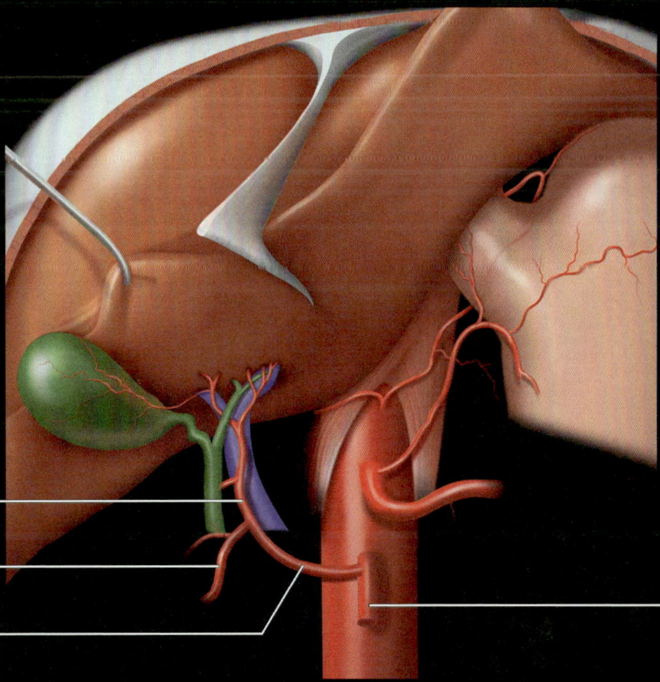

(Top) *In > 40% of individuals, there are variations in the origin and course of the hepatic arteries that differ from the "conventional" depiction. In this graphic, the left hepatic artery arises from the common hepatic artery, proximal to the origin of the gastroduodenal artery. The gallbladder and extrahepatic common bile duct are supplied by the right hepatic artery, as usual. The hepatic artery courses parallel to the portal vein and lies between the vein and the bile duct.* **(Bottom)** *Graphic shows a completely replaced hepatic artery arising from the superior mesenteric artery. In this setting, the hepatic artery passes through or behind the head of the pancreas and the portal vein and may be inadvertently ligated during pancreatic surgery if this variant is not recognized.*

HEPATIC ARTERIAL VARIANTS

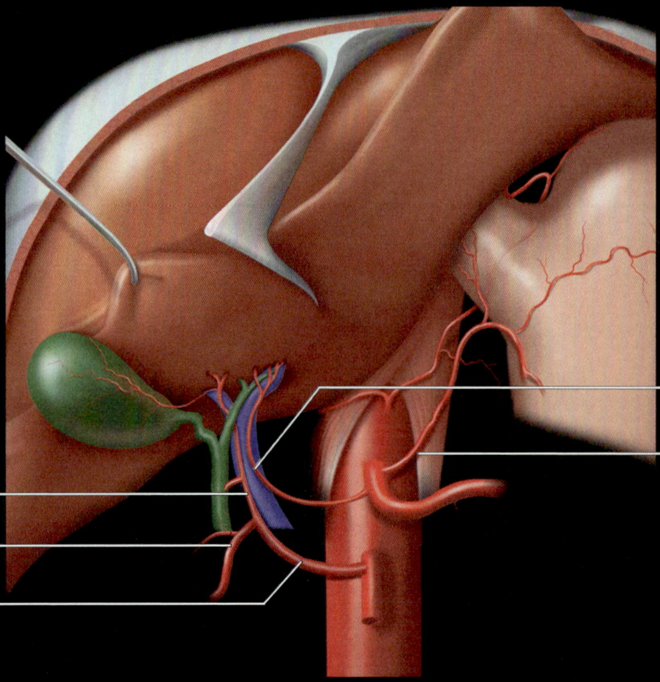

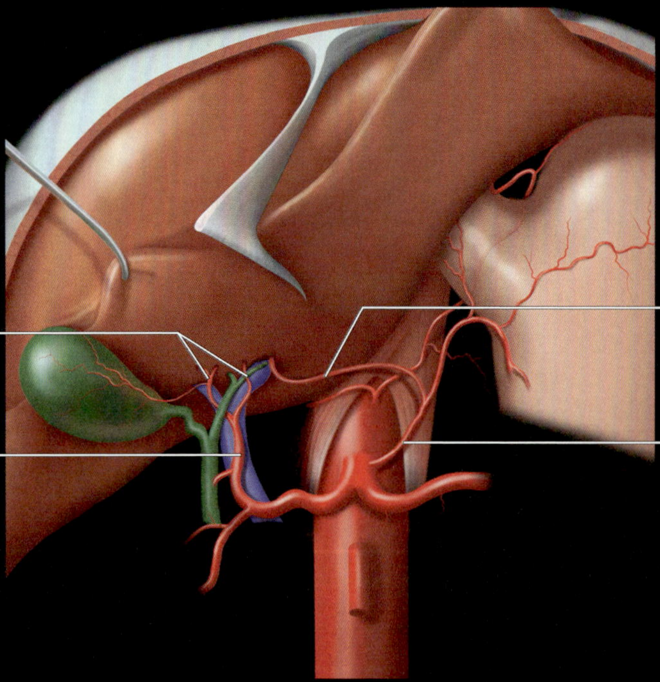

(Top) *Graphic depicts a separate origin of the left hepatic artery from the celiac trunk. In addition, the right hepatic artery is "replaced," arising from the superior mesenteric artery. The gastroduodenal and cystic arteries arise from the replaced right hepatic, as is common with this variation.* **(Bottom)** *Graphic depicts an "accessory" left hepatic artery arising from the left gastric artery. An accessory artery is a vessel in addition to those originating from the conventional depiction. In this case, there is a left hepatic artery arising from the proper hepatic artery as well. All of these variations are common and have major implications for patients undergoing any sort of upper abdominal surgery, especially partial hepatic resection or liver transplantation.*

REPLACED RIGHT AND LEFT HEPATIC ARTERIES

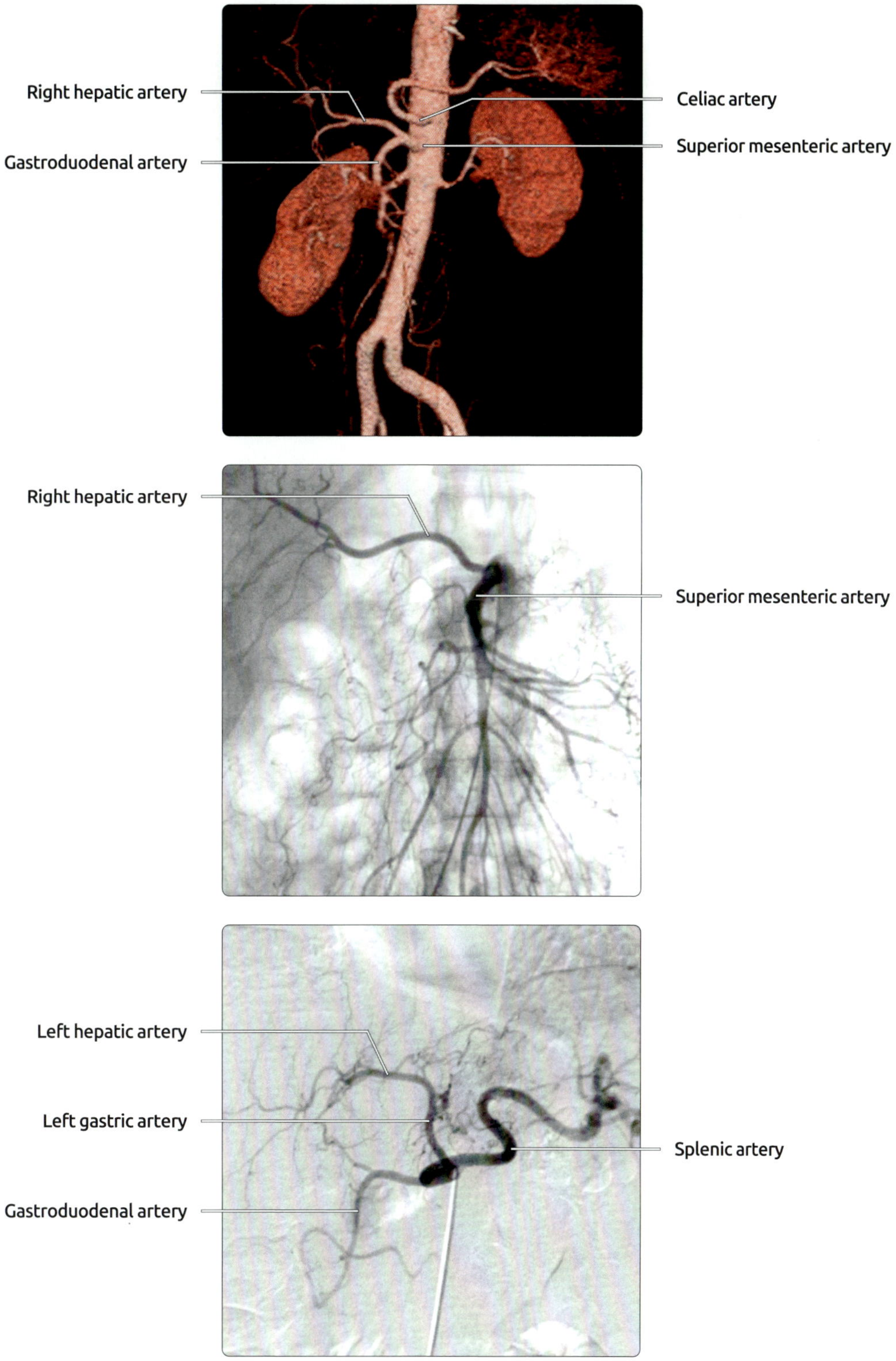

(Top) *First of 3 CECT images in a patient with separately replaced right and left hepatic arteries is shown. Coronal CECT with volume-rendered reconstruction shows the right hepatic artery arising from the superior mesenteric artery. The origin of the left hepatic artery is not shown clearly on this image.* **(Middle)** *A frontal film from a conventional catheter angiogram shows the replaced right hepatic artery arising from the superior mesenteric artery.* **(Bottom)** *Catheter injection of the celiac axis shows the left hepatic artery arising from the left gastric artery. In this subject, the gastroduodenal artery arises directly from the celiac artery. Variations of hepatic arterial anatomy are common and are important to recognize when hepatic transplantation, complex hepatobiliary surgery, or other intervention is being considered.*

HEPATIC ARTERY VARIANTS

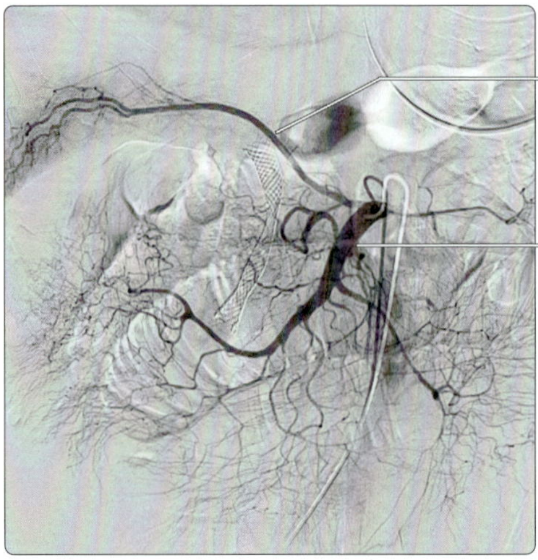

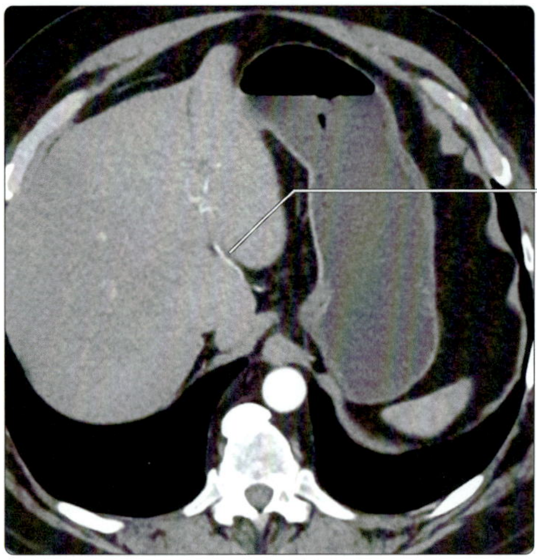

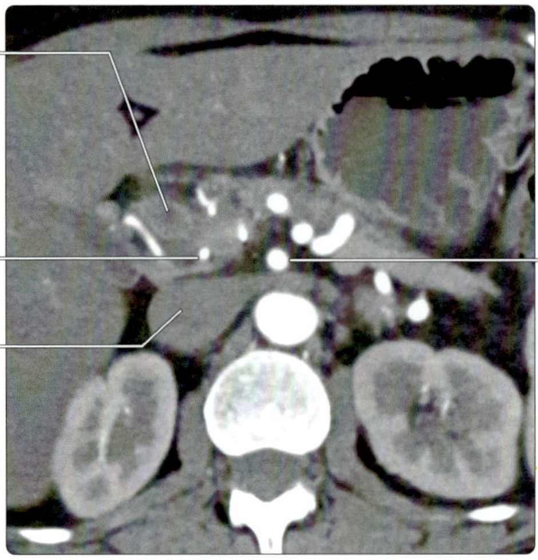

(Top) *Frontal angiogram demonstrates an aberrant right hepatic artery arising from the superior mesenteric artery and taking its typical course upward and to the right.* **(Middle)** *Axial CECT demonstrates a small vessel running in the fissure of the ligamentum venosum, a location where an artery should not normally be found. Any artery in this location must represent an aberrant (either accessory or replaced) left hepatic artery arising from the left gastric artery (replaced left hepatic artery in this case).* **(Bottom)** *Axial CECT demonstrates an artery extending upward toward the liver coursing between the main portal vein and the IVC, a location where the hepatic artery should normally not be visualized. This is the classic location of a replaced right hepatic artery arising from the superior mesenteric artery. The location of an aberrant artery can give away its site of origin, even if the artery is not traced out in its entirety.*

ACCESSORY RIGHT HEPATIC ARTERY

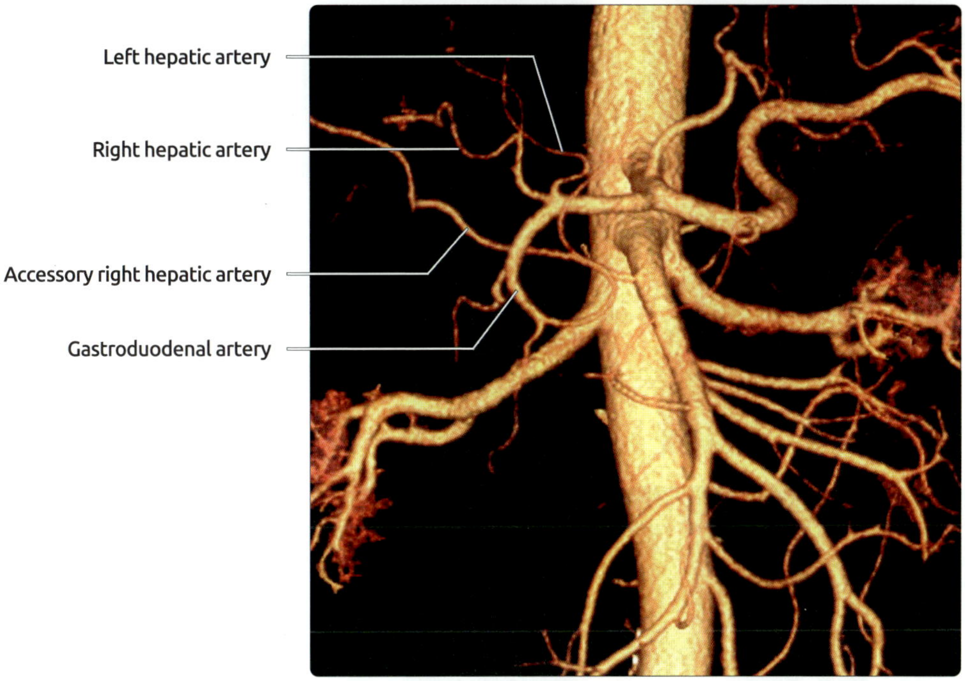

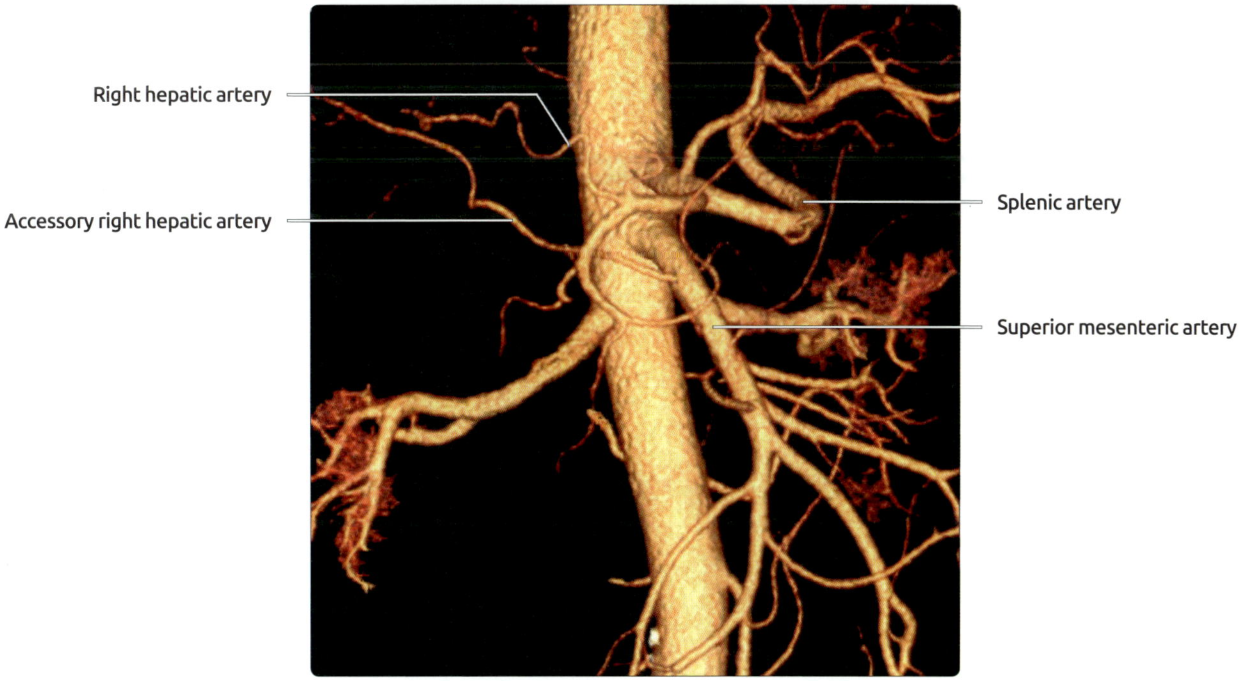

(Top) *First of 2 volume-rendered, coronal CECT images shows only a small right hepatic artery branch from the common hepatic artery. An accessory right hepatic branch arises from the superior mesenteric artery, a common variant. Note the anastomoses between the accessory right hepatic and gastroduodenal arteries.* **(Bottom)** *Oblique view of the CTA in the same patient helps to confirm the origin of the accessory hepatic vessel from the superior mesenteric artery.*

REPLACED HEPATIC ARTERY

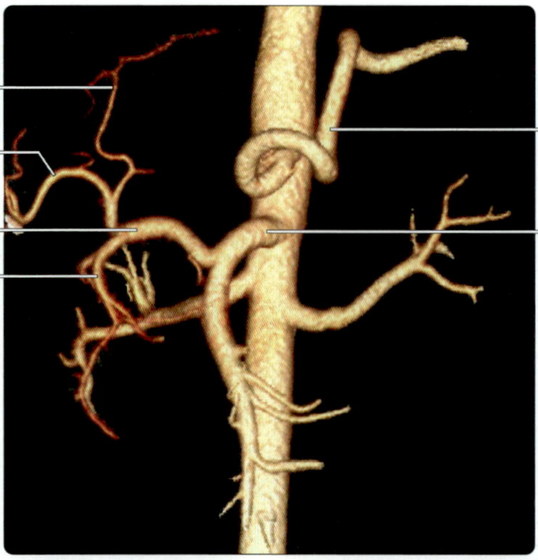

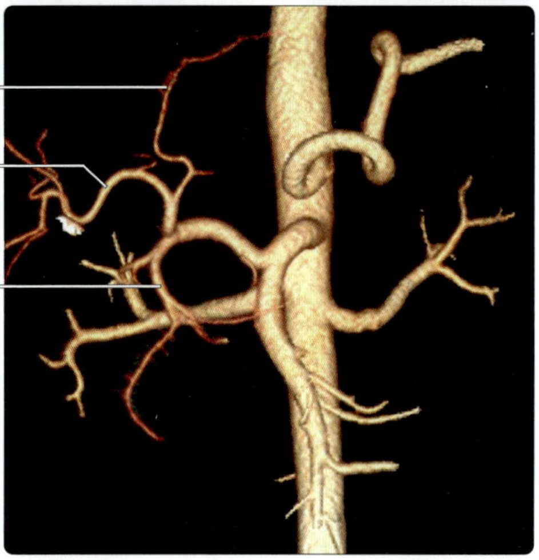

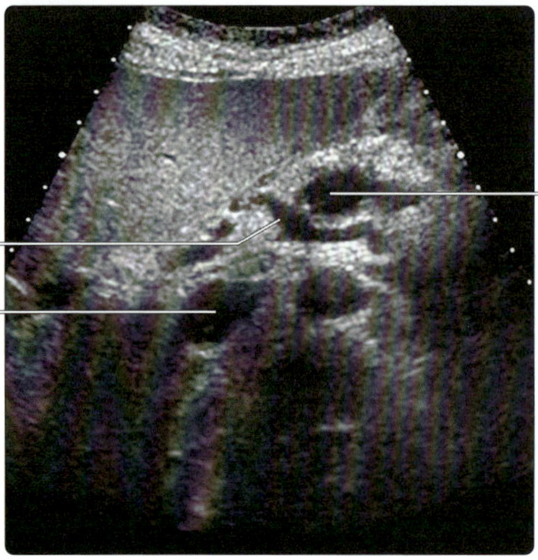

(Top) *Volume-rendered coronal CECT shows complete replacement of the common hepatic artery from the superior mesenteric artery. Both major hepatic arteries and the gastroduodenal artery arise from the superior mesenteric artery instead of the celiac artery.* **(Middle)** *Volume-rendered oblique CECT in the same patient helps to confirm the origin of the entire hepatic arterial system from the superior mesenteric artery.* **(Bottom)** *Transverse US demonstrates an aberrant hepatic artery coursing between the portal vein and the IVC, representing a replaced right hepatic artery arising from the superior mesenteric artery. With "conventional" anatomy, no artery should normally be seen in this space.*

PORTAL VEIN VARIANTS

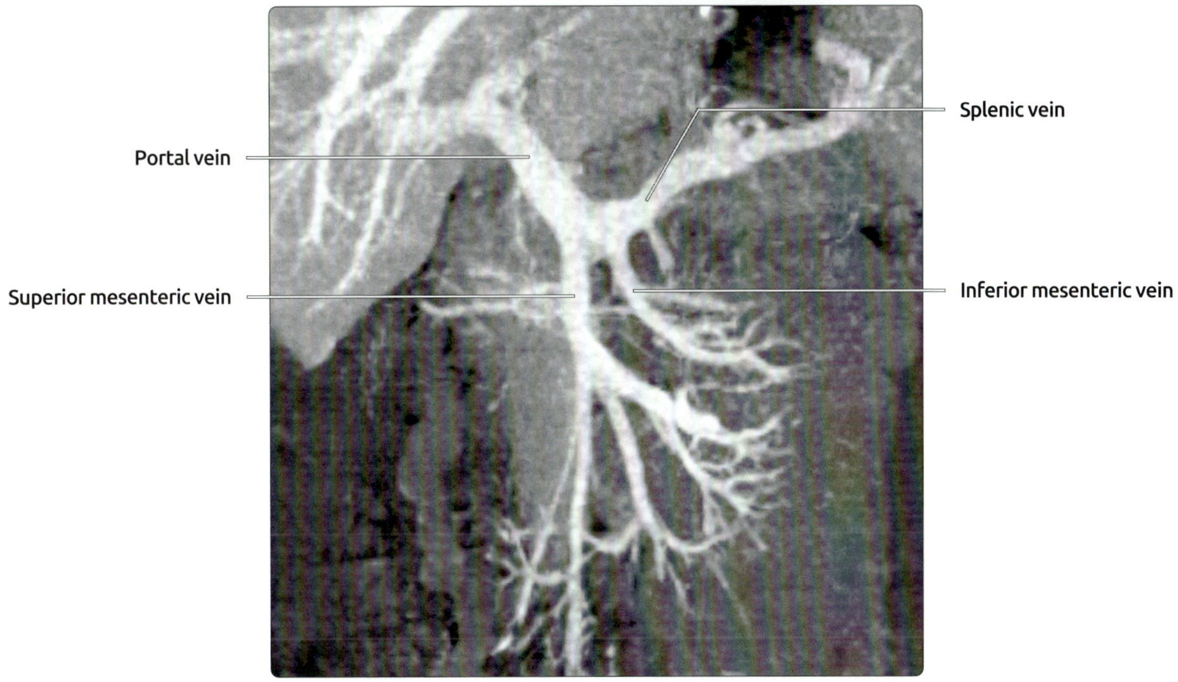

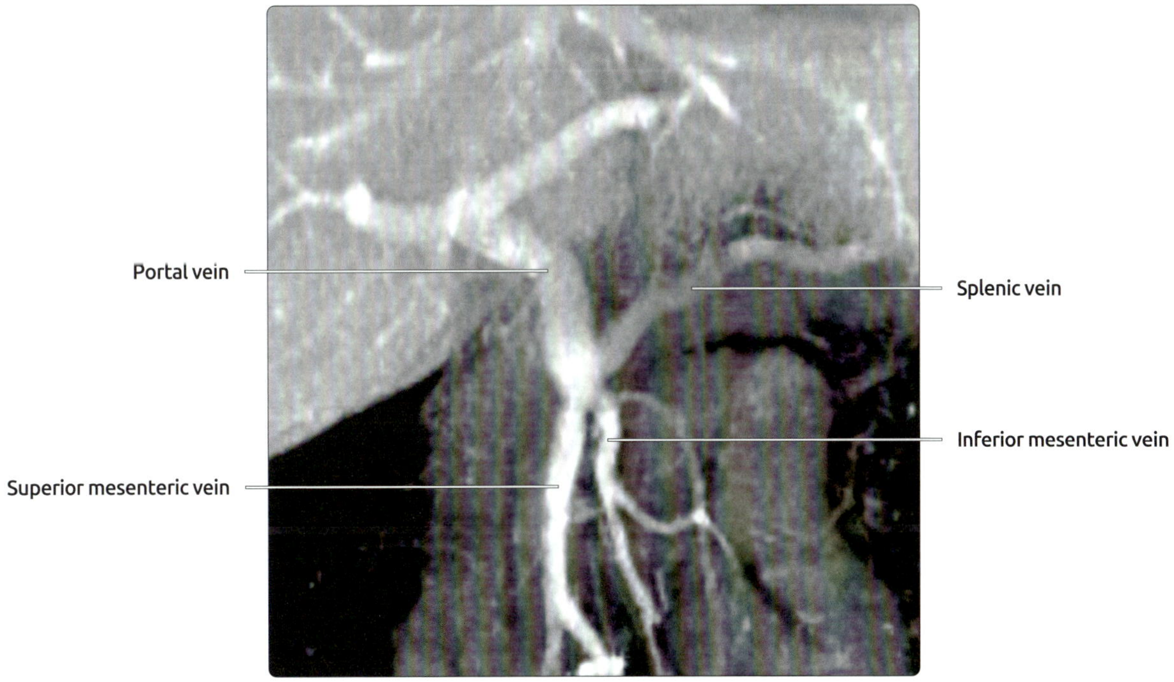

(Top) *MR angiogram (venous phase following IV injection of contrast) shows the most common arrangement of the major tributaries of the portal vein in which the inferior mesenteric vein joins the splenic vein just prior to its confluence with the superior mesenteric vein.* **(Bottom)** *In this patient, the MRA demonstrates a variant with all 3 major tributaries (portal vein, superior mesenteric vein, and inferior mesenteric vein) coming together at a trifurcated confluence.*

LIVER VESSELS, UNFAVORABLE FOR TRANSPLANTATION

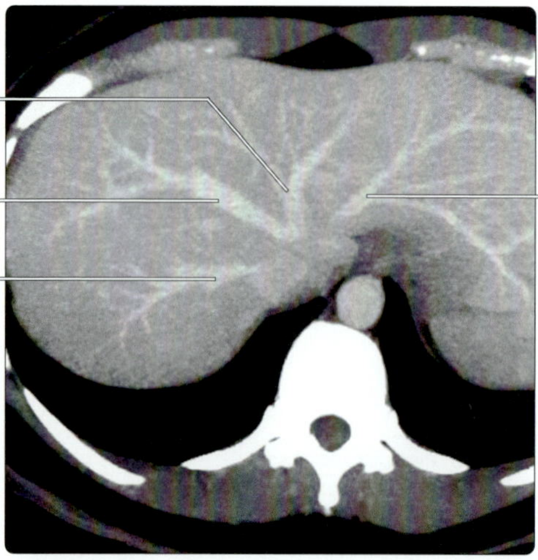

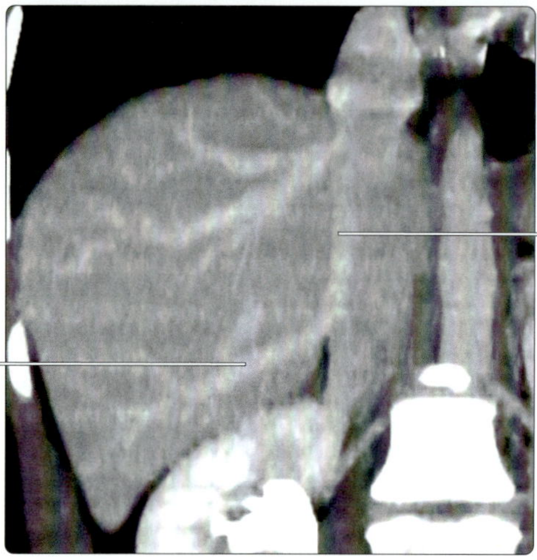

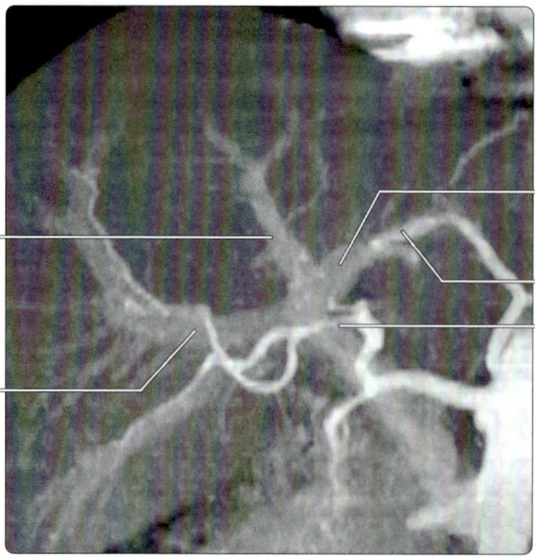

(Top) *Axial CECT in a potential partial liver donor (from whom the right hepatic lobe would be donated) shows a branch from segment VIII draining into the middle hepatic vein. In the usual incision used for harvesting the right lobe, the liver is divided just to the right of the middle hepatic vein, which would sever the segment VIII branch and require a separate anastomosis to the recipient IVC.* **(Middle)** *Coronal CECT shows a large accessory right hepatic vein draining segments V and VI directly into the IVC. If the right lobe were to be transplanted, this would require still another separate anastomosis to the recipient IVC.* **(Bottom)** *Coronal MIP shows a trifurcation arrangement of the portal vein in which the anterior and posterior branches of the right portal and the left portal vein join at the same point. This variant makes it difficult to transect the right portal vein without jeopardizing the left portal vein. Due to the various vascular anatomic variants demonstrated by CT in this patient, he was judged to not be a candidate for living donation.*

COLLATERAL FLOW THROUGH LIVER

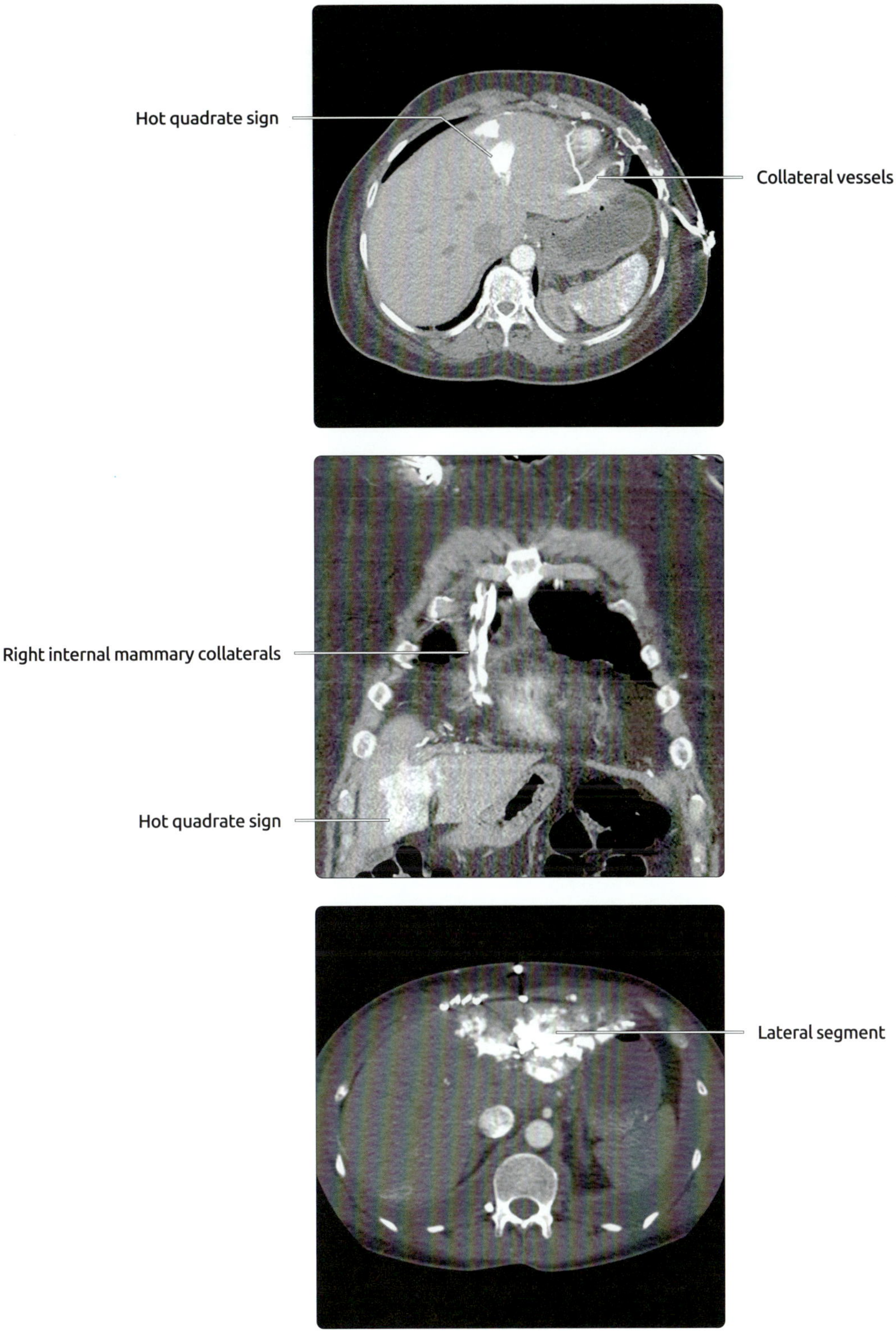

(Top) *Axial CECT demonstrates the classic hot quadrate sign in the liver in a patient with known superior vena cava syndrome as a result of collateral flow. This phenomenon is due to collateral blood flow through peridiaphragmatic collateral veins that communicate with hepatic veins and return blood to the IVC. On radionuclide and PET studies, this phenomenon may appear as a hot spot and be confused for a liver tumor.* **(Middle)** *Coronal CECT in the same patient nicely demonstrates the right internal mammary collaterals, which bypass the patient's occluded superior vena cava and subsequently enter collaterals in the liver, resulting in this very characteristic perfusion anomaly.* **(Bottom)** *Axial CECT demonstrates another example of an extensive liver perfusion anomaly resulting from collateral flow in a patient with superior vena cava occlusion.*

CIRRHOSIS WITH PORTAL HYPERTENSION

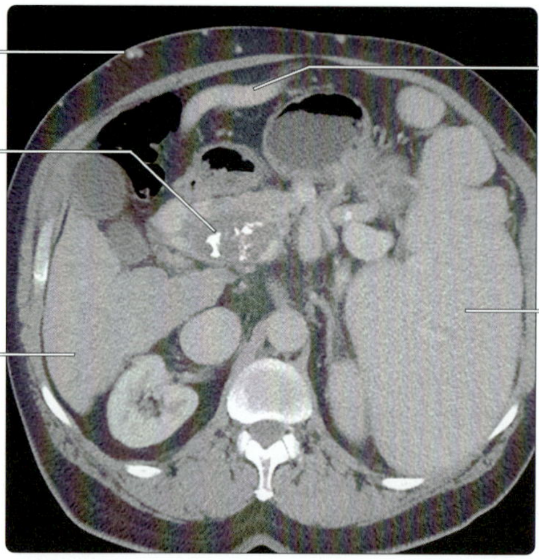

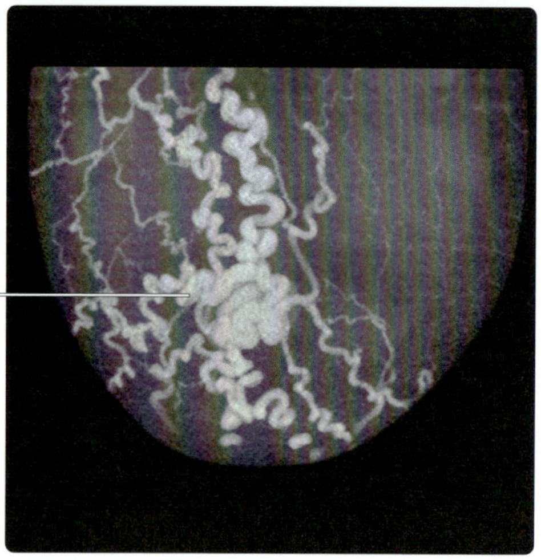

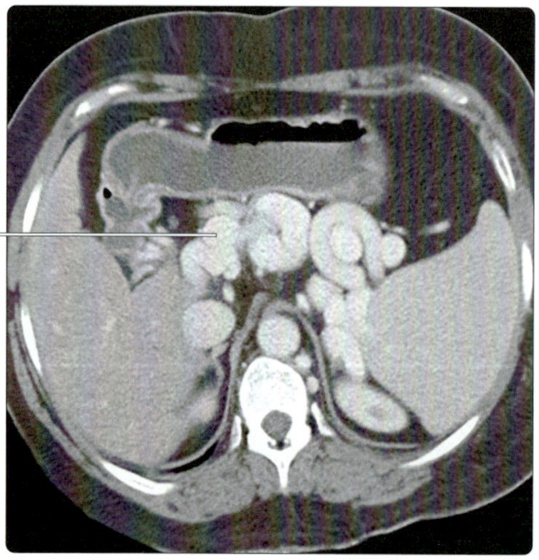

(Top) *Axial CECT demonstrates a cirrhotic-appearing liver as well as several stigmata of portal hypertension, including marked splenomegaly and varices. There is chronic-appearing portal vein thrombus with associated calcification, a common feature in patients with portal hypertension.* **(Middle)** *Coronal CECT with volume-rendered reconstruction in a known cirrhotic patient demonstrates extensive abdominal wall varices, a finding which has often been described as a caput medusa due to its appearance on clinical examination.* **(Bottom)** *Axial CECT in a cirrhotic patient demonstrates prominent varices in the upper abdomen. Prominent varices can sometimes be confused for a mass or tumor on noncontrast or arterial-phase imaging, although the diagnosis is obvious in the venous phase.*

DIFFUSE HEPATIC STEATOSIS

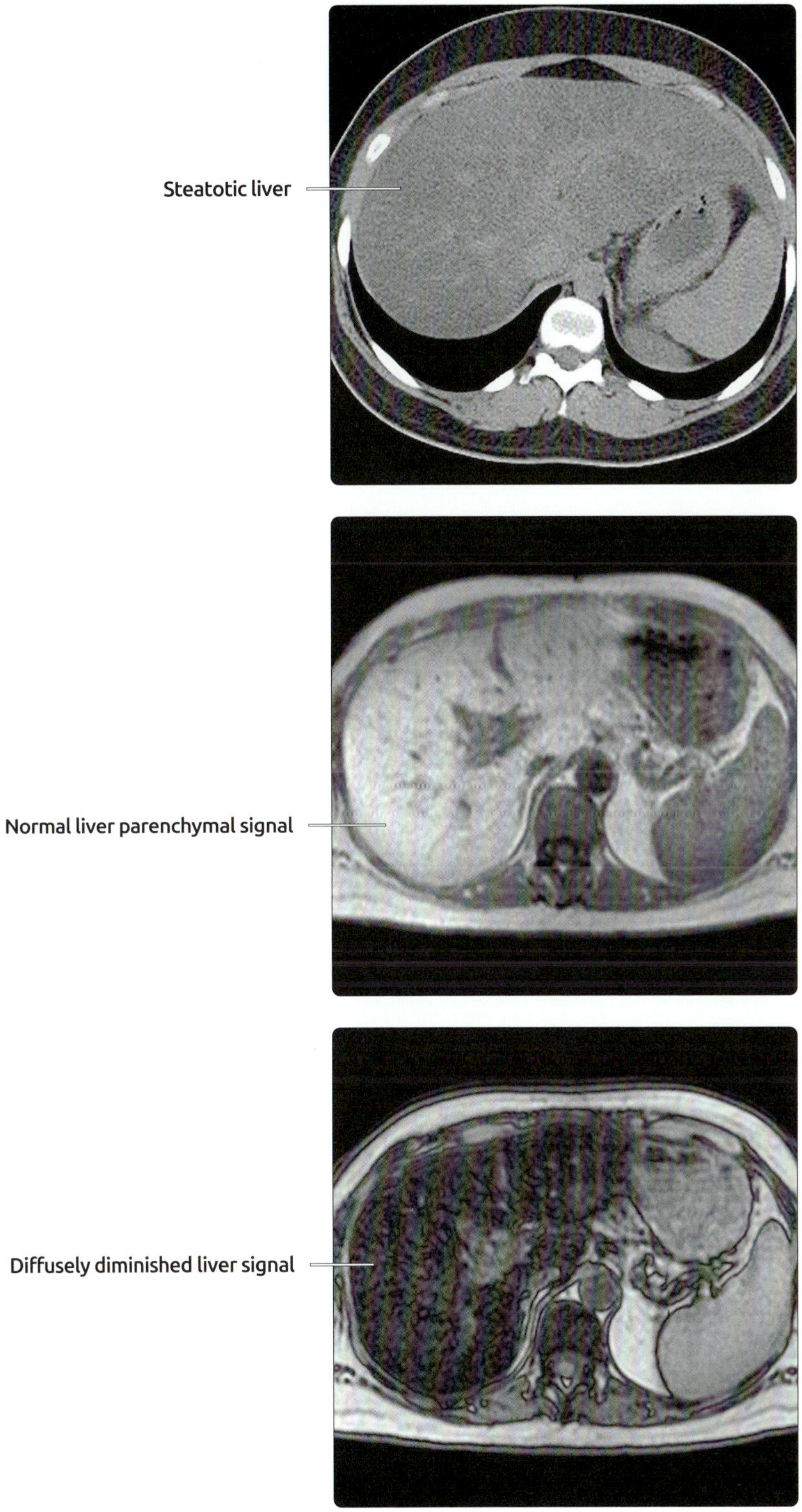

(Top) *Axial NECT in a patient with hepatic steatosis ("fatty liver") shows that the liver appears darker, or less dense, than the spleen rather than being slightly more dense (higher in attenuation), as would be normal. Generally speaking, a liver parenchymal attenuation < 40 HU on a NECT suggest suggest underlying hepatic steatosis.* **(Middle)** *Axial GRE in-phase MR demonstrates homogenous signal throughout the liver in a patient with elevated liver enzymes.* **(Bottom)** *Axial GRE out-of-phase chemical-shift image in the same patient demonstrates diffuse signal loss throughout the liver compatible with diffuse hepatic steatosis.*

MULTIFOCAL STEATOSIS

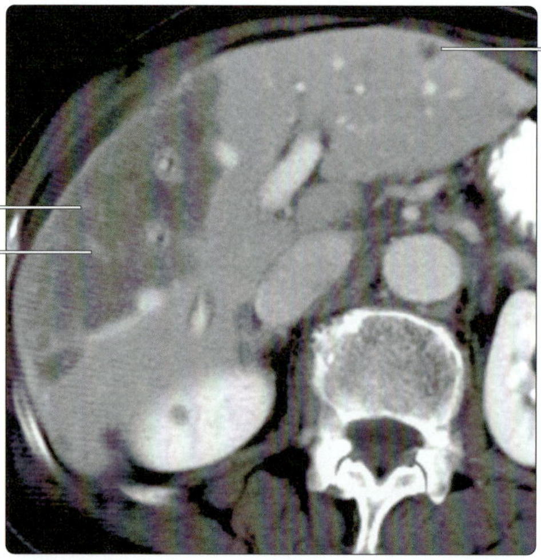

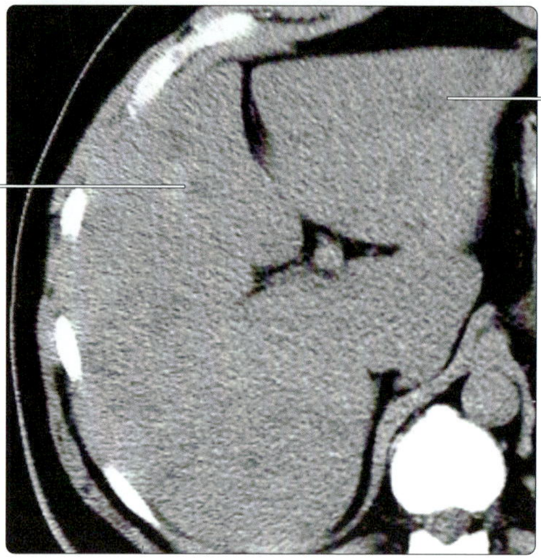

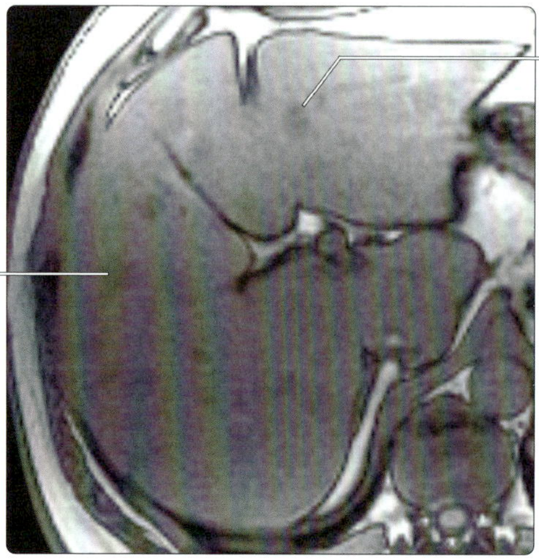

(Top) *Axial CECT shows a geographic area of low attenuation throughout the central and right hepatic lobes. In addition, there are spherical and oval lesions in other segments of the liver. Notice that hepatic vessels course through the low-density lesions without being displaced or occluded. Also note that the low-density lesions are not associated with mass effect on the liver contour nor capsular retraction. Percutaneous biopsy of the right lobe revealed steatohepatitis.* **(Middle)** *Axial NECT shows innumerable small hypodense focal liver lesions. Although the patient did not have a known history of malignancy, these findings initially raised concern for potential hepatic metastatic disease.* **(Bottom)** *Axial out-of-phase GRE MR in the same patient shows selective dropout of signal from each of the lesions, indicating lipid content and essentially confirming the diagnosis of multifocal fatty infiltration (steatosis). Focal steatosis can simulate a primary or metastatic liver tumor, although these entities can generally be easily distinguished using chemical-shift imaging MR.*

BENIGN LIVER MASSES

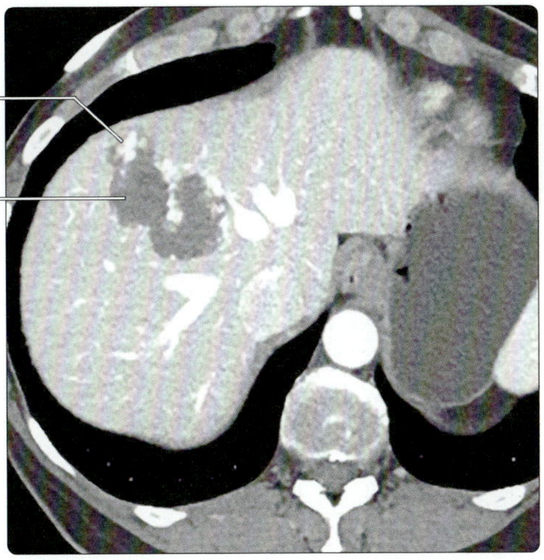

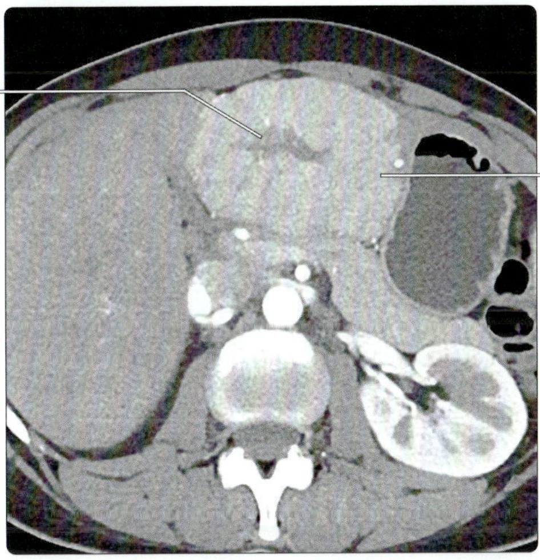

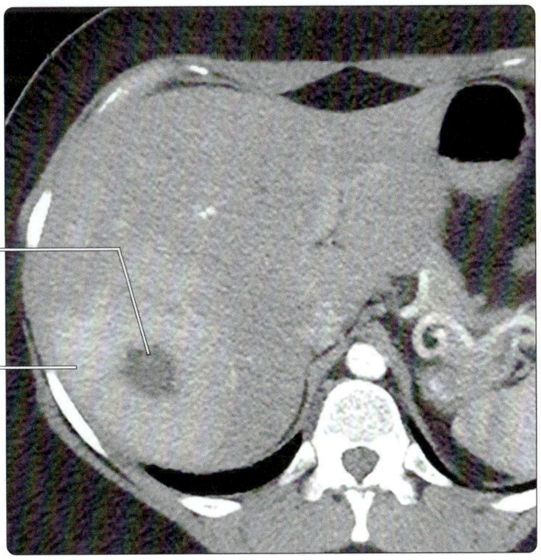

(Top) *Axial CECT in an asymptomatic patient demonstrates a mass in the right hepatic lobe with peripheral, nodular enhancement, compatible with a benign hemangioma. This appearance is so classic and typical that no further confirmatory imaging or biopsy is needed.* **(Middle)** *Axial CECT demonstrates a markedly hypervascular mass with a central scar. The mass was virtually invisible on the venous-phase images (not shown). This constellation of findings is typical for a benign focal nodular hyperplasia.* **(Bottom)** *Axial CECT demonstrates a hypervascular mass in the right hepatic lobe with central necrosis in a young female patient with a history of oral contraceptive usage. While difficult to make a definitive diagnosis on this single image, this lesion actually demonstrated washout on the delayed-phase images, and there was some clinical concern for potential malignancy. This lesion was ultimately biopsied and shown to represent an hepatic adenoma.*

LIVER METASTASES

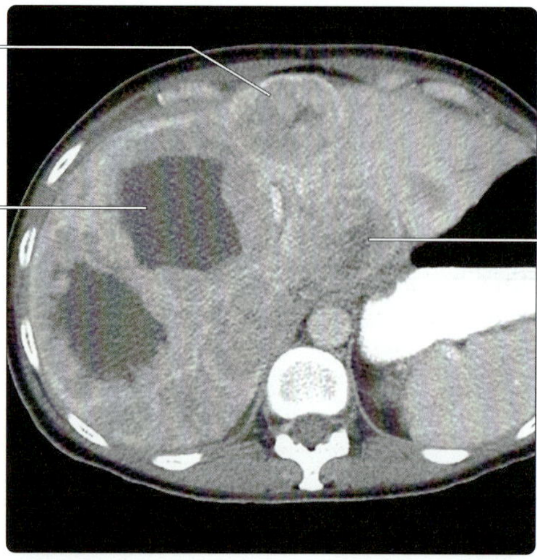

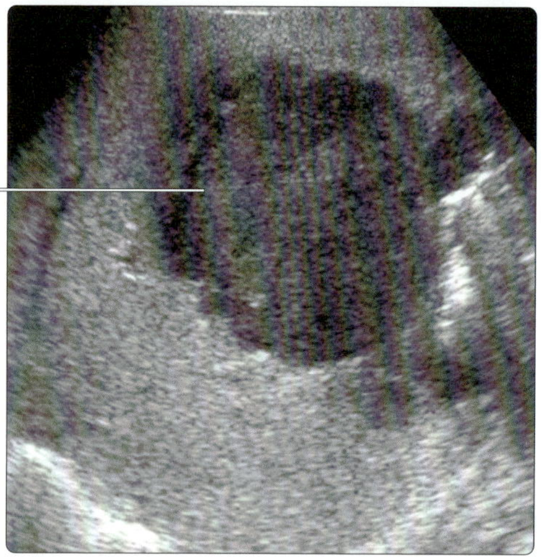

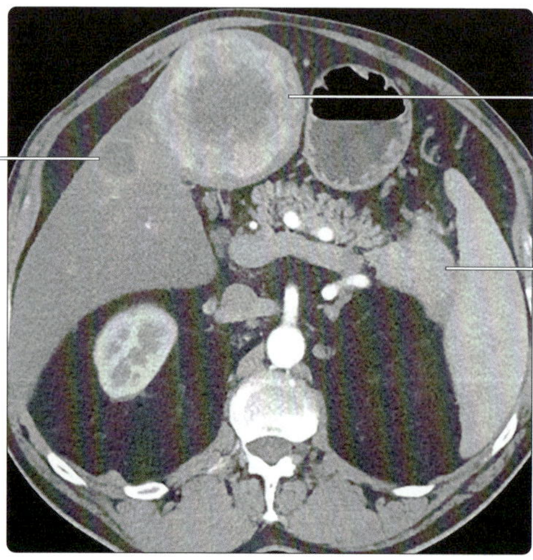

(Top) *Axial CECT shows numerous heterogeneous masses within all segments of the liver, representing metastases from the patient's primary gastrointestinal stromal tumor. Some of these have a significant central cystic component, consistent with central necrosis.* **(Middle)** *Transverse grayscale US demonstrates a solitary ill-defined hypoechoic mass in the liver. While a number of entities could have this appearance (including primary liver tumors), the patient's history of a known lung cancer was the key to the correct diagnosis of a metastasis in this case.* **(Bottom)** *Axial arterial-phase CECT demonstrates a hypervascular mass in the pancreatic tail representing a primary pancreatic neuroendocrine tumor. There are multiple similarly enhancing lesions in the liver, nicely visualized in this arterial phase of imaging, compatible with pancreatic neuroendocrine tumor metastases. The arterial phase of imaging can be critical in highlighting and detecting hypervascular metastases, particularly when small or subtle.*

HEPATOCELLULAR CARCINOMA WITH TUMOR THROMBUS

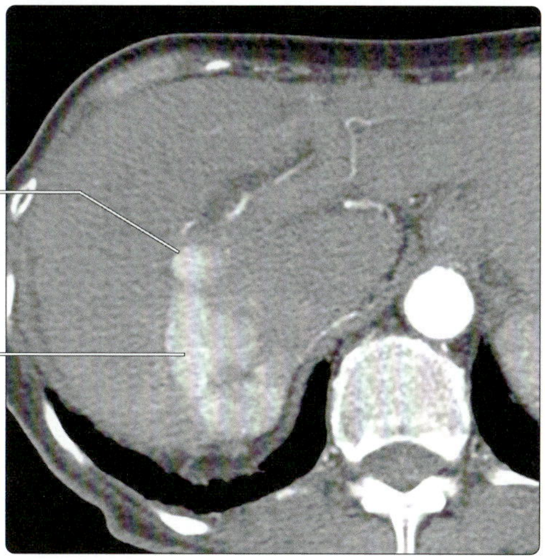

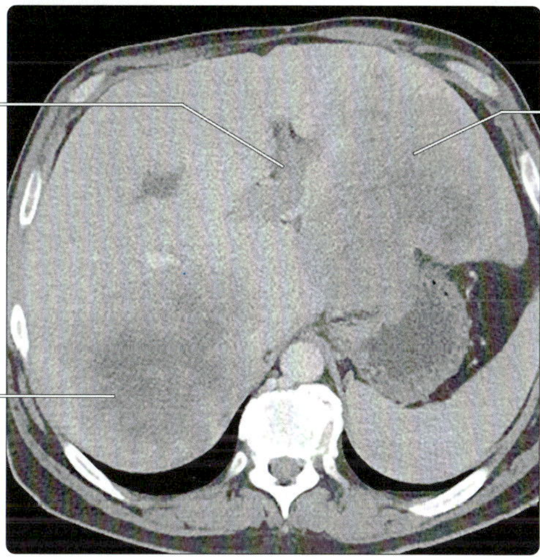

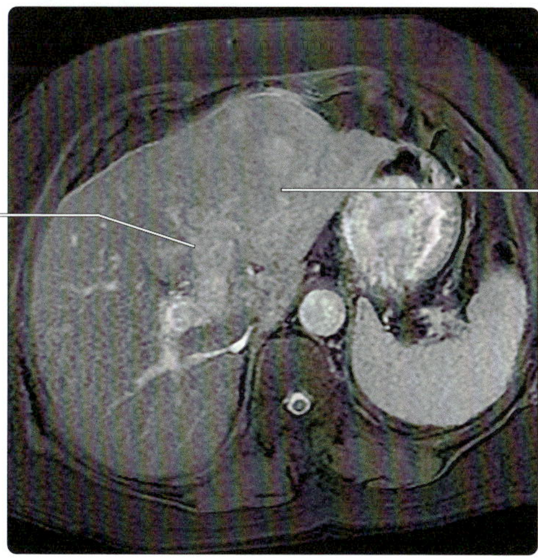

(Top) *Axial arterial-phase CECT demonstrates a typically hypervascular hepatocellular carcinoma in this cirrhotic patient. Note the direct extension of tumor into the right portal vein, a feature that usually precludes liver transplantation in these patients.* **(Middle)** *Axial CECT demonstrates extensive infiltrative hepatocellular carcinoma involving both hepatic lobes in a known cirrhotic patient. Note the presence of directly contiguous thrombus in the left portal vein, which demonstrates soft tissue attenuation suggestive of tumor thrombus (rather than the typical low density of bland thrombus).* **(Bottom)** *Axial T1 C+ arterial-phase MR demonstrates extensive infiltrative hepatocellular carcinoma involving much of the left hepatic love. There is directly contiguous tumor thrombus in the left portal vein, which demonstrates arterial enhancement on these early-phase images.*

IMAGING ANATOMY

Overview

- **Biliary ducts** convey bile from liver to duodenum
  - Bile is produced continuously by liver, stored and concentrated by gallbladder (GB), and released intermittently by GB contraction in response to presence of fat in duodenum
  - Hepatocytes form bile → bile canaliculi → interlobular biliary ducts → collecting bile ducts → **right** and **left hepatic ducts** → **common hepatic duct** (**CHD**) → **common bile duct** (**CBD**)
    - Right hepatic ducts drain right lobe (segments V-VIII) with right posterior duct draining segments VI/VII and right anterior duct draining segments V/VIII
    - Left hepatic duct drains segments II-IV via segmental ducts
    - Bile from segment I drains directly into junction of right and left hepatic ducts
  - Intrahepatic bile ducts follow branching pattern of portal veins
    - Usually lie immediately anterior to portal vein branch
    - Confluence of hepatic ducts just anterior to confluence of right and main portal veins
- **CBD** forms in free edge of lesser omentum by union of cystic duct and CHD
  - Length of CBD: 5-15 cm, depending on point of junction of cystic duct and CHD
  - CBD descends posterior and medial to duodenum, lying in groove on dorsal surface of pancreatic head
  - CBD joins pancreatic duct to form **hepaticopancreatic ampulla** (**of Vater**) (dilated short segment)
  - Ampulla opens into duodenum through **major duodenal** (**hepaticopancreatic**) **papilla** (surrounded by smooth muscle)
  - Distal CBD is thickened into **sphincter** (**of Boyden**) and hepaticopancreatic segment is thickened into **sphincter** (**of Oddi**)
    - Contraction of these sphincters prevents bile from entering duodenum and forces it to collect in GB
    - Relaxation of sphincters in response to parasympathetic stimulation and cholecystokinin (released by duodenum in response to fatty meal)
- **Vessels, nerves, and lymphatics**
  - Arterial supply
    - Intrahepatic bile ducts supplied by hepatic arteries
    - CBD blood supply divided into 3 segments
      - **Cystic artery** supplies proximal common duct
      - **Right hepatic artery** supplies middle 1/3
      - **Gastroduodenal** and **pancreaticoduodenal arcade** supply distal CBD
  - Venous drainage
    - From intrahepatic ducts → hepatic veins
    - From common duct → portal vein (in tributaries)
  - Nerves
    - Sympathetic via celiac ganglion and plexus
    - Parasympathetic via branches of vagus nerve
    - Sensory: Right phrenic nerve
  - Lymphatics
    - Same course and name as arterial branches
    - Collect at celiac and omental foramen nodes
- **GB**: Pear-shaped intraperitoneal sac that stores, concentrates, and releases bile (after eating)
  - ~ 7-10 cm long and holds up to 50 mL of bile
  - Lies in shallow fossa on visceral surface of liver
    - Vertical plane through GB fossa and middle hepatic vein divides left and right hepatic lobes
    - Touches and indents duodenum
  - Fundus is covered with peritoneum and is relatively mobile, while body and neck are attached to liver and covered by hepatic capsule
  - 3 major anatomic divisions
    - **Fundus**: Wide, rounded distal tip of GB that (usually) projects below liver edge
    - **Body**: Largest portion of GB that contacts liver, duodenum, and transverse colon
    - **Neck**: Tortuous portion of GB that narrows to join cystic duct
      - **Hartmann pouch**: Mucosal fold in GB neck where stones are commonly lodged
    - **Cystic duct**: 3- to 4-cm duct that connects GB to CHD and helps regulate bile flow to and from GB
      - Marked by **spiral folds** (**of Heister**)
  - Arterial supply: Cystic artery (usually arises from right hepatic artery but highly variable and can arise from left or common hepatic artery)
  - Venous drainage: Cystic vein (drains directly into portal vein) and small veins that drain directly into liver and hepatic veins
  - Lymphatics: Nodes draining GB are prominent in porta hepatis (e.g., cystic node near GB neck) and around pancreatic head
  - Innervation: Autonomic supply via vagus nerve (anterior vagal trunk) and sympathetic supply via celiac plexus
    - Contraction of GB and relaxation of biliary sphincters are caused by parasympathetic stimulation, but more important stimulus is from hormone cholecystokinin

CLINICAL IMPLICATIONS

Clinical Importance

- **Many common variations of biliary arterial** and **ductal anatomy** result in challenges to avoid injury at surgery
  - **Calot triangle** is bounded by CHD, cystic duct, and inferior liver surface
  - Replaced right hepatic artery and aberrant right anterior or posterior ducts run in Calot triangle and can be mistaken for cystic duct during cholecystectomy
- Close apposition of GB to duodenum can result in fistulous communication due to chronic cholecystitis or erosion of gallstones into duodenum
- **Obstruction of CBD** is common due to CBD stones, pancreatic or CBD malignancies, etc.
  - Result is **jaundice** due to backup of bile salts into bloodstream
- Cystic artery is end artery, placing GB at higher risk of ischemia and necrosis

Embryologic Events

- Abnormal embryologic development of fetal ductal plate can lead to spectrum of liver and biliary abnormalities, including polycystic liver disease, congenital hepatic fibrosis, biliary hamartomas, Caroli disease, and choledochal cysts

GALLBLADDER IN SITU AND INTERIOR

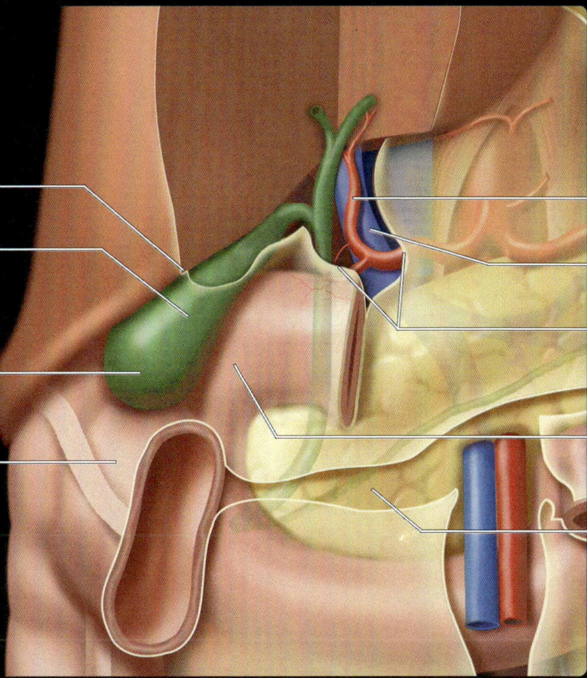

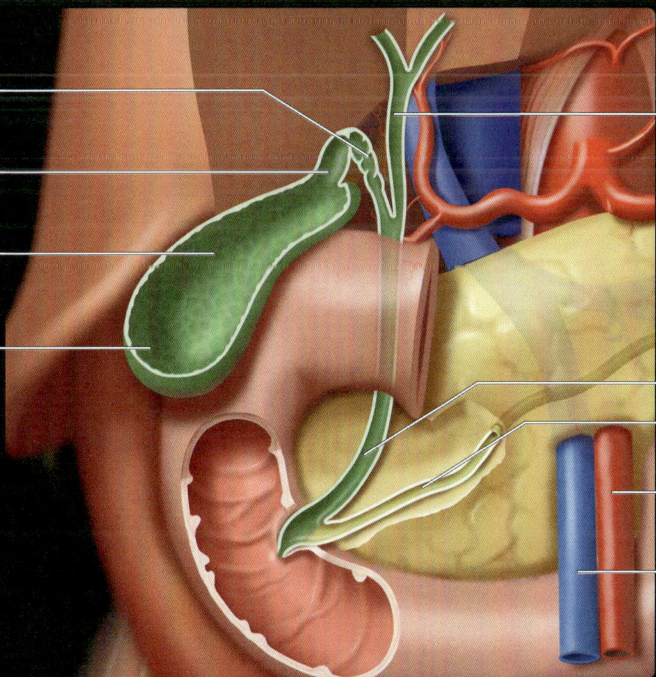

(Top) *The gallbladder is covered with peritoneum, except where it is attached to the liver. The extrahepatic bile duct, hepatic artery, and portal vein run in the lesser omentum. The fundus of the gallbladder extends beyond the anterior-inferior edge of the liver and is in contact with the hepatic flexure of the colon. The body (main portion of the gallbladder) is in contact with the duodenum.* **(Bottom)** *The neck of the gallbladder narrows before entering the cystic duct, which is distinguished by its tortuous course and irregular lumen. The duct lumen is irregular due to redundant folds of mucosa, called the spiral folds (of Heister), that are believed to regulate the rate of filling and emptying of the gallbladder. The cystic duct joins the common hepatic duct to form the common bile duct, which passes behind the duodenum and through the pancreas to enter the duodenum.*

VARIANT EXTRAHEPATIC DUCTS

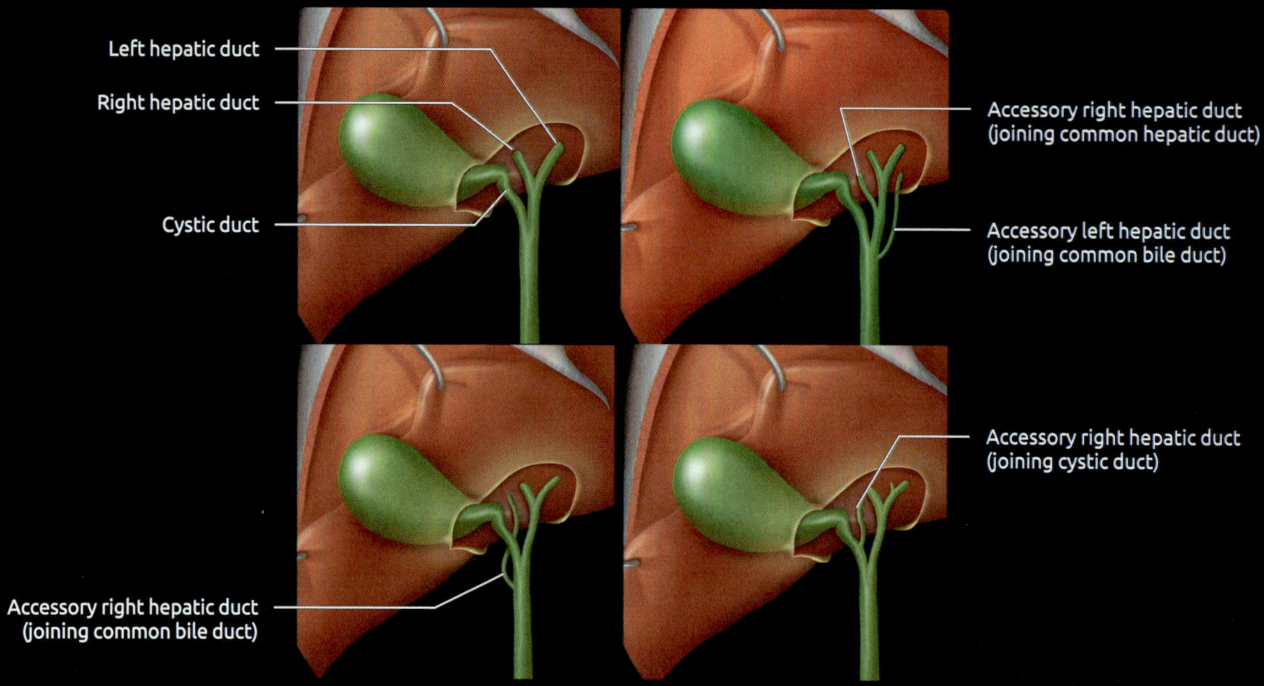

(Top) *Graphic shows the conventional arrangement of the extrahepatic bile ducts (left upper image), but variations are common (20% of the population) and may lead to inadvertent ligation or injury at surgery, such as cholecystectomy, in which the cystic duct is clamped and transected. Most accessory ducts are on the right side and usually enter the common hepatic duct but may enter the cystic duct or common bile duct. Accessory left ducts enter the common bile duct. While referred to as "accessory," these ducts are the sole drainage of bile from at least 1 hepatic segment. Ligation or laceration can lead to significant hepatic injury or bile peritonitis.* **(Bottom)** *The course and insertion of the cystic duct are highly variable, leading to difficulty in isolation and ligation at cholecystectomy. The cystic duct may be mistaken for the common hepatic or common bile duct.*

BILIARY TREE

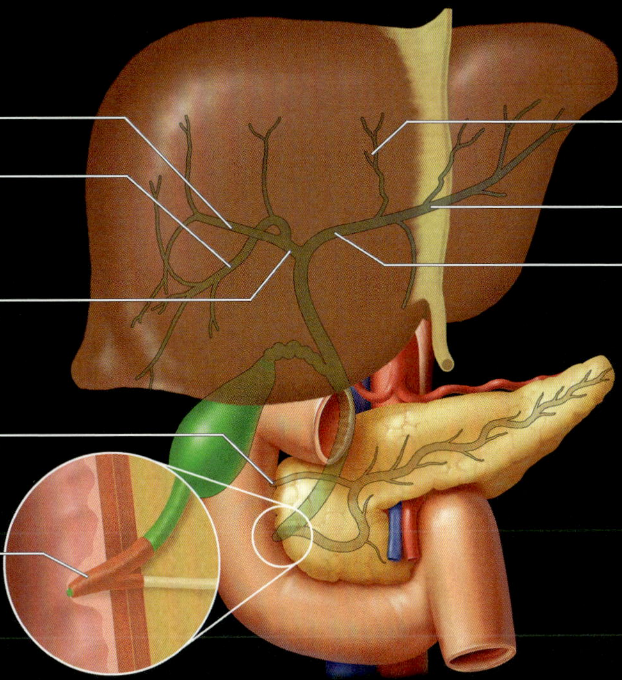

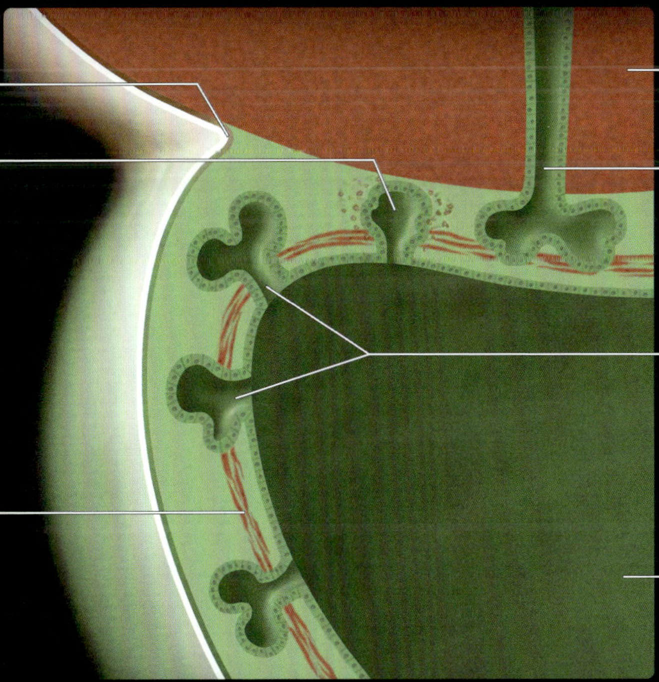

(Top) *Note the distribution of the larger intrahepatic bile ducts. The common bile duct usually joins with the pancreatic duct in a common channel or ampulla (of Vater) but may enter the major duodenal papilla separately. The distal bile duct has a sphincteric coat of smooth muscle, the choledochal sphincter (of Boyden), which regulates bile emptying into the duodenum. When contracted, the sphincter causes bile to flow retrograde into the gallbladder for storage. The common hepaticopancreatic ampulla may be surrounded by a smooth muscle sphincter (of Oddi).* **(Bottom)** *The gallbladder body and neck are adherent to the liver and may be bridged by aberrant bile ducts (of Luschka). Mucous glands are found in the gallbladder neck. Rokitansky-Aschoff sinuses are pseudodiverticular that extend into the wall and may collect debris, becoming inflamed.*

CHOLANGIOGRAM, BILIARY TREE

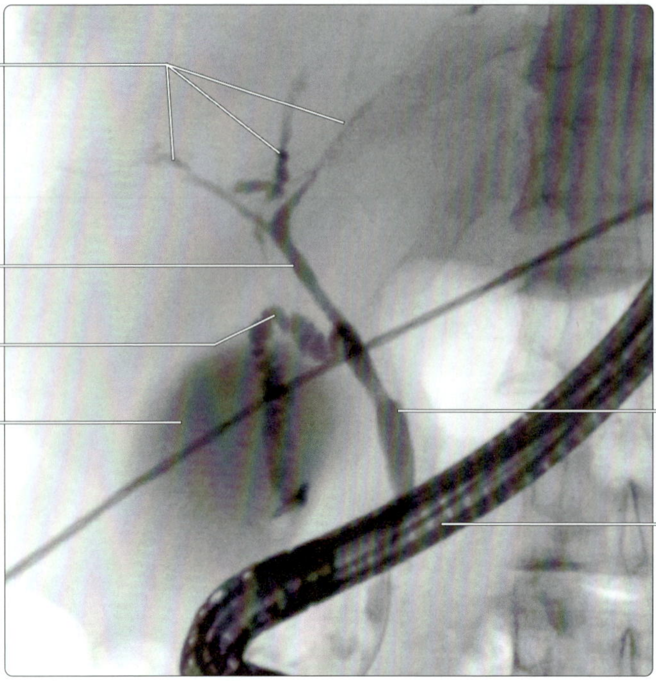

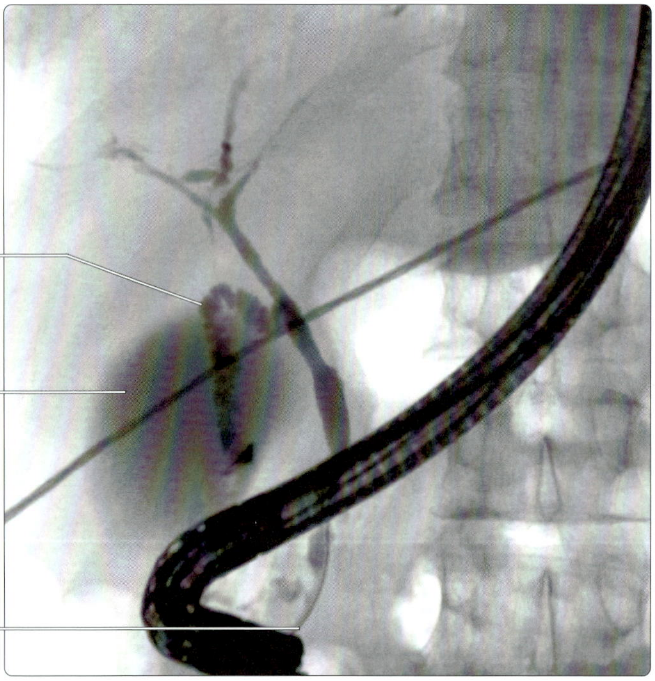

(Top) *First of 2 images from an endoscopic retrograde cholangiopancreatogram (ERCP) shows normal intrahepatic and extrahepatic bile ducts. Note the irregular contour of the cystic duct caused by the spiral folds of Heister. The gallbladder is opacified by retrograde passage of contrast material with the tip of the endoscopic cannula inserted into the hepaticopancreatic ampulla (of Vater).* **(Bottom)** *Portions of the intra- and extrahepatic ducts appear to have an irregular contour, but this is due to incomplete filling. ERCP studies are monitored in real time by x-ray fluoroscopy, and multiple films are obtained in different obliquities to distinguish artifacts, such as incomplete filling and air bubbles, from ductal pathology.*

MRCP, BILIARY TREE

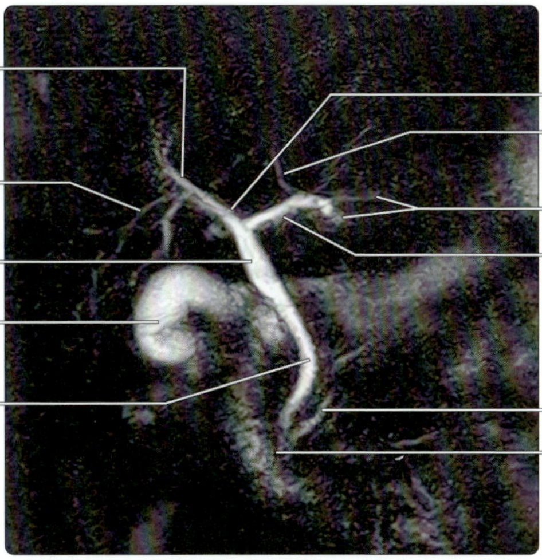

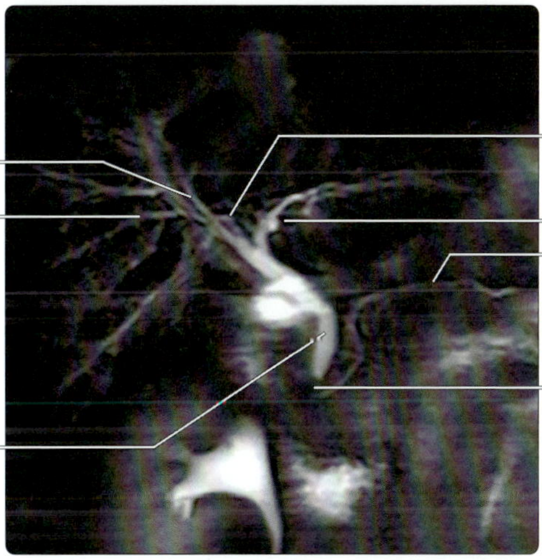

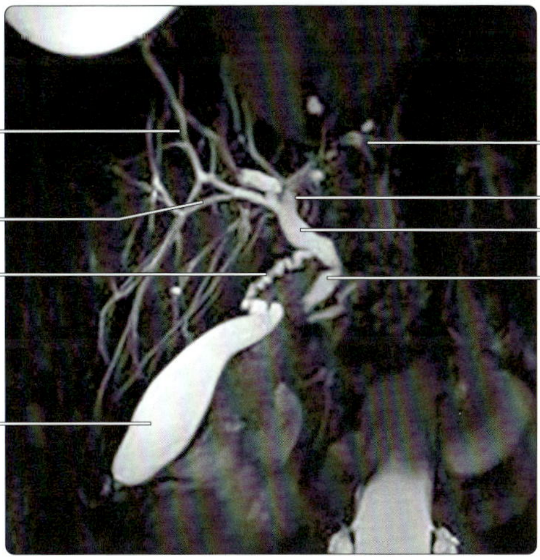

(Top) *Coronal MR cholangiopancreatogram (MRCP) with MIP reconstruction demonstrates a normal MRCP with conventional anatomy. Notice the different courses of the right posterior and anterior ducts with the anterior duct extending more vertically, while the posterior duct courses horizontally. The right anterior and posterior ducts join to form the right hepatic duct, which then combines with the left hepatic duct to form the common hepatic duct.* **(Middle)** *Coronal MRCP with MIP reconstruction again demonstrates conventional anatomy. The distal common bile duct and pancreatic duct are nicely seen to join together at the level of the ampulla.* **(Bottom)** *Coronal MRCP with MIP reconstruction nicely demonstrates the cystic duct inserting into the common hepatic duct to form the common bile duct. Note the corrugated spiral valves of Heister, which are seen throughout the cystic duct. The left hepatic duct receives flow from segmental ducts originating in segments II-IV, usually only vaguely visualized on MRCP.*

MRCP, BILIARY TREE VARIANTS

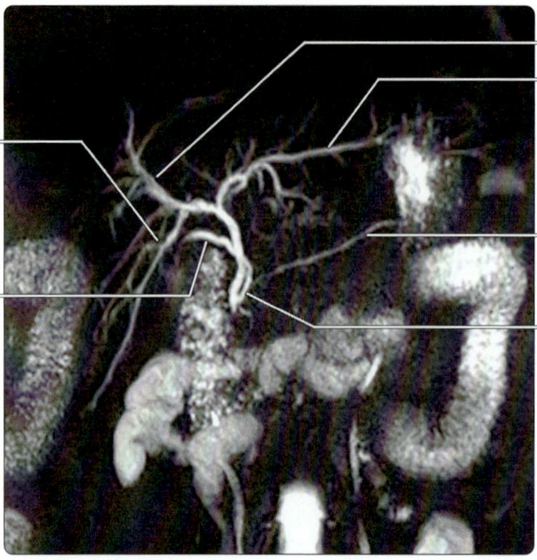

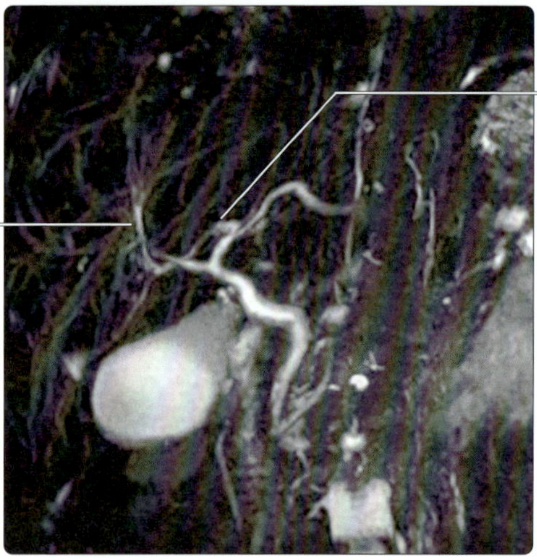

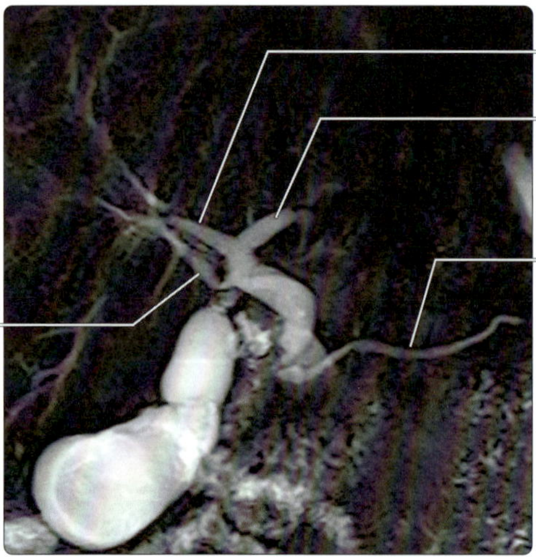

(Top) *Coronal MRCP with MIP reconstruction demonstrates a cystic duct variant with abnormally low insertion of the cystic duct onto the common hepatic duct. Several other cystic duct variants are possible, including high insertion and a parallel course of the common hepatic duct and cystic duct, and these can have some importance when the patient undergoes cholecystectomy.* **(Middle)** *Coronal MRCP with MIP reconstruction demonstrates a common biliary tree variant with the right posterior duct draining into the left hepatic duct (rather than normally combining with the right anterior duct to form the right hepatic duct).* **(Bottom)** *Coronal MRCP with MIP reconstruction demonstrates variation of the biliary tree with the right posterior duct arising from the proximal aspect of the left hepatic duct. Such variants are usually of little clinical consequence unless the patient undergoes hepatic surgery or transplantation.*

ULTRASOUND, BILIARY TREE

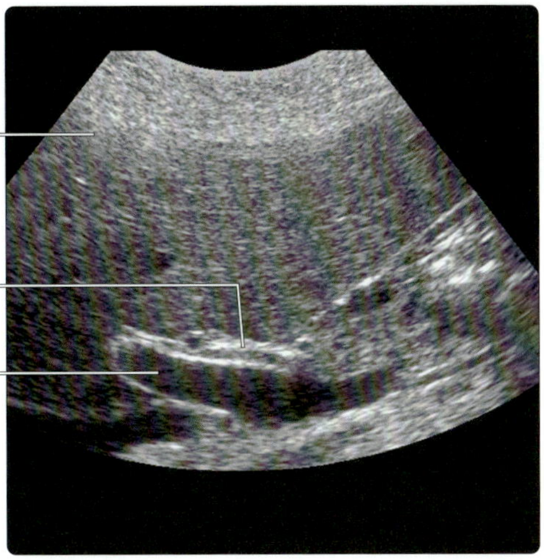

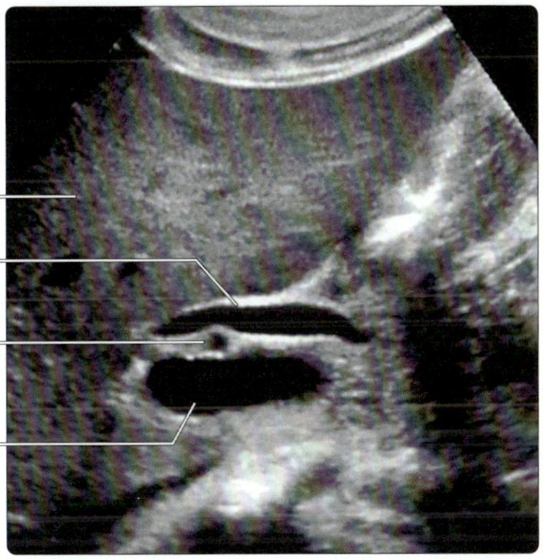

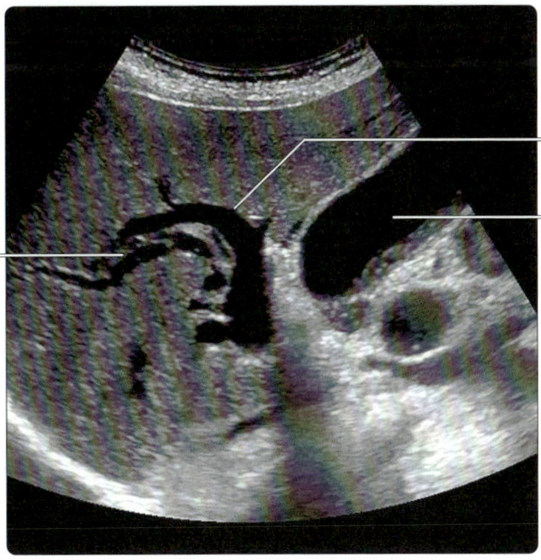

(Top) *Sagittal US demonstrates the normal relationship of the common bile duct relative to the main portal vein with the common bile duct typically lying anterior to the portal vein in the porta hepatis.* **(Middle)** *Sagittal US demonstrates the conventional appearance of the porta hepatis on US with the common bile duct lying anterior to the main portal vein. The right hepatic artery extends vertically upward between the common bile duct and the portal vein.* **(Bottom)** *Sagittal US demonstrates dilated intrahepatic bile ducts. While normal-caliber intrahepatic ducts can sometimes be difficult to visualize on US, this image with dilated ducts (due to a distal tumor) nicely demonstrates how the bile ducts run alongside the portal veins.*

ULTRASOUND, GALLBLADDER

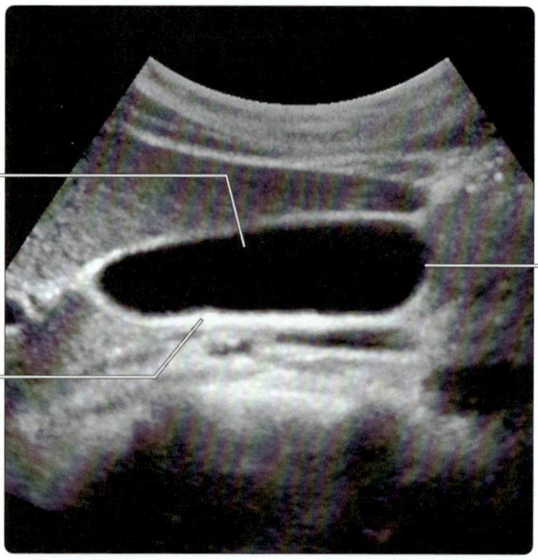

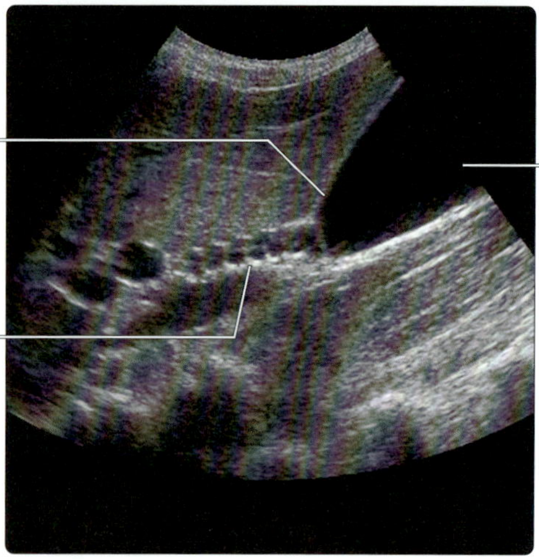

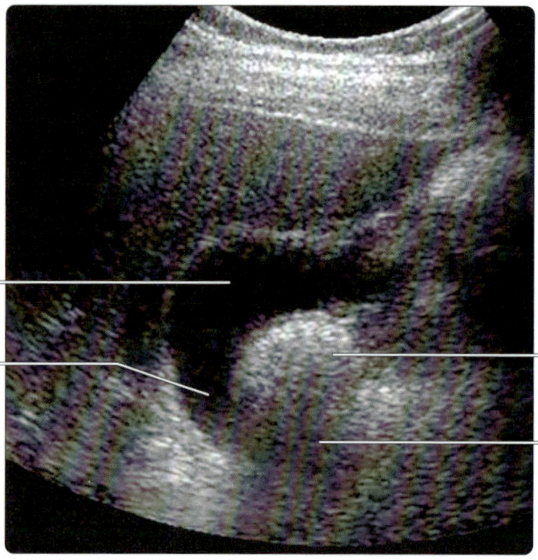

(Top) *US demonstrates a normal appearance of the gallbladder fundus and body. The fundus represents the distal tip of the gallbladder, whereas the body represents the largest central portion of the gallbladder. Notice the normal-appearing gallbladder wall (which should normally measure ≤ 3 mm) as well as the clear, anechoic normal bile filling the gallbladder lumen.* **(Middle)** *The gallbladder neck represents the proximal portion of the gallbladder, which tapers toward the gallbladder's junction with the cystic duct. The cystic duct is not routinely visualized well on US but is nicely visualized here as a tubular, tortuous structure. Note the ridge-like striations in the cystic duct on this image, representing the spiral valves of Heister (which are not always seen well on imaging).* **(Bottom)** *Sagittal US demonstrates a large echogenic reflector with posterior "dirty" acoustic shadowing abutting the gallbladder body and neck. While this could be mistaken for a stone or even gas in the gallbladder (i.e., emphysematous cholecystitis), this is merely bowel gas adjacent to the gallbladder. The gallbladder abuts the duodenum and colon, and bowel gas should not be confused for true pathology.*

ANGIOGRAM, CYSTIC ARTERY

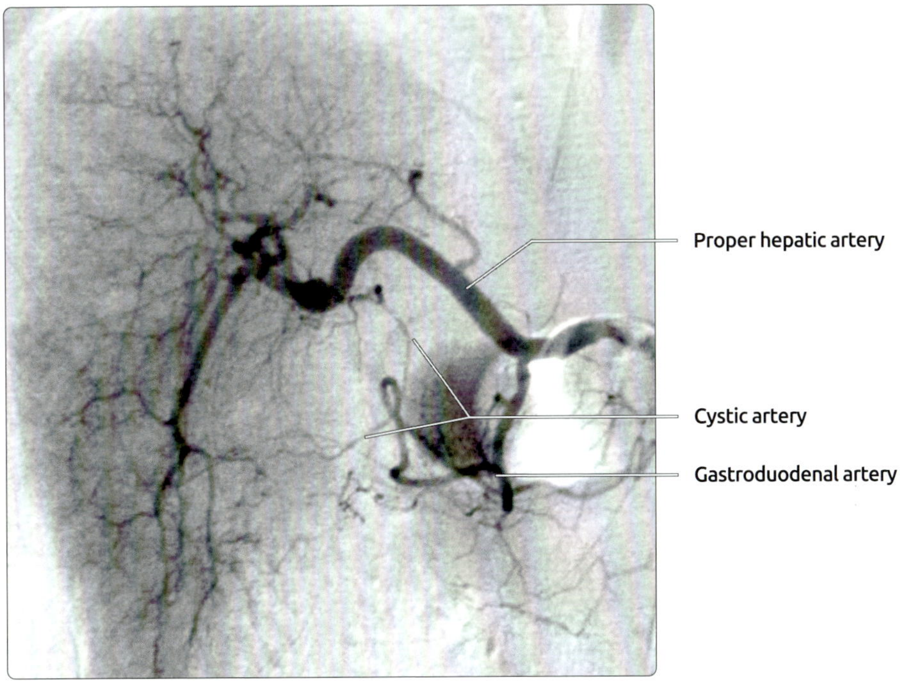

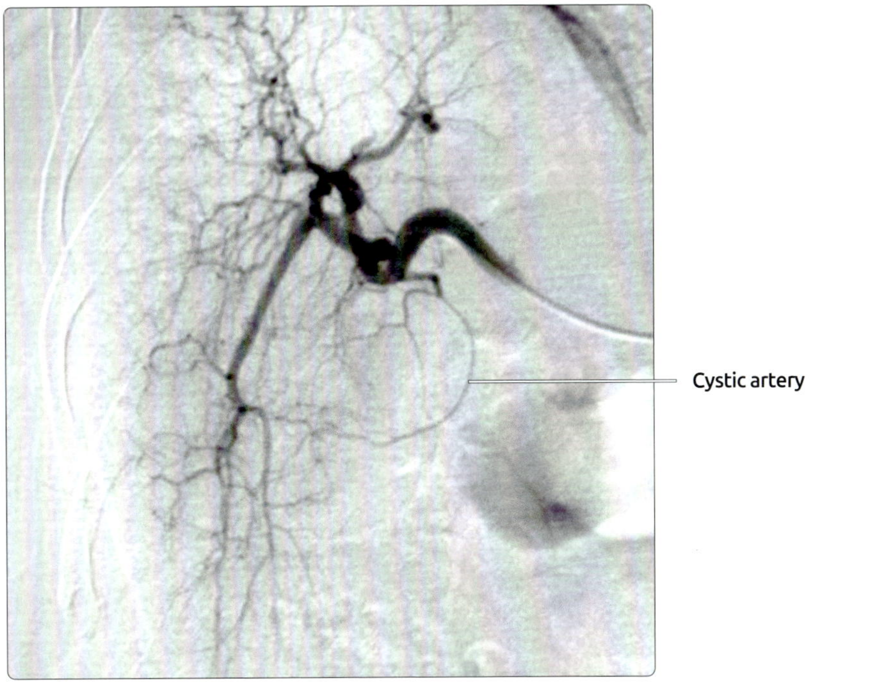

(Top) *Frontal catheter angiogram shows the cystic artery stretched along the surface of a distended gallbladder. The cystic artery usually arises from the right hepatic artery, though this is quite variable.* **(Bottom)** *Selective injection of the proper hepatic artery again shows the course of the normal cystic artery.*

CHOLANGIOGRAM, BILIARY VARIANTS

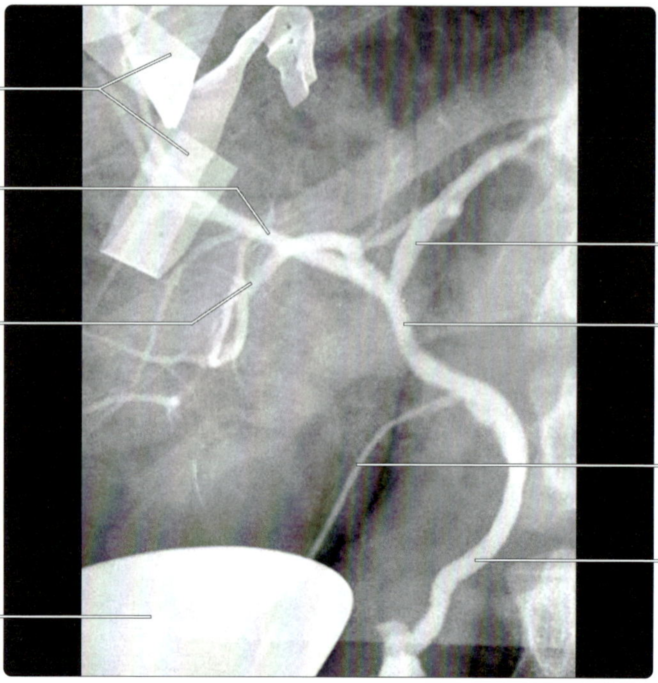

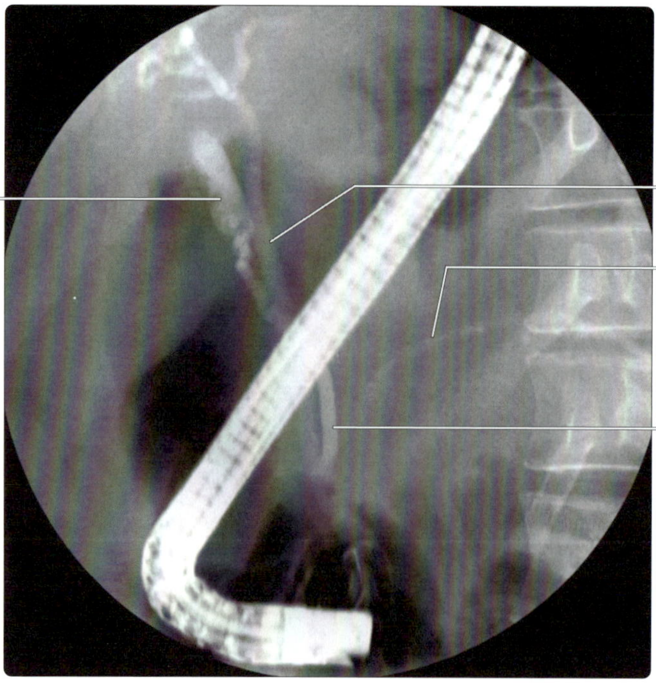

(Top) *Intraoperative cholangiogram was performed in a subject who was about to undergo right hepatic lobectomy as a living donor to a relative with liver failure. The cholangiogram shows a trifurcation pattern of the main ducts draining the right and left lobes. This is an unfavorable anatomic variant in this setting because the surgical plane of section will cut across both the anterior and posterior segmental bile ducts, requiring 2 separate anastomoses to the recipient's biliary (or enteric) system.* **(Bottom)** *Frontal ERCP shows a low insertion of the cystic duct into the distal common duct, likely within its intrapancreatic portion. Note the valves of Heister that impart a corrugated appearance of the cystic duct near its junction with the gallbladder. The cystic duct appears to spiral around the common duct and is likely enclosed in a common sheath with the common duct. All of these features have clinical implications, such as the likelihood of Mirizzi syndrome in case of a cystic duct stone.*

CHOLELITHIASIS AND CHOLEDOCHOLITHIASIS

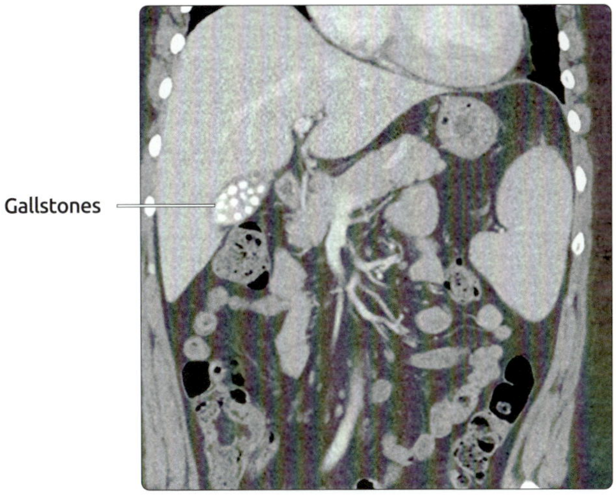

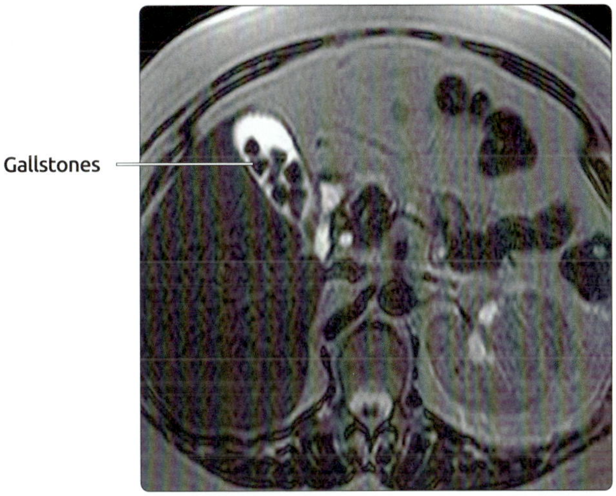

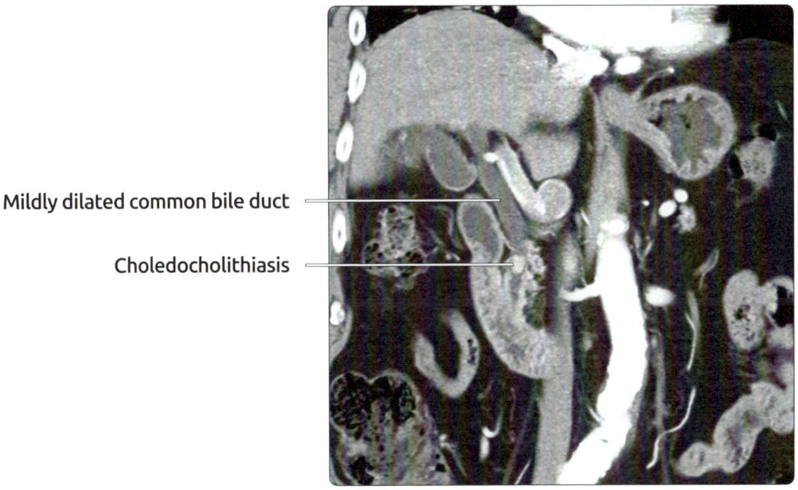

(Top) *Coronal CECT demonstrates multiple discrete high-density calcified gallstones in the gallbladder. CT is generally less sensitive for gallstones compared to either MR or US, particularly with regards to cholesterol stones.* **(Middle)** *Axial T2 MR depicts numerous low-signal gallstones in the gallbladder. With only a few rare exceptions, gallstones tend to be low in signal on all MR pulse sequences.* **(Bottom)** *Coronal CECT demonstrates a calcified gallstone in the common bile duct. The distal common bile duct stone is causing mild obstruction, as evidenced by the mildly dilated common bile duct proximal to the stone.*

CHOLELITHIASIS AND CHOLEDOCHOLITHIASIS

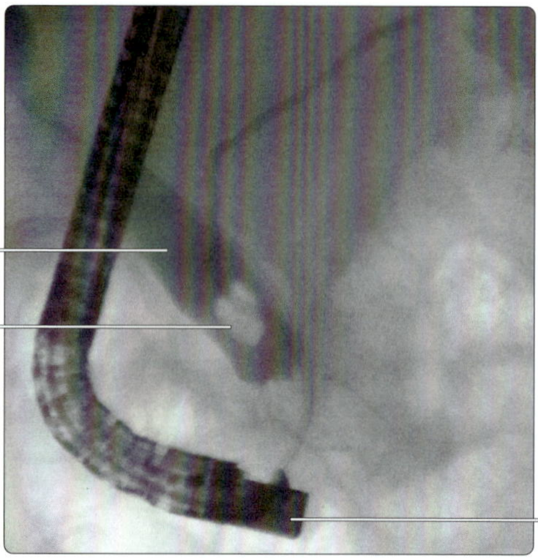

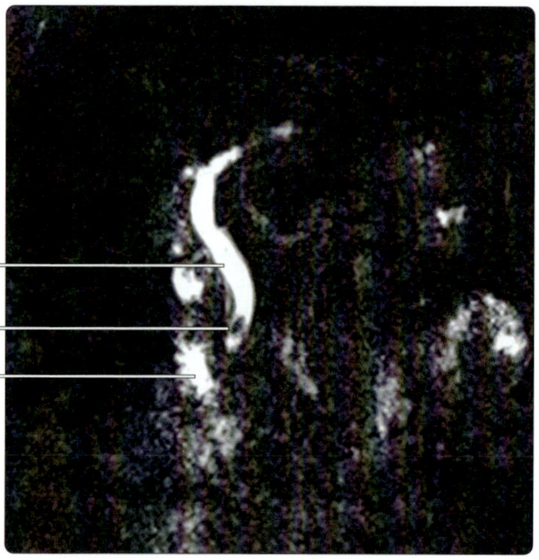

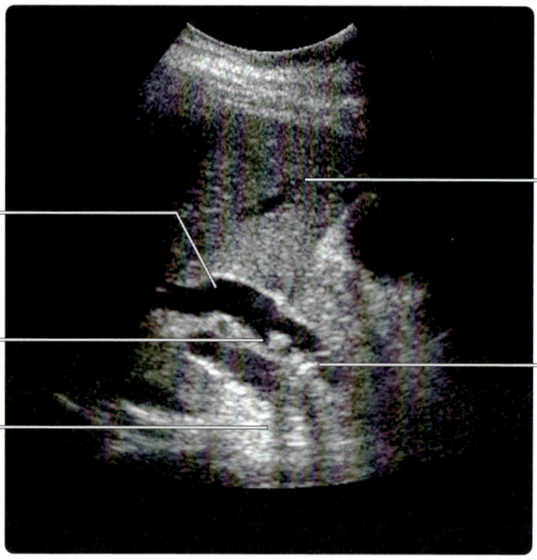

(Top) *Frontal ERCP in a patient with right upper quadrant pain shows a filling defect within the common bile duct, representing a ductal stone.* **(Middle)** *Coronal MRCP demonstrates a low-signal nodular filling defect in the distal common bile duct without evidence of substantial proximal biliary obstruction. MRCP is the most sensitive imaging test for identifying stones in the common bile duct, but it is important to utilize thin-section 3D MRCP (rather than simply viewing the MIP reconstructions) to maximize sensitivity.* **(Bottom)** *Longitudinal US demonstrates calcified gallstones with posterior acoustic shadowing in the common bile duct. US often struggles in the identification of common duct stones, as a result of frequent overlying bowel gas obscuring the duct.*

ACUTE CHOLECYSTITIS

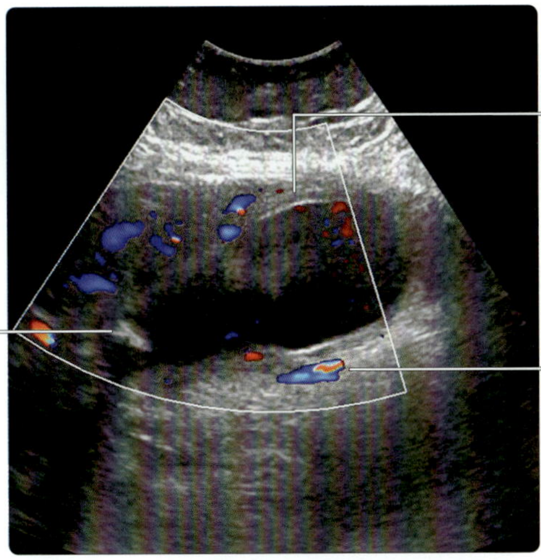

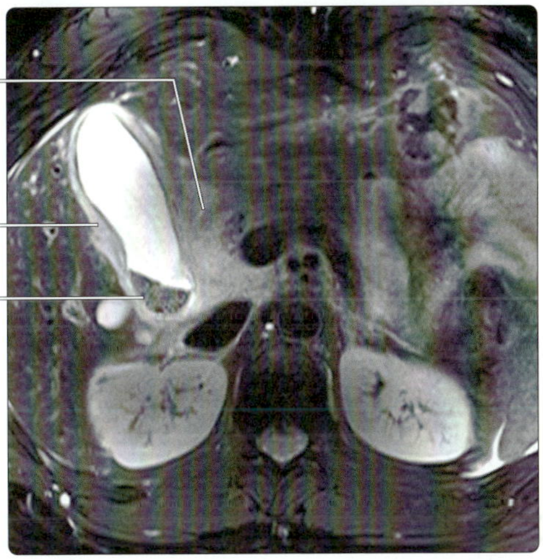

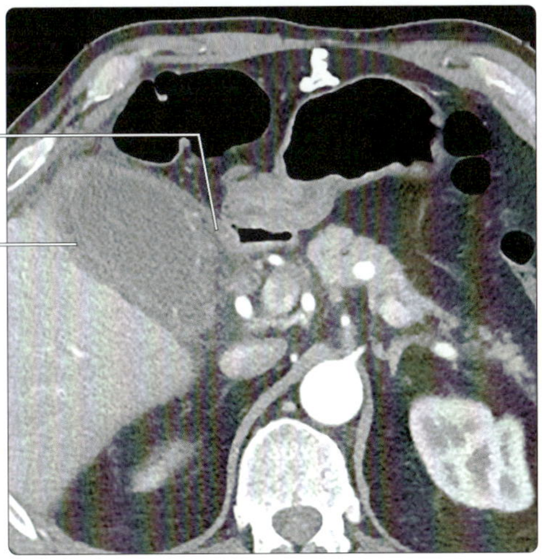

(Top) *Sagittal color Doppler US demonstrates a nonmobile gallstone lodged in the neck of the gallbladder with mild diffuse gallbladder wall thickening and wall hyperemia. The patient was tender in the right upper quadrant, and the diagnosis of acute calculous cholecystitis was confirmed at resection.* **(Middle)** *Axial T2 FS MR in a patient with right upper quadrant pain demonstrates diffuse gallbladder wall edema with adjacent stranding and soft tissue edema as well as multiple layering low-signal gallstones, compatible with acute cholecystitis.* **(Bottom)** *Axial CECT demonstrates gallbladder wall thickening and mucosal hyperemia with adjacent stranding. The diagnosis of acute cholecystitis was confirmed at surgery.*

MALIGNANT BILIARY OBSTRUCTION

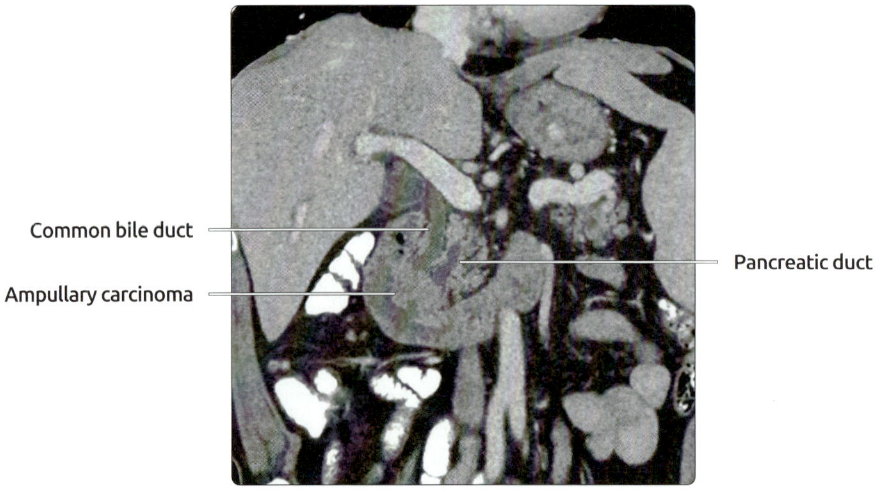

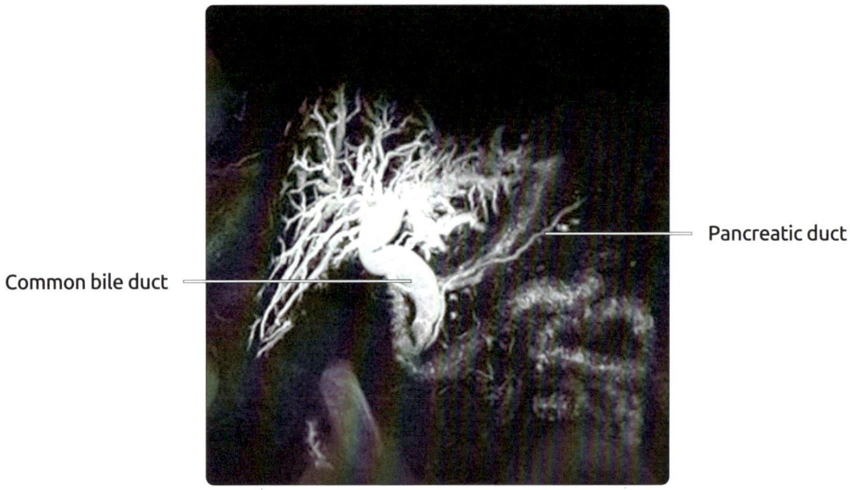

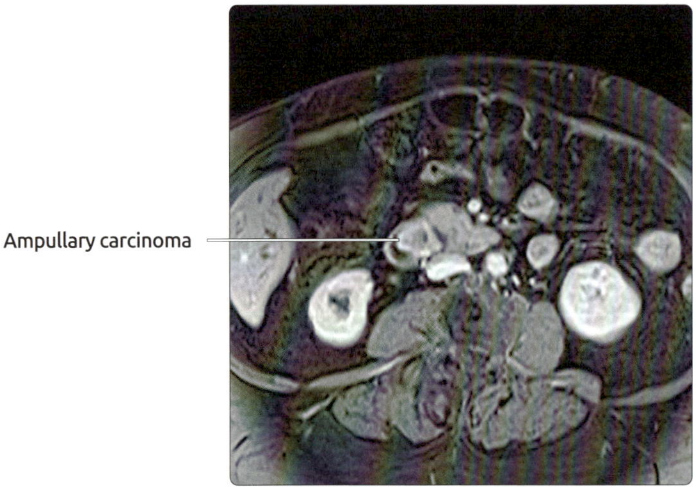

(Top) *Coronal CECT demonstrates a periampullary mass resulting in mild dilatation of the upstream common bile duct and pancreatic duct. The mass was found to represent an ampullary carcinoma at resection. The prognosis of these tumors can vary depending on whether they arise from pancreaticobiliary epithelium (worse prognosis) or intestinal epithelium (better prognosis).* **(Middle)** *Coronal MRCP with MIP reconstruction demonstrates marked biliary dilatation without substantial pancreatic ductal dilatation.* **(Bottom)** *Axial T1 C+ FS MR in the same patient demonstrates a small enhancing mass at the ampulla, representing the cause of the patient's biliary obstruction, ultimately shown at surgery to represent an ampullary carcinoma.*

MALIGNANT BILIARY OBSTRUCTION

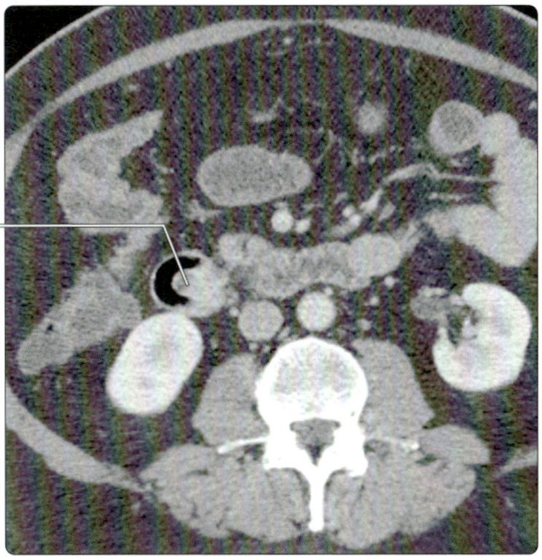

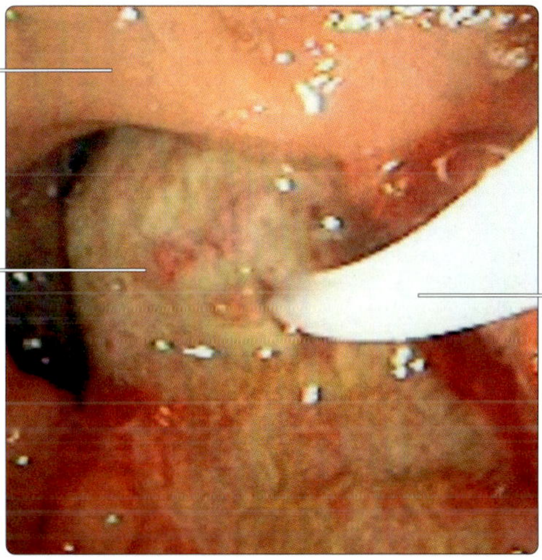

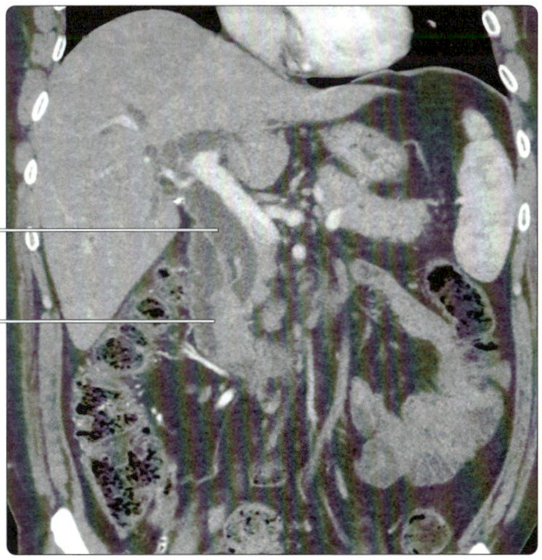

(Top) *Axial CECT demonstrates a soft tissue density mass that distorts the medial wall of the duodenum, representing an ampullary carcinoma.* **(Middle)** *Endoscopic view of the interior of the duodenum in the same patient shows the periampullary mass projecting into the lumen. The endoscopic probe or canula has been inserted through the tumor to opacify the bile duct and pancreatic duct. The mass was resected and proved to be an ampullary carcinoma.* **(Bottom)** *Coronal CECT in a patient with advanced HIV/AIDS demonstrates a periampullary soft tissue mass resulting in upstream biliary dilatation. This was ultimately found to represent an unusual manifestation of Burkitt lymphoma.*

MALIGNANT BILIARY OBSTRUCTION

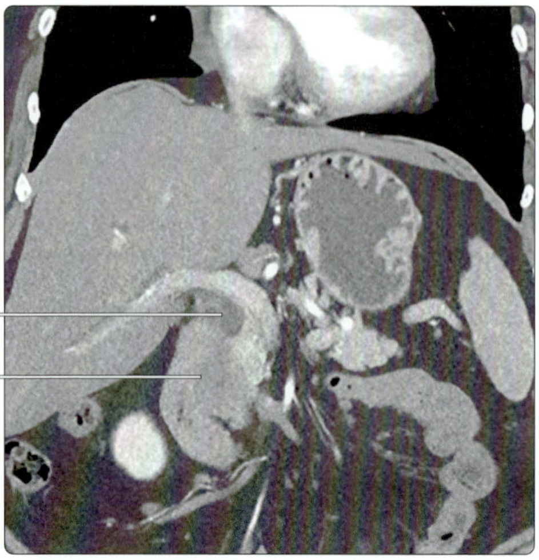

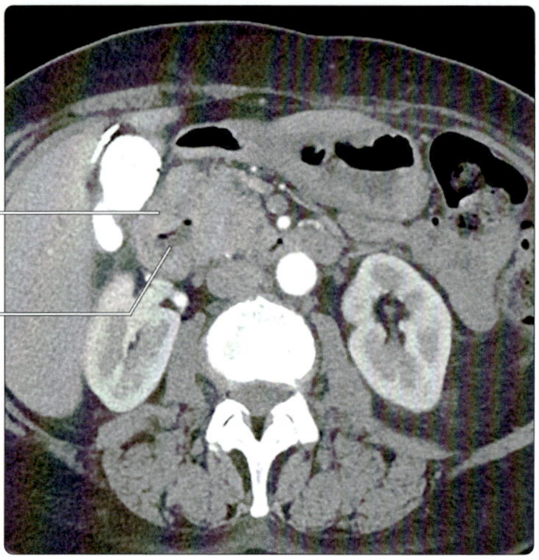

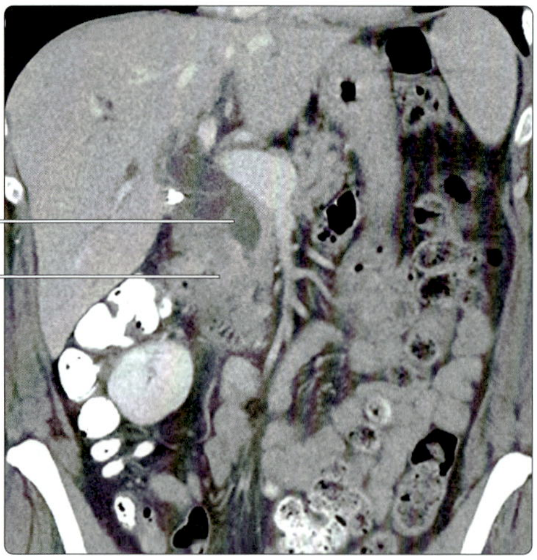

(Top) *Coronal CECT demonstrates a soft tissue mass at the ampulla causing mild upstream biliary dilatation. Indistinguishable on imaging from an ampullary carcinoma, this is a periampullary cholangiocarcinoma arising from the distal common bile duct.* **(Middle)** *Axial CECT demonstrates an annular, ulcerated mass centered in the 2nd/3rd portions of the duodenum near the ampulla.* **(Bottom)** *Coronal CECT in the same patient demonstrates that the periampullary duodenal mass results in obstruction of the biliary tree as a result of ampullary involvement. This was ultimately found to represent a periampullary duodenal adenocarcinoma.*

MALIGNANT BILIARY OBSTRUCTION

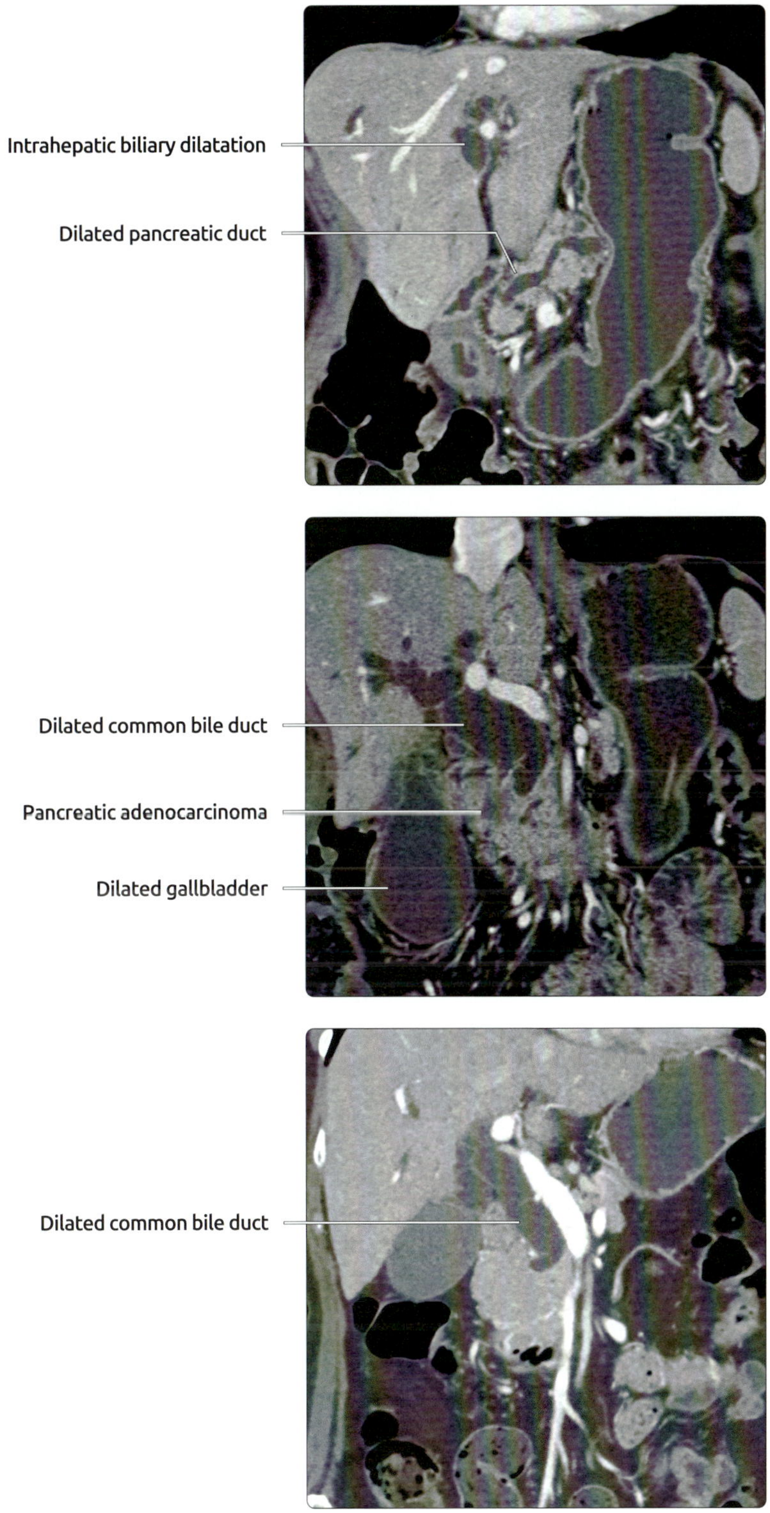

(Top) *Coronal CECT demonstrates a dilated pancreatic duct along with partially visualized dilated intrahepatic ducts.* **(Middle)** *Coronal CECT in the same patient demonstrates a hypodense, poorly marginated mass near the ampulla, ultimately found to represent a pancreatic adenocarcinoma, causing obstruction of the biliary tree with resultant dilatation of the common bile duct and the gallbladder.* **(Bottom)** *Coronal CECT demonstrates a dilated common bile duct with abrupt narrowing near the ampulla. Although an obstructing lesion is not clearly visible, a very tiny obstructing mass was found at endoscopy, which ultimately turned out to represent a pancreatic adenocarcinoma.*

BILIARY HAMARTOMAS

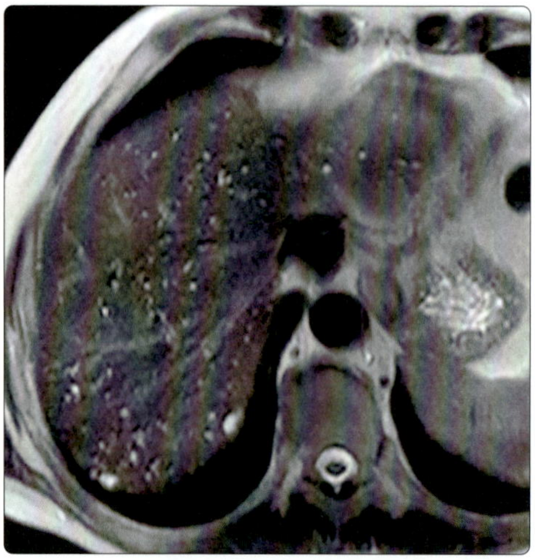

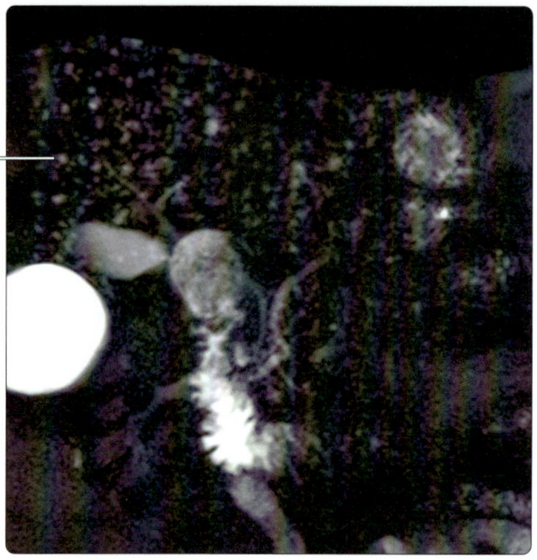

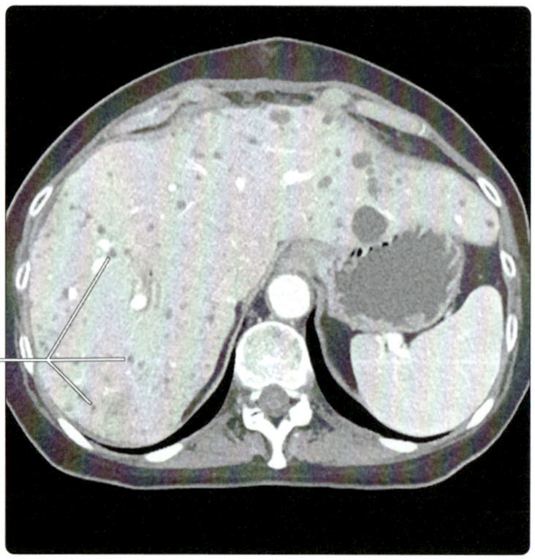

(Top) *Axial T2 HASTE MR demonstrates innumerable tiny T2-bright foci scattered throughout the liver, a common appearance for benign biliary hamartomas.* **(Middle)** *Coronal MRCP with MIP reconstruction in the same patient nicely demonstrates the innumerable tiny T2-bright foci in the liver without evidence of any biliary dilatation. The T2-bright foci are not in any apparent continuity with the larger bile ducts. The large number and small size of the cystic lesions indicate that these represent biliary hamartomas, which are developmental anomalies due to failure of involution of embryonic bile ducts that fail to connect with the normal biliary tree. This is an asymptomatic condition but may be mistaken for more serious conditions, such as polycystic disease or liver metastases.* **(Bottom)** *Axial CECT demonstrates the characteristic appearance of biliary hamartomas on CT with innumerable tiny hypodense foci that are distinguished from cysts by their small size, large number, and slightly irregular contours.*

CAROLI DISEASE

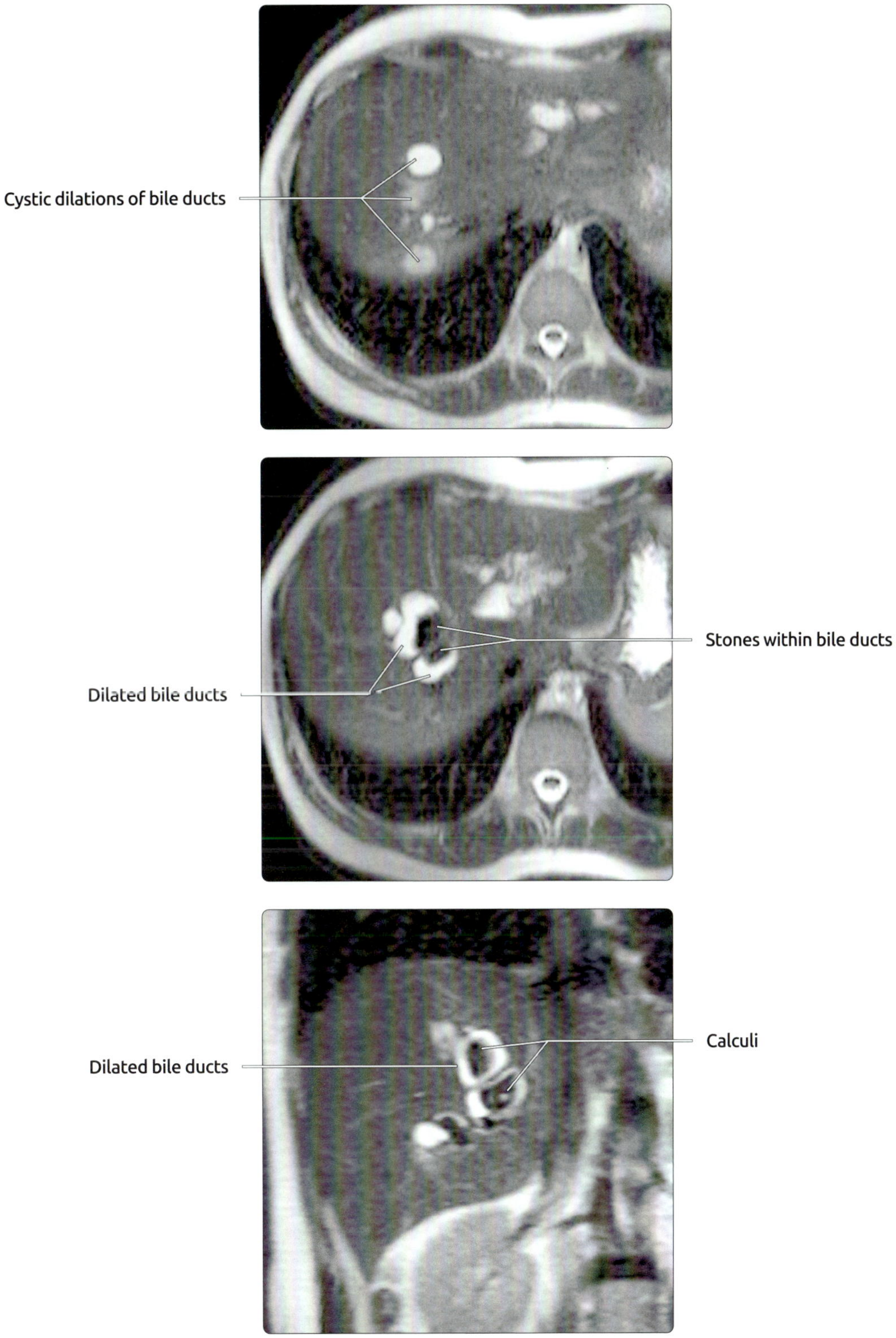

(Top) *First of 3 T2 HASTE MR images from a patient with Caroli disease (communicating cavernous biliary ectasia), a congenital anomaly that results in multifocal, saccular dilation of the intrahepatic bile ducts, is shown. Axial T2 MR shows multiple high-intensity spherical lesions within the liver, representing dilated bile ducts.* **(Middle)** *A more caudal image shows large calculi as low-signal nodules within the massively dilated bile ducts. The bile ducts do not "arborize" or branch in a uniform pattern; they are irregularly dilated in a cystic or fusiform pattern.* **(Bottom)** *Coronal T2 MR section shows the extent of the ductal abnormalities and the large calculi within the dilated ducts.*

CHOLEDOCHAL CYST

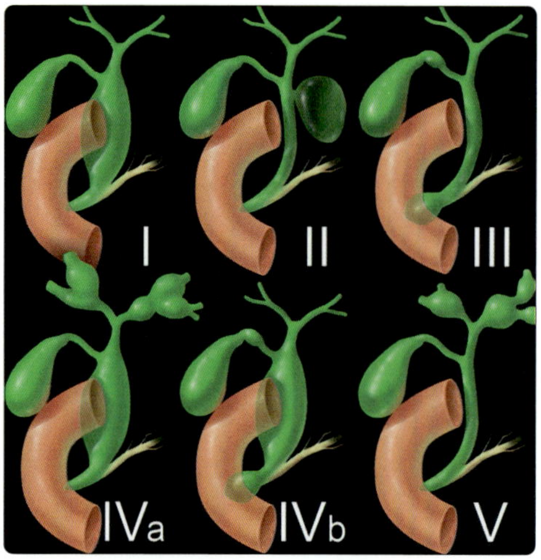

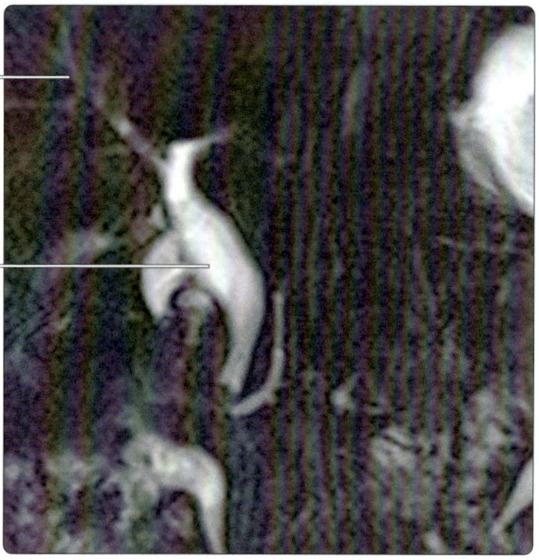

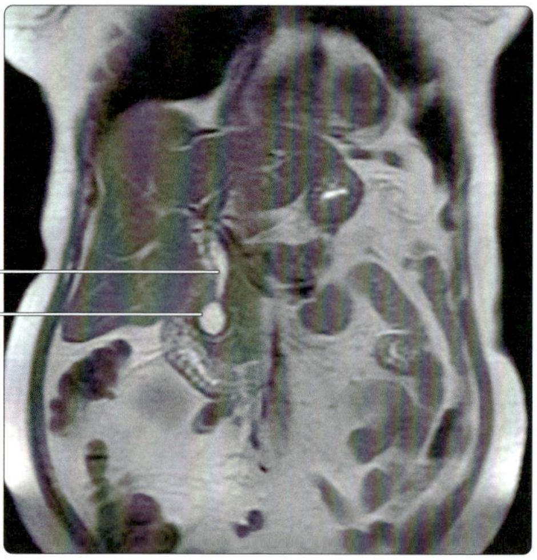

(Top) *Graphic illustrates the Todani classification of choledochal cysts. Type I is the most common and represents fusiform dilation of the extrahepatic bile duct. Type II is a diverticulum of the extrahepatic bile duct (rare). Type III is also rare and is a choledochocele, a diverticular expansion of the distal bile duct within the duodenal wall. Type IV is the 2nd most common type and is cystic dilation of the intrahepatic and extrahepatic bile ducts. Type V is cystic dilation of the intrahepatic bile ducts and is synonymous with Caroli disease.* **(Middle)** *Coronal MRCP with MIP reconstruction in an asymptomatic patient demonstrates diffuse, fusiform dilatation of the extrahepatic bile duct with normal-caliber intrahepatic ducts, characteristic of a type I choledochal cyst.* **(Bottom)** *Coronal T2 HASTE MR demonstrates a diverticulum (choledochocele) arising from the distal common bile duct at its insertion into the duodenum, compatible with a rare type III choledochal cyst.*

CHOLEDOCHAL CYST

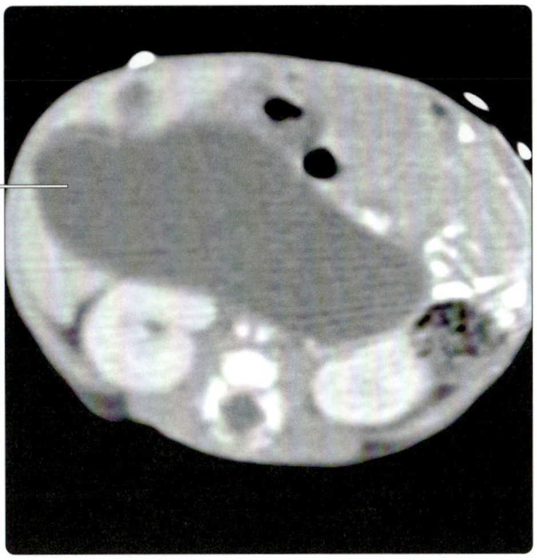

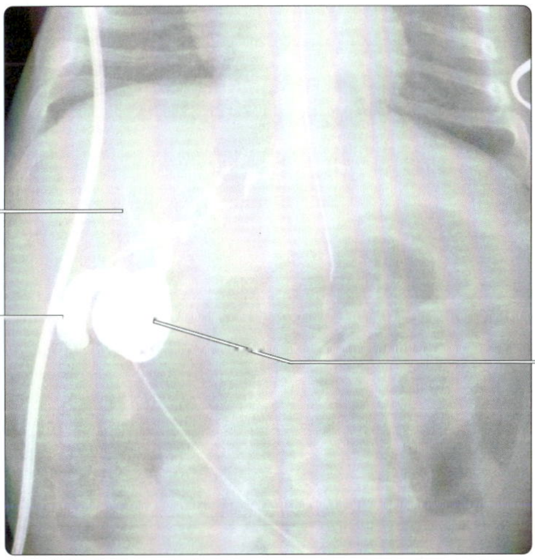

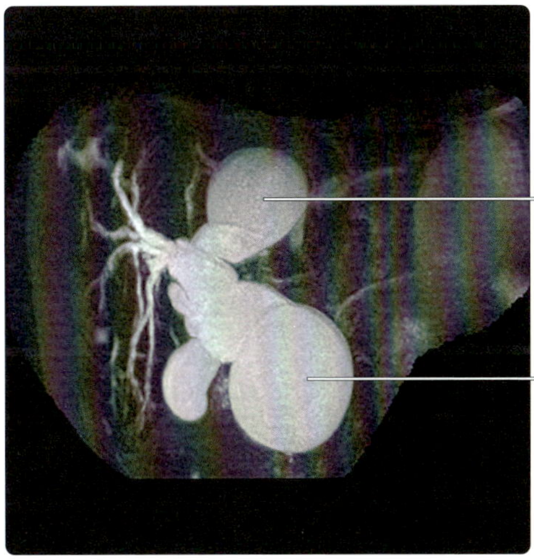

(Top) *Axial CECT in a newborn demonstrates a large cystic mass in the right upper quadrant, ultimately found to represent a type II choledochal cyst communicating with the common bile duct.* **(Middle)** *Frontal ERCP in a young pediatric patient shows fusiform dilation of much of the common bile duct, representing a type I choledochal cyst.* **(Bottom)** *Coronal MRCP with MIP reconstruction demonstrates a type IV choledochal cyst with cystic and fusiform dilation of both the intrahepatic and common bile ducts.*

GROSS ANATOMY

Overview

- Pancreas: Accessory digestive gland lying in retroperitoneum posterior to stomach and anterior to upper lumbar spine, which has both exocrine and endocrine function
  - Exocrine function: Pancreatic **acinar cells** secrete pancreatic digestive enzymes into duodenum via pancreatic duct
  - Endocrine: Pancreatic **islet cells** (of **Langerhans**) secrete insulin, glucagon, and other polypeptides into portal venous system

Divisions

- **Head**: Thickest portion of pancreas that lies to right of superior mesenteric vessels [superior mesenteric artery (SMA) and superior mesenteric vein (SMV)] and is attached to duodenal C loop (retroperitoneal 2nd and 3rd portions of duodenum)
  - **Uncinate** process is posterior/medial extension of pancreatic head, which lies posterior to SMV
  - **Common bile duct** lies along posterior surface of head and joins with **pancreatic duct** (**of Wirsung**) to form **hepatopancreatic ampulla** (**of Vater**)
  - Main pancreatic duct and common bile duct empty into **major papilla** in 2nd portion of duodenum
  - Hepatic artery located immediately superior to pancreatic head and gives rise to gastroduodenal artery (GDA), which runs along anterior/lateral aspect of pancreatic head
- **Neck**: Narrow portion of pancreas that lies immediately anterior to portal vein-splenic vein confluence
  - SMV joins splenic vein behind pancreatic neck to form **main portal vein**
  - Anterior border of pancreatic neck covered by gastric pylorus and lesser sac
- **Body**: Accounts for majority of pancreatic parenchyma and lies anterior to aorta, splenic vein, and left adrenal gland/kidney
  - **Splenic vein** lies in groove along posterior surface of pancreatic body
  - Anterior surface is covered with peritoneum, which forms posterior surface of **omental bursa** (**lesser sac**) separating stomach and pancreas
- **Tail**: Lies between peritoneal layers of **splenorenal ligament** in splenic hilum
  - Space immediately posterior to pancreatic tail is part of peritoneal space and is potential pathway of spread for pancreatic diseases into peritoneum
  - No defined dividing point between body and tail visible on imaging

Internal Structures

- **Pancreatic ductal anatomy**
  - **Main pancreatic duct** (**of Wirsung**) runs length of pancreas (from tail to head), eventually turning posteriorly and inferiorly through head to join common bile duct at major papilla
  - **Accessory pancreatic duct** (**of Santorini**) opens into duodenum at **minor duodenal papilla**
    - Usually (but not always) communicates with main pancreatic duct
  - Variations in pancreatic ductal anatomy are common, including dominant accessory duct draining most of pancreas
  - Duct should typically taper from head to neck (narrowest in tail segment)
- **Vessels, nerves,** and **lymphatics**
  - Arterial blood supply to pancreas derived from branches of celiac artery and SMA
    - Arterial supply to pancreatic head provided by anterior and posterior **pancreaticoduodenal arcades** (with majority of blood supply via GDA)
      - Anterior (1st branch from GDA) and posterior superior pancreaticoduodenal arteries arise from GDA
      - Inferior pancreaticoduodenal artery (with anterior and posterior branches) usually arises from right aspect of SMA
      - Dorsal pancreatic artery provides some blood supply to head/neck but variable origin (usually from proximal splenic artery)
    - Arterial blood supply to body and tail primary via **splenic artery**
      - Most important branches of splenic artery are **dorsal pancreatic artery** and **pancreatic great** (**magna**) **artery**, which arise from proximal and midportions of splenic artery, respectively
      - Small **caudal pancreatic artery** arises from more distal splenic artery and provides some minimal blood supply to body/tail
  - Venous drainage parallels arterial supply with draining veins (pancreatic veins, pancreaticoduodenal veins) serving as tributaries of SMV and splenic vein (eventually flowing into portal vein)
  - Autonomic nervous supply from **celiac** and **superior mesenteric plexus**
    - Parasympathetic stimulation of pancreatic secretion, but pancreatic juice secretion is mostly under hormonal control (**secretin** from duodenum)
  - Lymphatics follow blood vessels
    - Collect in **splenic**, **celiac**, **superior mesenteric**, and **hepatic nodes** with head drainage usually flowing into celiac and superior mesenteric nodes (and ultimately into cisterna chyli), while tail drains into splenic hilar lymph nodes

ANATOMY IMAGING ISSUES

Key Concepts

- Normal multimodality imaging appearance
  - **CT**
    - Normal pancreas demonstrates attenuation of 40-50 HU on NECT
    - Should demonstrate avid, rapid, homogeneous enhancement after administration of IV contrast with peak enhancement typically slightly earlier than standard portal venous phase
      - Peak enhancement often seen at roughly 40-45 seconds (pancreatic phase)
      - Inhomogeneous enhancement increasingly common in older patients due to fatty infiltration

- Main pancreatic duct typically measures < 3 mm and should gradually taper
- **MR**
  - Normal pancreas should demonstrate T1 signal, which is ≥ to liver
    - Pancreas generally shows highest T1 signal of any parenchymal organ in abdomen
  - Variable parenchymal signal on T2WI (generally low to intermediate signal), limiting utility of this sequence for detection of tumors
  - Enhancement kinetics after administration of gadolinium similar to CECT
  - Diffusion-weighted imaging may be helpful for lesion detection
- **Ultrasound**
  - Pancreas can be quite difficult to visualize completely on ultrasound due to overlying bowel gas (particularly distal body/tail segments)
    - Visualization may be improved by filling stomach with water immediately prior to scan, with patient in sitting/upright position, or utilizing Valsalva maneuver
  - Generally tends to be hypoechoic (compared to fat) in younger patients (similar to liver echogenicity) and hyperechoic in older patients (due to fatty infiltration)
  - Size thresholds frequently described for different portions of pancreas but generally of little clinical utility given great variability

- Shape, size, and texture of pancreas are quite variable and will typically change over patient's lifetime
  - Pancreas appears largest and most "full" in young patients
  - Global (not focal) atrophy is common in older patients (especially > 70 years) with increasingly evident lobulations, mild dilatation of pancreatic duct (usually still < 3 mm), and rare punctate parenchymal calcifications
  - Atrophy and fatty infiltration can also be normally seen in patients with obesity or diabetes
  - Any focal bulge or mass effect is abnormal and should prompt careful evaluation for mass
  - Focal fatty infiltration of pancreatic head is common variant and is not uncommonly mistaken for "mass"
- Pancreas lies in anterior pararenal space (APS) immediately posterior to lesser sac and does not have true capsule
  - APS defined by parietal layer of lesser sac anteriorly, anterior layer of Gerota fascia posteriorly, and by lateroconal fascia laterally
    - Other constituents of APS are duodenum, ascending/descending colon, and blood vessels
    - APS extends inferiorly into pelvis and has connection with transverse mesocolon and small bowel mesentery
  - Inflammation (e.g., from pancreatitis) easily spreads to duodenum and descending colon (which also lie in APS)
  - Inflammation easily spreads into mesentery and mesocolon (roots of these lie just ventral to pancreas)
  - Acute pancreatitis often results in lesser sac fluid (which should not be misinterpreted as pseudocyst)
- Important anatomic clues to pathology
  - **Dilated pancreatic duct**
    - Focal dilatation of pancreatic duct with abrupt narrowing (often in conjunction with upstream atrophy) should raise concern for obstructing mass (especially pancreatic adenocarcinoma)
    - Chronic pancreatitis may manifest as dilated, beaded pancreatic duct, often in conjunction with parenchymal and intraductal calcifications
    - Intraductal papillary mucinous neoplasm with main duct involvement can result in diffuse or segmental dilatation of duct
  - **Obstruction of splenic vein**
    - Common result of pancreatic tail mass (especially pancreatic adenocarcinoma) or chronic pancreatitis
    - Results in dilated venous collaterals, including **short gastric** and **left gastric veins** (gastric varices)
      - Gastric varices without portal hypertension almost always on basis of splenic vein occlusion
  - **Loss of peripancreatic fat planes**
    - Most commonly secondary to acute pancreatitis (commonly due to gallstones or alcohol)
    - Can also be secondary to adjacent inflammatory processes in stomach/duodenum or tumor spread (pancreatic adenocarcinoma, lymphoma)

CLINICAL IMPLICATIONS

Clinical Importance

- **Pancreatic ductal adenocarcinoma** is 4th leading cause of cancer death in USA and 12th most common cancer
  - Very poor prognosis with < 20% of patients candidates for surgery at time of diagnosis due to distant metastatic disease or tumor involving critical vascular structures
  - Imaging plays vital role in preoperative staging and determining which patients are potentially resectable
  - Classically presents as hypodense, poorly marginated mass with encasement of vasculature
- **Pancreatic neuroendocrine tumor**
  - Although classically considered relatively rare, small neuroendocrine tumors are increasingly being identified incidentally as result of improvements in imaging
  - Divided into syndromic and nonsyndromic types depending on whether they produce clinical syndrome secondary to hormone secretion
  - Classically presents as hypervascular, circumscribed mass
- Pancreatic anatomic variants are relatively common and can be mistaken for pathologic process
  - **Pancreas divisum**
    - Results from embryologic failure of fusion of ventral and dorsal pancreas
    - Pancreatic head and uncinate process drained by ventral duct, while body and tail drained by dorsal duct, neither of which communicates
      - Can predispose patients to recurrent bouts of acute pancreatitis
  - **Annular pancreas**
    - Congenital anomaly resulting in pancreatic parenchymal tissue either partially or completely surrounding duodenum
    - Can appear as "fullness" or enlargement of pancreatic head, which can be mistaken for mass

PANCREAS AND ITS RELATIONS

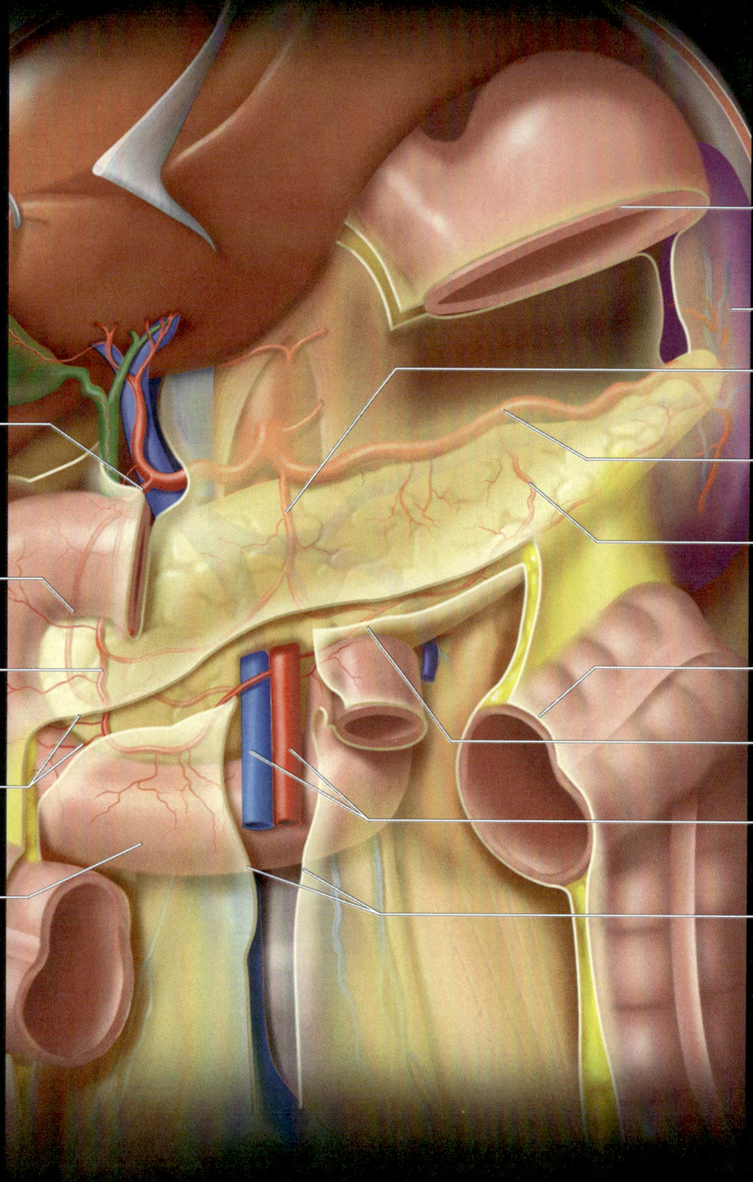

The arterial supply to the body and tail of the pancreas is through terminal branches of the splenic artery, which are variable in number and size. The 2 largest are usually the dorsal (superior) and great pancreatic arteries, which arise from the proximal and distal splenic artery, respectively. The arteries to the pancreatic head and duodenum come from the pancreaticoduodenal arcades that receive flow from the celiac and superior mesenteric arteries (SMAs). The superior mesenteric vessels pass behind the neck of the pancreas and in front of the 3rd portion of the duodenum. The root of the transverse mesocolon and small bowel mesentery arise from the surface of the pancreas and transmit the blood vessels to the small bowel and transverse colon. The splenic vein runs along the dorsal surface of the pancreas. The splenic vessels and pancreatic tail insert into the splenic hilum.

AXIAL VENOUS-PHASE CT

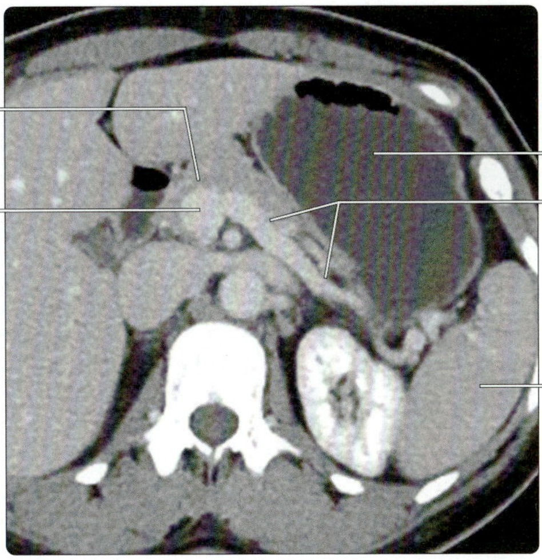

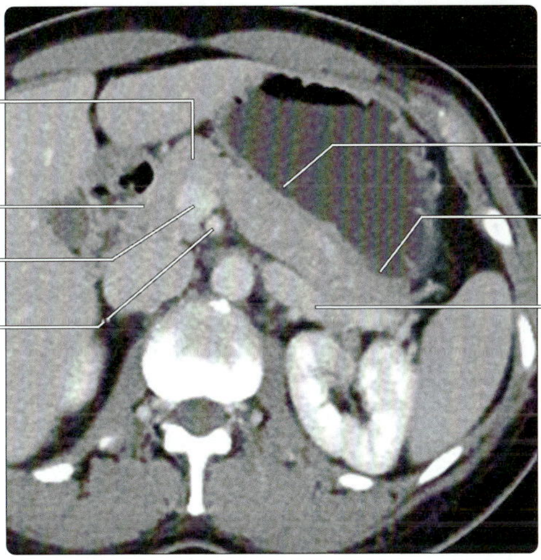

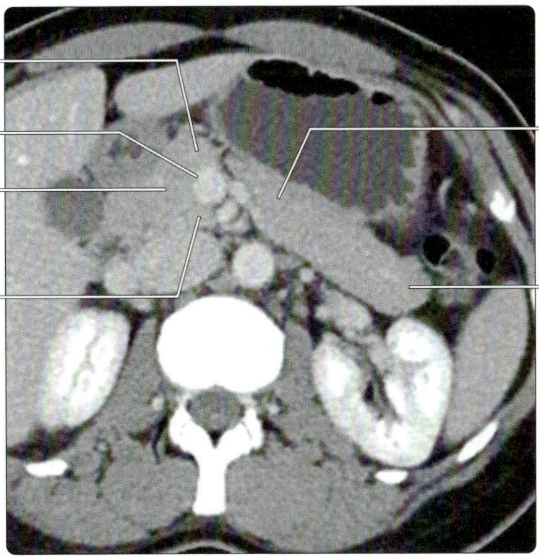

(Top) *Most cranial image from a series of axial venous-phase CECT images illustrates the splenic vein, which consistently runs in a groove along the dorsal aspect of the pancreatic body and tail. The pancreatic neck is seen on this image, lying immediately anterior to the portal vein-splenic vein confluence.* **(Middle)** *A slightly caudal image now illustrates the pancreatic body and tail, which lie immediately anterior to the splenic vein as well as the pancreatic head to the left of the portal confluence. Note the sheath of fat surrounding the SMA, a feature that allows rapid differentiation from the superior mesenteric vein.* **(Bottom)** *Another more caudal image demonstrates the pancreatic tail extending into the splenic hilum, where it is enveloped in a double layer of peritoneum, which serves as a common portal for pancreatic processes to extend into the peritoneal cavity. The uncinate process, a medial extension of the pancreatic head, is seen extending posterior to the superior mesenteric vein.*

AXIAL VENOUS-PHASE CT

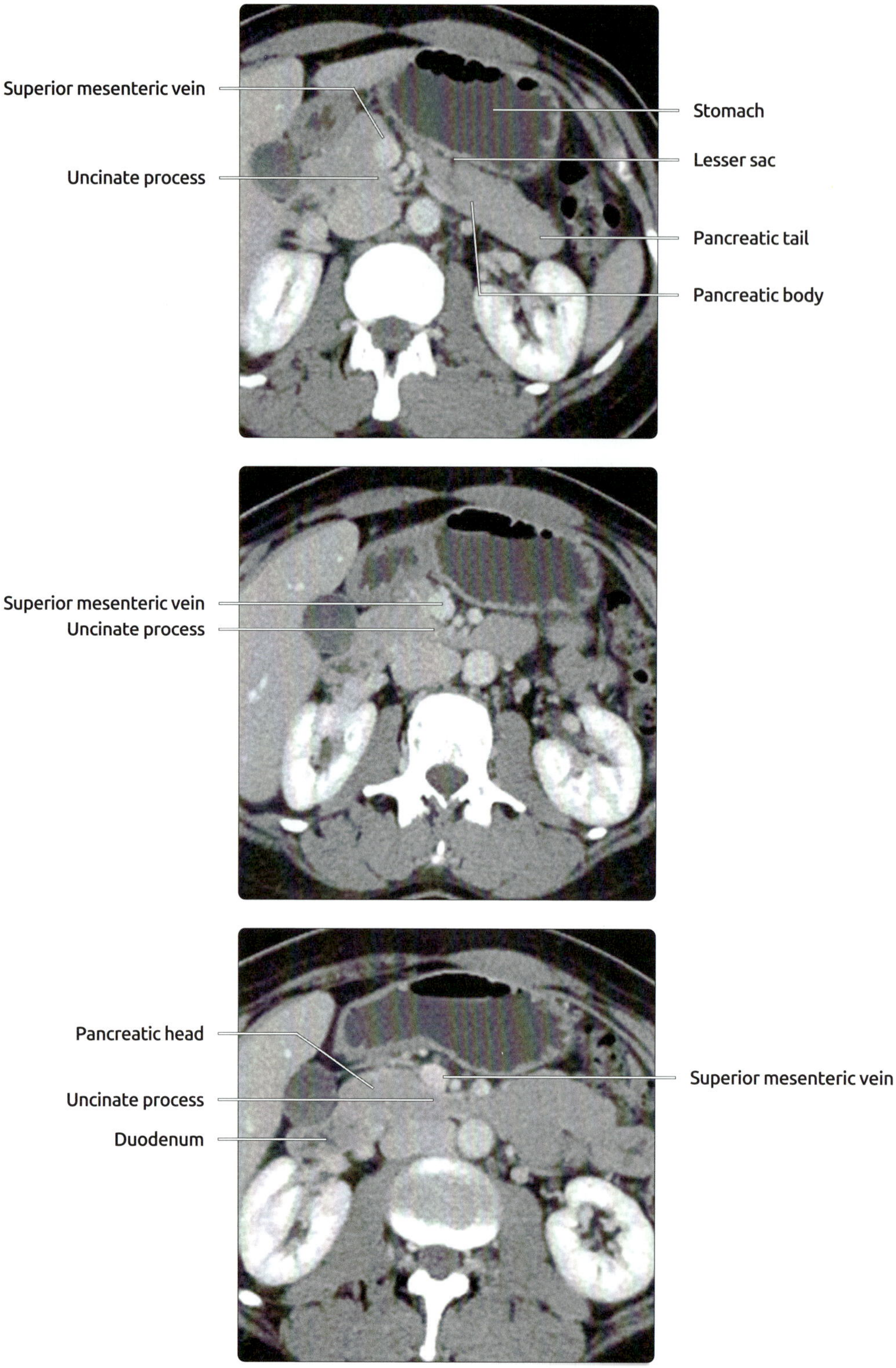

(Top) *The stomach lies immediately anterior to the pancreatic body and tail. The space between the pancreas and stomach represents the lesser sac, a space in continuity with the peritoneal cavity, which can be involved with pancreatic processes (such as pancreatitis).* **(Middle)** *The uncinate process extends immediately posterior to the superior mesenteric vein but does not typically extend as far medial as the SMA. The uncinate process can be quite variable in its shape/morphology.* **(Bottom)** *The pancreatic head and uncinate process are intimately associated with the retroperitoneal 2nd and 3rd portions of the duodenum.*

PANCREATIC ARTERIAL ANATOMY

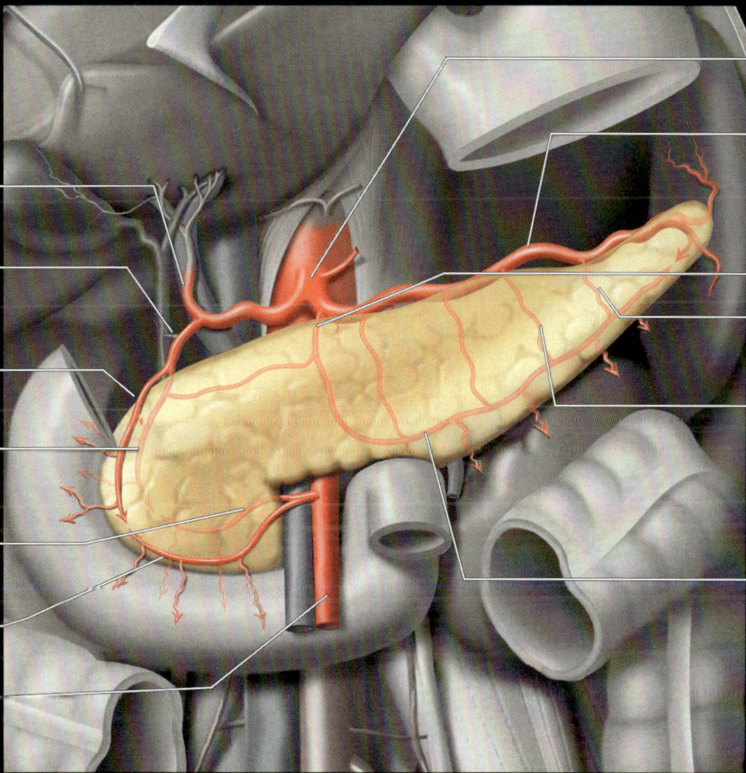

Illustration demonstrates the arterial supply to the pancreas. The pancreatic head is primarily supplied by the anterior and posterior pancreaticoduodenal arcades, including anterior and posterior superior pancreaticoduodenal arteries arising from the gastroduodenal artery (GDA) and anterior and posterior inferior pancreaticoduodenal arteries arising from the SMA. The blood supply to the body and tail segments is primarily via the splenic artery with the 2 biggest branches, including the dorsal pancreatic artery and the pancreatic great (magna) artery, which arise from proximal and midportions of the splenic artery, respectively.

CATHETER ANGIOGRAPHY, PERIPANCREATIC VASCULATURE

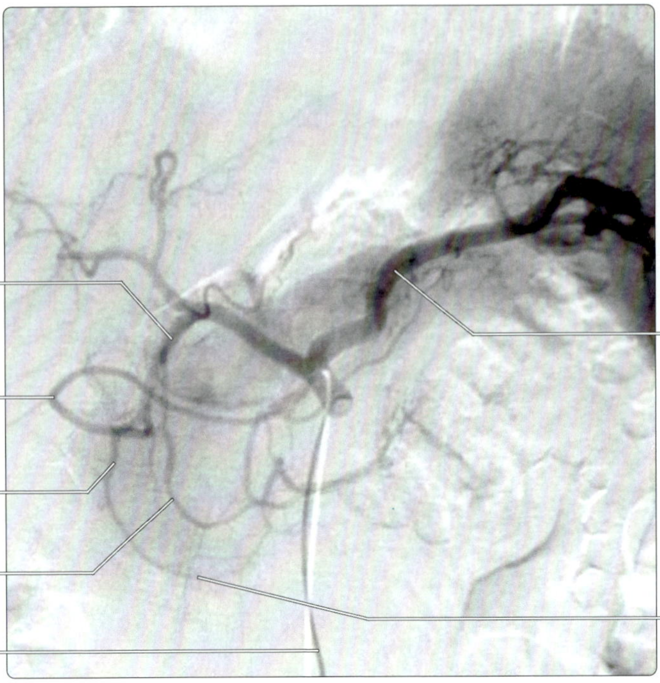

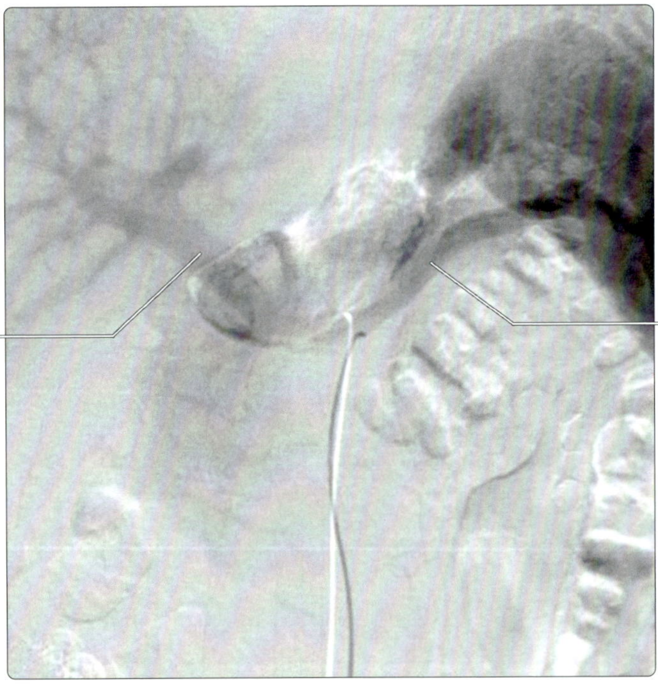

(Top) *First of 2 images from a celiac angiogram shows the arterial supply of the pancreas and adjacent organs. Arterial blood supply to the pancreatic head and duodenum is derived from the anterior and posterior pancreaticoduodenal arcades, which are formed by anastomosis of the anterior and posterior superior pancreaticoduodenal arteries arising from the GDA and branches of the inferior pancreaticoduodenal artery (arising from the SMA). The body and tail portions of the pancreas are supplied by branches of the splenic artery, which are quite variable in size and distribution. Some are tiny branches without specific names, while 2 are generally larger, the dorsal (superior) and great pancreatic arteries, which originate from the splenic artery proximally and distally, respectively. The GDA courses downward behind the 1st portion of the duodenum and anterolateral to the head of the pancreas.* **(Bottom)** *The venous drainage of the pancreas is to the splenic, superior mesenteric, and portal veins.*

CATHETER ANGIOGRAPHY, PERIPANCREATIC VASCULATURE

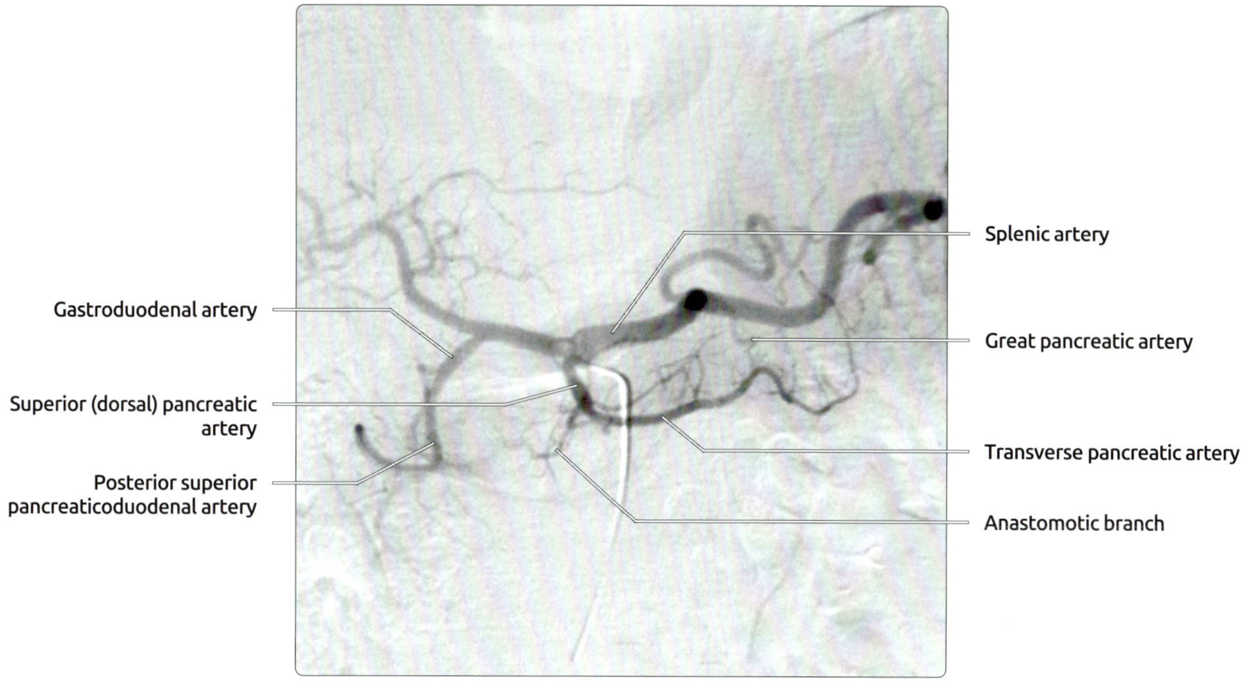

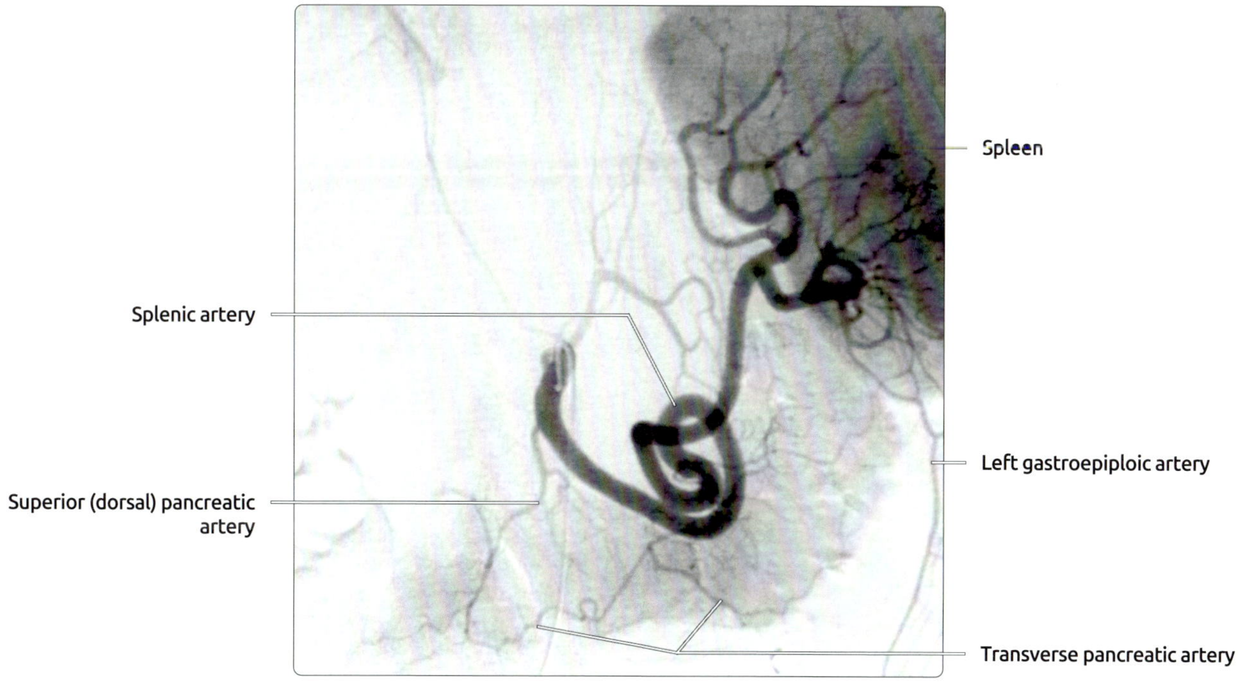

(Top) *Selective catheter injection of the celiac artery shows a prominent superior (dorsal) pancreatic artery arising from the celiac artery itself or the proximal splenic artery (its usual source). The transverse pancreatic artery runs along the main axis of the pancreas and receives branches from the superior and great pancreatic arteries and from a separate branch of the SMA, the inferior pancreatic artery. In spite of the name, the "great" pancreatic artery may be relatively small, as in this patient. A branch of the superior pancreatic artery anastomoses with the pancreaticoduodenal arcade.* **(Bottom)** *In this subject, the splenic artery is very tortuous, a common finding. The pancreatic arterial supply is mostly through unnamed terminal branches of the splenic artery. Neither the superior nor the great pancreatic artery are substantial vessels in this subject.*

CT ANGIOGRAPHY, VESSELS

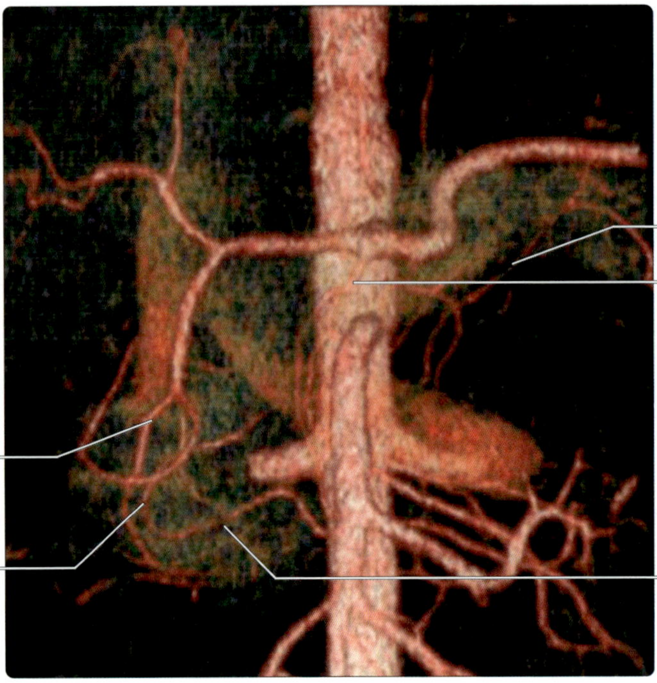

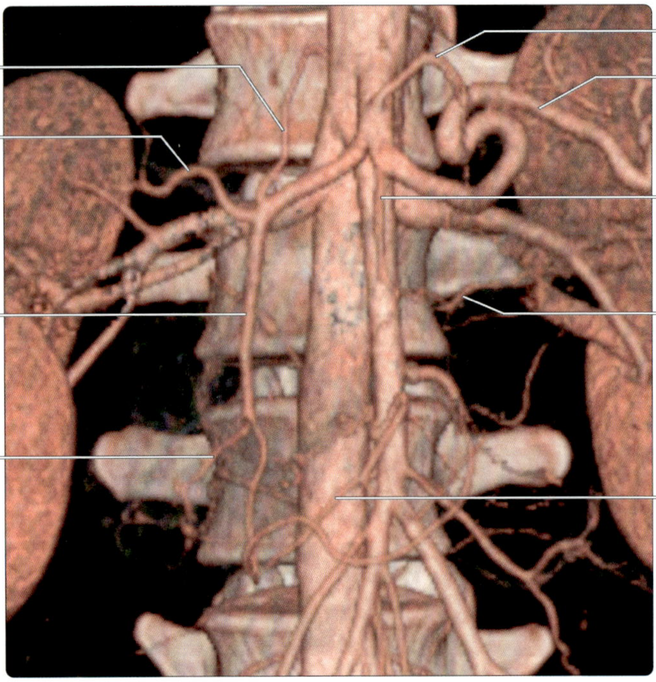

(Top) *Coronal volume-rendered image derived from a CT angiogram demonstrates conventional upper abdominal arterial anatomy. The anterior and posterior pancreaticoduodenal arteries arise from the GDA and their inferior extensions anastomose with branches of the inferior pancreaticoduodenal artery (arising from the SMA). The dorsal (superior) pancreatic artery arises from the proximal splenic artery. The arterial supply to the pancreas is highly variable, and all of the branches anastomose to a considerable degree. The transverse pancreatic artery runs parallel to the splenic artery along the long axis of the pancreas, also sending numerous small, anastomotic branches to the pancreas.* **(Bottom)** *Coronal volume-rendered CT angiogram demonstrates conventional pancreatic arterial anatomy, but variant hepatic arterial anatomy, with the left hepatic artery arising from the common hepatic and the right hepatic from the GDA.*

AXIAL ARTERIAL-PHASE CT, PANCREATIC ARTERIES

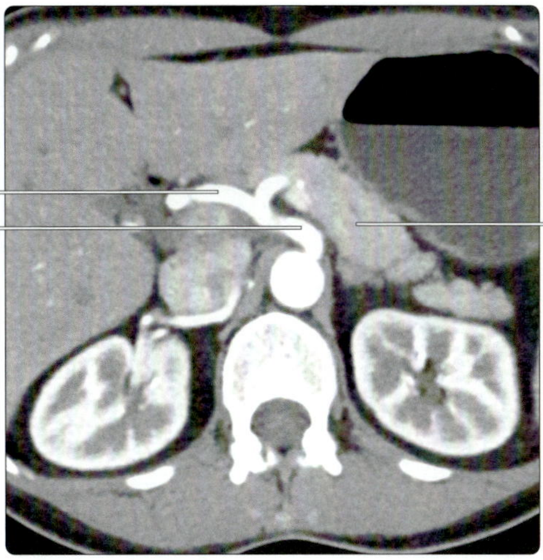

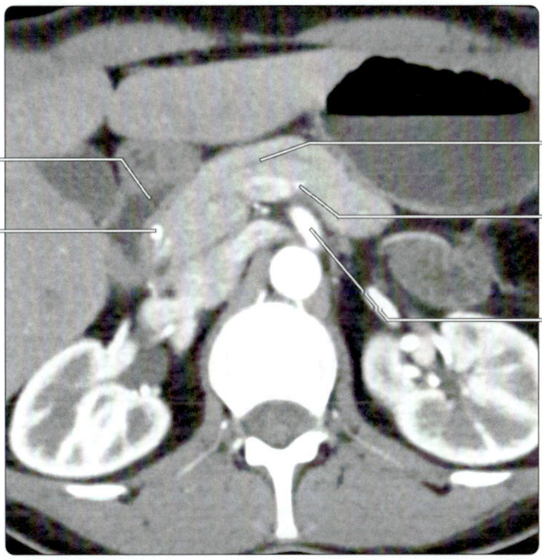

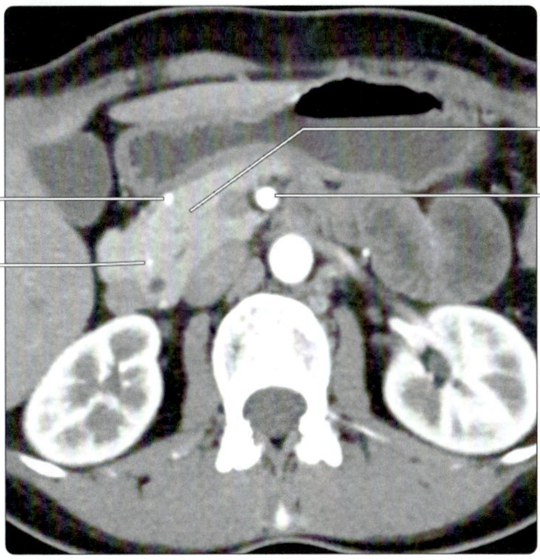

(Top) *First of 5 axial arterial-phase CT images shows a normal pancreas and conventional arterial anatomy. The splenic and common hepatic arteries arise from the celiac trunk and supply most of the blood supply to the pancreas through many branches with a highly variable pattern.* **(Middle)** *The dorsal (superior) pancreatic artery courses along the posterior surface of the pancreas in a vertical orientation and is usually the 1st branch of the splenic artery. The GDA is the 1st branch of the common hepatic artery and passes behind the gastric pylorus and duodenal bulb and anterolateral to the pancreatic head.* **(Bottom)** *The 3 major branches of the GDA are the right gastroepiploic artery and the anterior and posterior superior pancreaticoduodenal arteries.*

AXIAL ARTERIAL-PHASE CT, PANCREATIC ARTERIES

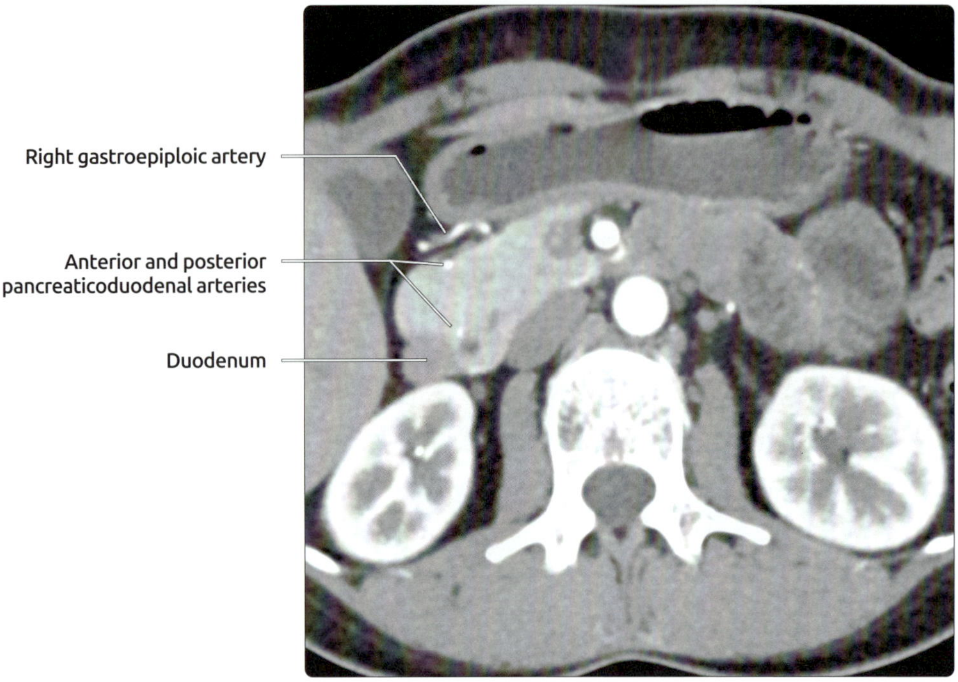

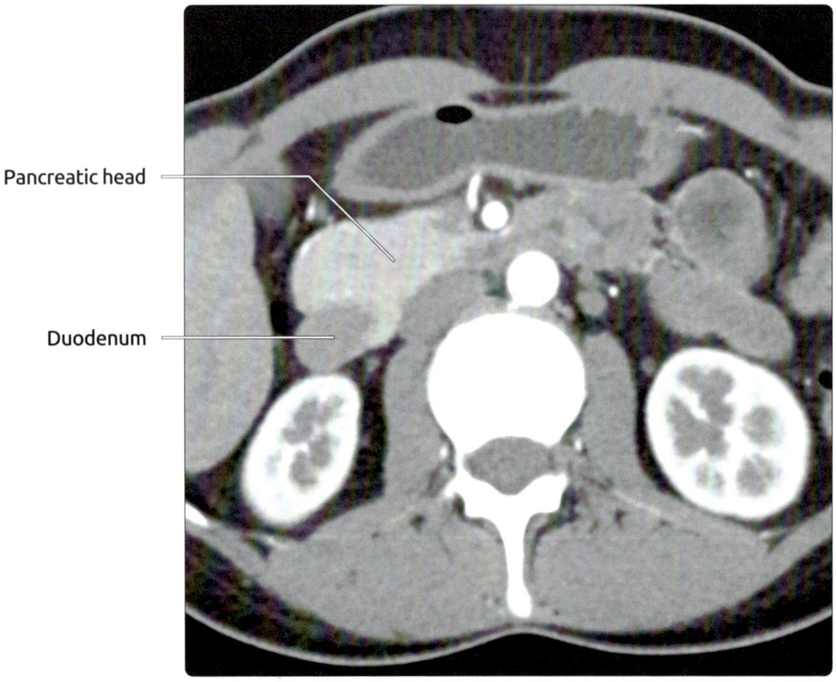

(Top) *The anterior and posterior pancreaticoduodenal arcades surround the head of the pancreas and anastomose with the SMA through the inferior branches of these vessels. Note the presence of the right gastroepiploic artery, one of the terminal branches of the GDA, which then extends along the greater curvature of the stomach to anastomose with the left gastroepiploic artery.* **(Bottom)** *The pancreaticoduodenal arcades send branches laterally to the duodenum and medially to the pancreas.*

AXIAL VENOUS-PHASE CT, PARENCHYMA AND VEINS

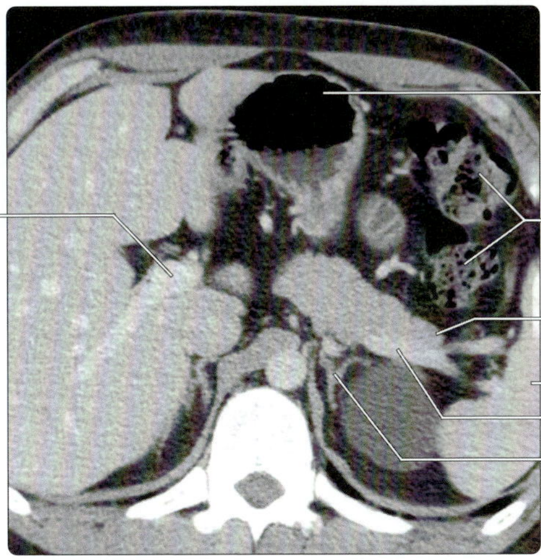

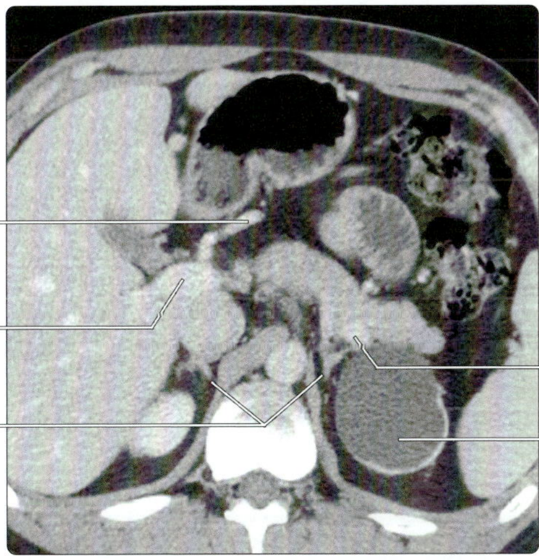

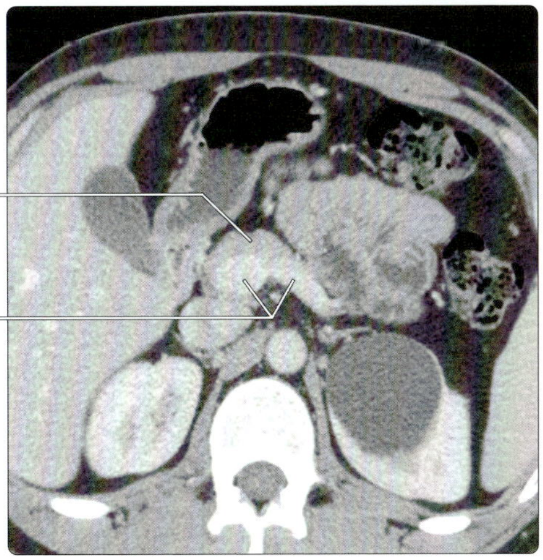

(Top) *First of 5 axial CT sections shows the body and tail of the pancreas lying in a nearly transverse plane with the splenic vein running in a groove along its dorsal surface. The body of the pancreas lies just ventral to the splenic vein and the adrenal just dorsal to it.* **(Middle)** *There are no reliable measures of "normal" pancreatic dimensions. The normal gland tapers smoothly from the head to the tail with some thinning through the neck region. Focal areas of enlargement are suggestive of tumor or focal inflammation.* **(Bottom)** *The neck of the pancreas lies just ventral to the confluence of the splenic and portal veins.*

AXIAL VENOUS CT, PARENCHYMA AND VEINS

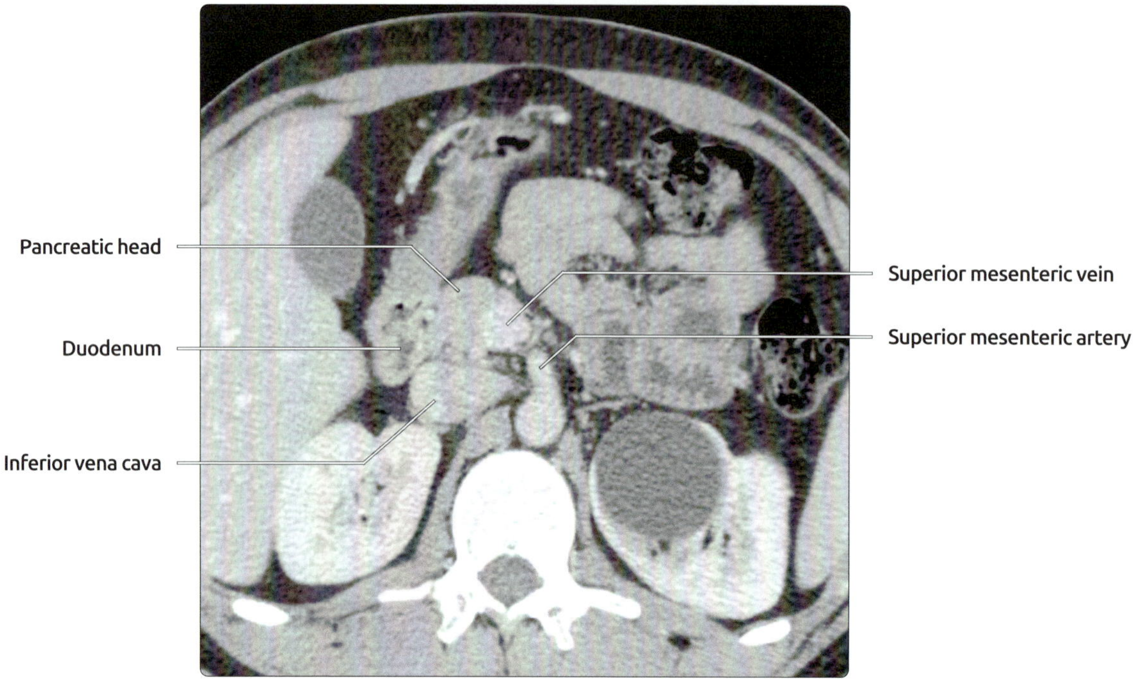

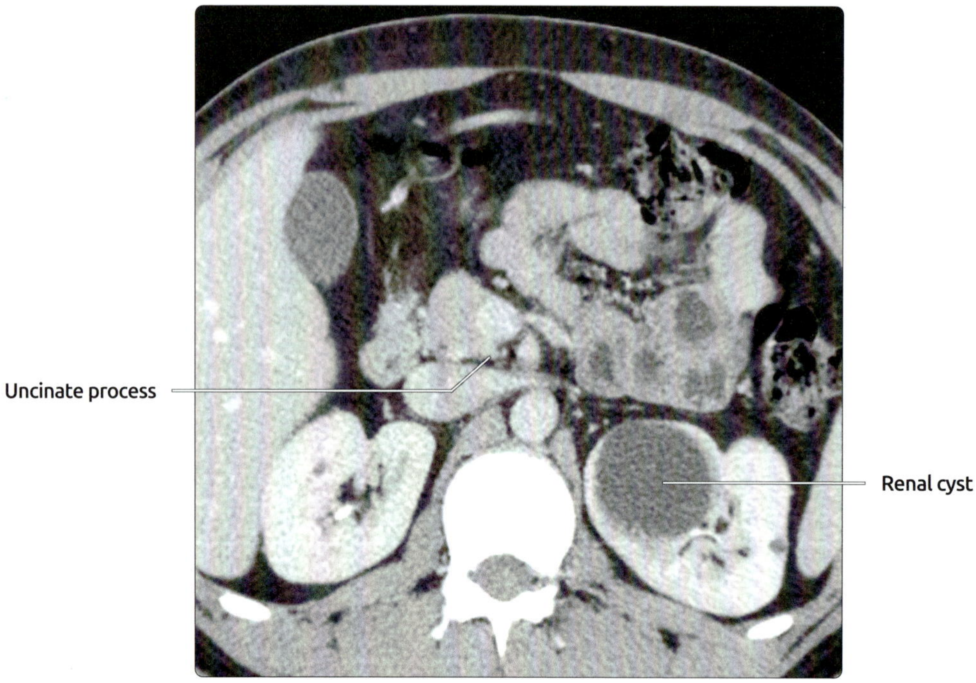

(Top) *The head of the pancreas lies between the 2nd portion of duodenum and the superior mesenteric vessels. The inferior vena cava lies just behind the head of the pancreas.* **(Bottom)** *The uncinate process is usually seen slightly inferiorly, and lies just behind the superior mesenteric vein and in front of the inferior vena cava.*

CORONAL CT

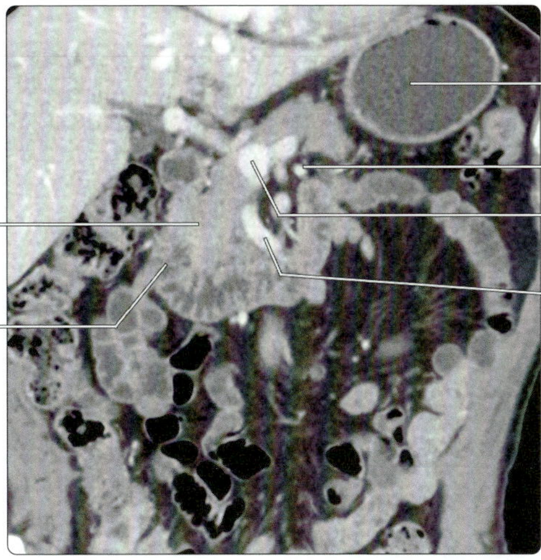

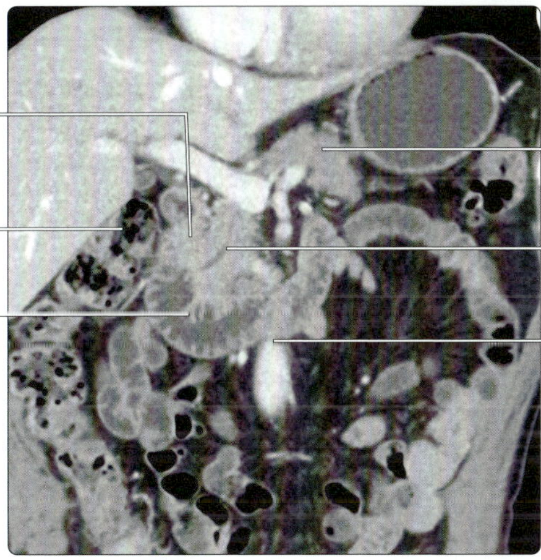

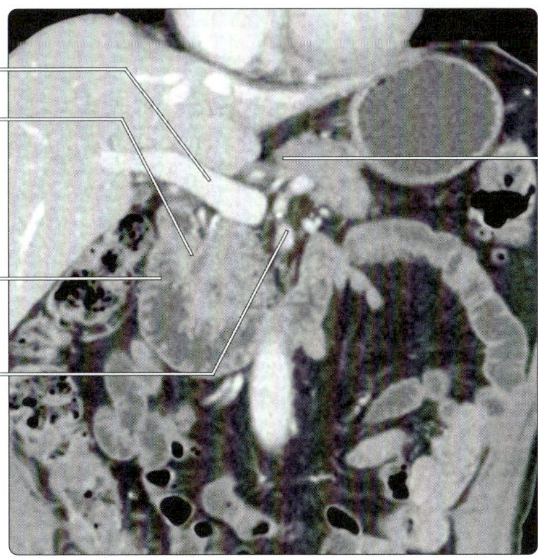

(Top) *First of 6 coronal CT images shows the normal pancreas and its anatomic relationships. This most ventral section demonstrates the intimate relationship of the pancreatic head with the 2nd and 3rd portions of the duodenum. The pancreatic head is located medial to the portal-superior mesenteric vein confluence. Also note the inferior mesenteric vein, which has variable origin, coursing inferiorly and toward the left.* **(Middle)** *A slightly more caudal section once again demonstrates the immediately adjacent pancreatic head and duodenum. In this image, the normal pancreatic duct is seen coursing toward the ampulla. The latest generations of CT scanners can easily visualize the normal pancreatic duct, which should normally taper along its course.* **(Bottom)** *A more caudal section demonstrates the pancreatic neck immediately posterior to the superior mesenteric vein and SMA as well as the normal-sized common bile duct coursing toward the ampulla.*

CORONAL CT

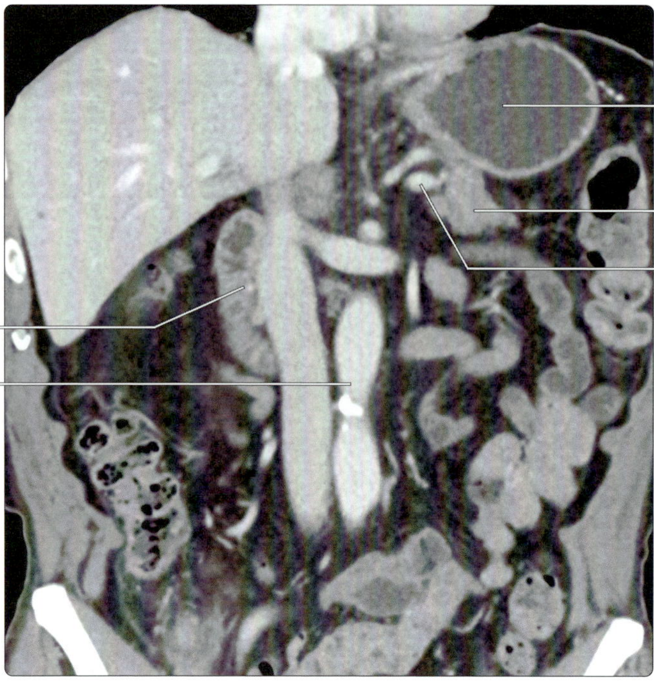

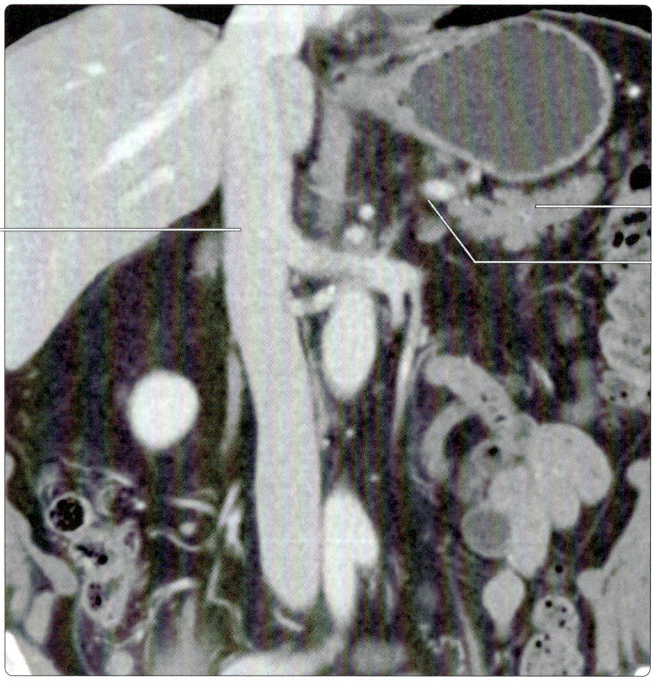

(Top) *Another more caudal image demonstrates the duodenal "C" loop as well as the pancreatic body extending into the left upper quadrant.* **(Bottom)** *This final, most caudal image demonstrates the pancreatic tail extending near the splenic hilum, intimately associated with the splenic vein, which runs in a groove along its posterior aspect. The pancreatic tail is enveloped loosely by 2 peritoneal layers, serving as a common site for pancreatic inflammatory or neoplastic processes to gain access to the peritoneal cavity.*

AXIAL T2 MR

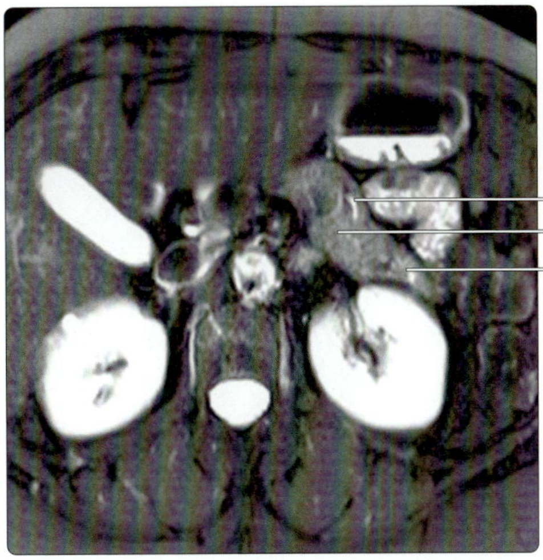

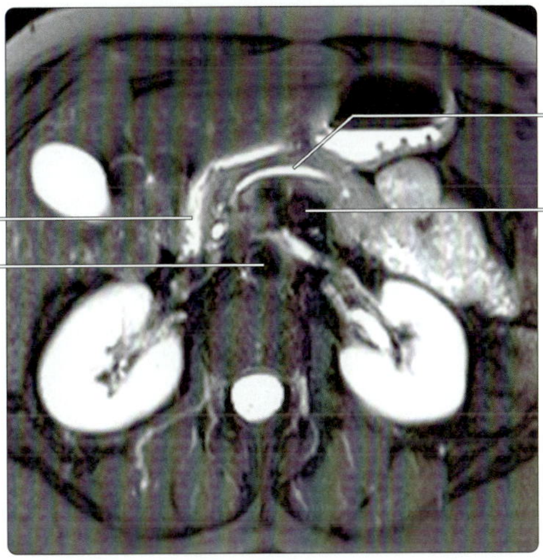

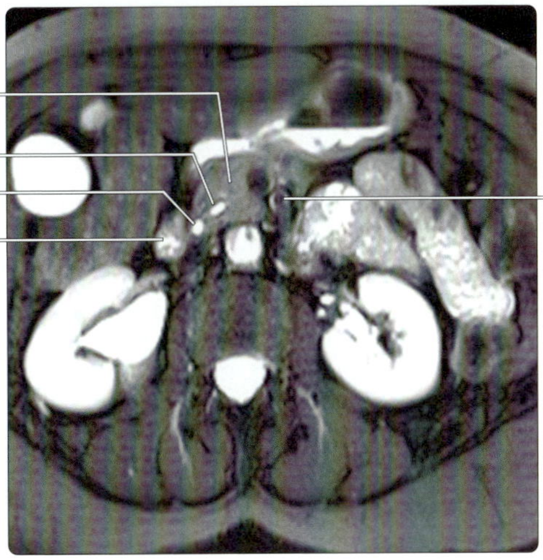

(Top) *Axial T2 FS MR through the pancreatic body and tail demonstrates the typical intermediate to low signal of the pancreas on this pulse sequence. While this sequence can be helpful to identify pancreatic and peripancreatic inflammation (i.e., pancreatitis) and cystic lesions, solid tumors can be relatively inconspicuous on T2 MR. Notice the T2-bright fluid within the normal-sized pancreatic duct.* **(Middle)** *Axial T2 FS MR demonstrates the pancreatic body with nice visualization of the normal-caliber pancreatic duct. Flowing blood on this sequence, as with the aorta and SMA, is visible as flow voids. The duodenum demonstrates internal high signal due to intraluminal fluid.* **(Bottom)** *Final axial T2 FS MR demonstrates the pancreatic duct in the pancreatic head coursing toward the ampulla to join with the distal common bile duct.*

AXIAL T1 C+ ARTERIAL-PHASE MR

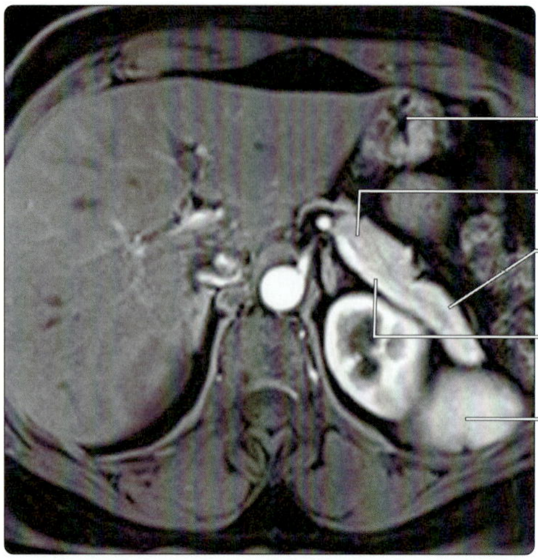

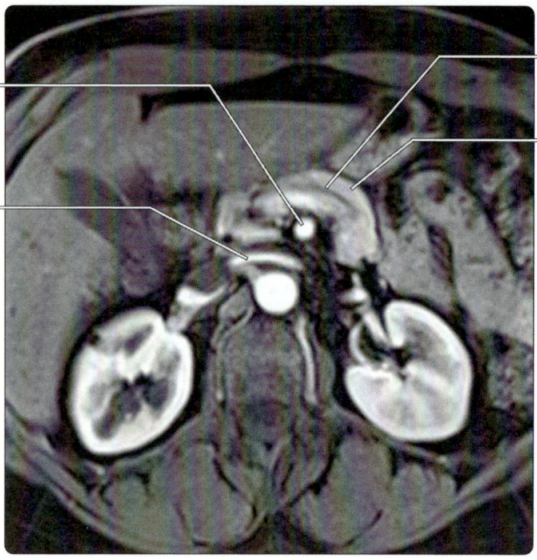

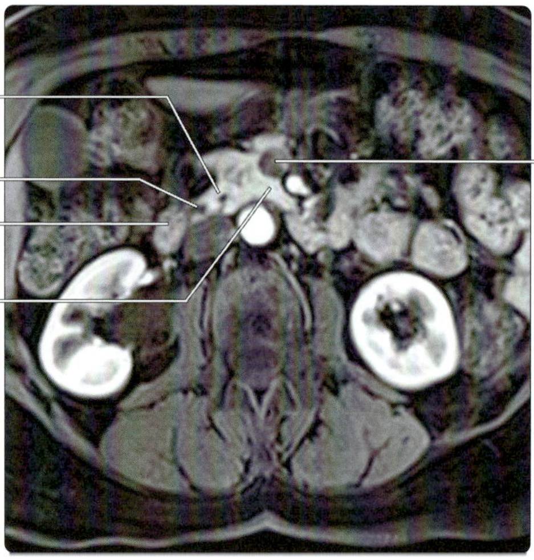

(Top) *First of 3 axial T1 C+ MR images demonstrates the relationship of the body and tail of the pancreas to the splenic vein and the spleen. The splenic vein runs in a groove along the posterior aspect of the pancreatic body and tail.* **(Middle)** *The pancreatic duct is seen as a thin fluid-filled structure at the center of the pancreas. The pancreatic body is located to the left of the portal/superior mesenteric vein confluence.* **(Bottom)** *The uncinate process lies behind the superior mesenteric vein (which is not yet well opacified on this arterial-phase image). Note the common bile duct and pancreatic duct as low-signal dots passing vertically through the head of the pancreas.*

AXIAL T1 C+ VENOUS-PHASE MR

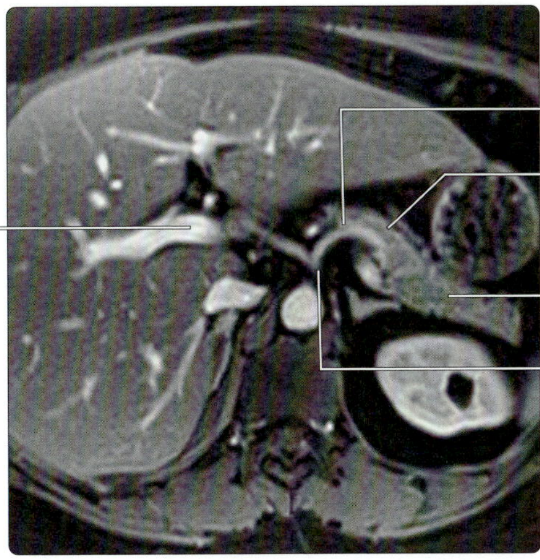

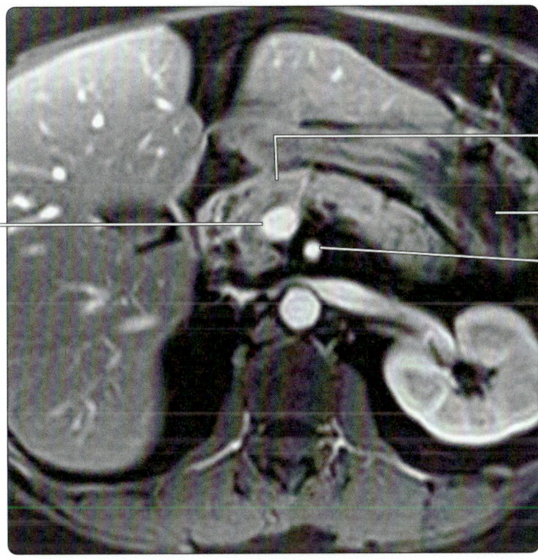

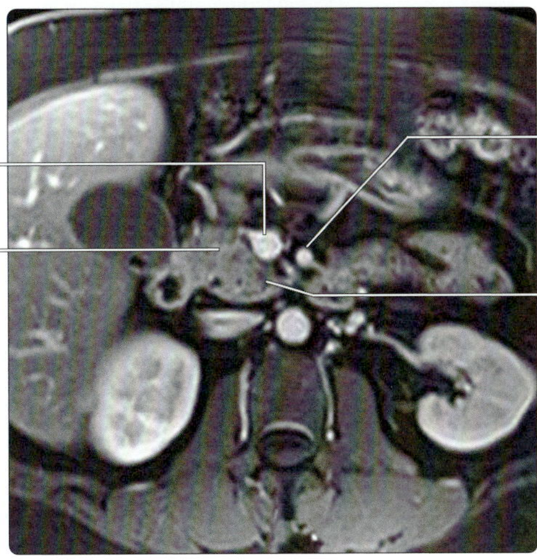

(Top) *First of 3 axial MR sections obtained during the portal venous phase of contrast enhancement shows the body and tail of the pancreas in relation to the celiac and splenic arteries.* **(Middle)** *The neck of the pancreas lies just anterior to the junction of the splenic vein and portal vein.* **(Bottom)** *The uncinate process lies posterior to the superior mesenteric vein but does not usually extend behind the artery. Another structure that lies in this space is the 3rd portion of the duodenum, which lies ~ 2 cm more caudal than the uncinate and renal vein.*

MRCP, PANCREATIC AND BILE DUCTS

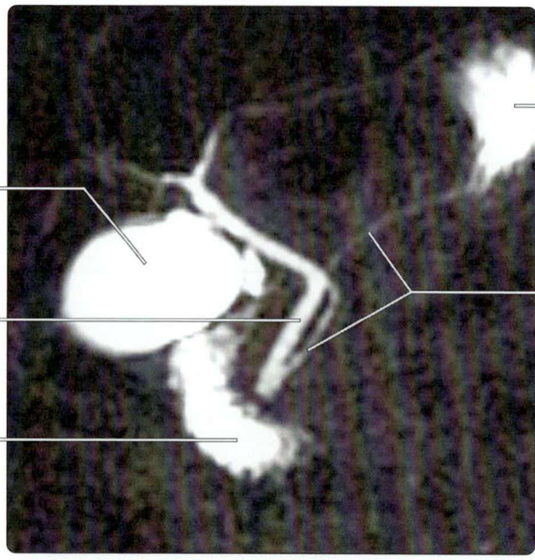

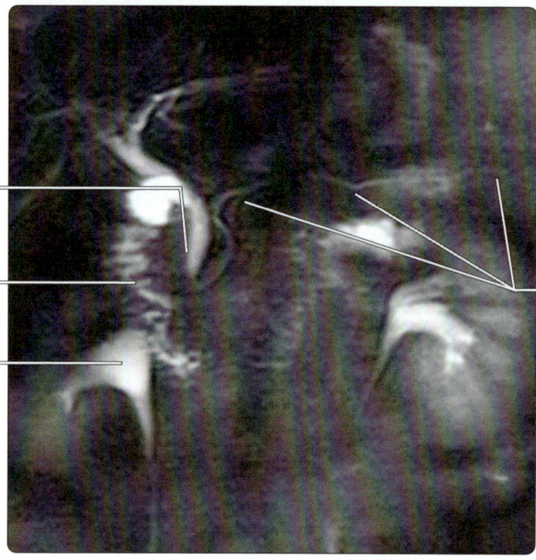

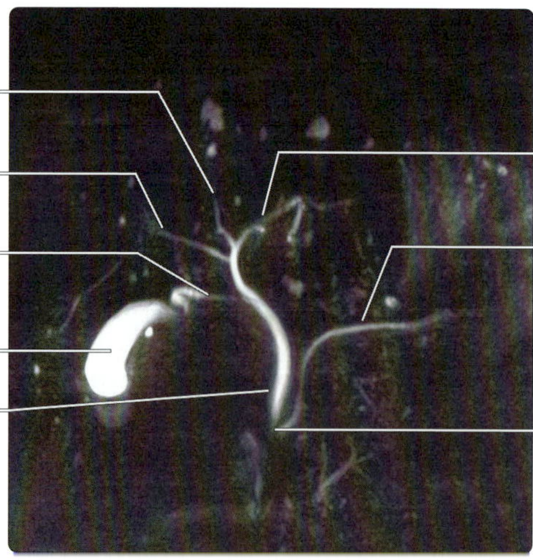

(Top) *Coronal MR cholangiopancreatogram (MRCP) with MIP reconstruction shows normal-caliber pancreatic and bile ducts. The common bile duct and pancreatic duct run a parallel course within the head of the pancreas and join together just prior to emptying into the duodenum at the hepaticopancreatic papilla (of Vater).* **(Middle)** *Coronal MRCP demonstrates the normal appearance of the pancreatic duct, which is normal in caliber, extending toward the ampulla to join the common bile duct. MRCP images are heavily T2 weighted and accentuate fluid-filled structures, such as fluid in the duodenum, renal collecting systems, biliary tree, and pancreatic duct.* **(Bottom)** *Coronal MRCP with MIP reconstruction demonstrates a normal appearance of the pancreatic duct, which joins the common bile duct at the ampulla. Note the smooth, thin appearance of the pancreatic duct without dilatation or ectatic side branches. Incidentally, the right anterior hepatic duct arises from the left hepatic duct, a normal variant.*

SENESCENT CHANGE

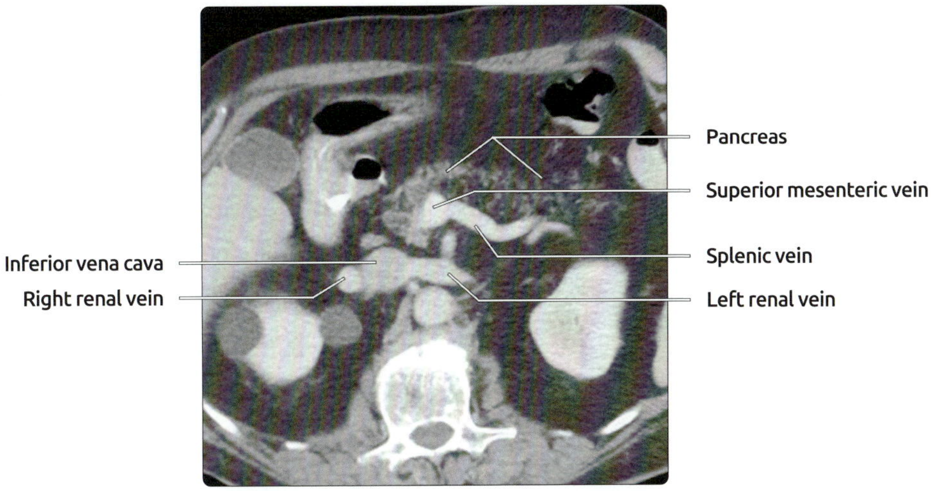

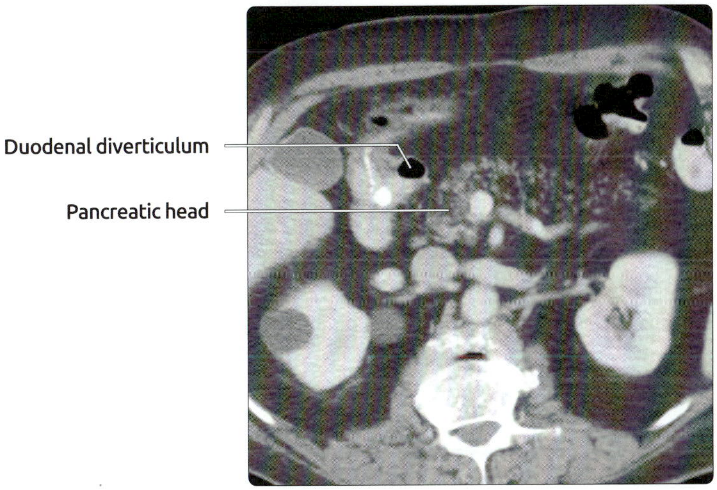

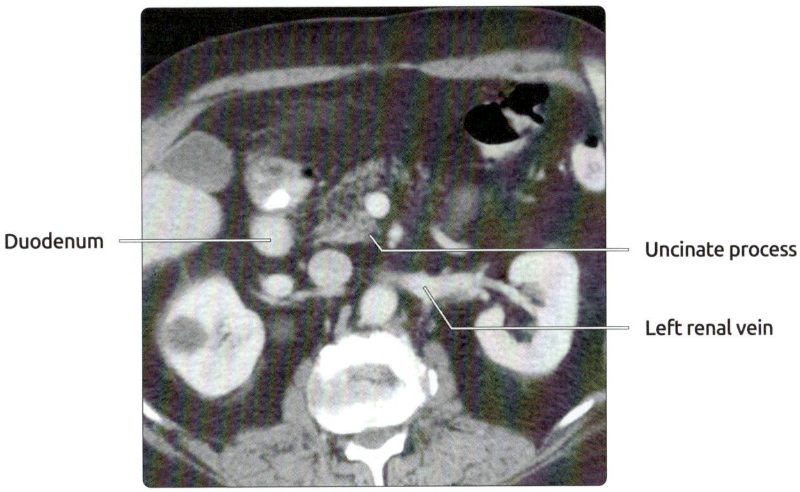

(Top) *First of 3 axial CECT images in an older adult patient shows extensive fatty infiltration of the pancreas. While this can be associated with chronic pancreatitis, diabetes, and obesity, it is often a normal finding in older patients of no clinical significance. The pancreatic parenchyma begins to atrophy, and the pancreatic duct may mildly dilate in older subjects (particularly beyond the age of 70) without symptoms or signs of pancreatic insufficiency.* **(Middle)** *In patients with extensive fatty infiltration of the pancreas, as in this patient, the pancreas can be difficult to visualize on ultrasound, as it may demonstrate similar echogenicity to the surrounding retroperitoneal fat.* **(Bottom)** *In this patient, the degree of fatty infiltration is relatively uniform throughout the gland. As with hepatic steatosis, however, this is not always the case, and fatty infiltration may be heterogeneous and potentially mimic a hypodense mass.*

VARIANT, ASYMMETRIC FATTY INFILTRATION

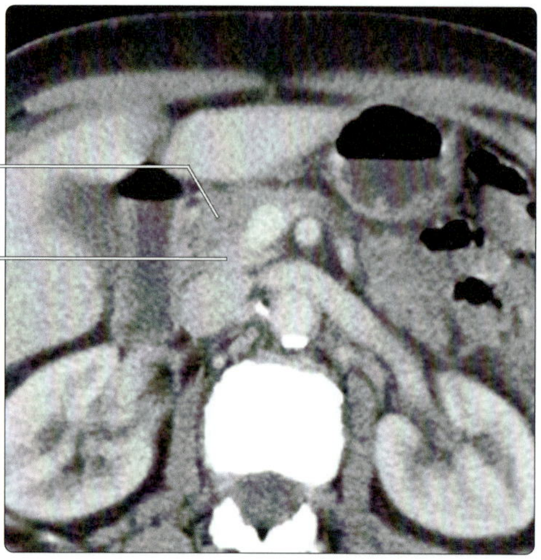

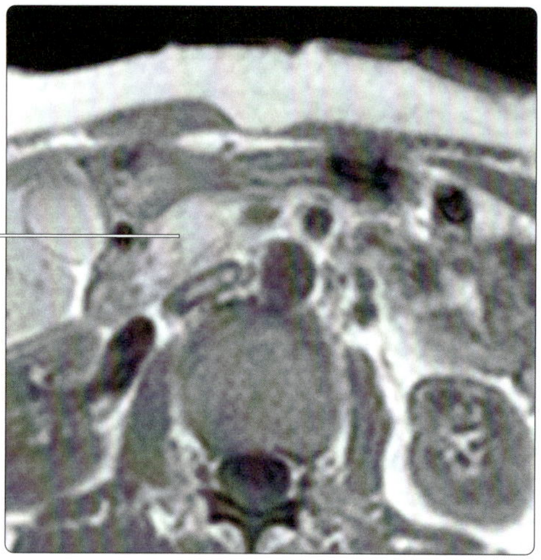

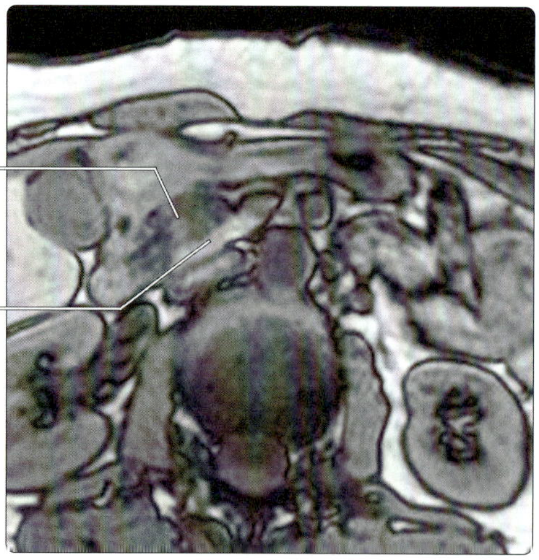

(Top) *Axial CECT demonstrates abnormal low density in the anterior aspect of the pancreatic head sharply demarcated from the more normal density of the posterior pancreatic head. The low-density area demonstrates no mass effect and does not appear to be a true mass lesion. This appearance is classic for focal fatty infiltration of the dorsal pancreatic anlage.* **(Middle)** *Axial in-phase GRE in the same patient demonstrates homogeneous high signal throughout the pancreatic head.* **(Bottom)** *Axial opposed-phase GRE in the same patient demonstrates signal loss in the anterior pancreatic head corresponding to the area of low density seen on CT and confirming the presence of fatty infiltration.*

ACUTE PANCREATITIS

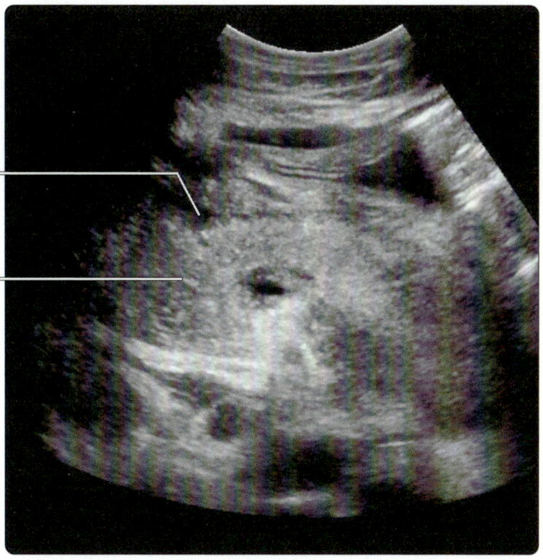

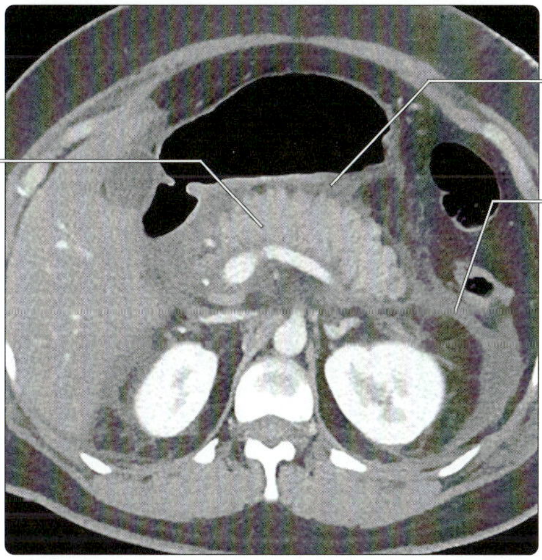

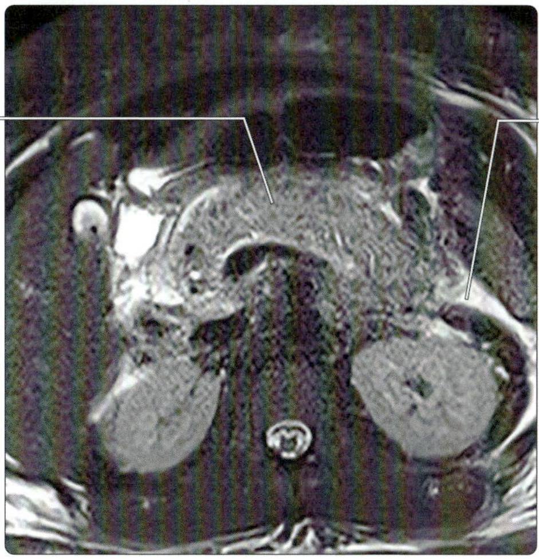

(Top) *Ultrasound of the pancreas in a patient with abdominal pain demonstrates the pancreas to be abnormally enlarged and "full" in appearance with a small amount of peripancreatic free fluid, features suggestive of acute pancreatitis.* **(Middle)** *Axial CECT demonstrates the pancreas to be enlarged, edematous, and inflamed with fluid tracking from the pancreas along the left anterior pararenal space. Also note that some inflammation spreads anteriorly into the lesser sac between the pancreas and stomach.* **(Bottom)** *Axial T2 FS MR in the same patient demonstrates edema and enlargement of the pancreas with surrounding peripancreatic fluid, compatible with acute pancreatitis.*

ACUTE PANCREATITIS

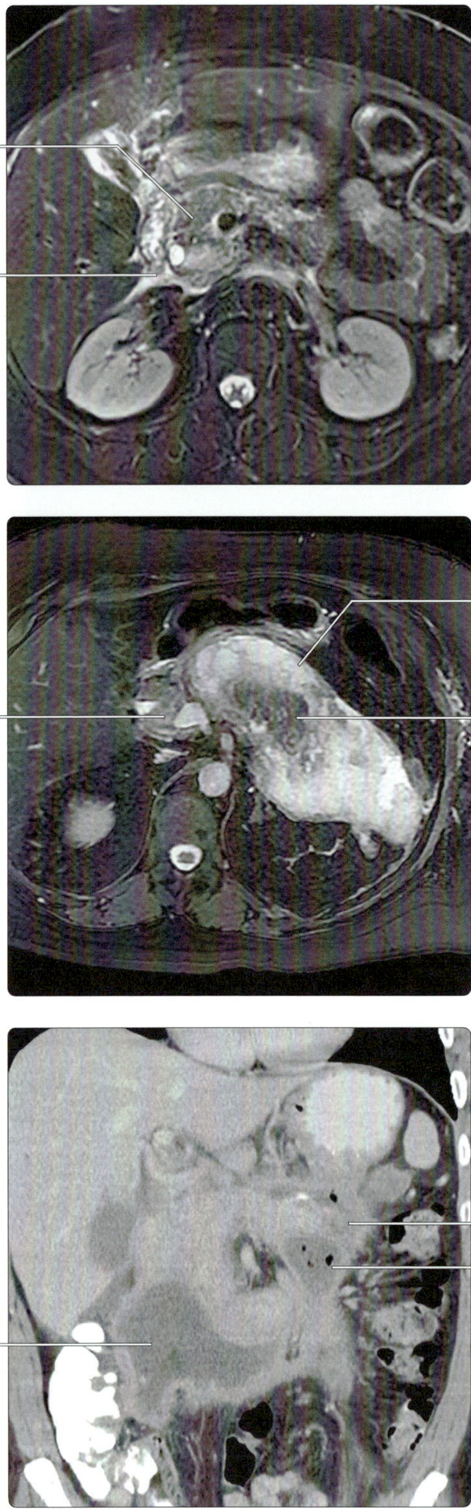

(Top) *Axial T2 FS MR demonstrates high T2 signal in the pancreatic head parenchyma with T2-bright free fluid surrounding the pancreatic head and tracking along the retroperitoneum, compatible with acute pancreatitis.* **(Middle)** *Axial T2 FS MR demonstrates absence of the pancreatic body and tail with only a small amount of residual enhancing pancreatic head, compatible with necrotizing pancreatitis. The body and tail have been replaced with a large fluid collection in the anterior pararenal space with extensive internal debris, compatible with walled-off necrosis.* **(Bottom)** *Coronal CECT in a patient with prior necrotizing pancreatitis demonstrates multiple thick-walled, heterogeneous fluid collections surrounding the pancreas, representing walled-off necrosis. Please note that a small amount of gas within one of the collections was secondary to placement of a drainage catheter (not shown), rather than superinfection.*

PANCREATITIS WITH PSEUDOCYST

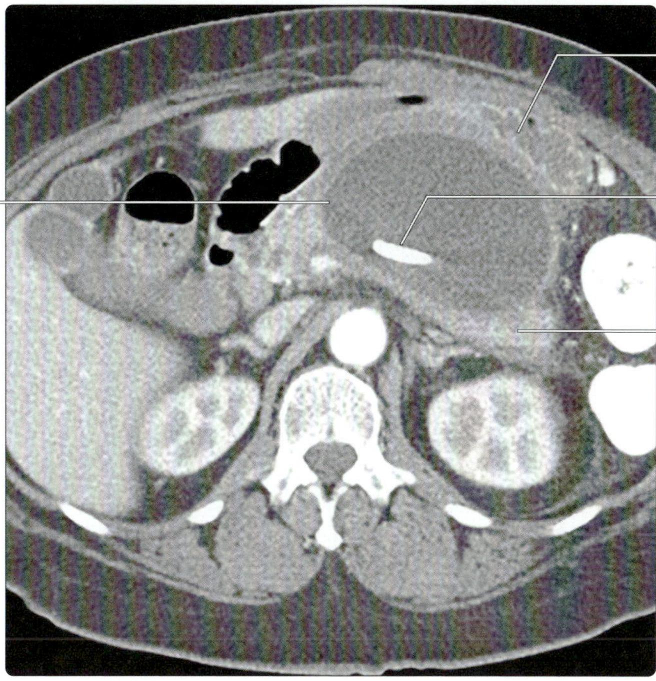

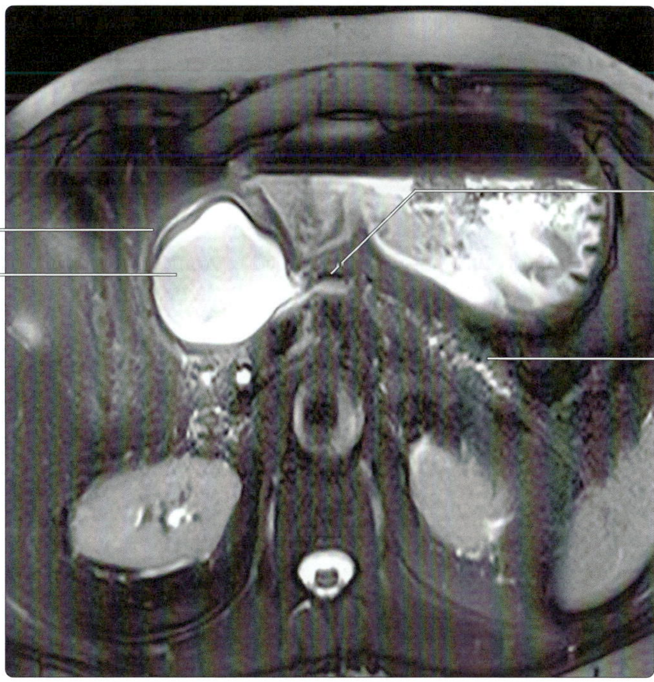

(Top) *Axial CECT demonstrates a loculated fluid collection centered in the lesser sac between the pancreas and stomach, representing a pseudocyst related to prior bouts of pancreatitis. As in this case, pseudocysts in this location can cause symptoms as a result of massive effect on the stomach or duodenum. Please note the presence of a drainage catheter within the collection.* **(Bottom)** *Axial T2 FS MR demonstrates a pseudocyst related to prior pancreatitis in the right upper quadrant adjacent to the pancreatic head and duodenum. The pseudocyst, in this case, appears to directly communicate with the adjacent pancreatic duct.*

PANCREATIC ADENOCARCINOMA WITH VASCULAR INVOLVEMENT

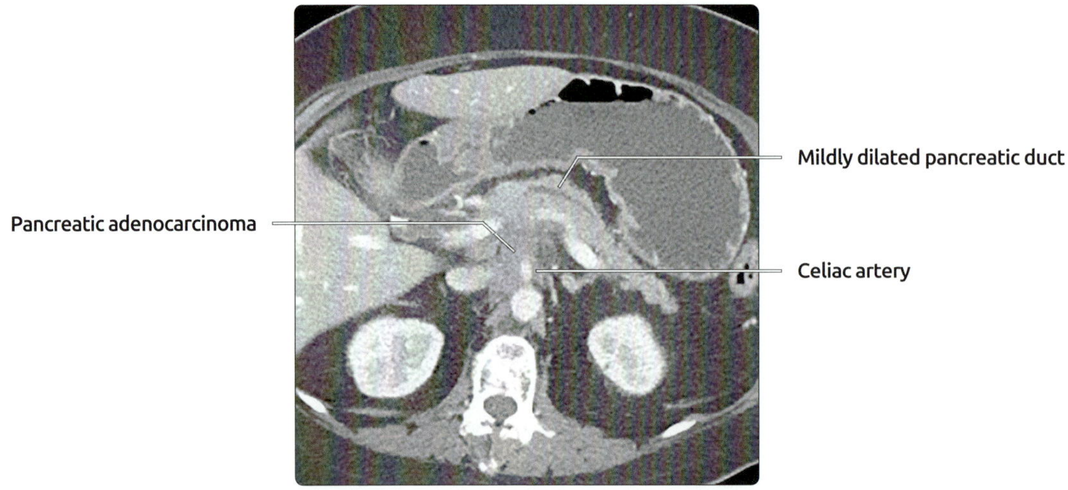

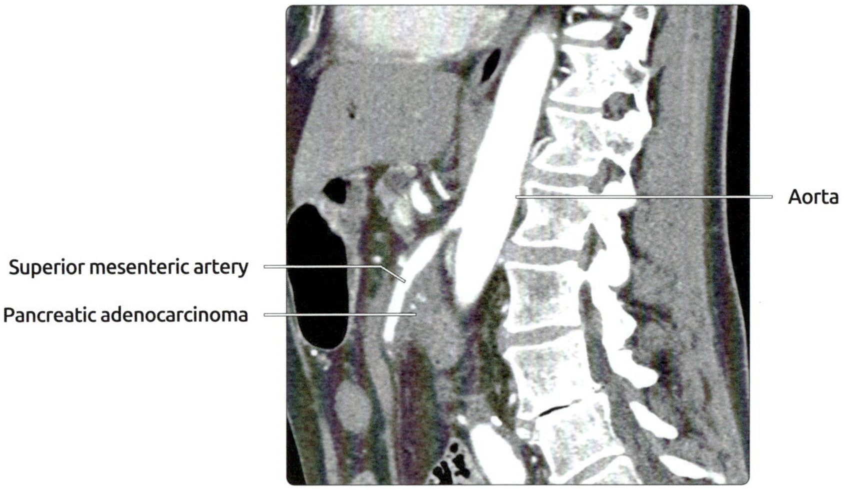

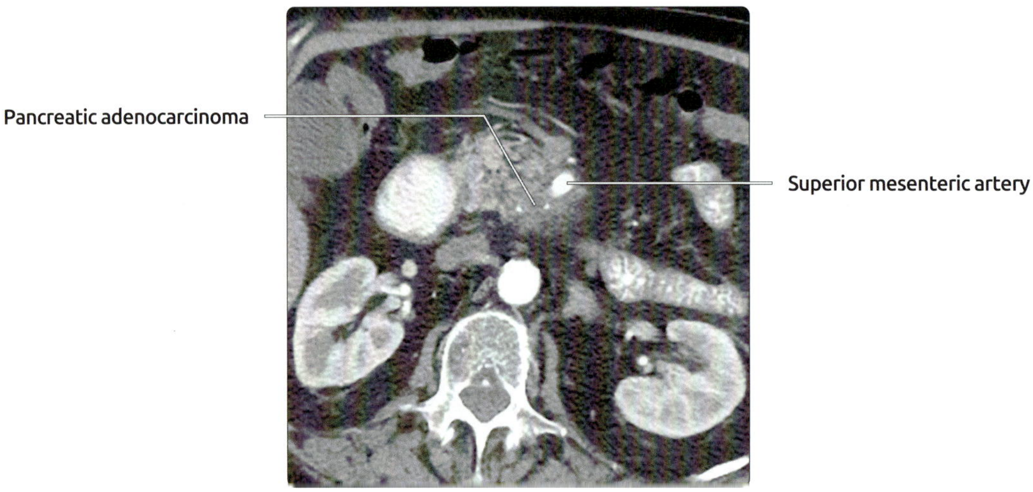

(Top) *Axial CECT demonstrates a hypodense mass, representing pancreatic adenocarcinoma, encasing the celiac artery with 360° involvement of the vessel circumference. Note that the mass results in mild upstream pancreatic ductal dilatation.* **(Middle)** *Sagittal CECT in a patient with pancreatic adenocarcinoma demonstrates an ill-defined hypodense mass surrounding the SMA with ~ 360° encasement. This patient was determined to be unresectable based on the extent of tumoral involvement of the SMA.* **(Bottom)** *Axial CECT demonstrates a pancreatic adenocarcinoma arising from the uncinate process completely encasing the SMA with 360° involvement. In most cases, this degree of SMA involvement by tumor precludes surgical resection.*

PANCREATIC ADENOCARCINOMA WITH DUCTAL OBSTRUCTION

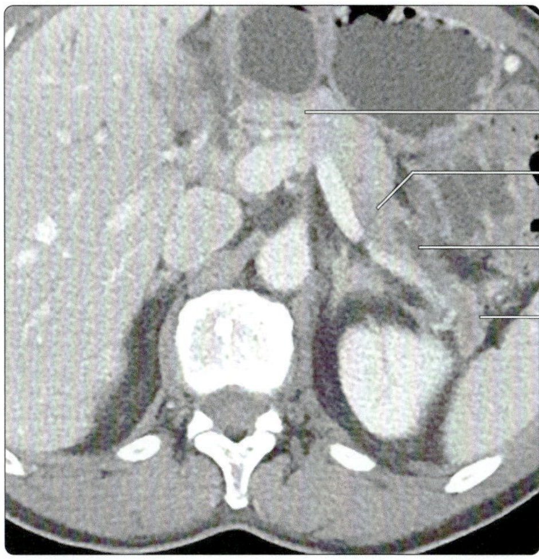

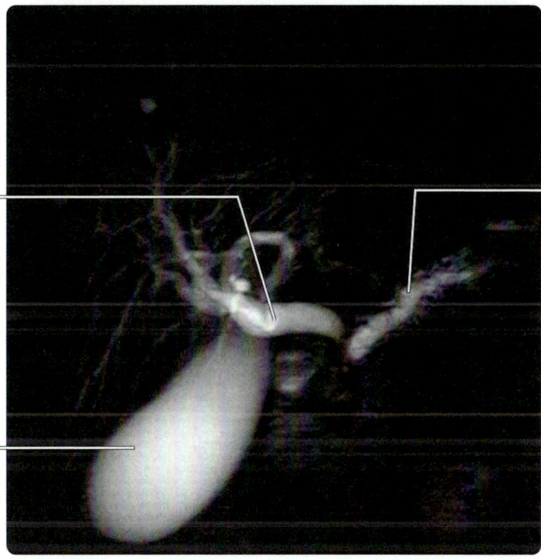

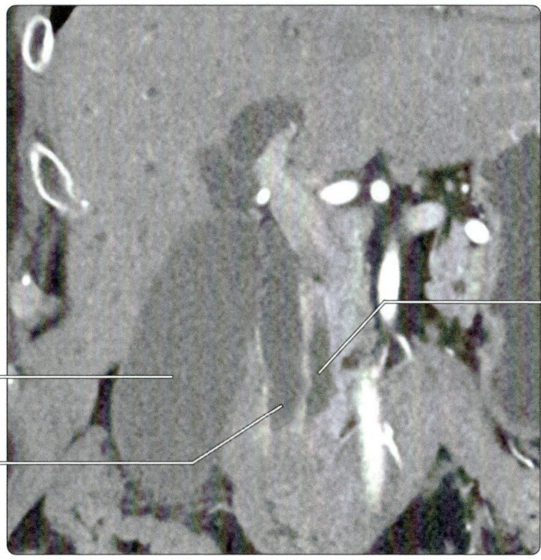

(Top) *Axial CECT demonstrates an ill-defined, hypodense mass in the pancreatic body/tail, resulting in obstruction and dilatation of the upstream pancreatic duct, a classic appearance for pancreatic adenocarcinoma. The pancreatic parenchyma upstream from the mass appears atrophic, a common feature with these tumors.* **(Middle)** *Coronal MRCP with MIP reconstruction demonstrates a classic double-duct sign with severe dilatation of the pancreatic duct and biliary tree (along with distension of the gallbladder), both of which narrow and are obstructed in the region of the pancreatic head in this patient with pancreatic adenocarcinoma.* **(Bottom)** *Coronal CECT demonstrates a subtle hypodense mass in the pancreatic head, found to be a pancreatic adenocarcinoma, causing severe upstream obstruction of both the pancreatic duct and biliary tree.*

PANCREATIC NEUROENDOCRINE TUMOR

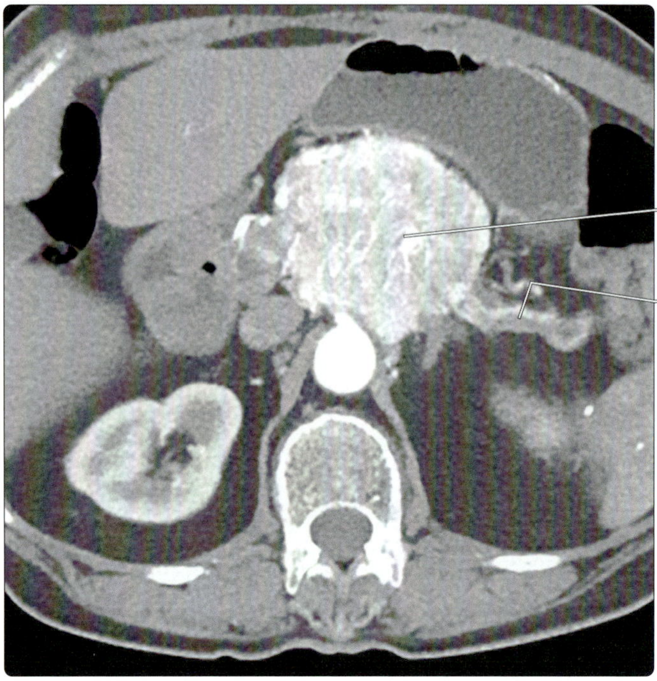

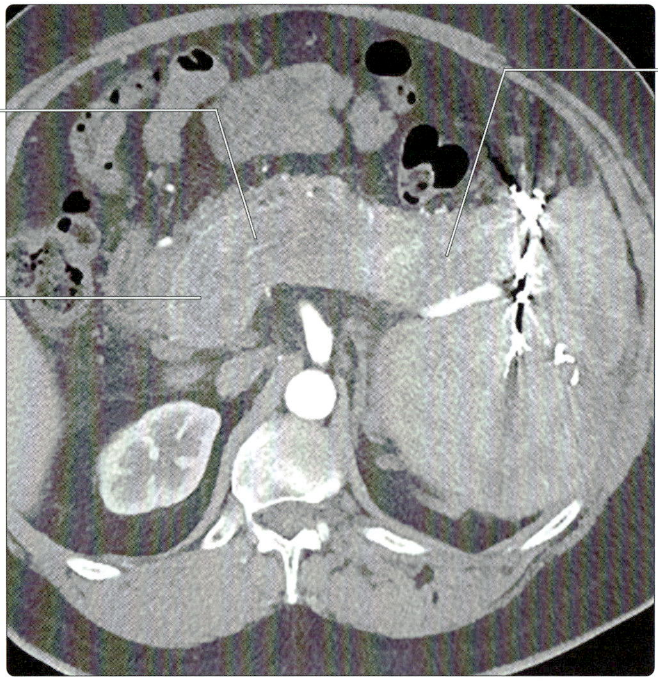

(Top) *Axial CECT in the arterial phase demonstrates a large, markedly hypervascular mass in the pancreas, a typical appearance for a neuroendocrine tumor. Note, however, that the mass in this case results in upstream atrophy with mild upstream pancreatic ductal dilatation, usual for neuroendocrine tumors, but a feature that can sometimes be seen with larger tumors or a subset of tumors that secrete serotonin-like hormones.* **(Bottom)** *Axial CECT demonstrates a neuroendocrine tumor of the pancreatic tail with extensive tumor thrombus filling the entirety of the splenic vein and extending into the portal-splenic confluence. Venous tumor thrombus is a common feature of pancreatic neuroendocrine tumors and is extremely rare in pancreatic adenocarcinomas.*

PANCREATIC DUCT VARIATIONS

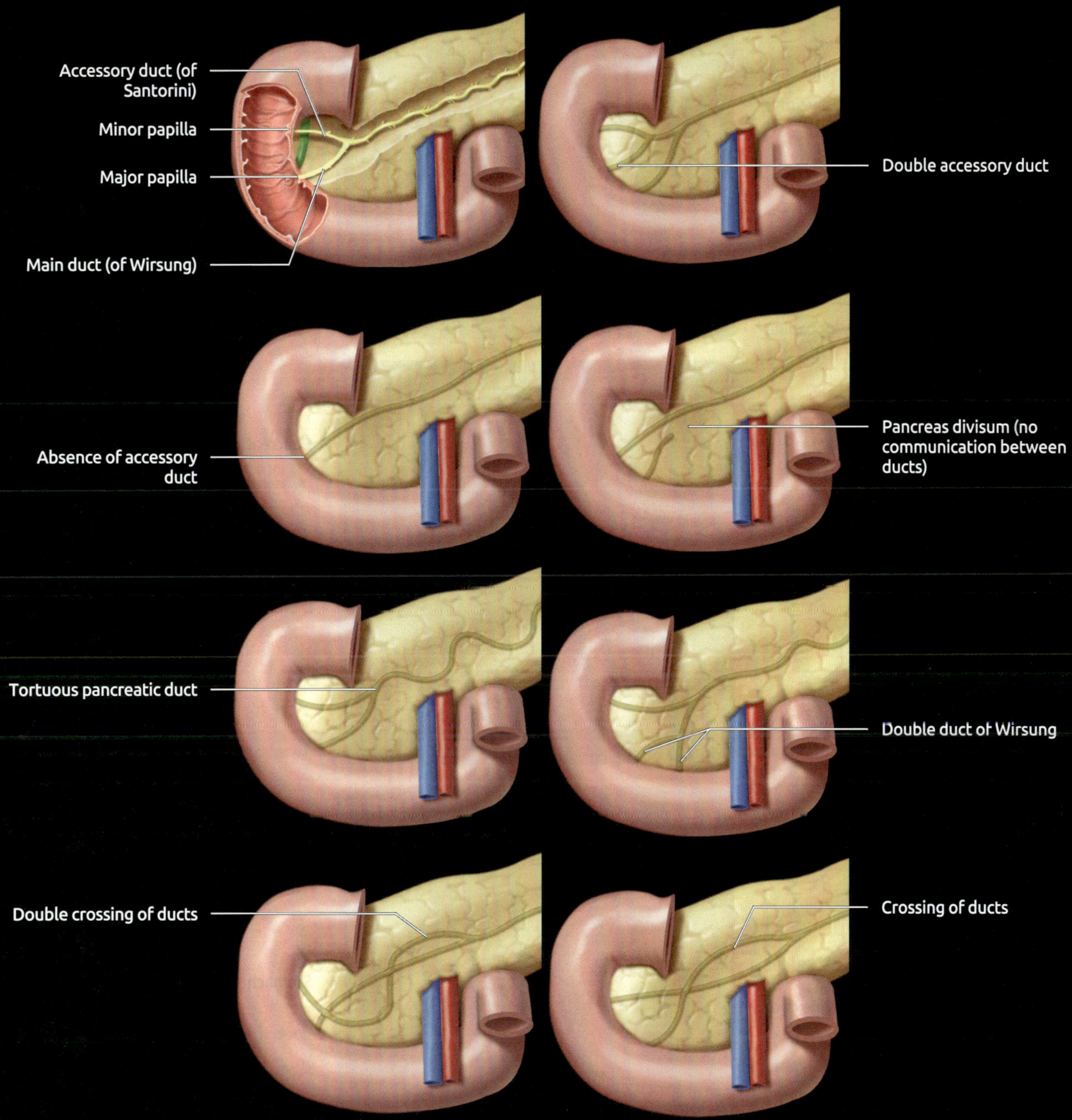

The accessory duct (of Santorini) originates with the dorsal pancreatic anlage, which is the larger bud from the embryologic foregut, composing the pancreatic body and tail. The main duct (of Wirsung) originates with the ventral, smaller anlage that develops into the pancreatic head and uncinate process. Usually, the main and accessory pancreatic ducts fuse, and the main duct becomes the primary conduit for drainage of secretions into the duodenum. The pancreatic duct courses through the center of the gland and is joined by tributaries that enter it at right angles. In the head, the duct turns caudally and dorsally and runs parallel to the common bile duct before joining it at the ampulla of Vater and entering the major papilla. The accessory duct usually enters the duodenum more proximally through the minor papilla.

TERMINOLOGY

Abbreviations

- Perirenal space (PS)
- Anterior pararenal space (APS)
- Posterior pararenal space (PPS)

IMAGING ANATOMY

Overview

- Retroperitoneum encompasses all abdominal contents in between **parietal peritoneum** and **transversalis fascia**
- 2 well-defined fascial planes (**renal** and **lateroconal fasciae**) separate retroperitoneum into 3 compartments
  - **PS** contains kidneys, adrenal glands, proximal ureters, and abundant fat
    - Enclosed by renal fascia
    - Does not extend across abdominal midline
  - **APS** contains pancreas, duodenum, colon (ascending and descending), and small amount of fat
  - **PPS** contains no organs and some fat (continuous with properitoneal fat)
- Renal fascia join and close PS, causing it resembles inverted cone with tip in iliac fossa
  - Caudal to PS, APS and PPS merge to form single **infrarenal retroperitoneal space**
  - Infrarenal retroperitoneal space communicates directly with pelvic **prevesical space** (**of Retzius**)

Anatomic Relationships

- Renal and lateroconal fasciae are major fascial planes, which divide retroperitoneum into discrete compartments
  - **Parietal peritoneum** separates peritoneal cavity from APS
  - **Anterior renal fascia** separates PS from APS
  - **Posterior renal fascia** separates PS from PPS
  - **Lateroconal fascia** separates APS from PPS and marks lateral extent of APS
- Renal and lateroconal fasciae are laminated planes (i.e., can split into layers and form potential spaces that can fill with fluid, etc.) and can form pathways of spread for rapidly expanding fluid collections or inflammatory processes (e.g., hemorrhage or pancreatitis)
  - Diseases originating in APS (e.g., pancreatitis) can extend posterior to kidney via interfascial plane
  - Anterior renal fascia can "split" into **retromesenteric plane**, which is continuous across midline abdomen
  - Posterior renal fascia splits into **retrorenal plane**
  - Lateroconal fascia splits into **lateroconal plane**
  - All 3 of these planes can be called **interfascial planes** and all communicate at lateroconal and renal fasciae junction
  - Interfascial planes communicate across abdominal midline and extend into pelvis caudal to PS
- **Great vessels** (**aorta and inferior vena cava**)
  - Occupy poorly defined prevertebral portion of retroperitoneum caudal to mediastinum
  - Share space with paraspinal nerve chains (sympathetic), major lymphatic trunks (including cisterna chyli and thoracic duct), and ureters

CLINICAL IMPLICATIONS

Clinical Importance

- **Disease within APS** is common
  - Pancreatic disease > duodenal > colon
  - **Pancreatitis** often spreads inflammation and fluid throughout APS to affect duodenum and descending colon
  - Perforated duodenal ulcer is most common cause of inflammation, fluid, and ectopic gas in right APS (may also result in intraperitoneal inflammation and ectopic gas since duodenal bulb is intraperitoneal)
- **Disease in PS** is common
  - Any traumatic, inflammatory, or neoplastic process involving kidney or adrenal
  - Renal fascia is strong and effective at containing most primary renal pathology within PS and excluding most other external processes from entering PS
    - PS is closed and does not communicate across midline or into pelvis but can decompress into interfascial planes that do communicate
    - PS is divided irregularly by **perirenal bridging septa**, which can result in loculated perirenal fluid collections that simulate subcapsular collections
      - Septa may act as conduits for fluid or infiltrative disease (including tumor) to enter or leave PS
- Traumatic or spontaneous **retroperitoneal hemorrhage** is common
  - Causes include anticoagulation therapy, ruptured abdominal aortic aneurysm, or bleeding from renal neoplasm (usually renal cell carcinoma or angiomyolipoma)
  - Bleeding commonly spreads along fascial planes to involve multiple retroperitoneal and pelvic extraperitoneal compartments, in addition to rectus and iliopsoas muscle groups (in coagulopathic patients)
- Neoplastic involvement of retroperitoneum
  - Primary carcinomas may arise from kidney, pancreas, colon (ascending or descending), adrenal glands, and duodenum
  - Primary retroperitoneal sarcomas (most commonly **liposarcoma**) can attain huge size by displacing abdominal viscera before being diagnosed
  - **Nerve sheath tumors** and **paragangliomas** can arise in retroperitoneum
  - **Metastases to** retroperitoneal nodes are very common (e.g., **lymphoma** and primary tumors arising in pelvis, including testicle, prostate, uterus, and ovary)
- **Retroperitoneal fibrosis** (**RPF**)
  - Chronic inflammatory process in lumbar retroperitoneum, which results in encasement of aorta, inferior vena cava, and ureters in fibrotic mantle of tissue
  - Classically results in medial deviation and obstruction of ureters

RETROPERITONEAL PLANES

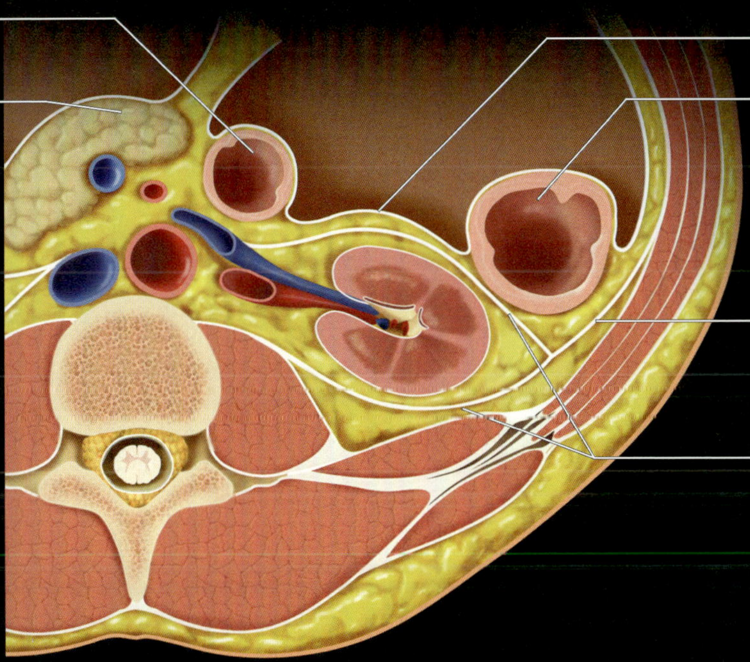

The renal and lateroconal fasciae divide the retroperitoneum into 3 compartments: Anterior pararenal space (APS), perirenal (perinephric) space (PS), and posterior pararenal space (PPS). The APS contains the duodenum, pancreas, and the ascending/descending colon. The APS compartment is limited anteriorly by the peritoneum, which reflects over the colon to form the paracolic gutter, an intraperitoneal recess. Lateral to the peritoneum covering the lateral abdominal wall is retroperitoneal fat, sometimes called the properitoneal fat stripe. The renal fasciae enclose the PS, and the anterior renal fascia forms the posterior boundary of the APS. The lateroconal fascia forms the lateral margin of the APS and the medial margin of the PPS. The posterior renal fascia usually joins with the fasciae of the psoas or quadratus lumborum muscle.

RETROPERITONEAL DIVISIONS

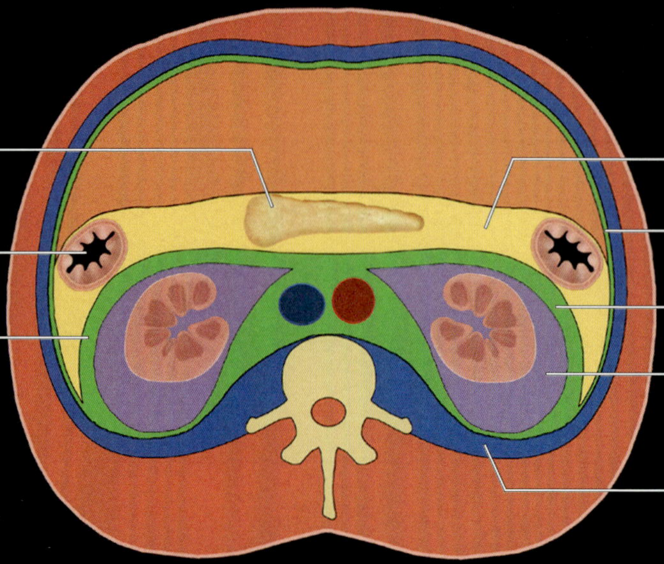

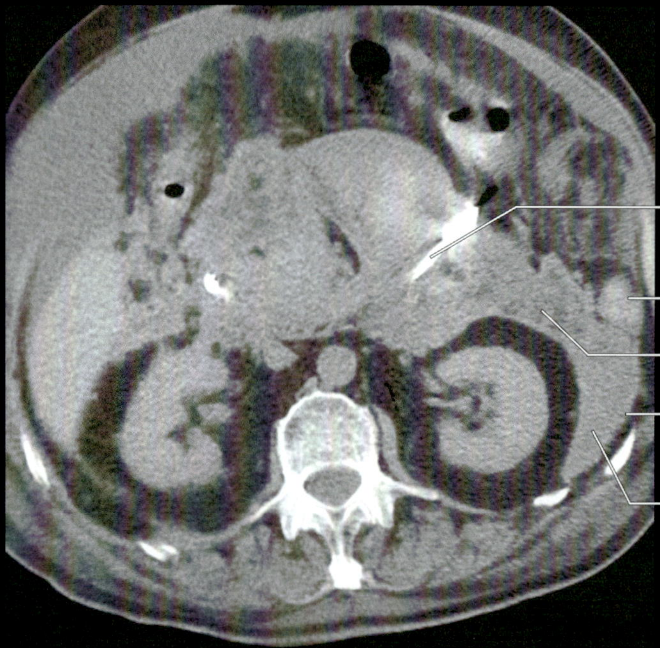

(Top) *The main divisions of the retroperitoneum are the APS (in yellow), perirenal space (PS) (purple), and PPS (blue). In addition, the interfascial planes are shown (green). These are the potential spaces created by inflammatory processes that separate the double laminated layers of the renal and lateroconal fasciae. The PPS contains no organs and is synonymous with the properitoneal fat that extends along the lateral and anterior abdominal wall. The PS do not communicate across the midline, but the APS and the interfascial planes do so.* **(Bottom)** *First of 3 NECT images of a patient with acute pancreatitis shows fluid distending the APS and interfascial planes. The pancreas, duodenum, and vertical colon segments all lie within the APS.*

CORONAL AND SAGITTAL REFORMATIONS

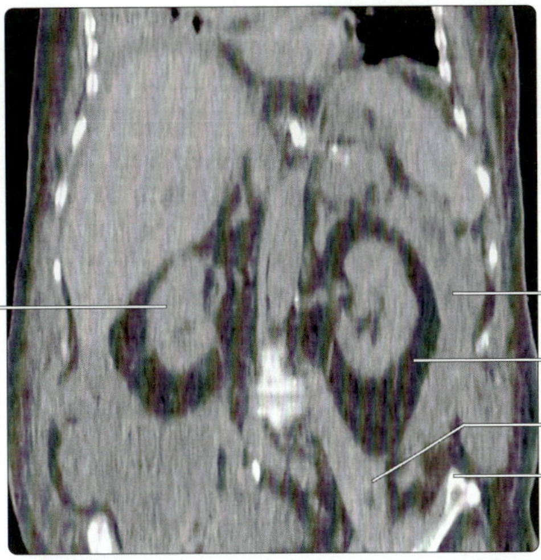

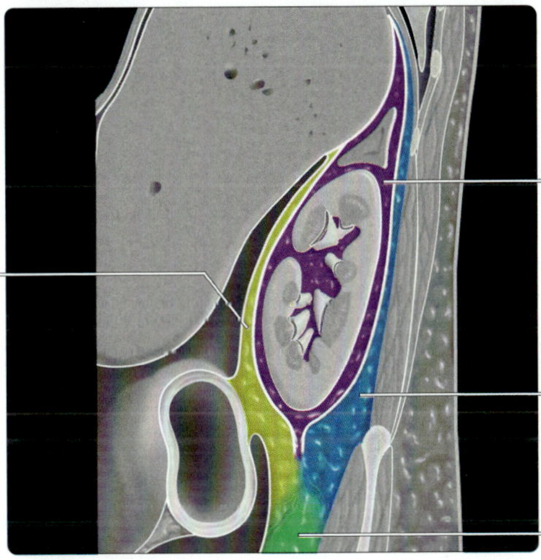

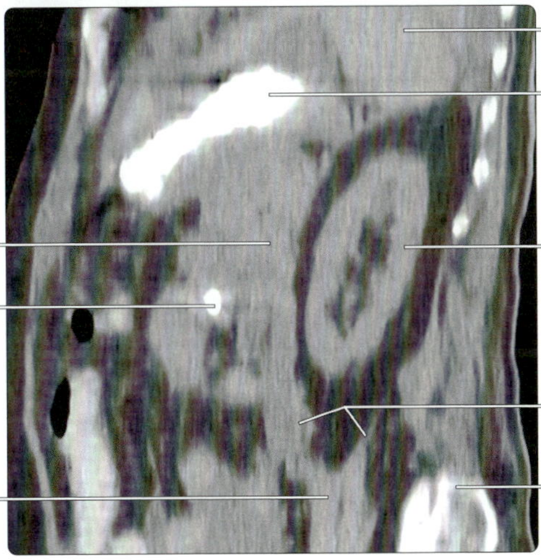

(Top) *A coronal reformation in the same patient shows sparing of the PS, which is protected by the renal fascia from the extensive inflammation that fills the APS, interfascial planes, and infrarenal retroperitoneal space (caudal to the termination of the renal fascia).* **(Middle)** *This sagittal graphic through the right kidney shows the 3 retroperitoneal compartments. Note the confluence of the anterior and posterior renal fasciae at about the level of the iliac crest. Caudal to this, there is only a single infrarenal retroperitoneal space.* **(Bottom)** *Sagittal reformation through the left kidney shows the shape of the PS, which is protected in this patient by the renal fascia.*

AXIAL CT, NORMAL FASCIAL PLANES

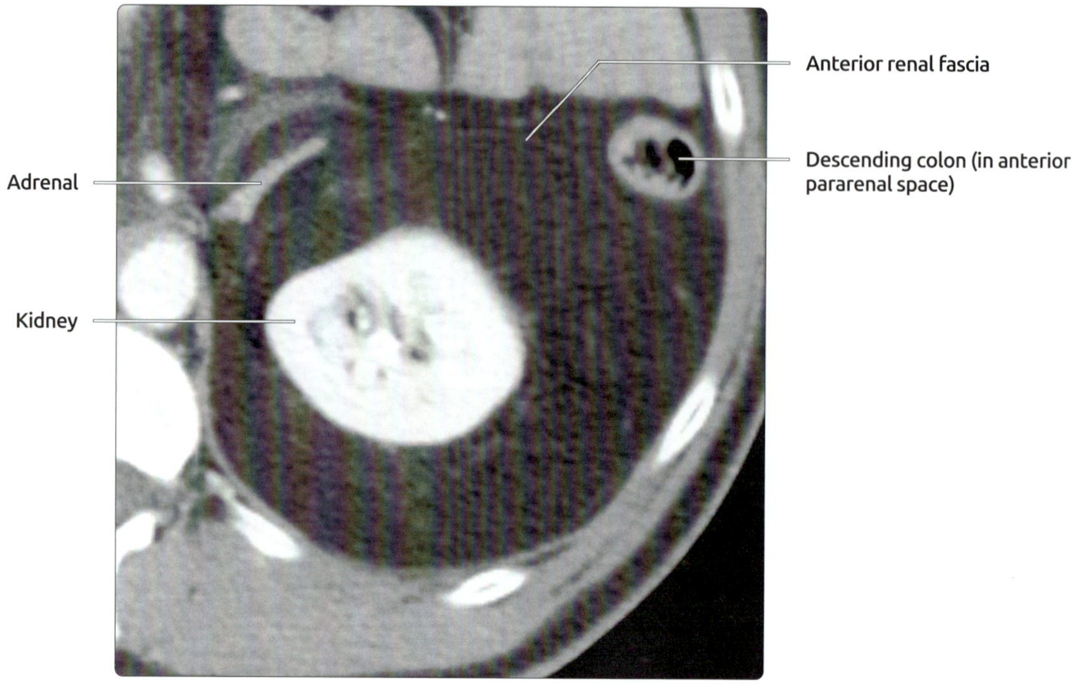

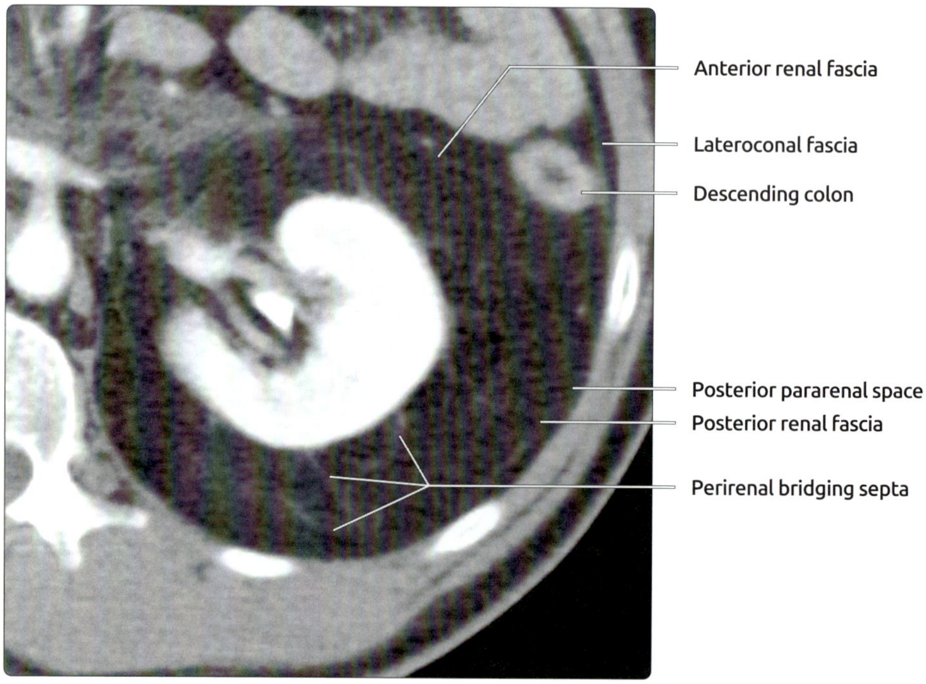

(Top) *First of 5 axial CECT sections show the renal and lateroconal fasciae that define the retroperitoneal spaces. The renal fascia encloses the PS, which includes the kidney, adrenal gland, and abundant fat.* **(Bottom)** *The posterior parietal peritoneum cannot be visualized when normal but forms the anterior margin of the APS whose posterior margin is the anterior layer of the renal fascia. The descending colon can be seen within the APS on this image. Note the perirenal bridging septa that potentially divide the PS into multiple smaller compartments.*

AXIAL CT, NORMAL FASCIAL PLANES

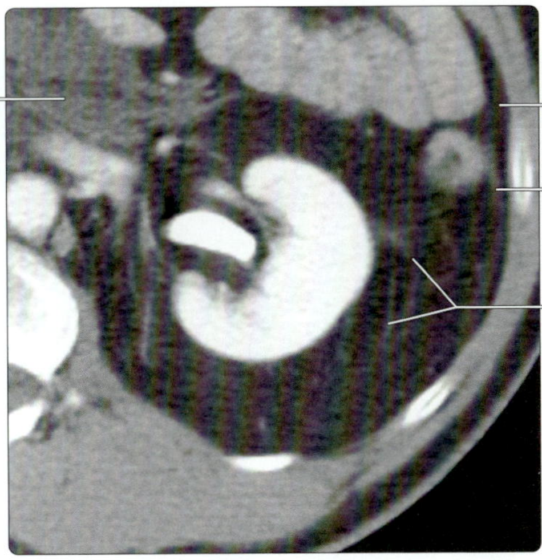

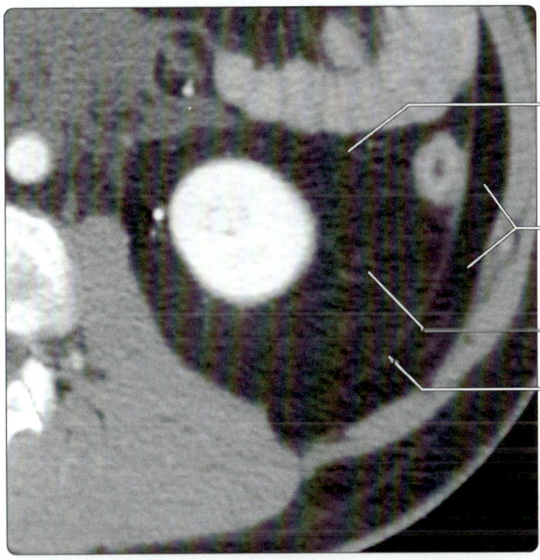

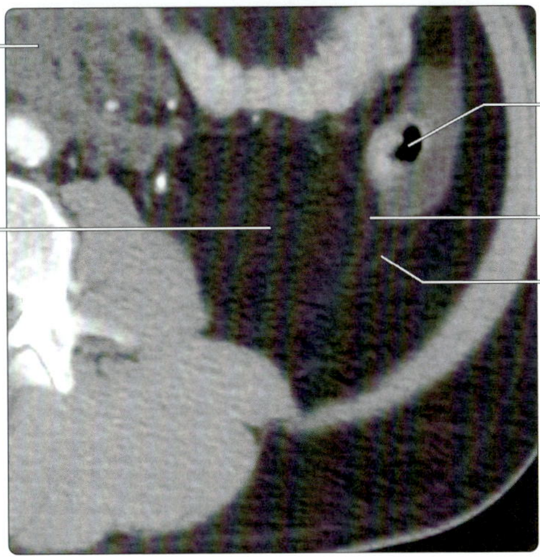

(Top) *More caudal CECT shows slight thickening of the lateroconal fascia, which forms the lateral margin of the APS. The fat plane between the lateroconal fascia and the transversalis fascia (which is not evident on this image) is known as the properitoneal fat (but also as the PPS). No organs or major structures lie within this space, but it is often involved secondarily by inflammation or bleeding that originates elsewhere in the retroperitoneum or abdominal wall.* **(Middle)** *The PPS is broader and more evident on this section.* **(Bottom)** *Most caudal section below the kidney shows the layers of renal fascia coming together as the PS narrows similar to an inverted cone.*

PANCREATITIS DEFINING PLANES AND SPACES

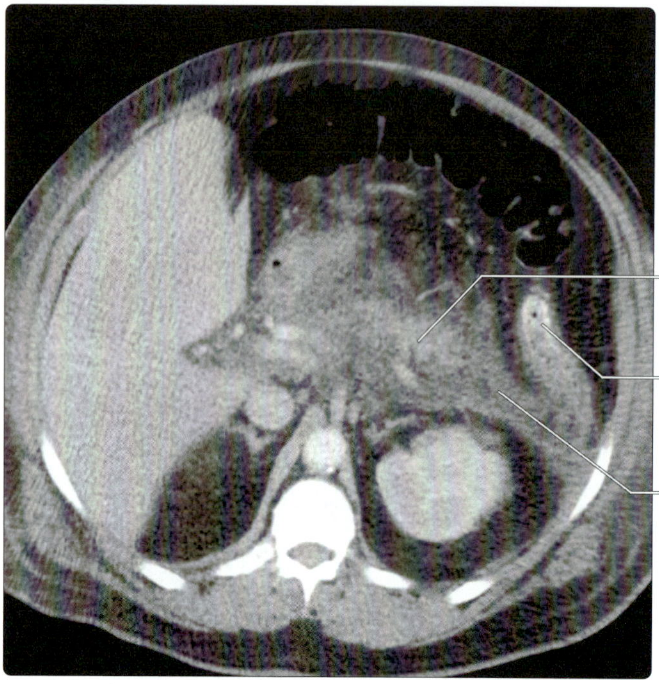

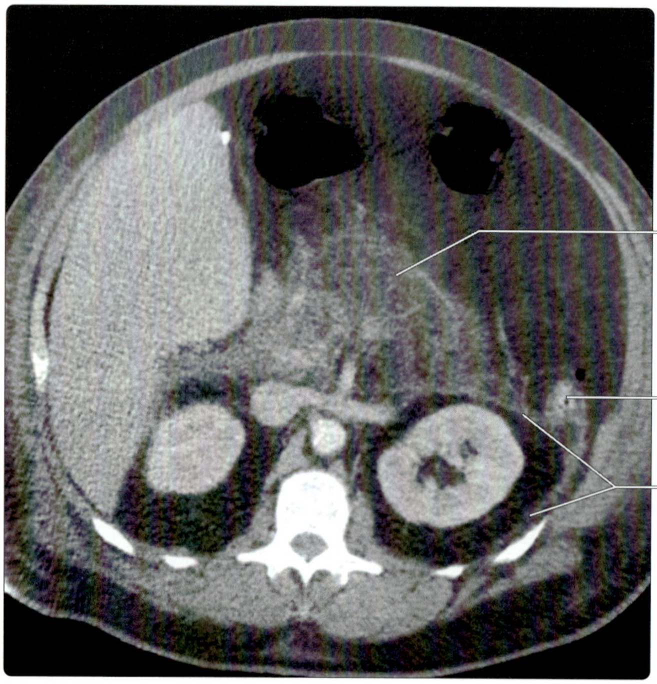

(Top) *First of 4 axial CECT images from a patient with acute pancreatitis and inflammation that help define the retroperitoneal spaces is shown. This section shows infiltration of the fat planes surrounding the pancreas, primarily within the APS. The inflammation spreads laterally to the anatomic splenic flexure where the colon leaves its mesocolon to enter the retroperitoneum as the descending colon.* **(Bottom)** *The inflammation also spreads caudally and ventrally to enter the mesentery. Recall that the roots of the mesentery and transverse mesocolon arise from just in front of the pancreas, and inflammation easily enters the fatty space between the peritoneal reflections.*

PANCREATITIS DEFINING PLANES AND SPACES

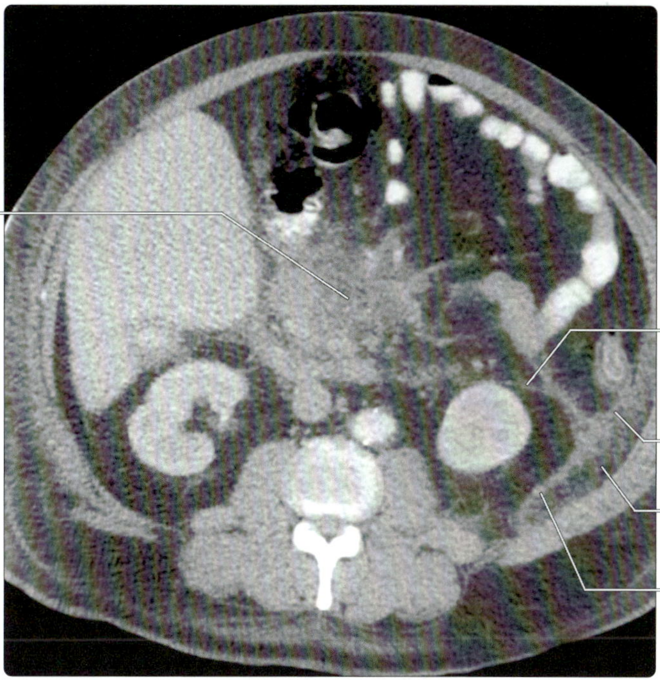

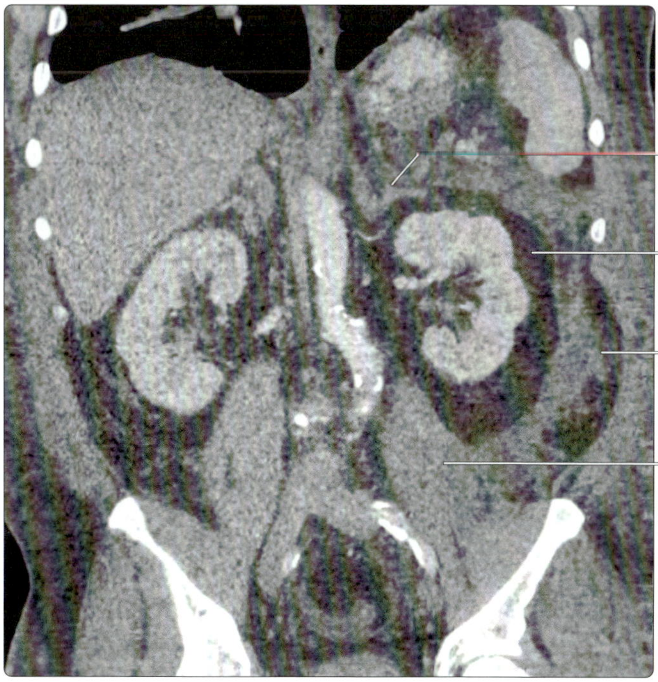

(Top) *The renal and lateroconal fasciae are thickened by inflammation, which has also spread into the PPS lateral to the lateroconal fascia.* **(Bottom)** *A coronal reformation in the same patient shows normal (spared) fat in the PS, while the fat in the APS and PPS is inflamed. Note how the layers of renal fascia close off the PS at about the level of the iliac crest. Caudal to the PS, there is only a single retroperitoneal space, since the APS and PPS come together. This helps to explain how the PPS becomes involved in inflammatory or hemorrhagic processes that originate in the APS.*

PANCREATITIS INVOLVING OTHER ORGANS

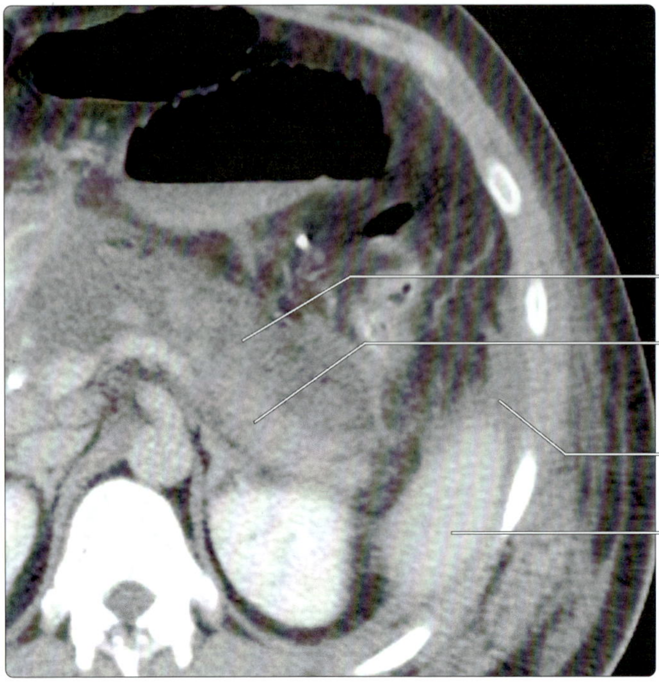

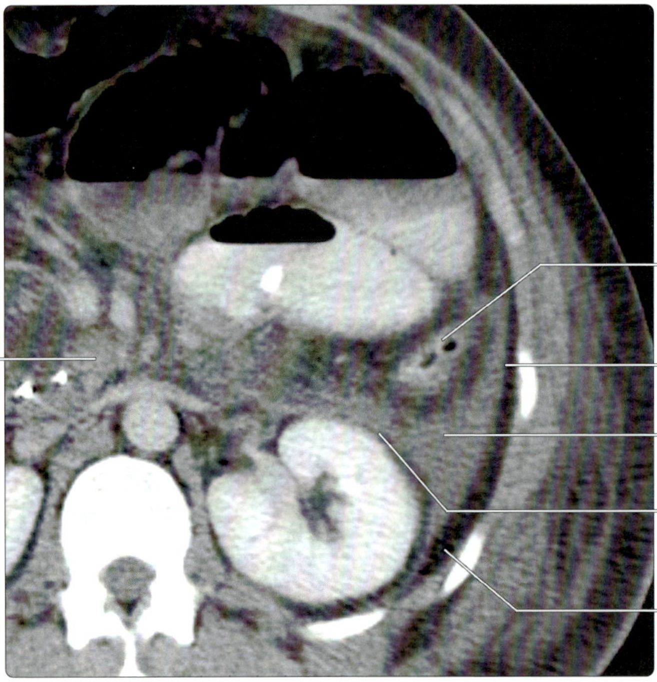

(Top) *First of 5 axial CECT images of a patient with acute pancreatitis shows extensive infiltration of the fat planes surrounding the pancreas. Fluid around the spleen is ascites, indicating that the inflammatory process has extended through the posterior parietal peritoneum to enter the peritoneal cavity.* **(Bottom)** *The inflammatory process has spread laterally through the APS but is prevented from entering the PPS by the lateroconal fascia. The layers of the renal and lateroconal fasciae are expanded by the inflammation at the point that they join, resulting in fluid tracking through interfascial planes. The descending colon, also residing in the APS, is contacted by the inflammation.*

PANCREATITIS INVOLVING OTHER ORGANS

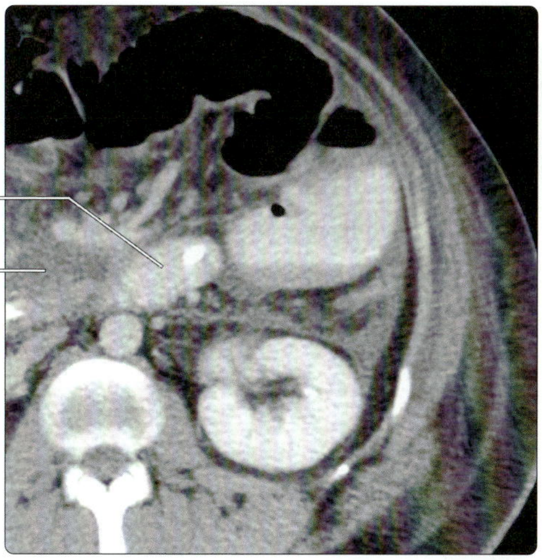

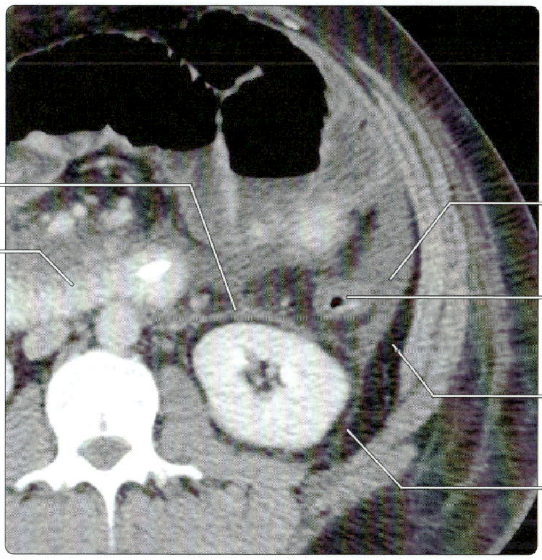

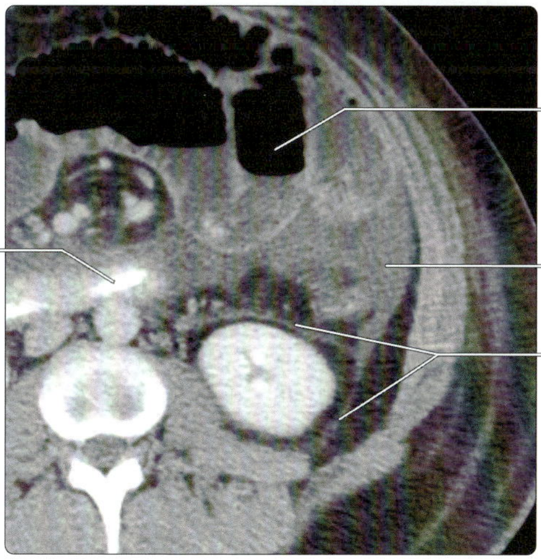

(Top) *A more caudal image shows the duodenum surrounded by inflammation as it too lies in the APS.* **(Middle)** *The parietal peritoneum reflects over the descending colon to form the paracolic gutter, an intraperitoneal recess. The peritoneum is a thin structure, not normally visualized on imaging, and is easily breached by inflammatory and hemorrhagic processes. No clear demarcation can be seen between the retroperitoneal inflammation and the ascites on this section.* **(Bottom)** *The contiguous involvement of the retroperitoneal and intraperitoneal spaces is evident on this section. The retroperitoneal duodenum is encased by inflammation, as are the intraperitoneal small bowel segments, resulting in an ileus (dilated bowel with diminished peristalsis).*

PERIRENAL BRIDGING SEPTA

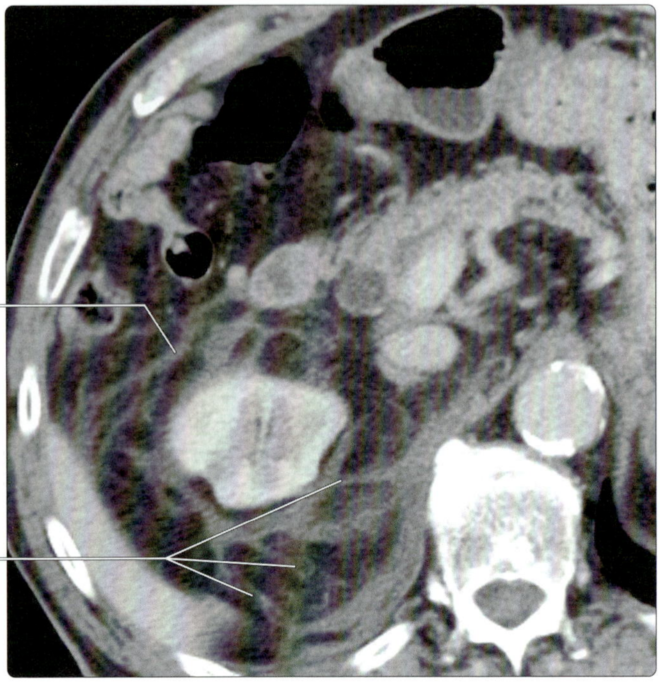

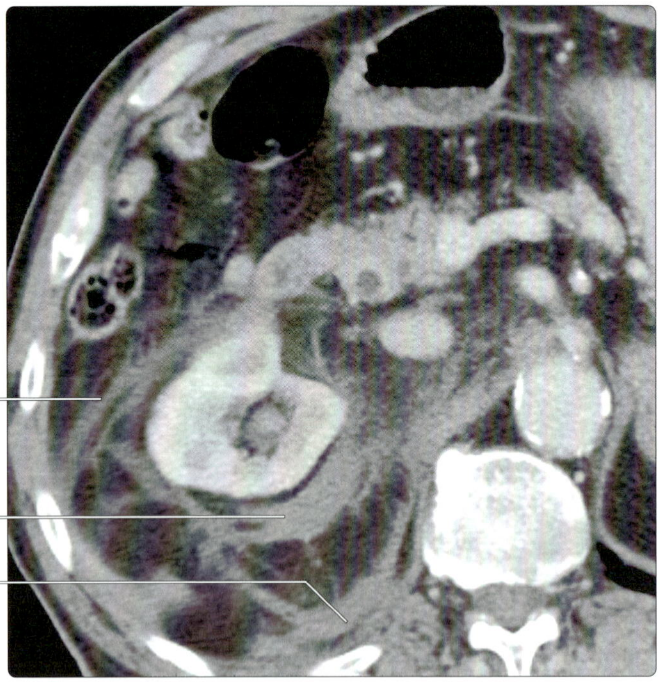

(Top) *First of 2 axial CT images of a patient with perirenal hemorrhage shows numerous curvilinear septations within the PS caused by hemorrhage tracking along the septa that divide the PS into multiple compartments. Blood and inflammatory infiltrates can decompress along these planes and spread into other retroperitoneal spaces.* **(Bottom)** *Loculated fluid within the PS can mimic subcapsular fluid. Note the intact fat plane between the loculated fluid and the surface of the kidney, indicating that the fluid is confined within the PS, but it does not spread diffusely to fill the space because of the perirenal septa.*

PERIRENAL SPACE WITH BLOOD

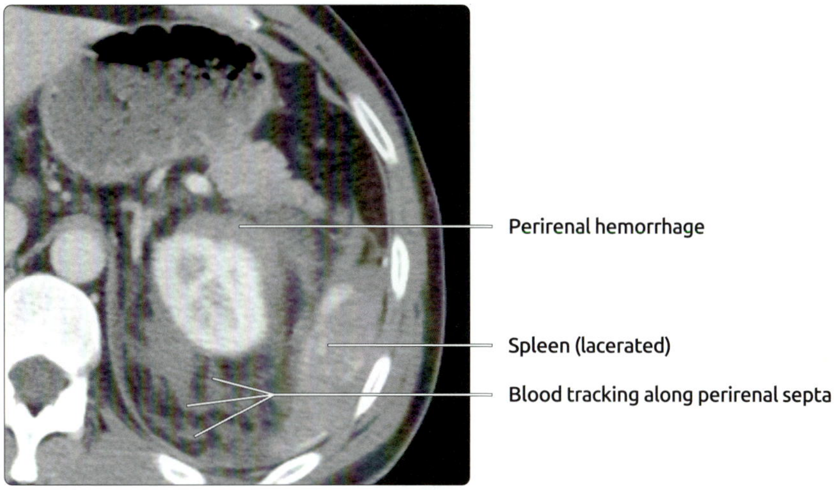

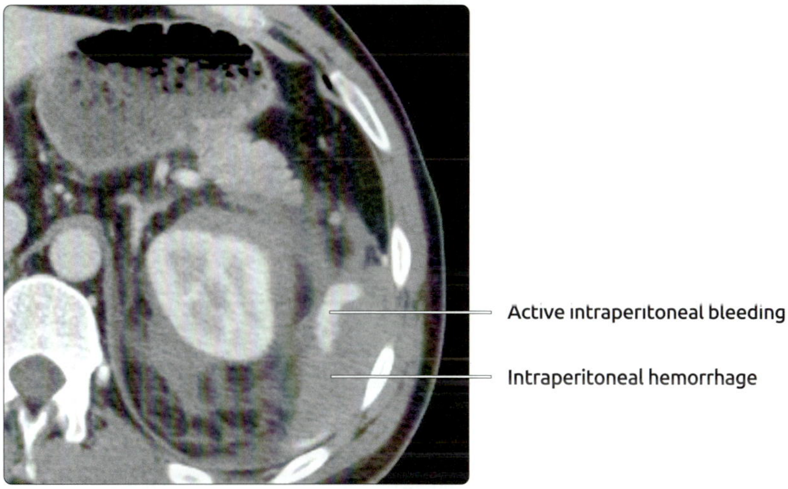

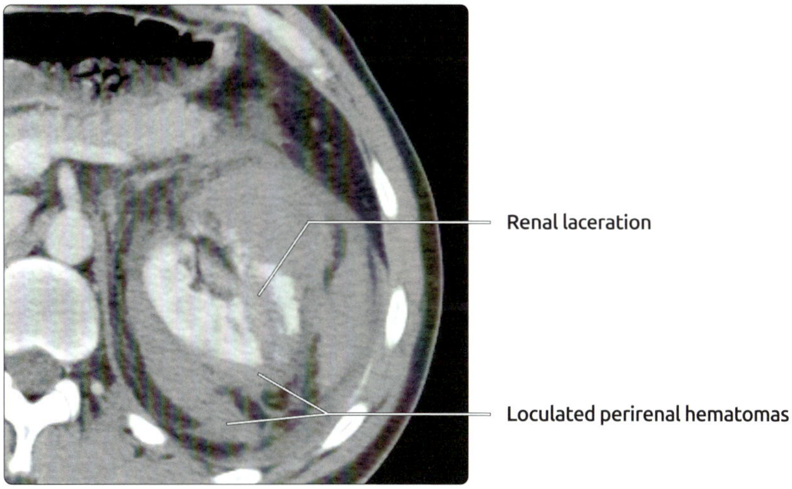

(Top) *First of 3 axial CECT images from a patient injured in a motor vehicle crash shows heterogeneous enhancement of the spleen due to parenchymal laceration. Coexisting renal injury has resulted in perirenal hemorrhage that is loculated by perirenal septa. The collection ventral to the kidney simulates a subcapsular hematoma but spreads along the surface of the kidney without causing compression of the parenchyma.* **(Middle)** *A more caudal image just below the spleen shows intraperitoneal hemorrhage with a collection that is isodense to the contrast-opacified aorta, indicating active arterial bleeding from the splenic injury. Active hemorrhage of this sort requires urgent intervention with surgery or angiographic embolization to occlude the bleeding vessel.* **(Bottom)** *A deep renal laceration is seen along with loculated perirenal hematomas.*

HEMORRHAGE INTO EXTRAPERITONEAL SPACES

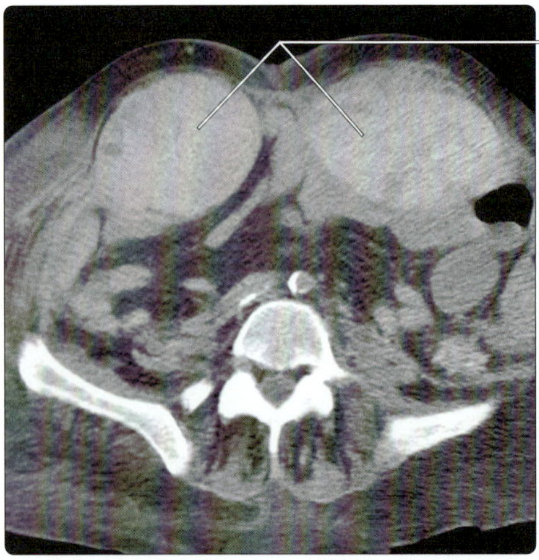

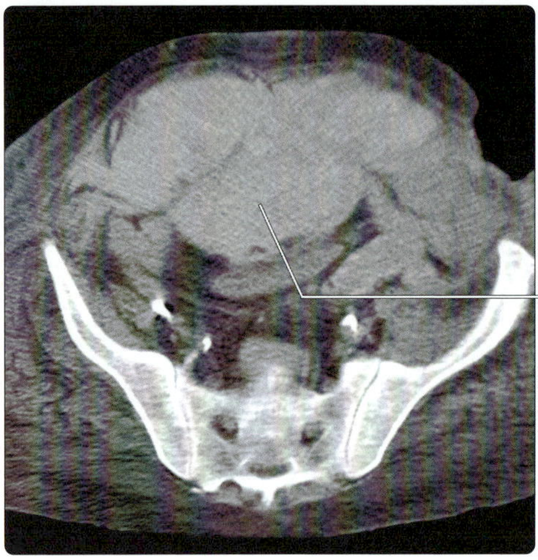

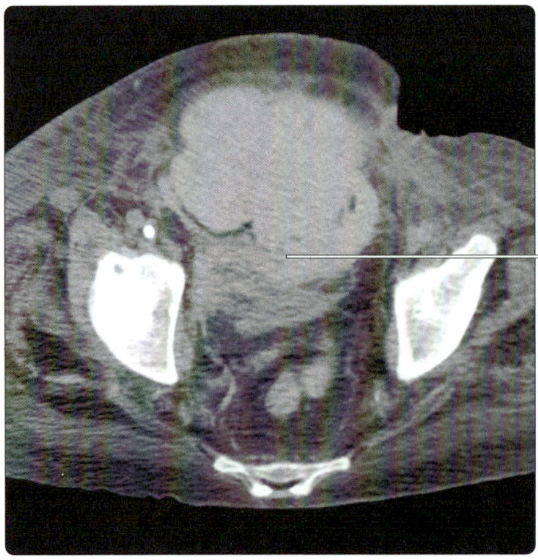

(Top) *First of 5 axial CECT images from a patient with spontaneous bleeding from overanticoagulation is shown. This section shows high-density, heterogeneous hemorrhage within the rectus sheath bilaterally, a common site for spontaneous bleeding in patients with coagulopathy.* **(Middle)** *Because the rectus sheath is incomplete posteriorly along the lower 1/3 of the muscle, the bleeding is no longer confined within the rectus sheath but extends into the adjacent extraperitoneal spaces, including the prevesical space (of Retzius). The prevesical and perivesical spaces communicate superiorly with the infrarenal retroperitoneal space, and fluid or inflammation may extend cephalad to involve all 3 retroperitoneal compartments.* **(Bottom)** *A more caudal image shows hemorrhage in the anterior abdominal wall (rectus sheath) and extraperitoneal pelvis (prevesical space).*

HEMORRHAGE INTO EXTRAPERITONEAL SPACES

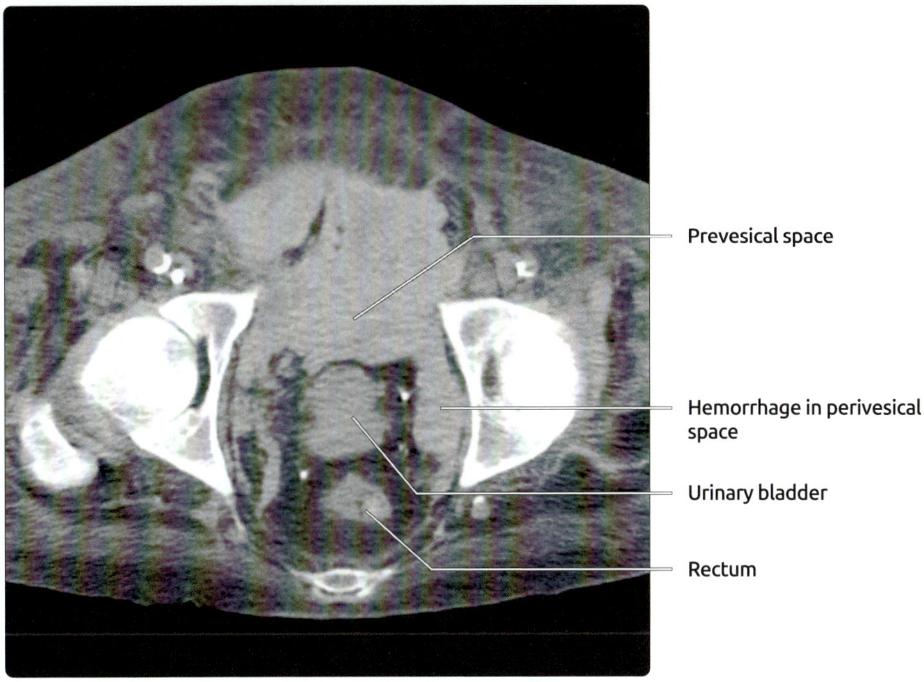

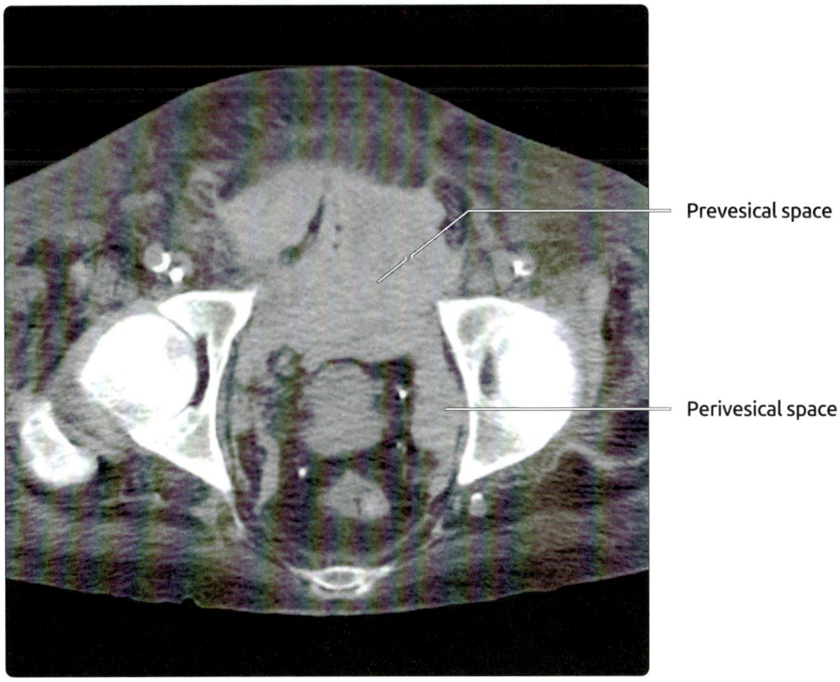

(Top) *The prevesical space continues caudally and laterally to involve the perivesical space lower in the pelvis.* **(Bottom)** *Hemorrhage spreads extensively through the pelvic extraperitoneal spaces. Involvement of the extraperitoneal pelvic spaces tends to assume a molar tooth appearance on axial CT sections, as in this image. The roots of the tooth are the extensions into the perivesical space. Extraperitoneal pelvic fluid also commonly extends posteriorly to the presacral space, which is minimally involved in this patient.*

HEMORRHAGE ACROSS MIDLINE

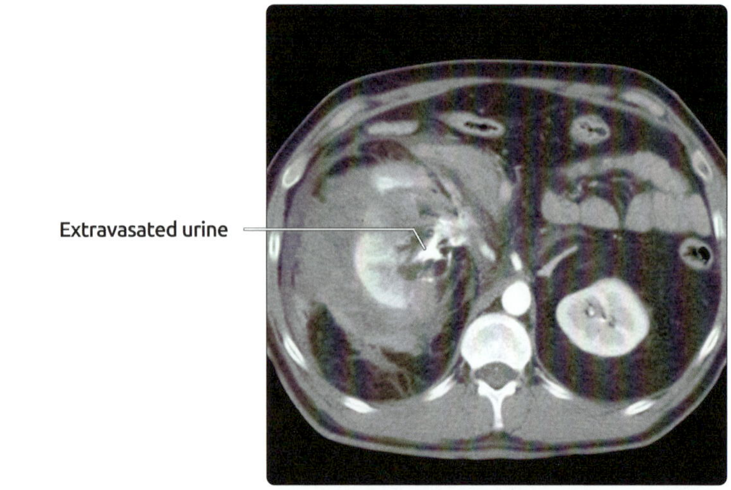

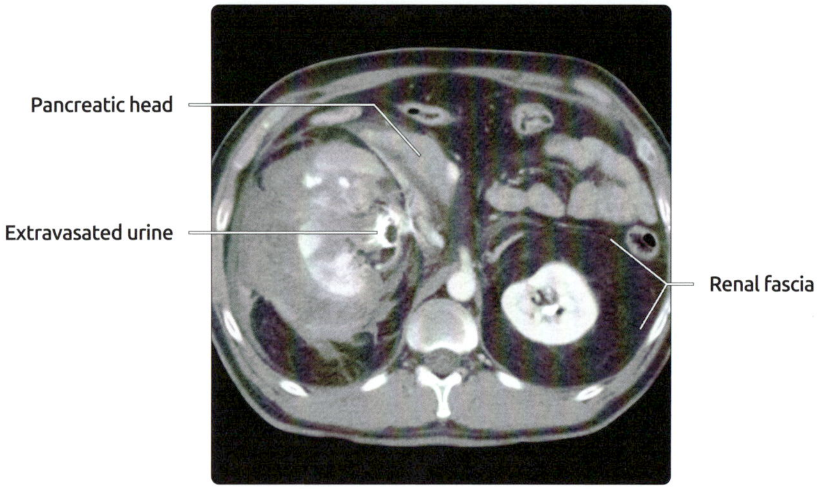

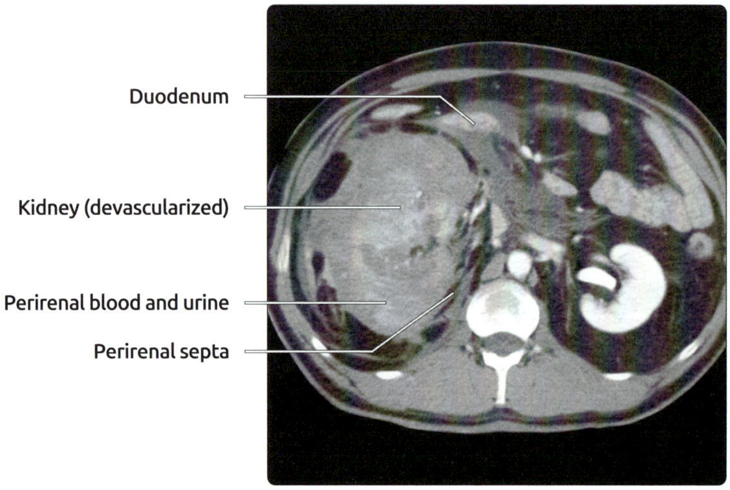

(Top) *First of 6 axial CECT images of a patient with a traumatic renal laceration shows a large perirenal hemorrhage that fills the PS. The renal parenchyma appears to be fragmented.* **(Middle)** *The deep renal lacerations have extended into the collecting system, evident as extravasation of contrast-opacified urine. The pancreatic head and duodenum are displaced by the large perirenal hematoma/urinoma.* **(Bottom)** *The lower pole of the right kidney shows no contrast enhancement, indicating that the renal artery to this segment has been occluded or avulsed. The large perirenal collection of blood and urine is spreading along the perirenal septa to decompress into the interfascial planes. The PS does not extend across the midline; however, the interfascial plane and APS do cross the midline.*

HEMORRHAGE ACROSS MIDLINE

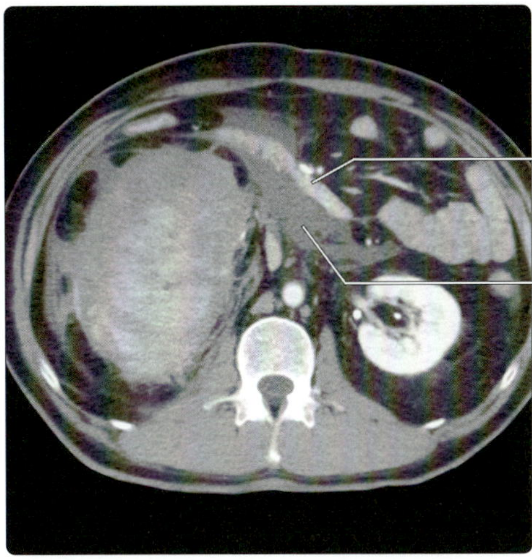

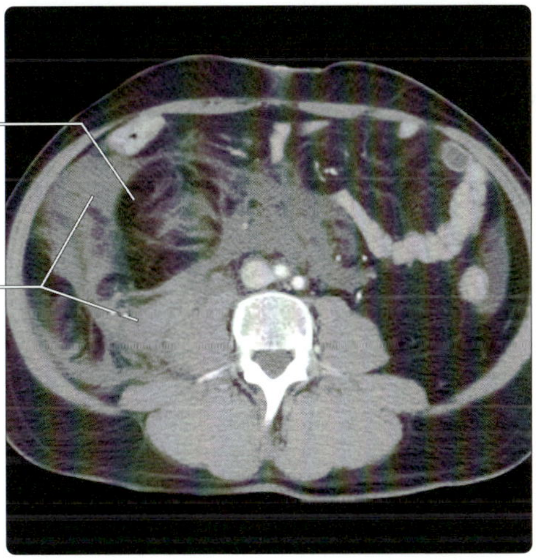

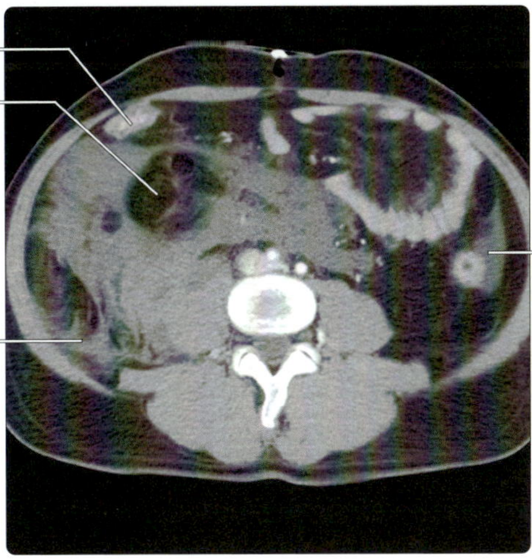

(Top) *The fluid crossing the abdominal midline with the duodenum is in the APS and the interfascial plane. These spaces have become involved secondarily after the renal injury resulted in extensive perirenal fluid collections that spread along the perirenal septa and separated the 2 layers of the renal fascia to form an interfascial plane.* **(Middle)** *On this more caudal image, there is little fluid within the PS but extensive fluid distention of the interfascial planes. Note the spread of fluid into the contiguous psoas compartment and PPS and the spread across the abdominal midline in front of the aorta and inferior vena cava (IVC).* **(Bottom)** *The right APS is relatively spared, as indicated by intact fat planes around the ascending colon. Blood has leaked through the peritoneum into the paracolic gutter.*

COMMUNICATION AMONG EXTRAPERITONEAL SPACES

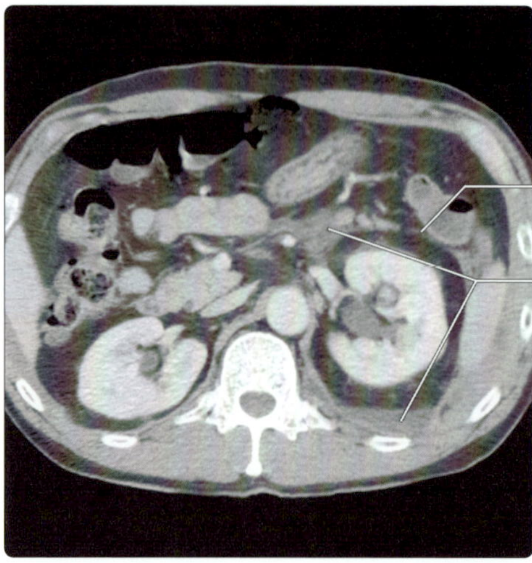

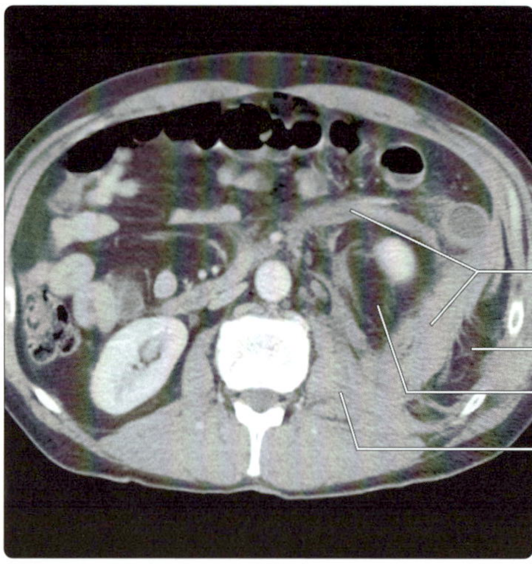

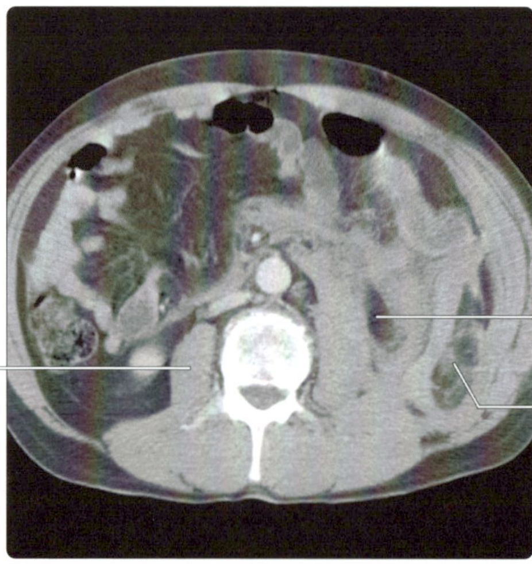

(Top) *First of 8 axial CECT images in a patient with a spontaneous retroperitoneal hemorrhage related to anticoagulation medication is shown. This section shows only subtle thickening of the left renal fascia and interfascial planes.* **(Middle)** *Extensive hemorrhage distends the interfascial planes on the left with contiguous bleeding into the psoas compartment and pararenal spaces. The PS is spared.* **(Bottom)** *The hemorrhage is spreading along curvilinear planes, corresponding to the interfascial planes and the fasciae covering the left psoas, quadratus lumborum, and transverse abdominal muscles. Coagulopathic hemorrhage often originates in the abdominal wall muscles, especially the rectus and iliopsoas, with subsequent dissection into the retroperitoneal spaces.*

COMMUNICATION AMONG EXTRAPERITONEAL SPACES

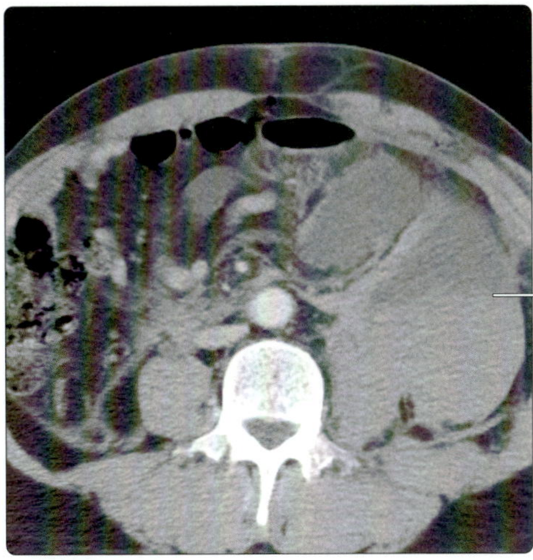

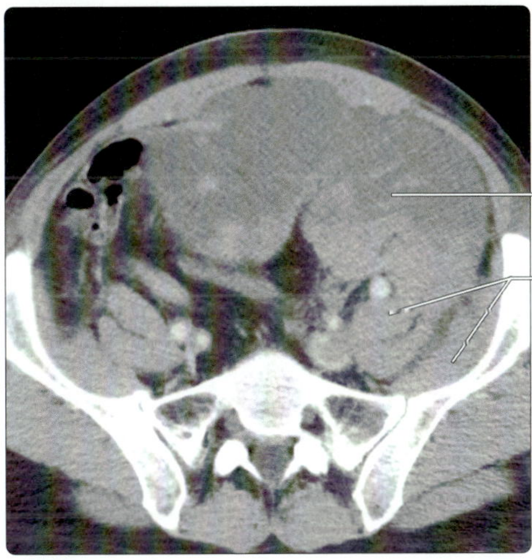

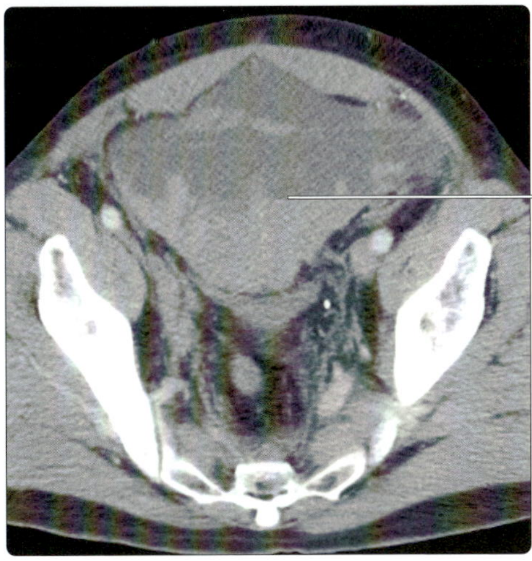

(Top) *A more caudal image shows massive hemorrhage into the psoas compartment and adjacent retroperitoneum. Note the cellular/fluid level, referred to as the hematocrit sign, that is almost pathognomonic of a coagulopathic hemorrhage. It indicates gravitational settling of blood cells below a serum level without clot development.* **(Middle)** *Hemorrhage has spread into the infrarenal retroperitoneal space and continues caudally to involve the pelvic extraperitoneal spaces, including the prevesical space. The blood remains contiguous and in communication with the left iliopsoas compartment.* **(Bottom)** *The hemorrhage distends the prevesical space in this more caudal axial CECT.*

COMMUNICATION AMONG EXTRAPERITONEAL SPACES

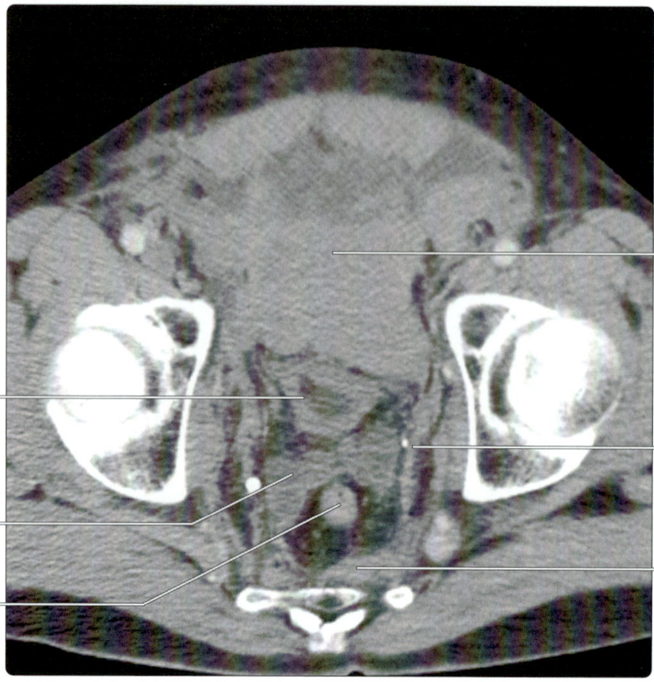

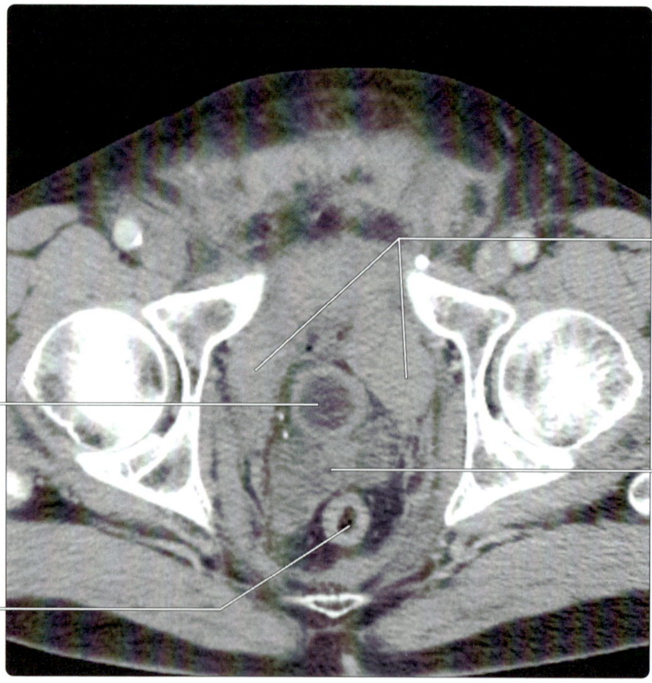

(Top) *Blood tracks from the prevesical space into the perivesical and presacral spaces. A small amount of blood has entered the peritoneal cavity and has settled into the rectovesical space (or pouch of Douglas). In this patient, blood has descended from its primary site of origin in the abdominal wall and retroperitoneum to the pelvic extraperitoneal spaces. The opposite can also happen, as with a pelvic fracture leading to extensive pelvic extraperitoneal bleeding that extends up into the abdominal retroperitoneum.* **(Bottom)** *Note the distinction between the rectovesical space (pouch of Douglas), an intraperitoneal recess, and the perivesical and presacral spaces, divisions of the pelvic extraperitoneal spaces.*

PRIMARY RETROPERITONEAL TUMORS

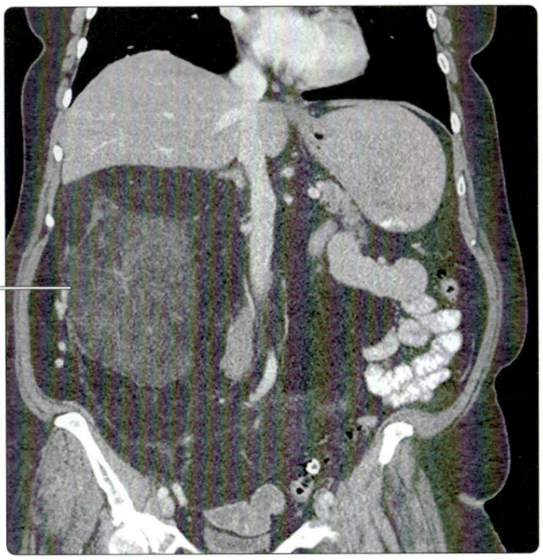

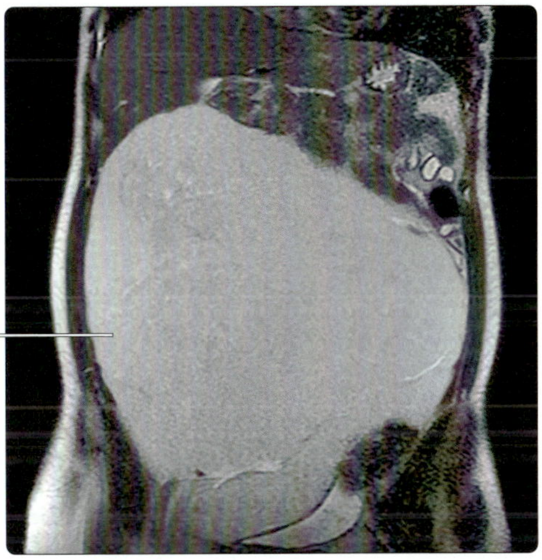

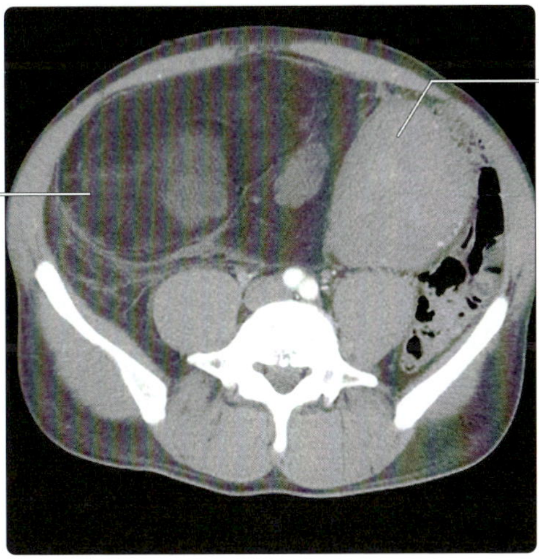

(Top) *Axial CECT demonstrates a large fat-containing mass in the right aspect of the retroperitoneum composed of extensive macroscopic fat along with some additional slightly higher density complex components, compatible with a liposarcoma.* **(Middle)** *Coronal T2 HASTE MR demonstrates a huge mass occupying a large portion of the abdomen, which has identical signal to the subcutaneous fat, representing an large retroperitoneal liposarcoma.* **(Bottom)** *Axial CECT demonstrates a large fat-containing liposarcoma composed both a macroscopic fat as well as some additional higher density and enhancing components. In general, the presence of complex higher density and enhancing components usually suggest greater degrees of dedifferentiation within these tumors.*

PRIMARY RETROPERITONEAL TUMORS

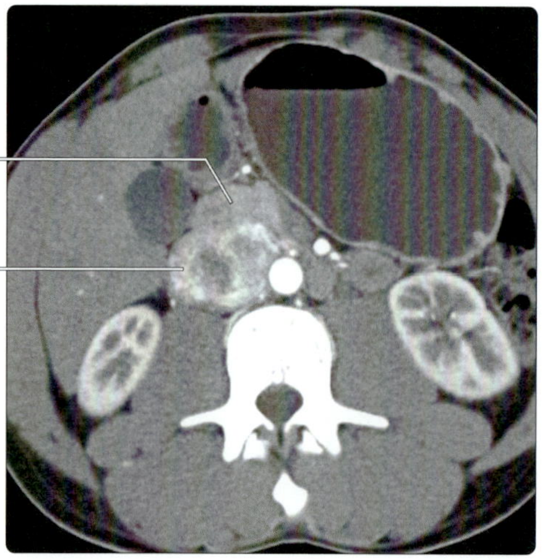

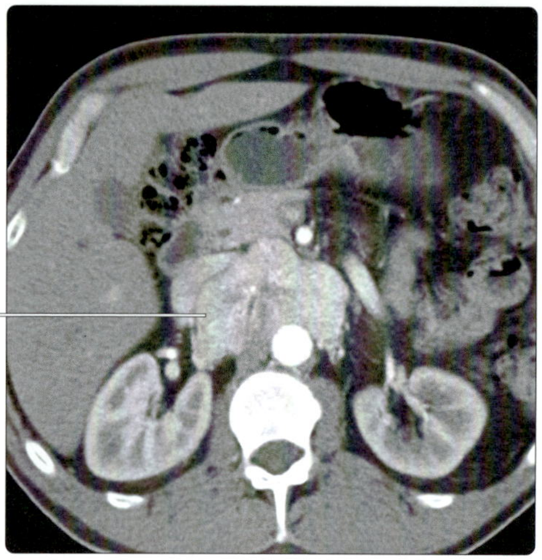

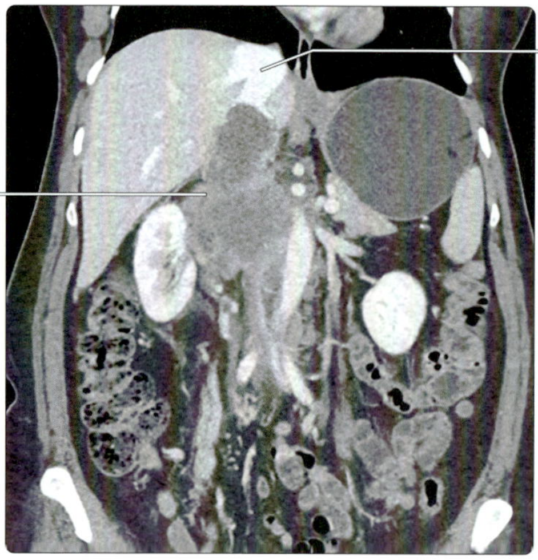

(Top) *Axial CECT demonstrates a markedly enhancing, hypervascular mass in the retroperitoneum. The mass does not arise from the duodenum or pancreas and was found to be a primary retroperitoneal paraganglioma. These tumors, as in this case, are often dramatically hypervascular, particularly in the arterial phase.* **(Middle)** *Axial CECT demonstrates a hypervascular mass in the retroperitoneum with marked enhancement, compatible with a paraganglioma. Paragangliomas in this location are often confused for pancreatic neuroendocrine tumors, duodenal carcinoid tumors, or duodenal gastrointestinal stromal tumors.* **(Bottom)** *Coronal CECT demonstrates a large mass in the right aspect of the retroperitoneum, with direct involvement of the IVC, found at resection to be an IVC leiomyosarcoma. While liposarcoma is the most common retroperitoneal sarcoma, a variety of other tumor types are possible. In this case, the location of the mass and the intimate involvement of the IVC should suggest the correct diagnosis.*

PRIMARY RETROPERITONEAL TUMORS

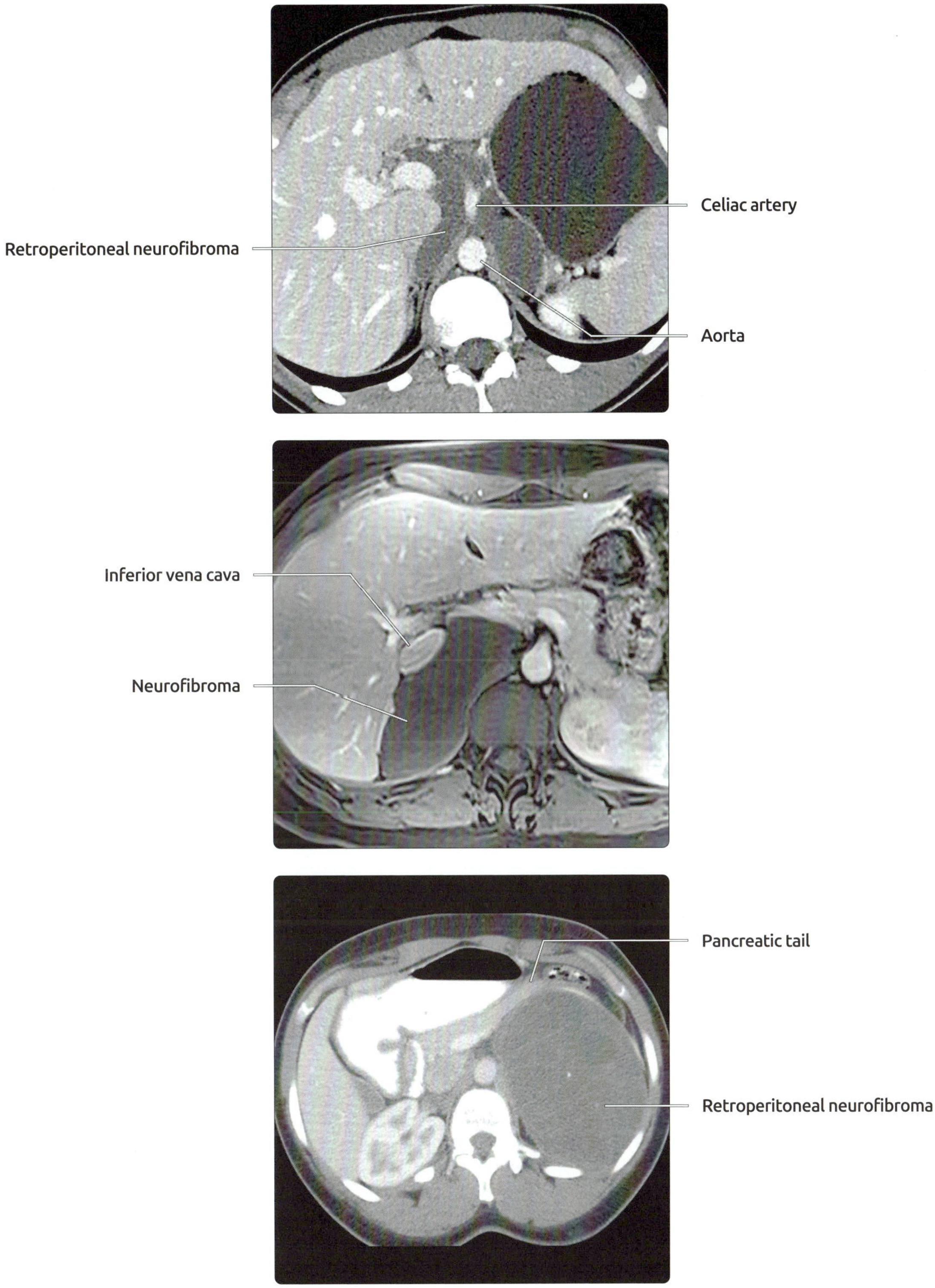

(Top) *Axial CECT in a patient with neurofibromatosis demonstrates a low-density mass in the retroperitoneum. The tumor surrounds the vasculature, including the celiac artery, without appreciable narrowing or attenuation. The appearance of this lesions is fairly typical of a retroperitoneal neurofibroma, particularly given the patient's history.* **(Middle)** *Axial T1 C+ FS MR demonstrates a poorly enhancing mass in the retroperitoneum, which was completely asymptomatic. The mass is very well circumscribed, conforms to the shape of the retroperitoneum, and has no significant internal complexity. The mass was ultimately resected and shown to represent a retroperitoneal neurofibroma.* **(Bottom)** *Axial CECT demonstrates a large, well-circumscribed hypovascular mass in the retroperitoneum, resulting in anterior displacement of the pancreatic tail. This mass was resected and found to represent a large neurofibroma.*

RETROPERITONEAL FIBROSIS

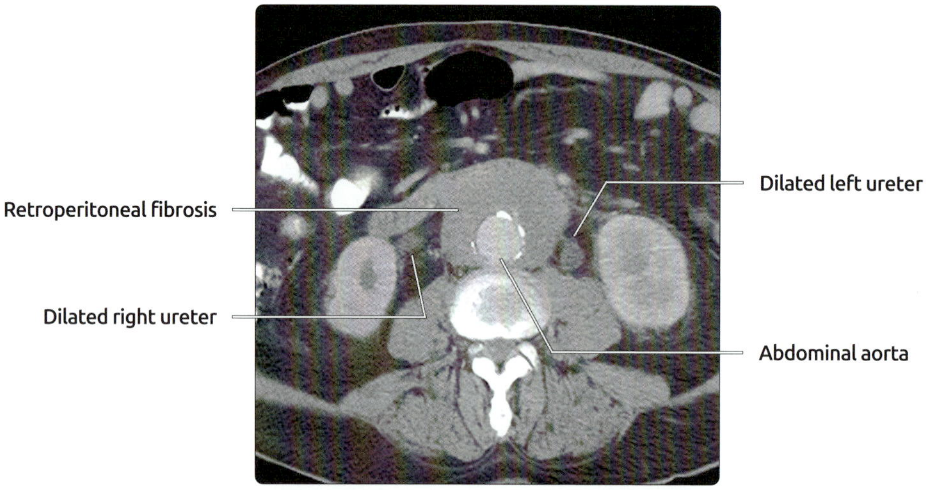

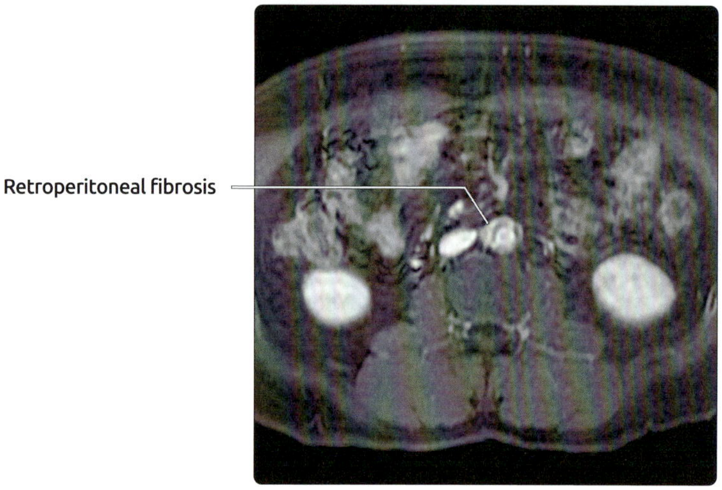

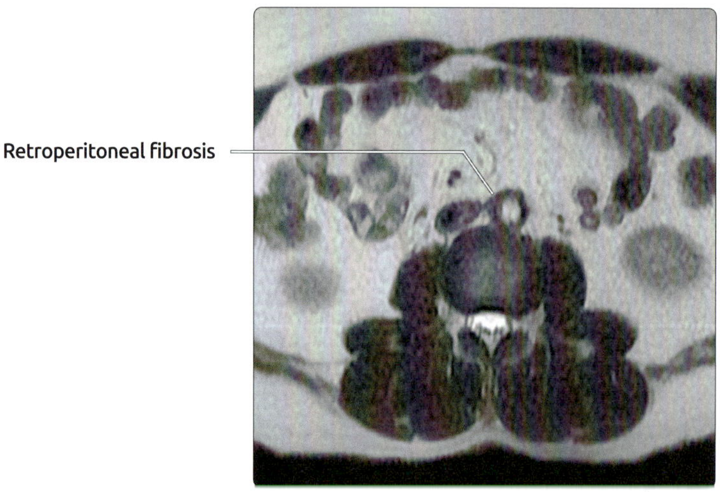

(Top) *Axial CECT demonstrates a large "rind" of soft tissue around the abdominal aorta with involvement of both ureters, resulting in bilateral hydronephrosis/hydroureter, ultimately found to represent retroperitoneal fibrosis.* **(Middle)** *Axial T1 C+ MR in the delayed phase demonstrates a "rind" of enhancing soft tissue around the abdominal aorta, compatible with retroperitoneal fibrosis. The soft tissue in this case demonstrates increasing enhancement on the delayed-phase images compared to the earlier phases (not shown) as a result of its fibrotic component.* **(Bottom)** *Axial T2 HASTE MR in the same patient demonstrates how the soft tissue "rind" around the aorta demonstrates marked T2 hypointensity, again secondary to its fibrotic component.*

RETROPERITONEAL FIBROSIS

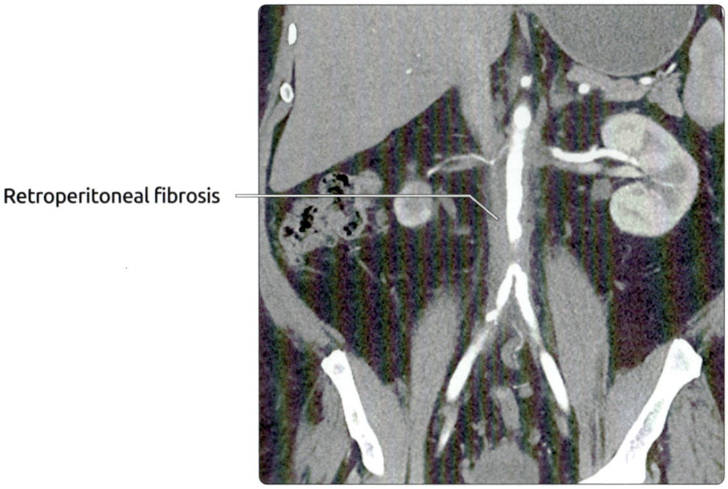

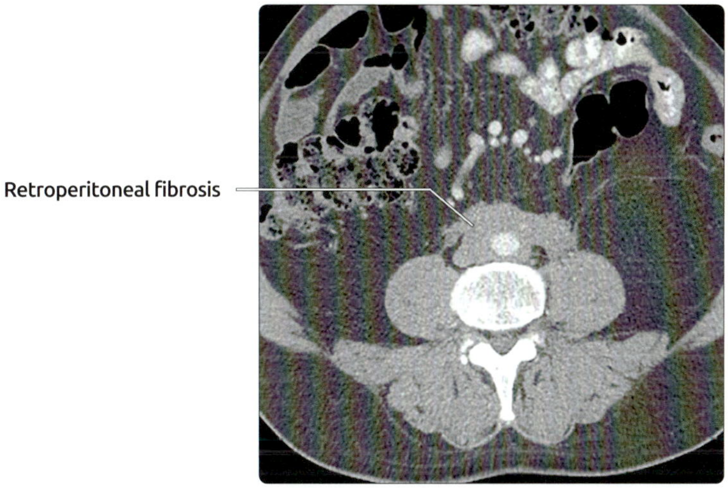

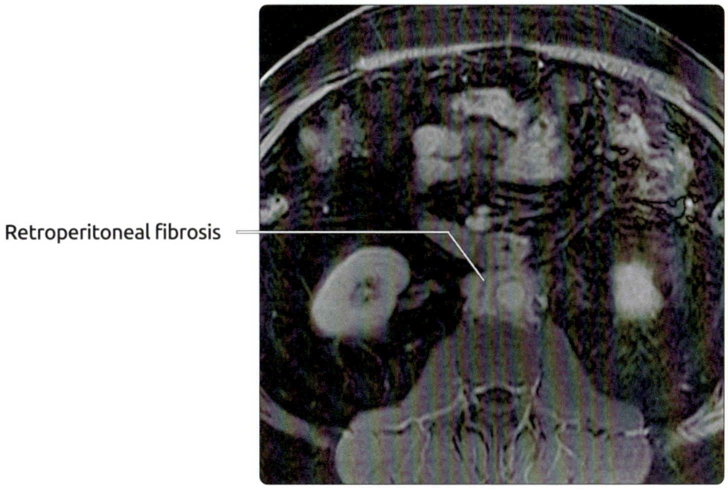

(Top) *Axial CECT demonstrates a mantle of soft tissue encasing the aorta, ultimately found to represent retroperitoneal fibrosis.* **(Middle)** *Axial CECT demonstrates soft tissue in the retroperitoneum surrounding the aorta, a classic appearance for retroperitoneal fibrosis.* **(Bottom)** *Axial CECT demonstrates a "rind" of soft tissue encasing the aorta, a classic appearance for retroperitoneal fibrosis. Retroperitoneal fibrosis can demonstrate relatively variable enhancement, depending on degree of fibrotic and inflammatory components, although the soft tissue in this case is clearly demonstrates a moderate degree of enhancement.*

LYMPHOMA WITH RETROPERITONEAL INVOLVEMENT

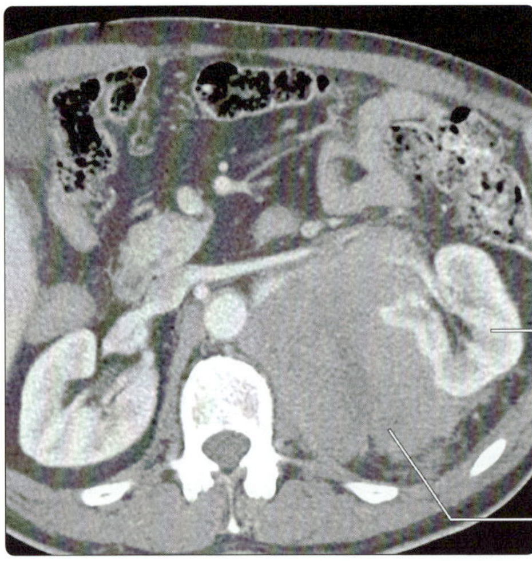

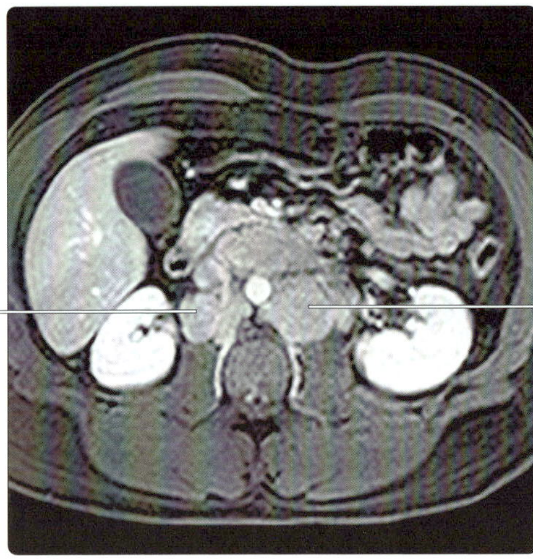

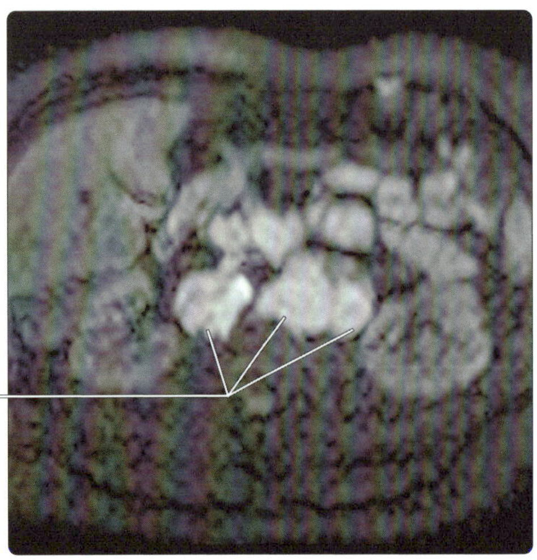

(Top) *Axial CECT demonstrates a large perinephric soft tissue mass extending medially in contiguity with retroperitoneal lymphadenopathy, ultimately proven to represent non-Hodgkin lymphoma.* **(Middle)** *Axial T1 C+ MR demonstrates multiple enlarged lymph nodes throughout the retroperitoneum. The most common entities to present with this appearance would be lymphoma or metastatic lymphadenopathy related to pelvic malignancies.* **(Bottom)** *Axial DWI MR in the same patient demonstrates many of these enlarged retroperitoneal lymph nodes show restricted diffusion. This was ultimately shown to represent non-Hodgkin lymphoma.*

LYMPHOMA WITH RETROPERITONEAL INVOLVEMENT

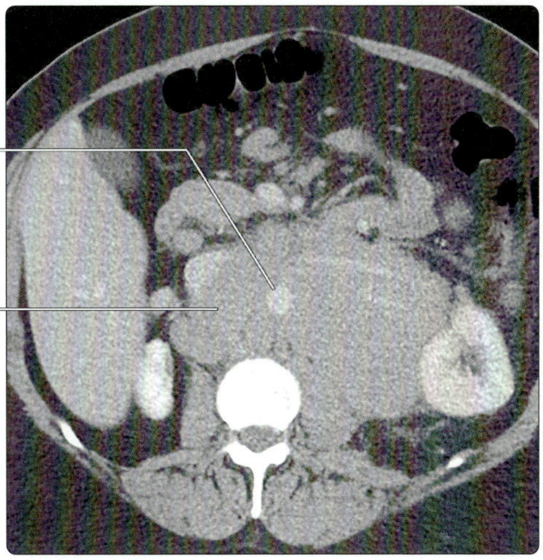

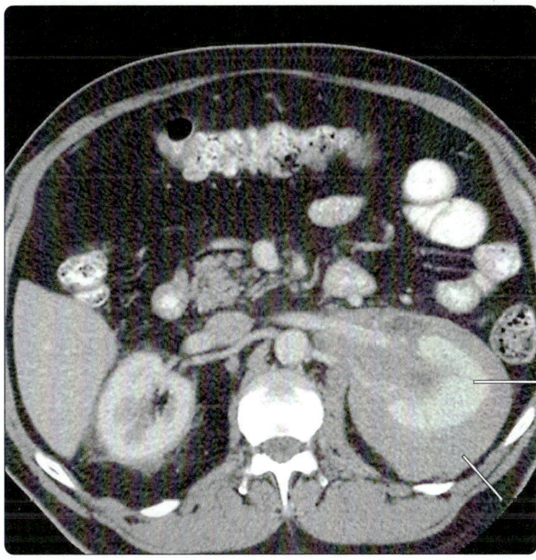

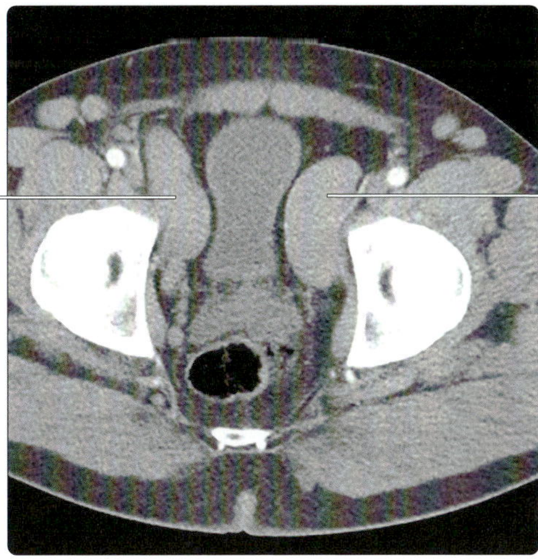

(Top) *Axial CECT demonstrates a large soft tissue mass encasing the aorta in the retroperitoneum, ultimately proven to represent non-Hodgkin lymphoma.* **(Middle)** *Axial CECT demonstrates extensive soft tissue encasing the left kidney in the retroperitoneum, a fairly typical appearance for lymphoma (although there are a few other more rare entities that can produce a similar appearance).* **(Bottom)** *Axial CECT demonstrates massive enlargement of the external iliac lymph nodes bilaterally, with resultant compression of the bladder, representing non-Hodgkin lymphoma. Lymphoma is the most common malignancy to cause widespread lymphadenopathy involving multiple abdominal and pelvic nodal chains. Primary pelvic malignancies may also spread to retroperitoneal nodes.*

GROSS ANATOMY

Overview

- Adrenal (**suprarenal**) glands are part of endocrine and neurological systems
  - Essentially different organs within same structure, composed of thick outer cortex and thin inner medulla
  - Lie within **perirenal space** bilaterally, bounded by **renal** (**perirenal**) **fascia**, above/medial to kidneys
  - Composed of "body" and 2 limbs (medial and lateral)

Anatomic Relationships

- Right adrenal is usually more apical in location
  - Lies anterolateral to right crus of diaphragm, medial to liver, and posterior to inferior vena cava (IVC)
  - Often pyramidal in shape with inverted V shape on transverse section
- Left adrenal is usually more caudal and lies medial to upper pole of left kidney, lateral to left crus of diaphragm, and posterior to splenic vein and pancreas
  - Often crescentic in shape with λ or triangular shape on transverse section

Divisions

- **Adrenal cortex**
  - Embryologically derived from mesoderm
  - Divided into 3 distinct zones (zona glomerulosa, zona fasciculata, and zona reticularis)
  - Secretes **mineralocorticoids** (aldosterone) from zona glomerulosa, **glucocorticoids** (cortisol) from zona fasciculata, and **androgens** from zona reticularis
- **Adrenal medulla**
  - Embryologically derived from neural crest
  - Part of sympathetic nervous system
  - **Chromaffin cells** secrete **catecholamines** (mostly epinephrine) into bloodstream
- **Vessels**, **nerves**, and **lymphatics**
  - Arteries
    - **Superior adrenal arteries**: Typically 6-8; from inferior phrenic arteries
    - **Middle adrenal artery**: 1; from abdominal aorta
    - **Inferior adrenal artery**: 1; from renal arteries
  - Veins
    - **Right adrenal vein** drains into IVC
    - **Left adrenal vein** drains into left renal vein (usually after joining left inferior phrenic vein)
  - Nerves
    - Extensive sympathetic connection to adrenal medulla
    - Presynaptic sympathetic fibers from paravertebral ganglia end directly on secretory cells of medulla
  - Lymphatics
    - Drain to **lumbar** (**aortic** and **caval**) **nodes**

ANATOMY IMAGING ISSUES

Multimodality Imaging Appearance

- No consensus on "normal" size or thickness of adrenals but average thickness of ~ 3 mm for medial/lateral limbs
  - While not based on any strong evidence, > 10-mm thickness can be used as threshold for hyperplasia
- MR: Generally isointense to liver on T1 MR and isointense to slightly hyperintense to liver on T2 MR
- Ultrasound: Easiest to visualize in newborns (as result of physiologic enlargement) and become progressively more difficult to visualize with age
  - Right adrenal gland easier to visualize than left (due to lack of liver as acoustic window and overlying bowel gas)

Key Concepts

- **Adrenal** (**cortical**) **adenomas**
  - Very common (at least 2% of general population) but usually cause no symptoms
    - Mostly "nonfunctioning" but identical to "functional" adenomas that cause Cushing/Conn syndrome
  - Most adenomas contain abundant lipid (precursor to steroid hormones), allowing definitive diagnosis using CT/MR sequences that highlight lipid
    - Lipid is intracellular/intercellular (not macroscopic deposits of fat)
  - Best CT technique: Nonenhanced CT with nodule measuring < 10 HU; or multiphase-enhanced CT with nodule demonstrating "washout" kinetics
  - Best MR technique: Chemical-shift MR with signal dropout within nodule on opposed-phase images
- **Pheochromocytoma** (tumor of adrenal medulla)
  - Signs: Headache, palpitations, excessive perspiration
  - 90% arise in adrenal, 90% unilateral, 90% benign
    - Similar tumor arising in other chromaffin cells of sympathetic ganglia is called **paraganglioma**
  - More common with multiple endocrine neoplasia, neurofibromatosis, and von Hippel-Lindau
  - Often markedly hypervascular in arterial phase
- **Adrenal myelolipoma**
  - Uncommon benign tumor (usually incidental finding) composed of mature adipose and hematopoietic tissue
  - Characterized by presence of **macroscopic fat**
- **Adrenocortical carcinoma**
  - Highly aggressive malignancy with poor prognosis
  - Large, heterogeneous mass (often with necrosis, hemorrhage, or calcification) with frequent local invasion, vascular invasion, and distant metastases
- **Cushing syndrome** (excess cortisol)
  - Signs: Truncal obesity, hirsutism, hypertension
  - Causes: Pituitary tumors (→ adrenal corticotrophic hormone), exogenous (medications) > adrenal adenoma > carcinoma
- **Conn syndrome** (excess aldosterone)
  - Signs: Hypertension, hypokalemic alkalosis
  - Causes: Adrenal adenomas > hyperplasia > carcinoma
- **Addison syndrome** (adrenal insufficiency)
  - Signs: Hypotension, weight loss, altered pigmentation
  - Causes: Autoimmune disease > adrenal metastases > adrenal hemorrhage > adrenal infection

CLINICAL IMPLICATIONS

Clinical Importance

- Rich adrenal blood supply due to endocrine function
  - Results in adrenal glands being common site for hematologic **metastases** (lung, breast, melanoma, etc.)
- Adrenal glands respond to stress (trauma, sepsis, surgery, etc.) by secreting ↑ cortisol and epinephrine
  - Overwhelming stress may result in **adrenal hemorrhage** or acute adrenal insufficiency (addisonian crisis)

ADRENAL VESSELS AND RELATIONS

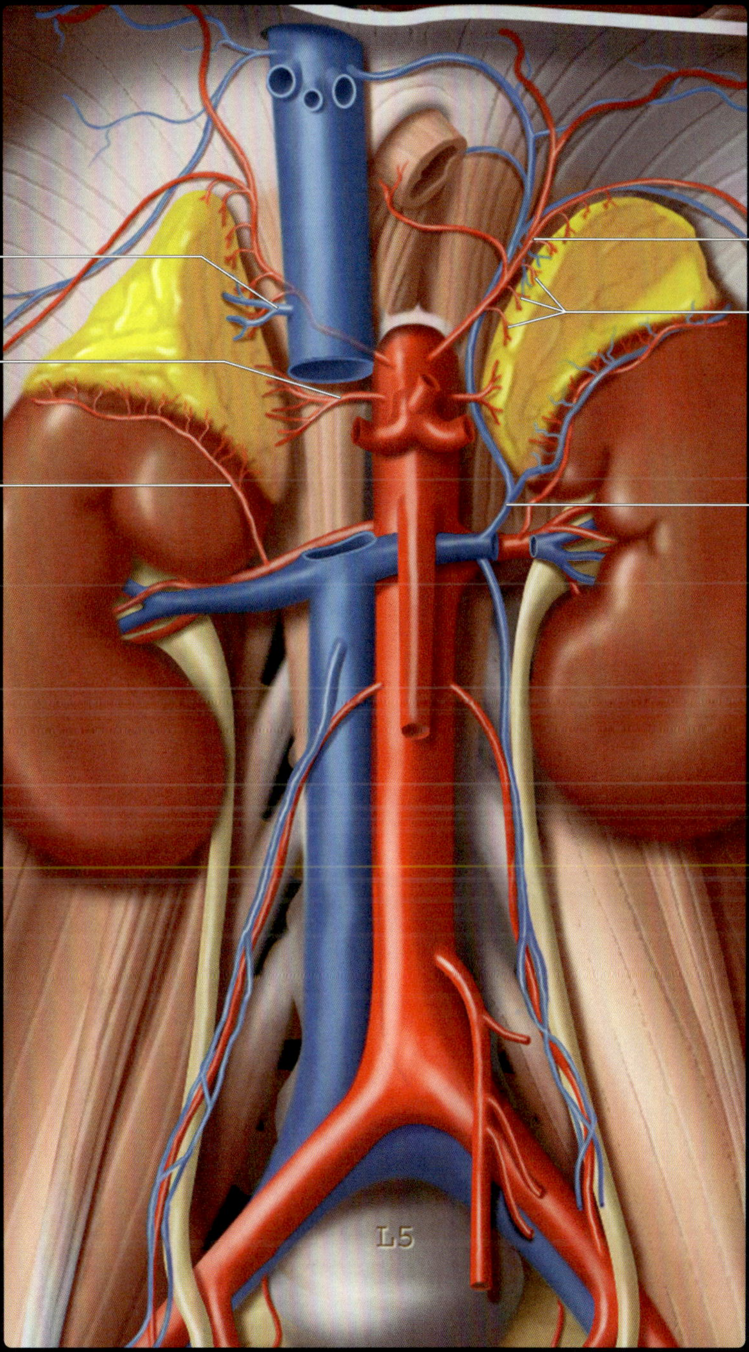

The adrenal glands rest atop the kidneys with an interposed layer of fat. Reflecting their critical role in maintaining homeostasis and responding to stress, the adrenal glands have a very rich vascular supply. The superior adrenal arteries are short branches of the inferior phrenic arteries bilaterally. The middle adrenal arteries are short vessels arising from the aorta. The inferior adrenal arteries are branches of the renal arteries. The left adrenal vein drains into the left renal vein, while the right adrenal vein drains directly into the inferior vena cava (IVC). Note the size of the adrenal glands is somewhat exaggerated in this graphic to facilitate demonstration of the vascular anatomy.

ADRENAL AXIAL ANATOMY AND RELATIONS

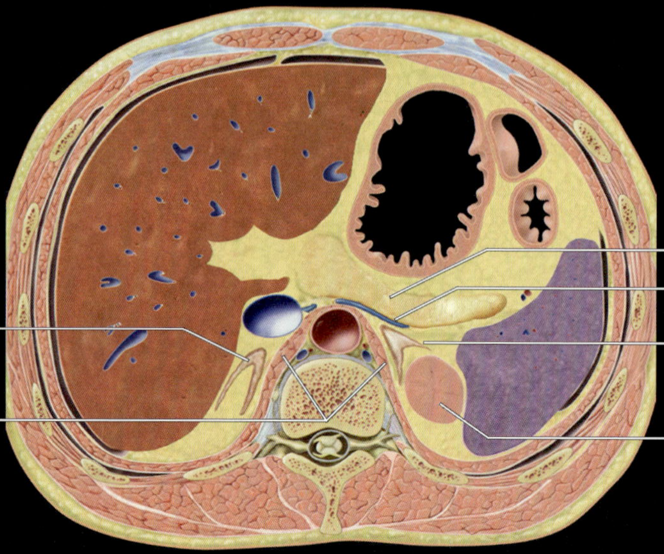

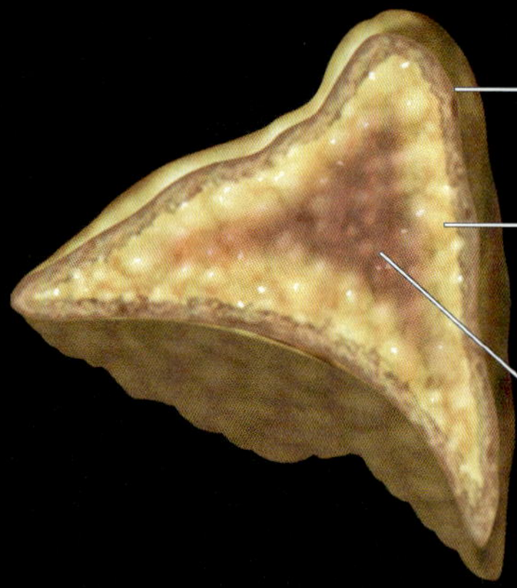

(Top) *The right adrenal is often more cephalic in location and lies above the right kidney, while the left adrenal lies partly in front of the upper pole of the left kidney. The left adrenal lies directly posterior to the splenic vein and body of pancreas and lateral to the left crus of the diaphragm. The right adrenal lies lateral to the crus, medial to the liver, and directly behind the IVC.* **(Bottom)** *The adrenal gland is essentially 2 organs in a single structure. The cortex is an endocrine gland, secreting primarily cortisol, aldosterone, and androgenic steroids. All of these hormones are derived from cholesterol, which imparts the characteristic lipid-rich appearance and imaging characteristics of the gland. The adrenal medulla is part of the autonomic nervous system and secretes epinephrine and norepinephrine.*

AXIAL CT, NORMAL ADRENAL ANATOMY

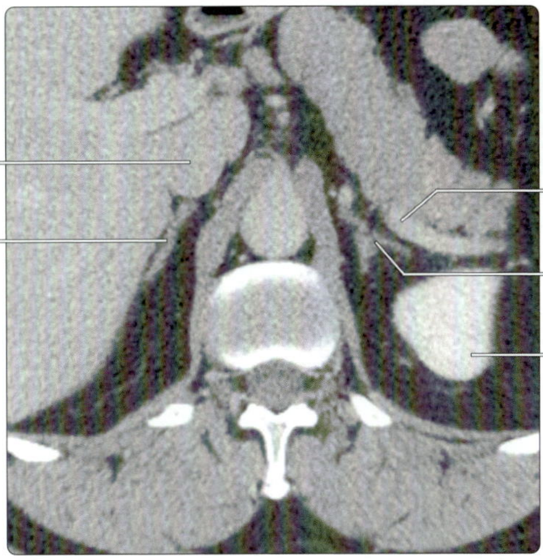

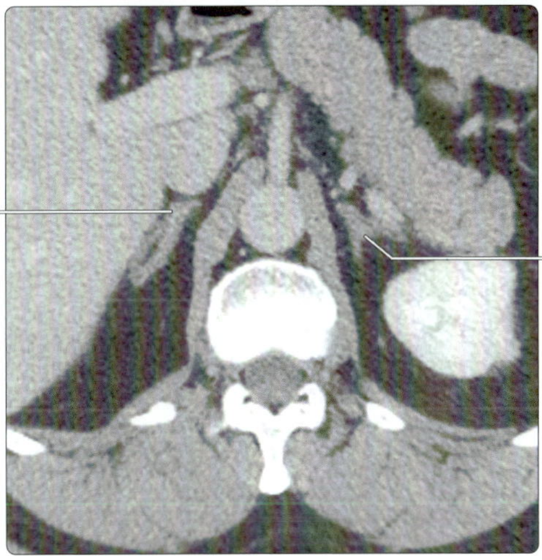

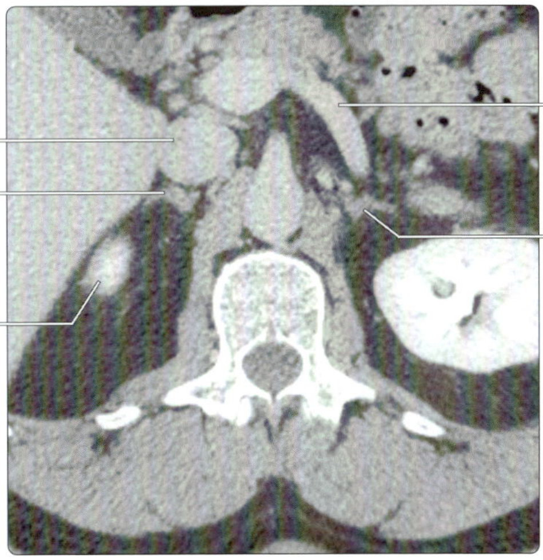

(Top) *First of 3 axial CECT images shows normal adrenal glands bilaterally. The right adrenal is usually suprarenal, touches the back of the IVC, and lies lateral to the right crus and medial to the liver. The left adrenal usually lies ventral to the upper pole of the left kidney and behind the splenic vein. The left adrenal often appears as an inverted Y shape, while the right is more like an inverted V.* **(Middle)** *Both limbs of the right adrenal are seen on this section.* **(Bottom)** *The lowest portions of the adrenals are seen on this section.*

CORONAL CT, NORMAL ADRENAL ANATOMY

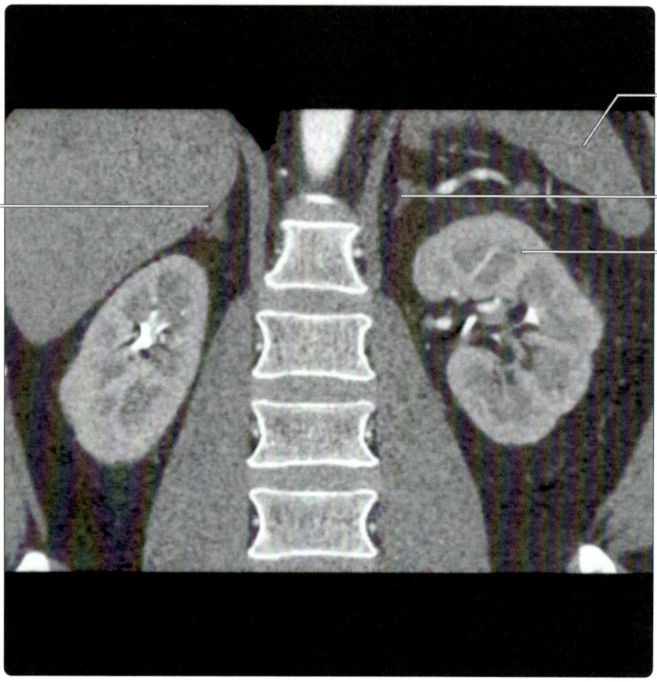

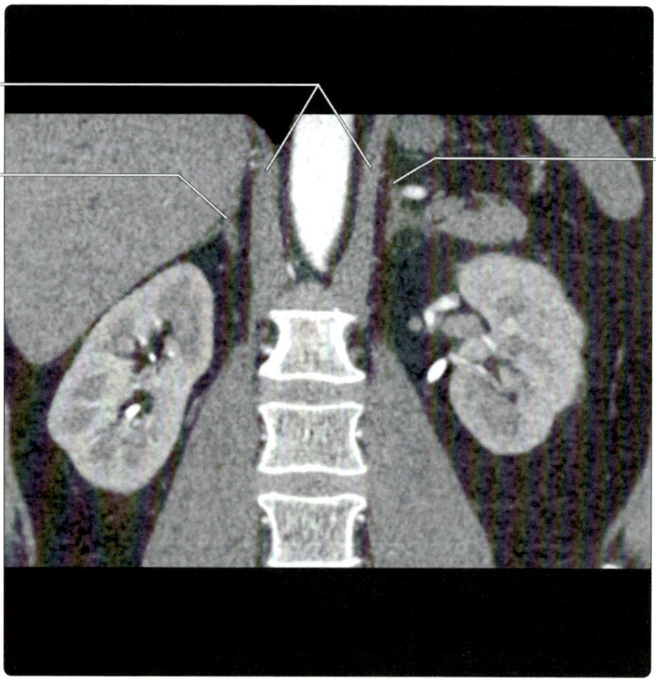

(Top) *First of 2 coronal CECT images shows the adrenal glands in their suprarenal location, which accounts for their alternate name of "suprarenal glands."* **(Bottom)** *More anterior coronal CT shows the adrenal glands and their relation to adjacent structures, such as the crura of the diaphragm.*

AXIAL AND CORONAL, NORMAL ADRENAL ANATOMY

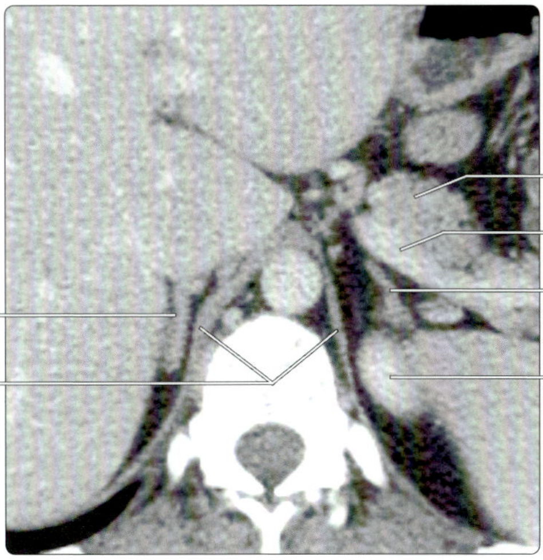

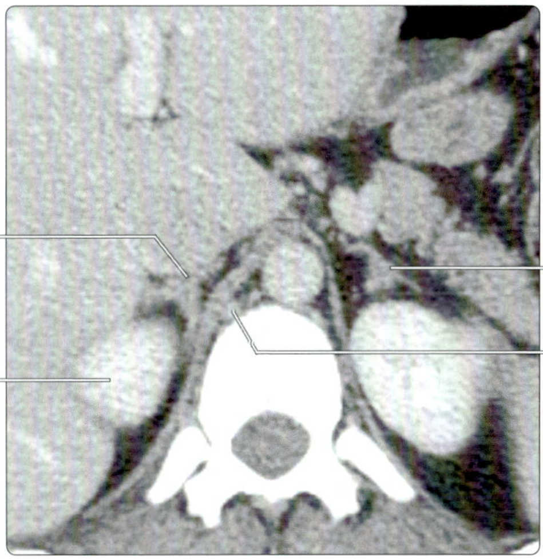

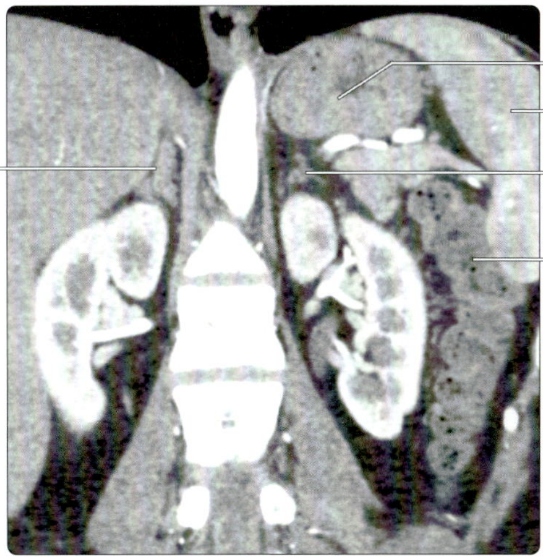

(Top) *First of 3 CT images of a subject with normal adrenal glands shows conventional anatomy.* **(Middle)** *More caudal section shows both adrenal glands with an inverted Y appearance. There is mild thickening of the left adrenal gland at the confluence of the medial and lateral limbs, a normal finding.* **(Bottom)** *Coronal view demonstrates the relationship between the adrenals and adjacent organs.*

AXIAL MR, NORMAL ADRENAL ANATOMY

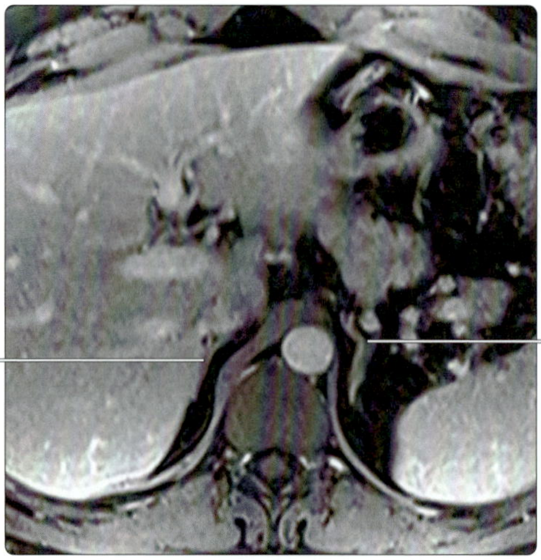

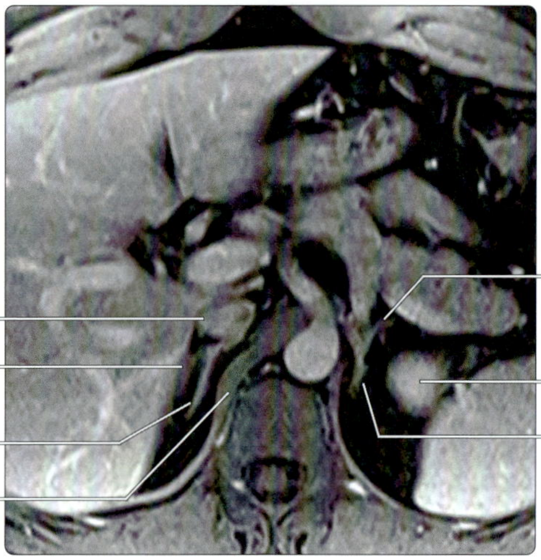

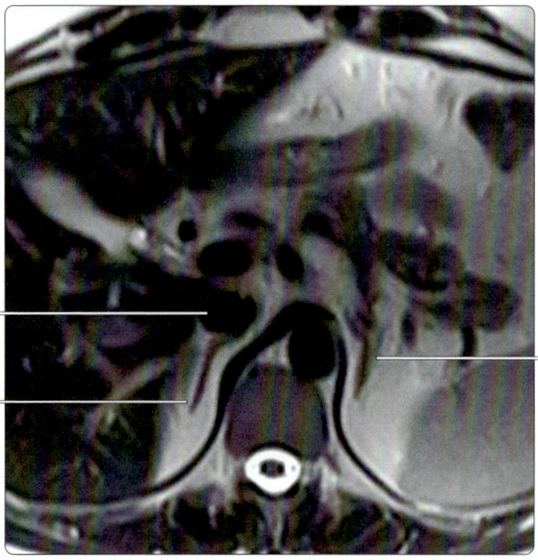

(Top) *First of 2 axial T1 C+ MR images demonstrates the bilateral adrenal glands. At this level, the 2 limbs of each adrenal gland are not yet discretely visible.* **(Middle)** *Axial T1 C+ MR at a slightly more caudal level now brings into view the twin limbs of each of the 2 adrenal glands. Each limb is thin and regular (without thickening or a nodule). Note that the right adrenal gland contacts the posterior margin of the IVC and is medial to the right diaphragmatic crus. The left adrenal gland is located slightly ventral and above the left kidney.* **(Bottom)** *Axial T2 HASTE MR demonstrates a normal appearance of the adrenal glands on these images acquired without fat saturation. The adrenal glands in this image appear relatively isointense to the liver (although the relative signal can vary depending on application of fat saturation).*

ADRENAL VENOGRAM

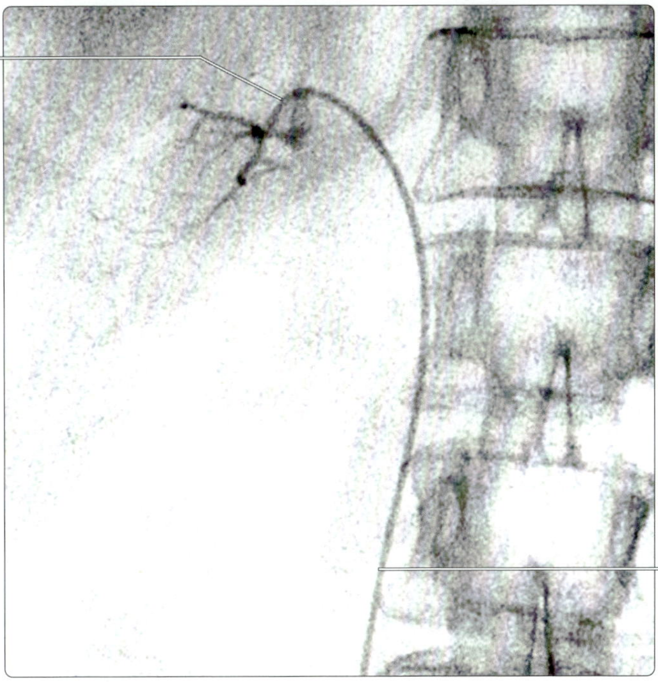

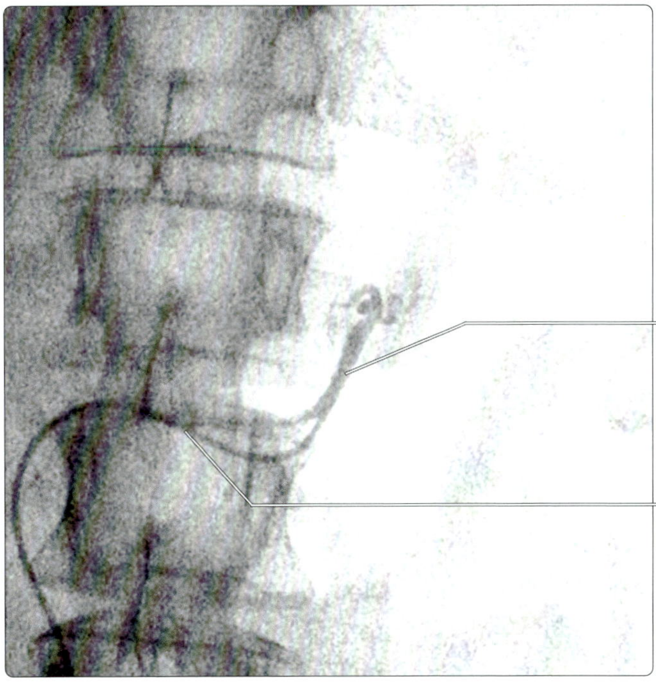

(Top) *First of 2 images shows selective catheterization of the adrenal veins in a young woman with hyperaldosteronism, but no definite mass is seen on CT. Selective adrenal vein sampling was requested to assess unilateral excess aldosterone secretion. A catheter has been inserted through the right femoral vein, and the tip was advanced into the opening of the right adrenal vein. The adrenal veins are very fragile and could be easily ruptured by a forceful injection of contrast medium. The angiographer must know the vascular anatomy and gently probe the venous orifice, confirming the location with a small bolus injection of contrast medium, as shown here.* **(Bottom)** *A subsequent image shows the catheter repositioned. The tip has been advanced through the left renal vein to enter the left adrenal vein. No attempt is made to opacify the smaller venous tributaries.*

FETAL ADRENAL AND KIDNEY

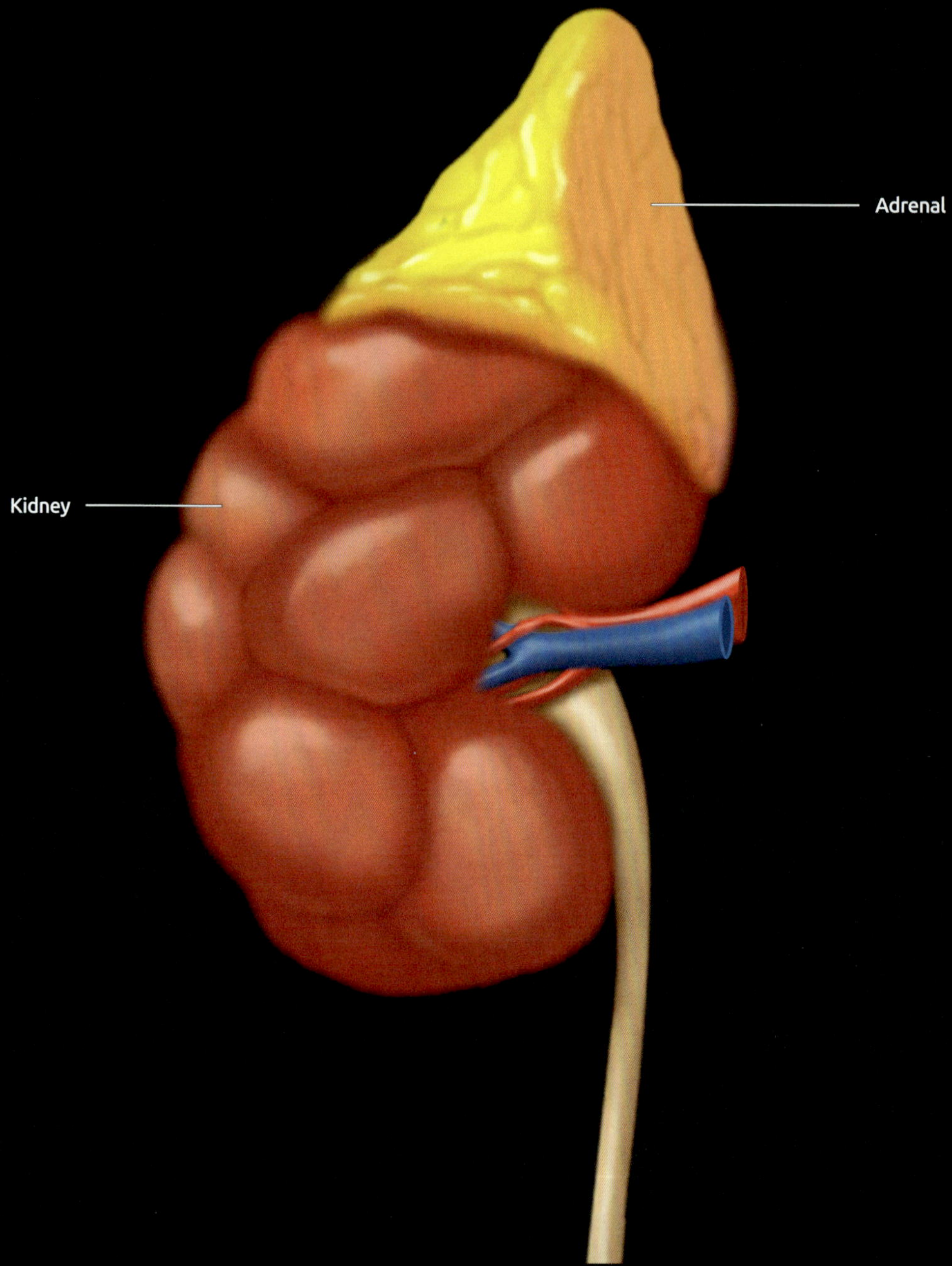

Graphic shows the appearance of the adrenal and kidney in the fetus and neonate. The adrenal is much larger, relative to the kidney, than in an adult. The kidney has a lobulated appearance reflecting the ongoing fusion of the individual renal lobes, each composed of 1 renal pyramid and its associated renal cortex.

NEONATAL ADRENALS AND KIDNEY

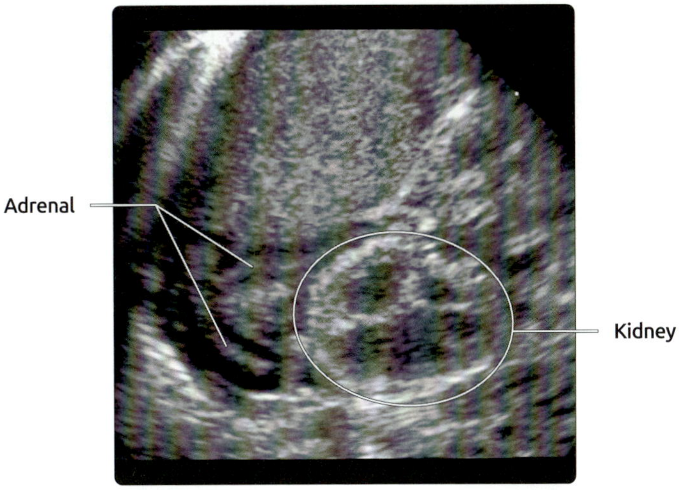

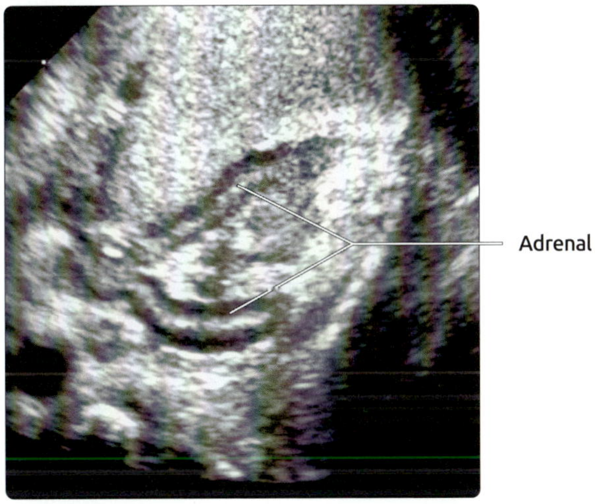

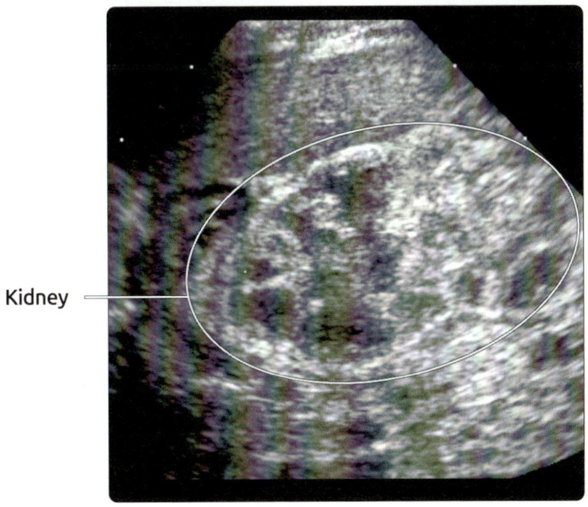

(Top) *First of 3 ultrasounds of a neonate shows the characteristic prominence of the adrenal gland and the lobation of the renal surface in early infancy. This sagittal image shows the large adrenal gland adjacent to the upper pole of the kidney.* **(Middle)** *Sagittal ultrasound shows the prominent limbs of the right adrenal gland.* **(Bottom)** *Sagittal ultrasound of the kidney shows its lobulated contour, a normal finding in the fetus and neonate.*

ADRENAL ADENOMA CT

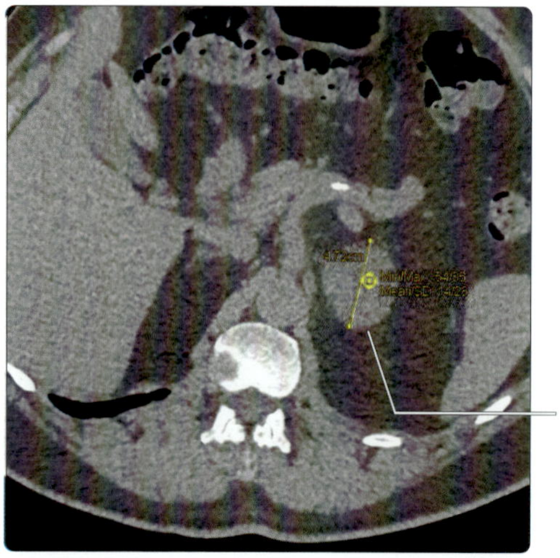

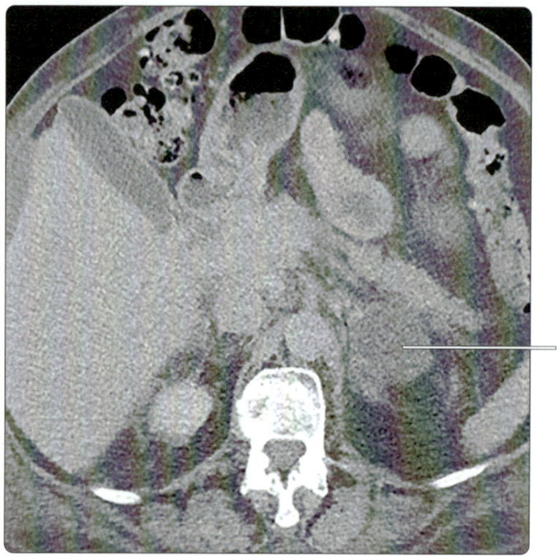

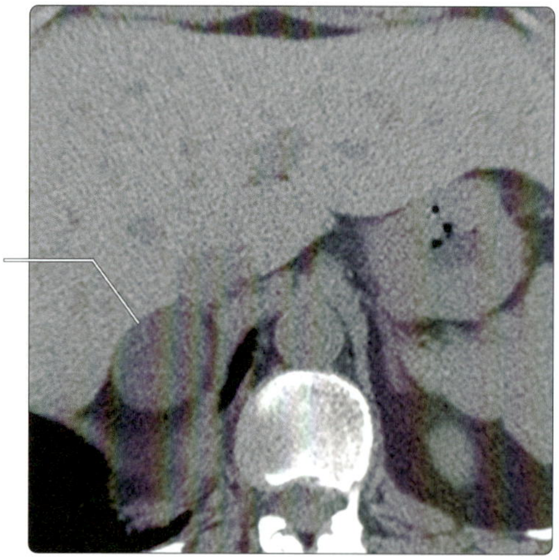

(Top) *Axial NECT demonstrates a low-density nodule within the left adrenal gland with a Hounsfield attenuation of 14 HU, compatible with an adrenal adenoma. Patients with adenomas may be symptomatic (e.g., signs of excess cortisol or aldosterone) or asymptomatic. Most subjects have the adrenal lesion discovered incidentally on a CT performed for some other reason and have no clinical symptoms or signs. In this setting, the adenoma is said to be nonfunctional.* **(Middle)** *Axial CECT demonstrates a low-density left adrenal nodule that demonstrated a noncontrast attenuation of 3, an absolute washout of 77.3%, and a relative washout percentage of 74.4%, features compatible with a benign adrenal adenoma (calculated based on a formal multiphase adrenal protocol CT).* **(Bottom)** *Axial NECT demonstrates a low-density, lipid-rich adenoma with Hounsfield attenuation of 6-8 HU, characteristic of a lipid-rich adenoma. Adenomas can be either lipid rich (common between 70-90%) or lipid poor. Lipid-rich adenomas, due to their high lipid content, are of low density on NECT.*

ADRENAL ADENOMA MR

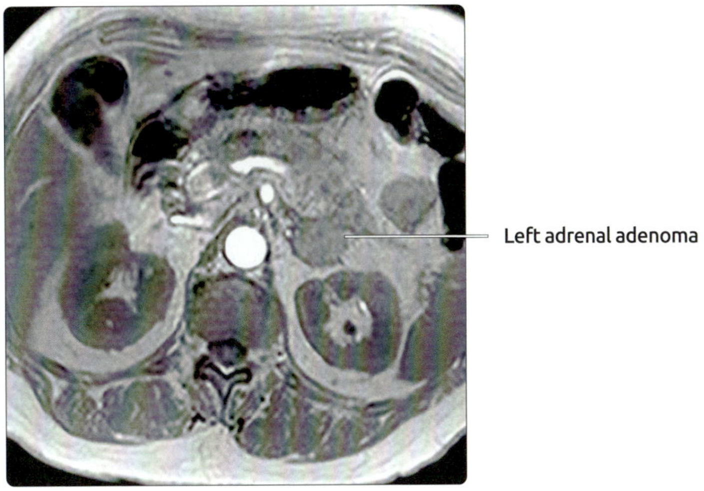

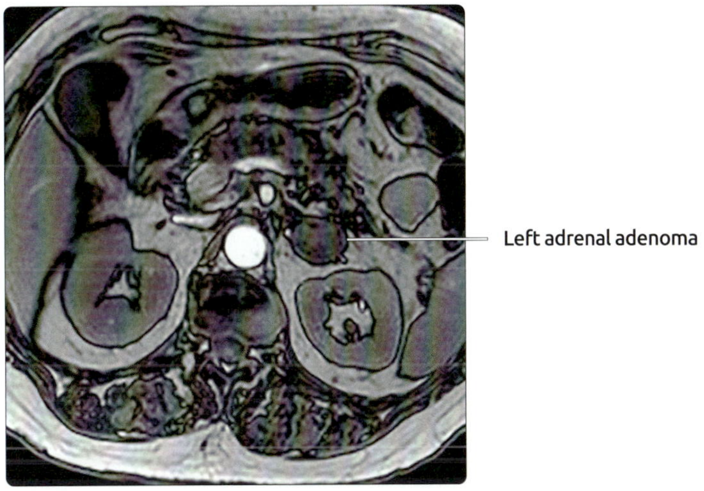

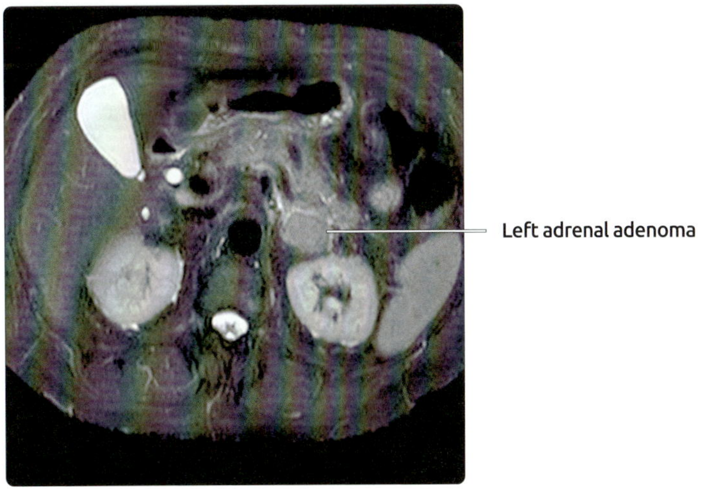

(Top) *Axial GRE in-phase MR demonstrates a homogeneous rounded left suprarenal mass.* **(Middle)** *Axial GRE opposed-phase MR through the same level shows marked loss of signal within the adrenal mass, confirming the presence of intravoxel/intracellular lipid. This finding is diagnostic of an adrenal adenoma.* **(Bottom)** *Axial T2 FS MR demonstrates that the adrenal adenoma is relatively low in signal, similar in signal to the normal-appearing left adrenal gland. Note that this is unlike most malignant adrenal masses, which tend to be T2 hyperintense.*

ADRENAL HYPERPLASIA

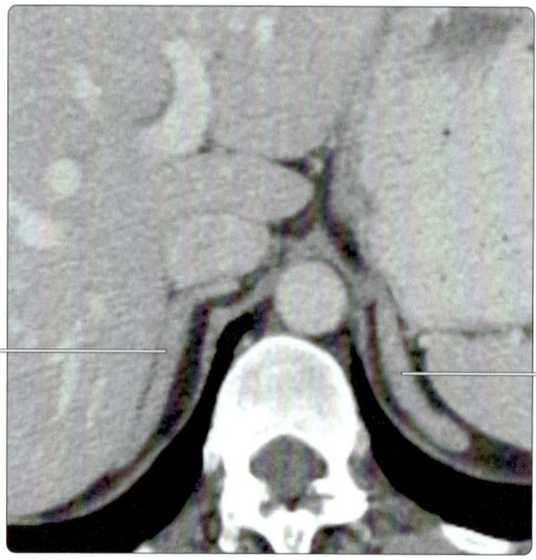

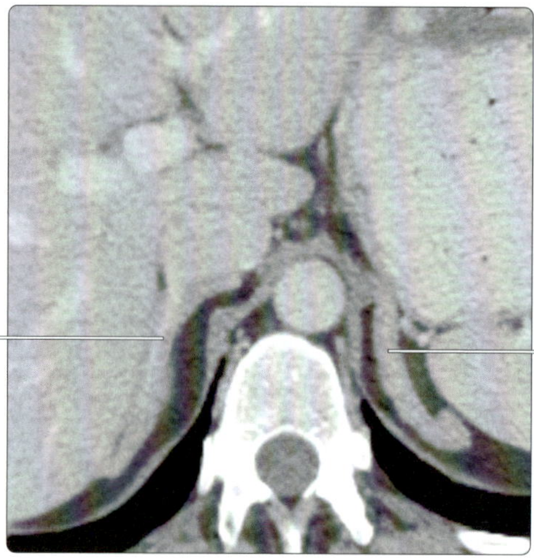

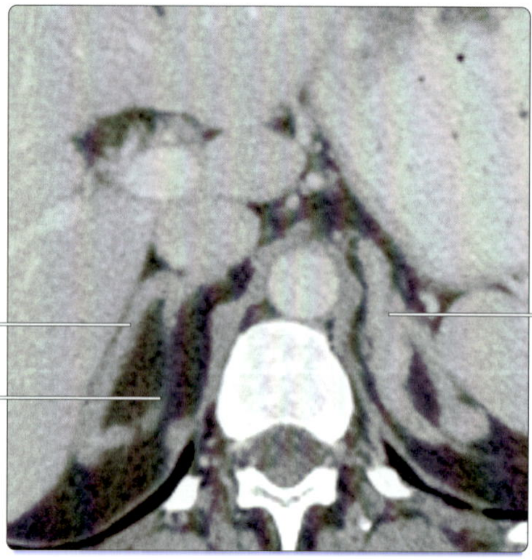

(Top) *First of 3 axial CECT images of a 40-year-old woman with congenital adrenal hyperplasia shows diffuse enlargement of both adrenal glands but preservation of their normal shape.* **(Middle)** *Each limb of the adrenal is in excess of 1 cm in diameter, one criterion used to diagnose or suggest adrenal hyperplasia. Most patients with adrenal hyperplasia have less markedly enlarged glands due to pituitary (or ectopic) production of excess adrenal corticotrophic hormone. In many cases the adrenal glands may appear normal by imaging.* **(Bottom)** *The striking enlargement of the adrenals is again evident on this image.*

ADRENAL MYELOLIPOMA

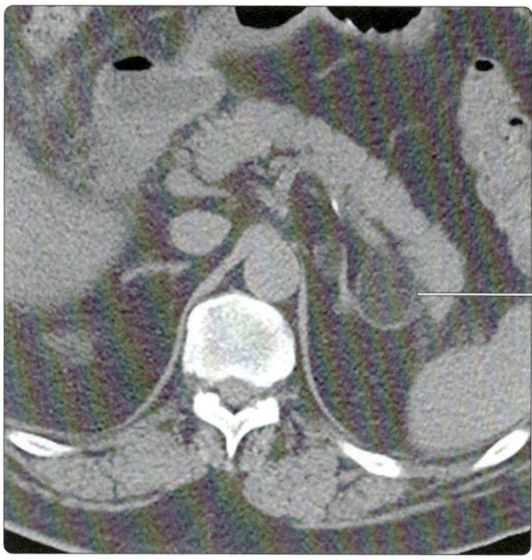

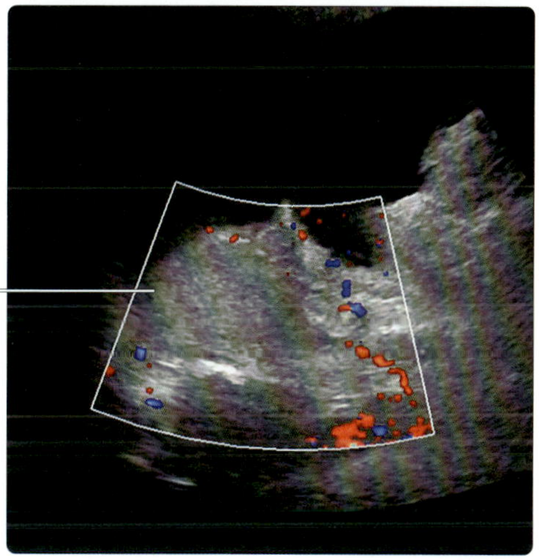

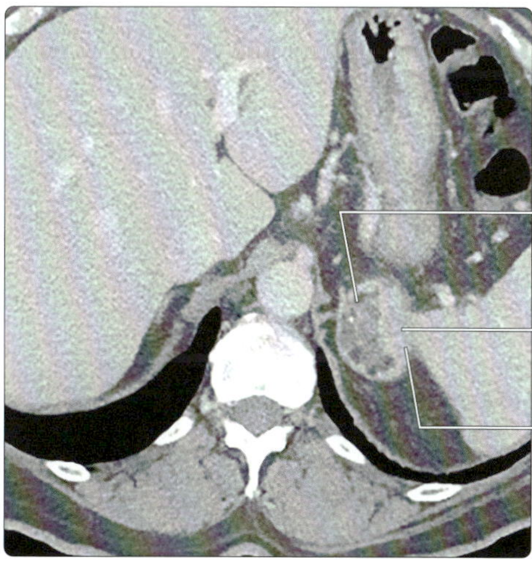

(Top) *Axial NECT demonstrates a fat-density mass arising from the left adrenal gland. The presence of macroscopic fat within an adrenal mass is virtually diagnostic of an adrenal myelolipoma.* **(Middle)** *Transverse ultrasound demonstrates a left adrenal mass, which is markedly echogenic and without internal color flow vascularity, representing a myelolipoma.* **(Bottom)** *Axial CECT demonstrates a left adrenal mass with internal solid enhancing components, but with clear evidence of macroscopic fat, compatible with a myelolipoma. Myelolipomas can demonstrate enhancing components and calcification, but the presence of macroscopic fat is essentially diagnostic of this entity.*

PHEOCHROMOCYTOMA

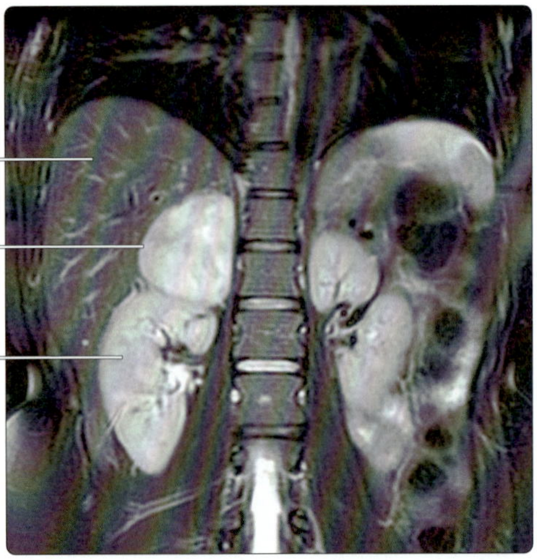

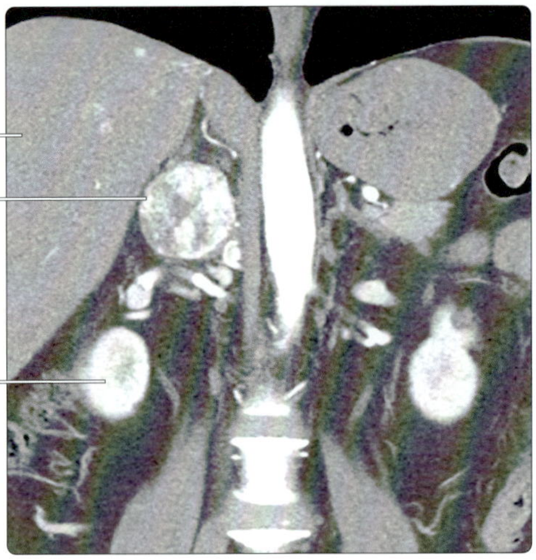

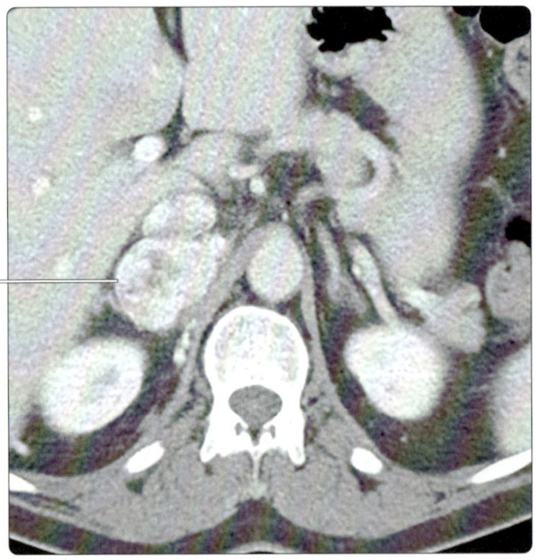

(Top) *Coronal T2 FS MR demonstrates a T2-hyperintense mass arising from the left adrenal gland in a patient with symptoms of malignant hypertension. Pheochromocytomas are classically said to be light bulb bright on T2 MR. While not true in every case, the majority of pheochromocytomas are T2 hyperintense.* **(Middle)** *Coronal arterial-phase CECT demonstrates a markedly hypervascular right adrenal mass, an appearance that is very suggestive of a pheochromocytoma. These lesions tend to show marked enhancement on arterial-phase imaging, making multiphase imaging (either CT or MR) critical in their imaging diagnosis.* **(Bottom)** *Axial CECT demonstrates an avidly enhancing mass with subtle central necrosis arising from the right adrenal gland, compatible with a pheochromocytoma, in this patient with malignant hypertension and palpitations.*

PHEOCHROMOCYTOMA

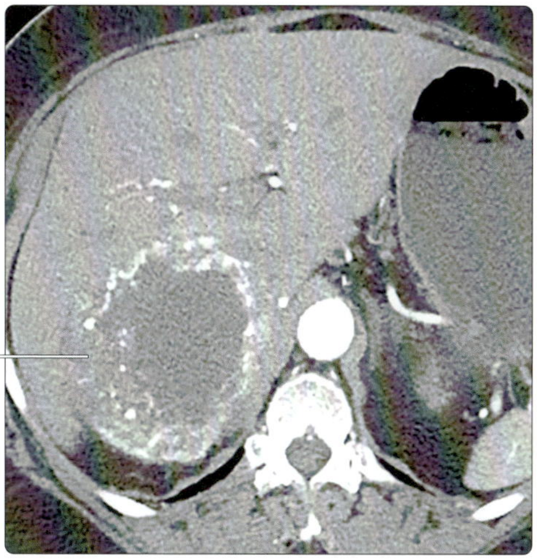

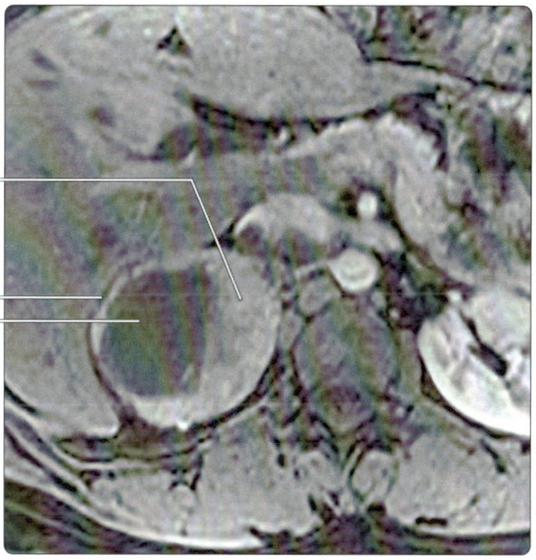

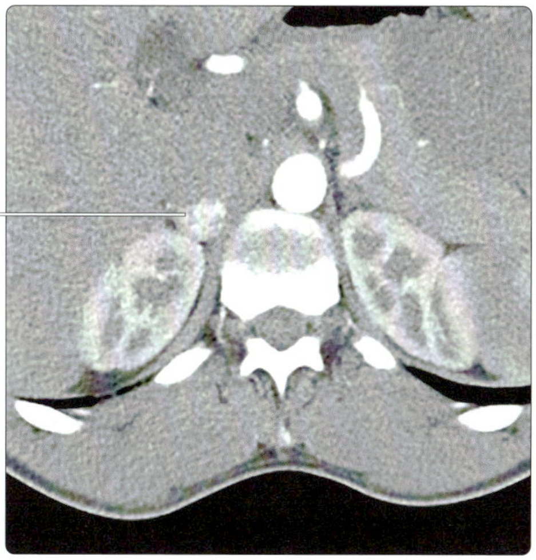

(Top) *Axial CECT in the arterial phase demonstrates a right adrenal mass, which is markedly hypervascular around its margins and centrally necrotic. The presence of markedly hypervascular soft tissue in an adrenal mass should strongly suggest the diagnosis of pheochromocytoma. This mass was surgically resected and confirmed to represent a pheochromocytoma.* **(Middle)** *Axial T1 C+ MR demonstrates a large right adrenal mass with a sizable central necrotic component and an avidly enhancing peripheral solid component, found to represent a pheochromocytoma upon surgical resection.* **(Bottom)** *Axial arterial-phase CECT demonstrates a small nodule in the right adrenal gland, which is markedly hypervascular with a Hounsfield attenuation of 150 HU. Adrenal nodules with attenuation values > 110 HU in the arterial phase are highly likely to represent pheochromocytomas, confirmed in this case at surgery.*

ADRENOCORTICAL CARCINOMA

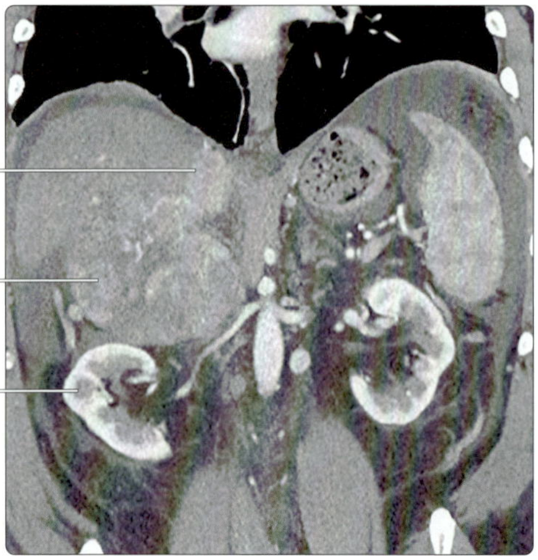

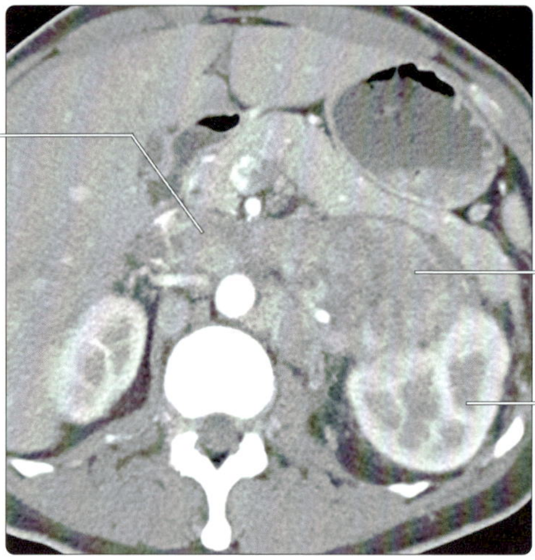

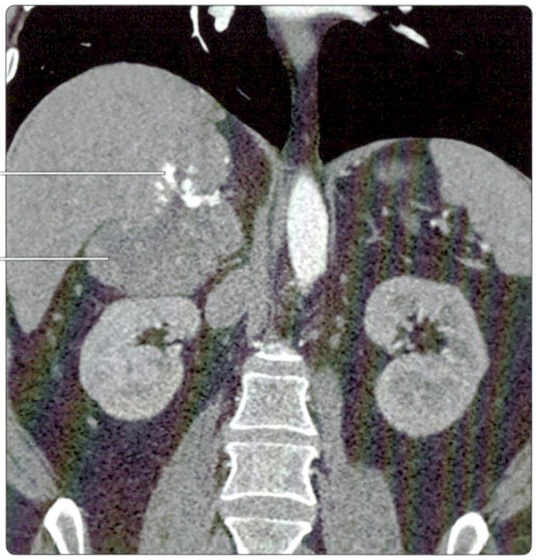

(Top) *Coronal CECT demonstrates a large, heterogeneous, enhancing mass arising from the right suprarenal fossa with tumor directly extending into (and distending) the IVC. Adrenocortical carcinoma commonly invades surrounding venous structures, including the renal vein or IVC.* **(Middle)** *Axial CECT demonstrates a large right suprarenal mass with direct extension to involve the left kidney as well as extensive tumor thrombus involving the left renal vein. This was ultimately confirmed to represent an adrenocortical carcinoma at resection.* **(Bottom)** *Coronal CECT demonstrates a right-sided adrenocortical carcinoma that directly invades the right hepatic lobe. Note the presence of extensive calcification within the mass, a common feature with these tumors.*

ADRENAL HEMORRHAGE

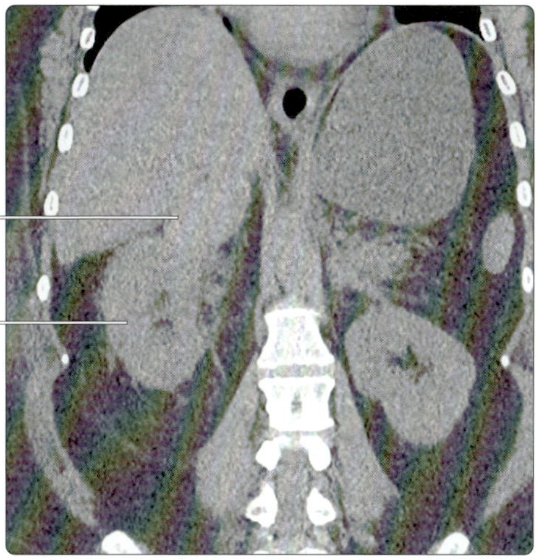

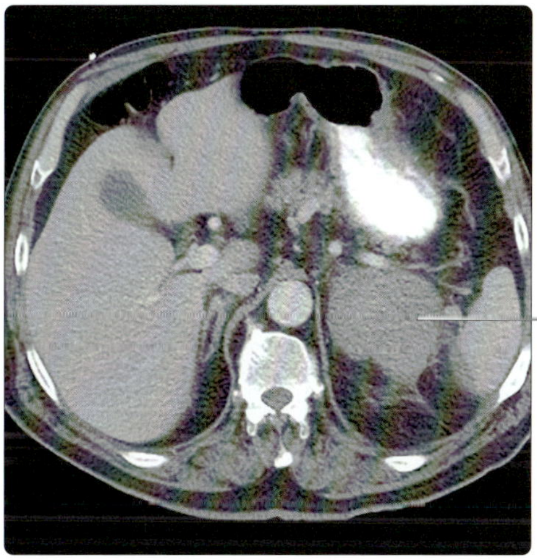

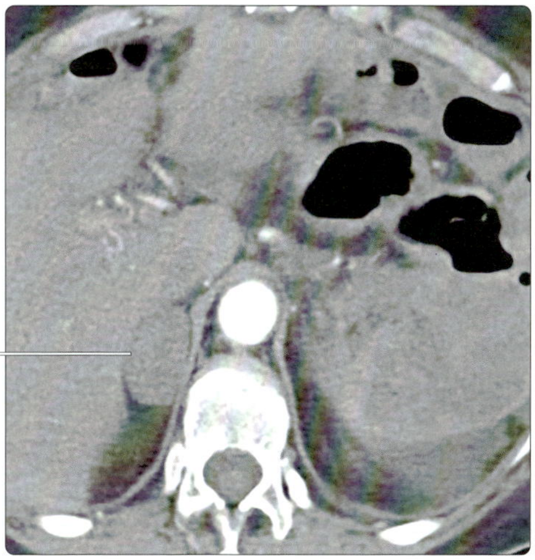

(Top) *Coronal NECT in a patient who was hypotensive due to blunt abdominal trauma demonstrates enlargement of the left adrenal gland with heterogeneous, high-density material, characteristic of acute hemorrhage. This "nodule" was not present on a prior CT from a few months earlier, aiding in the diagnosis.* **(Middle)** *Axial NECT in a septic, hypotensive patient in the ICU setting demonstrates a large left adrenal hemorrhage. Notably, a study performed a few days prior had shown a completely normal left adrenal gland.* **(Bottom)** *Axial CECT in a patient who had recently undergone surgery for ovarian cancer demonstrates a new intermediate-density nodule arising from the right adrenal gland. This nodule was new since a recent prior study, compatible with a small adrenal hemorrhage.*

GASTRIC DIVERTICULUM SIMULATING ADRENAL MASS

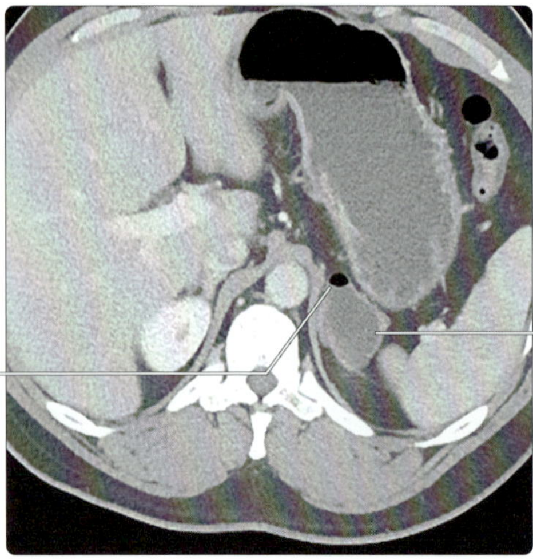

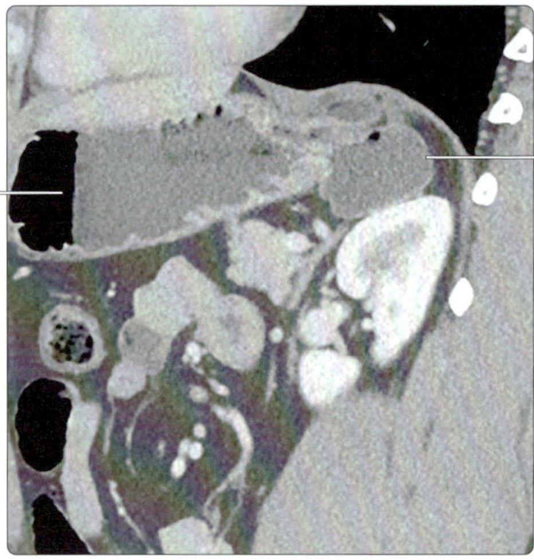

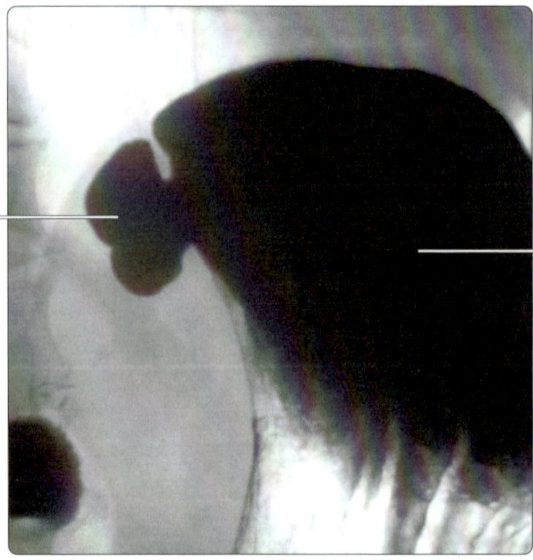

(Top) *Axial CECT shows a cystic-appearing "lesion" in the left suprarenal region, simulating an adrenal mass. The mass has the same density as the water-filled stomach, but note the presence of internal gas.* **(Middle)** *Sagittal CECT in the same patient nicely shows that this "lesion" is actually a gastric diverticulum directly communicating with the adjacent stomach. Gastric diverticula, which often arise in this location, are not infrequently mistaken for adrenal nodules.* **(Bottom)** *An oblique film from a barium upper GI series shows a barium-filled gastric diverticulum projecting posteriorly from the gastric cardia.*

ADRENAL INSUFFICIENCY

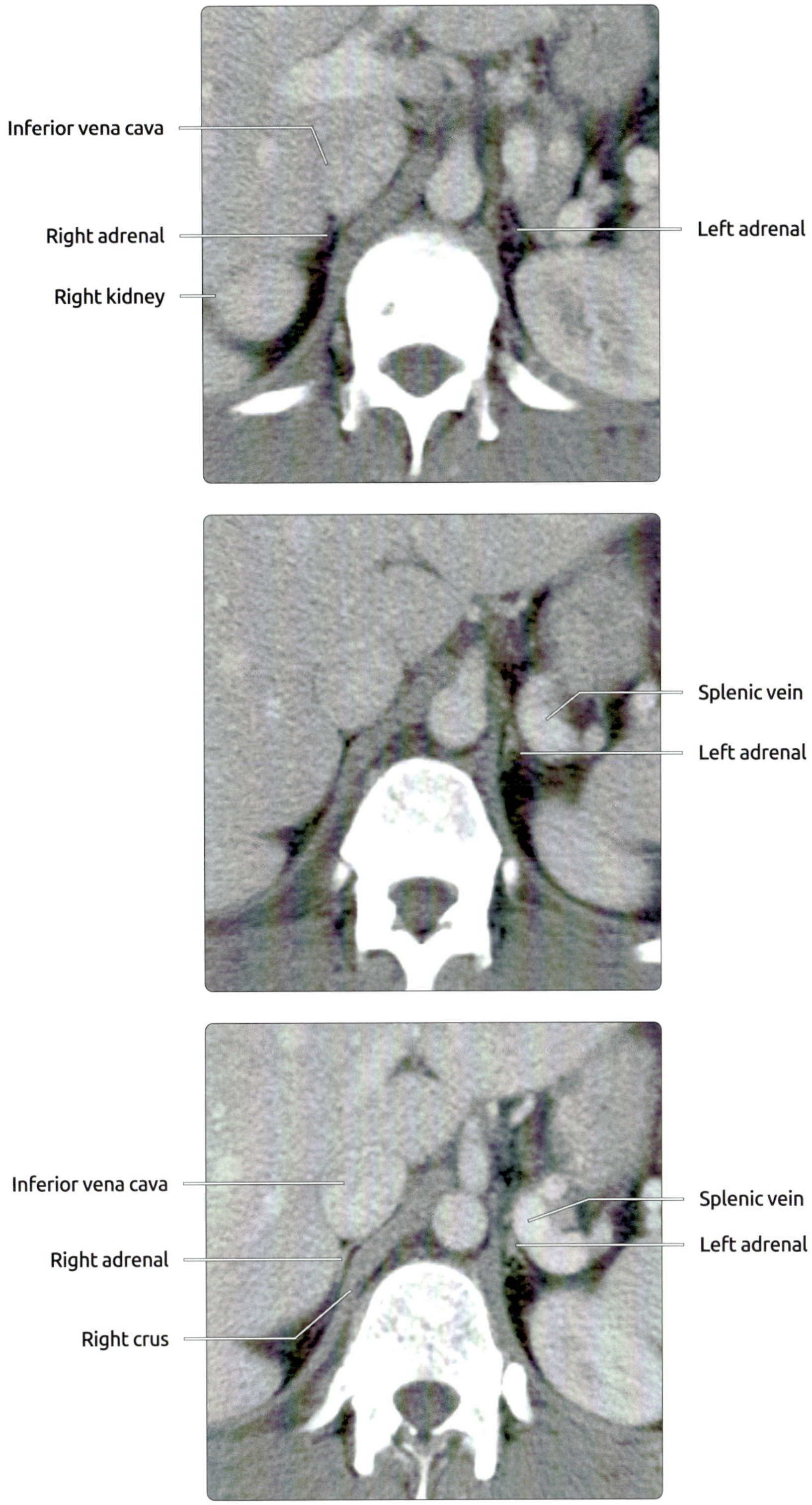

(Top) *First of 3 axial CT sections in a patient with adrenal insufficiency (addisonian syndrome) due to autoimmune disease is shown. The adrenal glands are extremely small and difficult to visualize.* **(Middle)** *The small adrenals are again evident. If adrenal insufficiency were due to adrenal tumors, bleeding, or infection, the glands would be enlarged.* **(Bottom)** *The adrenal glands have a normal shape and morphology but are extremely small.*

GROSS ANATOMY

Overview

- Kidneys are paired, bean-shaped organs located in retroperitoneum, which remove excess water, salts, and wastes from blood and maintain water/electrolyte balance

Anatomic Relationships

- Kidneys lie in retroperitoneum within **perirenal space** surrounded by **renal fascia** (**of Gerota**)
  - Thick renal capsule (directly covering kidney) separated from renal fascia by layer of perirenal fat
  - **Pararenal** fat is found surrounding renal fascia
- Kidneys typically found at T12-L3 levels with left kidney usually slightly more cephalad than right (due to liver)
  - Both kidneys lie on **quadratus lumborum muscles** and lateral to **psoas muscles**
  - Each kidney roughly 9-13 cm in length and 5 cm in width

Internal Structure

- Renal parenchyma divided into cortex and medulla with cortex extending centrally to divide medulla into pyramids
  - **Renal cortex**: Outer aspect of renal parenchyma, which contains renal corpuscles (glomeruli, vessels), proximal portions of collecting tubules, and loops of Henle
  - **Renal medulla**: Inner portion of renal parenchyma, which contains renal pyramids, distal parts of collecting tubules, and loops of Henle
  - Apex of each renal pyramid is renal papilla, which consists of collecting tubules that excrete urine
    - Path of urine flow: Papilla → minor calyx → major calyx → renal pelvis
    - Each papilla has 1 minor calyx, while multiple minor calyces (2-4) drain into each major calyx (infundibula), and multiple major calyces (2 or 3) drain into pelvis
  - **Renal pelvis**: Funnel-shaped expansion of upper ureter
- **Renal hilum**: Medial aspect of kidney where renal artery enters and renal vein and ureter leave renal sinus
- **Vessels, nerves, and lymphatics**
  - Renal artery
    - Normally 1 for each kidney (arising below level of superior mesenteric artery)
    - Arise from aorta at ~ L1-L2 vertebral level with right renal artery extending posterior to inferior vena cava (IVC)
    - Main renal artery → segmental branches (at hilum) → interlobar arteries (run along margin of pyramids) → arcuate and interlobular arteries → afferent arterioles
    - Branches of renal artery also supply proximal collecting system (pelvic/superior ureteric branches)
  - Renal vein
    - Normally 1 for each kidney (draining into IVC) with left renal vein passing anterior to aorta
    - Lies in front of renal artery and renal pelvis
  - Nerves
    - Sympathetic autonomic supply from **celiac/aorticorenal ganglia** and parasympathetic supply via vagus nerve (CNX)
  - Lymphatics
    - Drain to **left lateral aortic** and **right lateral caval nodes** (near renal artery origins)

ANATOMY IMAGING ISSUES

Key Concepts

- Accessory renal vessels (arteries and veins) and vascular variants are very common
  - Accessory arteries arise from aorta or common iliac
  - Left renal vein has several common variants (retroaortic or circumaortic renal veins)
  - Renal vascular anatomy must be considered when planning surgery (e.g., resection, transplant)

CLINICAL IMPLICATIONS

Clinical Importance

- Kidneys and urinary tract often etiology for patients with abdominal pain or fever
  - **Renal calculi** ("stones") may form within and obstruct renal calyces or ureter
  - **Pyelonephritis** often visible as heterogeneous renal parenchymal enhancement ("striated nephrogram")
- Renal malignancies
  - Most common tumors include parenchymal masses (**renal cell carcinoma**, **oncocytoma**) and tumors of collecting system (**urothelial carcinoma**)
  - Strong renal fascia usually prevents direct tumoral invasion of adjacent organs and body wall

EMBRYOLOGY

Embryologic Events

- Congenital anomalies of renal number, position, and structure are very common and often accompanied by other anomalies
  - **VATER** acronym
    - Vertebral, anorectal (e.g., atresia), tracheoesophageal (e.g., T-E fistula), radial (e.g., hand and wrist anomalies or absence), renal (e.g., agenesis, ectopia)
  - Congenital absence of kidney
    - Commonly associated with ipsilateral anomalies of genital tract (e.g., seminal vesicle absence or cyst)
  - Anomalies of position (ectopia) are common
    - Often due to failure to ascend from pelvic location where fetal kidneys lie close together
    - May be accompanied by abnormal fusion of kidneys, such as **crossed fused ectopia** and "**horseshoe kidney**" (1 in 400 adults)
    - Accompanied by anomalies of vessels and ureters
  - Anomalies of structure
    - Congenitally large **septum/column of Bertin** (**lobar dysmorphism**): Asymptomatic normal variant, which may simulate mass on imaging
    - **Fetal lobulations** (**lobation**): Single or multiple indentations of lateral renal contours, which represent persistent clefts between renal lobules (distinguish from cortical scarring due to infection or ischemia)
    - **Duplicated collecting system**: Variety of partial or complete duplications possible due to incomplete fusion of upper/lower pole moieties
  - **Autosomal dominant polycystic disease**: Common hereditary disorder characterized by multiple renal cysts, progressive renal failure, and various systemic manifestations (such as cerebral aneurysms)

KIDNEYS IN SITU

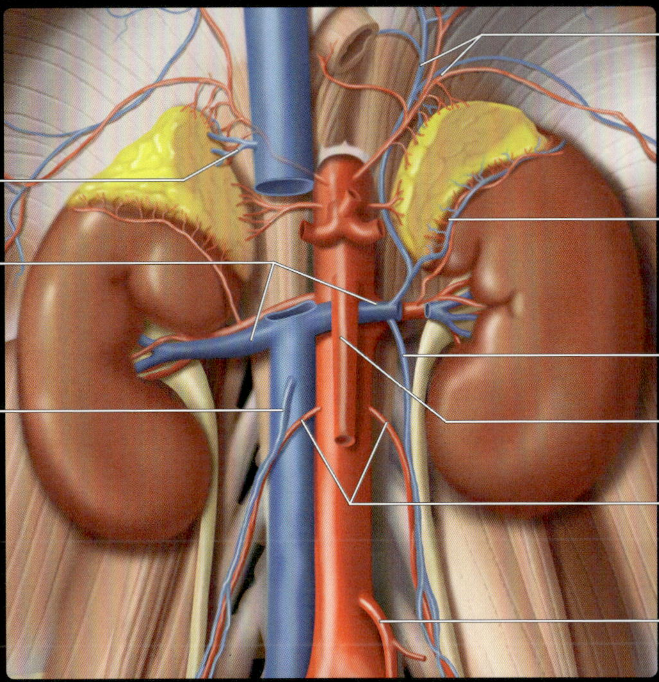

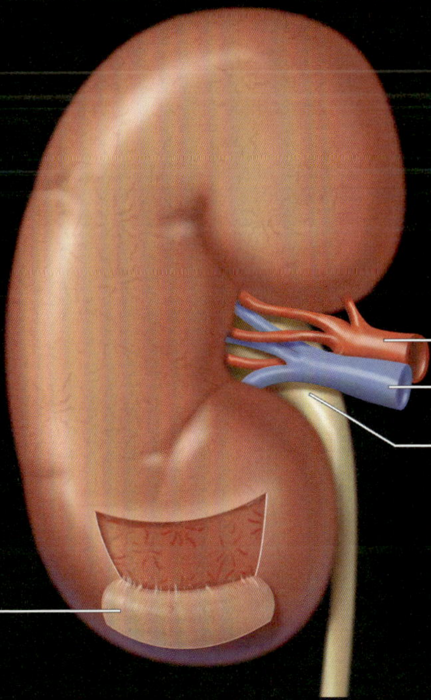

(Top) *The kidneys are retroperitoneal organs that lie lateral to the psoas and "on" the quadratus lumborum muscles. The oblique course of the psoas muscles results in the lower pole of the kidney lying lateral to the upper pole. The right kidney usually lies 1-2 cm lower than the left due to inferior displacement by the liver. The adrenal glands lie above and medial to the kidneys separated by a layer of fat and connective tissue. Peritoneum covers much of the anterior surface of the kidneys. The right kidney abuts the liver, hepatic flexure of the colon, and duodenum, while the left kidney is in close contact with the pancreas (tail), spleen, and splenic flexure.* **(Bottom)** *The fibrous capsule is stripped off with difficulty. As a result, subcapsular hematomas do not spread far along the surface of the kidney but instead result in compression of the renal parenchyma, unlike most perirenal collections.*

KIDNEY ARTERIES AND INTERIOR ANATOMY

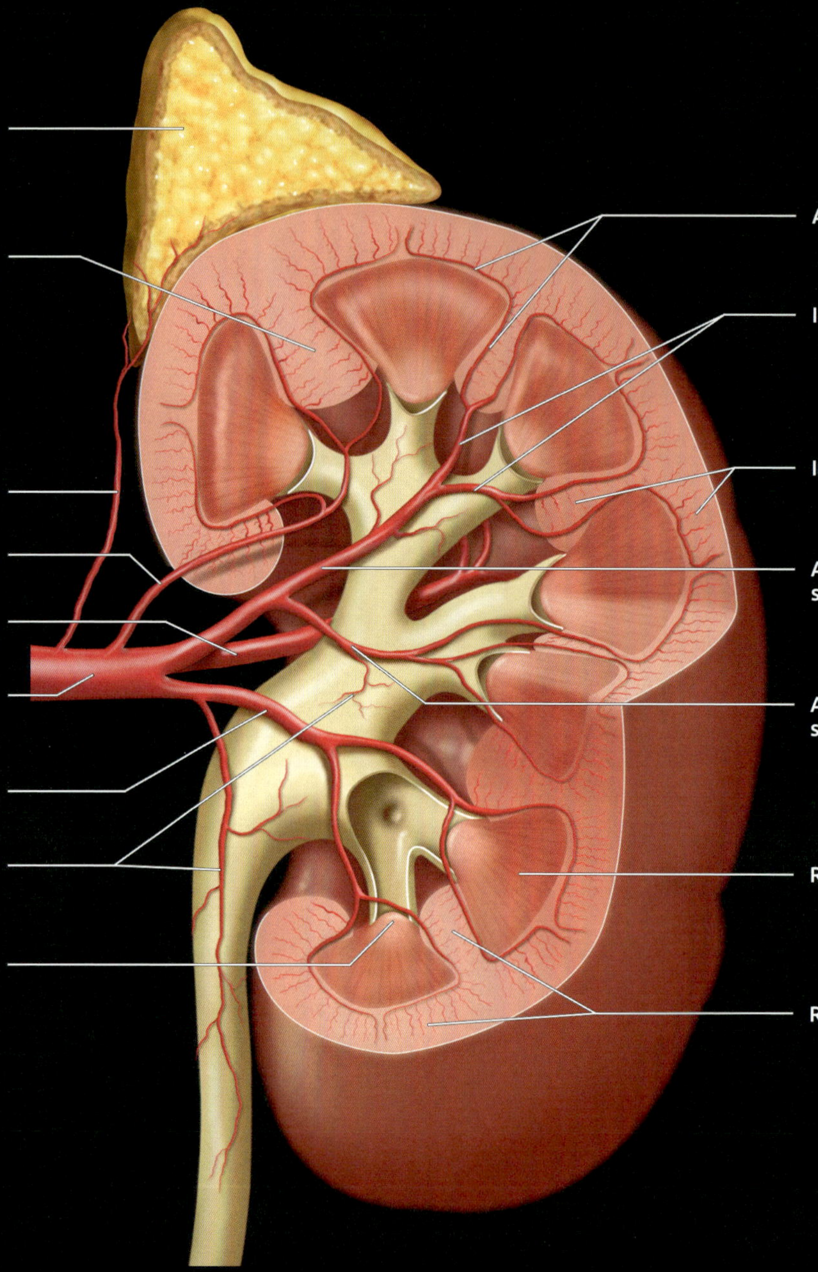

The kidney is usually supplied by a single renal artery, the 1st branch of which is the inferior adrenal artery. It then divides into 5 segmental arteries, only 1 of which, the posterior segmental artery, passes dorsal to the renal pelvis. The segmental arteries divide into the interlobar arteries that lie in the renal sinus fat. Each interlobar artery branches into 4-6 arcuate arteries that follow the convex outer margin of each renal pyramid. The arcuate arteries give rise to the interlobular arteries that lie within the renal cortex, including the cortical columns (of Bertin) that invaginate between the renal pyramids. The interlobular arteries supply the afferent arterioles to the glomeruli. The arterial supply to the kidney is vulnerable, as there are no effective anastomoses between the segmental branches, each of which supplies a wedge-shaped segment of parenchyma.

CT UROGRAM

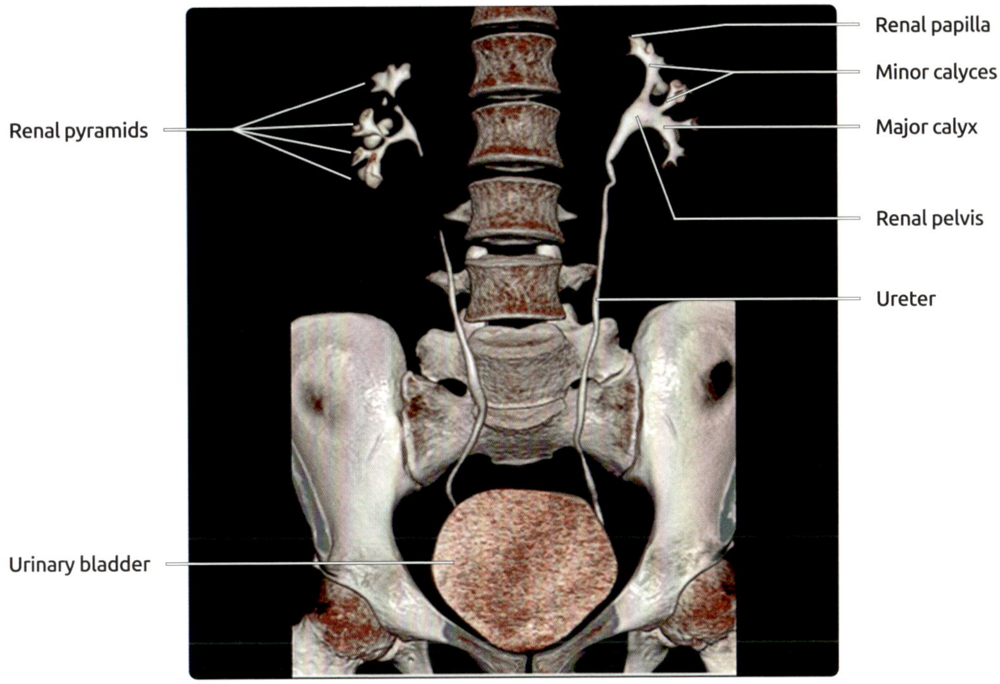

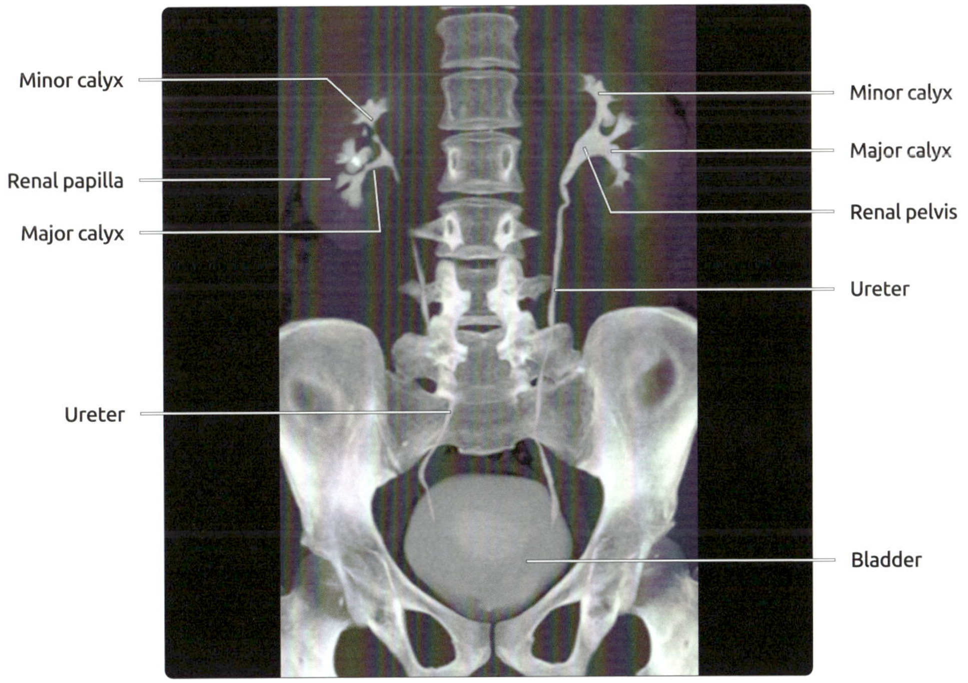

(Top) *A series of axial CT images can be viewed as a volume-rendered 3D image to simulate an excretory urogram. The window levels and workstation controls have been set to optimally display the renal collecting system with opacified urine displayed as "white" and less dense urine within the renal tubules in the pyramids and the diluted urine within the bladder displayed as "red." This CT was obtained in deep, suspended inspiration, resulting in caudal displacement of the kidneys.* **(Bottom)** *Coronal CECT in the excretory-phase image with MIP reconstruction provides a nice comparison with the previous images with this reconstruction providing the closest CT approximation of a traditional excretory urogram. This technique relies on a projection of the highest attenuation voxels in a dataset, resulting in an image that highlights the high-density contrast within the collecting systems and ureters in the delayed excretory phase.*

RENAL FASCIA AND PERIRENAL SPACE

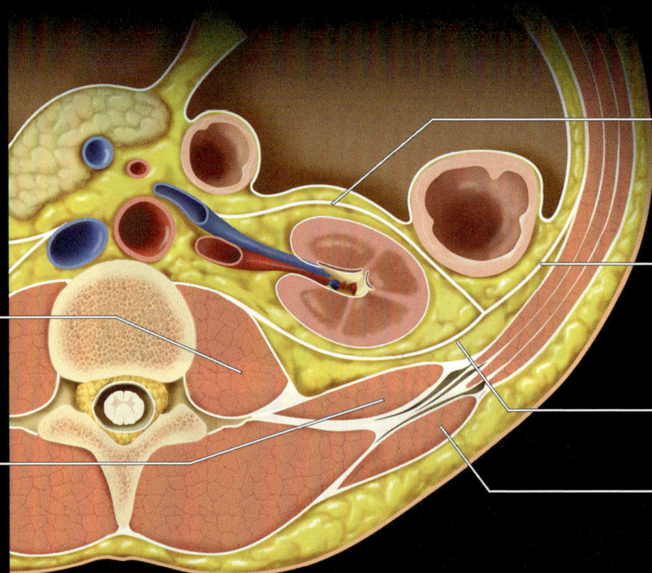

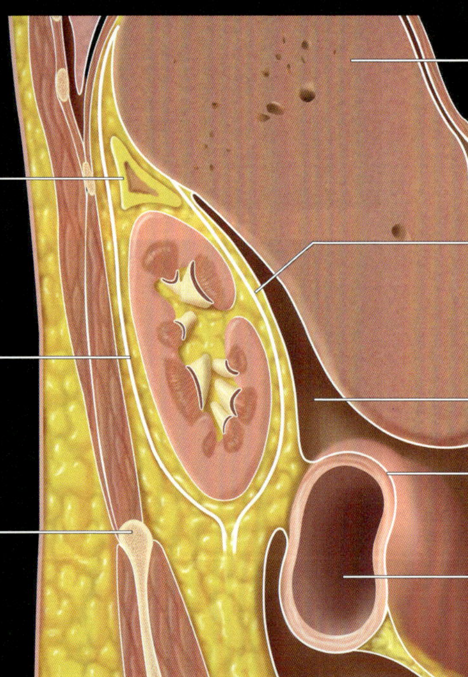

(Top) *The anterior and posterior layers of the renal fascia envelop the kidneys and adrenals along with the perirenal fat. Medial to the kidneys, the course of the renal fascia is variable (and controversial). The posterior layer usually fuses with the psoas or quadratus lumborum fascia. The perirenal spaces do not communicate across the abdominal midline. However, the renal and lateroconal fasciae are laminated structures that may be distended with fluid collections to form interfascial planes that do communicate across the midline and also inferiorly to the extraperitoneal pelvis.* **(Bottom)** *Sagittal section through the right kidney shows the renal fascia enveloping the kidney and adrenal. Inferiorly, the anterior and posterior renal fasciae come close together at about the level of the iliac crest. Note the adjacent peritoneal recesses.*

PERIRENAL PLANES

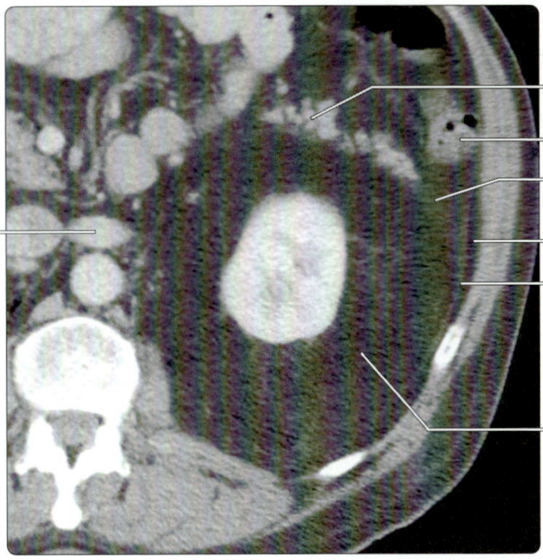

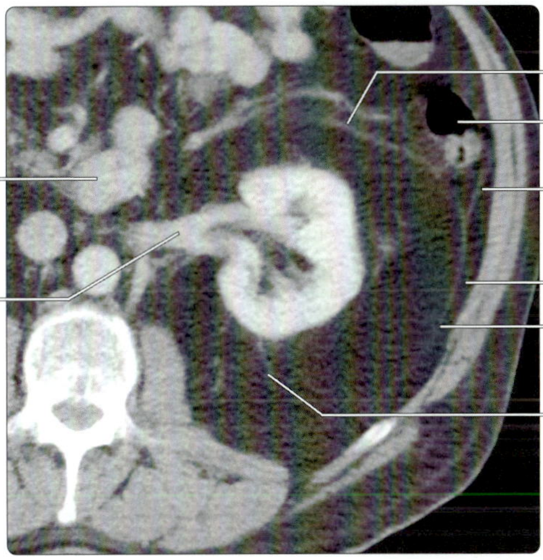

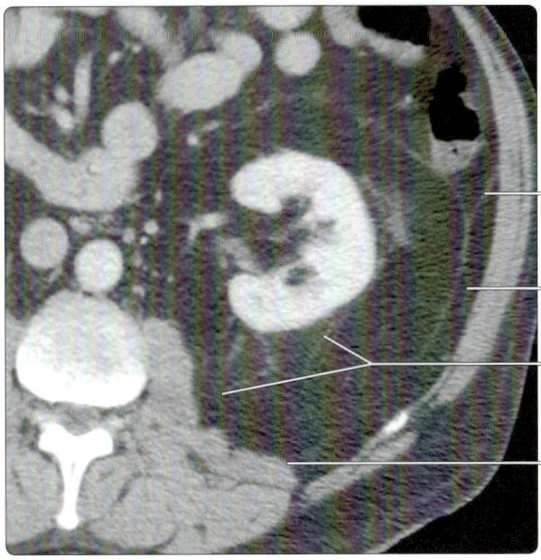

(Top) *First of 6 axial CECT images shows the upper pole of the left kidney surrounded by fat within the perirenal space. The perirenal space is the fat-containing space between the renal capsule and renal fascia. The pararenal spaces lie outside of the renal fascia. The pancreas, duodenum, and ascending/descending colon lie anterior to the renal fascia within the anterior pararenal space.* **(Middle)** *The perirenal space is bounded by the anterior and posterior renal fasciae, and within this space lie the adrenal and kidney, with a variable number of vessels, nerves, and lymphatics. The anterior renal fascia (also known as Gerota fascia) is well demonstrated on this image.* **(Bottom)** *The posterior pararenal space lies lateral to the lateroconal fascia and posterior renal fascia and is synonymous with the "properitoneal fat stripe" seen on radiographs. Note the fusion of the posterior renal fascia with the quadratus lumborum fascia. Several perirenal bridging septa are visible within the perirenal space that divide this space into multiple poorly communicating spaces.*

RENAL FASCIA, PERIRENAL PLANES, AND PERIRENAL SPACE

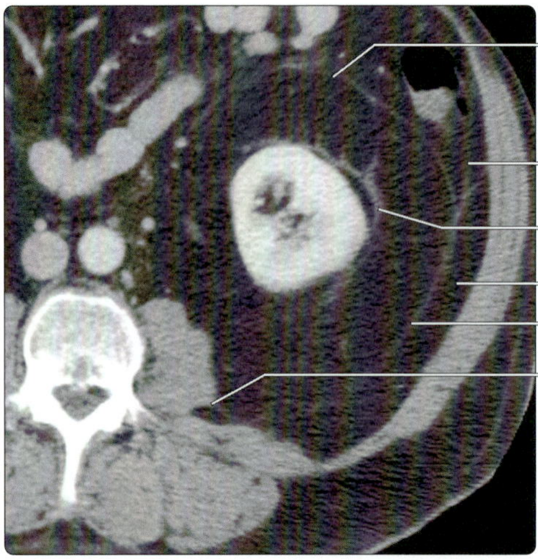

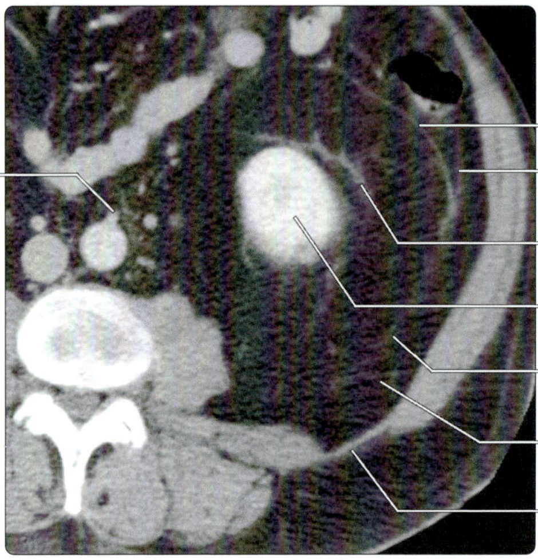

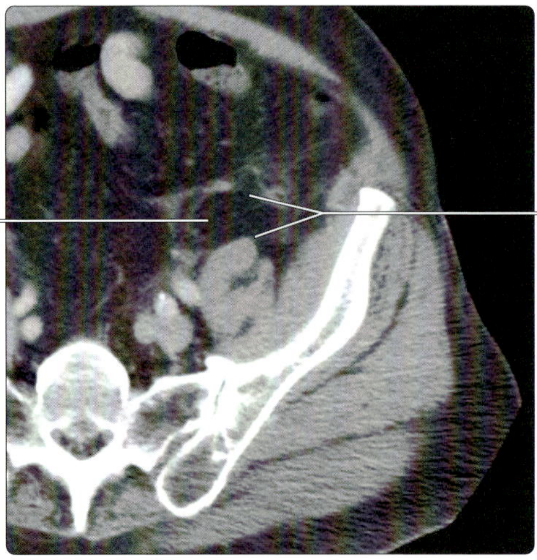

(Top) *The renal and lateroconal fasciae are key planes that define the 3 divisions of the retroperitoneum. The descending colon, duodenum, and pancreas lie in the anterior pararenal space. The lateroconal fascia is the lateral margin of the anterior pararenal space and the medial margin of the posterior pararenal space. On this section, the renal fascia courses medially to fuse with the psoas fascia.* **(Middle)** *Some perirenal bridging septa course parallel to the renal capsule and renal fascia and could be mistaken for the renal fascia itself. A flank incision to approach the kidney usually involves dividing the aponeurosis of the transverse abdominal muscle to enter the retroperitoneum. First encountered is the fat within the posterior pararenal space, sometimes referred to as the pararenal fat body. The renal fascia is next seen within which lies the perirenal space, also known as the pararenal fat capsule.* **(Bottom)** *At a level just caudal to the iliac crest, the layers of the renal fascia come close together to almost seal off the lower portion of the perirenal space.*

CATHETER ANGIOGRAM

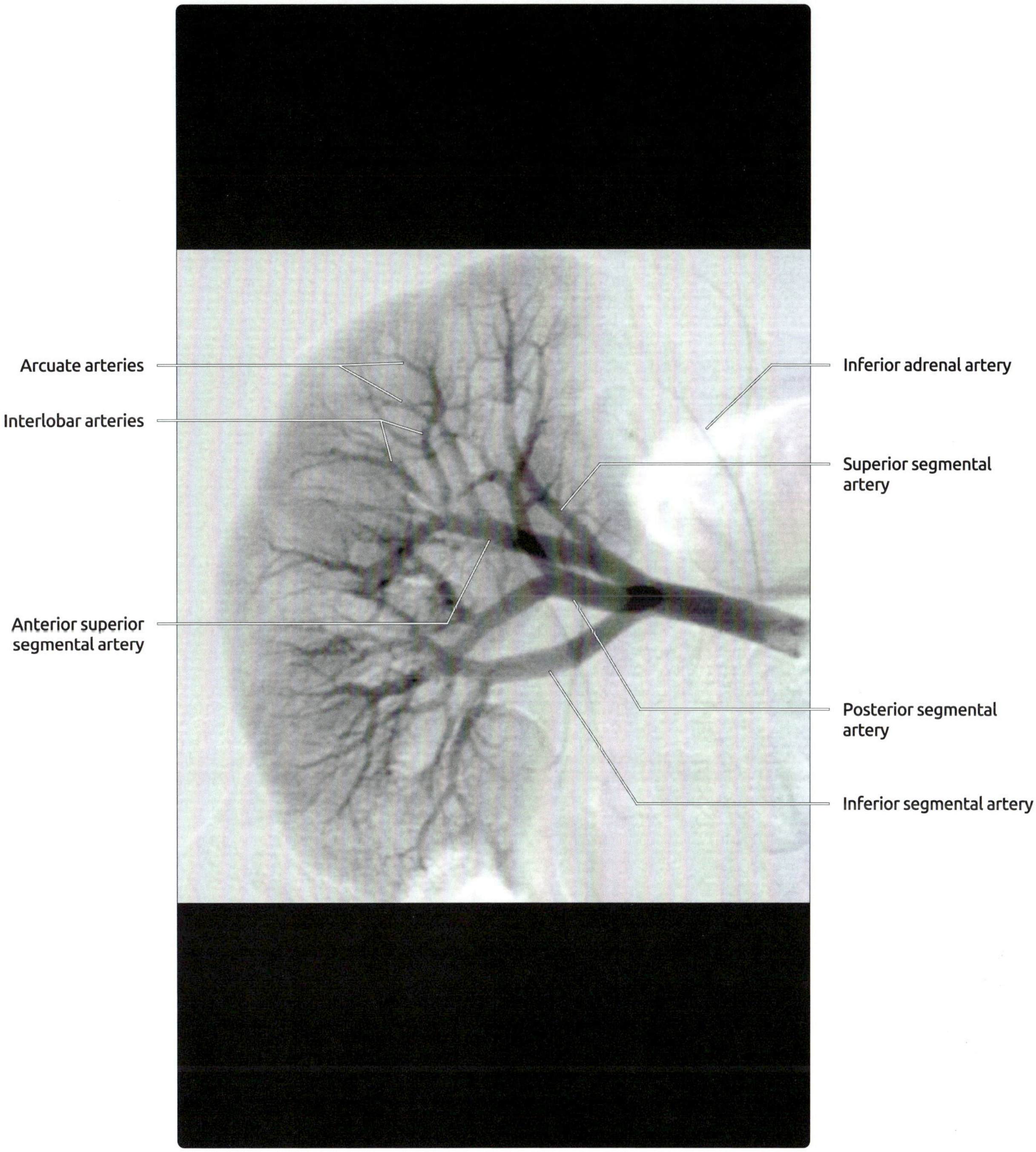

A selective catheter injection of the right renal artery depicts the vessels to the level of the arcuate arteries, which course along the convex surface of each renal pyramid.

RENAL ARTERY VARIANTS

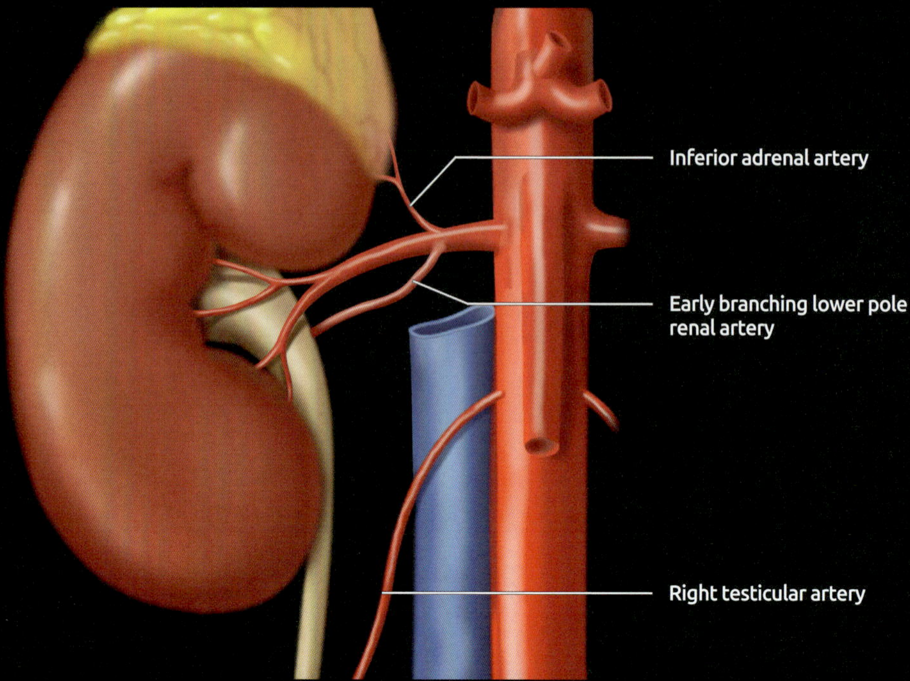

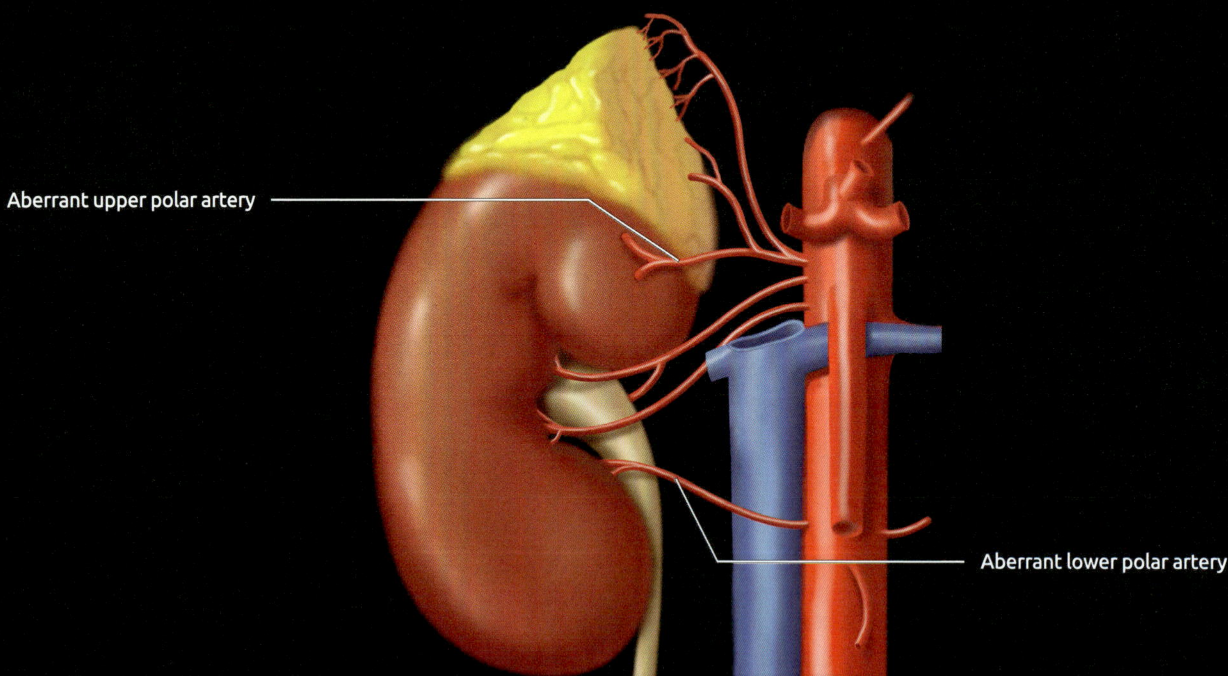

(Top) *Graphic depicts proximal ramification (early branching) of the renal artery. This may have important implications if renal surgery is being considered. For instance, a person with this vascular anatomy might be considered a poor candidate as a potential living renal donor, as "harvesting" this kidney might jeopardize the inferior adrenal and lower pole renal arteries.* **(Bottom)** *Graphic depicts supernumerary renal arteries arising directly from the aorta. Some of these enter the kidney at locations other than the renal hilum, close to the renal poles. These "polar" or "extrahilar" arteries may be ligated or transected unintentionally during renal, aortic, or other retroperitoneal abdominal surgeries. These are sometimes referred to as "accessory" renal arteries, but each is an end artery and the sole arterial supply to a substantial portion of the renal parenchyma.*

MULTIPLE RENAL ARTERIES

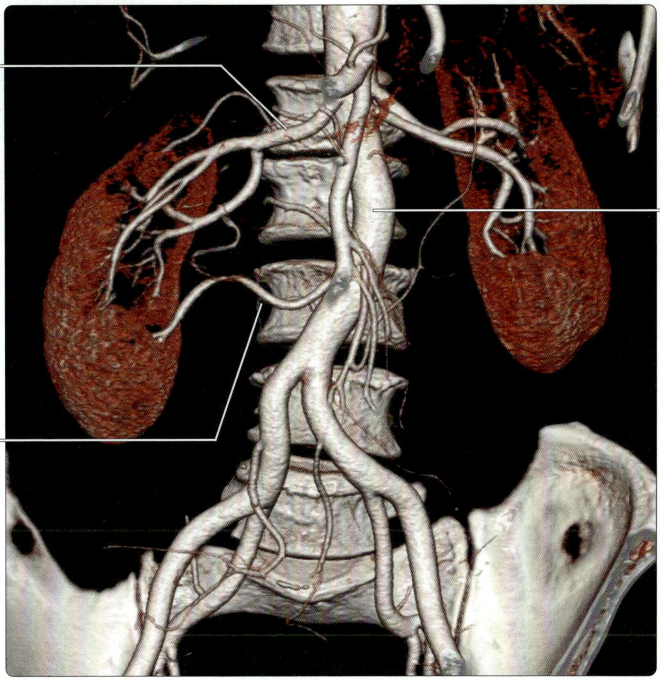

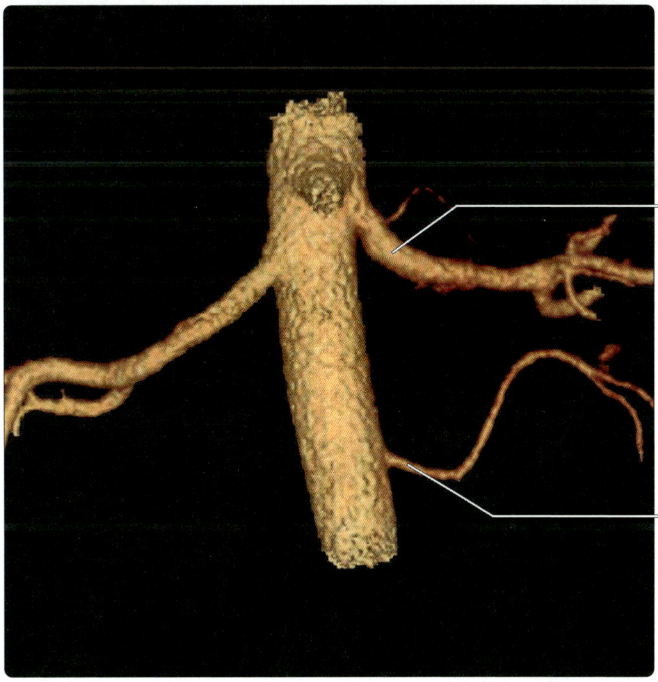

(Top) *Volume-rendered coronal CECT demonstrates 2 discrete renal arteries supplying the right kidney, including a dominant right main renal artery arising at roughly the L1 level and a smaller accessory renal artery arising at the L3 level. Identifying accessory renal arteries, critical in patients undergoing renal surgery or transplantation, is often much easier using 3D techniques (such as VR or MIP) than it is on standard-source axial images.* **(Bottom)** *Volume-rendered coronal CECT with subtraction of extraneous overlying vessels reveals a substantial aberrant left inferior polar artery arising directly from the aorta in a patient being evaluated for kidney donation. The transplant surgeons elected to instead harvest the right kidney for transplantation.*

RENAL VEIN VARIATIONS

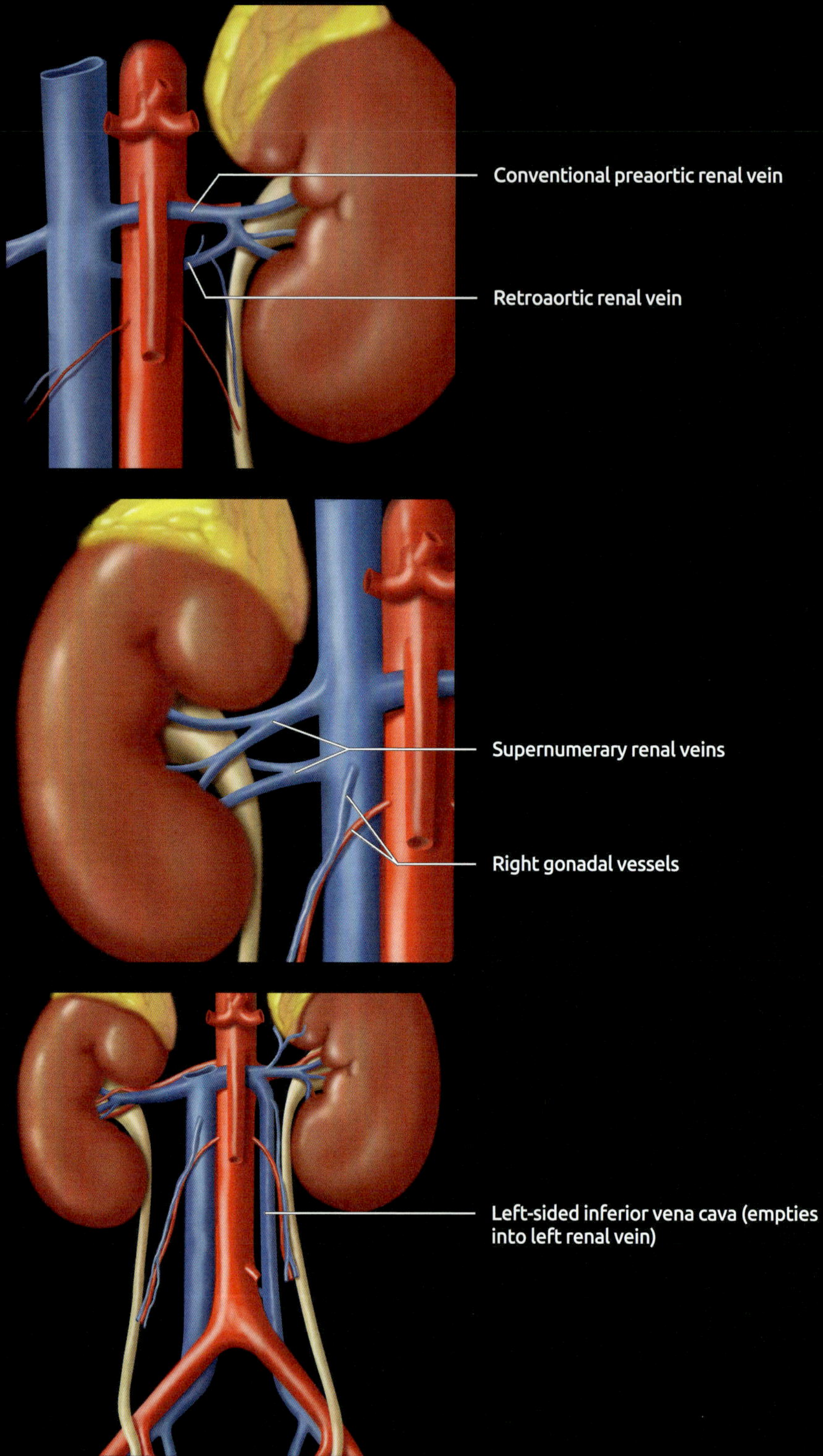

(Top) *Anomalies of the renal veins are less common than those of the arteries but are encountered in clinical practice and may have important implications. All anomalies are variations of the embryologic development and persistence of portions of the paired longitudinal channels, the subcardinal and supracardinal veins, which form a ladder-like collar around the aorta. Normally, only the anterior components persist, becoming the renal veins, which course anterior to the aorta. Persistence of the whole collar results in a circumaortic renal vein, which is depicted in this graphic. This anomaly is more common than an isolated retroaortic renal vein.* **(Middle)** *Persistence of the collar of veins on the right results in supernumerary right renal veins that encircle the renal pelvis.* **(Bottom)** *Persistence of the left supracardinal vein below the kidney results in a "duplicated" inferior vena cava.*

CIRCUMAORTIC LEFT RENAL VEIN

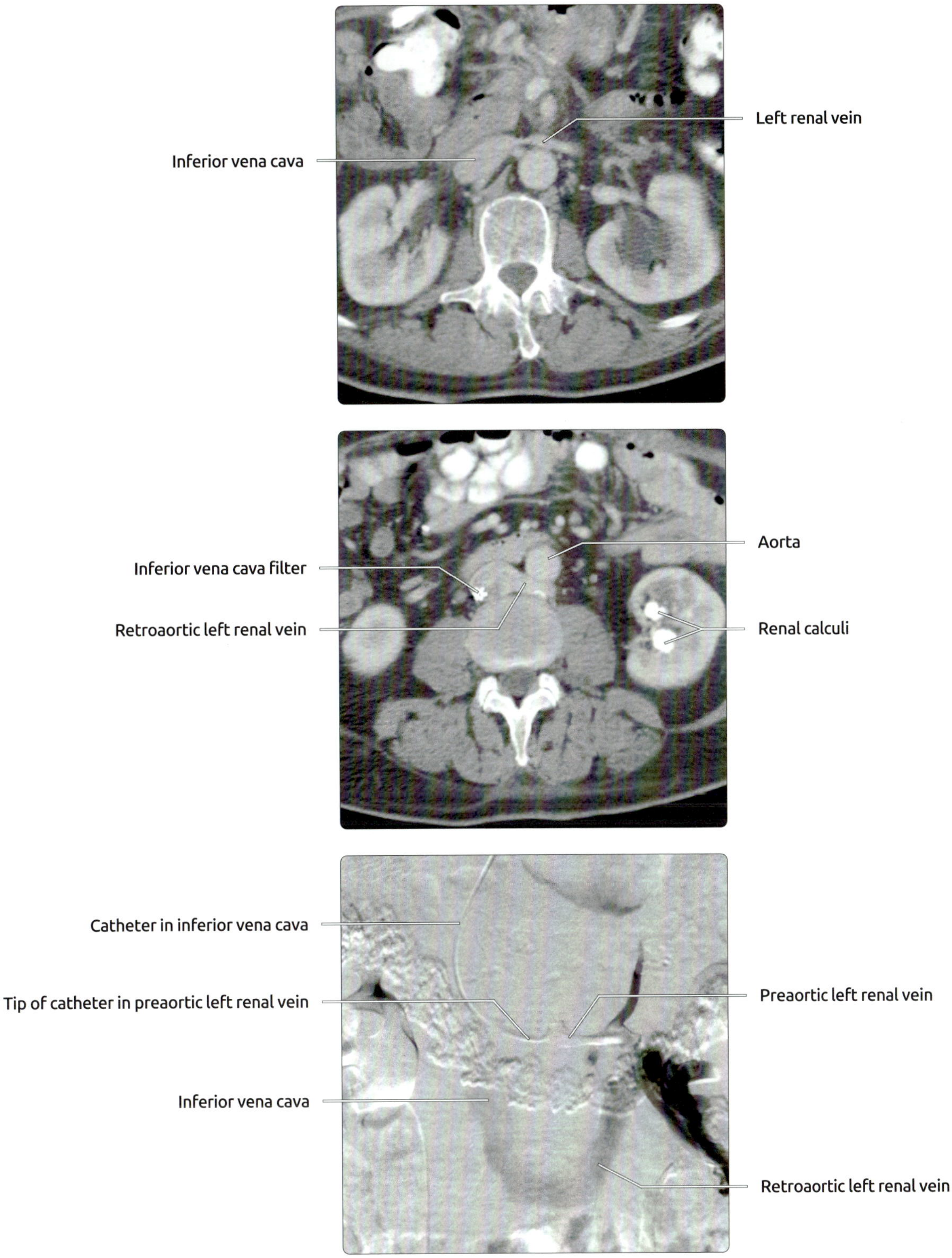

(Top) *First of 3 axial CECT images demonstrates a circumaortic left renal vein. This section shows a small preaortic (conventional) left renal vein as it joins the inferior vena cava.* **(Middle)** *A more caudal section shows a larger retroaortic left renal vein joining the inferior vena cava. Also note the tip of an inferior vena cava filter placed to prevent propagation of blood clots into the lungs (pulmonary embolism). Knowledge of renal venous anomalies is important to avoid inadvertent injury to the renal vein during interventional or surgical procedures involving the inferior vena cava or aorta.* **(Bottom)** *Frontal film from inferior vena cavagram preceding the CT is shown. A catheter has been inserted through an arm vein and advanced so that its tip is in the left renal vein. Injection of contrast has passed retrograde through the preaortic left renal vein and is flowing antegrade through the larger, more caudal retroaortic renal vein.*

LEFT RENAL VEIN COMPROMISED BY SUPERIOR MESENTERIC ARTERY

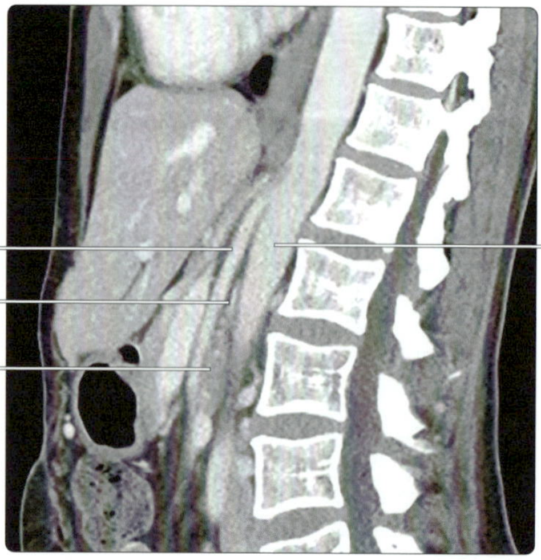

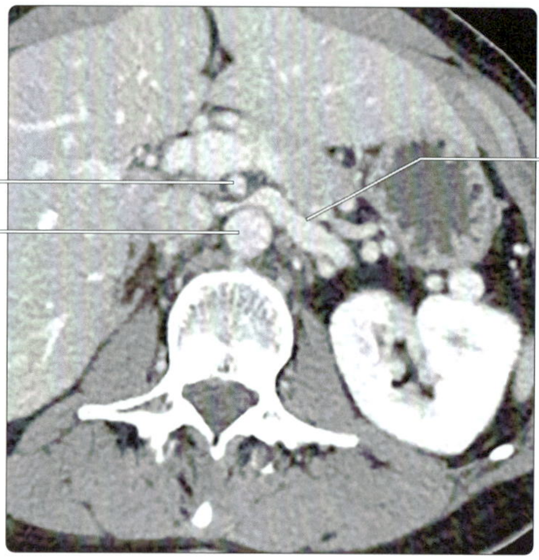

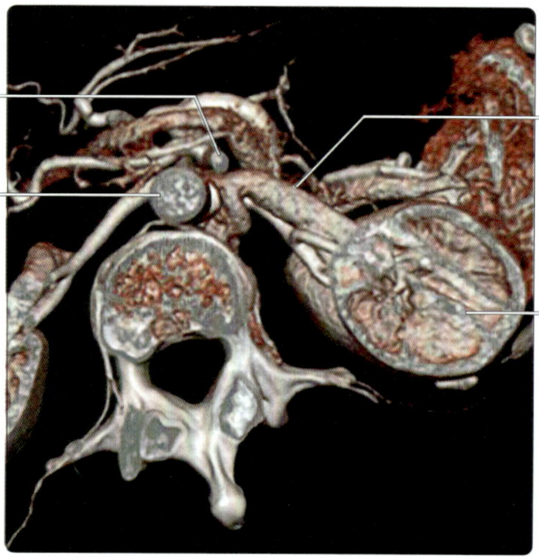

(Top) *Sagittal CECT demonstrates an abnormally narrow distance between the superior mesenteric artery (SMA) and aorta with narrowing of the normal "aortomesenteric angle." Notice the left renal vein and the duodenum compressed between the aorta and SMA.* **(Middle)** *Axial CECT in the same patient shows narrowing of the left renal vein between the aorta and SMA with distention of the renal vein "upstream" from the compressed portion. This may result in increased pressure within the renal vein with hematuria and flank pain. This patient was thought to have symptoms secondary to SMA syndrome and underwent corrective surgery.* **(Bottom)** *Volume-rendered CECT in a patient who underwent surgery for SMA syndrome demonstrates the left renal vein compressed between the aorta and SMA with poststenotic dilatation of the vein distal to the compression. Notice the abnormally narrow distance between the aorta and SMA. This "nutcracker" phenomenon (compression of left renal vein) is an important secondary sign in the diagnosis of SMA syndrome.*

STAGHORN CALCULI

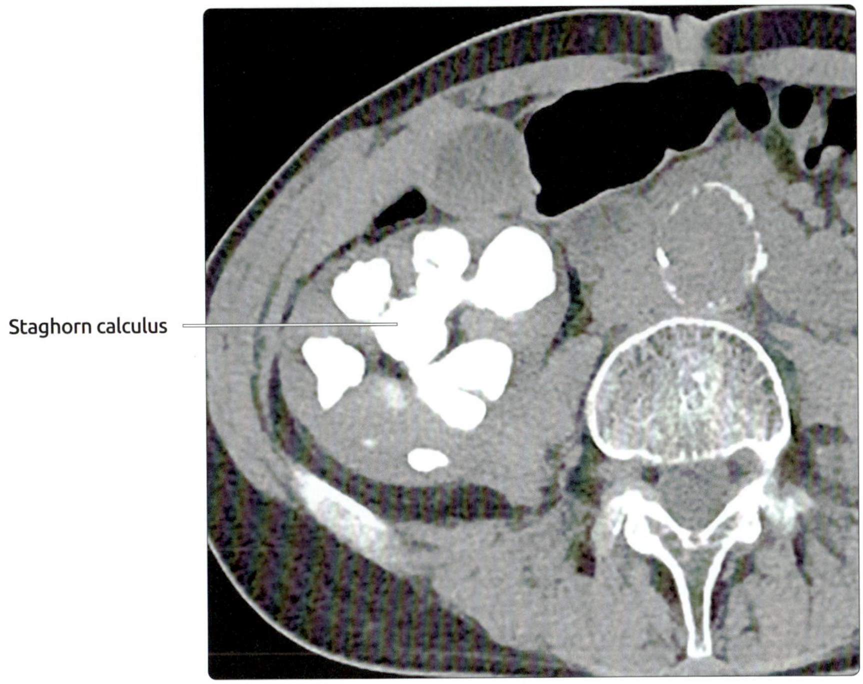

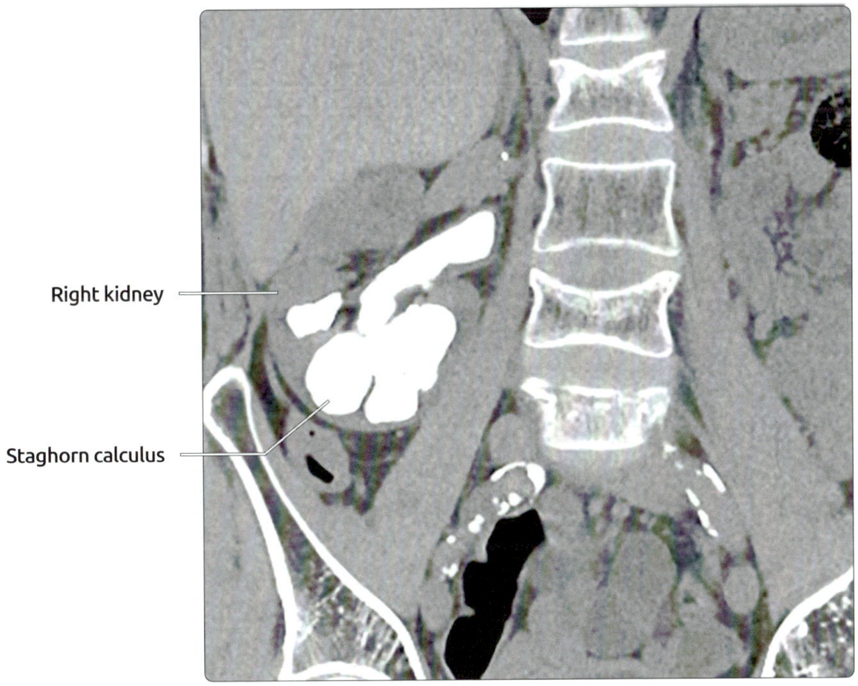

(Top) *First of 2 axial NECT images showing renal calculi demonstrates a large hyperdense staghorn calculus filling much of the intrarenal collecting system within the right kidney.* **(Bottom)** *Coronal NECT in the same patient shows that the renal calculus fills and conforms to the shape of the right renal calyces, resembling the horns of a deer ("staghorn calculus"). Essentially all renal calculi are dense and easy to recognize on CT, while many stones are too small or insufficiently opaque to diagnose on plain abdominal radiographs.*

RENAL CELL CARCINOMA

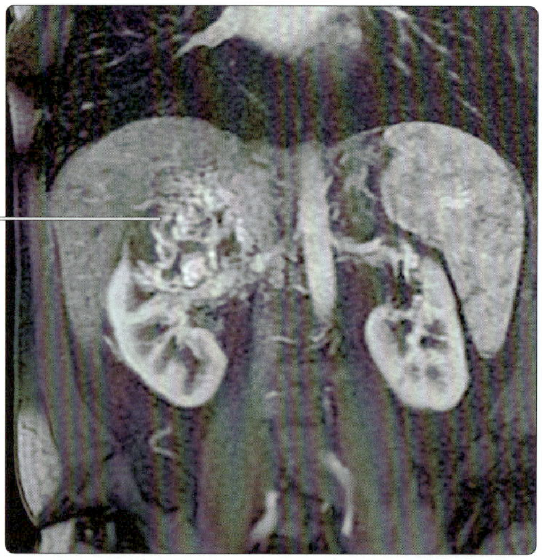

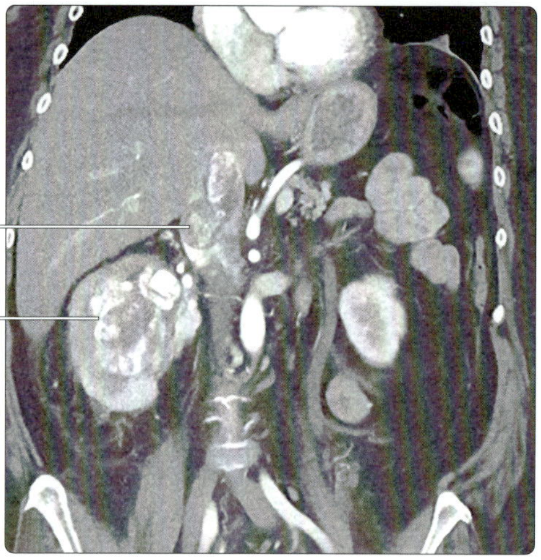

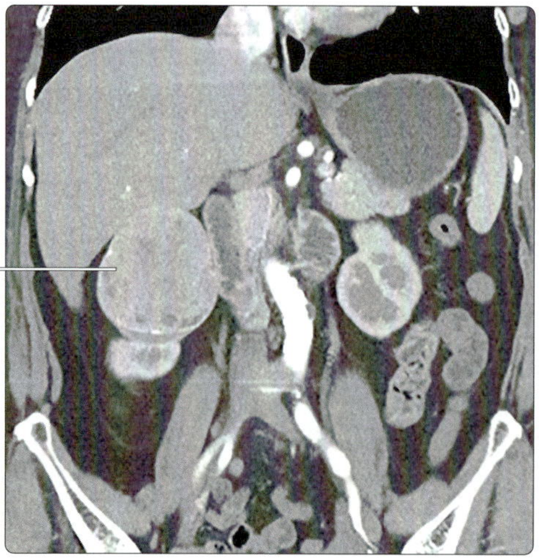

(Top) *Coronal T1 C+ FS MR demonstrates a large, heterogeneously enhancing mass with areas of internal necrosis arising from the right upper pole kidney found to represent a clear cell renal cell carcinoma (RCC) at resection.* **(Middle)** *Coronal CECT demonstrates a large mass with a substantial hypervascular soft tissue component in the right kidney compatible with a clear cell RCC. The mass directly invades the right renal vein and inferior vena cava with tumor thrombus also demonstrating similar hypervascularity, features that are not uncommon with RCC and which likely increase the risk of lung and other systemic metastases.* **(Bottom)** *Coronal CECT demonstrates a relatively hypoenhancing solid mass in the right kidney found to represent a chromophobe RCC at resection. Chromophobe RCC has a much better prognosis compared to the more common clear cell subtype.*

RENAL CELL CARCINOMA

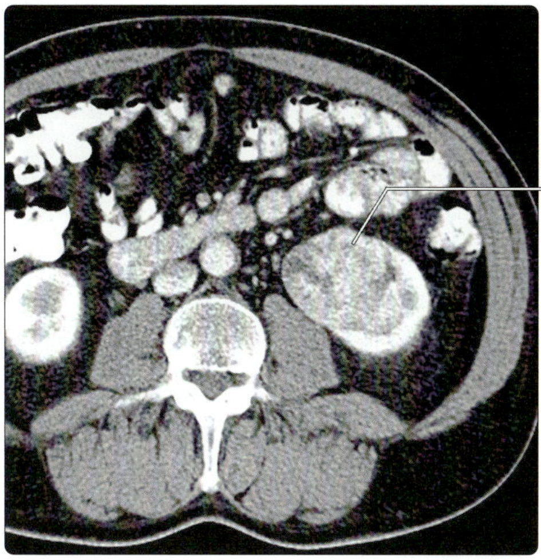

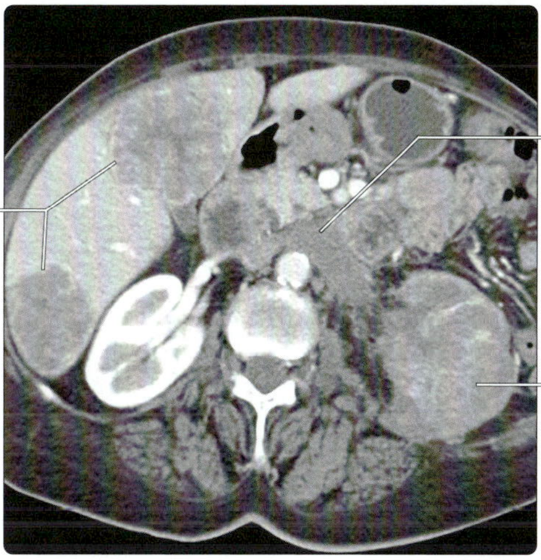

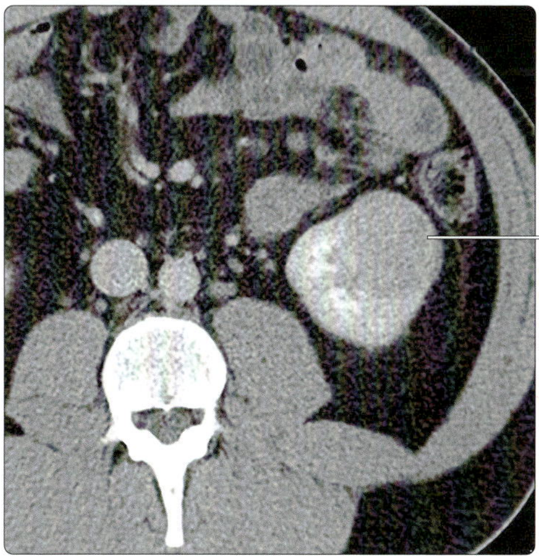

(Top) *Axial CECT demonstrates an avidly hypervascular solid mass arising from the right upper pole kidney, compatible with RCC. RCCs with this degree of enhancement are most likely to represent the clear cell type of RCC, a finding confirmed at resection, while papillary and chromophobe RCC typically demonstrate lesser degrees of enhancement.* **(Middle)** *Axial CECT demonstrates a mass in the left kidney, which replaces much of the renal parenchyma, directly contiguous tumor thrombus within the left renal vein, as well as several liver metastases, compatible with metastatic RCC.* **(Bottom)** *Axial CECT demonstrates a rounded, circumscribed mass in the left kidney found to represent a papillary RCC at resection.*

OTHER RENAL MASSES

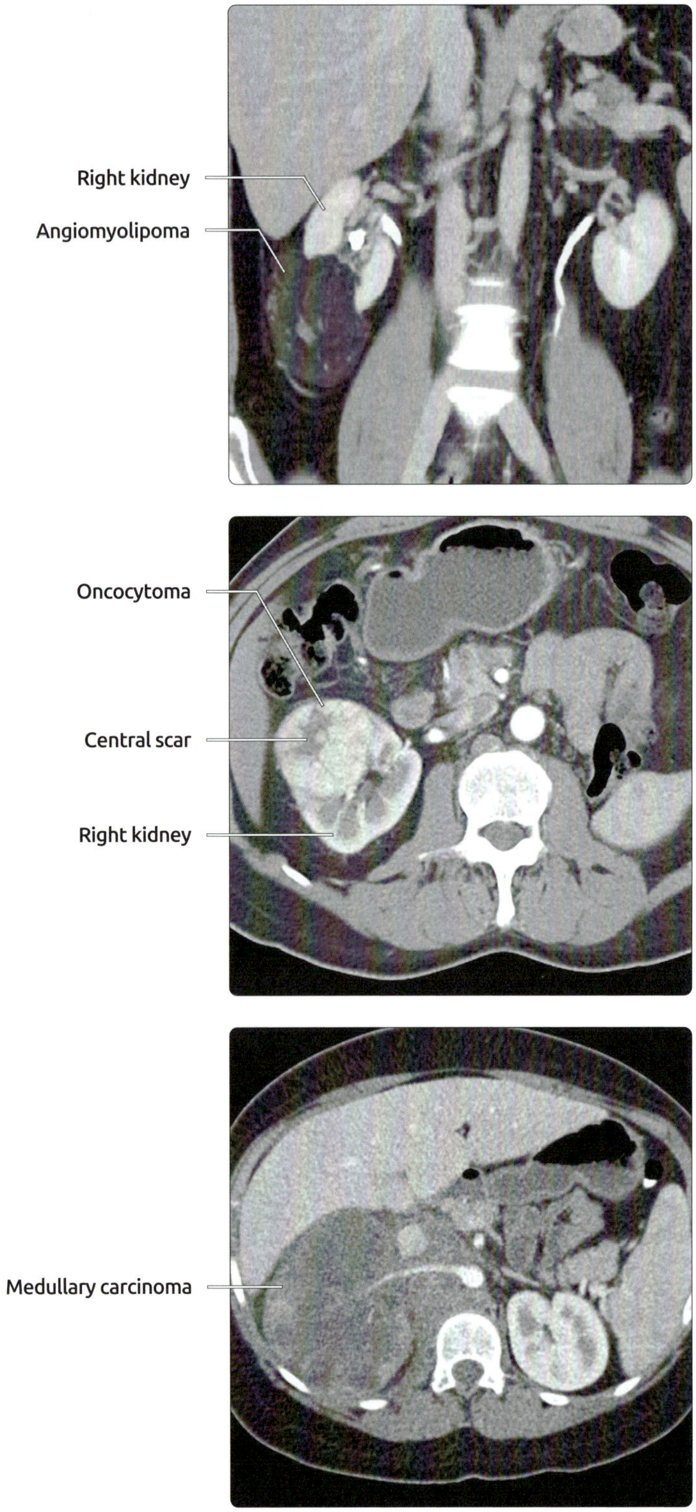

(Top) *Coronal CECT demonstrates a fat attenuation mass arising exophytically from the right kidney. The presence of macroscopic fat within a renal mass is virtually diagnostic of a benign renal angiomyolipoma (AML).* **(Middle)** *Axial CECT demonstrates an enhancing solid right renal mass with a central scar. While this mass cannot be reliably distinguished from an RCC, thereby requiring surgical resection, the presence of a central scar can suggest the possibility of a benign renal oncocytoma, a diagnosis confirmed at surgery.* **(Bottom)** *Axial CECT in a 10-year-old patient with sickle cell trait demonstrates an aggressive, infiltrative mass replacing much of the right kidney. The clinical history of sickle cell trait is the key to the correct diagnosis in this case of a medullary carcinoma. Medullary carcinoma is an extraordinarily aggressive malignancy with a very poor prognosis, as patients often live only a matter of a few months after diagnosis.*

TRANSITIONAL CELL CARCINOMA

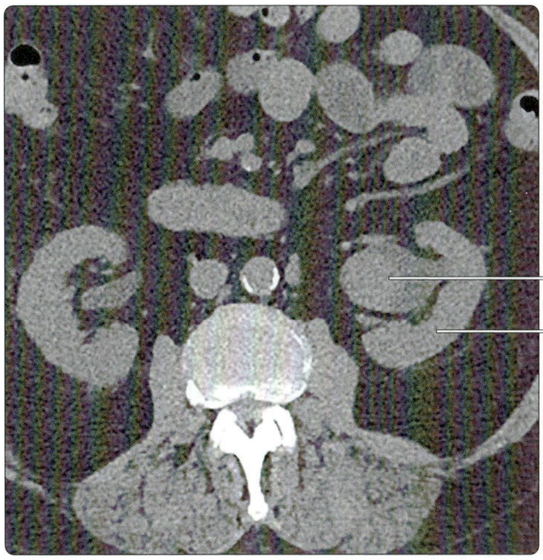

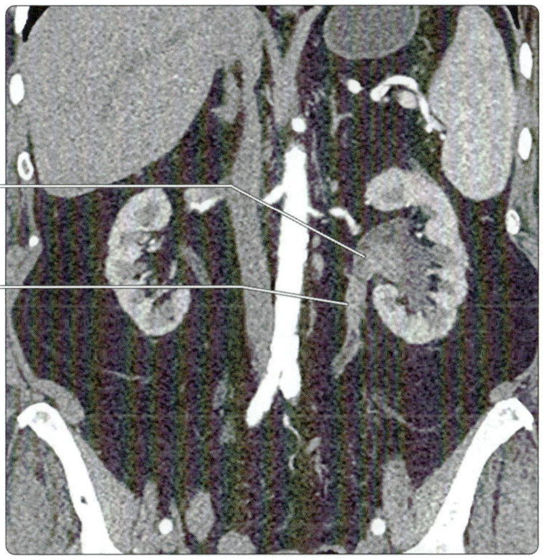

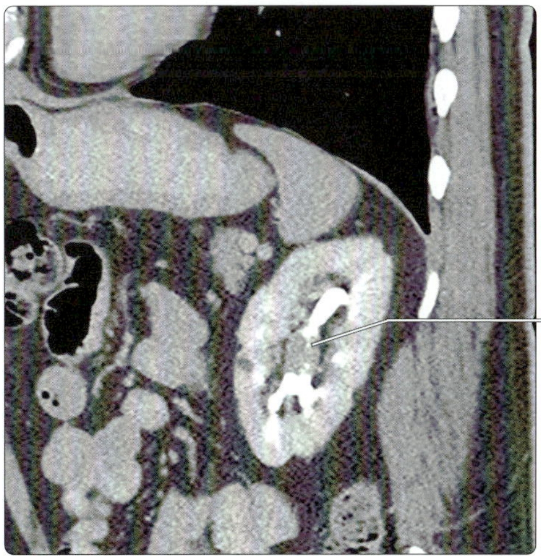

(Top) *Axial NECT demonstrates a high-density filling defect within the left renal pelvis. Note the presence of mild asymmetric left-sided hydronephrosis proximal to the filling defect.* **(Middle)** *Coronal CECT in the same patient demonstrates that the filling defect seen on NECT corresponds to an enhancing mass within the left renal pelvis with extension into the proximal ureter, compatible with urothelial cancer.* **(Bottom)** *Sagittal CECT in the delayed excretory phase demonstrates focal soft tissue thickening in the intrarenal collecting system, ultimately confirmed to represent a transitional cell carcinoma at resection. While diffuse urothelial thickening is usually infectious, focal urothelial thickening should always raise concern for malignancy.*

TRANSITIONAL CELL CARCINOMA

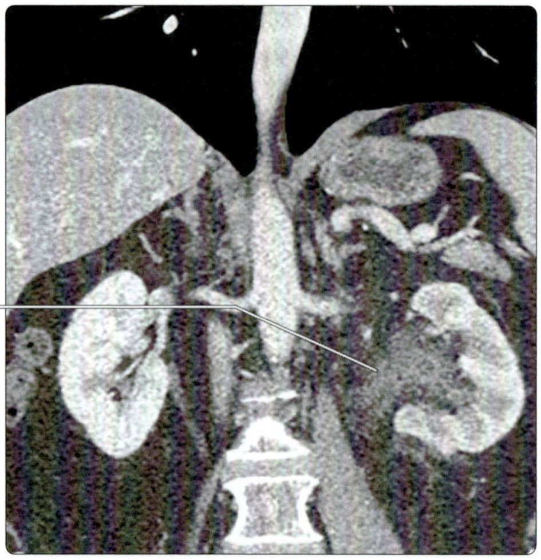

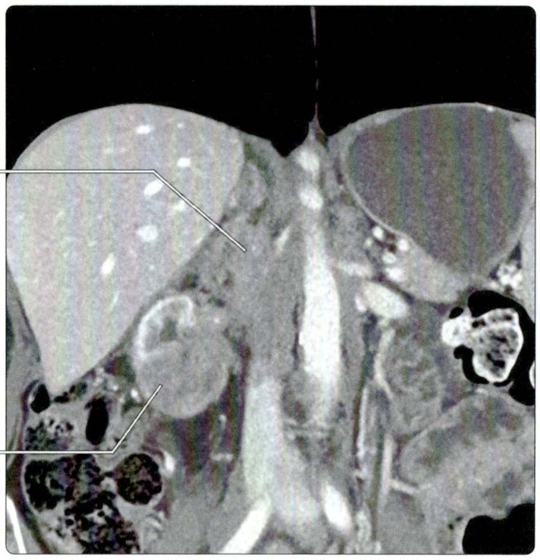

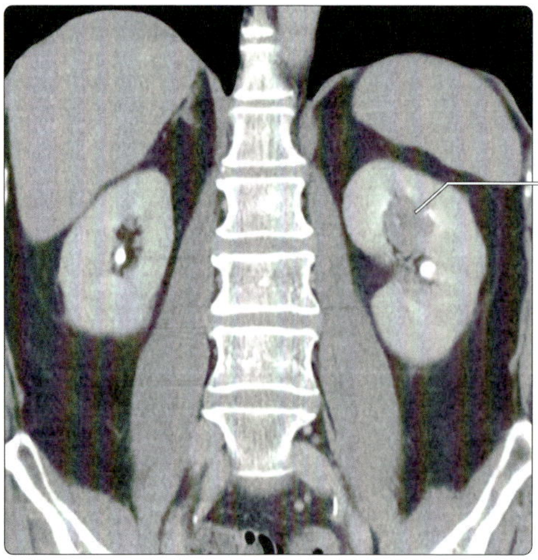

(Top) *Axial CECT in the excretory phase demonstrates severe focal left-sided urothelial thickening with extensive adjacent fat stranding. This was found to be a transitional cell carcinoma at resection with the stranding likely corresponding to extramural spread of malignancy.* **(Middle)** *Coronal CECT demonstrates an infiltrative mass replacing much of the right kidney with extensive contiguous lymphadenopathy in the retroperitoneum. Note that this mass, unlike RCC, is very poorly marginated and infiltrative, findings that are strongly suggestive of transitional cell carcinoma.* **(Bottom)** *Coronal CECT demonstrates a transitional cell carcinoma filling the left upper pole renal calyx.*

SIMPLE CYST

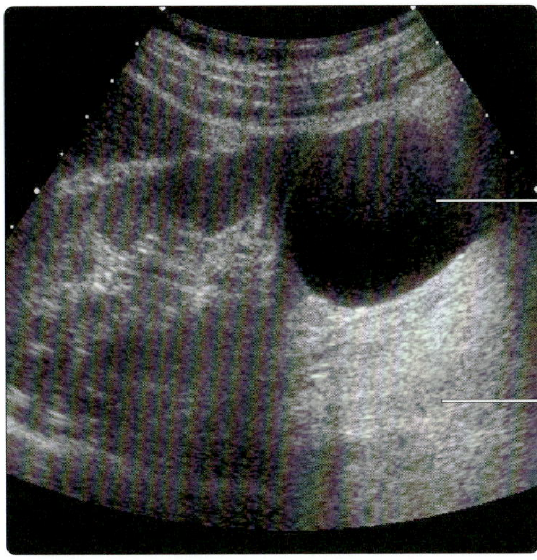

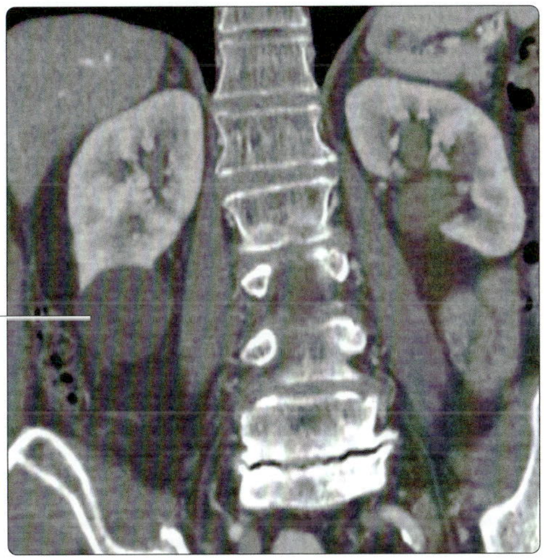

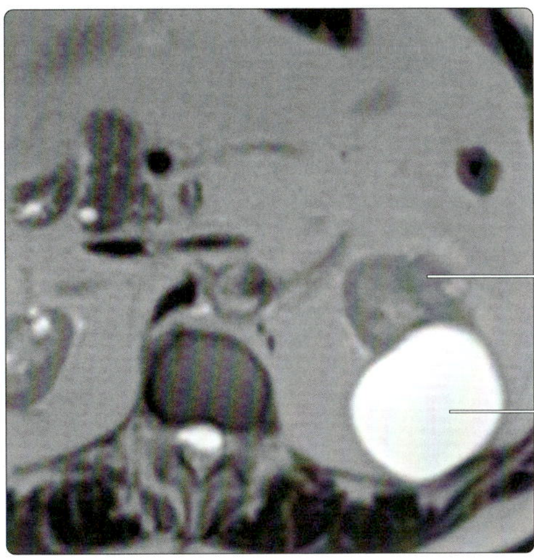

(Top) *Sagittal ultrasound shows a cystic lesion arising from the lower pole of the right kidney. The lesion is totally anechoic with no mural nodularity or septations, and there is posterior acoustic through transmission deep to the lesion.* **(Middle)** *Coronal CECT in the same patient shows the spherical cyst with nonenhancing contents that measured near water density (0 HU).* **(Bottom)** *Axial T2 MR demonstrates a classic simple cyst arising from the upper pole of the left kidney with simple T2-hyperintense fluid signal within the cyst and no evidence of any appreciable internal complexity.*

CONGENITAL ABSENCE OF KIDNEY

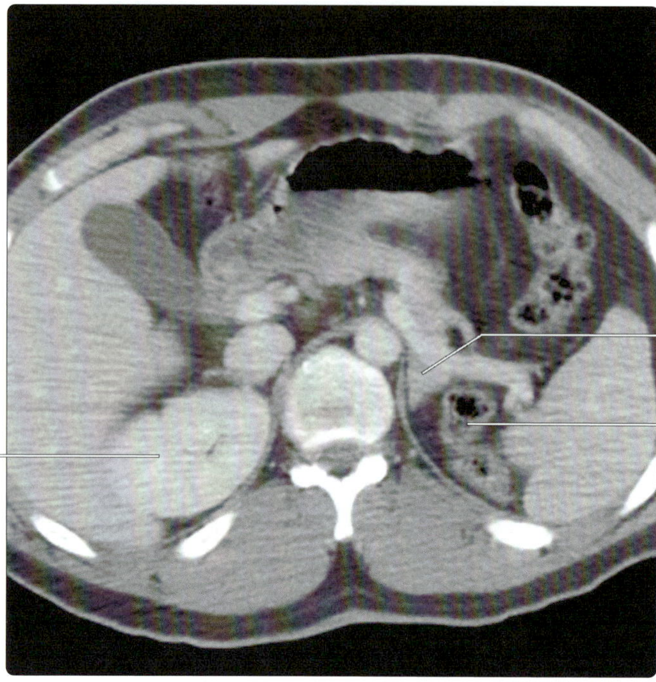

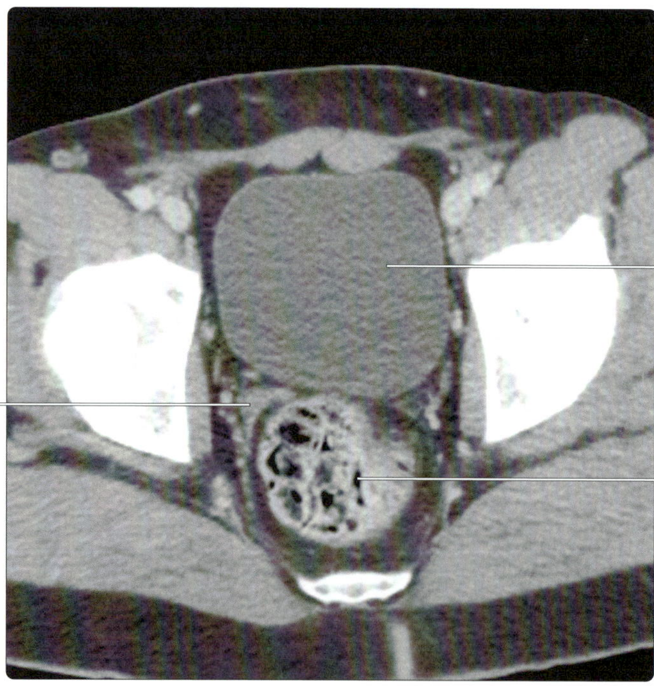

(Top) *First of 2 images of a patient with congenital absence of the left kidney is shown. An axial CECT demonstrates a normal right kidney and left adrenal gland but no left kidney. No kidney was found in an ectopic location.* **(Bottom)** *A more caudal image shows a normal right seminal vesicle but no left seminal vesicle. Congenital absence of the kidney is often accompanied by anomalies in other organ systems, including musculoskeletal, cardiovascular, and genital. Congenital anomalies of the kidney and genital organs usually occur ipsilaterally (on the same side).*

PELVIC KIDNEY

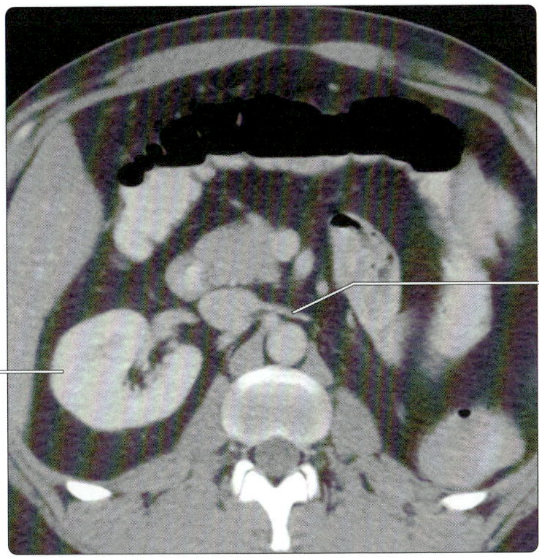

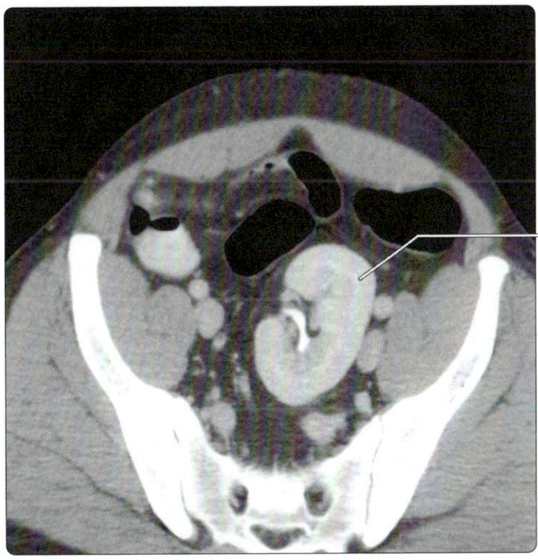

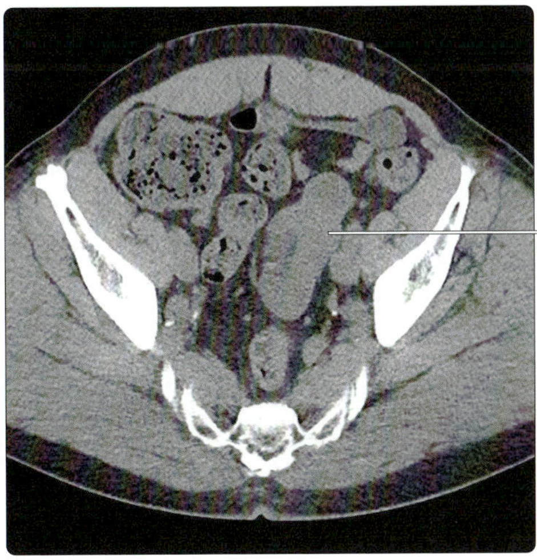

(Top) *First of 2 axial CECT images shows a normal right kidney in its usual location. The left kidney is not seen, though the left renal vein is present.* **(Middle)** *CECT through the pelvis in the same patient shows an ectopic, malrotated, pelvic kidney. In the early embryo, both kidneys lie in the pelvis. As the fetus grows, the kidneys usually "ascend" to their normal abdominal position, successively recruiting more proximal arterial and venous branches from the aorta and inferior vena cava. Ectopic kidneys are invariably low in position and usually are malrotated with aberrant blood supply from the distal aorta or iliac arteries.* **(Bottom)** *Axial NECT demonstrates a pelvic kidney in the left hemipelvis in this patient with absence of the left kidney in the left renal fossa (not shown).*

CROSSED FUSED ECTOPIA

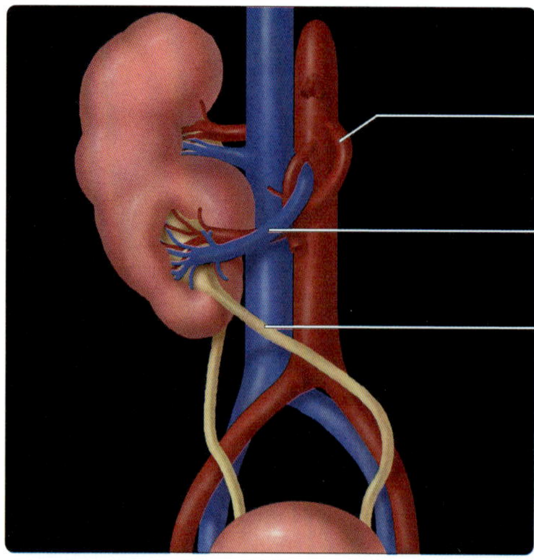

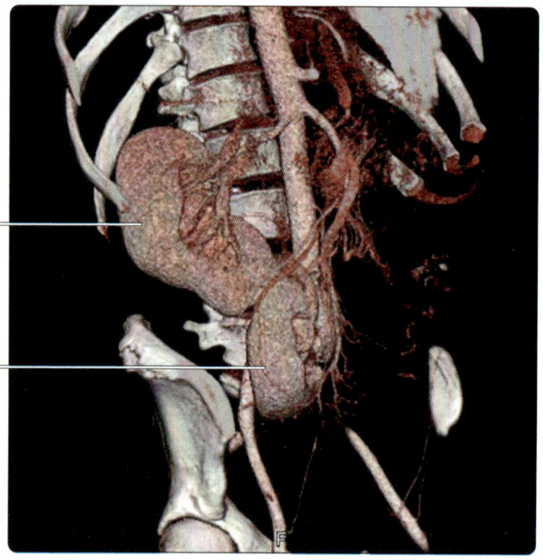

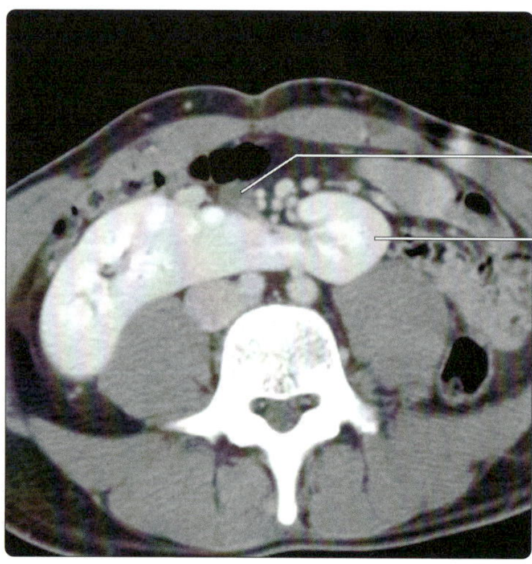

(Top) *Graphic illustrates the typical appearance of crossed fused ectopia. The right renal moiety develops normally with conventional vascular anatomy. The left renal moiety is fused to the lower pole of the right moiety. The left renal vessels and ureter have their normal origins and insertions but cross the midline to the left renal moiety. Ectopic kidneys are more vulnerable to trauma, calculi, and hydronephrosis.* **(Middle)** *Coronal volume-rendered CECT demonstrates the typical appearance of crossed fused renal ectopia, with the right renal moiety located superiorly in relatively normal position, while the left renal moiety is fused to the lower pole of the right kidney and is abnormally rotated.* **(Bottom)** *Axial CECT demonstrates that the left renal moiety is fused to the lower pole of the right kidney, and its hilum is directed anteriorly. This appearance is characteristic of crossed fused renal ectopia, although this entity can sometimes be easier to appreciate on coronal or 3D reformations compared to the axial plane.*

HORSESHOE KIDNEY

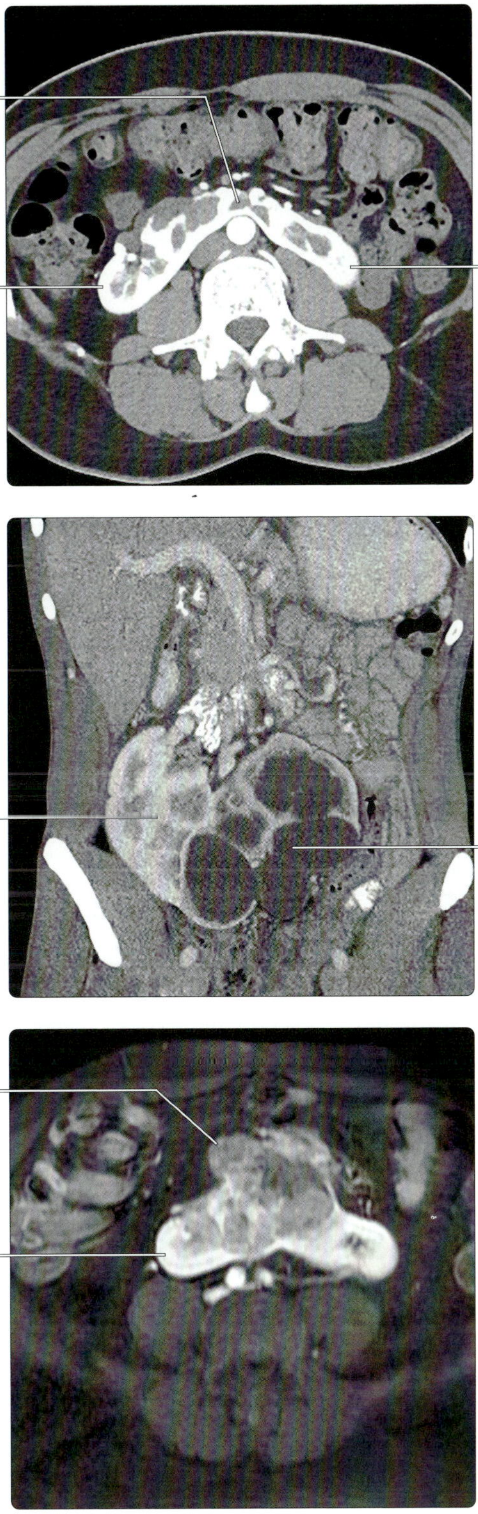

(Top) *Axial CECT demonstrates the classic appearance of a horseshoe kidney with both kidneys abnormally oriented and fused together near the midline. Notice the markedly abnormal orientation of both collecting systems, which are oriented laterally in this image.* **(Middle)** *Coronal CECT shows a U- or horseshoe-shaped kidney with the lower poles joined across the midline. There is severe left-sided hydronephrosis of the left renal moiety. Hydronephrosis is common in this anomaly and may be caused by aberrant arteries that compress the collecting system, aberrant and multiple ureters, renal calculi, or even tumor.* **(Bottom)** *Axial T1 C+ FS MR demonstrates a horseshoe kidney with a large enhancing renal mass arising from the midline, found to represent an RCC. Although a matter for debate, there is little evidence that horseshoe kidneys are at greater risk of developing malignancy.*

HORSESHOE KIDNEY

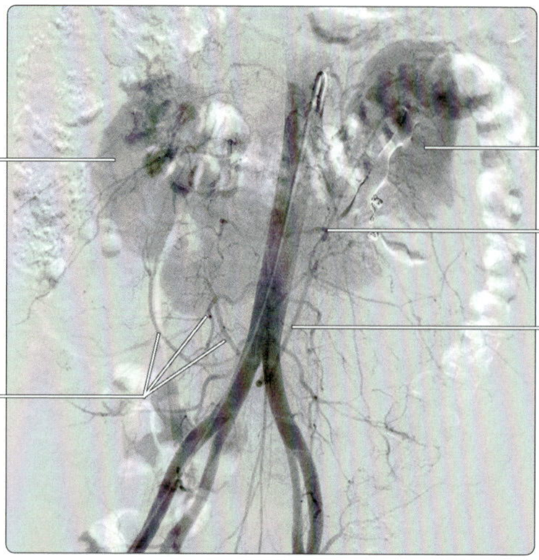

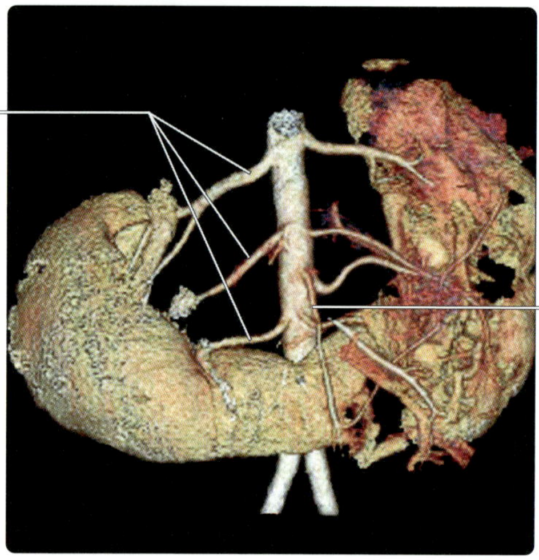

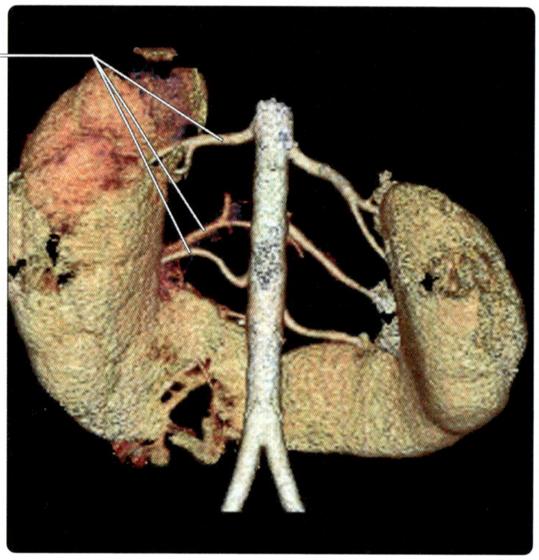

(Top) *Frontal angiogram demonstrates a horseshoe kidney with several bilateral renal arteries. Note that some of these renal arteries arise not just from the aorta but the common iliac arteries as well. The arterial supply to horseshoe kidneys can be extremely complex and variable in many of these patients.* **(Middle)** *Coronal volume-rendered CECT shows a horseshoe kidney crossing the midline between the aorta and the inferior mesenteric artery. Multiple renal arteries are evident. The left renal moiety appears moth eaten on this 3D image as a result of its hydronephrosis and consequent decreased parenchymal enhancement relative to the right side.* **(Bottom)** *Posterior 3D volume-rendered view of the horseshoe kidney in the same patient is displayed. Again, note the presence of multiple renal arteries supplying both the right and left renal moieties.*

HYPERTROPHIED COLUMN OF BERTIN

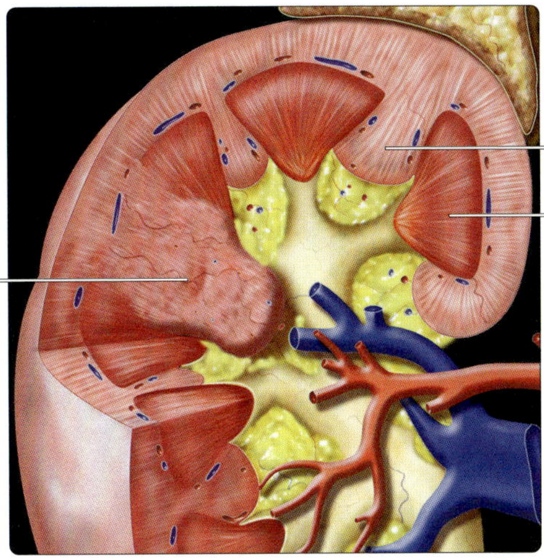

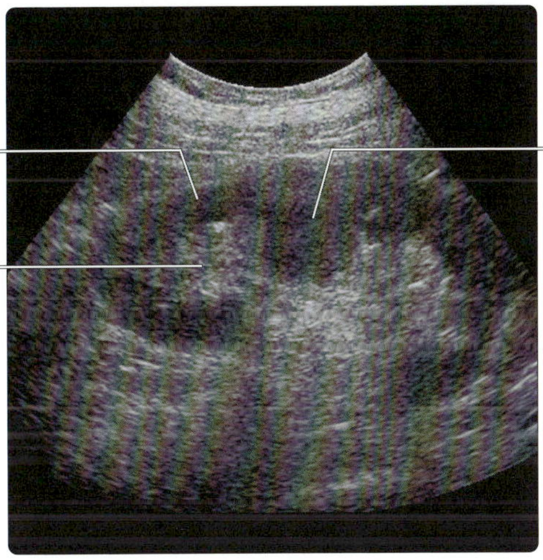

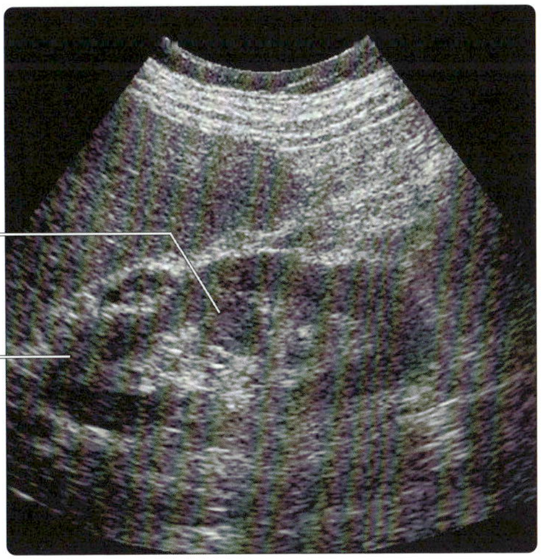

(Top) *Graphic depicts a hypertrophied column of Bertin, which is a rounded enlargement of the septal cortical tissue that separates the renal pyramids. This is normal tissue with the same imaging features as other parts of the renal cortex, but it may protrude into the renal sinus fat and be mistaken for a renal mass. This generally tends to be more of a diagnostic dilemma on ultrasound relative to either CT or MR.* **(Middle)** *Sagittal ultrasound shows a rounded "mass," a hypertrophied column of Bertin, projecting into and displacing renal sinus fat. This "mass" has the same echogenicity as other cortical tissue, allowing one to distinguish this benign entity from a true mass lesion.* **(Bottom)** *Sagittal ultrasound demonstrates a rounded, prominent "mass" of cortical tissue projecting deep into the renal sinus. Notice that this "mass" has identical echogenicity to normal renal cortex, typical for a benign column of Bertin. The location near the midpole kidney is fairly typical for this entity.*

FETAL LOBATION

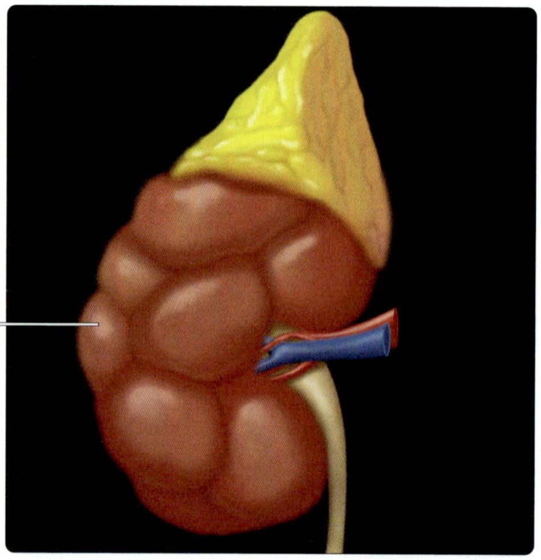

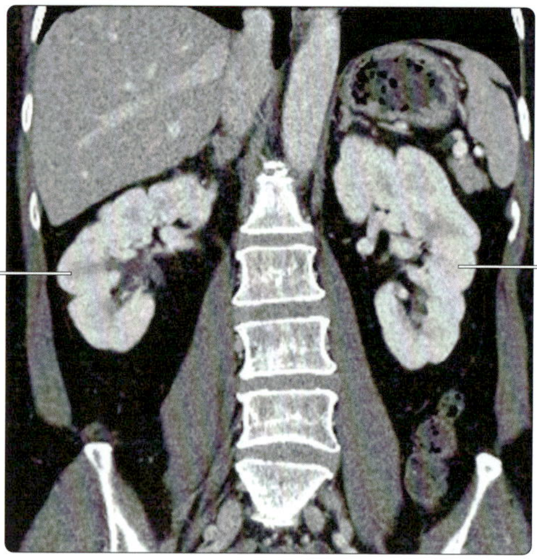

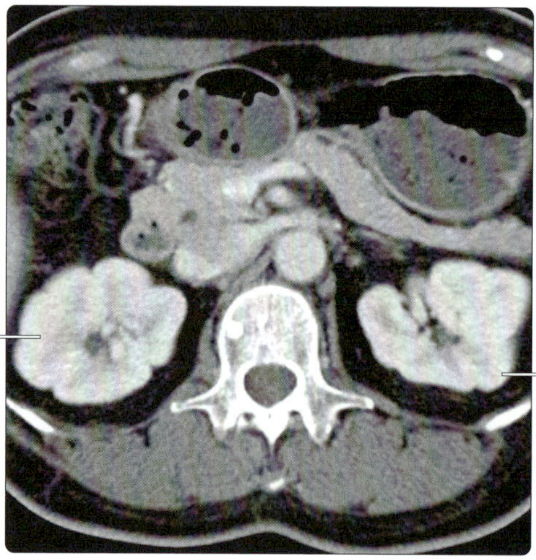

(Top) *Graphic depicts the typical lobulated appearance of the kidney in fetal life reflecting the development of the kidney from numerous lobes, each consisting of a renal pyramid and its associated cortex. This may persist into infancy and occasionally into adult life, though to a lesser degree. Note that the adrenal gland is relatively large compared with the kidney in fetal life and childhood.* **(Middle)** *Coronal CECT of an asymptomatic adult shows a lobulated surface of each kidney representing persistent fetal lobation (lobulation). This must be distinguished from cortical scarring in which renal tissue is lost as a result of ischemia or inflammation.* **(Bottom)** *Axial CECT in the same patient shows the same lobulated contour of the renal surface bilaterally with shallow clefts partially separating segments of the kidneys. Note that there is no cortical scarring or thinning and that the cortex is invaginated into each of these clefts. This is a normal variant and represents persistence of the more prominent lobulation of the fetal kidney, which is formed by coalescence of multiple renal lobes, each consisting of a renal pyramid and its associated cortex.*

DUPLICATION OF COLLECTING SYSTEM

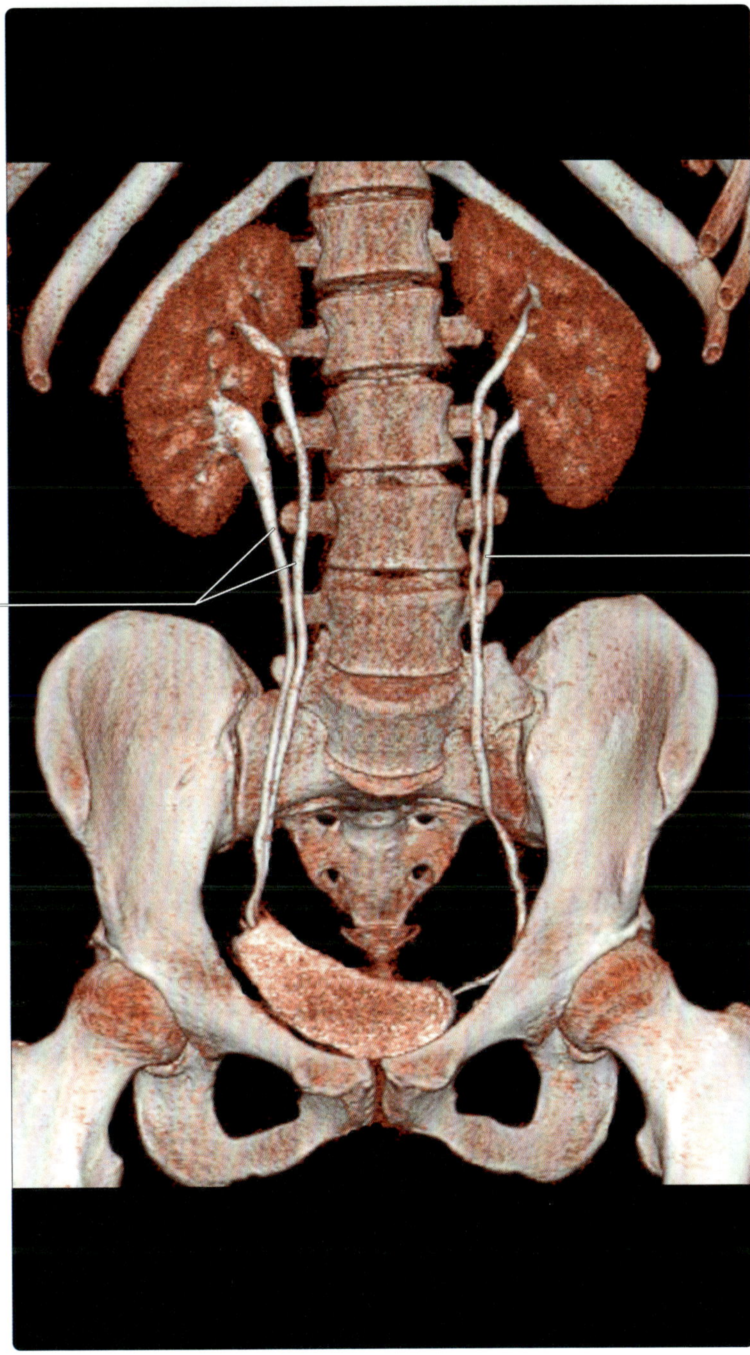

Coronal CT urogram with volume-rendered reconstruction shows a bifid left ureter with 2 separate ureters leaving the kidney but joining distally to form a single ureter. The entire right ureter is duplicated with separate ureteral orifices into the bladder. Each kidney in this subject had 2 hila with supernumerary arteries and veins.

DUPLICATED COLLECTING SYSTEM WITH URETEROCELE

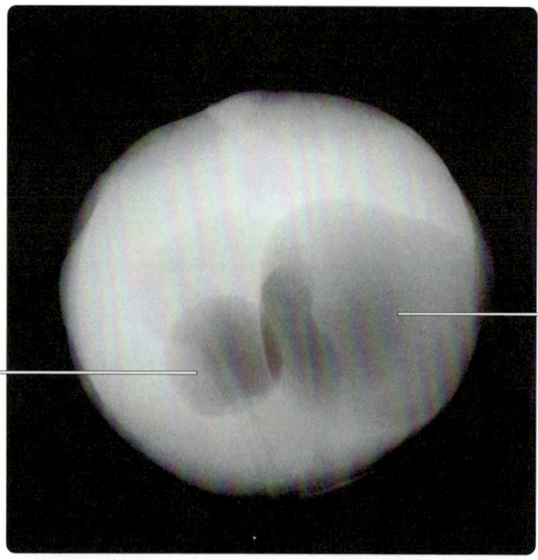

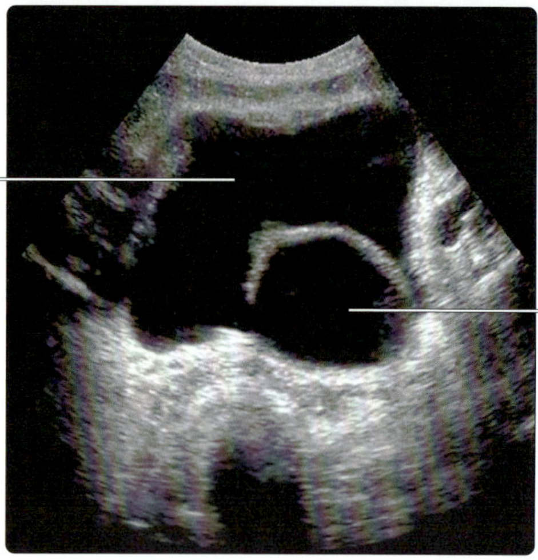

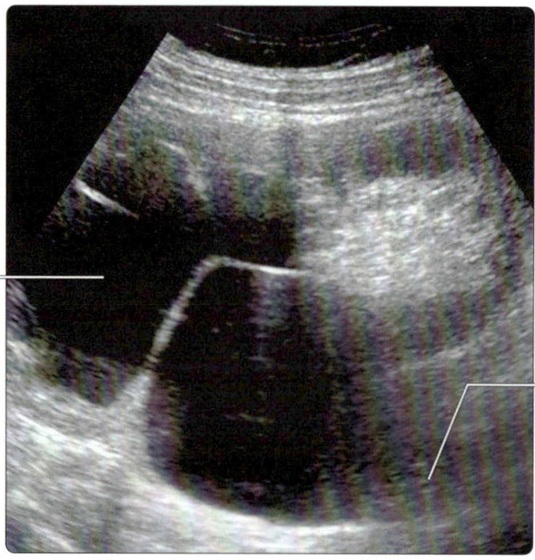

(Top) *Frontal film from a cystogram demonstrates a large, rounded filling defect in the left aspect of the bladder (adjacent to the Foley catheter balloon).* **(Middle)** *Ultrasound of the bladder in the same patient demonstrates that the rounded mass seen on the cystogram corresponds to a cystic mass in the bladder compatible with a ureterocele.* **(Bottom)** *Sagittal ultrasound in the same patient demonstrates marked severe hydronephrosis and hydroureter isolated to the left upper pole kidney (without hydronephrosis in the lower pole). These 3 images together nicely illustrate the Weigert-Meyer rule of duplicated collecting systems with the upper pole moiety frequently inserting ectopically medial and inferior to the lower pole and terminating in a ureterocele that causes obstruction.*

POLYCYSTIC KIDNEY DISEASE

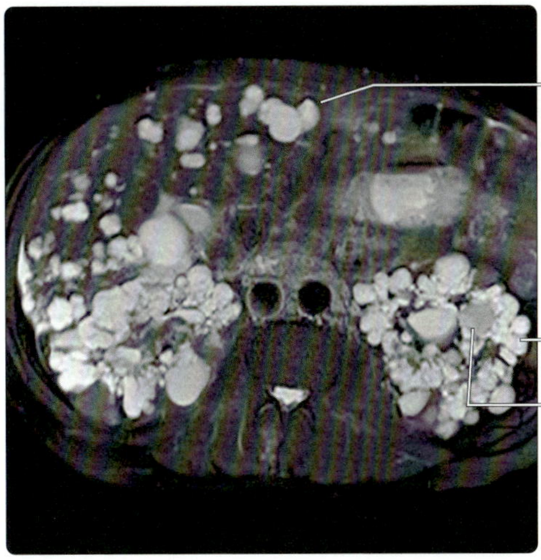

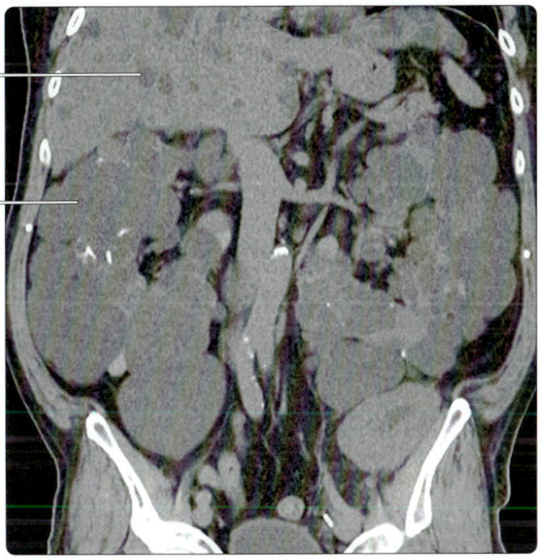

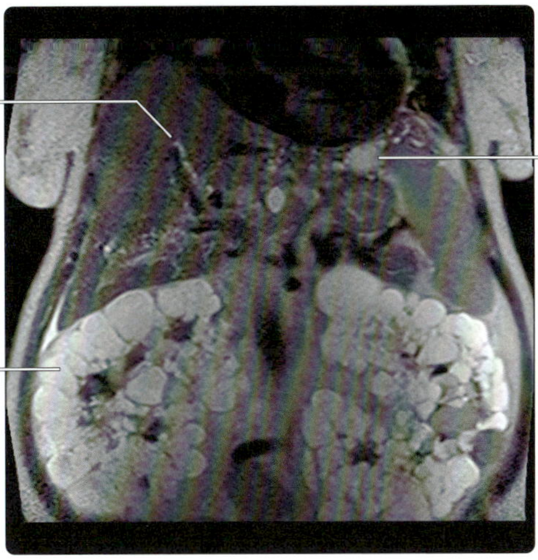

(Top) *Axial T2 FS MR demonstrates the classic imaging features of polycystic kidney disease, including innumerable cysts replacing and enlarging the bilateral kidneys and liver. Some of the cysts demonstrate slightly decreased T2 signal as a result of proteinaceous or hemorrhagic content.* **(Middle)** *Coronal NECT in a patient with polycystic kidney syndrome demonstrates extensive cystic enlargement of both kidneys as well as multiple hepatic cysts. In spite of the multiple liver cysts, hepatic function is normal here and in most patients with polycystic liver disease. Conversely, the polycystic renal disease is progressive and usually results in renal failure by the 6th decade.* **(Bottom)** *Coronal T2 MR in a patient with polycystic kidney syndrome demonstrates innumerable cysts within both kidneys. Note, in this case, the presence of cystic dilatation of the bile ducts in the liver represents superimposed Caroli disease (an entity that can be associated with polycystic kidney disease).*

GROSS ANATOMY

Ureter

- Muscular tubes (roughly 25-30 cm long) that carry urine from kidneys to bladder
    - Abdominal ureters course along posterior abdominal wall in retroperitoneum (behind parietal peritoneum)
        - Ureters arise from renal pelvis at **ureteropelvic junction** and lie in **perirenal space**
        - Ureters course along anterior surface of psoas muscles in lower abdominal retroperitoneum
    - Pelvic ureters enter pelvis at level of sacroiliac joints where they course anterior to iliac artery bifurcation and near ovaries (in women)
        - Course downward along lateral pelvic walls near internal iliac vessels
        - Near ischial spines, ureters curve anteromedially to enter bladder at level of seminal vesicles/cervix
        - **Ureterovesical junction** (**UVJ**): Ureters pass obliquely through muscular wall of bladder, creating valve effect that prevents reflux of urine
    - 3 points of physiological narrowing: **Ureteropelvic junction**, pelvic brim, and **UVJ**
- **Vessels**, **nerves**, and **lymphatics**
    - Arterial branches are numerous and variable (from aorta, renal, gonadal, internal iliac, vesical, and rectal arteries) and anastomose along length of ureter
        - Arterial supply to abdominal ureter primarily from renal artery and testicular artery/ovarian artery
        - Arterial supply to pelvic ureter from superior and inferior vesical arteries
    - Venous branches follow arteries with similar names
    - Nervous supply
        - Autonomic innervation from adjacent plexuses (renal, hypogastric, testicular/ovarian) causes peristalsis
        - Ureters carry pain/stretch receptors: Stone in abdominal ureter causes back/flank pain, while stone in pelvic ureter causes lower abdominal/groin pain
            - Sensory fibers enter spinal cord at T11-L2, causing ureteral pain to refer to that dermatomal level
    - Lymphatics to **external** and **internal iliac nodes** (pelvic ureter) and **aortocaval nodes** (abdomen)

Bladder

- Hollow, distensible viscus with strong muscular wall, which temporarily stores and expels urine
    - Bladder divided into **apex** (lies near symphysis and connects to median umbilical ligament), **body**, **base** (posterior aspect of bladder), and **neck** (joins urethra and fixed to pelvic floor by pelvic fascia and ligaments)
    - Wall of bladder composed mostly of detrusor muscle
        - **Trigone of bladder**: Triangular structure at base of bladder with apices marked by 2 **ureteral orifices** and **internal urethral orifice**
        - Internal urethral sphincter (in men only) and external urethral sphincter (men and women) are formed of muscle and located in urethral orifice
- Lies in **extraperitoneal** (**retroperitoneal**) pelvis with peritoneum covering dome of bladder
    - Reflections of peritoneum form deep recesses in pelvic peritoneal cavity
    - **Rectovesical pouch** is most dependent recess in men (and in women, following hysterectomy)
    - **Vesicouterine pouch** and **rectouterine pouch of Douglas** are most dependent recesses in women
- Bladder surrounded by retroperitoneal spaces (containing connective tissue/fat), which can expand to contain large amounts of fluid (e.g., extraperitoneal bladder rupture)
    - **Perivesical space** contains bladder and urachus
    - **Prevesical space of Retzius** between bladder and symphysis pubis communicates superiorly with **infrarenal retroperitoneal compartment** and posteriorly with presacral space
- **Vessels**, **nerves**, and **lymphatics**
    - Arterial supply primarily from internal iliac artery via **superior vesical arteries** and other branches of internal iliac arteries in both sexes
    - Venous drainage
        - Men: **Vesical** and **prostatic venous plexuses** → internal iliac and internal vertebral veins
        - Women: **Vesical** and **uterovaginal plexuses** → internal iliac vein
    - Nervous supply
        - Parasympathetic innervation from pelvic splanchnic and inferior hypogastric nerves (causes contraction of detrusor muscle and relaxation of internal urethral sphincter to permit emptying of bladder)
        - Sympathetic innervation via hypogastric nerve (relaxes bladder musculature and leads to urinary retention)
        - Somatic nervous supply via pudendal nerve (provides voluntary control over urination via external sphincter)
        - Sensory fibers follow parasympathetic nerves
    - Lymphatics to iliac chain and obturator nodes

CLINICAL IMPLICATIONS

Clinical Importance

- Ureters are often injured inadvertently during abdominal or gynecological surgery due to traction
- **Ectopic ureter**
    - **Weigert-Meyer** rule: Ureter from upper renal pole inserts inferior and medial to lower pole ureter
    - Upper pole ureter often becomes obstructed and inserts ectopically (not at trigone)
- **Ureterocele**: Cystic dilation of distal ureter can be simple (orthotopic) at trigone or ectopic (inserts below trigone)
- **Extraperitoneal bladder rupture**
    - Most often due to trauma with pelvic fractures
    - Urine distends prevesical space (molar tooth sign) and may track posteriorly into presacral space and superiorly into retroperitoneal abdomen
- **Intraperitoneal bladder rupture**
    - Usually due to blunt trauma with distended bladder
    - Urine flows up paracolic gutters into peritoneal recesses and surrounds small bowel
- Fetal **urachus** forms conduit between umbilicus and bladder and usually becomes obliterated, forming **median umbilical ligament**
    - May persist as "cyst" or diverticulum, which can rarely become infected or develop carcinoma
- **Bladder diverticula** can lead to infection, stones, tumor
    - Congenital diverticula typically near UVJ, while acquired diverticula usually due to bladder outlet obstruction

URETERS, BLADDER, AND VESSELS

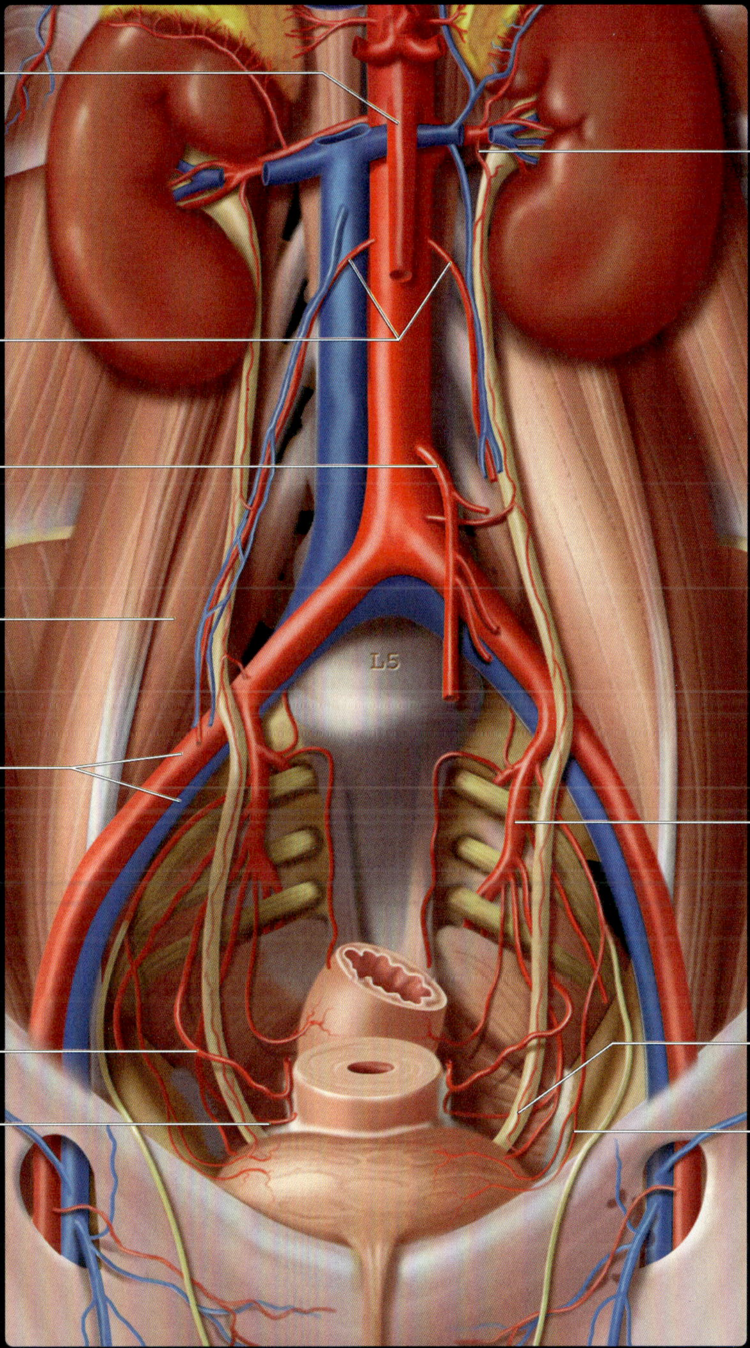

The ureters receive numerous and highly variable arterial branches from the aorta and the renal, gonadal, and internal iliac arteries. These vessels are short and can be easily ruptured by retraction of the ureter during surgical procedures. The arterial supply to the bladder is also quite variable. Both sexes receive supply from the superior vesical arteries and from various branches of the internal iliac arteries. Branches to the prostate and seminal vesicles (in men) also send branches to the inferior bladder wall. In women, branches to the vagina send arteries to the base of the bladder. Note how the ureters deviate anteriorly as they cross the external (or common) iliac vessels and pelvic brim. This may constitute a point of relative narrowing where the passage of ureteral calculi (stones) may be impeded. In the abdomen, the ureters course along the psoas muscles.

FEMALE AND MALE BLADDERS

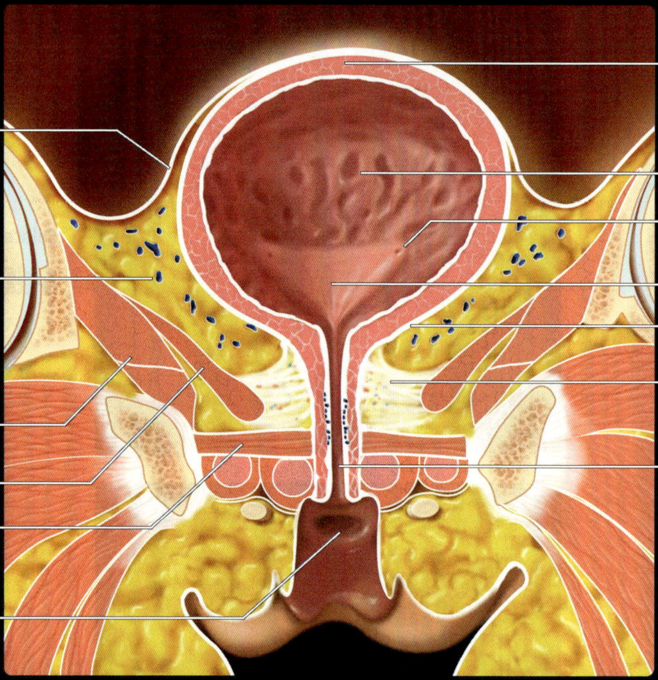

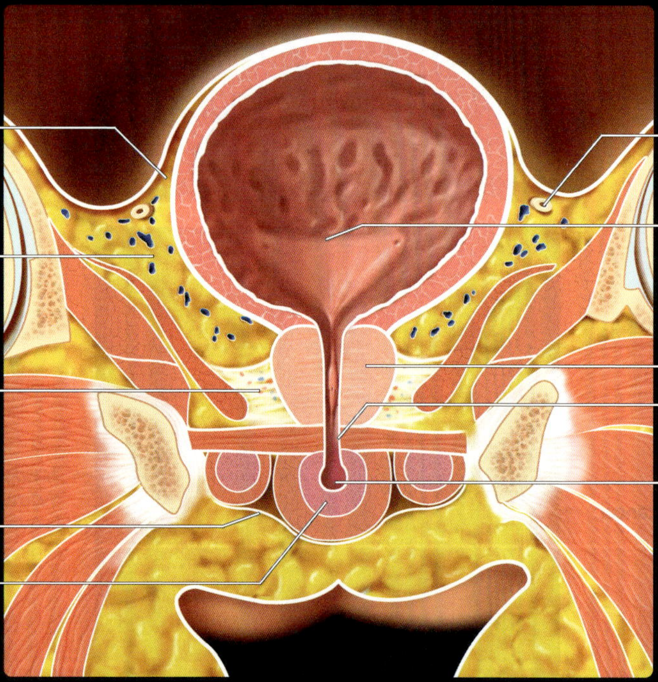

(Top) *Frontal (coronal) section of the female bladder shows that it rests almost directly on the muscular floor of the pelvis. The dome of the bladder is covered with peritoneum. The trigone is the distinct triangular base of the bladder whose apices are formed by the ureteral and urethral orifices. The bladder is surrounded by a layer of loose fat and connective tissue (the prevesical and perivesical spaces) that communicate superiorly with the retroperitoneum.* **(Bottom)** *Coronal section of the male bladder shows that it rests on the prostate, which separates it from the muscular pelvic floor. The bladder wall is muscular, strong, and very distensible. The ureters enter the bladder through an oblique anteromedial course that helps to prevent urinary reflux into the ureters. The mucosal surface of the ureters is continuous with the bladder and is the same transitional cell type.*

INTRAPERITONEAL BLADDER RUPTURE

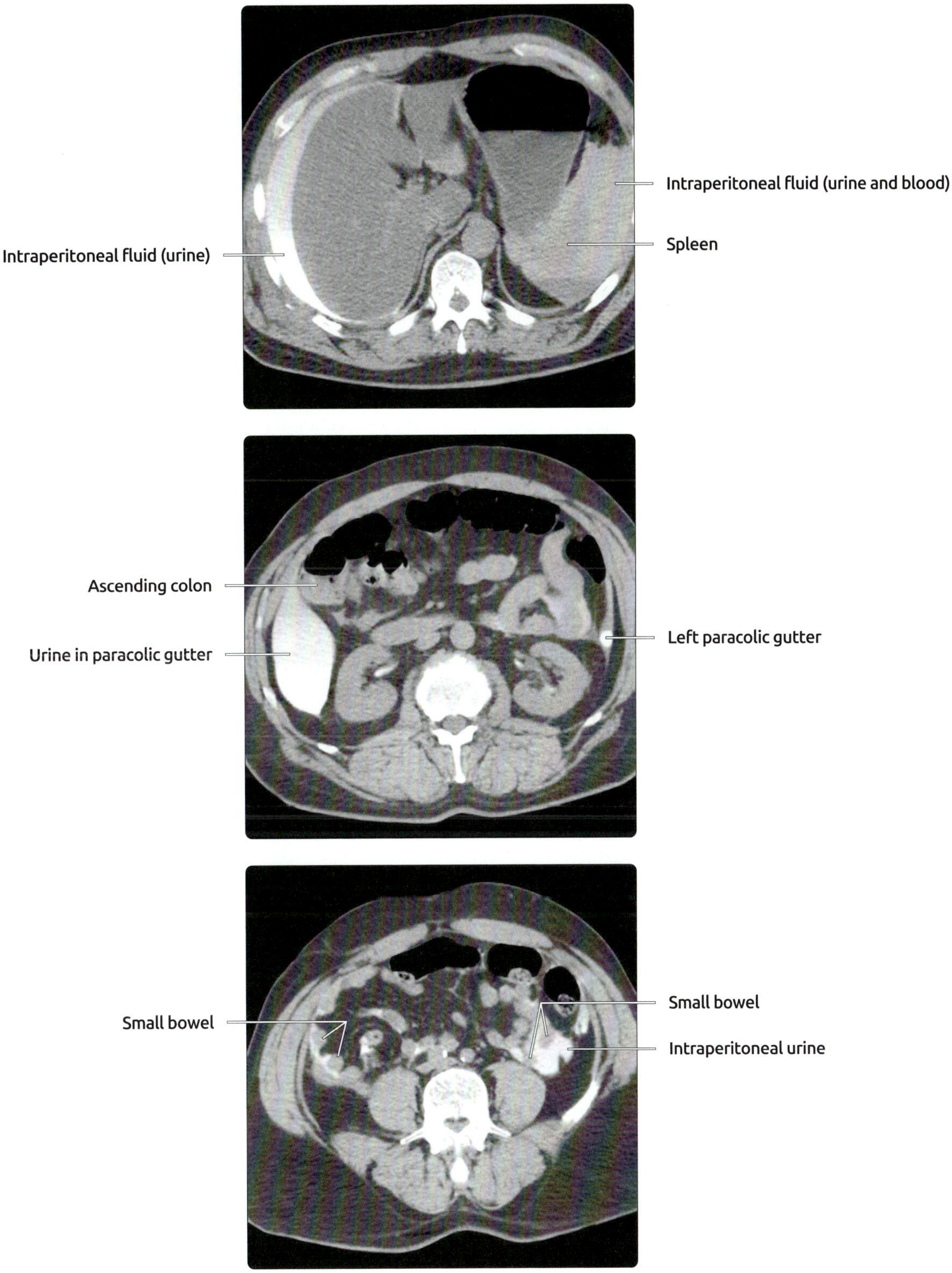

(Top) *First of 6 axial CT cystograms shows intraperitoneal rupture of the urinary bladder. The CT was obtained after instillation of contrast medium into the bladder. This section shows dense intraperitoneal fluid (urine) surrounding the liver in the right subphrenic space. The fluid surrounding the spleen is diluted by blood.* **(Middle)** *The right paracolic gutter is distended with intraperitoneal urine. Note that the right paracolic gutter is more dependent and more voluminous than the left.* **(Bottom)** *A more caudal section shows dense urine surrounding small intestinal loops, confirming the intraperitoneal location of the fluid.*

INTRAPERITONEAL BLADDER RUPTURE

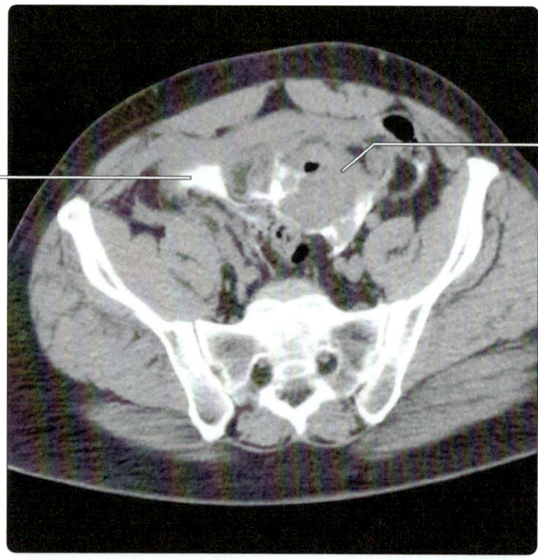

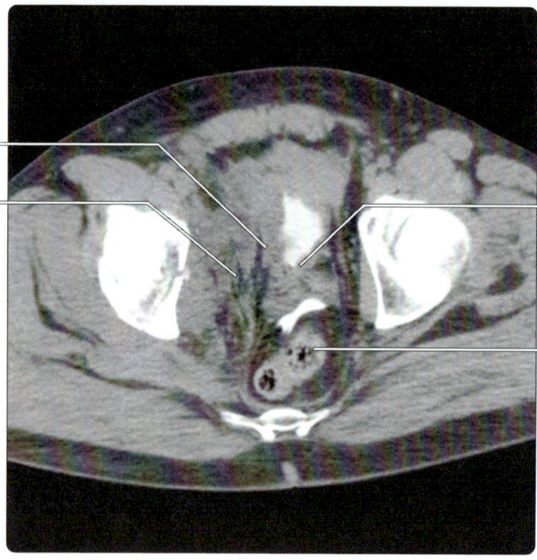

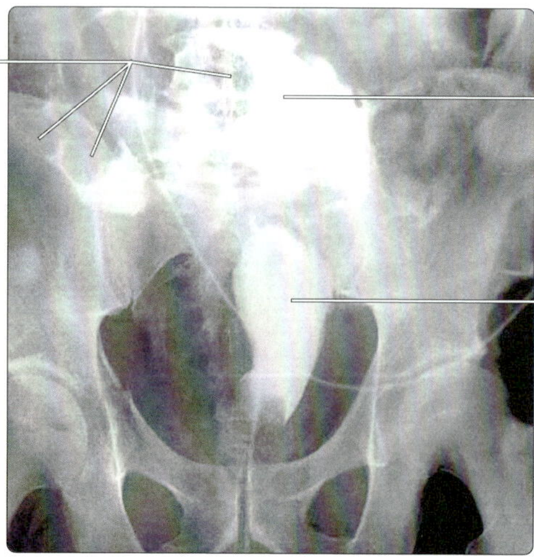

(Top) *A section near the dome of the bladder shows a hematoma surrounded by extravasated urine.* **(Middle)** *Densely opacified urine settles into the rectovesical recess, the deepest intraperitoneal recess. The bladder is displaced to the left by the pelvic hematoma.* **(Bottom)** *A frontal film from the cystogram that preceded the CT shows rupture of the dome of the bladder with intraperitoneal spill of opacified urine/contrast medium. Note the absence of contrast medium in the perivesical and prevesical (extraperitoneal) spaces and the contrast medium outlining small intestinal loops, seen as "filling defects" in the pool of contrast medium. The urinary bladder is displaced to the left by a pelvic hematoma.*

INTRAPERITONEAL BLADDER RUPTURE

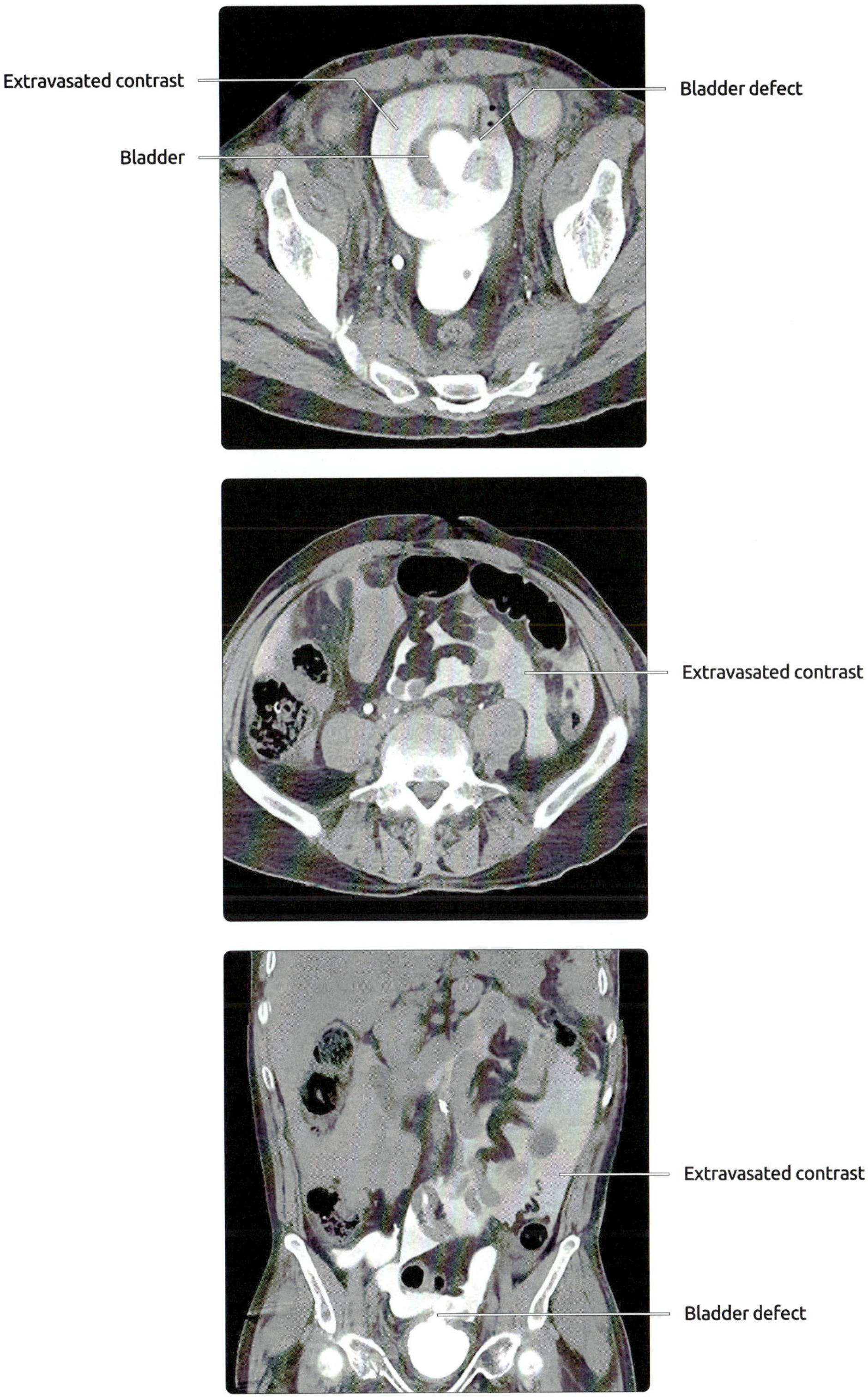

(Top) *Axial CT cystogram in a trauma patient demonstrates a defect in the wall of the bladder dome with extravasated contrast extending outward.* **(Middle)** *Another more superior image from the same patient now demonstrates extravasated high-density contrast surrounding multiple loops of small bowel, confirming the intraperitoneal location of the contrast.* **(Bottom)** *Coronal CT cystogram from the same patient nicely demonstrates the features of intraperitoneal bladder rupture in this case, including the defect in the bladder dome (most common location in intraperitoneal bladder rupture) and contrast extending upward into the intraperitoneal space to outline multiple loops of small bowel. Intraperitoneal bladder rupture is most common in the setting of blunt trauma to a distended bladder.*

EXTRAPERITONEAL BLADDER RUPTURE

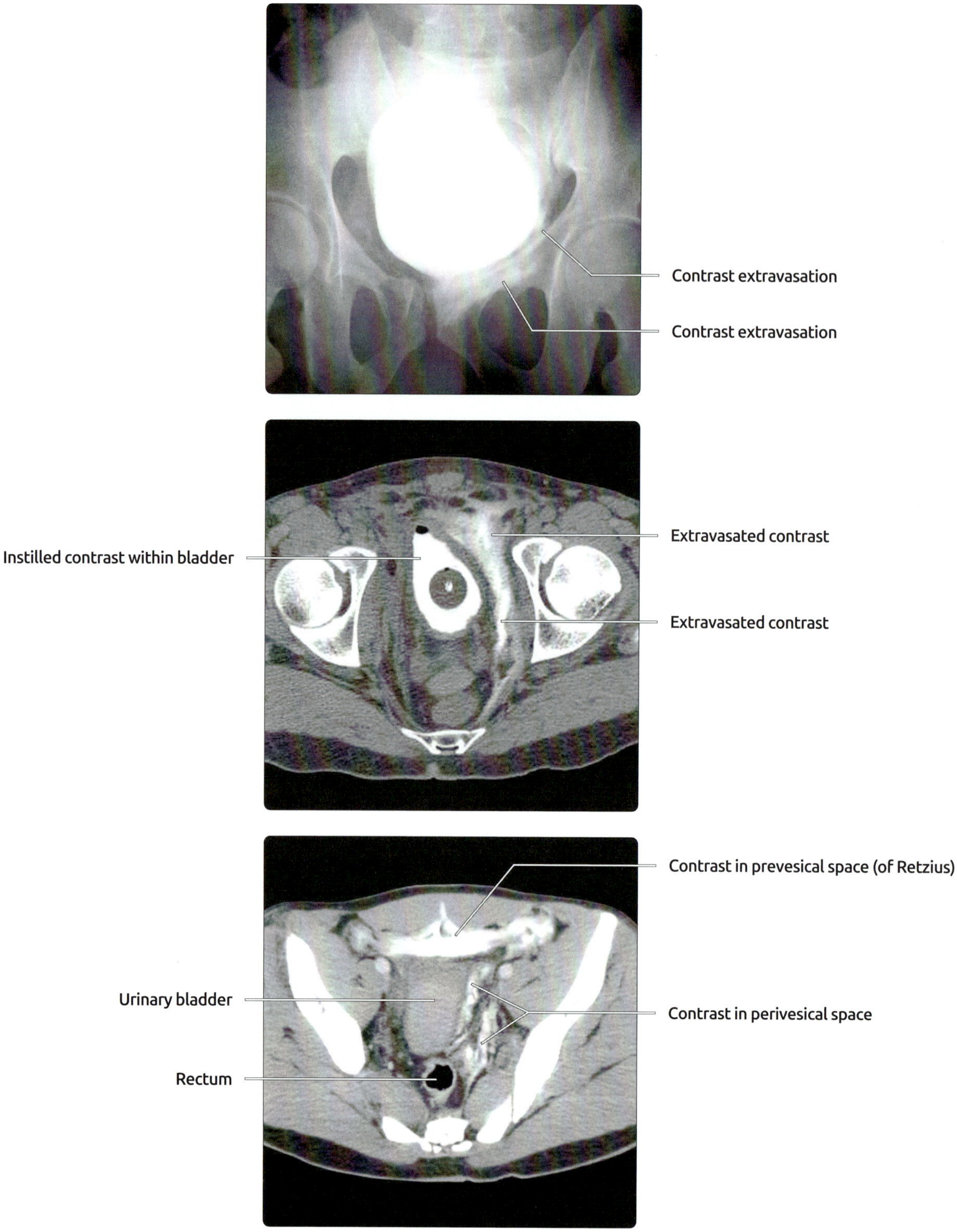

(Top) *Frontal cystogram in a trauma patient demonstrates extraluminal contrast in the extraperitoneal space surrounding the bladder.* **(Middle)** *CT cystogram in the same patient demonstrates extravasated contrast in the extraperitoneal space surrounding the bladder, compatible with extraperitoneal bladder rupture.* **(Bottom)** *Axial CT cystogram in another trauma patient with extraperitoneal bladder rupture demonstrates contrast within the prevesical space of Retzius, the extraperitoneal compartment that separates the bladder from the anterior abdominal wall and symphysis pubis. Contrast material is also present in the perivesical space, immediately surrounding the bladder.*

EXTRAPERITONEAL BLADDER RUPTURE

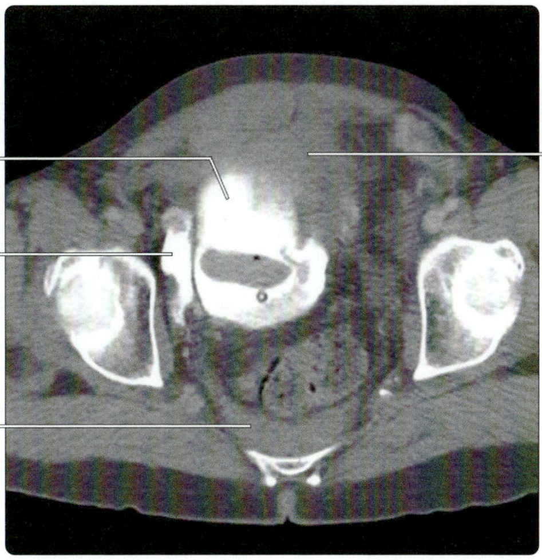

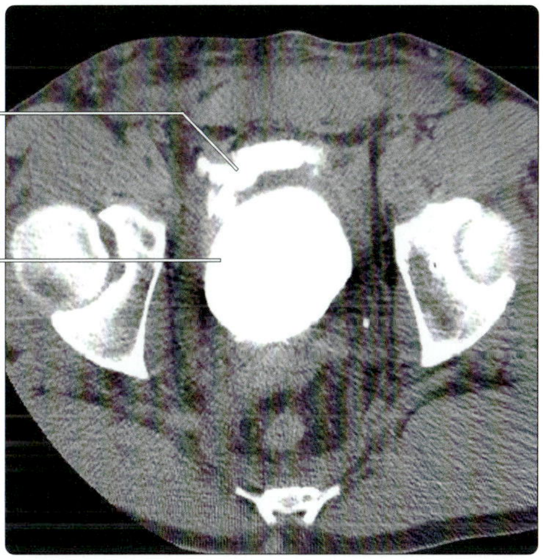

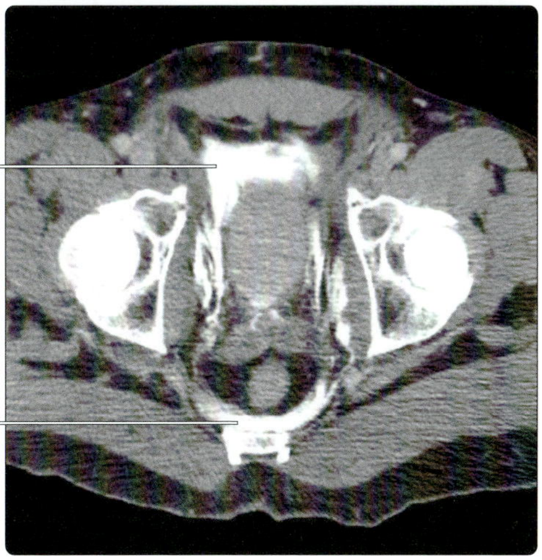

(Top) *Axial CT cystogram in a trauma patient with pubic symphysis diastasis (not shown) demonstrates extraperitoneal extravasation of urine. Unopacified urine or blood is also seen in the prevesical and presacral spaces.* **(Middle)** *Axial CT cystogram in a trauma patient demonstrates direct extravasation of contrast into the extraperitoneal space, in this case, the retropubic space of Retzius. Extraperitoneal bladder ruptures are the most common type of bladder rupture and are often associated with pelvic bone injuries. Note that the extravasated contrast material is irregular and flame-shaped or feathery in nature. Conversely, in intraperitoneal ruptures, the extravasated contrast material usually outlines bowel loops and the paracolic gutters.* **(Bottom)** *Axial CT cystogram in a patient with extraperitoneal bladder rupture demonstrates the classic molar tooth sign with contrast surrounding the bladder in the retropubic space of Retzius and along the pelvic side walls in the shape of a molar tooth. Note the additional extravasated contrast in the extraperitoneal presacral space.*

URACHAL CYST

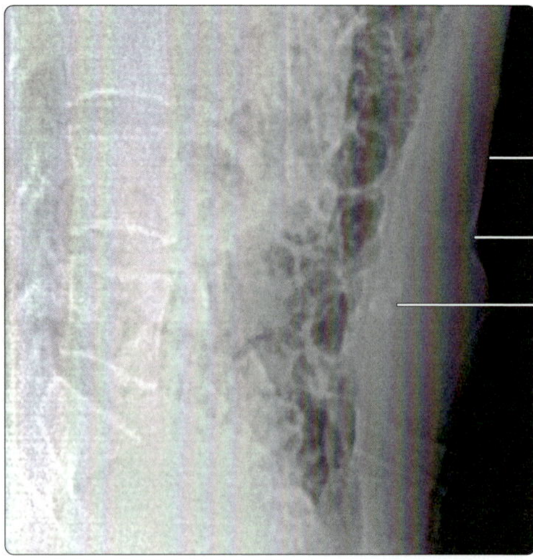

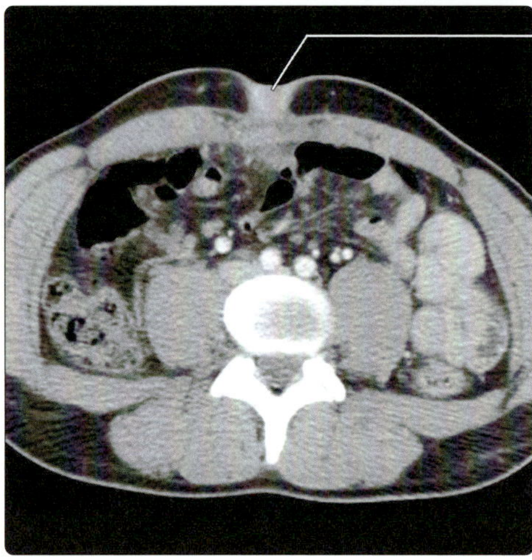

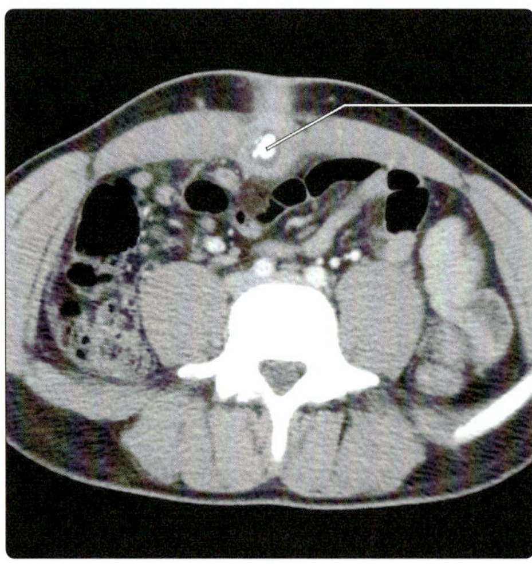

(Top) *Lateral plain radiograph in a patient who presented with infraumbilical pain and a palpable, tender mass demonstrates a faint calcification in the abdominal wall just inferior to the umbilicus.* **(Middle)** *Axial CECT in the same patient shows the umbilicus and subtle infiltration of the abdominal fat just deep to it.* **(Bottom)** *More caudal axial CECT in the same patient shows a calcified lesion within a rounded focus of inflammation. This is a typical site for the urachus. Chronic inflammation, which can occur within a urachal cyst, often develops calcification.*

URACHAL CYST

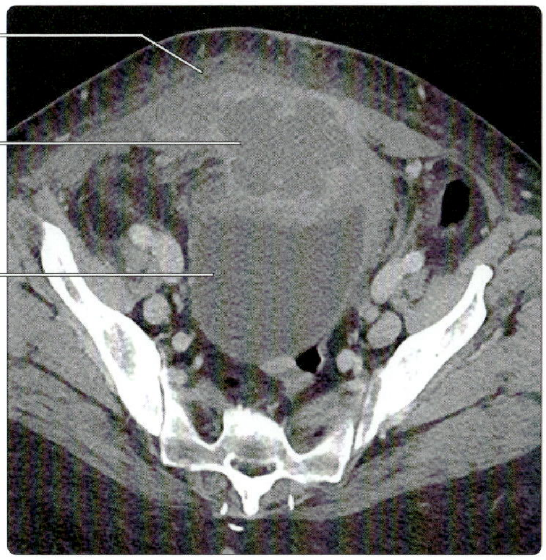

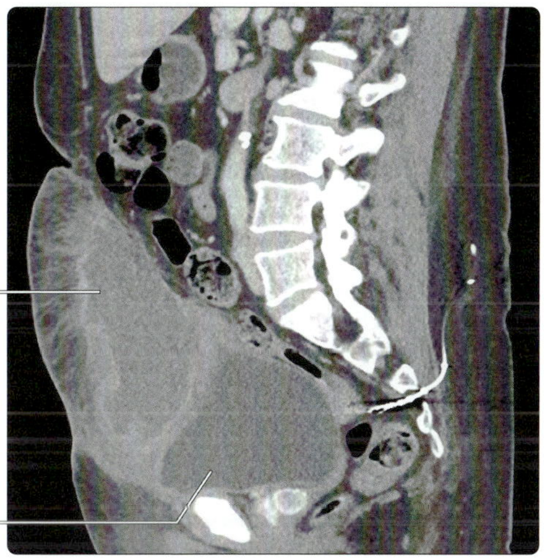

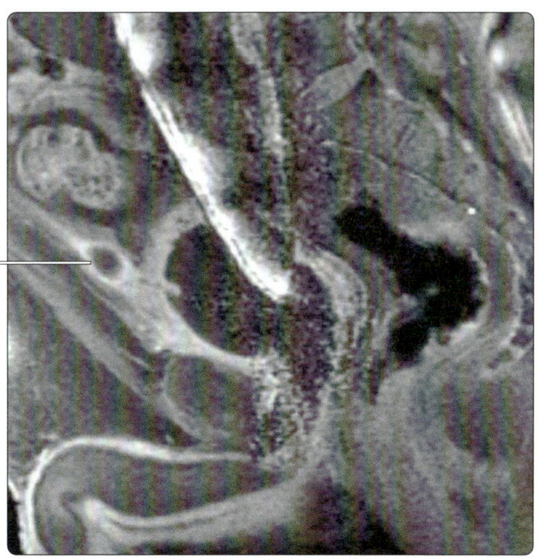

(Top) *Axial CECT in a patient with signs of infection demonstrates a large complex fluid collection in the midline anterior abdomen extending down toward the bladder with associated marked inflammation and edema in the adjacent anterior pelvic wall.* **(Middle)** *Sagittal CECT in the same patient demonstrates that this large fluid collection extends from the umbilicus downward to the anterior aspect of the bladder dome, in the expected location of the urachus. The patient was taken to surgery, as the possibility of a urachal adenocarcinoma could not be excluded, and this was found to be a large abscess arising from a urachal cyst.* **(Bottom)** *Sagittal T2 FS MR shows a midline cystic structure above the bladder dome, characteristic for the relationship of a urachal cyst to the bladder dome. A urachal cyst results when the bladder and umbilical ends of the fetal urachus close, but the central part of the tract remains open, as in this case.*

URACHAL CANCER

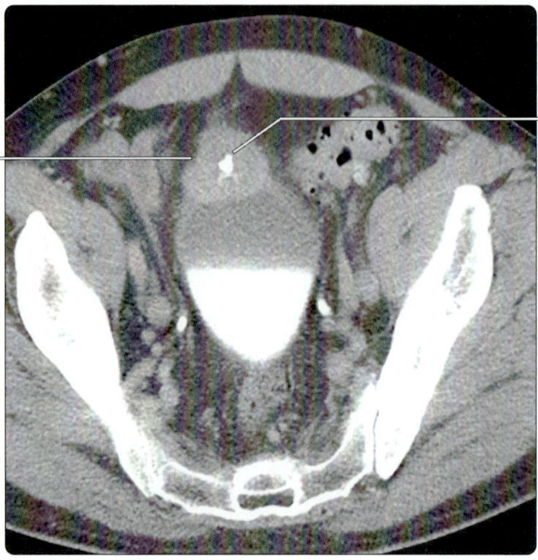

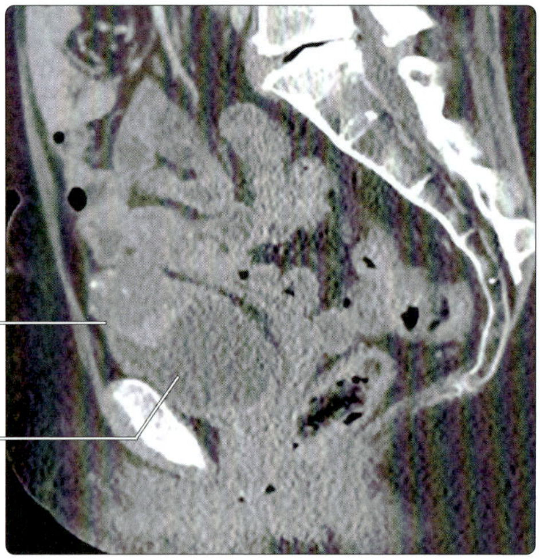

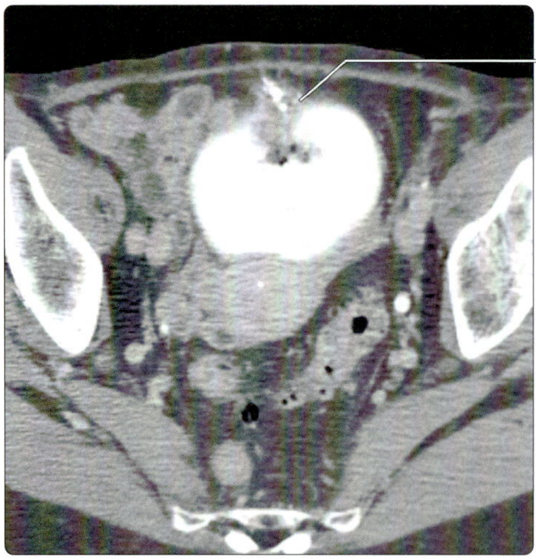

(Top) *Axial CECT demonstrates a mass that arises from the dome of the urinary bladder with internal calcification, a typical location and appearance for a urachal malignancy.* **(Middle)** *Sagittal CECT demonstrates a mass arising from the anterior dome of the bladder in close proximity to the expected location of the urachus, compatible with a urachal cancer. There is subtle speckled calcification within the lesion, a fairly typical feature of these tumors.* **(Bottom)** *Axial CECT demonstrates a mass involving the anterior superior wall of the bladder with internal calcification. A urachal carcinoma was excised at surgery along with a portion of the bladder.*

BLADDER DIVERTICULA AND STONES

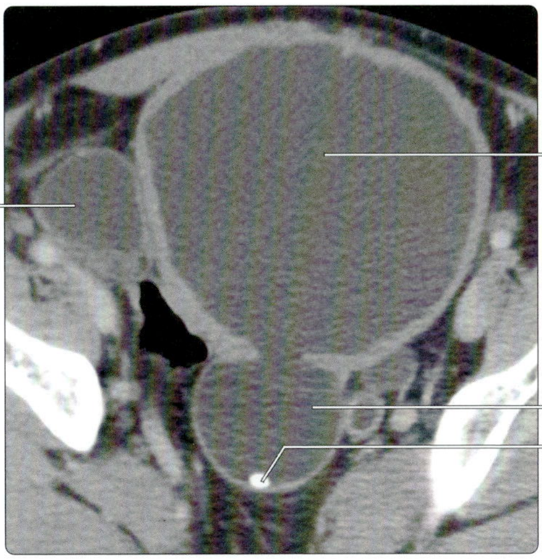

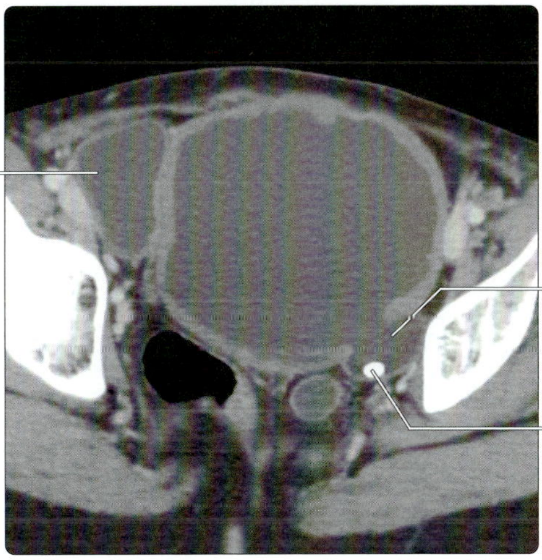

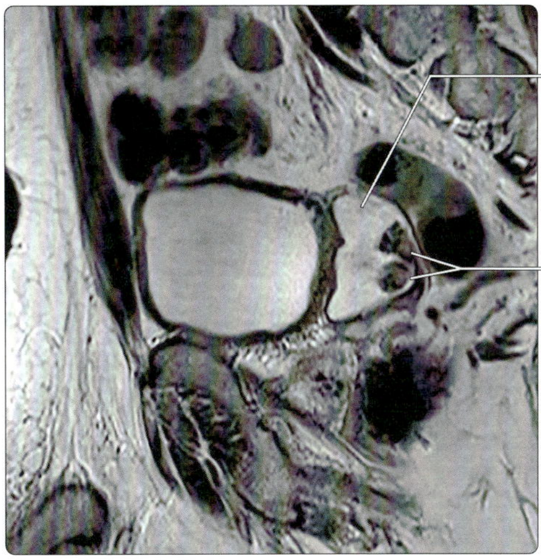

(Top) *First of 2 axial CECT images of an older man with hematuria and dysuria demonstrates multiple outpouchings from the urinary bladder that represent bladder diverticula. A calculus is present in the dependent position within one of the diverticula.* **(Middle)** *Another axial CECT in the same patient demonstrates additional diverticula and stones. Bladder diverticula can be either congenital or secondary to longstanding bladder outlet obstruction or neurogenic bladder.* **(Bottom)** *Sagittal T2 MR demonstrates a large "Hutch" bladder diverticulum extending posteriorly from the bladder with 2 internal T2-hypointense stones. Stasis of urine within bladder diverticula increases the chances of developing internal stones.*

BLADDER DIVERTICULUM WITH CARCINOMA

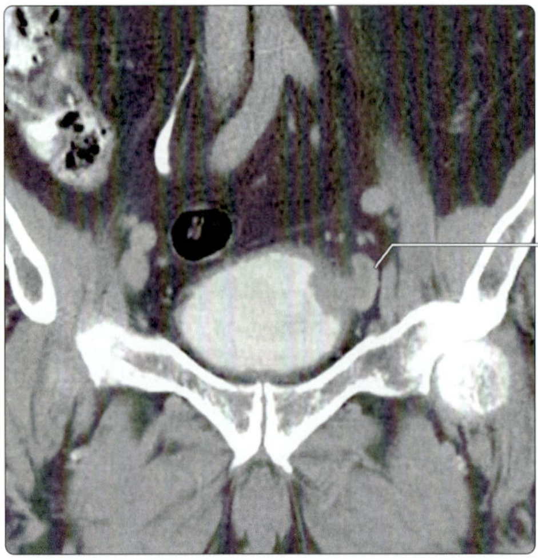

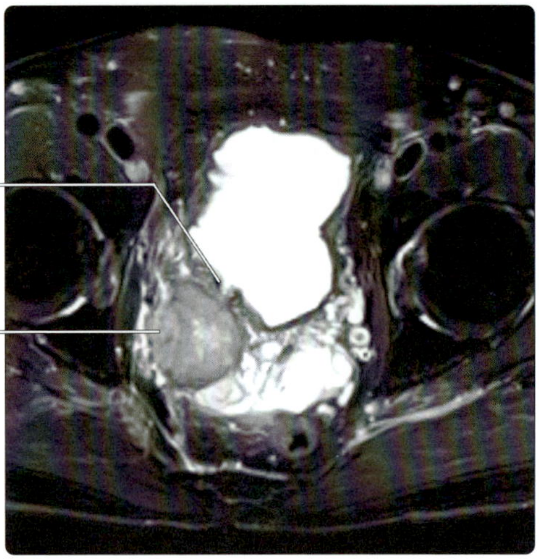

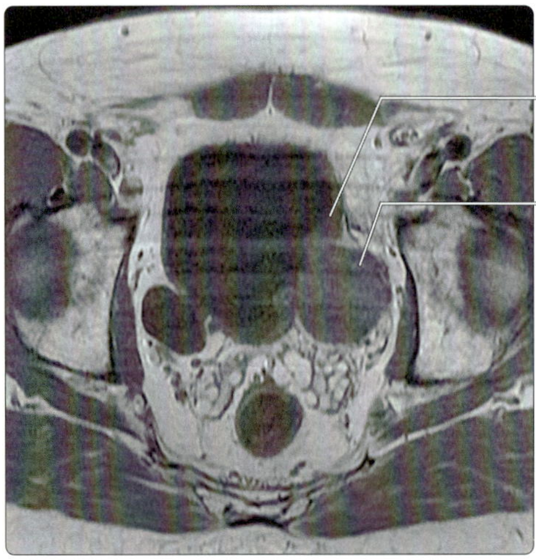

(Top) *Coronal CECT demonstrates a left-sided diverticulum with an internal soft tissue mass nodule representing a bladder diverticulum with internal urothelial cancer. Stasis of urine within bladder diverticula does increase the risk of developing a malignancy.* **(Middle)** *Axial T2 FS MR demonstrates a large mass arising from (and filling) a right-sided bladder diverticulum. While this could initially be confused with a mass arising outside of the bladder, the neck of a bladder diverticulum is visible, making it clear that this is a bladder cancer arising within a diverticulum.* **(Bottom)** *Axial T1 MR demonstrates a soft tissue mass filling a left-sided bladder diverticulum with additional extension into the main bladder itself.*

BLADDER CANCER

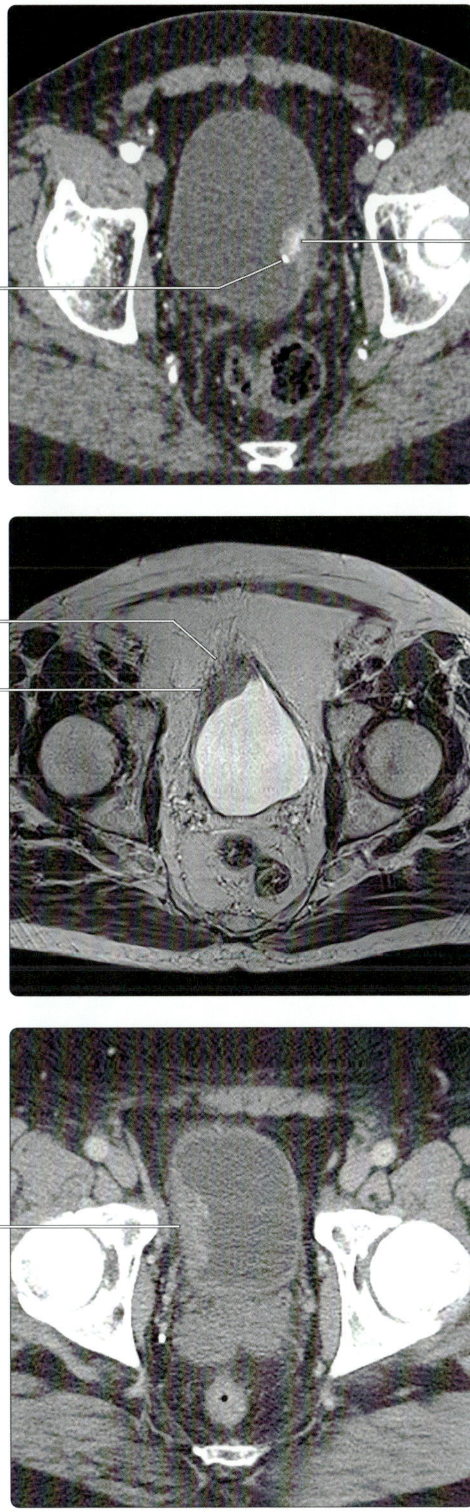

(Top) *Axial CECT in a patient with hematuria demonstrates an enhancing polyploid nodule with evidence of calcification along its surface, compatible with a bladder cancer. Although cystoscopy is certainly a better examination, CT can often identify many bladder cancers as sites of polyploid nodularity or wall thickening, particularly when the bladder is well distended.* **(Middle)** *Axial T2 MR demonstrates a bladder cancer along the anterior margin of the bladder with intermediate T2 hyperintensity. Note the infiltration of the tumor into the soft tissues anterior to the bladder, compatible with extramural spread.* **(Bottom)** *Axial CECT demonstrates a site of focal, mass-like wall thickening along the right lateral margin of the bladder. Any focal thickening in the bladder, particularly when the bladder is well distended, should raise concern for malignancy and prompt cystoscopy.*

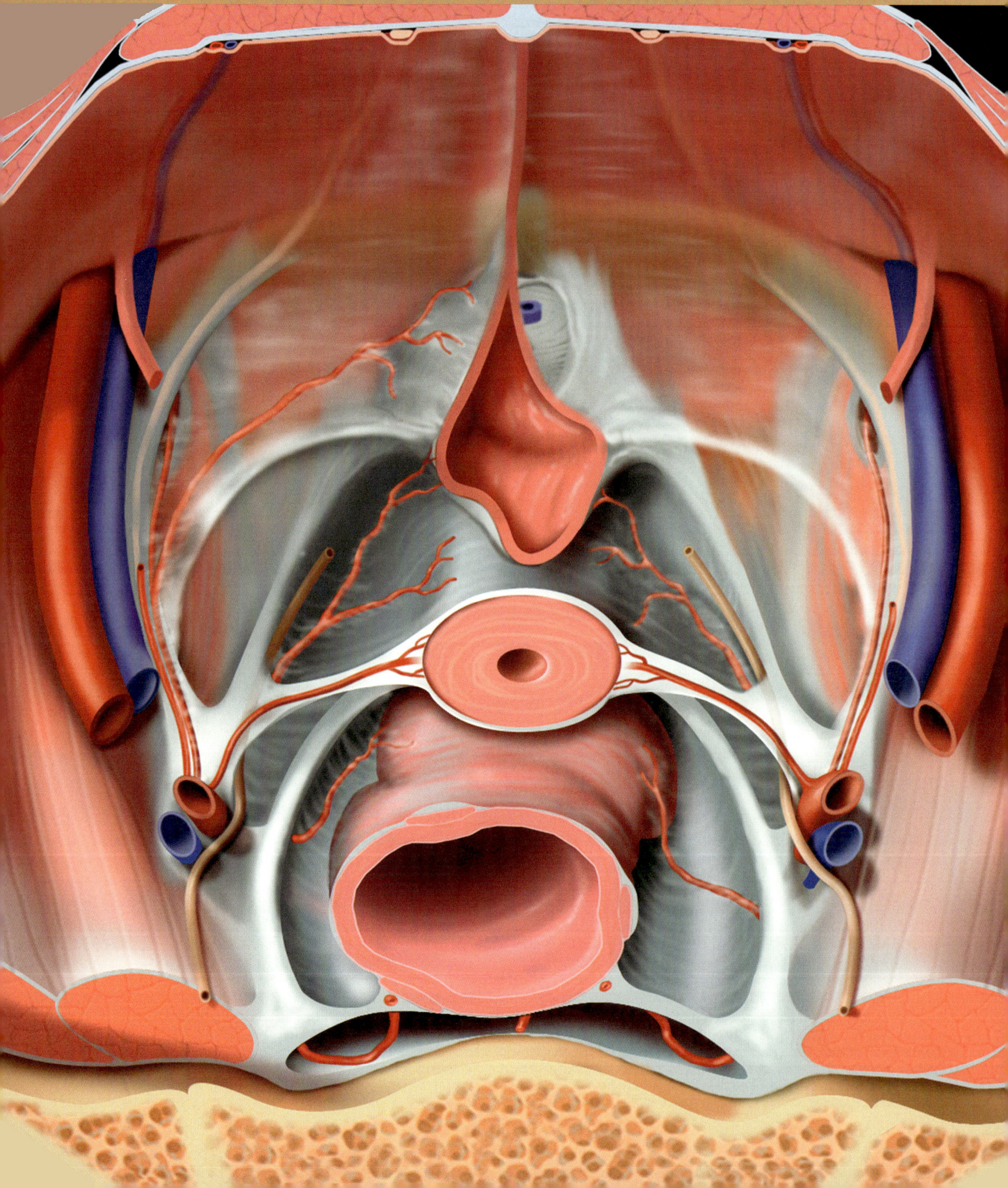

SECTION 3

Pelvis

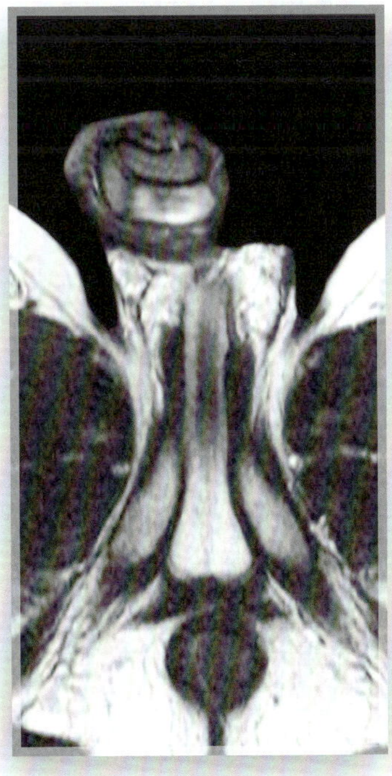
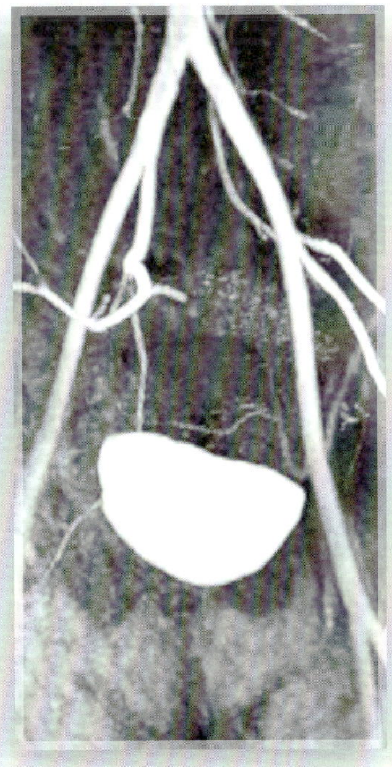
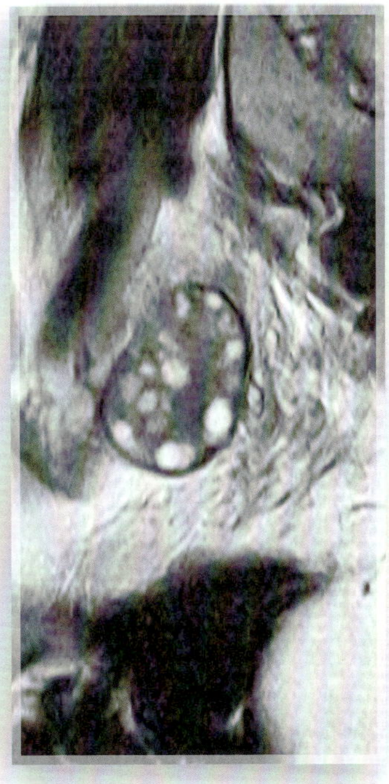

IMAGING ANATOMY

Arteries

- **Abdominal aorta**
  - **Testicular** and **ovarian arteries** originate below renal arteries
  - **Median** (**middle**) **sacral artery** is small, unpaired branch from posterior aspect of distal aorta
  - Divides into common iliac arteries at L4-5
- **Common iliac arteries**
  - Run anterior to iliac veins and inferior vena cava (IVC)
  - Usually no major branches
    - Rarely gives off aberrant iliolumbar or accessory renal arteries
  - ~ 4 cm long
- **External iliac artery**
  - No major branches
  - Exits pelvis beneath inguinal ligament
  - Larger than internal iliac artery
  - **Inferior epigastric** (medial) and **deep iliac circumflex** (lateral) arteries demarcate junction between external iliac and common femoral arteries
- **Internal iliac** (**hypogastric**) **artery**
  - Principal vascular supply of pelvic organs
  - Divides into anterior and posterior trunk
    - Anterior trunk to pelvic viscera
    - Posterior trunk to pelvic musculature
- **Anterior trunk** of **internal iliac artery**
  - **Branching pattern quite variable**
  - **Umbilical artery**
    - Only pelvic segment remain patent after birth
    - Remainder becomes fibrous **medial umbilical ligaments**
  - **Obturator artery**
    - Exits pelvis through obturator canal to supply medial thigh muscles
  - **Superior vesicle artery**
    - Supplies bladder and distal ureter
    - Gives off branch to ductus deferens in males
  - **Inferior vesicle artery** (male)
    - May arise from middle rectal artery
    - Supplies prostate, seminal vesicles, and lower ureters
  - **Uterine artery** (female)
    - Courses through cardinal ligament, at base of broad ligament
    - Passes over ureter at level of cervix ("water under bridge")
    - Anastomoses with vaginal and ovarian arteries
  - **Vaginal artery** (women)
  - **Middle rectal artery** runs above pelvic floor and anastomoses with superior and inferior rectal arteries to supply rectum
    - Superior rectal artery continuation of inferior mesenteric artery
    - Inferior rectal artery is branch of internal pudendal artery and runs below pelvic floor
    - Also anastomoses with inferior vesicle artery to help supply prostate and seminal vesicles
  - **Internal pudendal artery**
    - Supplies external genitalia (penis, clitoris) and rectum
    - Long complex course
    - Exits pelvis through greater sciatic foramen
    - Curves around sacrospinous ligament to enter perineum through lesser sciatic foramen
  - **Inferior gluteal** (**sciatic**) **artery**
    - Largest and terminal branch of anterior division of hypogastric artery
    - Passes between either 1st and 2nd or 2nd and 3rd sacral nerves
    - Exits pelvis below piriformis muscle
    - Supplies muscles of pelvic floor, thigh, buttocks, and sciatic nerve
- **Posterior trunk of internal iliac artery**
  - **Iliolumbar artery**
    - Ascends laterally to supply iliacus, psoas, and quadratus lumborum muscles
  - **Lateral sacral artery**
    - Runs medially toward sacral foramina to anastomose with middle sacral artery
  - **Superior gluteal artery**
    - Largest and terminal branch of posterior division
    - Exits pelvis through greater sciatic foramen, superior to piriformis muscle
    - Supplies piriformis and gluteal muscles

Veins

- **External iliac vein**
  - Upward continuation of femoral vein at level of inguinal ligament
  - Receives inferior epigastric, deep iliac circumflex, and pubic veins
- **Internal iliac vein** begins near upper part of greater sciatic foramen
  - Gluteal, internal pudendal, and obturator veins have origins outside pelvis
  - Pelvic viscera drain into multiple, deep pelvic venous plexuses
    - These drain into veins, which roughly parallel pelvic arteries
- **Right gonadal vein** drains into IVC, **left gonadal vein** drains into left renal vein
- **Common iliac vein** is formed by union of external and internal iliac veins
  - Unites with contralateral side to form IVC

Lymphatics

- **Superficial inguinal nodes**
  - In superficial fascia parallel to inguinal ligament, along cephalad portion of greater saphenous vein
  - Receive lymphatic drainage from superficial lower extremity, superficial abdominal wall, and perineum
  - Flow into deep inguinal and external iliac nodes
- **Deep inguinal nodes**
  - Along medial side of femoral vein, deep to fascia lata and inguinal ligament
  - Receive lymphatic drainage from superficial inguinal and popliteal nodes
  - Flow into external iliac nodes
  - Deep inguinal node in femoral canal is called gland or node of Cloquet
- **External iliac nodes**

- Along external iliac vessels
- Primary drainage from inguinal nodes
- Flow into common iliac nodes
- **Internal iliac nodes**
  - Along internal iliac vessels
  - Drainage from inferior pelvic viscera, deep perineum, and gluteal region
  - Flow into common iliac nodes
- **Common iliac nodes**
  - Along common iliac vessels
  - Drainage from external and internal iliac and sacral nodes
  - Flow into lumbar (lateral aortic) chain of nodes

Nerves

- **Sacral plexus**
  - Coalescence of lumbar and sacral nerves
    - L4 (minor branch), L5
    - S1-S3 (ventral rami), S4 (minor branch)
- **Sciatic nerve**
  - Upper band and major branch of sacral plexus
  - Runs on ventral surface of piriformis muscle
  - Broadest nerve in body (2 cm)
  - **Exits pelvis through greater sciatic foramen**, below piriformis
  - Innervates capsule of hip joint, posterior thigh, and leg muscles
- **Pudendal nerve**
  - Lower band of sacral plexus
    - Ventral rami of S2-S4
  - Follows internal pudendal artery
    - Exits pelvis through greater sciatic foramen
    - Curves around sacrospinous ligament to enter perineum through lesser sciatic foramen
    - **Pudendal canal** on lateral wall of ischiorectal fossa
  - Major nerve supply to perineum and external anal sphincter
- **Obturator nerve**
  - Anterior branches of L2-L4 ventral rami
  - Follows obturator vessels, exiting pelvis through obturator foramen
  - Innervates hip adductors
- **Femoral nerve**
  - Posterior branches of L2-L4 ventral rami
  - Descends through psoas muscle into iliopsoas groove
  - Exits pelvis beneath inguinal ligament through **femoral canal**
    - Most lateral structure in femoral canal
    - Acronym for **femoral canal** contents from lateral to medial = NAVL (nerve, artery, vein, lymphatic)
  - Branches innervate anteromedial thigh

ANATOMY IMAGING ISSUES

Imaging Recommendations

- CT angiography (CTA) and MR angiography (MRA) are imaging modalities of choice to evaluate pelvic vessels
  - 3D reconstructions and maximum intensity projections (MIP) rotated in 360° for complete evaluation
- Delayed imaging (180 seconds) for evaluation of pelvic veins
- Conventional angiography for therapy (embolization)

CLINICAL IMPLICATIONS

Arterial

- **Collateral circulation**
  - Rich, complex collateral circulation helps ensure delivery of blood to pelvic organs and lower limbs in event of proximal obstruction
  - **Common iliac artery obstruction**
    - Ipsilateral lumbar and contralateral lateral sacral → internal iliac → retrograde into external iliac artery
  - **Internal iliac artery obstruction**
    - Lumbar → iliolumbar arteries
    - Median sacral → lateral sacral arteries
    - Superior rectal → inferior rectal arteries
    - Deep femoral artery → femoral circumflex branches → inferior gluteal arteries
  - **External iliac artery obstruction**
    - Posterior trunk of internal iliac → deep iliac circumflex artery → common femoral artery
    - Anterior trunk of internal iliac → circumflex femoral branches of deep femoral artery

Venous

- **Rectal plexus** is major site of portosystemic collaterals in patients with portal hypertension
  - Hemorrhoids represent engorged collaterals
- **Ovarian vein thrombosis**
  - Postpartum septic thrombosis of ovarian vein
  - 90% occur on right side
- **May-Thurner syndrome**
  - Compression, thrombosis, and eventually occlusion of upper left common iliac vein as it passes between right common iliac artery and spine
- **Rectal carcinoma**
  - Hematogenous spread of metastases depends on location of primary tumor
  - Lower 2/3 → internal iliac veins → lungs
  - Upper 1/3 → inferior mesenteric vein → liver

Lymphatic

- **Testicular carcinoma**
  - Lymphatics follow venous drainage
  - Right-sided tumors first spread to infrarenal, precaval, and aortocaval nodes
  - Left-sided tumors initially spread to left paraaortic nodes near renal hilum
- Ovarian lymphatics follow same course; however, ovarian cancer more often spreads by direct peritoneal seeding

Nerves

- **Piriformis syndrome**
  - Piriformis muscle may irritate sciatic nerve, causing pain in buttocks and referred pain along course of sciatic nerve
- **Pudendal nerve block**
  - Not as frequently utilized secondary to increased use of epidurals
  - Performed either transvaginally or perineally
  - Needle directed toward ischial spine where pudendal nerve curves around into pudendal canal

PELVIC ARTERIES

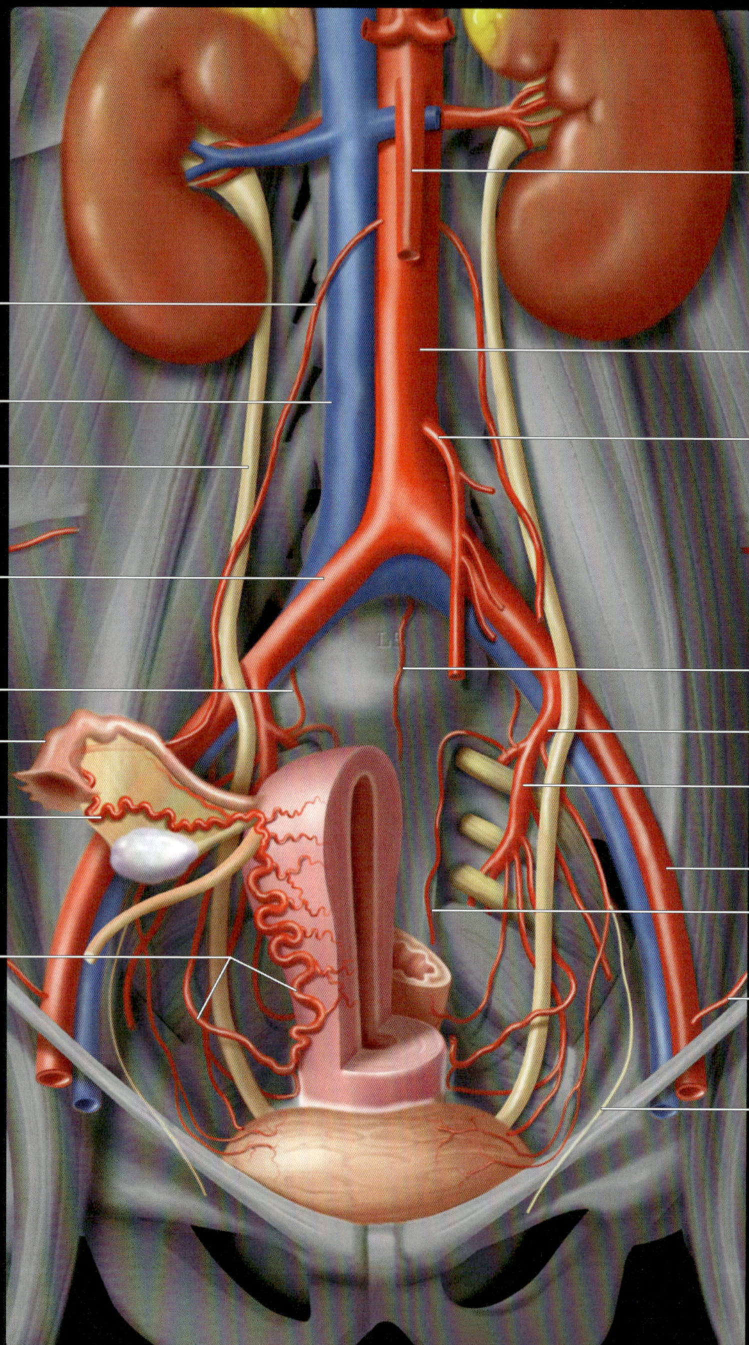

Frontal graphic shows the abdominal aorta, inferior vena cava (IVC), and iliac vessels in a woman. The inferior mesenteric artery is the smallest of the anterior mesenteric branches of the aorta and continues in the pelvis as the superior rectal artery. The paired ovarian arteries arise from the aorta below the renal arteries and pass inferiorly on the posterior abdominal wall to enter the pelvis. The ureters cross anterior to the bifurcation of the common iliac arteries on their way to the urinary bladder. The common iliac artery divides into the external iliac artery, which supplies the lower extremity and the internal iliac (hypogastric) artery, which supplies the pelvis. The internal iliac artery divides into an anterior trunk for the pelvic viscera and a posterior trunk for the muscles of the pelvis.

PELVIC ARTERIES

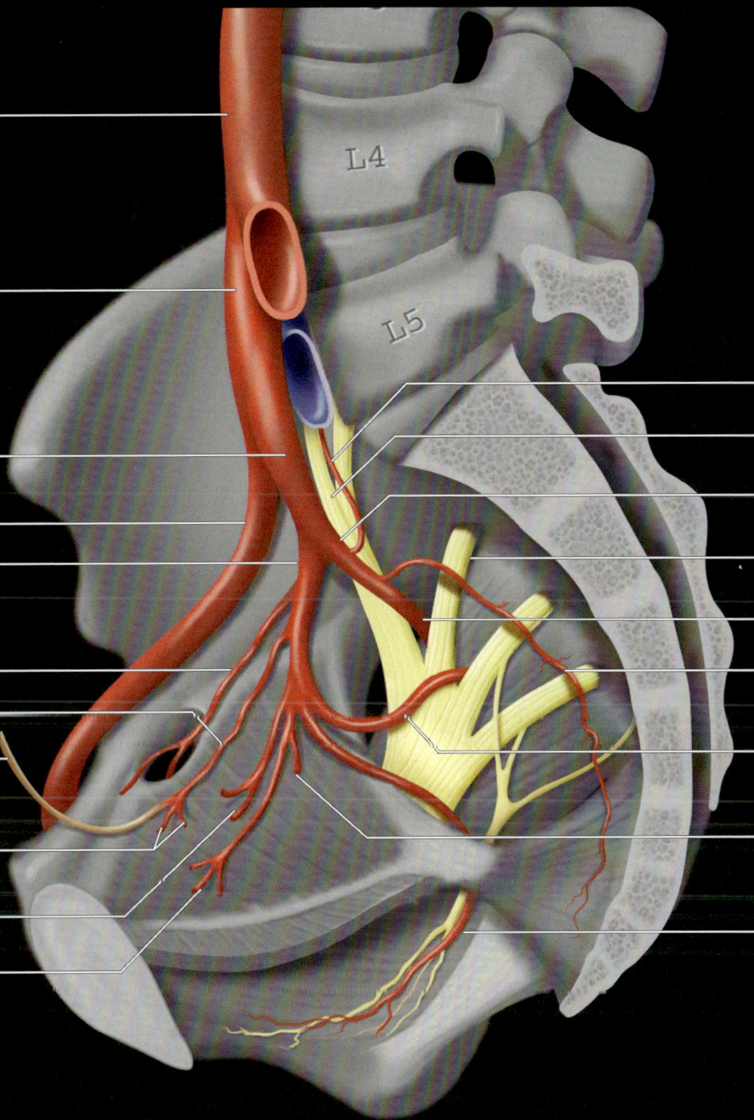

Graphic shows the pelvic arteries and their relation to the sacral nerves. The superior gluteal artery passes posteriorly and runs between the lumbosacral nerve trunk and the anterior ramus of the S1 nerve, whereas the inferior gluteal artery usually runs between the S1-S2 or S2-S3 nerve roots to leave the pelvis through the inferior part of the greater sciatic foramen. Only the proximal portion of the umbilical arteries remains patent after birth, while the distal portion obliterates, forming the medial umbilical ligaments. Arteries to the deep pelvic viscera include the superior and inferior vesicle, uterine, middle rectal and internal pudendal. The individual branching pattern is quite variable.

PELVIC VEINS

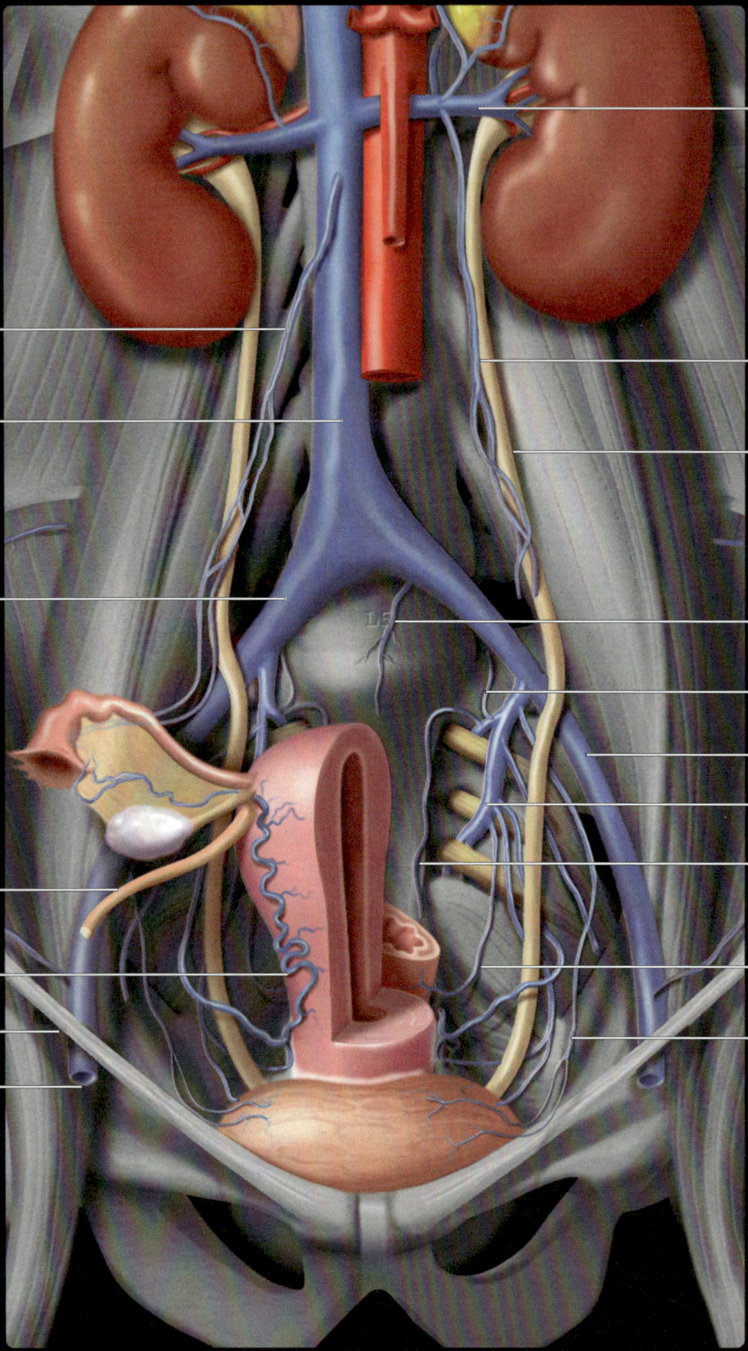

Graphic shows the veins of the pelvis in a woman. The left ovarian vein drains into the left renal vein, whereas the right ovarian vein drains directly into the IVC (the same is true for the testicular veins). Multiple intercommunicating pelvic venous plexuses (rectal, vesicle, prostatic, uterine, and vaginal) drain mainly to the internal iliac veins. There is a communication between the pelvic veins and the intraspinal epidural plexus of veins through the sacral venous plexus.

AXIAL CECT, PELVIC VESSELS

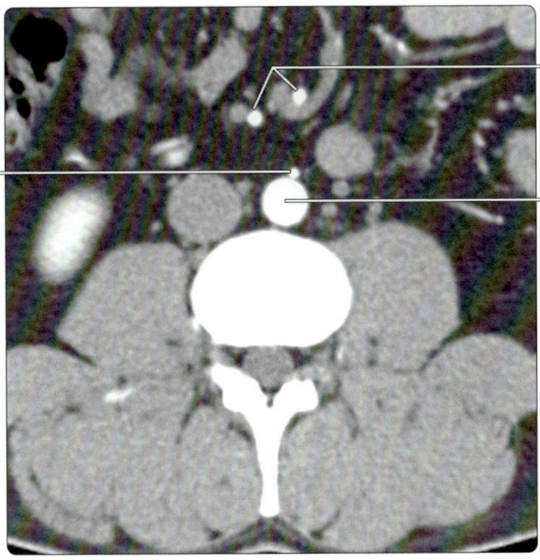

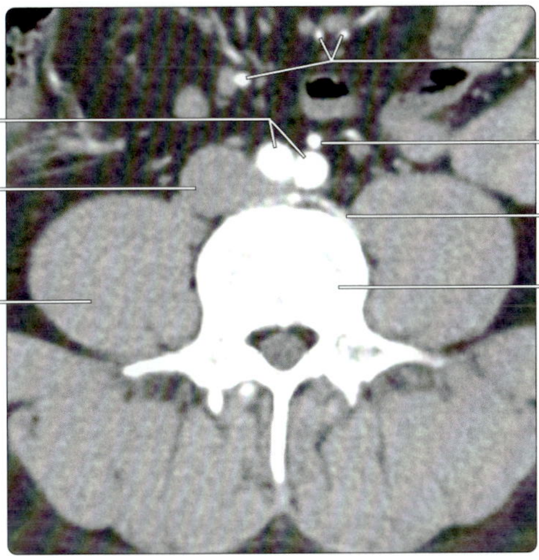

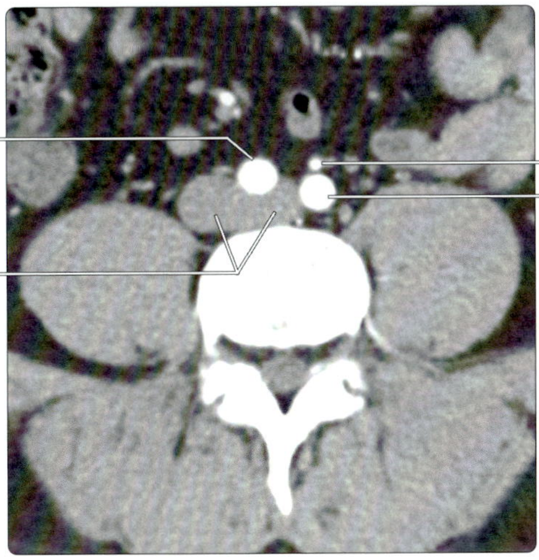

(Top) *First of 12 CECT images of the pelvic vessels is shown. The inferior mesenteric artery, the smallest of the mesenteric arteries, is an anterior branch of the abdominal aorta. It continues in the pelvis as the superior rectal artery. It supplies the distal transverse colon, splenic flexure, descending and sigmoid colon, and the upper part of the rectum.* **(Middle)** *The aorta bifurcates at the level of L4 into the 2 common iliac arteries.* **(Bottom)** *The IVC is formed below the aortic bifurcation by the joining of the common iliac veins. The left iliac vein passes between the left common iliac artery and the L5 vertebral body. Chronic iliac vein compression syndrome, or May-Thurner syndrome, is deep vein thrombosis resulting from the chronic compression of the left common iliac vein by the overlying right common iliac artery. Patients frequently present with left lower extremity swelling or pain.*

AXIAL CECT, PELVIC VESSELS

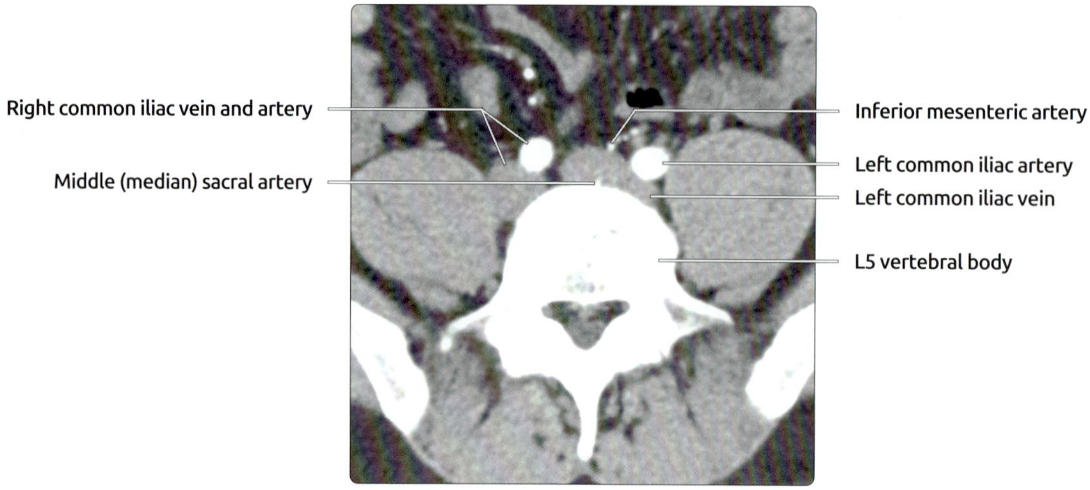

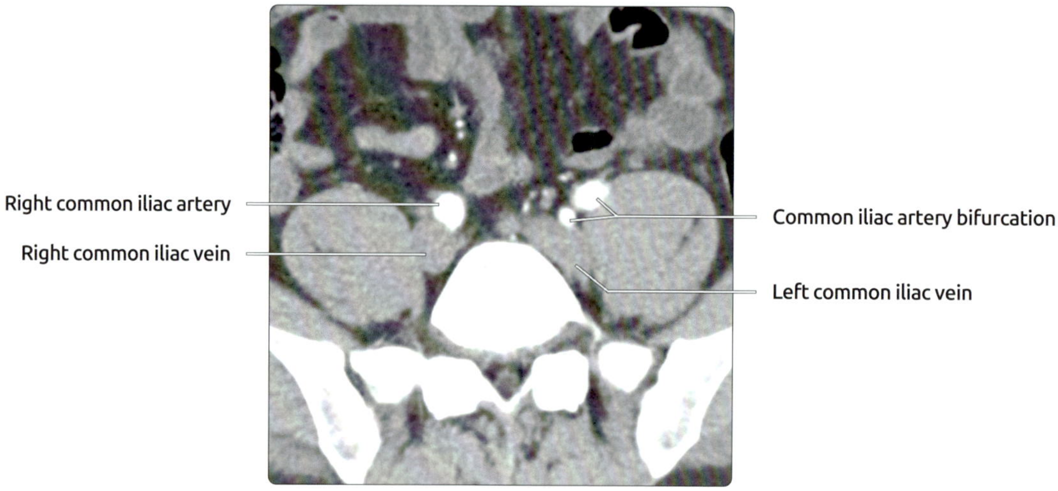

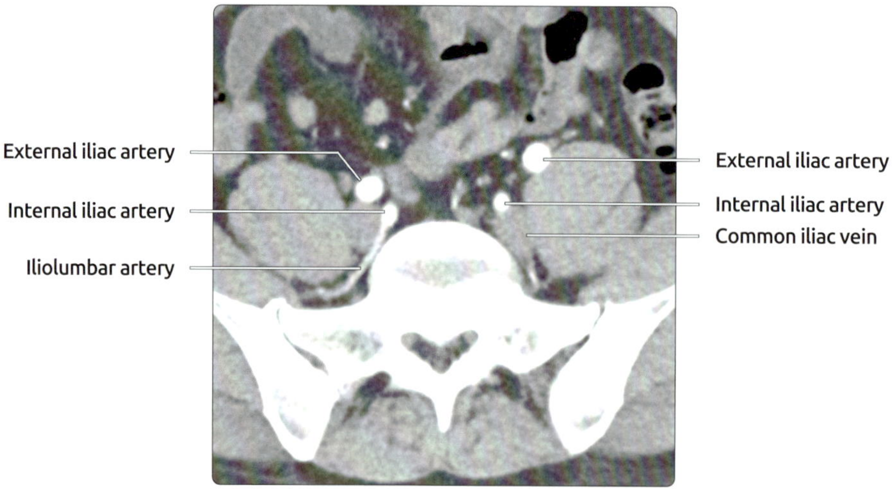

(Top) *The middle or median sacral artery is a small unpaired vessel, which comes off the posterior aorta and descends in the midline to the coccyx. The common iliac arteries usually give off no visceral branches. They may, however, give origin to accessory renal arteries.* **(Middle)** *The common iliac arteries divide, at the level of L5-S1, into 2 branches, the external and internal iliac (hypogastric) arteries. The external iliac artery supplies the lower extremity, while the hypogastric artery supplies the pelvic viscera and muscles of the pelvis.* **(Bottom)** *The iliolumbar artery, usually a branch of the posterior trunk of the internal iliac artery, arises in this patient from the main internal iliac artery.*

AXIAL CECT, PELVIC VESSELS

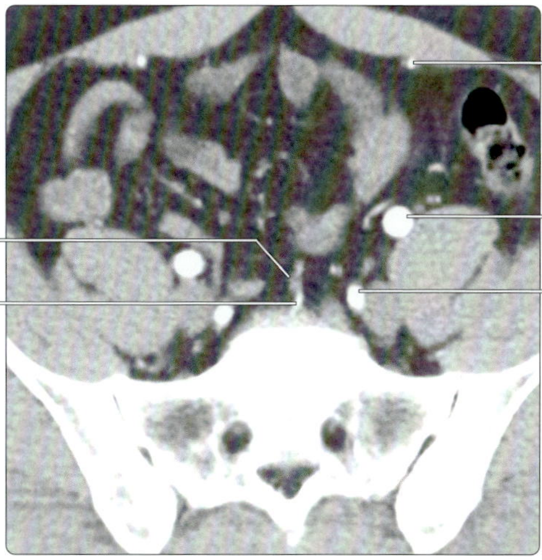

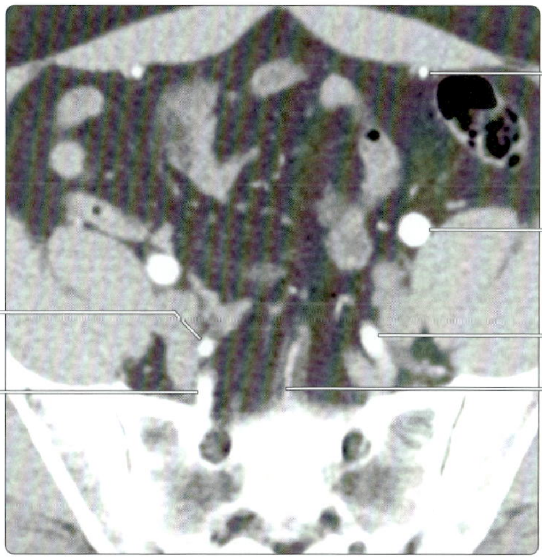

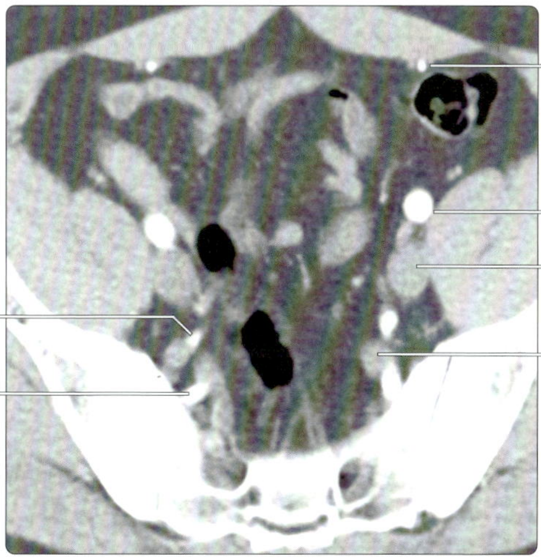

(Top) *The inferior mesenteric vein, a continuation of the superior rectal vein, accompanies the inferior mesenteric artery and usually drains into the splenic vein. The rectum is an important site of portosystemic anastomoses. In patients with portal hypertension, blood is shunted from the high-pressure portal venous system through the inferior mesenteric vein to the middle and inferior rectal veins, which drain into the internal iliac vein.* **(Middle)** *The internal iliac artery divides into an anterior and posterior trunk. The anterior trunk mainly supplies the pelvic viscera, whereas the posterior trunk supplies the pelvic musculature.* **(Bottom)** *The internal iliac vein receives blood from multiple pelvic venous plexuses. It unites with the external iliac vein, a continuation of the common femoral vein, to form the common iliac vein.*

AXIAL CECT, PELVIC VESSELS

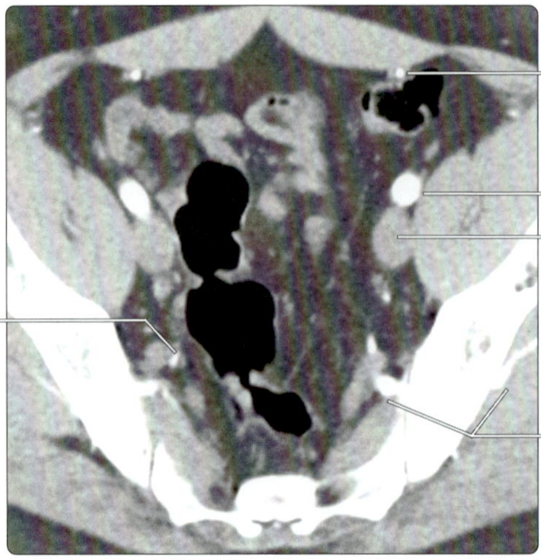

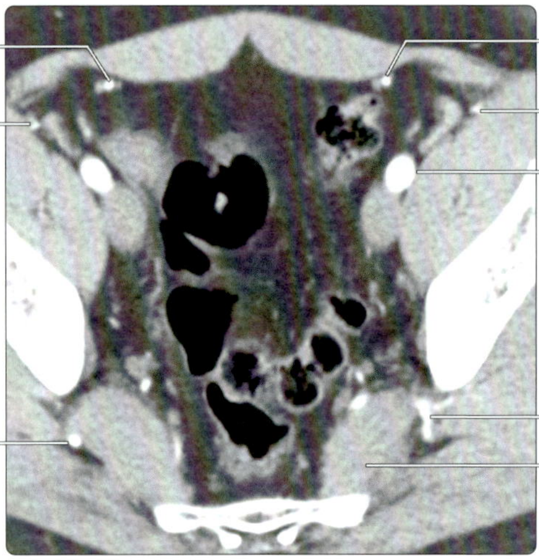

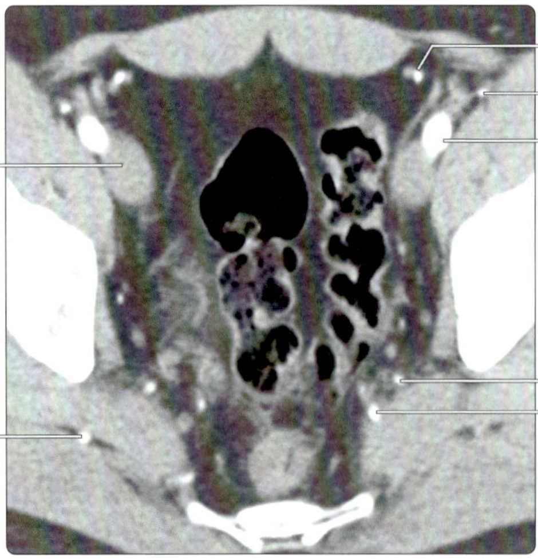

(Top) *The superior gluteal artery runs posteriorly between the lumbosacral trunk and the 1st sacral nerve. It exits the pelvis through the greater sciatic foramen, above the piriformis muscle.* **(Middle)** *The inferior epigastric and deep iliac circumflex arteries arise from the external iliac artery just above the inguinal ligament. These are important landmarks on angiography to determine the point where the external iliac artery becomes the common femoral artery.* **(Bottom)** *The internal pudendal artery and inferior gluteal arteries are the terminal branches of the anterior trunk of the internal iliac artery. They exit the pelvis through the greater sciatic foramen beneath the piriformis muscle. The internal pudendal artery then curves around the sacrospinous ligament to enter the perineum through the lesser sciatic foramen.*

PELVIC ANGIOGRAM, ARTERIES

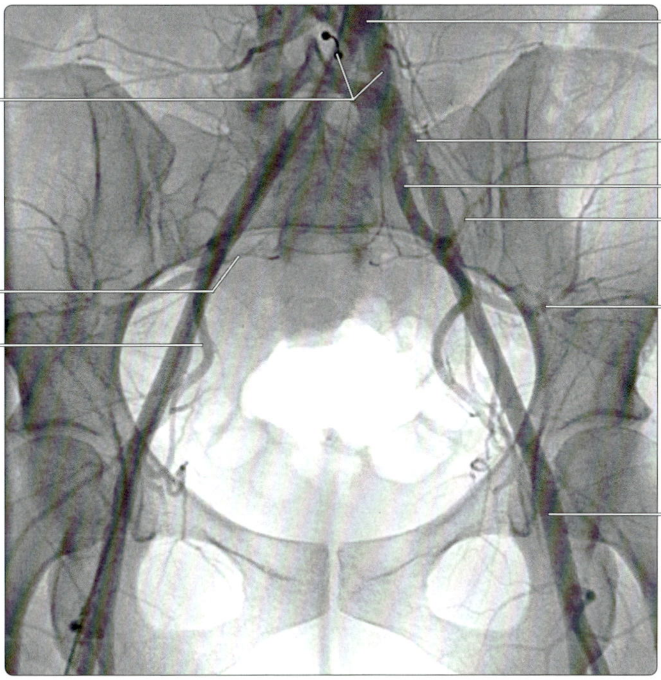

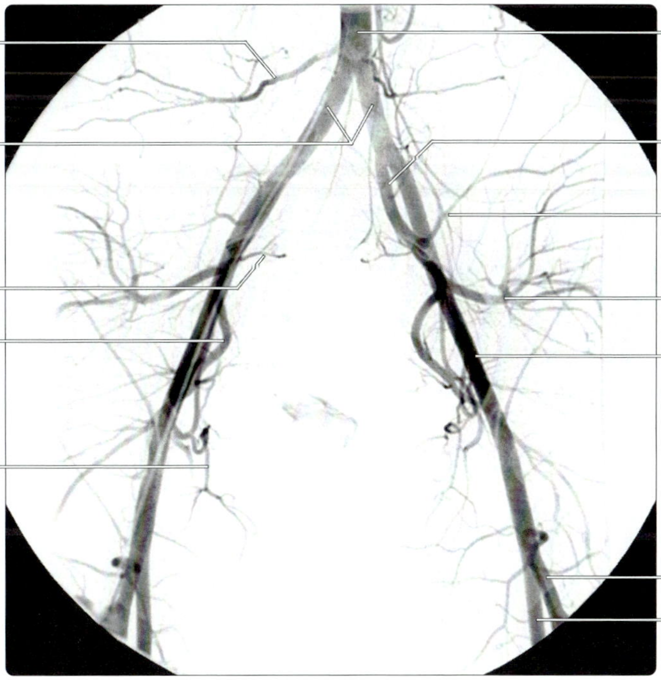

(Top) *Conventional angiogram shows the pelvic arteries. Note the overlap between the various visceral branches of the internal iliac artery.* **(Bottom)** *Digital subtraction angiography allows better visualization of the pelvic arteries.*

3D CT, ARTERIES

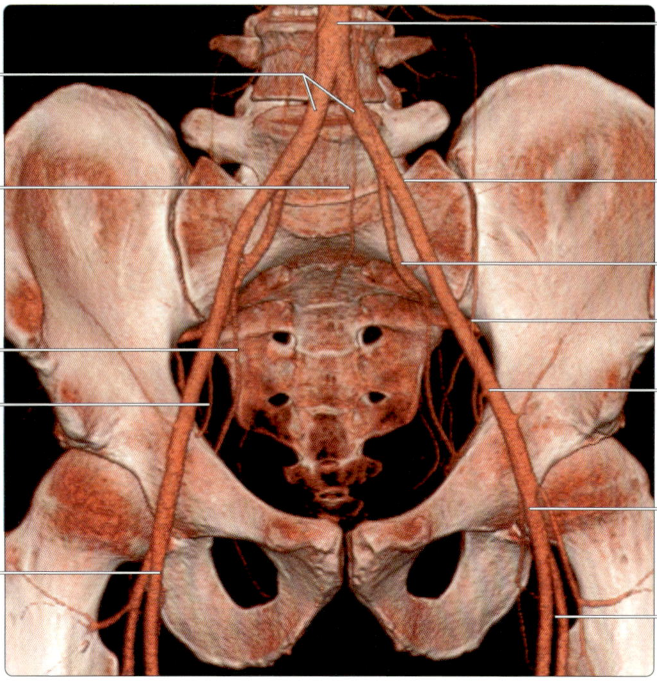

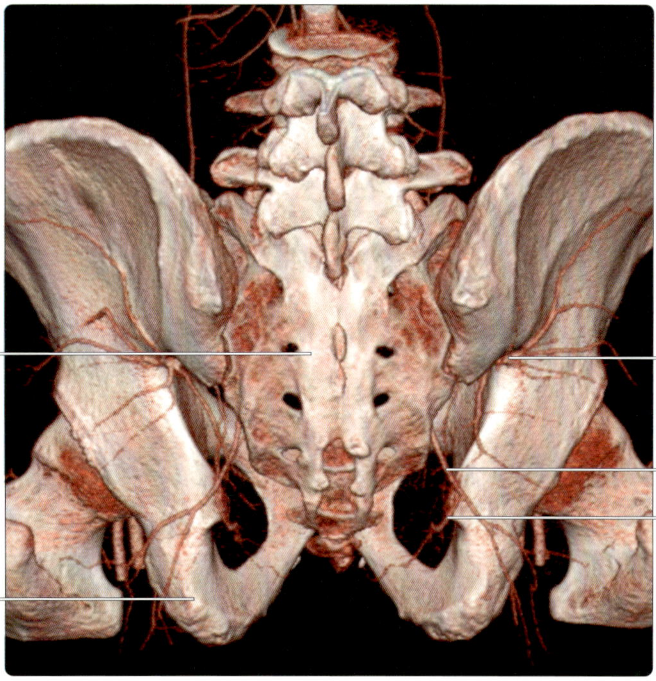

(Top) *3D volume rendering frontal image shows the iliac arteries. The aorta bifurcates at the level of L4 into the common iliac arteries, and the common iliac arteries bifurcate at the level of L5-S1 into the external and internal iliac arteries.* **(Bottom)** *3D volume rendering posterior image of the pelvis shows the superior gluteal artery, the continuation of the posterior trunk of the internal iliac artery, and the inferior gluteal artery, a branch of the anterior trunk. Note the proximity of the superior gluteal artery to the iliac bone, making it vulnerable to injury in cases of pelvic fractures.*

3D CT, ARTERIES

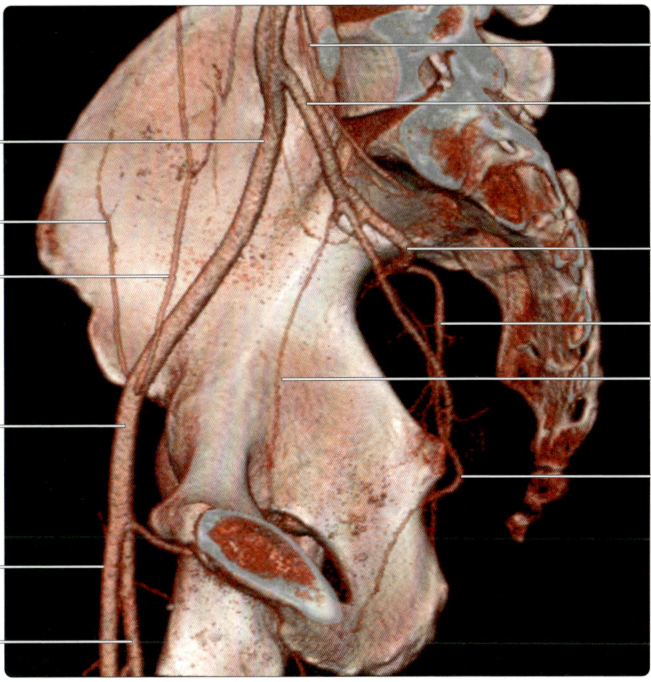

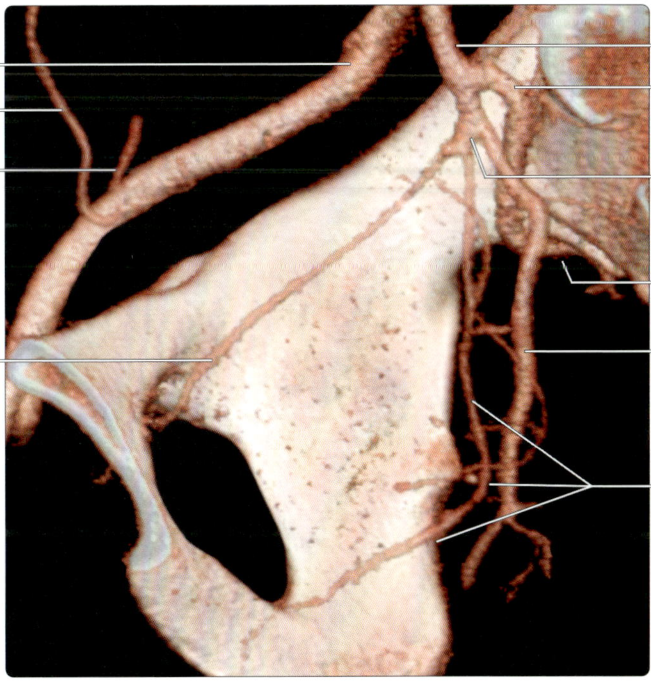

(Top) *3D volume-rendered side view of the iliac arteries is shown. The deep iliac circumflex artery arises opposite the inferior epigastric artery just above the level of the inguinal ligament. They demarcate the external iliac artery above, from the common femoral artery below.* **(Bottom)** *3D volume-rendered oblique side view of the iliac arteries in a woman shows a large uterine artery. The obturator artery, a branch of the anterior trunk of the internal iliac artery, runs along the pelvic side wall and leaves the pelvis through the obturator canal. Occasionally, an aberrant obturator artery can come from the inferior epigastric artery. Note the tortuous course of the internal pudendal artery, as it curves around the sacrospinous ligament to enter the perineum through the lesser sciatic foramen.*

AXIAL CECT, THROMBOSED RIGHT OVARIAN VEIN

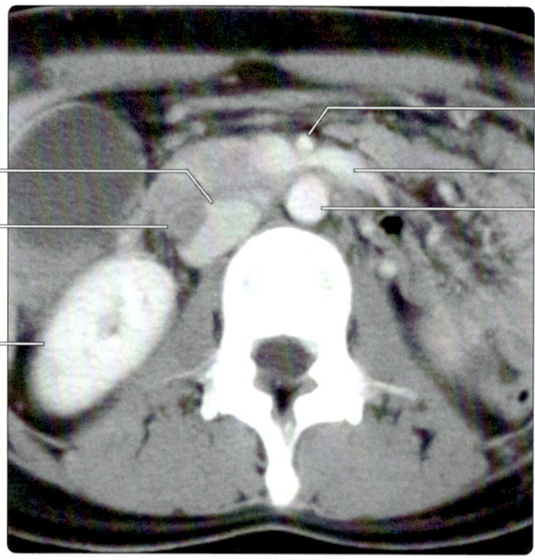

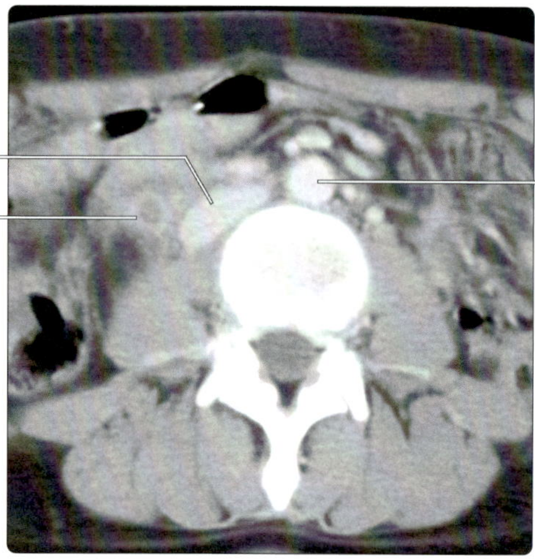

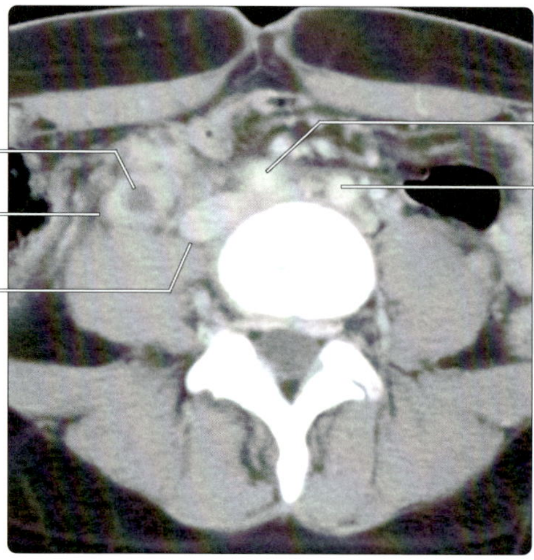

(Top) *First of 3 axial CECT images in a postpartum woman with fever and right-sided ovarian vein thrombosis is shown. On this most cephalad image, the thrombus is seen extending to the junction of the right ovarian vein into the IVC.* **(Middle)** *The right ovarian vein is dilated and filled with thrombus.* **(Bottom)** *Thrombosis usually occurs secondary to pelvic infection. In addition to the thrombus itself, the vessel wall is thickened, and there is a surrounding inflammatory reaction. Following the normal course of the vein up toward the IVC allows differentiation from other pelvic pathologies, such as appendicitis and hydroureter.*

AXIAL CECT, THROMBOSED LEFT OVARIAN VEIN

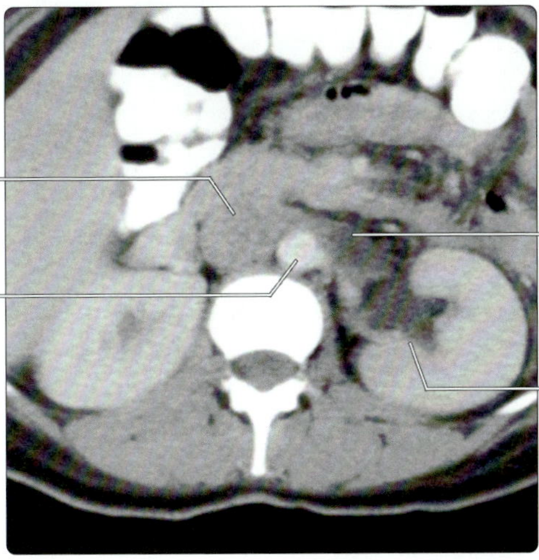

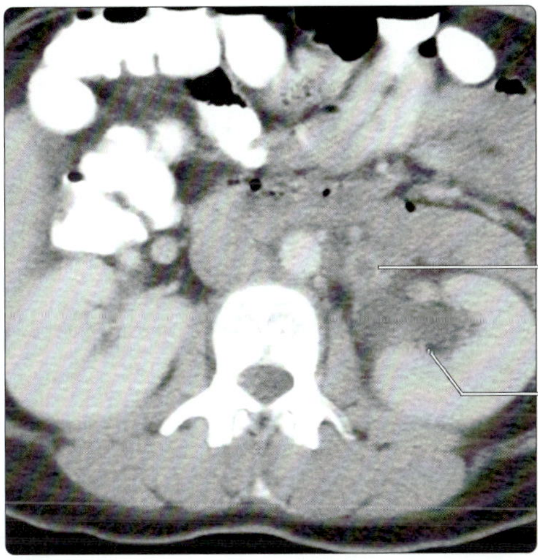

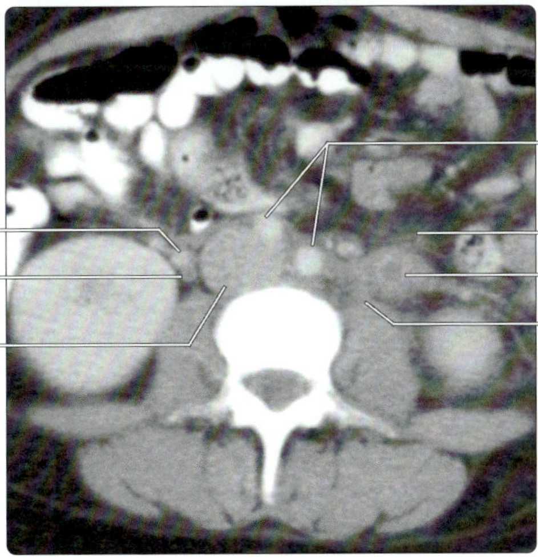

(Top) *First of 3 axial CECT images in a woman with postpartum thrombosis of the left ovarian vein is shown. This image shows the thrombus extending into the left renal vein.* **(Middle)** *Clot is present within the left ovarian vein, and the wall is thickened with surrounding inflammatory changes.* **(Bottom)** *The ovarian veins ascend along the psoas muscle and cross anterior to the ureters, near the confluence of the common iliac veins. Inflammation from ovarian vein thrombosis may also affect the ureter and result in hydronephrosis, as is seen in this case.*

PELVIC LYMPH NODES

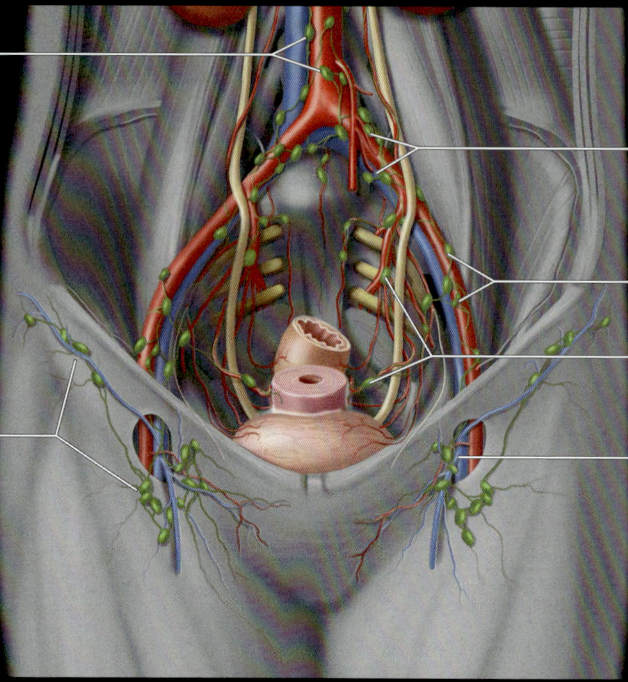

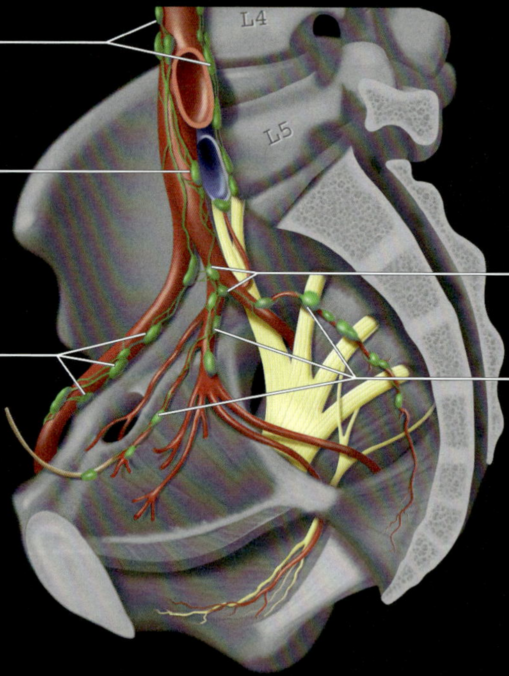

(Top) *Graphic shows the lymph nodes of the pelvis. Generally, the lymph nodes are located along, and named after, the iliac vessels. Two groups of inguinal lymph nodes are described: The superficial and deep groups. The superficial inguinal nodes lie along the saphenous vein, superficial to the cribriform fascia that overlies the femoral vessels. There are ~ 10 superficial lymph nodes. The deep inguinal lymph nodes are located medial to the femoral vein and under the cribriform fascia. There are ~ 3-5 deep nodes. The most superior inguinal node is located under the inguinal ligament and is called the gland or node of Cloquet.* **(Bottom)** *Graphic of the side wall of the pelvis shows the lymph nodes along the corresponding iliac vessels.*

CORONAL MR, PELVIC LYMPH NODES

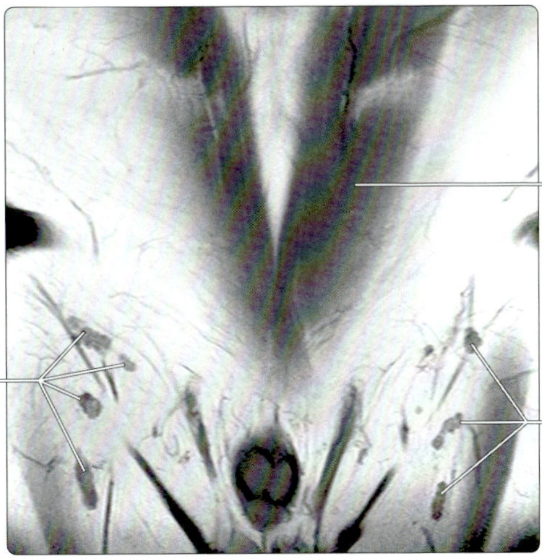

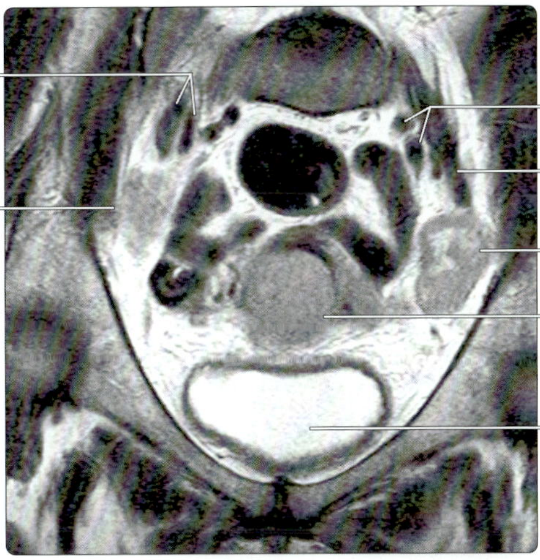

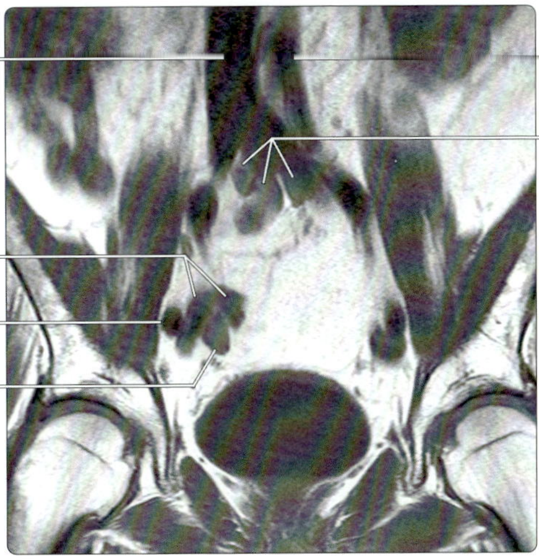

(Top) *Coronal T1 MR shows the distribution of normal-sized inguinal lymph nodes along the inguinal ligament. The superficial inguinal nodes drain the skin of the penis, lower abdomen, perineum, scrotum, and part of the buttock area.* **(Middle)** *Coronal oblique T2 MR of a woman with cervical cancer shows enlarged external iliac lymph nodes from metastatic disease.* **(Bottom)** *Coronal T1 MR shows multiple enlarged pelvic lymph nodes in a patient with lymphoma.*

PET/CT, PELVIC LYMPH NODES

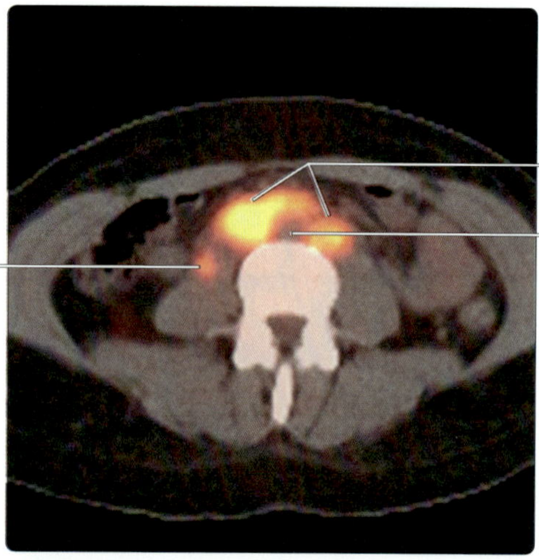

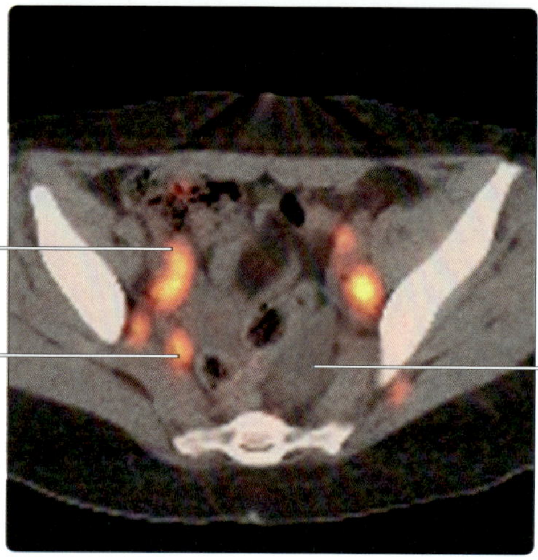

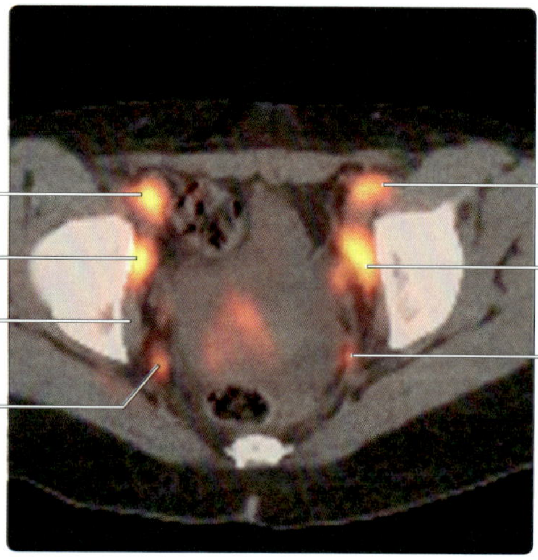

(Top) *First of 3 axial PET/CT images, after injection of F-18 FDG, shows increased metabolic activity in multiple enlarged groups of lymph nodes in a young patient diagnosed with non-Hodgkin lymphoma. Image shows involvement of the paraaortic lymph nodes. Increased activity is seen in the right ureter due to excretion of F-18 FDG in the urine.* **(Middle)** *Increased metabolic activity in the internal and external iliac lymph nodes on both sides is shown.* **(Bottom)** *Obturator nodes lie adjacent to the obturator internus muscle and are a frequent site of metastases for pelvic malignancies. The deep inguinal nodes are the main lymphatic drainage from the lower extremities, but they also receive small branches from the penis and efferent vessels from the superficial inguinal nodes.*

PET/CT, PELVIC LYMPH NODES

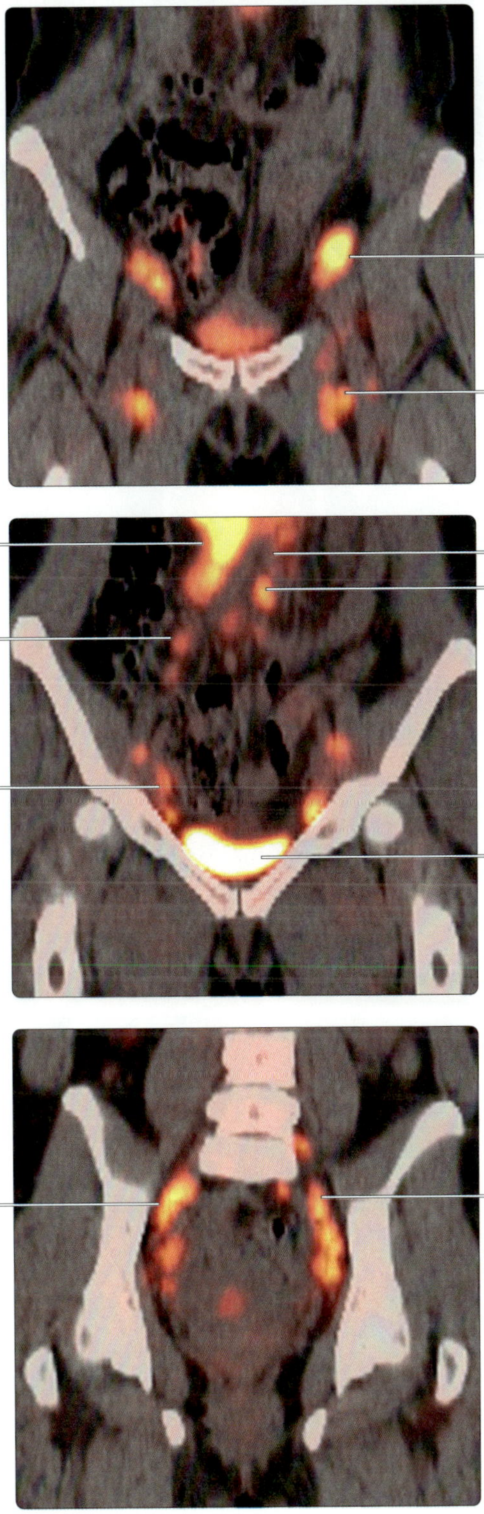

(Top) *First of 3 coronal PET/CT images in the same patient diagnosed with lymphoma shows multiple enlarged metabolically active lymph nodes. The external iliac nodes are above the inguinal ligament and are the primary drainage for the inguinal nodes.* **(Middle)** *More posteriorly, metabolic activity is seen in internal iliac, common iliac, and paraaortic lymph nodes. Paraaortic lymph nodes receive lymphatic drainage from the common iliac lymph nodes. In addition, the paraaortic nodes receive drainage from the gonads. Typical patterns of spread in testicular cancer patients occur according to the side of the primary tumor. Right-sided testicular tumors spread initially to the aortocaval nodes, right paraaortic nodes, and precaval nodes. Left-sided primaries metastasize to preaortic nodes and left paraaortic nodes just below the level of the renal vein.* **(Bottom)** *The common iliac nodes receive lymphatic drainage from the internal and external groups of lymph nodes and drain into the paraaortic lymph nodes.*

PELVIC NERVES

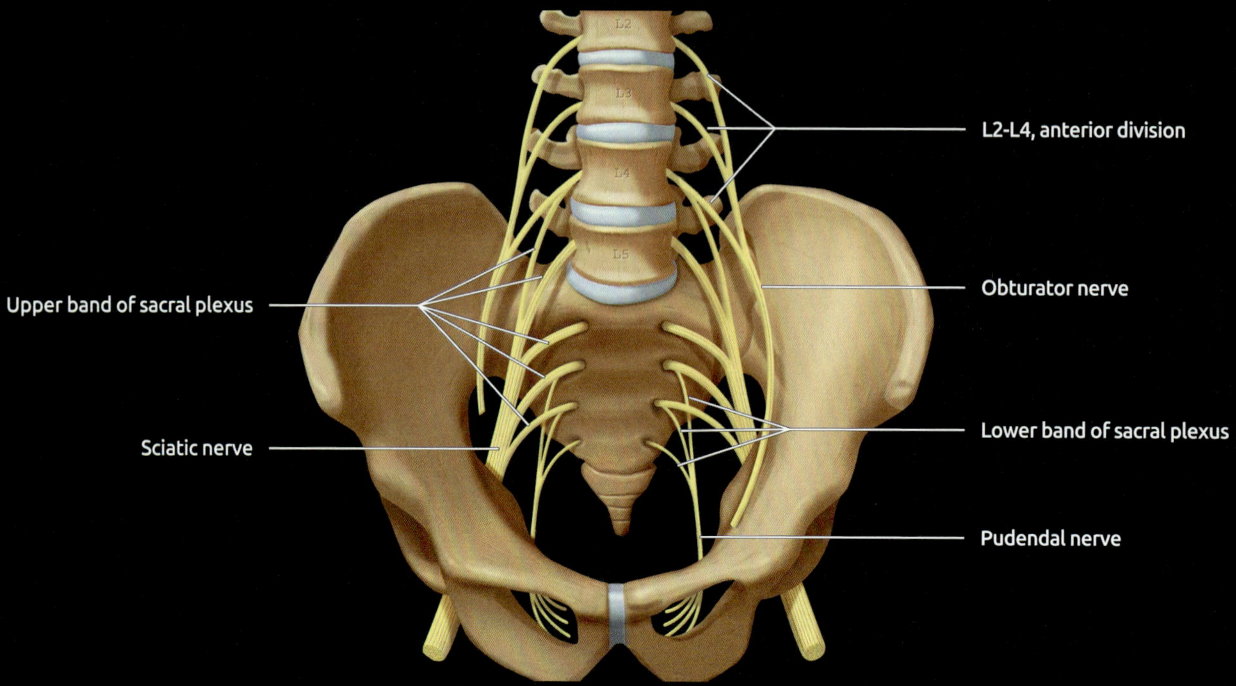

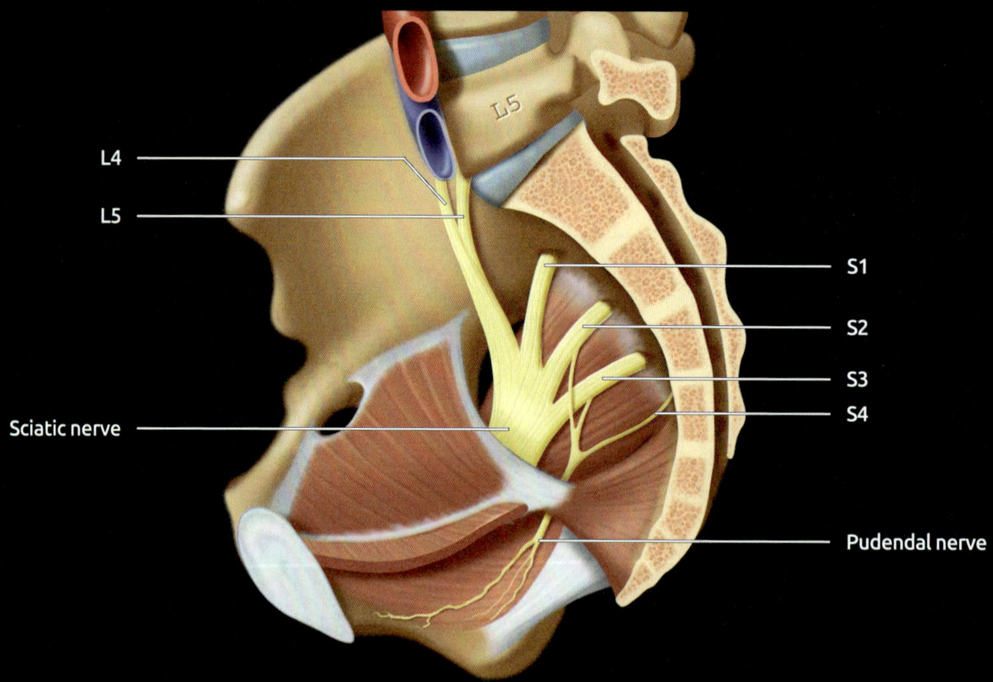

(Top) *The ventral rami of L2-L4 split into 2 divisions: A larger posterior division, which forms the femoral nerve (not shown), and a smaller anterior division, which forms the obturator nerve. The obturator nerve follows the obturator vessels, leaving the pelvis through the obturator foramen. The sacral plexus has a larger upper band (L4, L5, S1-S3), which becomes the sciatic nerve and a lower band (S2-S4), which becomes the pudendal nerve.* **(Bottom)** *The upper band of the sacral plexus coalesces into the sciatic nerve on the ventral surface of the piriformis muscle. The lower band forms the pudendal nerve, which exits the pelvis through the greater sciatic foramen. It then curves around the sacrospinous ligament to enter the perineum through the lesser sciatic foramen. The pudendal nerve is the primary innervation for the perineum and external anal sphincter.*

PELVIC NERVES

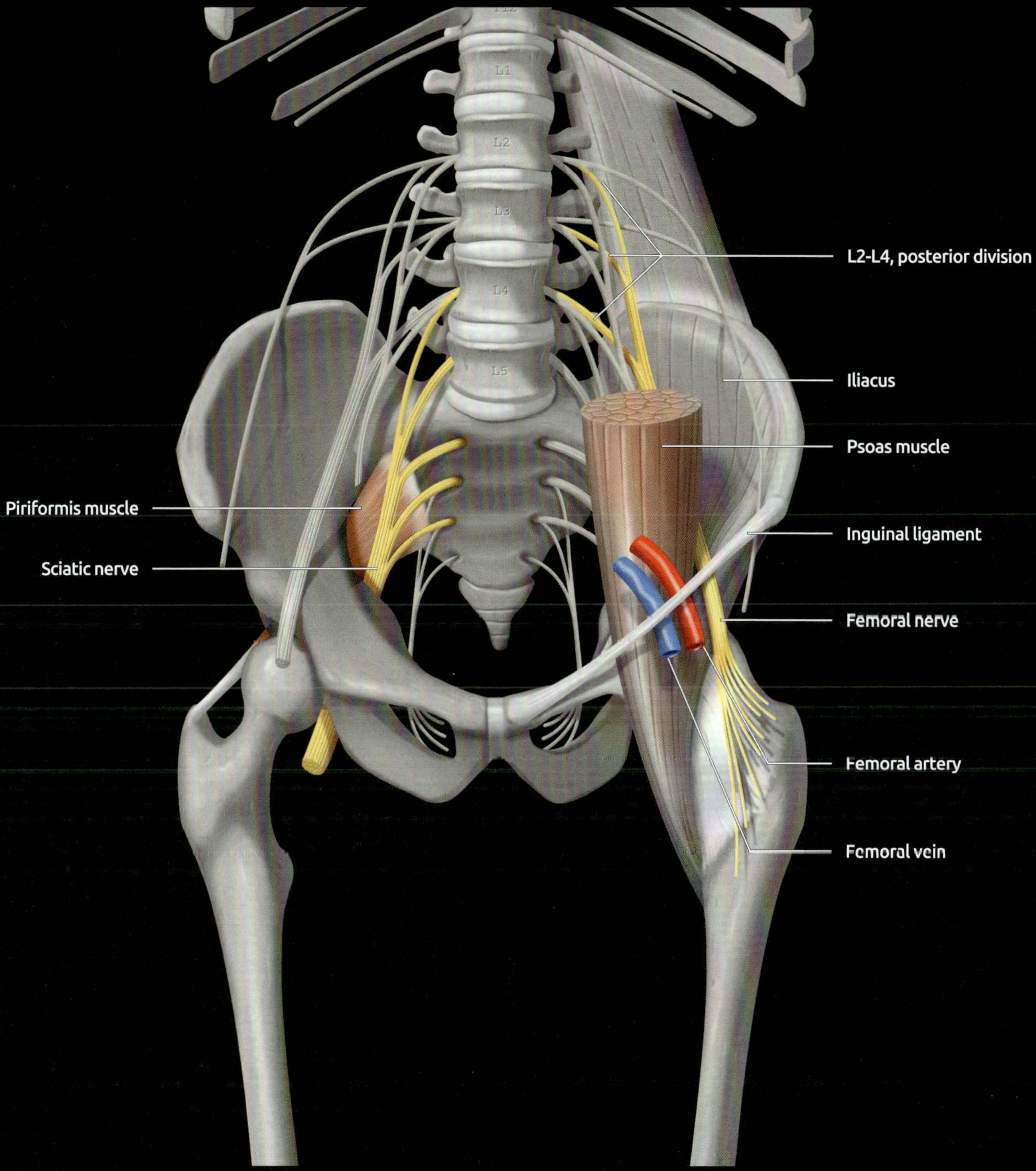

The femoral nerve is formed from the posterior division of the L2-L4 ventral rami. It descends through the psoas major muscle and into the iliopsoas groove. It exits the pelvis beneath the inguinal ligament through the femoral canal. The well-known acronym NAVL (nerve, artery, vein, lymphatic) describes the order of the femoral canal contents from lateral to medial. The femoral nerve innervates the anteromedial thigh. The sciatic nerve exits the pelvis through the greater sciatic foramen below the piriformis muscle to innervate the posterior thigh and leg.

SCIATIC NERVE

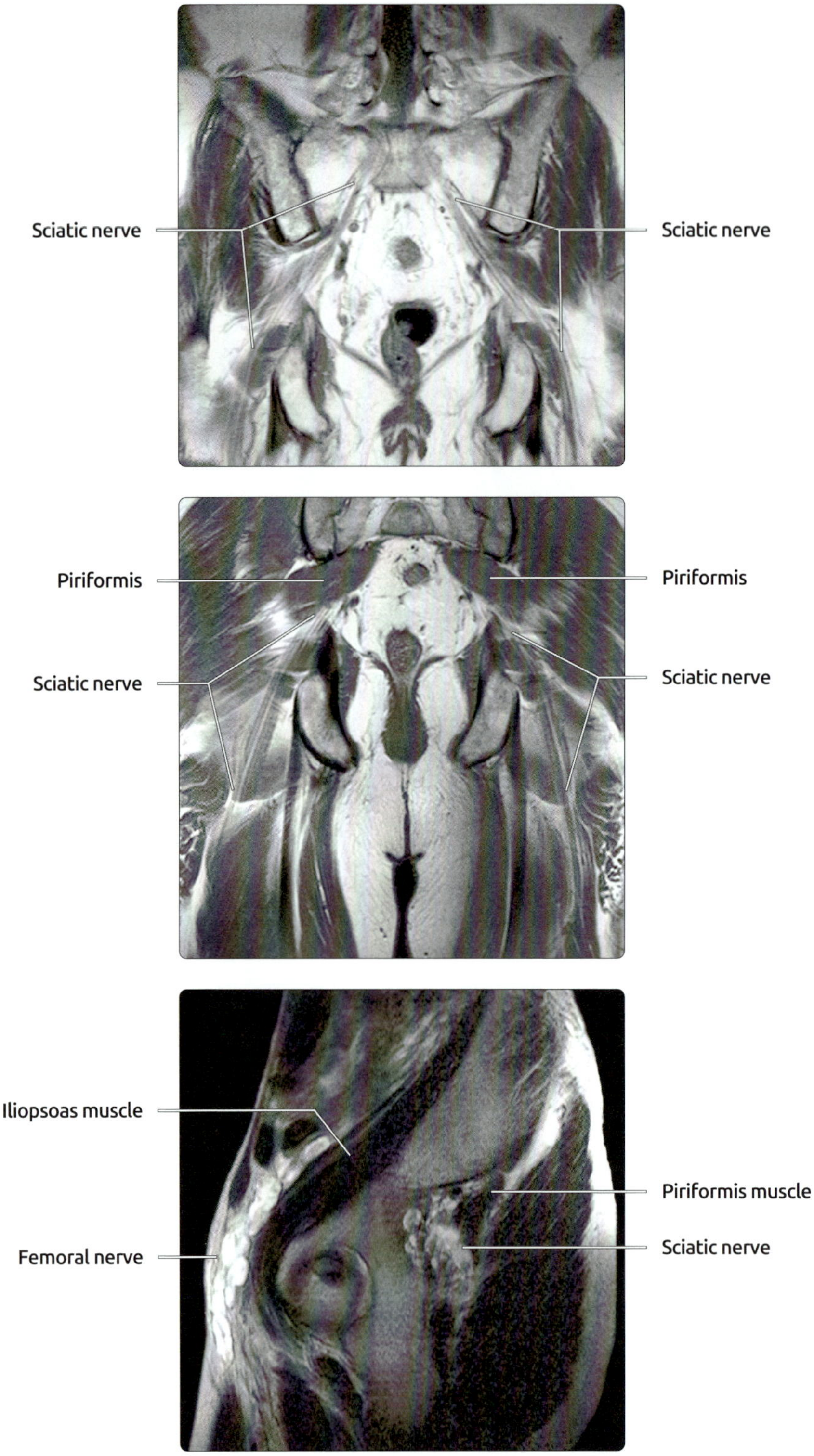

(Top) *Coronal T1 MR shows the nerves from the sacral plexus coalescing to form the sciatic nerve. The long fascicles composing this large nerve can be easily recognized.* **(Middle)** *Coronal T1 MR shows the sciatic nerve passing beneath the piriformis muscle into the posterior thigh.* **(Bottom)** *Sagittal T2 MR in a patient with neurofibromatosis type 1 is shown. The sciatic nerve is dramatically enlarged as it passes through the greater sciatic foramen beneath the piriformis muscle. The femoral nerve is also grossly enlarged.*

SCIATIC NERVE

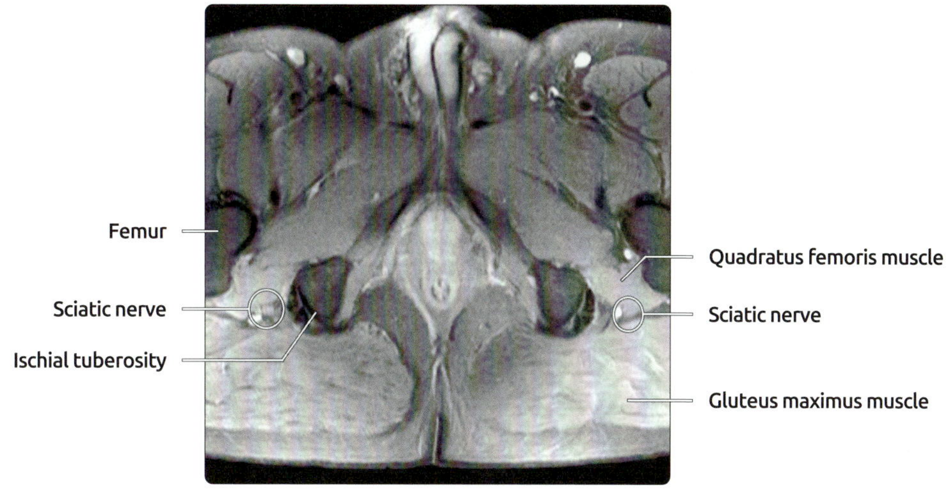

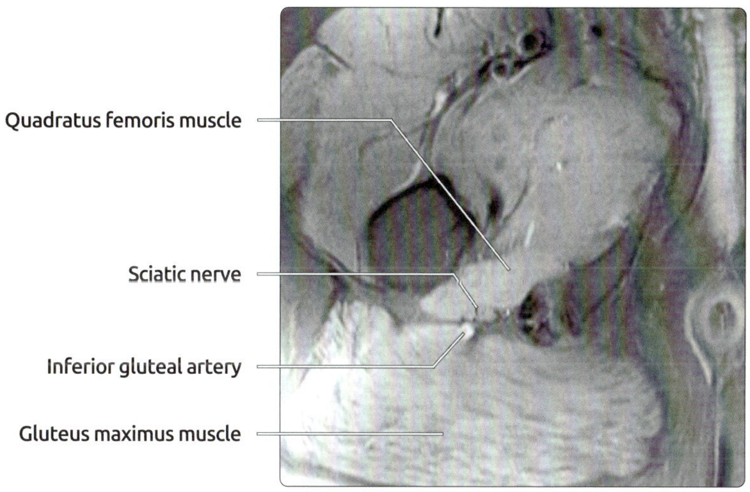

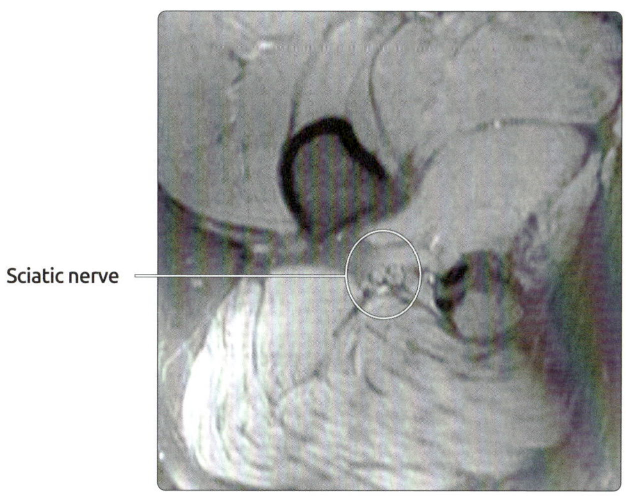

(Top) *First of 3 axial proton density, fat-saturated images shows the sciatic nerve. After it has left the pelvis, the sciatic nerve can be seen near the ischial tuberosities, between the gluteus maximus and quadratus femoris muscles.* **(Middle)** *As the sciatic nerve passes beneath the gluteus maximus muscle, it is accompanied by the inferior gluteal artery.* **(Bottom)** *The fascicular architecture of the nerve is very distinctive and permits differentiation from other soft tissues and vessels.*

FEMORAL NERVE

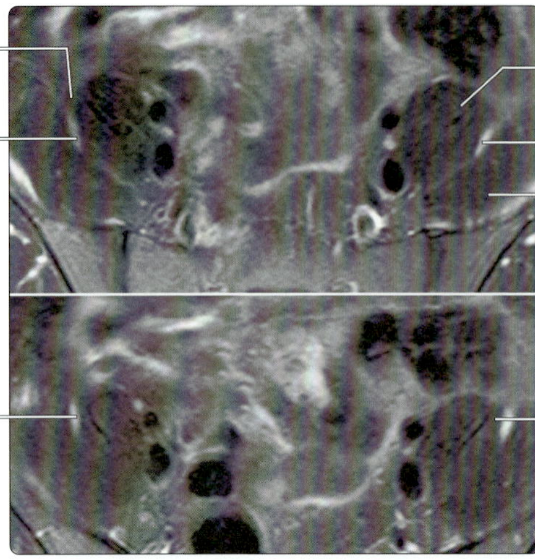

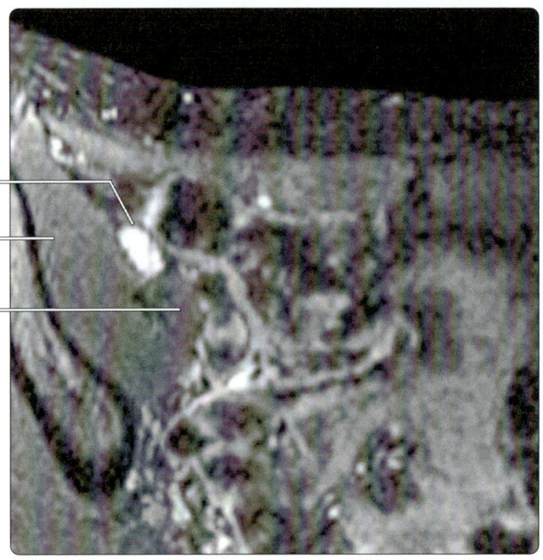

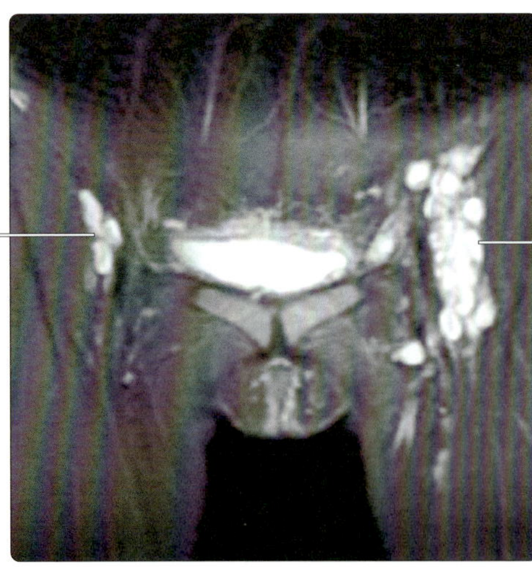

(Top) *T2 fat-saturated MR in a patient with bilateral femoral nerve inflammation shows the course of the pelvic portion of this nerve. The femoral nerve is formed from the posterior division of the L2-L4 ventral rami. It descends through the psoas muscle into the groove created between the psoas and iliacus muscles (iliopsoas groove). It maintains this position until it exits the pelvis beneath the inguinal ligament.* **(Middle)** *Axial STIR MR shows an abnormal, inflamed, hyperintense right femoral nerve traveling in the iliopsoas grove. The characteristic fascicular architecture of peripheral nerves is well shown. The femoral nerve had been inadvertently sutured during herniorrhaphy.* **(Bottom)** *Coronal STIR MR in a patient with neurofibromatosis type 1 shows involvement of both femoral nerves (left > right).*

FEMORAL, OBTURATOR, SACRAL NERVES

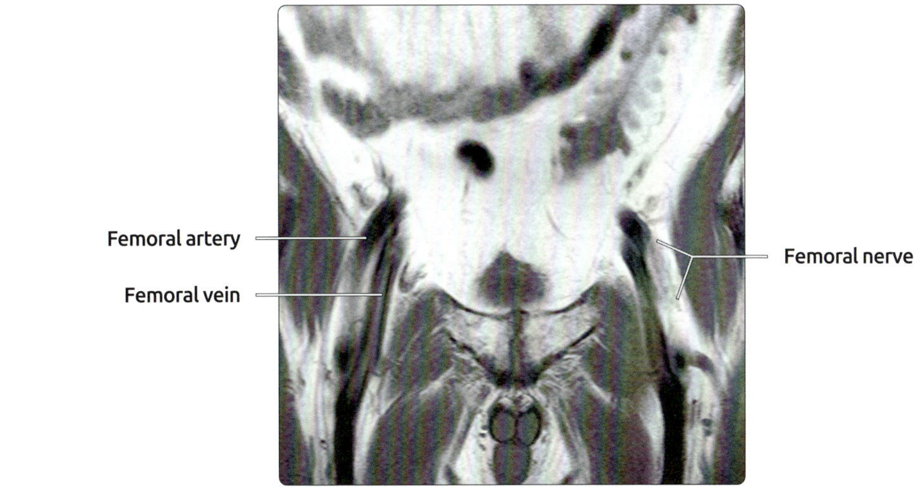

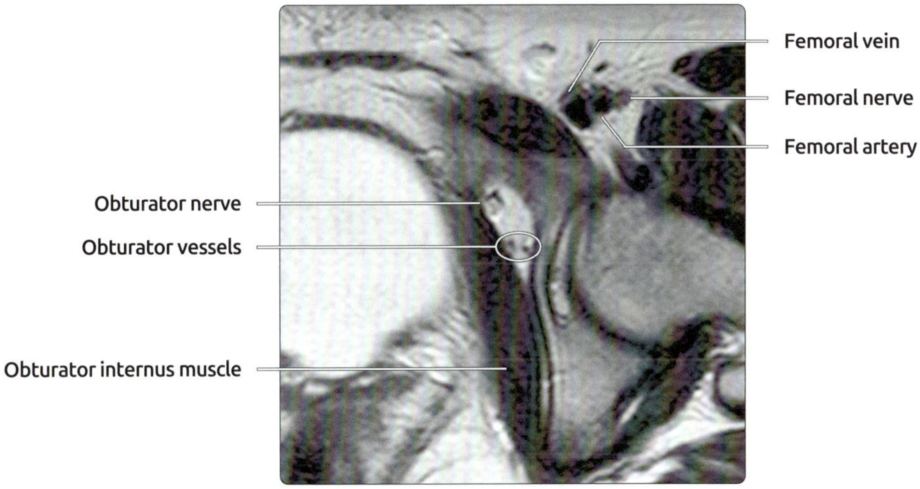

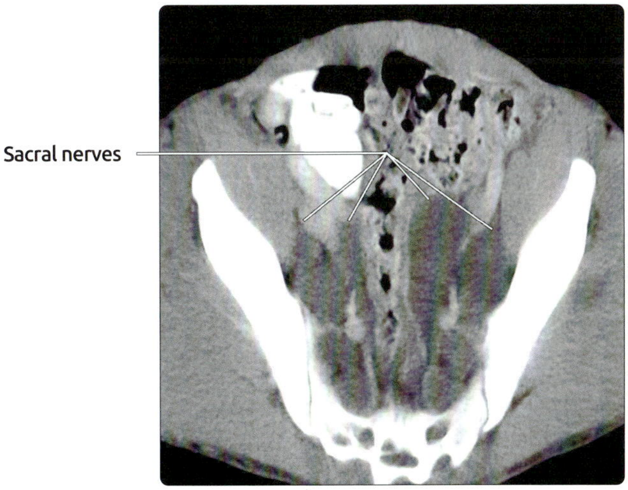

(Top) *Coronal T1 MR shows the femoral nerve, which exits the pelvis beneath the inguinal ligament where it is the most lateral component within the femoral canal. Branches of the femoral nerve innervate the anteromedial thigh.* **(Middle)** *Axial T2 MR shows the obturator nerve, which runs anterosuperior to the obturator vessels to enter the upper part of the obturator foramen. Here, it divides into an anterior and posterior division as it enters the thigh.* **(Bottom)** *Axial CECT in a patient with neurofibromatosis type 1 is shown. The sacral nerves, which form the sciatic nerve, are dramatically enlarged.*

IMAGING ANATOMY

Anatomic Boundaries

- Iliac crest to perineum
- Divided into **false** (**major**) **and true** (**minor**) **pelvis**
  - Dividing point from sacral promontory, along arcuate and iliopectineal lines to pubic crest
    - Arcuate line divides flared iliac fossa from inferior, narrowed portion of iliac bone
  - False pelvis wide "bowl" above line
  - True pelvis below arcuate line
    - **Pelvic inlet** upper border of true pelvis (tilts forward ~ 60° from horizontal)
    - **Pelvic outlet** formed by ischiopubic rami, ischial spines, inferior symphysis pubis, sacrotuberous ligaments, and coccyx

Skeletal Anatomy

- 4 components
  - Paired innominate (pelvic) bones
    - Composed of Ilium, ischium, and pubis
  - Sacrum
  - Coccyx
- Key ligamentous attachments and foramina
  - **Sacrospinous ligament**: From sacrum to ischial spine
    - Divides **greater sciatic foramen above** ligament from **lesser sciatic foramen below**
  - **Sacrotuberous ligament**: From sacrum to ischial tuberosity
    - Posterior boundary of lesser sciatic foramen
  - **Inguinal ligament**: From anterior superior iliac spine to pubic tubercle
    - Formed from fibers of external oblique aponeurosis
  - **Obturator foramen**
    - Formed by body and rami of pubic and ischial bones
- **Male vs. female pelvis**
  - Male pelvis
    - Pubic arch (angle formed by inferior pubic rami at symphysis pubis) < 90°
    - Heart-shaped pelvic inlet with narrow outlet
    - Round obturator foramen
    - Narrow sciatic notch
    - Larger, thicker bones
  - Female pelvis
    - Flattened pubic arch (> 90°)
    - Oval pelvic inlet with larger pelvic outlet
    - Oval obturator foramen
    - Wide sciatic notch
    - Thinner, more delicate bones

Anterior Pelvic Wall

- Continuation of upper abdominal muscles
  - **External oblique muscle**
    - Largest and most superficial of 3 flat muscles
    - Fibers course inferomedially
  - **Internal oblique muscle**
    - Middle flat muscle
    - Runs superomedially at right angles to external oblique muscle in abdomen
    - Muscle fibers become more horizontal in pelvis and become inferiorly directed more caudally
  - **Transversus abdominis muscle**
    - Innermost flat muscle with horizontal course
  - **Rectus abdominis muscle**
    - Vertically directed paired broad muscle
- **Linea alba**
  - Midline fibrous raphe formed by aponeuroses of flat muscles
- **Rectus sheath**
  - Strong fibrous sheath, which **invests rectus abdominis muscles** and superior and inferior epigastric vessels
  - Formed by aponeuroses of external oblique, internal oblique, and transversus abdominis muscles
  - **Internal oblique aponeurosis splits** in upper portion of abdominal wall
    - Anterior portion joins external oblique aponeurosis to form **anterior rectus sheath**
    - Posterior portion joins transversus abdominis aponeurosis to form **posterior rectus sheath**
  - Lower 1/3 of abdominal wall (below anterior superior iliac spine), all aponeuroses join and course anterior to rectus abdominis muscles
    - Creates **arcuate line** on posterior surface of abdominal wall
    - Wall below arcuate line covered only by **transversalis fascia**, which is separated from parietal peritoneum by extraperitoneal fat

Inguinal Canal

- Passage through anterior abdominal wall
  - Conveys spermatic cord in males and round ligament in females
  - Formed embryologically by evagination of processus vaginalis through anterior abdominal wall
- **Internal or deep ring**
  - Opening through transversalis fascia
  - Located above midportion of inguinal ligament
  - Lateral to inferior epigastric artery
- **External or superficial ring**
  - Triangular opening of external oblique aponeurosis
  - Just lateral to pubic tubercle
- **Hesselbach triangle**
  - Area of weakness in pelvic wall
    - Site of direct inguinal hernias
  - Medial border: Lateral edge of rectus sheath
  - Lateral border: Inferior epigastric artery
  - Inferior border: Inguinal ligament

Posterior Pelvic Wall (False Pelvis)

- Psoas and iliacus muscles fuse caudally to form iliopsoas muscle
  - Inserts on lesser trochanter and is powerful hip flexor

Pelvic Floor

- Curved bowl so distinct walls not clearly defined
- Posterior wall
  - Sacrum and coccyx
  - Piriformis and coccygeus muscles
- Lateral wall
  - Ilium and ischium
  - Piriformis, obturator internus, and levator ani muscles
- Anterior wall

- Symphysis pubis
- Anterior portion of obturator internus muscle, levator ani muscle
- Inferior boundary
  - **Pelvic diaphragm** (levator ani and fascia)
- Key muscles
  - **Obturator internus**
    - Runs from inner surface of obturator membrane to greater trochanter
    - Covered by thick fascia (**arcus tendineus**), which is origin of pelvic diaphragm
  - **Coccygeus**
    - Runs from ischial spine to sacrum and coccyx
  - **Levator ani forms pelvic diaphragm**
    - **Formed by 3 muscles**: **Iliococcygeus**, **pubococcygeus**, and **puborectalis**
    - All muscles covered by fascial sheath
    - Pierced by urethra, rectum, and (in females) vagina
    - Primary **support for pelvic organs**
    - Critical for maintaining continence as well as normal micturition and defecation

Perineum

- **Space inferior to levator ani**
- **Central tendon of perineum** (**perineal body**) formed in midline
  - Behind urethra in males
  - Behind vagina in females
- Diamond-shaped area bordered by pubic symphysis, ischial tuberosities, and coccyx
  - **Subdivided into 2 spaces** (urogenital and anal triangles) by line drawn slightly anterior to ischial tuberosities, along superficial transverse perineal muscle
- **Urogenital triangle** contains urethra and external genitalia
  - Muscle development greater in male
    - Paired muscles: Superficial and deep transverse perineal, bulbocavernosus, and ischiocavernosus
  - **Urogenital diaphragm**: Deep transverse perineal muscle and fascia, urethral sphincter
  - Male: **Penile bulb**
  - Female: Bartholin glands (greater vestibular glands) and vestibular bulbs
- **Anal triangle** contains anus
  - Identical in both sexes
  - Circular **external anal sphincter** surrounds anus and maintains continence
    - 3 muscle components: Subcutaneous, superficial, and deep
    - Fibers extend posteriorly to form **anococcygeal ligament**
  - **Ischiorectal fossa** located in posterior and lateral portion of triangle
    - Filled with adipose tissue
    - Communicates posteriorly, forming U shape around levator ani and anus
    - Bounded posteriorly by sacrotuberous ligament and gluteus maximus

CLINICAL IMPLICATIONS

Clinical Importance

- **Pelvic wall musculature** protects and supports pelvic viscera, stabilizes pelvis for ambulation, increases abdominal pressure for defecation, micturition, and, in females, parturition
  - Laxity or diastasis of musculature results in hernias
    - Incarceration may lead to bowel obstruction
- **Pelvic floor musculature** stabilizes lower pelvis, controls defecation and micturition, and maintains continence
  - Laxity results in pelvic floor descent with cystoceles/rectoceles, urinary incontinence, fecal incontinence, or constipation (more common in females)

Inguinal Hernia

- May be either indirect or direct
- **Indirect inguinal hernia**
  - Passes **through internal inguinal ring**, traverses inguinal canal to external ring
    - May extend into scrotum in males and labia majora in females
  - Passes **lateral to inferior epigastric vessels** and has oblique inferior course
  - Considered congenital defect and is associated with patent processus vaginalis
  - 5x more common than direct inguinal hernias
- **Direct inguinal hernia**
  - Protrusion through Hesselbach triangle
    - Generally does not extend into scrotum
  - Passes **medial to inferior epigastric vessels**
  - Considered acquired defect
- Inguinal hernias 5-10x more common in males

Femoral Hernia

- Begins posterior to medial portion of inguinal ligament
  - Traverses **femoral canal to fossa ovalis**
- Herniated contents are **below inguinal ligament**, lateral to pubic tubercle, and medial to femoral vessels
- More common in older women

Obturator Hernia

- **Herniation through obturator canal** (superolateral aspect of obturator foramen)
- Herniated contents may lie between obturator muscles or between pectineus and obturator externus
- More common in older women

Sciatic Hernia

- Herniation through greater sciatic foramen laterally into subgluteal region

Perineal Hernia

- Posterior: Through levator ani and coccygeus muscles
- Anterior: Through urogenital diaphragm

3D CT, BONY PELVIS

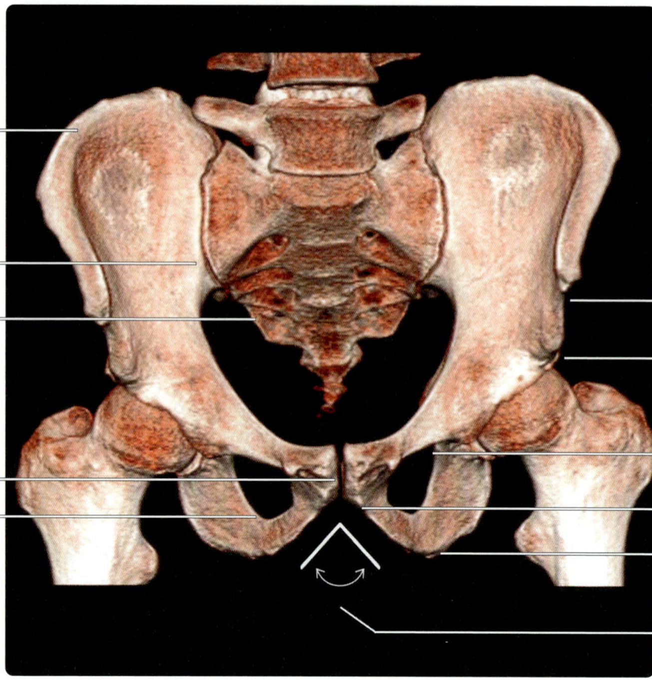

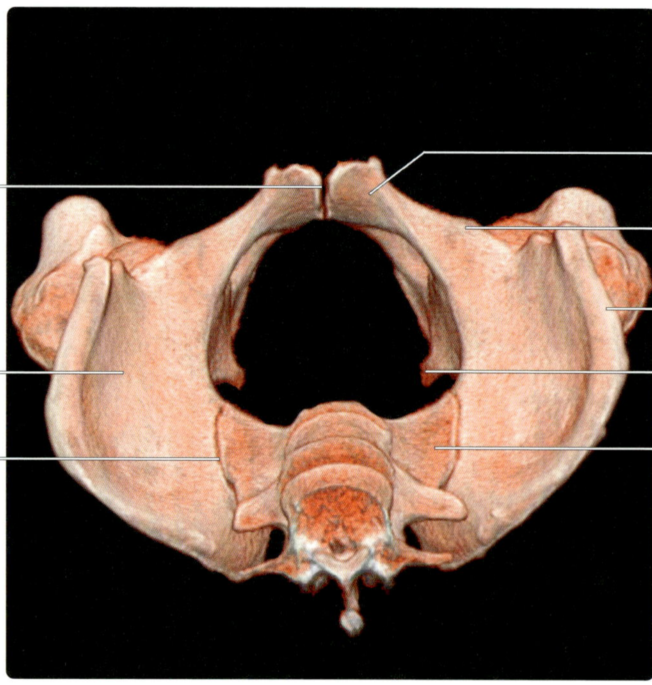

(Top) *3D bone CT reconstructions of the male pelvis are shown. There are several striking differences between the male and female pelvic bones. The bones are thicker and heavier in the male. The pubic arch is higher with a narrower angle. The pelvic inlet is more heart-shaped, and both the inlet and outlet are much narrower compared to the female pelvis.* **(Bottom)** *Superior view of the pelvic inlet is shown. Note the ischial spines are easily visible as the pelvis continues to narrow at the outlet.*

3D CT, MAJOR LIGAMENTS AND FORAMINA

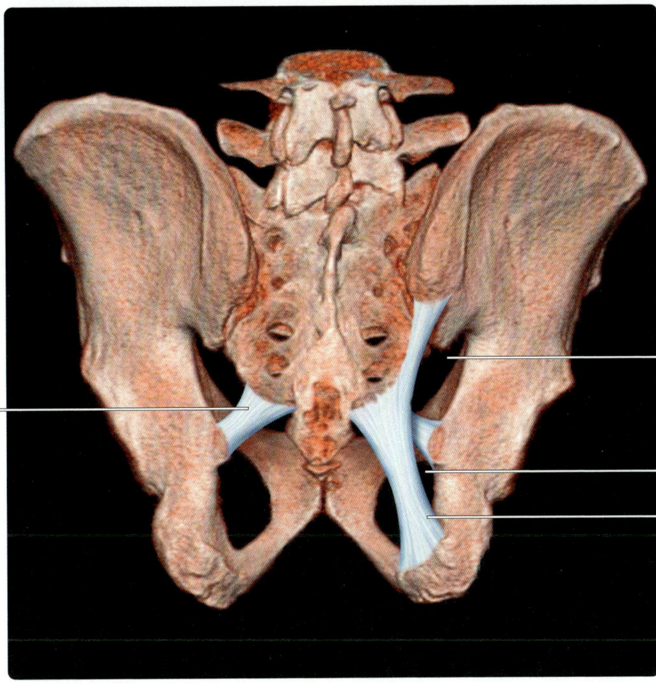

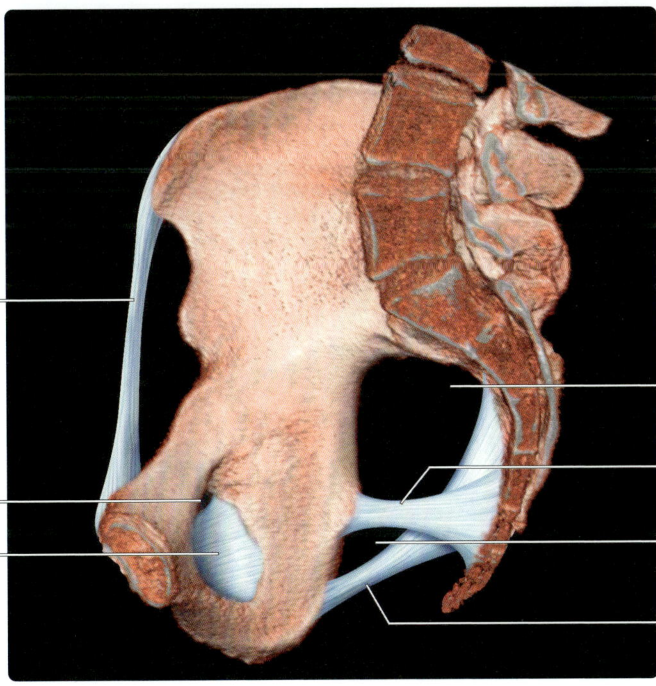

(Top) *3D CT reconstruction of the pelvis is shown, graphically enhanced to show the key ligaments. The sacrospinous ligament extends between the sacrum and ischial spine. The greater sciatic foramen is above the sacrospinous ligament, and the lesser sciatic foramen is below it. The sacrotuberous ligament has a somewhat broader attachment to the sacrum and extends to the ischial tuberosity. It forms the posterior boundary of the lesser sciatic foramen.* **(Bottom)** *The inguinal ligament extends from the anterior superior iliac spine to the pubic tubercle. A membrane covers the obturator foramen, except for the obturator canal, which is in the superolateral aspect of the foramen. The obturator vessels and nerve pass through this opening.*

ANTERIOR PELVIC WALL

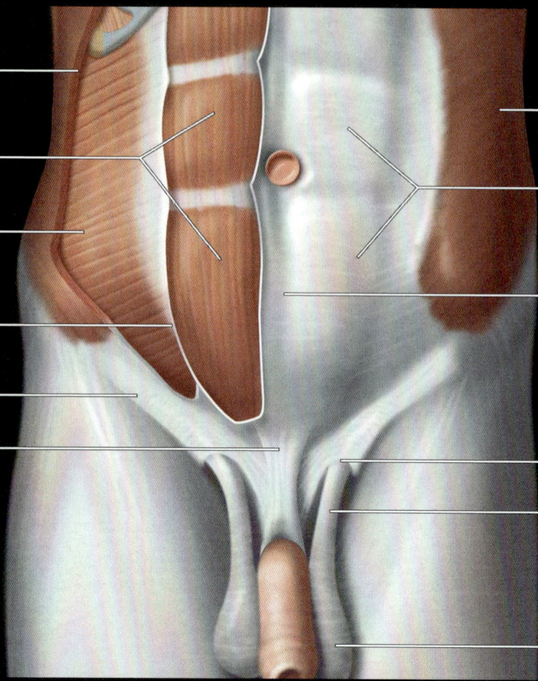

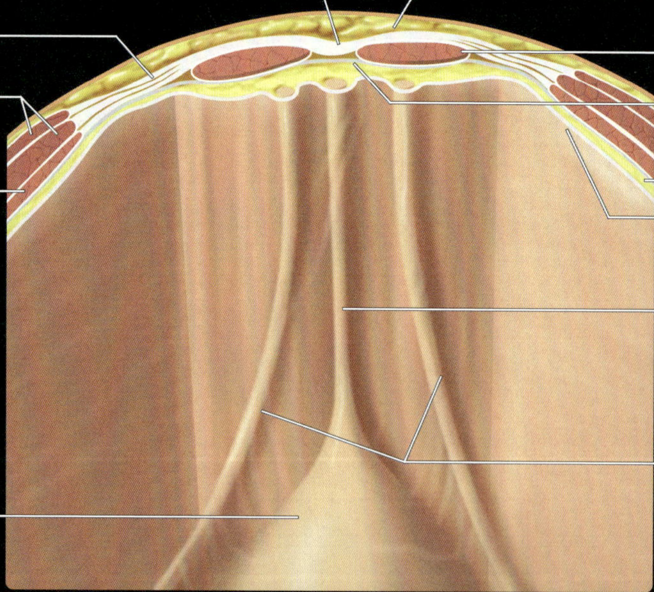

(Top) *Anterior graphic shows abdominal wall muscles.* **(Bottom)** *The inner aspect of the anterior abdominal wall, below the level of the arcuate line, is shown. The rectus sheath, which is formed by the aponeuroses of the external and internal oblique and transversus abdominis muscles, encloses the rectus muscle both anteriorly and posteriorly in the upper abdomen. In the pelvis, at the level of the anterior superior iliac spine, the sheath is present only anteriorly. The transition where this occurs is called the arcuate line. Only the transversalis fascia covers the posterior rectus muscle below this point, as depicted in this graphic. Extraperitoneal fat separates the transversalis fascia from the parietal peritoneum.*

POSTERIOR PELVIC WALL

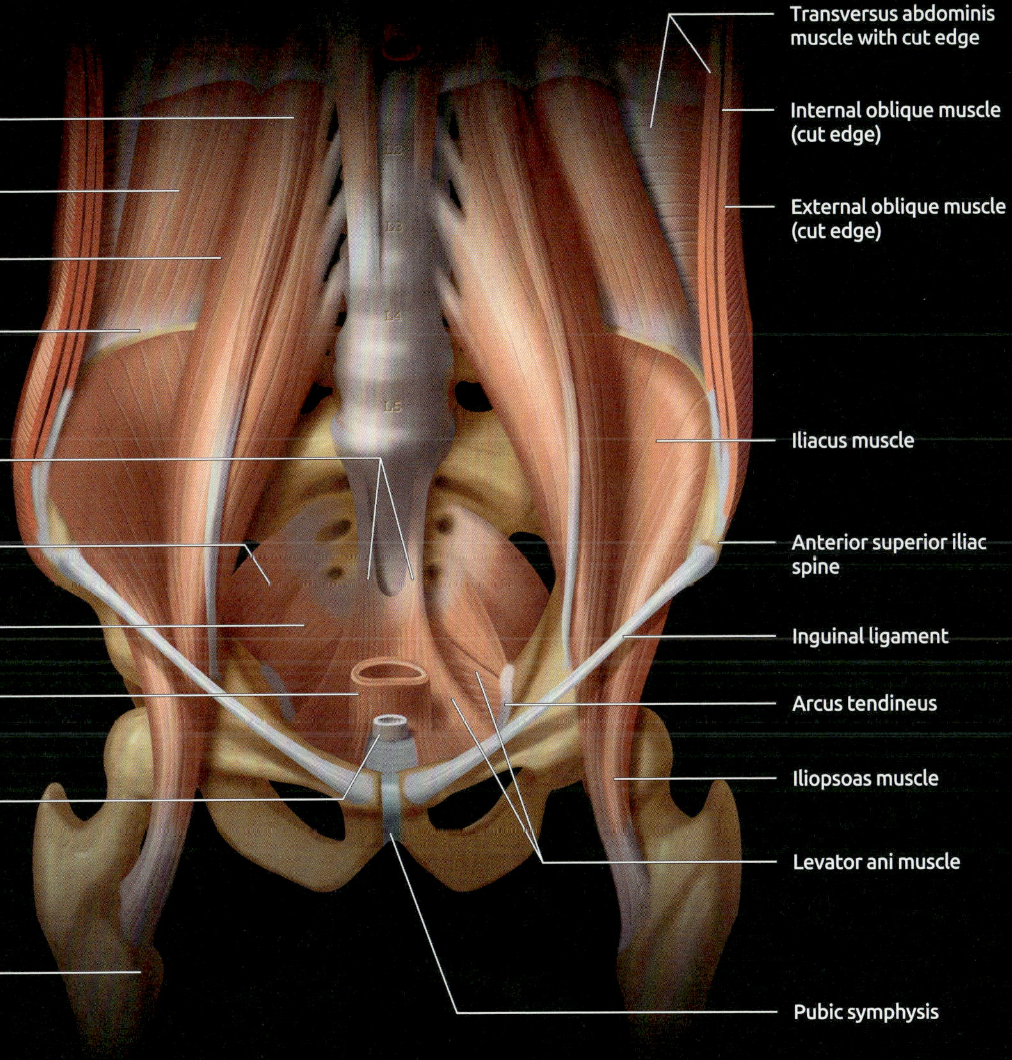

The iliac bones, sacrum, and iliacus and psoas muscles form the posterior wall of the false pelvis. The iliacus and psoas muscles fuse caudally to form the iliopsoas muscle, which passes anterior to the hip joint to insert onto the lesser trochanter of the femur. The sacrum, coccyx, and piriformis and coccygeus muscles form the posterior wall of the true pelvis. The inguinal ligament is formed by the external oblique aponeurosis and is continuous with the fascia lata of the thigh.

PELVIC FLOOR AND PERINEUM

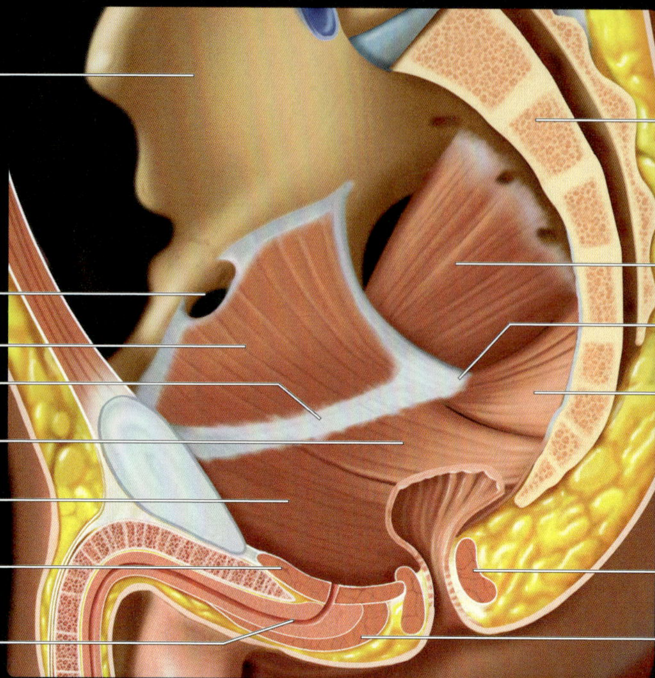

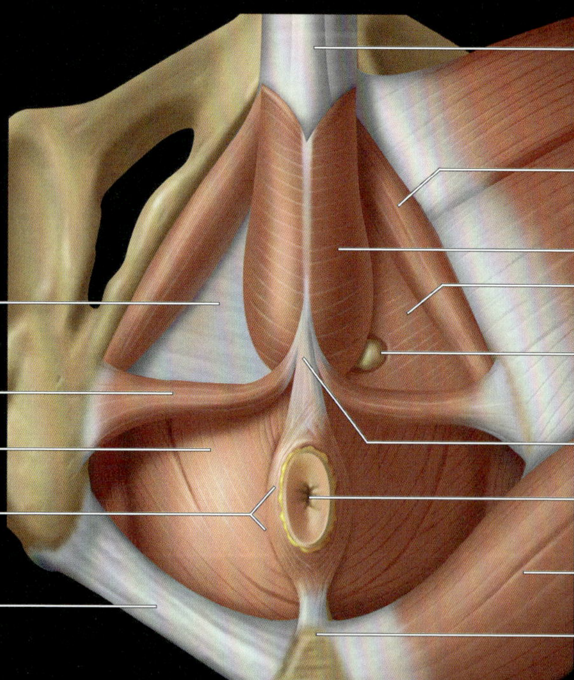

(Top) *The coccygeus muscles and pelvic diaphragm (levator ani muscles and fascia) form the pelvic floor. The levator ani is composed of 3 separate muscles: Pubococcygeus, ileococcygeus, and puborectalis. The levator ani is attached to the pubic bones anteriorly, the ischial spines laterally, and the arcus tendineus (thickening in the obturator fascia) between the bony attachments. The pelvic diaphragm separates the pelvic cavity from the perineum. The obturator internus and piriformis muscles form the lateral and posterior walls of the true pelvis.* **(Bottom)** *A thin sheet of tough, deep fascia called the perineal membrane covers the deep transverse perineal muscle. The urethra pierces the membrane (as it does in the vagina in females) and branches off the pudendal neurovascular bundle. The deep transverse perineal muscle and membrane form the urogenital diaphragm and provide an attachment for the external genitalia. No such membrane exists posterior to the transverse perineal muscles (anal triangle).*

AXIAL CT

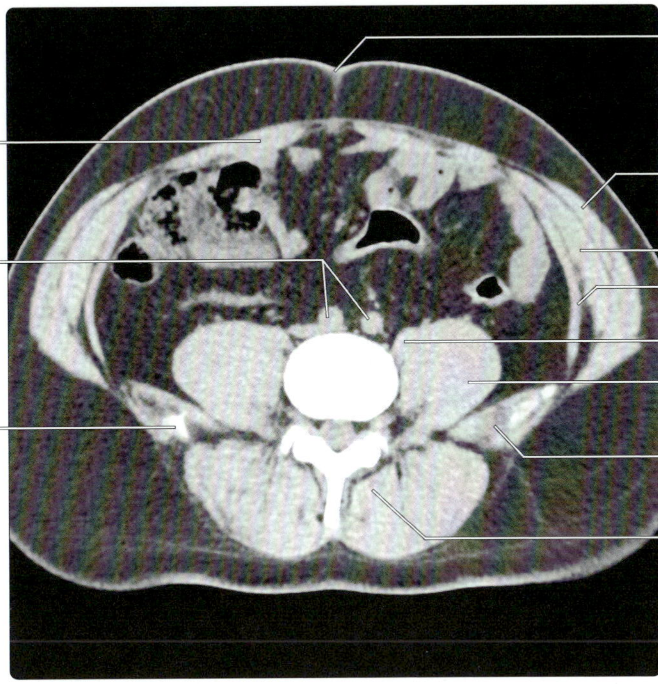

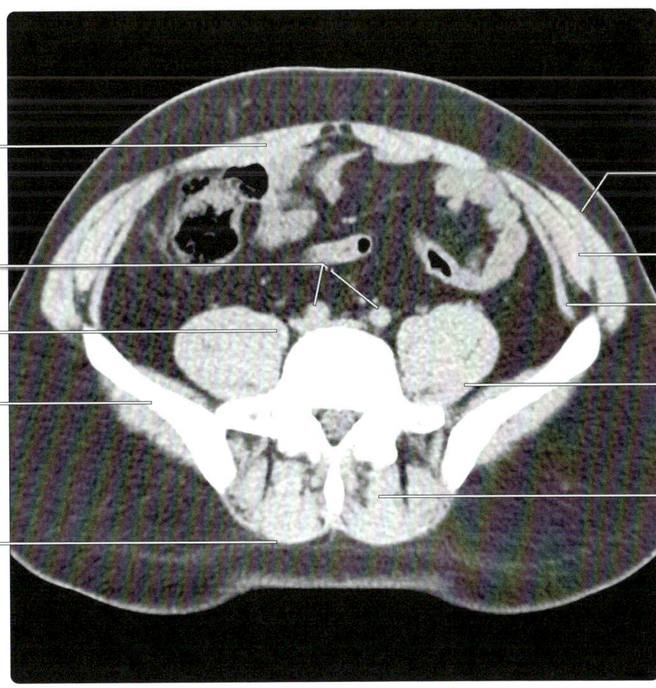

(Top) *First of 14 axial NECT images through the pelvis is shown. The pelvis begins at the top of the iliac crest and extends to the perineum.* **(Bottom)** *The flat muscles that form the lateral anterior abdominal wall include (from external to internal) the external oblique, internal oblique, and transversus abdominis muscles. The rectus abdominis muscles are paired, vertically oriented, strap-like muscles running on either side of the midline.*

AXIAL CT

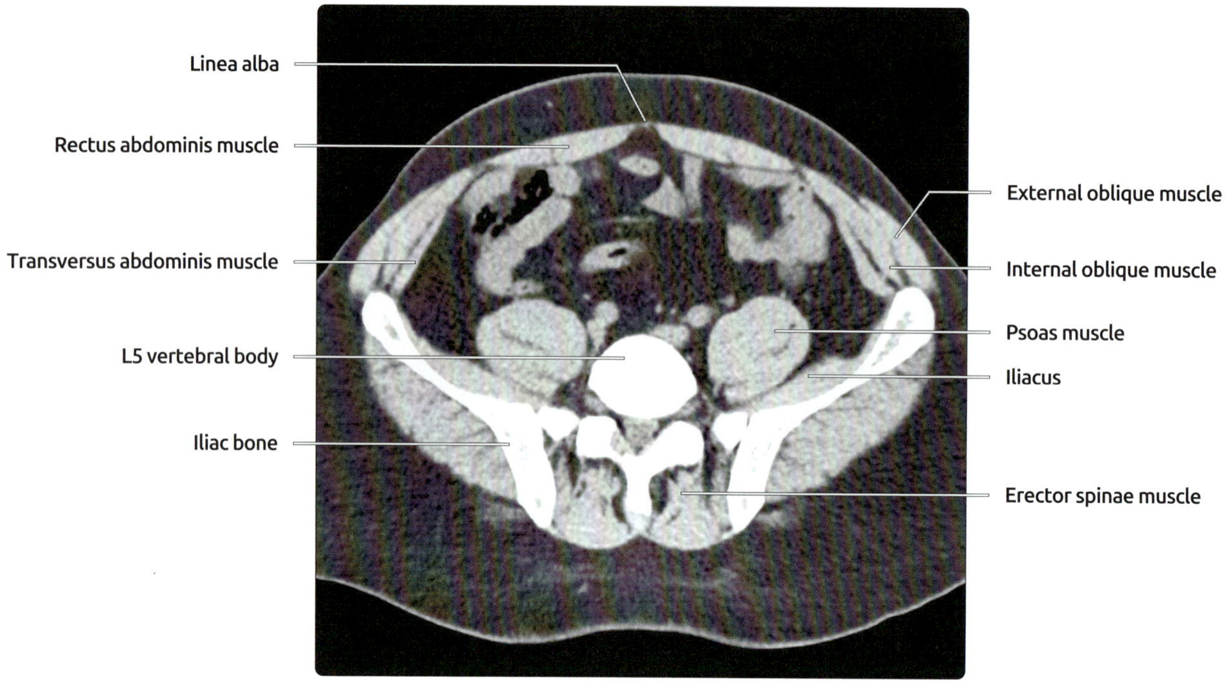

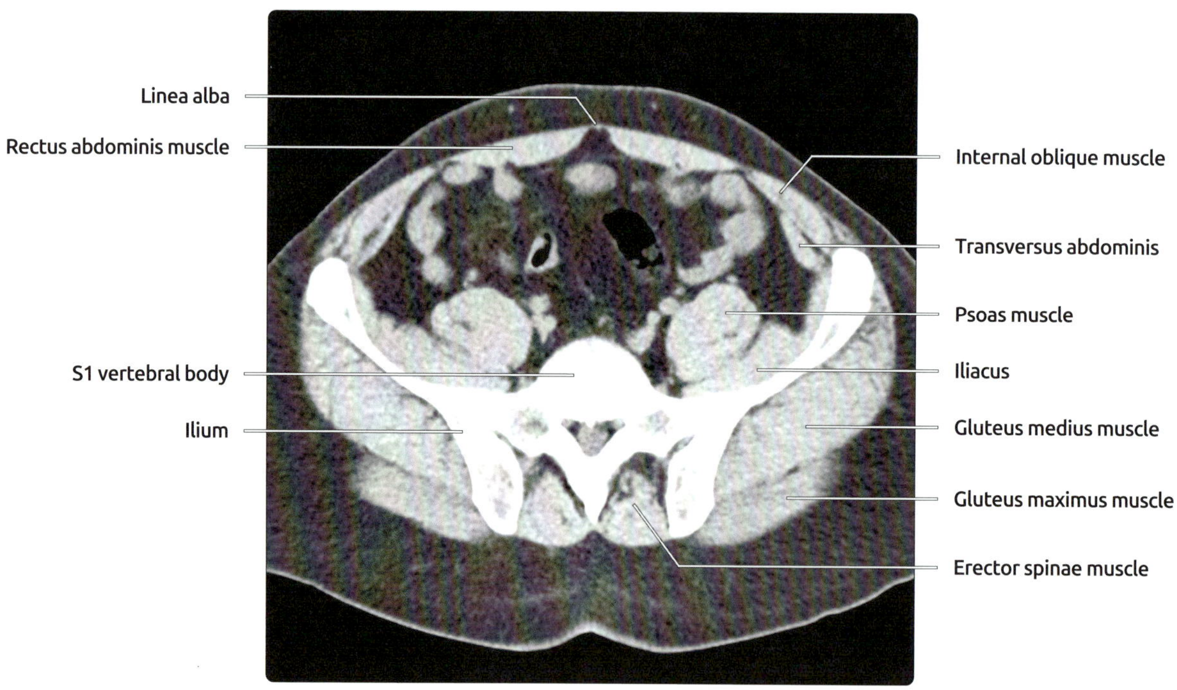

(Top) *The linea alba is a fibrous, midline raphe between the rectus muscles. It is formed by the aponeuroses of the flat muscles (external oblique, internal oblique, and transversus abdominis muscles).* **(Bottom)** *The arcuate line (not radiographically visible) occurs at roughly the level of the anterior superior iliac spine; above this line, the aponeurosis of the internal oblique splits. The anterior portion joins the external oblique aponeurosis to form the anterior rectus sheath. The posterior portion joins the transversus abdominis aponeurosis to form the posterior rectus sheath.*

AXIAL CT

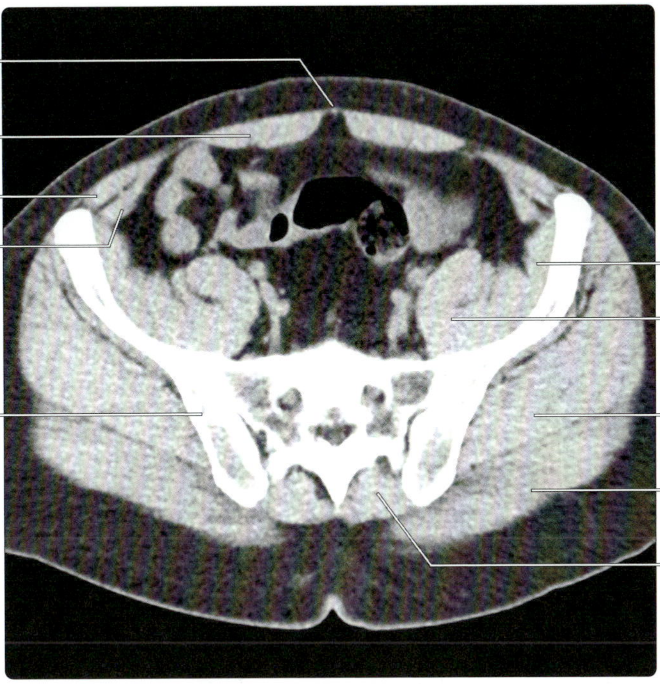

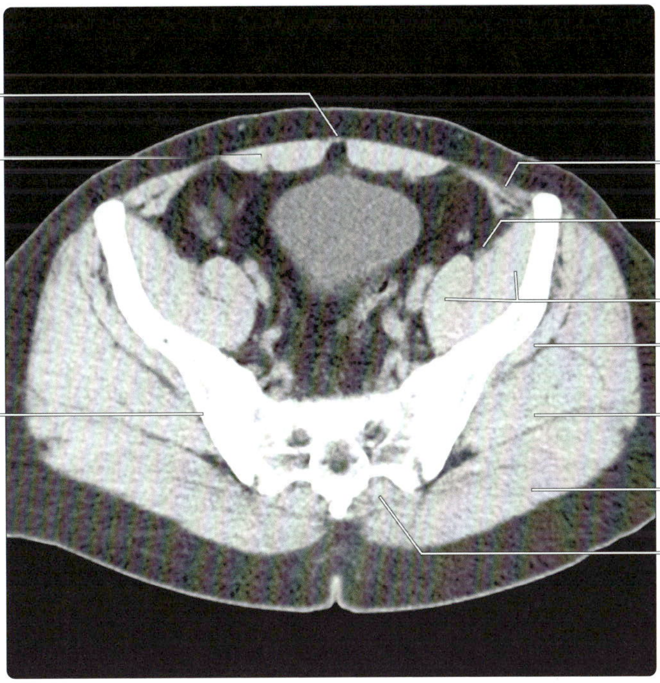

(Top) *Below the arcuate line, all 3 aponeuroses join and course anterior to the rectus muscle, making the caudal portion of the posterior rectus sheath incomplete. In this region, the posterior rectus abdominis muscle is covered only by the transversalis fascia, which is separated from the parietal peritoneum by a layer of extraperitoneal fat. Because the rectus sheath is incomplete in this area, a rectus hematoma or abscess may extend into the extraperitoneal space.* **(Bottom)** *The psoas and iliacus muscles merge to form the iliopsoas muscle, which continues inferiorly to insert on the lesser trochanter and serve as a powerful hip flexor. The femoral nerve lies in the iliopsoas groove on its way to the inguinal canal.*

AXIAL CT

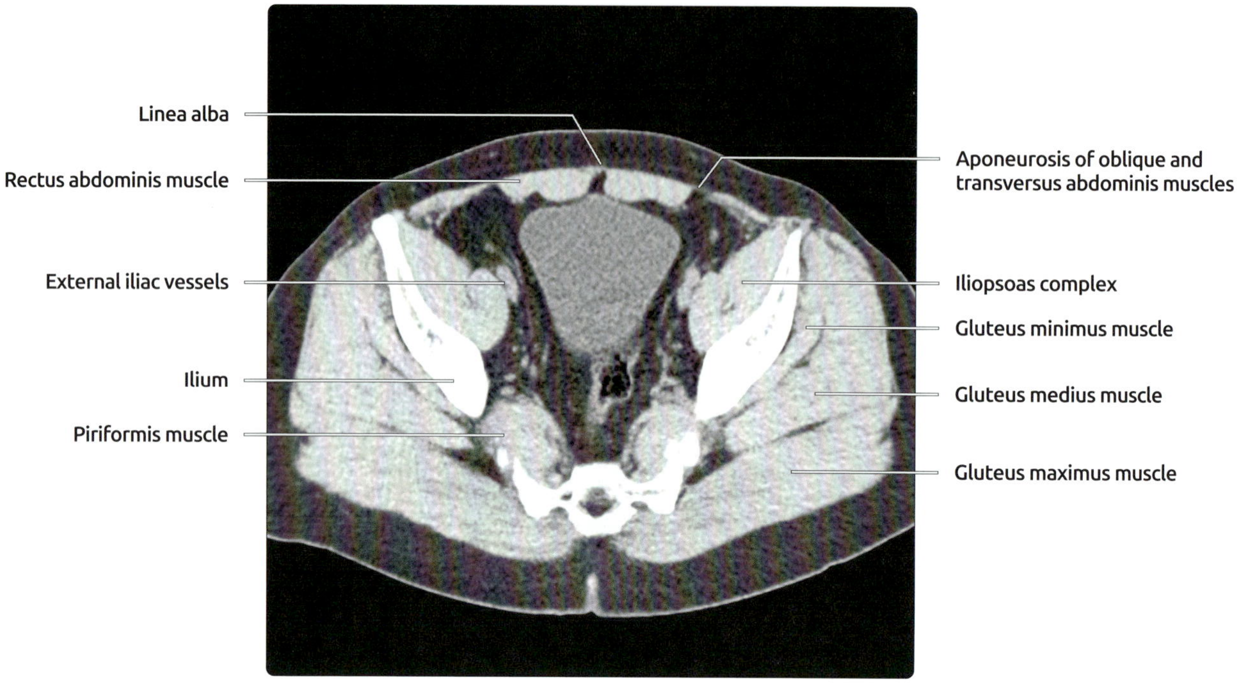

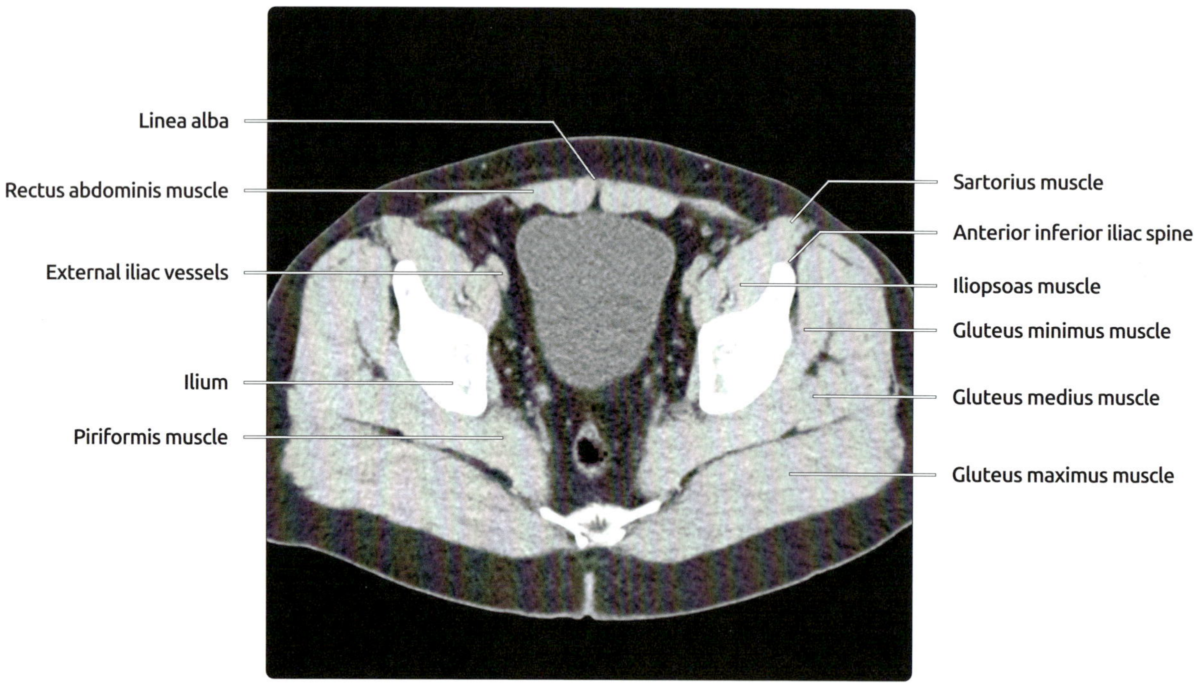

(Top) *Note the thinning of the abdominal wall lateral to the rectus muscle. This is the aponeurosis of the internal oblique, external oblique, and transversus abdominis muscles. This is a point of inherent weakness; a Spigelian hernia may occur through this area. The piriformis muscle forms an important part of the posterior and lateral wall of the true pelvis.* **(Bottom)** *Note how the pelvis narrows inferiorly as it transitions from the false pelvis to the true pelvis.*

AXIAL CT

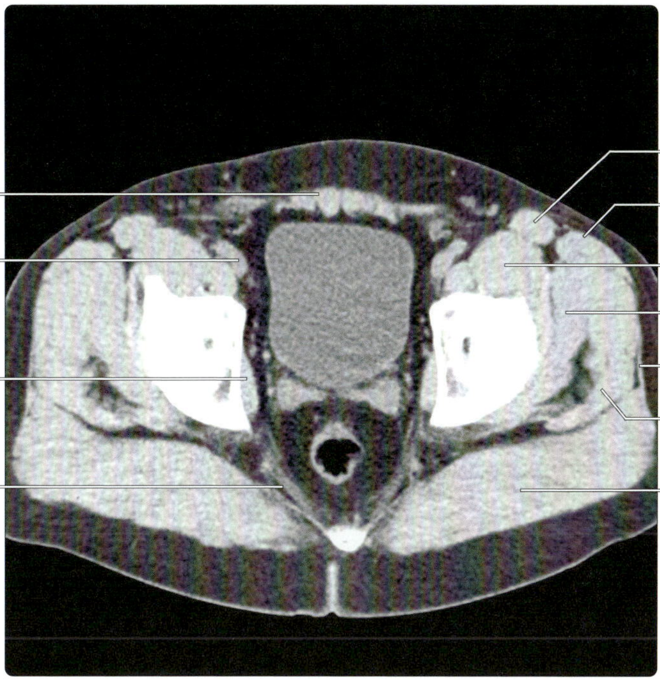

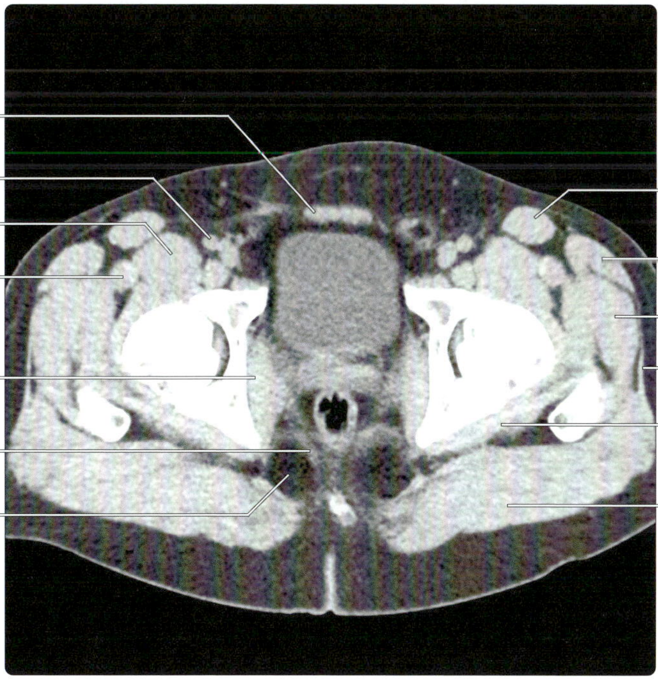

(Top) *The sacrospinous ligament is an important anatomic landmark with the greater sciatic foramen being above this ligament and the lesser sciatic foramen below it.* **(Bottom)** *The obturator internus muscle, an important part of the lateral pelvic floor, exits the pelvis through the lesser sciatic foramen. The tendon of the obturator internus fuses with the gemellus muscle before inserting on the medial surface of the greater trochanter. Within the pelvis, the fascia of the obturator internus forms the arcus tendineus, which serves as the origin for the levator ani muscles. The obturator internus also forms the lateral wall of the ischiorectal fossa.*

AXIAL CT

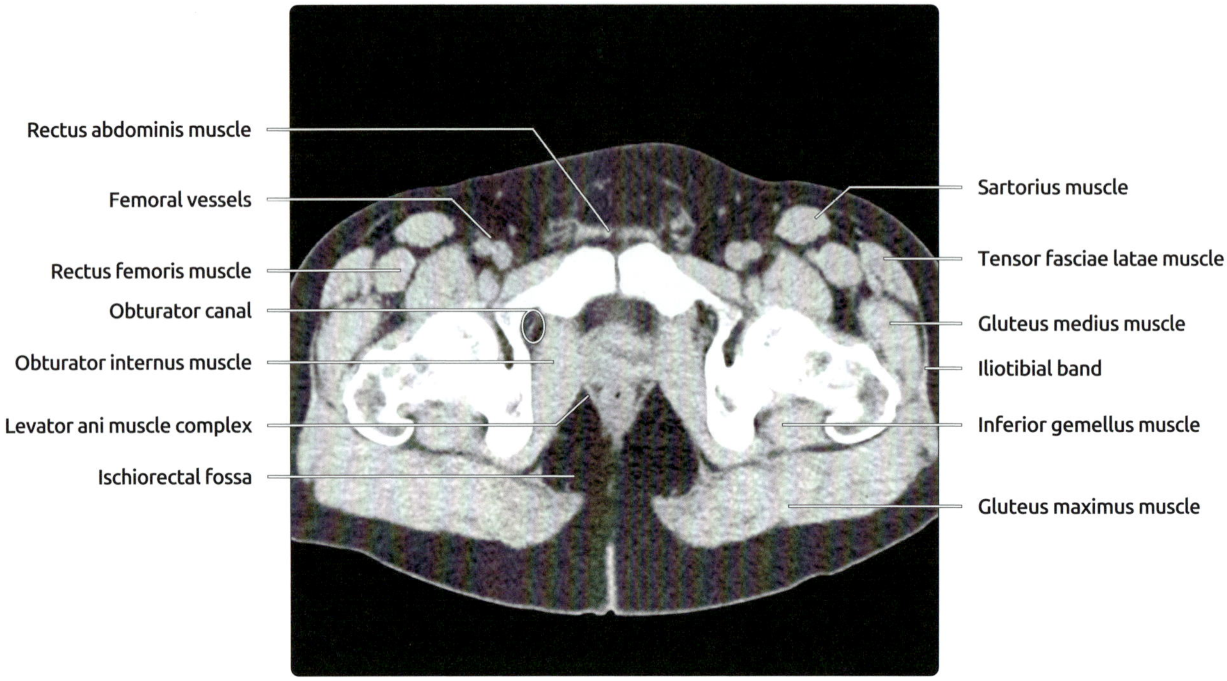

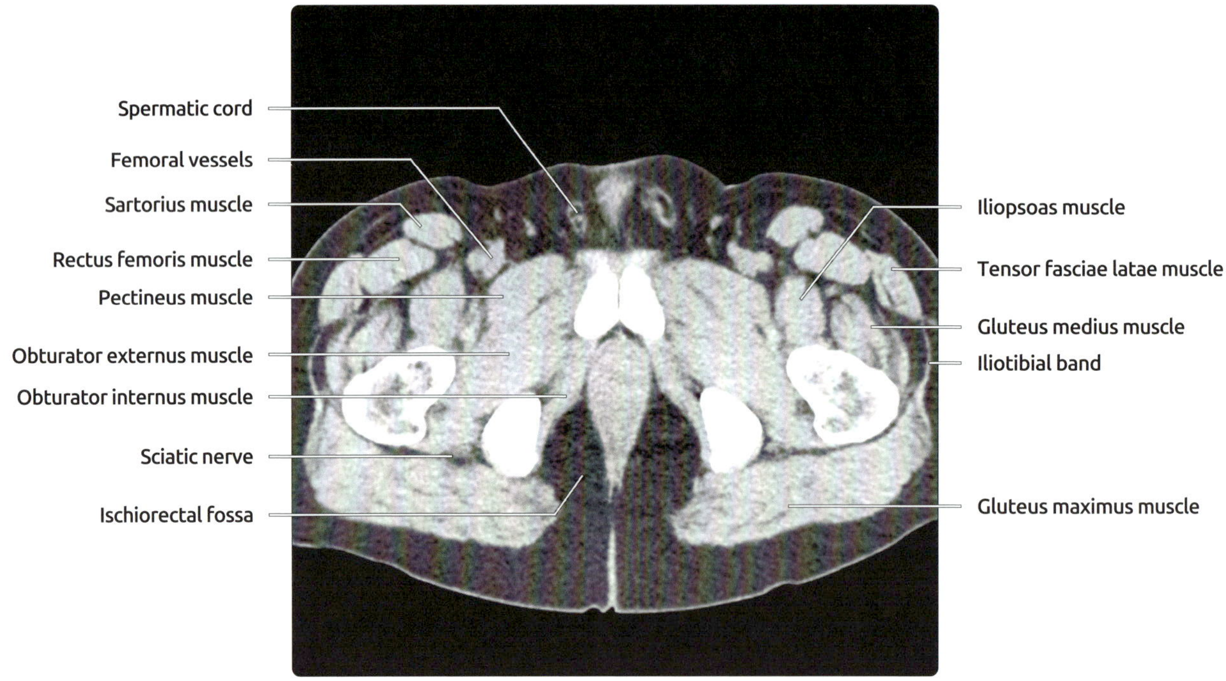

(Top) *Inferiorly, the rectus abdominis muscle thins and inserts on the pubic symphysis.* **(Bottom)** *The ischiorectal fossa is filled with fat and communicates across the midline, forming a U shape around the levator ani and anus. A perirectal abscess may extend into this space.*

AXIAL CT

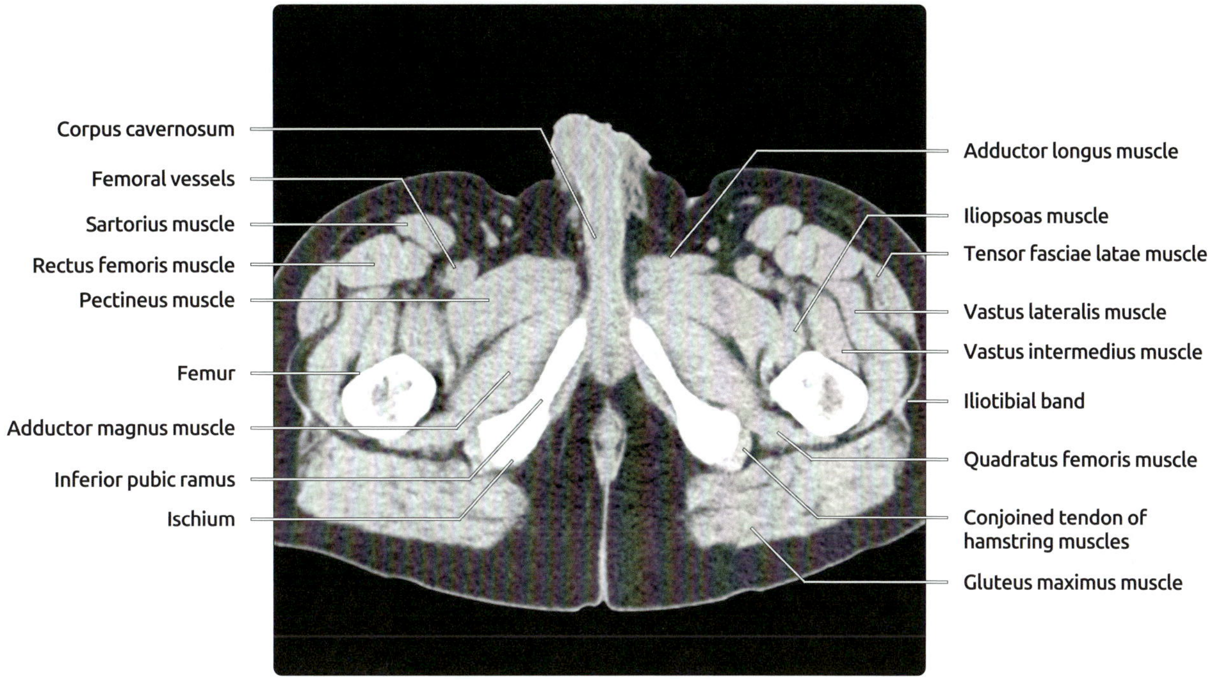

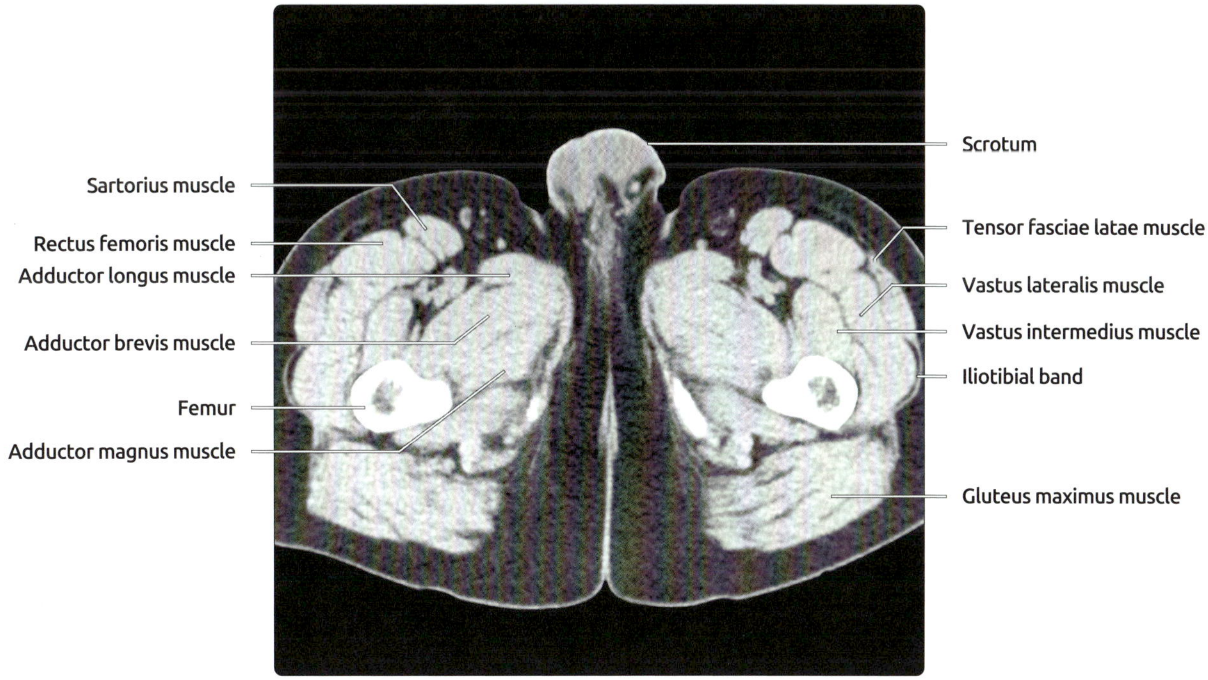

(Top) *The perineum is the space below the levator ani and includes the external genitalia, urethra, and anus.* **(Bottom)** *More inferiorly, the powerful adductor muscle group is seen.*

CORONAL T1 MR

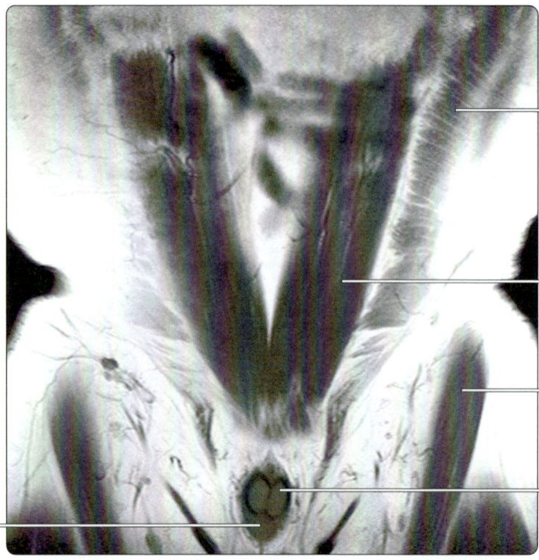

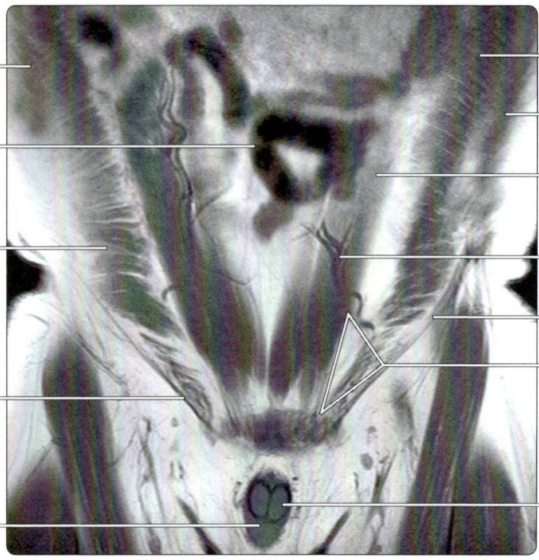

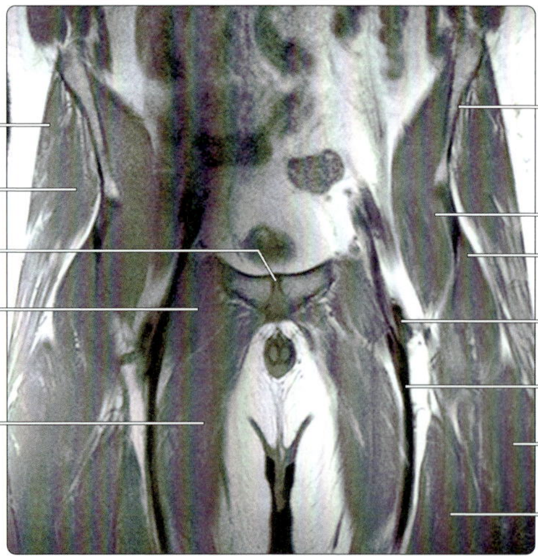

(Top) *First of 12 coronal T1 MR images through the abdominal wall is shown. The rectus abdominis muscle originates from the pubic symphysis and crest, coursing superiorly and inserting onto the anterior surface of the xiphoid process and medial portion of the 5th-7th costal cartilages.* **(Middle)** *The fibers of the external oblique muscle (partially seen) are directed inferomedially. The internal oblique muscle runs superomedially in the abdomen but becomes more horizontal and eventually courses inferiorly, low in the pelvis. The inguinal ligament extends from the anterior superior iliac spine to the pubic tubercle. The triangular area bounded by the lateral aspect of the rectus sheath medially, the inferior epigastric vessels laterally, and the inguinal ligament inferiorly is the Hesselbach triangle, an area of weakness of the anterior abdominal wall.* **(Bottom)** *The femoral vessels and nerves pass underneath the inguinal ligament through the femoral canal. Contents of the femoral canal include (from lateral to medial) the femoral nerve, artery, and vein.*

CORONAL T1 MR

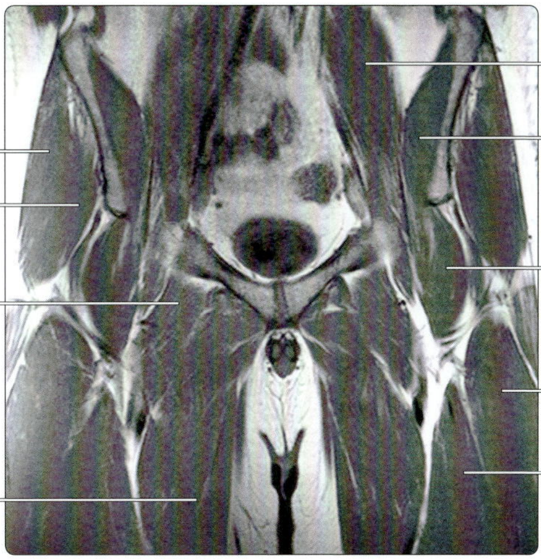

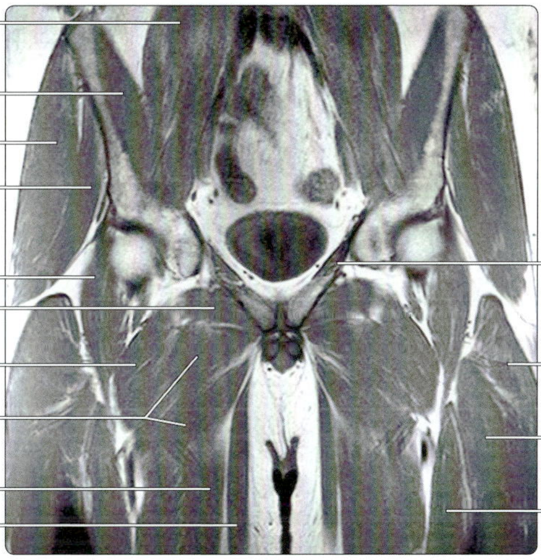

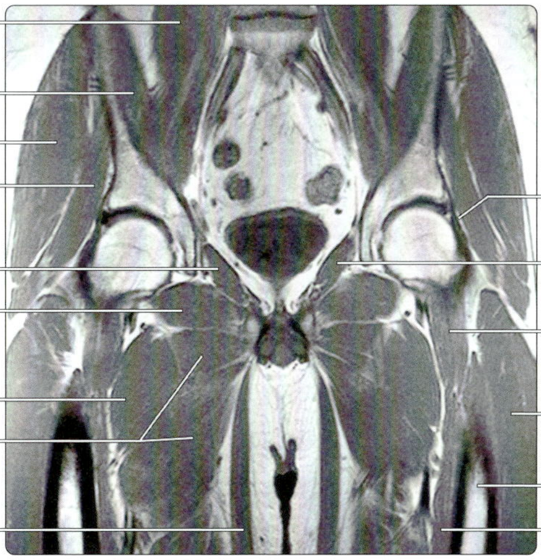

(Top) *The iliacus and psoas muscles, the largest muscles within the pelvic cavity, fuse caudally to form the iliopsoas muscle, a powerful hip flexor. After leaving the pelvis, it inserts on the lesser trochanter of the femur. The pectineus is a flat, quadrangular muscle in the femoral triangle. It arises from the pectineal line of the pubis and passes posterolaterally to insert below the lesser trochanter.* **(Middle)** *The superior pubic ramus forms the superior border of the obturator foramen, which is covered by a strong membrane. The obturator externus muscle arises from the outer surface of the obturator membrane.* **(Bottom)** *The obturator internus arises from the inner surface of the obturator membrane and the adjacent inner surfaces of the pubic bone and ischium. It passes posterolaterally along the lateral wall of the pelvis to form a tendon that traverses the lesser sciatic foramen. Inferior to the ischial spine, it turns laterally to insert into the medial aspect of the greater trochanter of the femur.*

CORONAL T1 MR

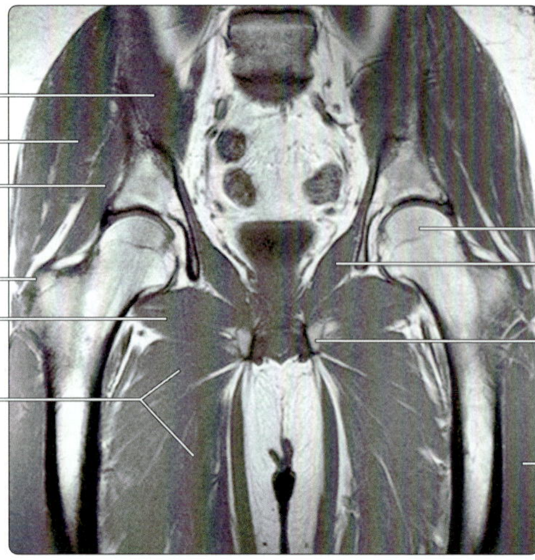

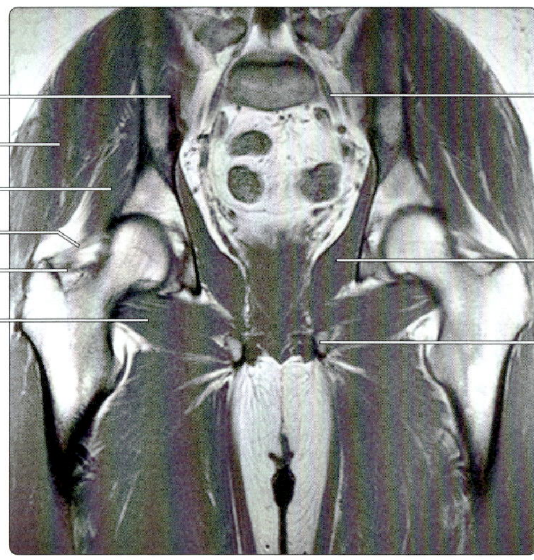

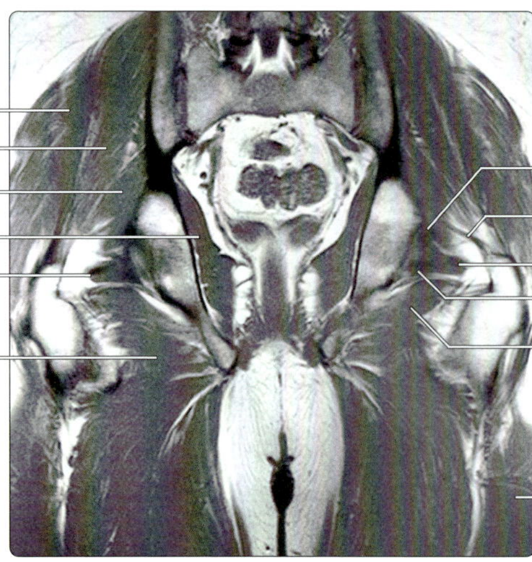

(Top) *More posteriorly, the gluteus medius and minimus muscles are seen inserting onto the greater trochanter of the femur. Both muscles are powerful abductors of the thigh. The obturator externus muscle passes behind the neck of femur to insert onto the greater trochanter.* **(Middle)** *The piriformis and obturator internus tendon are seen inserting onto the greater trochanter of the femur. Both muscles are lateral rotators of the hip.* **(Bottom)** *Slightly posteriorly, the obturator externus tendon is shown inserting on the medial aspect of the greater trochanter of the femur. It too rotates the hip laterally.*

CORONAL T1 MR

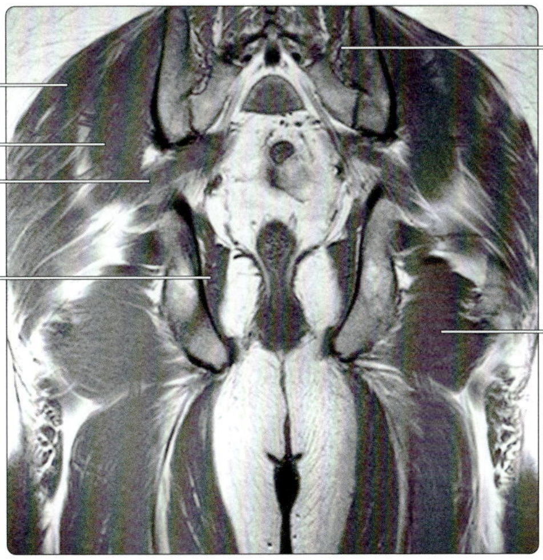

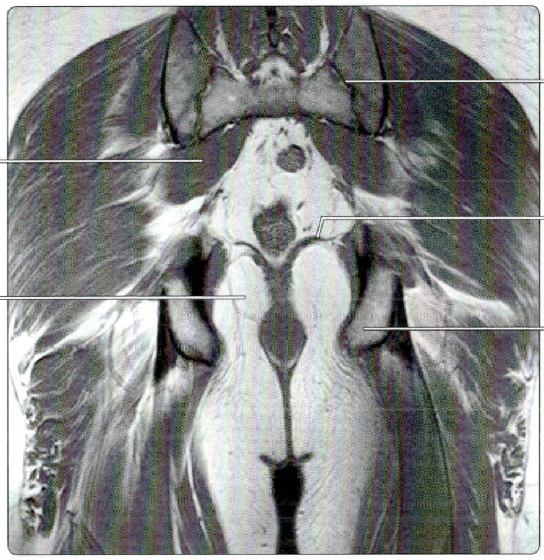

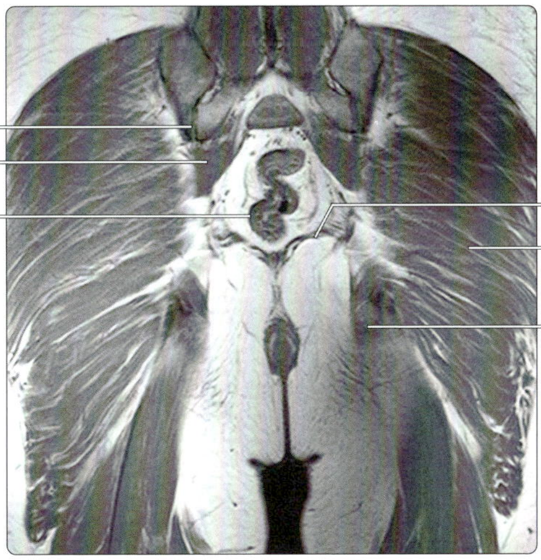

(Top) *The piriformis muscle is seen exiting the pelvis through the greater sciatic foramen. Once outside the pelvis, it runs inferior to the gluteus medius and may fuse with it.* **(Middle)** *The ischiorectal fossa is a pyramid-shaped space with its base toward the perineum and its apex at the origin of the levator ani muscle. The ischiorectal fossa lies between the ischial tuberosity and the obturator internus muscle laterally and the external anal sphincter and the levator ani muscle medially. The space contains fat, fibrous septa, and pudendal and rectal nerves and vessels.* **(Bottom)** *The piriformis muscle arises from the anterior surface of the sacrum and gluteal surface of ilium. Note the close relationship to the rectum, especially on the left.*

SAGITTAL T2 MR

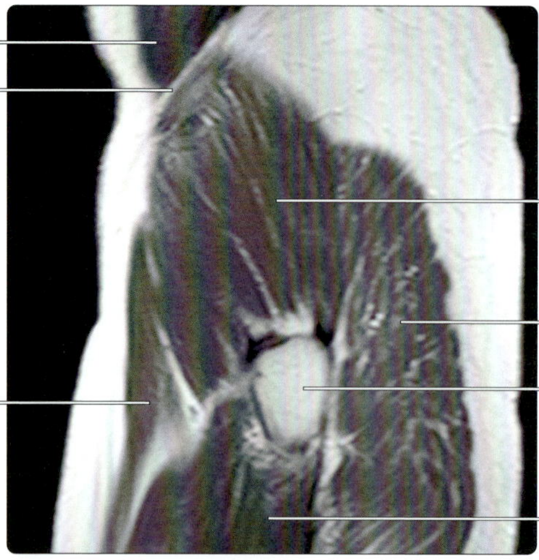

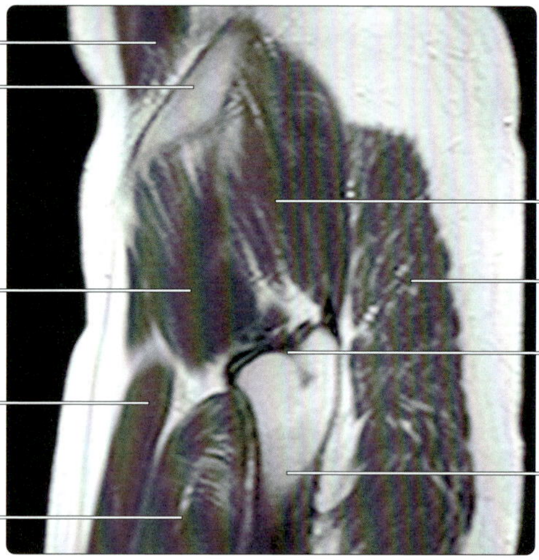

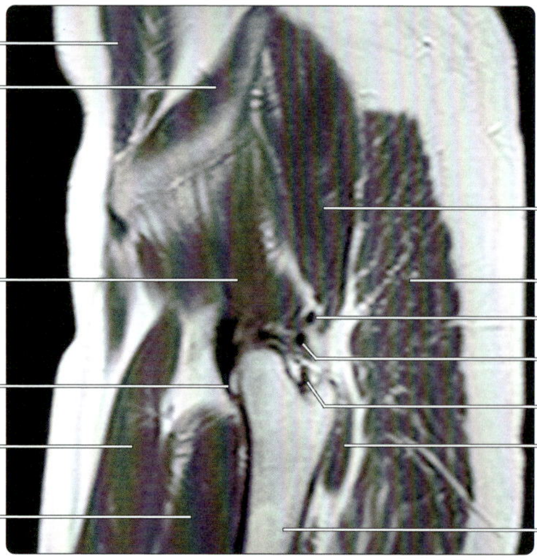

(Top) *First of 12 sagittal T2 MR images through the pelvic wall presented from lateral to medial is shown. Both the gluteus medius and minimus muscles arise from the dorsal ilium inferior to the iliac crest. The gluteus medius muscle inserts onto the lateral and superior surfaces of greater trochanter, while the gluteus minimus muscle attaches to the anterior surface of the greater trochanter. Both muscles are hip abductors.* **(Middle)** *The gluteus maximus is the largest muscle in the body. Its origin includes the posterior gluteal line, portions of the medial iliac crest, sacrum, coccyx, and sacrotuberous ligament. It inserts on the iliotibial tract of the fascia lata and gluteal tuberosity of the femur. It is the principle extensor of the thigh.* **(Bottom)** *The tendons of the piriformis, obturator internus, and obturator externus muscles are seen coursing to their insertions on the greater trochanter. This group of muscles helps to laterally rotate and abduct the thigh.*

SAGITTAL T2 MR

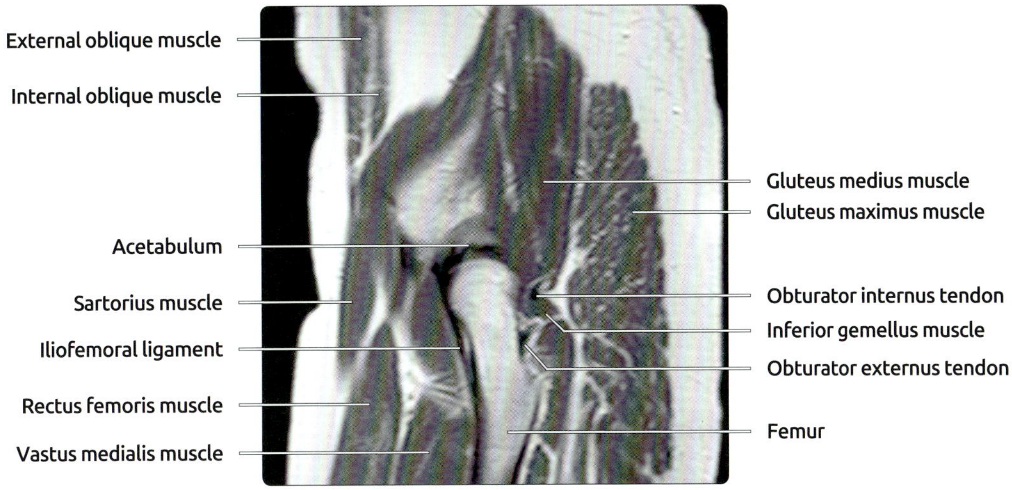

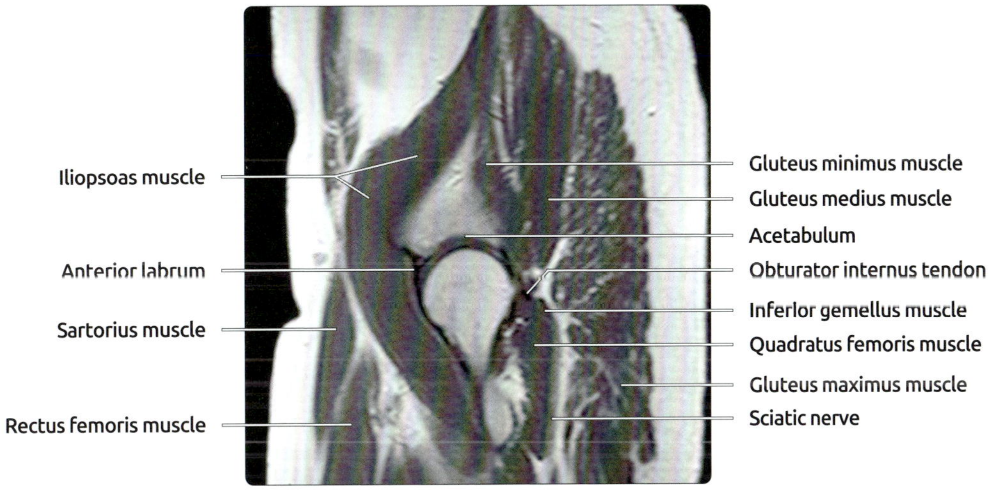

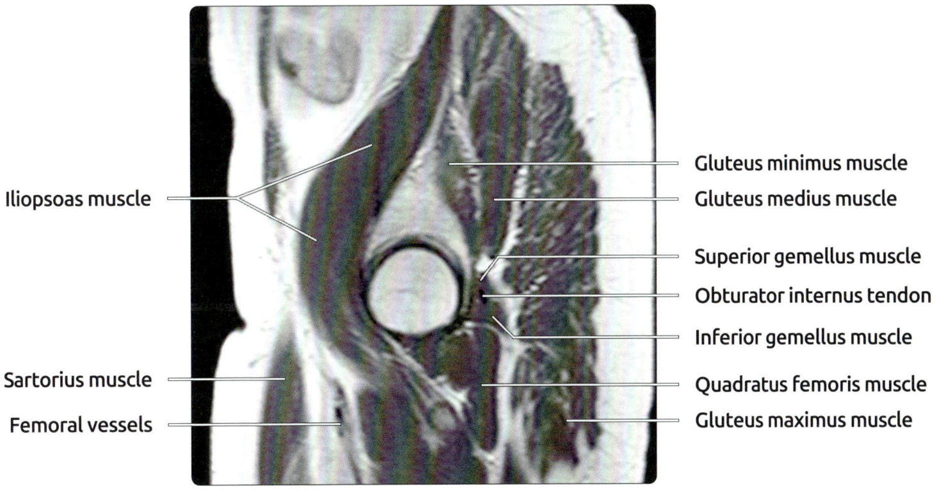

(Top) *The sartorius muscle arises from the anterior superior iliac spine and courses inferiorly to insert on the medial surface of the tibial shaft, near the tibial tuberosity.* **(Middle)** *The sciatic nerve is seen posterior to the quadratus femoris muscle, where it descends after leaving the pelvis through the greater sciatic foramen, below the piriformis muscle.* **(Bottom)** *The superior gemellus muscle arises from the ischial spine, while the inferior gemellus muscle arises from the ischial tuberosity. Both muscles insert on the medial surface of the greater trochanter of the femur, along with the obturator internus tendon. All 2 muscles rotate the hip laterally.*

SAGITTAL T2 MR

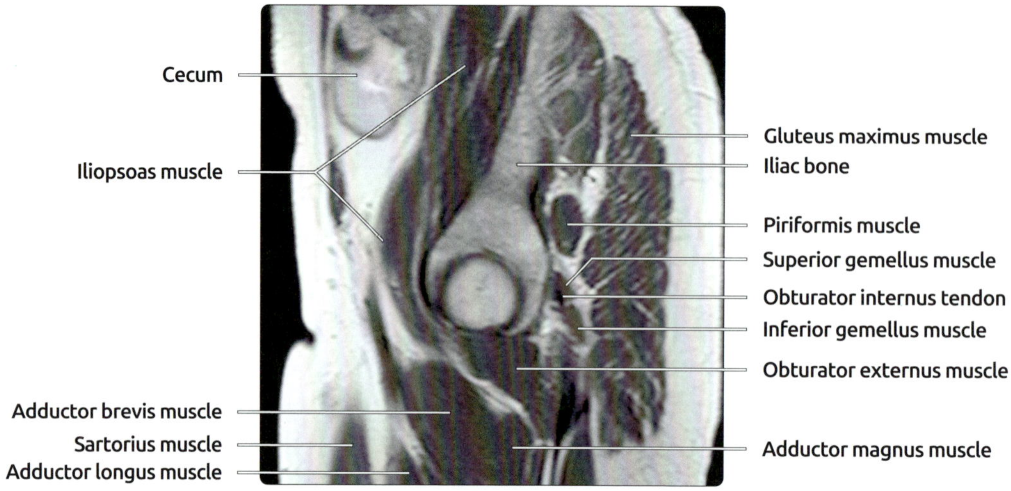

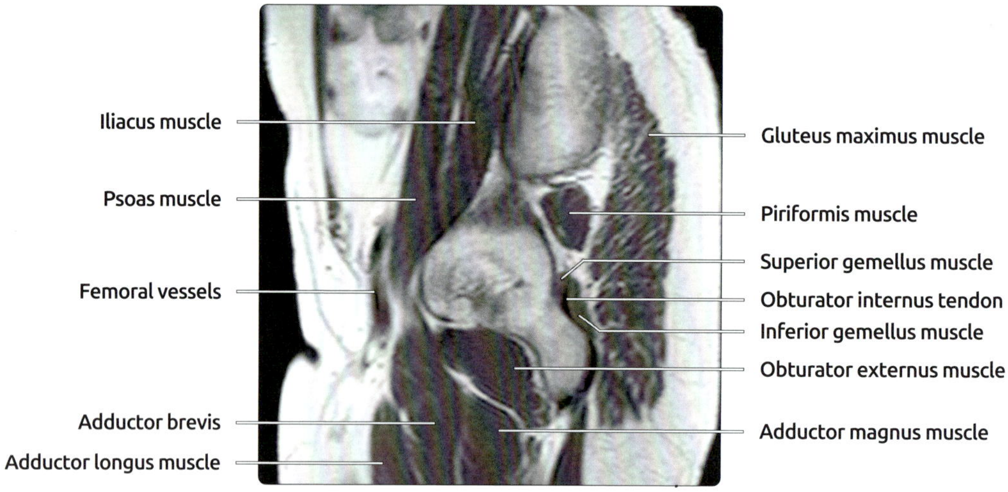

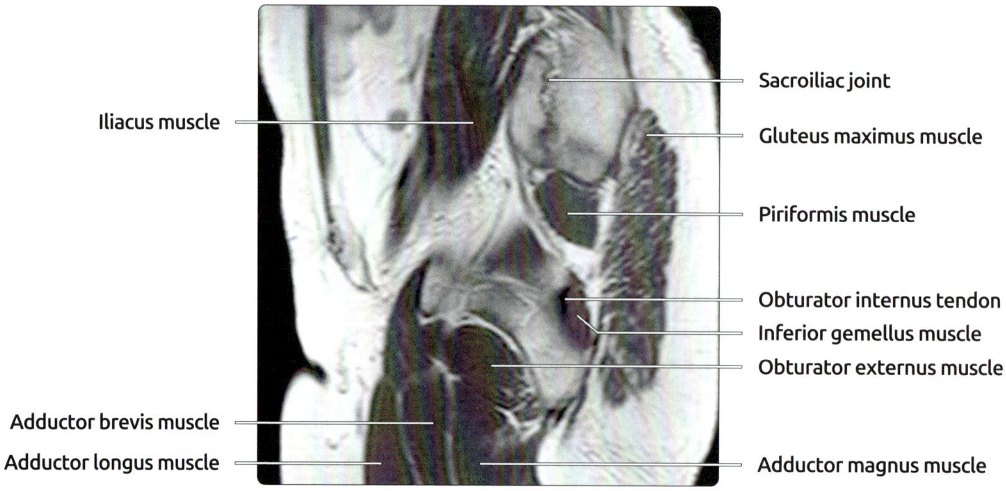

(Top) *The close relation between the iliopsoas muscle and the cecum explains the psoas sign: Pain on extension of the right thigh in patients with an inflamed retrocecal appendix. Inflammation in this area may irritate the iliopsoas muscle and cause spasm.* **(Middle)** *The piriformis muscle exits the pelvis through the greater sciatic foramen and is anterior to the gluteus maximus muscle.* **(Bottom)** *More medially, the powerful group of adductor muscles (longus, brevis, and magnus) is seen anterior and inferior to the obturator externus muscle.*

SAGITTAL T2 MR

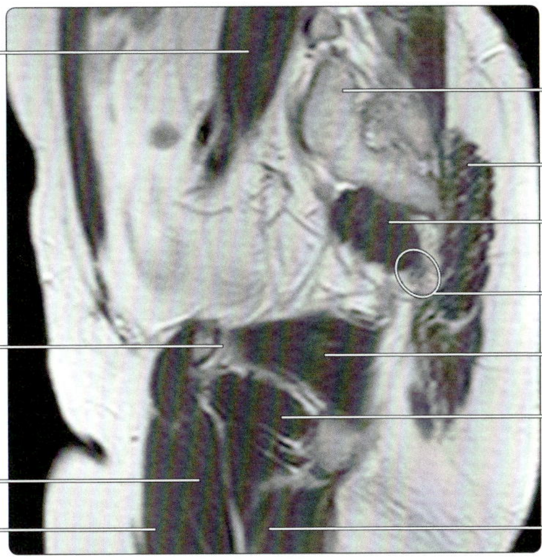

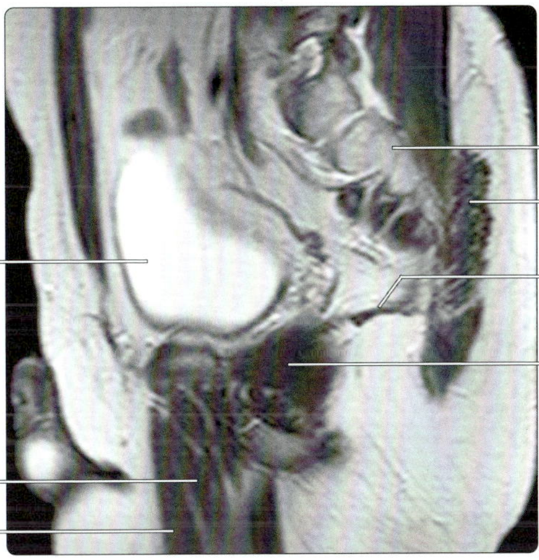

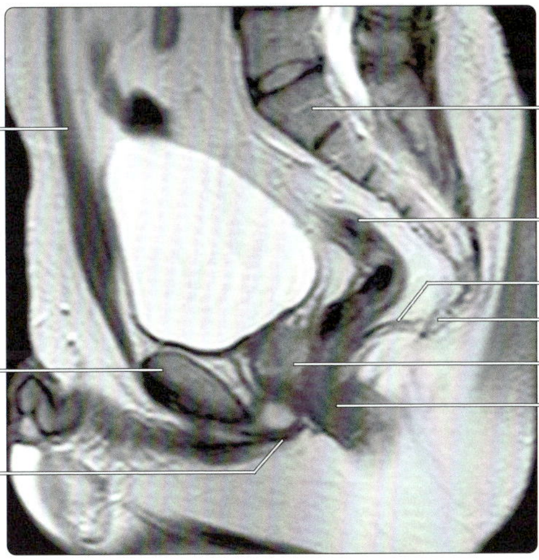

(Top) *The sciatic nerve exits the pelvis through the greater sciatic foramen, inferior to the piriformis muscle. Piriformis syndrome is a condition in which the piriformis muscle irritates the sciatic nerve, causing pain in the buttocks and referred pain along the course of the sciatic nerve.* **(Middle)** *The levator ani muscle constitutes the pelvic diaphragm and provides support to the pelvic organs. It is critical for maintaining continence and aiding in normal micturition and defecation.* **(Bottom)** *The bladder neck, prostate, and rectoanal junction are above or at the level of the pubococcygeal line, which extends from the last joint of the coccyx to the lower border of the symphysis pubis.*

PELVIC FLOOR

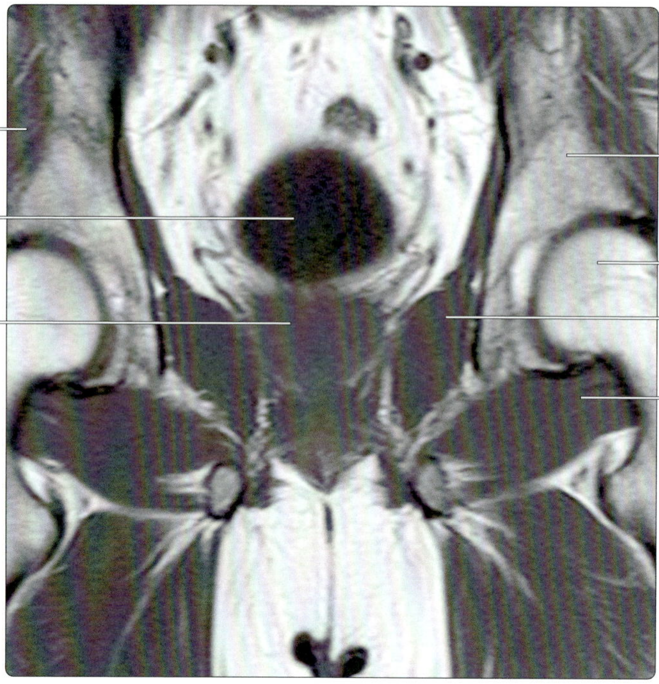

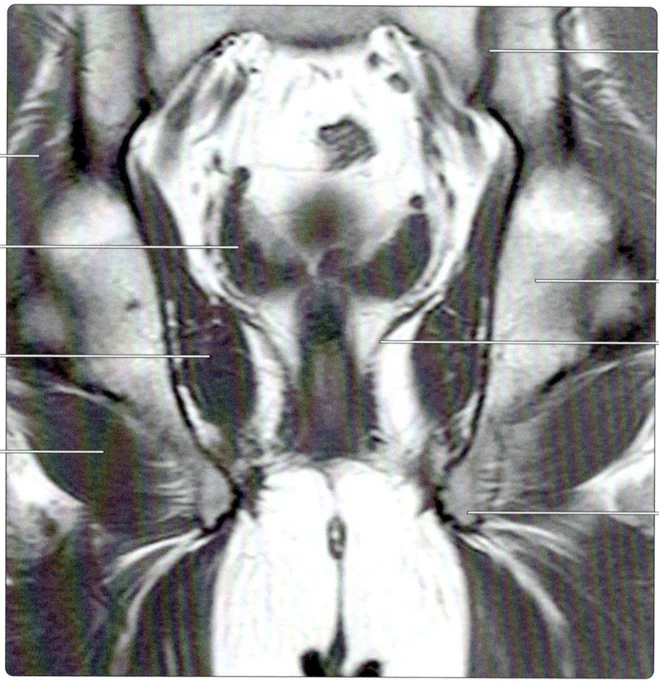

(Top) *First of 4 coronal T1 MR images of the pelvic floor is shown. The true pelvis in a male is much narrower than that of a female.* **(Bottom)** *The bowel-shaped nature of the levator ani is well appreciated on the coronal view. Although it cannot be discerned as a separate muscle by imaging, the most anterior portion of the levator ani is the pubococcygeus muscle.*

PELVIC FLOOR

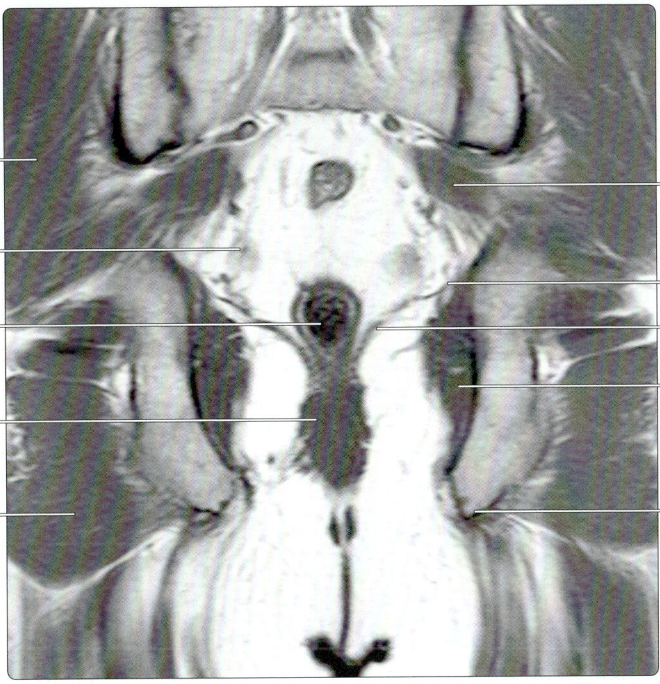

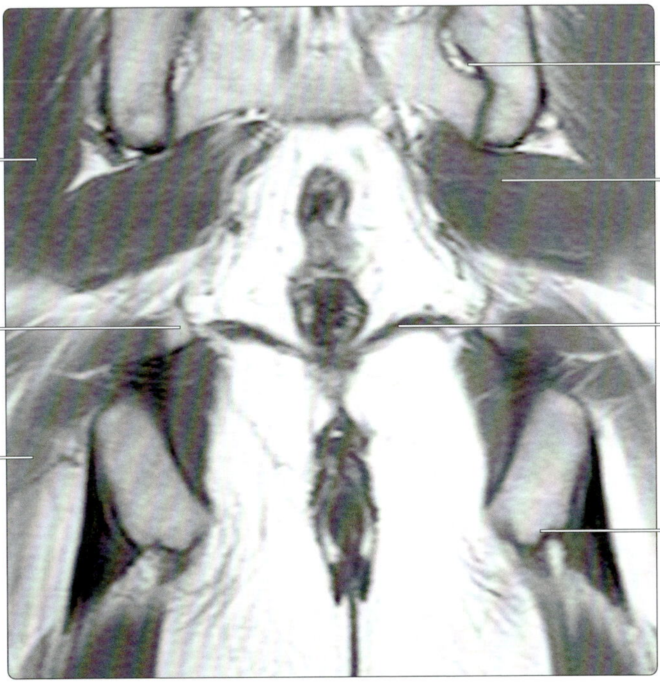

(Top) *The arcus tendineus is an important source of attachment for the levator ani muscle. It is formed from a strong fascial membrane covering the obturator internus muscle.* **(Bottom)** *The iliococcygeus forms the most posterior portion of the levator ani muscle. It arises from the arcus tendineus and ischial spines and inserts on the sacrum and coccyx. The coccygeus muscle also arises from the ischial spine and likewise inserts on the sacrum and coccyx. The coccygeus forms the most posterior support of the pelvic floor. It may not be seen as a discrete muscle, but note the thickening of muscle fibers in this area.*

INGUINAL CANAL

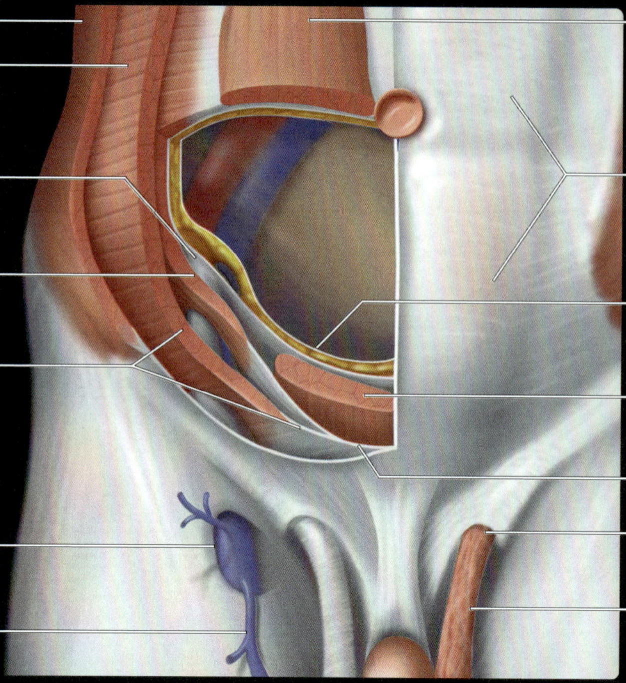

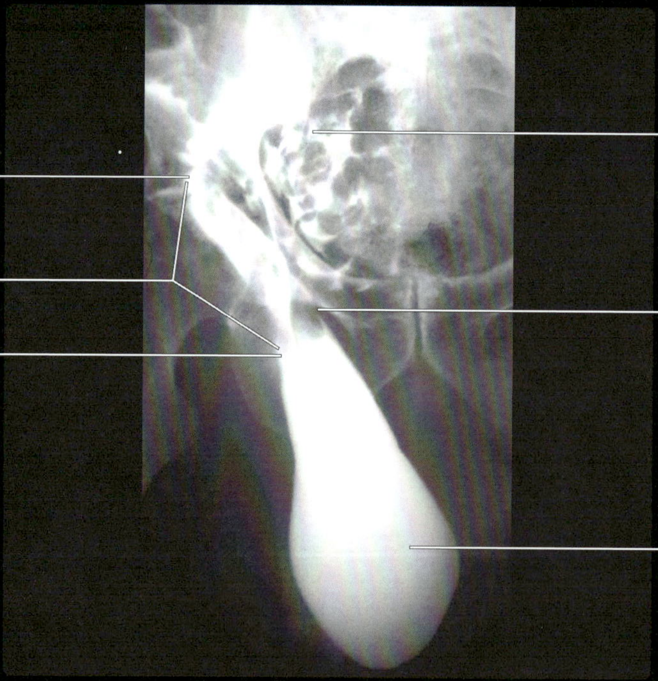

(Top) *The muscle layers of the anterior pelvic wall have been separated to show the inguinal canal. The inguinal canal extends from the internal (deep) inguinal ring, an opening through the transversalis fascia, to the external (superficial) inguinal ring, a triangular opening in the external oblique aponeurosis. The canal courses inferomedially and transmits the spermatic cord in males and the round ligament in females. It is formed embryologically by the processus vaginalis, an evagination of the peritoneum through the abdominal wall. Failure of closure of the processus vaginalis (patent processus vaginalis) puts the patient at risk for an indirect inguinal hernia.* **(Bottom)** *Herniogram shows an indirect inguinal hernia. There is a right hydrocele communicating with the peritoneal cavity through a patent processus vaginalis. A loop of bowel is seen as a filling defect within the inguinal canal. Herniograms are an antiquated examination, but it nicely demonstrates the course of the inguinal canal, which roughly parallels the superior pubic ramus.*

INDIRECT, DIRECT, AND FEMORAL HERNIAS

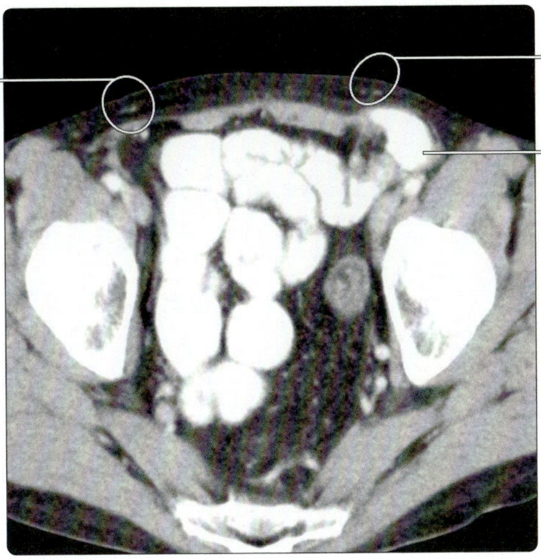

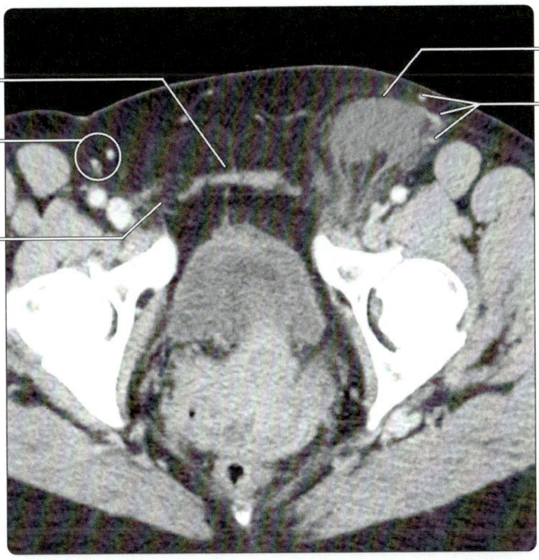

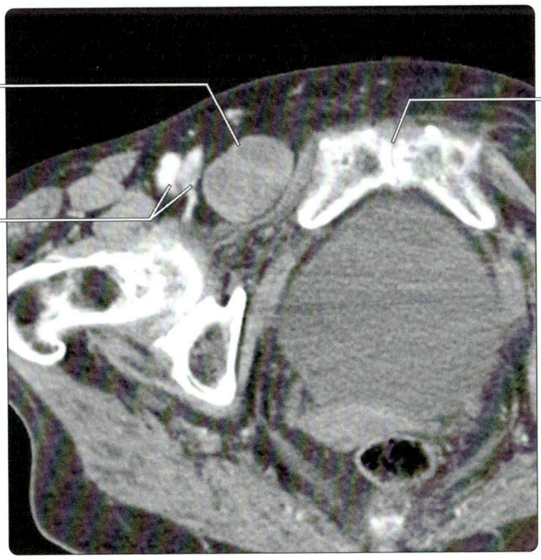

(Top) *Axial CECT shows an indirect inguinal hernia. The initial bulge through the abdominal wall is lateral to the inferior epigastric vessels. This represents the area of the internal inguinal ring. The course of an indirect inguinal hernia is lateral to medial, following the course of the spermatic cord (round ligament in females). Herniated contents may extend into the scrotum in males or labia majora in females.* **(Middle)** *Direct inguinal hernias occur through the Hesselbach triangle, a weakness in the lower pelvic wall. The triangle is formed by the inferior epigastric artery laterally, the lateral edge of the rectus sheath medially, and the inguinal ligament inferiorly.* **(Bottom)** *A femoral hernia in an older woman is shown for comparison. Femoral hernias traverse the femoral canal, which begins posterior to the medial portion of the inguinal ligament and ends at the fossa ovalis. Herniated contents are below the inguinal ligament, medial to the femoral vessels, and lateral to the pubic tubercle.*

EMBRYOLOGY AND HISTOLOGY

Testes and Scrotum

- **Testes form from genital ridges**, which extend from T6-S2 in embryo
- Composed of **3 cell lines**
  - Germ cells
  - Sertoli cells
  - Leydig cells
- **Germ cells**
  - Form in wall of yolk sac and migrate along dorsal mesentery to genital ridge
  - Form spermatogenic cells in mature testes
    - Mature from basement membrane to lumen of seminiferous tubule
      - Spermatogonia → spermatocytes → spermatids → spermatozoa
- **Sertoli cells**
  - Supporting network for developing spermatozoa
  - Form tight junctions (**blood-testis barrier**)
  - Secrete müllerian inhibiting factor
    - Causes **paramesonephric** (**müllerian**) **ducts** to regress
    - Embryologic remnant may remain as **appendix testis**
- **Leydig cells**
  - Principal source of testosterone production
  - Lie within interstitium
  - Causes differentiation of **mesonephric** (**wolffian**) **ducts**
    - Each duct forms epididymis, vas deferens, seminal vesicle, and ejaculatory duct
    - Embryologic remnant may remain as **appendix epididymis**
- **Scrotum** derived from labioscrotal folds
  - Folds swell under influence of testosterone to form **twin scrotal sacs**
    - Point of fusion is median raphe, which extends from anus, along perineum to ventral surface of penis
  - **Processus vaginalis**, sock-like evagination of peritoneum, elongates through abdominal wall into twin sacs
    - Forms anterior to developing testes
    - Aids in descent of testes, along with **gubernaculum** (ligamentous cord extending from testis to labioscrotal fold)
    - Results in component layers of adult scrotum
- **Testicular descent**
  - Between 7-12th week of gestation, testes descend into pelvis
    - Remain near internal inguinal ring until 7th month, when they begin descent through inguinal canal into twin scrotal sacs
    - Testes **remain retroperitoneal throughout descent**
    - Testes intimately associated with posterior wall of processus vaginalis
  - Component layers of spermatic cord and scrotum formed during descent through abdominal wall
  - Transversalis fascia → **internal spermatic fascia**
    - Transversus abdominis muscle is discontinuous inferiorly and does not contribute to formation of scrotum
  - Internal oblique muscle → **cremasteric muscle and fascia**
  - External oblique muscle → **external spermatic fascia**
  - **Dartos muscle** and **fascia** embedded in loose areolar tissue below skin
  - Processus vaginalis closes and forms **tunica vaginalis**
    - Mesothelial-lined sac around anterior and lateral sides of testis
    - Visceral layer of tunica vaginalis blends imperceptibly with tunica albuginea

GROSS ANATOMY

Testis

- **Tunica albuginea** forms thick fibrous capsule around testis
  - **Tunica vasculosa** forms vascular plexus along inner surface of tunica albuginea
- Densely packed seminiferous tubules separated by thin fibrous septa
  - **200-300 lobules** in adult testis
    - Each has 400-600 seminiferous tubules
    - Each tubule 30-80 cm long
    - **Total length of seminiferous tubules 300-980 m**
- Seminiferous tubules converge posteriorly to form larger ducts (**tubuli recti**)
  - Drain into **rete testis** at testicular hilum
- Rete testis converges posteriorly to form 15-20 **efferent ductules**
  - Penetrate posterior tunica albuginea at mediastinum to form head of epididymis
- **Mediastinum testis** thickened area of tunica albuginea where ducts, nerves, and vessels enter and exit testis

Epididymis

- Crescent-shaped structure running along posterior border of testis
- Efferent ductules form head (**globus major**)
  - Unite to form single, long, highly convoluted tubule in body of epididymis
- Tubule continues inferiorly to form epididymal tail (**globus minor**)
  - Attached to lower pole of testis by loose areolar tissue
- Tubule emerges at acute angle from tail as **vas deferens** (a.k.a. ductus deferens)
  - Continues cephalad within spermatic cord
  - Eventually merges with duct of seminal vesicle to form ejaculatory duct

Spermatic Cord

- Contains vas deferens, nerves, lymphatics, and connective tissue
- Begins at internal (deep) inguinal ring and exits through external (superficial) inguinal ring into scrotum
- Arteries
  - **Testicular artery**
    - Branch of aorta
    - Primary blood supply to testis
  - **Deferential artery**
    - Branch of inferior or superior vesicle artery
    - Arterial supply to vas deferens
  - **Cremasteric artery**
    - Branch of inferior epigastric artery

- Supplies muscular components of cord and skin
- Venous drainage
  - **Pampiniform plexus**
    - Interconnected network of small veins
    - Merges to form testicular vein
    - **Left testicular vein** drains to left renal vein
    - **Right testicular vein** drains to inferior vena cava
- **Lymphatic drainage**
  - Testis lymphatics follow testicular veins
    - Right side drains to interaortocaval chain
    - Left side drains to left paraaortic nodes near renal hilum
  - Epididymis may also drain to external iliac nodes
  - Scrotal skin drains to inguinal nodes

IMAGING ANATOMY

Ultrasound

- **Testes**
  - Ovoid, homogeneous, medium-level, granular echotexture
  - **Mediastinum testis** may appear as prominent echogenic line emanating from posterior testis
  - **Blood flow**
    - Testicular artery pierces tunica albuginea and arborizes over periphery of testis
    - Multiple radially arranged vessels travel along septa
    - May have prominent transmediastinal artery
    - Low-velocity, low-resistance wave form on Doppler imaging with continuous forward flow in diastole
- **Epididymis**
  - Isoechoic to slightly hyperechoic compared with testis
  - Best seen in longitudinal plane
  - Head has rounded or triangular configuration
  - Head: 10-12 mm; body and tail often difficult to visualize
- **Spermatic cord**
  - May be difficult to differentiate from surrounding soft tissues
  - Evaluate for varicocele with color Doppler

MR

- T1 MR
  - Testes uniform, intermediate signal
  - Epididymis isointense or slightly hypointense compared to testis
- T2 MR
  - Testes moderately high signal
    - Tunica albuginea, low-signal capsule around testis
    - Mediastinum testis appears as linear band of low signal along posterior aspect of testis
    - Thin, low-signal septa can be seen radiating back toward mediastinum
  - Epididymis hypointense relative to testis

CLINICAL IMPLICATIONS

Hydrocele

- Fluid between visceral and parietal layers of tunica vaginalis
- Small amount of fluid is normal
- Larger hydroceles may be either congenital (patent processus vaginalis) or acquired

Varicocele

- Idiopathic or secondary to abdominal mass
  - Idiopathic more common on left
- Vessel diameter **> 3 mm abnormal**
- Always evaluate with provocative maneuvers, such as Valsalva

Scrotal Calculi

- Free-floating calcifications within tunica vaginalis
- May result from torsion of appendix testis or epididymis

Cryptorchidism

- Failure of testes to descend completely into scrotum
- Most lie near external inguinal ring
- Associated with **decreased fertility and testicular carcinoma**
  - **Risk of carcinoma is increased for both testes**, even if other side is normally descended

Torsion

- Occurs most commonly when tunica vaginalis completely surrounds testis and epididymis
  - Testis is suspended from spermatic cord-like "**bell clapper**," rather than being anchored posteriorly
- Normal grayscale appearance with early torsion
  - Becomes heterogeneous and enlarged with infarction
- Doppler required for diagnosis
  - **Some flow may be seen even if torsed** but will be decreased compared to normal side
  - Venous flow compromised first, then diastolic flow, finally systolic flow

Dilated Rete Testes

- Tubular ectasia of rete testis posteriorly by mediastinum
  - May also see tubular ectasia in epididymis
- Form secondary to obstruction in epididymis or efferent ductules
  - Epididymal cyst common associated finding
- May be confused with mass, but careful scanning shows series of dilated tubules

Testicular Carcinoma

- Most common malignancy in young men
  - 95% are germ cell tumors
    - Seminoma (most common pure tumor), embryonal, yolk sac tumor, choriocarcinoma, teratoma
    - Mixed germ cell tumor (components of 2 or more cell lines) most common overall
  - Remainder of primary tumors are sex cord (Sertoli cells) or stromal (Leydig cells)
  - Lymphoma, leukemia, and metastases more common in older men
- Most primary testicular tumors metastasize via lymphatics in predictable fashion
  - Right-sided 1st echelon nodes: Interaortocaval chain at 2nd vertebral body
  - Left-sided 1st echelon nodes: Left paraaortic nodes in area bounded by renal vein, aorta, ureter, and inferior mesenteric artery

TESTIS

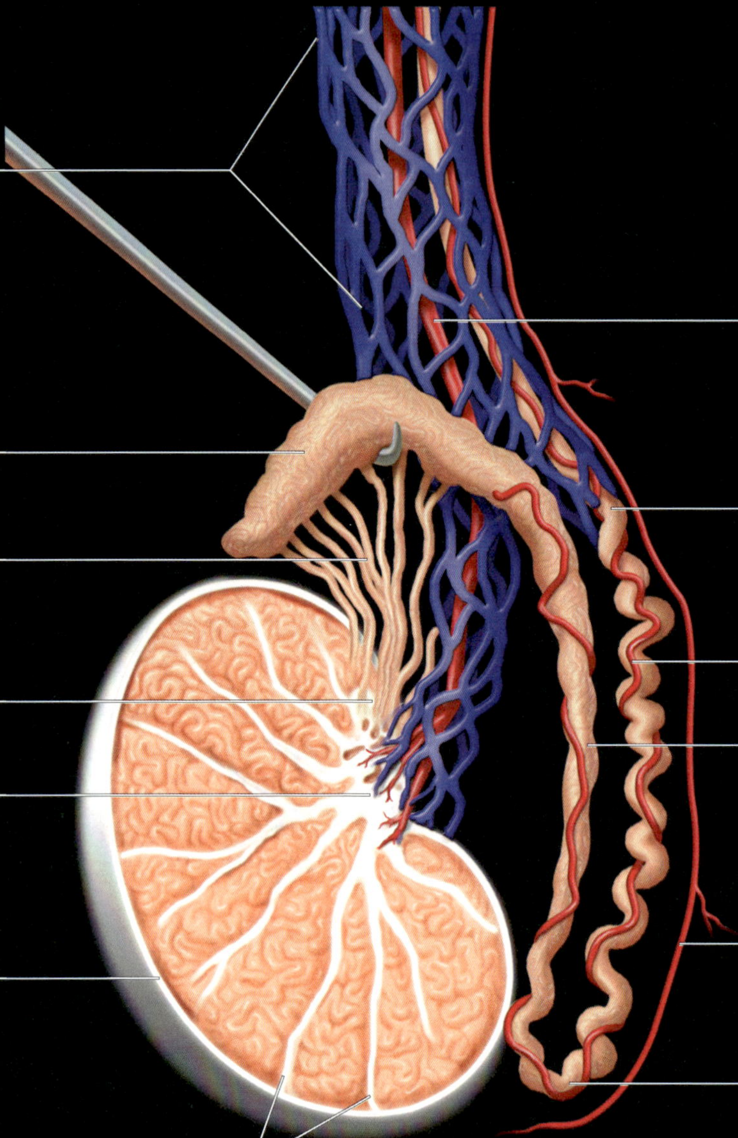

The testis is composed of densely packed seminiferous tubules, which are separated by thin fibrous septa. These tubules converge posteriorly, eventually draining into the rete testis. The rete testis continues to converge to form the efferent ductules, which pierce through the tunica albuginea at the mediastinum testis and form the head of the epididymis. Within the epididymis, these tubules unite to form a single, highly convoluted tubule in the body, which finally emerges from the tail as the vas deferens. In addition to the vas deferens, other components of the spermatic cord include the testicular artery, deferential artery, cremasteric artery, pampiniform plexus, lymphatics, and nerves.

VAS DEFERENS, SPERMATIC CORD

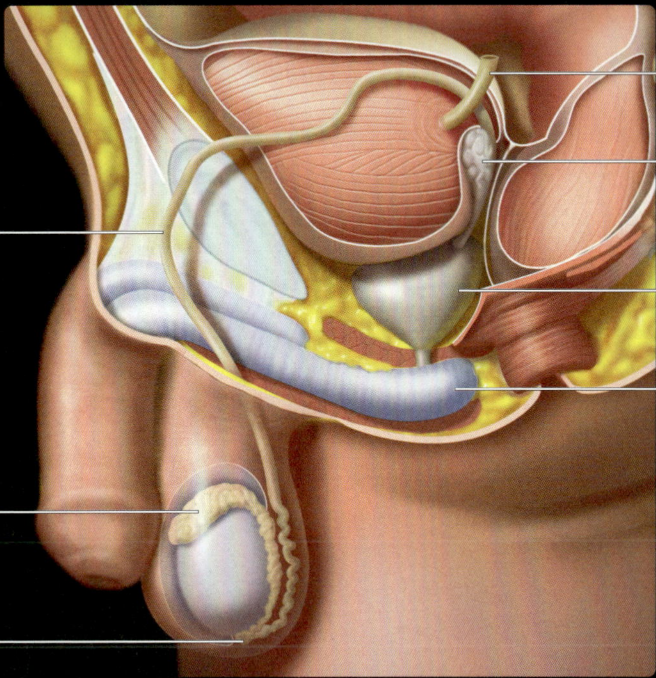

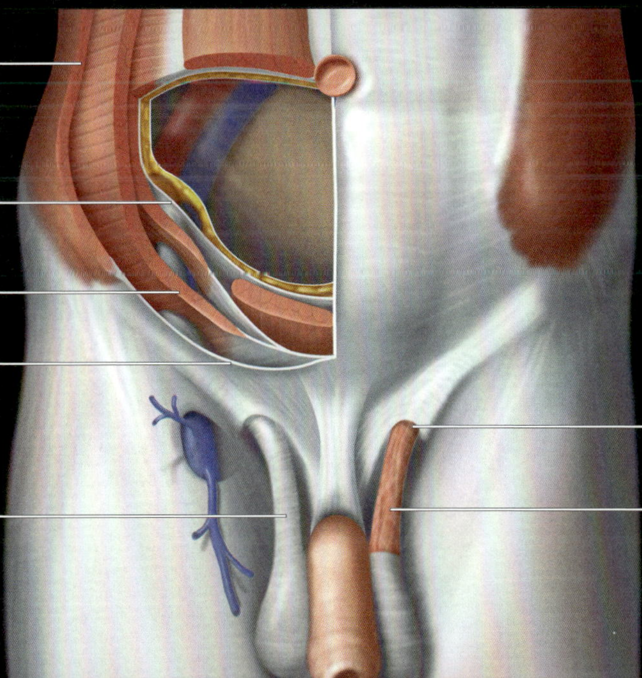

(Top) *The tail of the epididymis is loosely attached to the lower pole of the testis by areolar tissue. The vas deferens (also referred to as ductus deferens) emerges from the tail at an acute angle and continues cephalad as part of the spermatic cord. After passing through the inguinal canal, the vas deferens courses posteriorly to unite with the duct of the seminal vesicle to form the ejaculatory duct. These narrow ducts have thick, muscular walls composed of smooth muscle, which reflexly contract during ejaculation and propel sperm forward.* **(Bottom)** *The muscle layers of the pelvic wall have been separated to show the spermatic cord as it passes through the inguinal canal. The cremasteric muscle is derived from the internal oblique muscle, while the external spermatic fascia is formed by the fascia of the external oblique muscle.*

SCROTAL DEVELOPMENT

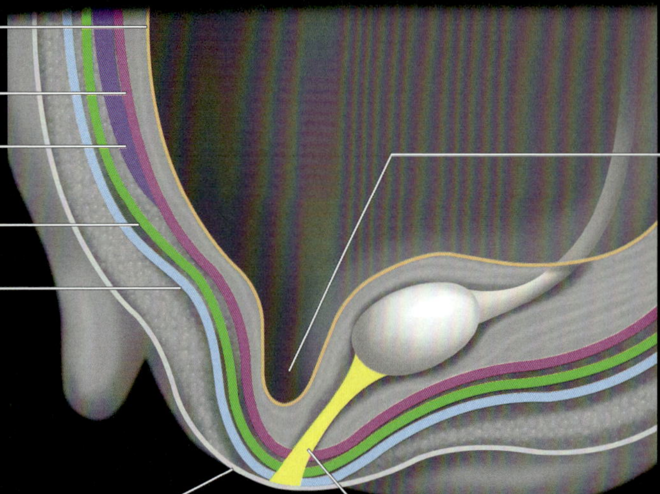

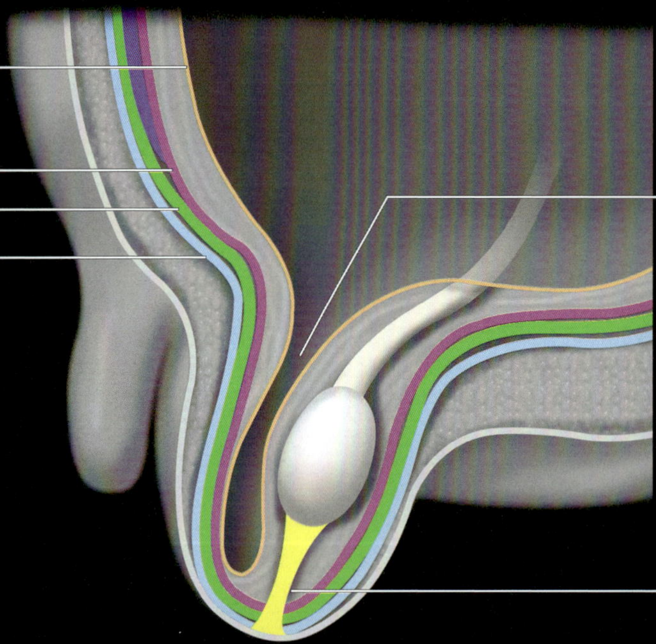

(Top) *The processus vaginalis is a sock-like evagination of the peritoneum, which elongates caudally through the abdominal wall. It forms on each side of the lower abdomen just anterior to the developing testes and, along with the gubernaculum (a ligamentous cord extending from the testis to the labioscrotal fold), aids in their descent.* **(Bottom)** *As the processus vaginalis evaginates, it becomes ensheathed by fascial extensions of the abdominal wall, which ultimately form the layers of the scrotum and spermatic cord. The transversus abdominis muscle is discontinuous inferiorly and does not contribute to the formation of the scrotum.*

SCROTAL DEVELOPMENT

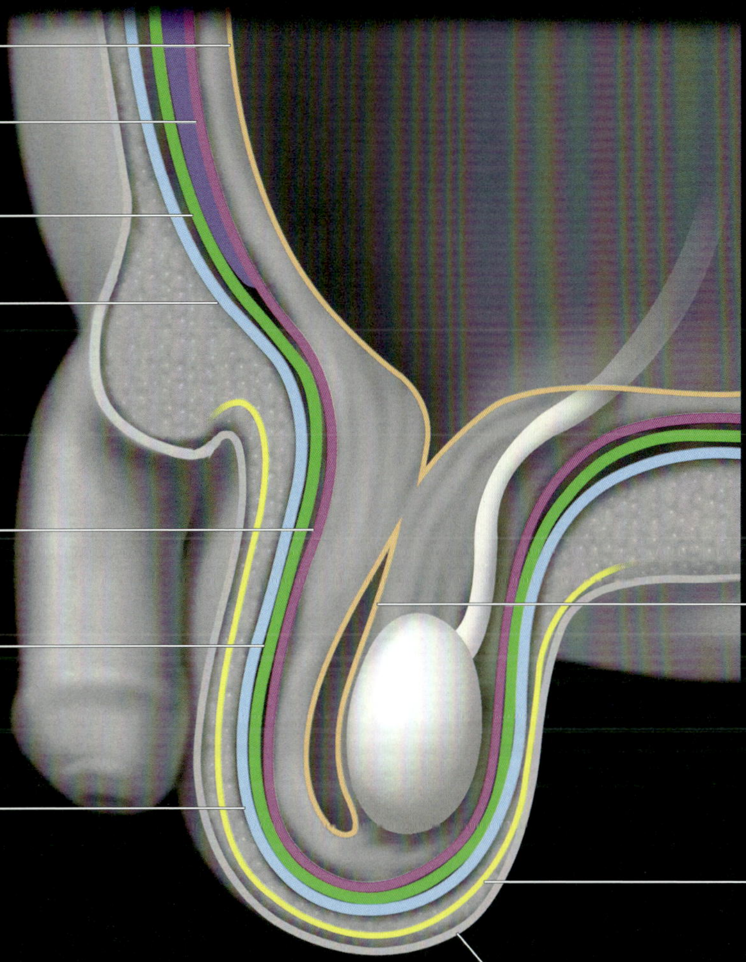

The abdominal wall derivative layers of the scrotum are as follows: Transversalis fascia → internal spermatic fascia, internal oblique muscle → cremasteric muscle and fascia, external oblique muscle → external spermatic fascia. The dartos muscle is embedded in the loose areolar tissue and is closely associated with the skin. Its primary function is to contract the skin and elevate the testes in response to cold. The various layers of the scrotum cannot usually be discerned with imaging. The superior portion of the processus vaginalis closes and forms an isolated mesothelial-lined sac, the tunica vaginalis. Failure of closure may result in a congenital hydrocele and is a risk factor for an inguinal hernia.

ULTRASOUND, TESTIS

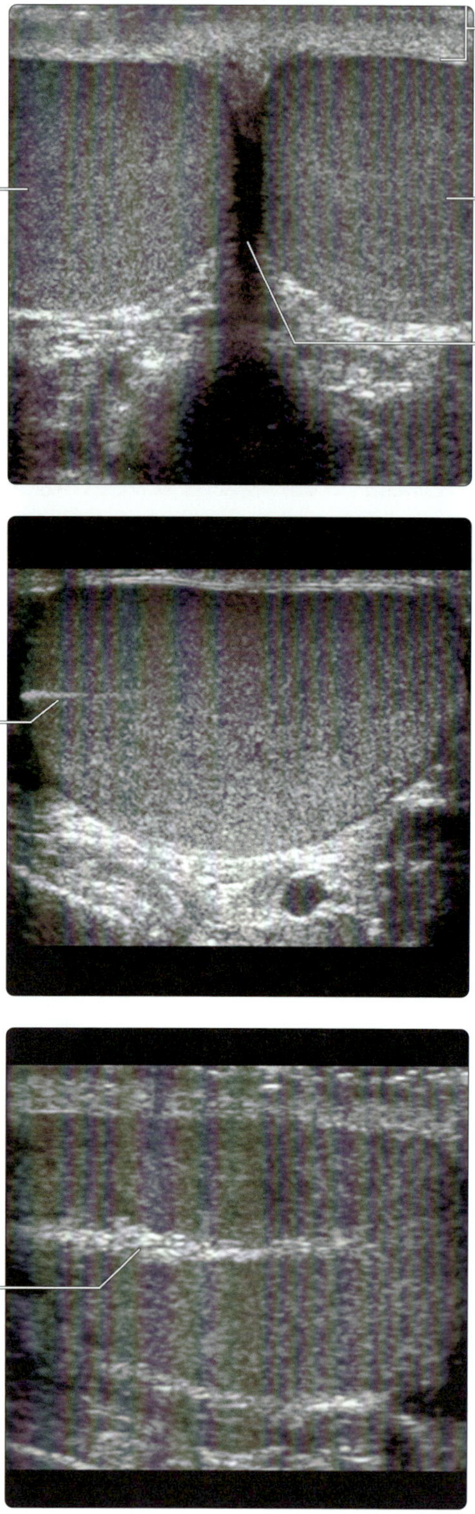

(Top) *Transverse ultrasound shows both testes. The testes have a homogeneous, medium-level, granular echotexture.* **(Middle)** *Longitudinal ultrasound shows the ovoid shape. The tunica albuginea may form an echogenic linear band where it invaginates at the mediastinum testis. The mediastinum testis has a craniocaudal linear course and is where the efferent ductules, vessels, and lymphatics pierce through the capsule.* **(Bottom)** *Longitudinal ultrasound shows a very prominent mediastinum testis.*

ULTRASOUND, EPIDIDYMIS

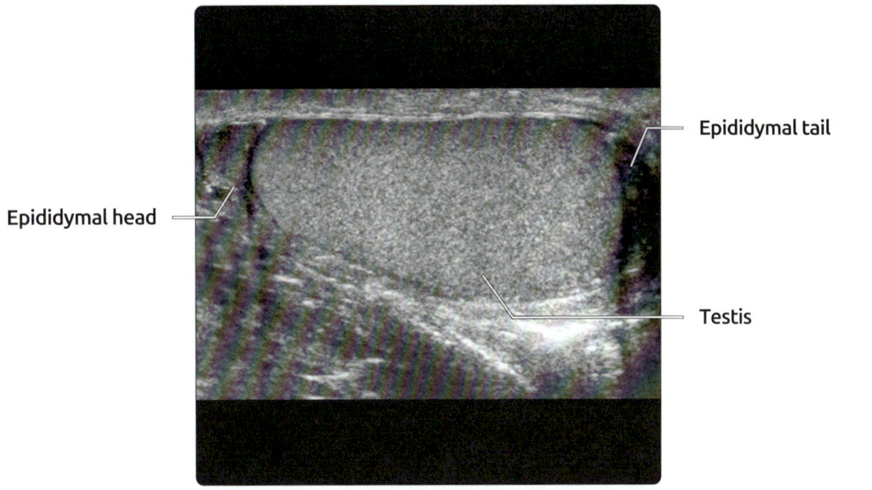

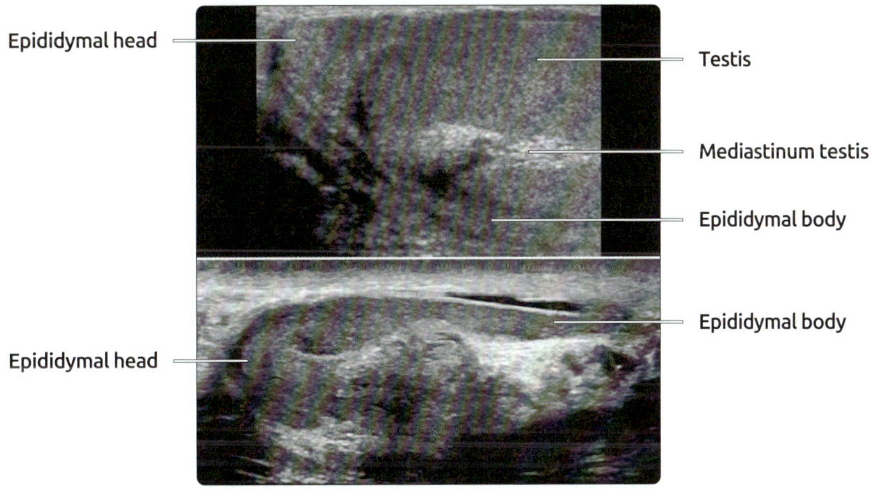

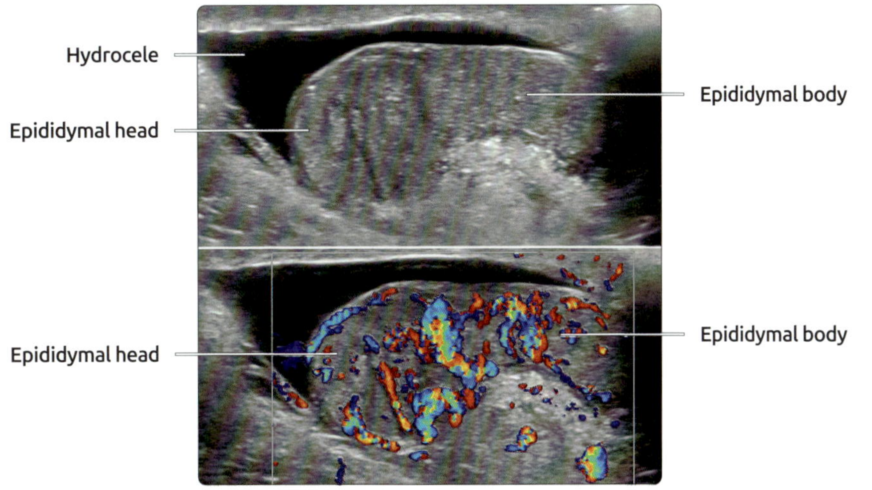

(Top) *Longitudinal ultrasound shows both the epididymal head (globus major) and tail (globus minor). The epididymal head measures ~ 10-12 mm and is iso- to slightly hyperechoic compared to the testis. The body and tail are often more difficult to visualize and may be slightly less echogenic than the head.* **(Middle)** *Composite image shows the normal epididymal head and body. The body may be difficult to visualize from an anterior scanning plane (above), but by moving the transducer more posteriorly (below), it is often better visualized.* **(Bottom)** *Composite image of a longitudinal grayscale ultrasound (above) and a color Doppler ultrasound (below) in a patient with acute epididymitis shows enlargement and hyperemia of both the epididymal head and body.*

TESTICULAR DOPPLER

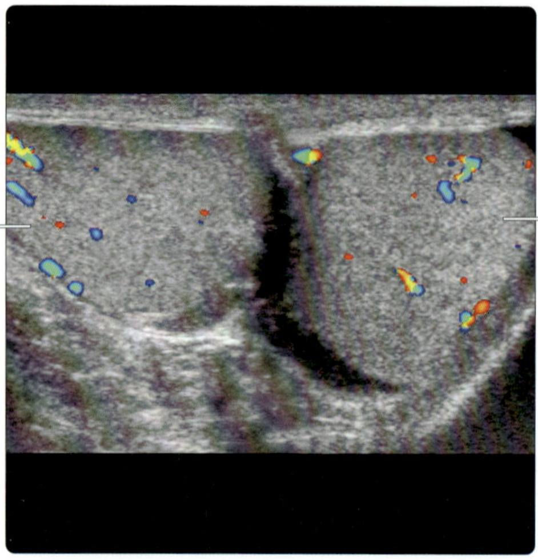

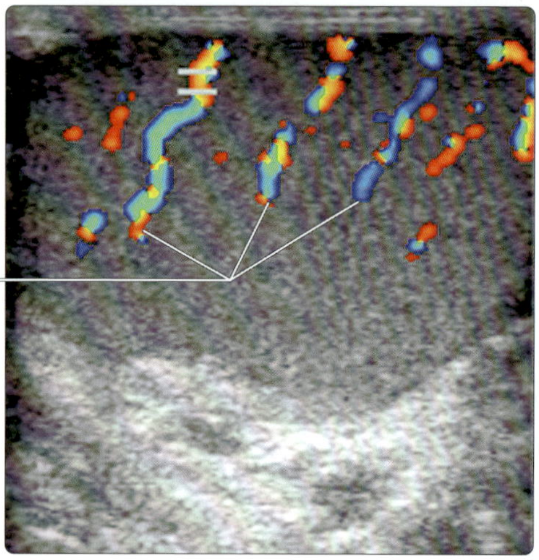

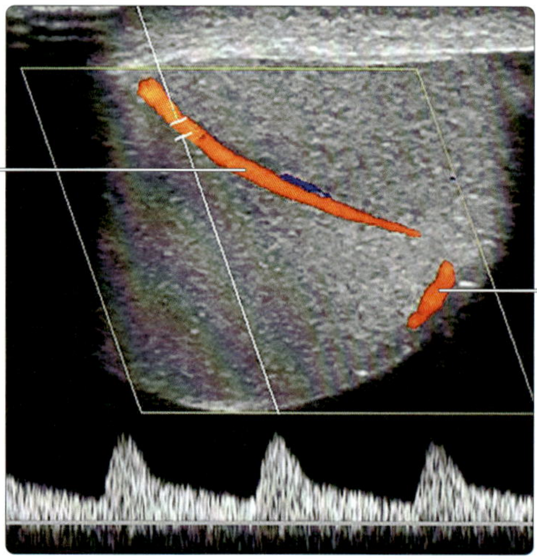

(Top) *Transverse color Doppler ultrasound shows normal symmetric blood flow in the testes. It is important to compare flow between testes to determine if the symptomatic side has increased or decreased flow when compared to the asymptomatic side.* **(Middle)** *Longitudinal color Doppler ultrasound shows prominent, radially arranged vessels within the testis.* **(Bottom)** *Color Doppler ultrasound shows a prominent transmediastinal artery with pulsed Doppler showing normal low-resistance arterial flow. Also shown is the tunica vasculosa, a vascular plexus that runs beneath the tunica albuginea.*

TESTICULAR AND EPIDIDYMAL APPENDAGES

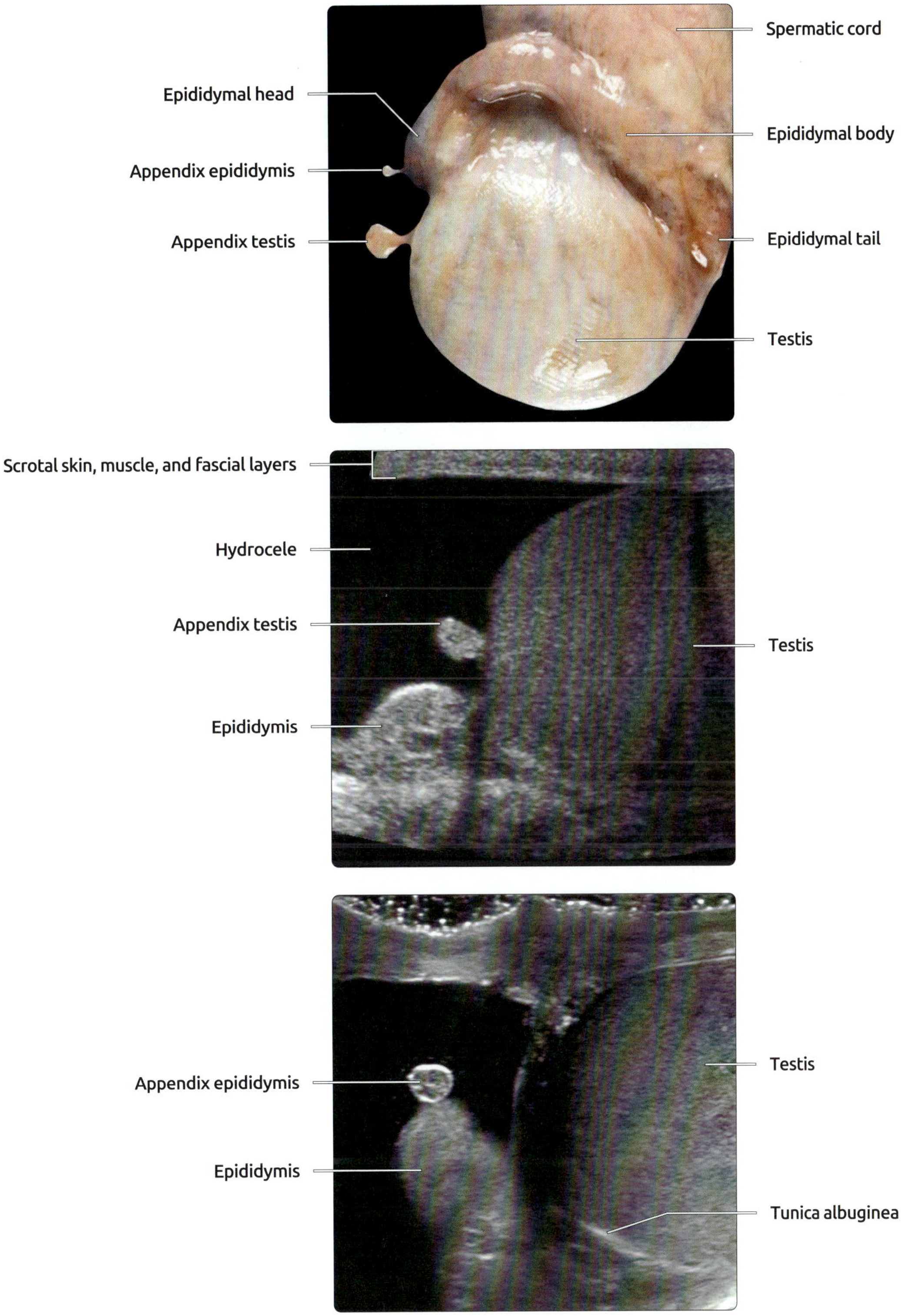

(Top) *Gross specimen shows the the testis, epididymis, and spermatic cord. Two embryologic remnants form small appendages from the testis and epididymis.* **(Middle)** *Ultrasound of the testis in a patient with a hydrocele shows a small, nodular protuberance from the surface of the testis. This is the appendix testis, which is a remnant of the müllerian system.* **(Bottom)** *Longitudinal ultrasound of the upper testis and epididymis shows a small, cystic "tag" of tissue projecting from the epididymis. This is an appendix epididymis, which is a remnant of the wolffian system. Both the appendix testis and appendix epididymis are usually not visible sonographically unless there is a hydrocele. They are usually of no clinical significance; however, they can rarely torse and be a cause of scrotal pain.*

CORONAL MR

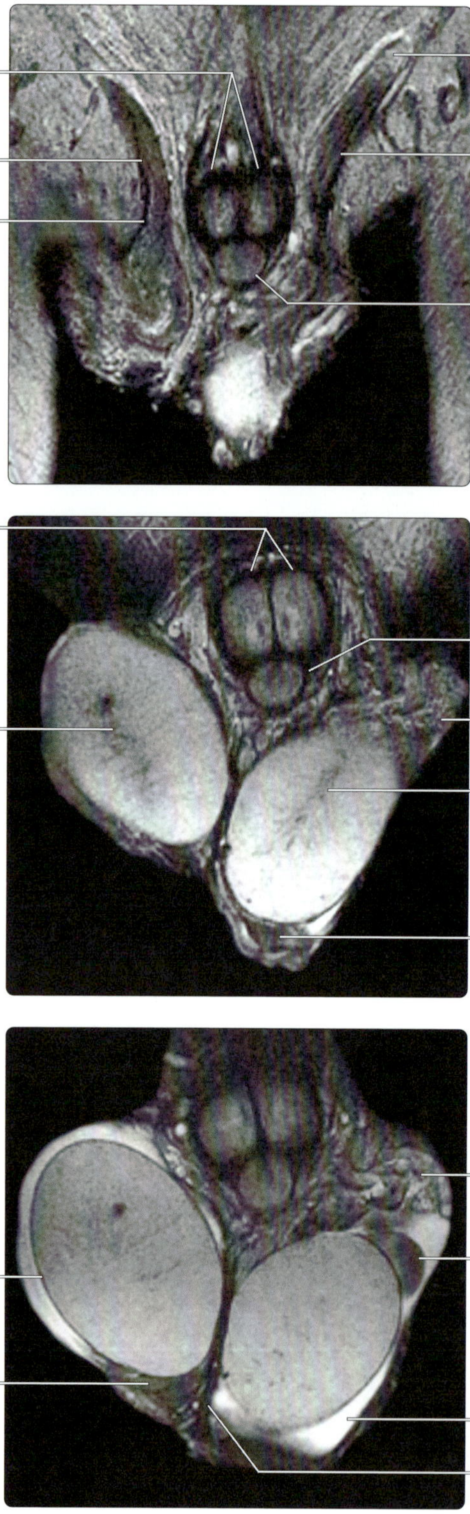

(Top) *First of 3 coronal T2 MR images, presented from back to front, shows normal scrotal anatomy. The spermatic cord begins at internal (deep) inguinal ring and exits through external (superficial) inguinal ring into scrotum.* **(Middle)** *The testes have a moderately high signal intensity. The mediastinum testis is an invagination of the tunica albuginea along the posterior aspect of the testis. It is low signal intensity with thin low-signal septa radiating away from it. The pampiniform plexus is a meshwork of interconnected veins. These generally have slow flow and will appear as an area of serpiginous high signal above the testis.* **(Bottom)** *The epididymis is relatively hypointense when compared to the testis. The tunica albuginea forms a distinct low-signal capsule around the testis. The median raphe separates the twin scrotal sacs and extends in a linear fashion from the anus to the ventral surface of the penis.*

CRYPTORCHIDISM

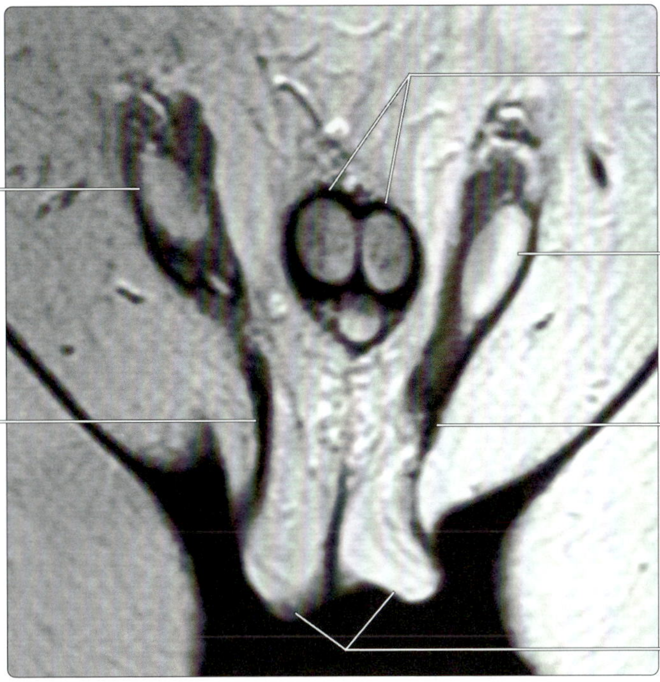

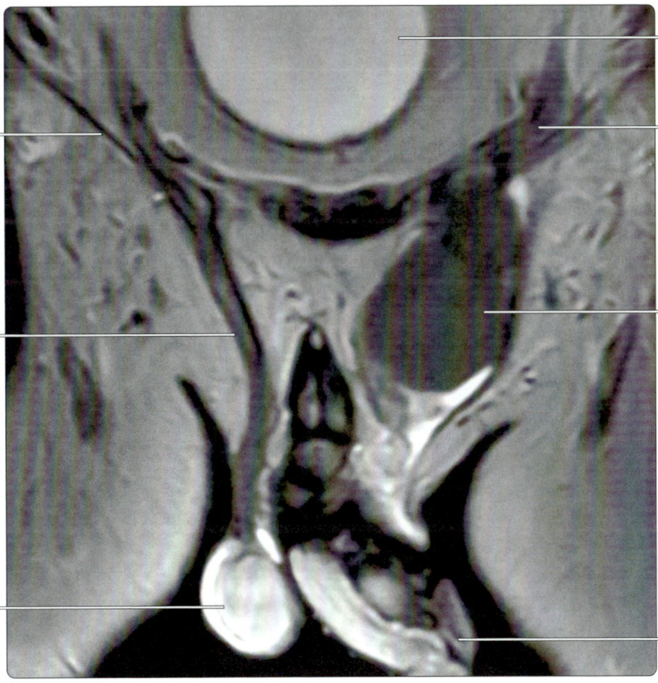

(Top) *Coronal T2 MR of bilateral cryptorchidism is shown. Most undescended testes lie within the inguinal canal near the external ring but can be anywhere along the path of testicular descent. Cryptorchidism is associated with testicular carcinoma, decreased fertility, and other congenital genitourinary abnormalities.* **(Bottom)** *Coronal T2 MR shows a tumor (seminoma) in an undescended left testis lying within the inguinal canal. It is important to be aware that the increased risk of carcinoma does not just apply to the undescended side but also to the normally descended testis. The increased risk may be from a generalized embryogenesis defect that results in bilateral dysgenetic gonads and a predisposition to tumor formation.*

AXIAL MR

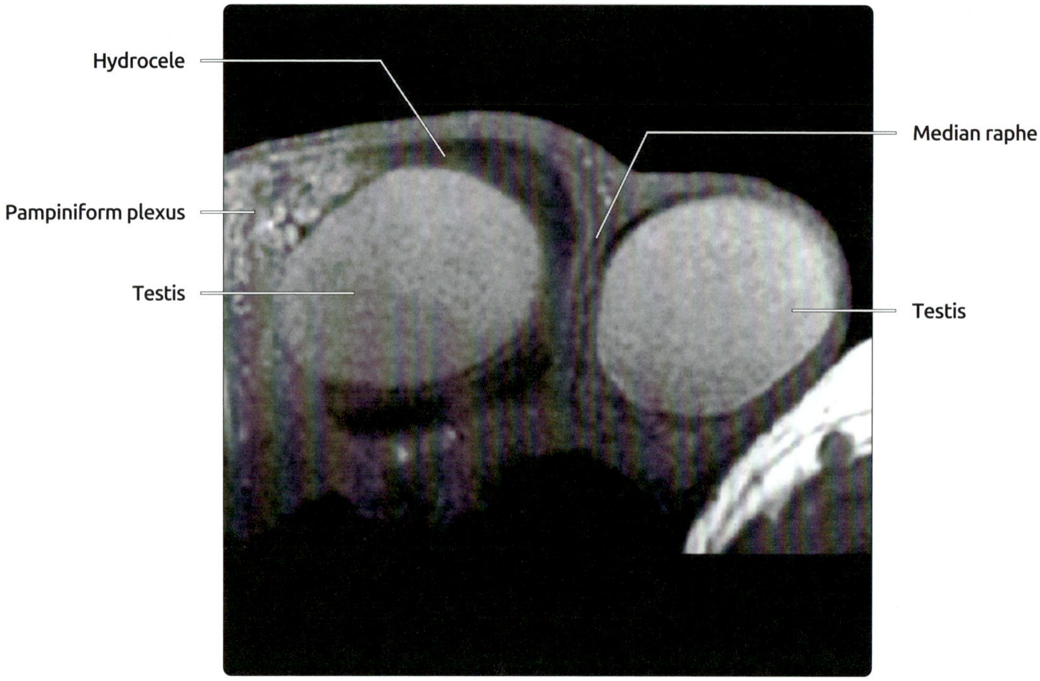

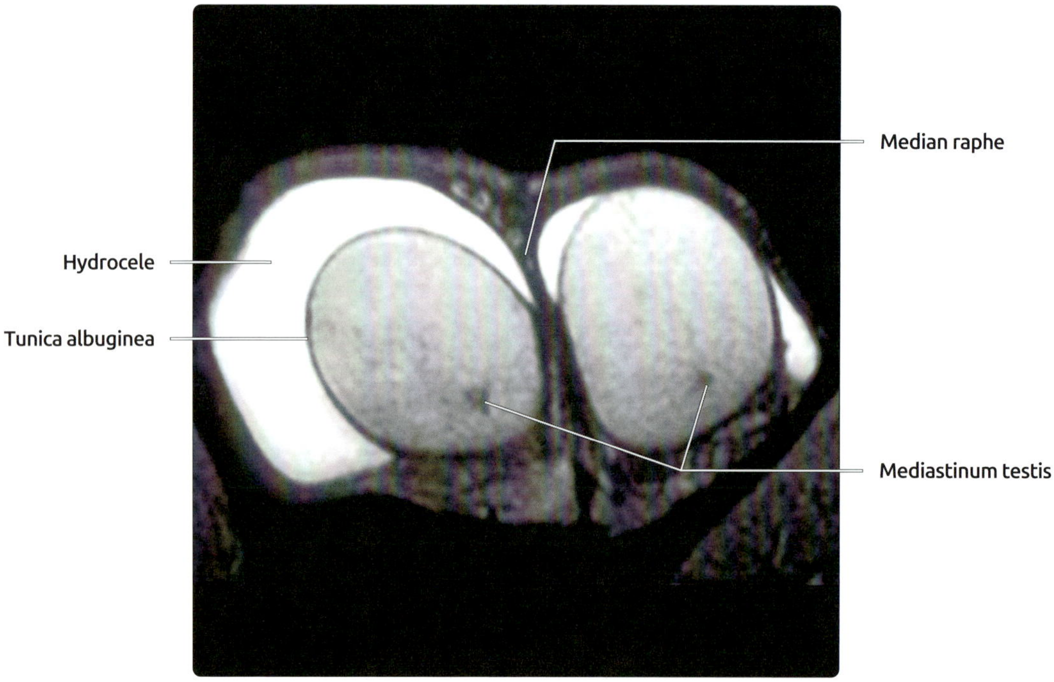

(Top) *Axial T1 MR of the scrotum shows the intermediate-signal testes and low-signal hydrocele.* **(Bottom)** *Axial T2 FS MR shows bilateral hydroceles (right > left). A hydrocele is an accumulation of fluid between the parietal and visceral layers of the tunica vaginalis. The tunica vaginalis surrounds all but the posterior aspect of the testis, which is attached to scrotal skin. If the tunica vaginalis completely surrounds the testis, it will be freely mobile ("bell clapper" anomaly) and is predisposed to torsion.*

RETE TESTIS

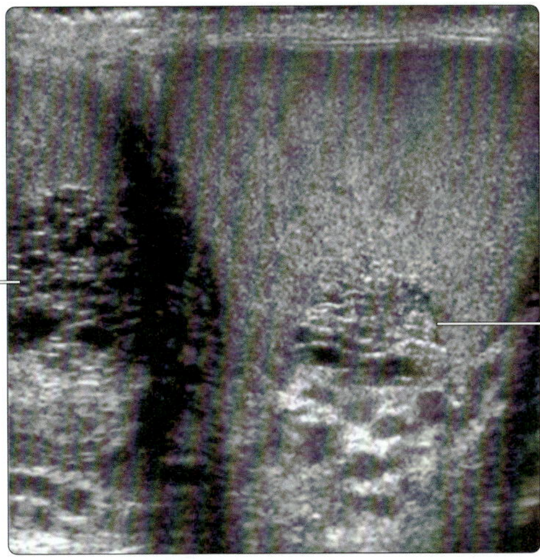

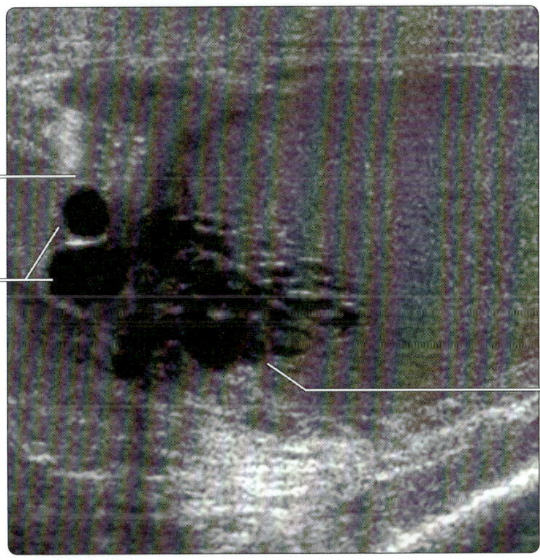

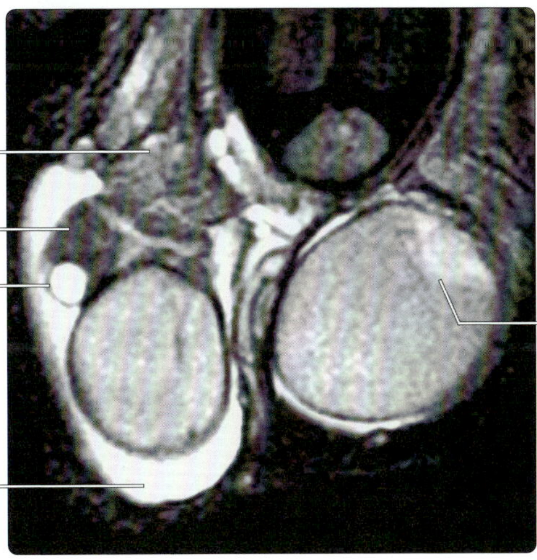

(Top) *First of 3 ultrasounds in a patient with tubular ectasia of the rete testis is shown. Dilation of the rete testis is thought to occur secondary to obstruction in the epididymis or efferent ductules. Tubular ectasia is located posteriorly by the mediastinum and is frequently bilateral. It may give an impression of a mass, but careful scanning shows the "mass" is actually a series of dilated tubules.* **(Middle)** *Longitudinal ultrasound of the right side shows 2 epididymal cysts. These are frequently associated with dilatation of the rete testis.* **(Bottom)** *Coronal T2 FS MR shows a triangular area of increased signal within the superior aspect of the left testis. This appearance helps to further differentiate tubular ectasia from a tumor, which is typically a round, low signal intensity mass on T2 MR.*

SPERMATIC CORD AND PAMPINIFORM PLEXUS

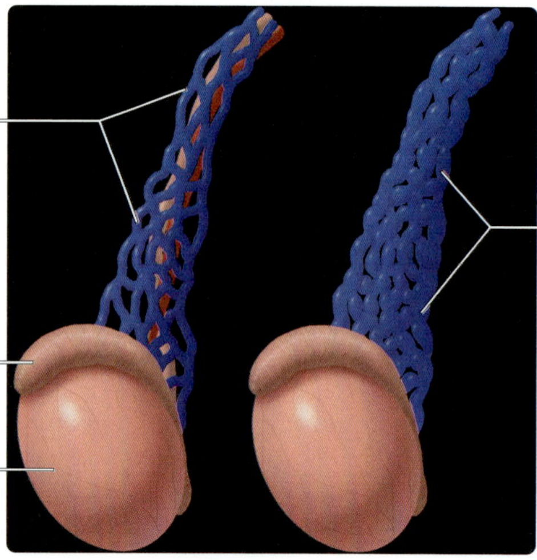

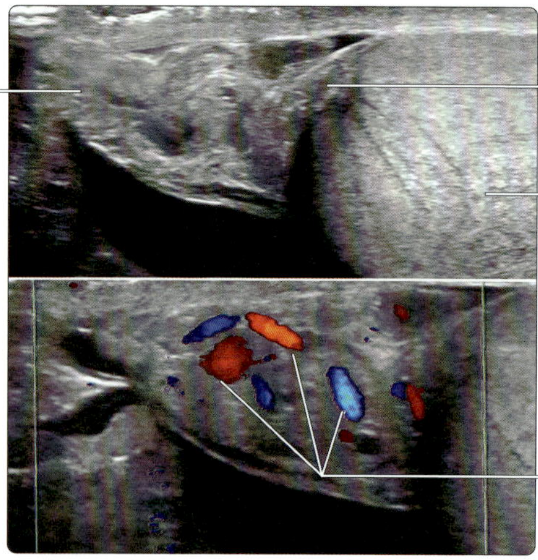

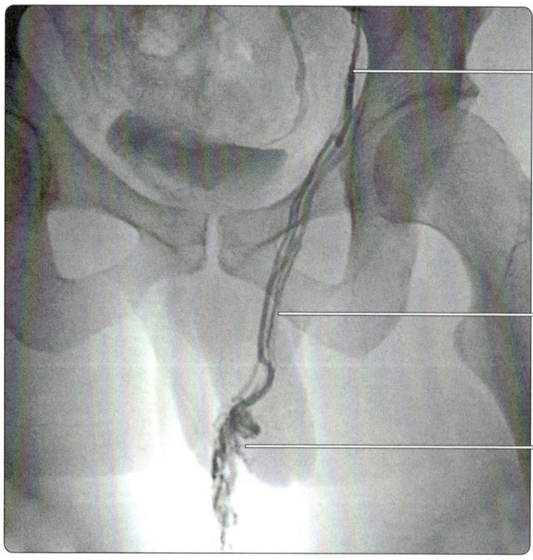

(Top) *The pampiniform plexus is an interconnected network of small veins, which merge to form the testicular vein. The left testicular vein drains to left renal vein, and the right testicular vein drains to inferior vena cava. The veins within the pampiniform plexus may dilate, forming a varicocele. Varicoceles are often idiopathic, particularly on the left side. Several anatomic factors predispose the left side: (A) The left testicular vein has a longer course than the right; (B) it has a perpendicular insertion into the left renal vein, while the right flows obliquely into the inferior vena cava; and (C) the left renal vein passes beneath the superior mesenteric artery, creating a nutcracker effect.* **(Middle)** *Longitudinal grayscale and color Doppler ultrasounds near the external inguinal ring show a normal spermatic cord and pampiniform plexus.* **(Bottom)** *This image from a left testicular venogram shows the pampiniform plexus and a varicocele, which was subsequently embolized.*

VARICOCELE

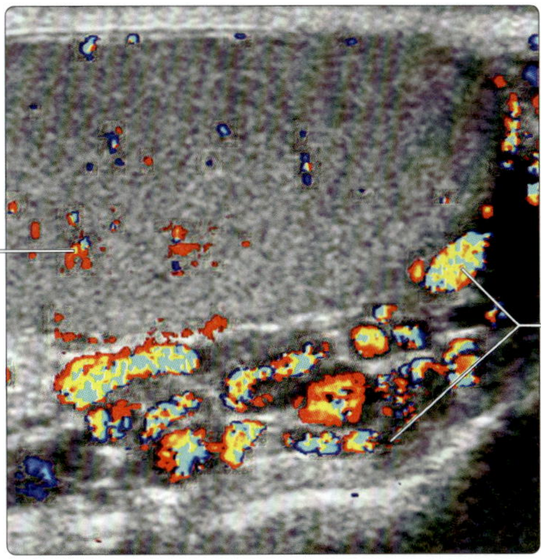

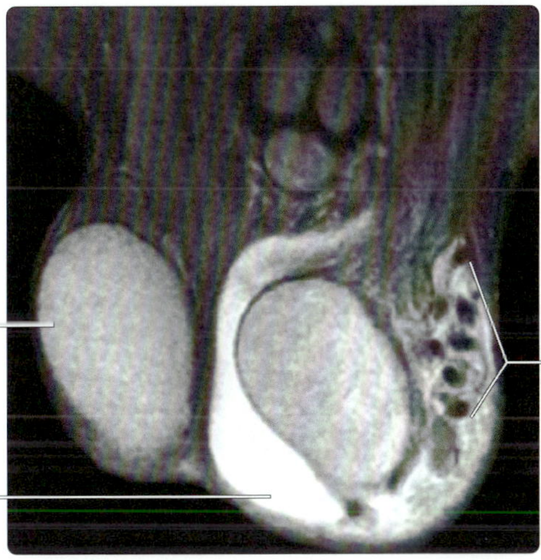

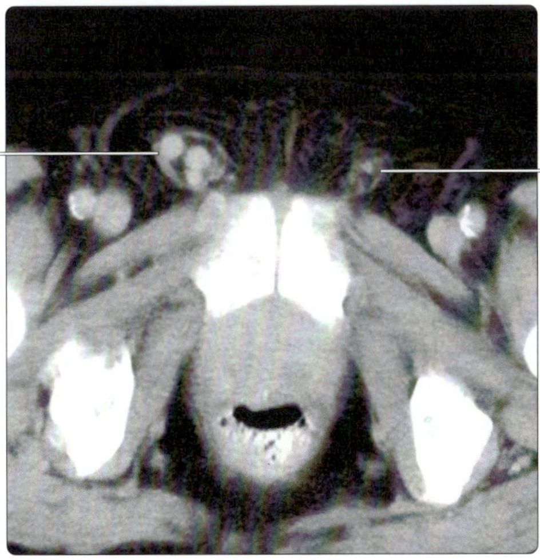

(Top) *Color Doppler ultrasound shows filling of a varicocele with the Valsalva maneuver. Normal vessel diameter should not exceed 3 mm. Color Doppler ultrasound at rest shows parallel channels often with little to no flow. Other provocative maneuvers, such as scanning in the upright position, can also be used to help make the diagnosis.* **(Middle)** *Coronal T2 MR shows a left-sided varicocele. The signal intensity may vary according to blood flow velocity. Fast-flowing varicoceles have a signal void, while slow-flowing ones will be intermediate to high signal.* **(Bottom)** *CECT of a right-sided varicocele in a patient with a right renal tumor is shown. Varicoceles may be caused by compression or invasion of the renal vein or inferior vena cava.*

RIGHT-SIDED LYMPHATIC DRAINAGE

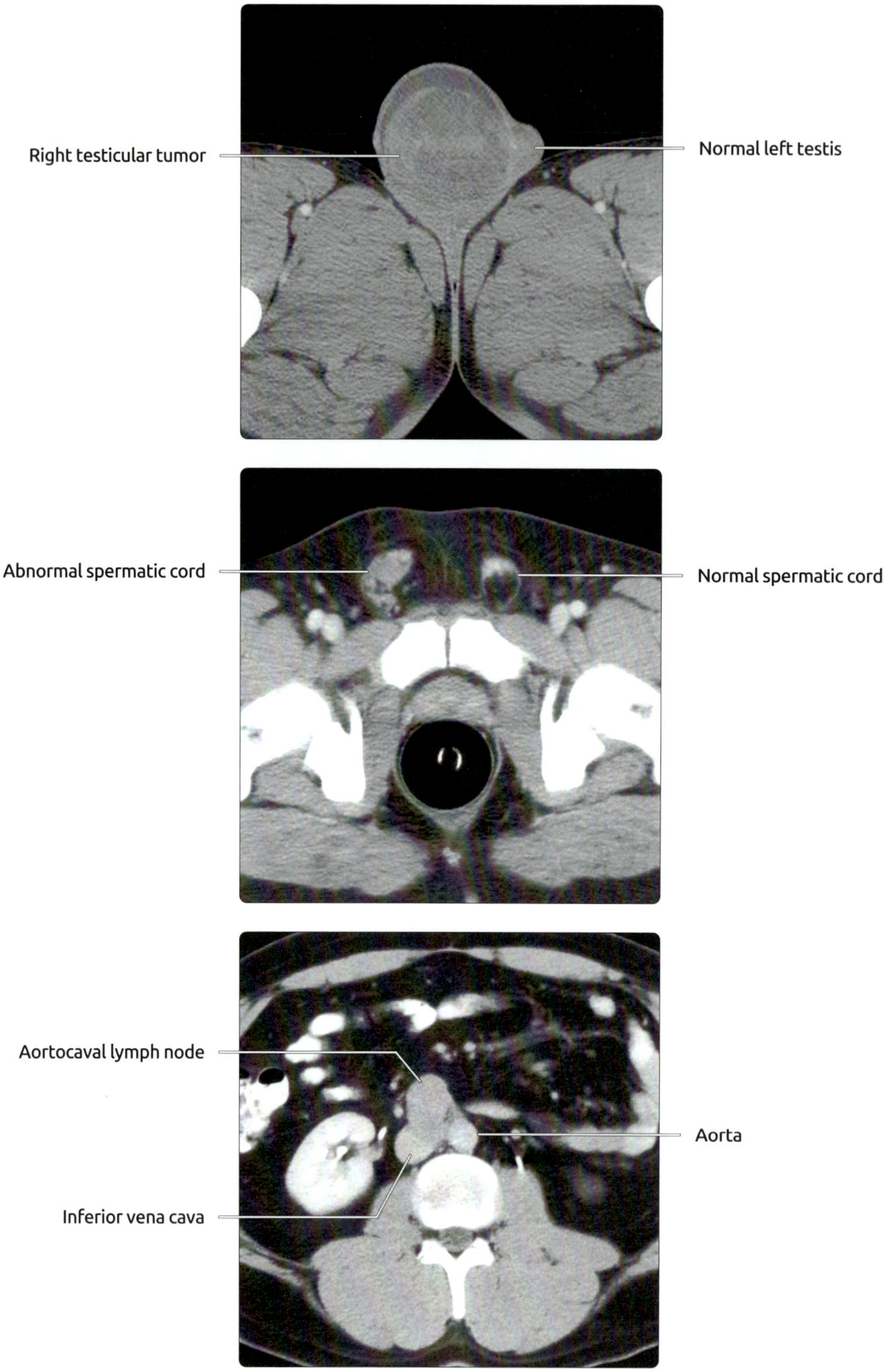

(Top) *First of 3 CECT images in a patient with a right-sided mixed germ cell tumor shows a large right testicular mass.* **(Middle)** *The right spermatic cord is enlarged and was invaded by tumor.* **(Bottom)** *A single enlarged aortocaval node is present. Most testicular tumors metastasize in a predictable fashion. Lymphatic drainage follows the testicular veins with the right side, first metastasizing to the interaortocaval chain at the level of the 2nd lumbar vertebral body.*

LEFT-SIDED LYMPHATIC DRAINAGE

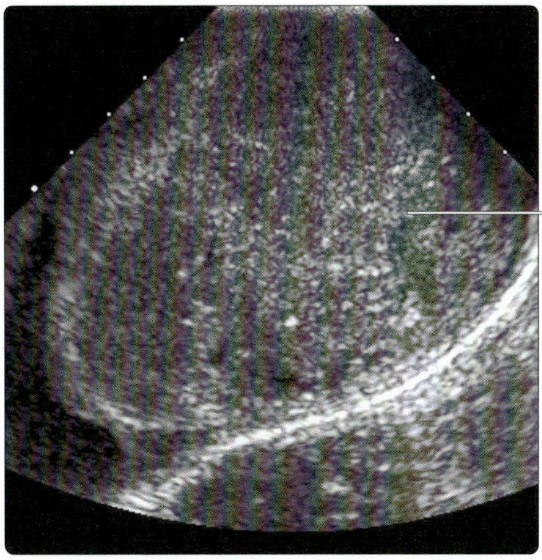

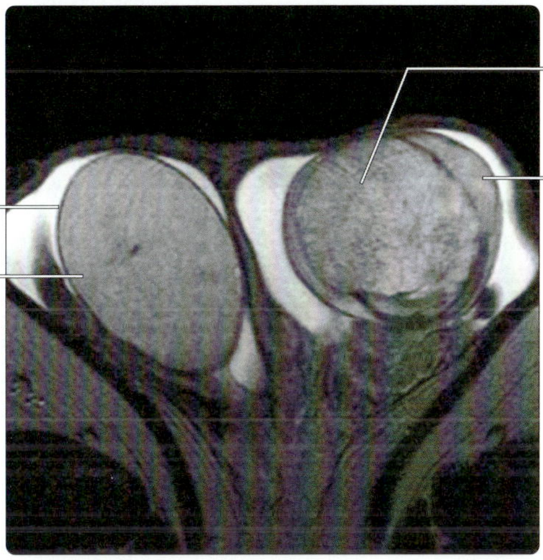

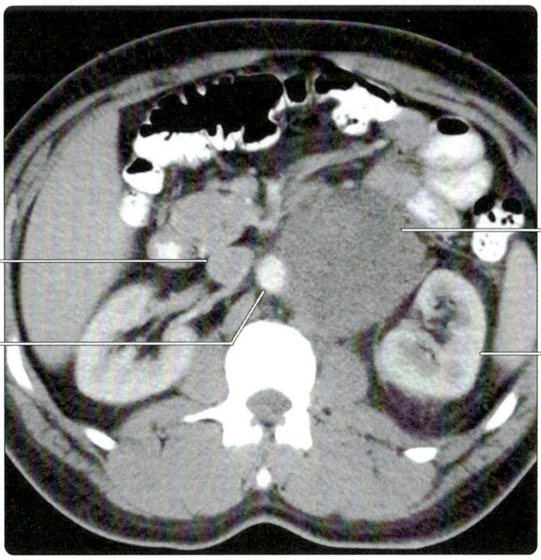

(Top) *First of 3 images of a patient with a left-sided seminoma is shown. Longitudinal ultrasound shows a large, hypoechoic mass.* **(Middle)** *Axial T2 MR shows a large, lobular mass essentially replacing the left testis. A small amount of normal parenchyma remains in the periphery.* **(Bottom)** *CECT shows significant left-sided adenopathy. The 1st echelon nodes for a left-sided testicular tumor are in an area bounded by the left renal vein, aorta, ureter, and inferior mesenteric artery.*

Prostate and Seminal Vesicles

TERMINOLOGY

Abbreviations

- Anterior fibromuscular stroma (AFMS)
- Central zone (CZ)
- Peripheral zone (PZ)
- Transition zone (TZ)
- Neurovascular bundle (NVB)
- Benign prostatic hyperplasia (BPH)

GROSS ANATOMY

Prostate

- Walnut-sized gland located beneath bladder and in front of rectum
  - Lies above superior fascia of urogenital diaphragm
  - Surrounds uppermost part of urethra
  - Normal prostate ~ 5 cm craniocaudal x 3 cm transverse x 3 cm anteroposteriorly
  - Normal volume ≤ 25 mL
  - From superior to inferior
    - **Base** refers to wider, superior 1/3
    - **Midgland** refers to middle 1/3
    - **Apex** refers to narrower, inferior 1/3
- Multiple points of contact with surrounding structures
  - Base closely related to inferior surface of bladder
  - Apex in contact with superior fascia of urogenital diaphragm
  - **Posterior surface** separated from rectum by rectovesical septum (**Denonvilliers fascia**)
    - 2 **ejaculatory ducts** enter prostate through posterior surface
    - Posterosuperiorly, seminal vesicles lie between bladder base and rectum
  - **Anterior surface** separated from symphysis pubis by extraperitoneal fat and plexus of veins
    - Connected to pubic bone on either side by puboprostatic ligaments
  - **Inferolateral surfaces** separated from levator ani by periprostatic plexus of veins
- **Prostate capsule**
  - Surgically and radiographically, capsule-like structure is seen; however, prostate lacks true capsule
  - Pathologically, capsule is made up of loose smooth muscle fibers and interspersed connective tissue
    - For this reason some authors have therefore termed it pseudocapsule
  - Capsule surrounds posterior and lateral portions of prostate
    - Anteriorly, it merges with AFMS
- **Prostatic urethra**
  - Prostatic utricle, prostatic ducts, and ejaculatory ducts enter prostatic urethra
  - **Verumontanum** (colliculus seminalis)
    - Median elevation of urethral crest below its summit
    - Openings of prostatic utricle and ejaculatory ducts
  - **Prostatic utricle**
    - Small vestigial blind pouch of prostate gland, ~ 6 mm long
    - Developed from united lower ends of atrophied müllerian ducts
- **NVBs**
  - Lie posterolaterally to prostate
  - Separated from prostate by Denonvilliers fascia
  - Carries nerve and vascular supply to corpora cavernosa
    - Critical for normal erectile function
  - Increasing evidence that nerves important for erectile function and continence surround prostate and are not just present posterolaterally at bundles
- **Arterial supply** derived from internal pudendal, inferior vesicle, and middle rectal arteries
  - All branches of internal iliac arteries
- **Prostatic venous plexus**
  - Receives blood from dorsal vein of penis
  - Drains into internal iliac veins
  - Prostatic venous plexus communicates with internal vertebral venous plexus (**Batson plexus**)
- **Nerve supply**
  - Parasympathetic fibers from pelvic splanchnic nerves (S2-S4)
  - Sympathetic fibers from inferior hypogastric plexuses
- **Lymphatic drainage** chiefly to internal iliac and sacral lymph nodes
  - Some drainage from posterior surface joins with bladder lymphatics and drains to external iliac lymph nodes

Zonal Anatomy (McNeal)

- Prostate is histologically composed of glandular (acinar) and nonglandular elements
- 2 nonglandular elements: Prostatic urethra and AFMS
  - AFMS is contiguous with bladder muscle and external sphincter
  - Proximal 1/2 of prostatic urethra is surrounded by cuff of smooth muscle ("**preprostatic sphincter**")
- **Glandular prostate** consists of outer and inner components
- **Inner prostate** has 2 parts: Periurethral glandular tissue and TZ
  - **Periurethral glands**
    - < 1% of normal glandular prostate
    - Small glands confined to submucosal layer of proximal prostatic urethra above verumontanum and deep to preprostatic sphincter
    - Ducts drain into proximal posterolateral wall of prostatic urethra in 2 rows
  - **TZ**
    - ~ 5% of normal glandular prostate
      - Enlarges in setting of BPH
    - Surrounds anterior and lateral aspects of proximal urethra
- **Outer prostate** has 2 parts: CZ and PZ
  - **CZ**
    - ~ 25% of normal glandular prostate, becomes much smaller with aging and BPH
    - Surrounds ejaculatory ducts
    - Funnel-shaped with its widest portion making up majority of prostatic base
    - Tapered tip extends inferiorly to level of verumontanum
    - Encloses both periurethral glands and TZ
    - Ducts of CZ drain to region of verumontanum, clustered around entry of ejaculatory ducts

- **PZ**
  - ~ 70% of normal glandular prostate, becomes much smaller with aging and BPH
  - Surrounds both CZ and distal prostatic urethra
  - Ducts of PZ drain exclusively to distal prostatic urethra
  - Ducts are arranged in 2 vertical lines along posterolateral urethral wall
- **Prostate pseudocapsule** ("surgical capsule")
  - Visible boundary between TZ and PZ

Seminal Vesicles and Ejaculatory Ducts

- **Seminal vesicles**
  - Sac-like structures located superolaterally to prostate
    - Between fundus of urinary bladder and rectum
  - **Secrete fructose-rich**, **alkaline fluid**, which is major component of semen
    - Secretions are energy source for sperm
    - Do not store sperm
- **Ejaculatory ducts**
  - Located on either side of midline
  - Formed by union of seminal vesicle duct and ductus deferens
  - Start at base of prostate and run forward and downward through gland

IMAGING ANATOMY

Prostate

- MR
  - T1 MR: Homogeneous intermediate signal intensity
    - Often best sequence for NVB, which are 5 and 7 o'clock positions
    - Best depicts areas of hemorrhage (usually related to recent biopsy) and T1-hyperintense cysts
  - T2 MR depicts zonal anatomy best
    - AFMS is low signal intensity band of tissue, of variable thickness, along anterior gland
    - PZ is high signal intensity, ≥ periprostatic fat
      - PZ is posterior, wraps laterally and extends anteriorly in midgland to apex in some men
    - PZ is surrounded by thin, low-signal "capsule" (though there is no true capsule)
    - CZ and TZ are of similar T2 signal intensity, < PZ
    - TZ easily recognized by variable-sized BPH nodules in most men > 50 years of age
    - CZ easily recognized in older man with BPH as symmetric dumbbell-shaped T2-hypointense areas at base surrounding paired ejaculatory ducts
    - Tumor is typically homogenously T2 hypointense in both TZ and PZ
  - Diffusion-weighted imaging (DWI)
    - High b values (≥ 1,400 sec/mm²) give best contrast between normal prostate and cancer
      - Tumor is markedly bright on DWI
      - Tumor is markedly dark on apparent diffusion coefficient (ADC) map
  - Dynamic contrast enhancement
    - Normal PZ typically shows low-level slow enhancement; however, in some men, diffuse avid enhancement is seen and may be due to chronic inflammation or fibrosis
      - TZ often shows avid early enhancement; BPH nodules show variable enhancement from low level to avid (mimicking tumor)
    - Tumor in both PZ and TZ enhances early and often washes out rapidly over minutes

Seminal Vesicles and Vasa Deferentia

- MR is excellent modality to evaluate seminal vesicles
  - Fluid-containing structures are therefore of low signal intensity on T1 MR, high signal intensity on T2 MR
  - Evaluation of seminal vesicles is important during staging of prostate cancer
  - Recent ejaculation results in seminal vesicle collapse, and ↓ T2 MR signal and can limit interpretation

CLINICAL IMPLICATIONS

Zonal Distribution of Prostatic Disease

- Prostatic diseases have zonal distribution
  - **Prostate carcinoma**
    - 70% of adenocarcinomas arise in PZ
    - 20% in TZ
    - 10% in CZ
  - **BPH**
    - Originates in TZ
    - Compresses CZ and PZ
    - May extend into bladder base (intravesical prostatic protrusion)
      - Associated with bladder outlet obstruction
      - Not to be confused with bladder mass
    - May extrude into PZ as focal heterogenous encapsulated nodules

Prostate Carcinoma

- Predictors of extracapsular extension of tumor
  - Asymmetry of NVB
    - Nerve-sparing procedure can leave positive margins if NVB invaded
  - Obliteration of rectoprostatic angle
  - Irregular bulge in prostatic contour
  - Tumor-capsule interface > 1.5 cm
- Patterns of spread
  - Direct invasion of seminal vesicles common
  - Tumor at base may spread to bladder neck and change surgical approach
  - Rarely spreads posterior to seminal vesicles across Denonvilliers fascia to involve rectum
  - Lymph nodes: Pelvic sidewall > common iliac > retroperitoneal
    - Perirectal nodes increasingly recognized as harboring disease
  - 90% of prostatic metastases involve pelvic bones and spine usually in ascending fashion
    - Lumbar spine affected 3x > cervical spine

PROSTATE GLAND

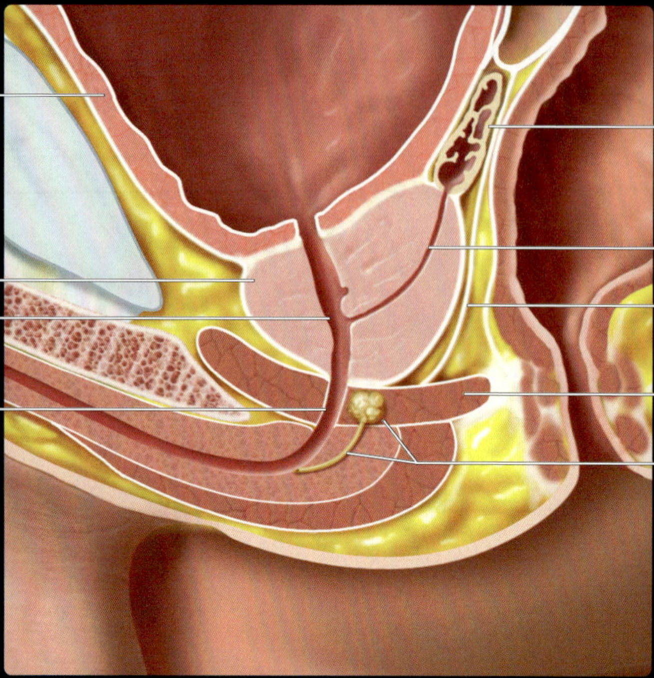

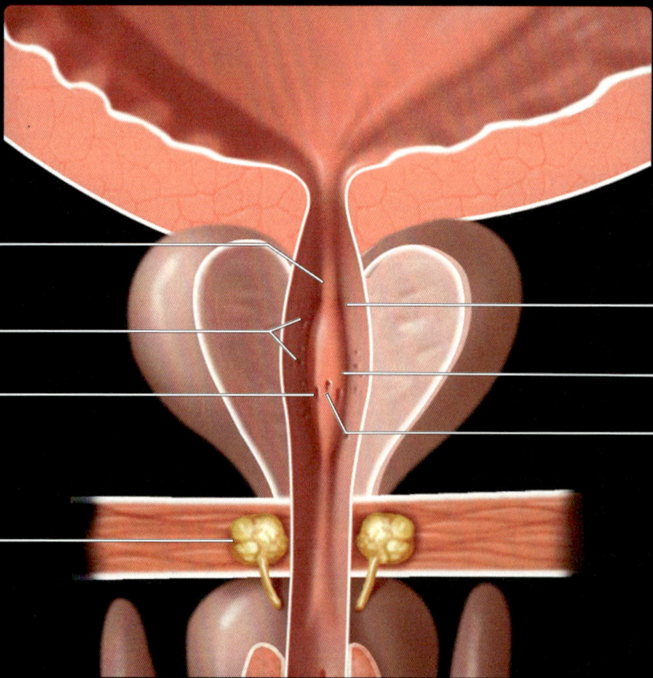

(Top) *Graphic illustrates the relationship between the prostate and the male pelvic organs. The prostate surrounds the upper part of the urethra (prostatic urethra). The base of the prostate is in direct contact with the neck of the urinary bladder, and its apex is in contact with the urogenital diaphragm. The posterior surface is separated from the rectum by rectovesical septum (Denonvilliers fascia).* **(Bottom)** *Graphic illustrates the topography of the posterior wall of the prostatic urethra. The urethral crest is a mucosal elevation along the posterior wall with the verumontanum being a mound-like elevation in the midportion of the crest. The utricle opens midline onto the verumontanum with the ejaculatory ducts opening on either side. The prostatic ducts are clustered around the verumontanum and open into the prostatic sinuses, which are depressions along the sides of the urethral crest.*

VAS DEFERENS AND SEMINAL VESICLES

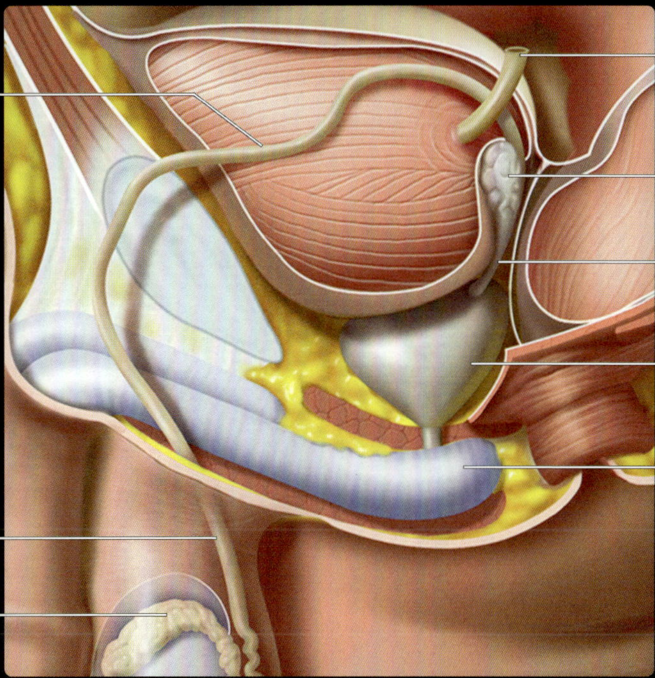

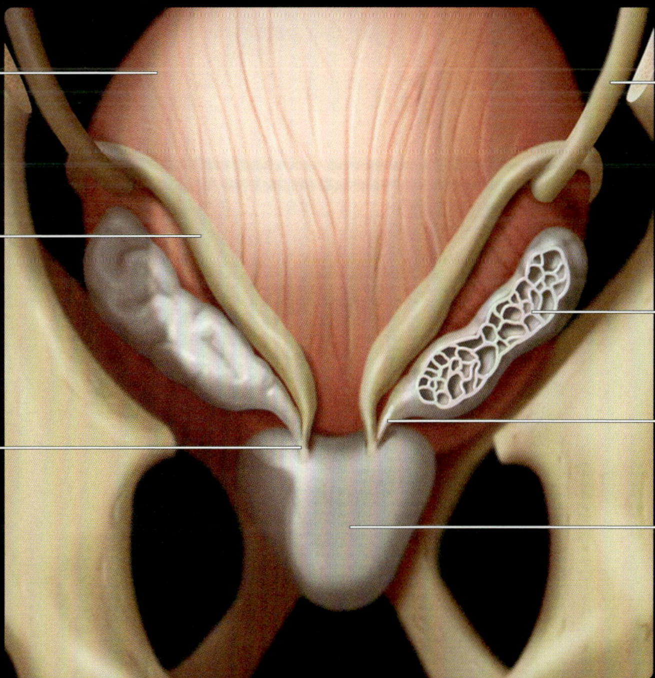

(Top) *Lateral view shows the position of the prostate deep in the pelvis. The vas deferens leaves the scrotum as a component of the spermatic cord, which courses through the inguinal canal into the pelvis.* **(Bottom)** *Posterior view of the prostate gland and seminal vesicles is shown. The cut surface of the seminal vesicle shows its highly convoluted fold pattern. The vas deferens crosses superior to the ureterovesical junction and continues along the posterior surface of the urinary bladder medial to the seminal vesicle. In the base of the prostate, it is directed forward and joined at an acute angle by the duct of the seminal vesicle to form the ejaculatory duct. The ejaculatory ducts course anteriorly and downward through the prostate to slit-like openings on either side of the orifice of the prostatic utricle.*

NORMAL PROSTATE AND BENIGN PROSTATIC HYPERPLASIA

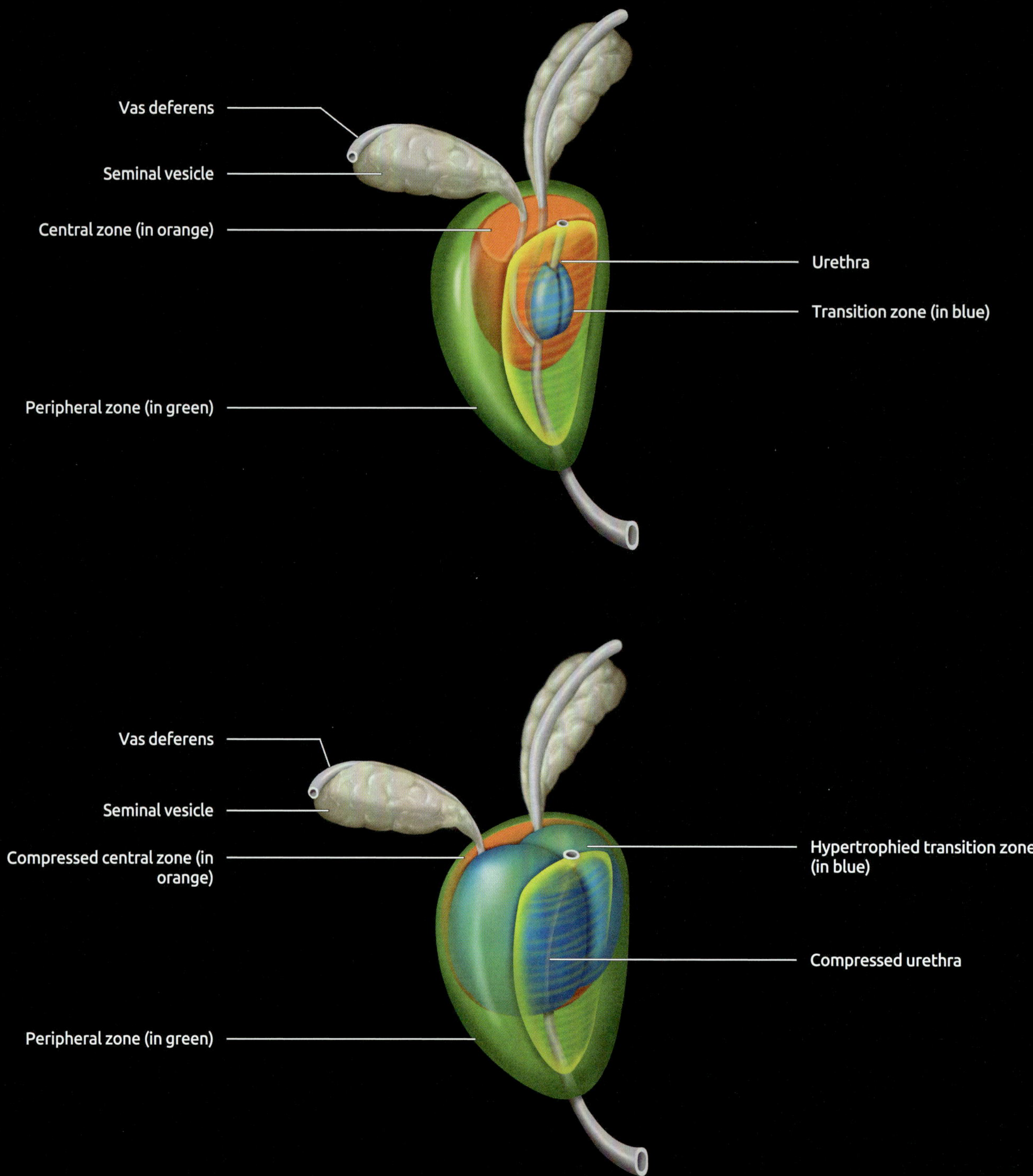

(Top) *First of 2 graphics comparing zonal anatomy in a young man to an older man with benign prostatic hyperplasia (BPH) is shown. The TZ in young men is small in size, composing 5% of the volume of the glandular tissue of the prostate. It surrounds the anterolateral aspect of the urethra at the level of verumontanum in a horseshoe fashion.* **(Bottom)** *With the development of BPH, there is enlargement of the TZ. This causes enlargement of the prostate and compression of the CZ and PZ. BPH mainly involves the TZ, though other zones may also be involved. The enlarged TZ causes compression of the prostatic urethra, the primary reason for development of urinary obstructive symptoms in patients with BPH.*

GROSS PHOTOGRAPHS, PROSTATE

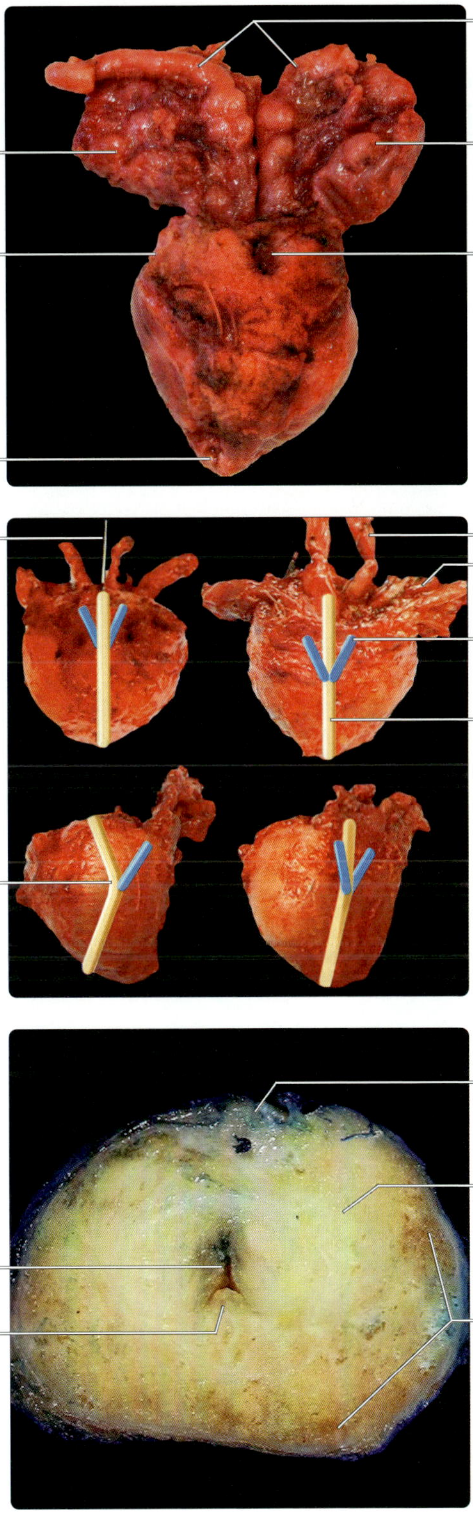

(Top) *Anterior-superior view of a resected prostate shows the urethra at the base. Paired bilateral seminal vesicles are attached to the posterior aspect of the base, lateral to the tubular vas deferens. (From DP: Genitourinary.)* **(Middle)** *These different views of the prostate depict the course of the prostatic urethra (yellow), which has a 35° anterior angulation halfway between base and apex, at the level of the verumontanum; this divides the prostatic urethra into proximal and distal segments. The ejaculatory ducts (blue) have a straighter course from the base of the seminal vesicles to the verumontanum. (From DP: Genitourinary.)* **(Bottom)** *Cross section of the prostate at the verumontanum shows the PZ, which has a more spongy appearance, surrounding the hypertrophied, nodular TZ. (From DP: Genitourinary.)*

T1 MR, PROSTATE

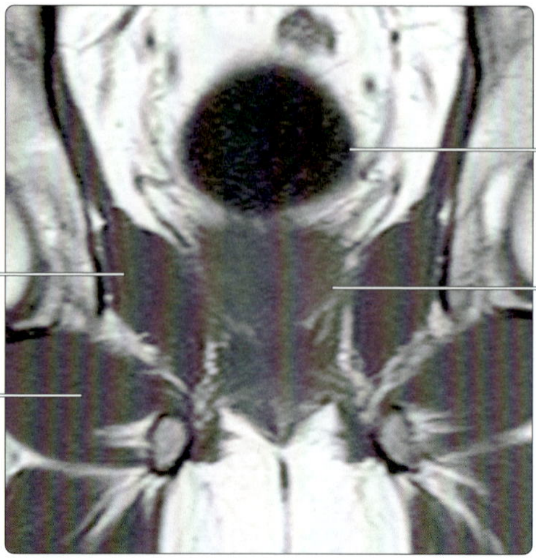

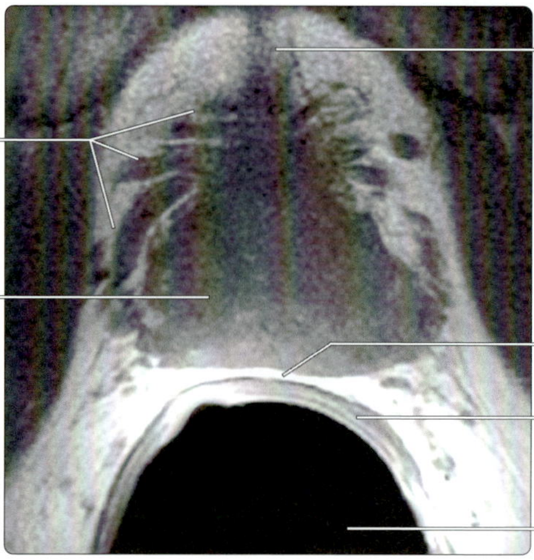

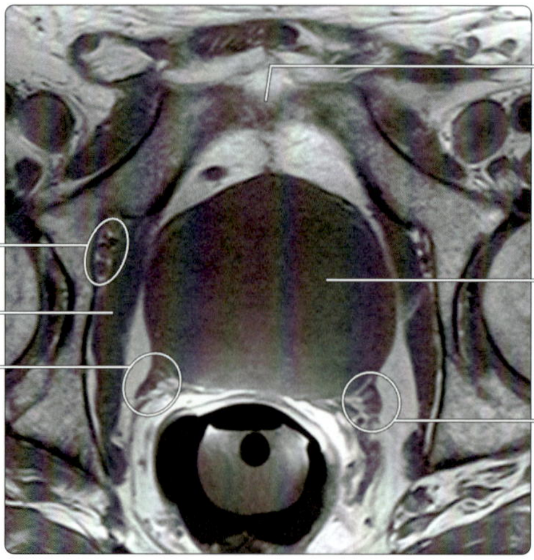

(Top) *Coronal T1 MR of the prostate in a normal young adult shows the conical-shaped prostate with the wide base in contact with the urinary bladder neck and the apex in contact with the urogenital diaphragm.* **(Middle)** *Image is immediately below the bladder neck, just entering the base of the prostate gland. The anterior surface of the prostate is separated from the symphysis pubis by extraperitoneal fat and a venous plexus. Posteriorly, it is separated from the rectum by the rectovesical septum (Denonvilliers fascia).* **(Bottom)** *Slightly inferiorly, the complete contour of the prostate is well seen. It is homogeneous and intermediate to low signal intensity, similar to the pelvic floor muscles. T1 images are not helpful in depicting prostate zonal anatomy, but the neurovascular bundles are usually well seen at the 5 and 7 o'clock positions.*

AXIAL T2 MR, PROSTATE

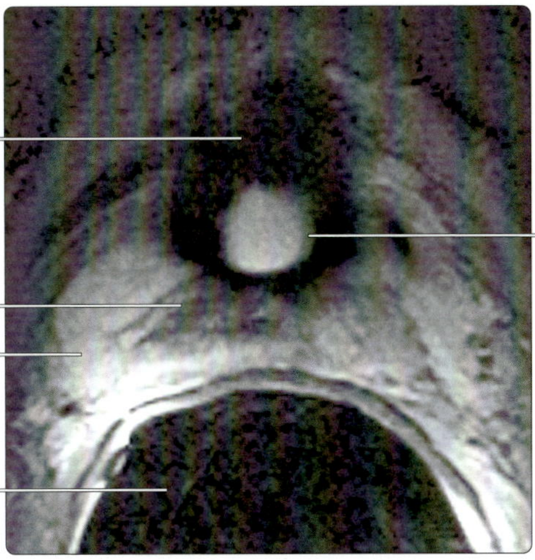

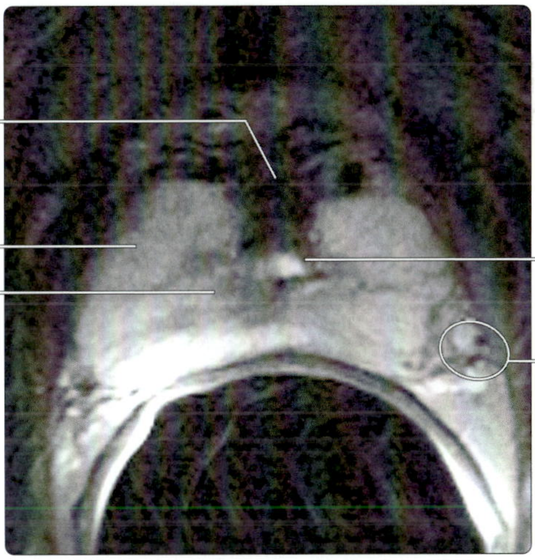

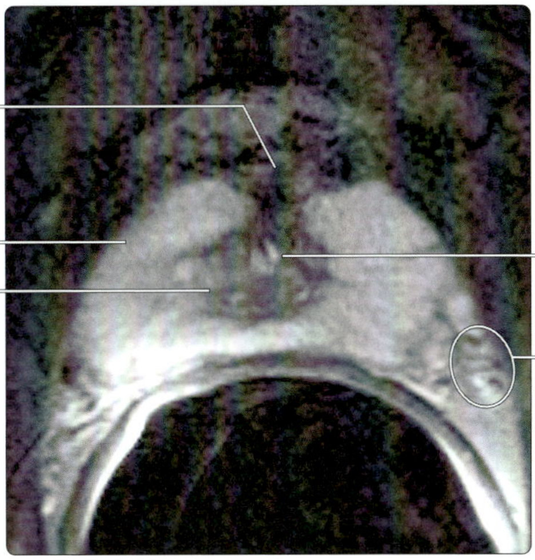

(Top) *First of 3 axial T2 MR images of the prostate of a 45-year-old man with mild prostatic obstructive symptoms is shown. This image, at the level of the prostatic base, shows the AFMS continuous with the bladder muscle and external sphincter. The PZ encloses the CZ, which forms a large part of the prostatic base.* **(Middle)** *The tapered end of the funnel-shaped CZ extends to the level of verumontanum. The neurovascular bundles lie posterolaterally to the prostate, at roughly the 5 and 7 o'clock positions. They supply the corpora cavernosa and are critical for normal erectile function. The periurethral glands are not discernible on imaging.* **(Bottom)** *At the level of the verumontanum, the slightly prominent TZ can be identified as an area of heterogeneous low signal intensity with scattered, small areas of high signal.*

DIFFERENT GRADES OF BENIGN PROSTATIC HYPERPLASIA

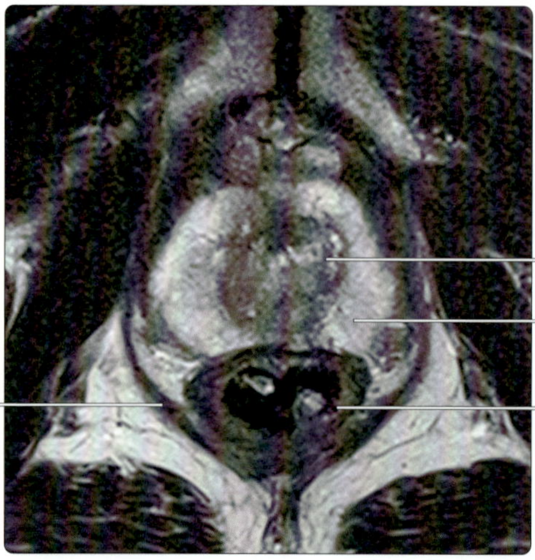

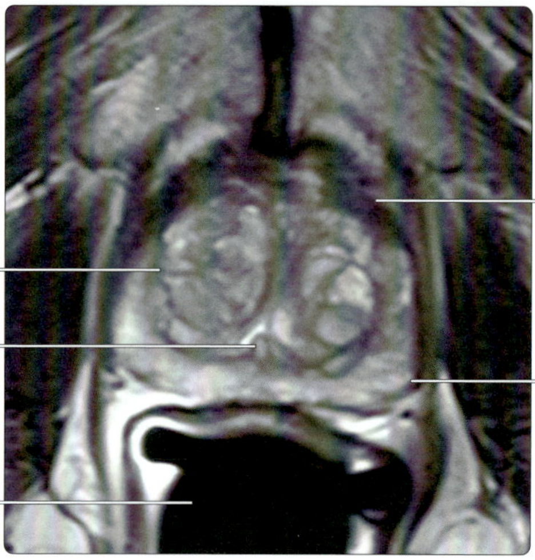

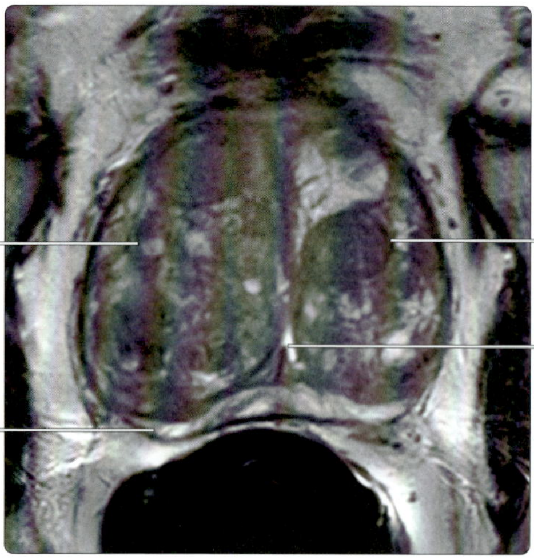

(Top) *Axial image of the prostate shows mild BPH. The hypertrophied TZ is heterogeneous with areas of high and low signal.* **(Middle)** *Axial T2 MR shows moderate BPH with an enlarged, heterogenous, and nodular TZ. The gland measures 52 mL.* **(Bottom)** *Axial T2 MR shows a markedly enlarged, 147-mL prostate due to BPH. Notice the compressed and thinned PZ. Areas of T2 dark signal in the TZ represent stromal elements of BPH, whereas bright signal represents glandular elements. BPH is often arranged into nodules, which are typically heterogenous and show encapsulation.*

DIFFERENT GRADES OF BENIGN PROSTATIC HYPERPLASIA

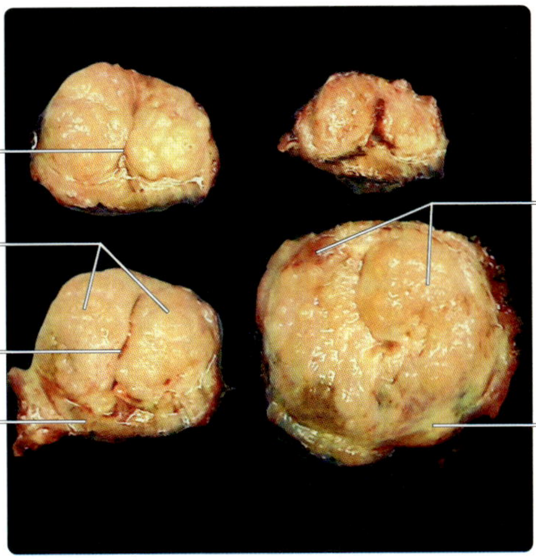

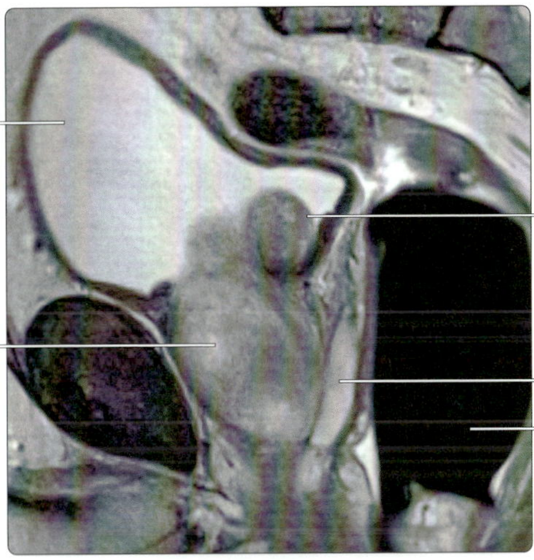

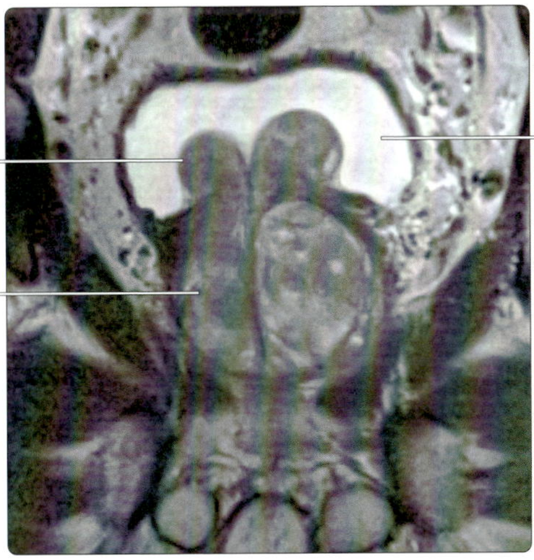

(Top) *Consecutive cross sections of a prostate gland with BPH shows marked nodular enlargement of the TZ. It is compressing the prostatic urethra, which was the cause of the patient's bladder-obstructive symptoms. The PZ is compressed and attenuated by the hyperplastic TZ. (From DP: Genitourinary.)* **(Middle)** *Sagittal T2 MR of the prostate shows BPH with a lobulated midline structure protruding into the bladder neck. This represents hyperplastic periurethral glands. The periurethral glands are less commonly involved with BPH but, when enlarged, can form what is termed intravesical prostatic protrusion (formerly known as median lobe hypertrophy).* **(Bottom)** *Coronal T2 MR of the prostate shows intravesical prostatic protrusion extending from BPH. This can cause severe obstructive voiding symptoms by creating a ball-valve mechanism of obstruction of the urethra. This is distinguished from a bladder mass by its typical appearance and location.*

ENDORECTAL ULTRASOUND, PROSTATE

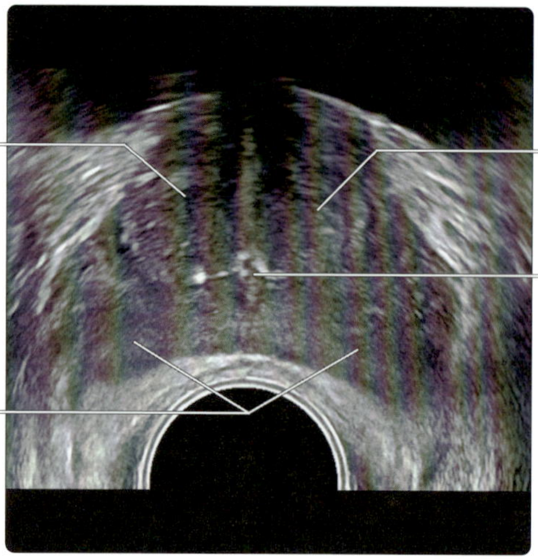

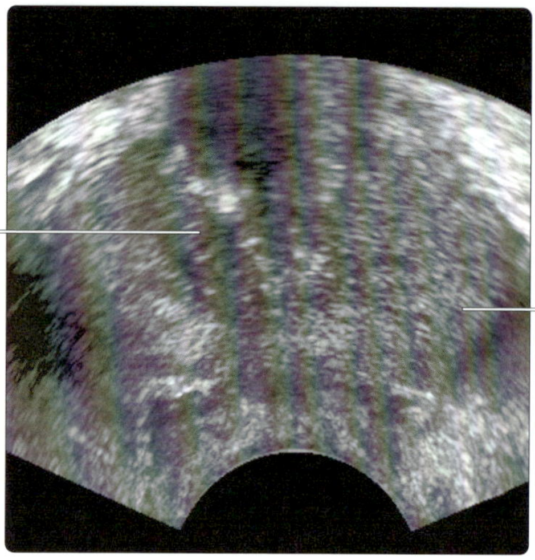

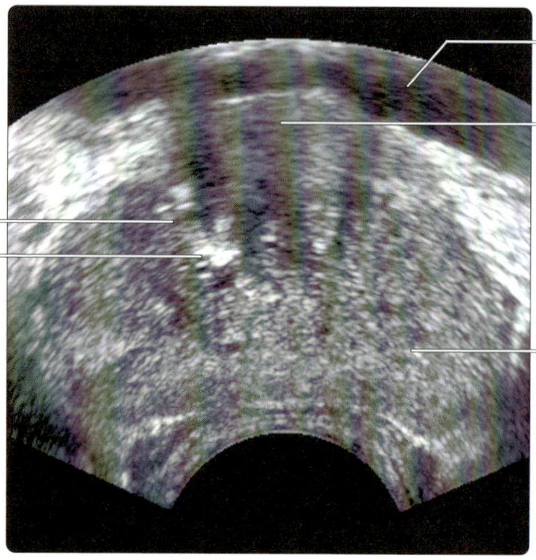

(Top) *Transverse transrectal ultrasound (TRUS) at the level of the midprostate shows the 2 lobes of the TZ on either side of the urethra. The more homogeneous PZ is along the posterolateral aspects of the prostate. Note the periurethral calcifications.* **(Middle)** *First of 2 transverse TRUS images shows a heterogeneous, slightly enlarged TZ. As BPH begins to develop, the TZ becomes distinguishable by ultrasound. It has a heterogeneous echogenicity with visible cysts, nodules, and calcifications.* **(Bottom)** *Focal areas of calcifications with shadowing are seen in this patient.*

ENDORECTAL ULTRASOUND, SEMINAL VESICLES

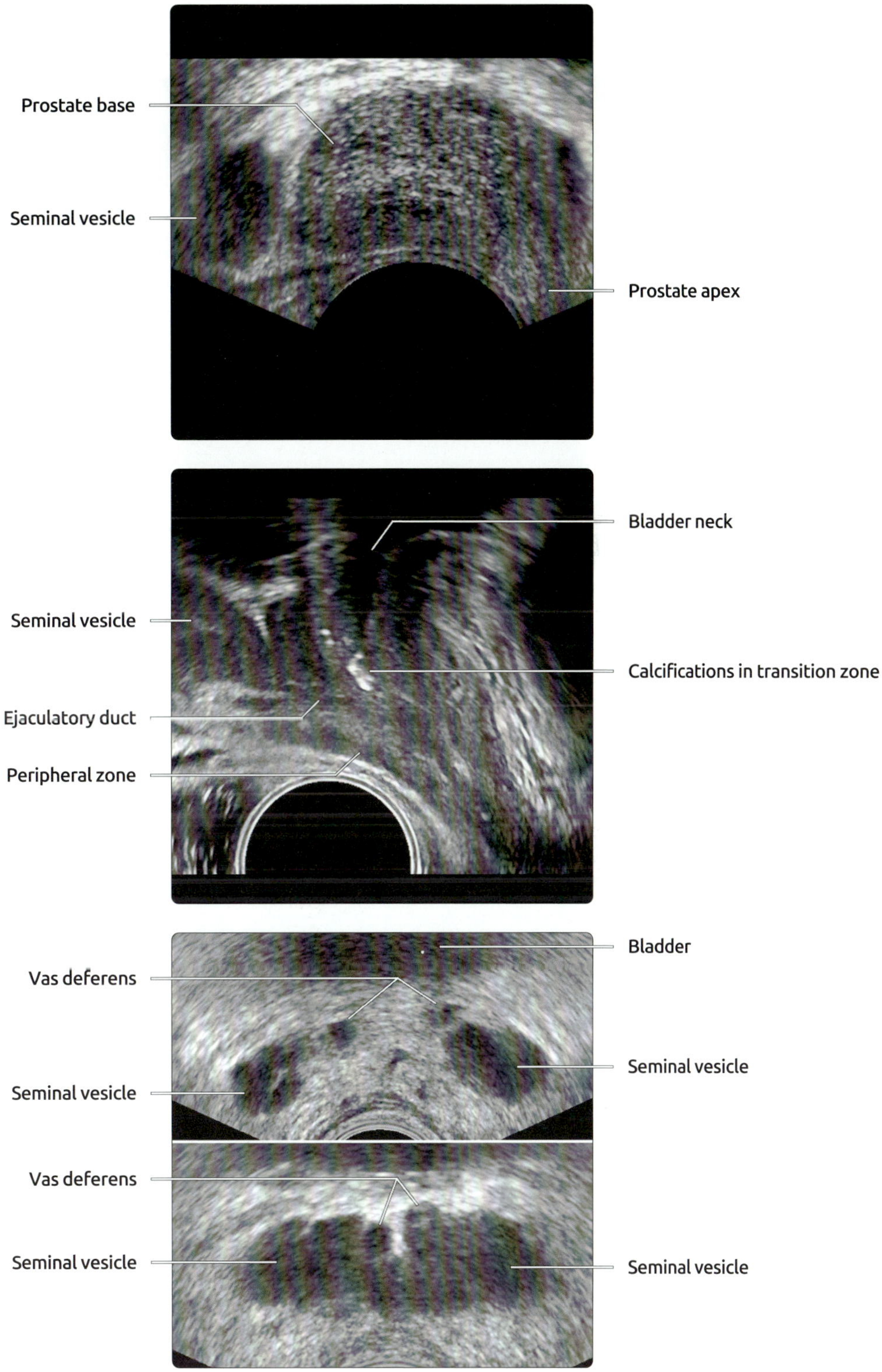

(Top) *Parasagittal TRUS shows the seminal vesicle adjacent to the base of the prostate.* **(Middle)** *Parasagittal TRUS shows the ejaculatory duct entering the prostate base. The ejaculatory ducts are surrounded by the CZ, which is not readily distinguishable on TRUS. Calcifications are seen within the more anterior TZ. The PZ runs along the posterolateral aspect of the prostate.* **(Bottom)** *Two consecutive transverse TRUS show the vas deferens converging with the seminal vesicles. Their union will form the ejaculatory ducts, which enter the prostate base and course within the prostate enclosed within the CZ. The ejaculatory ducts empty into the urethra at the verumontanum.*

DIFFUSION-WEIGHTED IMAGING

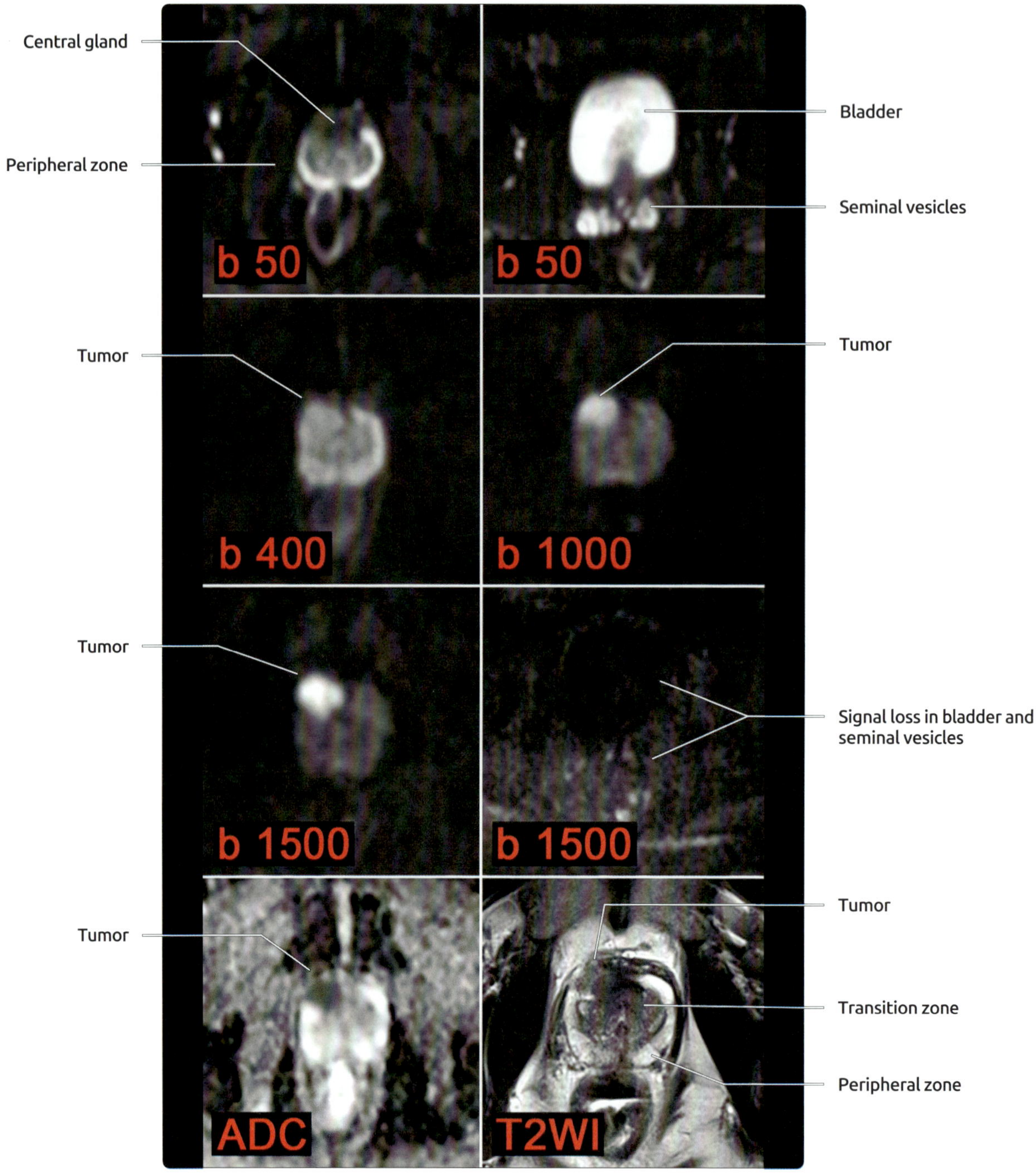

This sequence of diffusion-weighed images (DWI) shows a tumor near the apex of the gland. With low b values (strength of diffusion sensitizing gradient), the signal is similar to a T2 MR with high signal observed in the PZ, bladder, and seminal vesicles. As the b value increases, there is a diffusion of water molecules except in highly cellular areas, such as cancer, which has restricted diffusion. With a b value of 1,500, the tumor is well seen with loss of signal in the surrounding structures. The apparent diffusion coefficient gradient (ADC) map shows low signal (lack of diffusion) in the tumor.

DIFFUSION-WEIGHTED IMAGING

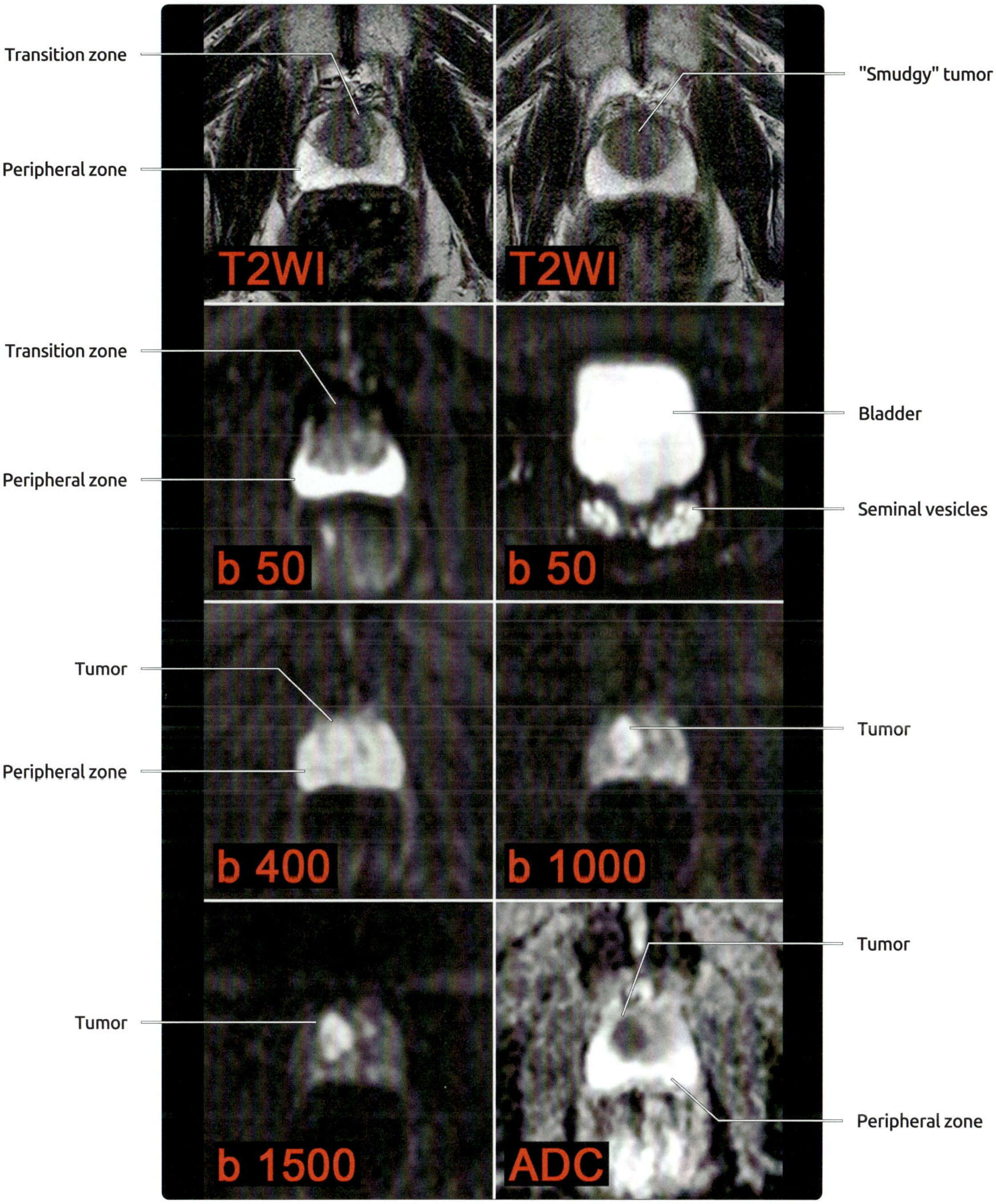

This sequence of DWIs shows a tumor in the TZ of the prostate gland. These tumors can be particularly difficult to diagnose, as they can be masked by surrounding BPH. One of the features described on T2 MR is a "smudgy," charcoal-like appearance in comparison to the more well-defined, encapsulated nodules seen in BPH. With increasing b values, the tumor becomes evident. It is important to note that at low b values, the tumor may be isointense to the rest of the prostate (b 400 in this case); therefore, high b values (> 1,200 sec/mm²) are recommended for prostate cancer imaging.

IMAGING ANATOMY

Penis

- **Composed of 3 cylindrical shafts**
  - **2 corpora cavernosa**: Main erectile bodies
    - On dorsal aspect of penis
    - Diverge at root of penis (**crura**) and are invested by **ischiocavernosus muscles**
    - Chambers traversed by numerous trabeculae, creating sinusoidal spaces
    - Multiple fenestrations in septa between corpora, creating multiple anastomotic channels
    - US
      - Intermediate echogenicity
      - Homogenous echotexture
    - MR
      - T2: Intermediate to hyperintense
      - T1: Intermediate to hypointense
      - Enhance gradually in centrifugal fashion
    - CECT: Enhance to variable degree
  - **1 corpus spongiosum**: Contains urethra
    - On ventral aspect, in groove created by corpora cavernosa
    - Becomes **penile bulb** (urethral bulb) at root and is invested by **bulbospongiosus muscle**
    - Forms **glans penis** distally
    - Also contains erectile tissue but of far less importance
    - US
      - Higher echogenicity than cavernosa
    - MR
      - T2 hyperintense, T1 intermediate
      - Enhances early and often avidly
- **Tunica albuginea** forms capsule around each corpora
  - Thinner around spongiosum than cavernosa
  - Also extends in between corpora (penile septum)
  - US: Thin linear echogenic band
    - 1- to 2-mm thickness
    - Dense shadowing from vertical portions and septum due to fibrous nature
  - MR: T2 dark thin band surrounding corpora
- All 3 corpora surrounded by deep fascia (**Buck fascia**) and superficial fascia (**Colles fascia**)
  - These are poorly seen by imaging
- **Suspensory ligament** of penis (part of fundiform ligament) is inferior extension of abdominal rectus sheath
- Main arterial supply from internal pudendal artery
  - Highly variable branching patterns
  - **Cavernosal artery** runs within center of each corpus cavernosum
    - Gives off **helicine arteries**, which fill trabecular spaces
    - Primary source of blood for erectile tissue
    - US: Double echogenic line in center of corpora
    - Color Doppler US shows flow in center of corpora
  - **Paired dorsal penile arteries** run between tunica albuginea of corpora cavernosa and Buck fascia
    - Supplies glans penis and skin
    - Well identified with color Doppler US
  - Multiple anastomoses between cavernosal and dorsal penile arteries
  - Bulbourethral artery supplies spongiosum at penile bulb
- **Venous drainage** of corpora cavernosa
  - Emissary veins in corpora pierce through tunica albuginea → circumflex veins → deep dorsal vein of penis → retropubic venous plexus
  - Superficial dorsal vein drains skin and glans penis
  - US: Color Doppler shows superficial and deep veins well
  - MR: Dorsal unpaired flow voids
- Primary innervation from terminal branches of internal pudendal nerve

Normal Erectile Function

- Neurologically mediated response eliciting smooth muscle relaxation of cavernosal arteries, helicine arteries, and cavernosal sinusoids
- Blood flows from helicine arteries into sinusoidal spaces
- Sinusoids distend, eventually compressing emissary veins against rigid tunica albuginea
  - Venous compression prevents egress of blood from corpora, which maintains erection
- **US** can be used to evaluate erectile function
  - Doppler evaluation of cavernosal arteries performed after injection of vasodilating agent
  - In flaccid state, there is little diastolic flow
  - At onset of erection, there is dilatation of cavernosal arteries
    - Increase in both systolic and diastolic flow
  - At maximum erection, venous drainage is blocked
    - Waveform changes to high resistance with reversal of diastolic flow
  - **Normal measurements**
    - Peak systolic velocity (PSV) > 30 cm/sec
    - End-diastolic velocity (EDV) < 5 cm/sec
    - Resistance index (RI): > 0.9

Urethra

- Divided into 4 sections: Prostatic, membranous, bulbous, and penile
  - **Posterior urethra**: Prostatic and membranous
  - **Anterior urethra**: Bulbous and penile
- **Prostatic urethra**
  - **Verumontanum** is 1-cm ovoid mound along urethral crest (smooth muscle ridge on posterior wall)
  - Prostatic utricle, prostatic ducts, and ejaculatory ducts enter in this segment
  - Lined by transitional cells
    - Remainder of urethra lined by columnar cells, except glans, which has squamous epithelium
- **Membranous urethra**
  - Short course through urogenital diaphragm
    - Level of external urethral sphincter
  - Contains bulbourethral glands (**Cowper glands**)
    - Ducts travel distally ~ 2 cm to enter bulbous urethra
- **Bulbous urethra**
  - Below urogenital diaphragm to suspensory ligament of penis at penoscrotal junction
- **Penile urethra**
  - Pendulous portion, distal to suspensory ligament
  - **Fossa navicularis**: Widening at glans penis
  - Both penile and bulbous urethra lined by mucosal urethral glands (**glands of Littre**)

- **2 points of fixations**: Urogenital diaphragm (membranous urethra) and penoscrotal junction

CLINICAL IMPLICATIONS

Erectile Dysfunction

- Complex and often multifactorial
  - Major causes include vascular, neurogenic, hormonal, drug-induced, anatomic, and psychologic etiologies
  - Vascular etiology most common
- Arteriogenic impotence effects inflow
  - Usually internal pudendal and penile arteries
  - Blockage may be as high as distal aorta (**Leriche syndrome**)
- Venogenic impotence effects outflow
  - Ineffective venoocclusion with continuous outflow of blood from sinusoids
- Doppler US with measurements in flaccid and erect state
  - Pharmacologic-induced erection with intracavernosal injection of vasoactive drug
  - Evaluation of PSV, EDV, RI, and waveforms can diagnose arterial vs. venous etiology
  - Less commonly performed in era of erectile dysfunction medication

Trauma

- **Urethral trauma** most commonly involves **membranous urethra**
  - High association with pelvic fractures
  - Retrograde urethrogram: Leakage of contrast at membranous urethra level above &/or below pelvic diaphragm
  - Stretch injury of posterior urethra with elongation of membranous segment and elevation of prostate
  - Direct blow to penile urethra shows leakage of contrast into spongiosum ± intravasation into veins
- **Straddle injuries** most commonly involve bulbous urethra
  - Urethral injury from compression against inferior pubic symphysis
- **Penile fracture**
  - Mainly occurs during sexual intercourse
  - Tear in tunica due to excessive bending of erect penis
  - US: Hypoechoic defect in tunica with adjacent collection
  - MR: Thin or absent corpora ± overlying hematoma

Peyronie Disease

- Idiopathic plaque formation on tunica albuginea
- May be fibrous or calcified
- Causes painful erections with shortening and curvature of penis
- Imaging useful to identify nonpalpable plaques (generally in septum)
- US: Focal thickening of tunica ± shadowing calcification
- MR: T1-/T2-hypointense thickening of tunica ± enhancement

Priapism

- Low-flow priapism is most common and results in ischemia
  - Medical emergency and does not require imaging
  - Seen in sickle cell disease, hematologic dyscrasias, and spinal cord conditions
- High-flow priapism may be due to arteriovenous fistula
  - Imaging may be used to identify vascular fistula

Penile/Urethra Cancer

- Most cancers are squamous cell carcinoma and arise from glans or prepuce
- Urethral cancers occur and may be squamous, urothelial carcinoma, or adenocarcinoma
  - Associated with urethral strictures
- MR has role in local staging if clinically unclear
  - Infiltrative T1-/T2-hypointense mass with enhancement
- Nodal spread is stepwise from inguinal to pelvic nodes

Penile Thrombophlebitis (Penile Mondor Disease)

- Thrombosis of superficial or deep dorsal vein of penis
- Most often associated with vigorous or prolonged sexual intercourse
- Presents as palpable cord along dorsal surface ± pain and swelling
- US: Echogenic noncompressible filling defect with no flow on CDUS

Urethral Stricture

- Location and length may be clue to etiology
- Iatrogenic: Penoscrotal junction and bulbomembranous junction
- Traumatic: Bulbar segment due to straddle injury or membranous due to pelvic fracture
- Infectious/inflammatory: Multifocal, long segment
- Malignant: Irregular margins

SELECTED REFERENCES

1. Albasha H et al: Imaging review of penile pathologies encountered in the emergency department. Emerg Radiol. 29(1):147-59, 2022
2. Galgano SJ et al: The male urethra: imaging and surgical approach for common pathologies. Curr Probl Diagn Radiol. 50(3):410-8, 2021
3. Wongwaisayawan S et al: Imaging spectrum of traumatic urinary bladder and urethral injuries. Abdom Radiol (NY). 46(2):681-91, 2021
4. Tu LH et al: MRI of the penis: indications, anatomy, and pathology. Curr Probl Diagn Radiol. 49(1):54-63, 2020
5. Varela CG et al: Penile Doppler ultrasound for erectile dysfunction: technique and interpretation. AJR Am J Roentgenol. 214(5):1112-21, 2020
6. Parker RA 3rd et al: MR imaging of the penis and scrotum. Radiographics. 35(4):1033-50, 2015

PENIS AND URETHRA

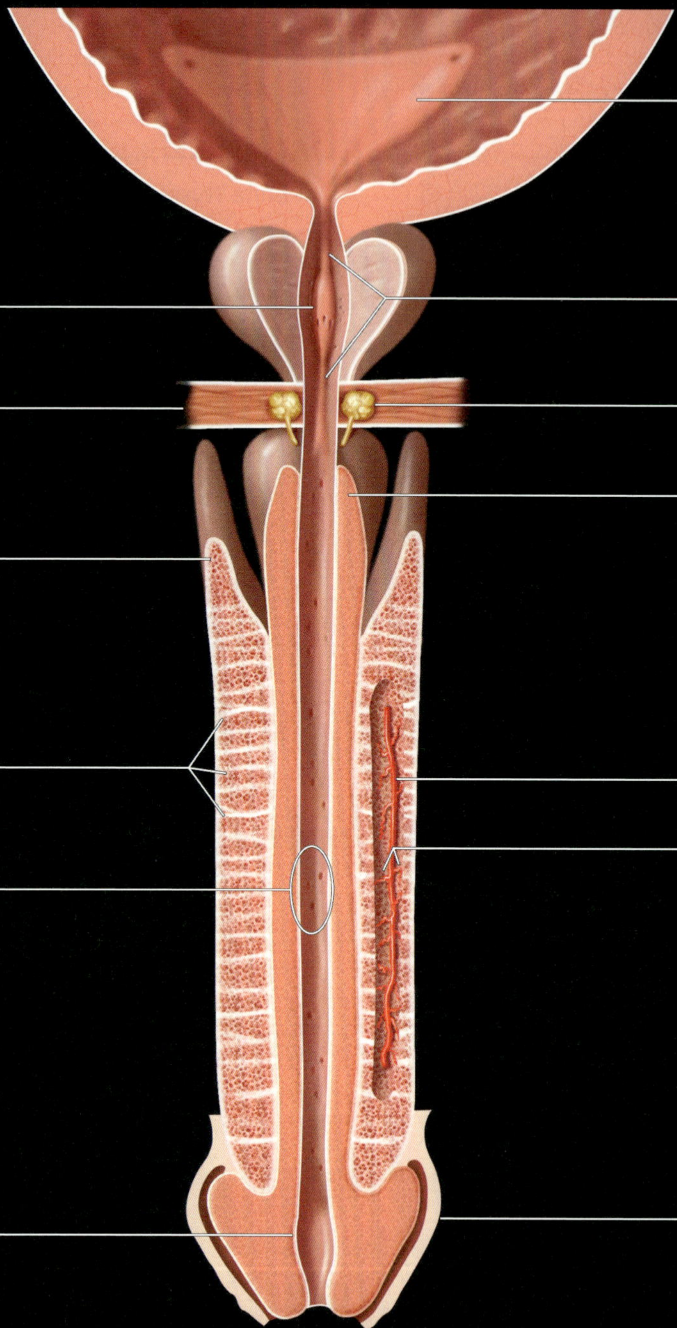

Graphic shows the posterior wall of the urethra from the bladder base to the fossa navicularis. The verumontanum is the most prominent portion of the urethral crest. The prostatic utricle (an embryologic remnant of the müllerian system) enters in the center of the verumontanum, along with the ejaculatory ducts, which are just distal on either side. Cowper glands are within the urogenital diaphragm, but their ducts course distally ~ 2 cm to enter the bulbous urethra. Multiple small, mucosal glands (glands of Littre) line the mucosa of the anterior urethra. The corpora cavernosa are the primary erectile bodies. They are traversed by numerous trabeculae, creating sinusoidal spaces. The cavernosal artery runs within the center of each corpus cavernosum and gives rise to the helicine arteries, which flow into the sinusoids.

PENIS AND PERINEUM

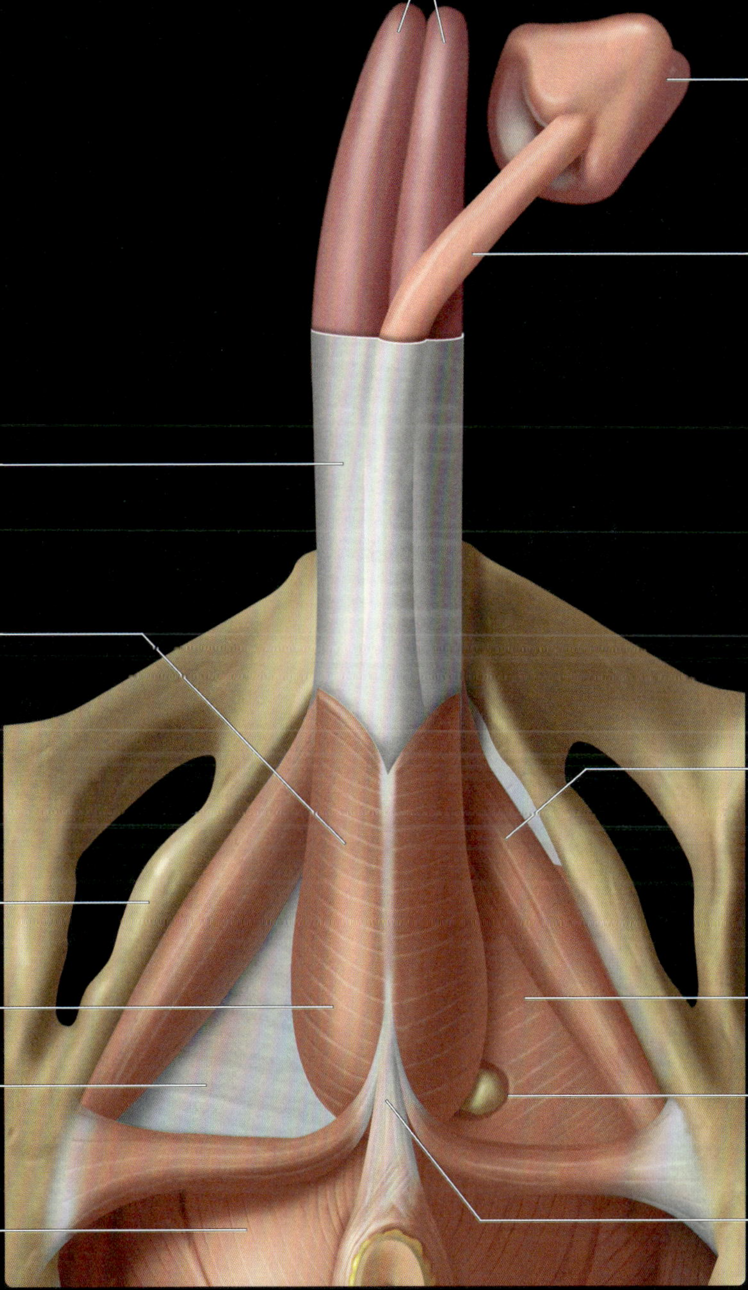

At the root of the penis, the corpora split into a triradiate form. The crura of the corpora cavernosa diverge, and each crus is invested by an ischiocavernosus muscle. The corpus spongiosum remains midline and is invested by the bulbospongiosus muscle. It widens at its attachment to the urogenital diaphragm and is called the penile or urethral bulb. The root of the penis is firmly anchored between the urogenital diaphragm above and Colles fascia below (cut away to show perineal muscles). Additional supports include fascial attachments to the medial surface of the pubic rami and and pubic symphysis.

PENIS CROSS SECTION

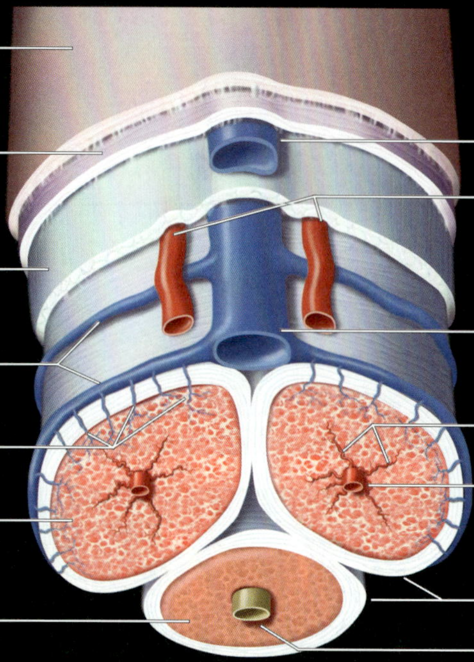

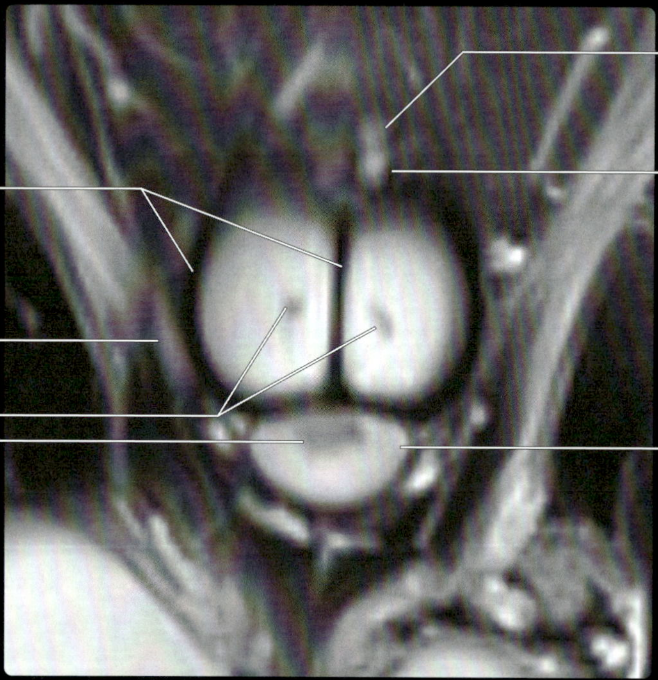

(Top) *Cross section of the penis viewed from above shows the 3 corpora, each of which is invested by a tunica albuginea. The 3 corpora are surrounded by Buck fascia. The cavernosal arteries run within the corpora cavernosa and give rise to the helicine arteries, which fill the sinusoids. Normal drainage is via the emissary veins → circumflex veins → deep dorsal vein of penis. With an erection, the emissary veins are compressed against the strong tunica albuginea, occluding venous drainage. The dorsal penile arteries and deep dorsal vein run beneath the Buck fascia. The penile arteries supply the glans and skin, but there are significant anastomoses between both the arterial and venous systems.* **(Bottom)** *Coronal T2 FS MR through the base of the penis shows the low-signal tunica albuginea surrounding each of the corpora. The cavernosal arteries and the urethra are seen in cross section within the corpora. Buck and Colles fascia are often not well seen by imaging.*

NORMAL ERECTILE FUNCTION

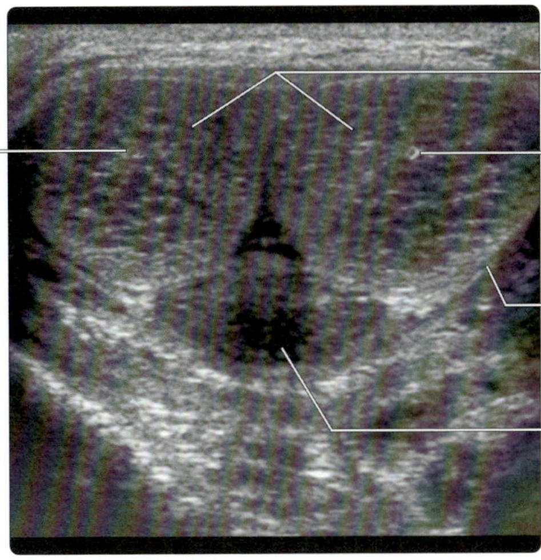

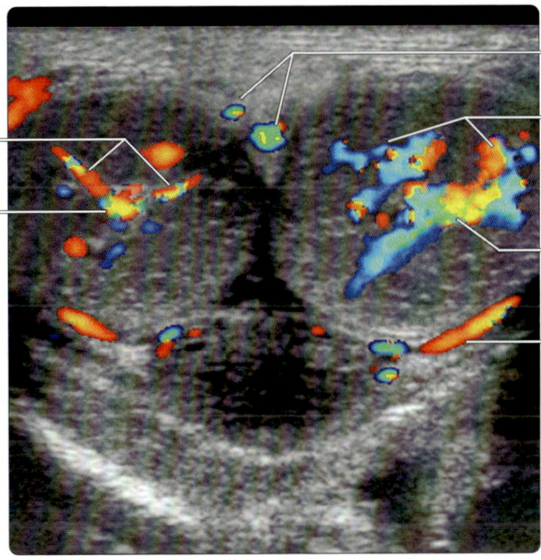

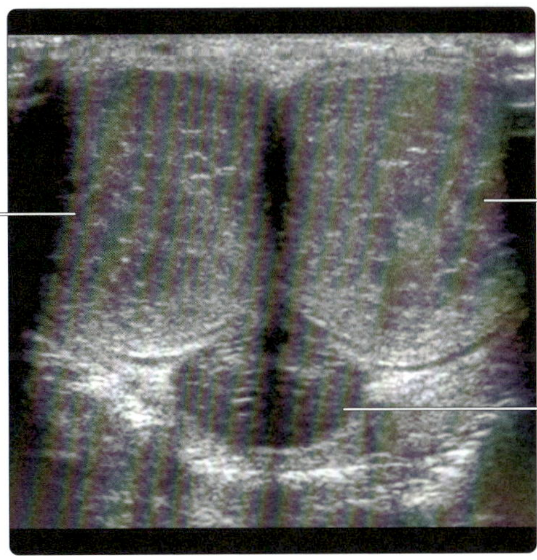

(Top) *First of 3 cross-sectional US images of the penis shows the changing appearance during a normal erection. The paired corpora cavernosa are the primary erectile bodies. They are composed of a complex network of trabeculae, which create the sinusoids and give a sponge-like appearance on US. Each corpus is surrounded by a tough tunica albuginea, which appears as a thin echogenic line. Buck fascia, which surrounds all 3 corpora, is intimately associated with the tunica and cannot be distinguished as a separate structure.* **(Middle)** *Color Doppler US at the onset of an erection shows arterial inflow. There is dilation of the cavernosal and helicine arteries as they begin to fill the sinusoidal spaces.* **(Bottom)** *With continued arterial inflow and compressed venous outflow, the corpora become maximally distended and rigid.*

PEYRONIE DISEASE

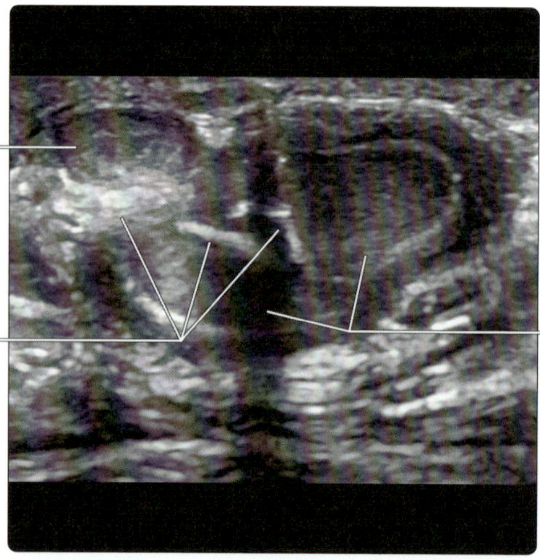

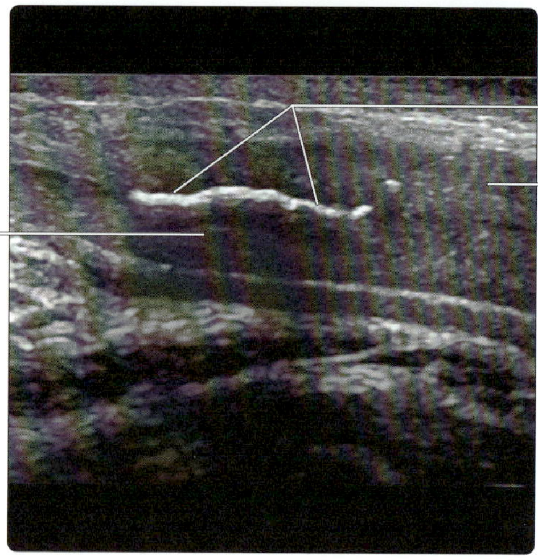

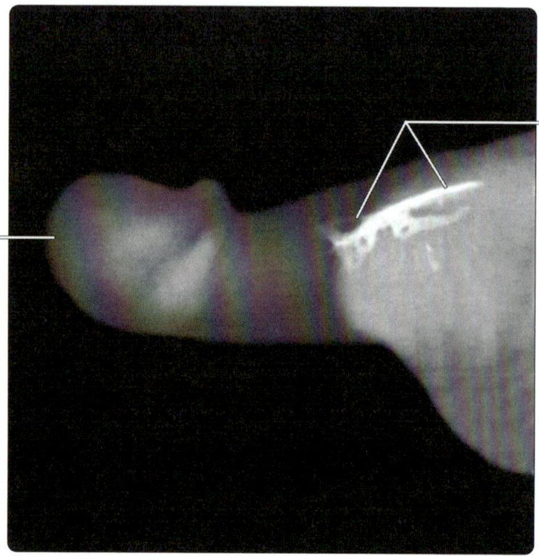

(Top) *First of 3 images in a patient with Peyronie disease shows ventral cross-sectional US of calcified plaques within the tunica albuginea of the corpora cavernosa involving the penile septum and corpus spongiosum.* **(Middle)** *Longitudinal US shows densely calcified plaques with significant posterior shadowing between the corpus cavernosum and spongiosum.* **(Bottom)** *Lateral radiograph of the penis shows not only the calcified plaques but also shortening and upward curvature of the penis. Peyronie disease is an incompletely understood inflammatory process with fibroblastic proliferation and calcification involving the tunica albuginea. It manifests as pain and deformity with erection. The penis may shrink in size with disease progression.*

PENILE INJURY AND MALIGNANCY

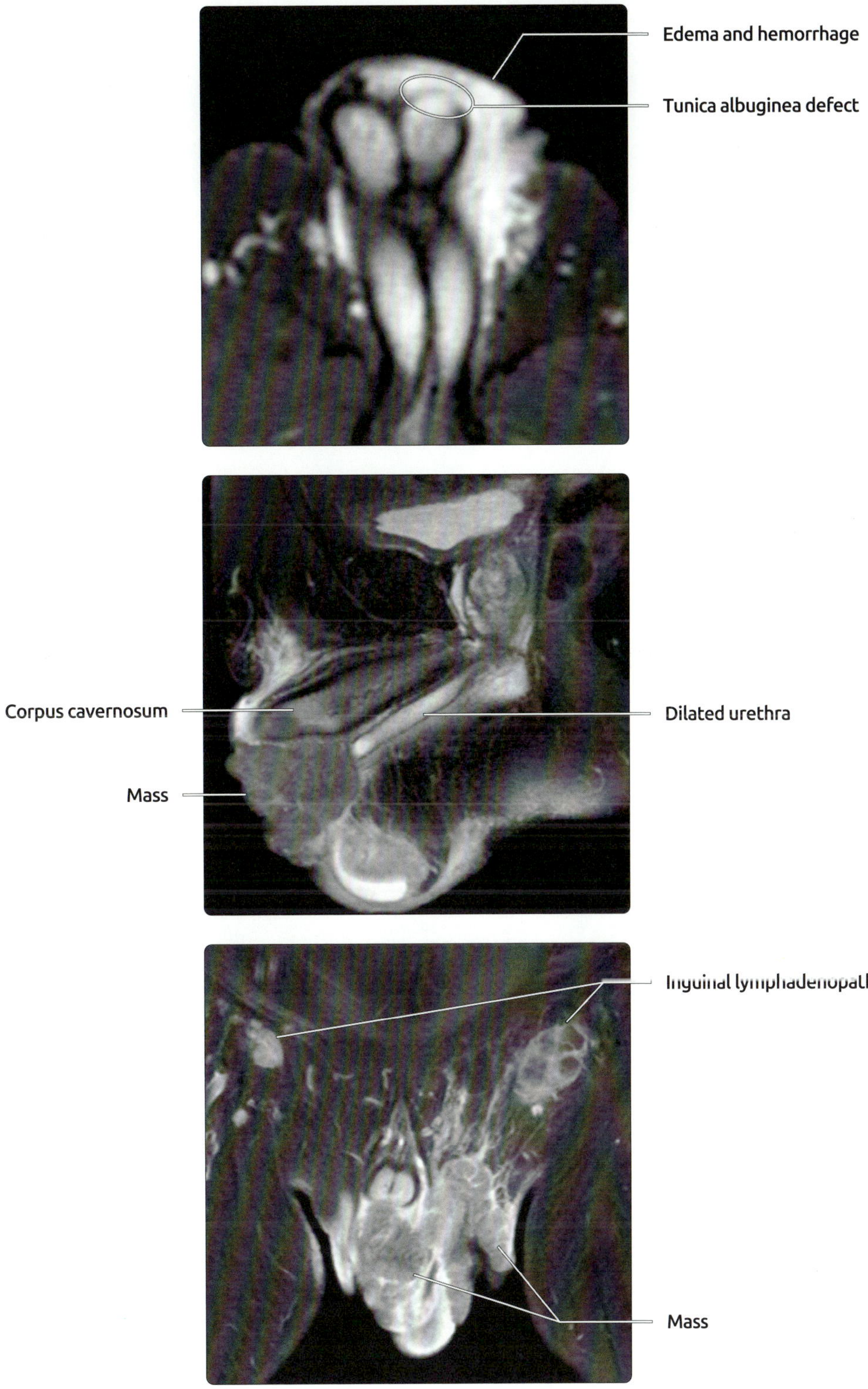

(Top) *Axial T2 FS MR shows a defect in the tunica albuginea of the left corpus cavernosum with surrounding edema and hemorrhage. Patient felt a pop during sexual intercourse and had immediate pain. Findings are consistent with penile fracture.* **(Middle)** *Sagittal T2 FS MR shows an infiltrative mass at the tip of the penis invading the corpus spongiosum. Biopsy showed squamous cell carcinoma, the most common form of penile cancer.* **(Bottom)** *Coronal T2 FS MR in the same patient shows the infiltrative penile mass involving the corpus spongiosum and extending to the scrotum and groin. Notice bilateral inguinal lymphadenopathy, the 1st nodal drainage pathway of spread of penile cancer.*

PENIS AND URETHRA

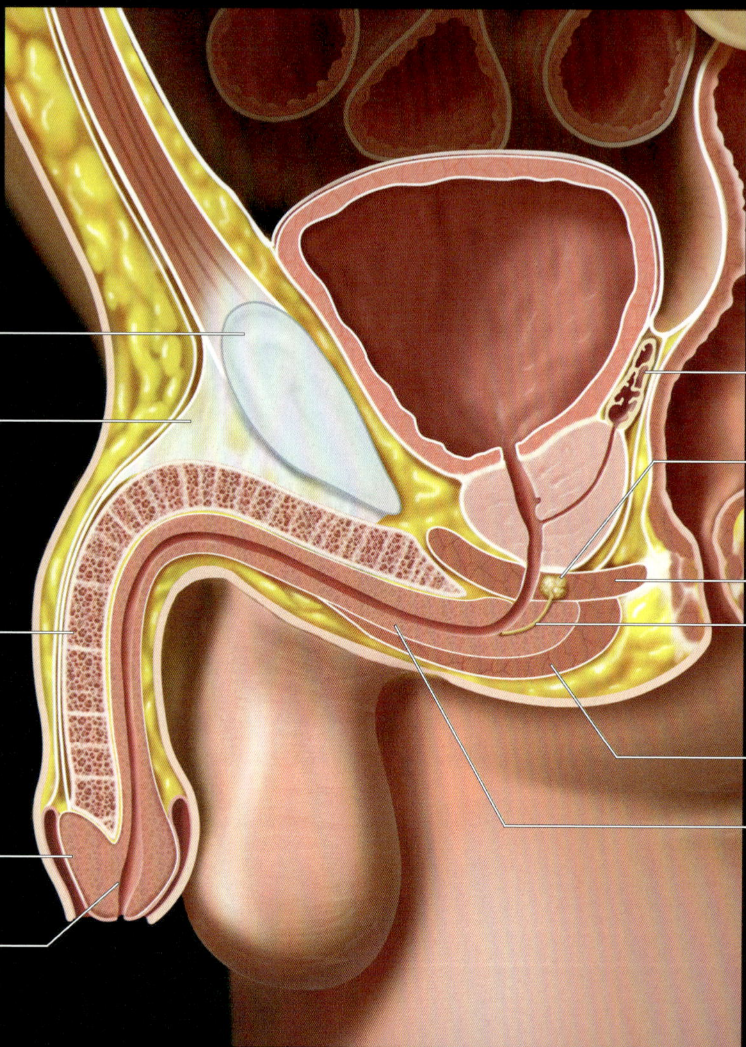

Sagittal midline graphic shows the course of the urethra. The anterior urethra runs within the corpus spongiosum. The external (pendulous) portion of the penis begins below the pubic symphysis at the penoscrotal junction. It is fixed at this position by the suspensory ligament of the penis, which is part of the fundiform ligament, an inferior extension of the rectus sheath. The proximal (internal) urethral sphincter extends from the bladder neck to above the verumontanum. The distal (external) sphincter has both intrinsic and extrinsic components. The intrinsic component is composed of concentric smooth muscle extending from the distal prostatic urethra through the membranous urethra. The surrounding extrinsic component is composed of striated muscle and is under voluntary control.

URETHRAL ANATOMY

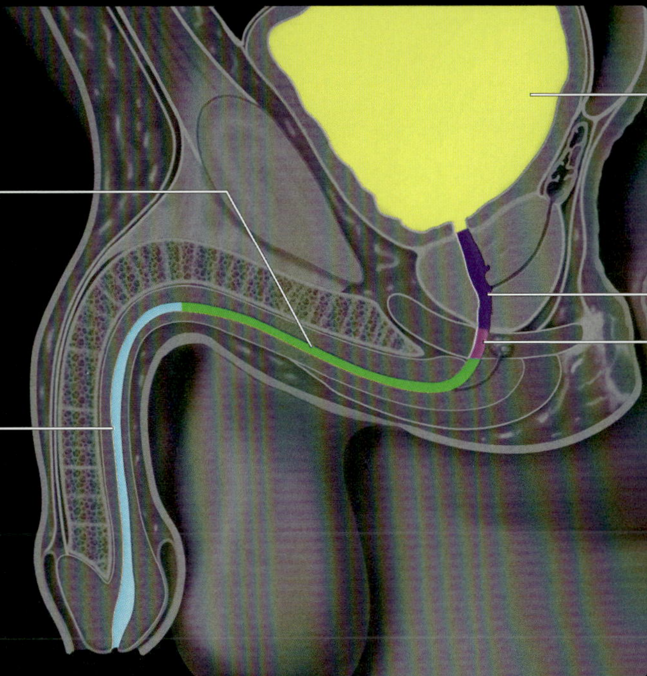

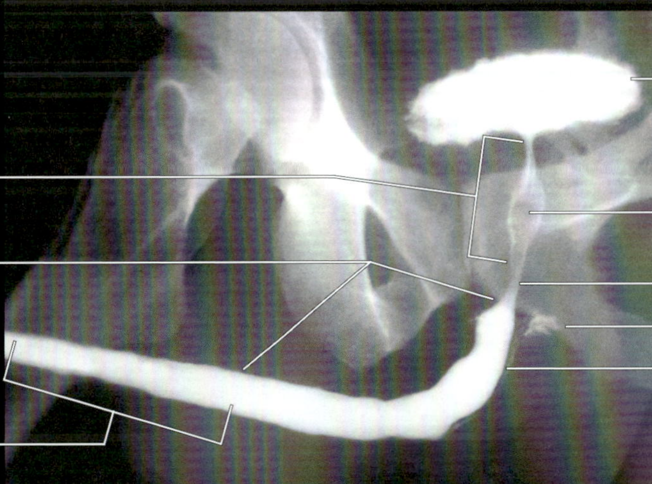

(Top) *The urethra has 2 major divisions: The anterior and posterior urethra, each of which has 2 parts. The posterior urethra is composed of the prostatic and membranous portions, and the anterior urethra is composed of the bulbous and penile portions. The prostatic urethra begins at the bladder base and extends to the apex of the prostatic gland. The membranous urethra traverses the urogenital diaphragm. It is the shortest portion of the urethra but the area most vulnerable to injury. The bulbous urethra extends from the bottom of the urogenital diaphragm to the suspensory ligament of the penis. The penile urethra is distal to the suspensory ligament and travels through the pendulous portion of the penis. It widens into the fossa navicularis at the distal glans.* **(Bottom)** *The verumontanum is a smooth filling defect along the posterior wall of the prostatic urethra. The paired Cowper glands lie within the urogenital diaphragm, but their ducts extend ~ 2 cm to enter the bulbous urethra.*

- Crossed by urethra and vagina; membrane is attached medially to lateral vaginal walls
- Trilaminar structure, which includes
  - Deep transverse perineal muscles
  - Superior and inferior fascia
  - Perineal muscle, which is sandwiched between superior and inferior fascia
- Perineal body
  - Fascial condensation posterior to vagina
  - Insertion site of perineal muscle and external anal sphincter

PERINEUM

Location and Description

- Perineum is superficial soft tissues below pelvic diaphragm
- When seen from below with thighs abducted, perineum is diamond-shaped
  - Bounded anteriorly by symphysis pubis, posteriorly by tip of coccyx, and laterally by ischial tuberosities

Divisions

- Perineum is divided by line between ischial tuberosities into urogenital triangle anteriorly and anal triangle posteriorly
- **Urogenital triangle**
  - Bounded anteriorly by pubic bone and pubic arch and laterally by ischial tuberosities
  - Contains urethra, vagina, perineal membrane, and external genital muscles
- **Anal triangle**
  - Bounded anteriorly by line between ischial tuberosities, laterally by sacrotuberous ligament, and posteriorly by tip of coccyx
  - Contains anal canal in midline with ischiorectal fossa on either side

SUPERFICIAL EXTERNAL GENITAL MUSCLES

Location

- Most superficial of 4 layers of pelvic floor
  - Includes superficial transverse perineal, bulbospongiosus, and ischiocavernosus muscles

Muscles

- Superficial transverse perineal muscles
  - Spans posterior edge of urogenital diaphragm
  - Inserts into perineal body and external anal sphincter
- Bulbospongiosus muscle originates from clitoris, courses along vestibule and inserts on perineal body
- Ischiocavernosus muscle covers crus of clitoris, then has posterolateral course and terminating at ischiopubic ramus
- Bulbospongiosus and ischiocavernosus play role in sexual function, compressing venous return of clitoris

PELVIC FLOOR MUSCLE AND ENDOPELVIC FASCIAL INTERACTION

Normal

- Muscles give active support of pelvic floor, whereas ligaments and endopelvic fascia give passive support to hold pelvic organs in place
- When levator ani muscles function properly, pelvic floor is closed, and ligaments/fascia are not under tension

Abnormal

- When pelvic muscles relax or are damaged, they cannot close levator hiatus, and ligaments/fascia are under strain
- Connective tissue can sustain high abdominal pressure loads for shorter periods, ultimately leading to prolapse

BONY PELVIS AND LIGAMENTS

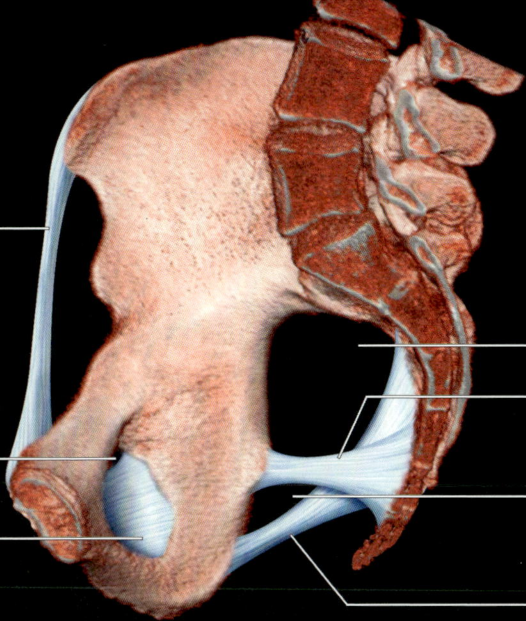

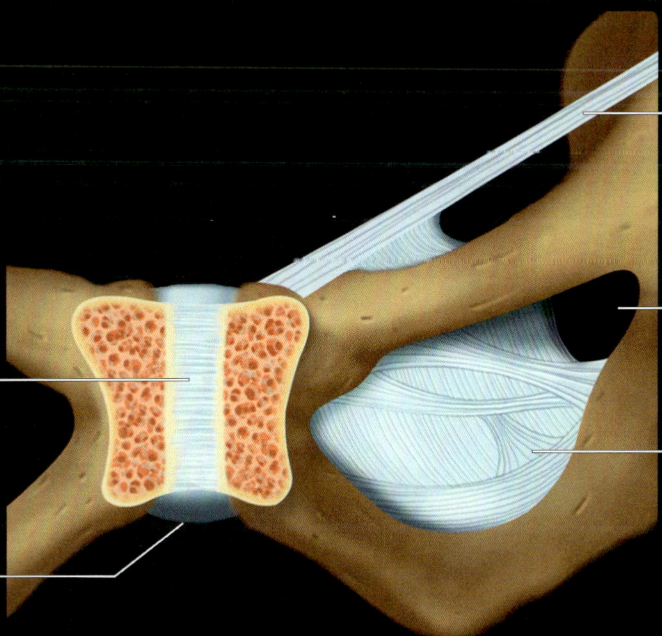

(Top) *3D CT reconstruction with graphic enhancement of the ligaments shows a medial view of the pelvic sidewall. The sacrospinous ligament extends between the sacrum and ischial spine. The sacrotuberous ligament extends from the lateral part of sacrum, coccyx, and posterior inferior iliac spine to insert on the ischial tuberosity. The greater sciatic foramen is above the sacrospinous ligament, and the lesser sciatic foramen is below it.* **(Bottom)** *Graphic of the pubic symphysis shows the arcuate ligament. It blends with the interpubic disc and extends laterally, attaching to the inferior pubic rami. This is an important landmark when assessing the urethral supporting ligaments. The interpubic disc is strengthened anteriorly by several interlacing collagenous fibrous layers passing obliquely from bone to bone. The obturator foramen is covered by a membrane, except for the obturator canal, through which the obturator nerve and vessels pass.*

POSTERIOR PELVIC WALL

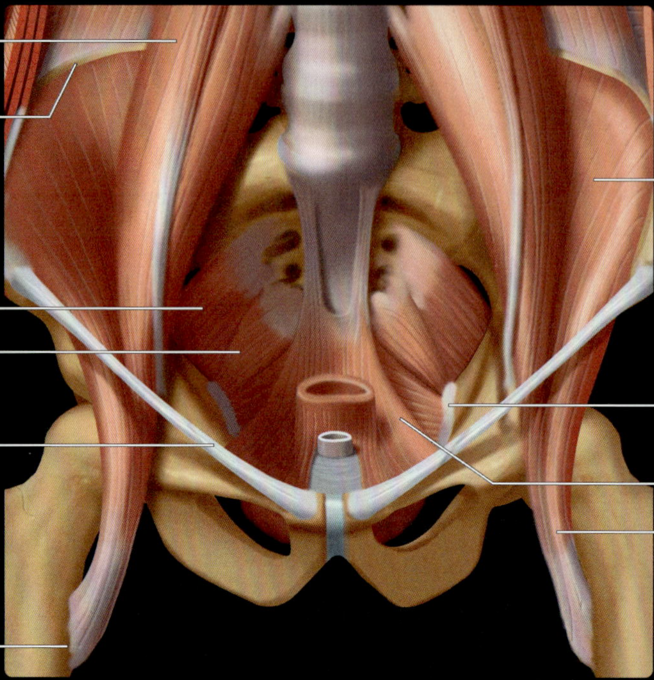

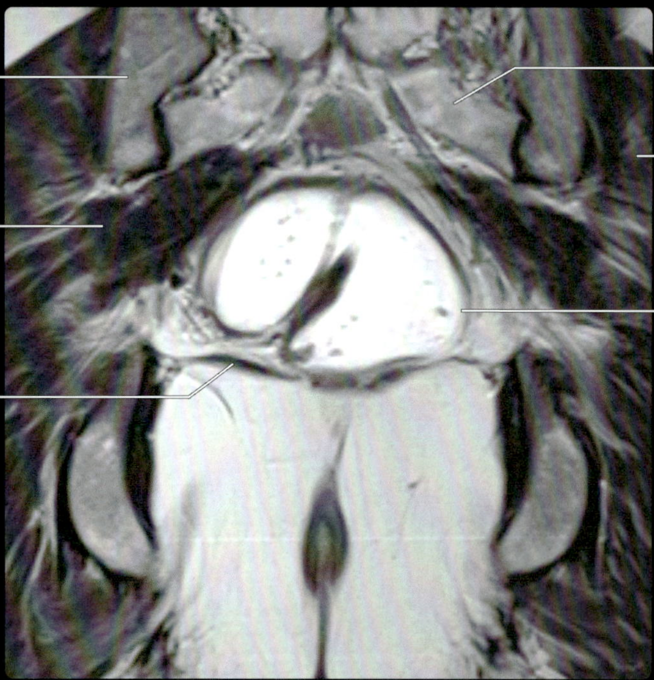

(Top) *The anterior pelvic wall is the shallowest wall formed by the posterior surfaces of the bodies of the pubic bone, symphysis pubis, and pubic rami. The posterior wall of the false pelvis is formed by the iliac bones, sacrum, and the iliacus and psoas muscles. These 2 muscles fuse caudally to form the iliopsoas muscle, which passes anterior to the hip joint to insert onto the lesser trochanter of the femur. The posterior wall of the true pelvis is formed by the sacrum, coccyx, and the piriformis and coccygeus muscles. The inguinal ligament is formed by the external oblique aponeurosis and is continuous with the fascia lata of the thigh.* **(Bottom)** *Coronal T2 MR shows the piriformis muscle originating from the anterior sacrum. The piriformis muscle (along with the sacrum, coccyx, and covering fascia) forms the posterior pelvic wall. The piriformis muscle then leaves the pelvis through the greater sciatic foramen and inserts onto the upper border of the greater trochanter of the femur.*

LATERAL PELVIC WALL

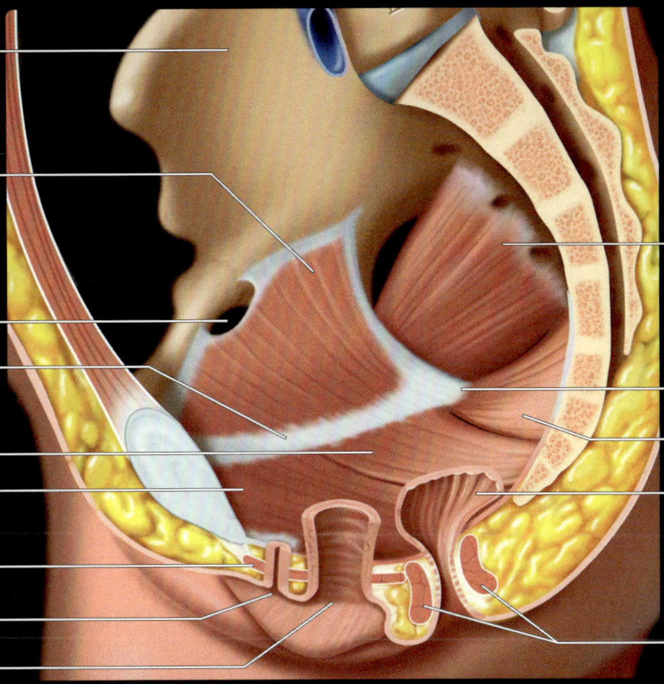

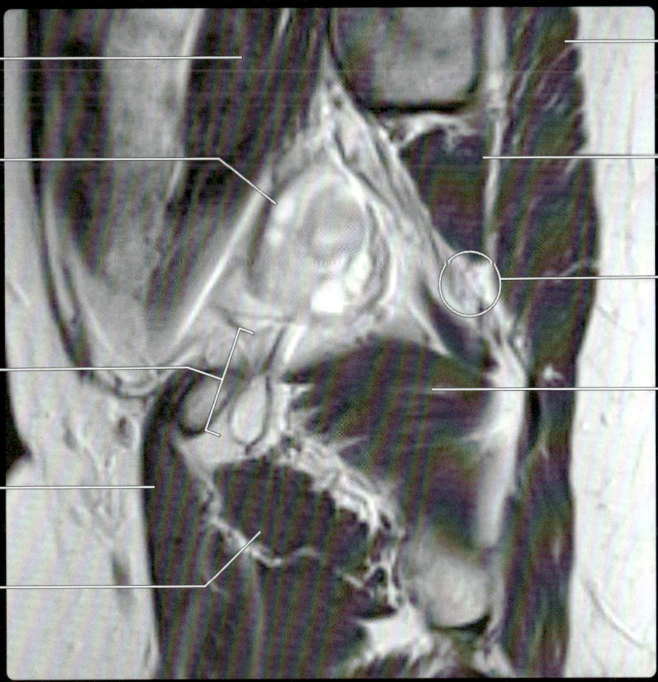

(Top) *The true pelvis is bowl-shaped; therefore, the designation of walls is somewhat arbitrary. The lateral wall of the true pelvis is formed by part of the ilium and ischium below the pelvic inlet, the obturator internus muscle and its covering membrane, and the sacrotuberous and sacrospinous ligaments. The pelvic floor is formed by the pelvic diaphragm (coccygeus and levator ani muscles and fascia). The levator ani is composed of 3 separate muscles: Pubococcygeus, iliococcygeus, and puborectalis. The levator ani is attached to the pubic bones anteriorly, the ischial spines laterally, and to the arcus tendineus levator ani (ATLA) (thickening in the obturator fascia) between the bony attachments. The pelvic diaphragm separates the pelvic cavity from the perineum.* **(Bottom)** *Sagittal T2 MR of the lateral pelvic sidewall shows the piriformis and obturator internus muscles.*

1ST LAYER OF PELVIC FLOOR: ENDOPELVIC FASCIA AND LIGAMENTS

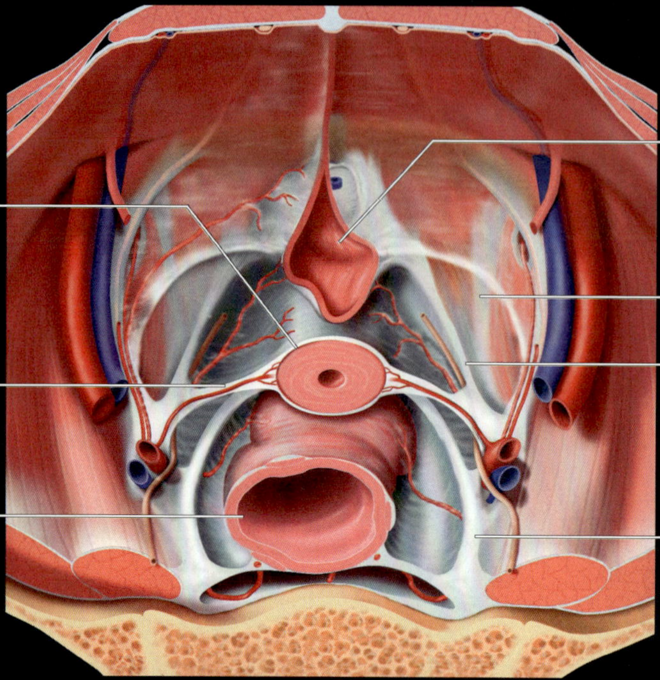

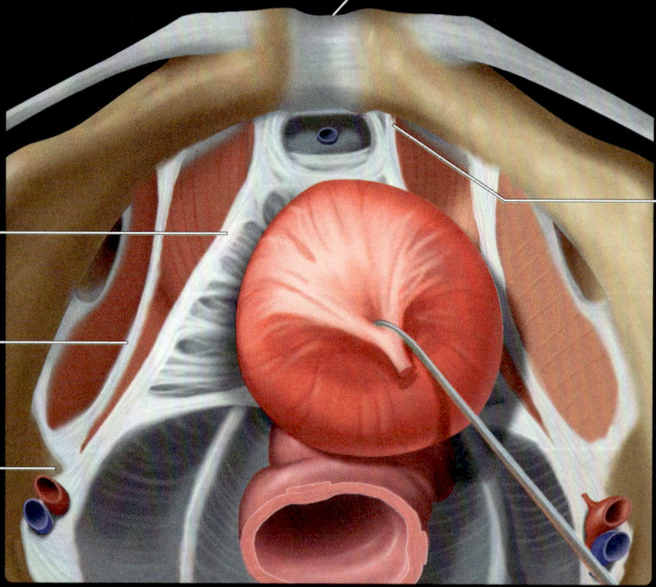

(Top) *Graphic looking down into the pelvis shows the fascia with the bladder, cervix, and rectum cut away. Endopelvic fascia is a continuous adventitial layer, covering the pelvic diaphragm and viscera. It is a complex network of connective tissue composed of collagen, fibroblasts, elastin, smooth muscle cells, and neurovascular bundles. Ligaments are a more well-defined aggregate of connective tissue.* **(Bottom)** *Graphic looking down on the pelvic floor with the bladder in situ shows the ATLA and the arcus tendineus fascia pelvis (ATFP). The ATLA extends from the pubic ramus anteriorly and attaches posteriorly to the ischial spine. It is a condensation of the obturator fascia and provides an important function in pelvic floor support as most of the muscles arise from it. The ATFP has a more inferior and medial course than the ATLA, attaching to the pubis close to the pubic symphysis. It provides a lateral anchoring site for the anterior vaginal wall and supports the urethra.*

1ST LAYER OF PELVIC FLOOR: ENDOPELVIC FASCIA AND LIGAMENTS

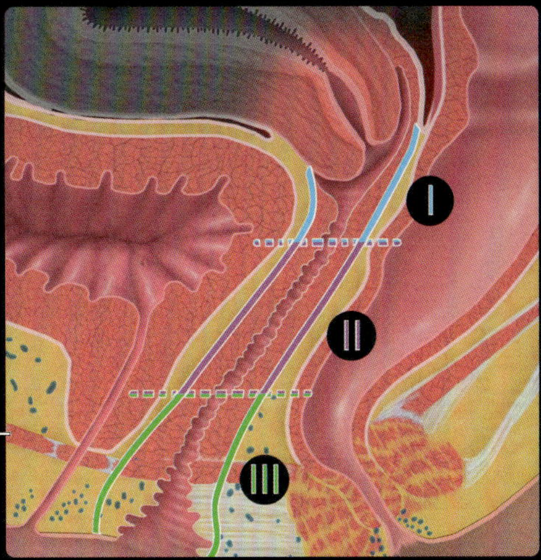

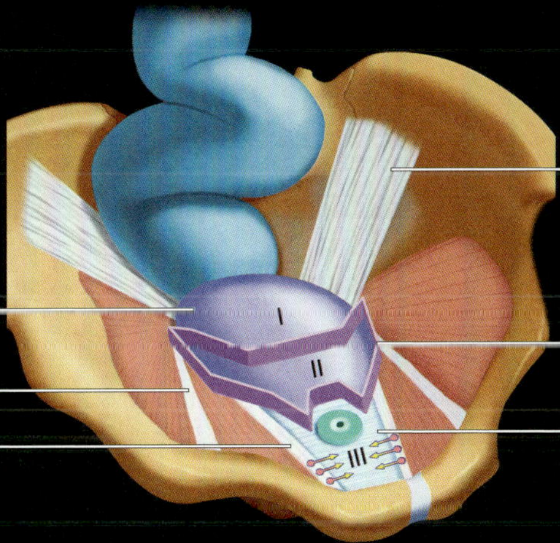

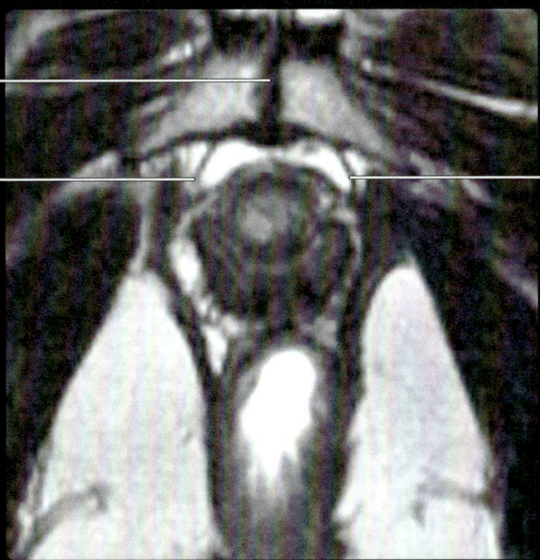

(Top) *There are 3 levels of endopelvic fascia support. Level I is the upper 2-3 cm of the vagina adjacent to the cervix. This gives support to the upper vagina and uterus. Level II is the midportion of the vagina. This is the level of support for the urinary bladder. Level III is the distal 2-3 cm, which extends to the introitus. At this level, there is no intervening paracolpium, and the anterior vaginal wall is fused with the lower posterior urethra. Level III provides urethral and vesical neck support.* **(Middle)** *This schematic diagram (uterus removed) shows the type of support the vagina receives at each level. In level I (suspension), the paracolpium suspends the vagina from the lateral pelvic walls. Fibers of level I extend both vertically and posteriorly toward the sacrum. In level II (attachment), the vagina is attached to the arcus tendineus fasciae pelvis and the superior fascia of levator ani. In level III (fusion), the vagina, near the introitus, is fused laterally to the levator ani.* **(Bottom)** *Axial oblique T2 TSE MR shows the ATFP on either sides of the symphysis pubis.*

2ND LAYER OF PELVIC FLOOR: PELVIC DIAPHRAGM (COCCYGEUS)

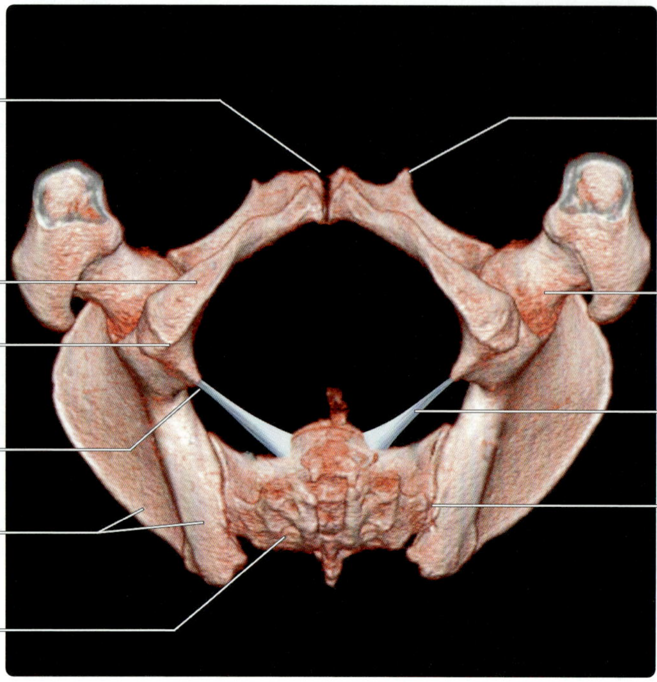

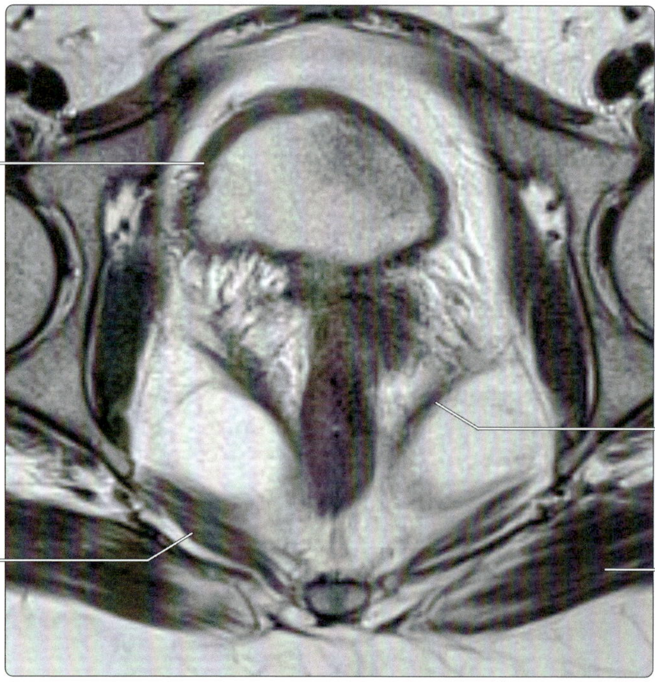

(Top) *3D CT reconstruction shows the pelvic outlet graphically enhanced to show the sacrospinous ligament. The pelvic outlet is formed by the ischiopubic rami, ischial spines, inferior symphysis pubis, sacrospinous ligaments, and coccyx.* **(Bottom)** *Axial oblique T2 MR shows the coccygeus muscle. It arises from the ischial spine, and the fibers fan out and insert into the lateral side of the coccyx. The sacrospinous ligament is at the posterior edge of the coccygeus muscle. The coccygeus muscle, along with the levator ani muscles, forms the pelvic diaphragm. The primary function of the coccygeus muscle is to give support to the pelvic wall, but it does not have a direct role in function.*

2ND LAYER OF PELVIC FLOOR: PELVIC DIAPHRAGM (LEVATOR ANI)

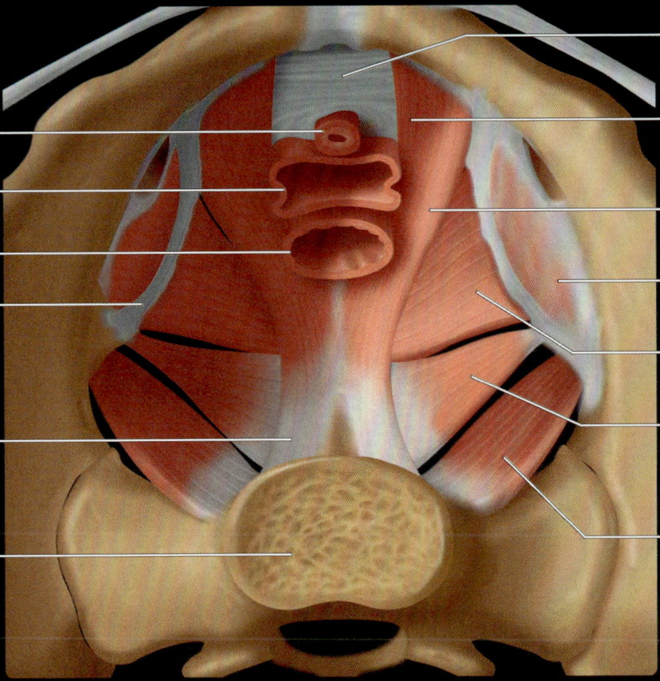

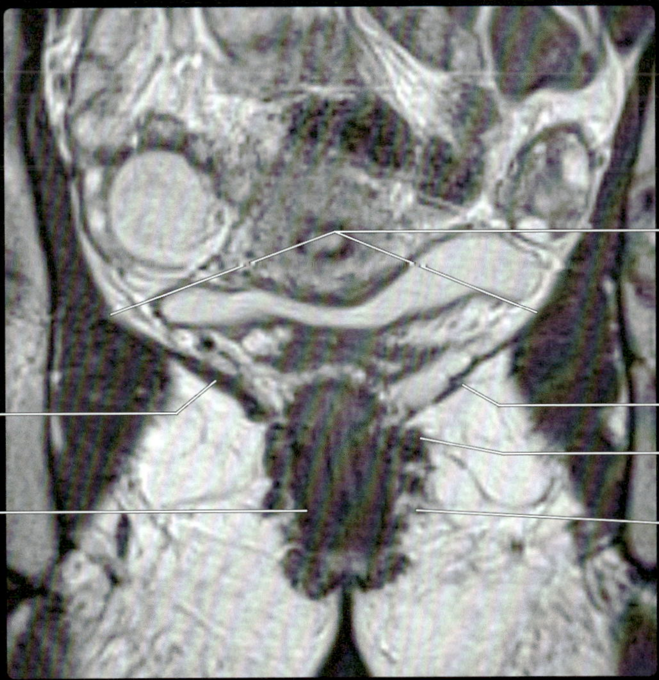

(Top) *Graphic provides a superior view of the pelvic floor. The levator ani is formed by the puborectalis, pubococcygeus, and iliococcygeus muscles. The obturator internus is covered by a fascial layer, which forms a thick band, the ATLA. This is a crucial area of attachment for the levator ani. The levator ani muscle with the coccygeal muscles form the pelvic diaphragm (floor). The piriformis muscle contributes to the posterior wall.* **(Bottom)** *Coronal T2 MR shows the levator ani extending from the ATLA to form the pelvic diaphragm.*

3RD LAYER OF PELVIC FLOOR: UROGENITAL DIAPHRAGM

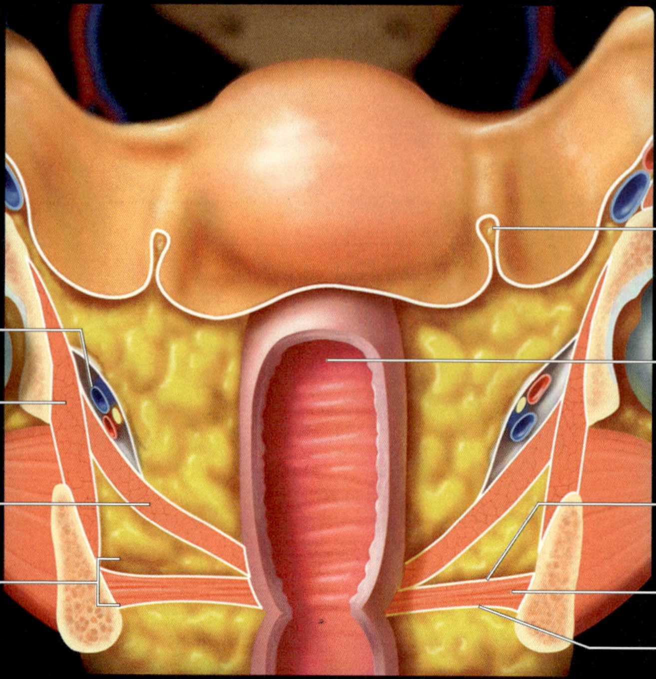

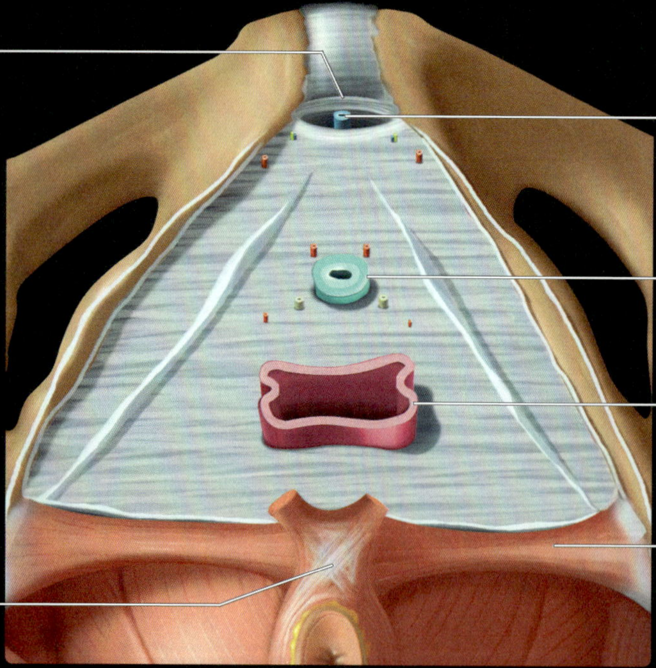

(Top) *Coronal graphic of the pelvic floor shows the urogenital (UG) diaphragm. The UG diaphragm is the fibromuscular layer directly below the pelvic diaphragm (levator ani muscles). It is a trilaminar structure with the deep transverse perineal muscle sandwiched between superior and inferior fascial layers. It is part of the perineum, which is located below the levator ani and includes the external genitalia.* **(Bottom)** *Graphic shows the inferior view of the UG diaphragm. It is triangular in shape and attaches laterally to the pubic bones. At the most anterior (ventral) aspect of the perineal membrane (covering fascia), the base of the arcuate ligament is separated from the anterior border of the UG diaphragm by an opening for the deep dorsal vein of the clitoris. Both the urethra and vagina pass through the UG diaphragm.*

3RD LAYER OF PELVIC FLOOR: UROGENITAL DIAPHRAGM

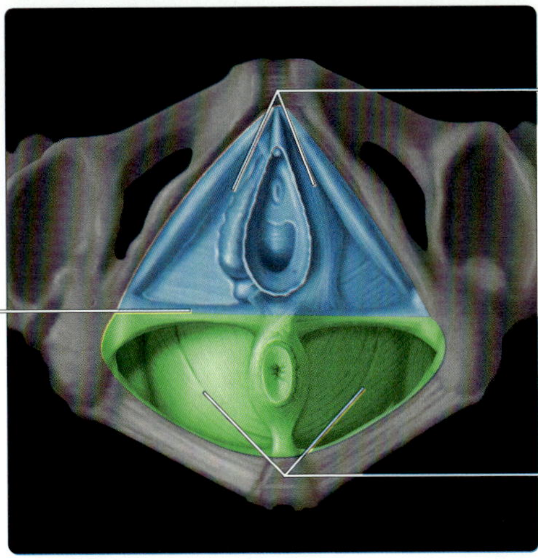

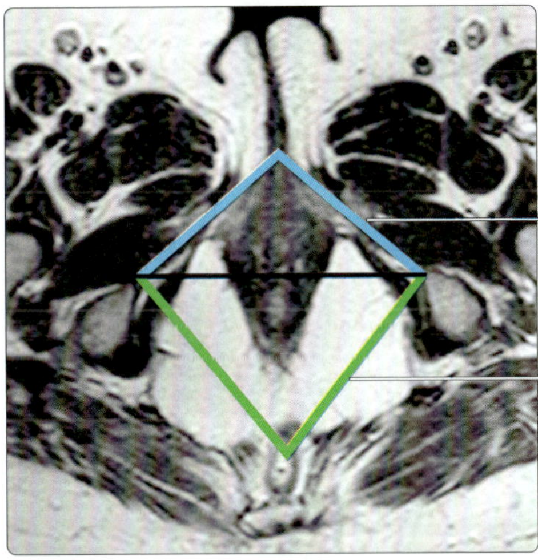

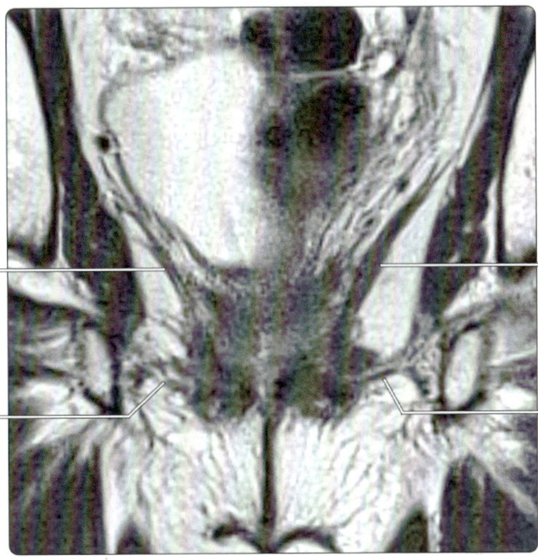

(Top) *The perineum is bordered by the symphysis pubis, ischial tuberosities, and coccyx, creating a diamond shape. It can be subdivided into 2 triangular compartments by a line drawn slightly anterior to the ischial tuberosities along the superficial transverse perineal muscle, creating the UG triangle anteriorly and the anal triangle posteriorly.* **(Middle)** *Axial oblique T2 MR shows the diamond-shaped perineum. The UG triangle contains the urethra, vagina, perineal membrane, and external genital muscles. The anal triangle contains the anus and ischiorectal fossa.* **(Bottom)** *Coronal T2 MR shows the UG diaphragm inferior to the pelvic diaphragm.*

4TH LAYER OF PELVIC FLOOR: SUPERFICIAL EXTERNAL GENITAL MUSCLES

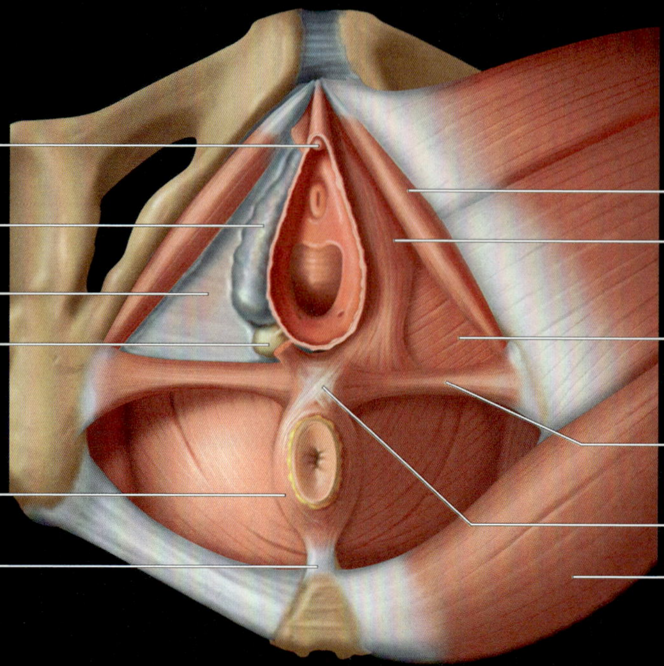

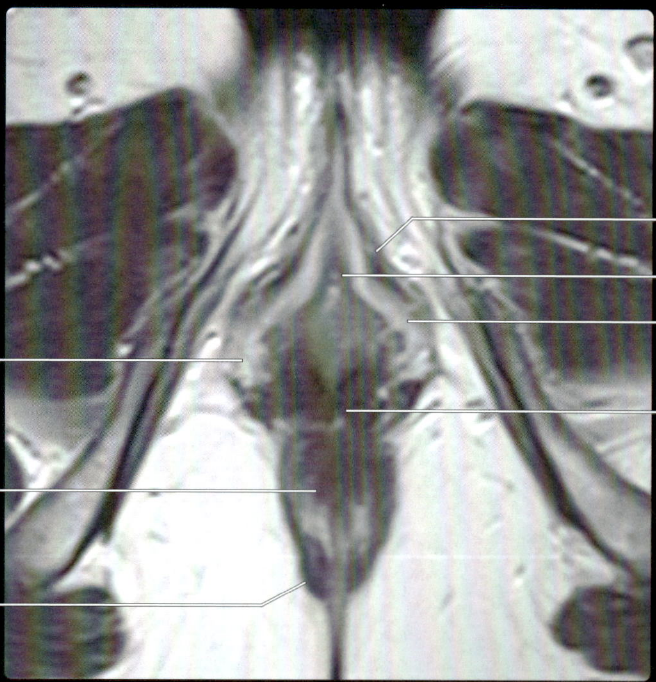

(Top) *The external genital muscles are located anteriorly in the UG triangle, whereas the anal sphincter complex and perineal body are in the anal triangle. The perineal body is a thickened, midline condensation of fibrous tissue at the midpoint of a line joining the ischial tuberosities. At this point, several important muscles converge and are attached: The external anal sphincter, paired bulbospongiosus muscles, paired superficial transverse perineal muscles, and fibers of the levator ani.* **(Bottom)** *Axial T2 MR in a female patient at the level of the superficial external genital muscle shows the convergence of structures at the perineal body, which appears as a thickening of tissue in the midline.*

4TH LAYER OF PELVIC FLOOR: SUPERFICIAL EXTERNAL GENITAL MUSCLES AND URETHRA

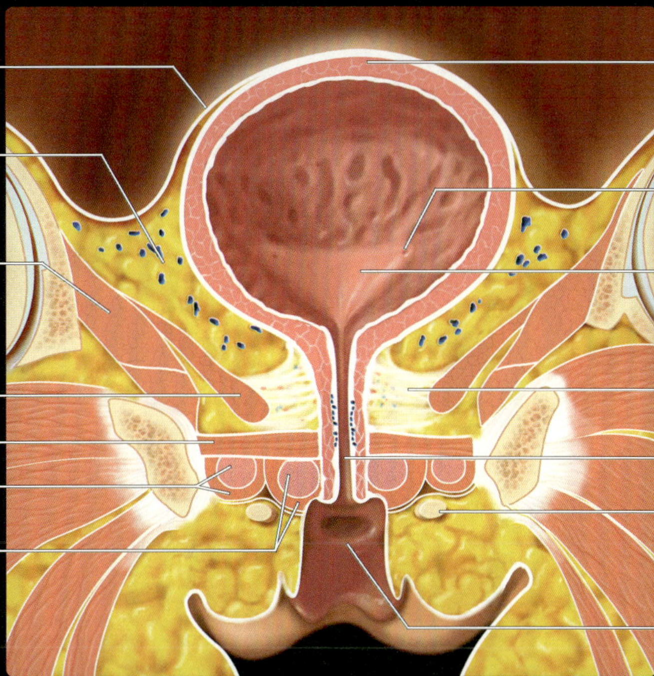

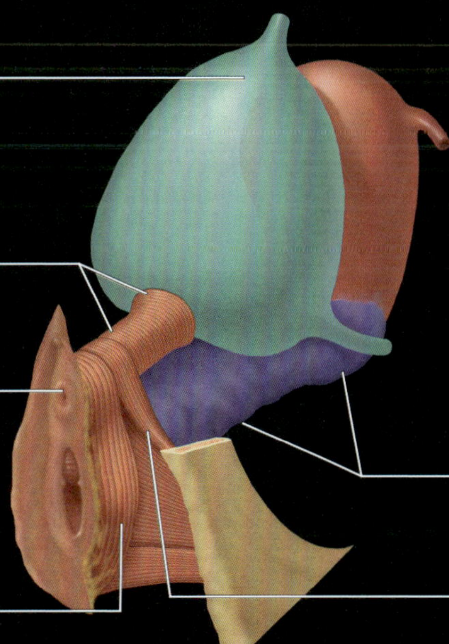

(Top) *Frontal graphic of the female bladder shows that it rests on the muscular floor of the pelvis and is supported by the endopelvic fascia suspended between the ATFP (1st and 2nd layers of pelvic floor). The urethra pierces the UG diaphragm (3rd layer of pelvic floor) to the external meatus, which is surrounding by the muscles of the external genitalia (4th layer of pelvic floor).* **(Bottom)** *The urethra has both an inner smooth muscle sphincter and an outer striated muscle sphincter complex, which is shown in this graphic. The circumferential upper external urethral sphincter, arch-like muscular bands, compressor urethrae, and urethrovaginal sphincter act as 1 functional unit to maintain continence.*

PASSIVE AND ACTIVE COMPONENTS OF PELVIC SUPPORT

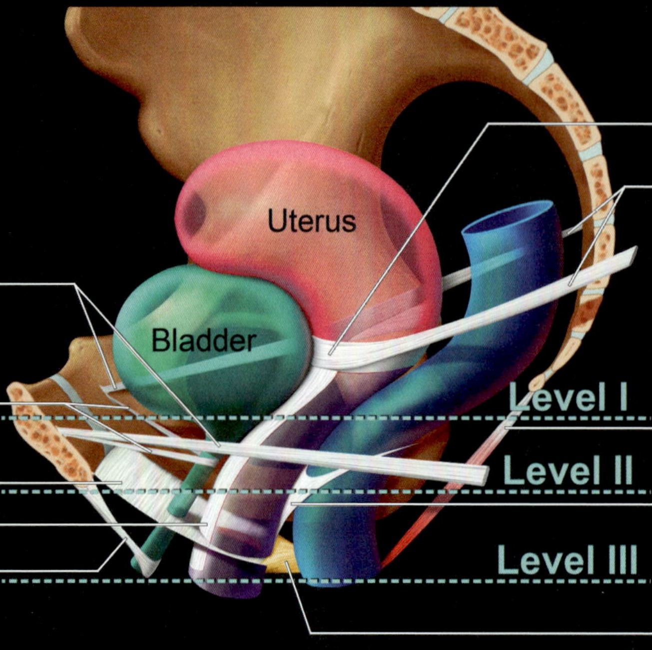

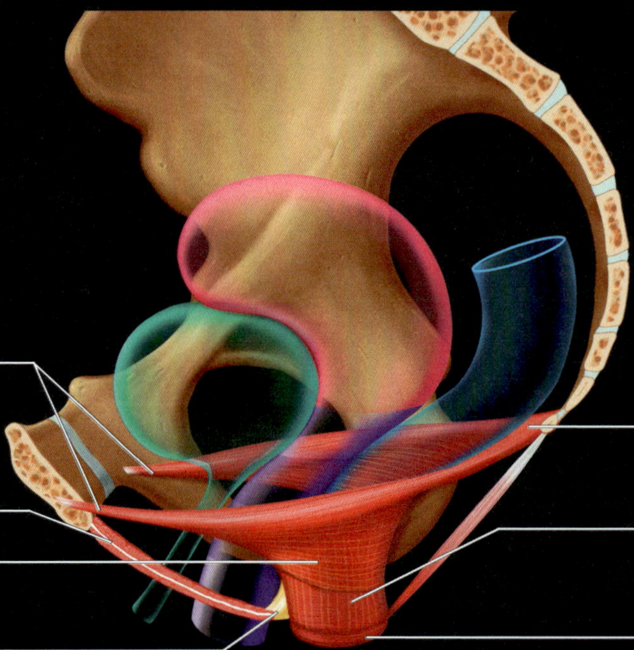

(Top) *Series of 2 graphics illustrates the passive and active conceptual approach to the pelvic floor. Passive components include the bony pelvis and supportive connective tissue. The supportive connective tissue is either in the form of a diffuse ill-defined layer (the endopelvic fascia) or as well-defined specialized aggregations of connective tissue (ligaments). The 3 endopelvic fascial levels include level I (upper vagina adjacent to the cervix), level II (midportion of the vagina), and level III (from the introitus to 2-3 cm above the hymenal ring).* **(Bottom)** *Graphic shows the main active component of pelvic support system, the levator ani muscle. The levator ani muscle is a wide sheet of muscle whose main components are the puborectalis muscle (PRM), which forms a sling around the junction of the rectum and anal canal, the pubococcygeus muscle (PCM), which passes posteriorly to insert into the anococcygeal body, and the iliococcygeus muscle, whose fibers fuse to form the levator plate and insert on coccyx.*

PASSIVE AND ACTIVE COMPONENTS OF PELVIC SUPPORT

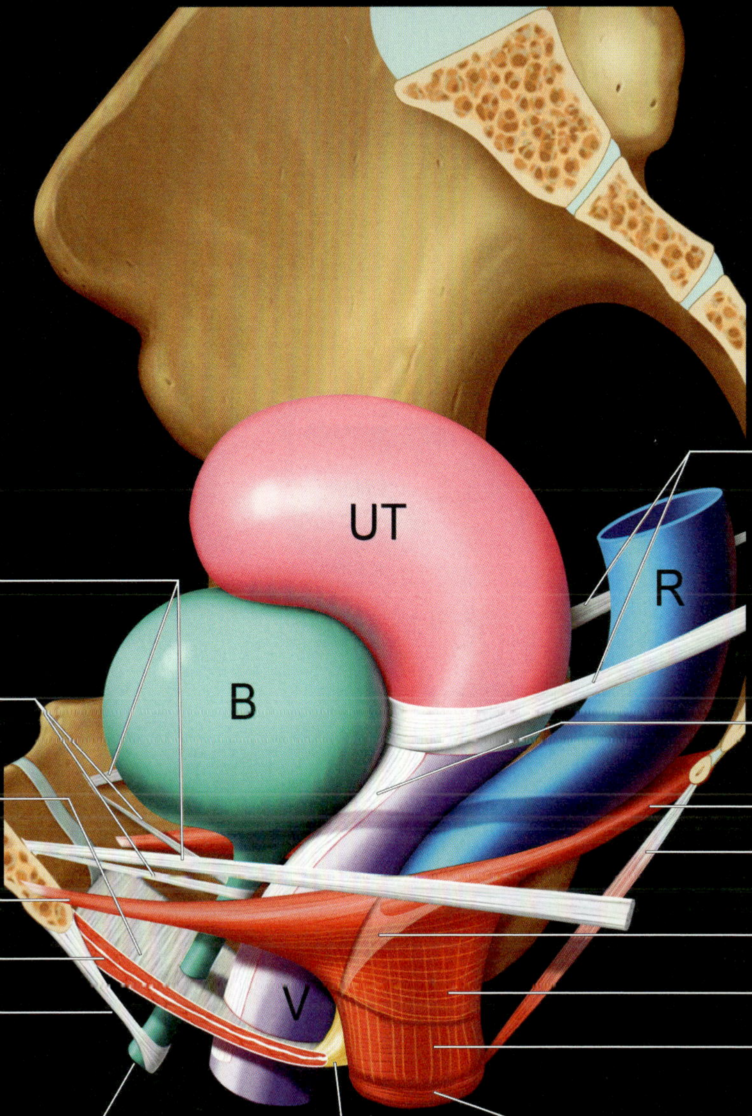

Graphic of the pelvis illustrates the multilayered system approach that considers the passive and active components of pelvic floor as an integrated multilayer system. From cranial to caudal, the pelvic support system consists of endopelvic fascia, pelvic diaphragm, perineum, and the external genital muscles. The muscles (levator ani) give active support to the pelvic floor, whereas the ligaments give passive support holding organs in place. When the levator ani is functioning properly, the pelvic floor is closed, and the ligaments and fasciae are under no tension. When the musculature is damaged and cannot close the levator hiatus, ligaments are put under strain and will eventually fail resulting in pelvic organ prolapse.

INTERACTION BETWEEN PELVIC FLOOR MUSCLES AND ENDOPELVIC FASCIA

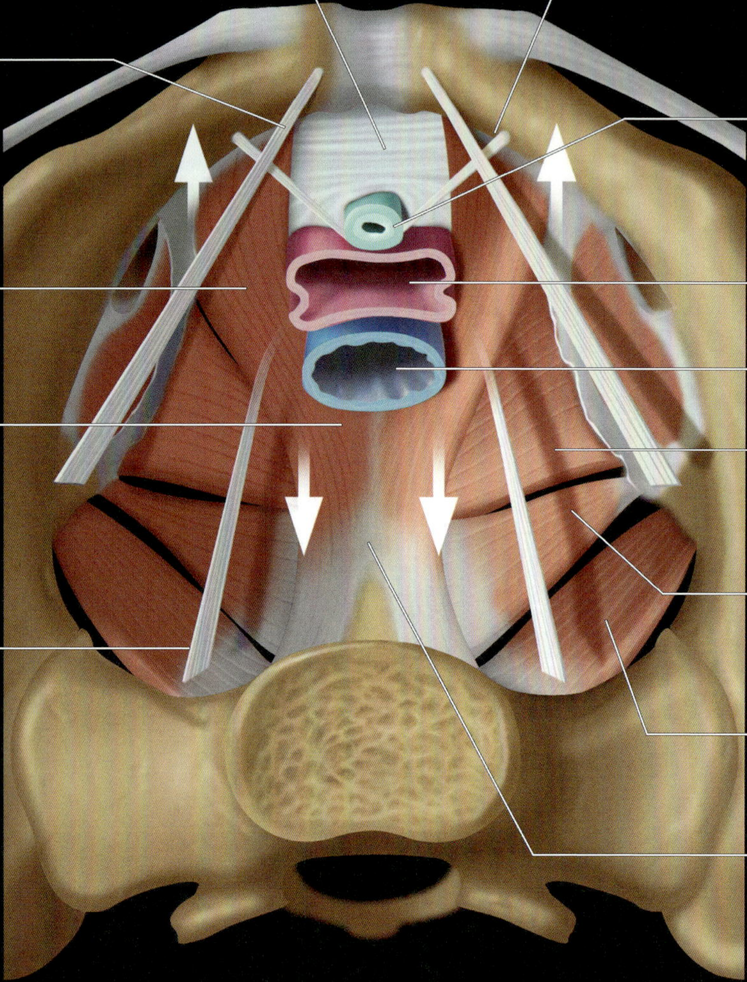

Graphic viewing into the pelvis from above shows the upper layer of pelvic floor muscles and its interaction with the endopelvic fascial condensations. The arrows indicate the direction of muscle forces. The anterior portion of the PCM is attached to the lateral vaginal wall and constitutes the forward force. The levator plate, attached to the posterior wall of rectum, constitutes the posterior force. The net result of these opposing forces is a tightly closed UG hiatus. As a result, the endopelvic fascia is not exposed to any tension. Muscle weakness or tear disrupts this balanced situation leading to pelvic organ prolapse.

FUNCTIONAL 3-PART PELVIC SUPPORT SYSTEM

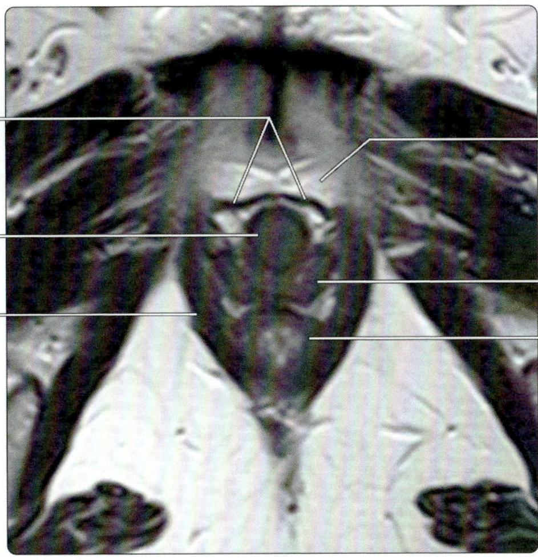

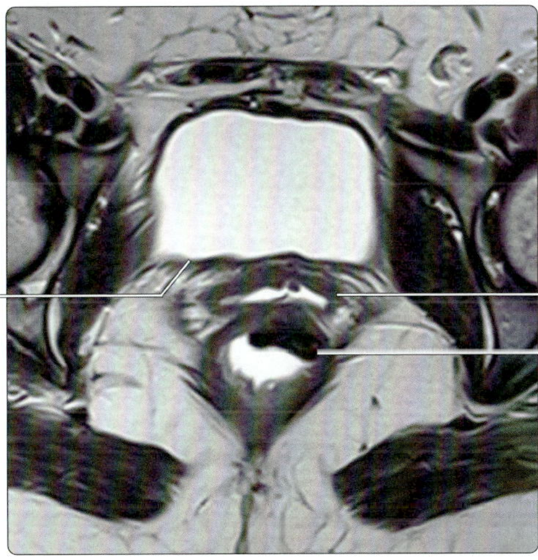

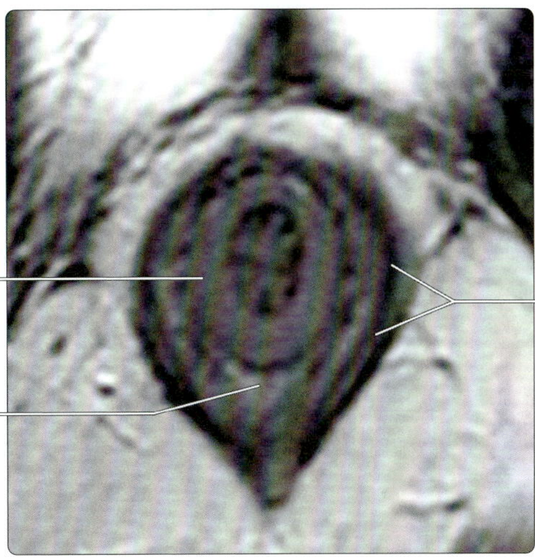

(Top) *Axial T2 MR shows a small symmetric prevesical space (space of Retzius). The vagina has a normal H shape, and the urethra is midline. The periurethral ligaments are seen ventral to the urethra.* **(Middle)** *Axial T2 MR shows the normal vaginal support system. Levels I and II endopelvic fascial attachment are evaluated at the level of the uterine cervix and bladder base, respectively. When normal, the posterior urinary bladder wall is more or less straight, as seen here.* **(Bottom)** *Axial balanced fast-field echo (BFFE) MR shows the anal sphincter complex, which along with pelvic floor muscles, is responsible for the 3rd functional component, anal continence. The deep external anal sphincter has a characteristic teardrop appearance.*

LEVATOR ANI

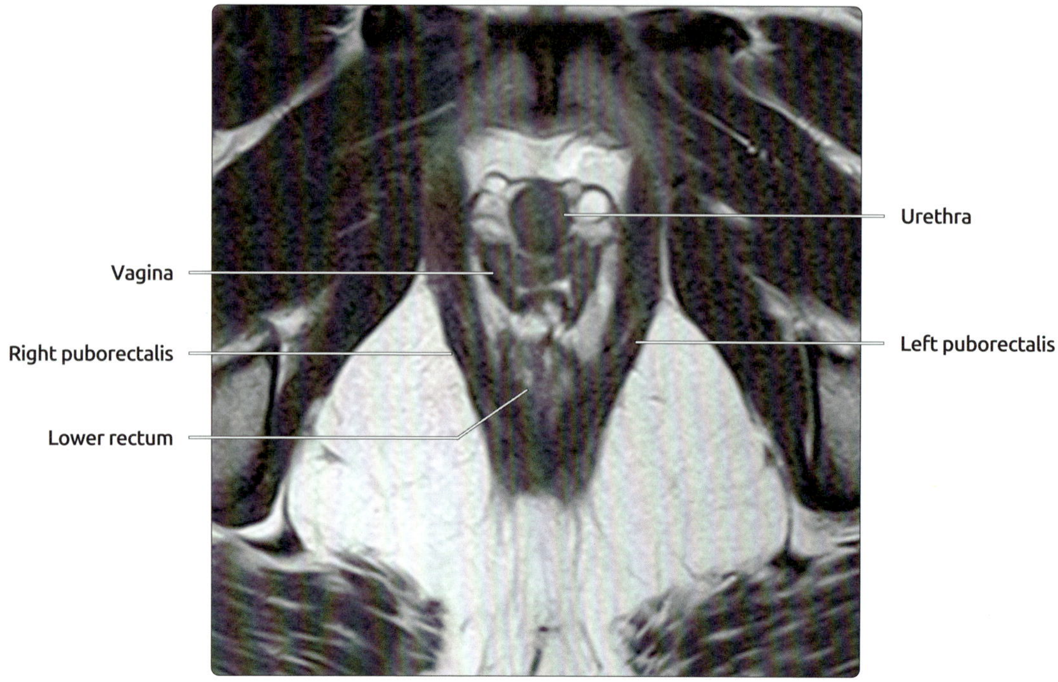

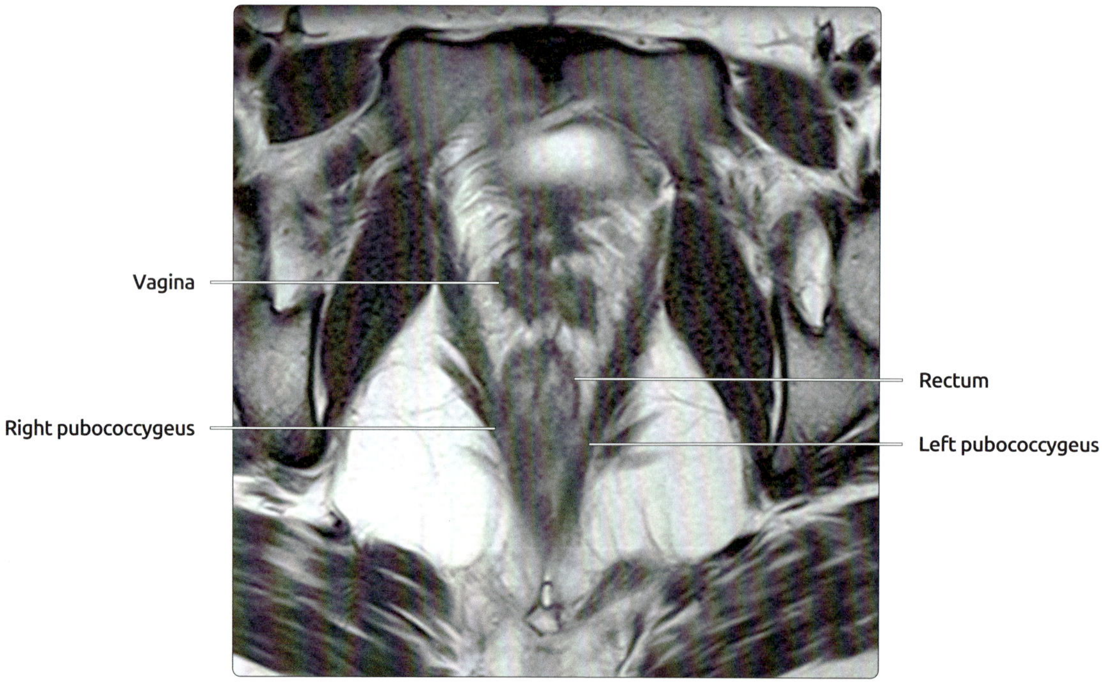

(Top) *Two consecutive T2 MR images show the relationship of the puborectalis and pubococcygeus. The puborectalis can be differentiated on axial plane as it forms a sling around the rectum and does not insert into any skeletal structure.* **(Bottom)** *At a higher slice, axial T2 MR shows the muscle fibers passing the rectum to have an osseous attachment on the coccyx, representing the PCM.*

LEVATOR ANI

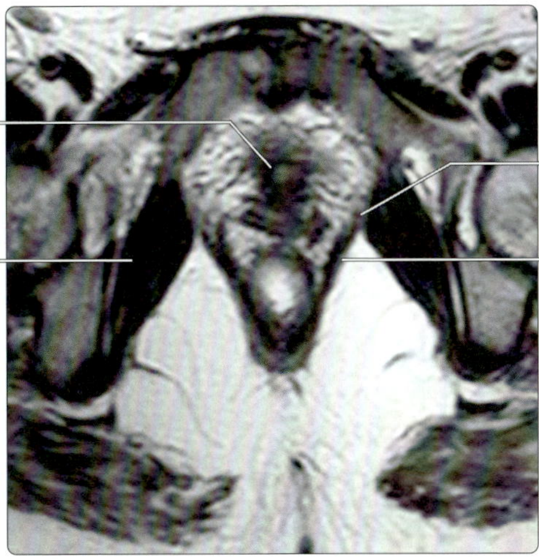

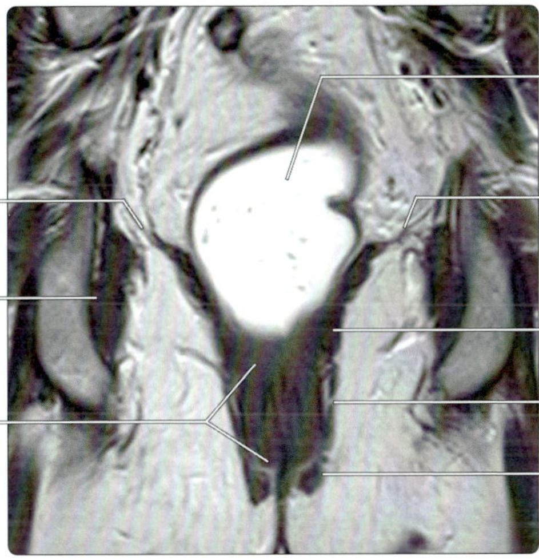

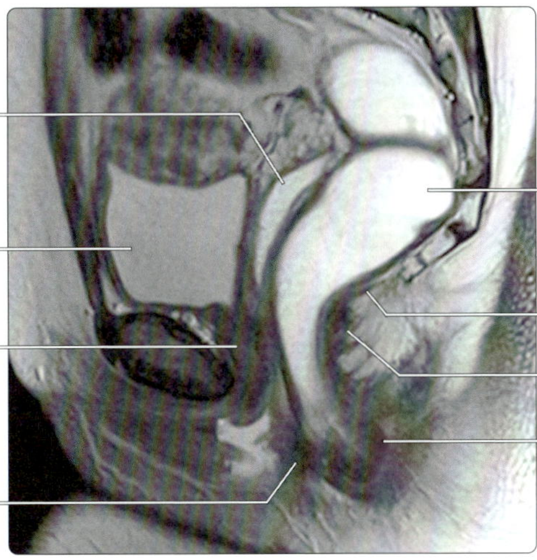

(Top) *Axial T2 MR in the same woman shows the origin of the iliococcygeus muscle from the obturator internus.* **(Middle)** *The attachment site of the IC muscle with the obturator fascia is shown in the coronal plane. Fibrofatty tissue separates muscle bundles at their origin and should not be mistaken for a tear. During straining, there is descent of ischiocavernosus muscle, which assumes a more vertical orientation.* **(Bottom)** *The PRM forms a sling around the rectum, while the iliococcygeus extends posterior to the coccyx, forming the levator plate.*

LEVATOR ANI (ABNORMAL)

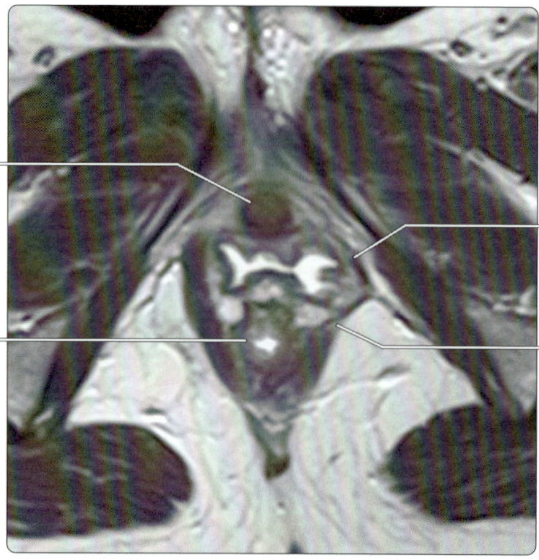

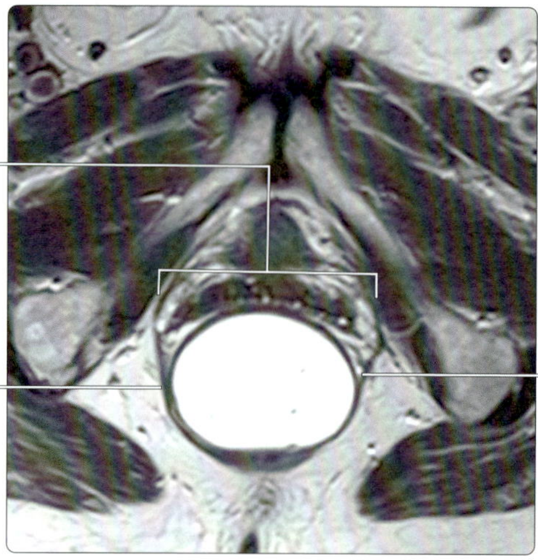

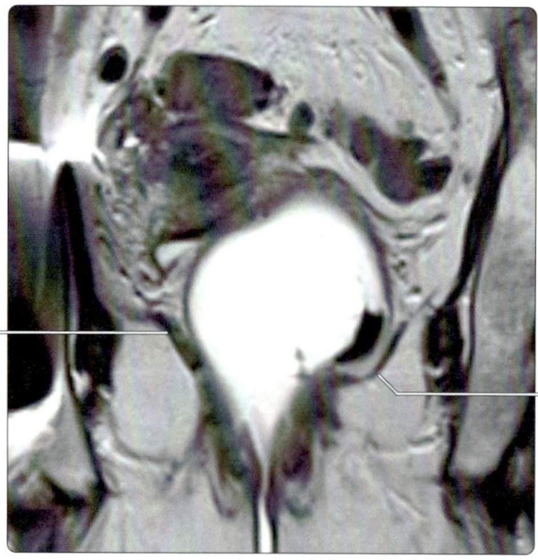

(Top) *Axial T2 MR shows a tear/detachment of the left aspect of the PRM with lateral bulging of the vaginal wall, which loses its normal H shape.* **(Middle)** *Axial T2 MR shows thinning of the PRM bilaterally with widening of the levator hiatus.* **(Bottom)** *Coronal T2 MR shows normal right levator ani muscle and thinning/laxity of the left levator ani muscle.*

AXIAL MR OF PELVIC FLOOR MUSCLES

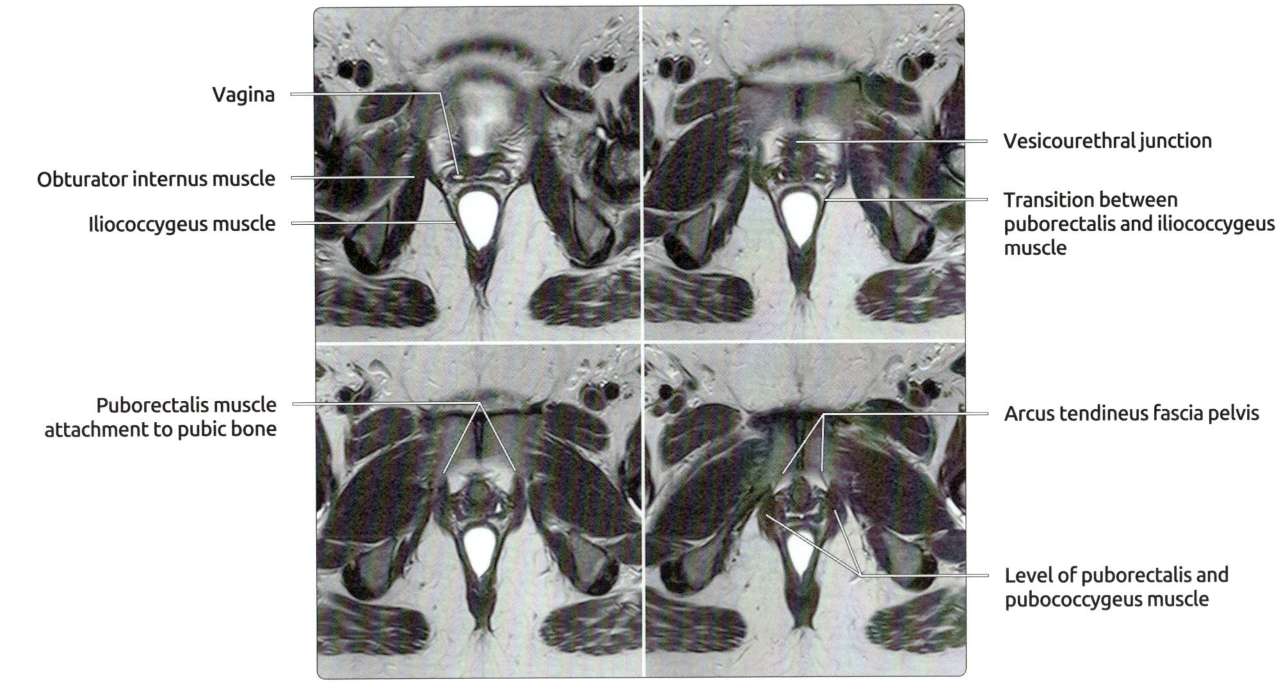

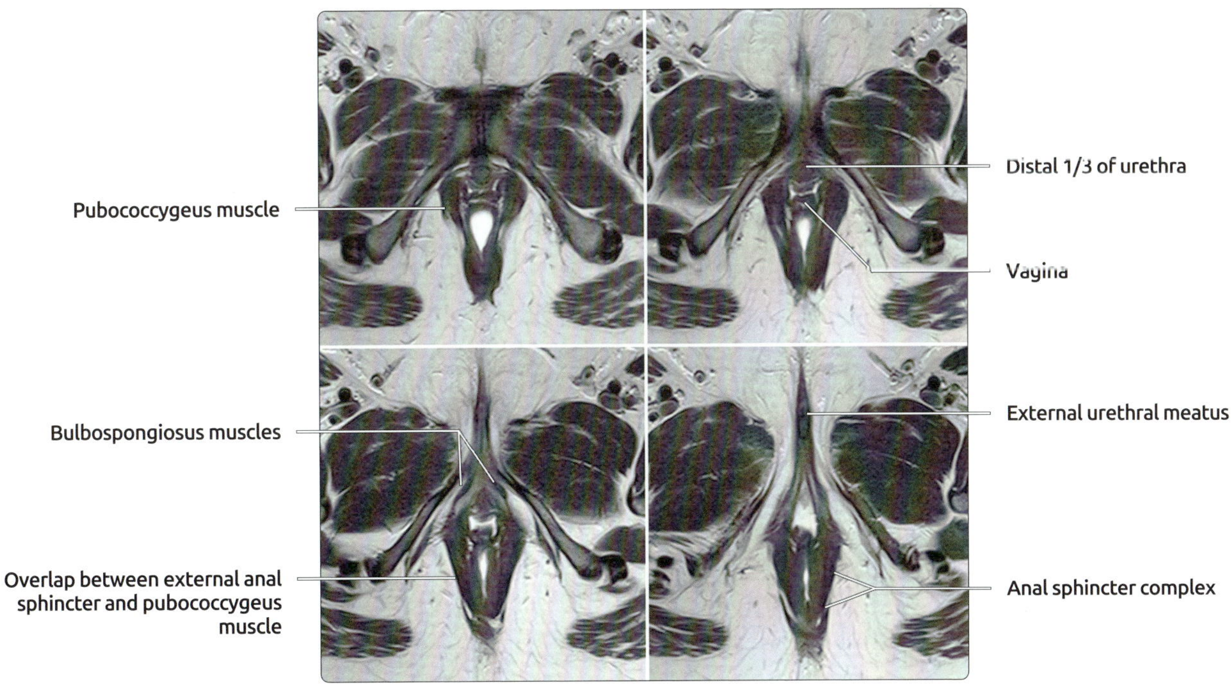

(Top) *Consecutive axial T2 MR images of a female volunteer are displayed superior to inferior. The sections are plotted perpendicular to the longitudinal axis of the pelvis. The images show subdivisions of the pelvic floor muscles.* **(Bottom)** *Consecutive axial T2 MR images of the same female volunteer at a more caudal level show subdivisions of the PCM and external anal sphincter. At the level of the perineum, the external genital muscles are seen anterior to the anal canal. At the level of the inferior pubic rami, the PCM is seen.*

CORONAL MR OF PELVIC FLOOR MUSCLES

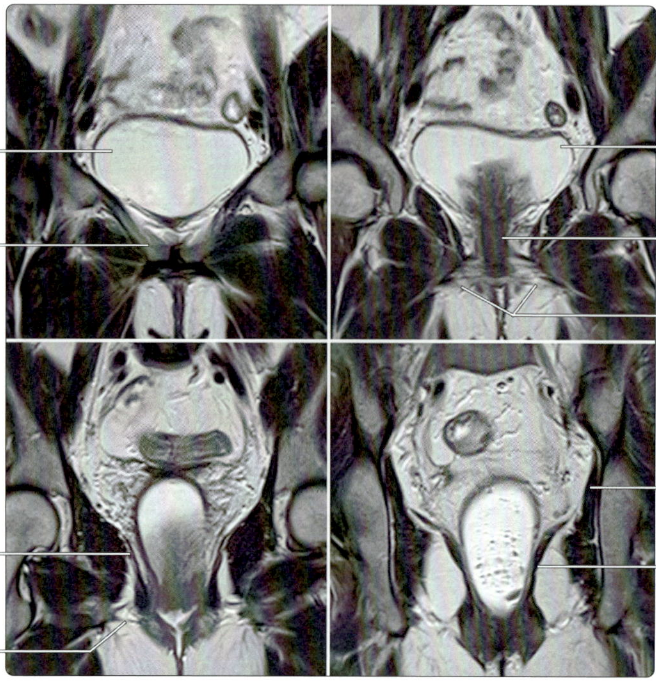

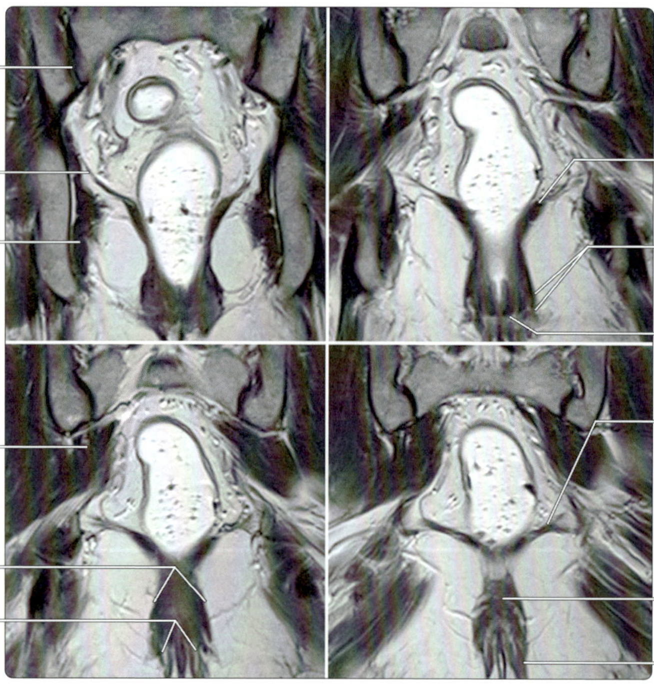

(Top) *Serial coronal images obtained in a female volunteer are displayed from anterior to posterior. These 4 images show the anatomy of the perianal structures and the anterior part of the ischiocavernosus muscle. The urethra is seen passing through the UG diaphragm and layers of the perineum.* **(Bottom)** *Serial coronal images obtained in the same woman at a more posterior level show the anatomy of the posterior part of the ischiocavernosus muscle and anal sphincter complex.*

SAGITTAL MR OF PELVIC FLOOR MUSCLES

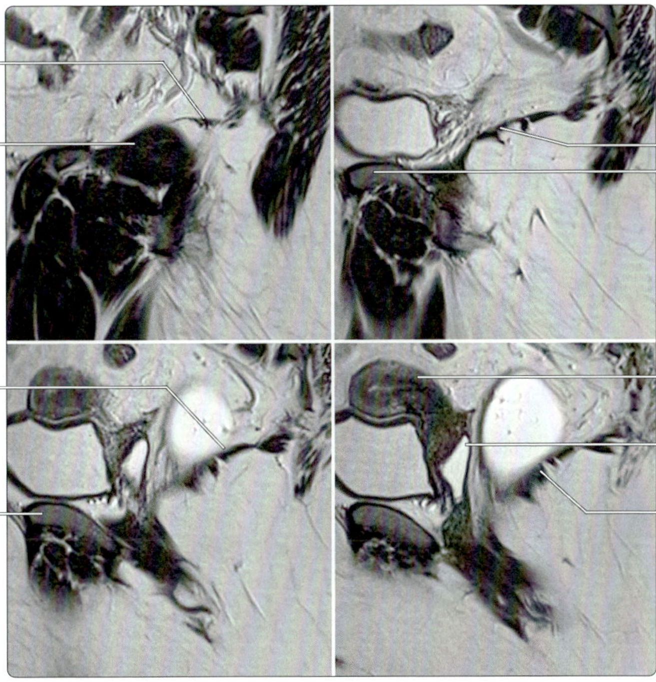

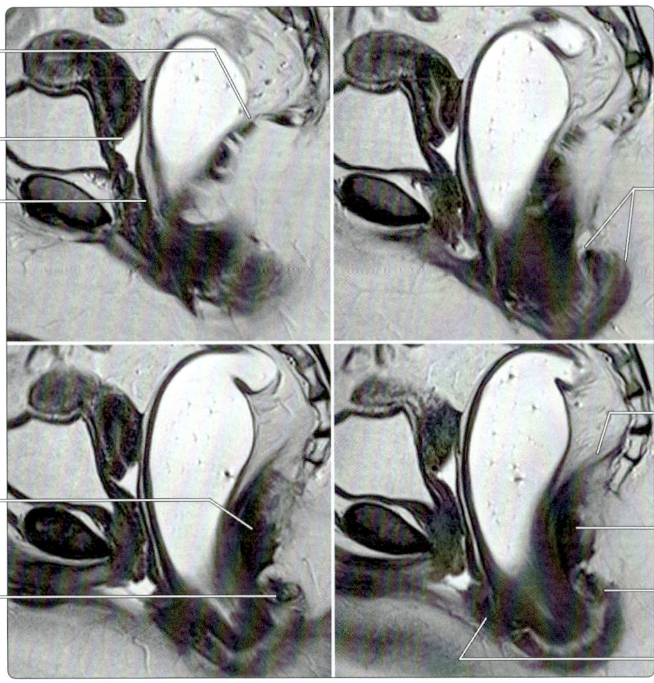

(Top) *Serial sagittal images obtained in a female volunteer displayed from lateral to medial show the lateral part of the ischiocavernosus muscle and the more curved slope of the muscle away from the midline at its origin from the obturator internus muscle.* **(Bottom)** *Serial sagittal images obtained in a female volunteer at a more medial level display the normal MR anatomy of anal sphincter and of the levator plate.*

URETHRAL SUPPORT, NORMAL AND ABNORMAL

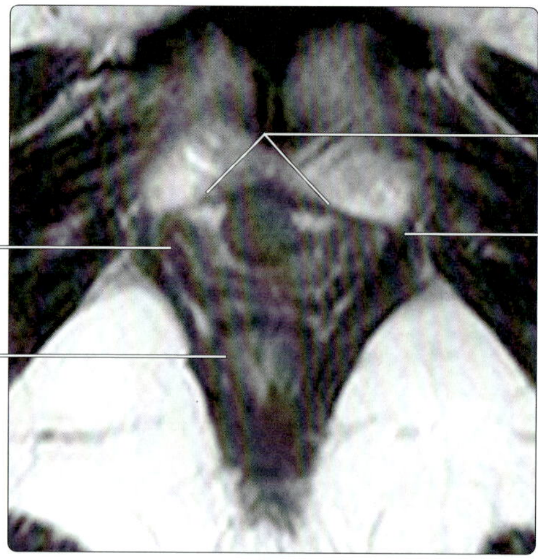

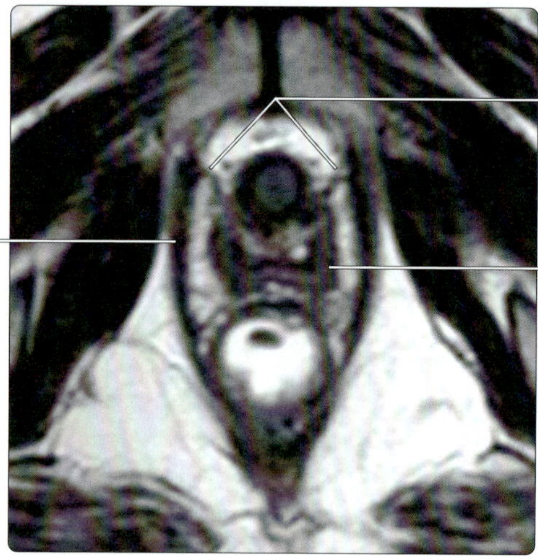

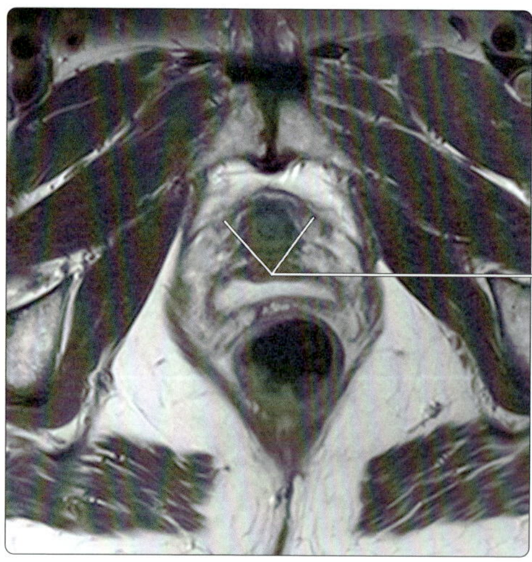

(Top) *Axial T2 MR of a female patient with normal level III endopelvic fascia shows a small symmetric prevesical space (space of Retzius). The vagina has a normal H shape, and the urethra is midline. The periurethral ligaments are seen ventral to the urethra coursing between the medial borders of the PRM.* **(Middle)** *Axial T2 MR of a female patient with a normal urethral support system shows the suburethral ligament.* **(Bottom)** *Axial T2 MR shows disruption of the periurethral ligaments with the drooping mustache sign and enlargement of the space of Retzius.*

NORMAL VAGINAL FASCIAL SUPPORT

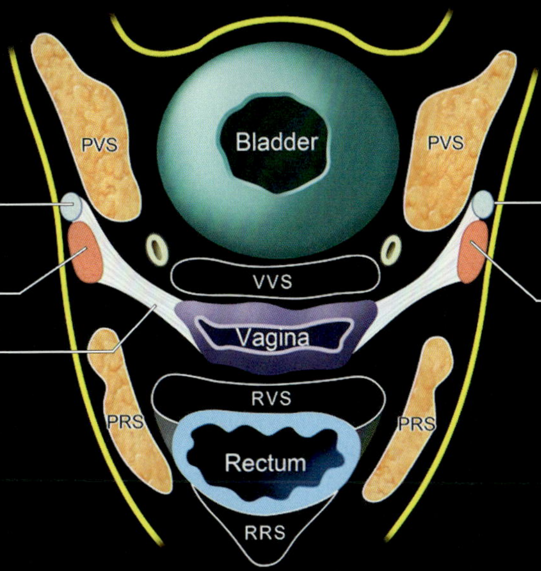

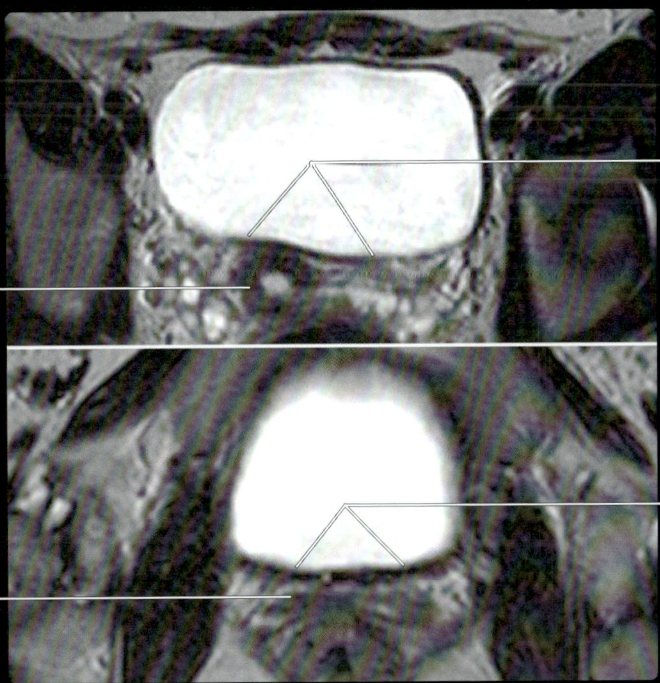

(Top) *Oblique axial graphic shows the endopelvic fascia and spaces. The vagina is suspended between the 2 ATFP ligaments by lateral fascial extensions. These lateral extensions fuse with the pubocervical fascia superiorly and the rectovaginal fascia inferiorly. Pelvic organs are separated from each other by spaces that allow organs to move independently from each other. Please note the following: PVS = paravesical space, VVS = vesicovaginal space, RVS = rectovaginal space, RRS = retrorectal space, PRS = pararectal space or ischiorectal fossa.* **(Bottom)** *Axial T2 MR composite shows normal level I and level II endopelvic fasciae. The landmark to define level I is the cervix (upper image), and level II is the midvagina at the level of the bladder base (lower image). While some ligaments are visualized, the fascia is not; however, the integrity of the fascia can be inferred by the appearance of surrounding organs. The posterior bladder wall is seen as a straight line, indicating that the level I and level II endopelvic fascia is intact.*

PARAVAGINAL FASCIAL DEFECT

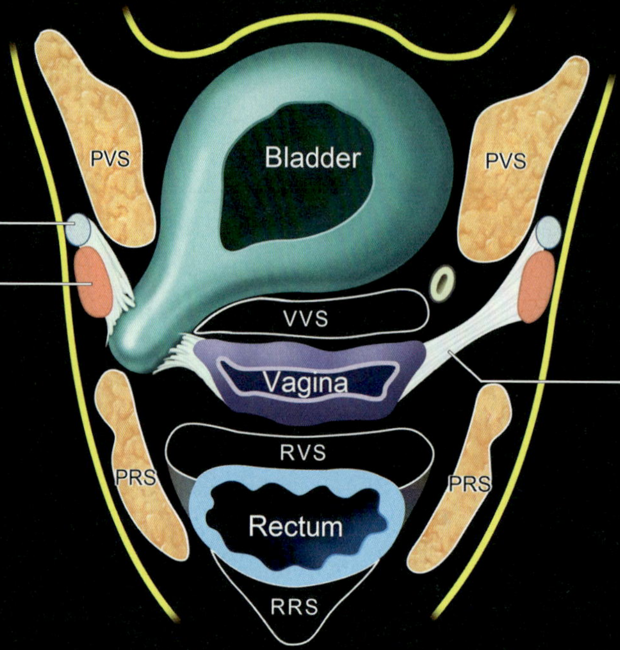

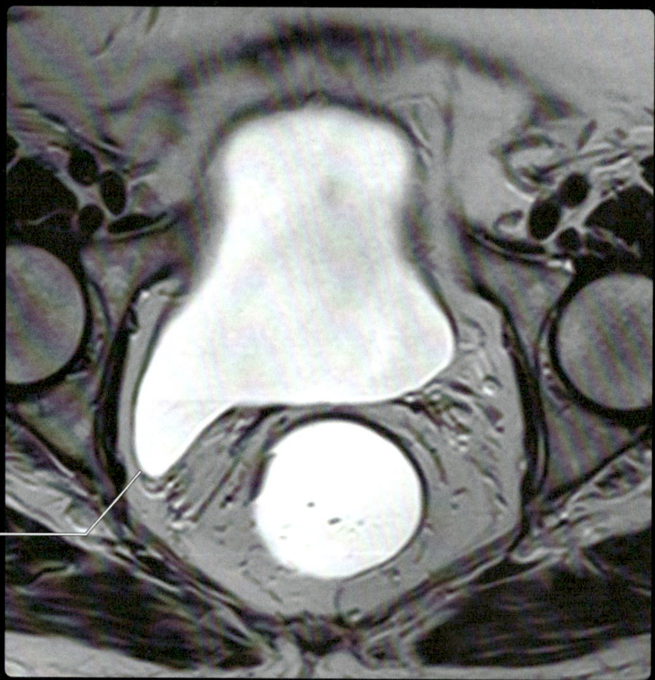

(Top) *In this oblique axial graphic, there is right-sided endopelvic fascial detachment causing a paravaginal defect. Because of the defective support mechanism, there is sagging of the right posterolateral wall of the urinary bladder to fill the resulting defect. Please note the following: PVS = paravesical space, VVS = vesicovaginal space, RVS = rectovaginal space, RRS = retrorectal space, PRS = pararectal space or ischiorectal fossa.* **(Bottom)** *Axial T2 MR obtained in a female with pelvic organ prolapse shows the consequences of fascial detachment with sagging of the posterior vaginal wall (saddlebag sign). The degree of sagging of the bladder wall corresponds to the size of the fascial defect.*

CENTRAL FASCIAL DEFECT

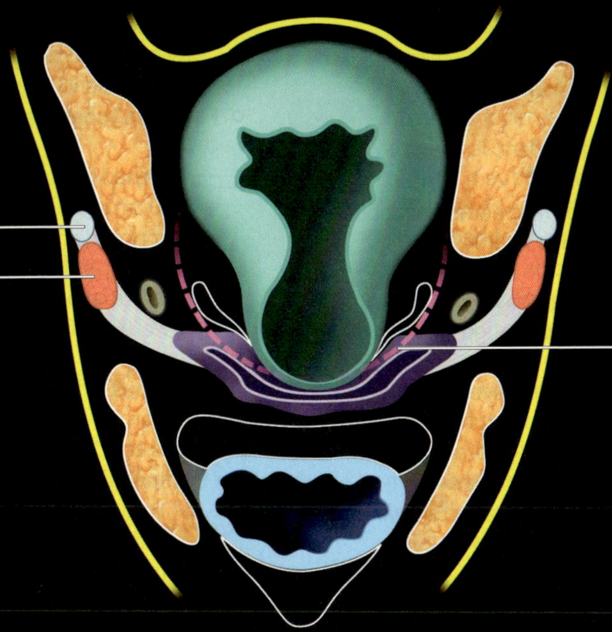

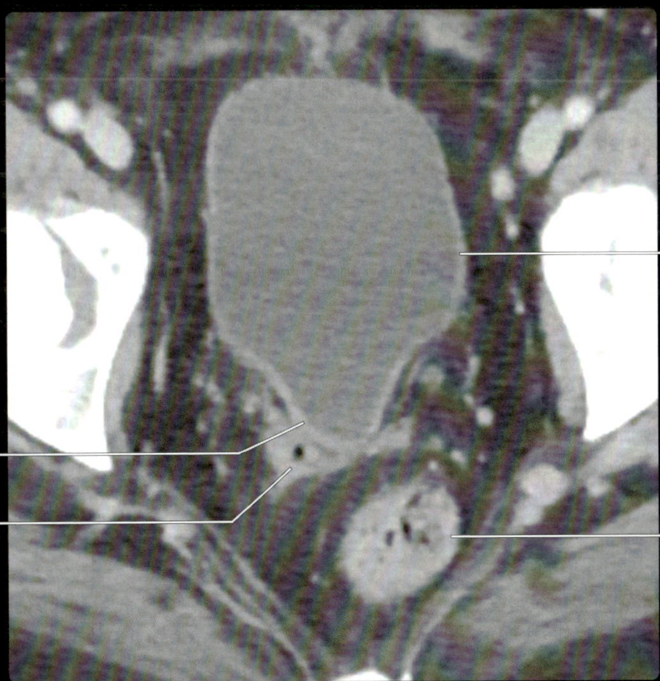

(Top) *Oblique axial graphic shows loss of central fascial support. This results from weakness and stretching of the midline pubocervical fascia (depicted by the dotted pink line), allowing the central portion of the bladder wall to bulge posteriorly (distension cystocele).* **(Bottom)** *Axial CECT shows posterior midline herniation of the urinary bladder caused by a central endopelvic fascial defect in a patient who had a hysterectomy.*

- During first 6 weeks of development, both male and female fetuses have paired mesonephric (wolffian or male genital) ducts and paramesonephric (müllerian or female genital) ducts
- In males, testes produce müllerian-inhibiting factor (MIF), suppressing ipsilateral müllerian development
- Absence of MIF in female fetus → growth of paired müllerian ducts + regression of mesonephric ducts
- Interruption of müllerian duct development → aplasia or hypoplasia of vagina, cervix, &/or uterus

- **Ductal fusion**
  - Müllerian duct growth is accompanied by midline migration and fusion of these paired ducts to form uterovaginal primordium
    - Fusion forms uterovaginal canal (uterus and upper vagina); unfused portions remain as fallopian tubes
    - Lower vagina forms from urogenital sinus
  - Interruption of müllerian duct fusion process → bicornuate or didelphys uterus
- **Septal reabsorption**
  - Between 9- and 12-weeks gestation, fused müllerian ducts undergo process of reabsorption of intervening uterovaginal septum
    - Reabsorption process occurs in both cranial and caudal directions
  - Interruption of septal reabsorption → septate or arcuate uterus

CLINICAL IMPLICATIONS

Müllerian Duct Anomalies

- Failure of müllerian duct development, fusion, or septal resorption leads to spectrum of congenital uterine anomalies
  - Although these are broadly categorized, it is important to think of them as a spectrum because fusion and resorption abnormalities may occur in combination and to varying degrees
  - Presence of deep external uterine fundal cleft > 1 cm defines presence of müllerian fusion abnormality
  - High association with **renal anomalies**
  - **Class I: Agenesis or hypoplasia**
    - Early developmental failure of müllerian ducts → agenesis or hypoplasia of proximal 2/3 of vagina, cervix, &/or uterus
  - **Class II: Unicornuate uterus**
    - Normal development of one müllerian duct and near-complete to complete arrested development of contralateral duct
    - Rudimentary horn ± endometrium may be present
      - If noncommunicating, often associated with adenomyosis
  - **Class III: Uterus didelphys**
    - Complete failure of müllerian duct fusion
    - Each duct develops fully with duplication of uterine horns, cervix, and proximal vagina
  - **Class IV: Bicornuate uterus**
    - Incomplete or partial fusion of müllerian ducts
  - **Class V: Septate uterus**
    - Normal external contour with variable degree of reabsorption of uterovaginal septum
    - Higher risk for recurrent miscarriage
  - **Class VI: Arcuate uterus**
    - Near-complete reabsorption of uterovaginal septum
    - No significant increase in rate of infertility
  - **Class VII: Diethylstilbestrol (DES) uterus**
    - Dysmorphic or T-shaped uterus occurs in women exposed to DES in utero

Deep Pelvic Endometriosis

- Fibrotic plaques from deep pelvic endometriosis most commonly develop along ridge formed by uterosacral ligaments, which meet posteriorly at torus uterinus
  - This leads to adhesions between uterus and anterior rectum and lack of sliding during transvaginal ultrasound
  - 2nd most common location is dependently in posterior cul-de-sac
  - Adhesions between uterine fundus and rectum may be associated with uterine retroflexion &/or retroversion

Ureteral Injury During Surgery

- Ureter courses along lateral aspect of uterosacral ligaments into base of broad ligament, lateral to cervix and under uterine artery within cardinal ligament, over anterior vaginal fornix, and into bladder trigone
- Most common site of ureteral injury during gynecologic surgery is within cardinal ligament
  - Close proximity to cervix and uterine artery

C-Section Scar

- C-section scar and potential associated wedge-shaped niche are associated with multiple possible complications
  - Accumulation of blood/menstrual products in niche with intermenstrual/irregular vaginal bleeding
  - Placenta accreta spectrum
  - C-section scar ectopic pregnancy
  - Uterine rupture

Ectopic Pregnancy

- Most common location is in ampullary portion of fallopian tube
- Uncommonly, interstitial ectopic pregnancy occurs within intramural portion of fallopian tube
  - Difficult to diagnose and associated with higher risk of rupture and life-threatening hemorrhage
  - Look for eccentric gestation (3D ultrasound is helpful) with < 5-mm surrounding myometrium and interstitial line sign
- Rarer ectopic pregnancy locations include C-section scar, within rudimentary uterine horn, and intraabdominal
- While IUDs do not increase risk for ectopic pregnancy, higher proportion of pregnancies in presence of IUD are ectopic

Postablation Tubal Sterilization Syndrome

- Functioning endometrium may be present in patients with prior tubal occlusion and endometrial ablation
- Obstructed outflow leads to fluid distention in interstitial portions of fallopian tube, which may cause severe pain

UTERUS AND CUL-DE-SACS

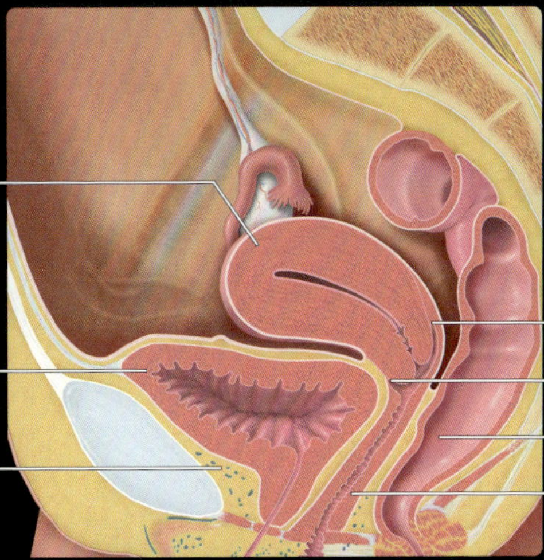

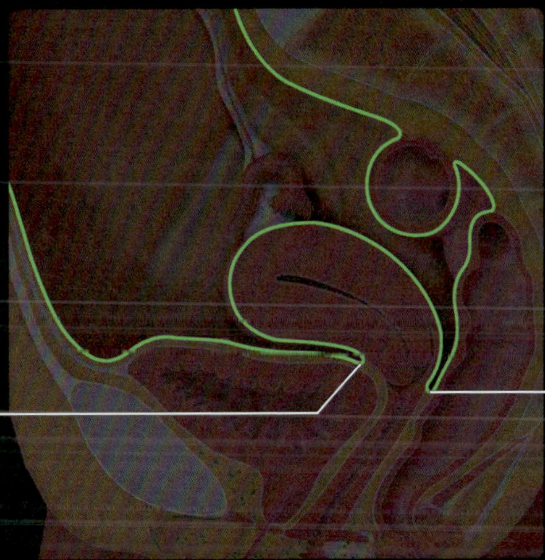

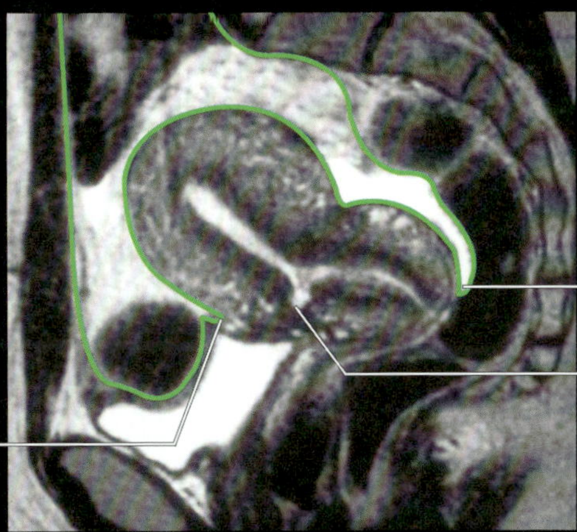

(Top) *Sagittal graphic of the female pelvis shows the bladder, uterus, and rectum. The peritoneal cavity is a complex potential space; peritoneal lining covers portions of these organs, including the bladder dome, the uterine corpus extending posteriorly to the posterior vaginal fornix, and reflects obliquely off the upper rectum. Note the posterior vaginal fornix extends more cephalad than the anterior vaginal fornix.* **(Middle)** *The peritoneum has been highlighted to show the cul-de-sacs. Posteriorly, the peritoneum extends along the posterior vaginal fornix, creating the posterior cul-de-sac (pouch of Douglas), the most dependent portion of the pelvis. Anteriorly, at the level of the lower uterine segment, the peritoneum is reflected over the dome of the bladder, creating the anterior cul-de-sac.* **(Bottom)** *Sagittal T2 MR with the peritoneum outlined is shown. The peritoneum over the bladder is loosely applied, allowing for filling of the bladder.*

UTERINE LIGAMENTS

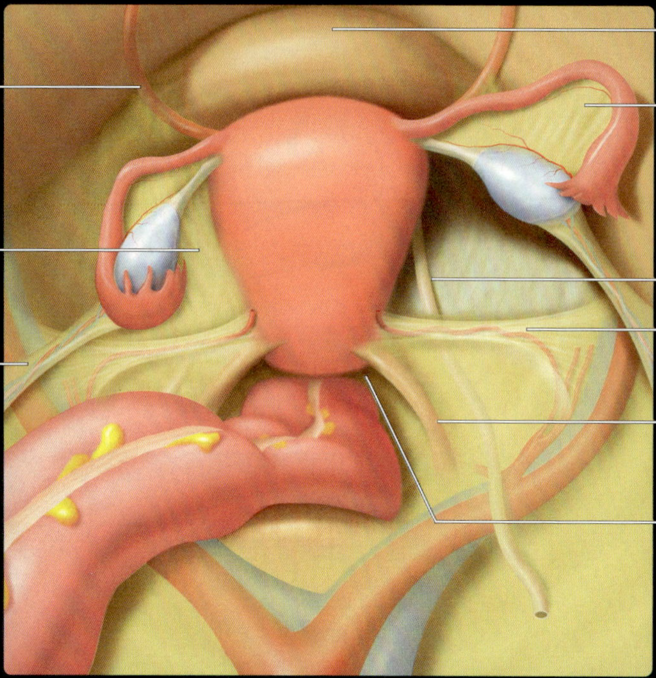

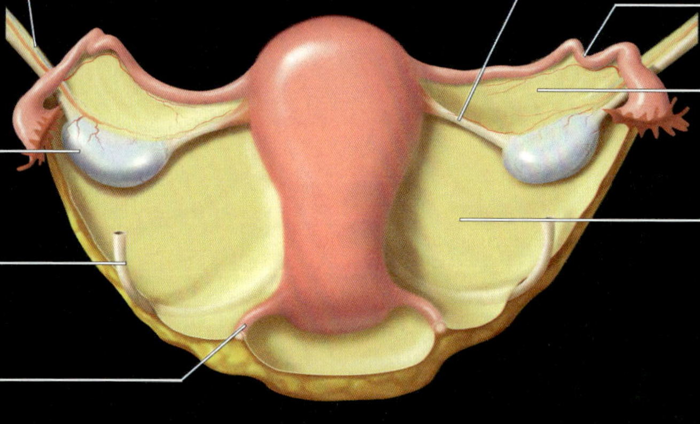

(Top) *Uterus viewed in situ from above and behind shows its positioning and major ligaments. The uterus is covered by a sheet of peritoneum, creating a double layer (the broad ligament), which sweeps laterally to attach to the pelvic wall. Areas of thickening at its base are the cardinal ligaments, which attach to the lateral pelvic wall, and the uterosacral ligaments, which attach to the sacrum. The uterosacral ligaments form the lateral borders of the posterior cul-de-sac (rectouterine pouch or pouch of Douglas). The round ligaments arise from the cornu of the uterus and course anteriorly to pass through the inguinal canal and insert on the labia majora, offering little support for the uterus.* **(Bottom)** *Graphic shows the posterior aspect of the uterus and ovaries with the other organs removed. The ovary is suspended from the pelvic side wall by the suspensory ligament of the ovary and from the uterus by the proper ovarian ligament. These ligaments separate the mesosalpinx above from the broad ligament below.*

UTERUS AND UTERINE ARTERIES

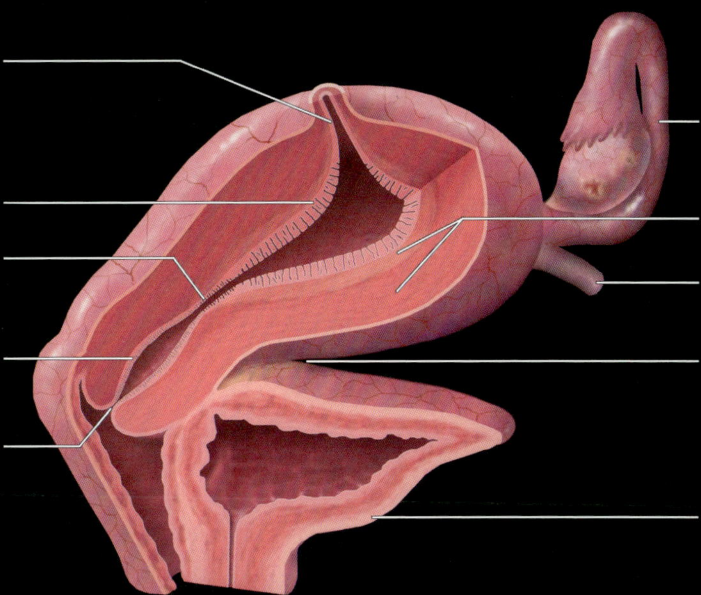

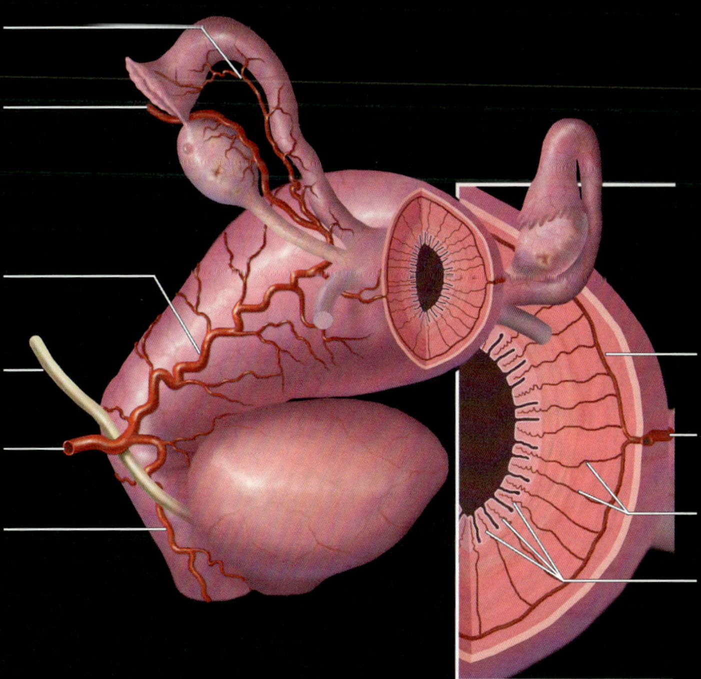

(Top) *The uterus is composed of a glandular endometrium and muscular myometrium. The smooth muscle within the inner portion of the myometrium (junctional zone) is more compacted with less water content and is relatively hypovascular.* **(Bottom)** *The uterine artery arises from the anterior trunk of the internal iliac artery. It crosses over the ureter ("water under the bridge") as it courses medially within the cardinal ligament to the lateral wall of the uterus. At the level of the cervix, it bifurcates into ascending and descending trunks. The ascending trunk forms the major blood supply to the uterus. It courses superiorly along the lateral wall of the uterus where it anastomoses with the ovarian artery, a branch of the aorta. Arcuate arteries course circumferentially in the outer of the myometrium and give rise to the radial arteries and finally the spiral arteries, which supply the endometrium.*

UTERINE VEINS

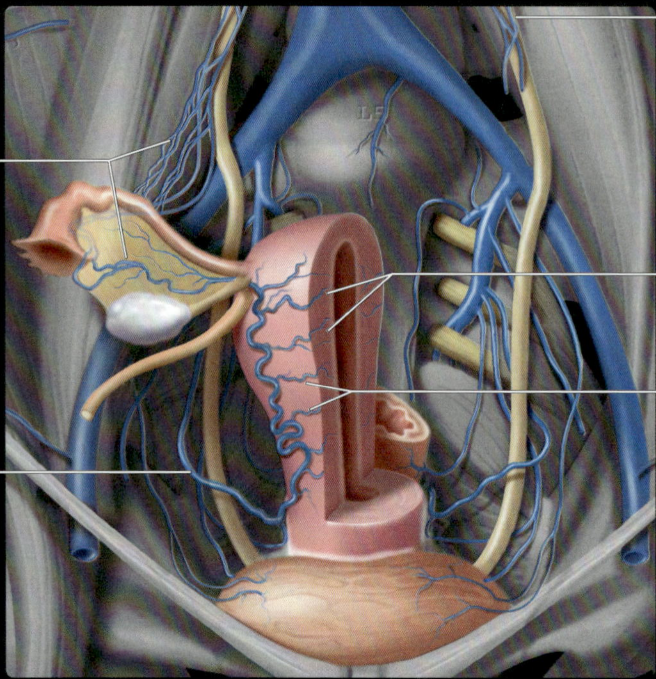

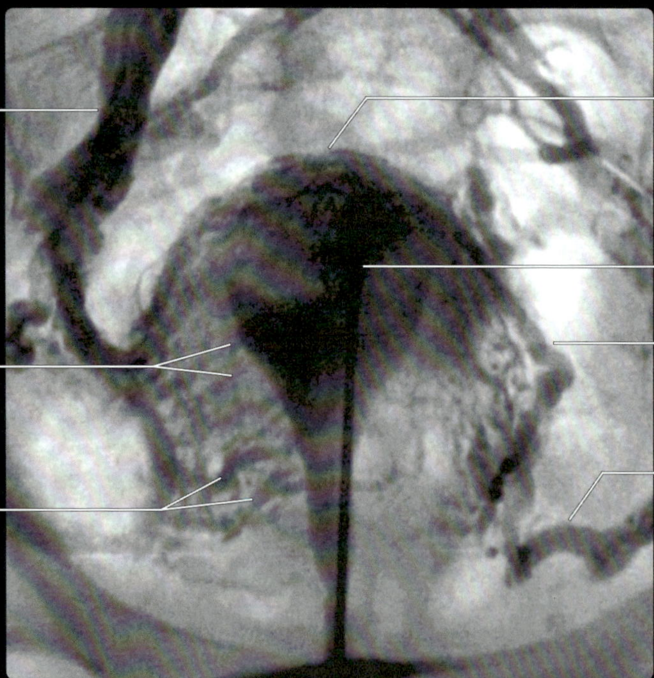

(Top) *Graphic shows the venous drainage from of the uterus, which follows the arterial system. There is venous drainage via both the uterine and ovarian veins.* **(Bottom)** *Pelvic venogram with the needle tip within the fundal myometrium shows very fine radial veins extending from the endometrium and larger arcuate veins, which course around in a circular fashion in the outer 1/3 of the myometrium. Note the presence of numerous small surrounding veins. These are part of the extensive venous plexus network within the pelvis, which forms a rich collateral venous drainage system. Pelvic venous congestion can cause development of varices and, in some people, chronic pelvic pain.*

MR UTERINE ANATOMY

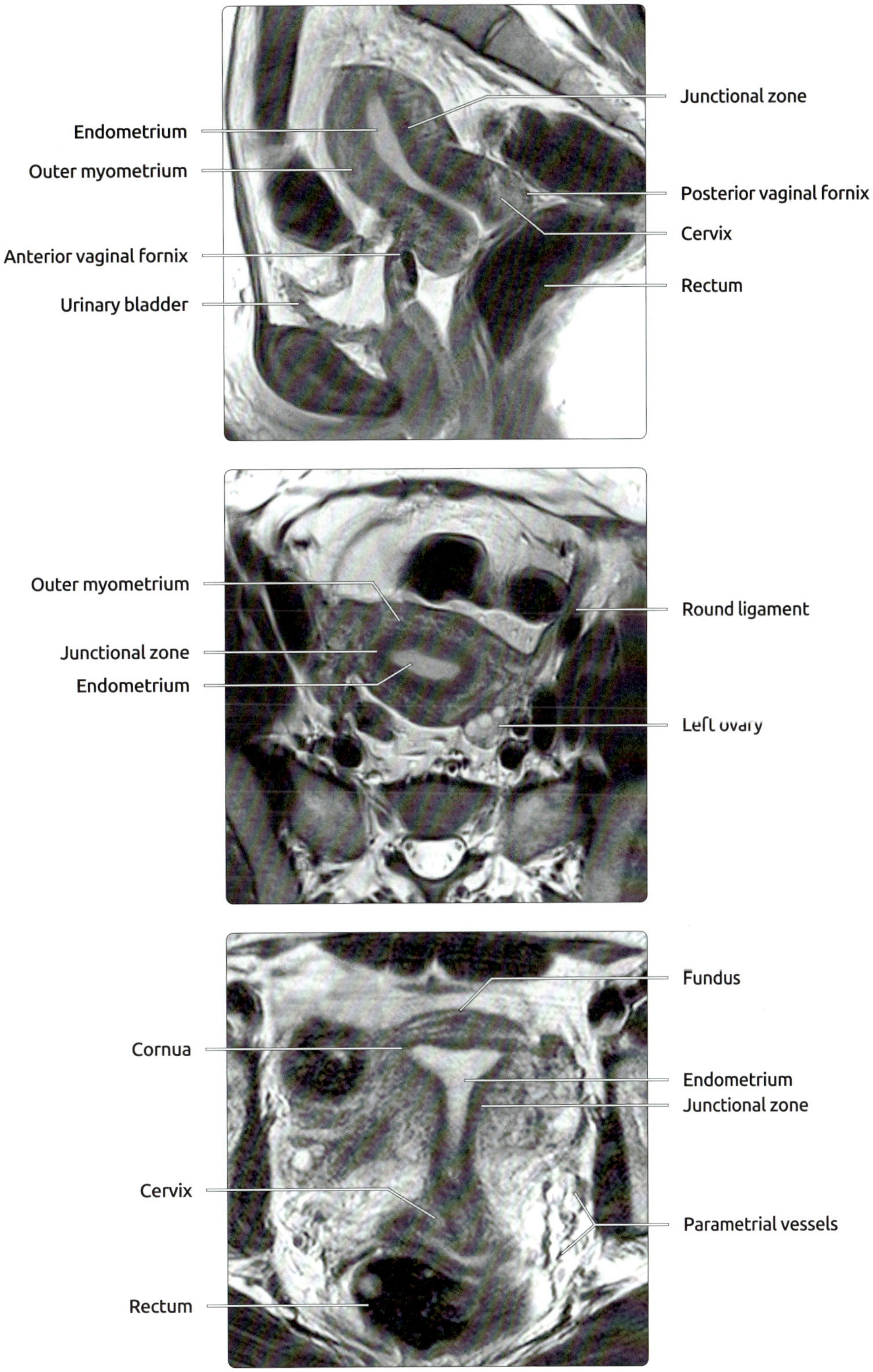

(Top) *Sagittal T2 MR of the uterus shows the typical zonal anatomy with a high-signal endometrium (which should not be mistaken for hydrometra), low-signal junctional zone, and intermediate-signal outer myometrium. The junctional zone represents the innermost portion of the myometrium, which has denser smooth muscle and lower water content, resulting in the lower signal. Junctional zone thickness > 12 mm is highly correlated with underlying adenomyosis.* **(Middle)** *Axial oblique T2 MR obtained orthogonal to the plane of the uterine corpus is shown. The uterus and endometrium have an elliptical contour at this level.* **(Bottom)** *Coronal oblique T2 MR obtained orthogonal to the uterine corpus allows simultaneous viewing of the endometrial cavity and uterine fundus. The fundus is located above the level of the cornua.*

MR AND ULTRASOUND CERVICAL ANATOMY

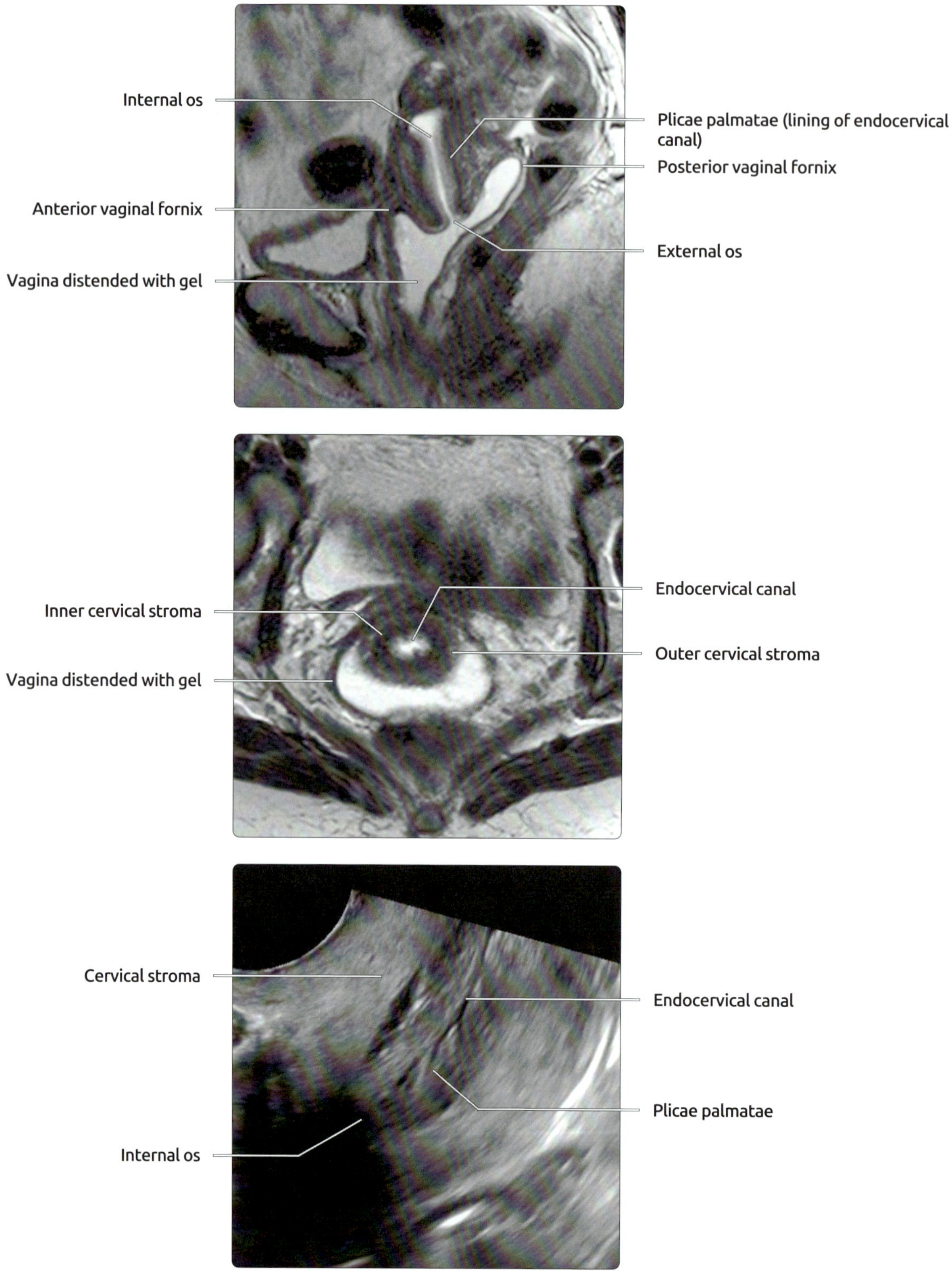

(Top) *Sagittal T2 MR of the cervix shows the feathery appearance of the endocervical canal, which is lined by mucus-secreting, columnar epithelium. This epithelium is arranged in small, V-shaped folds, giving a frond-like appearance, known as plicae palmatae. The posterior vaginal fornix extends more cephalad than the anterior vaginal fornix.* **(Middle)** *Axial T2 MR of the cervix shows the high signal intensity endocervical canal surrounded by low signal intensity cervical stroma. The T2-hypointense cervical stroma is largely fibrous with a high proportion of elastic fibers interwoven with smooth muscle. The outer layer of smooth muscle is often intermediate signal; this has variable thickness and is not always seen.* **(Bottom)** *Transvaginal ultrasound shows a longitudinal view of the cervix. The endocervical canal is lined with V-shaped folds known as plicae palmatae.*

UTERINE POSITION

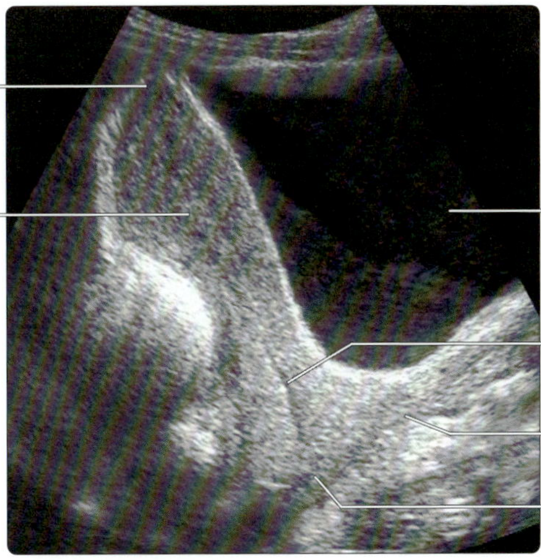

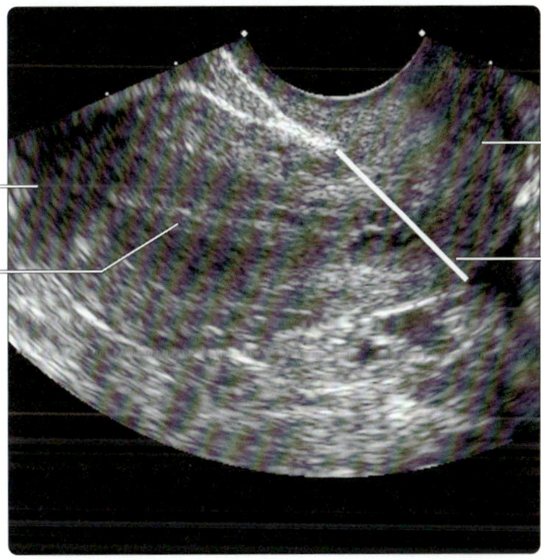

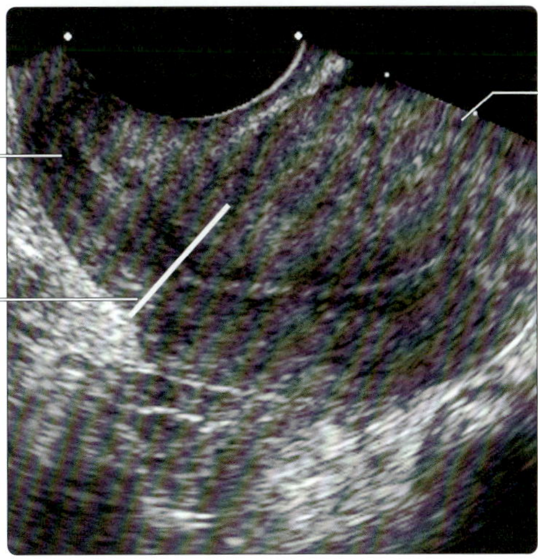

(Top) *Transabdominal ultrasound shows an anteverted uterus. Version refers to the angle the cervix makes with the vagina. In this case, the cervix is angled anteriorly relative to the vagina (anteversion), and the uterine corpus continues in a straight line with the cervix (neutral flexion). Uterine position can change with the degree of bladder filling.* **(Middle)** *Transvaginal ultrasound shows an anteflexed uterus. Flexion refers to the angle the uterine corpus makes with the cervix (a line has been drawn across the uterine isthmus as a reference point). The uterine corpus is angled anteriorly with respect to the cervix at this junction.* **(Bottom)** *Transvaginal ultrasound shows the cervix angling posteriorly with respect the probe, which is in the vagina (retroversion). Additionally, the uterine corpus is posteriorly angled with respect to the cervix (retroflexion).*

ENDOMETRIUM CYCLICAL VARIATION

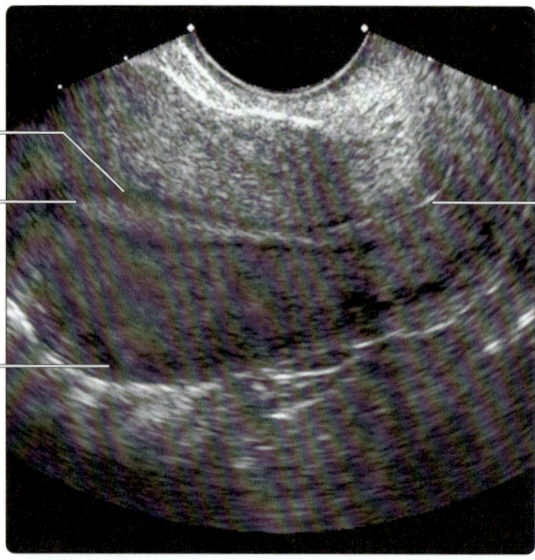

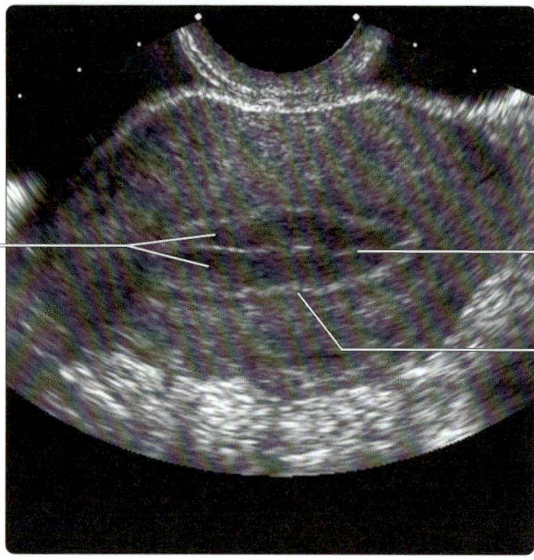

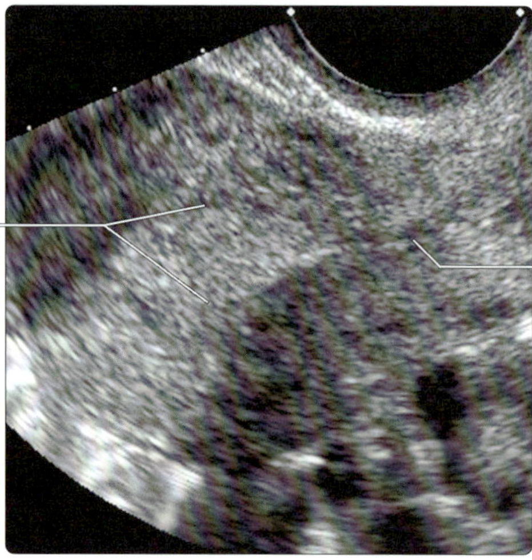

(Top) *In the early proliferative phase, the endometrium is thin and echogenic. The smooth muscle within the inner band of myometrium is more compacted and relatively hypovascular, giving it a more hypoechoic appearance (subendometrial halo). The majority of the myometrium has a homogeneous echogenicity with the outer portion (peripheral to arcuate vessels) sometimes slightly less echogenic.* **(Middle)** *In the late proliferative phase, the functionalis layer of the endometrium thickens and becomes hypoechoic (the basalis layer remains echogenic). A hyperechoic central line is created where the endometrial walls coapt. This gives the endometrium a trilaminar appearance.* **(Bottom)** *During the secretory phase, the endometrium becomes thickened and progressively echogenic. The increased echogenicity is the result of the endometrial glands becoming enlarged/tortuous and filled with glycogen, mucus, and other substances necessary to sustain a pregnancy. Secretory endometrium may be confounding for underlying pathologies, such as polyps, hyperplasia, or neoplasia. Consider rescanning a suspected endometrial abnormality earlier in a later menstrual cycle to see if it persists.*

SALINE-INFUSED SONOHYSTEROGRAM

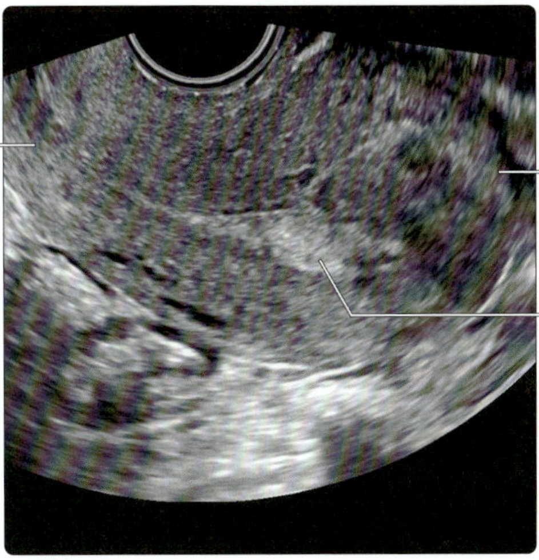

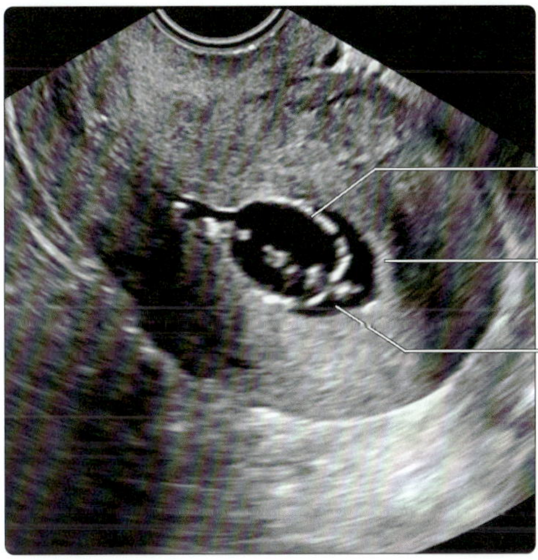

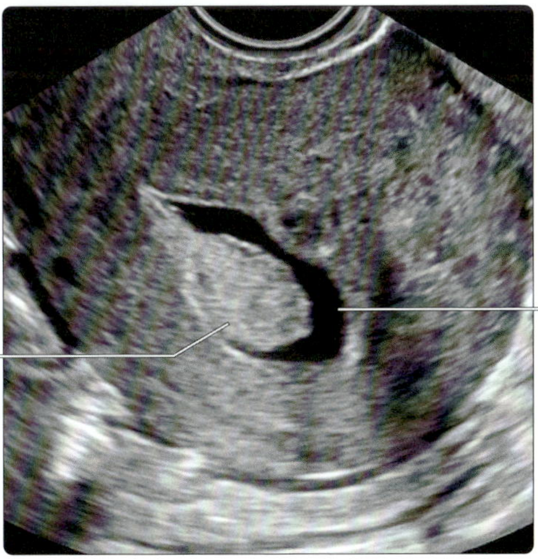

(Top) *Transvaginal ultrasound shows a longitudinal view of a retroverted and retroflexed uterus. There is a region of endometrium that appears relatively thickened and more echogenic; therefore, saline-infused sonohysterogram was recommended to further evaluate.* **(Middle)** *A balloon catheter is inflated in the lower uterine segment, and sterile saline is infused while scanning. As the cavity distends, the uterine walls begin to separate, and the endometrium can be evaluated. This longitudinal image at the beginning of the infusion shows the inflated balloon in place and saline beginning to expand the endometrial cavity.* **(Bottom)** *In this case, distending the endometrial cavity with fluid outlined a discrete endometrial polyp, which previously appeared as nondescript endometrial thickening.*

3D ULTRASOUND

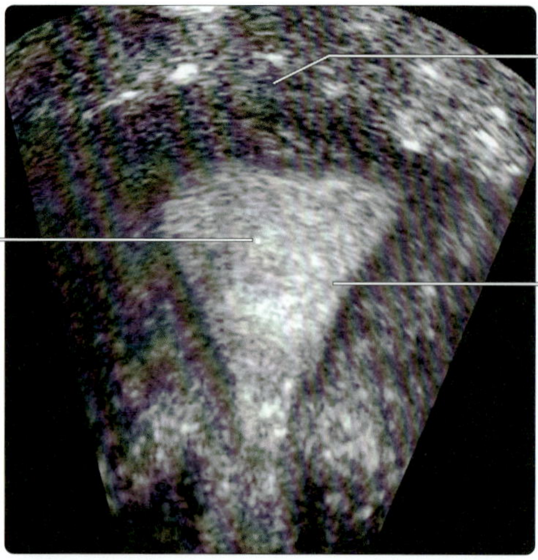

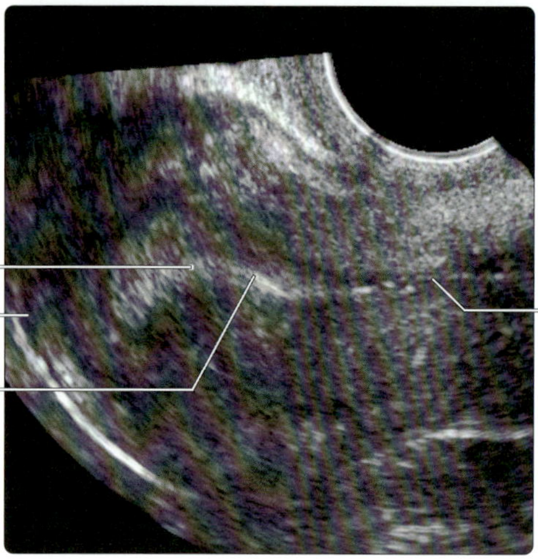

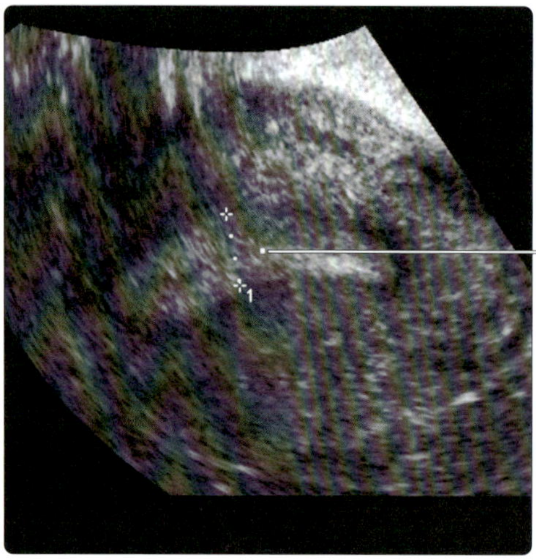

(Top) *First of 3 simultaneously acquired 3D images is shown. A dedicated 3D probe provides automated acquisition of ultrasound volume data. This is displayed in 3 simultaneous orthogonal planes. Multiplanar and rendered images can be rotated and sliced through like CT and MR images. A central localizer point on each image allows the operator to know the precise location in all 3 planes. In this coronal image, the entire endometrial and fundal contour are displayed. This has particular utility in evaluating müllerian duct anomalies.* **(Middle)** *The same 3D data set projected in the longitudinal plane shows both the endometrial cavity (bright line where the endometrial walls coapt) and endocervical canal.* **(Bottom)** *The same 3D data set projected in the transverse plane is shown. The cursors are measuring the endometrial thickness.*

UTERINE LIGAMENTS

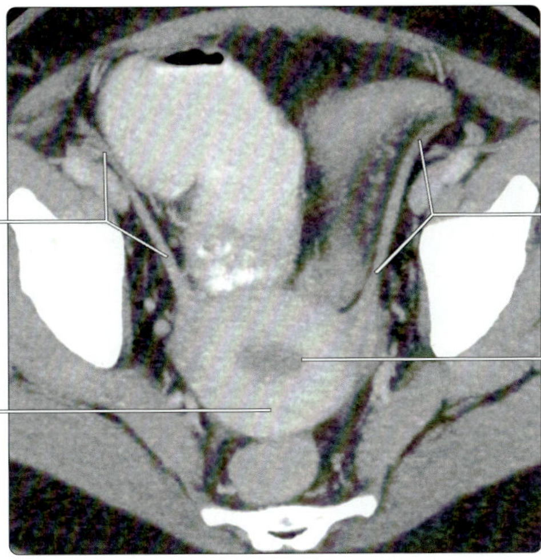

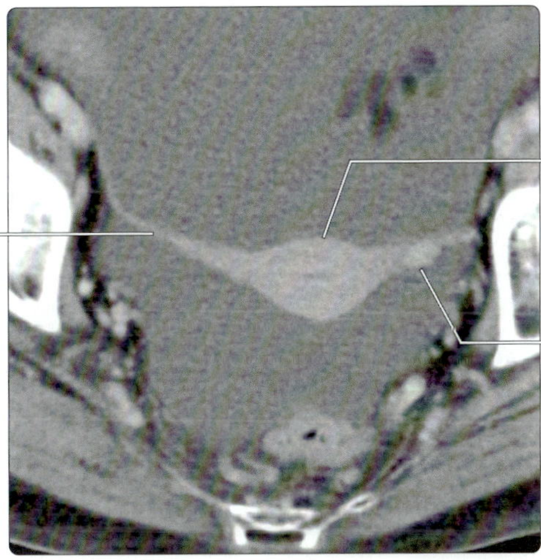

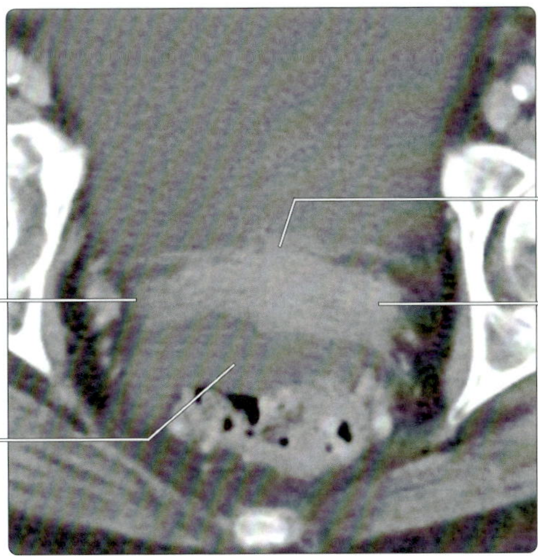

(Top) *Axial CECT shows the round ligaments, which arise anterior to the fallopian tubes near the cornua and extend anteriorly. They pass through the inguinal canal and insert on the labia majora. They are the embryologic homologue to the gubernaculum in the male, and offer little support to the uterus. This may be the site of varices in pelvic venous congestion, and canal of Nuck hydroceles usually occur along the round ligament insertion within the inguinal canal, which is normally obliterated.* **(Middle)** *Axial CECT in a female with ascites shows the broad ligaments extending to the pelvic sidewalls. The broad ligaments represent dual-layered peritoneal reflections that cover the uterus and adnexa. They offer little structural support to the uterus and are normally difficult to see without an adjacent peritoneal abnormality.* **(Bottom)** *Axial CECT in the same patient with ascites shows connective tissue thickening inferior to the broad ligaments, which represent the cardinal ligaments. These are subperitoneal, attaching the lateral cervix/vagina to the pelvic sidewall, and contain the uterine arteries.*

UTEROSACRAL LIGAMENTS

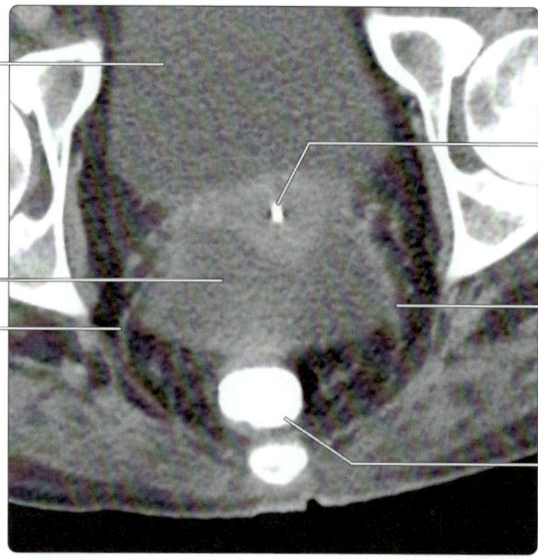

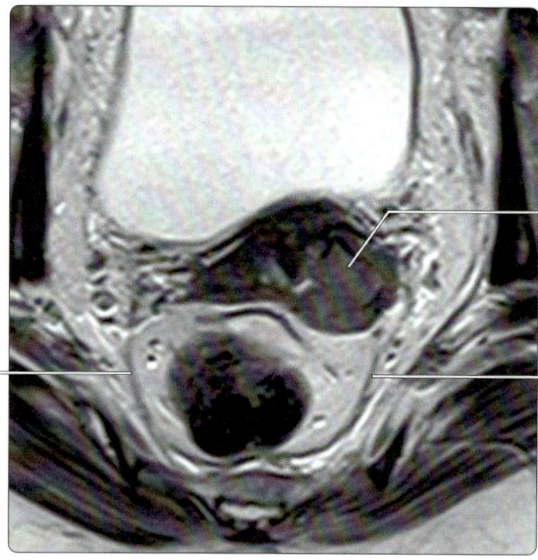

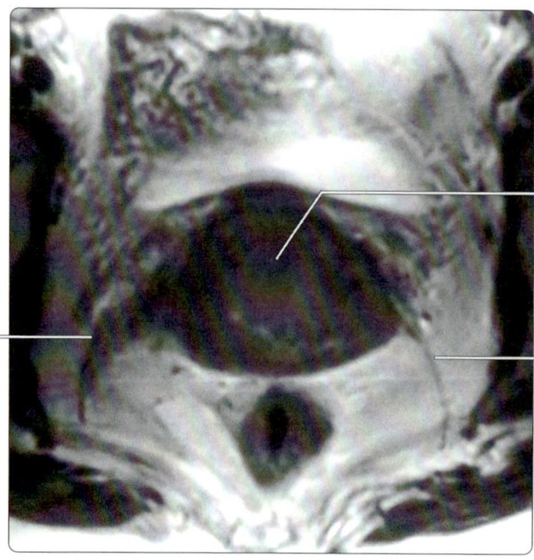

(Top) *Axial CT with rectal contrast in a patient with ascites shows fluid extending into the posterior cul-de-sac. The uterosacral ligaments outline the lateral borders of this space. They are composed predominately of smooth muscle, and convey the autonomic nerves to the pelvic organs. They, along with the cardinal ligaments and levator ani, are the main support for the uterus and cervix.* **(Middle)** *Cervical carcinoma commonly invades into the parametrium, which includes both the uterosacral and cardinal ligaments. These are very important areas to evaluate when staging cervical carcinoma. This axial T2 MR shows normal uterosacral ligaments without invasion.* **(Bottom)** *Axial T2 MR shows invasion of the right uterosacral ligament by cervical cancer.*

DEEP PELVIC ENDOMETRIOSIS

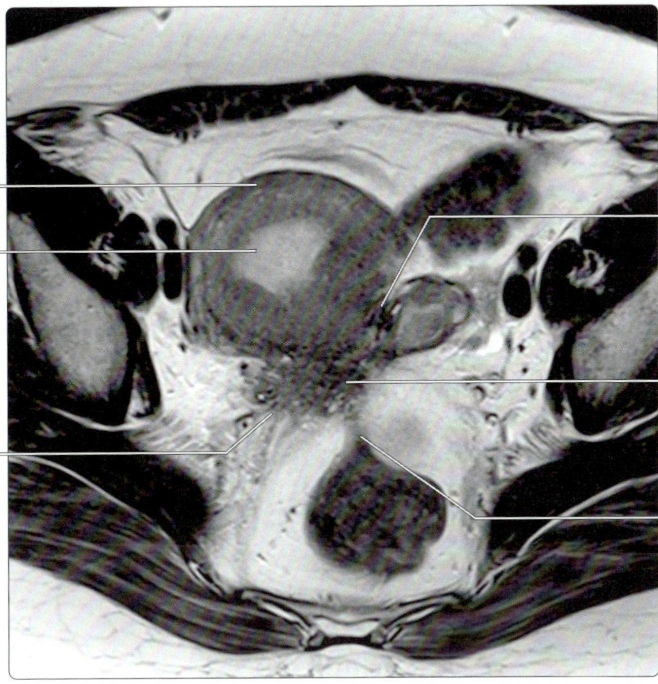

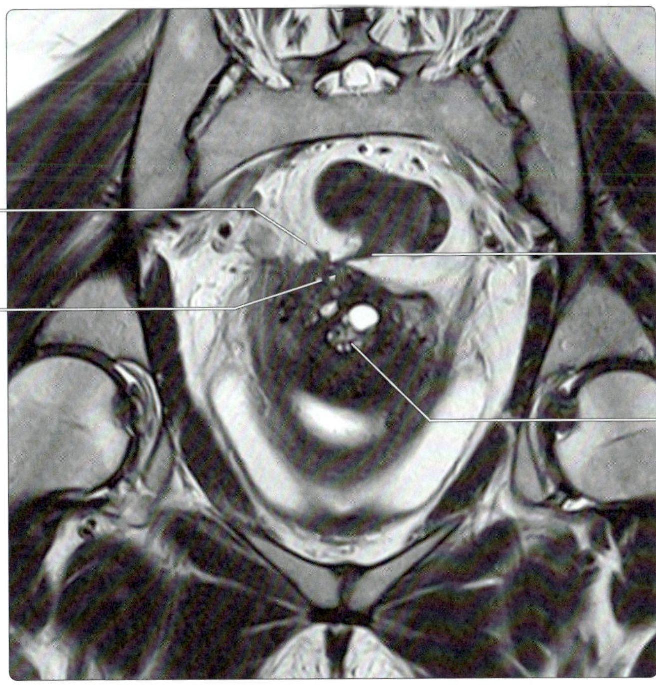

(Top) *The uterosacral ligaments meet to form a small ridge posteriorly on the cervix, known as the torus uterinus. This forms a natural anatomic platform where pathologies, such as malignancy, infection, or endometriosis collect. Axial T2 MR shows T2-hypointense endometriosis fibrotic plaque with T2-hyperintense glandular foci that tethers the anterior rectum, left adnexa, and extends along the right uterosacral ligament.* **(Bottom)** *Coronal oblique T2 MR shows T2-hypointense endometriosis plaque tethering the anterior rectal wall and extending along the right uterosacral ligament. The torus uterinus is the most common location for deep pelvic endometriosis, followed by the posterior cul-de-sac.*

FALLOPIAN TUBE ANATOMY

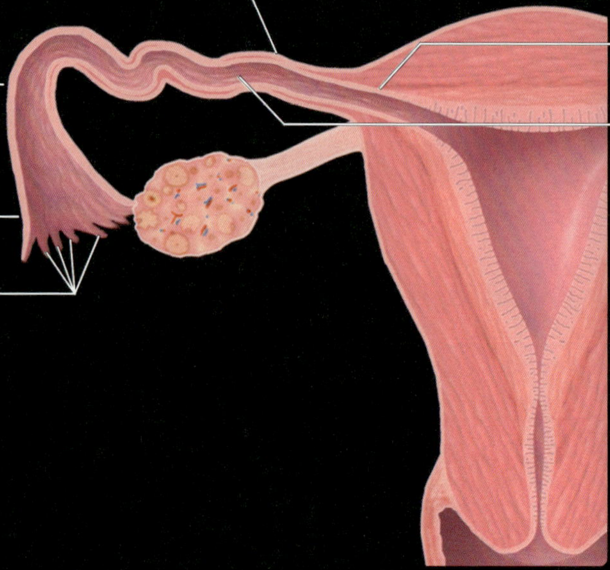

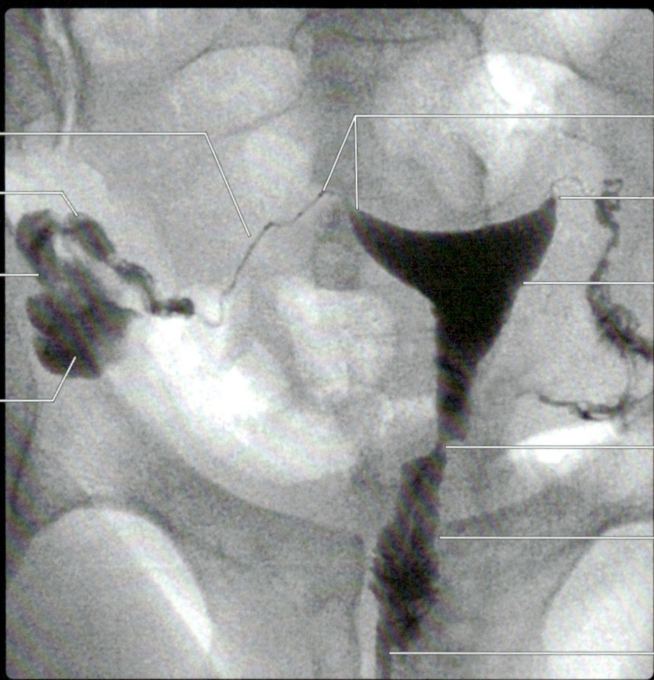

(Top) *Graphic of the fallopian tube shows the 4 segments, including the interstitial (intramural) portion, isthmus, ampulla, and infundibulum, which is ringed by the fimbriae.* **(Bottom)** *Hysterosalpingogram shows the endocervical canal with a feathered appearance created by the endocervical glands (plicae palmatae). The uterine cavity has a triangular configuration. The interstitial portion of the tube traverses the myometrial wall at the cornua (~ 1 cm in length). An acute angulation can often be seen between it and the isthmus. The fallopian tube widens at the ampulla and infundibulum before opening into the peritoneal cavity adjacent to the posterior surface of the ovary.*

HYDROSALPINX

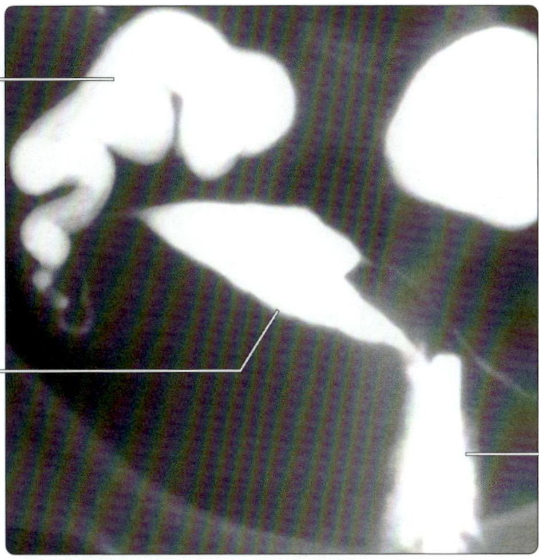

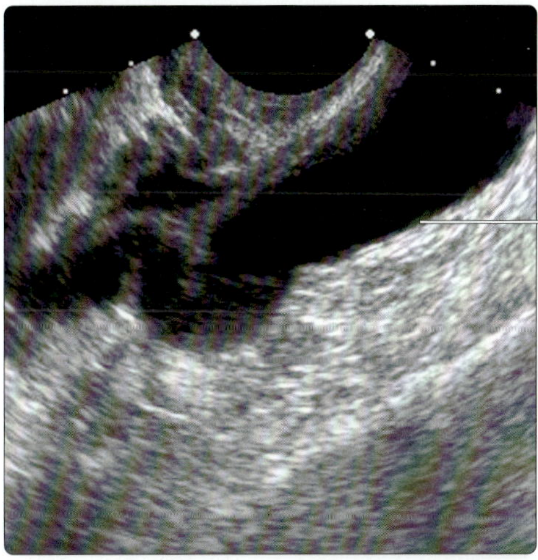

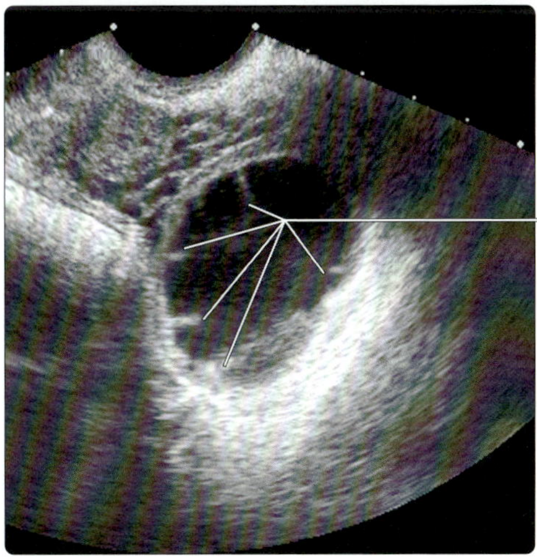

(Top) *Hysterosalpingogram shows hydrosalpinx. The right ampullary portion expands dramatically, and there is no free intraperitoneal spillage of contrast.* **(Middle)** *Transvaginal ultrasound shows an elongated tubular structure in the left adnexa, indicating hydrosalpinx. Obtaining this elongated view helps differentiate from a cystic ovarian mass.* **(Bottom)** *In the cross-sectional plane, a hydrosalpinx will often display a cogwheel appearance with mucosal folds projecting into the lumen. As the hydrosalpinx becomes more chronic, these folds may become thick and nodular.*

POSTABLATION TUBAL STERILIZATION SYNDROME

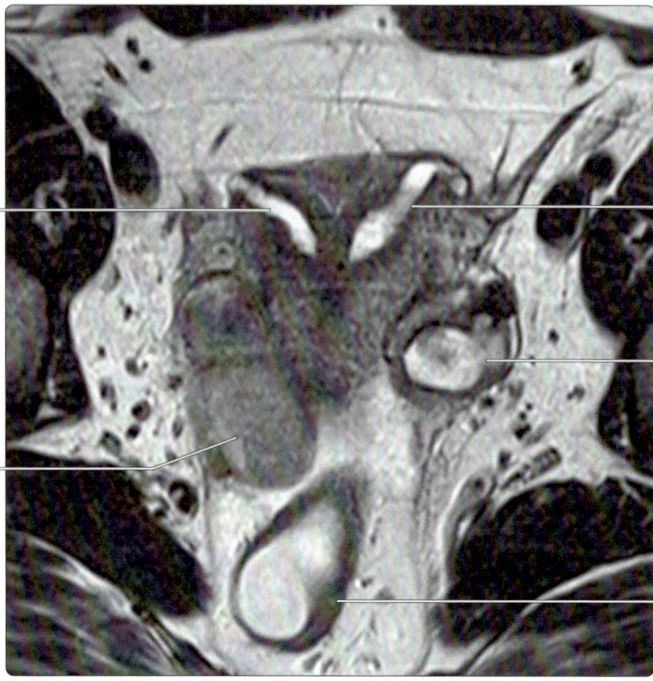

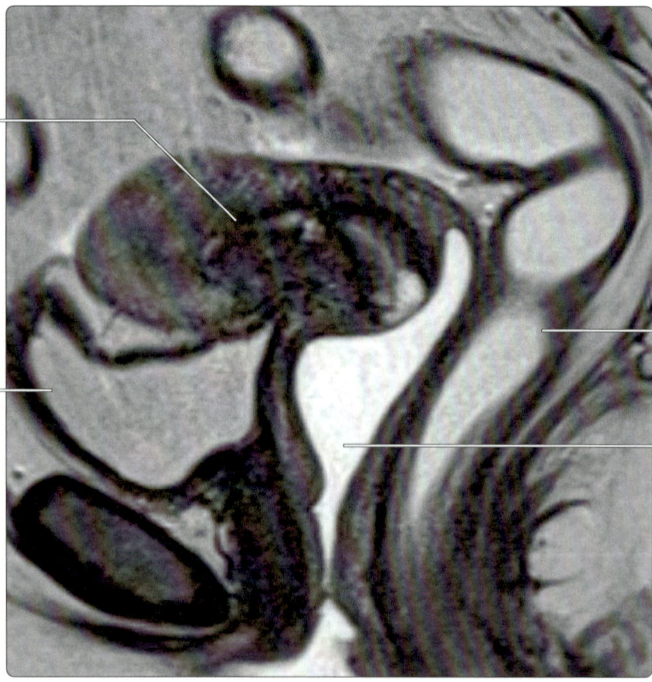

(Top) *Axial T2 MR in a female patient with prior endometrial ablation and tubal occlusion shows accumulation of fluid with distention of the interstitial portions of the fallopian tubes bilaterally. This occurs when there is residual functional endometrium near the fundus, and there is no way for the fluid/blood it produces to drain. Even a small amount of tubal distention can cause severe pain.* **(Bottom)** *Sagittal T2 MR in the same patient shows T2-hypointense obliteration of the endometrial cavity, a typical appearance after endometrial ablation.*

TUBAL ECTOPIC PREGNANCY

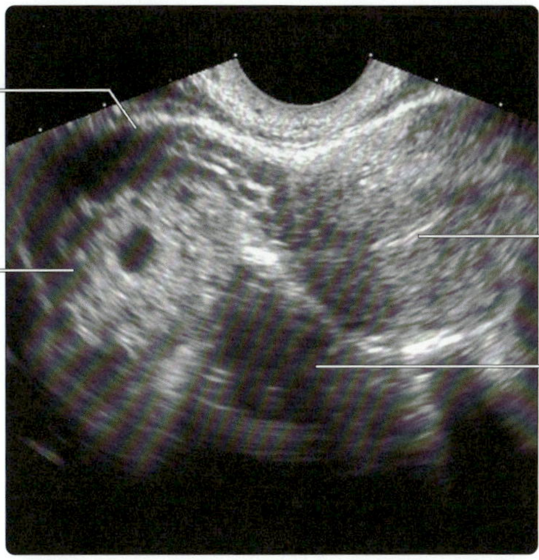

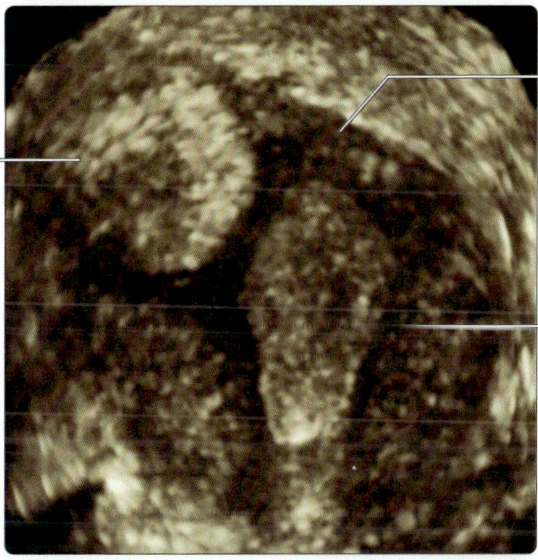

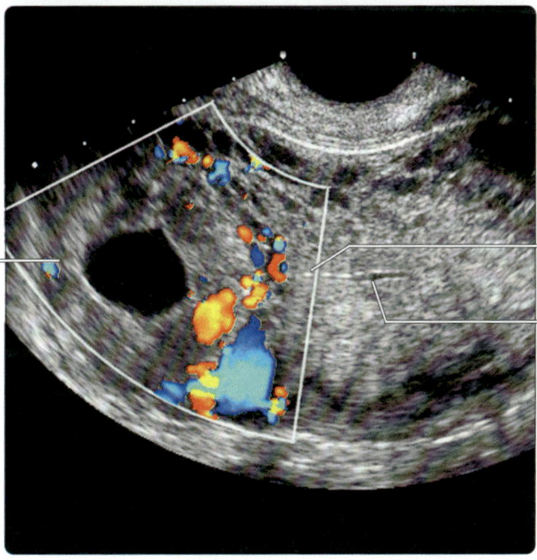

(Top) *Transverse transvaginal ultrasound shows a "probable" ectopic pregnancy (thick echogenic ring) within the ampullary portion of the right fallopian tube. This is the most common location for ectopic pregnancy, although precise tubal localization is usually not possible on imaging.* **(Middle)** *Coronal 3D ultrasound shows a gestational sac implanted in the interstitial (intramural) portion of the fallopian tube. These can grow to large size before rupturing and are higher risk for life-threatening hemorrhage.* **(Bottom)** *Transvaginal color Doppler ultrasound shows an interstitial ectopic pregnancy and an empty endometrial cavity. At the right cornua, the proximal interstitial portion of the tube can be followed to the ectopic gestational sac, known as the interstitial line sign. The ectopic pregnancy has dramatically enlarged the interstitial portion of the tube, and there is virtually no surrounding myometrium.*

UTERINE EMBRYOLOGY

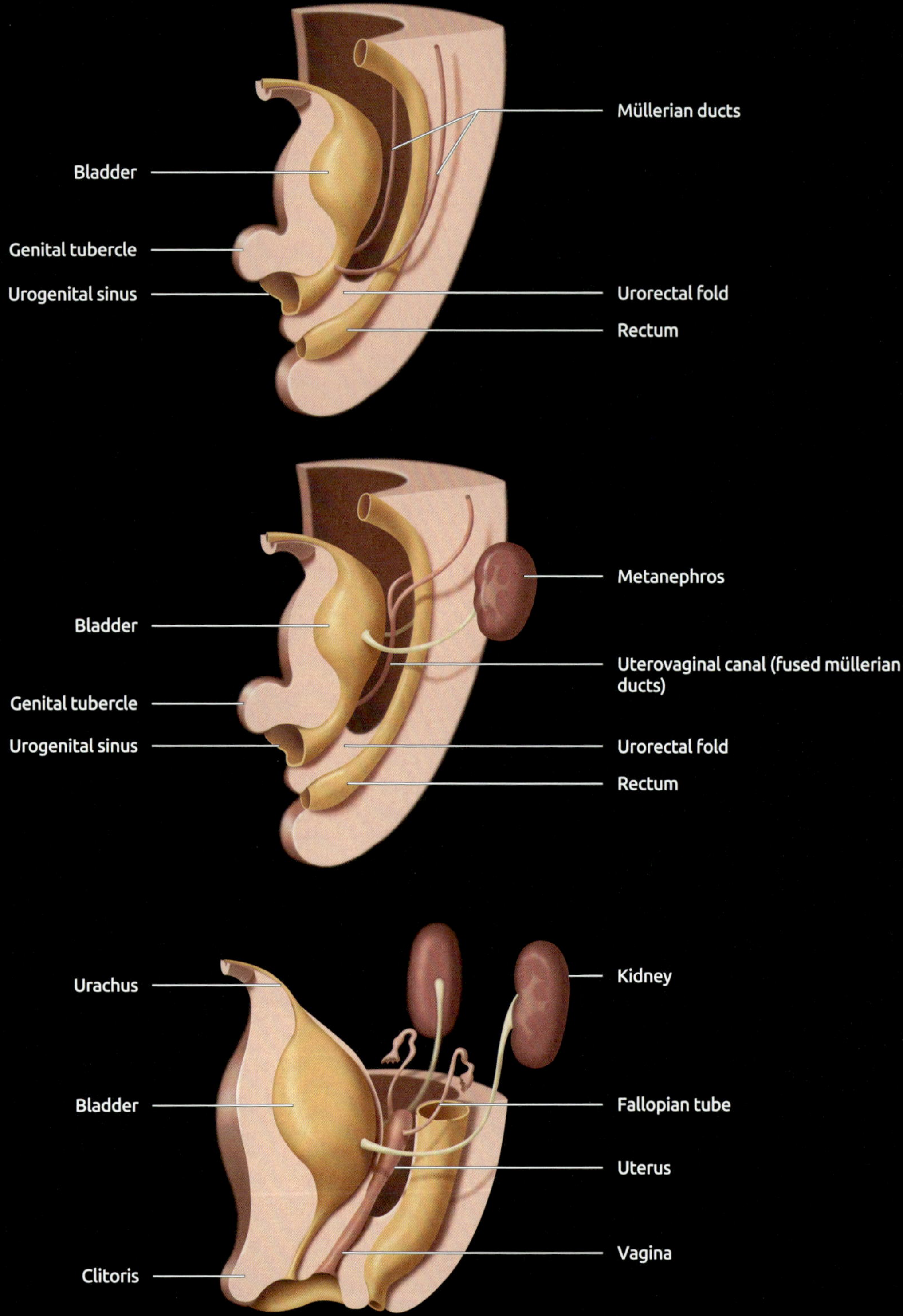

(Top) *The fallopian tubes, uterus, and upper vagina form from the paired müllerian (paramesonephric) ducts, which develop on either side of the midline. The lower vagina originates from the urogenital sinus.* **(Middle)** *The müllerian ducts must meet in the midline and fuse to form the uterus and upper portion of the vagina (uterovaginal canal). The unfused portions will form the fallopian tubes. The development of the kidney (metanephros) is closely related to uterine development, and coexistent renal and müllerian duct anomalies are common.* **(Bottom)** *The distal portion of the vagina is formed from the urogenital sinus, which splits to form the bladder and urethra anteriorly and the vagina posteriorly.*

UNICORNUATE UTERUS

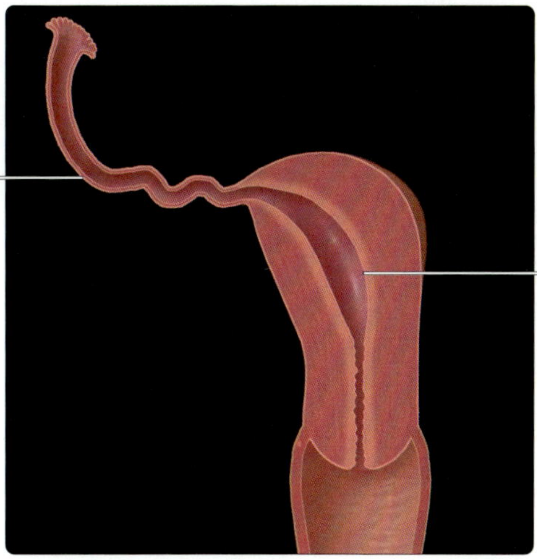

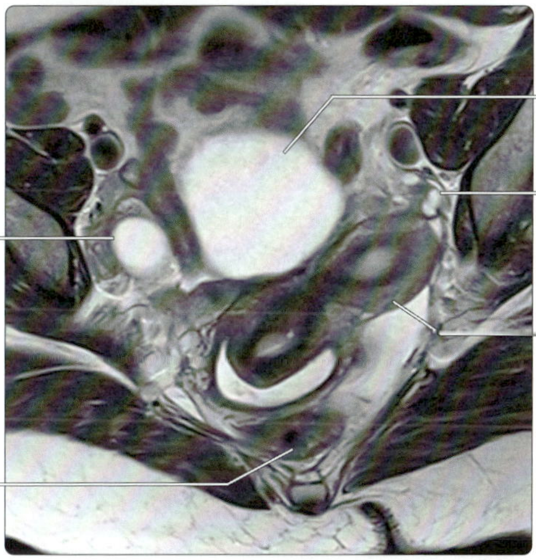

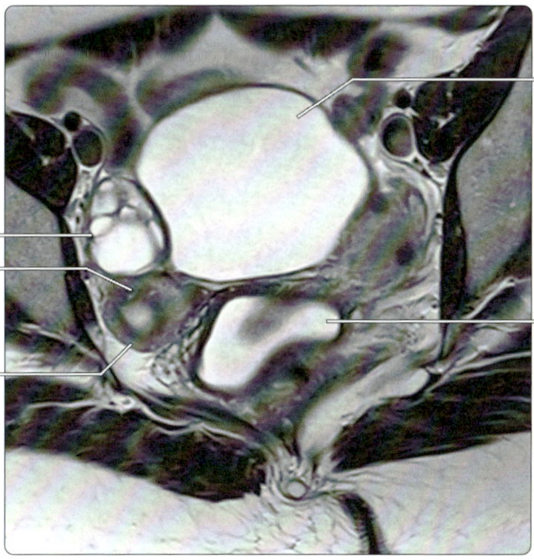

(Top) *Graphic shows a unicornuate uterus, which occurs if only 1 paramesonephric duct forms. The absence of the left uterine cornua may be a useful imaging clue.* **(Middle)** *Axial T2 MR shows a left-sided unicornuate uterus. This is associated with congenital renal anomalies, particularly renal agenesis.* **(Bottom)** *Axial T2 MR in the same patient shows a noncommunicating right-sided rudimentary horn, which are prone to adenomyosis. Pregnancy may rarely occur within a noncommunicating rudimentary horn via transperitoneal migration of sperm or zygote.*

UTERUS DIDELPHYS

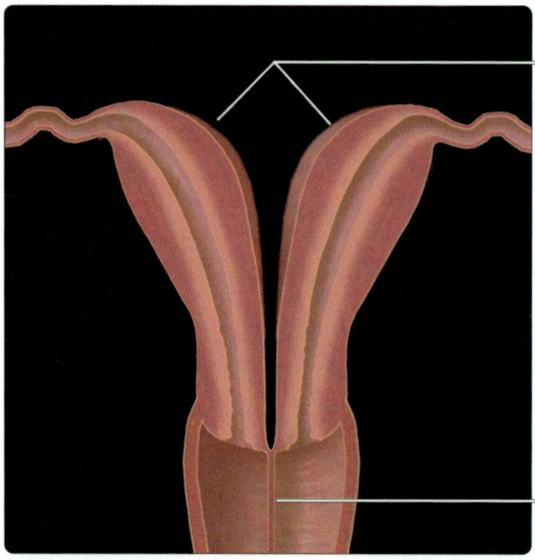

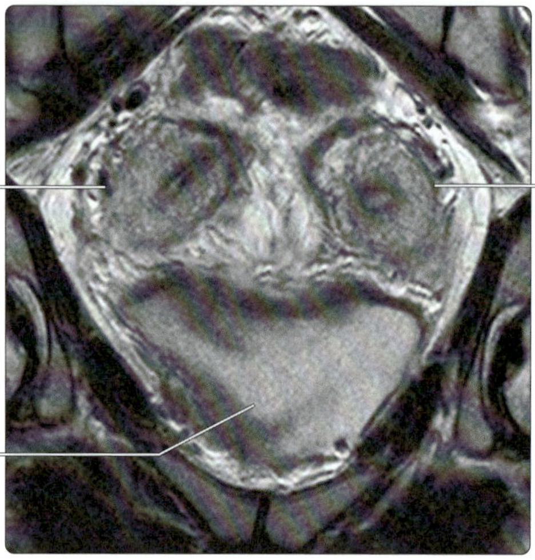

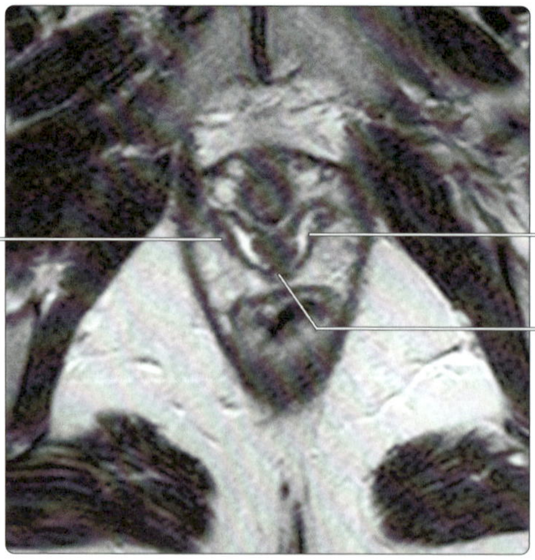

(Top) *Coronal graphic of uterus didelphys is shown. If both paramesonephric ducts form but fail to fuse, the result is uterus didelphys with the appearance of 2 unicornuate uteri side by side. A vaginal septum is seen in ~ 75% of cases.* **(Middle)** *Coronal T2 MR shows a uterus didelphys with 2 completely separate uterine horns.* **(Bottom)** *Axial T2 MR in the same patient shows a longitudinal vaginal septum; these are usually due to müllerian anomalies. Transverse vaginal septa are usually caused by failure of fusion between the müllerian duct and urogenital sinus.*

BICORNUATE UTERUS

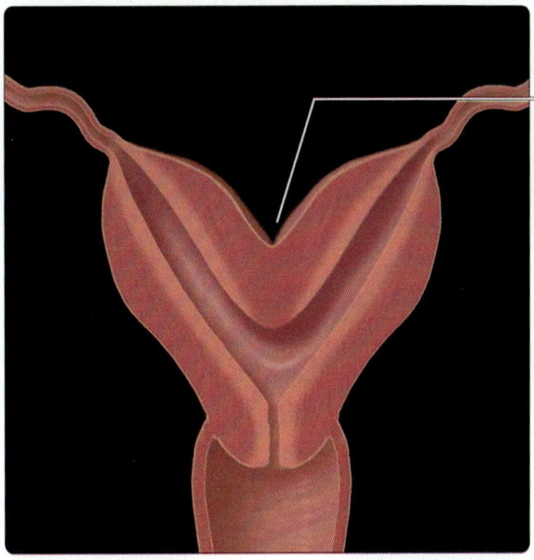

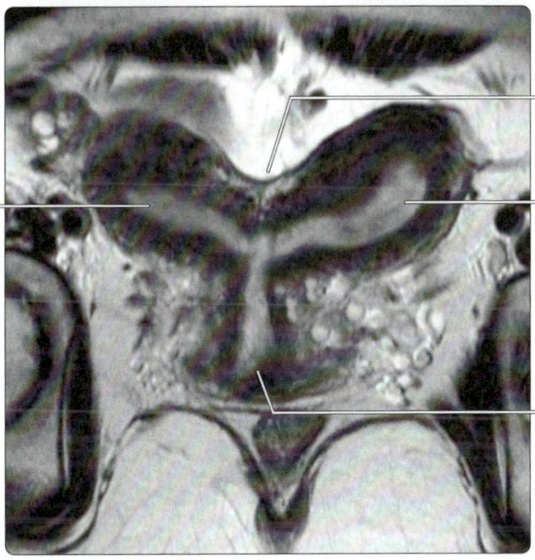

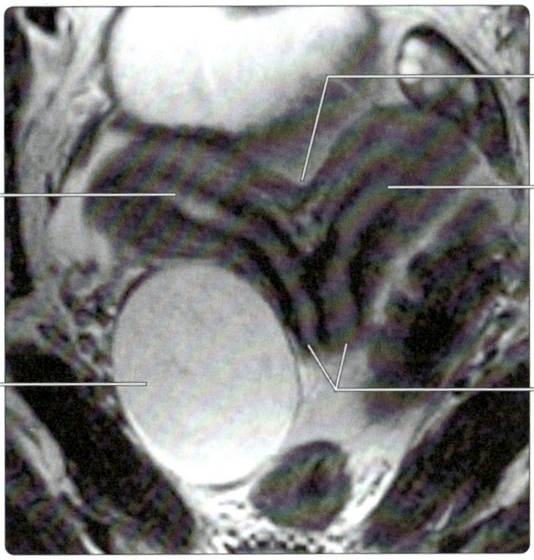

(Top) *Coronal graphic shows a bicornuate uterus. If the paramesonephric ducts only partially fuse, the external contour of the uterus will be concave. There may either be a single cervix (unicollis) or 2 cervices (bicollis), depending on the degree of fusion and septal resorption.* **(Middle)** *Coronal oblique T2 MR shows the classic concave appearance of a bicornuate uterus with 2 endometrial cavities communicating inferiorly, and there is a single cervix (unicollis).* **(Bottom)** *Coronal oblique T2 MR shows a concave uterus with 2 endometrial cavities, indicating a bicornuate uterus. In this case, the endometrial cavities do not communicate inferiorly, and there are 2 cervices (bicollis).*

SEPTATE UTERUS

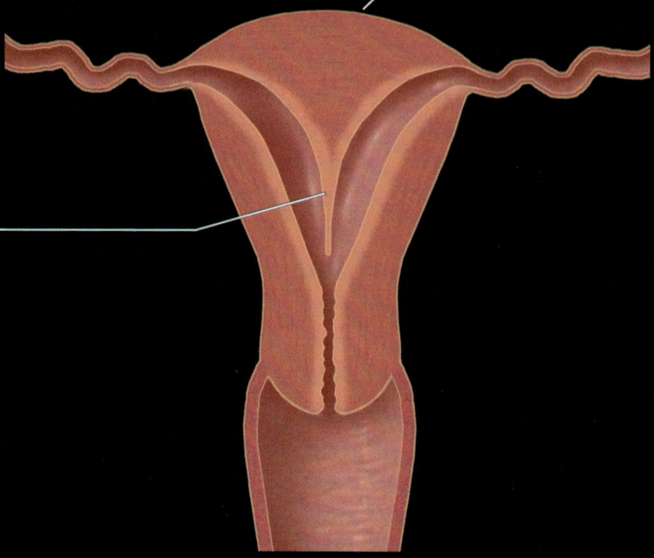

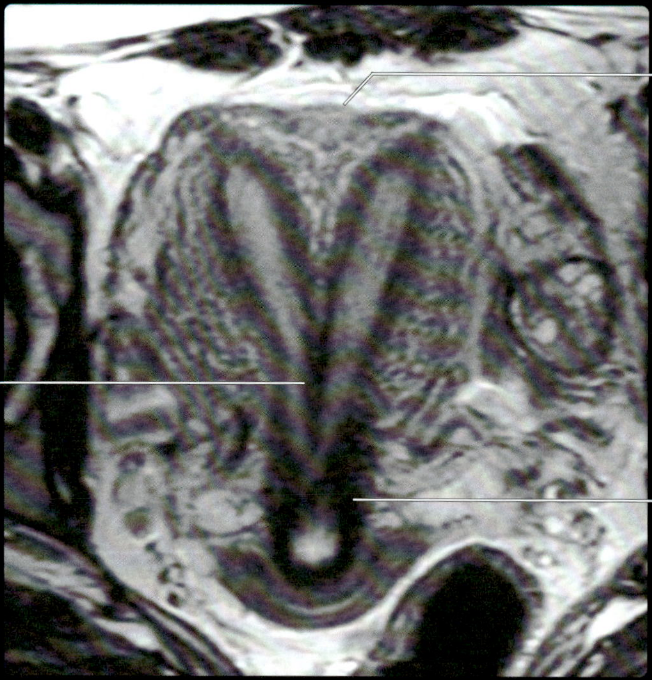

(Top) *Coronal graphic shows a septate uterus. If the paramesonephric ducts fuse appropriately, but the wall between them is incompletely resorbed, the result is a septate uterus. Unlike a bicornuate uterus, which is due to partial failure of paramesonephric duct fusion, the fundal contour is normal (not concave). The septum itself can be quite variable and extends only partially into the endometrial cavity or all the way to the cervix &/or vagina.* **(Bottom)** *Coronal oblique T2 MR of the uterus shows a normal fundal contour with a muscular septum extending to the level of the cervix. Septate uterus is the müllerian abnormality most associated with recurrent pregnancy loss.*

CESAREAN SECTION ABNORMALITIES

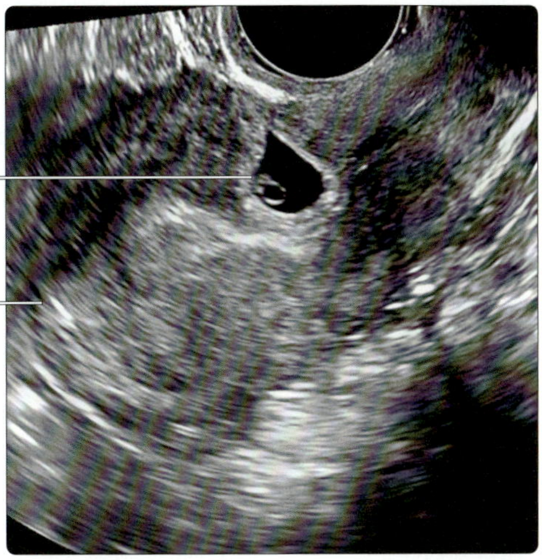

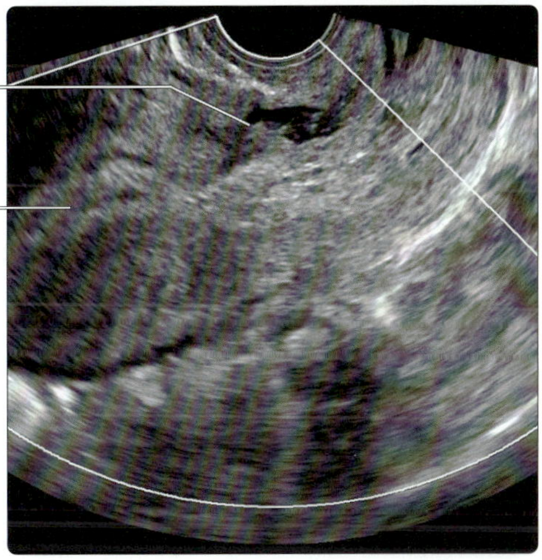

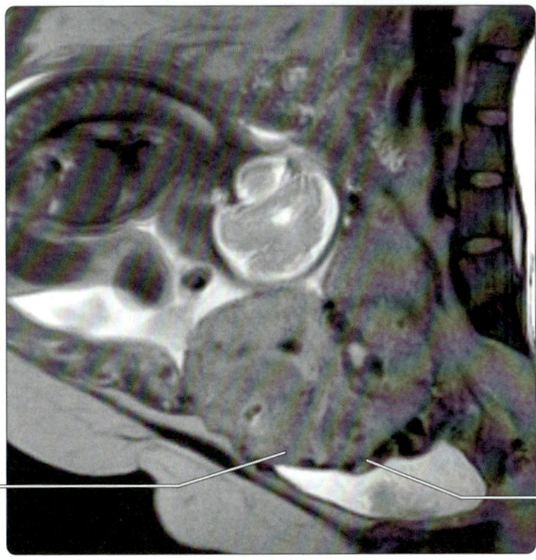

(Top) *Longitudinal transvaginal ultrasound shows a gestational sac with a yolk sac within the C-section scar niche.* **(Middle)** *Longitudinal transvaginal ultrasound shows accumulation of heterogeneous menstrual blood products within a prominent and triangular C-section scar niche, which may be a source of intermenstrual bleeding.* **(Bottom)** *Sagittal T2 MR shows placenta accreta spectrum. The placenta bulges through a prior C-section scar defect to involve the bladder dome with associated extensive vascularity. This condition is extremely high risk for life-threatening hemorrhage.*

GROSS ANATOMY

Overview

- **Size**
  - Ovaries are ovoid structures that vary in size depending on age, hormonal status, and stage of menstrual cycle
    - Normally ~ 8 mL in menstruating women, and < 3 mL before menarche and after menopause
    - Ovarian volume is estimated using formula for prolate ellipse (0.523 × length × width × thickness)
- **Position**
  - Ovaries are located in true pelvis, although exact **position is variable**
    - Location affected by parity, ovarian size, and uterine size/position; ligamentous laxity allows some mobility
    - Usually within **ovarian fossa** in nulliparous women
      - ☐ Lateral pelvic wall below bifurcation of common iliac artery
      - ☐ Anterior to ureter and internal iliac artery
      - ☐ Posterior to broad ligament
    - More variable position in parous women because ovaries are pushed out of pelvis during pregnancy
  - Fallopian tube drapes over ovarian surface, which is partially covered by fimbriated end
- **Structure**
  - Ovary is covered by specialized peritoneum composed of simple squamous/cuboidal cells called **surface or germinal epithelium**
  - **Tunica albuginea** is deep to surface epithelium, composed of short connective tissue fibers
  - Outer ovarian **cortex** contains follicles (usually < 25) surrounded by connective tissue
    - Includes maturing, regressing, and dominant follicles; corpus luteum or corpus albicans
  - **Medulla** is central vascularized stroma; vessels, lymphatics, and nerves enter/exit though hilum
    - Ovarian ligament attaches at **hilum**, where vessels, lymphatics, and nerves enter/exit
- **Ligamentous support**
  - **Suspensory ligament** (infundibulopelvic ligament)
    - Peritoneal fold that attaches ovary to pelvic wall
    - **Contains ovarian artery/vein** and sympathetic/parasympathetic plexuses
  - **Mesovarium**
    - Peritoneal fold attaching anterior ovary to posterior surface of broad ligament
  - **Ovarian ligament**
    - Fibromuscular band along superior margin of broad ligament extending from ovary to uterine cornua
  - **Mesosalpinx**
    - Connects ovarian ligament and fallopian tube
  - **Broad ligament**
    - Peritoneal fold connecting uterus to pelvic floor/sidewall, terminating superiorly with ovarian ligament
- **Blood supply**
  - **Ovarian artery** is infrarenal branch of aorta at ~ L2 level (~ 5-mm diameter)
    - Descends to pelvis and enters suspensory ligament, continuing through mesovarium to enter ovary
    - Anastomoses with uterine artery, source of **dual blood supply**
  - Drainage via venous plexus into ovarian veins
    - Right ovarian vein drains to inferior vena cava, left ovarian vein drains into left renal vein
  - Arteries and veins markedly enlarge during pregnancy
- **Lymphatic drainage**
  - Ovaries have 3 pathways of lymphatic drainage
    - Along ovarian veins → paraaortic and paracaval nodes at level of renal hilum
      - ☐ Accounts for majority of lymphatic drainage
    - Through broad ligament → internal/external iliac and obturator nodes
    - Along round ligament → inguinal nodes

PHYSIOLOGY

- **Hypothalamus-pituitary-ovarian axis**
  - Gonadotropin-releasing hormone (GnRH) from hypothalamus stimulates anterior pituitary to produce follicle-stimulating hormone (FSH) and luteinizing hormone (LH)
  - FSH acts on ovarian granulosa cells to induce follicle maturation
    - ~ 400,000 follicles present at birth, but only 0.1% (~ 400) mature to ovulation
  - LH acts on theca cells, which produce estrogen precursors
    - Increase at puberty leads to development of secondary sex characteristics
- **Menstrual cycle**
  - **Follicular phase** (days 0-14)
    - Several follicles begin to develop (range: 1-11 days; mean: 5 days)
    - By 8-12 days, dominant follicle develops, while remainder regress
  - **Ovulation** (~ day 14)
    - Egg extruded from ovary (caused by "LH surge")
    - Dominant follicle typically 2.0-2.5 cm at ovulation
  - **Luteal phase** (days 14-28)
    - LH induces formation of **corpus luteum** (temporary endocrine structure)
      - ☐ Luteinization of theca and granulosa cells surrounding dominant follicle that underwent ovulation
      - ☐ Corpus luteum secretes progesterone, estradiol, and inhibin A, which decidualizes endometrium, preparing it for pregnancy
      - ☐ If egg is not fertilized, corpus luteum decays into corpus albicans after ~ 10 days
      - ☐ If fertilization occurs, blastocyst secretes human chorionic gonadotropin (hCG), signaling corpus luteum to continue progesterone production

IMAGING ANATOMY

Ultrasound

- Ovarian medulla is typically mildly hyperechoic compared to cortex
- **Developing follicles** are anechoic
  - Immediately before ovulation, tiny curved line (**cumulus oophorus**) may be visible at periphery of dominant follicle

- Represents layer of granulosa cells around mature ovum

- **Corpora lutea** have variable appearance
  - Classically have thickened, vascular wall with crenulated inner margin
  - Majority contain fluid/hemorrhage
  - Variably associated with adjacent stromal edema
- Small echogenic foci may represent specular reflectors from walls of tiny unresolved cysts
- Doppler shows low-velocity, low-resistance arterial wave form

CT

- Ovoid soft tissue attenuation structures frequently present at either side of uterus
- May contain areas of fluid attenuation representing follicles
- Sometimes difficult to identify ovaries on CT, particularly in premenarchal or postmenopausal females
  - Ovaries are smaller, devoid of visible follicles
- **Techniques for identifying ovary on CT**
  - Ovary typically located in ovarian fossa, inferior to common iliac artery bifurcation and anteromedial to pelvic portion of ureter
  - Follow ovarian veins within suspensory ligament along anterior surface of psoas muscle to ovarian hilum
  - If ascites is present, broad ligament may be traced laterally to ovary

MR

- T1
  - Intermediate stromal signal with low-signal follicles, unless hemorrhage is present
- T2
  - Multiple high-signal follicles (in premenopausal females)
  - Low signal intensity capsule
  - Medulla higher signal intensity compared to cortex
- Postmenopausal ovaries usually have homogeneous intermediate to low T1 and T2 signal

CLINICAL IMPLICATIONS

Clinical Importance

- **Pelvic pain**
  - Physiologic ovarian processes are extremely common source of pelvic discomfort
    - Follicle rupture (ovulation) with variable amounts of bleeding irritate peritoneum
      - In rare cases, uncontrolled bleeding requires surgical intervention
    - Hemorrhage within ovarian follicles/cysts causes enlargement and pain
    - Corpus luteum may be associated with ovarian edema (CLOE), which causes pain and is occasionally mistaken for ovarian edema from adnexal torsion
- **Ligamentous implications**
  - Adnexa may undergo **torsion**, commonly along suspensory ligament
    - Low-pressure venous and lymphatic outflow are initially compromised → ovarian edema and enlargement
    - Over time, arterial circulation is compromised → thrombosis → ischemia and hemorrhagic infarction
  - More likely with ovarian mass serving as lead point
- **Vascular implications**
  - Ovaries have **dual blood supply** due to anastomosis of ovarian and uterine arteries
    - Up to 60% of adnexal torsion cases have detectable blood flow with Doppler ultrasound
  - **Pelvic congestion syndrome**
    - Dull/heavy aching pain exacerbated by prolonged standing, walking, coitus, and menstruation
    - Caused by venous dilation during pregnancy with persistent valve incompetence, "nutcracker" phenomenon, May-Thurner syndrome, or compression of retroaortic left renal vein between aorta and spine
    - Difficult to diagnose due to **extremely common overlap of imaging findings with asymptomatic patients**
    - Imaging features include dilated ovarian vein (> 8 mm), dilated/tortuous parauterine veins (> 4 mm) communicating with dilated uterine arcuate veins, slow venous blood flow (< 3 cm/s), and communication with other pelvic varices
    - Retrograde ovarian venography is gold standard, demonstrating reflux of contrast into ovarian vein and pelvic varices, often crossing midline
- **Lymphatic implications**
  - Lymphatics are mode of spread for malignancy involving ovary
    - Most commonly spreads via suspensory ligament/ovarian vein to paraaortic nodes
    - Spread via broad ligament to pelvic sidewall nodes and via round ligament to inguinal region also possible
- **Physiologic implications**
  - **Polycystic ovarian syndrome (PCOS)**
    - Failure of ovarian follicles to mature; associated with hyperandrogenism, metabolic syndrome, and abnormal menstrual cycles
    - Imaging criteria for polycystic ovary is enlargement (> 10 mL) with numerous small peripheral follicles (> 25)
      - Patients without PCOS commonly meet these criteria; this must be correlated with clinical features
  - **Theca lutein cysts**
    - Excessive hCG stimulates development of multiple large cysts
    - May be seen with infertility treatment, gestational trophoblastic disease, triploidy, and multiple gestations
- **Histologic implications**
  - Ovary does not contain epithelium and rather is covered by specialized peritoneal surface epithelium
    - Most epithelial "ovarian" cancers actually spread from other sources (serous adenocarcinoma originating from fallopian tube, mucinous tumors from gastrointestinal tract)

OVARY AND LIGAMENTS

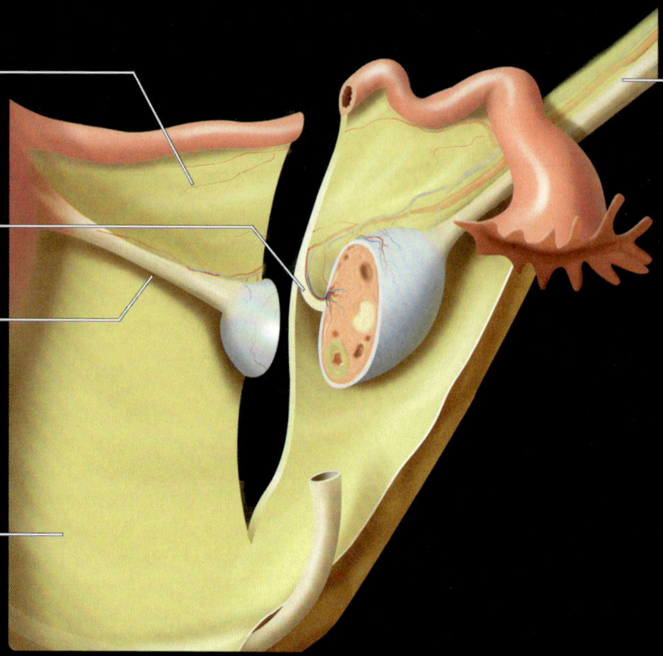

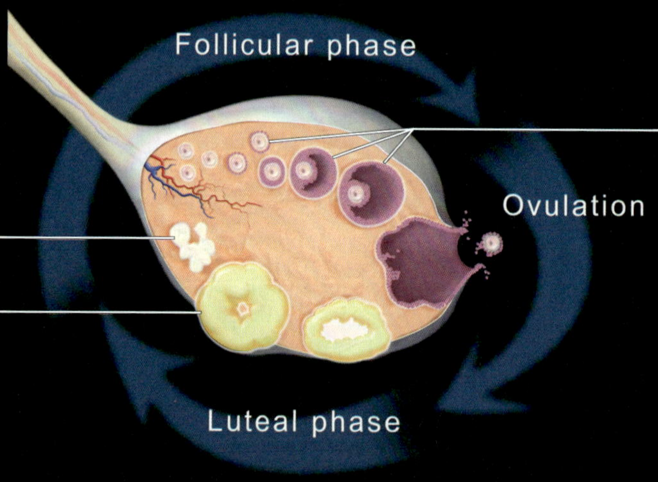

(Top) *Posterior view of the ligamentous attachment of the ovary is shown. The ovary is attached to the pelvic sidewall by the suspensory ligament (infundibulopelvic ligament) of the ovary, which transmits the ovarian artery and vein. These vessels enter the ovary through the mesovarium, a specialized ligamentous attachment between the ovary and broad ligament. The ovary is attached to the uterus by the proper ovarian ligament, which divides the mesosalpinx above from the broad ligament below.* **(Bottom)** *During the follicular phase of the menstrual cycle, several follicles begin to develop, but by days 8-12, a dominant follicle has formed, and the remainder begin to regress. On ~ day 14, the follicle ruptures, and the egg is extruded. After ovulation, a corpus luteum forms, and if fertilization does not occur, the corpus luteum degenerates into a corpus albicans.*

OVARY AND LIGAMENTS

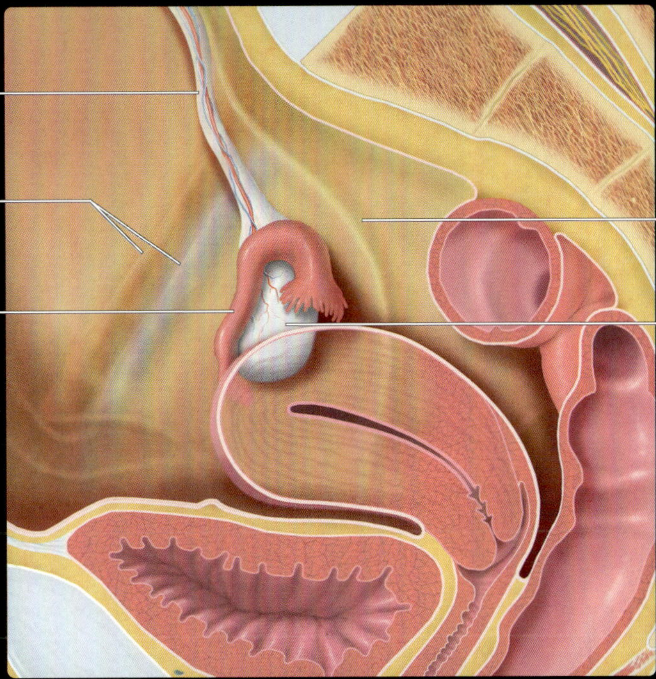

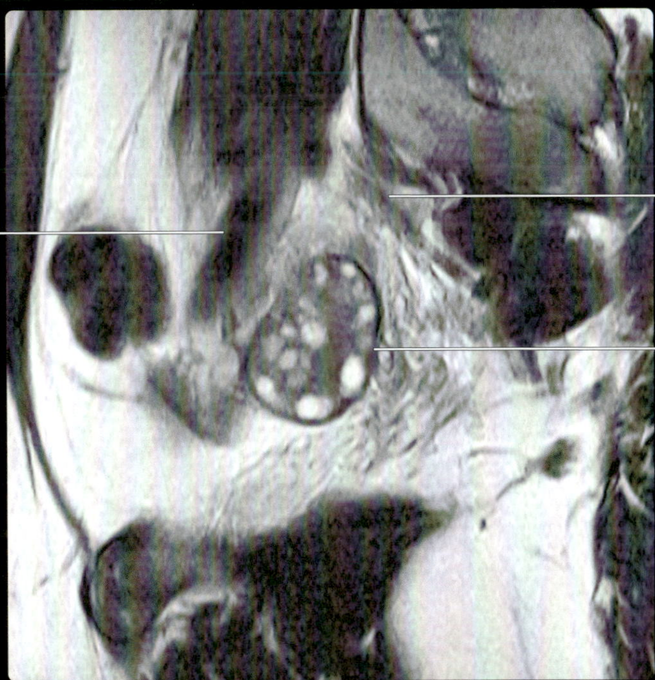

(Top) *Graphic shows the location of the ovary in a nulliparous woman. It lies against the lateral pelvic wall within the ovarian fossa, which is the area below the iliac bifurcation, posterior to the external iliac vessels and anterior to the ureter. With pregnancy, the ovaries are pushed out of the pelvis and seldom return to the same spot.* **(Bottom)** *Sagittal T2 MR of the lateral pelvic wall shows the ovary within the ovarian fossa, inferior to the iliac bifurcation.*

OVARY AND LIGAMENTS

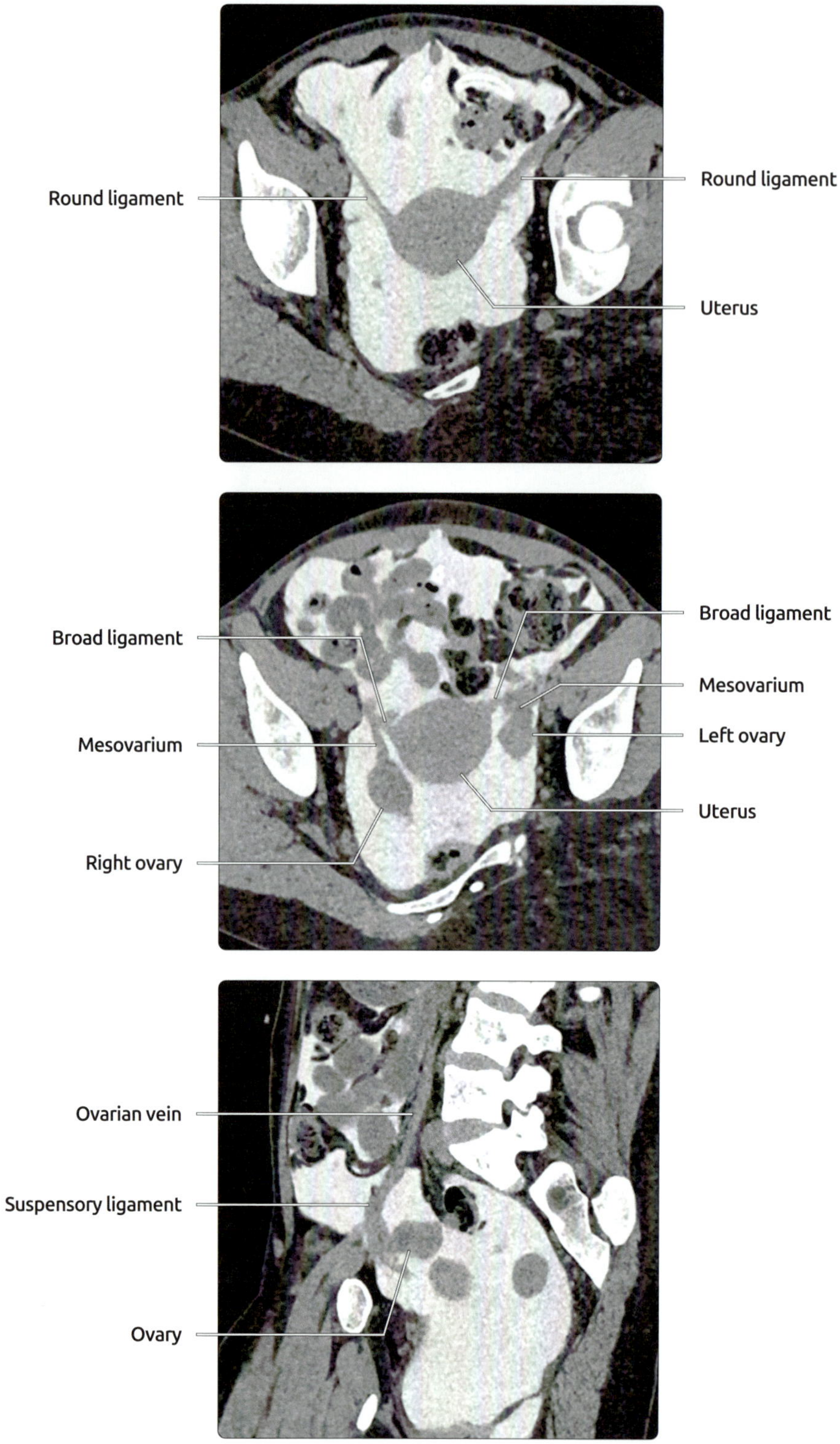

(Top) *Axial CT with intraperitoneal contrast outlining the pelvic structures shows the round ligaments extending from the uterine cornua to the inguinal regions.* **(Middle)** *A higher axial slice outlines the uterus, ovaries, and their ligamentous attachments.* **(Bottom)** *Sagittal multiplanar reformat shows the ovarian vessels extending from the retroperitoneum into the suspensory ligament, proceeding into the ovarian hilum.*

ULTRASOUND, NORMAL OVARY

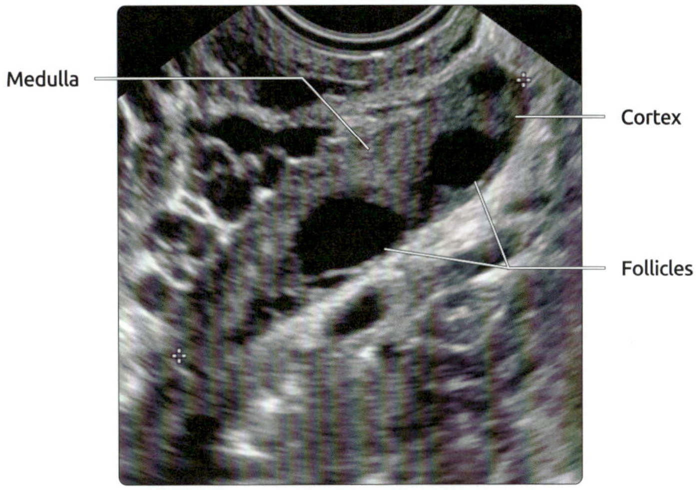

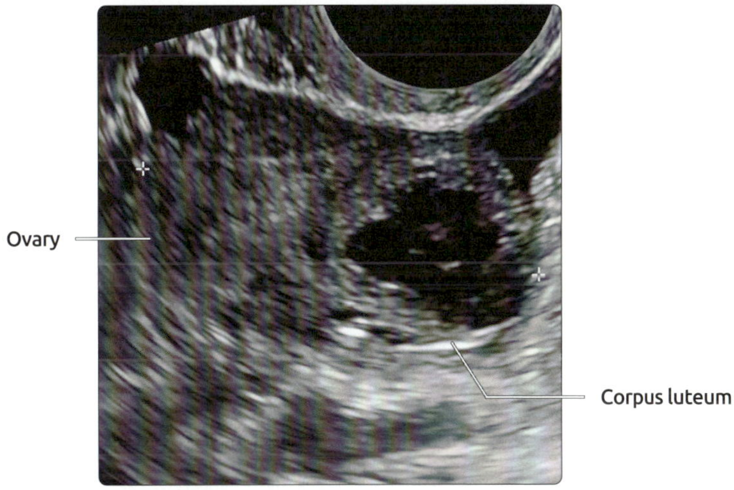

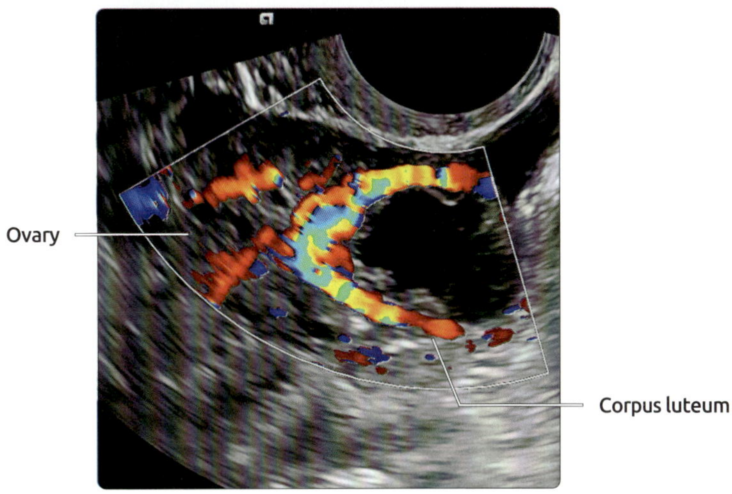

(Top) *Transvaginal ultrasound shows a normal ovary with maturing follicles peripherally within the cortex. The medulla is typically slightly more hyperechoic compared to the cortex.* **(Middle)** *Transvaginal ultrasound shows the classic appearance of a corpus luteum, including a thickened wall with crenulated inner margin, and internal fluid/hemorrhage.* **(Bottom)** *Color Doppler ultrasound shows hypervascularity within the wall of the corpus luteum.*

ULTRASOUND, NORMAL OVARY

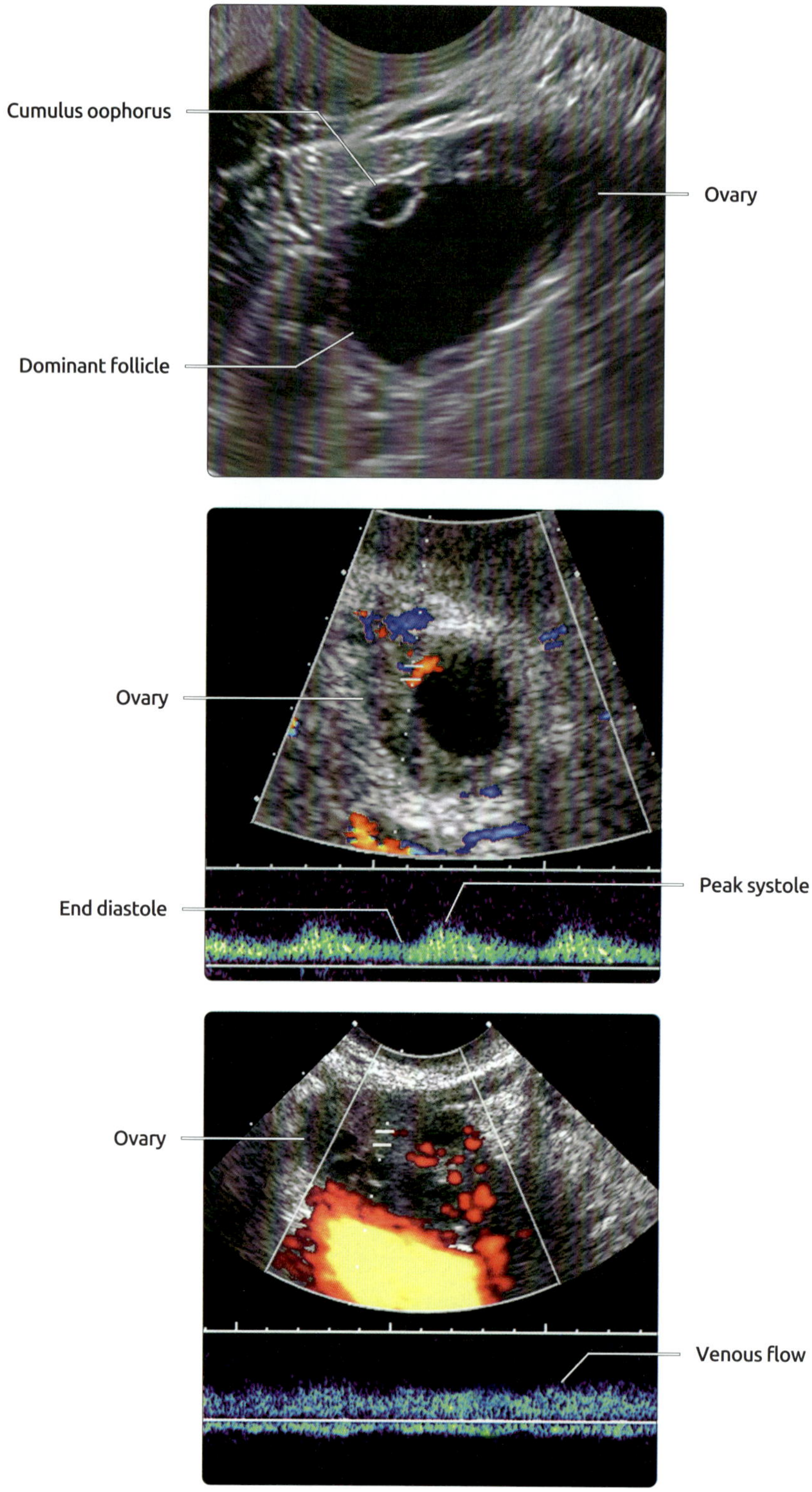

(Top) *Transvaginal ultrasound shows an ovary containing a dominant follicle with a peripheral cumulus oophorus, which is composed of a mature ovum surrounded by granulosa cells and is indicative of eminent ovulation.* **(Middle)** *Transvaginal color Doppler ultrasound in a patient on day 20 of the menstrual cycle shows normal ovarian arterial blood flow. The ovary normally has a low-resistance, low-velocity wave form.* **(Bottom)** *Transvaginal power Doppler ultrasound shows normal venous flow. Venous flow is the first to be compromised in adnexal torsion and should be documented in cases where torsion is suspected.*

MR, NORMAL OVARY

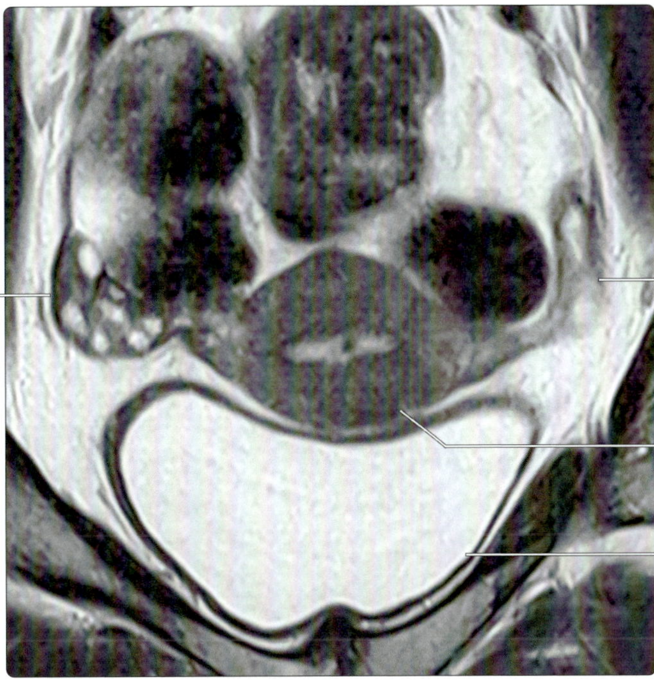

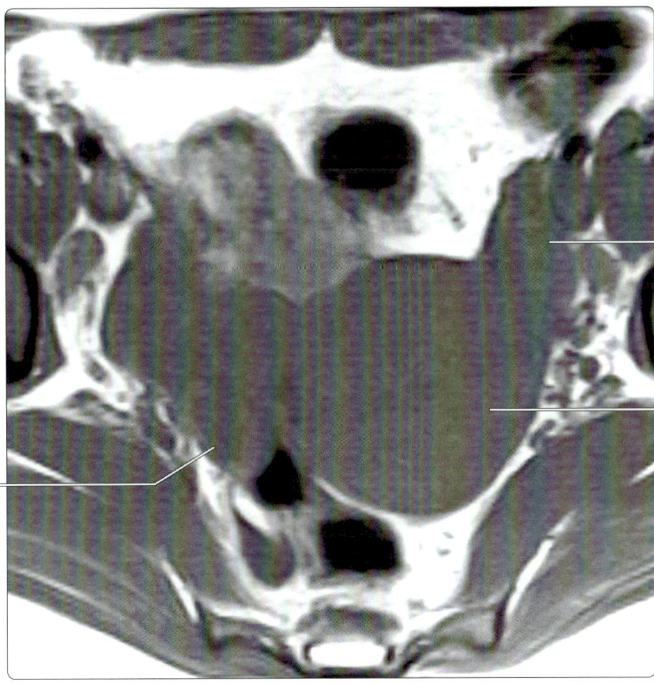

(Top) *Coronal T2 MR shows normal ovaries, which are nonenlarged and have a T2-hypointense rim and T2-hyperintense follicles.* **(Bottom)** *Axial T1 MR shows normal ovaries and uterus, which are intermediate signal intensity, similar to skeletal muscles. Fluid-containing ovarian follicles have low T1 signal intensity.*

ULTRASOUND, EVOLUTION OF OVARIAN HEMORRHAGIC CYST

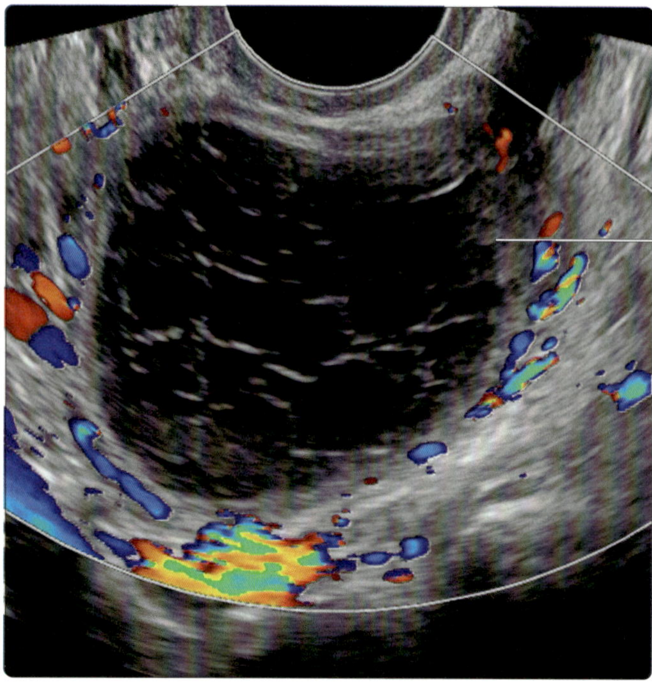

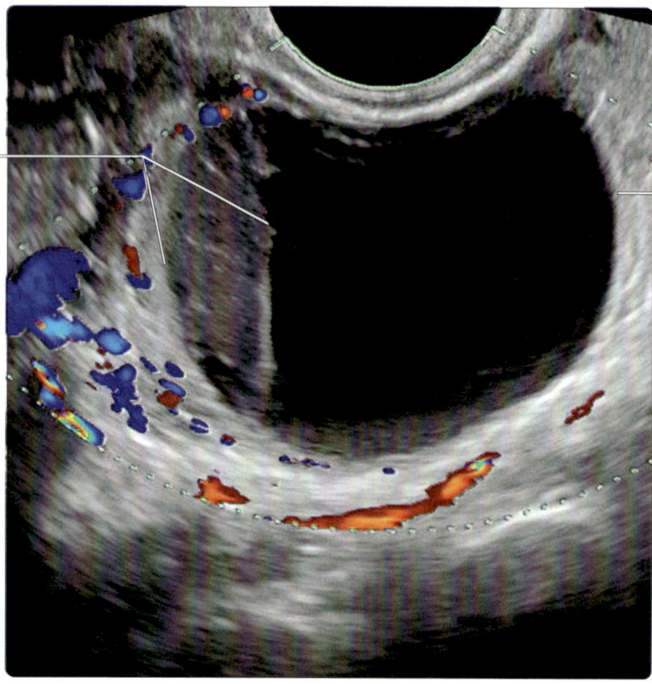

(Top) *Transvaginal color Doppler ultrasound shows the typical appearance of a hemorrhagic ovarian cyst, which contains lace-like internal septations. There is no internal flow on Doppler imaging.* **(Bottom)** *Transvaginal color Doppler ultrasound in the same patient obtained 3 weeks later shows the expected evolution of a hemorrhagic ovarian cyst. The cyst becomes predominantly anechoic with a peripheral avascular retracted clot.*

ADNEXAL TORSION, TWISTED VASCULAR PEDICLE

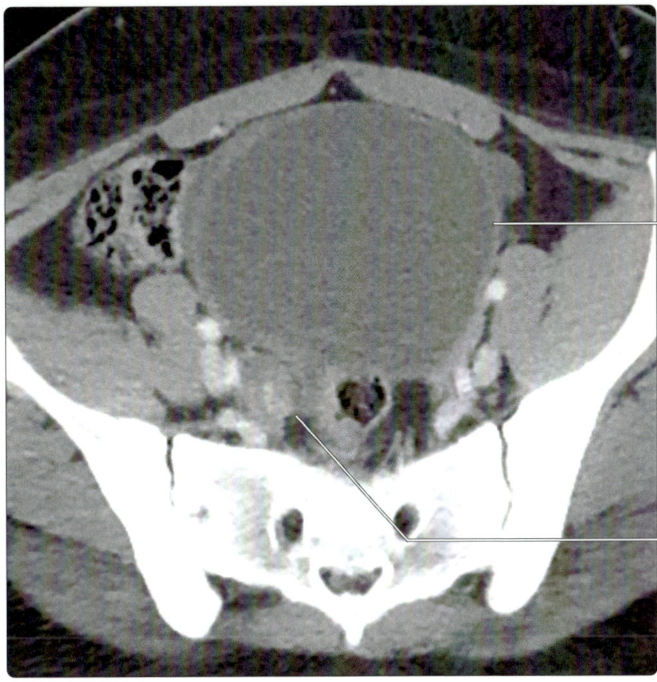

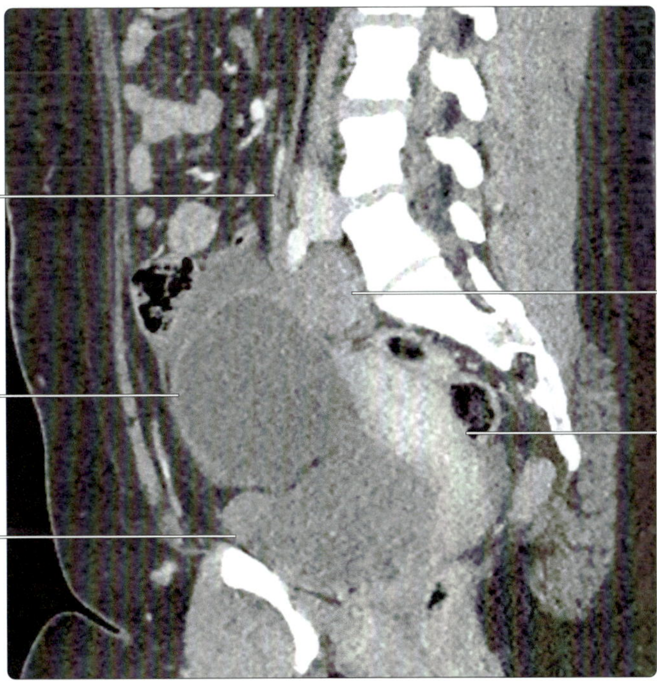

(Top) *Axial CECT shows an enlarged right ovary due to a lead point cyst and twisting of the suspensory ligament.* **(Bottom)** *Sagittal multiplanar CECT shows the ovarian vessels extending into the suspensory ligament, which is enlarged and twisted due to the lead point right ovarian cyst, which is anteriorly displaced.*

POLYCYSTIC OVARIAN SYNDROME

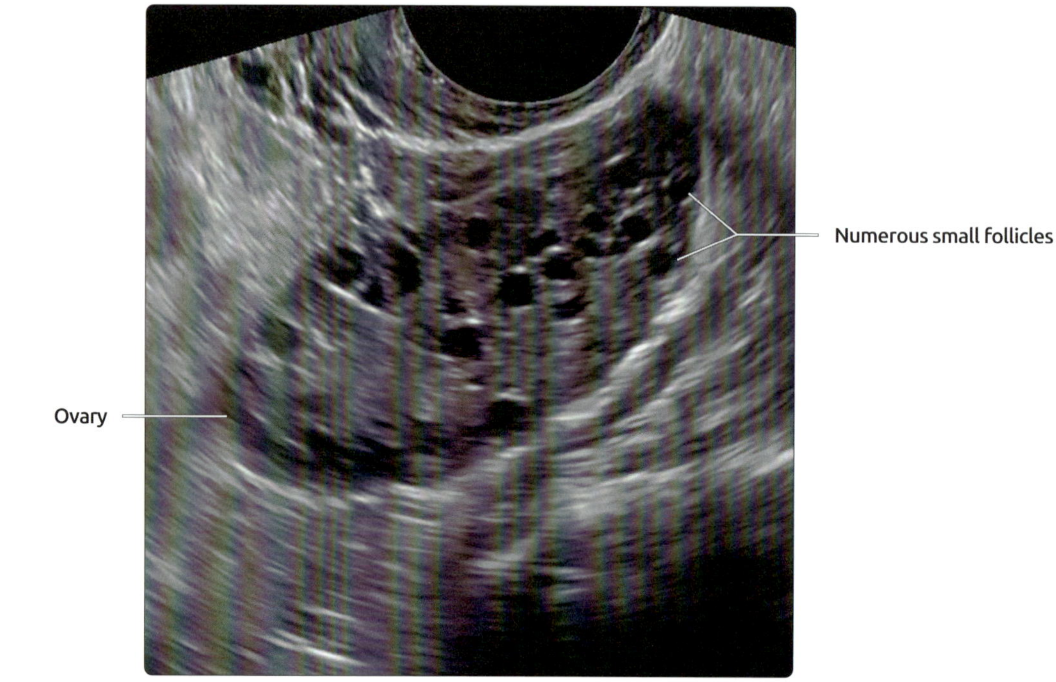

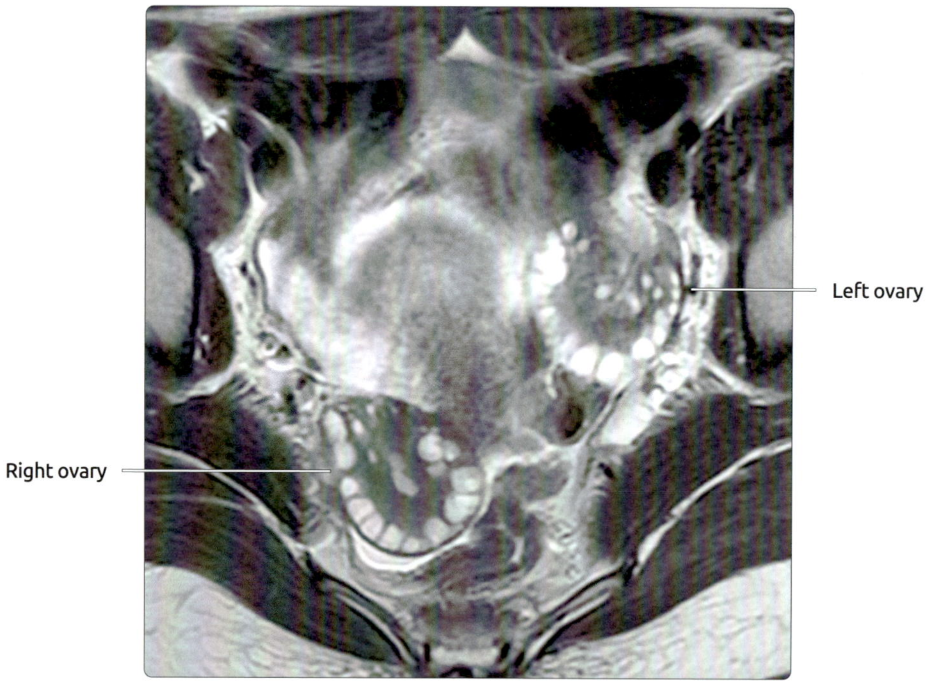

(Top) *Transvaginal ultrasound shows an enlarged ovary with numerous small follicles.* **(Bottom)** *Axial T2 MR shows bilateral ovarian enlargement with numerous small, peripherally located T2-hyperintense follicles. These follicles fail to mature. The lack of dominant follicle leads to anovulatory cycles.*

THECA LUTEIN CYSTS

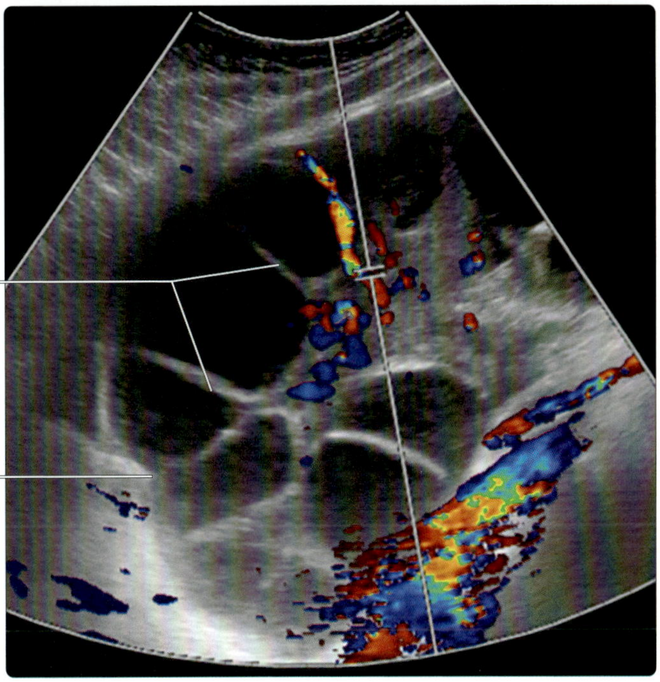

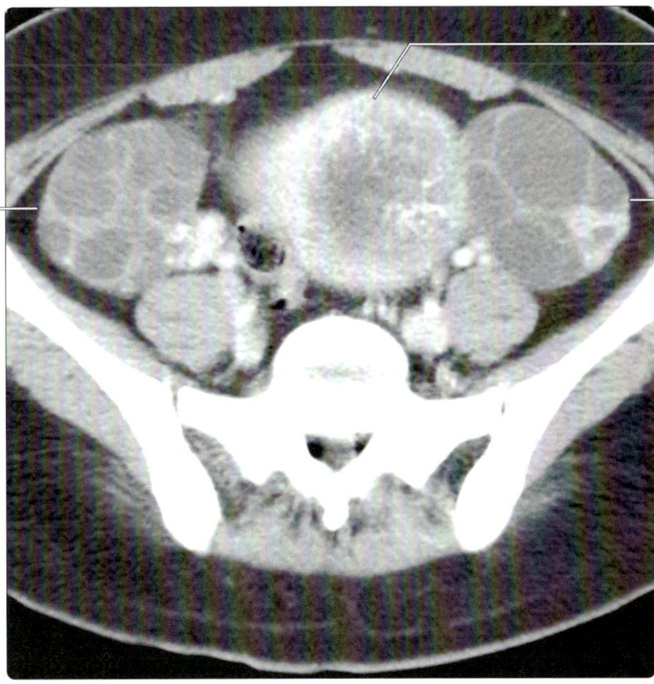

(Top) *Transvaginal ultrasound with color Doppler shows multicystic ovarian enlargement. Theca lutein cysts are associated with increased risk for torsion and cyst rupture.* **(Bottom)** *CECT shows bilateral ovarian enlargement due to theca lutein cysts. A mole is present in the uterus, which secretes human chorionic gonadotropin, leading to ovarian hyperstimulation. This can be seen with infertility treatments, triploidy, and multiple gestations.*

INDEX

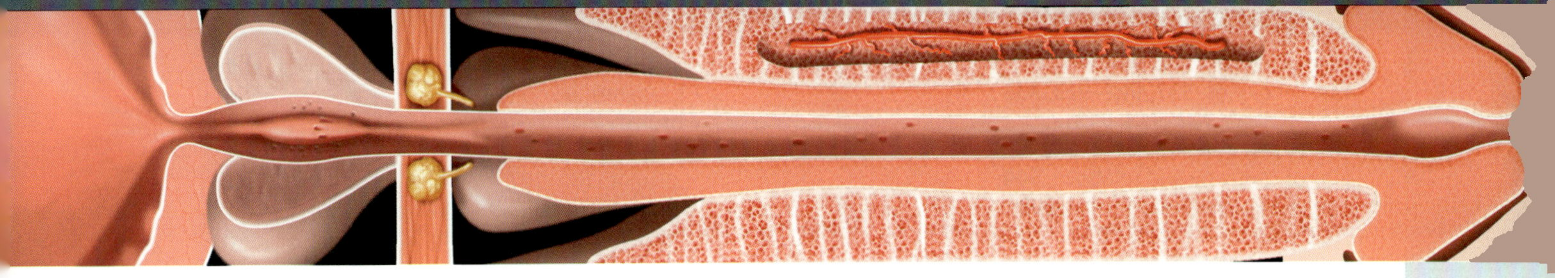

A

INDEX

INDEX

INDEX

B

C

INDEX

INDEX

INDEX

D

E

INDEX

F

G

H

INDEX

INDEX

INDEX

J

K

INDEX

M

INDEX

INDEX

N

O

P

Q

R

S

INDEX

T

U

V

INDEX

W

X

Y

Z